SME MINING ENGINEERING HANDBOOK

THIRD EDITION

VOLUME TWO

EDITED BY PETER DARLING

PUBLISHED BY
SOCIETY FOR MINING, METALLURGY, AND EXPLORATION, INC.

Society for Mining, Metallurgy, and Exploration, Inc. (SME)
(303) 948-4200 / (800) 763-3132
www.smenet.org

SME advances the worldwide mining and minerals community through information exchange and professional development. With members in more than 70 countries, SME is the world's largest association of mining and minerals professionals.

Copyright © 2011 Society for Mining, Metallurgy, and Exploration, Inc.

Supported by the Seeley W. Mudd Memorial Fund of AIME.

All Rights Reserved. Printed in the United States of America.

Information contained in this work has been obtained by SME, Inc., from sources believed to be reliable. However, neither SME nor its authors and editors guarantee the accuracy or completeness of any information published herein, and neither SME nor its authors and editors shall be responsible for any errors, omissions, or damages arising out of use of this information. This work is published with the understanding that SME and its authors and editors are supplying information but are not attempting to render engineering or other professional services. Any statement or views presented herein are those of individual authors and editors and are not necessarily those of SME. The mention of trade names for commercial products does not imply the approval or endorsement of SME.

No part of this publication may be reproduced, stored in a retrieval system, or transmitted in any form or by any means, electronic, mechanical, photocopying, recording, or otherwise, without the prior written permission of the publisher.

ISBN 978-0-87335-264-2
Ebook 978-0-87335-341-0

Library of Congress Cataloging-in-Publication Data

SME mining engineering handbook / edited by Peter Darling. -- 3rd ed.
v. cm.
Includes bibliographical references and index.
ISBN 978-0-87335-264-2
1. Mining engineering--Handbooks, manuals, etc. I. Darling, Peter, 1956- II. Society for Mining, Metallurgy, and Exploration (U.S.) III. Title: Mining engineering handbook.
TN145.S56 2011
622--dc22
2010050815

Contents

Foreword ix
Preface xi
About the Managing Editor xiii
Contributing Authors xv
Technical Reviewers xxi

PART 1: MINING: SETTING THE SCENE
1.1 Mining: Ancient, Modern, and Beyond 3
1.2 Current Trends in Mining 11
1.3 Future Trends in Mining 21

PART 2: MARKET ECONOMICS
2.1 Economics of the Minerals Industry 39
2.2 Pricing and Trading in Metals and Minerals 49
2.3 Market Capitalization 65
2.4 Investment Analysis 73

PART 3: EXPLORATION
3.1 Geological Features and Genetic Models of Mineral Deposits 83
3.2 Minerals Prospecting and Exploration 105
3.3 Geophysics Prospecting 113
3.4 Geochemical Prospecting 127

PART 4: DEPOSIT ASSESSMENT
4.1 Geological Data Collection 145
4.2 Geologic Interpretation, Modeling, and Representation 173
4.3 Sample Preparation and Assaying 187
4.4 Ore-Body Sampling and Metallurgical Testing 193
4.5 Mineral Resource Estimation 203
4.6 Valuation of Mineral Properties 219
4.7 Mineral Property Feasibility Studies 227
4.8 Cost Estimating for Underground Mines 263
4.9 Cost Estimating for Surface Mines 281

PART 5: MANAGEMENT AND ADMINISTRATION
5.1 Mine Economics, Management, and Law 297
5.2 Economic Principles for Decision Making 309
5.3 Management, Employee Relations, and Training 317
5.4 A Global Perspective on Mining Legislation 331

PART 6: MINING METHOD SELECTION
6.1 Evaluation of Mining Methods and Systems 341
6.2 Mining Methods Classification System 349
6.3 Selection Process for Hard-Rock Mining 357
6.4 Selection Process for Underground Soft-Rock Mining 377
6.5 Comparison of Underground Mining Methods 385
6.6 Comparison of Surface Mining Methods 405

PART 7: ROCK BREAKING METHODS
- 7.1 Mechanical Rock Breaking 417
- 7.2 Blasthole Drilling 435
- 7.3 Explosives and Blasting. 443

PART 8: GROUND MECHANICS
- 8.1 Introduction to Ground Mechanics 463
- 8.2 Soil Mechanics 471
- 8.3 Slope Stability. 495
- 8.4 Rock Mechanics 527
- 8.5 Geotechnical Instrumentation 551
- 8.6 Hard-Rock Ground Control with Steel Mesh and Shotcrete 573
- 8.7 Soft-Rock Ground Control. 595
- 8.8 Ground Control Using Cable and Rock Bolting 611
- 8.9 Mine Subsidence 627
- 8.10 Tailings Impoundments and Dams. 645
- 8.11 Waste Piles and Dumps 667

PART 9: INFRASTRUCTURE AND SERVICES
- 9.1 Electric Power Distribution and Utilization. 683
- 9.2 Compressed Air 705
- 9.3 Mine Communications, Monitoring, and Control 717
- 9.4 Mine Surveying. 731
- 9.5 Dewatering Surface Operations. 743
- 9.6 Dewatering Underground Operations 765
- 9.7 Physical Asset Management 781
- 9.8 Automation and Robotics 805
- 9.9 Mine Infrastructure Maintenance. 825
- 9.10 Systems Engineering. 839

PART 10: SURFACE EXTRACTION
- 10.1 Introduction to Open-Pit Mining. 857
- 10.2 Open-Pit Planning and Design. 877
- 10.3 Mechanical Extraction, Loading, and Hauling 903
- 10.4 Selection and Sizing of Excavating, Loading, and Hauling Equipment. 931
- 10.5 In-Pit Crushing 941
- 10.6 Design, Construction, and Maintenance of Haul Roads. 957
- 10.7 Surface Ore Movement, Storage, and Recovery Systems 977
- 10.8 Strip Mining 989
- 10.9 Strip Mine Planning and Design 1013
- 10.10 Highwall Mining. 1027
- 10.11 Quarrying. 1031

PART 11: HYDRAULIC AND PIPELINE MINING
- 11.1 Hydraulic Mining 1049
- 11.2 Placer Mining and Dredging 1057
- 11.3 Heap Leaching. 1073
- 11.4 Surface Techniques of Solution Mining. 1087
- 11.5 In-Situ Techniques of Solution Mining. 1103
- 11.6 Coal-Bed Methane Production. 1121

PART 12: UNDERGROUND DEVELOPMENT
- 12.1 Introduction to Underground Mine Planning. 1135
- 12.2 Hard-Rock Equipment Selection and Sizing 1143
- 12.3 Soft-Rock Equipment Selection and Sizing. 1157
- 12.4 Underground Horizontal and Inclined Development Methods. 1179
- 12.5 Subsurface Mine Development 1203
- 12.6 Construction of Underground Openings and Related Infrastructure 1223
- 12.7 Tunnel Boring Machines in Mining. 1255
- 12.8 Underground Ore Movement. 1271
- 12.9 Hoisting Systems. 1295

PART 13: UNDERGROUND EXTRACTION
- 13.1 Room-and-Pillar Mining in Hard Rock ... 1327
- 13.2 Room-and-Pillar Mining in Coal ... 1339
- 13.3 Shrinkage Stoping ... 1347
- 13.4 Sublevel Stoping ... 1355
- 13.5 Cut-and-Fill Mining ... 1365
- 13.6 Backfill Mining ... 1375
- 13.7 Cave Mining ... 1385
- 13.8 Longwall Mining ... 1399
- 13.9 Sublevel Caving ... 1417
- 13.10 Block Caving ... 1437

PART 14: MINERAL PROCESSING
- 14.1 Introduction to Mineral Processing ... 1455
- 14.2 Crushing, Milling, and Grinding ... 1461
- 14.3 Classification by Screens and Cyclones ... 1481
- 14.4 Gravity Concentration and Heavy Medium Separation ... 1507
- 14.5 Fundamental Principles of Froth Flotation ... 1517
- 14.6 Magnetic and Electrostatic Separation ... 1533
- 14.7 Dewatering Methods ... 1547

PART 15: HEALTH AND SAFETY
- 15.1 Mine Safety ... 1557
- 15.2 Health and Medical Issues in Global Mining ... 1567
- 15.3 Mine Ventilation ... 1577
- 15.4 Gas and Dust Control ... 1595
- 15.5 Heat, Humidity, and Air Conditioning ... 1611
- 15.6 Radiation Control ... 1625
- 15.7 Noise Hazards and Controls ... 1633

PART 16: ENVIRONMENTAL ISSUES
- 16.1 Site Environmental Considerations ... 1643
- 16.2 Mining and Sustainability ... 1665
- 16.3 Impacts and Control of Blasting ... 1689
- 16.4 Water and Sediment Control Systems ... 1705
- 16.5 Mitigating Acid Rock Drainage ... 1721
- 16.6 Waste Disposal and Contamination Management ... 1733
- 16.7 Closure Planning ... 1753

PART 17: COMMUNITY AND SOCIAL ISSUES
- 17.1 Community Issues ... 1767
- 17.2 Social License to Operate ... 1779
- 17.3 Cultural Considerations for Mining and Indigenous Communities ... 1797
- 17.4 Management of the Social Impacts of Mining ... 1817

APPENDICES
- A Web Sites Related to Mining ... 1827
- B Coal Mine Gas Chart ... 1833
- C Conversion Chart ... 1835

Index ... I-1

Dedication

*With deep appreciation for his contributions to the mining industry, we dedicate this
3rd edition of the SME Mining Engineering Handbook
to the memory of Richard E. Gertsch, 1945–2005.*

*Richard provided the initial leadership and direction for this edition.
His guidance allowed others to carry on with his vision.*

*Richard was a widely respected and admired mining engineer enjoying an illustrious career
spanning decades of work both in industry and academia.*

*Richard served on the SME Board of Directors and as the Chair of the M&E Division.
He was active on many committees working on important SME functions such as peer review,
programming, membership, publications, nominations, scholarships, and awards.
He received the Distinguished Service Award in 1991.*

Foreword

Mining engineers throughout the world are the salt of the earth, and this third edition of the *SME Mining Engineering Handbook* will be their bible. It builds on the grand tradition begun by highly respected leaders of the mining industry—Peele (1918, 1927, and 1941) and continued by Cummins and Given (1973) and then Hartman (1992).

Handbooks by their nature are often hard to read, but this one is a striking exception. Its readability immediately stands out and allows one to quickly absorb and comprehend its content—not only the text, but also the many tables, figures, and photographs. Artwork has been substantially upgraded and is especially appealing. The superb presentation reflects the managing editor's technical writing background, as well as the input and skills of Jane Olivier, SME's manager of book publishing, who spent a decade urging a new edition, and Diane Serafin, an editing specialist, who delved doggedly into the tedious details.

The bar of excellence in the quality and scope of material is well maintained and at a high level throughout. Often a work involving numerous experts can lead to conflicting views on countless aspects. However, the clashing of ideas is notably a hallmark of serious thinking. The clear beneficiary of the high standard is the reader, which in this case is the practicing mining engineer in the worldwide mining industry, as well as others in academia and government.

The strength of this handbook lies clearly in the quality of the chapter authors and peer reviewers. They are indeed talented experts in their specialized technical fields. A decided trend toward internationalism, paralleling similar trends in the supply and demand of mineral and energy raw materials, is evidenced by a cursory perusal of the lists of contributing authors and technical reviewers. Almost half of them are working and living outside the United States. In total, ten countries are represented—primarily English-speaking ones.

Such a diverse cadre of individuals offers wide-ranging views of the worldwide mining industry's multifaceted problems and potential solutions in a period of accelerated technological and social change. The broad scope is another strength for which no unanimity of thought can be expected or anticipated when taking a global perspective on the various issues.

Carryover material from previous editions is strictly limited, and when included, it is updated to be genuinely contemporary and purposeful. *Practical* and *useful* instead of merely *theoretical* and *interesting* are the watchwords throughout the work.

Comprehensiveness is an important feature not to be overlooked in a world-class handbook. This edition sets the standard and will be emulated far into the future. Much attention has been given to what can be instead of what is past. Productiveness received authoritative treatment. Specific emphasis has been given to broad topics that will continue to confront the industry in the years ahead, such as environmental issues, public concerns, health and safety matters, and sustainability. This handbook provides a pathway for the synthesis and solution of many of the complex issues and problems the mining industry is facing in the 21st century.

Raymond L. Lowrie
Editor of the *SME Mining Reference Handbook*
Recipient of the 2004 SME President's Citation
February 2011

Preface

It was Robert Peele who gathered a group of 46 specialists almost a century ago to write the now classic *Mining Engineers' Handbook*, published in 1918 by John Wiley & Sons, with second and third editions published in 1927 and 1941. When Wiley declined to publish further editions, SME began publication of the *SME Mining Engineering Handbook*. SME published the first edition in 1973 and a second edition in 1992.

From that initial publication in 1918, the handbook has been acknowledged as the repository of all essential information and useful, practical detail relating to virtually every aspect of mining.

From the beginning of my mining career, I have relied heavily on this book as a source of inspiration and information. With the honor of being asked to take on the somewhat challenging role of managing editor for this third edition, I owe a debt of professional gratitude to several generations of managing editors, authors, and technical reviewers who set the seemingly impossible high standards in the previous editions of the handbook.

From the onset, several objectives were established. These objectives included (1) to produce a book that would stand shoulder to shoulder as an equal alongside previous editions, (2) to maintain its tradition of being the "handbook of choice" for every practicing mining engineer, (3) to be practical rather than theoretical in its content and approach, and (4) to be international in its appeal and examples.

Mining is an international business, and the importance of sharing experiences, knowledge, and examples from around the world cannot be overemphasized. A water "problem" in one part of the world may be considered "normal" in another. A "deep" mine on one continent may connote "very deep" on another. Tailings dams in some countries need to be earthquake proof, whereas in other countries they must be free draining.

This edition attempts to take the best of the best from around the world and package it in a standard and logical format for the benefit of the global industry. This internationalism is shown by the subjects covered in each chapter as well as by the nationalities of the world-class authors and technical reviewers represented. It is noteworthy that most significant mining operations have graciously shared their knowledge, techniques, experience, and alternative viewpoints.

The handbook has moved with the times to cover the issues that are exercising the industry, the innovations that are exciting, and how the industry is dealing with changing attitudes toward a number of its constituents such as energy (both electrical generation and carbon/petroleum based), water management, resource maintenance, and the whole subject of sustainable development. One of the significant areas in which this handbook differs from previous editions is that it includes several chapters on both the social and environmental issues often associated with mining, and, more importantly, how these issues and their impacts can be mitigated and managed.

It is important to note that this is a handbook and not an encyclopedia of everything mining. Several subjects previously included as dedicated chapters are now covered in less detail within other chapters. This move was not designed to marginalize or dilute the importance of certain subjects but was done to acknowledge the significant shifts in the way mining has changed as a result of new technologies. It also reflects how the work and the responsibilities of today's mining engineer have changed and developed.

Attempts to encapsulate the essence of so diverse a discipline as mining engineering could not have been possible without the unselfish contributions of the hundreds of authors, coauthors, technical reviewers, and mentors who are very much the unsung heroes of this publication. Much gratitude and thanks are due to the many talented and world-class professionals who have given so freely, patiently, and enthusiastically of their time, hard-earned experiences, and masterful knowledge on a plethora of mining and related subjects to ensure that this handbook was produced on time and to the meteoric standards that the industry has come to expect.

One of the many delights of managing this project has been the acquaintance (usually electronically) of so many helpful, patient, enthusiastic, and friendly people from within the various mining associations, academic establishments, mining companies, and consultancies, as well as retired engineers and specialist editors—everyone an expert in their field. Without their cooperation, steady guidance, constructive comment, and encouragement, the managing of this edition would have been an impossible task. I cannot name them all, and to mention some but not others would be ungracious, but tremendous appreciation is due.

These experts are headed by the authors themselves, almost every one of whom is a world leader in their specialist field. Often with the briefest of guidance and without any form of remuneration, they have passionately undertaken their writing tasks. Their enthusiasm, commitment, and professionalism formed the bedrock on which the handbook has been based, and, as a result, they have produced superb and exceptional texts. I salute and thank you all.

Next, I acknowledge the many technical reviewers who, often at short notice and, as this project progressed, with an ever shortening lead time, have been called on to read through many drafts before articulating and listing their comments, suggestions, and observations. It typically was not an easy task and often may have been a marathon requiring skills, diplomacy, and knowledge that went well beyond their original brief. I apologize and thank you.

Thanks are extended to SME's book publishing team who have been thoroughly professional and efficient in their handling of this edition. This team has checked every comma, word, phrase, sentence, and illustration. They have ensured that copyright rules have not been flaunted in the quest for expedience, checked and rechecked references, standardized units of measure, and performed a hundred and one other tasks to ensure that this finished product is a source of information in a format that at best pleases and at worst does not annoy. Their attention to detail in this exacting task is very much appreciated.

I also thank SME for affording me this opportunity to repay some of the debt that I believe is owed to an industry that has kept me fed, watered, enthralled, and enthused for more than three decades.

Finally, I trust that any errors in opinion, facts, or perceptions in the handbook are few in number and will not overtly detract from the usefulness of and, I dare say, enjoyment of this third edition of the *SME Mining Engineering Handbook*.

Peter Darling
February 2011

About the Managing Editor

Peter Darling, managing editor of the third edition of the *SME Mining Engineering Handbook*, has more than 30 years of experience as a mining engineer and journalist.

A graduate of the Royal School of Mines, Imperial College, London, Darling worked as an oil industry engineer on offshore projects in Gabon, Congo, Angola, the Gulf of Guinea, the Gulf of Mexico, the North Sea, Tunisia, Egypt, and Abu Dhabi. He was also involved in underground platinum mines in South Africa and open-pit tin operations in Brazil.

Darling then embarked on a career as a technical writer and editor. His assignments took him to mines, quarries, tunnels, and construction sites stretching from Chile to China, Alaska to Australia, Peru to Papua New Guinea, and Russia to La Reunion. During this period he served as editor for a variety of respected industry publications, including *International Mining, Engineering and Mining Journal, Rock Products, Tunnels & Tunnelling International, International Construction, The Cement Edition, Construction Asia,* and *Coal* (North America). Darling also served as the press officer for Rio Tinto in London.

As a Royal Air Force Reserve officer, Darling was deployed to Gulf War II and twice to Afghanistan where he completed the final edit of the handbook. He is a Chartered Engineer and member of the Institute of Materials, Minerals and Mining and a member of the Society for Mining, Metallurgy, and Exploration.

Darling is currently a freelance technical editor, journalist, reporter, and speech writer specializing in mining, quarrying, oil and gas, tunneling, heavy construction, and engineering.

Peter Darling BEng (Hons), ARSM, CEng
Gothic House, Aylsham, Norfolk, England
www.peterdarling.co.uk
peter@peterdarling.co.uk

Contributing Authors

L. Adler
Professor
West Virginia University
Morgantown, West Virginia, USA

Tom Albanese
Chief Executive Officer
Rio Tinto Ltd.
London, United Kingdom

Breanna L. Alexander
Metallurgical Engineer
Lyntek, Inc.
Lakewood, Colorado, USA

Derek B. Apel
Professor, School of Mining & Petroleum Engineering
University of Alberta
Edmonton, Alberta, Canada

Odd G. Askilsrud
President
Tunnel Engineering and Applications, Inc.
Renton, Washington, USA

William F. Bawden
Pierre Lassonde Chair in Mining Engineering
University of Toronto
Toronto, Ontario, Canada

Scott Beer
Chief Operating Officer
Rajant Corporation
Malvern, Pennsylvania, USA

Eric N. Berkhimer
Senior Applications Engineer
P&H Mining Equipment
Milwaukee, Wisconsin, USA

Stephen L. Bessinger
Engineering Manager
BHP Billiton–San Juan Coal Company
Farmington, New Mexico, USA

Evelyn L. Jessup Bingham
Group Manager, Closure & Waste
BHP Billiton
Melbourne, Victoria, Australia

Martyn Bloss
Manager Long Term Planning, Olympic Dam
BHP Billiton
Adelaide, South Australia, Australia

Ernest Bohnet
Vice President of Mining and Geological Services
Pincock Allen & Holt
Denver, Colorado, USA

Richard K. Borden
Principal Advisor Environment
Rio Tinto
South Jordan, Utah, USA

Robert G. Boutilier
President
Boutilier & Associates
Vancouver, British Columbia, Canada, and Cuernavaca, Mexico

Charles A. Brannon
Manager Underground Planning
Freeport-McMoRan Copper & Gold Inc.
Phoenix, Arizona, USA

Ian Brown
National Production Manager
Lafarge Aggregates, Mountsorrel Quarry
Loughborough, Leicestershire, United Kingdom

Richard L. Bullock
Professor Emeritus, Mining & Nuclear Engineering
Missouri University of Science and Technology
Rolla, Missouri, USA

Jeremy Busfield
Principal Consultant
MineCraft Consulting Pty Ltd.
Queensland, Australia

Felipe Calizaya
Associate Professor
University of Utah
Salt Lake City, Utah, USA

Gordon K. Carlson
Chief Mine Engineer
Henderson Mine, Climax Molybdenum
Empire, Colorado, USA

Peter G. Carter
Manager of Mining Engineering
BHP Billiton
Melbourne, Victoria, Australia

Timothy P. Casten
Director, Underground Planning
Freeport-McMoRan Copper & Gold Inc.
Phoenix, Arizona, USA

Joe Cline
Lead Discipline Engineer
Cementation USA, Inc.
Sandy, Utah, USA

L. Graham Closs
Associate Professor, Dept. of Geology & Geological Engrg.
Colorado School of Mines
Golden, Colorado, USA

Mark Colwell
Principal
Colwell Geotechnical Services
Queensland, Australia

Ivan A. Contreras
Vice President
Barr Engineering Company
Minneapolis, Minnesota, USA

J. Alan Coope
(Deceased) Former Director of Geochemistry
Newmont Exploration Ltd.
Denver, Colorado, USA

Phillip C.F. Crowson
Honorary Professor & Professorial Research Fellow
Centre for Energy, Petroleum & Mineral Law & Policy
University of Dundee, Scotland

Michael J. Cruickshank
Consulting Marine Mining Engineer
Marine Minerals Technology Center Associates
Honolulu, Hawaii, USA

Kadri Dagdelen
Professor of Mining Engineering
Colorado School of Mines
Golden, Colorado, USA

Donald A. Dahlstrom
(Deceased) Former Professor Emeritus
University of Utah
Salt Lake City, Utah, USA

Diana Dalton
Lawyer and Independent Consultant in Mining Law
Halifax, Nova Scotia, Canada

Peter Darling
Managing Editor and Freelance Technical Writer
Aylsham, Norfolk, United Kingdom

J.A.J. (Barry) de Wet
Director and Consulting Geophysicist
BDW Geophysics Consulting
Brisbane, Queensland, Australia

Malcolm G. Dorricott
Principal Consultant
AMC Consultants Pty Ltd.
Melbourne, Victoria, Australia

Charles Dowding
Professor of Civil and Environmental Engineering
Northwestern University
Evanston, Illinois, USA

Geoff Dunstan
Mining Manager
Newcrest Mining Ltd.
Melbourne, Victoria, Australia

Gary Dyer
Manager Strategy
BHP Billiton Mitsubishi Alliance
Brisbane, Queensland, Australia

Erik Eberhardt
Professor of Geological Engineering
University of British Columbia
Vancouver, British Columbia, Canada

A.J. (Joe) Erickson Jr.
(Retired) Mining Geology Consultant
Anaconda, UPCM Co., AMAX, EXXON Coal & Minerals Co.
Houston, Texas, USA

Reinhold A. Errath
Technology Manager for Drives
ABB Switzerland Ltd.
Baden, Switzerland

Robin Evans
Senior Research Fellow, University of Queensland
Sustainable Minerals Inst., Centre for Social Resp. in Mining
Brisbane, Queensland, Australia

Brian Flintoff
Senior VP Tech. Dev., Equipment & Systems Business Line
Metso Mining & Construction Technology
Kelowna, BC, Canada

Frank H. Fox
Head of Occupational Health
Anglo American plc
Johannesburg, South Africa

Daniel M. Franks
Research Fellow, University of Queensland
Sustainable Minerals Inst., Centre for Social Resp. in Mining
Brisbane, Queensland, Australia

Russell Frith
Adjunct Professor, School of Mining Engineering
University of New South Wales
Sydney, New South Wales, Australia

Maurice C. Fuerstenau
Newmont Professor of Minerals Engineering
University of Nevada
Reno, Nevada, USA

Rajive Ganguli
Professor of Mining Engineering
University of Alaska Fairbanks
Fairbanks, Alaska, USA

Ginger Gibson
Adjunct Professor
University of British Columbia
Vancouver, British Columbia, Canada

Don Grant
Superintendent Life of Asset Planning, Olympic Dam
BHP Billiton
Adelaide, South Australia, Australia

Jedediah D. Greenwood
Senior Geotechnical Engineer
Barr Engineering Company
Minneapolis, Minnesota, USA

Ed Grygiel
Manager of Six Sigma Engineering
Jim Walter Resources
Brookwood, Alabama, USA

Bernard J. Guarnera
President and Chairman of the Board of Directors
Behre Dolbear Group, Inc.
Denver, Colorado, USA

John Hadjigeorgiou
Director, Lassonde Mineral Engineering Program and
Lassonde Institute for Engrg. Geosciences, Univ. of Toronto
Toronto, Ontario, Canada

Douglas F. Hambley
Associate
Agapito Associates, Inc.
Golden, Colorado, USA

Kenneth N. Han
Professor Emeritus
South Dakota School of Mines and Technology
Rapid City, South Dakota, USA

Jack Haptonstall
Mining Consultant
Pincock Allen & Holt
Lakewood, Colorado, USA

John P. Harrison
Lassonde Institute, Department of Civil Engineering
University of Toronto
Toronto, Ontario, Canada

Paul Harvey
President
Ekati BHP Billiton
Yellowknife, Northwest Territories, Canada

Zaher Hashisho
Assistant Professor, Dept. of Civil & Environmental Engrg.
University of Alberta
Edmonton, Alberta, Canada

Richard Herrington
Researcher, Economic Geology
Natural History Museum
London, United Kingdom

Ken Hill
Managing Director
Xenith Consulting
Brisbane, Queensland, Australia

Walter E. Hill Jr.
(Deceased) Former Chief Chemist
AMAX Exploration, Inc.
Lakewood, Colorado, USA

R. Anthony Hodge
President, International Council on Mining & Metals
Professor, Mining & Sustainability, Queen's University
Kingston, Ontario, Canada

Lok Home
President
The Robbins Company
Solon, Ohio, USA

John Hooper
Managing Director
Joem Promotions
Deal, Kent, United Kingdom

Paul B. Hughes
Research Assistant, Norman B. Keevil Institute of Mining
University of British Columbia
Vancouver, British Columbia, Canada

James D. Humphrey
Market Professional–Mining
Caterpillar, Inc., Global Mining Division
Decatur, Illinois, USA

David Humphreys
Independent Consultant
Former Chief Economist at Rio Tinto & Norilsk Nickel
London, United Kingdom

Partha V. Iyer
Consultant
Reading, Pennsylvania, USA

Jeffrey A. Jaacks
President
Geochemical Applications International, Inc.
Centennial, Colorado, USA

Andrew Jarosz
Associate Professor, Curtin University
Western Australian School of Mines
Kalgoorlie, Western Australia, Australia

Daniel W. Kappes
President
Kappes, Cassiday and Associates
Reno, Nevada, USA

S. Komar Kawatra
Professor of Chemical Engineering
Michigan Technological University
Houghton, Michigan, USA

Deanna Kemp
Senior Research Fellow, University of Queensland
Sustainable Minerals Inst., Centre for Social Resp. in Mining
Brisbane, Queensland, Australia

Thomas Kerr
President
Knight Piésold and Company
Denver, Colorado, USA

Philip King
Technical Director
Truro, Cornwall, United Kingdom

Charles A. Kliche
Professor of Mining Engineering
South Dakota School of Mines and Technology
Rapid City, South Dakota, USA

Peter Knights
BMA Chair and Professor of Mining Engineering
University of Queensland
Brisbane, Queensland, Australia

Abby Korte
Project Hydrologist
Lidstone and Associates, Inc.
Fort Collins, Colorado, USA

Ronald Kuehl II
General Manager, Vibrating Equip. & Systems Business Line
Metso Mining & Construction Technology
Columbia, South Carolina, USA

Uday Kumar
Professor of Operation and Maintenance Engineering
Luleå University of Technology
Luleå, Sweden

Mahinda Kuruppu
Senior Lecturer
Curtin University of Technology
Kalgoorlie, Western Australia, Australia

John I. Kyle
Vice President
Lyntek, Inc.
Lakewood, Colorado, USA

Dennis H. Laubscher
Mining Consultant
Bushmans River Mouth, South Africa

David Laurence
Acting Director, School of Mining Engineering
University of New South Wales
Sydney, New South Wales, Australia

G. Aubrey Lee
Senior Consultant
SESCO Management Consultants
Bristol, Tennessee, USA

José L. Lee-Moreno
Adjunct Professor, Dept. of Mining and Geological Engineering
University of Arizona
Tucson, Arizona, USA

Jennifer B. Leinart
CostMine Division Manager
InfoMine USA
Spokane Valley, Washington, USA

Paul Lever
Prof. and CRCMining Chair, Mech. & Mining Engrg. School
University of Queensland
Brisbane, Queensland, Australia

Christopher D. Lidstone
President
Lidstone and Associates, Inc.
Fort Collins, Colorado, USA

Braden Lusk
Assistant Professor, Mining Engineering Department
University of Kentucky
Lexington, Kentucky, USA

Alistair MacDonald
Environmental Assessment Specialist
SENES Consultants Limited
Edmonton, Alberta, Canada

Travis J. Manning
Metallurgical Engineeer
Kappes, Cassiday & Associates
Reno, Nevada, USA

John Marks
Consultant
Lead, South Dakota, USA

Michael D. Martin
Senior Associate
Behre Dolbear & Company (USA), Inc.
Denver, Colorado, USA

Douglas K. Maxwell
Senior Process Engineer
Lyntek, Inc.
Lakewood, Colorado, USA

Peter L. McCarthy
Chairman and Principal Mining Consultant
AMC Consultants Pty Ltd.
Melbourne, Victoria, Australia

John McGagh
Head of Innovation
Rio Tinto Ltd.
Brisbane, Queensland, Australia

John Mosher
Executive Vice President–Operations
PT Freeport Indonesia
Tembagapura, Papua, Indonesia

Michael G. Nelson
Department Chair, Mining Engineering
College of Mines & Earth Sciences, University of Utah
Salt Lake City, Utah, USA

Jerry M. Nettleton
Environmental Manager
Peabody Energy
Steamboat Springs, Colorado, USA

Antonio Nieto
Associate Professor, Energy and Minerals Engineering Dept.
Pennsylvania State University
University Park, Pennsylvania, USA

Alan C. Noble
Principal Engineer and Owner
Ore Reserves Engineering
Lakewood, Colorado, USA

Ciaran O'Faircheallaigh
Professor, Politics and Public Policy
Griffith University
Brisbane, Queensland, Australia

Marc Orman
Senior Geotechnical Engineer
Ausenco Vector
Grass Valley, California, USA

Jeffrey T. Padgett
Consulting Geologist
Monterey Coal Company
Carlinville, Illinois, USA

Rimas T. Pakalnis
Associate Professor, Norman B. Keevil Institute of Mining
University of British Columbia
Vancouver, British Columbia, Canada

Rich Peevers
Senior Engineer
Ausenco Vector
Grass Valley, California, USA

Paul R. Peppers
Supt. Central Maintenance & Projects, Sierrita Operations
Freeport-McMoRan Copper & Gold Co.
Green Valley, Arizona, USA

Fiona Perrott-Humphrey
Consultant to NM Rothschild (mining team)
Dir. of AIM Mining Research & PURE P-H Strategic Consulting
London, United Kingdom

Edwin V. Post
(Retired) Former President of Skyline Labs, Inc.
Wheat Ridge, Colorado, USA

Yves Potvin
Director of the Australian Centre for Geomechanics
University of Western Australia
Perth, Western Australia, Australia

Gavin Power
Director
Power Geotechnical Pty Ltd.
Brisbane, Queensland, Australia

Marc Rademacher
Director, Western Operations, Minerals Services Div.
SGS North America, Inc.
Denver, Colorado, USA

Marcus Randolph
Group Executive & Chief Executive, Ferrous & Coal
BHP Billiton
Melbourne, Victoria, Australia

Michael Rawlinson
Director and Head of Mining and Metals
Liberum Capital
London, United Kingdom

Mark Richards
Mines Manager
Imerys Minerals Limited
Cornwall, United Kingdom

Jamal Rostami
Assistant Professor, Energy and Mineral Engineering
Pennsylvania State University
University Park, Pennsylvania, USA

Cameron Routley
Superintendent Five Year Planning, Olympic Dam
BHP Billiton
Adelaide, South Australia, Australia

Ian Runge
Founder
Runge Ltd.
Brisbane, Queensland, Australia

Kristin Sample
Staff Engineer
Ausenco Vector
Fort Collins, Colorado, USA

W. Joseph Schlitt
President
Hydrometal, Inc.
Knightsen, California, USA

Ross Seedsman
Director
Seedsman Geotechnics
Wollongong, New South Wales, Australia

Ian Sherrell
Business Development Engineer
Outotec
Jacksonville, Florida, USA

Ernest T. Shonts Jr.
Senior Mining Engineer
Colorado Springs, Colorado, USA

Rod Stace
Associate Professor, Department of Civil Engineering
University of Nottingham
Nottingham, United Kingdom

Doug Stead
Professor of Resource Geoscience and Geotechnics
Simon Fraser University
Burnaby, British Columbia, Canada

Scott A. Stebbins
President
Aventurine Mine Cost Engineering
Spokane, Washington, USA

George Stephan
Senior Consulting Engineer
Stantec Mining
Tempe, Arizona, USA

Nimal Subasinghe
Associate Professor of Minerals Engineering
Curtin University (Western Australian School of Mines)
Kalgoorlie, Western Australia, Australia

Pramod Thakur
Manager, Coal Seam Degasification
CONSOL Energy, Inc.
Morgantown, West Virginia, USA

Roger J. Thompson
Professor of Mining Engineering
Curtin University (Western Australian School of Mines)
Kalgoorlie, Western Australia, Australia

S.D. Thompson
Assistant Professor
University of Illinois at Urbana–Champaign
Champaign, Illinois, USA

Ian Thomson
Principal
On Common Ground Consultants, Inc.
Vancouver, British Columbia, Canada

Jerry C. Tien
Department of Mining and Nuclear Energy
Missouri University of Science and Technology
Rolla, Missouri, USA

Peter Tiley
Consulting Engineer
G.L. Tiley & Associates Ltd.
Flamborough, Ontario, Canada

Michael A. Tuck
Associate Professor of Mining Engineering
University of Ballarat
Ballarat, Victoria, Australia

Bryan Ulrich
Senior Vice President
Knight Piésold and Company
Elko, Nevada, USA

Ronald W. Utley
Consultant
FLSmidth, Inc.
Bethlehem, Pennsylvania, USA

Klaas Peter van der Wielen
Research Assistant
Camborne School of Mines (University of Exeter)
Penryn, Cornwall, United Kingdom

Rens Verburg
Principal Geochemist
Golder Associates, Inc.
Redmond, Washington, USA

Joshua D. Wagner
Marketing Product Consultant
Caterpillar, Inc., Global Mining Division
Peoria, Illinois, USA

Gary E. Walter
Principal Consultant
Primo Safety and Health Services, LLC
Wilmington, Delaware, USA

Mark Watson
Technical Services Group
Alliance Coal LLC
Lexington, Kentucky, USA

Andrew Wetherelt
Senior Lecturer in Mining Engrg., Prog. Dir. BEng Mining Engrg.
Camborne School of Mines (University of Exeter)
Penryn, Cornwall, United Kingdom

David Whittle
Global Manager, Planning Leadership Program
BHP Billiton
Melbourne, Victoria, Australia

John Woodhouse
Chief Executive
The Woodhouse Partnership Ltd.
Kingsclere, Berkshire, United Kingdom

Paul Worsey
Professor, Mining and Nuclear Engineering Department
Missouri University of Science and Technology
Rolla, Missouri, USA

Sergio Zamorano
Technical Director of Conveyor Technologies
FLSmidth
Spokane, Washington, USA

Technical Reviewers

David M. Abbott Jr.
Senior Associate and Principal
Behre Dolbear & Company (USA), Inc.
Denver, Colorado, USA

Mark Adams
Chief Operating Officer
Barminco
Hazelmere, Western Australia, Australia

Hugh E.K. Allen
Mining Consultant
Allen Associates
Harrow-on-the-Hill, London, United Kingdom

Timothy D. Arnold
General Manager
General Moly, Inc.
Eureka, Nevada, USA

Doug Austin
Senior Vice President
M3 Engineering and Technology Corporation
Tucson, Arizona, USA

Peter Balka
Chief Mining Engineer
Tigers Realm Minerals Pty Ltd.
Melbourne, Victoria, Australia

John C. Barber
Technical Director and Principal Mining Engineer
AMEC E & C
Mesa, Arizona, USA

Michael Barber
Self-Employed Blasting Consultant
Parker, Colorado, USA

William F. Bawden
Pierre Lassonde Chair in Mining Engineering
University of Toronto
Toronto, Ontario, Canada

John Baz-Dresch
Manager–Technical Services
Cia. Minera del Cubo S.A. de C.V.
Guanajuato, GTO, Mexico

Douglas L. Beahm
Principal Engineer
BRS Engineering
Riverton, Wyoming, USA

Carmen Bernedo
Lead/Supervising Engineer
MWH Americas, Inc.
Denver, Colorado, USA

Stephen L. Bessinger
Engineering Manager
BHP Billiton–San Juan Coal Company
Farmington, New Mexico, USA

Patricia Billig
Environmental Toxicologist
Boulder, Colorado, USA

Christopher Bise
Professor and Chairman, Mining Engineering Department
West Virginia University
Morgantown, West Virginia, USA

Steve Boydston
Senior Consultant
DNV Business Assurance
Centennial, Colorado, USA

Bill Bradford
(Retired) Mineral Processing Consultant
Formerly with Selection Trust Ltd. & UN Dev. Prog.
Tring, Hertfordshire, United Kingdom

Wade W. Bristol
General Manager–Nevada Underground
Newmont Mining Corporation
Carlin, Nevada, USA

Richard L. Bullock
Professor Emeritus, Mining & Nuclear Engineering
Missouri University of Science and Technology
Rolla, Missouri, USA

Al Campoli
Vice President Special Projects
Jennmar Corporation
Pittsburgh, Pennsylvania, USA

Mark Chalmers
Chair
Uranium Council of Australia
Adelaide, South Australia, Australia

Rebecca Chouinard
Regulatory Specialist
Wek'eezhii Land and Water Board
Yellowknife, Northwest Territories, Canada

Phillip C.F. Crowson
Honorary Professor & Professorial Research Fellow
Centre for Energy, Petroleum & Mineral Law & Policy
University of Dundee, Scotland

James Davidson
General Manager
Urtek LLC
Adelaide, South Australia, Australia

Gary Davison
Managing Director
MiningOne Pty Ltd.
Melbourne, Victoria, Australia

Fred Delabbio
Gen. Mgr. Innovation–Underground, Technology and Innovation
Rio Tinto
Brisbane, Queensland, Australia

Phil Dight
Winthrop Professor of Geotechnical Engineering
Australian Ctr. for Geomechanics, Univ. of Western Australia
Perth, Western Australia, Australia

Mal G. Dorricott
Principle Consultant
AMC Consultants Pty Ltd.
Melbourne, Victoria, Australia

Sjoerd Rein Duim
Director, Principal Consultant (Mining)
SRK Consulting
Perth, Western Australia, Australia

W. Scott Dunbar
Professor, Norman B. Keevil Institute of Mining Engineering
University of British Columbia
Vancouver, British Columbia, Canada

Roger Ellis
Minerals Industry Consultant
Senior Geological Associate with ACA Howe International
Hertfordshire, United Kingdom

Richmond Fenn
Director Resource Development
Freeport-McMoRan Copper & Gold, Inc.
Oro Valley, Arizona, USA

Courtney Fidler
Department of Geography and Planning
University of Saskatchewan
Saskatoon, Saskatchewan, Canada

Russell Frith
Adjunct Professor, School of Mining Engineering
University of New South Wales
Sydney, New South Wales, Australia

Steven Gardner
President and CEO
Engineering Consulting Services, Inc.
Lexington, Kentucky, USA

Barry Gass
General Manager Asset Management
Rio Tinto Copper Projects
Salt Lake City, Utah, USA

Mike Gleason
Energy Superintendent
Climax Molybdenum Company
Empire, Colorado, USA

John Grieves
Project Manager
New Hope Group
Brisbane, Queensland, Australia

Douglas F. Hambley
Associate
Agapito Associates, Inc.
Golden, Colorado, USA

Jack Haptonstall
Mining Consultant
Pincock Allen & Holt
Lakewood, Colorado, USA

Alex Hathorn
Vice President Technical Services and Continuous Improvement
Peabody Energy Australia
Brisbane, Queensland, Australia

Bruce Hebblewhite
Professor & Head of the School of Mining Engineering
University of New South Wales
Sydney, New South Wales, Australia

Michael G. Hester
Vice President
Independent Mining Consultants, Inc.
Tucson, Arizona, USA

Ronald R. Hewitt Cohen
Professor of Environmental Science and Engineering
Colorado School of Mines
Denver, Colorado, USA

Evert Hoek
(Retired)
Vancouver, British Columbia, Canada

Kevin Holley
Principal Geotechnical Engineer, Director
SRK Consulting
Brisbane, Queensland, Australia

Steven Holmes
General Manager–Ray Operations
Asarco LLC
Hayden, Arizona, USA

David Hull
Chair, Educational Advisory Committee to the South Africa
Council for Professional and Technical Surveyors
Johannesburg, South Africa

Louie Human
Senior Consultant (Rock Mechanics)
SRK Consulting Australasia
Perth, Western Australia, Australia

David Humphreys
Independent Consultant
Former Chief Economist at Rio Tinto & Norilsk Nickel
London, United Kingdom

Conrad Huss
Chairman of the Board
M3 Engineering and Technology Corporation
Tucson, Arizona, USA

Nils I. Johansen
Pott College of Science and Engineering
University of Southern Indiana
Evansville, Indiana, USA

Daniel W. Kappes
President
Kappes, Cassiday and Associates
Reno, Nevada, USA

Patrick Killeen
(Retired) Research Scientist
Geological Survey of Canada
Ompah, Ontario, Canada

Linton Kirk
Senior Principal and Chief Mining Engineer
Coffey Mining
Belo Horizonte, Brazil

Charles A. Kliche
Professor of Mining Engineering
South Dakota School of Mines and Technology
Rapid City, South Dakota, USA

David Krizek
Principal
Tetra Tech
Tucson, Arizona, USA

Stanley T. Krukowski
Industrial Minerals Geologist IV
Oklahoma Geological Survey
Norman, Oklahoma, USA

Katherine Laudon
Vice President
Lidstone and Associates, Inc.
Fort Collins, Colorado, USA

Christopher D. Lidstone
President
Lidstone and Associates, Inc.
Fort Collins, Colorado, USA

Joe Luxford
Principal
Luxford Mine Management Services
Brisbane, Queensland, Australia

Michael D. Martin
Senior Associate
Behre Dolbear & Company (USA), Inc.
Denver, Colorado, USA

Bill McAuley
Former Managing Director
British Oxygen Corporation (BOC) Group
Camberley, Surrey, United Kingdom

Michael K. McCarter
Professor, Department of Mining Engineering
University of Utah
Salt Lake City, Utah, USA

Peter L. McCarthy
Chairman and Principal Mining Consultant
AMC Consultants Pty Ltd.
Melbourne, Victoria, Australia

Stefan Muller
Principal Hydrogeologist
SRK Consulting
Perth, Western Australia, Australia

Jan M. Mutmansky
Professor Emeritus of Mining Engineering
Pennsylvania State University
University Park, Pennsylvania, USA

Michael G. Nelson
Department Chair, Mining Engineering
College of Mines & Earth Sciences, University of Utah
Salt Lake City, Utah, USA

Dave Osborne
Group Manager (Coal Technology–Business Development)
Xstrata Coal
Brisbane, Queensland, Australia

Paul R. Peppers
Supt. Central Maintenance & Projects, Sierrita Operations
Freeport-McMoRan Copper & Gold Co.
Green Valley, Arizona, USA

Clyde Peppin
Consulting Engineer
Stantec – Mining (Formerly Mcintosh Engineering)
Tempe, Arizona, USA

Fiona Perrott-Humphrey
Consultant to NM Rothschild (mining team)
Dir. of AIM Mining Research & PURE P-H Strategic Consulting
London, United Kingdom

Gavin Power
Director
Power Geotechnical Pty Ltd.
Brisbane, Queensland, Australia

Jerry Ran
Manager of Geotechnical Engineering
Barrick Gold Corporation
Toronto, Ontario, Canada

Donald E. Ranta
President and CEO
Rare Elements Resources Ltd.
Golden, Colorado, USA

Larry Reimann
Manager Technical Services
Cameco Resources
Casper, Wyoming, USA

Abani Samal
Geologist/Geostatistician
Pincock Allen & Holt
Denver, Colorado, USA

Lee W. Saperstein
Dean Emeritus, School of Mines and Metallurgy
Missouri University of Science and Technology
Rolla, Missouri, USA

Scott A. Stebbins
President
Aventurine Mine Cost Engineering
Spokane, Washington, USA

Doug Stiles
Assistant General Manager
Carlota Copper Company–QuadraFnx Mining
Globe, Arizona, USA

Barton Stone
Chief Geologist
Pincock Allen & Holt
Lakewood, Colorado, USA

Nimal Subasinghe
Associate Professor of Minerals Engineering
Curtin University (Western Australian School of Mines)
Kalgoorlie, Western Australia, Australia

Edward Thomas
Professor (private practice)
Sydney, New South Wales, Australia

Tom Vandergrift
Senior Associate
Agapito Associates, Inc.
Golden, Colorado, USA

Gary E. Walter
Principle Consultant
Primo Safety and Health Services, LLC
Wilmington, Delaware, USA

Bill Warfield
Product Manager (Ground Engineering Products)
Atlas Copco Construction Mining Technique
Roseville, California, USA

Alan Weakly
Consultant
Innovative Mining Solutions
Story, Wyoming, USA

Andrew Wetherelt
Senior Lecturer in Mining Engrg., Prog. Dir. BEng Mining Engrg.
Camborne School of Mines (University of Exeter)
Penryn, Cornwall, United Kingdom

David Whittle
Global Manager, Planning Leadership Program
BHP Billiton
Melbourne, Victoria, Australia

Anne L. Williamson
Principle Scientist/Senior Project Manager
URS Corporation
Phoenix, Arizona, USA

Johnny Zhan
Senior Environmental Manager (Hydrology)
Barrick Gold Corporation
Salt Lake City, Utah, USA

# PART 11

Hydraulic and Pipeline Mining

CHAPTER 11.1

Hydraulic Mining

Mark Richards

The use of water plus energy to mine unconsolidated material has a long history. In recent years, however, the most popular techniques have been based on either water-jet cutting or dredging. Water-jet cutting is used for the mining of unconsolidated materials, alluvial deposits, freshly blasted ores, and for the recovery (or remining) of dewatered tailings dams. Dredging is used for the recovery of unconsolidated materials in rivers, lakes, in other extremely wet environments, and for the recovery (remining) of fluid tailings dams, see Chapter 11.2.

HISTORY

There are many claims for the invention of the hydraulic method. With the basic principles so obvious, however, it is thought that there was concurrent development of similar mining operations around the world in the 18th and 19th centuries. Certainly its use was heavily championed by the gold miners in North America; by gold, diamond, and possibly rutile miners in South Africa; and by china clay miners in Cornwall, United Kingdom. (Numerous sources regarding the history of hydraulic mining are available through the Internet.)

In North America, the heavy industrialization and scaling up of this mining method led to millions of tons of waste materials flowing down major rivers. The uncontrolled release of this debris led to one of the western world's first environmental action laws as the authorities forced the miners to capture the waste on the mine site and reuse the water.

Uncontrolled hydraulic mining and the use of rivers for high-volume waste removal continued in many countries well into the early 20th century and indeed still occurs today in some South American and Far Eastern countries. Now, in a more environmentally conscious age, this method is largely restricted to operations that can successfully contain the waste and reuse the water at an acceptable cost, such as the current china clay industry (where rheological parameters apply) and some tailings dam recovery operations.

The initial development of the technology concentrated on *monitors*—high-volume and relatively low-pressure output water jets. These were very effective but quite inefficient in energy terms. The monitors of the time had nozzles that ranged up to 200 mm (8 in.) in diameter, with feed pipe work being much bigger. This led to a fairly static operation, which was fine for the denudation of hillsides and thick alluvial operations. The technical considerations of balancing the resultant thrust vectors and steering the monitor were solved by "brute-force" applications using big counterweights, thrust vanes, and so forth.

At the end of the 19th century, the "easy pickings" were beginning to disappear in both the gold and china clay operations. The need to wash blasted material and less easily sorted material resulted in the need for a higher-pressure, more robust, and more transportable monitor. The technology for this was freely available in the parallel development of fire-suppression systems. The early 20th century saw the introduction of higher-pressure and lower-volume systems, loosely based on fire-industry designs. From that moment on, design diverged to match the mining method.

HYDRAULIC MINING TODAY

Today, after the factual features (i.e., grade, size, structure, depth, and location) of a deposit are taken into account, the energy costs associated with hydraulic mining are probably the most important factor influencing the viability of this mining method. To minimize the water pumping requirement, the liberation of ore from country rock using water must be carried out at the highest fluid density possible.

Low fluid density liberation means that a great deal of energy is required for

- Pumping the water to the production face,
- Pumping the resulting ore streams away from the production face,
- Recycling the water and separating the ore, and
- Setting aside a great deal of space for large settling ponds or investing in high-cost dewatering systems.

Mark Richards, Mines Manager, Imerys Minerals Limited, Cornwall, UK

Figure 11.1-1 Hydraulic monitor

Water-Jet Cutting
Water-jet cutting is normally achieved by using a water canon or hydraulic monitor (Figure 11.1-1).

Water-jet cutting technology has been developed regionally to match the specific mining requirements of the ore bodies. Alluvial gold mining, for instance, and china clay mining use quite different designs. The basic principles remain consistent, however, and there is starting to be a convergence of design based on significant step changes introduced by the fire-suppression industry.

Design Parameters
Conditions that will maximize the efficiency of hydraulic mining include the following:

- A reliable source of water, high above the washing faces, that minimizes pumping energy
- Pumps that are energy efficient and in the correct location
- A ready supply of large-diameter pipe to feed water to the monitors
- Monitors that do not need to be moved very often to maintain their optimum position at the mine face
- A mine face drifting into a hill with a gravity runoff that will direct the ore stream toward the processing plant
- No need to recycle the water

In today's operations, however, the mine designer may have to cope with many suboptimal parameters, probably the most crucial of which will be the requirement to recycle the water. This reflects on the energy use most of all and has been the reason that many hydraulic mining efforts have been replaced with mechanical make-down systems, such as washer barrels and water-sprayed vibrating screens.

Whatever these parameters may be, the task is exactly the same: to maximize the efficient use of the energy available.

SYSTEM DESIGN
Water-jet mining projects require a pumped water supply. It is essential that the pump and pipe-work system be designed to make the most efficient use of the electricity supply. More energy efficiency gains are possible from pump and pipe-work optimization than from monitor optimization.

Pumps
In an ideal project, a single pump would be preferred; however, most pressure pumps have a fairly narrow efficiency curve. If there is a requirement to move monitors between benches (thus varying the head requirement of the pump) or if there is a need to run varying numbers of monitors, then a multiple-pump setup should be considered. A single pump can be used for varying loads if the pump has a very flat efficiency curve and if it can be driven by an inverter-controlled motor. Large inverters (above 200kW) are also very expensive, and in certain circumstances they can cause unwanted harmonic effects back to the grid. For this reason, some electricity utilities place limits on their use. A complete cost analysis should be carried out, therefore, to determine which system is more appropriate.

The ideal pump arrangement is based on one pump per monitor, where the head and volume requirements are matched. If the water is free of suspended solids and has a neutral pH, consideration should be given to hydrophobic pump coatings which will definitely improve pumping efficiency, sometimes by up to 5%. If suspended solids are present, they should be minimized by the use of a large settling pond, which can also be used as a water storage lagoon. The presence of even small amounts of suspended solids or adverse pH conditions will significantly alter the choices regarding pump construction materials. In these cases, it is recommended that duplex stainless steel be used.

Modern pressure pumps need a net positive suction head in order to run efficiently, and therefore the pumps should be gravity fed with a minimum of least 1 m (3.2 ft) of head. Consideration needs to be given to stone traps, isolation valves, servicing room, crane access, and most importantly pipe friction; it is always good practice to oversize the pump pipe work on the suction side.

On start-up, a lot of damage to the pumps can occur through cavitation, especially if they are required to pump into an empty and unpressurized pipe-work system. It is therefore advisable that the pressure-side isolation valve be designed so that it can be partially closed during the start-up sequence, allowing the pump to be "throttled" during the start-up phase. Many installations have a timer sequence built into this valve for automatic throttling during start-up and shutdown.

It is advisable to have some facility for pump health checks that does not necessitate pump dismantling. Many modern diagnostic systems need pressure tappings before and after the pump, as well as on the pump casing, and also a facility to measure the bearing temperatures.

On high-power installations, it is always advisable, even with flexible couplings, to use laser alignment techniques between the motor and pump. Misalignment will mean heat buildup and bearing damage in the long term.

Finally, pressure pumps can have very high starting currents. It is therefore prudent to consider this when designing the power supply and also when designing the control panels. It is wise to have these control panels linked in some way to prevent operators from turning pumps on simultaneously.

Pipelines
Multiple-pump systems should be fed via a well-designed manifold for efficient distribution to the pipe-work system. If the manifold has not been well designed and properly aligned, a great deal of energy can be needlessly wasted at this stage. Ideally there should be a swept bend off each pump into the manifold, and the manifold itself should be at least large enough and long enough to reduce the turbulence of the water before it is fed into the pipe-work system. Manifolds without

Figure 11.1-2 A typical multipump arrangement in the Cornish clay industry

swept bends can be used if they are considerably oversized. The manifold will need drain and inspection points.

Figure 11.1-2 shows a four-pump (duplex stainless steel) arrangement with 355-kW motors. Of particular note are the oversized suction and delivery manifolds, the suction isolation valves, and the motorized delivery valves for throttling the start-up. The stone traps are out of view to the right.

It is all too common to build a system based on a "Christmas tree" approach to pipeline design, where sections are added over time with little thought for either isolation or friction losses. Where possible, the entire pipeline system should be designed to have a pipe friction of less than 1%. Many free on-line calculators for pipe friction and other pumping parameters are available on the Internet.

Because collisions between mobile plant and the pipeline network are inevitable, having the ability to isolate a section without losing production is a necessity. The use of *ring mains* with multiple pathways and isolation points will allow the friction losses to be kept to a minimum. Multiple pathways enable smaller-diameter pipes to be used, thus saving money and making pipe handling, maintenance, and movement considerably easier. The location and inclusion of both drain points and air bleed-off points should be considered during the design process.

If there are a variety of pipe sizes in the system, it is obviously wise to put the biggest diameters where the highest flows occur. Any large pipe, however, wherever it is located in the main feed system (before a branch occurs), will always reduce the friction losses.

It is normal to use steel pipes with rubber-lined steel wraps for pipe jointing; this allows a slight give for ease of fitting, movement under varying pressure, and some flexibility in alignment. Some industries have experimented successfully with high-density polyethylene pipes, which can be constructed in convenient lengths for dragging around the mine (typically 100 m [328 ft]). These longer lengths of pipe suffer proportionately less from the friction associated with the pipe joints and short bends.

MONITOR DESIGN

Historically, the major centers of hydraulic mining have tended to develop their own style of monitors with little concern about what was happening in other industries or areas. This has resulted in specialized designs that have given rise to an "evolutionary fit" for the hydraulic mining industry.

The basic principles are universal, and the overriding principle of jet cutting is to maximize the energy per unit area exerted by the water jet impacting the face. Therefore, the water jet should hold its shape and coherence for as long as possible, which means that the main efficiency driver for a monitor is the nozzle. To perform efficiently, the nozzle must be provided with nonturbulent water.

To steer the water jet in two dimensions, it is necessary to have two axes of maneuver and therefore at least two sets of bearings in the monitor. To balance the forces involved, a loop arrangement is needed where the bearings cross over. Figure 11.1-3 shows a monitor in operation in the Cornish clay industry. The thrust is in excess of half a metric ton (5,000 N), and therefore the forces must be balanced. Other approaches to this balanced-forces issue are shown in Figure 11.1-4.

A monitor is built of the following components:

- A base to counteract any overturning moment
- A stone trap
- A loop for balancing the forces as much as possible
- A multiple bearing assembly for two-dimensional movement
- Hydraulic rams, hydraulic ring motors, or electrical actuators for steering the jet
- A barrel that contains a streamformer, a stilling zone, and a nozzle

The nozzle needs to have a quick-change facility (normally using steel pipe wraps) for varying the volume and pressure in order to keep the pressure pump(s) on the peak efficiency point.

All of these elements must work together to minimize any turbulence before the water hits the nozzle. To facilitate this,

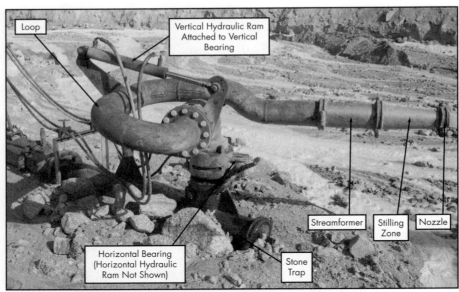

Figure 11.1-3 Close-up, showing bearing arrangement and hydraulics

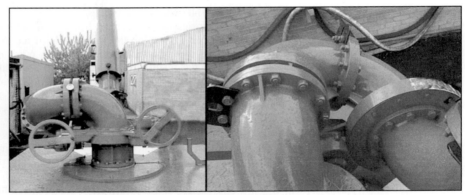

Note: The long barrel is for foam generation, which would not be used for hydraulic mining.
Figure 11.1-4 Latest designs from the fire-suppression industry

there must be minimum friction all the way through the unit, which calls for large-diameter pipes, swept bends, constant diameter, and a low-friction streamformer. There is a trade-off among all these elements as, for instance, large-diameter pipes will require much tighter bends, while an efficient streamformer will cause friction. Current design cues coming from the fire-suppression industry, however, indicate that large diameters are more important than tight bends. As shown in Figure 11.1-4, the latest thinking suggests that the bends be approximately two pipe diameters with the pipe indexed downward off the top flange to keep the turning moment as low as possible.

The streamformer is to some extent dependent on the operating conditions. Since even the most efficient streamformers can be blocked or damaged by foreign bodies in the water, a degree of pragmatism has to be used in this case. The Cornish china clay industry, for instance, uses a streamformer that has no center at all (Figure 11.1-5); the fins are mounted on the circumference of the barrel and point inward. The fire industry uses a much more delicate honeycomb design because their water tends to be much cleaner.

The nozzle construction material is dependent on the water quality and ranges from aluminium-backed polyurethane to brass and gunmetal (Figure 11.1-6). The industry is converging, however, on a fire-hose-type design of a simple cone with some variations in angle. Many other designs have been tried, but ultimately the original fire-hose type has tended to be the most efficient.

Of major importance is the surface roughness of the nozzle. It is essential that the nozzle be free of nicks and gouges. As a general rule, no roughness should be felt when a fingernail is passed over the surface. The water emerging from a good nozzle setup should appear as a straight bar for at least 50 nozzle orifice diameters and should degrade slowly but steadily on its path to the mining face. Any sign of instability, expansion, or pulsing in this critical area should be seen as a sign of excessive turbulence within the monitor.

Actuation

In most monitor designs, the actuation, or steering, is accomplished by the use of hydraulic rams, although some designs use hydraulic ring motors or electric actuators. In smaller

Figure 11.1-5 Self-cleaning streamformer

Figure 11.1-6 Polyurethane nozzle with aluminum backer

operations, or where the risk level is low, monitor steering may be a manual task by means of a yoke on the back of the monitor. In cases like this, it is normal to have a small tray on each axis where weights can be added to fine-tune the balance.

Respectively, motive power has to be supplied in the form of a hydraulic power pack or a generator. In all of these cases, it is common for a Pelton wheel to be used to supply the raw energy. The Pelton wheel is essentially a high-pressure water wheel that takes a small amount of water from the pressure line prior to the monitor to provide the motive force (Figure 11.1-7). Operation of the actuators is normally via pilot hydraulics from levers operated from the control hut (Figure 11.1-8).

In some instances, a secondary Pelton wheel has been used to supply lighting, although this is an inefficient use of pumping energy. In most cases, diesel-powered mobile lighting units or fixed lighting are preferred.

Control Hut Design

A demonstrable and significant increase in productivity, and therefore efficiency, can be achieved if the worker is able to maintain concentration on the production task, with eyes focused on the work face. Holding this attention span is difficult, however, if the worker is uncomfortable or having to work in a badly designed environment. Ergonomics, therefore, is the main driver for control hut design.

Obviously the prevailing climatic conditions will have a bearing on the control hut design; however the main design elements can be split into two categories: operator comfort and safety.

Operator Comfort

It is essential to treat the operator's seating position as if it were a workstation (similar to that for someone who uses a computer routinely) and to allow for the posture associated with operation of the levers. In other words, the controls should be designed to prevent repetitive strain injuries, and their positioning should prevent musculoskeletal stress. Other important considerations for operator comfort include the following:

- Soundproofing
- A draft-free operating environment
- A door design that can be kept open, shut, or anywhere in between, without being caught by the wind

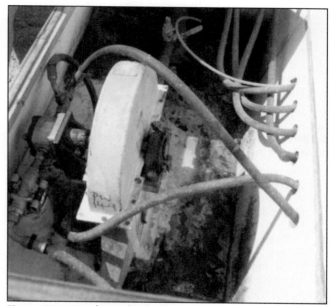

Figure 11.1-7 Pelton wheel

Figure 11.1-8 Monitor control hut

- An ability to control glare if working at night or when the sun is low on the horizon
- A means of preventing windscreen dirt or, failing that, an efficient windscreen cleaning system
- The need for the mobile plant drivers that move the huts (normally small wheeled loaders) to be able to see through the hut while carrying it

Safety Considerations

Stones can be flung a long way by a working monitor, and it is therefore necessary to have a basic understanding of this process:

- Stones get ejected when the monitor jet hits a reflector, such as a flat stone or a hole, where the majority of the jet is reflected back toward the monitor.
- The vertical angle of the monitor jet is crucial; when pointing down, it is much more likely to reflect.
- Stones falling through the reflected jet pick up acceleration only while they fall through the jet.
- The accelerated stones are self-limiting on the distance they can be thrown. The small ones are accelerated quite rapidly but slow down rapidly because of air resistance; the larger ones are difficult to accelerate and thus do not travel as far. Overall, the small stones travel the farthest but do the least damage.

It is wise, therefore, to have a parking rule and a distance rule for avoiding ejecta where possible and to have regard for the safety of others where it is not possible. Recommended precautions include the following:

- The monitor should be "parked" when the wheeled loader is operating in front of the monitor position.
- A minimum distance should be established (the Cornish clay industry recommends a minimum distance of 80 m [262 ft]) requiring that
  - People not involved in the act of jet washing be kept out;
  - All individuals use a safety helmet and eye protection within the zone;
  - Vehicles without safety guard glass (i.e., service vehicles) be advised to park outside the zone until they have the attention of the monitor operator, who will then carefully park the monitor to allow safe entry.
- Because it is obviously difficult to physically put a cordon around the monitor in an ever-changing mining environment, education and warning signs are vital.

If the host rock consists largely of stones, then some armoring of the glass in the hut may well be necessary to prevent breakage. This can be achieved through metal mesh guards or with a ballistic film (a polymer sheet coating similar to laminated glass). In the case of a hand-operated monitor, the energies involved are so low that ejected stones should not be a problem.

OPERATION

The following series of steps provide an overview of the water-jet mining process:

1. The monitor jet washes the material to separate the ore and fine contaminants from the host rock.
2. The resulting ore stream should be tightly channelled to ensure that there is no dropout en route to the processing area or to the gravel pumps which then transport the stream to the processing area.
3. A wheeled loader is normally employed to remove the washed host rock from the mining area.
4. Subsequent removal of the waste host rock can be by haul trucks being loading directly, or the waste can be stockpiled for later removal. Economies of scale may favor stockpiling.

Monitor and Control Hut Placement

The optimum distance from the monitor to the washing face is commonly considered to be 100 nozzle orifice diameters. This will maximize the energy per unit area of the water jet impacting the face. However, from an operational (health and safety) point of view, this is very close and, depending on the scale of the operation, may well compromise the reach in terms of the overall washing area. What is not in doubt is that the monitor should be as close as is safe and practical to the face, and that the control hut should be situated out of harm's way. This distance obviously has to be tempered by the fact that the operator needs a clear view of the jetting and washing process.

It is very easy to bury a monitor with debris runoff from the washing face. Although the operator should be able to prevent this by periodically cleaning the area in front with the monitor jet, it is sensible to place both the monitor and the control hut on small platforms made from local debris. This removes the equipment from the ore stream and makes it much more visible to any mobile plant working in the area.

Rewashing of consolidated tailings and other materials may result in a collapse of the face when liquefied. The operation must be risk assessed according to local conditions. Nevertheless, certain points are common to these operations:

- The control position must be substantially distant from the monitor.
- The control position should be placed on a high bank or mound of material that is not likely to be washed away.
- Washing face heights should be much lower than normal.
- Frequent checks of the washing face should be carried out by a suitably qualified person.

Monitor Operation

With due regard for safety, the operator should use the water jet in a horizontal action at the base of the ore pile in order to undercut the pile. Gravity will then bring unwashed material into the washing area. There is little need to raise the water jet into the higher reaches of the pile unless there is a need to dislodge a hanging rock. Washing from the higher reaches will mean that washed debris comes down to cover unwashed ore near the bottom, making the process inefficient. In addition, this practice tends to reduce the overall slope angle of the ore pile, which means that the optimum monitor position is quickly compromised as the ore pile is washed away.

To dislodge a hanging rock, a technique known as *doubling the stream* can be employed. The operator will direct the water jet to the top of the face above the rock for long enough that the face is covered with a falling cascade of water. The operator will then lower the water jet to the bottom of the hanging rock and take advantage of the weight of the cascade.

With the monitor centrally placed in front of the ore pile, the operator can wash one side until the density begins to drop off and there is a buildup of cleaned host rock or debris. At this point, the operator can move to the other side of the target

Figure 11.1-9 Wheeled loader removing washed host rock

area, thus allowing the wheeled loader access to load out the waste (Figure 11.1-9). When the wheel loader is forward of the monitor position, the jet should be parked so that flying stones are minimized.

It is at this point that good communication between the monitor operator and the wheeled loader driver is essential. At the beginning of their period together, they should agree on a set of signals or actions so that each can clearly understand the other's actions and intentions. It is normal, therefore, for each side of the ore pile to be worked in turn, allowing the wheel loader to be sufficiently available to remove the debris on demand and thereby keeping the density of the ore stream as high as possible.

Mobile Plant Operation

In most operations, the wheeled loader is sized to allow one machine to be fully occupied by one monitor. Additionally it is normal practice to stockpile the debris so that a fleet of trucks dedicated to waste removal can take advantage of the economies of scale.

The efficient interaction of the wheeled loader driver and the monitor operator is paramount to maximizing productivity and minimizing accidents. That the two must work as a team at all times cannot be stressed enough.

Ore Stream Pumping Versus Trucking

The method of mining used depends to a great extent on the debris ratio and the method of its removal. Across all industries the debris percentage can range from 90% down to 20%. Thus, it is wise to consider the cost of waste removal and the cost of ore stream removal as a whole.

If the debris ratio is very high, it may be appropriate to dispense with ore stream removal costs and carry the raw ore to a central washing point with a gravity run into the processing system. Conversely, if the debris ratio is very low, then electrically pumped ore stream removal is the most efficient. In addition, some ores degrade quite rapidly with too much energy input or shear. Thus, a high degree of pumping may do damage, and trucking may again be more appropriate.

In recent years, there has been a trend toward centralized hydraulic mining, and in some cases its abandonment, in favor of wet mechanical separation. In every new case, however, a comprehensive total costing of the ore stream *and* the waste stream will indicate the preferred process.

ACKNOWLEDGMENTS

The author thanks Angus Fire, Imerys Minerals Ltd., J.B. Thomas, Professor Dean Millar, and W. John Tonkin for their contributions to this chapter.

CHAPTER 11.2

Placer Mining and Dredging

Richard L. Bullock, Michael J. Cruickshank, and Ian Sherrell

PLACER MINING METHODS OTHER THAN BY DREDGING

The Spanish word *placer* means sand or gravel bank or shoal where gold can be obtained by washing. As it is defined for this chapter, a placer is any area of unconsolidated rock fragments or particles that may also contain minerals potentially of economic value. Placer mining involves the extraction of these unconsolidated materials without the use of explosives or any other significant means of rock-breaking force. This chapter contains a classification of placer deposits and describes the mining and processing of these deposits, which requires the application of mining and metallurgical engineering.

Many placers of economic value can be and have been mined by hand; indeed, the famous gold rushes of the American West and Klondike were in this category. Although it is fascinating history, such mining does not require the engineering methods described in this handbook. For individual prospecting and hand mining, as well as processing, see Wells (1973) and Peele (1945).

Classification of Placers

The many types of placer deposits described in this chapter all have four things in common:

1. A source of the original mineral existed.
2. Some mechanism released the mineral from its original location.
3. The valuable mineral must be resilient enough to survive transport and remain commercially recoverable.
4. A reason existed for the placer being located and concentrated where it was found.

Breeding (1985) describes the steps involved in the formation of a commercial placer as follows:

> The formation of a commercial placer requires a source of valuable minerals. Above primary deposits, eluvial deposits may exist that were formed by the erosion of gangue minerals and the concentration "in-situ" of valuable minerals. Down slope from these deposits are the hillside or colluvial deposits, and below them are the alluvial deposits of redeposited material. Most of the great placer fields of the world are the result of several generations of erosions and deposition. Well-known examples are in California and Colombia.

Gold is a very resistant and malleable matter, and gold placers may extend for 40 to 50 mi (64 to 80 km) along a river system. Platinum is less malleable, but is very resistant to disintegration. Diamonds are extremely hard, and (especially gem diamonds) may be found over great lengths of a river system. Cassiterite (i.e., tin crystals) are less resistant to disintegration, and such placers seldom extend more than 2 mi (3.3 km) without resupply from an additional source or sources of mineralization. Tungsten minerals are generally more friable, and within a few hundred yards (meters) of the source disintegrate to the point that they are uneconomical to recover. Rutile, ilmenite and zircon placers generally result from the weathering of massive deposits and may be encountered over extensive areas; most are fine grained and durable.

Source of Mineral

Sometimes the source material is a mineralized vein (lode) containing high-grade mineralization that is attacked by the release mechanism. Sometimes the source material is a more massive mineralized rock containing disseminated bits of low-grade mineralization. In both cases, the matrix rock and the mineral are broken down into smaller particles that may or may not then be transported by the release mechanism. Portions of the original vein or host rock may remain, or it may all have been removed by the release mechanism.

Richard L. Bullock, Professor Emeritus, Mining & Nuclear Engineering, Missouri University of Science & Technology, Rolla, Missouri, USA
Michael J. Cruickshank, Consulting Marine Mining Engineer, Marine Minerals Technology Center Associates, Honolulu, Hawaii, USA
Ian Sherrell, Business Development Engineer, Outotec, Jacksonville, Florida, USA

Release Mechanism

Physical release. The more common types of placers are those whose release mechanism is by flowing water. It may be normal stream movement, very rapid seasonal stream movement, flooding rivers or streams, or intermittent streams. When the water is moving fast enough to pick up the small particles, it will carry them to the point where a loss in velocity causes the stream to drop its load. Another release mechanism is the movement of ice as a glacier. This can be followed by streams flowing from the glacier, which then transport the material to a point where it will drop its load. Finally, wind erosion can also wear the rock particles down and separate them from the mineral particles.

Hydrothermal alteration and weathering. Both vein and massive types of mineral deposits can be broken down into smaller rock particles by hydrothermal alteration or by weathering. In these cases, the mineral usually stays in place and may have little chance of physical displacement other than by rolling down an adjacent slope, so there may be little chance for concentrating the mineral.

Deposit Type Formation

Several placer-building mechanisms can take place over long periods of geologic time and form the various types of deposits discussed in this chapter. The mechanisms are

- In-situ residual building,
- Eluvial and gravitational creep movement,
- Stream transport, and
- Steam transport and the wave action of a sea or ocean.

Residual Placers

Residual placers form when the remaining eluvials, consisting of mineralized rock, decompose in situ principally by chemical weathering or in some cases by hydrothermal alteration, often with little or no change in volume (Daily 1973).

Eluvial Placers

Eluvial placers, also termed *colluvial* or *hillside placers*, are transitional between disintegrated rock and stream gravel. They are formed on slopes where smaller particles are removed by rilling, wind, and dissolution (Daily 1973).

Various Stream Placers

Stream transport is a mechanism for forming placers.

Gulch placer. Creek placers, river placers, and gravel plain deposits (Wells 1973) are forms of the gulch placer. The differences are the volume, water force, and areas of deposit attributable to the water's action. McCulloch et al. (2003) classifies these deposits as colluvial, debris flow, pulse placer, and alluvial fans.

Debris flow, pulse placers. These deposits form when the original lode deposit has been sufficiently weathered so as to be weakened. Then, because of above-normal precipitation, the strong rush or pulse of water carries the debris from its original location down and across the valley. Coarse and fine material will flow with the pulse; fine material could be transported farther by normal precipitation.

Alluvial fan placers. When the normal stream flow carries its load down the narrow mountain valley to meet a wider valley or plain, the water fans out, whereupon it drops its load. It fills the current channel to the point that the stream is forced to find a new channel. In this manner a fan deposit is gradually created.

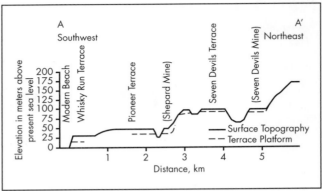

Source: Drew and Lessard 2009.

Figure 11.2-1 Schematic section depicting the relationship between terraces

Eolian placers. Eolian placers are concentrations resulting from the removal of lighter material by wind action (Daily 1973).

Bench Placers

When streams cut downward through a previously deposited alluvial deposit, a bench placer is developed. The economic bench placers are characteristically larger in volume than stream placers. These types of placers were typically mined in 1850 in California (United States) by hydraulicking.

Beach Placers

Where mineral-bearing streams reach the sea or ocean before they drop the transported mineral load, the wave action of the sea tends to concentrate the heavier minerals and forms bands or layers of the concentrated mineral or gem. These types of deposits are extremely important around the world for concentrating heavy minerals such as ilmenite, chromite, rutile, zircon, garnet, diamonds, and so on. It is important to understand that frequently such "beaches" are no longer by the seaside and are at elevations much higher than sea level. They may now be miles from the sea—for example, the Oregon chromite sands (Figure 11.2-1) or the deeply buried deposits of ilmenite/rutile sand east of Nashville, Tennessee (United States).

Desert Placers

Desert placers are found in arid regions where erosions and transportation of debris depend largely on fast-rising streams that rush down gullies and dry washes following summer cloudbursts that fall at higher elevations. These are typical in the southwestern United States and in other desert regions of the world. In these areas, the deposits can be quite large in extent, but erratic in location and mineral grade.

Glacial Placers

Glacial placers are mineral deposits that have been associated with a moving glacier that scraped off loose rock debris, soil, and sometimes minerals. The mineral is later concentrated either in glacial moraines or the streams forming from these glaciers. The gold deposits that have been worked along the foothills of the Sierra Nevada (United States) and the placer deposits of Fairplay and Breckenridge, Colorado (United States) are examples of glacier-related placer deposits (Wells 1973).

Marine Placers

Any of the placers just described may be found submerged offshore on the continental shelf, which extends at its outer edge to a water depth of about 150 m (500 ft) and as much as several hundred kilometers (nautical miles) from the current shoreline. All coastlines have been subjected to frequent changes in sea level over geological time and during periods when existing placers were subjected to high-energy surf or ocean currents. As the shoreline advanced or retreated, they may have been dispersed, reconcentrated, or buried. Occurrences have been reported of placer gold in Alaska (United States) carried off the shelf by turbidity currents in submarine canyons to depths of approximately 4,000 m (13,000 ft).

Placer Exploration and Sampling Techniques

Before an exploration program is undertaken, it is important to understand what type of placer deposit is being prospected for. The type of deposit will have to be addressed to judge how the mineral concentration took place, how far the mineral may have been transported, where the larger particles of the mineral are likely to be, and where the finer particles will probably be. The normal range of thickness of the deposit will also have to be judged.

In performing exploration and sampling of a potential placer deposit, it is necessary to be aware of the mining and recovery method likely to be used to extract and recover the valuable mineral. To perform exploration and sampling effectively, it is necessary to understand the limits of the mining and recovery methods. This same philosophy also applies to assaying: If the mineral cannot be recovered by the concentrating methods being applied in the mining and processing, then fire assaying of the total mineral content of the material sampled does not make much sense, since recovery may not be possible.

As to the sampling technique, sampling tools that are normally used for hard-rock minerals or coal are simply not adaptable to most placer deposits. What is needed is a sampling tool that extracts a sample large enough to be processed to determine what is recoverable and to assay that which is recoverable.

The many methods of extracting samples from a placer deposit vary from being highly prone to error to fairly accurate. They are

- Hand grab samples gathered from existing terrain,
- Hand-dug pits or trenches,
- Backhoe trenches,
- Bulldozer trenches,
- Churn drill holes,
- Auger drill holes,
- Reverse-circulation drilling,
- Vibratory drilling,
- Other types of drills that case and capture the entire sample within the hole, and
- Large-area bulk sampling.

One method of capturing a large sample for bulk sampling and still cover a wide area of a deposit is with a drill. Between 1979 and 1984, a subsidiary of St. Joe Minerals Corporation (Missouri, United States) combined a measuring hopper, a portable jig plant, and the drilling of large bulk samples for sample testing. The jig concentrate was panned and amalgamated to recover the gold. This system was used to prospect 28 holes in the tertiary gravels of the San Juan gold property of the North Columbia area of California. The drill first sank a 915-mm (36-in.) casing and used a cactus-grab bucket to muck out the sample. Where the gravels were cemented, they had to be broken with a chopping bit. The advantage of the total system was that it minimized the possibility of unintentional salting, and it only sampled and assayed that which could be recovered by the pilot processing.

If the gravels are re-cemented, it is also possible that a diamond drill can be used. One such case was reported (Stone et al. 1988) in which a 100×140 mm (4×5.5 in.) diamond drill bit was used very successfully on the cemented gravels described previously.

The problem with the majority of the drilling methods is that a typical alluvial placer deposit contains boulders that are too large to be broken down on the surface by a drill bit or auger flights. This is why for most placer deposits, a bulk sample extracted with mining tools that can handle the boulders is more accurate. However, for a desert placer deposit that is likely to have only a few boulders, an auger drill is the best option. Another problem with drilling holes for the sample is that unless the hole is cased, or unless the deposit is a cemented conglomerate, caving of the hole will occur. If the sampling horizon passed a high-grade lens, the samples below this lens are likely to be unintentionally salted. The resulting sample values will be elevated and thus inaccurate.

Probably the biggest problem is trying to sample the highly erratic nature of the precious-metal placer deposits. It will take many holes over a large area to statistically even out the nugget effect, which often occurs. In fact, with enough drilling and sampling, the outliers should be discarded from the sample collection.

Bulk Sampling Advantages and Disadvantages

The advantages of bulk sampling include a good view of the gravel in place and knowledge of the amount of force required to excavate it. Some of the same digging machines used in bulk sampling are used in production. The large samples obtained minimize the nugget effect and errors. On the other hand, a serious disadvantage is the fact that pits or trenches may not reach bedrock in all cases. Other disadvantages include the following. Curves often make it difficult to accurately measure a pit if the excavation is by a backhoe, because of sloughing from the sides. If groundwater is encountered, sample integrity is lost. Larger samples require larger transport and processing equipment, and the situation becomes more complicated. As unit costs of each sample are high, the exploration cost to delineate reserves is very expensive (Cope and Rice 1992).

Drill Sampling Advantages and Disadvantages

Historically, drilling has been the sample method of choice and has been proven by operation in most cases. Drilling gives greater coverage than bulk sampling in developing gravel grade and reserves at less unit cost, the profile of bedrock, and indications of enriched and barren areas. A disadvantage of drilling is that it gives a small sample, which aggravates the nugget effect. Other disadvantages are that at times boulders

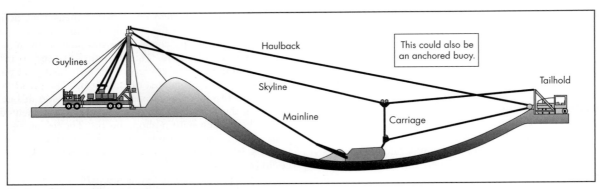
Courtesy of Ramsey Company, Inc.
Figure 11.2-2 Sand or gravel being mined by Ramsey scraper bucket

will necessitate abandonment of uncompleted drill holes, vibration from driving casing or from a reverse-circulation hammer drill may cause gold to migrate downward into the gravel, and drill rigs require at least minimal access roads and drill pads (Cope and Rice 1992).

Other Mining Methods

Although the most common method of large-scale mining of placers is by dredging, several other methods of placer mining have been used over the years, and they are discussed in this section.

Drift Mining

Frequently, the precious metals of a placer deposit are located in relatively thin lenses on or very close to bedrock or some other impervious material, and they may be covered by many feet of worthless gravels and debris. The economic value does not warrant the removal of all of the overburden; in such cases, if the mineral is above the valley floor, adits may be driven into the hillside and the high-grade placer mineral extracted by drifting wherever the deposit exists. Usually, these adits or drifts must be timbered to maintain stability. Occasionally, the rock has re-cemented itself into a conglomerate and drifting could take place without timbering. Typical deposits that have been mined extensively for both gold and platinum are remnants of ancient channels now above groundwater in California, British Columbia (Canada), and Alaska; frozen stream-valley and bench-type placers in the Yukon (Canada) and Alaska; and deep lead deposits in gutters in the bedrock below the groundwater table in southwestern Australia and British Columbia (Daily 1973).

Conventional Equipment for Dry Unconsolidated Placers

The term *conventional equipment* refers to the same type of equipment that is used in various types of standard earth moving in surface or open-pit mines. It includes

- Front-end loaders and trucks;
- Tread-mounted power shovels or hydraulic excavators and trucks;
- Bucket-wheel excavators and trucks;
- Draglines combined with some other materials-moving system;
- Diesel-powered scrapers, which are excellent for large-area and desert placer deposits; and
- Bulldozers and a truck-loading chinaman chute (a colloquial term for a loading ramp that carries material to a truck parked below it).

Also included are various types of wire rope scrapers (tuggers), which move loose material by bucket to a truck- or hopper-loading system. These systems having curved-type buckets are used for mining diamond placers off the coast of West Africa. Another example of a scraper that can be used in deep water is the Ramsey bucket, shown in Figure 11.2-2.

It is unnecessary to describe the many combinations of mining sequence for each of these pieces of equipment, since their use is current practice in hundreds of sand-and-gravel, quarry, and open-pit operations around the world. However, how to mine the placer and reclaim it at the same time is worth at least one example.

At the Oregon Resources Corporation Coos Bay mine, the material to be mined is a series of deposits of dry black sands containing commercial chromite, zircon, and garnets located on ancient beach terraces. In some ways, the mining will be more like a sand-and-gravel operation than an open-pit mine. Two mining systems were proposed (Figures 11.2-3 and 11.2-4), both ending in reclaiming the strip as shown in Figure 11.2-5.

After the small amount of waste overburden (averaging only 6 m [20 ft]) is removed, the mining method is simple front-end-loader/dozer, truck-operation strip mining. The process is similar to coal strip mining in the Midwestern United States, but on a much smaller scale. That is, they immediately mine and reclaim a series of parallel pits across the deposit. After the stripping has taken place for that pit for a length of 91 to 122 m (300 to 400 ft), the mining can commence. Most of the pits are not level, and they will be worked from the lower elevation upgrade.

The mining is a simple process of bulldozing the black sands down a slope to the pit bottom, where a front-end loader loads the material to a portable scalping-screening plant hopper, which removes the deleterious materials and transfers the heavy mineral sands to a stockpile. From the stockpile, the material is loaded into bottom-dump trucks and hauled to the plant in Coos Bay, approximately 33 km (20 mi) away. When developed, the mine is expected to produce an annual average of 977,810 t/yr (metric tons per year) (1,077,840 stpy [short tons per year]) for the first 8 years and an average of 850,000 t/yr (936,935 stpy) for at least the next 12 years.

The trucks returning from the plant carry cleaned and dewatered sand from which the ore has been extracted. This

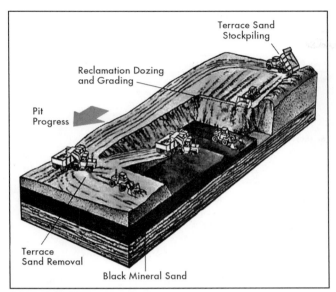

Source: Smith 2008.

Figure 11.2-3 Beach placer being mined and reclaimed within the same strip of mining

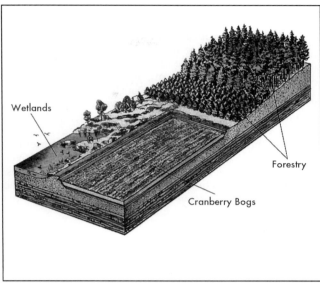

Source: Smith 2008.

Figure 11.2-5 Results of the mining and reclamation process shown in Figures 11.2-3 and 11.2-4

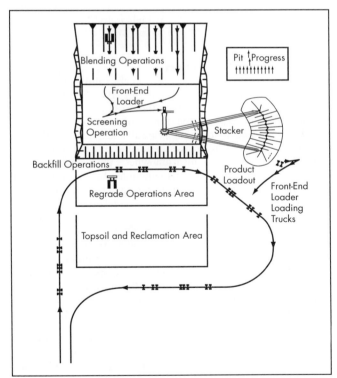

Source: Smith 2008.

Figure 11.2-4 Mining with dozers and front-end wheel loaders with reclamation being accomplished concurrently

is dumped at the edge of the pit, and a dozer pushes the material into the pit and fills it. The backfill pace keeps up with the mining extraction, so that no more than about 30 m (100 ft) of pit bottom is ever exposed. The disturbed swell of the material being brought back will fill the pit within about 81% of the original elevation.

The dozed backfill is then graded to the approximate contour of the original ground or is restored to a contour requested by the landowner to be reclaimed or to satisfy environmental regulations.

Frozen Placers

Mining the frozen placers of the earth's Arctic regions poses special problems. The following summarizes the problems connected with the mining of frozen placers (McLean et al. 1992):

> Practically all the auriferous gravel deposit north of the Alaska Range in Yukon and Alaska were frozen and had to be thawed for mining by all methods other than hydraulicking (Daily 1973). Gravel deposits to about 15 ft (5 m) in depth often could be thawed by solar action after removal of all vegetation and muck. The early drift miners used wood fires to sink shafts, and steam was delivered to pipe points driven into the face of drifts. During earlier years, ground was thawed for dredging by the use of steam introduced through pipes driven into the ground, which was inefficient and expensive. The fact that frozen gravel could be thawed best with cold water was discovered in 1918–1920, and that method was developed to a high level of efficiency as a result of research of the United States Smelting, Refining and Mining Co. The method consist of delivering water at ambient temperature, under low natural or pump pressure, through special pipe points either driven into the ground to bedrock or set in predrilled holes. Water temperature was seldom over 50°F (10°C). Flow was continued until the ground was thawed, which required six weeks [for] two or more seasons, depending on the depth of ground and character of deposit.
>
> The total volume of ground thawed with cold water at the three major properties in Alaska and

Yukon from 1924 until termination in 1964 was more than 360 million yd^3 (330 million m^3). Average water duty was approximately 11 yd^3 (10 m^3) per MID (miner's-inch-day) of water (1 MID = 1.5 cfm per 24 hours). Average rate of thaw in stream gravel to depths of about 45 ft (14 m), using ¾-in.- (19-mm-) diameter pipe points driven in an equilateral triangle spacing of 16 ft (4.9 m), was 1½ days of water under pressure.

In the mid-1970s, the University of Alaska performed a research project demonstrating that drifts could be driven through frozen ground using drilling and blasting techniques. Likewise, frozen fractured rock is often drilled and blasted in mine sites like Polaris on Little Cornwallis Island near the North Pole.

Hydraulic Mining of Placers

The term *hydraulicking* as applied to placer mining describes the excavation of gravel banks by streams of water under pressure from nozzles. Thrush (1968) defines it as follows:

> The process by which a bank of gold-bearing earth and rock is excavated by a jet of water, discharged through the converging nozzle of a pipe under a great pressure, the earth of debris being carried away by the same water, through sluices, and discharged on lower levels into the natural streams and watercourses below.

This practice was at one time (the late 1870s through 1984) a very efficient manner by which to mine the auriferous (gold-bearing) gravels in northern California, Alaska, and the Yukon. These early mining techniques brought about the first antimining environmental laws in the United States (the 1884 Sawyer decision), which prohibited any further discharge of waste material into the Sacramento and San Joaquin rivers. The mining method that was applied—well documented in Peele (1945)—has few applications in "enlightened" countries with environmentally restrictive regulations. However, hydraulic mining is also applied to nonplacer mining, namely coal and tailing reprocessing, neither of which is covered in this chapter.

Subgrade Hydraulic Mining

Places in Thailand and Malaysia for many years were mined by subgrade hydraulic mining (Daily 1973). The following conditions favor this type of mining:

- Many deposits of limited areal extent exist that are unsuitable for mining with dredges.
- An abundance of clay is pulped satisfactorily by water action.
- Extremely irregular limestone bedrock has deep channels in many places, which become exposed in the bottom of the pit.
- High-grade concentrations of mineral are often present in recesses in the top of the bedrock, which can be cleaned effectively only by handwork.
- Water is available in sufficient quantity during all or most of the year.
- Electric power is available at a reasonable rate.

Table 11.2-1 Estimated cost for dryland placer mining (2009 U.S. dollars)*

Daily Production	Stripping Ratio			
	1:1	2:1	4:1	6:1
Per metric ton of ore				
500	6.40	7.70	10.55	13.25
1,000	6.00	6.65	7.20	7.85
2,000	5.60	6.10	6.75	7.30
Per short ton of ore				
500	5.80	7.00	9.55	12.00
1,000	5.45	5.95	6.55	7.10
2,000	5.10	5.55	6.15	6.60

*Estimate includes wages/benefits, operation of equipment, fuel, power, and supplies. It covers stripping, mining, waste replacement, and minimum reclamation. Long haulage is not included. It does not include processing except for a scalping screen on-site. Assumes 1,672 kg/m^3 (104 lb/yd^3) material.

- Space is available for slime-settling ponds sufficient to permit compliance with regulations.
- An adequate supply of labor at acceptable cost is available; many operations are conducted by a type of cooperative or family organization.
- Security can be provided. Security is a constant problem, especially protection from roving individuals or groups at night.

Estimated Placer Mining Cost

The average cost of a timbered drift in consolidated ground is estimated to range from approximately $219 to $390/m^3 ($6 to $11/ft^3), depending on the size of the drift.

Table 11.2-1 shows the range of mining costs using wheeled front-end loaders or hydraulic shovels, bulldozers, and trucks to mine a dry placer deposit above the water table in mild terrain and in a location where the weather allows work year-round. The cost also includes in-pit scalping or screening but not mineral concentrating. The accuracy is no better than ±30%–40%.

PLACER MINING BY USE OF DREDGES

Placer mining is the removal of alluvial material of economic value. Traditional placers were gold deposits found in active or paleo and buried river courses. More recently, the term has come to include alluvial deposits containing economic values of gold, platinum, or silver (rare); heavy minerals of tin, titanium, rare earths, or iron; diamonds; and commercial aggregates. These deposits may be concentrated beneath an overburden of alluvial material some hundreds of meters thick, or they may be disseminated throughout the material or exposed on or near the surface. They may be in small concentrations or cover very large areas, and they may be deep beneath the oceans. The mining of unconsolidated materials such as manganese nodules and metalliferous muds from the deep oceans will be touched on only briefly.

Dredging is the recovery of mined materials from under water. The two basic types of dredge, mechanical and hydraulic, are distinguished by the method used to lift the material to the surface. Many variations and combinations of these types are available. The final selection of a dredge for a mineral deposit, however, is based on many factors including, but not limited to, the natural and social environments in which the deposit is located, the characteristics and current value of the deposit, the depth of the water, transportation distance, the efficiency of recovery, and the cost of the dredging system.

The decision to dredge or use other means of excavation depends on factors relating to the presence or availability of water and the characteristics and tenor of the deposit. With rare exceptions the dredge is used only in alluvial material, and the type of dredge chosen may be strongly influenced by the presence or absence of coarse materials, boulders, or other obstructions. Mining dredges for gold and heavy minerals are generally self-contained; the concentration plant is an integral part of the dredge or floats in close proximity. When there is a considerable overburden and little water, dry excavation employing large bulldozers, scrapers, or draglines may initially be used to remove the overburden and expose the ore, much like strip mining for coal. Offshore the depth of water may be paramount, but the nature of the deposit may call for specially designed systems for the platform and for excavation and recovery of the minerals.

Preparations and Planning the Dredging Operations

Dredging preparation begins with accurate evaluation of the area by sampling on an appropriate grid and preparing a three-dimensional representation of the deposit. Reverse-circulation or churn drills suitable for sampling are normally used. More modern methods of downhole sensors are sometimes employed. In lacustrine or coastal areas, continuous surveys with seismic or acoustic penetration may be preferentially used to develop the dimensions of the deposit in place and to select appropriate sites for drilling. The grade of the ore and characteristics of the deposit determined from the sampling data are used to select a dredge of the appropriate type and capacity.

The initial placement of large mineral dredges on land may typically involve the excavation of a pit in which the dredge is constructed. On completion, the pit is flooded and the dredging proceeds according to plan. In offshore or coastal areas, dredges may be towed directly to the site, but many more modern dredges have ship-type hulls and are self-propelled.

Crew sizes will vary considerably depending on the size and type of dredge and the social environment in which the deposit is located. The offshore tin recovery bucket ladder dredges in Indonesia typically had crews of 80 or more. Modern dredges are more automated but each shift needs a dredgemaster, an engineer, surface crews to maintain the control lines, and mechanical crews to conduct constant maintenance on all moving parts. Including shore, office, and transportation crew, the operation usually employs about 70 during the dredging season and 30–50 during the winter months. Fully automated systems require fewer personnel on-site (Garnett 1999a).

Dredge Location and Grade Control

Mining and grade control are determined from the sampling data. The sampling data will be treated in the same way in lakes as on land, although it may be necessary to adapt the sampling equipment to the environment. In northern latitudes, drilling may be carried out from an ice cover during the winter. Control in ocean beach environments is difficult, particularly where heavy surf or high tides are prevalent, and draglines may be used to mine in bulk without regard to whether the recovered material is high or low. In other situations, notably the diamond operations in southwestern Africa, the ocean may be held back by specially constructed massive berms or dikes so deposits can be worked in the dry.

Sampling equipment varies considerably in its methodology and cost. The more valuable the economic material, the more accurate and difficult the sampling will be. Sampling of beach sand may require only a determination of the depth of the sand to bedrock. Sampling of alluvial gold deposits requires extreme accuracy because of the value of the grains, and a "rich" deposit may contain only a few parts per million of gold. Because the specific gravity of gold is approximately 19, the strong tendency of gold grains is to migrate to or away from the sampling device when it penetrates and disturbs the ground. This effect can create significant sampling errors. An alluvial diamond deposit may contain only one part per billion of diamonds; these areas are generally sampled by very large bulk samplers that may be prototypes of the proposed mining device (Garnett 1999b).

An important aspect of determining the efficiency of the dredging operation is the R/E factor, a measure of actual recovery divided by the recovery estimated from the sampling data. This number may vary between 0.5 and 2.5 depending on the accuracy of reserve estimation and the recovery of the gold by the dredge and in the beneficiation process. A factor of 1.04 was recorded by the Alaska Gold Company, averaged over a period of 4 years (Garnett 1991).

Natural Hazards, Waste Disposal, and Environmental Controls

Natural hazards in minerals dredging may include severe winter conditions in Arctic environments, the inclusion of large boulders in river gravels, buried trees in tropical areas, or deeply gullied bedrock in other areas. The overcoming of these problems may be expensive and must be factored in to the unit costs of the operations. In the Arctic, it may be possible to keep a dredge pond open by sawing the ice and removing it from around the dredge, but more often the dredge may be put on a maintenance basis during the winter months. If large boulders are present, the dredgemaster may be required to mine around them, thus losing time and pay dirt, or it may be possible to lift them out of the way. Some large bucket-ladder dredges have such a capability. Buried trees, commonly found in the tin fields of Southeast Asia, may be circumvented in similar fashion. The problem with gullied bedrock can be quite severe if the rock is too hard to disintegrate with the dredge. In the diamond placers on the southwest African coast, massive berms have been built to keep out the sea, and the overburden and surficial waters are completely removed. The gullies where many of the diamonds are retained are swept by hand with small brushes. Offshore, new dredge heads have been developed that have the capacity to disintegrate the bedrock incorporating the gullies.

Discharge of tailings from a dredge may be the result of overburden removal, or the separation of the pay dirt from the gangue material. In either case the waste materials may be a significant part of the dredge output, and they must be disposed of in a manner that does not interfere with the recovery of the pay dirt. This may be done by hydraulic or mechanical diversion of the waste to an area normally located behind the dredge. If the mined ground must be restored, the soil cover may be removed separately and stockpiled away from the dredged area for later replacement over the mined area by earthmoving equipment.

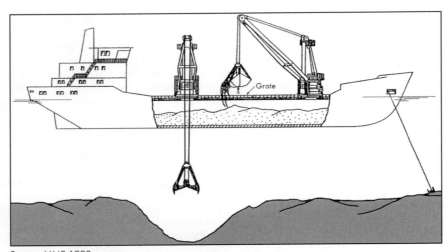

Source: MMS 1993.
Figure 11.2-6 Hopper dredge using clamshell for both loading and unloading

Environmental controls may preclude the use of certain areas for tailings disposal because of the runoff from the waste. However, most primary treatment at the dredge site involves only sizing and separation by physical properties such as specific gravity, color, or magnetic susceptibility. Chemicals are rarely added and the environmental problems may not be paramount. However, the issuance of permits is generally a political matter highly influenced by local and subjective feelings and can result in much frustration and delay. Waste-disposal permits must be secured at a very early stage in the planned development process. It should be recognized that all forms of placer mining are extremely cost sensitive. As such, dredging operations cannot withstand imposed expenditures on generic environmental research programs that are geographically located well outside the zone of influence of the dredging operation (MMS 1993).

Silting of watercourses can be prevented or lessened by containing and recycling the wastewater using cyclones or other dewatering devices. Containment or settling ponds may also be installed in appropriate areas. Offshore, some hydraulic dredges have used an antiturbidity system that allows the major solids to settle in a holding tank before discharge.

Types of Dredges

Mineral dredges may be generally classified as mechanical or hydraulic, depending on the means of transporting the dredged material from the dredge head to the dredge platform (Duane 1976).

Mechanical Dredges

Mechanical dredges transport the material from the deposit via moving buckets or moving belts. Major types of dredge buckets include the dragline, grab bucket, fixed arm, bucket ladder, bucket wheel, and continuous line.

Dragline. The dragline is a simple excavating bucket attached to a cable on a cranelike ladder structure. The bucket may be drawn back and forth until the required depth is achieved. Excavated materials are swung aside on the crane for dumping. Normally used in the mining of nonmetallics such as phosphates, aggregates, coal, or overburden, the dragline has the advantage of simplicity and comparatively low cost. Bucket size may vary from 0.76 m^3 (1 yd^3) or less to more than 38 m^3 (50 yd^3) for some mega-draglines.

Grab bucket. The grab bucket is commonly used for the mining of aggregates or overburden. The bucket or clamshell is manipulated by a crane-type ladder to dig vertically into the material. The open bucket is dropped onto the material and its two halves automatically close on the load when the bucket is raised and swung to the point of discharge. Normally limited in capacity and depth of excavation because of multiple cables, the grab bucket is, however, frequently used in coastal marine environments for reef coral sands (see Figure 11.2-6.)

Fixed arm. The fixed-arm bucket is a more rigid machine that can dig into a deposit using an upward and forward force applied to a toothed open bucket on the digging arm. It is normally used for dry excavation but has been adapted for underwater use on submersible platforms with the addition of a bucket wheel or a rotating cutter feeding in to a suction line.

Bucket ladder. The bucket ladder, a heavy and sometimes ponderous dredge, has been traditionally and widely used for the mining of gold and tin placers throughout the world. Bucket ladders are generally capable of handling heavy ground containing coarse materials and boulders. On land they may be constructed on a barge in a dry dredge pond that is flooded after construction is completed. Bucket ladders use a continuous rotating loop of heavy steel buckets on the ladder, cutting into the forward face of the pond and discharging into a washing and concentrating plant on the dredge. Waste materials are discharged well behind the dredge from chutes or belt conveyors as the bucket ladder moves forward. The concentrates are gathered onboard and shipped by local means to the smelters on shore.

There are two basic types of bucket-ladder dredge. The New Zealand type is moved forward on a headline anchored several hundred meters ahead of the barge and controlled laterally by anchored lines on each side. The more common California or spud type is propelled forward by a walking effect using two widely separated spuds at the stern of the barge and alternately raising and setting them as the ladder is swung from one side of the cut to the other, also using lateral anchored lines (see Figure 11.2-7). Bucket ladders are limited in the depth of cut by the length of the ladder. When

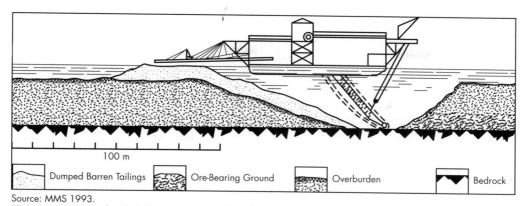

Source: MMS 1993.

Figure 11.2-7 Bucket ladder dredge with head line and anchors for estuarine or offshore operations

the bucket-ladder dredge is used offshore, the barge hull is modified to oceangoing standards, and headlines and multiple anchors are used for control. The dredge may be towed directly to the site, but many of the more modern ones have ship-type hulls and are self-propelled.

Bucket wheel. The bucket-wheel system consists of a rotating wheel with digging buckets on the periphery that discharge into a ladder-mounted belt conveyor or a hydraulic pipeline. The wheels may vary in diameter from a few meters to 10 m (30 ft). The bucket wheel is very adaptable, and with the major advantage of a powerful digging torque, it is now commonly preferred over the heavier and less mobile bucket ladders.

Continuous-line bucket. The continuous-line bucket is based on a system for unloading bulk granular materials from stockpiles or ship cargo holds (Figure 11.2-8). The bucket may be bell shaped or flat and attached to a continuously rotating flexible line. Its advantage when excavating loose materials such as sand is its capability to adapt to water depths. The system has been modified and tested for undersea mining of surficial deposits such as manganese nodules in water depths of 4,000 m (12,000 ft).

Hydraulic Dredges

Hydraulic dredges transport the dredged material from the dredge head to the surface in a slurry through pipes that may form, or be contained in, the dredge ladder. The dredge pump, usually a specially designed solids-handling centrifugal pump, may be on the dredge or submerged on the ladder. Submergence of the pump or addition of a submerged booster pump eliminates cavitation in the suction and may be necessary when the dredging depth exceeds cavitation depth (Figure 11.2-9). In special cases, air lifts or Venturi suction may be used in place of centrifugal pumps. The pump may discharge directly to a hopper, a treatment plant, or a sidecasting device on the dredge or it may be piped to a remote treatment plant or directly (in the case of a beach reclamation) to a fill area.

Variation in Dredge Heads

Straight suction heads are commonly used for removing sands or mud in harbor maintenance or for back-of-the-reef collection of carbonate sands for beach maintenance or cement manufacture. Small diver-operated suction devices are sometimes used for prospecting for river alluvials or in archaeological work for the recovery of artifacts.

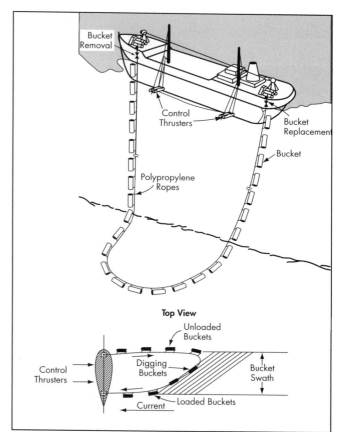

Source: MMS 1993.

Figure 11.2-8 Continuous-line bucket dredge tested in 4,000-m (12,000-ft) water depth. Note the bucket swath width controlled by the separation of the lines and the transverse movement of the vessel.

Trailing drag heads are commonly used from seagoing hopper dredges to mine commercial aggregates or for the clearance of river channels for commercial shipping. They are also adaptable for the mining of some seabed surficial mineral deposits such as heavy mineral sands or manganese nodules (Figure 11.2-10). They have a shallow but strong digging capability because of the momentum of the moving vessel.

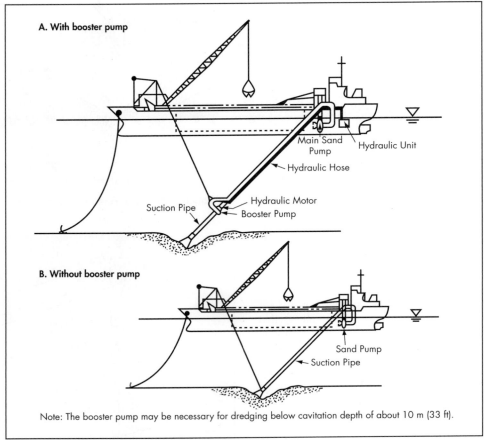

Source: MMS 1993.
Figure 11.2-9 Suction hopper dredges

Rotary-cutterhead dredges are commonly used in heavy ground where considerable energy is required to penetrate the material being mined. The material at the working face may be extensively disturbed, and use of these heads for the recovery of heavy minerals may result in significant losses of pay dirt, particularly if the underlying bedrock is very hard, rough, or gullied (Figure 11.2-11).

Horizontal auger heads are used in soft material that may be passed from a wide cut into the suction pipe opening. They are more often used for maintenance work in harbors than for mining. However, their use has been proposed for the recovery of surficial unconsolidated mineral deposits such as manganese nodules on deep seabeds.

Bucket-wheel cutterheads are a relatively recent innovation that has proven to be ideal for mining dredges. Digging buckets similar to those on a bucket-ladder dredge, but generally smaller, are mounted on the periphery of a rotating wheel attached to the digging end of the ladder. The buckets have tremendous torque and are capable of removing weathered bedrock. They can be designed to discharge directly into a suction pipe or conveyor without spillage of materials. For the mining of precious metals and heavy minerals, the bucket-wheel suction dredge has capabilities similar to those of the bucket-ladder dredge, and they generally cost less to operate. They have also been used in the form of remotely operated vehicles for sampling of gold placers and deepwater marine metalliferous sulfides.

Dredge Manufacturers

Mining dredges of all kinds are manufactured throughout the globe. *Dredgers of the World* (Clarkson Research Services 2009) lists 233 backhoe, dipper, grab, and clamshell dredges; 18 barge-unloading dredges; 53 bucket-ladder dredges; 412 cutterhead suction dredges (including bucket-wheel dredges); 32 types of special equipment including agitation, water-injection, and wormwheel dredges; and 39 suction and 481 trailing suction hopper dredges. The reader should refer to appropriate Web sites such as www.westerndredging.org or www.worlddredging.com for additional information. Manufacturers of pumps are listed on the Internet at www.dredgepumps.com.

Dredging Costs

Costs vary widely depending on many factors, as outlined previously. Economies of scale may provide a significant cost advantage for dredging over many other forms of earth moving and transportation (Correia 2005).

Other data have been presented by Garnett (1999a, 1999b) for gold placers and diamonds. These include costs of $1.10/m^3$ ($0.87/yd^3$) for Indonesian bucket-ladder tin dredges offshore and $7.00/m^3$ ($5.51/yd^3$) for gold dredging offshore in Nome, Alaska. The latter cost was similar for an experimental remote-controlled underwater miner bucket-wheel excavator with a higher R/E factor. Offshore diamond recovery in Southern Africa was reported in the same sources to be

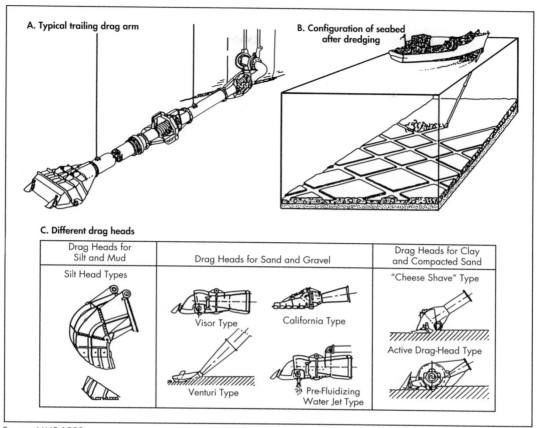

Source: MMS 1993.

Figure 11.2-10 Trailing suction hopper dredge and various configurations of drag head

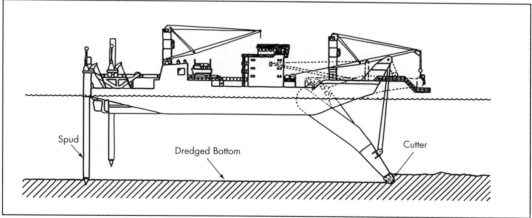

Source: MMS 1993.

Figure 11.2-11 Cutterhead suction dredge with rigid boom and spuds (cutterhead could be replaced by a bucket-wheel cutter)

about $1/m^3 ($0.79/yd^3) in the ore stratum and $1.25–$1.50/m^3 ($0.98–$1.18/yd^3) in the overburden. In the same location, costs for the underwater miner (Namsol) were reported to be $10–$20/m^3 ($7.87–$15.75/yd^3).

Recent costs for beach replenishment sands from immediately offshore at Waikiki Beach in Hawaii (United States) in 2007 were stated to be about 59/m^3 (47/yd^3) using barge-mounted flexible-line submersible pumps piped directly to shore.

The average costs for all dredging carried out by the U.S. Army Corps of Engineers for 2008 ranged from $3.00 to $17.85/m^3 ($2.37 to $14.06/yd^3), more than doubling their reported costs since 2000.

Obviously, dredging costs are highly variable, and extreme care should be taken in preparing cost estimates for any future operations.

Future of the Dredging Industry

The technology for placer mining on land will continue to improve as new materials and automation are applied to earth moving in general. This will bring down costs and improve the capability to work sustainably within the natural and social environments involved. The worldwide demand for heavy minerals is increasing, and operations are prevalent in newly industrializing countries such as India, People's Republic of China, and Mongolia (Yernberg 2006).

An even brighter future is likely to be associated with the deeper waters of the world's oceans, where applicable technology from other industries is already being applied. Placer diamond mining off the coast of southwestern Africa was instigated in the early 1960s (Cruickshank et al. 1968) and has since been developed into a significant industry involving several companies and countries. See also www.diamondfields.com/s/Namibia.asp and www.theartofdredging.com/peaceinafrica.htm.

In July 2008, Neptune Minerals (www.neptuneminerals.com), a leading explorer and developer of seafloor massive sulfide (SMS) deposits, lodged its first mining license application over the Rumble II West seamount within the exclusive economic zone (EEZ) of New Zealand. The application represented a key milestone in that company's progress toward commercialization and pilot mining by 2010. Neptune also successfully received prequalification proposals for its first seafloor mining contract. Leading international dredging and subsea engineering companies were invited to undertake a build-own-operate contract mining system for the commercial development of the deposits.

Nautilus Minerals, Inc. (www.nautilusminerals.com), has reported that the company was following the lead of the oil and gas industry to tap into the vast marine resources in deep waters offshore. Nautilus was the first company to commercially explore for deposits of SMSs and potential sources of high-grade copper, gold, zinc, and silver in water depths of approximately 2,000 m (6,000 ft). Its initial project, Solwara 1, under development in the archipelagic waters of Papua New Guinea (PNG), utilizes technologies from the offshore oil and gas, mining, and dredging industries, as well as current technologies from marine scientific exploration and positioning. In December 2009, the company reported eight new discoveries in the area, secure port capacity for its operations at Rabaul, and the grant of an environmental permit from PNG authorities. Farther afield, high-grade SMS systems were discovered during a Tongan exploration cruise. Long-term plans include an increase in tenement licenses and exploration applications in the EEZs and territorial waters of PNG, Fiji, Tonga, the Solomon Islands, and New Zealand.

Commercial dredging for deep-seabed nodules, crusts, or other mineral deposits has not yet been carried out in the United Nations area beyond the EEZs. Significant research has been conducted, however, on the potential technology available for their mining at water depths from 3,000 to 5,000 m (9,843 to 16,404 ft) and for the environmental sustainability of proposed operations. Extensive documentation may be accessed at the headquarters of the International Seabed Authority in Jamaica at www.isa.org.jm/en/home.

The future of placer mining by dredging is obviously changing and matching the global needs of the 21st century. Many other useful data may be accessed at www.OneMine.org.

PLACER MINING PROCESSING

When placer ores are mined, they must be upgraded to produce salable product(s) by either a wet or dry process. This section describes the most common upgrading processes and equipment. A heavy mineral sands example shows how processes can be combined.

Placer Ore Beneficiation Methods

Wet Versus Dry Beneficiation

Beneficiation of placer ores generally involves wet processing, with further downstream dry processing necessary for certain mineral assemblages and markets. A majority of the deposits (i.e., tin, phosphate, precious metals) can be upgraded purely by wet processes, but ilmenite, rutile, zircon, and other minerals found in heavy mineral sand deposits almost always require some sort of dry processing.

When compared with dry processing methods, wet processing offers the benefit of easy ore movement (pumping versus dry conveyance) and typically cheaper operating/maintenance costs (per ton treated or transported). For these reasons, combined with the common placer mining method of dredging, wet processing is used at the beginning of placer ore beneficiation flow sheets. The wet separation equipment may not always achieve the product grade and recoveries that are needed. In these instances, dry processing can be used.

In dry systems, the circuit always includes a dryer, which can require an equivalent of 8+ L/t (2+ gal/st) of diesel fuel per metric ton dried. The cost of drying, almost entirely made up of fuel costs, only allows further downstream equipment to achieve the desired separation. Drying does not increase the value of the ore itself and should always be minimized. When dry processing is required, a cost reduction may be obtained if it is preceded by a partial upgrade of the ore by wet processes. The partial upgrading will reduce the amount of material that must be dried and can drastically reduce drying costs. This process is decidedly different than minerals processing that requires flotation, in which the drying that takes place is primarily to meet customer specifications or reduce shipping weight of the concentrate.

Even when wet processing equipment obtains acceptable grades and recoveries, dry processing equipment may be able to achieve higher grades and recoveries. To determine if dry processing is required, lab testing should be performed. Each upgrading process has advantages and disadvantages, and specific mineral assemblage separations can only be determined by testing. The benefits and drawbacks of each type of processing must be weighed in full before a processing flow sheet can be finalized. The performance of the equipment and the overall capital and operating costs of the circuit should also be taken into consideration.

Sizing

Unlike most mineral processing, in which comminution is required to achieve a specific liberation size, particles generally have already been liberated by the time placer deposits are being processed. However, a wide range of particle sizes, from rocks (centimeters/inches in diameter) to clay (microns in diameter), must be properly handled. Trommel screens are used to remove the large waste rock and organic matter and are a last place to break up or remove large clay conglomerates. Cyclones are commonly used to deslime (remove clays and ultrafines) and remove the fine fraction of the ore that cannot be economically upgraded.

The nature of placer deposits may make it necessary to remove clay coatings on the minerals to allow proper separation and improve grade. These coatings are not typically seen in hard-rock operations in which comminution has been used to reach the liberation size. To clean the particle surface, attrition scrubbers are used. These scrubbers mix the slurry at high percent solids, which results in multiple particle–particle collisions. These collisions remove the softer surface coating of clay, which can then be easily washed away.

Vibrating screens are also used to either remove uneconomic size ranges of the ore or to separate the feed into multiple size fractions that can be upgraded using separate circuits. The resulting tighter size range will increase the separation efficiency of physical separation equipment since the size effect of the particles does not come into play as much. Generally wet screens are used on the higher-tonnage initial feed portions of the circuit, whereas dry screens can be found in later portions of the flow sheet or on final product sizing.

Hydraulic classifiers can also be used to size material. They can achieve excellent size separation given a monodensity feed, and they offer the benefits of high throughput, low operating cost, and very low maintenance cost when compared with wet vibratory screens.

Gravity Separation

In general, the greatest upgrading in placer ore beneficiation occurs with spirals. They are typically the first separation devices that the ore encounters after the initial sizing operations. The ore must be sized prior to spiral feeding, with oversize and slimes removed. Particles outside of a separable size range (that varies from mineral to mineral) do not separate well on spirals and can even hinder the separation of particles of more acceptable size. Initial wet concentrator spiral plants are generally installed on a barge that trails the mining dredge. Having the concentrator in the mining area reduces transportation cost of the raw ore and tailings.

In addition to their use as sizing devices, hydraulic classifiers can also be used as density separators. Denser particles report to the underflow and lighter particles report to the overflow. Because hydraulic classifiers combine density and sizing separations, coarser light particles may be misplaced to the underflow while finer heavy particles report to the overflow. To remedy the problem that this poses, both the underflow and overflow can report to separate spiral circuits. The misplaced coarse lights and fine heavies, the result of processing by the hydraulic classifier, must be separated from their respective streams, and feed that has undergone the sizing process lends itself to easier spiral separation than a raw mixed feed would allow (Benson 2005). Because of this phenomenon, newer flow sheets are incorporating hydraulic classifiers into spiral plants at the beginning and middle of a spiral circuit, whereas they traditionally were incorporated at the end (Elder et al. 2001).

Centrifugal jigs are another common gravity separator for placer ore. These enhanced gravity separators use a ragging material that can be varied to change the apparent specific gravity of the separating medium. Both product and tailings streams are screened to remove the ragging material, which can then be recycled. The adjustability of the apparent specific gravity of the separating medium and the centrifugal action of the jig allow for efficient separations. The drawbacks to these separators include the cost of the ragging material and the high maintenance required for the enhanced-gravity portion of the separator compared with other common placer ore gravity separators. Maintenance can include daily (or more frequent) shutdown and cleaning, and yearly rebuilds that can be costly and incur extensive downtime. Typically, centrifugal jigs are used in difficult specific gravity separations to recover higher-valued minerals. For example, they can be used in a scavenger stage of zircon wet gravity circuits.

Wet shaking tables can still be found in placer ore beneficiation, generally where near-specific-gravity separations are required at low throughputs. They provide excellent gravity separations but do have big drawbacks such as low feed rates and large size. Because of these drawbacks, wet tables are typically only used for higher-priced products (e.g., gold and zircon) that have already been at least partially concentrated. Their low feed rate (0.5–2 t/h [0.55–2.2 stph] per industrial-size table) and capital and operating expense must be economically justified before introduction into a flow sheet.

Magnetic Separation

Magnetic processing can be done both wet or dry, and so it lends itself to all aspects of a placer deposit flow sheet. In wet magnetic processing, a low-intensity magnetic separator (LIMS) is used to remove magnetite or other ferromagnetic minerals. In most placer deposits, the magnetite is usually a tailings stream, but in the case of iron sands, it can be a product. In general, medium-intensity wet magnets are not used for placer deposits. However, wet high-intensity magnetic separators (WHIMSs) are widely used. They offer the advantage of high throughput and wet processing (coinciding with the common wet mining methods), and they excel when used at the beginning of circuits where product grades are easy to achieve or extra tailings can be removed prior to drying, resulting in cost savings. As an example, in a typical ilmenite/rutile/zircon deposit, a WHIMS can be used to remove ilmenite that does not meet grade (e.g., low percentage of TiO_2).

Dry magnets have ranges of field intensities similar to those of wet magnets, and they are used to remove similar minerals. Some common dry magnets used in placer beneficiation include the rare earth drum (RED), induced-roll magnet (IRM), and rare earth roll (RER). REDs are medium-intensity dry magnetic separators and are commonly used to remove, as product or tails, more highly paramagnetic minerals such as ilmenite. Higher-strength dry magnets are available as both permanent rare-earth magnets, which are found within RERs, and electromagnets, which are used for IRMs; each has its advantages and disadvantages (Dobbins and Sherrell 2009). The higher-strength magnets are typically used to remove the middle to weakly paramagnetic minerals such as staurolite and leucoxene or to reduce the final %Fe_2O_3 grade of a non-magnetic product (e.g., zircon).

In general, wet magnets are a little less selective than dry because the liquid medium adds a restrictive drag force. On the other hand, dry magnets are not as tolerant of fine material and generally can only effectively treat down to 75 μm, whereas wet magnets routinely treat material finer than –45 μm (although that fine size is not very common in placer ore deposits).

In placer deposits where iron-bearing minerals are present, magnetite can be found. If enough ferromagnetic material is fed to higher-strength magnets, the ferromags cannot be effectively released, and wearing of shells and belts or plugging of matrices become issues. In these instances, low-intensity magnets can be used as scalpers for both wet and dry processes.

Electrostatic Separation

The typical separation equipment that coincides with the dry magnetic separators are the electrostatic separators. These can be found before (standard for heavy mineral sands) or after the magnets in placer ore beneficiation plants. They are mostly found in heavy mineral sand separation plants where the added physical property of conductivity is used to separate and upgrade the various products. Because the surface conductivity is so important for this type of separation, mineral surfaces must be clean (see information about attrition scrubbers previously in the "Sizing" section). All electrostatic separators either require minerals to be heated for a separation to occur or separate more effectively when minerals are heated. Some minerals, such as titanium-bearing sands, change conductivity at higher temperatures. The heat (+100°C [+212°F]) also removes the surface moisture that may be present, which can mask the true conductivity of a mineral.

The two common types of electrostatic separators are high-tension rolls (HTRs) and electrostatic plates (ESPs). The HTR has a higher capacity per unit length than the ESP (e.g., for heavy mineral sands 2 t/h/m of roll length versus 1 t/h/m [2.2 stph/m versus 1.1 stph/m]) and are generally the first stage of electrostatic separators. Because of the nature of HTRs, coarser nonconductive particles can report with the conductive material, and finer conductive particles can report with the nonconductive product stream. The opposite is true for ESPs. The misplacements associated with these machines make it possible to place ESPs within or after an HTR circuit to achieve effective separation of all particle sizes.

A much less common third type of electrostatic separator is the triboelectric separator. These can separate two nonconductive minerals based on the pre-charge that the minerals receive before they are introduced to the separation chamber. These may be used in place of flotation where water or environmental concerns are an issue. The greatest potential for these within placer ore processing plants is upgrading of phosphate by removal of silica sand and fine zircon cleaning by removal of undesirable nonconductors such as sillimanite and kyanite.

Other Processing

Flotation is not that prevalent in most placer ore upgrading but is frequently found in phosphate operations to separate silica from phosphate. The phosphate flotation circuits are generally large, with feed rates in excess of 1,000 t/h (1,100 stph). The flotation circuits are generally located after hydraulic classifiers that split the feed into tighter size ranges to allow for better separation efficiencies of the flotation cells. Circuits can be set up as rougher-cleaners, where the rougher flotation uses fatty acids as collectors for the phosphate. Because the phosphate floats very quickly, silica is usually entrained within the phosphate concentrate. To clean up the phosphate further, a cleaner flotation is used with amines as collectors to float the silica.

A placer ore beneficiation process generally limited to zircon sand is hot acid leaching. This process removes iron staining on sand grains with the addition of a high-concentration sulfuric or hydrochloric acid. This would only be incorporated where low %Fe_2O_3 grades are needed for final product specifications that cannot be accomplished by other means and where the value of the treated mineral can pay for the addition of this type of circuit. A hot acid leach circuit would include a dryer for heating, reactor vessels, rinsing equipment, acid storage and distribution systems, and an effluent treatment plant. Effluent treatment plants are generally simple and include addition of a base to neutralize the acidic effluent and a thickener to remove precipitated solids from the effluent system. The effluent water would be recycled back to the acid leaching circuit so that no off-site discharge would have to occur.

Tailings Handling

Tailings from the floating concentrators are immediately pumped off the barge to backfill the pond or cut (in the case of offshore tin mining). In pond-dredging operations, clays can be pumped into the operating pond when they quickly drop out of suspension or are not much of an issue. If they build up in concentration and hinder separation, clays can be pumped to settling ponds where they dewater and are eventually mixed with the top sand layer during reclamation.

With dry mining there is no working pond, so all tailings from the concentrator must be deposited into settling ponds. These ponds allow reclamation of water as well as settling of clays. In some instances, mixing the clay and sand (which come from different tailings streams within the concentrator) speeds overall dewatering rates.

When mineral separation plant tailings streams are not pumped directly back to a mining pit, they are generally sent to a stacker. The stacker includes a dewatering cyclone with some type of distribution mechanism for the underflow of high-percentage solids. The distribution mechanism (e.g., swinging chute or movable cyclone) allows the underflow tailings to stack because of their high percentage of solids and to become partially dewatered (5%–10% moisture) over time. The dewatered solids can then be trucked back to the mine to backfill mining areas or to create berms for dewatering ponds. Clays report back to process water ponds or thickeners, where they are allowed to settle.

General Heavy Mineral Sands Flow Sheet

Heavy mineral sands operations use nearly all of the processes mentioned previously. For this reason, an example of a typical heavy mineral sands processing flow sheet is given to illustrate how the previously mentioned equipment would be incorporated into one flow sheet (Figure 11.2-12). The overall flow sheet is broken up into a wet concentrator section that produces a heavy mineral concentrate and a mineral separation plant that produces the final products. The wet concentrators are mostly found on floating barges located behind the mining dredge. Mineral separation plants (MSPs), on the other hand, are always land based and are located away from the mine site so concentrate can be easily transported from any location within the mine plan and so that multiple mine sites can feed a single MSP. The central-location concept can move the MSP far from the mine site(s); 50+ km (31+ mi) can still be economical.

When feed is introduced to a wet concentrator, it is first sized by trommel to remove oversize and then treated with hydrocyclones to remove clays and fines. The feed is then sent to a gravity circuit that includes spirals (Reichert cones can be found but are rare) and may include hydraulic classifiers. Typically, the classifiers would be placed at the end of the spirals circuit to help remove any misplaced silica and any fine material that would not be recoverable within the MSP. An attritioning step would be included between the spiral circuit and classifier to clean the sand surface and allow washing of clays if they are present. Spiral circuits have multiple stages; the final product usually passes through four spiral stages

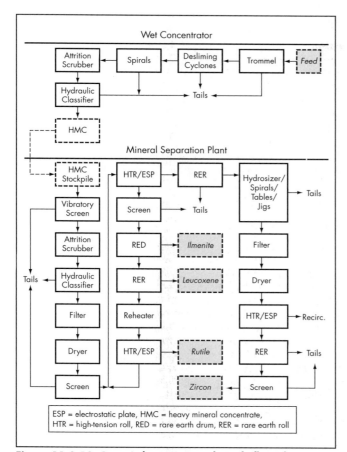

Figure 11.2-12 Generic heavy mineral sands flow sheet

Table 11.2-2 Heavy mineral beneficiation cost example

Area	Ore, US$/t	Heavy Mineral Concentrate, US$/t
Mining (dredge)	0.60	5.10
Wet concentrator	1.40	13.45
Mineral separation plant	0.60	6.50
Product handling and delivery	0.95	10.95
General and administrative	0.35	3.75
Concurrent reclamation	0.10	1.05
Total operating cost	4.00	40.80

Courtesy of Outotec (USA), Inc.

including at least one scavenger step. The final product from the wet concentrator is referred to as heavy mineral concentrate (HMC).

After the HMC is produced, it is transported to the MSP. Blending may occur from multiple mine sites to achieve consistent product grades and mineral assemblages in the feed. Consistent mineral assemblages can reduce bottlenecks in the MSP that may occur at low feed rates if too much of one mineral is present in the feed.

The first circuit of an MSP would generally be a feed-preparation step that may include wet screens to remove oversize as well as attrition scrubbers and hydraulic classifiers to guarantee clean grain surfaces and minimal silica. The feed would then be filtered and dried. High-tension separators are used after the drier to utilize the inherent heat from that process to ensure efficient separation of conductors from nonconductors. Conductor minerals would then be run through various magnets (RED and RER magnetic separators) to remove ilmenite and leucoxene as products. The resulting conductor/nonmagnetic minerals would be reheated and passed through more electrostatic separators to produce a rutile product.

The nonconductor stream from the first high-tension separator circuit is run through a separate magnet circuit to remove nonconductive/magnetic minerals such as monazite and staurolite. Depending on market conditions, monazite and staurolite may be marketable products. The resulting nonconductor/nonmagnetic minerals have a high percentage of zircon but also an upgraded silica content. To remove the lighter silica, this stream is rewetted and run through wet gravity separators that include a combination of spirals, hydraulic classifiers, wet tables, and/or centrifugal jigs. The heavy concentrate from the MSP wet gravity circuit has a high zircon content but generally does not meet product specifications on %TiO_2, %Fe_2O_3, and potentially %Al_2O_3. To clean up this zircon stream, it is dried and passed through high-tension separators and RER magnetic separators. Based on customers' product specifications, a final screening stage may be needed to remove any oversize or to size the zircon appropriately.

Cost for Mineral Processing of Placer Deposits

Table 11.2-2 shows typical costs for beneficiation of a heavy mineral sands placer deposit. The data assume a 1,000-t/h (1,100-stph) feed rate to the wet concentrator and ore containing an average total heavy mineral (HM) grade of 12.5%. The heavy mineral concentrate produced by the wet concentrator contains 95% heavy minerals; the valuable heavy minerals are ilmenite (65% of HM), rutile (5% of HM), and zircon (5% of HM). Recoveries in the wet concentrator are on the order of 95% for ilmenite, 90% for rutile, and 95% for zircon. Recoveries in the mineral separation plant are 98% for ilmenite, 90% for rutile, and 80% for zircon.

REFERENCES

Benson, S.J. 2005. Solving a conundrum: Why a conventional wet gravity circuit was abandoned in flowsheet design for a garnet project in central Australia. In *Proceedings of the 2005 Heavy Minerals Conference*. Littleton, CO: SME.

Breeding, W.H. 1985. Geology—Its application and limitation in the selection and evaluation of placer deposits. In *Applied Mining Geology: Problems of Sampling and Grade Control*. New York: SME-AIME. pp. 77–81.

Clarkson Research Services. 2009. *Dredgers of the World*, 7th ed. London: Clarkson Research Services.

Cope, L.W., and Rice, L.R. 1992. *Practical Placer Mining*. Littleton, CO: SME. p. 169.

Correia, K.G. 2005. Optimization of Millenium Chemical's Paraiba Guajú mine. In *Proceedings of the 2005 Heavy Minerals Conference*. Littleton, CO: SME. pp. 27–30.

Cruickshank, M.J., Romanowitz, C.M., and Overall, M.P. 1968. Offshore mining present and future with special emphasis on dredging systems. *Eng. Min. J.* 169(1):86–91.

Daily, A.F. 1973. Placer mining. In *SME Mining Engineering Handbook*. Edited by I.A. Given. New York: SME. pp. 17-151–17-161.

Dobbins, M., and Sherrell, I. 2009. Significant developments in dry rare-earth magnetic separation. SME Preprint No. 09-004. Littleton, CO: SME.

Drew, J., and Lessard, T. 2009. Developing a new process design for the southwestern Oregon industrial mineral bearing placer system via practical study of the unique deposition, mineralogy, and dry tailing requirements. SME Preprint No. 09-038. Littleton, CO: SME.

Duane, D.B. 1976. Sedimentation and ocean engineering: Placer mineral resources. In *Marine Sediment Transport and Environmental Management*. Edited by D.J. Stanley and D.J.P. Swift. New York: Wiley.

Elder, J., Kow, W., Domenico, J., and Wyatt, D. 2001. Gravity concentration—A better way (or how to produce heavy mineral concentrate and not recirculating loads). In *International Heavy Minerals Conference*, Freemantle, Western Australia, June 18–19. Carlton South, Victoria: Australasian Institute of Mining and Metallurgy.

Garnett, R.H.T. 1991. Dredge recoveries in North American placer gold dredging. In *Alluvial Mining*. London: Elsevier Science Publishers.

Garnett, R.H.T. 1999a. Marine placer gold, with particular reference to Nome, Alaska. In *Handbook of Marine Mineral Deposits*. Edited by D.S. Cronan. Boca Raton, FL: CRC Press. pp. 67–101.

Garnett, R.H.T. 1999b. Marine placer diamonds with particular reference to Southern Africa. In *Handbook of Marine Mineral Deposits*. Edited by D.S. Cronan. Boca Raton, FL: CRC Press. pp. 103–141.

McCulloch, R., Lewis, B., Keill, D., and Shumaker, D. 2003. Applied gold placer exploration and evaluation techniques. Special Publication 115. Butte, MT: Montana Bureau of Mines and Geology.

McLean, C.A., McDowell, W.K., and McWaters, T.D. 1992. Dredging and placer mining. In *SME Mining Engineering Handbook*, 2nd ed. Edited by H.L. Hartman. Littleton, CO: SME. p. 1459.

MMS (Minerals Management Service). 1993. *Synthesis and Analysis of Existing Information Regarding Environmental Effects of Marine Mining*. OCS Study MMS 93-0006. Washington, DC: U.S. Department of the Interior.

Peele, R. 1945. *Mining Engineers Handbook*, 3rd ed., Vol. 1. New York: John Wiley and Sons.

Smith, D. 2008. Presentation at the 16th Annual Bay Area Economic Outlook Forum, Bay Area Chamber of Commerce, Oregon Resources Corporation, Coos Bay, OR.

Stone, J.G., Mejia, V.M., and Newell, G.T. 1988. Using diamond drilling to evaluate a placer deposit: A Case Study. *Min. Eng.* 40(9):875–880.

Thrush, P.W., ed. 1968. *A Dictionary of Mining, Minerals and Related Terms*. Washington, DC: U.S. Bureau of Mines.

Wells, J.H. 1973. *Placer Examination Principles and Practice*. U.S. Department of the Interior, Bureau of Land Management, Technical Bulletin 4. Washington, DC: U.S. Government Printing Office. pp. 73–84.

Yernberg, W.H. 2006. Heavy minerals activity on the rise worldwide. *Min. Eng.* 48(1):33–36.

CHAPTER 11.3

Heap Leaching

Travis J. Manning and Daniel W. Kappes

HISTORY

The method of heap leaching has a long history in the mining business. In the mid-1500s, mines in Hungary recycled solutions bearing copper through waste rock for additional copper recovery. Around that same time, Georgius Agricola wrote about piling rocks in heaps and sprinkling them with water for alum recovery in his book *De Re Metallica* (Agricola 1556). At the original Rio Tinto mine in Spain, acid solutions were added to large heaps of copper oxide ores circa 1750. In the 1950s, heap leaching was used for uranium extraction from ore. It was not until 1967 that heap leaching was developed for precious metals by the U.S. Bureau of Mines; the first large-scale gold heap leach project was at Cortez, Nevada. Fewer than 20 years after heap leaching of gold was first developed, it accounted for 30% of all gold production in the United States. In parallel with gold heap leaching, the last 40 years have seen a tremendous expansion of copper dump and heap leaching (Scheffel 2002). Uranium ores have been heap leached with both acid and alkaline solutions since the 1950s. A list of current and historical uranium heap leach operations can be found on the Web site for the World Information Service on Energy: Uranium Project (WISE 2007). Heap leaching of nickel laterites is still in its infancy with only one project near production, the Çaldag project in Turkey.

HEAP LEACHING EXPLAINED

In a simplistic sense, heap leaching involves stacking of metal-bearing ore into a "heap" on an impermeable pad, irrigating the ore for an extended period of time (weeks, months, or years) with a chemical solution to dissolve the sought-after metals, and collecting the leachant ("pregnant solution") as it percolates out from the base of the heap. Figure 11.3-1 is an aerial photograph showing the typical elements of a heap leach operation: an open-pit mine is shown on the left; on the right is a 2-Mt heap of crushed, conveyor-stacked ore on a plastic pad. Pregnant and barren solution storage ponds are located downslope from the heap. Buildings include a solution process facility for recovering metals from the pregnant solution, a laboratory, a maintenance shop, and administration offices. For a small operation such as the one illustrated here, very limited infrastructure is required.

In a more complex sense, heap leaching should be considered as a form of milling. It requires a nontrivial expenditure of capital, and a selection of operating methods that trade cost against marginal recovery. Success is measured by the degree to which target levels and rates of recovery are achieved. This distinguishes heap leaching from dump leaching. In dump leaching, ores are stacked and leached in the most economical way possible, and success is measured by any level of net positive cash flow.

WHY HEAP LEACHING?

Options for recovering metals from ore are many, including agitation leaching, gravity separation, magnetic separation, flotation, and vat leaching. If the ore can be heap leached (which requires good metals recovery by leaching and a rock type that allows the construction of permeable heaps), this technique offers some significant economic benefits. Recent large-scale, very simple heap leaches (40,000+ t/d) have been constructed for a total project capital cost as low as $3,000 per daily metric ton of ore processed, whereas the total capital costs for an agitation leaching plant are often $15,000–$35,000 per daily metric ton. The capital costs of a flotation plant are $10,000–$25,000 per daily metric ton. Another advantage of heap leaching is that it is a chemical process and the product of the leach operation is usually a metal. For gold and silver ores, the primary product at the mine for a leaching operation is an impure gold/silver (doré) bar that can be sold for 95% or more of the quoted price for the metals recovered from the ore. For copper ores, the primary product is usually copper metal that can also be sold for a high percentage of its value. In comparison, flotation plants, which are the primary competitor to the chemical leaching processes, produce a concentrate of the main value mineral(s), which often carries lower-value components and must be sold to a smelter so that the realizable value of gold, silver, and copper is often only 60% of the quoted metal price of the recovered metals.

A number of factors will determine whether heap leaching is the best fit for a project:

Travis J. Manning, Metallurgical Engineeer, Kappes, Cassiday & Associates, Reno, Nevada, USA
Daniel W. Kappes, President, Kappes, Cassiday and Associates, Reno, Nevada, USA

Source: Kappes 2002.
Figure 11.3-1 Heap leach installation at Mineral Ridge, Nevada

- **Risk.** An increasing number of mining projects are being developed in exceedingly remote locations or places with political and social upheaval. It can be a wise choice to select heap leaching in these situations where a mill, with higher capital costs and more complex equipment systems, would be harder to maintain (or stop and restart) as social and political changes occur.
- **Lack of sufficient reserves.** In many underground and some open-pit mining situations, it is difficult (occasionally even impossible) to develop a large ore resource early in the project. Heap leaching offers a very quick, low-cost avenue for treating such ores. As an example, the Sterling mine in Beatty, Nevada (Imperial Metals Corporation) began life as an underground mine with a reserve of 100,000 t (metric tons) of ore at a grade of 11 g Au/t but eventually heap leached nearly 1 Mt.
- **Differential recovery is not sufficient to justify the added investment.** The operating cost plus debt service for a medium-sized mill (3,000 t/d) will typically be $34/t of ore treated, whereas the comparable cost for the heap is $12/t (year 2009 cost basis). Since the heap leach typically gets 25% less recovery than the mill, the gross value of the ore would typically have to be greater than $85/t in order to justify the mill.
- **Capital is very difficult or expensive to raise.** Heap leaching can give a small company a project that can catapult it into the "big league" from a relatively small investment. Often early-stage capital can be raised only by excessive equity dilution or acceptance of onerous bank covenants. The heap leach also gives the management of a small company more time to build up an experienced operational staff before it tackles more complex processes.

HEAP LEACHING CONFIGURATIONS
Configurations of heap leaching available include dynamic heaps, permanent single-lift and multilift heaps, and valley-fill heaps.

Dynamic Heaps
Dynamic heaps reuse the same lined area by loading the ore onto the pad, leaching, washing, and then removing the ore from the pads. They are also referred to as "on/off heaps." The advantage of an on-off heap is that a large amount of ore can be leached in a limited area in relatively thin layers. For ores where the permeability of thicker layers cannot be maintained, which is typical of oxidized copper ores, dynamic heaps often offer the only option. One factor often overlooked is that the leached ore is still an environmental liability and must be moved to another lined area. However, the water balance is easily manageable because of the limited area of the pads. The small area also makes it possible to cover the heap if necessary. Dynamic heaps are applicable to low-grade, precious metal ores containing sulfides that will start to oxidize to form acid within 1 to 3 years. The benefit of using a dynamic heap for this type of material is that the values can be recovered during the short leach time associated with the dynamic heap and the generation of acid would occur in the waste dump, not affecting the leach process.

The pad below a dynamic heap needs to be of sturdier construction than that for a permanent heap because there is more traffic at elevations close to the pad. Pads for dynamic heaps are frequently made of concrete or asphalt for this reason, though with careful ore removal a geomembrane (plastic liner) could be used. Slowly leaching ore can be re-treated by simply restacking it along with the new ore, but this will reduce capacity. Typically the maximum recovery is not reached with dynamic heaps because of the limited time frame for leaching. Before the ore can be removed from the pad, it is usually washed to recover any dissolved values and detoxified if necessary.

Permanent Heaps
In permanent heaps, the ore is stacked on a low-permeability surface and never removed. While this definition also applies to valley-fill heaps, the term *permanent heap* is usually reserved for a heap that is stacked on a relatively tabular ground surface where solution can exit at multiple points across the face of the heap. To expand the available volume for stacking ore, the pad is either expanded or additional lifts are added on top of the ore already stacked. The primary advantages of permanent heaps are that operations are less expensive than dynamic heap operations, and they allow for much longer leach times (up to several years).

Multilift heaps have been built as high as 200 m with successive lifts stacked on top of previously leached lifts. Single-lift heaps are appropriate for some ore types or leaching situations where, for example, the ore is high grade and not very permeable. For single- or multilift permanent heaps, the initial capital cost is relatively low because the liner does not need to be as robust as for a dynamic heap, and a small area can be lined at first with expansion in subsequent years. As multilift heaps grow taller, there are added operating costs for pumping solutions and transporting ore to higher elevations. Permanent heaps require a large area with gently changing topography. The large area required often leads to issues with the water balance in high-rainfall environments, but permanent heaps can be effectively used where rainfall exceeds 2.5 m/yr.

Valley-Fill Heaps
A valley-fill heap leach is just as it sounds: the ore is dumped at the bottom of a valley and is built up, "filling" the valley. Valley-fill heaps are built in areas that do not have enough level terrain to build an expanding permanent heap. Valley-fill

heaps can accommodate long leach times well. A valley-fill expands upward and outward as the valley widens near the top. The necessity for multilifts means that the ore should be strong and maintain high permeability under high loads. The pregnant solution is often stored inside the bottom of the heap and a small pond is used to collect the solution for recovery. This has advantages of less cost in liner, less precipitation collection, and it can keep the solution from freezing in cold climates. If solution is stored within the heap, a better liner needs to be in place because of hydraulic head on the liner. The front slope of the heap needs to be thoroughly designed and very stable to prevent any catastrophic failure. Frequently a retaining structure, or dam, is used to support the toe of the heap.

GEOLOGY OF GOLD DEPOSITS

Carlin-Type Sedimentary Ores
These ores consist of shales and "dirty" limestones containing very fine (submicroscopic) gold. Oxidized ores leach very well, with low reagent consumption and production recovery of 80% or better being achieved. Ores are typically coarse-crushed (75 mm) but may show recovery of 70% or better at run-of-mine (ROM) sizes. The largest of northern Nevada's heap leaches (Carlin, Goldstrike, and Twin Creeks) treat this type of ore. Unoxidized ore contains gold locked in sulfides (typically 1%–3% pyrite) and also contains organic (carbonaceous) components, which absorb the gold from solution. This ore shows heap leach recovery of only 10%–15% and is not suitable for heap leaching. Because of the different ore types, the northern Nevada operations (for instance, Barrick's Goldstrike mine) may employ roasters, autoclaves, agitated leach plants, and heap leaches at the same mine site. Crushing is usually done in conventional systems (jaw and cone crushers) and ores are stacked by trucks.

Low-Sulfide Acid Volcanics or Intrusives
Typical operations treating this type of ore are Round Mountain (Nevada) and Wharf mine (South Dakota, United States). Original sulfide content is typically 2%–3% pyrite, and the gold is often enclosed in pyrite. Oxidized ores yield 65%–85% recovery but may have to be crushed to below 12 mm. Usually the trade-off between crush size and percent recovery is a significant factor in process design. Unoxidized ores yield 45%–55% gold recovery and nearly always need crushing. At Round Mountain, approximately 150,000 t/d of low-grade oxide ore is treated in truck-stacked ROM heaps, 30,000 t/d of high-grade oxide ore is treated in crushed (12 mm) conveyor-stacked heaps, and 12,000 t/d of unoxidized ore is treated in a processing plant (gravity separation followed by leaching in stirred tanks). Primary and secondary crushing is done using jaw and cone crushers; finely crushed ore contains enough fines that conveyor stacking is preferred over truck stacking.

Oxidized Massive Sulfides
The oxide zone of massive sulfide ore deposits may contain gold and silver in iron oxides. Typically, these are very soft and permeable, so crushing below 75 mm often does not increase heap leach recovery. The Filon Sur ore body at Tharsis, Spain (Lion Mining Company) and the Hassai mine in Sudan (Ariab Mining Company) are examples of successful heap leaches with this type of ore. Because the ore is fine and soft, the ore is agglomerated using cement (Hassai uses 8 kg cement/t), and stacking of the heaps is done using conveyor transport systems.

Saprolites/Laterites
Volcanic- and intrusive-hosted ore bodies in tropical climates typically have undergone intense weathering. The surface cap is usually a thin layer of laterite (hard iron oxide nodules). For several meters below the laterite, the ore is converted to saprolite, a very soft water-saturated clay, sometimes containing gold in quartz veinlets. Silver is usually absent. These ores show the highest and most predictable recovery of all ore types, typically 92%–95% gold recovery in laboratory tests, 85% or greater in field production heaps. Ores are processed at ROM size (which is often 50% minus 2 mm or 10 mesh) or with light crushing. Ores must be agglomerated and may require up to 40 kg of cement/t to make stable, permeable agglomerates. Many of the western African and central American heap leaches process this type of ore; good examples are Ity in the Ivory Coast and Yatela in Mali. When crushing is required, one or two stages of toothed roll crushers (Stamler-type feeder-breaker or a mineral sizer) are usually employed. Conveyor systems are almost always justified; ore can be stacked with trucks if operations are controlled very carefully.

Clay-Rich Deposits
In some Carlin-type deposits and volcanic-hosted deposits, clay deposition or alteration occurs along with gold deposition. These ores are processed using the same techniques as for saprolites, except that crushing is often necessary. Because of the mixture of soft, wet clay and hard rock, a typical crushing circuit design for this type of ore is a single-stage impact crusher. Truck stacking almost always results in some loss of recovery. Agglomeration with cement may not be necessary, but conveyor stacking is usually employed.

Barneys Canyon (Utah, United States) uses belt agglomeration (mixing and consolidation of fines as it drops from conveyor belts) followed by conveyor stacking. The La Quinua operation at Yanachocha, Peru, employs belt agglomeration followed by truck stacking.

Silver-Rich Deposits
Nevada deposits contain varying amounts of silver, and the resulting bullion may assay anywhere from 95% gold and 5% silver to 99% silver and 1% gold. Mexico also has multiple silver-rich deposits, like the Dolores and Ocampo projects. Silver leaches and behaves chemically in the same way as gold, although usually the percentage of silver recovery is significantly less than that of gold. Examples of nearly pure silver heap leaches are Coeur Rochester and Candelaria mines in Nevada, and Comco in Bolivia.

GEOLOGY OF COPPER DEPOSITS
Copper occurs in three basic assemblages: oxides, secondary sulfides, and primary sulfides. Porphyry deposits produce the bulk of the world's copper, and if these outcrop, they usually contain all three mineral assemblages (Bartlett 1998). The secondary sulfides are often the highest grade because of the oxidized cap leaching naturally and depositing copper in the secondary sulfide zone. If acidic conditions are not sufficient, the copper is not transported and an oxide deposit is formed.

Normally, there is no distinct zoning and there is frequently a transition zone from oxides to sulfides. The varying chemistry of copper makes the chemistry of copper heap leaching more difficult and complex than that of gold or silver. Adding to the chemical issues, feldspar minerals in copper ores often break down over several weeks of leaching to form clays, which reduce the permeability of the heap. Dump leaches—where uncrushed low-grade mine waste is leached with dilute acid—are fairly common at copper mines because extra copper can be recovered at a very low cost.

Oxide Copper Ores
There is a wide range of copper oxide minerals. With the range of minerals comes a range of metallurgical responses. Most copper oxide minerals leach relatively quickly although some, especially the silicates (like chrysocolla), may be only partially soluble because of the formation of impermeable coatings.

Sulfide Copper Ores
There are primary sulfides, like enargite and chalcopyrite, and secondary sulfides, like chalcocite, bornite and covellite. The secondary sulfides usually leach faster and more completely than the primary sulfides. All sulfides need an oxidant present for dissolution, and the slow introduction of oxygen or oxidizing chemicals can result in a significant lengthening of leach times over what was predicted by laboratory tests. Within the heap, oxygen is usually converted by bacteria into oxidized (ferric) iron, and it is this ferric iron that reacts to oxidize the sulfides, so the heaps must be built in such a way as to encourage the growth of bacteria.

GEOLOGY OF URANIUM DEPOSITS

Sedimentary Uranium Deposits
Oxidized uranium is readily soluble and may be transported from granites or tuffs by groundwater to another location where it is concentrated. The uranium may precipitate in an area for a number of reasons, including lower temperatures and pressures, reduction, ion exchange, neutralization, and chemical replacement. Generally, for ore bodies, the uranium is deposited by reduction from the soluble, hexavalent uranium dioxide to an insoluble tetravalent state. The classic uranium deposits of the western United States are "roll front" deposits in which uranium is dispersed from sandstone and reprecipitated in a continually moving zone.

Granitic Uranium Deposits
Uranium may also be concentrated as large bodies of granite solidify. The initially low concentrations of uranium are continually forced into the remaining solution in the granite until small concentrated amounts of uranium-bearing solutions are left. The fluid can get into fissures in the surrounding rock and form veins of ore.

GEOLOGY OF NICKEL DEPOSITS
Nickel laterite deposits are created when a parent rock is weathered by high rainfall and elevated temperature, typically in a tropical or subtropical climate. The minerals that dissolve easily are removed by the rainfall, leaving behind higher concentrations of the less-soluble elements like iron and aluminum. Nickel is somewhat mobile and concentrates near the base of the leached zone.

CHEMISTRY OF GOLD DEPOSITS
The chemistry of leaching gold and silver from their ores is essentially the same for both metals. A dilute alkaline solution of sodium cyanide (NaCN) dissolves these metals without dissolving many other ore components (copper, zinc, mercury, and iron are the most common soluble impurities). Solution is maintained at an alkaline pH of 9.5–11. Below a pH of 9.5, cyanide consumption is high. Above a pH of 11, metal recovery decreases. A pH above 11 can also result in dissolving silica, which can cause problems with scale control and blinding of carbon.

Many heap-leachable ores contain both gold and silver. Deposits in western Africa and Australia tend to be very low in silver, while those in Nevada and Mexico are highly variable, ranging from pure gold to pure silver. Silver is usually not as reactive with cyanide as gold. This is because gold almost always occurs as metal, whereas silver may be present in the ore in many different chemical forms, some of which are not cyanide soluble. The consequence is that heap recoveries for gold typically range from 50% to 90% while silver typically ranges from 25% to 60%.

Recovery of gold from leach solutions is usually accomplished by adsorption of the gold cyanide complex onto granular activated carbon by pumping the solution through columns filled with the carbon. The leach solutions are recycled to the heap. Gold-rich solution from the heap usually contains only a few parts per million gold, whereas the carbon loads up to about 5,000 ppm. Gold is removed from the carbon by contacting it with a hot chemical solution that is then sent to an electrolytic cell for production of gold metal. This is usually melted on-site to produce a doré bar for sale to refineries (Marsden and House 2006).

The recovery of silver, and the recovery of both gold and silver from solutions high in silver, is usually accomplished using the Merrill–Crowe process. In this process the metal-rich solution from the heap is prepared by filtration and then vacuum treated to remove oxygen from solution. Zinc dust is added to precipitate the metals, and the precipitate is collected on filters. This precipitate is melted on-site to produce a gold/silver doré bar.

The level of cyanide in the heap leach on-flow solution typically ranges from 100 to 600 ppm NaCN, although some ores may require more than 1,000 ppm. Cyanide consumption via complexation, volatilization, natural oxidation, or oxidation by ore components typically ranges from 0.1 to 1.0 kg/t of ore. Cement and/or lime consumption ranges from 0.5 to 70 kg/t of ore. Several operations use cement for alkalinity control (instead of or in addition to lime) as well as for agglomeration.

Other leaching agents—thiosulfate, thiourea, hypochlorite, bromine—have been experimented with as an alternative to cyanide, but cyanide is by far the most effective and the most environmentally friendly leaching agent.

CHEMISTRY OF COPPER DEPOSITS
Copper ores are nearly always leached with a dilute sulfuric acid solution. Copper in heap pregnant solution is typically several grams per liter, whereas for gold/silver ores the solutions contain only a few parts per million of metals. With copper oxides it is not uncommon for the solution availability to the metals in the ore to limit the leach rate of copper. If a slow leaching rate is simply due to the avpplication rate, this can be remedied, but it is often due to the capillary action in the ore

particles, and this is difficult to speed up. Wetting agents have been used to overcome this, with limited success.

Depending on the ore, the natural production of sulfuric acid can be sufficient for the leaching process. The pH of the leach solution needs to be in the range of 2.0 to 2.8 or lower to prevent hydrolysis of the ferric ions. If other oxidizing sulfides (such as pyrite) are present, the acidity can drop to the point where gangue minerals react quickly to form clays, which plug the heap.

Recovery of copper from leach solutions is usually accomplished by processing the copper-rich solution through a solvent extraction plant where copper is extracted into an organic liquid and then back-extracted into concentrated acid for concentration and purification of the copper solution. The barren solution is recycled to the heap. The concentrated solution is sent to electrolytic cells for production of metallic copper. Copper cathodes produced from this process are normally of marketable purity and are sold directly to copper end users (Jergensen 1999).

CHEMISTRY OF URANIUM DEPOSITS

Uranium ores are typically leached with dilute sulfuric acid, often naturally generated, or ammonium carbonate. The tetravalent uranium in oxide minerals needs to be oxidized to the hexavalent state. This is normally done in an acid heap by natural bacterial oxidation, but use of chemical oxidants may be employed. For an alkaline heap, hydrogen peroxide, H_2O_2, is usually employed as an oxidant. The roll of the ferric ion in the oxidation in an acid uranium heap is the same as that in copper heaps. As a result of similar chemical processes, acidic uranium and copper heaps have the same issues to overcome (Merritt 1971).

If lime is present in the host rock, this can cause permeability issues as the leach solution dissolves and reprecipitates the calcium as gypsum. It is common to begin the leach process at a low sulfuric acid concentration and increase it after the gypsum has had a chance to reprecipitate throughout the heap instead of in a single layer, as would occur if higher initial acid concentrations were used. If the lime in the host rock causes excessive acid consumption, alkaline ammonium carbonate $[(NH_4)_2CO_3]$ leaching can be used.

Ammonium carbonate leaching is selective to uranium and as a result the uranium bearing minerals need to be exposed to wetting, possibly requiring finer crushing.

Recovery of uranium from both acid and alkaline leaching is typically accomplished by upgrading solutions with solvent extraction or ion exchange resins followed by a caustic treatment to precipitate "yellow cake" uranium, which can be refined to uranium dioxide (UO_2).

CHEMISTRY OF NICKEL DEPOSITS

Heap leaching of nickel has been proposed but rarely applied. Either hydrochloric (HCl) or sulfuric (H_2SO_4) acid may be used as lixiviants. Recovery of nickel from recycled heap solution requires expensive neutralization of the acid, and this is a barrier to application of the process.

LABORATORY WORK

As with any processing method, it is very important to base the design on the results of a comprehensive program of laboratory testing. For a proper heap leach laboratory program to be developed, it helps to know early on, preferably in the exploration stage, that heap leaching may be an option. To conduct representative column tests, the samples need to be coarse rocks that cannot be produced from reverse circulation drilling. Either large-diameter core diamond drilling (100–200 mm) or bulk sampling (tons) is required. It is common practice to drive a drift into the ore body to take a sample, but as the resource gets larger, this sample becomes less representative. The initial ore samples are seldom representative of the entire ore body; therefore, laboratory tests, including column leach tests, should be continued on a regular basis during mining.

After heap leaching is selected, there is a range of variables that need to be tested. These include crush size, heap stability, permeability versus heap height, solution application rate, reagent strength and consumption, the need for agglomeration, and the type (usually portland cement for gold/silver heaps) and the amount of agglomerating agent required, leach time, and percent recovery.

Heap leaching has inherent risks that can be largely eliminated if the design and operating practices follow the results of the initial and ongoing laboratory testing. The risks result from the nature of the operation. The results of the process are usually not known for several weeks or even months after the ore is stacked, and at this point it is not economical to reprocess the ore. Mistakes made in the initial plant design or incorrect operating practices (e.g., not crushing finely enough or not agglomerating or stacking properly) can result in cash-flow issues that might persist for up to a year after the problem is solved.

An on-site laboratory is an important part of the infrastructure at a heap leach operation. It should include an analytical section (for ore control) and a metallurgical testing section that regularly runs column leach tests on production samples. In addition to the standard production column leach tests, periodic test programs should be run to check the effects of chemistry, crush size, and agglomeration. For a small operation processing fewer than 5,000 t/d of ore, staffing should involve two or three technicians for sample preparation and assaying, and one metallurgist to conduct process tests. Large operations may have a laboratory staff of 10 to 15 people and around-the-clock operation.

HEAP PERMEABILITY AND FLOW EFFICIENCY

The key element in a successful heap leach project is a heap with high and uniform permeability. In any heap there are three zones of different flow regimes:

1. Coarse channels, which allow direct short-circuiting of solution from top to bottom
2. Highly permeable zones, in which solution is efficient at contacting the rock and washing the values downward in "plug flow"
3. Zones of low permeability where high-grade solution or unleached ore may be trapped

Efficiency of Solution Displacement

If the heap was "ideal" (i.e., moving in true plug flow), then when one displacement volume of solution was placed on top of the heap, it would fully replace the solution in the heap. This would be 100% wash efficiency. In practice, the "best" heap leaches exhibit a wash efficiency of about 70%. At 70% per displacement, three displacement washes are required to achieve a recovery of 95% of the dissolved metals. A fourth "displacement" is required, initially, to saturate the ore. Since a typical heap contains 20% moisture, 95% recovery (of the dissolved value content) requires that 0.8 t of solution must be applied to each ton of ore. In gold heaps, typical practice is to apply 1.3 t of solution per ton of ore during a 70-day primary leach cycle. This suggests two things: (1) most heap leach operations are able to maintain reasonably good permeability characteristics, yielding at least 50% wash efficiency; and (2) a high percentage of the recoverables is solubilized early in the 70-day leach cycle.

Drainage Base

A drainage base of crushed rock and embedded perforated pipes should be installed above the plastic leach pad and below the ore heap. The importance of this drainage base cannot be overemphasized. Solution should percolate vertically downward through the entire heap and then enter a solution removal system with zero hydraulic head. If the drainage base cannot take the entire flow, solution builds up in a stagnant zone within the heap and leaching within this stagnant zone can be very slow.

To put this in context, a "typical" heap might run 500 m upslope from the solution collection point. All of the on-flow solution in a 1×500-m strip must flow out at the downslope edge of the heap through the drainage base, which is typically 0.65 m thick. The design of the horizontal percolation rate through the drainage base is therefore nearly 800 times the design rate of the heap itself. This is not a difficult engineering accomplishment since flow is carried in pipes within the base.

At one Australian copper heap leach operation, three adjacent leach panels were built. The two flanking panels had a good installed drainage base, but the center panel did not. Recovery in the center panel was depressed 20% by the poor drainage base. A similar effect has been seen but not quantified at some gold heap leach operations.

Recovery Delay in Multilift Heaps

As subsequent lifts are stacked, the lower lifts are compressed and the percentage of low permeability zones increases. The first solution exiting an upper lift may have a values concentration of up to three times that of the ore. If impermeable zones have developed in a lower lift, high-grade solution may be trapped, causing a severe reduction in recovery rate and possibly in overall recovery percentage. The highest heap leaches currently in operation are 200 m high, with about 10 lifts of ore. Hard ore, crushed or ROM, can withstand the resulting pressure without significant permeability loss. Many softer ores can be agglomerated with enough cement so that they can perform under a load of 30 m; some agglomerated ores perform satisfactorily to 100 m. These properties can be properly evaluated in advance in laboratory column tests, which are run under design loads.

The delay in recovery as lifts are added to the heap is partly a function of the impermeability of the lower lifts and partly a function of the wash efficiency discussed earlier. The net effect is that average recovery is delayed as the heaps get higher and overall pregnant solution grade decreases (requiring more solution processing capacity).

Intermediate Liners

If the permeability of lower lifts becomes a serious problem, it is possible to install intermediate liners, though this is not recommended. Two problems occur with installing an intermediate liner: (1) the heap below the liner is compressed as the upper lift is placed, resulting in differential settlement and possible tearing of the liner; and (2) the ore below the liner cannot be washed with water, which is sometimes required as part of final heap closure.

SOLUTION APPLICATION RATE AND LEACH TIME

With regard to sprinkling rate, the timing for metals recovery is a function of the following factors:

- The rate at which the metal dissolves. Coarse particles dissolve very slowly, and may not fully dissolve for several months in a heap leach environment.
- The percentage of the ore minerals that exist as free or exposed particles
- The rate of diffusion of the solution into rock fractures, and dissolved metal back out of the rock fractures. Where ore minerals occur on tight fractures or in unfractured rock, the rock must be crushed into fine particles to achieve target rates and levels of recovery.
- The effect of chemical reactions within the heap, or within rock particles, which consume the reagents needed for leaching
- The rate of washing the values off the rock surfaces and out of the lift of ore under leach. This is a complex issue that depends on the overall permeability of the lift and the local permeability variations due to segregation and compaction as the lift is being constructed.

These factors cause wide theoretical differences in the response of various ores to leaching. However, in practice, most heap leach operations apply solution to crushed-ore heaps within a fairly narrow range of flows. The typical range is from 8 to 12 $L/h/m^2$, though some heaps range far above or below these rates (Van Zyl et al. 1988). Some operations start at higher solution application rates to saturate the ore and decrease the application rate with time. Other operations will start solution application very slowly until the heap is thoroughly wetted, to prevent the formation of preferential channels of water flow and then increase the application rate.

Laboratory columns always respond much faster than field heaps. Generally the cause of this problem is reduced reagent-to-values contact in a production heap. Two major reasons for the reduced contact are: (1) the ore is placed in the laboratory column much more uniformly so that percolation is more effective; and (2) the solution-to-ore ratio (tons of solution per ton of ore in a given time frame) is generally higher in laboratory columns than in field heaps. Both small- (150-mm) and large-diameter (1,000-mm) column tests tend to leach similarly. For some field heaps, notably where the ore is fine crushed and the ore leaches quickly, the solution/ore ratio is a more important factor than overall leach time. However, for the majority of heap leaches, time seems as important as specific application rate. The general target for the solution/ore ratio is between 1:1 and 1.5:1.

For ores with very slow leaching characteristics, an intermediate pond and a recycle stream may be added to the circuit, so that each ton of ore sees 2 t of leach solution during an extended leach period. The process plant treats only the final pregnant stream—1 t of solution/t of ore.

The use of multiple cycles is good operating practice for single-lift heaps of high-grade ore. However, for multilift heaps this is not the case. Heap modeling indicates that after the heap attains a height of three lifts, the intermediate solution contains almost as much metal as the pregnant solution. Recycling results in a significant buildup of dissolved values

Figure 11.3-2 Heap leach installation at Brewery Creek. The open-pit mine is shown in the background. In the center of the photograph is an operating heap leach completely covered with snow.

within the heap, causing a slight overall recovery loss and a cash-flow delay. For multilift heaps, it is often possible to justify an increase in the size of the recovery plant so that only fully barren solution returns to the heap.

It is extremely important to design a heap leach system so that the ore can be leached for a very long time. Unlike an agitated leach plant where the ore can be ground to a fine powder and intensively mixed, heap leaching is not a very energy-intensive process. After a heap is built, one of the most significant variables the operator can employ to solve design or production problems is the leach time. Some operations use on-off leach pads to achieve rapid first-stage recovery and then transfer the ore to long-term heaps to complete the process (Kappes 2002).

SOLUTION APPLICATION

The primary goal in the irrigation of a heap is to apply the solution as uniformly as possible. Solution distribution to a heap is done by flood, spray, or drip irrigation. The choice of irrigation can vary greatly depending on specific heap conditions. In recent years, flood irrigation has become a rarity. Spray and drip irrigation are widely used and frequently both are used on the same heap, depending on the season. The following equipment has become standard in heap leaching.

Drip Irrigation

Drip emitters, which issue drops of water from holes every 0.5–1.5 m across the heap surface, are very common. Drip emitters are small in-line devices spaced along a tube that distributes the solution evenly by forcing the water through a complicated path at each emitter, creating an equal pressure loss for the first and last emitter in a line. Drip emitters are easy to maintain and minimize evaporation and cyanide loss. In very cold climates it is possible to bury the emitters to prevent freezing, as shown at Brewery Creek (Alaska, United States) in Figure 11.3-2. In the winter months, the solution flow is piped directly to the process facility on the lower right. The barren leach solution is heated prior to application on the heap and solution pipes are heat traced. The ponds in the foreground are frozen over but, if necessary, excessive flow may be directed into them. Drip emitters can be advantageous for some copper and uranium heaps because they conserve heat that may be required for biological activity. The main drawback to drip emitters is that they do not provide continuous drip coverage. Thus the top 1 m of the heap may not be leached very well until it is covered with the next lift. Other drawbacks are that emitters, due to their small channels, require intensive (and expensive) use of antiscalant, and the use of in-line filters.

Wobbler Sprinklers

Wobbler sprinklers are used at a large number of operations. Their main advantages are that they issue coarse droplets, which control but do not eliminate evaporation, and they deliver a uniform solution distribution pattern that ensures uniform leaching of the heap surface. The coarse droplet size has another advantage in gold heaps. Cyanide is readily oxidized by air and sunlight, and the wobbler-type sprinkler minimizes this loss (but not as well as drip systems). Wobblers are typically placed in a 6 × 6-m pattern across the heap surface. A disadvantage of all sprinklers is that they require continual servicing, and personnel spend extended periods working in a "rainstorm." Occasional skin contact with cyanide solution does not pose a health problem, but an environment that encourages repeated skin/solution contact is not recommended. Sprinkler maintenance personnel, especially on acid copper heap leaches, wear full rain gear to eliminate any exposure problem, but the working environment (especially in cold weather) is not as pleasant as with drip emitters. Because of the impact of the water on the surface of a heap, wobblers, or any sprinkler system, can lead to the breakdown of delicate agglomerates and migration of fines into a heap. In a tropical climate it may be necessary to use sprinklers because of evaporation issues, but they can cause permeability problems. The influence of the sprinklers on the breakdown of agglomerates and migration of fines can be minimized by placing screen material over the heap to dissipate the impact energy of the water droplets.

Reciprocating Sprinklers

Reciprocating sprinklers shoot a stream, typically 5–8 m long, of mixed coarse and fine droplets. They are not considered ideal for heaps but often find application for sprinkling side slopes since they can be mounted on the top edge to cover the entire slope. If emitters and wobblers are used on side slopes, they must be installed on the slope, which is a difficult and sometimes dangerous place for service personnel to operate from.

High-Rate Evaporative Sprinklers

High-rate evaporative sprinklers typically operate at high-pressures with an orifice designed to produce fine droplets and shoot them in a high trajectory. Evaporative blowers using compressed air to atomize and launch the droplets can also be used. The drifting of the fine particles with wind can cause a concern with chemicals entering the surrounding environment. This type of equipment is not normally used at heap leach operations, but it will become more common as more heaps enter the closure mode where rapid evaporation is needed.

Regardless of the systems used for solution application and management, capital and operating costs for solution handling are usually small.

WATER BALANCE

Since many heap leach operations occur in desert areas where water is scarce, and others occur in environmentally sensitive areas where water discharge is not acceptable, the balance between water collection and evaporation is important. Fortunately, by adjusting the method and scheduling of solution application, it is usually possible to meet the local conditions.

The evaporation of water, regardless of its mechanism, requires a heat input of 580 cal/g of water evaporated. A heap leach gets this heat input from three sources: direct solar heating on heap and water surfaces, latent heat in the shroud of air within the "sprinkler envelope," and latent heat in the air that is pulled through the heap by convection.

The average 24-hour incident solar radiation on a flat horizontal surface ranges from 3,000 $kcal/m^2/d$ in the central United States to about 7,600 $kcal/m^2/d$ in the equatorial desert, which could theoretically evaporate 5–12 L of solution/d/m^2. With a typical heap application rate of 10 $L/m^2/h$, incident solar radiation could account for an evaporation rate of 2%–5% of applied solution when using sprinklers. Evaporation would be somewhat less when using drip irrigation (1%–4%) because some of the solar energy is reradiated from dry areas on top of the heap. This same heat input would result in pond evaporation of 5–13 mm/d.

Overall evaporative losses include the sprinkler losses, convective loss from air flowing through the heap, and losses due to heating/evaporation from ponds and from other areas not sprinkled. These have been determined at several Nevada operations to be up to 20% of the total solution pumped in summer months, but averaging 10% annually. Thus, direct sprinkler loss accounts for about 60% of the total evaporation. Use of drip irrigation can reduce but not eliminate evaporative loss.

In tropical climates, noticeable losses occur even during the rainy season. On several tropical heap leach projects where rainfall is seasonal and up to 2.5 m/yr, the overall annual evaporative loss from all sources, when using wobbler-type sprinklers operated 24 h/d, is about 7% of the solution pumped. A typical heap application rate is 10 $L/m^2/h$, or 88 $m/yr/m^2$. Thus, evaporative loss of 7% is equal to 6.2 $m/yr/m^2$ on the areas actually being sprinkled. If the heap and pond systems are properly designed, the active leaching area can be up to 40% of the total area collecting rainfall; it is therefore possible to operate in water balance when rainfall is 2.5 m/yr. For these operations, very large solution surge ponds are required.

Where rainfall is high and evaporation rate is low, some operations cover the side slopes with plastic to minimize rain collection. Others have tried to cover the entire heap during the rainy season, but this has not worked very well because of the mechanical difficulties in moving the covers.

In western Africa and Central America, it is acceptable practice to treat and discharge excess solution during the rainy season. Typically, excess process solution is routed through a series of ponds where cyanide is destroyed using calcium hypochlorite [$Ca(ClO)_2$] or H_2O_2, followed by adjustment of the pH to near neutral. The INCO SO_2 system, using copper-catalyzed hyposulfite to destroy cyanide, is also employed for this purpose. Cyanide-free solution is further treated in controlled wetlands (swamps) to remove heavy metals prior to discharge.

The worst water balance situation occurs in cool, damp climates such as in high-altitude operations. In such climates, rainfall and snowfall may be significant and evaporation is minimal. Generally such heaps can stay in water balance with an aggressive program of summer sprinkling. Arctic heap leaches have been able to stay in water balance because precipitation is lower than the total water requirement needed to saturate the ore (Kappes 2002).

LEACH PADS

The leach pad below the heap is a significant element of a heap leach design. The ideal location for the heap is a nearly flat (1% slope), featureless ground surface. Usually some earthwork is required to modify contours, but it is not necessary to eliminate all undulations. It is only necessary that all the solution will flow across the surface toward the collection ditches on the base or the sides of the heap. Where the slope exceeds 3%, the front edge of the heap (30–50 m) should be graded flat to provide a buttress to prevent heap failure. In the western United States, where the water table is often far below the surface, the current practice is to construct the leach pad of 1.5-mm thickness high-density polyethylene or very-low-density polyethylene on a 30-cm-thick layer of compacted clay. A leak detection/water seepage system of pipes is installed below the liner.

HEAP HEIGHT

As discussed previously, for multilift heaps, there is a delay in the recovery and grade. This is not only true for multilifts but also for single lifts as the lift gets taller. A lower recovery grade means that either a larger recovery plant needs to be installed or a lower cash flow will have to be tolerated. The delay in cash flow caused by delayed recovery also needs to be taken into account for additional lifts or height. Depending on the permeability of the material and lift height, the delay per lift can range from 3 to 30 days. It is common to see a delay of 7 days per lift. As a single lift gets taller, there is the added concern of particle size distribution. As the ore is stacked, it rolls down the side slope of the heap. This cascading action naturally segregates the fine particles near the top of the lift, and the large particles near the bottom. The taller the lift the more stringent the controls on this problem need to be. On the other hand, the advantage of taller lifts is that there is less high-grade solution flowing through already-leached ore beneath it. The larger the quantity of high-grade solution flowing through leached ore, the more opportunity there is to lose values. For copper heaps there are additional factors that must be considered when choosing lift height. Copper oxide heaps are frequently limited in height to maintain the pH in the range that keeps the copper soluble. Copper sulfide heaps will create acid within the heap that may cause a pH problem, and these can also easily become oxygen deficient. Because of these issues and permeability problems, copper ores may be leached in "thin layer" heaps where the ore is stacked only 3–4 m high and leached for a relatively short period (weeks instead of months).

With harder ore and quality agglomeration, permeability is not always the driving factor in the ultimate heap height. It may be good operating practice to reduce the heap height in order to maintain a high, consistent solution grade, which is important for planning purposes, especially at very large operations.

MINING, ORE PREPARATION, AND STACKING

Mining of ore for heap leaching employs the same techniques and equipment as mining of ore to feed any other process

Figure 11.3-3 Agglomerating drum and conveyor stacking system with 6-m-high heap

method. Where uncrushed (ROM) ore is placed on the leach pad, ore may be blasted very heavily in order to reduce rock size and improve gold recovery. In high-rainfall environments when processing clay-rich material, it is very important to practice a mining routine that minimizes the amount of rainfall absorbed by the ore.

Ore preparation varies widely. ROM ore may be hauled from the mine and dumped directly onto the heap. At the other extreme from ROM leaches, some heap leaches crush the ore and dry grind it to more than 50% passing 150 μm (100 mesh) and agglomerate the fines (AusIMM 1991). At times, the high-grade ore will be ground and subsequently reblended with the coarse low-grade ore going to the heap. This process is called "pulp agglomeration" and is practiced at Ruby Hill in Nevada.

Ores high in clay (such as saprolites) are typically processed by two stages of crushing using toothed roll crushers, then agglomerated in drums and stacked using a conveyor stacking system. Many ores are crushed and then either truck-stacked or conveyor-stacked without agglomeration. For these harder ores, crushing is usually achieved by a jaw crusher, followed by one or two stages of cone crushing.

Agglomeration

The term *agglomeration* means different things to different operators.

The simplest application of agglomeration is practiced where the ore is hard but contains a large percentage of fines. Agglomeration means simply wetting the ore with water so the fines stick to the coarse particles and do not segregate as the heap is built.

A more involved form of agglomeration is practiced where the ore contains amounts of clay or fines that may begin to plug a heap of untreated ore. Belt agglomeration may be employed. In this technique, cement and water are mixed with the ore at a series of conveyor drop points, and the mixture tends to coat the larger rock particles. The primary goal is stabilization by mixing and contact. A typical conveyor stacking system involves 10 or more drop points, so belt agglomeration may occur as a normal part of the process.

Where ores are nearly pure clays, such as the laterite/saprolite ores in tropical climates, drum agglomeration is usually employed. Figure 11.3-3 shows a typical agglomerating drum. The ore is first crushed finely enough (typically 25–75 mm) to form particles that can be a stable nucleus for round pellets. For gold and silver ores, cement and water are then added and the ore is sent through a rolling drum. The fines and the cement form a high-cement shell around the larger particles, and the rolling action of the drum compacts and strengthens the shell. Copper ores cannot use cement because the sulfuric acid will break it down. Concentrated sulfuric acid is used as a binding agent instead, although the binding effect of sulfuric acid is not very good. Where copper ores are very high in clay or in minerals that decompose in the heap to clay, heap leaching may not be very effective. Drum size and throughput are a function of several factors, but typically a 3.7-m-diameter 10-m-long drum can process 750 t/h. A 2.5-m-diameter drum can process 250 t/h. At the Tarkwa mine in Ghana, two 3.7-m drums were installed to process up to 20,000 t/d of ore. For multilift gold heaps, it is often necessary to use a higher cement addition on the lower lifts, and this is usually determined by laboratory tests in which the ore is leached under the full heap load. Maintaining permeability of the lower lifts is extremely critical to the success of a multilift heap (Kappes 2002).

Truck Stacking

Where rock is hard and contains very little clay, it is possible to maintain high permeability even when ore is crushed and dumped with trucks (Figure 11.3-4). Truck dumping causes segregation, of the ore as it cascades down the slope. To control the degree of this segregation, the ore may be partially agglomerated (wetted to cause the fines to stick to the coarse material) prior to placing in the trucks.

Truck dumping can also result in compaction of roadways on top of the heap. Several studies have indicated large trucks noticeably compact ore to a depth of 2 m. To mitigate this problem, most operations rip the ore after stacking but prior to leaching. The number of ripper passes is important; usually

Figure 11.3-4 A truck is dumping an upper lift onto a lift that is actively under leach. The road formed by haul truck traffic on the heap can be seen on the right.

it is four passes in a crisscross pattern. Some operations practice building an elevated truck roadway that is later bulldozed away. However, this requires substantial bulldozer traffic on the heap surface, which can lead to permeability problems.

Stacking the ore with trucks can result in the tie-up of a large tonnage of ore below the truck roadways. This is a bigger problem for small operations than for large ones, because the roadway width is usually the same regardless of the daily production rate. For a heap leach of 5,000 t/d, the roadways on the heap can tie up one month's ore production. A conveyor system that stacks ore from the base of the lift can reduce the unleached inventory to a few days' production. Because of this inventory reduction, at smaller operations where the ore is crushed, it is usually less capital intensive to install a conveyor stacking system. Conversely, for operations of 100,000 t/d or more, truck stacking is more flexible and may be less capital intensive than a conveyor system.

Conveyor Stacking

Two major types of conveyor stacking systems are used on heaps: radial stacking systems and spreader conveyor systems.

Radial conveyor stacking systems commonly include the following equipment:

- One or more long (overland) conveyors that transport the ore from the preparation plant to the heap. Typically these consist of conveyors up to 150 m or longer.
- A series of 8–15 "grasshopper" conveyors to transport the ore across the active heap area. Grasshoppers are inclined conveyors 20–50 m long, with a tail skid and a set of wheels located near the balance point.
- A transverse conveyor to feed the stacker-follower conveyor
- A stacker-follower conveyor, typically a horizontal mobile conveyor that retracts behind the stacker
- A radial stacker 25–50 m long, with a retractable 5–10-m conveyor at its tip called a stinger. Wheels, discharge angle, and stinger position are all motorized and are moved continuously by the operator as the heap is built.

Radial stackers are usually operated from the base of the lift but may be located on top of the lift, dumping over the edge. Figure 11.3-5 shows a heap that is not only stacked with conveyors on top of the heap but is also stacked in a novel manner: a spiral of one continually climbing lift instead of multiple individual lifts. Inclined conveyors can be installed up the sides of the lower lifts, and the stacking system can be used to build multilift heaps. Stackers for this purpose should have very-low-ground-pressure tires and powerful wheel drive motors to cope with soft spots in the heap surface.

Radial stacking systems can be used for heaps processing up to 50,000 t/d of ore, but beyond that, the size of the stacker (and the bearing pressure that is exerted by the wheels) becomes prohibitive. For operations stacking very high tonnages, large stackers can be mounted on caterpillar tracks to reduce ground pressure.

Spreader conveyors can be used on very-high-tonnage heaps. On on-off (dynamic) heaps, spreader conveyors span the entire width of the heap and continually travel back and forth, distributing the ore on the heap. A bucket-wheel excavator can be installed to remove the ore after leaching is complete. Several recent dynamic heap installations have been built in the form of a ring or oval with a moving "slot" from which the ore has been removed. Ore is removed on the advancing face of a slot and new ore is placed on the trailing edge. The "slot" can be seen in Figure 11.3-6 between the new ore stacked by a spreader conveyor on the right and leached ore removed by a bucket-wheel excavator on the left. The slot between old and new material with a short distance into each face of the heap for the two conveyors to work is the only area of the heap that cannot be under leach.

Mobile spreader conveyors up to 400 m long, mounted on multiple caterpillar tracks so they can travel in any direction, are used on multilift, very-high-tonnage heaps, with notable examples being in the Chilean copper industry.

Figure 11.3-5 Valley-fill heap leach at Ocampo, Mexico

SOLUTION COLLECTION

After the values have been dissolved, it is necessary to collect the pregnant solution. A wide variety of arrangements are used to collect the solution and direct it to the pregnant pond. On some valley-fill heap leaches, all of the solution is drained to a single point on the pad and directly into the pregnant liquor pond. As mentioned earlier, a portion of the solution may even be stored inside the void spaces in the heap, thereby minimizing the need for a large pregnant pond. Pregnant solution is usually collected in lined ditches that run adjacent to the heap. More than one inlet to the ditch may be used if it is desired to observe the recovery of different areas on the heap. To further increase the ability to watch the recoveries of different parts of the heap, small ridges may run the length of the leach pad, perpendicular to the collection ditch, to separate the heap into "cells." It is common for a weir to be placed at the point where the pregnant solution exits from each cell, or from the entire pad, to measure the flow draining from the heap. With a heap designed with cells, it is possible to segregate solutions into an intermediate pond. Multiple ditches or pipes run parallel to the heap, one running to each pond. Figure 11.3-7 illustrates this process at the Tarkwa heap leach in Ghana. The three large gray pipes on the right distribute the barren and intermediate leach solutions to each cell of the heap while the black pipes collect the intermediate and pregnant heap runoff solutions. The box at the base of the cell is a solution control point to select which solution will be used for leaching and to which pond the runoff will go. There is also a weir in the distribution box for an instantaneous measurement of solution from the heap. The pad below the heap is commonly a single plastic liner above compacted clay, but the ditches and ponds (and flooded areas of valley-fill heaps) are usually constructed with a double plastic liner with intermediate leak detection.

METALLURGICAL BALANCE

If the values available are not well tracked, the amount of recovery will never be known. The following are some of the values that are important to obtain to be able to calculate a good metallurgical balance:

- Sampling of the ore to be stacked
- Knowing when, where, and how much ore is stacked
- Conducting quality column leach tests to monitor the continually changing ore, and modeling the test results to apply to the heap
- Sampling and flow measurement of the barren and pregnant solutions to and from the heap
- Good assay procedures employing fire assay of the ore samples for gold ores, and total and soluble copper assays for copper ores
- A good model that accounts for delays caused by adding lifts
- Patience

It is important to collect good grade, tonnage, and column test data from day one. It cannot be assumed that the entire ore body will act the same as the initial sample taken for design. To maintain a representative model, the recovery curves will change with the grade, ore type, variations of dilution from waste, and other factors that may affect the recovery. It is difficult to produce a stable and accurate metallurgical balance early in the life of a heap because of the slow reaction time. As a heap ages, the intricacies of how the permeability and leach rates of that particular heap will become easier to predict. A metallurgical balance is a continually changing tool for managing a heap and predicting the cash flow. A good model will keep bankers, stockholders, and your managers happy. A poor one can put a mine in dire straits when the predicted cash flow does not materialize.

DESIGN CONSIDERATIONS FOR RECLAMATION AND CLOSURE

After the heap leaching operation is completed, the facility must be closed in accordance with local environmental requirements. Closure activities are highly variable depending on the environmental sensitivity of the site and on the regulatory regime. In general, heaps are washed for a short period of time (commonly 3 years), during which time 1 t of wash water or recycled treated process solution per ton of ore is applied. Heaps are then capped, and ponds are filled and covered with suitable materials.

The easiest heaps to reclaim are single-lift heaps because the older heaps are abandoned early in the life of the operation and can be washed while production operations continue. In

Courtesy of FLSmidth RAHCO, Inc.
Figure 11.3-6 Slot between new ore (right) and leached ore (left)

Source: Kappes 2002.
Figure 11.3-7 Solution collection at the Tarkwa heap leach

valley-fill heap leaches, nearly all the ore ever placed on the pad is situated directly under active leach areas. Thus, washing of the entire heap must wait until operations are completed. Larger operations may have two or more valley-fill leach areas and can therefore appropriately schedule closure activities.

Environmental regulations usually applied in the United States call for reasonably complete washing of the heap to reduce pH, to remove cyanide, and to partially remove heavy metals. Cyanide is fairly easy to remove because it oxidizes naturally, but pH and heavy metals are more difficult to control. Regulators are recognizing that a better approach is to conduct a "limited" washing program and then to cap the heap with a clay cover and/or an "evapotranspiration" cover of breathable soil with an active growth of biomass. These covers are designed to prevent infiltration of water into the heap. After several years of active closure activities, the flow rate of the heap effluent decreases to a manageable level (or to zero in arid environments). After the flow rate has reached an acceptably low level, heap closure is accomplished by installing a facility for recycling collected effluent back to the heap. A relatively small "cash perpetuity bond" is maintained such that the interest on the bond covers the cost of maintaining and operating the intermittent pumping facility for as long as necessary.

Worldwide practices range from simple washing and abandonment to the more complex procedure described above. Environmental design is an industry unto itself, and the simplistic concepts discussed here may not be universally applicable. Heap closure needs to be addressed in the feasibility stage of the project.

TROUBLESHOOTING

As with most things in life, the best troubleshooting for a heap is to do it right the first time. One of the primary challenges in operating a heap is the delay in metallurgical response. With long leach times, the results of a change in operations will not be seen for months. A common mistake people make is to watch a heap on a day-by-day basis, which is an insignificant amount of time. Arguably the most critical time in the construction of a multilift heap is the base lift of the heap, when operators are least familiar with the equipment and the characteristics of the ore. If the first lift is built poorly and has poor permeability or chemistry, this can have a lasting effect throughout the life of the heap.

Permeability
When permeability problems arise, there are few options to improve it. Often the first sign of low permeability is puddles of standing water on the surface of the heap, called ponding. Apart from ripping the top couple of meters, there are few viable options, although several have been tried. Some heaps are rehandled to improve permeability, which may be profitable on a higher-grade heap as pockets of ore will not have leached. Heaps have also been blasted to try to "fluff up" the ore, though with limited success. Allowing the water to pond might help to force the solution through. Efforts to improve the permeability of already-stacked ore are typically cost prohibitive.

Stability
Low permeability can also cause the ore to become saturated with water and lead to stability issues. The primary cause of stability issues in a heap is the complete saturation of a structurally significant portion of the heap. At the first signs of solution discharge from the side of the heap (more than 1 m above the base) or small blowouts, it is best to stop leaching that area immediately and assess the problem. The problem may be a general issue of poor permeability, or it may be that during stacking of an upper lift, the top surface of the lower lift was made impermeable. If this problem occurs, it usually implies that there will be an overall loss of recovery in the entire heap. To solve the immediate problem, the only practical solution is to reduce leaching near the heap edge, but it is possible to drill the area for installation of vertical drains. The problem indicates that the operation must do a better job of agglomerating new ore, or of ripping and then limiting traffic on the surface below a new lift.

Liner Leaks
Most pads have leak detection pipes below the pad, and leaks will report to the secondary containment sump. Some leakage through the primary liner is generally permitted, but excessive leakage will require abandonment of the heap. Several geophysical technologies have been developed to help locate a hole in a liner, but their accuracy is limited to a circle on the pad with a diameter equal to the vertical height of the heap. It is expensive to repair a liner under an operating heap, as the simple act of exposing it can complicate the issue by creating more holes. It is critically important to prevent liner leaks during construction.

Poor Recovery
A common problem that a heap leach will run into is that the ore takes longer to leach than predicted by the laboratory tests. To eliminate this risk, heap leach design must allow for total flexibility to extend leach times. Many heaps, especially in the copper industry, have achieved economically acceptable performance even though leach times were very long.

It is common practice to minimize the use of cyanide on a heap to save money. In reality this can lose money if done incorrectly. Initial laboratory tests can indicate starting concentrations and operators may periodically explore different cyanide levels to see the effects on recovery. The problem with this approach is that the results of any changes in a heap take a long time to notice. A better approach is to maintain an active production support laboratory and to periodically run a series of parallel column leach tests at different cyanide addition levels to find the best addition concentration. The cyanide concentration in a heap is controlled by the addition of cyanide to the leach solution, but the cyanide in the pregnant solution coming from the heap should also be monitored.

Acid Production/pH
For cyanide heap leaches, lime is usually used to maintain a pH above 10. For ores that generate acid slowly, limestone can be blended in with the ore as well. Caustic can be used but is normally not economic and can result in chemical reactions that plug the heap or are detrimental to the recovery process. If the pH of the pregnant solution flowing from the heap drops because of the oxidation of pyrite and other sulfides, there is generally no economic remedy. Such acid zones should be watched carefully as they can cause gold loaded in upper lifts to precipitate. That area of the heap may have to be abandoned. Future ore should be stacked with the correct amount of pH modifier.

For copper heaps, H_2SO_4 can be generated by the ore itself, but in most heaps, extra acid must be added. However, it may be needed in such quantities as to make the availability of this reagent the key economic issue in evaluating the project.

CAPITAL AND OPERATING COSTS
Heap leaching normally has significant capital cost advantages over agitation leaching. The capital costs for a heap leach can vary widely, from $2,500 to $8,000 per daily metric ton of ore treated depending on many variables including, but not limited to, process rate, location, infrastructure, and company policy. The capital costs of a typical agitated leach circuit per daily ton of ore treated is up to eight times that of a typical heap leach.

Operating costs for an agitated leach circuit may be several times that of a heap. In other cases where a large amount of cement is required for agglomeration or where the ore needs to be fine-crushed, the operating costs of agitation leaching are not necessarily higher than for heap leaching. Heap leach operating costs are not very sensitive to the size of the operation because as some things get less expensive (i.e., general and administrative expenses), other things get more expensive (recovery plant operation). This relationship can be seen in Table 11.3-1.

CONCLUSION
Although the concepts of heap leaching are simple, the practices have substantially evolved over the past 40 years. Early choices for pad materials, sprinkler systems, and stacker designs have been discarded under the pressure of operating experience and cost-reduction factors. Overall operating costs have continually declined as "superfluous" activities and controls have been eliminated.

In spite of the apparent simplicity of the heap leach process—or perhaps because of it—there were many failures in the early years. Now there is a large number of successful operations from which to draw the experience needed to optimize the process. Heap leaching is expected to maintain its

Table 11.3-1 Heap leach operating cost distribution

Operation	3,000 t/d	15,000 t/d	30,000 t/d
Mining, %	27.8	26.0	24.2
Crushing, primary, %	3.7	2.6	2.9
Crushing, second plus third stage, %	4.6	5.1	2.9
Crushing, fourth stage, %	7.4	10.4	11.4
Agglomeration/stacking, %	1.9	1.3	1.4
Leach operations (including sprinkler supplies), %	4.6	3.9	2.9
Recovery plant operations, %	13.9	16.9	20.0
General site maintenance, %	5.6	3.9	4.3
Cement for agglomeration (10 kg/t), %	9.2	13.0	14.3
Cyanide, lime, other reagents, %	2.8	3.9	4.3
Environmental reclamation/closure, %	4.6	6.5	7.1
General and administrative, support expenses, %	13.9	6.5	4.3
Total operating costs, %	100	100	100
Total operating costs, $	10.80	7.70	7.00

place as one of the principal tools for extracting gold, silver, and copper from their ores for both large and small deposits. The challenge for the future will be to remember and apply the experiences of the past.

REFERENCES

Agricola, G. 1556. *De Re Metallica*. Translated by Herbert C. Hoover and Lou Henry Hoover. 1950 edition. New York: Dover Publications.

AusIMM (Australasian Institute of Mining and Metallurgy) 1991. *World Gold '91: Gold Forum on Technology and Practice*. 1991. Second AusIMM-SME Joint Conference. Victoria, Australia: AusIMM.

Bartlett, R.W. 1998. *Solution Mining: Leaching and Fluid Recovery of Materials*, 2nd ed. Amsterdam: Gordon and Breach Science Publishers.

Jergensen, G.V. 1999. *Copper Leaching, Solvent Extraction, and Electrowinning Technology*. Littleton, CO: SME.

Kappes, D. 2002. Precious metal heap leach design and practice. In *Mineral Processing Plant Design, Practice, and Control*, Vol. 2. Edited by A.L. Mular, D.N. Halbe, and D.J. Barratt. Littleton, CO: SME.

Marsden, J., and House, I. 2006. *The Chemistry of Gold Extraction*, 2nd ed. Littleton, CO: SME.

Merritt, R. 1971. *The Extractive Metallurgy of Uranium*. Golden, CO: Colorado School of Mines Research Institute.

Scheffel, R. 2002. Copper heap leach design and practice. In *Mineral Processing Plant Design, Practice, and Control*, Vol. 2. Edited by A.L. Mular, D.N. Halbe, and D.J. Barratt. Littleton, CO: SME.

Van Zyl, D., Hutchison, I., and Kiel, J. 1988. *Introduction to Evaluation, Design and Operation of Precious Metal Heap Leaching Projects*. Littleton, CO: SME.

WISE (World Information Service on Energy). 2007. Uranium heap leaching operations. www.wise-uranium.org/uhl.html#OPS. Accessed November 2009.

CHAPTER 11.4

Surface Techniques of Solution Mining

W. Joseph Schlitt

Most categories of solution mining represent alternatives to mechanical excavation of ore. For example, dredging is simply mechanical excavation from the bottom of a river or other body of water. Hydraulic mining uses a high-pressure jet of water to dislodge material from a working face and even transport it to the processing plant. In-situ mining extracts values chemically or by simple dissolution, leaving the host rock in place and essentially undisturbed. In contrast, surface (leaching) techniques are forms of ore processing not unlike milling and flotation. All rely on mechanical excavation of ore to provide the feed to be treated.

Most types of solution mining represent the primary method of production, including mine-to-leach operations. However, surface leaching can also provide a secondary or supplemental form of production, with leaching limited to sub-mill-grade mine waste. Usually such material is of too low a grade to justify the cost for some degree of crushing, even though crushing could improve recovery or at least the rate of recovery. Hence, leach material is typically truck-hauled in run-of-mine (ROM) condition and dumped in lifts on the leach pile. Supplemental leach operations are fairly common in the copper industry, where higher-grade ore is usually milled and floated. They also exist in the gold industry, where higher-grade ores are typically ground- and agitation-leached.

Cost accounting usually differs significantly for primary and secondary leach operations. In a primary mine-to-leach scenario, the leach operation carries the full cost of mining. This is not unlike typical milling and flotation operations, which also carry the full cost of mining. Conversely, secondary operations typically carry no mining costs, as the leach material must be mined anyway to expose mill-grade ore. In effect, secondary operation lowers the strip ratio and provides better utilization of the resource, as some production is derived from material that is otherwise wasted. However, some incremental costs may be assigned to secondary leaching. An example is the extra cost needed to construct one or more leach-lined dumps with profiles that optimize recovery, rather than simply dumping material at the shortest haul location.

Both primary and secondary operations must meet two general sets of criteria, one that focuses on the characteristics of the ore and one that involves the development of efficient operating practices.

Consideration must first be given to the characteristics of the deposit and the leach material itself. Parameters include location of the resource, extent of the mineral inventory, mineralogy and mode of occurrence of the metal values, and chemical and physical nature of the host rock. Nature must provide a sufficient tonnage of material with adequate recoverable metal values per ton; otherwise the leach project never reaches the point where operating practices are of any consequence. For a proposed leach operation, the same type of exploration and development program is required as for any other mining project. The difference is that metallurgical testing is directed toward leaching instead of ore comminution and beneficiation.

This chapter describes the establishment of leach operating practices for treating whole ore, as opposed to some type of concentrate. Coverage begins with a section on ore–lixiviant systems for the different commodities, as well as bioleaching and metallurgical testing. This is followed by a section on materials handling that includes mining, leach systems, ore preparation, and emplacement. The next section covers solution management, including environmental considerations. The chapter concludes with shorter sections on downstream operations and infrastructure requirements.

Throughout this chapter, the following terms are used:

- *Heap leaching* refers to a mine-to-leach operation involving crushing, possible pretreatment, and careful stacking to optimize metal recovery.
- *Dump leaching* refers to an operation in which low-grade ROM material is placed in mine dumps and leached to obtain additional metal values.

ORE–LIXIVIANT SYSTEMS

One of the first decision points in a proposed leach operation is to select the optimum lixiviant with which to treat the ore. Clearly, if the mineralization is too refractory, leaching is not viable and other process options need to be considered.

The two key requirements for any lixiviant are high selectivity and rapid leach kinetics. The kinetics should always be

W. Joseph Schlitt, President, Hydrometal, Inc., Knightsen, California, USA

such that dissolution at the surface of the ore is the most rapid step, so that the overall leach rate is controlled by transport of reagents or products to or from the mineral surface. Other desirable characteristics for the lixiviant include low price, low toxicity, ease of handling, and capacity for regenerating or recycling. The lixiviant should also allow for easy concentration and recovery of the metal values.

In spite of many years of research on possible lixiviants, only a few are used widely in commercial operations. The principle reagents used to extract each major commodity are discussed in the following subsections.

Gold and Silver

Alkaline sodium cyanide solution is used almost exclusively to extract gold and silver during heap and dump leaching. Studies on the mechanism of dissolution indicate that two reactions are involved in gold dissolution (Dorey et al. 1988). Most of the gold dissolves according to the following reaction:

$$2Au + 4CN^- + O_2 + 2H_2O \rightarrow 2Au(CN)_2^- + H_2O_2 + 2OH^- \quad (11.4\text{-}1)$$

However, a small but still significant portion of the gold dissolves according to the Elsner reaction:

$$4Au + 8CN^- + O_2 + 2H_2O \rightarrow 4Au(CN)_2^- + 4OH^- \quad (11.4\text{-}2)$$

In both cases, the balancing cation is sodium (not shown, for simplicity), and oxygen is required. For any given ore, the gold dissolution rate depends on cyanide concentration, oxygen content, and solution alkalinity. The optimum pH is about 10.3 (Dorey et al. 1988), but many operators tend to run at somewhat higher values as a margin of safety. Because most ores are not so basic, alkaline reagent must also be added to the system to raise the pH. This is generally done by blending lime (CaO) with the ore, or adding milk of lime [$Ca(OH)_2$] or caustic (NaOH) to the solution.

In theory, nothing should prevent dissolution of virtually all of the gold in a heap or dump. In actual operations, extraction seldom exceeds 85% and recoveries as low as 40% occur. The problem is caused by poor accessibility of the gold to the lixiviant. One cause for this is poor heap construction that leaves portions of the ore blinded off so that solutions do not wet the entire rock mass. Another cause is gold encapsulated in silica or other refractory minerals such as pyrite or arsenopyrite. The processing of these so-called refractory ores is an area of technology that is still developing but is often applied to concentrates rather than whole ore.

Silver, although commonly extracted along with gold, is generally not recovered as effectively. Its dissolution reactions with cyanide are analogous to those for gold, but a lesser chemical driving force is involved. In addition, silver is much less noble than gold and tends to associate with various refractory minerals in addition to pyrite and arsenopyrite.

In spite of its almost universal usage on low-grade material in heaps or dumps, cyanide has a number of inherent problems, including the following:

- High potential toxicity
- Environmental concerns such as aquifer contamination
- Relatively slow kinetics for dissolution of both gold and silver
- Generally poor extraction of silver
- Requirement for alkaline reagents when leaching acidic ores
- High reagent consumption due to poor selectivity when leaching complex ores containing various base metals

The presence of copper is particularly troublesome, as most oxide and secondary sulfide minerals leach well and form highly soluble cyanide complexes. In addition, copper usually reports with gold and silver as they are stripped from the leach solution. The presence of copper then complicates the final parting and refining of the precious metals.

In spite of the problems with cyanide leaching, operators have learned to deal with issues such as cyanide destruction in effluent streams and detoxification of heaps and dumps at closure. Nonetheless there remains a strong impetus to find an alternate lixiviant. Most attention has been focused on thiourea [$(CS(NH_2)_2]$ (Hiskey 1981, 1984). Its potential benefits, compared to cyanide, include the following:

- Up to 12 times faster leach times, resulting in leach cycles of hours rather than days or months
- Functional in an acidic rather than an alkaline medium
- Better selectivity for gold and silver
- Lower toxicity

Unfortunately, however, after three decades of development, thiourea has not lived up to its potential, and its use generally remains confined to high-grade materials. The main problems are high unit costs and high reagent consumption.

Copper

Only three acidic lixiviants have been used in commercial copper leaching operations on ores and mine wastes: (1) acidic sulfate solution is the dominant choice and can be used on both oxide and sulfide mineralization, including ores that contain a mix of the two types; (2) acidic chloride solution has also been used; examples involve the use of sea water or saline groundwater for the leach solution and do not involve additions of hydrochloric acid; (3) sulfuric acid is used for leaching, even with a chloride solution, and actually produces a mixed sulfate–chloride lixiviant. Of these, sulfuric acid offers the following advantages:

- It is relatively benign with regard to materials of construction, with well-known handling characteristics.
- It is compatible with recovery of copper by conventional solvent extraction–electrowinning (SX-EW) technology.
- It is inexpensive and commonly available as a by-product from copper smelters that are often located nearby.

In addition to acidic systems, an alkaline ammonia-based lixiviant has also been used commercially in specialized cases, including the leaching of both ores and concentrator tailings containing malachite and azurite (copper carbonates). For many years, the Calumet and Hecla Mining Company ran ammonia leach plants to recover native (metallic) copper from mill tailings (Benedict and Kenny 1924). Today, no copper ores or tailings are treated in ammonia leach plants. The technology is therefore not covered further in this chapter; for more information, see Chase (1980).

The copper mineralization dictates how the ore should be leached. Oxidized ores, including most oxides, silicates, carbonates, hydroxides, and chlorides, can be dissolved effectively with a nonoxidizing acid—that is, sulfuric acid. The general reaction is a straightforward, irreversible dissolution that consumes acid. The specific reaction for the common mineral chrysocolla is

$$CuSiO_3 \cdot nH_2O + H_2SO_4 \rightarrow CuSO_4 \\ + SiO_2(H_2O)_{n+1} \quad (11.4\text{-}3)$$

However, unoxidized ores, mainly the various sulfide minerals and native copper, do require an oxidant. As described below, the chemistry is more complex and relies on natural oxidation processes to accomplish leaching. The oxidant may be either dissolved oxygen or an ionic species such as ferric iron in a sulfate-based system or ferric or cupric ions in a chloride system. Regardless of the lixiviant, the leach solution must be either acidic enough or basic enough to carry copper in solution, as copper salts tend to precipitate from near-neutral solutions.

In the absence of chloride input, the natural lixiviant for copper sulfide minerals is acidic ferric sulfate. An example for the leaching of chalcopyrite is

$$4Fe^{+3} + CuFeS_2 \rightarrow Cu^{+2} + 5Fe^{+2} + 2S° \quad (11.4\text{-}4)$$

Here the balancing anion is sulfate (not shown). Although small amounts of elemental sulfur are detected on partially leached mineral surfaces, no accumulation of elemental sulfur is observed in actual operations. Thus, any elemental sulfur is transitory and progressively oxidizes all the way to sulfate in the presence of bacterial catalysts and oxygen in the air within rock-pile voids.

The ferric iron in Equation 11.4-4 can come from several sources, including acid-soluble iron in various gangue minerals. Without going into detail, much of the iron comes from the oxidation of pyrite (Hiskey and Schlitt 1982), which plays an important role in sulfide leaching. Pyrite is typically 3 to 10 times more abundant than the copper species, so it is important simply on a mass basis. Pyrite oxidation is analogous to the chalcopyrite reaction shown in Equation 11.4-4. In addition to supplying iron, pyrite oxidation is an in-situ source of acid generation that helps maintain the acid balance and low pH needed to keep copper in solution (Templeton and Schlitt 1997).

These sulfide oxidation reactions are also exothermic. Heat released during reaction boosts the temperature in the heap or dump and accelerates the leach kinetics. It also produces a buoyancy effect and promotes convective airflow through the rock pile, which provides the oxygen that is needed for reoxidation of ferrous iron back to the ferric state. This reaction is typically bacterially catalyzed, as described later in the "Bioleaching" section.

Uranium

Commercial heap leaching of low-grade uranium ores was first practiced in the United States during the uranium boom in the 1950s. More recent examples include the Los Gigantes acid heap leach in Argentina and the bacterial heap leach on pyritic ore at Urgeiriça in Portugal (now shut down). Commercial operations currently exist in Brazil, China, Niger, and Spain, and both acid and alkaline leach projects are being planned in Australia, Botswana, and Namibia. The list seems likely to grow, given the renewed interest in nuclear power and the economics of heap leaching.

As is also the case for copper, both mine-to-leach and secondary operations can use sub-mill-grade feed. Most of these projects currently produce no more than a few thousand tons of yellow cake per year, and total output is small compared to in-situ operations. In addition to primary production, uranium has also been produced as a by-product of copper leaching operations in a few cases (George et al. 1968).

Regardless of the leaching method, the same lixiviant chemistry is applicable to uraninite (UO_2) and coffinite [$U(SiO_4)_{1-x}(OH)_{4x}$] ores that are commonly treated. Both acidic and alkaline lixiviant choices exist (Grunig 1981). To minimize reagent losses in solution mining, the lixiviants have been tailored to avoid excessive gangue reactions. Thus a sulfuric acid–ferric sulfate leach was developed for sandstone ore, and a sodium bicarbonate or ammonium carbonate leach was developed for uraninite in calcareous ore. The oxidant in the latter system is either hydrogen peroxide or oxygen dissolved in the leach solution.

For both acid and alkaline circuits, dissolution requires oxidation of tetravalent uranium to the hexavalent state. In dilute sulfuric acid (pH 1.6 to 2.0), uranium is oxidized by ferric ion and goes into solution as the very stable uranyl sulfate complex. Similarly, sodium or ammonium bicarbonate lixiviants form uranyl carbonate.

The overall reactions in an acid circuit are (Merritt 1971)

$$UO_2 + Fe_2(SO_4)_3 \rightarrow UO_2SO_4 + 2FeSO_4 \quad (11.4\text{-}5)$$

$$UO_2SO_4 + 2H_2SO_4 \rightarrow [UO_2(SO_4)_3]^{-4} + 4H^+ \quad (11.4\text{-}6)$$

The overall reactions in an alkaline circuit are (Merritt 1971)

$$2UO_2 + 2H_2O_2 \rightarrow 2UO_3 + 2H_2O \quad (11.4\text{-}7)$$

$$UO_3 + Na_2CO_3 + 2NaHCO_3 \rightarrow Na_4\,UO_2(CO_3)_3 \\ + H_2O \quad (11.4\text{-}8)$$

Both types of circuits are well established and have seen extensive commercial application in both solution mining and conventional mills using agitation leaching.

Nickel

Nickel and by-product cobalt are the most recent commodities to be recovered by heap leaching. The target ore types are laterites, which contain more than 70% of the world's nickel but account for only ~40% of the world's production (McDonald and Whittington 2008a, 2008b). Unlike gold, silver, copper, and uranium, virtually no discrete nickel or cobalt oxide minerals are found in laterite ores. The occurrence of both nickel and cobalt is quite variable, as they substitute for iron in goethite (limonitic ore) and smectite group clays or for magnesium in saprolite (serpentine) ore. A typical example is the hydrous magnesium silicate falcondite [$(Ni,Mg)_4Si_6O_{15}(OH)_2 \cdot 6H_2O$].

As of this writing, the technology for heap leaching of laterites is still developing and competes with the more established high-pressure acid leaching. Key issues for heap leaching have recently been reviewed (Steemson and Smith 2009). For both types of leaching, sulfuric acid is the overwhelming choice of lixiviant. For heap leaching, the acid concentration is typically 50–75 g/L. Because nickel and cobalt substitute in the structure of various complex minerals, the entire mineral must be solubilized to release the metal values. As a result, sulfuric acid has poor selectivity, and mass loss can be as high as 30%. Acid consumption is also high and varies from 10 to 75 kg/kg (lb/lb) of nickel or 100 to >500 kg/t (200 to 1,000 lb/st) of ore.

Because of high acid consumption, heap permeability can be a problem. As a result, laterite heap leaching is typically done in shallow, single-lift heaps run in on/off mode, as described later in the "Leach Systems" section.

Other Metals

Other than the commodities covered previously, zinc is the only metal that is currently produced commercially by heap leaching. Teck Cominco (now Teck Resources) reportedly pilot-tested the heap leaching of low-grade sphalerite (ZnS) ore, but results have not been reported, nor has a commercial project been announced. However, a zinc silicate heap leach is reportedly operating in Namibia with sulfuric acid as the lixiviant.

Low-grade manganese ores should also be amenable to solution mining, although this has never been done commercially. The lixiviant would be aqueous sulfur dioxide (SO_2), which reacts with manganese dioxide (MnO_2) as follows:

$$MnO_2 + SO_2 \text{ (aq)} \rightarrow MnSO_4 \text{ (aq)} \quad (11.4\text{-}9)$$

Less-oxidized forms of manganese require an oxidant such as dissolved oxygen, hydrogen peroxide, or perhaps ferric sulfate. Tests reported by Potter et al. (1982) suggest pregnant-leach-solution (PLS) grades ranging upward of 10% to 12% $MnSO_4$, implying that manganese can be recovered as the sulfate by evaporation or as the carbonate or hydroxide by neutralization and precipitation.

Polymetallic Systems

Other than gold/silver production, leach operations recovering more than one metal have been rare. In addition to the by-product uranium mentioned previously, efforts have also been made to extract cobalt from copper leach liquors in the United States. Cobalt recovery from some African copper leach operations seems certain, as does recovery of cobalt from nickel laterite heap leaching. Another possibility is dual-circuit recovery of copper and zinc. Recovery of cobalt, copper, manganese, and zinc was planned for the mine-to-leach operation at the Sanyati mine in Zimbabwe; however, the current status of this project is uncertain. The technology for cobalt recovery already exists for both copper and nickel circuits. However, when multiple recovery circuits operate in series, problems always exist with coextraction of metal values and carryover of reagents from one circuit to the next.

Bioleaching

Only the leaching of sulfidic ore and mine waste has a bacterial component; the leaching of nonsulfide materials is based on straightforward chemical control. Years of development work have been spent on bacterial or bioleaching, but commercialization has generally been limited to high-grade ores and concentrates.

The exception is Newmont Mining Company's two-stage heap-leach process for treating refractory (pyritic) gold ore. In the first step, bacterial leaching with forced aeration partially oxidizes the pyrite, liberating or at least exposing the gold inclusions. When at least 40% of the pyrite is oxidized, the heap is thoroughly rinsed to remove acid that was formed. In the original plan, the material was to be off-loaded, blended with lime, restacked on a second pad, and leached with cyanide to recover the gold. In actual operation, the first heap is off-loaded and hauled to the carbon-in-leach mill, where bio-oxidized material is neutralized in a semiautogenous grinding mill. This process has not found widespread use, but details are available in Logan et al. (2007).

Other commercial leach operations having a bacterial component all rely on indigenous microbial species for the necessary activity. The natural bacterial oxidation provided by the chemoautotrophic bacterial strains *Thiobacillus ferrooxidans* and *Thiobacillus thiooxidans* has been well known for years. *Thiobacillus* are generally active at temperatures below 40°C (104°F). Large sulfidic leach dumps typically operate at higher temperatures, with indigenous moderate or extreme thermophiles playing the same roles.

The exact mechanisms of bacterial action are still in dispute, but the accepted theory is that the bacteria obtain energy for growth from enzymatic oxidation of ferrous to ferric iron and elemental sulfur to sulfate.

T. ferrooxidans bacteria catalyze the oxidation of ferrous to ferric iron (Brierly 1978; LeRoux et al. 1973):

$$2Fe^{+2} + \tfrac{1}{2}O_2 + 2H^+ \rightarrow 2Fe^{+3} + H_2O \quad (11.4\text{-}10)$$

Similarly, *T. thiooxidans* bacteria catalyze the oxidation of sulfur to sulfate:

$$S° + 3/2 O_2 + H_2O \rightarrow 2H^+ + (SO_4)^{-2} \quad (11.4\text{-}11)$$

The rates of these reactions can increase by several orders of magnitude when catalyzed bacterially. With the addition of carbon, nitrogen, oxygen, phosphorous, potassium, and other elements as nutrients, the bacterial metabolism can be maintained to sustain a stable population.

The key factor to keep in mind in such an operation is that microbial populations are living organisms. Actions that cause sudden changes to their preferred environment should be avoided—otherwise the change can render the population inactive or even kill it outright, negatively impacting production. An example is sudden introduction of chloride ion via use of magnesium or calcium chloride as a dust suppressant or deicer on haul roads.

Metallurgical Testing

Regardless of the values to be recovered, a proper metallurgical test program is required for any new leach project. Test results establish expected operating parameters and set the design criteria needed to engineer and construct leaching and recovery facilities.

The most important single factor in any test program is the selection of samples to be tested. Great care must be taken to ensure that samples are representative of the material to be leached. If a deposit contains more than one type of ore, in terms of either mineralogy or lithology, then each type must be tested thoroughly. When a mine plan is available, samples representing various mining periods should also be tested to avoid surprises in recovery or reagent consumption. A common approach is to test quarterly composites for the first 3 years of operation and annually thereafter.

Test Levels

At least two and sometimes three levels of testing must be undertaken (McClelland 1988):

1. Preliminary tests: small-column percolation or bottle roll tests
2. Detailed tests: larger column tests to optimize feed size and other parameters
3. Large- (pilot-) scale tests: large-column or field tests

Preliminary tests accomplish the following: (1) provide insight into the amenability of the material for leaching; (2) allow selection of the preferred lixiviant, should the choice not be obvious; (3) provide information on the chemical aspects of the proposed ore–lixiviant system; (4) indicate whether or not agglomeration is needed to avoid permeability problems with clays or fines. With respect to the chemical aspects of the proposed system, chemical parameters include a first estimate of reagent consumption and the rate and ultimate level of extraction of the one or more metals. Differences in the behavior of the various ore types become evident.

Care must be taken in extrapolating preliminary test results to predict design parameters in the eventual operating system. Preliminary results are almost always misleading in the following ways:

- Leach rates are higher than in commercial operation. Material is crushed much more finely in small-scale tests than in a full-sized system. Finer crushing maximizes liberation of the metals and minimizes diffusion distances to any mineralization that is not exposed. Thus metals are solubilized rapidly and completely, assuming that results are not impacted by complete reagent consumption in typical batch-type tests.
- Reagent consumption is higher than in commercial operation. This reflects not only the high extraction of metals but also the high host-rock surface area. In particular, fine rock size tends to maximize reagent consumption when the lixiviant is not highly selective for the metals and attacks host-rock (gangue) minerals as well. Such behavior is particularly common for acidic lixiviants, which tend to be less selective than alkaline reagents. For example, in acid leaching of copper, on a mass basis there is generally much more reaction with gangue constituents than with copper minerals.

Detailed tests are run in larger columns. These tests are generally intended to determine whether or not the ore should be leached in ROM conditions, crushed to some smaller size, or crushed and then agglomerated. Extraction-rate curves and reagent-consumption figures are determined as a function of particle size and degree of agglomeration.

Detailed tests can be run in either of two ways: single- or multiple-feed sizing. In the former, medium-sized material is column-leached until the extraction rate levels off to near zero. The contents of the column are screened and assayed to determine recovery as a function of particle size. Generally a second confirmatory test is run on material crushed to the selected size. This approach is less costly but more time-consuming than running parallel tests on multiple samples crushed to a range of sizes. With either approach, the last step is to balance projected recoveries and revenues against the cost of progressively finer crushing and/or agglomeration. The most economical operating scenario can then be selected.

Similar detailed tests can be done to optimize other operating parameters as well. These parameters include agglomerating conditions, heap height, leach-solution irrigation rate, reagent concentration, pH, and, in some cases, oxidation potential. Along with crush size, these parameters define the operating conditions.

Large-scale tests are done less frequently now than in the mid- to late 1900s. They typically involve large columns or square cribs that represent a slice of the expected commercial heap. Cribs are frequently loaded in the same way that the commercial heap will be stacked, replicating ore emplacement with its natural angle of repose and size segregation with fines at the top. Typical large-scale column tests are run with 18–36 t (20–40 st) of gold/silver ore (McClelland 1988) and as much as 180 t (200 st) of copper ore (Murr 1980). In many cases, large-scale column tests are more meaningful than tests performed on small field heaps because of edge effects caused by the large percentage of material contained under the sloping heap faces.

Large-scale tests are often omitted, and test programs stopped at the detailed-test stage, for the following reasons:

- Considerable expense and an extended schedule are required to complete the tests and evaluate the results.
- Obtaining a representative sample can be difficult. Even driving an adit into the middle of the ore body may be insufficient if the deeper ore has different characteristics.
- For deposits that have only been drilled, there may simply not be enough core samples available for large-scale tests.
- Permitting and environmental requirements are likely to be as onerous as for full-scale operation.
- Extensive databases for actual projects, together with advanced computer-modeling techniques, now make projection of expected operating performance much more accurate than it has been in the past. However, this effort still requires a comprehensive test program that accurately replicates the proposed operating parameters.

Logistical considerations and the availability of representative leach material often limit really large-scale field tests to existing mine operations. Use of a large-scale heap or dump is required only when assessing the impact of a parameter that cannot be replicated in smaller-scale testing. A good example is the million-ton ROM test heap that Kennecott Utah Copper Corporation built at the Bingham Canyon mine (Utah, United States) (Schlitt 2006). The primary objective was to assess the impact of severe winter weather on performance of the heap and the 6 t/d (6.6 st/d) SX-EW plant used to recover the leached copper. A secondary objective was to provide a facility for operator training. By the end of the program, the project had produced 771 t (850 st) of cathode copper that was sold as a credit against the cost of the project.

Test Results

Regardless of the test scale, at the completion of metallurgical testing, results are typically presented as plots of cumulative metal extraction or reagent consumption versus a test variable.

One obviously useful plot is cumulative extraction versus time under leach (Figure 11.4-1). The leach curve has a pseudo-parabolic shape, indicating that the reaction rate is fast initially but slows progressively with time. The slope of the tangent to the curve at any given time gives the instantaneous leach rate at that time. For the curve shown in the figure, the rate is in units of percent extraction per day. A linear initial leach rate implies that the initial rate is controlled by either

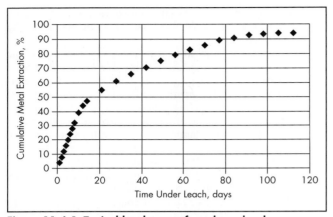

Figure 11.4-1 Typical leach curve for a heap leach

reagent availability or reaction rate at the mineral surface. As leaching continues, metal-bearing minerals exposed on the rock surfaces are consumed and the rate decreases as diffusion to or from less-accessible mineralization becomes rate-controlling. The rate continues to decrease until all accessible mineralization has reacted and the ore is leached to exhaustion.

Another useful plot is recovery versus cumulative solution flux (total quantity of leach solution applied per metric ton of ore under leach). Such a plot has a normalizing effect, enabling comparison of, for example, results for columns of different heights. Using solution flux also compensates for periods when flow to the heap is interrupted. Other plots may be appropriate for special situations. For example, for sulfuric acid lixiviant, a plot of recovery versus acid flux may be informative.

In addition to the general test criteria outlined previously, there are many commodity-specific aspects to metallurgical testing. These cannot be addressed in detail here but do require that a well-planned research program be conducted by knowledgeable metallurgists and engineers to get meaningful results. To cite just one commodity-specific example, bottle roll tests on cyanide extraction of gold should be done in open bottles. If the bottles are sealed, poor results may be due to oxygen deficiency rather than the presence of refractory mineralization.

MATERIALS HANDLING

This section covers everything from mining of the leach material to its emplacement on the heap prior to leaching. This includes a description of the leach systems, with alternatives to conventional heap and dump leaching. The section concludes with a description of ore preparation, which covers crushing, agglomeration, and the various methods of emplacement.

Mining

Surface heap and dump leaching operations are almost invariably associated with open-pit mining operations. These require essentially the same planning functions and use the same unit operations as those described in Chapter 10.2. However, two areas may be impacted by leaching operations: mine planning, and drilling and blasting.

If heap or dump leaching supplements conventional milling, mine planning can become quite complex. One complexity lies in the need for more segregation of mined materials. Material must be divided into not just two categories, mill ore and waste, but three: mill ore, leach material, and waste.

Both leach heaps and waste dumps need to be designed, and movement of all three types of material must be scheduled. In addition, water distribution and collection systems must be designed and installed for the leach system.

Supplemental production of leach copper also affects calculation of the ore cutoff grade and ultimate pit limits. Since leaching permits partial recovery of copper from material that would otherwise be wasted, the effect of a leaching operation is always to raise the cutoff grade on ore going to the mill. This, in turn, lowers unit costs for the mill output. Also, since leaching boosts total recovery of values from the mineral resource, it also increases the tonnage of material that can be profitably mined. Stated another way, a supplemental leach operation expands the ultimate pit limits. The exact impact of leaching is site specific and depends on the relative costs and output of leach production and mill production.

Drilling and blasting is another area that may be impacted by leaching, particularly when a mine-to-leach operation uses ROM ore. The drill-and-blast pattern must be designed to give a well-fragmented product with minimal quantities of oversized rock, about plus 200–250 mm (8–10 in.) top size. The powder factor also requires control to avoid generation of excessive amounts of fines that could cause permeability problems in the pile. The well-fragmented product should stack and percolate well while providing good mineral exposure needed for effective leaching. Because rock characteristics are very site specific, costs for an optimized drill-and-blast program that gives the desired as-mined leach product may need to be compared with costs for a low-cost drill-and-blast program that includes primary crushing to produce material for leaching.

Leach Systems

Heaps and dumps are by far the most common types of surface leaching operation. There are two types of heap leach systems to consider: a permanent multilift system and an on/off single-lift system. Each has advantages and disadvantages.

Permanent Multilift Leach System

An advantage of the permanent multilift system is that it minimizes material-handling costs. After a lift is leached, fresh ore is stacked over the spent ore. If the original lift is leached to exhaustion, an interlift liner can be used to separate the old and new ore. This prevents high-grade leach liquor from percolating down into the old material where it will be diluted by the residual moisture in the heap. However, in most cases, the original lift still contains residual metal values that can be extracted. In this case, no interlift liner is used and the solution percolating down through the new ore continues through the original lift, recovering additional metal. Stability issues and installation costs require additional consideration when employing interlift liners.

A disadvantage of the permanent multilift system is that the base of the heap must be impervious to solution flow to avoid contamination of the environment and avoid loss of metal-bearing solution. As a result, the lined-pad area must be expanded as the heap footprint grows. At the end of operation, the lined area must be sufficient to contain the entire leach ore output. A related problem is that loading will develop at the base of the heap as additional lifts of ore are added. This may compact the lowest ore and cause permeability problems at the bottom of the heap. The result can be development of a perched water table within the heap, causing solution to break out at intermediate heights or even heap instability. Even in the absence of permeability problems, the amount of residual moisture in the heap system increases as ore tonnage in the heap increases. This residual moisture typically varies from 6% to 12% of the ore's dry weight, increasing as ore size becomes finer. This moisture contains soluble metal values that were extracted from the ore but are not available for recovery. Thus, they represent in-process inventory and must be taken into account when forecasting production. These in-heap metal values will not be recovered until the operation is closed and the heap system is given its final rinsing as part of the closure exercise.

On/Off Single-Lift Leach System
The on/off single-use leach pad is typically used for ores that leach quickly or have serious permeability problems. The size of the lined pad area is fixed and is much smaller than that for a permanent multilift system. Ore is stacked on the pad, leached to exhaustion, rinsed, and removed to a permanent impoundment. Thus the pad must be large enough to hold material being stacked, leached, and off-loaded, and some additional area must be available for loading or to provide some flexibility in the leach cycle. If the residual ore is not acid-generating, the impoundment area may not require a liner. Because of the rinsing and small active ore volume, in-heap inventory is not a serious issue.

An on/off pad has two major disadvantages. One is the cost of double-handling the ore, once to stack it and once to remove it to the residual impoundment area. The other is the lack of opportunity to recover any values remaining when the leach cycle is complete and the spent ore is removed.

Alternate and Variant Leach Systems
Some alternatives or variations on these systems are appropriate for special circumstances.

Dump leaching is equivalent to a permanent multilift system but involves low-grade material handled in ROM condition. It is subject to the same issues as is the permanent multilift system.

Sand–slime separation is an alternate means for handling ore with a high fines content. Separation typically is done at 2 mm (10 mesh). The coarse fraction can be heap- or vat-leached and the fines can be agitation-leached in some type of countercurrent system. This circuit is currently in use at the Tintaya mine in Peru, where the coarse fraction is heap-leached and the fines are agitation-leached. It was previously used at the Ray mine (Arizona, United States), where the coarse fraction was vat-leached and the fines were leached in air-agitated Pachuca tanks.

In-place systems represent a special case of materials handling, intermediate between surface heap and dump leaching and in-situ mining. Broken or rubblized material is leached more or less in place. In-place leaching clearly involves no loading or hauling of broken ore to heaps or dumps. However, unlike in-situ mining, in-place leaching still relies on downward percolation of solution through unsaturated leach material. (In contrast, in-situ mining is characterized by pressure-driven solution flow in a saturated medium).

In-place systems are so site specific that discussion is limited to brief descriptions of a few commercial operations. Although broken uranium ore has been leached in underground settings, copper appears to be the main target to date. Probably the largest-scale operation was undertaken by Ranchers Exploration and Development Corporation when they drilled and blasted the entire upper portion of the Old Reliable mine (Arizona) ore body and then acid-leached the rubblized zone in place (Longwell 1974). Technically they were able to fracture and rubblize the ore, but leaching yielded less copper and produced a solution that was lower in copper concentration or tenor than expected. Thus the project was not the commercial success that was hoped for (Catanach et al. 1977).

Pit-wall leaching has yielded greater success. It involves progressive rubblization and leaching of abandoned pit walls. Since most porphyry ore bodies contain a halo of low-grade mineralization around the economically mineable zone, pit-wall leaching often recovers a portion of the copper that is otherwise left behind. Examples of pit-wall leaching include the Mineral Park and Silver Bell mines in Arizona. Examples of underground leaching of broken, remnant ore left behind in block-caving operations include the El Teniente mine in Chile and the San Manuel mine in Arizona.

Ore Preparation
Material scheduled to go to heaps or dumps can be handled in ROM condition or prepared for leaching by crushing, with or without agglomeration. Size reduction should be undertaken when the increase in production achieved with finer product generates sufficient incremental revenue to cover the capital and operating costs for crushing. In general, low-grade mine waste that must be removed to expose mill-grade ore does not merit much extra material handling or preparation. Such material is often dumped in ROM condition at locations that minimize haulage costs rather than maximize leach output. However, there has been a move to optimize at least the dump profiles to enhance recovery. Then the dump is leached to recover some of the values that would otherwise be lost.

Agglomeration
Agglomeration is widely applied in precious metals operations. It can provide at least two benefits:

1. Agglomeration provides strong, porous, and coarsely sized material that otherwise would be unleachable because of poor percolation characteristics when placed in heaps. This is particularly true when at least 10% of the material is finer than 420–500 μm (35–40 mesh). Such conditions usually lead to low metal extraction because of slow and uneven percolation (channeling) of solution and development of impermeable (dead) zones within the heap or dump. Clays are probably the leading source of fines, although fines produced by poor crushing practices are undesirable as well. Even tailings can be agglomerated and subsequently heap-leached, as was done for uranium–vanadium recovery from salt-roasted tailings at Naturita (Colorado, United States) (Scheffel 1981).
2. Agglomeration uniformly blends the lixiviant into the crushed ore. Leach reactions have a chance to start before the ore is even placed on the heap, which provides for quick release of metal values and high PLS grades. However, experience indicates that agglomeration is likely to have little or no effect on the final extraction level.

Agglomeration follows one of two pathways: it causes either adherence of fines onto coarser particles or conversion of fines into stable pellets.

For gold and silver, important process parameters include quantity of binder (typically 5 kg/t or 10 lb/st), moisture content (10% to 20% of total), and agglomerate curing period (≥8 hours). Since the subsequent cyanide leach requires alkaline conditions, binders are typically lime or portland cement. These binders generally have two important properties: (1) clay permeability is improved by the exchange of sodium ions in the clay with calcium ions in the binder, and (2) the binders have a cementing or pozzolanic effect that strengthens the agglomerates. Leach extraction rates can also be accelerated by using a barren cyanide solution as the source of moisture.

For copper ores and other materials that use an acidic lixiviant, the purpose of agglomeration is often primarily to make the ore acidic. The acid-cure dosage is normally less than half the amount of acid consumption expected for the particular ore. A higher dosage does little to improve extraction but raises the amount of acid consumed by gangue mineralization. In addition to accelerating the initial release of copper, acid agglomeration prevents a high pH in the initial drainage from the heap—that is, the PLS. If the pH is higher than 3.5 to 4.0, copper can precipitate from solution and be difficult to resolubilize.

Other process parameters include feed characteristics (particle size and presence of fractures), process control, and equipment selection (Milligan 1984; McClelland and van Zyl 1988). For gold and silver, the most popular types of agglomeration equipment are belts, drums, and pans. For copper, the most popular type is the drum, which thoroughly blends the acid and water throughout the ore charge. To save on capital costs, however, acidulation is sometimes achieved by spraying acid on the ore while it is transported by conveyor belt, although significant blending occurs only at the conveyor drop points and transfer stations.

Alternate Methods

As an alternative to agglomeration, some operations perform sand–slime separation as described in the previous section. For dump leaching, size reduction is generally not accomplished unless primary (in-pit) crushing is performed so that the material can be transported and stacked by conveyors. For heap leaching, primary, secondary, and tertiary crushing are often used to produce a final product that is nominally 25 to 12.5 mm (1 to 0.5 in.). Screening requirements depend on the nature of the ore, extraction sensitivity to crush size, and equipment manufacturer recommendations. Modeling the crushing circuits based on the various rock properties can enable selection of the optimum equipment configuration.

Ore Emplacement

How material is stacked for leaching—how it is emplaced—is a major factor in the success of a surface leaching operation. The approach to emplacement represents a principal difference between dump leaching and heap leaching.

For dump leaching, dumping practices reflect topographic constraints and efforts to minimize haulage costs. Consideration of leach recovery is usually limited to practices such as waste segregation, modification of the heap profile, and dump-surface preparation. Segregation usually involves efforts to avoid mixing barren or mineralogically unleachable materials into the leach dumps. Surface preparation before addition of the next lift is typically limited to leveling and deep ripping to break up the compacted zone associated with the truck traffic pattern. However, dump profiles can be adjusted to enhance leach conditions. For example, sulfide leaching normally requires some type of aeration, and dump dimensions can often be adjusted to enhance natural convective airflow through the rock mass. Optimization of the profile for different conditions has been modeled extensively and provides a guide to dump design for a particular ore.

Waste dumps are often high to minimize haulage costs; elevation differences of 90 to 230 m (300 to 750 ft) between toe and crest are common. This profile exacerbates two types of segregation:

1. Size segregation, which occurs when ROM material is dumped over the crest. This allows the coarsest material to slide or roll to the bottom while the fines tend to remain at or near the crest. As a result, the least permeable and most easily compacted zone of material is found at the top. Leach solution then percolates slowly and unevenly at the top of the dump and tends to channel and short-circuit through this layer. These poor percolation patterns normally persist at depth so that a significant volume of material in the dump is leached poorly if at all. The presence of leaching material below the sloping face may also be a problem.
2. Layer segregation, which creates alternate layers of coarse and fine material that lie parallel to the angle of repose. The coarse zones act as conduits and carry the bulk of the lixiviant flow. Hence they are usually well leached in spite of their large particle size. In contrast, the fine material has a low permeability and so undergoes little solution penetration or leaching unless proper solution management practices are followed.

These factors in waste emplacement are responsible for the generally poor extractions observed in waste dumps. For ROM material, recovery of metal values seldom exceeds 50% after many years of leaching. Only 10% to 15% extraction during the first 1 or 2 years is not uncommon, at least for copper sulfide dumps.

For heap leaching, careful emplacement of rock is important to maximize extraction of values. This generally involves construction of much shallower lifts, typically 3 to 10 m (10 to 35 ft) high to minimize size segregation. A further way to minimize size segregation is to use crushed or agglomerated ore. In turn, solution percolation is more uniform and solution-rock contact is improved, resulting in faster and more complete metal recovery.

Several additional approaches to heap construction, described by Muhtadi (1988), warrant consideration due to their widespread use. Foundation requirements are covered later in the "Environmental Aspects" section.

Truck dumping and dozing. Emplacement by means of truck dumping and dozing is essentially as described previously for waste dumps. However, more care must be given to emplacement of material. Typically, a ramp of waste is constructed to the height of the first lift. Ore is dumped from the end of the ramp down onto the pad or foundation of the heap. As ore builds up, the heap extends away from the ramp. To minimize compaction, an elevated roadway can be extended as the heap expands, with haulage trucks restricted to this path. When the heap is completed, the roadway is dozed off and the surface of the heap is graded and ripped to ensure good permeability and uniform percolation of solution.

In general, this method of heap construction is restricted to ores that do not undergo compaction or generate fines. This requires a strong, competent material handled either in ROM or crushed condition. Agglomerated materials cannot generally be handled by truck dumping, as they are too soft and friable. In addition, truck dumping still causes coarse/fine segregation.

Plug dumping. Emplacement by plug dumping also involves trucks but is not characterized by the rough handling noted previously. Hence it is suitable for softer ores and agglomerates. The first step is to bed the liner or foundation with crushed rock as a protective measure to ensure liner integrity. Then each haul truck dumps its load as closely as possible to the previously dumped pile until the entire pad is covered with overlapping mounds. The maximum heap height is about 2 to 3 m (7 to 10 ft), slightly less if a dozer is used to level the tops of the mounds.

A variation of this method is to replace the trucks with large rubber-tired front-end loaders. These are very efficient and can stack material to heights of ≥5 m (16 ft) without the need to first bed the liner.

Conveyor system. Emplacement by conveyor stacking has gained widespread acceptance in heap leaching. The concept is not new, however. It started early in the 20th century when high-grade copper ores were bedded and removed from leach vats by means of a conveyor-tipper and a clamshell bucket on a traveling gantry crane, respectively. Indeed, one of the first stacking systems in gold heap leaching (Gold Fields Mining Corporation, Ortiz, New Mexico, United States) used a traveling bridge that spanned a reusable pad and used a moving tripper to continuously add ore to the face of the heap. Spent ore was removed from the asphalt pad by front-end loaders.

Now conveyor hauling and stacking of ore on heaps has evolved into a highly mobile operation. The front end of the system usually consists of a short fixed conveyor that receives prepared ore from a crusher, agglomerator, or a stockpile via a feed hopper. The conveyor transfers material to a mobile, radial-arm conveyor/stacker by means of a system of movable intermediate conveyor sections. The stacker has great flexibility and can bed the heap uniformly and without compaction to almost any height desired. Thus conveyor systems can handle anything from primary crushed ore to agglomerated tailings. For on/off systems, front-end loaders or bucket-wheel excavators are used to remove the leach residue.

SOLUTION MANAGEMENT

Materials handling and solution management are interactive, with rock preparation and stacking having a substantial effect on solution flow. For example, maximum recovery of metal values can be achieved only if the heap/dump design permits good distribution and collection of leach solution. The design must also provide for movement of air into the piled rock, as oxygen is needed as a reactant in solubilizing many values. Likewise, the rock must be broken and stacked so that the entire rock mass is accessible to solution percolating down from the surface of the heap or dump.

Nevertheless, after the rock has been prepared and put in place, there is little that an operator can do except optimize the way the leach solution is managed. Doing so accomplishes two things: (1) it transports the lixiviant to the mineralized values throughout the piled rock, and (2) it washes the solubilized values from the heap or dump to the recovery plant.

Fluid Flow Phenomena

Fluid flow phenomena include both static and dynamic aspects. The latter covers the optimum application rate and the potential benefits of leach/rest cycles and pulsed leaching.

Static Considerations

The first step in understanding solution flow through rock mass is to categorize the space within the dump or heap. There are at least three distinct regions:

1. Solid rock (including isolated pores). By far the largest, this completely stagnant space typically involves about 60% of the total volume.
2. Space associated with openings created by rock blasting and handling. This space exposes pores and creates fractures or cracks on the faces of the rock fragments. Generally only 2% to 4% of the total volume, nevertheless, this space greatly increases the exposed surface area and makes the contained mineral values more accessible to the lixiviant.
3. Void space that occupies the remaining volume within the rock pile. This space results from the swelling that is a consequence of rock fracturing and stacking during mining and handling. For newly prepared heaps, void space normally constitutes about 40% of the volume of the rock pile. This value decreases as weathering and leaching consolidate the material and typically drops to <30% if a mine dump is leached over a period of years.

Obviously, void space can be filled with either leach solution or air. However, because the water–rock interface has a lower surface energy than does the air–rock interface, all the pores within the rock fragments fill with solution. The rocks themselves are covered with water under conditions of percolation. The water-filled void space represents the area for active water flow. In contrast, mass transport within the water-filled fractures or pores occurs only by diffusion. A consequence of the difference in interfacial energies is the location of the air-filled void space. Under conditions of percolation, only the larger voids and channels contain air, and the smaller openings are flooded.

Except under conditions of flooding, competition between air and water for available void space is almost independent of the rate at which leach solution is applied. Instead, the controlling factor is the effective particle size of the rocks in the pile. This is a consequence of the limitations imposed by capillary forces. As shown in Figure 11.4-2, capillary rise or drain-down height is related to effective rock size. The figure shows the correlation between rock size and percentage of the total void space that remains filled with water after leaching ceases and drain-down is complete. At this point, capillary rise is balanced by the force of gravity, and only evaporation can further reduce the water-filled void volume.

The experimental results shown in Figure 11.4-2 are significant when ingress of air is needed to help solubilize the metal values. For material that is finer than ~320 μm (48 mesh), the pile does not drain down at all, and little or no air can penetrate the pile. For material that is coarser than ~2.0 mm (10 mesh), the pile drains down almost completely, allowing air to permeate throughout the heap or dump.

Dynamic Considerations

When leaching begins at a low application rate on a thoroughly drained dump, some additional air-filled void space

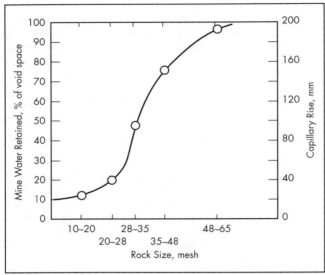

Source: Schlitt 1984.
Figure 11.4-2 Solution retention and corresponding capillary rise as a function of rock size

Table 11.4-1 Effect of solution application rate on water-filled void space

Application Rate		Water-Filled Void	
mm/h	gal/ft²/h	Space, %*	Comment
0	0	8.0	Drained
1.2	0.03	9.6	Capillaries filled
6.0	0.15	9.7	Typical range
24.0	0.60	10.1	of rates
120.0	3.0	10.3	
3,050.0	75.0	10.5	
4,070.0	110.0	27.6	Flooded

Source: Schlitt 1984.
*Expressed as a percentage of the total space within the rock pile.

becomes water-filled at the solution–air interfaces. Flow then begins. As the application rate increases, the only change is to thicken the film of flowing water in the larger voids and channels where there is a solution–air interface. No substantial change in the volumes occupied by air and liquid occurs until the application rate increases to the point where the air-filled voids begin pinching off. When such bottlenecks occur, the open voids flood upward from the pinch points and the air-filled void spaces begin to disappear.

Experimental column data demonstrating this phenomenon are shown in Table 11.4-1. At typical application rates (4–40 mm/h or 0.1–1.0 gal/ft²/h), there is considerable air-filled void space. Thus, adequate aeration should not be a problem unless the material being leached is very fine (e.g., flotation mill tailings).

Regardless of the application rate, the velocity of solution flowing through a water-filled void is controlled by the pressure gradient associated with the hydrostatic head and the permeability at that location. This relationship is expressed by Darcy's law:

$$V°_L = k/\mu \, (dp/dx) \quad (11.4\text{-}12)$$

where

$V°_L$ = superficial velocity of the leach solution
k = permeability
μ = fluid viscosity
dp/dx = pressure gradient or hydrostatic head of the leach solution

Optimum Application Rate

From the standpoint of the leach operator, the maximum effective application rate (q_{max}) is limited by the zone of minimum permeability in the system under leach. This may be the bulk permeability of the rock pile. However, care should be taken to ensure that the solution application rate does not exceed the surface acceptance rate of the heap or dump. This is an important consideration because several factors can combine to make surface permeability lower than the bulk value. These factors include the following:

- Accelerated decrepitation or weathering of the surface layer
- Breakup and compaction of the surface rock due to equipment movement, particularly haulage truck traffic
- Segregation of fine material at or near the top of the heap where dumping occurs
- Formation or accumulation of chemical precipitates such as gypsum or hydrated iron oxides due to evaporation and crystallization of the leach solution

Where surface permeability is lower than the bulk value, the application rate should be reduced to avoid ponding (surface flooding), which prevents air circulation into or through the rock. Ponding also provides ideal conditions for continued chemical precipitation that further seals the rock-pile surface.

To select the maximum effective application rate, the leach operator needs some knowledge of the permeability range in the system. This information is probably best obtained by performing one of the standard water permeability tests. Unfortunately, there is little published information on heap and dump permeability to serve as a guide. Limited field measurements on copper leach dumps suggest a value of q_{max} in the range 20–40 mm/h (0.5–1.0 gal/ft²/h). Higher flows cause increasing amounts of water simply to short-circuit through the rock without recovering additional values.

The importance of controlling application rates is further demonstrated in Kennecott data for copper dump leaching (Jackson and Ream 1980). In a comparison of sprinkler leaching at 20 mm/h (0.5 gal/ft²/h) and trickle leaching at 80 mm/h (2 gal/ft²/h), sprinkler leaching produced 11% more copper and a 15% higher copper content in the PLS. Overall, 13% less water was pumped per pound of copper produced at the lower application rate, significantly reducing power costs per unit of copper produced.

Leach/Rest Cycles

Leach/rest cycles are a factor in solution management when the active lixiviant (such as ferric iron) is generated in situ. However, such cycles have little value when the lixiviant is delivered externally (such as by acid in the irrigating solution). For lixiviants generated in situ, when a volume of rock has been thoroughly wetted with solution, the lixiviant distributes throughout the rock mass. Then, during the rest cycle, dissolution of the values continues without further solution application. For lixiviants delivered externally, as soon as the externally supplied reagent is consumed, leach reactions stop. Flow must resume supplying reagent before reactions can continue.

For appropriate ores, rest cycles can extend the economic life of older heaps and dumps, as they permit maximum recovery of metal values from material that was mined. Rest cycles also permit an operator to reduce costs by maintaining a limited flow of high-grade PLS to the plant. In both cases, the key factor is switching application of water from one leach area to another. While standing fallow, chemical and diffusional processes dissolve metal values and cause their concentrations to build up in the water-filled void space. Then a relatively short period of active solution application washes the solubilized values from the heap or dump. As an example, a 1-week leach cycle might be followed by a 4-week rest cycle. On older material, a longer rest cycle may be required to build up an economic concentration before washing.

Similarly, recycling barren solution to the oldest material first, then advancing the outflow to fresher material, reduces the volume and increases the grade of the pregnant solution being treated in the plant. This is sometimes referred to as *solution stacking*.

A third option is to reduce the irrigation rate progressively, which minimizes the solution flow when extraction is slow and helps to maintain a high-PLS metal concentration.

Thus, leach/rest cycles should not be overlooked as a way of optimizing a leach operation and minimizing pumping and facility costs.

Pulsed Leaching

Recent developments in vadose (unsaturated) zone flow (Breckenridge and Henderson 2005) may have application in situations where a heap or dump has a wide particle size distribution that includes a high fines content. Typically a leach dump would likely form a low-permeability zone that restricts leaching at higher flows. Under this condition, the capillaries are all filled and the strongest force is gravity, which leads to the familiar downward percolation of solution through the larger partially water-filled voids. However, as flow rate decreases, the point is reached where matric forces become dominant and most of the flow is via capillaries in the finer material.

Thus, for material with both coarse and fine sizes, leaching by cycling the irrigation rate between high and low values may improve recovery of metal values from the full spectrum of particle sizes, leading to higher overall production. For perspective, Breckenridge and Henderson (2005) give an example for a gold ore heap for which the high and low flows are 1.45 and 0.48 mm/h (0.036 and 0.012 gal/ft^2/h).

Production Aspects

The production aspects of solution management include the pumping systems and the pipeline requirements. Also covered are the types of equipment used to irrigate the heaps and dumps: either spray systems or drip emitters.

Pumping Systems and Pipeline Requirements

Selection of a pumping system depends on several factors: hydrostatic head on the system, line pressure needed to drive the solution distribution equipment on the leach surface, flow volume, and corrosion potential of the lixiviant. Because there are so many site-specific factors involved, only general guidelines are provided here.

In many operations, the recovery plant is located at a lower elevation than are the heaps or dumps. Thus, PLS generally flows to the plant by gravity in lined ditches or low-pressure pipelines. Any pumping from a heap or dump outflow or from a plant feed pond generally requires only a low-pressure horizontal pump, which can be mounted beside the pond with the pump intake suspended from a float in the pond to readily accommodate rising or falling water levels.

For the situation noted previously, recycling of barren (tail) solution back to the heaps or dumps requires a high-head pumping system, typically involving multistage vertical pumps. Pumps can be mounted on a fixed structure such as in a pump house where pumping is from a sump. Alternatively, they can be mounted on a barge and allowed to float in the barren solution pond.

If system pressure is not too great, flexible pipe or couplings can be used to tie the pumps into the shore-based pipeline system, allowing them to rise and fall with the pond level. The type of pipe is dictated by the elevation difference between the plant and the leach surface, plus the line pressure needed to drive the solution-distribution equipment on the heap or dump. The elevation difference can be as little as 30 m (100 ft) or as much as 600 m (2,000 ft). The latter requires sophisticated designs starting with heavy-walled metal pipe, shifting gradually to lighter schedules. The entire pipe system must be well anchored, yet allow for expansion and contraction due to temperature changes. In addition, the entire header system must almost certainly be higher than the 70–345 kPa (10–50 psi) pressure needed on leach areas, requiring installation of outlet stations on the main header line to serve as pressure-reduction points as well. When the pipeline profile and other flow parameters are defined, sophisticated computer programs can be used to aid in pipeline design.

Pipe materials must be selected with the chemical nature of the lixiviant in mind. For higher line pressures that dictate metal pipe, 316 L stainless is a frequent choice for acidic copper leach liquors that contain ferric and cupric ions. Polyvinyl chloride (PVC) or other liners have been tried on black iron or steel pipe, but the linings tend to deteriorate with time, particularly at joints, leading to corrosion and eventual failure at the joint. For alkaline solutions or concentrated sulfuric acid, black iron may be satisfactory. For lower line pressures, high-density polyethylene (HDPE) pipe has become the standard of choice. This is rugged, easy to cut and fuse in the field, resistant to scaling, and essentially corrosion-proof. PVC and other rigid plastic pipes may be less expensive but are not as rugged and tend to shatter. One of the few problems with HPDE pipe is its lack of rigidity: it twists and turns, and may require stakes or anchor blocks to keep it in place. However, this flexibility can also be an asset when routing the pipe during initial installation.

Solution Application Systems

As pointed out previously, careful distribution of solution on the heap or dump surface is required to optimize the leach operation. Development of such a solution management system requires a clear understanding of three concepts:

1. *Application rate,* calculated by dividing the leach solution volumetric flow rate by the surface area to which the solution is actually being applied
2. *Surface acceptance rate,* the maximum application rate that can be employed without encountering noticeable ponding on top of the heap or dump
3. *Irrigation rate,* a measure of the intensity of leaching, calculated by dividing the total volumetric flow rate by

the total area planned for leaching, including areas under active leach and areas that are being allowed to rest

For sulfidic copper minerals, typical irrigation rates are only 20%–25% of the application rate, and may be only 10% on old areas needing a long rest cycle. For gold, silver, and oxide copper, the percentages are much higher because the extraction rates are higher.

Only two classes of irrigation system are in widespread use today: spray or leach sprinklers and drip emitters. Pond leaching and high-intensity trickle systems have generally been abandoned, because their application rates normally exceed the optimum levels, leading to short-circuiting of solutions and dilution of values in the PLS. The desired application rate should be based on q.

Spray or sprinkler leaching devices include the following:

- **Sprinklers.** These are of two types: (1) rotating impulse sprinklers, which are much like typical "rain bird" agricultural sprinklers except that they are constructed of acid-resistant plastic or stainless steel for all wetted parts; and (2) Senninger wobblers, which are off-center rotary-action sprinklers that provide a coarse spray pattern to minimize evaporation. However, the off-center design creates a thrust problem, and wobblers must be staked down.

  Both designs provide a pattern of reasonably uniform solution coverage at an application rate of 20 mm/h (0.5 gal/ft^2/h), which is safely below q. With their own pressure regulators, wobblers are sophisticated enough to provide even lower rates (such as 12 mm/h or 0.3 gal/ft^2/h), the level typically cited as optimum for gold. Ranges of operating parameters for several types of sprinklers are shown graphically by Schlitt (1984) and are available in vendor literature.

- **Pressure emitters.** These are equivalent to drip-irrigation devices used in agricultural applications. They operate at low gauge pressure (100–140 kPa [15–20 psi]) and produce a steady seepage of leach solution that gives about the same application rate as sprinklers.

In spite of their similar delivery rates, each system has a number of relative advantages or disadvantages with respect to the following considerations:

- **Droplet impact.** Pressure emitters minimize droplet impact, reducing particle decrepitation and fines migration. This is particularly valuable when leaching soft, friable ore or agglomerates.
- **Winter weather.** Pressure emitters are effective even when buried. Thus they can be used in areas that experience severe winter weather and can cause sprays to ice up.
- **Wind.** Pressure emitters are immune to wind drift. Under extreme conditions, sprays may have to be shut down to avoid spreading acid or other chemicals beyond the boundaries of the leach system.
- **Evaporation rate.** Pressure emitters have low evaporation rates, only 1% to 2% of total flow, whereas spray losses have rates in the range of 5% to 10%. Which type has the advantage depends in part on a site's water balance. In dry areas where makeup water must be purchased, emitters are favored. In net precipitation areas, even on a seasonal basis, sprays are favored, since a high evaporation rate can benefit the water balance. In fact, special sprays called "foggers" are designed to maximize water removal by evaporation.
- **Blockage.** The major drawback to pressure emitters is the likelihood of blocking, or plugging, the small flow channels with precipitated salts. To avoid plugging, conditioning of the recycle solutions to prevent scale formation or remove suspended solids must receive careful attention. Conditioning can be done by means of settling ponds, filtration systems, or chemical additives.
- **Coverage.** Sprinklers give complete coverage of the top surface. In contrast, pressure emitters, which are actually arrays of point sources of water, create a conical volume of wetted ore below each emitter and leave some near-surface material unleached. For a permanent multilift system, this is not a serious issue, because any unleached ore will be wetted when the next lift is added and placed under leach. However, for a shallow on/off heap, as much as 10% of the ore can remain unleached.

Environmental Aspects

At least in the United States, obtaining the necessary environmental permits is probably the single most time-consuming activity in the development of a heap or dump leaching project.

Permitting requirements vary from state to state and may differ depending on whether the project is to be sited on private, state, or federal land. Because of these differences and constantly changing regulatory requirements, a detailed guide to the appropriate permitting process cannot be presented here. The best approach is usually to hire knowledgeable environmental consultants who practice in the area where the project will be located. These consultants should be quite familiar with applicable regulations and procedures and perhaps even know the local and state regulatory personnel.

Although specifics may differ from one site to another, regulatory agencies generally have two major concerns regarding surface leaching projects: (1) ensuring the safe handling of toxic chemicals and reagents, and (2) preventing contaminated runoff or seepage to groundwater. Detoxification of the heaps or dumps may also be a concern for all future operations and is already having an impact on the gold heap-leaching segment of the industry. However, groundwater restoration, a major factor for in-situ operations, has little impact on surface operations unless leaks have contaminated the subsurface aquifer.

Other environmental concerns are common to most mining operations. These include control of particulate emissions (dust) from the mining and any crushing or processing operations as well as topographic restoration to preexisting profiles and revegetation of mine and leach areas. Restoration activities usually have some sort of bonding requirements. In terms of costs, solution containment and heap or dump stability are the major considerations. Each is discussed in the following subsections.

Geotechnical Design

The design of piled rock structures that are stable under conditions of water flow, especially in seismically active areas, is a complex undertaking. Entire books have been written that address the necessary design features for both heaps (van Zyl 1987) and dumps (McCarter 1985). Because of this complexity, a detailed analysis of the problem cannot be presented here.

The first consideration is the method of heap or dump construction. Alternative methods involve use of reusable pads, expanding pads, and valley dumping. Major requirements and considerations in each case are discussed by Dorey et al. (1988).

Reusable pads are best in terms of both containment and stability. This method involves a sloping base that is highly impermeable and can easily support the weight of a single shallow lift of ore. In the past, the base has been asphalt or even concrete. However, today a multilayer base with plastic liners over a clay foundation is more common. From an environmental point of view, the biggest concern is the integrity of the final impoundment area for the leached ore. This requires a properly designed retaining structure and one or more liners to prevent seepage losses of residual reagents. A monitoring system is also required to detect any leakage. Backfilling an abandoned pit area should not be overlooked for creating a possible containment site.

Expanding pads and valley dumping do not entail the material rehandling required with reusable pads. Rather, use of expanding pads involves periodic extension of the prepared heap foundation and liner system, which is then loaded with an initial lift of ore. This method is characterized by a gently sloping impervious liner that is overlain with a network of drainage pipes and a coarse drainage layer. The intent is to drain off the PLS as soon as it percolates down through the ore. Rapid removal minimizes the hydrostatic head on the liner to help minimize leakage. As more ore lifts are added, loading on the liner system can become an issue.

Valley dumping uses a natural or constructed retaining structure such as a ridge or dam located at the bottom of a canyon or valley. Ore is stacked upstream behind the structure in progressive lifts. Total land area required is usually less than that required for expanding pads, but lining the underlying surface can be difficult because of slope and terrain features. Solution containment for this type of system is addressed in the following subsection, while stability is covered here. In a true valley-fill system, such as that of Carlota Copper Company in Arizona, the PLS is retained within the heap system as a saturated zone. Submersible pumps then advance the collected PLS to the recovery plant.

Stability depends on a number of factors, including strength and slope of the retaining structure, load imposed by the stacked ore, and buildup of a saturated PLS zone at the foundation or elsewhere in the rock mass. The latter is critical, as more heap and dump failures have been caused by buildup of excessive pore pressure than by any other cause (Caldwell and Moss 1985). Without going into mathematical details, the reason for this is that pore pressure within a saturated zone reduces the shear strength and frictional resistance to movement of the stacked material. As pore pressure builds, especially near a free face, it can cause shear strength to drop below that needed to withstand the loading imposed by the rock mass. At this point, the structure fails, often catastrophically.

Solution Containment
An effective liner is the principal barrier to seepage loss. This is true not only for heap or dump foundations but also for catchment basins, flow channels (including ditches and trenches), and holding ponds. In addition, no containment system is perfect. Some seepage always occurs and can be measured by Darcy's law (Equation 11.4-12). The factors that control seepage are liner permeability and hydrostatic head. The engineer must design the system so that the expected seepage level is within environmentally acceptable limits.

Strachan and van Zyl (1988) cover liners in detail, including liner selection, system design, and installation. Selection is based on the type of application, nature of the solution to be contained, physical loading, and exposure to the elements. Liner materials include clays, clay or chemically amended soils, and geomembranes (PVC, Hypalon, HDPE, asphalt, etc.), used alone as single liners or in combination. When used in combination, the upper liner is usually a geomembrane and the lower one is a soil material. On double or triple liners, a leak-detection system is often installed in a porous zone between the layers. The topmost liner is overlain with drainage-collection piping and a coarse rock drainage layer. If a forced aeration system is to be used to boost the sulfide leach rate, then under-heap aeration piping should also be installed during construction of the leach-pad overliner system.

The use of systems that minimize seepage loss not only meet regulatory requirements, but also make good sense from an operational perspective. Even a small leak, if undetected, can cause a major loss in metal values and reagents over time. Such losses are equivalent to lowering the ore grade or getting poorer metallurgical recovery. In many cases, the dollar value of lost production can equal or exceed the cost of a good solution containment system.

COMMODITY RECOVERY SYSTEMS
Systems used to recover metals from heap and dump leaching operations are the province of hydrometallurgists. Although the recovery systems themselves depend on the lixiviant used in leaching, they are essentially independent of ore-emplacement or solution management practices. Thus the unit operations are commodity specific and covered in standard metallurgical texts and reference volumes. Current recovery systems are briefly described in the following subsections.

Copper
Historically, copper has been recovered as an impure metallic mud or "cement" by cementation (precipitation) on scrap iron. However, today, virtually all copper is recovered by means of SX-EW, a process that produces directly salable cathode copper at the mine site.

SX involves two stages. In the extraction stage, copper in the leach solution is transferred to the lean organic liquid at a low acid concentration (≤ 10 or 15 g/L) and the two immiscible liquids are allowed to settle. The less-dense organic phase rises to the top and is drawn off. In the strip stage, the acid level in the lean electrolyte is much higher (≥ 150 g/L) and copper transfers from the organic phase to the electrolyte. The rich electrolyte is then subjected to electrowinning to produce copper cathodes.

Gold/Silver
Pregnant solutions from gold/silver heaps and dumps are currently treated by either of two methods.

One method involves absorption on activated carbon granules packed in columns. The gold and silver cyanide complexes load on carbon, which is periodically stripped into a concentrated cyanide solution. The gold and silver are then recovered by electrolytic deposition from the strip liquor.

The other method involves stripping the gold and silver from solution as a sludge by means of zinc precipitation (the

Merrill–Crowe process). The precipitate is dried and smelted to produce doré metal for refining. Because carbon is less effective for silver than for gold recovery, high silver solutions are usually treated by the Merrill–Crowe process.

Uranium

The relatively low-grade leach liquors from uranium heaps and dumps are usually concentrated by ion exchange, then purified and concentrated further by solvent extraction. Yellow cake (U_3O_8 [triuranium octoxide]) is precipitated from the solvent-extraction strip liquor.

Nickel

As for copper, nickel and any by-product cobalt are normally recovered by SX-EW.

Solution Conditioning

The only potentially unusual feature of the various recovery systems is conditioning of the barren solution before recycling it to the heaps or dumps. Because most of these facilities must operate under conditions of zero discharge, use of a bleed stream to control buildup of undesirable impurities may not be possible. Thus, conditioning may be needed to control scale-forming constituents, remove potentially harmful impurities, replenish reagents that have been consumed, adjust solution Eh or pH, or simply make up the water lost to evaporation.

Conditioning is highly site specific and depends on the ore–lixiviant system, gangue reactivity, and even chemical makeup of the available process water.

Overall conditioning requirements for heap and dump leaching are less than for in-situ mining. This is due to the sensitivity of the in-situ formation to loss of permeability caused by suspended or precipitated solids building up and blocking the flow paths between the injection and recovery wells.

ANCILLARY AND INFRASTRUCTURE REQUIREMENTS

Ancillary and infrastructure requirements for surface leaching are much the same as for any open-pit mining operation. This is because leaching is merely the ore-processing operation associated with mining. The main requirements are for electric power, water, a transportation system, and access to a sufficient labor force. Fuel (e.g., natural gas) is not a concern, as there are generally no large furnacing or other heating operations. Building heat, small precious-metal furnaces, and driers can easily be fired electrically or with bottled gas if a natural-gas line is not available.

In general, ancillary and infrastructure requirements are lower for primary heap and dump leaching operations than for mine-and-mill operations. Leaching plants are thus well suited to operations that are small or remotely located.

Site Requirements

A mine-and-mill operation requires a rail line, heavy-duty haul road, or both to move ore from the mine to the mill and concentrate from the mill to a smelter. In contrast, a leach operation requires merely an all-weather access road, since shipping is limited to receipt of reagents and other supplies and shipment of small tonnages of crude or refined product. Acid trucks or similar vehicles are likely to be the largest units entering or leaving the plant on a regular basis.

Labor Requirements

Labor requirements are lower for leaching operations than for conventional mills, as most workers perform actual mining activities. Exclusive of mining, labor requirements for a leaching operation should represent no more than 15% of the direct operating costs. Thus, much of the work force can be recruited locally, even in rural areas. However, a camp arrangement may still be required at very remote sites. Specialized services such as instrument and computer-system maintenance or even assaying may have to be done on a contract basis.

Power Requirements

Power requirements vary widely, depending on the type of operation.

Gold plants generally have low requirements, mainly for pumping relatively small solution flows and running small heaters and furnaces. Many remote gold plants use diesel generator sets or other types of on-site generation, as power so generated is cheaper and more dependable.

However, large copper leach operations are power-intensive. With flows as high as 3.2 m³/s (50,000 gpm), water-distribution systems may include multistage pumps with 37.5–110 kW (500–1,500 hp) motors. Copper electrowinning is also power intensive, requiring at least 2.2 kWh/kg Cu (1 kWh/lb Cu). Power is therefore usually purchased from a utility company. This purchased power is a big factor in the operating costs and the operation may incur a high capital cost if an extensive pole line with transformers, rectifiers, and substations is required.

Water Requirements

Water requirements vary according to operational stage. Start-up requirements are relatively high, principally because the various catchment basins and recovery-plant feed and tail-water reservoirs must be filled. The heaps and dumps must also be saturated with water before flow occurs. This represents a significant solution inventory, since the rock mass retains 8% to 12% moisture by weight, which is about 13 L/m³ (0.1 gal/ft³) of the heap or dump material.

Steady-state requirements are relatively low, because they cover only evaporation and other miscellaneous losses. Since most plants must operate under zero-discharge permits, no net makeup is required to compensate for bleed-off. The exact makeup requirement generally varies seasonally with the local evaporation rate. For copper in the southwestern United States, a typical annual average figure is 35 to 50 L/min/t (10 to 15 gpm/st) of daily production.

REFERENCES

Benedict, C.H., and Kenny, H.C. 1924. Ammonia leaching of Calumet and Hecla tailings. *Trans. AIME* 70:595.

Breckenridge, L., and Henderson, M. 2005. Heap leach pad optimization. Presented at SME Annual Meeting, Salt Lake City, Utah.

Brierly, C.L. 1978. Bacterial leaching. *CRC Crit. Rev. Microbiol.* 6:207–262.

Caldwell, J.A., and Moss, A.S.E. 1985. Simplified stability analysis. In *Design of Non-Impounding Mine Waste Dumps*. Edited by M.K. McCarter. New York: SME-AIME. pp. 47–61.

Catanach, C.B. et al. 1977. Copper leaching from an ore body blasted in-place. *In Situ* 1:283.

Chase, C.K. 1980. The ammonia leach for copper recovery. In *Leaching and Recovering Copper As-Mined Materials*. Edited by W.J. Schlitt. New York: SME-AIME. pp. 95–103.

Dorey, R., van Zyl, D., and Kiel, J. 1988. Overview of heap leaching technology. In *Introduction to Evaluation, Design and Operation of Precious Metal Heap Leaching Projects*. Edited by D. van Zyl, L.A. Hutchison, and J.E. Kiel. Littleton, CO: SME. pp. 3–22.

George, D.R., Ross, J.R., and Prater, J.D. 1968. Byproduct uranium recovered with new ion exchange techniques. *Min. Eng.* 20:73.

Grunig, J.K. 1981. Chemical aspects of the occurrence and extraction of uranium. In *Process and Fundamental Considerations of Selected Hydrometallurgical Systems*. Edited by M.C. Kuhn. New York: SME-AIME. pp. 17–26.

Hiskey, J.B. 1981. Thiourea as a lixiviant for gold and silver. In *Gold and Silver—Leaching, Recovery, and Economics*. Edited by W.J. Schlitt, W.C. Larson, and J.B. Hiskey. New York: SME-AIME. pp. 83–91.

Hiskey, J.B. 1984. Thiourea leaching of gold and silver—Technology update and additional applications. In *Practical Hydromet '83*. Edited by G.F. Chlumskey. New York: SME-AIME. pp. 95–99.

Hiskey, J.B., and Schlitt, W.J. 1982. Aqueous oxidation of pyrite. In *Interfacing Technologies in Solution Mining*. Edited by W.J. Schlitt. New York: American Institute of Mining, Metallurgical, and Petroleum Engineers. pp. 55–74.

Jackson, J.S., and Ream, B.P. 1980. Solution management in dump leaching. In *Leaching and Recovering Copper from As-Mined Materials*. Edited by W.J. Schlitt. New York: SME-AIME. pp. 79–94.

LeRoux, N.W., North, A.A., and Wilson, J.C. 1973. Bacterial oxidation of pyrite. In *Proceedings of the International Mineral Processing Congress*. Paper No. 45. London: Institution of Mining and Metallurgy.

Logan, T.C., Seal, T., and Brierly, J.A. 2007. Whole-ore heap biooxidation of sulfidic gold-bearing ores. In *Bio-Mining*. Edited by D.E. Rawlings and D.B. Johnson. New York: Springer-Verlag. pp. 113–138.

Longwell, R.L. 1974. In place leaching of a mixed copper ore body. In *Solution Mining Symposium*. Edited by F.F. Aplan, W.A. McKinney, and A.D. Pernichle. New York: American Institute of Mining, Metallurgical, and Petroleum Engineers. p. 233.

McCarter, M.K., ed. 1985. *Design of Non-Impounding Mine Waste Dumps*. New York: SME-AIME.

McClelland, G.E. 1988. Testing of ore. In *Introduction to Evaluation, Design and Operation of Precious Metal Heap Leaching Projects*. Edited by D. van Zyl, L.A. Hutchison, and J.E. Kiel. Littleton, CO: SME. pp. 61–67.

McClelland, G.E., and van Zyl, D. 1988. Ore preparation: Crushing and agglomeration. In *Introduction to Evaluation, Design and Operation of Precious Metal Heap Leaching Projects*. Edited by D. van Zyl, I.A. Hutchison, and J.E. Kiel. Littleton, CO: SME. pp. 68–91.

McDonald, R.G., and Whittington, B.I. 2008a. Atmospheric leaching of nickel laterites review, Part I: Sulfuric acid technologies. *Hydrometallurgy* 91(1):35–55.

McDonald, R.G., and Whittington, B.I. 2008b. Atmospheric leaching of nickel laterites review, Part II: Chloride and bio technologies. *Hydrometallurgy* 91(1):56–69.

Merritt, R.C. 1971. *The Extractive Metallurgy of Uranium*. Golden, CO: Colorado School of Mines Research Institute. pp. 60, 64, 84.

Milligan, D.A. 1984. Agglomerator design for heap leaching of gold and silver ores. Presented at First International Symposium on Precious Metal Recovery, Reno, NV, June.

Muhtadi, O.A. 1988. Heap construction and solution application. In *Introduction to Evaluation, Design and Operation of Precious Metal Heap Leaching Projects*. Edited by D. van Zyl, I.A. Hutchison, and J.E. Kiel. Littleton, CO: SME. pp. 92–106.

Murr, L.E. 1980. Theory and practice of copper sulfide leaching in dumps and in situ. *Miner. Sci. Eng.* 12:121–189.

Potter, G.M., Chase, C.K., and Chamberlain, P.G. 1982. Feasibility of in situ leaching of metallic ores other than copper and uranium. In *Interfacing Technologies in Solution Mining*. Edited by W.J. Schlitt. New York: American Institute of Mining, Metallurgical, and Petroleum Engineers. pp. 123–130.

Scheffel, R.E. 1981. Retreatment and stabilization of the Naturita tailings pile using heap leaching techniques. In *Interfacing Technologies in Solution Mining*. Edited by W.J. Schlitt. New York: American Institute of Mining, Metallurgical, and Petroleum Engineers. pp. 311–321.

Schlitt, W.J. 1984. The role of solution management in heap and dump leaching. In *Au and Ag Heap and Dump Leaching Practice*. Edited by J.B. Hiskey. New York: American Institute of Mining, Metallurgical, and Petroleum Engineers. pp. 69–83.

Schlitt, W.J. 2006. Kennecott's million-ton test heap—The Active Leach Program. *Miner. Metall. Proc.* 23(1):1–16.

Steemson, M.L. and Smith, M.E. 2009. The development of nickel laterite heap leach projects. In *Proceedings, ALTA Nickel-Cobalt Symposium*, May. Blackburn South, Victoria: ALTA Metallurgical Services.

Strachan, C., and van Zyl, D. 1988. Leach pads and liners. In *Introduction to Evaluation, Design and Operation of Heap Leaching Projects*. Edited by D. van Zyl, I.A. Hutchison, and J.E. Kiel. Littleton, CO: SME. pp. 176–202.

Templeton, J.H., and Schlitt, W.J. 1997. Method for determining the sulfuric acid balance in copper-leach systems. *Miner. Metall. Proc.* 14(3):1–7.

van Zyl, D., ed. 1987. *Geotechnical Aspects of Heap Leach Design*. Littleton, CO: SME.

CHAPTER 11.5

In-Situ Techniques of Solution Mining

John I. Kyle, Douglas K. Maxwell, and Breanna L. Alexander

INTRODUCTION
In-situ solution mining is the selective dissolution and recovery of target mineral(s) without the use of conventional excavating techniques. Instead, the minerals are dissolved where they reside within the ore body. The dissolved mineral solution is then pumped to a processing plant located on the surface, where the desired metals are produced for market. This method can be considerably more economical than conventional mining methods, because massive tonnages of gangue materials are simply left where they exist, resulting in elimination of waste handling efforts, tailing piles, and waste brines.

More importantly, low-grade deposits can potentially be economic, because solution mining typically is cheaper than conventional mining methods. Additionally, through careful engineering, drilling, and monitoring, in-situ mining can have a greatly reduced environmental impact compared to mining by conventional methods. Mineral commodities exploited by in-situ techniques include salts such as sodium chloride, potash, magnesia, soda ash, borates, sodium sulfate, and lithium, as well as metals such as uranium and copper.

Deposit Characteristics
The primary requirement for in-situ solution mining is that the ore body has amenable characteristics to allow the lixiviant (the dissolving fluid employed to extract the valuable mineral) to come into contact with the target mineral throughout the ore body in sufficient quantity that an economical recovery rate of the mineral can be achieved. Beyond this requirement, metallurgical and ore host issues must be acceptable so the mineral can be leached from the ore. The deposit may have characteristics that prevent efficient leaching in a sufficient time frame. The general category of minerals where in-situ solution mining has proven to be economically successful includes both salts and metals.

The two main types of deposits associated with soluble salts are evaporites and brines. Evaporites may be classified as either marine or alkali; both types accumulate as a result of concentration through the evaporation of inland brines. Natural brines may occur as surface or subsurface deposits and are often associated with playas.

Marine Evaporite Deposits
Marine evaporite deposits are typically impermeable and located in large plastic salt domes. Often, they are found in a series of horizontal beds and in anticline deposits. Some deposits are hard to mine from a conventional perspective because of the deposit depths and the plastic behavior of the salt under stress. These deposits, however, can be quite attractive for a solution mining application. Marine deposits mirror seawater composition to a degree, which has about 31% sodium, 4% magnesium, 1% calcium and potassium as cations, and 55% chlorides, 8% sulfates, and less than 1% bicarbonate as anions. Therefore, halite, sylvite, and sodium sulfate are commonly found in marine deposits.

Alkali Evaporite Deposits
Alkali evaporite deposits are typically found in inland seas such as the Great Salt Lake in Utah and Searles Lake in California (both in the United States), and the Dead Sea in Israel/Jordan. Evaporation of an inland sea leads to formation of a playa or salt bed. If the salt bed formation is followed by geological burial, then a subsurface alkali basin is formed. Typical deposits that are mined include halite, potash, trona, borates, sodium sulfate, lithium (as lithium chloride), and soda ash.

Natural Brine Deposits
Brine occurs naturally as surface brine, such as in the Great Salt Lake or the Dead Sea, and as subsurface brine, such as at the Hector deposit in California. The Searles Lake deposit has both surface and subsurface brines at various times of the year. Where groundwater or surface water runoff congregates in a depression, subsurface brines typically exist in the permeable matrix below the depression. The water percolates through the depression or playa, dissolving water-soluble salts along the way. The salt-laden water then flows to a location where it is stable, creating a surface or subsurface brine deposit.

John I. Kyle, Vice President, Lyntek, Inc., Lakewood, Colorado, USA
Douglas K. Maxwell, Senior Process Engineer, Lyntek, Inc., Lakewood, Colorado, USA
Breanna L. Alexander, Metallurgical Engineer, Lyntek, Inc., Lakewood, Colorado, USA

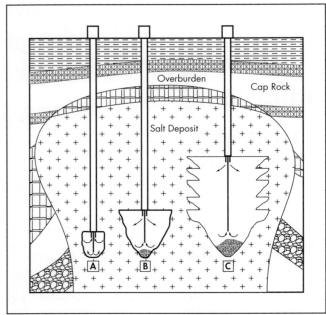

Figure 11.5-1 Leaching methods in impermeable salt deposit: (A) direct circulation solution mining—cavity establishment phase, (B) reverse circulation solution mining—brine production phase, and (C) reverse circulation—brine production continues

Metal Deposits

For metal deposits, in-situ leaching has been successfully employed in sandstone-hosted uranium deposits. The characteristics of these ore deposit types are discussed in depth later in the chapter, but in-situ recovery production for uranium requires ore within the water table so that a regime can be employed to move fluids through the sandstone in order to extract the uranium from the host rock.

Solution Mining Methods

Mining Impermeable Salt Deposits

Impermeable deposits are typically mined by introducing hot water or other solutions into the deposit to create a cavity. The formation of the cavity through dissolution of the target mineral produces a salt brine containing the mineral. The brine is pumped from the cavity to the ground surface where it is processed further in a plant to achieve the desired marketable product. Prior to the creation of the underground cavity, the potential for the collapse of overlying rock into the cavern must be considered. Salt-solution mining operations must be designed and carried out with the structural integrity of the resultant caverns in mind. A layer of rock or salt (beam) must be designed and left intact above the cavern to support the cavern. The minimum thickness and dimensions of this beam are designed based on the characteristics, strength, and cavern dimensions within the salt structure and the overlying geologic structure above the cavern. The cavern dimensions can be monitored with sonar, and its volume can be estimated from continuously recorded mass balance readings during production of the salt.

In order to develop a salt cavern, a vertical hole is drilled from the drill collar to a point near the base of the salt formation. Coaxial tubing is then inserted into the drill hole to provide for flow of lixiviant brine, product brine, and blanket fluid. Lixiviant brine chemically bonds with the target salt to make the product brine, which is extracted to harvest the target salt. The fluid blanket is used as an immiscible surface that rides on top of the salt brines to control vertical cavern creation. Alternately, tubing is inserted side by side rather than in a coaxial fashion. In either case, the depth of injection for each fluid is set at an appropriate vertical point for the leaching method and cavern design being implemented.

Using the direct circulation leaching method shown in Figure 11.5-1, an initial cavity (sump or sedimentation cavern) is formed at the bottom of the drill hole, which may extend below the base of the target salt bed. This sump creates an area designed to collect insoluble material as it drops out of the salt matrix in the cavern above, as the salt goes into solution. Lixiviant, as fresh water or brine, is pumped into the cavern at a designed elevation at a calculated flow rate. A blanket fluid, selected to float on top of the other liquids, or compressed air is injected at the top of the forming cavern to maintain the roof elevation of the developing cavern and drive the dissolution process horizontally rather than upward. As salt dissolves, a density gradient forms within the brine. During start-up, the lower density solution flows to the top of the cavern, becomes more saturated along the way, and is pumped out at the top of the cavern. As the overall brine saturation increases, the brine production flow rate is also increased. This method allows for the formation of a relatively wide lower cavity.

After the sump cavity is formed, the fluid injection levels are adjusted so the lixiviant is introduced near the top of the cavern. In some cases, heat may be added to the system to assist in facilitating dissolution. The new lighter-density lixiviant spreads horizontally under the blanket and removes the target salt in the ceiling of the cavern. The saturated brine, or pregnant solution, is pumped from the lower cavity region to the surface for processing. There is no need for a blanket to protect the cavern floor, because the brine at this level is already saturated. Periodically, the tubing is raised upward in successive steps, thereby controlling cavern diameter and mining of individual horizontal cuts upward through the height of the cavern. This reverse circulation leaching method is also illustrated in Figure 11.5-1.

The lixiviant can be tailored to selectively dissolve the target salt while leaving behind undesirable salts. However, the remaining salt matrix can drastically decrease dissolution rates and in some cases can *blind off*, stopping fluid flow altogether. Selective lixiviant can be prepared by recycling spent product brine from an established processing plant. The blanket can be any material that is immiscible with the lixiviant and brine, and has a specific gravity less than the lixiviant and brine—for example, compressed air or oil. Typical cavities can be 80 to 150 m (87 to 164 yd) in diameter.

Depending on the ambient surface temperature, *salting out* may occur when a change in brine concentration, due to temperature change or brine influx, causes premature crystallization. This is detrimental in a situation where, as the production brine nears the surface, the temperature drops enough to significantly decrease the solubility of a salt, causing crystallization of a potentially desirable product before it reaches the surface. The salt then comes out of solution prior to plan, which causes production problems—whether in the product brine line within the well, the pipeline on the surface, or in any facility accepting the brine prior to processing. Conducting dissolution and bench-scale studies prior to implementation of a plan in the field will be helpful in determining the overall design of the system to avoid (or deal with) salting out issues.

Mining Permeable Salt Deposits
Permeable salt deposits, such as occur below playas or partially dissolved salt deposits containing natural brine, are deceivingly simple to mine in situ. This method does not require a long period to establish saturated brine, because the natural brine is already at equilibrium. To extract the salt, a series of wells are placed within the salt deposit to inject brine being used as a lixiviant. This lixiviant flows through the salt deposit to a pump that is placed within a centrally located production well, so the product brine can be collected, extracted to the surface, and processed.

The flow rate of fluids extracted/injected must be carefully designed to achieve maximum production while avoiding short circuiting. This occurs when the recycle/injection brine is not allowed enough time to resaturate, thus not reaching the desired target salt content. This then results in a lower density than desired, and flow of the injection brine within the formation (or cavern in an impermeable deposit) is compromised.

It is important to understand the initial permeability of the salt deposit and the salt characteristics to ensure that the well field, well, and flow rates are properly designed for production operations. These types of production schemes can be employed to feasible depths based on economic conditions that are dictated by commodity price, mineral grades, capital costs, and production costs.

Mining Permeable Metal Deposits
Permeable deposits are those that have sufficient space between the particles or grains (pore space) to allow a lixiviant to be passed through the deposit so contact with the desired mineral can be accomplished. For example, where uranium is found in permeable sandstone, a lixiviant can be injected into the sandstone to allow the lixiviant access to the uranium particles so the uranium can be chemically attracted by the lixiviant for pumping to the surface and on to the processing plant. Often, sandstone of appropriate character will have a pore volume of 20% to 30%. To accomplish this, a well field is drilled with a number of injection wells and a much fewer number of production wells. Wells are drilled to surround a single production well that has been established within the ore body and are specifically designed and cased so that a lixiviant can be pumped into the injection wells. The lixiviant then flows through the ore body to the central production well, in which a pump is installed to control the movement of lixiviant from the injection wells to the production well.

In some cases, the deposit needs to be submerged within an aquifer so that solution flow can be controlled. Often, the aquifer will already be tainted because of the ore body's chemistry, but, nonetheless, there are groundwater issues to consider and manage. In more developed countries, regulations concerning in-situ mining are quite evolved, providing good guidance on acceptable operating practice. It is also of great value to have aquitards or aquicludes (an impermeable layer that prevents fluid flow) above and below the target ore body, which helps manage production solutions. Generally, the solutions are overproduced from the production well, so that an excess of solution is pumped from the production area of the ore deposit. This bleed is performed as a protective measure to capture all the production solution placed in the aquifer.

New innovations include modifications to standard vertical drilling methods. By employing practices perfected in the oil sector, inclined and horizontal drilling methods are now being considered. Production monitoring has also become much more sophisticated. This is achieved by using monitor wells to check for and identify possible loss in the production solutions within the aquifer. New process-control instrumentation and logic controllers can now be employed in the complete production system. These methods are of solid value in ensuring that the water supply and quality are properly managed at the site.

Mining Abandoned or Flooded Conventional Mines
When conventional underground mining methods have reached their feasible or economic limits, solution mining can provide for the additional recovery of minerals. These conventional mines are commonly established in bedded deposits that typically used a room-and-pillar (R&P) mining method. With this method, pillars of valuable ore are left to support the roof as the mining operation is conducted. Valuable ore left in the walls of the mining openings can be extracted by flooding the mine, thereby introducing a lixiviant to extract the ore. As solution mining is employed, it is essential to consider existing groundwater regimes, undesirable ore types, lixiviant flow paths throughout the mine openings, recovery of saturated brine, and any issues relative to induced subsidence of the natural ground layers if the pillars have already collapsed (or will collapse) as solution mining is implemented.

Whether the mine is flooded deliberately or by natural means, extraction wells are drilled into the lowest elevations of the mine, and injection wells are placed at higher elevations. Gravity does its work, with the more dense saturated brine collecting at the bottom of the mine from where the pregnant solution is pumped to the surface for harvesting.

Hydraulic Fracturing
Hydraulic fracturing is the process of expanding or creating artificial fractures around a well to enhance its permeability or the communication between adjacent wells. It is typically used where a section of the well is segregated, and a liquid such as brine, water, or oil is used to pressurize the target zone and induce high pressures, which will either extend fractures or generate new fractures. In the oil industry, acid has been used for decades to etch the fracture sides, while sand/glass beads have been included in the fracture fluid to prevent fracture closure when the hydraulic pressure is bled off. The fracturing system is dependent on inherent weaknesses, such as bedding planes, within the host structure; regional stress fields; and other geologic phenomena. Two of the primary difficulties of the technology are projecting the fracture propagation direction, intensity, and growth; and estimating the pressure requirements for successful fracturing. However, the technique has been successfully employed for several decades in the oil industry, and hydraulic fracturing has been shown to be a valuable technique.

SALTS EXPLOITED

Salt
Salt is mined utilizing three methods:

1. Conventional underground mining, which produces rock salt
2. Solution mining, which produces evaporated or vacuum pan salt
3. Solar evaporation of ocean or lake brine, which produces solar or sea salt

Desirable Geologic Conditions and Chemistry

Salt occurring in underground salt domes is typically exploited using a solution mining method. Many salt deposits have been found in a thick bed located 6 to 8 km below the earth's surface. The high temperature and pressure encountered at these depths allow the salt to flow in a plastic mass. With the salt having a lower specific gravity (2.168 for pure sodium chloride [NaCl]) than the surrounding rock, the salt naturally pushes upward to reach equilibrium near the surface. These salt domes are generally pure halite (the mineral form of NaCl), typically have a circular footprint in the horizontal plane, are 1.5 to 3 km (0.9 to 1.9 mi) in diameter, and extend for a vertical distance of 6,000 m (6,562 yd) or more. The deposits become more commercial when the top of the dome is within a few tens to a few hundreds of meters below the ground surface.

Mining Method

Salt dome deposits (impermeable-type deposits) are typically solution mined by employing multiple wells. Sometimes, hydraulic fracturing or lateral drilling is used to link several wells together within a well field, or *gallery*. Employing hydraulic fracturing technology has been successful and has greatly improved the competitiveness of mining operations by

- Enhancing the stability of underground openings by changing the cavity configuration (to eliminate the unstable "morning glory" shape) to a design based on hemispheres and ellipsoids, which increases the stability of underground cavities. This is quite beneficial in reducing the potential for subsidence at the surface;
- Cutting the costs related to brine generation by approximately 50%;
- Greatly increasing product extraction by roughly 40%;
- Reducing well maintenance costs by about 95%; and
- Increasing the well productive capacity by a factor of 10 (Feldman 2006).

Processing Method

Salt brine must be evaporated to extract the entrained salt values. With the proper climate and available land, this may be achieved through solar pond evaporation followed by further upgrading using flotation and centrifuges, much like the processes employed for potash. Alternatively, high-grade salt is obtained through vacuum pan evaporation. This process uses a series of three to five steam-heated multiple-effect evaporators in which a vacuum is pulled to lower the boiling temperature of the brine. Each successive evaporator has a lower pressure than the previous, and the superheated vapor proceeds through the series, helping heat the subsequent evaporators. In either case, before processing, the brine is often chemically treated to remove impurities that may yield a lower-grade product or cause scaling.

Production

Much of the brine produced through solution mining in the United States is never crystallized but is used directly by chlor-alkali plants in the manufacturing of such products as chlorine, caustic, and soda ash. When the desired product is food-grade salt, vacuum pan evaporation can yield fine-grained salt that is 99.8% to 99.95% pure.

Worldwide, the chemical industry is the largest user of salt, consuming about 60% of the total salt produced. Salt for human consumption takes up 30%, and the remainder goes to road deicing, water treatment, and other minor applications.

Potash

There is a long history of producing potash from brines collected from mixed salt deposits on the earth's surface. Commercial production at Wendover, Utah, dates from 1937 (Garrett 1996). Production of potash-bearing brines from underground deposits that had no prior conventional mining production began in the early 1960s. Potash has also been recovered from flooded conventional mines since the early 1970s. In 2008, construction began on another operation to deliberately flood and recover potash by solution mining from mines in New Mexico (United States) that were closed due to prior reserve exhaustion using conventional mining methods.

Desirable Geologic Conditions and Chemistry

Typically, the types of deposits exploited are natural brines and underground evaporite deposits. In many parts of the world, natural brine sources contain sufficient amounts of potash to justify commercial production. Some operations produce potash from brine lake deposits such as the Dead Sea, Searles Lake, and the Great Salt Lake. Natural surface evaporite deposits dominated by halite at Bonneville Salt Flats in Utah contain a brine with 2.5% (potassium chloride or potash [KCl]). Potash has been produced by solution mining from the generally flat-lying, bedded sylvinite (mixed KCl-NaCl) underground evaporite deposits found in North America, particularly in Saskatchewan in Canada, and in Utah and New Mexico. Some properties have been developed solely as solution mines, while others started life as conventional underground mining operations and have been converted to solution mining after cessation of traditional mining operations.

The chemistry of the salt deposits is important in both naturally occurring brines and in minerals. Naturally occurring brines containing potash are found in many locations around the world. The most common anion in those brines is chloride, but sulfate and other anions are known. The potassium levels are typically low (often less than 1% KCl), with sodium and magnesium being the most prevalent of the other metals.

The most commonly exploited mineral containing potassium is sylvite (KCl), which is found in the mineral sylvinite, a mixture of sylvite and halite. However, some leaching operations exploit carnallite ($KCl \cdot MgCl \cdot 6H_2O$) for its potash content.

Mining Method

Solution mining for potash employs three of the mining methods previously discussed: (1) mining of permeable salts; (2) mining of impermeable salts; and (3) mining abandoned or flooded conventional mines.

Mining of permeable salts. This includes naturally occurring brines containing potash that are recovered from brine lakes, salt beds, and wells. Brine lake operations, such as those at the Dead Sea and Great Salt Lake, pump brine from the lake into a solar evaporation pond system. The pond system is either created from a portion of the lake bed or uses nearby shore land that is separated from the main body of brine by human-made dikes. Initially, the brine can simply flood the solar ponds through holes in the dikes, but pumping arrangements are necessary for continued recharge of fresh brine into the system.

The Intrepid Potash operation at Wendover, Utah, gathers brine from the Bonneville Salt Flats via a system of collection canals. Intrepid has a network of canals that extends for miles into the salt flats. About 3 m (10 ft) deep and 5 m (16 ft) wide at the bottom, these canals are dug with hydraulic excavators to allow brine to flow to pumping stations near the solar evaporation ponds. Cantilever pumps are used to transfer brine from the pumping stations to the solar ponds. This system of canals is effective, but it requires constant and extensive maintenance. Intrepid Potash has also installed brine wells drilled into the salt beds. About 10 m (33 ft) deep, these wells provide a more constant baseload flow to the solar ponds than do the collection ditches.

Mining of impermeable salts. Underground evaporite deposits can be classified as impermeable-type deposits. Two operations, one in Saskatchewan and one in the Netherlands, exploit potash-containing salt beds by means of single hole— a combined injection and extraction well—without any other preparation. Large-diameter drill holes are drilled, and downhole, multistage extraction pumps are lowered to the bottom of the target leaching area. The leaching fluid is often pumped down the annulus, and pregnant brine is pumped up a pipe hanging in the center of the well. Some wells are designed to use parallel injection and production lines contained within a large casing.

Beds have been exploited at depths ranging from 1,400 to 1,700 m (1,530 to 1,860 yd) (Garrett 1996). Leaching begins at or below the bottom of the potash bed. These operations use a blanket or pad of an immiscible fluid floating on the brine in the cavity. The immiscible liquid limits the dissolution of minerals in the roof of the cavity, which causes the cavity to expand laterally. The cavity diameters typically grow to more than 100 m (109 yd) and may even reach 150 m (164 yd). Initial leaching is conducted with water to create an open cavity. Selective leaching, using saturated NaCl brine, leaves a matrix of salts that slows diffusion of the lixiviant into the bed and the brine out of the bed. The salt matrix also can blind off completely, preventing further dissolution.

Recovery of potash minerals from an impermeable type of deposit varies. Depending on overburden characteristics and thus well-field design, the mineable area becomes limited; extraction of minerals can be as low as 30% to 40% (strictly before processing.)

As reported by Garrett (1996), Kalium (now a subsidiary of the Mosaic Company) produces potash from a sylvinite bed about 9 to 15 m (10 to 16 yd) thick, at depths of up to 1,650 m (1,804 yd) in Saskatchewan. Formation temperature is 57°C (135°F), which increases solubility. Water is injected to dissolve both the potash and salt until the cavity is more than 100 m (110 yd) in diameter. As the cavity size increases, the large surface area allows selective leaching despite the slower diffusion rates. Annual production is more than 1,800,000 t/a (2,000,000 stpy).

BHP Billiton in the Netherlands exploits a deposit dominated by carnallite, with other salts that are bounded at the top and bottom with halite. The bed averages 100 m (328 ft) thick with depths of 1,400–1,700 m (1,531–1,859 yd). Directional drilling is used to create a *seven-spot* pattern of wells in the salt bed from a single drilling site on the surface. Both the leach solution composition and the flow rate must be carefully adjusted and maintained to ensure dissolution of the potash. Production is approximately 820,000 t/a (900,000 stpy).

The Potasio Rio Colorado project in Argentina, now owned by Vale Inco, proposes to use solution mining to produce up to 5.1 Mt/a (5.6 million stpy) of potash.

Mining abandoned or flooded conventional mines. There are many examples of such mines being exploited for solution mining operations. Texas Gulf developed a conventional underground potash mine near Moab, Utah, in the 1960s, which is now owned by Intrepid Potash. Before the solution mining was initiated, mining conditions were difficult because of methane problems, a weak roof, and an undulating ore body. After several years of R&P mining, a decision was made to flood the mine and operate it as a solution mine. Extraction wells were drilled into lower elevations of the mine, injection wells were placed at several higher elevations, and the mine was flooded with lixiviant through the injection wells. Saturated brine is pumped from the production wells within the mine to solar evaporation ponds for potash recovery.

The Potash Company of America's (purchased by Potash Corporation of Saskatchewan in 1993) Patience Lake mine in Saskatchewan was a conventional R&P operation that experienced repeated problems with major water inflows that closed the mine for 6 years early in its life. Eventually, this caused the company to abandon conventional mining techniques and convert the operation to a solution mining methodology in the late 1980s. A recovery well was located at the deepest point in the mine, and 10 injection wells were drilled into the old workings at higher elevations to establish the flow through the old workings.

In 2008, Intrepid Potash began a project to convert an abandoned mine in the Carlsbad, New Mexico, district to solution mining. The former Eddy Potash Company mine had been closed for many years after exhausting the resources available for R&P mining that it controlled on its leases. As with the mines previously described, extraction wells will be drilled into lower elevations of the old workings and injection wells drilled into the higher elevations. In this operation, because of the large surface area available in the old workings, the leach solution is proposed to be NaCl-saturated brine, to selectively leach the KCl from the pillars, floor, and roof of the old mine. Pregnant brine will be treated in a solar pond system.

Processing Method

Two basic approaches are employed to treat potash containing brines: solar evaporation ponds and evaporator-crystallizers. The low concentration of potash typically found in most naturally occurring brines makes solar evaporation pond treatment a necessity because of its low cost. The use of evaporator-crystallizers greatly increases the capital costs as well as the operating costs because of higher energy requirements. Solar evaporation ponds are limited to operations having locations with sufficient land area and suitable climate for the solar pond application. In northern climates, producers typically recover potash by crystallization rather than solar evaporation, because of low ambient temperatures.

The brines produced from the underground deposits may contain high enough concentrations of potash to allow the use of vacuum evaporators and crystallizers without pretreatment. This system has the advantage of being able to produce finished potash without need for further upgrading, but there is an increased cost in the form of additional energy requirements. When a potash product of sufficient grade is produced, regardless of the method employed, common operations include de-brining in centrifuges, drying, compacting, and screening to produce the various market products.

As described, naturally occurring brines are universally processed by solar evaporation as a concentration step. Some brines containing relatively high concentrations of sulfate will be treated with lime to remove most of that anion. The low-sulfate brine is pumped into a series of shallow, flat-bottomed ponds, which are often unlined and are simply lakes that have been divided or contained by dikes. The brine is sequentially transferred from pond to pond as the water evaporates and the concentrations of the contained salts increase. With careful monitoring, different salts will saturate and precipitate in different ponds. This allows operations, such as Searles Lake and the two Dead Sea plants, to make several different products. Evaporation rates vary with the environmental conditions, so solar pond operations are generally found in areas that enjoy warm, sunny, and dry conditions for long periods during the year. Rainfall is a risk, because high rainfall can slow production, and unusual rain events have even shut down operations for a year or more.

Generally, relatively pure NaCl is precipitated in one or more stages. Depending on location, this may be a marketable product, rather than a waste. When the potash approaches saturation in the brine, it is typically sent to harvesting ponds where the potash precipitates along with sodium chloride if there is little magnesium present. At the Israeli Dead Sea operation, the potash precipitates as carnallite. Harvesting ponds are typically constructed to have a 300-to-600-mm (12-to-24-in.) thick layer of hard NaCl deposited first to build a solid floor for the harvesting process and to protect the liners, if present.

A variety of methods are used to recover the precipitated salts from the evaporation ponds. Floating dredges are used in a few operations (Dead Sea Works), but dry harvesting is more common. In some dry harvesting operations, the salt mixture is plowed into windrows with graders, loaded with elevating scrapers, and hauled to a dump pocket. In others, specialized salt-harvesting machines dig up the salt or collect it from windrows and load it into dump trucks. The harvested solar salt typically contains 24% KCl and 75% NaCl. This material is commonly upgraded with froth flotation to produce a high-grade KCl concentrate. The concentrate is leached with a saturated KCl brine to remove the remaining NaCl and is then de-brined in centrifuges and dried. Potash for fertilizer or feed supplements is usually 93% or more KCl, which is also called muriate of potash.

Brines from underground mines are exemplified by the Intrepid Potash mine at Moab, where they are treated in a system of ponds similar to those for naturally occurring brines. However, the total area of the ponds at Moab is considerably smaller than the area at a natural brine operation with a similar production rate. This is due to the higher potash concentration in the brine from the mine. These ponds are also lined with synthetic material to prevent leakage into the local groundwater. Additionally, dye is also added to the brine in some ponds at Moab to increase evaporation rates. The dyed brine absorbs more light and heat than a clear brine over a reflective white salt bed.

The Patience Lake operation uses a pond to recover potash from the mine-produced brine, but by cooling rather than evaporation (Garrett 1996). The brine pumped out of the mine is warm and, when pumped to a pond near the plant, it cools by natural convection, especially during nighttime. Potash crystallization rates in this pond are highest in the wintertime as temperatures drop. This is one operation that recovers potash by dredging. The precipitation is fairly selective, and the dredged potash can be upgraded to market standards by leaching.

Some brines with high potash content from underground mines and some solar evaporation operations are treated by a combination of evaporators and crystallizers. Triple-effect evaporators remove much of the water from the brine and in many cases precipitate NaCl. The resulting slurry is separated and the salt is washed with potash brine to recover dissolved potash. The strong potash brine is fed to vacuum crystallizers where the potash is precipitated. Although these plants are compact, compared to the solar pond operations, the energy consumed in the evaporators and vacuum crystallizers results in higher operating costs. The potash from the crystallizers generally does not need further upgrading before it is dewatered, dried, compacted, and screened into final products. The brine from the crystallizers is recycled in the plant, but a bleed stream is generally required to eliminate deleterious elements.

The most prevalent waste product from potash solution mining is sodium chloride. In many natural brine operations, the sodium chloride is a co-product with the potash. In others, the sodium chloride can be returned to the brine source. Some of the salts associated with potash production, notably magnesium chloride, have virtually unlimited solubility in water. These will produce *bitterns,* which are brines of high concentration. Some bitterns can be treated and used as a source of magnesium chloride or even magnesium metal. In many cases, disposal of the bitterns is managed by returning them to brine lakes, deep-well injection, or storage in ponds.

Production
Worldwide consumption of potash was 52 Mt (57.3 million st) in 2005. Two-thirds of its production comes from Canada, Russia, and Belarus, with only nine other countries contributing to production. Primary consumers of potash include the United States, China, Brazil, and India. The most prevalent use of potash is as a fertilizer, primarily in the form of muriate of potash, KCl. Muriate of potash is also sold as an animal feed supplement.

Magnesium
Magnesium is found in about 60 minerals, of which only a few are of commercial value. The third most common element in seawater, it is found in deposits of magnesite and dolomite, and, in soluble form, in lake brines, mineral waters, and springs. Magnesium is produced from minerals, seawater, wells, and lake brines.

Desirable Geologic Conditions and Chemistry
Much like many other soluble salts of value, magnesium is produced from evaporites and brines. The brines are typically sourced from concentrated seawater, wells drilled into salt deposits, and concentrated inland salt lakes. In addition, brines are produced from saline aquifers, oil-field production brines, and mineral springs. In salt beds, magnesium is found in minerals such as carnallite found in New Mexico, kainite found in the Stassfurt mines in Saxony, Germany, or bischofite. The primary water-soluble magnesium minerals are shown in Table 11.5-1.

Mining Method
Solution mining of magnesium, practiced for decades, has been proven to have substantial advantages over conventional mining methods, given the right deposits. Heated water is used as the lixiviant and injected into impermeable deposits, as described previously in this chapter. Naturally occurring brine simply needs to be collected from the source. US Magnesium LLC uses

Table 11.5-1 Principal magnesium minerals

Mineral	Composition
Carnallite	$KMgCl_3 \cdot 6H_2O$
Kainite	$MgSO_4 \cdot KCl \cdot 3H_2O$
Bischofite	$MgCl_2 \cdot 6H_2O$

Table 11.5-2 Principal soda ash minerals

Mineral	Composition
Trona	$Na_2CO_3 \cdot NaHCO_3 \cdot H_2O$
Nahcolite	$NaHCO_3$
Dawsonite	$NaAlCO_3(OH)_2$
Thermonatrite	$Na_2CO_3 \cdot H_2O$
Natron	$Na_2CO_3 \cdot 10H_2O$

solar evaporation to concentrate natural brine from the Great Salt Lake prior to processing and production of magnesium metal.

Recovery is based on the ore deposit characteristics, the mining method employed, and the volume of nonproduct minerals such as alumina, silica, lime, and iron oxide.

Processing Method

The processing method varies by desired product. Magnesium hydroxide ($Mg(OH)_2$) is used to produce magnesia (MgO). Magnesium chloride ($MgCl_2$) can be used as is or further processed to produce magnesium metal. Magnesium can also be a by-product of other soluble salts such as salt, potash, and borates. Because of its very high solubility in water and other brines, magnesium is typically the last salt to precipitate in solar evaporation processes.

Magnesium hydroxide. Magnesium hydroxide is recovered from brines by precipitating the dissolved magnesium as magnesium hydroxide with the addition of dolime (calcined dolomite $CaO \cdot MgO$). After precipitation of calcium carbonate, the product-bearing solution is treated with sulfuric acid to remove any remaining calcium bicarbonate and then pumped to thickeners where concentrated magnesium hydroxide slurry is generated. This slurry is then filtered to produce a filter cake containing about 50% solids. The filter cake is either directly calcined to produce refractory or caustic-calcined magnesia, or it is calcined and pelletized before dead-burning to produce the desired size and density characteristics.

Magnesium chloride. To recover magnesium chloride from brines, the brine is pumped to a series of solar evaporation ponds where water evaporates, thus concentrating the brine. Generally, the first salt to precipitate as the water evaporates is sodium chloride. Double salts containing potassium and magnesium precipitate next and, dependent on other factors, kainite ($MgSO_4 \cdot KCl \cdot 3H_2O$), schoenite ($MgSO_4 \cdot K_2SO_4 \cdot 6H_2O$), and carnallite ($KMgCl_3 \cdot 6H_2O$) can be precipitated. Sodium sulfate is then precipitated, while the final product to be precipitated from the brine is magnesium chloride.

Although magnesium chloride is a directly marketable product, it can be used to produce magnesium metal by feeding high-quality, anhydrous magnesium chloride to a direct-current electrolytic cell with a very high amperage to reduce the magnesium chloride feed to magnesium metal and chlorine gas.

Soda Ash

Egyptian writings mention crude soda ash (Na_2CO_3) as early as 5,000 years ago. Natrun, crude trona, has been used as a valuable resource well before 3500 BC from the Wadi Natrun lakes in Egypt. Although natural deposits are rare, soda ash has been historically obtained by leaching wood or plant ashes. The uses of soda ash as a cleansing compound and as a flux in the making of glass were eventually discovered, and these are now its primary uses.

In 1792, the Leblanc synthetic soda ash process replaced the leaching of wood ashes. In 1865, the Solvay process made the Leblanc process obsolete because Solvay was simpler and more efficient and its by-products were less difficult to manage. In the late 1930s, the discovery of the large Green River deposit in Wyoming made natural soda ash less expensive in the United States than the synthetic soda ash obtained via the Solvay process.

Desirable Geologic Conditions and Chemistry

Using natural soda ash deposits rather than synthetic generation of soda ash from salt and limestone is preferable because the synthetic process is usually more expensive and produces more waste. About 60 natural sodium carbonate deposits are identified globally, only some of which have been assessed. Natural soda ash deposits can be defined in two manners: solid evaporites and brines.

Solid deposits of trona mineral are found in the extensive Green River Formation in Wyoming and at Lake Magadi in Kenya. Sodium-carbonate-rich deposits, in the form of nahcolite and dawsonite minerals, are located in the Piceance Creek basin in Colorado (United States). Other deposits occur as trona in dry lakes such as Searles Lake in California. The Rift Valley of northeast Africa has a series of lakes and dry lake beds that contain both solid trona and brines of various minerals, such as Lake Natron in Tanzania and the Sua Pan deposit in Botswana. The principal soda ash minerals are shown in the Table 11.5-2.

The Green River Formation is host to the world's largest deposit of trona. It is believed that about 45 billion t (49.6 billion st) of soda ash resources can be recovered through mining processes. Bed thicknesses are typically greater than 1.8 m (6 ft) thick and occur as beds that are primarily trona or with trona and halite intermixed. The Green River basin contains geologic formations created by inland lake sediments with many saline minerals.

The brines of the Searles Lake and Owens Lake deposits in California contain a series of dissolved minerals of sodium, calcium, magnesium, and sulfur, and are said to contain about 740 Mt (815 million st) of soda ash. The nahcolite deposit located in the Piceance Creek basin of northwestern Colorado contains reserves estimated at 26 billion t (29 billion st). Solution mining was practiced by American Soda for 4 years but was halted in 2004 because of economic conditions. Other deposits, such as those in northeastern Africa, have been created by the carbonatite leaching of volcanic ash or lava.

Mining Method

The conventional in-situ solution mining method most commonly employed is to create a cavity within the target ore formation and use a blanket on top of the lixiviant to control the shape of the cavity. Different methods of hydraulic fracturing of the beds are also used to increase permeability within

Table 11.5-3 Principal boron minerals

Mineral	Composition	B_2O_3, wt%
Boracite	$MgB_7O13 \cdot Cl$	62.2
Sassolite	$B(OH)_3$	56.4
Kernite	$Na_3B_4O_7 \cdot 4H_2O$	51.0
Colemanite	$Ca_2B_6O_{11} \cdot 5H_2O$	50.8
Hydroboracite	$CaMgB_6O_{11} \cdot 6H_2O$	50.5
Priceite	$Ca_4B_{10}O_{19} \cdot 7H_2O$	49.8
Probertite	$NaCaB_5O_9 \cdot 5H_2O$	49.6
Tincalconite	$Na_3B_4O_7 \cdot 5H_2O$	47.8
Meyerhofferite	$Ca_2B_6O_{11} \cdot 7H_2O$	46.7
Ulexite	$NaCaB_5O_9 \cdot 8H_2O$	43.0
Szailbelyite	$MgBO_2(OH)$	41.4
Inyoite	$Ca_2B_6O_{11} \cdot 13H_2O$	37.6
Kurnakovite	$Mg_2B_6O_{11} \cdot 15H_2O$	37.3
Borax (Tincal)	$Na_3B_4O_7 \cdot 10H_2O$	36.5

the formation. The average recovery, using solution mining techniques, is about 30%, whereas recovery with conventional mining techniques is about 45%. Techniques are now being investigated to improve this recovery percentage.

Mining operations began in the Green River basin in 1949 using conventional underground methods. As economic conditions changed, new methods were considered. For example, one operation, the Granger plant, was shut down as a conventional mining facility in 2001 because of poor markets but was reopened in 2005 employing a solution mining method.

Solution mining at the American Soda project at Yankee Gulch in the Piceance Creek basin injected adequately pressurized high-temperature (177°C to 216°C [351°F to 421°F]) water to thermomechanically fracture the nahcolitic oil shale and generate a bicarbonate liquor that was further processed on the surface. Ore-body depths were 670 to 800 m (733 to 875 yd), and the cavities were approximately 150 m (164 yd) tall with diameters up to 60 m (66 yd).

Trona and nahcolite ore are solution mined with aqueous sodium hydroxide and aqueous hydrogen chloride. After processing, the brine is regenerated in an electrodialysis cell before being returned to the solution mining cycle. After the trona is captured, it is clarified, filtered, recrystallized, and placed in a centrifuge to dewater the product. After this, it is dried and packaged or shipped in bulk to the market.

Processing Method
In cases where solution mining has been used, the pregnant liquor is a carbonate solution that has been carbonated to form sodium sesquicarbonate or sodium bicarbonate. In some instances, both carbonates are formed. These carbonates are then calcined to produce monohydrate or anhydrous sodium carbonate, depending on the required soda ash product. The extraction solution is then treated with slaked lime and water to produce a weak sodium hydroxide solution to recirculate back to the mine.

Borates
Although borates are commonly found, economic deposits only occur in the Alpine belt of south-central Asia, the Andean belt of South America, the eastern part of Russia, and the Mojave Desert in California. Most borate production is by conventional mining methods.

Several types of deposits are mined to produce specific products that typically have stringent standards. Approximately 70% of global borate supplies come from sodium borates produced in North and South America, while nonsodium borates are produced in Turkey. The primary borate minerals are shown in the Table 11.5-3. Borate deposits occur as natural brines and as crystallized underground deposits. Underground ore bodies may contain natural brine within void spaces.

Mining Method
Searles Lake employs permeable-type techniques to solution mine natural subsurface brines in addition to processing weak lake brines. The Fort Cady Minerals project in the Mojave Desert mines a deep permeable-type deposit consisting of colemanite ore beds permeated with naturally occurring brine. The deposit is situated more than 400 m (437 yd) below the surface, has an average thickness of 36 m (39 yd) that extends across 110 ha (272 acres) and appears to be completely encapsulated in clay. This encapsulation allows hydrochloric acid to be used as a lixiviant to further saturate the existing brine, allowing 50% boron content to be achieved. The operation is permitted for a production of 90,000 t/a (99,000 tpy) (Hartman 1996).

In-situ mining of an impermeable deposit was tested at the Kramer deposit in Death Valley, California, in 1948. This deposit was 72% borax with an 8.8-m (28.9-ft) thick layer at a depth of 107 m (351 ft). Water was injected at 102°C to 110°C (216°F to 230°F). The experiment ran for 38 days and leached 230 t (250 st) of borax (Garrett 1998).

Processing Method
Most commercial deposits have borax, colemanite, kernite, or ulexite as the primary boron mineral in the deposit. The chemical makeup of these minerals dictates the required different processing methodology. At Fort Cady, the leach liquor is sent to solar evaporation ponds where crude boric acid is crystallized or lime is added to precipitate synthetic colemanite, with the option to regenerate hydrochloric acid with sulfuric acid, for re-injection (Garrett 1998).

At the Searles Lake Trona plant, currently operated by Searles Valley Minerals, solvent extraction is used to recover boric acid from weak lake brines and produce anhydrous borax (Pyrobor) and borate decahydrate. The Searles Lake Westend plant manufactures anhydrous borax (Pyrobor), borax penta-hydrate (V-Bor), and sodium sulfate. Borax production methods can be quite varied when the brines contain a complex of soluble salts. At Searles Lake, two processing methods are used: evaporation and carbonation.

In the evaporation process, the brine is evaporated to crystallize sodium chloride, licons, and burkeite, thereby saturating the potash. Rapid cooling then precipitates filterable potash. The property of slow borate crystallization assists in separation of the potash from the borates. The liquor from the potash recovery operations is pumped to crystallizers where the boron is recovered as boron pentahydrate. The sludge from the crystallizer is thickened, filtered, washed, repulped, and redissolved to prepare a liquor, which is sent to polishing filters and vacuum crystallizers to make a high-purity deca- or penta-hydrate, which is then carefully dried and prepared for market. Boric acid and anhydrous borax are also made.

In the carbonation process, the brine with high borax and sodium carbonate is loaded into the top of carbonating towers. It is then subjected to compressed flue gas, which is

Table 11.5-4 Principal sodium sulfate minerals

Mineral	Composition
Thenardite	Na_2SO_4
Mirabilite (Glauber salt)	$Na_2SO_4 \cdot 10H_2O$
Glauberite	$Na_2SO_4 \cdot CaSO_4$
Astrakanite (bloedite)	$Na_2SO_4 \cdot MgSO_4 \cdot 4H_2O$
Burkeite	$2Na_2SO_4 \cdot Na_2CO_3$

controlled with very fine bubbles and rotating screens from distributors in the base of the towers. Carbon dioxide causes sodium carbonate (Na_2CO_3) to convert to sodium bicarbonate ($NaHCO_2$), and since the latter has very low solubility in the brine, it crystallizes, further purifying the brine. The liquor from the carbonation towers is then sent to the boron vacuum crystallizers, where borax seed crystals serve as nucleation points for growth.

Sodium Sulfate

Sodium sulfate has been used for more than 5,000 years. Records show that thenardite was used in ancient Egypt to preserve mummies while later uses included soap production and a mixture for glass. Its many applications and convenient location in playas and lake beds has made it a common industrial mineral for centuries.

Geologic Deposits

Sodium sulfate deposits can range from very large to quite small. Deposits in the Gulf of Kara-Bogaz in Turkmenistan, Searles Lake, Laguna del Rey in Mexico, Great Salt Lake, Lake Kuchuk in Russia, and the Aral Sea in Central Asia fall into the large-deposit category. Some of the smaller deposits of glauberite or thenardite in nonmarine deposits are less pure than the larger deposits. About 95% of the global sodium sulfate production comes from the larger deposits, which employ more sophisticated processing plants.

Many of the sodium sulfate deposits are found in playas, saline lakes, or brine deposits that have been developed along with a complex of other salts. One would think that sodium sulfate deposits would be formed within a mixture of salts, but, because of the steep solubility curve of sodium sulfate, mirabilite selectively crystallizes during cool evenings to form a more pure deposit that can be quite massive. In many deposits, mirabilite is then further enhanced to produce thenardite, astrakanite, burkeite, or glauberite. The chemical formulas for these minerals are shown in Table 11.5-4.

Mining Method

Although conventional mining methods have been used, solution mining is also being employed on suitable projects. Typical lixiviant is injected at moderate temperature (around 40°C [104°F]) with the mining operations determined by mineral solubility rather than by a chemical methodology of extracting the desired mineral.

One of the world's largest glauberite–thenardite deposits is located in Spain where 2.5 billion t (2.7 billion st) of equivalent sodium sulfate has been estimated. These resources are located in the central Madrid and Tajo basins along with the northeast Ebro basin.

Quimica del Rey S.A., a subsidiary of Industria Peñoles, produces sodium sulfate from subsurface brines, which are pumped from the subsurface to the surface and processed to produce sodium sulfate. The spent brine is further treated to produce magnesium oxide.

Processing Method

Equipment used to process naturally occurring sodium sulfate varies widely and includes air-cooled ponds, spray pipes, spray towers, polished-tube coolants within agitated tanks, scraped surface coolers, multicompartment vacuum crystallizers, and growth-controlled vacuum or heat exchange crystallizers. Generally, all operations are quite different up to the point where the evaporative process begins.

The production of a pure product requires that a minimal amount of insoluble material remains in the brine and that other soluble salts be minimized as well. To accomplish this, the feed brine is filtered, and an intermediate crystal sludge is prepared and then washed. The evaporators must be thermally efficient, have high capacity, and be able to operate in an economic manner in order to hold down costs. Multi-effect evaporators should be used to ensure lower operating costs, and heat sources should be selected in order to provide for lower costs. The design should reflect the fact that uniform crystals produce a higher-quality product.

Lithium

Lithium carbonate production from alkali brines started in 1966 at Clayton Valley, Nevada (United States), by Foote Mineral Company, which has been producing lithium carbonate from brines ever since. This is the only primary lithium carbonate-producing plant in North America.

Three types of brines are usually associated with lithium deposits: continental, geothermal, and oil-field. Continental brines occur where the lithium content is mainly derived from the leaching of volcanic rocks. These brines vary greatly in regard to lithium content, largely as a result of the extent to which they have been subjected to solar evaporation in concentrated solutions in lakes. They range from 30 to 60 ppm in the Great Salt Lake, where the evaporation rates are modest and dilution is constant because of the high volume of fresh water inflow. Examples of subsurface brines are Searles Lake (a former location of lithium production), Silver Peak, Nevada (a current source), and the high-altitude salars in Bolivia, Argentina, Chile, Tibet, and China, where lithium concentrations can be very high.

Geothermal brines occur as small quantities contained in brines at Wairakei, New Zealand (13 ppm Li), at the Reykanes Field (8 ppm) and other areas in Iceland, and at El Tatio in Chile (47 ppm). The most attractive known occurrences are in the Brawley area south of the Salton Sea in Southern California. Large tonnages of lithium are contained in U.S. oil-field, brines in North Dakota, Wyoming, Oklahoma, east Texas, and Arkansas where brines grading up to 700 ppm are known to exist. Other lithium brines exist in Paradox Basin, Utah.

Lithium salts are produced from brine processed in both North and South America. Lithium salts crystallize out of solution after the precipitation of sodium, potassium, and magnesium compounds. In marine evaporites, the Li:Na ratio is generally equal to or less than a 0.0001:1 ratio due to natural concentrations of natural seawater. In contrast, the average lithium concentration in seawater is about 0.174 mg/L, or 0.174 ppm. In nonmarine evaporites, the Li:Na ratio is substantially enhanced.

Lithium deposits with higher lithium and lower magnesium concentrations are more economic than other alkali

Table 11.5-5 Chemical composition of lithium brine lakes

	Salar de Atacama, Chile	Salar del Hombre Muerto, Argentina	Salar de Rincon, Argentina	Salar de Uyuni, Bolivia	Clayton Valley	Great Salt Lake	Zhabuye Salt Lake, Tibet	DXC Salt Lake, Tibet	Taijinaier Salt Lake, China	Seawater
Lithium content, %	0.15	0.062	0.033	0.035	0.023	0.004	0.12	0.04	—	0.000017
Mg:Li ratio	6.4:1	1.37:1	8.61:1	18.6:1	1.43:1	250:1	Low	0.22:1	—	7,000:1
Height, m (AMSL*)	2,300	3,700	3,700	3,653	1,300	1,280	4,422	4,475	—	0
Recoverable lithium carbonate, Mt	5.23	2.1	1.3	3.2	0.62	—	4.0	0.42	1.4	—

Source: Tahil 2007. Recoverable LCE updated based on Tahil 2008.
*AMSL = above mean sea level.

evaporites (Bartlett 1998). The ratio of magnesium to lithium is a critical factor in recovery of lithium from brines. If the level of magnesium in the brine is too high, the evaporation rate is slowed and the lithium yield is reduced. The average magnesium concentration in seawater is about 1,200 ppm, or a Mg:Li ratio of 7,000:1, making recovery of lithium from seawater uneconomical.

Mining Method
Subsurface brines are produced from wells, whereas lake brines are simply pumped to solar evaporation ponds. The brines have high concentrations of chloride ions, which are corrosive to most common metals, so materials of construction for pipe and pumps are generally plastic, polyethylene, and stainless steel. In the past, these solar evaporation ponds were not lined. The evaporation of the brine produced a layer of salt that sealed the ponds. However, present operations have to consider liners to reduce leakage of bitterns and loss of solution. Characteristics of several lithium-contained lake brines are shown in Table 11.5-5.

The depth of the ore deposit, thickness of the zone with high lithium concentrations, and the permeability of the material are of primary importance to the successful extraction of brine from the deposit. Typically, in production environments, the subsurface solutions flow by gradient into the extraction wells. Experience has shown that if brines are pumped into these systems to accelerate the capture of the lithium, they typically short-circuit, which causes the production system to fail.

A driving factor for the production of lithium from brines is the evaporation rate in solar evaporation ponds. High rainfall, high humidity, and high altitude are all detrimental to rapid evaporation of the water and subsequent concentration of the salts in the starting brine. Based on the measured lithium content of the brines, recovery of lithium carbonate is typically only 50%.

During mining, there is limited potential for gas generation, with the exception of working with oil-field brines, which underlie deposits of oil and natural gas. Because of the association with petroleum products, these brines must be carefully handled. Caution must be exercised during lithium extraction so that combustible gases are properly vented.

Most, if not all, solution mining and solar-evaporation-based lithium carbonate operations have yet to start production, so little history is available on the final reclamation of lithium production operations. However, reclamation requirements will be similar to most salt production operations and most likely will require removal of any residual sodium chloride, potash, or other materials in the ponds, removal of the dikes and liners, and leveling the area for natural drainage. The groundwater status in the immediate area will also have to be addressed and the site recontoured and revegetated to regulatory standards.

Processing Method
The process used to obtain lithium carbonate (Li_2CO_3) is known as the lime soda evaporation process. Brine is pumped from the evaporite deposits as a naturally occurring solution with little or no recharge or makeup with fresh water from the operator. The brine then flows through a series of evaporation ponds. Sodium chloride is the first to precipitate and is often harvested from the first set of ponds in order to avoid either raising the height of the pond dikes or adding ponds as they fill with precipitate. The sodium chloride may also be a by-product that can be sold, if there is a nearby market. The next series of ponds produce a mixed sodium/potassium chloride precipitate that may also be recovered for reprocessing to separate the potassium chloride (potash or KCl) as a by-product. At this point, the lithium content is 40 to 60 times the original concentration of the evaporite brine. The bittern is further processed to remove the soluble magnesium and boron, if present. Though the boron is removed by solvent extraction, the magnesium is precipitated using lime. The magnesium hydroxide and magnesium sulfate are then settled and filtered. The final bittern may be further concentrated by evaporation or, if in sufficient concentration, is precipitated with soda ash. The lithium carbonate precipitate is then filtered, dried, and prepared for market.

The primary product from in-situ mining of lithium, which is followed by solar evaporation and precipitation, is lithium carbonate. Other products that can be produced are lithium hydroxide, lithium chloride, and mixed lithium-magnesium sulfates.

Table 11.5-6 summarizes the main lithium resources, recoverable reserves, and potential chemical grade lithium carbonate production to 2020.

METALS EXPLOITED

Uranium
Because uranium was recognized in the 1940s as having significant value in several applications, underground and surface mining were employed to recover this material in quantity. As the industry was founded in the western United States, ores were treated in a number of conventional surface milling facilities. In the late 1960s in Wyoming and Texas, the uranium industry pioneered the commercial use of in-situ acid leaching. In the mid- to late 1970s and 1980s, in-situ leaching (ISL)

Table 11.5-6 Lithium carbonate production

Deposit Lithium Brines	Resource (Li metal), Mt	Reserve (Li metal), Mt	Current Li$_2$CO$_3$ Production, t, 2007	Probable Li$_2$CO$_3$ Production, t, 2010	Optimum Li$_2$CO$_3$ Production, t, 2015	Optimum Li$_2$CO$_3$ Production, t, 2020
Salar de Atacama	3.0	1	42,000	60,000	80,000	100,000
Hombre Muerto	0.8	0.4	15,000	15,000	20,000	25,000
Clayton Valley	0.3	0.118	9,000	9,000	8,000	8,000
Salar de Rincon	0.5	0.25	—	10,000	20,000	25,000
Salar de Uyuni	5.5	0.6	—	—	15,000	30,000
Zhabuye	1.25	0.75	5,000	10,000	20,000	25,000
Taijinaier	1.0	0.5	10,000	20,000	40,000	50,000
DXC	0.16	0.08	—	5,000	5,000	5,000
Salar de Olaroz*	0.32	0.16	—	—	5,000	5,000
Salton Sea	1.0	—	—	—	5,000	10,000
Smackover†	1.0	—	—	—	10,000	25,000
Bonneville	—	—	—	—	—	—
Searles Lake	0.02	—	—	—	—	—
Great Salt Lake	0.53	—	—	—	—	—
Dead Sea	2.0	—	—	—	—	—
Total			81,000	129,000	228,000	308,000

Source: Tahil 2008.
*Argentina
†Arkansas (United States).

techniques were largely perfected and proven to be economically feasible, and the technology became universally acceptable and employed in sandstone-hosted uranium deposits. At the same time, ISL techniques were also developed in the Soviet Union, utilizing similar approaches in slightly different geological environments. A few decades later, Australia also started to use the ISL technology for uranium recovery, and presently Kazakhstan is a major producer of uranium by ISL. The overall process of solution mining is now largely known as in-situ recovery (ISR).

Several types of uranium deposits are suitable for exploitation by ISL/ISR technologies. Of these, sandstone-hosted and paleochannel-type (as mined in Australia) deposits are amenable to ISR mining. These deposits typically occur as roll-front or tabular deposits and account for about 18% of known uranium resources. Ore bodies that can be economically recovered range in size from a few million pounds to hundreds of million pounds and are typically low- to mid-grade (0.04% to 0.15% U_3O_8), although higher grades are certainly possible. Typical uranium minerals in sandstone deposits include uraninite (UO_2), coffinite ($U(SiO_4) \cdot 0.5(H_2O)$) and carnotite ($K(UO_2)_2(VO_4)_2 \cdot H_2O$). Often, vanadium that occurs with the uranium can also be processed and recovered. Conversion factors for uranium species are given in Table 11.5-7.

In the United States, the ISR of uranium from sandstone deposits started in the 1950s with research into the heap leaching of run-of-mine ores. Eventually, it progressed to working subsurface deposits of uranium when mining conditions were favorable. Generally, flat-lying deposits with groundwater within confined ore zones and with good permeability are necessary prerequisites. Economical ore zones can range from a meter to tens of meters thick. Good permeabilities are typically considered to be better than 0.5 darcys, but deposits with lower darcy ratings have been successfully operated. After the ore deposit is drilled to determine thickness and uranium content, maps are generated to depict the grade multiplied by the thickness in order to prepare a grade-thickness map

Table 11.5-7 Uranium species conversion factors*

Species	Multiply by
U to U_3O_8	1.1792
UO_2 to U_3O_8	1.0395
UO_3 to U_3O_8	0.9814
UO_4 to U_3O_8	0.9294
UO_2SO_4 to U_3O_8	0.7667
$UO_2(SO_4)_2^{(2-)}$ to U_3O_8	0.6074
$UO_2(SO_4)_3^{(4-)}$ to U_3O_8	0.5028
$UO_2(CO_3)_2^{(2-)}$ to U_3O_8	0.7196
$UO_2(CO_3)_3^{(4-)}$ to U_3O_8	0.6237
$UO_4 \cdot 2H_2O$ to U_3O_8	0.8303
$UO_4 \cdot 2H_2O$ to U	0.7041

*For example, 1 kg U is equivalent to 1.1792 kg U_3O_8.

(in Kazakhstan, a similar method is used to develop a *productivity* map, which shows metal per square meter) that is used to guide evaluation and planning exercises. The minimum acceptable economic grade thickness is determined by evaluating economics, including the depth of ore zones, price variations, and establishing what ore can be mined by in-situ techniques at that time after considering overall project economics. After this minimum is established, the ore resources exceeding this amount, if the ore bodies are of sufficient size, are included in the mining plan.

Because it is critical to understand the leachability of the target uranium ore zones, metallurgical tests are conducted. Simple bottle roll tests, or column leaches, can be used to obtain an idea of potential optimal leachability, but this needs to be followed by complementary groundwater studies to establish groundwater quality, permeability, and other in-place characteristics. The overall geologic framework must also be fully defined to understand the structural geology, fracturing, definition of confining layers for the target groundwater regime, and, of course, both valuable and gangue mineral

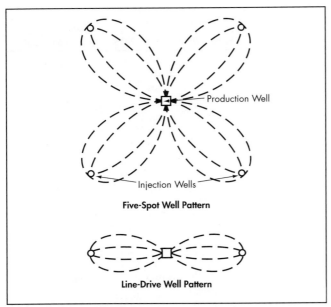

Figure 11.5-2 Well patterns

content. It may be necessary to conduct more in-depth mineralogical and metallurgical studies to define other constituents, types of clays, types of grain cement, and other items of interest. For example, titanium might be found, which is indicative of branerite; or clays or gypsum that create permeability problems; or highly reducing environments that must be overcome for effective in-situ leaching to be conducted.

Mining Method

The mining concept is to drill holes into the ore body and inject a lixiviant and then extract the pregnant lixiviant from a central production drill hole that is surrounded by the injection drill holes. Figure 11.5-2 provides a graphic of a five-spot and a line-drive production pattern, while Figure 11.5-3 shows the injection and production concept for a single target bed with a line-drive pattern. The type of pattern is dependent on the deposit shape, access, and the specifics regarding geohydrology. The specific design of the well-field pattern is dependent on the ore zone width, which can be tens of meters to hundreds of meters wide; the number of ore zones in the vertical ore zone to be mined; and the thickness of these ore zones. A third pattern, which is similar to the layout of the five-spot pattern, uses six injection wells so that a seven-spot pattern is created. This pattern is generally used in the United States when permeability of the ore zone, and therefore lixiviant flow, is low and more wells are needed for better lixiviant distribution throughout the ore zone. In Australia, seven-spot patterns are often used in very high permeability and/or high-grade zones. The additional wells help to ensure better contact and allow for the ability to reverse flow directions to stimulate production rates and obtain the highest uranium recovery possible. Figure 11.5-4 shows the transmission of pregnant and fresh/barren lixiviants to and from the ore body.

Economic ISR overall uranium recovery factors can range from 50% to 90%, depending on the deposit and time allowed for the mining process. Typically, recovery factors in the range of about 60% to 80% are common in the United States, while Kazakh well fields have often been operated with much higher recoveries, despite being uneconomic, and lower recoveries have historically been achieved in Australia. However, evaluation of recovery factors is not a simple task and requires in-depth knowledge of the ore calculation methodology to ensure that the ore in the margins of the roll front, the nose, and the tails, which may be included in the mine plan, can in fact be economically recovered. Some portions of a deposit might not be accessible to leaching because the uranium could be contained in lower-permeability units and/or clays. To further complicate this issue, detailed drilling is necessary to establish the most accurate location and characteristics of an ore body.

Both acidic and alkaline lixiviants are used to extract uranium from sandstone and paleochannel deposits. Alkaline lixiviants are prevalent in North America, while acidic lixiviants are common in Kazakhstan, other former Soviet Union countries, and Australia. Alkaline lixiviants place fewer metals into solution, and leave the well field easier to reclaim once the mining process has been completed. These lixiviants are generally composed of sodium carbonate and sodium bicarbonate, and many include oxygen or an oxygen-bearing compound that will oxidize the uranium to a soluble state.

Sulfuric acid is generally used as an acidic lixiviant, which is more powerful and places more metals into solution, generally leads to faster leaching rates, and can be necessary in specific types of deposits. The choice of lixiviant depends on the metallurgical characteristics of the deposit (specifically acid-consuming gangue), the economics associated with recovery rates, the groundwater regime, and the sociopolitical acceptability of using an acidic lixiviant. In addition, it is necessary to consider other groundwater uses and class of use within the region, as well as the potential for self-amelioration of the deposit once mining has ceased.

Processing Method

Uranium contained in the pregnant lixiviant extracted from the ore body is pumped to the processing plant where it first undergoes ion exchange to load the uranium onto the resin; the now barren lixiviant is recycled to the well field. The loaded resin is stripped of the captured uranium through an elution process. The eluate is subjected to precipitation and filtering processes, which generate yellow cake slurry that is dried in a low-temperature vacuum dryer, dehydrating the product while controlling emissions by capturing particulates. After drying is completed, the yellow cake is packaged and stored for shipment. In some cases, where salinity precludes the use of ion exchange, solvent extraction methods may be utilized.

The recovery of uranium within the processing plant is about 98%. The design of the plant is typically specific only to uranium, but other by-products can be produced. Uranium deposits typically contain other metals, such as molybdenum, vanadium, and copper, and it may be possible to economically extract these metals from lixiviant solutions. The overall processing plant can be designed to include modules to either remove or recover these metals as is economically feasible.

The uranium production process generates waste products. Under U.S. regulations, these solid waste products are classified as 11(e)(2) material and must be disposed of in a Nuclear Regulatory Commission–licensed disposal facility. The primary methods for liquid waste disposal are by deep disposal wells or by evaporating the liquids and disposing of the solids under the directives of regulatory requirements. In Australia, liquid wastes are disposed of in the mining zones

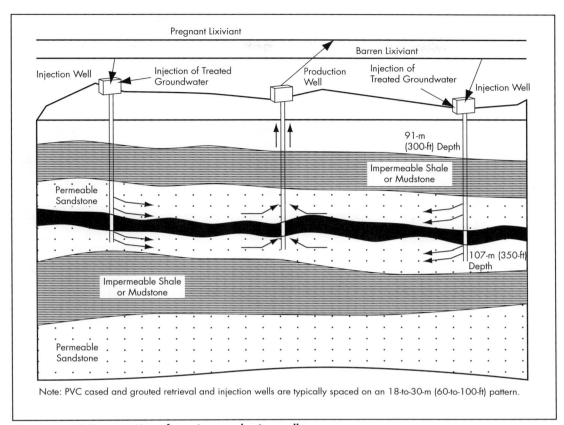

Figure 11.5-3 Cross section of uranium production wells

and solid wastes are disposed of in lined disposal facilities under the directives of the state or territory governments.

Groundwater Restoration

The ideal goal of groundwater restoration is to return the production zone (the area of an aquifer where uranium extraction occurs) to premining conditions. In the event this is not feasible, the Nuclear Regulatory Commission regulations allow for the achievement of restoration to the aquifer's quality of use class prior to mining operations. These standards have been deemed to be protective of public health and the environment.

The type of reagent used determines types of contaminants to the aquifer. ISR using acidic reagents causes a drastic decrease in pH and dissolution of other minerals in addition to the target uranium ores, concentrating most of the elements present in the surrounding rock to levels above the maximum admissible for drinking water supply. Microbes in the sulfide environment of the leach zone can produce a reducing environment, which may cause metals to precipitate in their sulfide forms. Meanwhile, bicarbonate reagents selectively dissolve the uranium ore, and contamination is noticeably lower. However, radium, selenium, and arsenic are still issues that must be addressed during remediation.

Two approaches can be used for groundwater restoration, direct cleaning and self-cleaning, as described in Table 11.5-8.

In Kazakhstan, the mining aquifers are allowed to attenuate naturally, whereas in Australia, some aquifers have been considered unusable for human or animal consumption and hence require no restoration. More recently, however, some of the proposed ISR projects are in areas that may have better quality and more valuable groundwater resources. In these areas, it will be necessary to perform some measures of restoration to at least initiate increases in pH and return to class of use.

Current Production

According to the World Nuclear Association, primary global uranium production for 2008 was approximately 52,000 t (114 million lb) of U_3O_8 (World Nuclear Association 2010). Primary producers include Canada at 10,600 t (23.4 million lb), Kazakhstan at 10,000 t (22.1 million lb), Australia at 9,900 t (21.9 million lb), Namibia at 5,200 t (11.4 million lb), Russia at 4,200 t (9.2 million lb), Niger at 3,600 t (7.9 million lb), and the United States at 630 t (1.4 million lb). Most of the production from the United States and Kazakhstan comes from ISR production, which is recognized as the most cost-effective and environmentally friendly method of uranium production.

Uranium is primarily used in nuclear reactors throughout the world. Total demand forecast, which for 2009 was 77,000 t (170 million lb) U_3O_8, is met by both primary production from mining operations and secondary supply, which includes drawdown from inventories and highly enriched uranium sources. Nonetheless, most researchers forecast that demand will continue to outstrip supply unless more uranium mines are placed into production or expanded.

FRASCH SULFUR MINING

The process by which elemental sulfur is produced from a geological formation is singularly credited to Herman Frasch.

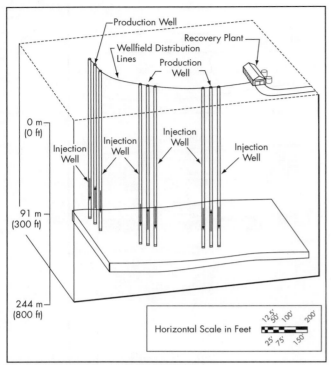

Figure 11.5-4 Schematic of ISR facility

Table 11.5-8 Groundwater restoration methods

Method	Description
Direct Cleaning	
Cleaning by reverse osmosis (RO)	• Varying number of pore volumes drawn to surface RO unit • Permeate recycled back underground • Brine sent to disposal • Most common method in United States
Washing with formation water	• 5–10 pore volumes drawn to surface • Depleted volume restored by adjacent natural groundwater
Cleaning by precipitation with reagents	• 5–10 pore volumes pumped to surface for uranium recovery and precipitation of undesirables with lime pulp • Saline residuals buried • Only recommended when other methods not suitable
Self-Cleaning	
Method of natural attenuation	• Given time—tens to hundreds of years—natural groundwater current restored to mined area • Rate dependent on natural flow rate, residual compositions, minerals present in rock, and acid versus alkaline reagent
Methods for accelerating natural attenuation	• Increase flow rates by pumping groundwater from beyond ore-body limits; introduce material to encourage natural bacteria growth; e.g., cheese whey, and molasses

Source: Data from IAEA 2001.

His concept of the mechanism of melting sulfur and then airlifting the molten liquid to the surface required a persistent effort in technical development and an expenditure of time and capital for a span of more than 10 years before realizing significant success.

The initial production of sulfur by the airlift (Frasch) process began in 1896 with the operation of three wells that produced 2,134 t (2,352 st) of sulfur (Haynes 1942). From a worldwide standpoint, however, the Frasch process is applicable to situations where fuel and water are available and the geological circumstances are favorable—for example, the Middle Eastern petroleum production areas, Poland, and Russia.

Basic Requirements for Frasch Sulfur Mine
Three basic resources are needed to develop a profitable Frasch sulfur mine:

1. Large deposit of ore of good grade (5% sulfur is suggested as a lower limit)
2. Adequate, reliable, and inexpensive source of water, preferably containing little dissolved solids
3. Low-cost source of fuel

Other factors such as market price, royalty payments, tax liabilities, and transportation costs also influence the economic feasibility of the project. Utilities that mine sulfur have been quoted requirements for

- Average water: 19.2 m^3/t (5,000 gal/long ton) of sulfur mined;
- Fuel: 20,000 to 22,300 Mm^3 of gas/m^3 (2,700 to 3,000 million ft^3 of natural gas/million gal) of mine water; and
- Power: 118 to 138 W/m^3 (600 to 700 hp/million gpd) of mine water (*Oil and Gas Journal* 1967).

Reliable and economic transportation systems are also essential for distribution of the bulk sulfur. In this regard, transportation and storage of molten sulfur has been growing, particularly where low-cost water transport can supply shore-based distribution terminals. Sophisticated production personnel are needed to operate and maintain the production and delivery systems.

Ore Deposit
The Frasch process was originally applied to removing sulfur from deposits associated with salt domes formed by the intrusive flow of deep-bedded salt that penetrates the upper layers of sediments. Typically, the sedimentary layers are mud, shales, limestone, gypsum, and anhydrite. The salt plug normally is covered by an insoluble mineral cap, below which the sulfur is found in the vuggy porosity of the limestone. The sulfur location in the cap rock of the domes is highly variable, and not all salt domes contain sulfur-bearing rock.

Sulfate-reducing bacteria in conjunction with hydrocarbons and water are clearly capable of producing elemental sulfur. Thus sulfur can be formed from the gypsum and anhydrite and be precipitated in the porous limestone with the progression of water circulation.

The west Texas deposits of sulfur are associated with layers of gypsum, anhydrite, limestone, and dolomite. These deposits are not associated with salt plugs, and the porosity and permeability of sulfur-bearing beds follow the natural layered porosity of the strata and fracture system (Zimmerman and Thomas 1969; Hentz et al. 1987).

Field Development
Efficient extraction of sulfur from the deposit depends on several interrelated factors. The hot water has to be introduced in such a way and at such a rate as to melt the sulfur so it will

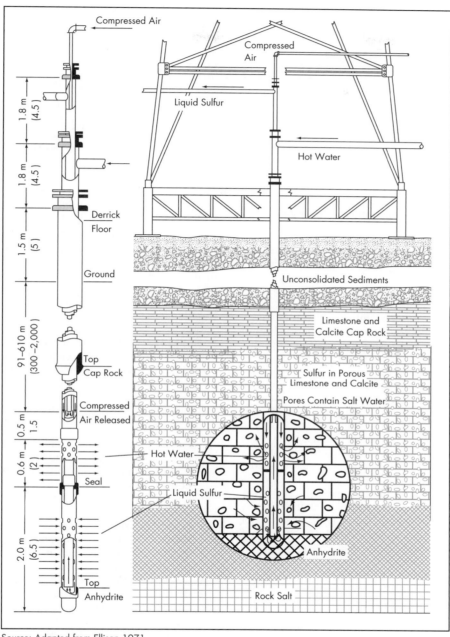

Source: Adapted from Ellison 1971.
Figure 11.5-5 Operation of a sulfur well

form a pool below the hot water, which is then airlifted to the surface. The piping system consists of concentric pipes which extend into the mineable formation (Figure 11.5-5). The operating strings are a series of these concentric pipes. The outer casing is set into the cap rock above the sulfur-bearing strata. The next string conducts the hot water and is set through the sulfur-bearing strata and is perforated at the bottom and also at some distance up the pipe. Another pipe is set concentrically within this pipe, and an internal seal is placed between the upper and lower perforations so that the flow of hot water is above the lower level. At a higher level, another pipe is run concentrically. The piping diagram in Figure 11.5-5 shows how hot water is introduced first, melting the sulfur, which is heavier, sinks to the bottom, runs into the lower pipe, and fills the second string. The molten sulfur is then airlifted with the compressed air and emerges as shown in the figure.

The molten sulfur is sent to storage, which can be handled in a number of ways. Since it may contain solid materials, or even small amounts of acid hydrocarbons, further processing may be required. Steam-heated leaf filters are commonly used for sulfur filtration. Precoating, usually with diatomaceous earth, is almost universal. The precoat should do the actual filtration to minimize blinding and will also neutralize some acid and reduce corrosion. However, small amounts of hydrated lime or ammonium bicarbonate may be added to the filter feed to neutralize acid. Such a system will produce

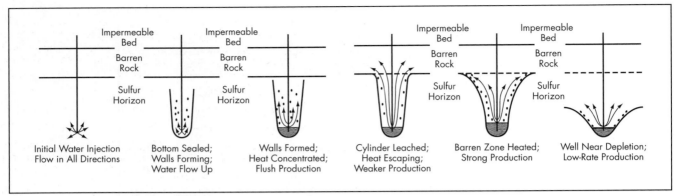

Source: Adapted from Donner and Wornat 1973.
Figure 11.5-6 Sequence during operation of sulfur well

sulfur that is free of ash and acid. If hydrocarbons must also be removed, activated clay may be added to bleach the dark sulfur. Other filter aids are also available to remove hydrocarbons. Filtration rates and precoat requirements will depend on the quality of the sulfur coming from the wells.

The sequence of the development of the well is shown in Figure 11.5-6. The initial injection melts a pool of sulfur and seals the bottom, as shown. Hot water is forced farther up the string, and a larger cylinder of rock is heated, melting more sulfur, which is airlifted to the surface. This yields a concentrated flush production. This scenario is simple in concept, but in practice, where variability of permeability, sulfur content, heat flow problems, and the influence of other wells exist, the actual operation is complex. It is necessary to reinject water produced in bleed wells to control flow, save heat, and dispose of wastewater. The mining process is highly variable because of the nature of the deposits, and the uncontrollable nature of the reservoir system makes the process something of an art that is gained by years of experience. Attempts have been made to apply reservoir engineering by substituting paradichlorbenzene as a meltable material for sulfur with some interesting insights (Rayne et al. 1956).

As mining proceeds, subsidence and collapse of overburden become a problem that can sometimes be helped by reducing the volume of hot water necessary, but may also shear tubing and cause many difficulties. Wells are often separated into groups to help control these problems and avoid total loss of field control. Corrosion due to molten sulfur contact is quite negligible. When the sulfur pool is drawn down to where the hot water and air come up together, the water flashes as steam, and a *blow* takes place. Hot water rising up the sulfur production string produces a corrosive environment, and, with a blow, it becomes appreciable. The salinity of the water influences the rate markedly. If seawater is used and a blow occurs, the corrosion rate is disastrous (Hackerman and Shock 1947; Shock and Hackerman 1949). One of the practical solutions to controlling corrosion has been to use specially produced cement-lined pipes as well as limiting the salinity of the water. Treatment of the water to minimize corrosion and scale problems is essential.

Bleed wells are required to balance the water pumped into the formation at the production wells. Location of the bleed wells will depend on such site-specific factors as the depth, permeability, and hydrology of the sulfur-bearing structure. Because of environmental concerns, bleed water should be recycled to the maximum extent possible. When required, disposal of bleed water needs to be carefully planned because of the high concentration of dissolved hydrogen sulfide and other sulfur-bearing compounds. Aeration in ponds, chemical treatment, and use of deep disposal wells are all methods of handling these waters.

Facilities Needed at Frasch Sulfur Mine

When a sulfur deposit has been located and defined by exploration drilling, a water supply located, and a fuel supply established, the design and construction of the needed facilities can proceed. Typical layouts showing both the well field and surface facilities are provided in the literature (*Oil and Gas Journal* 1970). Most mines will require some or all of the following:

- Pumps, pipelines, and reservoir for the water supply
- Power plant for heating and treating the water, compressing the air, and generating the electricity and steam
- Equipment for drilling and servicing wells
- Pipeline system for carrying the mine water and air to the wells and the sulfur to collecting stations
- Control stations for the producing wells
- Sulfur collecting, loading, and storage facilities
- Purification equipment needed to meet customer specifications
- Shipping facilities for solid and/or liquid sulfur
- Bleed-water disposal system
- Equipment needed for the maintenance of these facilities
- Employee transporting, dining, and housing facilities if the mine is remotely located
- Offices and buildings for material storage, laboratory, engineering, personnel, administration, and supervisory staff

ACKNOWLEDGMENTS

The authors recognize the work of D'Arcy A. Shock for his contribution of the section on Frasch sulfur mining, which has been reprinted (with minor reductions on history) from the previous edition of this handbook. Minor alterations have been made to convert to the International System of Units.

Additionally, coworker Edwin Bentzen, project manager at Lyntek, Inc., is thanked for his contribution of the section on lithium, as are the draftsmen turned illustrators who helped with the figures; Rachel Hawkins who offered moral support; and Nick Lynn, owner of Lyntek, who gave us the opportunity to complete the work.

REFERENCES

Bartlett, R.W. 1998. *Solution Mining: Leaching and Fluid Recovery of Materials*, 2nd ed. Netherlands: Gordon and Breach Science Publishers. pp. 373, 387.

Donner, W.S., and Wornat, R.O. 1973. Mining through boreholes–Frasch sulfur mining system. In *SME Mining Engineering Handbook*, 1st ed. Edited by A.B. Cummins and I.A. Given. New York: SME-AIME. pp. 21-60–21-67.

Ellison, S.P., Jr. 1971. Sulfur. In *Texas Handbook No. 2*. Austin, TX: Bureau of Economic Geology, University of Texas at Austin.

Feldman, S.R. 2006. Salt. In *Industrial Minerals and Rocks: Commodities, Markets and Uses*, 7th ed. Edited by J.E. Kogal, N.C. Trivedi, J.M. Barker, and S.T. Krukowski. Littleton, CO: SME.

Garrett, D.E. 1996. *Potash: Deposits, Processing, Properties and Uses*. London: Chapman and Hall.

Garrett, D.E. 1998. *Borates: Handbook of Deposits, Processing, Properties, and Use*. San Diego: Academic Press.

Hackerman, N., and Shock, D.A. 1947. Corrosion of steel in sulfur producing tubes—Frasch process [fundamentals]. *Ind. Eng. Chem.* 39:863–867.

Hartman, G.J. 1996. Fort Cady: Developing an in situ borate mine. *Min. Eng.* 48(8).

Haynes, W. 1942. *The Stone That Burns*. New York: Van Nostrand.

Hentz, T.F., Price, J.G., and Gutierrez, G.N. 1987. *Assessment of Native Sulfur Potential of State Owned Lands, Trans-Pecos, Texas*. Austin, TX: Bureau of Economic Geology, University of Texas at Austin.

IAEA (International Atomic Energy Agency). 2001. *Manual of Acid in Situ Leach Uranium Mining Technology*. IAEA-TECDOC-1239. Vienna, Austria: IAEA. p. 236.

Oil and Gas Journal. 1967. Oil industry background can pay off in sulfur production. *Oil Gas J.* 65(44):120–124.

Oil and Gas Journal. 1970. New sulfur region has modern plant. *Oil Gas J.* 68(18):125–129.

Rayne, J.R., Craft, B.C., and Hawkins, M.F. 1956. The reservoir mechanism of sulfur recovery. *Pet. Trans. AIME* 207:246–251.

Shock, D.A., and Hackerman, N. 1949. Corrosion of steel in sulfur producing tubes—Frasch process [field data]. *Ind. Eng. Chem.* 41:1974–1977.

Tahil, W. 2007. *The Trouble with Lithium: Implications of Future PHEV Production for Lithium Demand*. Meridian International Research. www.meridian-int-res.com/Projects/Lithium_Problem_2.pdf. Accessed February 2010.

Tahil, W. 2008. *The Trouble with Lithium 2: Under the Microscope*. Meridian International Research. www.meridian-int-res.com/Projects/Lithium_Microscope.pdf. Accessed December 2009.

World Nuclear Association. 2010. World Uranium Mining. www.world-nuclear.org/info/inf23.html. Accessed July 2010.

Zimmerman, J.B., and Thomas, E. 1969. *Sulfur in West Texas, Its Geology and Economics*. Bureau of Economic Geology Circular 692. Austin, TX: University of Texas.

CHAPTER 11.6

Coal-Bed Methane Production

Pramod Thakur

INTRODUCTION

Coal seams were formed over millions of years by the biochemical decay and metamorphic transformation of the original plant material. This process, known as coalification, produces large quantities of by-product gases. The volume of by-product gases increases with the rank of coal and is the highest for anthracite at about 765 m^3/t (Hargraves 1973). Most of these gases escape to the atmosphere during the coalification process, but a small fraction is retained in the coal. The amount of gas retained in coal depends on a number of factors such as the rank of coal, the depth of burial, the immediate roof and floor to the coal, geologic anomalies, tectonic pressures, and temperatures prevailing at the end of the coalification process. In general, the higher the rank of coal and the greater the depth of the coal seam, the higher the gas content. Gas contents of coal seams generally range from 0.1 to 25 m^3/t.

Methane is the major component of gas in coal, accounting for 80%–95% of the total gas content. The balance is made up of ethane, propane, butane, carbon dioxide (CO_2), hydrogen, oxygen, and argon. Because the coal-bed gas is mainly methane, it is generally called coal-bed methane (CBM). Methane recovered from active and abandoned coal mines is called coal mine methane.

Coal-bed methane production (CMP) from longwall gobs started in Europe in the 1940s, but methane drainage from virgin coal prior to mining began in the United States in the 1970s. The primary need for CMP was to make the coal mines safe from explosion hazards. The methane gas control techniques are discussed in Chapter 15.4 in detail. As the CMP process developed and became widespread, other benefits became apparent. In moderately gassy and very gassy mines, immediate improvement in coal productivity was realized. In some very gassy mines in the United States, the productivity improved from a low 15 t/worker-shift to 40 t/worker-shift. It significantly reduced the cost of mining by minimizing the gas delays caused by excessive gas emissions. In the 1980s, the volume of CBM produced prior to mining and postmining continued to increase, eventually leading to gas processing and commercial marketing. Extensive drilling of nonmineable coal deposits, such as CMP activities in the San Juan Basin of the United States, also produced a large volume of CBM. At present, nearly 10% (51 Gm^3 [billion cubic meters]) of U.S. natural gas production is from CMP. This number is likely to increase to 20% in the near future because CMP is less cost-intensive and financially less risky than deep drilling for natural gas production. The CBM, after processing to remove impurities, is exactly like natural gas with a calorific value of 35.8 MJ/m^3 (million joules per cubic meter).

Another financial incentive is driving CMP forward. Methane gas is a powerful greenhouse gas (GHG) with a radiative forcing that is 21 times higher than carbon dioxide (*radiative forcing* is a phrase used to measure the capacity of a gas to trap infrared energy). In many industrialized countries (e.g., the United States and Australia), a financial incentive is provided to capture CBM from active mines. The current U.S. incentive is $6.50/t of carbon dioxide. Thus, 1 t of CBM can qualify for a carbon credit of $136.50 if it is captured and marketed.

CMP today is a standalone, profitable enterprise. Besides saving lives and reducing the cost of coal production, it provides a clean-burning fuel and reduces GHG emissions. With adequate financial incentives, it can become a major supplier of natural gas in many countries such as the United States, China, India, and Australia.

COAL-BED METHANE RESERVES

As illustrated in Figure 11.6-1, coal deposits occur in 19 major basins around the world, and global reserves of coal and CBM are large. Coal has been a global energy source for a long time, but for the past 200 years it has played a vital role in the growth and stability of world economy. At present, 70 countries around the world annually mine about 6,500 Mt of coal. China, the United States, and India, the three largest coal producers, derive most of their electricity from coal. The total proven world reserve of mineable coal is 1 Tt (trillion metric tons), but indicated reserve to a depth of 3,000 m ranges from 17 to 30 Tt (Landis and Weaver 1993). Estimated reserves of CBM range from 78–959 Tm^3 and are shown in Table 11.6-1.

Pramod Thakur, Manager, Coal Seam Degasification, CONSOL Energy, Inc., Morgantown, West Virginia, USA

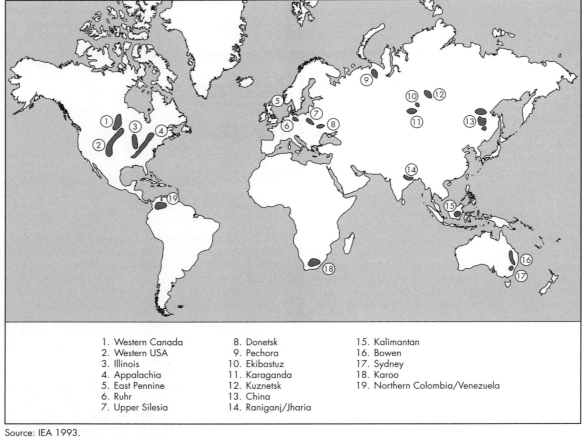

Source: IEA 1993.
Figure 11.6-1 Major coal basins of the world

Table 11.6-1 Global coal-bed methane reserve estimates*

Country	1990 Estimate, Tm³†	1997 Estimate, Tm³‡
Australia	8–14	—
Canada	6–76	92
China	20–35	31
Germany	3	2.83
India	1	0.7
Poland	3	0.4–1.5
Russia	20–166	720–790
South Africa	1	—
United States	11	30–41
Other countries	5–10	—
Total	78–320	877–959

*Recoverable reserve: 30%–60% of the total gas in place.
†Data from Kuuskraa et al. 1992.
‡Data from Cairn Point Publishing Company 1997.

CMP techniques were originally developed in the United States but are now prevalent in the rest of the world. At present, most CMP is realized from relatively shallow (<1,000 m deep) mineable coal deposits. However, coal deposits reach a depth of 3,000 m in many basins. Deep coal deposits are rich sources of CBM, but production techniques are not very successful yet because of very low permeability. New techniques are currently being developed (Thakur 2002). The vast global deposit of deep coal can be used in three ways:

1. CMP from deeper horizons (1,000–3,000 m)
2. Sequestration of carbon dioxide, a GHG. Production of CBM from deeper formation is assisted by CO_2 flooding. Coal seams have a much higher affinity for CO_2; hence, they absorb CO_2 and at the same time release methane.
3. Underground coal gasification (UCG). Coal seams not used for CO_2 sequestration can be burned in situ with a limited supply of oxygen to produce low-calorific-value combustible gases. In-situ combustion of coal also greatly enhances methane production by heating the coal formation.

Thus, there is a synergy in methane production from deep coal seams, CO_2 sequestration, and UCG. The subject will be discussed in the order that a typical CMP project is undertaken:

- Reservoir properties
- Wire-line logging
- Production technology
  - Shallow coal
  - Medium-depth coal
  - Deep coal
- Well servicing and maintenance
- Produced water disposal

- Gas gathering and processing
- CO_2 sequestration
- UCG

COAL-BED RESERVOIR PROPERTIES

The most important reservoir properties that not only influence the gas production rate but also determine the production techniques are

- Gas content of coal seams and their gas isotherms;
- Permeability;
- Reservoir pressure;
- Diffusivity of coal;
- Water content and quality of water; and
- Ground stresses and elastic properties of surrounding strata (e.g., compressive strength and Young's modulus of elasticity).

Coal porosity is generally low (1%–4%) and does not have a significant influence on gas production or production techniques.

Measurement of Gas Contents

The volume of gas contained in a ton of coal is termed *gas content* of the coal and is expressed in cubic meters per metric ton (m^3/t). It is generally accepted that the gas is stored in a monolayer on the micropore surfaces of coal. The volume of gas contained in coal is dependent on the rank, temperature, and pressure or depth of the coal seam. The micropore surface in coal is large; a ton of coal has a surface area of approximately 200 Mm^2. One cubic meter of coal can store two to three times the amount of gas contained in a typical sandstone reservoir for natural gas of the same volume.

For commercial gas production, it is best to core drill the entire field on a grid pattern and do a direct measurement of gas contents of all coal seams that comprise the gas reservoir. A typical spacing for core holes is one in every 200 ha.

Gas content measurement methods are classified as (a) conventional and (b) pressurized desorption techniques. In the conventional technique, coal cores or drill cuttings are retrieved from the core hole and immediately put in a sealed container to measure the desorbed gas. The method suffers from the uncertainty in the estimation of gas lost during sample retrieval and handling. To eliminate this problem, the pressured core desorption technique has been used. In this method, gas loss is minimized by sealing the coal sample down the core hole.

Both methods provide a positive proof of gas presence and data on the gas desorption rate that can be used to calculate diffusivity (to be discussed later). Desorbed gases are chemically analyzed to determine the composition and calorific value of the seam gas.

The industry standard for direct measurement of gas content is the U.S. Bureau of Mines method shown in Figure 11.6-2 (Kissell et al. 1973). Coal cores or drill cuttings are deposited in a sealed vessel and desorbed gases are measured periodically until the desorption rate is insignificant. Total volume of gas thus produced is known as the desorbed gas. The cumulative volume of desorbed gas plotted on the y-axis against the square root of time yields a straight line, and its intercept on the y-axis is a measure of the gas lost from the core before it was deposited in the sealed container. Subsequently, a portion

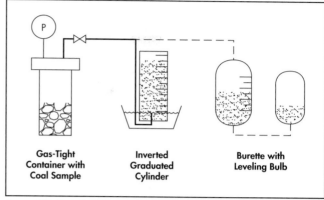

Source: Kissell et al. 1973.

Figure 11.6-2 Direct measurement of coal-bed gas content

of the core is ground in a hermetically sealed mill to release the residual gas. The sum of the three component volumes—desorbed gas, lost gas, and the residual gas—is the total gas contained in the sample. This total volume of gas is divided by the weight of the sample to obtain the gas content of the coal seam.

Total gas reserve (gas-in-place) for the coalfield must be established before any commercial undertaking is started.

$$\text{gas reserve, } G_r = \wp AHV$$

where

$\wp$ = density of coal, t/m^3
A = area of deposit, m^2
H = height of coal seam, m
V = gas content, m^3/t

If there is more than one coal seam in the reserve, as is the normal case, the total gas reserve is the sum of individual coal seam reserves.

Permeability

The permeability of a coal seam is the most important reservoir property next to gas content. Permeability of a porous medium is defined as 1 darcy when a fluid with a viscosity of 1 cP (centipoise) flows with an apparent velocity of 1 cm/s under a pressure gradient of 1 atm/cm (atmosphere per centimeter). The permeability is usually expressed in millidarcies (1 mD = 1/1,000 of a darcy).

Coal seams can have a permeability range of 1 microdarcy to 100 mD depending on the depth of the seam. Shallow coal seams have higher permeability, but deep coal seams show very low permeability.

Reservoir Pressure

The reservoir pressure can be calculated from the depth of the coal seam, but it is best measured directly. A vertical well is drilled and cased just above the coal seam. The coal seam is cleaned up with high-pressure water jets. A commercially available recording pressure gauge is installed in the well and the borehole is shut off at the surface. The gas pressure begins to build up, and in 4 to 6 days the final pressure is reached. The pressure gauge is retrieved and the final pressure is read from the chart. The coal seam reservoir pressure is typically less than the hydrostatic head (about 0.7 times).

Thus a 300-m-deep coal seam in the Northern Appalachian Basin of the United States has a reservoir pressure of approximately 2 MPa, whereas 600-m-deep coal seams in the Central Appalachian Basin have a reservoir pressure of 4–5 MPa. The reservoir pressure will rarely exceed the hydrostatic head, but if it does, it is a very fortunate situation for commercial gas production. The "fairway" area of the San Juan Basin in the United States is an example of an overpressurized basin. Many highly productive CBM wells have been drilled and produced in this area.

Diffusivity of Coal Seams
The diffusivity of a coal seam is a parameter that measures how fast the coal seam will desorb the gases when the confining pressure is reduced to the atmospheric pressure. It is commonly measured by a parameter called "sorption time." It is the time a coal seam needs to desorb 63% of the original gas content and is measured in days. Sorption time for some coal seams in the San Juan Basin is less than 1 day, whereas it can be as high as 900 days for coal seams in the Northern Appalachian Basin (Rogers et al. 2007).

Coal seams with high sorption time will take a long time to yield a high percentage of in-situ gas and should be drilled at a closer spacing and produced for a long time. On the other hand, coal seams with shorter sorption times should be drilled at a wider spacing (greater than one well per 20 ha) but produced for a shorter time (usually fewer than 10 years).

Water Content and Quality of Water
The coal seams are usually saturated with water. Actual fluid flow in coal seams is a two-phase flow. Invariably, the water phase is produced first. As the water content depletes, the gas production begins to increase, reaching a peak when water is almost drained out. A typical CBM well can produce from a few cubic meters per day to hundreds of cubic meters per day of water.

The amount of water produced and its quality have a serious impact on the economy of a CBM project and will be discussed later in the chapter.

Ground Stresses and Rock Properties
Every point in a coal seam lies in a stress field that comprises the vertical stress (σv), the major horizontal stress (σH), and the minor horizontal stress (σh). The comparative magnitudes of these stresses are the determining factors for a successful "hydrofracing" of a vertical well. For successful hydrofracing, the sequence $\sigma H > \sigma v > \sigma h$ must be true. This results in the most favorable vertical fracture because the fracture is always orthogonal to the least stress. It will also travel in the direction of σH; that is, the azimuth of the fracture is parallel to σH. Major directional permeability of a coal seam is also parallel to σH (refer to Chapter 15.4).

Rock mechanics properties such as compressive strength, elastic modulus, and Poisson ratios are also needed for a successful design of gas-production techniques.

WIRE-LINE LOGGING
The vertical boreholes drilled to establish the gas reserve and determine reservoir properties are also used for wire-line logging. The process involves lowering special sensors down the hole to measure various characteristics of the coal seam and sending the data to the surface using electrical wire line, hence the name. The most commonly used logging techniques to evaluate coal seams for CMP are

- Gamma ray logging (passive and active),
- Resistivity logging,
- Sonic logging,
- Neutron logging,
- Density logging, and
- Induced gamma ray spectrometry logging.

All these techniques had their origin in the oil and gas industry. They are only slightly modified, if needed, for coal-seam logging. Details of their application are described in literature (Scholes and Johnson 1993).

Figure 11.6-3 shows basic log responses for multiple coal seams in a vertical well. For coal seams, the gamma ray response is low, the resistivity response is high, the density response is high, and the neutron density response is also high. Most commonly, only gamma ray logging is used to precisely locate coal seams in a borehole. Some tools measure the natural gamma ray emissions from the formation, whereas others use an active gamma source and measure the scattered gamma rays.

The only other logging technique commonly used in CMP projects is sonic log, particularly the cement-bond log technique. Several acoustic sources mounted on a tool are lowered into the borehole. The sound wave is reflected by the steel casing and the cement in between the casing and the formation. The velocity of sound and/or the amplitude of sound waves indicate whether the cement is there or only a void is present.

These two logs are highly recommended to locate the coal seam and to make sure the steel casings are well set. All other logging techniques are optional. New logging tools that measure the gas contents, gas isotherms, and directional permeability in situ are being developed, but they are not yet commercially available as of 2010.

PRODUCTION TECHNOLOGY
The CBM resource as shown in Table 11.6-1 is huge. Production strategy varies from basin to basin, and even in the same basin based on local reservoir properties. Since all coal seam reservoir properties are depth dependent, the CBM resources can be classified into three broad categories to identify the best production technique for different depth ranges: shallow, medium-depth, and deep. The main characteristics of these reservoirs are summarized in Table 11.6-2.

Shallow Reservoirs
Shallow reservoirs are characterized by high-permeability, low gas content; low reservoir pressures; and low diffusivity. Vertical stresses (<3 MPa) are smaller than the horizontal stresses. Hydrofracing of a vertical well in these reservoirs results in horizontal fractures that are inefficient for gas production.

The following are preferred production techniques:

- Mine shafts with multiple laterals (horizontal boreholes) at the bottom of the shaft. A plan view of this drilling pattern is shown in Figure 11.6-4. A 3-m-diameter shaft is drilled in the target coal seam and the coal seam is reamed out to a larger size to accommodate drill rigs. In active coal mines, these shafts can be 6–8 m in diameter and are later used as ventilation shafts. The entire mine property may need six to eight shafts to recover coal. They can be sunk 10 to 20 years ahead of mining for CBM production. Depending on the price of gas, the marketed gas can

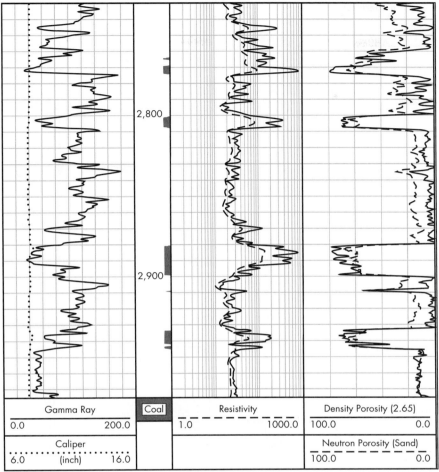

Figure 11.6-3 Basic responses to various logs in coal

defray all costs associated with shaft sinking. Specific gas production from these shallow coal seams ranges from 10–25 m³/d/m of the borehole. The specific gas production for a coal seam is a characteristic of the coal seam and is a measure of the rate of gas production per unit length of a horizontal borehole drilled for gas production. Assuming six horizontal boreholes drilled to 1,000 m, an initial production of 60,000–150,000 m³/d can be realized.

- Another production technique uses surface-based rigs to drill horizontal laterals (see Chapter 15.4 for details). The property is drilled with "production wells" on a grid pattern. Typically, four to six horizontal laterals are drilled from access wells 1 km away to intersect the production wells to realize commercial CMP (Figure 11.6-5).
- A special case of gas production from a shallow reservoir is the Powder River Basin of Wyoming (United States). Coal seams in this basin are very thick (30–60 m) and shallow; the maximum depth is <200 m. The gas is of biogenic origin (unlike deeper seams where gas is of thermogenic origin) and gas content is quite low (<3 m³/t). Permeability and water content are quite high. The production technique used here is a vertical well with hydrojetting but no hydrofracing. A typical well completed with a cost of $50,000 can yield gas production rates of 7–10 km³/d, making it a profitable venture. More than 10,000 wells have been drilled and completed this way in the Powder River Basin so far. The technique can work successfully in similar deposits in Montana (United States) or other places in the world.

Medium-Depth Reservoirs

Medium-depth reservoirs are characterized by low to medium permeability, high gas contents, relatively higher reservoir pressures, higher diffusivity, and the coal tends to be of higher rank. Vertical wells with well-designed hydrofracing are ideally suited to such reservoirs for gas production. Specific gas productions for vertical fractures created in these seams range from 4 to 8 m³/d/m. The hydraulic fracturing process basically involves pumping a fluid (such as water, nitrogen foam, or a high-viscosity gel) with fine sand (0.1–2 mm diameter) at a pressure high enough to hydrofrac the coal seam. Under ideal circumstances ($\sigma h < \sigma v < \sigma H$), a vertical fracture is created in the coal from the floor to the roof (Figure 11.6-6). Depending on the volume of the frac fluid pumped, the total frac length can be 300–600 m. The fracture will typically have two wings of 150–300 m that travel in the opposite directions but on the same azimuth that is parallel to the direction of σH.

For commercial gas production, all coal seams with a thickness >1 m are hydrofraced in the vertical interval 500–1,000 m. Multiple coal-seam fracing techniques are well

Table 11.6-2 CBM reservoir characteristics

Reservoir Type	Approximate Depth, m	Permeability, mD	Remarks*
Shallow	≤500	10–100	v < h < H
Medium-depth	500–1,000	0.1–10	h < v < H
Deep	≥1,000	<0.1	h < v < H

*v is the vertical stress; h and H are the minor and major horizontal stresses.

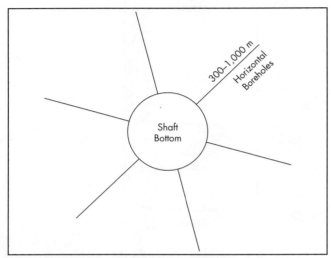

Figure 11.6-4 Mine shaft with horizontal boreholes

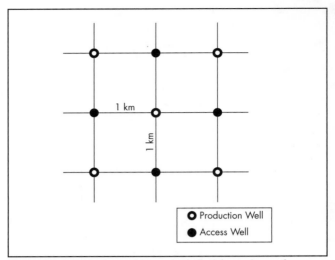

Figure 11.6-5 Horizontal laterals drilled from the surface

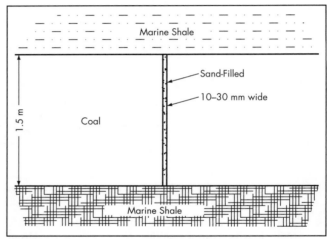

Figure 11.6-6 Sectional view of sand-propped hydraulic fracture

established but beyond the scope of this chapter. A typical gas production from each horizon of 1.5–2-m-thick coal seam is 3,000 m³/d. If three or four horizons of this thickness are hydrofraced in a well, gas productions of 9,000–12,000 m³/d can be realized.

Observations of nearly 200 hydrofraced wells and mapping of the subsequent mining of the fracture created can be summarized as follows:

1. The fracture volume (length × width × height, or LWH) is always proportional to the fluid volume, Q, pumped in:

    $LWH = CQ$

    where C is the leak-off coefficient; it is typically 1 or 2 orders of magnitude higher than that for sandstone or limestone reservoirs.
2. The fracture plane is always orthogonal to the least stress (σh or σv, whichever is smaller).
3. The fracture azimuth is always parallel to the major horizontal stress, σH.
4. The fracture always travels up into the roof but not in the floor, following the law of physics that the fluid will always flow down the pressure gradient.
5. When a coal seam bounded by shale or sandstone is fraced, the formation with the smallest modulus of elasticity, E, will hydrofrac first. In fact, strong shale and sandstone provide a confining barrier containing the fracture in coal.
6. The fracture width is proportional to (fluid viscosity/E) raised to the power of 0.25.

Figure 11.6-7 shows two methods of completion, namely (coal-bed A) open-hole and (coal-bed B) through perforations.

If the coal seam has a high compressive strength, the lowest coal seam can be fraced open-hole. The upper coal seams are fraced through perforations in the steel casing. If the coal seam has low compressive strength, it is preferable to hydrofrac it with perforated casings.

The choice of frac fluid varies from basin to basin. In the depth range of 500 to 750 m, gelled water or nitrogen foam, 70% quality, has given good results in the Central and Southern Appalachian Basin of the United States. On the deeper end (750–1,000-m depth) and for thick coal seams (>10 m), cross-link borate gels with very high viscosity have produced better results. An example is the San Juan Basin of New Mexico and Colorado. Sand concentrations in the frac fluid vary from 0.1 t/m³ to 0.5 t/m³ depending on the viscosity of the fluid. Fluids with higher viscosity can carry higher concentrations of sand.

A special case of CMP from deep and thick coal seams is the cavitation technique. Figure 11.6-8 shows a completion technique called cavitation through a perforated casing. If the coal seam has high compressive strength, cavitation can be

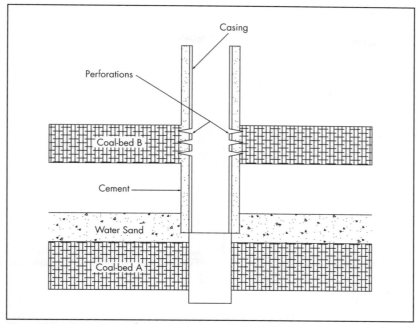

Figure 11.6-7 Two completion procedures for vertical wells in thin coal seams

done in an open hole. In the cavitation technique, the borehole is pressurized with water and the pressure is released. The process is repeated several times in a quick sequence with the hope that coal will cavitate in the depressurizing phase and enhance permeability leading to good gas production. Cavitation completions have worked well yielding 30,000–60,000 m^3/d of gas production in 10–20-m-thick coal seams.

Deep Coal Reservoirs

The bulk of the worldwide CBM reserves are in deep coal seams. These reservoirs are characterized by high gas content and high reservoir pressures but extremely low permeability (a few microdarcies). Coal is a viscoelastic material. An extreme reduction in permeability with increasing depth (and hence pressure) is a quite natural phenomenon. Because of depth, no horizontal drilling from a mine or from the surface has been done. As such, specific production of horizontal boreholes is not known but it is likely to be low, <1 $m^3/d/m$.

Mine shafts are obviously too expensive for these seams. Even vertical wells with hydrofracing are very inefficient. Actual hydrofracing efforts in Virginia (United States) at a depth of 1,400 m and in eastern France at a depth of 1,800 m ended in failure.

The only feasible technique at present is to drill long horizontal laterals from the surface. Figure 11.6-9 shows a plan view of 12 horizontal laterals drilled in four groups from the same surface location. The outlying horizontal laterals in each subgroup are hydraulically fractured at intervals of 300 m. Hydrofracing should be designed such that the fractures can extend at least 300 m on each side of the vertical well.

In the beginning, all hydrofraced horizontal laterals will have a high gas production rate, but after the gas production rate has significantly declined, the middle lateral in each subgroup should be injected with carbon dioxide. Coal seams have a much higher affinity for CO_2. It will be preferentially adsorbed by the coal matrix releasing methane at the same time (Thakur 2002).

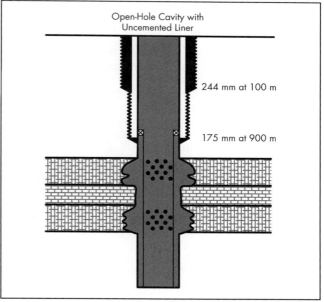

Figure 11.6-8 Cavitation completion procedure for thicker coal seams

WELL SERVICING AND MAINTENANCE

After drilling and completion, all CBM wells will produce water and gas, in that order. All wells need some servicing and maintenance. Figure 11.6-10 shows a typical gas and water production profile. Depending on the water content of the coal seam, a CMP well can produce moderate to large quantities (1–100 m^3/d) of water. As the reservoir pressure is reduced and more water is produced, the water production declines and gas production rises, reaching a peak. After that, the gas production rate also declines. Horizontal and vertical wells need different kinds of servicing and maintenance.

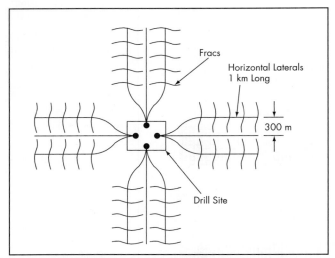

Figure 11.6-9 Gas production from deep coal seams

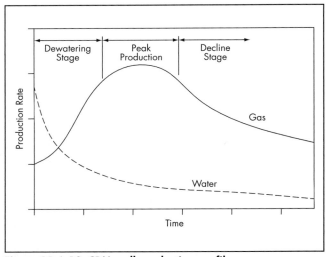

Figure 11.6-10 CBM well production profile

Horizontal Laterals Drilled from the Surface

These wells do not need much maintenance. The produced water collects at the bottom of the production well. If several horizontal laterals discharge water in one production well, the volume is large enough to warrant the use of a progressive cavity pump. As water production declines, the pump can be put on a timer. The gas pressure in the reservoir is enough to clean up the borehole.

This method of well completion still suffers from two drawbacks: an inability to stay in the coal seam and an inability to completely dewater the well. The first problem is due to drilling-instrument limitations. The borehole exits the coal seam and often enters the roof or floor, limiting gas production and the ability to completely dewater. Hence the specific production (cubic meters per day per meter) is generally only 50% or less compared to in-mine drilling, which is more precise (see Chapter 15.4). It is also difficult to reenter a horizontal lateral for maintenance, especially if multiple laterals are drilled from the same vertical well.

Vertical Hydrofraced Well

The servicing procedure and equipment for these wells is borrowed from the oil and gas industry.

After hydrofracing, a well is generally swabbed clean to remove any debris, and a dewatering pump is installed. Depending on the depth, gas and water production rates, and reservoir pressure, one of the following systems is used:

- Sucker rod pump
- Progressive cavity (Moyno) pump
- Submersible electric pump
- Gas lift

A sucker rod pump is most often used when water flow is less than 20 m^3/d. For heavier flows, a progressive cavity pump is used. In the dewatering phase of gas production, the pump is usually kept slightly above the coal seam and a 30–60-m water column is maintained above the coal seam to prevent a blowout (i.e., a sudden discharge of gas and proppant sand). Submersible electric pumps are used when the water production is excessive and progressive cavity pumps are unable to handle the flow. Gas lift is used when no electric power is available to operate a dewatering pump.

PRODUCED WATER DISPOSAL

All produced water must be metered to determine the flow rate. If the water is clean, a turbine meter is used to precisely measure the water flow. If the produced water contains some solids, a positive displacement water meter is used. The latter is more durable and commonly used. Where the water flow is intermittent and low, a 20-L bucket is used to collect the pump discharge, and the time of discharge is measured by a stopwatch from which the flow rate is calculated.

Gathering water and its proper disposal is often a significant part of the CMP. The following are the possible options for water disposal:

- Surface disposal
- Deep-well injection
- Evaporation in a pond or a boiler
- Land application
- Off-site disposal in abandoned mines
- Reuse for hydraulic stimulations

The choice of the technique depends on the volume of water produced per day and its quality. Permitting from national and state authorities is often needed and it is a prolonged process. Most commonly, only the first two choices are used, and they are the only ones discussed here. Figure 11.6-11 shows a layout of various options.

Surface Disposal

The simplest method of surface disposal comprises pH (a measure of water acidity) adjustment of water to meet the permit requirements, aeration of water with compressed air, and addition of some polymers to facilitate settling. The clean water is used for land applications or simply discharged into a local stream. The total dissolved solids (TDS) and chloride concentrations are monitored to match the quality of treated water with that of water in the stream.

If the produced water has a higher concentration of contaminants, it is pretreated, and the reverse osmosis process is

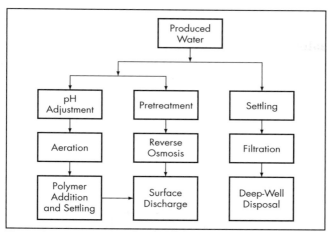

Figure 11.6-11 Options for treating produced water

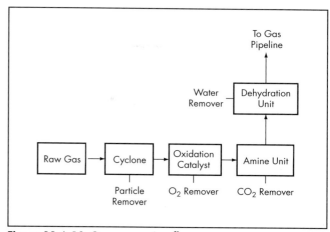

Figure 11.6-12 Gas processing flow

used to remove the TDS. Average efficiency of TDS removal is in excess of 95%. Only in the presence of bicarbonates does the efficiency go down to 50%–80%. The main limitation of this process is membrane fouling caused by suspended solids or oils in the produced water.

Deep-Well Injection

If there are no streams in the neighborhood of a CBM field or the necessary permits to discharge into the local streams cannot be obtained, the treated water can be injected into a deep well. Deep-well injection also requires a permit but it is easier to get. Typically these injection wells are old gas wells with a depth of 3,000 m or so. They are often hydrofraced to create good permeability.

Produced water must be properly treated before injection to keep the injection pressure below the permitted limits. Thick sandstone or limestone reservoirs with high porosity and permeability are ideal for this purpose. Preinjection filtration is almost always essential. Scale and corrosion inhibitors and bactericides are often added to the injected water to increase long-term use of the injection well. Finding or even drilling an injection well in the vicinity is crucial because the cost of gathering and hauling water can be quite substantial.

GAS GATHERING AND PROCESSING

CMP must be gathered to a central location for processing and commercial marketing.

Gas Gathering and Measurement

Gas flows from individual wells must be measured. The primary measurement instruments are either an orifice meter or a turbine meter. The orifice meter measures the differential pressure upstream and downstream of a measured orifice. The pressure drop is a function of the flow rate, gas pressure, and temperature. The pressure differential is continuously recorded on a chart that is calibrated to give the gas production. This is the most common method of measuring individual well production.

The turbine meter measures the flow rate and total volume of gas produced. Gas flows over rotary blades, and magnets in the meter housing pick up signals that are proportional to gas velocity. The area between the blades has fixed volumes so that the number of turns equates to the volume of gas produced.

The turbine meter is more accurate than the orifice meter, but it is more expensive and needs frequent maintenance.

Gas from each producing well is gathered using polyethylene pipes and compressors. Usually a cluster of wells is connected to a small compressor. These small compressors deliver the compressed gas to a central compressor facility where it is processed for marketing. In the design of the gas-gathering system, the main object is to maintain a low wellhead pressure, usually <3–5 kPa. Pipe diameters and compressor sizes are designed to minimize line losses. In general, one may start with 50-mm-diameter pipe at a single wellhead, but progressively the pipe diameter may increase to 250–300 mm at the main gas-gathering and processing plant.

Gas Composition and Processing

For commercial marketing, the CBM composition must meet the pipeline specification. All major pipeline companies require a minimum calorific value and put a limit on oxygen, moisture, and noncombustible concentrations. CBM composition varies from basin to basin, but generally it contains 85%–95% methane. Impurities must be removed to raise the calorific value to about 37 MJ/m^3. Other requirements are that the noncombustible content of gas should be <4%, moisture should be <0.11g/m^3, and oxygen is restricted to 20 ppm. In many applications, CBM is mixed with coal mine methane that contains higher oxygen and nitrogen, in which case excess nitrogen is removed by a cryogenic process.

Figure 11.6-12 shows a typical gas processing flow diagram. First, the raw gas is passed through a cyclone to knock solid particles out. Next, the gas is passed over an oxidation catalyst where methane is oxidized that consumes all oxygen and produces water and carbon dioxide (CO_2). Next, the CO_2 is removed by passing the gas through an amine unit. Finally, the moisture in gas is removed by dehydration units containing glycol. If there is excess nitrogen in the gas stream, it is generally removed by a cryogenic process. The clean, pipeline-quality gas is compressed to pipeline pressure, typically 7–8 MPa for delivery into a commercial gas pipeline.

Small rotary compressors are used for primary gathering because they are more efficient in compressing somewhat dirty but large volumes of gas. The main station compressors are generally reciprocating piston-type compressors that are more efficient for duty. The prime mover for compressors can be electric, diesel, or even a CBM engine to keep the cost low.

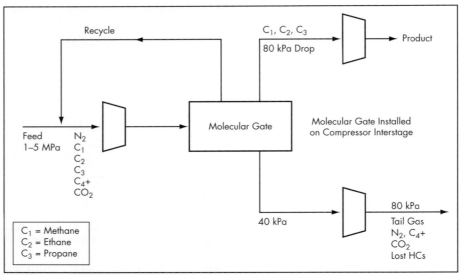

Figure 11.6-13 Molecular gate separation

Molecular Gate Separation System

The molecular gas separation system is a recent breakthrough for processing gases where gas molecules differ in size by only a fraction of an angstrom (Å). A new family of titanium silicate molecular sieves has a pore size that can be controlled down to 0.1 Å (0.01 nm).

In upgrading CBM and coal mine methane, the desired product, methane, has a molecular diameter of 3.8 Å whereas the impurities such as nitrogen (3.6 Å), carbon dioxide (3.3 Å), and oxygen (3.5 Å) are all smaller in size and easily fit within the pore of an adsorbent, where they can be adsorbed and removed in one step from the product stream (Mitariten 2001).

Figure 11.6-13 illustrates the basic flow sheet for a molecular gate separation process. The feed gas is compressed to 1–5 MPa and fed into the molecular gate system. Several repetitions/cycles are made to achieve high purity. The product gas comprising methane, ethane, and propane is recovered at practically the original pressure, which avoids recompression. The tail gas contains most impurities but often has enough methane that it can be used for steam production. The process has a methane recovery efficiency of 95% at 2.4 MPa, but it becomes more efficient at higher pressures. It can remove CO_2, N_2, and O_2 in one step. It is available in sizes of 30,000–900,000 m^3/d of methane. The cost of gas processing declines significantly with increase in the flow rate.

CARBON DIOXIDE SEQUESTRATION

It was mentioned earlier that coal has a great affinity for CO_2 and it can displace methane stored on the micropores of the coal matrix. This property of coal has been used to produce methane from deep coal seams and sequester CO_2 in it simultaneously. Laboratory experiments to displace methane from coal using nitrogen, helium, and carbon dioxide show that the latter is the most effective flooding agent. The storage capacity of coal for CO_2 is at least twice the capacity it has for methane (Collins et al. 2002).

Figure 11.6-9 shows an ideal layout for CO_2 sequestration. When gas production begins to decline, CO_2 injection can start. The optimum parameters of CO_2 sequestration have

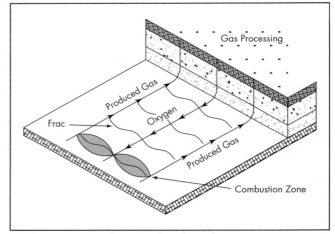

Figure 11.6-14 Underground coal gasification

not yet been established, but many research projects are in progress to determine the following:

- Optimum storage capacity for CO_2 in various coal seams
- Rate of injection and travel velocity of CO_2 in coal
- Optimum pressure of injection; generally the reservoir fracturing pressure should not be exceeded
- Ideal mix of gases; CO_2 by itself is slow moving and nitrogen is often added to accelerate the process
- Economics of CO_2 sequestration

CO_2 sequestration research is encouraged by many countries because it is considered a GHG. In many countries, there are financial incentives for sequestering CO_2. The current carbon finance instrument in the United States is fixed at $6.50/t of CO_2. Two sources of CO_2 can be related to coal: (a) power plants that burn coal, and (b) CBM itself that may contain 10%–50% CO_2 in some coal seams. CO_2 can be stripped using gas-processing techniques discussed earlier and its sequestration may become a profitable venture.

UNDERGROUND COAL GASIFICATION

CMP from deep coal seams can also be enhanced by heating the coal matrix. The drilling layout shown in Figure 11.6-14 is a schematic for UCG. Heating of the central horizontal boreholes can be done with steam or radio frequency devices, but the most synergistic approach is to ignite the coal with a limited supply of air/oxygen-combustible gaseous blends. The UCG process is, in fact, already well established, but it has not been done in conjunction with methane recovery. In this process, the outer horizontal laterals are used for gas production. When most methane is driven out, the coal matrix is combusted in situ to produce a lower calorific value gas. The process is also called in-situ gasification of coal.

The product gas composition depends on coal composition and process parameters such as operating pressure, outlet temperature, and gas flow. These parameters are not only monitored but constantly adjusted to optimize the process.

The UCG process is quite flexible. It can be used in coal seams with a thickness of 1–60 m, and calorific values of 10–30 MJ/kg including lignites and other low-quality coal. The coal deposits shown in Figure 11.6-1 have a reserve of 17–30 Tt. Only 1 Tt is amenable to actual mining. Thus, the remaining reserve can be economically exploited by the UCG process.

UCG experiments carried out in the 1970s by the U.S. Department of Energy in West Virginia and Wyoming (United States) produced only low calorific value gases (10–20 MJ/m^3), but new processes are being developed to improve the calorific value. Products of different UCG projects can be mixed with gases with higher calorific values to meet a consumer's specific need.

CMP and UCG processes supplement conventional mining by recovering energy from coal seams that are abandoned or cannot be economically mined. The gaseous product of the UCG process can be used in a variety of ways, such as

- Boiler fuel to produce steam for electric power generation,
- Feed to chemical plants to produce Fischer–Tropsch liquids or fertilizers, and
- A clean gaseous fuel after the raw gas is processed using techniques discussed earlier in the chapter.

The U.S. Environmental Protection Agency is making a great effort to export CMP technology to all major coal-producing countries. This is mainly being done to make mining safer and more productive. But it also creates a clean source of energy that supplements the natural gas supply. Additional incentives such as the carbon credit can provide a great impetus to develop this valuable resource even further. The combination of CMP with CO_2 sequestration can be used to enhance CMP from deeper coal seams and eventually, using the UCG process to convert coal into natural gas, can greatly increase the world supply of natural gas. Further research is needed to make the UCG process more efficient. In the future, the in-situ coal may also be converted into liquid fuel to increase the petroleum supply.

REFERENCES

Cairn Point Publishing Company. 1997. *The International Coal Seam Gas Report*. Queensland, Australia: Cairn Point Publishing.

Collins, R.C., Pilcher, R.C., and Marshall, J.S. 2002. Modeling CO_2 sequestration in abandoned mines. In *Proceedings of AIChE Spring Meeting*, March 10–14, New Orleans, LA. New York: American Institute of Chemical Engineers.

Hargraves, A.J. 1973. Planning and operation of gaseous mines. *CIM Bull*. (March): 119–128.

IEA (International Energy Agency). 1993. *Major Coal Fields of the World*. IEA Coal Research Report 51. London: IEA Clean Coal Centre.

Kissell, F.N., McCulloch, C.M., and Elder, C.H. 1973. *The Direct Method of Determining Methane Content of Coalbeds for Ventilation Design*. Washington, DC: U.S. Bureau of Mines.

Kuuskraa, V.A., Boyer, C.M., and Kelafant, J.A. 1992. Coalbed gas 1: Hunt for quality basins goes abroad. *Oil Gas J*. 90:49–54.

Landis, E.R., and Weaver, J.W. 1993. Global coal occurrences: Hydrocarbons from coal. *AAPG Stud. Geol*. 38:1–12.

Mitariten, M.J. 2001. One-step removal of nitrogen and carbon dioxide from coal seam gas with the molecular gate system. In *The 2nd Annual CBM and CMM Conference*, Denver, CO.

Rogers, R.E., Ramurty, M., Rodvelt, G.D., and Mullen, M. 2007. *Coalbed Methane: Principles and Practices*. Starkville, MS: Oktibbeha Publishing.

Scholes, P.L., and Johnson, D. 1993. Coalbed methane application of wireline logging. *AAPG Stud. Geol*. 38:287–302.

Thakur, P.C. 2002. Coalbed methane production from deep coal seams. In *Proceedings of AIChE Spring Meeting*, March 10–14, New Orleans, LA. New York: American Institute of Chemical Engineers.

PART 12

Underground Development

CHAPTER 12.1

Introduction to Underground Mine Planning

Richard L. Bullock

GENERAL BASIS OF MINE PLANNING

Many details go into the planning of a mine. The information gathered must come from several sources. First is the geological, structural, and mineralogical information, combined with the resource/reserve data. This information leads to the preliminary selection of potential mining method and sizing of the mine production. From this the development planning is done, the equipment selection is made, and the mine work force projections are completed, all leading to the economic analysis associated with mine planning.

Planning as just described, however, will not necessarily guarantee the best possible mine operation, unless the best possible mine planning has been done correctly. Any sacrifice in mine planning introduces the risk that the end results may not yield the optimum mine operation. Planning is an iterative process that requires looking at many options and determining which, in the long run, provide the optimum results.

This chapter addresses many of the factors to be considered in the initial phase of all mine planning. These factors have the determining influence on the mining method, the size of the operation, the size of the mine openings, the mine productivity, the mine cost, and, eventually, the economic parameters used to determine whether the mineral reserve should even be developed.

PHYSICAL AND GEOTECHNICAL INFORMATION NEEDED FOR PRELIMINARY MINE PLANNING

Many pieces of engineering and geologic data must be gathered before mine planning can take place. These are covered in the sections that follow. This section draws heavily from Chapter 2 of *Underground Mining Methods* (Bullock 2001).

Technical Information

Assuming that the resource to be mined has been delineated with prospect drilling, study will be based primarily on information supplied through exploration. The results of the exploration are recorded in a formal report for use in project evaluation. The exploration report should contain the following information with appropriate maps and cross sections:

- Property location and access
- Description of surface features
- Description of regional, local, and mineral-deposit geology
- Review of exploration activities
- Tabulation of geologic resource material
- Explanation of resource calculation method, including information on geostatistics applied
- Description of company's land and water position
- Ownership and royalty conditions
- Mining history of property
- Rock quality designation (RQD) values and any rock mass classification work that has been done
- Results of any special studies or examinations the exploration group has performed (metallurgical tests, geotechnical work, etc.)
- Report on any special problems or confrontations with local populace of the area
- Any other pertinent data such as attitude of local populace toward mining, special environmental problems, availability of water and hydrologic conditions in general, and infrastructure requirements

This critical information should be established to assist the mine planning. If the exploration project is one that has been drilled out by the company exploration team, this information should have been gathered during the exploration phase and passed to the mine evaluation team or the mine development group. More information on each of these subjects may need to be obtained, but if the inquiry can be started during the exploration phase of the project, time will be saved during the feasibility/evaluation and development phases of the project. If this is an ongoing mine operation, then most of the technical information will be available from other mine planning projects.

Geologic and Mineralogic Information

Knowledge of similar rock types or structures in established mining districts is always helpful. In developing the first mine in a new district, there is far more risk of making costly errors than in the other mines that may follow. The geologic and mineralogic information needed includes

- The size (length, width, and thickness) of the areas to be mined within the overall operations area to be considered, including multiple areas, zones, or seams;

Richard L. Bullock, Professor Emeritus, Mining & Nuclear Engineering, Missouri University of Science & Technology, Rolla, Missouri, USA

- The dip or plunge of each mineralized zone, area, or seam, noting the maximum depth to which the mineralization is known;
- The continuity or discontinuity noted within each of the mineralized zones;
- Any swelling or narrowing of each mineralized zone;
- The sharpness between the grades of mineralized zones within the material considered economically mineable;
- The sharpness between the ore and waste cutoff, including
  - Whether this cutoff can be determined by observation or must be determined by assay or other means,
  - Whether this cutoff also serves as a natural (physical) parting resulting in little or no dilution, or whether the break between ore and waste must be induced entirely by the mining method, and
  - Whether the mineralized zone beyond (above or below) the existing cutoff represents submarginal economic value that may become economical at a later date;
- The distribution of various valuable minerals making up each of the potentially mineable areas;
- The distribution of the various deleterious minerals that may be harmful in processing the valuable mineral;
- Whether the identified valuable minerals are interlocked with other fine-grained mineral or waste material;
- The presence of alteration zones in both the mineralized and the waste zones;
- The tendency for the ore to oxidize once broken; and
- The quantity and quality of the ore reserves and resource with detailed cross sections showing mineral distribution and zones of faulting or any other geologic structure related to the mineralization.

Structural Information
Required physical and chemical structural information includes the following:

- The depth of cover
- A detailed description of the cover, including the
  - Type of cover;
  - Approximate strength or range of strengths;
  - Structural features in relation to the proposed mine development; and
  - Presence of and information about water, gas, or oil that may have been encountered
- The quality and structure of the host rock (back, floor, hanging wall, footwall), including the
  - Type of rock,
  - Approximate strength or range of strengths,
  - Noted zones of inherent high stress,
  - Noted zones of alteration,
  - Major faults and shears,
  - Systematic structural features,
  - Porosity and permeability,
  - The presence of any swelling-clay or shale interbedding,
  - RQD throughout the various zones in and around all of the mineralized area to be mined out,
  - The host rock mass classification (Rock Mass Rating [RMR] or Barton's Q-system),
  - Temperature of the zones proposed for mining, and
  - Acid-generating nature of the host rock
- The structure of the mineralized material, including all of the factors listed previously, as well as the
  - Tendency of the mineral to change character after being broken (e.g., oxidizing, degenerating to all fines, recompacting into a solid mass, becoming fluid, etc.);
  - Siliceous content of the ore;
  - Fibrous content of the ore;
  - Acid-generating nature of the ore; and
  - Systematic fault offsets.

PLANNING RELATED TO PHYSICAL PROPERTIES
The physical nature of the extracted mass and the mass left behind are very important in planning many of the characteristics of the operating mine. Four aspects of any mining system are particularly sensitive to rock properties:

1. The competency of the rock mass in relation to the in-situ stress existing in the rock determines the open dimensions of the unsupported roof unless specified by government regulations. It also determines whether additional support is needed.
2. When small openings are required, they have a great effect on productivity, especially in harder materials where drill-and-blast cycles will be employed.
3. The hardness, toughness, and abrasiveness of the material determine the type and class of equipment that can extract the material efficiently.
4. If the mineral contains or has entrapped toxic or explosive gases, the mining operation will be controlled by special provisions within government regulations.

In countries where appropriate regulations do not exist, best-practice guidelines must be sought.

THE NEED FOR A TEST MINE
From this long list of essential information required for serious mine planning, it becomes evident that not all of this information can be developed from the exploration phase. Nor is it likely that it can all be obtained accurately from the surface. If this is the first mine in this mining area or district, then what is probably needed during the middle phase of the mine feasibility study is development of a test mine. While this may be an expense that the owners were hoping to avoid up front, the reasons for a test mine are quite compelling. They include opportunities to

- Confirm, from a geologic point of view, the grade, ore continuity, ore configuration, and mineral zoning;
- Confirm, from the engineering viewpoint, rock strengths and mass rock quality; verify mining efficiencies; confirm water inflows; and demonstrate waste characteristics;
- Pilot-test the metallurgical process;
- Enhance the design basis for cost estimates, improve labor estimates, build more accurate development schedules; and
- Lower the mining investment risk.

Having a test mine in place will shorten the production mine development and will serve as a training school for the production mine.

LAND AND WATER CONSIDERATIONS

Information needed about the property includes

- Details on the land ownership and/or lease holdings, including royalties to be paid or collected (identified by mineral zones or areas);
- Availability of water and its ownership on or near the property;
- Quality of the water available on (or near) the property;
- Details of the surface ownership and surface structures that might be affected by subsidence of the surface;
- The location of the mining area in relation to any existing roads, railroads, navigable rivers, power, community infrastructure, and available commercial supplies of mine and mineral processing consumables (drill steel, bits, explosives, roof bolts, steel balls and rods for the mills, chemical additives for processing, steel liners for all mining and processing equipment, etc.); and
- The local, regional, and national political situations that have been observed with regard to the deposit.

OTHER FACTORS INFLUENCING EARLY MINE PLANNING

Many operational decisions must be made in planning the mining operation. None is more important than sizing the mining operation. However, it is not an easy or obvious decision to make and several factors must be considered.

Sizing the Production of a Mine

A considerable amount of literature is available on the selection of a production rate to yield the greatest value to the owners, including works by Tessaro (1960), McCarthy (1993), Runge (1997), and Smith (1997). Basic to all modern mine evaluations and design concepts is the desire to optimize the net present value (NPV) or to operate the property in such a way that the maximum internal rate of return is generated from the discounted cash flows. Anyone involved in the planning of a new operation must be thoroughly familiar with these concepts. Equally important is the fact that any entrepreneur who is planning a mining operation *solely from the financial aspects of optimization*, and who is not familiar with the issues associated with maintaining high levels of production at low operating cost per metric ton over a prolonged period, is likely to experience disappointment in years with low (or no) returns.

Other aspects of the problem for optimizing mine production relate to the effect of NPV. When viewed from the purely financial side of the question (i.e., producing the product from the mineral deposit at the maximum rate), optimization that yields the greatest return is the case often selected. This is due to the fixed costs involved in mining, as well as to the present-value concepts of any investment. Still, there are "...practical limitations to the maximum intensity of production, arising out of many other considerations to which weight must be given" (Hoover 1909). There can be many factors limiting mine size, some of which are listed here:

- Market conditions
- Current price of the product(s) versus the trend price
- Grade of the mineral and the corresponding reserve tonnage
- Time before the property can start producing
- Attitude and policies of the local and national government and the degree of stability of existing governments and their mining policies, taxes, and laws
- Availability of a source of energy and its cost
- Availability of usable water and its cost
- Cost and method of bringing in supplies and shipping production
- Physical properties of the rock and minerals to be developed and mined
- Amount of development required to achieve the desired production related to the shape of the mineral reserve
- Amount and complexity of mineral processing required
- Availability of nearby smelting options (if required)
- Size and availability of the work force that must be obtained, trained, and retained
- Availability of housing for employees (in remote locations)
- Potential instability of the government in the future, which might cause a company to develop a smaller, high-grade mine in the beginning until they have received their objective return, then use the income from the existing property to expand and mine out the lower-grade ores

Although all of these factors must be taken into consideration, another approach to sizing the mine is to use the Taylor formulas (Taylor 1977). Taylor studied more than 200 mining properties and then used regression analysis to determine the formulas for sizing a mine. Taylor notes, however, that the formulas are not very applicable to steeply dipping mineral reserves or when mining from deep shafts. The formulas are

$$\text{life of mine} = 0.20 \text{ \#} \text{ (expected ore metric tons [tons])}^{0.25}$$
$$= \text{life of mine (in years)} \pm 1.2 \text{ years} \quad (12.1\text{-}1)$$

$$\text{daily production} = \text{reserve ore metric tons (tons)} / \text{expected life/operating days/year} \quad (12.1\text{-}2)$$

Assume 37.2 Mt (41 million tons) of mineable resource:

$$\text{life of mine} = 0.20 \times (37,200,000)^{0.25}$$
$$= 15.6 \text{ yr } (\pm 1.2 \text{ yr or } 14.4 \text{ to } 16.8 \text{ yr})$$

$$\text{daily production} = 37,200,000/16 \text{ yr}/365$$
$$= 6,525 \text{ Mt/d } (6,059 \text{ Mt/d to } 7,068 \text{ Mt/d})$$

or

$$\text{life of mine} = 0.20 \times (41,000,000)^{0.25}$$
$$= 16 \text{ yr } (\pm 1.2 \text{ yr or } 14.8 \text{ to } 17.2 \text{ yr})$$

$$\text{daily production} = 41,000,000/16 \text{ yr}/365$$
$$= 7,020 \text{ tpd } (6,530 \text{ tpd to } 7,590 \text{ tpd})$$

The two formulas tend to overestimate the production of small-vein-type mining deposits where a lot of vertical development must be completed compared to the daily tonnage that can be extracted. For mineral resources that are steeply dipping, and for deeper mining, the shape of the resource must be considered, as well as how much development is necessary to sustain the desired production. According to McCarthy (1993), for Australian underground narrow-vein mines, approximately 50 vertical meters (165 vertical feet) per annum is currently economically appropriate for modern mechanized mines. Thus for example, if given certain reserves for mining 10,000 metric tons per vertical meter, then the production would be 500,000 t/a. Properties that were above this "best fit" trend line, McCarthy says, are usually overly capitalized or have higher-than-average operating costs.

Not only does the resource's tonnage affect the mine size, but the distribution of ore grade can certainly affect the mine planning. Unless a totally homogeneous mass is mined, it may make a considerable economic difference as to which portion is mined first or later. Furthermore, no ore reserve has an absolute fixed grade-to-tonnage relationship; trade-offs must always be considered. In most mineral deposits, lowering the mining cutoff grade means that more metric tons will be available to mine. But the mine cutoff must balance the value of each particular block of resource against every type of cash cost that is supported by the operation, including all downstream processing costs, as well as the amortization of the capital that was used to construct the new property.

Even in bedded deposits such as potash or trona, the ability or willingness to mine a lower seam height may mean that more ore can eventually be produced from the reserve. In such cases, the cost per unit of value of the product generally increases. Similarly, narrow-seam mining greatly reduces the productivity of the operation compared to high-seam and wide-vein (or massive) mining systems.

In evaluating the economic model of a new property (after the physical and financial limits have been considered), all of the variables of grade and tonnage, with the related mining costs, must be calculated using various levels of mine production that, in the engineer's judgment, are reasonable for that particular mineral resource. At this point in the analysis, the various restraints of production are introduced. This will develop an array of data that illustrate the return from various rates of production at various grades corresponding to particular tonnages of the resource. At a later stage, probability factors can be applied as the model is expanded to include other restraining items.

Timing Affecting Mine Production

For any given ore body, the development required before production start-up is generally related to the size of the production, as well as the mining method. Obviously, the necessary stripping time for most large porphyry copper deposits is very large compared to the stripping time for even a large quarry. For an underground example, very high production rates may require a larger shaft or multiple hoisting shafts, more and larger development drifts, the opening of more mineable reserves, as well as a greater lead time for planning and engineering all aspects of the mine and plant. The amount of development on multiple levels for a sublevel caving operation or a block caving operation will be extensive compared to the simple development for room-and-pillar operation. In combination, all of these factors could amount to considerable differences in the development time of an operation. In the past, the time for mine development has varied from 2 to 8 years. Two indirect economic effects could result from this:

1. Capital would be invested over a longer period of time before a positive cash flow is achieved; and
2. The inflation rate-to-time relationship in some countries in the past has been known to push the costs upward by as much as 10% to 20% per year, thereby eliminating the benefits of the "economy of scale" of large-size projects.

To aid the engineer in making approximations of the time it takes to develop shaft and slope entries related to the size and depth of the mine entry, the reader is referred to Bullock (1998). For example, sinking and equipping a 6.1-m (20-ft) diameter shaft ranging in depth from 305–914 m (1,000–3,000 ft) will take approximately 64–113 weeks, depending on the production capacity that the shaft must be equipped for. Driving a slope to the same depths will take approximately 52–151 weeks, respectively.

For mines that are primarily developed on one or two levels and have extensive lateral development, the speed of the development can vary considerably. The lateral development on each level of a room-and-pillar mine opens up new working places, and the mine development rate can be accelerated each time a turnoff is passed, provided that there is enough mining equipment and hoisting capacity available. This is in contrast to a vein-type mine that has a very limited number of development faces, as the vein is developed for each level and each level requires deepening the main hoisting shaft or slope. Experience has shown that the mine's development only progresses deeper at the average rate of about 50 m/a. Chapter 4.7 of this handbook discusses several examples of the timing of mine developments.

The timing of a cost is often more important than the amount of the cost. In a financial model, timing of a development cost must be studied in a sensitivity analysis. In this respect, any development that can be put off until after a positive cash flow is achieved without increasing other mine costs should certainly be postponed.

Government Considerations Affecting Mine Size

Government attitudes, policies, and taxes generally affect all mineral extraction systems and should be considered as they relate to the mining method and the mine size. One can assume that a mine is being developed in a foreign country and that the political scene is currently stable but is impossible to predict beyond 5–8 years. In such a case, it would be desirable to keep the maximum amount of development within the mineral zones, avoiding development in waste rock as much as possible. This will maximize the return during a period of political stability. Also, it might be desirable to use a method that mines the better ore at an accelerated rate to obtain an early payback on the investment. If the investment remains secure at a later date, the lower-grade margins of the reserve might later be exploited. Care must be taken, however, that the potential to mine the remaining resource, which may still contain good grade ore, is not jeopardized, whatever the stability situation. There is merit in carefully planning to mine some of the higher-grade portions of the reserve while not impacting the potential to mine the remaining reserve.

Some mining methods, such as room-and-pillar mining, allow the flexibility of delaying development that does not jeopardize the recovery of the mineral remaining in the mine. In contrast, other mining systems, such as block caving or longwall mining, might be impacted by such delays.

Similar situations might arise as a result of a country's tax or royalty policies, sometimes established to favor mine development and provide good benefits during the early years of production; in later years the policies change. Again, as in the preceding case, the flexibility of the mining rate and system must be considered, but not to the extent of jeopardizing the remaining resource.

PREPLANNING BASED ON GEOLOGY AND ROCK CHARACTERISTICS

In the previous section, the influence of geology from a macro point of view was considered. In this section, the geology from a micro point of view of the rock and mineral characteristics must be considered in mine planning.

Geologic Data

Using geologic and rock property information obtained during preliminary investigations of sedimentary deposits, isopach maps should be constructed for all potential underground mines to show the horizons to be mined and those that are to be left as the roof and floor. Such maps show variances in the seam or vein thickness and identify geologic structures such as channels, washouts (called "wants"), and deltas. Where differential compaction is indicated, associated fractures in areas of transition should be examined. Areas where structural changes occur might be the most favored mineral traps but are usually areas of potentially weakened structures. Where possible, locating major haulage drifts or main entries in such areas should be avoided; if intersections are planned in these areas, they should be reinforced, as soon as they are opened, to an extent greater than that necessary elsewhere. Further details on development opening in relation to geologic structures and other mine openings appear in Spearing (1995), reproduced by Bullock (2001).

Again referring to flat-lying-type deposits, extra reinforcing (or reducing the extraction ratio) may also be necessary in metal mines where the ore grade increases significantly, since the rock mass usually has much less strength. Where the pillars were formed prior to discovering the structural weakness, it will probably be necessary to reinforce the pillars with fully anchored reinforcing rock bolts or cables. It is advisable to map all joint and fracture information obtained from diamond-drill holes and from mine development, attempting to correlate structural features with any roof falls that might still occur.

Characteristics of Extracted Material

The hardness, toughness, and abrasiveness of the material determine whether the material can be extracted by some form of mechanical cutting action, by drilling and blasting, or by a combination of both methods.

The mechanical excavation with roadheaders of the borate minerals from the Death Valley (California, United States) open-pit mine operated by Tenneco, and the later use of the same type of roadheader at the nearby underground Billie mine (California) illustrate the point. Bucket-wheel excavators have been used extensively in Germany and Australia for stripping overburden from the brown coalfields.

Technological advances in hard-metal cutting surfaces, steel strengths, and available thrust forces allow increasingly harder and tougher materials to be extracted by continuous mining machines. The economics of continuous cutting or fracturing, as compared to drill-and-blast, gradually are being changed for some of the materials that are not so tough or abrasive. However, for continuous mining—other than tunnel boring machines—to be competitive with modern high-speed drills and relatively inexpensive explosives, it appears that the rock strengths must be less than 103,400–124,020 kPa (15,000–18,000 psi) and have a low abrasivity. Rock that is full of fractures, however, is also a great aid to mechanical excavation. In one case that was covered in an article about the gradual trend toward mechanical excavation in underground mining (Bullock 1994), a roadheader was being used in a highly fractured, welded volcanic tuff, even though the rock strength was well over 137,800 kPa (20,000 psi).

At times, reasons other than the cost of extraction favor one mining system over another. Using a mechanical excavation machine is nearly always advantageous in protecting the undisturbed nature of the remaining rock where blasting might be prohibited. Likewise, the continuous nature of mechanical excavation can be used to speed mine production. This was seen in the openings driven by Magma Copper in developing the Kalamazoo ore body in Arizona, United States (Chadwick 1994; Snyder 1994) and by Stillwater Mining Company in developing their original ore body as well as for their East Boulder ore body in Montana, United States (Tilley 1989; Alexander 1999). A continuous boring tool may also be desirable for extracting an ore body without personnel having to enter the stoping area. Where their use is applicable, continuous mining machines are certainly much easier to automate than the cyclical drill-and-blast equipment. The automation of the 130 Mobile Miner at the Broken Hill mine in Australia is one case in point (Willoughby and Dahmen 1995). Another more dynamic case is the complete automation of the potash mines of Potash Corporation of Saskatchewan, Canada (Fortney 2001).

PLANNING THE ORGANIZATION

The amount of equipment or numbers of personnel required to meet the needs of all mines cannot always be given in absolute and precise terms. The purpose of this discussion is to mention some of the general problems that may be encountered and actions to mitigate those problems.

Work Force and Production Design

When planning the details of a mining work force, it is necessary to consider several factors:

- Is the supply of labor adequate to sustain the production level dictated by other economic factors? If not, can the needed labor be brought in, and at what cost?
- What is the past history of labor relations in the area? Are the workers accustomed to a 5-day work schedule, and if so, how will they react to a staggered 6- or 7-day schedule?
- Are the local people trained in similar production operations, or must they be trained before production can achieve full capacity?
- Will a camp have to be built and the workers transported in on weekly schedules?
- If long commutes are required to reach the property daily, are company-contracted busses for all shifts an option?
- If long commutes are required, are shifts longer than the normal 8 hours an option?
- Can people with maintenance skills be attracted to the property, or will the maintenance crew have to be built up via an apprenticeship program?

Apprenticeship programs are very slow in terms of producing results. Accordingly, some state laws in the United States restrict the annual number of people who can be trained in such programs. This one item could cause a mine designed and equipped for a very large daily production to fall far short of its desired goals.

Work-force issues and local community resources need to be investigated at the same time that the property is being evaluated and designed. This will provide adequate time for specialized training, minimize unexpected costs, and also prevent economic projections based on policies that, if implemented, could negatively impact employee morale or community relations. The productivity and profitability difference between an operation with high morale and good labor relations and an operation where such parameters are poorly rated can be drastic. Such matters can make the difference between profit or loss. Of all the items involved in mine design, this one is the most neglected and can be the most disastrous.

Equipment Selection

Field-Tested Equipment
The equipment selected should be produced by manufacturers that field-test their equipment for long periods of time before they are marketed to the industry. Too many manufacturers build a prototype machine and install it in a customer's mine on the contingency that they will stand behind it and make it work properly. Eventually, after both the end user and the manufacturer redesign, rebuild, reinforce, and retrofit this model, a workable machine is obtained. The cost in lost production, however, is imposed on the mine operator, not the manufacturer. The manufacturer can then proceed to sell the "field-tested" retrofitted model to the entire industry, including competitive mines. Beware of taking on equipment in underground mines that has not been thoroughly field tested in severe applications by the manufacturer, unless it is going to be used for an isolated research application.

Equipment Versatility
The equipment selected for the mining operation should be as versatile as possible. Normally, for large surface mines, this is not much of a problem. For small quarries and for most underground mines, however, it can be a problem if the equipment can only be employed on a limited number of mining operations. For example, in one room-and-pillar mining operation (Bullock 1973), the same high-performance rotary percussion drilling machines used for drilling the bluff or brow headings were then mounted on standard drill jumbos for drilling holes for burn-cut drifting and stoping rounds and for slabbing rounds in the breast headings. Because the drills on these jumbos have high penetration rates, they were also used to drill the holes for roof bolts and, in some cases, the holes for the reinforcing pillars. The same front-end loader was used to load trucks in one stope and to perform as a load-haul-dump unit in another stope. By switching working platforms, the same forklift tractors served as explosive-charging vehicles and as utility service units for handling air, water, and power lines. They also served as standard forklifts for handling mine supplies. This equipment philosophy results in the following advantages:

- Less equipment must be purchased and maintained.
- Less training is required for operators and maintenance personnel. In addition, all personnel have a better chance of becoming more efficient at their jobs.
- Having fewer machinery types means having a smaller equipment inventory.

Possible disadvantages of this approach are the following:

- A more efficient machine may be available to do the task currently being performed by the versatile—but less efficient—machine.
- The mine may become too dependent on a single manufacturer to supply its equipment needs.

Equipment Acceptance
Equipment selected should have a very broad acceptance and must be in common use throughout both the mining and construction industries. However, since underground mines impose a headroom restriction not encountered on the surface, this is not always possible. Nevertheless, where headroom is not an issue, selecting a standard piece of equipment means that the components will have endured rigorous testing by the construction industry. Furthermore, equipment parts are normally off-the-shelf items in distributors' warehouses.

Application Flexibility
The selected equipment should be flexible in application. For example, the equipment should be able to accelerate and move rapidly, have good balance and control at high speeds, be very maneuverable, and have plenty of reserve power for severe applications. Both trucks and loaders should have ample power to climb every grade in the mine and be able to accelerate quickly to top speed on long, straight hauls.

PLANNING THE UNDERGROUND MINE SERVICE FACILITIES
Underground facilities such as underground pumping stations, power transfer and transformer stations, underground shops, storage warehouse space, ore storage pockets, skip-loading stations, lunchrooms, refuge chambers, shift and maintenance foremen's offices, and central controlled computer installations and communications all extract a strong influence on engineering study and design. More detailed information appears in Chapter 12.5, where these types of facilities are described and their designed size is discussed.

REFERENCES
Alexander, C. 1999. Tunnel boring at Stillwater's East Boulder project. *Min. Eng.* 51(9):15–24.

Bullock, R.L. 1973. Mine-plant design philosophy evolves from St. Joe's New Lead Belt operations. *Min. Cong. J.* 59(5):20–29.

Bullock, R.L. 1994. Underground hard rock mechanical mining. *Min. Eng.* 46(11):1254–1258.

Bullock, R.L. 1998. General mine planning. In *Techniques in Underground Mining*. Edited by R. Gertsch and R.L. Bullock. Littleton, CO: SME.

Bullock, R.L. 2001. General planning of the noncoal underground mine. In *Underground Mining Methods: Engineering Fundamentals and International Case Studies*. Edited by W.A. Hustrulid and R.L. Bullock. Littleton, CO: SME.

Chadwick, J. 1994. Boring into the lower Kalamazoo. *World Tunnelling* (North American Tunnelling Supplement) (May).

Fortney, S.J. 2001. Advanced mine-wide automation in potash. In *Underground Mining Methods: Engineering Fundamentals and International Case Studies*. Edited by W.A. Hustrulid and R.L. Bullock. Littleton, CO: SME.

Hoover, H.C. 1909. *Principles of Mining*. New York: McGraw-Hill.

McCarthy, P.L. 1993. Economics of narrow vein mining. In *Proceedings, Narrow Vein Mining Seminar*, Bendigo, Victoria, June. Victoria, Australia: Australasian Institute of Mining and Metallurgy.

Runge, I.C. 1997. *Mining Economics and Strategy*. Littleton, CO: SME.

Smith, L.D. 1997. A critical examination of the methods and factors affecting the selection of an optimum production rate. *CIM Bull.* (February): 48–54.

Snyder, M.T. 1994. Boring for the lower K. *Eng. Min. J.* (April): 20ww–24ww.

Spearing, A.J.S. 1995. *Handbook on Hard-Rock Strata Control*. Special Publication Series SP6. Johannesburg, South Africa: South African Institute of Mining and Metallurgy. pp. 89–93.

Taylor, H.K. 1977. Mine valuation and feasibility studies. In *Mineral Industry Costs*. Spokane, WA: Northwest Mining Association. pp. 1–17.

Tessaro, D.J. 1960. Factors affecting the choice of a rate of production in mining. *Can. Min. Metall. Bull.* (November): 848–856.

Tilley, C.M. 1989. Tunnel boring at the Stillwater mine. *Proceedings, Rapid Excavation and Tunneling Conference (REtc.)*. Edited by R. Pond and P. Kenny. Littleton, CO: SME. pp. 449–460.

Willoughby, R., and Dahmen, N. 1995. Automated mining with the mobile miner. Presented at Mechanical Mining Technology for Hard Rock Short Course, Colorado School of Mines, Golden, CO, June.

CHAPTER 12.2

Hard-Rock Equipment Selection and Sizing

Peter L. McCarthy

INTRODUCTION

Whereas underground equipment should be selected to suit the mine, it is equally true that the mine should be designed to suit the equipment. It is impossible to begin mine planning and scheduling without a good idea of the type and size of equipment available and how it might be applied. Design of stopes or panels, mine accesses, development cross sections, ramp gradients, electric power reticulation, ventilation circuits, and so on must be based on an assumed type and fleet of equipment.

GENERAL APPROACH

In practice, equipment selection for a new hard-rock mine starts with the stoping operation. Stoping equipment should be appropriately sized for the characteristics of the ore body and the stope dimensions. Stope development openings should be sized to suit the stope dimensions and the stoping equipment capabilities and requirements.

Equipment should then be sized to mine the required size of stope development. Separate development fleets for access and stope development may be required. If this logic is not applied, big "efficient" development equipment may be acquired for access development, and then used in the stoping areas. The result is that the stope development is sized to fit jumbo drills, rock-bolter booms, or even the haul trucks, and must be far larger than the optimum, causing inefficiencies in stope design, poor geotechnical conditions, or excessive overbreak and development cost.

Of course, it is often necessary to select new equipment for an existing mine, where the constraints including access dimensions, ventilation, and so forth have to be identified and managed. In a shaft mine, for example, the machine size may be limited by the dimensions of shaft compartments and the component sizes in which the machine can be broken down for transport.

Proposed or possible changes to the mine plan, such as a move from a 5-day working week to continuous working, or a change in the decline gradient, will affect the equipment duty and must be considered. Sometimes a new machine may allow a new way of working. For example, replacement of rear-wheel-drive trucks with four-wheel-drive trucks would allow ramps to be steepened. This could reduce overall costs, and it is this larger context that must be analyzed.

Machines should be compared in terms of capital cost, operating cost, specifications, performance, availability, ease of maintenance, service and parts backup, and service life.

Fleet Description

The first step is to describe the key elements of the equipment fleet. For example, in a hard-rock ramp-access mine, this might include

- Two-boom jumbo drills,
- One-boom jumbo drills,
- Rock-bolting machines,
- Long-hole drills,
- Large load-haul-dump units (LHDs),
- Small LHDs,
- Haulage trucks,
- Integrated tool carriers, and
- Road graders.

Many other machines such as personnel transporters, shotcrete machines, air compressors, raise drills, and others may be needed, but the items in the list will be considered in determining stope and development layouts and dimensions.

Equipment Matching

Individual items of equipment, considered together, form a development system or a production system, and thus they must be physically compatible. Jumbo drills must be capable of reaching the limits of the largest excavation required, usually determined for the production drills and the trucks. LHDs must be able to reach and fully load the tubs (boxes or trays) of the trucks in three to five passes.

Sometimes a workaround will enable incompatible units to be used. For example, a "step" in the floor may enable small LHDs to load large trucks.

Electric machines should all operate from the same electric supply voltage. A decision must be made about whether compressed air will be reticulated, and if not, each pneumatic machine must be self-sufficient for compressed air.

Peter L. McCarthy, Chairman and Principal Mining Consultant, AMC Consultants Pty Ltd., Melbourne, Victoria, Australia

When items are obtained from a single manufacturer, the manufacturer will be able to determine how well matched they are. If they come from different manufacturers, some research may be needed, including detailed study of the specifications and discussions with staff of other mines about using the units together.

Choice of Manufacturer or Supplier

Manufacturers of underground equipment range from large, international firms with worldwide representation to small manufacturers that generally serve a local market or a specialized niche. In general, the international firms offer robust, reliable, and well-proven equipment at the upper end of the price range. Equipment is usually priced in euros or U.S. dollars, which may appear prohibitive when converted to a local currency.

Small regional manufacturers can offer attractive pricing and may offer a viable alternative to the name brands. Because their level of research and development is limited, and because they are building to a price, machines from such manufacturers generally have lower availability and may have reduced performance. These factors should be assessed carefully in deciding whether lower capital cost outweighs higher operating costs, including the impact of servicing and breakdown delays on the mining system as a whole.

Other manufacturers offer specialist equipment that is not available from the major suppliers. Examples of the products of international niche manufacturers include narrow-vein mining equipment from France and Canada, rock-bolting platforms from Canada, backfill slinger trucks from Germany, raise climbers from Sweden, and so on.

Equipment with similar specifications from the major manufacturers will generally give similar performance. This is often compared with the choice of buying a car from Ford or from General Motors—both will do the job. It is worth enquiring at other mine sites that use the same equipment, because specific models sometimes have problems that are not really resolved during the model run, whereas competing machines are free of those problems.

Apart from operating performance, equipment analysis should include the ease of maintenance. Some equipment models are easier to maintain, which can have a positive impact on the availability of the equipment.

The most important factor in choosing between otherwise similar machines is the level of backup support provided in a region by the manufacturer. Appropriate questions include the following:

- What training will the manufacturer offer to operators and maintenance personnel?
- Does the manufacturer have service facilities with trained personnel within an hour or two of the mine by road? If not, is it willing to establish such facilities on the basis of an order?
- Will the manufacturer stock a comprehensive range of spare parts in-country for any specific model? What is the likely delivery time?
- Will the manufacturer provide spare parts in the mine store on a consignment basis, so that payment is made only when they are used?
- Which major components are not stocked in-country? How long would it take to obtain them, if needed, and what would be the cost of urgent air freight?
- Will the manufacturer offer maintenance and consumables on a fixed cost per hour or per meter basis, and establish its own maintenance work force on-site?
- What is the reputation of the manufacturer for service and spares backup in the region?
- Are components interchangeable between different pieces of equipment such as the same axles in LHDs and trucks?
- Are the driveline components from the same manufacturer?

After these questions have been addressed, it may be advantageous to source the key fleet items, such as trucks, LHDs, and jumbo drills, from a common supplier. It is also attractive to minimize the number of different models of equipment on the mine site, even if from the same manufacturer, to minimize duplication of stores inventories or maintenance skill sets. Mixed manufacturer fleets, especially trucks, may result in unexpected operational conflicts such as different truck speeds, with slower trucks delaying faster trucks, especially when hauling loads up a ramp.

Appropriate Technology

Any new machine should incorporate a level of technology that can be understood and supported by the operators and maintenance team. Although it may be appealing to engineers to acquire the latest technology, the potential performance will not be delivered if the machine spends much of its time waiting for repairs. At the crudest level, there are mines that operate compressed-air rock drills and do not have the workshop facilities or the skills to maintain hydraulic rock drills. Similarly, many mines do not have the electronics capability to maintain the latest fully computerized jumbo drills and prefer to buy the versions with manual controls.

In general, the longer the supply lines to the mine and the less educated or skilled the work force, the more robust and less innovative should be the technology. It is also difficult to mix different technologies within one operation. For example, a mine developed some years ago in southern Africa tried to combine handheld drills and slushers in the stopes with rubber-tired jumbo and truck development and haulage. The two systems were not compatible: large numbers of walking stope miners impeded the movement of trucks; the rock broken in development was too large for the conveyor feeders; and the maintenance workshops had to service large numbers of stope drills and scraper winches, with which they were familiar, and smaller numbers of hydraulic jumbos and diesel trucks, which were new to them. The mine failed for these and unrelated reasons.

Mine systems should be designed to accommodate the technology to be used. For example, if a conventional mine is to be mechanized, not only must the required infrastructure be provided, but materials-handling logistics need to be thoroughly planned and assessed. For example, this could include transportation and storage underground of diesel, spare parts, and hydraulic oils.

Innovative Models

Innovative new models of machines, even from the most reputable manufacturers, may present operational problems. It often takes 2 or 3 years of operation to resolve design problems in trucks, loaders, and jumbo drills, with trucks giving the greatest problems. Usually the machine gets a bad name in the industry and a revised model with a new code name is released after the problems are resolved. The problems are not predictable and could be as simple as shearing of wheel studs on a high-powered truck.

The promised performance of a new model may be appealing, and some mine has to be the first to try it. When considering such a machine, it is usually possible to negotiate a performance guarantee with the supplier. This should take the form of an operating availability guarantee, possibly with a minimum performance expressed as metric ton–kilometers per hour, meters drilled per hour, or similar. Financial compensation applies if the machine does not deliver the guaranteed performance.

It is usual for the manufacturer and supply agent to be very interested in the success of a new product, and they may provide high standards of operator training and assume responsibility for maintenance for an extended period. The duration of these above-normal levels of service should be ascertained.

Examples of new equipment that have proved disappointing in the past include

- Some high-powered 50–80 t (metric ton) capacity trucks;
- Some combined drilling and cable-bolting machines;
- Some rock-bolting jumbos;
- Many models of scaling machines;
- Undersized impact breakers; and
- Novel surface or soft-rock crushers and sizers, when applied underground in hard-rock mines.

Equipment Replacement

All equipment has a service life, after which it becomes uneconomical and should be replaced. For some items the service life may be equal to or exceed the mine life, in which case replacement is not necessary. Generally, major items such as headframes, winders (hoists), crushers, ball mills, and so on are designed to last the life of the mine and, in practice, may outlast several mines. Worldwide, a number of operations are still getting satisfactory performance from crushers and winders that were built in the 1930s.

Most portable and mobile equipment has a finite service life. Typically, new mobile equipment will operate at low maintenance costs for approximately 10,000 hours or 2 years, after which the cost increases steadily. If the use of the machine is lower than around 5,000 h/yr, this "honeymoon period" may be extended. This cycle may be repeated following one or two major rebuilds, although the subsequent maintenance costs are rarely as low as for a new machine.

If the working environment changes, a machine may be deemed obsolescent and need to be replaced. Examples include a change to the stoping method, requiring a larger machine, or a change in the ventilation standards that requires a cleaner engine. If there is a substantial improvement in the available technology, a machine may be replaced, even though it is still delivering to its original specification. For example, new rear-dump haul trucks with much-improved power-to-weight ratios may displace older, slower units.

A decision to replace an existing machine must consider the entire physical and financial environment. For example, in a struggling mining company, capital may simply be unavailable no matter how well a replacement can be justified. In such circumstances, the equipment life can be extended almost indefinitely but with a much-increased operating cost. To extend useful life, many operations cycle their equipment through two or more duties. For example, haul trucks may be converted to water trucks, or production LHDs become development LHDs, where they are used more intermittently.

A particular problem arises in shaft-access underground mines. The difficulty involved in dismantling and reassembling large mobile units for shaft transfer leads to inflexibility in maintenance, and major overhaul periods may be extended. Operators tend to retain old equipment as backup units long after they have reached the end of their economic life. This creates an environment of low usage and excessive maintenance costs. For example, one Australian shaft mine was able to reduce its mobile equipment fleet from 50 to 11 units after establishing a service decline from the surface.

Service life and maintenance costs also depend on a company's maintenance policy and the level of onboard monitoring installed.

Making Financial Comparisons

The benefit of investment in a new machine should be quantified in advance, and in most companies this is a requirement of the capital approval process. The economic benefit is best measured by comparing the net present values (NPVs, or usually net present costs) of the alternatives. If the expected service lives of the alternatives are different, the analysis becomes complicated. For example, two machines with service lives of 5 and 7 years would need to be compared (artificially) over a period of 35 years to get a meaningful comparison, or an accurate estimate be made of the residual value of the longer-life machine after 5 years. To overcome this problem, costs should be annualized using an approach known as the annual average cost or the equivalent annual value method.

Equipment replacement decisions are not exclusive to mining, but apply throughout all business activities. Detailed discussions of the methods can readily be found in textbooks on engineering economics and on the Internet, and suitable financial modeling software can be downloaded and purchased for a small sum.

Many mining companies require that the internal rate of return for the investment exceed a hurdle rate, a minimum acceptable rate of return that is set from time to time by the administration and is based on the company's cost of capital or desired rate of return on all of its investments. Although this may simplify decision making, it often leads to suboptimal decisions because it prohibits many investments that add some value to an operation, and, in particular, it ignores the many benefits that cannot be quantified in a simple financial analysis. Companies may also limit the timing of capital approvals within the annual budget cycle and take a dim view of "out of budget" applications.

Nonfinancial measures of performance may be important. For example, improvements in workplace safety or environmental quality may not be economically measurable but may be highly desirable. Provision of a new machine with an air-conditioned cab may improve industrial relations even if the machine specifications suggest there will be no increase in productivity.

At many mines, particularly at remote sites, there is a large benefit in reducing labor numbers. Increased mechanization will aid this, and can become a matter of policy even if it is difficult to quantify the gains from a particular decision on an NPV basis. Labor costs are typically >50% of total costs, but when the cost of extras such as airfares and site accommodation are included, they may be higher yet.

Conversely, the social–political climate in some areas may dictate that reduction in labor by introduction of more productive machines is politically undesirable. An example is the large amount of hand labor on ramp roadway maintenance seen in many Mexican mines when a grader would be more efficient.

Information Required for Financial Analysis

The capital cost of a new machine should be estimated including the cost of spares, which can be 20% of the purchase price for a mobile unit. This will depend on whether spares are available on consignment; that is, they can be held in the store and paid for when used. Estimators should use the final or likely negotiated price rather than the manufacturer's list price, which is usually somewhat higher.

If an item is sourced overseas, estimators should check what currency will be used for the transaction and what protection (if any) is available against exchange rate variations. Any payable import duties and the transport and unloading costs including in-country transport to the mine should be checked. For a shaft mine, the transport cost will include shaft transfer and reassembly, which can be substantial for bigger items.

Operating costs can be estimated from

- Detailed cost records and performance of similar equipment,
- Cost records for the new item from another operation,
- Manufacturer's cost and performance estimates, and
- A database of costs from a range of operations.

To support ongoing decision making, good record keeping and cost allocation are required over the life of each unit. The mine accounting system must be able to track each piece of major equipment and not report a combined cost for all LHDs, trucks, jumbos, and so on. Records should include engine hours and, for rock-drilling equipment, percussion or hydraulic pump hours, as well as outputs such as metric tons loaded, metric ton–kilometers hauled, or meters drilled.

Caution must be used when comparing costs from different operations, where different operating conditions may apply and the basis of reported costs may not be well understood. Engineers should be particularly wary of manufacturer's estimates, as they are inevitably optimistic. This is an area where consultants can be very helpful, as a good consultant should have an extensive cost and performance database.

The salvage value may sometimes be important. For mobile equipment it will depend on the age of the unit, its condition, and the operating hours since a major rebuild. A rebuilt 10,000-hour underground mobile item could be worth 60%–70% of new cost. The rebuild itself might cost 20%–30% of the original purchase price. An older unit will generally be saleable for about 20% of new cost, if it is operable. Machinery merchants are generally willing to provide an estimate of value for a used item, and for common items may not require an inspection.

The cost of removing the old item from the mine, particularly if it is a shaft mine, may negate the salvage value, in which case the unit should be pushed into a stope and backfilled.

Taxation effects and the method of financing the new item may affect the analysis.

Leasing and Contracting

Besides outright purchase, other forms of equipment finance are available:

- An operating lease is similar to rental, is usually for short-term requirements, and can be cancelled without penalty. It is relatively expensive, because the lessor assumes the risk of re-leasing and technological obsolescence, but is fully deductible as an operating expense. It does not affect the lessee's balance sheet.
- A finance lease is a contract, whereby payments are made over much of the useful life of the asset. During this time the lease cannot be cancelled, although it may be paid out if the item is disposed of or destroyed. The cost includes capital (less salvage) cost, interest, a risk premium, and the commercial cost of providing the service. Typically 60%–70% of the purchase price is financed, with the balance payable as a residual. The item appears on the lessee's balance sheet.

As an alternative to leasing, a contractor (and contractor's equipment) could be used. Applications include, for example, truck haulage, production drilling, surface crushing, and the entire mine development or production program. Contracting is often appropriate where a small additional increment of capacity is required for 1 or 2 years. It also makes sense where specialist skills such as cable bolting are required for a limited time, or where the necessary expertise is not available in the vicinity of the mine and there is insufficient time to train a work force.

Estimating the Average Annual Cost

The average annual cost is the sum of depreciation, interest, and operating cost. Consider a truck that has an initial cost of $1.0 million and a salvage value of $0.2 million after an estimated 4-year service life. The depreciation is thus $0.8 million over 4 years or $0.2 million/yr.

The interest expense arises, because owning an asset ties up capital. This is true whether all or part of the purchase price is borrowed or existing cash is used. In the latter case, there is an opportunity cost because that money could have been invested elsewhere. The average investment is

($1.0 million + $0.2 million)/2 = $0.6 million

If the company has a cost of capital (or cost of borrowing) of 10%/yr, the annual interest expense is $0.1 \times \$0.6$ million or $60,000 for each of the four years.

The annual operating cost may be calculated from the estimated hourly operating cost of $200/h (which includes operator, fuel, tires, maintenance, etc., and the cost of a major rebuild at 12,000 hours) and the expected use of 5,500 h/yr, which is a total of $1.1 million/yr. From this, the average annual cost is

$0.2 million + $0.06 million + $1.1 million
= $1.36 million

A similar calculation can be made for each of the trucks under consideration. The equipment selection decision can then be based on these results, together with all of the other important considerations outlined in this chapter.

Courtesy of Atlas Copco.
Figure 12.2-1 Articulated rear-dump truck

A more accurate calculation can be made, if required, to include consideration of tax effects. The capital tax factor (CTF) is used to adjust capital costs for tax effects. It is determined by

$$CTF = 1 - (td / (i + d))$$

where

t = corporate tax rate
d = annual depreciation rate allowable by the taxation authority (average)
i = cost of capital or interest rate

For example, if $t = 30\%$, $d = 20\%$, and $i = 10\%$, then

$$CTF = 1 - (0.3 \times .20)/(.10 + .20) = 0.8$$

The CTF is applied to capital costs and salvage values, whereas interest and operating costs are adjusted using $(1 - t)$ to get the after-tax cost. In the example,

annual depreciation = $0.2 million × 0.8
= $0.16 million

annual interest cost = $0.06 million × 0.7
= $0.042 million

annual operating cost = $1.1 million × 0.7
= $0.77 million

Hence, the average annual cost (after tax) is $0.972 million.

Of course, all the trucks must be evaluated in the same way, either before or after tax, and experience shows that the rankings rarely change. Although more-accurate spreadsheet-based evaluations are possible, taking into account the time value of money, the average annual cost is only one of the considerations in selecting the most-appropriate equipment.

HARD-ROCK EQUIPMENT SELECTION

The key items of mobile equipment are loaders, trucks, jumbo drills, and production drills, each requiring a different approach to selection. Specific selection criteria apply to each item of ancillary equipment. In all cases the considerations discussed in the previous sections should be addressed.

Selecting a Truck

Trucks used underground fall into three categories: rigid-body rear-dump trucks adapted from surface mining units; articulated rear-dump trucks that may be adapted from surface mining units but usually are purpose designed to have a low profile (Figure 12.2-1); and tractor–trailer units, some with separately powered trailers, usually side-dumping, which can be assembled as an underground road train (Figure 12.2-2). All have diesel engines except for trolley-electric trucks, which require special infrastructure.

Where trucks are required to travel through development openings, the truck cross section will fix the dimensions of the opening, or vice versa. Allowance must be made for clearance from the load to any pipes, ventilation duct, or other services; and clearance to side walls, which may be a statutory requirement. A typical design profile is shown in Figure 12.2-3.

The clearance to ventilation ducts is influenced by the following:

- The shape of the tunnel profile, whether flat, arched, or shanty-backed, which is determined from the geology and ground-support requirements.
- The duct diameter, which depends on the airflow required. Sometimes oval-shaped ducts are used to improve clearances but at the cost of greater resistance and fan power consumption.
- The type of duct, whether rigid or flexible. Flexible ducts may be more mobile and require greater clearances.
- The size, number, and position of ducts within the cross section. The position could be in the crown or in one corner of the profile. In wider headings, vent ducts could also be along a sidewall.
- The size and number of ducts depends on the planned distance of forced ventilation (before connecting to a return airway) and the ventilation volume required to service the heading. A forced ventilation heading may typically require one loader and one truck. The ventilation volume is usually a relationship based on a factor relative to the

Courtesy of Powertrans Pty Ltd.

Figure 12.2-2 Underground road train

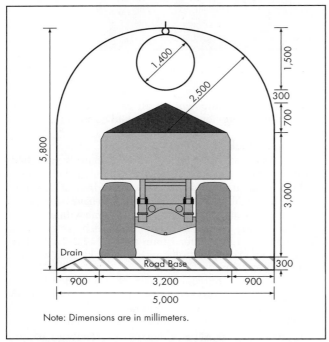

Note: Dimensions are in millimeters.

Figure 12.2-3 Typical mine development profile for haul truck

engine power. Duct diameter is then related to required ventilation volume and duct resistance. Duct diameter may become quite large (>1.2-m diameter) for significant forced ventilation lengths, which may require duplication or a mix of types (forced and push–pull systems).
- The expected load profile in the truck. Sometimes trucks are not fully loaded during the development phase, but can be fully loaded after the development is completed, when a ventilation circuit is established and the ventilation ducts have been removed.
- The initial thickness of road base material and any subsequent buildup of road base dressing material that is not removed by grading, and allowance for road spillage that a truck may accidentally ride over.
- The layout of turnouts from the main drive and the way ventilation ducts make the transition through corners.

Sidewall clearance is required to minimize impact damage and for safety reasons. There must be room for pedestrians to walk past a stationary truck and for recovery work to be done in the event of a breakdown or tire failure. It is not acceptable for trucks to pass pedestrians.

Trucks require a minimum turning radius, which generally will not be <20 m due to truck requirements (capability and maintenance) and development issues on grade (overbreak on development rounds). Widening of the heading and/or flattening the gradient may be necessary on bends to maintain required clearances or reduce longitudinal torsion of the truck frame.

Road trains require additional attention to radius and wall clearances due to the variable tracking path of the powered trailers. Development loops are also required for the road trains as they are not designed for reversing, and tipping is usually sideways.

The operating cost per metric ton–kilometer decreases as the truck capacity increases. Thus, there is an incentive to maximize truck capacity, with diesel trucks of up to 60-t capacity being available for underground applications. A trade-off study is required to assess the benefit of the low truck operating cost against the disadvantages that come with size. Factors to be considered include the following:

- The volume of waste rock broken increases with the cross-sectional area of the heading. For example, a 5.5 × 5.5 m drive will produce 53% more waste rock than a 4.5 × 4.0 m drive. If the cost of development breaking and disposal is $130/m^3 and 10 km of primary development is planned, the cost difference of more than $1.2 million must be recovered from a reduced trucking cost for all of the ore and waste rock from the mine.
- Ground-support costs generally increase with the square of span.
- Larger, more powerful trucks require more ventilation per truck but not per ton moved.

A larger mine opening with larger trucks will have a greater total production capacity in metric tons per day or per year.

Equipment manufacturers provide comprehensive manuals of specifications of their products, and many provide detailed approaches to equipment selection. The key considerations for an underground truck are

- Physical dimensions and load capacity;
- Exhaust gas and exhaust particulate quality;
- Safety features including falling object protection system and rollover protection system, and whether these comply with local standards;
- Air conditioned, enclosed cabin;
- Ergonomic cabin and seat;
- Power-to-weight ratio, and hence speed on grade;

- Engine braking capability or hydraulic retarder;
- Whether fitted with spring activated, hydraulic release fail-safe brakes;
- Whether the torque converter locks up to provide better power transfer and longer transmission life;
- Centralized lubrication; and
- Centralized fire protection system.

The truck load can be calculated from the truck body volume, the broken density of the rock to be carried, and a fill factor that typically varies from 85% to 95%. The stated tonnage capacity of the truck is typically quoted as "SAE heaped" (where SAE refers to the Society of Automotive Engineers), which has the top surface of the load rising one length unit vertically for every two length units horizontally from the top edge of all sides of the tray. This is rarely achieved in practice, particularly for up-ramp haulage. It is not unusual to find that a 50-t-capacity truck achieves an average payload of 43 t, particularly if there is any problem with back clearance on the haul route. Capacity is maximized by using a well-matched loader and with the use of an onboard weightometer.

The manufacturer's tables will provide torque-speed curves, speed-on-grade curves, and other information to allow cycle times to be estimated. In general, it is sufficient to determine the cycle as the sum of

- Loading time (including any wait for loader), which will depend on the loader capacity and the distance from the muck pile to the loading point;
- Travel time (at an average speed on grade) from the loading point to the dump point, usually including level and uphill segments;
- Dump time (including any delay waiting for other trucks to dump); and
- Travel time (at an average speed on grade) back to the loading point.

The average speed, when loaded, will depend on the condition of the roadway, including the presence of spillage or water, the clearance to the tunnel walls, the road gradient, the number of curves or bends, and any mine-imposed or statutory speed limit. The empty travel speed, if downhill, may be limited for safety reasons to less than the general mine speed limit. Similarly, there may be load or gradient limits for loaded travel downhill.

A typical cycle time calculation for a 50-t-capacity truck in an efficient decline-access mine might appear as follows:

Loading	2.5 minutes
Travel loaded	3.5 km at 7 km/h = 30.0 minutes
Dumping	1.0 minutes
Travel unloaded	3.5 km at 9 km/h = 23.3 minutes
Total cycle time	56.8 minutes
Efficiency factor	90%
Cycle time	63.1 minutes

Thus, if the truck can be loaded to an average 46-t payload, the productivity will be 43.7 t per effective operating hour. If traffic congestion is expected to interfere with the trucking cycle, then a further congestion factor may be introduced to increase the estimated cycle time. One simple way to estimate this factor is to assume that a truck returning empty must pass every other truck on its loaded haul out of the mine. If, as is typical (but not necessarily efficient) practice, empty trucks wait in passing bays for full trucks to pass them in the opposite direction, the cycle time could be increased by the time it takes for a full truck to travel half the average distance between passing bays multiplied by the number of trucks operating, less one.

After the cycle time has been determined for a particular haulage route, the tonnage to be moved, the effective time per shift, and the truck availability must be considered to determine how many trucks are required. It is almost always the case in underground mining that the best outcome is achieved by having an excess of loading capacity, so that truck use is maximized. In other words, it is better to have loaders waiting for trucks than have the trucks waiting for loaders. Often the loaders can spend the time between truck arrivals moving the muck up to a temporary stockpile at the loading point to minimize the loading time and turn the truck around quickly.

If there is in fact little or no queuing of trucks at loaders, the cycle time could theoretically be reduced by having slow full trucks wait for faster empty trucks to pass, rather than the more typical rule noted previously. If this were done, the average delay time in passing bays would be reduced. However, there are other practical operating reasons for why this would not usually be done.

In the example, assume a continuous 24-hour operation with two 12-hour shifts per day. Realistic allowances might include

Fuel up, safety meeting, and travel to job	1 h/shift
Meal breaks	1 h/shift
Job inspections	0.5 h/shift
Planned maintenance (8 h/week)	0.6 h/shift
Breakdowns/unplanned maintenance	0.5 h/shift
Job or efficiency factor	90%
Effective hours per shift (8.4 h × 0.9)	7.56
Effective hours per year (7.56 h × 730 shifts)	5,500
Annual truck production (5,500 h × 43.7 t)	240,350 t

If the target production is 1.5 million t/yr of material brought to surface, then the number of trucks required would be

$$1,500,000/240,350 = 6.2 = 7 \text{ trucks}$$

Traffic congestion occurs when the number of trucks required is high relative to the length of haul, so efforts should be made to provide effectives traffic management through the use of radio communications, block lights, and well-spaced, efficient passing bays. The problem is exacerbated for mixed-fleet systems (different capacity, different loaded speeds).

The difference between the efficiency factor introduced within the cycle time calculation and the job or efficiency factor discussed here is that the former increases the operating engine hours required and, hence, the annual operating cost, whereas the latter reduces the operating engine hours available. A useful check is to see that the annual hours required from the truck are consistent with experience at similar sites. There can be a large difference in annual operating hours for underground haulage trucks ranging from around 2,000 hours up to 5,600 hours depending on the management strategy and mine layout.

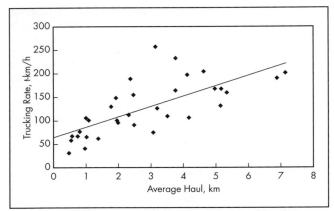

Figure 12.2-4 Truck productivity as a function of haul distance

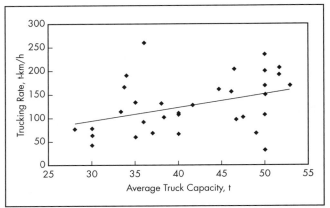

Figure 12.2-5 Truck productivity as a function of truck capacity

Because haulage routes vary, it is often useful to compare truck performance on a metric ton–kilometer basis. This assumes that hauling 1 t for 10 km requires about the same truck capacity as hauling 10 t for 1 km. In the previous example, a 50-t truck hauled 43.7 t/h over a distance of 3.5 km, an output of 153 t-km/h. Actual truck performance, as recorded from 32 operating mines, is shown in Figures 12.2-4 and 12.2-5.

The metric ton–kilometer approach should be used cautiously, because it ignores roadway gradient, which can vary substantially. Whereas some costs, such as tires, vary with load and distance traveled, others, such as fuel, are more closely correlated with engine hours and engine power setting.

When calculating the number of trucks required, allowance must be made for the haulage and disposal of waste rock from development. Often this rock will be backfilled into stopes, requiring a shorter cycle time than a haul to the surface. However, backfilling often causes delays while waiting for a loader or bulldozer to clear the dump point, so that the advantage may not be great. Where trucks must dump fill rock into stopes under a limited back height, ejector-tray trucks may be used. These have an ejector blade, something like a bulldozer blade, which pushes the load from the truck body without having to elevate it into the tipping position.

Ejector-tray trucks have a reduced payload compared to the equivalent standard model, and because of the increased weight in the truck body, they are less fuel efficient. They also have more moving parts to maintain and can suffer damage if large rocks wedge between the sidewall and the ejector.

Trolley-wire electric trucks have been available for many years but have limited application because

- The system has a high capital cost when the trolley-wire installation is considered;
- Although there is no diesel exhaust, they require substantial ventilation for cooling;
- If the mine generates its own power from diesel on the surface, there is no operating cost saving;
- A very high standard of road maintenance is required; and
- Without auxiliary power, trucks cannot travel beyond the fixed trolley wire into actively developing areas.

Despite these limitations, trolley-wire trucks may offer advantages for a mine using low-tariff grid power with a fixed loading point, such as the bottom of an orepass system.

Selecting a Load-Haul-Dump Machine

LHDs are known variously as loaders, muckers, or boggers and differ from front-end loaders in that they have a lower profile and are designed to carry a load in the bucket efficiently for up to 200 m between the muck pile and the truck or orepass that they are loading into. Thus, an LHD spends more time traveling than does a front-end loader, and is engineered accordingly.

A mine may use different-sized LHDs for stoping and for development. Often two or more sizes of LHD are used for stoping, to suit the ore-body width and mining method at different locations within the mine.

Standard models are available in bucket capacities ranging from around 3 to 11.6 m^3. Special machines designed for narrow-vein mining can be as small as 0.5 m^3. Corresponding payloads range from 1 to 25 t. Extremely low-profile machines are also available for working in flat-dipping narrow-vein ore bodies with as little as 1.6 m of headroom.

LHDs may be diesel or electric powered. The diesel units are versatile and can tram quickly from one location to another (Figure 12.2-6). The electric units (Figure 12.2-7) carry a cable drum and rely on a trailing electric cable, so they are tethered to a location during normal production. Electric LHDs have low noise levels and zero emissions, so they require less ventilation. They are highly productive in situations such as block caving, where ore is transported from a series of drawpoints to a fixed orepass location.

The ergonomic and safety considerations for selecting an LHD are similar to those for trucks, as listed previously. In the past, some operators sustained back injuries due to poor operator location and poor seating. The best operating position is midway between the front and rear wheels, and particular attention should be paid to the quality of the seat suspension.

LHDs often operate in situations with limited forced ventilation, so an enclosed, air-conditioned cab is desirable, and the engine should be the latest-generation, low-emission type. Fail-safe brakes are highly desirable. LHDs can operate effectively in headings as steep as 20% (1 in 5), although they perform best if the gradient is limited to around 14% (1 in 7) or less.

The size of a selected LHD must fit within the planned development and stope openings, and the bucket must be able to reach above and fill a truck efficiently.

Courtesy of Atlas Copco.
Figure 12.2-6 Diesel load-haul-dump unit

Courtesy of Atlas Copco.
Figure 12.2-7 Electric unit showing cable drum

The bucket load can be calculated from the bucket volume, the broken density of the rock to be carried, and a fill factor, which depends on the fragmentation of the rock and may be different for development rock and for broken ore in stope drawpoints. As with trucks, the nominal capacity quoted is typically for SAE-heaped loading, which is also typically difficult to achieve in practice. However, a high fill factor can be achieved with well-blasted rock. For example, an 8-m³ bucket with a 90% fill factor and broken ore of bulk density 2.2 t/m would carry 15.5 t.

The manufacturer's tables will provide speed on grade and other information to allow cycle times to be estimated. Some manufacturers provide estimates of productivity in metric tons per hour for various haulage distances.

In general, it is sufficient to determine the LHD cycle as the sum of the following:

- Loading time, which depends on the power of the LHD, the muck pile condition, the floor condition, and fragmentation of the broken rock. The presence of large rocks, which require individual attention, can slow the loading cycle considerably.
- Travel time (at an average speed on grade) from the loading point to the dump point.
- Dump time (including any delay waiting for trucks), which is usually faster when tipping into a pass because no attention to the shape of the load is required. However, depending on the fragmentation of the rock and the size of the pass, grizzlies or other methods of size restriction to limit the size of rocks in the pass may constrain the tipping rate into a pass.
- Travel time (at an average speed on grade) back to the loading point.

A more complex cycle may apply. For example, it is usual to cut stockpile (remuck) bays every 120–250 m along a development heading. After blasting, the LHD quickly moves the muck from the round into the stockpile bay to allow ground support and drilling to proceed, after which the LHD loads trucks from the stockpile. It may be necessary to consider the LHD performance to estimate times for the two loading cycles within the overall development cycle.

The average speed, when loaded, will depend on the condition of the roadway, including the presence of spillage or water, the clearance to the tunnel walls, the road gradient, and the geometry of the tramming route. A typical cycle time calculation for an 8-m³ LHD that is loading trucks might appear as follows:

Loading	30 seconds
Tramming loaded	200 m at 7 km/h = 103 seconds
Dump into truck	30 seconds
Travel unloaded	200 m at 7 km/h = 103 seconds
Delays, waiting	60 seconds
Total cycle time	5.4 minutes
Efficiency factor	90%
Cycle time	6.0 minutes
Cycles per hour	10
Metric tons per hour	155 (using 15.5 t/bucket from previous calculation)

As in the truck example, the effective hours must be determined. For a continuous 24-hour operation with two 12-hour shifts per day, realistic allowances might include the following:

Fuel up, safety meeting, and travel to job	1 h/shift
Meal breaks	1 h/shift
Inspect job, wash down, bar down	0.5 h/shift
Planned maintenance (8 h/week)	0.6 h/shift
Breakdowns/unplanned maintenance	1.0 h/shift
Job or efficiency factor	80%
Effective hours per shift (7.9 h × 0.8)	6.32
Effective hours per year (6.32 h × 730 shifts)	4,614
Annual LHD production (4,614 h × 155 t)	715,170 t

If the target production is 1.5 million t/yr of ore, the number of production LHDs required would be

1,500,000 / 715,170 = 2.1 = 3 production LHDs

The number of production LHDs required could be higher if

- LHDs are locked into production areas, or the tramming distance between production areas is excessive;
- More redundancy is required to ensure that production targets are met in the event of a major breakdown;
- There are periods of time in the stoping cycle when there are insufficient active drawpoints to use the available loaders; and
- LHDs are operated on remote control for stope cleanup, which reduces productivity.

A separate calculation is required to determine the number of development LHDs required.

Some LHD models are available with ejector buckets for discharging under limited headroom, which can reduce the need for back stripping at loading points, and can be useful for tight backfilling.

Remote Control and Automation of LHDs and Trucks

Remote control of loading units is desirable where conditions at the face or in the stope are hazardous, or where the ground overhead has not been secured. Radio remote control of LHDs has been available for many years, having evolved from cable-remote control of compressed-air-powered loading units. Line-of-sight remote control has proven to be hazardous and has fallen out of favor. Despite training and strict operating rules, operators tend to move into the path of the loading unit and hence are exposed to injury.

Teleremote control, using television cameras mounted on the LHD, allows the operator to be seated in a safe, air-conditioned environment underground or even to operate the machine from the surface. Current developments include employing the operator in a city office, remote from the mine, thus making the job more attractive while greatly reducing on-site costs.

The productivity per hour of an LHD is typically less under teleremote control than with an operator on board, because the teleremote operator does not have the full range of audible, visual, and physical inputs to enable optimum performance. However, most teleremote operations currently use two-dimensional video screens for providing visual information to operators. Such things as three-dimensional (3-D) visual presentation and virtual-reality helmets, available in other industries, have been trialed in mining operations but have not come into widespread use. It is conceivable that in the future, provision of real-time 3-D virtual-reality visual and audio information to remote operators may reduce the difference in productivity between remote control and direct operation. Also, teleremote operation offers the possibility of operating for a greater part of the working shift because delays such as travel time do not apply and because operators can be rotated in a surface office to keep the machines at maximum use.

The major manufacturers now offer autonomous LHDs that can operate with minimal or no human intervention. These machines have sophisticated onboard computers and typically use lasers to scan their environment and determine position. Such a machine can load, tram, and dump autonomously so that a human supervisor can manage two or more LHDs.

The loading part of the cycle requires the muck pile to be well broken. The current generation of autonomous machines is not as effective as humans at assessing the position of large rocks and maneuvering the bucket to displace and lift them. The LHD can be damaged if used autonomously with large rocks, so a common compromise is to load the LHD on teleremote and then switch to autonomous for the remainder of the cycle.

The use of surface teleremote and autonomous LHDs requires a good broadband communications backbone to be installed in the mine, and a high standard of training for maintenance personnel. For safety reasons these machines can only be used in an area that has been locked off from human entry, so the opportunities for their use are limited. They are suited to block caving and some sublevel caving operations where production activities can proceed for a significant time without the need for other activities around them. In many other mines, activities such as development, geological sampling, surveying, and so on interact with production loading, so that it is impractical to create a nonentry zone.

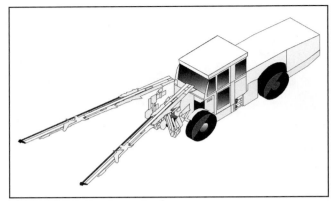

Figure 12.2-8 Two-boom jumbo

For autonomous operations, the mine must be designed to accommodate the level of technology. Where automation has been adapted to an existing mine layout, the results have generally been unsatisfactory. Design of operations needs to include interaction between activities; dealing with spillage, drainage, and road maintenance; blast damage; planned and unplanned maintenance activities; surveying; geological mapping and sampling; and so on.

Autonomous haulage trucks are also available but are not yet widely used. Autonomous trucks must operate on a dedicated haulage route that is locked off from other traffic and from pedestrian traffic, which can rarely be arranged in existing mine layouts. In most cases, duplicate development of a dedicated haulage roadway would be required and may not be justified.

Selecting a Jumbo Drill

A modern jumbo drill has an articulated diesel-powered carrier and an electrically powered hydraulic pump, which operates the boom rams and hydraulic rock drill(s) when the machine is in drilling mode. The jumbo may have one, two, or three booms, depending on the size of openings.

The diesel engine enables the jumbo to tram between working places. It is rare for the diesel engine to provide hydraulic power for drilling because it is inefficient, and because there is usually insufficient ventilation to operate the engine at the face. Some models have a "fast tramming" option that is useful in a large mine where the work places are far apart.

Single-boom jumbos are suited to face areas from 6 to 30 m^2, whereas twin booms may fit in an opening as small as 8 m^2 but can extend 100 m^2 or more. Two booms are sufficient to cover the face area in most mines, with three-boom jumbos being more popular in civil engineering works. Where a three-boom jumbo would, in theory, provide a faster drilling cycle, in practice, the operator is fully busy managing two. A valuable application of the center boom is drilling the large diameter reliever holes in the cut. Many jumbos have a third boom with a basket. A single boom is used in small headings or narrow veins, where there is insufficient room to operate a two-boom machine (Figure 12.2-8).

The manufacturer's data will show the capability of the jumbo to cover the proposed face area. The blast design will determine the hole diameter, generally in the range of 33 to 89 mm, although larger holes can be drilled at greatly reduced penetration rates. Depending on the duty, rock drills (drifters) with impact power in the range of 10 to 30 kW are available. The length of round to be drilled and fired will determine the length of the drill steel and the boom. In general, longer, heavier booms require a larger carrier.

It is important that the jumbo has adequate reach to enable back holes to be drilled parallel with the axis of the excavation. If an undersized machine is used to drill large profile excavations by angling the back holes out (not parallel to the direction of advance), the result may be a sawtooth profile, excessive overbreak, and unsatisfactory conditions for ground control.

High-powered drifters may require a high-voltage system (1,000 V) to deliver sufficient energy to the hydraulic pumps. The voltage also impacts the trailing cable diameter and, hence, trailing cable length and power extension requirements. This is of relevance for countries where lower-voltage systems (~450 V) are the norm.

An important consideration is whether the jumbo will also be used for rock bolting and meshing, either consistently or occasionally, and the type of rock bolts to be used. For rock bolting, it must be possible to turn the booms at 90° to the axis of the carrier, and to drill holes into the walls and back. The boom length cannot exceed the available width and height. In these situations a split-feed boom may be required, which effectively telescopes the boom to full length as required. This adds complexity, maintenance, and weight and may lead to inaccuracy as parts wear, so the benefit must be weighed against these considerations.

A jumbo may also be used for scaling or "rattling" the tunnel walls and back, as part of the development cycle, in lieu of barring down or using a dedicated scaling machine. This is never recommended by the drill manufacturers because it can cause a lot of damage to the booms from falling rock. Nevertheless, it is a common practice as operators have learned that the convenience and safety benefits can outweigh the increased maintenance. For this application, a robust and well-protected boom should be selected.

The face advance that can be achieved from a jumbo drill depends on the number of faces available to it. In a single-heading situation, such as during the development of a main decline from the surface, a twin-boom jumbo can typically average 45–50 m of advance per week, including the main heading and stockpile bays. This means that the main face will advance, on average, about 40 m/week. This rate can be exceeded for periods of several months if in good ground, but most mines present a range of conditions, including zones of poor ground or faults, which slows progress, so that over a 2-year development program the 40-m/week rate would apply.

A higher advance rate can be achieved using the latest "long round" technology. For example, during the development of Newcrest Mining's Cadia East project in New South Wales, a record for 6.0 m × 5.5 m mine development with full ground support was set with an advance of 283 m in a single month (including stockpile bay development) (Willcox 2008). Long round development has limitations on curve radius and turnout or crosscut development, so is not generally applicable to underground mine development. Higher development rates exceeding 400 m/month have been achieved in civil tunneling works, but the Cadia East performance is a good example for mine development.

The process of estimating the performance of a two-boom jumbo drill is set out in Tables 12.2-1 through 12.2-3, which are based on actual measurements from a mine where each jumbo had at least five faces available to it for drilling, and

Table 12.2-1 Physical development parameters

Parameter	Measurement
Width, m	5.0
Height, m	4.5
Area, m^2	22.5
Rock density (wet), t/m^3	2.8
Overbreak, %	5
Metric tons per meter (including overbreak), t/m	66.2
Drill factor, charged holes/m^2	2.20
Number of charged holes	50
Number of 100-mm cut holes	4
Hole depth, m	3.05
Pull, %	90
Advance per round, m	2.75
Metric tons per round, t	181.6
45-mm-diameter face holes, meters drilled, m	151.0
100-mm-diameter cut holes, meters drilled, m	12.20
Total face meters drilled, m	163.18
Rock bolt length, m	2.4
Friction bolt holes per round	5
Resin bolt holes per round	8
Rock bolt meters drilled per round, m	31.2
Total small-diameter meters drilled per round, m	182.2
Face meters drilled per meter advance, m	59.4
Total meters drilled per meter advance, m	70.8

so could be kept continuously at work. Causes of delays were recorded in detail on shift reports. The jumbo drills were conventional units without computer alignment and were not the latest generation of high-frequency drifters.

After this analysis is completed, using performance parameters for the jumbo drill under consideration, the required number of jumbos can be estimated. If the target is 800 m of development per month, then, in this example, three jumbos would be required. Of course, if sufficient faces cannot be made available to fully use the drills, this problem is not solved by adding more drilling capacity. Instead, the materials handling capacity must be addressed.

Selecting a Rock-Bolting Machine

Rock bolting may be done using a handheld rock drill from the muck pile, a work basket, or a scissor-lift platform. Purpose-built articulated, four-wheel-drive scissor lifts are available with or without a bolting boom attached. Alternatively, the development jumbo drill can be used to install rock bolts and mesh with minimal exposure of the operator to the area being supported.

Purpose-designed machines are available for drilling holes and installing both conventional rock bolts and cable bolts. A rock-bolting machine resembles a jumbo drill and may use the same carrier. The working boom carries a rock drill and a carousel that can handle a range of rock-bolt lengths and typically 10 rock bolts. A separate hydraulic arm may be fitted to maneuver and place screen mesh.

A rock-bolting machine may be capable of installing mechanical- or friction-anchored bolts as well as cement- or resin-grouted bolts. Machine selection should address the following points:

Table 12.2-2 Jumbo cycle time

Component	Measurement
Setup	
Tramming speed, k/h	4.0
Average distance between faces, km	1.0
Tramming time, min	15.0
Setup and markup time, min	25.0
Total setup time, min	40.0
Setup time (=engine hours) per cut, h	0.7
Drill meters per engine hour, m/h	272.0
Drilling Cycle	
Instantaneous penetration rate, 45-mm holes, m/min	1.40
Instantaneous penetration rate, 100-mm holes, m/min	0.7
Meters drilled, 45-mm holes, m	182.2
Meters drilled, 100-mm holes, m	12.2
Drilling time, 45-mm holes, min	130.1
Drilling time, 100-mm holes, min	17.4
Total drilling time, min	147.6
Total drilling time, h	2.5
Holes drilled (all types)	67
Boom setup time per hole drilled, s	55.0
Total boom set-up time, min	61.0
Total boom set-up time, h	1.0
Total set-up and drill time (2 booms), h	3.5
Average penetration rate per boom, m/min	0.71
Jumbo drilling time, h	1.74
Ground Support Using Jumbo	
Change centralizers, min	10.0
Rattle backs, min	15.0
Installation time per rock bolt, min	1.0
Total rock bolt installation time, min	5.0
Total ground-support time, min	30.0
Total ground-support time, h	0.5
Total jumbo time per face, h	2.9

Note: Calculations are given in seconds or minutes, and the results are rounded to hours. The final result of jumbo time per face in hours is the sum of setup time plus jumbo drilling time plus ground support time.

Table 12.2-3 Jumbo development performance

Component	Measurement
Roster	
Days per year	365
Days worked per week	5
Days worked per month	21.7
Shifts per day	3
Shift length, h	8
Delays	
Shift change, h	0.8
No face available, h	0.6
No power, h	0.5
Meetings, h	0.2
Face not scraped, h	0.2
No ventilation, h	0.2
Pump/water problems, h	0.1
Maintenance/breakdown, h	0.1
Total delays, h	2.7
Productivity	
Balance of shift available, h	5.3
Faces drilled out	1.82
Face advance per shift, m	5.01
Face meters drilled per shift, m	298
Total meters drilled (including rock bolts) per shift	354.7
Meters drilled per month, m	19,356
Face advance per working day, m	15.0
Face advance per month, m	325.6

- Can the desired type and length of rock bolt be installed?
- Can it handle screen mesh?
- Will it fit into the required minimum back height?
- Can it reach the required maximum back height?
- Is the cab enclosed to protect the operator from resin fumes?
- Is a remote control option available for hazardous situations?

The rock-bolting machine size should be matched to the required heading size, which in turn depends on the purpose of the heading. It is unfortunately quite common to find that limited availability of rock-bolting boom lengths forces headings to be larger than they need to be for the purpose for which they were designed. At best, this might merely produce more development waste than necessary. At worst, it can make the use of the heading for its intended purpose more difficult, more costly, or more time-consuming than it would have been if correctly sized.

Rock-bolting machines are subject to impact damage from falling rock and require tight tolerances on working parts, so that the rock bolt can be indexed to the drilled hole. It is useful to enquire from other operators about their experience with maintenance on the model being considered.

Rock-bolting machines tend to be slower and less mobile than jumbos. A check should be made to ensure that the rock-bolting machine can travel between development faces within the cycle time required.

Selecting a Production Drill

Production drills are used for drilling stope blastholes and service holes. Top hammer drills are generally available in sizes from 5 to 127 mm in diameter, whereas in-the-hole hammer (ITH) drills range from 95 to 178 mm in diameter. The drill-and-blast design will determine the optimum diameter and length of the hole. The selected drill must be able to quickly drill the hole, at minimum cost, with acceptable deviation. Deviation will limit the smaller hole sizes to around 12–15 m in length, whereas holes of around 100-mm diameter with tube drills can be up to 40 m long. Longer holes require larger diameters, favoring ITH drilling.

Skid-mounted, crawler-mounted, and trolley-mounted rigs have generally been replaced by highly mobile diesel-powered rigs. Top-hammer rock drills are hydraulic, using electric hydraulic power packs, whereas ITH drills use compressed air from a stand-alone or booster compressor. High-pressure water-powered hammers are also available.

For narrow-vein mining, production drills that can operate in vein widths of 3.0 m or slightly less are available. In the larger sizes, fully automated operation is possible so that a complete ring of blastholes can be drilled without human

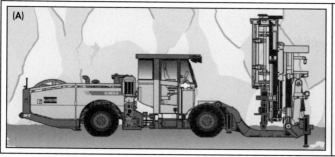 

Courtesy of Atlas Copco.
Figure 12.2-9 Production drill rigs

intervention. Considerations in selecting a production drill include

- The performance, including penetration rate, of the drifter at the desired hole diameter;
- Whether the desired hole inclinations can be drilled;
- Parallel hole coverage capability;
- Feed pivoting system (rigid with sliders—more accurate holes or totally variable—highly variable hole alignment);
- Whether reticulated compressed air or an onboard compressor will be used;
- Whether centralized lubrication is fitted;
- Whether automated rod handling is available;
- Whether carousel capacity is adequate for planned hole lengths;
- Whether remote control is available (umbilical cord system for misfire drilling management);
- What onboard surveying system is used for hole positioning;
- Whether cab is closed and air conditioned;
- The level of automation required (for increased use);
- Operator visibility of hole collar and lighting conditions; and
- The ability to collar the hole close to the wall, particularly in narrow-vein applications.

Drill string selection (rods, in-hole stabilizers, tubes) and feed stabilization (front- and/or rear-feed stingers) are also important considerations for managing drill-hole deviation.

Figure 12.2-9 shows production drilling rigs with hydraulic top-hammer rock drills. Both are designed for long-hole ring and parallel hole drilling upward or downward. In Figure 12.2-9A the drill unit faces the canopy and is suitable for holes in the range of 89 to 127 mm. The drill in Figure 12.2-9B is boom mounted for maximum reach and flexibility in both production drilling and bolt-hole drilling and is suitable for holes in the range of 51 to 89 mm.

Other Underground Equipment

The considerations outlined in the opening sections of this chapter apply to a wide range of underground equipment. Specific considerations include the following:

- Integrated tool carriers are based on wheel loaders for greater load-carrying capacity, versatility, and stability than conventional forklifts. Operator visibility and automated quick coupling are important features.
- Robust scalers are becoming more effective in headings between 4 and 6 m high. Backhoes fitted with a scaling pick or lightweight impact hammer and very large scalers have proved less successful. Some limestone mines use a rotary head on an excavator very effectively.
- Shotcrete (or Fibercrete) machines may be designed for wet or dry mix material. Mobility, boom reach, and remote-control capability are important in underground operations. The machine should have a remote-controlled boom nozzle to remove the operator from the unsupported area.
- Explosives charge-up machines may be used for cartridge, emulsion, or ammonium nitrate and fuel oil, and the charging process may be manual or mechanized. As with jumbo drills and rock-bolting machines, the charge-up machine must be sufficiently mobile to service the faces required within the development and production cycle. The reach and load capacity of the basket and the explosives carrying capacity must be suited to the proposed duty.
- Road maintenance equipment, such as graders (with or without rippers), bobcats, water trucks, and rollers need to be fully assessed for the suitability of the road surface medium and dimensions of the underground workings.
- A wide range of four-wheel-drive vehicles are suitable for underground mining, although many of the lower-cost models may not stand the rough conditions of a typical mine and may have a very high maintenance cost. It is a good strategy to buy the best rugged model available secondhand, with around 100,000 highway-kilometers. In corrosive or rough mine environments, the best vehicle life may be around 2 years. Estimators should not underestimate the cost of fitting a vehicle for underground operation, which might include crash bars, engine isolator, and removal of unwanted interior fittings.
- Handheld (airleg or jackleg) drills, rope scrapers (slushers), and rail equipment may still be used in some circumstances, although their use is rapidly diminishing. Proper selection of rail equipment, in particular, requires careful consideration based on cycle times, gradients, and locomotive capability. It is best to consult old textbooks including earlier versions of this handbook to gain an appreciation of the selection process.

REFERENCE

Willcox, P. 2008. Rapid tunnelling technologies at Cadia East. In *Proceedings of the Tenth AusIMM Underground Operators' Conference*. Melbourne: Australasian Institute of Mining and Metallurgy.

CHAPTER 12.3

Soft-Rock Equipment Selection and Sizing

Jeremy Busfield

INTRODUCTION

Soft-rock mining predominantly applies to coal mining but is also applicable to other bedded soft mineral deposits. Therefore, although this chapter primarily focuses on underground coal mining, certain aspects are relevant to mineral deposits such as potash.

Underground coal mining methods currently fall into the following principal categories:

- Longwall (full-extraction) operations
- Room-and-pillar (partial-extraction) operations
- Room-and-pillar (first-workings) operations

These methods are discussed in the following sections of this chapter. Figure 12.3-1 provides examples of the three categories.

LONGWALL MINING METHODS

Because of its mechanized process, longwall mining generally achieves the highest productivity and reserve recovery and is typically the primary choice for new mines being developed, unless geological, surface, or capital constraints dictate otherwise.

In longwall mining, a set of roadways is driven out from the main entries down each side of the longwall panel to block out a portion of reserves. The mechanized longwall equipment is then retreated through the panel to extract the entire panel of coal. The roads used for traveling and the roads used for coal clearance during panel extraction are termed the *maingate* (or *headgate*) entries, whereas the tailgate entries are generally used for return air. Where possible, a set of parallel adjacent panels is extracted sequentially to allow reuse of one gate road (i.e., the maingate for the first panel becomes the tailgate for the next panel, and so forth). Working in this way maximizes coal recovery, while keeping the quantity of roadway drivage to a minimum. The number of roadways that comprise a maingate varies around the world, with single entries being common in Europe, two roadways common in Australia and China, and three roadways common in the United States. The number of roadways influences the style of coal clearance and the mobile equipment used.

Longwall mining methods vary slightly according to the seam height, which can range from as low as 1.0 m to in excess of 15 m. The height influences the longwall equipment used. Single-pass seam extraction is undertaken in seam heights of up to approximately 5.5 m, for which conventional longwall equipment is used. For seams in excess of 5.5 m, either multislice or top coal caving methods can be used. Top coal caving requires a different style of roof support and an armored face conveyer (AFC) configuration. In very thin seams, the longwall plough system can be used, which uses a different coal-cutting machine (a plough as opposed to a shearer).

The longwall panel width can vary up to more than 400 m, and this influences the longwall equipment (quantity of roof supports, quantity of AFC pans, hydraulic power, electrical power, etc.). Panel length can be dictated by geology, surface subsidence restrictions, equipment limitations, or other mining constraints. Very short panel widths in the order of 25 to 50 m are typically termed *mini-wall longwalls*, and these use a specific type of AFC arrangement.

ROOM-AND-PILLAR MINING METHODS

Until the advent of high-production longwall techniques, the majority of underground coal extraction used room-and-pillar (R&P, also termed *development* or *bord-and-pillar*) mining techniques. In many parts of the world this is still the case, with many mines achieving high production rates from R&P methods. The use of the R&P method is often favored because of its low capital requirements compared to the longwall system. The R&P method is also used in small or irregular-shaped deposits, deposits with surface subsidence restrictions, and deposits where geological constraints (such as faulting) preclude the economic use of longwall mining.

R&P methods range from first workings, where only roadways are driven (forming pillars), to second workings, where the pillars are then later either fully or partially extracted. R&P methods are generally deployed in seam heights ranging from 1.5 to 4.5 m. In seam heights of less than 2.1 m, the roof is often excavated periodically or continually to provide an adequate working height. Low seam heights can restrict equipment options, particularly bolting equipment. With greater

---

Jeremy Busfield, Principal Consultant, MineCraft Consulting Pty Ltd., Queensland, Australia

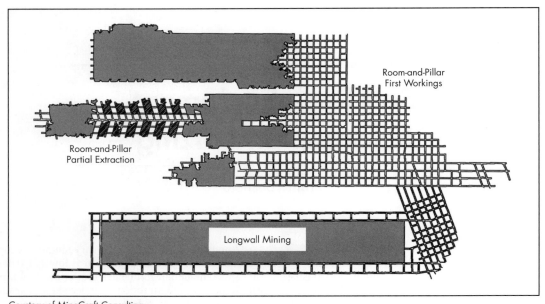

Courtesy of MineCraft Consulting.
Figure 12.3-1 Mining methods

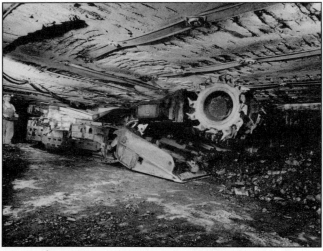

Courtesy of Bucyrus.
Figure 12.3-2 Continuous miner in a room-and-pillar operation

seam heights, some mines extract the floor as a second working, thus allowing up to 6-m seams to be partially extracted. In these cases consideration is required for roadway gradients and equipment capabilities in regard to negotiating steep grades. In first workings, a benefit can be gained by increasing the roadway width to up to 7.5 m (or more), depending on the spanning ability of the roof. In potash mines, roadway spans typically exceed what can be achieved in coal, with the roadways formed by wide-head continuous miners (CMs) taking several passes. A CM in an R&P operation is shown in Figure 12.3-2.

Secondary extraction methods generally fall into pillar-splitting or stripping methods or fall into split-and-fender methods, for example, Wongawilli extraction in Australia (Figure 12.3-3).

In the standard pillar-splitting technique, the main development roadways are driven toward the boundary of the mining area, forming coal pillars to support the roof. When the limit of reserves or the mining boundary is encountered, the coal pillars can either be split or stripped, thus forming smaller pillars that are then left to support the roof or that collapse in a controlled manner on retreat. This method of mining requires specific equipment that can extract coal and retreat in a nimble and rapid manner.

Split-and-fender mining involves the formation of main access roads into a particular district and the subsequent "blocking out" of an area of coal, typically 60 to 80 m in width and up to 1.5 km in length. A roadway or "split" is driven through to the end of the block using a CM, leaving a thin coal "fender" that is extracted as the CM retreats back toward the access roadways, allowing the roof to collapse behind. When the fender is fully extracted, the mining equipment is retracted, and the entire sequence is repeated until the whole coal block is mined out. Although similar equipment can be used for this method, the use of powered roof supports (mobile breaker line supports) is often used to increase safety and allow for a greater level of roof control.

LONGWALL MINING EQUIPMENT
This section begins with a discussion of standard longwall equipment, followed by discussions of longwall equipment variations, selection, specifications, and sizing.

Standard Longwall Equipment
The standard longwall face equipment consists of roof supports; a shearer; an AFC; a beam stage loader (BSL), including crusher and boot end; a monorail; a pump station; and system electrics. Figure 12.3-4 is a schematic of a longwall system showing the relative locations of each piece of equipment.

Roof Supports
Roof supports (sometimes referred to as shields or chocks) hold the exposed roof as the coal is mined, allowing the roof strata to cantilever over the support fulcrum and then break off into the goaf (or gob). In this manner, the operators are

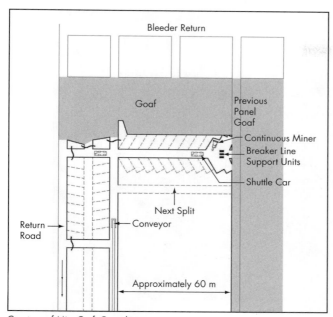

Courtesy of MineCraft Consulting.
Figure 12.3-3 Wongawilli extraction panel

shielded from falling goaf as the face is advanced and the immediate face area is destressed. As the shearer cuts along the face, the AFC is snaked across the new face line using the double-acting sequence rams fitted between the supports and the AFC pans. As the AFC is snaked, the roof supports are lowered and pulled across to the advanced position by the sequence ram and reset to the roof, referred to as the lower-advance-set cycle. In this way the face is automatically advanced as each longwall slice is taken through the panel. The immediate roof falls behind the support, forming a goaf.

Roof supports can be operated in either "conventional mode" or in "immediate forward support" (IFS) mode. Conventional supports cannot be advanced to support the roof until the AFC is advanced (i.e., the roof is exposed and unsupported for some distance behind the shearer), whereas in IFS mode the supports are advanced immediately after the shearer has passed. IFS support is generally needed in weak roof conditions and is the most popular mode of support currently used. IFS supports require longer canopies than "conventional" supports with corresponding larger leg cylinders and hydraulic systems.

Modern longwall supports have been built to operate to approximately 5.8 m in height. Support capacity depends largely on support configurations, although the largest supports available have a capacity of approximately 1,750 t (metric tons), equating to a support density of approximately 125 t/m^2 and weighing up to 65 t.

The roof supports used at the gate ends are configured slightly different to the "run-of-face" supports, in that they have longer canopies so as to extend over the maingate drives. In addition, the gate end supports have higher rated capacity relay bars to enable pushing the maingate drive frames and BSL. Typically, there are five gate end supports at the maingate and three at the tailgate.

Originally operated by valves fitted to each support (lower, advance, and raise), modern supports are controlled by electrohydraulic systems, whereby solenoids operate the valves and the supports can be operated automatically in sequence mode or from adjacent or nearby supports. This automatic sequence also can operate with a positive set mode, whereby each support is set against the roof to a minimum pressure (often between 70% and 80% of the yield pressure), thus guaranteeing effective support when in automatic mode.

When specifying roof supports, various options are available, including the following:

- Four-leg versus two-leg supports (because of advancements in cylinder technology and capacity and because of lemniscate design, two-leg supports are now more popular)
- Front walkway, rear walkway, or both (rear walkways are often used in thick seams to offer more protection against face spall)
- Side shields on one or both sides (used to push supports uphill)
- Face sprags or flippers (used to prevent face spall in thick seams)
- Base lift systems (used to raise shield prior to advance when in boggy floor)

A longwall roof support with face sprag and base lift fitted is shown in Figure 12.3-5.

Shearers

Modern longwall double-ended ranging drum shearers have two cutting drums, one at each end mounted on hydraulically raised ranging arms that are in turn connected to the shearer body (Figure 12.3-6). The shearer is driven by onboard traction units that drive a sprocket wheel that connects to rack bars mounted on the walkway side of the AFC. The face side of the shearer slides along the top (or toe) of the AFC sigma section. The shearer operators control the speed of the shearer and the position of the cutting drum using handheld controls (either radio remote or tethered cable) to maintain a suitable cutting horizon within the seam and minimizing dilution.

Modern heavy-duty shearers are fitted with up to 2,000 kW of available power at the cutting drums, enabling maximum cutting rates of more than 5,000 t/h to be obtained. Shearers can be up to 14 m in length and weigh up to 100 t.

Armored Face Conveyor

The longwall AFC is used to convey cut coal along the face to the maingate, where it is connected to the BSL and outby conveyor system. Steel AFC pans are constructed in sections typically 1.75 or 2.0 m long to match the width of roof supports. Flexible (dog bone) connectors are used to join the pans together. A clevice bracket is fitted to each pan, by which the roof support relay bar (hydraulic ram) attaches. As each slice is taken on the longwall, the AFC pans are advanced or "snaked" into the newly cut area behind the shearer by activating the relay bars in sequence.

The AFC supports the shearer body as it runs along the face, and the cut coal is conveyed along the AFC by flight bars attached to AFC chains driven by motors at both the maingate and tailgate face ends (Figure 12.3-7). Inspection pans that feature a removable decking plate will typically be fitted as every fifth pan to allow for access to the bottom chain. Transition pans are used at the gate ends to transition the AFC onto the drive frames and BSL. Maintaining tension in the AFC chain is important with a chain-tensioning device fitted at the tailgate drive frame. Often automated using an appropriate control

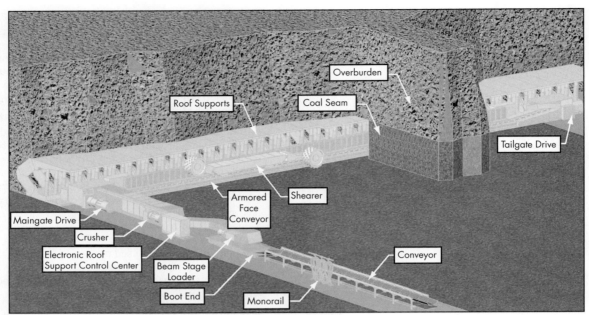

Courtesy of MineCraft Consulting.

Figure 12.3-4 Schematic view of longwall system

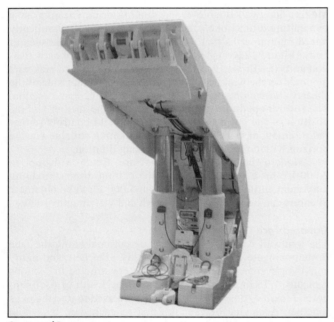

Courtesy of Bucyrus.

Figure 12.3-5 Longwall roof support with face sprag and base lift fitted

system (reed rods, etc.), the tensioner operates by extending a hydraulic ram (1-m stroke) fitted to the tail sprocket assembly. While the tensioner provides fine adjustment on a day-to-day basis, coarse adjustment is achieved by removing links from the AFC chain.

Important aspects of the AFC include the following:

- The drive motors must be able to start the AFC when fully loaded.
- The AFC speed must be sufficiently faster than the shearer's cutting rate to allow coal to be cleared when cutting from tailgate to maingate.
- The AFC chain must be of sufficient strength to handle the applied tensions.

Modern AFCs can be supplied with up to 3,000 kW of installed power in AFC pan sections with 1,300-mm raceways. Coal can be transported at rates in excess of 4,500 t/h. The current limiting factor for face length is the AFC (chain strength versus installed power). Modern chains are being developed so as to allow longer faces.

Currently, the two main AFC drive systems are controlled-slip transmission (CST) systems and fluid-coupling systems. CST systems incorporate clutch systems in the output stage of the AFC gearbox (Figure 12.3-8) with the motor directly coupled to the gearbox. Fluid-coupling systems are fitted between the motor and gearbox and incorporate valves to change the fill volume of the coupling, thus changing the level of torque. Both systems allow for controlled starting and load sharing between drives. An AFC maingate drive frame is shown in Figure 12.3-9.

Beam Stage Loader, Crusher, and Boot End

The BSL transfers the coal from the AFC to the conveyor belt and includes a chain conveyor running over deck plates, a crusher, a gooseneck, and a boot end. The BSL is capable of handling a higher volumetric capacity than the AFC to ensure that coal removal from the face occurs without accumulation at the maingate corner (AFC/BSL intersection).

The crusher sizes the cut coal to ensure that oversize lumps are not carried onto the conveyor, and they consist of a high-inertia rotating drum fitted with large "hammers" to break up large lumps of coal that may have slabbed from the face or large pieces of stone from the roof. The gooseneck is an elevated section that raises the coal so it can be transferred onto the belt.

Courtesy of Bucyrus.
Figure 12.3-6 Double-ended ranging drum shearer

Courtesy of Bucyrus.
Figure 12.3-7 AFC chain and pan

The boot end consists of the tail pulley for the conveyor belt fitted to a sliding frame so that the BSL can be retracted a certain distance (up to 3 m) without moving the tail pulley location. The boot end is fitted with tracks or pads for movement and features hydraulic rams so that the tail pulley can be leveled and adjusted for belt tracking.

Monorail
The monorail system provides a flexible services link between the operating longwall face and the pump station and transformer equipment. Services, including electricity, hydraulic fluid, water, compressed air, communications, and control, are carried along the monorail system as either hoses or cables. The purpose of the monorail is to allow the longwall to independently retreat typically 200 m (two pillars) before the need to relocate the services equipment further outby.

The monorail equipment comprises long-length runs of hosing and cables mounted to trolleys supported by roof-mounted monorail beams located adjacent the gate conveyor in the belt heading.

Pump Station
The pump station provides the high-pressure hydraulic fluid required to operate the longwall face equipment, including the hydraulic emulsion supply for the roof supports and water for shearer cooling, dust suppression, and other auxiliary systems as required. The pump station consists of sleds containing typically up to three pressure pumps, including one high-pressure set pump (if used), a high-pressure water pump for the shearer, a reservoir for the closed emulsion fluid system, and the pump control units. The pump station is typically located in the mine roadways (on a steel-framed sled) or in cut-throughs (trailer mounted to assist movement around the mine). Hydraulic fluid can be either fed from a surface-mounted emulsion-mixing farm (via poly pipe to the pump station) or from mixing tanks located on the sled.

Electrical System
A track-mounted transformer located outby (with the pump station) reduces the panel's incoming power to the required face voltage prior to supplying the distribution and control board (DCB) equipment, which is normally located on the BSL. The DCB provides isolation, control, and monitoring capabilities to the longwall face electrical equipment. The

Courtesy of Bucyrus.
Figure 12.3-8 AFC gearbox

face voltage is typically specific to the country (e.g., 3.3 kV in Australia and 4.4 kV in the United States).

The control and monitoring equipment, which is the electrical control center for the longwall system, is typically located on the maingate corner and provides an interface for operators with the longwall equipment.

Longwall Equipment Variations
Alternate equipment configurations are available for alternate mining methods and/or methods of operation.

Top Coal Caving
The roof supports for top coal caving feature a hinged rear canopy shield that, when hydraulically lowered, will allow coal that has fallen from the roof to flow onto a rear AFC and thus be collected (Figure 12.3-10). Although this feature can successfully allow for the recovery of coal from thick seams, the caving methodology and sequence (lowering of the canopy) requires careful integration into the overall longwall sequence to ensure that the productivity of the entire operation is not compromised.

Top coal caving requires a rear AFC, which is a second AFC located behind the roof supports along the edge of the goaf and protected by the hinged rear canopy. The rear AFC

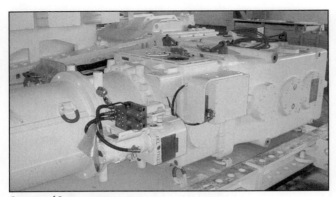

Courtesy of Bucyrus.
Figure 12.3-9 AFC maingate drive frame

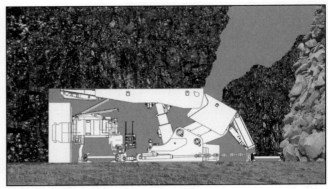

Courtesy of Bucyrus.
Figure 12.3-10 Top coal caving schematic

is connected to the roof support by a hydraulic relay bar and chain. The rear AFC transfers the coal onto the front AFC at the maingate end of the face. Hence, it requires a short transfer loader that crosses the main front walkway. In addition, the rear AFC drive unit must be located in the maingate area. To fit all this equipment into the maingate area requires a special roof-support configuration (Figure 12.3-11).

Longwall Ploughs
Longwall ploughs are used in place of shearers in thin-seam longwall applications. A plough consists of a vertical steel frame housing a series of cutting picks that sits on top of the AFC and is driven at high speed (2.5 to 3.6 m/s) the length of the face by a chain drive system (Figure 12.3-12). The plough cuts a thin slice of coal (150 to 250 mm) in either one or both directions with the coal falling onto the AFC to be cleared. The longwall supports and AFC are advanced after each cut in a staged manner. Because of the low seam height applications, plough faces are generally fully automatic with the equipment being operated and monitored from the maingate.

Longwall Automation
The longwall mining method is highly mechanized, which lends itself to various aspects of automation. This has been progressively developed over time. A longwall face in operation is shown in Figure 12.3-13.

Until approximately 2003, the level of longwall automation was limited and consisted primarily of inner-equipment automation (e.g., roof support sequencing) and intra-equipment automation (e.g., shearer initiated support advance). Recent Australian industry-funded studies have advanced the capability of longwall automation to now include the following capabilities:

- Face alignment control (use of an inertial navigation system)
- Horizon control (improved memory cut)
- Shearer-state-based automation (shearer speed control based on position)
- Automated fly and straightening cuts
- Automated gate road support operation
- Open communication systems (more user-friendly software systems)

Several mines are now testing these capabilities, and they have been reporting improved productivities.

Longwall Equipment Selection
The detailed specification of a longwall system needs to be determined at the time of tender preparation and is based on the results of detailed studies carried out during the feasibility studies. These studies include detailed mine planning, geotechnical characterization, caving studies, and engineering analysis. Key specification aspects include face length, web depth, nameplate capacity (NPC), and support density.

Face Length
Because of improved technology, longwall face lengths have progressively increased from 200 m in the mid-1980s to 440 m today. Long faces in the world include 400 m at Ulan in Australia, 385 m at Cumberland in the United States, and 440 m at Prosper in Germany. The current limiting factor for face length is the AFC (chain strength versus installed power). Hence, there is a trade-off between face length and NPC. Modern chains are being developed so as to allow longer faces. Current technology indicates that a 400-m face length with 4,500-t/h NPC is the current limit.

Web Depth
The depth of cut made by each pass of the shearer can range up to 1,200 mm and commonly ranges from 800 to 1,000 mm. There is a significant productivity benefit from having a deep cut, as this translates to more coal per pass of the shearer. However, the greater the web depth, the greater the required shield canopy length to achieve the required tip-to-face distance. Hence, the cautioning aspect to web depth is the risk of face instability. A detailed geotechnical study is required to establish an appropriate web depth that balances the need for high productivity with the need for face stability. Another approach is to use a half-web (or partial-web) system, whereby only half the web depth is cut during each pass, which can lead to better ground control in particular situations. A half-web system requires faster shearer speeds to maintain productivity. However, this has been very successful (e.g., Twenty Mile mine in the United States). If a half-web operation is envisaged, then the longwall specification will need to include partial push capability of the relay bars (half stroke).

Nameplate Capacity
NPC is the rated capacity of the AFC and is a measure of the coal output during the main cutting run of the shearer. The most important aspect of NPC is that it directly impacts the capacity of the entire underground coal clearance system. The longwall

Courtesy of Bucyrus.
Figure 12.3-11 Top coal caving roof support

Courtesy of Bucyrus.
Figure 12.3-12 Longwall plough

belt conveyor must match the longwall NPC, or the longwall will not be able to operate at designed capacity. This is commonly seen in practice for various reasons. Consequently, the shearer speed is commonly restricted to compensate.

Similar to face length, nameplate capacities have steadily increased in line with increasing technology up to the maximum of 4,500 t/h in Australia and 5,200 t/h in the United States. Although high NPCs can theoretically directly translate to higher productivities, this is not always seen in practice, with many of the highest-producing longwalls in the world rated at 3,500 t/h. This is because NPC is not the only aspect of the determination of productivity.

In general terms, an NPC of 4,500 t/h will require 6,000- to 6,500-t/h trunk belts (2,000- to 2,200-mm belt widths), while an NPC of 3,500 t/h will require 5,000- to 5,500-t/h belts (1,600- to 1,800-mm belt widths).

Support Density

Support density is the rated capacity of the roof shield and is expressed in units of metric tons per square meter of canopy area, usually after the shearer has made its pass (after cut). Again, improving technology has allowed this capacity to increase over time, with capacities up to 130 t/m^2 now available. Typical support densities are between 90 and 110 t/m^2. Variables such as support width (1.75 m versus 2.0 m), shield geometry, and leg cylinder diameter have an impact on support density. The required support density is determined using geotechnical techniques such as ground response curve modeling or Fast Lagrangian Analysis of Continua (FLAC) modeling. Modern supports typically have a support density rating between 95 and 115 t/m^2. Typically, the ratio of setting pressure to yield pressure will be specified at 80%. However, there is an increasing trend toward 90% set, which requires use of a high-pressure set hydraulic supply system.

Longwall Equipment Specifications

There are several longwall equipment suppliers throughout the world, many of which offer complete systems. The typical process of selecting suppliers is to conduct a formal prequalification process involving the following:

- **Expression of interest:** All suppliers are requested to formally express their interest in supplying the equipment.

Courtesy of Bucyrus.
Figure 12.3-13 Longwall face in operation

The expression of interest may be advertised in a suitable media. However, in most countries, equipment suppliers maintain close relationships with organizations, so they are aware of forthcoming tenders.
- **Technical prequalification:** An indicative technical specification is issued to interested suppliers who are requested to prequalify by submitting a response as to whether they can meet the equipment technical requirements, provide examples of where they have supplied before, and provide sufficient equipment description to satisfy the prequalification.
- **Commercial prequalification:** Often attached to the technical specification, the commercial prequalification requests answers to various questions of a commercial aspect relating to security of supply, financial strength of the supplier, warranty provisions offered, and a budget price for the supply of the equipment.
- **Short listing:** Following the prequalification process, a short list of two to four suppliers can be selected. These suppliers would then be subjected to a formal tendering procedure for the supply of the equipment.

Examples of a typical longwall requirements and specifications are included in Tables 12.3-1 and 12.3-2.

Table 12.3-1 Example of a longwall system's general requirements

Parameter	Requirement
Mining method	Conventional retreat longwall
Number of maingate roadways	Two
Panel width	305-m centers
Mining extraction range	Nominal 4.5 m, minimum 3.5 m, maximum 4.8 m
Face cross grades	Flat to 1:10 downhill
Panel lengths	1,500 to 4,000 m (total of 16 panels)
Longwall retreat	~50 km and ~100 Mt over the life of the mine
System cutting process capability	Bidirectional and unidirectional sequence mandatory; half-web optional
Automation capabilities	In accordance with the capabilities expected to be supplied at the time of ordering (approximately 2010)
Extraction web depth	Nominal 1,000 mm
Nameplate capacity	3,500 t/h
Process cycle capacity	3,000 t/h

Courtesy of MineCraft Consulting.

Table 12.3-2 Example of longwall equipment preliminary specifications

Equipment	Requirement
Roof supports	• Two-leg immediate forward support (IFS), 2.0-m nominal width roof support • Operating height range, 2.8 to 4.0 m • Single-piece rigid construction IFS-type canopy with single moveable side shield and automated face sprags • Support density, 100 t/m^2 • Tip to face distance of 400 mm before cut • Set to yield ratio of 80% • Rigid catamaran-type base fitted with base lift and telescopic relay bar • Full electrohydraulic system, including filtration • Closed height, 2,200 mm; maximum extension height, 300 mm higher than maximum extraction height • Run of face roof support typical mass, 35 t; gate end, 37 t • In-service life of 45,000 lower advance set (LAS) cycles without loss of function and 45,000 to 60,000 LAS cycles with up to 25% withdrawn from service for maintenance; maximum life, 80,000 cycles
Shearer	• Double-ended ranging arm/drum multimotor shearer • Ranging arms individually powered by electric motors • Haulage via indirect drive sprocket arrangement and track type to suit AFC pans • Drum diameter and arrangement to be optimized to provide acceptable coal loading and equipment undercut/cut-out capabilities; sprays incorporated • Shearer typical mass, 80 t without drums fitted
Armored face conveyor (AFC)	• Power rating to suit 3,500 t/h over full length of conveyor and volumetric capacity to suit 4,000-t/h short-term peaks • Drive units incorporating soft-start and load-sharing capabilities • Automatic chain-tensioning system • Capable of greater than 150-t reserve chain pull for face overload events • AFC pans of length to suit 2.0-m nominal width roof supports fitted with services trough and bretby* trough arrangement
Beam stage loader (BSL)	• Power rating and volumetric capacity to suit 4,000-t/h short-term peaks • Automatic chain-tensioning system • Fitted with ancillary equipment, including spray system, wet scrubber, and monorail removal platform
Crusher	• Power rating, motor inertia, and volumetric capacity to suit 4,000-t/h short-term peaks • Output material sizing to <300 mm
Boot end	• Skid-type advancing boot end with steering, side shift, and leveling capabilities • Travel to suit two BSL advances • Trough and skirting to suit 1,600-mm belt-width gate conveyor belting
Monorail system	• Nominal 200-m services retraction frequency, plus 40-m reserve compression (240 m total) • Located in belt road heading, with capacity to house all services required to support the longwall system • Materials management system to suit 100-m cut through spacing and 1,600-mm belt-width gate conveyor structure
Pump station	• Track-mounted pump and tank stations; cut-through located • Services to be supplied include roof support hydraulic supply, shearer water, roof support dust suppression, plus miscellaneous auxiliary systems • Output capacity to suit hydraulic power requirements plus a reserve of not less than 20% of maximum demand requirements • 350 bar standard pressure, 420 bar high-pressure set • Equipment transport configuration design to suit typical mine size and constraints
Electrical system	• Track-mounted transformer; cut-through located • BSL/monorail-mounted distribution and control board unit and AFC-mounted control and monitoring equipment unit • Incoming supply at 11-kV, 3-phase, AC 50 Hz; longwall motors typically 3.3 kV

Courtesy of MineCraft Consulting.
*Bretby is a trailing cable attachment, usually made from rigid, hinged plastic, that protects the cable from damage.

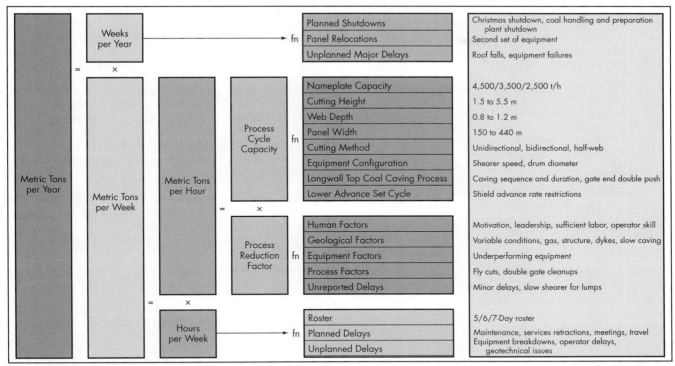

Courtesy of MineCraft Consulting.
Figure 12.3-14 Longwall productivity factors

Longwall Equipment Sizing

Longwall equipment size is generally referred to as its nameplate capacity. Selecting the longwall equipment size relates to the required levels of productivity. Hence, it is important to understand the relationships between the longwall system, its NPC, and its productivity. The relationship between the longwall system and annual productivity is shown in Figure 12.3-14. Two important productivity key performance indicators (KPIs) are the average metric tons per operating hour and the operating hours per week. For an operating mine, these statistics can be collected as part of the production monitoring data. For a new longwall operation, these KPIs need to be derived, which requires a combination of modeling and benchmarking.

To model longwall productivity, the following definitions are used:

- **Nameplate capacity (NPC):** NPC is the rated instantaneous capacity of the longwall (e.g., 2,500 t/h). This is the load that is carried by the AFC during its main cutting run and will occur somewhere in the cycle for between approximately 10 and 20 minutes. In some instances the NPC is determined (restricted) by the coal clearance system.
- **Process cycle capacity (PCC):** PCC is the average capacity achieved during one full cycle at maximum efficiency. This is equal to the metric tons produced in cutting maingate/tailgate and tailgate/maingate divided by the time taken (e.g., 1,800 t/h).
- **Actual productivity:** This is the actual operating rate achieved at the mine over an extended period, as determined from the statistical reporting system (e.g., 1,000 t/h). Depending on the analysis period, the actual productivity will fluctuate from very low levels up to, but not exceeding, the PCC.
- **Productivity reduction factor (PRF):** PRF is the difference between the actual operating rate and the PCC. It is a measure of efficiency and reflects various issues such as adverse mining conditions, operator skills, motivation, and organization. For new mines, this can be derived by benchmarking against similar operations and will generally range between 55% and 75%.
- **Operational availability:** This is a measure of the average number of hours in a shift that the longwall is producing divided by the total hours in a shift (e.g., 56%). It is calculated by dividing the actual operating hours recorded over an extended period by the planned operating hours. The difference between the two measures is the amount of unplanned downtime that occurs, either due to equipment breakdowns or process delays. For new mines, the operational availability can be derived by benchmarking against similar operations and will generally range between 50% and 70%.

The longwall model will calculate the PCC, assuming everything operates to its design capacity.

However, in reality, the longwall does not always operate in this manner. Numerous operational issues, such as operator skill and attention, the presence of geological anomalies, taking fly cuts (straightening cuts), slowing for lumps on the AFC, and gate end cleanups, all slow and impact the cycle. The PRF takes this into account and is a measure of how efficient the longwall cycle is performed. The PRF will fluctuate from very low levels to very high levels but can be measured over an extended period for use as a benchmarking tool.

In addition, numerous unplanned breakdowns (mechanical/electrical) and operational delays occur, which reduce the total amount of operational time available within a shift. This is termed *operational availability* (also called utilization).

PCC is calculated as follows:

- PCC equals the cycle metric tons per cycle time, where cycle metric tons equals metric tons cut from the maingate to the tailgate and the tailgate to the maingate. This is a function of face length, extraction thickness, web depth, and wedge cuts at gate ends.
- Cycle time equals the time taken for the shearer to cut from the maingate to the tailgate and the tailgate to the maingate. This is a function of NPC, shearer speed, shearer acceleration, snake length, cutting mode (bidirectional and unidirectional), and face length.

When calculating the PCC, it is important to conduct some operational reality tests regarding issues such as the following:

- Excessive shearer speed
- Sufficient shearer underframe clearance
- Excessive drum rotational speed
- Adequate ranging arm power
- Adequate AFC power

Typically, the engineer should have some operational experience with the longwall method to perform these calculations.

Longwall Top Coal Caving

Calculating the productivity of the longwall top coal caving (LTCC) method creates an additional level of complexity. However, the same principle can be applied. It is important to accurately model the LTCC process, particularly the caving cycle, as to when this occurs in the process. Typically, the main cut and the cave cannot occur at the same time as the BSL, and the section conveyor cannot handle the dual output. Therefore, the cave either occurs while the shearer is stationary at the gate end or while the snake is occurring. Therefore, the caving cycle is generally the rate-determining step for the process cycle, with several factors to consider, including the capacity of the rear AFC, the number of caving drawpoints, the time required to cave each support, and the method of caving (manual or automatic).

The key benefits of LTCC are resource recovery (more of the coal seam is recovered) and improved development ratio (more longwall metric tons per development meter). However, this can often be at the expense of productivity when compared to a conventional high-capacity longwall, and care is required when designing and modeling the process.

Other Longwall Equipment Considerations

Other engineering considerations when selecting a longwall system include the following:

- **AFC power demand calculations:** These calculations are performed to ensure that sufficient AFC drive power is installed so that one can be able to start a fully loaded AFC with sufficient reserve chain pull.
- **Electrical load flow study:** This study is performed to ensure that adequate electrical power is available to cater for both normal loads and for starting under adverse conditions.
- **Hydraulic flow simulations:** These simulations ensure that the hydraulic system is sufficient to meet the roof support requirements, including routine LAS cycle (up to three supports moving at once), high-pressure set, gate end advance, and to cater for adverse loading events.

ROOM-AND-PILLAR MINING EQUIPMENT

R&P mining can be divided into two categories: first workings (development) and second workings (extraction). Both processes require equipment to cut and gather the coal at the mining face, convey and discharge the cut coal onto a section conveyor, and install ground support at the mining face. Ancillary support equipment is required to provide ventilation and power. There are numerous options of equipment that can be used as shown in Figure 12.3-15.

Standard R&P Mining Equipment

Standard R&P equipment and systems include CMs, roadheaders, haulage units, feeder breakers, mobile bolters, electrical power, ancillary equipment, and mobile roof supports.

Continuous Miners

A CM is a large electrohydraulic machine that extracts the coal to form a rectangular profile roadway or tunnel. It features a rotating cutterhead (drum) laced with rock picks at the front. The cutterhead is driven into the coal face, thus breaking out the coal. The broken coal falls to the ground and is loaded onto a centrally located chain conveyor using a loading apron and gathering arms, spinners, or east–west conveyor. The coal is conveyed through the body of the CM and loaded into coal haulage units (typically shuttle or ram cars), which come to the rear of the CM to be loaded.

Modern CMs are fitted with hydraulic drill rigs that drill and install the primary ground support as the roadway is formed (roof and rib bolts). Platforms are provided for the drill rig operators to stand on while the CM is cutting coal. Commonly, the drill platforms feature hydraulically operated temporary support mechanisms to protect the operators from roof or rib collapse. The CM is fitted with caterpillar tracks to allow it to be propelled forward and backwards (skid steering). CMs are also commonly operated by handheld radio remote-control units, although older units have pendant controls (cable-connected) or operator cabins.

The three main types of CMs are as follows:

1. **Simultaneous cut and bolt:** This type features a full roadway width cutting head (e.g., 5.2 m) that can sump into the coal face while the body of the CM remains stationary, resulting in the ability to use the drill rigs at the same time as cutting coal (Figure 12.3-16).
2. **Sequential cut and bolt:** This type features a full roadway width cutting head. However, the body of the CM moves forward when cutting coal, and thus drilling cannot occur at the same time.
3. **Place change (or cut and flit):** These CMs are not fitted with drill rigs, often have a narrower cutting head (e.g., 3.5 m), and are used for the place-changing method of mining. Other machines are used to install the ground support in these applications (Figure 12.3-17).

Roadheader

A roadheader can be used as an alternative to the CM. A roadheader is a large electrohydraulic machine that extracts the coal to form an arched profile roadway, often extracting a portion of the stone roof. Roadheaders are fitted with a pineapple-shaped

Soft-Rock Equipment Selection and Sizing

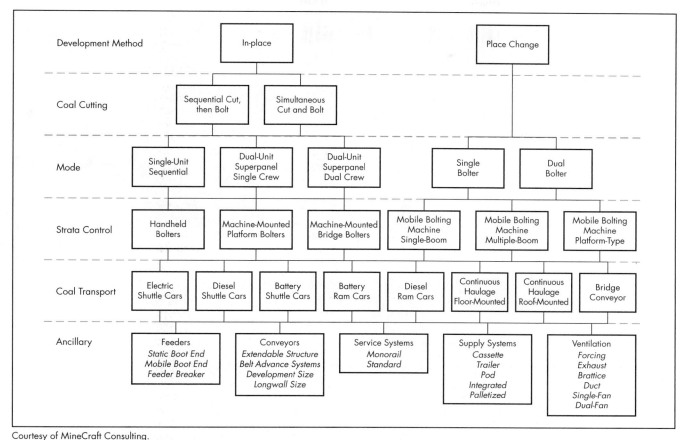

Courtesy of MineCraft Consulting.
Figure 12.3-15 Development process options flow chart

Courtesy of Sandvik.
Figure 12.3-16 Simultaneous cut-and-bolt continuous miner

Courtesy of Sandvik.
Figure 12.3-17 Place-change continuous miner

cutting head attached to a slewing boom. Similar to the CM, the cut material falls on a loading apron equipped with gathering arms or spinners that directs the material to a central chain conveyor that discharges from the rear of the roadheader.

Coal Haulage

Coal haulage units are required to convey the cut coal from the CM to the gate conveyor loading point. Key requirements include a large payload and rapid and flexible maneuverability. Three types of units are available, namely, electric-powered shuttle cars (SCs) with a tethered trailing cable, battery-powered ram cars, and diesel-powered ram cars.

Cable electric SCs (Figure 12.3-18) are commonly used for longwall roadway development and offer the benefits of being fumeless, do not require refueling or battery replacement, do not generate large amounts of heat, and do not require turning around (shunting) to discharge. Their disadvantages include attention to cable handling (cable damage) while in operation and a limited travel distance of approximately 200 m (the length of the trailing cable). Ram cars offer

Courtesy of Sandvik.
Figure 12.3-18 Electric shuttle car

Courtesy of Bucyrus.
Figure 12.3-20 Ram car being loaded at the face

Feeder Breaker

Feeder breaker units are typically required for large-capacity coal-haulage units. These consist of coal hoppers that accept the rapid discharge from the haulage units and then size and discharge the coal at a steady rate onto the section conveyor belt. Feeder breakers are fitted with crawler tracks so that they are self-advancing during panel extensions. However, they are typically left stationary during the mining process.

An advantage of feeder breakers, when using large haul units, is their ability to protect the gate conveyor boot end from knocks/collisions, which otherwise will misalign the conveyor. For a high-capacity development system, a basic requirement is the ability to quickly drive the haulage unit up to the discharge point and rapidly discharge the load. Feeder breakers can provide this requirement, whereas boot ends require a slower approach to avoid damage.

Some suppliers offer mobile boot ends that are a feeder breaker with a tail pulley and sufficient power to extend the conveyor belt (pull belt out of the loop take-up) when they advance. This can assist the belt-extension process. However, care is required to ensure the mobile boot end can withstand the belt tension as the panel extends. Staking props are often fitted to the mobile boot ends to anchor against the roof so that the belt tension does not pull them back.

Courtesy of Sandvik.
Figure 12.3-19 Battery ram car

unlimited travel distance. However, they require regular refueling and require shunting to discharge. If battery ram cars are used (Figure 12.3-19), then a battery charge station is required, which is commonly placed some distance from the working face because of their construction and ventilation requirements. A ram car in operation is shown in Figure 12.3-20.

Several SCs can be used behind each CM. However, most Australian mines use a single car in longwall gate roads because of the short shuttle distance, the congestion caused by two SCs, and the need for an additional person in the crew. If ram cars are used, then at least two units are required to provide service while one is being refueled. For pillar extraction methods, several SCs need to be used (between two and four) to rapidly clear the coal when a pillar split has commenced.

As an alternative to coal haulage units, various techniques of continuous haulage have been developed over time, including flexible conveyors, bridge conveyors, and ADDCAR systems. Careful panel design is required for continuous haulage to gain the benefits of this technology. Consideration is required for roadway width and height, cutting sequence, clearances, face ventilation, cut-through angles, access to the face by vehicle, and conveyor extension process. Continuous haulage has been successfully used in several non-longwall coal mines and potash mines throughout the world.

Mobile Bolters

When the place-changing (cut-and-flit) method is used, roof support is installed by an independent bolting unit that trams into the heading when the CM is flitted out. Modern mobile bolting units are electrohydraulic powered; are fitted with up to four roof bolting rigs, two rib bolting rigs, temporary roof, and rib protection canopies; and have provision for the storage of consumables.

Electrical Power

Electrical power to a development panel is reticulated at the country-specific mine supply voltage (e.g., 11 kV in Australia) to a section transformer via an isolator switch located at the start of the development panel. The transformer, located within 600 m of the working face, steps the voltage down to typically between 950 V (United States) and 1,000 V (Australia) and is then fed to a distribution control box located within 200 m of the face. The distribution control box features up to seven outlets to which trailing/flexible cables feed the various devices at the face (CM, SCs, fan, feeder, pump, etc.). Cable sizes reflect the various power demands of the equipment, with the CM cable typically 120 mm^2 and other cables typically 35 mm^2. The section feed cables are typically 150 mm^2.

Ancillary Equipment

Auxiliary fans are required to provide immediate face ventilation. Typically, these are centrifugal exhaust fans mounted on relocatable sleds or trailers placed in close proximity to the operating face in conjunction with either lightweight fiberglass or steel ducting (600 to 750 mm in diameter). These fans effectively place the fan's negative pressure suction inby of the CM operators, minimizing dust and methane buildup at the face. The fans commonly feature variable inlet vanes (to control airflow) and methane degassing valves (to prevent high gas levels from passing over the fan blades). The number of fans required for each working face depends on the number of roadways being developed, the method of working, and the level of activity. Commonly, between one and three fans are required.

An alternative method of face ventilation is by forcing ventilation in conjunction with dust scrubbers mounted on the CM. In this case, relocatable axial or centrifugal fans are located in close proximity to the face, with air directed using flexible or rigid ducting or brattice sheeting.

Panel dewatering is typically through pneumatic diaphragm pumps, as these pumps are able to operate in this arduous environment with minimal attention. Because of their relatively low output flow rate and their pressure-head capability, they normally report to a nearby electric pump pod. The electric pump pod units are capable of the greater pumping duties required to reach the central mine dewatering station before delivery to the surface.

Because of the high solids content associated with mine dewatering, system design and pump selection requires sound detail to avoid the settling of solids in the reticulation system (causing pipe blockages).

Mobile Roof Supports

For some of the secondary extraction methods (e.g., Wongawilli), mobile roof supports that are also called breaker line supports are used to provide a level of protection along the goaf (gob) edge. These are electric-powered, four-legged hydraulic roof supports mounted on caterpillar tracks with pendant controls. Typically, three units are used in a section. Breaker line supports are often used during longwall takeoffs to provide roof support as the roof supports are removed.

R&P Development Equipment Selection

The detailed specification of the R&P equipment needs to be determined at the time of tender preparation and is based on the results of detailed studies carried out during the feasibility studies. These studies include mining method analysis, panel configuration design, development process studies, productivity requirement studies, ground support specification, and engineering analysis.

Key specification aspects include mining strategy (development for longwall, first workings R&P, secondary extraction pillar splitting, etc.), face mining method (place change, in-place sequential, in-place simultaneous), type of CM, coal clearance method (continuous haulage or cars), bolting method (mobile bolters, CM-mounted rigs), and panel configuration.

Development Equipment Specification

The key aspects in selecting the most appropriate development arrangement and equipment specification include the following:

- Mining method: development or extraction
- Panel design: number of headings, pillar width, angle of cut-throughs, and roadway width
- Cutting method: in-place or place-change
- Cutting machine: roadheader or CM
- Type of CM: sequential, simultaneous, or place-change
- System of coal haulage: SC, ram car, or continuous haulage
- Number of SCs: one or two
- Number of bolting units: one, two, or three
- Method of face ventilation: method used

Table 12.3-3 summarizes a typical development arrangement for a longwall mine, and Table 12.3-4 summarizes a typical general specification for the development equipment for a first workings development system.

R&P Development Equipment Sizing

Development productivity is fundamentally measured in meters advanced per week or in metric tons per week (for which the conversion is relatively simple). Therefore, development equipment is typically sized to match a desired advance rate (e.g., 8 m/h) or output (e.g., 250 t/h). Unlike a longwall, where the equipment is completely integrated into a highly mechanized system, development is a collection of equipment that requires the mining engineer to integrate it into a process. The key process requirements are cutting, coal clearance, and ground support, and these are conducted by separate and independent items of equipment.

Development productivity can be calculated in a similar manner as longwall productivity, as shown in Figure 12.3-21. However, for development calculating the process advance rate is more complex, as it needs to take into account that the CM is continually moving from a fixed conveyor discharge point.

Two example development productivity cases are given in the following discussion. The first example, shown in Table 12.3-5, is for a conventional two-heading gate road entry for a longwall panel using a simultaneous miner bolter with a single shuttle car (100-m cut-through spacing). The second example, shown in Table 12.3-6, is for a three-heading development using the place-change method (60-m cut-through spacing). This example compares three mines with slightly different plunge depths.

Definition of the terms used in development productivity modeling include the following:

- **Nameplate capacity (NPC):** NPC is the peak cutting and loading rate of the CM (e.g., 950 t/h). It is often quoted by the equipment supplier and can be used to determine the time taken to load a shuttle car or to load onto a continuous haulage system.
- **Nameplate advance rate (NAR):** NAR is a measure of the peak advance rate of the CM. It is the time taken to cut and load one car, drive the car to the boot end, discharge its load, and then return, divided by the cut-out distance of the CM (e.g., 0.5 m). NAR is a useful factor when sizing the section conveyor belt.

Table 12.3-3 Development system—general arrangement

Parameter	Requirement
Roadway cross section	Nominal 5.2 m wide by 3.4 m high; rectangular profile
Working grade	Up to 1 in 10 (maximum)
Quantity of development panels	One main (up to seven headings) and two gate roads (two headings)

Courtesy of MineCraft Consulting.

Table 12.3-4 Development equipment—general requirements

Equipment	Requirement
Continuous miner	• Single-pass wide-head track mounted; one per panel; either simultaneous or sequential cut-and-bolt configuration • Fully remote operation with manual override • Four roof and two rib automatic-type bolters; to suit standard or long tendon bolts • Data communications, including real-time bidirectional between equipment and mine communication system • Onboard ventilation system • Dust-suppression system fitted; preferably with wet head design • Hydraulic system with capacity to suit simultaneous operation of all bolting equipment • Onboard roof support materials storage-and-handling equipment • Methane-detection monitors fitted • Fitted with integrated temporary roof-and-rib support systems
Coal haulage	• Electric shuttle car with trailing cable; one per panel • Nominal 15-t capacity; with feed-out rate of 1,400 t/h • Capable of travel speeds up to 8 km/h
Feeder breaker	• Track-mounted mobile feeder breaker; one per panel • Controlled via remote umbilical cable system • Nominal 22-t capacity with a feed-out rate 300 to 1,000 t/h (coal sizing to <200 mm)
Ancillary equipment	• Auxiliary fan; typical output 23 m³/s • Ventilation ducting; nominal 720-mm diameter; range 610 to 760 mm • Pumping: pneumatic diaphragm face pumps and electric pump panel pods

Courtesy of MineCraft Consulting.

- **Process advance rate (PAR):** PAR is a measure of the maximum advance rate during the cycle, assuming that all activities are completed at their peak rate with no delays. The importance of this factor is that it can be determined by modeling and using actual measurements observed at the mine (or from benchmarking). Therefore, it is a scientific means of calculating productivity. The PAR will be less than the NAR, as it takes into account bolting, extending the ventilation ducting, and any other process-related activity such as repositioning the miner and lowering the canopy. The PAR will only be evidenced in practice over short durations at the mine (e.g., 2 to 6 hours), as it is rare that the process flows at their peak rate over an extended time in the development process.
- **Actual advance rate (AAR):** AAR is the advance rate achieved at the mine over an extended period (e.g., 3.6 m/h). AAR is measured from the mine statistical reporting system for an operating mine. For a new mine, this would be calculated from the modeled process advance rate (AAR = PAR × PRF).
- **Process reduction factor (PRF):** PRF is the difference between the AAR and the PAR. It is a measure of efficiency and reflects various issues such as adverse mining conditions, operator skills, motivation, and organization, and the complexity of the mining process. For new mines, this can be derived by benchmarking against similar operations and will generally range between 45% and 70%.
- **Operational availability:** This is a measure of the average number of hours per week that the development section is operating divided by the total planned operating hours in a week (e.g., 56%). It is calculated by dividing the actual operating hours recorded over an extended period by the planned operating hours. The difference between the two measures is the amount of unplanned downtime that occurs, either due to equipment breakdowns or process delays. The time to extend the panel is not included in planned operating hours. For new mines, the operational availability can be derived by benchmarking against similar operations and is generally between 50% and 70%.

COAL CLEARANCE EQUIPMENT

The coal clearance system provides the means for transporting run-of-mine (ROM) coal from all mining units operating in the underground mine to the surface ROM stockpile. The coal clearance system consists of various types and sizes of belt conveyors, which are designed specifically for the various duty requirements associated with each mining unit. The loading pattern from each production face varies depending on the mining method. For example, longwall mining will produce a fairly constant flow for 15 to 20 minutes while the main cut is taken, punctuated by 10 to 20 minutes of irregular flows as the snake is taken. First workings development will produce very short duration high flows of 1-to-2-minute duration as each SC is discharged with a frequency between cars of 10 to 20 minutes. Occasionally, underground surge systems are used (bins, bunkers, etc.) to smooth the flow and allow for lower-capacity outby conveyors to be used.

The required duty of a coal clearance system depends on the mining system. For example, if the mine employs a 3,500-t/h NPC longwall system and up to three development units operating simultaneously (i.e., one unit completing mains development and two units completing gate road development), then the coal clearance system volumetric capacity duty would be as follows (simple method):

- Longwall gate conveyor: 3,500 t/h (matched to NPC of longwall)
- Development gate conveyors: 1,000 t/h (loading rate reduced to approximately 500 to 600 t/h by use of feeder breaker)
- Main headings trunk conveyor: 5,000 t/h (to allow for longwall plus three development units)

Detailed conveyor calculations are required to refine these duty requirements, which are undertaken during the mine feasibility study. These calculations take into account the panel gradients and panel lengths over the mine life and provide key specification data such as belt rating, power requirements, and the need for tripper drives (mid-panel booster drives). Specialist conveyor duty software is available to perform

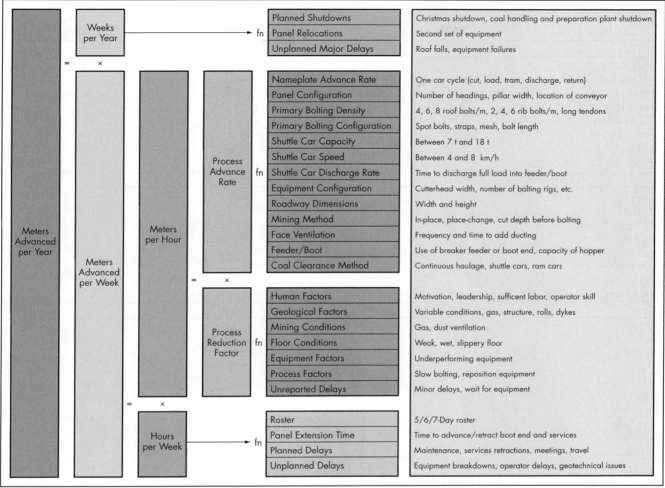

Courtesy of MineCraft Consulting.
Figure 12.3-21 Development productivity factors

these calculations. However, some practical experience is required to audit the outputs and to ensure that realistic specifications are produced.

Where conveyors are required to run downhill, braking units are required. Because gate road conveyors are extended as the panel is developed and then retracted as the panel is extracted, care is required in the calculations to ensure that the maximum duty case is calculated. This is especially prevalent where undulating seams are encountered (both up- and downhill sections).

Because the development conveyor duty is commonly significantly lower than the longwall conveyor duty, the following options are available when developing the gate roads:

- Install a small development conveyor (e.g., 1,050 mm) for development and then fully replace this conveyor prior to longwall extraction.
- Install the longwall conveyor drive head and structure (e.g., 1,600 mm), but with a narrow belt (1,200 mm) and minimal drive units, and then replace the belting prior to longwall extraction and fit additional drive units.
- Install the longwall conveyor drive head, structure, and belting (e.g., 1,600 mvm), but twith minimal drive units, and then fit additional drive units prior to longwall extraction.

Preliminary Specification

Table 12.3-7 is an example of the general requirements for the coal clearance equipment for a new longwall mine. Specific details would also be required in regard to requirements for related aspects, including the following:

- Dust control and belt scrapers
- Transfer chute design
- Detection systems for belt wander, belt alignment, belt tear, chute blockage, belt slip, and belt tension
- Belt monitoring, emergency stop, communication systems, gas/fire monitoring (e.g. carbon monoxide detectors)

ANCILLARY EQUIPMENT

Ancillary equipment used in underground coal mines must comply with the local statutory regulations, which are intended

Table 12.3-5 Development productivity model for in-place mining

Process (two-heading gate road)	Australia Mine Calibration 90 m Up Travel Road	Australia Mine Predicted for Belt Road	Australia Mine Predicted Average for Pillar
Nameplate			
Shuttle car speed, km/h	7.1	7.1	7.1
Time to load car, min	3.40	3.40	3.40
Car cycle time, min	4.60	3.53	3.94
Total car cycle, min	8.0	6.9	7.3
Total cars per hour	7.5	8.7	8.2
Metric tons per car	15.50	15.50	15.50
Metric tons per hour	116.2	134.2	126.7
Nameplate advance rate, m/h	3.8	4.3	4.1
Process			
Roof bolting cycle time, min	2.80	2.80	2.80
Rib bolting time, min	3.60	3.60	3.80
Lower canopy time, min	0.20	0.17	0.17
Advance time, min	0.60	0.60	0.60
Load mesh time, min	0.20	0.20	0.60
Raise canopy time, min	0.20	0.20	0.20
Prep tube time, min	0.00	0.00	0.00
Install tube time, min	12.50	12.50	12.50
Process Advance Rate			
Bolting Cycle 1			
Install roof bolts, min	2.80	2.80	2.80
Install rib bolts, min	3.60	3.60	3.80
Lower canopy, min	0.20	0.17	0.17
Load mesh mat, min	0.20	0.20	0.20
Advance 1 m, min	0.60	0.60	0.80
Raise canopy, min	0.20	0.20	0.20
Cutting Cycle 1			
Cut and load car 1, min	3.40	3.40	3.40
SC1 cycle time, min	4.60	3.53	3.94
Cut and load car 2, min	3.40	3.40	3.40
SC2 cycle time, min	4.60	3.53	3.94
Test if still bolting	Wait for SC	Wait for SC	Wait for SC
Cycle 1 duration, min	16.00	13.86	14.68
Bolting Cycle 6			
Install roof bolts, min	2.80	2.80	2.80
Install rib bolts, min	3.60	3.60	3.80
Lower canopy, min	0.20	0.17	0.17
Load mesh mat, min	0.20	0.20	0.20
Advance 1 m, min	0.60	0.60	0.80
Raise canopy, min	0.20	0.20	0.20
Install vent tubes, min	12.50	12.50	12.50
Cutting Cycle 6			
Cut and load car 1, min	3.40	3.40	3.40
SC1 cycle time, min	4.80	3.53	3.94
Cut and load car 2, min	3.40	3.40	3.40
SC2 cycle time, min	4.60	3.53	3.94
Test if still bolting	Wait for bolting	Wait for bolting	Wait for bolting
Cycle 6 duration, min	20.10	20.07	20.07
Total duration of cycles 1 to 6, min	100.10	89.36	93.46
Total cars	12	12	12
Total metric tons	186	186	186
Total advance, m	6.0	6.0	6.0
Process advance rate, m/h	3.60	4.03	3.85
Process reduction factor, %	67	67	67
Actual advance rate, m/h	2.4	2.7	2.57

(continues)

Table 12.3-5 Development productivity model for in-place mining (continued)

Process (two-heading gate road)	Australia Mine Calibration 90 m Up Travel Road	Australia Mine Predicted for Belt Road	Australia Mine Predicted Average for Pillar
Pillar Cycle			
Meters per pillar	237.2	237.2	237.2
Duration, h	98.9	88.3	92.3
Operational availability, %	72	72	72
Actual duration, h	137.9	123.1	128.7
Panel extension duration, h	34.0	34.0	34.0
Total pillar duration, h	171.9	157.1	162.7
Roster			
Rostered days per week	7	7	7
Scheduled production time per day, h	17.6	17.6	17.6
Number of days per pillar	9.8	8.9	9.2
Number of pillars per week	0.7	0.8	0.8
Advance per week, m	170.2	186.2	179.8

Courtesy of MineCraft Consulting.

to ensure that these vehicles are safe to operate in potentially hazardous environments. The following comments relate to Australia, which has strict vehicle requirements.

Mobile Equipment

Mobile equipment refers to the diesel vehicles that are typically used in underground coal mines. They are categorized as personnel carriers, material transporters and utility vehicles, or special-purpose vehicles. Typically, these vehicles were developed over the years into purpose-designed vehicles to suit the rugged environment of the underground coal mine. Statutory regulations require that these machines comply with diesel engine exhaust-emission limits, be flameproof, have wet braking systems, and be fitted with various shutdown monitors (high temperature, low scrubber water, etc.). Recently, nonflameproof personnel carriers (modified Toyota Land Cruisers) have been successfully introduced into some Queensland mines. However, these suffer some restrictions in their ability to travel into working faces, depending on methane levels and statutory zones.

Personnel Carriers

Transport of personnel from the surface to the work site and back is through the use of personnel carriers having typical carrying capacities of up to 14 people, including the driver. Several models are available from suppliers, with some companies requiring additional safety features such as forward facing seats and seatbelts. Some personnel carriers are converted into maintenance utility vehicles with a rear flat tray for carrying tools and spares.

Material Transporters and Utility Vehicles

Material transporters and utility vehicles include a range of vehicles for duties such as equipment and materials transport and other work duties around the mine. Typical configurations include the following:

- **Load-haul-dump vehicles (LHDs):** LHDs are used with attachments including fork tines, loading plates, buckets, jib cranes, elevated work platforms, augers, hydraulic drilling rigs, stone dusters, pipe installation platforms, cable reelers, and trailers. LHDs are typically the workhorse of the modern mechanized mine and are commonly fitted with a quick-detach system. Typically, there are one or two of these vehicles, with a capacity between 7 and 10 t, at every panel. The mine will also have some higher-capacity LHDs, with capacities of between 10 and 15 t. These are used for roadwork and other heavy-duty requirements. A utility loader with a 10-t capacity is shown in Figure 12.3-22.
- **Multipurpose vehicles:** Multipurpose vehicles are specifically designed to pick up and carry modular pods and tubs.
- **Bobcats:** Occasionally, mines will use small LHDs such as bobcats for light cleanup duties.

Special-Purpose Vehicles

These are purpose-built vehicles that perform only one or a few specific duties on a periodic basis, with typical configurations including the following:

- Heavy-duty LHDs (40 to 50 t): Used for transport of longwall equipment and some belt conveyor equipment (Figure 12.3-23)
- Roadway graders and rollers: Used for road construction and maintenance
- Roof support trailers: These are used in conjunction with an LHD to transport longwall roof supports
- Specific shield haulers: Used for larger roof supports
- Shearer transporter: Used to transport the longwall shearer
- Mine dozer: Diesel powered, heavy-lift machine for moving longwall equipment
- Mule: Electric-powered, heavy-lift machine for moving longwall equipment
- Mobile generator: Used for CM flits
- Mobile bolters: Used for outby roof support
- Drilling rigs: Used for gas drainage and/or exploration

Some of this equipment is used infrequently, and there are often a number of leasing companies available to service the mine requirements. The majority of mines in Australia rent this equipment during longwall moves, although if the mine purchases equipment that is not typical (e.g., roof supports weighing more than 35 t), then the mine may be required to purchase its own longwall move equipment. The general

Table 12.3-6 Development productivity model for place changing

Process (three-heading gate road)	Australia Mine Example Four Bolts + Strap at 1.0-m Spacing, 10-m Plunge Depth	U.S. Mine Predicted Four Bolts + Mesh at 1.0-m Spacing	
		10-m Plunge Depth	15-m Plunge Depth
Nameplate			
Time to load car, min	1.0	1.0	1.0
Car cycle time, min	1.5	1.5	1.5
Total car cycle, min	2.5	2.5	2.5
Total cars per hour	24.0	24.0	24.0
Metric tons per car	15	12	12
Metric tons per hour	360.0	288.0	288.0
Nameplate advance rate, m/h	13.5	13.6	13.6
Mining Cycle			
Number of cars advanced	18	17	26
Metric tons mined	270	204	312
Plunge depth, m	10.1	9.6	14.7
Duration of mining cycle, min	45.00	42.50	65.00
Miner flit time, min	28.00	20.00	20.00
Bolting Cycle			
Number of rows bolted	17.0	10.0	15.0
Bolting time per row, min	3.50	3.50	3.50
Bolting duration, min	59.50	35.00	52.50
Vent tube duration, min	0.00	0.00	0.00
Stone dust duration, min	0.00	0.00	0.00
Duration of bolting cycle, min	59.50	35.00	52.50
Bolter flit time, min	15.00	20.00	20.00
Longest cycle, min	74.50	62.50	85.00
Total cars	18	17	26
Total metric tons	270	204	312
Total advance, m	10.1	9.6	14.7
Process advance rate, m/h	8.1	9.2	10.4
Process reduction factor, %	44.3	70	70
Actual advance rate, m/h	3.60	6.5	7.3
Pillar Cycle			
Meters per pillar	324.5	501.0	501.0
Duration, h	90.1	77.6	69.0
Operational availability, %	28.5	70	70
Actual duration, h	316.3	110.9	98.6
Panel extension duration, h	46.0	16.0	16.0
Total pillar duration, h	362.3	126.9	114.6
Roster			
Rostered days per week	7	7	7
Scheduled production time, h/d	2.7	18.0	18.0
Number of days per pillar	16.0	7.0	6.4
Number of pillars per week	0.4	1.0	1.1
Advance per week, m	142.3	497.6	550.9

Courtesy of MineCraft Consulting.

requirements for the mobile equipment are summarized in Table 12.3-8.

ELECTRICAL EQUIPMENT AND DISTRIBUTION

Electrical power is typically reticulated in and about an underground mine at the mine supply voltage, which will be derived from the surface main mine substation. Locally positioned substations are then used to transform the electrical power to a voltage suitable for utilization by the various electrically powered equipment. This includes the equipment used in the following areas:

- Development panels
- Longwall panel
- Conveyor systems

Table 12.3-7 Coal clearance equipment—general requirements

Equipment	Requirement
Construction conveyors	Conveyors are installed during the construction phase of the mine, catering for the lower outputs from the development units, and installed in a suitable location to allow construction of the permanent conveyors. The construction conveyors include a trunk and surface radial stacker conveyor. General specifications of the construction trunk conveyor include the following: • Conveyor length to suit mains development progress to a final length of approximately 1,250 m • A volumetric capacity of 1,800 t/h and a power capacity of 600 t/h, 1,200-mm belt width operating at 3.0 m/s, and solid woven 8,000 belting • Head end drive unit (1 × 500 kW) with transformer/starter unit • Electric auto winch take-up system General specifications of the construction radial stacker conveyor include the following: • Same volumetric capacity and belting width as the construction trunk • Nominal 11-m vertical height discharge rotating through 45°, providing ~5,000-t stockpile capacity On installation of the permanent conveyors, the temporary conveyors are decommissioned and removed.
Ramp conveyor	Conveying ROM coal to surface ROM stockpile. General specifications include the following: • Conveyor length of 710 m with a lift of 108 m from tail to head • Volumetric capacity of 5,900 t/h and a power capacity of 5,550 t/h • 1,800-mm belt width operating at 4.5 m/s, EP2500/5-ply belting • Head end drive station (3 × 750 kW) with transformer/starter unit • Nominal 35-m vertical height gantry structure for ROM stockpile discharge to provide ~100,000-t conical stockpile capacity • Gravity take-up system The complete conveyor is installed during the project phase construction.
Trunk T1A conveyor	Used for conveying ROM coal to the ramp conveyor tail end. General specifications include the following: • Conveyor length of 1,575 m with a lift of 151 m from tail to head • A volumetric capacity of 5,900 t/h and a power capacity of 5,000 t/h • 1,800-mm belt width operating at 4.5 m/s, EP2000/4-ply belting • Head end (2 × 750 kW), tripper 1 (2 × 750 kW), and tripper 2 (1 × 750 kW) drive stations with transformer/starter units as required • Electric auto-winch take-up system The complete length conveyor is installed during the project-phase construction, including both tripper stations. The tripper 2-drive unit and associated transformer/starter is not installed during the project phase because the additional power is not required until panel 2 longwall coal is being produced.
Trunk T1B conveyor	Used for conveying ROM coal to the Trunk T1A conveyor tail end. General specifications include the following: • Conveyor length of 2,000 m with a lift of 136 m from tail to head • A volumetric capacity of 5,900 t/h and a power capacity 4,600 t/h • 1,800-mm belt width operating at 4.5 m/s, EP2000/4-ply belting • Head end (2 × 750 kW) and tripper (2 × 750 kW) drive stations with transformer/starter units as required • Electric auto-winch take-up system A nominal 300-m conveyor length is installed during the project-phase construction, in line with expected mains progress at that time. Only one of the head end 750-kW-drive units is installed due to the low loading requirements, accepting coal from mains and gate road MG103 development units. The second head end 750-kW-drive unit and tripper drive stations, plus remaining structure and belting, is installed in line with mains development progress and conveyor duty requirements.
Trunk T2 conveyor	Used for conveying ROM coal to the trunk T1B tail end. General specifications include the following: • Conveyor length of 1,360 m with a lift of –38 m from tail to head • A volumetric capacity of 5,900 t/h and a power capacity of 5,050 t/h • 1,800-mm belt width operating at 4.5 m/s, EP2000/4-ply belting • A head end (1 × 500 kW) drive station with transformer/starter unit as required • A conveyor braking station located at the conveyor head end is required to ensure suitable conveyor stopping times • Electric auto-winch take-up system This conveyor is installed during the project phase construction.
Gate conveyor—longwall duty	Used for conveying ROM coal from the longwall system to the trunk conveyor system. General specifications include the following: • Conveyor lengths ranging from 1,620 to 6,200 m with lifts ranging from 77 to 103 m from tail to head • A volumetric capacity of 4,200 t/h and a power capacity of 4,000 t/h • 1,600-mm belt width operating at 4.0 m/s and solid woven 10,000 belting • Power requirements up to tripper 1 (2 × 500 kW), tripper 2 (2 × 500 kW), and head end (3 × 500 kW) drive stations, with transformer/starter units as required • A conveyor braking station located toward the tail end of the conveyor is required to ensure suitable conveyor stopping times • Electric auto-winch take-up system This arrangement of power and braking is for the worst-case longwall gate conveyor. Other conveyors with shorter length and less lift may require less drive units and no braking requirements.
Gate conveyor—development duty	Used for conveying ROM coal from the development system to the trunk conveyor system. General specifications include the following: • Conveyor lengths in accordance with "Gate Conveyor–Longwall Duty" section of this table • A volumetric capacity of 550 t/h and a power capacity of 200 t/h • 1,600-mm belt width operating at 2.0 m/s and solid woven 10,000 belting • The longwall duty head end and take-up equipment is used for development duties; however, belt speed is reduced to 2.0 m/s with a single 500-kW-drive unit and a transformer/starter being employed • A conveyor braking station located on the return belt is required to ensure suitable conveyor stopping times

Courtesy of MineCraft Consulting.

Courtesy of Sandvik.
Figure 12.3-22 Utility loader with a 10-t capacity

Courtesy of Bucyrus.
Figure 12.3-23 Heavy lift LHD

- Dewatering pump stations
- Area lighting

Table 12.3-9 provides an example of the various power supplies utilized in the underground electrical power distribution system for an Australian mine. For other countries, the voltage and frequency will differ (e.g., the United States uses 60 Hz and 950 V for the development face and 4.2-kV for the longwall face). The applicable standard may revert to international standards as opposed to Australian standards.

All underground power circuits operate using an impedance-earthed system (IT earthing system code in accordance with AS 3007.2-2004). The purpose of this is to limit the magnitude of an earth fault and, hence, the resultant touch-potential, thus reducing the risk of electric shock from indirect contact. Also, because most electrical faults occur as a result of cable damage, cables used underground are generally constructed with an earthed screening around each phase conductor, so that the initial type of fault is a low-energy earth fault rather than a high-energy short circuit. Controlling the energy dissipated during an electrical fault in this way reduces the risk of injury to personnel, fire, ignition of methane or dust, or damage to equipment as a result of an arc flash.

The performance of the underground electrical power-distribution system has a significant impact on the performance

Table 12.3-8 Mobile equipment—general requirements

Equipment	Requirement (Nominal Quantities)
Personnel carriers	• Three production vehicles (14 seat) for longwall and development crews • Three support vehicles (14 seat) for maintenance and outby support crews • Two spare vehicles (14 seat) that are used when the others are in maintenance • Two utility vehicles (6 seat) for maintenance crews
Material transporters	• Three-panel supply LHDs (7 to 10 t) for longwall and development • Three outby LHDs (7 to 10 t) • Two spare LHDs (7 to 10 t) • Two roadwork and outby LHDs (15 t)
Special-purpose vehicles	• One grader—roadwork • One mobile generator

Courtesy of MineCraft Consulting.

Table 12.3-9 Underground electrical power distribution*

Purpose	Voltage	Number of Phases and Frequency
Underground reticulation	11,000 V	3 Phase, 50 Hz
Development equipment	1,000 V	3 Phase, 50 Hz
Longwall equipment	3,300 V	3 Phase, 50 Hz
Underground conveyors	Drive type and size dependent	3 Phase, 50 Hz
Dewatering pumps	1,000 V	3 Phase, 50 Hz
Underground lighting	240 V	1 Phase, 50 Hz

Courtesy of MineCraft Consulting.
*IT earthing system code in accordance with AS 3007.2-2004.

and reliability of the electrically powered mining equipment. Consequently, the design for the underground power supply system needs to account for the following key power supply performance parameters:

- Fault level
- Thermal capacity
- Voltage regulation
- Transient motor starting capability (particularly with respect to the large longwall AFC motors)

Other factors that may influence the quality of the power supply and require detailed study include transient stability and harmonic distortion.

All electrical equipment must comply with the relevant parts of AS/NZS 4871:2002. In addition, equipment for use in an area designated as an explosion risk zone must be certified as being explosion-protected by complying with the relevant parts of AS/NZS 60079.10:2004.

Underground Reticulation

The underground reticulation system is typically comprised of the following components:

- Power supply (11 kV) from the surface to underground (via either cables in the mine entry roadway or via cased boreholes)
- Underground switchboard (11 kV)
- Underground reticulation cables (11 kV) (mounted along the roof of the development roadways)
- Section circuit breaker (isolators)

For a new mine, the power supply from the surface to underground typically occur in the following two stages: the initial development phase and the operational phase.

The power supply for the initial underground development is typically via the mine access portals and is achieved by means of an overhead power line installed overland from the surface main mine substation (e.g., 132/11 kV) to a site in close proximity to the portals (this could be to the top of a highwall if the entry is via an open cut or trench). A cable then connects to switchgear located at the portal entry switch room, which will control the power supply entering the mine and will be interlocked with the mine ventilation fans, to trip the underground power in the event of a ventilation failure.

This feeder will initially provide a power supply for the following loads:

- Initial underground power supply
- Development equipment
- Temporary drift conveyor
- Temporary ventilation fans
- Pumps and other ancillaries

In the longer term, this feeder will also provide the permanent power supply for the drift conveyor, pumps, and other loads located in the portal area.

Prior to the commencement of the longwall, the power supply to the underground operations will require upgrading, typically by using borehole feeder cables originating from the surface main mine substation (132/11 kV). Two 300-mm^2, 12.7/22-kV, cross-linked polyethylene (XLPE), single-point suspension borehole cables provide the power supply to an underground switchboard (11 kV) located in the main headings at the bottom of the boreholes. The underground switchboard is configured with two incomers, one bus tie, and four outlets for the control and distribution of the power supply to the following loads:

- Longwall feeder
- Development and conveyor feeder
- Pit bottom feeder
- One spare

The underground-cable selection process is a balance of the following considerations:

- Achieving acceptable performance parameters
- Minimizing the number of different types and sizes used underground
- Practical sizing for the purposes of handling the cables underground

All 11-kV reticulation cables must comply with Australia/New Zealand safety standards (AS/NZS 1972:2006), while cable couplers must comply with AS 1300-1989.

Load-flow modeling of the underground reticulation system is required to accurately specify the required cable types and sizes. A typical specification is listed in Table 12.3-10.

Although paper-insulated lead-covered (PILC) cable has traditionally been used for underground reticulation cabling, the use of XLPE cable is often recommended because it offers superior current-carrying capacity for the same size cable (approximately 50% increase in capacity over the PILC type). This is important to provide adequate current rating, while keeping the cable size as small as possible for cabling handling purposes.

Table 12.3-10 Underground 11-kV reticulation cable sizes

Application	Cable Type	Cable Size, mm^2
Development/conveyor feeder in the main headings	12.7/22 kV, XLPE	150
Development feeder in the gate roads	12.7/22 kV, XLPE	95
Longwall feeder in the mains headings	12.7/22 kV, XLPE	240
Longwall feeder in the gate roads	12.7/22 kV, XLPE	150
Longwall conveyor tripper drives in the gate roads	12.7/22 kV, XLPE	95

Courtesy of MineCraft Consulting.

Section circuit breakers are used to sectionalize the underground reticulation system (11 kV) and to control the power supply entering the various areas of the mine. Section circuit breakers are required at each gate road entry to control the power entering that development or longwall panel. The section circuit breakers incorporate the following features:

- Incoming and through-going supplies
- Two switched outlets
- Circuit breakers and protection relays for each outlet
- Isolation and earth switches for each outlet
- Programmable logic controller (PLC), network, and control equipment
- IP56-rated (minimum) enclosures (IP56 is an International Electrotechnical Commission weatherproof standard)
- Arc fault rated enclosures (using arc fault control methods)
- Base frame, wheels, draw bar and stabilizing/leveling legs

Development Panels

The electrical equipment for the development panels includes the following items: panel substations, panel distribution and control boxes, and panel trailing cables.

The power supply equipment up to the input connection of the development substation, which includes the 11-kV cables and section circuit breakers, forms part of the underground reticulation system.

The panel substations transform the incoming power supply (11 kV) to a 1,000-V (nominal) supply for distribution to the development machines and ancillary equipment. The current growing trend in the industry is for 2-MVA- (megavoltampere) rated substations to accommodate the increased power requirements of the development equipment. Generally, these substations are not flameproof, as they are located far enough outby of the working face to be in a nonhazardous zone. The panel substations typically incorporate the following features:

- Incoming and through-going high-voltage supplies
- High-voltage circuit breaker and protection relays
- Isolation and earth switches for the transformer section
- 2.0-MVA, 11/1.05-kV transformer
- Four-outlet low tension end
- 1,000-V switchgear and protection relays
- PLC, network, and control equipment
- Communication components for the CM
- IP56 (minimum) rated enclosures
- Arc-fault-rated enclosures (using arc-fault-control methods)
- Base frame, wheels, and draw bar

The panel DCB receives the 1,000-V power supply from the panel substation via a 150-mm^2 trailing cable and provides

Table 12.3-11 Panel trailing cable requirements

Application	Cable Type (AS/NZS 1802:2003)	Cable Size, mm²	Cable Length, m
Panel 1,000-V feeder cable	241.1	150	110
Continuous miner cable	241.1	120	150
Shuttle car cable	275.1	35	275
Auxiliary cable (fan/feeder/pump)	241.1	35	110

Courtesy of MineCraft Consulting.

separate outlets for the distribution control and protection of the 1,000-V power supplies to the various development panel machines. The panel DCB incorporates the following features:

- Flameproof enclosures certified to the relevant parts of AS/NZS 60079.10:2004
- Incoming and through-going supplies
- Six to seven 1,000-V outlets with plug receptacles
- 1,000-V switchgear and protection relays
- Arc-fault-rated enclosures (using arc fault containment methods)
- PLC, network, and control equipment
- Communication components for the other development equipment
- Base frame with either wheels or fork tine boxes for transport

In Australia, all trailing cables must comply with AS/NZS 1802:2003, while the cable plugs must comply with AS 1299-1993. The plugs must be fitted and the cable assemblies tested in accordance with AS/NZS 1747:2003. Table 12.3-11 summarizes the recommended cable requirements for each development panel. All cables are normally supplied assembled complete with plugs, labels, and dust covers. Back-to-back couplers are required for the connection of the individual lengths of cable.

Longwall Panel

The electrical equipment for the longwall panel includes longwall substations; longwall DCB(s); monorail and face trailing cables; and control, monitoring, and signaling equipment. The power supply equipment up to the input connection of the longwall substation, which includes the 11-kV cables and section circuit breakers, forms part of the underground reticulation system. The longwall electrical equipment is normally included in the scope of supply for the longwall equipment package.

Underground Conveyors

The electrical equipment for the conveyor systems includes conveyor substations, conveyor starters, power and control cables, and control and signaling equipment. The power supply equipment up to the input connection of the conveyor substation forms part of the underground reticulation system (11 kV). The conveyor electrical equipment is normally included in the scope of supply for the respective conveyor system package.

Dewatering Pump Stations

The electrical equipment for the dewatering pump stations includes substations, pump station starters, power and control cables, and control and monitoring equipment. Generally, the dewatering pump stations will obtain their power supply from a substation conveniently located nearby, such as a conveyor substation or starter. However dedicated substations may be required for more distant pumping locations such as the pit bottom area to provide a power supply for pumping, an equipment tramming facility, and general power for the area.

Underground Lighting

Underground lighting is typically required in the following locations:

- Electrical equipment locations such as switchboards, section circuit breakers, substations, and DCBs
- Conveyor drives, loop take-ups, and transfer points, including the walkways in between
- Pump stations
- Transport access portal
- Other areas where a particular risk exists and which can be mitigated by lighting
- Initial 50 m (approximately) of the mine access drifts to assist the transport vehicle operators in the transition from daylight to darkness (and vice versa)

REFERENCES

AS 1299-1993. *Electrical Equipment for Coal Mines—Flameproof Restrained Plugs and Receptacles.* Sydney: Standards Australia.

AS 1300-1989. *Electrical Equipment for Coal Mines—Bolted Flameproof Cable Coupling Devices.* Sydney: Standards Australia.

AS 3007.2-2004. *Electrical Installations—Surface Mines and Associated Processing Plant.* Sydney: Standards Australia.

AS/NZS 1747:2003. *Reeling, Trailing and Feeder Cables.* Joint Australia/New Zealand Standard. Sydney: Standards Australia; Wellington: Standards New Zealand.

AS/NZS 1802:2003. *Electric Cables—Reeling and Trailing.* Joint Australia/New Zealand Standard. Sydney: Standards Australia; Wellington: Standards New Zealand.

AS/NZS 1972:2006. *Electric Cables—Underground Coal Mines—Other Than Reeling or Trailing.* Joint Australia/New Zealand Standard. Sydney: Standards Australia; Wellington: Standards New Zealand.

AS/NZS 4871:2002. *Electrical Equipment for Coal Mines.* Joint Australia/New Zealand Standard. Sydney: Standards Australia; Wellington: Standards New Zealand.

AS/NZS 60079.10:2004. *Electrical Apparatus for Explosive Gas Atmospheres.* Joint Australia/New Zealand Standard. Sydney: Standards Australia; Wellington: Standards New Zealand.

CHAPTER 12.4

Underground Horizontal and Inclined Development Methods

Michael A. Tuck

INTRODUCTION

Two types of development exist in mining: primary and secondary. The main difference between these two types is the life expectancy of the development.

Primary implies long life and includes such items as shafts, declines, and other mine accesses; main tunnels, haulages, crosscuts, ore- and waste passes, crusher chambers, and other large chambers. Thus primary development is the development of the more permanent features of a mine. These do not have to be developed early in a mine life, but they need to have a relatively long life span.

Secondary development is more temporary in nature and tends to be associated with the needs of a particular production unit or units. In many cases this type of development is consumed as production proceeds. Thus a typical lifetime for this type of development is 1 or 2 years and is often less than this.

Development is a necessary part of mining as it provides the infrastructure with which production of ore can be undertaken. Development is essential but represents an additional cost. In most situations this cost is not recoverable until mining operations—such as preparing to excavate secondary development undertaken within the ore body itself—begin. However, in some situations (e.g., coal mine development of revenue-generating material and some payback is obtained during development. Primary development is normally almost entirely undertaken in waste external to the ore body; the exception to this is the mining of bedded deposits where some ore in primary development can sometimes be taken.

Development requirements are influenced by a number of parameters; the following list is far from exhaustive:

- Ore-body type
- Ore-body depth
- Mining method selected
- Economics
- Ore-body extent

The mining method selected is a significant influence, especially with regard to secondary development but also with respect to primary development, as it is important to the layout and size of the primary mine infrastructure to be able to handle the production of waste and ore over the lifetime of the mine. It is also important to account for possible variations in production over the mine life as well. Given the cost of development and the usual need for mining companies to generate income relatively rapidly, due to the long lead times between exploration and eventual production, the aim in most cases is to bring production on stream as quickly as possible. However, the long-term capabilities of a mine are dictated by the capabilities of the primary mine infrastructure, the need for accurate geological information, environmental considerations, and numerous other factors. During the mine planning stage, the primary infrastructure and secondary development are investigated. It is essential at this stage that a thorough economic, technical, and sensitivity analysis of all possible mine production systems be undertaken to evaluate the project on a logical basis.

Mine development serves a number of purposes, which include

- Accessing the mineral deposit;
- Provision of accurate geological information to aid the production process; and
- Providing accurate information on other technical needs such as geotechnical characteristics of the rock, water inflow rates, ventilation pollutant inflow rates, and so on.

Figure 12.4-1 illustrates some of the openings whose construction will be discussed in this chapter.

EXCAVATION METHODS

Development openings in mines can be excavated in one of two ways, either by drilling and blasting or by mechanical excavation. Both techniques are used and are discussed in the following sections for various types of development openings. Both have their advantages and disadvantages; mechanized mining has the following advantages compared to drilling and blasting methods:

- Higher advance rates/development rates
- Smooth development profile and superior rock stability (tunnel boring machine and mobile miner circular profile in particular)

Michael A. Tuck, Associate Professor of Mining Engineering, University of Ballarat, Victoria, Australia

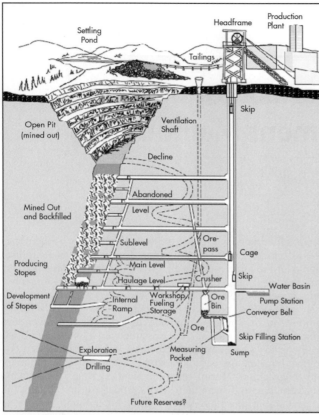

Source: Atlas Copco 2010.

Figure 12.4-1 Basic infrastructure of an underground mine

- Minimized ground support
- Reduced overbreak
- Less surface disturbance
- Less noise
- Lower running costs
- Continuous haulage system, such as conveyor
- Automated haulage systems
- Less labor required
- Automation
- Precise direction control (laser guidance)
- Safety (no explosives)

However, there are disadvantages also, including

- Cost,
- Difficultly dealing with variable ground conditions, and
- In hard, abrasive rocks, high rates of wear on the cutters can lead to high operating costs and long periods of downtime for cutter replacement and other maintenance.

Shafts

According to *The Mining Glossary* (AusIMM 2007), a shaft is defined as "a primary vertical or non-vertical opening through mine strata used for ventilation or drainage and/or for hoisting of personnel or materials; connects the surface with underground workings." Although this is a good definition, shafts can also be internal within a mine connecting vertically separate workings as well as levels.

In some ways, a better definition of a shaft is a vertical or subvertical opening in a mine that is excavated from the top downward, as opposed to a raise, which will be discussed in a later section of this chapter.

Shafts are the most important capital openings for deep mines. Shafts provide all the services for underground operations including

- Fresh air,
- Transport of ore and supplies,
- Personnel transport,
- Power,
- Communications,
- Water supply, and
- Drainage.

The depth of the shaft has a direct influence on the total time to develop a mine prior to first production; in some cases, shaft sinking can account for up to 60% of the total development time. Consequently, proper selection of the shaft-sinking method to minimize sinking time and ensure uninterrupted operation is of great importance.

Determining the shaft diameter and hoisting depths requires consideration of future mining needs. It is generally better to overdesign the shaft in the initial stages of the project life span than prevent feasible increases in production later by the creation of a bottleneck.

Designing the excavation requires data to be gathered and solutions to elements within the design to be determined. In order to design the excavation the following need to be taken into account:

- Site geology, including a description of the geological column, usually in a table form identifying the rock strata, their geotechnical parameters, groundwater levels (including water heads, inflow rates, and chemical contamination, if any)
- Determination of the shaft diameter
- Choice of shaft-sinking technology
- Description of the shaft lining (should include a list of lining sections with their thicknesses detailed)
- Shaft foundations, including their location and dimensions
- Shaft collar depth, foundations, thickness, and materials to be used (the type and number of openings with a description of their function, dimensions, and elevations should also be detailed)
- Shaft sump, including depth, structural characteristics, pumping arrangements, and cleaning system
- Surveying data
- Calculations, including ground and water pressure acting on the shaft lining, the resulting lining thicknesses, shaft insets and their dimensions, and airflow capacity
- Timetable of construction
- Cost specifications
- Engineering drawings

Shaft Shape and Diameter

Numerous factors regarding the shape and diameter of a shaft need to be considered, such as

- Shaft duties and purpose, such as rock hoisting, ventilation, and worker and materials transport. Is the shaft to be used for a single purpose or for a variety of purposes? What are the quantities involved?
- The amount of water to be handled, number of pipes, and the diameter of the pipe are important parameters, as is the method of hanging these within the shaft.
- Ground conditions
- Size of mining equipment to be handled by the shaft. Is the equipment to be handled in single or multiple parts? For transportation of oversize elements is suspension underneath the shaft conveyance possible?
- The expected lifetime of the shaft

The majority of shafts in modern mining are circular in shape; however, rectangular cross sections were common in the past. Although rectangular shafts were more popular within metalliferous mines, most modern shafts tend to be circular. Other shapes are used such as square or elliptical. Circular shapes have inherent strength and have the advantage that for a particular cross-sectional area the minimum perimeter results. Elliptical shafts tend to be used in areas where a significant horizontal stress field acts in a particular direction; the major axis of the ellipse is aligned to this direction to enhance the shaft strength.

Shaft Linings

Shaft linings basically serve to support the shaft equipment and the walls of the excavation. The minimum lining usually required to support shaft equipment is about 200 mm. Thicknesses required for wall support are generally higher. In modern projects where round or elliptical shafts are employed, concrete lining is used almost exclusively. Such shapes reduce airflow resistance, make sinking easier and allowing the full advantages of concrete to be employed. The placement of concrete can be mechanized, allowing high sinking rates as well as lower costs. The strength of the concrete can be adjusted as required and watertightness can be ensured within aquifers of moderate head. Where heads exceed the strength of concrete, cast iron linings are employed.

In order to calculate the required thickness of the lining material, a number of factors need to be considered. First, the ground pressure acting on the circular shaft needs to be determined.

The vertical or hydrostatic stress, σ_v, is proportional to the weight of the overburden and can be calculated from

$$\sigma_v = \sum \gamma \times h$$

where
γ = density of the rock strata multiplied by the acceleration due to gravity
h = thickness of the rock stratum

The next stage is to evaluate the horizontal stresses. Horizontal stresses vary substantially from one location to another. The minimum value can be calculated using elastic theory employing Poisson's ratio to a maximum value derived from active or residual tectonic stresses. A good average ratio to employ at 500-m depth would show that the horizontal stress is two to three times the vertical stress.

For conventional shaft sinking, the technical limit for shaft water inflow is approximately 500 L/min. However, even the smallest inflow of water can cause problems and will require special arrangements to control it. Water seepage into a finished shaft may cause corrosion of steel installations such as buntons and reinforcing, and will lead to costly and troublesome replacement. Therefore, water inflow to the shaft should be controlled; this is especially true for the buildup of hydrostatic pressure behind a shaft lining. Numerous methods can be employed as described by Unrug (1992). Examples include the use of polyethylene or other shaft liners installed at the rock–lining interface.

The thickness of a shaft lining can be determined using

$$d_i = a\left(\sqrt{\frac{R_c}{R_c - np\sqrt{3}}} - 1\right)$$

where
d_i = shaft lining thickness (m)
a = shaft radius
R_c = allowable stress in the concrete (MPa) (see Figure 12.4-2)
n = coefficient of lining work conditions (see Figure 12.4-3)
p = calculated outside pressure acting on the lining (MPa)

Shaft Collars

The shaft collar is the upper portion of a shaft, which is anchored to the first footing as a minimum and to the bedrock (if required). Overall, the dimensions of the collar depend to a large degree on its depth, cross section, thickness, the function of the shaft, the overburden rocks, hydrologic conditions, the ground pressure (on a diurnal and annual basis), sinking method, and additional loading conditions. A simple collar design is illustrated in Figure 12.4-3.

The size of the shaft collar depends to a large extent on the loads in the immediate surrounding areas from the shaft and associated equipment. Examples of these loads include winding equipment and its associated dynamic loading, loading impacting on the shaft collar area from other surface facilities, and so on. The shape, thickness, and choice of shaft collars depend on numerous factors, but in essence depend on structural factors. Some typical collar shapes are illustrated in Figure 12.4-4.

Shaft Insets and Sumps

Shaft insets should be designed to minimize airflow resistance. Model and/or computational fluid dynamics models can be used to ensure this.

Sumps at the base of a shaft and at intermediate levels to minimize pumping heads must be of sufficient depth or size to cope with the water flow requirements. Obviously, suitable dewatering arrangements must be provided. In addition, periodic cleaning must be designed into the system to account for sedimentation, which will reduce the holding capacity of the sump.

Technology of Shaft Sinking

The main classification between traditional shaft sinking and other specialist methods is based on the engineering and hydrogeological conditions of the individual site.

The conventional method of sinking is summarized in Figure 12.4-5 and involves repeating cycles of face advance and lining erection without any previous ground stabilization. Such methods are applicable to good- to fair-quality rocks with limited water inflow into the shaft (<500 L/min).

Symbol	Type	Magnitude	Remarks
n_o	Influence of tectonic forces	In-situ measurements	n_o
n_1	Influence of shaft diameter D	$n_1 = \frac{1}{2}\sqrt[3]{D+1}$	
n_2	Influence of the shaft station	$n_2 = 1.5$	n_2
n_3	Influence of the strata dip x	$n_3 = 1.0$ if $x < 30°$ $n_3 = 1.25$ if $x > 30°$	n_3
n_4	Influence of the weak strata of thickness h	$n_4 = 1.0$ if $h > 1.3$ m $n_4 = 0.7$ if $0.8 < h < 1.3$ $n_4 = 0.3$ if $h < 0.8$	n_4
n_5	Influence of the water filtration through the lining	$n_5 = 1.0$ for watertight lining $n_5 = 0.1$ for formation grouting $n_5 = 0.1$ to 0.2 for water drainage $n_5 = $ for preliminary lining	n_5

Remarks:

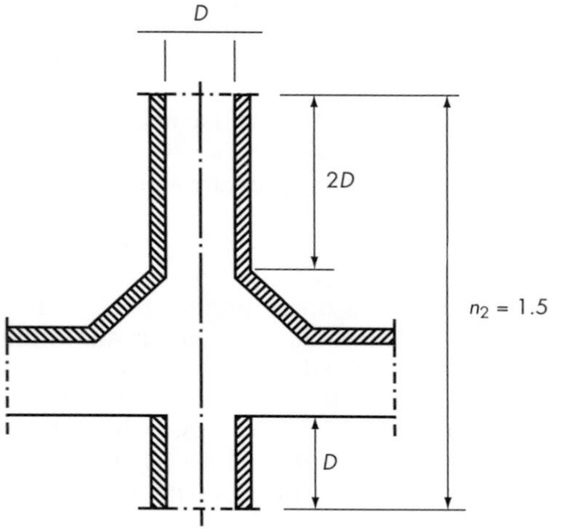

n_o—apply coefficient if $\sigma_h/\sigma_v > 1$

n_3 applies for cohesive soil and rock without water
n_4 applies for calculation of $H_{critical}$ in rocks
n_5 applies for calculation of water pressure

Source: Adapted from Roesner et al. 1984.

Figure 12.4-2 Determination of stress coefficient, n

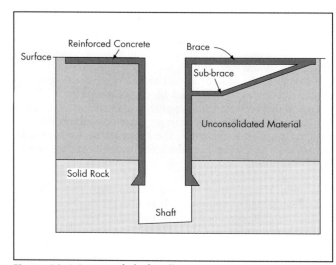

Figure 12.4-3 Typical shaft collar

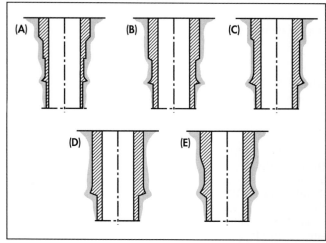

Source: Unrug 1992.

Figure 12.4-4 Typical shapes of shaft collars

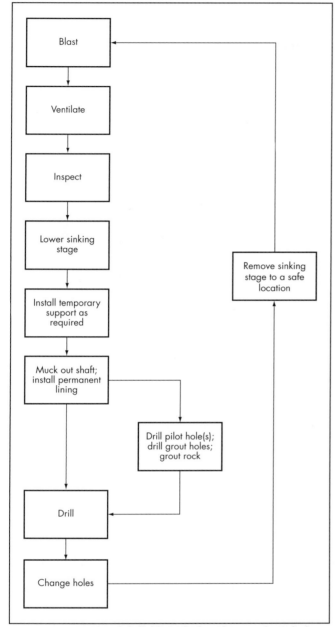

Figure 12.4-5 Typical drill–blast shaft-sinking cycle including a possible grouting sequence for rock strengthening

Specialized methods of shaft sinking include those applied to

- Strong rocks that have an extensive fracture pattern resulting in a strong water inflow (such as the South African goldfields);
- Coherent, weak, plastic flowing ground such as permafrost areas; and
- Loose water-bearing grounds.

Conventional Drill-and-Blast Shaft Sinking

The main components within a traditional drill-and-blast shaft-sinking operation are illustrated in Figure 12.4-5. Additionally, the following classifications of traditional shaft sinking can be made based on the order of the main components of the operation:

- *Series system*, in which sinking occurs with nonsimultaneous face advance and erection of the permanent lining. The shaft is sunk in sections. The length of the section depends on the rock quality. In a particular section of the shaft, the face is advanced first with a temporary lining. When sinking is suspended, the permanent lining is constructed. Fast sinking rates are not ensured; however, high capital expenditure is not required. This method is commonly applied to the sinking of small-diameter shafts at shallow to moderate depths.
- *Parallel system*, in which sinking with simultaneous face advance and lining erection at certain distances behind the face occurs. In this method, a temporary lining is required as with the series system. It is applied to the sinking of large-diameter and deep shafts where the high capital cost can be paid off over the life of the project.
- *Simultaneous system*, in which advancing the face and erection of the lining in the same shaft section occurs. This method can be further subdivided into series, simultaneous, and parallel simultaneous methods. The series simultaneous system involves face advance and lining completed in one cycle with each being done one after the other. The lining is erected either downward or upward and temporary lining is eliminated. The parallel simultaneous system involves simultaneous face advance and lining erection.

Shaft buntons, beams, guides, and so on are usually installed following the advance of the permanent lining; however, in some methods of shaft sinking (e.g., shaft boring) they are erected following the completion of the shaft lining.

Shafts can be sunk to the final depth or to depths required for certain stages of the mining process. Shaft extensions to subsequent levels need to be synchronized with reserve depletion in the upper levels to ensure uninterrupted production. Shaft extensions can be undertaken by either sinking or raising.

Before sinking a shaft, preparatory work needs to be done. This includes

- Determination of geological and hydrological conditions,
- Preparation of technical documentation, and
- Construction work associated with site preparation and construction of the shaft head.

The shaft site should be secure from flooding. From a design perspective, this would be based on a 100-year flood level. Numerous activities associated with the preparation of a shaft site include but are not limited to

- Construction of access roads and materials storage areas,
- Grading,
- Provision of a water supply,
- Provision of electrical power and other power sources, and
- Temporary and permanent buildings.

Drilling and Blasting

When undertaken correctly, drilling and blasting will ensure

- Correct size and shape of the excavation,
- An even face surface for simpler mucking and drilling of the next round,

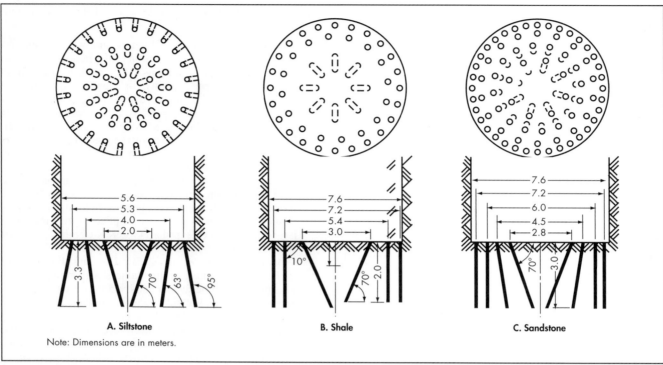

Figure 12.4-6 Typical circular shaft drilling patterns for different rock types

Source: Unrug 1992.

- Uniform sizing of the broken rock for effective mucking, and
- Safe and economic operation.

Shaft blasting uses a variety of high explosives. The correct product has to be used given site-specific conditions such as explosive gas emissions, reactive ground conditions, and the presence of water. Electric detonators with millisecond delays are also employed. The drilling pattern has a strong influence on sinking performance. It should ensure

- Accuracy in the shape of the excavation,
- Clean separation of rock from the bottom of the excavation,
- Even and sufficient comminution of rock,
- Avoiding cutoffs of neighboring active holes, and
- Avoiding damage to nearby shaft installations.

The drilling pattern depends on such factors as

- The shape of the shaft section,
- Rock strength,
- Cleavage,
- Dip of strata,
- Water inflow, and
- Hole-loading structure.

Typical drilling patterns for circular shafts are shown in Figure 12.4-6. With small and moderate strata dip, holes are placed on concentric circles drawn from the center of the shaft. The number of circles is three to five, depending on the shaft diameter. The central hole is vertical and shorter, about two-thirds the length of the other holes. This hole initiates conical excavation by cut holes and is especially recommended in strong rocks. Holes of the first (central) circle ("wedge holes") are inclined toward the shaft center and are fired first to open an additional free surface and to make the work of the remaining external holes easier and more efficient. They are drilled with an opposite (outward) inclination of 65°–75°. Four to ten holes are drilled for the first circle. The ends of four holes are intended to create a square with sides of 180–300 mm. For a larger number of holes, a 400–600-mm circle is used. In general, the harder the rock, the more first-cut holes are required. Holes of the next two, three, and sometimes four circles crush and lift the excavated material.

The inclination of holes varies from 65° to 85°; the farther they are from the center of the shaft, the more vertical they are. The distance between adjacent holes on a circle should not exceed 0.8 to 0.9 m for 32-mm and 36-mm cartridges; for a 45-mm cartridge, it should lie in the range of 1 to 1.2 m. The last circles of perimeter holes are supposed to leave the shaft wall smooth. They are holes drilled toward the shaft wall, 65°–85° (less inclination for weak rocks and more for stronger ones). In very weak rocks, they can be slightly inclined toward the center of the shaft. Perimeter holes are placed in a circle with a diameter of 0.6 m less than the shaft diameter in rocks with a coherence factor of $f = 2–8$ and 0.3 m for rocks with $f = 9–20$. The distance from the shaft wall has to be such that in rocks with $f < 8$, holes should not project from the design shaft section, and in rocks with $f > 8$, they should protrude about 100–200 mm outside the shaft contour. The distance between rib holes in moderate-strength rock is 0.9–1.2 m and in strong rocks 0.7–0.9 m.

Drilling of Shot Holes

Blastholes can be drilled with sinker percussive drills, hand-held or mounted on a special shaft jumbo. Sinkers are

Table 12.4-1 Factors influencing mucking efficiency

Depth, m (factor)	<150 (1.0)	150–300 (0.9)	300–500 (0.85)	>500 (0.8)
Rock strength f (mechanical mucking factor)	12–15 (0.85)	10–12 (0.89) 6–10 (0.92)	3–6 (1.0) 2–3 (1.1)	0.5–2 (1.4)
Shaft section area, m² (factor)	<4.0 (0.65)	4–8 (0.75) 8–12 (0.85)	12–16 (0.9) 16–20 (0.9)	20 (1.0)
Water inflow, L/min (factor)	<100 (1.0)	100–200 (0.9)	200–300 (0.8)	300 (0.75)

Source: Unrug 1992.

self-rotating, driven by compressed air or hydraulically with air or water flushing.

Two systems exist:

1. *Series*, in which the drilling starts from the shaft center toward its perimeter after completion of mucking
2. *Parallel*, where mucking and drilling are done at the same time. Drilling starts from the perimeter toward the shaft center.

A shaft jumbo can be used only in the series system. A series system is used in smaller shafts when a Cryderman or similar type of mucker is used. This system ensures precision in blasthole drilling.

The parallel system is sometimes chosen in large cross-sectional shafts where cactus-grab cabin muckers start to clean the sides of the shaft moving toward its center. Drilling is done in the cleaned area. Owing to parallel drilling and mucking, some time saving can be achieved; however, it is at the expense of drilling accuracy. Blastholes can be drilled individually or by groups of drillers. In the first case, approximately 25% of the time is lost in changing bits. In the second, one shaft area is divided according to the number of drillers; each group drills in one sector holes of a certain depth only. Bits are changed only when they get dull. Then the drilling rod is changed (lengthened), and work is done in the next sector. Wooden sticks are used to prevent holes from being plugged by muck.

Drilling equipment is delivered in a special bucket (known as a kibble) or basket to the shaft bottom. An elastic hose with quick coupling is connected to the main steel-compressed air tubing. At the end of the elastic hose, there is a distributor with multiple outlets and valves for particular drills. Drilling is a labor-intensive operation that takes about 16% of the cycle time.

Shaft-drill jumbos have been developed to mechanize blasthole drilling. These permit the drilling of larger-diameter holes more precisely, faster, and deeper.

Mucking in Shafts

Mucking in shafts accounts for as much as 50%–60% of cycle time. Table 12.4-1 details the different factors influencing mucking efficiency. Figure 12.5-2 in Chapter 12.5 shows a schematic of the sinking phases with an Eimco 630 loader.

Mucking efficiency is also dependent on the degree of rock fragmentation and the height of the bucket. For mechanical mucking, medium-sized fragmentation, approximately 125 mm, is most efficient. Generally, in a muck pile after a shot, 70%–80% of the material is fine fragmentation, and the rest is composed of larger fragments.

Several types of muckers are in use depending on shaft shape, size, depth, and location:

- Scrapers, used in rectangular shafts and very large-diameter shafts (underground bunkers)
- Overshot loaders on caterpillars (e.g., Eimco 630)
- Cryderman muckers
- Cactus grabs (light)
- Cactus grab (heavy, on central pivot)

A variety of systems are used. Some are driven by an operator (e.g., Eimco), some are hand-guided at the face when hanging from a platform, and some are attached to the shaft wall (Cryderman) and operated from a control cabin. Heavy cactus grab units have a rotating arm on a central pivot below the stage. Muckers can be electric, pneumatic, or hydraulically driven.

To determine the capacity of the mucking unit, the following may be used. The theoretical mucking capacity is

$$P_t = 3{,}600 \times k \times \frac{V_b}{t}$$

where

P_t = mucking capacity (m³/h)
k = fill factor (0.8–0.9 for sandstones and 1.1–1.2 for shales)
V_b = the bucket volume (m³)
t = mucking time (seconds)

The mean actual efficiency is found by calculating the actual mucking time and all time losses for unproductive activity such as mucker lowering time to work position, time of temporary support erecting when mucking has to be suspended, and time for final cleaning.

Most shafts intersect one or more water-bearing horizons, so shaft dewatering is an essential part of sinking. The characteristic features of shaft dewatering consist of changing water inflow rates, changing the pump head, filtering water to eliminate rock and mud particles, and periodic changes in positioning dewatering devices.

Dewatering can be considered in two ways, either as limiting water inflow into the shaft or pumping water out of the shaft. Water inflow into the shaft can be diminished by initial drainage of the rock mass, sealing of the water flow routes, or by tapping the water. The following are used to dewater the face:

- Rock kibbles and special water kibbles
- Face pumps
- Combination of face and stationary pumps

- Combination of hanging and stationary pumps
- Submersible pumps
- Airlifts

Shaft Lining Construction

The type of lining is determined by the following factors:

- Hydrogeological conditions
- Shaft function
- Planned lifetime
- Shape of shaft section and its depth
- Availability of construction materials
- Construction cost

In choosing a shaft lining, the geotechnical properties of the rock and hydrologic conditions are decisive in most cases. The chemical activity (corrosiveness) of the water can also be an important consideration.

Since modern new shafts often have automatically operating hoisting gear (moisture sensitive), they should be dry. Main shafts are usually planned for the entire mine life span, so they should be constructed to minimize repairs and maintenance time.

Temporary support is placed to protect the crew and equipment against falling rocks from the exposed shaft wall before a permanent lining is placed. Numerous systems of temporary lining are in use. Forms of temporary lining include rock bolts and mesh and other traditional methods of temporary support used in other mining operations.

Depending on the shaft design and environmental conditions, several different permanent-lining systems can be used:

- Timber
- Brick or concrete blocks
- Monolithic concrete
- Reinforced concrete
- Tubing
- Shotcrete
- Anchor bolts

Shaft-Sinking Stage

Sinking stages are an important element of any shaft-sinking program; they provide a working platform from which to perform the sinking operations. They also ensure the safety of the personnel working at the shaft bottom. An example of a sinking stage can be found on the Andra Web site (2009). This Web site details the construction of test facilities in France for nuclear waste emplacement, with access by shaft. It shows the sinking stage being installed in the shaft following construction of the collar and gives technical details of the shaft-sinking process as well as the development of the test galleries underground. These can vary considerably in design from a simple single-deck working platform to complicated multi-deck arrangements with facilities for drilling, mucking, shaft concreting or timbering, and storage of hoses, pumps, and all other equipment associated with shaft sinking.

Sinking stages have been used since the earliest shaft sinkings and have developed to maximize efficiency of the operations while minimizing delays. The primary purposes of sinking stages are to enhance safety and maximize efficiency.

Stages are suspended in the shaft by wire ropes and winches that are independent of the main hoisting system. Stages are raised prior to blasting and then lowered to their new position at the end of each sinking cycle. Sinking cycles are usually based on a certain height of concrete pour or timbering.

Specialized Methods of Shaft Sinking

Special methods of shaft sinking are used when the following conditions are encountered:

- Incompetent, water-bearing rocks (e.g., quicksand)
- Weak, unstable, soil-type rocks
- Competent fractured rocks with high water inflows above 0.5 m^3/min

Numerous special methods exist that encompass many variations in method and application because of environmental conditions and range of depths. The characteristics of these are given in Table 12.4-2. The following methods will be described in subsequent paragraphs: depression of groundwater level, grouting methods, freezing method, and shaft drilling. An alternative to these methods can be applied in some cases (e.g., moving the shaft to another site).

Depression of groundwater level. Prior to designing a method for depressing the groundwater, the following decisions must be made:

- Choice of the most suitable method for depression of the water table under site conditions
- Determination of the number, spacing, diameter, and depth of the wells
- Calculation of the radius of the depression cone, amount of water to be pumped out, and the height of the remaining wet portion of the water-bearing strata
- Evaluation of equipment, hardware, and pumping facilities
- Choice of the auxiliary method to go through the wet remainder of the water-bearing strata
- Choice of a suitable water discharge plan

To lower the water table in the area of shaft sinking, the following systems can be applied:

- Pumping water from three to six depression wells drilled around the planned shaft perimeter, placed on a circle about 4 m larger than the diameter of the shaft excavation (this system is used most often)
- Draining water through drill holes to existing openings in the mine below
- Draining water through drill holes to water-absorptive strata (e.g., dry sands under water-bearing strata)

Depression wells are drilled with 300–500-mm diameters, without drilling mud. They are spaced uniformly around the future excavation. Within the range of the water-bearing strata, they are equipped with filters perforated by round or elongated holes 5–10 mm in diameter, spaced in a checkered pattern, with an outside screen having eyes 0.4–0.9 mm in diameter, sized against entry of sand particles. Generally, the construction of the filter must be suitable for the character of strata to be drained. Submersible drainage pumps have a wide-capacity range, varying up to 350 m^3/h. Calculations for well efficiency are based on principles of flow through a porous media. The objective of the design is to determine a pumping regime to achieve a dry-out zone surrounding the shaft. This is created by the combined effect of the depression cones around particular wells. The movement of water through a porous media is given by the Darcy formula:

Table 12.4-2 Special methods of shaft sinking and their applicability

Method	Rock Mass	Thickness of Water-Bearing Strata, m	Water Inflow	Range of Depth, m	Approximate Rate of Advance, m/month	Comments
1. Pile wall • Wooden • Sheet pile wall (steel profiles or pipes)	Quicksands, soils (saturated with water without stones and rock fragments)	2–6 6–10	Limited	6–15 Down to 30	2–5 6–10	Required presence of impermeable strata of clay, into which the piles could be driven to lock water in above water-bearing strata.
2. Concrete pile wall drilled or driven piles	Quicksands, soils, loose rocks (sands, gravels), clays saturated with water (also with stones)	10–20	Any	25–30	8–10	Concrete pile wall can be used as a foundation for a shaft headgear.
3. Caisson method • Without pumping water in flow rocks • With steady pumping in weak and loose water-bearing rocks	Uniform flowrocks or loose rocks (sands, gravel), clays, without intrusion of hard rock or boulders	10–15	Not limited Up to 0.3 m^3/min	Down to 30; more in special cases when vibrators or compressed air can be applied	10–20	Flat-lying strata required or with little dip. Presence of the clay strata below water-bearing zone 2–3-m thick is necessary.
4. Caisson with compressed air	Loose or noncoherent water-bearing strata	8–30	Moderate	Down to 40	20–30	Maximum air pressure in the face chamber 3–3.5 atmospheres.
5. Depression of groundwater level	Mid and coarse solids and gravels with coefficient of filtration 10–100 m/d. Fractured competent rocks.	20–30	Great	Around 100–150	15–20	Not supposed to be applied in clay rocks, quicksands, and alternating packages of permeable and nonpermeable rock.
6. Chemical grouting	Sands with clay content not exceeding 10%–15%. Coefficient of filtration 2–80 m/d. Also competent rocks with small tissue.	A few	Moderate	Unlimited		Usually used as an additional interventional method.
7. Grouting (cement)	Competent rocks, with tissues and cracks filled with water (sandstones, shales, mudstones, limestones), and coarse sands and gravels.	Not limited	Mid and large, up to about 10 m^3/min	Any	10–15; 20–30 in good conditions	Not recommended in flow rocks or quicksands, with water flow velocity more than 80 m/d, or in presence of caverns and large voids.
8. Grouting with clay	Water-bearing rocks with large cracks and voids.	Not limited	Not limited	Not limited	10–15	Can be used in presence of aggressive waters. Usually is used combined with grouting (cement).
9. Bitumination	Competent rocks with large cracks and voids; also gravels	20–30	Mid and large	Down to about 40		Used as an auxiliary method for local sealing of the rocks with high flow velocity and aggressive waters.
10. Freezing of the rock mass	All types of water-bearing rocks	10–1,000	Not limited	Down to 1,000	Differs depending on depth, sinking frozen rocks up to 50 m overall less due to length of freezing method	Most reliable method, however, costly and time-consuming because of preparation work (drilling of the freezing holes and the time taken for an ice wall to form after freezing commences.)
11. Shaft drilling	Weak, water-bearing strata	Several hundred	Not limited	To 600		Thickness of competent hard rocks not supposed to exceed 10% of overall thickness of water-bearing strata.

Source: Unrug 1992.

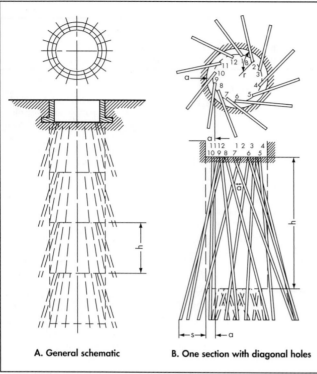

A. General schematic   **B. One section with diagonal holes**

Source: Unrug 1992.

Figure 12.4-7 Grout hole distribution when drilling from the face

$$V = K\frac{dh}{dl} = K_i = K\frac{h}{l}$$

where
- V = mean velocity of flow or mean filtration velocity (m/s)
- K = permeability or coefficient of filtration (m/s)
- d = equivalent diameter of the flow routes (m)
- $h = h_2 - h_1$ = difference of water levels in the ground or the depression created by the pumping of water
- l = distance in which the head loss appears (m)
- $i = \frac{h}{l} = \frac{h_2 - h_1}{l}$ = hydraulic gradient (dimensionless)

Application of grouting. Based on geologic data, the following decisions have to be made prior to the start of the project:

- Choice of the grouting system, including the decision as to whether grouting will be done from the surface or from the bottom of the shaft; determination of the grouting depth; and determination of the length of the grouting sections
- Determination of the number of grouting holes and their respective distances, diameters, depth and angle of dip, and applicable drilling method
- Construction and calculation of the grouting plug
- Grout type to be used, its density and admixtures
- Injection system and required equipment
- Determination of working pressure to be applied in particular sections, and approximate calculations of grout consumption
- Number of control holes and their locations
- Set of drawings corresponding to this list of considerations

Necessary hydrogeological information should contain

- Number and characteristics of the water-bearing water horizons,
- Chemical analysis of water flows and determination of their degree of corrosiveness,
- Flow velocity, and
- Coefficient of filtration.

Properly performed grouting seals the rock mass in the area of the shaft, making sinking possible by a conventional method. Grouting is performed from the surface when the water-bearing strata are relatively thick and close to the surface, or from the shaft face when the strata to be grouted are located at such a depth that it is not feasible to drill long holes from the surface. Grouting is most successful in water-bearing, competent but fractured rocks. Grout fills the fractures and cracks, creating a sealed zone around the shaft to be sunk and protecting it against water inflow.

Grouting for sealing is often applied to improve tightness of the lining in existing shafts and to eliminate the voids behind the lining, as well as to seal rocks in the immediate vicinity of the lining (Petterson and Molin 1999).

When grouting from the surface, grouting holes with an initial diameter of 75–150 mm are drilled vertically around the future shaft on a circle with a radius 3–4 m larger than the excavation diameter. Holes should be straight, and with their greater depth, any deviation should be recorded.

Grouting from the shaft face is illustrated in Figure 12.4-7 and is recommended when water-bearing strata occur below 100 m and when their thickness does not exceed 70 m. The length of sections varies between 12 and 25 m.

Holes are drilled along the length of the grouting section, usually by means of heavy drill hammers or small rigs when space allows. Holes are directed at an angle in the vertical plane to get the hole bottoms 1–2 m away from the future excavation. To increase the probability of encountering fractures by the hole, they are also tilted in a tangential direction as shown in Figure 12.4-7. The distances between the hole entries are 0.8–1.5 m. Spacing between the circles of holes is 1.5–3 m. The number of holes drilled depends on the rock character, shaft diameter, and hydrogeological conditions. Practically, it varies from 10 to 30. The initial drill hole size is 50–75 mm; the end size is 30–50 mm. The sequence of drilling and grouting is as follows:

1. The first one-half of the total number of holes is drilled and grouted; during hardening of the cement, the second half is drilled and grouted.
2. All holes are drilled and grouted, one after another, starting with the hole with the largest water inflow.
3. A pair of holes lying on opposite sides is drilled and grouted; then the pair lying in the plane perpendicular to the first one is processed, and so on.

This system is the one most often used, being recognized as the best, because the initial maximum distance between the holes lowers the probability of grout penetration from one hole to another. Grouting plugs are put in the shaft bottom to prevent water and grout inflow into the shaft. Entrance pipes to the holes are secured firmly by grouting them in the plug. The plug can be either natural or artificial.

A natural plug is composed of a protective layer of grouted rock 3–5 m thick, depending on the strength of rock and expected pressure (water and grout).

Artificial plugs are used more often and are constructed of concrete or reinforced concrete. They are constructed on the shaft bottom, 2–4 m from the water-bearing strata. When grouting is completed, the plug is demolished with explosives, and sinking proceeds through the cemented section, but again is suspended 2–4 m above the end of the grouted zone. The shaft lining is constructed and the procedure with the plug is repeated until the shaft passes the whole water-bearing zone. Details of the use of grouting to sink a shaft through water-bearing ground at the Longos gold mine in the Philippines are described by Ayugat et al. (2002).

Freezing method. This is a method commonly applied where a relatively large thickness of water-bearing strata has to be sunk through. Some good examples are presented in Harris (1995). It is commonly applied when shaft sinking to the coal measures in the United Kingdom due to the thickness of the water-bearing Bunter sandstone.

For a freezing process, a suitable design procedure requires solution of the following problems:

- Calculation of the wall thickness of the frozen rock cylinder
- Calculation of the diameter of the circle where freezing holes are located, their number, depth and spacing, and the number and distribution of surveying and control holes
- Selection of the drilling system for freezing holes, specification of equipment, choice of freezing pipes and their diameter, and assessment of time required for drilling
- Design of a temporary shaft head with a freezing basement
- Planning of the freezing procedure in active and passive periods, with specifications of freezing temperature, heat balance, required efficiency of freezing installation, and freezing time
- Determination of the system and time of thawing and liquidation of freezing holes
- Initial planning of freezing technology, shaft sinking and lining, and complementary grouting

An artificially created cylinder of frozen rock or an ice wall around the shaft creates a shaft lining, thus shielding the inner space against water and unstable rock. The thickness of the frozen rock cylinder, e_c, in meters, for shallow and moderate depths is calculated for external pressure, p, in megapascals, from the Lamé formula (in metric units):

$$e_c = r\left(\sqrt{\frac{k_c}{k_c - 2p}} - 1\right)$$

where

r = shaft radius of the excavation (m)
k_c = allowable stress of the frozen rock (mPa)

The allowable stress of the rock is 50% of its compressive strength in the mean range of freezing temperatures. For calculation of the thickness of the cylinder of frozen rock, the mean temperature of the rocks should be taken. The temperature within the wall of frozen rock changes from a minimum value just beside the freezing hole to a maximum around 0°C on the outer wall limit.

The cylinder of frozen rock, when properly created, allows the sinking of a shaft of moderate depth even in long sections using temporary lining. This is desirable because there are fewer seams between lining sections.

The most important problem during the drilling of freezing holes is to ensure verticality of holes. Every hole is equipped with two columns of tubing as follows:

1. **Outside:** The freezing pipe itself has a diameter of 102–152 mm and is constructed of steel with inside connectors so it is smooth on the outside and has a welded bottom.
2. **Inside:** The inner tubing has a smaller diameter of 32–51 mm, is made of steel or polyethylene pipe, and has inside diameter and wall thickness of 75 × 43 mm, respectively. Polyethylene pipes are used because they have several advantages over steel. They are cheaper and easier to install (by unwinding from a reel), have lower thermal conductivity, and the specific gravity of the pipe is close to the specific gravity of the brine, resulting in less suspended weight. Freezing brine flows at high velocity down the inner pipe. It rises more slowly in the annulus, absorbing heat from the surrounding rocks.

The collars of freezing holes with pipes connecting them with the distributing ring, shut-off valves, and thermometers are installed all together in a so-called freezing cellar. A freezing cellar is designated either as (1) temporary, to speed up construction time (it will be demolished when the permanent shaft headframe is installed); or (2) permanent, considered a part of the permanent shaft headframe, requiring a longer construction time.

Freezing media used as refrigerants are

- Ammonia (NH_3): Ensures a freezing temperature of –20°C with one stage and –40°C with two stages of compression, and is used in most cases.
- Carbon dioxide (CO_2): Provides a freezing temperature of –50°C.

A schematic of an ammonia freezing station is shown in Figure 12.4-8. Two main periods of freezing can be identified: the active period and the passive or maintaining period.

During the active period, a frozen collar of a predetermined thickness is created around the future shaft. To achieve the designed capacity of the freezing station, two or three aggregates are installed together to provide the required heat efficiency. During an active period, all units work at full capacity. The passive period starts after the frozen collar is created and lasts through the sinking, lining, and assembling of the shaft installations. During the passive period, usually only one unit works, while the others are on standby as a reserve.

Methods of excavation in frozen ground are similar to those under normal conditions; however, when explosives are used, special care is required because of the proximity of the freezing holes and the potential danger of their destruction. When the core of a shaft is not frozen, a cactus grab or other mechanical mucker can dig the shaft bottom. When the rock is partly frozen or too hard to be excavated, handheld drill hammers with chisel picks are used; however, these working conditions are very unhealthy. In hard-frozen rock, the drill-and-blast method is used with limitations. The *Hard Rock Miners Handbook* (de la Vergne 2009) provides some good examples of freezing calculations and "rules of thumb" for ground freezing, as shown in Table 12.4-3.

Shaft drilling. Shaft drilling is a commonly applied technique for the development of small-diameter ventilation shafts in the United States. It has also been applied with moderate

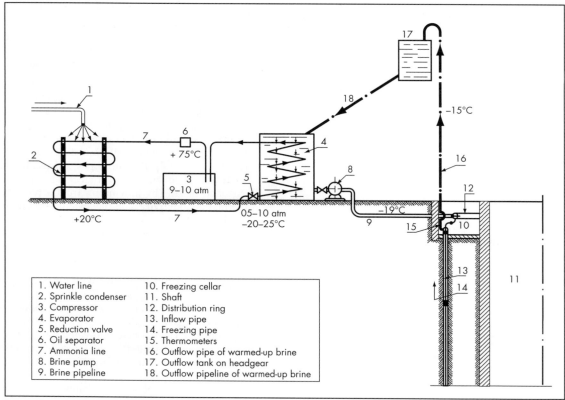

Source: Unrug 1992.
Figure 12.4-8 Ammonia freezing station

success in Australia. Shaft boring uses a mechanical borer similar to a large-diameter drill bit to excavate the shaft. A typical 4.25-m cutterhead is illustrated in Figure 12.4-9 and further details of shaft drilling technology including technical details on equipment, methods of shaft drilling, and some case studies are provided on the Shaft Drillers International Web site (n.d.).

Zeni and Maloney (2006) identify the following hazards with a conventional drill-and-blast shaft-sinking operation:

- Falling of persons
- Falling objects
- Methane and other gases
- Hoisting and communications failures
- Equipment operating in a small work area
- Respirable dust
- Water inundation
- Back injuries
- Ignitions in the shaft
- Concrete burns
- Blasting
- Escape in an emergency
- Noise

Zeni and Maloney also state that many of these hazards can be eliminated using blind boring of shafts, because blind boring is probably safer than conventional sinking. Shafts in the range of 4.3–4.9 m in diameter are routinely bored using this technology, with larger diameters achievable. The blind-boring machine is rotated from the surface using a rotary table. In addition, the shaft is also completely filled with a fluid such as water; therefore, the boring element is completely submerged. The cuttings from the boring operation can be lifted from the bottom of the shaft in a number of ways; however, hydraulic and airlift systems are most common.

The most common problem associated with blind drilling is lining the shaft. The most common lining for shafts fewer than 4.6 m in diameter is the use of steel linings placed from the surface with concrete or grout used to bind the lining with the rock by injection in the annulus between the steel and rock. Other techniques do exist and are discussed by Zeni and Maloney (2006). These include remote slip-forming concrete, lightweight composite linings made from extruded fiber-reinforced polymer, lightweight concrete combined with steel or polymer, and totally waterproof cast-in-place concrete. The future of shaft drilling depends on technological advances in fabricating and installing linings remotely, which would avoid the need for humans to work in a shaft environment.

Another good example of the method is presented by Bessinger and Palm (2007) for the installation of a ventilation shaft at the San Juan coal mine in New Mexico (United States). The paper particularly discusses the options available for construction of the shaft, the reasons why boring was chosen, and a number of the technical aspects of the project such as collar construction, pilot-hole drilling, shaft reaming, lining installation, and fitting of the ventilation and escape hoist system.

Table 12.4-3 Rules of thumb for shaft sinking using ground freezing

- For a ground-freezing project, the lateral flow of subsurface groundwater in the formation to be frozen should not exceed 1 m/d. (Source: Khakinkov and Sliepcevich)
- To determine the diameter of a proposed circle of freeze pipes around a shaft collar, 60% should be added to the diameter of the proposed excavation. (Source: Sanger and Sayles)
- When ground freezing is employed for a shaft collar, the area of the proposed collar excavation (plan view) should not be greater than the area to remain inside the circle of pipes (area that is not to be excavated). (Source: B. Hornemann)
- The minimum practical thickness for a freeze wall is 1.2 m. (Source: Derek Maishman)
- The maximum practical thickness for a freeze wall with a single freeze circle is 5 m. Concentric circles of freeze pipes should be employed when a thicker freeze wall is required. (Source: Derek Maishman)
- The radiation (heat transfer) capacity of a freeze pipe containing brine may be assumed to be 165 kilocalories/m^2 of pipe surface. However, if the brine velocity is too slow (laminar flow), this capacity will be reduced by 40%. (Source: Jack de la Vergne)
- The capacity of the freeze plant selected for a ground-freezing project should be 2–2.5 times the capacity calculated from the radiation capacity of the total length of freeze pipes installed in the ground. (Source: Berndt Braun)
- Groundwater movements over 3 to 4 ft/d are significant in a ground-freezing operation. (Source: U.S. National Research Council)
- If the drill casing is left in the ground after installing the freeze pipes, it will cost more but the freeze pipes will be protected from blast damage or ground movement and the heat transfer will be increased due to the greater surface area of the steel casing. (Source: Jim Tucker)
- The heat gain from circulating brine is equal to the sum of the friction losses in the pipes plus the heat generated due to the mechanical efficiency of the brine pump. The value calculated for the heat gain should not exceed 10% of the refrigeration plant capacity. (Source: Jack de la Vergne)
- The amount of liquid nitrogen (LN) required to freeze overburden at a shaft collar is 1,000 lb LN/yd^3 of material to be frozen. (Source: Weng Jiaje)
- Due to the heat of hydration, the long-term strength of concrete poured against frozen ground will not be affected if the thickness exceeds 0.45 m. Below this thickness, designers will sometimes allow a skin of about 70 mm. (Source: Derek Maishman)
- Ground freezing with liquid nitrogen should only be considered for temporary or emergency conditions (1–2 weeks). For the longer term, conventional freezing with brine is less expensive. (Source: George Martin)
- The freezing method may be simply applied to a small problem by laying bags of solid carbon dioxide (dry ice) against the problem area and covering it with insulation. In this manner, ground may be frozen to a depth of up to 1.5 m. (Source: C.L. Ritter)
- When making the calculations required for ground freezing, it is necessary to have a value for the natural temperature of the ground. If this temperature is not available, the average or mean annual surface temperature at the project may be used for this value without sacrificing accuracy. (Source: Jack de la Vergne)
- The brine circuit employed for a ground-freezing operation requires a surge chamber. The best one consists of an elevated tank (open to the atmosphere) in a short tower. It provides simple visual observance of the pumping head and, if equipped with a sight gauge, gives precise changes in the volume of brine in the system and early notice if a leak should occur. Slight losses in brine volume will occur as it becomes colder (coefficient of volumetric thermal expansion is 0.00280 per degree Fahrenheit). (Source: Leo Rutten)
- When freezing a shaft collar with a conventional brine system, one can tell when the freeze wall cylinder has built up to the point of closure by noting a sudden rise of the water level in the pressure relief hole drilled in the center. Excavation may normally commence soon afterward, since the ground pressure against the freeze wall is smallest near the surface. (Source: Derek Maishman)
- When freezing a shaft collar with a conventional brine system, the freeze pipe headers must be bled of any air accumulation once a day. (Source: Leo Rutten)
- When freezing a shaft collar with a conventional brine system, thermometer instruments should be recalibrated once a week by inserting a thermocouple into a container filled with water and ice (temperature exactly equal to the freeze point of water). (Source: John Shuster)
- If the freeze wall does not close when calculated, the problem may be due to a flow of groundwater. If the return temperature of the brine from holes on opposite sides of the freeze circle is slightly higher than the average, there is very likely a significant lateral flow of groundwater. The elevation of this flow in the ground strata may be determined by stopping the brine flow to one or more of the "problem" freeze holes, waiting 2 hours, and then measuring the standing brine temperature at vertical intervals (usually 0.6 m). The water is flowing where the brine temperature is slightly higher than the rest of the measurements. If the return temperature of the brine from only one hole is higher than average, it can mean that groundwater is flowing from the bedrock where it is penetrated by the freeze pipe. In either case, the default remedy is to grout from holes drilled to the vicinity of the problem area(s). If the return temperature from one hole is lower than average, it usually means there is a short circuit in that hole. (The inner freeze pipe, usually plastic, was not installed to the bottom of the hole, or it has come apart at a splice.) The proof is to pull the inner freeze pipe from the hole so it can be measured and examined. (Source: Jack de la Vergne)
- When a liability rider is incorporated in the contract documents for a ground-freezing project, the contractor will often extend the ground-freezing cycle. For a shaft collar, this often means that the excavation will be frozen solid inside the freeze circle before releasing it for excavation. Although this procedure provides added safety, it increases the costs and delays the project because excavation is much more difficult in frozen ground than it is in a soft core. (Source: Derek Maishman)
- An acoustic instrument has been developed to measure the thickness of a freeze wall under construction. (The speed of sound is different in frozen and unfrozen ground.) This instrument is not reliable (and its use may lead to great confusion) unless it is employed by a single experienced operator and is properly calibrated before the freezing process begins on a particular project. (Source: Jack de la Vergne)

(continues)

Table 12.4-3 Rules of thumb for shaft sinking using ground freezing (continued)

- If a brine freeze plant does not cool to the expected final temperature [at least –250°C], the most likely causes in order of greatest frequency are as follows:

Cause	Remedy
Low level of refrigerant	Top up
Clogged water spray holes	Remove and clean piping over condenser
Oil buildup in refrigerant	Bleed off at bottom
Air in refrigerant	Bleed off at top
Vacuum feed to compressor	Adjust to positive feed
Faulty valves in brine line	Replace
Faulty expansion valve	Replace
Undersize brine mains	Replace
Oversize brine pump	Replace

(Source: Jack de la Vergne.)

- If the freeze wall ruptures during excavation of a shaft collar, it will occur near the bottom of the excavation. Normal practice is to immediately clear personnel and equipment from the bottom and dump a truckload(s) of fine gravel or coarse sand in the excavation to prevent further inflow of soil through the freeze wall. It is then typical practice to drill from the surface and chemical grout in the vicinity of the breach. Then drill and install another freeze hole in the vicinity of the breach to accelerate and reinforce its closure, after which the excavation may resume. The new freeze hole may be frozen with liquid nitrogen to minimize the delay. (Source: Leo Rutten)

- If a freeze wall rupture occurs in a section of the excavation that is in very compact granular soil, undisturbed glacial till, or in the bedrock, normally only water will flow into the excavation. In this case, it is typical procedure to allow the excavation to flood. When the level of the water in the shaft collar has peaked, ready-mix concrete is tremied or pumped down through the water to form a concrete pad on the bottom. After the concrete has set and sufficiently cured, the shaft collar is pumped dry and grout is injected through the pad. When the grouting is completed and the freeze wall restored, the pad is broken up and the excavation can safely resume. (Source: Jack de la Vergne)

Source: Adapted from de la Vergne 2009.

WINZING

Winzing is the term for small-scale shaft sinking. Usually it is a form of internal mine development used, for example, to connect levels. Winzing has the following characteristics:

- Usually small diameter
- No room for mechanization, although in modern systems raise-boring machines can be used to develop winzes; however, most are still sunk by hand
- The possible range of sizes and depths are limited due to problems with muck removal
- Muck removal is usually by hand shoveling
- Use of pilot holes makes winzing easier as deviation can be avoided if levels are connected by a pilot hole
- Ventilation is highly important, but usually poorly done
- Secure ladders are essential for access
- Small winches are used for raising and lowering small kibbles
- Typically a bench cut is used
- Water can be a problem, but its impact can easily be overcome by using a pilot hole
- All services need to be removed prior to a blast (ventilation ducting, water pumps and hoses, compressed air lines)
- Hard physical work, especially in hot conditions
- Vacuum-lift devices have been used successfully for muck removal

Winzing is always a slower and more expensive undertaking than anticipated. In modern mining it is avoided wherever possible and raising is used in preference.

RAISING

A raise is a vertical or steeply inclined opening excavated from a lower level to a higher level (i.e., inverse winzing). The inclination is such that a ladder is required for passage. Raises are used to gain access to ore, as traveling and supply routes, and as ventilation routes. Raises are also used as orepasses, waste passes, or as slot raises to provide free faces for initial blasting of certain types of stopes.

Raises can be constructed through conventional raising, cage or gig raising, Alimak raising, longhole raising, and raise boring.

Conventional Raising

This method is restricted to shorter raises of up to 100 m in length and involves drilling and blasting using handheld drills. An important requirement in this type of raising is the provision of a working platform at the face to allow drilling and charging operations to occur. This platform is dismantled prior to blasting. Raising operations can be dangerous if correct procedures are not followed. The advantages of raising include the following:

- Raising is independent of mining production, so its input on current mine performance is minimal.
- Loading of muck into buckets is avoided.
- Blasting effectiveness is greater.
- No water pumping is required.

The disadvantages of raising include the following:

- The height of a raise is limited to between 100 and 120 m.
- Ladder climbing is an inconvenience for personnel and the transport of materials to the face.
- Potential hazards of rock falling from the face exist.
- Potential threats of muck jamming in the chute compartment exist.
- Surveying is difficult (deviation is possible).
- Time and cost implications are involved in preparing the opening for raising purposes.

Courtesy of Shaft Drillers International.
Figure 12.4-9 Blind boring cutterhead

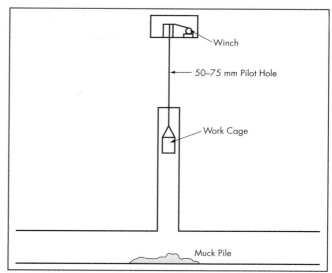

Figure 12.4-10 Cage or gig raising

Cage or Gig Raising

Cage or gig raising is a mechanized raising technique and is illustrated in Figure 12.4-10. A pilot hole is drilled from the upper level to the lower level. A hoist rope is passed down the pilot hole and attached to a cage (or work platform) that can be raised or lowered in the raise to the face, allowing work to proceed as normal by hand methods.

Alimak Raising

Alimak raising is a development of cage raising using a rigid steel guide rail attached to one wall of the raise. No pilot hole is required and all the controls are mounted on the cage or at the base of the raise. The design of the system incorporates provision for air and water hoses, drilling equipment, blasting requirements, and accessories. Alimak raising has the capability of driving longer raises at faster rates than conventional methods.

The machine runs on a guide rail that incorporates a pin rack. The segmented guide rail is fastened to the side of the excavation with rock bolts. The Alimak method has the following advantages over conventional systems:

- It allows the driving of very long raises, vertical or inclined, straight or curved.
- When traveling to the head of the raise, personnel are well protected in a cage under the platform.
- The miners work from a platform that is easily adjusted for convenient height and angle.
- Risks involving gases are reduced, as the raise is ventilated after blasting with an air–water mixture supplied in the raise climber.
- All equipment and material required can be taken to the head of the raise in the raise climber.
- Timbering is avoided, and if needed, rock bolts, with or without mesh, are used for support.
- The raise climber can be used for areas up to 8 m^2.

The disadvantage is the cost of the raise-climber system, which cannot be justified for short raises; therefore, the most frequent application is in mines where driving raises in hard rock is part of the mining method. The Australian Contract Mining Web site (2009) provides some excellent examples of the use of the Alimak climber as well as many images of Alimak climbers.

Longhole Raising

In this method all operations are undertaken from the top level. Long holes are drilled from the top level to the level at which the raise is to start. Holes need to be drilled accurately and can be reamed out. Charging is undertaken from the top level and blasting carried out in a similar manner to that used in vertical crater retreat mining. Figure 12.4-11 illustrates the longhole raise method.

Raise Boring

Raise boring is the most common method of raising in modern mining. It is a safe method for the driving of long raises, but it requires a high capital expenditure. Numerous methods of using raise borers are available such as blind-hole raising, pilot-hole methods, and so on.

Raise boring is one of the newest technologies in shafts and raises and has matured through improvements made during the decades since its initial application. Large-diameter holes or shafts can be drilled vertically or inclined. The costs of raise boring have been substantially reduced due to the improvement of cutters.

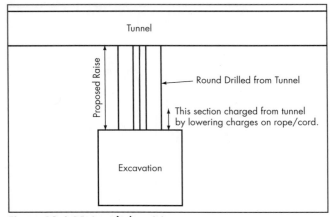

Figure 12.4-11 Longhole raising

Raises are used as a part of multilevel mining systems for orepasses and ventilation, as coal-mine ventilation shafts, and also as full-sized shafts, primarily in coal mining.

Excavating rock by a continuous boring system has several advantages over the conventional drill-and-blast methods, especially in vertical excavations such as shafts and raises, where it eliminates the hazards associated with the presence of workers at the face, similar to shaft boring. The continuous action of raise boring usually ensures faster completion of the project than conventional blast-and-drill methods. Boring does not disturb the surrounding rock, and as such, often there is no need for support or, alternatively, only minimum support is required for the finished excavation. A perfectly round shape makes the hole inherently stable, and its smooth walls reduce resistance for ventilation, as well as for the passage of ore with little prospect of hangups. Provided conditions are amenable, the cost advantage over other systems is clearly apparent for long raises and for shafts.

The disadvantages of a continuous boring system include a general requirement that rock conditions be similar along the length of the raise bore. Another disadvantage is that drilling in very hard rock is slow, and costs rapidly increase. Also, access to upper and lower levels is necessary. A raise-bored rock mass is preferably dry. The initial cost of the drilling rig is also high.

A large variety of machines are available that are designed for certain sets of rock conditions and nominal sizes and lengths of the finished hole. Typical characteristics of the machine specify the pilot-hole diameter, drill-pipe diameter, reaming torque, type of drive (AC, DC, or hydraulic), its rpm and power (hp or kW), and type of gear reducer. Additional specifications are the pilot-hole thrust and feed rates, reaming pull, size of drill pipe, base plate and derrick dimensions with characteristics (especially important for underground application), the type of transporter/erector, if applicable, and the water and air consumption.

Raise-boring machines operate on the principle of first drilling a small pilot hole and then reaming the hole, in one or more stages, to the desired size. The modes of operation are shown in Figure 12.4-12. The most frequently used is a conventional scheme with the pilot hole down and reaming upward. However, in some mining systems, box-hole drilling upward in one stage is chosen.

The raise-boring machine consists of several parts. The rig is composed of a rigid plate and a structure on which are assembled drilling, hydraulic, and electrical equipment. The rig has its own crawler driven by compressed air, or it is mounted on rail wheels for haulage. The rig can also be disassembled and transported. A pilot hole is drilled through the stem with stabilizers and a conventional drilling bit.

Stabilizers are used to ensure directional accuracy of the pilot hole. Several types are in common use. They are of a larger diameter than drill rods, with strips of tungsten carbide welded vertically or spirally on the outside, bringing the outer dimension close to the hole diameter. These stabilizers are used for reaming up. Other types, for pilot holes and reaming down, have more clearance, allowing cuttings to pass.

After the pilot hole penetrates to the target opening, the pilot bit and stabilizer are removed. The stabilizer is mounted on the drill stem and the reamer is put in the place of the small pilot-hole bit. Starting the hole (collaring) is a critical operation, and damage to the cutters or bit body can occur because of the uneven surface. The tension on the stem has to be adjusted to allow starting the hole without highly variable, dynamic loads acting on the drilling components. The stem is the longest at this stage of drilling and is, therefore, prone to maximum elongation and twist. The design of the reamer depends on the final size of the hole and the number of reaming stages.

In the past, the primary body of the reamer was often supplemented with stages mounted below, resulting in a "Christmas tree"-shaped tool, as illustrated in Figure 12.4-13. Because of large torque requirements, this system is presently used only for shaft-sized holes, together with large nonrotating stabilizers. In another configuration, extensions are attached to the primary body. By doing this, an increase of the final diameter is achieved by adding more cutters in the same planes as the cutters on the primary body itself.

The accepted measure of drilling performance is the rate of penetration, which depends primarily on bit design. The cutting action has to match the rock properties to optimize the drilling rate. During the reaming operation, in addition to the cutters, the design of the reaming head also influences the raise-boring performance. If no other source of information is available, an analysis of the wear of cutters can help in the design of an improved cutting structure. Independent variables when drilling are rock properties, which are usually characterized by compressive, tensile (indirect), and shear-strength measurements.

One of the most important technological aspects of boring is the removal of cuttings. Standard drilling technology with compressed air, air and foam, or drilling fluid is used for cuttings removal during pilot holing. During the reaming operation, cuttings simply fall down the hole.

The access point usually determines whether to drill raises from the top or bottom. If only one end of the raise has adequate room, the machine must be put there. The extent to which raise boring will interfere with production must also be evaluated. Locating the borer where fewer disturbances will be caused is preferred. Cost is also of major concern. Generally, drilling down and reaming up is cheaper than other configurations. Rock quality and its predicted response to drilling can influence the choice of direction as well.

In strong rocks without fractures, usually no support or lining is needed. If support is required, shotcrete, rock bolts (with or without wire mesh), or steel lining can be applied. Shotcrete can be placed remotely or by workers from a cage.

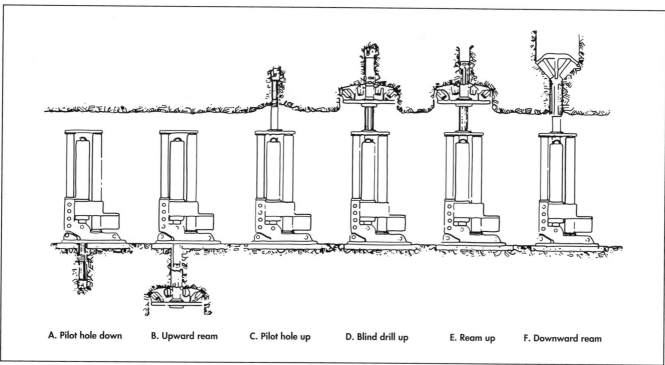

Source: Unrug 1992.

Figure 12.4-12 Typical modes of operation for raise boring

A. Pilot hole down   B. Upward ream   C. Pilot hole up   D. Blind drill up   E. Ream up   F. Downward ream

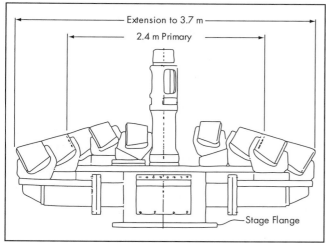

Source: Unrug 1992.

Figure 12.4-13 Raise head with extensions

Rock bolts must be installed manually. Steel lining, required in unstable formations, is installed remotely, directly behind the reamer.

Table 12.4-4 compares the factors that influence the selection of vertical/subvertical construction methods used in mining.

DECLINES

A decline is an inclined tunnel from a higher level to a lower level, usually but not necessarily from the surface. The purpose of declines is to access ore bodies or coal seams and provide use at a later stage as haulage routes. In coal mining, declines are known as drifts. These either provide access from the surface or from seam to seam; in the latter case they are known as cross-measure drifts. Drifts in coal mining operations can be driven at steep grades, up to 1 in 2.5 (40%), as the predominant mode of haulage of coal is by belt conveyor, which can handle such steep grades. Steep grades shorten the length of the drift, saving costs and giving early access to the coal seam(s).

In metalliferous mining, declines also allow access and may be used for truck haulage. Grades of 1 in 7 (14%) are common; however, steeper grades are possible in modern mining as more powerful underground trucks continue to develop. Various cutoff depths for declines can be found in mining literature as shown in Table 12.4-5. Beyond these depths, shaft systems should be employed. Over the years, the economic cutoff depth for declines has increased; for individual mines, the cutoff depth of declines is a purely economic decision and should be treated as such (McCarthy and Livingstone 2009). Medhurst (2001) shows that other factors also need to be considered in the determination of a cutoff depth including advances in technology such as the development of larger underground haul trucks, production rate, proposed mine life, potential for mine life being increased, and ventilation flow. Tables 12.4-6 and 12.4-7 compare and contrast decline and shaft access.

Declines are excavated and supported in a similar manner to horizontal tunnels, but drainage may require more attention. Details of tunneling techniques applied are described in the following sections.

TUNNELING

After the site of the surface connection(s) has been selected, the sites of tunnels or drives are often limited by the position

Table 12.4-4 Comparison of factors influencing the selection of vertical and subvertical construction methods

Factor Influencing Selection	Blind Shaft Borer	Blind Drilling	Raise Drilling	Vertical V-Mole	Conventional Shaft Sinking
Design Considerations					
Safety	Uses underground operators working under cover and behind shields. No operation of large moving equipment required.	Does not require miners to work underground	Only requires underground labor setup and mucking. No labor required in the shaft prior to shaft lining.	System uses underground operators working in a controlled environment. Considered safer than conventional shaft sinking as operators are located remotely from the working face.	Requires equipment operation in confined environment. Considered to be the most dangerous of the five shaft construction methods.
Shaft size	Depth is limited by skip-hoist rope capacity.	Limited by required depth, available equipment, and cost. Drilled shaft diameters range from well size to 20+ ft.	Nominally limited by available machine torque. Short (300-ft) shafts have been raised at 20-ft diameter.	Limited by available equipment from 16 to 23 ft in diameter.	Required to be larger than 10–12 ft for most applications. Upper limit not controlled by method.
Shaft depth	Depth is limited by skip-hoist rope capacity.	Limited by required diameter, available equipment, and cost.	Raise depths up to 3,200 ft have been reported.	—	—
Shaft verticality	Verticality can be controlled within extremely tight tolerances. Equal to or better than conventional.	Deviation can be estimated from geological data. Difficult to maintain an absolutely vertical shaft in subvertical structure. Suggested tolerance for design purposes = 0.25°–0.5°.	Shaft verticality is controlled by pilot borehole. Pilot-hole accuracy controlled by directional survey and careful drilling practice. Design tolerance should be specified based on use requirements.	Shaft verticality is controlled by pilot hole. However, offset reaming can be controlled by the operator providing additional control when following a deviated pilot borehole.	Verticality can be controlled within extremely tight tolerances. The most accurate shaft construction method with regard to verticality.
Ground disturbance	Minimal mechanical disturbance of shaft wallrock.	Provides minimal mechanical disturbance of shaft wallrock.	Provides minimal mechanical disturbance of shaft wallrock.	Provides minimal mechanical disturbance of shaft wallrock.	Conventional (drill-and-blast) excavation can result in deep-seated blast damage.
Timing/schedule	Setup time is 1 month or more. Advance rate should be between 20 and 40 fpd. In general, muck haulage and lining limit advance.	Typically much faster than conventional shaft construction methods.	Reaming range from 15 fph (5-ft dia.), 3 fph (10-ft dia.), to 1 fph (15-ft dia.).	Reported rates between 30 fpd (16-ft dia.) and 60 fpd (23-ft dia.).	Generally restricted to one round per shift (e.g., 1 fph).
Operational Considerations					
Groundwater	Same techniques as conventional sinking.	Method provides superior control of groundwater during excavation.	Groundwater controlled by pretreatment where necessary.	Large projected inflows require pretreatment (e.g., using grouting or freezing).	Large projected inflows require pretreatment (e.g., using grouting or freezing).
Support during excavation	Temporary or final support can be installed a short distance behind advancing face.	Support provided by hydraulic pressure and impermeable polymer skin permitting excavation in very poor ground conditions.	Not possible.	Temporary/final support can be installed a short distance behind advancing face.	Temporary/final support can be installed a short distance behind advancing face.
Final lining	Installed during excavation.	Steel/concrete composite lining typically used in weak ground. May incorporate bitumen layer for groundwater control and be designed for full hydrostatic, bituastatic, or lithostatic loading conditions.	None is usually required. Many rapid lining systems available	Final lining may be installed during excavation.	Final lining may be installed during excavation.
Miscellaneous	—	—	Requires existing underground access.	Requires existing underground access.	—
Other Considerations					
Shaft outfitting	—	If a final steel lining is installed, considerable time can be saved through surface installation of guides, brackets, and pipes. All furnishings can be aligned prior to welding to the downhole liner assembly.	Pilot-hole deviation usually prevents raises from being used for worker or materials winding.	Outfitting in line with excavation and final lining if required.	Outfitting in line with excavation and final lining if required.
Initial costs	—	High	Medium	High	Low
Operating costs	—	Medium	Medium	Medium	High

Source: Breeds and Conway 1992.

Table 12.4-5 Cutoff depths for declines compared to shafts over time

Maximum Decline Depth, m	Comments	Source
350	Average value	Northgate and Barnes (1973)
500	Based on a maximum haul length of 3,500 with a 1 in 7 (14%) gradient	Thomas (1978)
1,000	Average value	Medhurst (2001)
1,400	Stawell gold mine, Victoria, Australia	Atlas Copco (2008)

Table 12.4-6 Shaft/decline impact on mining operations

Operation	Decline Truck Haulage	Shaft Hoisting
Truck haulage	• Direct to surface • Long haulage cycles from depth • Expanding fleet size • Rising insurance for fleet • Increasing fleet maintenance	• Underground haulage to shaft • Additional trucking or conveying at surface • Shorter underground haulage cycles
Ore fragmentation	• Popping of oversize at surface	• Stope firing to suit hoisting systems • Underground crusher and/or grizzlies required to limit ore size at shaft loading station
Decline	• Increasing decline congestion for haulage access • Increasing road maintenance in decline	• No decline haulage between the shaft production level and the surface
Production shaft and underground crusher	• Continued surface crusher operation	• Hoisting system operation and maintenance • Underground crushing station operation and maintenance (if applicable) • Fixed plant insurances required
Workshops	• All truck servicing at surface workshops • Surface refueling for trucks generally	• Routine truck servicing underground • Truck refueling underground • Major truck service/overhaul on the surface
Site crew	• Increasing crew size for drivers, service staff, grader operation, support staff, and management • Expanding fly in/fly out routine	• Crushing and hoisting system maintenance staff required • Overall smaller site crew
Change of shift	• Inefficient shift change with large truck fleet and long haulage cycle times (approximately 1-hour cycle from 800-m depth)	• Organized crew transport for shift change of underground truck and maintenance staff
Energy source and greenhouse effect	• Diesel fuel required for haulage (generally) • Increasing fuel usage with depth (30%–50% higher than fixed plant)	• Hoisting system may be supplied from the grid (if available) • Remote power plant may use diesel fuel or natural gas • Overall lower energy use
Operating costs	• Haulage costs increase steadily as depth increases • Higher cutoff grades required for economic mining	• Reduced total haulage/hoisting costs at depth • Lower grades may be mined economically, extending reserves and mine life • Deferral of costs associated with mine closure
Health and safety	• Increased risk of truck fires in decline	• Safety systems and procedures required to ensure safe operations and maintenance in production shaft

Source: Adapted from Medhurst 2001.

of the ore body in relation to the surface connection(s). If choice is available, the most competent ground should be chosen. As with surface connections, the drive size is determined by duties such as ventilation, access, haulage, and so on. Another prime consideration is the mining method, as this controls the size of mining equipment required.

In modern mining, drives smaller than 2 × 2 m are not common. A 2 × 2 m drive will accommodate a small loader and allows easy passage for people. Medium-sized drives have a cross-sectional area in the range 10–25 m² and represent the most common size range of mining drives. Such drives can accommodate locomotives, conveyors, and rope haulage systems as well as allowing ventilation flow and access.

From a mining viewpoint, large drives lie in the range 25–40 m². This drive size is becoming increasingly popular, especially in mines employing trackless equipment and mines employing decline access. This drive-size range allows for the use of off-highway trucks and other larger items of trackless equipment.

Drives have a limited range of sizes due to the need to maintain a level floor for haulage and access reasons. Common shapes are square, rectangular, and arched. Arched roofs have better strength characteristics than square or rectangular cross sections; they are used for drives where long-term stability is required. Square and rectangular section drives are very popular in mining, especially in areas of competent ground or in production areas where the drives have a limited life span.

Two methods of excavation exist: traditional drill-and-blast and mechanized methods (which are described in subsequent sections). Drill-and-blasting is a cyclical operation, whereas mechanical methods offer a more continuous system of excavation.

Table 12.4-7 Shaft/decline impact on mine infrastructure

Infrastructure	Decline Truck Haulage	Shaft Hoisting
Truck fleet	• Larger truck fleet for decline haulage at depth • Upgraded truck capacity (subject to decline limitation) • Surface haulage fleet (optional)	• Modest truck fleet for underground haulage to shaft • Surface fleet for haulage from shaft to plant (if required)
Mechanized ore handling	• Continued use of surface crushing and conveying facilities	• Introduction of hoisting system, underground crushing, ore bins, and associated plant • Local mine development to integrate with shaft
Mine ventilation	• Expansion of mine ventilation for decline haulage from depth	• Hoisting shaft acts as main airway from surface • Reduced ventilation demand
Decline	• Additional passing bays and/or duplication of the decline to address decline congestion • Upgrade to traffic control system in the decline • Upgrade decline fire protection and associated refuge bays	• No significant impact
Workshop	• Upgrade/expansion of surface workshops to suit larger truck fleet • Increasing inventory of truck spares	• Expansion of underground workshops • Upgrade electrical maintenance facilities
Power supply	• Upgrade for ventilation fans and surface facilities	• Upgrade for demand of winder drive, underground crusher, and associated conveyor drives
Fuel	• Upgrade diesel storage for truck fleet • Expand fleet refueling bays for larger truck fleet	• Upgrade fuel storage for power station demand (if applicable) • Provide underground fuel storage and refueling bays
Water supply	• Upgrade for larger crew and truck fleet, and reticulation for decline fire protection	• Upgrade underground water reticulation for maintenance, sprays, and deluge systems
Second means of egress	• Extend/upgrade emergency egress facilities at depth	• Hoisting shaft provides emergency egress above loading station
Surface facilities	• Upgrade/extend due to larger crew	• Minimal upgrade
Underground services: rising mains, raw water, potable water, power cable(s), fiber-optic cable, leaky feeder, blasting control cable	• Route mechanical and electrical services via the decline and/or boreholes • Establish rising mains in boreholes	• Most services can be routed directly via the hoisting shaft
Mine control	• Mine operations generally controlled from surface	• Main control center usually located at winder house with unstaffed satellite stations underground

Source: Adapted from Medhurst 2001.

Drilling and Blasting

The majority of mining drives are excavated by traditional drill-and-blast; however, mechanical mining is becoming much more popular outside its traditional area of coal and other soft-rock mining. Drilling can be done by hand using air legs. The number of air legs in use will depend on the drive size (e.g., in a 2 × 2 m drive, only one drilling machine can be accommodated). In larger drives, drilling is mostly undertaken using mechanized drill jumbos with twin or three booms common. The most common form of drill jumbo is electrohydraulic, but other forms of power are used. In dry holes, ANFO (ammonium nitrate and fuel oil) is the most commonly applied explosive, whereas in wet holes, explosive emulsions or slurries are increasingly being used. Figure 12.4-14 shows the basic cycle of operations for a typical drill-and-blast tunnel.

A good example of a typical cycle of operations is provided by Argyle Diamonds (2007). Argyle is developing a new block cave operation to mine diamonds in Australia. The cycle of operations during the mine development prior to establishing caving uses typical drill-and-blast tunneling, with one of the key design elements being that at no time does the work force work under an unsupported roof, a common element in Australian mining. For each heading under development, the cycle takes about 18 hours to complete for a nominal face advance of 4 m/cycle and an average heading advance rate of 32 m/week.

In small drives, mucking may be carried out by hand, shoveling into mine cars. This is really only a viable method when cheap labor is available. In the past, rail-mounted shovel loaders were used extensively. These were almost universally powered by compressed air and operated as a front-end loader, "slinging" the muck over the body of the loader into mine cars located behind. Generally, bucket capacity was limited to 0.5 m^3.

The real revolution in mucking operations came with the development of rubber-tired loaders, which lead to the development of load-haul-dump units (LHDs). The advantage of these units is their mobility and flexibility. Bucket sizes vary from 1 to 10 m^3; the size selected depends on drive dimensions and production rate required. Motive power is either by diesel engine or in some cases electrical power (either battery or trailing cable). LHDs have a limited economic hauling distance range; if haul distances become too large, a more economic intervening haulage system such as trucks needs to be added between the LHD and the transfer point to the main mine haulage system. Numerous other methods of mucking also exist such as front-end loaders and gathering-arm loaders.

Numerous methods of haulage exist. In small drives, hand tramming of mine cars may be used. Other commonly applied methods include the following:

- Locomotive (rail) haulage: Motive power can be provided in numerous ways such as diesel engine, battery locomotive, electric trolley wire systems, or rope haulage.
- LHD with or without truck.
- Shuttle cars with or without trucks/conveyor.
- Belt conveyor.

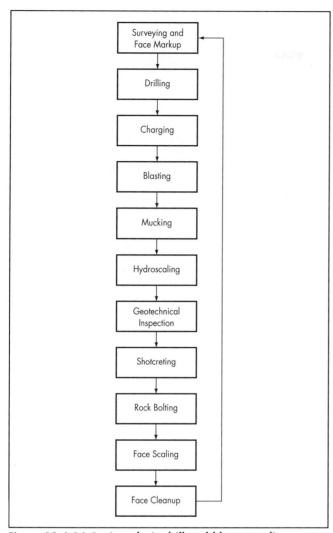

Figure 12.4-14 Basic tasks in drill-and-blast tunneling

Several methods of support are possible; the support required depends on local conditions. The main methods of support include

- Rock bolts,
- Cable bolts,
- Steel sets,
- Timber sets,
- Shotcrete,
- Concrete,
- Combinations of methods,
- Mechanical shields, and
- Natural support.

To design a drive blast, the method outlined by Persson et al. (1994) can be employed. The method can be broadly summarized as follows:

- Determination of the burden and spacing for the cut holes, to provide the relief space required such that the main blast does not freeze, for both angled and parallel hole cuts
- Determination of the burden and spacing for the main blastholes or production holes, also called easers, for the holes that blast downward and sideways into the cut area
- Determination of the burden and spacing for the main blastholes or production holes, also called easers, for the holes that blast upward into the cut area
- Spacing of the trimming or perimeter holes in the backs and sides of the tunnel for both conventional and/or perimeter blasting
- Spacing of the lifter or floor holes

Mechanized Tunneling

Three basic rapid mechanical tunneling systems exist: full-face tunnel boring, mobile miners, and roadheaders or boom-type machines.

Full-Face Tunnel Boring Systems

Full-face boring systems, or tunnel boring machines (TBMs), have been in common use in civil tunneling for many years but are used less frequently in mining projects. Nevertheless, TBMs are viable for mine development. Some good examples of TBM application and other mechanized tunneling techniques are shown in the video *The Burrowers* (Harris 1994).

Constant development and improved tooling have resulted in machines that are capable of advancing large-diameter openings in strong igneous and metamorphic formations at rates that compete favorably, and in many cases exceed, conventional drill-and-blast methods. In addition to high advance rates, TBMs leave a smooth profile, and usually minimal ground support is required. A disadvantage of TBMs is their wide turning circle, although a range of mini-full-face machines are available that have smaller turning radii. The high initial cost of these machines is balanced by low running costs compared to drill-and-blast excavation systems.

Full-face boring machines consist of a rotating cutting head fitted with disk cutters, drag bits, button bits, or various combinations of these. Advanced machines are available on which the tool type can be changed and tool spacing varied. The cutting head may be an open structure with spoke-like cutting arms, or it may completely conceal the face except for muck-removal openings and accessways for tool maintenance. The open-type head gives better access to the face and tools, and can be used with a forepoling arrangement. Cutting forces are provided by the head rotation, whereas normal forces are provided by the thrust of the machine against the tunnel face. Reaction to this thrust is provided by grippers mounted on the TBM body, which in turn react against the tunnel sidewalls. Mucking is performed by buckets mounted on the periphery of the cutterhead, and muck is removed via a central conveyor system.

Preparations for TBM excavation typically involve construction of a portal, placement of a concrete pad on which the TBM is assembled, and installation of support services and equipment as shown in Figure 12.4-15. Careful excavation, including the use of controlled blasting techniques, is usually required to mine the setup area to the tight tolerances required. Placement of a thin concrete layer against the start-up face is recommended to reduce out-of-balance loads.

TBM excavation is a continuous process, with cutting, mucking, and support installation proceeding concurrently. As the cutting head rotates, it moves forward, reacting against the grippers. The grippers are repositioned periodically when they reach the limit of their travel. The grippers are also used to steer the machine in some cases; in others, a floating main

Courtesy of The Robbins Company.
Figure 12.4-15 Tunnel boring machine

beam is used to steer while boring. Mucking in the immediate vicinity of the face is done by buckets located on the head periphery, and a central conveyor system moves the muck through the body of the machine to a bridge conveyor. When the rock is wet or where water inflow is a problem, earth pressure balance systems are used and the muck is removed in a coagulated form in a batch process. The bridge conveyor allows access for track laying and service installation without disrupting the mucking operation. A variety of mucking systems can be used to haul muck to the surface, but usually conveyors or shuttle trains are used. Adequate muck removal rates are critical to optimum face advance rates.

Rock support requirements for hard-rock TBM drives are generally minimal, and estimates have suggested that the savings in rock support costs (compared to drill-and-blast) can offset the cost of the machine in as little as 7 km of tunnel. Hard-rock TBMs are commonly equipped with a partial or slotted shield, and when support is required, conventional rock-support methods are used. Both soft-rock and hard-rock TBMs can be equipped with a full shield and segmental linings installed. This equipment enables hard-rock TBMs to cope with localized occurrences of soft ground.

Mobile Miners
Mobile miners are continuous hard-rock mining machines. A prototype Robbins mobile miner was used in 1984 for the development of a 1,150-m decline at Mt. Isa in Australia. Advance rates of 3.66 m/shift were attained mining a decline 3.66-m-high × 6.4-m-wide section through high-strength quartzite (Boyd 1987).

Roadheaders
Roadheaders have been used in mining and tunneling for a number of years. They are known by a number of names including boom headers, boom-type tunneling machines, and selective tunneling machines. Figure 12.4-16 shows commonly used terminology for the various machine components.

Originally developed as a means of advancing roadways in underground coal mines, early roadheaders were limited to cutting relatively low-strength strata. Development of these machines has greatly extended the range of applications, and they are now used in a wide variety of mining and civil tunneling work. Improvements in cutterhead design and the increasing use of water-jet-assisted cutting will result in a further extension of the range of roadheader applications in the coming years.

Roadheaders offer a number of advantages over TBMs, chiefly related to flexibility. They can cut a variety of cross sections, limited only by the basic dimensions of the machine, and are able to cut tight curves or corners.

These machines also offer advantages over conventional drill-and-blast methods. One of the most important advantages of roadheaders is the avoidance of blast damage to the rock and the consequent savings in ground-support costs. In addition, because mechanical excavation is a continuous process, shift time is more effectively used.

Roadheader cutting assemblies consist of a cutterhead on a movable, hydraulically powered boom, mounted on a rotatable turret attached to a track-driven chassis. In addition to the cutterhead, the machine also incorporates either a gathering arm or chain conveyor mucking system to remove broken rock from the face. The machine may be controlled by an operator seated on the machine or located some distance away, inside a shield, or beneath a supported roof section. To improve machine stability during cutting, many machines are equipped with hydraulically powered steering rams that are used to brace the machine off the excavation sidewall.

Two types of cutterheads are available from several manufacturers and are interchangeable on specific machines. These heads are termed *transverse* (or "ripping") heads and *in-line* (or

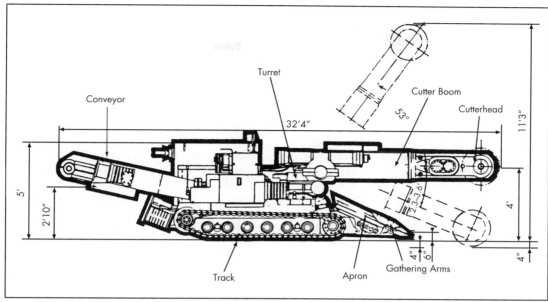

Source: Breeds and Conway 1992.
Figure 12.4-16 Main components of a roadheader

"milling") heads. In-line heads rotate coaxially with the boom. These heads are best suited to cutting rock with an unconfined compressive strength of 80 MPa or less. In-line heads require less thrust when sumping, and the head shape allows greater selectivity in cutting specific beds or bands. Transverse heads rotate at right angles to the boom axis. Advances in cutter booms have resulted in machines equipped with telescopic booms, or booms with extended length for cutting high backs or crowns. Telescopic booms are useful for cutting on steep gradients or on weak floor materials where the thrust from the machine's travel system would not be adequate.

Roadheader face-mucking systems generally consist of an apron with either a scraper chain or gathering arm system. Without adequate mucking capacity, the face will become muck bound and excavation will be delayed. In low-strength formations in which high cutting rates are possible, the mucking system may become the limiting factor in controlling advance rates. Roadheader cutting booms are generally mounted on a custom chassis with track propulsion. Track dimensions control ground pressures and should be carefully assessed in conditions in which the invert rocks are weak or prone to slurrying.

In the local area around working roadheaders, high concentrations of airborne dust, generated during both cutting and transport of the muck, and high ambient temperatures commonly occur. In addition to exceeding statutory respirable dust limits, excessive dust may completely obscure the face, resulting in inefficient excavation and increased overbreak. High temperatures and humidity result in labor inefficiency and overheating of electrical motors. The use of water-jet-assisted cutting leads to a reduction in dust levels, but does not entirely eliminate the problem, and may actually increase humidity at the face.

One of the primary advantages of using a roadheader excavation system over drill-and-blast methods is the elimination of blast damage and the consequent savings in rock-support costs.

All types of rock support can be adopted for use in conjunction with roadheaders. Because it can be relatively difficult to reverse a roadheader away from the face, the machine must be covered prior to shotcrete application.

ACKNOWLEDGMENTS
The author acknowledges the works of Unrug (1992) and Hood and Roxborough (1992) who wrote two excellent chapters in the previous edition of this handbook that were used extensively in this chapter.

REFERENCES
Andra. 2009. www.andra.fr. Accessed January 2009.
Argyle Diamonds. 2007. *Block Cave Mining: Construction and Operation*. DVD. London: Rio Tinto.
Atlas Copco. 2008. The truck race. www.youtube.com/watch?v=hi_Tatc4bAE. Accessed September 2009.
Atlas Copco. 2010. Underground mining. www.atlascopco.com/rock. Accessed March 2010.
AusIMM (Australasian Institute of Mining and Metallurgy). 2007. *The Mining Glossary: The Essential Handbook of Mining Terminology*. Innaloo City, WA: RESolutions Publishing and Media.
Australian Contract Mining. 2009. www.australiancontractmining.com.au/. Accessed December 2009.
Ayugat, A.V., Aquino, J.H., and White, P. 2002. Grouting and shaft sinking through water-bearing ground. www.multiurethanes.com/PDFfiles/pdf4.htm. Accessed December 2009.
Bessinger, S.L., and Palm, T.A. 2007. Installation of a ventilation shaft at San Juan coal. *Min. Eng.* 59(1):35–38.
Boyd, R.J. 1987. Performance and experimental development of the mobile miner at Mt. Isa. In *Proceedings of the Rapid Excavation and Tunneling Conference*, Vol. 2. New York: SME-AIME.

Breeds, C.D., and Conway, J.J. 1992. Rapid excavation. In *SME Mining Engineering Handbook*, 2nd ed. Edited by Howard L. Hartman. Littleton, CO: SME.

de la Vergne, J. 2009. *The Hard Rock Miners Handbook*, 3rd ed. www.mcintoshengineering.com/HardRockMiners Handbook/tabid/76/Default.aspx. Accessed January 2010.

Harris, J.S. 1995. *Ground Freezing in Practice*. London: Thomas Telford.

Harris, N. 1994. *The Burrowers: People and Their Tunnels*. Video. London: London Television Service.

Hood, M.C., and Roxborough, F. 1992. Rock breakage: Mechanical. In *SME Mining Engineering Handbook*, 2nd ed. Edited by H.L. Hartman. Littleton, CO: SME.

McCarthy, P.L., and Livingstone, R. 2009. Shaft or decline? An economic comparison. www.amcconsultants.com.au. Accessed January 2010.

Medhurst, G. 2001. Time to examine mine options more deeply. *Aust. Min. Mon.* (April): 72–76.

Northgate, G.G., and Barnes, E.L.S. 1973. Comparison of the economics of truck haulage and shaft hoisting of ore from mining operations. In *Papers Presented at the Symposium on Transportation, October 1973*. Sydney: Australasian Institute of Mining and Metallurgy.

Persson, P.A., Holmberg, R., and Lee. J. 1994. Rock blasting and explosives engineering. Boca Raton, FL: CRC Press.

Petterson, S., and Molin, H. 1999. *Grouting and Drilling for Grouting: Purpose, Applications, Methods and Equipment with Emphasis on Dam and Tunnel Projects*. Stockholm: Atlas Copco.

Roesner, E., Poppen, S., and Konopka, J. 1984. Stability during shaft sinking. In *Proceedings of the 2nd International Conference on Stability in Underground Mining*. New York: SME-AIME.

Shaft Drillers International. n.d. www.shaftdrillers.com/. Accessed January 2010.

Thomas, L.J. 1978. *An Introduction to Mining*. Sydney: Methuen of Australia.

Unrug, K.F. 1992. Construction of development openings. In *SME Mining Engineering Handbook*, 2nd ed. Edited by Howard L. Hartman. Littleton, CO: SME.

Zeni, A.J., and Maloney, W.J. 2006. Blind shaft drilling as a safer alternative to conventional sinking. In *Proceedings of the 11th U.S./North American Mine Ventilation Symposium*, University Park, PA, June 5–7. London: Taylor and Francis.

CHAPTER 12.5

Subsurface Mine Development

Richard L. Bullock

INTRODUCTION

To prepare a deeply buried ore body for extraction by underground mining methods, the mine must be developed. This means driving all of the underground openings for entry into the mine as well as all of the openings needed for exploration, exploitation, and services for the mine. Many openings must be driven before the mine can begin production and many more ongoing openings must be progressed as the ore body is extracted. For some block caving mines, this means driving an average of as much as 25 km/yr (14 mpy). This chapter discusses the types of openings that must be developed, the purposes for developing them, what they are called, and how they relate to various mining methods.

A development in a hard-rock mine is any initial opening driven to provide an accessway anywhere in the mine. It may not necessarily be in the ore. Such accessways usually are driven as single openings and provide space for roads, ventilation, power, air and water lines, drainage, storage areas, shops, dump stations, offices, and so on. They also expose the ore so that stope exploitation can begin at the points of exposure. The term *development* implies only that there is an opening for some mine service or where production might begin.

In contrast, in coal mining the term *development* refers to that phase of the mining cycle during which multiple entries are advanced through a virgin seam, forming pillars by breaking crosscuts or breakthroughs between the entries. When the development has reached the extent of the advance, the coal-extraction process removes all or part of the pillar on the retreat, or a longwall panel is set up for the extraction. The only area of obvious similarity between the two meanings is where the development opens more extractable ore or coal.

All underground mining development openings can be classified into four categories:

1. To open production accessways
2. To obtain information
3. To construct service facilities
4. Combinations of the first three categories

For each development, decisions must be made pertaining to each of the five physical variables:

1. Length
2. Direction
3. Inclination
4. Size
5. Method of ground control

Each development must serve the purpose for which it is intended and usually named. Predevelopment studies should include a discounted cash-flow analysis and ensure that all elements of each alternative are included and are indeed comparable.

Spacing and Alignment of Development Excavations

There are other general rules for recommended or preferred location and direction of mine developments. The following is taken from the works of Spearing (1995) and has been distilled from many years working in South African deep mines. While it may be true that all of this information applies to rock under considerable stress, it applies equally to weaker ground under low to moderate stress.

These rules are based on the theory of stress concentrations around underground openings and the interaction of those stress concentrations. The usefulness of these guidelines has been proven by experience obtained underground. Stress interaction between excavations can obviously be controlled by an increase in the installed support, but costs will also increase significantly. If there is adequate available space, it is generally more cost-effective to limit stress interaction between excavations.

Flat and Vertical Development

In flat development, the spacing between two square cross-sectional openings is

- Spaced horizontally at three times the combined width of the excavations; and

- Spaced vertically at three times the width of the smaller excavation, provided the area of the larger excavation is less than four times the area of the smaller.

The spacing between a rectangular cross section is

- Spaced horizontally at three times the combined maximum dimensions of the excavations; and
- Spaced vertically at three times the maximum dimension of the smaller excavation, provided the height-to-width ratio of either excavation does not exceed either 2:1 or 1:2.

The spacing between two circular cross-sectional openings is

- Spaced horizontally at three times the diameter of the larger excavation; and
- Spaced vertically at three times the diameter of the smaller excavation, provided the area of the excavation is less than four times the area of the smaller.

In vertical development (e.g., shafts and raises), the spacing of

- A square cross section is three times the combined widths of the excavation,
- A rectangular cross section is three times the combined diagonal dimensions of the excavations, and
- A circular cross section is three times the diameter of the larger excavation.

Successive Turnouts (Breakaways)
Multiple turnouts from a given intersection should be avoided to reduce dangerously large roof spans at intersections. The rule is that turnouts should be spaced at six times the width of the excavation between successive tangent points. (This rule does not apply to room-and-pillar stoping. In this case the roof spans are safely designed.) Acute turnouts (less than 45°) should be avoided because these result in pointed "bullnoses," which are unstable. Fracture and failure of a bullnose result in large, unsupported hanging wall spans. The breakaways for inclines are often brought too close to the connecting crosscut. The length of the connection between an incline and the flat drifts should be three times the diagonal dimension of the flat end.

Orientation of Adjacent Excavation Developments
The most highly stressed part of an excavation is the corner. Therefore, positioning a development such as a decline in an unfavorable orientation, too close to the corner of an adjacent opening, should be avoided.

Development Influenced by Geology
In all cases, the geology of the area should be taken into consideration in planning the development. Known weak geological horizons should be avoided even at the expense of longer developments. In permanent excavations (sumps, settlers, hoist chambers, etc.), the position of faults and the orientation of joint sets are critical to the stability of the development. Layouts must therefore cater to such geological features. All excavations should be kept away from high stress dikes where possible. When a dike is traversed, the most direct route should be used. Breakaways should not be sited in dikes, even at the expense of extra development. Haulages (drift opening) should not be positioned in fault zones, and development should not occur alongside a fault. A fault should always be intersected at an angle as near to 90° as possible.

Development Influenced by Stress Condition
When haulages are positioned beneath mined-out areas, consideration should be given to the 45° destressing guideline. An overstoping angle of 45° is generally required to destress a haulage. At an angle greater than 45°, stress concentrations are higher. This rule applies to reefs of 0° to 20°. Dips greater than this often call for computer modeling to show the extent of overstoping needed. Haulages should not be laid out too close to remnant pillars, which are very highly stressed. Another case of this application is where the production level for block caving is developed beneath and well ahead of the extraction level to avoid the high-stress conditions that will come once the extraction level is opened.

OPENING PRODUCTION ACCESSWAYS

Initial Entry into the Underground Mine
Figure 12.5-1 shows various types of entries into an underground mine. The initial entry may be for further exploration, a test mine, production openings, or eventual service openings. The opening may be a drift entry (adit) into the side of a hill, a decline (slope), a vertical shaft, an incline shaft, or a multiple-stage hoisting shaft. The type of opening will depend on many factors: the purpose of the opening; the elevation of the terrain at the surface of the mine in relation to the elevation of the ore body; the type of ore body or shape of the ore body; how material, workers, and ore/waste will be moved in and out of the mine; the depth needed to eventually develop openings; and the mining method that will be used to exploit the ore. Each of the types of initial openings is discussed in the following subsections.

Drift Entry (Adit)
An adit, or drift entry, is the most economical approach when the extractable material is above the floor elevation of a nearby valley. Literally thousands of Appalachian room-and-pillar coal mines have been opened this way to mine coal seams exposed on the hillsides; these are known as "punch mines." Most of the early western-U.S. hard-rock mines used this type of development, as did most of the limestone and dolomite mines throughout the Missouri–Mississippi river basin. The exposed rock could be opened through horizontal adits into the sides of the mountain bluffs.

A few of the advantages of the drift entry are

- Mine drainage by gravity,
- Level or downhill haulage for mined material,
- Easy access for equipment in and out of the mine, and
- Rail haulage can be used if desired.

Adits are usually driven by drill-and-blast methods using trackless jumbos, rail-mounted jumbos, jackleg handheld drills, or with a mechanical excavation machine such as a tunnel boring machine (TBM) or roadheaders. The grade of the adit should be driven at a +2% to 3% grade for good drainage, but probably limited to +1% to 2% if rail haulage is to be used. Almost never should a drift be driven by drill-and-blast methods on a flat grade if the adit goes any significant length, because there is no way to drain it unless the water rises enough to top the roadbed. For nearly flat grades, laser-guided equipment should be used, and if there is a lot of water, it will likely be necessary to overexcavate the bottom, build a flume-type ditch (U-shaped) on the side, and build a roadbed up to the grade that is needed.

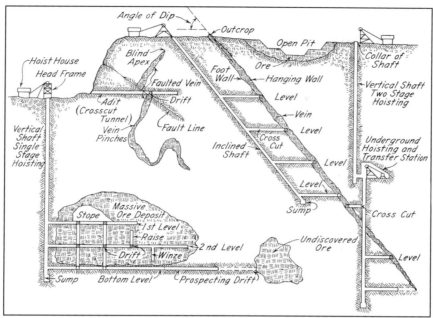

Source: Lewis and Clark 1964.
Figure 12.5-1 Underground mine showing names of mine openings

When the mineable material goes to a depth lower than the surrounding terrain, either a slope (decline) or a shaft must be developed. Many mines start out as adit mines and then a shaft is sunk or a decline is driven to exploit the lower portions of the ore body. The ore is then brought to the level of the adit, where it is hauled out of the mine. The same is true for ore lying above the adit. The shaft or ramp will be excavated above the adit and the material will be dropped down to adit level for haulage to the surface. Some refer to these as "upside down" mines. A few examples can be found at the Pine Creek tungsten mine (Bishop, California, United States), the El Soldado copper mine (Chile), and the Stillwater and East Boulder platinum mines (Montana, United States).

Decline (Slope) Entry

Again, there is a difference between what this development is called in hard-rock mining and coal mining: coal mines term such development as *slopes*, and these slopes usually contain a production conveyor belt; hard-rock miners call them *declines*, and unless it is a very high production mine, more than about 4,540 t/d (5,000 tpd), they probably will not contain a conveyor belt. Slopes/declines are usually limited to relatively shallow mines because for a given vertical depth, declines require approximately five to six times the linear distance compared to a shaft. Thus, to reach a depth of only 460 m (1,500 ft), an 18% slope would be approximately 2,540 m (8,333 ft) long.

The declines will normally be driven by using drill jumbos and trackless loaders and trucks. Roof bolting and shotcrete will normally be used for support. It is common practice to drive "muck bays" at regular intervals as the decline is being driven so the blasted round can be moved very quickly from the face to the muck bays, leaving the face available for another drill round sooner. The development muck can then be transferred to the surface while the next round is being drilled and charged. The usual distance between muck bays is about 120–180 m (400–600 ft). Declines can, in theory, be excavated by TBMs but rarely are.

There is an ongoing debate regarding what the proper decline grade for truck haulage should be. However, it depends on the length of the haul, power of the truck's engine, and drivetrain versus the size of the load and the surface condition of the decline. Normally, for good, efficient haulage, the truck should be able to maintain a speed of 12–13 km/h (7–8 mph) hauling a full load up the ramp. As an example, hauling a 48-t (53-ton) load up a straight, well-maintained road surface at 15% grade, a truck with a 485-W (650-hp) engine should be able to maintain 12 km/h (7 mph) for the full length of a 4-km (2.4-mi) decline. Many operators try to drive haulage declines at 18%–20% to save initial cost with a shorter decline, but this usually increases the operating cost by increasing the cycle time of the haul. This can be remedied by employing engines and drivetrains that are much more robust than those used for normal mine haulage. Many suppliers of haulage equipment today can supply the needed drivetrain for very steep grades, even for large-haul trucks.

Gradients steeper than 15% are not practical for main production areas due to difficulties maintaining the road surface to an adequate standard. The road surface tends to undulate and creep, particularly when wet. There are also issues with accelerated transmission wear on trucks.

The slope driven for conveyor-belt haulage for most crushed ore and coal can be up to 30°; however, if rubber-tired mobile equipment must also operate on the decline, this may limit the grade to less than 20% for safe operation.

The advantages and disadvantages of declines are discussed later in this chapter (see the "Slopes Versus Shafts" section).

Vertical Shaft Entry

In North America, for most deep hard-rock mines over 450 m (1,500 ft) deep, vertical shafts are still the preferred method of opening the mine for production. The purposes of vertical shafts

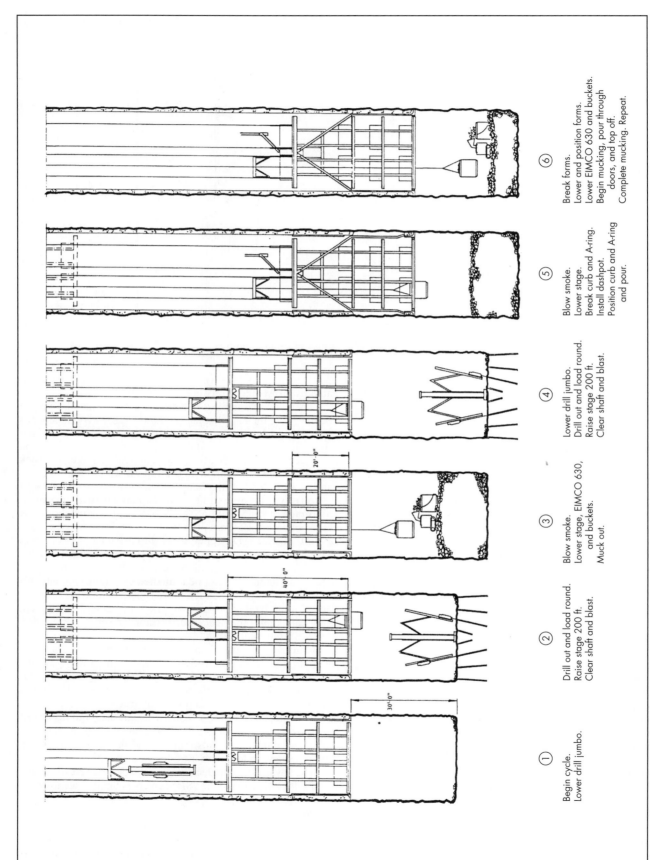

Figure 12.5-2 Cycles of the sinking operation

are distributed as: production, 22%; exploration and test mine development, 18%; service (including ventilation), 27%; and production and service combination, 33% (Dravo Corporation 1974). These proportions have probably not changed since the 1980s for North America. However, in Australia, where declines are used more than shafts for production, the primary use for vertical shafts is ventilation.

Basic design considerations. Like all underground openings, vertical shafts demand considerable study, planning, and design; indeed, this becomes even more important with shafts. The detailed engineering includes seven specific areas that must be considered:

1. **Size:** In the United States, shaft size varies from only 1.2 m (4 ft) to 8.5 m (28 ft), depending primarily on the purpose of the shaft.
2. **Configuration:** While most new shafts that are sunk from the surface today are circular, many older shafts and those sunk from underground are rectangular.
3. **Ground support:** For deepening old shafts, timber supports are still used. However, semimonolithic concrete lining is the most important shaft lining used today. Concrete linings have many advantages:
   - Good bond between the lining and the rock
   - Decreased labor intensity for permanent linings
   - High strength-resisting rock movement
   - In some cases, complete mechanization of lining placement

   The disadvantages of concrete linings are
   - The inability to take load immediately after placement,
   - Lower resistance to some corrosive waters, and
   - Higher sensitivity to mass rock movement and more difficult to repair.
4. **Internal arrangement of services:** Various compartments for hoisting ore and waste—which may or may not include timber or steel conveyance guides, workers, materials, and utility services—all have to be designed. Sometimes the utilities are incased within pipes embedded in the concrete lining.
5. **Depth (length of shaft):** For hoisting ore and waste rock, the shaft needs to be sunk to a depth where there is adequate room below the haulage level for
   - A large storage bin (orepass) between the rock dump and the feeder system that feeds the crusher,
   - The crusher system,
   - Feeding the crushed material into the skip loading system,
   - Accommodating the length of the skips, and
   - Accommodating some sort of shaft-spill rock cleanout equipment.

   The practical depth for shafts sunk for hoisting in one pass is about 1,220–1,500 m (4,000–5,000 ft); beyond that depth, an underground hoisting and ore transfer system should be considered (see Figure 12.5-1). The notable exception to this is what is being done in South Africa. The constraints on the unique deep gold mines in South Africa have led to improvements in the development of deep-shaft hoisting. These followed an industry research and development program undertaken in the last decade. Changes in legislation, winder control, and operating practice has allowed the depth limits of single-lift hoisting to be extended beyond a previous practical limit of about 2,400 m (7,900 ft). Several shafts have been commissioned where the winding depth is beyond 3,000 m (9,850 ft).
6. **Shaft linings:** Concrete lining strength is usually between 19 and 33 MPa (2,800 and 4,200 psi) but may be as high as 50 MPa (7,200 psi). Type III cement is usually used where the concrete is mixed on the surface and can be placed either by dropping batches down the shaft in a container and then distributing to the forms (shutterings), or by using slick lines and pots that distribute the concrete to the forms. In either case, the concrete must be compacted in place with pneumatic vibrators. Usually, the shutterings are collapsible steel forms that can be quickly erected and torn loose. Curb ring forms are placed and poured first, then the shuttering forms are placed on top of them.

   Shotcrete linings can be placed in a few variations depending on the shaft wall stability. The shotcrete may be applied directly on the wall by itself; it can be sprayed over woven wire fabric or the walls can be rock-bolted before or after the shotcrete is placed. Some operators prefer using steel fibers in the wet shotcrete for the added yield strength rather than shotcreting over woven wire fabric. Shotcreting of very small shafts is particularly desirable where steel forms are difficult to work with. Using shotcrete over wire mesh is also fairly common in European coal mines for internal shafts where the permanent lining is only roof bolts, wire mesh, and shotcrete.
7. **Shaft collars:** All hoisting shafts require a massive block of reinforced concrete designed to anchor the headframe and secure the top portion of the shaft in the unconsolidated material through which the shaft collar was excavated. They must be designed to allow all of the mine utilities to pass through the collar into the shaft.

Basic shaft-sinking equipment. Most deep shafts are sunk using a multistage platform that is installed in the shaft and stays in the shaft until it is completed. Some call this platform a Galloway stage. The drill jumbos can pass through the opening in the stage, the concrete forms can be stored and assembled on the stage, and mucking-out the bottom of the shaft can take place with the stage in place while buckets (kibbles) can pass through the opening.

Most shafts use one of the following mucking (lashing) systems:

- Cactus-type air-operated grab
- Pneumatic or hydraulic Cryderman mucker
- Backhoe-type machine
- Riddle clamshell-type machine
- Crawler-type, overshot loader

Also, the first portion of any shaft (the "presink") may be excavated with a clamshell or cactus grab bucket operated from a surface crane, before the Galloway stage is placed in the shaft.

The shaft-sinking schedule using a multistage platform is shown in Figure 12.5-2. Shaft sinking is normally performed by specialized contractors.

Another concept is blind boring or upreaming the shafts. Literally hundreds of blind-bored shafts have been completed in the United States and elsewhere around the world.

Slopes Versus Shafts

No general rule governs whether a shaft or a slope is the proper type of entry to be developed; each has advantages

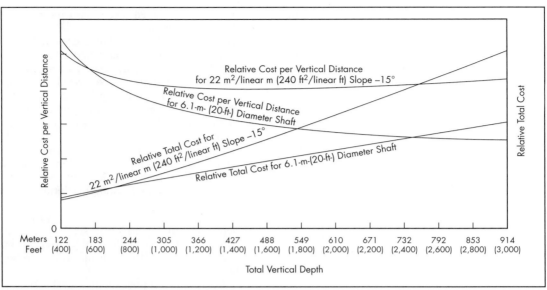

Source: Adapted from Dravo Corporation 1974.

Figure 12.5-3 Relative comparison between shafts and slopes of cost in vertical distance versus total depth

and disadvantages. Every major development decision should include a complete trade-off study combined with an accurate discounted cash-flow analysis that considers net present value (NPV) and internal rate of return (ROR) on capital investment alternatives.

Although the cost per unit distance for a decline may be somewhat lower than for a vertical shaft, the overall cost usually is higher and the development time is usually longer as a result of the greater length required. Thus the cost per vertical meter (foot) for the development, except for very shallow depths (e.g., 150 m [500 ft]), is considerably higher for slopes than shafts. Figure 12.5-3 compares relative costs per vertical unit of distance with total depth for declines and shafts. This chart was taken from work done many years ago on U.S. mines that produced between 1.6 Mt/a and 6.6 Mt/a (1.8 million tpy and 7.3 million tpy), and while the actual cost is no longer accurate, the relative costs are still approximately comparable.

Declines (slopes) have specific advantages over shafts:

- The technique of driving a decline does not require as much specialized technique as sinking a shaft.
- Declines can be used for trackless, track, or conveyor-belt haulage. When used for conveyor-belt haulage, the tonnage can be very large.
- The ground support cost for roof bolts and shotcrete is usually less expensive than the concrete liner normally required in a shaft.
- Large equipment can be moved in and out of the mine without disassembly.
- Maintenance is easier and less costly.
- If the production is from a fairly shallow depth and greater than about 5,000 t/d (5,500 tpd), the decline with a conveyor is likely to be lower cost.
- No visible head gear is required (i.e., no visual pollution).
- No winding gear is required, which in a remote location may need to be transported considerable distances along an unsuitable road network.
- No surface winding facilities need to be constructed, maintained, or manned.
- For major wrecks within the opening (shaft or decline), there are fewer production delays in declines. For an inclined ore body, and if only the top portion is initially developed, declines incur a much-lower initial cost than a shaft that would go to the depth to mine the entire ore body.

The disadvantages of declines over shafts are

- For an equivalent depth, the slope will be four to six times longer. The total cost and the time to construct for very deep declines will be greater (but for inclined ore bodies, production may be able to start early in the upper ore body).
- If there is a thick, very weak zone of rock to penetrate, the shorter distance of the shaft to penetrate through this zone will probably be much easier and less costly than for a decline where substantial ground support may be required.
- If the development opening is to be used for ventilation, the length of the opening will be increased four to six times, and the wall rock will be rougher than the concrete-lined shaft. Both of these issues will require more power to overcome the increased resistance for the same size development opening.

From depths of about 250 to 350 m (800 to 1,100 ft), mines producing from 3.3 to 6.6 Mt/a (3.6 to 7.3 million tpy) have been developed with declines. At these tonnages, at depths of more than about 350 to 500 m (1,100 to 1,600 ft), a shaft is probably more economical than a decline, except for very high tonnage mines using conveyor belts. However, many factors must be taken into consideration before making this decision. For example, if developing from the bottom of a mined-out open pit at a depth of 300 to 400 m (980 to 1,300 ft), the decline down another 800 m (2,600 ft) would probably make more sense than a shaft from the surface down to 1,200 m (3,900 ft). Likewise, if there is only proven ore down to a depth of 400 to 500 m (1,300 to 1,650 ft), and the annual tonnage was to be between 1 and 2 Mt/a (1.1 and

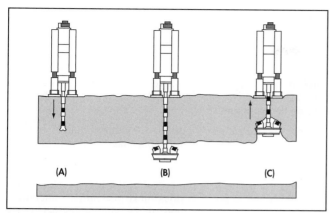

Source: Bullock 1997, © The McGraw-Hill Companies, Inc.

Figure 12.5-4 Raise boring by upreaming. (A) The machine drills a pilot hole. (B) The drill bit is replaced by a reamer. (C) The raise drill lifts the reamer, and tailings fall down the shaft and are removed.

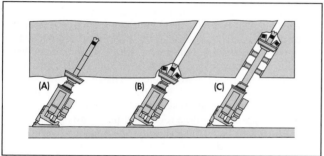

Source: Bullock 1997, © The McGraw-Hill Companies, Inc.

Figure 12.5-5 Raise boring using a blind "box hole" drill. (A) The machine drills a pilot hole. (B) The drill bit is replaced by a reamer. (C) The reamer bores into rock, and tailings fall down the shaft and are deflected by the drill.

2.2 million tpy), then in all probability the decline would have the advantage. However, there is a strong caveat even to this general statement: if the surface material is very much unconsolidated and/or is below the water table, the cost of developing through this material is much higher for a decline rather than a shaft. For example, if the ground must be frozen in order to penetrate safely, a shaft would be the obvious choice. This is one of the reasons that shafts in North America are sometime favored over declines, particularly in the plains area where large regional aquifers occur.

Incline Shaft Development

An incline shaft is an opening driven steeper than a decline but not vertical. They were often used in the past for maintaining a somewhat constant haul distance between the dipping ore body and the shaft. However, with today's trackless haulage equipment this advantage is minimized. Hoisting with incline shafts is less efficient and requires more maintenance than vertical shafts. Thus they are used very little today.

Multiple-Stage Hoisting

When the hoisting depth reaches approximately 2,400 m (8,000 ft) in depth, a second hoist may be placed underground and continued with multiple-stage hoisting. Usually these facilities are not placed one beneath the other as shown in Figure 12.5-1, but rather the lower shaft is placed wherever it makes more engineering sense to optimize both hauling to the shaft and hauling from the lower shaft to the loading facility of the upper shaft. Also, ground conditions could play a role in where to locate the lower shaft. Many times with today's powerful diesel or electric trucks, the choice may be to haul up an incline rather than sink a second deeper shaft.

Types of Raise Developments

Raises are needed for many reasons, mainly for the transfer of ore, workers, and materials. They also provide for a void space for stope blasting, ventilation, escapeways, and combinations of these. There are many ways to excavate a raise:

- **Conventional raise drilling and blasting:** Here the miner enters a bald raise and drills and blasts the drill round. A work floor is created by drilling pin holes in the sides of the raise.
- **Raise boring by upreaming:** This system requires two levels, between which the raise is bored (Figure 12.5-4). First, a pilot hole is drilled from the top down, then a large reaming bit is put on the drill string and the hole is reamed upward.
- **Blind-hole raise drilling with a "box hole" drill:** In this case there is only one level, and the blind-hole boring machine usually drills a small pilot hole and then reams out the blind raise (Figure 12.5-5). This is sometimes called a "box hole."
- **Drop raise by predrilling and blasting the raise in stages from the bottom up:** The drillers use long-hole drill machines and very accurately drill small blastholes and larger, parallel-cut open holes. The holes are charged and blasted in sequence from the bottom up, making sure that each sequence blasts clear before the next round is blasted.
- **Raise climber:** The raise climber is a mechanical machine that climbs on a cog-wheel track installed in the raise (Figure 12.5-6). The miner drills out the raise round, charges the round, then moves the raise climber back out of the raise before he shoots the round. The miner then runs the machine back up the raise, scales the loose rock under the protection of a canopy, installs another section of track, and repeats the cycle.
- **Boliden raise cage:** As raise-boring technology has become increasingly efficient, the Boliden raise cage is not used as much as in the past. Still, it is a viable method. Where there are two levels, a pilot hole is first drilled between them, and a small hoist is set at the top of the hole. A hoist cable is dropped down to the lower level and a two-deck drill cage is hoisted to the top of the raise. The upper deck has a canopy, to protect the miners while scaling and drilling. After drilling and charging, the cage is lowered and moved from the bottom of the raise for blasting. The cycle is then repeated.

Other Types of Mine Developments

Many other types of developments shown on Figure 12.5-1 need some explanation. The terms *drift* and *crosscut* have unique meanings in a vein type of ore body. Drift usually means the opening is driven parallel to the strike of the vein, whereas a crosscut typically is an opening driven perpendicular to the

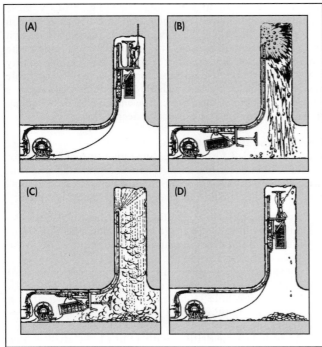

Source: Svensson 1982.

Figure 12.5-6 Raise boring using a raise climber. (A) The raise climber track is installed; the climber goes up and drills and charges a raise round. (B) The raise climber travels back down the track and the raise round is blasted. (C) The raise is ventilated. (D) The raise climber goes back up the raise with the miner under the protection of the steel cage. The miner, from the top of the raise climber, scales the loose material in the raise and extends the new track up the raise. Drilling can again commence.

strike of the ore body. However, crosscuts can also mean any opening driven between two other openings.

The term *raise* is commonly used for an opening driven at a high angle upward from the existing opening, and a *winze* is aan opening driven at a high angle downward from an existing opening. However, with today's technology, it is very common to drill and blast a drop raise from the opening in the downward direction and then blast it from the bottom up; this would still be called a raise.

Ramps are similar to declines in that they are driven at a steep grade that can still be navigated by trackless equipment. However, they are usually driven between mine levels, more for the purpose of traveling between production stopes and dumping points or to develop a new stoping entry. Ramps are better developed in an oval or figure-eight shape than in a true circular configuration. Ramps are more efficient when they are limited to a 15%–18% grade.

Stope Developments for the Various Mining Methods

For those who are not familiar with the various underground mining methods, the following discussion should be studied along with the chapters on mining methods in Part 13 of this handbook. It is not the intent of this chapter to explain or describe how the mining methods work, but rather to describe the differences in the developments that set up the various mining methods.

Hard-Rock Room-and-Pillar Mine Development

The configuration of a room-and-pillar mine generally follows the shape of the mineralized ore body to be mined. Therefore, the basic shape of the mine should start by determining the path of the main haulageways (entries), branching from these to reach the lateral extent of the reserve. Wherever possible, a few basic principles should be followed in planning the production developments:

- The central gathering point for hoisting should be as near to the center of the extractable mass as practical. That keeps the haulage, personnel, and material movement distances to a minimum. Although this may mean a large shaft pillar is left in the center of the mineral reserve of a flat-lying deposit, it does make the mining operation more efficient. At the end of the mining project, most of the pillars can probably be extracted, providing that proper mine planning has been used in the design and proper pillar removal systems have been established.
- Normally, room-and-pillar mines without ground stability problems should be developed as symmetrically as possible to allow the maximum number of working areas to be opened with a minimum of development time. In trackless mines, it is advisable to cut the main haulage opening as much as 50% larger than the minimum size needed to operate the existing equipment. Large main haulageways improve productivity, ventilation, and road conditions, and they are much more efficient to drive. They also allow productivity to be maintained by permitting efficient operation of larger equipment. The haulage drifts in several Mississippi Valley–type room-and-pillar mines are driven 5.5 × 9.8 m (18 × 32 ft), and the mines use very large equipment such as 36-t (40-st) or larger trucks.
- Haulage roads and railroads should be kept as straight as possible, with steep grades other than conveyor slopes being kept as short as possible. The grade limits depend on the type of haulage equipment being used. For main haulageways, railroad grades should be less than 2%, and for rubber-tired diesel equipment it is advisable to keep the grade less than 8%. However, for very short distances where the equipment is allowed to gain mass momentum, 15%–18% grades can be negotiated quite efficiently. If a 20% decline is to be driven for a conveyor system or service slope, rubber-tired, diesel-powered loaders are the best equipment for loading and hauling the material from the development. As mentioned previously, for such steep declines, very high horsepower engines are required for efficient haulage.
- The radius of a curve that can be negotiated by trackless equipment is somewhat dependent on the size of the equipment in relation to the size of the drift and the turning radius of the equipment. Normally, a radius of 15 m (50 ft) is handled easily by any trackless mining equipment. A road grader usually requires the greatest turning radius. For rail-mounted haulage, a radius of 24 m (80 ft) is adequate for stope development, but a limited-service main line requires a minimum radius of 38 m (125 ft). However, a radius of 60 to 150 m (200 to 500 ft) would be appropriate for main-line haulage, depending on the speed of the train.

The preceding discussion has outlined the need for operating flexibility in relation to the size, shape, and thickness of the mineral reserve in the room-and-pillar mine plan. For

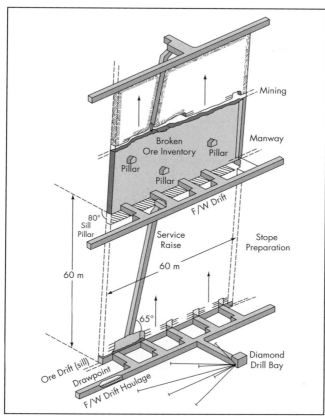

Source: Adapted from Marchand et al. 2001.

Figure 12.5-7 Typical shrinkage stope mine development

example, it is often advantageous to incorporate a lower level of uniform grade below the main ore horizon to create a ventilation circuit, providing mine drainage that removes water from the operating headings and roadways to make ore movement more efficient. The Fletcher, Brushy Creek, Buick, and Sweetwater mines—all located in the United States—of Doe Run Company incorporated multiple-level planning in the early years of operations, with all of the planning features described. These mines have been extremely efficient over their 45 years of operation.

Shrinkage Stoping Development for Production

Shrinkage stopes for extraction can be developed in many ways. Several years ago, William Lyman (1998) wrote on the subject, describing thirteen ways to develop the shrinkage stope and variations of the ventilation and service facilities. In Chapter 13.3 Haptonstall describes several of these methods. What is described here is a very common method that can be adapted to developing a trackless or a rail haulage shrinkage stope as depicted in Figure 12.5-7 (Marchand et al. 2001). The mine described here is Cambior's Mouska mine in Quebec, Canada.

An ore drift (sill) is driven and kept in the ore at about 30 m (100 ft) ahead of a footwall haulage drift. It is important that the sill drift width stays as near the same width of the ore vein as possible. The sill drift and the haulage drift are kept about 10 m (30 ft) apart. In this case the vertical distance between levels is 60 m (200 ft) with a stope width of 60 m (200 ft). At 10-m (30-ft) intervals, crosscuts are driven between the haulage drift and the ore drift to serve as drawpoints to shrink the broken ore from the stopes. (*Shrink* refers to the process of shrinkage stoping where the amount of broken ore out of the stope is decreased; most of it is left temporarily as a working platform and ground support.) The crosscut size will depend on what type of equipment is used to remove the ore. In this case small overshot pneumatic muckers were used and trammed back to the railcars for dumping, and the crosscuts were only 2.7 × 2.9 m (8.9 × 9.5 ft). For load-haul-dump (LHD) haulage from a drawpoint to an orepass or dump pocket, the crosscuts would have to be sized at least 1.2 m (4 ft) larger than the dimensions of the LHD. From the lower sill drift, a service raise is driven to the haulage drift of the level above. This service raise is equipped with a ladderway, service pipes, and a slide (with air tugger) all connecting to the level above. The upper level also provides the ventilation circuit. Shown also is a service manway from the bottom of the stope, which is, in this case, timbered as the stope advances upward, thus completing the ventilation circuit and maintaining two access points to the stope at all times. Other mines sometimes leave pillars between the stopes and place the second access in these pillars, which would require small doghole drifts between the raises and the stope. (A *doghole* is a very small drift driven to serve as an access between a vertical manway and the stope. In some cases it may be only 1.2-m (4-ft) high, where one must crawl on all fours to access the stope.) These pillars can later be removed and capture the ore in the drawpoint below when the stope is completed.

Sublevel Open-Stope Development

Most sublevel open-stope mines are applicable to dipping-type deposits, but the thickness of the deposit between the hanging wall and footwall normally varies from a few meters to tens of meters wide. The development access from the main decline or hoisting shaft is gained by level developments, usually in the footwall, that vary from 46 to 122 m (150 to 400 ft) depending on the vertical extent of the ore body, the eventual stope heights, and the number of production stopes that may be excavated per level. These levels usually become the haulage drifts and approach the bottom of each stoping block running parallel to the block with connecting access crosscuts running over and into the ore. Between the main levels, ramps are usually driven for trackless haulage transport between levels. These ramps also give access to the sublevels, which are developed at intervals to remove blocks of ore. The sublevel drifts are located in the footwall for haulage, with crosscuts into the ore for connecting to the drifts in the ore, which first become undercut access drifts and later draw drives. These sublevel drifts in the footwall can also be used for accurate delineation of the vein (reef) by drilling prospect holes from the sublevel drifts. Up above at the next sublevel, these drifts driven in ore are used for drilling the long holes that blast the ore. For wide ore bodies, there will probably be a hanging wall drift and a footwall drift. Usually at the widest part of the ore for that block there will be a slot crosscut drive and a slot raise drive between the sublevels, which opens to the undercut of the lower level. This slot raise is then slabbed (slashed) out to the full width of the ore body, and when the broken slot ore falls to the undercut opening, the slot allows room for the vertical slabs of ore to be drilled and blasted from the upper sublevel drill drifts. The ore then falls down to the draw drifts, which are accessed by drawpoint crosscuts for loading out the ore. When the ore is completely drilled and blasted from the first sublevel block,

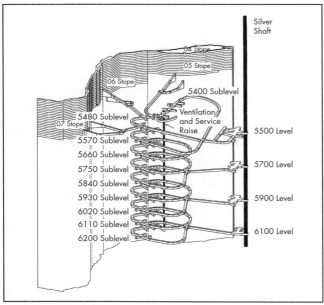

Source: Peppin et al. 2001.

Figure 12.5-8 Lucky Friday undercut-and-fill mine development infrastructure

the drilling and blasting moves up to the next sublevel and the cycle is repeated until the next main level is reached. An ore sill pillar is usually remaining at each main level.

Sometimes the ore body is narrow enough that a horizontal room can be opened and, instead of drilling fan holes from drill drives, vertical holes can be drilled for the long-hole blasting. This also applies when vertical crater retreat stoping is used. In the sublevel extraction method, termed "longitudinal mining" in Chapter 13.4 (Figure 13.4-21), the sublevel sill drifts serve as both the top cut for the downhole drilling and as a bottom cut for remote mucking. Ventilation raises can be placed in the ore or close to the stope accessway from the ramp. Raises are usually opened with raise boring machines, raise climbers, or drop raises. Rarely in today's modern mechanized mines are raises drilled and blasted by conventional methods.

Cut-and-Fill Stoping

Because cut-and-fill mining is so versatile and can be used to mine any shape of an ore body, it is impossible to generalize a specific development approach that fits all types of cut-and-fill systems. Normally the essential criteria for using cut-and-fill stoping is that the excavated stope needs support; the remaining rock is not competent enough to stand open to allow further mining of the area. However, where the ore is so high-grade that it is planned to extract nearly 100% of the ore and leave no pillars, there is a need for immediate backfill, even in competent rock. Consideration must be given to the general types of deposits that are to be mined by one of the many forms of cut-and-fill mining.

Steeply dipping, narrow ore bodies. Steeply dipping ore bodies are most commonly mined by cut-and-fill systems. Such mining occurs throughout the Silver Valley (in Idaho's Coeur d'Alene district), the Stillwater mines and the Cannon mine in Washington (all in the United States), to name but a few. The major infrastructure developments are normally developed in the footwall, but not necessarily, since nothing will be allowed to cave—even the hanging wall should remain structurally competent for developments. A classic example of infrastructure development is seen in Figure 12.5-8 of Idaho's Lucky Friday mine, illustrating the main shaft, the haulage levels coming off the shaft, the oval-shaped ramp contacting all of the sublevels, and the service and ventilation raise. Because this is an undercut-and-fill mine, there are 15% stope attack ramps to the top of each sublevel stope block and working down, slice by slice, to the bottom of the sublevel stope block. In an overhand cut-and-fill mine, the sublevel attack ramps would start at the bottom of the stope block and work up the ore block in slices, establishing a new attack ramp for each slice. Not shown on this sketch are orepasses from the sublevel ramp down to the next haulage level.

Flat, thick, and wide ore bodies. A few years ago only steep vein deposits were worked by cut-and-fill mining, but with today's technology wide and thick ore bodies are also mined in this manner. Almost all of the underground mines of the Nevada (United States) gold district use cut-and-fill mining, and more specifically, undercut, drift-and-fill, or ramp-and-fill. In this type of mining, the upper-level drifts should be driven and filled at a different direction azimuth than the drifts immediately below. This can be seen in Chapter 6.5, Figure 6.5-6.

Sublevel Caving

Sublevel caving is an underhand method where all of the blastholes are drilled upward. Gravity moves the ore down to the extraction and drilling drift. The main haulage drifts on the major levels will probably be longer from the hoisting facility to the production stoping than in noncaving mining systems, since this mining system allows for the overburden to cave. When the angle of the caving surface is taken into consideration, the hoisting facilities must be well removed from the zone of potential caving.

For transverse sublevel development, the haulage drift for a particular level is driven down the strike of the ore body, probably in the footwall waste, with production crosscuts turned off at regular intervals crossing the ore body. This divides the ore body into geometric blocks that will correspond to the geometric spacing of the sublevels. These drifts are driven to the hanging wall, and a slot raise up to the cave above is developed at the end. These slot raises must be slabbed out to the full dimension of block and will provide the opening for the initial blasting at the end of the stope. There are many development schemes to connect the sublevels, but all of them involve ramps that connect to the various sublevel haulage drifts. Also connecting to the sublevel haulage drifts are major orepasses that transfer the ore to a main haulage level.

Development for a sublevel cave stoping method is extensive. This is because so much of the ore must be mined by development to set up the drilling and blasting of the ore so that the surrounding waste rock will cave uniformly. In 1990, about 25% of the ore was extracted by drifting in the sublevels; now this can be as little as 6% (Bullock and Hustrulid 2001). The reason for the drastic decrease in development is because the better drills and drilling systems today are more efficient, drilling larger, longer, and much more accurate holes for blasting; thus the spacing between sublevels has more than doubled.

As indicated, ore is recovered both through drifting and through stoping. Because the cost per ton for drifting is several times that for stoping, it is desirable to maximize stoping and minimize drifting. This has meant that through the years, the height of the sublevels has steadily increased until today they are up to 30 m (98 ft). The sublevel intervals have changed from 9 m up to nearly 30 m (30 to 98 ft). A number of factors determine the design. The sublevel drifts typically have dimensions (width × height) of 5 × 4 m, 6 × 5 m, or 7 × 5 m (16 × 13 ft, 20 × 16 ft, or 23 × 16 ft) to accommodate LHDs.

After the sublevel vertical interval has been decided, it is necessary to position horizontal dimensions of the sublevel drifts. As an example, the drifts might be placed so that the angle drawn from the upper corner of the extraction drift to the bottom center of drifts on the overlying sublevel is 70°. This is approximately the minimum angle at which the material in the ring would move to the drawpoint. The resulting center-to-center spacing is 22 m (72 ft).

Longitudinal sublevel development is commonly used in ore widths of 18 m (60 ft) or less but may also be used to advantage in wider ore bodies. For longitudinal sublevel caving developments, almost all of the stope development is in ore, but for irregular ore bodies, there is a greater chance of leaving ore undrilled and blasted.

Mine Development for Block Caving (Panel Caving)

In this section, the term *block caving* will be used to represent both block and panel caving, suggesting the mining of individual blocks, and panel caving, indicating a laterally expanding extraction. The time and cost of developing a block caving mine is very significant compared to other types of mining.

Since block caving methods are associated with extremely high production rates, the main haulage developments are likewise designed for large high-volume/high-speed haulage. These initial developments must also accommodate very large ventilation capacities to handle all of the equipment needs of a modern massive caving mine. At the same time, the major permanent development facilities, including shafts and permanent ventilation raises, must be located far enough from the caving material so as not to be engulfed within a zone of high stress or rock-mass deterioration. An example of this infrastructure development in relation to the caving ore body is shown in Chapter 6.1 (Figure 6.1-2).

The most important development elements of a caving system are the undercut level, which removes the support for the overlying ore column; the funnel, trough, or bell through which the ore is transported downward to the extraction level; and the extraction or production level. In some cases, the undercutting can take place on two levels rather than one (see Chapter 6.5, Figure 6.5-24C). There are three types of systems for drawing the ore from the caving ore body: the LHD system, the slusher system, and the gravity system (in order of importance). Much of the main development infrastructure is similar for the three methods, but the stope developments are somewhat different.

Stope development for LHD caving systems. In Figure 12.5-9 the undercut level is the upper level where the long drill holes of the undercut ring are shown. The production level is the level below where the LHD is loading ore from the bell draw holes. Note the intake and exhaust drifts and raises that carry the ventilation and the orepasses and load-out chutes on the mail haulage level drift.

For more extensive discussion and illustrations on LHD caving systems development, see Chapter 13.10.

Stope development for slusher caving systems. In the top level shown in Figure 12.5-10—the undercut level—longhole drilling will open the troughs that the ore will collapse

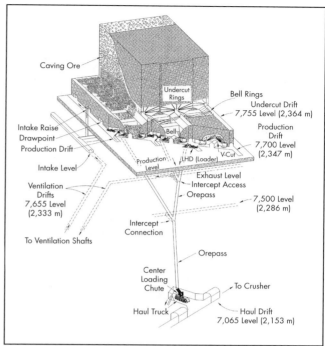

Source: Adapted from Rech 2001.

Figure 12.5-9 Underground development needed for LHD block caving

into and then flow down into the multitude of finger raises, where it flows by gravity to the grizzly (haulage) level where it is slushed to the main haulage level. In some cases, instead of slushing directly into the ore trains, the ore may be slushed into another orepass leading to an even lower main haulage level. The ventilation drifts and raises are not shown in the diagram. This mining system is labor intensive and would only be used where labor costs are extremely low.

Stope development for gravity caving systems. Figure 12.5-11 illustrates how the development has much less horizontal development, but many more raises for orepasses and draw extraction bells, in a 14 × 7.6 m (76 × 25 ft) grid pattern. When the ore is being drawn from a specific area, a miner must stand and assist the ore flow with a mining bar or explosive bombs in each of those grizzly chambers, as the ore flow becomes blocked with boulders. Because this method of block caving is so labor intensive, it is seldom used in new mines being developed in countries where labor costs are high.

PLANNING FOR EXCAVATION

Rapid-Development Incentives

Very large initial capital expenditure is required to construct a block cave mine and bring it into production. The eventual NPV of the investment is dominated by the time taken to start production, typically 5–8 years. The rate of advance is a key factor, and cost per meter, although important, tends to be secondary. One very large, multinational mining company that is currently operating or developing four large block caving mines, plans to develop 1,000 km (600 mi) of development and 12,000 drawpoints over the next 11 years (Moss 2009). Over many years and projects, this mining company's advance rate has stayed at around 4 m/d (13 ft/d), and the company states this has been a critical factor for bringing any block cave projects in on schedule.

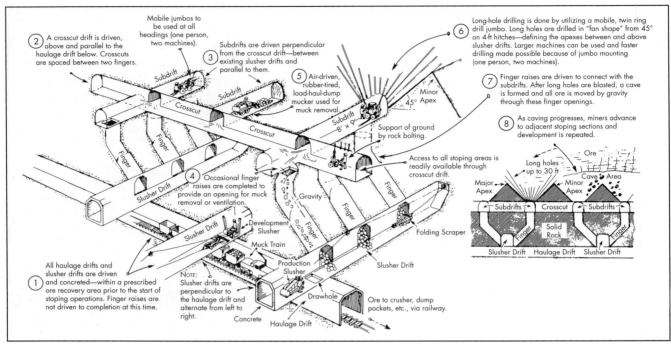

Source: Gould 1998.

Figure 12.5-10 Development needed for slusher-type block caving

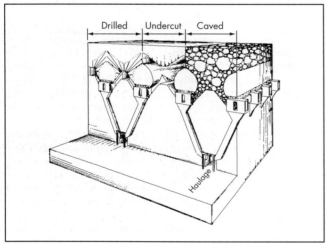

Source: Lacasse and Legast 1981.

Figure 12.5-11 Stope development required for gravity draw block caving

While rock-drilling efficiency has increased by fivefold over the years between 1960 and 2000, the advance rate in meters per day of development has decreased by threefold; thus, the cost per meter has gone up by tenfold. The primary reason for this is the new emphasis on safety at the development face. These protocols have now evolved in the mines to a point where most development requires that all ground is at least fully bolted and in most cases has skin support in the form of shotcrete or wire mesh. This means that currently most of the ground control functions must be carried out within the development cycle. Yet a few mines and many tunneling operations get much higher advance rates, some more than 10–15 m/d (31–49 ft/d). How is this accomplished? It must be planned, equipped, and executed as if it were a civil tunnel contract.

What Must Be in Place?

From the very beginning, there cannot be any compromise of worker safety; management must develop procedures whereby no worker is exposed to ground that has been inadequately secured. By reviewing many studies of rapid excavation, it is apparent that the following steps must be set in place to achieve these advance rates:

- The mining operation must know the best drill round to achieve 95% of the drilled distance for the type of ground at the mine location.
  - A thorough geotechnical study must be ongoing to define rock types related to blast design and ground control problems, and detailed planning must precede each step of the mine development.
  - Drill rounds must be designed by qualified engineers, and the execution of drilling and blasting must be followed according to the design. In a recent survey by the National Institute for Occupational Safety and Health (NIOSH), 60% of mines allow drillers to decide what type of a drill pattern and how much explosive to use in the round.
  - Miners think in terms of blasting or bombarding the rock with as much force as possible, and every hole is the same; blast designers tend to think of cutting the rock, particularly in the cut area and the periphery of the drift.
  - Oxygen-balanced emulsions should be used to minimize ventilation time, and electronic blasting should probably be used to ensure perfect hole timing and minimize vibration and peak particle velocity where it may be required.

- Since 50% of the cost and about 25%–40% of the face time are now spent in ground control, controlled smooth wall blasting should be used. Damage to the remaining rock must be limited and overbreak minimized. NIOSH recently conducted a research study showing that the total cost of drifting using controlled blasting is about 4%, or $59/m ($18/ft) less than conventional drilling and blasting. No credit was given for less ground-control cost (Camm and Miller 2009). The ground-control savings could easily be $150/m ($45/ft).
- Driving the largest faces consistently with geotechnical issues of ground control and ground stress conditions allows for the most drifting efficiency, since larger equipment can be used and, in some cases, there is room for multitasking at the face. In one case the development advance went from 1.7 m/shift (6 ft/shift) to 3.4 m/shift (11.1 ft/shift), even though the size of the drift tripled (Bullock 1961).
- The most efficient and reliable equipment should be acquired for driving the developments, particularly the critical path developments, including computerized drilling controls and the use of improved explosives systems designed for controlled blasting. The initial, up-front cost may be millions of dollars, but many millions more will be saved in the NPV of the operation.
- The entire mine development operation must focus on getting the men and supply logistics and maintenance planned, so there will be little or no time lost at the face between cycles. In one Rio Tinto study (Moss 2009), 39% of the time was spent "waiting" for the next step in the 15-hour cycle.
- Procedures and metrics must be put into place where all parameters of the drilling, blasting, ground control, and loading portions of the drifting cycle can be defined in quantitative forms and are measured for development projects. If quality is not measured, it cannot be managed.
- To achieve the desired results, every project needs to establish a quality control group that has the responsibility and authority to enforce the drilling, blasting, and ground control procedures that have been put in place. This group should report to the mine manager. In one project observed by the author, where quality control was completely ignored during the "controlled" blasting of poor ground, the result was that the tunnel had a slab of more than 1.5 m (5 ft) of overbreak for at least 15 m (50 ft) of tunnel, which the contractor (on a cost-plus contract) was allowed to simply fill with layers of shotcrete until the correct tunnel dimension was achieved.

The following are a few recent examples of high-speed development.

Norwegian tunneling technology opened a new haulageway for a coal mine, which averaged 103 m/week (338 ft/week) driving a 38.5-m^2 (414-ft^2) tunnel 5,630-m (18,471-ft) long (Nilsen 2009). Much of the tunnel was in permafrost, making the advance more difficult. The key is to use the face to advance development. To achieve this, some Norwegian tunneling operations use

- Side-tipping loaders;
- Trucks sized to the tunnel, enabling drivers to perform a three-point turn in the tunnel;
- Drilling equipment with navigation that is fully instrumented to implement a plan of the drill layout, including the survey of the hole collar and hole toe;
- Use of 5.8-m- (19-ft-) long drill steels (where ground is appropriate), logging of the drilling performance, blasting using emulsion, and rapid-setting shotcrete (one-half hour in freezing conditions); and
- Most importantly, a dedicated, highly skilled work force regularly achieving more than 100 m (328 ft) per week of development in openings up to 38.5 m^2 (414.4 ft^2). The best week achieved 150.1 m (492.5 ft) of advance.

The world's best decline advance rate is that currently achieved at the Cadia East project in New South Wales, for Newcrest Mining Limited, which was 260 m/month (853 ft/month) for 9 months ending in May 2008 and included a single best month of 311 m (1,020 ft). This project also shows a cost of A$4,500/m (US$3,894/ft), which is 31% lower than the industry average for Australia (Lehany 2008).

In general, tunnel contractors do a better job of rapid advance rate excavations than mining companies do.

What Is the Payoff for Rapid Excavation?

Assume one is developing a 2% copper, block caving mine to produce 100,000 t/d (91,000 tpd). One can also assume a typical cost per ton of copper production, a copper sales price of about $0.90/kg ($2.00/lb), a $360 million capital cost ($45 million for 8 years versus $72 million for 5 years), a 10% discount rate, and a regular development rate of 4 m/d (13.1 ft/d) compared to 8 m/d (26.2 ft/d). Based on this data, the mine would come into full production in the sixth year rather than the ninth year. This would result in a difference in NPV in the eighth year of $260 million for the regular development compared to $86 million for the accelerated development, or a difference in NPV of $154 million, favoring the accelerated case. However, by the 13th year, the accelerated development case shows a positive discounted cash flow and NPV compared to the regular development case, which does not show a positive discounted cash flow until approximately the 35th year. This theoretical example illustrates that with a difference of $154 million, considerably more capital dollars might be spent to put the proper equipment and practices in place, resulting in the doubling of development advance per shift.

Developments Driven for Obtaining Information

Exploration
Often developments must be driven out ahead of the rest of the mine infrastructure to identify both the specific location and grade of the ore to be mined. There are no set rules as to how these developments must be driven, since so much depends on the type of deposit. In some cases, the development needs to be driven directly into the ore to assess its character, and in other geologic settings the developments need to be driven a specific distance from the deposit so a series of diamond drill holes can be used to sample those deposit characteristics. The primary objective should be to get these developments completed well ahead of the need to mine the area. This means getting the developments done many months before the information will be needed for mining operations.

Information for Mine Planning Obtained by Development
During the very early stages of evaluation and planning, a lot of information is needed to make the correct decisions concerning the property. Therefore, the initial opening might be driven to test a particular aspect of the mineral reserve that

is considered critical for successful exploitation. More often, this is termed a *test mine*. This could be from an adit, ramp, or shaft (see Chapter 4.6). In any case, permitting is usually required that may delay the project.

A large amount of the technical information required can be obtained from diamond-drill holes, particularly where the mineralized zone is fairly shallow, continuous, and flat. Unfortunately, the true economic potential of many mineral resources cannot be determined until after an exploration development or test mine is driven. From the development opening, underground prospecting can further delineate the materials present, a bulk sample can be taken for metallurgical testing, and/or true mining conditions can be revealed. This procedure involves considerable financial exposure and, unfortunately, the delineated resource may turn out not to be economically viable. Nevertheless, there may be little choice about driving a test development opening if the potential of the resource is to be proven and the risk to larger downstream capital expenditures is to be minimized. This is another case where the decision must be made between a decline or a shaft as well as the size of the opening.

Regardless of the reason for driving a preliminary development, three basic principles should be observed:

1. The development should be planned to obtain as much of the critical information as possible without driving the opening in a location that will interfere with later production. For most mining methods, it is usually possible to locate at least a portion of the exploration development where it can be used in the developed mine that may come later, even if it becomes necessary to enlarge openings to accommodate production equipment at a later date.
2. The preliminary development should be restricted to the minimum needed to obtain the necessary information.
3. The development should be completed as quickly as possible. Most exploration drifts for metal mines are only large enough to accommodate small loading and hauling units (from 2.4 × 3 m to 3 × 4.3 m [8 × 10 ft to 10 × 14 ft]). If geologic information is needed, it is often possible to locate the developments in proximity to several of the questionable structures without actually drifting to each one. With modern high-speed diamond drills having a range of 213 to 305 m (700 to 1,000 ft), these structures can usually be cored, assayed, and mapped at minimum cost.

After the preliminary development has been completed to satisfy a specific need, the development should be used to maximize all other physical and structural information to the extent that will not delay the project. As much geotechnical information on the rock mass quality should be obtained as possible, and should be just as important to the company as resource information.

No rules regulate the size of the shaft or the entry for exploration development. The size depends on many factors, including the probability of the mine being later developed to the production stage. If the probability favors follow-up with full mine development immediately after exploration development, and if the shaft depth is not excessive (e.g., 244–366 m [800–1,200 ft]), it would be logical to sink a shaft sized for later use in production development. This is particularly true of operations normally using small shafts for production. Using a small shaft of 3.7 to 4.0 m (12 to 13 ft) in diameter is not uncommon in hard-rock mining in the United States. For example, most of the Doe Run Company's shafts in the Viburnum Trend area of Missouri are of this size, as well as many of the shafts in the Tennessee zinc district—the Elmwood mine, originally opened by New Jersey Zinc, is a good example of this practice. Initially a 3.7-m- (12-ft-) diameter conventional shaft was sunk for exploration. After the continuity and quality of the ore were proven by drifting and underground prospecting, the shaft was used for further development of the mine and was eventually used as the production shaft.

The Savage Zinc Company's mine at Carthage, Tennessee (United States) is another example of a 3.8-m shaft originally sunk as a test mine and then later opened as an operating shaft. In contrast, some companies choose smaller-drilled shafts to speed shaft sinking and minimize early expenses. Such shafts are usually completed considerably faster than conventional shafts, but not as fast as a raise-bored shaft (i.e., where an opening already exists, and the hole is reamed upward, and the cuttings fall back into the mine). Because this section is directed toward initial mine openings, removal of cuttings must be upward through the hole and out from the shaft collar. Various methods involve the use of air or mud, with either direct or reverse circulation. A large number of blind-bored shafts exist in the industry. Although bored shafts may stand up very well without support, most of these shafts contain either a steel or concrete lining.

Developing Service Facilities

Shop and Storehouse Developments

The efficiency of a modern mining system depends heavily on the productivity and availability of the equipment used to extract the material. Most of the energy to move the material horizontally must be provided. Since the material may be very heavy and/or abrasive, and because the environment of an underground mine imposes adverse operating conditions, underground mining equipment requires a great deal of preventive maintenance and repair. One of the most serious and most prevalent errors made in designing a mine is the failure to provide adequate space and equipment for necessary maintenance and repair work. The amount of service, if any, provided underground depends on several factors:

- The degree of difficulty in moving equipment into or out of the mine is a major consideration. If the property is a limestone mine with adit entrances and a good shop on the surface, it would probably not be advantageous to duplicate the facilities and personnel underground. As a generalization, most decline mines do not need an underground workshop; they require only service and refueling bays. Maintenance is done on the surface.
- Underground shops should provide a safe and clean working environment. In mines that are gassy or carry a gassy classification, building underground shop facilities may not be practical.
- If the active working area will be totally abandoned after a fairly short life, it would not pay to invest in an extensive underground shop area. Mines such as the punch mines in Appalachia (United States) are typical examples of this situation.

Most underground mines with shaft entries should develop good underground shop facilities. In the Viburnum Trend mines, which are developed with shafts 152–396 m (500–1,300 ft) deep, all of the mines rely on large mining equipment and have well-developed underground shop areas. The total area usually used as a shop area can range from 465 to 1,394 m² (5,000 to 15,000 ft²). The important criteria for shop design are to provide

- One or two large bridge cranes over the motor pits;
- Easy access from several different directions for access to and around the cranes when they are in service for extended periods of time;
- A motor pit and service area for scheduled lubrications performed as part of the preventive maintenance program;
- A separate area for welding operations;
- A separate area and equipment for tire mounting and repair;
- An enclosed separate area for recharging batteries;
- Various work areas equipped with steel worktables or benches;
- Close proximity to the main supply house;
- An area for washing and steam-cleaning the equipment;
- An office for the shop foreman, records, manuals, catalogs, drawings, and possibly a drafting table. The main pit areas should be visible from the office windows; and
- If rail haulage is used, a separate locomotive and car repair shop must be developed, with provisions for at least one motor pit.

The size of the underground supply room also depends on several factors. Some mines have no supply room, whereas others have very large rooms providing as much as 1,115 m² (12,000 ft²). Factors influencing the inventory policy include the frequency and ease with which the mine receives supplies, whether a supply system is installed at the collar or portal of the mine, whether there is a separate access to obtaining repair parts without disrupting the mine production, and the dependability of the parts suppliers for the equipment being used.

Sump Area and Pump Station

In mines below surface drainage, areas must be provided to store the water before it is pumped to the surface and discharged. No general rule exists to determine the capacity for water storage. The water flow into mines in the United States has varied from 0 to 2,208 L/s (0 to 35,000 gpm). Adequate pumping capacity is the only permanent solution to removing water from a mine. However, fluctuations in the water inflow and/or the time periods when some or all of the pumps are inoperable must be handled by having an adequate sump capacity.

Many mines use "dirty water" positive displacement pumps. The challenge is to keep the mud suspended until it enters the pumps, screening out the big lumps. This is economical up to about 80 L/s (1,320 gpm). Beyond that, settlers with centrifugal pumps will be necessary. Sumps are also needed in wet trackless room-and-pillar mines; the rubber-tired vehicles traveling on the roadways create fine material that eventually collects in the water ditches if the roadways have drips or continuous streams of water on them. Although this situation should be corrected to keep water off the roadways, it invariably occurs or recurs. A place must be provided for the fines to settle out of the water so they do not damage the clear-water high-head pump impellers. This can sometimes be done effectively in a small catch basin that is frequently cleaned out. However, if the water flows are substantial, the main sump receives the bulk of this fine material. As a result, provisions must be made to clean the sump with conventional or special equipment, divert the water into another sump while one sump is being cleaned, provide a place to put the fines ("soup") removed from the sump, and transport the fines to the surface.

One of the most common methods of cleaning a sump is to use conventional front-end loaders to remove the material. However, a ramp must be provided down to the sump-level floor. If deep-well impeller sections are suspended from a pump station down into the sump, considerable care must be taken to avoid bumping the impellers with the loader. Other systems using slurry pumps, scrapers, or diverting the material into skips have been tried, but there still is no single, efficient, low-cost way to clean a sump.

When cleaning a sump, diverting the water flow into another sump can also be a problem unless at least twice the normally required pumping capacity is provided. Half of the pumps must handle all of the inflowing water while one sump is being pumped down and cleaned. Mines that make large amounts of water also produce large amounts of fine material that fills the sumps much faster than anticipated. The sump of a new mine in the development stage cannot be cleaned too soon; if the mine is not producing much water at a particular stage of development, the sump should still be kept clean. The next shift may bring in more water than can be handled with half the pumps, and when that point is reached, there will be no way to reach the muck in the bottom of the sump for cleaning.

Two methods are often used to divide the sumps for cleaning. One is to locate a concrete wall between the two areas and provide a means of closing off one side and making the water flow to the other side. With this method, the wall should be anchored properly on the top and bottom with reinforcing steel into the rock. This provides a safety factor in preventing the wall from collapsing into the side being cleaned. The other system is to develop two physically separate sumps. Both sumps must be provided with pumps, a method of stopping the water inflow and diverting it to the other sump, and a means of access for cleaning.

Many mines locate the pump station below the sump chambers, using horizontal centrifugal pumps instead of deep-well turbine pumps. These mines avoid the installation of a vacuum priming system because the overflow feeds the main pump. Although this system provides easier access to the sumps and makes cleaning easier, the risk of losing the pump station by flooding is increased because of its lower elevation.

If the power for the pumps originates from a source not maintained by the mine (i.e., public power with or without "tie lines"), the mine may be without power for extended periods of time. Therefore, every protective measure should be provided. Using deep-well pumps on one of the upper or middle levels helps protect the facility for the maximum length of time. If the pumps are in a room or chamber with the power lines coming in and the water pipes going out, it may be impossible to access the pumps for replacement without a small overhead crane on a rail above the pumps. If deep-well pumps are used, sufficient headroom is needed to remove the multistage impellers.

Storage Pocket (Skip Pocket) Size

For mines that extend over a large vertical distance, there is usually enough storage capacity in the orepasses to accommodate any mismatch between production schedule and hoisting schedule. But for mines that are on a single level, there is often a need to build a large storage capacity between the production of the mine and the hoisting system.

The correct size of the skip pocket depends on what it is intended to accomplish. For example, if management decided to try to hoist for 20 shifts per week but stoped two shifts per day in a mine producing 4,536 t/d (5,000 tpd) between midnight Friday and 7:00 AM Monday, the minimum pocket size would be found from

$$C = [S_m/S_h] \times S_{md} \times T_{ms} \qquad (12.5\text{-}1)$$

where C is the capacity needed in metric tons or short tons, S_m is the number of shifts per week the mine operates, S_h is the number of shifts per week the hoist operates, S_{md} is the number of shifts the mine is down, and T_{ms} is the number of metric tons produced per mine shift.

For this example, the capacity would be calculated as

$$C = [10/20] \times 6 \times 2{,}268 \text{ t/shift} = 6{,}804 \text{ t} \qquad (12.5\text{-}2)$$

or in English units

$$\begin{aligned} C &= [10/20] \times 6 \times 2{,}500 \text{ st/shift} \\ &= 7{,}500 \text{ st} \end{aligned} \qquad (12.5\text{-}3)$$

It is invariably found that a much larger pocket or alternative surge capacity will improve the performance of the mine. The loading station and shaft hoist are often a bottleneck on production.

Other Service Facility Developments

A variety of other rooms or drifts must also be developed. A lunch area (or room) is essential around an underground shop. In many underground mines the lunchroom doubles as a mine rescue chamber. Office space should be provided for the privacy of the underground supervisor and the security of records and equipment.

Another underground facility that must be developed is a fuel storage area. For supplying all of the diesel-powered equipment used in the mines, it is usually advisable that a system be developed for centralized fuel storage. Where the mines are shallow, it is common to put the fuel oil underground through boreholes. However, where the mines are deep and have multiple levels, palletized storage containers are common. In either case, a retaining wall must be maintained that will contain the fluids in case a tank ruptures.

A new consideration is that of storing biodiesel fuel underground. Such fuel has the same requirements as diesel fuel in that it is constantly needed in the refueling of the trackless equipment, but there is also the added advantage that biodiesel is low-temperature sensitive and storing it underground will keep it fluid.

There must also be a space for explosive storage. Of course this should be located in a safe location, away from main haulageways and high-voltage electrical lines. Equally important is to construct the explosive magazine with complete security and management control. In addition, a separate storage and secure location must be built for the detonator systems.

MAKING MINE DEVELOPMENT DECISIONS

Many ways exist to develop a mine. However, there is only one "least-cost" method for any specific mine, and it is up to the mine engineering staff to develop trade-off studies to determine the optimum method to develop the mine. A classic example was recently completed for Stillwater Mining Company (SMC) for the development of mining of ore below the 3,200 Level down to the 2,000 Level. The mine production is at a nominal capacity of 4,200 t/d (4,600 tpd) of combined ore and waste, and is projected to increase to 4,900 t/d (5,400 tpd) over the next 10 years.

The 3200 Haulage System Trade-Off Study (C. Jacobs and B. LaMoure, personal communication) at the Stillwater mine was initiated to evaluate several life-of-mine muck-handling systems, each having the potential to provide for long-term muck-handling requirements for the operation while meeting common operating criteria and long-range goals. Each system evaluated was unique with respect to mine development, timing, required infrastructure, and expenditure profile.

The objective of the study was to evaluate alternate long-term production from the deeper levels of the mine for ore and waste handling systems of various hauling options in lieu of the proposed belt conveyor system, and to make a recommendation for the best haulage option. The proposed conveyor system was listed as an option to provide a "base case" comparison with other options. The primary options in this study included

- A decline with a conveyor option, with diesel trucks (base case);
- Deepening of the existing shaft with rail haulage;
- An internal shaft or winze with rail haulage and secondary hoisting;
- A decline with trucks using only diesel fuel; and
- A decline using electric trucks.

All options consider a common 20-year mining plan, with a maximum total throughput of 4,900 t/d (5,400 tpd).

Economic evaluation of the trade-off study options considered incremental differences in capital and operating expenditures within SMC's gross mining model (GMM). This precluded a pure incremental analysis of the different options but allowed SMC Technical Services to construct 20-year cash flows for each different haulage option, which consider all gross mining costs with respect to each plan, the revenue stream generated for each plan based on metals price assumptions, and downstream processing and commercial cost combined with applicable credits.

It is important to evaluate both annual cash flow and cumulative cash flow. The option with the highest cumulative cash would logically be the best option but may not be the desired option if annual cash flow is negative for 1 or more years.

Net cash flow (NCF) from each option was taken directly from option-specific cash-flow models and compared to perform an economic analysis considering mutually exclusive projects. The NPV for each cash flow is considered. Incremental analysis is then applied to derive incremental NPV values and ROR. Based on the analysis, the following results are relevant:

- NCF in all options are positive values; there are no negative years. This is due to full consideration of revenues and costs via the GMM, applied to what are relatively small capital costs and incremental capital costs of the haulage options.
- Using a discount rate of 10%, the electric trucks option is the best economic choice based on NPV. The incremental NPV for the cash-flow comparison confirms this with a

cash-positive incremental NPV, and a positive incremental ROR. Present value ratio is also favorable.
- In the absence of the electric trucks option, the shaft-deepening option becomes economically favorable. Increasing the discount rate in the analysis to 20% indicates that the base-case option is the economic choice based on NPV, where the shaft-deepening option has more cost in forward years. The diesel truck option was eliminated because of ventilation constraints and the logistics of operating eight trucks in the ramp system.

MECHANICAL EXCAVATION METHODS OF DEVELOPMENT

Because of the many advantages of mechanical mining when the physical properties of the rock and the tools are right for cost-competitive excavation, hard-rock mining companies keep trying to push the edge of technology so they, too, can benefit from mechanical mining. While it has been a very slow process of change, there have been a few notable successes. The principal advantages to mechanical excavation are

- Improved personal safety;
- Minimal ground disturbance;
- Reduced ground support needed;
- Minimum overbreak, thus less material to move;
- Minimal ground vibration or air blast;
- Uniform muck size;
- Reduced ventilation requirements;
- Continuous operations;
- Conducive to automation of system;
- Where the system is applicable, higher production rates; and
- Where the system is applicable, reduced tunneling costs.

The disadvantages to mechanical excavation are

- In hard rock, the system may have a poor penetration rate;
- In hard or abrasive rock, the system may have a high cutter cost;
- Some mechanical systems are not easily adaptable to changing ground conditions;
- It may lack flexibility of operations (TBMs can't cut sharp curves); and
- It often has a very high initial cost for the mechanical excavating equipment.

For designing a TBM for mine developments as compared to civil applications, Friant (2001) makes the following design recommendations:

- Multiple-speed cutterhead—An infinitely variable drive with high/low-speed torque is needed for broken ground changing to solid ground. A variable frequency drive is available.
- Closed or shielded-face cutterhead—Recessed cutters combine with radial bucket openings that extend toward the center of the head. Cuttings should not have to drop all the way to the invert through the slot between the rock face and cutterhead shield to be picked up. Back-loading cutters should be mandatory. There is no excuse for a fatality due to rock falls while changing cutters.
- Adequate shielding for rock support—Shielding, sometimes covering a 360° area, should extend up nearly to the gauge cutters. Adequate space must be provided near the cutterhead to install bolts and ring beams. Shields should not have wide gaps that allow small rocks to rain in. Shields should be active (i.e., with hydraulic adjustments) so if the machine gets squeezed, it can release itself.
- Soft-touch grippers—Grippers, if applied with too much pressure, can break up the tunnel wall and cause massive cleanup jobs after each push. Putting pressure on the crown should be avoided.
- A small length/diameter ratio—Either the whole machine length or its segment lengths dictate the radius curve through which it can bore. Often a 30-m (100-ft) radius is desirable on a midsize machine. A short stroke is an acceptable way to achieve the short turn radius desired. Continuous boring systems are desirable (i.e., with two sets of grippers), but only if they do not compromise other requirements of shielding and grippers.
- Mobility—A machine needs the ability to back out of its own tunnel, break down into manageable size subassemblies, and come apart and go together quickly. This may be achieved by the machine itself or through accessories for the machine. It must be kept in mind that in a mine there will often be many short tunnels as opposed to one continuous drive.
- Access for drills—In addition to rock-bolt drilling, the machine must supply access for drilling for grout curtains, spiling, soil stabilization, probe holes, and drain holes.
- Design simplicity—The simplest methods of thrust, steering, and control must be used, not those that are the most clever. Human-engineered controls and safety systems should be incorporated.

In soft material, such as coal, borax, potash, and trona, the applications of mechanical excavation are too numerous to mention. However, the successful hard-rock applications in mining are covered by a few examples:

- Stillwater Mining Company—Three different TBMs have been used to drive very long lateral drifts that are used first for ore-body delineation drilling, and then for main-line haulage. The rock has an unconfined compressive strength (UCS) ranging from 87 to 165 MPa (12,500 to 24,000 psi). The first machine used had trouble following the curvature of the ore strike, but the latter two machines were built for underground mine applications and were more adaptable. The biggest benefit for using the TBMs was time saved in the development schedule (Bullock 1994).
- Magma Copper Company—To expand the mining from the San Manuel (Arizona, United States) mine, a 4.5-m- (15-ft-) diameter TBM was used to drive 3,156 m (10,353 ft) of development during preparation for panel caving of the Kalamazoo ore body. The UCS of the rock ranged from 150 to 180 MPa (21,755 to 26,105 psi). The tunneling involved some 150-m- (500-ft-) radius curves and a decline of 5.5% (Bullock 1994).
- Heinrich Robert mine—At this coal mine in Herringen, Germany, a Voest-Alpine AM 105 roadheader was used to drive a drift 38.5 m^2 × 3.9 km (421 ft^2 × 2.3 mi). The rock is a massive quartzite sandstone with a UCS of 119 to 167 MPa (17,259 to 24,235 psi). During a period of 1 year, the machine averaged 5.6 m/d (18.5 ft/d), which included arch support on 1-m (3.3-ft) centers (Bullock 1994).

Table 12.5-1 Typical underground development estimated costs in 2009 U.S. dollars*

Development Item	Dimensions in Metric Units	$/m (Accuracy ±25%)	Dimensions in English Units	$/ft (Accuracy ±25%)
Drifts and crosscuts	3.7-m w × 4.0-m h, bolted	2,790–3,100	12-ft w × 13-ft h, bolted	850–950
Stope crosscuts	3.4-m w × 4.0-m h, bolted	2,550–2,900	11-ft w × 13-ft h, bolted	780–880
Stope attack ramps	3.0-m w × 3.0-m h, bolted	1,950–2,050	10-ft w × 10-ft h, bolted	600–630
Decline (contracted)	4.9-m w × 5.5-m h, bolted, and 152-mm shotcrete	6,900	16-ft w × 18-ft h, bolted, and 6-inch shotcrete	2,100
Ventilation drift	4-m w × 4.6-m h, bolted, and 76-mm shotcrete	3,850–4,050	13-ft w × 15-ft h, bolted, and 3-inch shotcrete	1,175–1,230
Ventilation raise	2.4-m diameter, raisebored and lined	6,250	8-ft diameter, raisebored and lined	1,900
Pump station, shops, stores, etc.	10.7-m w × 6.1-m h × length needed for m^3, bolted, and shotcrete	19.0–20.70/m^3	35-ft w × 20-ft h × length needed for ft^3, bolted, and shotcrete	5.80–6.30/ft^3

*Fairly fractured and weathered ground located in central Nevada requiring regular bolting with split sets on 1.2-m (4-ft) centers.

Other Types of Mechanical Excavation

Several other types of tools have been tried in mechanical excavation (Bullock 1994):

- Impact breakers (tunnels in Italy)
- Machines with disks on rotating drums (mobile miners)
- Machines with disks on rotating, programmable arms (continuous mobile miner)
- Various microboring machines

DEVELOPMENT COST ESTIMATES BY DRILL-AND-BLAST

It is not extremely useful to present cost estimates for developments, because the many variables of the rock types, rock-mass conditions and strengths, water condition, size, inclination, and use will affect cost. However, a few examples are presented in Table 12.5-1.

In another example, the Stillwater mine contracted two Alimak 3.35-m- (11-ft-) diameter raises, driven an average of 521 m (1,705 ft) for a cost of $5,612/m ($1,700/ft) in 2005 and 2006. When escalated to 2009 costs, this approximates to $6,600/m ($2,000/ft). There were two reasons why the raise climber was used for these ventilation raises, rather than using upreaming with a raise-boring machine: (1) there was no access to the top of the mountain in this isolated area and everything would have had to have been transported by helicopter; and (2) permitting to build a heliport and raise-boring site would have been difficult to obtain in this pristine area.

In the end, designers and consultants must remember that in planning a mine project, unlike a civil project, the product is not the hole: the hole is only a means to get at what you came for, the ore. Versatility, capability, flexibility, and at times raw power are the features that count.

ACKNOWLEDGMENTS

The author acknowledges the assistance of Curt Jacobs, Brent LaMoure, and the Stillwater Mining Company for their assistance in furnishing both development trade-off study information and the Alimak raise mining costs used in this chapter. The assistance of Dave Linebarger, a mining consultant from Elko, Nevada, is also acknowledged for furnishing typical underground development estimated costs.

REFERENCES

Brechtel, C.E., Struble, G.R., and Guenther, B. 2001. Underhand cut-and-fill mining at the Murray mine, Jerritt Canyon Joint Venture. In *Engineering Fundamentals and International Case Studies*. Edited by W.A. Hustrulid and R.L. Bullock. Littleton, CO: SME.

Bullock, R.L. 1961. Fundamental research on burn cut drift rounds. *Explos. Eng.* 39(1-2).

Bullock, R.L. 1994. Underground hard rock mechanical mining. *Min. Eng.* 46(11):1254–1258.

Bullock, R.L. 1997. Mechanical excavation applied to underground hard rock metal mining. In *McGraw-Hill Yearbook of Science and Technology*. New York: McGraw-Hill. pp. 466–468.

Bullock, R.L., and Hustrulid, W.E. 2001. Planning the underground mine on the basis of mining method. In *Engineering Fundamentals and International Case Studies*. Edited by W.A. Hustrulid and R.L. Bullock. Littleton, CO: SME.

Camm, T.W., and Miller, H. 2009. An economic analysis of controlled drilling and blasting techniques in metal drift development. Presented at the Controlled Blasting and Rapid Excavation Session at the SME Annual Meeting, Denver, CO, February 24.

Dravo Corporation. 1974. *Analysis of large scale non-coal underground mining methods*. Publication PB 234 555. Springfield, VA: U.S. Bureau of Mines. p. 605.

Friant, J.E. 2001. Mine tunneling using TBMs and their cost. Class presentation. University of Missouri, Rolla.

Gould, J.C. 1998. Climax panel caving and extraction system. In *Techniques in Underground Mining*. Edited by R. Gertsch and R.L. Bullock. Littleton, CO: SME.

Lacasse, M., and Legast, P. 1981. Change from grizzly to LHD extraction system. In *Design and Operation of Caving and Sublevel Stoping Mines*. Edited by D.R. Stewart. New York: American Institute of Mining, Metallurgical, and Petroleum Engineers.

Lehany, T. 2008. *Diggers and Dealers: Newcrest Executive General Manager's Operations Report*. www.newcrest.com.au/upload/506_5x08x200843531PM.pdf. Accessed September 2009.

Lewis, R.S., and Clark, G.B. 1964. *Elements of Mining*. New York: John Wiley and Sons.

Lyman, W. 1998. Shrinkage stoping: An introduction. In *Techniques in Underground Mining*. Edited by R. Gertsch and R.L. Bullock. Littleton, CO: SME.

Marchand, R., Godin, P., and Doucet, C. 2001. Shrinkage stoping at the Mouska mine. In *Engineering Fundamentals and International Case Studies*. Edited by W.A. Hustrulid and R.L. Bullock. Littleton, CO: SME.

Moss, A. 2009. Rapid construction a key to success. Presented at the Controlled Blasting and Rapid Excavation Session at the SME Annual Meeting, Denver, CO, February 24.

Nilsen, F. 2009. Norwegian high-speed tunneling experience. Presented at the Controlled Blasting and Rapid Excavation Session at the SME Annual Meeting, Denver, CO, February 24.

Peppin, C., Fudge, T., Hartman, K., Bayer, D., and DeVoe, T. 2001. Underhand cut-and-fill mining at the Lucky Friday mine. In *Engineering Fundamentals and International Case Studies*. Edited by W.A. Hustrulid and R.L. Bullock. Littleton, CO: SME.

Rech, W.D. 2001. Henderson mine. In *Engineering Fundamentals and International Case Studies*. Edited by W.A. Hustrulid and R.L. Bullock. Littleton, CO: SME.

Spearing, A.J.S. 1995. *Handbook on Hard-Rock Strata Control*. Johannesburg, South Africa: South African Institute of Mining and Metallurgy.

Svensson, H. 1982. Raise climber. In *Underground Mining Methods Handbook*. Edited by W.A. Hustrulid. Littleton, CO: SME.

CHAPTER 12.6

Construction of Underground Openings and Related Infrastructure

Joe Cline

INTRODUCTION

The extraction of deep mineral deposits usually requires the development of significant underground infrastructure in the host rock. Construction of these openings constitutes a major aspect in the development of an underground mine. Depending on their use, development openings will assume various sizes, shapes, and orientations. The openings must be large enough to provide ample space and perform the intended functions to support the underground mining process with allowances for the geologic and geotechnical characteristics of the rock mass.

Mine excavations are categorized into two distinct groups: excavations that are used primarily for rock handling and excavations that are used for service. Development openings that fall in the latter category are considered permanent or semipermanent, usually with an operating life equal to or exceeding that of the operating mine. Thus the locations of these excavations are particularly critical in the overall mine infrastructure plan; they must be designed and constructed to meet rigorous performance standards while maintaining their operating role.

In the development of underground mine openings, the design and construction aspects of the excavations must be considered concurrently. The design aspect will address the size, location, and ground support requirements of the openings with respect to the geotechnical characterization of the host rock to ensure the associated equipment can be installed and any unfavorable geologic/geotechnical conditions will be accommodated. The construction aspect will address the excavation methods and equipment, excavation sequencing, and integration of the development with the associated construction activities.

A significant portion of the underground development schedule is normally dedicated to the construction of major mine service facilities such as pump rooms, crusher rooms, underground maintenance shops, ore bins, orepasses, loading pockets, ventilation and haulage drifts, and so on. In hard-rock mines, the openings for most of these facilities are normally excavated using drill-blast-muck cycles. Vertical and inclined openings, including raises and orepasses, can be excavated using either drill-and-blast or semicontinuous methods (such as raise boring). Accessibility requirements are critical as well in order for mobile equipment and workers to gain access and construct the associated facilities.

Each underground facility has a significant role in the continuous operation of the underground mine. The decisions that are made from the development of the design (such as operational redundancy) to the operation and maintenance phase (such as implementation of control systems and periodic maintenance) are critical in ensuring constant, uninterrupted mine operations. For instance, the pumping system for an underground mine facility with high water inflows would be designed and implemented with enough redundancy to ensure that in the event of a system failure, auxiliary equipment is present that can be commissioned immediately to augment the primary system until such time that repairs are completed and the primary pumping system can be put back into operation. The ultimate reasoning behind designing and implementing redundancy in most underground mine service facilities is to ensure (with economic considerations for capital expenditure) a safe working environment and eliminate or significantly reduce the potential for catastrophes that could result in the eventual loss of revenue.

UNDERGROUND OPENINGS AND RELATED INFRASTRUCTURE

A number of underground openings and related infrastructure are common to most underground mine operations. They include developments such as access drifts, ventilation drifts, sumps, refuge stations, battery rooms, and so on. Major infrastructure such as conveyor galleries, underground crusher chambers, pump rooms, sumps, and underground maintenance shops normally involve significant capital expenditure but are necessary facilities in underground mining operations.

Sumps and Pump Rooms

Sumps and pump rooms are used in underground operations primarily to collect and manage water in and around the mine

---

Joe Cline, Lead Discipline Engineer, Cementation USA, Inc., Sandy, Utah, USA

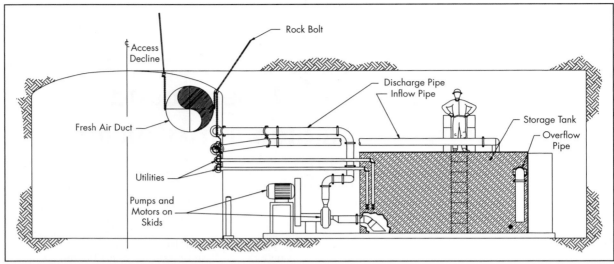

Courtesy of Kennecott Eagle Minerals Co., Miller Engineering, and Stantec – Mining.
Figure 12.6-1 Unitized storage tank and pump arrangement

workings. Water inflow in a mine can occur from the following primary sources:

- Drilling and other associated activities during underground development and production (service water)
- Groundwater inflows

In either case, the water must be effectively handled by locating, designing, selecting, constructing, and maintaining a dependable storage and pumping system. The size and configuration of the pumping system depends largely on the predicted amount of water inflow to the mine. Depending on the predicted water inflows, the water handling system arrangement may consist of

- A series of drainage ditches, channels, and pipes at different working levels that direct water to sumps; the water is then stored and ultimately pumped to the surface either with a single high-efficiency pump or by a series of pumps; or
- Drainage galleries or drifts used to transfer water to a central sump or storage facility and then pump it to the surface.

The first arrangement is best suited for small to medium water inflows (up to several hundred cubic meters per hour) where large inflows not anticipated in the hydrologic study could overwhelm the pumping system. The second arrangement is suited for underground operations with large water inflows (several thousand cubic meters per hour) where hydrologic studies indicate there is a potential for flows to peak beyond those predicted.

Common characteristics of underground sumps and pump rooms include the following:

- Excavation and concrete structures for the sump capacity and pump room size
- Concrete bulkhead(s) for the sump containment, equipment foundations, and pipe support structures
- Access platforms and walkways to provide for equipment inspection and maintenance
- Mechanical equipment (pumps, motors, piping, and associated appurtenances)
- Power, electrical, instrumentation, and control equipment (switchgear, mine load center, drives, etc.)
- Lifting devices (monorails, cranes) for removing major equipment and reassembling after service

Figures 12.6-1 through 12.6-3 illustrate three varieties of underground pumping system arrangements that are common in underground mine operations.

In the majority of cases, underground mine pumping systems are designed to handle suspended solids that can be abrasive (depending on the size and distribution of the particles in suspension) on the pump components that are in direct contact with the carrier liquid. Loss of pump operating efficiency due to wear of pump components is a major concern in underground mine operations. The pumps specifically designed for this type of application are referred to as slurry pumps.

Centrifugal pumps and positive displacement pumps are the two most common types of pumps used in underground mine applications. Other pumping arrangements, such as those employing multiple-stage vertical or horizontal pumps, have been employed in underground mining for pumping clear water. These do not handle suspended solids well and thus require the excavation of settling sumps. Vertical deep-well pumps are unpopular in conventional mine dewatering because the entire pump column needs to be removed to access the pump for repair or maintenance; however, they are typically used when dewatering a flooded mine.

The pumping unit illustrated in Figure 12.6-1 is one of several skid-mounted packaged units that comprise the entire pumping system. This type of pumping system is commonly used during the early development of a mine. The pumping arrangement can consist of multiple units of similar design connected in series and located at specific elevations within the mine to collect and transfer water upstream from the lowest extremities of the mine workings. Each unit can incorporate single or multiple centrifugal pumps with motors driving the pumps integrally or via V-belt drive systems. Typically, each unit discharges water upstream into the storage tank of the successive unit in the pumping circuit. The setup in Figure 12.6-1 can potentially handle flows up to several

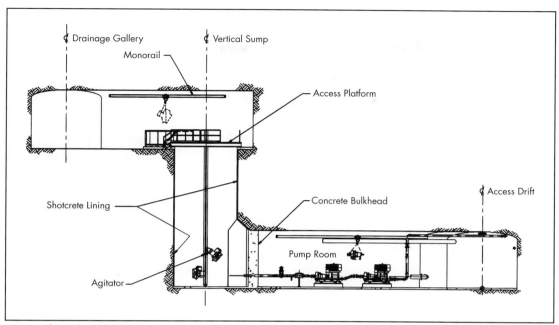

Courtesy of Barrick Gold and Stantec – Mining.
Figure 12.6-2 Combined vertical sump and pump-room arrangement

hundred cubic meters per hour; the actual performance is contingent on the total head to overcome, the number and size of pumps per unit, and the total number of units within the pumping circuit.

Figure 12.6-2 (also shown in plan view later in Figure 12.6-21) incorporates a vertical sump that is located directly upstream of a pump room. A concrete bulkhead separates the sump compartment from the pump room. In this type of arrangement, particular attention is required to locate, design, and construct the bulkhead to ensure stability and provide a watertight seal around the rock–concrete interface and at penetrations through the concrete bulkhead. The sump walls are typically lined with shotcrete or concrete to limit and/or eliminate any seepage into the surrounding rock and to maintain the integrity of the rock wall.

Pumps are usually erected on a concrete pump base that is raised several centimeters above the pump room floor. This arrangement prevents equipment from coming into contact with water in the event of a pipe leak or other malfunction. An overhead crane or monorail is usually provided to aid initial installation and for removing and replacing pump components. The pumping arrangement shown in Figures 12.6-2 and later in 12.6-21 consists of multiple banks of pumps, each having multiple centrifugal pumps connected in series. The number of pump stations that are required to make up the complete pumping system depends on the pressure rating of the pumps.

Contingent on the quantity and specific gravity of suspended solids in the carrier liquid, agitators may be installed inside the sump to keep solid particles in suspension at all times. This type of layout provides the following advantages:

- Can be used to provide redundancy in a mine's pumping system where only one bank of pumps operates at any time while the other is on standby

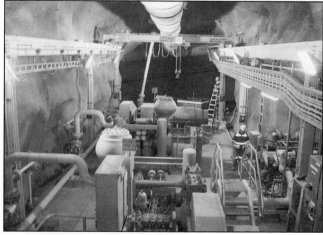

Courtesy of Northparkes Mines.
Figure 12.6-3 Remote sump and pump-room arrangement

- Allows for the possibility of installing additional banks of pumps should there be a future demand for more pumping capacity
- Supports the ability to perform equipment maintenance or repair without a complete system shutdown

The primary pumping equipment shown in Figure 12.6-3 comprises multiple piston diaphragm positive displacement pumps, complete with electronics and other accessory features. Each unit is capable of operating at up to 1,000 m of head at low flow rates, and up to about 90 m^3/h at lower heads. They are efficient in handling high concentrations of solids. The figure also illustrates an arrangement where the pump room is located

Courtesy of Northparkes Mines.
Figure 12.6-4 Remote sump at an underground pump station

in a separate excavation that is completely independent of the storage reservoir/sump cutout (see Figure 12.6-4 for sump illustration). Water is fed to the pumps by pipes running from the sump to the main pump-room intake. Unlike Figure 12.6-2, allowance is made immediately above the concrete bulkhead to access the agitator (the agitator keeps solids in suspension) and other equipment in the sump. Pumping arrangements of this type offer several operating advantages, including

- Pumping of dirty water,
- Operating under high static heads,
- Pumping in a single lift without the need for multiple intermediate pump stations,
- Operating at a high efficiency, and
- Eliminating the need for large settling sump cutouts.

Crusher Rooms

Underground crushers are used to reduce run-of-mine ore or waste to a size that is suitable for the materials-handling system to transport to the surface. Crusher chambers can be the largest single excavation and infrastructure development in an underground mine. Substantial capital expenditure normally goes toward the planning, design, procurement, and construction of these underground facilities. An underground gyratory crushing station is illustrated in Figure 12.6-5.

Critical information necessary for making a decision regarding plant suitability and final underground crushing arrangement includes

- Location,
- Access,
- Mining method,
- Ground conditions,
- Ore properties, and
- Required support facilities.

The underground ore transport network must be integrated with the crushing plant location and design, as the mined ore must be transported and fed into the plant for crushing before being transported to the surface.

Courtesy of Northparkes Mines.
Figure 12.6-5 Gyratory crusher

Underground crushing operations can be classified as single-stage or multistage (i.e., primary crushing or primary–secondary crushing). The size and complexity of the facility depend on the number of crushing stages. Because of advancements in belt conveyor equipment and other associated materials-handling systems (to handle larger and more abrasive materials), most modern underground mines only need to install a primary crushing system. There is also an added capital and operating cost–savings benefit that can be realized through elimination of a stage in the overall crushing circuit. Some underground mines in the past have implemented multistaged crushing operations. In the 1970s, ore produced from underground operations at the Climax mine in Colorado, United States (currently abandoned) was crushed by primary and secondary crushing at their No. 4 crusher station.

In hard-rock mines, gyratory and jaw crushers are most commonly used because of their proven capability in breaking hard rock. Sizer technology is advancing to the point where they are now suitable for size reduction of hard rock. Feeder breakers are more successful in softer ores. In coal mines where rock fragmentation activities are not as aggressive, other crushing plant alternatives (such as roll and impact crushers) are more predominant. When developing underground crusher rooms, the following must be considered:

- Location (remote or central) with respect to ore delivery and conveyance after crushing

Courtesy of Northparkes Mines.

Figure 12.6-6 Underground shop service bay with overhead crane

- Crusher chamber access (primarily for maintenance and supply of spare parts)
- Excavation size, stability, ground control, geotechnical instrumentation, and monitoring of the rock mass around the chamber
- Overhead cranes of suitable capacity and other localized lifting devices within the chamber
- Dump pocket (truck or rail dump) and coarse ore surge bin
- Oversize handling arrangement at truck and rail dumps (grizzly, rock breaker, etc.)
- Crusher, drives, mountings, control room
- Concrete structures supporting the crusher, drives, and motors
- Crushed ore loading pocket beneath the crusher or remote surge bin
- Pan feeders, impact feeder belts, screens, and belt conveyors
- Access platforms and walkways
- Dust collectors, scrubbers, and crusher room ventilation
- Power, electrical, instrumentation, and control equipment (substation, switchgear, drives, etc.)

Underground Maintenance Shops

When no direct ramp access exists between the underground operation and the surface, or when the access decline is too long to tram equipment to the surface in a reasonable time frame, a shop and maintenance facility may be constructed underground to service and repair mobile and stationary equipment. In mines with only shaft access, major equipment is normally transported, assembled, and maintained underground.

After commissioning and operating, major equipment often only leaves the underground mine for replacement or major rebuilding in a surface or off-site shop. An underground shop is a major lifeline for the mobile fleets in an underground mine. The main advantage of locating a shop facility underground is for close proximity to the operations so scheduled maintenance and emergency repairs can be completed promptly with less equipment downtime.

Whether fleet maintenance is contracted out to a separate entity or performed by the mine's own personnel, it is essential to establish an ergonomic layout for the shop plan that is safe and maximizes service to equipment with easy access to supplies, spares, and consumables so that equipment times in the shop are limited. For this reason, multiple entry and exit points in the shop must be created to ensure that immobilized equipment undergoing repairs will not hinder the exit or entry of other equipment. When an underground mobile fleet is comprised of both rubber-tired and track equipment, underground maintenance of both types of equipment must be available. The maintenance requirements for track equipment (such as work on engines, braking systems, and other critical locomotive components) are fundamentally different from those of rubber-tired equipment. For these reasons, it is essential to have a dedicated repair shop for track equipment that is independent of the primary maintenance shop. Underground shops are often equipped to replace components such as engines, transmissions, axles, buckets, and so forth.

An essential component of any underground shop is the overhead crane and other associated lifting devices (see Figure 12.6-6) used to move heavy components around the shop floor. Other items of prime importance in an underground shop include

- Ventilation inside the shop;
- Fire protection;
- Access in and around the shop;
- Number of bays and their function;
- Types and level of services to be performed;
- Supply and management of consumables;
- Management of spent hydrocarbons, lube, and gray water;
- Shop lighting and illumination;
- Compressed air supply;
- Storage of tires, spare parts, and tools; and
- Equipment cleaning.

Ore Bins

Ore bins are used primarily for temporary storage of ore underground before final transport to the surface. Depending on the location in the ore circuit and the materials handling methodology, ore bins are generally referred to as run-of-mine or coarse-ore bins when used to store ore that has been sized with a grizzly or sizer, or that discharges from the primary crusher.

Because of the cyclical nature of ore extraction from the mining face, an underground ore bin provides the surge capacity necessary to maintain a constant flow of the ore stream. The material to be stored is normally fed into the bin via trucks, rail, belt conveyor system (typical for remote bins), directly from a system of orepasses, or directly from the crushing plant (if located directly beneath the crusher). The content of the bin is later discharged from the bin bottom through a hopper-chute arrangement for further ore transport by a belt conveyor, rail, or truck system.

Ore bin design and construction depend on a variety of conditions, but, most importantly, on the type and condition of the host rock. In massive hard rock, ore bins are normally excavated in and integrated with the host rock (Figure 12.6-7). In other rock environments (such as medium, soft, or sedimentary rocks), ore bins are predominantly constructed as independent structures in order to limit the effect of any

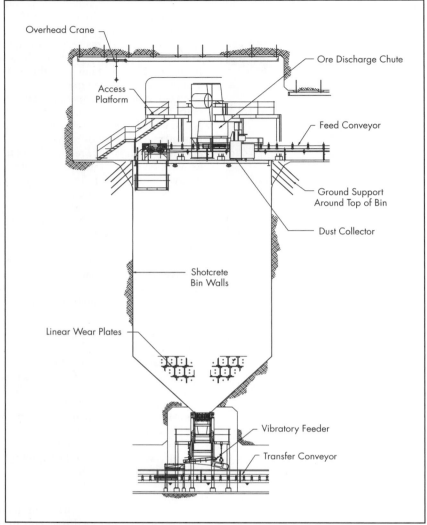

Courtesy of Stantec – Mining.

Figure 12.6-7 Ore bin and associated infrastructure in a hard-rock environment

potential ground movement on their ongoing performance (see Figure 12.6-8). Following are other important decision factors to consider with underground ore bins:

- Permanent access
- Access during bin development and construction
- Surrounding rock type and ground support requirements
- Location (above or beneath crusher, remote, close to hoisting plant, etc.)
- Size and capacity (based on production rate, scheduled, and/or nonscheduled shutdowns). Refer to the "Storage Pocket (Skip Pocket) Size" section in Chapter 12.5.
- Shape (circular shape most common, can also be rectangular)
- Associated materials handling infrastructure (feeders, chutes, conveyors, other materials handling equipment) to aid in transporting the material effectively
- Bin wall lining and other materials of construction (depending on abrasiveness of the stored material and characteristics of the host rock)

Ore- and Waste Passes

Passes are vertical or subvertical raises that are excavated in the host rock for transporting ore or waste by gravity flow from the upper levels in a mine to a predetermined lower level. Orepasses can be used to improve underground equipment productivity because they minimize overall travel distances from the ore loading point to the final ore dump point. The location and design of the orepasses are therefore critical in mines that primarily haul waste and ore with trackless equipment. Orepasses commonly run subvertically and may connect production levels by means of finger raises, as shown in Figure 12.6-9A. The top of the pass and finger raises, are usually equipped with a rock-sizing mechanism (such as a grizzly) to ensure that oversize materials which can potentially cause hang-ups in the pass are reduced in size. A load-out control mechanism (such as a discharge chute, chain controls, and press frame) is usually installed at the lowest point of a drawpoint, as illustrated in Figure 12.6-9B. Following are some major items to consider in underground orepasses:

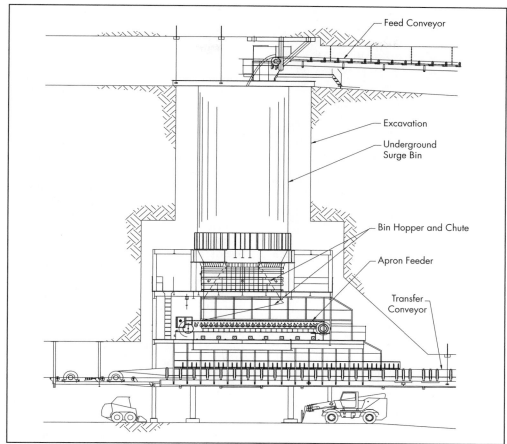

Courtesy of Potash Corporation of Saskatchewan–Allan Division.

Figure 12.6-8 Ore bin and associated infrastructure in an underground potash mine

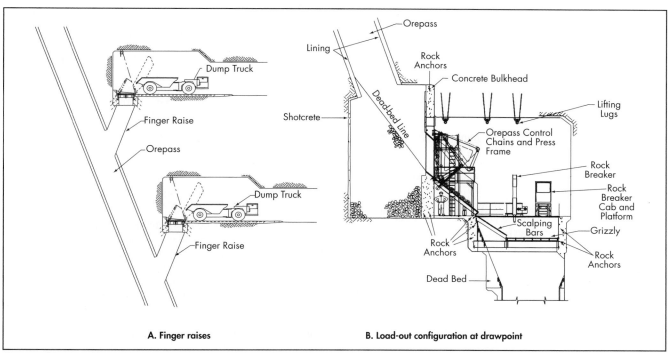

A. Finger raises

B. Load-out configuration at drawpoint

Courtesy of Stantec – Mining.

Figure 12.6-9 Orepass and finger raises

Courtesy of Newcrest Mining Ltd.

Figure 12.6-10 Access ladderway looking from top to bottom

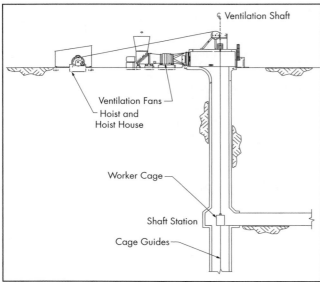

Courtesy of Stantec – Mining.

Figure 12.6-11 Emergency hoist and escape pod

- Shape and size of the orepass, enough to prevent material hang-ups
- Stability of the orepass
- Maximum interval spacing between successive passes
- Orepass inclination
- Daily and total tonnages anticipated
- Type and orientation of ore body
- Material characteristics (fines, moisture content, particle size, etc.)
- Load-out configuration at bottom
- Static pressures on load-out mechanism
- Effect of impact on load-out mechanism (from release of material hang-up or material falling down an empty pass)

Emergency Access

A secondary means of egress is required in all underground mine operations and normally governed by local mining regulations. An emergency access (worker access) is a secondary means of egress that is provided primarily for miner exit from the underground workings in the event of a fire, temporary loss of electric power, or any other occurrence warranting worker evacuation. It is important that emergency exits are located in an area where the ground is stable and the serviceability of the facility can be guaranteed and will not be hampered if the primary means of egress is blocked. A worker access is a vertical or subvertical shaft that is constructed by drill-and-blast or raise bore and is normally located within the mine's fresh-air intake network to ensure the presence of fresh air in the shaft at all times when in use. During the development phase of the mine, rather than install a separate worker access shaft, one of the compartments of a hoisting or ventilation shaft could be dedicated to emergency access. (After a U.S. mine is in the operating phase, the Mine Safety and Health Administration [MSHA] requires a separate entry as an escapeway.) A worker access is usually equipped with one of two evacuation systems, illustrated in Figures 12.6-10 and 12.6-11: an access ladderway, or an emergency hoist and escape pod.

If an emergency hoist and escape pod system is implemented, provision must be made for standby electric power (provided the emergency hoist is not powered by an internal combustion engine) to operate the hoist in the event of an electric power failure. Following are some major items to consider in designing and constructing a worker access (emergency exit):

- Depth of mine operations from the surface
- Type of primary access (shaft or decline)
- Type of evacuation system
- Mine environment (corrosive, may require galvanized steel components)
- Cost

Loading Pockets

Loading pockets are used to feed ore or waste into skips in a production shaft system. Crushed rock or run-of-mine ore is conveyed and discharged into flasks at the loading pocket by rail, belt conveyors, or mobile equipment. The flasks at the loading pocket serve as a temporary underground repository for rock waiting to be fed into production skips. Several factors, including the following, govern the design and construction of loading pockets:

- Shaft production rate
- Shaft layout and design
- Type and capacity of the skips
- Method of ore (or waste) delivery to the loading pocket
- Construction methodology
- Ground support requirements

Loading pockets consist primarily of steel structures and usually must be integrated with the shaft steelwork at the skip's loading level. It is important that the whole infrastructure around the loading pocket be designed to effectively sustain impact loads from falling rock without inducing undue stresses and movement that could hinder its performance. The design of the loading pocket excavation is therefore important to ensure the structure's long-term stability and must incorporate adequate ground support. Ore delivery to the loading pocket can be semiautomated (with a belt conveyor system feeding from an ore bin or crushing facility) or by means

Construction of Underground Openings and Related Infrastructure

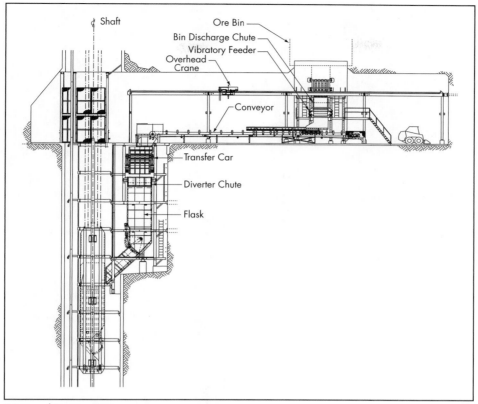

Courtesy of Stantec – Mining.
Figure 12.6-12 Loading pocket with belt conveyor feed

of mobile equipment (such as trucks and load-haul-dumps [LHDs]). Figures 12.6-12 and 12.6-13 illustrate typical loading pocket arrangements for both methods of ore handling.

Special Purpose Drifts

Conveyor, ventilation, and access drifts are three primary types of major lateral development that belong to the category of special purpose drifts. Common characteristics among all three underground openings include size (normally constructed to the smallest practical cross section to meet the service requirement), stability of the final excavations (usually over the life of the mine), and slope of the drift (to maximize equipment productivity in tramming).

Conveyor drifts are normally constructed with a minimum cross section to accommodate the associated equipment while allowing for access, maintenance, and material spill cleanup. They may or may not be inclined, depending on the vertical distance the conveyor has to travel. The belt conveyor equipment is normally supported from the back of the drift, as shown in Figure 12.6-14, usually with rock-bolted chain supports or on the sill of the drift with steel supports on concrete pedestals as shown in Figure 12.6-15. The former method of support is more cost-effective and can be installed faster. Depending on the size, capacity, and layout of the conveyor equipment, the drift may be overexcavated at the head and tail ends or transfer points to accommodate additional equipment such as drives, motors, take-ups, transfer tower structure, chutes, and access platforms (see Figure 12.6-16).

Access drifts have similar characteristics to conveyor drifts except that they are sized and constructed to accommodate

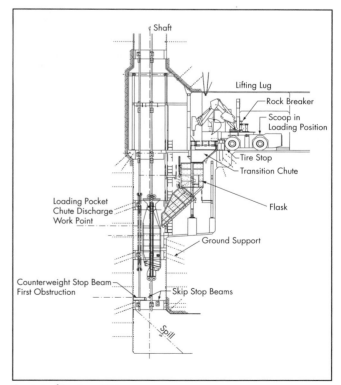

Courtesy of Stantec – Mining.
Figure 12.6-13 Loading pocket fed by rubber-tired equipment

Courtesy of Northparkes Mines.
Figure 12.6-14 Belt conveyor supported from back of drift

Courtesy of Northparkes Mines.
Figure 12.6-15 Belt conveyor supported on sill of drift

mobile equipment and utility lines where applicable. Passing bays (to accommodate two-way vehicular traffic), muck bays, and safety bays (where personnel and vehicular traffic interact) may be incorporated into the construction depending on the use of the drift, as in Figure 12.6-17. The layout and construction of an access drift must take the following into consideration: type and size of haulage equipment to be used (trucks and LHDs or rails); clearance requirements around equipment for passage of mine personnel; routing of ventilation ducts, pipes, electric cables, and other utilities as shown in Figure 12.6-18; vertical distances between successive mining levels; presence of spiral ramps; maximum slope of invert to minimize tramming effort and maintain equipment productivity; type of sill construction; and drainage requirements.

Ventilation drifts are essential for moving air into and from the underground mine workings. These drifts are the primary link between the ventilation shafts or ventilation decline and the underground mine environment, and are classified as follows:

- Intake drift—for taking fresh airflow from a fresh-air intake facility to the mine workings
- Exhaust (return) drift—for taking the mine's return air from the underground workings to an air-exhaust facility

Most mine operations normally use the primary and secondary accesses for ventilation. However, large underground operations with large ventilation airflow requirements may require dedicated ventilation drifts, shafts, and raises. These ventilation routes are usually not accessible to personnel and equipment traffic because of the high velocity of air being moved.

Primary considerations in ventilation drift design and construction include the ventilation fan size and airflow velocity, total length of the ventilation circuit, excavation cross-sectional dimensions, roughness of the excavation wall (to limit friction losses in airflow), and stability of the excavation. The fan and associated equipment can be installed underground in the drift, at the ventilation shaft collar, or at the portal, depending on the system design and operating requirements. Figures 12.6-19 and 12.6-20) illustrate a ventilation decline and an underground ventilation drift, respectively. The setup in Figure 12.6-19 represents a ventilation system used

Courtesy of Northparkes Mines.
Figure 12.6-16 Underground conveyor drift and equipment at transfer point

in the interim (i.e., during initial mine development) until the required lateral development and permanent ventilation raise is completed. Fresh air is heated and, with the aid of booster fans located downstream of the heaters, drawn through heating plants and discharged through vent ducts that extend down the decline to the working face at all times. Return air is exhausted up the decline and at the portal.

DESIGN, CONSTRUCTION, AND MAINTENANCE
Development openings and their associated infrastructure are unique in terms of design approach and design methodologies. The interrelationship between various components of the infrastructure usually requires multiple design iterations. This can be further complicated by the intrinsic variation of the mechanical properties of the host rock in which the development occurs. Design methods are established on sound engineering principles combined with empirical concepts, research, and extensive practical experience. Because the design process can sometimes be quite involved, only the salient points concerning design criteria and design methodologies are addressed in this section.

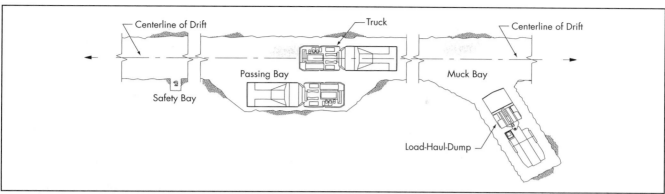

Courtesy of Northparkes Mines.
Figure 12.6-17 Mine access drift with passing bay, muck bay, and personnel safety bay

Courtesy of Northparkes Mines.
Figure 12.6-18 Mine access drift showing utility lines with ample clearance for mobile equipment

Courtesy of Rio Tinto.
Figure 12.6-19 Ventilation decline with surface heating plant fan and duct

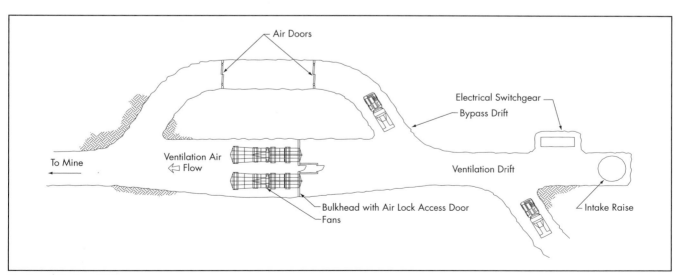

Courtesy of Stantec – Mining.
Figure 12.6-20 Underground ventilation drift with fan

Underground mine infrastructures are normally designed and constructed with four primary materials of construction: steel, concrete/shotcrete, wood, and rock. In most modern mines, wooden structures have been completely replaced by steel and concrete components. Other types of construction materials (such as polymeric materials and carbon-fiber reinforced polymer) have recently been introduced in the underground mine environment but have not gained widespread acceptance because of limited performance data. Steel, concrete, and the host rock will continue to be the primary materials of construction in underground mine development for the foreseeable future because of their proven durability and the ability to accurately predict their behavior under service.

Design Aspects

The following key aspects must be addressed at various stages of the design of underground openings and related infrastructure:

- Scope of work—Including, but not limited to, the extent of work to be performed, the different engineering disciplines that will be involved, and battery limits
- Functional requirements—Including the intended purpose and performance requirements, end results to be achieved, effect of the underground environment on performance, size of excavation necessary, and location within the global infrastructure plan
- Design criteria—Including the key factors and information that govern the design development, size, shape, capacity, and ultimate performance of the completed facility
- Design development—The initial and final design of the facility must consider the following:
  - Duty requirement of the excavation
  - Maximum size of excavation needed to satisfy duty requirement
  - Geotechnical characteristics of the surrounding rock
  - Access limitations
  - Constructability
  - Materials of construction to be used
  - Impact of the facility's construction and operation on the mine environment
  - The underground environment's impact on the facility's serviceability
  - Possible methods of assembly of the various equipment
  - Supporting structures underground
  - System operating costs
- Planning and engineering—Including the potential impact of the new facility on existing operations, resource requirements (e.g., can engineering be completed in-house or must it be contracted out to a consultant?), and schedule for completion of engineering

Construction Aspects

The techniques, installation methods, and equipment normally employed during construction and installation depend on the type and size of the overall facility. The construction phase normally starts with planning of the various stages of construction and installation of the complete facility. The primary stages during construction and installation are

- Project preplanning,
- Procurement,
- Construction and installation,
- Schedule and resource management, and
- Closeout and commissioning.

Construction and installation aspects of the underground development must address the following:

- Safety—Implementation of a comprehensive and practical safety program. There are fundamental differences between the protocols of construction project safety and the safety protocols implemented by the mine operation group.
- Resource requirements—Ascertaining whether the work should be outsourced to an independent development contractor. (The mine may or may not have the equipment and/or skilled personnel available to self-perform construction.)
- Contractual—Preferred method of outsourcing (competitive bidding, single source, alliance, etc.).
- Construction resources—Materials, equipment, and services that can be supplied by the mine during construction (i.e., electric power, compressed air, water, concrete, shotcrete, etc.).
- Planning—Emphasis on minimizing interference with other ongoing mine operations.
- Infrastructure—Underground and surface infrastructure necessary to support the construction and installation work.
- Safeguards—Adequate lighting around stationary operating equipment, guards around rotating equipment, and so on.
- Access requirements and interface with existing mine operations.
- Construction equipment—Various pieces of equipment needed for each phase of the development, size, and capacities of equipment with respect to underground access.
- Waste handling—Removal of rock generated from excavation.
- Ventilation—Ensuring adequate ventilation of the construction area.
- Constructability—Construction methods and approach for each phase of the development.
- Installation sequencing—Consideration of construction equipment limitations.
- Construction management—Determination of whether the mine has the resources to dedicate toward management and procurement of supplies and equipment during construction and installation.
- Concrete—A versatile component in underground mine construction. It is used in almost all facets of underground development, from ground support to foundations for structures.

Blasting is a routine activity in underground hard-rock mines, and the resulting blast-induced vibrations, if left uncontrolled, can have detrimental effects on concrete structures. The level of impact of blast vibrations on a concrete structure is dependent on variables such as the distance of the structure from the blast initiation, the age of the structure, size of the structure, and type and amount of explosives used. Because it is common practice for concrete construction to occur in close proximity or adjacent to underground excavation development, blasting around newly constructed concrete structures should be considered during the design and installation phases of underground mine development. The hydration and crystallization process in concrete normally occurs within 24 hours of placement. This period is very critical in all concrete structures because it is when the hardening and strength-gain process starts to occur. Normal concrete usually attains about 30%–40% of its compressive strength within 3 days and about 60% within 7 days of placement. Thus, blasting in close proximity to major new concrete structures is not advisable before 3 days following placement. Between 3 to 7 days following the concrete installation, blast vibrations adjacent to major concrete structures should be limited to that which produces a peak particle velocity of not more than 25 mm/s. After 7 days following concrete placement, a peak particle velocity of 50 mm/s or less should be used as a guide to limit blast vibrations until the concrete reaches its maximum strength around 28 days.

Project Preplanning

Tasks that are normally accomplished during preplanning include the following:

- After the contract has been awarded, become familiar with and understand the contract documents and general conditions of the contract. Arrange for independent services such as inspections and nondestructive testing, training, and so forth. This is particularly applicable to underground mine development contracts that are outsourced to an independent contractor.
- Determine and secure the various construction equipment (drill jumbo, LHD, haul truck, access platforms, mobile cranes, etc.) required for completing the work.
- Understand the ground conditions and plan the excavation, muck handling, and ground support sequence in accordance with the excavation designs.
- Establish procurement protocols and determine the long-lead items. Review their manufacture and delivery schedules and verify when they should be procured in order to avoid schedule disruptions.
- Establish availability of services and supply of all construction materials.
- Implement subcontracts for various fabrications (such as for miscellaneous steel structures, pipe spools, etc.).
- Plan the site set-up and administrative aspects of the site office, office equipment requirements, required project personnel, and so on (depending on the contractual arrangement).
- Arrange for transportation and schedule mobilization of resources to site.

Procurement

Procurement involves soliciting of suppliers and services; leasing of equipment; and purchasing and marshaling of permanent equipment such as pumps, electrical gear, overhead cranes, crushing equipment, lifting devices, and any other special item associated with the permanent installations. One aspect of the procurement process that must be addressed in the initial stage of planning is the on-site management of materials and equipment that are delivered for the construction of mine infrastructure projects. Planning for such undertakings (involving either a single major project or multiple small- to medium-size construction activities) must consider the logistics of transportation, sequencing of material delivery, and the need for adequate lay-down space and warehousing. It is advisable to manage construction project–related procurement separate from that of mine operations. A list of critical spare parts should be developed during the procurement stage. The procurement process can become lengthy when dealing with multiple vendor selection, whether prequalification-based or otherwise. Mine operations normally establish service agreements with suppliers and contractors that can be capitalized on in order to fast-track the procurement process.

Construction and Installation

Excavation methods. Raises, orepasses, and other vertical or inclined openings can be excavated by either drilling and blasting or semicontinuous methods such as raise boring, raise climbing, or drop raising. Ground conditions and rock type ultimately influence the excavation method chosen.

The most commonly employed excavation method for underground service facilities in hard-rock mines is drilling and blasting. Each drill-and-blast cycle follows a repetitive pattern consisting of blasthole drilling, charging and blasting, ventilation and dust removal, scaling, muck handling, and ground support, as described in the following paragraphs.

Work at the face is normally started by the mine survey crew that sets controls for the excavation and demarcates the excavation limits and layout of the blasthole drill pattern. The blasthole drilling sequence is normally arranged in a manner that maximizes drilling productivity.

Blasthole lengths are selected to complement the length of each round, which in turn are dictated by ground conditions and the maximum unsupported roof span allowed. Short rounds require short drilling depths, accomplished with a single-boom feed during fully mechanized drilling. All subsequent activities (face advancement, etc.) can be completed quickly because there are no equipment-related delays for handling drill rod extensions. In contrast, longer rounds require longer drilling depths, accomplished with multiple-boom feeds. Longer rounds can result in blasthole drilling inaccuracies, which can lead to increased drilling times, increased charging and blasting times, longer muck handling times (due to larger muck volumes), and larger roof spans to be supported.

After drilling is complete the holes are flushed, loaded with explosives, and prepared for the blast initiation.

Upon completion of charging and blasting, the blastholes are flushed with water (or other flushing agent, depending on the underground requirements). Out-of-reach blastholes are normally accessed with a work platform or basket. All equipment at the face is receded to a safe location prior to initiating the blast.

After the charging and blasting phase is complete, the face is ventilated to remove dust and fumes. Depending on the distance to the new working face, the vent duct and fan system may need to be extended to ensure that adequate fresh air reaches the working face.

Any loose rock on the excavation walls are removed by scaling. This process can be completed manually (by a worker in a basket with a scaling bar) or mechanically with a mechanized scaler. Manual scaling can be time-consuming and difficult. Mechanized scaling, on the other hand, is quicker and does not expose mine personnel to difficult working positions.

The muck handling phase commences after it has been determined that no potential rockfall hazards are imminent. Muck removal can be accomplished by one of the pieces of muck handling equipment listed in Table 12.6-1, or a combination thereof. For short rounds, high mucking rates may be attained with only LHDs, depending on the muck volume and haul distance. For long rounds or where a large volume of muck is involved, other equipment combinations (such as trucks and LHDs) may be necessary in order to attain the required mucking speeds.

Hauling the muck away as soon as possible frees up the face for advancement and ultimately improves work cycles. As the distance to the face increases, the haulage distance and time increase as well. To free up the face for subsequent rounds, muck can be temporarily moved into muck bays excavated at regular intervals from the working face. The muck can later be hauled from the muck bay to the final dump location (as a separate operation off the critical path of other face-development activities) while the face is being further developed.

Ground support normally commences after the muck has been hauled away from the face. The type of required ground support depends on the excavation size and ground conditions. In poor ground conditions, it may be necessary to perform multiple ground support activities (initial and final) instead of a single activity at the end of each round. If the ground is fractured and roof stability becomes critical, initial ground support (such as with shotcrete and mesh) can be installed immediately after scaling to stabilize the excavation before mucking out completely. Final ground support (bolting, final shotcrete, etc.) can be installed after the muck has been completely removed and the excavation fully exposed for the next round.

Consideration should be given to the benefits of using the same drill jumbo for drifting and ground support installation; however, it may prove to be more cost-effective overall to obtain a separate jumbo for the ground support exercise. With a single jumbo, the bolt holes could be drilled while the muck is being hauled away. In large excavations and where the excavation completion schedule is critical, separate jumbos for blasthole drilling and ground support installation may be warranted.

Excavation sequence. Ideally, underground openings are excavated full face, depending on the excavation layout, ground conditions, and operating range of available equipment. In openings where the cross-sectional dimensions exceed the operating reach of available equipment (particularly in fractured rock), the face may be advanced by multiple headings in order for the following to occur:

- Limit the unsupported roof span of the excavations.
- Prevent collapse.
- Maximize equipment operating range per round.
- Reach the back and promptly install ground support.

Large excavations for pump rooms, crushers, and similar facilities can be developed by driving a top heading followed by a single bench or multiple benches several rounds behind the top-heading development. Other excavation techniques for large, regular-shaped openings include driving a pilot drift and slashing the back of the excavation on retreat. Factors that influence the excavation sequence include rock properties, ground conditions, equipment operating range limits, and imposed limits on blast vibrations.

Major construction equipment. Several types of major equipment are used for the various work phases. Examples of the major equipment used in the construction of underground

Table 12.6-1 Work task, major equipment, and selection criteria

Excavation Development

1. Blasthole Drilling
   - Equipment—Drill jumbo (with a two-feed boom plus one with a work basket), air-leg drill
   - Criteria for selection—Access drift dimensions; equipment operating range or excavation dimensions (face area, maximum height); excavation layouts and turning radii requirements; length of each round, number, length and diameter of drill holes per face; maximum single drill rod length that can be handled; construction schedule; and number of working shifts per day

2. Charging and Blasting
   - Equipment—Examples include ammonium nitrate and fuel oil (ANFO) loader or ANFO cassette on cassette carrier truck, and emulsion loading units.
   - Criteria for selection—Fixed-frame or articulated-frame vehicles must meet underground access requirements; integrated units that can be used with available mine utility vehicles; type and size of mining operations

3. Ventilation and Dust Removal
   - Equipment—Ventilation fans and ducts
   - Criteria for selection—Volume of fresh air required at the face; system resistance to deliver air to the face
   - Comments—Flexible vent ducts (vent bags) can be installed quicker and are best suited for temporary applications; rigid vent lines have less resistance and allow the system to operate on negative pressure.

4. Scaling
   - Equipment—Work basket with scaling bar or mechanized scaler
   - Criteria for selection—Size of excavation; equipment reach

5. Muck Handling
   - Equipment—LHD, LHD-haul truck, or excavator-haul truck combinations
   - Criteria for selection—Clearance and access requirements; desired mucking productivity; haul distance; tramming surface; and haul road condition and grade
   - Comments—Hydraulic excavators may be used in medium to large excavations where ample room exists for maneuverability.

6. Ground Support
   - Equipment—Drill jumbo, handheld drills, work platform, work basket, scissor lift, rock-bolting jumbo, shotcrete equipment/truck or shotcrete cassette on cassette carrier truck
   - Criteria for selection—Similar to that of blasthole drilling
   - Comments—Ground support may be staged (initial and final) depending on ground conditions.

Fit-out and Installation of Permanent Equipment

1. General Lifting and Erecting
   - Equipment—Mobile rough terrain crane(s) with telescopic boom, work platform (telescopic boom), scissor lifts, heavy-duty forklift, portable light towers, portable welders, portable generators, portable air compressors
   - Criteria for selection—Available working room, equipment reach, maximum single load to be picked, etc.

2. Concrete Work (including bulkheads, dams, equipment foundations, slabs, etc.)
   - Equipment—Concrete truck(s), concrete pumps, vibrators
   - Criteria for selection—Equipment productivity; available access to the work area; distance from concrete source to the final placement location

3. Steel Structures (all major steelwork, access platforms, ladders, etc.)
   - Equipment—Similar to General Lifting and Erecting
   - Criteria for selection—Similar to General Lifting and Erecting

4. Mechanical and Electrical Equipment (pumps, piping, valves and fittings, switchgear, cable trays, and cable runs, etc.)
   - Equipment—Similar to General Lifting and Erecting
   - Criteria for selection—Similar to General Lifting and Erecting

service facilities and the criteria by which they are selected are listed in Table 12.6-1.

Schedule and Resource Management

The excavation and ground support schedule normally drives the overall construction schedule. If the round length and advance rates used in the project schedule are met, the excavation and ground support phase will probably be completed on or ahead of schedule. If the excavation phase is advancing at a slower rate than planned, it may become necessary to adjust the resources, round length, and working cycles in order to attain the required development rates. When certain unavoidable circumstances (such as unforeseen ground conditions and unpredicted water inflow) occur that prevent the crew from attaining the required advance rates during all or part of the excavation phase, the time lost due to such circumstances may never be recovered during excavation, resulting in adjustment/ extension of the overall construction schedule. In some circumstances it may be possible to totally or partially recover lost time during the fit-out and installation phase. Ultimately, managing the schedule effectively depends on the skill set of the crew, the condition of the equipment, knowledge of the ground conditions, and overall practical experience.

Equipment and personnel management are directly affected by the unique set of conditions that exist underground (bearing in mind that underground working environments are different from other construction activities of comparable size). Limited space, humidity, flowing water, and minimal lighting are some of the conditions that affect underground productivity. These and other factors should be taken into account during development of the excavation schedule. The following should be considered for effective management of the resources:

- Equipment interruption during operation should be limited and daily maintenance should be performed during break or standby periods.
- Equipment capabilities and productivities should be observed and the schedule adjusted accordingly based on actual site performance.
- Fleet size should be appropriately matched to the size of operations and construction activities.
- Construction materials should be inventoried on a regular basis to ensure adequate supply is on-site.
- The construction crew should be adequately trained.

Closeout and Commissioning

Precedence for project record-keeping information should be established at the onset of construction. All survey records of the actual excavation dimensions, ground support, installed services and permanent equipment, construction inspections, testing, and equipment commissioning records are assembled together with the operation and maintenance (O&M) manuals for each piece of installed equipment and submitted to mine operations personnel at completion of the project. All major equipment should undergo trial runs and be commissioned according to the manufacturers' specifications. All construction and installation deficiencies should be recorded on an ongoing basis by a punch-list system for tracking to ensure that all defects are corrected. A list of critical and common spare parts that should be maintained in inventory should be provided to mine purchasing and maintenance personnel.

Maintenance Aspect

Maintenance of the completed underground facility is normally the responsibility of mine operations personnel. The effect of ongoing O&M on the infrastructure's general arrangement must be considered during the preconstruction stage. Any potential impact on design and construction decisions should be addressed; among the items to consider are

- Maintenance and repair of equipment
  - Type of maintenance and repair
  - Surface versus underground repair
  - Need for lifting devices (monorail hoist or overhead cranes, depending on size of largest piece of equipment to be handled) to remove and reassemble major equipment
- Access requirements during major equipment overhaul (e.g., need for additional supporting equipment)
- Expendable parts change-out
  - System for changing out expendable parts on equipment (wear plate replacement, crusher mantle change-out, pump rebuild, conveyor belt splicing, etc.)
  - Method for incorporating maintenance and repair requirements into the design and construction decision making
- Preferred methods for system monitoring
  - System performance monitoring (computerized, remote, or central monitoring system)
  - System malfunction notification (visual or audible alarms)

Sumps and Pump Room

In the design and construction of sumps and pump rooms, several important criteria are considered. Some of the design considerations that are central to the engineering decision-making process are discussed here.

Design Considerations

Table 12.6-2 provides a list of factors and recommendations that should be considered in the design of sumps and pump rooms.

Construction and Installation

Construction usually starts with the procurement of services, materials, and equipment. In a new mine, the sump and pump-room development may constitute one of several packages in the global underground construction contract for which the various activities and resources should have been preplanned. For an existing mine, construction and installation of a pump station may be atypical of the daily production and development activities that occur underground (for reasons mainly pertaining to labor and equipment resource requirements) and may warrant the decision to outsource the work to an independent contractor. The process of planning and executing construction is somewhat similar whether mine management decides to perform all or part of the construction (using available crews and equipment) or have a development contractor perform the work.

Preplanning. Because several major tasks are usually involved in the construction stage, the work should be preplanned similar to the outline previously presented under "Project Preplanning" in the "Construction Aspects" section. Some critical items of importance during preplanning include understanding the contract documents and scope of work, the

Table 12.6-2 Factors and recommendations for sump and pump-room design

Factors	Recommendations for Decision Making
Location of facility	Conduct study to determine preferred location for excavations and associated infrastructure. Give primary consideration to geotechnical characteristics of the rock mass and how excavations will be supported. Locate excavations in host rock, away from active mining areas and major geological features (faults, shear zones, etc.) that could cause complications during excavation or require a higher level of support maintenance. Locate as close as practically feasible to active mine access.
Excavation design and ground support	Complete geologic characterization of the rock mass and excavation design with input from the geotechnical team. Geotechnical characteristics of the surrounding rock coupled with excavation sizes dictate type(s) and complexity of ground support. Final plant layout influences overall size of excavations. Large excavations may require extensive ground support (cable bolting, concrete/shotcrete lining and wire mesh, etc.) depending on ground conditions. Smaller openings may only require one of the mine's preestablished ground support designs (pattern bolting, wire mesh/screen, shotcrete, etc.). Final excavation designs should ultimately be based on the excavation service life and required size of openings to accommodate the total facility.
Hydrology and water inflow estimates	Perform study to estimate the amount of water flowing into the mine. Identify presence of aquifers. Specify hydraulic conductivity (K, in cm/s) of seepage zones, expected peak and minimum amount of water inflow over specific duration (12-month period, for instance), and amount of water produced from the mining operations. Specify water quality (pH, chlorides content, hydrocarbons, sulfates, quantity and sizes of suspended solids, specific gravity of the pumped product, etc.). Develop criteria for differentiating between clear water and dirty water. Grey water and surface runoff from active mining zones, drill stations, maintenance facilities, backfill regions, and so on may contain traces of contaminants that warrant pumping and treatment methods separate from the mine's ordinary groundwater inflow.
Plant layout and general arrangement	Develop plant layout and general arrangement based on types and configuration of pumps, piping and valve arrangement, sump capacities, electrical gear, and other supporting equipment (see examples in Figures 12.6-1 through 12.6-3, and Figure 12.6-21). Indicate clearances around fixed equipment for access during maintenance, miscellaneous steel and concrete structures, sump locations, and how the facility will be accessed within the underground mine layout. Locate electrical equipment preferably in separate cutouts within the pump room. Have mine operations personnel review overall plant layout and provide input to ensure that fundamental operational requirements are met.
Detail engineering and design	Complete final engineering and design of the facility. Design critical structures to withstand all possible loading conditions—bulkheads, dams, and other retaining structures to support the full height of the fluid being retained, plus any allowance for load amplification that may result from blasting or earthquake. Size and design access platforms to support personnel and equipment loads imposed during operation and maintenance activities. Methodology to clean out and maintain sumps should be incorporated into the design development. Locate pumps, motors, and electrical gear on raised foundations to prevent contact with water in the event of leakage. Prepare engineering drawing and specifications. Ensure design input and reviews from mine operations personnel at all stages.
Engineering and procurement	Complete final contract documents (drawings, specifications, etc.) for tender or construction. Certain long-lead items (pumps, cranes, valves, and other specialty equipment, fabrication services, etc.) can be procured at the mine's discretion, prior to completion of final design of the facility. Certain aspects of the final engineering and design that are dependent on specific equipment (especially with vendor-specific items such as pumps and cranes) may not be completed until such time that complete information on that equipment is made available. This procedure usually provides a beneficial reduction in the overall facility development schedule. Mine operations and maintenance personnel may influence decisions to select certain permanent equipment in order to take advantage of preestablished supply service agreements with suppliers, maintain consistency in spare parts inventory, and use the operating and maintenance knowledge already acquired on similar installed equipment.

ground conditions and excavation design, procurement of long-lead items, and resource requirements.

Procurement. Important items to procure include

- Pumps and pump motors;
- Flow control valves;
- Cranes or other lifting devices;
- Miscellaneous steel structures;
- Piping, miscellaneous pipe supports, and fittings;
- Electrical equipment (switchgear, load centers, substations, etc.);
- Rock bolts and other ground support consumables (these items should be present on-site at start of excavation);
- Fabrication services for steel structures, platforms, pipe spools, and other equipment; and
- Any specialty equipment necessary to facilitate the construction.

Construction activities. Depending on contractual arrangements, the initial construction activities normally involve provision of access to the sump and pump-room location and installation of services (electric power, service water, compressed air, discharge line, etc.) necessary for commencing excavation. Extension of these services, as required to further the construction, is normally the responsibility of the development contractor and usually done as an independent activity. An existing mine may decide to undertake all or part of the excavation development and ground support installation based on some of the circumstances outlined previously in the "Construction Aspects" section. In either case, the construction activity commences with development of access drifts to the facility location, followed by the various activities listed earlier in the "Excavation Methods" section.

After establishing access to the face, excavation development begins as previously outlined under "Construction and Installation" in the "Construction Aspects" section. The excavation sequence depends on ground conditions and size and type of facility. In hard rock, full-face drilling and blasting can be used for small to medium facilities where the sump and pump room are integrated into a single excavation. In medium to large facilities where the sump and pump rooms are located in separate excavations, full-face drilling and blasting may not be practical based on equipment capabilities, excavation dimensions, and ground conditions. In such cases, sequential excavation techniques such as top heading and benching are applicable.

For vertical sump and pump-room arrangements similar to Figure 12.6-21, the vertical sump section can be excavated from the bottom up by raise boring or drop raising, followed

Construction of Underground Openings and Related Infrastructure

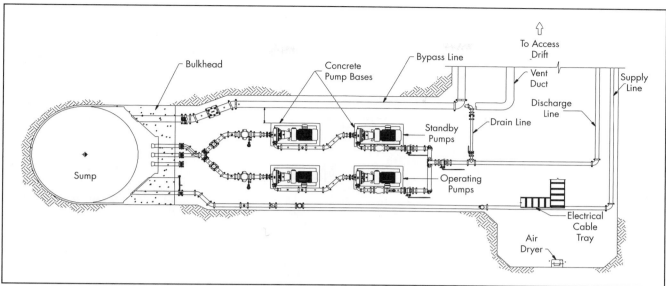

Courtesy of Barrick Gold and Stantec – Mining.

Figure 12.6-21 Typical pump room arrangement showing equipment layout and clearances

by slashing out the final excavation limits. Other methods previously described under "Construction and Installation" in the "Construction Aspects" section may be applicable, but ground conditions, access, potential hazards and risks, depth of excavation, required accuracy, and overall costs must all be considered when selecting the preferred excavation method. Submersible pumps and flexible hoses should be available to handle drill water or groundwater inflow at the face.

Fit-out and installation of permanent equipment normally commences after completion of the excavation and permanent ground support. This phase of work usually starts with installation of various formwork to construct reinforced concrete structures including concrete bulkhead, dams, retaining structures, equipment bases and foundations, foundations for overhead crane structures, and others. Concrete bulkheads, dams, and similar retaining structures must be constructed as monolithic structures in order to prevent the formation of cold joints where leakage could occur. The rock around the perimeter of reinforced concrete bulkheads and dams in the sump can be prepressure and postpressure grouted to seal any cracks present at the rock–concrete interface to enhance the sump's water tightness.

Another technique frequently used to achieve a watertight sump is to provide a continuous thread of hydrophilic rubber-based waterstop at the rock–concrete perimeter prior to placing concrete. The waterstop helps reduce leakage by expanding on contact with water. Overhead crane structures or other lifting devices normally comprise the first of major permanent equipment to be installed because they are used to set other major equipment in place (see Figure 12.6-22) and complete the fit-out. After all major equipment is in place, the piping and electrical installations commence. Normally, the piping and miscellaneous mechanical installations precede the electrical and instrumentation work.

Closeout and commissioning. At the closeout phase, all construction data and O&M manuals for the pumps, valves, fittings, cranes, electrical gear, and so on are submitted to mine operations. A list of important spare parts, developed during procurement, is also submitted upon closeout. Inspections and

Courtesy of Northparkes Mines.

Figure 12.6-22 Overhead service crane used for pump installation

testing that may be required during commissioning include the following:

- Pipe integrity testing
- Weld inspections
- Motor rotation
- Instrumentation verification and calibration
- Verification of valve positions (i.e., open/close position)

Facility Maintenance

After commissioning, operation and ongoing maintenance becomes the responsibility of the mine's O&M personnel. Common situations to monitor during regular operations include

- Drive belt failure,
- Bearing operating condition,
- Motor operating condition,

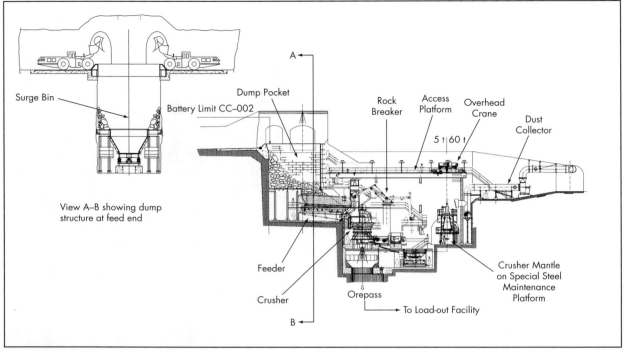

Courtesy of Northparkes Mines.
Figure 12.6-23 Typical underground crusher room layout

- Clogged strainers (especially if pumping dirty water),
- Amount of fines collecting in the sumps,
- Leaky valves, and
- Valves that do not open and close properly.

The mine should ensure that critical spares and consumables are present on-site at all times in case of a system malfunction or failure. Important parts that may require replacement include

- Drive belts and sheaves,
- Pump impellers,
- Pump motors,
- Valves,
- Pipe couplings and gaskets,
- Strainers, and
- Instruments (such as flowmeters, pressure gauges, etc.).

Crusher Room

Crusher rooms are among the largest service excavations in an underground mine. Excavation stability is paramount throughout the service life of the facility because of the considerably large roof and wall spans normally created. It is important that crusher rooms are located in areas remote from the influence of mining-induced stresses. In general, crusher rooms must be situated within the overall underground ore handling circuit to provide quick access by ore transport equipment. The overall crusher room design and construction can be a major undertaking, incorporating a dump structure at the feed end of the crusher, the crushing plant, and miscellaneous infrastructure in the crusher room and a load-out structure at the discharge end (see Figure 12.6-23). The level of detail normally involved in the design calls for coordination among several engineering disciplines and specialized equipment suppliers.

Design Considerations
A suggested list of important factors to consider during the design stage is provided in Table 12.6-3.

Construction and Installation
The construction and installation phase of underground crusher rooms is usually a major undertaking. The excavation of the crusher room and adjoining underground openings is normally more complex than other underground excavations such as a conveyor drift.

Preplanning. There are several major tasks associated with crusher room construction. The preplanning effort outlined under "Project Preplanning" in the "Construction Aspects" section should assist in developing a comprehensive excavation method, excavation sequence, ground support installation methods, and geotechnical instrumentation; identifying and selecting the required construction equipment; and procuring permanent equipment.

Procurement. Specific items to procure include

- Crusher and supporting equipment,
- Overhead crane and other lifting devices,
- Dust collector system,
- Feeders,
- Rock breaker (if specified as auxiliary to the crusher), and
- Miscellaneous steel structures and platforms.

Construction activities. Construction usually starts by establishing access to the various levels in the crusher room. In an existing mine, access to the crusher location may have already been established with the required services for construction (such as electric power, compressed air, service water, ventilation, etc.). This may not be the case in a new mine where the access development must be planned and

Table 12.6-3 Factors to consider in crusher room design

Factors	Recommendations for Decision Making
Location	The crusher room should be located away from major geological features such as faults and shear zones. The location must be remote from active mining areas, high-stress zones, and areas affected by mine-induced stresses. In mines employing skip hoisting, the preferred location is near the shaft or beneath the ore body.
Plant layout	The plant layout comprises three main areas: dump structure, crusher room, and load-out structure (see Figure 12.6-23). The layout must consider the following in developing overall excavation limits and general arrangement: • Access to and from the crusher room for maintenance • Repair and replacement of mechanical/wear components • Production capacity • Type and size of crushing plant • Appurtenant infrastructure in the crusher room (such as service crane, dust collector) • Dump and load-out structures • Method of ore transport at both ends of the crusher • Type of dump arrangement • Load-out arrangement The dump pocket above the crusher should be sized and dimensioned to accommodate the largest combined dump load from the largest anticipated haulage equipment (trucks, LHDs, or others). The general layout should indicate continuity between the three main areas and how the different areas and infrastructure relate to one other. Input from specialist equipment suppliers is mandatory at the plant development stage.
Excavation design	The duty requirements of the excavation must be satisfied. The final sizes of the underground openings should be established before completing the excavation design. Existing ground conditions, rock type, and geotechnical properties are some of the variables that will govern the excavation design. Accesses to the crusher room, excavation methods and sequences, ventilation, broken rock removal, ground support, geotechnical instrumentation, and monitoring must be established during the excavation design.
Overall plant design	The overall plant design addresses detailed design of the crushing equipment and other supporting structures. Design should be based on life-of-mine requirements so that the maximum anticipated excavation size over the mine life can be established and incorporated into the final excavation design. Access to and from the crusher room for installation and maintenance must be established. If a service crane is part of the infrastructure, allowance should be made for it and its support structures. Reinforced concrete foundation design for the crusher and supporting infrastructure is critical at this stage. Other important design features that should be considered include dust control devices at the dump, inside the crusher room, and at the crusher. A dust collector system is preferred for dust control inside the crusher room. Designs from suppliers of specialized equipment for crushers and other equipment must be complete and incorporated into the overall plant design. Input from mine operations personnel on the design should be highly sought. Preparation of engineering drawings and specifications for tender or construction is completed during this phase.
Engineering and procurement	Procurement of long-lead items (such as crusher components, overhead cranes, steel structures, etc.) should begin at this phase, at the mine's discretion. Completion of final engineering drawings and specifications for construction in most situations will need to progress in tandem with final equipment selection and design because of design information normally required from equipment suppliers.

executed prior to completing the crusher room excavation. The access drifts driven to the crusher room location must be of adequate size to accommodate the single largest piece of permanent equipment or mobile equipment required for the construction of the crusher.

The excavation design should outline the ground conditions, rock excavation sequence (normally in multiple lifts), types of ground support, and ground support installation schedule in accordance with the geotechnical recommendations. The excavation method and sequence will closely follow the procedures outlined under "Construction and Installation" in the "Construction Aspects" section. A typical excavation process normally involves developing multiple headings from the upper access into the upper region of the crusher room by sequentially drilling and blasting predetermined rock panels and providing ground support as the excavation progresses (see Figure 12.6-24). After the entire span of the crusher room is excavated and the roof supported, the remaining rock panels are excavated by sequentially benching and installing ground support in predetermined lifts until the excavation reaches a predetermined level in the crusher room (see Figure 12.6-25). The excavated lifts above the upper access could be mucked out by LHDs and trucks or a hydraulic shovel/truck combination, as long as mobile equipment can tram in and out of the cut. The final excavation lifts at the lower portion of the crusher room can then be drilled, blasted, and mucked out from the lower access.

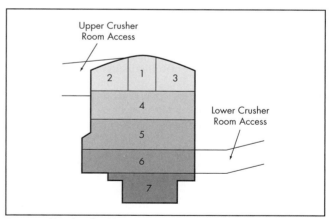

Courtesy of Northparkes Mines.

Figure 12.6-24 Sectional view through a crusher room showing numbered excavation sequence

Ground support for such large excavations normally includes rock bolts, cable bolts, shotcrete, and wire mesh or a combination thereof. The roof in the crusher room is normally excavated with an arched profile in the least cross-sectional dimension to enhance ground stability. In a situation where the selection of the crusher room location supersedes the requirement for competent ground, or when competent rock is

Courtesy of Northparkes Mines.

Figure 12.6-25 Underground crusher room excavation and ground support

inexistent within close proximity, steel arches could be used to support the excavation if bolting and shotcrete become insufficient. The excavation development phase should include comprehensive geotechnical instrumentation and monitoring programs. Borehole extensometers, tape extensometers, and strain gauges are some of the common geotechnical instrumentation that can be installed inside the excavation to monitor roof and wall displacements. Wall movements are expected during the excavation and are usually predicted from the geotechnical investigations and design. Actual wall and back displacements should be monitored throughout the excavation stages and during operation of the facility for comparison with those predicted from the geotechnical design to determine the need for any remedial action (i.e., additional ground support, etc.) that may be necessary prior to final crusher room fit-out. After the excavation is complete, preparations are made to commence the fit-out of the crushing facility. This stage normally involves extensive scaffolding, falsework, and concrete construction to build out foundations for the crusher and other miscellaneous infrastructure (Figure 12.6-26).

The service crane or other lifting device (if specified as part of the infrastructure) will possibly be one of the initial permanent infrastructures to be installed because it will be used extensively for the permanent installation of the crusher and associated components (see Figure 12.6-27). If a crane is not part of the permanent facility, an allowance must be made for the crusher room to be accessed by a mobile crane to facilitate the installation. All components of the crushing facility must be sized and manufactured so they can fit in the respective drifts for transportation to their final location.

Closeout and commissioning. The closeout and commissioning process for underground crusher rooms involves trial runs and equipment testing to ensure they operate smoothly.

Facility Maintenance

Maintenance of underground crushing facilities depends on the type of crusher and usually concentrates on replacing crusher wear parts, replacing wear liners on impact-resistant

Courtesy of Rio Tinto.

Figure 12.6-26 Extensive falsework set up for construction activity in crusher room

Courtesy of Northparkes Mines.
Figure 12.6-27 Crusher room fit-out in progress with aid of overhead service crane

surfaces, and resurfacing major crusher components. For the most popular types of crushers used in underground hard-rock crushing, the following is a partial list of maintenance activities that may occur:

- Replacement of toggle plate, jaw plate, and jaw wedges (for jaw crushers).
- Resurfacing of crusher mantle and replacement of concave segments in the crushing chamber, rim liners, spider arm liners, spider cap, and main shaft sleeves (for gyratory crushers). Resurfacing of crusher mantles and other associated work on the main shaft of gyratory crushers normally requires removal of the parts with a service crane or other mobile lifting equipment. Special steel access platforms can be provided in some cases to facilitate this particular maintenance activity (Figure 12.6-23). In order to limit the crusher's overall downtime, a spare mantle and shaft should be provided and located in the crusher room for replacement while resurfacing is ongoing on the worn mantle and shaft.
- Replacement of bearings, bushings, dust seals, oil seals, drives, and wear liners in dump pockets and other rock impact surfaces.

Underground Maintenance Shop

The design of underground maintenance shops must take into consideration shop ergonomics, access, and the types of services that will be performed underground. The size of the shop (and subsequently, the extent of services performed) is normally larger in mines accessed by a shaft only than for one with direct drive-in access. Compared to a similar facility on the surface, an underground maintenance shop has the advantage of close proximity to the underground operations and the mobile equipment fleet. It is essential that the shop layout and equipment are functionally adequate to satisfy the duty requirements and maximize the types of repairs that are performed underground.

Design Considerations
The following paragraphs discuss the factors that should be considered in the design of an underground maintenance shop.

Some of the major items to consider when designing an underground maintenance shop are listed in Table 12.6-4.

Construction and Installation
Figure 12.6-28 shows a typical mine shop layout. The construction methods employed in underground shops are similar to those previously described under "Construction and Installation" in the "Sumps and Pump Room" section. The preplanning process is also similar to that previously outlined under "Project Preplanning" in the "Construction Aspects" section. Procurement is normally for equipment, structures, and specialty services required to equip the shop. This includes purchase of the service crane or other lifting device (if available), other miscellaneous steel structures, fuel and lube storage and dispensing equipment, and furniture for the administration areas.

Construction activities. The excavation method and excavation sequence are similar to the procedures outlined previously in the "Sumps and Pump Room" section. Build-out of the shop commences after the excavation and ground support phases are completed. Construction of the permanent structures usually begins with concrete construction for floor slab structures, foundations for the service crane, and containment facilities. Equipping the shop and installation of compressed air, electrics, water supply, and other services normally happen in tandem.

Closeout and commissioning. All construction record information and equipment manuals should be compiled and handed over to the mine's operations personnel. A record of all construction deficiencies should be kept and all corrective actions must be documented and incorporated into the project handover documents. All installed equipment should be tested and commissioned according to the manufacturer's specification.

Facility Maintenance
The primary maintenance activities (other than mobile equipment maintenance and repair) that occur in underground shops focus on handling and disposal of spent hydrocarbons (lube and oils); cleanout of mud and other materials that collect in sumps, floor drains, and inspection pits; changing burnt-out electric light bulbs and batching of fuel; restocking of consumables and spare parts; and periodic maintenance on the various permanent equipment in the shop. Mines with only a shaft access normally batch fuel with a remote setup at the surface consisting of a batch tank of equal or slightly lesser capacity than the fuel tank located underground. The batching exercise must establish a procedure that guarantees complete gravity drainage of the underground fuel supply line after completing each batch.

Other items that require regular attention include

- Safety awareness in the shop and when working on heavy equipment;
- Fire prevention;
- Regular warehouse inventory;
- Regular maintenance of shop operating equipment;
- Disposal of waste products;
- Regular clean-out of the wash bay and other housekeeping activities;
- Monitoring of ventilation, especially around fuel and bulk lube storage areas; and
- Proper signage, guarding, and covering of floor openings.

Table 12.6-4 Factors to consider when designing underground maintenance shops

Factor	Recommendations for Decision Making
Shop location	The preferred location is in a dedicated service area of the permanent facility, with direct access to a main travelway and the ventilation exhaust routing. Excavation should be located in competent rock.
Required maintenance services	The essential services performed should include all periodic maintenance activities such as changing out oil and other essential fluids, filter change, lube, and the various mechanical inspections. Repairs to engines, transmissions, and other mechanical components; bucket and bucket teeth replacement; and hose and drill repair should be performed underground as well. Other major repairs not performed regularly, such as those involving special cutting and welding procedures (air arc, etc.), major equipment rebuild, and others should be done on the surface or off site, where space, equipment requirements, and any noise from associated activities can be accommodated. Overall, the type and extent of services performed in the shop will depend highly on whether or not there is direct access to the surface, and where applicable, the time or distance it takes equipment to be moved to the surface.
Shop layout and design	The underground shop should be laid out based on the following considerations: the number and types of equipment in the fleet, the type of maintenance or repair to be accomplished, the type of surface-to-underground access available, and cost. Minimum facility requirements for shop floor layout include • Service bays (two minimum, one equipped with a service crane), • Wash bay (one minimum), • Fuel/lube storage bay (one designed to meet emergency spill provisions of local jurisdiction), • General administration area (for shop foreman's office with visual window viewing the main shop area, warehouse, etc.), • Storage area, • Major parts marshaling area, and • Welding area. A shop layout with individual bays similar to Figure 12.6-28 may be equipped with a fire-rated access door and personnel door at each service bay for equipment and worker access, respectively. In other shop layouts such as that shown in Figure 12.6-6, or where the individual service bays are all located within one main complex, the shop itself will be fitted with one main fire-rated access door. All areas within the shop facility should have a concrete floor so that any spills can be contained and controlled. Containment must be provided around hydrocarbon products. Diesel and other hydrocarbon products should be stored in double-walled tanks. Proper ventilation is required in and around fuel storage areas. Wash bays should be equipped with an oil–water separator to remove hydrocarbons from wash water before it is finally discharged. Equipment is also available in truck wash systems that can filter and recycle water in wash bays. Other important items to consider include provision of a floor inspection pit or alternate structure in the service bay for inspecting vehicle undercarriage; adequate shop lighting, electric power, compressed air, water, and other services; design for containment of hydrocarbons; floor drainage and sumps; vehicle traffic control; personnel safety; fire suppression system; and first aid.
Excavation design	The excavation sizes are governed by the physical dimensions of the largest equipment that will be serviced underground plus allowances for personnel traffic around equipment. The service bay that is equipped with an overhead crane should be overexcavated at the back to accommodate the overhead crane trolley and the equipment used for its installation. The prevailing ground conditions, rock type, size of excavations, pillar sizes between openings, and geotechnical properties are factors that will govern the excavation and ground support design. Sufficient height should be allowed to permit a truck bed to be raised while maintaining enough clearance to operate an overhead crane. The walls of shops should be shotcreted in a very light color and painted white or a very light color to enhance visibility in the shop.
Engineering and procurement	Procurement of long-lead items (such as overhead cranes, tanks, dispensing equipment, steel structures, etc.) can begin prior to completion of the final engineering drawings. Some of the for-construction documents will normally progress concurrent with final equipment selection and design because of the delay in receiving detailed design information required from the respective suppliers.

Ore Bin

Ore bins are essentially used to provide surge capacity in the underground ore circuit.

Design Considerations

The following paragraphs discuss the primary considerations when designing these facilities.

Location and access. Ore bins can be located in several areas within the underground mine ore stream. Depending on the layout of the materials-handling system, ore bins can be located above or below a crusher or they can be remotely situated and fed by mobile equipment or belt conveyors. In mines that use skip hoisting for material transport, the ideal location for ore bins is near the shaft. In a block cave mine, ore bins should be particularly located outside the caving zone to reduce the effect of mining-induced ground movement on the structure. The general consensus in underground mine operations is to locate the bin in an area that will not be adversely affected by the mining activity. Access is crucial, especially during the development stage. Depending on the excavation techniques adopted, a temporary haulage drift may be required at the bottom of the bin for muck removal during the excavation.

Shape, size, and capacity of bin. The most common shapes for underground ore bins are rectangular and circular. Sometimes a hybrid of both shapes is used above and below the bin mainly to accommodate infrastructure layouts. The selected overall bin shape should be based on the existing rock conditions, rock geotechnical properties, and a geotechnical stress analysis of the rock mass around the ore bin location. The size of the bin is a function of its capacity. The capacity of the bin is dictated by the mine's operating requirements and factors such as daily production tonnages, amount of stockpile needed in the event of an emergency shutdown, expected duration of an emergency shutdown, and cycle times of mobile equipment feeding material into the bin. The horizontal dimensions and vertical depth between the dump and load-out structure are the two variables used to establish the overall bin size.

Excavation and ground support design. The in-situ stresses (at the time of initial excavation and during operation of the mine) around the region of the bin excavation should be taken into consideration when establishing the final excavation sizes and orientation. In a nonhydrostatic stress field (depending on the ground conditions), it is usually beneficial to align the maximum dimension of the bin in a direction

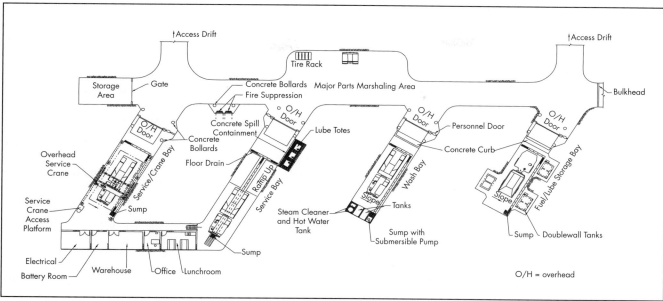

Courtesy of BHP Billiton and Stantec – Mining.

Figure 12.6-28 Typical underground mine shop layout

parallel to the maximum stress. Initial ground support design normally consists of rock bolts unless the geotechnical analysis dictates a more stringent program. If additional ground support (such as grouted rock bolts or cable bolts) is required, they are usually installed as a separate operation because of their longer installation times and must be completed prior to installing the bin wall lining. The ground support design should be continuously updated with rock properties and wall movement data obtained in the field as construction progresses. Grouted steel rebar bolts or point anchor bolts are generally not recommended in bin walls because, as the inside walls wear and deteriorate from abrasion, the exposed rock bolts tend to vibrate when they come in contact with the flowing rock, causing the host rock to loosen further. Fully grouted fiberglass bolts are a preferred alternative because the fiberglass tends to wear away at the same rate as the rock wall while maintaining full contact support. Also, steel bolts can damage equipment on the downstream side of the ore bin if they break off and fall into the ore stream.

Infrastructure. Underground ore bins are commonly used in conjunction with a conveyor haulage system. In mines with production shafts and a truck haulage system, they can be located close to the shaft where the material can be discharged directly into skips, thereby eliminating the need for conveyor belts. The critical support infrastructure in an ore bin is located at the feed and discharge ends. Depending on the material transport methodology, the feed end will comprise a truck dump, rail dump, or chutes and associated steel structures for conveyors. A service monorail or other permanent lifting device above the bin is not mandatory, but it can be useful when provided for erecting the permanent equipment. The bin walls should, at a minimum, be lined to maintain the integrity of the ground support and to resist abrasion from material drawdown. The lining is normally installed after the ground support is in place. Depending on the surrounding rock conditions and particularly the abrasiveness of the bin content, the lining design can vary from several inches of shotcrete to fully reinforced cast-in-place concrete panels with or without embedded steel liner plate on the exposed face. Abrasion-resistant concrete without embedded steel liner plates has been used successfully in many applications. The general arrangement at the discharge end of the bin is also dependent on the material transport methodology. If the bin discharges directly into skips in a production shaft, the load-out mechanism will resemble that used in a loading pocket, as illustrated in Figures 12.6-12 and 12.6-13 with a flow control mechanism. For bins discharging onto conveyors, a feeder is usually provided between the bin discharge point and the conveyor belt (Figures 12.6-7 and 12.6-8). The essence of the feeder is to reduce impact loads from falling rock on the conveyor structure and to distribute material evenly over the length of the conveyor belt. Special consideration should be given to the design of the bin bottom and discharge opening to minimize dead-bedding of the material and prevent hang-up inside the bin.

Construction and Installation

The excavation phase of ore bins is quite similar in many ways to that of underground crusher rooms but on a much smaller scale. The excavation size, level of intricacy, existing ground conditions, rock type, and rock properties will dictate the preferred excavation methods.

Preplanning. Planning the work is essential for successful completion of the construction. The specific tasks will vary depending on the type of host rock (i.e., hard rock or soft-medium rock), but overall the preplanning should adhere to the recommendations previously discussed under "Construction and Installation" in the "Construction Aspects" section to establish the following for the excavation and ground support:

- Excavation method for ore bin—Raise boring, pilot raising with slashing, Alimak raising, drop raising, and so on, depending on depth, geotechnical data, and geotechnical recommendations

- Access provision—Top and bottom drifts, and schedule for completion of access in conjunction with start of ore bin excavation, also depending on the bin excavation method
- Excavation sequencing—Full face or pilot raise and slash depending on ground conditions, diameter, depth, and geotechnical recommendations
- Muck removal—Type of equipment to be used, establishment of dump location
- Ground support—Types, procedure, and schedule of installation for bolt and mesh, shotcrete, final lining, and so on

Procurement. The main items to procure ahead of construction are for permanent equipment in the dump pocket above the bin, the bin structure, and the load-out structure below the bin. Examples of specific items include the following (see Figures 12.6-7 and 12.6-8):

- Overhead crane or monorail hoist at the dump pocket
- Dump structure and related appurtenances, depending on the materials transport method (i.e., trucks, conveyor belt, or rail)
- Wear liners and other associated steel or concrete furnishings in the bin
- Load-out structure and appurtenances below the bin

Construction activities. Construction activity starts with the access development to the ore bin location. The type of access depends on the excavation method and sequencing. The size of the access drift must consider the overall size of the largest piece of gear that will be transported to the bin location and the various sizes of the mobile equipment to be used during the construction and installation.

If a raise-boring technique is used, an upper drift will be required to set up the raise drill and drill the pilot hole. A lower access drift will also be required to recover the pilot bit, launch the reamer, and remove the drill cuttings. If excavation is done with a pilot raise, followed by subsequent slashing out of the ore bin, an upper and lower drift will be required for similar reasons. However, the bottom drift will only be required for recovering the pilot bit and hauling the muck. The bin excavation could be slashed out from the top down with the length of each lift limited to the maximum allowable unsupported rock span. The excavated rock can then be mucked toward the center of the pilot raise and hauled from the lower drift. Successive drilling and charging of blastholes can commence on each working bench. The pilot hole is covered with a temporary bulkhead as a safety precaution whenever people are working on the active bench. Depending on the ground conditions, ground support may be required immediately after completing the mucking cycle followed by final lining of the ore bin as a separate operation. Installing the ground support immediately after excavation is imperative before allowing workers to proceed with operations, maintain the integrity of the bin walls, and prevent unraveling of the rock face.

Other applicable excavation techniques include full-face drilling and blasting, Alimak, and long-hole methods with subsequent slashing out of the ore bin. These methods are suitable for ore bins with uniform cross sections, located in rock with good self-supporting characteristics because the ground support and installation of the final bin wall lining is normally performed as a separate activity after all excavation is complete.

Installation of the permanent equipment and other support structures follows after the excavation and ground support phase is completed. If a service crane is provided at the dump pocket location, it can be advantageously used during the erection of the permanent equipment; otherwise, a mobile crane will have to be brought in. The final lining of the bin can be installed in conjunction with the initial ground support or as a separate activity several stages behind the development of each lift. Cast-in-place concrete and a steel composite structure with embedded steel liner plate on the exposed surface is a popular method for lining ore bins. However, recent advancements in shotcrete technology have made it possible to replace such labor-intensive methods with the installation of specially formulated impact- and abrasion-resistant shotcrete lining. This specialty shotcrete can attain a compressive strength of 25 MPa in as little as 8 hours and more than 50 MPa after 28 days of installation. Vale Inco's Creighton mine in Canada used this shotcrete technique to replace the lining in their 5,280-level ore bin. The 100-mm-thick shotcrete lining used to replace the conventional concrete-encased steel wear liners have performed successfully with insignificant wear after installation.

Depending on the schedule, construction of the dump pocket and load-out structures can occur concurrently. Even if work is not progressing simultaneously at both locations, steps must be taken to protect personnel at all times. The top of the bin should be covered with a temporary bulkhead to protect personnel working below. Access doors should be provided in the bulkhead for transporting personnel, equipment, and supplies.

Commissioning and closeout. Handover documents at the end of construction and installation should include all construction record information and equipment manuals, compiled and handed over to the mine's operations personnel. A punch list of all construction deficiencies should be generated during construction and all corrective action must be documented and incorporated into the handover documents. All installed equipment should be tested and commissioned according to the manufacturer's specifications. A critical spare parts list should also be provided as part of the project closeout documentation.

Facility Maintenance

Replacing wear liners is a common maintenance activity in ore bins. This activity can be both time-consuming and labor intensive because it normally involves removing worn-out liners, repairing concrete surfaces, and drilling and installing new wear liners. The bin and all supporting equipment must be decommissioned for the duration of the liner replacement. Depending on which part of the bin needs repair, the top of the ore stockpile must be lowered several meters below the working level to fully expose the repair area. A temporary work platform set up inside the bin as a staging area and secured to the walls or top of the bin is preferred. Using the top of the bin stockpile as a work platform is highly discouraged because the presence of any voids within the stockpile could result in subsidence of the working surface. The top of the bin should be temporarily covered with a bulkhead and access doors of adequate size provided to transport personnel and supplies. A temporary safety ladder or other means of egress must also be provided from the working level to the top of the bin. Some jurisdictions may treat the interior of a bin as confined space, in which case special precautions (such as provision of adequate ventilation, lighting, emergency personnel evacuation, etc.) must be taken before mine personnel can enter to perform work. Liner replacement should concentrate on implementing alternative repair techniques (such as specialty shotcrete in lieu of the conventional concrete-encased wear liner) that will increase bin wall service life and reduce the frequency of repair shutdowns.

Other items that constitute part of the long-term maintenance program include the repair of feeders at the bottom of the bin, cranes, monorail hoists, drives, and motors, and other operating gear, and replacement of parts such as bearings, pulleys, drive belts, and conveyor belts.

Ore- and Waste Pass

Ore- and waste passes are used primarily for transporting broken rock by gravity flow from one level to another in an underground mine. They can be used as temporary storage bins with a flow control mechanism at the bottom. Orepasses can also be situated above a crusher to supply feed and below a crusher to transfer crushed rock to other materials handling equipment (such as feeders and impact belts). The former is a common underground arrangement while the latter is uncommon and usually incorporates a shorter orepass than the conventional layout. An inclined orepass that is located above a crusher is seldom equipped with any additional flow control mechanism other than a chain curtain. The main characteristics that differentiate a pass from an underground bin are the cross-sectional area, length orientation, and bottom discharge configuration. The height-to-diameter ratio of a pass is larger than that for an underground ore bin.

Design Considerations

The following paragraphs discuss the factors to consider in the design of ore- and waste passes.

Location and orientation. The preferred location for ore- and waste passes is in an area not adversely affected by mining-induced rock stresses. Sometimes this may result in a location that creates longer equipment travel distances. For a mine with production shafts as the primary means of material transport, passes are usually located as close to the shaft as practical. If underground material transport involves trucks and LHDs, passes should be located in a manner that will reduce equipment travel distances in order to optimize production. The ultimate decision on the location of the pass should take into account both economic considerations and rock mechanics stability.

In most operating hard-rock mines the general rule for ore- and waste pass inclination is around 70° from horizontal. Vertical orepasses are not uncommon, and in some instances a dogleg is introduced at the bottom of the pass to reduce the impact effect that results from falling rock. The inclination is governed by a variety of factors including the maximum rock size in the muck. A good guideline is that an orepass diameter should be a minimum of four times larger than the largest anticipated rock fragment to be transported in the pass. The flow characteristics of the material going through the pass, the amount of fines, and moisture present will also play an important role in determining the optimal inclination and cross-sectional area. Other considerations include impact load limitations on the flow control mechanism, ground conditions inside the raise, and the induced ground stresses around the excavation. A grizzly is usually installed at the top of a pass to control the rock sizes going through. The presence of wet fines in the muck may exacerbate hang-ups inside a pass due to the resulting cohesion. Steeper inclinations can be used in such situations. The impact load on the flow control gate at the bottom of a steeply inclined pass is significantly higher than that for a pass with a shallow incline, and could be as much as four times in some cases (Blight 2006).

Size and shape of opening. The most common cross sections are circular, rectangular, and square. Oval and elliptical shapes have been used in areas with a nonhydrostatic stress field, with the major axis of the pass aligned parallel to the direction of maximum stress. From a rock mechanics perspective, a circular cross section is more stable than a rectangular one where there is potential for stress concentration at the corners.

The size of an ore- or waste pass is a function of the largest rock size and amount of fines content in the muck. The optimum size of a pass has in the past been estimated by three main approaches: traditional guidelines, standards established at other mines, and empirical methods. All three methods are prone to unreliable and/or overconservative results. Numerical modeling of the excavation for ground support optimization and material flow prediction inside the pass can yield a more accurate design and can be used to refine any preliminary designs that result from the preceding three approaches. Another advantage of numerical modeling is the ability to incorporate staged excavation and updated structural and rock data as construction progresses in order to see the effect of any changed ground conditions on the final excavation and ground support design.

Excavation and ground support. The most common excavation methods used for ore- and waste passes are Alimak raising, drop raising and slashing (long-hole drilling and blasting), pilot raising–slashing, and full-face raise boring. The full-face raise boring technique is only suitable for circular cross sections, while all the other methods are applicable to any cross section.

Ground support is normally provided as the excavation progresses. A circular pass excavated by full-face raise boring is the most stable and is normally not ground supported. However, linings are commonly used to prevent deterioration of the wall from abrasion caused by muck flow. Ground support (similar to that for ore bins) is conventionally installed on other cross-sectional profiles and usually consists of friction bolts and fully grouted fiberglass rock bolts used primarily to maintain the integrity of the walls during development. The dimensions of the pass must be large enough to allow installation. In some cases a final lining (and any additional ground support) will be installed as a separate operation.

Lining. The most common lining methods employed in modern mines are cast-in-place concrete (with or without embedded wear plate), shotcrete, and grouted/concreted steel lining. When steel linings are used, they can also act as the exterior formwork for the concrete pour in the annulus.

Muck flow control. Passes are normally equipped with a flow control mechanism at the bottom to control the rate of discharge of rock. Material flow at the bottom is normally controlled by a chain curtain with or without a press frame, depending on the equipment setup and haulage method, as illustrated in Figure 12.6-9B. When a conveyor haulage system is employed, a vibratory feeder is normally installed below the chain curtain/press frame mechanism. The withdrawal and extension action of the press frame is controlled by a pneumatic cylinder. In practice, the throat height of the

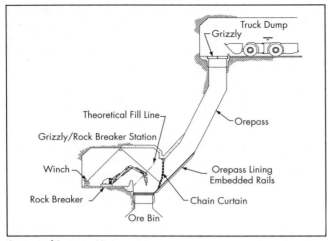

Courtesy of Stantec – Mining.
Figure 12.6-29 Orepass with chain curtain

pass is determined from the maximum rock size in the muck and should not be fewer than three times the maximum lump size. The natural angle of repose of the muck in the orepass chute is altered by the weight of the chain curtain, thereby restricting material flow. Muck is released by lifting the chain curtain from the hinge point (see Figure 12.6-29). The weight of the chain curtain also determines the rate of muck flow. The flow control mechanism should be designed to resist the static pressure due to the height of the muck in the pass plus any potential impact loads that may be generated from the release of a hang-up.

Construction and Installation
Drop raising–slashing, Alimak raising, pilot raising–slashing, and full-face raise boring are common excavation methods. The excavation method and sequencing is governed by various factors including the orepass configuration, ground conditions, rock type, economic considerations, and so on. A circular orepass is a good candidate for a full-face raise bore excavation method but will require a top and bottom access drift to set up the raise bore machine, drill the pilot hole, recover the pilot bit, launch the reamer, and remove the drill cuttings. A pilot raise, followed by slashing, also requires a top and bottom access drift. Slashing should be done from the top down and mucked through the pilot raise to the bottom access drift. The top-down procedure has the advantage of using the working bench as staging for ground support and other installations following each completed round. Excavation by an Alimak raise requires a bottom access drift and is applicable to all regular-shaped passes. With the drop raise method, slashing is normally done from the bottom up. This method is best suited for rock with good stand-up time where ground support can be delayed until the excavation is complete. Ground support is installed immediately after each mucking cycle (see Figure 12.6-30).

If the pass is to be lined, the lining is normally installed as a separate operation after the excavation is complete. Depending on the type of lining, length of the pass, and construction technique, the lining can be installed as a single unit or in sections (see Figure 12.6-31). The appurtenant equipment at the top and bottom are normally installed after the pass is completed. During filling (and anytime the pass is drawn empty), crushed and screened gravel should be dumped into the bottom portion of the pass to protect the headblock and other attached flow control mechanisms. The crushed rock acts as a buffer against direct impact from falling rock on the control gate and associated equipment.

Facility Maintenance
Repairing linings, releasing hang-ups, and installing rock reinforcement are the primary maintenance activities done inside a pass. Maintenance issues are similar to those encountered in ore bins. The top of the pass must be protected when repair work is in progress, and safety precautions similar to those required during repair work in ore bins should be implemented as well. Other areas that may require periodic assessment and maintenance include replacement of liners in the discharge chute, the flow control mechanism, and the feeder at the bottom of the pass.

Emergency Egress
The design and construction of subvertical emergency egresses is similar to ore- and waste passes, except that they differ in size and the equipping requirements. An emergency access can be constructed in a vertical or subvertical shaft depending on the preferred method of access/egress and equipment requirements. When a shaft is dedicated solely for emergency egress, its overall size will be limited to the minimum required to accommodate the access equipment and any installed services. If an existing shaft is partitioned to accommodate an emergency egress in one of the compartments, only the specific partition needs be sized accordingly for the emergency access equipment.

Design Considerations
The following paragraphs discuss some of the factors to consider in the design of an emergency access system.

Size and cross section. The size and cross section of the excavation depends on the selected evacuation system, ground conditions, and excavation method. Common cross sections are circular, square, or rectangular in shape. These cross sections can be constructed by methods similar to those previously outlined under "Construction and Installation" in the "Ore- and Waste Pass" section. The sizing and layout of evacuation systems are governed to some extent by local regulations. With a cage and hoist system, the cage may be sized to accommodate one or several mine personnel at a time, depending on various regulatory factors. A minimum factor of 0.14 m^2/person is normally used in North America to size the floor area of each deck in the cage. Allowance should also be incorporated for the cage guides, tolerances, and other appurtenances similar to a cage hoisting shaft. An access ladder system is normally designed to accommodate a single mine worker at any level along the ladder. Where a side-step platform is provided as a rest stop during a climb, it is normally designed to accommodate one individual at a time as well. Each rung on the ladder and side-step platforms is designed for a minimum specified imposed load. In the United States, these and other design requirements for emergency access ladders and platforms are regulated by the Occupational Safety and Health Administration and MSHA.

Location and layout. An emergency egress should be located in stable ground where the serviceability can be guaranteed and accessibility maintained in case the mine's primary means of egress is blocked. In the United States, an escapeway that is more than 100 m in height and inclined more than

30° from the horizontal must be provided with an emergency hoist. The layout must connect all primary working levels in the underground mine and should also be situated within the fresh-air intake network to guarantee the presence of fresh air at all times when in use.

Evacuation method. The evacuation method is normally by an installed access ladder system or a cage/escape pod with a small single-drum hoist or winch. The access ladder system is normally installed in a steeply inclined shaft. The inclination assists in reducing the climbing effort of personnel but ultimately with a longer access development. In inclined escapeways with fixed-ladder systems, sublevel rest platforms must be provided at specific intervals (9.1 m is the MSHA requirement in the United States). With a cage and hoist system, it is common practice to use a vertical shaft because it is more adaptable to the conveyance's guides, lower maintenance costs, shorter access development, and faster development time.

Construction and Installation

Usually a top and bottom access drift will exist when the emergency egress is excavated. Therefore, almost all the excavation and ground support techniques used to construct ore- and waste passes (as long as they fit the particular purpose) can be used in constructing emergency egresses. Ground support normally consists of mechanical bolts combined with wire mesh where there is potential for areas of the walls to slough over time. Fully grouted rebar rock bolts or cable bolts are common for facilities with longer service life. Friction bolts (such as Split Sets) are normally used in facilities with a shorter life span because of the possibility for corrosion. In exceptional situations, a shotcrete lining may be added as secondary ground support.

Equipping of the emergency egress starts after the excavation and ground support is completed. If an access ladder system is used as the method of evacuation, it is normally made of steel sections with a surrounding safety cage, as illustrated in Figure 12.6-10, and of modular design so that it can be erected in sections. In a moist and/or corrosive environment, the entire ladder system can be constructed of galvanized steel. The lowest section is first assembled and lowered down the raise to the bottom where it is secured in place. Successive sections are assembled and lowered in a similar manner with each preceding section used by the installation crew as a working platform to complete successive sections of the ladder. The process is repeated until the access ladder reaches the collar. The entire ladder system is attached to the footwall of the raise with rock bolts to provide added stability. A landing platform is normally provided at specific intervals along the entire length of the access ladder system. In the United States, the MSHA requirement is that the platform landing should be at no more than 9.14-m intervals.

If a worker cage is used as the method of evacuation, it will consist of a small cage or torpedo on guides and sized to carry a small number of personnel at a time. The cage or torpedo is lifted by a small single-drum hoist or winch, as shown in Figure 12.6-11. Installation of this type of system bears similarity to shaft-hoisting systems.

Facility Maintenance

Because emergency egress systems are not used on a daily basis, it is important to ensure they are functioning properly in case they have to be used in an emergency. Proper functioning of the system must be checked at regular intervals as

Courtesy of Northparkes Mines.

Figure 12.6-30 Orepass development: (A) installing initial ground support, and (B) excavating subsequent round with hydraulic excavator (ground support from previous round of excavation is already in place)

Courtesy of Northparkes Mines.

Figure 12.6-31 Lowering first lift steel lining into pass

part of the maintenance schedule. Periodic inspection, testing, and maintenance are more critical for a system with a worker cage and hoist than for one with only an access ladder. Some jurisdictions specify the required intervals for inspection and testing of emergency equipment. With access ladder systems, items to check for on a periodic basis include proper anchoring (loose rock bolts and connection bolts, etc.), corrosion of steel members, and connections. With an emergency hoist and worker cage, inspection and testing should include a trial run of the hoist and inspection of the hoist ropes, rope attachments, and guides. Periodic nondestructive testing of the ropes is normally specified by regulatory authorities and should be conducted when the system is initially commissioned and on a regular basis.

Loading Pocket

Loading pockets are used for transferring ore and waste onto skips in a shaft-hoisting facility. Following are the main components of the infrastructure that constitutes the loading pocket:

- The excavation required to accommodate the entire loading pocket infrastructure.
- The loading pocket chute and flow control mechanism. This is the first chute and control gate structure at the loading pocket. The layout and design vary depending on the materials handling design but are normally sized based on the size of materials being handled. The floor of the chute is also inclined at between 45° and 55° to the horizontal in order to promote material flow into the measuring pocket. The sidewalls and floor of the chute that are in contact with flowing material are usually lined with abrasion-resistant wear plates. The types of flow control gates normally used are the chain curtain, radial, or guillotine. The chain curtain–type mechanism is not suited for conditions where fines and water are present in the feed. Flow control gates can be operated by either a pneumatic or hydraulic cylinder.
- The measuring flask or pocket. The size of the measuring flask or pocket should match that of the skip. The normal arrangement underground is to provide one measuring pocket for each skip. Irrespective of the measuring system adopted (i.e., volumetric or weight), each load from the measuring pocket should completely load the empty skip. Measuring pockets are fabricated of steel with internal abrasion wear liners on all surfaces. The bottom of the pocket is usually inclined at about 60° to the horizontal to initiate quick and complete load-out into the skip. The two types of measuring pocket designs normally used underground are the fixed body and swing-out body. The former type is preferred in most underground mine environments because it operates faster, is easily automated, and the structure is fixed and generally incurs lower maintenance costs. The pocket is equipped with a flow control gate. The arc-type gate is very common with fixed pockets and is controlled either hydraulically or pneumatically. An access hatch is normally provided on one of the vertical walls for periodic inspection of the condition inside the pocket.
- The feed or discharge chute and control mechanism. The feed chute is used to direct material from the measuring pocket into the skip. Its construction is similar to loading pocket chutes with abrasion wear liners provided on the interior walls. It spans from the bottom lip of the measuring pocket to the edge of the skip. Normally some clearance is provided between the feed chute and the skip to account for misalignment and movement so that the feed chute is not hit by the skip during travel. The amount of clearance depends on the type of measuring pocket used and the width of the skip measured parallel to the width of the chute. The angle of inclination of the feed chute should match that at the bottom of the measuring pocket in order to provide complete load-out into the skip.

Loading pockets can comprise a simple materials transfer system involving truck or rail haulage as in Figure 12.6-13, or a more complex arrangement involving automated belt conveyors as seen in Figure 12.6-12. The decision to adopt either system depends on the scale of the underground operation and the mine's preference. In the general case, any loading pocket arrangement will contain the four primary components listed previously.

Design Considerations

The following paragraphs discuss the basic criteria to be considered in the design of loading pockets.

Operating criteria. The operating criteria should consider the time it takes to load the measuring pocket and the travel time of the skip to ensure that the pocket is fully charged and ready to load the skip upon return. The accuracy of the measuring devices is also critical in determining the size of the measuring pocket. The material in the pocket can be measured volumetrically or by weight. A volumetric measuring system is best suited for situations with uniform material density. In situations where material density is expected to vary, a weight measuring system should be adopted. In either case, the intent is to prevent overloading the skip and hoisting system.

Loading and materials-handling systems. Two types of layouts are normally used for materials loading systems. Material can either be loaded into the pocket from a surge bin or orepass situated directly above the loading pocket, or by conveyor belt feeding material from a remote crusher, surge bin, or other temporary underground material repository. The steel structure supporting the chutes and measuring pocket is usually designed to withstand impact loads from the material being dumped into the measuring pocket plus the static and drawdown load effects due to the full height of the rock before discharge into the skips.

O&M costs. The fixed body–type measuring pocket normally will incur less O&M cost than the swing-out body type because the additional gear needed to move and reposition the pocket is absent, and fewer operating parts exist that will require periodic servicing. Wear liner replacement costs will normally be similar for both systems.

Spill control. Material spillage during skip loading is expected at the feed chute/skip interface. In some loading pocket arrangements, spill can be minimized by limiting the clearance between the feed chute and the skip wall. Two methods are commonly used in skip hoisting to manage material spillage below the loading pocket. The first method is the installation of a spill pocket below the loading station location for collecting material that spills from the loading station. The spill pocket is periodically unloaded after it is full. The second method uses an access ramp developed to the bottom of the shaft to provide access for mobile equipment to periodically clean out material that spills to the bottom of the shaft.

The second method is the most popular and may incorporate a diversion berm at the shaft bottom to deflect all spillage toward the access ramp.

Construction and Installation
The initial aspect of construction normally covers the excavation of the loading station. Since the loading pocket constitutes part of the loading station, its excavation should be done during the loading station development. The loading station is normally excavated during the shaft sinking operation. The excavation and concrete work can be done during the shaft sinking and lining phase by slashing out the loading pocket and installing ground support and the concrete structures before continuing downward with the shaft development. The fit-out of the loading pocket can be done during the equipping phase of the shaft, and the primary loading pocket equipment can be transported into place by rigging from underneath the shaft conveyances or installing before the skips are set in place.

After the excavation and ground support is complete, concrete formwork is installed and the various reinforced concrete structures that form the foundations and walls of the loading pocket are installed. The measuring pocket, control gates, and other associated structures are usually designed to be supported by steel structures that form part of the shaft steelwork at the loading station. All the loading pocket steelwork should be transported and installed together with the shaft steelwork around the loading station.

The measuring pocket can be prefabricated off-site and installed as a single unit (depending on its size) after all the steelwork is in place. Depending on their sizes, the various components of the loading pocket can be transported down the shaft with or without the skips and set in position. Tall measuring pockets can be prefabricated in modular sections that can be transported down the shaft and installed in sections.

Facility Maintenance
The critical items that may need attention in a loading pocket are steel wear liner replacement in the chutes and measuring pocket, and lubrication of moving parts. When these items wear out, they normally expose the head of the connection bolt holding them in place. The bolts eventually shear and release the liner into the skip load. Periodic inspection is normally done after a certain amount of operating hours to assess the conditions of the wear liners. Access into the measuring pocket is normally through hatches located on one or more of the vertical walls. Access stairs and platforms are normally provided for this purpose around the loading pocket as part of the shaft steelwork. Inspections and repairs inside the loading pocket are normally scheduled during the shaft and hoist down times and require lock-out on all operating equipment.

Special Purpose Drifts
Drifts are used to access underground mine workings, connect two or more shafts together, for ventilation and drainage purposes, haulage, and emergency egress. One common characteristic among drifts is that the cross section is restricted to the minimum size that will satisfy duty requirements.

Design Considerations
The following paragraphs discuss the factors that should be considered in the design of drifts.

Location. Factors that govern the location of the drift include the intended use of the drift, the underground mine layout, and geotechnical characteristics of the rock. Ideally the drift should be located where ground conditions and rock geotechnical properties can accommodate the required opening sizes without incurring excessive capital and operating costs.

Purpose. Three primary functions of drifts in underground mine operations are ventilation, haulage, and general mine access. It is common for mines to use drifts for a combination of these functions, except in situations such as those involving large ventilation airflow requirements. Underground operations with large ventilation requirements normally have dedicated ventilation drifts in which personnel and equipment are not allowed because of the high air velocity through those drifts.

Layout and design. This is normally a repetitive process, starting with initially defining the purpose of the drift, followed by laying out the internal cross-sectional arrangement, then refining each iteration of the plan as necessary. For haulage and general access purposes, the type and physical dimensions of the fixed and mobile equipment in the drift, (during construction or production) will directly impact the cross-sectional dimensions of the drift (see Figure 12.6-32). The drift cross-sectional area, wall roughness, and overall length are critical parameters that affect airflow resistance in a ventilation drift. A rectangular cross section is the most common shape used in underground mine development, sometimes with an arched back modification, as shown in Figure 12.6-32A, to improve roof stability. With a rectangular cross section, the invert (sill) of the drift can easily accommodate a roadbed for mobile equipment without adversely reducing headroom.

After the basic sizes and lengths are established, the temporary (during development) and permanent (during production) cross-sectional arrangements are used to determine the largest excavation sizes required. The construction setup may produce the largest excavation cross section depending on the types of equipment and services provided in the drift. Typical services installed in the drift include pipes for service water, compressed air and discharge, ventilation ducts, and electrical cables.

Truck haulage drifts normally have a finished roadway surface to enhance equipment travel. The roadway normally has a 1%–2% slope toward a drainage ditch located close to the drift walls to collect surface runoff. In rail haulage drifts the track is sometimes offset from the drift centerline to accommodate personnel traffic and a drainage ditch on the opposite side. Conveyor drifts usually have the conveyor offset from the drift centerline as well to accommodate personnel traffic or small rubber-tired equipment (such as a skid steer loader or grader) for spill cleanup. Conveyors can either be supported on the sill with steel frames on concrete footings or hung off the back of the drift on chains or cables attached to fully grouted rock bolts. The latter support method is normally less expensive and can be installed quickly while providing easy access for spill cleanup underneath the conveyor. The fully grouted rock bolts for conveyor support can be installed during the ground support program or as a separate activity after the ground support installation is complete. The slope of the drift is another important design criterion to consider. The following is a guide for establishing the slope of a drift:

- Haulage drifts for mobile equipment—Ideally, under 8%, but not more than 10%–15%, depending on use
- Conveyor drifts—Up to 25%, depending on conveyor design
- Rail haulage drifts—Ideally flat to 0.25%, but not more than 2%

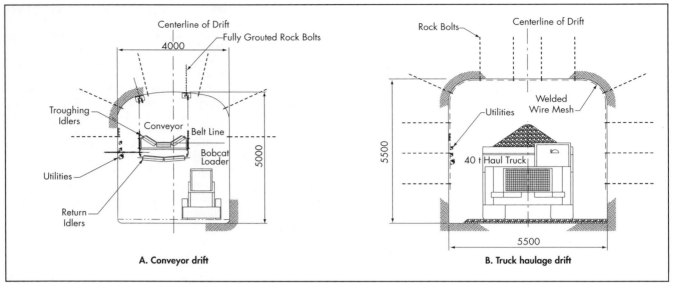

Courtesy of Stantec – Mining.

Figure 12.6-32 Typical drift cross sections for different haulage scenarios

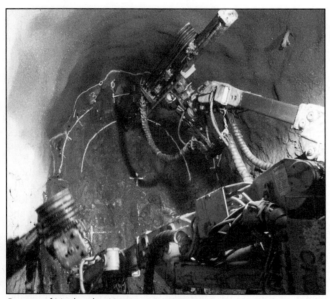

Courtesy of Northparkes Mines.

Figure 12.6-33 Access drift face markup and drilling with jumbo

Excavation and ground support. The excavation design can be completed after the excavation sizes are finalized. Ground support design is normally not as complicated compared to other underground openings and will usually range from no support necessary to roof bolting and other combinations. Excavations in blocky ground normally will require roof bolts set in a pattern. Depending on the rock type and ground conditions, other ground support combinations may be required (such as with wire mesh, shotcrete, or higher-capacity fully grouted rock bolts). When severe ground conditions exist that cannot be supported by bolting and/or shotcrete, steel sets with lagging or shotcrete can be used for ground support. Excavation is conventionally carried out by drill-and-blast. Muck can be removed with various equipment combinations depending on the size of the drift. LHD and LHD/truck combinations are the most common mucking equipment used nowadays.

Construction and Installation

Preplanning. The preplanning efforts for drifts are quite straightforward compared to other underground service facilities such as crusher rooms. Preplanning should include establishing drill-and-blast pattern designs for the various drift cross sections, sizing and location of any temporary cutout such as muck bays, installation of services, roadbed construction, and dumping of muck.

Procurement. Items to procure include pipes, ventilation ducts, conveyors, electrical cables and cable trays, and other permanent equipment. Specific materials and equipment to facilitate construction (rock bolts and other ground support consumables, mobile equipment, concrete construction equipment, etc.) will also be required and must be procured as well. Rock bolts and similar ground support items should be procured ahead of time and present on-site at the time the excavation phase commences.

Construction activities. As previously mentioned, the most common excavation method is by drill-and-blast with full-face advancement typical for small- to medium-drift headings. For larger drift cross sections, it may be necessary to use multiple headings for face advancement because of limitations in the operating range of equipment or ground conditions. Each face is marked up according to the blasthole pattern design followed by drilling with a jumbo or other available equipment (see Figure 12.6-33). The excavation and ground support phase should be completed before beginning installation of services or permanent equipment for various reasons including

- To avoid crowding in a limited underground work space for multiple construction activities,

Courtesy of Rio Tinto.
Figure 12.6-34 Rail haulage drift: (A) wet conditions in drift resulted in corrosion of rails and rotting of wood ties, and (B) ballast and wood ties were replaced with rails cast in concrete slab and provision of drainage ditch

- To eliminate construction schedule conflicts between different crews and workplace traffic congestion that can result in a safety incident,
- To reduce the risk of damage to installed equipment from ongoing construction activities involving mobile equipment, and
- To maintain the amount of airflow required per personnel and operating equipment at any time underground.

The next important task that usually follows completion of the excavation and ground support is the permanent roadbed construction. Because of the irregular surface left behind due to the look-out on the contoured blastholes, dressing of the sill followed by the roadbed construction is mandatory to improve equipment travel and productivity. The drift roadbed can comprise either a rigid construction (with a concrete slab construction) or flexible (constructed with at least a 150-mm depth of open-graded gravel). A truck haulage drift normally only requires a gravel roadbed, graded in place with a grader or similar equipment. Compaction of the gravel roadbed is normally attained by the travel of mobile equipment. Regular regrading and maintenance is a requirement with gravel roadbeds. Other drifts such as for conveyors with permanent equipment are sometimes equipped with a concrete roadbed that incorporates the foundations of adjoining equipment (see Figure 12.6-16). Rail haulage drifts are constructed with a gravel subgrade and ballast that supports the ties and rail track. Alternatively, the rail track could be continuously embedded in a concrete slab when the prevailing conditions underground dictate the need for such a method (see Figure 12.6-34). Embedding the rail track in a concrete slab may be necessary where there is a possibility for the following situations to occur:

- Poor subgrade conditions beneath the ballast
- Difficult drainage or surface runoff in the drift resulting in corrosion and rotting of wood ties
- Contamination of the ballast with fines deposited from the wheels of rolling stock
- Lack of suitable ballast material or production is too costly
- Where rubber-tired mobile equipment is used in conjunction with track equipment

Allowance should be made in the construction for expansion and contraction of the rail tracks where extreme seasonal variations in temperature occur.

The centerlines of the drift should be established in the field as the working point for installation of permanent structures and equipment. All permanent facilities can be accurately located and installed underground by their horizontal and vertical offset distances from the drift centerlines.

Facility Maintenance

The maintenance activities associated with drifts are minimal and normally concentrated around the permanent facilities installed within the drifts. Following are some of the routine maintenance activities that should be anticipated:

- Realigning and leveling rails and ties
- Replacing worn and damaged rail track
- Removing ballast that has been contaminated with fines and replacing with clean open gravel
- Cleaning out sumps and drainage ditches
- Resurfacing and grading roadbed
- Cleaning up spill beneath conveyors
- Repairing and replacing conveyor belt

Of all three drift types, ventilation drifts tend to require the least maintenance. Maintenance efforts for conveyor and rail haulage drifts can be similar under certain conditions and depending on the type of construction.

REFERENCE

Blight, G. 2006. *Assessing Loads on Silos and Other Bulk Storage Structures.* London: Taylor and Francis Group.

CHAPTER 12.7

Tunnel Boring Machines in Mining

Lok Home and Odd G. Askilsrud

A tunnel boring machine (TBM) is a device used to excavate tunnels in a circular cross section through a variety of soil and rock strata. TBMs can bore through hard rock, sand, and almost anything in between. Tunnel diameters can range from ~1 to about 16 m (3.3 to 52.5 ft); tunnels smaller than that are typically created by horizontal directional drilling rather than by TBM.

This chapter describes mechanical excavations in soft/moderate to very hard rock and in rock-mass conditions ranging from marginally self-supporting to massive. TBM types discussed include state-of-the-art open, shielded, and other types of machines that incorporate specific features to suit particular applications, all using disc cutters. Mechanical tunnel excavators that create noncircular cross sections, such as mobile miners, mobile tunneling machines, roadheaders, and drum miners, are not discussed in this chapter, although they also serve important roles in civil and mining applications. These machines are discussed elsewhere in this handbook.

TBM TYPES
TBMs are normally classified into three groups:

1. Open-type machines: main beam, Kelly drive, and open gripper
2. Shielded-type machines: single shield and double shield
3. Soft-ground machines: hybrid, earth pressure balance, mixed shield, and slurry

A feature common to all TBM types is a rotating cutterhead that does the following: (1) presses forward and against the rock face under high pressure; (2) uses individual disc cutters placed strategically on the cutterhead; (3) performs muck removal by means of buckets; and (4) is supported by a main bearing with the structure being moved forward by a thrust system composed of cylinders. All TBMs have a backup system of power units and muck-removal systems.

Shielded-type machines are seldom used in mine applications because of the difficulty in removing the shields. A third shielded-type machine, the gripper shield, was designed to actively support the ground from the cutterhead to behind the TBM where rock support can be installed and has been used in a wide range of geological conditions. However, it is no longer available because of limited acceptance in the industry.

Because most mining applications take place in moderate to hard rock, soft-ground machines are not discussed in this chapter.

GENERAL TBM APPLICATIONS
Compared to their use in civil-engineering applications, TBM use in mining has been relatively modest since the first successful hard-rock TBM was introduced by James S. Robbins in 1957.

TBMs in mining have ranged in diameter from ~1.7 to 7.6 m (7 to 25 ft). They have been used predominantly in soft to moderately hard formations. Mine applications at times place severe restrictions on component size and weight so that equipment can be lowered down existing shafts, transported to a designated assembly and launching area, and, after tunnel completion, removed. TBM systems in some mines must include explosion-proof equipment. Such equipment has been used for access to coal mining in Germany, England, and Canada.

Detailed study is required to select the proper TBM system for a project. Such study involves reviewing geological design reports, geological profile drawings, and baseline design or equivalent reports prepared by the mine owner or a consultant depicting anticipated conditions along the drive and required design parameters. Important factors to consider include the following:

- Rock types
- Rock strength parameters from a testing laboratory
- Rock mass characteristics, including stand-up time at the actual tunnel depth, profile, and alignment of the proposed tunnel
- Tunnel diameter
- Temporary and permanent tunnel-support requirements
- Water and gas
- Geological profile
- Mucking requirements

Lok Home, President, The Robbins Company, Solon, Ohio, USA
Odd G. Askilsrud, President, Tunnel Engineering and Applications, Inc., Renton, Washington, USA

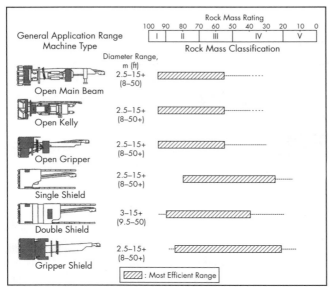

Source: Askilsrud and Moulton 2000.

Figure 12.7-1 General application ranges for various TBM types

Courtesy of The Robbins Company.

Figure 12.7-2 Main-beam TBM

Figure 12.7-1 can be a useful tool for discussion purposes in the planning stage. The chart shows typical TBM types and their most efficient application ranges based on rock mass rating.

The estimated TBM advance rate is based on rock strength and fracturing characteristics, in-situ condition of the rock mass, machine characteristics, and operating condition in the tunnel. The actual TBM advance rate is equal to the product of the average instantaneous penetration rate in millimeters per cutterhead revolution (inches per revolution), cutterhead speed (rpm), shift hours per day, and machine utilization (percent).

TBM DESIGN FEATURES

Important design features to consider for achieving efficient rock chipping include effective thrust load on the cutterhead and cutters, layout and spacing of cutters, and diameter and tip width of the cutter rings.

The nominal load capacity of a single-disc cutter has evolved from 89 kN (20,000 lb) for the 300-mm (12-in.) cutter of the 1950s to >312 kN (70,000 lb) for the 500-mm (20-in.) largest-size cutter of today. Today's state-of-the-art hard-rock TBM typically has a flat-face cutterhead with single tracking center, face and gauge cutters, each of which provides concentric kerfs on the rock face during excavation. Kerf spacing (center-to-center spacing of cutter tracks) is nominally ~90 mm (3.5 in.) for the face cutters, decreasing into continuously tighter spacing for the gauge-cutter positions. A flat-type cutterhead normally consists of nine cutters in the curved gauge area. A typical 4.6-m (15-ft) TBM hard-rock cutterhead with 432 mm (17 in.) back-loading cutters is shown in Figure 12.7-2. Single-disc cutters (483 mm or 19 in.) in wedge-lock housings are shown in Figure 7.1-4 in Chapter 7.1.

Selection of cutter diameter and cutter characteristics is determined by the application study and/or specific requirements of the application. State-of-the-art cutter rings for hard-rock applications are made of tool steel, specifically heat-treated so as to reduce the number of cutter changes and downtime for the anticipated ground conditions during the drive. The cutter hub assembly includes high-capacity tapered roller bearings and seals capable of supporting multiple cutter-ring changes.

Early during TBM design, it must be decided whether the cutterhead should have front- or back-loading cutters. Front-loading cutters are faster to change but require miners to enter and work beneath unsupported ground in front of the machine. Back-loading cutters can be changed from behind the cutterhead and under the protection of a shielded structure. The largest individual cutters 500 mm (20 in.) weigh up to 190 kg (425 lb). Cutters are moved and installed by means of a cutter-handling system normally designed for the specific TBM.

Other design features vary by TBM type. For further information on TBM cutters, see Chapter 7.1.

Main-Beam TBM

The main-beam TBM (Figure 12.7-3) is suited for high-strength rock and short faulted zones. Ground support methods can include ring beams, bolting, and shotcreting.

A main-beam TBM consists of four main elements: (1) cutterhead, (2) cutterhead support and main beam, (3) gripper and thrust assembly, and (4) conveyor assembly.

The gripper assembly with one set of grippers forms the stationary anchoring section of the TBM. It transmits thrust and torque to the tunnel wall during boring and carries part of the machine weight. The anchoring force is approximately 2½ to 3 times the total forward-thrust force.

The machine is forced forward a stroke at a time by hydraulic cylinders connected to the anchoring section and main beam as the cutterhead rotates against the rock face. At the end of a stroke and after the cutterhead stops rotating, the rear support is lowered to the invert, the grippers are retracted from the tunnel wall and moved forward equal to the stroke of the thrust cylinders, the grippers are again energized against the tunnel wall, and rear support is retracted so that boring can start again.

The front supports, support shoe, and extendable side and roof supports provide ground contact immediately behind the rotating cutterhead to stabilize the TBM during tunneling. The

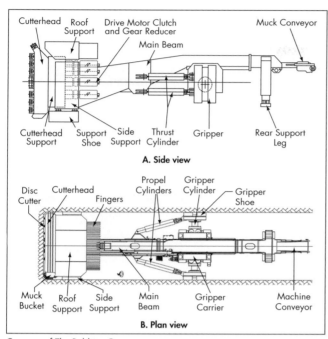

Courtesy of The Robbins Company.
Figure 12.7-3 Main-beam TBM

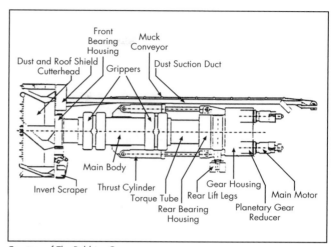

Courtesy of The Robbins Company.
Figure 12.7-4 Kelly-drive TBM

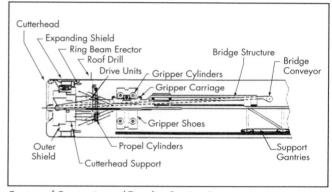

Courtesy of Construction and Tunneling Services, Inc.
Figure 12.7-5 Open-gripper TBM

cutterhead is typically driven by water-cooled electric motors located on the cutterhead support, each directly mounted to a gear reducer with output pinions. Drive pinions engage with the large ring gear, which is connected to the cutterhead. The cutterhead itself is supported by the main bearing.

The main-beam TBM has essentially become the industry standard open TBM mainly because of its ease of operation and ample room around the cutterhead area for ground support.

Kelly-Drive TBM

The Kelly-drive TBM (Figure 12.7-4) is suited for high-strength rock and short fault zones. Ground support methods can include ring beams, bolting, and shotcreting.

The Kelly-style TBM consists of two main elements: the anchoring section and the working or moving section.

The gripper assembly with two sets of grippers forms the stationary anchoring section. It transmits thrust and torque to the rock during boring and supports the weight of the machine. Four grippers in the radial direction are arranged horizontally in pairs and controlled individually.

The working section includes the cutterhead, front bearing housing, torque tube, rear bearing housing, and drive train. Cutterhead rotation is controlled by motors at the rear of the machine coupled to gear reducers and pinions. A common ring gear turns the drive shaft through the center of the torque tube to transmit rotational power to the cutterhead. Hydraulic pulling cylinders develop thrust that is transmitted from the main body through the rear bearing housing and drive shaft.

Use of this type of TBM has declined, mostly because of the limited access around the TBM for installing ground support.

Open-Gripper TBM

The open-gripper TBM (Figure 12.7-5) is suited for high-strength rock and short fault zones. Ground support methods can include ring beams, bolting, and shotcreting.

The TBM has expanding supports close to the excavation face and a simpler gripper design (no main-beam or Kelly-drive structures) than do the other types of open-type TBMs described. It typically uses an arrangement of lattice-type (diagonal pattern) thrust cylinders to react to cutterhead torque and correct roll. Advantages can include low cost, short overall length, and tight turning radius.

The machine consists of a cutterhead, forward expanding shield, and gripper assembly. The cutterhead and cutterhead drive are the same as typically seen on main-beam and gripper-shield TBMs. This configuration provides cyclic advance of the machine in the following sequence: bore, regrip, bore.

A bridge structure spans from the TBM cutterhead support to the first platform car. The bridge carries the grippers and also provides space to install muck-haulage rail and a mounting for the forward conveyor, material-handling equipment, exhaust duct, and tunnel-support installation equipment.

Single-Shield TBM

A single-shield TBM (Figure 12.7-6) performs boring and lining installation in sequence. Lining, typically concrete segments, is used as an anchoring station to propel the machine during boring. The machine can be used in hard as well as soft

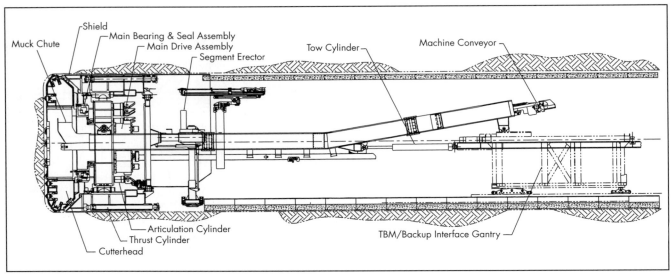

Courtesy of The Robbins Company.
Figure 12.7-6 Single-shield TBM

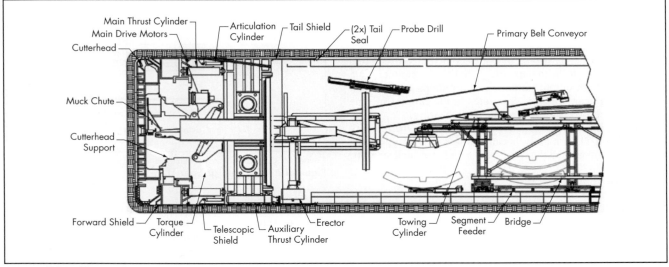

Courtesy of The Robbins Company.
Figure 12.7-7 Double-shield TBM

ground, and is well suited for tunnels in mixed ground and loose formations with low stand-up time.

The main section consists of a cutterhead, cutterhead support with integrated ring gear, main bearing, and drive units. Thrust cylinders are located in the rear part of the machine and act against the tunnel lining. Complete lining rings are installed inside the shield by a segment erector.

Steering is done by individual control of the oil volume supplied to separate groups of thrust cylinders. Steering can be enhanced by the articulation cylinders, which bias cutterhead support in relation to the shield.

Double-Shield TBM

The double-shield TBM, also called a telescopic-shield TBM (Figure 12.7-7), can be used in hard as well as soft formations. It is used when there is a risk of unstable ground conditions. Normally, the tunnel walls are used to anchor the machine, which allows installation of segmental lining during boring. If the walls cannot take the gripper pressure, the machine can push off the lining instead.

The working section at the front of the machine has a cutterhead, cutterhead support with integrated ring gear, main bearing, and drive units. A telescoping section between the anchoring and working sections includes hydraulic cylinders for thrust and steering. The anchoring section at the rear of the machine has two horizontal grippers that transmit thrust and torque. Auxiliary cylinders transmit these reactions to and propel the machine from the lining. A segment erector is installed inside the tail shield.

Steering is done by individual control of the oil volume supplied to separate groups of thrust cylinders. Steering can be enhanced by the articulation cylinders, which can off-angle

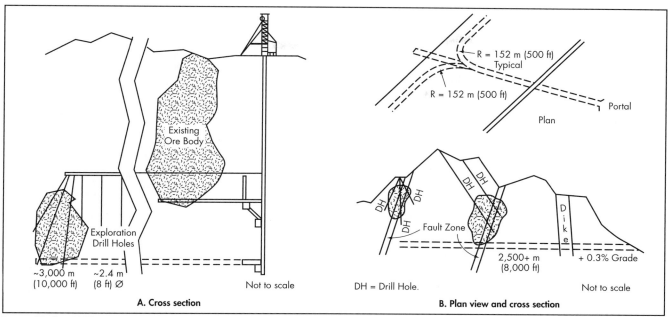

Source: Moulton 1997.
Figure 12.7-8 Mine openings for exploration

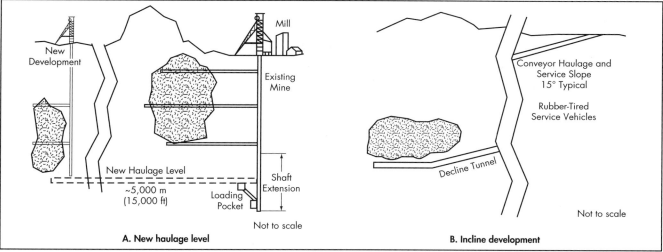

Source: Moulton 1997.
Figure 12.7-9 Mine openings for haulage by rail, rubber tire, or conveyor

the working section in relation to the anchoring section to bias the cutterhead in the proper direction.

If segment lining is being placed, an electronic guidance system is mandatory to ensure correct line and grade for installation of the segment ring within the tail shield.

TBM MINING APPLICATIONS

Underground openings in mines can be associated with exploration, access to and development of ore bodies, haulage (by train, conveyor, or rubber tire), ventilation, drainage/dewatering, movement of supplies and labor, safety, or a combination of these. Figures 12.7-8 through 12.7-13 show examples of some underground openings in mines.

A key consideration for TBM mining applications is the type of muck-removal system. Tunnels under consideration for TBM application are generally close to mine boundaries, because only long-distance tunnels could justify capital investments in TBM equipment. Tunnel cross sections are generally relatively small, because muck must be transported and hoisted without interrupting full-time mining.

Haulage of muck has historically been done by rail. Rail haulage is often the lowest option in terms of capital and operating costs and still reasonably flexible, but conveyor haulage is also common today, especially for longer tunnel drives. However, conveyor systems are not always suitable for TBM applications, because their minimum turning radius is ~200 m

(660 ft), while a custom-designed TBM may be able to bore a curve radius of ~100 m (330 ft).

Haulage by rubber-tire vehicles may be considered for larger tunnels with diameters of ~7 m (23 ft) or more. The invert is then typically backfilled with concrete or good-quality rock materials to provide a roadbed. Vehicle passing stations may be required at intervals to maintain a reasonable TBM utilization.

Another consideration for TBM mining applications is incline and decline. Declines have been limited to a maximum grade of ~15°, however, because a typical muck-removal system uses conveyor belts. Inclines can be quite steep since mucking can take place as safely contained gravity flow along the invert. In civil works, inclines have exceeded 45°.

Alignments are further determined by the end use of the excavation, geologic conditions, and other management or owner constraints such as property boundaries.

TBM VERSUS DRILL-AND-BLAST EXCAVATION METHODS

TBM excavation methods have both advantages and disadvantages compared with conventional drill-and-blast methods.

TBM Advantages

High Rate of Advance

Instantaneous excavation rates of 1–6 m/machine hour (3–20 ft/machine hour) are regularly achieved by TBMs with diameters of 3.5–5.5 m (11.5–18 ft). Overall advance rates per day, week, and month depend on the overall TBM system utilization and work schedules. Faster ore-body development shortens the time period of capital use and capital cost.

With TBM use, fewer headings are operated at one time, making supervision and planning easier and management effort more concentrated and focused.

Some types of geological environment can be more effectively supported if support is installed relatively quickly after excavation.

No Blast Vibration or Blast Fumes

TBMs enable work to be carried out continuously in the heading without delays for clearing gases from blasting from the heading or ventilation system. No workers need to be removed from other headings or work places to safety refuges for a blasting cycle. However, other drill-and-blast work in the mine can cause interruption in the TBM heading.

Reduced Ground Disturbance and Support

Mechanical excavation is well known to be less destructive to the structural integrity of material surrounding the excavated opening than is drill-and-blast excavation. Less ground support is required with significant reduction in material costs and installation time.

Roof bolts and steel ribs are the two predominant support systems used with TBM excavation in mines. Yieldable steel sets have commonly been employed in deep German coal mines for TBM drifts. Installation of ground support in larger tunnel cross sections can largely be automated.

Reduced Lining Costs

Where a completed excavation must be lined with shotcrete or concrete, the smooth uniform surface created by TBM excavation reduces costs of the lining material and raises the possibility that lining installation can be automated.

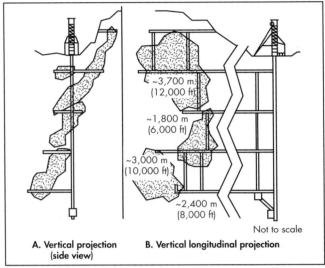

Source: Moulton 1997.

Figure 12.7-10 Mine opening drifts for development access and haulage

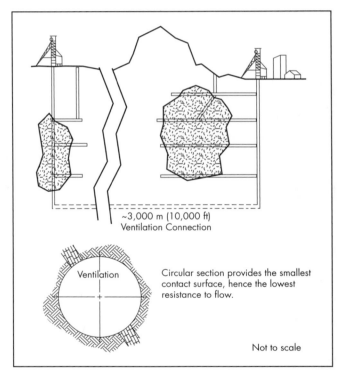

Source: Moulton 1997.

Figure 12.7-11 Mine openings for ventilation

Uniform Muck Size

Mechanically excavated materials essentially all pass through a 100-mm (4-in.) grizzly used for coarse screening.

Suitable for Automation

The uniform size of mechanically excavated materials makes material handling ideal for automation. Less wear and tear occurs on transport equipment such as rail cars, trucks, or conveyors in the tunnel.

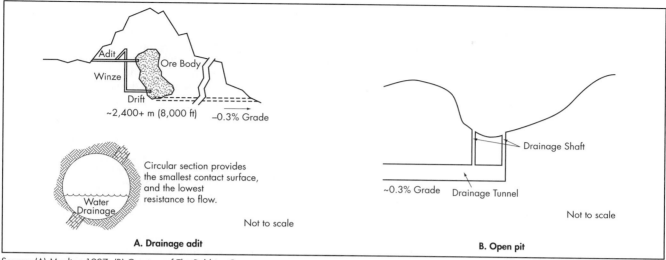

Source: (A) Moulton 1997; (B) Courtesy of The Robbins Company.
Figure 12.7-12 Mine openings for water drainage

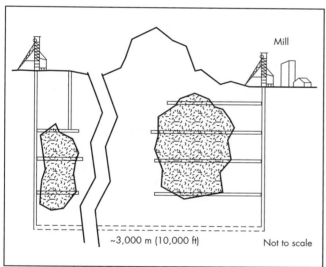

Source: Moulton 1997.
Figure 12.7-13 Multiple-use mine openings for ventilation, safety, drainage, haulage, and supplies

Vertical transport by means of loading pockets and hoist conveyances can be automated, minimizing maintenance costs. Ore that is encountered in a heading can be partially segregated and sent to the concentrator, eliminating primary crushing costs.

Reduced Labor Costs
A rapid rate of advance reduces the unit cost of excavation. In most western countries, both crew size and rate of advance are larger for TBM than for drill-and-blast operation. Consider the following typical average progress for a tunnel of small cross section, ~10 m² (108 ft²), each with two 10-hour work shifts per day:

- TBM operation:
  - Crew size: 4 to 7 (assume 6 for the following calculation)
  - Rate of advance: ~140 m (460 ft) per week
  - Productivity: 140 m/week ÷ 72 shifts/week = 1.94 m (6.4 ft) per shift
- Drill-and-blast operation:
  - Crew size: 2 to 3 (assume 3 for the following calculation)
  - Rate of advance: ~38 m (125 ft) per week
  - Productivity: 38 m/week ÷ 36 shifts/week = 1.06 m (3.5 ft) per shift

Good Safety Record
TBM excavations are considered relatively safe compared with drill-and-blast excavations because the latter incur greater risk to workers from falls or ground-related accidents, even in supported ground. Most TBM systems have protective canopies for workers in ground support installation work zones.

Both excavation methods involve some heavy lifting, which, if not properly performed, can lead to back injuries. A typical 432-mm (17-in.) single-disc-cutter assembly weighs about 136 kg (300 lb). Back-loading cutters normally required in mining applications are best changed with slings and hoists using proven cutter-handling systems. Back-loading cutters can normally be used for TBMs with diameters of ≥4 m (13 ft).

Lower Manning's Factor
TBM excavation in hard rock leaves a smooth circular profile with a roughness coefficient that is lower than that for a typical drill-and-blast operation. The cross section of an unlined TBM tunnel therefore needs to be only about two-thirds of the area of an unlined drill-and-blast tunnel to carry the same water flow.

Circular Cross Section
A circular profile typically gives the most stable cross section. TBM excavation requires substantially less ground support in fairly massive to massive hard-rock conditions.

Improved Ventilation Characteristics
The smoother rock surface of a TBM excavation means reduced friction and improved possibility of excavating longer drives from a single heading.

Less Concrete When Lining Is Required
A TBM tunnel has substantially less overbreak than does a blasted tunnel. Large savings in concrete and concrete placement may be realized.

No Muck Bays or Passing Bays
No muck bays or passing bays are needed for TBM tunnels with diameters of >3.5 m (11.5 ft) because all advance is centerline advance.

TBM Disadvantages

High Capital Cost
TBM costs vary widely with size, supply, and power requirements. The cost of a typical open-type TBM excluding roof drills, ring erectors, and other ground support or probing equipment is about US$1.7 million per meter of diameter (2009 price levels). A simple backup system in a heading with a diameter of 3–4.8 m (10–16 ft) varies from US$1.2 to $2 million, depending on facilities, storage, muck-transport systems, and so on. A system with a diameter of 3.6 m (12 ft), with spares and cutters, thus costs about US$8 million. A used, fully reconditioned, technically comparable system costs about 65%–75% of this amount.

Repair and maintenance work are necessary during the life of a machine. The life of a TBM system has been suggested at times to be ~15 km (50,000 ft) of tunnel excavation, although this varies widely depending on technical obsolescence, operating and maintenance practices, and application. Conditions in the marketplace external to the mine can have a large impact on equipment life and salvage value. A number of TBMs have bored more than 25 km (80,000 ft) of tunnel. Generally at least 2.5 km (~8,000 ft) of tunnel are necessary to justify the use of a TBM system; this includes equipment salvage value.

Large Value of Spares and Cutter Inventory
The recommended inventory of major spare parts, including main bearing, ring gear, drive unit (motor, clutch, gear reducer, and pinion), propel cylinder, and gripper cylinder are in the range US$600,000 to $800,000 (2009 price levels). Inventory of minor spare parts adds another US$60,000 to $100,000 of capital cost. US$300,000 can be spent for initial cutter mounting, cutter spares, and cutter repair tools and fixtures.

Long Lead Time for Delivery
Design and manufacturing time for a new TBM with a diameter of 3.6–5.5 m (12–18 ft) is typically a minimum of 10 months. The time for a smaller TBM, with a diameter of ~2.6–3.0 m (8–10 ft), may be as low as 8 months. Delivery time ex-works varies with actual supply, clarity of product definition, and subsuppliers' workloads.

Reconditioning and modification times for existing equipment is ~65%–95% of these values, depending on the particular equipment, its location, and the scope of work required.

Large Components That Are Difficult to Transport
To reach the high thrust and torque levels required for TBM excavation, components must be large and heavy, and are therefore difficult to move through existing mine shafts or mine drifts. TBMs are also often difficult to remove from completed tunnels.

Boring of TBM drifts from shafts can be challenging. Transport from the surface and assembly underground, as well as removal from the mine, require careful planning and preparation. At times transport and assembly must take place without interrupting full-time mining. Plans and methods for slinging and lowering heavy components down the shaft must be devised early during machine design. Machine components and support systems must be designed to be small and light enough for accommodation in the shaft and by the hoist. The existing hoist configuration, controls, or brakes may also require modification, and special transport or handling devices may be required to lower large components down the shaft or move them along the level.

Limited Range of Application in Varying Ground Conditions
TBM design is determined by the requirements of the particular application—the more uniform or consistent the geologic environment, the easier and more reliable the design.

Competent moderate hard rock, requiring moderate ground support and with no major water inflows, is ideal. However, mines are normally associated with geologic nonconformities that result in mineral deposits, and it is to be expected that all or part of an underground excavation will require some type of rock support. Machine design can, within limits, accommodate hard rock to caving or squeezing ground.

For mining applications, a high degree of flexibility is generally desirable. Today's machines can operate in varying ground conditions because of their variable-frequency drive motors for controlling cutterhead rpm and torque. They can be equipped with effective systems for probing, pre-excavation grouting, shotcreting near the cutterhead, rock bolting, ring beam installation, and application of mesh and lagging. In addition, they can be furnished with the McNally system, a longitudinal roof-support system for operation in bad ground conditions (McNally 2010).

Consumable Costs Can Be High in Hard, Abrasive Rock
Cutter costs can be a very significant part of the overall TBM costs. In soft to medium-hard sedimentary rocks such as are associated with limestone and other nonabrasive rock types, cutter costs are typically as low as US$1.00/m^3 ($0.75/yd^3) or less. However, in hard, massive quartzites, they can reach US$20/m^3 ($15/yd^3) or more due to the high abrasiveness of the rock combined with low instantaneous penetration.

When cutter costs are high, machine utilization drops because of downtime for cutter changes and for cutterhead maintenance and repairs.

Several laboratories can test rock samples needed for TBM application studies. Most TBM manufacturers and some TBM consultants can provide estimates on instantaneous penetration and cutter consumption based on these lab test data, anticipated rock-mass characteristics, and proposed machine and cutter specifications.

Limited to Circular Cross Section
When a heading is intended for haulage, the circular cross section produced by a TBM may limit the choice of haulage method and equipment. This is particularly true when rubber-tired haulage trucks or load-haul-dump units are to be used. The diameter of the excavation must be increased and the invert may require a concrete cast floor as a roadbed.

In some rock burst conditions, a TBM-produced circular cross section, although usually advantageous for ground support, can also be a detriment. In highly stressed and brittle rocks, the surface stress of the circular opening can be quite high compared to that in rock that has been deeply fractured and stress-relieved by use of drill-and-blast methods.

Limited Turning Radius
A typical TBM has a minimum turning radius of ~200 m (660 ft), although custom-designed TBMs may be able to bore a radius of ~100 m (330 ft). Backup equipment, conveyors belts, mining cars, mucking solutions, and so on must all be able to negotiate a curve so created. In contrast, a drill-and-blast operation has basically no limitations regarding curve radius.

Substantial Preparation at Work Site
Site preparation for most mine projects that involve adits is not unusually difficult. However, for a TBM application, there must be ample room at the portal site, cranes for lifting large pieces, and adequate supply of power. Other requirements are basically similar to those for drill-and-blast headings.

When an underground work station is planned, adequate hoist capacity and ample length for assembling the trailing gear are needed. A typical overall length for a TBM with a diameter of 3–5 m (10–16 ft) can be 100–140 m (320–450 ft). The underground assembly chamber for a 3.66-m (12-ft) TBM is usually sized as follows (Figure 12.7-14):

- Width and height: ≥1.5 times the TBM diameter, depending on the type of hoist
- Length: Length of the TBM, ~20 m (65 ft)

For safety and cost reasons, it is desirable that the assembly chamber be constructed in good ground, close to the launch point for the drive.

Requires Substantial Crew Training and High Mechanical and Electrical Skill Levels
A TBM system is essentially a crude moving processing plant or factory. Material is removed and sized in a cutting process from its in-situ state. The process from that point on involves several electrical, mechanical, hydraulic, lubrication, compressed-air, cooling-water, wastewater, ventilation, material-handling, and storage systems along with monitoring and control of these systems. These systems must be maintained in a hostile environment of water, dust, vibration, and varying temperature. Some of these systems may not be in common use with other machines or systems in the mine operation.

Mine operating personnel must be trained in the use and maintenance of these systems and how they relate to the overall TBM production process. Training is largely accomplished in the TBM manufacturer's shop during floor testing and machine disassembly, and at the work site during final assembly and full-load testing. TBM operators receive training by visiting or temporarily working at other sites and by working with the TBM supplier's field service personnel.

Instead of training mining crews, a better solution can be to use contractors who specialize in TBM work. A TBM contractor must be carefully selected and must understand mine priorities and operations requirements.

Requires Large Power Supply
Most TBM systems today use total power in the range 1,000–2,600 kVA or more for TBMs with diameters of 3–5 m (10–16 ft). Total power includes power for cutterhead, hydraulics, ventilation, conveyor, lighting, pumping, cooling, and other systems.

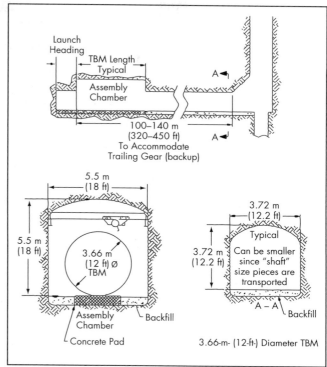

Source: Moulton 1997.

Figure 12.7-14 Example of an underground TBM assembly chamber

The power required to operate a TBM system may not have been planned in the mine's initial design stage, as many mines are more than 25 years old. Even today, mine infrastructure planning does not usually include provision for the use of mechanical excavation equipment in major headings. The large power requirement usually necessitates addition of high-voltage distribution cables in the shaft and addition of underground substations.

Vary Excavation Diameter
Normally, it is not practical to change tunnel diameter during a drive, because of costly downtime. The TBM design itself may allow a change of ~1 m (3.3 ft) or more from a minimum established diameter. Possibilities with regard to diameter increases should be based on the original TBM design criteria and the upcoming project application, including desired diameter, length of drive, and rock-mass and rock-strength characteristics.

DESIGN AND OPERATING PROBLEMS AND SOLUTIONS
Historically, design and operating problems have arisen with TBM applications in the mining industry. Samples of past problems and their solutions are given in Table 12.7-1.

Since 2000, several important improvements have been implemented in TBMs and auxiliary equipment that decrease downtime and increase TBM utilization:

- Water-cooled motors and hydraulics
- Larger-diameter cutters
- Higher-capacity cutter bearings and cutter ring with improved abrasive resistance characteristics

Table 12.7-1 TBM design and operating problems and their solutions

Problem	Solutions
Ground fall on personnel between cutterhead and face while changing cutters	• Developed back-loading cutters that can be changed from inside the cutterhead • Replaced bucket lips from inside the cutterhead
Undercutting of an unstable face with dome cutterheads, causing collapse of the tunnel crown	• Developed flat-face cutterheads with smooth surfaces
Front-mounted cutter housings rip blocks of material from the face, blocking cutterhead rotation	• Developed flat-face cutterheads with smooth surfaces
Large distance from initial cutting of face to provision of adequate ground support, causing fallout and caving. Massive ground support effort required	• Developed flat-face cutterheads with short standoff distance and narrow rim bar to reduce the distance to the first adequate support of the excavation
Methane fire	• Implemented early detection of gas and shutdown of equipment • Increased ventilation • Implemented better sensing of hot materials, particularly cutters • Interlocked water sprays with cutterhead rotation
Probe drilling in advance of boring limits advance	• Implemented state-of-the-art high-capacity hydraulic drills that enable probe drilling during cutter changes and/or TBM system maintenance
Additional machine waste heat in deep mines detrimental to work-crew efficiency	• Upgraded to water-cooled electric motors, gearboxes, and cutterheads
Low utilization in hard abrasive rock in deep mines	• This is not a good application for TBMs. More research and development is needed, particularly for cutters
Entrapment by squeezing ground	• Implemented adjustable shield designs • Avoided stoppages or delays for weekends or holidays in zones of squeezing
Inadequate support for gripper pads	• Increased gripper shoe area to reduce ground load • Implemented timber cribbing or shotcreting in gripper-pad areas as necessary
Running ground cave-in	• Used higher torque at low speed on cutterhead with ground treatment • Applied shotcrete early to stabilize ground
High repair and maintenance of open-type cutter mounting	• Added breasting plates to smooth rock-contact surface • Designed and built new cutterhead
Equipment not suitable for geology	• Modified TBM to meet conditions
Small size of TBM limits workspace for ground support installation	• Installed shield for temporary support and placed permanent support behind TBM
Cannot hold curve-alignment radius	• Implemented modern laser guidance systems • Designed a short TBM • Reduced TBM stroke • Implemented steering by means of "crabbing"
Large blocks of rock have damaged TBM muck chute and conveyor system	• Installed grill bars • Decreased size of bucket openings • Replaced bucket lips from behind • Lowered speed and increased torque
Plugging of muck-flow passages in the cutterhead	• Removed grill bars • Increased size of bucket openings • Smoothed the contact surface • Used foam injection to lubricate muck
Ground caving above cutterhead	• Stabilized the ground by grouting and filled the void with quick-setting grout • Installed spiling or combinations of the above • Installed a McNally roof-support system

Source: Adapted from Moulton 1997.

- High-efficiency variable-frequency electric drives
- Improved hydraulic probe-drill schemes and pre-excavation grouting solutions
- Faster-setting grout

OPERATIONAL AND MANAGEMENT PROBLEMS AND SOLUTIONS

The following typical challenges can arise in a TBM mining application:

- Failure on the part of mine managers, due to categorization of the TBM as a "mine research project," to commit fully to the decision to use a TBM
- Failure on the part of direct supervisors and miners to commit fully to the success of the TBM system
- Failure to obtain acceptance of new technology due to unrealistic expectations
- Failure on the part of machine suppliers to fully understand the unique mine operating conditions and priorities related to design challenges, operating costs, performance, and the high skill levels required of personnel
- Lack of jobsite preparation and coordination with other mine activities
- Lack of training of operating and maintenance personnel

Several TBM projects have been considered failures when original schedules, costs, and full desired end benefits were not achieved. However, most had acceptable results when compared to realistic alternates. The solution to these

Table 12.7-2 Summary of TBM projects, listed by original machine manufacturer

Mine	Year	Model	Bore Diameter	Length, m (ft)	Geology
TBM Manufacturer: Robbins*					
Thetford mine asbestos, Quebec, Canada	1957	103-105	2.74 m (9 ft 0 in.)	30 (100)	Greenstone
Step Rock iron mine, Quebec, Canada	1957	103-105	2.74 m (9 ft 0 in.)	300 (1,000)	Limonite
Homer-Wauseca mine, Iron River, MI, USA	1963	71A-109	2.13 m (7 ft 0 in.)	355 (1,163)	Limestone & iron
White Pine mine, White Pine, MI, USA	1968	181-122	5.48 m (18 ft 0 in.)	2,590 (8,500)	Shale
Milchbuck mine vent, Switzerland	1969	104-120-1	3.22 m (9 ft 11 in.)	91 (300)	Greenstone
Chingola, Zambia	1970	125-135	3.65 m (12 ft 0 in.)	3,200 (10,500)	Granite
Minster Stein Coal, Germany	1971	163-136	5.4/5.1 m (16 ft 9 in. / 16 ft 0 in.)	6,797 (22,300)	Shale & sandstone
Schelklingen mine Addition, Germany	1971	111-117	3.5 m (11 ft 6 in.)	1,240 (4,068)	Coal measures
Libanon gold mine, Gold Fields, South Africa	1974	114-163	4.5 m (15 ft 2 in.)	Not applicable	Quartzite
Mt. Lyell mine, Australia	1975	133-146-1	3.93 m (12 ft 10 in.)	3,048 (10,000)	Andesite & schist
Monopol Coal, Kamen/Westfalen, Germany	1976	163-136	5.4/5.1 m (16 ft 9 in. / 16 ft 0 in.)	12,802 (42,000)	Schist & sandstone
Blyvoor mine, South Africa	1977	61-176	1.84 m (6 ft 6 in.)	305 (1,000 multiples)	Quartzite
East Driefontain mine, South Africa	1977	61-177	1.84 m (6 ft 6 in.)	305 (1,000 multiples)	Quartzite
Westfalen Coal, Germany	1979	202-201	6.1 m (20 ft 0 in.)	12,701 (41,670)	Sandstone, shale
Blumenthal mine, Germany	1979	214-202	6.5 m (21 ft 4 in.)	10,600 (34,780)	Sandstone, shale
Selby mine road, York, UK (decline)	1981	193-214	5.8 m (19 ft 0 in.)	13,600 (44,620)	Sandstone, mudstone, coal seams
Sotiel mine, Huelva, Spain (decline)	1981	122-126-1	3.81 m (12 ft 6 in.)	1,774 (5,820)	Shale & phyrites
Stillwater mine, Nye, MT, USA	1987	146-193-1	4.1 m (13 ft 6 in.)	9,673 (31,737)	Gabbro and Norites
San Manuel copper mine, AZ, USA	1995	156-275	4.62 m (15 ft 2 in.)	9,754 (32,000)	Granodiorite, quartzmonzonite dykes
Stillwater mine, Nye, MT, USA (East Boulder Project)	1998	156-275	4.62 m (15 ft 2 in.)	5,650 (18,550)	Gabbro, anorthosite, norite
Da Tong coal mine, Shanxi, China	2007	154-273-1	4.90 m (16 ft 1 in.)	2,800 (9,186)	Sandstone, siltstone
OTM mine, Papua, New Guinea (dewatering tunnel)	2008	1410-251-3	5.56 m (18 ft 3 in.)	5,444 (17,861)	Metamorphic, sandstone, siltstone
TBM Manufacturer: Wirth[†]					
Sophia Jacoba, Aachen, Germany	1972	TB V-540	5.30 m (17 ft 5 in.)	137 (450)	Sandstone
Consolidation, Ruhr, Germany	1972	TBII-300HT TBE 300/530	3.00 m (9 ft 10 in.) 5.30 m (17 ft 5 in.)	1,600 (5,250) 1,600 (5,250)	Sandstone
Anglo American Free State Geduld, Republic of South Africa	1976/1979 1982/1984	TBS II-340H	3.40 m (11 ft 2 in.)	1,150 (3,772) 2,280 (7,480)	Quartzite/granite Quartzite/sericite
Lohberg I, Osterfeld Dinslaken, Germany	1982 1983/1985 1986	TBS V-650	6.50 m (21 ft 4 in.)	5,455 (17,897) in total	Coal measures
Lohberg II, Osterfeld Dinslaken, Germany	1988/1990		6.50 m (21 ft 4 in.)	3,770 (12,369)	Coal measures
Lohberg III (used the same TBM as Loberg II at a different diameter), Germany	1992	TBS V-650	6.80 m (22 ft 4 in.)	2,630 (8,629)	Coal measures
TBM Manufacturer: Società Esecuzione Lavori Idraulici SpA (SELI)[‡]					
Los Broncos copper mine, Chile	2009–2010	Universal Compact Double Shield	4.5 m (14 ft 9 in.)	8,500 (27,887)	Quartzmonzonite, andesite, ryodacitic porphyry, breccias

(continues)

Table 12.7-2 Summary of TBM projects, listed by original machine manufacturer (continued)

Mine	Year	Model	Bore Diameter	Length, m (ft)	Geology
TBM Manufacturer: Jarva*					
Adirondac mine, 27% incline, Mineville, NY, USA	1967	MK 11-4	3.05 m (10 ft 0 in.)	234 (768)	Magnetite, granitic gneiss
Mather "B" mine, MI, USA	1967	MK 14-1	4.03 m (13 ft 3 in.)	325 (1,065)	Graywacke, shale
Hecla Mining Co. development drift, Wallace, ID, USA	1968	MK 8-3	2.74 m (9 ft 0 in.)	152 (500)	Quartzite
Oak Park mine, incline, Cadiz, OH, USA	1970	MK 14-1	4.27 m (14 ft 0 in.)	549 (1,802)	Sandstone shale
R&P Emilie No. 4 Shelochta, Indiana, PA, USA	1976	MK 12-13	4.27 m (14 ft 0 in.)	227 (746)	Shale, coal, sandstone, limestone
Bethlehem Steel, Grace mine, Morgantown, PA, USA	1976	MK 8-2	2.49 m (8 ft 2 in.)	183 (600)	Chlorite magnetite
R&P, Urling No. 3, Shelochta, PA, USA	1976	MK 12-13	9.27 m (14 ft 0 in.)	745 (2,443)	Shale, sandstone, limestone
Mine development drift, Pachuca, Mexico, USA	1978	MK 11-4	3.66 m (12 ft 0 in.)	1,524 (5,000)	Shale
Westmoreland Coal 30% Decline, Eccles, WV, USA	1979	MK 8-2	3.05 m 10 ft 0 in.)	319 (1,045)	Shale sandstone
R&P coal conveyor haulage tunnel, Indiana, PA, USA	1979	MK 12-13	8.27 m 14 ft 6 in.)	427 (1,400)	Shale, sandstone
United Pocahontas Coal, Thurman, WV, USA	1980	MK 8	3.05 m (10 ft 0 in.)	945 (3,070)	Coal, sandstone
R&P coal trans tunnel, Shelochta, PA, USA	1980	MK 12-13	4.27 m (14 ft 0 in.)	335 (1,100)	Coal, sandstone
Kiena mine, North Bay, Ontario	1986	MK 6-19	1.98 m (6 in.)	1,375 (4,511)	Basalt, peridiotite diorite
TBM Manufacturer: Demag§					
Niederrhein, Germany	1973–1979	TVM 54	6.0 m (19 ft 8 in.)	21,085 (69,176) in total	Schist, sandstone, shale
Victoria, Germany	1977	TVM 54	6.0 m (19 ft 8 in.)	1,700 (5,577)	Schist, sandstone
Petrosani, Germany	1979	TVM 45H	4.8 m (15 ft 9 in.)	13,000 (42650)	Sandstone, sandy shale, schist
Saarbergwerke, Germany	1979	TVM 55H	6.0 m (19 ft 8 in.)	16,000 (52,493)	Schist, sandstone, sandy shale
Franz Haniel, Germany	1980	TVM 54-58	6.0 m (19 ft 8 in.)	8,000 (26,247)	Schist, sandstone, shale
Hans Aden, Germany	1981	TVM 65	6.5 m (21 ft 4 in.)	19,000 (62,336)	Schist, sandstone
TBM Manufacturer: Calweld†					
Climax mine, Climax, CO, USA	1969	TBM 40	3.96 m (13 ft 0 in.)	607 (2,000)	Silver plume, granitic gneiss
TBM Manufacturer: Ingersoll-Rand†					
Magma Copper, Superior, USA	1974	008	4.27 m (14 ft 0 in.)	Not applicable	Not applicable
TBM Manufacturer: Boretec Equipment*					
Falconbridge Fraser mine, Sudbury, Ontario, Canada	1988	Cub 3000	2.29 m (7 ft 6 in.)	300 (984)	Ultra-hard granite
TBM Manufacturer: Construction & Tunneling Services (CTS)**					
Stillwater mine, Nye, MT, USA (East Boulder Project)	1997	460-12	4.57 m (15 ft 0 in.)	5,650 (18,550)	Igneous & sedimentary
TBM Manufacturer: Lovat Tunnel Equipment†					
Donkin-Morien, Cape Breton, Nova Scotia, Canada	1983	RM300RL	7.6 m (24 ft 11 in.)	2x4,000 (2x13,123) Maximum declines at 20%	Sandstone with coal seams
Port Headland, North Western Australia	1999	RME 200SE	5.08 m (16 ft 8 in.)	1,037 (3,402)	Soft soil, limestone

*Data courtesy of The Robbins Company.
†Data from Moulton 1997.
‡Data courtesy of SELI.
§Data courtesy of Mannesmann Demag.
**Data from Moulton 1997 and Alexander 1999.

problems is to keep expectations realistic, fully identify mine requirements, and possess adequate design experience.

A summary of mine projects in the Western world is presented in Table 12.7-2 by date and original TBM manufacturer. The summary may not be complete, but the large number of projects listed should indicate that there is enough experience in the industry to know what to do and what not to do when applying TBMs to mining applications.

Many market studies, in attempting to identify the worldwide need for mines, have estimated the total annual drive lengths of horizontal openings. The total number is very large. Of this total, much is not really suitable for TBM excavation. However, when drive lengths are sufficiently long and when mine owners, operators, equipment suppliers, and contractors focus on a common result that benefits each of them, they can commit to their mutual success and achieve it. A good example of such commitment is the case history that follows.

TBM MINING CASE STUDY

The Magma Copper Company's San Manuel mine in Arizona and the Stillwater Mining Company's platinum mine in Montana, both in the United States, are examples of successful TBM use in mining. Many comprehensive reports concerning these projects are in the public domain. TBM use in the San Manuel mine is discussed here in detail.

Mine Location and Layout

The San Manuel mine is located about 72 km (45 mi) north of Tucson, Arizona. In order to secure future production, it became necessary to develop the Lower Kalamazoo ore body 1,050 m (3,450 ft) below surface for a modern block-caving operation. The base of the mass to be caved would be accessed by drifts on two levels from the existing No. 5 shaft.

Based on a feasibility study, it was decided to use a TBM rather than traditional drill-and-blast methods because of the very tight schedule for starting production in the new ore body. Because of a lack of qualified labor and the short construction time, it was determined that an experienced TBM contractor was needed.

The development layout was designed specifically for TBM use (Figure 12.7-15). The TBM tunnel route consisted of 9,826 m (32,235 ft) of tunnel circling the ore body on three levels: (1) the 3,440 grizzly level, (2) the 3,570 haulage level, and (3) the 3,600 conveyor-gathering drift level.

The grizzly level consisted of 3,963 m (13,000 ft) of TBM tunnel with slight grade changes to provide two permanent dewatering sumps. The level has five curves, with radii in the range 152–107 m (500–350 ft) and a total curve length of 1,031 m (3,383 ft).

The haulage and conveyor-gathering drift levels are accessed from the grizzly level by means of two parallel decline tunnels with grades of ≤5.7%, which accounts for the remaining 5,863 m (19,235 ft) of the TBM tunnel route. These decline tunnels have two curves, with radii in the same range as for the grizzly level and a total curve length of 1,255 m (4,119 ft).

Mine Geology

The geology of the San Manuel mine is quite complex, with rock structure ranging from very weak to reasonably strong. The area geology is described briefly as follows (Atlas Copco Robbins 1996).

The ore bodies resulted from porphyritic intrusion of granodiorite at the end of the Cretaceous period into pre-

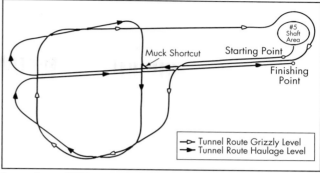

Source: Janzon 1995.

Figure 12.7-15 TBM tunneling scheme in the San Manuel mine

Cambrian quartz monzonite. Mineralization along the contact zone was shaped originally like a hollow ellipse, which later was split and displaced by faults and intersected by numerous dikes.

The TBM's excavation route included the stable quartz monzonite of ~150–180 MPa (22,000–26,000 psi) unconfined compressive strength, two mineralized haloes, and a mineralized core. The bore path also crossed the San Manuel fault six times and the Virgin fault five times. The San Manuel fault was flat, dipping in an ~1-m- (3-ft-) wide clay zone that followed the bore some 30 m (100 ft) at a time, but it influenced the drive for a total of about 190 m (600 ft). The Virgin fault dipped steeply but a series of related minor faults produced poor rock conditions for about 500 m (1,650 ft) of the tunnel.

In addition, where dacitic, andesitic, and rhyolitic dikes contacted the granodiorite and quartz monzonite, weak zones 0.2–0.6 m (0.5–2 ft) wide affected the bore for approximately 180 m (600 ft). Along most of the TBM's path, hydrothermal metamorphosis had weakened the rock further by veining, fracturing, and jointing. Unsupported wall stability ranged from <30 minutes to months, and the critical zone with the shortest stand-up time was anticipated to be ~170 m (560 ft) in length.

Although none of the rock masses could be considered an "excellent" TBM environment, comprehensive studies and the 1992 geotechnical report (Janzon 1995) concluded that the rock conditions would permit mechanical excavation and that stability, even in the weakest sections, would be adequate for TBM passage and ground support installation. The report actually favored the TBM approach because the TBM machine's low-level vibrations would destabilize hydrothermally altered rock far less than would shock waves from blasting. The report concluded that TBM tunneling would be substantially faster than drill-and-blast excavation.

Mine Equipment Selection

TBM equipment design for the project was based on three parameters: geology, logistics, and tunnel geometry. It was decided to use a hard-rock main-beam TBM.

The TBM was ordered in 1993, before the TBM contractor was selected, in order to meet the construction schedule. Mine management conducted detail studies and held discussions with outside consultants, machine manufacturers, and all potential contractors, then placed the TBM order with The Robbins Company.

The machine (shown assembled at the factory in Figure 12.7-2) is relatively short to allow boring of tight

Table 12.7-3 Specifications for Robbins TBM

Parameter	Value
Boring diameter	4.62 m (15 ft 2 in.)
Cutterhead	
Installed power	1,260 kW (1,680 hp)
Two speed	12.2/4.1 rpm
Maximum recommended thrust	7,350 kN (1,650,000 lb)
Cutters	
Number	33 back-loading/front-loading type
Diameter	432 mm (17 in.)
Maximum recommended load	222.7 kN/cutter (50,000 lb/cutter)
Boring stroke	1.5 m (5 ft)
Minimum turning radius	105 m (345 ft)
Electrical system	
Primary power	12.870 V/60 Hz
Secondary power	600 V for main motors
	120 V for control circuit/lighting
	480 V for backup equipment
Transformer power	2 × 1300 kVA
Weight	225 t (248 st)

Source: Data from Van Der Pas and Allum 1995.

curves. It has back-loading cutters, roof drill fixtures, a ring beam erector for installing W6×20 ring beams, invert thrust system, finger shield, soundproof operator station on the backup, including a closed-circuit TV for monitoring selected areas along the TBM and backup. Specifications for the TBM are listed in Table 12.7-3.

The TBM for the San Manuel mine was custom designed to be lowered down the mine's access shaft piece by piece and assembled underground. The cutterhead and main beam were built in two pieces to meet the space and weight limitations of the access shaft and its 24-t (26.5-st) hoist (Figure 12.7-16). Components were limited in physical dimension to 2 × 4 × 12 m (6.75 × 13 × 39 ft). More than 65 components were lowered down the shaft. Large components were suspended beneath the service cage; other parts fit inside the cage.

A further requirement was that all transport and assembly operations had to take place without interrupting full-time mining. Assembly of the machine was essentially complete in 114 shift hours, which was proof of excellent planning and a dedicated work crew.

The TBM contract was awarded to Frontier Kemper Constructors, Inc./Daniel Haniel GmbH (FK/DH). The contractor designed and manufactured the trailing gear and support equipment to fit the same logistical criteria and requirements as specified for the TBM. The 143-m- (470-ft-) long backup equipment consisted of 16 frame cars to house all electrical, hydraulic, ventilation, cooling, and muck-handling equipment. Because of the many short-radius curves, the contractor selected rail haulage using Mühlhäuser 10-m^3 (13-yd^3) muck cars and Brookville 27.2-t (30-st) locomotives. A California switch was selected and moved; permanent switches were installed, two on the grizzly level and one on the 3570 haulage level.

Mining Setbacks and TBM Modifications

TBM progress in the initial 1,830-m (6,000-ft) tunnel was not as expected. A number of problems plagued the operation. For example, clay plugged the cutterhead; a large cavity (h × w × l = 9 × 6 × 5 m [30 × 20 × 16 ft]) developed above the tunnel

Source: (A) Atlas Copco Robbins 1996; (B) Courtesy of Gomez International.

Figure 12.7-16 Two TBM components being lowered down the mine's No. 5 access shaft: (A) part of main beam; (B) split cutterhead

profile; loose material fell through the finger shield; steering problems developed because of failure to maintain proper alignment in the curves; and some machine problems developed. In addition, stand-up times in highly fractured areas were not as predicted, because rock often turned to rubble immediately or within minutes in the crown or along the tunnel sidewalls.

To achieve the planned rate of advance, the TBM needed to be modified quickly. An evaluation team consisting of Magma Copper, the TBM manufacturer, the contractor, and two outside consultants recommended the following modifications:

- Replace two of the gauge cutters with two additional muck buckets and close the peripheral side buckets off to reduce ingestion of rock behind the gauge cutters.
- Extend the roof support forward by a canopy to limit cutterhead exposure.
- Add fingers to the side supports to increase the surface area of the support.
- Improve starting and breakout torque by reducing cutterhead rpm from 12.2 to 9.3.
- Replace the two 1,300-kVA transformers with two 1,600-kVA transformers.

Some of these modifications are shown in Figure 12.7-17.

Early on, because of the highly variable ground conditions, the use of rock bolting had been abandoned, roof drill fixtures had been removed, and the operation had moved to exclusive use of ring beams to support the rock. Doing so not only saved labor but also provided more room in the congested work area immediately behind the cutterhead support.

The modifications greatly improved TBM performance in all areas. Cutterhead starting problems and ingestion of muck from above the cutterhead were eliminated, fallout of material from the sidewall was reduced to acceptable levels, and bucket plugging occurred only in areas of high clay content. A curve with a radius of 109 m (350 ft) was accomplished by means of "crab steering" without plowing the sidewalls. A beneficial side-effect of lowering the cutterhead speed was more efficient transfer of muck from the cutterhead buckets to the muck chute and machine conveyor belt. The results of the

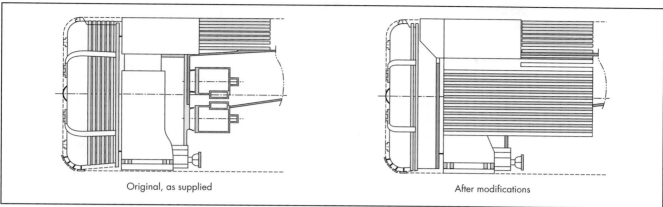

Source: Van Der Pas and Allum 1995.
Figure 12.7-17 Modifications to the main-beam TBM to improve ground control and steering

Table 12.7-4 Measures of advance before and after TBM modifications

Measure of Advance	Length of Advance, m (ft)	
	Before Modifications	After Modifications
Average daily advance	6.46 (21.19)	22.6 (74.15)
Best shift	19.2 (63)	21.6 (71)
Best day	37.5 (123)	44.5 (146)
Best week	141.7 (465)	263 (863)
Best month	333.5 (1,094)	831.2 (2,727)
Best quarter	882.4 (2,895)	2,142.2 (7,028)

Source: Van Der Pas and Allum 1995.

modifications, measured by comparing the lengths of advance before and after modifications, are shown in Table 12.7-4.

The Magma Copper Company concluded the following (Van Der Pas and Allum 1995):

> Advance rates achieved after completion of the modifications have proven that the decision to utilize a TBM for the initial development of the Lower Kalamazoo was the correct one. Even though the TBM was designed to the best interpretation of conditions likely to be encountered, as so often happens in underground construction work, the conditions encountered did not exactly match the assumptions that were the initial basis for the project. The correct analysis of the problems by TBM operations personnel, the TBM Evaluation Team, The Robbins Company and the consultants resulted in the successful modifications. Careful planning, scheduling and teamwork by operations personnel allowed the modifications to be completed on schedule and as planned....
>
> Many individuals and individual organizations spent a great deal of time and effort to determine what was required to make the project a success. It was the ability of those groups of people to focus on solutions to the problems instead of who to blame that built the foundation to success. The formation of a joint management team enabled both Magma and FK/DH to function side by side while solving the daily coordination problems as created by both organizations. Commitment to a successful project by the owner, the contractor, and the equipment manufacturer created a true working team. That team collectively was able to generate the results needed to prove that TBM technology in a deep underground copper mine can be a success.

CONCLUSIONS AND RECOMMENDATIONS

Excavation of strategic mine openings by TBM methods is safe, fast, and economically feasible when

- Geological conditions are suitable for the intended mechanical excavation method;
- Length and alignment of the heading is amenable to the equipment selected;
- Operating and maintenance personnel are properly trained and committed to the success of the method;
- Mine management is committed to the priority and success of the work;
- Proper equipment, methods, and support supplies are selected; and
- The scope of work justifies the capital expenditure.

TBM manufacturers are prepared to make the following modifications to standard civil-engineering TBMs in order to expand successful use in mines:

- Lower machine capacity to enable consistent daily advances of 10 to 15 m (33 to 50 ft). This results in smaller, lighter TBM components and more room around the cutterhead for ground support.
- Tighten the turning radius to <100 m (330 ft).
- Create a total system that enables continuous concrete invert installation and access by standard mine vehicles, and muck removal by conveyor.
- Facilitate modern ground support techniques that allow loose rock to be held in place and rock support to be installed while boring.

With a good working relationship between the mine and suppliers, as took place at the San Manuel mine, beneficial use of TBMs in mines can be achieved.

ACKNOWLEDGMENTS

This chapter is based largely on work by Bruce G. Moulton, a retired mining engineer and former president of Construction and Tunneling Services, Inc., who has kindly allowed use of his paper, "Tunnel Boring Machine Applications In Mining" (Moulton 1997). The information herein reflects technical information, project summaries, and price levels as of October 2009.

REFERENCES

Alexander, C. 1999. Development of Stillwater Mining Company's East Boulder Project using tunnel boring technology. Presented at the 1999 SME Annual Meeting, Denver, CO, March 1–3.

Askilsrud, O.G., and Moulton, B.G. 2000. Tunnel Boring Machines for Hard Rock and Mixed Ground Applications. Unpublished.

Atlas Copco Robbins. 1996. TBM Excavates Drifts for Early Ore Production. Project Report 4.96.

Janzon, H. 1995. Atlas Copco Robbins: TBM for Development in Block Caving Mine, San Manuel, Arizona.

McNally, M.P. 2010. The McNally System—Increasing production of main beam TBMs in difficult rock through reliable roof support. Presented at the International Tunnelling and Underground Space Association (ITA-AITES) World Tunnel Congress General Assembly, Vancouver, May 14–20.

Moulton, B. 1997. Tunnel boring machine applications in mining. Presented as part of a mechanical mining short course at the Colorado School of Mines, Golden, CO, March 19–21.

Van Der Pas, E., and Allum, R. 1995. TBM technology in a deep underground copper mine. In *Proceedings of the Rapid Excavation and Tunneling Conference*, Vol. 1. New York: SME-AIME.

CHAPTER 12.8

Underground Ore Movement

Martyn Bloss, Paul Harvey, Don Grant, and Cameron Routley

INTRODUCTION

This chapter provides an overview of the common ore-handling system components available to underground mining operations, including loaders, trucks, orepasses, crushers, conveyors, rail haulage, hydraulic haulage, and other systems. Shafts are described in more detail in Chapter 12.4.

Context Within the Underground Mine

The ore-handling strategy has considerable impact on the value of any underground mining operation. The definition of *ore handling* is any process that moves or transforms ore from one state in the overall value chain to the next more valuable state. In this context, every major process in the mine can be considered ore handling, including those that do not physically move ore. These include resource drilling, mine planning, mine development, production drilling, production blasting, and backfilling. As such, although this chapter mainly considers the physical movement of ore, it is only one component of the total ore movement. It is, therefore, important to consider a more holistic and integrated view when developing the physical ore-handling strategy, because this strategy is most likely to be influenced by (and complementary to) all value-adding processes within the mine, whether physical or not.

Poor ore-handling decisions can lead to underperformance of the mine, high production variability, declining performance over time, and unexpected costs. In addition, opportunities could be lost if little consideration has been given to future changes in mining strategy or expansion options. Value of the business could also be reduced if the ore-handling system is not flexible enough to respond to external influences such as changes in metal prices, technology, and license-to-operate conditions. A successful ore-handling system will demonstrate

- Planned variability within limits over the expected life of the operation,
- Practical and effective maintenance programs to maintain high equipment availability without creating long-term throughput bottlenecks,
- Good cost management, and
- Optionality to respond to changes in business strategy and external influences.

Because an ore-handling system can be the most inflexible component of the overall mining system, a mine's ability to adapt to future business conditions is linked directly to its ore-handling system's ability to respond to such changes. Planning for the future should therefore be a significant consideration in any ore-handling system design process.

Waste handling also needs to be considered as an integrated part of the ore-handling strategy. Because business value impacts (e.g., costs, destination, downstream processing) associated with handling ore and waste will be different, the two material streams cannot be considered in isolation. For example, if the financial cost of disposing waste separately exceeds the cost of including it in the ore stream, then the waste should be included in the ore stream.

Strategy

With reference to Figure 12.8-1, each step in a mining system value chain consists of an upstream buffer (stockpile) of ore, an ore-handling process that upgrades the ore to another value state, and a downstream buffer. The downstream buffer then becomes the upstream buffer to the next ore-handling process. A real system may be more complex in some parts than the simple series value chain shown in Figure 12.8-1, but the linkage between buffers and ore-handling processes remains unchanged.

Bottlenecks

Based on the theory of constraints (Goldratt and Cox 1992), any manufacturing process contains a single bottleneck, or a

---

Martyn Bloss, Manager Long Term Planning, Olympic Dam, BHP Billiton, Adelaide, South Australia, Australia
Paul Harvey, President, Ekati, BHP Billiton, Yellowknife, Northwest Territories, Canada
Don Grant, Superintendent Life of Asset Planning, Olympic Dam, BHP Billiton, Adelaide, South Australia, Australia
Cameron Routley, Superintendent Five Year Planning, Olympic Dam, BHP Billiton, Adelaide, South Australia, Australia

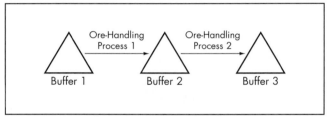

Figure 12.8-1 Linkage between ore-handling processes and supporting buffers

process that controls the output rate. By definition, the bottleneck has the lowest rate of production. Controlled systems have a clearly identified bottleneck that is stable, meaning it does not migrate from one process to another over time. Because of their predictability of performance, controlled systems are more amenable to cost efficiency and continuous improvement, whereas uncontrolled systems have high throughput variability and significant unplanned interruptions.

The process of developing a controlled ore-handling system is to

- Identify where the bottleneck is (or should be); and
- Establish capacities and rates for upstream and downstream buffers and ore-handling processes, respectively, in order to stabilize the bottleneck at the designed location in the value chain.

When this is achieved, the overall performance of the underground mine can be managed to achieve a desired outcome. If throughput is fixed, then cost efficiencies can be achieved by ensuring that only the minimum work required to maintain the buffers at the desired levels is undertaken. If throughput is to be increased, efforts should be focused on maximizing the rate of ore movement through the bottleneck, because this is the rate-limiting step in the overall value chain. Resources allocated elsewhere in the mine will not impact overall throughput. As the rate through the bottleneck increases, its location may migrate to another ore-handling process. At this stage the exercise is repeated at this new location, with the establishment of new buffer capacities and so forth, and an ongoing drive to increase the rate through the new bottleneck. This is continued until a new throughput level is achieved.

For an operation to run effectively there must be a single, well-defined bottleneck. Attempting to match all rates of the process will only lead to poor control of the overall operation, as the bottleneck will shift in response to short-term variability in the performance of each individual process.

The choice of the most appropriate bottleneck location depends on many factors. Foremost is whether the mining operation is currently in an operating state or the design phase. For a new operation, there is an opportunity to optimally locate the bottleneck from the start of the project. Ultimately, the constraint on mine throughput is the resource footprint and the rate at which this resource can be exploited. However, from a valuation perspective (e.g., net present value), sensitivity analyses for various operational configurations should be undertaken to establish the best location for the bottleneck. As a guide, the process in which the capital cost per ton of incremental ore is the greatest should be considered as optimal for locating the bottleneck. This is because over time it is likely that the bottleneck will naturally migrate to this location through continuous improvement of lower capital cost processes, so the sooner this is established, the greater the opportunity to bring forward (in time) positive cash flows to the business. This will not always be the case, however, because a good business case may be made for installing latent capacity initially in high capital items if expansion is planned in the short to medium term and the cost (in present value terms) of expanding these capital items is far greater at a later date. Essentially, the appropriate choice of bottleneck needs to factor in both initial and future plans for the operation.

If the operation is currently in production, an initial choice for the bottleneck is the most practical location to stabilize production in the short term. This allows control to be established and an effective process of debottlenecking to commence. Without this initial control, frequent shifts in the bottleneck will make any attempt at continuous improvement difficult to establish and progress. At the same time, a vision of the bottleneck's long-term location should be established, together with a plan for gradually migrating the bottleneck to this location. Once again, sensitivity analysis of various configurations should be undertaken to establish this optimal position, with a likely outcome being the process that has the highest capital cost per ton of incremental ore.

By definition, because the bottleneck is the only process that is 100% utilized at any given time, nonbottleneck processes cannot be 100% utilized, unless buffer capacity is unlimited and high-cost stock buildup was not being effectively managed. Therefore, nonbottleneck processes must cycle through periods of high and low utilization, according to the buffers' capacities. This is a difficult concept to manage in a mining operation where high utilization is demanded for all equipment and people. One option is to manage buffer levels between high and low trigger values, by deliberately under-resourcing with permanent equipment and people, and employing contractors for fixed periods of time when required. In that way, utilization of the mobilized resources is always high. This concept is described in Figure 12.8-2.

Bloss (2009) describes a practical example where the theory of constraints was applied to the underground mine at Olympic Dam in South Australia to increase throughput by 18%.

Buffers

As described in Figure 12.8-1, an effective ore-handling strategy needs to include the application of buffers between all processes in the ore-handling stream. Buffers exist in a number of forms; for example, as stockpiles, storage bins, orepasses, broken ore stocks within stopes, drilled and developed ore stocks, ore reserve, and mineral resource. They exist between two processes in the value chain in order to provide surge capacity so that variability in performance of either process can exist with minimal impact on the other process. Buffers are particularly important in order to protect the bottleneck, as, by definition, lost production through the bottleneck is lost to the system and cannot be recovered. As a default starting point, it is good practice to assume that a buffer is required between every pair of processes and that any decision not to provide a buffer is justified explicitly as part of the mining strategy. In that way, inadvertent omissions of critical buffers are minimized.

Some key aspects that need to be addressed during designing any buffer include the following:

- What is the size of buffer required?
- What is the required life?

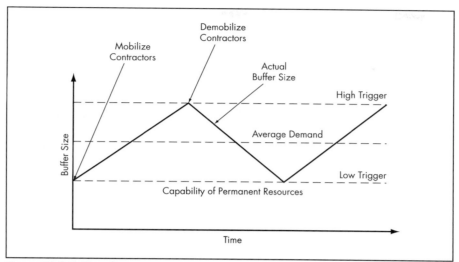

Figure 12.8-2 Use of contractor resources to manage buffers in nonbottleneck processes

- What options are available to provide a buffer of the required size and life span?
- What are the capital and ongoing operating costs of the first three options?
- How can it be designed to minimize operational downtime (e.g., blockages)?
- How can it be effectively maintained in a serviceable state (including replacement options) without adversely impacting future mine throughput?
- How effectively does the buffer integrate into upstream and downstream processes?
- What optionality exists for expanding the capacity of the buffer (enlarging, duplicating) to meet future strategic scenarios?
- Are there requirements for separating the ore into high/low/waste grade or according to other specifications?

Maintenance, Contingency, and Optionality Plans

In order to plan for future uncertainties, management strategies for each buffer and ore-handling process should include maintenance, contingency, and optionality plans. Maintenance plans protect throughput during ongoing operations and discrete and planned periods of major maintenance work. Contingency plans manage the risk of reduced performance as a result of unforeseen events. Optionality plans consider the potential for either a change in mining strategy or future capacity expansions. In many cases, these management plans are complementary. For example, provision of a redundant orepass can be an effective strategy to manage both maintenance and operational issues but can also be built into an optionality strategy in which the additional pass can be used, for example, to increase throughput in the mine or to segregate ores of different grades.

Table 12.8-1 describes some examples of maintenance, contingency, and optionality plans that could be employed to various components of an ore-handling system.

Selection Process

Materials handling is generally a big investment and can be the largest single investment in any new mine. The selection process needs to balance the knowns (capital cost, current operating conditions) against the unknowns (future operating conditions). This is not a trivial exercise, as the future operating condition has many dimensions, most of which are difficult to quantify with certainty. These include the following.

- **Safety:** Is the ore-handling system capable of adopting ongoing improvements in safety systems, processes, and technology (e.g., automation)?
- **Revenue:** Will changing metal prices affect business decisions that could impact the choice of ore handling?
- **Cost:** Will the cost of handling ore increase excessively?
- **Throughput options:** Does the ore-handling system provide the flexibility to increase throughput if required, without significant financial hurdles (including business interruptions)?
- **Geographic footprint:** Will the changing mining footprint over time adversely impact production cost and/or throughput?
- **Ground conditions:** What will be the impact of deteriorating ground conditions (directly relating to ore handling) on production cost and/or throughput?
- **License to operate (company strategy and regulations, government regulations, etc.):** Will mandated changes to operating practices (e.g., health, safety, environmental) impact the ore-handling system such that production costs and/or throughput are adversely affected?
- **Availability and reliability:** Will maintenance performance decline excessively and will the cost of sustaining performance be excessive? Can major maintenance works be carried out without adversely affecting mine performance?
- **Maintenance support:** Will the maintenance support be adequate? This can be an issue if new and/or specialized technology has been employed, where maintenance support can be problematic in terms of either speed of response or quality of service. A risk–reward evaluation is recommended to determine whether the benefits of the technology are outweighed by the reduction in equipment availability as a result of substandard maintenance support.
- **Replacement and/or major maintenance:** Is there scope for replacing/ maintaining key infrastructure, if required, without significant business interruptions?

Table 12.8-1 Examples of maintenance, contingency, and optionality plans

Component of Ore-Handling System	Maintenance Planning	Contingency Planning	Optionality Planning
Loaders/trucks	• Good availability track record • Fast response of spare parts • Duty/standby equipment • Major component change-out plan • Flexible fleet configuration • Good maintenance support—large skills base developed and maintained	• Capability to increase/decrease resources on demand • Flexible fleet configuration • Remote and automation capability	• Capability to increase resources on demand • Haulage routes capable of being expanded to account for changing mine's geographic footprint or mining strategy • Capability to adopt latest technology into fleet • Highly flexible to respond to different mine configurations
Orepasses	• Designed in best possible ground conditions, based on rock quality and both virgin and induced stresses over time • Good access for regular inspection and remedial work • Good monitoring and reconciliation systems (tons in versus tons out) implemented • Good understanding of long-term deterioration due to stress conditions and planned maintenance program • Artificial support options • Provision of backup pass • Simple design • Management of rock size going into pass (sizing)	• Provision of backup passes • Provision to duplicate passes if required due to failure of existing passes—passes have limited life, and, if small in number, premature failure could be a major issue • Good access for regular inspection and remedial work • Good access to manage hang-ups	• Provision for additional passes • Limited optionality for each individual pass—long lead time to design and install
Chutes	• Simple, robust, and reliable design • Integrated planned maintenance schedule	• Effective plan for unblocking chutes	• Fully automatable
Trains	• Duty/standby equipment • Good maintenance support—specialist skills required • Integrated planned maintenance schedule	• Latent instantaneous capacity • Critical spare parts inventory • Good maintenance support • Fully automatable	• Provision for additional capacity • Provision to expand rail system • Designed and constructed for long life and optionality—no option for long shutdown periods • Control of systems to optimize performance
Storage facilities (bins, stockpiles)	• Provision of backup storage facility—shutdowns are long and costly	• Adequate capacity storage (buffer) between all ore-handling processes • High-capacity buffers supporting bottleneck	• Provision for additional storage facilities • Optionality to split into different material streams (ore, waste, etc.)
Conveyors	• Latent instantaneous capacity • Good maintenance support • Integrated planned maintenance schedule	• Latent instantaneous capacity • Critical spare parts inventory • Good maintenance support	• Provision to increase belt speed • Provision to duplicate belt system • Provision to extend system to new mining areas
Crushers	• Latent capacity • Good maintenance support • Integrated planned maintenance schedule	• Latent capacity • Critical spare parts inventory • Good maintenance support	• Latent capacity • Provision to duplicate crusher
Shafts	• Latent instantaneous capacity • Good maintenance support (noting the extreme skills shortage in this field) • Integrated planned maintenance schedule	• Latent instantaneous capacity • Critical spare parts inventory • Good maintenance support	• Provision to increase skip weight/speed • Provision to duplicate shaft • Automation

- **New technologies:** Can the system take advantage of new technologies to add value (e.g., reduce costs, increase throughput)?
- **System robustness:** Is the system capable of effectively responding to unplanned variability in key input characteristics and maintain planned throughput? Some examples are
  - Particle size changes;
  - Presence of undesirable material (e.g., rock bolts, cables);
  - Presence of backfill material, clay, or other sticky waste material (e.g., fine, high moisture content potentially causing significant blockages in chutes and transfer points);
  - Water content changes;
  - Major unplanned failures of internal, upstream, or downstream processes; and
  - Changes in spatial distribution of ore supply.
- **Change in mining method:** Will potential future changes in the mining method be adequately serviced by the ore-handling system?
- **Change in ore-handling system:** Can possible additions, changes, or enhancements to the ore-handling system over time be adequately integrated into the existing system and implemented without adversely affecting system availability?
- **Continuous improvement:** Is the ore-handling system capable of facilitating ongoing improvement in order to increase operational efficiency?

- **Short-term operational variability:** Can the system adequately manage short-term operational variability through effective use of upstream and downstream buffers? This is particularly important if the ore-handling system is the business bottleneck.

A common selection process, described in Table 12.8-2, is based on the development of a matrix in which the various ore-handling options are assessed and ranked against key design and operational criteria. Each criterion is assigned a weighting based on the importance of that criterion to the selection. For each ore-handling option, a performance score is then assigned to each criterion. The weighted scores are then totalled, and the highest score is used to determine the preferred option.

Although this process is relatively simple and efficient, and can achieve good stakeholder buy-in, it has a number of potential shortcomings:

- The ore-handling options will be based on experience and may not include a broad enough choice.
- All relevant selection criteria, particularly those that relate to future uncertainty, may not be considered.
- The weighting assigned to each selection criterion may be subjective.
- The score assigned to some selection criterion for each ore-handling option may also be subjective.
- The summing of weighted scores for each ore-handling option to determine an overall score may not reflect the true relative impact of each selection criterion.

There is a risk that the selection outcome will be biased toward those options that prioritize currently known aspects of the design (e.g., capital and operating costs) over future unknown aspects. If, for example, two options have similar appeal in the current time frame, but the less attractive has greater optionality in the future, this option should be the preferred option. For this reason, it is recommended that this process be used as only one component of an overall selection study.

Comparison Between Ore-Handling-System Options

The following discussion compares the three most common ore-handling systems: shaft, conveyor, and truck. These systems usually contain other supplementary ore-handling components, including loaders, orepasses, and crushers. In this comparison, it is assumed that a significant vertical lift is required to bring the ore to the surface, therefore limiting the application of rail as a primary ore-handling system.

In assessing the relativity of these systems, notably, other issues not related to materials handling may often affect the choice of system. For example, if a decline system is required for access and provision of routine services to the underground, the incremental cost to also utilize this access for trucking, for example, may be much lower than the total cost. Hence the decline option may be cost-effective relative to other ore-handling options.

Cost
Shaft systems usually attract the highest capital cost; however, as depth increases, the alternatives become progressively more expensive, and the capital cost difference relative to shaft systems decreases. Once installed, however, shafts attract a relatively low operating cost.

Table 12.8-2 Example of a common ore-handling selection process*

Ore-Handling Options	Selection Criterion				Total Weighted Score
	1	2	3	4	
Weighting	40	30	20	10	
Option 1	2.0	6.0	1.0	8.0	3.6
Option 2	7.0	7.0	3.0	1.0	5.6
Option 3	8.0	2.0	9.0	5.0	6.1
Option 4	6.0	6.0	7.0	7.0	6.3

*All scores are from a total of 10.

Conveyor systems have recently been demonstrated to be cost-effective relative to shaft systems. Pratt and Ellen (2005) demonstrated negligible capital and operating cost differential between the two systems during a selection study at the Telfer operation in northern Western Australia. The shaft option was preferred due to other factors such as physical impact on the orebody, footprint required, and flexibility to service future economic ore zones.

Trucks are usually lower in capital cost but have higher operating costs. Recent improvements in trucking technology have increased the depth to which trucking costs remain economically attractive. Pratt and Ellen (2005) suggest that trucking is not economically viable over a range of mining rates from 1.5 Mt/a (million metric tons per annum) to 6 Mt/a below a depth of 650 m.

Figure 12.8-3 describes the capability of established underground mine haulage systems at the Telfer and Ridgeway gold mines in Australia. The choice of system is largely based on the cost efficiency as a function of depth and throughput.

Safety
All options have safety risks that need to be managed. For shafts, risks are associated with working at heights during inspection and maintenance work, and interactions with personnel during operations. For conveyors, fires and personnel interaction with the moving belts are of particular concern. For trucks, fires and vehicle-to-vehicle and vehicle-to-pedestrian interactions are the highest risk issues, all of which can be managed effectively through the application of adequate design and appropriate procedures. Some examples are presented in more detail later in the chapter.

Timing
Truck haulage via a decline offers a particular advantage in the ability to access ore progressively as the decline is sunk. In this way, positive cash flow can be generated earlier in the mine life compared to alternative systems where installation must be achieved to full depth prior to ore production. In cases in which a surface access decline is being established in a mine where the ore will ultimately be hoisted using a shaft, it may be advantageous to utilize truck haulage up the decline early in the mine life, prior to commissioning of the shaft. Examples of this application include the BHP Billiton Cannington mine in northwest Queensland, Australia, and Newmont Mining's Leeville mine in northern Nevada, United States.

Optionality
Because mines typically evolve over time, the ability of a mine to vary its operating strategy, in order to optimize production

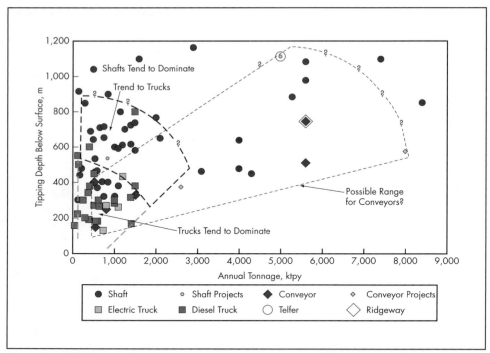

Source: Pratt 2005. Reprinted with permission from the Australasian Institute of Mining and Metallurgy.
Figure 12.8-3 Capability of established underground mine haulage systems

in response to changes, can often be a function of the optionality embedded within the ore-handling system. Although shaft and conveyor systems can be inflexible because of the limited number of fixed feed points, trucking systems are flexible because generally trucks can travel to most locations in the underground mine. Where fixed infrastructure ore-handling systems are employed, the tendency is to introduce hybrid systems (e.g., trucks feeding shaft) as the mine evolves in order to respond to changes.

In situations where the underground mine capacity needs to expand to increase business value, the potential is often related to the configuration and current utilization of the ore-handling system. If the existing system is based on trucks, the expansion can typically be achieved incrementally by adding trucks as required until the capacity (e.g., ventilation, traffic congestion) of the decline system is reached. Following this, additional access would be required at significant cost. Increasing throughput in fixed systems such as shaft hoisting and conveyors is relatively cheap up to a point where the systems' utilization is at the industry benchmark. Following this, higher utilization will probably require duplication of the existing system at significant cost. This is likely to be justified financially only if there is a step increase in throughput or the geologic footprint has evolved substantially enough that a duplicate system is justified to maintain or create a step change in expected efficiencies and associated operating costs.

LOADERS

Loaders move ore and waste from the development face, stope, pass, or stockpile to the next step in the ore- or waste-handling system. The types of loaders include

- Rubber tired,
- Rail mounted,
- Track mounted, and
- Shuttle loaders.

Loaders are most commonly a single-function unit but in certain applications can be part of a multifunction unit. The most common form of the latter is a continuous miner, which can cut coal from a coal face, load the coal into a transport vehicle, and also install roof support (when equipped to do so).

Rubber-tired loaders in metalliferous mines are referred to by many different names around the world, according to local preference. These include loader, load-haul-dump (LHD), scooptram, scoop, bogger, digger, mucker, and underground loader. An underground loader is shown in Figure 12.8-4.

Underground loaders are typically the first component of the ore-handling system. Loaders extract ore from the stope and either dump (tip) directly into an orepass or load into a truck that hauls the ore to an orepass or surface stockpile. The size of the loader fleet will be determined according to the following factors:

- Mining method
- Geometry of the stope/ore working face
- Size of underground excavation
- Geometry of the ore-handling system
- Production rate based on planned strategic objectives
- Productive operating time or availability and utilization per shift
- Loader productivity
- Geometry and condition of the tram/haul route
- Capital and operating costs

The mining method will determine the relative proportions of stope ore, development ore, and waste. The geometry of the stope or ore working face may limit the size of loader that can be used. Ground conditions will determine underground

Source: Caterpillar 2009.
Figure 12.8-4 2900G Elphinstone loader

excavation dimensions, which in turn will impact fleet size and equipment selection. The geometry of the ore-handling system will determine the proportion of material required to be rehandled. Sweigard (1992) provides a good overview of the equipment selection and the fleet size estimation process. This information allows mapping of loader tram segments to determine quantities of ore movement required to achieve the nominated production rate. Typical units of measurement are metric tons per operating hour (t/h) and ton-kilometer per operating hour (t·km/h), which is based on the one-way length of the tram. Preliminary estimates of total operating hours per year or total ton-kilometers per year for various sizes of loaders allow for an initial estimate of loader fleet size.

Further details on loader productivity are available in Sweigard (1992) and the various underground equipment manufacturers' specifications (e.g., available on the Caterpillar, Sandvik, and Atlas Copco Web sites). Gradeability/speed/rimpull charts similar to Figure 12.8-5 and equipment dimensions (Caterpillar 2004) provide useful specifications for determining loader performance in differing conditions. The calculation for estimating truck fleet performance is performed using a similar methodology.

Loaders tend to be as large as development allows, and tram distances are minimized to maximize productivity per loader. Loader trams greater than 250 m become less effective. To allow production to continue when the stope brow becomes open, the loader requires line-of-sight remote or tele-remote capacity. The loader bucket may also use different tooth configurations to assist digging in the stope drawpoint.

Safety Considerations

Safety issues that need to be considered during the design phase and implemented during the operations phase include the following.

- Light vehicle, heavy vehicle, and pedestrian interaction: Some strategies that can be employed to manage this issue include separating heavy vehicles from light vehicles and pedestrians, detection and collision avoidance systems, traffic rules, and visual and audible signalling systems.
- Remote loading at the stope drawpoint (operator interaction): The most effective method is to isolate the operator from the machine by using tele-remoting systems.
- Loading at a stope brow or orepass and initiating a mud rush or air blast: To manage this situation, a good understanding of the stope's material characteristics is required, together with strict controls and procedures regarding the methods of extraction from the drawpoint to ensure predictable behavior of material flow from the drawpoint.
- Fire on the loader: Minimizing risks and consequences of loader fires requires quality equipment design, effective preventive maintenance programs, well-trained operators, reliable automated onboard fire prevention and fire-fighting systems, and sound procedures for managing fires.
- Equipment failure during operation: Minimizing risks and consequences of equipment failure requires quality equipment design, an effective preventive maintenance system, well-trained operators, and controls that minimize personnel exposure to loader operations.
- Open holes (including tipples): Effective hard and soft controls are required to minimize risks associated with operating around open holes, including solid barriers, good lighting and signage, and procedures.
- Operator ergonomics: The well-being of personnel during loader operation and maintenance is paramount to sustainable operations. High-quality, fit-for-purpose design, both within the operator's cabin and wherever personnel work on the loader, is critical to achieving this objective.

Design and Construction Issues

Underground access development needs to be purposely designed to accommodate selected loaders and trucks. As a general rule, all rubber-tired equipment should operate in development with gradients of at least 1:50 to minimize water ponding and to provide good road conditions. Key development components are

- Drawpoints,
- Tramming drifts (drives),
- Tipple accesses,
- Stockpile bays, and
- Truck-loading bays.

These components are described in detail in the following sections.

Drawpoints

Drawpoints provide access to the stope for loaders to extract ore from the stope. The design of a drawpoint needs to consider horizontal and vertical clearances relative to the access walls and back, and gradient of the access. Ideally, the drawpoint should be angled relative to the tramming drive (e.g., 60°) to facilitate ease of entering and exiting. It also needs to be of sufficient length to accommodate the loader while digging the muckpile, without being too long to trigger poor ventilation conditions. At mines where remote loading from the stope is required, a safe operating position needs to be designed into the drawpoint area to protect the loader operator. Figure 12.8-6 shows a tele-remote setup in a Canadian mining operation.

Tramming Drifts (Drives)

Similar to drawpoints, the main design considerations for tramming accesses are clearance, gradient, and orientation geometry.

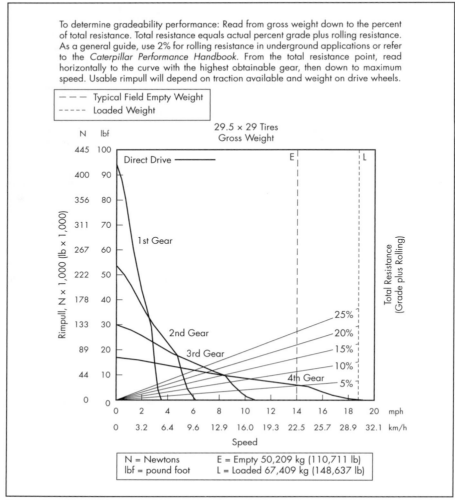

Source: Caterpillar 2004.

Figure 12.8-5 Caterpillar gradeability/speed/rimpull chart for Elphinstone 2900G

Tipple Accesses

Once again, clearance, gradient, and orientation of tipple accesses are the main design considerations. Back heights above the orepass and in the approach to the orepass are critical to facilitate the dumping (tipping) of the loader bucket and truck tray. If back height becomes problematic, trucks equipped with hydraulic ejector beds and plates should be considered. A steep upward gradient into the orepass reduces the potential for mine water to accumulate and flow into the orepass. Orientation of the tipple access to the main tramming drift (drive) can be perpendicular or angled. A perpendicular access allows loaders to approach from both directions without the requirement for a turnaround bay.

Stockpile Bays

Stockpile bays provide surge capacity and can improve overall ore-handling system utilization. They allow loaders to keep producing while waiting for trucks or full orepasses to clear. Stockpile bays need enough capacity to minimize delays to other parts of the development or production cycle.

Courtesy of Atlas Copco.

Figure 12.8-6 Tele-remote station at a stope drawpoint in Canada

Truck-Loading Bays

Truck-loading bays need to consider back height for loading and localized gradient and geometry constraints. Detailed access development design criteria can be obtained from the different manufacturers' equipment handbooks. Examples of the information available in the equipment handbooks are illustrated in Figure 12.8-5. Generally, maximum efficiency and minimum cost are achieved if the loader is sized to load the truck on the same level, rather than from an elevated loading ramp.

Operational Aspects

Stope Mucking

The productivity of a loader is dependent on a number of factors that facilitate the removal of ore from the drawpoint. Ore size distribution should be relatively homogeneous with minimal oversize or high percentage of fines. Oversized rocks slow the production rate because they must be resized either in the drawpoint or in nearby bomb bays. Processing oversized material in the drawpoint also introduces safety issues, which need to be managed. A high percentage of fine material (e.g., from the failure of a backfill mass) can create difficult operating conditions throughout the ore-handling system, particularly if water is present. Both oversize and fines issues can be better controlled through drill-and-blast design, geotechnical design, and the placement of quality, engineered backfill.

Good physical operating conditions include good ventilation for visibility and water sprays to help suppress dust. Drawpoint brows need to be supported to minimize wear over their anticipated lives. Brow wear can lead to a reduction in effective drawpoint length, making loading difficult, and increases the tonnage of ore that needs to be mucked by remote control.

The ventilation system needs to be monitored, because changes in airflow can occur when the stope brow opens. Additional consideration is required in uranium mines where air transit time becomes a factor in limiting radon residence time. A ventilation quantity of 0.06 m^3/s/kW of engine power is a good standard to meet acceptable environmental conditions within typical mining operations.

Tramming

Loader operators generally prefer to tram with the loader bucket at the rear when loaded and the front when empty. This reduces the occurrence of loaders driving over rocks that fall from the bucket while tramming.

The following key aspects need to be considered and implemented in a practical manner to achieve the best cycle time results:

- **Road maintenance:** Well-maintained roadways are critical for achieving maximum loader speeds and reducing equipment maintenance issues associated with continuous vibration of critical loader components.
- **Drainage:** The establishment of well-drained roadways will also increase achievable tramming speeds, reduce road and vehicle maintenance, and reduce collision with submerged objects that are not visible.
- **Ventilation:** A well-ventilated roadway will provide good visibility of the tramming route, allowing higher loader speeds and reducing the potential for vehicle collision.
- **Lighting:** The installation of lighting on the tramming route can allow higher loader speeds; however, lights should only be considered on a tramming route that is in use for a significant time period.
- **Tram geometry:** This includes the drift (drive) size or available clearance, the gradient (uphill or downhill loaded) and length of tram, and the number of corners or bends in the tram.

Loader Costs

Budget capital cost estimates can be provided from equipment manufacturers. The maintenance change-out program will determine ongoing sustaining capital requirements, depending on the frequency of rebuilds and major component change-outs. Loader operating costs are dependent on labor, fuel and lubricants, tires, bucket wear, maintenance consumables, and equipment damage.

A remote loading operation typically has a higher operating cost resulting from reduced productivity. The ability to estimate the cost of maintenance depends on the relative proportion of predictive and preventive maintenance to breakdown maintenance. Equipment suppliers can provide budget estimates for the majority of consumable operating costs. Equipment damage is difficult to estimate because it is related to a number of factors, including road conditions, visibility, operator experience, equipment clearance dimensions, mining method, tram geometry, and stope drawpoint conditions. A loader performing remote operation into an open stope will have a higher exposure to rockfall damage from unsupported areas of the stope.

Loader productivity and utilization will depend on the activity; for example,

- Development loaders: <4,000 h/yr;
- Sublevel open stoping production loaders: ~4,500 h/yr; and
- Block cave and sublevel cave production loaders: >5,000 h/yr.

Technology

Production Management Systems

A number of commercially available production management systems are used in underground mining operations, including Micromine's PITRAM and Modular Mining's Dispatch. Numerous operations have also developed their own in-house systems, which capture data and allow managers to monitor performance to determine whether operational targets are being achieved. The data are collected using radio communication or tag readers located throughout the mine.

Automation

Automation and remote control technology have been available for underground loaders for more than 20 years. Numerous automation systems are being developed by equipment manufacturers, mining companies, research organizations, technology companies, and various partnerships between these groups. The level of advancement with these automation systems and their ability to operate practically and cost-effectively in the challenging underground environment are still under debate. Substantial progress has been made, and cost-effective, practical automation solutions will be an important part of the future for improved loader productivity and operator safety.

Maintenance

All maintenance programs can be broken down into preventive and breakdown components. Both are required to minimize the risk of fatalities, injuries, and incidents arising from the use of mobile equipment underground and to maximize equipment productivity.

An effective predictive and preventive maintenance program that is prioritized over breakdown maintenance will reduce the potential for personnel injury and increase equipment utilization as a result of the following aspects:

- Replacement of components before they fail and while the equipment is not operating prevents potentially serious consequences, including injury to personnel and damage to associated components.
- More predictable scheduling of maintenance activities allows for better matching between supply of, and demand for, maintenance resources. Breakdown maintenance tends to create boom-and-bust work cycles that cannot be effectively managed using a fixed level of resources.
- More predictable availability of equipment allows for more efficient utilization, as a result of better matching between availability of equipment and supply of operators.

Best practice equipment availabilities are generally achieved by reducing the frequency of breakdown maintenance and increasing mean time between equipment failures. To maintain high availability throughout the useful life of equipment, a well-planned component change-out program is essential. The operating life and timing of major equipment rebuilds will have an impact on capital and operating cost profiles for the mine. Costs of preventive maintenance and major component change-outs need to be consolidated with other operating costs to determine an overall cost per ton-kilometer. This measure can be used to determine the effectiveness of the maintenance program.

TRUCKS

Trucks are used for long or inclined hauls to orepasses or directly to the surface. Commonly available trucking options include

- Diesel or electric articulated dump trucks, or ADTs (Figure 12.8-7);
- Diesel or electric rigid dump trucks;
- Diesel road trains (Figure 12.8-8); and
- Diesel, electric, and battery shuttle cars (Figure 12.8-9).

Trucks can be side loaded near the ore source (e.g., stope drawpoint), or side loaded or chute loaded at the bottom of an orepass.

Safety Considerations

Safety issues that need to be considered during the design phase and implemented during the operations phase include

- Light vehicle, heavy vehicle, and pedestrian interaction;
- Chute loading (the key to minimizing risks is to ensure that the design of the chute fully accounts for the material flow characteristics and the loads that the chute will be subjected to, so there are no unexpected failures in the chute);
- Equipment failure;
- Fire on the truck;

Courtesy of Atlas Copco.
Figure 12.8-7 ST1520 loader side-loading an MT5020 truck

Source: Findlater and Robertson 2004.
Figure 12.8-8 Ninety-ton road train exiting Sirius mine portal (Western Australia)

- Tipping into open holes; and
- Operator ergonomics.

Selection Criteria

Selection of the appropriate size and type of truck fleet depends on the following key factors:

- Ore reserve tonnage
- Production rate
- Haulage distance
- Depth below surface
- Lateral extent of the mine
- Mining method used
- Integration with upstream and downstream materials-handling processes

Truck haulage is generally the preferred option where the ore reserve is not large enough to justify the capital for a shaft or conveyor hoisting system. For hoisting shaft or inclined conveyor systems, the lateral extent of the mine, distribution of ore, and throughput requirements determine the most effective solution for upstream delivery to a centralized hoist, from a choice of trucking, rail haulage, or conveyor.

ADTs have traditionally been used because of their flexibility and ability to maneuver throughout the ore body in variable conditions. Road trains, electric articulated, and rigid-body trucks have been utilized with varying degrees of success depending on road conditions (grade, vehicle clearance, and road maintenance) and environmental conditions (temperature and visibility).

Source: Joy Mining Machinery 2008.
Figure 12.8-9 Shuttle car in operation in a potash mine

Courtesy of Joy Mining Machinery.
Figure 12.8-10 Coal shuttle car being loaded

Underground coal and potash mines have a unique application for trucks, given the typically narrow seams that are mined. Specialized shuttle cars are used to transport the coal or potash. These are low-profile AC-electric trucks designed to interact effectively with high-capacity continuous miners or roadheaders. They remove material that has been cut by the mining machine and transfer it directly to a main conveyor, which creates an efficient ore-handling system. Shuttle cars can operate in seam heights down to 1.2 m with a capacity of 8 t (e.g., Figure 12.8-10).

Design and Construction Issues

Access design considerations discussed in the "Loaders" section can also be applied to trucks (equipment clearance, gradient, and ground conditions). Additional considerations include

- Side-loading bays,
- Orepass dump point (tipple) configuration,
- Chute-loading systems, and
- Decline haulage access.

Side-Loading Bays

Side-loading bays facilitate the efficient loading of a truck. Side loading of trucks can be performed on level ground (Figure 12.8-7) or with the loader elevated above the truck, utilizing a perpendicularly configured side-loading bay. The latter configuration can be used when the loading point is fixed over an extended time frame and truck spotting times have a significant impact on truck productivity. To facilitate efficient truck loading and reduced spillage, the loader can be elevated 1.5 m above the truck.

Orepass Dump Points (Tipples)

Orepass tipple configuration for trucks requires additional height and length of back stripping compared with a loader configuration. It is also necessary to be able to reverse into the tipple. Additional development may be required to achieve efficient truck tipping.

Road trains dump (tip) to the side and therefore require a significantly different layout for an underground tipple. To accommodate side dumping (tipping), a truck loop is generally required. Consideration for a turning radius and a straight section at the dumping point can add a significant amount of underground development as compared with other truck dumping layouts.

Chute-Loading Systems

Truck loops and truck-reversing bays are two commonly used configurations for loading trucks from chutes. When sufficient tonnage is available (e.g., 3 to 4 million t), a truck loop configuration can achieve higher productivity. For lower tonnages, a chute that requires trucks to reverse provides an option with less capital. To further reduce capital costs in low-tonnage, short-term situations, ore can be side loaded into trucks at the bottom of orepasses. To facilitate chute loading of multiple trailers, road trains require a truck loop configuration.

Decline Haulage Access

Decline (or ramp) haulage provides access for trucks to haul ore from underground to a surface stockpile. Design considerations include clearance, gradient, geometry, and frequency and geometry of passing bays. Declines can be designed as a series of long straight sections or as a circular spiral. Passing bays can be designed parallel with or perpendicular to the decline, depending on ground conditions. In most Australian decline trucking operations, stockpile bays are developed every 150 m during construction and then utilized for passing bays. Block light systems are also commonly used to control traffic flow.

ADTs, large rigid-chassis trucks, and road trains with power trailer are effectively used to haul ore long distances to surface stockpiles. Figure 12.8-8 shows a road train exiting a decline portal, and Figure 12.8-11 shows design information for drift (drive) size and road train turning requirements. Robertson et al. (2005) provides a good overview of the future direction of trucking. Because decline haulage trucks usually have to share the decline with other vehicles, numerous traffic control systems are available to reduce the hazard of vehicle interaction.

Trucking Costs

Many factors that influence trucking costs are similar to those for loaders (see the "Loaders" section). The differences are primarily due to the specific functions of the equipment. Loaders are generally more influenced by loading conditions, whereas trucks are more influenced by haulage conditions (road conditions and haul distances).

The two technologies described under "Technology" in the "Loaders" section have similar application to underground trucks and should be considered if justified cost-effectively.

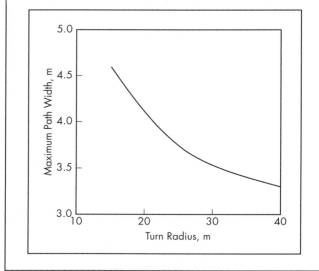

Source: Findlater and Robertson 2004.

Figure 12.8-11 Underground road train vehicle sweep paths for Powertrans underground truck and Powertrailer

OREPASSES

Orepasses provide a low-cost method to move ore and waste downward between operational horizons in a mine. Initial capital is the dominant cost associated with an orepass system, with lower ongoing sustaining capital and operating costs.

Orepasses can be constructed as one long section or as multiple shorter sections depending on the geometry and/or timing of access to the top and bottom of the orepass and the relative operational and safety aspects during construction.

Flow of material in the orepass is critical to its success. A number of factors need to be considered to maintain continuous flow conditions. Early origins of material flow were analyzed by Janssen (1895), who determined that the weight of the material in the bin or orepass was actually transferred to the walls rather than the bottom of the orepass. This produces the potential for arching, where broken material stabilizes in the pass and prevents material flow. Strategies for avoiding this are discussed in later sections, for example, when considering orepass width.

Safety Considerations

The safety issues that need to be considered during the design phase and implemented during construction and operation include the following.

- **Risks during construction:** The main risks are people working below vertical openings from which material can fall and people falling into open holes. Thorough risk assessments need to be undertaken to ensure that key risks are identified and managed.
- **Clearing of blockages:** Clearance in orepasses involves drilling hang-up holes into the immediate vicinity of the blockage and feeding explosives into the pass to free the blockage.
- **Fall prevention:** Stop (bumper) blocks, orepass covers, signs, barricades, and flashing lights are used to prevent people from falling into or driving equipment into orepass open holes.
- **Mud rushes:** These hazards occur when material containing a high proportion of fines is dumped (tipped) into an orepass and the orepass becomes blocked. When mine water or natural groundwater mixes with the fines in the orepass, there is the potential for a mud rush. The severity of a mud rush depends on the amount of fines in the orepass and the proximity of personnel and equipment to the bottom (discharge point) of the orepass during a mud rush. Management of mud rushes includes understanding material flow properties and developing procedures for managing undesirable combinations of materials that report to the drawpoint. Typically, pedestrians are isolated from any orepass loading activity, and, in cases where mud rushes are likely, remote loader operations are common.

Design and Construction Issues

Orepass design is largely dependent on the geometry of the ore body and the strength of the rock mass in the mine. Hadjigeorgiou et al. (2005) provide a good review of orepass practice in Canadian mines.

Inclination Angle

Orepass inclination angles are generally greater than 60° and should have at least a small amount of inclination from vertical to allow the dumped material's energy to be dissipated on the orepass footwall and therefore prevent free fall of the material to the orepass bottom. Orepass fingers that connect to chutes can be as low as 50° in order to reduce the amount of energy transfer from the ore in the orepass onto the chute.

Length

The length of an orepass is dependent on the vertical distance between stope extraction level(s) and the common ore-handling level. Orepasses are sometimes extended upward to allow more cost-effective dumping (tipping) of development material from upper levels of the stope. The orepass system can be constructed using one or two longer sections or a larger number of shorter sections, depending on access available for construction. The length of each orepass section generally determines the excavation method. Typically, Alimak raising and raise borers can construct longer orepasses, whereas drilled and blasted drop raises and conventional raises are used for shorter-length holes.

Orientation

The orientation of an orepass can be a critical design consideration in rock masses with dominant structures. The most favorable orientation is as close to perpendicular to the dominant structure as practical. In situations where the orepass needs to be oriented parallel with the dominant structure, additional support, liners, or a redundant orepass should be considered.

Shape

The shape of an orepass is dependent on the excavation method. Raise boring will ream a circular shape, while Alimak, drop raising, and conventional raising typically excavate a square or rectangular shape. If high stress is believed to be a stability issue for the orepass, circular shapes can be excavated with drill-and-blast methods.

Width

The width of an orepass is dependent on size distribution of the broken material, characteristics of the wall rock, and capacity requirements of the orepass. It generally needs to be no greater than five times the dimension of the largest particle that enters the orepass, as described in Table 12.8-3.

Recent work by Hustrulid and Changshou (2004) indicates that the ratios in Table 12.8-3 are conservative. Their work recommends considering size distribution and orepass wall conditions in the design, which will usually lower D/d ratios being practically achieved. Rock mass strength of the orepass wall relative to the predicted induced stress levels and orientation also needs to be considered when sizing the orepass.

Quality of the rock mass in orepass walls will partially determine what orepass size is achievable. Empirical design methods can be used to determine stable spans, and, if high stress fields exist in the mine, numerical modeling should be considered to understand possible future states of stress (magnitude and orientation) in the orepass walls.

Surge capacity requirements of the ore-handling system need to be considered when designing each individual orepass. When the ore-handling system downstream of the orepasses is taken out for maintenance, the loader and truck operation will stop when the orepasses are filled. Effective surge capacity in the ore-handling system allows the components of the system to be less dependent on each other. Surge capacity is generated by expanding the dimensions of the orepass and/or developing strategically located stockpile bays. Enough surge capacity needs to be designed into the orepasses to increase operational flexibility of the total ore-handling system. How much surge capacity can be installed has practical and cost limitations, and the amount of surge required will depend on the benefit realized by smoothing out interruptions in ore flow due to downstream process variability. Surge capacity from several hours to a week should benefit any mining operation if this surge supports the operational bottleneck. Where downstream process variability is significant, consideration should be given to installing the maximum possible surge capacity.

Material may become captured in certain areas of the orepass, and this reduces the effective capacity of the orepass. Sections of the orepass that have changes in direction or angle are more likely to create areas for captured ore and are more susceptible to blockages.

Construction

Orepasses can be constructed using a raise boring machine (Figure 12.8-12), drop raising (long-hole winzing), Alimak raising (Figure 12.8-13), or conventional handheld raising. The most common methods for longer passes are a raise bore machine or Alimak raising. The preferred construction method depends on the specific requirements of the orepass, the geometrical layout of the ore-handling system, and local preference and experience. In the past, Alimak raises were more popular in North America but are becoming less common because the Alimak platform operators are exposed to vertical openings. Alimak raises have the advantage of providing

Table 12.8-3 Orepass design to prevent interlocking arches

Ratio of Orepass Dimension (D) to Maximum Particle Dimension (d)	Relative Frequency of Interlocking
$D/d > 5$	Very low; flow almost certain
$5 > D/d > 3$	Often; very uncertain flow
$D/d < 3$	Very high; no flow almost certain

Source: Hambley et al. 1983.

access to install ground support during construction, which can provide long-term orepass stability. The Alimak raise climber shown in Figure 12.8-13 runs on a special guide rail using the Alimak rack-and-pinion system. The guide rail contains three pipes providing compressed air and water for ventilation, scaling, and drilling. Drop raises are commonly used for short orepasses (60 m or less) as blasthole deviation becomes greater at 60 m. Conventional handheld raises are generally limited to short finger raises that connect chutes to the orepass.

Support Systems

Ground support requirements for orepasses depend on rock mass and induced stress conditions, orientation of the orepass relative to geological structures, size and inclination of the orepass, planned mode of operation (choke fed or open), and planned life span of the orepass.

The type and method of ground support depends on the excavation method. For example, if an Alimak raise is used, it will need to be supported with rock bolts and mesh to protect operators during the construction phase. Ground support systems include metal cans or rings, concrete liners, specialized wear-resistant sprayed-on shotcrete, mesh, and rock reinforcement systems such as rock bolts and cable bolts. In some cases the cost of ground support can exceed excavation, and therefore the decision can be made to leave the orepass unsupported. This option is justified if life expectancy of the orepass in an unsupported state is sufficient. A replacement plan and monitoring program needs to be well defined and communicated to management. Rehabilitation of an orepass, sometimes considered in order to extend its life, can be a time-consuming and costly exercise, which can be avoided by installing ground support at the construction stage or by planning to construct replacement orepasses. Hadjigeorgiou et al. (2004) provides additional details of the different orepass support systems.

Ventilation

Good ventilation at orepass tipples and the extraction horizon is essential to maintaining a productive ore-handling system. Exhaust air from the orepass needs to be removed quickly from operating areas. Good ventilation can be provided by designing and constructing a parallel return air system, which provides effective exhaust capacity at each tipple and the extraction horizon. Spray bars can also be used to suppress dust. A number of ventilation design systems for orepasses are reviewed in Calizaya and Mutama (2004).

Operational Aspects

Key considerations for safe and efficient operation of an orepass are

- Controlling material size entering the orepass,
- Controlling material flow,

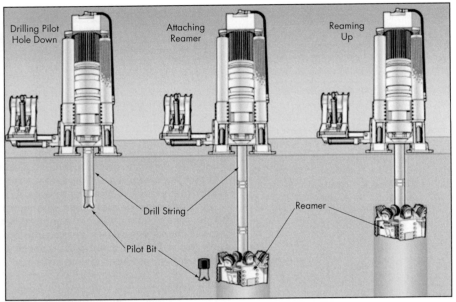

Courtesy of Atlas Copco.
Figure 12.8-12 Raise bore machine drilling a pilot hole and reaming an orepass

A number of monitoring techniques are available, including quantitative physical surveying of the orepass using probe hole drilling to determine the orepass profile (costly and time-consuming), a cavity monitoring system (e.g., Optech's), cavity autoscanning laser system (e.g., Measurement Devices Ltd.'s), or other laser scanning devices. Other qualitative methods include orepass camera surveying to obtain a video of the orepass, production data to determine a tonnage reconciliation of material movements, and other data captured by operations (i.e., number of hang-up blasts per time period and type of material blasted down during hang-up blasts). Orepass wear is largely caused by structural failure, stress failure, material impact, and abrasion and blast damage (Hadjigeorgiou et al. 2005). The orientation of an orepass relative to dominant structures can have a significant impact on its long-term stability.

The control of material size in an orepass is facilitated using grizzlies, scalpers, mantles, and orepass rings at the entry point. Material flow in the orepass can be controlled by chutes with chains or gates, or a plate feeder at the orepass exit point. When the orepass bottom has no control, a loader is used to control material flow. When oversize material (producing interlocking arches) and/or fines (producing cohesive arches) are introduced to the orepass, the probability of hang-ups increases and can cause delays. This either reduces throughput or increases the cost to maintain throughput by implementing an alternate ore-handling option.

Oversize can be introduced to the orepass by loaders dumping (tipping) elongated, shaped material through the size controls at the entry point of the orepass or from internal fall-off within the orepass. Reestablishment of material flow in the orepass usually requires some form of blasting to provide vibration or sometimes water flow to initiate material flow.

A comprehensive review of methods for releasing blockages is discussed in Hadjigeorgiou et al. (2004). Depending on the severity of the hang-up, it can be a long process to reestablish material flow, and redundant or alternate orepasses should

Courtesy of Australian Contract Mining.
Figure 12.8-13 Alimak raise system

- Continuously monitoring the orepass for wear and stability, and
- Maintaining good ventilation.

Over time, wear of the sidewalls can make an orepass inoperable. With good condition monitoring, orepass refurbishment or replacement can be planned in a proactive and controlled manner. If wear in an orepass goes undetected and makes it unserviceable, the ore-handling system can be disabled for an unacceptable period of time.

Orepass monitoring is an effective tool that can be used to understand the condition and potential life of an orepass.

be considered at the design stage. Choke feeding of material in the orepass can limit internal fall-off by providing confinement to reduce the effect of high-velocity material impacting the orepass walls. In choke-fed orepasses, the material should be continually drawn to reduce the risk of hang-ups.

Fine material can come from backfill falloff within the stope, raise borer cuttings, or highly fragmented development material. The cohesive nature of fine material can cause blockages, especially when the material is not periodically moved in the orepass. The issues with fine material can be magnified if water is simultaneously introduced into the orepass, which increases the potential for a mud rush at the orepass exit point.

Cost Considerations

Cost estimations need to consider the expected life of an orepass and future cost of replacement. Budget capital cost estimates can be obtained from mining contractors or in-house project teams that specialize in raise boring or Alimak raising. Initial capital cost of orepass controls (chutes, feeder, and chain controls) and material sizing systems (grizzlies, orepass rings, rock breakers, etc.) can also be estimated by specialized mining contractors. Ongoing sustaining capital costs for replacement of chutes, plate feeders, grizzlies, orepass rings, liners, and rock support systems need to be factored into overall cost estimates.

Orepass operating costs include mechanical and electrical maintenance on control systems and rock breakers, and drilling and blasting costs associated with clearing orepass blockages. A good orepass design, potentially at a higher capital cost and followed up with disciplined operating practice, will minimize future operating costs and increase productivity of the orepass system.

Advantages and Disadvantages

Advantages of utilizing orepasses include

- A high-capacity system for vertical transport of material,
- Low operating cost, and
- Buffer capacity.

Disadvantages of utilizing orepasses include

- High capital cost,
- Good technical design required;
- Subject to ongoing operational challenges—for example, hang-ups;
- Performance subject to ongoing ground conditions; and
- Difficulty to maintain if ground conditions deteriorate.

CRUSHERS

Material size reduction in many operations is often required for productive and reliable materials handling. By crushing ore underground, an operation can reduce wear on equipment, reduce clearances required in materials-handling processes, increase hoist skip or other conveyance capacity (reduce material void ratio), reduce oversize handling issues such as hang-ups or blockages, and increase secondary crushing and processing productivity.

Crushers are typically required ahead of conveyors and shafts. Where the material is transported to the surface by trucks, underground crushing is generally less justified because a greater size range is capable of being transported in this manner. Although the material needs to be crushed prior to being processed, in most cases the crushing process is more cost efficient when undertaken on the surface.

In operations where underground crushing occurs, uncrushed material is usually delivered to the primary crusher by dump trucks, wheeled loaders, trains, or orepasses. A feeder device such as a plate feeder or apron feeder can be used to control the rate at which this material enters the crusher. The feeder often contains a preliminary grizzly or screening device, which ensures an appropriate material feed size is received by the crusher.

Safety Considerations

The modern crushing plant includes the following safety features:

- Safety guards around all moving equipment (Boyd 2002).
- Emergency stop facilities on components where personnel access is required (Boyd 2002).
- Dust abatement, suppression, or collection equipment: Dust emissions must comply with the latest regulations for the jurisdiction. Designers must make provisions for the installation of equipment to control these emissions.
- Spillage prevention and collection facilities: Spillages from feeders, chutes, and conveyors should also be minimized.
- Noise abatement devices and controls: Because crushers, screens, and dust-collection fans all contribute to high noise levels, strategies need to minimize such exposure to personnel (e.g., engineering solutions, restricted access, or personal protective equipment).
- Fire protection systems: Fires are major hazards in any underground mine, and the potential for fires on conveyors is high. Protection through the use of automatic sprinklers or deluge systems is important to managing this risk.

Design and Construction Issues

Crushing plant design can be split into three main components: process design, equipment selection, and layout. The first two are often dictated by production requirements and design parameters. The layout can reflect input preferences and operational experience of a number of parties in order to provide a balanced, workable, safe, and economic plant design. Because many underground crushing facilities are located within purposely built rock excavations, it is vital that access and maintenance of all components is sufficiently allocated during the design phase. This should include suitable lifting equipment for major component maintenance or replacement, as large mobile crane access in many underground locations is not possible.

The fundamental goal for a crushing plant design is to provide an installation that

- Meets the required production rates and material sizing,
- Can be constructed within capital constraints,
- Operates at competitive cost, and
- Complies with safety and environmental regulations.

The following critical design parameters should be considered:

- Ore characteristics
- Geographical location

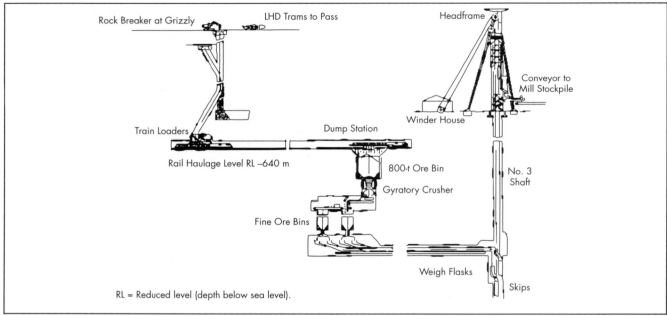

Courtesy of BHP Billiton Olympic Dam.

Figure 12.8-14 Ore-handling system at Olympic Dam mine

- Expected operational life
- Expansion potential
- Safety and environment
- Operability and maintainability

Figure 12.8-14 describes the ore-handling system at the Olympic Dam mine, including the primary cone crusher.

Underground Crusher Location
In underground operations, the excavations required for the installation of crushers and associated infrastructure can be substantial and are often the largest (unfilled) excavations within the mine. When considering an underground crusher layout and location, consideration should be given to

- Local and regional ground conditions,
- Proximity to planned and potential future production areas,
- Current and future stress conditions within the excavation, and
- Proximity to planned and potential future mine infrastructure.

Good geotechnical information is essential in the selection of a crushing plant location and the extent of ground support required.

Life of Mine/Expansion Plans
Planning for expansion should be considered in the design stage for underground crushing facilities. Expansion plans for most crushing plants can be incorporated in the early planning stages at much lower cost than waiting until the mine is up and running, particularly if the cost includes an interruption to normal production activities.

Crusher Selection
Crusher selection is dependent on many operation-specific variables including throughput rates, planned utilization, ore characteristics, material feed size, and required discharge product size. Many of the factors influencing underground crusher selection are common to surface crusher selection and are not covered in detail in this section.

As a guideline, the following can be applied: For a hard-rock mine application below 600 t/h feed rate into the crusher, a jaw crusher is the preferred option as the primary crusher. For feed rates more than 1,000 t/h, a gyratory crusher is the preferred option. Between these capacities, the choice will be dependent on other factors (Ottergren 2003). An example of a primary cone crusher is presented in Figure 12.8-15.

Equipment Overview
The major equipment in a primary crushing circuit usually includes a crusher, feeder, and conveyor. Secondary and tertiary crushing circuits have the same basic equipment items, along with screens and surge storage bins.

Grizzlies
A grizzly is a sizing device used to ensure a maximum passing size of material and to protect the crusher. Square mesh grizzlies get a more accurate maximum size of material than other shaped grizzlies but can be prone to plugging. Rectangular and nongrided grizzlies are less prone to plugging but can allow material of larger size to pass. Many grizzlies are rigidly mounted so that material must flow through them or be pushed through with equipment. Some grizzlies vibrate to encourage material on the grid to reorient and pass through. Grizzlies can be fed using a loader, truck, or conveyor.

Grizzlies also act as a metal scavenger to trap rock bolts, wire mesh, drill steels, and cables, removing them from the ore stream periodically by a loader from the top of the grizzly.

Rock Breakers
Fixed or mobile rock breakers are often used in conjunction with a grizzly. A modern rock breaker uses a hydraulic arm

Courtesy BHP Billiton Olympic Dam.
Figure 12.8-15 Primary cone crusher and rock breaker at Olympic Dam mine

with a pneumatic impact tip to move material around on the grizzly and impact oversize until it passes. Fixed rock breakers can be operated from a remote operator's cabin or even from the surface to reduce the installation size and ventilation needed underground. Before the advent of the modern rock breaker, oversize material feeding the primary crusher was a greater problem, and primary crushers were designed largely on the basis of gape (minimum dimension of feed opening) rather than capacity.

Additional and Optional Equipment

Other equipment items in crushing circuits can include the following:

- Overhead crane
- Freight elevator
- Service air compressor
- Sump pumps
- Air vacuum cleanup systems
- Rock grapple
- Conveyor belt magnets
- Conveyor belt metal detectors
- Belt monitoring systems
- Belt feeders
- Screw feeders
- Bin ventilators
- Apron feeder to the primary crusher
- Dust collection/suppression system
- Eccentric trolley removal cart
- Personnel elevator
- Air cannons
- Water booster pumps
- Service trolleys
- Conveyor gravity take-up service winch
- Conveyor belt rip detector
- Conveyor belt weigh scales
- Vibratory feeders
- Sampling station

Underground crushing installations are high-capital-cost items, due to fixed equipment costs and large-scale excavations required to house the system. Crusher installation capital-cost estimates should include equipment costs plus the following direct and indirect construction costs:

- Excavation and ground support
- Mechanicals
- Concrete
- Electrical
- Structural steel
- Instrumentation

Indirect costs can fall within a range of 40% to 60% of the direct costs.

Integration Aspects

As mentioned in the introduction, each step in a mining system value chain consists of an upstream buffer, an ore-handling process that upgrades the ore to another value state, and a downstream buffer. The upstream buffer in most operations will be the storage capacity of uncrushed material available for crusher feed. Within typical feed crushing configurations this may be in the form of one or a number of stockpiles, or live capacity within one or a number of orepasses. Downstream buffer is the capacity of crushed material storage between the crushing discharge and the next step in the materials-handling system (e.g., truck, conveyor or skip load, and transport). Where such upstream and/or downstream buffer is of limited capacity or nonexistent, operational variability around the crusher may adversely affect crusher uptime. Because of the high capital costs of large underground storage facilities for uncrushed or crushed material, it may not be possible to install large buffer capacity at this step of the materials-handling process. Therefore, crushing should not be the operation bottleneck by design, and the crushing and immediate upstream and downstream processes should have capacity in excess of the bottleneck process.

Technology

For the most part, advances in crusher design have moved slowly. Jaw crushers have remained virtually unchanged since about 1950. More reliability and higher production have been added to basic cone crusher designs that have also remained largely unchanged. Increases in rotating speed have provided the largest variation. Production improvements have come from speed increases and better crushing chamber designs.

The largest advance in cone crusher reliability has been in the use of hydraulics to protect crushers from being damaged when uncrushable objects enter the crushing chamber. Foreign objects, such as steel, can cause extensive damage to a cone crusher and additional costs in lost production. Advancement in hydraulic relief systems has greatly reduced downtime and improved the life of these machines.

CONVEYORS

Mine conveyor systems have been utilized in underground mass mining for many years. Conveyor systems are commonly found in underground continuous coal mining operations, which use longwall and room-and-pillar methods. More recently, underground conveyor systems have been employed in large-scale, long-life metalliferous mines. Conveyors are more commonly used in operations where mine geometry and production flow allow conveyor haulage to be more efficient or cost-effective than other methods.

Inclined, troughed belt conveyors are more frequently being selected from a range of alternatives, which include shafts and trucks, for ore haulage in underground mass mining projects. Belt conveyor haulage systems are being operated with lifts exceeding those normally associated with truck haulage systems and approaching the limits of shaft haulage systems (Spreadborough and Pratt 2008).

Continued development of conveyor technology has resulted in increased confidence in and reliability of these systems. Reliability of conveyor-based haulage systems as a whole is impacted by the complexity of larger configurations constructed of multiple units or lifts. Modern conveyor installations have proven high availabilities (>85% are achievable in underground operations), which has added confidence to their use in recent times.

Figure 12.8-16 shows a typical conveyor in an underground metalliferous mine operation.

Safety Considerations

In the selection of conveying options for an underground operation, several risk areas must be addressed, including

- Belt fires managed by self-extinguishing belt covers and auto detection and suppression systems;
- Belt failures addressed by belt-rip detection systems, monthly belt scans, and concrete bulkheads;
- Guarding requirements to prevent injury to personnel;
- Audible start-up alarms where personnel work near the belt;
- Emergency stop provisions in areas where access by personnel is required; and
- Dust management.

Selection Criteria

The selection of haulage systems for underground mass mining has historically focused on shaft haulage, trucks, and belt conveyors. The application of these alternatives in the Australian mining industry is summarized in Figure 12.8-3. Troughed belt conveyors are shown in this figure to be applied in the range up to 8 Mt/a production rate and 1,200-m lift.

Selection criteria for underground conveyor haulage include

- Inherent safety;
- Capacity (steady state and surge flow);
- Simplicity in design and operation;
- Dimensions (length, height, and width);
- Maneuverability/adaptability to various layouts;
- Cost (capital and operating);

Courtesy of BHP Billiton Olympic Dam.

Figure 12.8-16 Typical conveyor in an underground metalliferous mine operation

- Reliability/availability;
- Operating life;
- Size of product handled, spillage, and carry-back;
- Dust and noise generation; and
- Automatic, remote operations.

The advantages of conveying material include high automation, minimal operational labor requirements and lower operating costs, and high reliability/availability.

The disadvantages of conveying material include high capital cost (therefore lower probability of duplication), a large footprint, and limited flexibility.

Belt conveyors for underground service usually have a more rugged design and operate at slower speeds than a comparable overland conveyor. Hard-rock mine belt conveyors normally require the ore to be crushed before it is conveyed or at least sized through a grizzly. Reasons for this include increased belt life due to reduced impact from lumps and elimination of tramp material, including rock bolts, rebar, drill steels, and scaling bars, which can seriously affect the integrity of the belt system.

Belt conveyor systems are less flexible than truck haulage and require a high initial investment. This generally means that belt conveyors are the economical choice when the production rate is relatively high and the transport distance is significant. In certain applications, belt conveyor systems are selected for other reasons. For example, short conveyors are employed underground to optimize feed control to a loading pocket and prevent a run of fines from reaching the shaft.

Elements in a conveyor haulage subsystem can include

- Feed chute of dimensions appropriate for the material received,
- Belt to support the material and to transfer tractive forces to the material,
- Idlers to support the belt,
- Pulleys to resist belt tensions at changes of direction,
- Drives to provide driving and braking force,
- Take-up rollers to provide belt tension for no slip at the drives and for belt sag control, and
- Discharge or transfer chute.

Maintenance equipment at the haulage subsystem will typically include

- Monorails and/or cranes at drives and pulleys for change-out or repair,
- Belt clamps to resist belt tensions when installing or maintaining the belt (spring applied, hydraulic release, tested for holding capacity),
- Belt reel handling facilities, and
- Belt splicing/repair facilities.

Integration Aspects

When considering the conveyor components of the materials-handling system, consideration should be given to the upstream and downstream buffers. The upstream buffer in most operations will be the storage capacity of crushed material available for conveyor feed. In typical crushing/conveying configurations this may be in the form of fine ore stocks within dedicated stockpiles or storage bins. The downstream buffer may be underground storage at transfer points to the next step in the materials-handling system or surface stockpile storage capacity.

Technology

Advances in belt technology and particularly the development of stronger belt carcasses support the potential for further increases in capacity of inclined conveyor haulage systems. Vertical conveyors and hydraulic hoisting represent two potential technologies for haulage from underground mines.

The attraction of vertical conveying is due to the combination of a small footprint, continuous process, and overall energy efficiency. Limiting factors for vertical conveyors include tensile strength of the steel cored belts, safety of belt splices, and required production rate (Pratt 2008). Current precedence for the application of vertical or high-angle conveyors in a mining context is limited to a few hundred meters.

Other special types of belt conveyors include extendable systems, cable belts, and high-angle belts. To date, these special conveyor types have had few applications in hard-rock mines.

Conveyor Costs

Conveyor haulage systems are considered high in capital and low in operating cost compared with truck haulage. Generally, shaft haulage options attract higher capital costs and marginally lower operating costs for a similar-scale operation.

Operating costs will rise for conveyors as the complexity of the system increases with depth as a result of the addition of lifts and transfer points. This does not necessarily favor selection of shaft haulage over a conveyor system because other factors must be considered. Examples of aspects to be considered are horizontal transportation distances, confidence in geotechnical prediction for shaft and conveyor decline options, and other functional requirements of the mine. The actual selection of a haulage strategy for a large, deep mine will require evaluation of the trade-offs in risks and opportunities identified by the various options available.

Limits of Application

The slope of a high lift conveyor group is limited by the slope of the decline/incline development in which the conveyor is installed. This is normally in the range of 1:5.3 to 1:5.4 (10.5° to 10.7°) where the development is carried out using conventional rubber-tired equipment. Length and lift are limited by the belt carcass construction and the strength of the belt splice. The choice of belt construction is also limited by the troughability of the belt. Troughability is the ratio of the cross-belt sag to belt width and generally decreases with increasing carcass strength.

Future Applications

Advances in future designs will be fundamentally linked to the load-carrying capacity of available belt constructions and splice designs. Advances in conveyor technology continue to ensure success of conveyor systems in mass underground mining applications, including improved drive control equipment, belt condition monitoring, improved splice designs for higher strength and longer life, and improved chute designs.

RAIL HAULAGE SYSTEMS

Although belt conveying remains the dominant underground haulage system in coal mines, rail haulage continues to be favored in large tonnage (>5 Mt/a), long-life underground metalliferous operations.

Haulage by rail can be limited to ore and/or waste haulage, or can include integrated rail systems for transportation of personnel, materials, and supplies throughout the mine. Although the following discussion focuses only on the haulage of ore and waste from production areas to the hoisting system, it includes the consideration of interactions with other rail system users where required.

Prior to the emergence of trackless or rubber-tired mobile equipment in underground mining in the 1960s and 1970s, rail or tracked haulage systems were predominant. Many of today's rail haulage mining operations remain a legacy of that period. With improved productivity and performance of diesel mobile equipment and belt conveyors into the 1980s, this equipment has grown in popularity and application as significant methods of underground haulage. However, rail haulage continues to offer competitive operating cost, productivity, and throughput benefits over a range of applications in many underground mining settings.

Rail haulage systems consist of three primary types defined by their power supply: diesel, battery electric, and overhead trolley electric. A diverse range of systems are used around the world, from small-scale selective mines producing less than 100 ktpa to the largest bulk mining operations producing greater than 20 Mt/a.

Health and Safety Considerations

A fundamental consideration in haulage method selection and design is to ensure a safe and healthy workplace. Some of the key potential hazards that require consideration in rail haulage design and potential mitigating actions within a health and safety management system follow:

- High-speed equipment
  - Traffic management plan
  - System access control
  - Adequate lighting
  - Speed limits
  - Comprehensive equipment operating procedures
- Interaction between people and equipment
  - Equipment control systems
  - Physical barriers and guarding
  - Design for visibility
- Unmanned/automated equipment
  - Access control
  - Fail-to-safe control systems
  - Communication systems
- Overhead exposed conductors
  - Permit-to-work process
  - Isolation systems
  - Electrical protection systems
  - Entry height barriers
  - Water and dust management program
- Maintenance hazards
  - Custom-designed and fit-for-purpose maintenance facility
  - Regular inspection and maintenance schedules
  - High standard maintenance equipment
  - Isolation systems
  - Regular track cleaning program
- Mud rushes from chutes and orepasses
  - Water controls at all orepass entry points
  - Orepass monitoring
  - Regular chute inspections and/or continuous remote closed-circuit television monitoring
- Fire
  - Automatic fire suppression systems
  - Emergency response teams and training
  - Good housekeeping

Common and paramount in all the these circumstances are the underlying skilled workplace supervision, customized training programs, skills assurance process, good safety signage, and proactive maintenance programs that meet OEM (original equipment manufacturer) requirements at a minimum.

Selection Criteria

Selection of the preferred haulage system for a mine must consider a wide range of factors, including health and safety, economic, technological, reliability, operability, and flexibility.

Rail haulage can exist either as a stand-alone lateral haulage system or as part of an integrated system. In the case of the former, for example, rail provides for all lateral movement of personnel, materials, ore, and waste in the deep-level South African gold mines and northern Idaho (United States) silver mines. In the case of the latter, rail haulage can operate with mobile equipment, as at the Olympic Dam mine. Ultimate selection needs to match the shape, size, nature, and productive capacity of the ore body within the framework of an economically feasible operation.

The optimal haulage system for a mining operation is dependent on a variety of factors and their interrelationships, the more significant of which are discussed in the following paragraphs.

Health and Safety

Of fundamental importance are the design, installation, and operation of a safe rail system, as discussed previously.

Capital Cost

Rail haulage systems typically have a high capital cost and long construction time compared with alternative haulage systems. This is because economically competitive rail systems are often large scale by nature and require

- Long distances of tunnel development, which takes years to excavate;
- Orepasses and chute systems for train loading;
- Long-term ground support systems;
- High upfront equipment capital purchase costs for locomotives, rail, rolling stock, tipplers, trolley line system, and power reticulation equipment (where installed); and
- Large construction labor and installation costs.

Operating Cost

Converse to capital cost, rail haulage operating costs are among the most efficient per ton of material moved. This is driven by comparatively lower maintenance costs due to a more highly engineered and thus reliable system compared with diesel mobile equipment, requirement for a low number of operators, and better energy efficiency. Spare parts inventories may also be reduced.

Reliability and Operability

Well-engineered rail haulage systems can achieve high system availability (greater than 85% compared to diesel truck fleets, which range from 50% to 75% of total available clock time). The fixed nature of rail systems and repeatability of processes makes it highly attractive for automation, with the potential to deliver increased utilization. State-of-the-art systems, such as LKAB's 26-Mt/a rail system at the Kiruna mine in Sweden, CODELCO's 48-Mt/a Teniente 8 railway at the El Teniente mine in Chile, and BHP Billiton's 8-Mt/a rail system at Olympic Dam in South Australia, are highly automated, requiring high levels of technological support while delivering substantial unit operating cost and productivity benefits (Figure 12.8-17). Due to the fixed nature of rail systems, maintenance can be readily planned and scheduled, spare locomotives can limit production interruptions, while a smaller work force is typically required per ton hauled. Rail haulage systems require specific skill sets to maintain locomotives, high-voltage reticulation systems, communication systems, automation, and control systems.

Production Rate and Scale of Operations

Given the higher capital cost of rail haulage systems, they are typically used in operations with high production rates or long lives, in order to pay back the high upfront costs. Likewise, the high productivities and low unit operating costs offered by rail haulage favor their selection in mines with high throughput rates and/or long lives. However, smaller-scale rail systems

A. LKAB, Kiruna mine, Sweden B. BHP Billiton, Olympic Dam mine, Australia C. CODELCO, El Teniente mine, Chile

(A) Courtesy of LKAB; (B) Courtesy of BHP Billiton Olympic Dam; (C) Courtesy of P. Crorkan-Tait.

Figure 12.8-17 Underground rail haulage systems

are also competitive in mines where the reduced flexibility of a fixed horizon haulage system can be managed. This is particularly relevant in large flat-lying ore bodies where long-haul distances may exist, where larger openings for other forms of haulage cannot be considered, or where diesel mobile haulage systems are avoided because of environmental conditions (e.g., heat, gases, and fire risk).

Ore-Body Geometry
Rail haulage systems are by nature horizontal installations, and their selection must consider the lower flexibility of their fixed nature. In general, larger ore bodies are more suited to rail haulage as production is sourced from one or a few (albeit large) areas as opposed to disseminated, narrow-vein, or dispersed ore bodies where production comes from numerous locations in differing amounts at different elevations. However, rail haulage can be a competitive haulage option in such mines if the fixed nature of rail systems can be managed and matched to other aspects of the operations design.

Production Growth
Although diesel mobile equipment has the flexibility to travel wherever tunnel access is provided (flat or inclined), rail haulage is a constrained system. In order to support growth of operations to new areas or greater depths, rail haulages need to be extended or replicated to support mine growth or extension, or to sustain production from new areas. The 25-Mt/a rail haulage system of LKAB's Kiruna sublevel cave iron ore mine is a good case study, where major rail haulage horizons have been replicated at increasing depth in 300-m vertical increments on 10-to-15-year intervals as the mine has progressively moved its production areas downdip. Consideration at the design stage can enable rail systems to increase their productive capacity, if required, through measures such as adding more rolling stock, upgrading locomotive capacity, or increasing haul speeds.

System Life
The operational life of haulage systems and equipment is a key criterion in selection of haulage systems. Rail systems generally have a longer equipment life than mobile haulage equipment, which requires replacement nominally every 5 to 7 years. High throughput rail systems require longer operational life in order to provide an acceptable return on capital invested.

Mine Plan
Selection of haulage systems must be aligned and complementary to the mine plan. Rail haulage is more typically suited to deeper shaft hoisting operations where ore and waste need to be collected centrally for crushing, transport to, and removal from a single elevation, typically at depth. Development of ore bodies in areas of high topographic relief offers the potential for adit access and efficient rail haulage directly from the mine to the surface processing facility (such as Freeport-McMoRan Copper and Gold's deep ore zone mine at Grasberg in West Papua, Indonesia, and CODELCO's El Teniente mine).

Design Issues
Inclusive in the economic analysis of selecting a cost-effective haulage system are four key aspects for rail haulage design. Inherent to this analysis are capital and operating costs. The following factors need to be evaluated in a fully integrated design approach:

- Capacity
- System layout and geometry
- Materials properties and materials handling
- Equipment

Capacity
The required ore production rate is the primary driver of rail haulage system design. This will determine such key design parameters as tunnel layouts and size, equipment capacity, operating philosophy, the number of units in the fleet, integration with upstream and downstream production process capabilities, and haul speed. Consideration also needs to be given to whether the system is dedicated to ore haulage or is to be shared with movement of personnel and/or materials, and haulage of waste as well as ore.

Due to the fixed nature of rail haulage system infrastructure, consideration should be given at the design phase for the potential and likelihood of future increased production or extension of mining operations to new areas. In such circumstances, engineering designs should consider such key items as the adequacy of tunnel size and extent, maintenance facilities, power supply systems and infrastructure, track quality, and the number and configuration of orepasses and dumping (tipping) points.

System Layout and Geometry

The layout and extent of a rail haulage system requires economic trade-off studies of numerous factors that impact the geometric design, including

- Physical dimensions and depth of the ore body;
- Degree of coverage required by the haulage system beneath the ore-body footprint;
- Primary mining excavation costs (including ground support, final preparation, and drainage);
- Trade-off between optimal rail haulage design and operation, and optimal production mining systems in the mine workings above;
- Single or dual direction of travel;
- Haulage speed (a key driver of productive capacity);
- Radii of curves, single or double rail, tunnel dimensions, gradient, maximization of straights, and super-elevation of curves;
- Number and location of ore and/or waste delivery points into the rail system;
- Number and location of ore and/or waste dumping (tipping) points from the rail system;
- Future operating cost impacts and benefits; and
- Suitability for future expansion.

With an inherent need for wide radius curves to accommodate rail equipment and a preference for straight sections, rail haulage horizons can lend themselves to potential mechanical excavation techniques such as tunnel boring machines. These are commonly applied in civil engineering rail projects and warrant consideration in the preliminary phase of mine rail haulage system design.

The schematic presented in Figure 12.8-18 demonstrates the preference for geometric rail loop layouts, wide radius curves, and long straights.

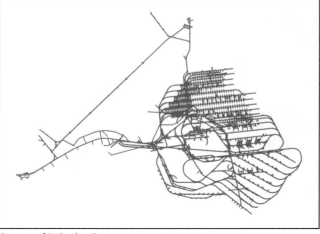

Courtesy of P. Crorkan-Tait.

Figure 12.8-18 Schematic of Teniente five-rail haulage design layout at El Teniente mine in Chile

Material Properties and Materials Handling

Rail haulage systems must be designed to manage the physical attributes of the material(s) to be hauled. For example, the physical properties of the material will determine the strength of mechanical equipment against wear and tear, and flow properties will impact orepass and chute design and determine the potential need for lining of passes, bins, or chutes for longevity.

The key physical properties to be considered include

- Moisture content;
- Size distribution, proportion of fines (mud-rush hazard), and maximum lump size;
- Specific gravity/material density;
- Abrasiveness;
- Dust generation;
- Presence of deleterious materials, such as tramp steel, cemented backfill, and waste dilution; and
- Changes in material properties over time.

Careful management of these properties has a significant downstream impact on system operability through aspects such as

- Pass and chute maintenance;
- Spillage, which can cause safety hazards, derailments, and significant downtime or equipment damage;
- Train hopper wear and tear;
- Downtime due to blockages, hang-ups, fines rushes, etc.; and
- Maintainability and maintenance cost.

Equipment

The selection of rail haulage equipment is a key component that must be fully integrated with all the previously discussed aspects of system design. Locomotive prime movers are typically powered by either battery electric, diesel, or overhead electric energy sources. The majority of large-scale, high-production underground rail haulage systems today are overhead DC electric inverted to AC-drive locomotives. Examples of such systems are shown in Figure 12.8-17.

Ore/waste hoppers and tipplers are available in various configurations, with capacities ranging from 1 t to more than 70 t per hopper. Hoppers can be either side or floor dumping (tipping), with a range of chassis, side-door, and floor opening designs available depending on various parameters. Train combinations can be sized to match production and rail system capability with rakes of up to 30 hoppers and total train payloads in excess of 1,300 t.

Rail haulage systems lend themselves to significant levels of automation. Design of remote control systems is critical to operating productivity and reliability, and needs to consider locally available skill sets, technical support, spare parts availability, mine communications systems, and environmental aspects such as water, humidity, dust, radiation, and gases. Train control systems must be fully integrated with the proposed operating philosophy for the haulage system, with the aim of preventing downtime and production delays, and minimizing maintenance.

Other equipment aspects to consider in system design include

- Rail gauge and installation quality;
- Control and communications systems;
- Switch design;
- Train loading points, passes, and chute design;
- Tippler design;
- Drainage and lighting; and
- Signaling.

The maintainability and operability of all equipment requires close attention. This includes access to all parts of the rail system for maintenance personnel and equipment in the event of derailment or breakdown; specialized equipment for rail maintenance; standby equipment, including spare locomotives and hoppers/ore cars; critical/insurance spares; and regular spare parts management.

HYDRAULIC HAULAGE SYSTEMS

Hydraulic ore transport is an often cited but rarely implemented ore transportation system in underground mining applications. While slurry pumping engineering is mature in a multitude of diverse surface applications, its serious consideration in underground environments is hampered by risks associated with

- The need to pretreat and prepare suitable material underground, beyond typical crushing and sizing arrangements;
- Pump and column wear considerations;
- Capital costs and equipment technologies for anything more than very small tonnage rates; and
- Lack of courage to lead change.

One well-established application of underground hydraulic transportation and hoisting is at the McArthur River uranium mine in Saskatchewan, Canada, operated by Cameco (H. Goetz, personal communication). McArthur River produces approximately 47,000 tpa of high-grade (average 17% U_3O_8 [triuranium oxtoxide]) uranium-bearing ore, from a depth of 640 m below the surface. The operation includes full ore preparation underground through a 600-t/d semiautogenous grinding mill, hydrocycloning, and 2×13-m-diameter thickeners. The thickener underflow feeds into a 200-m^3 thickener underflow holding tank that supplies the Wirth triplex piston/diaphragm positive displacement pumps (13 million Pa discharge pressure) for pumping up 2×100-mm-diameter columns in a single stage to the surface.

The physical characteristics and health and safety aspects associated with handling relatively low quantities of very high-grade uranium ore, with attendant high radiation levels, were a key driver in the selection of hydraulic transportation ahead of more conventional methods.

Hydraulic systems for transporting backfill are described in a separate chapter.

OTHER HAULAGE SYSTEMS

A small number of less common haulage systems are used in underground mines. These are typically selected because of highly specific geological environments, particular material properties, or unique operating systems and processes. These include the following:

- Rope scrapers/slushers—commonly used for ore haulage from operating stopes to orepasses in South African longwall gold and platinum mining operations. Slushers are also used for ore collection, haulage, and delivery into orepasses beneath established block caving operations.
- Monorails—traditionally a materials-handling system. Monorail systems can be either electrohydraulic friction or rack-and-pinion drive, and have been used in moderately dipping applications (up to 14°) in South African gold mines.
- Floor-mounted, rope-driven trap rail haulage systems—applied for haulage of equipment and ore/waste up or down steep inclines of up to 40°.

REFERENCES

Bloss, M. 2009. Increasing throughput at Olympic Dam by effective management of the mine bottleneck. *Trans IMM Sec. A* 118(1):33–46.

Boyd, K. 2002. Crusher plant design and layout considerations. Vancouver, BC: AMEC Mining and Metals.

Calizaya, F., and Mutama, K.R. 2004. Comparative evaluation of block cave ventilation systems. In *Proceedings of the 10th US/North American Mine Ventilation Symposium, Anchorage, Alaska, May 16–19*. Edited by R. Ganguli. Netherlands: Balkema. pp. 3–13.

Caterpillar. 2004. *Caterpillar Performance Handbook*. Peoria, IL: Caterpillar, Inc.

Caterpillar. 2009. R2900G Underground mining loader. Photograph AEHQ5608-01(8-07). www.cat.com. Accessed May 2009.

Findlater, G., and Robertson, A. 2004. Road trains underground—Trucking into the future. Presented at the Innovative Mineral Developments Symposium, Sydney, Australia, October 6.

Goldratt, E.M., and Cox, J. 1992. *The Goal: A Process of Ongoing Improvement*. Great Barrington, MA: North River Press.

Hadjigeorgiou, J., Lessard, J.F., and Mercier-Langevin, F. 2004. Issues in selection and design of ore pass support. In *Fifth International Symposium on Ground Support in Mining and Underground Construction*. Edited by E. Villaescusa and Y. Potvin. Perth, WA: Balkema. pp. 491–497.

Hadjigeorgiou, J., Lessard, J.F., and Mercier-Langevin, F. 2005. Ore pass practice in Canadian mines. *J. South Afr. Inst. Min. Metall.* 105:809–816.

Hambley, D.F., Pariseau, W.G., and Singh, M.M. 1983. *Guidelines for Open-Pit Ore Pass Design*, Vol. I and II: Final report. Washington, DC: U.S. Bureau of Mines.

Hustrulid, W.A., and Changshou, S. 2004. Some Remarks on Ore Pass Design Guidelines, Quantification and Analysis of Ore Pass/Draw Point Flow Problems, Leading to the Development of Design Rule of Thumbs—Final Report. Laboratory Ore Pass Studies in Support of Design Guidelines, Department of Mining Engineering, University of Utah, Salt Lake City, UT. Graham Mustoe Division of Engineering, Colorado School of Mines, Golden, CO. Unpublished.

Janssen, H.A. 1895. Experiments regarding grain pressure in silos. (Translated by W.A. Hustrulid.) *Z. Ver. Deut. Ing.* 39, 1045.

Joy Mining Machinery. 2008. Haulage systems product overview. Bulletin No. HS06-0208. www.joy.com. Accessed June 2009.

Ottergren, C. 2003. Crushers and rock breakers. In *Hard Rock Miners Handbook: Rules of Thumb*, 3rd ed. North Bay, ON: McIntosh Engineering.

Pratt, A.G.L. 2005. Application of conveyors for underground haulage. In *Proceedings of the Ninth Underground Operators Conference*. Melbourne, Australia: Australasian Institute of Mining and Metallurgy. pp. 179–188.

Pratt, A.G.L. 2008. Mine haulage—Options and the process of choice. In *Proceedings of the 10th Underground Operators Conference*, Hobart, Australia. Melbourne, Australia: Australasian Institute of Mining and Metallurgy.

Pratt, A.G.L., and Ellen, P.J. 2005. Selection of an ore haulage system for Telfer deeps. In *Hoist and Haul Conference*. Melbourne, Australia: Australasian Institute of Mining and Metallurgy. pp. 131–140.

Robertson, A., Ganza, P.B., and Noack, C.J. 2005. Underground trucking into the future. In *Proceedings of the Ninth Underground Operators Conference*. Melbourne, Australia: Australasian Institute of Mining and Metallurgy. pp. 285–288.

Spreadborough, J.C., and Pratt, A.G.L. 2008. Inclined troughed belt conveyor systems for underground mass mining operations. In *Proceedings of the Ninth Underground Operators Conference*. Melbourne, Australia: Australasian Institute of Mining and Metallurgy. pp. 179–188.

Sweigard, R. 1992. Materials handling: Loading and haulage. In *SME Mining Engineering Handbook*, 2nd ed. Edited by H.L. Hartman. Littleton, CO: SME.

CHAPTER 12.9

Hoisting Systems

Peter Tiley

INTRODUCTION

When it has been determined that a mine will include an underground section, it is necessary to review what kind of system will be used to transport workers and material to and from the surface to the underground, as well as to haul ore and waste from the mine to the surface.

This chapter deals with the selection, design, operation, and maintenance of hoisting plants that operate in vertical or near-vertical shafts. The term *hoisting plant* is used to collectively describe the openings and the equipment being considered. Information on the design and technical considerations to be reviewed when selecting a particular component or arriving at an arrangement of the hoisting plant is provided.

SHAFT HOISTING SYSTEMS

A shaft hoisting system consists of the following major sections, starting from the bottom of the shaft:

- A loading station for ore or a service station for workers and material
- Shaft conveyances called skips for ore transport and cages for transporting workers and material
- Ropes that suspend the conveyances
- A shaft that connects the underground to the surface and that is equipped with a system to guide the conveyances as they move in the shaft
- A headframe, located on the surface, that supports either the hoist itself or supports head sheaves that direct the headropes from the top of the conveyance to a ground-mounted hoist and that provides the tipping arrangements for the rock
- A hoist and hoist room

The headframe also typically includes an ore storage area into which ore and waste are temporarily discharged before being transferred to some other surface facility.

HOISTS

Two basic types of hoists are commonly used today: drum hoists, in which the hoist rope is stored on a drum, and friction hoists, in which the rope passes over the wheel during the hoisting cycle. Within each category are several variations.

Drum Hoists

Figure 12.9-1 shows the various arrangements of drum hoists. Drum hoists are usually located some distance from the shaft and require a headframe and sheaves to center the hoisting ropes in the shaft compartment and maintain a rope fleeting angle of less than 2°.

Single-Drum Hoist

The single-drum hoist may be used for balanced or unbalanced operation. When used for unbalanced hoisting, the cost of the electric drive may become quite high for long hoisting distances and high tonnages. This is because the motor must be large enough to produce sufficient torque to handle the combined weight of the rope, conveyance, and payload in the absence of a counterweight.

These costs are, however, offset by the reduced size of the installation because there is one less compartment, conveyance, head sheave, and head rope to accommodate, as well as a smaller hoist foundation and building.

There is virtually no difference in the power costs for balanced versus unbalanced operation, because power costs are for the energy consumed, which is derived from the number of tons hoisted.

As the shaft gets deeper, the difference in the size of the electrical drive for balanced versus unbalanced operation becomes less significant because the mass of the unbalanced rope, which is the same whether there are one or two conveyances, begins to dominate the total unbalance experienced by the hoist.

Single-drum production hoists are unusual because their hoisting efficiency is halved for the same skip size, but they are fairly common when used for service duty in which the hoisting cycle time is not as significant.

In a balanced hoisting system, one rope winds off the drum as the other winds on. When used with a skip or cage in balance with a counterweight, a single-drum hoist can service one or more levels because the location of the counterweight is not important. When used with two skips in balance, the single-drum hoist is only suitable for single-level hoisting. Any rope adjustments to locate the conveyance must be done manually. For shallow shafts with one layer of rope, no dividing of the

---

Peter Tiley, Consulting Engineer, G.L. Tiley & Associates Ltd., Flamborough, Ontario, Canada

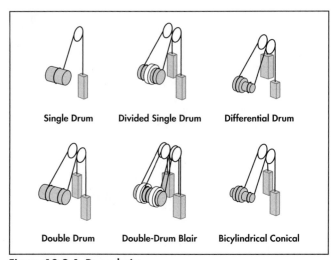

Figure 12.9-1 Drum hoists

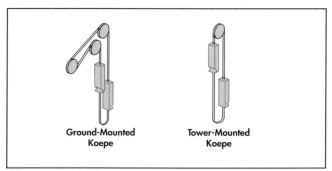

Figure 12.9-2 Friction hoists

drum is required. For deeper shafts, the drum must be divided to wind several layers of rope.

Divided Single-Drum Hoist
This type of hoist is used for deeper shafts with balanced hoisting when several layers of rope must be stored on the drum. Peak horsepower is less than with unbalanced hoisting because the skip weights are balanced. Because the payload and weight of the rope is not balanced, the maximum unbalanced load occurs when the loaded conveyance is at the bottom of the shaft.

Split Differential Diameter Drum
A third, though uncommon, type of single-drum hoist is the split differential diameter drum. In this arrangement, each of the drum compartments is of a different diameter. This type of hoist is used with a conveyance and counterweight in balance or with a skip and cage in balance. If, for example, the counterweight rope is wound on the smaller-diameter drum, it moves less than the main conveyance, and rope adjustment problems are reduced.

Double-Drum Hoist—One Drum Clutched
Although more expensive than a single-drum hoist, the double-drum hoist with one drum clutched has certain advantages. With this type of hoist, it is possible to make quick adjustments to the ropes due to initial stretch. As a service hoist with cage and counterweight, this type of hoist can serve several levels efficiently. As a production hoist with two skips, the ropes can be adjusted to maintain balanced hoisting at any level in a multilevel operation.

Double-Drum Hoist—Both Drums Clutched
The double-drum hoist with both drums clutched has the added feature of allowing hoisting to continue in one compartment should something happen to the other compartment. This is an excellent feature if there is only one hoist available. This type of hoist is also favored during shaft-sinking operations.

Multiple-Rope Hoist—Blair Type
With this type of hoist, each conveyance is suspended from two hoist ropes that are each coiled onto the same drum. The advantage of this is that by doubling the number of ropes per conveyance, the conveyance payload can be increased. This type of hoist was developed in South Africa for deep mines, typically more than 1,800 m (5,900 ft). Another advantage of this arrangement is its ability to dispense with safety dogs on the cage (which are only required for single-rope hoists).

Two such hoists are now operating in Canada, and several more are scheduled to be installed in the near future.

Bicylindrical Conical Drum Hoists
This type of hoist is no longer manufactured, but there are still a few in operation in North America.

The drums on these hoists, both single- and double-drum configurations, are made in two steps with a conical section in between.

The hoists were developed before multilayer winding was practical. Prior to multilayer winding, drum diameters simply increased to match depth. This practice leads to very large diameters (10 m [33 ft] or more), resulting in very high starting torque requirements. By beginning the acceleration on a small diameter, the high starting torque requirement could be avoided.

Friction Hoists
Figure 12.9-2 shows the two possible arrangements of friction hoists. The Koepe or friction hoist was developed by Frederick Koepe in 1877. It consists of a wheel with grooved liner or liners made from a friction material to resist slippage. The hoist rope is not attached or stored on the wheel. In early installations, the hoist was mounted on the ground, and a single rope was wound around the drum and over the head sheaves to the two conveyances in a balanced arrangement. In addition, a tail rope of the same weight per unit length as the headrope was suspended in the shaft below each conveyance. Thus the only out-of-balance load was simply the payload. Friction hoists can be either tower mounted or ground mounted.

Tower Mounted
The main reason for selecting tower mounting is the reduction in cost of the structure, because the rope loads are confined internally. This arrangement also provides a very small footprint that allows the structure to be supported by the headframe's collar concrete in the slip-formed design or by localized piled foundations in the structural-steel design of the headframe.

Because the Koepe wheel diameter is generally larger than the conveyance compartment centers, deflection sheaves must also be installed in the headframe below the hoist to move the ropes over to match these centers.

Tower mounting is also preferred in cold climates because the conveyance ropes are isolated from the weather.

Ground Mounted

A headframe for a ground-mounted friction hoist must be designed to resist the overturning loads produced by many headropes. These buildings are generally of a structural steel design and require the use of a pair of cluster sheaves to locate the rope centers over the conveyance compartments. Because they must support the entire suspended rope load, rather than simply deflecting the ropes by 10° to 15°, the cluster sheaves are much more robust and are, therefore, significantly more costly than the deflection sheaves used in tower-mounted applications.

Comparison of Friction and Drum Hoists

A friction hoist system differs from a drum hoist system in performance as well as components. Therefore, when deciding which type of hoist to use, it is necessary to compare the two complete systems rather than the two hoists alone. In addition to comparing the total capital costs of the hoist, headframe, ropes, conveyances, and shaft, it is necessary to consider operating costs, maintenance costs, reliability, power supply system, local custom, and individual preference.

The following general statements help distinguish between these two hoisting systems:

- Double-drum hoists are the preferred hoist for shaft sinking.
- Double-drum hoists are the best choice for hoisting in two compartments from several levels.
- Drum-type hoists are best suited for high payloads from shallow depths.
- The limitation on a drum hoist employing a single rope is the ultimate strength of the rope, because large ropes are difficult to manufacture and handle.
- The depth capacity of drum hoists can be extended by using two ropes per conveyance (Blair-type hoist), and with this arrangement, Blair hoists can be used for depths exceeding those of either single-rope drum hoists or friction hoists.
- Friction hoists with multiple ropes can carry a higher payload and have a higher output in tons per hour than drum hoists within a range of depths from 460 to 1,520 m (1,500 to 5,000 ft).
- Friction hoist mechanical operation is very simple, has a low rotational inertia, and is less costly than a drum hoist.
- Friction hoists have a lower peak power demand than drum hoists with the same output.
- The friction hoist can operate on a relatively light power supply.
- Rope maintenance is more intensive and difficult for friction hoists.

Hoist Components

The major components that make up a hoist are the drum or wheel, bearings, gearing, brakes, clutch motor, power conversion, and controls.

Drum

The drum of a drum hoist must be designed to store the amount of rope required for the shaft depth and accommodate the stresses caused by the rope tension loads. The prediction of these rope loads is well documented in the literature.

Although originally provided with no machined grooves (i.e., they were smooth), drums today are generally supplied with parallel grooving, which includes a half-pitch crossover twice per revolution. This grooving pattern was developed by LeBus International, Longview, Texas, to achieve stable spooling in multilayer winding. Drums can also be supplied grooved in a helical pattern, which is a continuous spiral from one end of the drum to the other. This pattern is limited to three layers of rope, because beyond this number of layers, the rope-winding pattern becomes unpredictable. LeBus International's grooved drums, in contrast, have been used to wind as many as 15 layers of rope without a problem.

Rope resonance must be considered in the selection of the grooving pattern. This is discussed in a later section of this chapter.

Wheel

The wheel of a friction hoist is not subjected to crushing forces. As a result, the drums are much lighter and are made wide enough to accommodate the rope-end attachment dimensions, as well as the space for the deflection sheaves.

The linings for the rope treads on Koepe hoists were originally made of wood or leather, which were then replaced by modern plastic.

Bearings

As shown in Figures 12.9-3 and 12.9-4, the hoist drum or wheel is supported on a shaft that rides in two or more bearings. Bearings provide the mechanism that supports the shaft loads, and bearings allow it to rotate at the same time with very little friction.

There are two types of bearings: sleeve bearings, which use a film of oil to keep the shaft out of contact with the fixed part of the bearing, and roller bearings. Sleeve bearings are being supplanted by spherical roller bearings that have the following advantages:

- Low coefficient of friction
- Economy of space
- Simplicity of lubrication
- Practical elimination of wear due to roller point contact
- Maintenance of accurate alignment
- Consistency in design and manufacture

The main disadvantages of the latter are the fact that they are not split, which complicates their replacement, and long order times are involved when bearings larger than about 500 mm (19.7 in.) in diameter are required.

Gearing

Hoists can be driven either directly by large-diameter, relatively low-speed AC or DC motors (90 rpm or less) or through a gear-reduction system by small-diameter, high-speed (400 to 1,800 rpm) motors. With the change in power conversion to variable-frequency drives, the motors have altered to AC machines operating in the 900-to-1,800-rpm range.

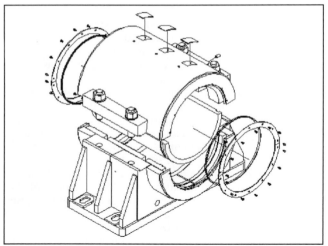

Figure 12.9-3 Sleeve bearing

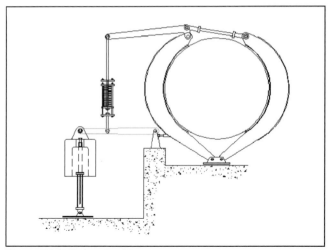

Figure 12.9-5 Jaw braking system

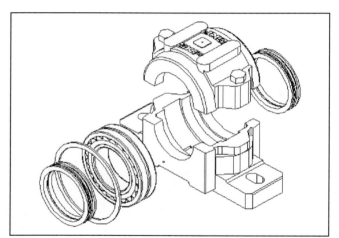

Figure 12.9-4 Spherical roller bearing

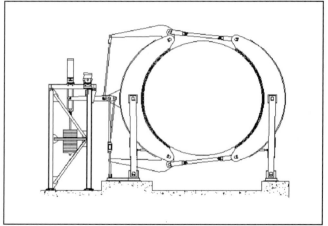

Figure 12.9-6 Parallel motion braking system

Gear reduction was commonly a single-stage system that incorporated a large gear fixed to the drum shaft and driven by one or more pinion gears. This cannot be used practically for the sealed, self-lubricating, multistage reducers now for high-speed AC motors. Reducers are now fully enclosed, multistage, and equipped with oil coolers as well as bearing vibration-monitoring systems. These utilize flexible couplings between the motors and high-speed input shafts as well as between the low-speed output shaft and the hoist drum shaft.

Brakes

A hoist must be equipped with a mechanical braking system to decelerate, stop, and hold the load. Although the electrical motor, drive, and controls can initially decelerate and stop the hoist, this equipment cannot hold the load because of thermal limitations inherent in the drive. Electrical systems also require a source of energy to function. For these reasons, hoists must be provided with mechanical braking systems that ultimately rely on either gravity or spring force to operate.

Drum brakes and disc brakes are the two main types of braking systems. Until the mid-1990s, drum hoists were exclusively equipped with drum brakes in jaw or parallel motion configuration, mainly because they were part of the hoist speed-control system rather than solely for holding or emergency braking functions. Most new hoists, both drum and friction, are equipped with disc brakes.

Figures 12.9-5, 12.9-6, and 12.9-7 illustrate the designs of jaw, parallel motion, and disc braking systems, respectively. Brake control systems are now required to control deceleration rates to very high accuracy to protect passengers from injury, limit the dynamic stresses that arise in the hoisting ropes during emergency stops, and prevent rope slip on Koepe hoists. This level of accuracy is readily achievable with the higher operating pressures and lower inertias associated with disc brakes.

In addition to accurate control of deceleration rates, braking systems must be capable of dissipating the energy absorbed during emergency stops. Disc brakes have lower mass than drum brakes and run at higher surface pressures. Therefore, they tend to heat up more quickly. By manufacturing brake discs in several segments and bolting the segments to the hoist drum or wheel, a disc brake can be designed to expand freely under thermal load and, therefore, avoid distortion or failure.

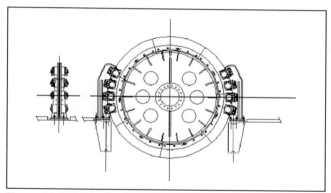

Figure 12.9-7 Disc braking system on a Koepe hoist

Clutch

Double-drum hoists used for unbalanced hoisting, such as when sinking or when operating from several levels, must have at least one of the drums clutch-connected to the drum shaft. This assembly consists of a clutched or floating drum that is supported on the shaft through a sleeve bushing and a clutch arm, which is used to transfer the torque to the drum. The clutch arm consists of a spider or gear that slides axially along the shaft supported by a hub driven by keys or splines or driven by a hexagonal section of the shaft. The drum also incorporates a clutch ring, which has teeth matching the clutch arm on its entire periphery, bolted to the side of the drum. As the spider moves axially, these two sets of teeth engage or disengage. The tooth pitch determines how accurately the rope lengths on each drum can be matched. Driving torque is then transmitted from the clutch spider to the drum through the bolts used to attach the clutch ring to the drum.

As the drum must be prevented from rotating when it is unclutched, it is necessary that the clutch's operating mechanism and the brake holding the drum be interlocked both electrically and mechanically.

Drive Motor and Power Conversion

Hoists have used energy sources such as steam, compressed air, and internal combustion engines to drive them. These have largely been replaced with electrical power.

Electrical drives for hoists originally employed AC wound-rotor induction motors using variable resistance in the secondary for speed control. These systems were then supplanted by DC machines that utilized motor-generator sets to provide DC power. The motor-generator sets were replaced by so-called static drives in the late 1960s. The name *static drive* referred to the fact that the power was converted from AC to DC using thyristors, which are made from silicon wafers and involve no moving parts. The DC technology began to disappear in the late 1980s when it was replaced with AC technology that utilized synchronous motors and variable-frequency sources called cycloconverters. The variable-frequency sources have now been replaced with pulse width-modulated, digitally controlled equipment that utilize transistors (injection-enhanced gate resistors or isolated gate bipolar transistors) for power conversion. The AC motors are now a mixture of synchronous and squirrel-cage designs, with the squirrel-cage design being the preferred option because of its simplicity and ruggedness.

This new generation of variable-frequency static drives also provides AC power with low harmonic content and ensures operation at unity power factor. These drives consist of two inverters back to back. The first converts line frequency AC to DC; the second, DC to variable frequency AC.

Hoist Control Systems

The hoist control system consists of two components: active and passive. The active component starts and stops the hoist, controls its speed throughout the hoisting cycle, and displays the state of the hoist to the operator. The passive component monitors the speed and position of the hoist itself as well as the state of its subsystems, such as the brakes, hydraulic power units, motor ventilation, and similar auxiliary equipment. The passive component also signals the active component when the system is operating outside of its normal range and thereby initiates a stoppage of the system until the fault that raised the alarm is cleared.

All of the required functions are now fulfilled by digital devices call programmable logic controllers. These devices utilize redundancy to a large degree to ensure that they are reliable.

The operator's interface has also evolved to what is called a human–machine interface, which is a graphical display provided on a computer monitor.

HOIST SELECTION

The capacity of a hoist depends on the drum or wheel dimensions, number and size of ropes, speed, and horsepower.

The mechanical components, such as drum, shaft, bearings, gearing, and brakes, are svized to match the rope loading; the electrical components are sized to match the torque required to handle the loads and the speed at which the hoist will run.

The rope is selected to handle the required payload while operating at a reasonable or legislated safety factor.

Selection Procedure for Production Drum Hoists

Sizing the hoist runs through an iterative process, which requires the following initial information:

- Hoisting depth
- Required annual production
- Hours available for hoisting

The process of selection is as follows:

1. Using the depth, calculate the hoisting cycle time. The cycle time will depend on the maximum speed, acceleration and deceleration rates, creep speeds and distances (arriving at and departing the loading pocket and dump), and the loading/dumping time when the hoist is stopped.
2. Varying these items will enable the designer to optimize the selection. Good guidelines are
   - a maximum speed of 5 m/s (1,000 fpm) per 300 m (1,000 ft) of depth up to a maximum of 20 m/s (4,200 fpm);
   - acceleration and deceleration times of 15 seconds;
   - a creep speed of 1.5 m/s and a creep time of 5 seconds; and
   - load and dump times of 10 to 15 seconds, depending on density of the load.
3. Use the cycle time to calculate the cycles per hour.

4. Use the required annual production and available hoisting times per year to calculate the skip payload.
5. Use the skip payload and a reasonable skip tare mass based on 70% of the skip payload to determine the total attached load.
6. Multiply the attached load by 8.5 to determine the required rope strength.
7. Select a rope with this strength.
8. Check that the rope strength also meets the required safety factor (typically 5 but may also have a factor that is depth dependent) when the rope weight is added.
9. Determine the drum diameter based on 80:1 drum-to-rope diameter (this ratio depends on the rope size).
10. Estimate the motor power using kW = (payload × speed)/102, where the payload is measured in kilograms and the speed is measured in meters per second.
11. Estimate the inertia of the rotating parts based on historical data.
12. Carry out a duty cycle calculation to verify the required motor power.

A typical double-drum production hoist cycle, starting with empty skips and loading pocket, no power on the hoist motors, and with the hoist brakes applied, is as follows:

1. Open the loading pocket door and load the empty skip.
2. Close the loading pocket door.
3. Apply power to the hoist.
4. When power is established (torque proving), release the brakes.
5. Accelerate to creep speed, about 1 m/s, and creep away from the end zones, about 2 m (6 ft), to clear dump scrolls.
6. Accelerate to full speed in 15 to 20 seconds.
7. Run at full speed.
8. Decelerate to creep speed in 15 to 20 seconds.
9. Enter dump scrolls at creep speed, about 0.75 m/s (150 fpm), and creep through dump scrolls until the full skip is in fully dumped position or empty skip is in position at the loading pocket.
10. Stop the hoist.
11. Repeat the cycle.

It is good practice to stop the cycle when the empty skip stops at the loading pocket rather than the full skip arriving at the dump point. Headframes are designed to allow for rope stretch and rope length mismatch, which means that the full skip can stop anywhere in about a 1- to 2-m (3- to 6-ft) zone when it is dumping. This is not possible at the loading pocket, because the headrope will stretch when the skip is being loaded and the loading window in the skip is limited in height.

Selection Procedure Drum Hoist for Service

The selection procedure for a service hoist is much simpler than a production hoist because service hoists are sized to handle a maximum payload at a speed that is largely independent of depth. In this case, the following guidelines apply:

- The maximum speed is 8 m/s (1,500 fpm).
- The acceleration and deceleration times are 20 seconds.
- The creep speed is 1.5 m/s (300 fpm), and the creep time is 10 seconds.
- The waiting time is 30 seconds.

The rope size is determined as before, except the cage weight is generally 50% of the payload.

When using a counterweight, it should be sized as the cage mass plus one-half the cage payload, which ensures that the maximum average unbalance for the hoisting cycle will never exceed one-half the payload.

Selection Procedure for Koepe Hoists

The selection procedure for a Koepe hoist is identical to that of a drum hoist, except that the number of ropes must be defined. A good starting point is to use four ropes, as this provides a reasonable limit on complexity while minimizing the size of the ropes.

One item that is different is the mass of the conveyances. As the ratio between the tension in the ropes on the loaded side of the wheel and the tension in the ropes on the empty side of the wheel must be kept below about 1.5, the conveyances must be heavy. A good guideline is to have the conveyance weight equal to the payload and then increase it until the ratio is 1.5 or less.

Koepe hoists often run on rope guides, so provision must be made in the hoisting cycle to slow the conveyance down as it enters the ends of travel where it is guided by fixed guides. These guides are required to keep the conveyances' lateral motion in check when it is traveling in the narrow confines of the headframe and shaft bottom. The guides are tapered to allow for the uncertainty in the conveyance position and are often called "spear" guides as a result.

The conveyance speed when entering these guides is usually kept below 3 m/s (600 fpm), but it can be as high as 6 m/s (1,200 fpm) if the system is properly designed.

The production hoisting cycle for the Koepe hoist is identical to the drum hoist, except for a situation when the skips are running on rope guides. This means that a high creep speed section will occur as the skip slows down to enter the spear guides and then runs in them until it reaches the dump point at which time it will decelerate to a low creep speed.

In addition to determining the hoist's capacity, consideration of the following items is critical for achieving efficient and trouble-free operation of friction hoists: T_1/T_2 ratio, rope tread, deflection sheaves, differential rope loads, and tail ropes.

T_1/T_2 Ratio

The T_1/T_2 ratio is the ratio of the static rope tensions on each side of the drum, where T_1 is the weight of loaded conveyance plus weight of ropes, and T_2 is the weight of unloaded conveyance plus weight of ropes. The maximum recommended value for a particular installation is a function of the coefficient of friction of the tread material, the acceleration and deceleration required, and the angle of lap of the rope on the drum. A typical maximum static value of T_1/T_2 is 1.5. It is also necessary to consider the dynamic loads that occur when the hoist is accelerating and decelerating, both under the normal and braking conditions. Because acceleration will decrease the rope load and deceleration will decrease it, T_1/T_2 can become too high, causing the rope to slip on the drum. When this happens, the coefficient of friction drops and control is lost. There will also be tread damage due to friction temperature rise and abrasion.

The maximum coefficient of friction that should be used when calculating the allowable T_1/T_2 ratio is 0.25. This number is achievable even with wet ropes, as long as they are not overlubricated.

The tension increase during acceleration is given by the factor $(1 + a/g)$; the tension decrease during deceleration is give by the factor $(1 - a/g)$. Here, a is the acceleration/deceleration and g is the acceleration due to gravity (9.81 m/s² [32.17 ft/s²]). Thus, the dynamic tension ratio is the static tension ratio $T_1/T_2 \times (1 + a/g)/(1 - a/g)$.

The amount of unbalance that the friction can support is given by

$$T_1/T_2 = e^{(\mu \times \theta)}$$

where

e = base of the natural logarithm (2.713)
μ = coefficient of friction
θ = angle of wrap measured in radians (typically π or 3.14 radians, equivalent to 180°)

As an example of the effect of dynamic variation in the rope tension on the margin to rope slip, assume that the actual static tension ratio is 1.5 and that the system has been designed with an angle of wrap of 180°. A coefficient of friction of 0.25 will produce an allowable static tension ratio, T_1/T_2, equal to 2.2, well above the actual static T_1/T_2 of 1.5.

If the hoist is subjected to a deceleration rate of 3 m/s², the dynamic T_1/T_2 increases as follows by the factor $(1 + 3/9.81)/(1 - 3/9.81) = 1.88$, giving a dynamic T_1/T_2 of $1.5 \times 1.88 = 2.82$. This is greater than 2.2, indicating that the ropes will slip if the coefficient of friction is only 0.25.

This example shows why braking systems for Koepe hoists must be carefully designed.

Rope Tread

Selection of an appropriate tread material for the drum rope grooves and tread pressure is important. As the rope passes over the drum, the load and thus the strain in the rope changes. This change in strain tends to cause the rope to creep (as opposed to slip) on the drum. A resilient tread material such as polyurethane tends to flow with the change in strain and reduce creep.

Typically, the average tread pressures (i.e., the sum of the rope tensions/projected rope area on the drum) of 2,069 kPa (300 psi) with lock coil ropes, 1,724 kPa (250 psi) with flattened strand ropes, and 1,379 kPa (200 psi) with round strand ropes on plastic treads will give good life.

More recently, nonrotating flattened strand ropes have replaced lock coil ropes in North America. To compensate for the greater tendency of these ropes to wear the tread liners, the liners have been replaced with harder plastics. Although these materials lack the resilience of polyurethane, good rope life can still be obtained provided the hoisting plant is properly designed in the first place. This is explained in greater detail in the sections that follow.

Deflection Sheaves

Deflection sheaves are used when the pitch circle diameter of the friction hoist pulley wheel (sitting above the shaft) is greater than the center-to-center distance of the compartments. In addition to positioning the hoisting ropes, the use of deflection sheaves has the advantage of increasing the angle of contact of the ropes on the wheel, permitting a higher T_1/T_2 ratio before slippage will occur.

However, deflection sheaves have the disadvantage of requiring additional torque during the hoisting cycle, increasing the height of the headframe and putting reverse bending into the ropes, which can reduce their life.

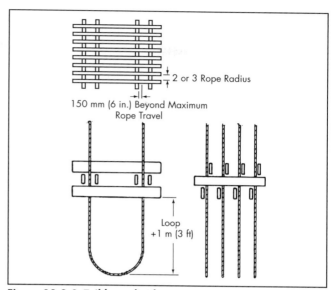

Figure 12.9-8 Tail loop divider

Differential Rope Loads

A great concern with multirope friction hoists is the equalization of loads between the ropes. The two most common causes of differential loading are a difference in rope length or in circumferential lengths of each rope groove on the wheel.

Analysis of this problem and procedures to identify and correct the effects are well covered in the technical literature. In the initial design stage, it is important that steps be taken to reduce its effects. Because there are no hard and fast rules, the following guidelines have been developed in the industry.

The ratio of rope lengths (skip at loading pocket to hoist wheel)/(skip at dump to hoist wheel) should be greater than 50. The total number of revolutions of the drum per trip should be less than 80.

The number of ropes should be kept to a minimum. There will be less difficulty maintaining equal tension on a four-rope hoist than on a six-rope hoist. In addition, a larger wheel will be required for a four-rope drum having fewer revolutions per trip, lower rope bending stresses, and lower sensitivity to tension variations. This requirement may conflict with rope size choice, however.

Rope attachments should have both fine and coarse length adjustment. Rope attachments that incorporate a hydraulic cylinder to balance the loads equally should always be used.

Tail Rope Loop Dividers

To avoid mishap and possible mechanical damage through entanglement, the balance rope loop area should be well controlled in the sump. One method is by constructing a loop divider. The balance ropes should be restricted quite closely in the direction of their lateral movement.

Early contact with the dividers between individual ropes will provide good control in the sump area and also assist in ensuring that adequate stability through the shaft is maintained. Longitudinal movement, on the other hand, should not be overly restricted. The dividers should be designed to avoid trapping any material from spillage. Some operators have found that a wire rope equipped with numerous ultra-high molecular

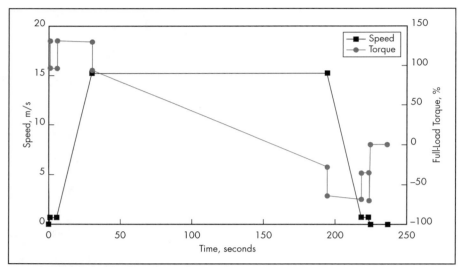

Figure 12.9-9 Double-drum hoist duty cycle

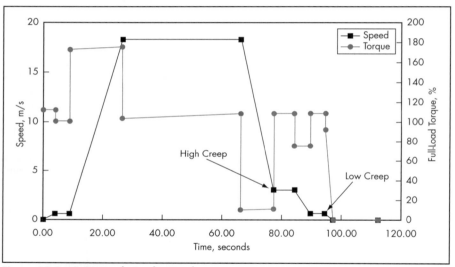

Figure 12.9-10 Koepe hoist duty cycle

weight polyethylene rollers, about 150 mm (6 in.) in diameter and placed in the configuration shown in Figure 12.9-8, has proven to be most successful in this connection. The elevation of the loop divider for good results is one loop diameter plus approximately 1 m (3 ft) above the bottom of the loop. Some operators have placed such dividers at several elevations in the sump area to ensure good contact over the height of the tail loop.

DUTY CYCLES

A duty cycle is a determination of the root mean square (RMS) thermal and torque load on the electrical equipment. The calculations assume that motor current is directly proportional to motor torque.

Duty cycles are usually carried out assuming the motor is force-ventilated. If the machine is self-ventilated, then factors have to be applied to account for the amount of time the motor spends accelerating and decelerating and at rest. These factors differ for AC and DC machines because of the differences in their construction.

The heating effect on each segment of the duty cycle is calculated using

$$\text{heat generated} \sim \text{torque}^2 \times \text{time}$$

The rate of cooling on each segment of the duty cycle is calculated using

$$\text{equivalent cooling time} = \text{cooling factor} \times \text{actual time}$$

The thermal load over the whole duty cycle is the square root of the sum of each segment, that is

$$\text{RMS torque} = \sqrt{\frac{\sum \text{torque}^2 \times \text{time}}{\sum \text{cooling time}}}$$

These results are reasonably accurate for force-ventilated motors but can be misleadingly low for modern self-ventilated machines.

A typical speed-and-torque-versus-time chart corresponding to a duty cycle for a double-drum hoist is shown in Figure 12.9-9 and for a Koepe hoist in Figure 12.9-10.

The drum hoist usually has a significant regenerative portion in its cycle because the large unbalance caused by the rope changes from a positive load to a negative one when the conveyance passes the shaft midpoint. Because Koepe hoists utilize balance ropes, this reversal of load is absent from the cycle.

HOIST UTILIZATION

The previous section outlined the procedure for sizing the hoisting system based on the production or service duty required. Inherent in this procedure is a step for determining how much production time is available each year.

Many mines have multipurpose shafts, that is, the shaft will provide both production and service capacity for the mine. Production and service can proceed at the same time if the shaft is equipped with both a service and a production hoist, as long as the compartments are physically separated by brattice or the personnel are not transported in the shaft while it is in production. It is also possible to use the same hoist for both production and service duty, but this requires a much higher degree of planning and leaves little room for any hiccups in the mining process.

Table 12.9-1 demonstrates the daily examination and testing required on a hoist in North America. In addition to examination and testing, time must also be allotted for maintenance of the hoists, conveyances, shaft, and shaft furnishings. The amount of time required will depend somewhat on the shaft configuration and hoist type, but all of the items listed in Table 12.9-2 are required to some degree.

Experience has shown that the average time available for production hoisting can vary from 12 to 20 hours per day when spread out on an annual basis. For most mines, the production time available will be between 16 and 18 hours per day.

The total hoisting time will also depend on the shift system in practice at the mine. Although most production hoists are automatic, they may have to be shut down during shift change if proper isolation between the cage hoist and production hoist is unavailable or if there is only one combined production/service hoist. Bearing all of this information in mind will permit the mine operator to determine how many hours of actual hoisting is available on an annual basis. Most hoisting plants today will achieve about 95% availability, as long as time has been allotted for their proper maintenance.

CONVEYANCES

The transportation of ore, waste, personnel, and material requires the use of a conveyance. Conveyances for handling ore and waste are termed *skips*; those used to carry personnel and material are termed *cages*.

Skips

The skipping cycles discussed above have implicit assumptions as to what type of skip is being used. There are three main types of skips in use today: the overturning (Kimberley), the swing-out body, and the fixed body, as shown in Figures 12.9-11, 12.9-12, and 12.9-13, respectively.

Overturning Skip

The overturning or Kimberley skip consists of a bail frame, a shaft to support the skip, and the skip body itself. The

Table 12.9-1 Daily hoist examination and testing required in North America

Procedure	Time, minutes
Run through shaft	5
Test overwind and underwind protection	10
Test brakes (dynamic drag)	10
Verify safety devices	20
Examine head- and tail ropes	30
Examine rope attachments	15
Examine hoist and deflection sheaves	10
Total	100

Table 12.9-2 Annual shaft maintenance summary for a shaft with two hoists

Activity	Hours per Activity	Annual Total, hours
Daily hoist examination and testing*	1.5	433.5
Daily rope examination	4	1,196
Weekly hoist, rope, and conveyance examination*	8	208
Weekly shaft equipment and guide examination*	4	208
Monthly rope adjustment and examination	8	96
Monthly shaft equipment repairs†	8	96
Annual rope electromagnetic testing (4 ropes)	32	32
Annual hoist testing	16	16
Annual brake testing	8	8
Biannual rope change	48	24
Total	—	2,317.5
Annual hoisting time with 1-week shutdown	357×24	8,568
Available hoisting time	8,568–1,121.5	6,250.5
Average per day	7446.5/357	17.51
Shaft availability	—	73%

*Will require one crew per hoist.
†May be done on special shutdown.

skip body rests on the shaft. The skip is dumped through the action of a bull wheel mounted on each side (at the top) of the skip body. Each bull wheel engages a scroll, causing the skip body to rotate around the shaft as the skip is being hoisted through the dump. The design of the tipping path is complex because the payload is being discharged as it moves along the path. The horizontal and vertical forces that need to be considered in designing the scrolls and their support steel are dependent on the speed at which the skip enters the dump, the inertia of the skip plus payload, the acceleration the skip produced by the curvature of the dumps scrolls, and the gravity forces.

The advantages of Kimberley skips are that there is less shaft spillage when handling fine or wet material and that they can handle larger pieces for a fixed cross-sectional area. The disadvantages are that the capacity is limited because of the height-to-width ratio; the payload ratio is usually lower, particularly with sticky material that tends to build up in the bottom; they generally impose higher stresses on the headframe and bins; and they require more headframe height to dump.

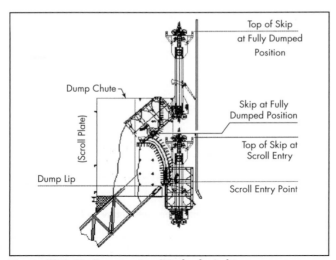

Figure 12.9-11 Overturning (Kimberley) skip

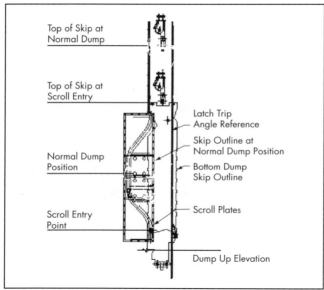

Figure 12.9-12 Bottom-dump skip (the most popular type of swing-out-body skip)

Swing-Out-Body Skip

The swing-out-body skip also consists of a bail frame, a shaft, and the skip body. In this case, the skip body is suspended from the shaft and the skip is dumped by means of rollers mounted on each side (at the bottom) of the body. The rollers engage a scroll and cause the body to pivot away from the bail into the dump. At the same time, a discharge door, which forms the bottom of the skip, opens and the material is discharged. The skip can be dumped by using either a mechanical operator or scroll or by using a hydraulic operator that pulls the skip body out of the bail. These options are illustrated in Figures 12.9-14 and 12.9-15.

For bottom-dumping skips in general, the cross-sectional area is partially determined by the size of the material to be handled. The door opening is sized to be at least three times the dimensions of the largest piece to be handled to prevent possible bridging.

The advantages of swing-out-body skips are that they have a more favorable payload ratio than the overturning type, the tipping path is less complex than for the overturning skips, the dumping distance in the headframe is less than for the overturning type, and they can be built to handle much larger payloads because dumping does not involve lifting the entire mass of the skip body and its payload. The main disadvantage is that muck buildup at the hinge point in the door can prevent the door from closing and thereby place large forces on the headframe scroll or skip body during the closing cycle. In addition, with wet material, there could be spillage or dribble from the door lip; and the bottom door mechanism is more complex than in the fixed-body type.

Fixed-Body Skip

In fixed-body skips, the body of the skip remains fixed and does not move out of the confines of the skip compartment during dumping. Instead, a door at the bottom of the body swings open to form a chute and allows the material to discharge into the dump bin. This type of skip may or may not have a bail.

There are several types of fixed-body skips. The most common are the Rolachute, Arc-Gate, and Sala-type skips. The difference among these types is the mechanism involved with the opening and closing of the door. In the Rolachute skip, the door is held closed by means of a door latch. When the skip enters the headframe, a latch track releases the latch, allowing the door to open and the material to discharge. With the Arc-Gate skip, a roller is mounted on each side of the door. When the skip enters the dump area, the roller engages a scroll, causing the door to swing open. In the Sala-type skip, the door is operated by means of air cylinders mounted on the skip body. Compressed air to operate the cylinders is provided through a yoke that is permanently mounted in the headframe and picked up by the skip as it enters the dump. The Sala design, with its onboard air cylinders, adds additional complications brought about by the requirement to interface with an external source of air.

The cross-sectional area of fixed-body skips is determined by the size of the material to be handled. Because fixed-body skips can be built without a bail, the cross-sectional area of the skip can be increased by 10% to 15% when compared to skips with a bail for the same sized compartment.

The advantages of fixed-body skips are as follows:

- The skip can load and dump through the guides.
- The skip has no bail, thus allowing a larger cross-sectional area to be used for the skip body (this can reduce the overall length of the skip for a given capacity).
- The only moving part is the door that forms its own dump chute.
- The skip requires less travel distance to open the door (none at all with the Sala-type skip).
- This type of skip places less stress on the headframe, skip body, and ropes during the dumping operation.

Both the Arc-Gate and Sala types have the added advantage that no part of the skip, particularly the dump door, protrudes beyond the skip body, thereby avoiding any possibility of hanging up or impacting shaft furnishings should the door fail to close completely.

The disadvantages of this type of skip are as follows:

- Problems can occur with sticky muck, as there is no shaking action when dumping.

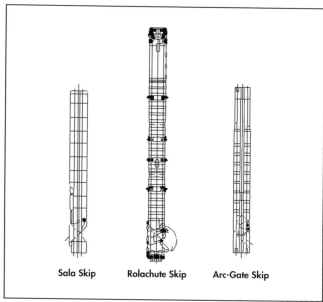

Figure 12.9-13 Fixed-body skips

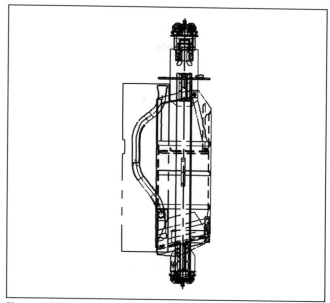

Figure 12.9-14 Dump with mechanical scrolls

- The body necks down at the transition to the dump portion of the assembly, and this can cause hang-ups.
- The entire skip must be removed for replacement of wear plates, not simply the body.
- There is always some dribble into the shaft because the chute does not extend beyond the projection of the skip's body (except with the Rolachute design).

Skip Design Considerations

When selecting a particular skip, the components to be considered during the evaluation process include design methods, body, doors, crosshead, bail, guide rollers, and guide shoes.

Design Methods

Skip components may be designed based on experience and practice or through analytical methods that define the loading being placed on the various components of the skip. The types of loading to be considered are dynamic loading (caused by filling, hoisting, and dumping the skip) and static loading. Factors of safety are usually specified by legislation.

Body

The skip body is designed as a rectangular box with stiffeners. The type of material, its thickness, and the stiffener size and spacing are the variables to be examined. Particular attention must be given to the area on the backside of the skip that receives impact and abrasion from loading.

Skip bodies are usually lined with abrasion-resistant material to increase overall life. Two types of steel and elastomers have been used for skip liners. Abrasion-resistant steels are members of the high-strength, low-alloy family of steels. Manganese steels are used because they have the ability to work-harden during service. Elastomers have the advantage of reduced weight, less ore sticking and buildup, corrosion resistance, and ease of handling during replacement.

The use of aluminum is also increasing, particularly in drum hoist applications. Such skips have a lower mass and allow a higher payload to be hoisted.

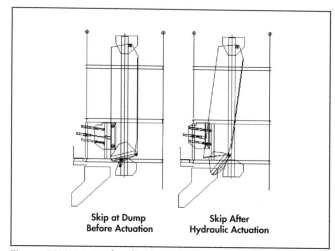

Figure 12.9-15 Hydraulic dump before and after actuation

The impact areas are often lined with heavy rubber materials that were originally designed for use inside of rotating grinding mills. These liners are quite thick, typically 75 mm (3 in.), so they must be accounted for when estimating the live volume of the skip.

Doors

Both swing-out-body and fixed-body skips have doors at the bottom of the skip. Doors must be designed to withstand the impact of large pieces falling on the floor during loading, the static loading of a full skip, and abrasion during unloading.

With swing-out-body skips, the door is hinged at the back, and several problems are associated with leakage paths between the door and the body. These door leakage problems may be solved through the use of rubber sheets, flexible materials, or special modification of plates at the door. Fixed-body skips generally have fewer problems with leakage or door closure. Problems may occur with both types when they are loaded on the side opposite the dump site. In this case, a vertical stop may be placed for the dump roller to react against. Doors are made and lined with the same material as used in the body.

Crosshead

The crosshead fastens the bail to the rope. It must be designed to take all the skip loads, including acceleration, impact, and service factors. The location of the crosshead above the skip is particularly important during loading. It should not interfere with the flow of muck from the loading pocket. The rope stretch in deep mines can be several feet and must be considered.

Bail

During hoisting, the bail supports the body pivot, prevents the body from moving, and supports the door. At the dump, the bail is subjected to horizontal loading and supports the door. It is thus subjected to both bending and axial loading.

Guide Rollers

The purpose of guide rollers is to keep the guide shoe off the guides. This in turn results in a smoother ride, reducing dynamic loading between the skip, guide, and buntons when traveling.

When selecting guide rollers, there is usually a conflict between a large-diameter tire, adequate shaft clearance, and the cross section of the skip. Left unresolved, the final design usually results in a small-diameter tire. Such tires need frequent replacement. Tire sizes range from 150 to 350 mm (6 to 14 in.) in diameter. Tires are rated on a load-speed basis. High-speed hoists require special guide rollers that are typically made from wire-reinforced rubber. Guide rollers should be part of an assembly comprising springs and shock absorbers in order to work effectively. Systems in which the guide roller is simply fixed to a bracket bolted to the skip are of little use.

Guide Shoes

Guide shoes are used on vertical conveyances to keep the main structure from being worn away by the guides. They are considered as expendable items to be replaced when they are too worn for further use. Clearance between the guide and guide shoe varies from 6 to 12 mm (0.25 to 0.5 in.). Guide shoes can be made from many materials, such as cast iron, mild steel, hardened steel, abrasion-resistant steel, brass, and bronze. The amount of wear to be expected is directly proportional to the load on the guide shoe and inversely proportional to the hardness of the material.

Cages

Cages are used primarily to move personnel and materials entering into and exiting from the mine. In some instances, cages are used to hoist cars loaded with ore and waste. The design and construction of cages are similar to those used for skips.

A cage is merely an enclosed box opened by a door or doors and suspended from a hoisting rope. The dimensions and capacity of the cage are determined by the quantity, volume, weight, and dimensions of the materials and items to be transported.

Cages are much lighter in construction than skips because they are not exposed to the impact loading and abrasion produced when loading and dumping the mine's ore and waste.

Counterweights

In some balanced hoisting systems, a counterweight may be used with either a cage or skip. Because counterweights serve no functional purpose other than to provide for balanced hoisting, their dimensions are quite variable. They are usually designed to fit the space available.

One configuration that has been used for a counterweight consists of a cylinder running in a steel pipe. These systems, although apparently simple, have been found to be hard to maintain because inspection of the pipe is very difficult and correct alignment of the pipes is complicated to determine and monitor.

When used in conjunction with skips on a Koepe hoist, counterweights allow the system to operate from several loading levels, which is not possible with a balanced skipping configuration.

Safety Devices

In North America, cages must be equipped with safety devices that are designed to prevent the cage from falling down the shaft in the event of rope breakage or serious overwind. Different types of safety devices are used for drum hoists and friction hoists.

Drum Hoists

With drum hoists, both excessive overwind and rope breakage can occur. Of these two types of accidents, rope breakage is considered the most serious, and provisions to bring the conveyance to a safe stop in the event of a rope break is required by legislation.

The most common safety device used is the safety dog. A pair of dogs is installed at the top of the conveyance, one pair for each guide. During normal operation, the tension in the hoisting rope keeps the dogs open. Should the rope break, or become slack, a large spring and a series of levers cause the dogs to bite into the guide and stop the conveyance from descending. The dogs dig into the guide to produce a retarding force.

Although dogs are available for use on both wooden and steel guides, only one design, the so-called Wabi single-tooth dog (Wabi Iron and Steel Corp.), which operates on wooden guides, has been demonstrated to produce consistent and acceptable deceleration rates. The dogs used on steel guides are prone to self-energizing, which results in extremely high deceleration rates causing serious injury to personnel riding in the cage.

Friction Hoists

In multiple-rope friction hoist installations, the probability of all hoisting ropes breaking is small. Therefore, overtravel is considered to be more serious, and provisions to bring the conveyance to a safe stop in the event of overtravel are required.

The most common device used is an arrestor gear, consisting of steel frames supported in the hoisting compartment and transported by the conveyance as it travels through the arrestor zone. These devices are located in the headframe and at the shaft bottom. They rely on friction or mechanical deformation of steel to decelerate the hoist.

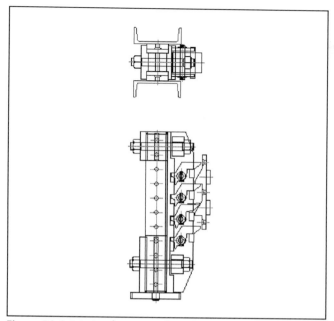

Figure 12.9-16 Catch-gear rack and catch plate

These devices are designed to arrest an upcoming skip at a g-force of 0.9 g's (8.8 m/s² or 29 ft/s²). This level of deceleration ensures that personnel on board will not "float" and thereby injure themselves and ensures that the balance ropes will not overtake the cage, fall back, and likely fail.

In addition to the arrestor gear, the headframe is also equipped with fallback protection. This can be as simple as tapered wooden guides that will jamb the conveyance and hold it until personnel can be evacuated or, more typically, by sets of ratchets that consist of a series of teeth on pivots. This arrangement allows the cage to move up the shaft but prevents it from reversing. Figure 12.9-16 is an illustration of this mechanism.

Loading Pockets

At the loading pocket, ore and waste are transferred from the mine to the skip. The equipment and systems installed can vary from simple manually operated to complex and automated. Two basic types of loading systems are available. These involve measurement by weight or by volume.

The criteria to be considered in the design of all systems include response time, loading time, accuracy of measuring devices, operating and maintenance costs, and amount of spillage.

Although details can vary, each system has the following common elements: a pocket chute with control gates, a measuring pocket, and a skip feed chute with a control gate.

Pocket chutes are generally fabricated from steel with liners. They have an angle of inclination of 45° to 55°. Chute gates can be pneumatically or hydraulically controlled and may be guillotine, ball-and-chain, or undercutting radial arc types.

Skip Loading

Skips are loaded directly from the orepass and are loaded using a measuring flask or a conveyor.

Direct Loading from the Orepass

This method is only used in small operations, typically less than 4 t (4.4 st). The orepass terminates close to the shaft and is equipped with a heavy chain gate, which is raised to allow the ore or waste to fill the skip via a chute. In double-drum installations, the chute will include some kind of diverter mechanism to select between the two skips, as shown in Figure 12.9-17.

Loading via a Measuring Flask

This is the most common arrangement with the flask or flasks installed in a cutout adjacent to the shaft and being fed via a conveyor or sometimes a pan feeder. The flasks, sized to match the skip volume, are filled from the top and are equipped with a door and chute to control and direct the discharge of material into the skip. Loading the flasks is typically controlled by load cells supporting the flask with a volume override. When the material is of a consistent density, such as crushed salt or potash, the systems are often controlled by volume only. The material flows into the flask until the flask is full. At this point, an undercutting arc gate will close the bottom of the storage bin that feeds the flasks. The loading flask is furnished with pneumatically or hydraulically controlled gates. The bottom of the flask is usually inclined at an angle of 60° or greater to ensure rapid loading and complete clean out. Examples of these systems are shown in Figure 12.9-18. With balanced skip hoisting, two loading flasks are provided, one for each skip.

Loading via Conveyor

One of the problems associated with the previously mentioned systems is the difficulty in controlling the loading rate. After the gate is raised to allow the material to flow, there is very little that can be done to regulate the flow, and, as a result, the entire load drops into the skip. This can produce very high dynamic loads on the ropes that support the skip. These loads will reduce the life of the ropes.

To overcome this problem, skips can be fed directly from a conveying system, which can be programmed to fill the skips at a constant rate over a period of 10 to 20 seconds. The conveyors can be equipped with load cells to measure the load accurately. Figure 12.9-19 shows a layout of such a system.

Skip Feed Chute

A feed chute is required to direct the material from the measuring pocket or conveyor to the skip. The angle of inclination of the feed chute is usually 60° or greater.

The clearance between the chute and the skip is approximately 75 mm (3 in.). It has been suggested that the ratio of the chute discharge width to the skip width should not exceed 0.7 for skips up to 1.5 m (5 ft) wide and 0.8 for skips up to 1.8 m (5.9 ft) wide. To reduce spillage, the sides of the chute must be narrower than the opening in the side of the skip body.

Automatic Loading

Automatic loading of skips is used to improve the efficiency of hoisting systems and to reduce labor requirements. Efficiency is improved by ensuring that each skip receives the required load without overloading or underfilling. Labor is reduced when automatic loading is used in conjunction with automatic hoisting. So that automatic loading can be accomplished, a minimum of three control systems must be included. There must be a means of measuring the weight and/or volume of

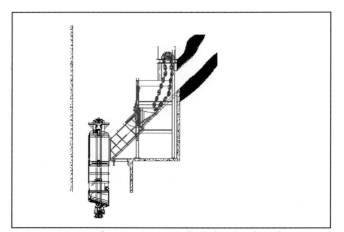

Figure 12.9-17 Chain-gate-controlled skip loading from orepass

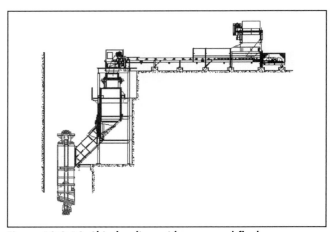

Figure 12.9-18 Skip loading with measured flask

material in the measuring pocket, a means of opening and closing feed chutes to fill and empty the measuring pocket, and a logic system to ensure that the above operations are carried out in the correct sequence to prevent such system failures as double loading and dumping into an empty shaft. With automatic hoisting, the logic system extends to the operation of the hoist.

Spill Pockets

Despite the precautions taken to prevent spillage during the loading and dumping of the skips, a certain amount is inevitable below the loading pocket. This spillage may be confined to the hoisting compartment by lining these compartments down to a spill pocket. According to data available from a number of mines in South Africa, the amount of spillage, expressed as a percentage of tons hoisted, varies from 1.5% to 5%. The larger amounts of spillage are associated with manually operated systems handling uncrushed material.

Two methods of handling spillage in modern hoisting systems are commonly used. One method consists of installing deflectors in the shaft below the underwind position of the skip. These deflectors direct spillage into a small flask similar to the loading pocket. The spillage pockets are then cleaned as required. With drum hoisting, this system can be used at any elevation in the shaft. With a friction hoist, the spillage pocket must be located below the tail rope loops, and a second conveyance, such as the cage, must be used to handle spillage. The second method of handling spillage consists of developing a ramp from the lowest horizon in the mine to the shaft bottom. In this system, any spillage falls directly to the shaft bottom, where it can be mucked out. Regardless of the system used, particular care is required with friction hoist systems to ensure that the tail ropes do not become entangled or affected by the buildup of spill in the shaft and to ensure that the guide rope tensioning weights do not end up resting on the accumulated spill. Most mines today provide ramps to the shaft sump for cleaning the spill.

ROPES

The basic purpose of the hoisting rope is to connect the conveyance to the hoist. It is selected primarily based on safety, compatibility, life, and costs. Safety requirements when hoisting personnel, or where persons may be endangered by hoists and their appurtenances, are usually determined by legislation.

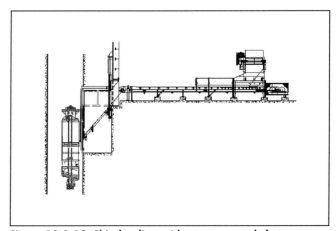

Figure 12.9-19 Skip loading with conveyor and chute

The life of a rope is usually expressed as the number of trips it will make; it is affected by rope construction, hoist and sheave dimensions, type of loading, shaft atmosphere, and maintenance. Costs of initial purchase, maintenance, and rope replacement, including lost production, are normally considered. In addition to hoisting ropes, the hoisting systems designer must be concerned with balance ropes (for friction hoist application), guide ropes, rubbing ropes, and rope attachments.

The types of rope construction commonly used for various hoisting applications are as follows.

Drum hoist ropes

- Production duty: Lang lay flattened strand 6 × 30 fiber, or wire metal or plastic core
- Shaft sinking duty: Multistrand, nonspin, 34 × 7, wire metal core

Friction hoist ropes

- Head ropes:
  - Full locked coil
  - Multistrand, nonspin, plastic cushion, 29 × 7, plastic core
  - Round strand, regular lay, plastic core

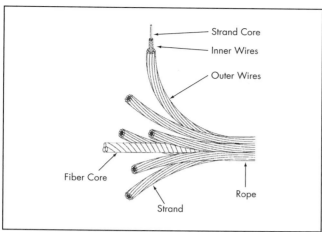

Source: Anon. 1980.
Figure 12.9-20 Exploded view of a rope

- Balance ropes: Multistrand, nonspin, 34 × 7, plastic core
- Guide and rubbing ropes: Half-lock coil, wire metal core

In this section, a general description of the various types of ropes, plus the factors to be considered to ensure correct application and selection, is presented.

Rope Construction
In the construction of most wire ropes, a number of individual wires are wound around a core to form a strand. The strands are then wound around a core to form the rope. Figure 12.9-20 is an exploded view of a rope showing its components.

Some ropes are constructed without strands by winding consecutive layers of wire around the inner layers. The factors to be considered in rope construction are the wire, strands, cores, and lay. By varying these factors, ropes with different characteristics can be constructed.

Hoist Ropes
The four types of ropes generally considered for hoisting applications are round strand, regular lay; flattened strand, Lang lay; multiple strand, nonspin; and locked coil. Typical properties for these types of rope are given in Table 12.9-3.

Round Strand, Regular Lay Ropes
The wires in these ropes are laid up (i.e., twisted together) to forms strands in the opposite sense to which the strands are laid up to form the rope. The wires on the outside surface lie approximately parallel to the axis of the rope.

Under tension, the wires tend to untwist in one direction and the strands tend to untwist in the other direction. With careful design, the rope will show little tendency to twist when loaded.

Flattened Strand, Lang Lay Ropes
The strands in these ropes are formed around a triangular core and have the wires laid up to form strands in the same sense as the strands are laid up to form the rope. The wires on the outer surfaces of the strands lie at approximately 45° to the axis of the rope. This construction was developed specifically for multilayer winding on hoist drums.

When compared to a round strand, regular lay rope, the flattened strand, Lang lay rope

- Is more flexible because the wires lie parallel to the rope near the rope center, not on the surface;
- Has wires at the outer surface of the rope that are less subject to "plucking" by adjacent turns as the rope coils on the drum; and
- Is designed so that contact loads at the surface of the rope and between strands are distributed over many wires, thereby reducing internal nicking of wires.

One disadvantage of a Lang lay rope is that it tends to untwist when loaded. The two types of lay ropes are illustrated in Figure 12.9-21.

Multiple Strand, Nonspin Ropes
These ropes were developed to overcome the torsion forces developed by Lang lay ropes. They are more susceptible to the crushing forces associated with multilayer winding on a drum and have a lower strength/weight ratio than Lang lay flattened-strand ropes.

This type of rope is used on drum hoists for shaft sinking, where the required service life is relatively short and the conveyances run on rope guides and, thus, low-torque ropes are required. More recently, their design has evolved to include plastic cushioning between layers to reduce internal nicking. The resulting ropes are used as head ropes and balance ropes on Koepe hoists.

Full Locked Coil Ropes
This type of rope is completely different from both round and flattened strand rope. The center or core of the locked coil rope consists of a concentrically laid strand of round wires. Around this core lies one or more layers of X-shaped cross-sectional wires, the outer layer always being Z-shaped and interlocking. The lay of the various layers can be arranged so that the rope does not have a tendency to untwist when loaded. In addition, the structure is compact and stable and better able to withstand crushing stresses than any other rope, and the line contact seals between the wires in the outer layer keeps lubricant in and corrosive environment out.

On the debit side, the specially shaped outer wires cannot be made as strong as round wires, and the rope is less flexible than stranded rope. Larger sheaves and drums are required for ropes of equivalent strength.

The main use of these ropes is on Koepe hoists running on rope guides in deep mines. Figure 12.9-22 shows the cross section of a full locked coil rope.

Rope Selection
For the initial selection, the following four requirements should be considered: strength, resistance to fatigue failure, abrasion resistance, and resistance to crushing or distortion. The choice should be made after correctly estimating the relative importance of each of the above requirements. Naturally, strength is the overriding concern.

Rope Selection for Drum Hoists
This segment discusses the selection of the size of rope to meet operating conditions for drum hoists used in vertical and slope applications. In the United States, the Code of Federal Regulations (CFR), CFR Title 30, Part 56, specifies minimum requirements for the following items that are relevant to drum hoisting: factor of safety, ratio of drum/sheave diameter to rope diameter, and fleet angles.

Table 12.9-3 Typical properties of rope types

Rope Type	Characteristics and Applications	Comparative Strengths, kN (for 38-mm diameter)	Unit Mass, kg/m
Round strand, regular lay	Used on sinking and Koepe hoists	980	6.1
	Tend to nick in multilayer applications		
	Available from wires up to 1,960 MPa (284,000 psi)		
Flattened strand, Lang lay	Used on drum hoists and shallow Koepe hoists	1.135	5.9
	Best for multilayer applications		
	Very high strength, easy to perform nondestructive testing		
	High torque under load		
	Available from wires up to 2,000 MPa (290,000 psi)		
Multiple strand, nonspin	Used on sinking and Koepe hoists	1.000	5.7
	Low torque, internal nicking unless plastic fillers incorporated into the design, low elastic modulus, and high initial stretch		
	More difficult to perform nondestuctive testing		
	Available from wires up to 1,960 MPa (284,000 psi)		
Locked coil	Used on drum and Koepe hoists	1.400	8.4
	Near-zero torque, lower unit strength, high elastic modulus		
	Available from wires up to 1,570 MPa (228,000 psi)		

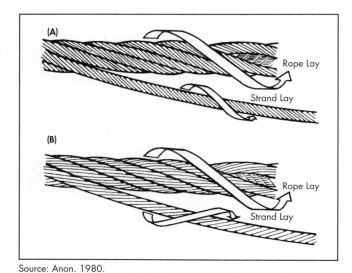

Source: Anon. 1980.

Figure 12.9-21 Rope types: (A) regular and (B) Lang lay

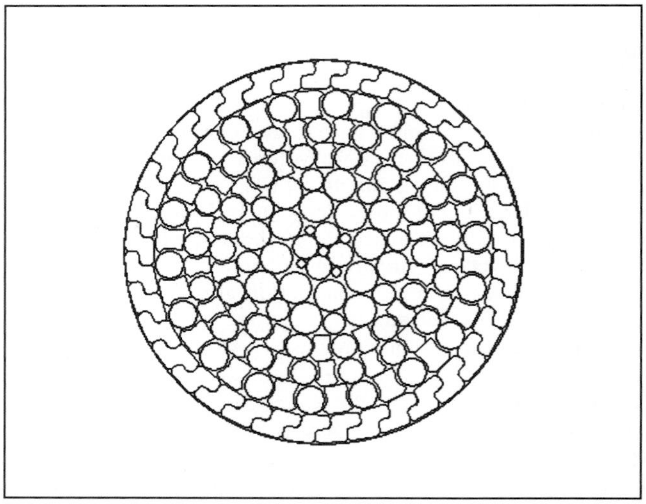

Figure 12.9-22 Cross section of full locked coil rope

Factor of safety. Mandatory requirements by the Mine Safety and Health Administration (MSHA) are covered in CFR Title 30, Parts 55 and 56, and Section 57.19-21 for metal and nonmetal underground mines and underground coal mines, which require the following minimum rope strengths, where L is the maximum suspended rope length in meters:

- For rope lengths less than 914 m (3,000 ft):
  minimum strength = static load × $(7.0 - 0.0033*L)$
- For rope lengths in excess of 914 m (3,000 ft):
  minimum strength = static load × 4.0

Drum/sheave diameter. Mandatory requirements for drum and sheave diameters have now been removed from the regulations, but insofar as the old regulations are still common practice in the United States, as well as Canada and Europe, they are repeated here.

Drum and sheave tread diameters should be

- Not less than 60 times the hoist rope diameter for slope or inclined shaft applications,
- Not less than 80 times the hoist rope diameter if the hoist ropes are 25 mm (1 in.) in diameter or greater or not less than 60 times the hoist rope diameter if the hoist ropes are less than 25 mm (1 in.) in diameter for vertical shaft applications, and
- Not less than 100 times the hoist rope diameter for locked coil ropes.

Fleet angles. The fleet angle refers to the angle made by the headrope between the drum and sheave. If this angle lies outside of fairly small limits, the rope will tend to either pile up at the drum flange or jump several pitches when it starts to return across the drum. Practice has found that the fleet angle should be between 0.5° and 1.5° for ungrooved drums but can be between 0.25° and 2.0° with drums grooved in the parallel, so-called LeBus pattern.

South African Code of Practice
The most recent codes of practice relating to the selection of headropes as well as assessment of rope condition are those produced by the South African Bureau of Standards (SABS). These are called SANS 10293 (1996) and SANS 10294 (2000) and are worth obtaining to better understand all facets of mine hoisting ropes.

Rope Selection for Koepe Hoists
A friction hoist differs from a drum hoist in that the rope is not wound on and off the drum but passes around the hoist drum, with the friction between the rope and drum providing the force necessary to move the rope. Friction hoists may be mounted in the headframe tower directly over the shaft, or they may be mounted in a hoist house at ground level. The friction hoist drum is often referred to as a wheel.

Regulations and other calculations and considerations involved in rope selection for friction hoists are discussed here.

Factor of safety. Mandatory requirements by MSHA are covered in CFR Title 30, Parts 55 and 56, and Section 57.19-21 for metal and nonmetal underground mines and underground coal mines, which require the following minimum rope strengths, where L is the maximum suspended rope length in meters:

- For headropes with lengths less than 914 m (3,000 ft): minimum strength = static load $\times$ (7.0 – 0.00164*L)
- For headropes with lengths in excess of 914 m (3,000 ft): minimum strength = static load $\times$ 5.0
- Balance or tail ropes = 7.0 on weight of rope (static load)
- Guide or rubbing ropes = 5.0 on weight of rope (static load)

Bending ratio. The required bending ratios are the same as for drum hoists when stranded construction ropes are use. The bending ratio for locked coil ropes must, however, be larger because these ropes are much stiffer in bending than stranded ropes.

Wire Rope Industries of Canada recommends the following minimum bending ratios for locked coil ropes, although all such specifications have been removed from CFR requirements:

- Less than 25 mm (1.0 in.)   100:1
- 25 to 32 mm (1.0 to 1.25 in.)   100:1
- 38 mm (1.5 in.)   110:1
- 51 mm (2.0 in.)   120:1
- Greater than 51 mm (2.0 in.)   130:1

Rope Types
Lang lay ropes work satisfactorily in installations where the hoisting distance is less than 750 m (2,500 ft); at greater depths, the spin upsets the rope structure, resulting in misplaced strands, broken wires, and so forth.

Ropes of nonrotating design, such as multistrand, nonspin, or full lock coil, should be used if the hoisting depth is greater than 750 m (2,500 ft).

Tail Rope Orientation
Any object falling down a mine shaft is affected by the angular velocity of the earth and will move toward the east wall. A tail rope behaves in a similar way. If the shaft compartments are arranged in an east–west direction, the tail rope loops will open and close freely. However, if the compartments are north and south, the ropes could foul the separators at the shaft bottom. Because tail ropes are required for friction hoist installations, the orientation of compartments should be considered. The above described action results from a physical principle known as the Coriolis effect and is a function of the hoist speed and the longitude of the mine.

Tail Rope Loop Diameter
When the tail rope hangs in a shaft beneath a pair of conveyances, there is a curved segment at the shaft bottom which is called the tail loop. The diameter of this loop is determined by the spacing of the conveyances. The rope, however, will twist if the conveyance spacing is less than the natural loop diameter of the rope. This natural loop depends on the rope construction and will vary between 30 and 60 times the rope diameter. The stiffer the rope construction, the bigger the loop.

Guide and Rubbing Ropes
Ropes used as guides and rubbing ropes should be of half-locked coil construction. These ropes are manufactured with a core of concentrically laid round wires covered with one or two layers of shaped wires. The outer layer of locked coil construction is always composed of interlocking wires of Z shape.

Some of the advantages of half-locked coil ropes as guides and rubbing ropes are as follows: their smooth surface reduces friction, they have a greater degree of rigidity, and they use wires with large cross sections on their outer layer. This is critical because one broken wire on the outer layer is cause for the rope to be replaced. A broken wire will catch in the guide rope bushing on the skip, which could lead to a major mishap.

Rope Sag
A rope between a ground-mounted hoist and a head sheave sags and takes the form of a catenary. With a heavy rope and a light load, the sag can be significant and cause problems with the rope impacting the sheave deck or the hoist building front wall, if not considered.

The shape of the catenary is given by

$$y = 2*H/q*\sinh(\bar{r} + a)*\sinh(\bar{r})$$

where

y = vertical coordinate along catenary
H = horizontal tension in the cable
q = unit mass of the catenary
$\sinh$ = hyperbolic sine function
$\bar{r}$ = horizontal coordinate along the catenary
a = horizontal spacing of catenary end points

Inclined (Slope) Haulage
The selection of ropes for an inclined shaft differs little from selection for a vertical shaft. The three major differences are calculation of rope tension, sizing of sheaves, and abrasive wear.

Calculation of Rope Tension

The tension in the rope is the sum of the gravitational, frictional, and acceleration forces. The gravitational force is the sum of the weights of the car or cars, payload, and the rope multiplied by the sine of the angle of the incline. The friction force is commonly taken as 2.5% of the weight of the car and payload plus 10% of the rope load.

Sizing of Sheaves

Because of the space restrictions underground and the need for portability, deflection sheaves are commonly smaller than used elsewhere for hoisting ropes. Minimum drum diameter to rope diameter D/d ratios are presented in Table 12.9-4.

Abrasive Wear

Abrasive wear is more likely to be a problem in a slope than in a vertical shaft, and this potential problem should be addressed in rope selection. One approach is to select ropes with thicker wires, such as 6×7. Some designers increase the factor of safety to allow more wear on the rope before it needs to be replaced.

Rope Operating Practices

The correct wire rope properly installed on well-designed equipment maintained in good working condition provides the foundation for satisfactory rope performance. To attain and sustain the required rope performance, it will be necessary that the rope be properly operated and adequately maintained. Well-developed rope-operating practices should consider the following: rope storage, rope installation, rope changing procedures, rope cuts, lubrication, and inspection procedures.

Rope Storage

Whether the rope must be stored temporarily or indefinitely, it should be indoors, where it can be shielded from weather, corrosive fumes, and excessive heat that might dry the lubricant or fiber core. The rope reels must be turned 180° every 4 months so that lubricants do not settle to the lowest point, thereby allowing unprotected steel to rust.

Rope Installation and Handling

Ropes should be wound onto the drum under load. The direction of coiling should be determined by the accepted rule-of-thumb method, as shown in any rope handbook. Before cutting, the rope supplier or a technical handbook detailing correct cutting procedures should be consulted. This typically requires seizing of the rope on both sides of the cut with hose clamps and cutting the rope with a grinding disc.

Drum Hoist Rope Changing Procedures

Rope changing procedures for drum hoists are relatively easy and are described in rope handbooks. Precautions must be taken, however, to ensure that adequate tensions are applied to provide correct spooling during installation and to pretension the dead turns. Pretensioning the dead turns is particularly important in multilayer winding. Failure to adequately pretension will lead to loose dead turns and catastrophic rope wear.

Adequate pretensioning can be provided by using a special rope-tensioning winch or by doubling down the headrope using a loaded skip.

Table 12.9-4 Minimum D/d ratios for sizing of sheaves

Rope Construction	Minimum D/d
6×26	30
6×31	26
6×36	23
6×41	20
8×36	18

Friction Hoist Rope Changing Procedures

There are several options for changing head and balance ropes on a Koepe hoist. Procedures involve careful planning and special equipment. These procedures are beyond the scope of this document.

Rope Cuts

The end of the rope adjacent to the conveyance suffers premature strength reduction as a result of corrosion and local bending. These ends must be cut off at regular intervals.

In multilayer winders, wear occurs on those areas of the rope that meet at the crossovers, that is, the region of the drum where the rope changes layer. These locations need to be moved regularly (every 3 or 4 months) to avoid premature retirement of the rope.

Lubrication

Atmospheric moisture, shaft water, and chemical fumes are a few of the agents that will cause corrosion. This frequently leads to wire breakage and to the premature removal of mine hoist ropes. In mines with very wet and corrosive shafts, it is distinctly advantageous to apply a lubricant that will penetrate the interior of the rope and that contains a corrosion inhibitor. These must be applied during fabrication of the rope. It is impossible to apply lubricants that can penetrate the rope strands after the ropes are manufactured. Ropes made with fiber cores soaked in lubricants will extend rope life in wet shafts.

Inspection Procedures

To achieve optimum rope performance, an inspection procedure that requires findings to be recorded, fully diagnosed, and translated into appropriate action should be instituted.

The presence of the foregoing not only allows the operator to estimate the remaining safe working life of the rope but can also pinpoint the source causing such damage to the rope.

Ropes should be visually examined daily and tested regularly using electromagnetic devices to detect loss of cross section and broken wires. When the rope's strength is determined to have been reduced below 90% of the original value, the rope must be replaced. Other retirement criteria also exist and are stipulated in CFR Title 30, Section 57.19024.

Rope Attachments

Wire rope end attachments are as important as the rope to which they are fastened. Therefore, the selection of the correct type of attachments should be based on an understanding of how they affect ultimate rope serviceability and efficiency.

Forming a loop in the rope or attaching a fitting to the rope are the only ways of attaching something to a wire rope.

Loops are made either by splicing the rope to itself or by the use of clamps or wedges. Fittings secured directly to wire rope can be applied by cold forming (snagging) of the metal in the fittings, either by pouring a liquid material, such as molten zinc, or by a wedging arrangement. In this section, some of the most commonly used attachments are described and are shown in Figure 12.9-23.

Factor of Safety

The factor of safety for rope attachments is not specifically covered in CFR Title 30, Part 57. However, it is considered good engineering practice to design the rope attachments to be stronger than the rope, that is, the rope will fail before the attachments. British codes require a factor of safety equal to 10 based on static loads.

Cappels

Cappels are rope fittings attached directly to the rope. These types of fittings consist of a pair of wedges grooved to suit the particular rope diameter and interlocked to ensure complementary movement. A number of bands driven over the diverging exterior surfaces of the cappel limbs provide the initial compressive force to ensure that the wedges grip the rope. The limbs' internal surfaces have been machined to fit the external surfaces of the wedges. A safety block is fastened on the end of the rope, protruding beyond the bottom of the wedges. Several types of cappels are used for hoist ropes, including the wedge and loop.

Thimbles

In North America, many mine hoist ropes on drum hoists are terminated with a wire rope thimble and U-bolt cable clamps. With this type of attachment, a loop is formed in the rope and the thimble placed within this loop. The free end of rope is then clamped to the long end to secure the thimble.

Thimbles are made of cast steel, whereas clamps, shackles, and pins are made of 1.5% manganese steel.

Thimble Cappels

These attachments represent a significant improvement over the basic thimble without the complication of installing bands to secure the wedges. These devices are self-locking, using the force on the rope to drive a wedge into a housing that incorporates an extension with a pin to connect the cappel to the conveyance drawbar.

Swivels

When winding with balance ropes, the lay length tends to shorten as each conveyance in turn approaches the bottom landing. It is for this reason that swivels are provided to allow the ropes to spin and regain their normal length. Shaft conditions and loadings influence the type of swivel used.

Attachment Maintenance and Inspection

If thimbles and clips are to be used, then the manufacturers' recommendations for the correct number of clips, amount of turnback, clip positions, and correct torques should be followed closely. If a wedge-type connection is being applied, then care should be taken to properly clean the rope surface. Frequent applications of cement dust and subsequent steel brushing will remove all the surface lubricant.

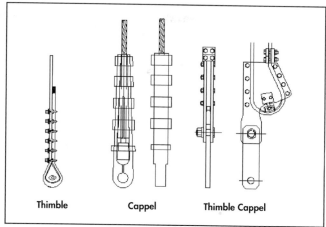

Figure 12.9-23 Rope attachments

In some jurisdictions, statutory mining regulations require that all types of attachments be inspected both visually and nondestructively on a regular basis. In other parts of the world, removal criteria for attachments have been established.

SHAFT AND GUIDE SYSTEMS

A shaft is a vertical or inclined primary opening in rock that provides access to and serves various levels of a mine. Primary openings are those that are considered to be permanent and require a high degree of safety. Chapter 12.4 discusses design and construction details.

Shaft Design Procedure

A suggested procedure for the design of shafts follows. It is important that the steps be followed in the given sequence. The process is iterative and involves working through the process several times before an optimum design is achieved. The shaft design procedure involves the following steps:

1. Define the purpose of the shaft.
2. Identify the location and determine the inclination.
3. Determine the number of hoists required.
4. Determine the size of the conveyances and compartments.
5. Determine the arrangement of the compartments.
6. Determine the exterior shape.
7. Design the interior members (guides, buntons, etc.).
8. Design the shaft lining.
9. Check the ventilation characteristics.
10. Determine the ground stabilization and temporary ground support.
11. Determine the shaft collaring method.
12. Determine the shaft sinking method.
13. Evaluate and modify this procedure by repeating these steps.

Purpose of Shaft

One of the first items to be examined when designing a shaft is to identify its intended purpose. Shafts usually fall into one of the following categories: production (ore and waste handling), service (personnel and materials handling), ventilation (upcast or downcast), exploration (for defining mineral deposits), and combinations of the above.

Location

Generally, the location of a new shaft at the mine site is determined after establishing the following:

- Mine surface layout
- Location, dip, and extent of the ore body
- Number of working levels to be considered
- Location of ore and waste handling facilities
- Water collection sump requirements
- Safety and stability of the shaft pillar
- Future planned expansion of the shaft

The bottom of the shaft, where possible, must also be in a stable formation, and it should be able to facilitate any future deepening. The location of the shaft must also avoid adverse surface features and should satisfy surface layout plus logistics of surface and underground ore and waste handling facilities.

Inclination

The dip of the ore body is the main factor involved in deciding to sink either a vertical or inclined shaft. A secondary factor is the relative ground strength and geologic formations to be encountered by the proposed shaft.

The main advantages associated with vertical shafts are that hoisting speeds are greater, shaft maintenance costs are lower, sinking can be carried out faster, and sinking can be achieved in almost any type of ground.

Inclined shafts (slopes) are generally associated with inclined (dipping) ore bodies, where the length of crosscutting to reach the ore body from a vertical shaft becomes longer with the increase in depth of the shaft. They have the advantage of minimizing development to reach the ore from the shaft and are frequently used in coal mines.

Number of Hoists

The number of hoists (and conveyances) required to meet hoisting demands has a major impact in the design of a shaft. The number of hoists will be determined using the procedures described previously.

Compartment Size

The cross-sectional area of a particular compartment (horizontal area) is dependent on its use. To determine the most appropriate size, it is necessary to list the items to be transported in the compartment, determine their approximate dimensions and weight, indicate their direction of flow, and indicate their approximate quantities. Another important factor to be considered is the size of the shaft-sinking bucket. After having determined the size of compartments for cage, counterweight, and skips, the remaining area is then divided to accommodate ventilation, manway, and pipe facilities.

Size of Skip Compartment

The skip capacity is determined by such variables as the hoisting capacity, material density, lump size, and vertical height available in the headframe. Generally speaking, the skip dimensions and cross-sectional area are determined by the skip supplier. The amount of clearance between the skips and shaft equipment is typically 300 mm (12 in.) for rope-guided skips and 75 mm (3 in.) for skips on fixed guides.

Size of Cage Compartment

Cages are generally used to transport personnel and materials. The following clause of CFR Title 30 (Section 57.19-66 Mandatory) applies in the United States: "In shafts inclined over 45°…each person shall be provided a minimum of 0.14 m^2 (1.5 ft^2) of floor space."

The cage capacity is determined by the cage supplier based on a knowledge of the following factors: the amounts of material to be transported, the dimensions and weights of the major pieces of equipment to be lowered and raised, the number of personnel to be lowered and raised per shift, and the type of conveyance preferred (i.e., single deck, double deck, combination skip/cage, etc.)

After establishing the cross-sectional area for the cage, the cage compartment is sized the same way as the skip compartment to provide similar clearances.

Size of Manway Compartment

From CFR Title 30 (Section 57.11-37 Mandatory): "Ladder ways constructed after November 15, 1979, shall have a minimum unobstructed cross-sectional opening of 610 × 610 mm (24 × 24 in.) measured from the face of the ladder."

From CFR Title 30 (Section 57.11-41 Mandatory): "Fixed ladders with an inclination of more than 70° from the horizontal shall be offset with substantial landings at least every 9.1 m (30 ft) or have landing gates at least every 9.1 m (30 ft)."

In addition to these mandates, it is necessary to provide adequate clearance at the landing platform to allow a person to move from one ladder to the next offset ladder. Generally, an additional floor area of 914 × 610 mm (36 × 24 in.) is provided.

Size of Counterweight Compartment

The counterweight compartment can be of any size or shape as long as it provides sufficient space and clearance for the safe passage of the counterweight with the same clearances as for skips and cages.

Size of Service Compartment

There are no hard and fast design rules for sizing the service compartment except that the cross-sectional area of the compartment should be adequate to provide the following: separate locations for communications (signaling) and power lines; space for locating air, water, pump, and backfill lines; clearance for pipe fittings; and space for pipe column brackets. The compartment must also be accessible from a shaft conveyance, preferably a cage, to allow maintenance of the services in the compartment.

Compartment Arrangement

The arrangement of the compartments in relationship to one another depends on the surface layout, the underground layout, and the type and size of hoist to be used. Because these design details are not known precisely during the initial design stages, one must first make a "best estimate" and be prepared to change the arrangement as the final design evolves.

On surface, the orientation of the compartments in relation to the hoist and headframe is of major importance. Good rope practice for drum hoists requires that the location of the hoist in relation to the shaft and headframe be in line to provide acceptable rope fleet angles.

For friction hoists, the center-to-center distance on the compartments is a major consideration for determining the size of the drum, the necessity for deflection sheaves, and the type of rope construction for the tail ropes.

Exterior Shape

The exterior shape of a shaft is established by considering ground-stability factors. A circular shape provides better resistance to deformation by lateral pressure. In fair to competent formations, an opening excavated in this form is generally self-supporting. In weak formations, the circular shape is adaptable to a variety of lining materials: concrete, steel segments, or cast-in-place concrete. Competent formations can be excavated to a rectangular shape and be generally self-supporting. An elliptical shape offers better support than a rectangular shape but not as good as a circular one. The geotechnical considerations leading to the selection of an exterior shape are beyond the scope of this discussion, but their importance in the overall design process cannot be overstated (see Chapter 12.4). Also shafts may have different exterior shapes at different depths, reflecting the different strata encountered.

Interior Members

Guides and buntons are the main vertical and horizontal structural elements in a mine shaft conveyance system. The primary function of these members is to facilitate running of the shaft's conveyances. Their characteristics influence not only speed of operation and amount of maintenance required, but also production and ventilation costs. Therefore, the choice of sets and guides is of great importance to achieving operational cost savings. To minimize construction costs and the time to install guides, bunton spacing is maximized. The current trend is to install buntons at 5-m (16.4-ft) spacing. This results in flexible guides and requires more sophisticated designs and simulation of the conveyance motion to avoid so-called conveyance slamming events, which can lead to destruction of the guides and conveyances.

Shaft Guides

Shaft guides are used in vertical shafts to keep the skips, cages, and counterweights in proper shaft position. The two types of guides in use are fixed guides (wood, rail, and structural steel), and rope guides.

Wooden guides. Wooden guides are supported by the horizontal sets installed in the shaft. They are favored where conveyances require safety devices. This is because the decelerating characteristics of safety dogs on wood are more reliable than on steel or rail. When hoisting on wooden guides, a speed of 11 m/s (36 fps) with a medium-sized skip is the practical upper limit. To achieve speed of this order with large conveyances and longer guides, it is necessary to install wooden guides supported by structural steel backers. These backers, typically wide-flange structural steel, are bolted between the steel buntons, and the wooden guides are then bolted to them. This arrangement avoids the problems associated with fastening wood guides directly to wood or steel buntons. Wooden guides are subject to changes in dimensions due to changes in moisture content. They are also subject to wear to a greater extent than steel guides. Wooden guides generally are not used in shafts where fire hazards exist, such as in coal mines.

Until recently, the wood used for these guides was either British Columbian fir or Australian Karri wood. Both of these products are now in scarce supply. Wooden guides are now available in composite wood materials, akin to plywood. The only disadvantage of the composite guide is its tendency to swell when wet; therefore, it is not suitable for wet shafts.

Rail guides. Rail guides are common in inclined shafts. These guides vary in size, with the larger rails giving a wider bearing surface for the wheel. The steel rail provides a harder surface but presents a smaller wear area; this creates more wear and maintenance on conveyance guide shoes and wheels. They have also been used to guide conveyances (usually in shafts equipped with Koepe hoists) in vertical shafts but suffer from their higher lateral flexibility, which leads to excessive motion of the conveyance as it moves in the shaft.

Steel guides. Structural steel tube sections have virtually replaced wood as the material of choice for guides in production shafts, where higher speeds and heavy loads require a fairly rigid and stable section. With steel guides, the conveyances are usually equipped with adequately sized guide roller assemblies to minimize wear between the guides and shoes on the conveyance. When hoisting on steel guides, a speed of 18 m/s (3,500 fpm) is a practical upper limit.

Rope guides. Rope guides are also used for guiding conveyances in vertical shafts. These are generally used with multirope hoisting systems where the effect of hoist-rope torque is minimal or nonexistent. Half-locked coil ropes are used for this service. Rope guides do not require any intermediate support, thereby eliminating the steel sets required for fixed guides. Because rope guides are not able to withstand the horizontal forces of loading and unloading, the conveyance must be supported by other means in these areas. Fixed guides are used at the extremes of travel, and a retractable guiding system can be considered at intermediate levels. There are several schemes by which rope guides are suspended and tensioned, and the selection of any one of these will depend on the particular installation. These methods will be reviewed in a later section.

Because rope guides eliminate the need for horizontal buntons, the ability of the shaft to handle airflow is enhanced. In a large shaft, with multiple conveyances traveling at the same time, the aerodynamic effects can cause unexpected rope deflections. These effects should be studied during the design stage. Because rope guides are the smoothest of all guides, hoisting speeds up to 20 m/s (3,900 fpm) are feasible. Discussion of the detail design of rope-guided systems is provided in a following section.

Shaft Buntons

Shaft buntons are horizontal members used to divide the shaft into compartments. They also support and carry the shaft guides, pipes, and cables. Buntons can be of any cross-sectional shape and can be installed in a shaft at any spacing. However, the correct design and layout of buntons and their installation at a suitable spacing minimizes the air resistance of shafts and, hence, reduces ventilation operating costs. Steel buntons are available with aerodynamic cross sections that minimize the resistance to airflow.

Guide and Bunton Structural Design

The design of guides and buntons has progressed to the point where their design is no longer empirical. The design requires the use of computer-based simulation of the motion of a conveyance as it moves up the shaft. This allows the designer to optimize the design and reduce the amount of steel that is required for this equipment.

In addition to using simulation software, the designer can also use guidelines detailed in SABS 0208-4 (2001) to provide a check on a proposed design before submitting it to the rigor (and expense) of a computer simulation.

Guide Alignment

One of the required inputs to either the guideline or computer simulation is the degree of alignment expected for the installation. This means how far off of plumb, or straight line, one expects, or allows, the guides to be.

A practical limitation for this has been found to be ±3 mm (±0.125 in.) over 5 m (16 ft). To verify that this has been achieved, it is necessary to either install two plumb lines for each guide the entire length of the shaft or to utilize sophisticated instrumentation developed in South Africa for this purpose. This latter option is a more reliable method, as it does not rely on measurement and recording by individuals with its inherent inaccuracy.

Guide Rope Design

Guide ropes are common in high-speed shafts operating large conveyances from a single horizon. This situation is common in North America in coal, potash, trona, and salt mines.

It is first of all necessary to calculate what resistance to lateral motion the guide ropes provide. The lateral restoring force produced by a guide rope is termed its *stiffness*. This term refers to the force produced by moving the guide rope laterally in the shaft as opposed to bending the rope in an arc.

Calculation of this stiffness in kilograms is depicted in Figure 12.9-24. As this figure shows, the restoring force at any point along the guide rope is simply a function of the tension in the guide rope at that point and the distance the force is from the end of the rope.

The more difficult problem comes in calculating what forces arise as the conveyance moves up or down the shaft. These forces can be put into one of the following categories:

- The Coriolis force, which is the result of the conveyance moving radially through the earth
- Impulsive forces from the buffeting produced as conveyances pass at mid-shaft or cross a ventilation opening
- Steady aerodynamic forces caused by the shape of the conveyance
- A twisting moment arising from the head- and tail ropes

Coriolis force. When a body descends in a shaft, it moves toward the east wall; when it ascends, it moves toward the west wall. This motion is a result of the different circumferential velocities that exist between points along any radius of a rotating sphere. The *apparent* lateral acceleration of the object, as it moves at a constant speed in the radial direction, results in a force on any structure, such as a guide rope, trying to restrain it. This force is called the Coriolis force and is proportional to twice the product of the object's radial and angular velocities. The force is required to increase or decrease the object's lateral velocity to match that of the guide rope, and all other matter at the same radial location. The magnitude of the Coriolis force is given by

$$F = 2\Omega V$$

where
Ω = rotational speed of the object about the earth's axis
V = radial speed of the object in the shaft

Aerodynamic and buffeting forces. Conveyances will experience short-term impulsive forces called buffeting forces when they pass one another as well as steady aerodynamic forces as they travel the entire length of the shaft. The buffeting forces have been found to be inconsequential and do not need to be considered when designing a shaft. The aerodynamic forces that arise as conveyances move through a shaft are, however, very significant and need to be determined. They arise because the flow of the air around the conveyance is asymmetrical and will cause lift to occur. The lift generally acts toward the closest walls because the air velocity is higher in the reduced space between the conveyance and shaft wall. It is difficult to determine the lift coefficient. Some recent research has provided a means for calculating this coefficient and has thereby enabled the designer to determine reasonable estimates for the level of force that a conveyance might encounter. More accurate determination of the lift coefficient requires the use of computational flow dynamics to model the flow around the conveyance in the shaft or scale-model wind-tunnel testing of the shaft.

The force on the conveyance is calculated from

$$F = \tfrac{1}{2} C_L \rho V_r^2 A_c$$

where
C_L = lift coefficient
ρ = air density
V_r = relative velocity of the conveyance and ventilating air
A_c = the area of the skip facing the section of shaft wall

Forces are typically in the range of 100 to 500 N (20 to 100 lb).

Rope Torque

Both headropes and tail ropes induce torque and, therefore, twisting of the conveyance because of the natural tendency of a rope to untwist when under tension. The data provided above shows that headropes are available with very low torque constants. The ropes used for balance, or tail ropes, are typically designed to minimize the torque that arises as load is applied to them. This minimum is still significant and is the reason why the ropes are attached to the bottom of the conveyance using swivels. This arrangement reduces the torque to a function of the friction in the swivels. The swivels are basically a rolling contact bearing that has a static friction, or stiction, of about 1.5 to 2 times the rolling friction. Based on the data provided by the manufacturer of the swivels, each tail rope will typically produce a torque of the order of 50 N·m (40 lb-ft). This will be developed when about 250 m (800 ft) of tail rope is suspended below the skip.

Guide Rope Location

Figure 12.9-25 illustrates the two most common arrangements of guide ropes. Option A has all four guide ropes along one side of the conveyance, while Option B has one rope at each of the four corners of the conveyance.

Although both options offer the same resistance to lateral motion, Option B provides significantly better resistance to

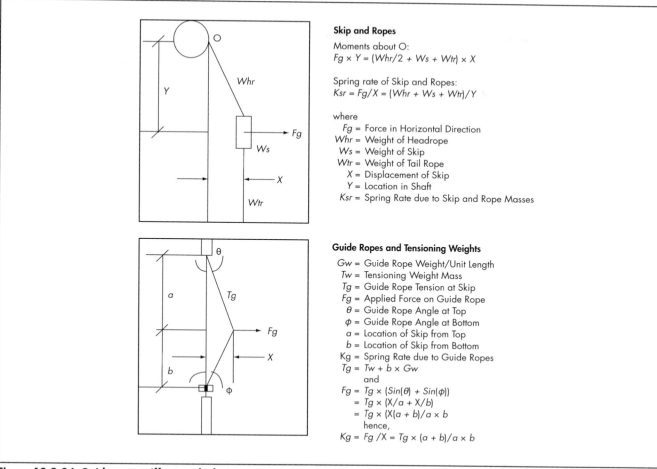

Figure 12.9-24 Guide rope stiffness calculation

twist. In fact, Option A will cause twist for any force that acts in a direction parallel to the plane of the guide ropes.

Guide Rope Tensioning
To keep conveyance motion to a minimum, guide ropes should be tensioned to the maximum amount permitted by local regulations. This typically permits operation with a minimum safety factor of 5.

Tensioning of ropes is possible by the following two means: hanging heavy weights (called cheese weights) from the bottom of ropes, and anchoring the bottom of the rope to a beam in the shaft and tensioning the rope in the headframe using either large springs or hydraulic cylinders. The use of hydraulic cylinders simplifies rope changing and reduces the need for shaft excavation. The most reliable method is still the use of tensioning weights, because this approach automatically compensates for rope stretch and temperature fluctuations.

If cheese weights are used, it is necessary to clean out the shaft bottom on a regular basis, or the weights will become buried in spill and drop their tensioning force.

Shaft Lining
Shaft lining may be comprised of shotcrete, cast-in-place concrete, or steel and cast-iron tubbing. When a lining is required, the selection depends on a number of design considerations that relate to the strata being excavated. For further discussion, see Chapter 12.4.

Shotcrete
Sprayed concrete, 150 mm (6 in.) or more in thickness, may be used to control immediate raveling and weathering of the shaft walls. To date, shotcrete has not been used extensively to provide long-term support nor has it been used as the only means of lining in shafts.

Concrete
The use of concrete for permanent support is common in most new shafts. Shafts may be fully lined or partially lined with "concrete rings" with areas of open ground between the rings. Common spacing used for ringed shafts is 1.2-m- (4-ft-) high concrete rings with a 1.2-m (4-ft) open area between, resulting in a 2.4-m (8-ft) set interval. Concrete is generally placed with a minimum thickness of 300 mm (12 in.), but this will increase with depth for linings designed to withstand the hydraulic head encountered in formations that include aquifers.

Steel or Cast-Iron Tubbing
Although the use of concrete as a lining material is becoming more common, there are certain geologic conditions where steel or cast-iron tubbing is a more appropriate

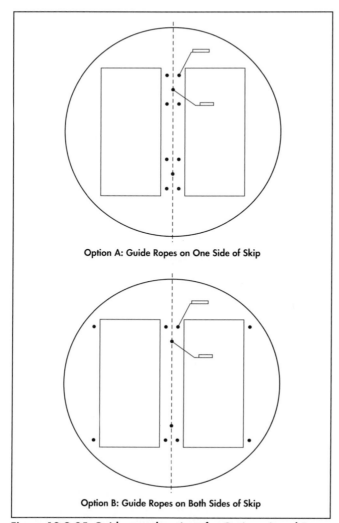

Figure 12.9-25 Guide rope locations for Options A and B

material. Concrete is suitable when water pressures are less than 1,700 kPa (250 psi) and when the stress field is uniform and compressive in nature. Steel is more suitable for a completely watertight lining and is capable of withstanding tensile stresses. These stresses could arise if the shaft is subjected to bending loads caused by subsidence or buckling loads caused by a nonuniform horizontal stress field.

Steel lining can be butt-welded to make it permanently watertight, and this has a higher buckling safety factor than tubbing. Steel is particularly useful as a lining for shafts sunk by drilling methods. Cast-iron tubbing consists of several segments bolted to form a ring and requires a lead gasket to be placed between segments. This type of tubbing is much stronger than the steel liner and is typically used to provide support when crossing aquifers are encountered at considerable depth and, hence, high pressure.

Design Considerations for Shaft Lining
The design requirements for lining a structure are governed by the required duty of the shaft and environmental conditions in which it is constructed. Careful consideration of the following points should be made in the design of structural lining: the precise behavior of the rock mass in situ, the self-supporting action of the rock itself, the effect of separation of the lining due to certain vertical and horizontal movements of the surrounding rock formation, the construction of watertight linings, and the economics of the lining structure.

The design of steel or cast-iron linings is a much more complex process, because the buckling characteristics of an elastically embedded hollow cylinder with external pressure must be evaluated.

Ventilation Characteristics
The shaft is generally the largest single contributor to total mine air resistance when it is equipped with rigid guides for hoisting purposes. The resistance consists of the following two components: frictional resistance caused by viscous drag in the thin air layer at the shaft walls and the periphery of the vertical elements (guides, pipes, cables, ropes, etc.) of the shaft; and shock resistance due to the transverse sets (buntons and dividers) completely immersed in the air stream, thereby obstructing the flow and recurring at intervals throughout the shaft.

These transverse elements create turbulence in the airstream with a corresponding loss of flow energy. The turbulent waves below buntons and conveyances often contribute much more to the total shaft resistance than the skin friction of the shaft walls. The forces caused by the buntons are parallel to the airstream and are proportional to velocity pressure, frontal area, surface texture, and shape. Because the frontal area of all the buntons in a shaft is likely to be more than the frontal area of the cages and skips, the aerodynamic design of the bunton is of importance in the design of a mine shaft.

A reduction in shaft resistance can be achieved through improving the aerodynamic effect of horizontal obstructions such as buntons, minimizing the frontal area of the set members, and increasing the spacing between the horizontal obstructions. Practical experience has indicated that there is little ventilation advantage to be gained from spacing greater than 5 m (16 ft).

A general treatment of mine ventilation is offered in Chapter 15.3.

HEADFRAME
The basic purpose of a headframe installed over a shaft is to support the sheave wheel over which a hoisting rope passes for raising or lowering conveyances or, in the case of a tower-mounted Koepe hoist, to support the hoist and all of its suspended loads. The construction of a headframe is also necessary to allow dumping of hoisted materials aboveground. Structurally, there are two types of headframes: steel headframes, which are typical for drum hoists, and slip-formed concrete headframe structures, which are typical for tower-mounted Koepe hoists. These designs are shown in Figures 12.9-26 and 12.9-27, respectively.

Construction Materials
Headframes may be constructed of timber, steel, or concrete. Modern trends in high-capacity hoisting have necessitated the construction of very large headframes, which effectively preclude the use of timber as a construction material.

Steel Versus Concrete
Both concrete and steel can produce economically viable, structurally compatible, and visually appealing headframes

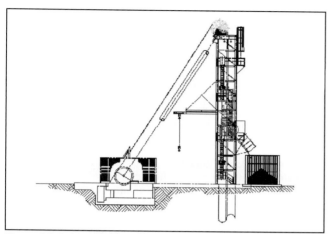

Figure 12.9-26 Steel headframe

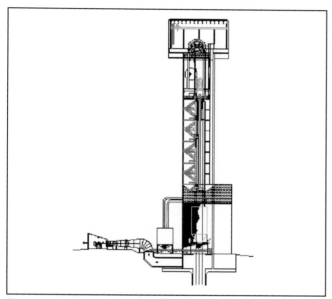

Figure 12.9-27 Concrete headframe

For steel structures, advantages include the following:

- Foundation loads are lighter.
- They are adaptable to minor changes.
- They offer no salvage value.
- Capital-cost saving in remote areas is possible.
- The structure can be prefabricated into transportable lengths for on site assembly.
- Construction is speedy.

The main disadvantage of steel is that it is prone to corrosion and, thus, is not suitable as a production headframe for corrosive ore (e.g., salt and sulfides) unless the steel is treated and regularly maintained.

Design Considerations

A successful headframe design must be based on an overall integrated hoisting system design, which involves a critical examination of the impact of at least the following items: loads, foundation, mine services, sinking provisions, equipment monitoring facilities, conveyance and rope handling, heating and ventilation, and others such as local mining regulations.

Loads

All headframes should be designed to withstand a combination of loads, comprised of the following:

- Dead load consisting of the weight of the headframe, sheave wheels, ore bins, and contents
- Live load from hoisting at maximum capacity
- Breaking loads of the ropes when the conveyance is stopped by the crash beams or if it is jammed in the shaft (breaking stresses introduced by an overwind are usually transferred to the structural system, often through an arrestor gear, and thus the supporting structure must be critically analyzed with a thorough understanding of the arrestor mechanism)
- Wind load, the intensity of which depends on the location, height, and shape of the structure
- Snow load
- Temperature and seismic stresses, taken from published data relating to the geographic area of the shaft location

Foundation

The foundation should be capable of withstanding and absorbing stresses imposed by the structure and associated loads. Foundation design in areas where the ground must be frozen by artificial means before the shaft is sunk require much more consideration than normal because the ground will heave when the freeze is removed, which can lead to tilting of the headframe.

Mine Services

The design of the headframe must suit surface layouts for providing the essential mine services. When possible, it should be located close to the shaft collar house, waiting area, lamproom, first-aid room, hoist house, maintenance shop, changing house, and administrative offices. The type and design of the headframe must also take into account the shaft's internal arrangement, position of the hoist with respect to the shaft, and the position of the sheave steel.

Facilities for the following key concerns must be considered and adequately developed during the conceptual design stage:

for any new development. The advantages of concrete structures include the following:

- Effective use can be made of space within the structure for tower-mounted friction hoists.
- When used in conjunction with ground-mounted drum hoists, the hoist may be located at any position around the headframe.
- Maintenance costs are low.
- A high damping effect of the massive concrete structure occurs on shock and vibration.
- The collar layout and the interconnection with service buildings is simple.
- Future expansion is possible.
- The structure is durable and resistant to corrosion.

The disadvantages of concrete structures include the following:

- Erecting in remote areas is difficult.
- Supporting the structures in poor ground is difficult.
- Concrete offers no salvage value.

- Arrangement of skip dumping area in relation to ore and waste storage and handling facilities
- Clearance required for installation, maintenance, and removal of conveyances
- Lifting equipment required for maintenance of conveyances
- Access required for maintenance of conveyances, dump areas, and so forth

In the case of headframes for drum hoists, provision must be made for

- Lifting equipment and access for maintenance of sheaves;
- Access and anchorage facility to support the ends of hoist ropes when doubling down for pretensioning drum end coils, and
- Subcollar loading access for double-deck cages.

The design of headframes for friction hoists should address the following:

- Supporting facility at predetermined elevations for hoist or balance ropes and conveyances or counterweights during installation and changes
- Clearance for installation of conveyance arrestor gear
- Provisions for handling shaft ventilation air and for sealing machinery rooms
- Arrangement of power conversion equipment adjacent to hoist motors to minimize runs of heavy bus bars
- Lifting equipment access during changing of headropes and tail ropes
- Support, accommodation, inspection, removal, and replacement facilities for rope guides where used
- Ventilation to specific machines and areas
- Controlling moisture, dust, and temperature where required
- Accommodation of sinking arrangements
- Accommodation of all services for air, water, communications, and power lines to the shaft and to the headframe-mounted equipment
- The accommodation of stairwells and elevators
- Lighting and lightning protection

Sinking Provisions

After the decision has been made to sink a shaft, considerations should be given to the type of surface sinking plant (such as headframe and hoist) that is to be used for shaft sinking. In particular, it should be determined whether the sinking plant is to be temporary or is to make use of the permanent headframe and/or hoist.

Temporary headframe. Sometimes a temporary headframe is used during shaft sinking by the contractor. These headframes are generally made in modular form from structural steel. Headframes of this type are portable and can be reused many times.

Design criteria for a temporary sinking headframe are the same as for a permanent headframe. Its height depends on the method of rock disposal and size of bucket. The sinking headframe should embody features for dumping buckets or skips, for protecting personnel on the surface and in the shaft from falling pieces of rock while dumping, and for minimizing the work of top personnel in dumping buckets and removing broken rock. The design of temporary headframe structures should accommodate an adequately sized storage bin and should provide sufficient clearance for a rock disposal vehicle.

Permanent headframe. Current trends favor installation of a permanent headframe for the sinking operation, as it tends to be more economical and time saving. The permanent main skeleton can be designed to facilitate erection and installation of the sinking sheaves at a suitable elevation. The structure can then be finished and equipped during the shaft sinking operation. The use of the permanent structure as a sinking headframe can make the permanent hoisting facilities available at an earlier date following completion of the shaft-sinking program.

Equipment Monitoring Facilities

Two main hoisting incidents, which can be detrimental to personnel and equipment and require monitoring, are overtravel at the shaft extremities (i.e., sump, headframe) and overspeed during the duty cycle. To protect against these incidents, it is mandatory that monitoring and safety devices be installed in the headframe.

Operation of all hoists is monitored by digital equipment designed for this purpose and often termed *safety controllers*. These are driven by the hoist drums and they check hoisting speed, overtravel at each end of the wind, and deceleration in the end zones. In addition to these dynamic checks, track limit switches must be incorporated to protect against a safety monitor losing its synchronism with the hoist. The function of track limit switches is to provide independent overtravel protection in the headframe.

Ultimate protection against overwind is provided for all Koepe hoists by conveyance arrestors. These are installed at the ends of travel in each hoisting compartment. The headframe arrestors are designed to retard an ascending conveyance from the speed of a person riding at a rate of 0.9 g's. Arrestors are also installed at the shaft bottom and are designed to retard a full descending conveyance at a rate of at 1 g or higher. It is not practical to install arrestors in the headframe to stop a hoist from full speed. For example, a hoist traveling at 18 m/s (3,600 fpm) would require a stopping distance of 18.5 m (60 ft) when decelerating at 0.9 g's. This amount of overtravel provision would be very expensive to provide and would still not guarantee absolute protection because a runaway hoist can easily reach speeds in excess of 50 m/s (10,000 fpm).

Where applicable, the headframe should accommodate installations of a crash beam and catch gear. With drum winders, the crash beam is located below the sheaves in the headframe. In the event of an excessive overwind, it stops the conveyance from traveling beyond this point and is designed to withstand the breaking strength of the headrope. Should a conveyance on a Koepe hoist be involved in an overwind situation in which the hoist ropes break, the conveyance and tail ropes will fall down the shaft, causing damage. To prevent this, fallback protection should be provided in the headframe to "catch" the conveyance.

Conveyance and Rope Handling

In the design and layout of a headframe, provisions must be made for the handling and changing of conveyances, hoisting ropes, and rope guides (if applicable). Friction hoisting often requires the use of very large skips and cages. Because of their size, the service and handling of these conveyances, and the tying off of the headropes and tail ropes associated with them, considerable problems can arise. In the friction hoist headframe, these can be facilitated through the installation of an adequately sized overhead traveling crane. More

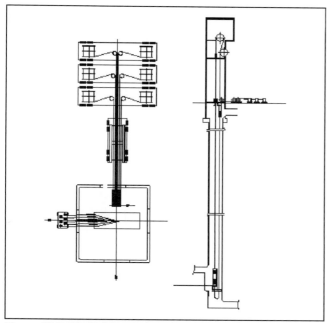

Figure 12.9-28 Multirope changing reeving diagram

recently, special equipment has been designed to permit the removal or installation of all of the headropes at the same time. The process requires a special winch that is designed to pull all of the headropes at once and transfer them on or off of numerous rope-holding winches. A reeving diagram for this arrangement is shown in Figure 12.9-28. In the drum hoist headframe, handling and changing of conveyances and ropes are relatively easy and are usually achieved through the use of mobile cranes, rope blocks, and tugger hoists.

Heating and Ventilation

In warm climates, the headframe may be left open. In colder climates, however, it is necessary to design the headframe as an enclosed structure with adequate provisions for a heating system. In tower-mounted friction hoist systems, with all of the hoists, motors, controls, and so forth enclosed in the headframe, it is necessary to provide a ventilation system to supply filtered air for both cooling and heating.

Some headframes located over downcast shafts are pressurized and must be equipped with air locks to isolate the movement of personnel and equipment into and out of the collar house.

Mining Regulations

Mining operations, including the design and construction of the headframe and shaft facilities, are subject to local, state, and federal regulations that relate to, among other things, safety of permanent structures and environmental protection. These regulations affect the design and construction of mine facilities.

Height

Because of increasing public awareness of the aesthetic impact of industrial development, the height of the headframe should be considered. This is especially true for locations in flat terrain where any large structure can be seen for long distances. The height of a headframe is generally based on the storage bin capacity requirements for ore and waste, the clearance required for the loading facility under the storage bin, the requirement that the resultant force from headframe loadings falls within the backleg post, and the need for providing minimum clearance for conveyance overtravel beyond its normal travel to the head sheaves. Of the above, the determination of the minimum clearance for overrun distance is the principal design consideration. The determination of the minimum clearances required in headframes is described in the following paragraphs.

Operating Allowance

An operating allowance is required to allow for variations in repeatability of the final stop position of the conveyance in the headframe. In multilevel hoisting with drum hoists, this tolerance compensates for permanent rope stretch, differences in drum and rope diameter, and clutching accuracy. In the case of Koepe hoists, the allowance compensates for creep during resynchronization and also allows a reasonable gap for rope adjustment due to permanent rope stretch.

Rope Stretch Allowance

This is included to avoid contact with the track limit switch or encroaching on the safety controller's overwind protection limit when empty conveyances are hoisted.

Stopping Allowance

A stopping allowance is needed to ensure that clearance remains between the top of the conveyance and the first obstruction in the headframe should the conveyance exceed its normal operating limits and require the use of emergency braking to stop.

Rope Tension Equalization

The ropes on a Koepe hoist operate at slightly different diameters because of uneven wear of the rope treads. To minimize the effects of this variation, it is recommended that the ratio of the maximum to minimum suspended rope length should not be more than 50.

Spacing of Koepe Wheel and Deflection Sheave

When a headrope is bent in an arc, the wires that make it up move relative to one another to accommodate the different lengths at the outer and inner radii of the surface of the rope. When the rope straightens again, the wires do not assume their original positions immediately, primarily because there is friction between them. The headrope passing over the hoist wheel is bent one way, straightens when it leaves the wheel, and then bends the other way when it passes over the deflection sheave. If the time taken to pass from the wheel to the sheave is less that 0.5 seconds, the wires will not have time to realign between these reverse bends and will eventually produce distortion and premature failure of the rope. Allowance for this spacing must be made when laying out the headframe.

Rope Harmonics and Vibrations

Vibrations in hoist ropes can be observed as "yo-yo-ing" at the collar, and, in the case of drum hoists, as the rope whips between the hoist and the head sheave. The causes of these vibrations are jerking operations, such as uneven braking, and the horizontal and vertical movement of the rope at the crossover points on the drum. These vibrations cannot be avoided and normally do not cause problems. However, there have

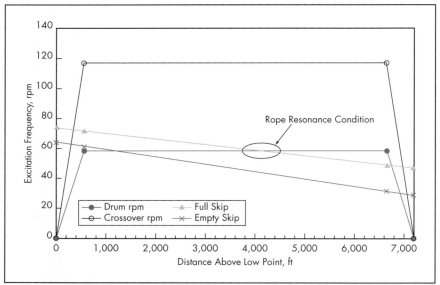

Figure 12.9-29 Rope resonance plot

been cases where the frequency of the crossover motion is a harmonic of the natural frequency of the flight rope (i.e., the length of rope between the drum and head sheave), and severe and dangerous whipping has occurred. Such a situation is most likely with a long, flat span of rope between the hoist and head sheave.

The hoist designer should analyze the harmonic behaviors of the ropes and try to select the type of drum grooving that will avoid harmonic vibrations. This selection requires two calculations. The first calculation determines the natural frequency of the flight rope (which varies with rope tension and therefore decreases linearly as the conveyance ascends the shaft). The second calculation determines the frequency spectrum excited by the crossover spacing. If the excitation spectrum includes harmonics that match the natural frequency of the flight rope, alternative crossover geometries will have to be considered. If suitable alternatives are not found, the hoist speed will have to be altered or idler towers will have to be added between the drum and the head sheave.

A typical plot of excitation frequency and flight rope natural frequency is illustrated in Figure 12.9-29. This plot illustrates that a rope resonance condition would exist if the rope crossover occurred once per revolution but not if it occurred twice per revolution with the crossovers spaced 180° apart.

Calculations for Minimum Clearance

The following details show how the required clearances are to be calculated.

Elastic Rope Stretch for Skip

The headropes stretch when a skip is loaded. This stretch is locked into the rope as it coils on the drum when the loaded skip is raised and is relieved as it uncoils when the empty skip is lowered. The drum will, therefore, make more revolutions when raising a full skip than an empty one. Because the skip distance is measured by revolution of the drum, the hoisting cycle is only accurate for a loaded skip. The headframe will usually be fitted with a "clutching-mark" magnet switch located at a distance above the "skip-at-dump" magnet switch equal to the elastic rope stretch. The stretch (S) can be calculated from

$$S = P*L/(A*E)$$

where
 P = payload
 L = rope length from the drum to the skip at the loading pocket
 A = rope area based on its nominal diameter
 E = rope elastic modulus in tension

A good guideline for elastic rope stretch is 300 mm (1 ft) per 300 m (1,000 ft) of shaft depth.

Clutching Accuracy

A clutch includes a ring with a finite number of teeth. The clutching accuracy is, therefore, limited to plus or minus the tooth pitch. The ability for the operator to compensate can be determined from

$$C = \pm\pi*D/(2*N)$$

where
 C = clutching accuracy that can be achieved
 D = pitch circle diameter of the headrope
 N = number of teeth in the clutch ring

Unequal Drum and Rope Diameters

Inequalities in drum or rope diameters between two drums of the same hoist will be amplified by the number of revolutions the hoist makes as it travels up the shaft. This difference is given by

$$\Delta T = \pi*N*(\Delta Dd + 1.7*\Delta Rd)$$

where
 ΔT = difference in travel
 N = number of revolutions of the hoist
 ΔDd = difference in drum diameters
 ΔRd = difference in rope diameters

Unequal Rope Lengths

If the ropes on two drums of a hoist are not exactly the same and the drums wind more than one layer of rope, then the drum with the longer rope will operate at a larger average diameter. As a result, this drum will take fewer revolutions to raise a skip than the drum with less rope. This difference is proportional to the number of extra turns times the constant π.

For example, consider the case where the rope and drum diameters are equal but there is one more turn on one drum than the other. Assume also that there are four layers of rope, the PCD of the first layer is $d = 4,653$ mm, and the PCD of the fourth layer is $D = 4,915$ mm. One drum has 6 turns on the first layers when the skip is at the loading pocket and 18 turns on the fourth layer when the skip is at the dump, and the other drum has 7 turns on the first layer when the skip is at the dump and 19 turns on the fourth layer when the skip is at the dump.

$$\text{difference in rope lengths} = \pi*d = \pi*4.653 = 14.6 \text{ m}$$

$$\begin{aligned}\text{difference in skip travel} &= \pi*(D - d) \\ &= \pi*(4.915 - 4.653) = 0.8 \text{ m}\end{aligned}$$

As these calculations demonstrate, when one rope is damaged and needs to be shortened, the same length must be cut off the other rope.

Stopping Distance

If a conveyance travels beyond the track limit switch (which must be checked at least daily) it will initiate an emergency stop using the brakes to stop and hold the hoist. The allowance to be made for the conveyance overtravel during this event is given by

$$S = 2*V^2/(2*D)$$

where

- S = stopping distance
- V = hoisting speed permitted by the safety controller in the end zone (typically 0.5 m/s or 100 fpm)
- D = deceleration rate achieved by the braking system

The factor 2 is included in the equation to account for the dead time of the brakes.

REFERENCES

CFR (Code of Federal Regulations). 2009. 30 CFR Parts 55, 56, 57. Mineral Resources. Available from www.gpoaccess.gov/cfr. Accessed January 2010.

SABS 0208-4. 2001. *Design of Structures for the Mining Industry. Part 4: Shaft System Structures.* South African Bureau of Standards.

SANS 10293. 1996. *Condition Assessment of Steel Wire Ropes on Mine Winders.* South African Bureau of Standards.

SANS 10294. 2000. *The Performance, Operation, Testing and Maintenance of Drum Winders Related to Rope Safety.* South African Bureau of Standards.

PART 13

Underground Extraction

CHAPTER 13.1

Room-and-Pillar Mining in Hard Rock

Richard L. Bullock

Room-and-pillar (R&P) mining is a mining method whereby a series of rooms (horizontal openings) is extracted, leaving pillars of ore, rock, or coal in place between the rooms. Figure 13.1-1 shows a plan view of an ideal R&P mine. The square blocks are pillars; the spaces between the pillars are rooms. In hard rock, pillars are usually much smaller horizontally than rooms. In soft rock and coal mines, they are usually much larger.

The R&P method has been used widely and often—for example, in lead/zinc/copper ore bodies of mines in southeast Missouri (United States) for 150 years; in mines in Charcas (Mexico) for more than 100 years; in copper mines in Poland for nearly 50 years; and in underground rock quarries and salt mines worldwide for more than 100 years. The proliferation of this method for such a long period of time suggests that it is low cost, versatile, and safe.

ORE-BODY CHARACTERISTICS OF R&P MINING

The R&P method is normally employed on fairly flat seams or ore bodies, but may in some cases be used in seams that angle up to 45°. Ideally, the ore body or seam to be mined should be very large laterally and fairly uniform in thickness. However, many R&P mines in hard-rock ores have nonuniform ore horizons. For example, in some of the R&P mines in the Old Lead Belt of Missouri, the ore horizon varies in thickness from 3 to 91 m (10 to 300 ft). In the Viburnum Trend mines of Missouri, ore varies in thickness from 3.5 to 30 m (12 to 100 ft).

The strength of both the back/roof and the ore are important. Obviously, roof strength must be sufficient to stand open for the width of the room when the room is opened. Similarly, ore strength determines the size of the pillar required to support the open span of the back. Thus pillar size is not fixed but rather is determined by the sum of (1) stress in the pillar due to the weight of the overburden above the opening, (2) additional stress in the pillar due to removal of the ore that originally supported the current back, and (3) tectonic stress-related forces that may still remain in the rock. The ore in the pillar must be strong enough to compensate for these collective stresses.

Farmer (1992) and Zipf (2001) described the design of stope pillars and barrier pillars normally used for a new R&P

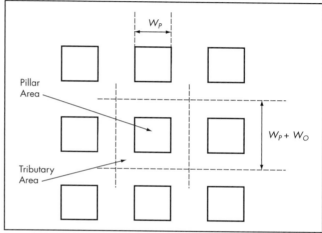

Source: Zipf 2001.

Figure 13.1-1 Plan view of a room-and-pillar mine

mine in a new mining district. The caveat "new mine in a new district" is important, since, in opening a mine in an already-mined district, decisions regarding pillar size and room size for a given stope depth and ore strength have already been made. Yet it is nevertheless important to understand how to design stope (panel) pillars and barrier pillars for a new mine.

TRADITIONAL STRENGTH-BASED PILLAR DESIGN METHODS

This section is adapted from Zipf (2001). Significant research has been done on pillar design for both R&P and longwall mining. Many ground-control and pillar-design experts have contributed (e.g., Holland 1964; Salamon and Munro 1967; Hardy and Agapito 1977; Mark 1999). Their studies include verification of pillar-design formulas and formula constants by statistical verification with a large database of both failed and nonfailed coal mine incidents from Australia and South Africa (Galvin et al. 1999; Gale and Hebblewhite 2005). However, with the exception of the work of Hardy and Agapito, these studies

---

Richard L. Bullock, Professor Emeritus, Mining & Nuclear Engineering, Missouri University of Science & Technology, Rolla, Missouri, USA

concern coal, and there has been no similar verification for hard rock or even other softer minerals such as trona, salt, oil shales, or borates. The purpose of the following discussion is to extend this work to the design of pillars for all minerals, not just coal.

R&P mines may cover only a hundred square meters and contain just a few pillars, as is typical for small zinc deposits. Or they may cover many square kilometers, as is typical for coal, trona, and limestone mines. In an R&P mine, large arrays of pillars are typically grouped into panels, which in turn are surrounded by barrier pillars. The small pillars within a panel are sometimes called panel pillars. In developing layouts, the mining engineer must develop appropriate dimensions for room spans, panel pillar widths, panel sizes, and barrier pillar widths. Developing these dimensions requires evaluation of not just pillar strength but also the consequences of pillar failure, which can happen anywhere in the layout at any time.

Traditional strength-based pillar design requires estimates of pillar stress and pillar strength. The safety factor for the pillar is then calculated by dividing pillar strength by pillar stress. What constitutes an acceptable safety factor depends on the tolerable risk of failure. Safety factors of 2 are typical for pillars in main development headings or panels during advance mining. Safety factors of 1.1 to 1.3 are typical for panel pillars after retreat mining. Safety factors much less than 1.0 are possible for panels where pillar failure is the eventual intent.

Pillar Stress

All equations in this section can be dimensioned as preferred by the user, as long as dimensions are consistent with the equation constants.

According to traditional R&P design methods, in-situ vertical stress is (Farmer 1992)

$$\sigma_s = \lambda z \qquad (13.1\text{-}1)$$

where

λ = unit weight of rock
z = depth of mining horizon

The tributary area method then provides a first-order estimate of the average pillar stress. For the square R&P system shown in Figure 13.1-1, the average pillar stress is

$$\sigma_{pa} = \sigma_z \left(\frac{W_p + W_o}{W_p} \right)^2 \qquad (13.1\text{-}2)$$

where

W_p = pillar width
W_o = opening width

For rectangular or irregular-shaped pillars, where R is the extraction ratio, the average pillar stress is

$$\sigma_{pa} = \sigma_z \left(\frac{1}{1-R} \right) \qquad (13.1\text{-}3)$$

$$R = \frac{A_M}{A_T} = \frac{A_T - A_P}{A_T} \qquad (13.1\text{-}4)$$

where

A_M = area extracted (i.e., total room area)
A_T = total area of the ore body
A_P = pillar area

The tributary area method assumes that the mined area is extensive and all pillars have the same dimensions. It also ignores the deformation properties of the surrounding rock mass relative to the pillar rock. However, pillar stress is generally higher at the center than at the edge of a panel. Coates (1981) partially solved this problem by developing the following relation for average pillar stress that accounts for the width and number of pillars across a panel and the relative mechanical properties of the pillar and rock mass:

$$\sigma_{pa} = \sigma_z \left\{ \frac{\left[2R - K_0 \dfrac{H}{L} \dfrac{(1-2\upsilon_{rm})}{(1-\upsilon_{rm})} - \dfrac{\upsilon_p}{(1-\upsilon_p)} K_0 \dfrac{H}{L} \dfrac{E_{rm}}{E_p} \right]}{\dfrac{H}{L} \dfrac{E_{rm}}{E_p} + 2(1-R)\left(1 + \dfrac{1}{N}\right) + 2 \dfrac{RB}{L} \dfrac{(1-2\upsilon_{rm})}{(1-\upsilon_p)}} \right\} \qquad (13.1\text{-}5)$$

where

K_0 = ratio of horizontal to vertical stress
H = mining height
L = lateral extent of the mined area
E, υ = elastic constants
rm = rock mass
p = pillar mass
B = individual opening width
N = number of pillars across the panel

Although this equation is based on two-dimensional elasticity theory and therefore applies only to long, narrow rib pillars, it illustrates the behavior of average pillar stress. As the E_{rm}/E_p ratio increases, average pillar stress decreases due to the bridging effect of the stiff rock mass. Similarly, as panel width L decreases and the H/L ratio increases, the average pillar stress decreases.

As with the tributary area method, the solution offered by Equation 13.1-5 gives only average pillar stress for all panel pillars and does not give changes in pillar stress across the panel. Boundary-element method programs such as ExamineTAB (Zipf 2001), MULSIM/NL (Zipf 1992a, 1992b), or LAMODEL (Heasley 1997, 1998) are needed to calculate changes in pillar stress across a panel or within a pillar.

Pillar Strength

Over the past several decades, a large amount of rock mechanics literature has addressed pillar strength in both coal and metal/nonmetal mines (Obert and Duvall 1967; Hoek and Brown 1980; Bieniawski 1992; Brady and Brown 1993). Much of this work is empirical and addresses two issues: (1) the size effect, whereby rock strength decreases as specimen size increases; and (2) the shape effect, whereby rock strength decreases as the width-to-height ratio decreases. Using energy considerations, Farmer (1985) developed theoretical expressions relating strength to size. When failure occurs in a brittle manner, as it does in most rocks, strain energy within the specimen transforms to fracture surface energy, which is constant for a particular rock. Based on energy conservation,

$$SED * V = FE * A \qquad (13.1\text{-}6)$$

where

SED = strain-energy density
V = volume
FE = fracture-surface energy, which is a material constant
A = fracture surface area

Table 13.1-1 Values for constants in empirical pillar-strength formulas

Source	a*	b*	α	β	Comments
Bunting (1911)	0.7	0.3	—	—	Pennsylvania anthracite
Obert and Duvall (1967)	0.78	0.22	—	—	Laboratory rock and coal
Bieniawski (1968)	0.64	0.36	—	—	South Africa coal
Skelly et al. (1977)	0.78	0.22	—	—	West Virginia coal
Greenwald et al. (1939)	—	—	0.5	0.83	Pittsburgh seam mines†
Holland (1964)	—	—	0.5	1	U.S. coal mines†
Salamon and Munro (1967)	—	—	0.46	0.66	South Africa coal mines‡
Hardy and Agapito (1977)	—	—	0.60	0.95	U.S. oil shale mines†

Source: Zipf 2001.
*a and b are dimensionless.
†Use with English units only.
‡Use with metric units only.

Rearranging gives

$$SED * \frac{V}{A} = SED * L = FE = \text{constant} \quad (13.1\text{-}7)$$

where L is the characteristic dimension or length of the rock specimen.

Assuming that laboratory-scale failure is mechanistically similar to full-scale pillar failure, then

$$SED_S * L_S = SED_P * L_P \quad (13.1\text{-}8)$$

where

S = subscript denoting laboratory-scale specimen
P = subscript denoting full-scale pillar

Since strain-energy density at failure is proportional to the square of stress at failure,

$$\sigma_S^2 * L_S = \sigma_P^2 * L_P \quad (13.1\text{-}9)$$

or

$$\frac{\sigma_P}{\sigma_S} = \left(\frac{L_S}{L_P}\right)^{1/2} = \left(\frac{V_S}{V_P}\right)^{1/6} = \left(\frac{V_S}{V_P}\right)^{0.17} \quad (13.1\text{-}10)$$

where

σ_P = strength of the full-scale pillar
σ_S = strength of the laboratory specimen
V = volume, which is proportional to L^3

This theoretical relationship accounts for the size effect in observed rock strengths. The following empirical strength formula, proposed by Hardy and Agapito (1977) for oil shale pillars, follows this general theoretical form and provides some experimental confirmation. It also includes an additional term for pillar shape.

$$\frac{\sigma_P}{\sigma_S} = \left(\frac{V_S}{V_P}\right)^{0.118} \left[\frac{W_P H_S}{H_P W_S}\right]^{0.833} \quad (13.1\text{-}11)$$

where

W_P = width of pillar
H_P = height of pillar
H_S = height of specimen
W_S = width of specimen

Classic empirical pillar-strength formulas usually follow one of two general forms:

$$\sigma_P = \sigma_{S'}\left(a + b\frac{W}{H}\right) \quad (13.1\text{-}12)$$

$$\sigma_P = K\frac{W^\alpha}{H^\beta} \quad (13.1\text{-}13)$$

where

σ_S = strength of a cubical pillar ($W/H = 1$) at or above the critical size
K = constant characteristic of the pillar rock
a, b, α, β = constants that account for the shape factor

Formulas by Obert and Duvall (1967) and Bieniawski (1968) follow the first form; formulas by Salamon and Munro (1967) and Holland (1964) follow the second. In these forms, the size effect is accounted for directly by the unit pillar strength σ_S or the rock constant K. The constants a, b, α, and β, which account for the shape factor, show reasonable agreement, as shown Table 13.1-1.

Several methods exist for estimating σ_S and K in Equations 13.1-12 and 13.1-13, respectively. One is to consider the strength of a cubical pillar ($W/H = 1$) at or above the critical size, where critical size is the size beyond which rock mass strength remains relatively constant. For coal pillars, the critical size is widely recognized to be about 0.9 m (36 in.). For coal mines in the United States, the recommended strength value for a cube of coal this size is σ_S = 6.2 MPa (900 psi) (Mark and Chase 1997). This unit strength for a full-scale cube of coal can then be adjusted for shape effects using the Mark–Bieniawski relation (Mark 1999):

$$\sigma_P = \sigma_{S'}\left(0.64 + 0.54\frac{W}{H} - 0.18\frac{W^2}{HL}\right) \quad (13.1\text{-}14)$$

which reduces to the original Bieniawski relation when pillar width W equals pillar length L.

If laboratory-scale strength data (σ_S) is available, the full-scale strength of a cube of rock mass ($\sigma_{S'}$), from Equation 13.1-10, is

$$\sigma_{S'} = \sigma_S\left(\frac{V_S}{V_{S'}}\right)^{0.17} \quad (13.1\text{-}15)$$

This unit strength for a full-scale cube of the rock mass can then be adjusted for the shape effect using the Obert–Duvall relation:

$$\sigma_P = \sigma_{S'}\left(0.78 + 0.22\frac{W}{H}\right) \quad (13.1\text{-}16)$$

Finally, the Hoek–Brown failure criterion can also provide an estimate of the strength for a full-scale cube of the rock mass. For most rock masses with good to reasonable quality,

$$\sigma_1' = \sigma_3' + \sigma_c\left(m_b\frac{\sigma_3'}{\sigma_c} + s\right)^a \quad (13.1\text{-}17)$$

where
m_b, s, a = constants that depend on the rock mass quality
σ = uniaxial compressive strength of the intact rock pieces (equivalent to σ_S)
σ_1' = axial principal stress
s_3' = confining principal stress

The constants m_b and s are estimated using a rock-mass classification index called GSI, or Geological Strength Index, which is equivalent to Bieniawski's rock mass rating (RMR) system, assuming a dry rock mass. For a complete discussion of the Hoek–Brown failure criterion for rock masses, see Hoek et al. (1995).

Barrier Pillar Design

The previous formulas for pillar stress and pillar strength apply mainly to pillars within a large array, or so-called panel pillars. The barrier pillars surrounding this array of panel pillars also require sizing. The tributary area method can provide a conservative estimate of barrier pillar stress, if we assume that the panel pillars have all failed or are all mined and therefore carry no overburden stress. Alternatively, numerical methods such as MULSIM/NL or LAMODEL can estimate barrier pillar loads without need for conservative assumptions inherent to the tributary area method.

The empirical pillar strength formulas shown earlier apply to pillars with width-to-height ratio $W/H < 5$. For barrier pillars and other pillars in coal mines with $W/H > 5$, Salamon's "squat" pillar formula applies:

$$\sigma_P = \sigma_{S'}\left(\frac{5^b}{V_P^a}\right)\left(\frac{b}{e}\left(\left(\frac{W/H}{5}\right)^e - 1\right) + 1\right) \quad (13.1\text{-}18)$$

where
$b = 0.5933$
$a = 0.0667$
$e = 2.5$

For hard-rock and other noncoal mines, an equivalent squat pillar formula does not exist. It is necessary to extrapolate the Obert–Duvall relation to high W/H ratios or use numerical methods and the Hoek–Brown failure criterion to estimate the required barrier pillar size.

Koehler and Tadolini (1995) reviewed no fewer than 10 empirical and observational approaches to barrier pillar design in coal mines. Most approaches calculate minimum barrier pillar width as a function of depth, and some include seam thickness. Coal strength is generally neglected. Equivalent empirical approaches to estimate barrier pillar size in noncoal mines do not appear to exist.

Summary of Traditional Strength-Based Pillar Design Methods

For R&P coal mining, the Analysis of Retreat Mining Pillar Stability (ARMPS) method and program developed by Mark and Chase (1997) applies. This method incorporates the tributary area method and empirical strength formulas for sizing coal pillars during advance and retreat mining, and includes adjustments to pillar stress for factors such as side-loading from previously mined panels. It can also determine barrier pillar size. Again, an equivalent program does not appear to exist for noncoal R&P mining, but the ARMPS method should apply with suitable adjustments to the input parameters.

With traditional strength-based pillar design methods, the mining engineer can determine panel pillar size and barrier pillar size, but not panel dimensions. Operational considerations such as equipment and productivity determine the panel width and usually set it as large as possible. Based on strength considerations alone, a narrow panel requires a narrow barrier pillar and a wide panel requires a wide barrier pillar, and rock mechanics factors do not affect the panel-width determination. Maximum panel width can be determined by the size of air blast that an operation can withstand, but this is not a rock mechanics factor. Other rock mechanics considerations are needed to determine maximum panel width as well as panel pillar and barrier pillar sizes.

Typical R&P Sizes for Metal Mines

Within the bounds of roof and pillar safety, larger rooms enable use of larger equipment and increase the productivity of the stoping operation. Case studies completed many years ago showed that the average productivity of three mines with average room size of 41 m² (440 ft²) was 21.7 t, or metric tons (23.9 st, or short tons) per worker-shift; in contrast, the average productivity of three mines with average room size of 158 m² (1,700 ft²) was 71.5 t (78.8 st) per employee shift (Bullock 1998a). While safety should never be compromised, larger rooms are more productive, providing that they do not get too high to safely maintain the back and the loose rock on high pillars and ribs. Many limestone producers typically have rooms 12.2 m (40 ft) high in a single pass.

Examples of typical R&P mines with very high rock-quality designations (RQD = >85) are the mines of the Viburnum Trend (Carmack et al. 2001). Room widths are held to 11 m (36 ft), but are typically 9.8 m (32 ft); final heights after multiple passes are preferred to be <15 m (50 ft). Roof bolting is done typically in every intersection and elsewhere as needed. Historically, pillar sizes were 8.5 × 8.5 m (28 × 28 ft) on 20-m (60-ft) centers; then panel pillar sizes of 11.6 × 23 m (38 × 76 ft) were adopted for wider areas of mining where pillar extraction is planned upon retreat from the area. Relative pillar strengths in the mines are monitored using a rating system of 1 to 6, where 1 denotes no visual stress and 6 denotes pillar failure (Figure 13.1-2). This classification system is important not only for pillar recovery but also for pillar maintenance. Pillars whose rating reaches 3 must be reinforced with grouted rebar bolts to halt further deterioration if they are to maintain their integrity for a long period. The rock strength of samples in this area depends considerably on mineral content, but varies from ~70 to 200 MPa (10,000 to 30,000 psi). A similar but less defined and seemingly more conservative pillar rating system was developed by Westmin Resources Ltd. (Lunder and Pakalnis 1997).

Pillar Rating	Pillar Condition	Appearance
1	No indication of stress-induced fracturing. Intact pillars.	
2	Spalling on pillar corners; minor spalling of pillar walls. Fractures oriented subparallel in walls and are short relative to pillar height.	
3	Increased corner spalling. Fractures on pillar walls more numerous and continuous. Fractures oriented subparallel to pillar walls and lengths are less than pillar height.	
4	Continuous, subparallel, open fractures along pillar walls. Early development of diagonal fractures (start of hourglassing). Fracture lengths are greater than half of pillar height.	
5	Continuous, subparallel, open fractures along pillar walls. Well-developed diagonal fractures (classic hourglassing). Fracture lengths are greater than half the pillar height.	
6	Failed pillar, may have minimal residual load-carrying capacity and be providing local support to the stope back. Extreme hourglassed shape or major blocks have fallen out.	 

Source: Carmack et al. 2001.

Figure 13.1-2 System for rating pillar strength

Some resources exist for which mineral continuity is so consistently inconsistent that no amount of surface prospecting can accurately identify ore continuity, and thus pillar and stope layouts cannot be preplanned. The Mississippi Valley type zinc deposits of central and east Tennessee (United States) are of this type. If this is the case and it is the first mine in a new mining district, an underground test mine is required so that better geotechnical information can be obtained on which to base plans for room widths and pillar widths and height, all of which greatly affect mine operating cost. Structural information can then be seen and accurately mapped, data needed to determine joint and fracture information can be accurately measured, in-situ stress measurements can be taken, and large core samples can be gathered for laboratory analysis.

ORIENTATION OF ROOMS AND PILLARS

Pillar orientation is affected by in-situ stress. The orientations of both rooms and pillars are affected by the extent of dip.

Pillar Orientation Due to In-Situ Stress

As in all mining methods, mining engineers who are planning R&P operations must be aware of probable in-situ stress within the rock before mining. Any significant horizontal stress in a particular direction must be taken into account by orienting the room advance and the direction of rectangular pillars to give the most support in the direction of stress. In the early phase of development, research should be done to determine the magnitude and direction of inherent stress. However, very shallow R&P operations may have very little horizontal stress, in which case the orientation of rectangular or barrier pillars is not a concern, particularly if a mine has been opened from a hillside with adits where nature could have relieved horizontal stresses eons ago.

In sharp contrast to this condition, some deep R&P mines have tremendous problems, not only with horizontal stress but also with rock that absorbs a large amount of energy and then fails violently. In such operations, not only is pillar orientation important, but the sequence of extraction and how it takes place is also important. Such an operation is documented as a case study by Korzeniowski and Stankiewicz (2001).

R&P Orientation Due to Dip

R&P mining is usually done on fairly flat strata but need not be limited to that. Shallow dips, of up to ~8%, have little effect on the layout of the R&P stope. If the mine is developed from a shaft, rooms should be laid out so that haulage of the load goes downgrade. If the mine is developed from an adit or decline, of course, this would not be the case. As to the orientation of rectangular pillars, it probably makes sense to orient the long side of the pillar in the direction of the dip.

Steep dips of 35% to 40% (~20°) are best mined in a series of parallel slices in steps, each following a level contour of the dipping ore body. As each round is blasted, much of the rock cascades down to the next level, where it can be loaded. This so-called cascade mining method was first reported for the Mufulira copper mine in Zambia (Anon. 1966). However, in this case the pillars were removed almost immediately in a second cycle of mining, leaving the hanging wall to cave as mining retreats along the strike.

Very steep dips of >20° do not allow operation of trackless equipment. Manufacturers have proposed back-mounted, cog-wheel-driven jumbos that can drill stope rounds under these conditions, but the ore so mined does not flow by gravity and must be removed by scraper, conditions being too steep for any other type of loading equipment. Thus many operators over the years have resorted to drilling with handheld jackleg drills and using rope scrapers (slushers) to move ore from steep slopes to the haulage equipment (Christiansen et al. 1959).

R&P MINE HAULAGE DEVELOPMENT

For hard-rock R&P mines, the production shaft is normally developed somewhere near the centroid of the ore body (unless the production opening is by adit or decline, or the ground will be allowed to cave). The objective of production development is to minimize the haul cost for the ore to the shaft. Rail haulage should be kept as straight as possible, with long-haul grades of <2%, although within the rooms themselves grades of up to 5% are permissible. In a trackless haulage operation, however, these numbers may differ and the following apply:

- Keep grades as low as possible and long hauls at <8.0%.
- Keep haul roads as straight as possible. Locate pillars in such a way that roads need not deviate around them. Lay out all main haul roads before mining, and lay pillars out from this plan.
- Keep haul roads in excellent condition. Use adequate crushed stone and keep them well graded and dry. Water

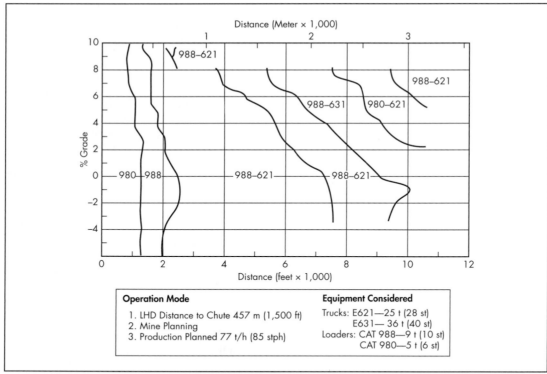

Source: Bullock 1975.
Figure 13.1-3 Equipment and operations planning from a materials-moving simulation program

not only causes potholes, but also lubricates rock to the point that it can cut tires.

In laying out a mine development plan, Spearing's rules of thumb should be kept in mind (Spearing 1995; Bullock and Hustrulid 2001):

- Keep intersections off of the main drift six widths apart.
- Avoid acute-angle turnouts that create sharp bullnose pillars.
- Separate near-parallel ramps and declines from the main drift by at least three times the diagonal width of the main drift.

In planning a trackless/rubber-tire haulage system for an R&P mine, consider the various methods of moving rock from the working face to the crushing/hoisting facility. The versatility and flexibility of modern hydraulic excavators and powerful rubber-tired equipment can make it difficult to choose the optimal method. The following guidelines can help:

1. When the mine starts operation from the bottom of a decline or shaft, the haul distance is short. Unless hauling is up a decline to the surface, a rubber-tired load-haul-dump (LHD) unit is probably optimal.
2. As the mine works out from the dump point, the LHD unit should eventually start loading a truck. Just one truck suffices as long as the loader can load the truck, then load itself and follow the truck to the dump point by the so-called load-and-follow (LAF) method.
3. As the distance between the mine and the dump point continues to increase, eventually the LHD unit should stay in the heading and load multiple trucks, enough to keep itself busy. At this point, the LHD unit is no longer optimal, and a front-end loader (FEL) may be required. As distance or haulage grade increases, larger trucks and FELs may be also required.

Examples exist (Bullock 1975) where true wheeled loaders have been successfully modified and adapted for both LHD and FEL operation, and perform efficiently at all three stages.

Figure 13.1-3 shows a graph generated by a materials-moving simulation program used in the late 1970s in the Viburnum Trend lead/zinc/copper mines. The program optimizes equipment selection based on least cost. At that time, the equipment manufacturer charged less capital cost per ton of capacity for the 25.4-t (28-st) truck than for the 36.2-t (40-st) truck. When it took this into account, the simulation found it cheaper to run a fleet of smaller trucks at the greater distance and grade. This is not the normal expected result, and thus the simulation proved its effectiveness.

If a mining zone has more than one level and the main haulage is on the bottom level, then various combinations of LHD units to the orepass or FELs with trucks hauling to the orepass are possible, and at the bottom of the orepass, automatic truck-loading feeders can transfer the ore to trucks or rail-mounted trains. If the upper ore body is small and an automatic ore chute cannot be justified, then the ore can be allowed to fall to the ground and an LHD unit or FEL can load it into another truck. Figure 13.1-4 shows all of the possibilities for moving material from an R&P face to the final ore pocket at the shaft. For a complete explanation of how to optimize these variable conditions, see Bullock 1975 or Bullock 1998b.

Clearly, then, in an expanding R&P mine, the optimal method or combination of methods for moving ore constantly changes. But for every condition and distance there

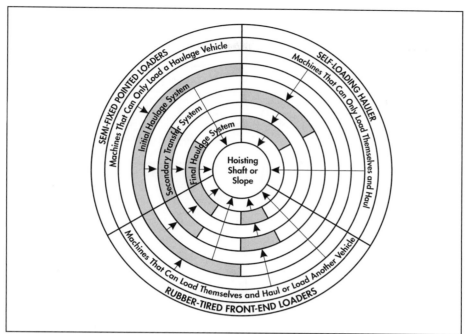

Source: Bullock and Hustrulid 2001.

Figure 13.1-4 Multiple methods of moving ore from a stope face to a hoisting conveyance

is one optimal least-cost method, which can be determined by computer simulation of the underground mine environment (Bullock 1975; Gignac 1978).

Another R&P concept is worth noting. In some older Tennessee mines, the haulage ways along which ore is transported are somewhat narrow. As haul distance increased, it became desirable to use larger-haulage trucks, but such trucks could not be accommodated on the narrow haulage ways. To solve the problem, Savage Zinc, in their Gordonville mine, hooked side-dump, powered truck-trailers (built by Gulf Transport of Australia) together into a mini-truck train, thus increasing the payload per trip and optimizing production for load-haul conditions.

If a conveyor system of haulage is used for the main distance haulage, then the preceding advice for mine layout still applies to the general layout of the mine, except conveyance starts from the decline of the mine entrance where the main conveyor carries mine production to the surface. In these cases, a hard-rock mine usually has a semimoveable crusher or breaker underground before the feeder to the conveyor system. The system of haulage, either LHD unit or FEL/truck, hauls broken rock from the faces to the central receiving point at the crusher. The crusher then feeds the conveyor. The crusher must be moved periodically—about 6 to 12 months—depending on how fast mine production moves the faces away from the central hauling point.

At one time, rail haulage was the principal method of gathering ore from the faces to the production shafts. In the Old Lead Belt of Missouri, more than 556 km (300 mi) of interconnected railroad was used to bring ore into two main shafts from what was originally about 15 mines. Today in the United States, very few R&P metal mines use rail haulage, but many coal mines still do, and in other countries many metal mines still do as well.

SINGLE-PASS OR MULTIPLE-PASS EXTRACTION

Two approaches exist to mining ore body or valuable rock: (1) taking the entire ore-body thickness in a single pass or slice, or (2) removing the ore body by multiple passes or slices. The choice of approach of course is related to the overall thickness of the ore body to be extracted and to concerns for safety and efficiency.

Normally, in mining bedded deposits for aggregate, the total thickness of the desired horizon is known and the decision concerning slice thickness can be made in advance. However, in metal ore deposits, particularly in Mississippi Valley or collapsed Breccia type deposits, most of the time the thickness of the total mining horizon is not known except where the formation has been penetrated by diamond drill holes and identified, although a few meters away from the hole the formation may very well be different. For both aggregate producer and metal-ore producer, the best approach is first to mine what is thought to be the top slice through the ore body. The thickness of this first mining step depends on the available equipment and on the height of ground that can be mined and maintained safely and efficiently. The advantage of first mining what is thought to be the top slice is that whatever back and rib/pillar scaling and reinforcement is required can be reached easily and safely, probably at distances not more than 6.1 m (20 ft). However, some aggregate producers mine 12.2 m (40 ft) or more in one pass by using high-mass jumbo and extendable-boom roof-scaling equipment.

R&P Rock Breaking for Single-Pass Stoping

In most hard-rock R&P mines, extraction is done by drilling and blasting the face. The initial pass of drilling and blasting is usually done by drilling either burn-cut or V-cut drill patterns. However, only about 40% of the rock should be broken with drilled swing patterns (rounds), breaking to only one free face.

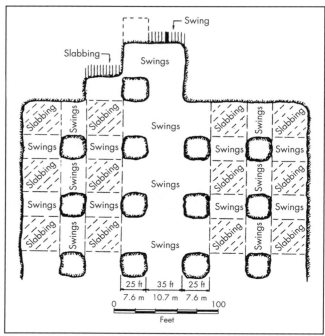

Source: Bullock and Hustrulid 2001.

Figure 13.1-5 Optimizing the drilling and blasting of R&P stoping using 40% pattern swing drilling and 60% slabbing

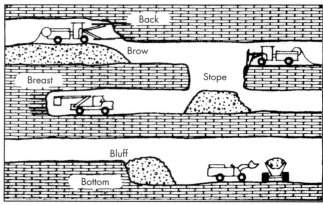

Source: Adapted from Bullock and Hustrulid 2001.

Figure 13.1-6 Various types of stoping action involved in R&P metal mines

The other 60% of the rock should be broken by slabbing—that is, by drilling holes parallel to the second free face as it is exposed (Figure 13.1-5), which minimizes the cost per ton of rock broken and maximizes productivity.

Particularly in the R&P mining of aggregates and metals, extraction is often done by mechanical means. In many trona and potash mines, and some salt mines, all extraction is done mechanically.

R&P Rock Breaking for Multiple-Pass Stoping

After the first pass is completed in a metal mine for a given stoping area, the back and floor should be sludge-sampled with either pneumatic or all-hydraulic drills. Some operations do this with a stope rock-breaking jumbo and some with an all-hydraulic jumbo designed for this role. The purpose of this sludge-sampling prospecting is to determine what ore remaining in the back and floor needs to be removed by additional slices. If ore is found in the back and floor, then the ore in the back should be removed first. After it is removed and additional back-prospecting reveals no more ore, the back should again be made secure and safe, and ore in the floor can be taken.

Methods of mining ore in the back vary depending on the thickness of the ore yet to be mined and the height of the original stope. If the original stope height was no more than 7.6 m (25 ft) and the back slice thickness was no more than ~2 m (7 ft), most extendable-boom jumbos can reach high enough to drill the brow with breast (horizontal) drilling. Drilling horizontal holes increases the chance of leaving a smooth back, which requires less maintenance than when holes break to the free face of the brow. This is especially true for smooth-wall blasting or flat-seam deposits. A jumbo feed should not be tilted up for drilling near-vertical holes and for drilling uppers to break a brow, particularly in bedded deposits that exist in so many R&P mines.

If ore thickness in the back is greater than the conditions cited previously or if room height is already at the maximum that can be safely maintained, then the back slice should be mined by first cutting a slot in the back at the edge of the entrance to the stope (Figure 13.1-6). This slot should be between the pillars and reach the height of the next slice or the top of the ore body. Care must be taken not to damage the rock that will form the top of the new pillars.

Smooth-wall blasting can be used to advantage in this area. A mine dozer (usually a small dozer of D-4 or D-6 size) begins by pushing rock up and making a roadway for the jumbo to travel on the top of the rock pile. If a dozer is unavailable, an FEL can also do a reasonable job of building the roadway. The jumbo then drills breast (horizontal) holes in the brow and follows the ore zone wherever it goes throughout the stoped area. Even if it goes into the solid beyond the original stoped area, this is not a problem, except that a loader now must load the ore as it is broken. About 75% of the ore must remain in the stope rock pile until back-mining is complete, which can be a disadvantage if the mine must produce immediately. When the top of the ore is finally reached, the back should be made safe since miners are still working close to the back. This type of mining has been practiced in the Tennessee zinc and Viburnum Trend lead/zinc districts. It is not uncommon for several passes of ore to be mined from broken rock piles. In these cases, either ore must be loaded from the bottom edges of the rock piles to make room for new broken rock or a loader must go up on the rock pile and load out the excess rock.

If after the first slice is taken through the stope it is then discovered that a thick, continuous ore zone lies above this first slice—a zone of perhaps 15 to 20 m (50 to 65 ft) or more—then an entirely different approach is required. It may be preferable to create a development ramp to what is now the top of the ore, mine the top slice from this new ramp, and then put through a slot raise from the bottom to the top level that can be slabbed out to make room for long-hole drilling and blasting the ore body down to the level below. Again, the pillars should be presplit or smooth-wall-blasted to protect them. If the eventual height of a pillar can be anticipated, the size of the pillar at its base should be determined taking this additional height into account. If it is not taken into account and safety factors in the design are insufficient, then the pillars may need reinforcement with grouted rebar bolts to give them the required strength and stiffness.

When all back ore is removed from the stope and the final back is made completely secure, bottom ore can be removed from the floor. This is best done by first cutting a ditch or short decline in the floor, at the entrance to the stope, to the depth of the desired bluff or bench. Bluffs can be carried very thick and are limited only by the height required for the loading equipment to work safely. It is common practice to carry bluffs up to 9 m (30 ft) in the lead mines of the Viburnum Trend area using 7–11-t (8–12-st) loaders. Beyond this height, the safety of the loader operator may become an issue.

Bluffs at least 4 m (13 ft) thick are usually drilled and blasted with downholes by small-surface quarry-type drills. Bluffs of less than this thickness can be drilled out with face jumbos, drilling breasting, or horizontal holes (sometimes called lifters and splitters). These procedures can be repeated over and over until the bottom of the ore is reached.

When it is known beforehand that bluff mining will be done, for the initial pass the pillar width must be left large enough to accommodate considerable height should it be needed. If additional ore is not found, then the pillars can be slabbed down to smaller size. However, again the precaution: in both removal of back ore and taking up of the bottom, the pillars most be protected with presplitting or smooth-wall blasting. Pillar design is discussed in more detail in the following section; the point here is that, unless the original mine plan specifically allows for taking down the back and taking up the floor or specifies a very large safety factor in pillar design, the width-to-height ratio required for safe operation will be exceeded by multiple-pass mining. It cannot be overemphasized that everything must be done to protect pillar integrity during the first of what may become multiple passes. Unless caution during presplitting or smooth-wall blasting is observed, blasting fractures may extend into the pillar as far as 2 m (6.6 ft) or more. As more slices are removed, these fractures begin to open, reducing pillar size and subjecting the pillar to failure.

METHODS OF PILLAR EXTRACTION IN HARD-ROCK MINES

The mining engineer should anticipate some percentage of pillar removal beyond what is cited in the original mine plan and stope layouts. If future pillar extraction is planned—whether by partial slabbing, removal of a few high-grade pillars, or complete removal incorporating some system of backfilling—then what is left after the first pass greatly influences what can be done in the future. Too often, too much is extracted on the first pass, so that the entire area is weakened and must be reinforced just to keep the mine open, much less enabling the selective robbing of pillars.

The first and most important thing to do is to install a complete network of convergence stations throughout the affected area. Experience shows that regular monitoring of such a network over time can reveal what rate of back–floor convergence is acceptable and what rate will lead to massive failure. Convergence in some mines can be as high as 1.25 mm per month without the back breaking up and failing (Parker 1973). Convergence in other mines, such as those of the Old Lead Belt of Missouri and the Viburnum Trend, can be as low as 0.1 mm per month and still cause problems. However, the following are good general guidelines:

- Convergence of 0.0254 mm (0.001 in.) per month is not significant.
- Convergence of 0.0762 mm (0.003 in.) per month indicates a serious problem, but is controllable with immediate action.
- Convergence of 0.1778 mm (0.007 in.) per month indicates that acceleration is getting out of control and the area may be lost.

Several methods exist for pillar removal of R&P stoping:

- Slab some ore off of each pillar containing the high-grade portion of the ore during retreat from the area.
- Completely remove a few of the most valuable pillars, but leave enough pillars untouched to support the back.
- In narrow stopes, completely remove all pillars in a controlled retreat.
- Use massive backfill methods to remove some or all pillars.

Placing cemented rock fill around pillars, all the way to the back, to properly support the stope between the solid and the backfill creates a pressure arch over the intervening pillars, which can then be removed. Economics permitting, the area can be backfilled if necessary, and the pillars that were encapsulated or trapped in the original fill can be mined from a sublevel beneath the pillar by long-hole blasting of the pillar into the sublevel area. (Lane et al. 1999) The total backfill prevents any future subsidence.

All of the previous methods (except sublevel extraction) were used in the final mining of the Old Lead Belt of Missouri in its last 25 years, and much more intensely in its last 10 years. This R&P mining district was mined for 110 years before finally shutting down.

Sometimes when pillars are partially removed without proper planning, or when pillars that are too small are left after first-pass mining, the pillars begin to break up and serious convergence accelerates uncontrollably. One of two things must be done to save the area: (1) massive pillar reinforcement, if there is time before collapse; or (2) massive backfill in the entire area.

Among examples of massive pillar reinforcement, fully grouted rebars were placed in more than 300 pillars in the one of the R&P mines of the Viburnum Trend. Weakly (1982) has documented the method employed, reinforcing pattern, cement grout mixture, convergence instrumentation, and results. In the Old Lead Belt areas, pillars to be reinforced were wrapped with used hoist cable, with a load of 5.5 t (6 st) placed on each wrap (Wycoff 1950), but this reinforcement method is not as fast, economical, or effective as use of fully grouted rebars in the pillars.

Among examples of massive backfill are the Leadwood mines of the Old Lead Belt, where this method was used only about 132 m (425 ft) below surface. However, the back was thin-bedded dolomite, interbedded with shale and glauconite, and badly fractured and leached. The roof bolt was originally developed in these St. Joe mines in the late 1920s as a means of tying the layers together to act as a beam (Weigel 1943; Casteel 1964); the roof bolts were used with channel irons to form a crude truss. Even though the rock in the pillars provided poor support because it contained bands of high-grade galena, for economic reasons pillar removal and slabbing took place over a period of 25 years. Occasionally local cave-ins would occur after an area had been pillared, but since the cave-ins were beneath uninhabited St. Joe–owned land, they were of no real concern. However, when slabbed pillars between two smaller cave-ins began to fail and a third and fourth cave-in occurred in more critical areas, there was a considerable amount of concern. Initial extraction of some of the area involved multiple-pass mining and room heights were mostly 6.1 to 12.2 m (20 to 40 ft). Final mining of the area had resulted in ore extraction of approximately 95%. To stop caving in the third and fourth areas, in around 1962 more than a million tons of uncemented, cycloned sand tailings were put into the mines, filling the rooms nearly to the back. The backfill was very successful in controlling the converging, failing ground.

The end result of use of these methods is comparable to the overhand-cut-and-fill practice of deliberately mining pillars very small and immediately filling in around them. This method, known as postpillar mining, was used in the Falconbridge nickel mines (Canada) (Cleland and Singh 1973), the Elliot Lake uranium mines (Canada) (Hedley and Grant 1972), and in some mines in San Martin and Niaca (Mexico) and San Vicente (Peru). The end results for the two mining methods, with very small pillars encapsulated in sand tailings, are similar.

VENTILATION

Most of the literature on ventilation design for R&P mining is written for coal mines. However, a few considerations are unique to metal and aggregate R&P mines. R&P metal or aggregate mines differ from R&P coal mines in the following respects:

- Everything is larger. It is not uncommon to have entry drifts of 9 × 9 m (30 × 30 ft) and rooms stoped out to 12 × 15 m (40 × 50 ft). A lot more air or a stream of high-pressure directed air is required to meet the required minimum velocity across the working face.
- Stopings are difficult to build and the total force against them can be enormous. Many R&P mines rely on auxiliary fans to pick up air from the main ventilation drifts and carry enough air through vent tubing to serve the needs of the active face.
- Ventilation doors are similar in design to airplane hangar doors. Again, the total force against these doors, which must be controlled automatically, is enormous.
- Air stratification in large stopes can be a problem.
- Diesel equipment is used extensively, and the exhaust must be diluted to meet particulate standards.
- Ventilation fans can be placed underground, if it is beneficial to do so.

THE EFFECT OF CHANGING MARKET CONDITIONS ON R&P MINE PLANNING

Market conditions are important in planning for R&P mining, even during the stages of feasibility planning. An advantage of R&P mining for metal or any commodity that is gradational in value is that mining methods and approaches are flexible and can be changed quickly in response to changing market conditions—more quickly, in fact, than is possible for other mining systems.

Another advantage of R&P mining is that, if the ore body is continuous, new faces are continuously being opened. A stope only four pillars wide can have as many as 15 faces open simultaneously for drilling and blasting. Imagine how many faces would be exposed to drilling and blasting if a stope were 10 pillars wide. For metal mines, this offers a great deal of flexibility; one can mine the grade of ore that is most desirable at the current market price. For at least short periods of time in each stope unit, it is usually possible to work only the higher-grade or lower-grade faces of a particular mineral, although doing so usually has a drastic effect on the grade within a few days. During initial mining of the high-grade Fletcher mine in southeastern Missouri (United States), after operation for only 3 to 4 years, approximately 50 to 70 faces were open for mining on any given day although only 10 to 12 were actually worked. One can then select the faces to be mined so as to maintain the grade of lead, zinc, or copper ore that can best be handled by the concentrator and still optimize the financial objective of the mine.

Spare equipment can be put into reserve stopes to increase production if the remaining materials flow can handle the added capacity. However, if this is done too often or for too long, mine development must also be accelerated. Old stopes, if maintained, can be reactivated quickly for mining of lower-grade minerals that have become financially worth mining.

Lower-grade ore can be left in the floor or the back of stopes until price rises, at which point the reserves are readily available for quick mining. This technique is often overlooked by those unaccustomed to planning R&P metal mining, where mineral values are gradational. The optimal technique is first to mine the better areas of the mineral reserve and maintain a grade of ore that satisfies the economic objectives at that time. Then when economics change, the lower-grade areas left as remnant ore reserves can be mined.

In spite of these considerations, however, even in R&P mining, drastic changes in the rate of mining (momentum) cannot be assumed to be free. It often takes several months with an increased labor force to regain a production level that seemed easy to maintain before a cutback in mine production. If spare equipment is used to increase production, maintenance will probably convert from a previous preventive-maintenance on-shift schedule to a breakdown overtime schedule, at least until permanent additional equipment can be obtained. Nevertheless, the necessary changes can be made.

As discussed in the section on pillar robbing, another method for achieving economic flexibility with R&P mining is to slab or remove high-grade pillars. Then, even in the later years of the mining operation, some "sweetener" is left to blend with the lower grade. Although not unique to R&P mining, this technique is easier to accomplish in an R&P operation than in other more complex mining systems.

CAPITAL AND OPERATING COSTS FOR HARD-ROCK R&P MINING

Capital and operating costs for hard-rock R&P mining are usually considerably less than for other hard-rock mining systems at the same daily production. However, panel caving is not comparable because R&P mines are not normally built to produce the equivalent lower tonnage per day.

Table 13.1-2 shows estimated preliminary capital costs for R&P mines. The original estimates (Stebbins and Schumacher 2001), based on 2000 data, included cost contingencies of 7.2%

Table 13.1-2 Estimated preliminary capital costs for R&P mines developed from shafts (mid-2009 dollars)

Daily tonnage, t (st)	1,200 (1,323)	8,000 (8,819)	14,000 (15,432)
Total capital cost	$39.8 M	$125.0 M	$152.8 M
Total capital cost/daily tonnage	$33,200	$15,600	$10,900

Source: Adapted from Stebbins and Schumacher 2001.

Table 13.1-3 Estimated direct operating costs for hard-rock R&P mining (2008 dollars)

Activity	Cost, US$ Per metric ton	Cost, US$ Per short ton
Drill & blast labor	0.47	0.43
Loading & hauling	0.73	0.67
Maintenance labor, repair, parts, & tools	1.84	1.67
Indirect labor	1.14	1.04
Overtime	0.15	0.14
Other materials & tires	0.64	0.58
Sundry supplies	0.49	0.44
Diesel fuel	1.60	1.45
Drill steel & bits	0.26	0.24
Grease & oil	0.76	0.69
Roof supplies	0.13	0.12
Tire recapping	0.18	0.16
Other outside services	1.10	1.00
Blasting supplies	1.11	1.00
Underground prospecting	0.38	0.35
Fringes on direct labor	0.55	0.50
Total direct operating cost	11.53	10.46

to 7.6%. The contingencies were removed, the percentage of the remaining total was calculated for each element of capital cost, elemental capital costs were escalated using the proper indexes for the time period and totaled, and a 25% contingency was added to the total. These costs include all equipment, preproduction development, mine-surface facilities, engineering and project management, working capital, and cost contingencies.

Table 13.1-3 shows estimated direct operating costs for R&P mines. The estimates are based on 1998 data from six highly productive R&P mines that vary in daily production from about 2,300 to 6,300 t (2,500 to 7,000 st) per day. The estimates were escalated to 2008 using the proper mining-unit inflation index for each cost element, and a 20% contingency was added. It is noteworthy that the six mines have very large rooms, are highly mechanized, and are considerably more productive in tons per worker-shift than is average for the industry (the latter is about 40–50 t [44–55 st] per worker-shift).

REFERENCES

Anon. 1966. Cascade mining. *World Min.* (June): 22–23.
Bieniawski, Z.T. 1968. The effect of specimen size on compressive strength of coal. *Int. J. Rock Mech. Min. Sci.* 5:325–335.
Bieniawski, Z.T. 1992. Ground control. In *SME Mining Engineering Handbook*, 2nd ed. Edited by H.L. Hartman. Littleton, CO: SME. pp. 897–937.
Brady, B.H.G., and Brown, E.T. 1993. *Rock Mechanics for Underground Mining*. Boston, MA: Kluwer Academic Publishers.
Bullock, R.L. 1975. Optimizing underground, trackless loading and hauling systems. Ph.D. dissertation, University of Missouri, Rolla, MO.
Bullock, R.L. 1998a. General mine planning. In *Techniques in Underground Mining*. Edited by R.E. Gertsch and R.L. Bullock. Littleton, CO: SME. pp. 87–127.
Bullock, R.L. 1998b. Production methods of noncoal room-and-pillar mining. In *Techniques in Underground Mining*. Edited by R.E. Gertsch and R.L. Bullock. Littleton, CO: SME. pp. 171–214.
Bullock, R.L., and Hustrulid, W.A. 2001. Planning the underground mine on the basis of mining method. In *Underground Mining Methods: Engineering Fundamentals and International Case Studies*. Edited by W.A. Hustrulid and R.L. Bullock. Littleton, CO: SME. pp. 29–48.
Bunting, D. 1911. Chamber pillars in deep anthracite mines. *Trans. AIME* 42:238–245.
Carmack, J., Dunn, B., Flack, M., and Sutton, G. 2001. The Viburnum Trend underground—An overview. In *Underground Mining Methods: Engineering Fundamentals and International Case Studies*. Edited by W.A. Hustrulid and R.L. Bullock. Littleton, CO: SME. p. 98.
Casteel, L.W. 1964. The first century of research by St. Joseph Lead Co. *Min. Eng.* (July): 16:111–111D.
Christiansen, C.R., Calhoun, W.A., and Brown, W.F. 1959. Mining and milling methods and costs at the Indian Creek Mine, St. Joseph Lead Co., Washington County, MO. Information Circular 7875. Washington, DC: U.S. Bureau of Mines.
Cleland, R.S., and Singh, K.H. 1973. Development of post-pillar mining at Falconbridge Nickel Mines Ltd. 66(732):113–130.
Coates, D.F. 1981. *Rock Mechanics Principles*. CANMET Monograph 874. Ottawa, ON: Canada Centre for Mineral and Energy Technology.
Farmer, I. 1985. *Coal Mine Structures*. London: Chapman and Hall.
Farmer, I. 1992. Room and pillar mining. In *SME Mining Engineering Handbook*. Edited by H.L. Hartman. Littleton, CO: SME. pp. 1681–1686.
Gale, W., and Hebblewhite, B. 2005. *Systems Approach to Pillar Design, Module 1: Pillar Design Procedures*. Final Report, Vol. 1. Australian Coal Association Research Program Project C9018. Brisbane: ACARP.
Galvin, J.M., Hebblewhite, B.K., and Salamon, M.D.G. 1999. University of New South Wales coal pillar strength determinations for Australian and South African mining conditions. In *Proceedings of the 2nd International Workshop on Coal Pillar Mechanics and Design*, Vail, CO, June 1999. NIOSH Information Circular 9448. Pittsburgh: National Institute for Occupational Safety and Health. pp. 63–71.
Gignac, L.P. 1978. Hybrid simulation of underground trackless equipment. Ph.D. dissertation, University of Missouri, Rolla, MO.

Greenwald, H.P., Howarth, H.C., and Hartman, I. 1939. Experiments on the strength of small pillars of coal in the Pittsburgh bed. Technical Paper 605. Washington, DC: U.S. Bureau of Mines

Hardy, M.P., and Agapito, J.F.T. 1977. Pillar design in underground oil shale mines. In *Proceedings of the 16th U.S. Symposium on Rock Mechanics*, University of Minnesota, Minneapolis, MN. New York: American Society of Civil Engineers. pp. 257–266.

Heasley, K.A. 1997. A new laminated overburden model for coal mine design. In *Proceedings: New Technology for Ground Control in Retreat Mining*. Information Circular IC-9446. Pittsburgh: National Institute for Occupational Safety and Health. pp. 60–73.

Heasley, K.A. 1998. Numerical modeling of coal mines with a laminated displacement-discontinuity code. Ph.D. dissertation, Colorado School of Mines, Golden, CO.

Hedley, D.G.F., and Grant, F. 1972. Stope and pillar design for the Elliot Lake uranium mines. *Can. Int. Metall. Bull.* 65(723):37–44.

Hoek, E., and Brown, E.T. 1980. *Underground Excavations in Rock*. London: Institution of Mining and Metallurgy.

Hoek, E., Kaiser, P.K., and Bawden, W.F. 1995. *Support of Underground Excavations in Hard Rock*. Rotterdam: A.A. Balkema.

Holland, C.T. 1964. Strength of coal in mine pillars. In *Proceedings of the 6th U.S. Symposium on Rock Mechanics*. Rolla, MO: University of Missouri. pp. 450–466.

Koehler, J.R., and Tadolini, S.C. 1995. *Practical Design Methods for Barrier Pillars*. Information Circular IC-9427. Washington, DC: U.S. Bureau of Mines.

Korzeniowsk, I.W., and Stankiewicz, A. 2001. Modifications of room and pillar mining method for Polish copper ore deposits. In *Underground Mining Methods: Engineering Fundamentals and International Case Studies*. Edited by W.A. Hustrulid and R.L. Bullock. Littleton, CO: SME. pp. 103–110.

Lane, W.L., Yanske, T.R., and Roberts, D.P. 1999. Pillar extraction and rock mechanics at the Doe Run Company in Missouri 1991 to 1999. In *Proceedings of the 37th Symposium on Rock Mechanics*. Rotterdam: A.A. Balkema. pp. 285–292.

Lunder, P.J., and Pakalnis, R.C. 1997. Determination of the strength of hard-rock mine pillars. *Can. Int. Metall. Bull.* 90(1013):51–55.

Mark, C. 1999. Empirical methods for coal pillar design. In *Proceedings of the 2nd International Workshop on Coal Pillar Mechanics and Design*. Information Circular IC-9448. Pittsburgh: National Institute for Occupational Safety and Health. pp. 145–154.

Mark, C., and Chase, F.E. 1997. Analysis of Retreat Mining Pillar Stability (ARMPS). In *Proceedings: New Technology for Ground Control in Retreat Mining*. Information Circular IC-9446. Pittsburgh: National Institute for Occupational Safety and Health. pp. 17–34.

Obert, L., and Duvall, W.I. 1967. *Rock Mechanics and the Design of Structures in Rock*. New York: Wiley.

Parker, J. 1973. How convergence measurements can save money, Part 3 of Practical rock mechanics of the miner. *Eng. Min. J.* 174(August): 92–97.

Salamon, M.D.G., and Munro, A.H. 1967. A study of the strength of coal pillars. *J. S. Afr. Inst. Min. Metall.* 68:55–67.

Skelly, W.A., Wolgamott, J., and Wang, F.D. 1977. Coal pillar strength and deformation prediction through laboratory sample testing. In *Proceedings of the 18th U.S. Symposium on Rock Mechanics*. Golden, CO: Colorado School of Mines Press. pp. 285-1–285-5.

Spearing, A.J.S. 1995. *Handbook on Hard-Rock Strata Control*. Special Publications Series SP6. Johannesburg: South African Institute of Mining and Metallurgy. pp. 89–93.

Stebbins, S.A., and Schumacher, O.L. 2001. Cost estimating for underground mines. In *Underground Mining Methods: Engineering Fundamentals and International Case Studies*. Edited by W.A. Hustrulid and R.L. Bullock. Littleton, CO: SME. pp. 70–72.

Weakly, L.A. 1982. Room and pillar ground control utilizing the grouted reinforcing bar system. In *Proceedings of the 17th U.S. Symposium on Rock Mechanics*. Littleton, CO: SME. pp 87–92.

Weigel, W. 1943. Channel irons for roof control. *Eng. Min. J.* 144(5):70–72.

Wycoff, B.T. 1950. Wrapping pillars with old hoist rope. *Trans. AIME* 187:898–902.

Zipf, R.K. 1992a. *MULSIM/NL Theoretical and Programmer's Manual*. Information Circular IC-9321. Washington, DC: U.S. Bureau of Mines.

Zipf, R.K. 1992b. *MULSIM/NL Application and Practitioner's Manual*. Information Circular IC-9321. Washington, DC: U.S. Bureau of Mines.

Zipf, R.K. 2001. Pillar design to prevent collapse of room and pillar mines. In *Underground Mining Methods: Engineering Fundamentals and International Case Studies*. Edited by W.A. Hustrulid and R.L. Bullock. Littleton, CO: SME. pp. 493–496.

CHAPTER 13.2

Room-and-Pillar Mining in Coal

Jerry C. Tien

INTRODUCTION

After the coal seam is reached by means of a slope or shaft, the room-and-pillar (R&P) mining method can be used to extract coal. Room-and-pillar (also called bord-and-pillar) mining is an "unsupported" mining method used to extract mineral deposits that are roughly tabular and generally flat-lying such as coal. Using this system, the coal seam is extracted across a horizontal or nearly horizontal plane while leaving "pillars" of unmined coal to support the overburden and leaving open areas or "rooms" underground.

This method is termed *unsupported* because no artificial pillars are used to support openings. The roof is supported by natural pillars left standing—usually in a systematic pattern—and generous amounts of roof bolting and localized support are also usually used. The key to successful R&P mining is selecting the optimum pillar size, which varies depending on many factors, varying between 40% and 60% of the seam volume and proper pillar layout. With proper planning, this method provides a very effective means for extracting coal.

The productivity of R&P mining operations experienced a distinct step-change in the latter part of the 20th century as a result of increasingly powerful mining equipment, technical progress, and rationalization of efforts, and is now an indispensable method in underground coal mining. Worldwide today, it is estimated that at least 60% of coal mined underground uses the R&P mining method, although the exact amount of coal mined using the R&P mining method is difficult to obtain—the 60% figure is estimated by examining major coal-mining countries. For example, the world's number one coal-producing country, China, has at least 85% of its coal mined using R&P mining; more than 95% of China's coal is mined underground. More than 60% of U.S. coal mined underground comes from the R&P method (including 10% coal from longwall operations). It is estimated that less than half of Australian underground coal is mined using the R&P method. South Africa produced 43% of that country's total coal production using R&P mining in 2008. It is also estimated that at least 60% of India's coal comes from R&P mining.

In the mid-1980s, R&P mining using either drilling and blasting or a continuous miner was the dominant method in underground coal mining, producing about 87% of all the coal mined underground in the United States. Following the growth of longwall methods, this dropped to approximately 60% in 2007 (including the coal from longwall gate road development). Even longwall mining operations require highly efficient R&P units for timely entry development, producing an estimated 20% of total coal production. It probably will be difficult for R&P to compete costwise with longwall operations in all circumstances, but with its modest capital requirement and versatile operation, the R&P method has been able to compete head to head in terms of cost per ton of coal extracted with longwall mining in many cases.

ROOM-AND-PILLAR MINING

Early mines using the R&P method were developed more or less at random, with pillar sizes determined empirically and headings driven in whatever direction was convenient. For example, in Australia in the early days, the rooms were laid out systematically to suit hand shoveling and endless rope haulage. In modern coal mines, however, rooms and pillars are typically laid out orthogonally in a regulated pattern (Figure 13.2-1).

In addition to support, coal pillars are also expected to restrict surface movement, protect critical airways from the effect of high load distribution and abutment stresses, form ventilation partitions for airflow distribution, act as yield (sacrificial) pillars to provide temporary support in high lateral-stress situations, and as shaft pillars to protect air shafts and other critical mine structures. Barrier pillars can be used to form a load-bearing framework to minimize the impact of ground movement from adjacent workings, to isolate major faults and local joints, to protect critical roadways, to separate adjacent mine leases, and most importantly, to provide a barrier to large masses of water-filled old workings.

R&P mining has two basic operations: entry development and coal production. The former is to develop mine entries to access coal seams and construct the necessary support infrastructure to facilitate coal production. Although in inclined coal seams it might be necessary to excavate either above or below strata to access the coal seam, most entries are in the

---

Jerry C. Tien, Department of Mining and Nuclear Energy, Missouri University of Science & Technology, Rolla, Missouri, USA

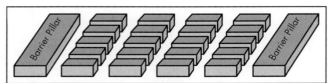

Figure 13.2-1 Room-and-pillar layout with five entries and long barrier pillars

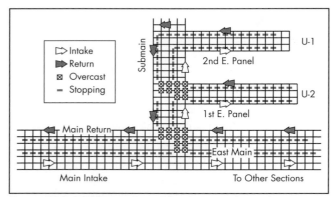

Figure 13.2-2 Mine section with two operating units, seven-entry main, five-entry submains, and four-entry panels

coal seam itself, which also produces coal in the process. Both horizontal and inclined R&P operations require other support operations such as roof bolting, ventilation, face cleaning, and transportation.

The mine layout has to be carefully planned to accommodate all functions for maximum efficiency. Development openings (entries) and production entries (rooms) closely resemble each other—both are driven parallel and in multiples, and connected by crosscuts to form pillars. Driving multiple entries simultaneously is necessary for high production and operational efficiency; it also facilitates ventilation, roof bolting, and coal transportation routes in the face area.

Usually, a hierarchy of entries exists in underground coal mines. Main entries are driven mainly to divide the property into major areas and usually serve the life of the mine for ventilation as well as for worker and coal transport. Submain entries can be regarded as feeders from the mains that subdivide each major area. From the submains, panel entries take off to subdivide a block of coal into panels and rooms for orderly coal extraction. It can be seen in Figure 13.2-2 that the East Main has a 3-2-2 or intake-neutral-return arrangement, and the submain off the East Main has the same arrangement (intake-neutral-return) but in 2-2-1 layout. The number of intake entries exceeds the number of return entries indicating it is a blowing system (with a fan on top of an intake air shaft) where neutral air travels away from working faces.

The number of development entries in any area can vary between three and seven, or as many as eleven in some cases; the exact number is often a balance of the following factors:

- Expected load on the pillar due to its depth or other geotechnical concerns
- Coal production
- The system that supports the expected production (number of pieces of mining equipment that are operating, labor and material transport, ventilation, etc.)

It is critical that a proper balance be achieved between the optimum coal extraction ratio and the size of pillar necessary to support the overburden.

Geotechnical Design for R&P Mining

Because of the complicated and varying nature of rock strata, it is customary to start the basic pillar design by assuming (1) rock strata to be homogeneous, uniform, and horizontally bedded deposits; (2) a uniform pillar layout; and (3) the rock materials are competent. Other factors such as pillar shape (uniformity of distribution and cross section and height–width ratio), true pillar geometry (outside dimensions minus an outer shell of distressed materials that in effect support very little load), and true mechanical behavior of pillar materials also have to be considered.

Bieniawski (1987) suggested the following eight-point general pillar design approach:

1. Data collection to (a) obtain a uniaxial compressive strength of roof rock and coal (σ_c); (b) spacing, condition, and orientation of geological discontinuities; and (c) groundwater condition.
2. Determine roof span and support methods.
3. Estimate pillar strength (σ_p) by considering size effect using

$$\sigma_p = \sigma_1 [0.64 + 0.36 (w/h)]$$

where σ_1 is the strength of a cubical specimen of critical size or greater (e.g., about 3.3 ft for coal), w is pillar width, and h is pillar height.

4. To obtain pillar load using

$$S_p = 1.1 H \left(\frac{w+l}{l}\right)^2$$

where S_p is pillar load (psi), H is the depth below surface (ft), w is pillar width (ft), and l is the roof span (ft).

5. Select a proper safety factor, F (ranges between 1.5 and 2), $\sigma_p = F S_p$, and solve for pillar width.
6. Determine extraction ratio (ER) using

$$\text{ER} = \frac{(w+l)^2 - w^2}{(w+l)^2} = 1 - \left[\frac{w}{w+l}\right]^2$$

If the ER is not acceptable and needs to be increased by decreasing the pillar size, select a new width and repeat Steps 3 through 6. Figure 13.2-3 shows that a 50 × 50 ft pillar with 20-ft width will result in nearly 50% ER.

The optimum ER must be a compromise between the safety of the structure and the cost of mining and conservation of resources. Considering all factors, the extraction ratio for R&P mining ranges from about 50% to 60% for underground coal mining, as shown in Figure 13.2-3.

7. Cross-check results by using the Holland formula ($\sigma_p = \sigma_1[0.778 + 0.222 w/h]$) and the Obert–Duvall formula ($\sigma_p = \sigma_1 \sqrt{w/h}$). The Holland formula is valid for pillar width-to-height ratio and the recommended safety factor is between 1.8 and 2 (Figure 13.2-4).
8. Exercise engineering judgment by considering a range of mining and geological parameters to assess the various options for mine planning in accordance with the local mining requirements.

For example, a coal mine in the Pittsburgh seam (United States) is currently mining at 500-ft deep with stable pillars and roofs of w = 40 ft, h = 10 ft, l = 18 ft, and k = 5,580 psi

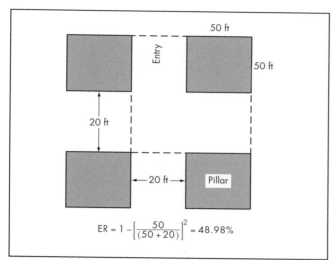

Figure 13.2-3 Extraction ratio calculation for a 50 × 50 ft pillar with a 20-ft-wide entry

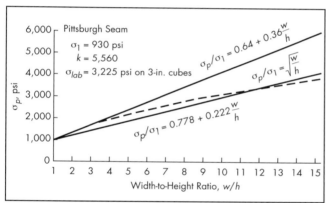

Source: Bieniawski 1987.

Figure 13.2-4 Pillar strength in the Pittsburgh seam versus w/h ratio as predicted by three formulas

Courtesy of Joy Mining Machinery.

Figure 13.2-5 Continuous miner at work

for the Pittsburgh seam. K is a constant for actual pillar material, defined by $k = \sigma_c \sqrt{D}$, where σ_c is the uniaxial comprehensive strength of rock specimens tested in the laboratory having a diameter or cube size dimension of D (in.). Mining is planned at 1,000 ft. Calculation shows that pillar stress (S_p) = 1,150 psi. For a 1,000-ft depth, using proper pillar strength (σ_p = 1,934 psi), safety factor (F = 1.67), and extraction ratio (0.52), the resulting pillar width will be 65 ft.

After coal is mined, it is hauled away using shuttle or ram cars to a properly located breaker station (usually 200–250 ft away from the working face) where it is crushed and transported to the surface by conveyor.

Practical limitations exist on R&P mining in coal operations. That is, at depths greater than 1,300–1,600 ft, this method may become very difficult or uneconomic to practice, owing to excessive roof pressure and the larger pillar sizes that are required. However, depending on other parameters, there are always exceptions. For example, KGHM is mining copper in Poland by the R&P method below 3,300 ft.

Continuous Mining Method

With a continuous mining method, separate unit operations of drilling, cutting, blasting, and loading are replaced by a single high-performance continuous miner (CM). As the steel drum (cutterhead) spins in a circle, the tungsten carbide steel teeth or cutting bits break the coal from the face. The coal falls onto a steel apron where moving arms load the coal onto a short conveyor on the back of the CM and transport it to a shuttle car waiting at the rear (Figure 13.2-5).

The shuttle car is driven to a feeder breaker three to four breaks (180–280 ft depending on pillar sizes) away and loads the panel conveyor, which carries the coal to the surface. After one working face is completed, the CM then moves to the neighboring face and the process is repeated.

A typical CM measures approximately 33 × 11 × 4.5 ft (length × width × height, respectively). The depth of each cut in a working face has increased today to 30–40 ft ("deep cut") because of more powerful and reliable machines, but this requires multiple passes and resetting of the miner to complete cutting in one face. Efficiency of the miner operator and coordination of other auxiliary functions such as face cleaning, establishing ventilation, and roof bolting are all key factors in determining the face advance rate and productivity. Remote-controlled CMs are now the norm in the industry, and they have significantly enhanced face productivity and safety.

A typical (standard) continuous mining section uses one set of equipment with one crew operating in a panel of three to seven entries. The section is equipped with one CM, two to three shuttle cars or ram cars, one roof bolter, and a section scoop for auxiliary jobs. In addition to a section foreman, a CM section would have nine to ten workers: two CM operators (one operating and one assisting/relieving), two roof bolters, three shuttle car operators, one mechanic, and one general laborer).

In addition to the standard unit, U.S. coal mines have also been using either a double or a super unit arrangement: The former uses two sets of equipment and two or nearly two crews operating side by side in a panel with 8–13 entries; the latter uses two sets of equipment and one crew plus two extra workers operating side by side in a panel with 4–13 entries. Depending on the mine's specific condition, either method has been used successfully in developing long-distance mine mains with multiple entries.

Since the late 1970s, mining machines progressed steadily from 440 V to 950 V with an increase in productivity. By 1990 virtually all CMs were powered by 950 V. The demand for larger, more-powerful mining equipment prompted the need for increased voltage even further for a CM. Since input voltage varies directly with mining rate and motor torque varies with the square of the voltage, higher voltages for a given size cable also lead to a lower power voltage drop. With further improvement in shielding technology, high-voltage miners have become available. For example, Joy Mining Machinery's 14 CM series high-voltage (2,300 V) continuous miner can mine up to 38 t/min, depending on the seam thickness. Special design, use, and maintenance precautions are needed to ensure an equivalent level of safety when high-voltage systems are used in permissible areas. In the United States, mining equipment operating above 1,000 VAC requires a petition for modification including an approval examination by the Mining Safety and Health Administration for each high-voltage CM deployed inby (the distance between where one stands and the face). CMs capable of cutting 19.7 ft of coal seam have also been produced.

Ongoing improvement in CM performance in both productivity and reliability has brought about a dramatic increase in coal production. Under certain conditions, CM units have been able to deliver more than 3,000 t/shift on a regular basis.

Conventional Mining (Drilling and Blasting) Method

Conventional mining is cyclical, employing mobile mechanized mining equipment to carry out production unit operations in five distinct steps:

production cycle = cut + drill + blast + load + haul

A cutting machine cuts a 5–7-in. slot or kerf at about 10 ft in length to improve breaking during blasting. A drill is then used to drill holes into the coal face in a predetermined pattern, which are then loaded with permissible explosives. After the coal is shot and fragmented by a qualified shot firer (following strict safe-blasting procedures), it is scooped up and loaded into a waiting shuttle car that transports the coal to a conveyor-belt loading head (feeder breaker) for transport to the surface. In some smaller mines, the coal is hauled in rubber-tired trailers drawn by battery-powered tractors. After the first working area is completed (depending on specific practice, this could be the first heading on the right in a panel if the mine works from right to left direction), the crew moves to the adjacent workplace and the steps repeat, as shown in Figure 13.2-6. Coal mined at the face is transported using a shuttle car or ram car to the breaker where it is transferred to the conveyor belt.

Several other auxiliary functions must also accompany production functions: face cleaning, roof control, proper face ventilation, cables and equipment advances, and movement of materials and supplies. As a result, multiple entries are necessary such that all functions can be performed at separate faces without delays.

Depending on the face advance rate and traveling distance, as well as the change-out points due to multiple hauling units, the location of the dumping point (breaker) needs to be located in a position that minimizes haulage travel time and congestion.

In the United States, haulage entries where coal is transported are separated from both intake and return airways, thus

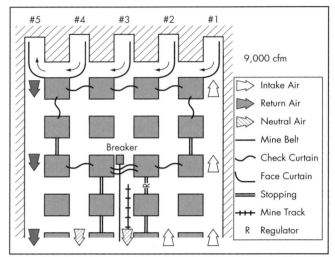

Figure 13.2-6 A five-entry unit with mining progressing from right to left

the introduction of "neutral" entry airways, which should not be used in the face area but be channeled directly into the return airway.

Increased safety concerns, efficiency demand, and the need to operate several pieces of mining equipment simultaneously have caused this conventional mining method to fall out of favor, and it is slowly being replaced by the continuous mining method. In 2008, only 0.6% of all coal mined underground in the United States was mined using this method.

KEY AUXILIARY FUNCTIONS

To provide a highly productive mining environment while maintaining safety, a well-coordinated mining plan that incorporates several key support functions is critical. The top two such functions are effective ground control and proper face ventilation.

Ground Control in Coal Mining

Ground control is concerned with both the short- and long-term stability of mine accesses and entries as well as subsidence control. Although no method of roof control yet devised has proven to be universally acceptable for the wide range of strata conditions, perhaps the most significant development in coal mine ground control during the 20th century was the introduction of roof bolting in the late 1940s and 1950s (Figure 13.2-7). Being able to provide effective and efficient ground control is essential to ensure safety and productivity in underground coal mines.

The theory of the interaction between roof bolts and the rock strata is complex and several models are available: beam theory (roof bolts bind together several weak strata into one stronger "beam"), suspension theory (weak members of the strata are suspended from a strong anchor horizon), and the keying-effect theory (roof bolts act much like the keystone in an arch). All seem to be able to explain at least part of the support phenomenon under certain conditions.

From an engineering standpoint, roof bolts are inherently more effective than the wooden timbers they replaced, they greatly accelerated the transition to trackless, rubber-tired face haulage, and they significantly facilitated coal production in underground coal mining.

Figure 13.2-7 Roof bolting has become an integral part of the mining cycle

Source: Barczak 2005.

Figure 13.2-8 Cribs and steel plates used to reinforce roof support

Steel bolts, usually 4–6 ft long and 0.6–1 in. in diameter, are inserted in holes drilled into the roof strata by a roof bolter and are secured as primary roof support by friction or resin. Various types of bolt systems are commercially available (frictional, tensioned, nontensioned, fully-grouted tensioned) and accepted by the United States and world mining industry.

The tensioned bolt system (also referred to as an active system) includes the conventional point-anchored mechanical bolt, a tensioned combination bolt, partially grouted tensioned rebar bolt, mechanically anchored resin-assisted bolt, partially grouted cable bolt, and so on. The nontensioned "passive" bolts include the fully grouted resin bolt and resin-anchored cable bolt. The fully grouted tensioned bolt combines the advantages of both the tensioned and nontensioned bolt. Roof channel, roof mat, roof pan, wire mesh, and shotcrete have also been used in combination with roof bolts for surface restraint. They are used to support small pieces of loose rock or as reinforcement for shotcrete, and the latter to provide support to the rock surface between bolts. Shotcrete provides the added advantages of reducing the rock exposure, thereby minimize weathering effects.

Today, all major mining countries have mandatory safety and ground control standards for underground coal mines. In the United States, specific regulations on roof support standards are set forth in Subpart C, Title 30, Part 75 of the Code of Federal Regulations (CFR 2010). Regulations stipulate that "no person shall work or travel under unsupported roof" underground, and "the roof, face and ribs of areas where persons work or travel shall be supported or otherwise controlled to protect persons from hazards related to falls of the roof, face or ribs and coal or rock bursts; no person shall work or travel under unsupported roof." Additionally, there are detailed regulations dealing with roof bolting, including installations, conventional roof support, pillar recovery, and automated temporary roof-support systems.

Additional supporting systems, either temporary or permanent in nature to meet specific geological conditions, are commonly used. Wooden cribs or posts, trusses, cable bolt, steel set, and wire mesh or shotcrete have all been successfully applied in underground coal mines. Wooden cribs (Figure 13.2-8) are prop timbers or ties laid in alternative cross-layers (i.e., log-cabin style) to effectively provide significant reinforcement capacity over a large deflection range. However, crib support systems restrict airflow, generate floor heave, and permit large roof deformations that can lead to failure. In other words, they have to be used carefully.

Roof truss as a means of secondary or supplemental mine roof support has been receiving increasing acceptance in recent years, particularly in bituminous coal mines. In those entries where long life is required or doubtful performance from other forms of support seems likely, trusses in many cases have provided the necessary protection for continuing operations. Steel trusses employ a combination of steel rods anchored into the roof to create zones of compression and provide better support for weak roofs and roofs over wide areas.

The aforementioned roof bolts used to reinforce the roofs of mines, although effective, are relatively expensive and by definition inflexible. Cable bolts use strand cable coupled with properly designed couplings. These are not only more economical, they provide relative flexibility as compared to the steel bar. This support system has been recognized as an effective ground control system in underground coal mines.

Steel sets are also commonly used as long-term support of underground openings such as inclined mine entries, main travelways, belt entries, and so on. The operator generally connects the steel set with tie rods, inserts wood blocks between the arch sets as lagging, and backfills the voids between the steel set and roof. Under some extreme geological conditions, yieldable arches are used to provide effective ground support in areas of excessive ground movement or faulted or fractured rock. They are designed such that when the ground load exceeds the design load of the arch as installed, yielding takes place in the joints of the arch permitting the load to settle into a natural arch of its own and thus bring all forces into equilibrium. The arch is often stronger after yielding than before due to increased joint overlap.

Ventilation

Proper ventilation is critical in any underground coal operation because of the presence of methane and coal dust as well as diesel particulate matters from diesel-powered equipment. Control of these hazardous pollutants has become increasingly challenging because of increasing mine sizes, coal production, and the use of diesel-powered mining equipment in recent years.

Engineering control techniques commonly used for ventilation purposes range from simple dilution with sufficient air quantity, water sprays and water additives, and local dust collectors to complex drainage systems to remove the gases prior to mining in many gassy mines. Nevertheless, dilution remains the most common method used. In general, the reduction in pollutants is roughly proportional to the increase in airflow; therefore, providing sufficient airflow underground remains the first priority in controlling the underground atmosphere.

However, providing the necessary additional air quantity can be difficult and expensive because, for a given fan pressure (H, in. water), the quantity of air (Q, ft³/min) delivered is inversely proportional to mine resistance (R):

$$Q = \left(\frac{H}{R}\right)^{1/2}$$

where R is determined from the physical characteristics of mine airways: airway frictional factor, airway perimeter, air-traveling distance, and average airway cross-sectional area shown as

$$R = \frac{KOL}{5.2A^3 n^2}$$

where K is airway frictional factor (lb·min²/ft⁴); O is airway perimeter; L is traveling distance, which increases with increasing mine size (ft); A is average cross-sectional area (ft²), and n is the number of entries.

In general, R&P mining will have low mine resistance because of multiple (parallel) entries. However, this can be both advantageous as well as challenging; the air needs expensive channeling systems (continuous stopping lines and overcasts) to effectively "push" fresh air to the working faces.

In the United States, mines using the R&P method have always been laid out in a structured and hierarchical fashion—intake, return, and neutral (belt and track) airways are always clearly defined and grouped by stopping lines in all mains, submains, and panel layouts, as shown in Figure 13.2-2 (where all three intake, two neutral, and three return airways are separated by stopping lines.) A great number of overcasts (or undercasts) are used to separate the intake, neutral, and return airflows. The number of entries at an intersection will determine the number of overcasts (or undercasts) for effective air distribution. For example, as many as 16 overcasts at one single intersection may be necessary to fully separate a 7 × 5–entries intersection.

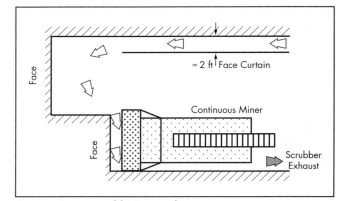

Figure 13.2-9 Scrubber at work

The power (AHP, or air horsepower) required to circulate a certain amount of air quantity (Q) under H (in. water) can be calculated using the following equation:

$$AHP = \frac{HQ}{6,345} = \frac{(RQ^2)Q}{6,345} = \frac{RQ^3}{6,345}$$

Alternatively, it is proportional to the cube of air quantity delivered. In other words, double the air quantity quadruples the mine pressure, which requires eight times the amount of energy compared to that needed before the change.

In the face area, U.S. federal regulation (CFR 2010) requires that 9,000 cfm of fresh air is to be delivered to the last open crosscut, properly installed face curtain (or brattice) in each individual heading used to push air to the deepest penetration, and check curtains must be designed to isolate the dusty air generated in the feeder breaker area from being blown to the face area, as shown in Figure 13.2-6.

Perhaps the most prolific success story in face ventilation since the 1980s has been the use of a machine-mounted flooded-bed scrubber (Figure 13.2-9). Flooded-bed scrubbers are devices installed on CMs to collect dust-laden air through several inlets near the front of the continuous mining machine. Inside the scrubber, the dust-laden air passes through a knit wire-mesh filter panel and is wetted with water sprays, where the dust particles are captured by the water. This filtered air then enters a demister where water is removed before the treated air is discharged at the back end of the miner.

Today, scrubbers are widely used on CMs in the United States and are highly successful in reducing coal dusts in the face area. Efficiencies for removal of respirable dust typically run from 90% to 95%, provided that screens are washed relatively frequently.

Many factors influence scrubber efficiency: scrubber capacity, face ventilation method (blowing or exhausting curtain or tubing), amount of fresh air reaching the beginning of the face curtain or tubing (a minimum of 3,000 cfm by regulation), ventilation setback distance (distance from the end of the face curtain or tubing to the deepest penetration), and CM location. Water-spray locations will also affect scrubber capture, with studies showing that by placing water sprays outby (away from the face), the scrubber inlets between the spray inlets and the rear end of the machine make these water sprays act as small jet fans. This arrangement will potentially induce additional airflow alongside of the machine and provide best results.

Another way to reduce dust includes displacement, which uses the airflow in a way that confines dust and methane sources and keeps them away from workers by forcing the dust and/or methane downwind of the worker. In addition, modifying the CM cutting cycles by leaving roof coal in place until it can be cut to a free face generates less dust and sometimes provides better extraction control.

OTHER ESSENTIAL AUXILIARY SERVICES

Other auxiliary operations are essential to the performance of an operating unit: coal haulage, power supplies, pumping, materials and supplies, stopping lines, and overcast construction.

A CM is not truly continuous but often only operates for less than half of the time, with the other half either employed for tramming (from location to location) or waiting for haulage vehicles to unload coal. Different methods have been used to minimize this nonproductive time in order to realize CMs' true production potential in the face area. The proper number of haulage vehicles for the face and an optimum distance between the face and the feeder breaker are the most commonly used measures to minimize a CM's waiting time. Face haulage uses either an electric-powered vehicle with an umbilical cable, rubber-tired shuttle cars, or battery-powered ram cars. Today it is common to use three haulage vehicles with well-planned travel routes to maximize efficiency, a compromise between optimum coal haulage and minimum congestion.

Flexible conveyor trains (FCTs) have also been used in some mines to minimize the CM's waiting time. They are electric-powered, self-propelled conveyor systems that provide continuous haulage of material from a CM to the main mine belt. The FCT uses a rubber belt similar to a standard fixed conveyor. The FCT's conveyor belt operates independently from the track chain propulsion system, allowing the FCT to move and convey material simultaneously. Available in lengths of up to 570 ft, the FCT is able to negotiate multiple 90° turns in an underground mine infrastructure, thereby greatly facilitating face haulage.

In addition to face haulage, proper planning is also critical to ensure an efficient haulage system to transport coal from the face area to the panel belt and ultimately to the surface through panel and main conveyor belts.

To prevent explosions from settled coal dust on the floor, rib and roof areas of mine entries are covered with pulverized limestone (inert) dust or rock dust. Studies show that during an explosion the rock dust disperses, mixes with the coal dust, and prevents flame propagation by acting as a thermal or heat sink (i.e., the rock dust reduces the flame temperature to the point where combustion of the coal particles can no longer occur). Under U.S. federal law (CFR 2010), all underground areas of a coal mine, except those areas in which the dust is too wet or too high in incombustible content to propagate an explosion, must be properly rock-dusted to within 40 ft of all working faces. Rock dusting is usually carried on a third or idle shift to not interfere with routine production.

In coal, methane drainage is still an art, but some of the drilling technology (e.g., directional drilling from both surface and underground) borrowed from the petroleum industry has been practiced at some gassy mines since the mid-1990s and has obtained quite satisfactory results; its performance is still continually being assessed.

PILLAR EXTRACTION

Only after careful review of pillar performance to date and pillar-extraction plans, and upon approval from regulatory agencies can coal pillars can be systematically removed after primary mining to maximize coal production. Some of the commonly used extraction methods are the "Christmas tree" or "split-and-fender" extraction methods. The former is favored because it does not require place changes and bolting. A third system is the "pocket-and-wing" method, which can fully extract large pillars. The general procedure is to extract one row of pillars at a time, leaving the mined-out portion, or gob, free to subside. Whereas extraction of all the coal in a pillar is a desirable objective, partial pillar-extraction schemes are more common. These are usually accomplished by CMs (or hydraulic mining with augers) driving into pillars according to predetermined extraction plans (Figure 13.2-10).

Pillar-extraction plans must be properly designed according to site-specific geotechnical and operational parameters (equipment and timber availability and cost, pillar dimensions, major geological features/structures, roof competency, seam thickness, extraction span or cavability, gob edge, and depth of cover) to ensure pillar stability during extraction. Case histories attributed to roof falls have found that both weak rock and competent immediate roof strata cause pillar failures. Hazards associated with pillar extraction also tend to intensify with depth; failures include both squeeze (nonviolent gradual pillar failures) and bumps, both of which are generally more severe at depth. Design of pillar-extraction schemes requires geotechnical input from specialists.

SYSTEM RATIONALIZATION AND OPTIMIZATION

Traditionally, coal productivity has always been measured by tons per workshift or similar measures based on total tonnage. Other factors such as size distribution, dust generation, and waste management also need to be considered at the planning stage to meet specific contractual demands and processing capacities at the surface preparation plant.

In the face area, CM cutting, loading and tramming capabilities, miner coal haulage, face layout, and so on must all be carefully planned and matched in order to achieve optimal overall efficiency. Materials handing still represents a significant component of the operational cost. A study on coal mining production costs in the United States (Chugh et al. 2002) shows that, of the total production cost, labor (with benefits) constitutes 47%, mining-related costs (supplies, maintenance, power, and other direct costs) uses another 29.5%, with the remaining designated to depreciation, taxes, and other expenses. Of this total production cost, half goes to face production, outby construction uses another 25%, coal processing and waste management uses 15%, and the remaining 10% is allocated to transportation. The same study indicates that face haulage is still the major bottleneck to productivity despite the use of more powerful high-voltage CMs. The study compares a Joy 14CM15 miner (995 V) with a high-voltage 14CM27 miner (2,300 V) in the Illinois coalfields. Results show that the high-voltage miner outperforms its counterpart in cutting and loading rate (by 29.89%), tons per car (by 8.64%), and tramming rate (7.84%), but they perform equally in tons per shift. (True productivity gains of the high-voltage miner were not fully realized because of the batch haulage system delays.) In most countries, the major impediment to productivity is still the time required to install roof supports.

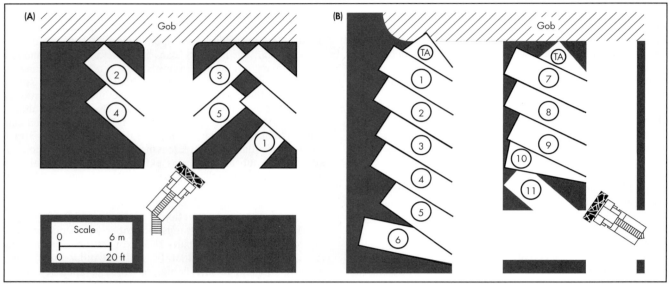

Source: Mark and Zelanko 2001.

Figure 13.2-10 Commonly used pillar extraction methods

SUMMARY

With its low capital investment and great flexibility, R&P mining continues to be an indispensible method in underground coal mining. In some cases it is the only method to access remnant parcels of coal that would otherwise be abandoned using longwall mining. It is also essential in gate road development in longwall mining, and under special circumstances, with proper planning, it may (in cost terms) even compete with longwall mining.

Although the basic methods of operation in R&P mining remain similar, there are a variety of systems with minor variations from location to location depending on local specifics. It is all a matter of system optimization—what would be an optimal number of entries or haulage method in one location may not be so in a different location depending on advance rates, ground control concerns, and dust and methane emissions that dictate ventilation practices. The optimum plan for a site is a trade-off of such factors. For example, the more entries, the higher the coal production, although at the expense of advance rate. On the other hand, fewer entries will cause congestion but will provide faster advances. Modern high-speed computers and simulation packages are ideal for modeling these scenarios for optimum performances.

A CM must be matched with a compatible haulage system in order to realize its full potential. Options could be a high-voltage miner matched with a continuous haulage system, a batch haulage system, or a surge car. Another benefit is that the high-voltage miner generates 43% less dust; using surfactants reduces that by an additional 6%.

Underground coal mining has always been a cost-competitive industry and will continue to be so in the foreseeable future. Results from continuing efforts to reduce production costs and/or increase productivity to stay competitive have been quite impressive. According to the U.S. Energy Information Administration (2010), in the period from 1949 to 1999 the underground coal productivity per worker-hour rose from 0.68 t/h (short tons per hour) to 3.99 t/h, a 587% increase, and to 4.04 t/h in 2004, a 594% increase from 1949 and a historical high. This rate decreased to 3.17 t/h in 2008.

The continuing improvement in equipment technology, system monitoring, automation, and system optimization in R&P mining make this method more productive and indispensible in underground coal mining.

REFERENCES

Barczak, T.M. 2005. An overview of standing roof support practices and developments in the United States. In *Proceedings of the Third South African Rock Engineering Symposium*. Johannesburg: South African Institute of Mining and Metallurgy.

Bieniawski, Z.T. 1987. *Strata Control in Mineral Engineering*. Rotterdam: A.A. Balkema.

CFR (Code of Federal Regulations). 2010. Title 30—Mineral Resources, Part 75, Mandatory Safety Standards. www.msha.gov/30CFR/CFRINTRO.HTM. Accessed April 2010.

Chugh, Y.P., McGolden, M., and Hirschi, J. 2002. Identification of cost-cutting strategies for underground mines in Illinois. Final Technical Report (November 1, 2000, through October 31, 2002). Office of Coal Development and Marketing and the Illinois Clean Coal Institute.

Mark, C., and Zelanko, J.C. 2001. Sizing of final stumps for safer pillar extraction. In *Proceedings of the 20th International Conference on Ground Control in Mining*. Edited by S.S. Peng, C. Mark, A.W. Khair, and K.A. Heasley. Morgantown, WV: West Virginia University.

U.S. Energy Information Administration. 2010. Table 7.6: Coal Mining Productivity, 1949–2008. www.eia.doe.gov/emeu/aer/coal.html. Accessed April 2010.

CHAPTER 13.3

Shrinkage Stoping

Jack Haptonstall

INTRODUCTION

Shrinkage stoping is a vertical overhand stoping method in which most of the broken ore remains in the stope to provide a working floor for the miners. The broken ore also provides wall support until a stope is completed and ready for drawdown of the remaining ore.

In general, the method is used in steeply dipping (>50° dip) narrow ore bodies with regular boundaries. Ore and waste (both the hanging wall and the footwall) should be strong, and the ore should not be affected by storage in the stope. The method is labor intensive and cannot readily be mechanized. It is usually applied to ore bodies on narrow veins or to ore bodies for which other methods cannot be used or might be impractical or uneconomical. The method can be easily applied to ore zones as narrow as 1 m (3.1 ft), but it can also be successfully used for extraction of much wider stopes. The method is most efficient when drilling of the ceiling or back is done with uppers instead of horizontally.

Shrinkage stopes are mined upward in horizontal or inclined slices. Usually about 35% of the ore derived from the stope cuts (the swell) can be drawn off (shrunk) as mining progresses. As a consequence, no revenues can be obtained from the ore remaining in a stope until it is finally extracted and processed for its mineral values.

Logically, ore broken in a shrinkage stope should be free flowing, and broken ore should not pack or re-cement itself together. Neither the ore nor the adjacent country rock should contain undue amounts of clay or other sticky material that would cause the broken ore to hang together in the stope and make it difficult or impossible to draw. The ore should not readily oxidize, which may cause the broken ore pile to re-cement itself and consequently "hang up." Also, oxidation of the ore minerals may have an adverse effect on recoveries of values in subsequent beneficiation of the ore. To avoid mining excess amounts of waste as dilution from the stope back, ore developed for shrinkage stoping should be fairly continuous along the strike of the vein or ore body. In many cases, however, it may be possible to mine around waste areas and leave them in place as random pillars.

In designing for shrinkage stoping, consideration must also be given to the plunge or rake of the ore body, especially where an entire ore body or ore shoot may be mined as a single stope rather than as preestablished stope panels with defined vertical endlines. A stope with a shallow plunge or rake (<50°) may be difficult to mine by shrinkage methods because the ore moves away too quickly from the predeveloped extraction system, occasioning the need for additional development of the system. Further development, especially raising, is also often required when the ore abruptly extends beyond stope endlines.

Following is a summary of the parameters for the application of shrinkage stoping to a particular ore body (Boshkov and Wright 1973; Lucas and Haycocks 1973; Morrison and Russell 1973; Lyman 1982):

- **Ore characteristics:** The ore should be strong and non-oxidizing, should not pack or stick together, and should not spontaneously combust.
- **Host rock characteristics:** The host rock should be characterized by moderately strong walls that are free of clay or geologic structures such as faults, joints, and so on.
- **Deposit shape:** The deposit can be almost any shape, but it should have uniform, definitive boundaries.
- **Deposit dip:** The dip should be greater than the angle of repose (>39°) and preferably steeper than 50°.
- **Deposit size:** Width of the ore deposit should be narrow to moderate; the deposit can be of any length and height.
- **Ore grade:** Average ore grade should be moderate to high.
- **Production rates:** Production rates should usually be small to moderate.

Vertical longitudinal sections of typical shrinkage stopes are shown in Figures 13.3-1 through 13.3-3.

DEVELOPMENT AND PREPARATION

Areas selected for shrinkage stoping are generally developed by drifting in the vein or ore body on two levels that are spaced vertically at predetermined intervals that customarily

---

Jack Haptonstall, Mining Consultant, Pincock Allen & Holt, Lakewood, Colorado, USA

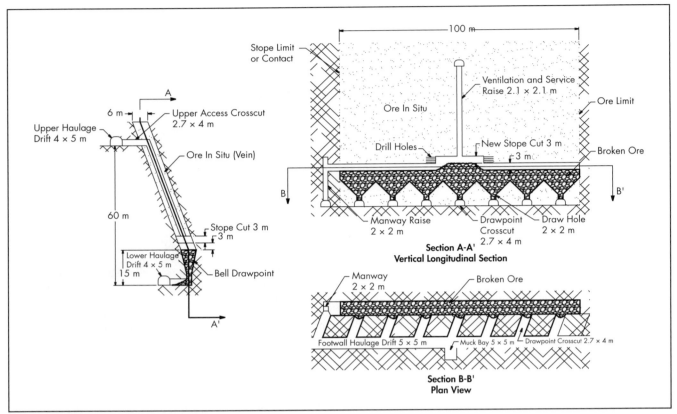

Figure 13.3-1 Idealized shrinkage stope with drawpoint extraction

vary between 30 and 180 m (100 and 600 ft). After a viable ore body has been established, the next phase is to develop one or more raises to establish vertical ore continuity and provide a ventilation opening and access to the stope. In some cases the raises are not developed, and access is only from the sill level below the stope.

Raising may be done conventionally with platform-type raise climbers or mechanically with raise-boring equipment. Drifting is normally done using conventional drill-and-blast track or trackless methods.

Stopes may be prepared with extraction raises on 7.6- to 9-m (25- to 30-ft) centers the full length of the stope block. Each raise is fitted with a chute (except for manways and/or service ways) that is normally of timber construction. Extraction raises are belled out and hogged over (mined over or excavated) to form the undercut for the start of the first stope cut (see Figure 13.3-2).

Another method used for preparing a stope is to blast down at least two cuts from the ore zone, clean up the broken ore, and install stull timbers or timber sets in the sill drift below the stope. Timber chutes or chinaman chutes are installed at approximately 7.6- to 9-m (25- to 30-ft) intervals as part of the timbering. A chinaman chute is a timber platform constructed below the caps of the timber sets; ore is shoveled from the platform into cars spotted below it.

Both of the foregoing types of stope preparation are still used, but on a very limited basis. A more common method for preparing shrinkage stopes in modern operations is to drive a sill drift along the vein or ore body as well as an extraction drift in the footwall of the ore zone and parallel to it.

The footwall drift is normally situated about 7.6 to 15 m (25 to 50 ft) from the ore body. Subsequently, draw-hole extraction crosscuts are driven from the footwall drift into the sill drift on 7.6- to 15-m (25- to 50-ft) centers. The back of the ore body sill drift is then drilled and blasted down, and the swell is extracted via the draw holes, either with rail-mounted mucking machines or with load-haul-dump (LHD) units. See Figures 13.3-1 and 13.3-4).

Obviously, other variations for the preparation of shrinkage stopes have been used successfully. These may include preparation with short crosscuts into the waste footwall from the ore sill drift, driving finger raises back up to the ore and preparing the first stope cut from the finger raises. Another variation might be the elimination of the construction of chutes or drawpoint crosscuts and drawpoints, electing instead to extract ore from the ore sill drift with remotely controlled LHD units.

STOPING OPERATIONS

When an ore block has been developed for shrinkage stoping, a manway is usually constructed in the raise to the next level. One or more manways and service ways are usually constructed with timber on one or both ends of the stope. Often a timber slide is installed in one of the manways for hoisting and lowering supplies and equipment into and out of the stope; hoisting and lowering are often done with an air or electric single-drum (tugger) hoist, which is installed in the level below the manway. After the manways, ventilation raise(s), and service ways have been established for a stope, mining can commence.

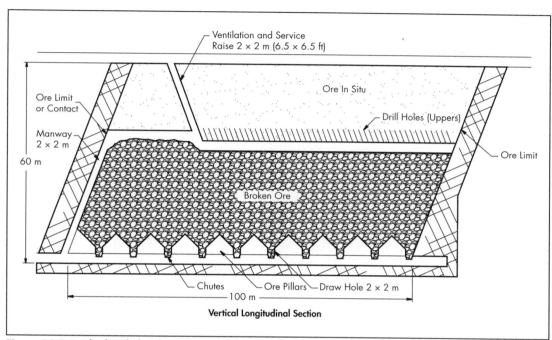

Figure 13.3-2 Idealized shrinkage stope with chutes

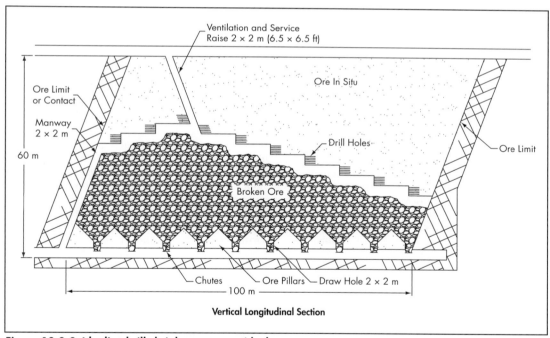

Figure 13.3-3 Idealized rill shrinkage stope with chutes

Drilling of a shrinkage stope back is usually done with handheld stoper or jackleg drills, although mechanized drill wagons or jumbos have been used in wide (>5.0 m [>16.4 ft]) stopes. Back stoping (vertical or near-vertical drilling) is the preferred mode of operation, but breasting down (horizontal drilling) is also common. Upholes are generally 1.5 to 2.4 m (5 to 8 ft) in length, and usually all drill holes are loaded and a complete cut is blasted simultaneously. Breasts are drilled with 1.8- to 3.0-m (6- to 10-ft) holes or normally are blasted once per shift. Drill holes are loaded with ANFO (ammonium nitrate and fuel oil) products, water gels, and slurry blasting agents. Primers are usually water gel sticks or dynamite, and holes are initiated using nonelectric, electric, or fuse-blasting systems.

After a cut or round has been blasted in a stope, drawdown (emptying out) of the 35% swell is necessary. Leveling of the muck pile is required as a next step to provide a work platform and facilitate drilling the next vertical or horizontal cut. Leveling of the ore in the stope can be done with hand

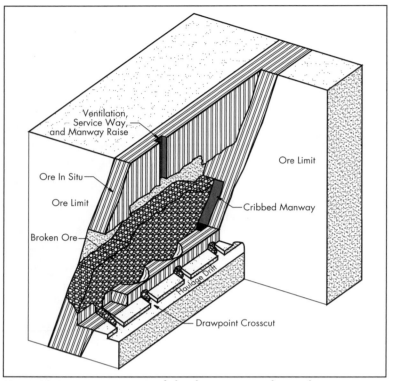

Figure 13.3-4 Isometric view of shrinkage stope with LHD drawpoints

shovels for small stopes, with two- or three-drum air or electric slushers for stopes that are longer and/or wider, and with LHD units for large stopes. After leveling, construction of manways and service ways and drilling of the next stope cut are done to continue the mining cycle.

Variations for establishing openings for manways, ventilation raises, or service ways may include strategically cribbed timber raises, steel culverts or rings, or even timber sets within the confines of the stope. These installations may be desirable during the mining phase but may create safety problems or nuisances during ore drawdown due to the collapse of the structures for these facilities. Pinning, stulling, or wedging these installations to the stope walls may prevent their destruction during drawdown. However, if the construction materials are dragged down with the broken ore, they can cause hang-ups in the extraction drawpoints or chutes.

A stope should have strong self-supporting walls to permit the application of shrinkage stoping. Excessive dilution through scaling of walls can preclude use of the method. Good mining practice and government regulations dictate a minimum ground support program for stope extraction. Shrinkage stoping has inherent dangers, merely because workers must be in the excavated area to drill, load rounds, and install ground support and stope services—all while standing on a broken ore pile. Many serious accidents have occurred when voids in the broken ore pile suddenly collapsed, dragging miners and equipment down into the void. In other cases, a chute or draw hole over which miners were working had mistakenly been pulled, with the same result. The requirement that miners use safety belts in a stope may reduce the danger, but it also reduces the workers' flexibility.

Ground support is usually minimal in shrinkage stopes, especially when the method is applied in areas of a vein or ore body that are steeply dipping, have strong ore and wall rocks, and have few structural features (e.g., faults, fractures, or joints) in the walls or within the vein or ore body. However, modern mining practice is to err on the side of safety, and so in most cases ground control is a primary function. Rock bolting of stope backs and ribs with mechanical or grouted rock bolts is the preferred method of ground support in shrinkage stopes. However, in very narrow stopes (e.g., <2.0 m [<6.6 ft] wide), correct installation of bolts in the walls is quite difficult because the bolts must be placed perpendicular to the planes of the hanging wall and footwall. More often than not, bolts in narrow stopes are installed at acute angles to the walls, which diminishes the effectiveness of the installations. Stulling of the ribs and back is still practiced; recent South African innovations in stull installation have made it a much more attractive option.

Stulling is simply the installation of timber props from wall to wall or floor to back. Typically, a board headpiece is inserted between the head end of the stull and the rock hanging wall, and the stull is secured by pounding wedges between the rock footwall and the butt end of the stull. In the past, the butt ends of stulls were almost always installed in a hitch cut from the rock with a hammer and chisel. The invention of the Jackpot headpiece has resulted in efficiency improvements in the installation of stulls. The Jackpot is a hollow steel can that fits over the butt-end of a timber stull and is pumped up hydraulically to pressurize the stull against the ribs or back of the stope. The unit eliminates the need to cut a hitch in rock to accommodate the stull, cut the stull to a precise length, and then try to tighten the installation with wooden headboards and wedges. Leaving pillars in areas of low-grade ore or waste within the stope boundaries is a very acceptable approach to supplementary wall support. The pillar should be shaped in such a way that ore does not remain on top of it on drawdown of the stope.

Sampling of narrow shrinkage stopes is usually done by taking either channel or chip samples at systematic intervals (e.g., 1.5 m [5 ft]) from the stope back, ribs, or face. Samples are usually taken by hand; sometimes mechanical means can be applied in the process. When high sulfides are present, X-ray analyzers can be used for sampling. In wider stopes, drill sampling of the back and ribs can be done. The drill sample holes may crisscross the stope back according to a predetermined pattern and may also be drilled into the stope walls. Drill cutting samples are collected in a canvas sample bag or other container through a hose and funnel or other device.

STOPE DRAWDOWN

Shrinkage stope drawdown usually results in a consistent and steady supply of ore to the process plant and/or mine stockpiles. If a mine has several shrinkage stopes in various phases of the mining cycle, there is no reason the ore flow to the plant or stockpiles should ever diminish after completion of the initial stopes in the mine plan. Maintaining a steady ore flow from shrinkage stoping is largely dependent on engineering the optimum production rate for a given operation.

Completed stopes can be drawn down from strategically placed chutes and from drawpoint crosscuts, or they can be remotely mucked with LHD units. In some mines, slusher trenches developed below the stopes have been used effectively to remove broken ore from shrinkage stopes. Haulage from the stope extraction points can be done using trackbound or trackless equipment. Chutes should be robustly designed and well constructed to ensure that they will not be destroyed during the extraction phase.

Shrinkage stopes usually should be drawn down evenly so that if the stope walls do peel or slough, the waste material remains atop the pile and does not add dilution to the broken ore. After a stope drawdown has started, ground control of the walls, pillar recovery, ore hang-ups, and so on, is minimal. Reentry of miners into a stope during the drawdown phase is not recommended.

One of the most dangerous jobs in a mine is the drawdown of shrinkage stopes, especially where the ore contains blocky or sticky material that may cause the broken ore to hang up in the stope. Stope hang-ups are often removed by washing them down with water, by destroying them with explosives, or by blowing them down using high-pressure air (blowpipes). In the past, miners entered completed stopes to pick them down by hand, but as has been stated previously, this is unsafe and not recommended.

A summary of the advantages and disadvantages of the shrinkage stoping method is shown in Table 13.3-1.

VARIATIONS AND APPLICATIONS

Variations of shrinkage stoping include inclined or rill shrinkage and long-hole shrinkage. Large pillars or small isolated ore blocks may sometimes be recovered using shrinkage methods. Certain underground openings, such as short shafts, may also be excavated using shrinkage methods.

A mine that employed shrinkage stoping as a primary mining method was the now closed Homestake gold mine in Lead, South Dakota (United States). The Homestake ore bodies were mainly fracture fillings situated in the limbs of folds. "Bullpen" stopes (i.e., fixed-minimum-width shrinkage stopes laid out in a transverse direction to the long axis of a large ore body) 15 m (50 ft) wide were developed across the great Main Ledge ore body. Stopes were mined over timbered sills, where strategically located chinaman chutes were constructed for ore extraction. Stopes were mined overhand for about 21.5 m (70 ft) to within 9.1 m (30 ft) of the next level. Ore pillars 7.6 m (25 ft) thick were left between the primary stopes, and these, along with the crown pillars left over in the primary stopes, were subsequently extracted with square-set stopes. Homestake abandoned this type of shrinkage stoping just before World War II.

A second example of large-scale shrinkage stoping is the shut-down Idarado mine located between Ouray and Telluride in Colorado (United States). The Idarado ore bodies were mainly steeply dipping veins with excellent ore continuity. Shrinkage stopes were mined along the veins' full width, spans that varied between 1.5 and 7.6 m (5 and 25 ft). The consistent ore continuity permitted the mine planners to divide the veins into stope panels generally 122 m (400 ft) long and about 61 m (200 ft) high. Panels were prepared over a slusher trench situated about 6.1 m (20 ft) above the primary on-vein drift. Raises on 7.6-m (25-ft) centers, developed from the slusher trench, were belled out and interconnected above the slusher trench to form the opening for the first stope cut. Ore extracted from the stope was slushed from the extraction raises to a chute in the exact center of the stope. The Idarado operations were closed in 1979.

Variations on the Idarado system were practiced at the Morococha and Casapalca mines of the Cerro de Pasco Corporation in the Andes Mountains of central Peru in South America and at the Tayoltita mine in the state of Durango, Mexico. Stopes in these mines were prepared conventionally over the main on-vein development level by driving 7.6-m- (25-ft-) long raises on 7.6-m (25-ft) centers. The raises were belled out to form the opening for the first stope cut about 5 m (16 ft) above the main level. On completion of this step, each raise was fitted with a timber chute for ore extraction. (See Figures 13.3-1, 13.3-2, and 13.3-4.)

In all of these examples, a raise was first developed through each ore block or stope panel for ventilation, manway, and service way. Manways from the sill level were carried as timber-cribbed raises within the stope, as stulled and lagged raises on the ends of a stope, or—as in the case of the Idarado mine—as boreholes located 3 m (10 ft) inside the footwall of the stope. In the vein mines, drilling was done with pneumatic handheld stopers or jackleg drills, whereas in the Homestake mine, drilling was done with bar-and-column Leyner-type drills.

The terms *inclined shrinkage* and *rill shrinkage* refer to an application in which multiple faces or benches for drilling are carried along the back of the stope as it is mined upward (Figure 13.3-3). This system was common practice at the Pachuca mine in the state of Hidalgo, Mexico. Stopes were developed conventionally over pillars as described previously, and chutes on 7.6-m (25-ft) centers were installed on the extraction raises. Pachuca miners preferred to use jackleg drills rather than stopers, and with the rill system multiple faces were available for breast stoping in a given shift (see Figure 13.3-3).

Table 13.3-1 Advantages and disadvantages of shrinkage stoping

Advantages	Disadvantages
Shrinkage method is easily adapted to narrow veins.	Method requires steep dip (>60°).
Stopes do not require backfill.	Method requires strong ore and walls.
Stope extraction system is usually simple.	Stoping is labor intensive and efficiencies are low at about 3 to 9 t (3.3 to 9.9 st) per worker-shift.
Stope drawdown provides a steady, consistent supply of ore.	
Application of method requires a minimum investment in equipment and machinery.	Only 35% to 40% of ore is available for extraction during mining cycle.
	Very little mechanization of stoping is possible.
Stope development is uncomplicated and minimal.	Excessive clay or oxidation of ore may result in stope hang-ups or problems in ore dressing.
Reasonable selectivity is possible.	
Ground support of ore and walls is minimal.	Ground control is sometimes difficult to install.
Gravity aids drawdown of stopes.	Risk exists of losing stope if drawdown is not properly controlled.
Mining recovery is good (75% to 100%).	Mining costs are usually moderately high to very high.
Dilution is low (0% to 20%).	

Source: Morrison and Russell 1973; Lucas and Haycocks 1973; Hamrin 1982; Lyman 1982.

Long-hole shrinkage (Figure 13.3-5) is developed in the manner previously described. The method was applied with limited success at the Stillwater mine near Nye, Montana (United States). In this method, drilling of the stope block is done from vertical raises driven through the ore block or panel on 15- to 30-m (50- to 100-ft) centers. Raises in this application are preferably developed with raise climbers or with cage raising techniques. The raise climber platform or the cage serves both as the entry/exit vehicle for the raise and the platform from which to drill and blast the stope block. Long holes are drilled from the raise, parallel with the strike of the ore zone, and these are also loaded and primed from the climber or cage platforms. Initiation of the stope blast is done from a safe area on the service level below the stope.

Shafts, winzes, or other large underground openings may be developed through shrinkage methods. Often these excavations are done as described previously for the long-hole shrinkage application. In certain situations, conventional shrinkage of a shaft, winze, or other excavation may be necessary.

SAFETY, EFFICIENCIES, AND COSTS

Shrinkage stoping has largely been discontinued in most countries because it has numerous safety hazards, is very labor intensive and largely inefficient, and is generally expensive. Coupled with these reasons is that more than 60% of the mined ore remains in the stope until stoping is completed, which may take many months and sometimes even years. During this period, amortization of the investment required to develop and mine an ore block by shrinkage stoping may outweigh any advantage in applying the method to a particular operation (see Figure 13.3-5).

Typically, the efficiencies in a modern shrinkage stoping operation are in the range of about 0.3 to 1.2 t (0.3 to 1.3 st) per worker-hour. In western operations, very few miners have experience in the development and operation of shrinkage stopes, but the method is still widely practiced in Latin America and in developing countries elsewhere in the world.

Costs in a typical North American shrinkage operation are currently on the order of US$25 to US$50/t (US$55/st), with more than 50% of the costs assigned to labor. Operations in Latin America or Asia may enjoy lower costs because labor is less expensive, but efficiencies are generally not equal to those of North American operations.

CASE STUDY: EAGLE RIVER MINE

The Eagle River gold mine is owned and operated by Wesdome Gold Mines of Toronto, Ontario, Canada. The mine is located near the northeast corner of Lake Superior, 50 km (31 mi) west of the town of Wawa, Ontario. It is accessible by gravel roads connecting Trans-Canada Highway 17 near Wawa to the site.

The Eagle River mine has been in production for about 12 years. Historic production from the mine is about 22,900 kg (711,000 oz) of gold from mining and processing about 2.4 Mt (2.64 million st) of ore. The mine is currently being operated at a production rate of 100,000 t (110,000 st) per year.

Geology

The Eagle River ore deposit occurs in the Mishibishu greenstone belt, which is a broad arcuate belt about 60 km (37.2 mi) long and 18 km (11.1 mi) wide. The area in which the Eagle River deposit occurs contains magnesium-rich tholeite, calc-alkaline andesite, dacite, and minor rhyolite. The ore occurs within shear-hosted near-vertical quartz veins that have mineable widths of 1.5 to 7.5 m (4.9 to 24.6 ft). The ore zones of the Eagle River mine are of a series of quartz lenses that to date have been traced over a strike length of 2.5 km (1.5 mi) and a depth of 650 m (2,132.5 ft). At least four distinct ore zones or shoots have been developed, and these vary in length from 50 to 250 m (164 to 820 ft).

Ore Reserves and Mining Method

Proven and probable ore reserves at the end of 2007 were 253,000 t (278,300 st) at an average grade of 12.9 g gold/t (0.37 oz/ton). The mining method, formerly all shrinkage stoping, was changed to long-hole stoping in 2000.

Details regarding shrinkage stoping are as follows:

- Stope heights are 100 m (328 ft) with drawpoint scooptram/truck mucking and hauling.
- Two raises with doghole accesses are driven on the boundaries of ore blocks, which can be up to 100 m (328 ft) long.
- Two people work a stope, breasting toward the middle. Breasts are 2.5 m (8.2 ft) high and 3 m (9.8 ft) long and fill a void measuring 3.5 m (11.4 ft).
- Uppers at one end of a stope provide the start for the next lift.

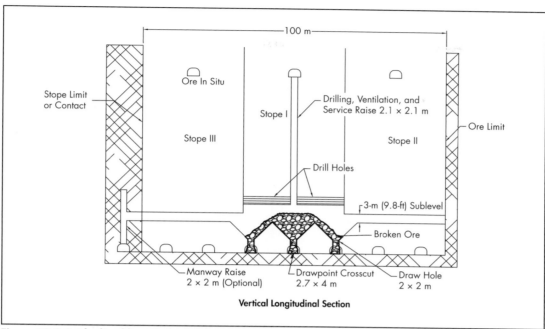

Figure 13.3-5 Idealized long-hole shrinkage stope

- Dogholes are staggered at 10-m (32.8-ft) intervals in both access raises. No stull timbering is required. Ladders provide access from dogholes, at least one of which is always available because of stagger.
- Productivity is 15 t (16.5 st) per worker-shift.

Details regarding long-hole stoping are as follows:

- Sublevels are spaced at 15 m (49.2 ft); jackleg and slusher subdrifts are driven from a central mill hole.
- Slot raises are developed as drop raises.
- Drill holes are 50-mm- (2-in.-) diameter downholes on a 3 × 3 m (9.8 × 9.8 ft) pattern.
- Productivity is 40 t (44 st) per worker-shift.
- Retreat mining is practiced with haulage and drawpoints on main levels spaced at 60-m (196.8-ft) intervals.

Equipment, Productivity, and Costs

Equipment includes 3.8-m^3 (5-yd^3) diesel LHD units, 25-t (27.5-ton) trucks, two-boom drift jumbos, scissors lifts, raise climbers, jacklegs, and stoper drills.

Productivity is about 15 t (16.5 st) per worker-shift in stopes. The shrinkage mining cost is unavailable. The long-hole mining cost is about US$40/t (US$44/ton).

SUMMARY

The labor intensity of shrinkage stoping as well as the limited mechanization possible for the method precludes its wide-scale application in most modern mining operations. However, in some applications, the method may be appropriate—for example, in a mine in which the ore occurs in very narrow veins and cannot be economically extracted using other methods.

REFERENCES

Boshkov, S.H., and Wright, F.D. 1973. Underground mining systems and equipment. In *SME Mining Engineering Handbook*, 1st ed. Edited by A.B. Cummins and L.A. Given. New York: SME-AIME. pp. 12.1–12.13.

Hamrin, H. 1982. Choosing an underground mining method. In *Underground Mining Methods Handbook*. Edited by W.A. Hustrulid. New York: SME-AIME.

Lucas, J.R., and Haycocks, C., eds. 1973. Underground mining systems and equipment. In *SME Mining Engineering Handbook*, 1st ed. Edited by A.B. Cummins and L.A. Given. New York: SME-AIME. pp. 12.1–12.262.

Lyman, W. 1982. Introduction to shrinkage stoping. In *SME Underground Mining Methods Handbook*. Edited by W.A. Hustrulid. New York: SME-AIME. pp. 485–489.

Morrison, R.G., and Russell, P.L. 1973. Selecting a mining method: Rock mechanics, other factors. In *SME Mining Engineering Handbook*, 1st ed. Edited by A.B. Cummins and L.A. Given. New York: SME-AIME. pp. 9.1–9.22.

CHAPTER 13.4

Sublevel Stoping

Rimas T. Pakalnis and Paul B. Hughes

INTRODUCTION

This chapter refers to the generic mining method of sublevel stoping. The most commonly used sublevel stoping mining methods are sublevel open stoping, long-hole open stoping or blasthole stoping, and vertical crater retreat (VCR). Variations of this method include vein mining, transverse stoping, Avoca, longitudinal, and other less-used methods such as slusher mining of uppers. The shrinkage stoping method is also a variation of sublevel stoping and is discussed in Chapter 13.3.

Definitions

Following are definitions of general terms for this mining method, also shown schematically in Figure 13.4-1.

- **Span:** The length of the stope along the strike.
- **Width:** The perpendicular distance between the footwall and the hanging wall.
- **Height:** The distance along the exposed hanging wall and not the vertical height between levels.
- **Longitudinal pillar:** A pillar aligned along the strike of the stope.
- **Rib pillar:** A pillar aligned transverse of the stope, perpendicular to the strike.
- **Sill pillar:** Horizontal pillars that separate levels or stopes.
- **Dilution:** The reduction of ore grade due to mixing of ore with barren rock.
- **Internal dilution:** Rock that must be mined because of the geometry of the ore body and the requirement to mine rectangular areas. The term is synonymous with *planned dilution*.
- **External dilution:** Dilution caused by sloughing or failure of stope walls and back and is outside the blasted stope boundary. External dilution is defined as the external waste tonnage divided by the ore tonnage. The term is synonymous with *unplanned dilution*.

Sublevel stoping, in the absence of consolidated fill, employs pillars to separate the individual stopes to reduce the potential for wall slough. Sublevel stoping requires a straight/linear layout of stope and ore boundaries. Inside of the stope,

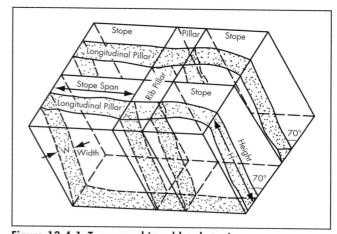

Figure 13.4-1 Terms used in sublevel stoping

everything is ore with no chance of recovering small mineralization in the wall rock. This method requires knowledge of the ore boundaries as shown in Figure 13.4-2.

Sublevel stoping with no fill is a mining method in which ore is mined and the stope is left empty. The result is a large void that requires individual pillars be placed to separate the stopes. Sublevel stoping is largely restricted to steeply dipping ore bodies (50°–90°) with a competent hanging wall (HW) and footwall (FW). Figure 13.4-3 shows the general approach to sublevel stoping whereby ring drilling is used from levels generally spaced ~20 m apart in a vertical dimension. Level spacing is largely limited by the length of the production holes, which range in diameter from 50–75 mm and maximum lengths of 25 m. This can be modified if in-the-hole (ITH) drills or top hammer tube drills are used. Characteristic with the sublevel stoping method are the intermediate levels, which largely differ from long-hole (blasthole) stoping as depicted in Figure 13.4-4 where the intermediate level has been removed. In sublevel stoping the mining is accomplished from individual levels at predetermined vertical intervals. These intervals are largely governed by

---

Rimas T. Pakalnis, Associate Professor, Norman B. Keevil Institute of Mining, University of British Columbia, Vancouver, British Columbia, Canada
Paul B. Hughes, Research Assistant, Norman B. Keevil Institute of Mining, University of British Columbia, Vancouver, British Columbia, Canada

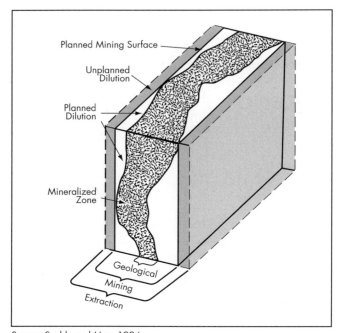

Source: Scoble and Moss 1994.
Figure 13.4-2 Defining dilution

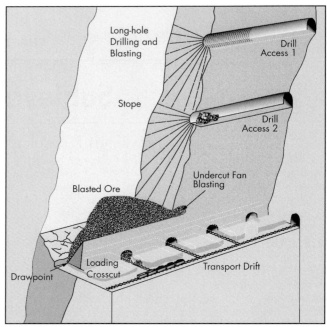

Source: Hamrin 2001.
Figure 13.4-3 Sublevel open stoping

- Ore geometry, in order to minimize internal dilution by enabling the extraction of irregular ore bodies,
- Rock mechanics constraints in terms of minimizing the external dilution through wall slough, and/or
- Operational restrictions such as drilling equipment constraints.

Sublevel and long-hole methods require blasting into a "vertical slot/free face," whereas a VCR, shown in Figure 13.4-5, differs in terms of blasting to a horizontal free face, which is largely confined due to the muck remaining within the stope as only the swell is drawn.

Variations of the sublevel method include narrow vein mining/Alimak, Avoca, longitudinal, sublevel retreat, and transverse stoping, as well as historical methods such as slusher and track mining (Haycocks and Aelick 1998).

SUBLEVEL STOPING REQUIREMENTS AND CONSTRAINTS

The following variables must be addressed in sublevel stope designs:

- **Size:** The minimum width generally ranges from 3 m to 6 m; however, in isolated cases it reaches 1.5 m (Clark and Pakalnis 1997) and lower (0.8 m). The width is governed by the production blast pattern, which, with the use of 50-mm blastholes, is typically 1.2 × 1.2 m, and the stope layout is based on this spacing (Figure 13.4-6).
- **Shape:** The shape is preferably tabular and regular in shape from level to level.
- **Dip:** The dip is preferably greater than the angle of draw, which typically is in excess of 50° in practice. The concern also is that a shallow hanging wall dip will result in a less-stable HW configuration because of gravity influences and increased wall exposures between vertical stope horizons, all resulting in increased potential for external dilution.

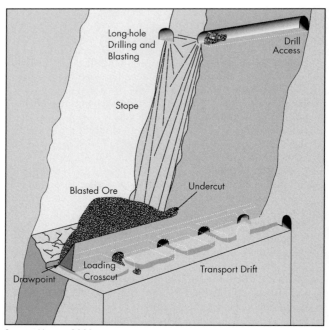

Source: Hamrin 2001.
Figure 13.4-4 Long-hole or blasthole stoping

- **Geotechnical:** This requires a moderate to strong ore strength and generally a competent HW-FW as these will be exposed and affect the level of external dilution. The ore will determine the potential pillar sizes, hole squeeze, and block size that affect production stope productivity (Pakalnis 2002).
- **Stope spans:** Since this is a nonentry method, stope spans can be larger. The span should be designed to control

Sublevel Stoping

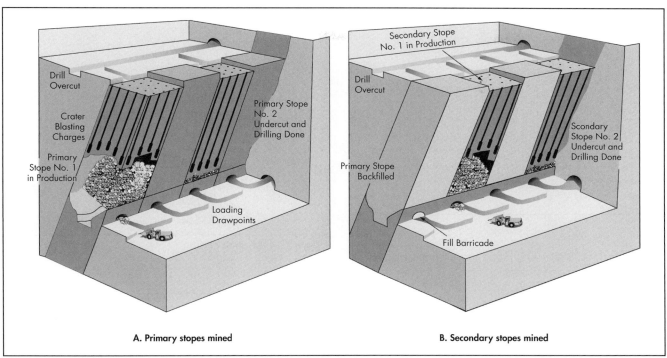

Source: Hamrin 2001.
Figure 13.4-5 Transverse stoping

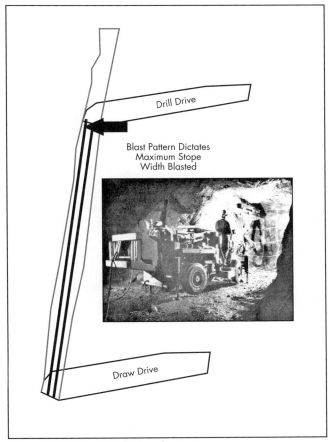

Figure 13.4-6 Sublevel stope layout

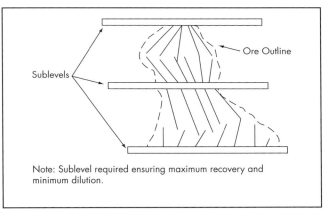

Note: Sublevel required ensuring maximum recovery and minimum dilution.

Figure 13.4-7 Geometric constraints

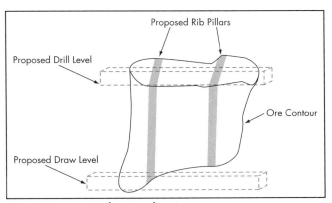

Figure 13.4-8 Initial stope planning

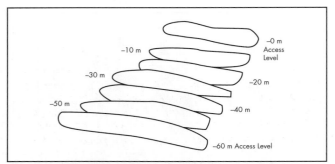

Figure 13.4-9 Composite plan section of ore body used to design blast layout

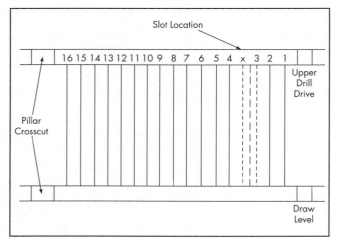

Figure 13.4-10 Longitudinal section of blasthole layout

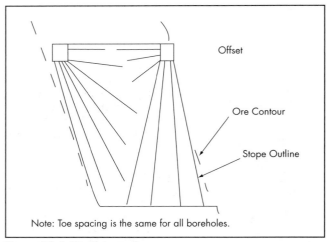

Figure 13.4-11 Ring section

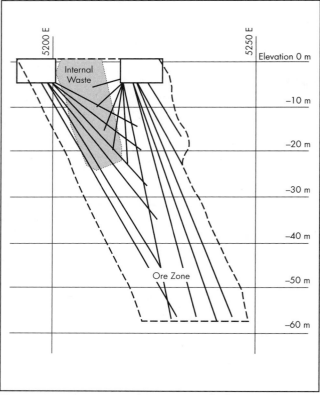

Figure 13.4-12 Engineered ring section, looking north (example)

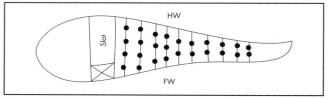

Figure 13.4-13 Location of slot raise (plan view)

external dilution and avoid stope collapse and air blast. Span length is governed by HW rock mass quality and generally is in the range of 30+ m with the stope height (inclined) in excess of 30 to 60+ m.
- **Pillar size:** The purpose of the pillars is to support the crosscuts and divide up the stopes. The size of the pillars is dependent on induced stresses, structure, rock mass, and operational constraints.
- **Selectivity:** Selectivity is limited because waste zones can be incorporated as pillars. Changes in ore-body geometry outlines are difficult to address unless the ore body narrows to the next pillar or sublevel where the drill pattern can be modified (Figure 13.4-7).

DESIGN CONSIDERATIONS FOR SUBLEVEL STOPING
General Design Guidelines
The design of a sublevel stope starts with an engineered layout that incorporates the geometry of the stope, stope span, stope height, pillar dimensions, drill levels, and draw levels (Figure 13.4-8). This layout is then superimposed upon the ore contours (plan) as defined from the upper drill drive to the lower draw level horizon. The example shown in Figure 13.4-9 is a schematic of a 60-m long-hole stope (151-mm blastholes) with geologic contour intervals shown every 10 m in the plan.

The resultant longitudinal composite is shown in Figure 13.4-10 employing a ring burden (distance between drill rings) of 3 m from pillar/stope boundary to pillar/stope boundary.

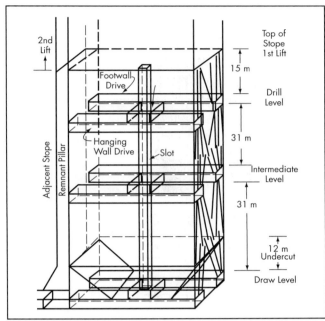

Figure 13.4-14 Sublevel open stope

The ring section is shown in Figures 13.4-11 and 13.4-12 with toe spacing (distance between toes of blastholes) of 4.2 m. The ring section incorporates the geological ore outline as defined by the geologic level contours with the stope outline coinciding with the drilled-and-blasted layout.

Development Considerations

Sublevel stoping uses long-hole drilling employing extension drill steels to achieve the appropriate blasthole depth. When ring drilling is used, the entire cross section of the stope is drilled with holes that radiate from the drill drive. The drilling pattern is matched to the shape of the ore body and location of the drill drift. Parallel holes are drilled when the drill drive can be silled out from the HW-FW, but this largely is constrained by the stability of the exposed working back. Two principal drill systems exist: top hammer and in-the-hole hammer. Both require long-hole rock drills equipped with extension steel in 1.2–1.8-m-long sections. Top hammer drills are more suited for narrower ore bodies (sublevel stoping), while ITH hammers are more suited for wider ore bodies (long-hole stoping). These will be discussed in a later section.

The blast layout for the individual rings will incorporate the ring number, hole number on that particular ring, the amount of explosive required (kilograms), delay interval, angle of hole to be drilled, length of hole to be drilled, and the depth of collar (stemming) to be used.

A slot raise must be developed in order to accommodate the swell of the blasted muck. It generally is developed at the extremity of the stope as shown in Figure 13.4-13, and subsequently the slot raise is enlarged FW to HW to open up the area for blasting. Generally, one cannot blast rings into a narrower void, so the slot should be located in the largest area of the stope.

Ore Handling Considerations

Ore handling in sublevel stoping involves removal of the ore at the bottom of the stope, and historically it involved track and/or slushers to remove the muck. This process is now

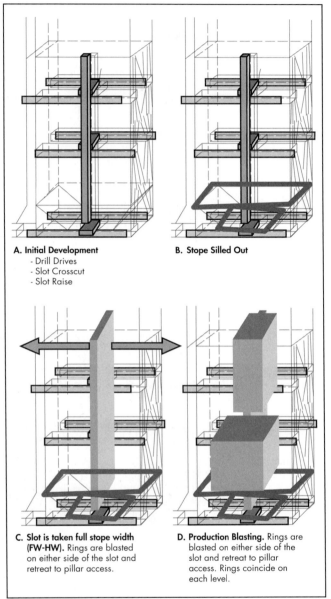

Figure 13.4-15 Sublevel open stope development

conducted largely by trackless mining equipment such as scoop trams used for drawpoint loading into mine trucks and/or orepasses as shown in Figures 13.4-3 and 13.4-4.

Sublevel Stoping

Sublevel stoping design is schematically shown in Figure 13.4-14, and the sequence of development and extraction sequencing is shown in Figure 13.4-15. The dimensions noted in the figures are typical of sublevel stoping dimensions and are employed solely to assist in the description of the method and not intended for design, as the dimensions of a stope are based on the geometry of the ore body and operational constraints.

This mining method employs sublevels located approximately 20–30 m apart. The distance between sublevels is largely governed by the length of hole that can be drilled with

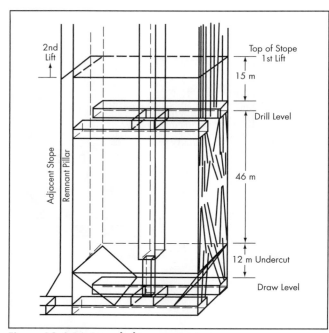

Figure 13.4-16 Long-hole open stoping

minimal drill-hole deviation (under 2%). The drill-hole diameter ranges from 50–75 mm using top hammer drills, which restricts the length of the hole to generally under 30 m with blasthole burden and toe spacing between approximately 1 × 1 m and 2 × 2 m (typically 1.2 × 1.2 m). Modern tube drills (top hammer) at 100 mm in diameter are able to drill 35–40-m-long holes.

Generally, if the stope width (HW-FW) is greater than 15 m, an FW and HW development drive as shown in Figure 13.4-14 is used; otherwise, only a center drive in the middle of the stope is developed.

The initial development is shown in Figure 13.4-15A, whereby the drill drives, slot x-cut, and raises are driven. The drill drives are comprised of the draw level, intermediate level, and the upper drill level. The undercut (Figure 13.4-15B) is silled out for a vertical height of approximately 12 m above the draw level. The height of undercut or void can be minimized through the use of programmable detonators, ensuring that sufficient void is created for the subsequent blast. The undercut serves the purpose as well to ensure breakthrough of the holes from the upper drill drive. A 2 × 2-m slot is bored/blasted to the full length of the level above the upper drill level (Figure 13.4-15B), which is subsequently slashed to 3.7 × 3.7 m for the full stope height and width from FW to HW (Figure 13.4-15C) to provide sufficient void space for the subsequent rings to be mined. Production blasting (Figure 13.4-15C) is comprised of individual rings blasting into the void for the full stope width on either side of the slot. This assumes pillar access exists on either side of the slot. Normally the production rings blasted from the intermediary level correspond to a similar set of rings on the upper drill level (Figure 13.4-15D) to ensure that a void from draw level to upper drill level exists. The geometry shown in Figure 13.4-14 employs a ring burden of 1.5 m and toe spacing of 2.1 m. The stope is normally drilled off prior to commencement of blasting, and only the holes that are scheduled for the blast are loaded. The upholes from the

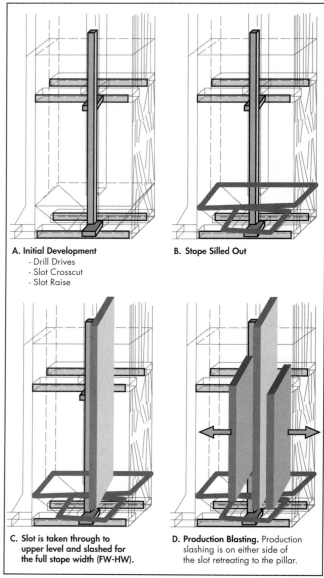

Figure 13.4-17 Long-hole open stoping development/mine sequence

intervening levels must ensure interleaf coverage of approximately 1 to 2 m. The example shown in Figure 13.4-15 uses 15-m-long drill holes with uppers and downholes, and 1–2-m interleaf coverage.

Long-Hole Stoping

Long-hole (blasthole) stoping development is shown in Figure 13.4-16 with the sequence of development and extraction shown in Figure 13.4-17. The subsequent examples given are typical of long-hole stoping dimensions and are employed solely to assist in the description of the method and not intended for design, as these dimensions change based on the geometry of the ore body and operational constraints.

Long-hole stoping largely eliminates the intermediary level with the draw and drill horizon interval governed by the length of hole that can be drilled with minimal drill-hole

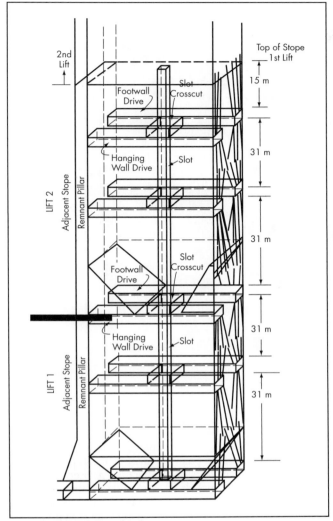

Figure 13.4-18 Piggyback stope

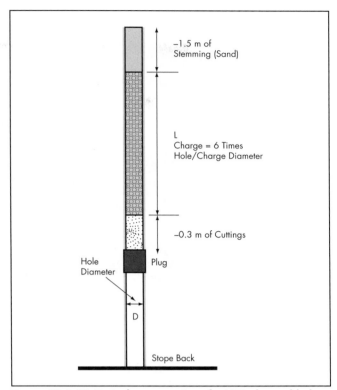

Figure 13.4-19 Typical cross section of a VCR-charged hole

150-mm-diameter blastholes. The stope is normally drilled off prior to commencement of blasting. The example shown in Figure 13.4-17 uses 46-m-long downholes and 15-m upholes.

A variation in the above sublevel and long-hole mining methods is to use nonconsolidated backfill above the upper drill drive of Lift 1 and subsequently drawing out from the level that serves as the draw horizon for the level above (Lift 2) as shown in Figure 13.4-18. This negates the need for cones in ore and consequently maximizes ore recovery as shown in Figures 13.4-15 and 13.4-17. The cones can be eliminated with the use of remote mucking equipment; however, the equipment will be traversing under potentially extended, unsupported spans (see Figure 13.4-5).

Vertical Crater Retreat

As shown in Figure 13.4-18, a VCR is a variation of long-hole open stoping where the "free face" is not a vertical slot but a "flat back" at the base of the block to be mined. Spherical charges are used to break the ore into slabs as shown in Figure 13.4-5 and have a length/diameter (L/D) ratio of 6:1. Field testing has shown that a ratio of explosive column length (L) to hole diameter (D) of 6 or less will behave similarly to a spherical charge. Blasting is carried out in horizontal slabs with only the swell being mucked at the drawpoint. This is a form of shrinkage stoping where the broken stope muck provides passive support to the stope walls. The ore is recovered at the base of the stope through drawpoints. Similar requirements and constraints to that of sublevel stoping exist except for the need for a competent HW-FW due to the option of maintaining the stope full of muck.

Development is similar to that of long-hole stoping, requiring an upper drill horizon and draw level, and it is

deviation (under 2%). The drill-hole diameter ranges from 75 to 150 mm using ITH hammer bits, thereby enabling the lengths to approach 30–60 m in length with blasthole burden and toe spacing approximately 3–4 m² (3 × 3 m). The development is as shown in Figure 13.4-17. Generally, if the stope width (HW-FW) is greater than 15 m, an FW and HW development drive is used as shown in Figure 13.4-16; otherwise, a center drive is driven in the middle of the stope.

The initial development is shown in Figure 13.4-17A, whereby the drill drives, slot crosscut, and raises are driven. The drill drives are comprised of the draw level and the upper drill level as the intermediary level has been removed. The undercut (Figure 13.4-17B) is silled out for a vertical height of ~12 m above the draw level. A 3.7 × 3.7-m slot is bored/blasted to ~12 m above the upper drill level (Figure 13.4-17B), which is subsequently slashed to 6.1 × 6.1 m for the full stope height and width from FW to HW (Figure 13.4-17C). Production blasting (Figure 13.4-17D) is comprised of individual rings blasting into the void for the full stope width on either side of the slot. This assumes pillar access exists on either side of the slot. The geometry shown in Figure 13.4-16 employs a ring burden of 3 m and toe spacing of 4.2 m with

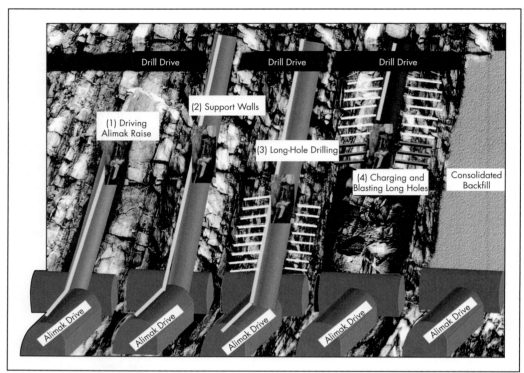
Figure 13.4-20 Alimak raise mining

generally recommended to sill out at the drill horizon to provide drill coverage for the entire block.

The vertical separation between drill and draw level is largely a function of the ore regularity and drill accuracy as detailed in general for the long-hole mining method. The dimensions are similar to that of long-hole mining where ITH drills are employed with heights ranging from 30–60 m and 75–150-mm drill diameters are used. A typical loaded blasthole for VCR is shown in Figure 13.4-19 employing a single deck charge.

Advantages of the VCR are the high productivity associated with this bulk mining method and the ability to mechanize. The ability to only muck the swell enables support to the stope walls. An advantage of this method over shrinkage is the nonentry and high mechanization associated with VCR.

Disadvantages of this method are the extensive pre-stope planning and development that is required prior to commencement of production mining, as the stope must be largely drilled off prior to bench blasting. Similar disadvantages to that of shrinkage mining exist in having the broken ore within the stope until the end of mining of the block.

VARIATIONS ON SUBLEVEL STOPING

The sublevel mining method has variations that have been implemented and will be discussed in the context of its similarity with sublevel stoping.

Vein Mining

Vein mining—also termed *Alimak mining*—has been employed within narrow vein ore bodies as detailed in the Namew Lake mine (Canada) case study (Madsen et al. 1991). Access to the ore is gained by a bored raise/Alimak such as that shown in Figure 13.4-20. The diameter of the raise is approximately 2–3 m and extends from draw level (Alimak drive) to upper drill drive as shown in Figure 13.4-20 (item 1 in the figure) and spans the length of the ultimate stope span with similar constraints as those detailed for sublevel stoping. Support may be in the form of cable bolts (item 2) in the HW providing the final wall support on stope extraction. The ore is drilled laterally by conventional drills, long-hole jumbos, or other methods and ranges from 5 m to 15 m in length to the adjacent stope (item 3) with blasting from the draw level vertically to the upper drill level (item 4). An intervening pillar may be left between stopes or the stope mined from one Alimak raise to the next depending on the geotechnical constraints. The major advantage is the ability to mine narrow ore bodies with minimal horizontal development. The vertical height of the Alimak is largely limited by operational and geotechnical constraints and reaches heights of 30 to 100 m. Blasthole sizes are generally 50–75 mm with burdens and spacing similar to that of sublevel stoping (1–2 m^2).

Transverse Open Stoping

Variations of sublevel stoping with delayed fill are shown in Figure 13.4-5. This mining method is largely used for stope widths in excess of 20 to 30 m or as dictated by geotechnical stable back spans; otherwise, conventional longitudinal or strike mining is used. Figure 13.4-5 shows the objective is to recover the secondary pillars between the primary stoping blocks, which can be excavated by sublevel stoping (general) and subsequently filled with consolidated fill that can be comprised of hydraulic fill, paste, or cemented rock fill. Mining of the secondaries occurs after curing of the primaries to a strength that is able to withstand minimal dilution. Generally, the binder content ratio is 30:1 to 20:1 (fill to cement by volume). Alternatively, a permanent pillar is left behind to confine

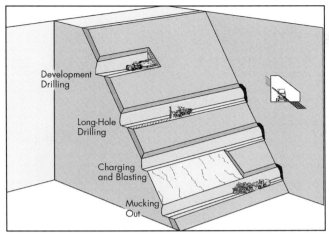

Source: Hamrin 2001.
Figure 13.4-21 Longitudinal mining without fill

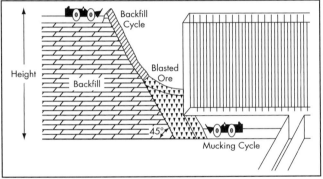

Source: Caceres 2005.
Figure 13.4-22 Longitudinal mining with fill (Avoca)

the unconsolidated fill with only primaries excavated along the strike. With this variation the secondaries are narrow pillars left behind (approximately 3–5 m). A disadvantage of this method is its inability to follow the variations of an irregular hanging wall dip.

Longitudinal Mining
Figure 13.4-21 shows sublevel extraction employing mucking along the strike (retreat). This is a variation of conventional FW drawpoints as shown in Figure 13.4-21. The stopes with no fill are as shown in Figure 13.4-21 and with delayed fill are as shown in Figure 13.4-22. The delayed fill method of longitudinal mining is also referred to as Avoca mining. Having longitudinal mucking access requires that remote load-haul-dump (LHD) equipment be used. This method is also referred to as sublevel benching.

CONCLUSION
Sublevel stoping accounts for more than 60% of all underground production in North America. This is largely due to the developments of extension steels, hollow tube and special long-hole rock drills, and ITH drilling techniques requiring less development and greater production capacities. Several variations exist; however, characteristic to this method is the development from a top drill drive and removal of muck from a draw level below for a steeply dipping stope. The variations of the method are selected to suit the ground conditions and operational requirements of the mine.

An essential part of sublevel stoping is the stope extraction sequence. The extraction sequence is governed by the development, rock mechanics, tonnage requirements, and, if applicable, fill cycle.

Sublevel stoping is a safer mining method because the operator is never within the stope under the unsupported back. Further, the mining method works on a retreat pattern where the equipment and operator work under a supported back. This mining method is suitable to modern hauling equipment including the use of remote LHD units where the operator is removed from any potential hazard associated with the stope.

An important safety consideration with open stoping is to ensure that drawpoints remain full above the brow of the stope. Adhering to this safety standard largely eliminates the risk of potential air blast due to hanging wall collapse.

The main advantage of sublevel stoping is the efficiency associated with drilling, blasting, and loading operations as they can be performed independently from each other. A high potential exists for mechanization with moderate to high productivities of more than 25 t per worker-shift.

The main disadvantage is the complicated and comprehensive development that is needed and the requirement for regular tabular ore geometries.

REFERENCES
Caceres, C. 2005. Effect of backfill on longhole open stoping. M.A.Sc thesis, University of British Columbia.

Clark, L., and Pakalnis, R. 1997. An empirical design approach for estimating unplanned dilution from open stope hangingwalls and footwalls. Presented at the 99th Annual General Meeting of the Canadian Institute of Mining, Metallurgy and Petroleum, Vancouver.

Hamrin, H. 2001. Underground mining methods and applications. In *Underground Mining Methods: Engineering Fundamentals and International Case Studies*. Edited by W.A. Hustrulid and R.L. Bullock. Littleton, CO: SME.

Haycocks, C., and Aelick, R.C. 1998. Sublevel stoping. In *SME Mining Engineering Handbook*. Edited by H.L. Hartman. Littleton, CO: SME.

Madsen, D., Moss, A., Salamondra, B., and Etienne, D. 1991. Stope development for raise mining at the Namew Lake mine. *CIM Bull.* 84:33–39.

Pakalnis, R. 2002. Empirical Design Methods—UBC Geomechanics. Presented at NARMS–TAC 2002, Toronto, July.

Scoble, M.J., and Moss, A. 1994. Dilution in underground bulk mining: Implications for production management. In *Mineral Resource Evaluation II: Methods and Case Histories*. Special Publication No. 79. London: Geological Society. pp. 95–108.

CHAPTER 13.5

Cut-and-Fill Mining

George Stephan

DEFINING CHARACTERISTICS

Cut-and-fill is the broad descriptive term applied to mining methods requiring that excavated voids be filled with barren material to facilitate the continuation of ore production. This fill is required to provide support for subsequent openings or to provide a working platform for further mining. Cut-and-fill methods may be applied in conjunction with other conventional methods, such as blasthole stoping, or they may be methods that stand alone, such as drift-and-fill.

Classic cut-and-fill mining involves the successive mining of horizontal or inclined slices upward through a relatively narrow and subvertical tabular ore body. This is followed by the placement of uncemented waste rock or hydraulically placed sand fill to create a new, higher working level to gain access to ore above and to support the ground below. Raises maintained up through the fill (and occasionally down from upper levels) provide for access, ventilation, ore removal, and the drainage of water from the fill material.

Modern variants may use paste fill and typically use ramp access to allow the use of mechanized mining equipment. These techniques preclude the need for raises.

None of the cut-and-fill methods described herein can match the productivity or the low mining cost of mass mining methods. Cut-and-fill methods are chosen because they are more economically attractive than other extraction methods available for the same situation.

FAVORABLE CONDITIONS

Cut-and-fill methods are applicable to a wide variety of situations, including the following:

- Ore zones are irregular in shape and orientation.
- Ore grade is high and dilution control is critical.
- The precise contacts between ore and waste are structurally critical but not readily visible.
- The waste rock is weak.
- Ore zones are large, but their rock quality is weak.
- Localized underground stability is required.
- Surface disturbance must be minimized.
- The value of the ore makes recovery of support pillars economically viable.
- The reduction of surface waste storage is important.
- The need exists to advance a working platform for the upward mining of the ore body.

Cut-and-fill methods are favored when the ore value is relatively high and ore recovery rate with minimal dilution cannot be satisfactorily accomplished by open stope mining or caving methods. If the openings are small enough and the ground conditions are competent, irregular ore boundaries can be mined with open stope methods without backfill. However, when the openings are larger and the ground conditions are less favorable, backfilling becomes necessary for safe economic production.

Fill methods are useful when regional stability is needed. This is the case where large empty stopes could adversely affect the stability of permanent mine access drifts, shafts, and raises. These stresses could also affect new mine production stopes and surface infrastructure.

There is also the advantage that fill can provide a relatively inexpensive and convenient way to advance a working platform to maintain access to the ore for drilling, blasting, and mucking. Another advantage is that ventilation air can be managed more efficiently because vacant and unused mine openings are closed by backfill.

LIMITATIONS

These factors limit the applicability of cut-and-fill:

- The availability of a sufficient quantity of a suitable fill material type
- The cost of binding agents (if required)
- Production, preparation, transportation, and placement cost of the fill material
- Storage and reclamation facilities to match the mining cycle demand
- Congestion and interruption of production mining activities

---

George Stephan, Senior Consulting Engineer, Stantec Mining, Tempe, Arizona, USA

The principal drawback for the cut-and-fill methods is the additional cost of producing, preparation, transporting, and placing the backfill.

Not all readily available fill materials are appropriate; the backfill material must match its intended use. The material that is most readily available and least expensive may not have characteristics that would allow its use as fill material. The material most readily available also may not be usable in the stope designs dictated by the size and shape of the ore body. Further, the least expensive delivery system may not be practical.

Assumptions about fill availability should be verified early in mine planning. Suitable aggregate for cemented rock fill may not be available near the mine site. Both sand fill and paste fill generally make use of tailings from an on-site mill. For sand fill, however, sufficient coarse fraction may not be available within the tailings, either because of the fineness of the grind or because of the need to supply coarse fraction for tailings dam construction. Conversely, the grind may be too coarse to provide sufficient fines for paste fill.

In addition to the added cost, backfilling operations can result in congestion, maintenance, and cleanup problems underground. Cut-and-fill imposes limitations on the sequence and timing of production that, if not carefully controlled, may take precedence over the maximization of production grade. Usually, conventional ground support in the form of rock bolts, wire mesh, and shotcrete still will be required to temporarily support the stope openings during production mining.

The cost of producing and placing fill requires consideration of the mining method trade-offs. Unless the deposit simply cannot be mined by any nonfill method, the additional cost and the advantages of using fill must result in additional ore production (increased recovery), reduced dilution, and/or cash cost reductions sufficient to justify the additional cost of the fill and its preparation and placement.

TYPES OF CUT-AND-FILL METHODS

There are numerous cut-and-fill methods and many hybrid combinations. The following is a basic set of descriptors:

- Vertical progression of the mining
  - Underhand (mining beneath the backfill)
  - Overhand (mining on top of the backfill)
- Mining method
  - Drift-and-fill
  - Postpillar
  - Bench-and-fill
  - Blasthole stoping
  - Uphole, flat-back slicing, or back stoping
- Type of fill
  - Cemented rock fill
  - Uncemented rock fill
  - Sand fill (hydraulic)
  - Paste fill
- Timing
  - Concurrent (placed during the cycle of stope advance)
  - Delayed (placed after the excavation of the stope is complete)

Cut-and-fill mining is extremely flexible, and various combinations of mining methods and fill materials can be used to cope with specific mining situations. However, some combinations are not feasible, such as underhand drift-and-fill with uncemented rock fill. Further discussion regarding possible combinations can be found in the "Cemented Fill Stopes" and "Uncemented Fill Stopes" sections.

METHOD SELECTION CRITERIA

Ore-Body Morphology

The size, shape, orientation, and degree of boundary uniformity of the ore body are the first considerations in selecting a cut-and-fill method. The use of fill for support and for provision of the necessary platform to access the ore provides the operator with great flexibility for adjusting the excavation to irregularities along the boundaries of the ore body. Alternative methods (that do not use fill) may result in unsafe working conditions, excessive dilution, or insufficient recovery. Cut-and-fill methods can be adapted to flat, vertical, undulating, or even en echelon zones and to situations in which the ore-body shape and attitude change quickly along the strike or dip.

Access, Infrastructure, and Development Cost

Typically, overall project cost for access and infrastructure to support cut-and-fill mining is less than the cost for bulk mining methods like block caving. Overall project cost is greater, however, than the cost for room-and-pillar or open stoping.

The use of fill for support, rather than leaving unmined pillars of ore, results in greater recovery of ore because fewer unrecoverable ore pillars are left behind at the conclusion of mine operations. In methods such as shrinkage stoping, much of the broken ore is left (albeit temporarily) in the stope to provide wall-rock support and a working platform from which to mine the whole block of ore. Although the shrinkage ore is withdrawn at the completion of the stope, the use of fill as a working platform and wall-rock support results in a faster return on the cost of mining.

Amenability to Mechanization

The mechanization of cut-and-fill mining has two phases: mechanization of the ore extraction (i.e., the cut) and mechanization of the fill process. The mechanization potential for ore extraction with cut-and-fill methods varies widely. The ground conditions that dictate the use of cut-and-fill mining may also limit drift dimensions or, alternatively, may require flexibility to adapt to varying stope conditions. This reduces the size of trucks and loaders that can be used. Options associated with production and delivery of fill are covered in Chapter 13.6. By itself, the use of fill seldom enables an increase in the level of mechanization.

Dilution

Cut-and-fill methods increase the extractable yield of ore bodies and reduce dilution from boundary overbreak. However, the effect of dilution from fill must be considered in any economic analysis. Fill dilution generally carries no recoverable value, whereas overbreak into country rock may carry at least some value because of its proximity to the ore. Dilution from fill can come from stope ribs where the adjacent stopes have been previously backfilled. If the stope in production is above a previously filled stope, the mucking operations will typically remove a layer of fill material from the floor to clean up as much ore as possible. Less fill will be taken if the floor surface is cemented fill. Conversely, ore can be lost in the floor of a stope if ore is used to build a roadway on top of the fill. Fine ore, which usually contains a significant portion of value, can be lost into the fill from the cut below if the fill is not well consolidated at its placement.

The dilution potential of various fill materials can differ significantly. Uncemented rock and sand fill typically dilute at the highest rates. Dilution is usually less with cemented rock, sand, and paste fills. However, dilution from cementing agents may have an adverse effect in the metallurgical plant. This difference in behavior must be included when evaluating the relative costs of using these fill materials in the mining operation.

Geotechnical Constraints

The in-situ geotechnical environment—including the rock quality of the ore and country rock, geologic structures, ore-body depth, water, and regional stresses—determines the size and shape of the stopes that can be excavated before placing fill. The ground support required to provide a safe working environment is also dictated by the geotechnical situation. Cut-and-fill methods frequently rely on the initial mining of stopes in virgin ore zones. These stopes are called *primary stopes*. They are subsequently backfilled to provide support for the mining of adjacent stopes, which are called *secondary stopes*. The size, shape, and mining sequence of the primary and secondary stopes are dictated by the geotechnical conditions of the ore and country rock and the mechanical properties of the planned fill material. Sometimes it is advantageous to place higher-strength, higher-cost fill in a primary stope and lower-strength, lower-cost fill in a secondary stope. The flexibility to control and differentiate fill quality and properties according to the application is crucial to mine operation, safety, and cost control.

Sill pillars and crown pillars are used to separate stoping zones vertically. The dimensions of these pillars are determined by the geotechnical environment and the stresses induced by mining. If these pillars contain valuable ore, it may be desirable to place highly engineered fill to facilitate the safe recovery of this ore. The objectives are to use cut-and-fill mine planning to maximize the extraction of the ore body and to avoid sterilizing any portion such that it cannot be safely recovered.

Availability of Fill Material

The availability of fill material (rock, sand, or paste) affects the selection of cut-and-fill methods. The quantity, engineering properties, and cost must be matched to the planned use. For example, if the on-site mill tailings are unsuitable, a nearby source of alternative material must be found. Similarly, if the price for cement is unreasonably high, the cost and availability of the various fill alternatives must be weighed against the engineering requirements of the selected cut-and-fill method. The safest and least expensive combination of fill material and cut-and-fill method must be found and continually reviewed.

The use of tailings for fill is often dependent on mill throughput. An unfortunate side effect of using mill tails, particularly in mines with numerous smaller stopes, is that when stopes stop producing to allow for fill placement, they must rely on continued production from other stopes to provide fill material. The design of the mine must include satisfactory storage capacity for fill because it is usually accumulated at a slow, relatively steady rate—typically as a by-product of the milling operation. When fill is required for placement, it must be delivered at a much greater rate, stope by stope. Careful planning is therefore critical to ensure that a balanced mining/filling cycle is maintained throughout the operation.

Failure to maintain a balanced cycle can result in fill shortfalls, leading eventually to loss of production until a proper cycle can be reestablished.

CEMENTED FILL STOPES

The inclusion of a binding agent such as cement in the fill is required when the fill must have sufficient strength to provide support for the excavation of stopes located above, below, or adjacent to the stope being mined. The term *delayed* is used to indicate that extraction of all ore in the stope has been completed before the stope is backfilled.

Occasionally, when the value of the ore in the pillars of a conventional room-and-pillar mine is considered high enough, cemented fill is placed around the pillars to permit recovery of the ore. Placement of cemented fill is also sometimes used to facilitate the later extraction of ore from sill pillars where the fill material is normally uncemented.

The use of hydraulic and cemented fill can cause significant problems with level cleanup, sumps, settlers, and pumps. This is usually worse with cut-and-fill stoping than with sublevel open stoping because there are many more points of drainage to manage.

Blasthole Stoping with Delayed Backfill

Blasthole stoping with delayed backfill is a method that can be applied to ore bodies that are generally tabular and vertical or near vertical. The rock quality of both the ore and the waste needs to be fair to good. Cemented fill for the primary stopes is required when a primary/secondary plan is pursued. Cemented fill allows secondary stopes to be mined subsequently. The secondary stopes may be backfilled partially or completely with uncemented fill.

Longitudinal Blasthole Stoping with Delayed Backfill

Where the ore zone is narrow and the long axes of the stopes are parallel to the strike, the stope orientation is called *longitudinal*. Occasionally the ore is contained in en echelon lenses: narrow parallel zones of ore separated by narrow parallel zones of waste. These conditions require sequences of longitudinal stopes followed by backfilling to prevent stability problems and dilution from the collapse of the waste material between ore lenses (see Figure 13.5-1). The width of the stopes is generally the distance between hanging wall and footwall. The maximum length is generally limited by geotechnical considerations. To minimize dilution from overbreak or slough when the hanging wall is weak, the installation of mechanical support (bolts, wire, shotcrete, or cable bolts) may be required.

Transverse Blasthole Stoping with Delayed Backfill

Where the ore zone has sufficient width, the long-axis orientation of the stopes may be perpendicular to the strike. These are *transverse stopes*. Geotechnical considerations dictate the transverse orientation as well as the stope dimensions. The sequence of lateral and vertical, primary and secondary stope mining, backfilling, and curing requires careful planning (see Figure 13.5-2). After mining begins, the plan seldom permits significant changes; it becomes difficult to adjust the production rate and grade.

Drift-and-Fill

Drift-and-fill is the term applied to projects in which the ore is extracted by excavating parallel drifts through the ore zone and backfilling them before mining the adjacent drift. The drifts are

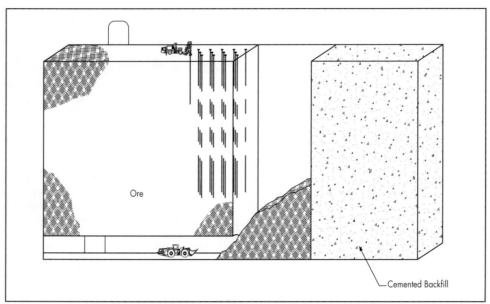

Figure 13.5-1 Typical longitudinal blasthole stoping with delayed backfill

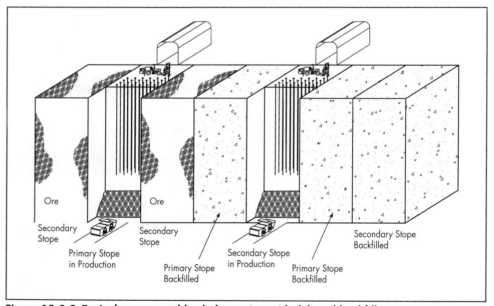

Figure 13.5-2 Typical transverse blasthole stoping with delayed backfill

usually alternately filled with cemented and uncemented rock fill. The orientation of the drifts mined on a level can be longitudinal, transverse, or diagonal to the strike of the ore body. For additional strength, the orientation is sometimes alternated between levels. Excavation of the ore by drifting is less productive and more costly than excavation by blasthole stoping. Therefore, drift-and-fill is usually employed where ground conditions are too weak and the ore boundaries are too irregular for blasthole stoping. Because of its higher associated operating costs, drift-and-fill is best used where the ore grade is high.

Overhand Drift-and-Fill
Overhand drift-and-fill is a bottom-up mining sequence (see Figure 13.5-3). This method is useful where the ore occurs in relatively thin, shallow dipping lenses with weak back or wall rock. The bottom-up sequence may also be desirable because of production grade considerations. Ideally, this method would be used where the rock quality of the ore in the back is not sufficiently high to permit postpillar stoping. This method permits backfilling of secondary stope drifts with uncemented fill, thus reducing costs. The first or lowest cut requires cemented fill in both primary and secondary drifts if mining blocks from below are scheduled to mine up to a previously mined block.

Underhand Drift-and-Fill
Underhand drift-and-fill is a top-down sequence (see Figure 13.5-4). The use of underhand drift-and-fill stoping

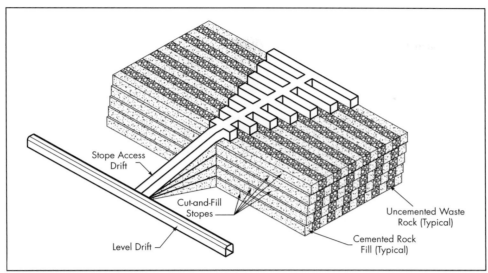

Figure 13.5-3 Typical overhand drift-and-fill

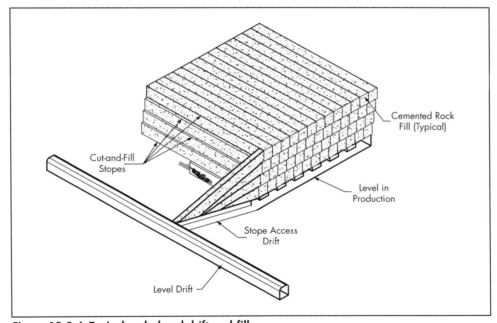

Figure 13.5-4 Typical underhand drift-and-fill

fills the critical need for a method that facilitates the mining of high-grade deposits with weak ore and weak country rock. This method results in the creation of a strong crown pillar of cemented fill beneath which subsequent mining can safely proceed. The initial level may be opened using small drifts that are easily supported. After the upper level is completed and backfilled, succeeding levels below can be opened to more efficient mining widths. Orienting the drifts on each level such that they are not parallel to the overhead drifts reduces the risk of ground falls. The fill material must be carefully engineered and the samples tested regularly. The method may involve the use of mats or tension cables to support and distribute the load above each open drift. The ore zone should be vertical or near vertical to maximize the benefit of employing this method.

UNCEMENTED FILL STOPES

The use of uncemented fill has the advantage of eliminating the cost for cement. Uncemented fill has limitations when used as the primary backfilling material because it lacks the strength of cemented fill. Therefore, it is seldom used where the mining of adjacent stopes or stopes below is planned. The use of uncemented fill dictates a bottom-up mining sequence or the use of substantial sill pillars between levels.

Usually, stope mining with uncemented rock fill has the advantage of being able to use any sort of readily available and inexpensive fill material. Frequently, waste muck from development mining can be used. This relieves the expense of transporting thve waste to the surface for disposal. Alternatively, the rock can be quarried on the surface and delivered to the

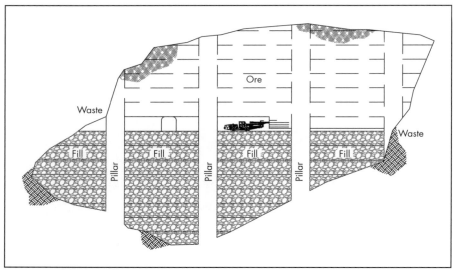

Figure 13.5-5 Postpillar mining

underground stopes via borehole or waste pass. Because of its coarse texture, rock fill is often capped with sand fill or cemented sand fill to provide a mucking surface that is less susceptible to ore loss and dilution.

Postpillar Mining

In horizontal or near-horizontal ore bodies of substantial thickness and width, the placement of uncemented fill serves the primary purpose of providing access to excavate successive slices of ore in a bottom-up sequence. The postpillar method can be described as a room-and-pillar excavation where backfill (usually uncemented rock fill) is placed on the floor so that successive layers can be mined upward through the ore zone (see Figure 13.5-5). The pillars, or at least the upper portions of the pillars, can be preserved and extended upward to continue supporting the back. The pillars can be much taller and more slender when they are surrounded by fill, and so recovery is higher than for nonfill room-and-pillar methods.

Vertical or Near-Vertical Cut-and-Fill Stoping (Conventional or Mechanized)

The use of uncemented fill in narrow-vein stopes that are inclined or vertical provides both stability and a working platform for the upward advance of the excavation of the ore. Hydraulic (sand) fill or rock fill with sand is superior to rock fill alone for this application because greater density is achieved. Frequently, a top layer of cemented fill is used to provide a stronger surface for machine operation and to minimize ore loss and dilution.

Using sand (usually coarse tailings) to backfill stopes was one of the earliest fill applications in mining. It was often used in conjunction with timber support, but the use of timber is now largely supplanted by bolts and shotcrete. The transport, distribution, and placement of fill are as hydraulic slurry via a pipe system. Less frequently, compressed air is used for the transport and pneumatic stowing of the fill. The pipe system prevents congestion of the haulageways and eliminates the need for trucks and loaders. With hydraulic fill, drainage of the excess water must be accommodated with designed filters, drainage raises, sumps, and pumping systems. These features impose substantial operational costs and scheduling limitations. Pneumatic fill systems have increased ventilation requirements and pipe wear as well as the physical limits associated with distribution and the potential health risks of increased airborne dust.

In conventional cut-and-fill stopes, access to the level of the stope cut was gained by timbered or steel-tubbed raises that were maintained and extended up through the backfill from the haulage level (see Figure 13.5-6). In the stope, drilling was accomplished with hand-operated stoper drills for vertical holes or jacklegs for horizontal holes where the cut would be taken by breasting down. Slushers were the principal means for mucking the blasted ore across the stope to the muck raise. From the bottom of the muck raise, chutes were used to transfer ore to muck cars for transport to the shaft by rail. In addition to ore transport, the raises provided manway access, hoisting of supplies, pipe and power reticulation, and drainage of the backfill water.

In the conventional system, it was difficult to bring equipment into or out of the stope; after a piece of equipment had been installed in a stope it remained isolated there, rising with the level of the cut. Low unit cost, light weight, simplicity, and durability were important equipment characteristics; slushers fit this description well. However, they were not as productive for mucking as low-profile LHDs (load-haul-dump units) with buckets.

Overshot muckers powered by compressed air were used at some mines to replace slushers. Overshot muckers are relatively simple to maintain, and ventilation requirements are no greater than those required for slushers.

Eventually, diesel-powered LHDs were placed into service (Figure 13.5-7). With their introduction, ventilation requirements increased and maintenance requirements demanded better access than raises could provide. The cost of LHDs forced the need to achieve higher utilization rates, which in turn made it necessary to allow stope-to-stope movement of LHDs. This change led to the shift from access by raises to access by ramp systems.

Resuing

Resuing is the term applied to the mining method whereby high-grade ore from very narrow veins is extracted selectively before stopes are widened in waste rock to obtain the additional

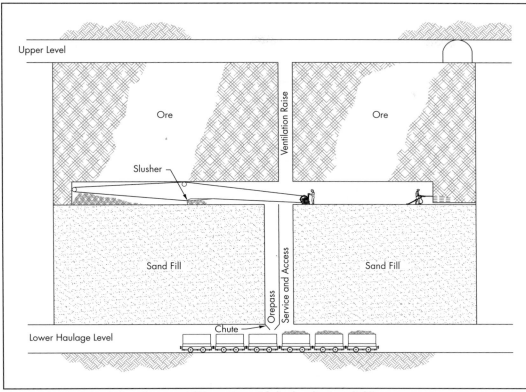

Figure 13.5-6 Conventional cut-and-fill stoping with hydraulic (sand) fill

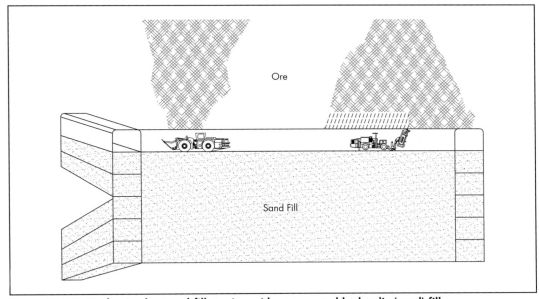

Figure 13.5-7 Mechanized cut-and-fill stoping with uncemented hydraulic (sand) fill

width required for equipment to operate (see Figure 13.5-8). Resuing requires a working platform of fill.

Bench-and-Fill

Bench-and-fill stoping is applied to vertical and near-vertical stopes that have sufficient width, length, and competency of wall rock to permit a system of laterally retreating blastholes followed by advancing waste rock fill (see Figure 13.5-9). This method allows variations in placement of the fill. The amount of waste rock exposed between the blasting and mucking of ore, and the advancing face of waste fill, can be adjusted to suit local conditions. In some cases, the waste fill is advanced such that all open space is completely eliminated. This lack of open space enhances support for weak hanging-wall and

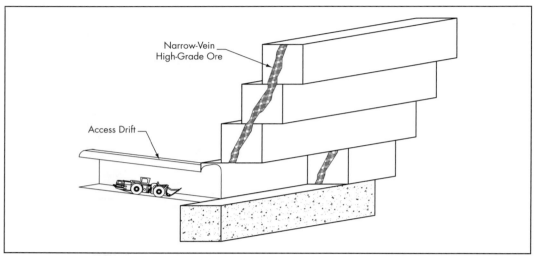

Figure 13.5-8 Resuing

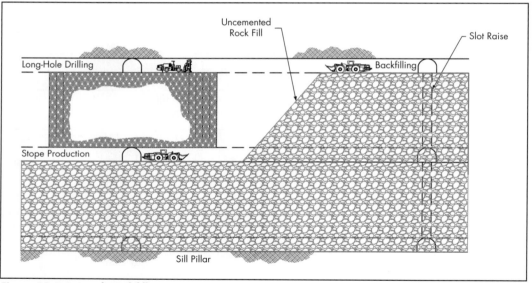

Figure 13.5-9 Bench-and-fill

footwall conditions, thus reducing dilution and ore loss. The use of some cemented fill to augment the uncemented fill can also increase recovery.

EXAMPLE MINES

The mines referenced in this section are using, plan to use, or have used the cut-and-fill methods described herein. Many of these operations use more than one method, and in many cases their specific working methods have evolved over time. For reference and research purposes, the lists include both currently operating and nonoperating mines.

Mines Using Cemented Fill

This group includes mines using longitudinal blasthole stoping with delayed cemented fill, overhand drift-and-fill with delayed cemented fill, overhand drift-and-fill with cemented fill, and underhand drift-and-fill with cemented fill:

- Longitudinal blasthole stoping with delayed cemented fill
  - Garpenberg—Bergslagen, Sweden
  - Zinkgruvan—Bergslagen, Sweden
  - Julietta—Far East Russia
  - K2—Republic, Washington, United States (closed)
  - Kensington—Juneau, Alaska, United States (planned)
- Transverse blasthole stoping with delayed cemented fill
  - Big Gossan—Papua, Indonesia (planned)
  - Cannon—Wenatchee, Washington, United States (closed)
  - Diavik—Northwest Territories, Canada (planned)
  - Greens Creek—Admiralty Island, Alaska, United States

- Kidd Creek—Timmins, Ontario, Canada
- Meikle—Carlin, Nevada, United States
- Mt. Isa—Queensland, Australia
- Overhand drift-and-fill with cemented fill
  - American Girl—Yuma, Arizona, United States (closed)
  - Golden Promise—Republic, Washington, United States (closed)
  - Rosebud—Lovelock, Nevada, United States (closed)
  - Strathcona deep copper mine—Levack, Sudbury, Ontario, Canada (closed)
- Underhand drift-and-fill with cemented fill
  - Apex—St. George, Utah, United States (closed)
  - Bullfrog—Beatty, Nevada, United States (closed)
  - Carlin East—Carlin, Nevada, United States
  - Deep Post—Carlin, Nevada, United States
  - Getchell/Turquoise Ridge—Golconda, Nevada, United States
  - Jerritt Canyon—Mountain City, Nevada, United States
  - Lucky Friday—Mullan, Idaho, United States
  - Rain—Carlin, Nevada, United States
  - Stratoni—Stratoni, Greece
  - Val Chisone—Italy

Mines Using Uncemented Fill

This group includes mines using vertical or near-vertical stoping with uncemented sand fill (hydraulic fill), resuing, postpillar mining with uncemented fill, and bench-and-fill stoping with uncemented fill:

- Vertical or near-vertical stoping with uncemented sand fill (hydraulic fill)
  - Bunker Hill—Kellogg, Idaho, United States (closed)
  - Galena—Wallace, Idaho, United States
  - Groundhog—Vanadium, New Mexico, United States
  - Homestake—Lead, South Dakota, United States (closed)
  - Star—Burke, Idaho, United States (closed)
  - Stillwater—Nye and Big Timber, Montana, United States
  - Sunshine—Kellogg, Idaho, United States (closed)
- Resuing
  - Golden Patricia—Kenora District, Ontario, Canada (closed)
  - Julietta—Far East Russia
- Postpillar mining with uncemented fill
  - Black Cloud—Leadville, Colorado, United States (closed)
  - Christmas mine—Winkelman, Arizona, United States (closed)
  - El Mochito—Las Vegas, Honduras
  - Mantos Blancos—Antofagasta, Chile
  - Milpillas—Sonora, Mexico
  - Rey de Plata—Guerrero, Mexico (closed)
  - Tizapa—Guerrero, Mexico
  - Madero—Zacatecas, Mexico
  - King Island Scheelite—Tasmania, Australia (closed)
- Bench-and-fill stoping with uncemented fill
  - El Peñón—Antofagasta, Chile
  - Ken Snyder—Midas, Nevada, United States (modified)
  - Mt. Isa lead mine—Queensland, Australia
  - Stillwater—Nye and Big Timber, Montana, United States

CURRENT TRENDS

Stope Mechanization

The widespread introduction of mobile equipment (diesel-powered loaders, trucks, drills, and service vehicles) since the 1960s and 1970s has largely replaced the use of jacklegs, stopers, and slushers, the primary underground mining machinery of the prior era. The maintenance and maximum usage of this mobile equipment precludes allowing it to become captive in a single stope. Therefore, access to stope levels is now almost exclusively by ramps rather than by raises.

The demand for ventilation has also been increased because of the increased usage of diesel equipment.

In recent years, production with remote controlled loader operation has been replacing the more costly development of gathering bells and chutes that have been commonly used in blasthole stope mines.

Paste Fill

Mining fill systems based on the use of paste for backfill are supplanting other fill systems for bulk backfill applications. The opportunity to transport and place the material by pipe reticulation substantially decreases congestion in drifts and ramps. The placement rate of paste fill is much faster than placement using trucks carrying fill from a surface or underground batch plant and then having to rely on LHDs and jammers to place the fill into the stope.

The resulting benefits (lower operating costs, lower ventilation demands, and a faster turnaround time for the filled stope) are similar to those achieved by hydraulic systems, but without the water drainage issues. However, paste plants are costly to construct and must be completed during preproduction development.

Cemented Rock Fill

The use of cemented rock fill continues to be a common choice where bulk material handling is not advantageous and where flexibility is important to maintaining operations.

CHAPTER 13.6

Backfill Mining

Douglas F. Hambley

INTRODUCTION
This chapter discusses the design of systems for backfilling underground mines, opening with a short history of backfilling followed by a listing of the types of backfilling systems to be covered in the chapter, and closing with a discussion of sources of further information. Stoping methods that use backfill are discussed in other chapters, especially Chapter 13.5.

Backfilling of mined openings may be performed for a number of reasons. The first is to provide support for walls of an opening in relatively weak rock so that ore immediately above, below, or adjacent to the opening can be mined. Another possible reason for backfilling is to contribute to the support of the roof and walls of a wide opening. In this case, the backfill will not prevent failure in the roof, but it will minimize the displacement of the broken rock and thereby minimize the potential for progressive upward failure, possibly reaching to the surface. This leads to another reason for backfilling, namely minimization of surface subsidence. A final important reason for backfilling is isolation and disposal of mining/milling wastes considered to be unattractive or an environmental hazard on the surface.

Unless otherwise noted, equations in this chapter are given in consistent units. The reader is expected to include units when performing calculations and provide any necessary conversion factors.

BRIEF HISTORY OF BACKFILLING PRACTICES
The backfilling of mined openings has a long history. It is only since about the 1980s that archeologists have recognized that backfilled bell pits in chalk in Europe that date from 2700 BC were actually mines for flint. The workings extended outward from the base of a shaft in all directions without support until it became unsafe, at which time the workings and shaft were backfilled to surface and another pit begun alongside. This method of mining apparently remained in use in near-surface ironstone and coal mines in the United Kingdom until the early 19th century (Shepherd 1980).

Stowing of wastes in workings to allow the mining of adjacent areas was a common practice in the metal mines in Cornwall (United Kingdom) and the coal mines in Belgium in the mid-19th century. Wastes generally consisted of barren or low-grade rock mined as part of the stoping process or in other areas of the mine away from the ore body or seam. However, if wastes obtained in this manner were insufficient, other materials such as sand or smelter slag could be brought into the mine to act as backfill (Callon 1876).

In the Kolar Gold Fields in India, quarried granite blocks were used as backfill/ground support in overhand stopes until the 1940s. The granite blocks were hand fitted to completely span the width of the stope. The advantage of this was that because the deformation modulus is similar to that of the wall rock, significant amounts of stress could be transferred to the blocks from the wall rock, thereby reducing the potential for rock bursts, which were a severe problem in those deep mines. The disadvantage of this method is that it is only feasible where labor is extremely cheap.

Hydraulic backfilling (fill placed hydraulically) has been used in the coal mines in Pennsylvania (United States) since 1884 (Peele 1941). With regard to metal mines, hydraulic placement of sand-sized tailings appears to have been first practiced in the gold mines on the Witwatersrand in South Africa in 1910, its initial adoption in North America occurring in Butte, Montana (United States) in 1917. As in Pennsylvania, the initial use at Butte was to seal off a mine fire in an area of broken ground where other sealing methods were impractical (Rahilly 1923). General rules were developed early from experience concerning the pipe velocities necessary to convey tailings hydraulically without having the pipes "sand out" (fill up from particle sedimentation due to insufficient flow velocity); however, criteria for flow and identification of the different flow regimes that operate with slurries have only been developed since the early 1950s beginning with Durand and his co-workers at SOGREAH (SOciete GRenobloise d'Etudes et d'Application Hydrauliques) in France and others working independently in the United Kingdom (Abulnaga 2002).

In the late 1950s, the International Nickel Company (INCO) of Canada (now Vale Inco) and Falconbridge Nickel Mines Ltd. (now Xstrata) began experimenting with the addition of portland cement to hydraulic backfill. The purpose was to produce a stronger fill that could be used for support in

---

Douglas F. Hambley, Associate, Agapito Associates, Inc., Golden, Colorado, USA

pillar recovery operations and could provide competent mucking floors, thereby eliminating the need for timber floors. In 1961, with financial support from the two aforementioned mining companies, Canada Cement Lafarge Ltd. began a comprehensive study to develop cemented fill technology that was completed in 1970 with the publication of their findings (Weaver and Luka 1970).

Pneumatic stowing was introduced in Polish coal mines in the 1920s and soon afterward adopted in other coal mining areas of Europe. In North America, the use of pneumatic stowing was promoted by the U.S. Bureau of Mines in the 1960s and 1970s but has not generally been adopted. The method was tested at the Sullivan mine in British Columbia (Canada) in the late-1960s and later used in uranium mines in Grants, New Mexico (United States), and in gold mines in South Africa.

The concept of thickened tailings, developed by Professor Robinsky of the University of Toronto and first applied at the Kidd Creek mine in Timmins, Ontario (Canada), in the mid-1970s, was extended by the late Dave Landriault and others at INCO Metals in the 1980s and later at Golder Paste Technology Ltd. to become paste fill when used for backfill underground. In 1975, pump-pack was developed by the British using broken coal and bentonite, which was a thixotropic material and pumped like a toothpaste but set like a low-strength concrete (Schmidt et al. 1976). Paste fill was developed independently in Germany at the Bad Grund mine in the early 1980s (Jewell 2006).

"Flowable fill" has been used in recent years by, among others, the Office of Surface Mining Reclamation and Enforcement (OSMRE) of the U.S. Department of the Interior to remotely backfill near-surface abandoned mines—especially coal mines—that represent a subsidence hazard. Flowable fill refers to a cementitious slurry consisting of a mixture of fine aggregate or filler, water, and cementitious material(s) that is used primarily as a backfill in lieu of compacted earth. This mixture is capable of filling all voids in irregular excavations and hard-to-reach places (such as under and around pipes), is self-leveling, and hardens in a matter of a few hours without the need for compaction in layers.

TYPES OF BACKFILL
As indicated in the history section, the different materials that have been used to backfill mine openings include the following:

- Dry sand and rock fill
- Uncemented hydraulic backfill
- Cemented hydraulic backfill
- Cemented rock fill
- Paste fill
- Pneumatic fill
- Flowable fill

In subsequent sections of this chapter, each of these alternatives will be discussed. For each, the method of placement, drainage requirements, if any, equipment requirements, pipe wear, and other relevant design and operating topics are considered.

Dry Sand and Rock Fill
Dry placement of stope fill materials is often the simplest method of fill placement and has a long history. In metal mines in North America and Australia that used cut-and-fill or square-set stoping, this method dominated until the 1950s and was still taught as a preferred method of backfilling in some mining schools until the late 1960s.

Advantages and Disadvantages
Although it has generally been superseded by other methods, there are still situations where dry placement of fill has advantages over other systems. For example, if a mine does not have a processing plant on site to provide mill tailings, other fill options must be considered. Such fills could include (1) natural surface sands (glacial till in North America, natural dune sands in Australia); (2) open-cut overburden, specifically quarried surface rock, or even rock generated in local civil works; and (3) underground development waste. Dry fill is also the preferred option in evaporite and gypsum mines where the ore or host rock is readily dissolved in water, or in mines in arid locations where water is scarce.

Dry placement of fill has two main disadvantages. First, it has a relatively low density as placed, which means it can undergo significant compression before it reaches its optimum density to provide stope wall support to prevent spalling of stope walls and minimize stope wall convergence. Second, depending on stope configuration, it may be difficult to place dry fill close to the roof of the opening across the full span of the stope roof.

Transport and Placement of Dry Fill
The first consideration, as with all types of backfill, is the means of transport of the dry fill underground. The most efficient means for transport of sand- or gravel-sized particles is a borehole from the surface. Selection of the size of the borehole depends on two considerations: the required throughput in metric tons per hour, and the size distribution of the particles. For coarse material such as gravel, the particle size will govern, because the diameter of the borehole should, as a general but not necessarily universally applicable rule, be a minimum of three to six times the maximum particle size of the gravel in order to avoid hang-ups. For finer material, the governing concern is the required throughput. In general, boreholes with diameters of 150 mm or 200 mm (6 or 8 in.) will be sufficient for most purposes. However, using this size of pipe, one must never let the pipe fill and possibly bridge, as it may plug.

The most common method of transporting dry fill to the workings is by haulage vehicles. In conventional, older mines, this usually meant railcars, which would dump the fill into a fill raise. More recently, haulage trucks have been used and belt conveyors are also being used for sand, gravel, or rock. All of these placement methods require good access to the workings that need to be filled.

Another important consideration with dry placement of granular fill is the method used to place and compact the fill within the stope. Historically, placement was done by hand, which is slow and costly. If the area to be filled is accessible to mobile equipment, spreading and compaction could be accomplished by means of a hydraulic ram and blade mounted on a haulage truck or load-haul-dump (LHD) vehicle. (The blade is pushed laterally against the fill pile by the hydraulic ram to compact it.) Another placement and compaction method that has been tried is a high-speed throwing belt, the main disadvantage of which is the limited distance that the fill travels beyond the discharge point.

In the case of rock fill, placement is not generally a problem if dumping is from the top of the working. Because of the mass of the rock pieces, gravity flow is usually satisfactory. However, large blocks contain a significant proportion of voids, and such fill may need considerable consolidation before it can provide significant stope wall restraint. This disadvantage is often resolved by filling the voids with cemented hydraulic fill, as discussed later in this chapter.

Uncemented Hydraulic Backfill System Design

Hydraulic transport and placement of backfill is the method most commonly used in metal mines but less commonly in coal mines. Deslimed and partially dewatered mill tailings are the materials most commonly transported underground hydraulically, but surface sand (e.g., glacial till in North America, natural dune sands in Australia) is also transported in this manner. The fine (slimes) portion of tailings is undesirable because of its slow drainage, but also because of the potentially disastrous consequences of liquefaction resulting from vibrations from, for example, production blasts.

Transport of Slurry Solids Content

Mill tailings generally consist of a heterogeneous slurry with approximately 30%–40% solids by mass and the rest water. The solids and the water represent two separate phases that will readily separate if the flow velocity is insufficient. When a heterogeneous slurry is discharged from a horizontal pipe, the solids will settle out with the coarsest particles settling first. As the solids are discharged, they will gradually form a cone around the end of the pipe, with the finer materials toward the toe of the slope. The slope of the cone varies with the solids content of the slurry; with very dilute slurries, the slope may not exceed 2°–3° and the sediments will be loosely packed, but at solids contents of 60%–70% by mass or more, the slope will approach the angle of repose of the particles and the solids will consolidate readily as they drain.

Hydraulic backfill systems typically operate at solids contents of 65%–70% solids by mass or more. Because, as stated previously, the tailings in the mill typically have a solids content of around 30% by mass, it is generally necessary to partially dewater the tailings prior to transport underground. Dewatering is usually accomplished using hydrocyclones or occasionally thickeners or rake classifiers. Hydrocyclones are vastly preferable because they most effectively deslime as well as dewater. One word of caution: The cut point commonly used by geotechnical engineers for drainage computations is the D_{10} size, whereas the D_{20} or D_{50} size is used by designers of hydrocyclones. Failure to appreciate this distinction could result in having a larger fines content than anticipated in the backfill. It is critical to maintain the design flow rate in a hydrocyclone if the cut point is to be achieved in practice.

Drainage requirements. After hydraulic backfill has been placed in a mined opening, it is necessary to drain away the water for a number of reasons. First, water-bearing sand/slime-sized material, especially if it contains excessive slimes and/or excessive water, can liquefy if subject to vibration caused by a seismic disturbance such as an earthquake, by a rock burst, or by production blasting. Second, drainage of the water is essential to consolidation of the fill. Because the fill is used as the working floor for the next lift in overhand cut-and-fill stoping, it is generally desirable to have it available for production traffic within 24 hours.

Prevention of the phenomenon of liquefaction requires essential consideration in the preparation and placement of hydraulic fill—particularly uncemented hydraulic fill—underground.

A general, but not absolute, rule is that the gravitational percolation rate of water through the fill should exceed 10 cm/h (4 in./h). Because drainage of water from the fill is by gravity with a hydraulic gradient of unity, this percolation rate corresponds to a hydraulic conductivity of approximately 3×10^{-3} cm/s (5.9×10^{-3} fpm), which is a typical value for a well-graded, fine-grained sand. To ensure this drainage rate is achieved, it is customary to remove the fines so that no more than 10% of the solids are finer than 74 μm (200 mesh). However, satisfactory backfilling has been achieved with as much as 10% of the solids finer than 43 μm (325 mesh). Also, hydraulic conductivity is temperature dependent, so as mine water temperature increases, fines content can increase and drainage rates remain satisfactory. Importantly, the converse applies. A comprehensive discussion of hydraulic fill drainage mechanisms is given by Thomas et al. (1979).

To facilitate fill drainage, it is customary to erect fences within the stopes being filled to contain the fill and to drain the water through ladderways or drain towers constructed up through the fill. To allow the fill water to drain without loss of fill fines it is customary to place filter fabrics on the inside of fill fences and outside drainage ladderways and towers inside the site to be filled. Historically, burlap (in North America) or hessian (in Australia) was used as the filter fabric. More recently, these materials have been largely replaced by monofilament geotextiles, though the traditional materials, where available, are still acceptable. To select the correct geotextile, the following filter selection criteria developed by the U.S. Army Corps of Engineers may be used:

- The ratio of the D_{85} (85% finer) size of the backfill to the opening size of the geotextile should exceed 1.0.
- The open area of the geotextile should be less than 40% of its total area.

For most tailings, the D_{85} size will range from 149 to 297 μm (0.0069 to 0.012 in.). Selection of the geotextile should also consider the required water flow rate per unit area of geotextile, which depends on the dimensions of the fill fences and drainage towers.

Backfill transport pipe size requirements. With two-phase (solids/fluid) heterogeneous slurries, the transport velocity in a pipeline must be high enough to prevent deposition of coarse particles. On the other hand, an excessive transport velocity in the pipe will result in excessive friction losses and pipe wear. A flow velocity of 1.5 m/s (5 fps) is often but not always considered the minimum necessary to provide satisfactory results. Conversely, a flow velocity of greater than 2.4 m/s (8 fps) may result in excess friction losses and pipe wear. Table 13.6-1 presents hourly placement tonnages that can be achieved at a pipe velocity of approximately 1.8 m/s (6 fps) and 65%–70% solids density by mass for a material with a specific gravity of 2.65.

Pipe friction losses and pump requirements. In mines accessed by a shaft, transport of backfill underground will be by means of either a pipe in the shaft or a borehole near the shaft (preferable to a pipe in the shaft, depending on shaft function) that connects the backfill plant to the workings. In such cases, the elevation difference is usually but not always sufficient to transport the fill to the working place without pumping. However, where access to the mine is by means of a horizontal adit, gravity flow will, with the occasional

Table 13.6-1 Hydraulic fill placement rate versus pipe diameter

Nominal pipe diameter, mm	50	80	100	150	200
Nominal pipe diameter, in.	2	3	4	6	8
Fill placement rate, t solids/h	15–20	35–40	65–75	145–165	255–290

Table 13.6-2 Fill tonnages handled before replacement by various types of pipe and fittings

Type	Standard (Schedule 40)	Heavy-Duty (Schedule 80)	Rubber-Lined
Pipe, t	250,000	800,000	>800,000
Bends and elbows, t	—	>250,000	>1,300,000
Tees, t	150,000	600,000	>600,000

Source: Adapted from Waterfield 1973.

exception, not be sufficient to transport the backfill slurry and pumping will be necessary. Where mine access is by decline, pumping may or may not be necessary, depending on the general mine layout. In all cases it is prudent to calculate the friction losses in the system. Friction losses are greater than those for water alone, but can be estimated from the conventional Darcy–Weisbach equation using the following adjusted friction factor (Abulnaga 2002):

$$f_m = f_L \{1 + 81 C_v [gD_i(\rho_s - \rho_L)/(V^2 \rho_L \sqrt{C_d})]^{1.5}\} \quad (13.6\text{-}1)$$

where

f_m = friction factor for the slurry mixture
f_L = friction factor for clear water under the same temperature and pressure conditions
C_v = solids concentration by volume
D_i = inside diameter of the pipe
ρ_s = solids density
ρ_L = water density
V = slurry velocity in the pipe
C_d = drag coefficient, which can be taken as 0.43 in most situations

If a pump is found to be necessary, a centrifugal slurry pump is the most likely selection, pump specification being based on head and flow rate requirements. These are relatively low-head pumps, and it may be necessary to install pumps in series to ensure that varying head requirements are met. However, material will flow through the pump impellers even if the head contribution is not needed and the motor not operating. To minimize wear in centrifugal pumps, the following guidelines should be followed (Paterson 2006):

- Discharge velocity less than 8 m/s (26 fps)
- Maximum impeller tip speed for all-metal pumps of 28 m/s (92 fps)
- Maximum impeller tip speed for rubber-lined pumps of 20 m/s (76 fps)
- Range of discharge flow/flow at best efficiency point of 60%–100%

Wear in pipes and line fittings and pipe material selection. Erosion of the walls of boreholes used to transport backfill underground is generally not considered a serious problem, though boreholes must be lined from the surface down to appropriately below the base of surface rock alteration, sometimes referred to as the base of complete oxidation. This is based on the usual situation where the oxidized/altered rock from the surface will be scoured by the fill flow, whereas the significantly more solid rock below will not. On the other hand, wear in elbows, tees, and other fittings and in pipes will result in the need for periodic replacement or selection of more appropriate lining materials. Consequently, the design engineer must consider whether to use

- Rubber-lined, heavy-duty (thick-walled), or standard pipes;
- Victualic, threaded, or flanged pipe couplings;
- Rubber-lined, built-up (heavy-duty), or standard bends, elbows, and tees;
- Heavy-duty rubber hose or rubber-lined pipe connections at horizontal takeoff points; or
- For long-term, high-tonnage applications, pipes and fittings with basalt or similar linings.

In general, wear is most severe at bends, elbows, and tees; however, wear will also occur in straight pipe, especially if the slurry is abrasive and contains angular particles. Wear rates depend on the density of the slurry, the diameter of the pipe, the flow velocity, and the maximum size and abrasiveness of the particles. Unfortunately, wear rates are rarely expressed in terms of these parameters but are more commonly given in terms of the tonnage handled by pipe prior to replacement. If line velocities are not excessive, the tonnages given in Table 13.6-2 can be expected before replacement becomes necessary.

The lack of data in Table 13.6-2 regarding the life of elbows and bends is because tees are much more commonly used for takeoffs because it is simpler to change the side of the tee that is blanked off than it is to uncouple a line to insert a bend or elbow. In stopes, distribution pipes can be either standard steel pipe or plastic pipe and should include tees and valves at 3-m (10-ft) intervals to facilitate even distribution of the fill. Wear is typically on the invert of the fill pipe, and pipe life can be extended significantly by rotating the pipe one-sixth of a revolution at regular intervals.

In multilevel mining, heavy-duty rubber hose connections are preferable to pipe connections at the levels. Only a single hose is required at a level even though there may be multiple branch lines. However, the connections on the ends of the hoses will be bolted flanges rather than victualic or threaded connections.

Backfill Plant Requirements

The first requirement in a typical backfill plant is a bank of hydrocyclones to remove excess water and fine particles from the tailings. The overflow from the cyclones (the fines and excess water) is typically transported to a surface tailings impoundment for disposal. Many mines once operated on a 5-day week, whereas concentrators typically operate on a 7-day week, and in such cases it became necessary to provide a storage silo capable of holding backfill pouring requirements for 1 or 2 days. This silo is known as a repulper, because the backfill is allowed to drain until needed, at which time water is added at high pressure to the base of the tank to refluidize the fill.

The fill plant will also need transfer pumps and valving, and care should be taken to ensure that the capacities of the transfer pumps are sufficient. Transfer pumps should be able to handle a flow rate approximately 20% greater than the design flow rate for the system (solids plus water), because if the flow rate decreases below the critical value for sedimentation, the inevitable result will be the complete filling of the pipe with solids. Cleaning out a pipe that has sanded up is time-consuming and costly, particularly if the fill contains portland cement.

Current practice is more likely to have both mine and mill operating around the clock, thus eliminating the need for fill storage tanks and the complexities they involve.

Cemented Hydraulic Backfill System Design

Correctly prepared uncemented hydraulic fill is predominantly sand-sized and must behave as a sand, albeit on occasions a slow-draining sand. Consequently, it has no true cohesion or, until drained, no unconfined compressive strength. For some applications, such as fill placed against unmined rock that will be recovered later, it is often desirable that the fill be able to stand unsupported upon exposure. For such applications, it is customary (but on occasions not at all necessary and possibly detrimental) to add cement to the backfill.

Because cemented fill is the same fill material as the uncemented fill except for the cement, several of the design parameters for a cemented hydraulic backfill system are the same as for an uncemented backfill system. However, slurry solids content must be more carefully controlled, though pipeline velocities remain essentially the same. As is the case for uncemented fill, the tailings are deslimed with a typical D_{10} size of 74 μm (200 mesh). However, the cement provides fines to the mix and the drainage rate is typically somewhat slower than that for uncemented fill. This potential problem is in part negated for the following reasons: (1) portland cement consumes fill water during hydration, reducing the volume of water to be removed; (2) portland cement produces heat during hydration, reducing the viscosity of fill water, tending to increase dewatering rate; and (3) portland cement lubricates the flow so higher placement densities can be used, reducing the amount of water to be removed. In pipeline flow, portland cemented fill flows more uniformly than uncemented hydraulic fill, and this tends to reduce pipeline wear for similar placement rates—coarse fill particles have reduced access to pipeline walls. Optimum solids content for pipeline flow generally ranges from 68% to 72% solids by mass, and closer attention is given to control of this solids content. Costs considered, after portland cement is added to hydraulic fill, the fill operation moves from underground disposal of mining waste to preparation of a quality-controlled engineering material.

Cement Content Versus Strength and Deformation Properties

The strength of a cemented fill depends on the percentage of cement in the fill. Figure 13.6-1 shows the relationship between cement content and strength for several cemented mine backfills in Canada and Australia. Although Figure 13.6-1 shows

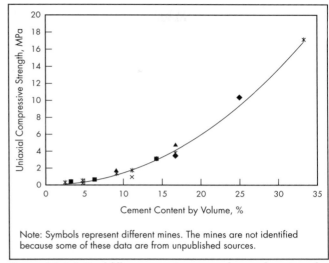

Figure 13.6-1 Backfill compressive strength as a function of cement content

the cement content as a volume percentage, it is best to express the cement content as a mass percentage of the total dry solids. The equation of the best-fit curve in Figure 13.6-1 is given by

$$C_o = 12.1\, P^{2.0448} \qquad (13.6\text{-}2)$$

where

C_o = compressive strength in kilopascals (kPa) (multiply result by 0.145 to obtain strength in psi)
P = cement content in volume percent

Some mines have experimented with replacing part of the cement with fly ash or ground slag and/or similar pozzolanic materials. In the case of fly ash, there does not appear to be a significant difference in compressive strength for fly ash contents up to 20% of the cement content in fill with cement/tailings ratios between 1:10 and 1:30 by volume. In the case of ground slag, the uniaxial compressive strength increases compared to portland cement only, with the increase approximately proportional to the slag content.

Cement Content Versus Use

The amount of cement required in a cemented fill depends on the strength requirement and the use for the fill. Some examples are given in Table 13.6-3. These should be used for guidance and not adopted as standard. For example, use of cement in mucking floors may have advantages for movement of mucking equipment, but it prevents almost totally the downward flow of fill drainage water.

Freestanding Height

The freestanding height of cemented backfill depends on the compressive strength and density of the fill as follows:

$$H = 2C_o/\gamma g \qquad (13.6\text{-}3)$$

where

H = freestanding height of the fill
C_o = uniaxial compressive strength of the backfill
γ = density of the fill

Table 13.6-3 Cemented fill cement/tailings ratio versus application

Application	Cement/Tailings Ratio
Bulk filling	1:30
Barrier against adjacent pillar, adjacent stope, or related structures	1:10
Mucking floors	1:6 to 1:12
Water-retaining bulkheads/barricades where no restriction of stope drainage must result—drain pipes through the strengthening plug	1:2

In practice, the known parameter in Equation 13.6-3 is the required freestanding height; the compressive strength necessary for the fill to stand unaided at that height is the unknown. For example, if the required freestanding height is 8 m and the specific gravity of the fill is 1.7, Equation 13.6-3 indicates that the required strength is 133 kPa (19.3 psi). From Figure 13.6-1, the strength corresponds to a backfill containing approximately 3% cement. Alternatively, for a compressive strength of 133 kPa, Equation 13.6-2 gives a cement content of 3.22%, which is equivalent to a cement/tailings ratio of 1:30.

Pump Requirements
Where there is extensive horizontal development on upper levels of a mine, the elevation head may be insufficient to support flow by gravity alone. In such cases, pumps will be necessary to transport the fill over all or part of the underground fill distribution system. Both positive displacement pumps and centrifugal pumps can be used for this application; however, rubber-lined centrifugal slurry pumps are considered a better choice. Pumps with dry glands, which minimize water handling requirements, are available. With centrifugal pumps, the slurry will flow through a pump even if it is not needed to supply head.

Backfill Plant Requirements
Backfill plant requirements are similar to those for uncemented fill, except for the need for a cement silo plus mixing tanks and feeders.

Cemented Rock Fill System Design
Cemented rock fill, as typical of modern practice, was developed at the Geco Division of Noranda Mines Ltd. in the early 1960s to provide support for stope walls that were prone to slough. By restraining stope wall rock, the fill reduced dilution and improved overall regional stability. A cemented rock fill system was developed at Mount Isa in Australia in the late 1960s, but the main reason for the rock fill component was to reduce the quantities of cemented hydraulic fill as portland cement was expensive in that part of Australia, being remote from cement suppliers and prone to disruption of supply during the wet season. In a cemented rock fill system, the main design function of the cemented hydraulic fill is to fill the voids in the broken rock.

Rock Fill Size Considerations
Rock fills should consist of coarse angular fragments of medium to high density, medium to high chemical stability, and high strength and abrasion resistance. Abrasion resistance can be tested using the Los Angeles abrasion test.

The rock fill can be modified to suit a particular application by optimizing the particle size distribution. A competent fill that will stand when exposed on a vertical face should have a low porosity. Fill should be appropriately graded with respect to particle size. The particle size distribution of a rock mass is often represented by the Gates–Gaudin–Schuhmann distribution:

$$P = (d/m)^n \qquad (13.6\text{-}4)$$

where
P = cumulative percent finer than size d
d = particle size
m = maximum particle size

In the concrete industry, a densely packed fill material is obtained when the exponent, n, in Equation 13.6-4 is 0.5.

A major concern with rock fill is the extent of rock degradation from impacts during freefall passing underground from the surface or from degradation from particles grinding against each other during choke-fed passing. An impact test that was used successfully at Mount Isa in Australia (Kuganathan 2005) consisted of dropping samples of up to 300-mm (11.8-in.) boulders from a height of 16.5 m (54 ft) onto a piece of 16-mm (⅝-in.) thick manganese steel liner plate. A test was considered successful if 75% of the material recovered was coarser than 25 mm (1 in.). The first full-scale drop tests at Mount Isa involved dropping several tens of metric tons of surface-quarried rock about 530 m (1,600 ft) down a vertical shaft, with particle size analyses before and after.

Binder Placement
The stiffener in a cemented rock fill can consist of deslimed mill tailings fill, cemented deslimed mill tailings fill, paste fill, portland cement, fly ash or other pozzolan, or a combination of two or more of these. All of these binders are transported most easily by pipeline to be added to the rock fill as it enters the top or some more-appropriate location in the stope. Standardization of such practice is difficult and large stopes must be considered on a case-by-case basis.

Paste Fill System Design
With paste fill, the tailings become a high-density slurry with a solids content of 75%–80% by mass. The slurry acts as a viscous, Bingham solid, for which a critical shear stress must be exceeded before the material will flow. The critical shear stress for paste fills typically ranges between 250 and 800 Pa (0.041 to 0.116 psi) (Boger et al. 2006).

Because of the viscous nature of the paste and the large amounts of energy required to transport paste in the turbulent regime, it is customary to transport paste in the laminar flow regime (Paterson 2006). The drawback to laminar flow, however, is the potential for sedimentation and blockage in the pipeline, though paste fill, because of its tightly controlled production and relatively high pulp density, is less prone to settling than conventional deslimed mill tailings fill. Two other important attributes of paste fill are (1) the water is bound into the slurry, so that little drainage is required; and (2) the fact that fines removal from tailings may not be required.

Paste Particle Size Requirements
At least 15% of the particles in the paste should be finer than 20 μm (0.00079 in.) (Hassani and Archibald 1998).

In addition, it is customary to add from 3% to 6% portland cement or a mixture of portland cement and fly ash as a binder. If the tailings contain insufficient fines, it may be necessary to add fly ash or additional cement.

Pump Requirements
The critical issue for pumping the paste from the backfill plant to the mine stopes is the initial shear stress that must be overcome before the paste will flow (Abulnaga 2002):

$$P_{st} = 4\tau_0 L/D_i \tag{13.6-5}$$

where

P_{st} = start-up pressure
τ_0 = initial shear strength of the paste
L = pipe length
D_i = inside pipe diameter

The initial shear stress to be overcome, or yield stress, can be measured by the vane shear test. Alternatively, it is becoming more common to estimate the yield stress from a slump test using the following relationship (Boger et al. 2006):

$$\tau_0/\rho gh = 0.5 + 0.5(s/h)^{0.5} \tag{13.6-6}$$

where

s = measured slump
h = height of the slump cylinder

Because of the viscosity and the density of the paste, positive displacement pumps are required. These pumps have the attribute that they can pump against any head up to the design head of the machine while the flow rate remains constant. These pumps should be fitted with pressure relief valves because they will continue to pump against blocked or closed pipes, which could result in a ruptured pipe or a damaged pump.

Pipe Wear
Few data are available regarding the wear of pipes transporting paste fill. However, because the paste moves as a viscous unit, it is expected that the wear rates are less than for conventional hydraulic fills. The wear can be checked in vertical sections of the pipelines using a three-legged caliper probe. It is recommended that readings be taken after 15,000 t (16,535 st) have been transported and thereafter at 50,000-t (55,116-st) intervals.

Backfill Plant Requirements
Components in a paste fill plant typically include tailings storage tanks with a 1-day live storage capacity; paste thickeners for reducing water content from 65% by mass to approximately 20% by mass; silos for cement (and possibly fly ash) storage; weightometer hoppers for tailings, cement, and fly ash (if used); mixers for blending tailings, fly ash, and cement (if used); and positive displacement pumps.

Pneumatic Stowing
Pneumatic stowing is often used in situations where hydraulic filling has significant drawbacks. Examples are situations where it is undesirable to handle the large quantities of water that drain off from the fill or where seams are thin and flat-lying so that hydraulic placement of fill tight to the roof is difficult to achieve. Pneumatic stowing has also been considered as a means of placing broken rock whose ore grade is too low for recovery by conventional means in order to extract a mineral by pressure or flood leaching.

Transport Considerations
For successful pneumatic stowing, the air velocity must be sufficient to move the solids. The air velocity at which the solids begin to move is the critical transport velocity. For vertical conveying, the critical transport velocity equals the terminal velocity, which is given by either Stokes' law or the Newton–Rittinger equation depending on whether the flow is laminar or turbulent. For pneumatic transport, the flow of both air and the solids is turbulent, and the terminal velocity is given by the Newton–Rittinger equation:

$$v_t = [4gD_i(\rho_s - \rho_a)/(3\rho_a C_d)]^{0.5} \tag{13.6-7}$$

where

v_t = terminal velocity
ρ_a = air density

For the fully turbulent regime, C_d has an essentially constant value of 0.44. The air density term in the numerator of Equation 13.6-7 can be ignored. For horizontal conveying, the critical velocity is approximately $0.7v_t$. For satisfactory flow, the air velocity should be four to five times the critical velocity.

Particle Size Distribution
The particle size distribution will affect the velocity necessary to transport the solids and the pipe size. To minimize the possibility of bridging of large chunks, the pipe diameter should be at least three times that of the largest chunk carried. As a practical matter, the maximum lump size should be 100 mm (4 in.) or less, with 75 mm (3 in.) preferred. Considered another way, typical stowing pipe diameters range from 150 mm (6 in.) up to 225 mm (9 in.). Consequently, the maximum lump size should preferably be less than 75 mm (3 in.), and ideally 50 mm (2 in.) or less. A smaller maximum particle size will also result in reduced velocity, power requirements, and pipe wear.

Feeder and Blower Considerations
A pneumatic stowing system has three basic components:

1. A hopper and feeder for the solids
2. An air supply consisting of compressed air from either a pipeline or compressor, or a rotary blower
3. An injection pipe and nozzle

The hopper should be large enough to contain a bucket load of broken material from a typical haulage vehicle. The discharge opening in the hopper should have a minimum dimension of at least three to six times the dimension of the largest piece being transported.

The feeder should be sized to handle the expected hourly throughput of the system. Three important considerations for the feeder, which is generally a rotary airlock type, are the number of vanes on the rotor, diameter of the airlock, and the rotor rpm. For a given flow rate in cubic meters per hour, better flow characteristics are obtained with a larger diameter airlock operated at a lower rpm. Six or eight vane models are available. In some applications, difficulties have been experienced with the eight-vane units.

Table 13.6-4 Estimated fill throughput through different types of pipe before replacement

Type of Pipe	Pipe Life, 1,000 m³ of material passed
Standard Schedule 40 steel pipe	8 to 30
Twin-wall pipe	20 to 150
Steel pipe with abrasion-resistant basalt lining	200 to 1,500

Source: Adapted from Jerabek and Hartmann 1966.

Except for very large tonnage operations, low-pressure systems using positive displacement blowers provide satisfactory air quantities for stowing. These blowers are noisy and should be fitted with silencers. Injection pipes are typically 150–225 mm (6–9 in.) in diameter. Air velocities in the stowing pipe should exceed 20 m/s (65 fps), and the exit velocity in the nozzle should be about 40 m/s (130 fps) for optimum results. Stowing machines are available to place up to 280 m³/h (367 yd³/h).

Solids Loading and System Pressure Drop

An important parameter is the solids-to-air ratio by weight. For most stowing applications this ratio will range from 10 to 40. However, the actual value of the ratio for a given stowing operation will depend on the quantity and size of the solids to be stowed, given that with coarse material the velocity required to keep the solids in the air stream will be higher than for fine material.

The system pressure drop consists of the pressure drop due to the air in the system and the pressure drop due to the solids. The air pressure drop is given by the Darcy–Weisbach equation in the following form:

$$\Delta p_a = p_a^2 - p_2^2 = f(L/D_i)(V_a^2/2g)\gamma_{a1} p_a \quad (13.6\text{-}8)$$

where

p_a = initial air pressure
p_2 = mine atmospheric pressure
f = friction factor = 0.016
L = pipe length
D_i = pipe internal diameter
V_a = pipe air velocity
γ_{a1} = specific gravity of mine air

Based on tests performed by Peter (1952), the total system pressure drop for a given solid/air ratio can be estimated from the following equation:

$$S = (0.2086 - 0.0001\ EQ)R_s + 1 \quad (13.6\text{-}9)$$

where

S = ratio of system pressure drop to air pressure drop
E = electrostatic charge
Q = airflow rate, m³/h
R_s = solids/air ratio by weight

Usually the equation shows the product of EQ in the direction of flow for determining the pressure drop. Thus, the presence of statically charged particles increases the pressure drop (Smeltzer et al. 1982). After the system pressure drop is known, the system power requirements can be determined.

Power Requirements

The power required to transport the solids is given by

$$P_b = RTQ\, \gamma_{a1} \ln((R_s\, \Delta p_a + p_2)/p_2) \quad (13.6\text{-}10)$$

where

P_b = blower power
R = universal gas constant
T = absolute temperature (degrees K in metric units or degrees Rankine in English units)
Q = airflow rate
γ_{a1} = mine air density

Additional power will be required for the feeder as follows:

$$P_f = Q\, \gamma_{a1}\, D_f \quad (13.6\text{-}11)$$

where

P_f = feeder power
D_f = feeder airlock diameter

It will be necessary in Equations 13.6-10 and 13.6-11 to include appropriate conversion factors to obtain the power in metric or English units.

Pipe Wear

Broken rock stowed pneumatically wears pipe rapidly, especially at bends. For this reason, standard Schedule 40 steel pipe should not be used. Heavy-duty (Schedule 80) pipe could be used; however, there are special multiwall pipes fabricated especially for stowing. Another option is pipes containing abrasion-resistant basalt pipe linings. Estimates of stowing pipe life for different types of pipe are presented in Table 13.6-4. Pipe life can be maximized by rotating the pipe one-sixth of a revolution at frequent intervals.

Flowable Fill System Design

Flowable fill is a self-consolidating, low-strength material used as an alternative to consolidated granular fill. It is an excellent solution for filling inaccessible areas such as abandoned mines and is used for this purpose by OSMRE and the Pennsylvania Department of Environmental Protection (PaDEP), among others. The mix proportions to provide 1 m³ (1.31 yd³) of the flowable fills used by these agencies are presented in Table 13.6-5.

Fly ash must conform in general to mineral admixture Class F but must have a moisture content not greater than 20% and a maximum loss on ignition of 6%. Black or brown fly ash will not be allowed unless certified lab results show the loss on ignition to be 6% or less. The sand must be 45%–80% finer than 1.19 mm (16 mesh) and approximately 10%–30% finer than 297 μm (50 mesh). The gravel must consist of hard, dense, durable uncoated gravel or crushed stone with a grain size distribution as follows:

Percent Finer, by Mass	Particle Size
100	19 mm (3/4 in.)
40–70	9.5 mm (3/8 in.)
0–15	4.76 mm (4 mesh)

Table 13.6-5 Flowable fill mixes for backfilling abandoned mines

Mix No.	Portland Cement, kg	Sand, kg	Gravel, kg	Fly Ash, kg	Water, L*
OSMRE No. 1	233	0	0	1,130	470
OSMRE No. 2	219	854	384	531	272
OSMRE No. 3	201	785	0	816	272
OSMRE No. 4	296	0	0	1,186	401
PaDEP	167	0	0	1,186	247

*Water contents may be modified from those shown to obtain sufficient slump for the mixture to flow.

Because flowable fill has a high slump by definition, gravity flow from the mixing tank will suffice for most applications. Other agencies that deal with abandoned mines may use other mixes.

Flowable Fill Equipment Requirements
Flowable fill is generally used to backfill abandoned mines. Consequently, it will generally not be possible to direct the flow of material from underground. Therefore, multiple access points from the surface are usually necessary, and storage silos and mixing equipment must be portable. Portable concrete batch mixers are readily available that can handle up to 153 m³/h (200 yd³/h). Portable silos for sand and fly ash storage can also be hired. Another option is to contract with a local concrete Ready Mix supplier, indicating the mix to be supplied.

OTHER CONSIDERATIONS

Remote Placement of Backfills
Backfilling of former mines as a subsidence-prevention measure is becoming increasingly common. In general, it will be necessary to provide multiple pour points to ensure the void will be as completely filled as possible. Similar to discharge from a stacker or conveyor, the solid discharge from a fill pipe will form a cone that is characterized by finer and finer particles as it moves away from the discharge point. The angle of repose of the cone depends on the solids percentage in the slurry—the higher the percent solids, the steeper the cone. Because a supply bore will become useless after the cone fills to the top of the opening, the spacing between supply borings depends on the height of the opening, as well as the percent solids in the fill.

Automated Systems
Because of the generally batch nature of filling and the need to change discharge locations during the course of backfill placement, not all facets of backfilling lend themselves to automatic operation. However, some portions of the cycle, such as starting and stopping of pumps and opening and closing of hopper gates, can be automated by means of interlocks, level controls, and other automatic controls. For example, the Manitoba Division of Vale Inco installed an automated fill pumping system at the Thompson mine in the 1960s. In theory, the valving on the discharge locations could be automated and interlocked with level switches placed below the discharges; however, the result would be a pouring system in the stope that would be significantly more difficult to install and dismantle.

Effect of Cemented Fill in Mill Feed on Processing Plant Recovery
When mucking on a fill floor, it is inevitable that some fill will be mixed with the broken ore. If the fill is uncemented, the only negative effect is some dilution. However, if the fill is cemented, there are likely to be unwanted effects on the concentrator circuit, especially in the case of a silver/lead/zinc ore. Specifically, the cement will significantly increase the pH of the process water because the pH of hydrous cement is approximately 10.5–11. In addition, the presence of calcium carbonate may chemically affect milling processes. Occasionally, the presence of cement will improve mill recovery percentages.

Heat Generation of Sulfide Backfills
Tailings containing pyrite and especially some forms of pyrrhotite readily oxidize, giving off heat and absorbing oxygen. As a consequence, it is not uncommon for the temperature in a stope to rise by 3°–5°C (5°–10°F) after the first lift of fill has been placed. However, the oxidation process may also result in additional cementing of the fill. Noranda took advantage of this at the Horne mine in Quebec when designing the hydraulic backfill system there in the 1930s (Patton 1957). One unfortunate consequence of this was loss of life by removal of oxygen from mine air. Also, it is reported that in the Kimberley mine in British Columbia, iron sulfides in placed rock fill increased fill temperatures to such an extent that fill behind bulkheads became molten.

SOURCES OF FURTHER INFORMATION
As indicated in the introduction, this chapter makes no claim for completeness. The following publications are aimed specifically at design of backfill systems:

- *Handbook on Mine Fill* (Potvin et al. 2005)
- *Paste and Thickened Tailings: A Guide*, 2nd ed. (Jewell and Fourie 2006)
- *Mine Backfill 1998* (Hassani and Archibald 1998)

These publications do not cover pneumatic stowing or flowable fill.

Numerous case histories of backfill systems can be found in the proceedings volumes of the International Symposia on Mining with Backfill that have been held approximately every 4 years since the first one was organized by Ed Thomas, the late Ken Mathews, Jack Barrett, and others at Mount Isa in Australia in 1973 under the auspices of the Australasian Institute of Mining and Metallurgy. Another source of case histories is the proceedings of the International Symposiums on Paste and Thickened Tailings that have been held since the 1990s, which have generally been published by the Australian Centre for Geomechanics.

ACKNOWLEDGMENTS
The author wishes to express his deep appreciation to Ed Thomas, an acknowledged authority on the design of backfill systems, for serving as technical reviewer for the chapter

and for providing input on backfill practices outside of North America. It has also been the author's great fortune to have known Ed since 1978 when Ed visited Elliot Lake, Ontario, where the author was involved in the design of a pillar extraction system using backfill.

REFERENCES

Abulnaga, B.E. 2002. *Slurry Systems Handbook*. New York: McGraw–Hill.

Boger, D., Scales, P., and Sofra, F. 2006. Rheological concepts. In *Paste and Thickened Tailings: A Guide*, 2nd ed. Edited by R.J. Jewell and A.B. Fourie. Nedlands, Western Australia: Australian Centre for Geomechanics.

Callon, J. 1876. *Lectures on Mining Delivered at the School of Mines, Paris*. Translated by C.L. Foster and W. Galloway. Paris: Dunod.

Hassani, F.P., and Archibald, J.F., eds. 1998. *Mine Backfill 1998*. CD-ROM. Montreal: Canadian Institute of Mining, Metallurgy and Petroleum.

Jerabek, F.A., and Hartman, H.L. 1966. Mine backfilling with pneumatic stowing. *Min. Cong. J.* 52(5):51–60.

Jewell, R. 2006. Introduction. In *Paste and Thickened Tailings: A Guide*, 2nd ed. Edited by R.J. Jewell and A.B. Fourie. Nedlands, Western Australia: Australian Centre for Geomechanics.

Jewell, R.J., and Fourie, A.B., eds. 2006. *Paste and Thickened Tailings: A Guide*, 2nd ed. Nedlands, Western Australia: Australian Centre for Geomechanics.

Kuganathan, K. 2005. Rock fill in mine fill. In *Handbook on Mine Fill*. Edited by Y. Potvin, E. Thomas, and A. Fourie. Nedlands, Western Australia: Australian Centre for Geomechanics.

Paterson, A. 2006. High concentration hydraulic transport systems. In *Paste and Thickened Tailings: A Guide*, 2nd ed. Edited by R.J. Jewell and A.B. Fourie. Nedlands, Western Australia: Australian Centre for Geomechanics.

Patton, F.E. 1957. Backfilling at Noranda. In *Mining in Canada*. Montreal: General Committee of the Sixth Commonwealth Mining and Metallurgical Congress, Canadian Institute of Mining and Metallurgy.

Peele, R. 1941. *Mining Engineers' Handbook*, 3rd ed. New York: John Wiley and Sons.

Peter, G. 1952. Messungen an blasversatzmaschinen zur klärung der zusammenhänge zwischen blasleistung, leitungslänge, luftmenge und luftdruck (Measurements on stowing machines to clarify the relationships between discharge capacity, pipe length, air quantity and air pressure). *Glückauf.* 33/34:807–819.

Potvin, Y., Thomas, E.G., and Fourie, A.B., eds. 2005. *Handbook on Mine Fill*. Nedlands, Western Australia: Australian Centre for Geomechanics.

Rahilly, H.J. 1923. Mine fires and hydraulic filling. *Trans. AIME* Vol. 102.

Schmidt, W.B., Moroni, E., Bullock, R., Sanford, C., and Murphy, J. 1976. Review of British and German coal mining technology. *Trans. AIME* Vol. 260.

Shepherd, R. 1980. *Prehistoric Mining and Allied Industries*. London: Academic Press.

Smeltzer, E.E., Klinzing, G.E., and Weaver, M.L. 1982. Pressure drop losses due to electrostatic generation in pneumatic transport. *Ind. Eng. Chem. Process Des. Dev.* 21(3):390–394.

Thomas, E.G., Nantel, J.H., and Notley, K.R. 1979. *Fill Technology in Underground Metalliferous Mines*. Kingston, ON: International Academic Services.

Waterfield, M.C. 1973. *A Review of Fill Mining Practice*. Internal Report 73/95. Ottawa: Mining Research Center, Mines Branch, Department of Energy, Mines and Resources.

Weaver, W.S., and Luka, R. 1970. Laboratory studies of cement-stabilized mine tailings. *CIM Bull.* (September): 988–1001.

CHAPTER 13.7

Cave Mining

Dennis H. Laubscher

INTRODUCTION

The expression *cave mining* will be used in this chapter to refer to all mining operations in which the ore body caves naturally after undercutting and the caved material is recovered through drawpoints. The term encompasses block caving, panel caving, inclined-drawpoint caving, and front caving. Caving is the lowest-cost method of underground mining, provided that drawpoint size and handling facilities are tailored to suit the caved material and that the extraction horizon can be maintained for the life of the draw.

The daily production from cave mining operations throughout the world (in 2008) was approximately 503 kt/d. Table 13.7-1 shows production broken down by layout.

Today, several open-pit mines currently producing in excess of 50 kt/d are examining the feasibility of implementing low-cost, large-scale underground mining methods. Several underground cave mines that produce high tonnages are planning to implement caving heights (or "dropdowns") of 500 m and at depths of 1,500 m. This will result in a considerable change in their mining environments. These changes will necessitate more detailed mine planning, rather than the simple projection of current mining methods to greater depths.

As more attention is directed to mining large, competent ore bodies with low-cost underground methods, it is necessary to define the role of cave mining. In the past, caving has been considered for rock masses that cave and fragment readily. The ability to better assess the cavability and fragmentation of ore bodies, the availability of robust load-haul-dump (LHD) machines, an understanding of the draw control process, suitable equipment for secondary drilling and blasting, and reliable cost data have shown that competent ore bodies with coarse fragmentation can be mined at a much lower cost using caving rather than with drill-and-blast methods. However, after a cave layout has been developed, there is little scope for change.

Aspects that have to be addressed are cavability, fragmentation, draw patterns for different types of ore, drawpoint or drawzone spacing, layout design, undercutting sequence, and support design. Table 13.7-2 shows some significant anomalies in the quoted performance of different cave operations.

It is common to find that old established mines that have developed standards during the course of successful mining in the upper levels of an ore body are resistant to change and do not adjust to the ground control problems that occur as mining proceeds to greater depths or as rock types change. Mines that have experienced continuous problems are more amenable to adopting new techniques to cope with a changing mining situation. Detailed knowledge of local and regional structural geology, the use of an accepted rock mass classification to characterize the rock mass, and knowledge of regional and induced stress environments are prerequisites for good mine planning. It is encouraging that these aspects are receiving more and more attention.

Table 13.7-1 Production per day obtained using different mine layouts in 2008

Mine Layout	Daily Production, kt
Grizzly	0
Slusher	3
Load-haul-dump machine	±500
Total	503

Table 13.7-2 Anomalies in quoted performance of caving operations

Quote	Explanation
96% of ore recovered for 100% mineral extraction	Underevaluation of ore body and dilution zone
Correct drawpoint spacing, but occurrences of 200% overdraw with 30% waste dilution entry	Highly irregular draw and underevaluation of dilution
15% dilution entry in spite of correct drawpoint spacing and uniform fragmentation	Drawpoints being drawn in isolation
Ore from lower 100 m of draw column still reporting in drawpoint, even though 260 m of ore has been drawn	Large range in fragmentation and irregular, high values in dilution zone

Dennis H. Laubscher, Mining Consultant, Bushmans River Mouth, South Africa

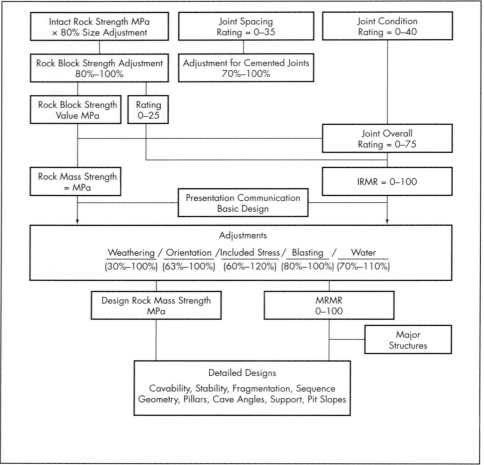

Figure 13.7-1 Flow sheet of MRMR procedure with recent modifications

The Laubscher rock mass classification system provides both in-situ rock mass ratings (IRMRs) and rock mass strength. Such a classification is necessary for design purposes. The IRMR defines the geological environment, and the adjusted or modified rock mass ratings (MRMRs) consider the effects of the mining operation on the rock mass. Figure 13.7-1 is a flow sheet of the MRMR procedure with recent modifications. The reader is encouraged to read the chapter "The IRMR/MRMR Rock Mass Classification for Jointed Rock Masses" by Laubscher and Jakubec (2001). The ratings describe in detail cavability, subsidence angles, failure zones, fragmentation, undercut-face shape, cave-front orientation, undercutting sequence, overall mining sequence, and support design.

FACTORS AFFECTING CAVING OPERATIONS

The 25 parameters that should be considered before implementing any cave mining operation are set out in Table 13.7-3. The parameters in bold are a function of the parameters that follow. Many parameters are uniquely defined by the ore body and the mining system and are not discussed further. The parameters considered later are common to all cave mining systems and need to be addressed if any form of cave mining is contemplated.

Cavability

Monitoring many caving operations has shown that two types of caving can occur—stress and subsidence caving. However, it is better to use the terms *vertical extension* to mean upward propagation of the cave and *lateral extension* to mean the propagation of the cave as a caved block is expanded.

Vertical extension caving occurs in virgin cave blocks when the stresses in the cave back exceed the strength of the rock mass. Caving can stop when a stable arch develops in the cave back. The undercut must be increased in size or the boundaries weakened to induce further caving. High horizontal stresses acting on steep dipping joints increase the MRMR. This was the situation in block 16 at the Shabanie mine in Zimbabwe. A stable back was formed when the undercut had a hydraulic radius of 28 and the MRMR was 64. When block 7 (adjacent to block 16) caved, the horizontal confining stress was removed, which resulted in a reduction in the MRMR to 56 in block 16. At this point, caving occurred.

Lateral extension caving occurs when lateral restraints on the block being caved are removed by mining adjacent to the block. This often results in a large stress difference, leading to rapid propagation of the cave and limited hulking.

Table 13.7-3 Parameters to be considered before implementation of cave mining

Cavability	Primary Fragmentation	Drawpoint/Drawzone Spacing
Rock mass strength (rock mass rating [RMR]/MRMR)	Rock mass strength (RMR/MRMR)	Particle size of ore and overlying rock
Rock mass structure-condition geometry	Geological structures	Overburden load and direction
In-situ stress	Joint and fracture spacing, and geometry	Friction angles of caved particles
Induced stress	Joint condition ratings	Practical excavation size
Hydraulic radius of ore body	Stress or subsidence caving	Stability of host rock mass (MRMR)
Water	Induced stress	Induced stress
Draw Heights	**Layout**	**Rock Burst Potential**
Capital	Particle size	Regional and induced stresses
Ore-body geometry	Drawpoint spacing and size	Variations in rock mass strength, modulus
Excavation stability	Method of draw—gravity or LHD	Structures
Effect on ore minerals	Orientation of structures and joints	Mining sequence
Method of draw	Ventilation, ore handling, drainage	
Sequence	**Undercutting Sequence (pre/advance/post)**	**Induced Cave Stresses**
Cavability: poor to good or vice versa	Regional stresses	Regional stresses
Ore body geometry	Rock mass strength	Area of undercut
Induced stresses	Rock burst potential	Shape of undercut
Geological environment	Rate of advance	Rate of undercutting
Rock burst potential	Ore requirements	Rate of draw
Production requirements	Completeness of undercut	
Influence on adjacent operations	Shape (lead, lag)	
Water inflow	Height of undercut	
Drilling and Blasting	**Development**	**Excavation Stability**
Rock mass strength	Layout	Rock mass strength (RMR/MRMR)
Rock mass stability (drill-hole closure)	Sequence	Orientation of structures and joints
Required particle size	Production	Regional and induced stresses
Hole diameter, lengths, rigs	Drilling and blasting	Rock burst potential
Patterns and directions		Excavation size (orientation and shape)
Powder factor		Drawpoint
Swell relief		Mining sequence
Support	**Practical Excavation Size**	**Method of Draw**
Excavation stability	Excavation stability	Fragmentation
Rock burst potential	Induced stress	Practical drawpoint spacing
Brow stability	Caving stresses	Practical size of excavation
Timing of support: initial, secondary, and production	Secondary blasting	Gravity or mechanical loading
	Equipment size	
Rate of Draw	**Drawpoint Interaction**	**Draw-Column Stresses**
Fragmentation	Drawzone spacing	Draw-column height
Method of draw	Critical distance across major apex	Particle size
Percentage hangups	Particle size	Homogeneity of ore particle size
Secondary breaking/blasting	Time frame of working drawpoints	Draw control
Seismic events		Draw-height interaction
Air blasts–drawpoint cover		Height-to-short-axis base ratio
		Direction of draw
Secondary Fragmentation	**Secondary Blasting/Breaking**	**Dilution**
Rock, block shape	Secondary fragmentation	Ore-body geometry
Draw height	Draw method	Mining geometry
Draw rate, time-dependent failure	Drawpoint size	Particle size distribution
Rock block workability and strength	Gravity grizzly aperture	Range of particles, unpay ore and waste
Range in particle size, fines cushioning	Size of equipment and grizzly spacing	Grade distribution of pay and unpay ore
Draw control program	Ore handling system, size restrictions	Mineral distribution in ore
		Drawpoint interaction
		Secondary breaking
		Draw control (techniques, predictions)
		Draw markers
Tonnage Drawn	**Support Repair**	**Ore/Grade Extraction**
Level interval	Tonnage drawn	Mineral distribution
Shut-off grade	Point and column loading	Method of draw
Drawpoint spacing	Brow wear	Rate of draw
Dilution percentage	Floor repair	Dilution percentage
Controls	Secondary blasting	Cutoff grade to lant
Redistribution		Ore losses
Subsidence		
RMR/MRMR		
Height of caved column		
Minimum and maximum spans		
Major geological structures		
Depth of mining		
Topography		

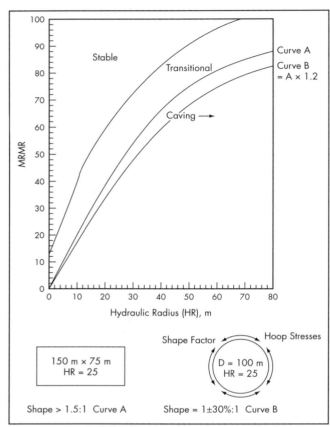

Figure 13.7-2 Stability diagram based on worldwide experience

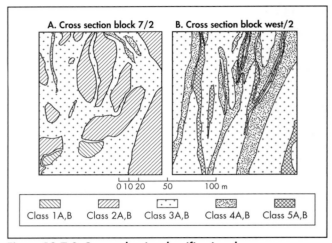

Figure 13.7-3 Geomechanics classification data

Figure 13.7-2, based on a worldwide experience base, illustrates caving and stable situations in terms of the hydraulic radius (area divided by perimeter) for a range of MRMR values. An additional curve has been added to account for the stability that occurs with equidimensional shapes.

All rock masses will cave. The manner of their caving and the fragmentation need to be predicted if cave mining is to be implemented successfully. The rate of caving can be slowed by control of the draw since the cave can propagate only if there is space into which the rock can move. The rate of caving can be increased by a more rapid advance of the undercut, but problems can arise if an air gap forms over a large area. In this situation, the intersection of major structures, heavy blasting, and the influx of water can result in damaging air blasts. Rapid, uncontrolled caving can result in an early influx of waste.

In conventional layouts, the rate of undercutting (RU) should be controlled so that the rate of caving (RC) is faster than the rate of damage (RD) due to abutment stresses. Thus, RC > RU > RD.

However, in areas of high stress, the rate of caving must be controlled to maintain an acceptable level/amount of seismic activity in the cave back. Otherwise, rock bursts can occur in suitably stressed areas (pillars and rock contacts). As advanced undercutting will be used in these situations, damage to the undercut and production levels will not be a problem.

The stresses in the cave back can be modified to some extent by the shape of the cave front. Numerical modeling can be a useful tool for helping an engineer determine the stress patterns associated with possible mining sequences. An undercut face that is concave toward the caved area provides better control of major structures. In ore bodies having a range of MRMRs, the onset of continuous caving will be in the lower-rated zones if they are continuous in plan and section. This effect is illustrated in Figure 13.7-3B, where the class 5 and 4B zones are shown to be continuous. In Figure 13.7-3A, the pods of class 2 rock are sufficiently large to influence caving, and cavability should be based on the rating of these pods. Good geotechnical information, as well as information from monitoring the rate of caving and rock mass damage, is needed to fine-tune this relationship.

Particle Size Distribution

In caving operations, the degree of fragmentation has a bearing on

- Drawpoint spacing,
- Dilution entry into the draw column,
- Draw control,
- Draw productivity,
- Secondary blasting and breaking costs, and
- Secondary blasting damage.

The input data needed to calculate the primary fragmentation and the factors that determine the secondary fragmentation as a function of the caving operation are shown in Figure 13.7-4.

Primary fragmentation can be defined as the size distribution of the particles that separate from the cave back and enter the draw column. The primary fragmentation generated by subsidence caving is generally more coarse than that resulting from stress caving. The reasons for this are (1) propagation of the cave occurs more rapidly, (2) the rock mass disintegrates primarily along favorably oriented joint sets, and (3) there is little shearing of intact rock. The orientation of the cave front or back with respect to the joint sets and the direction of principal stress can have a significant effect on the primary fragmentation.

Secondary fragmentation is the reduction in size of the original particles as they move through the draw column. The processes to which particles are subjected determine the size distribution that reports to the drawpoint. A strong, well-jointed

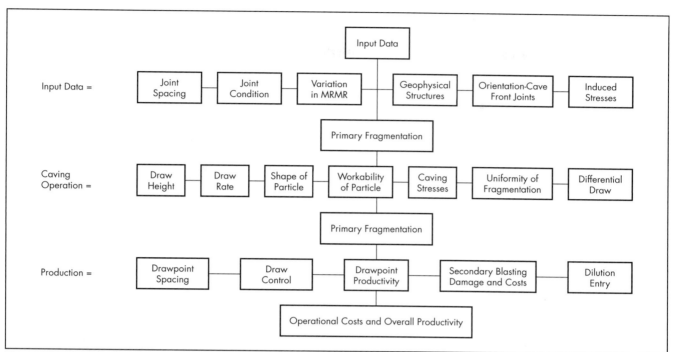

Figure 13.7-4 Input data for calculation of fragmentation

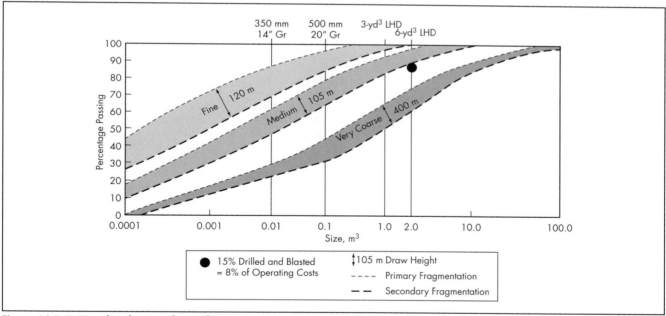

Figure 13.7-5 Size distribution of cave fragmentation

material can result in a stable particle shape at a low draw height. Figure 13.7-5 shows the decrease in particle size for different draw heights and coarse (less jointed) to fine (well jointed) rock masses. A range of RMRs will result in a wider range of particle sizes than that produced by rock with a single rating. This is because the fine material tends to cushion larger blocks and prevents further attrition of these blocks. This difference is illustrated in Figure 13.7-3B, in which class 5 and class 4 material is shown to cushion the larger primary fragments from class 3 material. A slow rate of draw results in a higher probability of time-dependent failure as the caving stresses have more time to work on particles in the draw column.

Fragmentation is the major factor determining drawpoint productivity. Experience has shown that 2 m^3 is the largest block that can be moved by a 6-yd^3 LHD machine and still allow an acceptable rate of production to be maintained. In Figure 13.7-6, the productivity of a layout using 3.5-, 6-, and

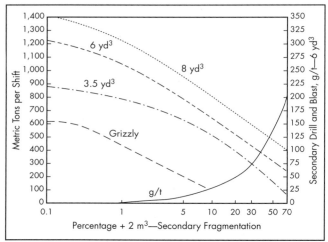

Figure 13.7-6 LHD productivity on a round trip of 100 m with 60% utilization

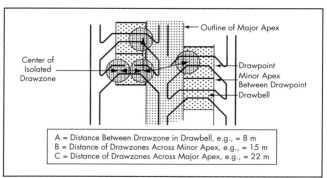

Figure 13.7-7 Maximum and minimum drawzone spacing (isolated drawzone = 10 m, area of influence = 225 m²)

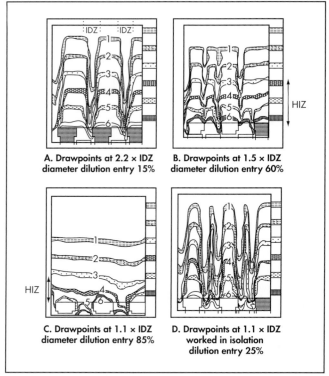

Figure 13.7-8 Results of three-dimensional sand-model experiments

8-yd³ LHD machines loading to a grizzly is related to the percentage of fragments larger than 2 m³. The use of secondary explosives is based on the amount of oversized rock that cannot be handled by a 6-yd³ LHD.

A computer simulation program has been developed for predicting primary and secondary fragmentation. The results obtained from this program are confirmed by underground observations.

Drawzone Spacing

Drawpoint spacings for grizzly and slusher layouts reflect the spacing of the drawzones. However, in the case of LHD layouts with a nominal drawpoint spacing of 15 m, drawzone spacing can vary from 18 to 24 m across the major apex (pillar), depending on the length of the drawbell. This situation occurs when the length of the drawpoint crosscut is increased to ensure that an LHD machine is straight before it loads. In this case, the major consideration (optimum ore recovery) is compromised by incorrect use of equipment or a desire to achieve ideal loading conditions. The drawzone spacings for 30-m center-to-center production drift spacings are shown in Figure 13.7-7, with the critical distance being across the major apex. The length of the drawbell can be increased from 8 to 10 m and the production drift spacing to 32 m without affecting the major apex spacing. Interaction in the drawbell will still occur.

Sand model tests have shown a relationship between the spacing of drawpoints and the interaction of drawzones. Widely spaced drawpoints develop isolated drawzones with diameters defined by the fragmentation. When drawpoints are spaced at 1.5 times the diameter of the isolated drawzone (IDZ), interaction occurs. Interaction also improves as drawpoint spacing is decreased, as shown in Figure 13.7-8. The flow lines and the stresses that develop around a drawzone are shown in Figure 13.7-9. The sand model results have been confirmed by observation of the fine material extracted during cave mining and by the behavior of material in bins. The question is whether this theory, which is based on IDZs, can be wholly applied to coarse material, where arching (spans) of 20 m have been observed. The collapse of large arches will affect the large area overlying the drawpoint, as shown in Figure 13.7-10. The formation and failure of arches leads to wide drawzones in coarse material, so that drawzone spacing can be increased to the spacings shown in Figure 13.7-10.

The frictional properties of the caved material must also be recognized. Low-friction material can flow greater distances when under high overburden load, and this can mean wider drawzone spacings. It is therefore logical to expect that when a line of drawbells is drawn, this creates a large low-density zone. Similarly, when lines of drawbells are drawn on alternate shifts, good interaction will result. The draw program at the Henderson mine in Colorado (United States) broadly follows this pattern.

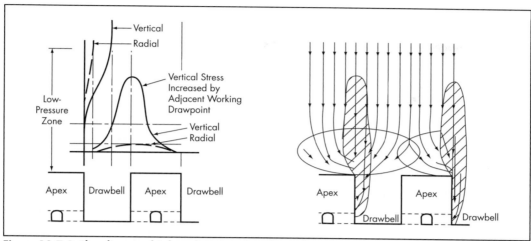

Figure 13.7-9 Flow lines and inferred stresses between adjacent working operations

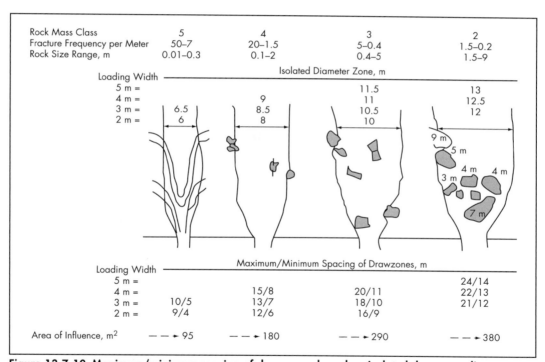

Figure 13.7-10 Maximum/minimum spacing of drawzones based on isolated drawzone diameter

Three-dimensional model tests need to be continued to establish some poorly defined principles, such as the interaction across major apexes when the spacing of groups of interactive drawzones is increased. Numerical modeling could possibly provide the solutions for the draw behavior of coarsely fragmented material.

Draw Control

Draw control requirements and applications are shown in Figure 13.7-11 and Table 13.7-4, respectively. The grade and fragmentation in the dilution zone must be known if sound draw control is to be practiced. Figure 13.7-12 shows the value distribution for columns A and B. Both columns have the same average grade of 1.4%, but the value distribution is different. The high grade at the top of column B means that a larger tonnage of waste can be tolerated before the shutoff grade is reached.

In LHD layouts, a major factor in poor draw control is the drawing of fine material at the expense of coarse material. Strict draw control discipline is required so that the coarse ore is drilled and blasted at the end of the shift in which it reported in the drawpoint.

It has been established that the draw will angle toward less dense areas. This principle can be used to move the material overlying the major apex by creating zones of varying density through the differential draw of lines of drawbells.

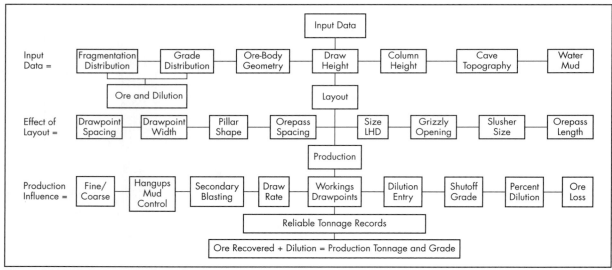

Figure 13.7-11 Draw control requirements

Table 13.7-4 Draw control applications

Objective	Application
Reduce dilution	Calculation of tonnage
Improve ore recovery	Recording of tonnages produced
Avoid damage to pillars	Controlling draw

Dilution

For the purposes of block caving, the *percentage of dilution* is defined as the percentage of the ore column that has been drawn before the waste material appears in the drawpoint. It is a function of the amount of mixing that occurs in the draw columns. Mixing is a function of

- Ore draw height,
- Range in fragmentation of both ore and waste,
- Drawzone spacing, and
- Range in tonnages drawn from drawpoints.

The range in particle size distribution and the minimum drawzone spacing across the major apex will give the height of the interaction zone (HIZ). This is illustrated in Figure 13.7-13. There is a volume increase as the cave propagates so that a certain amount of material must be drawn before the cave reaches the dilution zone. The volume increase, or swell factors, is based on the fragmentation and is applied to column height. Typical swell factors for fine, medium, and coarse fragmentation are 1.16, 1.12, and 1.08, respectively.

A draw control factor is based on the variation in tonnages from working drawpoints (Figure 13.7-14). If production data are not available, the draw control engineer must predict a likely draw pattern. A formula based on the above factors has been developed to determine the dilution entry percentage:

$$\text{dilution entry} = (A - B)/A \times C \times 100$$

where
A = draw-column height (H) swell factor
B = height of interaction
C = draw control factor

The graph for dilution entry was originally drawn as a straight line, but underground observations show that where the dilution occurs early, the rate of influx follows a curved line with a long ore "tail" (Figure 13.7-15). Figure 13.7-16 shows that dilution entry is also affected by the attitude of the drawzone, which can angle toward higher overburden loads.

Layouts

Eight different horizontal LHD layouts and two inclined drawpoint LHD layouts are used at various operating mines. An example of an inclined drawpoint LHD layout is shown in Figure 13.7-17. The El Teniente (Chile) and Henderson mine layouts are shown in Figures 13.7-18 and 13.7-19, respectively. Numerical modeling of different layouts showed that the El Teniente layout was the strongest and had several practical advantages. For example, the drawpoint and drawbell are on a straight line, which results in better drawpoint support and flow of ore. If there is brow wear, the LHD can back into the opposite drawpoint. The only disadvantage is that the layout does not have the flexibility to accommodate the use of electric LHD machines as does the herringbone layout.

A factor that needs to be resolved is the correct shape of the major apex. It is thought that a shaped pillar will assist in the recovery of fine ore. Also, in ore bodies characterized by coarse fragmentation, there is less chance that stacking will occur (Figure 13.7-20). The main area of brow wear is immediately above the drawpoint. If the vertical height of the pillar above the brow is small, failure of the top section will reduce the strength of the lower section and result in aggravated brow wear.

More thought must be given to the design of LHD layouts to provide a maximum amount of maneuver room within a minimum size of drift opening. Thus, larger machines can be used within the optimum drawzone spacings. Another aspect that needs attention is LHD design, that is, to reduce the length and increase their width. Although the use of large machines might be an attraction, it is recommended that caution be exercised and a decision on machine size be based on a correct assessment of the required drawzone spacing in terms of fragmentation. The loss of revenue that can result from dilution

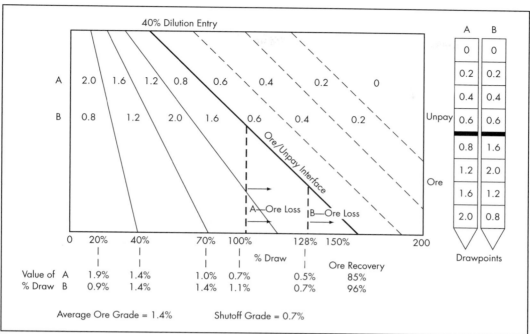

Figure 13.7-12 Grade analysis

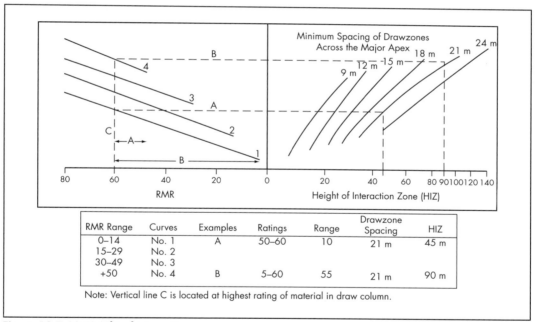

Figure 13.7-13 Height of interaction zone

far exceeds the lower operating costs associated with larger machines.

Undercutting

Undercutting is one of the most important aspects of cave mining because not only is a complete undercut necessary to induce caving, but the undercut method can reduce the damaging effects of induced stresses.

The normal undercutting sequence is to develop the drawbell and then to break the undercut into the drawbell, as shown in Figure 13.7-18. In environments of high stress, the pillars and brows are damaged by the advancing abutment stresses. The Henderson mine technique of developing the drawbell with long holes from the undercut level reduces the time interval and extent of damage associated with post-undercutting. To preserve stability, the Henderson mine has also found it

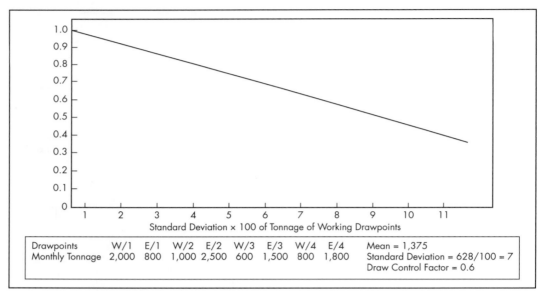

Figure 13.7-14 Draw control factor

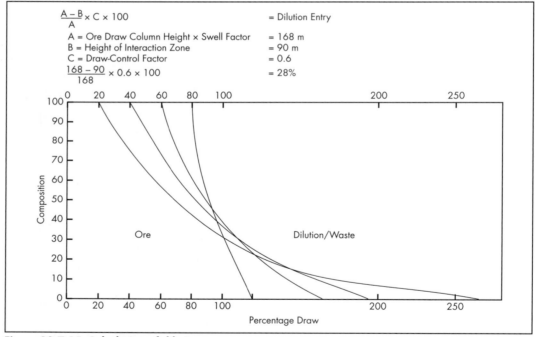

Figure 13.7-15 Calculation of dilution entry

necessary to delay the development of the drawbell drift until the bell must be blasted (Figure 13.7-19).

The damage caused to pillars around drifts and drawbells by abutment stresses is significant and is the major factor in brow wear and excavation collapse. Rock bursts also occur in these areas. The solution is to complete the undercut before development of the drawpoints and drawbells. The "advanced undercut" technique is shown in Figure 13.7-21.

In the past, the height of the undercut was considered to have a significant influence on caving and, possibly, the flow of ore. The asbestos mines in Zimbabwe had undercuts of 30 m with no resulting improvement in caving or fragmentation. The long time involved in completing the undercut often led to ground control problems. Good results are obtained with undercuts of minimum height, provided that complete undercutting is achieved. Where gravity is needed for the flow of blasted undercut ore, the undercut height needs to be only half the width of the major apex. This results in an angle of repose of 45° and allows the ore to flow freely.

Support Requirements

In areas of high stress, weak rock will deform plastically and strong rock will exhibit brittle, often violent, failure. If there is a large difference between the RMR and MRMR values,

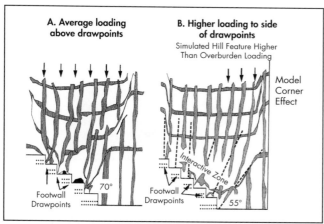

Figure 13.7-16 Inclined drawpoint layout showing effect of different overburden loading (three-dimensional sand-model experiments)

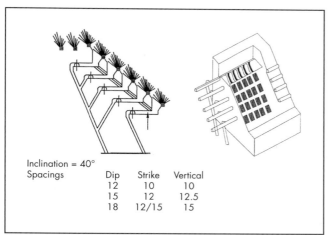

Figure 13.7-17 Inclined drawpoint LHD layout

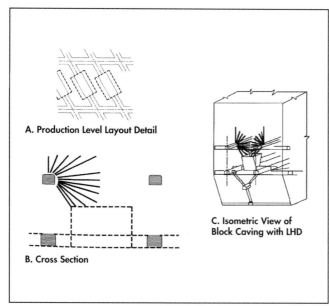

Figure 13.7-18 LHD layout at El Teniente

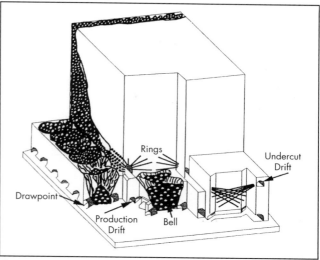

Figure 13.7-19 Isometric view of panel cave operation at the Henderson mine

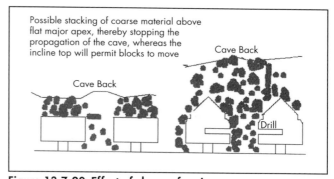

Figure 13.7-20 Effect of shape of major apex

yielding support systems are required. This is explained in Figure 13.7-22.

Prestressed cables have little application in underground situations unless it is to stabilize fractured rock in a low-stress environment. The need to constrain the rock laterally and for lining surfaces such as concrete cannot be too highly emphasized. Support techniques are shown in Table 13.7-5. The use of cable bolts in brows is common practice, but often these bolts are not installed in a pattern that takes into account joint spacing and orientation. Cable bolts should not be installed in highly jointed ground, as the bolts do not apply lateral restraint to the brow and serve only to hold blocks in place.

CONCLUSIONS

Cavability can be assessed provided accurate geotechnical data are available and geological variations are recognized. The MRMR system provides the necessary data for an empirical definition of the undercut dimension in terms of the hydraulic radius. Numerical modeling can assist an engineer in understanding and defining the stress environment.

Fragmentation is a major factor in assessing the feasibility of cave mining in large, competent ore bodies. Programs are being developed for predicting fragmentation, and even the less sophisticated programs provide good design data. The economic viability of caving in competent ore bodies

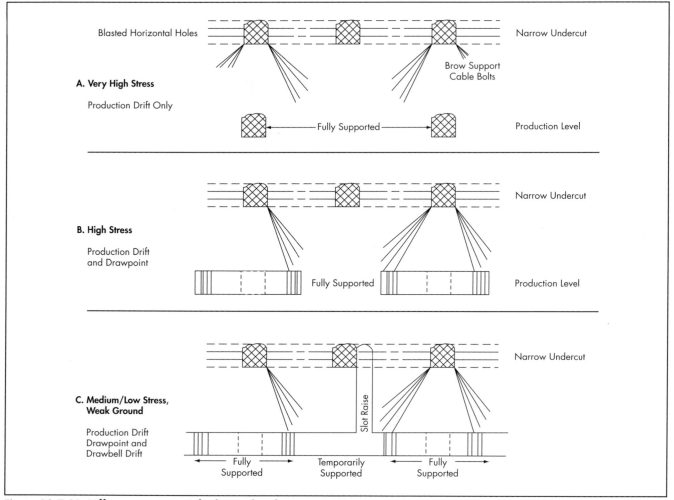

Figure 13.7-21 Different sequences of advanced undercutting RMR

is determined by LHD productivity and the cost of breaking large fragments.

Drawpoint and drawzone spacings for coarser material need to be examined in terms of recovery and improved mining environments. Spacings must not be increased to lower operating costs at the expense of ore recovery. The interactive theory of draw and the diameter of an isolated drawzone can be used in the design of drawzone spacings. Complications occur when drawzone spacings are designed on the basis of primary fragmentation and the secondary fragmentation is significantly different.

The drawpoint is the only control that the operator has once caving has started. The drawpoint should be commissioned as soon as possible after the ore body is undercut. It is therefore essential that every effort is put into ensuring the stability of the drawpoint. The support system must be designed for the rock mass strength and draw height. When it is planned to draw high column heights, the extra expense of ensuring drawpoint integrity is essential. Major drawpoint repairs in a horizontal layout tend to disrupt the draw pattern and can lead to column loading.

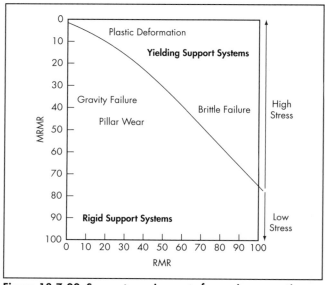

Figure 13.7-22 Support requirements for caving operations

Table 13.7-5 Support techniques

Support Element	Low Stress	High Stress
Bolts		
Length	1 m +(0.33 W × F)	1 m+(0.5 × F)
Spacing	1 m	1 m
Type	Rigid rebar	Yielding (e.g., cones)
Mesh	0.5 mm × 100 mm square	0.5 mm × 75 mm or 50-mm aperture
Deep-seated support	Cables, 1 m + 1.5 W	Steel ropes, 1 m + 1.5 W Long cone bolts
Shotcrete linings	Mesh-reinforced shotcrete	Mesh-reinforced shotcrete
Arches	Rigid steel arches Massive concrete	Yielding steel arches Reinforced concrete
Surface restraint	Large washers (triangles) Tendon traps	Large plate washers Yielding tendon traps
Corners	25-mm rope, cable slings	25-mm rope, cable slings
Brows	Birdcage cables from undercut level Inclined pipes	Birdcage cables from undercut level Inclined pipes
Repair	Grouting Extra bolts, cables, plates, straps, and arches	Grouting Extra bolts, ropes, plates, straps, and yielding arches

Notes: W = span of tunnel. F is based on MRMR = 0–20:F = 1.4; MRMR = 21–30:F = 1.3; MRMR = 31–40:F = 1.2; MRMR = 41–50:F = 1.1; MRMR = 51–60:F = 1.05; MRMR = >61:F = 1.0.

Block caving has advantages over large open pits in environmentally sensitive areas. The crater is much smaller and if necessary can be filled with waste rock even during drawdown. When the operation is completed, the rehabilitation costs will be considerably less than with a large open pit.

ACKNOWLEDGMENTS

This chapter, taken from SME's 2001 publication *Underground Mining Methods*, presents an update of the technology of cave mining. Although it is not possible to quote other references because the bulk of the data supporting the contents of this chapter have not been published, the basic concepts are known to mining engineers. However, it is appropriate to acknowledge the contributions from discussions with the following individuals in Canada, Chile, South Africa, and Zimbabwe: R. Alvarez, P.J. Bartlett, N.J.W. Bell, T. Carew, A.R. Guest, C. Page, D. Stacey, and A. Susaeta. The simulation program for the calculation of primary and secondary fragmentation was written by G.S. Esterhuizen at the University of Pretoria in South Africa.

REFERENCE

Laubscher, D.H., and Jakubec, J. 2001. The MRMR rock mass classification for jointed rock masses. In *Underground Mining Methods*. Littleton, CO: SME.

CHAPTER 13.8

Longwall Mining

Stephen L. Bessinger

INTRODUCTION

Longwall mining is a technique that evolved from a largely manual method in which rows of individual roof props supported the roof along a long face. The ore was broken from the face, first by drilling and blasting, and later by mechanical means powered by pneumatic, hydraulic, or electrical energy. Ore loading was done by hand shoveling into a conveyor or rail cars at the face. This method was labor intensive, lacked productivity, and exposed workers to significant hazards. Since then, technology evolution has produced spectacular improvements, motivated by the need for improved safety and productivity, as well as reduced labor and production costs. This evolutionary process has resulted in modern longwall systems, which are the focus of this chapter. The configuration and major components of a modern longwall are shown in Figure 13.8-1.

The shearer traverses the face, excavating the ore within a defined extraction height. The mined material is loaded onto the armored face conveyor (AFC) by the shearer and is transported to the main belt conveyor system via the stageloader. The stageloader normally has an integral crusher to provide suitably sized material for conveyor belts. The shields advance sequentially, following the shearer to hold up the roof directly above the face equipment and advance the AFC to repeat the cutting cycle. The excavated area behind the shields is allowed to collapse, which limits the roof load that the shields must support and starts a process that may create surface subsidence. High-capacity longwall mining systems depend on the breaking action of a machine to mine the ore. As such, modern longwalls are used to produce ores that can be mechanically cut or broken without explosives.

High-capacity longwalls mainly produce coal, but examples also exist in trona, potash, and phosphate rock. The potential is being investigated to apply longwall technology to platinum and gold-reef deposits with equipment adapted to the requirements of stronger rocks.

LONGWALL APPLICATIONS

Longwall mining has evolved mostly in response to coal-mining applications. Because many coal mines yield a relatively low-value product, particularly when coal is used for electric power generation, the output of the longwall must be a sustained stream, with low cost on a unit-of-production basis. The desirable image often suggested is that of a *coal faucet*, which can be opened or closed as needed. Embedded in the image is the ability to adjust the flow to desired levels. It should be the goal of those involved in the design, operation, and maintenance of a modern longwall to fulfill the goals of sustainability and productivity suggested by the coal faucet image. It is notable that health and safety considerations are intrinsically interwoven into longwall design and operating concepts because they are fundamental elements of sustainability and success. It is customarily accepted that longwall mining is the safest and most productive mining method that can be applied to soft rock deposits.

Longwall systems are serially dependent processes. This is because all of the associated equipment that supports longwall operation must function simultaneously, at the capacity of the longwall, to permit its operation. This dependence is a particular issue for the outby ore transport method, typically conveyor belts. Frequently shortfalls in outby conveyor belt performance are the most limiting constraint on longwall performance. The important point is that the success of the longwall system depends on the success of each of its constituent components, not just the machinery composing the longwall. This principle extends even to mine development activities that create the panel where the longwall will operate in the future.

Longwalls are typically designed for specific considerations related to the target deposit. Notable among these considerations are those described in the following sections.

Deposit Depth

Longwall mining has been applied to deposits with very shallow depth of cover, perhaps less than 50 m in a few cases and up to 1,600 m at the other extreme. Applications in the 100–600-m depth are common for longwall coal mines, especially in the United States and Australia. The loads on longwall roof supports (shields) are largely independent of depth of cover. Shield loads are more related to the extraction height and the

---

Stephen L. Bessinger, Engineering Manager, BHP Billiton–San Juan Coal Company, Farmington, New Mexico, USA

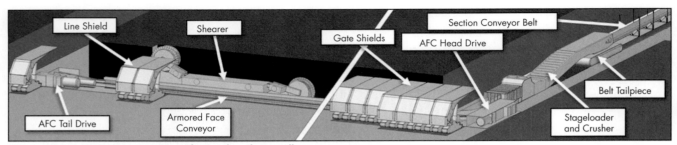

Figure 13.8-1 Major components of a modern longwall

strata overlying the ore bed (seam) than to any other factor. Mining-induced stresses on the face (front-abutment load) and gate road pillars (side-abutment load) are depth related. Not only must gate road pillars demonstrate expected stability, but the roof, in conjunction with the primary and secondary supports, must also be stable. Floor lift (floor heave) is frequently encountered at increased depth or with weak materials in the floor (i.e., under clay, fire clay, or claystone) (Peng 2006).

Deposit Dip

Most longwall mining systems are applied where the inclinations along the longwall face or the panel axis are low, less than 8.5°. Although longwalls are being used at much steeper inclinations, these are not consistent with high productivity. Steeply inclined longwalls also require special adaptations to stabilize the face against downhill creep, shield toppling, and product transport issues. Steep-seam, gate road development is also problematic. Steep, thick seams offer more opportunity for success but are still difficult.

As conveyor belt inclination increases, so do the issues in its operation, up to a practical limit and depending on the material being conveyed. For run-of-mine (ROM) coal, this limit is in the 15° range (CEMA 1997). This value can be even less, possibly 12°, if the material conveyed is wet. The wet condition can arise when production volumes diminish but water application is unchanged. In another case, large volumes of water can be transported by a submerged AFC onto the conveyor belt system. Importantly, operation of either conveyor belts or AFCs is easier when material flow is uphill. Numerous difficulties arise when conveying material down an appreciable grade, which far overwhelms any energy benefits that might be realized (Evans 1993).

The relative layout of the face and progression of panels in the mining sequence with respect to dip is important because of concerns about water migration, methane accumulation, and toppling of face and rib materials.

Deposit Thickness

Longwall systems are typically applied to laterally extensive tabular deposits with limited variation in ore thickness. After the ore thickness and its variation are characterized within the area to be mined, a nominal mining height will be defined. Practical minimum and maximum mining heights are defined by critical equipment sizes and operating envelopes. The longwall has little flexibility in extraction of material within the bounds of a panel and will cut through any area of the panel, extracting at least its minimum operating height. The only alternative is to bypass undesirable areas within the panel by withdrawing the longwall equipment and reinstalling it outside the undesirable area. This is a costly and time-consuming process, chosen only after considerable planning and reconnaissance. Where ore thickness is greater, the longwall can extract up to the maximum cutting height of the shearer or near the fully extended height of the shields. Caution must be used when approaching the fully extended height of the shields to avoid a condition where the shields fail to apply and maintain full loads into the roof without roof failure.

Mining equipment technology has the consequence of defining mining height in the following approximate intervals of deposit thickness (T).

- Thin: T < 1.75 m
- Moderate: 1.75 m < T < 3.75 m
- Thick: 3.75 m < T < 7.25 m
- Very thick: T > 7.25 m

Thin seam systems have historically applied longwall plow technology but have not achieved the productivity of thicker-seam, shearer-based applications.

Most modern, high-capacity longwalls operate in the moderate height category—from 1.75 to 3.75 m. Optimum height is generally agreed to be in the 2.5–4.0 m range for single-pass longwalls.

Above 3.75 m and up to approximately 7.25 m mining height, the thick category, is the upper limit of single-pass longwall capability. In this range, normal longwall equipment requires numerous adaptations to perform successfully until it reaches a practical limit at approximately 7.25 m.

Above 7.25-m extraction thickness, longwall technology is forced into several alternatives: multilift longwall or top coal caving. On a worldwide basis, as very thick coal seams extracted by surface mining extend to depths beyond economic limits, underground mining methods will be required that offer high resource recovery, such as longwall top coal caving.

Deposit Uniformity

Ideally, a tabular deposit well suited to longwall mining would exhibit uniformity in the ore bed/coal seam as well as the adjacent roof and floor. Unfortunately, examples of such ideal conditions are few. Generally, discontinuities result from structural or depositional origins but may also be the result of post-depositional events.

Structural geologic disturbances normally involve folding and faulting. Gentle inclinations induced by folding can be considered as previously described in the discussion on dip. However, if the gentle dips of a broad syncline, anticline, or monocline go over approximately 8.5° as a trend, this may limit applicability of the longwall. Expectations of high productivity may decline as the inclination steepens from 8.5° up to 15°.

Aside from overall inclination, the rate of change of inclination becomes a constraint, as the AFC, shearer, and shields

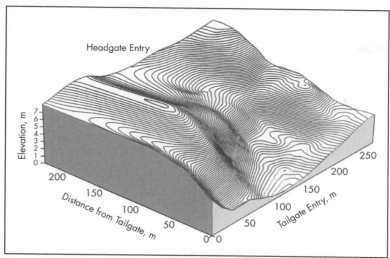

Figure 13.8-2 Disturbed seam topography

all have different concerns in this regard. As an example, if the radius of curvature of seam topography gets short, and the shearer is able to follow it, the AFC may have less vertical articulation than required to follow the changed topography and bridge, forming a span that is not supported by the floor. Thus, the AFC would only be supported by the connections that hold adjacent pan sections together (*dog bones* or *dumb bells*). By the AFC's own weight, or with the additional weight of the shearer, the AFC connections may fail or the pan structure may fail. This can cascade into an AFC system failure, such as a "chain wreck" where AFC chains break, flights are damaged, and main drive sprockets may fail. Such incidents may take hours to days to repair and cost in the hundreds of thousands of dollars in direct and lost-opportunity costs. They also create avoidable safety hazards.

In an alternate case, shields can lift their bases off the floor during cyclic advancement if their canopies become "iron bound" with adjacent shields because of a short radius of curvature in extraction profile. This condition can result in a suspended shield, which can fall to the floor unexpectedly and be a safety hazard. Of course, all of these problems are created when the equipment is asked to do something that was never intended by the original equipment designer or, conversely, was not identified as a performance requirement in the acquisition specifications. Figure 13.8-2. shows a disturbed area that was successfully mined with a longwall but not without impact to productivity.

Another structural geology concern is fracturing or faulting. An example might be fracturing along the axis of local bed flexure, which can adversely affect roof control or cause elevated gas or water inflows. Notably, fractures/faults parallel to hillsides exist in some areas and may complicate logical panel layouts beneath a ridge line. Faulting may be present with or without appreciable folding. Strata displacements may be normal faults, and total displacements may be distributed across one or more discrete fault traces within a short distance. This is generally a manageable issue so long as the local displacement is not too large. Because these features are preferably traversed at an acute angle to the face, grading roof, and floor into and out of the feature makes operating through local offsets feasible—up to double the mining height in displacement. This will introduce an adverse impact to ROM product quality and overall productivity. Obviously, proximity of faults on a single face is problematic, as is the case where faults, still at acute angle to the face, exhibit enduring persistence. Both of these cases may require revised ROM quality and productivity estimates.

Then there are those cases where the fault is at a shallow angle with the face line or possibly even parallel to the face at worst case. This scenario can have seriously negative impacts to productivity and may lead to premature face recovery or even threaten the loss of equipment. If such a feature is suspected, it should be approached with caution after careful risk assessment and with a plan of mitigating actions in place. If such a feature is first recognized as it emerges onto the face, it is certainly appropriate to make efforts to evaluate it and implement a mitigation strategy before exposing a large proportion of the face to the hazard. It has frequently been the case that a longwall could "outrun" a local adverse roof condition; however, that is not true for faults parallel or at shallow angle to the face.

In the case of most structural faulting, the ground stresses within one block of the fault may not be equivalent to the stresses in others. Although normal faulting is at some risk in this regard, it is particularly problematic in thrust faults. The stress differential between blocks in a thrust-faulted scenario can result in some blocks having extremely difficult ground control issues compared to others.

Depositional discontinuities develop in response to conditions present at the time that the mineral deposit formed. These discontinuities commonly arise in conjunction with sand channels above or below a coal seam or as scours (also termed *faults* or *wants* in some locales) where they erode and replace the coal seam. Figure 13.8-3 depicts sand channels as they may commonly occur in coal-bearing strata.

In near proximity to these features, it is possible to have elevated stresses and differential compaction surfaces. These glassy smooth discontinuities (slickensides) are weak in tension and can constitute a hazard to roof control. Where large thicknesses of coal occur over a short stratigraphic interval, compactional features can exist as faults, with displacements mostly confined in near proximity to the seam and having normally short lengths, perhaps less than 250 m. Compactional faults are not normally the source of great concern.

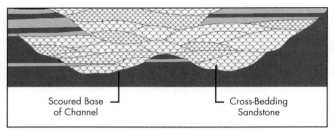

Source: Adapted from Papp et al. 1998.
Figure 13.8-3 Sand channel in coal strata

Sand channels above the mining horizon can lead to increased shield set load and yield load requirements and should be evaluated accordingly. Where these features might be encountered at a shallow angle to the face, opportunities to avoid the condition are initially preferred. When near-seam sand bodies are strong and stiff, in conjunction with significant depth, the possibility of sudden strain energy release in response to mining can be a concern.

Another variation of depositional discontinuity occurs when soft sediments are injected into or across the coal seam. These are not usually large or strong, so they represent more of a threat to ROM quality and local ground control as opposed to sustainability (Moore 1940).

In a similar fashion, it is occasionally the case that igneous materials can be injected into coal-bearing strata, forming lenticular or vertically crosscutting dikes. In various locations around the world, including South Africa, Australia, and the western United States, these igneous dikes can be a serious obstacle to longwall operation. Where these bodies are narrow, perhaps less than 1.0 m wide, they are often sufficiently contaminated by entrained debris that they do not achieve strength beyond the limits of mechanical cutting by a shearer or continuous miner. However, when encountered as thicker bodies, typically a stronger core requires other means to penetrate it. In South Africa, this material is identified as dolerite and forms dikes off of sills intruded into the coal-bearing strata. The dolerite dike material can be more than 140 MPa compressive strength (Papp et al. 1998). It is also the case that post-depositional events can lead to the formation of natural coke, or a natural burn can produce coal loss with only residual coal ash.

It is safe to assume that all of forms of rock-mass nonuniformity noted will have a negative impact on longwall operation. As such, it is useful to identify any geologic features at the planning stage. Although reconnaissance drilling and detailed geologic mapping are customary, various geophysical methods can be applied to search for the different discontinuity types. Satellite imagery, magnetic or seismic surveys, and electromagnetic tomography are a few of the methods used. Unfortunately, these methods are not universally useful, and choice of method deserves careful attention. It is common to find that none of the methods are particularly informative for any specific site.

Strata Gases

Characterizing the gases and quantities that might be encountered during longwall development or subsequent longwall mining is important when striving to achieve the coal faucet concept. If acceptable atmospheric conditions cannot be continuously maintained where people work and travel, the longwall is at risk of being idled during periods where excursions beyond acceptable limits occur. In the United States, concern is emerging for atmospheric safety in the mined-out areas (gob or goaf) behind the longwall or in sealed adjacent areas.

Several gases have prominent impacts on longwall design and operation worldwide. These include those described in the following paragraphs.

Methane
Methane (CH_4) is common in coal mines to varying degrees and may be encountered in most sedimentary strata. Coal seams have long been recognized as likely reservoir rocks for methane, but it is increasingly recognized that other porous and permeable near-seam rocks can be significant sources of methane liberation. This is particularly true when the in-situ rock mass is fragmented by mining-induced caving. Higher CH_4 liberation rates in new longwall gobs result from creation of smaller particles than existed in the undisturbed rock mass. Release of lithostatic stress also lends to increased production rates of methane in caved material. Not only does the methane fill the void space of rock porosity at some reservoir pressure, it can be chemically adsorbed by coals. It is also common for fractures or faults to be charged with fluids.

Occasionally the rock stress augmented with interstitial fluid (gas) pressure can be greater than the unconfined rock strength, which results in one mechanism of *outburst* when rock failure occurs. Not only can this create atmospheric inundations from the escaping gas, it can result in violent ground control hazards. Such events are known in a few U.S. and Australian coal mines.

Lastly, the most notable hazard of methane in the mining environment is its explosibility when mixed with atmospheric oxygen in certain ratios. As one step to avoid this hazard, most longwall face equipment is designed, tested, and certified to be explosion proof or intrinsically safe. Explosion-proof equipment will not allow an ignition within the enclosure to propagate outside it. Intrinsically safe components rely on the principle that methane–oxygen mixtures will not ignite below certain initiation energy thresholds, and they ensure that such energy is never available, even during component failure conditions.

Carbon Monoxide
Carbon monoxide (CO) can be adsorbed on coal and naturally occurring or it can be a product of partial combustion of hydrocarbon fuels or coal. This gas primarily represents a health hazard, but can be explosive in relatively high levels when mixed with atmospheric oxygen.

Aside from its human toxicity, it is often an indicator of combustion, spontaneous combustion in some coals, or may result from fire or explosion. Longwall mines with spontaneous combustion propensity may monitor the CO production rate from the longwall as an early indicator of an evolving thermal event.

Carbon Dioxide
Carbon dioxide (CO_2), a common strata gas, may be present along with methane or may be a dominant constituent. In some Australian cases, high-pressure CO_2 resides in localized areas of the coal seam and requires specialized control measures to manage the hazards of outbursts. CO_2 is also a product of full combustion of coal and hydrocarbons. As such, it can be

one of several gases monitored to detect evolving combustion events, spontaneous combustion in particular. Graham ratio, Jones-Trickett ratio, and CO/CO_2 ratio, as well as CO production rate, may be used to detect the onset or monitor progression of a combustion event (SIMTARS 1999).

Hydrogen Sulfide
Hydrogen sulfide (H_2S) is primarily recognized in the mining environment for its toxic effects on humans. It is also a concern in longwall mining for its corrosive effects on metals and its participation in unexpected steel failures.

H_2S can dissolve in still water and be released when the water is disturbed. The aqueous mechanism can bring it in direct contact with metals, such as bronze or steel in longwall components, and result in metal sulfide formation and the liberation of free hydrogen (H_2). Because H_2 is not normally a strata gas but is known to be an indicator gas in combustion events, this can explain its existence without the presence of a thermal event.

H_2S can be a metabolic product of sulfate-reducing bacteria. These bacteria are naturally occurring and can form microscopic colonies beneath which high concentrations of H_2S can exist, perhaps up to 4,000 ppm. This can create overwhelming corrosive or metallurgical effects while atmospheric levels remain low.

Other Gases and Vapors
On occasion, other gases and vapors may naturally occur. Low molecular weight hydrocarbons can exist in gas or liquid phases in sedimentary strata. Their combustibility and quantity may be reason for concern in some instances.

In most cases, the regulatory body that has jurisdiction over mining health and safety for a specific location will have statutory guidelines for safe and legal levels of the noted gases. In order to be successful and sustainable, a longwall system must be designed to accommodate safety and statutory compliance in order to avoid production outages associated with gas level excursions above allowable limits.

Water
In many areas, water is associated with the coal seam or near-seam strata. It may also be associated with overlying or updip workings and fracturing or faults. In any event, its management is reason for careful planning because of its generally negative impacts to longwall development and operation, as well as the typically negative impact it can have on ground control. Where swelling clay minerals are present in the rock, the matrix may decompose when the clay minerals hydrate and swell. The slake-durability test is one measure of a rock's behavior when exposed to moisture. When swelling clay minerals are present, the rock typically performs poorly in slake durability.

Siltstone, mudstone, claystone, and fireclays are often the operating floors for longwall development sections or the longwalls themselves. The often present water, coupled with the duty as a roadway for rubber-tired equipment, turns these floors to mud, sometimes hundreds of millimeters deep. In the immediate headgate or tailgate entries, this can mean that the belt tailpiece and stageloader are resting on mud and may even invite floor heave, which requires clearance before equipment advance. It also can mean that gate shields may unreliably advance or be unable to create the intended load into the roof, leading to a deterioration of roof control in the headgate or tailgate. This same principle can leave standing support such as props, cans, or cribs with an inadequate foundation upon which to create resistance to roof deformation. Water management, or the failure to manage it, is often a reason that longwall productivity falls short of its full potential.

Spontaneous Combustion
Spontaneous combustion is a condition where slow oxidation of a combustible substance leads to heat evolution and increasing temperature until the material reaches a point where combustion starts and spreads to adjacent fuel. Some coals, particularly those with lower rank, exhibit a propensity toward spontaneous combustion (Cliff et al. 2004). It is a particular concern when spontaneous combustion heatings or thermal events take place in a longwall gob (goaf), where they can be difficult to locate and identify, and even more difficult to access and extinguish. Such events typically cease production in the affected portion of the mine until, in the best case, the event ends or, in the worst case, causes loss of life and assets.

As serious as this scenario is, an even worse one is possible where methane is present in combination with a coal prone to spontaneous combustion. In that case, atmospheric oxygen in combination with explosive methane levels can be ignited by a small spontaneous combustion ignition source, resulting in an explosion and fire. In such an event, multiple fatalities and loss of assets are possible.

Because responsible mine operators are committed to a *zero-harm* philosophy, effective controls must be offered to preclude the possibility of a spontaneous combustion event or an explosion and fire. The mitigating strategy for all of these concerns is the long known *fire-triangle* concept for fuel, oxygen, and ignition source. Considering the fire triangle, it becomes immediately evident that in a spontaneous-combustion-prone coal mine, the only component that may be subject to control is oxygen. Thus, the choice of bleederless longwall ventilation, explained later, can effectively exclude oxygen from the gob (goaf) and mitigate the hazard.

As with any underground fire, the issue is only partially getting the fire extinguished, which usually entails oxygen deprivation, but also dissipating the residual heat resulting from combustion so that the event will not simply resume if oxygen becomes available again. This is also why early detection and intervention are key components to a spontaneous combustion control plan. But the most important element is a strong focus on prevention (Mitchell 1996).

Roof and Floor Strata
One of the advantages of a longwall mining system is that it can accommodate a wide range of roof conditions. Gate road stability issues aside, stabilizing the longwall roof is actively done by the shields. The shield must immediately stabilize the near-seam roof when it is initially set against the roof after it is advanced, referred to as *set load*. The set load required to stabilize the roof is a function of site-specific conditions and parameters of the face cross section. The set load produced by the shield is derived from the emulsion system pressure available to the shield during setting, the major stage diameter of the leg, and its resulting area. Set pressure applied to leg area and derated, to account for leg inclination (angle to vertical) functionally defines set load created by the shield. To be successful, set load required to stabilize the roof must be less than the vertical thrust created by the shield at setting. Subsequently, the shield must stabilize the roof throughout the time when the

shearer cuts past it, adding a web depth to the span, and ultimately as adjacent shields are lowered away from the roof as they are advanced. Loads also arise from face deterioration and span enlargement to the degree that it occurs in response to development of front-abutment loads from web to web. If the shield resists this load and the roof remains stable, the result is the end-cycle load required to support the ground. In concept, this should be at or below the yield capacity of the shield.

This yield load is calculated in a manner similar to that of set load, except that instead of system setting pressure, the opening pressure for the leg yield valve must be considered. With set load and yield load independently defined, it is possible to create some of the popularly discussed ratios such as set-to-yield ratio, yield capacity/linear length of face, and set or yield capacity/area of supported roof. Although these terms are frequently discussed as design factors, they are, in fact, only outcome statistics to the actual equilibrium that must be successfully established. They also have limited comparability over a range of possible operating heights, because only the vertical component of the leg load primarily contributes to roof stability. The horizontal force component is mostly resolved within the shield by the lemniscates links.

Because the interaction of the roof and the roof support (shield) has been described, and the need for stability is clear, the factors leading to loads arising from the roof must be characterized. Roof loads on shields increase with increasing distance along the face from either gate road up to a maximum load specific to any site. Distances greater than that effectively only expand the number of mid-face shields exposed to that maximum load. It is not prudent or practical to attempt to select shields in the design process for any load less than the maximum load experienced for the specific installation. Certainly, using outcome statistics or rules-of-thumb to size shield capacity is poor practice and risks the possibility that an entire longwall will fail to perform as intended. Should this be the case, available solutions may require premature equipment replacement or acceptance of undesirable operating performance.

Evaluating the factors that contribute to roof loading on the shields, it is clear that both site-specific and face cross-section factors are contributory. Site-specific factors include

- Seam thickness (extraction height),
- Near-seam strata (type and thickness),
- Seam and nearby strata material properties,
- Caving behavior of strata,
- Seam inclination,
- Natural and mining-induced fractures, and
- Previous mining above or below the seam.

In concept, the overlying rock collapses into the void created by mining. As the intact material fragments and caves, with particles rotating and translating into the void, its volume bulks to fill the void, largely in inverse proportion to particle size. This process continues until little enough void space remains between the rubble and the uncaved rock that further displacement of the overlying rock takes place through particle translation but with little rotation. Eventually, even the downward translation reaches equilibrium with the resistance created by the underlying compacted rubble. Thus, the weight of rock overlying the longwall can be supported across three areas:

1. Unmined face ahead of the longwall (front-abutment load)
2. Longwall roof supports
3. Rubble material of the gob

The area depicted in Figure 13.8-4 graphically shows the region of rock that must be stabilized by the roof supports (shields). In consideration of the loading process described, it is evident that the loading of longwall roof supports has little impact from mining depth. It also becomes evident that overlying roof materials that cave as small particles from thin layers can lead to smaller caving heights before particle rotation is diminished and consequently to lower shield loads and thus lower shield capacity requirements. Conversely, thicker and stronger overlying beds can impose much larger loads on roof supports, especially if they are close to the mining horizon. Another consideration is that stiff roof rocks, typically strong as well, can require high set loads to stabilize them. Further, they can lead to extreme cases of periodic weighting, where strong beds can act as cantilevered loads, imposing successively greater loads on the roof supports until the cantilever fails. Such strong beds can be multiple within the caving horizon and cause superimposed effects of periodic loading. Figure 13.8-5 shows an example of how this variation might look for an example longwall application (Barczak and Garson 1986). Wind blasts originating in the gob often coincide with the caving of thick, strong beds, or the first cave of a new longwall panel.

In some instances a cantilevered bed failure results in the release of stored energy and in a "bump" or "bounce." This mining-induced seismicity can be damaging to equipment and hazardous to personnel. In general, mining-induced seismicity is often associated with strong rock or coal and significant depth. High overburden stress coupled with mining layouts that may act as stress concentrators can lead to mining-induced seismicity originating at elevations ranging from beneath the mining horizon and extending upward to the surface. High-flow yield valves on shield leg cylinders can afford protection against equipment damage at moderate convergence rates in a periodic weighting event, but only purpose-designed rock-burst protection can prevent damage to shields during a bump where a large convergence can occur suddenly.

Face cross-section factors that affect loads imposed on the shields include

- Geometry of the shields, shearer, and AFC;
- Depth of cut (web); and
- Sequence state in cutting cycle.

Because of these factors, coupled with geologic variation that takes place on scales smaller than the area mined by a set of roof supports within their service life or even within the course of a panel, it is not surprising that the loads imposed on shields vary over a range at any installation. Figure 13.8-6 shows the cumulative probability distribution of loads required to stabilize the longwall roof.

These curves can be numerically modeled for a proposed longwall face cross section and strata section, and are site specific for every individual application. Models of this type have shown good correlation with data collected from pressure monitoring on equipment in service. As discussed previously, the set load curve is derived independently of the yield load curve.

It can be observed that in the region above approximately 96% cumulative probability (CP), providing additional support load into the roof offers diminishing returns. Using this method to select roof support capacity leads to selection of

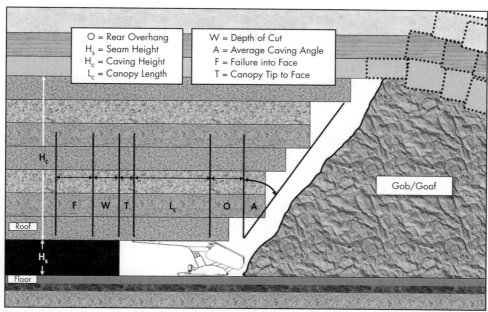

Figure 13.8-4 Conceptual loading model for longwall shield capacity evaluation

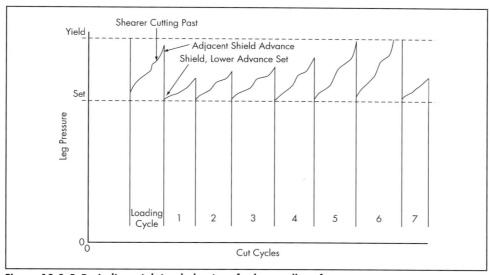

Figure 13.8-5 Periodic weighting behavior of a longwall roof

supports that control loads from the roof in the 91%–96% CP range. This leaves a 4%–9% incidence of yielding. In most cases, yielding events will be of brief duration, and equilibrium will be reached by sharing load onto adjacent shields. Selection of capacity in the higher end of the range, 93%–96% CP, allows for some deterioration in performance, as equipment ages, without creating performance problems.

Choice of supports larger in capacity than 96% CP of the roof loads that will arise offers higher cost, greater weight, reduced travelway dimensions, and elevated emulsion system demands with little perceptible benefit. However, because the probability distribution is nonlinear, selecting capacity much below 91% CP leads to an excessive occurrence of shield yielding and onset of roof damage companion to the more severe yielding events.

Notably, potential shields for any application should be evaluated at different heights (leg cylinder inclinations) to verify that they will be adequate for any mining height condition they may experience in service. Excess debris should not be allowed to accumulate on top of the roof canopy or beneath the base, as this material can be compacted and may allow excess roof convergence and damage to arise before enough support resistance is generated to create equilibrium. This is especially true for strong, stiff roof materials such as competent sandstones near the roof line in some coal mines.

Specific Energy of Cutting

The productivity of a mechanized cutting process for rock is largely a function of the material's specific energy of cutting, which depends on the type of cutting being applied. Shearers

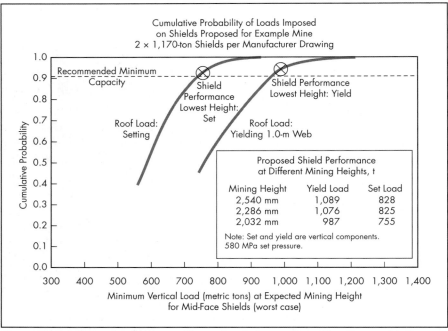

Figure 13.8-6 Example of site-specific stochastic longwall roof-loading evaluation

are different from drum-type continuous miners, which are yet different from longwall plows. While shearers and continuous miners work against forces related to compressive strength of the material being cut, plows primarily work against the tensile strength of the material, which is typically much weaker than the compressive strength. Particularly with respect to shearers, the cutter motor power used to produce a resultant quantity of ROM ore over a period of time defines the specific energy of cutting for the material (kilowatt-hour per metric ton [kW·h/t]). This assumes that the cut material can be cleared so that product accumulation does not impede the process. This measure is useful for the purpose of production process design and management but is not as pure a parameter as the name might suggest.

The load applied to the material to be cut can alter this value and is a benefit for the longwall system. When the front-abutment load has fully evolved, after the start of mining in a panel, this load is usually large enough to "soften" the face and reduce the energy required to produce a quantity of material. Until the front abutment fully develops, cutting will be more difficult than after it exists. The apparent softening can also be noted as abutment load varies in periodic weighting cycles and when production resumes after a stoppage of sufficient duration to allow significant stress redistribution.

The addition of work done by the abutment stress is also advantageous to creation of a larger product size distribution and lower fugitive dust production on a milligram/cubic meter/metric ton basis. Interestingly, the cutting bits deeper in the web do more work and produce a smaller size distribution of product than the bits shallower in the web. This smaller-sized product may be significant in dust production, which can become airborne. The possibility of improved machine utilization and proportionately reduced dust production exists with adoption of increased web depth. In the transition of web depths from 0.76 m to as much as 1.07 m, this appears to have been true.

Specific energy of cutting for various materials being cut by shearing machines ranges from 0.05 kW·h/t for easily cut coals to 1.3 kW·h/t or more for some very hard cutting sandstone and igneous dike materials.

Where very difficult cutting is expected for an extended duration, specially designed *rock drums* are used in lieu of normal production drums. These rock drums are designed with closer bit spacing and more bits per line in the drum lacing to accommodate the heavier cutting demands. The style of bits may also change between radial and conical cutting bits to offer improved longevity and performance.

Typically, high-capacity longwalls operate well when they encounter materials in the 0.05–0.45 kW·h/t range.

Abrasivity

When considering a longwall mining application, the abrasivity of the product to be mined and conveyed is of concern. Most wear on the AFC and beam stage loader (BSL) is related to the tonnage of material that has been conveyed across the systems. As an approximation to the wear that might be expected in a particular application, standardized laboratory tests have been developed that are also used by the electrical generation industry, which has similar concerns as mining about abrasive wear to equipment within its facilities. In high-level summary, sacrificial blades are rotated through a mass of the rock material of interest for a predetermined time. At the end of the test, the weight of material lost by the sacrificial blades is determined and reported. This weight loss ranges from only 8 mg Fe in a trona specimen to more than 1,300 mg Fe in a severely abrasive coal and sandstone ROM product. Typical values range from 200 to 900 mg Fe, which roughly correlates with wear of 0.5 to 2.5 mm of deck-plate thickness reduction per million metric tons of production. Unacknowledged in these values are the corrosion processes, which can certainly influence the outcome in service. Improved results can be achieved by use of abrasion-resistant steels where the corrosion environment will not provoke premature failures (Ludema 1996).

Mine Layouts

Having identified most of the important site-specific considerations that might have to be addressed by a successful mine design, the general mine layouts can begin to be discussed.

Historically, advancing and retreating longwall layouts have been the subject of much discussion worldwide. In the advancing system, the longwall setup room is close to the main or submain entry set, and the longwall mines away from these entries into undeveloped reserves. In the process, support is constructed to constitute one or more gate roads behind the longwall face. Mining continues until a designated extent is reached or conditions force face recovery. The fact that artificial roadways have to be maintained in the gob adds significant cost and labor to this technique. As cost and productivity benchmarks have evolved across the industry, the elevated costs and lower productivity of advancing faces have generally seen these operations cease or transition to retreating longwall layouts. Only in unusual circumstances or where historic precedent dictates are advancing faces operated. At present, all of the world-class longwall installations are retreating faces. This trend can be expected to continue into the future.

Retreating longwall layouts may be of two types: conventional or punch (highwall) type. Figure 13.8-7 shows a conventional retreating longwall layout with a mined-out panel adjacent to the active longwall. The entries of the previous panel's headgate (main gate) transition into becoming the tailgate for the active face. The active longwall face mines progressively from the setup room at the far (inby) end of the developed panel and is eventually recovered as it enters its intended recovery position. The recovery position of the face is shown by the line that coincides with the near (outby) extent of the mined-out panel. Optional recovery chutes may be used to aid free equipment recovery.

The punch (highwall) longwall is largely similar, except that instead of developing a main or submain entry set from which gate roads are driven, the gate roads are developed directly off a surface-mined box cut or highwall. Punch longwalling is attractive because it eliminates the need to develop and maintain long-term underground workings in the form of mains and submains entries. It also lends itself well to the transition from surface mining to underground mining that confronts many mine operators as the depth and production costs of remaining reserves increase.

Gate roads for retreating longwalls are typically formed with two- to four-entry developments. Although single-entry and dual-entry gate roads are common outside the United States, dual-entry gate roads are an exception in the United States. In the interest of worker safety in a fire or explosion emergency, U.S. mining law generally requires a minimum of three-entry developments in coal mines to accommodate separated intake (fresh air), return (exhaust), and conveyor belt entries.

In cases where it is safer to have a dual-entry development than the statutory minimum of three entries, a variance may be granted by regulatory enforcement authorities based on a demonstrated diminution-of-safety-argument. U.S. mine operators have succeeded in the diminution-of-safety argument where adverse geotechnical conditions threaten worse consequences than the reduction of number of entries. Companion to the grant of such a variance are additional controls to safeguard against the hazards envisioned during statute development.

Often, especially in the United States, three- or even four-entry developments are used for gate roads. The motivation for creation of more development length, in total, than is minimally required to sustain longwall operation should be carefully considered. It is ordinarily desirable to minimize the ratio of development costs per longwall metric ton produced. It is also desirable for time value of money considerations to associate as closely in time as possible the expenditure of costs that provide future benefits with realization of those benefits. In the pursuit of these principles, it is common for longwalls to be operated in geometric succession, with the headgate of the prior panel being used as the tailgate of the current panel and so forth, until some predefined extent is reached.

Often this extent is the aggregation of a group of panels to form a *district*. The number of panels in a district is variable, typically based on ventilation or geotechnical considerations. When managing the constraints of a large district eventually becomes difficult enough, mine operators prefer the development of the incrementally additional gate road, which is required to develop the first panel in a new retreating longwall district. Frequently, districts are sealed upon completion, removing them permanently from the worked and traveled portion of the mine. Mine operators are then largely relieved of the burden to maintain access for inspection and ventilation of such a sealed district, but access must be maintained to inspect the final seals where the district meets with active mine workings, and the seals themselves must be maintained in adequate condition.

Accumulations of water, explosive atmospheres, or spontaneous combustion events can be reasons for concern, even in sealed and otherwise abandoned areas. Relatively recent changes to U.S. regulatory requirements for seals and monitoring atmospheres in sealed areas address these concerns.

On occasion, it can be necessary to isolate overburden stress effects between the current longwall panel and the prior panel with a barrier pillar. Where panels are mined at depth with thick, stiff beds in the overlying strata, failure to adequately support the stiff strata can channel vertical stresses toward the current panel. This elevated stress can result in mining-induced seismic activity in the strata section, with possible events in excess of 3.5 on the Richter scale. These events can be damaging to equipment and hazardous to personnel, particularly when seismic activity originates at or near the face location. At a lesser scale, stiff coal subjected to large stresses can fail suddenly and result in the expulsion of material from the face toward the longwall travelway, creating a hazard to personnel.

Longwall gate-road development is characteristically "just in time" to avoid delaying production activity on the longwall. The degree to which the development is ahead of the needs of the longwall in time is often referred to as *float time*.

Some ventilation plans require the next adjacent longwall to be developed in order to implement the ventilation plan. More commonly, this is not the case, and float times are in the 6-to-50-week range. This allows adequate time to withdraw the development equipment and stage longwall-move supplies and equipment prior to the onset of a longwall move. However, it does not impose unnecessary "hang time" on the gate road, as deterioration of conditions can begin soon after development in some settings. It also associates the cost

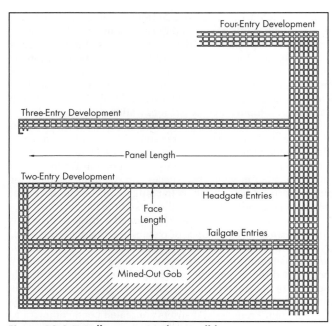

Figure 13.8-7 Fully retreating longwall layout

of such development in time, closer to the future benefits expected to arise from it. For many mine operators, the longwall move actually starts with emplacement of portions of the new face prior to completion of mining in the current panel. Such equipment might include: tailgate gate shields; AFC and BSL; a new or rebuilt shearer; electric power and control equipment; and emulsion-pumping equipment, controls, and tank. The amount of equipment preinstalled depends on arrangements specific to any mine operator. In general, shorter production outages occur with increased amounts of equipment preinstalled in the new longwall panel.

Gate road development itself is done with continuous miners and one of the several face haulage mechanisms available. Figure 13.8-8 shows the general layout of a three-entry gate-road development section with a continuous miner and battery coal haulers. The section depicted uses a place-changing method of operation where a continuous miner mines a cut and then changes places to mine a cut in another entry or crosscut. When the continuous miner has withdrawn, a separate machine installs roof bolts and possibly wire mesh or roof straps to roof and rib or roof trusses, if necessary.

Roof bolting progresses from where support was installed on the prior sequential cut in that working place to the face of the cut. After a cut is fully supported, the roof bolter is withdrawn and the cut is ready for auxiliary activities that may be necessary before the continuous miner returns to make its next cut in that working place. As this process is successively applied to entries and crosscuts, the entry set is advanced. Cuts may range from as little as 3.0 m to more than 12.0 m. In good roof conditions, the cut depth may be limited to the depth attainable without exposing personnel to unsupported roof or airborne contaminants.

Short cuts have a decidedly negative impact on the productivity of place-changing mining methods, as a disproportionately great amount of working time is lost to place changing. As depicted, ventilation is provided by section fans.

The coal haulers transport the mined product to a conveyor belt for ultimate clearance from the working section. It is common for the conveyor belt to be equipped with a feeder breaker. The feeder breaker has a limited amount of surge capacity, allowing coal haulers to unload quickly, but also meters product through a breaker and onto the conveyor belt at rates that optimize belt system performance. This is particularly the case where multiple sections all discharge onto a shared conveyor belt.

Place-changing operations are typically applied in relatively good roof conditions and often have better productivity than the alternative—in-place continuous miners. These in-place continuous miners, also called bolter miners, install roof bolts, mesh, straps, and even roof trusses in parallel with the mining cycle. Thus, roof support is installed close to the face, and the roof has minimal opportunity to deteriorate before additional support is implemented to augment its natural integrity. Also, this method minimizes the number of times that the bolter miner is moved between working places, thereby minimizing damage to potentially weak floors.

Historically, seriously flawed weight distribution on bolter miners led to worse floor damage than place-changing machines, even though machine travel was less. Recognition of this problem has allowed it to be mitigated on modern machines.

Aside from mining and bolting, section operations are much the same between in-place and place-changing systems. Some mine operators use loading machines between bolter miners and the coal haulers. The loading machine allows the bolter miner to discharge product into a small surge pile on the floor, behind the bolter miner. The machine loads from this surge pile into coal haulers as they become available and decouples the continuous miner from producing and loading only when a coal hauler is on station and ready to receive product. The use of a loading machine also caters to an often noted deficit of bolter miners, which is cleanup in-cut.

As a general principle, bolter miner productivity is less than place-changing operations when a reasonably long cut, perhaps greater than 6.0 to 7.5 m, is available. However, if cut depth falls below this range or if ground control is unacceptable in the cut depth available, a bolter miner system may become the preferred option. Current bolter miners have the ability to sump and bolt simultaneously. This has a notably favorable impact on productivity when compared to earlier style bolter miners, which could only cut as much as was available with the machine frame stationary when drilling and bolting activities were ongoing.

Additional concerns for gate road design are entry and crosscut span, crosscut angle, and spacing. Gate road entries must serve several purposes, including the following:

- Provide adequate cross section for ventilation
- Provide access of personnel and materials
- Transport and installation of longwall equipment
- Installation of the longwall section conveyor belt
- Temporary locations for emulsion-pumping equipment and electrical power and controls
- Retreat of the longwall in normal operation

Standing roof support is restricted to only those areas where it does not obstruct necessary machine travel. In the longwall conveyor belt entry, supplemental support may be installed as a precaution because there is little opportunity to add remedial support after the belt is commissioned.

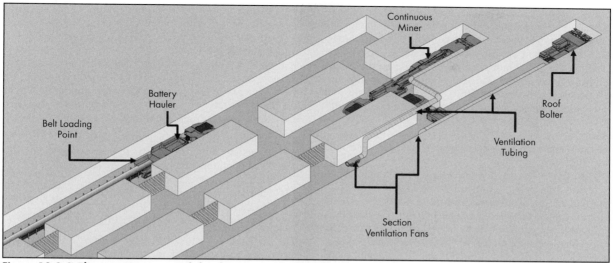

Figure 13.8-8 Three-entry gate road development with place-change mining

Four distinct regimes of increasing loading are applied to gate roads, aside from the first and last panels of a district:

1. Initial development
2. Adjacent to a single longwall gob (headgate)
3. Intermediate to adjacent longwall gobs (active tailgate)
4. Longwall gobs on both sides, well behind (inby) the face

The primary roof-support system controls the ground in the first and second stages, but supplemental support is commonly applied prior to use as a tailgate for the active longwall. This support is typically installed in the immediate tailgate entry and crosscuts before it reaches third-stage loading, and, where practical, it may even be installed at the time of development. Commonly used materials include wooden cribs (4-point or 9-point), cans, steel props, and pumpable cribs.

The alternative of right-angle intersection of crosscuts to main entries versus inclined is a balance of considerations between the convenience of access afforded by inclined intersections and the propensity of acute-angled corners to deteriorate with consequent enlargement of the intersection sum-of-diagonals (SoD). Increasing SoD often correlates with decreased stability of intersections. Floor stability and horizontal stress direction can influence the optimal intersection design.

The result of entry spacing and crosscut spacing is pillar size. The correct selection of pillar size for longwall gate roads is an involved discussion, beyond the scope of this chapter. In general, bigger pillars are more desirable from a stability perspective, but the unrecoverable reserves lost to gate road pillars, the increased development footage, and prolonged development time is not desirable. Long pillars are favorable to gate road development, but wider pillars are not. This is because development along panel length is useful to the purpose of completing panel development, but distance spent on crosscuts is not. Access to the longwall face for heavy maintenance is enhanced by shorter crosscut spacing.

Face Length and Panel Length

The choice of longwall face length is a balance of the initial capital cost committed to acquire longwall equipment versus the benefits of improved productivity, production cost savings, and additional reserve recovery for a wider longwall face. The decision is also limited by the capabilities of existing equipment, although capabilities have steadily improved over time. In subsidence-sensitive areas, wider faces can cause mine plans to be offset further from sensitive surface structures and sterilize additional reserves in the process, a cost weighed against recognized benefits.

As the industry-wide dedication to zero harm redefines acceptable hazards to ever diminishing severity and frequency, a conflicting prospect emerges as mining conditions tend to become increasingly adverse and present increased hazards. The cumulative result is an increased amount, and cost of development footage required to sustain longwall operations. Consequently, increased longwall face length is attractive. For most site-specific circumstances, an optimum longwall face length can be defined. Often there is a broad optimum in net present value terms—between 300 m of installed equipment length and up to perhaps 475 m in other cases. Present technology allows face length up to approximately 500 m with high productivity. Common face lengths range from 250 to 425 m. Achieving maximum lengths implies the use of the largest AFC chains and highest-rated electric motors and AFC gearboxes available.

Longwall panel length is an issue constrained by

- Ventilation system limitations,
- Timing considerations to avoid production outages,
- Longwall equipment longevity,
- Property constraints, and
- Reserve extent.

At some point, excessive ventilation pressure losses and insufficient air quantities dictate a practical limit to the length of a gate road development or may constrain longwall operation. Alternatively, the time required to complete a long development may compare unfavorably with the rate at which an active longwall is retreating. Yet another consideration is whether the longwall equipment can retain the required availability levels as it wears through continued service, although durability of the constituent equipment elements of the longwall has been improving through time. It is not unreasonable to expect shields, AFC drives, and frames to serve through the longest panels. AFC sprockets can be reversed or replaced during planned maintenance intervals, as can chains, flights, and chain connectors. Shearing machines have longevity on par with the longest panels (highest tonnages) but can also be repaired as subcomponents, or a replacement shearing machine can be implemented during mining of a panel more readily than undertaking a complete longwall move (panel transfer). With this, contemporary longwall panels range from 1,850 to 4,600 m in length, with contained panel tonnages up to 11 million metric tons of coal.

Thin Seams

Most of the discussion so far has been focused on the most common high-productivity longwall variation: single-lift longwall in the 1.75- to 3.75-m extraction height range. However, thin ore deposits, typically with higher value, such as coking coals, are of interest. Thin seams are also a topic where they overlay other coal seams but must be extracted first to prevent their sterilization by undermining. Historically, thin coal seams were extracted with low-productivity shearing machines and longwall plow systems. Low-seam shearing machines of the past are hopelessly obsolete compared to current expectations. Periodic attempts are made to make low-seam shearers practical, as a number of mining projects worldwide reside at the edge of viability and would be successful if only better productivity were available. Shearers in low seams are constrained by factors such as

- Motor size (diameter and power),
- Vane depth of drums and loading performance,
- Clearance (shearer tunnel and ranging arm),
- Length and height limits (traversing undulations), and
- Operator ability to travel at tram speed of the shearer.

While efforts continue to solve the low-seam shearer issues, progress has been made in the modernization of longwall plow systems. Figure 13.8-9 shows a longwall plow. Instead of cutting the coal with a shearer, the coal is cut by a plow body captivated on the AFC and drawn from gate to gate on a dedicated chain driven by motors on each face end. Since the plow cuts only a fraction of the web of a shearing machine, normally 100 to 140 mm but up to 250 mm on the highest-capacity installations, multiple transits of the face are required to achieve equivalent production to one pass with a shearer. In contrast, the plow speed may be more than 10 times that of typical low-seam shearers, so plows offer attractive potential in deposits too low for shearers, provided that conditions are favorable (Gluckauf GmbH 1985).

Favorable conditions for a plow-type longwall include

- Strong, continuous roof and floor rock;
- Ore that plows readily and parts cleanly from roof and floor;
- Ore horizon with little thickness variation and limited undulations; and
- Ore free of intrusions, faults, or dikes.

Roof falls in the tailgate or on the face can be serious obstacles to a plow. Replacement of a significant section of the extracted horizon with sandstone through faulting or washout can be equally problematic.

The history of plow-type longwalls is extensive in Europe, where almost every conceivable condition has been confronted by plowing, but not always to successful outcomes in modern terms. The seam conditions that make plow application attractive in a modern context are limited compared to the totality of potential low-seam reserves worldwide. At present, only a few plow-type longwalls compete with shearer-based longwalls for the same markets.

Thick Seams

Unlike thin seams, thick-seam longwall has great potential worldwide. Single-pass longwall designs up to 7.25 m high have been proposed with installations more than 6.5 m high in service. A variety of issues require adaptation of the designs of these high installations, including

- Shearer steering into/out of the face,
- Structural integrity of AFC to bear heavy shearers,
- Face operator travelways and safety,
- Slab development or persistence,
- Face spalling ahead of the shield tips,
- Shield capacity and resulting weight, and
- Potential worst-case AFC starting capability.

The production history of many of these installations bears testimony to the success of designers and mine operators in overcoming the numerous challenges arising from thick-seam operation. At any period of time, some of these high faces appear in the ranks of the world's most productive longwalls.

Interest is intensifying in longwall extraction of even thicker seams than 7.25 m, but there is an accompanying recognition that single-pass longwall is not the solution. To extract very thick coal seams, the longwall top coal caving (LTCC) technique is being developed. Two major variations of the LTCC method exist. In one, the face cross section is much like a conventional shearer-based system with a single AFC. An original pass is made as customary, at a convenient height, and then a special purpose caving chute integrated into the shield canopy is lowered so that material caving from above the shield is channeled into the AFC. When product quality or caving conditions dictate, the chute is retracted, and successive chutes are opened along the face. The process is repeated until the majority of the available product has been drawn. At this point, additional undercutting is conducted with the shearer in preparation for successive drawing of caved material.

Alternatively, another technique of LTCC uses an AFC, both ahead of the shields as expected but also behind them under a powered flipper used to allow or preclude caved coal introduction onto the rear AFC. Figure 13.8-10 shows a typical installation in cross section.

Potential exists for more new LTCC installations worldwide than the total number of modern high-productivity longwalls currently in service in North America. Today, LTCC installations exist in China and Eastern Europe with considerable expansion of the technique possible there, as well as significant prospects for introduction to Australia and possibly even the Americas. The LTCC technique could require adaptation to conform to statutory requirements in countries where the technique has not previously been practiced.

Multiseam Longwall

On occasion thick seams are extracted by single-lift longwall operations conducted in a coordinated sequence, with the upper longwall progressing ahead of the lower longwall in an interlocked sequence. Though this method has examples in the field, coordinated operation of independent longwalls can be difficult. Thin interburden is more problematic than thicker interburden, as are joints or fractures that transect the interval between the seams. Thin or low-shear-strength interburden, as well as jointed interburden, can transmit stress between

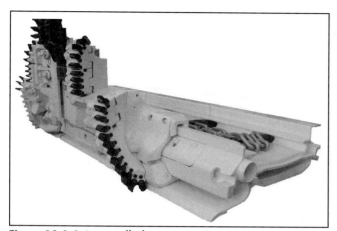

Figure 13.8-9 Longwall plow

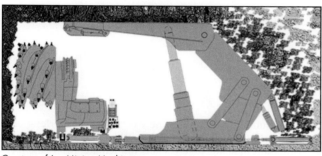

Courtesy of Joy Mining Machinery.
Figure 13.8-10 Top coal caving with rear AFC

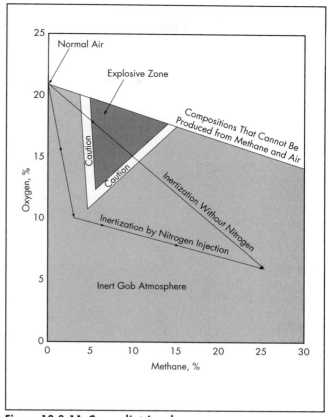

Figure 13.8-11 Coward's triangle

mining horizons more directly than horizons separated by even a modest thickness of strong, stiff interburden. High stresses are often transmitted from one horizon to another where strong structures surrounded by high-extraction workings in one horizon are approached by workings in the other horizon. In general, multiseam operations are more focused on extraction of independent seams at appreciably separated intervals of interburden and time. Often the earlier workings are not executed with anticipation of modern longwall mining in the future, thus making mine planning an exercise in mitigating interseam interactions.

Longwall Ventilation

Ventilation of longwall mining systems seems to exhibit nearly endless variety in the exact details of worldwide plans (AusIMM 1986; Kennedy 1999). Taken at a broader level, longwall ventilation falls into two general categories with very different mechanisms (MSHA 1996). To introduce these principles, some basic attributes of methane and air mixtures should be clearly understood. Figure 13.8-11 shows Coward's triangle (Coward and Jones 1952). Other representations of the explosion hazard are popular in various locations around the world, but the chemistry and physics of the matter are universal. Thus, other representations describe the same physical behavior.

Two principal regions are represented in Figure 13.8-11. The upper portion of the figure describes compositions that cannot be formed from methane and air and is therefore irrelevant. Below these naturally impossible compositions of methane and air is the region of naturally possible mixtures. These possible mixtures include two major subdivisions: inert mixtures and explosive compositions. The inert mixtures will not explode if exposed to a potential ignition source and occur as fuel-lean or fuel-rich compositions with respect to the explosible compositions. The explosible compositions exist within Coward's triangle shown in the approximately 5%–15% methane range and approximately 12%–21% oxygen range. The zone depicted around the triangle is an arbitrary safety margin. This physical behavior of methane and air mixtures being an absolute principal, two approaches to avoid the explosion hazard exist. One is to provide fresh air to dilute methane to a fuel-lean state. If such a state can be maintained without fail, it renders an acceptably safe condition. Paramount to this strategy is that the fuel-lean (oxygen-rich) atmosphere cannot transition from a respirable state, required where people work and travel, to a fuel-rich (oxygen-deficient) state, which is likely to occur in the unventilated gob. Where the respirable mine atmosphere transitions to an unventilated gob atmosphere, particularly after mine workings are sealed, it becomes impossible to guarantee that the reducing oxygen level and increasing methane level will not cross through the region of explosible compositions without active intervention. This case is described by the upper trajectory line in Figure 13.8-11. If even a small energy source above the intrinsic safety threshold is available to ignite the explosive composition, a violent event will result in proportion to the volume of gas mixture ignited (SIMTARS 2001). If coal dust becomes airborne secondary to a typical methane explosion, a much more energetic and violent event can arise (Burrows 1989). This mechanism has repeated itself many times worldwide, often with disastrous outcomes. Application of incombustible dust (rock dust or stone dust) to areas where

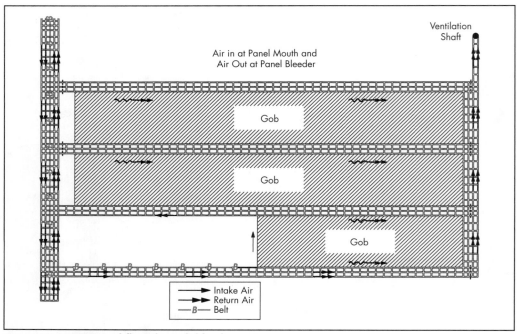

Figure 13.8-12 Typical flow-through bleeder system

coal dust can accumulate is effective in mitigating this dust explosibility hazard.

An alternative mitigation strategy for the explosion hazard involves the deliberate exclusion of oxygen from gob atmospheres, followed with forced inertization if necessary to depress oxygen levels below those where an explosion hazard can exist, regardless of methane concentration. This is represented by the lower trajectory line in Figure 13.8-11, which is actively controlled by inert gas injection and never crosses the region of explosion hazard, Coward's triangle. Periodic atmospheric monitoring, followed by application of the inertization process as necessary, can prevent explosible methane–oxygen compositions from being created by already inert atmospheres where introduction of new oxygen is possible.

With this in mind, the two major strategies of longwall ventilation are bleeder and bleederless ventilation systems. A conceptual flow-through bleeder longwall ventilation system is shown in Figure 13.8-12.

In this system, fresh (intake) air is supplied through the headgate entries and splits at the headgate to ventilate the longwall face. An adequate volume is allowed to flow behind (inby) the face at the headgate to ventilate the perimeter of the gob. Along the face and at the tailgate, air is allowed to "bleed" into the gob and tailgate to ventilate the methane to fuel-lean levels. This is the most common ventilation system used in the United States. It is notable that because atmospheric compositions are uncertain where they cannot be monitored and controlled in the bleeder-ventilated gob inby the longwall and in the adjacent abandoned gobs, any potentially persistent ignition source such as spontaneous combustion or intermittent source of ignition energy such as lightning is reason for concern.

The major alternative to bleeder ventilation of longwalls is bleederless ventilation. The bleederless longwall ventilation system is depicted in Figure 13.8-13.

The notable differences between the bleederless and bleeder ventilation systems are that in the bleederless system seals are built in headgate crosscuts immediately inby the face in order to progressively seal the gob and exclude oxygen as the longwall retreats. If concerns exist about atmospheric compositions in the gob or if oxygen invasion could stimulate spontaneous combustion, the added feature of forced inertization, as discussed previously, can be an effective risk-mitigation strategy. Gases such as nitrogen, CO_2, or even oxygen-depleted air can be used for safely inertizing mine atmospheres. The exhaust gas from combustion processes, such as gas turbine engines, can potentially be a source of oxygen-depleted gas mixtures for inertization.

To prevent oxygen inflow into sensitive gobs, pressure balancing is often applied to minimize pressure differentials across seal lines and minimize introduction of new oxygen to a gob. An alternative technique is the use of *balance chambers*, as depicted in Figure 13.8-14. In this method an initial seal is built to separate the atmospheres, and then another seal is built outby the initial seal, becoming primary. Inertization gas delivery pipes penetrate both seals to allow inert gas delivery to the sealed area and the space between the seals. If the seals have reasonable integrity, the space between the seals can be elevated to a low positive pressure, and the small leakage volume that occurs is either into the gob, the mine airway, or both. The net effect is to prevent oxygen ingress into the sealed area at low total inert gas expenditure. Sudden change in inertization gas consumption can be an early indication of deterioration in seal performance.

A valuable tool to monitor the composition of mine atmospheres in critical areas is a *tube-bundle* system. With this equipment, atmospheric samples are drawn from critical locations underground through dedicated individual tubes to analysis equipment located at the mine surface. This allows sampling of areas where electrical sensors might be prohibited

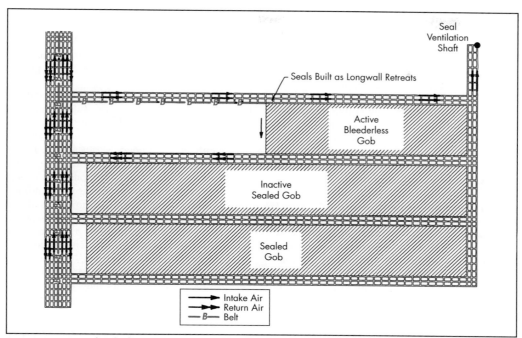

Figure 13.8-13 Bleederless longwall ventilation

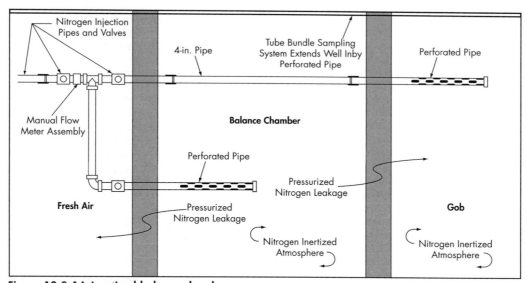

Figure 13.8-14 Inertized balance chamber

and sampling continuation in the event that electrical power is removed, as is the case in many mine emergencies.

An addition to the controls inherent in either of the ventilation systems described, gob ventilation boreholes (GVBs) can be used to assist the ventilation system by extracting gob atmosphere, predominantly methane, before it enters the ventilation system. GVBs are often centrally located in a longwall panel when the intention is to intercept methane evolved from broken strata before it enters a bleeder ventilation system. These holes, drilled from the surface in advance of mining, are often spaced at larger intervals than GVBs intended to control expansion volumes in a bleederless ventilation system. The location of GVBs is typically along the periphery of a panel, where void space is connected to workings at seam level, when the intention is to control barometric expansion volumes, as in bleederless ventilation systems.

Diurnal barometric pressure variation occurs daily, with the barometer falling as the atmosphere warms with duration and intensity of solar exposure and rises as night falls when the atmosphere cools and air density increases. Random barometric pressure variations are related to weather systems. The actual barometric pressure is a function of both effects and can be predicted for a surface location with good overall accuracy. Operation of GVBs in response to barometric pressure variation, alternating with inert gas injection, can control evolved methane and prevent oxygen invasion into progressively sealed gobs, as in bleederless systems, or other sealed areas.

Although GVBs are typically drilled vertically from the surface, an alternative method applies medium radius drilling technology to postmining methane drainage. The wells start with a vertical collar followed by a build section where the well is steered to a horizontal position. The horizontal lateral is located in a stable horizon above the longwall panel. Holes up to 1.6 km in length, producing approximately 1 m^3/s of methane, have been achieved.

An alternative to GVBs drilled from the surface is an array of cross-measure boreholes drilled at upward inclination from the seam horizon. Such cross-measure holes can drain methane prior to mining and be used later to a similar effect as GVBs when caving behind the retreating longwall opens a communication between the holes and the larger gob. Gas captured by the holes can be collected by an underground piping system and directed to a borehole to the surface without entering the mine ventilation system.

Horizontal boreholes, originating in-seam or turning into the seam from vertical wells to the surface, can be progressed in-seam for more than 1 km to drain methane in advance of mining. These holes have proven to be highly effective for methane drainage but must be planned carefully, as they can later present hazards that must be mitigated before mining progresses through them.

CONTEMPORARY ISSUES

Improving the health and safety of miners by reducing their exposure to hazards has been a major motivation in the steady progress of mining mechanization. Longwall mining as it exists today is the product of this ceaseless effort. Greater productivity and resource recovery are realized, while the incidence of injuries and accidents continues to decline as a trend. Few would dispute that longwall mining is, by far, the safest and most efficient underground soft-rock mining method.

Although many concerns have been addressed with great success, others have persisted. Coal workers' pneumoconiosis, more commonly called "black lung," is a chronic disease which investigators have correlated to respirable dust exposure in the coal industry. Black lung is often debilitating in its advanced phase and may lead to death. Most coal-producing countries worldwide acknowledge this hazard and have occupational health limits for respirable coal dust exposure intended to prevent workers from contracting the disease. Not all dust is deemed hazardous but only the submicrometer (<10-μm size particles) fraction of dust is believed to contribute to black lung. Because longwall mining results in large production, it is not surprising that increased amounts of dust can also result, with respirable dust being a fraction of the total.

Depending on the regulatory strategy applied, this can lead to different conclusions. One strategy is based on environmental exposure. This assumes that a worker will not benefit from any personal protective equipment and that safety is produced only by maintaining environmental levels below disease thresholds. This strategy does not give innovation or technology an opportunity to contribute solutions unless it impacts the environmental exposure as opposed to the respiratory exposure. However, personal protective equipment such as airstream helmets offers demonstrated effectiveness, and automation technology has the potential to remove workers from most dust exposure. Personal dust exposure management is now more feasible than ever, with the demonstration and introduction of personal dust monitors. Seemingly, the optimum solution to the problem of dust exposure would apply the best available technologies, crediting the benefits in realistic terms, to create incentive for continued progress in longwall technology.

Another issue of modern underground mining is related to surface subsidence. The most damaging aspect of mining-induced subsidence is the zone that connects undisturbed surface with areas subjected to near maximum vertical displacement. This boundary varies from a few meters to a few hundred meters wide and commonly wraps around the perimeter of the subsided area. The condition is exaggerated when subsidence takes place years after active mining and may continue for a protracted period of time. It is also worsened when small land areas are subsided, a few hectares per instance, leading to a greater proportion of subsided land area being subjected to the most damaging end outcome (Kratzsch 1983; SAIMM 1982). These situations, often acting in combination, make compensation for damage to surface assets difficult, because the mine operators may no longer have a corporate presence or the ability to pay for reparations years or even decades after the fact (Peng 1992).

The development of longwall mining has brought relief to these concerns. As modern longwall panels have gotten wider and longer, the proportion of subsided land area permanently subjected to the most damaging boundary conditions has decreased. Also, unlike the subsidence of smaller individual areas, the subsidence develops to its maximum level in close alignment with predictions and generally does so in a period of weeks to months after active mining by longwall methods. Thus, all stakeholders are in a better position, both in terms of damage and timing, to make such reparations as may be appropriate.

Longwall mining has been extensively applied to urban and suburban areas in Germany, Poland, and the United Kingdom to mine under a tremendous variety of structures, including public buildings and residences, industrial facilities, roads, bridges, railroads, pipelines, and power lines, to successful outcomes. Longwall mining has also been used worldwide to mine under rivers, lakes, and oceans successfully. With the worldwide knowledge base for mitigation of subsidence damage to surface structures, many alternatives can often be applied to minimize damage to surface assets or repair them effectively.

As longwall mining transitions from rural to suburban areas in the United States and Australia, it has vocal critics. The arguments put forward derive from many origins but generally do not suggest a better method to extract the resource but rather oppose the extraction all together or dispute the value of the damages that might be created or the ability to bring about acceptable reparations.

CONCLUSION

When longwall systems are properly selected and operated, the results can be spectacular and sustainable. These installations make longwall mining look easy. In reality, successful selection and operation of longwall systems are based on attention to detail and a commitment to continuous improvement. These pursuits require much effort and rely on a capable staff as much as the equipment composing a longwall mining system.

Success also depends on a close collaboration between mine operators and technology suppliers, often as a strategic alliance. Both parties must begin with the end goal in mind, which has to be attainment of safe production and profitability for both partners. Mine operators and technology suppliers must recognize their stake in the shortfalls as much as the successes, constantly being prepared to contribute investments in time and money in pursuit of continuous improvement. This is especially important in consideration of the long-term nature of the relationship created by long equipment service life and large capital investments.

Continuous improvement is necessary to meet the evolving expectations of stakeholders to the overall production process. These stakeholders expect the result of zero harm as it applies to health and safety of employees, local communities, and the environment. They also expect improving productivity and diminishing costs, even in consideration of future mining conditions that typically are more adverse than past or present conditions.

For all of these reasons, success at longwall mining is exclusively a result of organizations with aligned purpose, detailed planning, and controlled execution leading to sustainable results on the part of a wide diversity of involved stakeholders.

REFERENCES

AusIMM (Australasian Institute of Mining and Metallurgy). 1986. *Australasian Coal Mining Practice*. Monograph No. 12. Edited by C.H. Martin. Parkville, Victoria, Australia: AusIMM. pp. 328–363.

Barczak, T.M., and Garson, R.C. 1986. *Shield Mechanics and Resultant Load Vector Studies*. RI 9027. Washington, DC: U.S. Bureau of Mines.

Burrows, J., ed. 1989. *Environmental Engineering in South African Mines*. Johannesburg, South Africa: Mine Ventilation Society of South Africa. pp. 763–771.

CEMA (Conveyor Equipment Manufacturers Association). 1997. *Belt Conveyors for Bulk Materials*, 5th ed. Naples, FL: CEMA. pp. 29–43.

Cliff, D., Rowlands, D., and Sleeman, J. 2004. *Spontaneous Combustion in Australian Underground Coal Mines*. Redbank, Queensland, Australia: Safety in Mines Testing and Research Station.

Coward, H.F., and Jones, G.W. 1952. *Limits of Flammability of Gases and Vapors*. Bulletin 503. Washington, DC: U.S. Bureau of Mines.

Evans, M.P., ed. 1993. *Conveyor Belt Engineering for the Coal and Minerals Mining Industry*. Littleton, CO: SME-AIME. pp. 145–193.

Gluckauf GmbH. 1985. *German Longwall Mining*. Edited by R.H. Bachstroem. Essen, Germany: Verlag Gluckauf GmbH. pp. 57–61.

Kennedy, W.R. 1999. *Practical Mine Ventilation*. Chicago, IL: Intertec Publishing. pp. 305–337.

Kratzsch, H. 1983. *Mining Subsidence Engineering*. Berlin, Germany: Springer-Verlag. pp. 363–439.

Ludema, K.C. 1996. *Friction, Wear, Lubrication: A Textbook in Tribology*. Boca Raton, FL: CRC Press. pp. 189–200.

Mitchell, D.C. 1996. *Mine Fires: Prevention, Detection, and Fighting*, 3rd ed. Chicago, IL: Primedia Business Media and Magazines.

Moore, E.S. 1940. *Coal: Its Properties, Analysis, Classification, Geology, Extraction, Uses and Distribution*, 2nd ed. New York: John Wiley and Sons. pp. 231–248.

MSHA (Mine Safety and Health Administration). 1996. *Bleeder and Gob Ventilation Systems: Ventilation Specialists Training Course*. CL 42a. Beckley, WV: MSHA. pp. 125–145.

Papp, A.R., Ayers, W.B., Murray, D.K., and Shattuck, P.J. 1998. *Atlas of Coal Geology*. Vol. 1, Igneous Intrusions and Extrusions. Tulsa, OK: American Association of Petroleum Geologists.

Peng, S.S. 1992. *Surface Subsidence Engineering*. Littleton, CO: SME. pp. 67–151.

Peng, S.S. 2006. *Longwall Mining*, 2nd ed. Morgantown, WV: West Virginia University, College of Mineral and Energy Resources. pp. 45–91.

SAIMM (South African Institute of Mining and Metallurgy). 1982. *Increased Underground Extraction of Coal*. Monograph No. 4. Edited by C.J. Fauconnier and R.W.O. Kersten. Johannesburg, South Africa: SAIMM. pp. 92–110.

SIMTARS (Safety in Mines Testing and Research Station). 1999. *Interpretation of Mine Atmospheres*. Redbank, Queensland, Australia: SIMTARS.

SIMTARS (Safety in Mines Testing and Research Station). 2001. *Frictional Ignitions*. Redbank, Queensland, Australia: Safety in Mines Testing and Research Station.

CHAPTER 13.9

Sublevel Caving

Geoff Dunstan and Gavin Power

Sublevel caving is a top-down mining method allowing earlier production than sublevel stoping with less upfront development than traditional block caving. In large-scale operations it has the potential to be a *factory method* (meaning highly repetitive) with the cost and capacity benefits associated with high production, moderate development requirements, and maximum use of automated equipment (Bull and Page 2000). In smaller mines it can be used as a selective method with lower production rates, where the capacity benefits are less evident. Horizontal slices of in-situ ore are progressively blasted and extracted, with blasted ore and caved rock filling the void created by ore extraction (Figure 13.9-1). Dilution can be managed and minimized by disciplined draw control. As this process progresses, caving propagates upward to create surface subsidence.

Originally, sublevel caving was used in weak ground that would collapse when the timber supports were removed. After a supported drift was pushed across the ore body from an access or perimeter drift, the supports would be individually removed in a retreating fashion. As each support was removed, the ore would cave—thus the name sublevel cave—and be slushed out to the perimeter drift. The next support would not be removed until the amount of diluting material became excessive. Overall, the method was slow and combined poor ore recovery with high dilution.

Later applications of the method saw it applied in relatively competent ore bodies with weak hanging walls in order to facilitate the caving process and avoid the creation of large voids. The stronger ground required drill-and-blast to take the role of the support removal, so in this version of the method it is no longer the ore that caves but the host rock above and immediately adjacent to the ore. With the introduction of drill-and-blast, sublevel caving could be utilized in ore bodies with footprints too small for natural cave development and/or dipping geometry unfavorable for block caving. In recent times, the method has been applied to ore bodies with strong hanging walls where cave assistance techniques have been used as required.

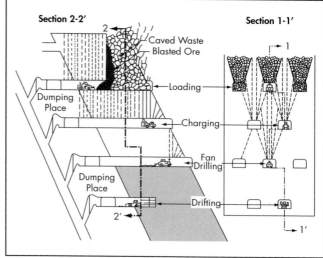

Source: Kvapil 1992.
Figure 13.9-1 Schematic of sublevel caving mining

Current applications of the method are accessed by decline(s) and have sublevels established on approximately 20- to 35-m vertical intervals, far greater than the early applications of the method, due largely to advances in drilling and blasting technologies. A perimeter drift is driven along (transverse layout) or perpendicular (longitudinal layout) to the strike of the ore body and is generally offset approximately 20 m from the ore–waste contact. Some mines develop the perimeter drift in ore and recover it after production is completed from the drifts. The perimeter drift is developed to access the production drifts for ore transportation, services, and ventilation. The production drifts are spaced between 12 and 30 m on horizontal centers. To enable optimal coverage for drilling and to allow for the downward flow of caved material, they are staggered between the levels (Figure 13.9-2).

Geoff Dunstan, Mining Manager, Newcrest Mining Ltd., Melbourne, Victoria, Australia
Gavin Power, Director, Power Geotechnical Pty Ltd., Brisbane, Queensland, Australia

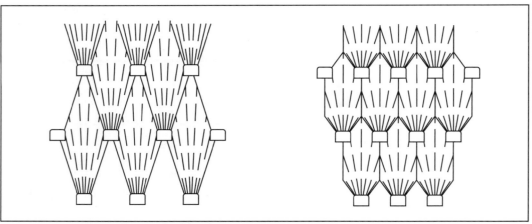

Figure 13.9-2 Sectional view of sublevel caving ring options

Over the years, the ratio of development-to-production ore has increased from approximately 1:3 to as high as 1:10.

The production ore extraction process commences by placing either a slot drift on the far ore–waste contact or placing individual slot raises at the end of each production drift. The slot drift will have a slot raise at one end and a series of parallel rings, generally on 2- to 3-m burden, the width of drift reaching up to the level above. Once the slot has been opened, to create adequate expansion room for the first production rings, some *slashing* rings will start the production rings in the drifts until the production rings reach full height. The production rings will usually be angled forward by 10° to 20° from vertical to assist the breakage and flow of the ore in the confined blasting environment and to help maintain geotechnical stability at the brow. The production ring burden is generally between 2 and 3 m. The production rings drill up through the pillar between the production drifts on the level above, as shown in Figure 13.9-2.

The layout of the method requires secondary ventilation to be forced into the production drifts throughout both the development and production phases of the level. Load-haul-dump (LHD) machines are used to extract the ore from the fired rings and transport it back to either stockpiles or ore-passes located on the perimeter drift.

As the initial level is opened, only the swell material from the blast (approximately 40%) is excavated, because caving will likely not be initiated until production on the level is well advanced. The extraction ratio of ore increases to a nominal value of approximately 80% to 100% over the subsequent two or three levels. The reduction in the early draw rates is designed to minimize the likelihood of air blast until the overlying cap rock and/or hanging wall starts to break up and flow. In addition, the reduced draw builds up a blanket of ore above the active drifts, which can assist in minimizing the impacts of dilution in a blind cave, as well as protecting the drifts when the sublevel caving is commenced directly below an open pit.

Operational sublevel caving practice varies widely, and consequently tonnage and mineral recoveries are also variable from operation to operation. Overall tonnage extraction rates for large-scale sublevel caving mines will be approximately 80% to 120% of the blasted tonnage for approximately 65% to 95% of the total ore-body mineral content. Of the rock drawn from each fired ring, approximately 15% to 25% will be dilution. More dilution comes at the end of each individual ring's life, and varying practices mean that dilution results also vary from site to site (Power and Just 2008).

When extracting the ore body, strict operational rules are required to ensure the method's success. Only after the upper level has been retreated a safe distance, somewhere between one and two times the sublevel interval, can the level below be commenced. Within the level, the production rings need to be brought back in a planned sequence (managing the lead–lag between adjacent production drifts) to minimize the impact of stress and explosive damage on neighboring drifts.

APPLICATION OF SUBLEVEL CAVING

The suitable ore-body characteristics that lead to the successful implementation of the sublevel caving method include

- Cavable overlying host rock,
- Relatively large ore-body footprint or a relatively weak hanging wall to initiate and propagate caving,
- Steeply dipping and relatively uniform ore body,
- Reasonably uniform grade distribution,
- Disseminated ore body with a low-grade zone around the sublevel caving production area to minimize the impact of waste dilution,
- Competent ore body to minimize the excavation support requirements and improve drift availability to enable continuous production, and
- Visually different ore and waste properties.

Although sublevel caving can be successfully applied to ore bodies missing some of these characteristics, any such omissions generally lead to increased operational complexity (Power and Just 2008).

GRAVITY FLOW

The sublevel caving method relies on the gravity flow process to get the ore to the drawpoint. As a result, gravity flow has been a focus of sublevel caving research for many years.

Gravity flow in the sublevel caving environment is unique in a number of ways:

- The effects of blasting produce differential fragmentation within the ring, both vertically and laterally.
- Draw occurs from drawpoints on multiple levels, laterally offset, resulting in multiple stages of flow.

Source: Power 2004b.

Figure 13.9-3 Large 3-D physical model used to study sublevel caving flow at the University of Queensland, Australia

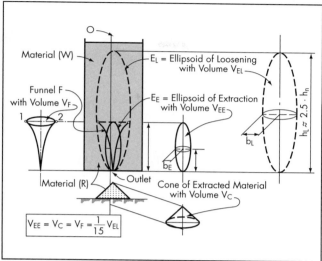

Source: Kvapil 1992.

Figure 13.9-4 Example of idealistic flow analysis derived from analysis of physical models

- Each individual ring is subject to unique factors, including geological structure, drill-and-blast inconsistencies, and differing draw patterns.

As in other fields of research, the gathering of information on flow has been a gradual process, with successive workers building on the data and theories of past workers. Early studies of gravity flow focused on small-scale, glass-fronted physical models and over time expanded to larger three-dimensional (3-D) models such as shown in Figure 13.9-3. These models involved the drawing of broken rock from geometrically scaled and well-controlled environments, so that the resulting movement and extraction zones could be studied in detail and accurately measured.

Research using these physical models resulted in the promulgation of influential theories and relatively precise mathematical formulas to describe concepts such as particle velocity profiles, the ellipsoid of loosening and of draw, and the relationship between these bodies (e.g., Figure 13.9-4). It was later determined from full-scale marker trials that the shapes of the flow and movement envelopes were not exactly elliptical (Janelid 1975); however, they are close enough for the ellipse to be used as a basic design concept in sublevel caves (Power 2004b).

In practice, these theoretical explanations of flow in bins and hoppers serve mainly to give a broad understanding of flow in a highly controlled environment. Taken in isolation, they comprise a highly idealized, single-level description of granular flow with only a general relationship to granular flow in real sublevel caving environments. However, results from these models were useful in guiding design of early sublevel caves and provided important insights into the mechanisms controlling flow. They also led to the development of some of the fundamental concepts underpinning the method, which are well established today (e.g., drawing drawpoints interactively and loading from side to side in drawpoints).

Although model studies provided a basic understanding of gravity flow mechanisms, full-scale field trials were needed to obtain flow data in operational sublevel caves. Marker trials involve the placement of markers (usually short lengths of steel piping or cable) into internal marker rings before the experimental production ring is blasted. These markers are collected as the ring is mucked, and the resulting data is analyzed to assess the flow patterns and ultimate recovery of the ring.

The earliest comprehensive trials were those carried out by Janelid at Grängesberg mine in Sweden (Janelid 1972). Janelid's marker trials showed gravity flow in the full-scale environment but only with reference to recovery from the level on which the rings were blasted. They were carried out on relatively small sublevel caving layouts (approximately 3-m-wide crosscuts on 7-m crosscut centers and 13-m sublevel heights). For the relatively small ring tonnages drawn, the trials showed high angles of draw and irregular draw patterns not quite elliptical in shape (Figure 13.9-5).

In recent times, marker trials have become more popular in response to significant changes in sublevel caving geometry, which are partly due to improvements in drilling and blasting technology. Figure 13.9-6 shows a comparison of changes in sublevel caving geometry at the Kiruna mine in Sweden between 1983 and 2008.

The Kiruna 2000 program of marker trials was also single-level trials, although they did not have the same density of markers as the Grängesberg trials and relied on visual identification of markers at the drawpoint for recovery of experiment information. This resulted in only about a third of the installed markers being recovered (Gustafsson 1998). A

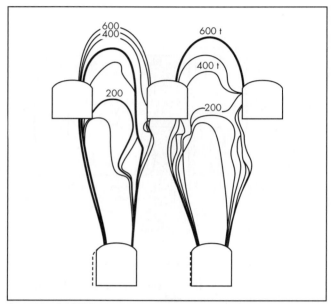

Source: Janelid 1975.

Figure 13.9-5 Example of analysis from Grängesberg marker trials

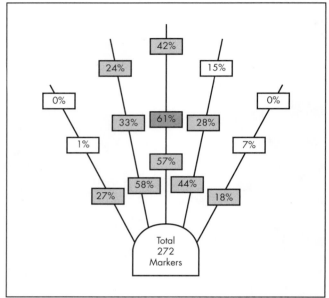

Source: Quinteiro et al. 2001.

Figure 13.9-7 Summary results from the Kiruna 2000 marker trial program, indicating percentage of markers placed in each location recovered

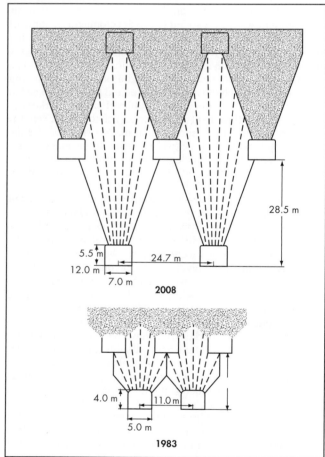

Source: Power and Just 2008.

Figure 13.9-6 Comparison of sublevel caving geometry at Kiruna mine in 1983 and 2008

diagram summarizing the results of these trials (Figure 13.9-7) indicates a total of 272 markers recovered and shows the percentage of markers recovered from various locations in the marker rings. This analysis indicates that most markers recovered were almost directly above the drawpoint. One important advantage of the Kiruna trials was that they gave an indication of recovery from greatly upsized sublevel caving rings, with crosscut widths, spacings, and sublevel heights all approximately double those used at Grängesberg.

The Kiruna marker trials showed that, although mining scale had dramatically increased because of improvements in equipment technology and a focus on cost reduction, granular flow properties had not changed, leaving large parts of the fired production rings unrecovered (if viewed only from a single-level perspective).

Marker trials carried out at the Ridgeway mine in New South Wales, Australia, were the first comprehensive multilevel marker trials concentrating on granular flow through an entire sublevel caving system (Power 2004a, 2004b). Carried out between 2002 and 2006, they involved more than 20,000 markers in more than 70 trial rings, with data on marker recovery collected up to six levels below the level on which the marker rings were fired. These trials involved the placement of markers in three planes of markers, evenly spaced within the burden of the production rings (nominally 0.65 m, 1.3 m, and 1.95 m from the next production ring to be fired) (Figure 13.9-8).

An important component of these trials was the confidence level gained from the systematic magnet-based marker recovery procedure, which removed some of the uncertainty associated with the need to visually identify markers at the drawpoint and allowed analysts to more confidently assess areas of recovery and nonrecovery in individual rings.

Analysis of data from these trials first introduced the concepts of primary, secondary, and tertiary recovery for the

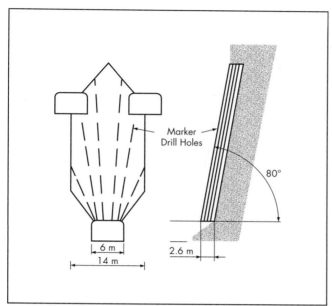

Figure 13.9-8 Layout of marker rings for Ridgeway marker trials. Drill holes in succeeding marker ring planes were offset to increase spatial coverage of markers. Markers were placed at 1-m intervals within marker drill holes.

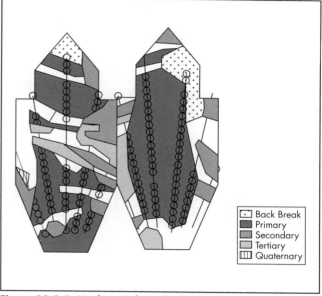

Figure 13.9-9 Marker trial results from Ridgeway showing typical width of draw for flow in adjacent crosscuts (indicating noninteraction between primary drawzones)

three main levels on which sublevel caving ore is recovered. *Primary recovery* is defined as ore reporting to the drawpoint on the level from which its sublevel caving ring was fired, whereas *secondary* and *tertiary* refer to ore that reports on successive levels below this. These trials also allowed testing of a number of theories that had gained traction in the absence of solid data and allowed other advances in understanding of granular flow in real sublevel caving environments.

Similar trials were later carried out at Perseverance mine in Western Australia (Hollins and Tucker 2004; Szwedzicki and Cooper 2007). These trials aimed to determine the precise time of marker arrival at drawpoints and therefore required manual collection of markers in order to determine exactly when, in the draw cycle, each marker appeared at the drawpoint. Although this made multilevel analysis and completion of a large, statistically significant trial program more difficult, important data largely comparable to previous field trial programs was collected.

More recently, research has led to the development of *smart* markers, which can transmit a radio frequency signal to receivers that record their extraction location and time (University of Queensland 2007).

The following sections discuss some of the new understanding that has been gained from full-scale marker trial programs carried out since 2000.

Width of Draw

Width of draw has consistently shown to be, on average, less than the width of the fired ring. The Ridgeway layout comprised 5–6-m-wide crosscuts on 14-m crosscut centers, as this mine was designed using a recovery rather than cost-reduction focus (Power and Just 2008). Here, a 6-m-wide drawpoint generally produced a primary recovery zone with a maximum width of between 8 m and 12 m. Therefore, despite the relatively close crosscut spacing, interaction between primary drawzones rarely occurred. Figure 13.9-9 illustrates this situation for sections taken through two adjacent crosscuts. All the different stages of recovery were recorded and are represented, and it is clear that the primary recovery zones do not interact.

Prior to the completion of these trials, a theory known as interactive draw (as proposed for sublevel caving) had been put forward (Bull and Page 2000) and influenced the design of some sublevel caving mines. This theory suggests that by drawing from adjacent sublevel caving crosscuts interactively or in close sequence, zones of low pressure can be created between the drawpoints, allowing the rock to flow at lower angles than in drawpoints that are drawn in isolation. Prior to this, physical modeling results had also been used to suggest that drawing from an interactive front, rather than in isolation, would decrease dilution ingress from the sides of the rings and reduce hang-ups (Janelid 1974).

In reality, interactive draw did not result in primary drawzones flowing to the full width of the ring and interaction with adjacent crosscuts (Power 2004a). Similar trials completed at Perseverance mine also indicated no evidence of interactive draw in any of the trial areas (Hollins and Tucker 2004). Summary results from these trials clearly show the episodic nature of flow from individual rings; those at relative narrow draw widths were similar to ones seen at Ridgeway (Figure 13.9-10).

Figure 13.9-5 also shows (albeit for reduced production tonnages) that recovery zones for rings spaced on 7–8-m centers at Grängesberg are not interacting, even though strict interactive draw controls were used in these experiments. Flow here was only allowed to develop to half the height seen at later mines (given the smaller sublevel spacing). Figure 13.9-7 also suggests no interaction between adjacent crosscuts at Kiruna, even though these results summarize the marker recovery results of all the trials completed in its program, rather than

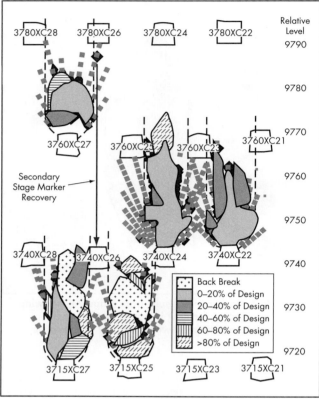

Source: Hollins and Tucker 2004.

Figure 13.9-10 Summary of the Perseverance marker trial results

of an individual marker trial. This is not surprising given the relatively large crosscut spacing at Kiruna mine.

In practice, every mine will have a range of differing input conditions that will affect marker trial outcomes, added to which are the significant differences seen between trials at individual sites. Even so, the results on width of draw at different sites are relatively similar.

Depth of Draw

In approximately 20% of cases, draw at Ridgeway never extended to the full 2.6-m depth of the ring. In more than 50% of cases, the primary flow zone only reached full ring depth late in the draw cycle for a small percentage of the ring height. Figure 13.9-11 shows an example of the primary draw, which is both narrow and shallow.

There is some conjecture about the cause for shallow draw as seen at Ridgeway. Visual observations indicated that unrecovered rock at the back of the ring sometimes retained the original rock-mass texture (Figure 13.9-12). However, if a drawpoint opened up, the less mobile ore at the back of the ring often collapsed into the drawpoint. This indicates that the rock was being fractured but the fragments were not being effectively separated from one another. However when hang-ups occurred and the LHD was used to dig farther into the base of the ring, greater depth of draw was sometimes achieved, with corresponding higher recoveries.

The shallow draw at Ridgeway shows a variation from the argument that early dilution entry originates behind the ring (from the waste side of the ring). Like other mines (Gustafsson 1998; Hollins and Tucker 2004), Ridgeway commonly measures early entry of rock from outside the ring (at approximately 20% draw). However, this rock has been shown to originate from the level above the ring, due to flow piping preferentially up the solid face of the ring.

One of the assumptions underlying the sublevel caving method is that the entire ring is fractured effectively, compacting the waste ahead of it and allowing the ore to be drawn relatively undiluted from the drawpoint. Some, however, have shown that the space available to allow adequate swell (and particle disassociation) for all portions of modern sublevel caving rings is insufficient (Hustrulid 2000). The Ridgeway results may support this hypothesis, indicating that the rock at the back of the ring is not being mobilized because of the combination of lack of opportunity to swell and cushioning of the ore at the free face of the ring. It is also possible that differing modes of dilution entry are occurring at different mines, where different blasting and draw control practices are used. Marker trials allow the opportunity to learn more about the sources of dilution entry and for mine operators to refine drill-and-blast and draw control practice to postpone and reduce dilution entry.

Ore Recovery over Multiple Levels

Before completion of the Ridgeway marker trials, assumptions about levels of primary recovery were significantly higher than those actually measured. However, although a lower than expected primary recovery is achieved on average (approximately 50% for 120% ring tonnage extracted), much of the remaining ore is recovered by draw from levels below. Even with this low primary recovery percentage, Ridgeway's closely spaced drawpoints, favorable ore-body geometry, and tightly managed draw control enable it to recover a high proportion of the ore fired, with relatively low levels of dilution (Power and Just 2008).

This indicates the importance of viewing the sublevel caving system as a multiple-level system, rather than making design decisions based purely on recovery from the primary level alone. Figure 13.9-13 illustrates how ring recovery is spread over many levels at Ridgeway. Figure 13.9-14 indicates that an equation can be fitted to this data with a good fit.

It is this general sense of order within an apparently chaotic system that has enabled the development of a number of forecasting tools that can be used with reasonable accuracy in operational mines for estimation of recovery, development of draw control strategies, and optimization of recovery through the targeting of residual grade.

Back Break

Back break is a phenomenon that is difficult to simulate effectively in physical models but can be seen in many marker trials and generally in sublevel caving production drifts. It occurs when a ring blast breaks rock away from the next ring to be fired in the crosscut. This back break often occurs at the brow but can also be present along the full height of the production ring, depending on ground conditions. This was noted in marker trials carried out by Janelid at the Grängesberg mine (Janelid 1972) and has also been seen at Ridgeway (Power 2004a) and Perseverance (Hollins and Tucker 2004). Back break is still ore recovered on the primary level and thus can be included, from a grade forecasting perspective, with primary recovery. Its effect on blasting and flow, however, can be significant. If a blast

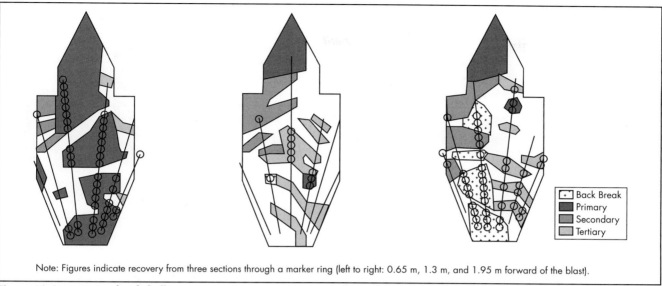

Note: Figures indicate recovery from three sections through a marker ring (left to right: 0.65 m, 1.3 m, and 1.95 m forward of the blast).

Figure 13.9-11 Example of shallow primary draw

Figure 13.9-12 Intact rock-mass structure visible at the back of a ring in which shallow draw has occurred

produces high amounts of back break, rock flows through this broken area, effectively changing the brow position of the blasted ring and reducing the likelihood of drawing to the full ring depth (Figure 13.9-15).

Figure 13.9-16 shows an example of a ring that has suffered significant back break, indicating that back break from the previous ring has broken almost to the full burden of the next ring to be fired. It is probable that this back break influenced the performance of the blast ring and possibly recovery. In this example, back break extends to the base of the ring; however, primary flow still does not reach the central marker plane for the lower half of the ring. This indicates that, even when ore is likely to have been well fragmented due to a high relative powder factor, shallow flow can still occur.

Parts of a ring that have been damaged by the previous blast can be significantly thinner than the rest of the ring. They will be blasted with a greater effective power factor, causing the blast to behave differently than intended. Generally, the zone of greatest explosive concentration produces most back break. Because this is also the area of greatest explosive concentration in the next ring, this accentuates changes in the actual blasting powder factor distribution for this ring.

Chaotic and Variable Flow

Figure 13.9-16 shows how back break in one ring can affect blasting and flow performance of the next ring to be blasted. Because the blasting and flow behavior of the vast majority of sublevel caving rings cannot be practically measured, it is not known what effects are being generated to other rings. If it is accepted that performance in one ring will affect performance of rings around it, and that these events cannot be predicted, it follows that sublevel caving flow should be treated as a chaotic system. Some of the parameters producing significant variability in sublevel caving systems include

- Rock mass properties,
- Blast design,
- Ability of charging crew to implement blast design (blocked holes, etc.),
- Hole deviation,
- Blasting performance of rings around the current ring,
- Hang-up occurrence impacting draw, and
- Grade variations impacting draw.

Although it is impossible to predict the flow outcome from a particular ring, the research program carried out at Ridgeway identified five common primary flow behaviors by which most of the trial results could be classified. These are shown in Figures 13.9-17 and 13.9-18. Although general primary recovery zones are shaded in these figures, it should be assumed that the rippling flow effects (discussed later) would result in portions of these drawzones being left for recovery on subsequent levels or not being recovered at all.

Figure 13.9-19 shows the relative frequency of these draw behaviors and that the standard and standard/shallow draw classes produce the highest ring recoveries (calculated after the secondary level). As depth of draw is reduced, width of draw increases. If apex mobility is low, width of draw is also likely to be low, and recovery is also likely to decrease.

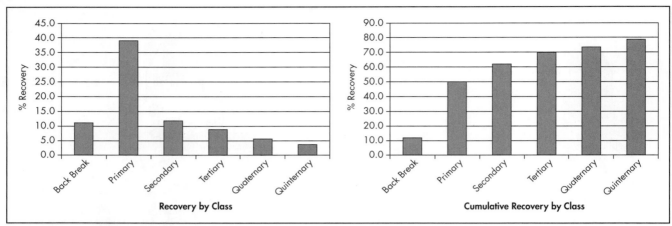

Figure 13.9-13 Recovery by class at Ridgeway

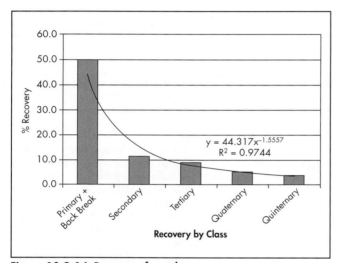

Figure 13.9-14 Recovery formula

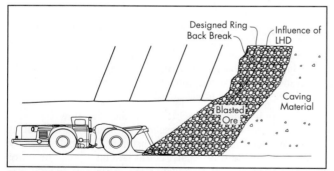

Figure 13.9-15 Impact of back break on ore mucking

Although not discussed in detail here, coupling full-scale flow assessment with analysis of the drill and blast factors producing these flow types is also important. Marker trials and other drill-and-blast research programs can be used to develop a greater understanding of drill-and-blast, and its effect on flow can be used to optimize operational drill-and-blast design. This research is also used as the basis for layout, blasting, and draw control design for new sublevel caving mines.

Many of the figures included here show examples of ripple patterns in ore recovery. These patterns produce regions of recovery interspersed with sections of nonrecovery, resembling ripple patterns produced in sand due to water flowing normal to the ripples. Researchers suggest a process whereby rock flowing toward the drawpoint encounters zones of lower mobility and continues around these on its original course. These ripple patterns indicate that, although a flow zone may progress through a certain region in the ring, parts of this region are often left unrecovered. Ripple patterns are often at slightly different orientations for different draw classes, suggesting that flow has progressed at different orientations for different recovery classes. The relatively narrow draw pipes progress subvertically from individual drawpoints to heights of at least 55 m, as illustrated in Figure 13.9-20, which shows the results of a secondary draw trial.

In this trial, no primary draw was taken in the rings on the upper level, allowing secondary draw attributable to the rings below to be studied. The primary recovery zone in the lower ring is produced by the same flow event as the secondary recovery zone on the level above. Flow from the ring below appears to have progressed into the center of the ring above (left). The secondary flow zone in the upper ring seems to have been preferentially broken in the blast, possibly due to more concentrated explosive density in the center of the ring. It is likely that low mobility in the apex of the ring below has caused this flow outcome (also resulting in very little secondary recovery from the upper ring on the right).

Episodic flow patterns from individual rings were also identified in the Perseverance marker trials (Hollins and Tucker 2004). Similarly, field experiments conducted at the Kiruna mine showed pulsations of waste entering the drawpoint, with increasing levels of waste appearing with tonnage drawn (Figure 13.9-21). These pulsations were believed to be related to differential mobility in the blasted ore from different parts of the ring and waste rock.

The results of the early model-based research into granular flow in sublevel caving led to advances in design and draw control practice and to a general understanding of granular flow. In many cases, the data recovered from marker trials now supplements or has replaced information previously gained from physical models. Such trials allow testing in theoretical thinking on sublevel caving granular flow in full-scale operational environments but must be designed and analyzed as part

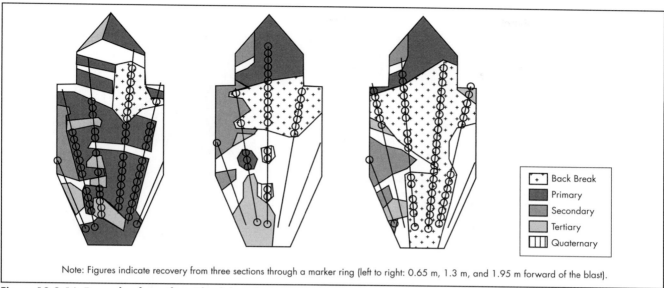

Note: Figures indicate recovery from three sections through a marker ring (left to right: 0.65 m, 1.3 m, and 1.95 m forward of the blast).

Figure 13.9-16 Example of significant back break

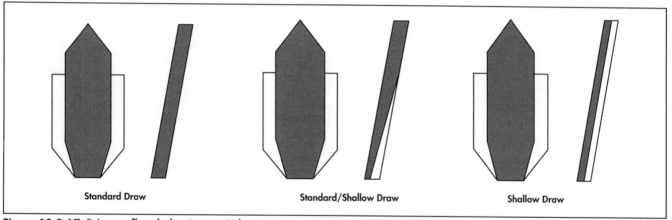

Figure 13.9-17 Primary flow behaviors at Ridgeway

of a systematic scientific program to be completely effective. They often produce results that lead researchers in totally new directions of thought.

Other results through full-scale research of the more recent understanding of flow (particularly in its multilevel nature) are improvements in computer-based forecasting and optimization capabilities. Several different methods of numerical flow modeling specifically designed for sublevel caving environments are now available. Some of these methods not only accurately forecast monthly production grade but also the targeting of high-grade zones and regions of the ore body where extraction has been low. In some cases, they allow shut-off simulation to produce detailed drawpoint-by-drawpoint production targets in situations where visual or weight-based draw control systems are ineffective.

With the continued development of new technology targeted at sublevel caving (e.g., smart markers and improved computing capacity for research into drill-and-blast effects and numerical modeling of granular flow), it is likely that sublevel caving will continue to be a productive and economically viable mass mining method.

MINE DESIGN

Layout selection is a critical success factor in all aspects of a sublevel caving operation. The layout directly relates to the ore recovery, drill-and-blast effectiveness, excavation stability, and the cost of mine development. Because it is a fundamental success factor, considerable attention needs to be given to selecting the layout prior to commencing mine establishment and determining the critical success factors for the mine. A mine focused on low-cost production will implement a different layout than a mine focused on maximizing recovery. Mine layouts will probably change over the operating life of the mine, so the original layout is often modified as a result of lessons learned under operational conditions.

Layout Orientation

Layout orientations are generally classed as either transverse or longitudinal. The transverse layout is generally applied in wider ore bodies, greater than approximately 50 m, where the parallel production drifts are driven across the ore body, according to Figure 13.9-22A. For widths below this, the production drifts are generally driven along the strike of the ore

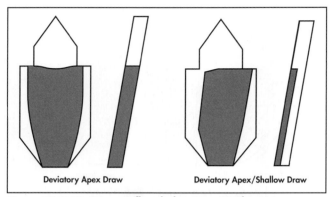

Figure 13.9-18 Primary flow behaviors at Ridgeway

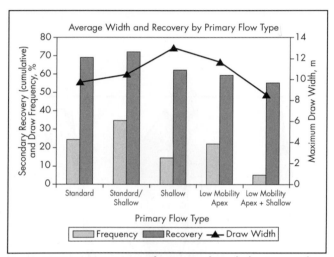

Figure 13.9-19 Frequency of primary draw behaviors and their effect on width of draw and recovery

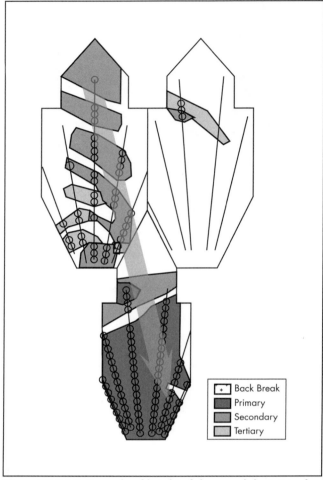

Figure 13.9-20 Example of height of draw and deviation due to low mobility in the apex, from secondary recovery trial

body in a longitudinal fashion, according to Figure 13.9-22B. The longitudinal layout may contain as few as one drift on a level depending on ore-body width. The choice of orientation is predominantly determined by the width of the ore body but can also be influenced by geotechnical and economic factors.

The transverse layout is generally more productive than the longitudinal because of the number of active production faces that can be made available. Some mines will operate with a mixture of layouts due to a variable ore-body geometry, although transition between the two styles is difficult in a vertical sense because of the criss-cross nature of the layouts. However, if transition is required between the layouts, it may be prudent to skip a level and recommence sublevel caving operations with the new layout and allow the skipped level to cave in rather than trying to perform complex production ring drilling and blasting. This technique was successfully employed at the Perseverance mine (Wood et al. 2000).

For either layout type, the supporting development is similar with a perimeter drift containing either orepass tipping points or truck-loading stockpiles. The perimeter drifts are also used to provide access to the primary ventilation circuit and enable the installation of secondary fans to allow airflow to be delivered to the production working areas on the level.

Layout Dimensions

Sublevel Height

The sublevel height is generally measured as the vertical floor-to-floor distance between the levels. It is determined by the dip of the ore body and the physical limitations of the production drill-and-blast equipment. If the ore body is vertical, then dip is not a constraint on sublevel spacing. However, as the ore body flattens, the sublevel height needs to be reduced to avoid incurring excessive amounts of hanging-wall dilution, as well as minimizing the amount of unrecoverable ore on the footwall.

Over the years, advances in drill-and-blast equipment have seen the sublevel height increase from less than 12 m to more than 30 m in some cases. The key criterion for drill-and-blast is to ensure that satisfactory fragmentation is achieved through the entire ring. To achieve this, the drill holes must be evenly spaced, straight, and adequately charged with the appropriate type and strength of explosive. The proposed drill pattern is an integral part of selecting the overall sublevel layout and impacts not only the sublevel height but also the drift dimensions and the drift and pillar spacing.

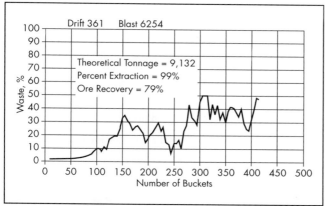

Source: Hustrulid 2000.

Figure 13.9-21 Pulsations of waste entering a ring at the Kiruna mine

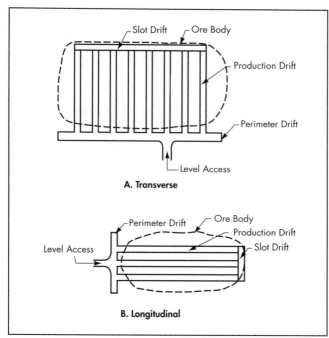

Figure 13.9-22 Transverse and longitudinal layouts

Production Drift Spacing

Important factors in selection of production drift spacing include

- Potential ring shape produced by the combination of sublevel height and production drift spacing (will it produce a ring shape that can be efficiently blasted?);
- Granular flow properties inherent to the mine (based on drill-and-blast procedures and rock-mass characteristics) and the likely recoveries to be achieved;
- Value of the mineral being mined and whether a cost-reduction or recovery-emphasis strategy will be followed;
- Geotechnical factors; and
- Production drilling and charging factors.

For maximum ore recovery, the entire level would theoretically be excavated, similar to a coal longwall operation, but this is impractical. Therefore, a trade-off is required between the amount of ore potentially recovered on its original level and the amount left in the cave for recovery, hopefully, on lower levels (with more associated risk to its ultimate recovery).

Thus, the spacing of the production drifts is strongly influenced by the operating philosophy of the mine. The more cost-focused a mine is, the wider the spacing between the production drifts, because this reduces the amount of expensive development required (Power and Just 2008). Figure 13.9-23 illustrates this for two different layouts, showing theoretical primary recovery zones only. Recovery-focused mines will try to place the production drifts as close together as practical in an attempt to maximize the amount of ore recovered on the primary level to minimize the effects of dilution, as shown on the left side of Figure 13.9-23.

An indicative estimate for the ellipsoid of draw width can be generated by either using a ratio of 1.4:1.7 of the drift width or the drift width plus 1.5 m on each side, according to the following formula:

$$W_{DE} = W_D + (1.5 \times 2)$$

where

W_{DE} = ellipsoid of draw width, m
W_D = drift width, m

Previous estimates for this W_{DE}/W_D ratio for the much smaller sublevel caving layouts also predicted this same ratio range of 1.4:1.7 (Hustrulid and Kvapil 2008).

Because sublevel caving relies on the extraction of ore from not only the level on which it is fired but subsequent lower levels, it is important to overlay the extraction envelopes over all levels to gain an understanding of likely system recovery rather than just recovery on a single production level. In extreme cases, significant overlapping of theoretical draw ellipses may result in excessive extraction of these areas, which can lead to early dilution entry. Conversely, too much of a gap between the extraction envelopes could lead to consolidation of material on the sides of the rings, which may cause poor firing conditions, due to the limited void available. Stress concentration areas may also result, which could lead to damage of the production drifts.

Production Drift Shape

The shape of the production drift impacts the mine's ore recovery and productivity. Ideally, the production drifts are as wide as possible to allow the maximum flow of ore. To further assist with flow, the drift roof should be as flat as practical while still providing adequate support to the roof and brow. A wide and flat roof also assists in improved spacing of the production drill holes, which limits the impacts of blasting on the brow, as well as reducing the amount of back break. If the drift roof is overly arched, the flow can be preferentially funneled to the center of the drift, leading to early dilution. In some cases, a trade-off will be necessary (i.e., an arch may be required to produce a wider stable drawpoint).

Greater drift widths also improve the productivity of the LHD machines. An LHD can tram faster in a drift where it has plenty of clearance. At the muck pile, the width is also an advantage for the LHD, because it is able to load from side to side, ensuring a more even flow of material into the production drift, as illustrated in Figure 13.9-24. In addition, the loader is able to extract material from deeper into the drawpoint,

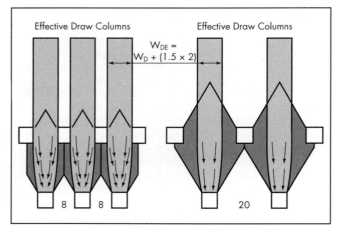

Figure 13.9-23 Effect of drift spacing on primary ore recovery

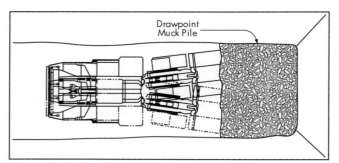

Figure 13.9-24. LHD side-to-side loading in a drawpoint

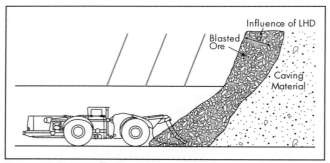

Figure 13.9-25. Influence of LHD on cave material

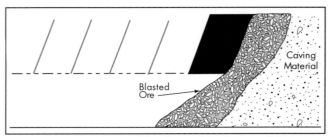

Figure 13.9-26 Unique sublevel cave-blasting conditions

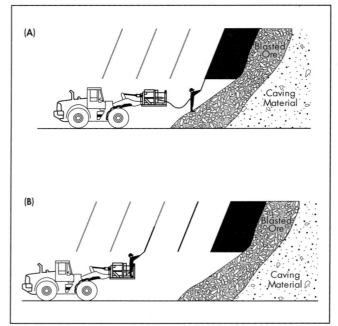

Figure 13.9-27 Charging option set-ups

because it is not extracting the material from the center of the drift, which has rilled the farthest out from the brow, as illustrated in Figure 13.9-25. This further improves ore recovery and influences the burden of the production rings, because the cave material loosens in front of the next ring to be fired.

Minimizing the height of the drift roof restricts the rock's ability to rill too far from the brow. As previously stated, if the depth of material excavation by the LHD can be increased, then the overall ore recovery will be increased. The drift height, however, is usually a trade-off between length of rill and size of machinery required to achieve the desired production rate. Although it is possible to obtain custom-made machinery to fit the optimal drive height, this strategy can have drawbacks if the machinery fails to meet production targets, as machinery replacements are generally not immediately available. In the authors' experience, it is generally the production drill that limits the drift height, and, as the large-scale operations push for longer and longer production holes, this is likely to remain the case. Low-profile LHDs are generally available to achieve the desired production rates, but the desire to drill with longer and longer drill rods for greater hole accuracy is forcing average drift heights higher.

Production Drill-and-Blast

The drill-and-blast requirements of a sublevel caving mine are unique because the blasting environment is always choked. This means that every production ring is blasted against the combination of previously blasted ore and waste sitting in and around the production drift (Figure 13.9-26).

Because of the environment in which the blasting occurs (i.e., the only free face is the roof of the production drift), the blast's function is different from nearly every other production drilling situation. With sublevel caving blasting, the intent is to initially compact the caved material in front of the blast to create a void area for the fired ring to access. After the fired ring is in the void area, the intent is to fragment the ore slice as much as possible so that it will preferentially flow when mucking commences. Although this is the intent, the reality is somewhat different. The nature of the fanned ring pattern used in sublevel caving and the reducing ratio of development-to-production work against the intended blasting criteria. With the reduction in this ratio, the only true swell available to the blasted rock is reduced and in some cases is <10%, while the blasted rock itself still wants to swell to >30% of its in-situ volume. This lack of initial swell results in the holes fired first, and those sections of the hole close to the drift have the opportunity to swell and move into the drift as desired, while the ends of the holes have limited opportunities to swell and move. In addition, the fan nature of the drill pattern ensures that explosives are concentrated in the center of the ring relative to the outer holes. This creates smaller, more uniform fragmentation in the center, which is thus more mobile than the material at the edges. These fragmentation and flow characteristics affect the gravity flow and ore recovery performance for the mine (Trout 2002).

With the diamond layout style of sublevel caving production drifts, a wide variety of production drilling patterns can be used. With the wider-spaced production drifts, a pattern involving larger and longer holes will generally be used, although, as the production drifts are brought closer together, the drill holes will interact. When this occurs, the production rings often need to be offset by approximately 1 m to reduce the chances of sympathetic detonation between the holes or damage to the neighboring ring when the rings are fired. Steeper side holes will lead to better funneling of ore to the drawpoint and smaller *dead zones* at the sides of the ring but also to higher apexes at the top of the ring (which can be more difficult to blast effectively).

The production drill rings are generally inclined forward by 10° to 20°. This inclination assists in feeding the ore preferentially into the drift by providing a small overhang to the diluting material directly above. The inclination also ensures that the holes are all drilled on a similar plane for improved blasting results and additional protection of the ring brow. Because prevailing structures in the rock mass can sometimes adversely affect rings inclined forward at similar angles, the amount of forward inclination may need to be adjusted in different parts of the mine in order to achieve optimal results.

The sublevel caving production rings can be charged with either ammonium nitrate fuel oil (ANFO) or emulsion products. Prior to any charging activities, the production holes should be *dipped* by the charging crew to ensure that the holes are clear of obstructions and are drilled to the correct depths. Often, charging personnel must stand on the muck pile to charge the ring, as the charging unit frequently cannot get the operators to the production ring because of the rill of broken rock in the drawpoint and the height of the drive (Figure 13.9-27A). This can lead to significant risk of injury and inferior drill-and-blast results. Safety hazards include the muck pile rilling unexpectedly or trip hazards resulting from the uneven nature of the muck pile on which the operators work.

To provide a safer charging environment, the preloading or precharging of multiple rings can be implemented (Figure 13.9-27B), although this is subject to the local regulators and may take time to get approval. *Precharging* relates to fully loading and priming the production rings, whereas *preloading* refers to the practice of only placing the explosive in the hole but not the detonator, which is added immediately prior to blasting. When utilizing precharging, the tails of the detonators are tucked back up the hole so they do not impede the LHD when it is mucking the rings in front; the tails are only pulled out when the ring is ready to be hooked up for blasting.

The implementation of precharging/loading in recent years at a number of large-scale operations has had significant benefits for those operations, including improved safety of personnel around the drawpoint brows, a decoupling of the charging and mucking activities, the removal of production disruptions caused by hole blockage and dislocation, the elimination or significant reduction in the number of production ring redrills, and improvements in draw control (as less mucking is needed to prepare the ring for firing).

Drawbacks to precharging/loading include the need to have a number of production rings predrilled in order to allow the charging and drilling processes to be independent. From a safety perspective, when a precharged/loaded ring is behind a misfire, the scope is limited to get a drill rig back to assist with the misfire unless the drill rig has a remote operation function or the explosives are washed from the precharged/loaded ring. For this reason, precharging is often easier to implement in mines with better geotechnical conditions, where charged holes are less likely to be dislocated or brows lost. If the ore body contains aquifers or seeping water, precharging/loading may be also unsuitable when using ANFO, because the explosive column can be desensitized or washed out by the time the ring is ready to fire (emulsion should overcome this issue).

Because of the choked blasting conditions of sublevel caving rings, a single production ring per drift is generally only fired at any one time. Firing of multiple rings may be undertaken in areas of very poor ground conditions, but this introduces the risk of misfires and leaves pillars within the cave, which can cause problems on the level below. It also leads to lower overall recovery, because most of the ore in the ring farther from the brow cannot flow to the drawpoint and can be occluded by waste flowing from above. Depending on infrastructure constraints and required production rate, some mines fire rings in multiple production drifts at the same time to ensure broken stocks are sufficient for the LHD fleets.

Hang-ups are a normal feature of the sublevel caving mining method, with up to 50% of rings hanging up at some stage of the mucking cycle. In mines with more closely spaced drawpoints, the hang-up can sometimes be cleared by removing additional material from the production drifts on either side of the hung-up drift. This material removal appears to disturb rock in the affected drawpoint by acting to release pressure on either side of the hung-up ring. Other hang-up retrieval methods include water cannons and (as a last resort) explosives, because they often damage the drawpoint brow. If a hang-up cannot be cleared within a specified period, the next ring in the production drift must be fired and the mining process continued.

Although production drill-and-blast design is an input into the overall layout of the sublevel caving mine, it can be readily changed as the characteristics of the ore body become more fully understood. The shape of the production ring may not vary significantly, but the drill-and-blast parameters may. Design parameters, including hole diameter, number of holes in the ring, drill-hole toe spacing, production ring burden and dump angle, and explosive column densities and lengths, can all be modified with relative ease. When adjusting the production drilling parameters, it is important to take a holistic view of the sublevel caving operation in order to ensure that the changes made will not dramatically and negatively affect other parts of the process.

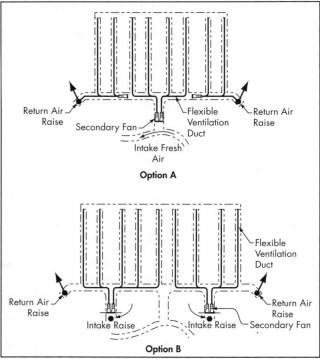

Figure 13.9-28 Variable ventilation layouts for sublevel caving operations

Services

The active ore production areas are all dead-end headings. A single fan may be used to supply up to four production drifts, as the ducting can be tied off in the headings that are not in use. If multiple secondary fans are required on the level, they may need to be set up in a cascading style (Option A, Figure 13.9-28). This means that a fan will be placed downflow of another fan and, in essence, will receive the used air from the upflow fan. This cascade style of secondary ventilation can lead to lost production time if there is a significant amount of dust in one of the drawpoints upflow of the secondary fan, because all the dust will be blown into the headings serviced by the downflow fan. Each level is set up as its own primary circuit so the levels do not interact with each other. Primary intake and return raises with fixed ventilation infrastructure are used to direct the appropriate amount of air throughout the mine. Figure 13.9-28 shows two styles of level ventilation: Option A is for mines using the main access decline as their fresh air source, whereas Option B is for mines that heat or refrigerate their intake air.

The cave itself may provide a path for airflow after it has broken through to the surface, though this will be determined by the porosity of the cave column and the pressure differences between the surface and the cave workings. Generally, movement of air through the caved rock will be in the form of leakage rather than a volume that can support mining activities. Individual mining circumstances will ultimately determine how much air passes through the cave zone area.

Water and compressed air are reticulated to each level to support the mining activities. After development of the level is completed, the compressed air line may be recovered unless required by the production drill. The water line is used for production drilling and to feed the drawpoint sprays mounted on the drift roof to keep the dust levels down during mucking. Drainage on the sublevels is important in ensuring that maximum production rates are achieved. The production drifts are usually driven on a slight incline of between 1 in 50 (2%) and 1 in 40 (2.5%) to remove the pooling of water along the drift or at the face. In addition to the introduced water for the mining process, water is also associated with aquifers that have been breached during mining. Sumps are generally located in the perimeter drift with a drain hole drilled to the level below. The production water can then be collected at a mine dam located lower in the mine and pumped out via the mine dewatering pump system.

Communications systems are run onto the production levels for radio communications, as well as to control any automated or semiautomated mining equipment, including production drill rigs and LHDs in some of the larger-scale sublevel caving operations. Such systems may also be utilized to transmit real-time production data to a central production management area. The larger-scale sublevel operations using automation or the tele-remote operation of equipment are doing so in an attempt to remove workers from hazards associated with drawpoint brows and in the general underground environment. The LKAB mines in Sweden are leading the way in this field. They use the automation of machinery to provide production increases by allowing the machines to work extra hours while the operators are located on the surface or an area of the mine not impacted by the mining process, especially the blasting and clearance cycle.

Geomechanical Factors

With sublevel caving, a number of geomechanical factors must be considered when either laying out the mine or making large-scale operational decisions. These include

- Quality of the ore body and hosting rock mass,
- Orientation and magnitude of the major and minor stress parameters, and
- Presence of large-scale and microstructures.

Prior to establishing a sublevel caving mine, an assessment of the likelihood of natural caving is required. The assessment is a specialized geotechnical task and is beyond the scope of this chapter. However, inputs for these techniques rely on a number of key geotechnical parameters, so it is imperative when diamond drilling into a deposit that may be applicable for caving that all of the relevant information is recorded. If stability assessments indicate that the ore body will not cave naturally, then sublevel caving may not be suitable, unless reliable techniques for cave preconditioning and/or caving assistance are available.

The orientation of the stress field will often determine the retreat direction of the sublevel caving production front. If there are no other influencing factors, it is preferable to retreat perpendicular to the major principal stress direction. By retreating in this manner, the effects of the stress field can be "shadowed out" after production has been initiated on a sublevel. This creates a distressed environment for the majority of production headings in the mine, because the stresses will be forced to wrap around, under, and over the production areas.

Stress that builds up around the periphery of the working areas is called abutment stresses, which can result in rock movement and bursting, depending on the rock strength-to-stress ratio. The ground support requirements are therefore greater in the perimeter drives than in the production drifts.

The bullnose section of the pillars at the turnout points of the production drifts from the perimeter drive are especially susceptible to stress damage, as shown in Figure 13.9-29, because they are reduced in size with the retreating of the cave front.

Production drive stability can also be impacted by the abutment stresses as production on the level above passes over the lower-level production drifts. The preferred ground support system in a sublevel caving operation is one that can yield slightly without failing when the abutment loads come on but is still sufficient for working at the brow in the distressed conditions. Some ground support combination examples include grouted rebar with 50 to 75 mm of fiber-reinforced concrete or split sets with wire mesh.

Identifying mine scale and minor structures is important when laying out the mine, because they can be used to assist in the propagation of the cave or in controlling the effects of the abutment stresses. Should the major structures not be favorable to the cave's progression, then considerable care needs to be taken to monitor caving progress, because it may be necessary to reduce or stop production in certain cave areas to reduce the impacts of these structures. The orientation of the minor structures is also important, because it can influence the inclination of the production rings. Such inclination on a plane parallel to the minor structures can reduce the number of production hole dislocations and therefore improve productivity.

Water-bearing aquifers or surface water courses—that is, streams and dams in the caving area—also need to be identified prior to the commencement of caving activities to ensure that there are no unexpected water or mud inrushes during the operation of the mine. Where an aquifer or water course cannot be dewatered prior to caving, extreme care should be taken and evacuation procedures implemented as the cave encroaches on the water-bearing area. Generally, as the cave propagates, it will ensure that a large number of fractures are created in the cap rock, which will breach the aquifer and potentially drain it before the main cave back moves through the water-bearing zone. However, in extreme cases, uncontrolled mud flows can be generated by excessive and poorly controlled draw in combination with high inflows of water (either from aquifers or channeling of precipitation through subsidence zones) and suitable clay mineralogy. When this is a risk, use of a robust inrush hazard management plan with appropriate triggers and controls is beneficial. Prior to commencing a recent sublevel caving operation at the Kiruna mine, engineers were forced to drain a small lake, while engineers at the Ridgeway mine identified an aquifer system below the surface and monitored it as the cave progressed, noting only minimal water reporting to the production drawpoints, as the majority of the water was absorbed in saturating the cave column.

PRODUCTION DRAW CONTROL

Draw strategy and production management are extremely important in sublevel caving operations to enable the accurate prediction of ore and waste flow from the cave after production commences. The draw strategies and production management plans used at each mine need to be variable and flexible in order to cope with unforeseen circumstances such as the early inflow of waste, hang-ups, geotechnical instability, and arrival of large oversize material at drawpoints.

Draw Strategy

The objective of a draw strategy is to maximize the overall ore recovery while minimizing the impacts of dilution. To achieve this, an effective sublevel caving mine relies heavily on a draw control system that is variable and adaptable to the constantly changing operational situations, including unscheduled equipment breakdowns and production incidents. Effective draw control practices also rely on an understanding of the cave's flow characteristics and the grade distribution in and above the ore body being mined.

Any adopted draw control system needs to include guidelines to allow the frontline management to make tactical decisions. Such guidelines will be grouped according to authority levels—ranging from the LHD operator to the general manager of the mine. In mines with a distinct visual or physically distinguishable difference between the ore and waste rock, guidelines will dictate the percentage of ore-to-waste in the drawpoint when production must stop for a particular drawpoint. To allow decisions to be made, mines without this distinction must rely solely on the tonnage drawn from the drawpoint followed up with a sampling system.

Regular monitoring of the drawpoint status is imperative to achieve overall efficient draw strategy and to ensure that underperforming or troublesome drawpoints are identified early so their potential impact can be minimized. Draw strategies can be classed as either *independent* or *interactive* and are determined by the mine layout and overall mine objective. These terms refer to drawpoints where tactical extraction targets are linked to those of surrounding drawpoints (interactive) and those where tactical production targets can be generated without reference to surrounding drawpoints (independent). Theories discussed previously concerning the widening of draw columns due to interaction through interactive draw do not have bearing on these draw strategies.

Independent draw is often used by operations with a limited number of production drifts available for mucking. Under these circumstances, each individual drawpoint is often depleted before moving onto the next one. The penalty for this type of draw strategy is a greater risk of dilution piping in from above and impacting adjacent production drifts. Interactive draw style operations rotate the LHD between a number of neighboring drawpoints in a fixed sequence, which is repeated until all of the drawpoints have been completed and the next set of rings fired. The interactive approach enables the blasted and caved material to be extracted over a larger lateral area, thus reducing the opportunity for dilution piping. Depending on the ore flow characteristics of the mine and the mine layout selected, either approach may maximize the overall mine performance.

As previously identified, drawpoints can be operated individually or in groups. In mines with substantial numbers of drifts on each level, the practice of grouping a number of drifts together and treating them as a single entity can substantially reduce the complexities of managing the operation on a short-term basis. The orientation and shape of the production front will be influenced by the stress field and major structures. No standard exists for the shape of the cave front, because all mines will have varying conditions and operating requirements. The three most widely used cave-front layouts are the flat, en echelon, and stepped, as shown in Figure 13.9-30.

When selecting a cave front, thorough consideration needs to be given to the lead–lag rules that will be used to operate the sublevel caving. The term *lead–lag* refers to the maximum and minimum spacing between either individual or groups of drawpoints being retreated across the levels. These rules are 3-D, because they also include the level above and the level below

Figure 13.9-29 Example of perimeter drive damage in sublevel caving operation

the specific level. When defining the lead–lag rules, the main influencing factors are stress field constraints, which relate to the minimum distance required between production fronts on adjoining levels to preclude stress damage. Damage can also occur when drawpoints retreat past their neighbors on the same level. The impacts of stress can be mitigated by increasing the amount of ground support in the sections of the drift, though additional support will be limited by the ability to cost-effectively change the lead–lag distances. Another operational limit to lead–lag distances is the interaction of drilling and charging activities within adjoining drifts. Most operations will have an exclusion limit around a charged hole, which must be considered when establishing lead–lag rules.

Compaction of the cave column is also another potential factor. If a level retreats too close to the one above, stability of the cave in front of the upper level may be compromised, and the ore blasted on that level may be more difficult to recover. Conversely, if the distance between the vertical levels is too great, it may take too much time for the lower level to pass under the upper level, and the cave may recompact or reconsolidate, depending on its characteristics, causing considerable production and ore losses.

Production Management

A number of factors impact the theoretical production rate of a sublevel caving operation, including

- Ore-body geometry;
- Cave fragmentation of both ore and waste;
- Mine layout and excavation size;
- Equipment employed, fleet size, and number; and
- Materials-handling capacity

As the sublevel caving method is now often used in more competent ore bodies, opportunities for larger, stable excavations exist. This is transforming production rates for sublevel caving operations as larger equipment is used. In most sublevel caving mines, the majority of activity occurs within the confines of the level with little activity in the access declines. This makes the method efficient, with some operations achieving >35,000 t/operator per annum, as relatively few operators are required underground once the mine is operational.

Overall, the aim of sublevel caving mining is to become as factory-like as possible. This approach is achieved by taking a holistic view of the mining method from the face through to the surface in the understanding that overall mine productivity is governed by the lowest-producing component. With the increase in equipment size and technology, larger fragmentation can be accommodated at the drawpoint, but this may cause downstream issues for the shaft/conveyor or truck fleet used to haul the material out of the mine. As such, it is imperative that the mine production system is integrated with the materials-handling system, as well as with the mill, to ensure overall optimum mine performance.

Recovery and Extraction

Recovery factors relate to the percentage of mineral recovered from the ore body relative to the amount blasted. Mineral recovery from sublevel caving operations can be above 85% of the quantity of in-situ mineral blasted. However, this recovery can be inflated by inflow of low-grade mineral contained in diluting rock from outside the mining boundary.

Extraction or tonnage factors relate to the percentage of tons drawn relative to the tons fired. This percentage varies from site to site but is generally between 80% and 125%. When the extraction rate is more than 100%, this is termed *overdraw*. Depending on the draw control strategies and grade distributions within the ore body and host rock, individual drawpoints may be drawn in excess of 1,000% and still be returning material above the shutoff grade (the grade at which drawpoints should theoretically be shut down). This generally occurs when previous levels of the mine have been significantly underdrawn because of geotechnical or mine design factors.

Some ore bodies will have physical differences between the ore and waste, which are readily identifiable—for example, color or weight. This allows mine personnel to assess in real time whether or not a drawpoint is still running at the required grade. In these situations the mines are able to more selectively recover ore left behind on previous levels. However, they must be careful to ensure that their grade control methods are accurately calibrated in order for their shutoff targets to be met.

Other operations with no readily identifiable physical difference between the ore and waste will often run to a tonnage limit and stop draw at a point independent of the drawpoint grade. This tonnage strategy can lead to a considerable amount of mineral being left in the cave, which these operations will then try to recover over subsequent levels by increasing their planned draw rates using a broader, less selective strategy. Under these systems, total tonnage factors for the mine can be slightly higher than situations where selective grade control can be applied.

Fines migration can also influence extraction strategies, especially when sublevel caving is applied beneath open pits with significant exposures of weak, weathered materials or in ore bodies with weak hanging walls. The cave column applies a significant amount of comminution to the rock as it flows, so weak material is often reduced to a small fragmentation size referred to as fines. Fines can also be created through the overcharging of the blastholes. Due to the properties of gravity flow, fines material will descend through the cave column faster than the coarser rock fragments, sometimes prematurely closing drawpoints, affecting ore-handling system efficiency or creating the potential for mud rush when combined with the introduction of large amounts of water to the cave column.

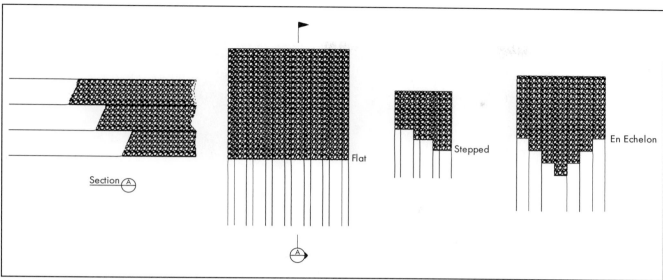

Figure 13.9-30 Examples of cave-front geometries

TECHNICAL COMPONENTS

Cavability

The caving of the host (or *cap*) rock above the sublevel caving operation is crucial to the success of the mine. Mine-scale structures and the general geotechnical conditions of the host rock influence the rate of cave propagation and the size of the footprint required to ensure caving.

There are two distinct phases to the caving of the cap rock: (1) initiation of caving and (2) the continuation of caving as the mine continues deeper (Figure 13.9-31). Should either of these stages not occur and the cave stalls or forms a stable arch, there are considerable safety risks relating to the creation of voids and the potential for air blasts. Cap rock failure may occur as either a time- or production-dependent failure. Time-dependent failure occurs when the cap rock yields and fails only at intermittent periods—a mode that can lead to significant cap rock failures depending on how much swell room is available. Production-dependent failure occurs when the cap rock unravels almost immediately after an air gap becomes available, largely controlled by extraction of the blasted material. Time-dependent failures are often accompanied by a small number of significant seismic events, whereas production-dependent failures tend to have a significant number of seismic events of a small magnitude. The mode of cap rock failure will determine the production ramp-up rate of the sublevel caving mine, because a number of safety hazards can be created if the initial production rates are not matched to the cave propagation rate.

The most significant safety risk is the creation of an excessive air gap, which can lead to a dangerous air inrush/air blast when the cave back yields. In order for rock to cave, some form of air gap has to be present. The geomechanical properties of the rock mass, the footprint of the production level, and the draw control strategy of the mine determine the size of the air gap that can form. Sublevel caving operators must manage their operations to ensure that the size of any air gap that forms is limited and does not constitute a hazard to mining personnel.

Void monitoring is a critical process when establishing a cave under any conditions. The results from the cave-monitoring system need to be cross-checked with the production figures to determine the type of cap rock failure mode that is in place and to ensure that there is a suitable coverage of blasted and caved material above the active production drawpoints to prevent a dangerous inrush of air into operational areas. It may be prudent to reduce or even stop production for a period of time in the sublevel caving to allow the cap rock failure a chance to catch up and return to the safety operating guidelines. Continuing to produce while creating a significant air void can have catastrophic results when the cap rock ultimately fails.

To maintain an understanding of how the cave is progressing, cave monitoring throughout the life of the operations is important. A cave-monitoring system should consist of a number of complementary measurement techniques to allow for the cross-referencing of results. Microseismic systems placed outside the expected cave column give an indication of the seismogenic zone, which exists just above the cave back (Duplancic and Brady 1999) and can be used not only until the cave breaks through to the surface but also as it relaxes over the life of the mine. Open holes into the cave back area can be used until the cave reaches a predefined distance from the surface where it is no longer safe for people to travel above the cave back. The open holes can be plumbed for depth using weighted wire rope and in some cases also provide access for borehole cameras. As the cave approaches the surface, the open holes can be used to place extensometers and time-domain reflectometers, which allow for remote monitoring of the failure of the cap rock.

Subsidence

As a caving method, sublevel caving will create a surface expression and a crater that deepens with time. This surface movement zone is referred to as the subsidence area. Understanding the likely size of this area is imperative when locating major infrastructure both on the surface and underground. Motion

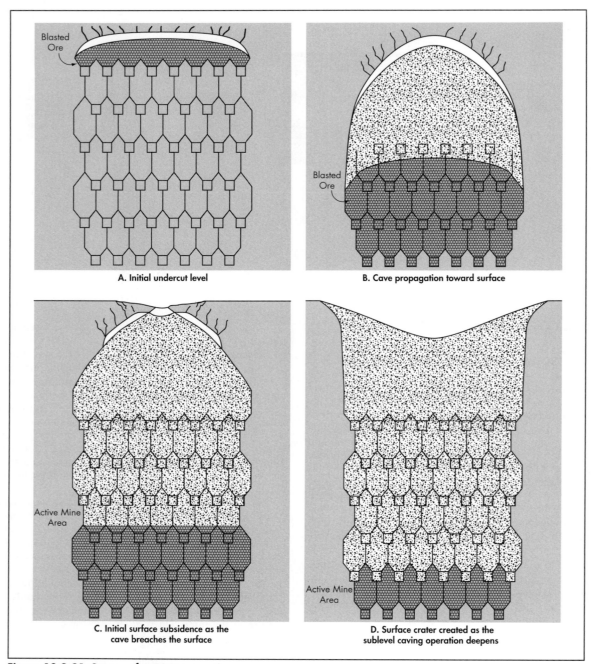

Figure 13.9-31 Stages of cave propagation

sensors and aerial surveys (Figure 13.9-32) can be used to monitor the growth of the subsidence area to determine if it correlates closely with the production being drawn from the mine. Should a discrepancy occur between the surface growth and mine production, a review of the potential void creation should be undertaken to ensure there is no risk to the work force.

As the subsidence zone creates a direct link from the surface to the underground workings, it is imperative that water inflow restrictions are placed around the subsidence zone where practical. For sublevel caving mines below an open pit or at the bottom of a valley, this can be a significant issue as these areas are catchments for rainfall. In these situations, monitoring of the rainfall and the saturation of the cave column may be utilized to determine the appropriate operational response. The flow of water into the subsidence area may not only result in additional pumping from the workings but can create a mud-rush issue when combined with the fines located within the cave column.

In the simplest case of a uniform vertical ore body, the initial cave crater will generally have vertical walls until the crater reaches a sufficient depth for slumping failure to occur. The surface expression will gradually expand as the crater deepens and an exclusion limit needs to be implemented. The size of the subsidence and fractured zone around a cave will vary due to the host rock conditions, major structures, and so forth, but a rule of thumb estimate is 65° from the lowest level, as shown in Figure 13.9-33.

A variety of methods can be used to estimate subsidence effects, wherein the reader is referred to other sections in the handbook.

ACKNOWLEDGMENTS

The authors acknowledge the efforts of all involved in advancing the understanding of sublevel caving mining, because this chapter is a synopsis of many presentations, research programs, and personal discussions.

REFERENCES

Bull G., and Page, C.H. 2000. Sublevel caving—Today's dependable low cost ore factory. In *Mass Mining 2000 Conference Proceedings*. Edited by G. Chitombo. Melbourne, Australia: Australasian Institute of Mining and Metallurgy.

Duplancic, P., and Brady, B.H. 1999. Characterisation of caving mechanisms by analysis of seismicity and rock stress. In *Proceedings of the 9th International Congress on Rock Mechanics*, Paris. Edited by G. Vouille and P. Berest. Brookfield, VT: A.A. Balkema.

Gustafsson, P. 1998. Waste rock content variations during gravity flow in sublevel caving: Analysis of full scale experiments and numerical simulations. Ph.D. thesis, Department of Civil and Mining Engineering, Luleå University of Technology, Sweden.

Hollins, B., and Tucker, J. 2004. Draw point analysis using a marker trial at the Perseverance nickel mine, Leinster, WA. In *Proceedings, Mass Mining 2004*, August 22–25. Santiago, Chile: Mineria Chilena.

Hustrulid, W. 2000. Method selection for large scale underground mining. In *Mass Mining 2000 Conference Proceedings*. Edited by G. Chitombo. Melbourne, Australia: Australasian Institute of Mining and Metallurgy.

Hustrulid, W., and Kvapil, R. 2008. Sublevel caving—Past and present. In *Mass Mining 2008 Conference Proceedings*, Luleå, Sweden, June 9–11. Edited by H. Schunnesson. Luleå, Sweden: Luleå University of Technology.

Janelid, I. 1972. Study of the gravity flow process in sublevel caving. International Sublevel Caving Symposium, Stockholm. Atlas Copco.

Janelid, I. 1974. "Rasbrytning." STU-rapport 73-3885. Stockholm, Sweden: Royal Institute of Technology.

Janelid, I. 1975. Study of the gravity flow process in sublevel caving. In *Proceedings of the Sublevel Caving Symposium*, Stockholm. Atlas Copco.

Figure 13.9-32 Surface subsidence at Ridgeway gold mine

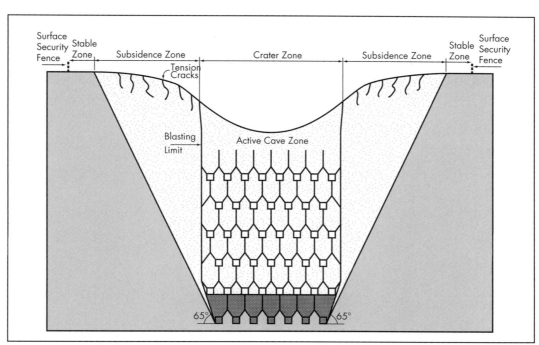

Figure 13.9-33 Subsidence terminology

Kvapil, R. 1965. Gravity flow of granular and coarse materials in hoppers and bins. *Int. J. Rock Mech. Min. Sci.*, Vol. 2, Part I, pp. 25–41; Part II, pp. 277–304.

Kvapil, R. 1992. Sublevel caving. In *SME Mining Engineering Handbook*. Edited by H.L. Hartman. Littleton, CO: SME.

Power, G. 2004a. Full scale SLC draw trials at Ridgeway gold mine. In *Proceedings, Mass Mining 2004*, August 22–25, 2004. Santiago, Chile: Mineria Chilena.

Power, G. 2004b. Modelling granular flow in caving mines: Large scale physical modelling and full scale experiments. Ph.D. thesis, University of Queensland, Brisbane, Australia.

Power, G., and Just, G.D. 2008. A review of sublevel caving current practice. In *Proceedings of the 5th International Conference and Exhibition on Mass Mining*, Luleå, Sweden, July 9–11, 2008. Luleå, Sweden: Luleå University of Technology.

Quinteiro, C.R., Larsson L., and Hustrulid, W.A. 2001. Theory and practice of very-large-scale sublevel caving. In *Underground Mining Methods: Engineering Fundamentals and International Case Studies*. Edited by W.A. Hustrulid and R.L. Bullock. Littleton, CO: SME.

Szwedzicki, T., and Cooper, R. 2007. Ore flow and fragmentation at Perseverance mine. In *Proceedings of the 1st International Symposium on Block and Sublevel Caving*. Marshalltown, South Africa: Southern African Institute of Mining and Metallurgy.

Trout, P. 2002. Production drill and blast practices at Ridgeway gold mine. In *Proceedings 2002 Underground Operators Conference, Townsville, Queensland*. Carlton South, Victoria, Australia: Australasian Institute of Mining and Metallurgy.

University of Queensland. 2007. *Mass Mining Technology Project*. Final Report. Brisbane, Australia: University of Queensland.

Wood, P., Jenkins, P.A., and Jones, I.W.O. 2000. Sublevel cave drop down strategy at Perseverance mine, Leinster Nickel Operations. In *Mass Mining 2000 Conference Proceedings*. Edited by G. Chitombo. Melbourne, Australia: Australasian Institute of Mining and Metallurgy.

CHAPTER 13.10

Block Caving

Charles A. Brannon, Gordon K. Carlson, and Timothy P. Casten

Block caving is a mass mining system that uses the action of gravity to fracture a block of unsupported ore, allowing it to be extracted through preconstructed drawpoints. By removing a relatively thin horizontal layer at the base of the ore columns using standard mining methods, the vertical support of the ore column above is removed and the ore then caves by gravity. As broken ore is removed from the ore column, the overlying ore continues to break and cave by gravity.

Although some relatively smaller block cave ore bodies are caved and mined as a single production block, most existing and planned block cave mines use either of the following:

- An extended block caving system that divides the deposit into discrete production blocks; or
- A single cave front (or series of fronts, or "panels") advancing forward through the ore body, continually opening up new production areas as the earlier caved areas become exhausted.

The block caving method typically allows for relatively large volumes of production after the mine has been developed and production ramp-up has been achieved. The preproduction development period can be significant (typically 5–10 years, depending on the length of time to achieve the initial access). The up-front capital required prior to any return on investment is very high because much of the production levels and infrastructure must be in place before caving can begin. After the mine has reached its sustained maximum production rate, the operating cost tends to be very low with minimal additional infrastructure required to maintain the high production volumes. Block caving is generally the least expensive of all underground mining methods, and can in some cases compete with open-pit mining in cost.

The proceedings from the MassMin conferences, held every 4 years, are outstanding sources of case histories as well as theoretical discussions of block caving. Many of the discussion points in this chapter are drawn at least in part from the many excellent papers in those proceedings. The two latest proceedings are from 2004 (Karzulovic and Alfaro) and 2008 (Schunnesson and Nordlund). An excellent reference book on the geotechnical design aspects of block caving is *Block Caving Geomechanics* (Brown 2003).

CURRENT AND PLANNED CAVING MINES

Figure 13.10-1 shows the worldwide distribution of active and planned block and panel caving mines. A number of the planned mines will be developed under very large open pits that will be exhausted in the next 10–15 years. Thus, many of the planned caving mines will be very large deposits, mined at greater depths.

As technology improves, the trend is for higher production rates in these deposits. The largest caving mines in the world in 2009 were producing in the range of 75,000 t/d (metric tons per day) from a single footprint. At the same time, mining complexes such as El Teniente in Chile were producing 130,000 t/d from a number of production blocks. Planned production rates from future mines are as high as 160,000 t/d, indicative of the evolution of caving mines into "rock factories" capable of these high production rates.

Another trend in the future is for increasing mining depths, as the new "super caves" are developed beneath the open-pit mines and other undeveloped deep deposits begin to come on-line. As technology improves our ability to manage issues in deep mines such as in-situ stress, temperature gradient, and more competent ore bodies, this trend is expected to continue. Currently, caving mines are being contemplated in excess of 2,000 m deep.

ORE-BODY CHARACTERISTICS

As for any mining method, determining the applicability of the caving method must consider many different aspects, including the location, type, size, geotechnical characteristics, and value of the ore body. No single formula can be readily applied to determine the suitability of an ore body to a particular caving method.

The caving method is typically applied to large, fairly flat-dipping ore bodies, or porphyry-type ore bodies, with rock

---

Charles A. Brannon, Manager Underground Planning, Freeport-McMoRan Copper & Gold Inc., Phoenix, Arizona, USA
Gordon K. Carlson, Chief Mine Engineer, Henderson Mine, Climax Molybdenum, Empire, Colorado, USA
Timothy P. Casten, Director, Underground Planning, Freeport-McMoRan Copper & Gold Inc., Phoenix, Arizona, USA

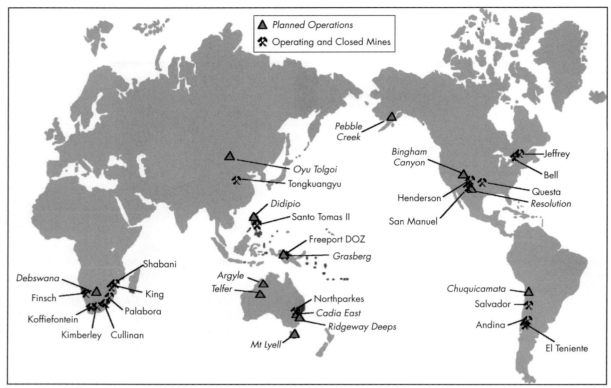

Source: Brown 2003.
Figure 13.10-1 Known operating and planned block and panel caving mines

mass characteristics that are amenable to sustainable, large-scale, rock mass failure (caving). The lateral extents of the ore body must be sufficient to both establish and sustain the cave. The required dimensions to achieve this will vary and depend on key characteristics such as rock strengths, fracture frequency and orientation (fragmentation), and in-situ stress conditions. The height of the ore columns must be sufficient to pay for the costs of developing and maintaining the production levels from which the ore is extracted. At the same time, the ground conditions must allow for development of the mine openings required to exploit the ore body, as well as for proximal location of required infrastructure.

GEOTECHNICAL CHARACTERISTICS
The geotechnical character of the rock mass to be caved is a key factor in the ability to effectively initiate and propagate the cave and successfully and efficiently produce ore for life of mine. An understanding of cavability, fragmentation, and stress are primary elements in designing and operating a caving mine.

Rock Mass Characterization
During the process of mine method selection, a thorough understanding of the rock mass behavior is required in order to assess cavability, fragmentation, mine production level design, mine infrastructure design, and surface impacts. For this purpose, topics that require a thorough understanding include

- Geology, including geometry, lithology, and structural features of the ore body as well as the surrounding country rock;
- Hydrology, including surface features and groundwater flow models;
- Geotechnical studies;
- Rock mass classifications;
- In-situ stresses; and
- Discontinuities of the rock mass (faults, joints, bedding, etc.).

Cavability
The cavability of the deposit is a fundamental feature of mine design. Initiating the cave, and then sustaining the cave throughout the mine life, is critical for achieving the desired productivity and thus directly impacts the economics of the project. The cavability may be established through empirical comparisons of the rock mass to other known caving operations, numerical modeling, or both.

A key aspect of the cave design is the hydraulic radius (HR), which is a calculation of the area of the caved zone divided by the length of the perimeter. Geotechnical studies will determine the required HR to initiate the cave, and also the HR required to sustain the cave. This ties directly into the mine design, as the production blocks must honor the required HR for that deposit. In general, caves tend to grow laterally in a circular fashion trying to reach equilibrium. In contrast, long laterals with narrow leads (extended rectangles) may satisfy the HR requirement but may not cave until more area has been opened.

Fragmentation
The fragmentation characteristics of the deposit are perhaps the single most important geotechnical feature. Fragmentation

impacts most aspects of mine design and operation, including drawpoint size and spacing, equipment selection, production rates, draw control requirements, dilution, hang-ups, and mill throughput. Thus an accurate prediction of fragmentation is a key to a successful mine design, and yet this prediction is one of the more difficult to produce with confidence.

The fragmentation of a deposit changes over time, and all of the stages below must be considered in the mine design:

- *In-situ fragmentation* refers to the natural, unaltered blocks before any caving takes place.
- *Primary fragmentation* is the distribution of block sizes during initiation of the cave.
- *Secondary fragmentation* is the fragmentation that occurs as the blocks move through the draw column during mining.

Monitoring

Monitoring of the progress of the cave development is important for many purposes, including

- Safety by maintaining a reasonable air gap between the cave back and the cave muck pile;
- Initiation and development of caving in order to validate the conceptual models, but also to allow modifications to the design as required by that information;
- Stability of the production and infrastructure areas in order to react to issues such as areas of excessive convergence; and
- Surface subsidence monitoring.

BLOCK CAVE MINE DESCRIPTION

Following is a discussion of block cave mine planning that offers additional details about the layout of a block caving mine.

Mine Layout

A block caving mine consists of several horizontal operating levels—the undercut level, extraction level, service/ventilation level, and haulage level. An additional level for drainage may be required, depending on water conditions in the mine.

Undercut Level

Undercut level designs depend on the method of undercutting chosen. Undercut methods are described in more detail later in this chapter. In the cases of pre- or advance undercutting, the undercut level is much closer to the extraction level and the major apex pillar is formed during the undercutting process. Post-undercutting usually places the undercut level drifts directly over the extraction level drifts and at the top of the major apex. Long-hole drilling and blasting is required for the formation of the drawbell in all cases but is sequenced differently according to method. Pre- and advance undercutting start with the extraction of the rings across the drawbell and create the angles of the major and minor apexes. This is followed by a final blast from the drawpoint to create the drawbell connection between the undercut and extraction levels. Post-undercutting has the opposite sequence, with the initial blast from the drawpoint, followed by a series of bell formation shots from the undercut level to create the drawbell. The final sequence of shot or shots will be the undercut rings, which will remove the remaining pillar material at the undercut inducing caving.

Extraction Level

Although there are a variety of possible extraction level layouts, there are two main drawpoint layouts employed in the majority of current and planned caving mines: the offset herringbone and the El Teniente (or "straight-through"). Both methods have advantages and disadvantages that must be considered when selecting the layout.

Herringbone layout. Figure 13.10-2 shows an example of an offset herringbone extraction level layout. Herringbone designs are best suited when electric load-haul-dumps (LHDs) are to be employed because the mining direction is the same on both sides of a production drift. Electric LHDs also help minimize the ventilation requirements. This option tends to be used in smaller-footprint operations such as Northparkes (Australia) and Palabora (South Africa) that do not require the ore body to be split into multiple production blocks. The drawbell of the herringbone layout is rectangular and lends itself well to single-shot drawbell blasting.

El Teniente (or "straight-through") layout. In the El Teniente layout, the drawpoint drifts are developed in straight lines oriented at about 56°–60° to the production drifts and the major apexes. Figure 13.10-3 is an example of the straight-through layout at the Henderson (Colorado, United States) mine. The minor apexes are short and inclined to the major apexes. The El Teniente layout is generally considered to be more robust than comparable herringbone layouts because it has a better pillar shape.

From the development perspective, the El Teniente layout has advantages over the herringbone layout, especially in a panel caving operation. The El Teniente straight-through design allows for the drawpoints to act as temporary crosscuts and accesses or "pseudo fringes," well before final boundary drifts must be completed. The straight line drifts in the El Teniente layout make development more efficient and cost-effective, and potentially create less overbreak, especially around the nose pillar.

Temporary fringes can be achieved with the herringbone layout, but this is more difficult as the drawpoints are staggered and therefore the temporary fringe drifts are not straight. In addition, the boundaries of the panels in a large deposit become problematic. Such problems occur because if a fringe drift is established between panels, the width of the pillar between the herringbone layouts on either side of the fringe drift is excessive, and merging the caves across that pillar is very difficult and likely to cause excessive weight problems.

From a scheduling perspective, the El Teniente layout also has advantages over the herringbone layout. This is because a line of drawpoints outside of the planned production block can be used as a "temporary or pseudo fringe," and in this way development meters can be deferred. This also postpones construction of drawpoints, grizzlies, and chutes. Overall this would delay capital spending as well as reduce the risk of slipping on the planned schedule.

A significant disadvantage of the El Teniente layout is that it is impractical for electric LHDs using trailing cables. With a herringbone layout, the LHD can produce from all drawpoints in a production drift while facing the same direction. But with an El Teniente layout, the LHD will have to change direction for either side of the production drift. This will require additional lost productive time to disconnect and reconnect the trailing cable at the other end of the production drift. The flexibility of having diesel equipment that can easily move to other areas or function in the mine is an advantage for El Teniente layouts. Of course, diesel equipment requires a higher ventilation airflow rate.

A safety concern in the El Teniente layout is with tramming the loaded LHD bucket-first. This restricts the operator's visibility 50% of the time because he or she has to operate the LHD in both bucket-first and engine-first positions in order to pull from all drawpoints. Secondary breaking units and pedestrians who access operating panels are examples of an increased risk that has to be managed in this case. Procedures are typically in place to protect pedestrians in a working panel and include lights, reflective personal protective equipment, and worker/machine proximity detection systems. A positive aspect of the El Teniente design is that the LHD operator always enters the drawpoints from his side of the LHD, giving him better visibility for safety and operational efficiency.

Another safety issue is the concern over the impact of a wet muck spill in an El Teniente layout. During a significant wet muck spill the straight-through configuration of the El Teniente layout can result in the LHD being pushed back into the adjacent drawpoint by the wet muck and being completely covered. This results in greater damage to remote-controlled LHDs and an extended time required for recovery of the gear from the muck spill. With the herringbone layout, the LHD is typically pushed down the drift and remains exposed. In either case, areas of wet muck would need to be worked by remote-controlled or automated LHDs, thereby eliminating any risk to the operators.

Service/Ventilation Level

The service/ventilation level is typical of a larger-footprint mine and is primarily a way to provide intake and exhaust ventilation across the production footprint for the undercut, extraction, and haulage levels. It also provides an access to these levels that is independent of production accesses and activities. Smaller-footprint mines will typically not require a separate ventilation level and will simply use the existing levels (extraction and haulage) to provide the required ventilation.

A large caving footprint is a very porous system from a ventilation point of view, with many accesses and cross connections in the preproduction and production areas. In order to provide reliable fresh air splits and exhaust connections, a series of ventilation plenums are required. This is usually via two ventilation levels (intake and exhaust) located between the extraction level horizon and the haulage level. This series of plenums are connected to vent raises that intersect the production levels in key points. The ventilation levels cover the extent of the footprint and are usually quite large-diameter drifts, which makes them good candidates for a service level. The service level can be used to access the haulage level in key areas as required for development as well as for construction and repair of haulage level infrastructure. This is especially important in a rail haulage system where independent rubber-tired access to the track is required for cleanup, construction, or maintenance work. The service level can also be used to assist with level drainage and waste haulage, and act as a conduit for services such as water, compressed air, and power.

Haulage Level

The haulage level design will be determined by the type of transportation method required—rail haulage, trucks,

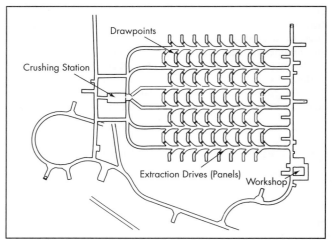

Source: Ross 2004.
Figure 13.10-2 Herringbone extraction level layout at the Northparkes mine

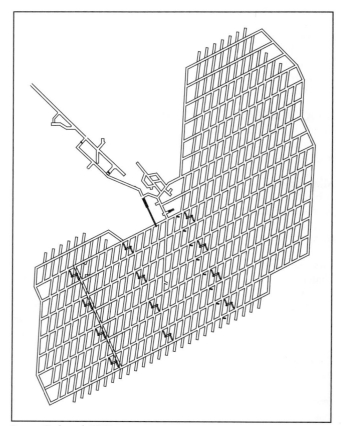

Source: Rech et al. 2000.
Figure 13.10-3 Extraction level plan at the Henderson mine

conveyors, or a combination of these. Each method has its advantages and disadvantages, which need to be carefully weighed in the context of that particular deposit before development begins because conversion after the fact is almost impossible.

The greatest advantage of rail haulage is its ability to move large tonnages in a single movement. This is especially important if the material requires long-distance travel to either crusher systems or to a surface milling facility. Some of the drawbacks to this system include large capacity bins to accommodate train sizes, grade and curve radii restrictions, increased power consumption (for electric locomotives), track maintenance, electrical capital and maintenance, and potential production shortfalls as a result of disabled trains in the loading drifts or main haulage lines.

The greatest advantage of truck haulage is its flexibility. Trucks have the ability to maneuver in smaller areas and can be reversed in traffic flow if the need arises. Road maintenance is continual but not labor intensive, and generally a single disabled truck can be removed from the traffic flow fairly easily, thereby reducing production shortfalls. Another advantage of a truck haulage level is the ability to modify designs easily over the course of time if new reserves or discoveries come into play. The main disadvantages of truck haulage include significantly smaller payloads, which require larger fleet sizing, larger ventilation requirements, increased maintenance costs, more labor, and the inability to efficiently move large daily tonnages long distances. In the case of the Henderson mine, for example, a fairly confined truck haulage route to a centralized crusher feeds a 24-km conveyor system to the mill and has proven to be very reliable and productive.

Several small-footprint caving operations (e.g., Northparkes and Palabora) use a system of conveyors to gather the production material from multiple crushing stations onto a central conveyor that feeds production shaft hoisting.

MINING METHODS

The selection of the methods used to undercut and extract the ore from a caving operation are fundamental to the success and efficiency of the operation. Selecting the appropriate methods at the design phase is key, as making major changes in methodology after the mine is in production can be difficult and costly.

General Mining Systems

The key to designing a suitable mining system is to have a good understanding of the geotechnical characteristics of the deposit. In particular, the fragmentation of the deposit will control the width of the "draw cone" or area of influence of the drawpoint as the material is mined. The area of influence then determines the required spacing of the drawpoints.

The three principal methods of block caving are the gravity, slusher, and LHD systems. Historically, the gravity system was commonly selected for ore bodies consisting of finer material essentially free-flowing to the drawpoint. For somewhat coarser material, the slusher system was employed, while the LHD system was employed for relatively coarsely fragmented deposits. Today mines are moving toward higher production rates with greater reliance on mechanization and automation, and moving away from the labor-intensive gravity and slusher mucking systems. The majority of existing and planned block caving operations now use some variant of the LHD (rubber-tired) system of mucking. Of course, this assumes that the ground conditions allow for development of stable production levels that can handle the size of the equipment. The majority of this section is therefore devoted to the LHD mining system.

Full-Gravity (Grizzly) System

The grizzly system consists of the haulage level, transfer raises, grizzly level, finger raises, and undercut level. The haulage and grizzly levels are driven across the ore block to be mined. This work can be done simultaneously because they are on different levels. Finger raises are driven from the grizzly level to the undercut elevation, and then short horizontal connections are driven between the tops of the finger raises forming pillars that will later be long-holed and blasted to initiate the caving action. Transfer raises are driven between the haulage and the grizzly level with a loading chute at the haulage level and grizzly rails for sizing at the grizzly level. After the undercut has been blasted, broken ore flows down the finger raises and is sized by the grizzly rail spacing, passes into the transfer chutes, and is loaded into rail cars. Sledgehammers or semi-mobile rock breakers are used to break up oversized pieces at the grizzly. Some large pieces may hang up in the finger raises, and these are usually broken up by secondary blasting with packaged explosives placed strategically against the oversized rocks. The Andina copper mine in Chile is an excellent example of the grizzly system.

LHD (Rubber-Tired) System

The use of LHD equipment usually requires that the drawpoints be spaced at greater intervals to allow room for the equipment to operate. The larger equipment requires larger drift/drawpoint excavations, so larger spacing is also needed for pillar strength. The wider spacing is suitable for coarser rock; finer fragmented deposits may require tighter drawpoint spacing and therefore smaller equipment. This system consists of the haulage level, ore-transfer raises, the production level, drawpoint entries, draw cones to the undercut level, and undercut drifts. Long-holing and blasting is used to form the undercut that promotes caving. The large draw cone will allow larger pieces of ore to move down near the drawpoint where small drills and explosives can be used to break up the larger pieces; high hang-ups practically never occur with this system. The haulage, production, and undercut levels can be developed simultaneously using separate accesses. The haulage and production levels should be separated by a substantial distance to provide adequate storage in the transfer raises, so that the loading of the trains is not delayed waiting for the LHDs. When the haulage and production drifts are complete, the transfer raises and drawpoint entries can be driven. Drilling of the draw cones can be done from the drawpoint entries or from the undercut level. Freeport-McMoRan's Henderson molybdenum mine in Colorado, the DOZ copper-gold mine in Indonesia, and CODELCO's El Teniente copper mine in Chile are examples of mines that use the LHD system.

Undercutting Methods and Issues

The undercutting strategy, design, and management are critical elements to a successful block cave mine. Some of the aspects of undercutting to consider include the following:

- Undercutting strategy (post-, pre-, or advance undercutting methods)
- Undercut design (initiation, direction, and shape of the undercut face, rate of undercut advance, and undercut heights)
- Shape of the undercut and extraction method (flat versus "crinkle" undercutting)
- Impact of the undercut design on stresses in the undercut and extraction levels
- Drill-and-blast designs for the undercut and drawbell

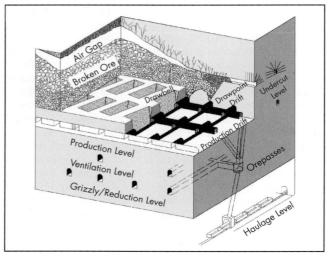

Courtesy of G. Flores.
Figure 13.10-4 Post- or conventional undercutting

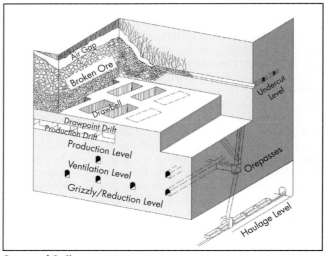

Courtesy of G. Flores.
Figure 13.10-5 Pre-undercutting

Post-Undercutting

Post- or conventional undercutting refers to undercut drilling-and-blasting after development of the underlying extraction level has been completed (Figure 13.10-4). Drawbells are also prepared ahead of the undercut blasting. Production blocks can be brought into production more quickly than with other undercutting methods, and this method has been most successfully used on smaller mining footprints. Other advantages are that this system requires about half of the development on the undercutting level as the other undercutting systems. In addition, no separate ore-handling facility is required on the undercut level, because the ore is transferred through the already-developed drawbells.

The primary disadvantage is that the mine openings are all subjected to the abutment stresses when the cave front passes through during undercutting. The resulting damage to the major and minor apex can be substantial and reduces the effective life of the drawpoints and production drifts (or at least often requires repair work to maintain drawpoint integrity). Selection of this method requires careful consideration of ground conditions and stress issues. Drawbell blasting can be easier to recover from because the undercut is not yet blasted, but conversely, undercut remnant pillars (stumps) can be difficult to deal with. The drawbell can be mucked clean and inspected prior to undercutting.

Pre-Undercutting

In pre-undercutting, the undercut is mined ahead of the extraction level development (Figure 13.10-5). This allows the development of the extraction level in a destressed environment. This method is typically seen in small-footprint mines and in mines where ground compaction is not deemed an issue.

The pre-undercut can occur quite some distance ahead of the last drawbell, with a considerable amount of development and construction work carried out in the cave shadow area. The footprint has to be small enough that the pre-undercut does not sit idle too long, allowing the abutment stress to refocus on the undercut level or the extraction level through compaction. In addition, facilities must be developed to handle the swell muck from the undercut level.

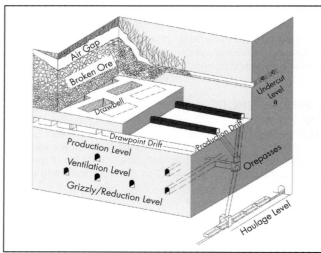

Courtesy of G. Flores.
Figure 13.10-6 Advance undercutting

Advance Undercutting

Advance undercutting is increasingly popular and many future caving mines are planning to use this technique. In this method, the undercutting is completed above a partially developed extraction level (Figure 13.10-6). The extraction drifts, and in some cases all or part of the drawbell drifts, are completed before the undercut passes. When the undercut has passed, the drawbelling is completed. The crucial drawbell blast is thus completed in a destressed condition. Because significantly less of the extraction level excavation is completed in this method, damage to the pillars can be greatly reduced. The cave is brought into production more quickly than for the pre-undercut method, and the probability of damage from compacted undercut material is reduced.

Typically an advance undercut only extends ahead by the length of one drawbell, thereby keeping the active cave close to the leading edge of the cave front. If the caved undercut is left to sit for too long before taking the drawbell beneath, it

Figure 13.10-7 Extraction level under construction using an advanced undercut, DOZ mine, Indonesia

Figure 13.10-8 Drawpoint construction-forming jumbo, Henderson mine

can begin to compact and "accept" the weight of the ground above. The abutment stress is redirected to the undercut level and far less damage is seen on the extraction level.

A separate level for undercutting is required and typically has twice the development as the other undercutting methods. Because development and construction occur ahead of the undercutting and drawbelling, this method requires relatively greater up-front capital and use of development resources. It can be difficult to recover a drawbell if it is not blasted properly. If this becomes necessary, a hung drawbell needs to be drilled from the adjacent drawbell to the unblasted drawbell through the minor apex.

PREPRODUCTION DEVELOPMENT AND CONSTRUCTION

When initial mine development is complete, and depending on the geotechnical requirements for the area, a significant amount of preproduction ground support may be required in the extraction level to maintain panel and drawpoint integrity for the duration of the caving and production periods. The production period can last from 3 to 5 or more years depending on the drawdown rate and column heights. The ground support may, for example, consist of grouted threadbar (or cable) bolts, wire mesh, and shotcrete throughout the extraction area and will vary depending on ground conditions.

Preproduction work is followed by drawpoint construction, which typically consists of two lintel steel sets placed at the mucking point of the drawpoint to protect the brow and pillar from erosion by the caving material. The goal is to maintain the opening for the life of the ore column with a minimum of repair. Figures 13.10-7 and 13.10-8 show typical drawpoint construction. The lintel sets are often concreted and bolted in place, with the posts being set into concrete in the floor. The drawpoint and panel floors are typically concreted. Some operations choose to add surface hardening products and steel wear bars (often this is simply used rail) to prevent early floor wear from both mucking and tire wear by the LHDs. If the area has particularly poor ground, a full series of steel sets and concrete may be applied to the drawpoint and panel section in addition to the two lintel sets.

A typical drawpoint and corresponding section of panel may cost in the range of US$250,000 to US$500,000, depending on the level of support required. Designing for adequate drainage in the panels is key as water can play a major role in early floor cracking and wear. A good general rule is to keep panel roadways at a minimum 3% gradient to ensure adequate drainage.

Some mines are small enough that a separate ore handling level is not required (e.g., Palabora and Northparkes). Where the footprint is larger and the haul distance to the crusher is prohibitively high, a separate haulage level is employed. As the panel and drawpoints are being constructed, the relevant orepasses, ore handling systems, and ventilation infrastructure must keep pace. Orepasses can be drop raised, raise climbed (mechanically or manually), or raise bored down to a handling level. The top of the orepass is usually fitted with a grizzly and often a rock breaker for oversize material. Grizzly dimensions vary depending on the orepass size and handling method, and can range from 400 × 400 mm up to 1,000 × 1,000 mm openings for mines with the larger equipment that can handle material of that size. Grizzlies should be sized to the estimated fragmentation parameters; otherwise, excessive choking can occur and clearing the obstructing material will decrease the LHD efficiencies. Ideally, as much of the rock reduction or breaking as possible should be done at the primary crusher to reduce secondary breaking requirements on the extraction level. This requires the largest-sized grizzly openings that can be delivered to by the LHD without causing hang-ups in the orepass or other downstream systems. With a 1,000 × 1,000 mm grizzly the orepasses would need to be at least 4,000 mm in diameter, with chutes and transport systems sized accordingly. The DOZ mine in Indonesia initially used 1,000 × 1,000 mm grizzlies on a 4,100-mm-diameter orepass. These fed into large chutes and 55-t-capacity haul trucks that dumped at a 1,370 × 1,955 mm gyratory crusher.

Figure 13.10-9 shows a typical grizzly installation from the DOZ mine. The grizzly openings are 800 mm, feeding into a 4.1-m-diameter bored orepass. The bumper is 300 mm high.

The orepasses can vary in length and angle depending on the haulage level elevation. Orepass diameter is a function of grizzly size and expected material types. A good general rule is to keep the orepass diameter greater than three to four times the size of a grizzly opening to minimize the potential for hang-ups. The base of the orepass is fitted with a typical lip and gate chute arrangement that is used to load either trucks

Figure 13.10-9 Grizzly installation example at the DOZ mine

or rail cars with ore. Where the broken ore is fine enough, a feeder may be used to load onto coarse ore belts. The haulage level delivers the ore to a primary crusher from which it is transferred to the surface by means of conveyor, hoists, or rail (as surface topography permits).

PRODUCTION

The production rate of a caving mine plays a major role in the required infrastructure, the time to ramp up to full production, and the drawpoint opening requirements. Selecting the production rate is one of the primary mine design variables required early in the process.

Capacity

A primary benefit of block cave mining is the high production capacity relative to other underground mining methods. Some of the larger caving mines in the world are producing up to 75,000 t/d from single mining blocks (the DOZ mine in Indonesia), and mines operating in multiple mining blocks are producing greater than 130,000 t/d (El Teniente, Chile).

Those upper limits will continue to grow as more and more of the large open-pit mines in the world transition to underground mining and the technology for large-scale underground production improves. Mines are being planned today that will exceed 150,000 t/d when at full production.

Setting the sustained maximum production rate for a caving operation is a key variable in the mine design process. The production rate is driven by several criteria: the rate of drawpoint opening, the rate of draw (t/d per drawpoint), the height of draw (HOD), the life of each drawpoint (calculated from the previous two variables), and ventilation and ore-handling constraints. The latter two variables typically act as a step function in production capacity, as the addition of either ventilation capacity or ore-handling infrastructure results in a significant step-up in production capacity for a mine. The goal is to ramp up production to achieve a production rate that is sustainable over a period of time, in the order of 7 to 10 years. That may be less than a potential maximum production rate, because it would not make economic sense to ramp up to some maximum rate only to have production immediately fall back as the mining block begins to become depleted. Commonly, the construction of drawpoints proves to be a primary limiting factor in the production ramp-up.

The maximum possible production may not be the economically best option, and an analysis of the economics of the mine plan needs to be considered. For example, should an inordinate amount of incremental capital be required to achieve a modestly higher production rate, cash-flow analyses would indicate that the optimal plan would be for the lower production rate.

Daily draw rates commonly range from about 150 to 600 mm. The allowable safe rate of draw depends on the rate at which the ore will cave. Drawing faster than the cave rate can result in an "air gap" that can produce conditions for a dangerous air blast should the cave fail suddenly. Coarsely fragmented material typically caves more slowly than weak, more finely fragmented material. In addition, coarsely fragmented ore will require extra effort in secondary blasting of oversize material and hang-ups, affecting the average rate of draw for those drawpoints.

Drawpoints at the leading edge of the cave are commonly mined at significantly lower rates to allow for new cave growth and to minimize opening of an air gap at the leading edge. In addition, draw rates can change over time for a given drawpoint; as the draw column is pulled, the material commonly becomes finer-fragmented through comminution in the column. Typically, early in the life of the drawpoint, very slow average draw is achieved due to coarse fragmentation and the resulting oversize and hang-ups that need to be dealt with. Most block cave planning tools use some sort of production rate curve that takes the change over time of the draw rate into account, so that the production capacity of the mine is not overestimated.

Another major consideration in determining the optimal production rate for a caving operation is the manner in which the cave is advanced in the ore body. Large deposits cannot be caved in a single pass because the advance of the cave front is slowed by the limits of the drawpoint opening rate, and the cave fronts themselves get excessively long. Long, slow-moving cave fronts generate a range of geotechnical issues that can be very damaging to the mine (Araneda and Sougarret 2007). Thus, the larger footprints must be divided into smaller mining blocks that can be caved and managed effectively, which limits the risk of excessive geotechnical issues. Larger deposits will require multiple caves that at times will be adjacent to or merged with other blocks. The issue then becomes one of production block design and drawpoint opening sequences that allow for an overall sustained production rate and at the same time are geotechnically achievable.

The maximum HOD varies from deposit to deposit and commonly ranges from 200 to 400 m, although column heights of >500 m have been achieved. The limit depends largely on

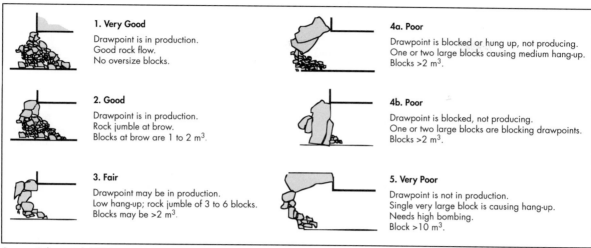

Source: Srikant et al. 2004.
Figure 13.10-10 Drawpoint rating system at DOZ mine

the ability to maintain the drawpoint and production drifts in good and safe operating condition. If a drawpoint is well constructed, excessive stress conditions avoided, and an even draw maintained, the life of the drawpoint can be maximized. It may be necessary to repair or rebuild a drawpoint for higher draw columns more than once, which is a common occurrence in mines with high column heights.

With a good estimate of the maximum sustained production rate and the life-of-mine production profile, the size and quantity of the major ore-handling and ventilation infrastructure can be designed.

Drawpoint Opening Rates

The rate at which an operation can open new drawpoints controls the production ramp-up as well as the full production rate that can be sustained. If the number of new drawpoints does not equal or exceed the number of drawpoints closed for depletion or due to damage, then production cannot be sustained. Commonly the development and construction of the drawpoint is the critical path to drawpoint opening rates. Undercutting and drawbelling can generally keep up with the pace of drawpoint construction.

For example, a typical maximum sustained rate of drawpoint opening for the DOZ mine is 10 drawpoints per month (usually referred to as five drawbells). The Henderson mine typically opens four to six drawbells a month, although the mine has peaked at higher rates.

Drawpoint Opening Sequences

Few aspects of the design of the block cave mine are more critical than developing a workable drawpoint opening sequence. The drawpoint opening sequence determines the order in which the ore is extracted. Where possible, preference is commonly given to the higher-value zones of the deposit. However, equally importantly, the sequence determines the shape and direction of the advancing cave and, for larger mines, determines how the production blocks interact with each other as the mine life progresses. It is critical to devise an opening sequence that is geotechnically sound; otherwise, the production goals, and perhaps even the mine itself, can be put at risk.

The mining sequence on a large-footprint cave will also drive the development and construction schedule. With a smaller footprint (e.g., Northparkes), the whole mine can be developed and constructed prior to caving. On larger footprints (e.g., DOZ, El Teniente, or Henderson) the development and construction of the levels takes place on an as-needed basis ahead of the caving front. This reduces early capital expenditures and smoothes development and construction activities. Overextending development in a large, active cave can often cause ground control problems in the open areas, especially if they sit for extended periods of time in stressed ground ahead of the cave front.

Production Issues

Cave fragmentation is a key issue, as the amount of ore that requires secondary breaking is a major factor in the ability of the mine to reach and sustain its maximum production level. Coarse fragmentation can give rise to high, medium, and low hang-ups in the drawpoints as well as excessive secondary breaking at the grizzly. Figure 13.10-10 shows the types of hang-ups that may occur.

Low hang-ups are dealt with using smaller drills and can be easily reached by the operators. Often nonexplosive breaking techniques are used to prevent damage to the drawpoint and lintel set area. The same mechanism can be used to break oversize material in the drawpoint or material too large to go through the grizzly. Mobile rock breakers and fixed rock breakers at the grizzly areas are also employed for secondary breaking.

When a drawpoint has been caved, accessing the drawpoint past the lintel set is too dangerous as it exposes personnel to falling rock. Longer-reach equipment is required to deal with intermediate and high hang-ups past the lintel set. An intermediate step is the use of a water cannon to "sound" the hang-up and often bring it down by dislodging key sections. These units consist of mobile water tanks with a high-pressure water nozzle fitted to a robotic boom.

Medium-height hang-ups are dealt with using single-boom jumbos and harness remotes to drill holes in the key rocks. In some cases high bombing will be used without drilling. A bundle of explosives is tied to the end of a series of blasting sticks and placed at key intersections of the hang-up to dislodge and bring the arch down. This process is high risk, however, as personnel spend considerable time around the front of the drawpoint and are exposed to potential falling and bouncing rocks.

In a high hang-up, the coarser fragmentation "jumbles" together inside the drawbell at a height above about 8 m. These are very rare occurrences and are handled using remote long-boomed drilling and charging rigs or by drilling through the major apex and blasting the larger rocks.

Ventilation
There are several components to ventilating a caving operation that bear particular attention. Planning these components into a general scheme early on will reduce operational trouble at a later stage. Because multiple LHDs will be operating in parallel galleries, sufficient air must be provided to support this. Regulation and control of the airflows becomes critical (especially with diesel equipment) in order to maintain appropriate levels for emissions and heat. Additional volume is also required in panel caving operations to support the development, drawpoint construction, and shotcrete/concrete operations that will be in the same panels, albeit upstream from the production.

Depending on the form of undercutting, more parallel paths on the undercut level will need to be planned for. This needs to be done years ahead of the actual caving. Henderson mine designs, for example, take several paths across the panel at set intervals to allow for the changing of the main intake as the caving advances. The intake level uses loop connections so that the airflow is equalized across the width of the panel and will not starve the last intakes to the production level at the end of a drive. Likewise, as the cave progresses over a long length, proper design can allow for the conversion of the intake paths to an exhaust path at the back end, thereby reducing the overall requirements to a boundary drift.

Another aspect that deserves attention is the orepass connections to the exhaust for dust mitigation. Every time a bucket of muck is dumped, dust will congregate at the orepass and potentially move downstream in the main drift. This not only causes visibility issues but also compromises the filtering and air conditioning systems on the LHDs. By connecting to the exhaust level, either at intercepts or through other channels, regulation of air can be accommodated to cause a slight drawdown, pulling the dust out of the production drift without compromising the overall airflow downstream of the orepass. In small-footprint caves, orepasses can be designed in the middle of the production drifts and, if connected to the main exhaust, will allow two LHDs to operate in the same panel from opposite ends without interference.

A final prudent design consideration is to have direct exhaust connections around any and all underground crusher stations. Often multiple accesses for high production rates are designed at these stations without proper assessment of the ventilation requirements to support them.

Automation
Caving mines lend themselves to automation due to the regular drawpoint layouts and fairly consistent nature of the production operations. Several caving mines employ automated LHDs on the extraction level (e.g., the DOZ and El Teniente mines) and in some cases automated trucks (e.g., the Finsch mine in South Africa) and rail (e.g., El Teniente) on the haulage level. Many aspects of rail haulage systems are automated and can typically be managed from a remote control center. The automated equipment operates in isolated areas to prevent worker/machine interface issues. Automation is employed to improve productivities and reduce exposure of operators to injury. In certain mines the risk of rock bursting or wet muck rushes is great enough to warrant the removal of the operator from the panel area. These areas have been typically operated using remote control stations, but automated technology provides better guidance for tramming than manual remote control. The automation uses scanning lasers mounted fore and aft on the LHD to scan and remember the tunnel profile on a single horizontal section. This allows for greater LHD speeds and less damage from side wall collisions. The buckets still require manual remote loading using cameras, as the muck type and size have been shown to be too variable to allow automation of the bucket-loading process. The isolation of areas due to remote and automated loading can cause productivity issues as all automated equipment is shut down when personnel are present for drawpoint sampling, inspection, secondary breaking, and so on.

Typical Equipment
Production in rubber-tired operations today is fairly standardized around the LHD equipment. These may be electric- or diesel-driven, both having their advantages and disadvantages. Small footprints and herringbone drawpoint designs favor the electric LHD, which also requires significantly less ventilation support. Diesel-driven LHDs have more versatility to move anywhere and can be used for more than production mucking. However, they do require significantly more ventilation. Newer regulations for dust, noise, and diesel exposure are also driving operators to use enclosed, air-conditioned cabs, which reduce some visibility but definitely give the operator a much-improved working environment. Capacities will vary, depending on ore characteristics and drift sizing, but generally range from 8 to 14 t in load capacity for most caving operations. Some of the super caves plan to use 17-t LHDs.

Production haulage is typically by truck or by train. In the case of trucks, most are end-dump designs in the 40–60 t range. These smaller trucks are designed more for development haulage but are being used in a production mode in many mines. PT Freeport Indonesia's DOZ operation employs 55-t-capacity trucks. The Henderson and Andina mines operate 80-t side-dump trucks, which have good haulage characteristics, but only on level ground. They are designed for production haulage only and cannot easily negotiate ramps when loaded.

Trains will be sized according to production requirements. The larger block cave mines using rail will typically use full-size rail equipment with a rail gauge of 1.4 m. For example, the El Teniente operation produced more than 130,000 t/d in 2008 using electric locomotives ranging from 72 to 130 t for ore haulage. Each of the five coarse ore trains hauls 18 bottom-dump railcars at 100 t per car. Another large caving operation plans for trains with two 40-t locomotives and 24 cars at 30 t per car. Rail systems lend themselves very well to automation and, in the right environment, can have much lower operating costs than haul trucks.

Several varieties of long-hole drills exist, and newer computerized versions are available. As with most other equipment, new regulations for operator exposures are pushing operators to use enclosed-cab drills. Undercut drilling requires these drills to have the capability of 360° rotation. Some drills have the enhanced ability to drill at forward angles from the perpendicular, so that the same machine is able to drill the V-cuts as well as undercut rings. Depending on the method of undercutting (explained in previous sections), other drilling equipment such as blind-bore machines or raise machines will be required for the bell relief slot raise.

In addition to the primary production equipment, a wide range of support equipment is used to develop and support production needs. In direct support are the secondary drills and rock breakers used to deal with hang-ups and oversize. Although these types of equipment may seem of lesser importance at first glance, they are probably among the most valuable tools for maintaining high production. The ability to efficiently clear choked or hung drawpoints has a huge impact on the ability of LHDs to meet draw-control parameters and production requirements. Storage of boulders for secondary blasting not only detracts from a LHD's primary function of production but requires removal of the LHD from mucking operations to load and blast this oversize. A variety of smaller drills and breakers are on the market, but no standardized methodology has arisen as yet to combat these issues. Some particular examples of this style of equipment were described in a previous section.

Also to be considered is the variety of equipment that supports the development and general operation of the mine. In particular, the concrete/shotcrete delivery systems are an integral part of maintaining appropriate development leads and ground support. Shotcrete placers, concrete pumps, drawpoint-forming equipment, and concrete transporters are critical components of the mobile fleet. Concrete/shotcrete batching systems, whether surface or underground, and the support needed for material storage are major considerations in infrastructure design to ensure a smooth operation.

Miscellaneous support equipment for moving workers, materials, and utilities throughout the mine should also be given careful thought. Forklifts, lube/fuel trucks, high-lift trucks for hanging utilities and ventilation fans/ventlines, water trucks for dust suppression, cable trucks for power runs, and service trucks for localized construction should all be understood and planned for. These types of units are as critical as production equipment for keeping the operation working efficiently and minimizing disruption to production.

Support equipment for rail haulage operations typically includes smaller (12-t) locomotives, often diesel-powered so as to be independent from issues with the electrical power supply. Functions handled include shift change, track repair, cleanup under ore chutes, switching and marshalling, electrical and control maintenance, and development support. In addition to the locomotives, a variety of flat cars, man-haul cars, and muck cars are used.

CAVE MANAGEMENT

The ultimate goal of the cave management functions is to safely maximize the recovery of the ore reserves. An effective cave management system is critical to the block cave mine operation and has an active role in virtually all phases of the mine. From the calculation of mining reserves and development of the production forecast, the drawpoint opening schedule is developed. The development drifting, construction, and undercutting plan are then based on that schedule. As caving proceeds, draw orders are developed that will pull the cave evenly to avoid early dilution, while managing issues such as drift convergence, rock bursts, areas of high stress, excessive dilution, water inflows, and the like. The cave must be monitored and plans altered to account for any issues that arise.

Reserves, Reporting, and Forecasts

The initial ore reserves are established during the feasibility and detailed design period, which sets the target production rate and guides the mine layout, design, and development and production schedules. Typically, mine planning and scheduling software is used.

Production forecasting is a critical process in cave management. The metal to be produced from the mine is the ultimate goal for the caving operation, and so as accurate a prediction as possible is required. Forecasting is an ever-evolving process. As mining proceeds and more is learned about how the cave is progressing, it is necessary to modify forecasting assumptions to match predicted to actual metal recovery as closely as possible.

As the ore body is mined, production data is tabulated and reported. This information must be reconciled back to the predicted tons and grade, as well as to the actual tons and grade received by the concentrator facility. These reconciliations provide key data on the modeling of the draw behavior of the cave, which in turn feeds into both the forecasting and reserve models.

Tracking the Progress of the Cave

How the cave progresses has a major impact on what can be expected from the drawpoints and so provides a constantly changing input into the forecasting process. Geotechnical monitoring provides input on the overall geometry of the cave. Geological observations at the drawpoints are reconciled back to the in-situ geological models that determine key parameters for modeling the draw such as the dilution rates and draw cone interaction.

Managing the Cave

When issues such as excessive weight problems or wet muck develop, modification of the draw in the problem areas can often alleviate or eliminate those issues. It is therefore important for cave management to understand and react to those issues. Special draw orders for specific drawpoints might be issued to address those issues. For example, a hung-up drawpoint might be causing excessive weight in a production drift, so orders are implemented to prioritize blasting of that hang-up. Extra tons might be pulled from that drawpoint and surrounding drawpoints for a period of time until the condition is stabilized. For a drawpoint with wet muck, a priority will be placed on pulling ore from that drawpoint, which helps to pull the water from the cave and prevents the water from spreading to additional drawpoints.

Draw Control

Draw control is an essential component in the caving process. Controlling the shape of the cave back and keeping the drawdown of the cave even and controlled will minimize cave issues such as early dilution, accelerated breakthrough to older workings, and excessive air gaps. Proper cave draw will improve the prediction of ore grades for the diluted block model. Draw control is managed by setting daily draw orders for each drawpoint in the mine. A record of draw is held in a database and regularly compared to the plan. Corrections in the short-term draw order are made to keep the draw control in line with the long-term plan. Systems to monitor draw from each drawpoint can range from a manual form filled out by the operator on an honor system or an electronic tagging system where every drawpoint has an individual tag that registers with a reader on the LHD every time a bucket is taken.

Drawpoint sampling is performed based on tonnage drawn and the HOD. Frequency of sampling is typically increased as a drawpoint nears the end of its life so the drawpoint can be closed as required to prevent overdilution. Sampling involves personnel manually sampling each drawpoint using some form of a sampling grid. Although this statistically tends to be quite a poor sample, it is a good indicator of drawpoint grades and can be compared back to the model to ensure a reasonable correlation.

CAPITAL COST CONSIDERATIONS

Caving operations typically require a significant commitment in up-front capital expenditures, and capital development for a caving mine requires a long lead time. The mine accesses and infrastructure, as well as sufficient portions of the operating levels, have to be in place before caving begins. This is particularly true for the current and planned super caves exceeding 100,000 t/d, Some of the deep deposits may take several years just to establish access and ventilation to the deposit, followed by multiple additional years developing the mine infrastructure. Once production is initiated, the time required to ramp up to full production is significant. As an example, a mine with a planned full production of 100,000 t/d may require 5 years after caving is initiated to reach that objective. Thus for a large caving operation, the lead time from initial access development to full production can be 10–15 years.

Capital estimates are typically divided into the following:

- *Capital development*, which includes all preproduction and production buildup development and infrastructure required to bring the mine to its sustainable full production rate; and
- *Sustaining capital*, work required on long-term development, facilities, or systems to sustain full production operations, but not direct production costs.

A typical breakdown of capital development and sustaining capital components follows. A number of the categories are the same as for the operating cost breakdown; these costs are classified as capital or operating, based on whether or not full production has been achieved.

- Drifting and excavating:
  - Rail haulage, conveyor drift, material handling excavation
  - Ramps and accesses, fixed facilities (shops, lunchroom, etc.)
  - Extraction and undercut fringe drift; production drift, drawpoint drift, drawbell drift development
  - Production and drilling drift on extraction and undercut levels
  - Ventilation drift and accesses, mass excavation, dewatering drifts
  - Shafts, raises, ventilation raises, orepasses, and waste passes
- Preproduction:
  - Ground support
  - Drawbelling and caving
- Ground repair (drawpoint rebuild and repair, and production drift repair)

Table 13.10-1 Development capital requirements for an 80-kt/d block cave mine (in 2009 U.S. dollars)

Capital Outflows	$, in millions
Drifting and excavating	260
Preproduction	10
Ground repair	10
Indirects	
Technical services and management	50
Electrical (including power cost)	130
Safety and miscellaneous	30
Mechanical shop (including fuel cost and rebuild)	170
Equipment and infrastructure	
Haulage and ore-flow infrastructure	190
Mobile equipment	250
Power generation and distribution	100
Fixed facilities infrastructure	100
Total development capital cash outflows	1,300

- Indirects (costs not directly allocated to production, such as ventilation, electrical, engineering, water control, and maintenance)
- Equipment and infrastructure:
  - Rail or truck haulage, or other ore-flow infrastructure
  - Drawpoint construction
  - Mobile equipment
  - Power generation
  - Fixed facilities (ventilation, pumping, batch plant, shops, offices, warehouses, compressors, etc.)

Table 13.10-1 presents a generic example of development capital costs for an 80,000-t/d block caving operation to illustrate the distribution of costs by major category. These expenditures occur over about a 10-year period from initial access development through to full production.

OPERATING COST CONSIDERATIONS

Block caving is typically the lowest-cost underground mining method. This is principally because the amount of drilling and blasting per metric ton of ore is much less than with any other system. Once the cave is progressing, blasting is limited to secondary breakage of oversize material within a drawpoint. In addition, because permanent production levels are established from which all ore tonnage is mined, total development drifting per metric ton of ore is substantially less than for other methods. Block caving lends itself to a high degree of mechanization, both in the production areas and in the ore-flow infrastructure. The overall result is lower costs due to reduced requirements for labor, mobile equipment, and consumables.

Table 13.10-2 is an example from an 80,000-t/d block cave mine operating in 2009, to illustrate the general distribution of costs by category. Of course, costs can vary widely from mine to mine because of many factors such as country of operation, ground conditions, infrastructure requirements, maturity of the mine, depth, and so on.

LABOR

Block caving mines can be highly mechanized and efficient operations, commonly achieving more tonnage per worker-shift than any other underground method, and potentially producing many times more tonnage per worker-shift. The initial labor required to develop a caving mine can be relatively high

Table 13.10-2 Annual operating costs breakdown for an 80-kt/d mine (in 2009 U.S. dollars)

Category	Direct Cost, $ in millions	per metric ton
Production	10	0.36
Development	16	0.59
Raises	5	0.17
Preproduction	12	0.43
Ground repair	3	0.11
Mine support services		
Drilling and water control	2	0.09
Underground technical services	9	0.31
Underground electrical	5	0.18
Management and administration	2	0.07
General services	3	0.09
Maintenance	49	1.80
Underground safety	5	0.17
Other	3	0.13
Construction	22	0.80
Total direct cost	145	5.30

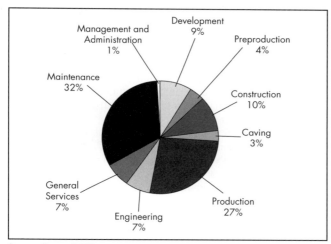

Source: Casten 2005.

Figure 13.10-11 Distribution of labor by category at the DOZ mine producing 50 kt/d

and is a factor in the early, high-capital expenditures typically required to bring a block cave into production. As major infrastructure items such as crushers, fixed facilities, ventilation, and ore handling systems are completed, labor levels will reduce down to a set of relatively specialized crews that focus on the repetitive activities required to achieve and maintain full production. These crews typically consist of (a) level development, (b) preproduction support, (c) construction, (d) caving, and (e) production. In addition, required support groups include management, engineering, geology, geotechnical, mine services, logistics, maintenance, and electrical crews. Most of these skill sets are common to most underground mining methods.

In a caving operation, the control and interaction of these different crews in a fairly constrained area can be a challenging task requiring clear management, good engineering, and seasoned supervision. Each of those key activities listed above requires careful scheduling to prevent the various crews from obstructing each other. In addition, the crews must work around interruptions such as scheduled and unscheduled blasting and shifting priorities to ensure that the cave is propagating well and being drawn correctly.

As the mine life progresses, activities such as development, construction, and undercutting will be completed before the mine is closed. Thus the number of different crews begins to decrease until finally the mine will be left with only production activities. In the case of small-footprint caves, all of the preproduction activities are managed by contractors, and production crews are quickly the only set of crews that remain. Northparkes Lift 1 was a good example of this with only 63 employees involved in production at 22,000 t/d as the mine had been fully constructed. In the case of the much larger-footprint DOZ mine, where development and preproduction activities continue on an actively producing and expanding cave front, labor levels remain higher for longer periods with approximately 1,800 personnel planned to produce at 80,000 t/d with only about one-half of the ore body developed. Figure 13.10-11 shows the labor distribution by category for the DOZ mine in 2005, when the mine was ramping up to a production rate of 50,000 t/d.

RISK CONSIDERATIONS

This section presents a review of typical risks in the block caving process. The first part discusses risks that present major operational hazards, and the second part discusses risks to consider when designing or operating a caving mine. A detailed discussion of the risk assessment process is beyond the scope of this chapter, but an informative section on operational hazards and risk assessment for block caving can be found in Chapters 10 and 11 of *Block Caving Geomechanics* (Brown 2003).

As with any mining project, a detailed risk assessment is required. Typically a *qualitative* risk analysis is initially completed to identify potential risks to the project, assess the likelihood and consequences of those risks, and develop an action plan to mitigate or eliminate the identified risks. A *quantitative* risk assessment, where a cost is calculated for each risk should its consequences be realized, could then be completed.

Major Operational Hazards

Major operational hazards that could result in loss of life or premature mine closure and are associated with caving methods include (1) issues with "inrushes" such as air blasts, mud rushes, or water; and (2) issues with stability of the underground workings, such as rock bursts or major excavation collapses.

Inrushes

An *air blast* may occur when the draw rates exceed the cave propagation rate, so that an air gap develops in the cave (prior to propagation of the cave to surface). Should the cave back then experience a large enough failure into the air gap, the compression of the air in that confined space can produce a rapid flow of air through exposed underground openings and can produce a violent and damaging air shock wave.

A *mud rush* is a sudden inflow of saturated fines from the drawpoint or other underground openings. A mud rush can occur when water builds up within caved material that contains sufficient potentially mud-producing materials. The inrush is then triggered by a mechanism such as production from the drawpoint, blasting, or seismicity. The sudden inflow puts at risk the personnel in the path of the mud rush. Mud rushes are a fact of life in caves with the appropriate conditions and must be managed to mitigate the risk to the people and to the operation. For example, automated or remote-controlled loaders can be used to produce from the wet areas. A key is to attempt to keep the material moving in the wet areas to minimize the buildup of the wet material and prevent the event from spreading further into the cave.

An external mud rush might occur when the cave breaks up into a saturated zone of material such as a tailings or slimes retention area, or perhaps a backfilled zone from an upper lift of the mine. Tailings or slimes ponds should never be sited over caving areas as catastrophic failures and inrushes can and have occurred that can claim hundreds of lives (e.g., West Driefontein in South Africa and Mufulira in Zambia).

Underground Stability Issues

A *rock burst* is an underground seismic event that, depending on its scale, can cause violent and significant damage to excavations in the mine. Rock bursts are the result of stored strain energy that is released suddenly into the rock mass. The induced stresses are high enough to overcome the strength of the rock mass and the released energy cannot be absorbed in slips or fractures. Rock bursts will usually occur only in strong, brittle rocks such as quartzite or granite. Effects include violent fracture and expulsion of rock from and near the boundaries of excavations, and the full or partial closure of mining excavations. Damage can range from localized spalling in a drift to complete destruction and closure of a part of the mine.

Current and planned caving mines are tending toward deeper, higher stress conditions, and in stronger and more brittle rock. Thus it is expected that the risk of rock bursts in future caving mines is an issue that will have to be addressed.

An additional type of collapse is stress-induced, but unlike a traditional collapse, it is generally progressive rather than sudden and violent. Collapses of this type are typically caused by the stress abutment produced ahead of the undercut; this is sometimes referred to as "weight" problems. This type of collapse involves failure of the pillars left on and above the extraction level. It may take weeks or months to develop and once initiated is difficult to arrest, although some mines have had success by overpulling from drawpoints within the stressed areas. Such collapses are exacerbated by the presence of faults or other discontinuities in the rock mass, by a change in rock type, or by incomplete blasting in the undercut level that leaves a "stump" or remnant that puts excessive stress (point loading) on the major or minor apex.

Major Risk Categories

The following subsections address major risk categories that must be considered during the planning and operation phases of a caving mine.

Geotechnical Issues

Most block caving operations, which tend to be large in size and occur at considerable depths below the surface, are difficult to explore completely in advance of having to make a development decision. Virtually all aspects of the mine design—from level designs, ground support criteria, ore-flow systems, equipment fleets, and so on—depend on a design that correctly incorporates the geological and geotechnical aspects of that deposit. Therefore, a considerable number of risks are present in the category of geotechnical information. A few of the potential risks include

- Cavability—An assumption that the ore will cave completely and in a predictable manner.
- Fragmentation—Underestimation of the fineness of the fragmentation that causes the drawpoint spacing to be too large, leading to isolated draw and loss of reserves. Underestimating the coarseness of the fragmentation can cause excessive hang-ups in the drawpoint, leading to excessive secondary blasting, and creating a bottleneck in an ore-flow system that was designed for fine fragmentation.
- Caving performance
  - **In-situ stress:** Underestimation of the stress levels encountered leads to inadequacies in ground support, mass excavation designs, and undercut sequencing.
  - **Undercut sequence design:** The geometry of the panels and the sequence of the undercutting are key components to the caving performance. Should the chosen design prove to be unstable, large portions of the production area could be lost and ground repair would be extensive, leading to lost or delayed production.
  - **Caving rate:** If lower than expected, ramp-up will be delayed.
  - **Cave shape:** If the developing cave fails to propagate as expected, excessive dilution could be introduced earlier than anticipated.
- Excavation stability
  - Panel excavations may be designed too large in areas of unanticipated poor ground conditions, resulting in too large a drawcone spacing and loss of reserves. The poor ground also leads to weaker pillars and excessive ground repair costs.
  - Location of the major underground fixed facilities relies on identifying adequate areas of good ground conditions in which to locate those facilities.
  - Orepasses are at risk as the cave front passes and must be designed to handle the abutment stresses.

Other Issues

The following are additional risk categories to be considered during planning and operation of a caving mine:

- **Fixed facilities design.** Infrastructure required to deliver workers and materials to the mine must be adequate for the production rates required. The ore-flow system must be capable of adequate ore delivery to the mill.
- **Environmental impacts and permitting.** If surface impacts exceed those anticipated, it could have an impact on the mine life or production capacity. For example, larger-than-expected subsidence areas that affect the surrounding area, or development of acid rock drainage exceeding that permitted, and so on.
- **Procurement.** Lead times for procurement of major components can be lengthy, leading to delays in mine development and production if not managed properly.
- **Labor.** Many future major block cave mines are scheduled to be developed and operated in similar time frames in the next few decades. Experienced labor, both in engineering and operations, may be limited when needed.
- **Safety.** For new mines, various aspects of the operation will comprise new types of facilities unfamiliar to the personnel, and will require training, familiarization, and supervision to minimize hazard exposure.

REFERENCES

Araneda, O., and Sougarret, A. 2007. Lessons Learned in Cave Mining: El Teniente 1997–2007. In *Proceedings of MassMin 2000*, Brisbane, Queensland, October 29–November 2. Carlton, Victoria: Australasian Institute of Mining and Metallurgy.

Brown, E.T. 2003. *Block Caving Geomechanics*. Brisbane: Julius Kruttschnitt Mineral Research Centre, The University of Queensland.

Calder, K., Townsend, P., and Russell, F. 2000. The Palabora underground mine project. In *Proceedings of MassMin 2000*, Brisbane, Queensland, October 29–November 2. Edited by G. Chitombo, T. Brown, and L. Hill. Carlton, Victoria: Australasian Institute of Mining and Metallurgy.

Casten, T. 2005. Development Plan—Manpower Forecast. Internal memorandum. PT Freeport Indonesia.

Karzulovic, A., and Alfaro, M., eds. 2004. *Proceedings of MassMin 2004*, Santiago, Chile, August 22–25. Chile: Mineria Chilena.

Rech, W.D., Keskimaki, K.W., and Stewart, D.R. 2000. An update on cave development and draw control at the Henderson mine. In *Proceedings of MassMin 2000*, Brisbane, Queensland, October 29–November 2. Carlton, Victoria: Australasian Institute of Mining and Metallurgy.

Ross, I. 2004. Northparkes Lift 2 development. In *Proceedings of the Innovative Mineral Developments Symposium*, Sydney, October 6. Carlton, Victoria: Australasian Institute of Mining and Metallurgy.

Schunnesson, H., and Nordlund, E., eds. 2008. In *MassMin 2008: Proceedings of the 5th International Conference and Exhibition on Mass Mining*, Luleå, Sweden, June 9–11. Sweden: Luleå University of Technology.

Srikant, A., Nicholas, D.E., and Rachmad, L. 2004. Visual estimation of fragment size distribution in the DOZ block cave. In *Proceedings of MassMin 2004*, August 22–25. Santiago: Mineria Chilena.

PART 14
Mineral Processing

CHAPTER 14.1

Introduction to Mineral Processing

Maurice C. Fuerstenau and Kenneth N. Han

The term *mineral processing* is used in a broad sense throughout this chapter to analyze and describe the unit operations involved in upgrading and recovering minerals or metals from ores. The field of mineral processing is based on many fields of science and engineering. In addition, environmental science and engineering have become inseparable components; the steps involved in mineral processing have to be founded not only on sound scientific and technological bases, but on environmentally acceptable grounds as well.

GOALS AND BASICS OF MINERAL PROCESSING

In the traditional sense, mineral processing is regarded as the processing of ores or other materials to yield concentrated products. Most of the processes involve physical concentration procedures during which the chemical nature of the mineral(s) in question does not change. In hydrometallurgical processing, however, chemical reactions invariably occur; these systems are operated at ambient or elevated temperatures depending on the kinetics of the processes.

The ultimate goal in the production of metals is to yield metals in their purest form. Mineral processing plays an integral part in achieving this objective. Figure 14.1-1 shows a generalized flow diagram for metals extraction from mining (step 1) through chemical processing. Steps 2 and 3 involve physical processing, and steps 5 and 7 involve low-temperature chemical processing (hydrometallurgy). All four steps are considered part of mineral processing. High-temperature smelting and refining (pyrometallurgy), steps 4 and 6, are not included under the heading of mineral processing.

Table 14.1-1 specifies processing routes from ore to pure metal for a number of metals. Processing routes can be quite different, and more than one route may be possible for many of these metals. For example, in the extraction of copper or gold from low-grade ores, dump or heap leaching is commonly practiced. The choice of this leaching practice is frequently driven by the overall economics of the operation. Because crushing and grinding of ores are quite expensive, leaching of ores in large sizes is attractive compared to the leaching of finely ground ores, even though the overall recovery of metals from the leaching of fine particles is, in general, much greater than that obtained with large particles. The introduction of this innovative leaching process has made feasible the mining of many mineral deposits that could not be processed economically through conventional technologies.

METALLURGICAL EFFICIENCY

One of the most important and basic concepts in mineral processing is metallurgical efficiency. Two terms are commonly used to describe the efficiency of metallurgical processes: *recovery* and *grade*. These phenomena are illustrated in the generalized process presented in Figure 14.1-2. In this example, 100 t/h (metric tons per hour) of ore are being fed into a concentration operation that produces 4.5 t/h of concentrate and 95.5 t/h of tailings. In upgrading this process, then, 1.0 t/h of the desired material, A, is introduced into the unit operation and 0.9 t/h (4.5 × 0.2) of this material reports to the concentrate, resulting in 90% recovery (0.9/1.0 × 100). The grade of the mineral, A, has been improved from 1% to 20%. The term *percent recovery* refers to the percentage of the valuable material reporting to the concentrate with reference to the amount of this material in the feed. Sometimes obtaining the highest possible recovery is not necessarily the best approach in a concentration process. High recovery without acceptable grade will lead to an unsalable product and is therefore unsatisfactory.

Mineral processing engineers are responsible for optimizing processes to yield the highest possible recovery with acceptable purity (grade) for the buyers or engineers who will treat this concentrate further to extract the metal values. To achieve this goal, economic assessments of all possible technological alternatives must be conducted.

ECONOMIC CONCERNS

Table 14.1-2 summarizes the total U.S. supply and recycled supply of selected metals in 1996. Although the data shown here are historic (reprinted from an earlier publication), the trends are still current. The total supply of iron and steel includes supply from primary and secondary sources as well as imports; these two metals represent by far the largest of commodities produced and consumed, followed by aluminum,

---

Maurice C. Fuerstenau, Professor of Metallurgy, University of Nevada, Reno, Nevada, USA
Kenneth N. Han, Professor Emeritus, South Dakota School of Mines & Technology, Rapid City, South Dakota, USA

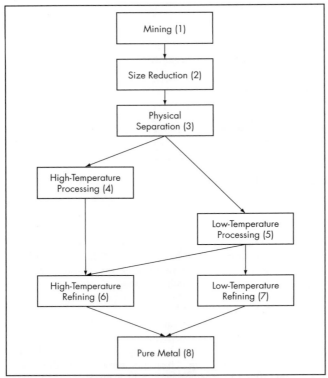

Figure 14.1-1 Generalized flow chart of extraction of metals

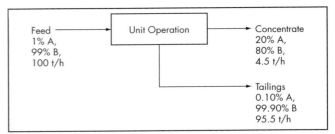

Figure 14.1-2 Simple material balance for a unit operation

Table 14.1-2 U.S. total and recycled supply of selected metals in 1996

Metal	Total Supply, million t metal content	Recycled Supply, million t metal content	% Recycled
Iron and steel	183	72	39
Aluminum	8.34	3.29	39
Copper	3.70	1.30	35.1
Lead	1.63	1.09	66.8
Zinc	1.45	0.379	26.1
Chromium	0.48	0.098	20.5
Magnesium	0.205	0.0709	35
Gold	516 t*	150 t*	29

Source: USBM 1997.
*Value for 1995.

Table 14.1-3 Abundance of various elements in the earth's crust compared to annual U.S. consumption

Element	Relative Abundance, %	U.S. Consumption, stpy
Fe	5.00	1.28×10^8
Al	8.13	5.4×10^6
Cu	7×10^{-3}	2.3×10^6
Zn	8×10^{-3}	1.0×10^5
Pb	1.5×10^{-3}	1.2×10^6
Au	1.0×10^{-7}	113
Ag	2.0×10^{-6}	4.52×10^3

Source: USBM 1990.

Table 14.1-1 Processing sequences for selected metals

Metal	Associated Major Minerals	Steps Involved in the Processing Route (Figure 14.1-1)							
		1	2	3	4	5	6	7	8
Iron	Hematite, Fe_2O_3; magnetite, Fe_3O_4	x	x	x	x		x		x
Aluminum	Gibbsite, $Al_2O_3 \cdot 3H_2O$; diaspore, $Al_2O_3 \cdot xH_2O$	x	x			x		x	x
Copper	Chalcopyrite, $CuFeS_2$; chalcocite, Cu_2S	x	x	x	x		x	x	x
Zinc	Sphalerite, ZnS	x	x	x	x		x		x
		x	x	x	x			x	x
Lead	Galena, PbS	x	x	x	x		x		x
Gold	Native gold, Au	x	x	x		x	x		x
		x	x*			x	x		x
Platinum	Native platinum, Pt; platinum sulfides	x	x	x			x	x	x
Silver	Native silver, Ag	x	x	x		x	x		x

*Only crushing is practiced; grinding is usually omitted.

copper, and lead. The recycled supply of these metals from processing scrap is strikingly high. In addition, the tonnage of precious metals consumed is rather small. However, because of the high prices of precious metals, their monetary value is substantial. For example, the monetary value of 516 t (metric tons) of gold was $12.8 billion in 1996, compared to $10.7 billion for 5.3 million t of copper and lead.

Table 14.1-3 lists the relative abundance of various metals in the earth's crust. Most metals are present in extremely small concentrations in nature, and none of these metals can be recovered economically at these concentrations. Rock that contains metals at these concentrations is not ore; ore is rock that can be processed at a profit. An average copper ore, for example, may contain 0.3% to 0.5% copper. Even this material cannot be treated economically at high temperature without prior concentration. There is no way that rock containing 4.5 kg (10 lb) of copper and 903 kg (1,990 lb) of valueless material can be heated to 1,300°C (2,372°F) and treated to recover this quantity of metal economically. Concentrating the ore by froth flotation to approximately 25% or more copper results in a product that can be smelted and refined profitably.

UNIT OPERATIONS

Numerous steps, called unit operations, are involved in achieving the goal of extracting minerals and metals from ores in their purest possible form. These steps include size reduction, size separation, concentration (froth flotation, gravity concentration, and magnetic and electrostatic concentration), dewatering, and aqueous dissolution.

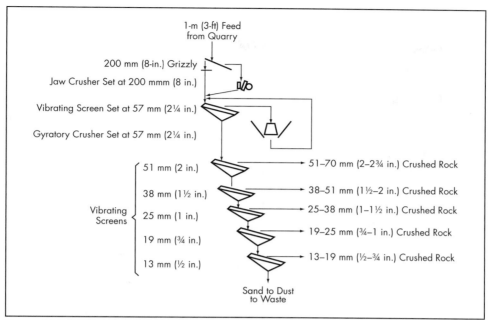

Figure 14.1-3 Flow sheet for crushing and grading rock

1. **Size Reduction.** The process of crushing and grinding ores is known as *comminution*. The purpose of the comminution process is threefold: to liberate valuable minerals from the ore matrix, to increase surface area for high reactivity, and to facilitate the transport of ore particles between unit operations.
2. **Size separation.** Crushed and ground products generally require classification by particle size. Sizing can be accomplished by using classifiers, screens, or water elutriators. Screens are used for coarse particulate sizing; cyclones are used with fine particulates.
3. **Concentration.** Physicochemical properties of minerals and other solids are used in concentration operations. Froth flotation, gravity concentration, and magnetic and electrostatic concentration are used extensively in the industry.
   - **Froth flotation.** The surface properties of minerals (composition and electrical charge) are used in combination with collectors, which are heterogeneous compounds containing a polar component and a nonpolar component for selective separations of minerals. The nonpolar hydrocarbon chain provides hydrophobicity to the mineral after adsorption of the polar portion of the collector on the surface.
   - **Gravity concentration.** Differences in the density of minerals are used to effect separations of one mineral from another. Equipment available includes jigs, shaking tables, and spirals. Heavy medium is also used to facilitate separation of heavy minerals from light minerals.
   - **Magnetic and electrostatic concentration.** Differences in magnetic susceptibility and electrical conductivity of minerals are utilized in processing operations when applicable.
4. **Dewatering.** Most mineral processing operations are conducted in the presence of water. Solids must be separated from water for metal production. This is accomplished with thickeners and filters.
5. **Aqueous dissolution.** Many metals are recovered from ores by dissolving the desired metal(s)—in a process termed *leaching*—with various lixiviants in the presence of oxygen. Following leaching, the dissolved metals can be concentrated by carbon adsorption, ion exchange, or solvent extraction. Purified and concentrated metals may be recovered from solution with a number of reduction techniques, including cementation and electrowinning.

EXAMPLES OF MINERAL PROCESSING OPERATIONS

Figure 14.1-3 shows a typical flow sheet for crushing and sizing rock in a quarrying operation. Run-of-mine ore can be present as lumps as large as 1.5 m (5 ft) in diameter. In this figure's example, 1-m (3-ft) lumps of rock are fed to a crusher that reduces the material to 200 mm (8 in.) or less in diameter. After screening to remove rock that is less than 57 mm (2¼ in.) in size, rock between the sizes of 57 mm (2¼ in.) and 200 mm (8 in.) is further reduced in size by a gyratory crusher. The product from this step is then classified by screening to the desired product for sale.

Figure 14.1-4 shows an integrated circuit demonstrating crushing, grinding, size separation, and gravity concentration of a tin ore. Initial size separation is effected with a grizzly set at 38 mm (1½-in.). Oversize material is fed to a jaw crusher set at 38 mm (1½-in.), and the crushed product is then further reduced in size to 20 mesh by ball milling. The minus-20-mesh material is classified by hydrocyclones set at 150 mesh, and the minus-150-mesh material is sent to shaking tables to concentrate the heavy tin mineral, cassiterite. The middlings in this process receive additional treatment. The concentrate from this operation is reground and sized at 200 mesh. Two-stage vanning is used to produce a fine tin concentrate.

The flow sheet describing the flotation processing of a copper ore containing chalcopyrite and molybdenite is shown

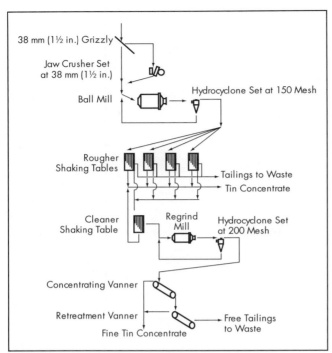

Figure 14.1-4 Flow sheet for the gravity concentration of a tin ore

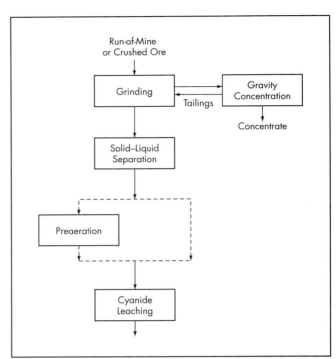

Figure 14.1-6 Flow sheet options for grinding and agitated leaching of free-milling oxidized gold ores

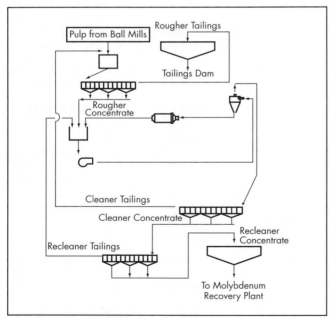

Figure 14.1-5 Flow sheet for the flotation of copper sulfide ore

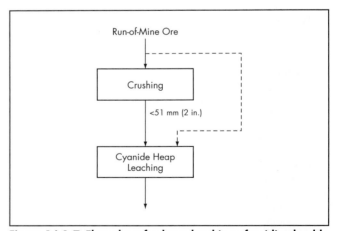

Figure 14.1-7 Flow sheet for heap leaching of oxidized gold ores

recovery plant for further processing. In this operation, the feed contains 0.32% Cu and 0.03% Mo. Rougher concentrate, cleaner concentrate, and recleaner concentrate contain 7% to 9% Cu, 18% Cu, and 25% Cu, respectively. Recleaner concentrate also contains 2% to 3% Mo.

Figure 14.1-6 depicts a flow sheet for processing free-milling oxidized gold ore. The kinetics of gold leaching is slow, and gold ores are frequently ground to less than about 76 mm (3 in.) before leaching. Even then, one day is usually required in the leaching step. In this process, run-of-mine ore is crushed and ground. The ball mill discharge is subjected to gravity concentration to recover the larger particles of free gold. The tailings from this operation are thickened, and the underflow from the thickeners is then subjected to cyanide

in Figure 14.1-5. After grinding and classification, pulp is fed to rougher flotation. The rougher tailings are thickened and sent to a tailings dam. The rougher concentrate is classified, and the oversize is reground. Cyclone overflow is fed to cleaner flotation, and the cleaner concentrate is recleaned. Cleaner tailings are recycled back to rougher flotation, and the recleaner concentrate is thickened and sent to the molybdenum leaching. In some instances, ores may contain oxygen-consuming minerals, such as pyrrhotite and marcasite, and a preaeration step may be conducted ahead of cyanide leaching.

Heap leaching has revolutionized the gold mining industry. Low-grade oxidized ores containing approximately 0.933 g gold per metric ton (0.03 oz gold per short ton) of ore can be processed with this technology, whereas they could not be processed by the higher cost of the grinding/agitation leaching (milling) process. Figure 14.1-7 presents a simplified flow sheet of heap leaching. As the figure shows, run-of-mine ore may or may not be crushed. If crushing is done, the ore is generally crushed to <51 mm (<2 in.) in diameter.

ACKNOWLEDGMENTS
This text is taken from the introductory chapter of *Principles of Mineral Processing,* edited by Maurice C. Fuerstenau and Kenneth N. Han and published by SME in 2003.

REFERENCES
USBM (U.S. Bureau of Mines). 1990. *Mineral Commodity Summaries*. Washington, DC: USBM.

USBM (U.S. Bureau of Mines). 1997. *Mineral Commodity Summaries*. Washington, DC: USBM.

CHAPTER 14.2

Crushing, Milling, and Grinding

John Mosher

Comminution is the collective term used to describe the progressive reduction in size of run-of-mine (ROM) ore. Size reduction is required either to prepare the ore for market (for iron ore and coal, for example) or, as in the case of base and precious metals, to allow liberation and separation of the valuable constituents. For ore requiring beneficiation, liberation and separation are the essential steps in mineral processing. The degree of liberation attained by comminution defines the grade–recovery curve for a given beneficiation process, as well as typically representing the largest category of mineral processing capital and operating expenses. As such, the comminution process has a large impact on an operation's bottom line. Proper circuit design and operation are critical to overall project success.

Crushing and grinding have been regarded as separate processes, with each step performed on a certain size range of material. This concept is rooted in the historical conventional circuit that entailed staged crushing, followed by open-circuit rod milling and closed-circuit ball milling. Modern equipment has blurred this concept, with autogenous grinding/semiautogenous grinding (AG/SAG) mills and high-pressure grinding rolls (HPGRs), in particular, spanning a broad range of comminution applications and particle size ranges.

This chapter addresses in general terms comminution circuit configuration, equipment types, and operation. Stages of comminution are discussed from ROM through fine grinding, with sections devoted to special grinding applications, ore characterization and circuit design, and mill liners and media. No attempt is made to cover all comminution machines or flow sheets; however, methods processing the largest tonnages of ore in the metals mining segment are summarized. Selected unit operations in broad use for each stage of comminution are presented in Table 14.2-1, with size ranges for the most commonly used unit operations depicted in Figure 14.2-1. Unit operations are discussed in more detail in relevant sections.

Examples of selected circuit configurations are presented in Table 14.2-2. This selection is not exhaustive; the summary presents both a range of grinding applications and demonstrates examples of selected circuit configurations. Throughout this chapter, commonly used references to describe feed and product sizes are employed; F and P represent feed and product, respectively, with the subscript indicating the percentage of the material passing a given size. For example, a P_{80} of 13 mm describes a size distribution for a material that has 80% of the material finer than 13 mm.

Individual comminution unit operations are well described in contemporary mineral processing textbooks (Wills 2006; Kelly and Spottiswood 1995). Classic works such as Taggart (1945) are still useful and provide perspective, and a number of comminution monographs provide detailed references (including Mular and Jergensen 1982; Napier-Munn et al. 1996). To keep up with changing practices, various forums addressing academic, engineering, and operational aspects of comminution were developed. Perhaps the most focused of these forums have been the International Autogenous and Semiautogenous Grinding Technology Conferences, hosted by the University of British Columbia. Proceedings from these conferences (1989, 1996, 2001, and 2006) are the broadest reference body for case studies on contemporary milling practices. Despite the name, a broad spectrum of mineral processing comminution circuits is addressed.

MINING AND PRIMARY CRUSHING

Because the process of comminution starts with mining, the degree of comminution achieved should be used as a performance indicator for the mining process. Correspondingly, the comminution circuit should also be developed in consideration of the selected mining process.

For surface mines that produce broken ore or muck via drilling and blasting, there are numerous case studies of optimizing drilling and blasting in conjunction with the overall economics of shovel productivity, primary crusher throughput, and downstream comminution efficiency. Efficient blasting is effective at both decreasing top size as well as improved fines generation. In general, the following conclusions can be drawn:

- Staggered patterns (versus square patterns) reduce the maximum distance between drill holes and typically reduces the ROM top size (boulders).

John Mosher, Executive Vice President-Operations, PT Freeport Indonesia, Tembagapura, Papua, Indonesia

Table 14.2-1 Unit operations for stages of comminution

Milling Stage	Typical Top Size	Common Unit Operations	Alternatives
ROM	F_{100}* = 1 m	• Primary crusher (gyratory/jaw) • Secondary pre-crush	• MMD Sizer • McLanahan log washer
Primary	F_{100} = 250 mm F_{80} = 50–100 mm	• AG/SAG/pebble crusher circuits • 2º/3º/4º crushing circuits • Crusher/HPGR circuits	• Rod milling • Impact mills • Dry AG/SAG
Secondary	F_{100} = 13 mm	• Ball milling • Pebble mills	• Attrition scrubbing • Dry ball mills
Tertiary (concentrate regrind, handling, pelletizing)	Nominally –0.2 mm	• Ball mill • HPGR • Vertical shaft mills (screw and pin) • Horizontal shaft mills (IsaMill)	• Roller mills • Vibratory ball mills • Centrifugal mills • Jet mills

*F = feed; subscript represents the percentage of the material passing a given size.

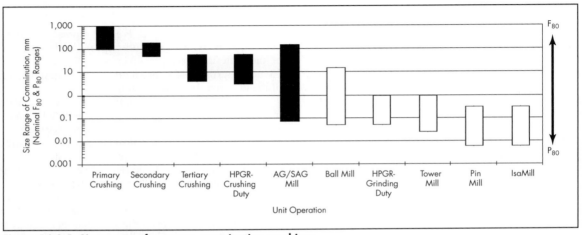

Figure 14.2-1 Size range of common comminution machines

- Quality of drilling in terms of the accuracy of holes in x-, y-, and z-axes improves fragmentation as well as shovel productivity.
- Customized blast designs by material type (burden and spacing, explosive type and amount, and initiation timings) can improve fragmentation and offer cost benefits.
- The effect of blast sequence (including presplitting, production, and trim blasting, where required for wall control), blast size, and pattern geometry should be considered for the effect on comminution.

Additional considerations for underground mines apply. As a consequence of mining through areas previously subjected to ground control, many underground mines produce extraneous metal in the mill feed. Additionally, block cave mines typically produce coarser material at drawpoints early in their development. With higher draw columns, greater AG breakage occurs, and drawn material becomes finer. Block cave mines usually employ secondary breakage to control material hang-ups in drawpoints, as well as mechanical rock breakers and grizzlies to ensure top size control prior to truck loading, before primary crushing is undertaken.

Typically, ROM material must undergo a stage of size reduction to facilitate conveyor transport. Questions to be addressed in circuit development include the size, number, and type of crusher used to produce suitable mill feed, where to place the crushers, and how to feed them. The choice of primary crushing equipment is typically between jaw and gyratory crushers. Jaw crushers can be selected for applications with low throughput, while throughputs greater than 1,000 t/h (1,100 stph) favor gyratory crushers. Gyratory crushers of up to 1.9 × 3.2 m (75 × 124 in.) and installed power of 1 MW or more are currently being marketed. A modern gyratory crusher is depicted in Figure 14.2-2.

The location of primary crushing stations is a critical question. Primary crushers can be permanently located, semimobile, or mobile. Even permanently located crushers can be disassembled and relocated during a mine's life, so the question as to which of the three options to select is one of trade-offs between the capital and operating cost of the incremental truck haulage fleet required versus mobile or semimobile crushing stations.

Fixed plants simply rely on the haulage fleet to bring ore to the crusher. Mobile and semimobile plants often used tracked crawlers, which have capacities of more than 1,000 t (1,100 st), to periodically relocate the crushing unit. In conjunction with movable conveyors, such systems are placed to shift the haulage burden from a mobile equipment fleet to conveyors and more economically transport material mined to either ore or overburden stockpiles. The evolution of larger truck sizes, along with a general trend of steeper and deeper pits, have tended to disfavor in-pit crushing and conveying. However, these can offer advantages in terms of reduced operating costs, increased availability, and reduced labor requirements compared to a truck fleet; a thorough review of the topic is provided by Boyd and Utley (2002).

Table 14.2-2 Selected contemporary circuit configurations (all circuits milling ROM ore)

Configuration	Operation*	Major Equipment, Dimensions and Installed Power, kW	Nominal Capacity, t/d	Total Installed Power, MW	P_{80}, μm
Crushing–milling	C1-2, PT Freeport Indonesia; Cu-Au (Indonesia)	Secondary crushing: 2 × 750; Tertiary crushing: 8 × 600; Quaternary crushing, HPGR: 2.0 × 1.5 m, 2 × 3,600/each; ball mills, 8 × 1,950	75,000	29	210
Crushing–dry milling†	Goldstrike roaster; Au (Nevada, USA)	Secondary crushing: 2 × 600; Double-rotator mills: 2 × 7,500	11,000	15.6 (excluding 5-MW air sweep fans)	74
Single-stage autogenous milling‡	Iron Ore Company of Canada, Carol Concentrator; Fe (Labrador, Canada)	One grinding line: AG mill, 9.8 × 4.2 m, 1 × 5,225 (closed with screens)	37,000	5.2	1,000
Single-stage SAG§	Henderson; Mo (Colorado, USA)	SAG mill, 8.5 × 4.3 m, 1 × 5,225 (closed with hydrocyclones)	7,500	5.2	105
AG pebble milling**	Empire IV; Fe (Michigan, USA)	AG mill, 9.8 × 5.0 m, 3 × 6,720; pebble crusher, 1 × 300; HPGR, 1.4 × 0.8, 1,400; pebble mills, 4.7 m × 9.9 m, 6 × 1,975	26,000	34	20
SAG ball milling	C4, PT Freeport; Cu-Au (Indonesia)	SAG mill, 11.1 × 6.1 m, 1 × 20,000; pebble crusher, 2 × 750; ball mills, 7.3 × 9.3 m, 4 × 10,400	120,000	64	175
HPGR ball milling††	Cerro Verde; Cu-Mo (Peru)	HPGRs, 2.4 × 1.6 m, 4 × 5000; secondary crushers, 4 × 750; ball mills, 4 × 7.3 × 11.4 m, 12,000, ABB VS drive	107,000	80	140

*For plants with multiple lines; data is by plant, concentrator, or individual grinding line (as indicated).
†Thomas et al. (2001).
‡Chung et al. (2001).
§Wood (1996).
**Greenwood and Rajala (1992).
††J. Monteith (personal communication).

Aside from selection of crusher and crusher location and mobility, design considerations include methods to feed the crusher and convey crusher product to the mill stockpile. Crusher feeding considerations include the number of truck dump locations and the size of the crusher feed hopper (and appropriate design for dead beds). Primary crushers can either be fed with direct truck dumping (as in Figure 14.2-3) or by using ore pockets and apron feeders. Direct dumping is the simplest, but it can be more difficult to sustain peak crusher utilization, and truck wait times can be incurred. Apron feeders offer the ability to partially disconnect the crusher from truck dumping (with the idea of both reducing truck queues and improving crusher performance) but introduce additional capital costs, as well as the substantial maintenance costs of operating apron feeders in the demanding duty. Crusher installations should also consider the requirement for mobile equipment to clear dump pockets in case of blockages, rock breaker installations to handle oversize inadvertently fed to the crusher, overhead craneage (fixed or mobile) for maintenance, and overall elevation required for an installation (controlled by crusher dump pocket and product bin dimensions).

Crusher product handling using a bin and apron feeders to place the material on a conveyor belt is effective and allows matching the operation of the semicontinuous, variable capacity crusher (depending on material type) with a conveying system of fixed capacity. Product bin sizing should consider peak flow-through rates observed when crushing fine ore to avoid bin overfilling. In cases with sufficient belt capacity relative to crusher throughput, product can be fed directly to belts via appropriate chutes.

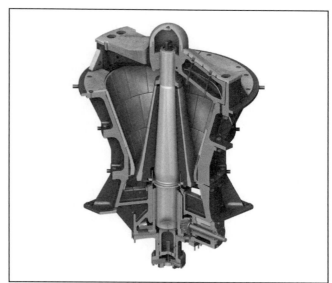

Courtesy of FLSmidth.
Figure 14.2-2 Primary gyratory crusher

Alternatives to conventional crushing prior to further processing include mineral sizers (whose manufacturers include Krupp and MMD). These machines operate dry, employ horizontal counter-rotating shafts equipped with wear tips, and are alternatives for size reduction of low- to moderate-strength materials with low-head-height devices. A somewhat similar

Courtesy of PT Freeport Indonesia.
Figure 14.2-3 Direct truck dumping to a gyratory crusher

device, the McLanahan log washer, operates wet and serves to deagglomerate coarse from fines rather than affect true comminution. These devices are well proven and work effectively in appropriate applications. There can be substantial capital cost savings with reduced civil and structural work for these units.

All of these machines are vulnerable to damage through introduction of incompressible foreign objects (notably ground engagement tools and tramp metal). In cases where direct dumping is used, protection is limited to tools such as cameras and image analysis software (useful for detection of the loss of shovel teeth), spotters, and the installation of safety clutches to limit damage for unavoidable events. Where material is reclaimed or dumped to feeders, metal detectors or magnets can be used in some applications.

PRIMARY MILLING

A critical linkage point between the mining and milling operations is the mill stockpile. Most comminution plants reclaim from a stockpile of primary crushed material, and stockpile management is essential to maintain consistent throughput. There is generally some degree of stockpile segregation, with coarse material preferentially accumulating toward the outside of stockpiles. Maintaining a live stockpile and balancing multiple reclaim feeders will result in the highest average (and most stable) throughput. Feed size to a primary milling circuit can have a substantial impact on throughput, particularly for AG and SAG circuits. This underscores the importance of evaluating mining technique and practice, primary crushing, and the downstream comminution circuit. It is also important in terms of nonsteady-state operation, such as stockpile drawdown conditions. As the proportion of material drawn down from the stockpile increases, the feed to the comminution circuit typically becomes coarser, correspondingly requiring a higher unit energy input for comminution. The often observed effect was well quantified by Morrell and Valery (2001); that paper also discusses broadly the effect of SAG feed on circuit performance.

Maintaining reclaim stockpiles at reasonable levels minimizes the effect of load and haul equipment shift changes on downstream operations and decouples the mine and mill from maintenance activities associated with shovel and mill maintenance periods. Stockpile size should be based on anticipated interruptions due to primary crusher maintenance, load-and-haul asset maintenance, mill maintenance downtime requirements, and normal fluctuations due to mine sequencing. Depending on climate, ore moisture, surrounding facilities, dusting characteristics, and regulatory requirements, covering stockpiles may be desired or necessary.

The following sections outline three circuit configurations in broad use as the first stage in a comminution plant (primary milling): (1) standard crushing plants, (2) AG/SAG mills, and (3) the combined application of cone crushers and HPGRs. The latter circuit is now broadly viewed as an alternative to SAG circuits.

Crushing Circuits

The classic crushing circuit prepares primary crushed ore for secondary milling through two stages of crushing. Most typically, stockpiled primary crusher product is reclaimed and fed to a secondary crusher. In turn, the product of a secondary crusher operating in open circuit is screened, with oversize fed to a tertiary crushing circuit operating in closed circuit with screens. The product of a standard tertiary crushing plant is nominally 100% passing 13 mm. The basic design of a tertiary crushing plant (secondary crushers in open circuit with tertiary crushers in closed circuit with screens), along with variations, has been widely used. Such a circuit, inclusive of HPGRs in a quaternary role and follow-on single-stage ball milling, is depicted in Figure 14.2-4; the crushing plant and concentrator depicted has a nominal capacity of 75,000 t/d (83,000 stpd). The plant depicted shows wet primary screening followed by dry tertiary screening. In some applications, this is reversed, and some circuits (particularly for heap leaching) are fully dry. Selection of wet or dry screening is a function of material type, ROM moisture, efficiency of classification required, and the type of follow-on material handling or processing.

The product of such a typical crushing plant can be directly fed to a fine-crush heap leach operation. In milling operations, the product of a crushing plant was historically further reduced in an open-circuit rod mill prior to comminution in a ball mill. However, rod mills are limited to nominally less than 5 m (16.4 ft) in diameter and 6 m (19.7 ft) in length due to practical constraints on mill size before rod tangling occurs—such a mill has a power draw of less than 2.0 MW. As such, economies of scale have largely led mill circuit designers to send crushing plant product directly to ball mill circuits. Doing so may introduce a degree of grinding inefficiency (based on the coarseness of feed to a ball mill or a high reduction ratio in one stage of grinding), but this has been widely accepted as being favorable to a requirement

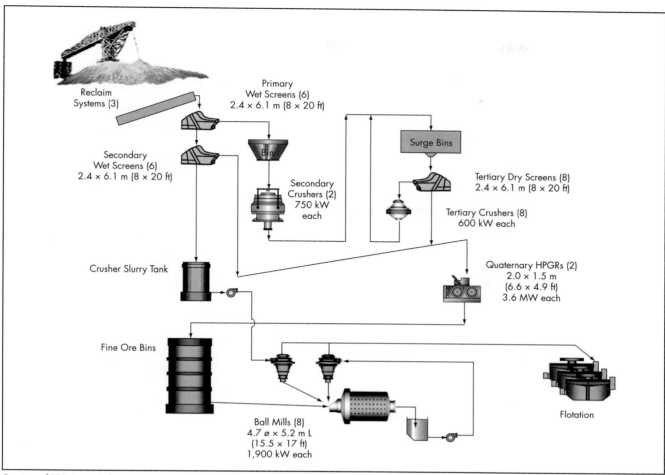

Courtesy of PT Freeport Indonesia.
Figure 14.2-4 Crushing ball mill plant

for a large number of rod mill lines and an additional unit operation. As will be discussed in a following section, HPGRs in a quaternary role perform top-size reduction admirably, as well as doing an excellent job of fines generation. Depending on circuit configuration, large single-line secondary/tertiary crushing lines have a nominal installed power of 3 MW and the capacity to crush 25 to 40 kt/d (28,000 to 44,000 st).

Crushers are generally of two types of mechanical design. In the Symons design (Figure 14.2-5), a threaded bowl liner is adjusted to modify the crusher gap. The bowl is threaded to the adjustment ring, which is secured flush against the mainframe through hydraulic accumulators that release in the event of a tramp metal event. In the Allis design (Figure 14.2-6), the crusher mantle is shaft supported against a fixed bowl liner (analogous to a gyratory crusher's concaves) to maintain either a fixed gap or a fixed pressure setting. In the event of tramp metal (or other uncrushable item) entering the crushing chamber, the hydraulically supported shaft drops, allowing the material to pass.

Because crushing plants are well understood, the skill sets required for plant operation and maintenance can be readily sourced or developed. Throughput for the machines used is largely volumetrically controlled; as such, plant throughput is relatively insensitive to changes in feed hardness or feed size distribution (though the amount of comminution work and power drawn by the machines is clearly affected). Power input is a direct result of comminution effort required to effect breakage, unlike tumbling mills, where it is the outcome of the potential energy transmitted to the mill charge.

Crushing plants have some drawbacks relative to other plant configurations. First, particularly for larger plants, they normally require many more units of equipment (bins, crushers, screens, conveyors, chutes, etc.) for an equivalent throughput than other major circuit types. This is due to the fact that the current upper limit for installed power for a conventional crusher is 750 to 1,000 kW, smaller than the average unit installed power for mills. Partially as a result of the number of units in series, and partially due to liner change-outs, crushing plant run time is lower than that of rotating mills. Crushing plants do not produce fines to the same degree as other primary milling options, which increases the amount of comminution effort required in a second stage of milling. The mode of breakage is dominated by compressive fracture, which produces fewer fines, and screening (wet or dry) at finer sizes is not practical at large tonnages; screen area required escalates nonlinearly with decreasing cut size. Finally, crushing plants have material-handling difficulties with sticky ores, significant downtime due to metal detector trips, and potentially a resulting loss in production with ores that contain a high degree of tramp metal or magnetite.

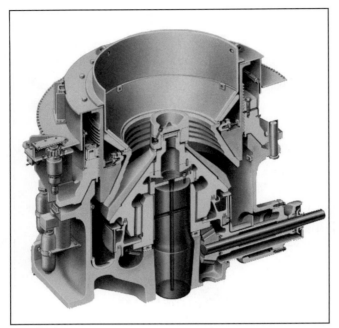

Courtesy of Metso.
Figure 14.2-5 Symons-style crusher (Metso MP-series)

Courtesy of Sandvik.
Figure 14.2-6 Shaft supported crusher (Sandvik CH-870)

AG/SAG Circuits

Since their early development, AG and SAG mills have become a mainstay of comminution in mineral processing applications. Although crushing-rod mill-ball mill circuits were referred to as "conventional" plants until relatively recently, AG/SAG mills are now the foundation unit operation for most large grinding circuits. The variety of AG/SAG circuits is substantial. The following duties are all well represented:

- Single-stage autogenous/semiautogenous milling
- AG/SAG mills as one stage in a larger comminution circuit
- Inclusion of pebble crushing circuits in the AG/SAG circuit
- Employment of a crushing stage after primary crushing (such that SAG feed has been through two stages of crushing)

Single-stage AG circuits have been installed in relatively high-tonnage, coarse-grind applications such as Iron Ore Company of Canada's (IOC's) Carol Concentrator, and in lower-tonnage, finer-grind applications (closed with hydrocyclones) like Noranda's Brunswick operation in Quebec, Canada. Very high single-line capacities have been demonstrated, with some circuits having up to 70 MW of installed grinding power in a single plant. PT Freeport Indonesia's 11.6-m- (38-ft-) diameter SAG mill in Concentrator 4 (Papua, Indonesia), presented in Figure 14.2-7, has registered monthly tonnages averaging more than 125,000 t/d (138,000 stpd), with single-day performances in excess of 150,000 t/d (165,000 stpd).

Although common convention generally refers to high-aspect-ratio mills as SAG mills (with diameter to effective grinding length ratios of 3:1 to 1:1), low-aspect-ratio mills (generally, a mill with a significantly longer length than diameter) can also grind autogenously or semiautogenously. Such mills are common in South African operations and are sometimes referred to as tube mills or ROM ball mills. Many of these mills operate at higher mill speeds (nominally 90% of critical speed) and often use grid liners to form an AG liner surface. These mills typically grind ROM ore in a single stage, either autogenously or semiautogenously.

The broad application of AG/SAG mills lies in the large single-line capacity of such plants, the ability of circuits to mill a broad range of feed types, and the variety of circuit configurations. The application of gearless drives has allowed mills to grow beyond sizes allowed by twin-pinion drives; a nominal limitation of 12–16 MW (based on a limit of 6–8 MW per pinion on a twin-pinion drive) results in a restriction of nominally an 11-m (36-ft) diameter SAG mill (based on typical diameter/length ratios) if using geared drives. Many large mills around the world (including Cadia, Collahuasi [New South Wales, Australia] with 12.2-m (40-ft) mills, and Antamina [Peru], Escondida Ph IV [Antofagasta, Chile], PT Freeport Indonesia, and others with 11.6-m [38-ft] mills) have installed SAG mills of nominally 20 MW with gearless drives. A number of mills with 25 MW or higher gearless drives are either under construction or in engineering. Some of these grinding lines have design capacities exceeding 100,000 t/d (110,000 stpd). The process flow sheet for the large SAG installation (with pebble crusher product combining with SAG discharge and feeding screens) depicted in Figure 14.2-7 is presented in Figure 14.2-8.

AG/SAG mills comminute ore through impact breakage, attrition breakage, and abrasion of ore serving as media. AG circuits require an ore of suitable competency (or fractions within the ore of suitable competency) to serve as media. SAG circuits may employ low to relatively high ball charges (ranging from 2% to 22%, expressed as volumetric mill filling) to augment AG media. Higher ball charges shift the breakage mode away from attrition and abrasion breakage toward impact breakage; as a result, autogenous milling produces a

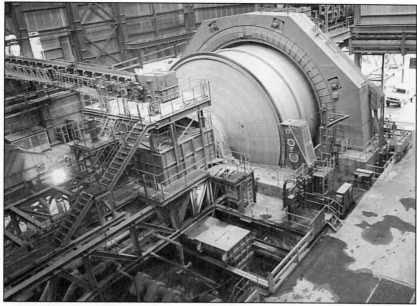
Courtesy of PT Freeport Indonesia.
Figure 14.2-7 Large SAG installation (20-MW gearless drive)

finer grind than semiautogenous milling for a given ore and otherwise equal operating conditions.

With a higher-density mill charge, SAG mills have a higher installed power density for a given plant footprint relative to AG mills. With the combination of finer grind and a lower installed power density (based on the lower density of the mill charge), a typical AG mill has a lower throughput, a lower power draw, produces a finer grind, and has a higher unit power input (kilowatt-hours per ton) than a SAG circuit milling the same ore. In the presence of suitable ore, an AG circuit can provide substantial operating cost savings due to a reduction in grinding media expenditure and liner wear. In broad terms, this makes SAG mills less expensive to build (in terms of unit capital cost per metric ton of throughput) than AG mills but more expensive to operate (as a result of increased grinding media and liner costs). SAG circuits tend to be less susceptible to substantial fluctuations, because of feed variation, than AG mills and are more stable to operate. AG circuits are more frequently (but not exclusively) installed in circuits with high ore density, with iron ore being the classic example. A small steel charge addition to an AG mill can boost throughput and result in more stable operations, typically at the consequence of a coarser grind and higher operating costs.

AG or SAG mills with low ball charges are often used in single-stage grinding applications. Conversely, SAG mills produce coarser grinds than AG mills in equivalent applications. Based on their higher throughput and coarser grind relative to AG mills, SAG mills are more likely to be used as the primary stage of grinding, followed by a second stage of milling. AG/SAG circuits producing a fine grind (particularly single-stage grinding applications) are often closed with hydrocyclones. Circuits producing coarser grinds often classify mill discharge with screens. For circuits classifying mill discharge at a coarse size (coarser than approximately 10 mm), trommels can also be considered to classify mill discharge. Trommels, however, are less favorable in applications requiring high classification efficiencies and can be constrained by available surface area for high-throughput mills. Regardless of classification (hydrocyclone, screen, or trommel), oversize material can be returned to the mill or directed to a separate comminution stage.

The addition of pebble crushing is the most common variant to closed-circuit autogenous/semiautogenous milling (instead of direct recycle of oversize material). The potential efficiency benefits, both in terms of grinding efficiency and in capital efficiency through incremental throughput, has long been recognized, but the challenges of metal removal (primarily grinding media passing mill grates) were perceived to be a substantial hurdle. With time, and after successful pioneer operations, operators and circuit designers became comfortable with the required magnetic separation and metal detectors/bypass chutes required to protect pebble crushers from tramp steel.

Even after steel ball removal had proved to be reliable, pebble crushers installations were still thoroughly scrutinized at the design stage because of the additional cost and circuit complexity. Instead of foregoing pebble crushing completely, though, many operations chose instead to leave provisions in the design for future pebble crushing expansions. Today, every AG/SAG flow sheet conceived is likely to seriously consider the inclusion of a pebble crusher circuit. For certain ore types (particularly those with a chert, andesite, or other hard component that develops a critically sized material that constrains milling rates), a pebble crushing circuit is almost an imperative for efficient circuit operation. For other cases, it is simply a technique to improve circuit capacity. Construction of an AG/SAG circuit without provision for later pebble crusher installation should have a persuasive economic justification. Important aspects of pebble crusher circuit design include

- Preparation of a clean, well-sized, and dry feed;
- Metal removal (with additional protection via metal detectors and bypass);
- Surge capacity (through bins or, more costly, a pebble stockpile);

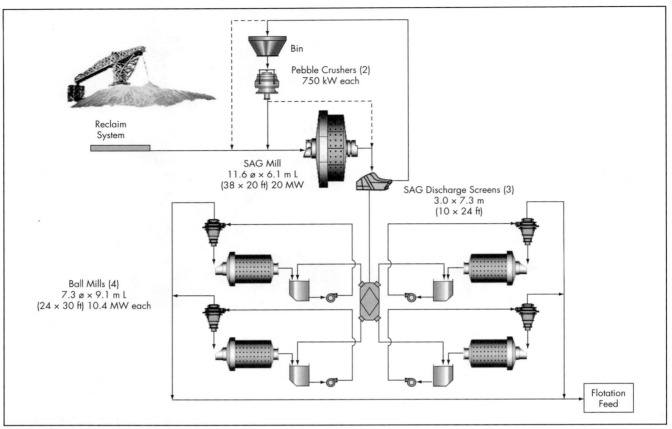

Courtesy of PT Freeport Indonesia.
Figure 14.2-8 SAG pebble crusher–ball mill circuit

- Sufficient capacity (primarily a concern of large circuits where multiple pebble crushers are required to serve one grinding line);
- Design for bypassing crusher(s) during maintenance; and
- Evaluation of where to reintroduce the crushed pebbles back to the grinding circuit.

The standard destination for crushed pebbles has been to return them to SAG feed. Doing so is often convenient based on conventional plant layouts. However, open circuiting the SAG mill by feeding crushed pebbles directly to a ball mill circuit is often considered a technique to increase SAG throughput. PT Freeport Indonesia's operation has the ability to return crushed pebbles to SAG feed per a conventional flow sheet, or to SAG discharge. By combining with SAG discharge and screening on the SAG discharge screens, top size control to the ball mill circuit feed is maintained while still unloading the SAG circuit (Mosher et al. 2006).

Aside from parameters fixed at design (mill dimensions, installed power, and circuit type), the major variables affecting AG/SAG mill circuit performance (throughput and grind attained) include

- Feed characteristics in terms of ore hardness or competency;
- Feed size distribution;
- Selection of circuit configuration in terms of liner and grate selection and closing size (screen apertures or hydrocyclone operating conditions);
- Ball charge (fraction of volumetric loading and ball size); and
- Mill operating conditions including mill speed (for circuits with variable-speed drives), density, and total mill load.

The effect of feed hardness is the most significant driver for AG/SAG performance, which is the case to a much greater degree than for (compressive) crushing plants; variations in ore hardness result directly in variations in circuit throughput. With AG circuits, the variations can be substantial and occur suddenly. The effect of feed size is also marked, with both larger and finer feed sizes having a significant effect on throughput. A number of case histories of AG mills have failed to consistently meet throughput targets because of a lack of coarse media. Compounding the challenge of feed size is that, for many ores, the overall coarseness of the primary crusher product is correlated to feed hardness. Larger, more competent material consumes mill volume and limits throughput.

To counteract this effect, many circuits (including Troilus [Quebec], Kidston [Queensland, Australia], Ray [Arizona, United States], Porgera [Papua New Guinea], Granny Smith [Western Australia], Geita Gold [Tanzania], St. Ives [Western Australia], and KCGM [Western Australia]) currently use or have used a secondary SAG precrush circuit to further comminute primary crusher product prior to feeding to a SAG mill. The objective of SAG precrush is to increase SAG throughput. Occasionally, secondary crushing is included in the original design but more often is introduced as an additional circuit to account for harder ore (that proves to be harder than expected or becomes harder as the deposit is developed) or as a capital efficient mechanism to boost throughput in an existing circuit. Such a flow sheet is not without drawbacks. Not surprisingly, some of the advantages of semiautogenous milling are reduced in terms of increased liner wear and increased maintenance costs. Also, precrushing can lead to an increase in midsize material, thereby overloading pebble circuits, and in challenges controlling recycle loads. In certain circuits, the loss of top size material can lead to decreased throughput. It is now widespread enough to be considered a standard circuit variant and is often considered an option in trade-off studies. The concept of feeding AG mills with as coarse a primary crusher product as possible in order to maximize AG mill throughput is obviously at the other end of the spectrum.

To a more significant degree than in other comminution devices, liner design and configuration can have a substantial effect on mill performance. In general terms, lifter spacing and angle, grate open area and aperture size, and pulp lifter design and capacity must be considered. Mill liner designs have moved toward more open shell lifter spacing, increased pulp lifter volumetric capacity, and tailored grate design to facilitate maximizing both pebble crushing circuit utilization and SAG mill capacity (Mosher 2005).

AG/SAG circuits can handle a broad range of feed sizes, as well as sticky, clayey ores (which challenge crushing plants). A range of circuit configurations have been developed, and large single-line plants can be designed with significant economies of scale and streamlined maintenance shutdowns and practices. Compared to crushing plants, wear media usage is reduced, and plants run at higher availabilities. Circuits, however, are more sensitive to variations in circuit feed characteristics of hardness and size distribution. Unlike crushing plants, for which throughput is largely volumetrically controlled, AG/SAG throughput is defined by the unit power required to comminute the ore to the closing size desired. A higher degree of operator skill is typically required of AG/SAG circuit operation, as is more advanced process control to maintain steady-state operation, with different operator/advanced process control regimes based on different ore types. Additionally, very hard ores can severely constrain AG/SAG mill throughput. In such cases, the circuits can become capital inefficient (in terms of the size and number of primary milling units required) and, in some cases, require more total power input compared to alternative comminution flow sheets.

Crusher–HPGR Circuits

Recently, HPGRs have made broad advances into the nonferrous metals mining. Long established in the cement industry, HPGRs first made inroads into diamond processing (where rock fracture along grain lines favored a reduction in diamond breakage during comminution) and in the iron ore industry. Penetration to the hard-rock mining industry was slow and hampered by high maintenance requirements both for wear surfaces in general and high wear on the edge of rolls. The intensity of the maintenance effort (relative to other comminution machines) required for a trial unit at the (then Phelps Dodge) Sierrita operation (Arizona) in 1996 was high.

Significant product improvements since then, however, have increased wear life, improved availability, and decreased the overall maintenance effort required. Incorporation of studs on the surface of rolls to allow formation of AG wear surfaces, as well as implementing edge blocks of a long-wearing material for edge protection, have allowed HPGRs to break into the mainstream of mineral processing. These wear-retarding innovations were the focal point of a full-scale trial at Lone Tree in Nevada, United States (Seidel et al. 2006). Successful completion of this trial marked somewhat of a turning point for interest in HPGRs for hard-rock applications. Manufacturers have also paid special attention to the bearings, wear surfaces, and the handling of tramp metal through the rolls to improve operational reliability, reduce maintenance, and obtain longer service lives. Availabilities are now such that aside from rolls change-outs (akin to a mill liner change), the unit rarely controls circuit availability.

The wear on a roll's surface is a function of the ore's abrasiveness. Increasing roll speed or pressure increases wear with a given material. Studs allowing the formation of an AG wear layer, edge blocks, and cheek plates are depicted in Figure 14.2-9. Development in each of these areas continues, including profiling of stud hardness to minimize the "bathtub" effect (wear of the center of the rolls more rapidly than the outer areas), low-profile edge blocks for installation on worn tires, and improvements in both design and wear materials for cheek plates.

In hard-rock-metals mining applications, HPGRs are currently used in tertiary and quaternary crushing applications, as well as in secondary pebble crushing. In some respects, HPGRs replace crushers as a unit operation. However, from a process standpoint, HPGRs produce a product with substantially more fines (for a given P_{80}) than a crushing circuit. In this regard, the size distribution of an HPGR circuit is much more similar to the product of a SAG circuit than a conventional crushing circuit, reducing the required amount of power in the ball mill circuit (relative to a crushing circuit). Also, an HPGR represents a much larger installation of power in a given footprint relative to tertiary crushers: An HPGR has up to a 5-MW power installation per unit compared to 750 kW for a single large cone crusher. HPGRs have also been cited as crushing more efficiently than rotating mills, as well as generating residual cracks that improve efficiency in subsequent milling operations. This effect is material- and operating-condition specific.

As a result, the number of material handling units (feeders, conveyors, screens, chutes) can be sharply reduced compared to a conventional crushing plant, and plant layouts are also much more straightforward. Freeport-McMoRan's (Phelps Dodge at the time of construction) Cerro Verde operation in Peru was a groundbreaking installation because the combination of secondary crushing (using MP1000s), tertiary crushing using HPGRs, and screens replaced what would have been more typically a large SAG mill feeding a multiple ball mill circuit (Vanderbeek et al. 2006). This was a significant step, as it presented an alternative to conventional crushing plants or autogenous/semiautogenous milling for primary milling applications (see Figure 14.2-10). Newmont Mining Corporation's consideration and ultimate selection of HPGR comminution for its Boddington project in Western Australia, widely documented in the literature, has now been commissioned with a similar comminution flow sheet to the Cerro Verde flow sheet (and also employing four units measuring 2.4×1.7 m).

Although the primary focus of this section is to discuss HPGRs (as they have been combined with crushers and screens) in primary milling roles, the versatility of the HPGRs in grinding applications must be noted. Though limited in top size they can accept (based on nip angle and resultant stud

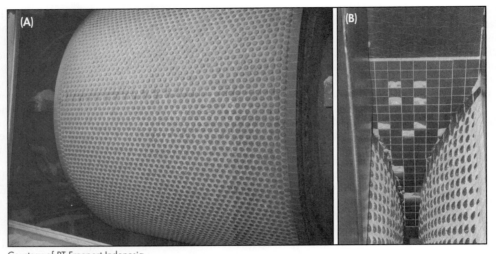

Courtesy of PT Freeport Indonesia.
Figure 14.2-9 HPGR roll (2.0 × 1.5 m diameter) with (A) studs and edge blocks and (B) cheek plates

breakage, leading to selection in tertiary rather than secondary roles), the machines are quite versatile because they have been installed in tertiary, quaternary, second-stage pebble crushing roles, and a number of finer grinding applications. In the fine-grinding role, HPGRs are widely used in pelletizing operations. One area not currently applicable to HPGRs is secondary milling in a wet process. In the cement industry, it is common practice to use HPGRs in closed circuits with a dry air classifier for finish grinding (S. Kirsch, personal communication). However, wet classification and subsequent dewatering does not compare favorably to the use of ball mills for wet processes most commonly used in mineral processing. A summary of the broad range of comminution applications of HPGRs is presented in Table 14.2-3.

SECONDARY MILLING

For those circuits requiring secondary grinding (most circuits aside from heap leach operations and single-stage AG/SAG circuits), the use of wet ball mills has remained the standard unit operation. Most such circuits use ball mills in closed circuit with hydrocyclones. However, closing circuits with fine screens is becoming more common, particularly in the iron ore industry (Valine et al. 2009). The majority of applications grind from a typical range of F_{80} values of 1–13 mm to P_{80} values of 300–37 µm. Pebble mills are also used in a number of applications.

Ball mills as large as 8.5 m (28 ft) in diameter are now being constructed. As with SAG mills, twin pinion and girth gears can be considered for mills with power input up to at least 15 MW (generally mills with a diameter of 7.3 m [24 ft] or less). Wraparound drives are the standard for larger mills.

In general, given a mill of suitable power and operated with an appropriate feed size, reduction ratio, media size selection, and the use of efficient classifiers, accurately predicting grinding circuit response is relatively straightforward. These principles apply for either wet or dry milling.

Aside from the widely used Bond equation (discussed later in the "Circuit Design" section), Bond used eight factors to "correct" for the power required for grinding in different operating scenarios (Rowland 2002). Bond's work to develop criteria for use in designing grinding circuits has proved to be enduring and useful, even with the contemporary availability of powerful simulation tools. Although the utility of correction factors for quantitative contemporary use in formal design is debatable, using the factors for qualitative evaluation of aspects affecting grinding efficiency can have value. Two of the eight correction factors apply strictly to rod milling, with the remaining six to ball mill grinding. Of greatest interest for current applications are corrections for coarse feed (with the definition of what is excessively coarse being a function of ore hardness), open-circuit milling (vs. closed-circuit), grinding to a P_{80} finer than 75 µm, and excessively low reduction ratios. One correction factor was developed to correct for inefficiency of large ball mills; many mill designers today believe that this factor had more to do with inefficiencies of classification in large hydrocyclones in early mill installations than in any actual milling inefficiency. The last factor for ball milling applies to dry grinding. In summary, although using these factors as originally developed for design may not be relevant, using them qualitatively to assess efficient operating conditions can be a useful tool.

By far the largest tonnages processed through secondary milling today are classified using hydrocyclones. Specialized low-tonnage operations may use screens, selected contemporary operations still employ screw classifiers, and dry operations generally use cyclones. A detailed discussion of the appropriate design for recirculating load is beyond the scope of this chapter. Milling efficiency typically increases with closed-circuit operation and will often increase to a certain level with increasing circulating load. After a certain point, the grinding efficiency declines with increasing recirculating load. Also, at a certain point, the power required for pumping for increased recirculation is not offset by improved grinding efficiency.

Some circuits perform grinding in one step, others in multiple stages (though rarely exceeding two), and occasionally with beneficiation integrated into the grinding circuit. The use of gravity concentration integrated into the mill recirculating load is common for the recovery of gold, flash flotation is often seen in sulfide mills, and recovery efforts after each stage of grinding are common for recovering platinum group metals and certain iron ore operations relying on magnetic separation.

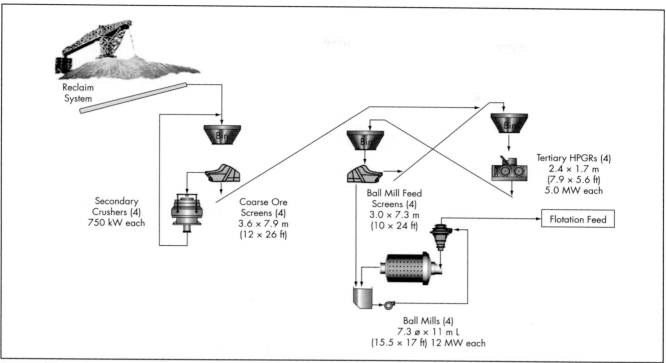

Courtesy of Minera Cerro Verde.
Figure 14.2-10 Crusher–HPGR–ball mill circuit

Table 14.2-3 Summary of range of HPGR applications

		Size				
Operation	Feed Application	Tire Diameter, m	Tire Width, m	Power, MW	Tons per Hour	Top Size, mm
Cerro Verde, Peru	Cu/Mo, tertiary, closed circuit	2.4	1.7	5.0	1,250	<50
CMH, Chile	Iron ore, tertiary, closed circuit	1.7	1.8	3.7	1,800	<60
Premier, South Africa	Diamond, tertiary, open circuit	2.85	0.5	1.2	330	<32
Empire, Michigan (USA)	Iron ore, second-stage pebble, open circuit	1.4	0.8	1.4	440	<30
Argyle, Australia	Diamond, quaternary, closed circuit	1.7	1.4	1.9	800	<15
PT Freeport Indonesia	Cu/Au, quaternary, open circuit	2.0	1.5	3.6	1,300	<10
Samarco, Brazil	Iron ore, pellet preparation regrind, open circuit	2.0	1.5	2.9	1,200	<0.2
Vale Tubarao, Brazil	Iron ore, pellet preparation regrind, open circuit	1.4	1.6	3.5	715	<0.2
Vale Sao Luis, Brazil	Iron ore, pellet preparation regrind, open circuit	1.7–2.4	1.2–1.4	2.0–3.6	650–1,200	<0.2

Source: Adapted from Mular and Mosher 2006.

TERTIARY MILLING

In the context of this paper, the term *tertiary milling* is used to describe concentrate regrind, grinding conducted for materials handling (iron ore pelletizing and pumping of mineral concentrates, for example), and fine grinding for specialized metallurgical applications. The border between a typical tertiary milling application and fine grinding is not distinct. However, as the methodologies for sizing change substantially at nominally 37 μm (400 mesh, and switching from conventional sieving practice toward subsieve analysis and alternate techniques at finer sizes), designating grinding to a P_{80} <37 μm as fine grinding is a useful designation.

Most grinding circuits are developed with the concept that the mill feed will be ground to an economic optimum such that further grinding expense (in terms of energy and media wear) is not justified based on the additional liberation and recovery of additional value units. In many cases, grinding 100% of the bulk mill feed to this size represents the optimum in terms of overall project economics and simplicity. In some circuits, it is desirable to have multiple stages of liberation and beneficiation. Since physical mineral separation processes have an optimum size range for efficiency, and essentially all lose efficiency at very fine particle sizes, this sequence can allow recovery of value minerals as they are liberated and before they are overground. Particularly in flotation, value minerals are often recovered to rougher concentrates when only partially liberated. Further grinding of such concentrates is employed to increase liberation and facilitate production of concentrate at a salable grade. Aside from liberation and recovery considerations, material handling of concentrates may dictate further size reduction. Examples include grinding of concentrates to allow pumping without pipeline wear or grinding of iron ore concentrates prior to pelletizing operations.

The field of options for fine grinding is diverse, with many originally developed to fulfill a specific need and subsequently marketed. Although not all operations require this stage of grinding, for those that do it is often a critical unit operation. Although the mills are smaller in size than mills for primary and secondary grinding, unit energy input (ranging from 5 to 40 kW·h/t) for tertiary mill feed often exceeds the total unit energy requirement for the bulk feed. In broad terms, the equipment field can be classified into rotating mills, stirred mills, and a general category of specialized mills. In general, the stirred mill category has the ability to grind finer with a higher efficiency than rotating mills. Specialized mills can cover a broad range of sizes but tend to be more focused to specific applications. Options for fine grinding are summarized:

- Conventional rotating mills (ball mills)
- Stirred mills
  - Vertical shaft mill (screw type). Also referred to generically as tower mills, after an early manufacturer, the Japan Tower Mill Company. Similar mills are now made by several manufacturers, with perhaps the most common being the descendant of the original mill design (by Nippon Eirich Co. Ltd.) by the trade name of Vertimill (marketed by Metso) (Lynch and Rowland 2005).
  - Vertical shaft mill (pin type). One trade name is the Metprotech mill (marketed by Bradken); another with a higher tip speed is the stirred media detritor (SMD) (marketed by Metso).
  - Horizontal shaft mill. The IsaMill has made broad inroads into fine grinding.
- Other specialized mills
  - Jet mills
  - Vibratory ball mills
  - Roller mills
  - Centrifugal mills

These machines are classified primarily by their mechanical configuration. Lichter and Davy (2002) provide another useful distinguishing feature for stirred mills—the charge motion. They cite a distinct difference between mills in which the charge is stirred (tower mills and some pin mills) and those in which the charge is fluidized (some pin mills and IsaMills), indicating that stirred mills are more efficient with coarse, hard feeds and that fluidized mills are more efficient with relatively finer, softer feeds.

Conventional rotating (tumbling) ball mills are often used to grind concentrates down to sizes of nominally –37 μm. Common applications include tertiary milling with multiple stages of grinding, in regrind applications, and in preparation of iron ore concentrates for pelletizing.

Stirred mills are often used for fine grinding. Mills in this category are sometimes more commonly referred to by their trade names. Stirred mills offer a higher power density (kilowatts per cubic meter), which, combined with the fact that they are filled with media (as opposed to tumbling mills, where the charge volume is nominally 35% of the total mill volume), means that these mills are compact in terms of installed power. Tower mills are occasionally used as way to retrofit additional grinding power with a small footprint and low capital for civil works; in this capacity they are often used for additional secondary or for tertiary grinding. They are widely used in lime slaking and in concentrate regrind applications. Pin mills (of either nonfluidizing or fluidizing tip speed) are generally the next mill considered at progressively finer grinds to a nominal grind size with a P_{80} of 5 μm. IsaMills are established as a versatile fine grinding mill, having been commercialized after being developed at Mount Isa Mines (Queensland, Australia). The IsaMill has demonstrated an ability to grind sizes with a nominal P_{80} of 5 μm. IsaMill models with up to 3 MW of installed power have broadened the mill's usage to regrind applications where higher tonnages dictate larger mill sizes (Burford and Niva 2008).

Grinding energies required (sometimes judged by the calculation of an operating work index, Wi_o, at coarser sizes, and at finer sizes by the specific energy input required to attain the target grind) often increase substantially as grinds get progressively finer. This can be due to a combination of factors, including higher energy required as the mineral grain size is approached, the number of micro-imperfections in the particle structure to exploit decreases, and, not insignificantly, the efficiency of classification is reduced. The required energy inputs can vary substantially based on mill type, media type, and the efficiency of classification employed. Tests should be conducted with a variety of machines and with a variety of media types to measure the specific energy input required to attain the target grind. Media types and sizes are summarized in a subsequent section ("Media and Liners"), but it is worth noting that the energy efficiency of fine grinding is strongly influenced by the media type selected (Gao et al. 2001). Additionally, based on the high unit power input, the integrity of media is also critically important. At very fine grind sizes, and depending on the material ground, the energy input can modify the reactivity of the material, and this can be a consideration in the chosen grinding methodology.

There are number of specialized fine-grinding applications, particularly in the fields of industrial minerals, powders, and the like. Fine grinding in specialized applications is often conducted dry and using either ceramic media or autogenously to avoid iron contamination. Examples include jet milling, centrifugal mills (including the nutating mill), vibratory ball mills, and roller mills. The reader is directed toward manufacturers and the body of literature for detailed coverage of these applications.

No discussion of ultrafine grinding would be complete without touching on the method of particle size analysis. At extremely fine sizes, particle size determination is not always straightforward. The measurements used are often indirect measurements from which particle size is estimated, with different techniques resulting in a different size being inferred. Additionally, the intense energy input can change the surface chemistry of the ground material, which can have effects including agglomeration or increased galvanic activity. For example, the use of specialized electroformed nickel subsieves (which offer precisely sized apertures for sieves at sizes less than 20 μm) is restricted for copper concentrates because of rapid galvanic action with the sieve itself.

Techniques often used include conventional sieves, specialized subsieves (generally available at 15, 10, and 5 μm) made of textile or electroformed and photo-etched nickel, and indirect methods, which employ material properties. Techniques rely on measurements of settling velocity (by means of elutriation, sedimentation, or centrifugal sedimentation), light dispersion characteristics, or both. Examples include the Warman Cyclosizer, the Malvern Mastersizer laser analysis, Microtrac's laser unit, Quantachrome's Microscan units, as well as others including the Coulter counter and surface area techniques. Regardless of technique, it is important to use the same one throughout a study. It can also be worthwhile to confirm sizing calculated via a certain technique using microscopy, either direct or in combination with image analysis software. Though this is unlikely to have any benefit for a circuit in operation, it can be useful in comparing size distributions of different machines. Comparing numbers derived using different techniques should be performed only with great caution.

SPECIAL APPLICATIONS

This chapter has focused largely on the comminution means employed by the majority of mineral processing operations (as judged by the tonnages processed). A number of special considerations apply either to specific industries, minerals, or focused applications. Comminution to develop a specific size distribution (rather than the simpler approach of assessing the size distribution through the use of a single data point) and a certain shape of particle (rather than using liberation as the goal), along mineral boundaries, and beneficiation steps integrated in the comminution process are all often considerations.

Beneficiation during comminution is often employed. The concentration of heavier specific gravity minerals in hydrocyclone feed or more often in hydrocyclone underflows is well documented and is used as a preconcentration step in flash flotation and gravity concentration units in a number of grinding circuits. The addition of reagents and lixiviants to grinding mills is commonly practiced, either for conditioning (commonly in flotation) or to increase leach residence time.

In certain applications (typically aggregate or industrial minerals), the particle shape resulting from comminution can be important and is a criterion in the selection of certain types of comminution machines. Examples include vertical shaft impact crushers or Gyradisc (marketed by Metso) crushers, which are examples of machines sometimes selected for comminution to a specific particle shape. Similarly, in certain applications, minimizing fines generation while attaining a grind below a maximum particle size is important. When this requirement is overriding, certain grinding equipment (listed in order of coarser grinds and lower fines generation) can be used: center discharge rod mills, end-peripheral discharge rod mills, overflow rod mills, and grate discharge ball mills. In quarrying and certain other applications, maximizing yield in certain size fractions preferentially over other size fractions can add significant value. For specialized applications such as this, close consultation with vendors and thorough testing are recommended.

Dry grinding is an important, albeit niche, application. For any pyrometallurgical operation, dry grinding is an obvious consideration. Newmont's Carlin mill and Barrick's Goldstrike mill in Nevada both have roaster facilities fed by Krupp double-rotator dry mills. The double-rotator mills consist of a twin chamber mill (with sides for coarse and fine milling), with both sides discharging to a common air classification system. Operation of Goldstrike's system is well described by Thomas et al. (2001). Newmont's Minahasa gold operation in Indonesia employed an Aerofall SAG mill followed by dry ball milling prior to decommissioning. The IOC's Carol Concentrator in Labrador (Canada) formerly milled iron ore for gravity concentration using dry Aerofall AG mills; Chung et al. (2001) demonstrated substantially better energy efficiency in grinding using dry milling. Energy costs for drying, particularly for a wet downstream process, contributed ultimately to the full conversion of the Carol Concentrator to wet AG. Richards Bay Minerals uses an Aerofall mill for dry grinding of titanium slags. Dry grinding with air classification should be considered in applications where the follow-on process is pyrometallurgical and drying would be required anyway, minimization of fines during comminution is critical (with air-sweep and classification assisting in this regard), or an ore's aggressive soluble components make dry grinding more cost-efficient than construction of a wet milling circuit made from corrosion-resistant materials.

A number of other specialized grinding units are referenced in this chapter based on their comminution range (primary, secondary, or tertiary grinding). Media selection to avoid contamination is also a consideration, referenced later in the "Media and Liners" section. Finally, certain applications find that media selection can impact metallurgical performance (through the interaction of redox potential and dissolved oxygen); for certain applications, AG, high chrome, or inert media can be preferred.

ORE CHARACTERIZATION, CIRCUIT DESIGN, AND CIRCUIT SELECTION

Many ore bodies can be successfully processed using a range of comminution circuit configurations. In some cases, however, certain types of circuits are favored over others. Ore characteristics (competency, density, work index, abrasivity, chemical composition), plant capacity, downstream processing requirements, mine life, capital costs, operating costs, delivery time for major components, and plant location are all factors in process selection. Given ore characterization data, alternative circuit designs to accomplish the required comminution can be developed. Capital and operating cost estimates for each circuit alternative allow for trade-off studies. Ore characterization, circuit design, and circuit selection are summarized briefly, with each addressed in depth in the broader literature.

Before an effective study culminating in a circuit design can be commissioned, a sufficient knowledge of the deposit and the mining method and sequence must be defined to allow development of a sampling design. Sampling of a greenfield deposit via drilling can be an expensive undertaking, and collection of a bulk sample for pilot testing (if desired or required) often exceeds the cost of the testing itself. Knowledge of the overall range of comminution characteristics within a deposit can be useful, but, as a guideline, the testing effort based on ore type should be proportional to the representation in the overall reserves. If ore types have divergent characteristics and/or will be mined at different periods in the mine life, it may be prudent to consider construction of a modular plant.

In some cases, blending of samples may also be considered. If so, blending should be done in a fashion consistent with the planned mining sequence. For instance, blending two samples that will be milled together may make sense. However, compositing of an entire drill core consisting of different horizons almost never makes sense and typically results in misleading data. Many designers prefer to blend the results of testing (mathematically) of individual ore types rather than blending a sample for testing. Both approaches have advantages and disadvantages. Obviously, testing for circuit design should be performed on ore and not waste; too many projects have had initial characterization done on waste (with waste samples being more readily available). Efforts in reasonable ore-body definition and the planned mining sequence to allow development of a rational sampling program is almost always a good investment.

Ore Characterization
Bench-scale breakage characterization tests are the basis for circuit design. Such tests fall into three broad categories: (1) grindability, (2) single-particle breakage, and (3) standard materials. Pilot-scale testing has long been considered the most reliable way to perform test work. However, the cost of testing, the difficulty in getting representative samples, advances in design methodologies, and greater experience with newer comminution circuits have tended to decrease the importance of pilot testing.

Typically, grindability tests rely on (1) a known size distribution of feed, (2) the application of a known or measured amount of energy to (3) a measured feed rate, and (4) a measured product size distribution. Based on these four items, an index is then calculated. The traditional Bond work index is perhaps the best known, but similar indices are calculated by other tests. Grindability tests that reach a locked cycle equilibrium or steady-state continuous operation are a better indicator of actual mill performance than batch tests. The design methodology for HPGRs using a bench-scale test (such as Polysius' LABWAL test) is largely analogous to this approach, except that primary index is a specific throughput, with measurements for specific energy input and the comminution outcome of that energy input also measured.

Single-particle breakage tests typically endeavor to quantify breakage as a function of energy input. Regardless of the test, using a breakage energy regime as close to the actual milling conditions as possible is preferred. Single-particle breakage tests provide fundamental data critical for a simulation effort; traditional grindability tests are unable to generate this data. Unfortunately, tests of this type cannot reliably predict the steady-state mill load; for autogenous/semiautogenous milling, the composition of the mill load defines mill performance.

Some comminution machines use conventional civil engineering measurements of strength as a direct basis for design, such as unconfined compressive strength. This is particularly true for crushers. Aside from the comminution tests presented, other indicative data can be used to map the overall comminution profile of an ore body, particularly rock quality designation and point load testing. Such tests, though not directly usable for design, can prove beneficial in determining the distribution of ore regimes. In some cases, correlations to comminution tests using these (inexpensive and quick) test results can prove constructive.

In addition to breakage tests, because media and liner costs represent significant operating costs, tests to determine wear rates are often conducted. The Pennsylvania abrasion test (often called the Bond abrasion test) is the most widespread. However, plant correlation with test results is often not particularly good. Because the test is conducted dry, it does not measure corrosion, a significant source of media consumption in wet grinding circuits. Abrasion data can also be obtained using specialized equipment, such as Polysius' ATWAL test for HPGRs, or by direct measurements of wear in pilot testing. A good approach is to use these tests to determine the relative wear of an ore type to others, then to benchmark with similar circuits and similar ores in operation.

Pilot testing involves larger-scale, steady-state testing at a range of operating conditions (including a feed size similar to the anticipated design) to evaluate potential design and operating scenarios. The response is then either directly scaled from pilot-sized to full-scale equipment using either a volumetric or a power basis. Pilot testing offers a high degree of confidence but can be inadequate in actual utility because of the limited number of variables that can be reasonably tested, the amount of sample required, and the logistics and expense of sample acquisition.

A number of tests have been developed over the years with various strengths and weaknesses. This is partly a reflection of proprietary design approaches by various companies and designers, and perhaps also a reflection of the never-ending quest for a test that can fully define the comminution response of an ore type quickly, with little sample, and at low cost. A summary of selected ore characterization tests are presented in Table 14.2-4. Most of the tests summarized are conventional comminution tests relevant for primary and secondary grinding operations. Bench-scale and pilot-scale HPGR tests are included, with direct consultation with manufacturers recommended for tertiary grinding tests (most of which rely on direct scaling based on unit power input). All testing (to include the benchmark of pilot testing) presents trade-offs, in terms of data developed, with the time, effort, and cost of sample collection and testing. In many cases, factoring how representative the samples tested are of the ore body presents its own challenges. Fundamentally, a large number of poorly selected samples cannot overcome a lack of sample design. Data developed from testing samples that poorly represent the intended population are at best useless and at worst damaging in that they can contribute to incorrect designs or performance projections.

Ore testing should also consider the methodology of the subsequent circuit design. Traditional power-based design methods are generally based on work indices measured during grindability tests. Then these data are typically scaled based on power input. Simulation circuit design generally relies on the results of single-particle tests. A combination of computational mathematical modeling (circuit simulation) and use of traditional work indices are generally used in current practice. Additionally, to broaden the applicability of certain tests, some designers have developed correlations (some with defined bands of variance) that allow prediction of single-particle test parameters from grindability tests (particularly the SMC Test; Morrell 2006) and vice versa.

A sound approach for ore-body comminution is an integrated test program using both well-designed steady-state tests and single-particle breakage tests. These tests may be complemented with other tests to map a deposit and with equipment-specific tests (i.e., bench-scale tests for specific comminution machines) supplanting or replacing portions of the test program. Simulation of bench- and pilot-scale mill tests via discrete element modeling or other high-definition models is well within the scope of today's technology. Perhaps single-particle ore characterization in conjunction with high-definition modeling and scaling of steady-state bench-scale tests will become the norm for future design work.

Circuit Design
Given a design throughput, grind size target, and ore breakage characteristics for an ore body (most appropriately evaluated as a range and ideally based on a preliminary mine plan), a number of circuit designs can be prepared for evaluation. Based on circuit specifications, ore characteristics, downstream processes, location, operator preferences, and scoping

Table 14.2-4 Summary of ore characterization tests

Test	Mill Diameter, m	Top Size, mm	Core Size, Standard Reference	Peak Energy*, J/kg	Peak Energy, J	Sample Requested, kg	Sample Used†, kg	Closing Size, mm	Type
AG/SAG pilot	1.83	100–150	N/A‡	50	150	~10,000/test	Varies	Varies	Continuous
Media competency	1.83	165	N/A	18	100	750	400	N/A	Batch
Pilot HPGR	0.71‡	32	NQ	N/A	N/A	700	Varies	Varies	Continuous
Bond impact	N/A	75	PQ	500	200	Twenty 50–75 mm rocks	7.5	N/A	Single particle
Drop weight	N/A	64	HQ	1,400	450	75	24	N/A	Single particle and batch
MacPherson autogenous grindability	0.45	32	NQ	70	3.3	135	100	1.2	Continuous
SMC Test	N/A	32 or 22	HQ-NQ	12,600§	500	15	2.5	N/A	Single particle
SAG Power Index	0.305	25, $F_{80} = 13$	NQ	8	0.2	10	2	N/A, $P_{80} = 1.7$	Batch
Bench-scale HPGR (LABWAL)	0.25**	12.5	BQ	N/A	N/A	250	25	N/A	Continuous
Bond rod mill	0.305	13	Any	500	1.6	20	10	1.2	Locked cycle
Bond ball mill	0.305	3.3	Any	—	0.6	10	4	0.149	Locked cycle

Source: Adapted from Mosher and Bigg (2002); McKen and Williams (2006).
*Peak energy is estimated as the nominal maximum impact breakage energy input based on the highest energy event on the coarsest size range used in the test.
†Sample is the amount typically used in conducting the test and typically still usable for metallurgical testing if required. The amount is typically less than requested based either on the amount required for sampling purposes or to allow for repeat testing by the laboratory.
‡N/A = not applicable.
§S. Morrell (personal communication).
**Diameter for HPGR tests specifies rolls diameter.

and prefeasibility studies, a short list of comminution circuit types can generally be developed that merit development of a process design.

Fundamental to achieving size reduction is the application of power; applying power at peak efficiency reduces unit operating costs and ensures efficient utilization of capital. The machines most widely used to apply comminution energy have been discussed in previous sections. The machines described apply energy via two mechanisms:

1. Directly through devices that break particles through compressive force (crushers, for example)
2. Initially to impart motion to a grinding charge, a portion of which energy is then translated into particle breakage

For the first case, the machines are generally controlled by volumetric throughput. The amount of power drawn to impart compressive forces is a direct function of the hardness of the material being broken by compression. As a result, the unit power can vary based on material type at a fixed throughput. For a given material type and circuit/machine configuration, the size distribution of the product tends to be fairly consistent.

In contrast, application of power in tumbling mills is somewhat more complex. Once power is drawn in a rotating mill, a portion of the power is applied to the charge, resulting in breakage through the mechanisms of impact, abrasion, and attrition. With a steady-state charge, the power imparted is constant, regardless of material hardness or throughput. In overflow mills, the throughput through the mill can be readily varied, and there is a degree of control over the unit power input. A SAG mill's throughput, though, is volumetrically controlled based on the mill's discharge grates. As throughput decreases in a tumbling mill with a constant power draw, the unit power increases, and a finer grind results.

In both cases, though, the fundamental driver of the amount of breakage achieved is the amount of power delivered to each unit mass of material. Therefore, in designing for volumetric throughput machines, it is essential to have a firm understanding of throughput and how much power will be drawn under the operating conditions used. In tumbling mills, it is essential to understand the power draw that the mill will achieve and the range of throughput that can be expected under such conditions.

Historical articles on comminution circuit design have presented tabulations of power draw estimates for mills and machines of a given size. Indeed, substantial effort by mill designers went toward understanding how the power installed on a mill would be drawn. In numerous case studies, the installed motor power was inadequate for operation at the desired operating conditions. Conversely, motors with rated power outputs larger than could ever be drawn have been installed, leading to confusion on the part of designers and operators who believed that this power should be available for grinding.

Much of the art of power estimation was converted to direct computation through the development of a mathematical model by Morrell (1996). His model is an outcome of modeling the physics of the power required to impart charge motion and lift; fundamentally, a mill's power draw is the sum of the power (applied over time) required to impart kinetic and potential energy to the mill charge. The power draw of a mill is a function of (1) the charge shape (including the volumetric filling of the mill, the charge density, amount of slurry pooling, etc.), (2) charge position (particularly top [or shoulder] and bottom [or toe] position of the charge), and (3) the mill speed. It is evident to practitioners that these three fundamental variables of mill power draw are interrelated—for example, charge shape and position are impacted by changes in mill speed.

The critical question in design is the efficient application of power. As with ore characterization, there are a number of design techniques. The use of power-based, simulation-based, and direct-scaling benchmarking with existing plants or (preferably) a combination of techniques can be used to develop circuit designs for evaluation. Regardless of methodology, a key metric of the design process is the design power input (kilowatt-hours per ton) for each ore type. Designers often evaluate (and sometimes rank) circuits in terms of unit power input; mill operators more commonly reference circuit performance based simply on throughput at target operating conditions.

In power-based designs, the Bond equation continues to be a fundamental tool. The correlation between the power applied in a grinding operation and the amount of size reduction attained has long been intuitively known, with several relationships between the two proposed through the years. Over decades of application, F.C. Bond's third-theory equation has an established track record for both design and analyses of conventional grinding circuits. The standard Bond equation is presented in Equation 14.2-1 (Bond 1960). The equation can also be used to evaluate the relative power efficiency of simulated and operating circuits. By rearranging the formula to Equation 14.2-2, the operating work index (Wi_o) of an existing grinding circuit can be calculated, given known mill power draw, throughput, feed, and product size distributions. The use of operating work indices can be useful for forecasting future mill performance and as a benchmark to compare against laboratory measured work indices.

$$kW \cdot h/t = Wi * \left(\frac{10}{\sqrt{P_{80}}} - \frac{10}{\sqrt{F_{80}}} \right) \quad (14.2\text{-}1)$$

$$Wi_o = \frac{kW \cdot h/t}{\left(\frac{10}{\sqrt{P_{80}}} - \frac{10}{\sqrt{F_{80}}} \right)} \quad (14.2\text{-}2)$$

where

$kW \cdot h/t$ = power applied per ton of feed to the grinding unit
Wi = standard Bond work index of the ore
P_{80} = 80% cumulative passing size of the product, in μm
F_{80} = 80% cumulative passing size of the feed, in μm

Care must be employed in using the Bond equation for sizing individual unit operations when the feed and product particle size distributions are not parallel; this challenge is an artifact of describing a particle size distribution through the use of one number (represented by the P or F size). For example, SAG mills and HPGRs both produce substantially more fines than a traditional crushing plant for a given F size. As a result, an operating work index of a ball mill following a SAG mill or an HPGR is substantially lower than the operating work index of a ball mill following a rod mill or crusher. Similarly, the operating work index of the unit preceding the ball mill will be higher for those machines producing substantial fines than for crushers or rod mills. The apparent advantage in ball mill efficiency is a misnomer; the ball mill simply has less work to do with greater proportions of fines in the feed. This effect can be addressed in a number of ways, but recognizing it is essential in design (McKen et al. 2001).

Power-based designs rely on circuit configurations that will deliver the targeted energy input. As such, sound use of computational methods, tables, or benchmarking to ensure that the designed mill/motor configurations will draw and apply the target power is essential. Aside from proper equipment sizing, the most critical pitfall of a power-based design is ensuring that the design is achievable in terms of the assigned power split for each grinding stage. Many designs have relied on arbitrary power allocation between grinding stages. When commissioned, it was found that throughput constraints between units would either prevent the design throughput from being achieved, most often while attaining a finer grind than designed, or vice versa.

Simulation-based designs use a set of mathematical models to calculate the anticipated response of a circuit under a set of operating conditions. A number of commercial and proprietary simulation packages are available. Major equipment manufacturers often use in-house simulation models. Many of these packages are based on a combination of empirical and fundamental models, and correlations with observed plant performance.

Simulation is a powerful tool that offers the ability to evaluate circuit response for a range of operating conditions and of a given circuit for a range of feed types, and readily compare a number of circuit options. It can also provide a great deal of useful process information (in terms of the volumes and size distributions of intermediate flows) that is unavailable from power-based designs. Notably, simulation offers the ability to evaluate the impact of classification on the comminution circuit, something that cannot be done with power-based techniques. In addition to computer simulation to model steady-state comminution circuit performance, discrete event simulation can be useful in evaluating overall circuit response with respect to linkages with discontinuous (truck dumping) or semicontinuous (such as a crushing plant with periodic stops due to metal detects) processes.

When conducting a design exercise, it is critical to validate that the circuit being modeled is within the boundaries of the underlying model. Although this is generally the case for typical processing plants, the use of models outside of the operating window for which the model was developed can be misleading. Calculating the operating work index of circuits developed using circuit simulation can provide a cross-reference (reality check) and can be useful for evaluating the relative efficiency of circuits.

Direct scaling is employed for a number of comminution machines and, depending on the design approach, can be based on achieving a target unit energy input (kilowatt-hours per ton) or on a specific throughput determined in laboratory testing. For single unit operations, and when used in conjunction with a good database of existing plant operations, robust designs can be produced. However, examples frequently emerge of direct scaling and commissioning of plants that could not achieve the design throughput but are capable of grinding finer than target grind, or vice versa—that is, they exceed nameplate throughput but do not achieve the target grind.

Regardless of the design technique, benchmarking is a useful tool. Frequently, comparison of ore breakage characteristics and circuit design parameters can yield a list of relevant similar operations. In fact, although this approach may not be bankable, it is widely used for low-risk, self-financed expansion projects or when scaling the anticipated throughput of a new deposit in an existing mill. Using it as the sole basis for design, however, can introduce a higher degree of uncertainty in the design outcome.

An additional consideration is the question of how to integrate the stages in a comminution circuit. Primary grinding units, whether they are crushers, AG/SAG mills, or HPGRs, arrive at their unit power input largely based on the volumetric throughput at which they are operated. For AG/SAG mills at a given operating condition of speed and load, the power input is fixed, so the only variable is feed rate (which will change based on ore hardness and size distribution). For crushers or HPGRs, though, the feed rate is fixed based on volumetric throughput at a given aperture, and the power drawn varies as a function of the ore characteristics. Regardless, a design methodology that considers designing the primary milling stage based on the required throughput, with the secondary stage based on the amount of power required to finish grind the design throughput, is more successful than a methodology that counts on a set amount of power input in the primary grinding stage.

For power-based designs, determining the appropriate and efficient way to input the design grinding power is the core question. For simulation-based designs, determining the amount of power drawn under a given set of simulated conditions is required. Cross checking this value with calculated power requirements or benchmarking with operating circuits is wise. For scaled designs, it is essential to ensure that power input scales similarly to volumetric throughput, or the unit power inputs will not match, and consequently the desired size reduction may not be attained at the target throughput.

Most modern circuit designers consider the various design techniques to be complementary. Although various designers may have preferences, a robust design will consider several different approaches. A higher degree of confidence can be assigned to a design that is confirmed based on traditional power-based approaches, with various operating conditions investigated and evaluated by simulation, and which benchmarks and scales favorably with similar operations. Although any of the techniques can be used independently, the most robust design will eventuate through the combination of these techniques.

Circuit Selection

Staged crushing plants are favored in cases of low-tonnage operations where little fines generation is desired and a dry, coarse product is required (heap leaches), or in moderate tonnage operations with hard ore. Plants that will be fed sticky ores, with large amounts of tramp metal, or high-tonnage operations are generally less favorable for a traditional crushing plant. A benchmark for annual run time (including mechanical availability and all other factors) for a well-maintained and run modern crushing plant is 85%. Lower availabilities are typical in more-demanding applications.

Semiautogenous milling is favored for high-capacity plants, soft to moderately hard ore types, sticky or clay-bearing ores, and ores containing significant amounts of tramp metal or magnetite. Ores with a competent fraction, particularly higher-density ores, can be amenable to autogenous grinding. This is particularly true for small to medium throughputs with a long mine life. Because AG mills tend to have a lower unit power input for a given mill size, they can be more expensive to construct but much less expensive to run (when compared on unit cost basis) compared to a SAG mill. Circuits designed for autogenous grinding are often designed to accommodate semiautogenous grinding either for operational flexibility or with the anticipation of milling future ore bodies; a case study of an operation with a versatile single-stage AG/SAG circuit is Brunswick (Larsen et al. 2001).

When using the largest mills available, a single SAG mill often produces feed for secondary grinding two or more ball mills. In a typical SAG–ball mill circuit, the SAG circuits should generally be designed for the target throughput, with the ball mill circuit designed to achieve the target grind. Based on ore breakage characteristics, feed size, grate aperture, and circuit closing size, the ability to shift the transfer size substantially between the SAG and ball mill circuits is often limited. Though exposure to significant downtime for a failure in a single line is often considered a drawback, SAG plants have demonstrated annual run times of 95%.

HPGRs are considered favorably as a circuit alternate to autogenous/semiautogenous milling. Such a circuit must be designed in conjunction with secondary crushing to reduce ROM ore to a size suitable for HPGR feed. HPGRs, in closed circuit with screens, then produce ball mill feed. HPGRs were selected for the Cerro Verde Cu-Mo project commissioned in 2006 and have also been specified for the Boddington project. Particularly in the case of hard and abrasive ores, HPGRs can offer operating cost benefits relative to SAG. Depending on the delivery time for large SAG mills (and often its accompanying wraparound electric motor), delivery and construction of HPGRs may be executed more quickly than for a SAG mill circuit. The largest HPGRs can be reasonably sized with a single large crusher to feed one ball mill for typical circuits. A crusher and closed-circuit HPGR grinding line feeding a ball mill will typically have lower availability and more unit operations than a similar SAG/ball mill line.

Substantial progress has been made in terms of availability and reliability. Preventive maintenance periods are required for cheek plates and inspections, but HPGR maintenance is rarely the controlling factor in overall circuit maintenance. Aside from rolls changes (akin to a mill reline and typically occurring between 2,000 and 15,000 running hours, depending on the material and application), more often circuit availability is controlled by belt, chute, screen, and conventional crusher maintenance. Circuit availabilities between a traditional crushing plant and a SAG mill should be expected for circuits where a secondary crusher/screen/HPGR replaces a SAG mill, with 90% run time a nominal benchmark. Aside from the fact that the degree of ancillary equipment (belts, transfer chutes, additional pumps, etc.) can increase the maintenance demand relative to a SAG plant, the additional unit operations also drive up capital costs.

Circuit selection is typically based on the calculation of the net present value of various circuit options (implicitly incorporating both differences in capital and operating costs and in performance between plant types). Qualitative considerations include location, operator preference, equipment lead times, and the like.

Selection of motor and drive types for tumbling mills is a specialized topic in itself. For larger mills (nominally more than 7.2 m [23.6 ft] in diameter for ball mills and more than 9.8 m [32.2 ft] for SAG mills), gearless (wraparound motor) and geared (pinion/bull gear combinations) drives are normally compared. Gear manufacturers indicate that geared drives with up to 10 MW per pinion are feasible (Hankes 2001). Failures with both large wraparound motors and large gears have occurred in the industry, and engineering trade-off studies will likely continue for those mills where either geared or wraparound motor drives could be used.

MEDIA AND LINERS

Table 14.2-5 summarizes the typical media sizes and types in the broadest use for mills discussed in this chapter. Contemporary practice in tumbling mills is to add a single-size makeup ball mass to maintain constant power draw. Ball

Table 14.2-5 Typical media sizes and types

Media	Size*	Type
SAG mills	127 mm (100–150 mm)	Forged steel
Ball mills	50–65 mm (25–100 mm)	Forged or cast steel
Regrind/fine-grind ball mills	25 mm (9–40 mm)	Forged or cast steel
Tower mills	12 mm (6–25 mm)	Forged, cast, or other (e.g., shot) steel
Pin mills	12 mm (3–12 mm) (Metprotech) 3 mm (SMD)	Steel or ceramic Ceramic or sand
IsaMills	3 mm (1–6 mm)	Sand, slag, ceramic

*Common makeup ball sizes indicated, with a typical range indicated in parentheses; coarser or finer sizes may be found in certain applications.

charging should be done on a daily basis or, more ideally, on a shift-wise basis. Some plants with a mechanized ball charging system (typically for SAG mills) have adopted continuous charging (at a rate matching the mill's steel consumption) to facilitate operating at peak conditions. As steel media most often have a hardness profile with a softer center, once balls reach a certain size, they tend to wear very quickly.

The use of a mixed-size makeup ball charge has been repeatedly demonstrated in laboratory studies to offer incremental efficiency benefits, but the benefits have been difficult to document at plant scale. Selected operations have adopted mixed-sized makeup ball charging for SAG mills, but it remains uncommon for ball mills. There is little reason to believe that the steady-state media size distribution resulting from the wear rate of the makeup ball size corresponds to the optimum ball size based on the mill's feed and target grind.

The concept of selecting a makeup ball size based on the coarseness of feed has long been recognized (Taggart 1945). In general, a mixed makeup ball charging regime improves grinding efficiency, with greatest potential benefit for single-stage milling applications with large size reductions. McIvor and Weldum (2004) suggests that the benefits from graded charges extend to pebble mills and hypothesizes that this is a fundamental reason for improved efficiency (relative to laboratory-measured work indices) at the Empire mill. Nonetheless, most operations tend to use a single-sized makeup ball size for reasons of convenience. In practical terms, for grate discharge mills, there is essentially zero grinding media smaller than the grate aperture at any given time. For overflow mills, the lower media size is constrained by the size at which media flow out of the mill (which, in turn, is effected by volumetric throughput, pulp density, and media size and density).

The use of ore as AG media was reviewed previously in the "AG/SAG Circuits" section. Pebble milling (a form of secondary milling) most often uses oversize material extracted from the primary milling circuit as media. This is most often done in applications with higher ore densities and can represent a substantial savings in milling costs. With ore densities lower than steel densities, larger mills are often required, and some degree of flexibility may be needed when the generation and consumption rates of AG media do not match.

The use of recycled SAG scats (worn and broken media recovered via magnets from SAG mill grate discharge) in ball mills has grown in application. The balls are direct charged to ball mills as a replacement for new grinding balls. The larger soft core of SAG media (as a result of larger size and requirements for ball durability) means that the recycled media wear more quickly than a typical like-sized new ball mill ball. Discussion of the use of alloys, cast/forged or other types of media, or whether to effect improved wear life through surface hardness, hardness profile, or improved resistance to corrosion is outside the scope of this chapter.

Liner innovation has contributed substantially to comminution performance. Nowhere has this been more evident than in semiautogenous milling. After early work that tended to focus more on maximizing life, current best practice is focused on maximizing total productivity over the life of a liner set. Generally this means development of a shell lifter set that facilitates (within the constraints of original mill shell drilling) a quick startup period with little to no lifter packing and for which full power draw can be maintained over the planned life of the liner set. Critical aspects of shell lifter design include lifter spacing, height, and face angle. In general, common practice has evolved toward operation with liner sets that are more open (have greater spacing between lifter bars relative to lifter heights). In general terms, this philosophy leads to less liner packing, the shortening (or elimination) of a break-in period with new liners, and higher mill performance over the liner set. Going too far can lead to short liner life offsetting performance benefits and, in extreme cases, insufficient charge lift to achieve baseline mill performance. Design of pulp lifters and discharge grates has been an area of intense effort by mill operators, particularly large, high-volumetric throughput mills.

Because of the HPGR's role in replacing duties performed by mills, HPGR liner wear is often relevant in comparison with liner and media cost. Cost savings relative to the overall costs of liner and media consumption in a rotating mill can often be realized. Relative to mill liners and grinding balls, HPGR wear surfaces are much lower volume and a more highly engineered product. The unit cost of media wear, generally in terms of dollars per metric ton (at equivalent kilowatt-hours per ton) is a good metric for comparison.

The use of liner handlers to reline mills is now standard practice, and the use of larger machines has facilitated consolidation of liner parts. The world's current largest liner handler is in use at PT Freeport Indonesia; the machine's 7,250-kg capacity has allowed consolidation of parts and reduction in the number of units requiring handling during a reline. Progressive refinements, with each pushing the then-existing technology's capacity in terms of casting integrity and liner handlers, now facilitates the use of 40% fewer liner pieces than the original design in PT Freeport Indonesia's 10.4-m (34-ft) and 11.6-m (38-ft) SAG mills. The use of a sound, integrated, and well-resourced maintenance approach can lead to annual run times in the 95%–96% range in even the most challenging environments. Such techniques can also be applied to ball mill liners or to primary crusher concave relines (e.g., with larger castings, reducing the number of rows from a typical five-row installation to three-row).

LOOKING FORWARD

This chapter has summarized a range of comminution options for mineral processing applications. Although machine design, performance, maintenance, and other aspects have been touched upon, the focus has been on the process. Since the last comminution chapter in the previous edition, the range of equipment options for mineral processing comminution plants has broadened. AG/SAG circuits as well as circuits incorporating HPGRs are in wide use. In particular, AG/SAG mills and HPGRs have blurred the distinctions of using a certain machine for a certain feed size.

The low fraction of total power input in the crushing and grinding process that actually goes toward size reduction (with the balance being dissipated largely as heat and noise) has long been cited (Kelly and Spottiswood 1995). Despite intense research efforts with both public and private sponsorship, no viable commercial processes with markedly better efficiency are on the horizon. With increasing focus on energy input, from both high energy cost aspects but also considering the embodied energy and greenhouse gas impacts of various products, these efforts will continue. Interestingly in the case of energy input in mineral processing applications, the effects of increased energy input for liberation are offset by the benefits of improved resource recovery.

To the extent that the marketplace can accurately assign a value to the material produced and a cost to the negative impacts, assessing the benefits of increased comminution to affect increased recovery versus the costs of higher energy consumption should be clear. The focus on comminution as often the highest-cost category of mineral processing (both in terms of capital and operating) is likely to increase given continuing demand for resources and increasing awareness and attempts to mitigate the impact of the human carbon footprint. At first glance, decreasing the comminution effort may seem to be an easy target to achieve both cost and carbon footprint benefits. In many cases, though, the incremental comminution effort required for the incremental value recovery is less than the energy requirement to recover a like quantity of values from a freshly mined ton of ore. From an energy perspective, increasing comminution will in many cases be favorable to mining and milling new ore tons. Thorough study and cost quantification are required for a true optimization of the overall process.

REFERENCES

Bond, F.C. 1960. Crushing and grinding calculations. *Br. Chem. Eng.* 6. (Rev. 1961 by A/C Pub. 07R9235B).

Boyd, K., and Utley, R.W. 2002. In-pit crushing design and layout considerations. In *Proceedings of Mineral Processing Plant Design, Practice and Control.* Edited by A.L. Mular, D. Halbe, and D.J. Barrat. Littleton, CO: SME.

Burford, B., and Niva, E. 2008. Comparing energy efficiency in grinding mills. Presented at Metallurgical Plant Design and Operating Strategies (MetPlant 2008), Perth, Western Australia.

Chung, G., McInnis, D., and Hong, D. 2001. Autogenous grinding at Carol Concentrator. In *Proceedings of Semi-Autogenous Grinding Technology.* Vancouver, BC: University of British Columbia.

Gao, M., Young, M.F., Cronin, B., and Harbort, G. 2001. IsaMill medium competency and its effect on milling performance. *Miner. Metall. Process.* 18:2 (May).

Greenwood, B.R., and Rajala, G.F. 1992. Autogenous grinding circuit at the Empire Mine. In *Comminution—Theory and Practice.* Littleton, CO: SME.

Hankes, B. 2001. Recent advances in girth gear and pinion manufacture. In *Proceedings of Semi-Autogenous Grinding Technology.* Vancouver, BC: University of British Columbia.

Kelly, E.G., and Spottiswood, D.J. 1995. *Introduction to Mineral Processing.* Australia: Wiley.

Larsen, C., Cooper, M., and Trusiak, A. 2001. Design and operation of Brunswick's AG/SAG Circuit. In *Proceedings of Semi-Autogenous Grinding Technology.* Vancouver, BC: University of British Columbia.

Lichter, J., and Davey, G. 2002. Selection and sizing of ultrafine and stirred grinding mills. In *Proceedings of Mineral Processing Plant Design, Practice and Control.* Edited by A.L. Mular, D. Halbe, and D.J. Barrat. Littleton, CO: SME.

Lynch, A., and Rowland, C. 2005. *The History of Grinding.* Littleton, CO: SME.

McIvor, R., and Weldum, T.P. 2004. Fully autogenous grinding from primary crushing to 20 microns. In *Plant Operator's Forum 2004: Things That Actually Work!* Littleton, CO: SME.

McKen, A., and Williams, S. 2006. An overview of the small-scale tests available to characterise ore grindability. In *Proceedings of Semi-Autogenous Grinding Technology.* Vancouver, BC: University of British Columbia.

McKen, A. Raabe, H., and Mosher, J. 2001. Application of operating work indices to evaluate individual sections in autogenous-semiautogenous/ball mill circuits. In *Proceedings of Semi-Autogenous Grinding Technology.* Vancouver, BC: University of British Columbia.

Morrell, S. 1996. Power draw of wet tumbling mills and its relationship to charge dynamics. Part 1, A continuum approach to mathematical modelling of mill power draw; Part 2, An empirical approach to modelling of mill power draw. *Trans. Inst. Min. Metall.* 105:C43-53, C54-62.

Morrell, S. 2006. Design of AG/SAG circuits using the SMC test. In *Proceedings of Semi-Autogenous Grinding Technology.* Vancouver, BC: University of British Columbia.

Morrell, S., and Valery, W. 2001. Influence of feed size on AG/SAG mill performance. In *Proceedings of Semi-Autogenous Grinding Technology,* Vancouver, BC: University of British Columbia

Mosher, J. 2005. Comminution circuits for gold ore processing. In *Advances in Gold Ore Processing.* Edited by M. Adams. Amsterdam: Elsevier.

Mosher, J., and Bigg, A.C.T. 2002. Bench-scale testing and pilot plant tests for comminution circuit design. In *Proceedings of Mineral Processing Plant Design, Practice and Control.* Edited by A.L. Mular, D. Halbe, and D.J. Barrat. Littleton, CO: SME.

Mosher, J., Banini, G., Mular, M., and Supomo, A. 2006. SAG mill pebble crushing: A case study of PT Freeport Indonesia's Concentrator #4. In *Proceedings of Semi-Autogenous Grinding Technology.* Vancouver, BC: University of British Columbia.

Mular, A.L., and Jergensen, G.V. 1982. *Design and Installation of Comminution Circuits.* New York: SME-AIME.

Mular, M., and Mosher, J. 2006. A pre-production review of PT Freeport Indonesia's high pressure grinding roll project. In *Proceedings of Semi-Autogenous Grinding Technology.* Vancouver, BC: University of British Columbia.

Napier-Munn, T., Morrell, S., Morrison, R., and Kojovic, T. 1996. *Mineral Comminution Circuits, Their Operation and Optimisation.* Brisbane, Australia: Julius Kruttschnitt Mineral Research Centre.

Rowland, C.A. 2002. Selection of rod mills, ball mills, and regrind mills. In *Proceedings of Mineral Processing Plant Design, Practice and Control* Edited by A.L. Mular, D. Halbe, and D.J. Barrat. Littleton, CO: SME.

Seidel, J., Logan T.C., LeVier, K.M., and Veillette, G. 2006. Case study—Investigation of HPGR suitability for two gold/copper prospects. In *Proceedings of Semi-Autogenous Grinding Technology*. Vancouver, BC: University of British Columbia.

Taggart, A.F. 1945. *Handbook of Mineral Dressing, Ores and Industrial Minerals*. New York: Wiley.

Thomas, K.G., Buckingham, L., and Patzelt, N. 2001. Dry grinding at Barrick Goldstrike's roaster facility. In *Proceedings of Semi-Autogenous Grinding Technology*. Vancouver, BC: University of British Columbia.

Valine, S.B., Wheeler, J.E., and Albuquerque, L.G. 2009. Fine sizing with the Derrick® Stack Sizer™ Screen. In *Recent Developments in Mineral Processing Plant Design*, Tucson, AZ.

Vanderbeek, J.L., Linde, T.B., Brack, W.S., and Marsden, J.O. 2006. HPGR implementation at Cerro Verde. In *Proceedings of Semi-Autogenous Grinding Technology*, Vancouver, BC: University of British Columbia.

Wills, B. *Mineral Processing Technology*. 2006. Oxford, UK: Butterworth-Heinemann (Elsevier).

Wood, C. 1996. Single stage SAG grinding experience at Henderson. In *Proceedings of Semi-Autogenous Grinding Technology*, Vancouver, BC: University of British Columbia.

CHAPTER 14.3

Classification by Screens and Cyclones

Brian Flintoff and Ronald Kuehl II

INTRODUCTION

The classification of particles based on size is an important step in almost all mineral processing flow sheets. Although the technical focus in plant design, control, and optimization is most often on comminution (e.g., size control and mineral liberation) and/or separation equipment (e.g., recovery/yield and grade), classification plays a critical role in optimizing process efficiency. This chapter focuses on two major classes of classification unit operations: screens and hydrocyclones.

SCREENS

Matthews (1985) said, "Screening is defined precisely as a mechanical process that accomplishes a separation of particles on the basis of size and their acceptance or rejection by a screening surface. Particles are presented to the apertures in a screening surface and are rejected if larger than the opening or accepted and passed through if smaller." The screening process has been a part of mineral processing flow sheets for a long time, and like many other unit operations, screening has seen some interesting mechanical and process developments over the past few years.

Looking across all of the process industries, there are many different kinds of screens and applications. However, the authors have restricted the discussion to the most common applications in the mining industry. Figure 14.3-1 is a generic graphical representation of the major applications of screening in mining: scalping, size control, and sorting.

Table 14.3-1 provides more amplification on the type of screen one might find in given applications.

Screen types can generally be categorized as follows:

- **Fixed types:** grizzlies, riffles, sieve bends, and so forth
- **Moving (linear) types:** vibrating, reciprocating, and resonance
- **Moving (rotating) types:** cylindrical trommels

Figures 14.3-2 through 14.3-6 are photographs of a grizzly screen, vibrating screen, trommel screen, Stack Sizer, and sieve bend. In general, these are the workhorses of the mining industry, with the vibrating screen standing out as the most widely employed screening unit operation. It is for this reason that most of the chapter is devoted to the vibrating screen.

Vibrating Screen Basics

The vibrating screen is ubiquitous in mineral processing plants. One would imagine that with this breadth and history of application and its apparent simplicity, there would be few remaining unknowns in the screening process. However, the design and operation of screening systems remains a blend of art and science—a statement used by many other authors (e.g., Mathews 1985). It is interesting to note that the screening process is enjoying something of a renaissance today, as new simulation tools offer a closer examination of the physics of the process, providing quantitative insights into mechanisms and their relation to design variables. In addition, new instrumentation offers real-time modulation and measurement of mechanical and process condition. The following section deals with vibrating screen basics and introduces some of the latest tools and techniques in the design, analysis, control, and optimization of vibrating screens.

Principles of Operation

The vibrating screen is used here as a proxy for all other screens, as the basic operating principles are similar or the same. In the screening process, a feed material comprised of particles of varying sizes is presented to the screen media (collectively, the screen deck) in such a way that those particles finer than the screen aperture fall through to the undersize stream, and those that are larger than the screen aperture continue moving along the screen surface to eventually report to the oversize stream.

Screening can be carried out with wet or dry feeds, although coarser particle separation (greater than a 5-mm aperture) is usually performed dry (at the surface moisture of the feed). Wet screening usually involves either sticky/clayey feeds or finer particle sizes where the solid material has been slurried to facilitate transportation and processing. In dry applications, some combination of vibration and gravity are responsible for particle transportation. In wet applications,

---

Brian Flintoff, Senior VP Tech. Dev., Equipment & Systems Business Line, Metso Mining & Construction Technology, Kelowna, British Columbia, Canada
Ronald Kuehl II, General Manager, Vibrating Equipment & Systems Business Line, Metso Mining & Construction Technology, Columbia, South Carolina, USA

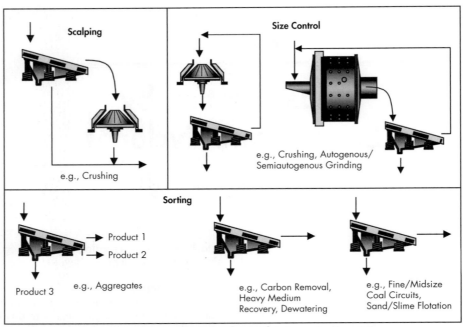

Figure 14.3-1 Examples of the application classes for screening in mining

Table 14.3-1 Types of screening operations

Operation and Description	Type of Screen Commonly Employed
Scalping: Scalping is strictly the removal of a small amount of oversize from a feed consisting predominantly of fines. Scalping typically consists of the removal of oversize from a feed with a maximum of 5% oversize and a minimum of 50% half size. Coarse scalping is typically smaller than 150 mm (6 in.) and larger than 50 mm (2 in.).	Coarse (grizzly), fine (same as fine separation), and ultrafine (same as ultrafine separation). *Example*: Grizzlies are usually used to scalp oversize material from primary crusher feed or are used to scalp fines from the feed to crushers or grinding rolls.
Course separation: Consists of making a size separation smaller than 50 mm (2 in.) and larger than 5 mm	Vibrating screens (banana, incline, and horizontal) and trommel screens. *Examples*: The removal of pebbles from a semiautogenous grinding discharge stream, sizing material for the next crushing stage, and sizing material for leach pile or product pile.
Fine separation: Consists of making a size separation smaller than 5 mm and larger than 0.3 mm (48 mesh)	Vibrating screens (banana, incline, and horizontal), which are typically set up with a high speed and a low amplitude or stroke; sifter screens; static sieves; and centrifugal screens. *Examples*: Sorting coal into ±28 mesh fractions for washing, and iron ore processing
Ultrafine separation: Consists of making a size separation smaller than 0.3 mm (48 mesh)	High-speed, low-amplitude vibrating screens; sifter screens; static sieves; centrifugal screens. *Example*: Size control on grinding circuit product
Dewatering: Consists of the removal of free water from a solid-water mixture and generally limited to 4 mesh and above	Decline vibrating, horizontal vibrating, inclined vibrating (about 10°), and centrifugal screens. *Example*: Wet quarrying operations
Trash removal: Consists of the removal of extraneous foreign matter. This is essentially a form of scalping operation, and the type of screen depends on the size range of the processed material.	Vibrating screens (horizontal or inclined), sifter screens, static sieves, and centrifugal screens. *Example*: The removal of extraneous organic and other matter from the leach feed in a gold plant
Other applications: Other applications include desliming (the removal of extremely fine particles from wet material by passing it over a screening surface), conveying (in some instances, the transport of a material may be as important as the screen operation), dense media recovery, a combination washing and dewatering operation, and concentration.	Vibrating screens (banana, incline, or horizontal), oscillating screens, and centrifugal screens. *Example*: Dense medium recovery in a coal washing plant, a wobbler feeder for conveying, and classifying sticky materials to a crusher

Source: Adapted from Mathews 1985.

Figure 14.3-2 (A) Static grizzly (scalping oversize from a primary jaw crusher feed) and (B) vibrating grizzly screen

Figure 14.3-3 Vibrating screen with a dust collection system

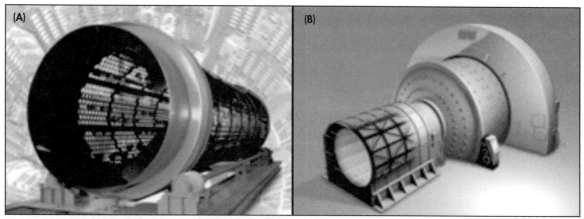

Figure 14.3-4 (A) Trommel screen and (B) mill discharge trommel screen

Courtesy of Derrick Corporation.
Figure 14.3-5 Derrick Stack Sizer

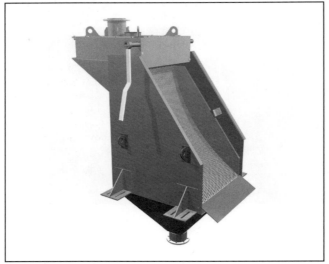

Courtesy of Multotec Process Equipment
Figure 14.3-6 Sieve bend

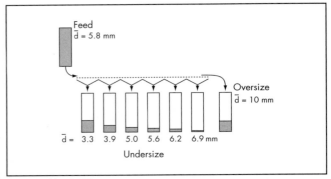

Source: Hilden 2007.
Figure 14.3-7 The relative mass flows and mean particle sizes for a horizontal vibrating screen

some form of vibrator mechanisms dictates the movement of material. However, hydrodynamic (drag) forces are introduced. In wet applications, deck water sprays or flood boxes are sometimes used to aid in classification. Spray selection is chosen based on two criteria: it must be decided whether the goal is to flush material through the deck or to wash off the fines. A general guideline is that water added to the feed is more effective than water sprayed on the deck in improving the classification.

Figure 14.3-7 is a schematic representation of the results of an experiment with a small horizontal vibrating screen having a screen aperture of 10 mm. (This particular screen and test [from Hilden 2007] is referenced throughout the text.) The shaded bars depict the relative mass flows, and the annotations show the mass mean size for each stream. As the feed is introduced onto the deck, it forms a bed that is usually several times the screen aperture in thickness. The bed moves under the influence of gravitation and vibrational motion, and in the process, the particles stratify as the finer materials move quickly through the interstices of the larger particles and find their way to and then through the screen apertures. In this initial zone on the screen, rapid segregation means there is always sufficient fine material in the layer next to the screen deck that the flow through the screen apertures is more or less constant. This is said to be typical of crowded or saturation screening. However, as the fines are depleted, the bed height is reduced, and at some point the particles begin to act more or less as individual entities. This is the zone of separated or statistical screening, that is, each time the particle approaches the deck it has a chance (or a statistical trial) to pass through the aperture. (This notion of repetitive trials underpins many semi-empirical screen models.)

Figure 14.3-7 shows that the greatest flow of material to the undersize stream occurs closest to the feed end, which is typical of crowded screening. Segregation ensures that these particles presented to the apertures are small enough that the probability of passage to the undersize stream is ~1.0. The mass flow to the undersize stream decreases along the length of the deck, which is typical of the transition to separated screening, more typical of the coarser particles (the so-called "nearsize," as it is near to the size of the aperture) that remain in the bed. For these particles, the probability of passage on any one trial is significantly less than 1.0. The increase in the mass mean size along the deck also reflects the fact that the remaining material on the deck becomes coarser as it moves from the feed end to the discharge. This helps to explain why screen length governs efficiency. On the other hand, screen width is related to the amount of material that can be fed to the screen and be effectively separated, and therefore this is the major factor governing capacity.

On this latter point, a guideline for good dry-screen operation is for the bed depth at the discharge point to be three to four screen apertures in thickness. A guideline for wet screening is to design for a bed depth of four to six times the opening. With anything larger, the efficiency of removal of the particles close to the aperture size is reduced, as segregation is more difficult. This suggests that multiple screen decks probably should be considered, using a larger aperture size on the top deck to scalp out much of the larger oversize and reduce the loading on the lower deck for the final classification. Anything less and these particles will probably bounce excessively, reducing the number of trials and the screen efficiency. (It is

Table 14.3-2 Design and operating variables for vibrating screens

Design Variables	Operating Variables
Screen area and open area	Particle size, shape, and distribution
Aperture size and shape	Solids feed rate, distribution, and bed depth
Slope of screen deck	Feed moisture content
Speed	
Magnitude of stroke	
Type of motion	
Feed arrangement	

Source: Bothwell and Mular 2002.

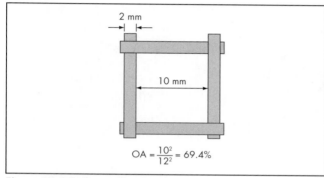

Figure 14.3-8 Computing open area

Figure 14.3-9 Blinding on a screen deck

also sometimes useful to install weirs on the deck's surface to hold back the flow as an aid to sustaining an active bed.)

Screen design and operation is about matching the screen characteristics to the (range of) feed conditions. Table 14.3-2 provides a summary of the important variables for each.

When focusing on design variables, screen area is important because capacity is proportional to width (W) and efficiency is proportional to length (L). It is common for $L \approx 2$ to $3\ W$. Because of the mechanical fittings on and around the screen decks, the effective area is often approximated as 90% to 95% of the actual area (LW). Screens are standard products, so the manufacturers dictate the actual sizes. It is therefore common to require more than one screen, operating in parallel, to effectively separate a feed.

Open area (OA) is expressed as the ratio of the total area of the apertures over the total active area of the screen deck. Figure 14.3-8 illustrates this for the screen in Figure 14.3-7 with a square aperture of 10 mm in a wire mesh screen with a wire diameter of 2 mm. From the explanations of screening mechanism, it is clear that the greater the OA, the greater the capacity of the screen. In practice, OA is limited by the need for sufficient mechanical strength and wear resistance of the deck and is a function of the media materials of construction, for example, wire mesh versus perforated plate.

This is a good point to introduce some common, and occasionally confusing, jargon associated with phenomena that act to reduce the OA and, hence, the capacity. The first is pegging (also called blinding or plugging), which occurs when a nearsize particle becomes firmly lodged in an aperture effectively eliminating it from the screening process. The second is blinding (also called bridging), which is the phenomena whereby smaller particles aggregate in such a way that they combine to block up the aperture, having the same overall effect as pegging, as shown in Figure 14.3-9. Blinding is more of a problem with moist or sticky materials and small aperture sizes.

The aperture size is, by convention, the minimum linear measurement in an opening; for example, the diameter for a round hole and the width (where width ≤ length) for a slot or a square. (Because screen performance assessments start with sieving samples of process streams and because the sieve has square apertures, the best correlations occur with square apertures in the screen media.) What is perhaps more important is the so-called throughfall size, h_T, which accounts for the change in effective opening as the screen is inclined. This is illustrated in Figure 14.3-10, where the angle of inclination is exaggerated to show the difference between the aperture size, h, and the throughfall size for a wire mesh screen surface. (There is a similar effect for rectangular slots.)

Although square apertures may be the most common, there is a multitude of aperture shapes. Some of the more common shapes are illustrated in Figure 14.3-11. Square or round shapes generally allow for better control of the separation size, although circular apertures are most often used in coarse-screening applications. Longitudinal slotted apertures allow slabby particles into the undersize stream. This increases capacity and reduces pegging and blinding. Transverse slots are mainly used in dewatering applications. (For steel or polymeric screen media, it is possible to design tapers and other forms of relief into the apertures to minimize pegging—see the insert in Figure 14.3-11 for an example.) Finally, the screen deck or panel is designed to withstand the load and maximize wear life. Therefore, in the case of a large spread in feed particle sizes and/or some large particles in the feed, it is common to use multiple screen decks to satisfy both mechanical and process requirements.

Inclining a screen deck increases the slope and causes the feed material to move more quickly on the deck, reducing the residence time and increasing capacity and sometimes efficiency. Vibration motion on a horizontal screen deck usually means that material velocities of 10 to 16 m/min (33 to 52 fpm), with an inclination in the neighborhood of 20°, will increase this into the range of 20 to 28 m/min (66 to 92 fpm). Generally, inclinations range from horizontal to as high as 30°, although ~20° is the usual limit. This led to the evolution of the multislope screen (also commonly known as a banana

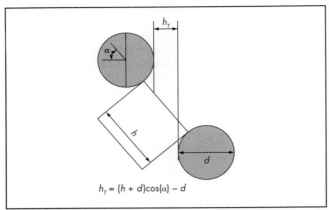

Figure 14.3-10 The relation between the throughfall size (h_T) and the aperture size (h)

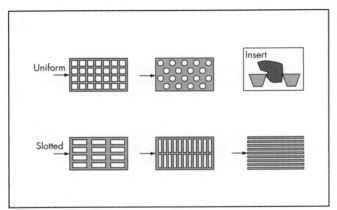

Figure 14.3-11 Common screen aperture shapes

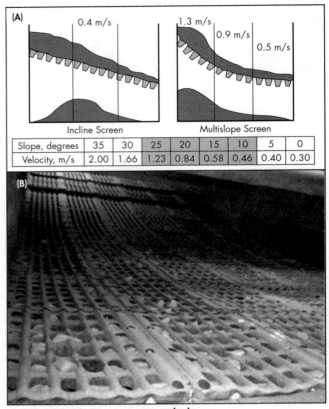

Figure 14.3-12 Banana screen deck

Table 14.3-3 G-force implications in inclined vibrating screens

G-Force	Comment
<1.5	Practically no movement on the deck, so it is of no use.
1.5–2.5	The stratification is limited and the relative velocity between the particle and the deck is low. These parameters are suitable for soft screening and where size degradation is problematic and there is a low tolerance to impact.
3.0–3.5	The particle is presented to the screen deck at its maximum relative velocity and offers the best screening condition for inclined screens. The stratification is adequate.
4.0–5.0	The stratification is good, but the relative velocity between the deck and particle is low, offering poor screening efficiency. Bearing life is shortened.
5–6	The stratification is good, but the particle projection is in excess of the pitch (particles are in flight too long) and screening opportunities are lost. The relative velocity between the screen and particle is low, combining to give poor parameters for efficient inclined screening. The high acceleration forces can act negatively on the screen structure, and bearing life is further shortened.

Source: Moon 2003.

screen). The banana screen (Figure 14.3-12) has a relatively high slope in the initial screening zone, which takes advantage of the fast kinetics of fine particle removal. The slope then decreases, which slows particle movement and offers an additional G-force to accelerate removal of the larger fine material. In the final section, the slope is relatively flat, yielding minimal velocity and maximum residence time, for finer nearsize particle removal to the undersize stream.

Vibrating screen deck motion is circular, elliptical, or straight line, with the vibrating mechanism rotating in the direction of flow or counter to the flow. Strokes are usually in the range of 3 to 20 mm (0.118 to 0.787 in.) and speed ranges from 650 to 950 rpm. A key parameter here is the G-force (G), which is computed by Equation 14.3-1.

$$G = \frac{N^2 S}{1,789,129} \qquad (14.3\text{-}1)$$

where
 G = the G-force (also known as the throw number)
 N = speed (rpm)
 S = stroke (mm)

Typical G-forces are in the 3.5-to-5.0 range, tending to higher values with heavy loads, sticky materials, or where pegging and blinding are issues. The G-force is usually higher in a horizontal screen, as it must provide the means for both stratification and material movement. (For example, with the horizontal test screen in Figure 14.3-7, the speed was 981 rpm and the stroke 5 mm [0.197 in.], giving a G of 2.69.) Table 14.3-3 amplifies on the conditions prevalent over a broad range of G-forces.

Figure 14.3-13 is a simplified graphic of particle motion with different trajectories, and Figure 14.3-14 shows how deck motion can be induced using weights on the shaft. Of

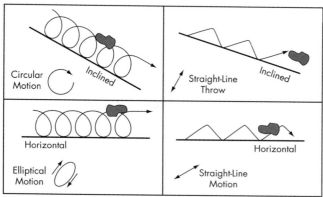

Figure 14.3-13 Particle motion as a function of the deck orbit

Table 14.3-4 Process matrix for a vibrating screen*

Design Variable	Screen Capacity	Screen Efficiency	Undersize Mean Size	Pegging/ Blinding
Screen area	+	0	0	0
Open area	+	+ or 0	0	0
Aperture size	+	+	+	–
Aperture shape (increasing nonuniformity)	+	+ or 0	+	–
Deck inclination	+	– or 0	–	– or 0
Speed	+	+	0	–
Stroke angle	–	+	0	– or 0
Stroke	+	+	0	–
Orbit (from linear toward circular)	–	+	0	+

*All other factors being equal, an increase in the design parameter will cause an increase (+), a decrease (–), or no change (0) in the performance metric.

The designer for the screening application chooses the orbit motion (circular, elliptical, or linear) based on a compromise between material movement (capacity), which emphasizes linear motion, and efficiency, which emphasizes circular motion. The ellipse is an optimized solution. All give good stratification with the correct stroke angle. For horizontal and banana screens, linear motion is the defacto standard as material movement is critical. For inclined screens, circular or elliptical orbits are more common.

Finally, there is the matter of feeding the screen. Here, there are the following three general rules:

1. It is important that the feed be uniformly distributed across the width of the screen for the unit to achieve rated capacity.
2. It is important that feed rate be modulated where possible (e.g., using belt feeders) to ensure that the screen works efficiently in the face of changing feed conditions.
3. Where possible, it is good to introduce the feed having a motion opposite to the eventual material flow, which helps in distribution and avoids flooding in the initial zone of the screen deck.

Table 14.3-4 provides a review of the major design variables of a vibrating screen and indicates usual design ranges where appropriate. This "one-variable-at-a-time" approach does not always take into account the myriad interactions among these variables.

Screen Selection

Many opinions and methods of screen selection suggest fertile ground for development and refinement. Three general selection method classes follow:

- *Class 1:* Empirical methods based on correlations derived from extensive databases of industrial experimental results
- *Class 2:* Lab screening tests with empirical model-based scale-up procedures
- *Class 3:* Fundamental model-based methods (these are relatively new and are described later in more detail)

Class 2 methods have been used sparingly in the past, but recent work (Hilden 2007) suggests there is good scope for improvement and further suggests there is room to combine Class 2 and Class 3 methods very effectively. Despite the

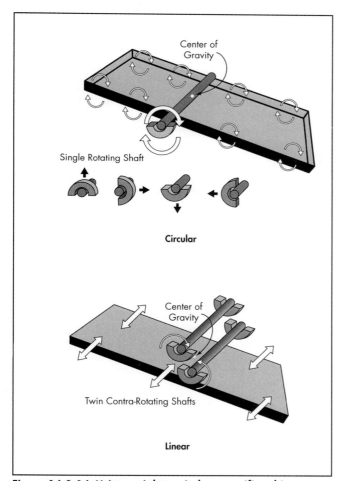

Figure 14.3-14 Using weights to induce specific orbits

course, there are other methods to induce motion and other, more complex orbits.

The stroke angle, measured relative to the length of the deck and in the direction of flow, can affect efficiency and capacity. Very low (~0°) or high (~180°) angles do little other than accelerate wear. For other angles, there is a component of vertical movement (aiding stratification) and a horizontal component aiding (<90°) or retarding (>90°) material flow. (For the screen in Figure 14.3-7, the stroke angle was 50°.)

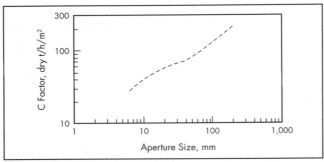

Source: Allis Chalmers, n.d.

Figure 14.3-15 Basic capacity factor (C) as a function of aperture size

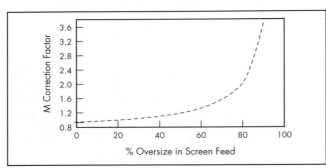

Source: Allis Chalmers, n.d.

Figure 14.3-16 Correction factor (M) as a function of percent oversize

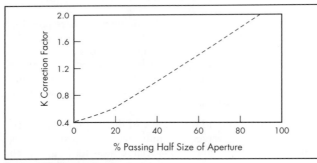

Source: Allis Chalmers, n.d.

Figure 14.3-17 Correction factor (K) as a function of percent half size

potential of using Class 2 and Class 3 methods, most industrial vibrating screens are still selected using Class 1 methods. Because in most cases these screens operate more or less as designed, it is understandable why these methods have survived with a few improvements for 70 years.

Class 1 selection methods employ empirical correlations to deduce screen area. The design process can begin once the application (e.g., a pebble screen on a semiautogenous grinding [SAG] mill discharge), the design feed rate of solids, and the typical feed particle size distribution are given. Early decisions on the number of decks, inclination, motion, lo-head or hi-head screens, and so forth, come from the design basis documents and past practices in similar applications. Once these parameters are set, the correlations can be used to compute screen area requirements, and the deck with the largest area determines the screen size. The task is then to match one of the standard products to the design. This often is an iterative design process.

Two usual approaches can be used to solve the Class 1 design problem—one based on the total solids feed rate to the screen and the other based on the solids feed rate of undersize material to the screen. Not surprisingly, this is also an area of debate, as the scientific community tends to prefer the latter method, yet the former is very commonly used. The former method will be described in an abbreviated way simply for illustrative purposes. To provide a real example, the data for the screen shown in Figure 14.3-7 is used.

The fundamental equation for determining the screen area required is given in the Equation 14.3-2 (Allis Chalmers, n.d.)

$$A = \frac{T}{CMKQ_1 Q_2 Q_3 Q_4 Q_5 Q_6} \qquad (14.3\text{-}2)$$

where

A = required screen area
T = mass flow of feed to screen
C = basic capacity factor
M = oversize factor (percent larger than aperture size in feed)
K = undersize factor (percent smaller that ½ aperture size in feed)
Q_1 = bulk density factor (usually ~60% of rock density)
Q_2 = aperture shape factor
Q_3 = particle shape factor
Q_4 = OA correction factor
Q_5 = wet or dry screening factor
Q_6 = surface moisture factor

Figures 14.3-15 through 14.3-17 provide the basis for estimating the factors C, M, and K, respectively. Table 14.3-5 provides the means for estimating the Q factors. (OA and aperture shape come from Figure 14.3-8.) Table 14.3-6 presents the summary data for the calculations, showing both the input data required and the empirical corrections, as deduced from the previously mentioned graphs and tables (which are adapted from Allis Chalmers, n.d.). The required area allows for a 6% loss due to mechanical fittings and so forth. Following the rough rule that $L = 3W$, this screen would have the approximate dimensions of 0.34×1.03 m (1.12×3.38 ft). In fact, the screen in Figure 14.3-7 is 0.34×1.67 m (1.03×5.48 ft). The width is the same, but the test screen has extra length, so a higher classification efficiency might be expected. This topic is discussed in the following subsection.

Performance Assessment

Mechanical and process performance can be distinguished, but this subsection only focuses on the latter.

A single parameter can be calculated in several ways that will quantify screen performance. Among these, the most common are E_u, which is defined as the efficiency of removal of undersize material from the oversize stream, and R_u, which is defined as the efficiency of undersize removal from the feed stream. Referring to Figure 14.3-18, Equations 14.3-3 and 14.3-4 are used to compute numerical values:

$$E_u = \frac{F(1-\hat{f}_u)}{O} = (1 - \hat{o}_u) \qquad (14.3\text{-}3)$$

Table 14.3-5 Q correction factors

Bulk Density Factor	Value
Bulk density, dry t/m³	Q_1
0.4	0.25
0.8	0.5
1.6	1
2.08	1.3
Aperture Shape Factor	
Opening shape	Q_2
Square	1
Rounded	0.8
Slotted	
2:1 Slot	1.15
3:1 Slot	1.2
4:1 Slot	1.25
Particle Shape Factor	
Particle shape	Q_3
Cubical	1
Slabby	0.9
Open Area Factor	
Open area	Q_4
Calculation	Q_4 = percent open area ÷ 50%
Wet or Dry Screening Factor	
Opening	Q_5
Dry	1
Wet	
0.8 to 3.2 mm	1.25
4.8 to 6.4 mm	1.4
8 to 12.7 mm	1.2
14.3 to 25.4 mm	1.1
Surface Moisture Factor	
Surface moisture	Q_6
≤3%	1
3% to 6%	0.85
6% to 9%	0.7
Wet screen	1

Table 14.3-6 Screen sizing example

Parameter	Value
Feed rate, dry t/h	15.7
Aperture (square), mm	10
Rock density, dry t/m²	2.7
Application	Dry
C Factor, dry t/h/m²	40.12
M Factor	0.972
K Factor	0.88
Q Factors	
Q_1	1.00
Q_2	1.00
Q_3	1.00
Q_4	1.39
Q_5	1.00
Q_6	1.00
Adjusted unit capacity, dry m/t/h/m²	47.64
Area required, m²	0.34

Size, mm	Feed Distribution, %
13.2	5.6
9.5	16.6
8	11.4
6.7	11.1
4.75	24.1
3.35	10.7
2.8	6.5
Pan	14.0
Percent +10 mm	18.7
Percent –5 mm	34.0

Source: Data from Hilden 2007.

$$R_u = \frac{U}{F\hat{f}_u} = \frac{(\hat{f}_u - \hat{o}_u)}{\hat{f}_u(1 - \hat{o}_u)} \qquad (14.3\text{-}4)$$

where

F = feed rate
O = oversize mass flow rate
U = undersize mass flow rate
$\hat{f}_u$ = cumulative mass fraction in the feed passing the aperture size
$\hat{o}_u$ = cumulative mass fraction in the oversize stream

To utilize these performance measures, one is required to perform a sampling experiment around the screen while it is running at steady-state conditions. This raises topics beyond the scope of this chapter (e.g., experimental design, sampling theory, sample preparation and analysis, and data massage). Fortunately, for the screen shown in Figure 14.3-7, the balanced data are already available and are summarized in Table 14.3-7.

Using these results, $E_u = 77.9\%$ and $R_u = 93.5\%$. The data suggest that there is a considerable amount of finer (less

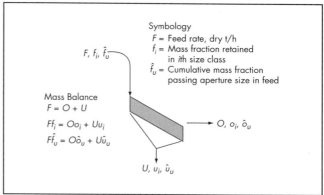

Figure 14.3-18 Screen performance parameters

than the aperture size of 10 mm [0.4 in.]) nearsize material in the oversize stream. In terms of assessing performance, the manufacturers have a curve that relates E_u to the rated capacity of the screen. This is shown in Figure 14.3-19, and the rated capacity is the ratio of the actual tonnage (or area) to the calculated tonnage (or area) from Equation 14.3-2. The respective area figures for the example are 0.35 and 0.57 m²,

Table 14.3-7 Sampled data summary for the horizontal screen in Figure 14.3-7

Size, mm	Feed, %	Oversize, %	Undersize, %
13.2	5.6	20.8	0.0
9.5	16.6	61.8	0.0
8	11.4	13.9	10.5
6.7	11.1	3.5	13.9
4.75	24.1	0.0	33.0
3.35	10.7	0.0	14.6
2.8	6.5	0.0	8.9
Pan	14.0	0.0	19.1
Mass fraction	1.00	0.27	0.73
Percent –10 mm	81.3	22.1	100.0
Mass flow, dry t/h	15.7	4.2	11.5

Source: Data from Hilden 2007.

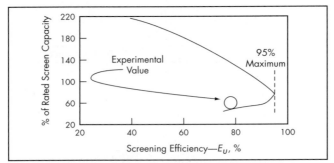

Figure 14.3-19 Percent of rated capacity versus E_u

which give a rated capacity of 59.9%. This point is plotted on the curve for reference.

This comparison indicates that the test screen in Figure 14.3-7 is not performing as well as would be expected at the given operating conditions, and further analysis of the situation would be required to validate the result and improve performance.

Another very common measure of efficiency is to use the data in Table 14.3-7 to compute an experimental partition curve, which can then be modeled with any number of simple functional forms. The model parameters are efficiency measures. Again, looking to the example in Figure 14.3-7, the experimental partition factors for the various size fractions are computed from Equation 14.3-5 and are defined as the mass fraction of material in a certain size fraction in the feed that reports to the oversize stream:

$$p_i = \frac{Oo_i}{Ff_i} \qquad (14.3\text{-}5)$$

where

p_i = observed partition factor for size class i
f_i = mass fraction retained in the ith class of the feed
o_i = mass fraction retained in the ith class of the oversize stream

A very common model and one that will be further discussed later is the so-called Rosin–Rammler curve, and

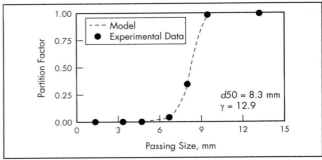

Source: Data from Hilden 2007.

Figure 14.3-20 Experimental and modeled partition function data

the estimated partition factors are computed according to Equation 14.3-6:

$$p_i = 1 - e^{-0.693\left(\frac{d_i}{d50}\right)^\gamma} \qquad (14.3\text{-}6)$$

where

p_i = calculated partition factor for size class i
d_i = the characteristic size for size class i
$d50$ = cut size—particles of this size have an equal chance (50/50) of going to the undersize or oversize
γ = sharpness of separation (larger values mean better separation)

The parameters ($d50$ and γ) are obtained by minimizing a least-squares objective function using nonlinear regression (e.g., Solver in Microsoft Excel). That is, the parameters are selected in such a way that the sum of the squares of the differences in the observed and calculated partition factors is minimized. Figure 14.3-20 presents the results for the screen in Figure 14.3-7.

One would normally expect the cut size to be close to the throughfall aperture size, and it is clear that the value is a little low here (in part because of the way the characteristic size was defined for this study). The sharpness of separation for screens cutting in the aperture size range is usually around 6, so a value of close to 13 suggests that efficiency is high. Taken together, the implication is that the screen is not processing the finer nearsize material very well, which would require further investigation.

Interestingly, the data acquired for the screen in Figure 14.3-7 permit the calculation of the cumulative partition curve as a function of the position along the length of the deck. The experimental results are shown in Figure 14.3-21. Given the explanation related to crowded and separated screening given previously, it is no surprise to see that the fine material moves quickly through the screen—collectively, particles do not have to progress far down the deck before all of the very fine material has passed through the apertures (and the corresponding $p_i = 0$). For the fine, nearsize material, it is clear that the process is a lot slower.

The periodic measurement of screen efficiency is as important to the maintenance of process performance as the periodic measurement of, for instance, bearing condition is to the maintenance of mechanical performance. Unfortunately, it is often the case that the process aspects of screens are

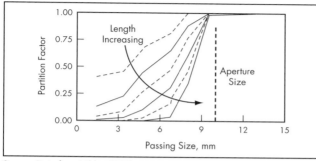

Source: Data from Hilden 2007.

Figure 14.3-21 Partition factors as function of position on deck

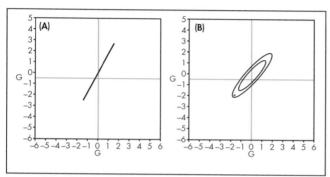

Figure 14.3-22 Feed end orbits in (A) normal and (B) abnormal operation

ignored. However, they are robust devices, and they often have a direct impact on the performance of comminution and separation equipment, for which a systems thinking approach must be applied.

Recent Developments
This subsection provides a brief introduction into some of the developments emerging in the vibrating screen market.

Real-time condition monitoring. Using vibration, thermographic, and occasionally other specialty sensors to automate the more usual manual inspections of predictive maintenance is becoming increasingly commonplace. Where it is done, this is still usually reserved for the typical applications of monitoring bearing health. Manufacturers such as Metso (Screen Security Package) have taken this a step further by using additional accelerometers to compute screen orbits, which can then be compared to expected values from run-in tests and/or to finite element modeling to analyze the mechanical and process health of the screen. Figure 14.3-22 is an example of normal orbits and the kinds of trends that develop when there is a problem, such as with the supporting springs. Loose media on the screen deck and other kinds of serious problems have their own orbit "signature."

In the case of the normal orbits (Figure 14.3-22A), both are linear and superimposed, showing the same stroke and angle. In the case of the abnormal conditions (Figure 14.3-22B), the orbits are now elliptical and the strokes are different.

The frequency spectrum of the accelerometers also provides useful information on the mechanical state of the deck, loading, and other interesting performance parameters. Moreover, when these signals are combined with others (e.g.,

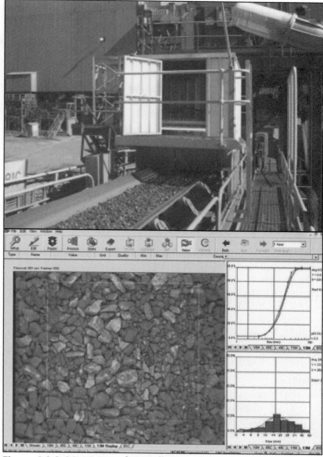

Figure 14.3-23 VisioRock installation and typical output display

from load cells and vision systems) in a kind of sensor fusion approach, the analytical detail and reliability can be dramatically increased.

Real-time deck motion modulation. As seen in the process matrix of Table 14.3-4, deck motion (stroke, angle, and orbit) all affect performance, and in light of the fact that screen feeds often change (mass flow and particle size distribution), there is a school of thought that says the ability to modulate these parameters on line would support real-time screen performance management. A number of manufacturers are implementing this kind of solution on selected screen models.

Instrumentation. One of the exciting areas of potential for performance monitoring is "machine vision," primarily as applied to measuring feed or product size. Typically, dry screening is very dusty and requires hoods covering the entire screening surface for effective dust control. Therefore, applications on the deck itself (oversize velocity, load, etc.) are challenging. However, in the case where the product streams are discharged onto belts, the standard technology, such as Metso's VisioRock, can be applied. For example, in the production of iron ore, lump size control is very important, and the early warning of a problem is essential if, for example, the screen media is broken, thereby producing off-spec material in the undersize stream. One operator is now using machine vision to detect and react to these types of quality-assurance events, and Figure 14.3-23 shows the VisioRock installation as well as the output from data processing. In this case, images are analyzed at a rate of approximately 10 per second, which is more than sufficient to monitor everything that passes.

Acoustics also offer interesting potential for analyzing screen performance, but it is very early for research in this area.

Population balance simulation methods. Population balance methods (PBMs) are not new, having been in use for at least two decades, but as the design and operations needs turn increasingly toward simulation to accelerate the activities, higher and higher fidelity is required. These are an element of the Class 2 methods previously mentioned in the "Screen Selection" subsection. Although the industry is moving in this direction, older approaches are still in use in most of the simulation packages. One of these is the model by Karra (1979), which is based on a Class 1 selection method, although in this case it is based on the method of sizing relating to the mass flow rate of undersize in the feed stream.

Using the model in Equation 14.3-6, Karra developed a correlation based on the empirical screen sizing factors used in the screen selection approach. The factors have been modified over time, and their estimation is described elsewhere (e.g., King 2001), but the equation for computing $d50$ is

$$d50 = \frac{G_c h_T}{\left(\frac{(T_u/H)}{ABCDEF}\right)^{0.148}} \quad (14.3\text{-}7)$$

where

$d50$ = cut size
G_c = nearsize correction factor
h_T = throughfall aperture (King 2001)
T_u = tons of undersize in the feed (dry t/h)
H = effective screen area (m²)
A = basic capacity factor
B = oversize factor
C = fine size factor
D = deck location factor
E = wet screen factor
F = bulk density factor

Because a number of the properties for the screen in Figure 14.3-7 have been given, Table 14.3-8 is a shortened summary of the Karra calculation of $d50$. The agreement with the experimental value in Figure 14.3-20 is good.

Karra assumes γ is a constant at 5.9, and in the authors' experience, numbers in the region are pretty typical for coarse screening. This number does not agree very well with the value in Figure 14.3-20, which highlights that Karra's simple two-parameter model approach probably requires additional parameters.

Nevertheless, using the model of Equation 14.3-7 and the feed data in Table 14.3-7, Karra's model can be used to predict the expected size distributions for the undersize and oversize products, and these are shown in Figure 14.3-24 along with the actual experimental results. The predictions are satisfactory, but the importance of γ is clear. The calculated fractional mass splits of 0.26 and 0.74 compare well with the product mass flows reported in Table 14.3-7.

Discrete element simulation methods. In the pursuit of higher-fidelity simulation of screen performance, the ultimate tool is probably discrete element modeling (DEM). Very briefly, this approach is based on fundamental models

Table 14.3-8 Computing Karra's $d50$ for the screen in Figure 14.3-7

Symbol	Factor	Units	Result
T_u	Feed	dry t/h	15.67
H	Effective screen area	m²	0.57
h_T	Throughfall aperture	mm	10
G_c	Nearsize correction factor	—	0.82
A	Basic capacity factor	t/h/m²	17.85
B	Oversize factor	t/h/m²	1.38
C	Fine size factor	t/h/m²	1.12
D	Deck location factor	t/h/m²	1.00
E	Wet screen factor	t/h/m²	1.00
F	Bulk density factor	t/h/m²	1.01
$d50$	Cut size	mm	8.34

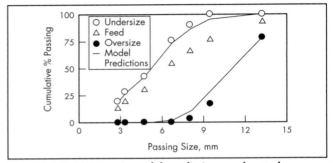

Figure 14.3-24 Karra's model predictions and actual cumulative size distribution curves for the screen in Figure 14.3-7

drawn from physics (e.g., Newton's laws of motion.) Using a mathematical description of the body and motion of the screen; some typical physical properties; and the feed size distribution, flow rate, and particle shapes, the DEM simulation can be run from only first principles (e.g., no correlations and no parameters drawn for experimental data). Although this sounds simple in concept, solving the problem is onerous, as there are millions of particles, each requiring several differential equations to describe translational and rotational motion in time and space. Because of the interactions of the particles among themselves and with the screen itself, the integration intervals are very short—on the order of one millionth of a second. Simulating several seconds of real-time operation can take days, weeks, or months, depending on the problem size, the computational capability, and the simulation code efficiency. It is for this reason that the goal is to combine DEM and PBMs, the former to work out the real mechanisms and the latter to abstract these and provide speedy calculations.

Because of the vast quantities (gigabytes or even terabytes) of numerical data generated in such simulations, it is common practice to use visualization techniques to "see the results." However, the numerical data must be used for any calculations. Figure 14.3-25 presents a snapshot of visualization, and Figure 14.3-26 shows the predicted (lines) and measured (bullets) particle size distributions from a simulation of the banana screen pictured in Figure 14.3-12B. The agreement in Figure 14.3-26 is very good, which provides some insight into the power this tool will bring to screen modeling.

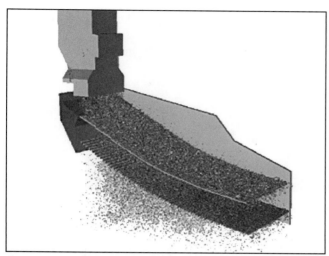

Figure 14.3-25 DEM visualization of screen operation

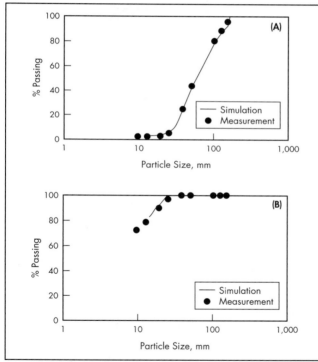

Figure 14.3-26 DEM predictions of (A) oversize and (B) undersize particle size distributions

Screen Media

Because screen media is often the single largest operational expense for a vibrating screen, this chapter would be incomplete without a brief discussion on this topic. The governing objective here is the absolute lowest total cost, which implies that availability, cost, wear life, downtime, and so forth must be considered in making the final media selection.

Three major materials of construction for screen media are steel (punched plate or wire mesh), rubber, and polyurethane. The polymeric compounds of rubber and polyurethane are usually more expensive, but they tend to exhibit superior wear performance in those applications that permit alternatives. Service life can be extended by factors of more than five with polymerics.

Figures 14.3-27 through 14.3-29 show wire mesh, rubber, and polyurethane media, respectively. Media specialists often break down the applications as scalping, standard production (size control and sorting), and washing (sorting).

Wire mesh can have standard weaves, as shown in Figure 14.3-27A, or slotted antiblinding or even custom weaves, as shown in Figure 14.3-27B. The wires are normal high-carbon or stainless steels.

Typical applications for rubber screen decks (Figure 14.3-28) include high-impact scalping type duty or situations where abrasion and/or blinding are a serious concern. The decks are manufactured by punching or molding.

Polyurethane decks (Figure 14.3-29) are usually found in standard production applications and in washing or dewatering applications. They are manufactured by casting or injection molding processes.

The polymeric materials lend themselves to modular installation or paneling. Table 14.3-9 weighs the pros and cons of steel versus the polymerics.

As a cautionary note, although there might be a tendency to choose thicker panels to increase wear life, this will reduce capacity and accuracy and will induce higher levels of pegging and blinding. Media suppliers familiar with best practices can be very helpful in these selections, and it is also recommended that the original equipment manufacturer (OEM) is consulted when changing the design of media.

Standard Screen Sizing and Relative Pricing

Table 14.3-10 provides a glimpse of a typical vibrating screen product line, and while the price data supplied here is somewhat out of date, the relative pricing of the screen equipment is more or less the same.

Fine Screening

In general, wet fine particle classification is achieved by mechanical/hydraulic means (e.g., rake or spiral/screw

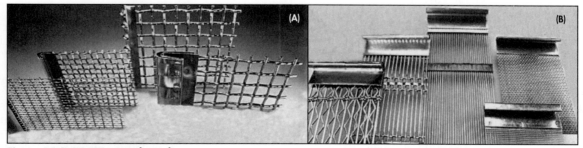

Figure 14.3-27 Wire mesh media

Figure 14.3-28 Rubber screen decks

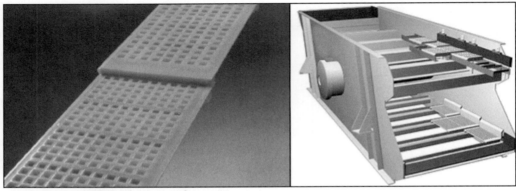

Figure 14.3-29 Polyurethane panels

classifiers), hydrocyclones (discussed later in this chapter), and fine screens. Although the mechanical devices still have their place, fine screens, and especially hydrocyclones, dominate fine and ultrafine classification. Fine screening is the term generally reserved for fine and ultrafine separations, as defined in Table 14.3-1. It is performed dry or wet, although in most mining or quarrying applications, this is a wet process. Consequently, the emphasis here is on wet fine screening.

The principles described previously for vibrating screens apply to most screening processes. One notable difference is that fine screening generally involves much higher rotational speeds (e.g., ~3,600 rpm). Therefore, much smaller strokes are required. Another major difference for fine screening is that the selection of the appropriate screen is as much an art (i.e., past practices in similar applications) as it is a science (i.e., laboratory testing of specific screens on the expected feed material and supplier databases). This complexity is due in large measure to the critically important aspect of fluid mechanics, as drag forces tend to govern wet fine and ultrafine separations. Suffice it to say, the expertise in fine-screen design generally lies with the major supplier organizations, such as Derrick Corporation, Conn-Weld Industries, Kroosh Technologies, Pansep, Multotec Group, and Wedgewire.

The two most common wet fine screening devices are the sieve bend (i.e., a stationary version of the banana screen, Figure 14.3-6) and the more conventional, high-frequency fine screen (i.e., having a more conventional geometry and represented herein by the Derrick product line, including the Stack Sizer [Figure 14.3-5], the multifeed screen, and the repulp screen). Wet fine screening is essentially the only flow-sheet

Table 14.3-9 Advantages and disadvantages of steel versus polymeric screen decks

Wire	Rubber/Polyurethane
Advantages • Price • Accuracy of cut • Open area is high (on installation) Disadvantages • Shorter life • More downtime • Pegging problems • Blinding problems • Reduces open area (in operation) • Noise levels	Advantages • Longer life • Less downtime • Shock resistance • Accuracy • Reduced pegging • Reduced blinding • Open area (in operation) • Noise reduction Disadvantages • Price • Open area (on installation) (e.g., although these screens tend to have lower open areas out of the manufacturing process than wire mesh, they retain open area much better)

option in a number of applications (see sorting class in Figure 14.3-1), for example, in coal preparation desliming and iron ore coarse silica removal. (A more complete review on this topic is available in Valine and Wennen (2002).

Many other applications provide wet screening as an alternative to hydrocyclones or other mechanical devices, as it offers some significant advantages in classification efficiency (as measured by γ or E_u). In some applications, both sieve bends and high-frequency screens are used in conjunction with hydrocyclones to exploit the physical and classification

Table 14.3-10 Standard screen sizing and relative pricing

Screen Size, W × L, m	Linear Motion, Low-Head, Horizontal, Multiflow Banana Screens*		Inclined Circular Motion Ripl-Flo Screens†	
	Single Deck, $	Double Deck, $	Single Deck, $	Double Deck, $
1.2 × 4.8	70,200	95,160	58,500	93,600
1.8 × 4.8	76,440	138,060	74,880	109,200
1.8 × 6.1	136,500	175,500	117,000	143,520
2.4 × 4.8	113,100	152,100	105,300	127,140
2.4 × 6.1	145,860	269,100	124,800	156,000
2.4 × 7.3	175,500	388,600	179,400	218,400
3.0 × 4.8	117,000	232,440	N/A‡	N/A
3.0 × 6.1	210,600	310,440	155,220	226,200
3.0 × 7.3	237,900	343,200	202,800	249,600
3.0 × 8.5	257,400	374,400	N/A	N/A
3.6 × 6.1	252,720	368,940	N/A	N/A
3.6 × 7.3	280,800	382,200	N/A	N/A
3.6 × 8.5	308,880	444,600	N/A	N/A
4.2 × 6.1	265,200	443,040	N/A	N/A
4.2 × 7.3	304,200	530,400	N/A	N/A

*Low-head and multiflow pricing includes motor and modular polyurethane screen surfaces.
†Ripl-Flo screens include motor or motors and woven wire screen surfaces.
‡N/A = not available.

attributes for each type of separator. In other instances, it may also be possible to feed a wet screen by gravity, rather than the pumping systems more commonly associated with hydrocyclone batteries. Nevertheless, in these kinds of trade-off studies, it generally comes down to the footprint advantages of hydrocyclones versus the process efficiency advantages of wet screening. For example, one study (Vince 2007) in coal indicated that purely from a capacity perspective, two to three Derrick Stack Sizers were the equivalent of one 900-mm (35.4-in.) classifying hydrocyclone. Interestingly, that same study came to the conclusion that the Stack Sizer should be considered for coal-preparation applications with cut points in the 100-to-350-µm range, while sieve bends could be considered for cut sizes in the 250-to-350-µm range. Of even greater interest was the observation that "similar" equipment made by different suppliers can have quite different performance characteristics—*caveat emptor* (buyer beware).

One area of growing interest for fine screening is close fine-grinding (ball mill or even stirred-mill) circuits (e.g., see Wennen et al. 1997). The opportunities to exploit improved efficiency to reduce specific energy and media requirements and to improve downstream metallurgical efficiency (e.g., by minimizing the overgrinding of valuable minerals) are very appealing, both technically and economically. In addition, while hydrocyclones separate based on particle mass (i.e., particle size and density are first-order effects), wet fine screens separate predominantly on size (i.e., density is a higher order effect, mainly related to particle stratification). For example, in iron-ore grinding, the cut size on a screen is the same for the iron minerals and silica, whereas in a hydrocyclone, the cut size for the iron minerals is about one-third of that for silica,

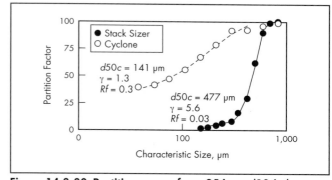

Figure 14.3-30 Partition curves for a 254-mm (10-in.) hydrocyclone and a Derrick Stack Sizer

which is to say that a lot of fine liberated iron mineral can be sent back to the grinding mill by a hydrocyclone.

Consider the partition curves shown in Figure 14.3-30. (The parameter Rf is defined later in Figure 14.3-49 and in Equation 14.3-17, but it effectively represents the very fine material hydraulically entrained in the oversize stream—the bypass.) Although there are some "apples to oranges" comparisons in terms of the cut sizes, these example are drawn from the same plant in similar applications (Aquino and Vizcarra 2007); the real interest comes from the comparison of the sharpness of separation (γ) and the efficiency (here inversely related to Rf). The benefits of wet screening are obvious, which makes it a technical option to cut sizes ranging down to 100 µm. For the hydrocyclone partition curve, the inflections

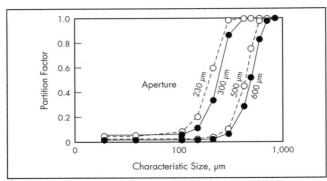

Figure 14.3-31 Partition curves for a Derrick Stack Sizer with different screen apertures

and rather poor sharpness of separation are a partial consequence of the existence of liberated heavy and light minerals in the feed (i.e., the density effect).

Finally, Figure 14.3-31 provides a glimpse of the ability of this fine screen to preserve efficiency over a range of aperture sizes (i.e., the sharpness in separation is good and the bypass is low).

CYCLONES

Cyclones (named for the spiraling motion within the body of the device) are one of a large number of classification devices aimed at separating fine particles in a fluid (water or air). For examples, see Kelly and Spottiswood (1982) or Svarovsky (2000). Cyclone classification is generally reserved for the separation of particles on the basis of size. For this reason, other applications for cyclones, such as thickening and dense-medium separation, are not considered here. Broadly speaking, cyclone classifiers can either treat dry feeds (the air classifier or air separator) or wet/slurry feeds (the hydrocyclone). The emphasis here will be on wet applications.

Unlike the vibrating screen, which employs a physical means (the screen media) to effect a separation, the cyclone classifier uses differential settling properties of particles of different mass. The cyclonic action within the body of the device develops G-forces significantly greater than gravity, which accelerate the separation process. In general, cyclones are robust, compact (small footprint), and relatively efficient size-separation devices.

Figure 14.3-32 shows the most common applications of wet cyclones in mining. Air classifiers are also commonly used in these applications, for example, as a size-control device in dry grinding (most notably in refractory gold, coal-fired power plant pulverizers, or cement clinker applications) and as sorters (usually in de-dusting applications).

Figure 14.3-33 shows the geometry of a standard hydrocyclone, the shape of which clearly resembles the natural phenomenon that gives rise to the name. The separator comprises a feed section, cylindro-conical section, discharge spigot, vortex finder, and overflow. In some cases, usually with larger diameters, each of these is supplied as a separate entity, whereas in others they may be cast into a single unit (i.e., with smaller diameters).

The capacity of a hydrocyclone is a function of its size and, subject to operational constraints, the rate and solids concentration at which the feed (slurry) material is supplied. In some cases, hydrocyclones are gravity fed from surge tanks,

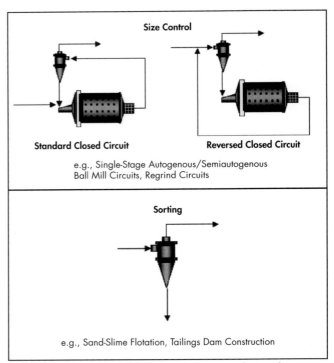

Figure 14.3-32 Examples of the application classes for hydrocyclones in mining

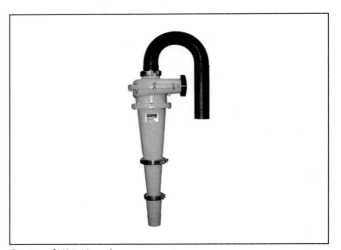

Courtesy of Weir Minerals.
Figure 14.3-33 Hydrocyclone

but the majority of applications use a feed sump and pump arrangement. Consequently, it is common to utilize a group of hydrocyclones operating in parallel to achieve the required process throughput capacity. As Figure 14.3-34 shows, the compact design of the units permits the use of a radial header to distribute feed to a number of hydrocyclones. This is commonly called a hydrocyclone cluster or battery, or a Cyclopac. Moreover, as Figure 14.3-34B highlights, the use of knife-gate valves to open or close a hydrocyclone is very convenient, both from an online process control point of view as well as for maintenance purposes. On the latter point, and given the relative cost of a hydrocyclone, it is typical to remove a

(A) Courtesy of Weir Minerals; (B) Courtesy of FLSmidth Krebs.
Figure 14.3-34 Hydrocyclone clusters: (A) side view and (B) top view

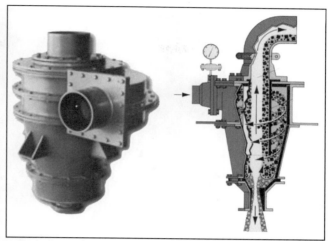

Courtesy of FLSmidth Krebs.
Figure 14.3-35 Water-only cyclone

hydrocyclone for maintenance and simply replace it with a refurbished spare.

There are a number of variants on the more typical hydrocyclone geometry of Figure 14.3-33. Figure 14.3-35 shows the water-only or automedium cyclone that is frequently used in washing fine (–0.6 mm) coal. (The truncated lower cone section is employed to set up a secondary density separation in a rotating particle bed.) The flat-bottomed hydrocyclone occasionally used in mineral processing applications has a very similar kind of body geometry.

Air separators tend to be designed to treat the whole feed flow, so they are larger units, as shown in Figure 14.3-36. These devices can be either static (such as the hydrocyclone) or dynamic (including moving blades, and so forth, to enhance separation).

Figure 14.3-36 illustrates a typical dry grinding circuit with the air separator used for classification and a cyclone used for solids recovery. The cutaway view on the right illustrates the internals and the drive for this dynamic air separator.

Hydrocyclone Basics

The hydrocyclone evolved into its current role in mineral processing in the period from approximately 1950 to 1970, by which time it had become a key element in the design of most large-scale grinding circuits. The hydrocyclone offers high capacities in small footprints and arguably enabled the ongoing development of very large grinding circuits. Although they still have their place, mechanical classifiers were effectively rendered obsolete in plant design by the hydrocyclone. Despite the numerous benefits of hydrocyclones, there have always been issues with efficiency, as introduced earlier in the discussion on fine screening. As today's operators look for ways to address energy efficiency and as competitive technologies emerge, hydrocyclone manufacturers have redoubled their development efforts. As is the case for screens, they are using the very latest simulation tools to help refine existing designs and develop new ones for specific applications, with good success. The following subsections provide a refresher on hydrocyclone basics and introduce some of the latest tools and techniques for the design, analysis, control, and optimization of hydrocyclones.

Principles of Operation

Figure 14.3-37 provides a sketch of a typical hydrocyclone showing the main design components. The principal dimensional variable is the diameter of the upper cylindrical section of the hydrocyclone body, D_c. The inlet (D_i) and discharge (D_o and D_u) dimensions usually lie within some ratio of D_c, (e.g., $0.2D_c \leq D_o \leq 0.45D_c$). Because the inlet is usually more rectangular in nature, to facilitate a reduction in turbulence on entry (involute designs—see Olson and Turner 2002), D_i is defined as the diameter of circle with an area equivalent to the actual inlet area. In the past, smaller hydrocyclones have had lower cone angles (~10°), but because of headroom considerations, larger hydrocyclones tend to have larger cone angles (~20°). Some typical dimensions for a hydrocyclone with D_c = 760 mm (30 in.) are shown in the figure.

The feed enters the hydrocyclone near the top of the cylindrical section, usually through some sort of involute inlet design, which is intended to introduce the slurry into the hydrocyclone with a minimum of turbulence, as this can have a deleterious effect on efficiency and wear. Figure 14.3-34A clearly shows this involute design. The rotational motion set up in the body of the hydrocyclone

Figure 14.3-36 A Gyrotor air separator in a typical dry grinding circuit

gives rise to the G-forces that accelerate the particle separation process. All of the particles are subjected to the centrifugal forces directing them to the wall of the hydrocyclone, where this material moves downward in a helical fashion to be discharged through the apex, as *underflow*. However, as the particles migrate to the outer wall of the hydrocyclone, they necessarily displace the fluid, which must move inward. The fluid exerts drag forces on the particles and retards their outward motion and, in the case of the finer particles, actually causes them to move inward. This stream moves upward in a helical path to be discharged through the vortex finder, as *overflow*. When the hydrocyclone is operating normally, the centrifugal forces acting on the slurry create low pressure along the central axis, establishing the conditions necessary for development of the characteristic air core.

To look more closely at the separation process, consider Stokes' law for laminar settling, given by Equation 14.3-8. (Whether it is Stokes' law, Newton's law, or something else—as experimental data would seem to indicate—the trends predicted by Stokes' law are seen in practice.)

$$v_t = \frac{(\rho_s - \rho_l)gd^2}{18\mu} \qquad (14.3\text{-}8)$$

where

v_t = terminal settling velocity
ρ_s = solids specific gravity
ρ_l = carrier fluid specific gravity
g = gravitational or centrifugal force acting on the particles
d = particle diameter
μ = carrier fluid viscosity

From the explanation given previously, anything that acts to increase the particle settling velocity increases the chances of particle reporting to the underflow stream. Clearly, the larger the particle, the more likely it is to go to the underflow. Similarly, higher G-forces (meaning higher feed flows or pressure drops) will also act to increase the probability that a particle will find its way to the underflow stream. Particles with higher specific gravities will also have a higher probability of reaching the underflow stream.

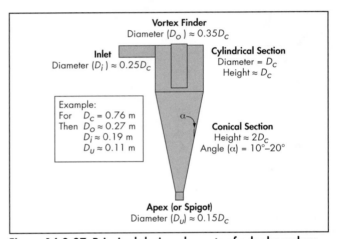

Figure 14.3-37 Principal design elements of a hydrocyclone

However, if the feed material contains a lot of very fine clay material, thereby increasing the effective viscosity, the particles will be less likely to reach the underflow stream. Similarly, with high-density hydrocyclone feeds, there is a significant concentration of fine particles in the carrier fluid, which increases the fluid specific gravity and makes it less likely for particles to reach the underflow stream.

To further complicate things, the size of the vortex finder also has an influence on the separation outcome. Smaller vortex finders will create a finer particle overflow stream, and vice versa. Although to a lesser extent, the apex size will also have an impact, as a smaller apex tends to limit the underflow capacity, diverting some coarser particles toward the upper regions of the hydrocyclone, the finest of which will then be entrained in the overflow stream.

Figure 14.3-38 (absent the air core) was developed by Renner and Cohen (1978) and shows roughly the main regions inside a cyclone. Region A is relatively narrow area in the vicinity of the inlet that is characterized by a size distribution similar to that of the feed. Most of the conical section is Region D, which is characterized by a coarser size distribution similar to the underflow. Region B is the finer material with a size distribution very similar to the overflow product.

Region C is perhaps the most critical region, as it was found to contain significant quantities of nearsize particles, which suggests that they tend to accumulate here until physical capacity constraints force them to move to one product zone or the other. Arguably, this is the region of active classification, and poor operation would be expected if it fails to form properly, possibly because of poor design or excessive feed solids concentrations.

Rather than produce a process matrix for the hydrocyclone, the authors have chosen to illustrate these concepts with the Plitt model (see Flintoff et al. 1987) given in Equation 14.3-9. This is the correlation for the corrected cut size, $d50_c$. For example, increasing the hydrocyclone diameter, D_c, would be expected to lead to a coarser cut size. Similarly, increasing the feed flow, Q, would lead to a finer cut size.

$$d50_c = \frac{CD_c^{0.46}D_i^{0.6}D_o^{1.21}\mu^{0.5}e^{0.063V}}{D_u^{0.71}h^{0.28}Q^{0.48}\left[\frac{\rho_s - \rho_l}{1.65}\right]^k} \quad (14.3\text{-}9)$$

where

$d50_c$ = corrected cut size
C = a calibration constant
D_c = hydrocyclone diameter
D_i = inlet diameter
D_o = overflow apex diameter
V = volume percent concentration of solids in hydrocyclone feed
D_u = underflow apex diameter
h = free vortex height
Q = feed volumetric flow rate
k = settling exponent (e.g., 0.5 in Equation 14.3-8)

A correlation exists for γ as well, but it is much weaker, generally indicating that γ will improve at lower flows (longer residence times) and lower feed solids volumetric concentrations, both of which are intuitively acceptable.

A couple of phenomena are worthy of some amplification. The first is the phenomenon known as roping (a name based on the appearance of the underflow stream in this regime). Roping occurs when the volumetric flow rate of the underflow stream causes the air core to collapse. When this happens, the usual helical spray from the apex alters appearance, as shown in Figure 14.3-39. Consequently, roping will increase the flow of coarser particles to the overflow stream (i.e., a coarser cut size). The impact on efficiency (as measured by γ in a partition curve) appears to depend on the hydrocyclone feed density—higher densities having a more deleterious effect. Roping does tend to maximize the underflow density. Thus, it minimizes the fraction of fine material that reports with the carrier liquid to this stream.

The other phenomenon is particle density. There are many cases (coal, iron ore, massive sulfide ores, and so forth) where liberated particles of widely different densities exist. Equation 14.3-8 indicates that, for a constant settling velocity, the particle size is related to solids specific gravity as $d \propto (\rho_s - \rho_l)^{0.5}$. In practice, and because of other phenomena such a secondary density separation occurring in the rotating bed, the exponent is usually much greater than 0.5, which means that density effects are exacerbated in the hydrocyclone. To illustrate, consider Figure 14.3-40, which shows coal slurry being processed in a hydrocyclone. Sink and float analysis

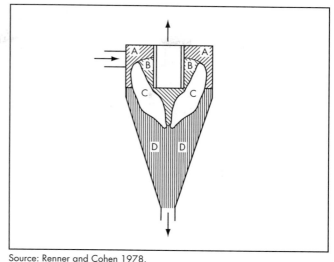

Source: Renner and Cohen 1978.
Figure 14.3-38 Regions in a cyclone

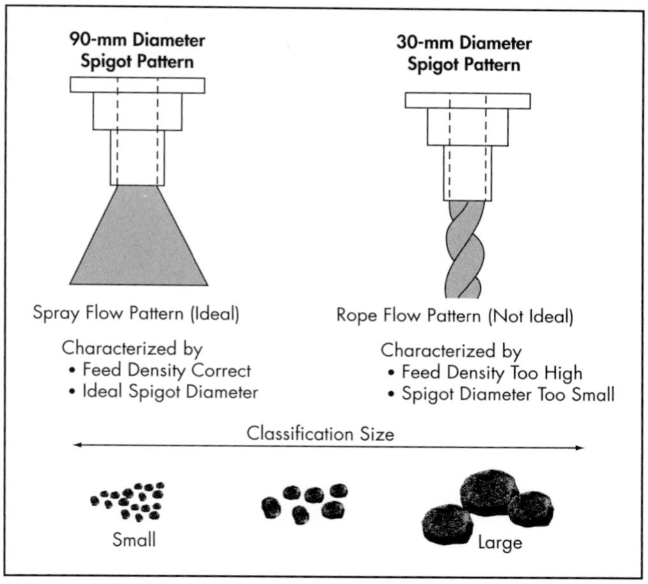

Courtesy of Weir Minerals
Figure 14.3-39 Spray and rope discharge

on the products allows the reconstruction of separate partition curves for each specific gravity (SG) class, all of which have a reasonable shape. However, looking only at the overall curve could suggest poor classification.

The last item of interest is hydrocyclone mounting. With the development of larger and larger hydrocyclones, mounting them in a vertical orientation puts a significant head on the underflow stream, which can influence wear and separation. However, tilting these hydrocyclones (usually at ~45°) allows operation at lower pressures while still producing a good underflow product, albeit at a coarser cut size. Maintenance-saving claims for wear in the lower region of the hydrocyclone are impressive (80%, e.g., Schmidt and Turner 1993), but it can be more difficult to perform in-situ maintenance for this kind of installation.

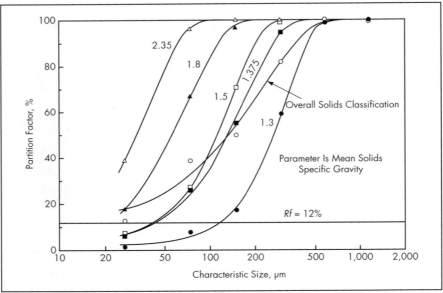

Source: Reprinted with permission of the Canadian Institute of Mining, Metallurgy and Petroleum. Originally published in the *CIM Bulletin*, Vol. 80, No. 905, p. 43.

Figure 14.3-40 Partition curves for coal slurry in a hydrocyclone

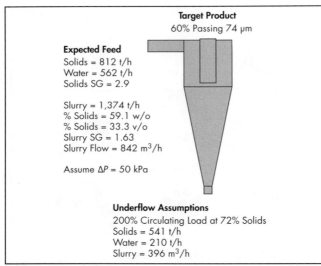

Source: Arterburn 1982.

Figure 14.3-41 Hydrocyclone design

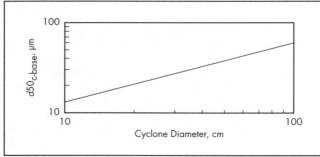

Source: Arterburn 1982.

Figure 14.3-42 For standard hydrocyclones, $d50_{c\text{-}base}$ versus hydrocyclone diameter

Hydrocyclone Selection

The selection of hydrocyclones is normally a task best performed with a supplier, as these firms possess the knowledge, expertise, databases, and tools to ensure that the best results are achieved. In most cases, it is possible to avoid pilot testing. However, there are some instances where this is preferable and even inevitable. One of the more challenging design problems is the closed-circuit grinding application, illustrated on the left in Figure 14.3-32. This is challenging because the hydrocyclone feed characteristics depend very much on mill performance, which in turn depends on the hydrocyclone underflow characteristics, which in turn depend on the feed. Experience has given most suppliers a very good understanding of this problem; nevertheless, circuit simulation tools can be useful adjuncts in developing a hydrocyclone design basis. (Public domain hydrocyclone models, e.g., Equation 14.3-9, include most of the basic design parameters, yet these are seemingly not commonly used in the design process, which underscores the experience factor necessary in this process.)

For the purposes of this chapter, the techniques of hydrocyclone design documented by FLSmidth Krebs are briefly reviewed (for more detail, see Arterburn 1982 and Olson and Turner 2002). The required information is summarized in Figure 14.3-41.

A set of standard tests under very specific design and operating conditions was performed to develop a relationship between what is called the base corrected cut size ($d50_{c\text{-}base}$) and the principal dimensional variable, the hydrocyclone diameter, D_c. The outcome of this work is captured in Figure 14.3-42, based on Equation 14.3-10, and is also given in the alternate form as Equation 14.3-11.

$$d50_{c\text{-}base} = 2.84 D_c^{0.66} \qquad (14.3\text{-}10)$$

$$D_c = 0.206 d50_{c\text{-}base}^{1.515} \qquad (14.3\text{-}11)$$

Table 14.3-11 Empirical correlation between the reference size and $d50_{c\text{-}actual}$

Required Overflow Size Distribution (Percent Passing) of Specified Micrometer	Multiplier (To Be Used to Factor the Specified Micrometer Size)
98.8	0.54
95	0.73
90	0.91
80	1.25
70	1.67
60	2.08
50	2.78

Not unlike the screen design methodology introduced earlier, the hydrocyclone design approach illustrated here relies on a number of empirical correction factors to scale the performance of the standard hydrocyclone to the expected industrial operation. In this case, the key scaling parameter is the $d50_c$, and the key design formula is given by

$$d50_{c\text{-}actual} = d50_{c\text{-}base}\frac{(CP1)(CP2)(CP3)(\ldots)}{(CD1)(CD2)(CD3)(\ldots)} \quad (14.3\text{-}12)$$

where

$d50_{c\text{-}actual}$ = the expected corrected cut size in industrial operation
$d50_{c\text{-}base}$ = the corrected cut size for the standard hydrocyclone
$CP1, CP2, CP3\ldots$ = process related correction factors (e.g., feed volume [~ viscosity] correction factor)
$CD1, CD2, CD3\ldots$ = design-based correction factors, (e.g., vortex finder diameter)

In practice, Equation 14.3-12 is rearranged and used in the form of Equation 14.3-13:

$$d50_{c\text{-}base} = \frac{d50_{c\text{-}actual}}{\frac{(CP1)(CP2)(CP3)(\ldots)}{(CD1)(CD2)(CD3)(\ldots)}} \quad (14.3\text{-}13)$$

With this short introduction, the basic steps in hydrocyclone design are as follows.

Step 1—Estimation of the $d50_{c\text{-}actual}$. Table 14.3-11 provides the relationship between the overflow size specification and $d50_{c\text{-}actual}$. For the overflow specification in Figure 14.3-41, 60% corresponds to a factor of 2.08, which when multiplied by the reference size of 74 μm gives the estimate for $d50_{c\text{-}actual}$ of 154 μm.

Step 2—Calculation of the correction factors and estimation of $d50_{c\text{-}base}$. Three important published process-related correction factors are the solids feed volume concentration ($CP1$), the operating pressure drop ($CP2$), and the solids specific gravity ($CP3$).

$CP1$ is the correction for the feed volumetric concentration, which is in effect a first-order approximation to viscosity. Of course, viscosity also depends on water temperature, clay concentrations, particle shape, and other factors—clays in fine tailings is what gave rise to the tailings curve in Figure 14.3-43. In this figure, Olson and Turner

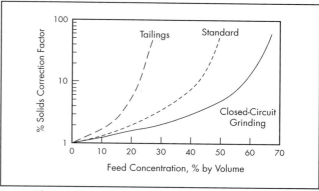

Source: Olson and Turner 2002.

Figure 14.3-43 Correction factors for hydrocyclone feed solids volumetric concentration

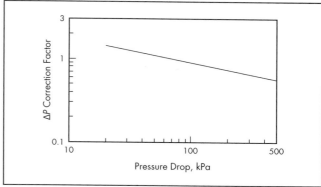

Source: Arterburn 1982.

Figure 14.3-44 Pressure drop correction factor

(2002) extended Arterburn's method, which was based on the standard curve in Equation 14.3-14.

$$CP1 = \left[\frac{53-V}{53}\right]^{-1.43} \quad (14.3\text{-}14)$$

In this example the standard curve is assumed, and with $V = 33.3\%$, then $CP1 = 4.09$.

$CP2$ is the correction factor for the operating pressure drop—usually the difference between the measured pressure in the radial header and atmosphere. Larger hydrocyclones tend to run at lower pressures, and vice versa. Figure 14.3-44 and Equation 14.3-15 show this relationship.

$$CP2 = 3.27\Delta P^{-0.28} \quad (14.3\text{-}15)$$

where ΔP is the pressure drop in kPa.

In this example, it was assumed that $\Delta P = 50$ kPa (7.25 psig), which should give relatively good hydrocyclone performance while minimizing wear. From Equation 14.3-15, $CP2 = 1.09$.

$CP3$ is the correction factor for the solids specific gravity and is shown in Figure 14.3-45 and Equation 14.3-16. For a solids specific gravity of 2.9, the correction factor $CP3 = 0.93$.

$$CP3 = \left[\frac{1.65}{\rho_s - \rho_l}\right]^{0.5} \quad (14.3\text{-}16)$$

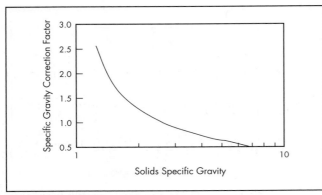

Source: Arterburn 1982.
Figure 14.3-45 Specific gravity correction factor

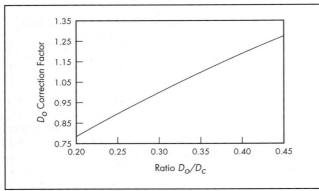

Source: Olson and Turner 2002.
Figure 14.3-46 Vortex finder correction factor

$CD1$ is the correction factor for the vortex finder diameter. In this example the default design of $D_o = 0.3D_c$ is used, but Figure 14.3-46 and Equation 14.3-17 illustrate the nature of the correction.

$$CD1 = \left[\frac{D_o}{0.3D_c}\right]^{0.6} \quad (14.3\text{-}17)$$

Equation 14.3-11 then solves as

$$d50_{c\text{-}base} = \frac{154\ \mu m}{(4.09)(1.09)(0.93)(1)} = 37\ \mu m \quad (14.3\text{-}18)$$

Step 3—Select size and number of hydrocyclones. From Equation 14.3-11 or Figure 14.3-42, the diameter of the hydrocyclone required to give this $d50_{c\text{-}base}$ is 49.1 cm (19.3 in.). Hydrocyclones, like screens, come in certain standard sizes, usually defined by D_c. As seen in Figure 14.3-47, the closest size to the "theoretical" value computed previously is a hydrocyclone with $D_c = 51$ cm (20.1 in.).

For a pressure drop of 50 kPa (7.25 psig), the capacity of a single 51-cm (20.1-in.) hydrocyclone will be approximately 140 m³/h (616 gpm). From Figure 14.3-41, the total feed volumetric flow is 842 m³/h (3,707 gpm), so approximately six hydrocyclones will be required in operation. In practice, two or three additional hydrocyclones would be added to accommodate maintenance as well as peaks in the feed flow rate. It is also not uncommon to provide an additional off-take valve in the radial header to be used for sample-collection purposes.

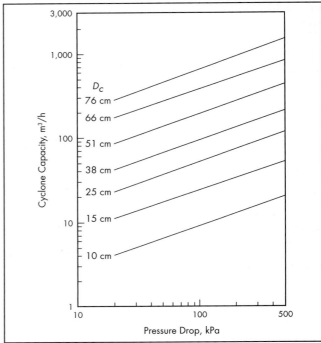

Source: Arterburn 1982.
Figure 14.3-47 Hydrocyclone capacity correlation

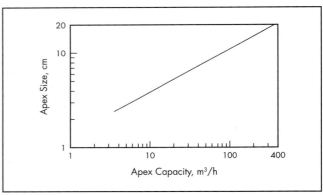

Source: Arterburn 1982.
Figure 14.3-48 Apex size correlation

Step 4—Select apex size. The last step is to determine the apex size, and that comes from Figure 14.3-48. For a total underflow rate of 396 m³/h (1,744 gpm), the flow through each apex would be ~66 m³/h (291 gpm), corresponding to an apex size of ~9.3 cm (3.66 in.). The manufacturer's closest size would be chosen, if possible tending to a smaller size in anticipation of wear.

Of course, after the hydrocyclones are installed and in operation, the reality of the application is known, and the optimization process (process variables and design variables such as D_o and D_u) can be modified as required.

Although the suppliers have undoubtedly added many refinements to their proprietary methods for hydrocyclone design, such as screens, the published methods for design have changed little since the 1980s. However, with the advent of some of the new, more efficient hydrocyclone technology

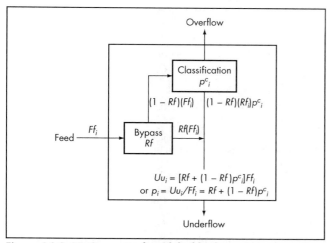

Figure 14.3-49 Conceptual model of hydrocyclone operation

Table 14.3-12 Experimental data for an inefficient hydrocyclone

Size, μm	Cyclone Feed	Cyclone Underflow	Experimental Partition
6,730	0.17	0.24	1.00
4,760	0.41	0.57	1.00
3,360	0.53	0.74	1.00
1,680	6.77	9.49	1.00
850	14.89	20.40	0.98
600	10.21	13.03	0.91
425	11.37	12.97	0.81
300	11.05	10.59	0.68
212	7.96	6.61	0.59
106	12.07	8.86	0.52
74	4.11	2.65	0.46
53	2.80	1.77	0.45
Pan	17.66	12.07	0.49
Total	100.0	100.0	—
Mass	1,237.3	880.9	—
Percent solids	65.1	73.6	—
Rf(water)	0.47	—	—

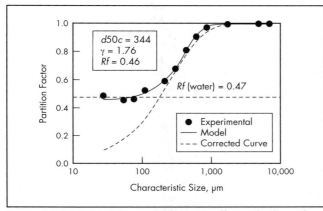

Figure 14.3-50 Partition curve for an inefficient hydrocyclone

to be discussed later, these published methods will require an adaptation.

Performance Assessment

Hydrocyclone efficiencies are usually assessed through the partition curve, and this implies a sampling experiment to acquire the necessary data. However, unlike screens, the fact that hydrocyclones are treating fine particles in slurries introduces another dimension to the simple modeling approach used for screening (Equation 14.3-6). The liquid phase reporting with the underflow stream carries with it fine particles that should have otherwise reported to the overflow stream. A common conceptual model is illustrated in Figure 14.3-49. It is assumed that a fraction of the feed material is bypassed directly to the underflow stream and the rest of the material is then subjected to the classification process (characterized by γ and $d50c$). (In this figure, upper case letters F and U are mass flows in the feed and underflow, while f_i, and u_i are the mass fractions retained in the ith size class. p^c_i is the corrected partition factor for the ith size class, usually given by Equation 14.3-6. Modeling convention is to number the size classes from 1 (coarsest size fraction) to N (the so-called pan fraction) from the sieve analysis. For example, in Table 14.3-12, the particles greater than 6,730 μm will be size Class 1, those falling into the interval 6,730 × 4,760 μm, will be size Class 2, and so on.

Thus, the basic partition curve model for hydrocyclones is given by

$$p_i = Rf + (1 - Rf)\left[1 - e^{-0.693\left(\frac{d_i}{d50_c}\right)^\gamma}\right] \quad (14.3-19)$$

where Rf is the the bypass fraction.

Table 14.3-12 provides some experimental data for an operating hydrocyclone. Figure 14.3-50 compares the experimental data with modeling results with Equation 14.3-19, and the fit is seen to be very good. (The inefficient hydrocyclone was chosen for this illustration as the data are used to explore other measures of efficiency.)

The corrected partition curve (the term in the braces in Equation 14.3-19) is shown as the dashed line in Figure 14.3-50. Removing the effect of Rf in this way permits a more objective assessment of $d50_c$ and γ. The $d50$ for the experimental data in Figure 14.3-50 is about 106 μm, not very close to the $d50c$ value of 344 μm. Of course, this finer cut is because of the large number of fine particles carried to the underflow by the water—the Rf factor.

To amplify on Rf, the very fine particles ($d_i \to 0$ mm) are assumed to be more or less trapped in the water by virtue of the extraordinary drag forces. Hence, it is expected that the Rf deduced from the partition curve analysis will be equivalent to the water recovered from the feed to the underflow stream. In a well-constructed and executed experiment, the Rf estimated from curve fitting and the Rf(water) calculated from the water split should indicate good agreement, as evidenced in Figure 14.3-50.

For completeness, not all experimental partition curves behave like those for a homogenous ore (i.e., a low-grade ore with a gangue matrix of approximately constant specific gravity). For example, it is not uncommon to see the so called "fish-hook" behavior at the fine end of a partition curve, where the p_i values actually dip significantly below the expected Rf(water) value, before coming back to this level at the

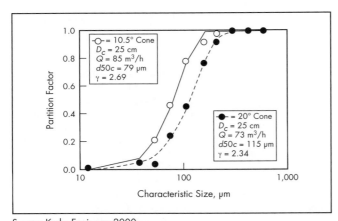

Source: Krebs Engineers 2000.
Figure 14.3-51 Effect of cone angle on hydrocyclone performance

ultrafine sizes. While beyond the scope of this work, there are ways to model this if necessary, but the physical explanations are not so straightforward. Similarly, there can be inflections at the coarser end of the curve, so called "humps" or even "dips," and these are usually related to the separate classification action of liberated heavy and light mineral particles. In this latter case, one must separate these fractions experimentally (or mathematically) and model the curves separately.

From an operational point of view, maximum efficiency is achieved when γ is maximized (values of ~2.5 or larger are considered "good") and Rf is minimized (values of 0.3 or less are considered "good"). In closed-circuit grinding, Rf is a clear indication of the amount of fine material needlessly sent back to the mill. In practice, there is always a compromise between efficiency and operational objectives. In general both γ and Rf will improve with a lowering of the solids volumetric concentration in the feed. Rf can be also reduced using flat-bottom or inclined hydrocyclones, both of which maximize the underflow solids content.

A final word of caution on partition curves is that there are numerous simple two-parameter functional forms used to characterize the corrected partition curve. They all have in common the parameters $d50c$ and γ. However, while the $d50c$ values will usually compare well, the same is not true for the γ values. It is important that one be aware of the underlying functional form in drawing any comparisons between plant results and those quoted by others. To very quickly illustrate, another commonly used model is given in Equation 14.3-18.

$$p_i = Rf + (1-Rf) \frac{\left[\exp\left(\gamma \frac{d_i}{d50_c}\right) - 1\right]}{\left[\exp\left(\gamma \frac{d_i}{d50_c}\right) + \exp(\gamma) - 2\right]} \quad (14.3\text{-}20)$$

When Equation 14.3-18 is fit to the data in Figure 14.3-47, the parameter values are $d50_c$ = 340 μm, Rf = 0.44, and γ = 2.13. For a given data set, γ values from the model in Equation 14.3-20 are always higher than those from the model given by Equation 14.3-19.

A number of other efficiency measures, such as imperfection and probable error, are sometimes used with hydrocyclones. Many of these are based on the partition curve. In addition, some people use the efficiencies defined earlier for screening (Equations 14.3-3 and 14.3-4). To illustrate, the partition curve in Figure 14.3-50 is seen to be rather inefficient, both in terms of γ and especially in terms of Rf. (This is a high-density feed with older model hydrocyclones.) The corresponding figures are E_u = 62.2% (fines removal from oversize) and R_u = 41.2% (fines removal from feed to overflow). Not surprisingly, these poor performance figures, especially R_u, have been the main drivers to explore fine-screening solutions and more efficient hydrocyclone designs.

Recent Developments
Design enhancement on existing hydrocyclones. Recently, both FLSmidth Krebs (gMAX) and Weir Minerals (Cavex) have introduced new, more efficient hydrocyclone designs. In both cases it appears that design improvements were made in the feed inlet to minimize turbulence when the new feed enters the body of the hydrocyclone, which impacts capacity, wear, and efficiency. Moreover, both OEMs made changes to the hydrocyclone body design as well. FLSmidth Krebs (Krebs Engineers 2000) exploited the potential for efficiency improvements associated with increasing residence time in the critical conical section of the hydrocyclone, as shown in Figure 14.3-51. An extensive redesign campaign yielded optimal combinations of cylinder and cone length, as well as cone design for each standard gMAX hydrocyclone diameter.

Mohanty et al. (2002) reported that a conventional hydrocyclone separating a fine coal feed (~45 μm cut target) yielded an efficiency of R_u = 37%, while the gMAX improved this to 68%. Adding the FLSmidth Krebs Cyclowash attachment (a means of injecting water into the lower cone section of a hydrocyclone to remove fines) gave a further increase to 70%. (Further technical details may be found on the supplier's Web site.)

New hydrocyclone systems. In addition to the work underway to improve traditional hydrocyclone designs, new designs are also under study by equipment manufacturers and researchers alike, although in general these are application-specific products. Two good examples of this are the three-product hydrocyclone (University of Cape Town) and the ReCyclone (Weir Minerals).

Briefly, the three-product hydroyclone (e.g., Mainza et al. 2004) is intended for use in applications where there are components of distinctly different densities. In conventional hydrocyclones, the heavier minerals concentrate near the hydrocyclone wall, which creates a heavy medium that selectively sends the larger, lighter particles toward the inside of the hydrocyclone, where they can be sent to the overflow. In the case of the platinum ores in South Africa, the lighter silica gangue contains the platinum group metals, while the heavier chromite is barren. Larger silica particles that report to the overflow represent potential recovery losses in the flotation process. The three-product hydrocyclone uses concentric vortex finders to produce two overflow products, and the inner vortex finder collects the fine particles, which are sent directly to flotation. The outer vortex finder collects a coarser middlings overflow, rich in coarse silica, which is then screened to ensure that the coarser material gets to the mill for further grinding, while the finer heavy material joins the flotation feed. Development work continues with this system, which would appear to have other applications in the mineral processing industry.

The ReCyclone (D. Switzer, E. Castro, and J. Lopez, personal communications) uses two-stage classification as a means to reduce Rf and increase γ. This concept has been studied and promoted by many academics and a few operators. It

Courtesy of Weir Minerals.
Figure 14.3-52 Schematic and installation of the ReCyclone

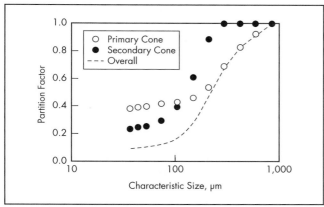

Courtesy of D. Switzer, E. Castro, and J. Lopez.
Figure 14.3-53 Partition curves for a ReCyclone

does find application in some coal and iron-ore flow sheets, but generally the extra footprint and the capital and operating expenses associated with such an approach has diminished interest among plant designers. One area where this has been extensively applied is in tailings dam construction, where it is critically important to move as much of the fines as possible from the sand product used to construct the dam. In these cases, two batteries of hydrocyclones (larger primaries followed by smaller secondaries) are common, to ensure that sand percolation rates are sufficiently high to avoid stability problems in the dam. The ReCylone performs these steps in a single unit.

Figure 14.3-52 is an illustration of the ReCylone unit and Figure 14.3-53 presents partition curves for the primary and secondary units and for the overall curve. Note the adjustment system on the primary underflow, which acts conceptually like a variable apex. The effects on Rf and, hence, fine particle content in the underflow are clear. All of the case studies reported by D. Switzer, E. Castro, and J. Lopez (personal communications) are on tailings. However, one can easily envisage other applications.

Instrumentation. Although little has been published thus far, the major hydrocyclone suppliers and some third parties are actively looking at technology value additions that would help to improve performance and minimize maintenance costs. For example, hydrocyclones are known to vibrate when they transition from normal underflow spray to a roping regime, and vibration and/or acoustics (e.g., Wedeles 2007) could be applied to diagnose this event or other events such as apex plugging (e.g., tramp grinding balls in the hydrocyclone feed). Another example is the elusive measurement of the relative wear of hydrocyclone liners, an area also under study by pump manufacturers. The interested reader should contact the suppliers directly on this topic.

Computational fluid dynamics (CFD). As was seen earlier in the discussion on screens, most major suppliers are rapidly embracing sophisticated simulation tools as a way to create and screen virtual machines for the purposes of design and optimization. Although this does not preclude the necessary step of field testing prototypes, it does help in the selection of the most appropriate candidate designs, reducing research and development costs and technical risk, and shortening the time to market.

CFD is the simulation technique of choice for fluid systems such as the hydrocyclone. However, at this stage it is impossible to simulate real slurries, and the standard approach is to simulate the fluid and then to use these results with some "tracer particles" (within CFD or in the DEM environment) to gain insight on classification efficiency. This one-way coupling (the fluid influences the particles, but the converse is not true), albeit with dilute suspensions, provides quantitative predictions based on first principles.

Suffice it to say, CFD played a major role in the development of both the Cavex and gMAX hydrocyclone designs described earlier. For the interested reader, the paper by Delgadillo and Rajamani (2007) provides a current vision of the benefits of CFD in advancing the hydrocyclone state of the art.

Media

Briefly, hydrocyclones are usually constructed of urethane or steel. Steel hydrocyclones are lined with wear-resistant materials such as rubber, ceramics, or urethane, and they can sometimes employ special metallic-based compounds. Again, the OEM is in a very good position to provide expert application advice on media choices.

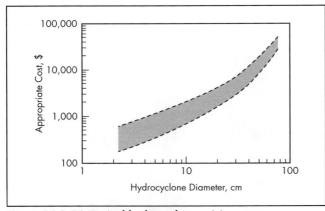

Figure 14.3-54 Typical hydrocyclone pricing

Standard Hydrocyclone Sizing and Relative Pricing

Each hydrocyclone manufacturer has a specific standard product size range, as shown in Figure 14.3-47. Generalizing across several of these suppliers, the usual sizes are approximately 10, 15, 25, 40, 50, 65, and 85 cm (3.9, 5.9, 9.8, 15.7, 19.7, 25.6, and 33.5 in.).

From an investment point of view, the total installation cost depends on the number of hydrocyclones, the materials of construction and lining, the type of radial header, the level of automation (manual versus automatic valves), and so forth. Figure 14.3-54 simply indicates the approximate cost of a single hydrocyclone and demonstrates the usual spread depending on whether lesser or more expensive materials are employed.

ACKNOWLEDGMENTS

All figures and tables not specifically referenced are included courtesy of Metso Mining. The authors extend their most sincere thanks to Derrick Corp., Weir Minerals, FLSmidth Krebs, and our many colleagues at Metso Mining who provided much of the information contained herein, as well as some very helpful feedback.

REFERENCES

Allis Chalmers. n.d. *Theory and Practice of Screen Selection*. Allis Chalmers Technical Note.
Aquino, B., and Vizcarra, J. 2007. Re-engineering the metallurgical processes at Colquirjirca mine. Sociedad Mineral El Brocal S.A.A., Mineria, Peru, October. pp. 30–38.
Arterburn, R. 1982. The sizing and selection of hydrocyclones. In *Design and Installation of Comminution Circuits*. Edited by A.L. Mular and G.J. Jergensen. Littleton, CO: SME. pp. 592–607.
Bothwell, M.A., and Mular, A.L. 2002. Coarse screening. In *Proceedings of Mineral Processing Plant Design, Practice and Control*, Vol. 1. Littleton, CO: SME. pp. 894–916.
Delgadillo, J., and Rajamani, K. 2007. Exploration of hydrocyclone designs using computational fluid dynamics. *Int. J. Miner. Process.* 84:252–261.
Flintoff, B., Plitt, L.R., and Turak, A. 1987. Cyclone modeling: A review of present technology. *CIM Bull.* (September): 39–50.
Hilden, M.M. 2007. A Dimensional Analysis Approach to the Scale-up and Modelling of Industrial Screens. Ph.D. Thesis, University of Queensland, Australia.
Karra, V.K. 1979. Development of a model for predicting screening performance of a vibrating screen. *CIM Bull.* (April): 167–171.
Kelly, E.G., and Spottiswood, D.J. 1982. *Introduction to Mineral Processing*. New York: John Wiley and Sons. pp. 200–201.
King, R.P. 2001. *Modeling and Simulation of Mineral Processing Systems*. Boston: Butterworth-Heinemann. pp. 81–125.
Krebs Engineers. 2000. *Krebs gMAX™ Cyclones—For Finer Separations with Larger Diameter Cyclones*. www.krebs.com. Accessed September 2009.
Mainza, A., Powell, M., and Knopjes, B. 2004. A comparison of different cyclones in addressing challenges in the classification of the dual density UG2 platinum ores. In *Proceedings of the First International Platinum Conference "Platinum Adding Value."* Sun City, South Africa: SAIMM. pp. 341–348.
Mathews, C.W. 1985. Screening. In *Mineral Processing Handbook*. Edited by N.L. Weiss. Littleton, CO: SME. pp. 3E1–3E13.
Mohanty, M., Palit, A., and Dube, B. 2002. A Comparative evaluation of new fine particle separation technologies. *Miner. Eng.* 15(10):727–736.
Moon, D. 2003. Understanding screen dynamics for optimum efficiency. Presented at the 34th Conference of the Institute of Quarrying Southern Africa, March.
Olson, T., and Turner, P. 2002. Hydrocyclone selection for plant design. In *Mineral Processing Plant Design, Practice and Control*, Vol. 1. Edited by A.L. Mular, D.N. Halbe, and D.J. Barratt. pp. 880–893.
Renner, V., and Cohen, H. 1978. Measurement and the interpretation of size distributions within a cyclone. *Trans. Inst. Chem. Eng.* 87:C139–C145.
Schmidt, M., and Turner, P. 1993. Flat bottom or horizontal cyclones—Which is right for you? *World Min. Equip.* (September): 151–152.
Svarovsky, L. 2000. *Solids–Liquid Separation*, 4th ed. Boston: Butterworth-Heinemann. pp. 191–245.
Valine, S., and Wennen, J. 2002. Fine screening in mineral processing operations. In *Mineral Processing Plant Design and Control*. Edited by A.L. Mular, D.N. Halbe, and D.J. Barratt. Littleton, CO: SME. pp. 917–928.
Vince, A. 2007. *Improved Classification at 100–350 Microns*. Australian Coal Association Research Program (ACARP). www.acarp.com.au/abstracts.aspx?repID=C15051. Accessed September 2009.
Wedeles, M. 2007. Smart cyclone: A tool for optimizing cyclone operation. Presented at SAG 2007, Vina del Mar, Chile (also available from FLSmidth Krebs).
Wennen, J., Nordstrom, W., and Murr, D. 1997. National Steel Pellet Company's secondary grinding circuit modifications. In *Comminution Practices*. Edited by S.K. Kawatra. Littleton, CO: SME.

CHAPTER 14.4

Gravity Concentration and Heavy Medium Separation

Nimal Subasinghe

Gravity concentration is one of the oldest processes of separating minerals and is aimed at separating heavy minerals from lighter gangue. Gravity concentrators have wide applications in the heavy minerals industry for the recovery of heavy minerals, such as ilmenite, rutile, leucoxene, and zircon, from the lighter gangue. They are used to separate metal sulfides from ores where the sulfides are coarse grained. They have been used to exploit placer gold deposits over many centuries, and as an adjunct in modern mills to recover coarse gold early in the process. When the feed material is comprised of particles with a wider size distribution, gravity concentrators do not produce a clean concentrate and have been used mainly for preconcentration or as a rougher stage. Gravity concentrators are simple devices that are inexpensive to operate and cause comparatively much less environmental pollution than using harmful chemical reagents for the same purpose. Gravity separation may be achieved in either of two ways: relative gravity separation and absolute gravity separation.

In relative gravity separation, all particles settle in the same direction within the device, and large differences in specific gravity (density) between the minerals are needed for effective separation. As settling velocity of particles depends not only on density but also on size, it is important to prepare narrowly sized feeds to these devices. However, as particles of finer sizes settle under gravity at much slower rates, the effectiveness of these devices decreases considerably below about 50 μm. The heavy and light minerals that settle at different rates are made to report to different exit streams by applying a transverse force. There are many devices of this nature, such as jigs, spiral concentrators, Wilfley tables (shaking tables), air tables, Gemini tables, pinched sluices, riffled sluices, Reichert cones, and so on. To improve the efficiency of separation of finer material, centrifugal concentrating devices have been introduced within the past two decades, including Knelson concentrators, Falcon concentrators, Kelsey jigs, and water-only cyclones. These devices use centrifugal acceleration to enhance the movement of finer particles and have been extensively used in the gold processing plants to recover fine gold.

In absolute gravity separation, a fluid or a nonsettling slurry comprised of fine solid particles with a specific gravity (SG) intermediate to those of the two minerals to be separated is used as the medium of separation. Particles of higher and lower specific gravities than the medium move in opposite directions. This technique requires only a small difference in specific gravities of the minerals for effective separation. Heavy or dense medium separation methods are absolute separation methods. Devices of this sort include dense medium vessels and dense medium cyclones.

The amenability to gravity concentration is indicated by the concentration criterion, C, defined (Taggart 1951) as

$$C = \left(\frac{\rho_h - \rho_f}{\rho_\ell - \rho_f}\right) \qquad (14.4\text{-}1)$$

which has been deduced from Newton's law of settling, where ρ_h, ρ_ℓ, and ρ_f are the densities of heavy mineral, light gangue, and water, respectively. If C is greater than 2.5, the separation may be achieved with relative ease; separation is not commercially feasible for values below 1.25. Table 14.4-1 shows the effectiveness of gravity separation at various feed sizes.

GRAVITY CONCENTRATORS

Gravity concentrators may be classified according to the way particles are made to move relative to each other so that the denser particles will segregate from the lighter ones under the influence of the gravitational force. A gravity concentrator uses essentially one of three mechanisms to provide movement of particles within it: jigging, shaking, and flowing film devices.

Table 14.4-1 Values of concentration criterion required for effective separation

Value of Concentration Criterion, C	Significance for Separation
>2.50	Separation effective down to ~75 μm
2.50–1.75	Separation effective down to ~150 μm
1.75–1.50	Separation possible to 2 mm, but difficult
1.50–1.25	Separation possible to 6 mm, but difficult
<1.25	Relative separations impractical. Absolute separation may be possible down to C of 0.2.

Source: Aplan 1985.

Nimal Subasinghe, Assoc. Professor of Minerals Engrg., Curtin University (Western Australian School of Mines), Kalgoorlie, Western Australia

These gravity separation devices rely on a fluid medium, such as water, to transport the material through the device and also bring about a separation based on settling characteristics of heavy and light particles. The process requires the feed material to be uniformly sized and fed as a slurry. If the feed has a wide size distribution, the smaller heavy mineral particles will report to the same exit as larger light particles because their settling velocities are similar. As the particle size decreases, the differences in gravitational forces acting on heavy and light particles, which bring about differences in their movement, also diminish. Recent advances in the design of gravity separation devices have addressed this problem by introducing "multi-G" devices that impart higher accelerations on smaller particles through centrifugal forces than those due to gravity. These devices are known as centrifugal concentrators. Thus, each gravity concentrator has an optimal feed particle size range for effective separation. The magnitude of the applied acceleration is usually quoted in terms of multiples of the gravitational acceleration, G. Figure 14.4-1 shows the classification of gravity separation devices. Table 14.4-2 shows the range of feed sizes suitable for various gravity separators.

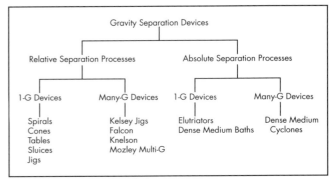

Figure 14.4-1 Gravity separation equipment

Table 14.4-2 Effective feed size ranges for typical gravity concentration devices

Coarse particle concentration [+6.4 mm]	
Jigs	Heavy media hydrocyclone
Heavy media separators	
Intermediate concentration [–6.4 mm + 0.150 μm]	
Jigs	Shaking tables
Heavy media hydrocyclone	Spirals
Water-only cyclone	Pinched sluices
Kelsey jig	Reichert cone
Pneumatic jigs and tables	Bartles–Mozley concentrator
Sluices	
Fines concentration [–0.150 μm]	
Shaking table	Falcon concentrator
Spirals	Knelson concentrator
Pinched sluices	Bartles–Mozley concentrator
Reichert cone	Mozley multigravity separator

Source: Adapted from Aplan 2003.

Jigging Devices

Jigging is one of the oldest methods of gravity concentration. It is commonly used in coal, cassiterite, gold, and iron ore industries. Generally jigs are used for separating coarser material from about 2 to 10 mm in diameter, although efficient separation of particles down to about 75 μm is possible with modified devices. They can handle large tonnages and operate on a continuous basis.

Jigging essentially uses alternating expansion and contraction of a bed of particles. The bed is generally formed over a screen through which the fluid is allowed to flow. Ragging comprised of heavy particles is generally placed over the screen to retain the bed and release the segregated heavies to the hutch during the suction stroke. The bed movement is activated by a pulsating current of fluid, usually water, or by the movement of the screen that holds the bed. The pulsation of the bed may be achieved by means of a plunger or by admitting compressed air at regular short intervals into a limb of a U-shaped tank. Light material is generally discarded through a gate positioned opposite to the feed end.

Gaudin (1939) suggested that three segregation mechanisms may take place within a jig: differential initial acceleration, hindered settling, and consolidation trickling. The extent of prevalence of these mechanisms depends on the properties of the material being processed, such as size and density and the manner in which the force is applied to dilate and contract the bed.

The differential initial acceleration between particles in a pulsating bed is a function of their density and is independent of size. That is, the initial acceleration conveyed to heavy mineral particles by the pulsating current of fluid is larger than that for lighter mineral particles. By applying a higher frequency of pulsation, this effect may be accentuated. This mechanism will segregate heavy minerals beneath the lighter gangue.

Also, during the pulsation stroke, heavy particles within the expanded bed will settle faster than the lighter ones. However, because of the presence of a large number of particles within the bed, hindered settling of particles takes place, the result of particle–particle interactions as opposed to free settling. The differences in hindered settling rates of larger particles are more pronounced than those of smaller particles. Hence, jigs perform better on coarser particles.

When the dilated bed collapses during the suction stroke, larger particles tend to interlock, leaving voids through which the finer particles may segregate. This process is known as interstitial or consolidation trickling.

Coal Jig

Baum jigs are commonly used for processing coarse coal whereas Batac jigs are used mainly for fine coal. Both these units use compressed air to pulsate the bed. In these units, water is typically pulsed at 60 to 80 pulses per minute and the units handle feed sizes up to 130 mm at capacities of up to 40 t/m^2/h.

Mineral Jig

Jigs that treat placer deposits generally have a diaphragm in the hutch area, which is placed underneath the screen that holds the bed. Typical jigs of this type are Cleveland and Pan-American placer jigs, which have a smaller footprint. Harz and Denver jigs have a wall that separates the hutch area and use an eccentrically driven plunger in one section to pulsate the bed. Schematic diagrams of a few jigs are shown in Figure 14.4-2.

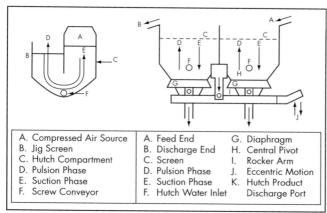

Source: Aplan 2003.
Figure 14.4-2 (left) Baum jig, and (right) Pan-American placer jig

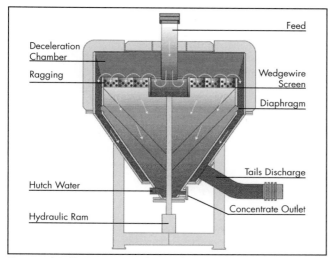

Courtesy of Gekko Systems Pty. Ltd., Australia.
Figure 14.4-3 In-line pressure jig

In-Line Pressure Jig

The in-line pressure jig (IPJ) combines a circular bed with a moveable sieve action. The screen is pulsed vertically by a hydraulically driven shaft. The length and speed of the up-and-down stroke can be varied to suit the application. Screen aperture, ragging dimension, and ragging material can also be altered for the application, which varies the recovery of solids to the concentrate stream. It uses less hutch water, compared to traditional jigs. The engineered ceramic ragging may be selected from a range of specific gravities from 1.6 to 5.0. IPJs have been used across many mineral types including gold, sulfides, tin, coal, and diamonds. Figure 14.4-3 shows a schematic diagram of an IPJ.

IPJs are available in a range of sizes handling 2 to 100 t/h of feed material. The solids recovery to concentrate may be varied from about 5% to about 50% by selecting the appropriate ragging density, ragging size, and screen aperture for a given feed size. Lower yields require tighter ragging sizing. The units are supplied with automatic controllers to control the feed rate, pulse frequency, stroke length, and air bleed rate for optimal performance.

Kelsey Centrifugal Jig

The Kelsey centrifugal jig (KCJ) imparts a centrifugal acceleration on the feed and ragging particles while a jigging action is being applied. It enables the effective separation of particles below about 500 μm down to fine sizes, and has found applications in tin, tantalum, gold, nickel, tungsten, chromite, base metals, iron ore, and mineral sands industries.

The centrifugal acceleration is conveyed by a spinning rotor, the speed of which may be varied and controlled. Inside the rotor, a screen of cylindrical shape is spun coaxially with the rotor. The ragging material is spread evenly on the inside of the screen because of the centrifugal acceleration. The feed enters through a fixed central pipe, then accelerates toward and continually rises up the ragging bed. Pressurized water is introduced into a series of hutches behind the screen, which is pulsed through the ragging to fluidize the bed. The pulsing of the bed is achieved via pulse arms connected to pulse pads, which push against flexible diaphragms at the back of each hutch, thereby pushing the water in the hutch through the screen and ragging bed.

KCJs can handle capacities up to 50 t/h solids and typically use about 30 to 50 m^3/h of hutch water. The rotor spins at speeds up to 200 rpm, generating accelerations of 40 G. The pulse rate is between 1,800 and 2,200 pulses per minute at a stroke length of 2 to 3 mm. A schematic diagram is shown in Figure 14.4-4.

The size of ragging used depends on the size distribution of the valuable mineral particles in the feed and should be chosen to avoid excessive blinding of the internal screen. Other operating variables include pulsation frequency, hutch water addition rate, ragging type, ragging SG, depth of bed, and so on. These units also have a number of design features, such as automatic screen cleaners and vibration detection. Programmable logic controller–based interlocks and sacrificial shear pins are provided to avoid major damage in the event of a malfunction.

Air Jig

Air jigs have been used in both coal and gold industries. In these devices, a low-pressure, high-volume air source is used to fluidize the particulate bed while a high-pressure, low-volume air source is used to impose the pulsation stroke for stratification. Recently, the air jig has found wider application in the coal industry as this method does not use process water, with the advantage of eliminating the need to dewater fines or manage slurry confinement. Particulate emissions have been minimized by the inclusion of fabric dust collectors. The new units use a punch plate instead of a fine wire mesh cloth to support a deeper particle bed. However, their application has limitations due to particle size, particle shape, moisture, and near-gravity content (Alderman 2002).

Flowing Film Separators

Flowing film separators have been used for centuries to accumulate heavy minerals such as gold and tin. These devices are cheap to make and operate. In flowing film concentrators, the feed slurry is made to flow over an inclined surface, and segregation of the heavy minerals beneath the lighter minerals takes place during the passage. At low flow rates, the thickness of the film is small and the particles generally settle out and roll over the surface. Heavier particles tend to accumulate in the lower regions closer to the surface and travel at lower velocities while the lighter materials travel closer to the surface and faster. At high flow rates the particles will be suspended in the flow due to turbulent suspension, resulting in a density gradient across the depth of the flow. The particles also experience shearing because of Bagnold forces that give rise to a dispersive pressure on the particles. Commonly used flowing film separators are pinched sluices, Reichert cones, riffled sluices, and spiral concentrators.

Pinched Sluice

A pinched sluice is a device comprised of an inclined plane bottom surface and tapering sides. The feed slurry is fed at the higher end and is made to flow down the inclined surface. The segregated heavy minerals flowing closer to the bottom of the surface are removed through slots placed across the bottom of the channel at the discharge end while the lighter minerals flow over the slot and report to tailings. The depth of flow increases toward the lower end because of the lowering of the area of cross section. These sluices are used by small-scale operators treating alluvial gold deposits and in tin and heavy minerals industries. Figure 14.4-5 shows a schematic diagram of a pinched sluice.

These units are operated at inclinations of between 15° and 20° to the horizontal. Subasinghe and Kelly (1984) demonstrated that at higher inclinations the recovery of coarser particles decreases because of the presence of Bagnold forces acting on the particles. At lower slopes, coarse and dense particles settle out on the deck resulting in poor segregation of heavies in the bed.

Reichert Cone

Reichert cones operate in a similar way to that of pinched sluices. The cone is designed as a series of pinched sluices joined in a circular pattern with their sides removed. An inverted cone is placed above the concentrating cone to distribute the feed slurry evenly. Heavy and light materials are removed through annular slots placed near the center column. Several cone pairs may be stacked on top of each other to increase the quality of the separation. Cones are operated continuously. Figure 14.4-6 shows a schematic diagram of a Reichert cone.

Riffled Sluice

Riffled sluices are commonly used by artisan miners to recover gold. The segregated gold or heavy minerals are recovered behind the riffles placed across the sluice. The feed flow rate, height, and spacing of the riffles determine the size of the recovered particles. If the feed contains fine gold, then a lower flow rate and smaller riffles are used. Sometimes instead of the riffles, corduroy cloth or coir mats are used to trap the fine gold particles. Sluices have been used to process fine cassiterite ores.

Spiral Concentrator

Spiral concentrators have found wide applications in the beach sand industry. In a spiral concentrator, the feed slurry is made to flow down a helical conduit of semicircular cross section. The slurry experiences a centrifugal force and has a tendency to move particles toward the outer rim. Because of the shape of the trough's cross section, these particles then travel inward, closer to the bottom of the trough. The heavy particles settle out of suspension and flow down the spiral closer to the trough's inner edge. There they are subjected to a smaller centrifugal acceleration and take up a trajectory closer to the

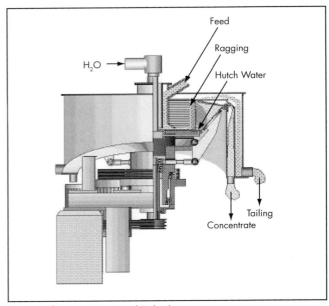

Courtesy of DownerEDi–Mineral Technologies.
Figure 14.4-4 Kelsey centrifugal jig

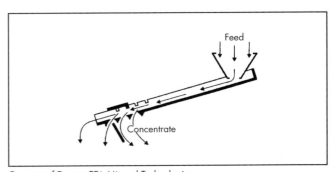

Courtesy of DownerEDi–Mineral Technologies.
Figure 14.4-5 Pinched sluice

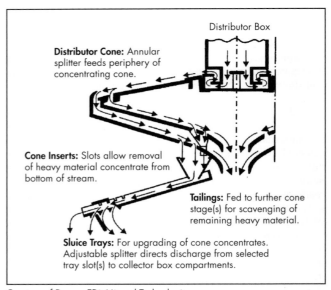

Courtesy of DownerEDi–Mineral Technologies.
Figure 14.4-6 Reichert cone

axis of the spiral. The lighter particles move closer to the free surface and are carried toward the outer rim as a result of the centrifugal acceleration. When the slurry reaches the bottom of the spiral, separation of the heavies has occurred, and the heavies may be collected through the inner ports while the lighter material is collected from the outer port. To save floor space, two or more spirals may be incorporated into one unit known as a twin, triple, or quad start, thereby increasing the throughput rate.

The shape of the trough varies with the type of mineral being separated. Coal spirals generally have a flatter profile than mineral spirals. The pulp density may vary between 20% and 40% solids. The feed sizes range from about 2 mm to about 75 μm. The capacity of a single-start spiral may be up to 7.0 t/h, depending on the material being processed. The cut specific gravity depends on the pulp density, feed rate, and size distribution of the feed. In the beach sand industry, spirals are commonly used as a preconcentration stage. Figure 14.4-7 shows a schematic diagram of a concentrating spiral.

Shaking Devices

The most common shaking device is the concentrating table or the Wilfley table. Concentrating tables are very efficient gravity separators generally used in cleaning concentrates obtained from other devices. Concentrating tables or shaking tables are comprised of a rectangular flat table inclined horizontally by about 15° to 20°. Riffles are placed along the length of the table and the table is made to vibrate in the longitudinal direction in a reciprocating manner, with a slow forward stroke and a rapid return. The feed slurry enters through a port at the top corner and is distributed over the flat surface because of the vibratory motion. Wash water is introduced along the top edge of the table surface, washing the particles down the slope. Heavy mineral particles get trapped between the riffles and travel across the table in a longitudinal direction. The lighter minerals flow over the riffles and are carried down the slope and report to the tailings stream, nearer to the feed end of the table. The heavy mineral concentrate is collected at the opposite end of the table with middling particles collected in between. The feed to tables may vary from about 1 mm down to about 50 μm. The feed pulp density is between 25% and 35% solids. Figure 14.4-8 shows a schematic diagram of a Wilfley table.

Many designs and riffle geometries exist, depending on the type of material being processed. Tables have been in use for processing of minerals such as coal, barites, beach sands, chromite, garnet, iron, manganese, tantalum, tin, tungsten, and zircon. Generally the variables that may be manipulated to increase the efficiency of separation are deck slope, wash water flow rate, and the frequency and amplitude of the vibratory motion. With larger particles, longer strokes and lower shaking frequencies are used. The separation is judged mainly by visual inspection, and modifications to operating variables may be made accordingly. These are low-capacity devices with about 2 and 0.5 t/h/m² for 1.5-mm and 100-μm particles, respectively. As they take up much floor space, multiple decks may be stacked up one above the other to save space.

Gemini tables have found wide application in the gold industry as a device for cleaning concentrates of centrifugal devices such as Knelson and Falcon concentrators. For finer feeds, Bartles–Mozley tables are generally used, in which the shaking mechanism is replaced by a smoother oscillator movement to avoid mixing of the bed due to turbulence. Air

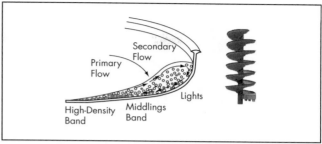

Courtesy of DownerEDi–Mineral Technologies.
Figure 14.4-7 Spiral concentrator

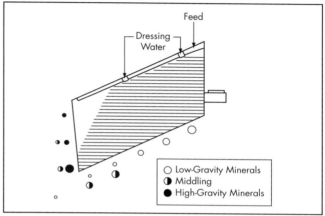

Source: Aplan 1985.
Figure 14.4-8 Wilfley table and distribution of table products

tables have also been used in the mineral industry to recover fine particles in which the bed is fluidized using air.

As an aside, air tables were originally developed for use in arid regions but have also been used in the separation of flat, fibrous materials of lower bulk densities such as asbestos and vermiculite. However, these devices demand that the feed be very narrowly sized for their efficient operation. As they also have only limited capacities, their use is limited only to special applications.

Centrifugal or Enhanced Gravity Concentrators

The movement of fine particles that settle under gravitational force within a fluid medium is slow. Even if there is a significant density difference between heavy and light minerals, the separations achieved are inadequate. However, by allowing the particles to traverse a circulatory motion at high speed, a centrifugal acceleration may be transmitted to them that may be several hundred times that of gravity. Devices that use this technique are called centrifugal concentrators. Two devices that have found wide applications in recovering fine gold and other minerals are the Knelson and Falcon concentrators.

Knelson Concentrator

Knelson concentrators, as shown in Figure 14.4-9, are comprised of a slightly conical-shaped rotating bowl with a series of grooves on the outer wall to capture the segregated heavy mineral particles. The feed material enters through a central feed inlet as a slurry. The slurry descends onto the base plate at the bottom of the rotating bowl and is thrown outward because of centrifugal acceleration. As a result,

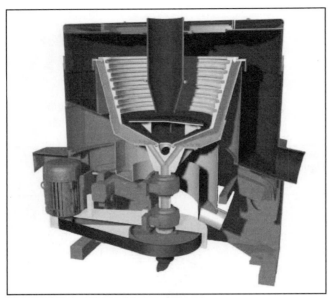

Courtesy of Knelson Gravity Solutions.
Figure 14.4-9 Knelson concentrator

some of the particles get trapped within the grooves of the bowl forming a bed while the excess get carried upward into the tailings stream by the rising current of water. Injection of water through fluidization ports located in the grooves of the bowl prevents compaction of the concentrate bed. This creates a fluidized bed that acts as a concentrating chamber for heavy mineral particles, under an enhanced separating force. Upon conclusion of the concentration cycle, concentrates are collected manually in batch units or flushed from the bowl into a concentrate launder through a multiport hub in continuous units.

As heavy particles experience a higher centrifugal acceleration, they tend to move through the fluidized bed to displace the light materials and produce a heavy mineral concentrate within the grooves of the bowl. The lighter particles are displaced into the tailings stream by the fluidizing water. Fluidizing water flow rate, bowl rotational speed, and particle characteristics of the feed, such as the size and size distribution, determine the optimal conditions for separation and the cycle time.

Falcon Concentrator
Falcon concentrators rely on a rotating bowl to generate high centrifugal accelerations similar to Knelson concentrators. The feed slurry rises upward over a slightly inclined smooth wall. Stratification of higher-density particles occurs toward the bowl wall, and lower-density particles move toward the center of the bowl. Different models vary in the way they collect and produce the concentrate.

In the semibatch unit, the upgraded product that has been forced against the bowl wall is made to move into a vertical, fluidized collection zone in the upper portion of the bowl. The collection zone consists of a number of rings that have pressurized water injected from the back, effectively elutriating and cleaning the concentrate that sits in these collection rings until a rinse cycle begins. These machines are known as semibatch devices because they continually accept feed during the run cycle but only produce concentrate during periodic rinse cycles. Run times range from 30 minutes to several hours depending on the application, while the rinse times are generally less than a minute. The throughput rates range from 1 to 400 t/h. The centrifugal accelerations range from 50 to 200 G. These units are mainly used in precious metal recovery plants.

In continuous devices, at the top of the bowl is a ring of specially designed concentrate flow hoppers, followed by pneumatically controlled variable orifice valves. These valves allow for a constant stream of concentrate to be produced. The continuous concentrator is able to vary mass pull to concentrate by adjusting valve orifice size through changes in air pressure. Because of their high mass pull rates of 5% to 40%, these units are used in the recovery and upgrade of tin, tantalum, tungsten, chrome, cobalt, iron, fine oxidized coal, and uranium ores. The ultrafine units are designed to concentrate material below 50 μm at less than 20% solids and may be typically used in deslime cyclone overflow streams. Figure 14.4-10 shows a schematic diagram of a Falcon continuous concentrator.

HEAVY MEDIUM SEPARATION
Heavy medium separation (HMS) is a very effective gravity separation technique in which a nonsettling heavy medium is used in place of a fluid. The heavy medium is generally

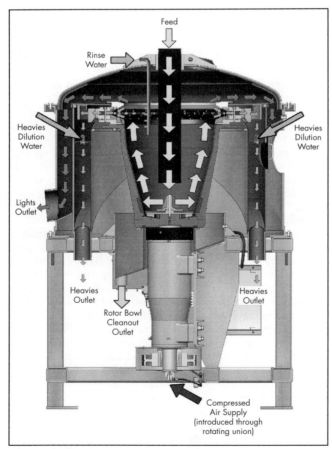

Courtesy of Falcon Concentrators Inc.
Figure 14.4-10 Falcon continuous concentrator

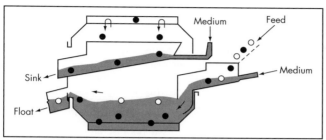

Source: Aplan 1985.
Figure 14.4-11 Heavy medium separator

comprised of a nonsettling suspension of fine heavy particles, the density of which is much higher than that of water. Thus, the concentration criterion described in Equation 14.4-1 increases considerably in this case, which indicates a more efficient separation. Also, since the medium density is between those of the heavy and light minerals to be separated, it gives rise to an absolute separation where the two products are separated into a sinks product and a floats product. This technique requires a much lower density difference between the minerals to be separated, compared to the relative separation techniques described previously.

With HMS it is possible to (a) produce a finished concentrate and a final waste product in one operation, (b) reject a waste product at a coarser size, thereby saving grinding costs, (c) achieve separation at a low operating cost with low maintenance costs, (d) make relatively sharp separations, (e) operate continuously, (f) tolerate feed with wide size distributions, and (g) produce a consistent product for further processing. The main industrial applications of HMS with coarse material feeds are in coal and iron ore industries. HMS has also been used in base metal industries such as copper, lead, and zinc. The most commonly used medium in coal separation is magnetite. Although silica, barite, and galena had been used in the past as the medium solid for ore separations, in recent times ferrosilicon (FeSi) has been used extensively. With finer feed materials, the viscosity of the slurry increases, and the separation efficiency decreases as a result. This situation has been rectified by the use of centrifugal separators such as heavy medium cyclones (HMCs). The water-only cyclones work similarly to an HMC that uses a loess medium instead of an external medium. See Figure 14.4-11 for a schematic diagram of a heavy medium separator.

Heavy Medium Solids

The solids used to make up the heavy medium depend on the type of material to be separated. The heavy medium suspension should be nonsettling, which requires the solid particles to be sufficiently small. For coal, the most commonly used are magnetite and barite, as the medium density should be less than the ash density. For ore separations, more dense solid materials such as ferrosilicon and galena have been used (Aplan 2003). Ferrosilicon contains about 15% Si. Ground ferrosilicon is used for separations at SGs up to 3.2; for higher SGs up to 3.8, atomized FeSi may be used. The main requirement in selecting a medium for a given operation is its ability to be separated from the products. Generally the products are washed and screened to remove the medium. Wet magnetic separators are also used extensively to recover magnetite and ferrosilicon. The medium recovered from the product streams needs to be thickened to the required consistency prior to recycling it back to the separating vessel.

Feed Preparation

It is important to prepare the feed prior to HMS, usually by wet screening. The main purpose is to size the ore into fractions that may be treated in various separating units. More importantly, screening is used to remove the slimes that would otherwise increase the viscosity of the medium, giving rise to inefficiencies.

Medium Recovery

Medium is generally removed from the products in drain and wash screens. Sieve bends are the preferred screen type for medium recovery. Prior to reusing, the medium needs to be cleaned using wet magnetic separators. Magnetic separators are fed at about 30% to 35% solids, and the rate is determined by the type of separator and magnetic susceptibility of the medium.

Heavy Medium Separators

HMS equipment must ensure the feed particles have sufficient time to report to the relevant discharge streams. They generally differ in the way the float and sink products are discharged. The float products will often discharge over a weir with or without the need for paddles or scrapers. The Wemco drum discharges the heavy mineral using lifters while rotating. The Drewboy washer uses paddles to elevate the heavies, while the Norwalt bath uses a bucket elevator. The Dutch State Mines bath uses a drag conveyor with a paddle to deliver the heavies and lights to different discharge ports.

The separations of finer sizes are carried out effectively in units that use centrifugal acceleration to aid in the separation and discharge of particles. HMCs and Dyna Whirlpool separators (DWPs) are the widely used equipment of this type.

Heavy Medium Cyclones

The HMCs are essentially a modified version of hydrocyclones used with a heavy medium as the separating fluid instead of water. In an HMC, the heavy mineral migrates to the outside wall and is discharged through the vortex finder. The shape of the HMC, however, differs from that of a classifying cyclone in that the height of the cyclone is comparatively shorter. HMCs treating coal operate at inlet pressures with a minimum head of about nine times the diameter of the cyclone. The diameters of these HMCs range from about 0.5 m to 1.4 m and have capacities up to 500 t/h. The relative density of the overflow stream should be between 3% and 12% lower than that of the feed stream.

Dyna Whirlpool Separator

The DWP also imparts a centrifugal acceleration to the feed particles. A DWP is essentially a sloping cylinder with cover plates and central openings to which the raw feed is fed centrally at the top end. There are also tangential orifices at either end. The medium enters tangentially at the lower end and migrates toward the top end, leaving the cylinder tangentially along with the heavy mineral. The lights gravitate to the bottom end and leave through a central discharge port. Generally, centrifugal devices such as the HMC and DWP experience higher media losses than static-type devices. A schematic diagram of a Dyna Whirlpool separator is shown in Figure 14.4-12.

PERFORMANCE EVALUATION OF GRAVITY SEPARATORS

Gravity separators aim to separate minerals at a desired separation density. However, because of the presence of middling particles, which have densities between those of heavy and light minerals, a sharp separation is not possible. Also, since these devices rely on differences between settling rates of the two minerals, fine feed material inherently gives rise to inefficiencies. The efficiency of a gravity concentrator is determined by two factors: material characteristics and machine characteristics. The degree of liberation of the feed material that affects the separation efficiency is represented by separability curves, whereas machine characteristics are evaluated in terms of performance curves (Kelly and Spottiswood 1982). The performance curve of an operating machine is represented by a plot of the fractional mass of particles reporting to the concentrate stream against the extent of the property being exploited for the separation. For gravity concentration, traditionally density (or SG) of the particles has been taken as the property being exploited. Such a plot is commonly known as the performance curve, partition curve, or the Tromp curve, which may be construed as a probability plot that indicates the probability of a particle of given SG reporting to the concentrate stream. A typical performance curve is shown in Figure 14.4-13.

A flatter curve represents a poor separation, whereas a steeper slope represents a sharp separation. The specific gravity of particles that have equal chances of reporting to either exit is known as the cut specific gravity (SG_{50}) or separation density. For static heavy medium separators the separation density is close to the medium density whereas for centrifugal separators, it is slightly higher.

To compare different operations with different separation SGs, several efficiency measures have been proposed. Measures of efficiency based on performance curves generally used for gravity concentrators are probable error (Ecart probable, E_p), imperfection (I), and the sharpness index (SI). These are defined as follows:

$$E_p = \tfrac{1}{2} (SG_{75} - SG_{25})$$

$$I = E_p / (SG_{50} - 1)$$

$$\text{and} \quad SI = \frac{SG_{75}}{SG_{25}} \qquad (14.4\text{-}2)$$

where SG_{25}, SG_{50}, and SG_{75} are the specific gravities at which 25%, 50%, and 75% of the material would report to the concentrate stream, respectively. The percentage of misplaced material measured as an area between curves has also been used as a measure of efficiency.

E_p represents the inverse of the slope in the middle part of the curve but does not take into account the shape of the two ends of the performance curve that represent most of the misplaced material. Generally, the cut specific gravity of

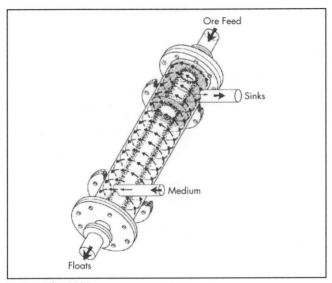

Source: Aplan 1985.
Figure 14.4-12 Dyna Whirlpool separator

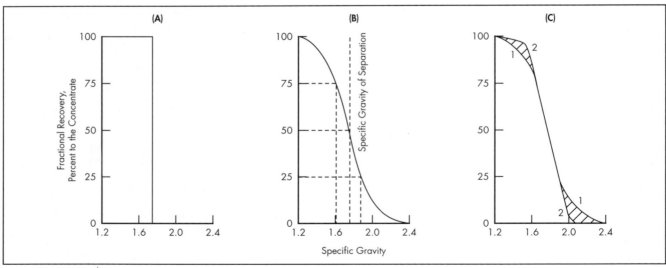

Source: Aplan 2003.
Figure 14.4-13 Performance curves of a gravity separator. Partition curves: (A) perfect separation, (B) actual separation (curve 1), and (C) same E_p as for graph B but with superior recovery of misplaced particles (shaded area between curves 1 and 2).

Table 14.4-3 Approximate E_p values for coal cleaning devices at 1.5 specific gravity

Coal Size and Appropriate Cleaning Device	Ep at Stated Size				Broad Size Range of Feed	
	+½ in. (+12.7 mm)	½ × ¼ in. (12.7 × 6.4 mm)	14 × 28 M* (1.7 × 0.6 mm)	20 × 200 M (830 × 74 mm)	Range	Ep
Course coal						
Baum jig	0.06	0.16	0.30	—	6 in. × 48 M	0.12
HMS, static bath	0.03	—	—	—	6 in. × ¼ in.	0.03
Intermediate to fine coal						
HMS, cyclone	0.02	0.03	0.05	—	¾ in. × 28 M	0.03
Shaking tables	—	0.07†	0.10	0.20	⅜ in. × 200 M	0.09
Water-only cyclone	—	0.15†	0.20	—	¼ in. × 200 M	0.28

Source: Aplan 2003.
*All mesh sizes (inches) are in Tyler Standard mesh.
†¼ in. × 14 M.

Table 14.4-4 Influence of near-gravity material on the difficulty of separation, process selection, and suggested equipment

Wt % ± 0.1 SG	Degree of Difficulty	Suggested Gravity Process	Suggested Devices
0–7	Simple	Almost any	Jigs, HMS, tables, spirals, sluices
7–14	Moderately difficult	Efficient process	
10–15	Difficult	Efficient process, good operation	
15–20	Very difficult	Very efficient process, expert operation	Heavy media separation
20–25	Exceedingly difficult		
>25	Formidable	Exceptionally efficient process, expert operation	Heavy media separation, close control

Source: Aplan 2003.

separations involving fine particles is higher. Thus, imperfection that takes into account this variation may be used to compare various equipment that process different-sized feeds. Table 14.4-3 shows typical E_p values for coal cleaning devices.

SELECTION OF GRAVITY SEPARATION EQUIPMENT

The selection of gravity separation equipment is based on the response of the ore to sink-and-float tests using heavy liquids. The commonly used heavy liquids are methylene iodide, tetrabromoethane, methylene bromide, and lithium sodium— tungstates whose densities are 3.3, 2.96, 2.48, and 2.85, respectively. For higher-density separations, Clerici solution is used. Generally an organic solvent may be used to dilute the reagents to a required specific gravity. Caution should be exercised in handling these reagents as most of them are toxic.

The data from these tests are presented as washability curves that essentially indicate the grade and recovery of the concentrate that can be achieved at various separation densities. The selection of gravity equipment is based on the amount of material present in the ore that is near the separation density. The amount of ore within ±0.1 SG units of the separation density is known as the tolerance value, which is a reflection of the degree of difficulty of separation. Table 14.4-4 shows the recommended equipment at various tolerance values.

The selection also depends on the feed size of the ore as each equipment type can handle material only within a specified size range.

For the selection of centrifugal separators for fine gold recovery, a gravity recoverable gold test has been proposed (Laplante et al. 2000). This test is a material characterization test that indicates the amenability of the ore to centrifugal gravity concentration. For the purpose of determining optimal operating conditions of a Knelson concentrator, a model based on a mechanistic approach has been proposed (Subasinghe 2007).

REFERENCES

Alderman, J.K. 2002. Types and characteristics of non-heavy medium separators and flowsheets. In *Mineral Processing Plant Design, Practice and Control Proceedings*, Vol. 1. Edited by A.L. Mullar, D.N. Halbe, and D.J. Barratt. Littleton, CO: SME. pp. 978–994.

Aplan, F.F. 1985. Gravity concentration, In *SME Mineral Processing Handbook*, Vol. 1. Edited by N.L. Weiss. New York: SME-AIME. pp. 4-1–4-55.

Aplan, F.F. 2003. Gravity concentration. In *Principles of Mineral Processing*. Edited by M.C. Fuerstenau and K.N. Han. Littleton, CO: SME. pp. 185–219.

Gaudin, A.M. 1939. *Principles of Mineral Dressing*. New York: McGraw-Hill.

Kelly, E.G., and Spottiswood, D.J. 1982. *Introduction to Mineral Processing*. New York: Wiley.

Laplante, A.R., Woodcock, F.C., and Huang, L. 2000. A laboratory procedure to characterise gravity recoverable gold. *Trans. SME* 308:53–59.

Subasinghe, G.K.N.S. 2007. Evaluating an alternative to the gravity recoverable gold (GRG) test. In *Proceedings of Ninth Mill Operators' Conference*, Perth, March 19–21.

Subasinghe, G.K.N.S., and Kelly, E.G. 1984. Modelling pinched-sluice type concentrators. In *CONTROL '84: Mineral/Metallurgical Processing*. Edited by J.A. Herbst. Littleton, CO: SME-AIME.

Taggart, A.F. 1951. *Elements of Ore Dressing*. New York: Wiley.

Fundamental Principles of Froth Flotation

S. Komar Kawatra

INTRODUCTION

Froth flotation is a highly versatile method for physically separating particles based on differences in the water-repellency characteristics of the surfaces of various rock and mineral species contained in aqueous slurry. The air bubbles generated within a separation vessel collect hydrophobic particles and ascend to the surface with the particles attached to be removed, while the particles that remain completely wetted stay in the liquid phase. Froth flotation can be adapted to a broad range of mineral separations, as it is possible to use chemical treatments to selectively alter mineral surface characteristics so that they have the necessary hydrophobicity for the separation to occur. Many diverse applications, both mineral and nonmineral, are currently in use. Examples of mineral applications include separating sulfide minerals from silica gangue (and selectively from other sulfide minerals), separating potassium chloride (sylvite) from sodium chloride (halite), separating coal from ash-forming minerals, removing silicate minerals from iron ores, and separating phosphate minerals from silicates. An example of a nonmineral application is the de-inking of recycled newsprint. Flotation is particularly useful for processing fine-grained ores that are not amenable to conventional gravity concentration because of their low density differentials that render gravity separation unacceptably ineffective.

The flotation system includes many interrelated components, and changes in one area will produce compensating effects in other areas, as shown in Figure 14.5-1 (Klimpel 1995). Therefore, it is important to take all of these factors into account when developing specific froth flotation applications. Changes in the settings of one factor (such as feed rate) will automatically cause changes in other parts of the system (such as flotation rate, particle size recovery, airflow rate, and pulp density). As a result, it is difficult to study the effects of any single factor in isolation, and compensation effects within the system can sometimes result in an unexpected and undesirable outcome (Klimpel 1995). This makes it difficult to develop predictive models for froth flotation; although work is being done to develop models that utilize readily measurable parameters such as solids recovery and tailings solid content (Rao et al. 1995). Such models incorporate fundamental physical and chemical reactions as well as empirical data that relate to the mineral components, and they are usually validated by various levels of test work (i.e., bench-scale, pilot-plant, and commercial plant trials).

PERFORMANCE CALCULATIONS

A universal method for expressing the effectiveness of a separation does not exist, but there are several useful methods for examining froth flotation processes.

Parameters

The following parameters are used to measure performance.

(a) *Ratio of concentration:* The ratio of concentration is the weight of the feed relative to the weight of the concentrate, which is expressed as follows:

ratio of concentration = F/C

where
F = total weight of the feed
C = total weight of the concentrate

One limitation with this calculation is that it uses the weights of the feed and the concentrate. Although these data can be obtained from laboratory experiments, in the plant it is likely that the ore is not weighed and only assays are available. However, it is possible to express the ratio of concentration in terms of ore assays. This can be done starting with the definition of the ratio of concentration (F/C) and using the following mass balance equations:

$$F = C + T$$
$$Ff = Cc + Tt$$

where
F, C, and T = percent weights of the feed, concentrate, and tailings, respectively
f, c, and t = assays of the feed, concentrate, and tailings, respectively

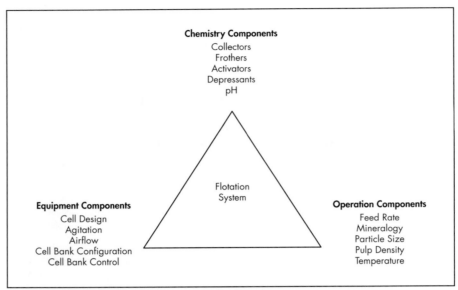

Figure 14.5-1 Interrelated components of a flotation system

So that F/C can be determined, one needs to eliminate T from these equations. Multiplying $F = C + T$ by t gives $Ft = Ct + Tt$, and subtracting this equation from $Ff = Cc + Tt$ eliminates T and gives $F(f - t) = C(c - t)$. Rearranging then gives

$$F/C = (c - t)/(f - t)$$

(b) *Percent metal recovery:* The percent metal recovery is the percentage of the metal in the original feed that is recovered in the concentrate. This can be calculated using weights and assays, for example, $(Cc)/(Ff) \times 100$, or because $C/F = (f - t)/(c - t)$, the percent metal recovery can be calculated from the assays alone using $100(c/f)(f - t)/(c - t)$.

(c) *Percent metal loss:* The percent metal loss is the opposite of the percent metal recovery and represents the material lost to the tailings. It can be calculated simply by subtracting the percent metal recovery from 100%.

(d) *Percent weight recovery:* The percent weight recovery is essentially the inverse of the ratio of concentration and equals $100C/F = 100(f - t)/(c - t)$.

(e) *Enrichment ratio:* The enrichment ratio is calculated directly from the assays as c/f. The weights are not involved in the calculation.

Example Calculations

A copper ore initially contains 2.09% Cu. After carrying out a froth flotation separation, the products are as shown in Figure 14.5-2. Using these data, parameters (a) through (e) are calculated as follows.

(a) *Ratio of concentration:* From Figure 14.5-2, the ratio of concentration can be calculated as $F/C = 100/10 = 10$. If only assays are available, the ratio of concentration equals $(20 - 0.1)/(2.09 - 0.1) = 10$. Therefore, for each 10 t (metric tons) of feed, the plant would produce 1 t of concentrate.

(b) *Percent metal recovery:* Using the example data from Figure 14.5-2, the percent Cu recovery calculated from the weights and assays is as follows:

$$\text{percent Cu recovery} = [(10 \times 20)/(2.09 \times 100)] \times 100 = 95.7\%$$

The calculation using assays alone is as follows:

$$\text{percent Cu recovery} = 100(20/2.09)(2.09 - 0.1)/(20 - 0.1) = 95.7\%$$

This means that 95.7% of the copper present in the ore was recovered in the concentrate, while the rest was lost in the tailings.

(c) *Percent metal loss:* The percent Cu loss can be calculated by subtracting the percent Cu recovery from 100% as follows:

$$\text{percent Cu loss} = 100 - 95.7 = 4.3\%$$

This means that 4.3% of the copper present in the ore was lost in the tailings.

(d) *Percent weight recovery or percent yield:* The percent weight recovery is equal to the percent weight of the concentrate in Figure 14.5-2. It can also be calculated as follows from the assay values given in the figure:

$$\text{percent weight recovery} = 100 \times (2.09 - 0.1)(20 - 0.1) = 10\%$$

(e) *Enrichment ratio:* The enrichment ratio is calculated by dividing the concentrate assay in Figure 14.5-2 by the feed assay as follows:

$$\text{enrichment ratio} = 20.0/2.09 = 9.57$$

This reveals that the concentrate has 9.57 times the copper concentration of the feed.

Grade–Recovery Curves

Although these calculated values are useful for comparing flotation performance for different conditions, it is most useful to consider both the grade and the recovery simultaneously using a "grade–recovery curve." This is a graph of the recovery of the valuable metal versus the product grade at that recovery, and it is particularly useful for comparing separations where both the grade and the recovery are varying. A set of grade–recovery curves is shown in Figure 14.5-3. If 100% of the

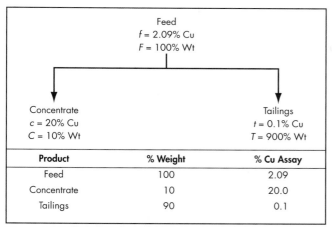

Figure 14.5-2 Grade–recovery performance of a hypothetical copper ore flotation process

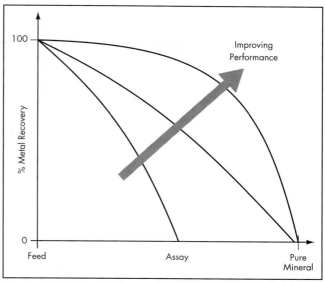

Figure 14.5-3 Typical forms of the grade–recovery curves for froth flotation

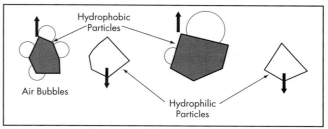

Figure 14.5-4 Selective attachment of air bubbles to hydrophobic particles

feed is recovered to the product, then the product will obviously have the same composition as the feed, and so the curve starts at the feed composition with 100% recovery. Similarly, if the purest mineral grain that contains the metal of interest is removed, this will be the maximum grade that can be produced by a physical separation, and so the zero percent recovery end of the curve terminates at an assay that is less than or equal to the assay of the purest grains available in the ore. In the curves shown in Figure 14.5-3, the points that are higher and to the right show better performance than the points that are lower and to the left.

HYDROPHOBICITY/HYDROPHILICITY

The basis of froth flotation is the difference in wettabilities of different minerals. Particles range from those that are readily wettable by water (hydrophilic) to those that are water-repellent (hydrophobic). If a mixture of hydrophobic and hydrophilic particles are suspended in water and air is bubbled through the suspension, then the hydrophobic particles will tend to attach to the air bubbles and float to the surface, as shown in Figure 14.5-4. The froth layer that forms on the surface will then be heavily loaded with the hydrophobic mineral, and it can be removed as a separated product. The hydrophilic particles will have limited or no tendency to attach to air bubbles, and so they will remain in suspension and be transported away in the slurry (Whelan and Brown 1956).

Particles can be naturally hydrophobic, or hydrophobicity can be induced by chemical treatments. Naturally hydrophobic materials include hydrocarbons and nonpolar solids such as elemental sulfur. Coal is a good example of a material that is normally naturally hydrophobic, because it is mostly composed of hydrocarbons. Chemical treatments to render a surface hydrophobic are essentially methods for selectively coating a particle surface with a monolayer of nonpolar oil.

The attachment of the bubbles to the surface is determined by the interfacial energies between the solid, liquid, and gas phases. This is determined by Young–Dupré equation as follows:

$$\gamma_{lv}\cos\theta = (\gamma_{sv} - \gamma_{sl})$$

where

γ_{lv} = surface energy of the liquid–vapor interface
γ_{sv} = surface energy of the solid–vapor interface
γ_{sl} = surface energy of the solid–liquid interface
θ = "contact angle"

The contact angle (θ) is the angle formed at the junction between vapor, solid, and liquid phases, as shown in Figure 14.5-5. If the contact angle is very small, then the bubble does not attach to the surface, whereas a very large contact angle results in very strong bubble attachment. A contact angle near 90° is sufficient for effective froth flotation in most cases.

PARTICLE–BUBBLE CONTACT

When particles are rendered hydrophobic, contact with gas bubbles must occur so that attachment to the surface can take place. Contact between particles and bubbles can be accomplished in a flotation cell such as the agitated cell shown in Figure 14.5-6.

In the agitated cell, the rotor draws slurry through the stator and expels it to the sides, creating a suction that draws air down the shaft of the stator. The air is then dispersed as bubbles through the slurry and comes in contact with particles in the slurry that is circulated through the stator.

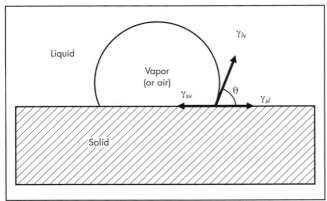

Figure 14.5-5 Contact angle between air bubble and solid surface immersed in liquid

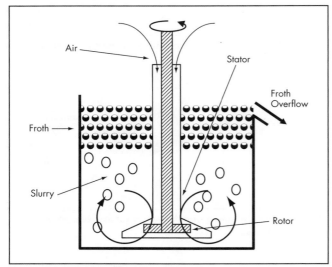

Figure 14.5-6 Simplified flotation cell

Particle–bubble collision is affected by the relative sizes of the particles. If the bubbles are large relative to the particles, then fluid flowing around the bubbles can sweep the particles past without them coming in contact. Therefore, it is best if the bubble diameter is comparable to the particle diameter to ensure good particle–bubble contact.

COLLECTION IN THE FROTH LAYER

When a particle and bubble have come in contact, the bubble must be large enough for its buoyancy to lift the particle to the surface. This is obviously easier if there are low-density particles (as is the case for coal) than if there are high-density particles (such as lead sulfide). The particle and bubble must remain attached while they move up into the froth layer at the top of the cell. The froth layer must persist long enough to either flow over the discharge lip of the cell by gravity or be removed by mechanical froth scrapers. If the froth is insufficiently stable, the bubbles will break and drop the hydrophobic particles back into the slurry prematurely. However, the froth should not be so stable as to become persistent foam, as foam is difficult to convey and pump through the plant.

The surface area of the bubbles in the froth is also important. Because particles are carried into the froth by attachment to bubble surfaces, increasing amounts of bubble surface area allows a more rapid flotation rate of particles. At the same time, increased surface area also carries more water into the froth as the film between the bubbles. Because fine particles that are not attached to air bubbles will be unselectively carried into the froth along with the water (entrainment), excessive amounts of water in the froth can result in significant contamination of the product with gangue minerals.

REAGENTS

The properties of raw mineral mixtures suspended in normal plant circulating water are rarely suitable for froth flotation. Chemicals are needed both to control the relative hydrophobicities of the particles and to maintain the proper froth characteristics. Therefore, many different reagents are involved in the froth flotation process, with the selection of reagents depending on the specific mineral mixtures being treated (Klimpel 1980; Thompson 2002).

Collectors

Collectors are used to selectively adsorb onto the surfaces of particles. They form a monolayer on the particle surface that essentially makes a thin film of nonpolar hydrophobic hydrocarbons. The collectors greatly increase the contact angle so that bubbles will adhere to the surface. Selection of the correct collector is critical for an effective separation by froth flotation. Collectors can be generally classed depending on their ionic charge; they can be nonionic, anionic, or cationic, as shown in Figure 14.5-7 (Glembotskii et al. 1963). Nonionic collectors are simple hydrocarbon oils, whereas the anionic and cationic collectors consist of a polar part that selectively attaches to the mineral surfaces and a nonpolar part that projects out into the solution, rendering the surface hydrophobic. Collectors can either chemically bond to the mineral surface (chemisorption) or they can be held on the surface by physical forces (physical adsorption).

Chemisorption

In chemisorption, ions or molecules from solution undergo a chemical reaction with the surface, becoming irreversibly bonded. This permanently changes the nature of the surface. Chemisorption of collectors is highly selective, as the chemical bonds are specific to particular atoms.

Physisorption

In physisorption, ions or molecules from solution become reversibly associated with the surface, attaching due to electrostatic attraction or van der Waals bonding. The physisorbed substances can be desorbed from the surface if conditions such as the pH or the composition of the solution changes. Physisorption is much less selective than chemisorption, as collectors will adsorb on any surface that has the correct electrical charge or degree of natural hydrophobicity.

Nonionic Collectors

Hydrocarbon oils, and similar compounds, have an affinity for surfaces that are already partially hydrophobic. They selectively adsorb on these surfaces and increase their hydrophobicity. The most commonly floated naturally hydrophobic material is coal. The addition of collectors such as no. 2 fuel oil and kerosene significantly enhances the hydrophobicity of the coal particles without affecting the surfaces of the associated ash-forming minerals. This improves the recovery of

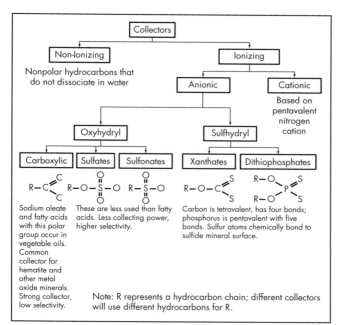

Figure 14.5-7 Basic collector types

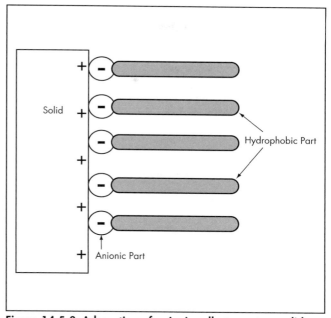

Figure 14.5-8 Adsorption of anionic collector onto a solid surface

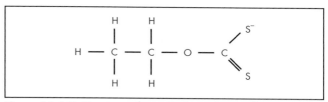

Figure 14.5-9 Structure of a typical xanthate collector (ethyl xanthate)

the coal and increases the selectivity between coal particles and mineral matter. Fuel oil and kerosene have the following advantages over specialized collectors for froth flotation:

- They have a low enough viscosity to disperse in the slurry and spread over the coal particles easily.
- They are very low cost compared to other compounds that can be used as coal collectors.

In addition to coal, it is also possible to float naturally hydrophobic minerals such as molybdenite, elemental sulfur, and talc with nonionic collectors. Nonionic collectors can also be used as "extenders" for other collectors. If another, more expensive collector makes a surface partially hydrophobic, adding a nonpolar oil will often increase the hydrophobicity further at low cost.

Anionic Collectors

Anionic collectors are usually weak acids or acid salts that ionize in water, producing a collector that has a negatively charged end that will attach to the mineral surfaces and a hydrocarbon chain that extends out into the liquid, as shown in Figure 14.5-8. The anionic portion is responsible for the attachment of the collector molecule to the surface, while the hydrophobic part alters the surface hydrophobicity.

Anionic collectors for sulfide minerals. The most common collectors for sulfide minerals are the sulfhydryl collectors, such as the various xanthates and dithiophosphates.

Xanthates are most widely studied and commonly used, and they have structures similar to what is shown in Figure 14.5-9. Xanthates are highly selective collectors for sulfide minerals, as they chemically react with the sulfide surfaces and do not have any affinity for the common nonsulfide gangue minerals (Crozier 1992). The $OCSS^-$ group attaches irreversibly to the sulfide mineral surface. Using xanthates with longer hydrocarbon chains tends to increase the degree of hydrophobicity when they adsorb on the surface.

There are a wide variety of other sulfhydryl collectors (Thompson 2002; Bulatovic 2007), which have somewhat different adsorption behaviors so they can be used for some separations that are difficult using xanthates. These include the following:

- *Dithiophosphates* have greater selectivity against pyrite than do xanthates, and some of these can collect liberated grains of gold. They are also useful in lead sulfide, zinc sulfide, and silver sulfide flotation.
- *Thionocarbamates* selectively float copper sulfides from pyrite at alkaline pH, and they work well for copper sulfides that tend to grind into the slime fraction, such as chalcocite.
- *Dithiophosphinates* are even more selective against pyrite than dithiophosphates, and they are particularly useful for complex polymetallic ores. Although they are selective, they are not particularly strong collectors (produce low levels of hydrophobicity), and so they are often used in combination with other collectors.
- *Trithiocarbonate* has a very high selectivity against pyrite in copper ore flotation.
- *Mercaptobenzothiazole* has improved the recovery of tarnished or semi-oxidized copper and zinc sulfides. It is often used in combination with xanthates.

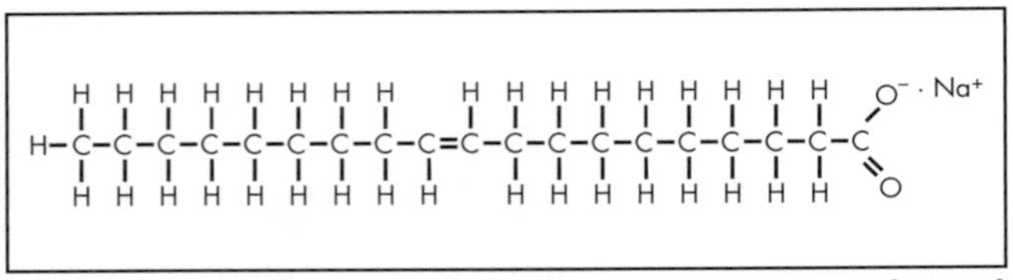

Figure 14.5-10 Structure of oleic acid, a very commonly used anionic collector

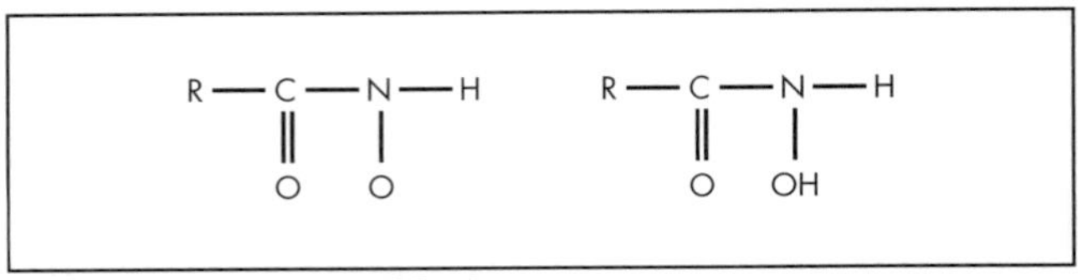

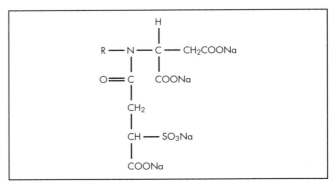

Figure 14.5-11 Typical hydroxamate structures

Figure 14.5-12 Active group of sulfosuccinamate

- *Mercaptans* can be used in copper flotation, but they are not very popular because of their strong, unpleasant odor.
- *Xanthogen formates* float both copper sulfides and metallic copper, and they have been used to recover both cement copper and precipitated copper sulfide from acid pulps.
- *Monothiophosphates* have strong collecting power for sulfide minerals at pH values as low as 2.0, and they can be used in combination with dithiophosphates at more alkaline pH.

Anionic collectors for oxide minerals. The collectors available for the flotation of oxide minerals (oxyhydryl collectors) are not as selective as the collectors used for sulfide mineral flotation because they attach to the surface by electrostatic attraction rather than by chemically bonding to the surface. As a result, there is some collector adsorption onto the minerals that are not intended to float.

The most widely used oxyhydryl collectors are the carboxylates, with a typical example being the fatty acid sodium oleate, the sodium salt of oleic acid, which has the structure shown in Figure 14.5-10. Numerous other fatty acids can also be used as collectors; these differ from oleic acid mainly in their chain length and their degree of saturation. The anionic group responsible for attaching all of these reagents to the mineral surface is the carboxyl group, which dissociates in water to develop a negative charge. The negatively charged group is then attracted to positively charged mineral surfaces.

Because particles that are immersed in water develop a net charge due to exchanging ions with the liquid, it is often possible to manipulate the chemistry of the solution so that one mineral has a strong positive charge while other minerals have a charge that is either only weakly positive or is negative. In these conditions, the anionic collector will preferentially adsorb on the surface with the strongest positive charge and render them hydrophobic.

Other oxyhydryl collectors include the following (Bulatovic 2007):

- **Alkyl sulfates:** Alkyl sulfates are derived from sulfuric acid and are essentially a sulfate ion with a hydrocarbon group attached to one of the oxygens ($R\text{-}O\text{-}SO_3^-$). They are not much studied as flotation reagents, and because of this their industrial application is limited. Alkyl sulfates are used for flotation of sulfate minerals such as barite ($BaSO_4$), celestite ($SrSO_4$), gypsum ($CaSO_4 \cdot 2H_2O$), and anhydrite ($CaSO_4$).
- **Sulfonates:** These collectors are similar to alkyl sulfates, except that the hydrocarbon group replaces one of the oxygens from the sulfate ion and attaches directly to the sulfur ($R\text{-}SO_3^-$). They are generally manufactured in a poorly controlled way by treating petroleum fractions with sulfuric acid and separating the sulfonates from the acid sludge produced. They are poorly chemically characterized, and they are not much studied as flotation reagents, although they have been available from Cytec Industries.
- **Hydroxamates:** Belonging to the group of "chelating collectors," hydroxamates have structures such as those shown in Figure 14.5-11. These are intended to attach to ions on mineral surfaces at multiple points, with the objective of being more selective than other oxyhydryl collectors. They have been investigated as collectors for malachite, bastnaesite, titanates, and pyrochlore.
- **Sulfosuccinamates:** Derived from carboxylic and succinic acids, the sulfosuccinamates can be considered to be a hybrid of the fatty acid collectors and the sulfonate collectors, with active groups such as shown in Figure 14.5-12. The presence of multiple carboxyl groups and additional sulfonate groups gives them greater collecting power, at some cost to the selectivity. They are used for flotation of cassiterite and monazite.
- **Phosphonic acid:** A typical structure of styrene phosphonic acid is given in Figure 14.5-13, which serves as a basis for a range of collectors that have recently been developed for the flotation of cassiterite, ilmenite, rutile, and pyrochlore.
- **Phosphoric acid esters:** This family of collectors consists of a phosphate core with one or more attached hydrocarbon groups, as shown in Figure 14.5-14. Collectors of this group have been developed for the flotation of apatite, scheelite, ilmenite, rutile, perovskite, zircon, cassiterite, and pyrochlore.

Cationic Collectors

Cationic collectors use a positively charged amine group (Figure 14.5-15) to attach to mineral surfaces. Because the amine group has a positive charge, it can attach to negatively charged mineral surfaces. Cationic collectors therefore have essentially the opposite effect from anionic collectors, which

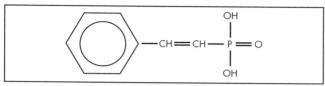

Figure 14.5-13 Styrene phosphonic acid structure

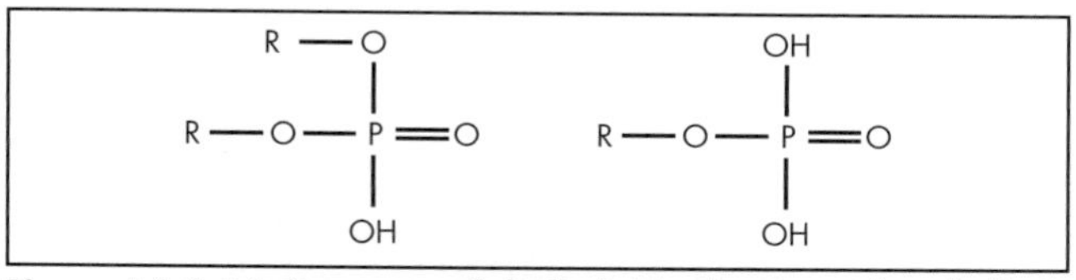

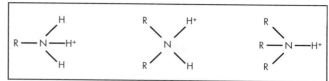

Figure 14.5-15 Primary, secondary, and tertiary amine groups

Figure 14.5-14 Structure of dialkylphosphoric acid and alkylphosphoric acid

attach to positively charged surfaces. Cationic collectors are mainly used for the flotation of silicates and certain rare-metal oxides and for the separation of potassium chloride (sylvite) from sodium chloride (halite).

Frothers

Frothers are compounds that act to stabilize air bubbles so that they will remain well dispersed in the slurry and will form a stable froth layer that can be removed before the bubbles burst. The most commonly used frothers are alcohols, particularly methyl isobutyl carbinol (MIBC), or 4-methyl-2-pentanol (a branched-chain aliphatic alcohol), or any of a number of water-soluble polymers based on propylene oxide, such as polypropylene glycols. The polypropylene glycols in particular are very versatile, and they can be tailored to give a wide range of froth properties. Many other frothers are available, such as cresols and pine oils, but most of these are considered obsolete and are not as widely used as they once were. Some work has also been done using saltwater (particularly seawater) as the frothing agent, and the process has been used industrially in Russia (Klassen and Mokrousov 1963; Tyurnikova and Naumov 1981).

Function of Frothers

Klimpel (1995) found that the use of different frothers produced changes in the flotation rate and recovery values in coal flotation and reached the following conclusions:

- When frother dosage was held constant while collector dosage was increased, it was found that the flotation rate increased to a maximum and then decreased. This was observed for all frother types and all particle size fractions. The difference between the frother families studied was that the collector dosage that produced the most rapid flotation rate was different.
- The finest (–88 μm) and coarsest (+500 μm) particles tended to float more slowly than the intermediate-size particles for all of the frother types.
- Changes in flotation rates were due to changes in the coal particle size and to changes in the frother/collector dosage. Although the contribution of particle size was generally more significant, the reagent dosage effect provides a useful means for adjusting the flotation rate in the plant.
- With aliphatic alcohol frothers, the flotation rate maximum was much more pronounced than for the propylene oxide (PO) and combined propylene oxide/alcohol (PO-alcohol adduct) frothers.
- Increasing the frother dosage to increase recovery always leads to less selective flotation, regardless of frother type.
- The PO and PO-alcohol adduct frothers are more powerful recovery agents than alcohol frothers, and they should therefore be used at lower dosages.
- Overdosing with alcohol frothers leads to a slower flotation rate, because excesses of these frothers will lead to eventual destabilization of the froth. This effect does not occur with the PO and PO-alcohol frothers, and therefore overdosing with these frothers leads to high recovery but poor selectivity.
- PO frothers with molecular weights of 300 to 500 are optimal for coal recovery.
- Alcohol frothers tend to be more effective for fine-particle recovery than for coarse-particle recovery. To recover coarse particles, the alcohol frother and the hydrocarbon collector dosages should both be high. The alcohol will still provide reasonable selectivity at these high dosages.
- The high-molecular-weight PO-based frothers are more effective for coarse-particle flotation than the alcohol or low-molecular-weight PO frothers, but they also have a lower selectivity. For both good coarse-particle recovery and good selectivity, the PO frothers should be used at low dosages, with low collector dosage as well. The PO-alcohol adduct frothers are even more effective for coarse-particle recovery, and they need to be used at even lower dosages.
- The optimal frother for high recovery with good selectivity will often be a blend of members of the various frother classes examined. It is reported that such frother blending will give enough benefit to be worth the effort in approximately half of all coal flotation operations.
- Neither the alcohol nor PO-based frothers will change the shape of the grade–recovery curve. Changes in frother type and dosage simply move the flotation results along the curve. Similarly, changes in hydrocarbon collector dosage also mainly move the performance along the grade–recovery curve.
- For medium- and coarse-coal size fractions (–0.5 mm), the total gangue recovered is linearly related to the total coal recovered. It is only for the finest particles (–50 μm) that the gangue recovery increases nonlinearly with increasing coal recovery.
- When floating coals with a broad particle size range, the majority of the gangue reaching the froth is from the finer particle size fractions.
- As the rate of coal flotation increases, the rate of gangue flotation increases proportionately. This is typical of a froth-entrainment process acting on the gangue.

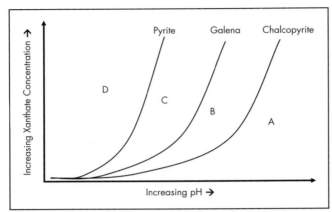

Figure 14.5-16 pH response curves for sulfhydryl collector adsorption on different sulfide minerals

adsorption is a function of pH. This makes it possible for sulfhydryl collectors to be used to progressively separate specific minerals. The pH where the xanthate ion wins the competition with OH⁻ depends both on the concentration of xanthate in solution and on the specific sulfide mineral present, as shown in Figure 14.5-16.

The curves in Figure 14.5-16 mark the boundaries where the given mineral becomes sufficiently hydrophobic to float. Both xanthates and dithiophosphates exhibit curves of this form, with different pH values and concentrations for each type of collector (Fuerstenau et al. 1985). For example, assume a mixture of pyrite (FeS_2), galena (PbS), and chalcopyrite ($CuFeS_2$). From Figure 14.5-16, if the pH and xanthate concentrations are in Region A, then xanthate will not be adsorbed on any of the minerals, and therefore none of the minerals will be floated. However, if the pH and xanthate concentrations are altered to move into Region B, only chalcopyrite becomes hydrophobic and consequently floats. In Region C, both chalcopyrite and galena will float, and in Region D, all three minerals will float. It is therefore possible to progressively lower the pH to float chalcopyrite first, then galena, and then pyrite, producing concentrates for each mineral and leaving behind any nonfloatable silicate gangue minerals.

Acids. The acids used are generally those that give the greatest control of pH at the lowest cost, with sulfuric acid being the most popular. A key point to bear in mind is that the anion of the acid can potentially have effects of its own that are quite separate from the lowering of the pH. Therefore, acids other than sulfuric acid are useful in some cases.

Alkalis. As with acids, the most popular alkalis are those that are cheapest, with the lowest-cost alkali generally being lime (CaO or $Ca(OH)_2$). However, the calcium ion often interacts with mineral surfaces to change the flotation behavior. In some cases the calcium ions have beneficial effects, while in other cases they change the flotation in undesirable ways. It may therefore be necessary to use sodium-based alkalis, such as sodium hydroxide (NaOH) or sodium carbonate (Na_2CO_3), because the sodium cation generally does not have any significant effect on the particle surface chemistries.

Activators

Activators are specific compounds that make it possible for collectors to adsorb onto surfaces that they could not normally attach to. A classic example of an activator is copper sulfate as an activator for sphalerite (ZnS) flotation with xanthate collectors (Fuerstenau et al. 1985). When untreated, xanthate cannot attach to the sphalerite surface because it forms a zinc-xanthate compound that quickly dissolves, as shown in the following reaction:

$$ZnS(s) + xanthate \rightarrow S(s) + Zn\text{-xanthate (aq)}$$

The surface of the sphalerite can be activated by reacting it with a metal ion that does not form a soluble xanthate, such as soluble copper from dissolved copper sulfate, as shown in the following reaction:

$$ZnS(s) + CuSO_4(aq) \rightarrow CuS(s) + ZnSO_4(aq)$$

This forms a thin film of copper sulfide on the sphalerite surface, which allows for stable attachment of the xanthate, rendering the sphalerite particle hydrophobic and floatable. Other metals such as silver and lead can also be used to activate zinc,

Synthetic and Natural Frothers

The frothers originally used were natural products such as pine oil and cresylic acid. These are rich in surface-active agents that stabilize froth bubbles and are effective frothers. As natural products, they are not pure chemicals, but instead contain components that are effective frother and others that are weak collectors. Although this can have the advantage of reducing the amount of collector that needs to be added separately, it introduces some problems with process control. If the frother is also a collector, then it becomes impossible to alter the frothing characteristics and the collecting characteristics of the flotation operation independently.

Synthetic frothers, such as the alcohol-type and polypropylene glycol-type frothers, have the advantage that their effectiveness as collectors is negligible. It is therefore possible to increase the frother dosage without also changing the quantity of collector in the system. This in turn makes the flotation process much easier to control.

Modifiers

Modifiers are chemicals that influence the way that collectors attach to mineral surfaces. They may either increase the adsorption of collector onto a given mineral (activators), or prevent collector from adsorbing onto a mineral (depressants). Importantly, just because a reagent is a depressant for one mineral/collector combination, it does not necessarily follow that it is a depressant for other combinations. For example, sodium sulfide is a powerful depressant for sulfide minerals being floated with xanthate, but it does not affect flotation when sulfide minerals are floated with the collector hexadecyl trimethyl ammonium bromide.

pH Control

The simplest modifiers are pH-control chemicals. The surface chemistry of most minerals is affected by the pH. For example, minerals in general develop a positive surface charge under acidic conditions and a negative charge under alkaline conditions. Because each mineral changes from negatively charged to positively charged at some particular pH, it is possible to manipulate the attraction of collectors to their surfaces by pH adjustment. Other, more complex effects are due to pH that change the way that particular collectors adsorb on mineral surfaces.

Sulfhydryl collectors such as xanthate ions compete with hydroxide ions (OH⁻) to adsorb on mineral surfaces, and so

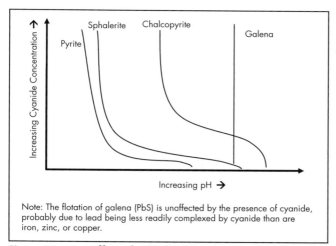

Figure 14.5-17 Effect of cyanide depressant on flotation of minerals as a function of pH

but the copper element is cheaper than silver and is less toxic than lead.

It is also possible to adsorb specific ions onto the surface that can promote attachment of the collector. For example, silica (SiO_2) normally has a strongly negative surface charge at approximately neutral pH, and therefore it has little affinity for anionic collectors such as oleic acid. However, calcium ions specifically adsorb onto silica surfaces, and the negative charge of the calcium ions can actually reverse the surface charge, making it positive. It is then possible for the anionic collectors to electrostatically attach to the calcium-activated silica surface.

Depressants

Depressants have the opposite effect of activators by preventing collectors from adsorbing onto particular mineral surfaces. Their typical use is to increase selectivity by preventing one mineral from floating while allowing another mineral to float unimpeded.

Cyanide. Cyanide (CN^-) is a particularly useful depressant in sulfide mineral flotation, as can be seen in Figure 14.5-17. Its activity is believed to be due to its ability to complex with, and in some cases dissolve, a number of metal ions, preventing them from attaching to the xanthate molecules. In particular, it is a strong depressant for pyrite (FeS_2) and can be used to "deactivate" sphalerite that has been activated by copper ions in solution (Fuerstenau et al. 1985).

Lime. Lime is added as either CaO or $Ca(OH)_2$, and when it dissolves it contributes calcium ions that can adsorb onto mineral surfaces. In combination with its strong alkaline nature, this makes it particularly useful in manipulating sulfide flotation. It is less useful in oxide mineral flotation because it can activate the flotation of silica by anionic collectors, causing it to float along with the other oxide minerals.

Organic depressants. A large number of organic compounds are useful as flotation depressants. These tend to be soluble polymers (such as starch polysaccharide) that selectively become adsorbed and coat mineral surfaces, thereby preventing a collector from attaching. An example of this is in the "reverse flotation" of silica from iron ore, where the silica tailings are floated using a cationic collector at a pH of 8.5 to 11, leaving behind the iron oxide minerals. Starch acts as a depressant for iron oxide in this process, preventing it from being floated by the cationic collector.

Recovery of Gangue Particles in Flotation: Entrainment or Hydrophobicity?

Some gangue material is always carried out of the slurry in the froth. For example, in the flotation of coal, a portion of the ash-forming minerals and pyrite will be carried into the froth along with the coal. It is commonly believed that this gangue mineral can often be prevented from floating if the correct depressant can be discovered. In the case of coal, more than 42 chemicals have been reported by investigators as being capable of depressing pyrite in coal flotation, and yet none of them have ever been useful industrially. This is because, in most cases, the pyrite was not truly hydrophobic (Kawatra and Eisele 1992, 2001).

In a froth flotation machine, there are two ways that a particle can reach the froth layer. It can be carried into the froth by attachment to an air bubble (true flotation), or it can be suspended in the water trapped between the bubbles (entrainment). Although true flotation is selective between hydrophobic and hydrophilic particles, entrainment is nonselective, and therefore entrained particles are just as likely to be gangue minerals as they are to be the valuable mineral. If particles are sufficiently coarse, then they settle rapidly enough that they are not carried into the froth by entrainment. As they become finer, particles settle more slowly and so have more time to become entrapped in the froth, and consequently they have less of a tendency to drain away. Clay particles in particular, which are only a few micrometers in size, are very easily entrained. For particles that are less than a few micrometers in size, their rate of recovery into the froth by entrainment is equal to the rate of recovery of water into the froth. For example, if 20% of the water entering a flotation cell is carried into the froth, then up to 20% of the fine particles entering the cell will probably be entrained. The entrainment of coarser particles will be less than 20% due to their greater ability to drain from the froth.

In addition to the entrained particles, gangue is carried into the froth by being physically locked to the floatable particles. In the case of coal, much of the pyrite consists either of submicrometer pyrite grains that are never liberated from the coal, or pyrite particles with surfaces consisting primarily of coal. These will therefore behave as if they were coal particles. The recovery of entrained particles can only be reduced by lowering the fraction of the water recovered into the froth.

The recovery of locked particles can only be changed significantly by either grinding to a finer size to improve liberation, or by rejecting the locked particles along with the least-floatable liberated particles, which sacrifices recovery of the valuable mineral. Depressant chemicals are not useful in either of these cases, as they are only intended for preventing the hydrophobic bubble attachment and true flotation of particles. Before deciding to use a depressant, it is therefore important to first determine whether the particles to be depressed are actually being recovered by true flotation in the first place or whether there are other causes.

EQUIPMENT

In addition to controlling the chemistry, flotation requires a machine for mixing and dispersing air throughout the mineral slurry while removing the froth product (Nelson et al.

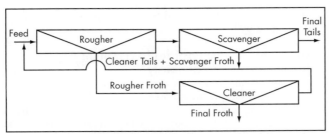

Figure 14.5-18 One configuration for a rougher/cleaner/scavenger flotation circuit

2009). These individual machines are then connected to form a flotation circuit to fully clean the product.

Conventional Cells

Conventional flotation cells consist of a tank with an agitator designed to disperse air into the slurry, as was previously shown schematically in Figure 14.5-6. They differ in the exact location and design of the rotor, the means of air induction, and the size/shape of the tank (Nelson et al. 2009). Considerable design effort has gone into optimizing the hydrodynamics of the system and ensuring that

- The suspended particles do not settle out into dead zones;
- The residence time of slurry throughout the cell is uniform;
- Solid particles are effectively brought in contact with air;
- Air is evenly dispersed throughout the cell;
- The froth layer is collected efficiently without excessive losses of attached particles back into the pulp;
- The pulp level can be accurately controlled;
- Maintenance of the machines is simplified; and
- Power consumption to achieve all of these goals is minimized.

The advances in understanding of flotation cell hydrodynamics has enabled the development of very large machines, with the recent "Supercell" design having volumes as large as 300 to 350 m^3 (FLSmidth Minerals 2009; Nelson et al. 2009).

Flotation cells are relatively simple machines with ample opportunity for particles to be carried into the froth along with the water making up the bubble films (entrainment) or for hydrophobic particles to break free from the froth and be removed along with the hydrophilic particles. Therefore, it is common for conventional flotation cells to be assembled in a multistage circuit, rather than using single flotation cells in isolation. Typically a "rougher" cell bank is used to make an initial, rapid separation of the valuable minerals from the gangue. The valuable mineral concentrate is often then reprocessed in the "cleaner" cells to remove gangue mineral particles that were misplaced into the concentrate. Similarly, "scavenger" cells are used to reprocess the gangue mineral product from the rougher to recover valuable minerals from the tailings. These different cell banks can be arranged in configurations such as the one shown in Figure 14.5-18, although, depending on the nature of the separation, there may be more or fewer stages of cleaners or scavengers added.

The use of a multistage circuit for froth flotation allows a great deal of flexibility, as the recirculation of material between cells can be adjusted to fit the specific needs of the material being processed. One concern is that a large number

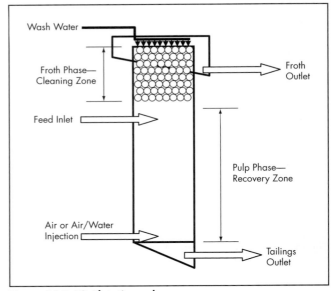

Figure 14.5-19 Flotation column

of flotation cells are typically necessary, which can use a great deal of the available plant floor space.

Flotation Columns

Flotation columns provide a means for improving the effectiveness of froth flotation using a single machine, which can be very useful if floor space is at a premium (Eberts 1986). A column essentially performs as if it were a multistage flotation circuit arranged vertically (Degner and Sabey 1988) with slurry flowing downward while the air bubbles travel upward, producing a countercurrent flow. The first flotation machine design to use a countercurrent flow of slurry and air was developed by Town and Flynn in 1919. It was not until the work of Boutin and Tremblay in the early 1960s that a new generation of countercurrent columns was developed that ultimately became industrially successful (Rubinstein 1995).

A typical flotation column schematic is shown in Figure 14.5-19. The basic principle of column flotation is the use of a countercurrent flow of air bubbles and solid particles. This is achieved by injecting air at the base of the column and feed near the midpoint. The particles then sink through a rising swarm of very small air bubbles.

Countercurrent flow is accentuated in most columns by the addition of wash water at the top of the column, which forces all of the water that entered with the feed downward to report to the tailings outlet. This flow pattern is in direct contrast to that found in conventional cells, where both the air and the solid particles are driven in the same direction. The result is that columns provide improved hydrodynamic conditions for flotation, and thus generally produce a cleaner product while maintaining high recovery and low power consumption. The performance differences between columns and conventional cells may best be described in terms of the following factors: collection zone size, particle–bubble contact efficiency, and fines entrainment (Kawatra and Eisele 1987).

The collection zone is the region where particle–bubble contact occurs, and it differs greatly in size between column and conventional flotation. In conventional cells, contact occurs primarily in the region surrounding the mechanical

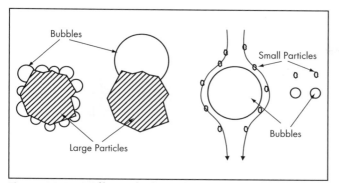

Figure 14.5-20 Effects of relative bubble and particle sizes on froth flotation

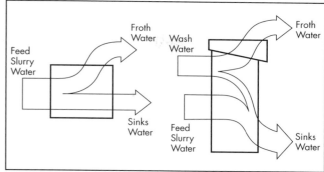

Figure 14.5-21 Effect of wash water on the entrainment of feed into the froth

impeller. The remainder of the cell acts mainly as a storage volume for material that has not yet been through the collection zone. This creates a bottleneck that keeps the flotation rate down. In contrast, flotation columns have a collection zone that fills the entire volume of the machine, so that there are more opportunities for particle–bubble collisions. The reduced level of turbulence needed to achieve a good rate of recovery in columns also reduces the tendency of coarse particles to be torn away from the bubbles that they attach to. Therefore, columns are more effective for floating coarser particles (Kawatra and Eisele 2001).

Columns exhibit higher particle–bubble contact efficiency than conventional machines due to the particles colliding head-on with the bubbles. As a result, the energy intensity needed to promote contact is less and, therefore, power consumption is reduced.

A second beneficial effect in certain types of flotation columns is the reduction of bubble diameter (Yoon and Luttrell 1986). As bubble diameter is reduced, the flotation rate of both the coarser and finer particles is improved. The effects producing the improvement are illustrated in Figure 14.5-20. Coarse particles can attach to more than one bubble if the bubbles are small and the chances of the particle being torn loose and sinking again are therefore reduced. For fine particles, the probability of collision with the bubble is improved if the bubble is small, as then the hydrodynamic forces tending to sweep the particle away from a collision are reduced. The reduction of bubble diameter has the added benefit of increasing the available bubble surface area for the same amount of injected air. Therefore, it is desirable to produce bubbles as fine as possible.

The entrainment of fine waste material in the froth product is a common failing of conventional flotation machines. It results from the phenomena whereby some water is transferred into the froth within the zone surrounding the air bubbles. As a result, fine suspended particles are swept into the froth with this water, even though they are not physically attached to the air bubbles. In most column flotation machines, the entrainment problem is addressed through the use of wash water, as shown in Figure 14.5-21. Where the conventional cell must allow a certain amount of feedwater to enter the froth, the wash water in the column cell displaces this feedwater to the tailings, thus preventing entrained contaminants from reaching the froth. The only drawback to the use of froth washing is that the demand for clean water is increased, which may cause problems in some situations.

In a coal flotation operation, the combined effects of the relatively gentle mixing, the countercurrent flow, and the use of wash water in columns is that, because there is a distance of several meters between the discharge points of the clean coal froth and the gangue tailings, there is a much reduced possibility of coal being misplaced into the tails or of gangue short-circuiting to the froth. The net result is that, depending on the design, a column is typically equivalent to between three and five stages of conventional flotation.

Bubble Generators

The impeller-type mixers that are used in conventional cells are not well suited for use in flotation columns, as they would either need excessively long shafts or rotating seals. In the original flotation column design, bubbles were produced using sintered ceramic air diffusers that produced very fine air bubbles. However, this was found to suffer from plugging problems, particularly in hard water, and for this reason cloth and perforated rubber sheeting alternatives were adopted (Boutin and Wheeler 1967; Dobby et al. 1985). These still tended to require excessive maintenance, and for this reason external bubble generators of various types have been adopted (McKay et al. 1988; Yoon et al. 1990; Davis et al. 1995; Rubinstein 1995). The external bubble generators combine a stream of water with air to produce a mixture of very fine bubbles in water. This mixture is injected into the column. This approach has a number of advantages, including the following:

- The bubble generator is accessible for adjustment and maintenance.
- The column has no porous elements inside that can become clogged or damaged, and so the dispersion of air in the column does not change.
- The bubble generator can be designed to tolerate particulate matter in the water, and so recycle water can be used in the generators.
- External bubble generators can consistently produce very small bubbles in the column.

Axial Mixing Effects

A key issue in flotation column operation is "axial mixing," which is mixing along the vertical axis of the column, as shown in Figure 14.5-22. As the air bubbles rise in the column, they also tend to transfer water to the base of the froth layer. The displaced water then descends again, setting up a strong mixing action. This tendency is greatest if the column is slightly tilted from vertical, as then the bubbles preferentially ascend on one side while the water descends on the other. This is why it is particularly crucial for flotation columns to be perfectly vertical. The tendency toward axial mixing is also increased if some large bubbles are present, and so performance is optimal if the bubbles are uniformly small in diameter.

Another approach to suppressing axial mixing is the introduction of horizontal baffles made from perforated plates (Kawatra and Eisele 2001). These baffles interrupt the flow of the liquid, as shown in Figure 14.5-22, preventing it from being rapidly carried to the surface or from short-circuiting directly to the tailings. Experiments with horizontal baffles in a flotation column have confirmed that they can greatly improve the operating characteristics of the column without sacrificing capacity, and a sufficiently open baffle design is highly resistant to plugging by coarse particles or debris. Overall, flotation columns generally have superior performance to conventional flotation cells. However, fundamentally, they require automatic control, especially for pulp level control, as they cannot utilize simple tailings overflow weirs, which are commonly used in conventional flotation cells.

APPLYING LABORATORY RESULTS TO PLANT EQUIPMENT SELECTION

Laboratory Reagent Screening

When the mineralogy of the ore for a new project has been determined, potential reagents for an ore can be based on a review of the relevant literature for similar ores. The sample to be used in laboratory test work should be prepared to minimize the quantity of slimes produced, with a combination of roll crushing and rod milling generally producing a good flotation feed. Because many flotation reagents have a limited shelf life, fresh reagents should be obtained (Thompson 2002).

Flotation experiments to evaluate reagent schemes are generally of two types. Single- and multiple-stage cleaning tests are used to evaluate the relative performance of different reagent types and dosages and show how cleanly the valuable minerals and gangue can be separated. Single-stage "rougher" kinetic flotation tests are used to establish grade–recovery relationships, which are necessary for calculation of residence times and equipment sizes.

Release Analysis

It is often useful to be able to determine the theoretical maximum amount of upgrading that can be achieved by froth flotation, given a specific ore and a particular set of reagents. One approach to this is "release analysis," which is carried out by progressively refloating froth products to collect only the particles that are fully hydrophobic. The procedure is illustrated in Figure 14.5-23.

Release analysis is used for determining the maximum grade of material that can be produced by a froth flotation process. Each stage of flotation removes more of the entrained and poorly floatable particles from the froth, until the final froth product consists only of the most strongly floatable material. It can be considered to be roughly analogous to a heavy-media washability test, except that instead of measuring the separating density that is needed to produce a given grade and recovery, it measures the number of flotation stages needed to achieve a given grade and recovery. It is therefore a valuable method for characterizing the inherent floatability of a particular ore and can be used to produce unbiased comparisons of the relative difficulty of upgrading different ores by froth flotation.

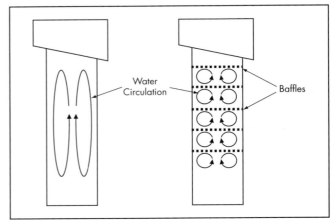

Figure 14.5-22 Comparison of axial mixing in regular columns and horizontally baffled columns

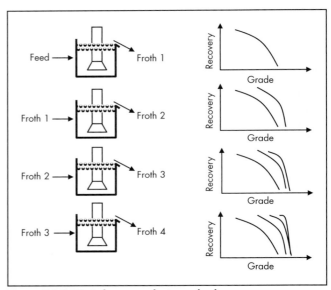

Figure 14.5-23 Release analysis method

Release analysis also provides a means for comparing the performance of conventional flotation with column flotation. The product from release analysis is typically much higher-grade than the product from a single stage of conventional flotation. On the other hand, a correctly operating flotation column will typically provide a product grade that is comparable to the grade of the product from the final stages of release analysis.

Laboratory and Plant Flotation Kinetics

Laboratory kinetic flotation experiments are best carried out in a flotation cell that allows froth to be collected in a reproducible manner over specific time intervals while maintaining a constant level of pulp in the cell. This can be accomplished by use of mechanical froth removal, and a mechanism to continuously provide makeup water to replace the water removed in the froth and prevent the pulp level from dropping and increasing the froth depth in the course of the experiment (Kawatra and Eisele 2001). The time intervals chosen for froth collection will obviously depend on the flotation rate of the mineral, as it is desirable to have at least five froth increments in order to be able to determine the kinetics properly.

To compare froth flotation experiments in the laboratory with operations in the plant it is necessary to take into account the differences in the way that the two types of cells are operated. This is most easily done by carrying out kinetic experiments to measure the recovery of material to the froth as a function of time.

The following is a simple kinetic equation (Arbiter and Harris 1962):

$$\ln(C_o/C) = kt$$

where
C_o = concentration of the valuable material at the start of the test
C = concentration of valuable mineral in the pulp at time t
k = rate constant

Assuming that all particles of a particular mineral have identical floatabilities, a plot of $\ln(C_o/C)$ versus t will produce a straight line with a slope equal to the rate constant. The higher the value of the rate constant, the higher the collecting power of the reagent used, which provides a means for comparing different reagents in the processing of a particular mineral.

The following equation is a sometimes more useful flotation kinetics model, and it includes a term for both flotation rate and ultimate flotation recovery and takes into account the fact that the particles exhibit a range of hydrophobicities and floatabilities (Klimpel 1995):

$$r = R\{1 - [1 - \exp(-Kt)]/Kt\}$$

where
r = total weight of component recovered at time t
t = time
K = rate constant
R = ultimate theoretical weight recovery at "infinite" time

This model takes into account the fact that the hydrophobic particles vary in size and degree of hydrophobicity, and it is therefore more appropriate than conventional reaction kinetics relationships that are intended to apply to systems of identical molecules. It is particularly useful for correlation of laboratory results with plant results. In conventional laboratory test work, it is common for R to be the most important one in determining the flotation performance, because laboratory tests are often run until all floatable material is recovered. In the plant, it is common for the rate constant, K, to be the most important factor. This is because it is too expensive to provide enough cell volume to recover all material that does not float in a reasonable time. Because of this difference, the results of laboratory studies can be very poor predictors of plant performance. To correct this, it is best to run timed flotation laboratory tests that can produce kinetic data, so that the R and K performances can be determined. Then, based on the residence time of the plant-scale units, it can be determined whether the plant performance is being dominated by kinetics, K, or by the ultimate recovery, R (Klimpel 1995).

Flotation Cell Selection

Flotation cells for use in a plant must be selected based on laboratory and pilot-scale data. Laboratory tests are usually carried out as batch experiments and are generally quite straightforward, although it is necessary to keep the following points in mind:

- The pulp must be agitated sufficiently to keep all particles in suspension.
- It is often necessary to condition the reagents with the minerals for a period of time to ensure good coverage with collector.
- In many cases, adding frother in stages along with makeup water may be necessary to keep the pulp level and froth depth constant.
- The capacity of the cell increases as the percent solids increases, and because of this the best process economics are achieved at the highest percent solids practical.

The following is the most important information obtained from this test work:

- **Optimum grind size of the ore:** This depends not only on the liberation characteristics of the ore but also on the floatability. Excessively coarse particles will be too large to be levitated by attached bubbles, and excessively fine particles will float poorly due to not striking the bubble surfaces, oxidation effects, or other problems. The coarsest material that can be floated is normally around 300 µm, whereas the finest particles are around 5 µm.
- **Reagent addition:** Quantity of reagents needed and the appropriate points in the circuit to add each reagent must be determined.
- **Optimum pulp density:** This is necessary for determining the size and number of flotation cells for a given capacity.
- **Flotation time needed to reach desired recovery:** It is important to be aware that a longer time is required for flotation in the plant than in the laboratory, mainly because of the increased time needed for selected particles to ascend through the large equipment. Typical comparative flotation times for different minerals are shown in Table 14.5-1.
- **Variability of the ore:** Because ore properties vary from one location to another in the mine, it is critical to run experiments with ore from several locations.
- **Corrosive and wear properties of the ore:** This information is needed for selecting appropriate materials of construction for the plant equipment.
- **Circuit design:** The type of circuit, the number cells per bank, the number of flotation stages, and the appropriate locations for recirculation of intermediate products must be determined. The number of cells per bank depends on the flotation characteristics of the material being floated. Typical practice can be as little as three or as many as 17 cells per bank, as shown in Figure 14.5-24.

Example Calculation

Problem. As an example of sizing flotation cells, consider the following copper ore:

- Specific gravity of dry ore = 2.7
- Optimum percent solids in laboratory flotation machine = 30%
- Optimum laboratory flotation retention time = 6.5 minutes

Table 14.5-1 Comparison of optimum flotation times in the laboratory and in the plant

Material Floating	Flotation Times in Industrial Rougher Flotation Cells, minutes	Usual Laboratory Flotation Times, minutes
Barite	8–10	4–5
Coal	3–5	2–3
Copper	13–16	6–8
Effluents	6–12	4–5
Fluorspar	8–10	4–5
Feldspar	8–10	3–4
Lead	6–8	3–5
Molybdenum	14–20	6–7
Nickel	10–14	6–7
Oil	4–6	2–3
Phosphate	4–6	2–3
Potash	4–6	2–3
Sand (impurities floated)	7–9	3–4
Silica (from iron ore)	8–10	3–5
Silica (from phosphate)	4–6	2–3
Tungsten	8–12	5–6
Zinc	8–12	5–6

Source: Metso Minerals 2006.

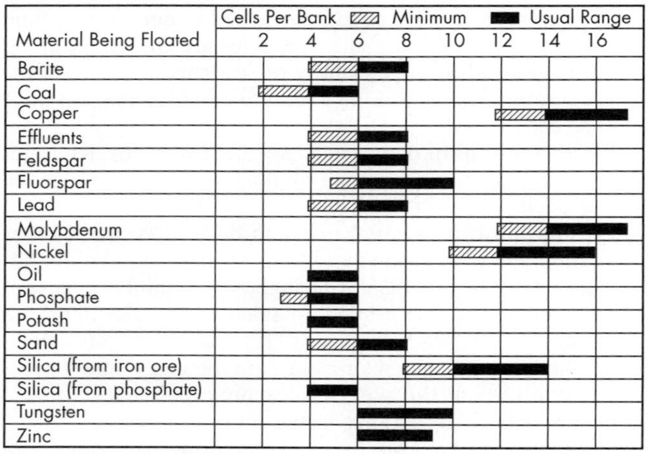

Figure 14.5-24 Normal numbers of cells per bank in froth flotation circuits

Source: Metso Minerals 2006.

Assume that the flotation facility will need to process 14,500 t/d of dry ore.

Solution. Using this information, the following can be calculated:

- From Table 14.5-1, the average laboratory flotation time for copper is 7 minutes, and the average plant flotation time is 14.5 minutes.
- The scale-up factor is 14.5/7 = 2.07, and therefore the time needed for flotation in the plant for this particular ore is (6.5)(2.07) = 13.5 minutes.
- A feed rate of 14,500 t/d of dry ore will require a flow of (14,500/1,440 min/d) = 10.07 t/min, and therefore for a 13.5-min residence time, there will need to be (13.5)(10.07) = 136 t of ore retained in the flotation bank.
- The slurry must be 30% (by mass) solids, and therefore, for 136 t of dry ore, there will be (136/0.3) = 453 t of slurry.
- The ore has a specific gravity of 2.7, and therefore the density of the slurry is calculated from the relationship $(100/\rho_{slurry}) = (X/\rho_{solids}) + ((100 - X)/\rho_{water})$, where X is the percent solids, and ρ_{slurry}, ρ_{solids}, and ρ_{water} are the specific gravities of the slurry, solids, and water, respectively.
- Therefore, $(100/\rho_{slurry}) = (30/2.7) + (70/1.0)$, and solving for ρ_{slurry} gives 1.23.
- The total volume of the 453 t of slurry that the cells must hold will be (453/1.23) = 368 m^3. However, when the pulp is aerated, it will consist of approximately 15% air and 85% slurry by volume, so the actual volume of pulp needed will be 368/(0.85) = 433 m^3.
- Cells of 14.15 m^3 (500 ft^3) are a standard size, and therefore the number required if these are used will be 433/14.15 = 30.6 cells (approximately 31 cells).
- Referring to Figure 14.5-24 for copper, between 12 and 17 cells per bank will be required, with 14 to 17 cells being optimum. Two banks of 16 cells would give 32 total cells, which would provide the 31 cells needed with some extra capacity. The cells would then be arranged in the individual banks in accordance with the manufacturer's recommendations.

Design Software

The design of flotation systems is simplified by the use of appropriate design software. Circuit-simulation packages can take data from laboratory test work and can predict the likely performance of a given design of a full-scale circuit. This allows the evaluation of many different arrangements before beginning pilot-scale studies (Collins et al. 2009). This type of software is under active development, so it is necessary to check the current literature to determine what is available at any given time. Such software packages are typically quite expensive and require training to use to their full potential, and they are most effectively used by specialist firms rather than by individual mining companies.

REFERENCES

Arbiter, N., and Harris, C.C. 1962. Flotation kinetics. *Froth Flotation 50th Anniversary Volume.* Edited by D.E. Fuerstenau. New York: American Institute of Mining, Metallurgical, and Petroleum Engineers. pp. 215–246.

Boutin, P., and Wheeler, D.A. 1967. Column flotation development using an 18 inch pilot unit. *Can. Min. J.* 88(3):94–101.

Bulatovic, S.M. 2007. *Handbook of Flotation Reagents.* Amsterdam: Elsevier.

Collins, D.A., Schwarz, S., and Alexander, D.J. 2009. Designing modern flotation circuits using JKFIT and JKSIMFLOAT. In *Recent Advances in Mineral Processing Plant Design.* Edited by D. Malhotra, P.R. Taylor, E. Spiller, and M. LeVier. Littleton, CO: SME. pp. 197–203.

Crozier, R.D. 1992. *Flotation Theory, Reagents, and Ore Testing.* Oxford: Pergamon Press.

Davis, V.L., Jr., Bethell, P.J., Stanley, F.L., and Lutrell, G.H. 1995. Plant practices in fine coal column flotation. In *High Efficiency Coal Preparation.* Edited by S.K. Kawatra. Littleton, CO: SME. pp. 237–246.

Degner, V.R., and Sabey, J.B. 1988. Wemco/Leeds flotation column develompent. In *Column Flotation '88, Proceedings of an International Symposium on Column Flotation*, Phoenix, AZ, January 25–28. Littleton, CO: SME. pp. 267–280.

Dobby, G.S., Amelunxen, R., and Finch, J.A. 1985. Column flotation: Some plant experience and model development. In *Proceedings International Federation of Automatic Control*. pp. 259–263.

Eberts, D.H. 1986. Flotation—Choose the right equipment for your needs. *Can. Min. J.* 107(3):25–33.

FLSmidth Minerals. 2009. Supercell. www.flsmidthminerals.com/supercell. Accessed December 2009.

Fuerstenau, M.C., Miller, J.D., and Kuhn, M.C. 1985. *Chemistry of Flotation*. New York: SME-AIME.

Glembotskii, V.A., Klassen, V.I., and Plaksin, I.N. 1963. *Flotation*. New York: Primary Sources.

Kawatra, S.K., and Eisele, T.C. 1987. Column flotation of coal. In *Fine Coal Processing*. Edited by S.K. Mishra and R.R. Klimpel. Park Ridge, NJ: Noyes Publications. pp. 414–429.

Kawatra, S.K., and Eisele, T.C. 1992. Recovery of pyrite in coal flotation: Entrainment or hydrophobicity? *Miner. Metall. Process.* 9(2):57–61.

Kawatra, S.K., and Eisele, T.C. 2001. *Coal Desulfurization: High-Efficiency Preparation Methods*. New York: Taylor and Francis.

Klassen, V.I., and Mokrousov, V.A. 1963. *An Introduction to the Theory of Flotation*. Translated by J. Leja and G.W. Poling. London: Butterworths.

Klimpel, R.R. 1980. Selection of chemical reagents for flotation. In *Mineral Processing Plant Design*, 2nd ed. Edited by A.L. Mular and R.B. Bhappu. New York: SME-AIME. pp. 907–934.

Klimpel, R.R. 1995. The influence of frother structure on industrial coal flotation. In *High-Efficiency Coal Preparation*. Edited by S.K. Kawatra. Littleton, CO: SME. pp. 141–151.

McKay, J.D., Foot, D.G., and Shirts, M.B. 1988. Column flotation and bubble generation studies at the Bureau of Mines. In *Column Flotation '88, Proceedings of an International Symposium on Column Flotation*, Phoenix, AZ, January 25–28. Littleton, CO: SME. pp. 173–186.

Metso Minerals. 2006. *Basics in Minerals Processing*, 5th ed. Helsinki: Metso Minerals.

Nelson, M.G., Lelinski, D., and Gronstrand, S. 2009. Design and operation of mechanical flotation machines. In *Recent Advances in Mineral Processing Plant Design*. Edited by D. Malhotra, P.R. Taylor, E. Spiller, and M. LeVier. Littleton, CO: SME. pp. 168–189.

Rao, T.C., Govindarajan, B., and Barnwal, J.P. 1995. A simple model for industrial coal flotation operation. In *High-Efficiency Coal Preparation*. Edited by S.K. Kawatra. Littleton, CO: SME. pp. 177–185.

Rubinstein, J.B. 1995. *Column Flotation: Processes, Designs, and Practices*. Basel, Switzerland: Gordon and Breach.

Thompson, P. 2002. The selection of flotation reagents via batch flotation Tests. In *Mineral Processing Plant Design, Practice, and Control*. Edited by A.L. Mular, D.N. Halbe, and D.J. Barratt. Littleton, CO: SME. pp. 136–144.

Tyurnikova, V.I., and Naumov, M.E. 1981. *Improving the Effectiveness of Flotation*. English edition translated by C.D. Zundorf. Stonehouse, England: Technicopy.

Whelan, P.F., and Brown, D.J. 1956. Particle–bubble attachment in froth flotation. *Bull. Inst. Min. Metall.* no. 591, 181–192.

Yoon, R.H., and Luttrell, G.H. 1986. The effect of bubble size on fine coal flotation. *Coal Prep.* 2:174–192.

Yoon, R.H., Luttrell, G.H., and Adel, G.T. 1990. *Advanced Systems for Producing Superclean Coal*. Final Report. DOE/PC/91221-T1. Blacksburg, VA: Virginia Polytechnic Institute and State University.

CHAPTER 14.6

Magnetic and Electrostatic Separation

Partha V. Iyer

The two methods of physical separation—magnetic and electrostatic separation—have developed from very early recognition of the forces that they utilize. Initially only low-field-strength permanent magnets and contact electrification were available, and hence the technologies were of limited value. However, as available magnetic field strengths increased, applications widened. Similarly, the introduction of safe high-voltage supplies opened up new electrostatic separation possibilities. Electrostatic separation is now something of a misnomer, as the most effective use of high-voltage electricity for separation involves small current flows.

Magnetic and electrostatic separation techniques have probably always been bracketed together because historically they were used for dry separation of liberated sand-sized mineral particles. They were mainly applied in the final stages of separation and/or upgrade of beach sand and alluvial tin concentrates (Gaudin 1939; Oberteuffer 1974; Knoll 1997; Svboda 1987; Venkatraman et al. 2003; Wills and Napier-Munn 2006).

Dry magnetic and electrostatic separation are most efficient on closely sized material. In the case of electrostatic separation, brief high-intensity drying can be applied to remove surface moisture and thus enhance differences in surface conductivity. In some cases, the magnetic or electrostatic properties of minerals can be modified by surface treatment at sustained high temperatures.

Wet magnetic separation has found major application in the treatment of fines at iron-ore operations. One of its attractions is its ability to handle a much wider and finer size range than dry systems. It is also applied in the cleaning of heavy media suspensions such as ferrosilicon.

Useful peripheral applications include removal of tramp iron from run-of-mine or coarse-crushed material (magnetic) and removal of dust from roaster and smelter off-gases (electrostatic). One particular attraction of both methods of separation is that they normally create few if any environmental problems and may in fact solve some.

Knowledge of magnetic and electrostatic forces dates back at least to the Greek philosopher Thales of Miletus, who lived about 600 BC. Thales knew some of the magnetic properties of the mineral lodestone, and he was also aware that when amber was rubbed with animal fur, the electrostatic charge produced on the amber (or fur) would attract light, nonconducting particles. The first record of magnetic separation of minerals appears to be a patent issued in 1792 to an English experimenter, William Fullarton, that described the concentration of iron ore. About a century later, in 1886, F.R. Carpenter obtained a U.S. patent for electrostatic concentration of ore.

The application of magnetic separators has extended well beyond either removing tramp iron or concentrating materials that are obviously affected by low-rate (and unsophisticated) magnetic fields. The latest developments in material science and magnet technology have allowed high-intensity and high-gradient industrial magnetic separators with field strengths as high as 6 T (60,000 G [gauss]) to be developed. Development of permanent rare-earth magnets and superconducting magnets has opened new markets for magnetic separators.

The electrostatic separator is still the most reliable and economic unit operation for processing beach-sand deposits rich in minerals such as ilmenite, rutile, leucoxene, zircon, and garnet. Increased environmental awareness has promoted the demand for unit operations that process secondary materials. A classic example is the successful use of electrostatic separators to remove plastics from metals. Triboelectrostatic separators, which can successfully separate two nonconductors, are being used in minerals and plastics separation.

REVIEW OF MAGNETIC THEORY

All materials can be classified as follows, based on their magnetic properties:

- *Paramagnetic* minerals are attracted along the lines of magnetic force to points of greater field intensity.
- *Diamagnetic* minerals are repelled along the lines of magnetic force to a point of lesser field intensity.
- *Ferromagnetic* minerals, a special category of paramagnetic materials, have a very high susceptibility to magnetic forces and retain some magnetism (remnant magnetism) after removal from the magnetic field.

Partha V. Iyer, Consultant, Reading, Pennsylvania, USA

Magnetic susceptibilities and electrostatic responses of minerals are provided in Table 14.6-1. In this section, the fundamentals of magnetic separation are introduced without detailed explanations of the physics involved.

Magnetic Force or Flux Density

In an electromagnet, an electric charge in motion sets up a magnetic field in the space surrounding it, and a magnetic field exerts a force on an electric charge moving through it. In fact, all magnetic phenomena arise from forces acting between electric charges in motion. Therefore, flux density B at point P, resulting from current element I of length dl, is as specified by Ampere's law:

$$dB = KI\left[\frac{dl(\sin\theta)}{r^2}\right] \quad (14.6\text{-}1)$$

where

B = magnetic flux density, weber/square meter (Wb/m^2)
K = constant of proportionality
I = current element, amperes (A)
l = length of current element, m
θ = angle between current element and radius vector to point P
r = distance to P, m

Magnetic flux density B, or force per pole, is defined as the number of magnetic flux lines per unit area normal to the lines. The unit for magnetic flux is the weber (Wb). The unit for term B is webers per square meter, called a tesla (T). The largest values of magnetic induction that can be produced in the laboratory are about 50 to 60 T (500,000 to 600,000 G).

Industrial high-intensity superconducting separators can reach fields of about 5 T (50,000 G). High-intensity induced-roll dry separators reach values as high as 2 T (20,000 G). Low-intensity wet drum separators for concentration of iron ore have drum-surface field strengths on the order of 0.1 T (1,000 G). However, the magnetic force acting on particles depends not only on the magnetic field B, but also on its gradient dB/dz, where z is the direction of the changing field.

Magnetization

A ferromagnetic material can be magnetized simply by being brought close to a permanent magnet or by passing current through a wire winding around the material. The magnetic state of a body can be defined by (1) stating the magnetization of all points within the body, (2) defining the strength of the magnetic poles, or (3) defining the magnitude of equivalent surface currents.

Coulomb's law for magnets is similar to that for electric charges; that is,

$$F = \frac{1}{4\pi\mu_o} * \frac{m_1 m_2}{r^2} \quad (14.6\text{-}2)$$

where

F = force, newtons (N)
μ_0 = permeability of a vacuum ($4\pi \times 10^{-7}$ H/m [henry/meter])
m_1 and m_2 = pole strength, A/m
r = distance, m

Magnetization (or, more completely, the intensity of magnetization M) is the total magnetic moment of dipoles per unit volume, in units of amperes per meter or pole strength m per unit area A.

$$M = \frac{m}{A} \quad (14.6\text{-}3)$$

Magnetic Field

The strength H of the magnetic field has the same units as does M (amperes/meter) and can be thought of as the cause of magnetization. It is defined as

$$H = \frac{B}{\mu} \quad (14.6\text{-}4)$$

where μ is absolute permeability.

In free space, a magnetic field produces a magnetic force given by $B = \mu_0 H$. However, in most applications the space is filled with some magnetic substance that causes an induced magnetization, $\mu_0 M$. Therefore, the total magnetic flux density B is the vector sum of the flux caused by the magnetic field H and the flux resulting from the magnetization M of the material. For ferromagnetic materials, however, the contribution of M usually dominates B:

$$B = \mu_0(H + M) \quad (14.6\text{-}5)$$

Permeability and Susceptibility

Magnetic flux density B, strength H of the magnetic field, and magnetization M can be used to compare the magnetic response of various materials. The ratio M/H is a dimensionless quantity called the *volume susceptibility* or simply *magnetic susceptibility*, χ. Similarly, the ratio B/H is called the *absolute permeability*, μ. The *relative permeability* μ_r is defined as

$$\mu_r = \frac{\mu}{\mu_o} \quad (14.6\text{-}6)$$

where

μ_r = relatively permeability
μ = absolute permeability
μ_0 = permeability in a vacuum ($4\pi \times 10^{-7}$ H/m)

The relative permeability of diamagnetic materials is slightly less than 1, that of paramagnetic materials is slightly greater than 1, and that of ferromagnetic materials is very high (e.g., for iron with 0.2% impurities, μ_r = ~5,000).

Summary

Paramagnetic minerals have magnetic permeabilities that are higher than the surrounding medium, usually air or water, and they concentrate the lines of force of an external magnetic field. The higher the magnetic susceptibility, the higher the field intensity in the particle and the greater the attraction up the field gradient toward increasing field strength.

Diamagnetic minerals have magnetic permeabilities that are lower than the surrounding medium, usually air or water, and they repel the lines of force of an external magnetic field. These characteristics cause the expulsion of diamagnetic

Table 14.6-1 Minerals and their magnetic and electrostatic responses

Mineral	Composition	Specific Gravity	Magnetic Response			Electrostatic Response	
			Ferromagnetic	Paramagnetic	Nonmagnetic	Conductive	Nonconductive
Actinolite	$Ca_2(Mg,Fe)_5(Si_4O_{11})_2(OH)_2$	3.0–3.2		x			x
Albite	$Na(AlSi_3O_8)$	2.6			x		x
Almandine	$Fe_3Al_2(SiO_4)_3$	4.3		x			x
Amphibole	$(Fe,Mg,Ca)_xSiO_3$	2.9–3.5		x			x
Anatase	TiO_2	3.9			x	x	
Andalusite	Al_2SiO_5	3.2			x		x
Andradite	$3CaO \cdot Fe_2O_3 \cdot SiO_2$	3.8		x		(2) ←	x
Anhydrite	$CaSO_4$	3.0			x		x
Ankerite	$Ca(Mg,Fe)(CO_3)_2$	2.9–3.1		x			x
Apatite	$(F_1Cl_1OH)Ca_5(PO_4)_3$	3.2			x		x
Aragonite	$CaCO_3$	3.0			x		x
Arsenopyrite	$FeAsS$	5.9–6.2		x →	(1)	x	
Asbestos	$Mg_3[Si_2O_5](OH)_4$	2.4–2.5			x		x
Augite	$Ca(Mg,Fe,Al)[(Si,Al)_2O_6]$	3.2–3.5		x		x →	(1)
Azurite	$Cu_3[CO_3]_2(OH)_2$	3.8			x		x
Baddeleyite	ZrO_2	5.6			x		x
Barite	$BaSO_4$	4.5			x		x
Bastnaesite	$(Ce,La,F)CO_3$	5.0		x			x
Bauxite	$Al_2O_3 \cdot 2H_2O$	2.6			x		x
Beryl	$Be_3Al_2[Si_6O_{18}]$	2.7–2.8			x		x
Biotite	$K(Mg,Fe)_3[Si_3AlO_{10}](OH,F)_2$	3.0–3.1		x			(4)
Bismuth	Bi	9.8			x	x	
Borax	$Na_2B_4O_7 \cdot 10H_2O$	1.7			x		x
Bornite	Cu_5FeS_4	4.9–5.0	(1) ←		x	x	
Brannerite	$(UO,TiO,UO_2)TiO_3$	4.5–5.4		x		x	
Brookite	TiO_2	4.1			x	x	
Calcite	$CaCO_3$	2.7			x		x
Carnotite	$K_2(UO_2)_2V_2O_8 \cdot 2H_2O$	5.0			x	(2) ←	x
Cassiterite	SnO_2	7.0			x	x	
Celestite	$SrSO_4$	4.0			x		x
Cerussite	$PbCO_3$	6.6			x	(2)	x
Chalcocite	Cu_2S	5.5–5.8			x	x	
Chalcopyrite	$CuFeS_2$	4.1–4.3	(1) ←		x	x	
Chlorite	$(Mg,Al,Fe)_{12}[(Si,Al)_8O_{20}](OH)_{16}$	2.6–3.2		x			x
Chromite	$(Fe,Mg)(Cr,Al)_2O_4$	4.6		x		x	
Chrysocolla	$CuSiO_3 \cdot nH_2O$	2.0–2.3			x		x
Cinnabar	HgS	8.1			x		x
Cobaltite	$(Co,Fe)AsS$	6.0–6.3		x		x	
Colemanite	$Ca_2B_6O_{11} \cdot 5H_2O$	2.4			x		x
Collophanite	$Ca_3P_2O_8 \cdot H_2O$	2.6–2.9			x		(3)
Columbite	$(Fe,Mn)(Ta,Nb)_2O_6$	5.2–8.2		x		x	
Copper	Cu	8.9			x	x	
Corundum	Al_2O_3	3.9–4.1			x		x
Covellite	CuS	4.7			x	x	
Cryolite	Na_3AlF_6	3.0			x	(2) ←	x
Cuprite	Cu_2O	5.8–6.2			x		x
Diamond (natural)	C	3.5			x		x
Diamond (synthetic)	C	3.5		x			x
Diopside	$CaMg[Si_2O_6]$	3.3–3.4		x →	(1)		x
Dolomite	$CaMg(CO_3)_2$	1.8–2.9			x		x
Epidote	$Ca_2(Al,Fe)_3Si_3O_{12}(OH)$	3.4		x			x

(continues)

Table 14.6-1 Minerals and their magnetic and electrostatic responses (continued)

Mineral	Composition	Specific Gravity	Magnetic Response			Electrostatic Response	
			Ferromagnetic	Paramagnetic	Nonmagnetic	Conductive	Nonconductive
Euxenite	(Y,Er,Ce,La,U)(Nb,Ti,Ta)$_2$(O,OH)$_6$U$_3$O$_8$	4.7–5.2		x		x	
Feldspar group	(K,Na,Ca..)$_x$(Al,Si)$_3$O$_8$	2.6–2.8			x		x
Ferberite	FeWO$_4$	7.5	(1) ←	x		x	
Flint	SiO$_2$	2.6			x		x
Fluorite	CaF$_2$	3.2			x		x
Franklinite	(Zn,Mn)Fe$_2$O$_4$	5.1–5.2	x			x	
Gahnite	ZnAl$_2$O$_4$	4.6			x		x
Galena	PbS	7.5			x	x	
Garnet complex	Ca,Mg,Fe,Mn silicates	3.4–4.3		x →	(1)	(2) ←	x
Gibbsite	Al(OH)$_3$	2.4			x		x
Geothite	FeO(OH)	4.3		x		(2) ←	x
Gold	Au	15.6–19.3			x	x	
Graphite	C	2.1–2.2			x	x	
Grossularite	Ca$_3$Al$_2$(SiO$_4$)$_3$	3.5			x	(2) ←	x
Gypsum	CaSO$_4$·2H$_2$O	2.3			x		x
Halite	NaCl	2.5			x	(2) ←	x
Hematite	Fe$_2$O$_3$	5.2		x		x	
Hornblende	Ca$_2$Na(Mg,Fe^{2+})$_4$(Al,Fe^{3+})[(Si,Al)$_4$O$_{11}$](OH)$_2$	3.1–3.3		x		(2) ←	x
Huebnerite	MnWO$_4$	6.7–7.5		x →	(1)	x	
Hypersthene	(Mg,Fe)SiO$_3$	3.4		x			x
Ilmenite	FeTiO$_3$	4.7		x		x	
Ilmenorutile	(Nb$_2$O$_5$,Ta$_2$O$_5$)$_x$TiO$_2$	5.1		x		x	
Ilvaite	CaFe$_2$(FeOH)(SiO$_4$)$_2$	4.0		x		x →	(1)
Kaolinite	Al$_2$Si$_2$O$_5$(OH)$_4$	2.6			x		x
Kyanite	Al$_2$O[SiO$_4$]	3.6–3.7			x		x
Lepidolite	K(Li,Al)$_3$(Si,Al)$_4$O$_{10}$[OH,F]$_2$	2.8–2.9			x		x
Leucoxene	FeTiO$_3$ TiO$_2$(alteration product)	3.6–4.3		x →	(1)	x	
Limonite	HFeO$_2$·nH$_2$O	2.2–2.4		x →	(1)	(2) ←	x
Magnesite	MgCO$_3$	3.0			x		x
Magnetite	Fe$_3$O$_4$	5.2	x			x	
Malachite	Cu$_2$CO$_3$(OH)$_2$	4.0			x		x
Manganite	MnO(OH)	4.3		x →	(1)	x	
Marcasite	FeS$_2$	4.6–4.9			x	x	
Martite (see Hematite)							
Microline	KAlSi$_3$O$_8$	2.6			x		x
Microlite	Ca$_2$Ta$_2$O$_7$ (see Pyrochlore)	5.5			x		x
Millerite	NiS	5.2–5.6		x		x	
Molybdenite	MoS$_2$	4.7–5.0			x	x	
Monazite	(Ce,La,Y,Th)PO$_4$	4.9–5.5		x			x
Mullite	Al$_6$Si$_2$O$_{13}$	3.2			x		x
Muscovite	KAl$_2$[AlSi$_3$O$_{10}$][F,OH]$_2$	2.8–3.0			x		(4)
Nahcolite	NaHCO$_2$	2.2			x		x
Nepheline syenite	(Na,K)(AlSi)$_2$O$_4$	2.6			x		x
Niccolite	NiAs	7.6–7.8		x		x	
Olivine	(Mg,Fe)$_2$[SiO$_4$]	3.3–3.5		x			x
Orpiment	As$_2$S$_3$	3.4–3.5			x	x	
Orthoclase	K[Al,Si$_3$O$_8$]	2.5–2.6			x		x
Periclase	MgO	3.6			x		x
Perovskite	CaTiO$_3$	4.0			x		x
Petalite	LiAl(Si$_2$O$_5$)$_2$	2.4			x		x

(continues)

Table 14.6-1 Minerals and their magnetic and electrostatic responses (continued)

Mineral	Composition	Specific Gravity	Magnetic Response			Electrostatic Response	
			Ferromagnetic	Paramagnetic	Nonmagnetic	Conductive	Nonconductive
Phosphate (pebble)	(see Collaphanite)						
Platinum	Pt	14.0–21.5		(1) ←	x	x	
Pyrite	FeS$_2$	5.0		(1) ←	x	x	
Pyrochlore	(Na,Ca..)$_2$(Nb,Ta..)$_2$O$_6$[F,OH]	4.2–4.4			x	x	
Pyrolusite	MnO$_2$	4.7–5.0		(1) ←	x		x
Pyrope	Mg$_3$Al$_2$(SiO$_4$)$_3$	3.5			x	(2) ←	x
Pyroxene	(Ca,Mg,Fe,Al)$_2$Si$_2$O$_6$	3.1–3.6		x →	(1)	(2) ←	x
Pyrrhotite	Fe$_{x-1}$S$_x$	4.6–4.7	x			x	
Quartz	SiO$_2$	2.7			x		(3)
Realgar	AsS	3.6			x	x	
Rhodochrosite	MnCO$_3$	3.7			x	(2) ←	x
Rhodonite	MnSiO$_3$	3.6–3.7			x	(2) ←	x
Rutile	TiO$_2$	4.2–4.3			x	(2)	
Samarskite	(Y,Er..)$_4$[(Nb,Ta)$_2$O$_7$]$_3$	5.6–5.8	(1) ←	x		x	
Scheelite	CaWO$_4$	6.1			x		x
Serpentine	Mg$_6$[Si$_4$O$_{10}$](OH)$_8$	2.5–2.7		x			x
Siderite	FeCO$_3$	3.9		x		(2) ←	x
Sillmanite	Al$_2$O[SiO$_4$]	3.2			x		x
Silver	Ag	10.1–11.1			x	x	
Smithsonite	ZnCO$_3$	4.1–4.5			x		x
Sodalite	Na$_8$[Al$_6$Si$_6$O$_{24}$]Cl$_2$	2.1–2.3			x		x
Spessarite	Mn$_3$Al$_2$[SiO$_4$]$_3$	4.2			x		x
Sphalerite	ZnS	3.9–4.0		x →	(1)	x →	(1)
Sphene	CaTi[SiO$_4$](F$_3$OH)	3.3–3.6			x	(2) ←	x
Spinel	MgAl$_2$O$_4$	3.6		(1) ←	x	x	(1)
Spodumene	LiAl(SiO$_3$)$_2$	3.1–3.2			x		x
Stannite	Cu$_2$FeSnS$_4$	4.3–4.5			x	x	
Staurolite	Fe^{2+}Al$_4$[Si$_4$O$_{11}$]$_2$O$_2$(OH)$_2$	3.6–3.8		x		(2) ←	x
Stibnite (Antimonite)	Sb$_2$S$_3$	4.6			x	x	
Struverite	(Ta$_2$O$_5$,Nb$_2$O$_5$)$_x$TiO$_2$	5.1		x		x	
Sulpher	S	2.1			x		x
Sylvite	KCl	2.0			x		x
Talc	Mg$_3$Si$_4$O$_{10}$(OH)$_2$	2.7–2.8			x		x
Tantalite	(Fe,Mn)(Ta,Nb)$_2$O$_6$	5.2–8.2		x		x	
Tapiolite	Fe(Ta,Nb)$_2$O$_6$	7.3–7.8		x		x	
Tetrahedrite	(Cu,Fe)$_{12}$Sb$_4$S$_{13}$	5.0		x		x	
Thorianite	ThO$_2$	9.7			x		x
Thorite	ThSiO$_4$	4.5–5.4			x		x
Topaz	Al$_2$SiO$_4$(F,OH)$_2$	3.5–3.6			x		x
Tourmaline	(Na,Ca)(Mg,Fe^{2+},Fe^{3+},Al,Li)$_3$(Al,Mg,Cr)$_6$ B$_3$Si$_6$(OH.O,F)$_4$	2.9–3.2		x →	(1)	(1,2) ←	x
Uraninite	UO$_2$	11.0		x			x
Vermiculite	Mg$_3$[Al,Si$_3$O$_{10}$](OH)$_2$·nH$_2$O	2.4–2.7		x			x
Wolframite	(Fe,Mn)WO$_4$	6.7–7.5		x		x	
Wollastonite	CaSiO$_3$	2.8–2.9			x		x
Wulfenite	PbMoO$_4$	6.7–7.0			x	x	
Xenotime	YPO$_4$	4.4–5.1		x			x
Zeolite	Hydrous aluminosilicate usually of Ca and Na	2.0–2.5			x		x
Zincite	ZnO	5.7			x	(1) ←	x
Zircon	ZrSiO$_4$	4.7			x		x

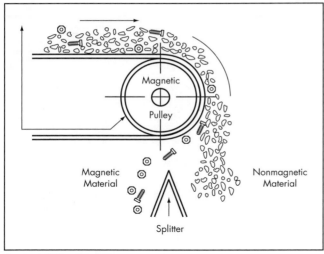

Courtesy of Eriez Magnetics.
Figure 14.6-1 Typical magnetic pulley

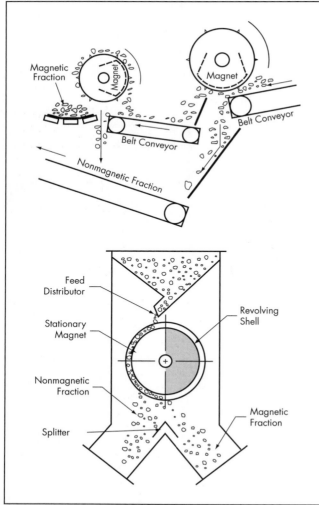

Courtesy of Eriez Magnetics.
Figure 14.6-2 Magnetic drum separators operating as lifting magnets

minerals down the gradient of the field toward decreasing field strength. This negative diamagnetic effect is usually orders of magnitude smaller than the positive paramagnetic attraction. Thus, a magnetic circuit can be designed to produce higher field intensity or higher field gradient, or both, to achieve effective separation (Falconer 1992).

CONVENTIONAL MAGNETS

Magnets are used in the mineral industry to remove tramp iron that might damage equipment and to separate minerals according to their magnetic susceptibility.

Low-Intensity Magnetic Separators

Low-intensity magnetic separators have flux densities up to 0.2 T (2,000 G). These separators are used mainly to remove ferromagnetic materials (such as iron bolts, bars, cables, and tools), to protect downstream unit operations (such as conveyor belts), or to scalp ferromagnetic materials to improve the performance of permanent or electromagnetic separators used to separate weakly magnetic materials. Low-intensity separators can treat wet slurry or dry solids.

Protective Magnets

The device most widely used to protect downstream operations from tramp iron is a magnetic pulley installed in the head of a conveyor (Figure 14.6-1). These devices remove tramp metals from dry solids. They contain either a permanent magnet or an electromagnet. Many types of magnets can be used—for example, plate magnets, cross-belt magnets, cobbing magnets, grate magnets, magnetic humps, and magnetic filters. An arrangement of magnetic drum separators is shown in Figure 14.6-2.

Wet Magnetic Separators

Low-intensity wet magnetic separators have been the workhorse of the iron-ore industry for several decades. Iron ore that is rich in magnetite has traditionally been enriched by these magnets. The coal industry uses these magnets to recover magnetite or ferrosilicon in a media-recovery circuit.

Several types of separator work on the same principle but have different design features. The common types are counter-rotation drum separators and concurrent-rotation drum separators (Figure 14.6-3).

High-Intensity Magnetic Separators

Separating paramagnetic or weakly magnetic particles requires a higher flux density. This higher density is achieved by means of electromagnetic circuitry that can generate a magnetic force of up to 2 T (20,000 G). For example, in a silica sand processing plant, these separators remove weakly magnetic iron-bearing particles.

Induced-Roll Magnetic Separator

Induced-roll dry magnetic separators are widely used to remove trace impurities of paramagnetic substances from feedstock such as quartz, feldspar, and calcite. The machine contains laminated rolls of alternating magnetic and nonmagnetic discs. A magnetic flux on the order of 2 T (20,000 G) is generated, and very high gradients are obtained where the flux converges on the sharp edges of the magnetic laminations.

A thin stream of granular material is fed to the top of the first roll (Figure 14.6-4). The magnetic particles are attracted to the roll and deflected out of their natural trajectory. Selectivity

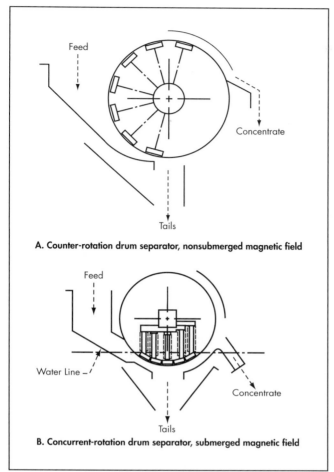

Figure 14.6-3 Wet magnetic separators

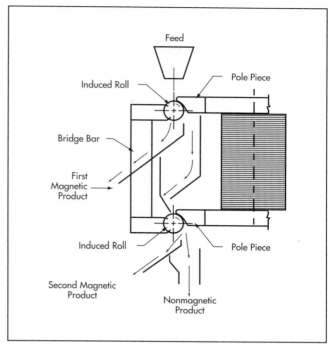

Figure 14.6-4 Induced-roll dry magnetic separator

Courtesy of Outotec USA, Inc.

Figure 14.6-5 Induced-roll dry magnetic separators treating silica sand

is obtained by varying roll speed and magnetic flux. A rather closely sized material must be treated if high selectivity is required. An industrial induced-roll magnetic separator consists of several rolls and can treat up to 10 tph (Figure 14.6-5).

Lift-Type Magnetic Separator

Lift-type magnetic separators are used on granular and powdered material that is dry and free-flowing. This type of separator produces a clean magnetic product because the magnetic particles are lifted out of the stream against the force of gravity, which minimizes entrapped particles (Figure 14.6-6). The selectivity of the lift-type separator is superior to that of induced-roll separators. Their main limitation is lower capacity. The cross-belt separator, a type of lift magnetic separator, has been used to some extent in processing ilmenite, garnet, and monazite in beach sands.

Jones Separator

The Jones separator is a wet high-intensity separator built on a strong mainframe made of structural steel (Figure 14.6-7). Magnet yokes are welded to this frame, and electromagnetic coils are enclosed in air-cooled cases. Actual separation takes place in plate boxes that are on the periphery of the one or two rotors attached to the central shaft.

Feed, which is a thoroughly mixed slurry, flows through the separator by means of fitted pipes and launders, and then into the plate boxes. The plate boxes are grooved to concentrate the magnetic field at the tips of the ridges. Feeding is continuous as a result of the rotation of plate boxes and rotors, with feed points at the leading edges of the magnetic fields. Each rotor has two symmetrically placed feed points. The feebly magnetic particles are held by the plates, while the remaining nonmagnetic slurry passes straight through the plate boxes and is collected in a launder. Before leaving the field, entrained nonmagnetic particles are washed by low-pressure water and collected as a so-called "middlings product." When the plate boxes reach a point midway between the magnetic poles where the magnetic field is essentially zero, the magnetic particles are washed out under high-pressure scour water sprays of up to 5 bars of pressure.

Field intensities of >2 T (20,000 G) can be produced in these machines. They are widely used to recover iron minerals from low-grade hematite ore. Other common applications include removing magnetic impurities from cassiterite

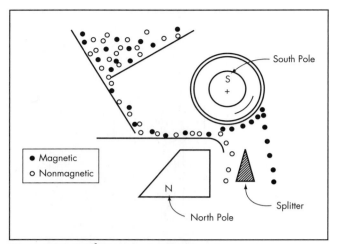

Figure 14.6-6 Lift-type magnetic separator

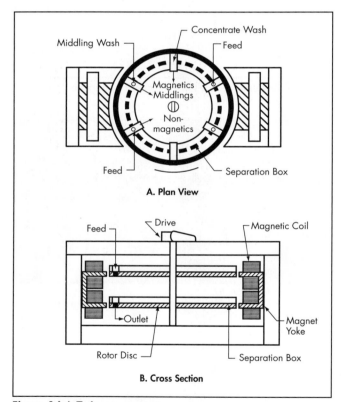

Figure 14.6-7 Jones separator

concentrate, removing fine magnetics from asbestos, and purifying talc.

Frantz Isodynamic Separator
The Frantz Isodynamic separator, introduced in the early 1930s, is the most efficient magnetic separator for separating minerals with field-independent magnetic susceptibilities. The isodynamic field, generated by a bipolar magnet with special pole-tip profiles, provides constancy of the product of the field and the field gradient. However, mineral separation in an isodynamic magnetic field is limited to minerals that have a constant susceptibility at the laboratory scale. Only this category of mineral then experiences a constant force throughout the isodynamic area.

PERMANENT MAGNETS
Most weakly magnetic minerals, such as garnet, ilmenite, and magnetic impurities in silica sand, can be effectively separated with a magnetic separator that has a flux density of >0.6 T (6,000 G). For nearly a century, induced-roll magnetic separators were the only economically viable unit operation in these applications. However, in spite of their considerable success, induced-roll separators have certain limitations in their selectivity and application. The development of permanent-magnet technology during the last three decades has reestablished the importance of magnetic separation and has increased the efficiency of fine-particle separations that were not successful with induced-roll magnets.

Principle and Design
In the last three decades, magnetic separation technology has undergone a revolution. Research in material science and ceramic technology has culminated in the development of new permanent rare-earth magnets and superconducting alloys. Successful adaptation of these magnetic materials, combined with knowledge of magnet geometry, has led to the design and development of a number of new magnetic separators that have opened niche markets that were previously considered beyond the realm of magnetic separation. These new separators are capable of

- Effectively removing magnetic impurities or reducing their concentration (even to parts-per-million levels);
- Producing high-grade mineral separates;
- Operating on virtually no energy, which makes them economical; and
- Generating higher magnetic flux levels, of up 2.1 T (21,000 G).

Dry Permanent Magnetic Separator
Recent improvements in magnet composition and design have led to the development of permanent magnetic separators. These improved rare-earth permanent magnets—for example, neodymium iron boron (NdFeB) and samarium cobalt (SmCo) magnets—have a magnetic attractive force that is an order of magnitude greater than that of conventional permanent magnetic circuits. The two main types of dry permanent magnetic separators that have found wide industrial applications are the rare-earth drum (RED) and the rare-earth roll (RER). They are widely used to separate weakly magnetic materials (such as garnet, ilmenite, and chromite) and to separate magnetic impurities present in low concentrations in silica sand (Dingwu et al. 1997; Dobbins et al. 2009; Wasmuth and Mertins 1997).

Rare-Earth Drum Separator
In a RED separator, NdFeB magnets are uniquely arranged to provide an intense (≤0.9 T or 9,000 G) and "deep" magnetic field perpendicular to the drum surface (Figure 14.6-8). Particles on the drum surface experience uniform flux density that minimizes misplacement of pinned particles to the middlings. Weakly magnetic particles pinned to the drum are carried to the region of no magnetic intensity and released as magnetics. The centrifugal force of the rotating drum throws those particles not influenced by the magnetic field into the nonmagnetic hopper.

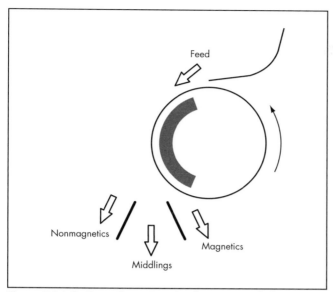

Figure 14.6-8 Operating principle of a rare-earth drum separator

An industrial-scale RED separator usually has three drums (Figure 14.6-9). The top drum is generally a low-intensity (≤0.2 T or 2,000 G) scalper magnet that removes ferromagnetic particles; the nonmagnetic fraction is subsequently treated on the REDs. The main purpose of the scalper is to protect the bottom two REDs as well as to increase their capacity. Some separators have a built-in internal air-cooling system to protect the magnets from overheating when feed is preheated, as in plants that process beach and silica sand.

Rare-Earth Roll Separator

An RER separator operates by means of a thin belt (usually 0.076–0.51 mm) that travels at very high velocity. The unique aspect of this separator is the way in which it is configured as a head pulley. Feed passes through the magnetic field, and magnetic (or weakly magnetic) particles are attached to the roll and separated from the nonmagnetic stream (Figure 14.6-10).

Although drum separators can effectively handle coarse particles (12.5 to 0.075 mm), roll separators are very effective in treating fine particles (<1 mm). The capacity of deeper-field drum separators is generally higher than that of roll separators at 7–9 t/h/m (400–500 lb/h/in.) for drum separators but 1.8–5.4 t/h/m (100–300 lb/h/in.) for roll separators. The major advantage of the drum separator is its low maintenance cost, because it does not have a belt that must be replaced frequently. However, both separators have their own niche markets. Drum separators can treat coarser particles (such as garnet, ilmenite, and iron ore) at a higher throughput. Roll separators can produce high-grade, high-purity glass sand products when feed is not preheated.

Considerable research has been conducted to increase the throughput of REDs and RERs without compromising product quality. Most of these improvements have come from improved magnetic-circuit design, increased drum diameter, higher-quality nonmagnetic stainless steel, better manufacturing processes, and more compact layout design.

Case study I: Performance evaluation of RED and induced-roll magnetic separators. A series of tests compared

Courtesy of Outotec USA, Inc.

Figure 14.6-9 Industrial-scale rare-earth drum separator

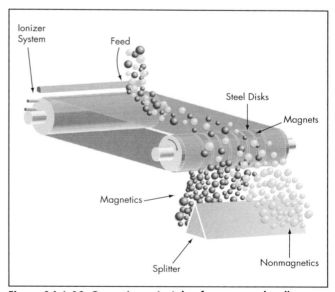

Figure 14.6-10 Operating principle of a rare-earth roll separator

the performance of a RED separator and an induced-roll magnetic separator to process a titanium-rich magnetic feed. The feed contained 76% TiO_2 and 1% Al_2O_3; it was obtained from a beach-sand processing plant. The goal was to produce high-grade nonmagnetic TiO_2 product (Table 14.6-2). Compared to the induced-roll magnetic separator, the RED separator gave product containing nearly twice the TiO_2, 45% rather than 24%, at a comparable product grade (TiO_2 grade of >90% and Al_2O_3 content of ~1%).

Case study II: Performance evaluation of RED and RER separators. A detailed study of the effect of feed rate on RED and RER separators used a garnet-rich heavy mineral sample (Figure 14.6-11). The RED separator was only minimally sensitive to feed rate. Garnet recovery decreased very marginally with increasing feed rate; it was 96.5% at a 2.6-tph feed rate and 95.1% at a 7.8-tph feed rate. In contrast, the RER separator seemed to be very sensitive to feed rate. Garnet recovery was 94.6% at a 2.6-tph feed rate, but fell to 81.3% at a 7.8-tph feed rate.

Eddy Current Separator

In the case of RED and RER magnetic separators, the outer drum or belt rotates; inside the shell, rare-earth magnets are mounted on a stator. In contrast, in an eddy current separator, not only does the nonmetallic outer drum rotate but the inside shell also rotates. The inside shell is a faster-moving rotor containing rare-earth magnets arranged in alternating polarity to induce eddy currents. An induced eddy current sets up a repulsive force in good conductors and thus separates nonferrous electrically conductive metals (such as copper and aluminum) from nonconducting materials.

Wet Permanent Magnetic Separator

A wet permanent magnetic separator (Figure 14.6-12) has permanent-magnet (NdFeB) bars positioned more or less horizontally inside a revolving drum made of stainless steel. This separator provides a field strength of 0.7 T (7,000 G) on the drum surface. The pulp tank is made of stainless steel with a concurrent or semi-countercurrent flow tank design and an adjustable discharge gap at the magnetics discharge end. The separator is equipped with an adjustable valve on the nonmagnetics discharge pipe to help control the flow rate and overflow level. This separator has found industrial applications in processing low-grade (martitic) iron ore.

SUPERCONDUCTING MAGNETS

High magnetic fields (up to 2 T or 20,000 G) are generated by passing current through a resistive coil or by permanent magnets. Development of newer computer models, through the use of finite-element-analysis techniques, has helped to achieve higher magnetic force. However, there is a logical maximum magnetic field for both the resistive coil and permanent magnet. Resistive coils are limited by the intrinsic resistance applied by the windings; the field strength of existing permanent magnets can be increased only marginally by modifying the magnet geometry. In the future, new magnetic materials may help to overcome this limitation.

Principle and Design

Currently, superconducting magnets are the only economically and technically viable way to achieve field strengths as high as 5 T (50,000 G). Most of the industrial superconducting magnetic separators designed and installed in the 1990s

Table 14.6-2 TiO_2 recovery and composition of nonmagnetic product*

Type of Separator	TiO_2 Recovery, %	Product Grade TiO_2, %	Product Grade Al_2O_3, %
Induced-roll magnet separator	24.2	92	0.96
RED-7000 gauss separator	44.7	92	1.03

*Treated by an induced-roll magnetic separator (feed contained 76% TiO_2 and 1% Al_2O_3; two passes; nonmagnetic retreat) and a RED-7000 gauss drum separator.

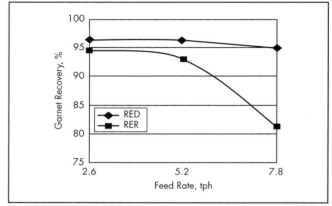

Figure 14.6-11 Performance of RED and RER separators

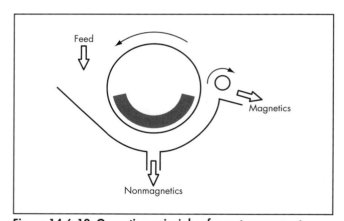

Figure 14.6-12 Operating principle of a wet permanent magnetic separator

and 2000s include expensive cryogenic systems that use liquid helium. This cooling system is essential to help maintain the temperature of the matrix, made of superconducting niobium/titanium alloy, below ~4 K, in order to achieve field strengths as high as 5 T.

Three types of cryogenic system are commonly used:

1. **Closed-cycle liquefier system.** In a closed-cycle liquefier system, the superconductor resides in a bath of liquid helium, and boil-off gas is recirculated through a helium liquefier. Although system installation is quite complex, system performance has been good and reliable, provided that there are no long-term interruptions in the supply of electrical power and cooling water.

2. **Low-loss system.** In a low-loss system, the superconductor windings reside in a reservoir of liquid helium. A very

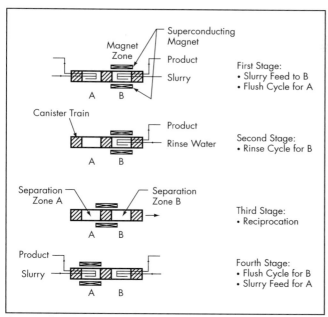

Courtesy of Outotec USA, Inc.

Figure 14.6-13 Operating principle of a reciprocating canister superconducting magnetic separator

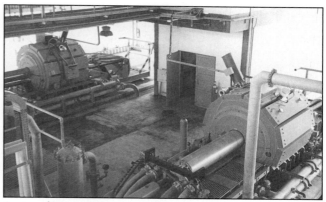

Courtesy of Outotec USA, Inc.

Figure 14.6-14 Industrial-scale reciprocating canister superconducting magnetic separator

efficient insulation system enables the magnet to operate for long periods, typically a year or more, between liquid-helium refills. The salient feature of this system is its relative immunity to short-term electrical failures. This feature allows this technology to be used where equipment is operated under difficult conditions.

3. **Indirect-cooling system.** The advent of heat engines based on the Gifford–McMahon cycle, which generate temperatures of ≤4 K, has made it possible to cool superconducting windings without the need for liquid helium. This technique offers great potential for small-scale systems in which the economics of helium supply or the cost of a liquefier cannot be justified. However, a constant power supply is essential for reliable operation.

In recent years, however, developments in cryogen-free cooling technology as well as advances in materials science have spurred design of more robust superconducting magnetic separators that do not require expensive helium refills.

In summary, superconducting magnets have two main advantages:

1. Low power consumption resulting from zero resistance of the magnet winding
2. Generation of high magnetic fields

Superconducting Wet High-Gradient Magnetic Separator

In a superconducting wet high-gradient magnetic separator (HGMS), magnetic particles are captured on a stainless-steel–wool matrix contained within the bore of a high-intensity magnet (Carpco SMS Ltd. 1993; D'Assumpção et al. 1995). The high intensity is generated using a superconducting coil. Because these coils have essentially zero resistance, little electrical power is required to energize the magnet. Furthermore, after the magnet is energized, the coil ends can be shorted, leaving the magnet in a fully energized state without any additional power supply. This practice is called operating the magnets in persistent mode.

Unloading trapped magnetic particles from the matrix is an essential step that determines the separation efficiency and capacity of the unit operation. Demagnetization is achieved either by deenergizing the magnet (a state commonly referred to as switch-mode) or by moving the matrix canister (referred to as a reciprocating canister HGMS). In reciprocating technology, captured magnetic particles are flushed using a ram to remove the trapping zone from the magnetic field regions. The ram operates on a magnetically balanced canister that houses a multisection separation region with unique and separate trapping zones (Figure 14.6-13).

Units that combine reciprocating canister technology and a low-loss cryogenic system have been used in kaolin processing throughout the world. Figure 14.6-14 shows a typical large-scale reciprocating canister HGMS.

Superconducting Dry Open-Gradient Magnetic Separator

In a conventional dry open-gradient magnetic separator (OGMS), the magnet structure is arranged to provide a region in open space with a highly divergent field. Thus, the magnet geometry supplies both the magnetic field and the field gradient. Paramagnetic material passing through this region experiences a force directly proportional to the field intensity and the magnitude of the field gradient. Compared with a conventional OGMS, a superconducting OGMS offers not only a higher magnetic force but also a deeper magnetic field, which translates to larger separation volumes.

ELECTROSTATIC SEPARATION

Almost all minerals show some degree of conductivity. The electrostatic separation process uses the difference in electrical conductivity or surface charge of the mineral species of interest. It has generally been confined to recovering valuable heavy minerals from beach-sand deposits. However, growing interest in plastic and metal recycling has opened up new applications in secondary material recovery (Knoll and Taylor 1985; Prabhu 1999).

When particles come under the influence of an electrical field, depending on their conductivity, they accumulate a charge that depends directly on the maximum achievable charge density and on the surface area of the particle. The

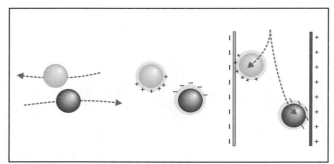

Figure 14.6-15 Work functions after particle–particle charging

charged particles can be separated by differential attraction or repulsion. Therefore, the important first step in electrostatic separation is to impart an electrostatic charge to the particles. The three main types of charging mechanism are contact electrification or triboelectrification, conductive induction, and ion bombardment. After the particles are charged, separation can be achieved by equipment with various electrode configurations.

Triboelectrification

Triboelectrification is a type of electrostatic separation in which two nonconductive mineral species acquire opposite charges by contact with each other (Alfano et al. 1988). The oppositely charged particles can then be separated under the influence of an electric field. This process uses the difference in the electronic surface structure of the particles involved. A good example is the strong negative surface charge that silica acquires when it touches carbonates and phosphates.

The surface phenomenon that comes into play is the *work function*, defined as the energy required to remove electrons from any surface (Figure 14.6-15). After particle–particle charging, the positively charged particle has a lower work function than does the negatively charged particle.

Over the years, several different types of triboelectrostatic separator have been designed and tested. A few tube-type and plate-type triboelectrostatic separators are now in commercial use processing minerals such as wollastonite, silica sand, and calcium carbonate. Unfortunately, in most applications, these separators suffer from low throughput and poor separation efficiency. The main challenge has been the inability to help particles acquire and retain charge for a desired period of time during the separation process.

Conductive Induction

When uncharged particles, whether conductors or nonconductors, contact a charged surface, the particles assume polarity and the potential of the surface. Electrically conductive minerals do so rapidly. However, in the case of nonconductors, the side that faces away from the charged surface is slower to acquire the same polarity as the surface. Hence, if both conductor and nonconductor particles are just separated from contact with a charged plate (Figure 14.6-16), the conductor particles are repelled by the charged plate, and the nonconductor particles remain unaffected by the charged plate—they are neither attracted nor repelled.

The most common industrial separators working on this principle are plate- and screen-type separators. Feed falls under gravity onto an inclined, grounded plate and into an

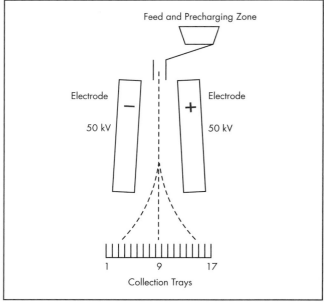

Courtesy of Integrated Mineral Technology, Ltd.
Figure 14.6-16 Operating principle of a plate-type electrostatic separator

electrostatic field induced by a high-voltage electrode. These electrodes are generally oval. Conductor particles acquire an induced charge from the grounded plate and move toward the oppositely charged electrode; that is, the particles experience a "lifting effect." Nonconductor particles are generally not affected by this field. Because the lifting effect depends on surface charge as well as particle mass, fine conductors are effectively separated from coarse nonconductors.

Ion Bombardment

When conductor and nonconductor particles placed on a grounded conducting surface are bombarded with ions of atmospheric gases generated by an electrical corona discharge from a high-voltage electrode, both conductor and nonconductor particles acquire a charge. When ion bombardment ceases, conductor particles rapidly lose their acquired charge to the grounded surface. Nonconductor particles react differently; the side that faces away from the grounded conducting plate is coated with ions of charge opposite in electrical polarity to that of the grounded conducting plate. Therefore, nonconductor particles remain pinned to the grounded plate because of electrostatic force (Figure 14.6-17).

An industrial high-tension electrostatic separator that uses the pinning effect to process plastics and metal scrap is shown in Figure 14.6-18. This separator consists of a rotating roll made from mild steel that is grounded through its supporting bearing. The electrode assembly consists mainly of two types of electrodes: a beam or corona electrode and a static-type electrode. The beam electrode, usually connected to a DC supply of ≤50 kV negative polarity, charges all particles and pin nonconductors to the roll. Feed is presented uniformly to the rotating roll surface by a velocity feed system. Both conductor and nonconductor particles are sprayed with ions. Conductor particles rapidly lose their charge to the grounded roll surface and are thrown off by centrifugal force. Nonconductor particles are pinned to and then brushed off the rolled surface.

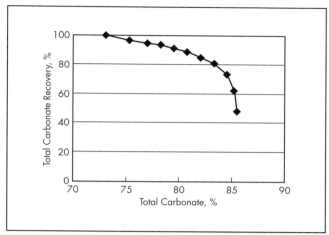

Figure 14.6-17 Operating principle of a roll-type electrostatic separator

Courtesy of Outotec USA, Inc.

Figure 14.6-18 Industrial roll-type electrostatic separator

Both conductor and nonconductor particles are collected in a partitioned product hopper at the bottom of the unit. The operating variables—roll speed, applied voltage, feed rate, splitter position, and electrode combination and position—can be adjusted to achieve effective separation.

After a long hiatus, there has recently been some significant development in designing new and efficient high-tension separators. OreKinetics has taken the lead and developed a new generation of high-tension roll and plate separators that have additional well-insulated static electrodes. This new design has helped improve separation efficiency and reduce the number of reprocessing steps.

ACKNOWLEDGMENTS

This chapter has been revised and updated from Chapter 7 in *Principles of Mineral Processing* (Venkatraman et al. 2003). Thanks go to previous coauthors Frank S. Knoll (retired) and James E. Lawver (deceased) for their contributions to that earlier chapter.

REFERENCES

Alfano, G., Carbinj, P., Ciccu, R., Ghani, M., Peretti, R., and Zucca, A. 1988. Progress in triboelectric separation of minerals. In *Proceedings of the XVI International Mineral Processing Congress*. Edited by K.S.E. Forssberg. Amsterdam: Elsevier.

Carpco SMS Ltd. 1993. Specification for Cryofilter HGMS. Model No. HGMS 5/460/10/S. Jacksonville, FL: Carpco.

D'Assumpção, L.F.G., Neto, J.D., Oliveira, J.S., and Resende, A.K.L. 1995. High gradient magnetic separation of kaolin clay. SME Preprint No. 95-119. Littleton, CO: SME.

Dingwu, F., Jin, S., Zhougyuan, S., and Shiying, P. 1997. Technical innovation and theoretical approach of a new type of permanent high gradient magnetic separators (PHGMS). In *Proceedings of the XIX International Mineral Processing Congress*, San Francisco, 1995. Littleton, CO: SME.

Dobbins, M., Dunn, P., and Sherrell, I. 2009. Recent advances in magnetic separator designs and applications. In *Proceedings of the 7th International Heavy Minerals Conference: What Next*, Drakensberg, South Africa, September 20–23. The Southern African Institute of Mining and Metallurgy.

Falconer, T.H. 1992. Magnetic separation techniques. File 4599. *Plant Eng.* (February): 85–87.

Gaudin, A.M. 1939. *Principles of Mineral Processing*. New York: McGraw-Hill.

Knoll, F.S. 1997. Solid–solid operations and equipment. In *Perry's Chemical Engineer's Handbook*. Edited by R.H. Perry and D.W. Green. New York: McGraw-Hill.

Knoll, F.S., and Taylor, J.B. 1985. Advances in electrostatic separation. In *Minerals and Metallurgical Processing*. New York: American Institute of Mining, Metallurgical, and Petroleum Engineers.

Oberteuffer, J.A. 1974. Magnetic separation: A review of principles, devices and applications. *IEEE Trans. Magn.* 10(2):223–234.

Prabhu, C. 1999. Design and testing of a triboelectrostatic separator for cleaning coal. Master's thesis, Virginia Polytechnic Institute and State University, Blacksburg, VA.

Svboda, J. 1987. Magnetic methods for the treatment of minerals. In *Developments in Mineral Processing*. Edited by D.W. Fuerstenau. Amsterdam: Elsevier.

Venkatraman, P., Knoll, F.S., and Lawver, J.E. 2003. Magnetic and electrostatic separation. In *Principles of Mineral Processing*. Edited by M.C. Fuerstenau and K.N. Han. Littleton CO: SME.

Wasmuth, H.D., and Mertins, E. 1997. A new medium-intensity drum type permanent magnetic separator and its practical application for processing ores and minerals in wet and dry modes. In *Proceedings of the XIX International Mineral Processing Congress*, San Francisco, 1995. Littleton, CO: SME.

Wills, B.A., and Napier-Munn, T.J. 2006. Wills' *Mineral Processing Technology*, 7th ed. London: Elsevier.

CHAPTER 14.7

Dewatering Methods

Donald A. Dahlstrom

INTRODUCTION

Liquid–solid separation is a relatively slow-moving field in terms of technological change. Improvements in this process have been evolutionary rather than revolutionary. Water is used to process the majority (between 80% and 90%) of the minerals and coal mined. (A small and decreasing percentage of coal is crushed and screened dry, and industrial minerals, such as diatomaceous earth, bentonite, cement raw materials and clinker, and base metal ores prior to roasting are often crushed and ground to a fine powder using dry processes.) Beneficiation processes usually use water because it allows for greater efficiency, higher recovery, and lower cost per unit of valuable product. In addition, it eliminates air pollution.

Where capital and operating costs are mentioned in this chapter, they should be regarded as order-of-magnitude only and not definitive.

Costs of Liquid–Solid Separation

In general, when considering the need to separate solids from water, as the particles to be separated decrease in size, the cost increases and the capacity per unit area decreases. When the solids are colloids (generally considered to be -10 μm), costs increase even faster. They are difficult to remove by filtration or centrifugation. Usually, a flocculant is added to the mixture to cause the colloids to form large flocculi or agglomerates. Accordingly, liquid–solid separation is a major cost in mineral processing, probably exceeded only by the cost of comminution, flotation, and endothermic reactions.

For example, the capital cost of a coal preparation plant increases by about 30% to 40% if the -28 mesh coal fraction is processed instead of discarded and the process water is recovered (by liquid–solid separation) for recycle and reuse. Operating costs per ton of -28 mesh coal also increase substantially when compared to coarser coals. These costs are due primarily to the use of flotation and the liquid–solid separation steps involved.

At the same time, liquid–solid separation by mechanical means (i.e., sedimentation, filtration, and centrifugation) is much less costly than thermal drying, primarily because those means consume less energy per unit of water removed.

To illustrate some highly efficient liquid–solid separations, consider the following example. A 30-m-diameter conventional gravitational thickener (at a normal design rate for tailings concentration and water reclamation of 0.1 m²/t of dry solids per day) will process more than 2,400 t (metric tons) of solids per day. With a feed of 15% (by weight) solids, an underflow concentration of 50% (by weight) solids or higher can usually be achieved if the solids contain 50% to 55% particles that are -200 mesh or coarser. This size means that 4.67 kg of water per kilogram of solids has been eliminated and that more than 82% of the water will report to the thickener overflow for reuse. The thickener drive head will be equipped with only a 3.7- or 5.6-kW motor.

In the processing of magnetite concentrates derived from the beneficiation of taconite, disk filters are used to dewater magnetite concentrates before the balling step. For a 1,800 cm²/g Blaine concentrate (approximately 85% to 90% -325 mesh), a filtration rate of 1,100 kg of dry solids/h/m² is used as a design basis (Wolf et al. 1971). The feed concentration would be maintained at 65% (by weight), and the vacuum level should be at 60 cm of mercury by using a vacuum pump capacity of about 1.8 m³/m² of filtration area. Power consumption will be very high because of the high vacuum and flow rate. Considering also the filter drive, compressed air requirements, filtrate pump, and agitator, a total of 26 kW would be required for 10 m² of filtration area. This power requirement is equivalent to 2.5 cal/m² of filtration area. However, 100 kg of dry magnetite solids per hour are dewatered to the 9% (by weight) moisture required for hauling. At the same time, 4,600 kg of water are extracted. Thus, only 25 kcal/kg are eliminated.

Thermal drying, on the other hand, requires around 4,900 kcal/kg of water evaporated. Obviously, mechanical methods of dewatering have relatively low energy consumption per unit of liquid removed.

Liquid–solid separation is also critical in water discharge, water reclamation and closed water circuits (Wolf et al. 1971). Water that has been used for processing and beneficiating minerals and coals will contain solids that can range in size from a fraction of a micrometer to 6 mm or more. Some streams will

---

Donald A. Dahlstrom, Former Professor Emeritus, University of Utah, Salt Lake City, Utah, USA

contain the valuable solids and others the refuse or tailings. In both cases, the solids must be separated out if the water is to be reused. Furthermore, the concentration of suspended solids in recycled water must be low enough so that the water does not contaminate the next product. In the case of iron ore processing, the return water must contain between 100 and 150 mg/L or less to minimize the percentage of silica in the final pellet. Coal requires a suspended solids concentration of less than 1% (by weight).

Water discharge requirements often necessitate even better solids–liquids separation than process requirements. In the United States, state and federal regulations generally require that effluent disposed in lakes, streams, or other public water sources contain no more than 10 to 50 mg/L of suspended solids.

Steps in Liquid–Solid Separation

Separating liquid and solids requires many steps, and these are treated extensively in the literature (e.g., Fuerstenau and Han 2003; Wills and Napier-Munn 2006).

In coal processing, centrifuges and screens are usually used. For minerals beneficiation, ore is usually ground much finer than is practiced for coal processing, often to 80% passing 75 μm or even 80% passing 44 μm (Henderson et al. 1957; Kobler and Dahlstrom 1979). Thus, base metals and iron ore concentrates, which are large-tonnage minerals, will usually be rated according to the percentage of –325 mesh or even –400 mesh solids. Because of their abrasive character, these and other minerals will be dewatered in thickeners and then on filters. Both continuous filters and centrifuges are used with crystalline solids such as potash sulfate. Gravity thickeners are used for both concentrate and tailings (Coe and Clevenger 1916; Roberts 1949; Terchick et al. 1975). Tailings are generally sent to tailings ponds without solids–liquids separation, as their tonnage and volume can be very large. However, more stringent regulation of the construction of tailings ponds ensures that mechanical dewatering methods will be increasingly used in the future (Chironis 1976).

Hydrometallurgical processing always creates abundant colloidal solids during the leach step. Proper liquid–solid separation enables the recovery of the maximum amount of pregnant liquor while minimizing its dilution. Thus, multistage countercurrent sedimentation, countercurrent washing filtration on a single filter, or two- or three-stage filtration with single washes per stage are practiced. Both continuous vacuum filters and pressure filters are involved (Nelson and Dahlstrom 1957; Osborne 1975; Kobler and Dahlstrom 1979).

Current practice at many locations requires "zero discharge," or 100% recycling of water streams. If solids are disposed of in a landfill, the discharged solids must be low enough in moisture (generally 80% solids or more) so that liquids do not leach out by gravity. Thus, in a modern mineral processing plant, efficient liquid–solid separation processes are critical to low-cost processing, and these processes can substantially affect capital and operating costs.

MAJOR INFLUENCES ON LIQUID–SOLID SEPARATION

Many factors external to the liquid–solid separation equipment itself influence its performance and productivity. The most common of these are as follows (Cross 1963; Robins 1964; Scott 1970; Bosley 1974; Silverblatt et al. 1974; Wetzel 1974; Hsia and Reinmiller 1977; Weber 1977; Dahlstrom 1978; Rushton 1978; Sakiadis 1984; Baczek et al. 1997; Lewis 2007), and of which only the first factor will be discussed in detail in this chapter:

- Particle size and shape
- Weight and volume percentage of solids
- Fluid viscosity and temperature
- pH and chemical composition of the feed
- Variation and range in feed quality (i.e., of the first four items)
- Specific gravity of solids and liquid
- Quality requirements of discharge streams from liquid–solid separation steps, particularly as they influence results upstream and downstream

Particle Size and Shape

Size distribution greatly affects liquid–solid separation rates. Stokes' law can be used to illustrate this fact. This law permits the terminal settling velocity (maximum velocity achieved during free fall) to be determined as follows:

$$v_t = \frac{(\rho_s - \rho)gD^2}{18\mu}$$

where

v_t = terminal velocity, m/s
ρ_s = particle density, kg/m^3
ρ = liquid density, kg/m^3
g = gravitational acceleration = 9.806 m/s^2 at sea level at 45° latitude. This value is used for g, the standard gravitational acceleration on this planet, and is usually symbolized by gc.
D = particle diameter, m
μ = fluid viscosity, kg/m × s

Viscosity is normally measured in centipoise. One poise = 1 g·cm^{-1}·s^{-1}. (All nomenclature is metric [International System of Units], but English units can be used as long as they are used consistently.)

This equation has some limitations. It assumes laminar flow (i.e., a Reynolds number of less than 0.1) and nonhindered settling. In nonhindered settling, a particle is not influenced by the presence of other particles or by the slurry's specific gravity, conditions that require very dilute slurries.

If one assumes a 20-μm particle and an 80-μm particle of the same density and in the same fluid, the terminal velocity is found from the equation to be directly proportional to the square of the diameter. In this case, the 80-μm particle will fall 16 times as fast as the 20-μm particle (see Table 14.7-1).

Unfortunately, mineral slurries contain a variety of particle sizes of different densities and agglomerating behaviors, so the application of the equations is usually difficult. The size distribution of particles in the slurry also has a major effect, so a simple characterization, such as 80% –200 mesh, is not very useful when designing equipment. Percentage of clay-like particles and type of clay are very important variables and can change throughout a mineral deposit. For most heavy metal mineral processing operations, the clay content is usually within manageable limits. For these plants, discharge streams from conventional thickeners are typically 55% to 60% solids, discharge steams from filters are typically 80% to 83% solids (dry enough so it can be stacked with a bulldozer), and discharge streams from high-density thickeners

Table 14.7-1 Drag coefficient and related functions for spherical particles

Reynolds Number, N_{Re}*	Drag Coefficient, C	CN_{Re}^2	C/N_{Re}
0.1	244	2.44	2,440
0.2	124	4.96	620
0.3	83.8	7.54	279
0.5	51.5	12.9	103
0.7	37.6	18.4	53.8
1	27.2	27.2	27.2
2	14.8	59.0	7.38
3	10.5	94.7	3.51
5	7.03	176	1.406
7	5.48	268	0.782
10	4.26	426	0.426
20	2.72	(1.09)(10³)	0.136
30	2.12	(1.91)(10³)	0.0707
50	1.57	(3.94)(10³)	0.0315
70	1.31	(6.42)(10³)	0.0187
100	1.09	(1.09)(10⁴)	0.0109
200	0.776	(3.10)(10⁴)	(3.88)(10⁻³)
300	0.653	(5.87)(10⁴)	(2.18)(10⁻³)
500	0.555	(1.39)(10⁵)	(1.11)(10⁻³)
700	0.508	(2A9)(10⁵)	(7.26)(10⁻⁴)
(1 × 10³)	0.471	(4.71)(10⁵)	(4.71)(10⁻⁴)
(2 × 10³)	0.421	(1.68)(10⁶)	(2.11)(10⁻⁴)
(3 × 10³)	0.400	(3.60)(10⁶)	(1.33)(10⁻⁴)
(5 × 10³)	0.387	(9.68)(10⁶)	(7.75)(10⁻⁵)
(7 × 10³)	0.390	(1.91)(10⁷)	(5.57)(10⁻⁵)
(1 × 10⁴)	0.405	(4.05)(10⁷)	(4.05)(10⁻⁵)
(2 × 10⁴)	0.442	(1.77)(10⁸)	(2.21)(10⁻⁵)
(3 × 10⁴)	0.456	(4.10)(10⁸)	(1.52)(10⁻⁵)
(5 × 10⁴)	0.474	(1.19)(10⁹)	(9.48)(10⁻⁶)
(7 × 10⁴)	0.491	(2.41)(10⁹)	(7.02)(10⁻⁶)
(1 × 10⁵)	0.502	(5.02)(10⁹)	(5.02)(10⁻⁶)
(2 × 10⁵)	0.498	(1.99)(10¹⁰)	(2.49)(10⁻⁶)
(3 × 10⁵)	0.481	(4.33)(10¹⁰)	(1.60)(10⁻⁶)
(3.5 × 10⁵)	0.396	(4.86)(10¹⁰)	(1.13)(10⁻⁶)
(3.75 × 10⁵)	0.238	(3.34)(10¹⁰)	(6.34)(10⁻⁷)
(4 × 10⁵)	0.0891	(1.43)(10¹⁰)	(2.23)(10⁻⁷)
(4.25 × 10⁵)	0.0728	(1.32)(10¹⁰)	(1.71)(10⁻⁷)
(4.5 × 10⁵)	0.0753	(1.53)(10¹⁰)	(1.67)(10⁻⁷)
(5 × 10⁵)	0.0799	(2.00)(10¹⁰)	(1.60)(10⁻⁷)
(7 × 10⁵)	0.0945	(4.63)(10¹⁰)	(1.35)(10⁻⁷)
(1 × 10⁶)	0.110	(1.10)(10¹⁰)	(1.10)(10⁻⁷)
(2 × 10⁶)	0.150	(6.00)(10¹¹)	(7.50)(10⁻⁸)
(3 × 10⁶)	0.163	(l.47)(10¹²)	(5.44)(10⁻⁸)
(5 × 10⁶)	0.174	(4.35)(10¹²)	(3.48)(10⁻⁸)
(7 × 10⁶)	0.179	(8.75)(10¹²)	(2.55)(10⁻⁸)
(1 × 10⁷)	0.182	(1.82)(10¹³)	(1.82)(10⁻⁸)

Source: Perry et al. 1984, with permission from The McGraw-Hill Companies, Inc.
*For values of Re less than 0.1, $C = 24/Re$.

are typically 68% to 73% solids (a thick paste but of pumpable consistency).

The designs for thickeners and filtration equipment are usually decided in a program of laboratory testing, and a large safety factor is included when designing equipment. In general, even very finely ground mineral slurry will respond to dewatering if the correct equipment is employed. Although equipment suppliers should be contacted early in the design process, it is important that the client take charge of this process (rather than leaving it up to equipment suppliers), because the results can be interpreted in a variety of ways, which might favor some equipment over others.

LIQUID–SOLID SEPARATION EQUIPMENT

Commonly used equipment for liquid–solid separation in the minerals processing industry can be categorized into the following unit operations: screens, cyclones, thickeners and clarifiers, filters, centrifuges, and tailings ponds.

Screens

Screens are often the first liquid–solid equipment encountered by the ore stream as it enters the processing plant. Screens serve the dual purpose of separating the solids into different size fractions and separating solids from liquids. Typical applications for the latter use are "trash screens," which remove nonmineral particles such as wood chips from the mineral slurry. While this is a somewhat mundane application, it can be critically important to the functioning of downstream equipment, so a trash screen is usually present following the grinding stage in most mineral processing plants. This screen is usually an inclined-deck vibrating screen. In the gold-processing industry, many plants employ coarse carbon particles (1,600 µm, 10 mesh) to adsorb dissolved gold from 75-µm ore slurry. The carbon particles are recovered from the slurry on a variety of screen types, typically inclined-deck vibrating screens (commonly called Dutch State Mines screens or DSM screens), which employ parallel bars/wires at right angles to the flow and a steeply dipping nonvibrated deck (Fuerstenau and Han 2003).

In 2005, vibrating screens with very fine screen decks came into limited use to make a very fine separation, typically at 200 µm (65 mesh), of the product fines fraction from slurry, which is recirculated to the grinding mills. Formerly this job was always done using cyclones or spiral classifiers.

Cyclones

Cyclones are typically configured as a straight-sided cylinder connected to a tapered pipe section. The ore slurry stream is introduced tangentially into the cylindrical section, and it rapidly spins to separate the solids from the liquids. Typically, the dimensions are 25 cm in diameter and 100 cm long, but they can be much larger or smaller. Cyclones are commonly used as size separation devices to separate the fine product slurry from the coarse recycle slurry in the grinding section of mineral processing plants. Size separation is not the subject of this chapter, but the cyclones also serve the dual purpose of making a dense underflow (high solids content, coarse material) from a dilute overflow. This effect is utilized by dewatering cyclones, which are used to separate dense solids slurries to form a nearly clear liquid overflow. In this application they substitute for thickeners.

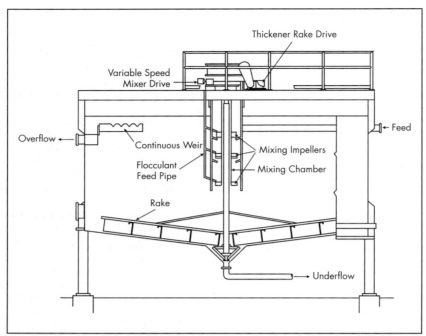

Figure 14.7-1 High-density thickener

Thickeners and Clarifiers

While cyclones employ centrifugal force, which can be many times the force of gravity, thickeners and clarifiers use gravity only to allow settling of mineral particles and separation of a dense slurry, which is pumped from the bottom of the thickener below a clear or nearly clear overflow liquid. Most mineral processing plants employ thickeners at multiple points in the process. Thickeners can range from 3 m to more than 100 m in diameter. Thickeners with diameters in the range of 20 to 50 m are applicable to settle the process stream in medium sized mineral processing plants treating 1,000 to 5,000 t of ore per day.

Where colloidal solids are present (which is usually the case), flocculants will improve operation of thickeners by causing the colloids to form agglomerates or flocculi of much larger size (which still contain probably 95% or more of liquid) that will settle at reasonable rates. In a typical mineral processing plant, flocculants are employed at multiple points, and the total reagent cost is between $1.00 and $5.00/t of mineral processed.

There are three basic types of thickeners: conventional, high-rate, and high-density. They all consist of a large, round, relatively shallow tank with a bottom that slopes toward the center, and they are equipped with a slowly rotating rake mechanism that moves the settled solids toward the central discharge. Overflow liquid is collected in a peripheral launder (trough) and is directed to a pipe outlet. Conventional thickeners can be very large and are sometimes run without the addition of flocculant. Capital cost is relatively high, but operating costs are low. High-rate thickeners can be much smaller in diameter than conventional thickeners for the same duty, and they rely on recycled slurry plus high usage of flocculants. Both conventional and high-rate thickeners discharge solids of the same density, usually 45% to 60% solids.

High-density thickeners have much deeper tanks with steep cone bottoms, and they rely on the auto-compression of the cake to increase density of the discharge solids up to 70% or more (see Figure 14.7-1). At this percent solids, the discharge is a thick paste solids that can be pumped only with special pumps. High-density thickeners are useful where water must be conserved or where a dense paste is needed for other applications such as mine backfill.

Although the overflow of a thickener must usually be of reasonable clarity, the emphasis is primarily on underflow concentration. Clarifiers are thickeners that operate on a dilute solids stream (often the discharge from thickeners) in which the flocculated stream is introduced below a blanket of flocs to capture all the suspended solids into the settled bed at the bottom of the clarifier. Clarifiers may have different internal configurations than the simple circular-rake thickener. For example, clarifiers may be rectangular tanks filled with submerged inclined plates (lamella thickeners), or they may be inclined tubes that reduce the settling distance of the particles to the vertical distance between plates.

The term countercurrent decantation (CCD) thickeners refers to the use of three to ten thickeners operated in a continuous train, in which slurry containing a process solution enters the first thickener and water enters the last thickener. Solids are pumped from the underflow of the first thickener into the next downstream thickener and so forth to the last thickener.

Solution is pumped from the overflow of the final thickener to the next upstream thickener, and it is mixed with the solids as they enter the thickener from the upstream direction. In this manner the solids slurry is continuously mixed and then settled, and the process solution (containing chemicals or dissolved products to be recovered) is gradually displaced by water. A five-stage thickener stream can often recover >95% of the dissolved values from the slurry using a volume of clean

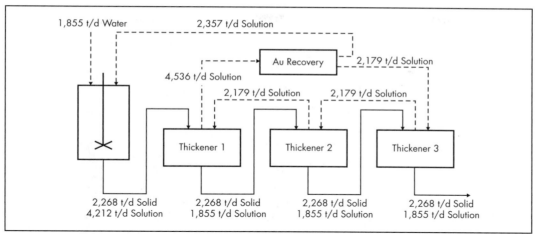

Figure 14.7-2 Three-stage countercurrent decantation circuit for gold ore

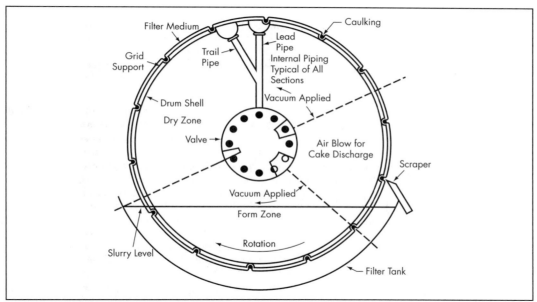

Source: Schweitzer 1979, with permission from The McGraw-Hill Companies, Inc.
Figure 14.7-3 Rotary drum vacuum filter

water equal to the volume of process solution entering with the slurry. CCD thickeners are the normal final stage in leach plants that chemically leach metals such as gold/silver, copper, nickel, or aluminum for their ores (see Figure 14.7-2).

Filters

Filtration can be divided into three modes: continuous, batch and semicontinuous, and clarifying. Each of these modes can be further subdivided. Ther are many books devoted to filtration and selection of filtration equipment, and the subject is discussed here in a very limited manner.

Vacuum Filters

Continuous vacuum filters can be rotating discs or drums, or traveling belts (Figure 14.7-3). They are usually used for dewatering mineral concentrates or for dewatering large volume, easy-to-filter materials such as the cleaned coal product of a coal washing plant. Traveling belt filters up to 6 m wide and 30 m long are employed for high-volume or slowly-filtering slurries. Large drum filters (with the filtration surface being the exposed outer surface of the drum) have been employed for the separation of solids slurry from process liquids, although this application has largely been replaced by thickeners.

Low-Tonnage Pressure Filters

Pressure filters are commonly employed in mineral processing plants to remove relatively low-tonnage, dilute solids from process streams. Although there are many types of pressure filters, in mineral processing plants these typically take the form of vertical-leaf filters with multiple parallel vertical leaves in a pressurized tank-like housing, or plate and frame filters with multiple plates sandwiched within a compression framework (Figure 14.7-4).

Typical applications are for achieving high clarifying process streams, where the prefiltration solids density is a few

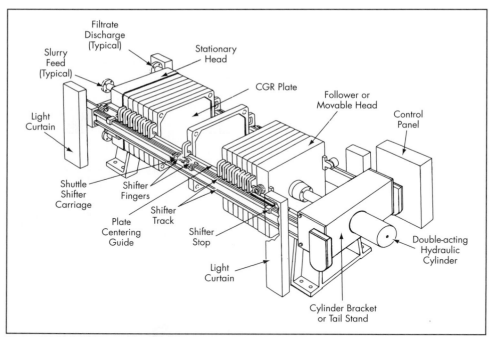

Figure 14.7-4 Plate and frame filter

hundred ppm up to 1%, and for the recovery of precipitated products such as gold and silver after these are precipitated in fine suspended solids form after the addition of zinc dust to a cyanide leach solution.

Clarifying filters typically have multiple parallel leaves consisting of a framework covered with filter cloth, and are usually precoated with diatomite to improve filtration capacity. Plate and frame filters for these low-solids applications are usually opened and closed, and cleaned manually.

High-Tonnage (Continuous) Pressure Filters

Continuous (actually semicontinuous) recessed plate filters are widely used to produce a dry tailings from a dilute slurry in high-tonnage process plants. A typical filter might contain 60 plates, each with a filter surface 1×1.5 m on each side of the plate. Such a filter might be able to process 1,000 t/d of slurry in a 12-minute cycle that includes 3 minutes of cake formation and other intervals for washing and blowing. The capital cost of a 1,000 t/d filter is around $2 million, but the final installed cost is around $5 million because the fitler is supported by a complex arrangement of pumps, piping, tanks, air compressors, and discharge conveyors.

Where a conventional tailings pond can be employed (in which the ore slurry is sent to the pond and clear solution is recovered), this is always preferable to tailings filtration, but in many cases topography or environmental considerations make tailings filtration necessary (see Table 14.7-2).

Centrifuges

Centrifuges are used with less-abrasive solids such as fine coal and crystalline solids. They can also beneficiate coal as the concentrate solids are usually appreciably higher in ash content than the feed.

Centrifuges, as their name implies, use a centrifugal force to substantially increase the settling rate of particles. This centrifugal force, created by pumping the feed into a cone-shaped unit, forces the feed to separate into two discharges: a solid-rich fraction and an aqueous fraction. There are no moving parts. However, a choke nozzle on the solids discharge allows adjustment and fine-tuning.

For mineral dewatering, the solid-bowl scroll centrifuge is frequently employed because it continuously discharges solids. The unit consists of a revolving horizontal cone-shaped casing inside which a screw conveyor of similar internal diameter revolves in the same direction but at a slower speed. The feed enters through a central tube of the screw conveyor from where it is subjected to high centrifugal forces that cause the solids to settle on the inside of the casing where a screw conveyor "scrapes" and transports them to the discharge port located at the tapered end of the unit. Bowl diameters are usually from 15 to 150 cm. They have a throughput of 0.5 to 50 m^3/h of liquid and 0.25 to 100 t/h of solids with the moisture content of the discharge varying from 5% to 20% (Wills and Napier-Munn 2006).

Tailings Ponds

Tailings ponds are widely employed at mineral processing plants. Mineral slurries are pumped to the ponds, the solids settle and consolidate to form a stable mass, and the clear liquid is recycled to the plant or discharged to the environment. For chemical containment, tailings ponds may be lined with plastic liners and leak detectors, but where the process involves only nonchemical processing (physical methods such as gravity concentration or flotation), the tailings pond may be a simple excavated area with a compacted soil base. The fine solids in the tailings usually serve as an effective seal to prevent most escapes of process water.

Tailings pond management and construction (which is a continuous process as more tailings are created) may be a significant part (5% to 8%) of overall plant operating cost.

Table 14.7-2 Typical equipment factors for continuous vacuum filters (standard designs)

Filter Type	Submergence, %*		Total % Under Vacuum	Maximum % Cake Wash	Maximum % Dewatering Only	% Required for Cake Discharge	Minimum Cake Thickness, mm
	Apparent	Effective					
Filters Forming Cake Against Gravity							
Drum filters							
Scraper	35	30	80	29	50–60	20	6.4
Roll 1	35	30	80	29	50–60	20	0.8
Belt	35	30	75	29	45–55	25	3.2–4.8
Precoat	35	35	100	29	50	0	0
Disk filters	35	28	75	NA†	45–50	25	9.5–12.7
Filters Forming Cake with Gravity							
Horizontal belt	as required		100*		As required	0	3.2–4.8
Horizontal table	as required		80		As required	20	19.1

*Horizontal belt filter is based only on effective area (area under vacuum).
†NA = not available.

Where zero discharge is practiced, excess solution may be evaporated from the tailings pond using natural solar evaporation or various types of evaporative sprinkler. Evaporators of the "turbo-mist" variety blow a stream of fine droplets high into the air after mixing the solution with compressed air to form the fine droplets. A typical equipment unit of this type might evaporate 2,000 m^3 of water per month while consuming 45 kW. Capital cost is of the order of less than \$150,000. Multiple units can be installed. The heat needed for evaporation can come from simple solar heating of the pond surface, but a larger contributor is the latent heat contained in the moving air mass above the tailings pond.

SUMMARY

Liquid–solid separation and the related problem of water balance is one of the major design issues for any mineral processing plant. Regardless of the mineral to be treated or the process employed, mistakes or design uncertainties in this area are one of the most common reasons for slow plant start-up or underperformance. This chapter has intended to identify some of the more common design practices, but a very broad body of published literature, supplier brochures, and Internet information exists on the subject of flow-sheet design and liquid–solid separation. After a liquid–solid application is identified, the engineer should invest substantial time in quantifying the design factors and limitations, and in researching the alternatives, before making the selection of specific equipment.

ACKNOWLEDGMENTS

This chapter is based on the "Liquid–Solid Separation" chapter by the same author in *Principles of Mineral Processing*, published by SME.

REFERENCES

Baczek, F.A., et al. 1997. Sedimentation. In *Handbook of Separation Techniques for Chemical Engineers*, 3rd ed. Edited by P.A. Schweitzer. New York: McGraw-Hill.

Bosley, R. 1974. Vacuum filtration equipment innovation. *Filtr. Sep.* 11:138–149.

Chironis, N.P. 1976. New clarifier/thickener boosts output of older coal preparation plant. *Coal Age* 81:140–145.

Coe, H.S., and Clevenger, G.H. 1916. Methods for dewatering the capacities of slime settling tanks. *Trans. AIME* 55:356–384.

Cross, H.E. 1963. A new approach to the design and operation of thickeners. *J. S. Afr. Inst. Min. Metall.* 63:271–298.

Dahlstrom, D.A. 1978. Practical use of applied theory of continuous filtration. *Chem. Eng. Prog.* 74(March):69.

Fuerstenau, M.C., and Han, K.N. 2003. *Principles of Mineral Processing*. Littleton, CO: SME.

Henderson, A.S., Cornell, C.F., Dunyon, A.F., and Dahlstrom, D.A. 1957. Filtration and control of moisture content on taconite concentrates. *Min. Eng.* March.

Hsia, E.S., and Reinmiller, F.W. 1977. How to design and construct earth bottom thickeners. *Min. Eng.* 29:36–39.

Kobler, R.W., and Dahlstrom, D.A. 1979. Continuous development of vacuum filters for dewatering iron ore concentrates. *Trans. SME-AIME* 266:2015–2021.

Lewis, S.M., ed. 2007. *Perry's Chemical Engineers' Handbook*, 8th ed. New York: McGraw-Hill.

Nelson, P.A., and Dahlstrom, D.A. 1957. Moisture content correlation of rotary vacuum filter cakes. *Chem. Eng. Prog.* 53:320–327.

Osborne, D.G. 1975. Scale-up of rotary vacuum filter capacities. *Trans. Inst. Min. Metall.* 84:C158–C166.

Perry, R.H., Green, D.W., and Maloney, J.O. 1984. *Perry's Chemical Engineers' Handbook*, 6th ed. New York: McGraw-Hill.

Roberts, E.J. 1949. Thickening, art or sciences. *Trans. AIME* 184:61.

Robins, W.H.M. 1964. The theory of the design and operation of settling tanks. *Trans. Inst. Chem. Eng.* 42:T158–T163.

Rushton, A. 1978. Design throughputs in rotary disc vacuum filtration with incompressible cakes. *Powder Technol.* 21:161–169.

Sakiadis, B.C. 1984. Fluid and particle mechanics. In *Perry's Chemical Engineers' Handbook*, 6th ed. Edited by R.H. Perry and D. Green. New York: McGraw-Hill.

Schweitzer, P.A., ed. 1979. *Handbook of Separation Techniques for Chemical Engineers*. New York: McGraw-Hill.

Scott, K.J. 1970. Continuous thickening of flocculated suspensions. *Ind. Eng. Chem. Fund.* 9:422–427.

Silverblatt, C.E., Risbud, H., and Tiller, F.M. 1974. Batch, continuous processes for cake filtration. *Chem. Eng.* 81:127–136.

Terchick, A.A., King, D.T., and Anderson, J.C. 1975. Application and utilization of the Enviro-Clear thickener in a U.S. steel coal preparation plant. *Trans. SME-AIME* 258:148–151.

Weber, F.R. 1977. How to select the right thickener. *Coal Min. Process.* 14:98–104.

Wetzel, B. 1974. Disc filter performance improved by equipment redesign. *Filtr. Sep.* 11:270–274.

Wills, B.A., and Napier-Munn, T.J. 2006. *Wills' Mineral Processing Technology*, 7th ed. Oxford: Butterworth-Heinemann.

Wolf, J., et al. 1971. Present methods and future needs in iron concentrate dewatering the process water reclamation. Paper presented at AIME Annual Meeting, March 1–4, New York.

PART 15

Health and Safety

CHAPTER 15.1

Mine Safety

David Laurence

The modern mine site is a different place than that which existed in 1989. Unlike the majority of mines that operated throughout the world at that time, many mines are no longer residential operations; instead they are staffed by a "fly-in, fly-out" work force. Workers at most mines are likely to be employed by contractors rather than the mining company itself and may not belong to a labor union. Many workers joined the industry during the boom times from 2004 and lack experience. All of these factors add to the challenges faced by mine managers to provide a safe place of work for all those employed on a mine site.

Even casual visitors to a mine are required to comply with the stringent safety requirements and the legislative responsibilities placed on mine management to ensure the health and safety of all people. The measures include

- Security procedures at the gate or entrance to the mining lease,
- Random or compulsory drugs and alcohol testing (many mines now have a zero-alcohol limit; many have no tolerance to any drugs, including analgesics),
- A safety induction program (for a contractor or employee, this can take a week to complete), and
- Continuous supervision by an employee.

One of the most significant issues for mine managers is achieving a safe working environment at their mine. "Zero harm" is the phrase that sums up the mission of many mining companies, contractors, and their employees. It is likely to be driven by one or all of the following points:

- The community no longer accepts that injuries are a necessary part of the mining process.
- If a fatality occurs at a mine site, there may be community angst, which grows to outrage if multiple fatalities occur.
- Politicians may act contrary to the mine's best interest but in their best interest.
- Regulator enforcement has strengthened, implying that
  - Mines can be shut down or
  - Managers and directors can be imprisoned or fined.
- The loss in human and financial terms can be immense.

- Remuneration and career progression are tied to safety performance.

Safety is one of the dimensions of sustainable mining practices, along with the other important dimensions of economy, environment, community, and efficiency (Figure 15.1-1). Three of these dimensions—economy, environment, and community—represent the pillars of sustainable development. In mining it is also important to consider safety and efficiency. Efficiency relates directly to the resource and encompasses optimization, technical excellence, nurturing the resource, and so forth (Laurence 2008).

Safety in the mining industry must encompass the full cycle of exploration, development, construction, operations, rehabilitation, closure, and final walkaway. Of course, associated activities such as mineral processing, smelting, and transport are all part of the industry, and excellence in safety needs to apply there as well.

In discussions of mine safety, the terms *dangerous* and *accidents* are not specific enough to engender clarity. Rather than using the word *dangerous*, the term *hazardous* is more meaningful when describing the working environment of mine sites. Hazards are present in the mining industry, but these can be identified and the resultant risks managed. The use of risk management techniques has been universally adopted in the

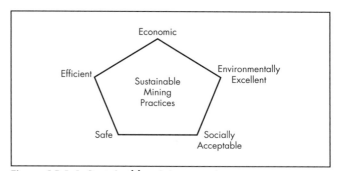

Figure 15.1-1 Sustainable mining practices

David Laurence, Acting Director, School of Mining Engineering, University of New South Wales, Sydney, New South Wales, Australia

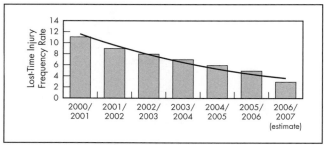

Figure 15.1-2 Australian mining industry lost-time injury frequency rate, 2000–2007

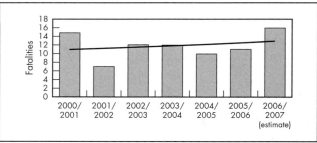

Figure 15.1-3 Mining industry fatalities in Australia, 2000–2007

Australian mining industry, and it has achieved a significant reduction in lost-time injuries and other safety performance measures. In a similar vein, the word *accident* connotes that uncontrollable natural forces may have been involved in an event. The industry prefers to use the term *incident*, with the knowledge that such events are preventable, whereas accidents are not.

Although the content of this chapter focuses on the Australian mining industry, the lessons can be applied anywhere.

DEFINING SAFETY

The Australian Standard 1470:1986, titled *Health and Safety at Work*, defines *safety* as "the provision and control of work environment systems and human behaviour which, together, give relative freedom from those conditions and circumstances which can cause personal damage." The Minerals Council of Australia (MCA), the peak lobby group for the mining industry in that country, defines *safety awareness* as "the state of mind where we are constantly aware of the possibility of injury and act accordingly at all times." On its Web site, MCA describes its vision for safety as "An Australian minerals industry free of fatalities, injuries and diseases." The MCA believes that

- All fatalities, injuries, and diseases are preventable;
- No task is so important that it cannot be done safely;
- All hazards can be identified and their risks managed;
- Everyone has a personal responsibility for the safety and health of themselves and others; and
- Safety and health performance can always be improved (MCA 2009).

Similar values and statements can be found on the Web sites of the major mining companies in Australia and throughout the world.

Exposure to dust, noise, and hazardous substances has long been an occupational hazard for mine workers. Cases of mesothelioma, asbestosis, silicosis, and pneumoconiosis are still appearing in mine workers many years after exposure. Although the health issues are at least as important as the safety issues and, in many cases, are more insidious and can take longer to manifest, the focus of this chapter is on mine safety rather than occupational health issues.

It is also important that workers are not subjected to injury through the use of machines. Mechanization and automation have helped to reduce any adverse interactions between humans and machines. Many mines still use handheld drilling techniques, however, and cases of repetitive strain injury in arms, hands, and fingers continue to occur. Similarly back injuries remain a major component of all workers' compensation claims in the mining industry. Part 15 of this handbook provides detailed information in this problem area.

MEASURING SAFETY IN MINING

Is mining a safe industry? Is *safety* a relative or absolute term? Some would argue that relative to other comparable industries, such as deep sea fishing, logging, and the construction and transport industries, mining is safe. Others would argue that, compared with the nuclear and chemical industries, it is not safe.

Defining how safety is measured is an important first step before attempting to compare mining with other industries. In mining, the two most commonly used measures are lost-time injuries and fatalities. Usually these measures are expressed as frequency rates, e.g., the lost-time injury frequency rate (LTIFR), which indicates the number of lost-time injury events that occur for a given number of hours worked. Figure 15.1-2 shows that on a lost-time injury basis, the Australian mining industry is performing in a continuously improving fashion. The use of this measure alone, however, leaves the industry open to the usual criticisms of reliability of this factor. It is well known that the LTIFR can be manipulated, particularly in this era of flexible work rosters.

One measure that cannot be manipulated is the number of fatalities that occur on mine sites. As shown in Figure 15.1-3, which covers the same period, 2000–2007 (financial years), an increasing number of deaths occurred. Although it can be persuasively argued that the number of employees in the mines increased, the numbers are far from the MCA's vision of "an industry free of fatalities, injuries and diseases." The risk of fatalities is measured by the fatal injury frequency rate (FIFR—the number of fatal injuries per one million hours worked). The Australian national FIFR in 2006–2007 was 0.05 per million hours worked, which was, according to the MCA, higher than the previous year at 0.04. Although an FIFR of 0.05 is below the 10-year average FIFR of 0.07, rates have fluctuated widely from year to year, and a consistent downward trend has not emerged.

According to the MCA, the Australian mining industry is performing relatively well compared with other countries. "The Australian minerals industry average FIFR for the ten-year period 1997–1998 to 2006–2007 was 0.07. Internationally the Australian industry compares favorably to South Africa on this indicator, which recorded an equivalent rate of 0.27 for this period, and the United States of America (USA), which recorded a rate of around 0.17 for this period" (MCA 2009). Comparisons must be made on equal terms, however, as the indices vary from country to country. For example, in Australia, an LTIFR is defined as the number of lost-time injury events per million hours worked; in North America, it is per 200,000 hours worked.

THE CONCEPT OF RISK

The minerals industry has approximately 10 fatalities per year, based on a 10-year rolling average (MCA 2009). The minerals industry work force is about 120,000 strong, which means that about 1 in 12,000 workers are killed each year.

Is this an acceptable risk? Although no government agency in the world has a prescribed acceptable rate of work-related fatalities, several studies have shown that the community is comfortable with industries that kill less than 1 employee per 100,000 per year, as indicated in Table 15.1-1. This means that fatality risk in the minerals industry is much too high.

It may be useful to begin by examining the mining industry's safety record in comparison with other industries. There is no doubt that many industries have a higher incidence of serious or fatal injuries than mining. It has been estimated, for example, that the construction industry kills one person a week in Australia. The transport sector also results in more deaths than the mining industry in Australia. In a single state, Queensland, 23 truck drivers died in work-related incidents in 2007. Recent safety data is notoriously difficult to obtain, but a 2003 study of fatal crashes involving articulated trucks in Australia indicated that the number of fatalities between 1991 and 2002 (the latest available data), was relatively stable at around 190 per year (Australian Transport Safety Bureau 2003).

In a final comparison and according to recent statistics, in Australia, between 5 and 10 people per year are killed by lightning strikes, approximately the same number as those killed in mines (Bureau of Meteorology 2009).

RISK MANAGEMENT

The mining industry in Australia has embraced the concepts of risk management in virtually every aspect of its business. Risk assessments are usually mandated by legislation and are carried out before a new (major) piece of equipment is introduced or a new mining method, process, or system is implemented. Risk management has been credited with markedly improving safety performance in Australian mines.

Figure 15.1-4 illustrates the steps usually followed in carrying out a risk assessment. First the hazard is identified. The risk presented by that hazard is then assessed. This can be done in a qualitative or a quantitative manner. Risk is generally assumed to be the product of the likelihood of a loss caused by that hazard occurring and the consequences of it occurring. After the risk is estimated, a decision is made to eliminate, mitigate, or control the risk to an acceptable level. Monitoring is the final step in the process to close the loop.

Hazards are a source of potential harm. In the mining industry, hazards are numerous, and the challenge for the work force, in the first instance, is to be aware of them. One of the reasons that hazards in mining are so great is because of the significant energies involved, be they gravitational, mechanical, chemical, electrical, or other types. The risk analysis allows the individual risks to be plotted on a matrix allowing management to allocate priorities to each.

Table 15.1-1 Estimates of risk

Activity	Risk of an Individual Dying in Any One Year
Smoking 10 cigarettes a day	1 in 200
Deep sea fishing	1 in 260
Influenza	1 in 500
Offshore oil and gas industries	1 in 600
Natural causes, 40 years old	1 in 850
Road accident	1 in 8,000
Road traffic accident	1 in 10,000
Playing soccer	1 in 25,000
Accident at home	1 in 26,000
Accident at work	1 in 43,500
Working in a "safe" industry	1 in 100,000
By fire or explosion at home	1 in 1,000,000
Struck by lightning	1 in 10,000,000

Source: British Medical Association 1987.

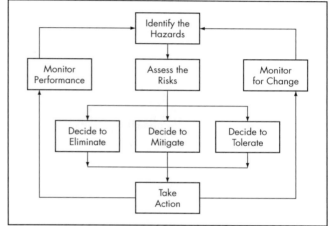

Source: Joy 1999a.
Figure 15.1-4 Basic risk-management process

MINE SAFETY LEGISLATION

In May 1970, the British government appointed a committee, chaired by Lord Robens, former chairman of the U.K. National Coal Board, to inquire into the provisions made for the safety and health of people in their workplace. The Robens Report was the major instigator to changing the philosophy and design of safety legislation in the United Kingdom and globally (Robens 1972). Up until that time, health and safety regulation consisted of a number of Acts of Parliament such as the Mines and Quarries Act (U.K. Office of Public Sector Information 1954), which had numerous sets of regulations covering specific risks or activities. Many of these had originated because of past accidents and disasters. Robens felt that this rule-based legislation was too complex, often inadequate, and had built-in obsolescence due to rapid technological changes.

Robens recommended a single broader and more flexible system of legislation based on more effective self-regulation by employers and workers jointly. Workplace health and safety should be effected by a tripartite approach involving workers, employers, and the regulator. The resultant Health and Safety at Work Act (U.K. Health and Safety Executive 1974) included the following features:

- The duty of the employer to provide safe and healthy working systems, premises, working environments, and equipment, and to ensure that people work safely through training, instruction, and supervision
- The duty on each employee to observe safety and health provisions and to act with due care
- Elimination of prescriptive detail, replacing it with goal setting or performance achievement
- An emphasis on hazard identification, risk management, and risk assessment
- The use of codes of practice and guidelines to assist industry in compliance

In Australia, the New South Wales (NSW) government incorporated the philosophy of Robens in its Occupational Health and Safety (OHS) Act (NSW Government Legislation 1983), which was updated in 2000 and applies to all industries including the mining industry. This is a performance standard or enabling act. That is, it states the standard of performance to be achieved, leaving it to those bound by the act to select for themselves the methods to be used in achieving that standard. It is thus, philosophically, standing apart from the traditional prescriptive legislation, which prescribed for operators how to achieve standards.

The OHS Act acknowledges that the best results will be achieved by influencing attitudes and creating a framework for better health and safety organization and action by industry itself. Key features of the approach are

- Replacement of a mass of legislation by a single statute of general application;
- Provision by legislation of a comprehensive set of duties relating to the basic and overriding responsibilities of employers, employees, manufacturers, and others, which should be simple, easily assimilated precepts of general application (the "general duty" provisions); and
- The use of approved and voluntary standards and codes of practice in order to provide progressively better conditions.

Thus, under this legislation, both employers and employees have duties of care. An employer must ensure a safe and healthy workplace. An employee must take reasonable care of the health and safety of persons (including themselves) while at work and to cooperate with the employer.

Tasmania has a similar act, the Workplace Health and Safety Act (Australasian Legal Information Institute Tasmanian Consolidated Acts 1995). Part 3 of the act describes the general duties and obligations of both employers and employees. Section 9 provides that an employer must ensure, so far as is reasonably practicable, that the employee is safe from injury and risks to health. Section 16 of the act requires an employee to take reasonable care for the employee's own health and safety and for the health and safety of other persons and to comply with any direction given by an employer with respect to any matter relating to health and safety. Under this legislation, all parties, including the employer (and its employees) and the contractor (and its employees), have a duty to contribute to a safe working environment at the mine.

Other countries have different styles of legislation. A feature of the Canadian regulatory framework is the internal responsibility system or IRS. This is a system in an organization where everyone has direct responsibility for health and safety as an essential part of the job (Ontario Ministry of Labour 2000). The position a person holds in the organization does not reduce the obligation; however, the chief executive officer has overriding responsibility for implementing the system. Each person takes the initiative on health and safety issues and works to solve problems and make improvements.

A company's Joint Health and Safety Committee has a key contributory role in making the IRS work well. The keys to a successful IRS involve fixing responsibility at the worker level. For instance, each person must

- Have a sincere wish to prevent accidents and illnesses,
- Accept that the causes of accidents and illnesses can be eliminated or greatly reduced,
- Appreciate that risks can be continually reduced so that the time between accidents and illnesses gets longer and longer,
- Make health and safety an essential part of the job,
- Have a clear understanding of their responsibilities,
- Be able to explain what proactive measures they have taken to ensure health and safety on the job,
- Have a clear understanding of their abilities and limitations,
- Strive to more than meet the minimum standards,
- Strive to cooperate with others to reduce risk,
- Understand and believe in the IRS process,
- Recognize that reprisals and incrimination are not part of the process (Ontario Ministry of Labour 2000).

The IRS is a system that drives responsibility for health and safety down to the operator level. If mine managers can obtain this level of responsibility from their workers, then there is a chance of ensuring the health and safety of employees, which is a general duty under most mining OHS legislation.

INFORMATION SHARING

One significant advance contributing to improved safety performance over the past 20 years has been a greater openness among mining companies and government regulators to share safety information. Although this openness is threatened in some jurisdictions because of the litigation that follows some incidents, there is a great deal of information available, particularly on the Internet. This is mainly in the form of safety alerts, significant incidents reports, and bulletins. Most of these are generated by regulators and designed to inform mine operators about significant incidents that occurred within their jurisdictions. They are also shared between regulators, however, so that an operator of an underground coal mine in Queensland, Australia, for example, is aware of a fatality at a similar mine in West Virginia in the United States. The fatality reports issued by the U.S. Department of Labor's Mine Safety and Health Administration (MSHA) contain detailed information and are designed to ensure that such tragedies do not reoccur. Reports of this nature can be found at the Web sites for MSHA; the Department of Mines and Energy, Queensland, Australia; and the Department of Primary Industries, New South Wales, Australia.

THE HUMAN FACTOR

One body of evidence suggests that despite the best systems and procedures, accidents and incidents do occur and many of these are caused by risk taking or human error. For example, "Recent history in the mining industry in Queensland has shown that incidents and accidents have occurred as a result

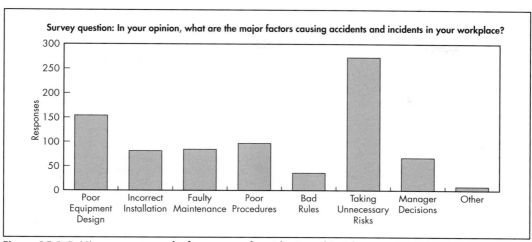

Figure 15.1-5 Miner survey results for causes of accidents and incidents

of unintentional errors or actions that do not recognise the personal risk present. Too often, the serious accidents are a result of conscious risk-taking, such as taking the easy way out, taking a short cut or not thinking about the risks in your workplace before starting work. It is not uncommon to find that the victims knew the potential risk of accidents before the incident" (Queensland Department of Mines and Energy 2002). And "very few people have accidents for which there is no procedure in place and we have a significant degree of accidents as a result of people breaking rules, ignoring rules or simply not knowing about them...." (Pitzer 2000).

A recent study of risky behaviors of continuous miner operators in underground coal mines found that more than 90% of employees interviewed "expressed misgivings" about safety rules and regulations. There was a clear distinction between operating within the rules and operating normally, with the latter referring to the quickest possible way to complete a task. "The issue is not whether the rules are practical or not, or whether the rules should be simply followed or not. A more serious issue is that possibly the large majority of employees (including deputies/supervisors) operate dangerous machinery every day in underground mines with a basic disposition that safety rules are 'irrelevant, superfluous, non-essential or excessive.' If this is the case, an important resource for limiting the risky behaviour of employees is critically deficient" (Pitzer 2000).

A further study of 39 multiple-fatality coal mine disasters, mainly from the United States, United Kingdom, and Australia, found that serious violations of mining laws were discovered among 33 (Braithwaite and Grabosky 1985). It was found that serious violations either caused the disaster, were components of the causes, or exacerbated the effect of the disaster.

These findings are supported by a 1991 study, commissioned by the NSW Department of Mineral Resources, which found that 83% of fatal accidents from strata movements in underground coal mines in the period 1972–1990 were associated with breaches of rules (Roylett et al. 1991). Human error was simultaneously present with breaches of rules in more than half the fatal accidents analyzed. It was concluded that fatalities would continue to occur unless management found strategies to focus on the development of better support rules, compliance with rules, and improvements in technology to "counter against human error." Subsequent investigations have suggested that, apart from rules breaches, a major contributory cause of these incidents was inadequate mine design, including undersized pillars (J. Galvin, personal communication).

A supplementary study on roof and rib fall incidents commissioned by the NSW Department of Mineral Resources in 1999 updated the findings of the Roylett et al. 1991 study. It found that "to effectively minimise roof fall fatalities the industry must control the behavioural causal factors involved in these fatalities. In particular, miners going out under unsupported roof and not working to the support rules" (Pereira 1999).

Another research project focused on the development of more effective mine safety rules and regulations (Laurence 2002). Responses from almost 500 mine workers were analyzed and some simple guidelines established, in particular regarding rule content. The recommendations were that management and regulators should not continue to produce more and more rules and regulations to cover every aspect of mining. Miners will not read nor comprehend to this level of detail. Detailed prescriptive regulations, detailed safe work procedures, and voluminous safety management plans will not "connect" with a miner. The aim should be to operate within a framework of as few rules as possible, but rules of the highest quality.

Of course, achieving more effective rules and regulations is not the only answer to a safer workplace. Possibly less emphasis should be on content and more about the process, in particular ensuring that a positive safety culture exists and that communication channels are open and working well. This was confirmed in the survey when the expressions "simply bad rules" or "poor rules" were rarely blamed for accidents and incidents, risk taking, or error making. More often the problems of implementation, communication, and learning were the main causal factors.

The mine workers were also asked what they considered to be the major causes of accidents and incidents. Taking unnecessary risks was considered by far to be the most significant response, as shown in Figure 15.1-5.

Human error is defined by Reason (1990) as "the failure of planned actions to achieve their desired ends—without the intervention of some unforeseeable event." Errors can be divided into those where the plan was adequate but actions

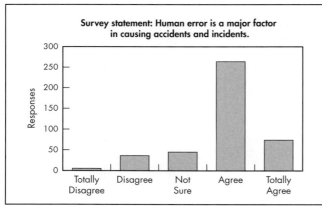

Figure 15.1-6 Miner survey results regarding human error and accidents

failed to go as planned, or those where the actions conformed to plan but the plan was inadequate. The former are known as slips, lapses, and trips, and the latter as mistakes.

There was general agreement among the mining work force that human error is a major cause of accidents and incidents, as illustrated in Figure 15.1-6. The major reason that people make errors is haste, according to the work force, as shown in Figure 15.1-7.

In his analysis of 25 incidents involving fatalities or serious injury, Joy (1999b) found that many were caused by management systems failures. Active human error was also a significant cause. The error was not generally intentional: "most active human error occurs where the person intended to do things correctly." Of the 25 accidents, 46% were violations; 31%, mistakes; 8%, slip/lapses; and 15%, no active human error. In some of these incidents, mine workers

- Were unaware of the rules,
- Were aware of but did not understand the rules,
- Mistakenly applied the rules,
- Ignored the rules,
- Deliberately violated the rules,
- Took a risk,
- Were unable to identify hazardous situations, or
- Were poorly trained or lacked sufficient educational background (Joy 1999b).

Further confirmation of the extent of the contribution of unsafe behaviors to accidents and incidents on mine sites in Australia is shown in Table 15.1-2. The table shows all published incidents over a 4-year period where a human factor can be identified. Such incidents are listed on the Web pages of each government mines' inspectorate in New South Wales, Queensland, and Western Australia. It can be seen that many of these accidents were caused, at least in part, by

- Lack of awareness of procedures or rules,
- Not complying with or ignoring rules,
- Lack of clear instructions,
- Poor communication generally,
- Production taking precedence over safety,
- Overriding or bridging out safety barriers,
- Inadequate training,
- Lack of familiarity with equipment, or
- Fatigue.

EXERCISING DUE DILIGENCE

It is important that mine managers are aware of the need for due diligence should an incident occur at their mine site. In the event of a serious incident, a regulator may wish to see

- A report of the incident;
- Whether it was reported to the authorities as required;
- Whether an accident report was filled out;
- What system of work was being used;
- What equipment was being used;
- The extent of the injured person's qualifications and experience in work of this type;
- Whether the injured person was inducted and trained;
- Who the supervisors and managers were, their qualifications and experience, and what instructions were provided to the injured person;
- Whether a risk assessment was carried out prior to the work;
- Whether a job safety analysis, safe operating procedure, or equivalent was in place; and
- Evidence of a safe system of work.

A SAFE SYSTEM OF WORK

An organization has implemented a "safe system of work" when it has employed and trained competent people, utilized fit-for-purpose equipment, and implemented effective systems and practices, all in a controlled work environment, as illustrated in Figure 15.1-8.

Competent People

To achieve safe production in a mine, competent people must be employed in senior management, in line management, and at the coal face. Evidence of competence includes inductions, training including refresher training, assessments, experience, formal educational qualifications, and so on. In most Australian states, mine managers have to demonstrate in a formal written and oral examination that they are competent to run a mine. They also need a degree or diploma in mining engineering or equivalent. Mine managers in New South Wales and Queensland are now required to hold the G3 certificate, which demonstrates competency in integrating risk management into the mining operation.

Equipment

Clearly the equipment and machinery used on a mine site needs to adequately carry out the task for which it was designed; it needs to be "fit for purpose." In recent times, manufacturers have worked in partnership with mines to continuously improve mining from a safety and ergonomic perspective.

Effective Systems and Practices

In an underground mine, examples of effective systems and practices include a safety management plan, ground control management plan, ventilation surveys, job safety analyses, formal risk assessments, and so on.

A Controlled Work Environment

Because of geological and geotechnical uncertainties, a controlled work environment is difficult to achieve, particularly in an underground mine. This can be overcome to a certain extent by ensuring that geological and geotechnical data is sufficient to enable sound decisions to be made for managing ground support, one of the principal hazards. Ground support measures are usually routine and can thus be considered as

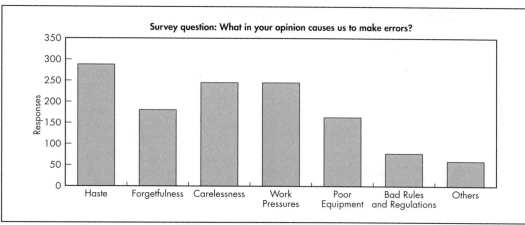

Figure 15.1-7 Miner survey results showing reasons for error

close to a controlled work environment as can be achieved underground.

AVOIDING MINE DISASTERS

In the past, there has been a tendency to focus on low-consequence, high-probability events, also known as "slips, trips, and falls." Doing this may certainly result in a lower injury rate and thus better safety statistics, which of course is one of the common key performance indicators of mine management. Experience shows, however, that if this aspect of safety is the only focus, opportunities may be missed for identifying potentially more damaging events, sometimes resulting in mine disasters.

Increasingly, the community is becoming less tolerant of any injury or human loss caused by mining and associated operations. Society is prepared to accept a certain amount of loss in mining, but community outrage may occur if fatalities are involved.

Reason (1997) argues that all organizational accidents arise from breaching a defense or series of defenses, barriers, or safeguards separating damaging hazards from vulnerable people or assets. Defenses should be present in layers, termed *defenses in depth*, and each defense would protect against the possible breakdown of the one before (Figure 15.1-9). Defenses should

- Provide alarms and warnings when danger is imminent,
- Restore the system to a safe state in an off-normal situation,
- Interpose safety barriers between the hazards and the potential losses,
- Contain and eliminate the hazards should they escape the barrier, and
- Provide the means of escape and rescue should hazard containment fail.

Many organizations harbor what Reason (1997) terms *latent conditions*, which resemble pathogens in the human body—they may be present for many years before they combine with local circumstances and active failures to penetrate the system's many defenses. Examples include poor design, gaps in supervision, manufacturing defects, maintenance failures, inadequate training, and impractical procedures. Latent conditions were present in the *Challenger* space shuttle tragedy, the *Herald of Free Enterprise* sinking, the *Piper Alpha* oil rig fire, and the underground fire at King's Cross Station in London, to name a few.

Using Reason's Swiss cheese example, an accident occurs when there is a connecting set of holes in successive defenses, allowing an accident trajectory to occur, as shown in Figure 15.1-9.

Hopkins (1999) analyzed multiple-fatality events at the Gretley mine in New South Wales and the Moura mine in Queensland, Australia, which resulted from water inrush and a coal dust explosion, respectively. He found the following indications:

- Normalizing of the evidence. In much the same way as the National Aeronautics and Space Administration's management and engineers came to regard O-ring failure in the space shuttle *Challenger* disaster of 1986 as normal due to the frequency of malfunction in earlier launches, Gretley supervisors and managers (and probably the employees as well) normalized the warning signs of water inflow.
- Management reluctance to withdraw workers from danger. Hopkins indicated that at both Moura and Gretley, common sense would dictate that workers not be placed in danger until the hazard had passed or been eliminated.
- A culture of denial. "New evidence appears reliable and informative if it is consistent with one's initial beliefs; contrary evidence tends to be dismissed as unreliable, erroneous or unrepresentative." There were strong production pressures at Moura, which may have resulted in a tendency to dismiss any contrary evidence. Hopkins calls this the culture of denial, "a set of beliefs which enabled management to deny that there was any immediate danger, no matter what the evidence."

Organizational and systemic factors are thus recognized as major contributors to mine disasters. Human factors also played a part in these mining disasters, namely

- Communication failures,
- Failure to recognize warning signs,
- Lack of knowledge about the hazards,
- Lack of training,
- Errors,
- Poor judgment, and
- Risk taking.

Table 15.1-2 Extracts from significant incident reports 1998 to March 2002, Queensland and New South Wales, Australia

Incident	Behavioral Contribution to Cause of Incident
Queensland Mines	
Highway trucks collide	Local **road rules were not clearly understood and followed.**
Light vehicle drives off high wall	Operators should ensure that they are provided with **clear work instructions**, particularly where they are working in an area of the mine with which they are not familiar.
Catastrophic failure of a dragline boom	Unconfirmed reports indicate that draglines are now being loaded above their rated suspended load, in **a drive for greater productivity**.
Dump truck with unbalanced load becomes unstable	**A standing instruction** that large rocks are not to be loaded but reblasted **was ignored** by the shovel operator.
Driller reports for work in an intoxicated state	The driller's offsider was disciplined and given a mine site reinduction where the **statutory requirement of not reporting for duty in an intoxicated state** was stressed. His workmate was also given the same mine site reinduction.
Methane monitors were bridged out of circuit in two mines while producing coal.	On each occasion the tailgate methane **sensor was bridged out of circuit** during calibration checks and not recommissioned after these routine checks were completed.
Dumping overburden near coal crew operations	A dump truck operator dumped overburden outside of the allotted dump area **without knowledge or consideration of the potential consequences.**
An operator had two fingers pinned and a third severely lacerated when a bolting rig autoretracted.	Operator was attempting to do multiple tasks consecutively and **pinch points were not guarded or identified.**
Collapse of bridged material in pugmill feed bin	The plant operator walked along the edge of the top of the feed bin with **no safety belt/harness attached.**
Rigger struck on head by 500-kg assembly	The crews dismantling the crusher **did not understand** the function of the component parts or **were unaware** of the condition of the component parts.
Dozer runs over light vehicle	The mine site **procedure** for entry into a work area **was not followed.**
Truck falls backward over tipping edge	**Ignorance of safe work practices**; the truck driver was **not familiar with a safe procedure** for tipping over the edge of a stockpile.
Rollover of rear-dump truck	The height and size of the rill was inadequate and **not identified as a risk** by the operators. Lack of handover between shifts led to hazards in the workplace **not being communicated**. The dozer operator on the previous shift **had not worked to the mine's standard** for a dump area by allowing the dump face to become misaligned.
New South Wales Mines	
Two large trucks out of control	Maintenance, testing, and inspection of the braking systems were inadequate. Drivers **had not been given training** in emergency response procedures.
Danger from a misfire	**Communications were not clear** between the day shift and the night shift about the result of the refiring.
Unstable structure crushes workman	For some time before the accident, both the injured person and his supervisor **had worked more than 11 hours** a day.
Quarry worker killed by front-end loader	Management **should have safe work procedures in place** for people who need to approach mobile equipment. Everyone at a mine, including contractors and visitors, **should be trained** in the procedures.
Operator crushed by overturned forklift	**No standard operating procedures** were available. **No formal, refresher, or emergency training** had been given. The vehicle appears to have been **traveling too fast** for the conditions; the driver was **not restrained by a seat belt.**
Electric shock from damaged circuit breaker	The electrical maintenance engineer and the electrical staff generally were **not aware of this hazard**, even though it was reported in previous safety alerts.
Continuous miner drill rig fatally crushes tradesman	Managers should also address **basic hazard recognition and control skills training** as a priority.
Vehicle brake failure while travelling in a decline	If the truck were driven in third gear in the decline, it would have eliminated the need to use the brakes.
The 415-V conductors shorted out, and the power tripped off on short circuit.	The electrical contractor was **not aware** of the site *Electrical Management Handbook*.

Source: Data from Queensland Department of Mines and Energy 1998–2002; NSW Department of Mineral Resources 1998–2002.
Note: The behavioral issue in each incident is emphasized in bold.

Disasters in the mining industry are usually followed by a political and/or bureaucratic response in the form of inquiries, inquests, changes to legislation or restructuring, and re-resourcing of mines departments. Some ideas for reducing these potential disasters include the following:

- Better communication in both directions at mine sites, both top-down and bottom-up, will reduce the opportunity for a high-potential incident.
- Relying solely on sophisticated safety management systems is not enough; it is necessary to ensure that employees understand and are trained in procedures.
- Failure to comply with existing regulations can contribute to the making of disasters, as can out-of-date regulations. On the other hand, strict adherence to rules also causes disasters, as was shown in the *Challenger* explosion.
- There is a need to avoid normalizing abnormal events and to be aware of incremental changes in processes.

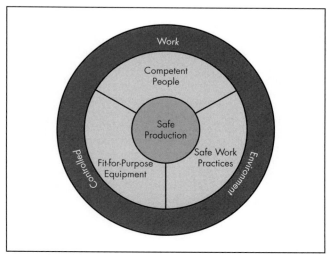

Source: Adapted from Nertney 1987 and Joy 1999b.
Figure 15.1-8 Work process model or Nertney wheel of safe work

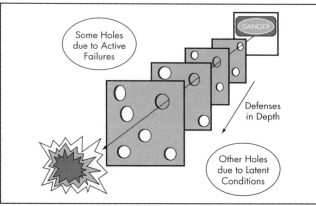

Source: Reason 1997.
Figure 15.1-9 Reason's Swiss cheese model

- The development of active foresight, or what Reason (1997) describes as constant vigilance, will also help to reduce disasters.

CONCLUSIONS
Advances in mine safety over the past two decades have been significant. Apart from the obvious advances in technology that have made tasks more efficient and less physically intensive, other initiatives include

- Changes to the regulatory framework in the form of goal or performance orientation rather than prescribing how to achieve safety;
- Stronger enforcement and higher penalties for mine operators if an incident occurs;
- Better training and standards for the work force;
- Better systems and practices;
- More rigorous due diligence;
- More awareness of the importance of the human factor;
- Extra emphasis on reducing the risk of a large event occurring, rather than focusing only on the less damaging slips, trips, and falls;
- More sharing of safety information among mining companies and regulators, particularly through governmental Web sites.

REFERENCES
AS 1470:1986. *Health and Safety at Work—Principles and Practices*. Sydney, NSW: Standards Australia International.

Australasian Legal Information Institute Tasmanian Consolidated Acts. 1995. Workplace Health and Safety Act. www.austlii.edu.au/au/legis/tas/consol_act/whasa1995250/. Accessed September 2009.

Australian Transport Safety Bureau. 2003. *Articulated Truck Fatalities*. Monograph 15. www.infrastructure.gov.au/roads/safety/publications/2003/Truck_Crash_5.aspx. Accessed May 2009.

Braithwaite, J., and Grabosky, P. 1985. *Occupational Health and Safety Enforcement in Australia*. Report to the National Occupational Health and Safety Commission. Canberra, Australia: Australian Institute of Criminology.

British Medical Association. 1987. *Living with Risk*. Chichester Sussex, UK: John Wiley and Sons.

Bureau of Meteorology. 2009. Warm season storms. www.bom.gov.au/weather/wa/sevwx/perth/storms_warm.shtml. Accessed May 2009.

Hopkins, A. 1999. A culture of denial: Sociological similarities Between the Moura and Gretley mine disasters. *J. Occup. Health Saf.—Aust. N.Z.* 16(1):29–36.

Joy, J. 1999a. An introduction to engineering risk management. Lecture presented at Minerals Industry Safety and Health Centre, Queensland, Australia. www.mishc.uq.edu.au/ugmodules/mod1.1ERMintro.pdf. Accessed May 2009

Joy, J. 1999b. A systems based analysis of 25 system safety accident investigations (SSAIs) of major mining events. In *Proceedings of the NSW Mining and Quarrying Industry OHS Conference*, Terrigal, New South Wales, Australia.

Laurence, D.C. 2002. Investigating the influence of the regulatory environment on the safety performance of the mining work force. Unpublished Ph.D. thesis, University of New South Wales, Sydney, Australia.

Laurence, D.C. 2008. Challenges and opportunities for sustainable mining practices in the Asia-Pacific region. Lecture presented at School of Mining Engineering, University of New South Wales, Sydney, Australia. Unpublished. www.mining.unsw.edu.au/lectures/KFL/UNSW_KennethFinlay_pp_2008.pdf. Accessed May 2009.

MCA (Minerals Council of Australia). 2009. *Safety Performance Report 2006/7*. www.minerals.org.au/_data/assets/pdf_file/0020/32384/2006_07_Safety_Performance_Report_v1.pdf. Accessed May 2009.

Nertney, R.J. 1987. *Process Operational Readiness and Operational Readiness Follow-On*. DOE 76-45/39, SSDC-39. Washington, DC: U.S. Department of Energy.

NSW (New South Wales) Government Legislation. 1983. Revised 2000. Occupational Health and Safety Act. www.legislation.nsw.gov.au/fullhtml/inforce/act+40+2000+FIRST+0+N. Accessed September 2009.

NSW (New South Wales) Department of Mineral Resources. 1998–2002. Minerals and Petroleum: Major Investigations. www.dpi.nsw.gov.au/minerals/safety/major-investigations. Accessed May 2009.

Ontario Ministry of Labour. 2000. *The Internal Responsibility System in Ontario Mines*. www.gov.on.ca/LAB/ann/00-75be.htm. Accessed March 2001.

Pereira, N. 1999. Report into and Management of Roof Fall Fatalities in NSW Coal Mines. New South Wales, Australia: Department of Mineral Resources. Unpublished report.

Pitzer, C.J., ed. 2000. *A Study of the Risky Positioning Behaviour of Operators of Remote Control Mining Equipment*. Report MGD-5004. New South Wales, Australia: Department of Mineral Resources.

Queensland Department of Mines and Energy. 1998–2002. Summary Statements of Accidents and Incident Reports. www.dme.qld.gov.au/mines/ssarchive.cfm. Accessed May 2009.

Queensland Department of Mines and Energy. 2002. *Mines Inspectorate Safety Newsletter*. January. www.dme.qld.gov.au/mines/publications___forms.cfm. Accessed May 2009.

Reason, J. 1990. *Human Error*. Cambridge, UK: Cambridge University Press.

Reason, J. 1997. *Managing the Risks of Organizational Accidents*. Hampshire, UK: Ashgate Publishing.

Robens, A. 1972. *Safety and Health at Work: Report of the Committee 1970–72*. Cmnd 5034. London: Her Magesty's Stationery Office.

Roylett, B., Russell, I., Raman, R., and Blyth, D. 1991. Analysis of Accidents from Strata Movements in Pillar Extraction in New South Wales Coal Mines. Internal report. New South Wales, Australia: Department of Mineral Resources.

U.K. Health and Safety Executive. 1974. Health and Safety at Work Act. www.hse.gov.uk/legislation/hswa.htm. Accessed September 2009.

U.K. Office of Public Sector Information. 1954. Mines and Quarries Act. www.opsi.gov.uk/RevisedStatutes/Acts/ukpga/1954/cukpga_19540070_en_1. Accessed May 2009.

CHAPTER 15.2

Health and Medical Issues in Global Mining

Frank H. Fox

INTRODUCTION

Mining normally takes place in remote areas, and as resources become increasingly scarce, new deposits are only found in the more inaccessible and inhospitable parts of the world. These are often in countries with poorly developed infrastructure and are characterized by extremes of terrain and climate, all of which contribute to potential medical and health challenges. Climatic extremes (heat, cold, high and low humidity, altitude, and hyperbaric conditions) increase the possibility of relatively minor medical conditions becoming major impediments to health, safety, and productivity and pose challenges to even the fittest individuals. Remote locations also bring psychological stresses that may manifest in the form of alcohol and substance abuse.

With a global shortage of skills in the mining industry it is likely that the work force will be a mixture of local and expatriate employees. Health problems may range from endemic diseases such as malaria to the psychological stress of isolation and continuous shift systems that can lead to wellness issues as well as alcohol and substance abuse.

Prevention of medical and health problems in mining operations (from exploration to production environments) starts with anticipation and recognition of the potential issues and proceeds with a thorough risk-assessment process inclusive of projected frequency and severity of incidents, development of risk-control measures, and allocation of adequate resources to implement the controls to manage the risk.

GENERAL APPROACH

The key to managing all potential health problems is prior risk assessment at the design stage of operations, before travel and before operations begin. During operations, continuous risk assessment is vital as things change. This is based on anticipation, recognition, evaluation, and control. Additionally, awareness of the differences between health and safety needs must be clear; too often, safety concerns are given priority over health issues. This is a short-term focus that must be corrected.

Global travel is so commonplace that travelers are often complacent about potential health risks. Yet the risks may be significant, even for local travel to mine sites within one's home country. And while mining and its associated processes have become safer over the years, and engineering design and equipment have improved, the extraction of minerals remains an inherently hazardous task. Mining requires close attention to detail to avoid these hazards.

Historically, mining has focused primarily on accident rates as the primary driver for performance measurements and safety awards. Improvements in engineering and processes have resulted in fewer accidents, but similar improvement has not been achieved for illnesses. In fact, some engineering improvements may have inadvertently contributed to an increase in health issues such as obesity, risk of cardiovascular disease, and cumulative trauma as more tasks have been engineered to reduce physical effort.

The following factors should be considered when planning for optimum health of mine workers: location of the site, climate, endemic diseases, local infrastructure (health care, transport, and accommodations), local legislation, the make-up of the work force (local, expatriate), the process (ore, products, and effluent), and shift system (continuous operation, fly-in/fly-out, etc.). If the site is an exploration base, travel to and from the site may pose a health risk and should be considered.

There are three ways in which health can be approached with regard to mining: those issues that affect personal health, those that concern occupational health, and those that relate to health in the community. These overlap in a number of areas but can be usefully approached in these subdivisions. A few of the more well-known conditions will be presented together with an approach for management from the mining engineer's perspective. Mining engineers will gain a basic understanding of the health issues that face them and an approach to dealing with them.

PERSONAL HEALTH

Personal health conditions and behaviors start at home, are taken to the mine site, and continue with residence there. A worker's lifestyle (diet, exercise, sleeping habits, stress level, substance abuse, etc.) has as important an affect on health and safety as conditions at the mine itself. Hazards at

---

Frank H. Fox, Head of Occupational Health, Anglo American plc, Johannesburg, South Africa

the mine site will include the location and local infrastructure of the mine and the occupational health hazards that may be encountered in the mining environment. In addition, there may be local climatic conditions that pose a risk: excessive humidity or low humidity, high altitude, and exposure to ultraviolet radiation.

Exposure to infectious diseases can occur both on and off the job. The specific diseases vary from country to country and cover the entire spectrum of bacterial, viral, protozoal, fungal, and parasitic infections. Exposure to disease occurs through personal contact, insect vectors, food sources, contact with animals, and water.

Malaria, tuberculosis (TB), and HIV/AIDS (human immunodeficiency virus/acquired immunodeficiency syndrome) are three prevalent diseases that are currently receiving worldwide attention. These diseases are not only important at a personal level but also as social issues in the communities surrounding mines in many developing countries. Ultimately they are an important factor in the productivity of the work force. These will be treated separately, as they illustrate the spectrum of involvement and intervention that may be needed in mining.

Traveling—locally and globally—presents its own unique set of health risks. Before traveling to exotic or unfamiliar locations, it is wise to seek advice from a health practitioner who is knowledgeable about worldwide medical issues, as well as from local medical practitioners who are experienced in dealing with health concerns specific to the area.

Health insurance is an important consideration when traveling and particularly when engaged in exploration work. Local health care facilities may be less than ideal, and suitable emergency evacuation arrangements should be in place before the work is undertaken. Many companies employ paramedical personnel to provide a basic level of primary and emergency care on-site with additional arrangements in place for evacuation in case of emergency. However, the key in all situations is prevention of the emergency in the first place, which means evaluation of fitness before placement, correct prophylaxis for endemic diseases while on-site, and careful attention to public health measures.

Although often taken for granted, some mention must be made of personal hygiene. Work in hot, humid conditions, underground or topside in tropical climates brings with it potential health problems. Fungal infections, insect bites, and bacterial skin infections are a common source of disease and can become serious problems if neglected. These are often prevented through simple measures and attention to personal hygiene.

OCCUPATIONAL HEALTH

Sir Thomas Legge, the United Kingdom's first Medical Inspector of Factories (appointed in 1897), wrote a simple axiom that remains relevant today:

> All workmen should be told something of the danger of the material with which they come into contact and not be left to find it out for themselves—sometimes at the cost of their lives (Robbins and Landrigan 2007).

This principle is enshrined in occupational health and safety law and various International Labour Organization conventions, specifically in Article 9 of Convention 176 (ILO 1995).

A full discussion of occupational health is beyond the scope of this chapter, but there are a few basic points that should be considered. Occupational health is defined as follows:

> Occupational health should aim at the promotion and maintenance of the highest degree of physical, mental and social well-being of workers in all occupations; the prevention among workers of departures from health caused by their working conditions; the protection of workers in their employment from risks resulting from factors adverse to health; the placing and maintenance of the worker in an occupational environment adapted to his physiological and psychological capabilities and; to summarize: the adaptation of work to man and of each man to his job.
>
> The main focus in occupational health is on three different objectives: (i) the maintenance and promotion of workers' health and working capacity; (ii) the improvement of working environment and work to become conducive to safety and health and (iii) development of work organizations and working cultures in a direction which supports health and safety at work and in doing so also promotes a positive social climate and smooth operation and may enhance productivity of the undertakings. The concept of working culture is intended in this context to mean a reflection of the essential value systems adopted by the undertaking concerned. Such a culture is reflected in practice in the managerial systems, personnel policy, principles for participation, training policies and quality management of the undertaking (International Labour Office 1998).

These objectives might imply that occupational health is a medical issue, whereas in reality the second and third parts of the statement can only be achieved through application of engineering and human factors and ergonomic solutions. Occupational health and safety is thus an engineering problem, not a medical one, so it is essential that the mining engineer has an understanding of the health hazards that are inherent in mining.

Management is responsible for operating the mine and its associated processes in such a way that it does not pose a risk to the health and safety of the work force and the surrounding community. The focus should be on control of exposure to health hazards through the application of the hierarchy of controls from the design stage onward (see Table 15.2-1). Utmost attention to monitoring and maintaining those controls must be observed. This control must be supported by a program of environmental monitoring and medical surveillance of exposed people. None of this can be effective without the active involvement of all employees. Everyone should be informed, trained, and involved in the control and prevention measures that are put in place.

HEALTH RISK ASSESSMENT

The health hazards associated with mining are numerous and cover the entire spectrum of physical, chemical, biological, ergonomic, and psychological stressors. The mining process itself exposes people to dusts, fumes, and vapors of various

Table 15.2-1 Hierarchy of controls

Elimination
Eliminate an unsafe process or substance (e.g., use ultrasound to clean components instead of chemicals).

Substitution
Substitute a process or chemical with one that is safer (e.g., use water instead of chemical solvents).

Engineering controls
Use physical controls that eliminate or reduce exposure (e.g., enclose the process, separate the worker, provide adequate ventilation [exhaust, directional, and natural], soundproof areas where there is a high noise level).

Administrative controls
Practice safe work procedures and good housekeeping (e.g., cover containers, clean up spills, rotate jobs, reduce shifts, restrict eating or drinking to specific areas, maintain washing facilities, provide training).

Personal protective equipment
Provide appropriate personal protective equipment for workers (e.g., gloves, hearing protection, overalls).

---

compositions and particle sizes; noise; vibration; radiation (ionizing and nonionizing); and ergonomic factors. Associated with this are ancillary engineering processes that present an additional set of hazards. Factoring in the location of the mine gives a complete picture of the occupational health risks of mining.

The first step is developing a program to ensure health and safety is risk assessment, followed by a program to control, monitor, and mitigate the residual risk. Health risk assessment (HRA) can be a complex process performed by a multidisciplinary team, but it can also be a basic process aimed at identifying the need for expert advice, intervention, and a more formal approach.

Typically, an HRA will take place in three stages: (1) a baseline or initial assessment to identify and prioritize potential risk; (2) an issue-based assessment to refine and quantify the risk identified in the initial assessment; and (3) continuous risk assessment. This final stage is integral to the ongoing risk management program and is a process of renewal that monitors the effectiveness of controls and continually assesses risk against process changes over time. Ultimately, an HRA should allow managers and workers in any industrial environment to answer these questions:

- What are the hazards in my operation or area of work?
- What am I doing to control exposure?
- What am I doing to monitor the controls and are they effective?
- Can I, and should I, do more?

HRA requires an understanding of the process, the hazards, and possible exposures inherent in the mining operation and an understanding of the potential health effects of any exposure. The team that undertakes an HRA therefore needs input from a process engineer, the medical staff, occupational hygiene personnel, line management, human resources, and the workers.

THE MEDICAL SIDE OF OCCUPATIONAL HEALTH

The medical component of occupational health is concerned with the contribution to HRA, assessment of fitness for work, medical surveillance, management of occupational diseases and injuries, and wellness of the work force.

Depending on the infrastructure in the area, the mine may also provide medical treatment for employees, their dependants, and, in some cases, the community. This may extend to public health issues such as disease control and management programs and, in remote sites where the mine provides all the infrastructure, sanitation and drinking water.

Fitness for Work

Assessment of fitness for work has a large impact on safety as well as productivity. It also is a means of preventing exposure of individuals with certain preexisting medical conditions to situations in the workplace that cause them undue risk. An example is the exposure of people with asthma to environments with irritant gases or sensitizing agents. In order for this process to be fair, it is essential that fitness criteria be based on the task and risks inherent in the particular position within the mine. This requires a job specification with fitness criteria derived from the HRA.

Fitness for work can also be extended into the wellness arena. Here the focus is on maintaining fitness through management of lifestyle (diet and exercise), stress, and chronic diseases. This may be achieved through specific wellness and employee assistance programs, or the company's in-house medical services. The athletic facilities provided in many mining towns support this approach to wellness.

Medical Surveillance

Medical surveillance is the process of screening for early signs of occupational disease and any signs of excessive exposure to hazardous agents in the workplace. The process involves recording an individual's occupational and medical history, undertaking a physical examination, conducting certain special tests (such as audiometry and spirometry), and, in some cases, biological monitoring (measurement of chemicals, metals, or metabolites in body fluids). In all cases this should be based on knowledge of the hazards and exposures in the workplace, with only necessary tests being done. In many countries, medical surveillance is a legal requirement, but regardless of location, all employees should be informed of the reason for testing and given the results with an explanation.

Occupational diseases are conditions that arise out of and as a consequence to exposure to a hazardous agent in the working environment. Typically there is a dose–response relationship between the exposure and the onset of the disease; that is, the exposure must be high enough and long enough to cause the condition. Examples of this are the dust diseases of the lung (pneumoconiosis) where an accurate diagnosis is based on exposure to a certain amount of dust over an appropriate period of time. In some conditions there may not be a clear dose–response relationship (such as allergies), and the condition is diagnosed on the basis of exclusion. Although the traditional occupational diseases have largely disappeared through better control of workplace exposures, there is still a risk from the many new chemicals and processes that are being discovered, and it is impossible to predict idiosyncratic reactions to chemical exposure. The mining industry needs to remain vigilant through continuous risk assessment with regard to exposure to potential health risks.

In most countries, exposure to hazardous agents in the workplace is regulated through the use of occupational exposure limits. These are typically based on a time-weighted average exposure over 8 h/d, 40 h/week. Management of exposure to health risks is through the application of engineering controls according to the hierarchy of controls.

The specific engineering approach to some of the health and safety issues falls into the disciplines of occupational (industrial) hygiene and ventilation engineering and is beyond the scope of this chapter.

Emergency treatment and management of occupational injuries is also a part of any medical service on a mine, and is too often the only consideration given. The scope will vary according to the site and the extent of the operation, but adequate preparation must be made for the emergency care and subsequent treatment of all possible injuries on the mine. Services may range from essential emergency care on-site with evacuation to a hospital to comprehensive on-site medical facilities that can provide sophisticated care quickly and efficiently.

Occupational Disease

Occupational disease may be acute or chronic, and despite the preventive measures outlined previously, occupational disease remains all too common, particularly in developing countries where legislation may not be as developed or effectively enforced. Acute conditions due to exposure to hazards in the workplace (such as muscular strains and sprains, dermatitis, and others) require timely diagnosis and treatment. Conditions such as silicosis and asbestosis (which occur after prolonged exposure) remain significant causes of both morbidity and mortality around the world, and despite having been almost eradicated in many countries, new cases of silicosis continue to be common in the developing world. Mining is an inherently dusty operation, and prevention of occupational disease requires a sustained long-term commitment to implementing and maintaining the hierarchy of controls. (Unfortunately, there is often a reliance on personal protective equipment as the main protection strategy.) The insidious onset of most occupational diseases contributes to the lack of a sense of urgency around prevention, and the overlap with some community diseases (e.g., asthma, skin disease, and TB) may contribute to the lack of recognition of the problem. The most common exposures throughout industry and mining include noise, heat, manual handling, vibration, and chemicals, leading to noise-induced hearing loss (NIHL), musculoskeletal conditions, and dermatitis as perhaps the most common diseases. Emerging risks include whole-body vibration, diesel particulate, nanoparticles, and psychological stress. Fatigue (often related to shift work) is also an important factor to consider, from both a safety and health point of view.

While this is not intended to be a text on occupational disease, a few conditions are worthy of brief mention, although this should not detract from the importance of any other condition.

Silicosis

Silicosis is one of the oldest recognized lung diseases due to dust exposure. In its typical form it is caused by exposure to crystalline silica dust at moderate levels over a period of 10 to 20 years. It is identified as a diffuse nodular pattern on the chest X-ray. This is due to the silica particles that provoke an inflammatory reaction and cause scarring of the lung tissue. In its milder forms it may not cause any appreciable symptoms, but with higher levels of exposure the scarring can be so severe that it interferes with the function of the lungs and the heart. In addition, there is an increased risk of TB in workers with silicosis, and silica has also been linked to lung cancer. There is no treatment, and the only solution is prevention through reduction of dust levels and avoidance of exposure.

Despite the large drop in the number of cases of silicosis in the developed world, there is no room for complacency. Surveillance in the United States showed a large reduction in the number of deaths attributable to silicosis between 1968 and 2005, but the incidence has leveled off since 1998 (Attfield et al. 2009). Although the incidence of silicosis has dropped significantly in developed countries through strict control, it cannot be assumed that current occupational exposure limits are entirely safe, and therefore it remains a risk (Park et al. 2002). In fact, the incidence of pneumoconiosis appears to be more closely correlated with the number of people employed in mining rather than the reduction in dust levels (Attfield et al. 2009). Silica is ubiquitous in soil and rock, and since it is not only hard-rock mining that poses a risk, it must be considered a hazard wherever mining is undertaken.

Noise-Induced Hearing Loss

Hearing loss from exposure to noise has been known since biblical times, and tinnitus (ringing in the ears) in blacksmiths is described in the book of Ecclesiastes. In more modern times it was described in coppersmiths and millers by Ramazzini (1713) in the 18th century. Despite its long history, NIHL remains one of the most common occupational diseases today, and it is a significant cause of morbidity.

Hearing loss can present in two ways: (1) acute, from exposure to very loud noise such as an explosion, and (2) chronic, from exposure to noise above 85 db for many years. The mechanism of damage is different between the acute and chronic types of hearing loss, with the acute form associated mostly with damage to the conduction system (the eardrum and small bones in the ear) and the chronic form associated with damage to the hair cells in the cochlea. Of the two forms, the chronic form of hearing loss is commonly referred to as NIHL. It typically presents with a slowly progressive loss of hearing, first in the higher frequencies and then in the frequencies used for speech. In many cases this is accompanied by tinnitus, which makes the disability even worse. The slow onset and the late age at which the condition occurs means that it may be confused with hearing loss due to aging. Occupational exposure to noise is not the only cause of NIHL, as any other source of loud noise (music, gunfire, etc.) can produce the same symptoms. In addition, some drugs and infections may cause a similar pattern of hearing loss. Modern hearing aids may provide some assistance with deafness, but in the presence of tinnitus this is doubtful. The best solution is prevention.

Heat Stress

Exposure to heat is a common problem in mining, whether on the surface or underground, and can lead to significant health and safety risks if not correctly managed. Any situation where there are high temperatures, high humidity, poor ventilation, or high radiant heat sources can result in heat stress.

There are four typical presentations of heat stress: heat fatigue, heat cramps, heat exhaustion, and heat stroke. Heat fatigue is a vague condition of impaired general performance and may be associated with mood changes. The behavioral effects of heat exposure should not be underestimated as they can lead to poor decision making that, in the mining and industrial setting, can have a significant impact on safety. Heat cramps occur in the muscle groups that are most used in work and are painful spasms that last several minutes but are usually self-limiting. Heat exhaustion is the condition that results from excessive fluid loss and presents with dizziness, sweating, headaches, weakness, nausea, and vomiting. Heat stroke is the most serious of the conditions and results from failure of the thermoregulatory center in the brain. It is almost always associated with prolonged physical exertion in a hot environment and presents with extreme hyperthermia (body temperature above 41.1°C or 106°F); confusion; collapse; and, typically, hot, dry skin (although this is not always the case). If not treated effectively, heat stroke can be rapidly fatal. Heat exposure can also cause a skin condition known as prickly heat, which is an acute itchy dermatitis.

Heat exposure is a combination of the ambient environmental conditions, the heat generated internally by physical effort, and the ability of the body to cool itself. There are four basic mechanisms of heat exchange involved: convection, conduction, radiation, and cooling through evaporation.

The management of risk from heat exposure starts with awareness of the problem, environmental controls where possible, increased fluid intake (200 mL/20 min), and control of physical exertion (increased work breaks) in hot environments. Salt tablets are often mentioned, but this may do more harm than good; simply increasing the amount of salt in the diet is all that is required. Acclimatization does occur after a period in a hot environment, and many deep-level mines have acclimatization programs for new workers. This reduces the risk, but there are some individuals who will remain at high risk for heat disorders. Age (over 50), obesity (body mass index >30), certain medical conditions, and a previous history of heat disorders all increase the risk of heat stroke and should be considered as barriers to working in hot environments. For a more complete review of the management of risk due to heat in the mining environment, see the chapter by Johan Kielblock in the *Handbook of Occupational Health Practice in the South African Mining Industry* (Kielblock 2001).

Fatigue
Mine work, with its physical nature, remote locations, and the need for continuous production, creates an environment in which fatigue is an ever-present risk. Fly-in/fly-out work rosters with 12-hour shifts over prolonged periods and sometimes long commutes make it difficult to avoid fatigue. Despite increased mechanization of the work, the rough terrain and long shifts make the work physically tiring.

The term *fatigue* is used in various ways to describe localized tiredness of muscle systems or in the general sense of tiredness. It can be acute or chronic, and in this chapter the term is only used in the general sense.

Fatigue manifests as increasing difficulty in performing physical or mental activities. This in turn results in poor productivity and poor decision making, which can be a safety risk. In its acute form, fatigue may simply present as sleepiness, but chronic fatigue may lead to more general effects such as mood disorders, stress, digestive problems, heart disease, and mental illness. The causes of fatigue are many, but put simply, fatigue in the workplace is the sum of the interaction between the shift system, the physical workload, and personal factors (such as diet, sleep, overall health, drug use, etc.). These can be combined into an equation that provides a useful reference for managing the risk from fatigue in the workplace (Schutte 2008):

$$F_t = F_{ss} + F_{ew} + F_{pf} \qquad (15.2\text{-}1)$$

where
F_t = total fatigue
F_{ss} = fatigue due to the shift system or roster
F_{ew} = fatigue due to ergonomic and work factors
F_{pf} = fatigue due to personal factors

The management of the risk from fatigue starts with recognition and assessment of the risk followed by implementation of an integrated plan to deal with the three components of the equation. Shift rosters should be designed to minimize sleep loss and the accumulation of sleep deficit. Shift lengths should be considered in conjunction with travel time as it is time awake that is important rather than just the length of the shifts. Tasks should be designed to reduce repetitive work, and work requiring sustained physical or mental effort should be broken up with short breaks. Personal factors must also be taken into account and are managed mostly through education, but some arrangements to improve quality of sleep through provision of decent accommodations and recreation facilities at the work site can be made. Where there are rotating shifts, education of the family about the social adjustments needed may also help in ensuring good quality rest for the worker. Diet, general health, and use or misuse of drugs and alcohol should also be addressed in the plan. Numerous guidelines are available regarding the prevention and management of the risks associated with fatigue. An Internet search for government documents on fatigue management will generate many reliable sources for up-to-date information on the subject.

Wellness
Employee wellness is a common part of many occupational health programs, particularly in developed countries, and is concerned with maintaining and improving the health and productivity of employees. This is achieved through targeted interventions on lifestyle management, stress management, and assisting employees in managing chronic illnesses. Employee assistance programs providing access to counseling services are a common intervention. The objective is a fit and healthy work force that has a direct impact on improved productivity and reduced health care costs to the company. In the context of a remote mine site, these programs may cross over into other aspects of community life such as recreational facilities and opportunities, provision of family accommodations, drug and alcohol programs, and many other interventions.

COMMUNITY HEALTH
The community surrounding any mine is intimately linked with the process through employment, emissions from the process, and the impact the mine has on the economy of the area. The influx of people seeking employment creates a demand and a strain on existing health care infrastructure and may provide the stimulus for the creation of additional or new infrastructure.

The mine may have some direct effects (both positive and negative) on the health of the community as well as some indirect effects. The health and welfare of the community also has an important influence on the mine's license to operate, and increasingly there is a call for health impact assessments to be carried out, both before and during operation of the mine.

These impact assessments should address the infrastructure needs, the potential health impact of the mine on the community, and the impact of any common diseases in the community on the mine. Malaria, tuberculosis (TB), and HIV/AIDS fall into this last category and there is a significant interaction in both directions.

Malaria, TB, and HIV/AIDS are perhaps the three most important communicable diseases in the world today; combined, they account for some 4 million deaths worldwide (WHO 2008; UNAIDS 2008; Global Malaria Partnership n.d.). The areas where they occur overlap considerably, and there is some interaction between the three diseases: HIV increases the risk of TB, TB hastens the onset of AIDS, and it would appear that malaria may increase the viral load in HIV infection (although this is not yet well understood). Exposure to silica dust in mining increases the risk of TB, and the in-migration generated by mining employment has a direct effect on increasing the prevalence of malaria and HIV. Since mining often occurs in countries where these diseases are prevalent and there is a direct relationship between mining and the risk of these diseases, it is important that the mining engineer has some basic understanding of the possible effects and interventions required.

The business case for managing these three diseases is compelling: companies that have good programs for managing these diseases in the workplace experience a range of bottom-line benefits that include

- Reduced employee turnover,
- Reduced absenteeism,
- Improved productivity,
- Decreased costs of health care,
- Access to international funding for mine development, and
- Enhanced company reputation and community relations.

The approach to managing these three diseases or any other disease that may be of concern to the mine should be based on a structured assessment of the risk leading to the development of a management program based on that assessment (see Table 15.2-2). Again, risk assessment is the starting point for management. It is beyond the scope of this chapter to describe the approach in detail, and only a brief overview of each disease is given. Because the three diseases have much in common, a case can be made for having an integrated approach to management. An extremely useful reference for an approach to these three diseases is the International Council on Metals and Mining's *Good Practice Guidance on HIV/AIDS, Tuberculosis and Malaria* (ICMM 2008).

Malaria

Malaria is an infectious disease caused by an intracellular parasite that is spread by the bite of the female *Anopheles* mosquito. The distribution of the disease is mainly tropical (see Figure 15.2-1), but it is also found in other parts of the world, carried by travelers returning from areas where the disease was picked up.

Four species of the plasmodium parasite cause malaria in humans: *Plasmodium falciparum, Plasmodium malariae, Plasmodium ovale,* and *Plasmodium vivax.* Of these, *P. falciparum* is the cause of the most severe form of malaria and has a significant mortality in nonimmune people. It is also becoming resistant to many of the more common drugs used to treat malaria.

Table 15.2-2 Management planning process for HIV/AIDS, TB, and malaria

Risk Assessment (Do we need the program?)	
1. Screening	Decide whether (or not) one or more of the diseases needs to be managed.
2. Scoping	Increase understanding of the nature and extent of the disease and the associated risks.
3. Situation analysis	Develop a health baseline.
4. Stakeholder analysis	Participate in a process of information sharing, problem solving, and identification of potential partners.
Risk Management (What do we need to do?)	
5. Priorities, goals, and objectives	Decide whether (or not) one or more of the diseases needs to be managed.
6. Scoping	Increase understanding of the nature and extent of the disease and the associated risks.
7. Situation analysis	Develop a health baseline.
8. Stakeholder analysis	Participate in a process of information sharing, problem solving, and identification of potential partners.

Source: Adapted from ICMM 2008.

The symptoms of malaria mimic those of influenza in the early stages and include headache, chills, fever, malaise, and vomiting. These usually start about 10–15 days after infection, and diagnosis and effective treatment is essential to prevent severe illness (which may have fatal complications).

The diagnosis is made through the use of tests similar to a pregnancy test and examination of blood under a microscope (looking for the parasite in red blood cells).

Treatment of uncomplicated cases is with combinations of a drug based on the Chinese plant *Artemisia annua* and another drug, lumefantrine. Complicated cases are usually treated with quinine. Depending on the parasite, further treatment to clear the parasite from the liver may also be required.

People who have not been exposed to malaria all their lives are regarded as nonimmune, whereas those who have been born in and grown up in malarial areas may develop some immunity. This does not mean they do not get malaria; it merely means they are less likely to get the severe forms of the disease.

Malaria Control Programs

Three components of malaria control are control of the vector, control of exposure to the vector, and chemical prophylaxis. The vector is the female *Anopheles* mosquito, which breeds in stagnant water. It also has certain habits (such as settling on window and door frames) that facilitate transfer to humans. Vector control therefore involves management of sources of stagnant water and spraying campaigns. This is an engineering solution, as drains, sewers, and other sites for stagnant water have to be actively managed. Spraying of buildings with residual insecticides is an important part of many programs, and the specific insecticide used varies depending on the time and place. DDT (dichlorodiphenyltrichloroethane) is still advocated in some areas as it has long residual effects on buildings, but must obviously be used with extreme caution because of its environmental impact.

Control of exposure involves behaviors such as wearing long-sleeved shirts and long trousers, avoidance of being outdoors at dusk when the mosquitoes are most active, the

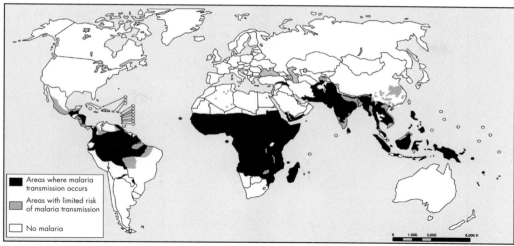

Source: WHO 2009.
Figure 15.2-1 Global distribution of malaria in 2009

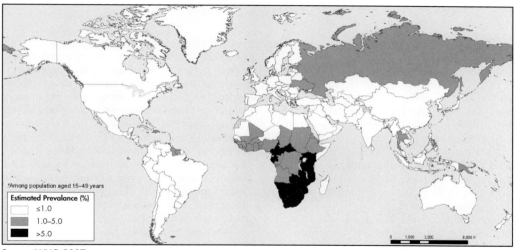

Source: WHO 2007a.
Figure 15.2-2 Estimated HIV/AIDS prevalence in 2007

use of mosquito repellents (both on the person and airborne), and barrier methods such as insecticide-treated bed nets. Chemoprophylaxis involves taking medication to suppress the parasite before it can grow in the bloodstream. These drugs should not be the same as those used to treat the condition, and medical advice should be sought on which is the most appropriate drug for the area and the period of use. It may only be practical for expatriate (nonimmune) employees, but there may be some use for general prophylaxis in the community surrounding the mine, which will reduce the reservoir of the parasite. Expert advice in this regard is essential.

A common myth surrounding malaria is that taking malaria prophylaxis hides the disease and makes the diagnosis more difficult. This is untrue and, in fact, dangerous as the complications are more likely in the nonimmune person who is not taking prophylaxis. It is true that no single prophylactic measure is 100% effective, but these all rely on human behavior. A high degree of suspicion and prompt diagnosis and treatment of malaria is essential.

HIV/AIDS

In 2007, UNAIDS reported that there were an estimated 33.2 million people living with HIV/AIDS, 2.5 million new infections, and 2.1 million deaths worldwide. More than 68% of people living with HIV/AIDS are in sub-Saharan Africa, and 76% of all deaths from AIDS occurred in this region (Figure 15.2-2).

AIDS (acquired immunodeficiency syndrome) is a condition caused by infection with the human immunodeficiency virus (HIV) and has collectively become known as HIV/AIDS. The virus is a lentivirus, one of a group of viruses known as retroviruses that attack the white blood cells of the body (specifically the CD4+ T-lymphocyte). The infection results in a gradual deterioration of the immune system, leaving the individual susceptible to a variety of opportunistic infections and some cancers. Transmission of the virus occurs through contact of a mucous membrane or the bloodstream with infected body fluids such as blood, semen, vaginal secretions, and breast milk. Contact can occur through sexual intercourse,

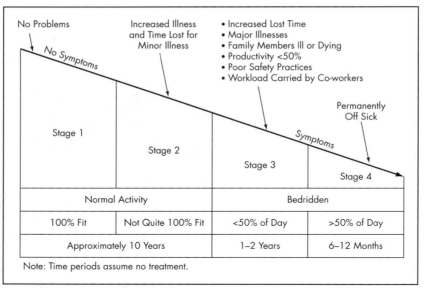

Figure 15.2-3 Stages of HIV infection

blood transfusion, use of contaminated hypodermic needles, and between mother and baby.

HIV infection carries no specific symptoms; in the primary infection it causes a short flu-like illness, but after that, the symptoms are those of the particular opportunistic infection that the person may have. The diagnosis of HIV infection is made by testing a blood (or saliva) sample for the presence of antibodies to the virus. There are various methods for antibody detection and usually the diagnosis is confirmed by a second test using a different test for the antibody. There are also tests for the presence of the virus, and viral counts are done using recombinant DNA techniques, normally used for monitoring the response to treatment.

HIV is no longer the death sentence that it was 15 years ago, and combination therapy with highly active antiretroviral drugs has made this just another chronic disease for many people. However, it is important that treatment is started before the disease has progressed too far. It is also important to prevent the spread of the virus from mother to child during childbirth and breastfeeding, and effective drug regimens exist for this purpose.

Workplace HIV Programs

HIV is preventable, and the primary goal of any workplace program should be to prevent new infections in employees. In addition to this, workplace programs should seek to provide care and support for those who are infected, and build capacity in senior managers to deal with the workplace issues of people living with HIV/AIDS. Any workplace programs developed should be aligned with and support the efforts of the local state health services.

A strong commitment and concerted efforts from management and unions, backed up by company policy, is essential for success. As a minimum, the policy should address the issues of stigma and discrimination, confidentiality, care and support, employment, and workplace voluntary counseling and testing.

The disease typically goes through four stages, and the intervention at each point is different. Figure 15.2-3 shows the stages from an employer's perspective. In the early stages of infection, there are no symptoms and relatively minor illnesses only. This may have little or no impact on productivity; however, this is perhaps the most dangerous stage as the individual is infected, infectious, and very often ignorant of their condition. They are also on a downward slope as far as health is concerned (as measured by the CD4 cell count). Voluntary counseling and testing (VCT) programs not only create an awareness of the disease among those who are HIV negative but also creates opportunities for those who are infected but still well enough to prevent further spread of the disease and maintain their health. Treatment with antiretroviral drugs can be started at the optimal time, and mother-to-child transmission prevented, but only if the infection status is known.

When the disease has progressed to the symptomatic stage, the only option is treatment and prevention of transmission. With early recognition of the infection through VCT and initiation of antiretroviral treatment at an appropriate point, Stages 3 and 4 can be almost indefinitely delayed. Unfortunately, many people leave the diagnosis of HIV and appropriate treatment to a late stage when treatment may not be very effective. However, even at this stage many people are able to return to productive employment and continue to support their families.

HIV/AIDS is seldom an infection that involves only one partner and it is essential that any management program involve the dependents of the employee. Education, access to counseling and testing, and possibly even provision of treatment should be considered. State health programs may provide all this but may need support. In addition, there are numerous donor agencies that can offer help and support in setting up and running HIV/AIDS management programs (ICMM 2008).

Tuberculosis

The global distribution of TB is shown in Figure 15.2-4. Note the similarity with both HIV/AIDS- and malaria-prevalent areas.

TB is a disease caused by a rod-shaped bacterium (bacillus) known as *Mycobacterium tuberculosis*. According to the

Source: WHO 2007b.

Figure 15.2-4 Global distribution of tuberculosis in 2007

World Health Organization (WHO), it infects about one-third of the world's population. Most of this is latent infection rather than active, but WHO estimates that in 2006, 1.5 million people died of TB, with another 200,000 dying of HIV-associated TB (Attfield et al. 2009). In latent infection, the body seals off the bacteria in small numbers and the disease remains dormant. However, this dormant infection can become active if the immune system is compromised by, for example, HIV. The estimated risk of an infected person becoming sick is between about 5% and 10% during their lifetime, but when HIV is a factor, the risk rises to between 10% and 15% per year.

Most commonly, TB causes a respiratory illness (pulmonary TB), but it can infect any organ in the body with devastating consequences. The disease is spread by droplet infection through coughing.

TB is a disease with a slow onset, and the symptoms may be insidious and difficult to diagnose, which means the diagnosis may often be late. Typically it will present with a cough, weight loss, and profuse sweating at night. The cough may be accompanied by bloodstained phlegm, but in the early stages this may not occur. The infected person, unaware, may infect many others before the diagnosis is made.

TB is a preventable and treatable condition. Prevention can occur at two places in the cycle of infection, the first being the interruption of the passage of the bacillus from an infected person to an uninfected person, and the second being the prevention of those who are infected (latent infection) from developing active disease. In the first instance, the focus is on finding and treating infectious cases, and in the second, the focus is on maintaining good health.

Treatment consists of a prolonged course (6 months) of multiple drugs (usually three), and good compliance to the treatment regimen is needed for a cure; this requires rigorous follow-up of cases by health care workers. The treatment program devised by WHO to assist with these requirements is known as "directly observed therapy short-course" (DOTS). It aims to ensure that treatment is taken as instructed and may involve family, friends, co-workers, and health care workers assisting the individual to comply with the treatment course. Despite this, poor compliance with the treatment remains a problem, and in resource-constrained countries, where medical staff numbers are low and drug supplies may not be continuous, drug-resistant strains of the bacteria have emerged. This has been complicated by the HIV epidemic, and multidrug-resistant (MDR) and extensively drug-resistant (XDR) cases are becoming commonplace in some countries. This is an extremely serious situation in some locations as XDR cases mostly have a fatal outcome and MDR cases are increasingly difficult to treat. Education, awareness, and an ability to detect cases is therefore essential for both prevention of the spread of TB and the treatment of those who are ill.

From a mining company perspective there are several things that can be done to manage the risk from TB. First, hostel-type accommodations should be avoided if at all possible. Second, education programs, together with active case-finding, should be undertaken. These can be accomplished through collaboration with local health authorities or through the mine's own health services. Where the risk coexists with HIV, the two programs can be combined.

For a detailed and thorough review of this subject, see the ICMM *Good Practice Guidance* document (ICMM 2008).

CONCLUSION

Although the focus may appear to be on HIV/AIDS, TB, and malaria due to their global importance, it must not be forgotten that there are numerous other infectious and noninfectious diseases that can and do affect the productivity and health of miners. These lend themselves equally to the approach advocated in this chapter, namely, risk assessment followed by a structured management program if the condition is found to be a significant problem to the mining operation. The same approach at an individual level is equally important for the management of personal risk due to the possible health risks that may arise.

REFERENCES

Attfield, M.D., Bang, K.M., Petsonk, E.L., Schleiff, P.L., and Mazurek, J.M. 2009. Trends in pneumoconiosis mortality and morbidity for the United States, 1968–2005, and relationship with indicators of extent of exposure. *J. Phys. Conf. Ser.* 151 012051.

Global Malaria Partnership. n.d. *Key Malaria Facts.* www.rollbackmalaria.org/keyfacts.html. Accessed July 2009.

ICMM (International Council on Mining and Metals). 2008. *Good Practice Guidance on HIV/AIDS, Tuberculosis and Malaria.* www.icmm.com/document/314. Accessed July 2009.

ILO (International Labour Organization). 1995. C176 Safety and Health in Mines Convention. www.ilo.org/ilolex/cgi-lex/convde.pl?C176. Accessed July 2009.

International Labour Office. 1998. *Encyclopaedia of Occupational Health and Safety*, 4th ed. Vol. 1, Part II. www.ilo.org/safework_bookshelf/english?d&nd=170000102&nh=0. Accessed July 2009.

Kielblock, A.J. 2001. Heat. In *A Handbook on Occupational Health Practice in the South African Mining Industry.* Johannesburg: Safety in Mines Research Advisory Committee.

Park, R., Rice, F., Stayner, L., Smith, R., Gilbert, S., and Checkoway, H. 2002. Exposure to crystalline silica, silicosis, and lung disease other than cancer in diatomaceous earth industry workers: A quantitative risk assessment. *Occup. Environ. Med.* 59(1):36–43.

Ramazzini, B. 1713. *De Morbis Artificum*. Translated by W.C. Wright as *Diseases of Workers*. 1964. New York: Hafner Publishing.

Robbins, A., and Landrigan, P.J. 2007. Safer, Healthier Workers: Advances in Occupational Disease and Injury Prevention. In *Silent Victories: The History and Practice of Public Health in Twentieth-Century America*. Edited by J.W. Ward and C. Warren. New York: Oxford University Press.

Schutte, P.C. 2008. Guidelines on fatigue management based on best practices for South African mines. Project Report SIM 060101. Pretoria, South Africa: Mine Health and Safety Council, Department of Minerals and Energy.

UNAIDS (Joint United Nations Programme on HIV/AIDS). 2007. AIDS Epidemic Update. http://data.unaids.org/pub/EPISlides/2007/2007_epiupdate_en.pdf?bcsi_scan_6E709EC7EBBC9476=0&bcsi_scan_filename=2007_epiupdate_en.pdf. Accessed July 2009.

UNAIDS (Joint United Nations Programme on HIV/AIDS). 2008. Global Facts and Figures. http://data.unaids.org/pub/GlobalReport/2008/20080715_fs_global_en.pdf?bcsi_scan_6E709EC7EBBC9476=0&bcsi_scan_filename=20080715_fs_global_en.pdf. Accessed July 2009.

WHO (World Health Organization). 2007a. Disease distribution maps. HIV Prevalence. http://gamapserver.who.int/mapLibrary/Files/Maps/Global_HIVPrevalence_ITHRiskMap.png. Accessed July 2009.

WHO (World Health Organization). 2007b. Disease distribution maps. Tuberculosis. http://gamapserver.who.int/mapLibrary/Files/Maps/Global_MDG6_TBprevalence_2007.png. Accessed October 2009.

WHO (World Health Organization). 2008. Global tuberculosis control—Surveillance, planning, financing. WHO Report 2008, WHO/HTM/TB/2008.393. www.who.int/tb/publications/global_report/2008/en/index.html. Accessed July 2009.

WHO (World Health Organization). 2009. Disease distribution maps. Malaria. http://gamapserver.who.int/mapLibrary/Files/Maps/Global_Malaria_ITHRiskMap.JPG. Accessed July 2009.

CHAPTER 15.3

Mine Ventilation

Michael A. Tuck

INTRODUCTION

Ventilation in underground mines is an essential element to ensure safe production of minerals. The basic objectives of any ventilation system are clear and simple: to provide airflows of sufficient quantity and quality to provide oxygen, and to dilute contaminants to safe concentrations and remove them from the mine. These basic requirements are enshrined in law in all countries; however, the quantity and quality requirements can vary greatly from country to country depending on a large number of factors as well as the situation being ventilated: coal, metalliferous, industrial minerals, or gassy/nongassy. The overall requirement is that workers must be able to live, work, and travel within an environment that is safe and provides reasonable comfort. Ventilation can also be employed in situations where human intervention is not required, but an adequate oxygen supply is required to operate and/or cool machinery or to provide oxygen to combustion processes.

Mine ventilation can be defined as the application of the principles of fluid mechanics and thermodynamics to the flow of air through mine openings. Air is introduced into the mine from the atmosphere, it flows around the mine, and it is expelled back into the atmosphere by the creation of a differential pressure between the intake and return openings of the mine. To achieve this, the mine ventilation system consists of interconnected airways and working zones within the mine, connections to the surface, fans to produce airflow, and control devices to ensure that the air courses around the mine as required. Controlling a ventilation system requires quantity as well as air quality control, and ventilation planning techniques are employed to achieve this.

This chapter concentrates on the fundamentals of mine ventilation engineering and its application to both coal and hard-rock mining. Other chapters in this handbook deal specifically with individual mine air pollutants.

PURPOSE OF MINE VENTILATION

The primary purpose of ventilation is to maintain an atmosphere that promotes a safe and effective work environment. This includes management of oxygen content and levels of contaminants including toxic gases, combustion products, and dust, plus maintaining workplace temperature conditions at levels where workers can achieve maximum efficiency. Legal limits relating to exact standards for the work environment vary depending on the purpose of ventilation as well as from country to country and state to state within some countries. Other than where prescribed by a definitive legal standard, a useful minimum oxygen concentration to apply to the design of any ventilation system is 19% oxygen by volume.

Pollutants enter a ventilation airstream from a variety of sources. The aim of any ventilation system is to dilute any pollutants to a safe level and then remove them. The range of pollutants that can be released into a ventilation system depends on the situation under consideration. Some of the more common types of pollutants encountered include

- Gases,
- Products of combustion,
- Dusts,
- Heat and humidity,
- Radioactive solids and by-products, and
- Diesel particulate matter.

Figure 15.3-1 shows the factors that contribute to hazards encountered in mine ventilation and the methods of control. The factors are divided into those imposed by natural conditions and those imposed by the design and engineering decisions regarding how the mine is to be developed and how ore is to be produced. The natural factors include geology, depth, surface climate, gases contained in the rock mass, ground water and other fluids, the physical and chemical properties of the rock, and the age of the workings. The engineering design of a mine impacts the hazards through the

- Mining method used,
- Method of fragmentation of rock,
- Rate of fragmentation of rock,
- Use of backfilling materials,
- Types of power used underground,
- Types and number of vehicles in the mine,
- Method of ore and waste removal,

Michael A. Tuck, Associate Professor of Mining Engineering, University of Ballarat, Victoria, Australia

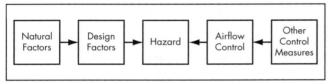

Source: Adapted from McPherson 1993; with permission from Springer Science and Business Media.

Figure 15.3-1 Factors contributing to hazards and methods of control in mine ventilation

- Types and quantities of materials stored and/or used in the mine, and
- Mine layout and the number and interconnectedness of the airways in the mine.

All mines are different, so the pollutants and their effects vary between mines, locations, and the cycle of operations within a mine (see Appendix B, Coal Mine Gas Chart). As an example, a deep mine in Canada may have issues with cold conditions in the shallower sections of the mine requiring mine heating during the winter months, while at the deeper levels the air has to be cooled by refrigeration. In addition to the ventilation pollutants, other hazards associated within a mine ventilation system are underground fires and explosions.

The main method of controlling the atmospheric conditions in an underground mine is by airflow. Fans generally induce the flow of air in mines, although natural ventilation due to air density changes can provide some airflow in modern mines. Main fans handle the whole mine airflow (singly or in combination) and are commonly sited on the surface, although at a number of mines they are sited underground. Legislation in some areas may require that main mine fans can only be located on the surface and in some cases may have to be reversible. While main fans generally handle all the mine airflow, other fans are used in mines to boost the airflow energy to allow air to be circulated over greater distances. These are called booster fans, and they generally serve specific areas of the mine. In mine development headings, a through-flow of ventilation may not be possible as the drive is single entry. In these cases an intake and return path for the air needs to be developed. In modern mining this is generally undertaken by using auxiliary fans to pass air through ducts to ventilate the blind entries. Alternatives include the use of brattice cloth and in some cases jet fans. Air doors, regulators, stoppings, seals, and air crossings control airflow distribution further.

While airflow is the primary method of achieving the goals of mine ventilation, in some situations the airflow required would become too costly to mine economically. In such cases other methods of control can be employed to reduce the load on the ventilation system so that the ventilation costs can be decreased. These methods include methane drainage, mine heating and cooling systems, dust suppression systems, and various other monitoring and control systems. These basic elements of a mine ventilation system are described in the video *Breathe and Live*, which can be viewed on the Mine Safety and Health Administration Web site (MSHA 1973). This video presentation details the hazards present from a ventilation perspective in underground mining, provides details of ventilation systems and devices, and includes an overview of equipment to provide ventilation and systems of auxiliary control such as heating and cooling systems. It is an old presentation, but it details the fundamentals of ventilation in an easy-to-understand format.

THEORY OF MINE VENTILATION

The background theory of mine ventilation comes from fluid mechanics and thermodynamics, specifically the Bernoulli equation, the steady-flow energy equation, and the Chezy–Darcy law for frictional pressure drop in a pipe. Atkinson (1854) developed the basis for the theory of mine ventilation that is still in use today. Taking fluid mechanics theory as a base and introducing a number of simplifying assumptions, he developed a series of equations that still form the basis of current ventilation theory and modern mine ventilation planning software. As described here, the theory uses an assumption of incompressibility. As modern mines progress to greater and greater depths, this assumption frequently becomes invalid due to changes in air density. Where changes in elevation and temperature produce changes in density greater than 5%, the assumption of incompressibility becomes flawed and must be taken into account.

Mine Airway Resistance (Atkinson's Law)

The Atkinson equation is based on the fundamental work undertaken by the French hydraulic engineers Chezy and Darcy. The Chezy–Darcy equation states the following for a pipe:

$$p = fL \frac{per}{A} \rho \frac{u^2}{2}$$

where

p = pressure drop (Pa)
f = friction coefficient
L = length (m)
per = perimeter (m)
A = cross-sectional area (m²)
ρ = density (kg/m³)
u = velocity (m/s)

For fully developed turbulent flow, Atkinson assumed that the coefficient of friction was a constant. Also assuming that the air density was a constant, he was able to combine the constants into a single factor known as the Atkinson friction factor, k (kg/m³),

$$k = \frac{f\rho}{2}$$

giving

$$p = kL \frac{per}{A} u^2$$

This is known as Atkinson's equation and can be written in terms of air quantity Q (m³/s), where $Q = uA$, giving

$$p = kL \frac{per}{A^3} Q^2$$

For any given airway, the length, L, perimeter, per, and cross-sectional area, A, are known. Ignoring the dependency on density, the friction factor varies only with the roughness of the airway lining for fully developed turbulent flow, so all these factors can be collected together to form a single variable or characteristic number, R, for that airway, where R is known as the airway resistance.

$$R = kL\frac{per}{A^3}$$

This gives the square law of mine ventilation:

$$p = RQ^2$$

where

p = pressure across airway (Pa)
R = resistance or Atkinson resistance (N·s²/m⁸)
Q = air quantity (m³/s)

The equation assumes a constant density; usually this is standard air density of 1.2 kg/m³. Values of R quoted in design tables usually refer to this standard density. For other air densities, ρ, the following can be used to give the frictional pressure drop and resistance:

$$p = k_s L \frac{per}{A^3} Q^2 \frac{\rho}{1.2}$$

$$R = k_s L \frac{per}{A^3} \frac{\rho}{1.2}$$

where k_s is the friction factor at air density 1.2 kg/m³.

From these calculations it is clear that the degree of roughness of an opening underground (k) has an important influence on its resistance and the cost of passing air through that opening. It also directly influences the heat transfer rate between the rock and the airstream. The friction factor, k, can be determined in a number of ways, details of which can be found in McPherson (1993):

- By analogy with similar airways, from ventilation surveys
- By direct measurement
- From design tables (see Table 15.3-1)
- From geometric data

Airway resistance (R) is an important concept in ventilation engineering. The square law states

$$p = RQ^2$$

This shows the resistance to be a constant of proportionality between the frictional pressure drop, p, and the square of the airflow, Q, at a specified air density. The parabolic form of the plot on a p, Q graph as shown in Figure 15.3-2 is known as the airway resistance curve.

The cost of passing any airflow through an airway varies directly with the resistance of that airway. This can be determined by evaluating the air power consumed in an airway using air power = $p \times Q$ or knowing $p = RQ^2$, resulting in air power = RQ^3.

Using a cost per kilowatt-hour for electricity, the cost to ventilate an airway can be simply determined. The equation also shows that doubling the airflow in a particular airway has the effect of multiplying the air power and hence the ventilation cost by a factor of 8.

Table 15.3-1 Friction factors (referred to an air density of 1.2 kg/m³) for a variety of mining situations

Type of Airway/Ducting	Friction Factor k, kg/m³
Rectangular airways	
Smooth concrete lined	0.004
Shotcrete	0.0055
Unlined with minor irregularities only	0.009
Girders on masonry or concrete walls	0.0095
Unlined, typical conditions, no major irregularities	0.012
Unlined, irregular sides	0.014
Unlined, rough or irregular conditions	0.016
Girders on side props	0.019
Drift with rough sides, stepped floor, handrails	0.04
Steel arched airways	
Smooth concrete all round	0.004
Bricked between arches all round	0.006
Concrete slabs or timber lagging between flanges all round	0.0075
Slabs or timber lagging between flanges to spring	0.009
Lagged behind arches	0.012
Arches poorly aligned, rough conditions	0.016
Shafts	
Smooth lined, unobstructed	0.003
Brick lined, unobstructed	0.004
Concrete lined, rope guides, pipe fittings	0.0065
Brick lined, rope guides, pipe fittings	0.0075
Unlined, well-trimmed surface	0.01
Unlined, major irregularities removed	0.012
Unlined, mesh bolted	0.014
Tubbing lined, no fittings	0.007–0.014
Brick lined, two side buntons	0.018
Two side buntons each with a tie girder	0.022
Longwall face with steel conveyor and powered supports	
Good conditions, smooth wall	0.035
Typical conditions, coal on conveyor	0.05
Rough conditions, uneven faceline	0.065
Ventilation ducting	
Collapsible fabric ducting (forcing systems only)	0.0037
Flexible ducting with fully stretched spiral reinforcement	0.011
Fiberglass	0.0024
Spiral-wound galvanized steel	0.0021

Source: McPherson 1993; with permission from Springer Science and Business Media.

Shock Losses

The theory outlined previously enabled the frictional pressure drop to be evaluated for an airway. In addition to the frictional or rubbing resistance, other losses of energy, termed *shock losses*, can be incurred. Whenever airflow is required to change direction, additional turbulent vortices are set up. These consume energy as they propagate and the resistance of the airway can increase significantly. These losses occur at bends, junctions, and changes in cross sections due to obstructions and regulators, and at points of entry/exit from the system.

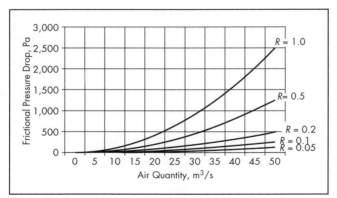

Figure 15.3-2 Mine airway resistance

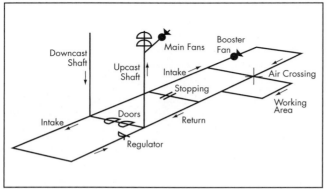

Source: McPherson 1993; with permission from Springer Science and Business Media.

Figure 15.3-3 Simple mine ventilation system

The effects of shock losses are the most uncertain of all the factors affecting resistance. Two methods are applied to the assessment of the additional resistance caused by shock losses. The head loss or pressure loss caused by a shock loss can be expressed as a shock pressure loss or p_{shock}:

$$p_{shock} = X\rho(u^2/2)$$

where
   X = shock loss factor
   ρ = air density (kg/m^3)
   u = air velocity (m/s)

This can be turned into a shock loss resistance, R_{shock}, using the square law:

$$R_{shock} = R_s = X\rho/2A^2$$

This can also be turned into an equivalent length due to shock loss using the Atkinson equation:

$$R_s = (kL_{eq} \, per/A^3)(\rho/1.2)$$

or

$$L_{eq} = 1.2 \, X \, A/(2k \, per)$$

Shock loss factors, X, are listed in McPherson (1993).

VENTILATION SYSTEMS

Every underground mine is unique in terms of its geometry/layout, extent/size, geology, and pollutants. As such, the pattern of airflows and pressure drops through the airways that make up the mine workings are also highly variable. However, certain features of all ventilation systems have some commonality, and classifications of ventilation systems and subsystems can be identified. The aim of this section is to discuss the essential characteristics of underground ventilation systems, describe the main elements of ventilation infrastructure, and introduce some of the technical terms employed by ventilation engineers. Ventilation systems can be broadly classified as follows:

- Mine systems
- District systems
- Auxiliary ventilation systems

Mine Systems

Figure 15.3-3 illustrates a schematic mine ventilation system showing a basic system layout and some of the ventilation control devices that can be encountered. Figure 15.3-3 shows the following:

- Fresh air enters via downcast shafts or other surface connections.
- Air flows along intake airways to working areas where most pollutants are added.
- Contaminated air returns back through the system along return airways.
- The return air passes back up to the surface via upcast shafts or other surface return connections.

However, ventilation is more complicated than this simplistic explanation. Air, like most fluids, will try to follow the path of least resistance from one point to another, and the path to the areas requiring ventilation usually lies along a high-resistance path. In order to get the air to the areas where it is needed, airflow control devices are required. These fall into two main categories: active devices that add energy to air to direct it to where it is needed, or passive devices that add resistance to flow routes to encourage the air to flow along other routes. Examples of active devices include main fans and booster fans. Examples of passive devices include air doors, regulators, air crossings, stoppings, and seals, which by adding resistance increase pressure drop. Both types are discussed in the following sections.

Active Devices
Main fans. Fans are the primary means of producing and controlling the ventilation airflow. Main fans (singly or in combination) handle all the air passed through the mine and are usually located at or near the surface. Main fan systems can be

- Forcing,
- Exhausting, or
- Push–pull (a combination of the two).

Surface fans have the following advantages:

- Simpler installation
- Simpler testing
- Simpler maintenance

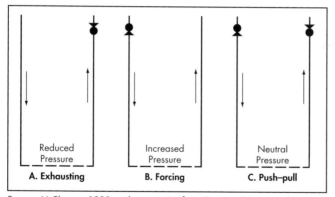

Source: McPherson 1993; with permission from Springer Science and Business Media.

Figure 15.3-4 Possible locations for main fans

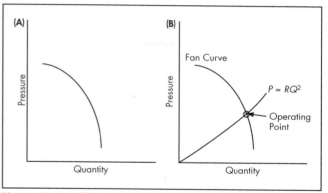

Figure 15.3-5 (A) Fan pressure–quantity characteristic curve, and (B) superimposed mine resistance characteristic and fan operating point

- Easier access
- Better protection in an emergency situation

Main fans can be located underground when fan noise is considered a problem or if shafts need to be completely free of airlocks. Underground fans incur additional problems associated with additional doors, airlocks, and leakage paths.

The choice of surface main fan location is illustrated in Figure 15.3-4. In Figure 15.3-4A, exhaust ventilation connects the main fan to the upcast shafts; in Figure 15.3-4B, forcing ventilation connects the main fans to the downcast shaft; and in Figure 15.3-4C, push–pull ventilation connects main fans to both the upcast and downcast shafts.

Most mines ventilate using exhaust ventilation (i.e., connecting the main fan to the exhaust shaft). The choice of whether to force or exhaust depends on four factors:

1. Gas control (control during falls of barometer and fan stoppage favors exhausting)
2. Transportation (selection depends on whether transporting mineral through intake or return airways)
3. Fan maintenance (exhaust fan subjected to air carrying dust, warmer air, and water droplets, thus the combined effects of impaction and corrosion are greater than in forcing case)
4. Fan performance (cooler air implies lower mass flow duty for a forcing fan and hence for a certain airflow, a reduced pressure and running costs). However, this effect is usually not great.

The combination or push–pull system of ventilation is most generally applied in metalliferous mining situations such as where caving techniques are employed and where the zone of fragmentation has extended to the surface. Maintenance of a neutral pressure underground minimizes the leakage of air between the workings and the surface. This is of particular concern where the fragmented rock is liable to spontaneous combustion. Push–pull systems are also employed in multishaft mines where multiple main fans are employed. In this case it offers the potential to achieve a better distribution of airflow, better control over pressure and leakages, greater flexibility, and reduced operating costs. However, multifan systems require skilled adjustment, planning, and balancing.

Fans generate energy and airflow by converting rotational mechanical energy into hydraulic energy. Numerous types of fans exist; the two most commonly applied fans in mine ventilation are centrifugal fans and axial fans, although in some situations mixed-flow fans are used. Every fan has its own characteristic set of performance or characteristic curves in terms of pressure–quantity, power–quantity, and efficiency–quantity curves. Figure 15.3-5A illustrates a typical pressure–quantity curve for a fan, while Figure 15.3-5B shows a mine system resistance curve superimposed on the fan curve. The point at which the two curves intersect is termed the *operating point* for the fan showing the fan pressure and associated flow. A typical main fan installation is illustrated in Figure 15.3-6.

Courtesy of Calizaya et al. 2000.

Figure 15.3-6 Main ventilation fans at PT Freeport DOZ mine

Booster fans. As air flows through a ventilation network, it loses energy. As mines become deeper and more extensive, the mine resistance increases and the reliance on a single main (i.e., its energy input) results in a reduction in the ventilation volume delivered to the extremities of the mine. Rather than upgrading the main fans to overcome this loss in energy, it is better to boost the energy of the airflow in sections of the mine by adding additional booster fans underground. This has numerous advantages including

- No need to upgrade existing main fan facilities, avoiding capital and operating cost increases; and
- Only a proportion of the whole mine airflow is boosted, saving operational costs compared to having to boost the whole mine airflow.

Courtesy of De Souza et al. 2003.
Figure 15.3-7 Booster fan installation

Figure 15.3-7 shows a typical booster fan facility. De Souza et al. (2003) provide a good example of the design and performance of a booster fan facility and the benefits that can be obtained. These include increased ventilation flows in remote mining sections, improved pressure control, improved energy use, and increased air velocities providing enhanced cooling power in hot mines.

Passive Devices

Air doors and airlocks. When access is still required between intake and return, air doors are employed to minimize short-circuiting of ventilation flow by increasing the resistance of a potential short-circuit route. Ventilation doors between main intake and returns are usually in the form of two or more sets of air doors and termed *airlocks*. Doors should always be hinged to open toward the higher pressure so that they are self-closing. Doors can be constructed with a variety of materials including wood or steel.

Regulators. A passive regulator is simply an air door fitted with an adjustable orifice, the purpose of which is to reduce the airflow in a particular section downstream to a defined airflow by increasing the resistance of the airway. When the airflow in a section of a mine needs to be higher than that normally obtainable, active regulation by the installation of a fan to boost the airflow is required. These are termed *booster fans* and can also be called "active regulation" as the energy of the airflow is increased.

Air crossings. Where intake and return airways need to cross over each other, the leakage between the two must be controlled by means of an air crossing. Air crossings can take the following forms:

- **Natural:** one of the airways is elevated above the other leaving a beam of strata separating the two airways.
- **Intersection:** this is the more usual method. Both airways intersect during construction, and the roof of one or the floor of the other is excavated to expand the zone of intersection. Masonry or concrete blocks or a steel structure with metal or timber shuttering separates the two airways. The high-pressure side is further sealed by the application of shotcrete or another sealant.
- **Prefabricated air crossings**

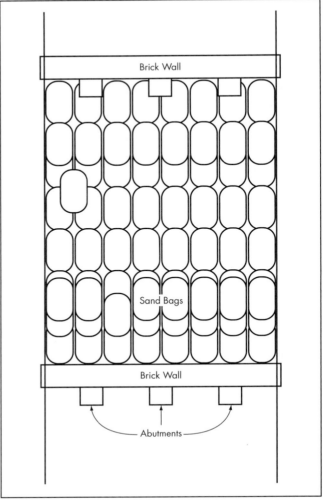

Source: Chalmers 2008.
Figure 15.3-8 Explosion-proof seal

In all cases the material from which the crossing is fabricated should be fireproof and capable of maintaining its integrity in case of a fire. Aluminum and other low-melting-point materials should not be used to construct an air crossing.

Stoppings and seals. In developing a mine, connections are necessarily made between the intake and return airways for a number of reasons including to provide quick access during mine development and to provide through ventilation. When these are no longer required for access they should be blocked by stoppings. These should be well constructed, keyed into solid rock, and leakage through should be minimized (e.g., by being coated with a sealant). As a short-term method, brattice curtains can be used when pressure differentials are low.

When required to isolate abandoned areas of the mine, seals should be used. Chalmers (2008) details the requirements for sealing abandoned coal mining areas, and many of the details discussed in his paper also apply to other forms of mining. Seals are constructed for a number of reasons, such as preventing unwanted air loss from a ventilation system and to preclude oxygen from entering mined-out areas to prevent spontaneous combustion. Regardless of the reason for the

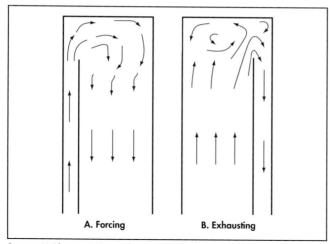

Source: McPherson 1993; with permission from Springer Science and Business Media.

Figure 15.3-9 Line brattices used in auxiliary ventilation

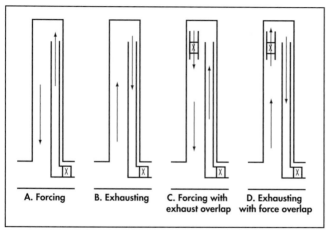

Source: McPherson 1993; with permission from Springer Science and Business Media.

Figure 15.3-10 Auxiliary fan and duct systems

seal, Chalmers presented a number of characteristics that a good seal should have:

- Highly resistant to airflow
- Strong
- Fire resistant
- Easily and rapidly deployed
- Ability to cope with strata convergence
- Insolubility
- Nonreactive
- Sampling points installed
- Stable, not easily blown over

The exact construction depends on local conditions and materials available; however, a distinction can be made between seals that are designed to purely stop air leakage and a seal that is required to be explosion proof. Figure 15.3-8 shows the typical construction of an explosion-proof seal.

Backfill is becoming more widely used, particularly in hard-rock mining. Backfilling worked-out stopes can also act as a very effective seal.

District Systems

District systems are discussed later in this chapter in the "Coal and Metaliferrous Ventilation" section.

Auxiliary Systems

Systems used to supply air to the working faces of blind headings, sublevels, and so on comprise three basic types: line brattices, fan and duct systems (forcing, exhausting, or overlap systems), and air movers and jet fans.

Line Brattices

Brattice cloth, hung from floor to ceiling, creates an artificial intake or return in a dead-end drive. Commonly employed in room-and-pillar operations, its major disadvantage is the resistance added to the mine ventilation network at the most sensitive point in the system. Also these systems are very leaky. The advantages are that they are cheap, require no power, and produce no noise. Figure 15.3-9 illustrates the use of line brattices.

Fan and Duct Systems

These systems consist of a duct or a multiplicity of ducts connected together to form a ventilation network with a fan or fans to provide the air movement. As such, fan and duct systems can be simple or complex ventilation networks. The main difference between these systems and the ventilation systems discussed previously is that they rely on open atmospheric connections in a factory, mine, or other context to complete the ventilation circuit.

Fan and duct systems basically come in four forms as shown in Figure 15.3-10:

1. Forcing systems
2. Exhausting systems
3. Forcing systems with an exhaust overlap
4. Exhausting systems with a force overlap

Within such systems the usual choice of fan is an in-line axial fan. In mining situations, if the dead-end heading is greater than 30 m in length, such systems are usually the only means of ventilation applicable. Care should be taken to ensure that the pressure and quantity characteristic of the fan is commensurate with the resistance offered by the duct and the airflow required. The latter of these depends on the pollutants to be removed. The resistance of the duct is established in the same way as for other ventilation systems by determining the frictional losses associated with the duct and adding extra losses incurred at bends, changes in the cross section, and entry and exit losses.

The choice of which system to choose depends on the pollutants of greatest concern. The higher velocity emerging from the exit of a forcing duct provides a scouring action, which effectively sweeps the area. This is good for removing gases by providing turbulent mixing. Where heat is the main concern, forcing systems provide high airflows and enhance air-cooling power. The main disadvantage of the forcing system is that pollutants added to the airstream pass back via the open atmosphere along the heading in mines, thus polluting the general atmosphere relatively slowly.

Where dust control is the primary hazard, the exhausting system is preferred as the dust is contained within the ducting and fresh airflows along the heading. The main disadvantage of the exhausting method is the lack of an effective scouring action as there is no high-velocity jet of air to achieve this. Another disadvantage of exhaust systems is that rigid or reinforced ducting needs to be used. These are more expensive than the flexible ducting types that can be used in forcing systems. The force/exhaust and exhaust/force overlap systems have been developed to overcome the disadvantages of the single forcing or exhausting systems. The selection of which of these methods to use depends on the pollutants under consideration.

Where long lengths of ducting are required (e.g., in tunneling work), the resistance of the duct may be such that multiple fans connected in series may be needed. As ducting leaks, it is not advisable to cluster fans at one end of the ducting system as the pressures developed will enhance leakage. The preferred method is to space the fans out along the duct to avoid excessive gauge pressures. Multiple fans must be interlinked electrically and airflow or pressure transducers used to detect accidental severing or blockage of the duct.

Air Movers and Jet Fans

Air movers are used to enhance and/or control air movement within localized areas. They are freestanding and can be powered by a number of sources including electricity, compressed air, and water. Jet fans, otherwise known as ductless, vortex, or induction fans, are common and come in a number of forms and sizes. A spray of water can also generate airflow.

VENTILATION SURVEYS

A ventilation survey is an organized procedure to acquire data to determine the quantity, pressure, and air quality distribution within a mine. The accuracy and required detail of the survey depend on the purpose of the survey. Surveys are undertaken for a variety of reasons, including

- Mandatory requirements,
- Ensuring that working areas of the mine receive the airflow required in an efficient and effective manner,
- Keeping ventilation plans and network models up to date,
- Verification purposes, and
- Planning purposes.

A major objective of ventilation surveys is to measure and obtain the frictional pressure drop, p, and corresponding airflow, Q, for each branch within the ventilation network. From these data the following may be calculated for both planning and ventilation control purposes:

- Distribution of airflows, pressure drops, and leakage
- Air-power losses and the distribution of ventilation operating costs around the network
- Volumetric efficiency
- Branch resistances
- Natural ventilation effects
- Friction factors

In addition to pressure and airflow measurements, other measurements are taken either as an integral part of the survey or separately. These will include airway area, airway perimeter, wet and dry bulb temperatures, barometric pressures, dust levels, radon levels, and concentrations of gaseous pollutants.

There are two main aspects to a ventilation survey: the determination of air quantity, and pressure drop.

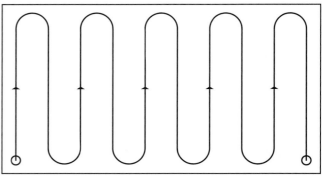

Source: Adapted from McPherson 1993; with permission from Springer Science and Business Media.

Figure 15.3-11 Path of an anemometer traverse

Air Quantity Surveys

The volume of air passing a given point in an airway or duct, Q, is normally determined as the product of the cross-sectional area of the airway or duct, A, and the mean air velocity, u:

$$Q = u \times A$$

Thus, most common techniques of determining Q are combinations of the methods available for determining area and mean velocity. Modern methods of determining air speed or velocity can be divided into three groups depending on

1. Mechanical effects,
2. Dynamic pressure, and
3. Thermal effects.

Mean velocity at a point can be determined either by a traversing instrument that integrates measurements as the cross section is traversed or mathematically combining fixed-point measurements to arrive at an average air velocity. Details of the methods used to undertake such measurements are described by numerous authors such as McPherson (1993) and are summarized in the paragraphs that follow.

Figure 15.3-11 shows a typical method used to undertake an anemometer traverse measurement. These are undertaken with a vane anemometer or similar type of integrating instrument. The anemometer should be attached to a rod about 1.5 m in length or greater to allow the whole cross section to be traversed. Anemometers are relatively insensitive to yaw; this is misalignment with the airflow, so long as the angle of yaw does not exceed about 5°.

The observer faces the airflow and holds the instrument at least 1.5 m away from his or her body by attaching the instrument to an extendable arm. The instrument is then started while another observer starts a stopwatch. The instrument is then moved as shown in Figure 15.3-11, the aim being to sweep equal areas of the traverse in equal times. The traverse should not take less than 60 seconds. After the whole area is swept, the anemometer is stopped, the total flow through the instrument noted and the time taken, allowing the average velocity to be computed. The process is repeated until three readings yield the same average velocity to within ±5%.

At least two sets of traverses are recommended to be taken at different locations within the airway. The measuring locations should be chosen to avoid potential problems wherever possible. Booking of results should include the following:

- Observers' names
- Location of station, time, and date
- Anemometer readings and corrections
- Air dry and wet bulb temperatures and barometric pressure
- Dimensioned sketch of the cross section
- Calculation of area
- Calculation of air volume flow

An anemometer traverse can also be undertaken at the ends of ducts but not if the duct diameter is less than six times the anemometer diameter. Anemometers can also be used for fixed-point measurements with corrections applied to obtain the real flow. These corrections are determined via traverses undertaken previously at the same location. When the duct/anemometer ratio of 6 precludes application of an anemometer survey, pitot tube techniques are most commonly employed, as described in McPherson (1993).

The measurement of a cross-sectional area can be done in a number of ways:

- Simple taping
- Offset method
- Profilometer
- Photographic methods

Pressure Surveys

The purpose of a pressure survey is to determine the frictional pressure drop corresponding to a certain flow in an airway. Basically there are two methods, the more accurate being the trailing hose or gauge and tube method, which is time-consuming but preferred where foot travel is relatively easy. The second method is the barometer or altimeter method, used where access is difficult such as in multilevel mines and in shafts. Details of these methods can be found in Hemp (1989) and are briefly described in the following paragraphs.

The principle of the trailing hose method is simple: A hose is laid along the length of the airway being surveyed and a manometer or other suitable differential pressure gauge is connected to one end of the hose. In the case of a horizontal airway, this gauge will indicate a differential pressure exactly equal to the pressure loss. When the airway is not horizontal, the differential pressure will differ from the pressure loss; however, methods are available to calculate the pressure loss derived from the airway equations, which are modified versions of the steady-flow energy equation for mining conditions.

In its simplest form, the barometer method involves the simultaneous measurement of barometric pressures and wet and dry bulb temperatures at either end of the airway. Airflow also needs to be measured. If the elevations of the two measuring stations are accurately known, the pressure and/or energy loss can be determined from the airway equations.

MINE VENTILATION NETWORK ANALYSIS

An essential component in the design of new mines is the quantified planning of the distribution of airflows, determining the location and duties of fans and other ventilation control devices to achieve the goal of providing acceptable environmental conditions throughout the ventilation system. In existing mines it is better to plan ahead to ensure that new fans, shafts, and other airways are available in a timely manner for efficient ventilation of the mine and extensions to the mine. Operating mines are dynamic operations with new workings continually being developed and older ones ending their productive life. Ventilation planning should, therefore, be a continuous and routine process.

The laws of fluid mechanics control the behavior of air in airways, ducts, and pipes. As shown previously, with regard to mining this involves using Atkinson's equation, the square law, and in some cases the laws of thermodynamics. Ventilation network analysis is concerned with the interactive behavior of airflows within the branches of a complete and integrated network. Ventilation network analysis problems can be addressed and formulated quite simply as stated by McPherson (1993):

- If the resistances of the branches of a network and how these branches are interconnected are known, how can the distribution of airflows for given locations and duties of fans be predicted?
- If the airflows in specific branches are known, how are the combination of fans and the structure of the network to provide these airflows determined?

Ventilation network analysis is a generic term for a number of techniques, a family, which enable such questions to be addressed. For a ventilation network there are theoretically an infinite number of combinations of fans, regulators, and airway resistances that will produce the desired flow distribution. Practical considerations will limit the number of acceptable alternatives. One essential aspect of any technique to be employed is that it must be easy to use, and be sufficiently rapid and flexible to allow investigation of multiple solutions.

Fundamentals of the Ventilation Network Analysis

The fundamental relationships that govern the behavior of electrical current within a network of conductors were first recognized by Gustav R. Kirchhoff (1824–1887). These same basic relationships are now known as Kirchhoff's laws and are also applicable to fluid networks.

Kirchhoff's first law (hereafter referred to as Kirchhoff 1) states that the mass flow entering a junction equals the mass flow leaving that junction (a consequence of the law of conservation of mass) and can be written as

$$\sum_j M = 0$$

where M is the mass flows, positive and negative, entering junction j. However,

$$M = Q\rho$$

where Q is the volume flow (m^3/s) and ρ is the air density (kg/m^3), thus,

$$\sum_j Q\rho = 0$$

If the density variation around the ventilation network is negligible (i.e., the flow can be regarded as incompressible), then

$$\sum_j Q = 0$$

This provides a means of checking the accuracy of airflow measurements taken around a junction when undertaking a ventilation survey.

Kirchhoff's second law (hereafter referred to as Kirchhoff 2) stated in its simplest form is that the algebraic sum of all pressure drops around a closed path, or mesh, in the network must be zero, taking into account the effect of fans and/or natural ventilation pressures, a consequence of the law of conservation of energy. The steady-flow energy equation for a single airway can be written as

$$\frac{\Delta u^2}{2} + \Delta Zg + W = \int VdP + F \text{ J/kg}$$

where
- u = air velocity (m/s)
- Z = height above datum (m)
- g = acceleration due to gravity (m/s^2)
- W = work input from fan (J/kg)
- V = specific volume (m^3/kg)
- P = barometric pressure (Pa)
- F = work done against friction (J/kg)

Considering a number of branches forming a closed loop or mesh within a network, the algebraic sum of all ΔZ must be zero and the sum of all the changes in kinetic energy, $\Delta u^2/2$, is negligible. Summing the remaining terms around the mesh, m, gives

$$\sum_m \int VdP + \sum_m (F - W) = 0$$

The summation of the $-VdP$ terms is the natural ventilation energy, NVE, which originates from thermal additions to the air, thus

$$\sum (F - W) - \text{NVE} = 0$$

This calculation is in joules per kilogram and can now be converted to pressure units by multiplying by density, ρ.

$$\sum (F - W)\rho - \rho\text{NVE} = 0$$

But if $\rho_F = p$, the frictional pressure drop, and $\rho W = p_f$, the rise in total pressure across a fan, and $\rho\text{NVE} = \text{NVP}$, the natural ventilation pressure, then Kirchhoff's second law becomes

$$\sum (p - \rho_f) - \text{NVP} = 0$$

Kirchhoff's laws can be applied to fluid networks that have either compressible or incompressible flow. In the case of compressible flow, the analysis is undertaken on the basis of mass flow; in the case of incompressible flow, the analysis is undertaken on the basis of volume flow.

From this point forward, incompressible flow will be assumed. The square law needs to be incorporated within Kirchhoff's laws for incompressible flow.

Kirchhoff 1

$$\sum_j Q = 0$$

Kirchhoff 2

$$\sum_m (RQ|Q| - p_f) - \text{NVP} = 0$$

where
- R = resistance (N·s^2/m^8)
- $|Q|$ = the absolute value of airflow

This ensures that the frictional pressure drop always has the same sign as the airflow and removes the complication of squaring or square-rooting negative numbers.

Deviations from the Square Law

The fundamental relationship of mine ventilation is the square law $p = RQ^2$. This is developed from the Chezy–Darcy relationship and is valid for fully developed turbulent flow. The resistance, R, is a function of the friction factor, k, which is itself related to the Darcy friction factor, f. If the flow regime falls into the transitional zone or the laminar flow zone, f, and, hence, k becomes a function of the Reynolds number, Re, the value of R varies with airflow. In these cases, the square law is inapplicable.

For the transition zone the following relationship applies:

$$p = RQ^n$$

Values of n have been reported to be in the range of 1.8–2.05. In the case of laminar flow, the following equation can be used:

$$p = R_L Q$$

where R_L is laminar resistance.

In the following sections it is assumed that the square law applies.

Methods of Solving Ventilation Networks

Basically there are two methods for solving ventilation networks:

1. Analytical methods formulate governing laws to produce equations that yield exact solutions.
2. Numerical methods iterate until a solution is found to a required degree of accuracy.

Analytical Methods

The following paragraphs describe analytical methods to solve ventilation networks.

Equivalent resistances. This is the most elementary method of network analysis. When two or more airways are connected together in series or parallel, each of these resistances can be combined into a single equivalent resistance in a similar manner to electrical circuits. A mine ventilation network is made up of a number of roadways connected together; these connections between branches can be broken down into roadways connected in series and in parallel.

- Series circuits—The common element is that the air quantity flowing in each branch is the same. The equivalent resistance of n branches connected in series is

$$R_{\text{total}} = R_1 + R_2 + \ldots + R_n \text{ (N·s}^2\text{/m}^8\text{)}$$

- Parallel circuits—The common element is that the pressure drop across every branch is the same. The equivalent resistance if n branches are connected in parallel is

$$1/\sqrt{R} = \sum_1^n 1/\sqrt{R}_n \text{ (N·s}^2\text{/m}^8\text{)}$$

The square root term in the equation results from the square law of mine ventilation rather than the linear Ohm's law for electrical circuits.

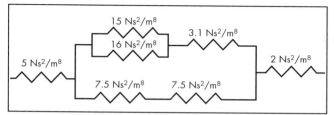

Figure 15.3-12 Ventilation network for equivalent resistance calculation (Example 1)

Example 1. A section of a mine ventilation system is connected together as shown in Figure 15.3-12. If the pressure drop across the system is 1.1 kPa and the resistances are as indicated, determine the flow in each airway.

Solution. First the equivalent resistance of the circuit needs to be determined. Taking the bottom line of the parallel portion of the circuit first, the equivalent resistance is $7.5 + 7 = 14.5$ N·s²/m⁸. The equivalent resistance of the top line now needs to be determined.

Taking the parallel portion first,

$$\frac{1}{\sqrt{R}} = \frac{1}{\sqrt{15}} + \frac{1}{\sqrt{16}} = 0.508$$

Thus $R = 3.875$ N·s²/m⁸, thereby showing the advantage of placing airways in parallel as the resistance of the combination is less than the individual resistances that make up the parallel combination.

This is in series with the 3.1 N·s²/m⁸ resistance, so the resistance of the top line is 6.975 N·s²/m⁸.

The equivalent resistance of the whole parallel section can now be evaluated:

$$\frac{1}{\sqrt{R}} = \frac{1}{\sqrt{14.5}} + \frac{1}{\sqrt{6.975}} = 0.641$$

Thus, $R = 2.433$ N·s²/m⁸.

Combining this in series with the intake resistance of 5 and the return resistance of 2 gives a total system resistance of 9.433 N·s²/m⁸.

The total inlet and outlet flow for the system can now be evaluated from the square law of mine ventilation, $p = RQ^2$.

$$Q = \sqrt{\frac{p}{R}} = \sqrt{\frac{1{,}100}{9.433}} = 10.8 \text{ m}^3/\text{s}$$

Knowing this, the pressure drop for the intake and outlet can be evaluated to determine the pressure drop acting over the parallel element of the circuit:

$$p = 7 \times 10.8^2 = 816.5 \text{ Pa}$$

So the pressure drop over the parallel element of the circuit is $1{,}100 - 816.5 = 283.5$ Pa. This acts across both legs so the flow in the bottom leg is

$$Q = \sqrt{\frac{p}{R}} = \sqrt{\frac{283.5}{14.5}} = 4.42 \text{ m}^3/\text{s}$$

and the flow through the upper leg is 6.38 m³/s.

To determine the pressure drop over the parallel portion of the upper leg, the pressure drop through the 3.1 N·s²/m⁸ resistance needs to be determined. This is

$$p = 3.1 \times 6.38^2 = 126 \text{ Pa}$$

So the pressure drop over the parallel portion of the upper leg is $283.5 - 126 = 157.5$ Pa. The flows can now be determined. Through the 15 N·s²/m⁸ resistance it is

$$Q = \sqrt{\frac{p}{R}} = \sqrt{\frac{157.5}{15}} = 3.24 \text{ m}^3/\text{s}$$

and through the 16 N·s²/m⁸ leg the flow is 3.14 m³/s.

This example shows the effect of placing resistances in parallel to reduce the overall mine resistance and the dominating effect that the entry and exit from the system has on the overall pressure drop in the system. In other words, the entry and exits from a mine ventilation system are primary consumers of air power in a mine ventilation circuit—not surprising as they carry the full system flow.

Direct application of Kirchhoff's laws. Kirchhoff's laws allow equations for each independent junction in a network and for each independent mesh. Solving these two sets of equations obtains the branch airflows that satisfy both laws. If b branches exist in a network, then b airflows need to be determined, and b independent equations are required.

If a network contains j junctions, Kirchhoff 1 can be written down for each in turn. But as each branch is assumed to be continuous with no intervening junctions, a branch airflow denoted by Q_i entering a junction will automatically imply that Q_i leaves a neighboring junction at the other end of that same branch. Therefore, by the last junction, all airflows will have been symbolized; thus, the number of independent equations from Kirchhoff 1 is $j - 1$.

This leaves $b - (j - 1)$ or $b - j + 1$, with further equations to be established from Kirchhoff 2. It is, therefore, necessary to choose $b - j + 1$ independent closed meshes around which to sum the pressure drop. The direct application of this method to mine circuits may result in hundreds of equations to be solved simultaneously. Computer assistance is required, as manual solution is limited to small networks.

Numerical Methods

Numerous numerical methods exist. The main differentiation between the methods revolves around mesh selection and numerical solving of the equations. The most robust—though not always the most efficient—method was devised by Professor Hardy Cross in 1936 at the University of Illinois (United States) for water distribution systems. This technique is the most widely used and was modified for use in mine ventilation systems by Professors F.B. Hinsley and D.R. Scott at the University of Nottingham (England) in 1951.

Example 2. Figure 15.3-13 shows the system resistance curve for a single representative branch of a ventilation network. If airflow Q is reversed, the frictional pressure drop, p, also becomes negative. When establishing a flow distribution, the true airflow in the branch, Q, is initially unknown. Assume a flow, Q_a, that is less than the true value by an amount ΔQ:

$$Q = Q_a + \Delta Q$$

Now find ΔQ. The square law can be written as

$$p = R(Q_a + \Delta Q)^2$$

More generally,

$$p = R(Q_a + \Delta Q)^n$$

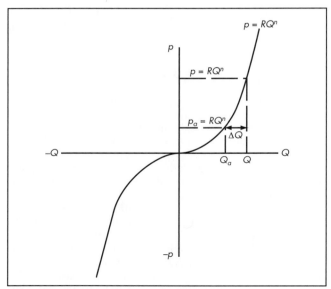

Source: Adapted from McPherson 1993; with permission from Springer Science and Business Media.
Figure 15.3-13 Iterative technique to solve ventilation networks

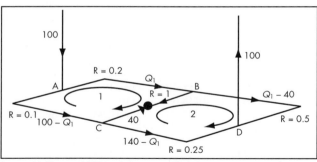

Source: McPherson 1993; with permission from Springer Science and Business Media.
Figure 15.3-14 Network for worked example of direct application of Kirchhoff's laws

Going back to the square law, this expands to

$$p = RQ_a^2 + 2RQ_a\Delta Q + R(\Delta Q)^2$$

The frictional pressure drop corresponding to the assumed airflow is

$$p_a = RQ_a^2$$

where the error in p is

$$\Delta p = p - p_a$$

Substituting from the equations gives

$$\Delta p = 2RQ_a\Delta Q + R(\Delta Q)^2$$

Assuming ΔQ^2 is small compared to $2RQ_a\Delta Q$, the approximation can be written as

$$\Delta p = 2RQ_a\Delta Q$$

Thus, $\Delta Q = \Delta p/2RQ_a$. The technique of dividing a function by its first derivative in order to estimate an incremental correction factor is a standard method of finding roots of functions and is usually called the Newton–Raphson method. The technique can also be applied to meshes as explained in McPherson (1993). For the general case where $p = RQ^n$, the following is true:

$$\Delta Q_m = \frac{-\sum(RQ_a|Q_a|-p_f-\text{NVP})}{\sum(2R|Q_a|+S_f+S_{nv})}$$

where

S_f = slope of the pressure/quantity characteristic curve for the fan
S_{nv} = slope of the pressure/quantity characteristic curve for the natural ventilation effects

The Hardy Cross technique can now be summarized as follows:

1. Draw a network schematic and choose at least $(b-j+1)$ closed meshes such that all branches are represented. Convergence to a balanced solution will be improved if each high-resistance branch is included in only one mesh.
2. Make an initial estimate of airflow, Q_a, for each branch.
3. Traverse one mesh and calculate the mesh correction factor, ΔQ_m, from the equation.
4. Traverse the same mesh in the same direction and adjust each of the airflows by the amount ΔQ_m.
5. Repeat Steps 3 and 4 for each mesh.
6. Repeat Steps 3, 4, and 5 until Kirchhoff 2 is satisfied to an acceptable degree of accuracy.

An example of this method is given in McPherson (1993).

Example 3. Figure 15.3-14 shows a network. The downcast and upcast shafts pass 100 m³/s and the resistance of each branch is shown. A booster fan in the central branch passes 40 m³/s. Determine the distribution of airflow and the total pressure developed by the booster fan.

Solution. Inspection of the figure shows that it cannot be solved by equivalent resistances. Therefore, it is a complex network. The given airflows are shown on the figure with flow from A to B denoted by Q_1. The flows in the other branches are expressed in terms of Q_1. The shafts can be eliminated, as their airflows are already known. Thus,

$$b - j + 1 = 5 - 4 + 1 \text{ independent meshes are required}$$

where

b = number of branches in the network
j = number of junctions in the network

Meshes along the closed paths ABCA and BDCB are selected. In order to apply Kirchhoff 2 to the first mesh, consider the pressure drops around that mesh branch by branch. Frictional pressure drops are positive in the direction of flow and p_b:

Branch	Frictional Pressure Drop (RQ^2)	Fan
AB	$0.2\,Q_1^2$	—
BC	1.0×40^2	$-p_b$
CA	$-0.1(100-Q_1)^2$	—

Summing these up according to Kirchhoff 2 gives

$$0.2\, Q_1^2 + 1{,}600 - p_b - 0.1(100 - Q_1)^2 = 0$$

Collection of like terms gives

$$p_b = 0.1 Q t_1^2 + 20\, Q_1 + 600$$

For the second mesh,

$$0.5(Q_1 - 40)^2 - 0.25(140 - Q_1)^2 - 1 \times 40^2 + p_b = 0$$

which reduces to, substituting for p_b from the previous equation,

$$0.35 Q_1^2 + 50 Q_1 - 5{,}100 = 0$$

Solution of this quadratic equation gives $Q_1 = 68.83$ or -211.69 m^3/s. The only practicable solution is 68.83 m^3/s, thus the flows are

Branch	Air Quantity, m³/s
AB	68.83
BD	28.83
BC	40
AC	31.17
CD	71.17

The required booster fan pressure is

$$p_b = 0.1(68.83)^2 + 20\,(68.83) + 600 = 2{,}450$$

Ventilation Network Simulation Programs

There are numerous ventilation network simulation packages on the market. These are mathematical models and are designed to simulate real systems. Use of these packages has revolutionized mine ventilation planning (Mine Ventilation Services 2007; Ventsim 2010).

MINE VENTILATION PLANNING

Planning of new mines and extensions to existing mines is a critical part of overall mine planning. Without adequate airflow provided in a timely and efficient manner, likely outcomes include poor environmental conditions for the work force, with associated health and safety issues, and reduced production. Ventilation planning is also essential in the testing of emergency and evacuation procedures.

A generalized approach to mine ventilation planning is shown in Figure 15.3-15. The figure has been simplified for clarity and indicates that a step-like approach should be taken with constant feedback between stages, similar to other mine design problems. All mines are different in terms of the pollutants to be diluted and removed, so generic rules are not applicable to all situations. Ventilation planning should be an integral part of overall mine planning, as well as production planning. With regard to ventilation planning, the following factors should be considered:

- The method should fit with the remainder of the mining operation.
- The method should be flexible.

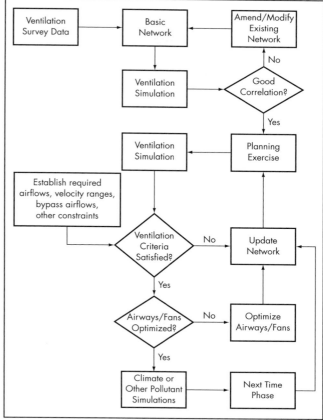

Figure 15.3-15 Generic mine ventilation planning

- The method should be economical.
- The method must be adaptable for extensions and future mining activities.

With reference to Figure 15.3-15, the initial stage is to create a simulation model of the mine. For an existing mine, this should be created using the results of ventilation surveys. For a new mine, typical design data such as friction factors should be combined with geometric data and layout data from other mine planning activities. For extensions to existing mines, a combination of survey data and design data will need to be used. After this has been achieved, the next stage is to test the ventilation network for existing mines to ensure that the results of the simulation correlate well with the real ventilation survey data. If a reasonable correlation (±5%) is not achieved, the ventilation network needs to be modified by examining assumptions and data against reality until a suitable correlation is achieved. Possible reasons why correlation is not achieved include compressibility occurring in the network while an incompressible analysis is being undertaken, airways not conforming to actual design specifications, shock losses not being accounted for, and numerous other reasons. For a new mine, after underground workings exist, ventilation surveys should be undertaken and compared against the simulated network to ensure that the simulations match the mining reality.

After a suitable network simulation has been developed, the next stage is to ensure that the predicted flows are reaching the production and other areas that require ventilation. To do this, the required airflows, air velocity limits, bypass

Table 15.3-2 Recommended maximum air velocities

Area	Velocity, m/s
Working faces	4
Conveyor drifts	5
Main haulage routes	6
Smooth-lined airways	8
Hoisting shafts	10
Ventilation shafts	20

Table 15.3-3 Diesel dilution requirements for selected locations

Location	Diesel Airflow Requirement	Comments
Western Australia	0.05 m^3/s per kW	
Queensland, Australia	None	Was 0.04 m^3/s per kW
Ontario, Canada	0.06 m^3/s per kW	
Manitoba, Canada	0.92 m^3/s per kW	Uses 100/75/50 rule
United Kingdom	None	
United States	0.032 to 0.094 m^3/s per kW	Typically design to 0.08 m^3/s per kW
Chile	0.063 m^3/s per kW	
South Africa	None	Exhaust/overlap system required for firing anytime. Typically design to 0.06 m^3/s per kW.
Indonesia	0.067 m^3/s per kW	

airflow, requirements to eliminate recirculation, and other requirements such as establishing limits for the concentration of pollutants need to be established. In some cases these limits are dictated by law, in other cases not. In some countries and states, best practice is mandated by health and safety legislation.

In the case of gases, the required airflow can be simply determined if the release rate of the gas is known and the required maximum concentration is known as detailed by McPherson (1984). For dust and some other requirements such as bypass airflows, air velocity limits such as those shown in Table 15.3-2 can be applied. Minimum velocity standards can also be fixed; a commonly applied standard is 0.3 m/s (McPherson 1993). For diesel dilution, a common method is to specify airflow per rated kilowatt of diesel power in a location. See Table 15.3-3 for a list of these requirements for several countries (Brake and Nixon 2008). For heat and humidity, heat stress indices need to be applied to define safe work, to reduce work and stop-work conditions, and to undertake climate simulations in addition to ventilation simulations (Tuck 2008).

Other limits can also be applied; for example, in wet exhaust shafts where condensation or water flow results in airborne droplets, the air velocity should not lie in the range from 7 to 12 m/s as water blanketing can occur between this range of velocities (McPherson 1993).

After the required air velocities, quantities, and other criteria have been set, the results of the ventilation simulation need to be carefully checked against these criteria. If the criteria are not satisfied, amendments need to be made to the network such as changing the duty of fans, changing airway lining or other inputs, and the simulation rerun until the criteria are realized.

The third stage involves an economic overview of the system. Mine ventilation systems can be major electricity consumers, so trying to optimize the system to minimize the total capital and operating cost is prudent.

Following this stage, if ventilation alone is not sufficient to dilute and remove the contaminants, the design requirements for auxiliary systems such as methane drainage, mine cooling, and mine heating can be investigated as required.

These considerations give a broad, simple overview of the ventilation design process; a fuller treatment can be found in numerous textbooks and conference proceedings (McPherson 1993). However, planning of mine ventilation is not simple, and a lack of care and attention at each stage of the process can result in a system that fails to deliver the outcomes required. A good illustration of this is provided by Brake and Nixon (2008), who examined the requirements to correctly estimate the primary airflow requirements in underground metalliferrous mines. Brake and Nixon found that airflows were underestimated for failure to

- Provide for leakage in the auxiliary ventilation ducts,
- Provide for leakage in the workings,
- Supply essential anti-recirculation bypass flows,
- Provide for diesel equipment mobility,
- Provide for ramps and other underground fixed plant and infrastructure travelways,
- Understand the relationship between airway dimensions and minimum wind speeds,
- Recognize which is the critical airborne contaminant to be diluted,
- Plan for reasonable capacity increases and other contingencies,
- Provide for likely changes in diesel technology,
- Provide for increased mine resistance,
- Understand the impact of both increased mine resistance and leakage on airflow requirements and fan performance,
- Understand that providing extra capacity in a ventilation system at the start of the project while it is an initial additional capital expenditure is less expensive than fixing the ventilation at a later date, and
- Properly assess ventilation planning and implementation lead times.

An important part of mine ventilation planning is that it should be a continual process and that the effects of time need to be carefully assessed. A good example of this has been reported by Sedlacek (1999). In this paper, four scenarios were investigated to enable the Kidd Creek mine (Canada) to develop to greater depths with the potential to extend to 2,987 m below the surface. The situation in 1998 involved the use of several fans arranged in a series of parallel arrangements to provide the required airflow. In total, 220 intake and return level exhaust fans were in use. In addition, the airflow within the system was seriously constrained by an inadequate return airway system. Sedlacek details four possible solutions to the long-term ventilation needs for Kidd Creek; of these, the preferred option involved the mine being separated into 15 distinct ventilation districts or circuits, each controlled by a single major supply fan. The overall volume of air is controlled by strategically located single exhaust fans placed at or close to the top of the major exhaust raises. This system reduced the number of main fans from more than 200 to 18 and enabled considerable savings in operating costs.

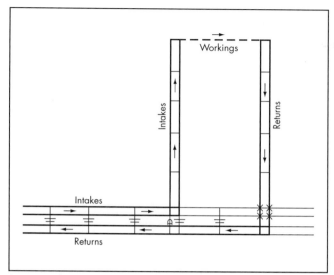

Source: McPherson 1993; with permission from Springer Science and Business Media.

Figure 15.3-16 U-tube ventilation

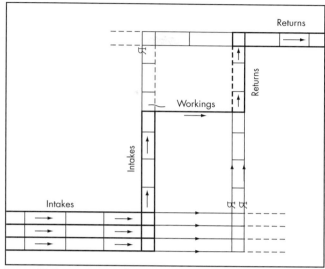

Source: McPherson 1993; with permission from Springer Science and Business Media.

Figure 15.3-17 Through-flow ventilation

A more modern approach to ventilation design is the potential to apply ventilation on demand to working stopes (Hardcastle et al. 1999). The principle can also be applied to mine cooling systems (Gundersen et al. 2005). Production stopes in mining do not require the same amount of air continuously; currently, ventilation design assumes worst-case conditions for specifying the airflow required—good from a safety perspective, but poor regarding economics. The principle of ventilation on demand is to match the air supply to the requirements of individual stopes depending on the point in the production cycle where they are currently operating. This has the potential to reduce overall mine airflows and to achieve considerable savings in power costs. The principle is still in the early stages of development. One main concern is that mine ventilation systems are highly inertial in nature and the changes induced to provide ventilation on demand for a specific stope could have a poor impact elsewhere in the system, even if only for a short, transient period of time. This could have serious safety implications.

COAL AND METALLIFEROUS VENTILATION

Before describing some of the main points of ventilating coal and metalliferous mines, a general overview of the two main forms of ventilation applied in both cases will be made. There are two broad classifications: U-tube systems (Figure 15.3-16) and through-flow systems (Figure 15.3-17).

Figure 15.3-16 shows the principle of U-tube ventilation. Air flows toward and through the workings, then returns along adjacent airways separated from the intakes by stoppings and doors. Room-and-pillar (R&P) layouts and advancing longwalls tend to be of this type. A problem with this system is the close proximity of the main intake and return airways; there can be a high pressure drop across the control devices separating these, such as doors and stoppings, which can lead to high leakage losses from the main intake to the main return before the working areas.

Figure 15.3-17 illustrates the through-flow system. In this system, intakes and returns are separated geographically, and adjacent airways are all intakes or returns. There are far fewer stoppings and air crossings, but additional regulators are required to control the airflow through the working area. Examples include the parallel flows from downcast to upcast across the multilevels of metal mines and the back-bleeder system of longwall retreating. In these systems, although leakage occurs, it is generally smaller and can be used to enhance flows as, for instance, at the return end of the working zone, which in a hot and humid mine is generally one of the main problem areas; the leakage flow increases the air velocity, enhancing the cooling power of the airflow.

Division can be made between stratified deposits (longwall or R&P systems) such as coal, gypsum, potash, and trona, and metalliferous ore-body deposits (also known as hard-rock deposits). Stratified deposits usually have a simpler two-dimensional system; ore-body deposits tend to be more complicated. This is due to their three-dimensional nature and the need to have more numerous working sections, not all of which are working at a particular time (in other words, air requirements are on a day-to-day basis, and air must be directed to where it is needed).

Stratified Deposits

Underground coal, potash, and other tabular forms of mineral deposits use either the longwall method of mining or the R&P (bord-and-pillar) mining method. Layouts can vary from country to country, however, the following sections highlight the modes of airflow distribution used. Other differences between these types of mines and hard-rock mines are the dominant pollutants that need to be accounted for. In coal mines, methane gas, dust, and in some cases diesel emissions are the dominant pollutants. In hard-rock mines, diesel pollutants, dust, and blast fumes are of concern. Because all mines will have different critical pollutants, assessment should be made on a mine-to-mine basis.

Longwall Systems

Two main features have influenced the design of longwall systems: methane and other gas control from the waste areas, and the high rate of rock breakage. Both have increased the

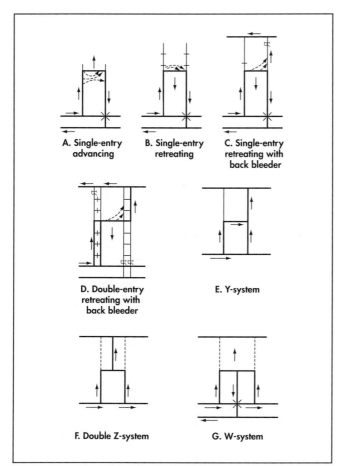

Source: Adapted from McPherson 1993; with permission from Springer Science and Business Media.

Figure 15.3-18 Longwall district ventilation systems

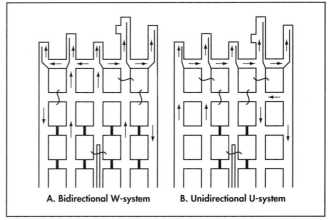

Source: McPherson 1993; with permission from Springer Science and Business Media.

Figure 15.3-19 Room-and-pillar development with line brattices (high face resistance)

production of dust, gas, heat, and humidity. A typical layout is shown in Figure 15.3-18.

With reference to Figure 15.3-18, single-entry systems are common in Europe where the deposits are mined at considerable depth and geotechnical issues would require very large pillars to be left if multiple access systems were used. With the advancing system, some leakage occurs through the waste area; control of this is achieved by roadside packing. This leakage air flushes gas out of the waste and can cause problems of high gas concentration at the return end of the face. Leakage of air through the waste can cause gob fires. In the retreating system in Figure 15.3-18B, the amount of air leaking through the waste is generally reduced, but the same problems can occur as in the advancing case.

The back-bleeder system employs a back or bleeder return to constrain the gas fringe back in the waste to prevent flushes of gas onto the face. The method illustrated in Figure 15.3-18C is a combination of U-tube and through-flow ventilation. Multiple opening systems such as those illustrated in Figure 15.3-18D are more common in the United States, South Africa, and Australia, where shallower coal seams are mined. Two or more entries are driven using R&P mining; these serve as the lateral boundaries of the longwall panel. If retreat mining is used, back bleeders can be used to control waste or gob gas. In some states, air from conveyor belt drives cannot be used to actively ventilate production areas; use of multiple entry systems enables this requirement to be achieved.

The types shown in Figures 15.3-18E through 15.3-18G are used when the gas is very heavy. An additional fresh air feed helps to maintain gas concentrations to safe levels by feeding fresh air directly to the return end of the face. These systems can also be applied in hot workings.

Room-and-Pillar Systems

In R&P (bord-and-pillar) mining, the potential for leakage is high because of the large number of interconnected entries as shown in Figure 15.3-19. The figure illustrates two methods of ventilating R&P development: a bidirectional or W-system in which intake air passes through one or more central airways with return airways on both sides, and a unidirectional or U-tube system in which intake and return airways are on opposite sides of the panel. In both cases the conveyor is sited in the middle airway and line brattices regulate the air through it.

The bidirectional system offers the advantage that the air splits at the end of the panel and each airstream only has to ventilate half the panel. In addition, the rib side gas emission is likely to be heavier in the outer airways. Its disadvantage, however, is that it requires double the number of stoppings, and hence has more leakage paths than the unidirectional system. Thus, the unidirectional system has a higher volumetric efficiency. In both cases the use of brattice cloth to control air distribution at the face end offers a high resistance. This problem can be overcome by employing auxiliary fans and ducts at the face end to force or exhaust air from the headings; the resistance of the face end effectively becomes zero. This is illustrated in Figure 15.3-20.

All the methods shown in the previous figures for longwall mining can also be applied to R&P workings. There are different strategies between R&P and longwall mining. The large number of interconnected airways and high leakage result in R&P layouts offering less resistance to airflow than longwall layouts. Thus, R&P layouts tend to require higher volume flows at lower fan pressures. Volumetric efficiencies of R&P layouts are very low.

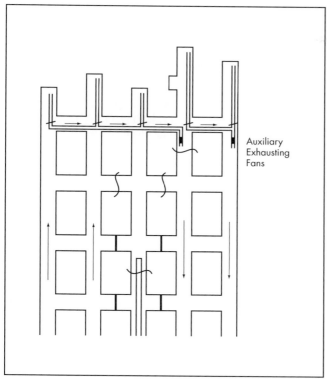

Source: McPherson 1993; with permission from Springer Science and Business Media.

Figure 15.3-20 Room-and-pillar development with auxiliary fans (zero face resistance)

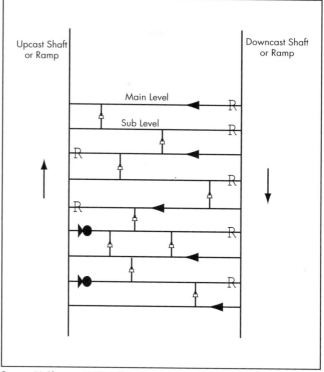

Source: McPherson 1993; with permission from Springer Science and Business Media.

Figure 15.3-21 Principle of through-flow ventilation applied to the levels of a metal mine

Ore-Body Deposits

The irregular geometry of metalliferous deposits, coupled with grade variation and fluctuating market prices, results in highly complex mine development in metal mining. Mine development can in a number of cases be related to chasing the grade, dictated by commodity prices at a particular point in time. These factors and the fact that drill-and-blast operations are cyclic in operation also necessitate many more stopes than would be required in a stratified deposit mine, some of which may not be worked on a particular shift. Consequently, metal mines are characterized by a three-dimensional aspect and require much more flexible ventilation arrangements.

Ventilation networks in metal mines are more complex than those for stratified deposits. Figure 15.3-21 illustrates the basic principles of ventilating a metal mine. Air moves in a through-flow manner from a downcast shaft or ramp, across the levels, sublevels, and stopes toward return raises, ramps, or upcast shafts. Airflow control is via fans and/or regulators (represented by R in the figure). Ascentional ventilation takes advantage of natural ventilation effects caused by temperature and humidity variations and avoids uncontrolled recirculation.

Airflow distribution in various stoping systems is highly variable. The guiding principles are as follows:

- Use ascentional ventilation.
- Through-flow ventilation should be used wherever possible; however, use of fan and duct systems is recognized as a need in certain systems.
- Series ventilation should be avoided.
- Leakage through orepasses is a problem; attempt to avoid large pressure drops across orepasses to minimize leakages.
- Wherever practical, every level and sublevel should be provided with its own through-flow of air between shafts, ramps, or raises.

REFERENCES

Atkinson, J.J. 1854. On the theory of the ventilation of mines. *N. Engl. Inst. Min. Eng.* 3:73–222.

Brake, D.J., and Nixon, T. 2008. Correctly estimating primary airflow requirements for underground metalliferous mines. In *Proceedings of the Tenth Underground Operators Conference*, Launceston, Tasmania, April 14–16. Carlton, Victoria, Australia: Australasian Institute of Mining and Metallurgy.

Calizaya, F., Casten, T., and Karmawan, K. 2000. Commissioning of a 750 kW centrifugal fan at PT Freeport Indonesia's deep ore zone mine. SME preprint 00-132. Littleton, CO: SME.

Chalmers, D.R. 2008. Sealing design. In *Proceedings of the 12th U.S./North American Mine Ventilation Symposium*, Las Vegas, NV, June 9–11.

De Souza, E., Rocque, P., and Sletmoen, L. 2003. Design and performance verification of a booster fan installation at the Red Lake mine. SME preprint 03-050. Littleton, CO: SME.

Gundersen, R.E., von Glehn, F.H., and Wilson, R.W. 2005. Improving the efficiency of mine ventilation and cooling systems through active control. In *Proceedings of the 8th International Mine Ventilation Congress*, Brisbane, July 6–8. Carlton, Victoria, Australia: Australasian Institute of Mining and Metallurgy.

Hardcastle, S.G., Gangal, M.K., Schreer, M., and Gauthier, P. 1999. Ventilation-on-demand—Quantity or quality—A pilot trial at Barrick Gold's Bousquet mine. In *Proceedings of the 8th U.S. Mine Ventilation Symposium*, Rolla, MO, June 11–17. Littleton, CO: SME.

Hemp, R. 1989. Pressure surveys. In *Environmental Engineering in South African mines*. Edited by J. Burrows. Johannesburg: Mine Ventilation Society of South Africa.

McPherson, M.J. 1984. Mine ventilation planning in the 1980s. *Int. J. Min. Eng.* 2:185–227.

McPherson, M.J. 1993. *Subsurface ventilation and environmental engineering*. London: Chapman and Hall.

Mine Ventilation Services, Inc. 2007. www.mvsengineering.com. Accessed February 2010.

MSHA (Mine Safety and Health Administration). 1973. *Breathe and Live*. www.msha.gov/streaming/wvx/Vintage/breathelive.wvx. Accessed February 2010.

Sedlacek, J. 1999. Optimization of the ventilation system at Kidd Creek mine of Falconbridge, Limited. In *Proceedings of the 8th U.S. Mine Ventilation Symposium*, Rolla, MO, June 11–17. Littleton, CO: SME.

Tuck, M.A. 2008. Ventilating deep mines—Time for a rethink of ventilation design. In *Proceedings of the Tenth Underground Operators Conference*, Launceston, Tasmania, April 14–16. Carlton, Victoria, Australia: Australasian Institute of Mining and Metallurgy.

Ventsim. 2010. www.ventsim.com. Accessed February 2010.

CHAPTER 15.4

Gas and Dust Control

Pramod Thakur

INTRODUCTION

The provision of an adequate air environment to promote the health, safety, and comfort of mine workers has always been and will continue to be a prime requisite for successful mining operations. Although the definition of an adequate environment varies from country to country, it generally means the provision of sufficient circulating air, often at specified velocities, to maintain at least 19.5% oxygen in the working areas; concentration of solids (dust) and gaseous pollutants such as methane, carbon dioxide, and respirable dusts below specified limits; and temperature and humidity below specified limits.

This chapter will deal only with the techniques used to control solid and gaseous contaminants in the mine air. Liquid contaminants, such as mists, are not an issue in the mining industry. Table 15.4-1 shows the maximum allowable concentrations of contaminants normally encountered in mine air (CFR 2003).

Methane gas is a big concern primarily in coal mines, but some salt and potash mines also have excessive methane emissions. Special techniques have been developed over the past 50 years to keep the mines safe. Respirable dust is a problem common to all mining and processing operations. In the past, diesel particulate matter was a health concern only in metal and nonmetal mines, but it is becoming a health concern in coal mines as the population of diesel equipment in coal mines is increasing. Diesel haulage is likely to replace all trolley wire haulage, leading to improved safety in mines.

Traditionally, all solid and gaseous pollutants in mines are rendered safe by dilution with adequate ventilation air. In deeper mines, methane emission is so high that it is drained prior to mining and postmining. Respirable dusts also need to be controlled before they become airborne. Diesel exhaust is often oxidized to remove carbon monoxide and unburned hydrocarbons, and the diesel particulate matter is collected on specially designed filters to keep its concentration in air below the threshold limit value.

GAS CONTROL STRATEGY

All gas emissions in mine airways must be controlled to keep their concentrations below the statutory limits. The control strategy depends on the type of contaminants, their sources, and rates of emission. It can range from simple dilution with ventilation air to complex procedures for removal of the contaminant prior to mixing with the mine air or suppression/elimination at the source. An overall strategy comprises the following steps (Hartman et al. 1997):

1. Prevention (e.g., proper blasting, internal combustion engine maintenance)
2. Extraction (e.g., methane from coal mines)
3. Chemical absorption—use of catalytic converter in diesel engines
4. Isolation—sealing of areas in coal mines to contain methane emissions
5. Ventilation—including local dilution at the source with auxiliary ventilation and disposal via the mainstream ventilation

Four cases will be discussed in greater detail to illustrate these strategies:

1. Methane control in coal mines
2. Respirable dust control
3. Diesel exhaust control
4. Radon gas control

Methane Control in Coal Mines

Coal seams and coal-bed methane are syngenetic in origin because they originate from the same plant material. All coal seams are gassy, but they vary in their degree of gassiness (gas contained per ton of coal). Typical gas contents of coal seams vary from negligible to 25 m^3/t (see Appendix B, Coal Mine Gas Chart). Methane and sometimes carbon dioxide are the major components of gas in coal comprising 90%–95% by volume. The remaining gas is primarily nitrogen, helium, argon, and hydrogen. Methane is released into coal mines from coal seams as mining proceeds. It is colorless, odorless, and combustible, and it forms an explosive mixture with mine air in the concentration range of 4.5%–15% by volume. The maximum permitted concentration of methane in mine air is restricted by law to between 1% and 2% in all major

---

Pramod Thakur, Manager, Coal Seam Degasification, CONSOL Energy, Inc., Morgantown, West Virginia, USA

Table 15.4-1 Permissible exposure limits for mine air contaminants

Contaminant	8-Hour Time-Weighted Limit	Ceiling Limit
Methane (fresh air)	—	1%–1.25%
Methane (return air)	—	2%
Carbon dioxide	—	0.5 %
Carbon monoxide	50 ppm	
Nitric oxide	25 ppm	
Nitrogen dioxide	—	5 ppm
Sulfur dioxide	5 ppm	
Hydrogen sulfide	10 ppm	
Radon (uranium mines)	1.0 working level	
Respirable coal dust	2 mg/m^3	
Respirable silica dust	0.1 mg/m^3	
Diesel particulate matter (coal mines)*	120 mg/m^3	
Diesel particulate matter (metal mines)	160 mg/m^3	
Asbestos	5 fibers/cm^3	

*Applicable only to Pennsylvania and West Virginia in the United States; it is the most stringent standard in the world.

Table 15.4-2 Recent methane explosions in U.S. coal mines

Year	Mine	Deaths
2006	Sago mine, Tallmansville, West Virginia	12
2001	Blue Creek No. 5 mine, Brookwood, Alabama	13
1992	No. 3 mine, Norton, Virginia	8
1989	William Station No. 9 mine, Wheatcroft, Kentucky	10
1983	McClure No. 1 mine, McClure, Virginia	7
1982	No. 1 mine, Craynor, Kentucky	7
1981	No. 21 mine, Whitwell, Tennessee	13
1981	No. 11 mine, Kite, Kentucky	8
1981	Dutch Creek No. 1 mine, Redstone, Colorado	15
1980	Ferrel No. 17 mine, Uneeda, West Virginia	5
1976	Scotia mine, Oven Fork, Kentucky	26
1972	Itmann No. 3 mine, Itmann, West Virginia	5
1970	Nos. 15 and 16 mines, Hyden, Kentucky	38

Table 15.4-3 Gassiness of coal seams

Category	Depth, m	Gas Content, m^3/t
Mildly gassy	≤200	<3
Moderately gassy	200–500	3–10
Very gassy	>500	10–25

coal-producing countries, but methane–air explosions are quite common even today. Table 15.4-2 shows a list of mine explosions in the United States from 1970 to 2006. In these explosions, lives were lost in spite of methane drainage in some of the mines. When methane drainage was not practiced and ventilation was the only method of methane control, mine explosions were much more disastrous with very high numbers of fatalities. To minimize the risk of methane explosion, mine ventilation must be supplemented by methane drainage from coal seams prior to mining and postmining when the gas content of coal exceeds 3 m^3/t.

The amount of gas contained in a coal seam depends on a number of factors, such as the rank of coal, the depth of burial, immediate roof and floor, geologic anomalies, tectonic stresses, and temperatures during the coalification process. For methane control, all coal seams can be divided into three categories: mildly gassy, moderately gassy, and very gassy, as shown in Table 15.4-3.

For mildly gassy and moderately gassy mines, ventilation is the primary technique for diluting and disposing of methane emissions. A large mine may have several fans circulating 50,000–100,000 m^3/min of air in the mine airways to keep methane concentration below statutory limits. Gassiness of a mine is also expressed as total methane emissions divided by the total daily coal production, and the ratio is called specific emissions. It is generally economically feasible to handle specific emissions as high as 30 m^3/t with a well-designed ventilation system, but with increasing methane emissions, a stage is reached when it becomes impossible to stay within statutory limits with mine ventilation alone. Methane drainage then becomes necessary. With a well-planned methane drainage scheme and a good ventilation system, even very gassy mines with specific emissions in excess of 100 m^3/t can be safely operated.

In gassy mines it is best if most of the methane is recovered by carrying the gas in pipelines without mixing it with mine air.

Properly designed and carried out, methane drainage has the following benefits:

- Enhanced mine safety
- Improved coal productivity leading to a lower cost of production
- Reduced number of entries in development sections, which leads to faster development and an improved longwall/development coal ratio
- Reduced number of ventilation shafts, size of fans, and power cost
- Controlled respirable dust due to boreholes drilled in coal seams providing a channel for advance exploration and water infusion
- Defrayed costs through marketing of produced gas and boosting energy production from coal mines

The Gas Emission Reservoir and Its Properties

Methane emissions during the mining process take place in two stages: during development and during actual longwall mining. In the development phase, all methane emissions are realized from the coal seam being mined. Methane emissions on the longwall face are also realized only from the coal seam being mined. In order to achieve a high rate of advance in longwall gate roads as well as the mains, the working coal seam must be properly degassed prior to mining. The most important reservoir properties that must be determined for designing an optimum degasification scheme are

- Gas content and gas isotherm of the coal seam;
- Water content of the coal seam and its chemical composition;
- Reservoir pressure;
- Directional permeability of the coal seam;
- Magnitude and directions of principal horizontal stresses; and
- Compressive strengths and elastic moduli of the coal seam, the roof, and the floor.

For most efficient mining, the gas content of very gassy coal seams must be reduced to <3 m^3/t and reservoir pressures should be reduced to <0.5 MPa. The amount of gas retained

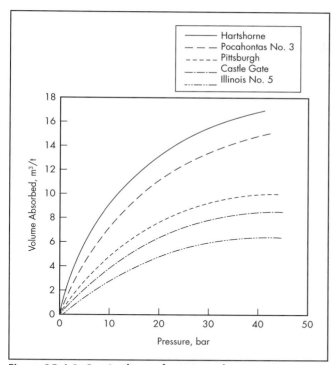

Figure 15.4-1 Gas isotherms for U.S. coal seams

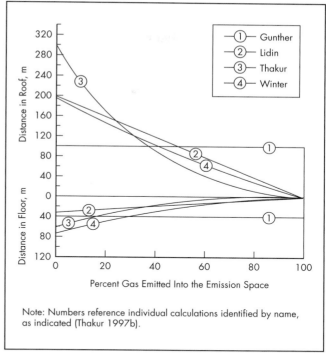

Figure 15.4-2 Vertical limits of the gas emission space

by coal seams at any given pressure can be determined from their gas isotherms. Typical gas isotherms for some U.S. coals are shown in Figure 15.4-1.

In the postmining phase, the gas emission reservoir is considerably expanded. Figure 15.4-2 shows the vertical extent of the gas emission space created by longwall mining and the percentage of gas contents released by various coal seams into the gas emissions space as a function of their distance from the mined coal seam. The vertical dimension of the gas emission space is highly dependent on the width of the longwall face. In general, the wider the longwall panel, the greater the vertical extent of the gas emission space and, consequently, the higher the specific gob gas emissions (volume of gas emitted per unit area of the gob). For optimum methane capture, the specific gob gas emissions for the longwall panel must be calculated from Figure 15.4-2 and subsequently verified by direct measurement of postmining methane emissions.

A very gassy coal seam is usually underlain and overlain by equally gassy, but generally nonmineable, coal seams. Longwall mining fractures them all, and their gas contents are released into the mine. Specific gob gas emissions can be as high as 2 million m^3/ha of the gob area, but it depends on the following conditions:

- Total thickness, gas contents, and proximity of other coal seams to the mined seam in the gas emission space
- Width of the longwall panel
- Presence of geological disturbances (e.g., lineaments, faults)
- Thickness and properties of noncoal strata above and below the mineable coal seam

Coal Seam Degasification

Coal seam degasification can be broadly classified into two phases: premining degasification and postmining degasification. Some of the premining degasification techniques can also be used for commercial coal-bed methane production (see Chapter 11.6). Premining degasification is usually done 3–5 years ahead of mining, but for commercial coal-bed methane production the coal seams should be drilled at least 10 years ahead of mining. This maximizes the return on investments.

Premining degasification. At present there are three techniques commonly used for premining degasification: in-mine horizontal drilling, vertical hydrofractured (hydrofraced) wells, and horizontal boreholes drilled from the surface. Depending on the gas content and depth of the coal seams, one or a combination of these techniques is used to reduce the gas content and reservoir pressure to <3 m^3/t and 0.5 MPa, respectively. Vertical hydrofraced wells and horizontal boreholes drilled from the surface can also be used for commercial coal-bed methane production.

In-mine horizontal drilling. Figure 15.4-3 (plan view of a longwall face) shows a commonly used pattern of drilling in a moderately gassy coal seam (with a gas content >3 m^3/t). The drilling rig follows the continuous miner driving the development section and drills ahead of the mining face to degas the section. Next, a cross-panel borehole is drilled in the longwall panel from the same site to degas it. This pattern is repeated approximately every 300 m. In a permeable coal seam (permeability ≥10 millidarcies), nearly 50% of in-situ gas can be removed prior to mining. A detailed description of this technique is available in published literature (Thakur and Davis 1977; Thakur and Zachwieja 2001). This technique of degasification is equally suited to the room-and-pillar type of mining.

Vertical hydrofraced well. In very gassy, deep coal seams (gas content > 10 m^3/t) it becomes necessary to first drill with vertical hydrofraced wells in order to reduce the reservoir pressure and partially degas the coal seam. Secondary

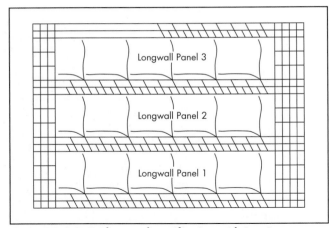

Figure 15.4-3 Coal seam degasification with in-mine horizontal drilling

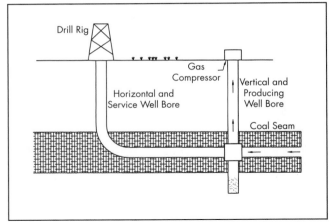

Figure 15.4-5 Horizontal boreholes drilled from the surface

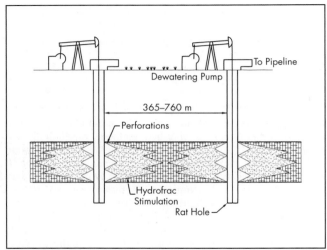

Source: Thakur 2005.
Figure 15.4-4 Vertical hydrofraced wells

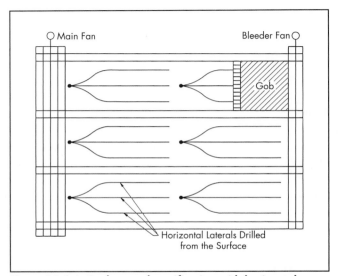

Figure 15.4-6 Coal seam degasification with horizontal boreholes drilled from the surface

degasification is performed by horizontal drilling, either from the surface or underground mine workings. Figure 15.4-4 shows a vertical section of hydrofraced wells. If drilling is done 3–5 years in advance, nearly 50% of in-situ gas can be drained by these wells prior to mining. A 300 × 3,000 m longwall panel in a very gassy mine may need 10–12 vertical hydrofraced wells to achieve this goal. When the reservoir pressure is reduced to <1.5 MPa, the longwall panel is drilled horizontally to remove an additional 25%–30% of in-situ gas content of the coal seam. Well-designed hydrofraced wells can create a vertical fracture in a coal seam that has a length of 300–600 m. Hydrofracing is typically done with nitrogen foam, but plain water, gelled water, and cross-linked gels have also been used depending on the specifics of the coal seam. This technique can be used for both coal seam degasification and commercial gas production.

Horizontal boreholes drilled from the surface. Figures 15.4-5 and 15.4-6 show the common design and application of horizontal boreholes drilled from the surface for coal seam degasification. For successful operation, water produced by the horizontal laterals must be removed. A production well with a sump (drilled below the coal seam to be degassed) is first drilled. Nearly 100 m away, a vertical well is drilled, and when the borehole approaches the target coal seam, it is deviated by 90° to intersect the coal seam and the production well horizontally. The borehole can be laterally extended to a distance between 1,000 and 2,000 m depending on the depth of the coal seam. A water pump is installed in the production well to remove water as well as coal fines, and gas production is maintained. Very expensive instruments are needed to articulate a horizontal borehole to intersect the production well and keep it in the coal seam for most efficient methane drainage. The cost of drilling horizontal boreholes from the surface is, therefore, much higher than the cost of drilling in-mine horizontal boreholes. However, this is the best technique for commercial coal-bed methane production from all coal seams, and its application is increasing (Thakur et al. 2002b).

Postmining degasification. Longwall mining creates a much larger gas emission space than the mined coal seam. Besides the specific gob gas emission of the longwall, the other factor that has the greatest influence on the postmining methane emission rate is the rate of longwall face advance.

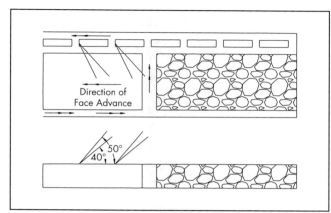

Figure 15.4-7 Methane drainage with cross-measure boreholes

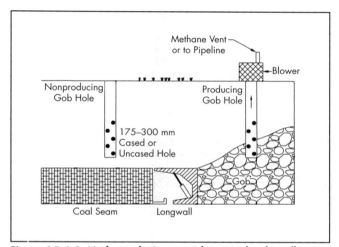

Figure 15.4-8 Methane drainage with vertical gob wells

Methane emission from the gob is linearly proportional to the rate of face advance in hectares per day. Approximate gob emissions in million cubic meters per hectare can be calculated from these two data points. At present there are only two viable techniques for postmining methane drainage: cross-measure boreholes and vertical gob wells.

Cross-measure boreholes. Figure 15.4-7 shows a typical layout of methane drainage from retreating longwall gobs using cross-measure boreholes. Inclined boreholes, 50–100 mm in diameter, are drilled from longwall gate roads into the roof (and sometimes into the floor). These boreholes are inclined at 40°–50° to the horizontal plane and are also slanted toward the gob; they are drilled roughly at 30-m intervals. All boreholes are manifolded to a 100–200-mm diameter pipeline and gas is conducted to the surface. Gas production is often assisted by applying negative pressure of 100–1,000 mm of water gauge. This system of methane drainage is best suited to slow-moving longwall faces in mildly gassy mines. It is popular in European mines but generally unsuitable for gassy and highly productive U.S. and Australian longwall faces.

Vertical gob wells. Figure 15.4-8 shows upward gas movement in the well over the gob. Steel casings of 175–300 mm diameter are anchored in competent rock strata nearly 100–300 m above the working seam, and the bottom 100–300 m is slotted to permit gas entry. Gas production is generally assisted with powerful blowers.

In U.S. coal mines, longwall panels in very gassy coal mines can be 240–400 m wide and can advance 15–30 m/d. The specific gob gas emission in the central and southern Appalachian basin of the United States can be 1–2 million m^3/ha. Hence, the gob gas emission rate can range from 0.5–1.2 million m^3/d. The bleeders with a limited number of entries and ventilation air (8,000–10,000 m^3/min) can capture only 0.1–0.2 million m^3/d of methane. The remaining gob emissions must be captured by vertical gob wells resulting in methane capture ratios of 65%–85%. Completion techniques for these gob wells are described elsewhere (Davis and Krickovic 1973; Mazza and Mlinar 1977). Although the spacing of these gob wells on a longwall panel can be calculated if good data are available, the best method is to go by direct experience in the local basin and the mine. Generally, a spacing of 3–6 ha per well should yield 70%–80% of methane capture if assisted with well-designed blowers.

This method of postmining degasification has also been successfully used in retreat mining of room-and-pillar sections.

Respirable Dust Control

When mineral (metal and nonmetal) and sedimentary deposits such as coal are mined, processed, or transported, fine dust is produced by the crushing of the original material. Dust has been defined as solid particles <100 μm in size that can be disseminated and carried by air in suspension. Dust clouds in mine airways constitute two well-known hazards, namely, dust explosions and dust-related lung diseases such as silicosis and coal workers' pneumoconiosis. Table 15.4-4 shows a list of different kinds of dust encountered in mines. They are classified as pathogenic or explosive dusts but in decreasing order of severity (Hartman et al. 1997). Some rate multiple listings.

Dust concentrations in air to sustain an explosion ranges from 20–500 g/m^3. Such high concentrations cannot be created or sustained under normal working conditions. Most dust explosions are preceded by a gas explosion that can kick up the deposited dust, creating a dangerous situation. Dust explosions have been made almost nonexistent by the process of

Table 15.4-4 Classification of dusts

Classification	Minerals
Fibrogenic dusts (harmful to respiratory system)	Silica (quartz, chert), silicates (asbestos, talc, mica), metal fumes (nearly all), beryllium ore, tin ore, iron ores (some), carborundum, coal (anthracite, bituminous)
Carcinogenic dusts	Radon daughters, asbestos, arsenic
Toxic dusts (poisonous to body organs, tissue, etc.)	Ores of beryllium, arsenic, lead, uranium, radium, thorium, chromium, vanadium, mercury, cadmium, antimony, selenium, manganese, tungsten, nickel, silver (principally the oxides and carbonates)
Radioactive dusts (injurious because of alpha and beta radiation)	Uranium ore, radium ore, thorium ore
Explosive dusts (combustible when airborne)	Metallic dusts (magnesium, aluminum, zinc, tin, iron), coal (bituminous, lignite), sulfide ores, organic dusts
Nuisance dusts (little adverse effect on humans)	Gypsum, kaolin, limestone
Inert dusts (no harmful effect or cellular response in lung)	None

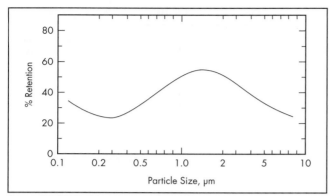

Source: ILO 1965, © International Labour Office.
Figure 15.4-9 Alveolar retention of particles of unit density

mixing inert dust such as gypsum or limestone dust with combustible dust in very high proportions. For example, mixing limestone dust with deposited coal dust so that the incombustible content is at least 60% will render the mixed dust nonexplosive.

Dust particles finer than 10 μm (unit density spheres) do not generally settle in the turbulent airflow in the mine airway. This size range is called respirable dust. The human respiratory system consists of a series of branching passages decreasing in size but increasing in number. Thus, starting with inhalation either through the nose or mouth, the air passes in succession through the trachea, bronchioles, alveolar ducts, and finally into alveolar sacs where gaseous exchange of oxygen takes place. The lungs act as both a filter and a scavenger for dust particles.

When dust-laden air is inhaled, only a small fraction of it is deposited in the lungs. The mechanisms of deposition are impaction, sedimentation, and diffusion, much the same as in the case of a simple paper filter. Figure 15.4-9 shows a typical deposition and retention curve for the portion of the lungs below the bronchioles, popularly known as the pulmonary deposition curve (ILO 1965). Thus particles >10 μm are deposited by impaction in the nose and the top of the trachea. Particles <10 μm are deposited by sedimentation in the upper respiratory region with progressively increasing efficiency as the size decreases. A maximum is reached around 2 μm, below which there is a decline. A minimum is reached around 0.3 μm, beyond which there is further increase in depositional efficiency due to diffusion forces.

The scavenging action of the lungs is equally important. Practically all dust particles deposited in the upper respiratory tract are trapped by cilia (hair-like structures lining the trachea) and transported to the mouth where they are spat or sneezed out or swallowed. Since the digestive system is sturdier than the lungs, the swallowed particles seldom do any harm. Dust particles deposited in the alveolar ducts and sacs are ingested by migratory cells, the phagocytes, and transported to lymph nodes with varying degrees of efficiency.

With continued exposure to high concentrations of dust, the lungs' parenchyma accumulates a critical mass and body tissues begin to react to the dust particles causing lung diseases. Depending on the type of dust inhaled, the lung disease is commonly called silicosis (for silica), asbestosis (for asbestos), silicatosis (for silicates), siderosis (for iron ore), and coal workers' pneumoconiosis (for coal dust, also known as black lung). Almost all industrialized nations have set up

Table 15.4-5 Respirable dust permissible exposure limits

Dust Type	PEL	Comments
Crystalline silica	0.1 mg/m^3	U.S. standard; other standards are less stringent
Silicates		
Asbestos	5 fibers/cc	>5 μm
Mica	706/million m^3	
Cement	1,056/million m^3	
Talc	706/million m^3	
Coal dust	2 mg/m^3 or 10/% crystalline silica	When silica <5% When silica >5%; U.S. standard is the most stringent
Metallic Dusts		
Mercury	0.05 mg/m^3	
Lead	0.15 mg/m^3	
Antimony, arsenic	0.5 mg/m^3	
Manganese	5 mg/m^3	
Iron, zinc molybdenum	5 mg/m^3	
Uranium (insoluble)	0.2 mg/m^3	
Vanadium (V$_2$O$_5$)	0.5 mg/m^3	

Source: Adapted from ACGIH 1972.

a permissible exposure limit (PEL) for the exposure to various dusts. The PELs for various respirable dusts are shown in Table 15.4-5. They are adapted from the general guidelines provided by the American Conference of Governmental Industrial Hygienists (ACGIH 1972).

These PELs were developed and implemented during the past 40 years. In general, they are more stringent than the old standards, and the impact on workers' health has been favorable. A fine example is the 70% decline of coal workers' pneumoconiosis (CWP) in U.S. coal mines during the past 40 years as shown in Figure 15.4-10.

In order to minimize, if not totally eliminate, the incidence and prevalence of any dust-related lung diseases, a holistic approach comprising the following requirements are necessary (the last two requirements will not be discussed in this chapter) (Thakur 1997a):

1. Better personal exposure monitoring (daily and cumulative)
2. Better mining and dust control technologies
3. Recognition of disease-causing characteristics of each dust, particularly epidemiological data on dust exposure (dosages) and related occurrence and progression of lung diseases
4. Lifestyle intervention program including cessation of smoking and medical surveillance

Personal Exposure Measurement
Earlier dust sampling instruments such as the impinger and konimeter were nonsize selective and simply measured the number concentration of dust in the air. The concentration of respirable dust was expressed as the number of dust particles per unit volume of air. Other instruments based on the same principle of particle counting were introduced later, namely, the membrane filter and the thermal precipitator. All of them took a long time to determine the dust concentration, in addition to which they were labor and laboratory time-intensive.

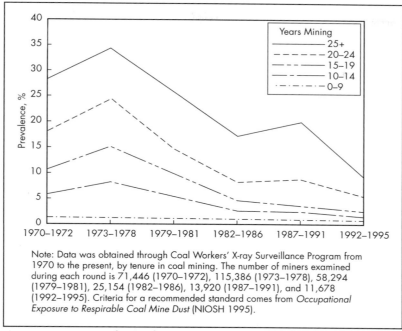

Figure 15.4-10 Prevalence of CWP Category 1 or higher

A simple but real-time dust measuring device was needed that could be used for evaluating dust conditions underground. Several dust concentration measuring instruments were developed in the 1970s that were based on light-scattering measurements. The tyndalloscope was the first instrument designed on this principle, but it did not perform very well compared to other instruments. It was very sensitive to the size distribution of dust particles and needed frequent calibration. Subsequently three instruments—the Tyndallometer, Simslin, and GCA Ram—were designed to measure scattered laser light in Germany, the United Kingdom, and the United States, respectively. They were also sensitive to the size distribution of dust particles. But their biggest drawback was their inability to distinguish between dust and water particles. Water sprays used to suppress dust significantly affected the dust concentration readings (Thakur et al. 1981). During the same time period, the dust measurement technology was required to be size selective. The new criteria for sampling and measurement of dust concentration were as follows:

- The sampling device should be size selective and it should mimic the human nose.
- The dust concentration should measure the "mass" of dust per unit volume of air.

Hartman et al. (1997) compared size-selective curves for respirable dust from the British Medical Research Council (BMRC)/Johannesburg and the U.S. Atomic Energy Commission. The BMRC/Johannesburg had 0% collection for 10-μm particles, 50% collection for 3.5-μm particles, and 100% collection for particles <2 μm. This was achieved by classifying the respirable dust with a horizontal elutriator. The U.S. Atomic Energy Commission developed a size-selection device using a cyclone that had 0% collection at 7.1 μm, 50% collection at 3 μm, and 100% collection at 2 μm or smaller particles.

The second criterion required that dust samplers should measure the mass of dust per unit volume in air. This conclusion is based on the findings of the National Coal Board's (United Kingdom) pneumoconiosis field research study (Jacobsen et al. 1971). Dust exposures of a large cohort of coal miners were measured and correlated with changes in their lung radiographs. Figure 15.4-11 shows a quantitative relationship between the coal mine dust exposure and the risk of developing simple CWP. The latter is characterized by the presence of small nodules in the lung and classified as Category 1, 2, or 3 according to the profusion of the nodules. Simple CWP Categories 1 and 2 can be reversed by moving the mine worker to an area with much less or no dust exposure. In Category 3, pulmonary massive fibrosis sets in, and it is generally irreversible.

MRE gravimetric dust sampler. The MRE gravimetric dust sampler was developed by the Mining Research Establishment (MRE) of the National Coal Board. It is the standard dust sampling instrument for most industrialized nations, including the United States, Canada, the United Kingdom, and Australia. It is based on the BMRC/Johannesburg size-selection curve and uses a horizontal elutriator for this purpose. The classified dust particles are collected on a membrane filter. The preweighted filter is weighed again after sample collection, and the differences in weight divided by the volume of air sampled yields the gravimetric dust concentration. The only drawback of this instrument is its size and weight. It measures the mass concentration of respirable dust but cannot be conveniently carried by a mine worker.

The U.S. personal gravimetric sampler. This sampler follows the U.S. Atomic Energy Commission size-selection curve and uses a nylon/metal cyclone to classify dust particles. It consists of two basic components: a cyclone assembly with a preweighed filter cassette, and a rechargeable battery-operated diaphragm pump with calibrated flow rates. The

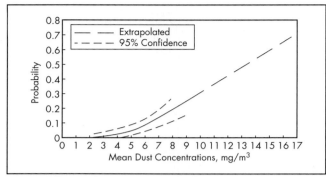

Source: Jacobsen et al. 1971.

Figure 15.4-11 Probability that a worker with no pneumoconiosis will be classified as Category 2 or higher after 35 years of exposure to various concentrations of coal mine dust

pump is worn on a miner's belt but the cyclone-filter assembly is clipped onto the miner's collar to be as close to his or her nose as possible. Thus, the sampler collects the dust particles that the wearer is likely to breathe and gives a better measure of mine worker's exposure.

The two instruments have excellent correlation. The personal sampler operating at a flow rate of 2 L/min yields a respirable dust concentration that is multiplied by a factor of 1.38 to make it compatible to the MRE sampler measurement. The 2.00 mg/m^3 dust standard for U.S. coal mines is based on the MRE sampler reading. For noncoal mines, the personal sampler draws air at 1.7 L/min and the measured dust concentration is the dust standard; that is, no correction is applied (CFR 2003).

Instantaneous dust monitors. Although the MRE sampler and the personal sampler are required instruments for compliance with dust standards, they are not able to give instantaneous dust readings that can be used for immediate engineering control. Two instantaneous reading instruments are currently in use in the mining industry, the MINIRAM and the TEOMB.

MINIRAM (miniature real-time aerosol monitor). The measuring principle is based on the detection of scattered electromagnetic radiation in the near-infrared region. The MINIRAM senses the radiation scattered by respirable dust particles by means of silicon-photovoltaic hybrid detectors, within a 1-m^3 volume of air. Air passes freely through the open sensing chamber induced by personal movement, ventilation, convection, or a combination of all three. No pump is required. The instrument measures particle diameter in the range of 0.1–10 μm and is independent of air velocity. It is controlled by a single-chip complementary metal-oxide semiconductor microprocessor that provides continuous real-time reading of airborne dust concentration. The instrument is light and small in size and can be easily used anywhere in the mine.

TEOMB (tapered-element oscillating microbalance). As the name implies, the instrument is a real-time mass monitor of respirable dust based on the principle of a tapered-element oscillating microbalance. This instrument consists of a replaceable filter cartridge mounted on the narrow end of a hollow, tapered tube (Figure 15.4-12). The wider end of the tube is fixed. Air is drawn through the filter and dust particles are deposited on it. The tapered tube oscillates but the frequency

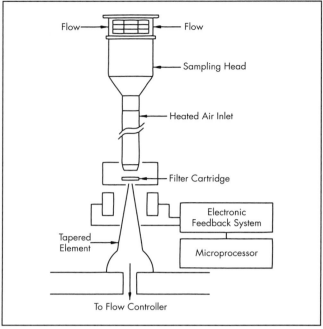

Source: Schmitz 1994, courtesy of the IEA Clean Coal Centre.

Figure 15.4-12 TEOMB ambient particulate monitor

of oscillation (initially 252 Hz) depends on the mass of the tapered element, including the filter. The frequency changes as the dust is deposited on the filter, increasing the total mass of the tapered element. The change in frequency translates into dust concentration. A microprocessor collects data on air volume sampled and corresponding dust mass collected, does the calculation, and displays dust concentration in real time in milligrams per cubic meter.

The instrument is sensitive to mechanical vibrations and should not be mounted on mining machines. It is best carried on a person. It is also sensitive to moisture. The intake air stream is maintained at 50°C to minimize this problem.

The TEOMB can operate with flow rates from 0.5 to 5 L/min. The data can be displayed instantaneously, hourly, or at the end of an 8-hour shift. It is highly recommended for engineering control of respirable dust in mines.

Engineering Control of Dust

Because the respirable dust in suspension behaves like a gas, the engineering control of respirable dust, in principle, is similar to that for gases discussed earlier. The sequence of control techniques (listed in descending order of importance and effectiveness) are

1. Prevention of generation of dust and its suspension in air,
2. Suppression of airborne dust on-site,
3. Collection of dust that could not be suppressed, and
4. Dilution with auxiliary and main ventilation.

Use of protective equipment/administrative control in special situations is the last resort, but it is not regarded as a dust-control technique. The following steps illustrate this strategy for dust control in underground coal mining:

1. Coal seams are infused with water and cut with very sharp bits to minimize dust generations.

2. Water mixed with surfactants is sprayed at high pressure to suppress dust on-site.
3. In many cutting and drilling operations, the airborne dust is collected and scrubbed before discharge.
4. Adequate ventilation is provided to dilute the remaining dust and keep the dust concentration below statutory limits.
5. For persons working in return airways, well-designed personal equipment, such as air helmets, is used to protect the workers.

Engineering control of dust is divided into the following types of mining and processing operations:

- Coal mining—continuous mining sections
- Coal mining—longwall sections
- Coal mining—surface
- Metal mining
- Mineral in transport
- Mineral/coal processing

Coal mining—continuous mining sections. All four techniques to control respirable dust have been used in continuous mining sections. Infusing the coal seam with water or water-soluble gels (e.g., sodium silicate and sodium bicarbonate) has been done on a very limited basis because of high cost. Using large and sharp cutting bits also reduces dust generation.

Most common dust control techniques comprise use of water sprays to suppress dust, use of water scrubbers in some cases, and use of ventilation air to dilute the respirable dust concentration below statutory limits.

Water spray systems. Water sprays installed on continuous miners deliver 120–240 L/min water at 0.5–1 MPa pressure at the nozzle. Besides suppressing dust, the sprays also wet the coal, keep the cutting bits cool, and serve as an emergency fire suppression system. Spray nozzles are strategically installed on the continuous miner to achieve these goals. A well-designed spray system can reduce respirable dust concentration by 60%; typical reductions range from 30% to 50%. Ideally, there should be a spray behind each cutting bit delivering 2–4 L/min at 0.3–0.8 MPa. Such machines are known as wet-head continuous miners and they achieve the best dust control.

Water scrubbers. Where seam height permits, the flooded-bed-type scrubber is becoming increasingly popular for use on continuous miners. It is mainly used with blowing-type face ventilation. The dirty air is drawn into a water scrubber by fans and collects the dust. The airflow in the scrubber must be properly matched with the airflow in the heading. A well-designed system with remote operation can cut down respirable dust exposure by 80% (Goodman and Listak 1999). Often, a demister is used to remove water from the scrubber discharge.

Ventilation air. The quantity of ventilation air at the face is perhaps the most important factor in controlling dust exposures. Average air velocity in the face area ranges from 20 to 150 m/min. Air is brought to the face using either line brattice or auxiliary fans and tubing. For best results the end of the line brattice or ventilation tubing should be kept within 4.5 m from the face. When face ventilation is done with auxiliary fans and tubing, it can be either a blowing or exhausting ventilation. Sometimes a combination of both blowing and exhaust ventilation is used for optimum results.

Table 15.4-6 Dust source contribution on longwall faces

Source	Average %	Concentration (median value)
Intake	9	0.33 mg/m^3
Stage loader–crusher	15	0.78 mg/m^3
Shields	23	1.80 mg/m^3
Shearer	53	3.50 mg/m^3

Coal mining—longwall sections. The sources of dust on a longwall face are the intake air, the stage loader–crusher, the shearer, and the shields.

The intake air is usually a minor source of dust unless the belt air is used on the face. Confinement and well-designed water sprays usually at <0.5 MPa pressure can generally keep the contribution of dust from the stage loader–crusher at a minimum. The dust created by shield movement is also contained by mounting water sprays on the shields. The biggest source of dust on longwall faces is the shearer. Table 15.4-6 gives a summary of data obtained from thirteen longwall faces (Kissell 2003).

Water infusion. Excellent dust control can be achieved by drilling the longwall for degasification, as discussed earlier, and using the horizontal boreholes for water infusion. Slow infusion at pressures of 0.5–2 MPa has given excellent results, often reducing dust generation by 80% (Nagy et al. 1957). Larger and sharper cutting bits also reduce dust generation. Use of very high pressure water (30–70 MPa) jets to assist coal cutting has also significantly reduced dust generation.

Water sprays. Well-designed water sprays are the next best way to control dust on longwall faces. A minimum of 400 L/min at 0.7 MPa is recommended (Colinet et al. 1997). Ideally, there should be a spray behind each cutting bit but, in general, the larger the number of sprays, the better the dust control. The water sprays, on and around a longwall shearer, can be so arranged that they direct the dust cloud away from the operators and hold it against the face (Figure 15.4-13). This arrangement of sprays is called a shearer–clearer spray system and it reduces shearer-generated dust by 30%–50% (Jayaraman et al. 1985).

Coal mining—surface. Respirable dust in surface mining is created by a number of operations such as drilling, mobile excavation equipment, and haul roads. The situation can become serious under dry and windy conditions.

Drill rigs. Two well-established practices for dust control on drill rigs are dry collection and wet suppression. In a dry collection system, an enclosure is created around the well bore being drilled, which is hooked up to a fan and a filtration system. The negative pressure created by the fan helps contain the dust within the enclosure. The filtration system collects most of the dust and clean air is exhausted to the atmosphere. In the wet collection system, water (often treated with some surfactants) is mixed with the compressed air used to bring drill cuttings out of the hole. Typically less than 4 L/min of water is used for this purpose. The collection efficiency appears to increase from a low rate of 0.8 L/min to a limit of approximately 4 L/min (Zimmer et al. 1987).

Mobile equipment. The operators of mobile equipment such as bulldozers, front-end loaders, and haulage trucks can be best protected by tightly enclosed cabs. In extremely dusty

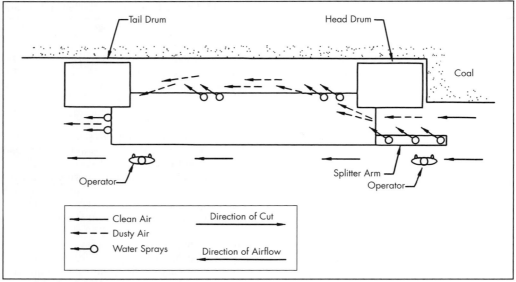

Source: Adapted from Kissell 1992.
Figure 15.4-13 Air currents when using the shearer–clearer system

situations, filtered air can be pumped into the driver's cab and a slightly positive pressure maintained within the cab.

Haul roads. Wetting the haul roads frequently with water is the most commonly used procedure. To improve the wetting capacity of plain water, hygroscopic salts such as calcium chloride, sodium silicates, or surfactants are often added. The other approach is to create a film of impervious material that can contain the dust. Most commonly used materials are portland cement and bitumen derived from coal or polymers. The best dust control method really depends on the type of road aggregate involved.

Metal mining. Main sources of respirable dust are drilling rigs; blasting; ore- and waste passes; and conveyor belts, transfer points, and roadheaders, if used. Fundamental principles of dust control in metal mines are similar to those for coal mining (i.e., prevention of generation, suppression on-site, collection of airborne dust, and dilution with ventilation).

Drilling rigs. Wet drilling is far superior to dry drilling and can reduce dust concentration by 95% (MSA Research Corp. 1974).

Blasting. Water can be important in controlling blasting dust. The area to be blasted should be thoroughly wetted with water spray beforehand. Good ventilation is also necessary for efficient control of blasting fumes and dust. However, perhaps the best technique to avoid exposure to blasting dust and fumes is to carry out the operation off-shift, when there are no miners in the section.

Ore- and waste passes. Ore- and waste passes create large quantities of respirable dust. Falling ores create a pressure wave discharging dusty air into mine workings. The first step in controlling dust is to design an orepass that is connected with other ore- and waste passes. This relieves the intensity of the pressure wave. The second step in controlling dust is to put the orepasses on exhaust with air velocity at about 60 m/min and discharge the air into a return airway (MAC 1980).

An alternative approach is to confine the dust in the orepass and use sufficient amounts of water to suppress the dust. Water sprays can be installed at the tipping site and the ore can be wetted prior to delivery to the tipping site. However, excessive use of water can create additional problems such as plugging of the chutes, excessive water accumulation in the orepass, and adverse milling conditions.

Conveyor belts, transfer points, and roadheaders. Conventional water sprays are used to suppress dust on conveyor belts, transfer points, and roadheaders. All dust control techniques discussed for continuous mining sections can be used to control dust from roadheaders.

Mineral in transport. Fugitive dust created in mineral transport can be successfully controlled by identifying the points of origin and installing proper control equipment. Major points of origin are loading, transportation, unloading, and stockpiles. Loading and unloading use silos, cranes, front-end loaders, and belt conveyors. For transportation, trucks, trains, and conveyor belts are used. Stockpiles are located at various points in the mineral chain, but primarily at the mines, ports, and the consuming plant. General principles of dust control in transportation are discussed in great detail elsewhere (Schmitz 1994), but they are similar to the guidelines discussed earlier for in-mine dust control.

Prevent dust generation. Every effort should be made to design loading, unloading, and transporting equipment to minimize dust generation. Stockpiles can also be designed to minimize total surface area for a given volumetric content.

Use dust suppressants. Increasing the moisture content of the mineral increases the force of cohesion. Water sprays are the most common devices used to control dust. Sometimes the water is atomized using compressed air or high-frequency sound to improve the efficiency of dust control. Adding surfactants to water also improves dust control. Besides water, many other liquids and chemical additives can also improve dust control. Oil and oil products are highly suited to use as coal dust suppressants because of their surface chemistry. Agglomeration agents such as polymer emulsions are also used for dust control. The main advantage of an agglomeration agent is its longevity of use. Effective dust control may be sustained over several transfer points. Coating agents such as polymers are mixed with water or are sprayed on bulk materials to form a water-insoluble crust on the material. This prevents the effects of weathering and wind erosion. Foams are used where the use of excessive water can create problems. Only water, compressed air, and foaming agents are needed to achieve this. Foaming agents are high-foaming surfactants, and they may also contain wetting and binding agents.

Contain and capture. Suppressants can successfully prevent dust from becoming airborne but can have high cost and some environmental concerns. They can be supplemented, or at times replaced, by containing and capturing fugitive dust. Capturing techniques require an exhaust fan, hoods, ducts, and fabric collectors/filters. Very high collection efficiency, up to 99%, can be achieved by proper design. The fabric collectors must be cleaned periodically. Wet scrubbers and electrostatic precipitators are also used for dust collection. In an electrostatic precipitator, the dust particles first receive a negative charge from high-voltage DC electrodes, and then they are collected by a positively charged collecting plate. The collecting plates are vibrated to release the dust into a discharge system. Electrostatic precipitators are not recommended for combustible dusts such as coal because of the risks of explosions or fire.

Mineral/coal processing. Sources of dust in a processing plant include belt conveyors, loading/unloading points, transfer chutes, crushers, bucket excavators, chutes, and stockpiles. A very good reference for additional information is the *Dust Control Handbook for Mineral Processing* (Mody and Jakhete 1987).

Personal Protection Equipment

In many countries, approved respiratory protection equipment is required for those working in an area where the respirable dust concentration exceeds the statutory limits. Mainly, they use a filtration system to clean up the air to be inhaled, or they provide filtered intake air with a positive pressure to the workers.

Replaceable filter respirator. A respirator consists of a mask that fits a miner's face and forms a seal against the dusty air. Dirty air is drawn through replaceable filters that can collect 80%–98% of all dust in the air. The filter-holding unit is often plastic but can also be made of metal or hard rubber. Although the respirator does an excellent job of respirable dust control, some personal discomfort may arise owing to breathing resistance and facial irritation caused by the face seal. It also tends to interfere with normal voice communication, as well as eyeglasses and goggles. To eliminate these difficulties, sometimes a much lighter, single-use respirator is used. The whole mask is made of the filter material and covers the entire nose and mouth area of the wearer. However, it does not form a perfect seal against the dusty air nor is it as efficient as the replaceable, heavier filters.

Air helmet. The air helmet is a special hardhat (somewhat bigger than the conventional hardhat) that contains a battery-powered fan, a filtration system for dust, and a face visor. It provides protection for head, eyes, and lungs in a single unit. Although the air helmet is much more expensive than mask filters and weighs about 1.4 kg, it is finding wearer acceptance in many mining operations.

In the air helmet, dust-laden air is drawn through the filtration system by a battery-powered fan and the clean air is directed to the full-face visor over the wearer's face. The air exits the wearer's face at the lower end of the visor. The battery is normally smaller than the miner's cap-lamp battery and is carried on the belt.

If air velocity is not high, the air helmet performs almost as well as the replaceable filter mask with collection efficiencies of up to 84%, but at high air velocity, the collection efficiency appears to go down to about 50% (Cecala et al. 1981). This may be due to excessive turbulence created by the high velocity of air forcing dusty air into the visor.

Personal protective devices are meant for situations when a miner enters and works in an area with dust concentrations above the statutory limits. They should not be required to be used for an extended period of time or (under normal operations) the entire working shift. Although personal protection devices can be effective in reducing dust exposures, they are vulnerable to human errors in their cleaning, maintenance, fitting, and use.

Diesel Exhaust Control

Mineral transport in metal mines has been done by diesel engines for a long time. It has many advantages, but the exhaust must be treated and diluted properly to keep the air environment safe.

Diesel Engines

Diesel engines were introduced in European, Asian, and Australian coal mines nearly 60 years ago, but more recently in U.S. coal mines. Currently, nearly 5,000 units are working in U.S. metal mines and another 3,000 units in U.S. coal mines. The driving force behind this phenomenal growth is mine safety and productivity. Diesel equipment improves safety by eliminating electric shocks, fire, and explosion hazards from trolley wires and cabled electrical equipment. It can also prevent personal injury by eliminating the need to carry small loads (up to 40 kg) by hand in work areas.

The main drawback of using diesel equipment in mines is the contamination of air with diesel exhaust emissions. The major components of diesel exhaust are carbon monoxide (CO), carbon dioxide (CO_2), unburned hydrocarbons (HCs), nitrogen oxides (NO_x), sulfur oxides, and diesel particulate matter (DPM). PELs for all gaseous components of diesel exhaust have been well established. Ventilation can dilute all these gases to below their respective PELs.

DPM consists of

- Solid carbon (includes elemental and organic carbons);
- Liquid and solid hydrocarbons, including polyaromatic hydrocarbons (PAHs);
- Sulfates; and
- Moisture.

The PAHs attached to solid carbon particles are generally believed to be potential carcinogens, but their concentrations are exceedingly low.

DPM Standards

At present, there are no DPM standards in the world that are based on sound epidemiological studies correlating sustained exposure to DPM and the resulting health effects. However, several countries have set arbitrary standards, mainly to minimize DPM concentrations in mine air as much as possible. Table 15.4-7 shows some of these standards.

DPM Control Strategy

The best strategy, by general concensus, to control DPM in all mines and tunneling is to take an integrated approach, as summarized by Thakur and Patts (1999):

- Reduce DPM generation
  - Clean engines
  - Clean (low sulfur) fuel
- Combust and collect generated DPM
  - A catalytic converter on all engines
  - Soot filters where needed
- Dilute DPM
  - Minimum ventilation air
- Monitor DPM emissions and maintain equipment
  - Routine engine performance check
  - Proper maintenance
  - Adequate training to operators and mechanics

Reduce DPM generation. The amount of DPM emitted in grams per brake horsepower is known as the specific DPM emission for the engine and is a measure for "clean" engines. Table 15.4-8 shows the specific DPM emission for some old and new diesel engines. The new engines with electronic ignitions and turbocharging are remarkably clean. Whereas the older engines had a specific DPM emission of about 0.3 g/bhp-h, the new engines have a specific emission of only 0.02 to 0.1 g/bhp-h.

The improvement is brought out by electronic ignition and turbocharging. Another technique to reduce the DPM emission is to derate the engine by restricting the maximum fuel rating. Even a 10% reduction in horsepower rating can cut down DPM emissions by 50% or so (Schnakenberg and Bugarski 2002).

Concurrent to the development of clean engines, much effort has been made to improve diesel fuel quality. Extensive research in the field (Cowley et al. 1993) has shown that fuel properties that most influence DPM and gaseous emission comprise sulfur content, fuel density, cetane number, aromatic contents, and oxygen content. Besides sulfur content, the other two most influential properties are cetane number and fuel density.

Sulfur in fuel creates both sulfur dioxide (gas) and sulfates (solids, counted as a part of DPM). When the current American Society for Testing and Materials (ASTM) D2 fuel (sulfur at 340–500 ppm) is used, nearly 50% of DPM is sulfate. Recently, the U.S. Environmental Protection Agency required all diesel fuels to have no more than 15 ppm sulfur, creating an ultra-low-sulfur fuel. The low sulfur level not only reduces the mass of DPM emitted by the engine but also considerably reduces the number of nanoparticles (<40 nm). In some countries the health risks of inhaling diesel exhaust is ascribed to these nanoparticles. Thus, the use of ultra-low-sulfur fuel is quite helpful in reducing the relative health risks.

Alternate fuels such as Fisher–Tropsch (F-T) diesel and biodiesel have received attention as additives to the petroleum diesel fuel. F-T diesel is produced from coal, biomass, or natural gas. It is a very high-quality fuel with a potential to further reduce exhaust emissions. It has a high cetane number (up to 70), a low sulfur content (<10 ppm), and very low aromatic content (<3%). The benefits are more pronounced in DPM emissions (Norton et al. 1999), owing to very small sulfur content. The only drawback of this fuel is low lubricity, and the addition of some lubricants to the fuel is usually necessary.

Table 15.4-7 World DPM standards

Country	PEL	Measurement Technique
United States (metal mines)	0.16 mg/m^3	NIOSH 5040 method for total carbon
United States* (Pennsylvania and West Virginia)	0.12 mg/m^3	Total DPM divided by nameplate air as specified by the Mine Safety and Health Administration
Canada	1.5 mg/m^3	Respirable combustible dust
Germany (metal and nonmetal)	0.3 mg/m^3	Elemental carbon (EC)
Proposed limits		
ACGIH	0.15 mg/m^3	All particles <1 μm
ACGIH	0.02 mg/m^3	EC particles <1 μm

Source: Thakur et al. 2002a; Schnakenberg and Bugarski 2002.
*U.S. federal regulations do not put a separate limit on DPM but combine it with the respirable dust standard of 2 mg/m^3. They do, however, put a limit on DPM emission rates (MSHA 2001):
- Light-duty outby engines: ≤5 g/h
- Heavy-duty outby engines: <2.5 g/h
- Heavy-duty (permissible) inby engines: ≤2.5 g/h

Table 15.4-8 DPM emissions from old and new diesel engines

Engine Type	Horse-power	DPM Emission, g/h	Specific Emission, g/bhp-h
Old engines			
Cat 3306	150	45.9	0.306
Cat 3304	100	29.7	0.297
Perkins 1006-60 T	152	20.4	0.134
New engines			
DD Series 40	250	5.10	0.020
Deutz F3L912W	40	4.3	0.108
Deutz F3L1011	42	2.6	0.062

Biodiesel is defined as mono alkyl esters of long-chain fatty acids derived from renewable biomass (e.g., corn, soya beans, sugarcane, and switch grass). Pure biodiesels have relatively low sulfur content (<50 ppm) and no aromatics. The cetane number is comparable to D2 diesel fuel (42–48). Since biodiesel is oxygenated (esters contain oxygen molecules), its combustion in diesel engines is more complete than that of petroleum fuels. As a result, there is a substantial reduction in ambient HC, CO, and DPM. A slight increase in the NO$_x$ emissions (caused by a significant increase in nitrogen dioxide, or NO$_2$) have been observed in neat biodiesels or biodiesel blends (Sharp 1998).

The synthetic diesel and biodiesel fuels can be used in existing engines and fuel injection systems without negatively impacting engine performance.

Combust and collect generated DPM. The second step in the control of diesel exhaust is to combust (oxidize) the gaseous and liquid portions of the exhaust and collect solids on a filtration medium. The process is also known as exhaust aftertreatment. It can broadly be divided into two processes: diesel oxidation catalysts (DOCs) and DPM filters.

The main function of DOCs is to oxidize CO, HC, and some liquids (PAHs) attached to solid particles in the DPM. The by-product is CO$_2$ and water. It has no impact on solids.

Although the designs of most DOCs are proprietary, they are usually made of either metal or ceramic monoliths with very high surface area/volume ratios. They are normally coated with proprietary catalysts made of noble metals or base metals that lower the temperature required to oxidize the volatiles in the exhaust. They can reduce the CO and HC concentration in the exhaust by approximately 90%. Even the diesel particulate mass is reduced by about 25% by driving out and oxidizing all liquids attached to solids.

The DOCs usually last the entire life span of the engine. They are relatively cheap, costing less than a few thousand U.S. dollars for the small engines used in mines.

The main reason for the failure of a DOC is deactivation of catalysts by very high temperature (>650°C) and poisoning by lubricating oil additives (phosphorus, zinc, and heavy metals) and sulfur in fuel. Overcatalyzation of a DOC can result in the excessive production of NO_2 and sulfates if the fuel has a high sulfur content. Most DOCs are designed to match a particular diesel engine and, by and large, perform well. They are universally recommended for all diesel engines working in underground/surface mines, especially in conjunction with ultra-low-sulfur fuels. They should be installed close to the engine for best results.

DPM filters are used downstream from DOCs and usually work in tandem. They can be broadly classified in two categories: a ceramic monolith or a fiber filter media.

The ceramic monolith is made of cordierites or silicon carbide. They are often catalyzed using noble or base metals. As the exhaust flows through the monolith, DPM is collected on catalyzed surfaces and eventually burnt to gases by the "auto regeneration" process. Cordierite monoliths are liable to thermal fractures, but silicon carbide monoliths perform better. They have lower thermal shock resistance but their higher melting point makes them more durable. They also create less backpressure because they have a high permeability for gases. The collection efficiency for DPM ranges from 80% to 95% depending on the manufacturer. The life of a ceramic/silicon carbide filter is normally in excess of 5,000 hours. Routine maintenance and periodic "complete regeneration" in an off-board electric kiln is needed. In some less-demanding operations such as diesel equipment working on flat rail tracks, off-board regeneration is becoming the standard and more reliable procedure. The ceramic filters also act as flame and spark arrestors.

When a fiber filter media is used on a diesel engine, it is necessary that the diesel exhaust be cooled in a water bath. The cooled gas goes through a replaceable paper or synthetic fiber filter where solid particles are collected. The collection efficiency ranges from 90% to 97%. Depending on the duty cycle, a paper filter can last one to five shifts in operation, after which it must be replaced. Intake air resistance (usually limited to 7–8 kPa) often warns when a replacement is due. Synthetic filters can be washed, cleaned, and reused. Since the temperature of all significant parts of a diesel engine is regulated to be below 150°C, the exhaust must be cooled to this temperature before being discharged into mine air. Water scrubbers not only cool the exhaust but also act as flame and spark arrestors. Their main drawback is the high maintenance. Recently dry scrubbers have been developed that do not require the exhaust to contact the coolant. The gases are conducted through tubes that are themselves cooled by the bath. In another design, the coolant is conducted in tubes to cool the exhaust. Both designs are commonly used in underground mines. The fiber filter system is usually large in size and quite expensive. They are, therefore, more suited to bigger equipment on the surface or on locomotives in the mines.

They would also be essential on permissible diesel equipment such as shuttle cars.

Dilute DPM. The earliest ventilation standard for diesel engines was developed by the United Kingdom's National Coal Board. It required 4 m^3/min of air for each brake horsepower of the engine. The latest standard for ventilation is the Pennsylvania and West Virginia standard. It requires the ventilation air to be the "nameplate ventilation" as specified by the Mine Safety and Health Administration (MSHA). This air is enough to dilute all gaseous components of diesel exhaust to or below their respective threshold limit values. The exhaust treatment system for the engine must reduce the DPM emissions to a level that, when diluted by the nameplate air, the resulting DPM concentration will not exceed 0.12 mg/m^3. With the advent of clean engines, clean fuels, and new DPM filters, it is not difficult to achieve this high standard anymore.

For multiple engines working in one air split, the old MSHA requirement was the 100-75-50 rule. This means that the ventilation air requirement for the first (usually the biggest) unit is the MSHA nameplate air, 75% of the nameplate air is needed for the second unit, and 50% of the nameplate air is needed for the third and each additional unit. The latest requirement is the total sum of all nameplate airs for the diesel units working in each air split (West Virginia Diesel Commission 1998). Ventilation air requirements for diesel engines working in mines in different situations are extensively described in Thakur (1974).

Monitor DPM emissions and maintain equipment. DPM emissions are actually based on the original eight-mode test of the engines in the laboratory. Two factors that may make the original emissions worse are poor maintenance and the age of the engine.

It is obvious to anyone working in a dieselized mine that a poorly maintained engine can aggravate the DPM levels in mine air. This can be avoided by preventive, routine maintenance of the entire diesel fleet. Such maintenance must be carried out by trained personnel on a periodic basis. Daily checking of gaseous emissions should be done to confirm the condition of diesel engines. Ambient air quality examination can also provide a warning that one or more engines are not working properly. All mine personnel should be properly trained in operating the equipment and monitoring its performance. Maintenance of a diesel engine is the key to good performance and cannot be overemphasized.

Unmaintained or poorly maintained diesel engines do show an increase in DPM emissions with age. However, a properly maintained diesel engine (including the change of injectors, if needed) should not suffer any deterioration in DPM emissions during its normal tenure in the mine. Routine maintenance as recommended by manufacturers should be carried out only by trained personnel.

Radon Gas Control

Radon is a chemically inert, gaseous, radioactive product of the disintegration of radium. It is primarily found in uranium mines, although it is present in trace amounts in other mines including coal mines. Both radium and radon are intermediate products in the disintegration of uranium-238 to become lead-206. Table 15.4-9 shows some important steps in the disintegration of uranium to lead with half-lives of the isotopes.

After radon is released into mine air, the decay process continues with formation of radium A, B, C, C', D, E, F, and G, which is lead. These products are commonly referred to

Table 15.4-9 Uranium disintegration process

Name	Isotope	Type of Radiation	Half-Life
Uranium	$^{238}_{92}U$	Alpha	4.49×10^9 years
Radium	$^{226}_{88}Ra$	Alpha	1,622 years
Radon	$^{222}_{86}Rn$	Alpha	3.825 days
Lead	$^{206}_{82}Pb$	—	Stable

as radon daughters. They are atoms of solid matter having relatively short half-lives. They emit alpha or beta particles and rarely gamma rays, and they are the main cause of health concern.

Radon gas is inhaled and exhaled before large amounts of alpha particles are emitted. The daughter products, however, attach themselves to respirable dust particles and get deposited in the alveolar sacs. Exposure to excessively high concentrations of radon and radon daughters have been linked to lung cancer.

The maximum exposure limit for radon daughters has been set at 1.00 WL (working level), with a yearly cumulative exposure of 4 WLMs (WL months). A working level is defined as that concentration of radon daughter products in 1 L of air that will yield 1.3×10^5 million electron volts (MeV) of alpha energy in decaying to radium C′ (U.S. Department of Health, Education and Welfare 1957).

The average WL for a given 8-hour shift is calculated by dividing the sum of exposure level multiplied by the number of hours by 8. Mathematically,

$$\text{average WL} = \frac{1}{8} \Sigma \text{WL}_i \times h_i$$

where
  WL = exposure
  h_i = hours

The WLM = {WL × 8} / 173. This is based on the assumption that there are 173 working hours in a working month.

Except for uranium mines, radon gas has not created a health hazard in coal and metal mines. However, it is desirable to check for its presence in all mines and to take action if necessary.

REFERENCES

ACGIH (American Conference of Governmental Industrial Hygienists). 1972. *Documentation of the Threshold Limit Values for Substance in Workroom Air*. Cincinnati, OH: ACGIH.

Cecala, A.B., Volkwein, J.C., Thimon, E.D., and Urban, C.W. 1981. *Protection Factors of the Airstream Helmet*. Avondale, MD: U.S. Bureau of Mines.

CFR (Code of Federal Regulations). 2003. *Title 30—Mineral Resources, Part 75. Mandatory Safety Standards—Underground Coal Mines*. Washington, DC: U.S. Office of the Federal Register, National Archives and Records Administration.

Colinet, J.F., Spencer, E.R, and Jankowski, R.A. 1997. Status of dust control technology on U.S. longwalls. In *Proceedings of the 6th International Mine Ventilation Congress*, Pittsburgh, PA, May 17–22. Littleton, CO: SME.

Cowley, L.T., Juene, A.L., and Lange, W.W. 1993. The effect of fuel composition including aromatics content on emissions from a range of heavy-duty engines. In *Fourth International Symposium on the Performance Evaluation of Automotive Fuels and Lubricants*, Birmingham, UK, May 5–7. Coordinating European Council for the Development of Performance Tests for Lubricants and Engine Fuels.

Davis, J.G., and Krickovic, S. 1973. *Gob Degasification Research: A Case History*. Information Circular IC-8621. Washington, DC: U.S. Bureau of Mines.

Goodman, G.V.R., and Listak, J.M. 1999. Variations in dust levels with continuous miner position. *Min. Eng.* 51(2):53–58.

Hartman, H.L., Mutmansky, J.M., Ramani, R.V., and Wang, Y.J. 1997. *Mine Ventilation and Air Conditioning*, 3rd ed. New York: Wiley.

ILO (International Labour Office). 1965. *Guide to the Prevention and Suppression of Dust in Mining, Tunneling and Quarrying*. Geneva: ILO.

Jacobsen, M., Rae, S., Walton, W.H., and Rogan, J.M. 1971. The relationship between pneumoconiosis and dust exposure in British coal mines. In *Inhaled Particles 3: Proceedings of an International Symposium Organized by the British Occupational Hygiene Society in London, 14–23 September, 1970*. Edited by W.H. Walton. Old Woking, Surrey, England: Unwin Brothers.

Jayaraman, N.I., Jankowski, R.A., and Kissell, F.M. 1985. *Improved Shearer-Clearer System for Double Drum Shearers on Longwall Faces*. Pittsburgh: U.S. Bureau of Mines.

Kissell, F.N. 1992. Gas and dust control. In *SME Mining Engineering Handbook*, 2nd ed. Edited by H.L. Hartman. Littleton, CO: SME.

Kissell, F.N. 2003. *Handbook for Dust Control in Mining*. Information Circular IC-9465. Pittsburgh: National Institute for Occupational Safety and Health.

MAC (Mining Association of Canada). 1980. *Design Guidelines for Dust Control at Mine Shafts and Surface Operations*, 3rd ed. Ottawa, ON: Mining Association of Canada.

Mazza, R.L., and Mlinar, M.P. 1977. *Reducing Methane in Coal Mine Gob Areas with Vertical Boreholes*. Final Report to U.S. Bureau of Mines, Contract No. H0322851.

Mody, V., and Jakhete, R. 1987. *Dust Control Handbook for Mineral Processing*. Pittsburgh: U.S. Bureau of Mines.

MSA Research Corp. 1974. *Control of Respirable Dust in Non-Coal Mines and Ore Processing Mills—Handbook*. Report to U.S. Bureau of Mines, Contract H0220030, NTJS No. PB 240646. Washington, DC: Government Printing Office.

MSHA (Mine Safety and Health Administration). 2001. Diesel particulate matter exposure of underground coal miners. *Fed. Reg.* 66(13):5526. January 19.

Nagy, J., Hartmann, I., and Mitchell, D. 1957. *Experiments on Water Infusion in the Experimental Coal Mine*. Washington, DC: U.S. Bureau of Mines.

NIOSH (National Institute of Occupational Safety and Health). 1995. *Occupational Exposure to Respirable Coal Mine Dust*. Cincinnati, OH: U.S. Department of Health and Human Services.

Norton, P., Vertin, K., Clark, N., Lyons, D., Gautam, M., Goguen, S., and Eberhardt, J. 1999. Emissions from buses with DDC 6V92 engines using synthetic diesel fuel. SAE Technical Paper Series No. 1999-01-1512. Society of Automotive Engineers.

Schmitz, J. 1994. *Control of Coal Dust in Transit and in Stockpiles*. London: IEA Coal Research.

Schnakenberg, G.H., and Bugarski, A.D. 2002. *Review of Technology Available to the Underground Mining Industry for Control of Diesel Emissions*. Information Circular IC-9462. Pittsburgh: National Institute for Occupational Safety and Health.

Sharp, C.A. 1998. Exhaust Emissions and Performance of Diesel Engines with Biodiesel Fuel. www.biodiesel.org/resources/reportsdatabase/reports/gen/19980701_gen-065.pdf. Accessed February 2010.

Thakur, P.C. 1974. Computer-aided analysis of diesel exhaust contamination of mine ventilation systems. Ph.D. thesis, The Pennsylvania State University, University Park, PA.

Thakur, P.C. 1997a. How to eliminate coal workers' pneumoconiosis. In *Proceedings of the 27th International Conference on Safety in Mines Research Institutes*. New Delhi: Oxford and IBH Publishing Company.

Thakur, P.C. 1997b. Methane drainage from gassy mines—A global review. In *Proceedings of the 6th International Mine Ventilation Congress*, Pittsburgh, PA, May 17–22. Littleton, CO: SME.

Thakur, P.C. 2005. Advancing mine safety and production through coalbed methane production. In *The 1st China International Conference Proceedings on Coal Mine Gas Control and Utilization*. Beijing: Coal Industry Publishing House.

Thakur, P.C., and Davis, J.G. 1977. How to plan for methane control in underground mines. *Min. Eng.* 10:415–422.

Thakur, P.C., and Patts, L.D. 1999. An integrated approach to control diesel particulate matter in underground coal mines. In *Proceedings of the 8th U.S. Mine Ventilation Symposium*. Rolla, MO: University of Missouri–Rolla Press.

Thakur, P.C., and Zachwieja, J. 2001. Methane control and ventilation for 1000-ft wide longwall faces. In *Proceedings of Longwall USA*, Pittsburgh, PA, June 13–15.

Thakur, P.C., Hatch, R., and Riester, J.B. 1981. Performance evaluation of machine mounted respirable dust monitors for U.S. coal mines. Presented at the International Symposium on Aerosols in the Mining and Industrial Work Environment, University of Minnesota, Minneapolis, MN, November 1–6.

Thakur, P.C., Chamberlin, E.S., and Holt, J.S. 2002a. Federal regulations vs. diesel exhaust control in U.S. coal mines. In *Mine Ventilation: Proceedings of the North American/Ninth U.S. Mine Ventilation Symposium*, Kingston, Ontario, June 8–12. Lisse, The Netherlands: A.A. Balkema.

Thakur, P.C., Statnik, R., and Cairns, G. 2002b. Coalbed methane production from deep coal seams. Presented at the AIChE Spring Meeting, New Orleans, LA, March 10–14.

U.S. Department of Health, Education and Welfare. 1957. *Control of Radon and Daughters in Uranium Mines and Calculations on Biologic Effect*. Edited by D.A. Holaday. Publication No. 494. Washington, DC: U.S. Department of Health, Education and Welfare.

West Virginia Diesel Commission. 1998. Rules for Operating Diesel Equipment in Underground Mines in West Virginia, Title 196, Series 1, 255–282.

Zimmer, R.A., Lucek, S.R., and Page, S.J. 1987. Optimization of overburden drill dust control system on surface coal mines. *Int. J. Surf. Min.* 1(2):155–157.

CHAPTER 15.5

Heat, Humidity, and Air Conditioning

Felipe Calizaya and John Marks

INTRODUCTION

In underground mines, heat and humidity can create adverse environmental conditions that can affect worker performance negatively. Such conditions can even limit the depth to which mining can be extended. Although the human body has a self-regulating mechanism against variations of external heat, under extreme conditions this mechanism can break down, resulting in occupational illnesses such as heat exhaustion and heat stroke.

In the United States, with a few exceptions, heat becomes an environmental problem in mines with workings at depths greater than 1,000 m (3,280 ft). In these mines, ventilation is not sufficient to control the heat loads and must be supplemented by some form of air conditioning. The Anaconda mine in Butte, Montana, and the Magma mine in Superior, Arizona, were the first to utilize air conditioning systems. Both mines are at least 1.5 km (0.9 mi) deep. Other U.S. mines that require air conditioning include the San Manuel mine in San Manuel, Arizona, which is 1,000 m (3,280 ft) deep; the Homestake mine in Lead, South Dakota, which is 2.4 km (1.5 mi) deep; the Sunshine mine in Kellogg, Idaho, which is 1.6 km (1 mi) deep, and more recently, the Meikle mine near Elko, Nevada, where heat became a environmental problem at depths ranging from 300 to 600 m (1,000 to 2,000 ft).

In Canada, heat is also an environmental concern in most deep underground mines. These include Inco's Creighton mine, which is 2,400 m (7,870 ft) below the surface; Falconbridge's Kidd Creek mine, which is 3,000 m (9,845 ft) deep; and Agnico-Eagle's Laronde mine, which is 2,000 m (6,560 ft) deep. Autocompression is the main source of heat in these mines. The ventilation systems in these mines are now being upgraded to include refrigeration plants.

This chapter presents summaries of the following topics:

- Psychrometric properties of the air
- Human heat stress indices
- Instrumentation for determining how thermal loads are estimated
- Heat sources
- Mine cooling
- The principles used to estimate airflow and refrigeration requirements

Sample problems are used to illustrate the various steps involved in determining ventilation requirements and the critical depths. The solution to each problem is found using CLIMSIM, a commercial mine climate simulation program from MVS Engineering, Clovis, California. This software is used to predict the properties of the air at various simulated workings, to evaluate the resulting parameters against the target values, and to determine a feasible solution to the problem.

Several books and articles have been written on ventilation and air conditioning. The references at the end of this chapter provide a selected list of publications on this topic. The ventilation engineer is strongly encouraged to study these references.

Definitions

Definitions specific to the control of heat and humidity in underground mines are as follows.

- **Autocompression:** This is the term given to a process by which a column of air moving downward is compressed and experiences a temperature increase. Autocompression is due to the additional pressure of air "stacked" on itself. In a dry shaft, autocompression raises the air temperature by about 1°C (1.8°F) for every 100-m (328-ft) increase in depth.
- **Heat:** Heat is a form of energy and, as such, it can never be destroyed. In mines, heat is produced by chemical reactions such as the oxidation of ores (slow) or the burning of fuels (fast), by friction (pulleys under a stationary belt), and in many other ways.
- **Heat stress:** Heat stress is a qualitative assessment of the work environment based on temperature, relative humidity, air velocity, personal clothing, and radiant energy (environmental factor). Several heat stress indices have been proposed. The most common ones are effective temperature, cooling power, and wet bulb temperature.

Felipe Calizaya, Associate Professor, University of Utah, Salt Lake City, Utah, USA
John Marks, Consultant, Lead, South Dakota, USA

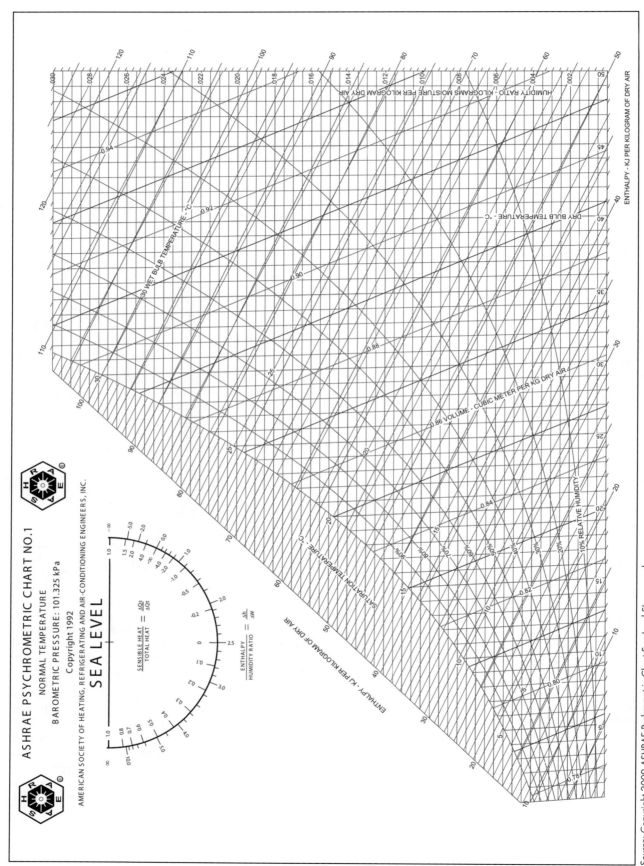

Figure 15.5-1 ASHRAE psychrometric chart

- **Heat strain:** The physiological response to heat stress (people factor) is called heat strain. Effects include dehydration, increased heart rate, fatigue, cramps, and heat stroke. Humans have different tolerance levels for heat. They can cope easier with heat than cold.
- **Sensible heat:** This is the heat that when added to or removed from a substance changes its temperature without changing its state.
- **Latent (hidden) heat:** This is the heat added to or removed from a substance that results in a change in state but does not increase or decrease its temperature.
- **Enthalpy:** Enthalpy is the total heat content of the air–vapor mixture per unit weight (sum of latent and sensible heat).
- **Temperature:** Temperature is the degree of hotness or coolness of a substance (measurement of a state).
- **Critical depth:** This is the depth in a mine at which the wet bulb temperature of the intake air exceeds the allowable wet bulb temperature, mainly due to autocompression. Beyond this depth, the mine will require some form of air conditioning.

PSYCHOMETRY AND AIR CONDITIONING

Psychrometry is the science of evaluating the thermal state of moist air. The psychrometric properties of interest to a ventilation engineer are dry bulb temperature (t_d), wet bulb temperature (t_w), dew-point temperature (t_{dp}), barometric pressure (P_b), specific volume (V), humidity ratio (W), and enthalpy (h). Because air is considered a mixture of dry air and water vapor, any three of these are necessary to be known, and the remaining properties can be calculated (McPherson 1993; Hartman et al. 1997). The easiest three to measure are t_d, t_w, and P_b. The determination of all of these properties is a prerequisite to solving air conditioning problems.

Three tools are available for estimating these properties: psychrometric tables, charts, and the fundamental equations. There are several computer programs available that make use of these tools and more advanced numerical solution methods. When tables are not available, the basic properties can be calculated using Equations 15.5-1 through 15.5-8 (McQuiston and Parker 1994).

Specific Volume, V:

$$V = \frac{RT_d}{P_b} \text{ m}^3/\text{kg} \qquad (15.5\text{-}1)$$

where
R = a gas constant, which is 287 J/(kg K)
T_d = the absolute temperature, which is equal to 273 + t_d, K
P_b = barometric pressure, Pa

Saturated Vapor Pressure, P_s:

$$P_s = 0.6105 \exp\left[\frac{17.27 t_d}{t_d + 237.3}\right] \text{ kPa} \qquad (15.5\text{-}2)$$

where t_d is the dry bulb temperature, °C.

It is evident in Equation 15.5-2 that the saturation vapor pressure is a function of temperature only.

Actual Vapor Pressure, P_v:

$$P_v = P_s - 0.000644 P_b (t_d - t_w) \text{ kPa} \qquad (15.5\text{-}3)$$

where
P_s = saturated vapor pressure, kPa
t_d = dry bulb temperature, °C
t_w = wet bulb temperature, °C

Relative Humidity, RH:

$$RH = \frac{P_v}{P_s} 100\% \qquad (15.5\text{-}4)$$

where
P_v = water vapor pressure, kPa
P_s = saturated vapor pressure, kPa

Absolute Humidity (or the "Humidity Ratio"), W:

$$W = 0.622 \frac{P_v}{(P_s - P_v)} \text{ kg/kg} \qquad (15.5\text{-}5)$$

where
P_v = water vapor pressure, kPa
P_b = barometric pressure, kPa

Air Density, d:

This equation gives kilograms of moisture per kilogram of dry air.

$$d = \frac{1}{V}(1 + W) \text{ kg/m}^3 \qquad (15.5\text{-}6)$$

where
V = specific volume, m³/kg
W = absolute humidity, kg moisture/kg dry air

Enthalpy, h:

$$h = 1.005 t_d + W(2501.3 + 1.86 t_d) \text{ kJ/kg} \qquad (15.5\text{-}7)$$

where
t_d = dry bulb temperature, °C
W = absolute humidity, kg moisture/kg dry air

The *2007 ASHRAE Handbook* gives the following values for standard air: sea level temperature = 15°C and barometric pressure = 101.325 kPa. The barometric pressure, P_b, depends mainly on elevation and can be approximated as follows:

$$P_b = 101.325(1 - 0.0000225577 Z)^{5.2559} \text{ Pa} \qquad (15.5\text{-}8)$$

where Z is the elevation in meters.

Example 1. Compute the humidity ratio and enthalpy of saturated air at 20°C and at a barometric pressure of 101.325 kPa.

Solution. Perform the following calculations:

- From Equation 15.5-2, for $P_v = P_s$ at 20°C, $P_v = 2.338$ kPa.
- Using Equation 15.5-5, $W = 0.0147$ kg/kg.
- Using Equation 15.5-7, for $W = 0.0147$ kg/kg and $t = 20°C$, $h = 57.56$ kJ/kg.

Quick estimates of all the psychrometric properties of air can be determined from a psychrometric chart. Each chart applies to air at a given barometric pressure. Figure 15.5-1 shows a psychrometric chart for $P_b = 101.325$ kPa. This is one of a series of charts produced by the American Society of Heating, Refrigerating, and Air-Conditioning Engineers (ASHRAE) for a range of pressures. Although all psychrometric properties can now be calculated using programs such as HotWork from

MVA (Mine Ventilation Australia, Sandgate, Queensland, Australia) (Brake 2008), charts provide a visual representation of process lines that represent changes in psychrometric conditions of the airstream. Figure 15.5-2 shows how the various processes can be followed from an initial state point (point A). Horizontal lines emerging from this point indicate changes in sensible heat, that is, cooling if the process proceeds from right to left or heating if the process proceeds from left to right. A vertical line indicates changes in latent heat—humidification if the process proceeds upward and dehumidification if the process proceeds downward. Combination processes run diagonally.

Example 2. A flow rate of 120 m³/s of air at $t_d = 30°C$, $t_w = 20°C$, and $P_b = 101.325$ kPa is to be chilled to $t_d = t_w = 10°C$ at the same pressure. First, calculate the following:

- The mass flow rate of dry air through the process
- The kW(R) required to chill the air
- The liquid water condensed in the air
- The relative humidity of the air entering the process

Solution. For $t_d = 30°C$, $t_w = 20°C$, and the specific volume of air is 0.873 m³/kg, calculate as follows:

- Mass flow rate: $m = Q/v = 137.3$ kg/s, where Q is the airflow rate in cubic meters per second
- Removed heat: $q = m(\Delta h) = 137.3 \times (57.2 - 29.4) = 3,817$ kW(R)
- Condensed water: $q_l = m(\Delta W) = 137.3 (10.6 - 7.7) = 398.20$ g/s (0.40 kg/s)
- Relative humidity: $RH = 40\%$ (from ASHRAE chart)

HEAT STRESS INDICES

These indices refer to the assessment of thermal conditions to which workers can be exposed repeatedly without developing any adverse health effect. They are expressed in terms of quantitative figures (maximum allowable limits) used by ventilation specialists for designing heat control systems. Since the introduction of the first heat stress index in 1923, more than 90 indices have been developed. Based on their application, these can be divided into two groups: empirical and rational. Empirical indices are expressed in terms of an external or environmental factor, such as the wet bulb temperature and the effective temperature. Rational indices are based on the responses of the workers to thermal stress, such as sweat rates and core temperature increases (internal factors). Only the empirical indices are applicable to underground mining conditions.

In most mining countries, these indices are regulated by code (Hartman et al. 1997; Hardcastle 2006). In U.S. mines, heat stress is not regulated. Management must set and enforce the applicable limits. The following are the most commonly used indices (Van der Walt et al. 1996; Del Castillo and Dawborn 2002; Mutama 2005; ACGIH 2008):

- Wet bulb temperature: $t_w \leq 27°C$ (81.6°F)
- Wet bulb globe temperature: $WBGT \leq 27°C$ for continuous work at moderate rate
- Effective temperature range: $ET = 28°C$ (81.6°F) for reduced work load and a maximum of 32°C (89.6°F) when work must be terminated
- Air-cooling power (ACP) for hard workload: 175 W/m² $\leq ACP \leq 275$ W/m²
- Air velocity: $v \geq 0.50$ m/s

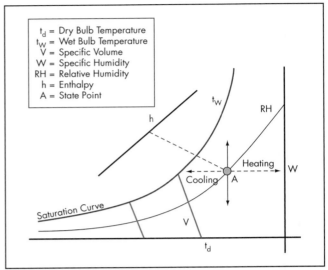

Figure 15.5-2 Psychrometric processes and state point

The wet bulb temperature is the simplest index that can be used by ventilation engineers to assess the work environment. It measures the evaporation of water from a wet bulb thermometer. It can be determined by a sling psychrometer after being whirled or vented at the rate of at least 3 m/s (10 fps).

The wet bulb globe temperature for an indoor environment is measured by a black globe thermometer or calculated using Equation 15.5-9 (ACGIH 2008):

$$WBGT = 0.7NWB + 0.3GT \tag{15.5-9}$$

where
  $WBGT$ = wet bulb globe temperature, °C
  NWB = natural (or unvented) wet bulb temperature, °C
  GT = globe temperature, °C

Effective temperature is an index of relative comfort determined from charts that require the knowledge of three parameters: dry bulb temperature, wet bulb temperature, and air velocity. It is still used in U.S. mines with negligible radiant heat.

Air cooling power is a rational heat stress index based on heat balance between workers and environment. For hard work rates, this index ranges from 175 to 275 W/m², depending on air velocity, wet bulb temperature, and type of clothing. For safety, vent systems should be designed for cooling power rates higher than the expected work rate.

Air velocity, as measured by an anemometer, is not an index, but it can be used in combination with the wet bulb temperature. A common lower limit for airways where personnel work is 0.50 m/s (1.6 fps).

The following are sample heat stress indices used in some hot mines:

- Homestake gold mine, South Dakota
  - Wet bulb temperature for acclimatized workers is 26.7°C $\leq t_w \leq$ 29.4°C (the lower limit indicates 100% worker efficiency, and the upper limit indicates that a corrective action is needed)
  - The air velocity range is 0.76 $\leq v \leq$ 3 m/s
  - The wet bulb globe temperature, $WBGT$, is $\leq$32.2°C

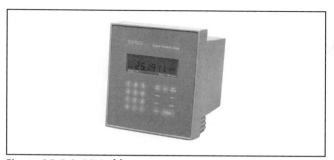

Figure 15.5-3 Digital barometer

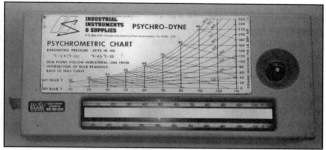

Figure 15.5-4 Portable psychrometer

- Meikle gold mine, Nevada
  - The wet bulb temperature, t_w, for acclimatized workers is ≤27.8°C
  - The effective temperature, ET, is ≤26.6°C
- Mindola copper mine, Zambia
  - The wet bulb temperature is 28°C ≤ t_w ≤ 30°C (depending on the air velocity, t_w is 28°C for $v = 0.77$ m/s and 30°C for 1.99 m/s)
  - Air-cooling power is $ACP \geq 300$ W/m^2

If two heat stress indices of a work environment exceed the design limits, the mine can take one or more of the following actions:

- Install ventilation and/or air conditioning systems (active engineering controls)
- Isolate the heat sources via insulation or blocking off heat sources (passive engineering controls)
- Enact administrative measures

Ventilation and/or air conditioning are generally the preferred control methods for long-term work areas, but they can be very expensive. Isolating heat sources can be done in certain cases, but if active or passive controls are uneconomical, the mine can consider administrative controls. These include

- Increasing workers' heat tolerance by heat acclimatization,
- Assigning workers to hot working areas selectively (on the average about 5% of relatively young workers are heat intolerant),
- Adjusting the work–rest regime by permitting workers to take short breaks to recover sweat rate and body core temperature, and
- Rotating personnel on hot jobs and providing workers with readily accessible cold areas, cold drinking water, and salt tablets. A half-teaspoon of salt per 2 L of water is sufficient.

In every case, the objective is to prevent the body core temperature from rising above 38°C (100°F).

INSTRUMENTATION

The following instruments are used to determine the heat load in underground mines: a barometer, a psychrometer, and thermocouples. Although most of these units are now digital, analog devices are still being used under harsh conditions and as primary standards for calibration.

Barometric Pressure

Barometers are used to measure the absolute pressure of mine air. This pressure is mainly a function of elevation of the working areas. Under standard conditions, the barometric pressure varies by about 500 Pa (0.07 psi) over a 24-hour period. Under special conditions, such as during major storms or cyclones, the pressure difference can be as high as 2,000 Pa (0.3 psi). Aneroid barometers, such as those manufactured by American Paulin System, Cottonwood, Arizona, are still being used for pressure measurement in underground mines. Because of their ruggedness and ease of operation, they are preferred by ventilation engineers. With the advent of electronic technology, digital barometers are gaining a wider acceptance in ventilation surveys, especially in hard-rock mining. They are based on pressure transducers that are highly sensitive to variations of pressure load. Figure 15.5-3 shows a capacitor-type barometer in a rugged enclosure. The instrument basically consists of two closely spaced metal surfaces, one of which is a diaphragm that flexes slightly under applied pressure and alters the gap between them (variable capacitor). This variation in distance is picked up electronically by the sensor. The instrument is highly accurate in the range of 60 to 110 kPa (8.7 to 16.0 psi).

Air Temperature Measurement

The dry and wet bulb temperatures of the air are measured by a psychrometer. This instrument consists of two identical thermometers: one to measure the true air temperature (dry bulb) and the other, covered by a water-saturated wick, to measure the evaporative cooling effect (wet bulb). The drier the surrounding air, the more rapidly the water will evaporate. This causes the wet bulb to depress in temperature until a balance takes places between the cooling effect and the heat gain from the ambient air. A requirement to measure the wet bulb temperature is the air velocity across the bulb. This velocity must be at least 3 m/s (10 fps). Without this velocity, one will always tend to measure a higher humidity level than actual.

Figure 15.5-4 shows a portable psychrometer. This instrument consists of two thermometers (dry and wet) and a small fan. Before any measurement, the wick on the wet bulb must be saturated with distilled water. While the fan is operating, the wet bulb temperature will begin to depress until it stabilizes at its lowest point. The procedure may take from 1 to 2 minutes depending on the dryness of the air. For better accuracy, the wet bulb temperature should be read first. The instrument can be used for temperature measurements in the –7° to 43°C (19° to 109°F) range. A humidity indicator (Figure 15.5-5) is another device used in underground mines. The instrument is based on a thin polymer film sensor that changes its capacitance as it absorbs water molecules. When equipped with a Vaisala HMP41 probe, it can be used to measure relative humidity, absolute humidity, dry and wet bulb temperatures, and dew-point temperature. The operating range of the instrument is from –20° to 60°C (4° to 140°F).

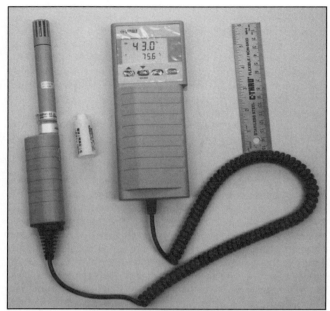

Figure 15.5-5 Digital humidity indicator

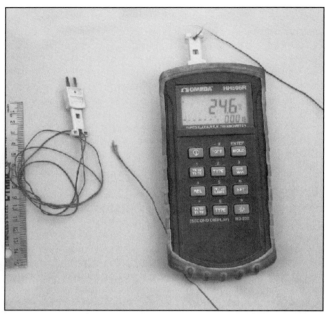

Figure 15.5-6 Thermocouple and digital thermometer

Rock Temperature Measurement

The rock temperature in a mine is usually measured by means of resistance thermometers and thermocouples. A resistance thermometer is constructed of a precious-metal conductor, such as platinum or silver, which exhibits increased resistance when heated. A thermocouple consists of insulated wire pairs with a junction point (bead) at one end and a male connector at the other. For underground applications, J-, K-, and T-type thermocouples are used. When a thermocouple is connected to a temperature indicator (reader), an analog or digital signal is obtained, making it suitable for temperature logging. Rock temperatures can be measured from boreholes that are drilled from underground openings or from the surface. In either case, it is good practice to allow the rock temperature to stabilize and eliminate the local heating and cooling effects. This is usually achieved 8 hours after the drilling is completed (Duckworth 1999). A plot of temperature verses depth is often used to determine the temperature distribution in the borehole and the virgin rock temperature (VRT). Figure 15.5-6 shows thermocouple wires and a digital thermometer. The fine wire thermocouples enable accurate temperature measurements without disturbing the base temperature of the body in which the installation is made. They are available in wire sizes ranging from 0.025 to 0.81 mm in diameter. The K-type thermocouple has an operating range of from −200° to 1,372°C (−328° to 2,502°F). The HH506R data logger (Omega Engineering) is a two-channel temperature-measuring device used with seven types of thermocouples. When recording mode is used, the instrument allows the user to store the maximum, minimum, and average temperatures during a 24-hour period.

HEAT SOURCES

Heat is emitted from various sources in underground mines, including autocompression, heat from wall rock, mining equipment, groundwater, oxidation of minerals, and metabolic heat. A study conducted on heat flow in seven underground mines in the United States has shown that wall rock, autocompression, and mining equipment account for about 80% of the total heat released. In shallow mines with depths of less than 1,000 m (3,280 ft), ventilation alone may be sufficient to remove the heat from the working sections. However, in deep mines with depths greater than 1,000 m (3,280 ft), the autocompression effect causes the wet bulb temperature to exceed the critical temperature of 28°C (82°F), and refrigeration becomes a necessity. The design of a refrigeration system requires a thorough evaluation of all sources of heat.

Heat from Rock

The heat flow from the rock depends on a number of variables, including the virgin rock temperature, rock type, size, shape, and depth of the openings and the presence of water on the rock surface. It also varies with the age of the opening. New and well-ventilated airways rapidly give off heat. After the airway has aged, the rock wall temperature will typically stabilize to within 2°C (3.6°F) of the dry bulb temperature. For rough estimates, Equation 15.5-10 can be used to approximate the radial heat flow, q, from the rock.

$$q = 2\pi \kappa L \frac{t_1 - t_2}{\ln(r_1/r_2)} \text{ W} \qquad (15.5\text{-}10)$$

where
κ = thermal conductivity, W/m °C
L = airway length, m
t_1 = rock wall temperature, °C
t_2 = virgin rock temperature, °C
r_1 = airway radius, m
r_2 = distance into the rock, m

Thermal conductivity is a measure of how well (or poorly) heat is conducted (or resisted) in a certain substance. It represents the temperature profile along each meter of drift length for a given heat flow. However, it does not measure how quickly the heat spreads along the drift. Thermal diffusivity is a measure of how quickly the heat spreads through the substance. Table 15.5-1 shows the thermal properties for

Table 15.5-1 Thermal properties of common substance

Substance	Conductivity, W/m °C	Specific Heat, kJ/kg °C	Diffusivity, m²/s 10⁻⁶
Water	0.60	4.19	0.14
Dry air	0.026	1.005	21.5
Coal	2.20	1.00	1.29
Copper	380	0.38	112
Gabbro	2.37	—	2.37
Granite	1.92	0.82	3.33
Pyritic shale	3.65	—	2.01
Quartz	5.50	0.88	2.32
Sandstone	1.97	0.71	1.68
Shale	2.39	—	0.90
Rhyolite	3.46	—	1.11
Gneiss	3.18	—	1.63
Concrete	1.20	0.88	0.59

Source: Adapted from ASHRAE 2007 and Brake 2008.

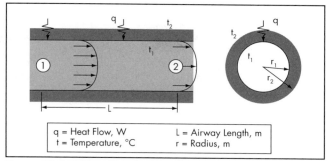

Figure 15.5-7 Heat flow from rock

a number of substances determined from laboratory or field tests (ASHRAE 2007; Brake 2008). Temperatures t_1 and t_2 are rock temperatures measured at radii r_1 and r_2, respectively. For rectangular openings, r_1 is approximated by the hydraulic radius (= 2 × area/perimeter). Radius r_2 is the distance into the rock from the center of the opening. The virgin rock temperature t_2 is determined from a borehole that is at least 4 m (13 ft) long and drilled perpendicular to the longitudinal axis of the opening.

A pair of t_i and r_i measurements, together with the length of the opening and the thermal conductivity of the rock, is sufficient to determine the heat flow into an opening. However, to improve the accuracy of the estimates, it is recommended to take a series of measurements from various boreholes drilled radially from the airway.

Example 3. A 400-m-long drift that is 4 m in diameter (r_1 = 2 m) is driven in quartzite rock (Figure 15.5-7). The VRT at r_2 = 10 m is determined to be 50°C. If the rock wall temperature is 30°C, how much heat will flow into the section?

Solution. For κ = 5.5 W/m °C, ($t_1 - t_2$) = 0°C and ln(r_1/r_2) = –1.61, Equation 15.5-10 yields q = 172 kW. For a given q (measured directly), this approach can be used to determine the thermal conductivity of the rock.

Autocompression

Autocompression is the process in which a column of air is compressed and experiences a temperature increase as it enters a downcast shaft. In dry shafts, the compression occurs adiabatically. On average, the dry bulb temperature increases by 1°C (1.8°F) per 100 m (328 ft). This occurs even when there is no external heat added into the shaft or external work done on the air. The wet bulb temperature also increases with depth, but at a lower rate. Typically, this temperature increases by about 0.6°C (1.1°F) per 100 m (328 ft) when the surface wet bulb temperature is 6°C (43°F) and 0.4°C (0.7°F) per 100 m (328 ft) when the surface wet bulb is 25°C (77°F) (Brake 2008). In an upcast shaft, the air decompresses as it travels up the shaft, resulting in a drop of both dry and wet bulb temperatures, a reduction in air density, and a decrease of atmospheric pressure. In locations with high humidity levels in the upcasting shafts and drifts, the decrease in temperature associated with decreasing pressure may result in the generation of fog as the conditions of saturation occur.

For a downcast shaft, the heat flow, q, can be estimated by

$$q = Q \frac{1}{V}(h_2 - h_1) \text{ kW} \qquad (15.5\text{-}11)$$

where
Q = airflow rate, m³/s
h_1 = enthalpy at state point 1 (shaft collar)
h_2 = enthalpy at state point 2 (shaft bottom)

Example 4. What is the heat load from autocompression when 100 m³/s of dry air of t_{d1} = 30°C and t_{w1} = 20°C at sea level drops down a 1,500-m shaft?

Solution. Estimate the wet bulb temperature at the bottom of the 1,500-m shaft (t_{w2} = 26.8°C by interpolation); determine the specific volume of the air at shaft collar (V = 0.873 m³/kg); and determine the enthalpies for t_{w1} and t_{w2} (h_1 = 57.2 kJ/kg and h_2 = 83.7 kJ/kg from Figure 15.5-1). Then, using Equation 15.5-11, q = 3,025 kW.

Electrical Equipment

Electrical motors, transformers, gearboxes, and other controls are not 100% efficient. Some of these inefficiencies are manifested in the form of sensible heat. All energy used in electrical equipment is transformed into two effects: work against gravity and heat. Conveyor belts, hoists, and pumps are examples where the electrical energy is transformed into potential energy and frictional heat. Other devices, such as rock breakers, fans, electric drills, and lights used on a horizontal plane, are examples where electrical energy is converted entirely into heat. Using Equation 15.5-12, for any given machine, the heat load, q, can be calculated as the difference in effective power and work done against gravity as follows:

$$q = PE - mgh \text{ kW} \qquad (15.5\text{-}12)$$

where
P = input power, kW
E = total efficiency, decimal fraction
m = mass flow rate, kg/s
g = 9.81 m/s²
h = elevation difference, m

Fans dissipate heat. However, this heat is detrimental only when the fan is located upstream of the main workings (either as blower or booster fans).

Example 5. A 1,500-m-long conveyor belt system transports ore at the rate of 720 t/h. The vertical lift is 280 m. The system consumes 2,500 kW of power at an efficiency of 85%. Determine the heat load along the belt line.

Solution. For an input power of 2,500 kW, an efficiency of 85%, $m = 200$ kg/s, and $h = 280$ m, Equation 15.5-12 yields $q = 1,576$ kW.

Diesel Engines

When diesel equipment is used in underground mines, in addition to smoke and fumes, a significant amount of heat is added to the air. The heat flow from a diesel engine is about 2.8 times the heat produced by an equivalent electric unit. About a third of this is added at the radiator, another third at the exhaust pipe in the form of smoke, and the remainder as useful work, which is finally also converted into heat. Each volume of fuel produces from three to ten volumes of water vapor. This results in increases of both latent and sensible heat of the air. The best way to determine the heat load is to estimate the fuel consumption, F, and the combustion efficiency, E. Using this information and the calorific value of the fuel ($C = 34,000$ kJ/L), the heat load, q, can be calculated using Equation 15.5-13.

$$q = FEC \qquad (15.5\text{-}13)$$

Alternatively, it can be approximated by multiplying the effective engine power by 2.8 ($q = 3PE$).

Example 6. A 30-t truck (298 kW) is operated at 80% of its nameplate specifications in a development heading. How much heat does this engine produce when fully loaded?

Solution. From Equation 15.5-13, $q = 2.8 \times 298 \times 0.8 = 668$ kW. This engine burns 668 kW of fuel to produce 298 kW of effective power.

Heat from Broken Rock

When freshly blasted rock is exposed to ventilation air and there is a temperature difference between them, there will be heat flow from the rock to the air. In most cases, the broken rock temperature will be less than the local VRT. However, the difference is small, and for rough estimations it can be ignored. Further, the rock temperature at the storage bin or exit point will be equal to the ambient air temperature. Based on these assumptions, the heat load from broken rock can be approximated using Equation 15.5-14 as follows:

$$q = mC_p(VRT - t_d) \text{ kW} \qquad (15.5\text{-}14)$$

where C_p is the specific heat of broken rock, kJ/kg°C.

Example 7. One hundred metric tons per hour of broken rock (granite) with a $C_p = 0.82$ kJ/kg°C is transported from a drawpoint where the rock temperature is 45°C to an ore bin where the rock has cooled down to 30°C. Estimate the heat transfer from the broken rock to the ventilation system.

Solution. For $m = 27.78$ kg/s and $\Delta t = 15$°C, Equation 15.5-14 yields $q = 342$ kW.

Heat from Groundwater

Mine water comes from two sources: natural fissures in the rock and human-induced sources from various applications, including water used for cooling. The fissure water has usually the same temperature as the VRT. However, the temperature of human-induced water can vary from below ambient temperature to more than 40°C (104°F). Ventilation air will pick up the heat from hot water in uncovered ditches. The heat flow from groundwater can be estimated by applying Equation 15.5-14, where the C_p(water) = 4.19 kJ/kg°C. Whenever possible, groundwater should be confined at the source and transported through pipelines or covered ditches.

Example 8. A mine produces 0.010 m³/s of fissure water at the local VRT of 55°C. If the water exits the discharge system at 30°C, how much heat is added to the ventilation air?

Solution. For $m = 10$ kg/s and $\Delta t = 25$°C, $q = 1,048$ kW.

Other Sources of Heat

Other sources of heat include those from the oxidation of timber, sulfide ores, and carbon-based minerals. Oxidation of timber usually begins with the growth of bacteria in poorly ventilated areas (with less than 2% oxygen). Coal and pyritic sulfides will oxidize at normal ambient temperatures. The reaction is exothermic. If the heat is not removed, this can initiate a self-heating process and can develop into fires. Except for mines with extensive caved areas, this heat is usually swept out by ventilation and ignored in the estimation of the heat load.

Total Heat Load

For a given section, the heat load is determined as the sum of the heat loads from different sources. Once evaluated, this load is used to determine the ventilation or refrigeration requirements. The requirements are calculated by equating the wet bulb temperature at the exit of the section to the reject temperature. Psychrometric charts can be used to determine the size of fans and cooling devices.

Example 9. Air enters a section of a mine at $t_d = 26$°C, $t_w = 20$°C, and 101 kPa of absolute pressure. The heat load in the section is estimated at 3,000 kW. If the wet bulb temperature of the air leaving the section is not to exceed 28°C, what is the required quantity of air?

Solution. For $V = 0.850$ m³/kg, $h_1 = 57$ kJ/kg, and $h_2 = 89.5$ kJ/kg, using Equation 15.5-11, $q = 78.5$ m³/s.

MINE COOLING

When actual or projected heat loads exceed the ability of a mine's ventilation system to remove the heat, cooling methods must be considered. In principle, a mine cooling system is a heat-removal system. A substance such as air, water, or ice is sent into the mine at a low enthalpy state and removed at a higher one. Planning for existing or proposed mines generally follows the methods listed in this section (McPherson 1993; Howes and Hortin 2005).

Airflow Increase

It is generally less expensive to increase airflow than to install refrigeration. Primary fans can be replaced or sped up, booster fans can be installed, or new airways can be driven. However, for older mines, airflow increases are often impractical because of the cube relationship between fan power and airflow increase through a given resistance. Nor does increasing airflows help mines at or below the critical ventilation depth, which is the depth at which the intake shaft wet bulb temperature exceeds the design reject temperature.

Chilling the Service Water

This technique was developed in South Africa. All metal mines require service water to suppress dust whenever rocks are drilled or broken and to wash sidewalls and the face for ground support and geologic mapping. Chilled service water is among the most flexible cooling methods available, because it is used where needed the most and then turned off and routed elsewhere. It intercepts rock heat before that heat enters the air. However, chilling the service water requires a refrigeration plant and insulated pipelines. Plate-and-frame evaporators can produce 1°C (34°F) water, whereas shell-and-tube heat exchangers are limited to about 3.3°C (38°F) because of potential tube rupturing. The effect can be wasted if chilled service water is used where cooling is not required. Chilled service water can be potable or nonpotable, but in any case it should be treated for microorganisms.

Bulk Cool Intake Air on Surface

This method is used at warm-climate mines to provide winter-like or better conditions year round. Surface refrigeration plants are almost always less expensive and easier to install and maintain than equivalent underground plants, are not limited in heat rejection, and are not subject to disruption following underground power outages. Ammonia machines can be used. Compressors are usually centrifugal or rotary screw. Multiple machines are often installed in parallel so that a single machine can be down for maintenance without disrupting cooling duty to any significant extent. Depending on conditions, evaporators and/or condensers on multiple machines can be installed in series to provide greater temperature difference (lift) to water temperatures. Mine refrigeration machines typically provide a lift of about 6° to 8°C (11° to 14°F). The lower positional efficiency and utilization over a year's time generally preclude surface bulk cooling from being used in cooler-climate mines.

Bulk Cool Air Underground

This practice has a higher positional efficiency and a higher utilization than do surface plants because it is closer to work areas. It also utilizes the full heat-removal potential of the ventilation system by rejecting heat to exhaust air via underground cooling towers. The refrigerant used must be nontoxic. Both centrifugal and rotary-screw compressors are used. As with surface plants, multiple machines are usually installed in parallel.

Installing Chillers

Also known as district cooling, midsize refrigeration machines of 350 to 1,250 kW(R) can chill water for a closed-loop system of cooling coils or spray chambers. These coils and chambers are located at the entrances of work areas. Coils can be located in a bank. Auxiliary fans draw air through the cooling units and blow it through a duct to work headings. Refrigeration machine condenser heat is preferably rejected to clean service water, if enough quantity is available. But heat is often rejected to mine sump water before being pumped out of the mine. Problems encountered with district cooling are coil-air side and condenser tube-water side fouling.

Spot Coolers

These units range from 70 to 350 kW(R) capacity. The units are portable and contain one or two direct-expansion cooling coils, one or two condensers, one or two compressors, and a fan. Cooled air is delivered through a duct to the heading. Condenser heat is usually rejected to the mine service water. This water typically discharges at temperatures between 35° and 45°C (95° and 113°F), and thus it should be piped to prevent it from contacting intake airflows. The service water flow rate is approximately 0.02 L/s per kW(R) of cooling at 21°C (70°F) water temperature, but can be less or more depending on temperature.

Spot coolers are popular with the work force. Although limited to moderate cooling duties, they are easily moved to positions close to advancing headings. Spot coolers cost more per unit of cooling but can be the only choice for development, exploration, or production workings on the periphery of operations. The service water system, mine dewatering system, and surface treatment or disposal system must be capable of handling the increased water flow required for condenser cooling. Some applications might employ air-cooled condensers. In the past, reciprocating compressors were used. More recently, scroll-type hermetic compressors have been employed because they are less expensive and can handle liquid slugging better than can reciprocating units. Larger spot coolers can use rotary-screw compressors. Current designs use non-ozone-depleting and non-global-warming refrigerants. As with water-cooling coils, direct expansion coils on spot coolers must be washed, the frequency depending on how much dusty air the unit receives.

Cooling Towers

Cooling towers utilize the full heat-removal capacity of a mine by rejecting refrigeration machine condenser heat to the exhaust air. Exhaust air typically exits the work areas at a temperature close to the design reject temperature, for instance, 28°C (82°F) wet bulb. A common misconception is that heat cannot be rejected to saturated air through evaporation, but the process in a cooling tower involves both sensible heating of the air by hot water followed by latent, or evaporative, cooling of the circulating condenser water. Latent heat is typically about 80% of total heat transfer. Water sprayed in the cooling tower is often about 45°C (113°F), which cools to about 35°C (95°F) before being sent back to the refrigeration room. Actual temperatures depend on the efficiency of heat transfer, both in tower and machine condensers, and on various design parameters, such as water flow rate and condensing temperature. Equation 15.5-15 calculates the total heat transfer on the water side, H_w, and Equation 15.5-16 calculates the total heat transfer on the air side, H_a, as follows:

$$H_w = M_w C_p \Delta t \text{ kW} \qquad (15.5\text{-}15)$$

where

M_w = mass flow of water, kg/s
C_p = specific heat of water, 4.18 kJ/kg°C
Δt = inlet water temperature minus the outlet water temperature, °C

$$H_a = M_a \Delta h \text{ kW} \qquad (15.5\text{-}16)$$

where

M_a = mass flow of air, kg/s
Δh = specific enthalpy of exit air minus the specific enthalpy of entering air, kJ/kg

Cooling tower analysis proceeds by equating H_w with H_a. Mines use the "factor-of-merit" method, developed in South Africa, for planning, designing, and analyzing cooling towers and spray chambers. The factor-of-merit for a direct-contact heat exchanger, such as a cooling tower or spray chamber, ranges between 0 and 1, where 0 implies no heat transfer takes place and 1 implies as much heat transfer as permitted by the second law of thermodynamics (Burrows 1982). In other words, the fluid with the lower thermal capacity, either water or air, heats up or cools down to the entering temperature of the other fluid.

Normally the designer knows the exhaust air quantity and temperature available for the cooling tower through measurements or the mine plan. For a new mine, the tower entering air wet bulb temperature can be assumed to be about 1.5°C (2.7°F) lower than the design reject. The lower temperature is caused by leakage airflows entering exhaust. At most deep and hot mines, the air can be assumed to be saturated. Very roughly, 19.8 to 40.3 kW of heat can be rejected to 1 m³/s (2,100 cfm).

Cooling tower analysis is a complicated procedure. The designer must know or assume the heat to be rejected (the evaporator duty times the condenser heat-rejection factor), the entering air quantity and temperature, the tower factor-of-merit (from experience, based on accepted design criteria), and a first guess on the condenser water flow rate. Condenser water flow depends on the heat to be rejected and the refrigeration machine specifications. The exiting air temperature, the entering and exiting water temperatures, and the water makeup requirements can then be calculated via an iterative procedure. An example of cooling tower design for a 3.5-MW underground refrigeration plant is given in the 2007 ASHRAE applications handbook.

Large Spray Coolers

Large spray chambers are used to bulk-cool ventilating airflows that are used for mining operations. These chambers can be located on the surface at the intake shaft or underground. Chambers are usually two- or three-stage horizontal openings equipped with Vee Jet spray nozzles lining the chamber. Wave-fin mist eliminators at the exit keep water from blowing out of the chamber. More details are given in the 2007 ASHRAE applications handbook.

Small Spray Chambers

Small spray chambers are typically portable, enclosed, direct-contact heat exchangers for small duties 35 to 350 kW. An auxiliary fan draws or blows air through the chamber, which is then ducted to the work heading. Wave-fin or mesh-type mist eliminators keep sprayed water in the chamber. On the water side, about 0.018 L/s (2.85 gpm) per kW of chilled service water is required for operation. This water is dumped into the ditch and into the mine dewatering system. Or, in a closed loop system, the water is collected and pumped back to a district refrigeration machine. These chambers are efficient and inexpensive for small duties, especially those requiring auxiliary ventilation. Limited air scrubbing takes place. But they require the chilled water and mine dewatering infrastructure to support them.

Integrated Refrigeration Systems

Integrated or combination systems can cool both intake air and/or service water. The proportion depends on the season, with more bulk cooling in the summer and more water chilling in winter. If service water is recycled after being pumped out of the mine, it is often sent first to a precooling tower before being refrigerated. Recycling mine water may reduce treatment costs by reducing water that is rejected to surface drainage streams or ponds. Integrated systems come in a wide variety of cooling and energy recovery methods that utilize both surface and underground components. Most deep, hot mines employ some type of integrated refrigeration system.

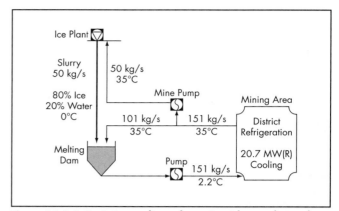

Figure 15.5-8 Basic ice cooling schematic with sample numbers

Ice Cooling

For deep, hot mines or mines already utilizing their full heat-removal capacity, ice cooling may be the answer. Ice can remove about 4.5 times the heat as an equivalent mass flow of chilled water. Ice is produced on the surface in chunk or slurry form and is sent underground to a melting dam, where the ice is melted by warm water returning from work areas. South African mines have been at the forefront of ice application. Problems with ice transport have been mostly solved. Currently, ice systems can compete on a cost basis with standard cooling techniques for deep mines. Figure 15.5-8 shows a basic ice cooling system.

Vests and Cabs

As mines become deeper, microclimate cooling should be considered for specific operations. Dry-ice (frozen carbon dioxide) or cold-pack cooling vests can be used by crews setting up ventilation in hot headings before permanent systems are in operation. These vests are heavy-duty "fire-fighter"-type jackets with internal pockets that are filled with ice slabs; the jacket is kept closed by a belt and/or straps, and the wearer is cooled as the ice melts. Vests are not popular with the work force—they are bulky and the cooling effect only lasts between 2 and 3 hours, then the vest must be recharged. Air-conditioned cabs on large load-haul-dump machines and trucks have proven themselves worthy of consideration. One of the biggest problems with diesel equipment is the heat produced. A diesel engine produces about 2.8 times the rated power in heat. Thus, a 100-kW loader operating at 80% utilization produces $100 \times 0.8 \times 2.8 = 224$ kW of heat. The design reject temperature can easily be exceeded in headings serviced by diesel vehicles. Air-conditioned cabs protect the operator from heat, noise, and dust. Although air-conditioned cabs are expensive, the cost is much less than trying to maintain the reject temperature in all headings serviced by diesel vehicles.

Energy Recovery

Most mines send water underground for various purposes, including wetting down muck piles and reducing dust. Hot mines also send down water solely for cooling. In deep, hot mines, water pressure must be periodically broken to prevent excessive hydrostatic heads in pipes. This is often accomplished by pressure-reducing valves or by using a cascading water-tank system. If service water flow is high enough, thought should be given to a system that recovers the water's potential energy and converts it to useful work. A turbine—either a centrifugal pump designed to run in reverse or a Pelton wheel—can reduce the pressure and recover up to 80% of the energy required to pump the water back to the surface. Another benefit is preventing the conversion of water potential energy into heat; that is, 1°C (1.8°F) for every 427 m (1,400 ft) of drop. Other energy-recovery devices include the hydrochambers or hydrotransformers (popular in Europe). These are underground chambers equipped with two pumps in two separate circuits. The chambers are alternately filled with chilled water delivered from surface and hot water returned from the mine (McPherson 1993).

Controlled Recirculation

This technique is used with bulk air-cooling underground to reduce the fresh-air quantity needed to ventilate the mine. A portion of the mine exhaust air is cooled, scrubbed, and recirculated back to the workings. Advantages include higher air velocities in work areas; lower circuit fan pressure, because less air is drawn through the primary circuit; and a lower heat load due to autocompression. Autocompression can account for half the heat load at some mines. Controlled recirculation is usually applicable to deep, hot, mechanized mines, but because of the risk involved with recirculating air, it has not been used widely. Controlled recirculation systems must be monitored and controlled very carefully (Marks 1987). Fresh-air flow from the surface must be high enough to remove dust, blasting fumes, and diesel emissions. Controlled recirculation can be shut down during blasting.

Thermal Storage Systems

This innovative technique has been used to good effect in cold-climate mines. In winter, intake air is drawn through surface rubble piles or ice stopes where it is heated by the rock or by spraying water, which then freezes, releasing its latent heat into the air. In summer, the process reverses, with intake air being cooled by the rock rubble or melting ice (Stachulak 1989). This system is being successfully applied at Vale's Creighton mine in Ontario, Canada.

Maintenance

Mine cooling is not an install-and-forget system. Adequate maintenance must be planned from the start. Any cooling coil, direct expansion, or chilled water is subject to air side fouling. If at all possible, coils should be placed upstream of blasting. Coils receiving dusty and fume-laden air must be washed with a high-pressure spray (but not too high of a pressure or the fins will bend) at least every other day. Cooling coils should have a fin spacing no tighter than one fin per 4.2 mm (6 fins per inch) to limit fouling. If sump water is used for district refrigeration machine condenser cooling, the condenser tubes must be cleaned, either by an automatic reversing brush system with a flow diverter valve or by shutting the machine down periodically (depending on the water-fouling factor) and cleaning the tubes manually. Mines with extensive refrigeration systems (e.g., a large chiller plant or more than 10 spot coolers) should employ a mechanic specializing in refrigeration. This person can be assisted by apprentice mechanics. Another viable approach is to retain the services of factory mechanics or a local high-volume air conditioning and repair shop. Any such mechanics must be certified to handle refrigerants.

Selecting a Cooling System

Selecting a cooling system takes place after the mine cooling and ventilation requirements have been specified. The relationship between mine cooling and ventilation requires special attention, especially if mine exhaust air is expected to remove heat via underground cooling towers. Cost-benefit is the most common analytical technique. Each alternative requires certain expenditure and each provides a certain benefit. But just as important are hardware dependability, maintainability, dependency on outside factors, flexibility, safety, and technological level. An attempt must be made to quantify some of the more intangible aspects. Factors influencing selection of a mine cooling system include the following.

- **Seasonal surface ambient conditions.** As mentioned previously, warm climate mines tend to bulk-cool air on the surface close to the intake shaft.
- **Ore-body and mining methods.** The more massive the ore body, the more ideal bulk cooling becomes. Scattered stoping or continuously advancing headings lend themselves more to spot or district cooling. Large ore bodies amenable to bulk mining may use larger airflows and yet less air per ton of production.
- **Mining rate.** A fast mining rate prompts a high instantaneous heat load. But ironically, less total heat (and therefore lower costs) might be drawn from the wall rock due to stopes being quickly filled and sealed. One should leave as much heat in the wall rock as possible.
- **Size and condition of major airways.** In older mines beginning to experience heat problems, airflow increases might be prohibitively expensive due to small airways. Refrigeration might be needed earlier than at larger mines.
- **Heat sources.** The percentages of heat flow from various sources influences the selection of cooling methods. If autocompression is a major source, bulk cooling at the bottom of an intake shaft might be called for. If large quantities of heat come from diesel equipment, cabs might be appropriate. If more work takes place in the stopes or in the development and exploration headings than in the haulways, spot cooling might be best.
- **Costs.** Knowing the costs of labor, water, supplies, and power is critical for determining the optimum capital expenditure to control operating costs. For example, if power cost is high, a higher coefficient-of-performance system may be warranted.
- **Governmental regulation.** Some governments regulate heat stress. There are also laws prohibiting toxic refrigerants or combustible insulation underground. These factors can influence design of cooling systems.

VENTILATION AND REFRIGERATION PLANNING

Example 10 is presented to illustrate the various steps involved in developing a ventilation-refrigeration system for a hot deep mine. Decision parameters such as the critical depth, total heat load, and airflow requirements to create acceptable work conditions are highlighted. Once these parameters are made available, ventilation and climate simulators can be used to represent the mine openings and to predict the ventilation and air-cooling requirements.

CLIMSIM, a climatic simulator based on radial heat transfer from or to an airway, is used to predict the temperature and air-cooling profiles for a set of new workings, especially

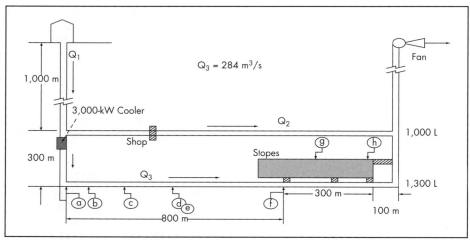

Figure 15.5-9 Mine ventilation schematic

for those subject to different types of heat loads (McPherson 1993). The profiles are then evaluated against preestablished standards, and the cooling requirements are determined. Wet bulb temperature and effective temperature are used as the target temperatures in the evaluation.

Example 10. A mining company extracts 1,200 t/d (1,320 stpd) of ore from an ore body located 1,000 m (3,280 ft) below the surface. Exploration work has shown that the ore body, a mineralized quartzite, extends to lower levels for at least 300 m (980 ft). This work has also shown a substantial increase in rock and groundwater temperatures with depth. Preliminary measurements have shown that, due to autocompression, the wet bulb temperature will increase by about 6°C (11°F) per 1,000 m (3,280 ft) of vertical depth and that the groundwater temperature will be 52°C (126°F) (VRT). The inflow of fissure water is estimated at 2.6 L/s (41 gpm). Currently, the company is preparing a prefeasibility study to deepen the mine to 1,300 m (4,265 ft) below the surface and to extract the remaining ore body at the rate of 1,500 t/d (1,650 stpd). The ore will be mined by conventional means of drilling and blasting. The required mining equipment includes seven jumbo drills (111 kW each), four production loaders (235 kW), and seven trucks (three 381 kW and four 135 kW), all diesel powered. In addition, the equipment includes a crusher-conveyor system (225 kW). The heat loads from the broken rock and groundwater are estimated at 352 and 262 kW, respectively. Figure 15.5-9 shows a schematic of the mine and the location of various heat sources. Table 15.5-2 summarizes the ventilation conditions that can be achieved with the current infrastructure, thermal properties of the rock, and heat loads for the resulting working areas. Based on this information, the problem is to determine the cooling requirement for this mine.

Solution. The problem was investigated using CLIMSIM for various heat load and ventilation conditions. During the analysis, two target temperatures were used: an average wet bulb temperature of less than 28°C (82°F) and a maximum wet bulb temperature of 32°C (90°F) at any working. When these targets were not met, the ventilation and cooling requirements were modified. Based on the psychrometric properties of the surface air, the depth of the openings, and an estimated VRT of 52°C (126°F), two scenarios were analyzed. The first simulated the mine before any air-cooling was installed and the second after an underground bulk air cooler was installed in the intake shaft.

Table 15.5-2 Airway, ventilation, and heat flow parameters for Example 10

Description	Parameter
Geometry of mine opening	
Length, m	1,200
Cross-sectional area, m^2	30.0
Perimeter, m	22.0
Airway friction factor, kg/m^3	0.04
Wetness factor	0.15
Ventilation at intake	
Airflow rate, m^3/s	284
Pressure, kPa	101.0*
Dry bulb temperature, °C	25*
Wet bulb temperature, °C	21*
Thermal properties (quartzite)	
Virgin rock temperature, °C	52.0
Geothermal step, m/°C	100
Thermal conductivity, W/m °C	5.50
Thermal diffusivity, $m^2\tau/s$	2.32
Heat sources†	
Ore bin, (x = 10 m), kW	352
Conveyor (x = 10 to 100 m), kW	100
Crusher (x = 100 m), kW	125
Drain water (x = 100 to 800 m), kW	262
Trucks (x = 100 to 800 m), kW	894
Loaders (x = 800 m), kW	940
Drills, controls (x = 1,000 m), kW	777
Booster fans (x = 1,100 m), kW	600

*Measured at intake shaft collar.
†x = Distance from intake.

For the first scenario, two sets of parameters were estimated: the required airflow rate to dilute the air contaminants and the heat generated at the workings. Based on the mining equipment, a flow requirement of 0.079-m^3/s (167 cfm) per kW of power, and a minimum quantity of 70-m^3/s (148,000 cfm) for two fixed facilities, the required airflow rate for the mine

Table 15.5-3 Airflow requirements for Example 10

Item	Utilization, %	Quantity, m³/s
Loaders (4 × 235 kW)	80	59.4
30-t trucks (3 × 381 kW)	80	72.2
Other trucks (4 × 135 kW)	80	34.1
Jumbo drills (7 × 111 kW)	50	30.7
Crusher system (225 kW)	100	17.8
Underground shops	—	40.0
Other fixed facilities	—	30.0
Total quantity		284.2

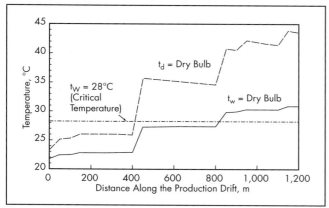

Figure 15.5-10 CLIMSIM output for Equation 15.5-10 (wet and dry bulb temperatures versus distance)

was estimated at 284 m³/s (600,000 cfm) (Table 15.5-3). The heat loads and their distances from the intake shaft are those shown in Table 15.5-2 and Figure 15.5-9. The conveyor belt and the drainage water were represented as linear sources of heat. Once these values were input to the model and the simulator was run, the following results were generated: the wet bulb temperature at the bottom of the intake shaft was 26°C (79°F) and averaged 33.5°C (92.3°F) at the workings, and the dry bulb temperatures at active workings varied between 36° and 52°C (97° and 126°F). Both temperatures are above the acceptable target values, indicating that to maintain safe working conditions, the mine would require some form of cooling system.

For the second scenario, a 3,000-kW(R) refrigeration plant was installed in the intake shaft. Physically, this could be installed on the 1,000-m (3,280-ft) level, just upstream the intake shaft. Once this parameter was added to the model (as a negative heat source), the simulator generated the following results: the average wet bulb temperature would be equal to 26.5°C (79.7°F), with a maximum of 30°C (86°F) in the stopes; and the dry bulb temperature would average 33°C (91°F). These results show that the 3,000-kW plant would provide acceptable work conditions in the mine. Figure 15.5-10 shows two temperature profiles for the 1,200-m- (3,940-ft-) long production drift. The 3,000-kW refrigeration plant would reduce the wet bulb temperature at the bottom of the shaft to 22°C (72°F). This temperature would remain below 26°C (79°F) for 800 m (2,620 ft) where four 235-kW loaders are operated in four production stopes and two development headings. The heat and water vapor generated by these and other diesel units would increase the wet and dry bulb temperatures to 31° and 42°C (88° and 108°F), respectively.

This example shows the manner in which CLIMSIM can be used to formulate alternate solutions to heat flow problems and determine the capacity of the cooling system. It also shows the effect of autocompression in the ventilation air and allows the user to determine the critical depth, that is, the depth at which the mine would require some form of external cooling. In this example, such a depth is located at about 900 m (2,950 ft) below the surface.

CONCLUSIONS

In the United States, heat becomes an environmental concern in mines where workings are located at depths greater than 1,000 m (3,280 ft). At such depths, ventilation alone is not sufficient to control the heat load, and the mine must be supplemented by some form of air conditioning.

Since the introduction of the first heat stress index in 1923, more than 90 indices have been developed in various countries. In U.S. mines, heat stress is not regulated. Management must set and enforce the applicable limits. The most commonly used indices are

- Wet bulb temperature for continuous work: ≤27°C (80.6°F)
- Effective temperature for reduced workload: ≤28°C (82.4°F)
- Air velocity: ≥0.5 m/s (100 fpm)

In underground mines, heat is emitted from various sources, including autocompression, wall rock, mining equipment, groundwater, the oxidation of minerals, and metabolic heat. The estimation of the total heat load is crucial to the design of an air-cooling system. Once the individual heat loads are quantified, mine climate simulators such as CLIMSIM can be used to represent the mine openings and to predict the ventilation and air-cooling requirements.

Mine cooling is a heat-removal process in which a substance such as air, water, or ice is sent down the mine at a low enthalpy state and removed at a higher one. Depending on the amount of heat to be removed, several cooling methods can be used to create acceptable work conditions. These can range from upgrading the primary fans (increased ventilation) to cooling the mine air (refrigeration).

Ventilation is the least-expensive method to remove heat from mine workings. However, it has a limit—the critical depth. The critical depth is the depth at which the wet bulb temperature of the intake air exceeds the allowable wet bulb temperature.

Refrigeration plants used in hot mines range from spot coolers (for localized heat loads) to bulk air coolers (for widespread heat flow problems). Bulk air coolers can be installed on the surface, near the intake shaft, or underground, near the workings. Deciding at which stage a refrigeration plant is required is a difficult question to answer. Economics and site-specific conditions will dictate the optimal choice.

REFERENCES

ACGIH (American Conference of Governmental Industrial Hygienists). 2008. *Documentation of the Threshold Limit Values and Biological Exposure Indices*. Cincinnati, OH: ACGIH. Available from www.acgih.org/home.htm.

ASHRAE (American Society of Heating, Refrigerating, and Air-Conditioning Engineers). 2007. Mine air conditioning and ventilation. In *2007 ASHRAE Handbook—HVAC Applications.* pp. 27.1–27.13.

Brake, D.J. 2008. *Psychrometry, Mine Heat Loads, Mine Climate and Cooling.* Brisbane, Australia: Mine Ventilation Australia.

Burrows, J. 1982. Refrigeration—Theory and operation. In *Environmental Engineering in South African Mines.* Cape Town, South Africa: The Mine Ventilation Society of South Africa. pp. 613–652.

Del Castillo, D., and Dawborn, M. 2002. Refrigeration and ventilation design for the deepening of Mindola copper mine. In *Proceedings of the North American/9th U.S. Mine Ventilation Symposium,* Kingston, Canada, June 8–13. Lisse, The Netherlands: A.A. Balkema. pp. 401–408.

Duckworth, I.J. 1999. Rapid evaluation of rock thermal parameters at the Lucky Friday mine. In *Proceedings of the 8th U.S. Mine Ventilation Symposium,* Rolla, MO, June 11–17. Rolla, MO: University of Missouri-Rolla Press. pp. 337–346.

Hardcastle, S. 2006. Controlling personnel heat exposure in Canada's deep and highly mechanized mines. In *Proceedings of the 11th U.S./North American Mine Ventilation Symposium,* University Park, PA, June 5–7. London, UK: Taylor and Francis. pp. 259–269.

Hartman, H.L., Mutmansky, J.M., Ramani R.V., and Wang Y.J. 1997. *Mine Ventilation and Air Conditioning.* New York: John Wiley and Sons.

Howes, M.J., and Hortin, K. 2005. Surface cooling at Kidd Creek mine. In *Proceedings of the 6th International Mine Ventilation Congress,* Brisbane, Queensland, Australia, July 6–8. Carlton, Victoria: Australasian Institute of Mining and Metallurgy. pp. 55–63.

Marks, J.R. 1987. Computer aided design of large underground direct-contact heat exchangers. In *Proceedings of the 3rd U.S. Mine Ventilation Symposium,* University Park, PA, October 12–14. Littleton, CO: SME. pp. 158–162.

McPherson, M.J. 1993. *Subsurface Ventilation and Environmental Engineering.* London: Chapman and Hall.

McQuiston, F.C., and Parker, J.D. 1994. *Heating, Ventilation, and Air Conditioning—Analysis and Design.* New York: John Wiley and Sons.

Mutama, K.R. 2005. Assessing new airflow requirements at the Meikle mine. In *Proceedings of the 8th International Mine Ventilation Congress,* Brisbane, Queensland, Australia, July 6–8. Carlton, Victoria: Australasian Institute of Mining and Metallurgy. pp. 481–487.

Stachulak, J.S. 1989. Ventilation strategy and unique air conditioning at Inco Limited. In *Proceedings of the 4th U.S. Mine Ventilation Symposium,* University of California, Berkeley, CA, June 5–7. Littleton, CO: SME. pp. 3–9.

Van der Walt, J., Pye, R., Pieterse, H., and Dionne, L. 1996. Ventilation and cooling at Barrick's Meikle underground gold mine. *Min. Eng.* 48:(4):36–39.

CHAPTER 15.6

Radiation Control

Derek B. Apel and Zaher Hashisho

RADIOACTIVE DECAY OF URANIUM SERIES
Uranium is a naturally occurring radioactive element. It is found in rocks, soils, rivers, and oceans, and traces can be found in food and in the human body. Uranium is a common substance that is found throughout the earth's crust. It is more common than all of the precious metals and is more common than many nonprecious metals. Uranium, for example, is 40 times more common than tin. More than 200 minerals contain uranium, of which uraninite is the most common. Uranium concentrations vary from substance to substance and place to place.

Uranium is the heaviest of the naturally occurring elements and is made up of a number of different forms known as isotopes. Natural uranium is primarily made up of two isotopes, ^{235}U and ^{238}U. About 99.3% of natural uranium is ^{238}U and approximately 0.7% is ^{235}U, the "fuel" component of natural uranium. Because it is present virtually everywhere, it contributes to what is called the natural background radiation. Almost every backyard contains traces of naturally occurring uranium.

Radon is derived from the radioactive decay of uranium to radium then radon (see Figure 15.6-1). During the decay process, the unstable elements emit radiation. In essence, three different types of radiation are emitted: alpha, beta, and gamma.

Alpha radiation occurs when an alpha particle (the nucleus of a helium atom, which contains two protons and two neutrons) is emitted from the atom's nucleus. An alpha particle carries a relative electrical charge of +2, and when this particle is emitted from an element, the atomic number of this element is reduced by 2 and its atomic mass is reduced by 4. Usually, alpha radiation is not able to penetrate clothing and human skin, but if the alpha-emitting material is inhaled, it will cause damage to the lung tissue.

In simple terms, the beta radiation occurs when an electron is ejected from the nucleus of the decaying element. Therefore, this radiation has a relative electrical charge of –1, which causes the decaying element to gain a positive charge. This is shown on the periodic table as an increase by 1 atomic number of the new element. Beta radiation may travel a couple of meters in the air and can penetrate human skin. If beta-emitting materials are left on the skin for a longer period of time, they will cause skin injury.

Like visible light, radio waves, and X-rays, gamma radiation is a type of electromagnetic radiation. Gamma radiation can travel for several meters in the air and several centimeters in human tissue. Gamma rays can penetrate most materials. Dense materials (e.g., lead) are needed for shielding against gamma radiation.

Radon is a colorless, odorless, tasteless, radioactive gas that generally lacks activity toward other chemical agents. However, radon occasionally forms clathrate compounds with some organic compounds, and it may form ionic or covalent bonds with highly reactive elements such as oxygen or fluorine. Radon is the heaviest noble gas and exhibits the highest boiling point, melting point, critical temperature, and critical pressure of all noble gases. Radon is highly soluble in nonpolar solvents and moderately soluble in cold water. Radon's isotopes, all of which are radioactive, include mass numbers ranging from 200 to 226. ^{222}Rn formed in the ^{238}U decay chain is the most important isotope because of its relatively long half-life of 3.82 days. The short half-life beta- and gamma-emitting decay products of ^{222}Rn achieve equilibrium with the parent isotope within several hours. The density of radon is 9.73 g/L, and it has solubility in water of 22.2 mL/100 mL at 20°C.

Units of Radiation
The activity (rate of decay) of ^{222}Rn is expressed in units called curies. The curie is based on the rate of decay of 1 g of ^{226}Ra or 3.7×10^{10} disintegrations per second. The International System of Units (SI) measure of activity is becquerels per cubic meter (Bq/m^3). One becquerel equals one disintegration per second. The radiation unit most widely used by the uranium industry is the working level (WL), which is based on the measurement of energy released by ^{222}Rn and its progeny. One WL means that the ^{222}Rn and its progeny released in 1 L of air will emanate 1.3×10^5 million electron volts (MeV)

Derek B. Apel, Professor, School of Mining & Petroleum Engineering, University of Alberta, Edmonton, Alberta, Canada
Zaher Hashisho, Assistant Professor, Department of Civil & Environmental Engineering, University of Alberta, Edmonton, Alberta, Canada

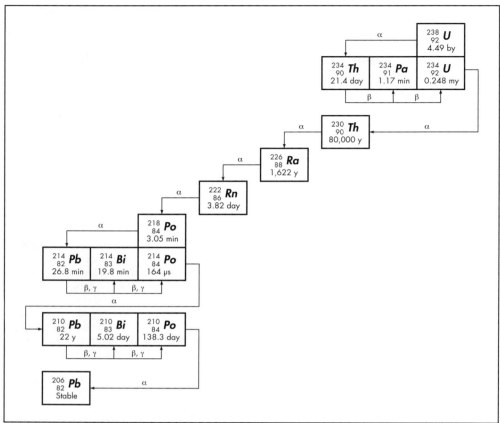

Figure 15.6-1 Uranium decay series (includes the half-life, atomic mass, atomic number, and emission type of each decay product)

of energy from alpha particles. The exposure to radiation at underground mines, which combines both exposure and time, is the working level month (WLM), which is defined as the exposure to one WL during 170 hours. For example, the WLM exposure for a miner who worked underground during a month for 120 hours and whose exposure was measured to be 0.9 WL has an exposure of 0.9 WL × 120 h/170 h = 0.64 WLM.

The joule-hours per cubic meter (Jh/m^3) is the SI unit of cumulative exposure to ^{222}Rn and its progeny. One WLM is equal to 3.5×10^{-3} Jh/m^3.

Occupational Exposure Limits

The exposure limit varies by regulating agency and type of worker. In the United States, the Mine Safety and Health Act regulates exposure to ^{222}Rn gas and ^{222}Rn progeny by underground miners. The act sets limits so that no employee can be exposed to air containing ^{222}Rn progeny in excess of 1.0 WL or 100 pCi/L (picocuries per liter) in active work areas. The act also limits annual exposure to ^{222}Rn progeny to less than four WLM per year (Field 1999).

EFFECT OF RADON PROGENY ON HUMAN HEALTH

The linkage between radon and the increased risk of cancer has been established throughout numerous studies (Frumpkin and Samet 2001). Even before the discovery of radon, many people blamed "mines air" for high mortality rates among underground miners. This was first described by Georgious Agricola (Agricola 1912) in his greatest work titled *De Re Metallica*, which was originally published in 1556. The study that most referenced in the literature linking radon exposure to lung cancer are findings summarized in a report titled *Health Effects of Exposure to Radon*, prepared by the National Research Council's Committee on the Biological Effects of Ionizing Radiation (BEIR) Report VI (NRC 1998). This report was based on data pooled from underground mines. The data show that of 60,000 miners who worked between 1941 and 1990 in different mines located in different countries, more than 2,600 developed lung cancer against only 750 cases based on background data. A study conducted by Kreuzer et al. (1999) reported that among 64,000 uranium miners in East Germany who worked a minimum of 180 days at these mines between 1946 and 1989, there were 1,436 deaths due to lung cancer.

The findings from these studies overwhelmingly show that radon progeny exposure causes lung cancer. Most of the radiation-hazard results from the inhalation of the short-lived radon progeny. Radon itself is of minor concern, because most of the inhaled radon is exhaled. Therefore, this is not radon itself but radon's progeny whose decay in lungs causes damage to the cells and DNA mutation that can lead to lung cancer. This was established by research conducted by Hursh et al. (1965). Additional work found that radon is removed from the body with a primary half-life of between 30 and 70 minutes (Crawford-Brown 1989).

EMANATION OF RADON AT UNDERGROUND MINES

The emanation of radon from uranium ore is the effect of radium decay. During this process, an alpha radiation is emitted from an atom of radium and its atomic mass is being reduced from 226 to 222. At the same time, the atomic mass drops from 88 to 86, and radium changes into radon. Radon can emanate to the mine atmosphere from the ore, rock walls, and mine water. The rate of radon emanation can be described as the number of radon atoms exhaled from a unit area of rock in a unit of time, and it is defined in units of picocuries per square centimeter per second ($pCi/cm^2/s$).

The emanation of radon from rock surfaces in underground mines (J) can be calculated (Bates and Edwards 1980) as follows:

$$J = C_\infty \sqrt{\lambda D \phi} \ pCi/cm^2/s \quad (15.6\text{-}1)$$

where

C_∞ = radon concentration at infinite distance into rock, pCi/m^3
λ = radon decay constant (0.00755 disintegrations per atom or 2.1×10^{-6} Bq)
D = diffusion coefficient, m^2/s
ϕ = rock porosity (0.50 = 50% porosity)

If the emanation of radon from a unit volume of rock (B) is known, Equation 15.6-1 can be modified to the following form:

$$J = B\sqrt{\frac{D}{\lambda \phi}} \quad (15.6\text{-}2)$$

where B is sometimes called the emanating power, $pCi/cm^3/s$.

The random movement of the radon gas atoms mixed in the air results in a net migration of the radon gas toward the direction of its decreasing concentration in the air. This phenomenon is called molecular or atom diffusion.

Table 15.6-1 shows examples of various media and their coefficients of diffusion. The table illustrates that the coefficient of diffusion decreases with the density of the material.

CONTROLLING RADON PROGENIES AT UNDERGROUND URANIUM MINES

The control of radon progenies at underground uranium mines should commence with the process of selecting the mining method, controlling water inflows, and designing a flexible ventilation system. In addition, each mine has to establish safety operating procedures specific for each operating mine. The latter is extremely important, as even the best ventilation system can malfunction because of a power outage, human error, or other unforeseen circumstances. When designing ventilation systems for underground uranium mines, deposits can be divided into two groups: low-grade deposits, usually ranging from 0.1% to 2% U_3O_8 (triuranium octoxide), and high-grade deposits, where the grade can exceed 20% U_3O_8.

Ventilation Systems

In the case of high-grade deposits, the radon emanation rate from the ore would make it practically impossible to dilute the radon daughters using flush-through ventilation, and in these cases the ore is mined using remote mining methods (e.g., raise boring or mining using water jets) while a minimum pillar thickness of 5 m of an inert rock is kept between the ore and the miners (Figure 15.6-2; Apel 2000).

Production areas where the emanation of radon is high can be ventilated using both flush-through ventilation and a secondary ventilation system that is used to capture radon at its source and then exhaust it to the restricted entry return air drifts. The new development headings shall be ventilated using a combination of positive and negative pressure ducting. The positive pressure ducting is used to flush the air at the heading, whereas the negative pressure rigid ducts are used to transfer the radon-contaminated air to the exhaust air drifts or shafts (see Figure 15.6-3).

All inactive areas must be sealed off by ventilation bulkheads, and the bulkheads should be periodically checked for radon leaks. In cases where a complete seal cannot be established by the ventilation bulkheads, it should be connected to the low-pressure negative ventilation system.

Experiments conducted in mines in the United States showed a reduction of radon concentration by using positive blowing versus the exhaust systems of up to 20% (Schroeder and Evans 1969; Franklin 1981). However, these experiments were conducted in relativly shallow mines, and it may not be feasible to achieve any decrease of radon production in deep mines where rock-in-situ stresses are much higher.

Water Control

Radon is soluble in water and is also readily released from it. Therefore, by limiting water inflows into mine openings, radon emanation into the mine air will be reduced. Controlling the water inflows can be achieved by using grout curtains ahead of the development or using ground freezing (Apel and Szmigiel 2006). The solubility of radon decreases with an increase in water temperature (e.g., 510, 230, and 169 cm^3/kg at 0°C, 20°C, and 30°C, respectively) (NRC 1998). Because not all the radon leaves water immediately after entering the underground openings, all the water sumps must be ventilated using negative pressure ventilation. The removal of radon from water can be done through aeration. Aeration simply forces radon from the water to the air and can be very effective. According to Dixon et al. (1991) and Kinner et al. (1993), bubble-plate aeration and diffused bubble aeration as point-of-entry units are capable of achieving removal efficiencies in excess of 99% at loading

Table 15.6-1 Coefficient of diffusion for radon in various materials

Material	Coefficient of Diffusion D, m^2/s
Dense rock	0.05×10^{-6}
Rock with porosity of 6.2%	0.2×10^{-6}
Rock with porosity of 12.5%	0.5×10^{-6}
Rock with porosity of 25%	3.0×10^{-6}
Unconsolidated soil material	3.0×10^{-6}
Compacted silty sands	3.2×10^{-6}
Compacted clayey sands	2.5×10^{-6}
Compacted inorganic clays	2.7×10^{-6}
Silty sandy clay	2.5×10^{-7}
Uranium mill tailings	6.0×10^{-8}
Concrete	$(1.7–3.0) \times 10^{-9}$
Air	$(10–12) \times 10^{-6}$
Water	1.13×10^{-8}

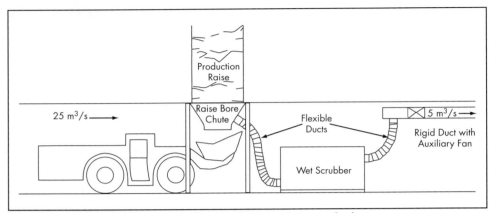

Figure 15.6-2 Ventilation setup for the raise-boring mining method

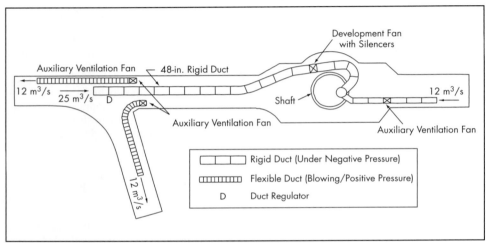

Figure 15.6-3 Ventilation setup during the development stage at a uranium mine

rates of 185 Bq/L or more, whereas the simple spray-jet aeration technique only removes between 50% and 75% of the radon.

Dust Control

Dust control is extremely important at uranium mines because radon progenies can get electrostatically attached to tiny particulates, such as dust, in air. If these particulates get inhaled, they will start decaying and releasing damaging alpha and beta particles. How much radon progenies get attached to the dust depends on factors such as the size and concentration of the airborne particles. The size and density of a particle determine whether or not the particle will reach the lung tissue. Particles with diameters of 10 μm are generally filtered by the nose and saliva during breathing and, therefore, should not reach the lungs. However, particles that range in diameters of from 2 to 10 μm have a good chance of being deposited in the bronchial region (Hinds 1999).

Obviously, selection of the mining method will impact how much dust will be created, but facts such as allowing longer periods after the blasts for dust to settle or spraying the muck with water after the blast will also limit the amount of air-bound dust created. Using personal protective equipment, such as an air respirator, is probably the best way to limit exposure to dust at uranium mines.

RADIATION MONITORING

Exposure of workers to radiation at underground mines can be controlled by adjusting the ventilation. The level of radiation at underground mines can change with further development of underground workings or changes in the uranium-ore production rates. Therefore, monitoring of radiation is very important in reducing the exposure of workers. Monitoring techniques at underground mines can be classified based on their ability to measure radon or its progeny. They can also be classified based on the measurement time frame into grab sampling, integrated sampling, or continuous monitoring and sampling. Radiation monitoring is not only limited to monitoring radon and its progeny, but it also includes monitoring of gamma-ray radiation.

Grab Sampling

Grab sampling consists of collecting samples over a short period of time, ranging from minutes to hours.

Radon Sampling with Scintillation Cells

A scintillation cell, such as the Lucas chamber, consists of a flask coated with silver-activated zinc sulfide on the inner surfaces, except for a clear window on one side to allow photons produced from alpha disintegrations to be counted. Sample air is admitted to the evacuated cell, which is typically 100 mL to 2 L in size (Sengupta 1989). Usually, the air sample is left to age for 4 hours to allow for equilibrium between radon and its progeny. Radon concentration is proportional to the photon count obtained with a photomultiplier tube. Correction factors are used to account for the decay during the time between sampling and counting and for decay during counting (McDermott and Ness 2004). Scintillation cells offer a reliable technique for radon sampling. For short counting periods, a sensitivity of 5 to 10 pCi/L is attained, whereas the limit sensitivity can reach 0.1 pCi/L for long counting periods (≥100 minutes) (Sengupta 1989).

Radon Progeny Sampling

Kusnetz method. The Kusnetz (1956) method involves drawing a known volume of air (100 to 250 L) through an air filter using a vacuum pump during a short time frame, typically 5 minutes (McDermott and Ness 2004). Airborne radon progeny attached to dust particles are collected on the surface of the filter. The decay of these progeny results in emission of alpha particles. The alpha particles are counted over a typical duration of 5 minutes after a delay period of 40 to 90 minutes. The delay time is the time in minutes from the end of sampling to the middle of counting. For a given delay time, t_k, an appropriate Kusnetz conversion factor, K, can be calculated using the following equations (CNSC 2003):

$$K = 230 - 2t_k \text{ for } 40 \leq t_k \leq 70 \quad (15.6\text{-}3)$$

$$K = 195 - 1.5t_k \text{ for } 70 \leq t_k \leq 90 \quad (15.6\text{-}4)$$

Knowing the volume of sampled air and the alpha particles count, radon progeny concentration can be determined as WL. The original method assumes perfect equilibrium, which results in greater than a ±25% error (Vutukuri and Lama 1986). Modifications of the Kusnetz method compensate for the assumption of equilibrium by taking alpha-activity count at several occasions during the growth of the progeny activity and making corrections for nonequilibrium (Vutukuri and Lama 1986).

The original Kusnetz method measures alpha particle activity with a rate meter, resulting in a practical lower limit of measurement of 0.3 WL. In the modified Kusnetz method, a scaler is used instead of the rate meter, resulting in a 10-times improvement in the sensitivity (Sengupta 1989).

The concentration in working levels, WL, can be obtained with the following equation (DCEP 2008):

$$\text{WL} = \frac{C_S - C_B}{E \times K \times V \times t_c} \quad (15.6\text{-}5)$$

where

- WL = concentration in working levels
- C_S and C_B = gross and background alpha counts, respectively, recorded over the interval t_c
- E = efficiency of equipment detection, fraction
- K = factor to convert from counts per minute to working level
- V = volume of air sampled, L
- t_c = counting time, minutes

For the Kusnetz method, K is the Kusnetz conversion factor determined from Equations 15.6-3 and 15.6-4.

The right side of Equation 15.6-5 can be multiplied by a correction factor to compensate for the absorption of radon progeny on the sampling filter. This factor will probably be less than 0.02 for a cellulose ester filter, but it may exceed 0.1 for a glass-fiber filter (CNSC 2003).

A disadvantage of the Kusnetz method is that it is not suitable for atmospheres containing both ^{222}Rn and ^{220}Rn progeny, because the delay period allows for the ingrowth of ^{220}Rn progeny, which will contribute to the gross alpha count (DCEP 2008). In Canada, the Canadian Nuclear Safety Commission inspectors in underground mines rely on the modified Kusnetz method for sampling, counting, and computational procedures (CNSC 2003).

Modified Tsivoglou method. The modified Tsivoglou method is very similar to the Kusnetz method, except that the filter is counted three times after collection to measure the concentration of the individual radon daughters. Filter counting occurs at the following intervals after completion of sampling: 2 to 5, 6 to 20, and 21 to 30 minutes (McDermott and Ness 2004).

Based on a half-life for ^{218}Po of 3.04 minutes, the concentration of each of the radon decay products, that is, ^{218}Po, ^{214}Pb, and ^{214}Po, can be determined using the following calculations:

$$C_1 = \frac{1}{FE}\begin{pmatrix} 0.16921\ G1 - 0.08213\ G2 \\ +0.07765\ G3 - 0.5608\ R \end{pmatrix} \quad (15.6\text{-}6)$$

$$C_2 = \frac{1}{FE}\begin{pmatrix} 0.001108\ G1 - 0.02052\ G2 \\ +0.04904\ G3 - 0.1577\ R \end{pmatrix} \quad (15.6\text{-}7)$$

$$C_3 = \frac{1}{FE}\begin{pmatrix} -0.02236\ G1 + 0.03310\ G2 \\ -0.03765\ G3 - 0.05720\ R \end{pmatrix} \quad (15.6\text{-}8)$$

where

- C_1, C_2, and C_3 = concentrations of ^{218}Po, ^{214}Pb, and ^{214}Po, respectively, pCi/L
- F = sampling airflow rate, L/min
- E = counting efficiency, counts per minute/disintegrations per minute
- $G1$, $G2$, and $G3$ = gross alpha counts at the 1st, 2nd, and 3rd intervals, respectively
- R = background counting rate (counts per minute)

The working level can be expressed using the following equation:

$$\text{WL} = 10^{-3}(1.028 C_1 + 5.07 C_2 + 3.728 C_3) \quad (15.6\text{-}9)$$

Using an alpha-scintillation scaler instead of a rate meter results in a 10-times improvement in the sensitivity (Sengupta 1989).

Rolle method. The Rolle method is used for the determination of ^{222}Rn progeny concentration (DCEP 2008). This method is similar to the original Kusnetz method, except that the timing for filter counting after sample collection is different. In the Rolle method, the delay period is 5 to 10 minutes, allowing for faster measurement compared to the Kusnetz method. In addition, the short delay time minimizes the error in the radon progeny ratio allowing better measurement of the concentration of ^{222}Rn progeny (Sengupta 1989). The working level of ^{222}Rn progeny can be calculated using Equation

Table 15.6-2 Conversion factor for Rolle method

Sampling Time, minutes	Delay Time, minutes	Counting Time, minutes	Conversion Factor	Error, %
4	6.46	5	213	12
5	6.06	5	213	11
10	4.35	5	213	11
15	1.36	10	212	10
20	2.12	5	210	10

15.6-5 (i.e., the same equation used for the Kusnetz method) but with a conversion factor selected from Table 15.6-2 (DCEP 2008).

Rock method. The Rock method is mainly used for the determination of ^{220}Rn progeny concentration because of the extended delay time allowing for the ingrowth of ^{220}Rn progeny. The method is similar to the Kusnetz method, but the delay period is typically between 5 and 17 hours. An advantage of this method is that the long delay period allows sampling of remote locations and sending the samples for laboratory analysis. The concentration of ^{220}Rn progeny can be calculated using Equation 15.6-5, but with the conversion factor selected from Table 15.6-3 (DCEP 2008).

Cote method. The Cote method allows for the determination of the concentration of ^{220}Rn and ^{222}Rn progeny based on the selection of the delay period for the alpha count. The method consists of 10-minute dust sampling followed by several 15-minute alpha counts. The first count, conducted 1.2 minutes after the completion of sampling, is used to obtain the concentration of ^{222}Rn progeny, because the contribution of ^{220}Rn progeny is negligible at this time. A conversion factor of 218 is used in Equation 15.6-5 to obtain the WL concentration of ^{222}Rn progeny. The second count, conducted 155 minutes after sampling, provides the combined contribution of ^{222}Rn and ^{220}Rn progeny using a conversion factor of 14.1 in Equation 15.6-5. The final count, conducted 225 minutes after sampling, is used to obtain the ^{220}Rn progeny contribution using a conversion factor of 14.2 in Equation 15.6-5. A disadvantage of the Cote method is the need for a counting instrument to be placed on or very close to the sampling location (DCEP 2008).

Integrated Sampling

Integrated sampling provides a single concentration value averaged over the duration of the sampling time, which can ranges from days to weeks, months, or up to a year.

Radon Sampling

Activated carbon canisters. Adsorption on activated carbon allows integrative sampling of radon gas over a period of up to 7 days. Because the adsorbed radon undergoes radioactive decays during the sampling period, this method does not uniformly integrate radon concentrations. Sampling can be active using a pump, or passive based on diffusion (Katz 1977; McDermott and Ness 2004). After sampling, the activated carbon is analyzed using a sodium iodide gamma-scintillation detector to count the gamma rays between 0.25 and 0.61 MeV emitted by the radon decay products on the activated carbon (Katz 1977; McDermott and Ness 2004). Alternatively, radon can be desorbed from the activated carbon and transferred to a counting cell for alpha counting (Katz 1977). For gamma counting of a diffusion-based activated carbon monitor, the

Table 15.6-3 Conversion factor Rock method

Time After Sampling, hours	Conversion Factor
5	13.2
7	11.3
9	9.8
11	8.6
13	7.5
15	6.7
16	6.3

lower limit of detection ranges from 0.1 to 0.2 pCi/L of radon for a sampling time of 4 days (George 1990). Radon sampling with activated carbon canisters is a simple and low-cost method. However, the results can be sensitive to humidity and temperature. Desiccant can be used to avoid interference from adsorption of water vapor.

Alpha track detectors. Alpha track detectors provide an integrated concentration of radon over the exposure period, which ranges from few months to a year. The detector consists of a plastic film, such as cellulose nitrate (LR-115, manufactured by Kodak-Pathé) or allyl diglycol carbonate (CR-39) films (Alter and Fleischer 1981). The films are enclosed in a container that allows only radon to diffuse through and prevent the penetration of radon progeny. Alpha particles emitted by the radon and its progeny in the container strike the detector and produce submicroscopic damage tracks. Following exposure, the detectors are treated with a caustic solution to visualize the damage tracks for manual counting with a microscope or automated counting with an optical counting system. Radon concentration is proportional to the net areal density of tracks per unit time. The advantages of alpha track detectors are their simplicity, cost-effectiveness, and the lack of need for external power sources. Their main disadvantage is the dependence of results on the type of film used. However, the variability in the results can be decreased by counting more tracks over a larger area of the detector (McDermott and Ness 2004).

Radon Progeny Sampling

Thermoluminescent dosimeter (TLD). TLD-based systems integrate radon progeny concentration over sampling periods ranging from few days to a few weeks. The system contains a detector assembly through which sample air is drawn. The detector assembly includes a filter and two TLDs. One TLD is used for measuring alpha particles emitted from radon progeny collected on the filter. The other TLD is shielded from the alpha particles emitted from the filter and is used for measuring background gamma radiation (Vutukuri and Lama 1986; McDermott and Ness 2004). Alpha particles emitted from the decay of the radon progeny are registered on a luminescent material such as calcium fluoride dysprosium or lithium fluoride. Following sampling, the detector assembly is sent for analysis using a TLD reader. Light emitted from the TLD as a result of heating in the TLD reader is detected with a photomultiplier tube and is proportional to the radon progeny collected on the filter paper during sampling (George 1990). A TLD is accurate and suitable for long sampling periods; however, they suffer from their relatively high cost and clogging of filters (Vutukuri and Lama 1986). Instead of TLDs, optically stimulated luminescence (OSL) dosimeters are used for measuring exposure of mine workers to radiation. OSL dosimeters use luminescent material such as Al_2O_3:C that responds to gamma radiation (Figure 15.6-4). The OSL dosimeter is typically exposed for several months before it is analyzed. Unlike the case of TLDs, the readout method for OSL dosimeters is all optical, eliminating the need for heating of the samples. OSL dosimeters also provide a high-sensitivity, fast-readout process, as well as the possibility of multiple readings (McKeever 2001).

Electret detector and alpha track detector systems. A filter-based system similar to the one used with a TLD system can be used with other types of detectors, such as an alpha track detector or an electret detector. An electret is an electrostatically charged disk positioned in an ion chamber. Alpha particles generated from the decay of radon progeny collected on the filter result in negatively charged ions that are collected on a positively charged electret, resulting in a reduction in its surface voltage. Radon progeny concentration is proportional to the voltage reduction (McDermott and Ness 2004). Short-term electrets integrate measurements up to 7 days. Long-term electrets are less sensitive but integrate measurements for 1 to 12 months. The lower limit of detection for a short-term electret is 0.2 pCi/L of radon for a 3-day exposure (George 1990).

Continuous Monitors

In continuous monitoring, sampling and analysis occur simultaneously and continuously.

Scintillation Cell

A variation of the scintillation cell used for grab sampling of radon can be used for continuous radon monitoring. One configuration consists of a flow-through scintillation cell in which air is continuously pumped through for a few minutes. Alpha particles from the decay of radon gas are counted as the gas is swept through. Another configuration consists of a periodic-fill cell, into which air is pumped once during each preselected time interval, and the alpha particles are counted in each interval. Passive scintillation cells in which radon diffuses through a filter area allow passive continuous monitoring of radon after a short diffusion time lag with a lower limit of detection of about 1 pCi/L for a 1-hour counting (George 1990; McDermott and Ness 2004).

Diffusion-Based Ionization Chamber

A diffusion-based ionization chamber allows radon gas to diffuse into the chamber while radon progeny are removed electrostatically. As is the case with a passive scintillation chamber, there is a short time lag with this method. Inside the ionization chamber, the emitted alpha particles from radon decay generate bursts of ions that are recorded by an electrometer. The lower limit of detection is about 0.5 pCi/L for a 1-hour counting (George 1990).

Courtesy of K. Messer.

Figure 15.6-4 An optically stimulated luminescence dosimeter

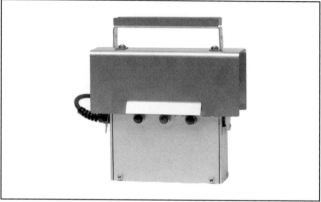

Courtesy of K. Messer.

Figure 15.6-5 A Continuous working-level monitor intended for operation as a station monitor

Continuous Radon Progeny Monitors

Continuous radon progeny monitors use filters to collect radon progeny and an alpha detector to measure alpha particles generated from the decay of the radon progeny. Hence, they are similar to integrated radon progeny samplers, except that automated sampling of air is done on a continuous basis and alpha counting is integrated over short-time intervals ranging form several minutes to an hour (George 1990). Gross alpha counting is also used to provide WL radon progeny. One commonly used system, alphanuclear prism, has four alpha-energy counting windows that indicate WL concentrations of <0.1 WL, 0.1 to 0.25 WL, 0.25 to 0.5 WL, 0.5 to 1 WL, and >1 WL (Figure 15.6-5).

ACKNOWLEDGMENTS

The authors thank Jean Alonso and Carolyn Ingram of Cameco Corporation for their input during the preparation of this chapter.

REFERENCES

Agricola, G. 1912. *De Re Metallica*. Originally published in 1556. Translated in 1912 by H. Hoover. London: Salisbury House.

Alter, H.W., and Fleischer, R.L. 1981. Passive integrating radon monitor for environmental monitoring. *Health Phys.* 40(5):693–702.

Apel, D.B. 2000. Designing the ventilation system for the McArthur River Mine—World's largest uranium deposit. In *Mining in the New Millennium: Challenges and Opportunities—Proceedings of the American–Polish Mining Symposium*, Las Vegas, NV, Oct. 8. Rotterdam: A.A. Balkema.

Apel, D.B., and Szmigiel, P. 2006. Mapping ground conditions before the development of an underground hard rock mine—McArthur River uranium mine case study. *Int. J. Rock Mech. Min. Sci.* 43(4):655–660.

Bates, R.C., and Edwards, J.C. 1980. Mathematical modeling of time dependent radon flux problems. In *2nd International Congress on Mine Ventilation*, Reno, NV. pp. 412–419.

CNSC (Canadian Nuclear Safety Commission). 2003. *Measuring Airborne Radon Progeny at Underground Mines and Mills*. Regulatory guide G-4. Cat. No. CC173-3/2-4E.

Crawford-Brown, D.J. 1989. The biokinetics and dosimetry of radon-222 in the human body following ingestion of groundwater. *Environ. Geochem. Health* 11:10.

DCEP (Department of Consumer and Employment Protection). 2008. *Managing Naturally Occurring Radioactive Material (NORM) in Mining and Mineral Processing—Guideline*. Western Australia: Resources Safety, DCEP. www.docep.wa.gov.au/ResourcesSafety. Accessed December 2009.

Dixon, K.L., Lee, R.G., Smith, J., and Zielinski, P. 1991. Evaluating aeration technology for radon removal. *J. Am. Water Works Assoc.* 83(4):141–148.

Field, R.W. 1999. Radon occurrence and health risks. In *Occupational Medicine Secrets*. Edited by R.M. Bowler and J. Cone. Philadelphia: Hanley and Belfus. pp. 85–92.

Franklin, J.C. 1981. Control of radiation hazards in underground mines. In *Radiation Hazards in Mining: Control, Measurement, and Medical Aspects*. Edited by M. Gomez. New York: SME-AIME. pp. 441–446.

Frumpkin, H., and Samet, J.M. 2001. Radon. *CA Cancer J. Clin.* 51(6):337–344.

George, A.C. 1990. An overview of instrumentation for measuring environmental radon and radon progeny. *IEEE Trans. Nucl. Sci.* 37(2):892–901.

Hinds, W. 1999. *Aerosol Technology: Properties, Behavior, and Measurement of Airborne Particles*, 2nd ed. New York: Wiley-Interscience.

Hursh, J.B., Morken, D.A., Davis, R.P., and Lovass, A. 1965. The fate of radon ingested by man. *Health Phys.* 11(465).

Katz, M. 1977. *Methods of Air Sampling and Analysis*. Washington, DC: American Public Health Association.

Kinner, N.E., Malley, J.P., Clement, J.A., and Fox, K.R. 1993. Using POE techniques to remove radon. *J. Am. Water Works Assoc.* 85(6):75.

Kreuzer, M.G.B., Brachner, A., Martignoni, K., Schnelzer, M., Schopka, H.J., Bruske-Hohlfeld, I., Wichmann, H.E., and Burkart, W. 1999. The German uranium miners cohort study: Feasibility and first results. *Radiat. Res.* 152(6):56–58.

Kusnetz, H.L. 1956. Radon daughters in mine atmospheres: A field method for determining concentrations. *Ind. Hyg. Q.* 17(1):85–88.

McDermott, H.J., and Ness, S.A. 2004. *Air Monitoring for Toxic Exposures*, 2nd ed. Hoboken, NJ: Wiley-Interscience. pp. 447–471.

McKeever, S. 2001. Optically stimulated luminescence dosimetry: Nuclear instruments and methods. *Nucl. Instrum. Methods Phys. Res., Sect. B.* 184(1-2):29–54.

NRC (National Research Council). 1998. *Health Effects of Exposure to Radon: BEIR VI*. NRC Committee on the Biological Effects of Ionizing Radiation. Washington, DC: National Academy Press.

Schroeder, G.L., and Evans, R.D. 1969. Some basic concepts in uranium mine ventilation. *Trans. SME-AIME.* 235:91–99.

Sengupta, M. 1989. *Mine Environmental Engineering*, Vol. I. Boca Raton, FL: CRC Press. pp. 147–155.

Vutukuri, V.S., and Lama, R.D. 1986. *Environmental Engineering in Mines*. New York: Cambridge University Press. pp. 200–201.

CHAPTER 15.7

Noise Hazards and Controls

Gary E. Walter

INTRODUCTION

Noise is an inherent health hazard in the mining industry. The mining process is highly mechanized from the earliest ore removal to final processing, and heavy equipment is essential at virtually every stage of operation. Exposure to noise is a concern for workers who drive mechanized equipment as well as those who operate or work near stationary equipment such as haulage belts or crushing equipment. Exposure to noise levels above regulatory or recommended limits can result in hearing loss that adversely affects the daily lives of workers and negatively impacts society as workers so affected access health care services.

Worker exposure to noise is regulated by many government agencies. In the United States, these agencies include the Mine Safety and Health Administration (MSHA) and the Occupational Safety and Health Administration (OSHA). In the United Kingdom, the regulatory agency for worker health and safety is the Health and Safety Executive (HSE). The European Union (EU) has also enacted a directive on noise that must be implemented by member countries. Another significant contributor to the assessment of risk and establishment of industrial hygiene standards in the United States is the American Conference of Governmental Industrial Hygienists (ACGIH), which is not a government agency but rather a professional society that publishes recommendations for exposures to various hazardous agents to workers.

It is important to emphasize that most hearing loss is preventable. Prevention can be achieved by eliminating noise sources, substituting quieter equipment, installing appropriate engineering controls, implementing administrative controls, using personal protective equipment, and adopting effective hearing-conservation programs. Regulations in many parts of the world require noise-control and hearing-conservation programs. Many of these regulations were recently updated to reflect new information regarding the levels of noise that can cause hearing loss. This has resulted in reduced permissible exposure levels and enhanced requirements to conduct risk assessments and implement control measures.

DEFINITIONS

Action level (AL)—The level, established by regulation, at which action must be taken to protect the hearing of workers exposed to noise. The action required is country-specific.

- **Action level (MSHA)**—The 8-hour time-weighted average sound level (TWA_8) sound level of 85 dBA, or equivalently a dose of 50%, integrating all sound levels from 80 to ≥130 dBA (A-weighted). (Note: An A-weighted setting measures sound similarly to the way the human ear responds to sound. A C-weighted setting measures sound in a flat or linear response.)
- **Lower exposure action value (HSE)**—The lower of the two levels of daily or weekly personal noise exposure or of peak sound pressure that, if reached or exceeded, requires specified action to be taken to reduce risk; that is, a daily or weekly personal noise exposure of 80 dBA.
- **Upper exposure action value (HSE)**—The higher of the two levels of daily or weekly personal noise exposure or of peak sound pressure that, if reached or exceeded, requires specified action to be taken to reduce risk; that is, a daily or weekly personal noise exposure of 85 dBA.

Criterion level—The sound level that, if constantly applied for 8 hours, results in a dose of 100% of that permitted by regulation.

dB—The unit of measure for sound pressure, defined in two ways:

1. For sound-pressure levels measured by sound level meter (SLM), the decibel is 20 times the common logarithm of the ratio of the measured sound pressure to the standard reference sound pressure of 20 µPa, which is the threshold of normal hearing sensitivity at 1,000 Hz.
2. For hearing-threshold levels measured by audiometer, the decibel is the difference between audiometric zero (reference pressure equal to 0 hearing-threshold level) and the threshold of hearing of the individual being tested at each test frequency.

Gary E. Walter, Principal Consultant, Primo Safety and Health Services, LLC, Wilmington, Delaware, USA

dBA—The level of noise, measured by SLM or dosimeter, A-weighted to represent the way the ear processes sound.

Dosimeter—A personal monitoring device that integrates noise exposure. Dosimeters must be set to match the criterion, threshold, and exchange rate established by the relevant regulation.

Exchange rate—The amount of increase in sound level, in decibels, that would require halving of the allowable exposure time to maintain the same noise dose. The MSHA and OSHA exchange rates are both 5 dB (which the American Industrial Hygiene Association recommends lowering to 3 dB). The EU and HSE exchange rates are both 3 dB. The ACGIH threshold-limit value (TLV) exchange rate is also 3 dB.

Noise dose—The integrated noise exposure that reaches a worker's ear, measured by personal noise dosimeter, including all sound that is >80 dBA. The permissible exposure limit or the recommended criteria is exceeded when this value is >100%.

Permissible exposure limit (PEL)—The noise-exposure limit that must not be exceeded, established by government regulation. The MSHA PEL is 90 dBA.

Sound level meter—An instrument designed to react to noise much as does the human ear in order to measure and compare against appropriate noise standards.

Threshold—The cutoff level of sound used to measure and calculate noise dose. Sound below this level is considered zero in noise–dose calculations.

Threshold-limit value (ceiling), or TLV—An exposure that should not be exceeded, even instantaneously.

Threshold-limit value (time-weighted average) for physical agents—The time-weighted average exposure for an 8-hour workday and a 40-hour workweek.

NOISE REGULATIONS AND EXPOSURE LIMITS

Noise-exposure regulations vary by country, resulting in the existence of differing exposure limits, measurement procedures, compliance activities, and required controls. In the United States, workplace noise-exposure standards are governed by MSHA and OSHA. Whether a mine operation is regulated by these two entities determines which regulations the operator must meet. Some operations may be covered by both agencies for different parts of the operation. In general, most mine operations are regulated by MSHA. Some states have additional noise regulations. Environmental or community noise requirements established by the U.S. Environmental Protection Agency or by state and local governments may also affect operations near residential or public areas. The MSHA (2006) regulation for noise in mines is in the Code of Federal Regulations (CFR) as Title 30, Part 62, *Occupational Noise Exposure*. For OSHA-covered operations, the OSHA (2004a, 2004b) regulations are in the CFR as Title 29, Part 1910.95, *Occupational Noise Exposure* and *Methods for Estimating the Adequacy of Hearing Protector Attenuation*.

The European Union adopted Directive 2003/10/EC, *Minimum Health and Safety Requirements Regarding the Exposure of Workers to the Risks Arising from Physical Agents (Noise)*, for application throughout Europe. Each member country must adopt this directive at a minimum but may adopt more stringent regulations. An example of a country adopting more stringent regulations is the United Kingdom, where the HSE issued *Control of Noise at Work Regulations 2005*, which applies to all U.K. industry sectors, including mining.

The ACGIH has for longer than 50 years adopted TLVs for chemical substances and physical agents including noise (ACGIH 2006). ACGIH members are industrial hygienists or other occupational health and safety professionals dedicated to promoting workplace health and safety. The ACGIH publishes guidelines regarding safe levels of exposure to chemical and physical agents found in the workplace. In many parts of the world where no regulations exist, ACGIH TLVs have become the standard reference for the protection of worker health.

Although present-day international standards for measuring noise and hearing are generally in agreement, there is some difference of opinion regarding what level of noise adversely affects human health. As a result, there are different acceptable levels of allowable exposure for workers and some differences in the actual application for regulatory standards. The following are MSHA, EU, HSE, and ACGIH standards for TWA_8 exposure (MSHA 2006; EU 2003; HSE 2005; ACGIH 2009).

- MSHA standards
  - AL: 85 dBA (50% dose); all sound >80 dBA is included
  - PEL: 90 dBA (100% dose); only sound >90 dBA is included
  - Threshold: AL is 80 dBA; PEL is 90 dBA
  - Exchange (doubling) rate: 5 dBA
- EU and HSE standards
  - Lower AL: 80 dBA
  - Upper AL: 85 dBA
  - PEL: 87 dBA
  - Threshold: 80 dBA
  - Exchange (doubling) rate: 3 dBA
- ACGIH recommendations
  - TLV: 85 dBA
  - Threshold: 80 dBA
  - Exchange (doubling) rate: 3 dBA

Note: A modern personal noise dosimeter can record and calculate both AL and PEL levels.

Using the EU (HSE) upper AL limit of 85 dBA and exchange (doubling) rate of 3 dB as the reference for worker exposure, the amount of time permitted at various noise levels can be calculated by adding 3 dBA to the 85 dBA limit and cutting the time in half from the permitted 8 hours. Table 15.7-1 shows the amount of time at various noise levels that is equivalent to 85 dBA for the time shown.

Table 15.7-2 shows the same information when the MSHA limits are used as the reference for worker exposure. The times are different partly because ACGIH uses the more conservative 3-dB exchange (doubling) rate whereas the MSHA standard uses 5 dB. The data in these tables can be used to calculate worker noise exposure using the formula given in Equation 15.7-1.

INSTRUMENTATION AND SURVEY TECHNIQUES

Noise levels must be measured before any control measures are undertaken. It is important to document worker exposures in order to meet regulatory requirements and make informed decisions about implementing hearing-conservation programs for exposed workers. These measurements are the basis for most decisions regarding the application of engineering controls and assist in determining the most effective control measures. Measurement of worker exposures as part of a risk-assessment process can help prioritize the application of engineering controls so that the greatest number of affected employees can be protected through implementation

Table 15.7-1 Noise-level equivalencies at various sound levels (EU reference standard)*

Noise Level, dBA	Maximum Permissible Time of Noise Exposure, h
85	8
88	4
91	2
94	1
97	0.5
100	0.25

Source: EU 2003.
*Maximum noise pressure level must not exceed 115 dBA. Peak level of sound pressure (in case of impulsive sound as well) must not exceed 140 dBC.

Table 15.7-2 Noise-level equivalencies at various sound levels (MSHA reference standard)*

Sound Level, dBA	Reference Duration T, h	Sound Level, dBA	Reference Duration T, h
80	32.0	100	2.0
85	16.0	101	1.7
86	13.9	102	1.5
87	12.1	103	1.3
88	10.6	104	1.1
89	9.2	105	1.0
90	8.0	106	0.87
91	7.0	107	0.76
92	6.1	108	0.66
93	5.3	109	0.57
94	4.6	110	0.50
95	4.0	111	0.44
96	3.5	112	0.38
97	3.0	113	0.33
98	2.6	114	0.29
99	2.3	115	0.25

Source: MSHA 2006.
*At no time shall any excursion exceed 115 dBA. For any value, the reference duration T in hours is computed by $T = 8/2^{(L-90)/5}$ where L is the measured A-weighted, slow-response sound pressure level.

Courtesy of Quest Technologies, a 3M company.
Figure 15.7-1 Sound level meter in use near a haul truck

of engineered solutions or administrative or other control procedures. A good basic reference for understanding the measurement of noise is the *Handbook of Noise Measurement* (Peterson 1980).

Three types of instruments are used for measuring noise: the SLM, personal noise dosimeter, and octave band analyzer. Each has a specific purpose. Any instrument used for the measurement of noise must meet minimum standards to ensure accuracy. Figure 15.7-1 shows an SLM being used to measure noise levels near a haul truck. Figure 15.7-2 shows a miner wearing a newer personal noise dosimeter.

In the United States, SLMs must meet the standards set forth in American National Standards Institute (ANSI) S1.4-1983. Personal noise dosimeters must meet the standards set forth in ANSI S1.25-1991. In other parts of the world, international standard ISO 1999:1990 is the generally recognized standard for measurement of noise. In underground mines where explosive gases or atmospheres may be present, electrical instruments must be rated intrinsically safe for use in explosive atmospheres. In the United States, MSHA maintains a list of permissible instruments and other equipment.

The SLM is the most commonly used instrument for identifying both stationary and mobile noise sources that may contribute to worker exposure. Most have both A-weighted and C-weighted settings. Readings are often used to assess machinery or community noise and to evaluate the attenuation of hearing protectors for use in high-noise environments.

A basic noise survey conducted with an SLM helps identify potential noise sources, determine sound levels in areas where workers may be exposed, and determine whether workers may be exposed to hazardous levels of noise. The entire area should be surveyed to identify all stationary sources so these sources and associated noise levels can be documented on a mine map or other plan. Mobile equipment must be measured while in service. SLM readings can help identify potential sources that require additional monitoring, often with more sophisticated equipment.

Although the noise dose that a worker receives can be calculated using SLM readings, in practice this is difficult to do with sufficient accuracy to determine compliance with regulations or standards. To use SLM readings, a worker must be followed throughout the work period to identify each source of noise, measure each source's noise level, and determine how much time the worker is exposed to each noise source and level. The worker's noise exposure is

$$\text{dose (\%)} = 100(C_1/T_1 + C_2/T_2 \ldots C_n/T_n) \quad (15.7\text{-}1)$$

where C is the total duration of noise exposure at a specific noise level and T is the total duration of noise exposure permitted at that level. If the sum of the fractions exceeds 100, then the combined exposure exceeds the PEL if comparing against a regulatory limit or the TLV if comparing against the ACGIH recommended limit.

The personal noise dosimeter overcomes some of the disadvantages of the SLM. It is worn by the worker throughout the day and records all sounds exceeding the regulatory or recommended limit. At the end of the work shift, the dosimeter gives a readout of the equivalent dose in dBA or percentage of allowed exposure. Most contemporary dosimeters have software for downloading the collected data for individual time periods recorded, often for periods as short as 1 minute. The data can be printed out in table or graph form. Comparison of this data with observations about the work performed throughout the day can reveal which tasks or noise sources contributed to the overall exposure, allowing prioritization of noise sources for further study or engineering control.

The octave band analyzer is a more sophisticated SLM. It is more accurate and has the asset of filters that measure noise at different frequencies. Table 15.7-3 lists the octave frequency bands that are commonly used for determining the frequency distribution of noise. Use of an octave band analyzer to obtain the frequency distribution of sound is important when designing engineering controls to reduce noise levels. As with other sound-measuring instruments, an octave band filter set and integrating SLM must meet specific standards (ANSI S1.43-1997 (R2002), ANSI S1.11-2004). Different frequencies may require different strategies for reducing worker exposure to noise. The information gained from this analyzer can also be used to calculate the protection factors of hearing-protective devices.

In general, noise surveys are of two types: (1) a health survey, to determine whether workers are exposed to noise levels above regulatory or recommended limits; and (2) an engineering survey, to identify sources of noise that contribute to worker exposure.

Health surveys, as noted previously, can be accomplished using either an SLM or a personal noise dosimeter. In conducting a health survey, it is important to ensure that an adequate number of workers are sampled so as to obtain a representative assessment of noise exposure. The best approach is to divide workers into homogeneous exposure groups (HEGs), each containing workers assigned similar tasks and thus expected to experience similar levels of exposure to noise. In deciding which workers to include in each HEG, the following must be considered: fluctuations in noise level, fluctuations due to working activities (variations in operating conditions, shutdowns, productions schedules), and variations in task performance among workers.

A number of samples for the same worker or multiple workers performing the same tasks over several different work periods is usually necessary to ensure that worker exposure is correctly measured. The number of samples required is based on statistical evaluation of the HEG and the variability of the noise levels. Six measurements for each HEG should be considered the minimum for a representative and statistically significant assessment. Samples should be as close to full-shift as possible and distributed over different days and shifts.

Software tools are available that can assist in obtaining statistically valid samples. After basic screening has been performed and potential workers and tasks that may be above

Courtesy of Occupational Health Dynamics.
Figure 15.7-2 Personal noise dosimeter worn by a miner

Table 15.7-3 Commonly used octave frequency bands (in Hertz)

Lower Band Limit	Center Band (Geometric Mean)	Upper Band Limit
22	31.5	44
44	63	88
88	125	177
177	250	354
354	500	707
707	1,000	1,414
1,414	2,000	2,828
2,828	4,000	5,656
5,656	8,000	11,312

Source: OSHA 1999.

regulatory or recommended limits have been identified, such tools can help determine whether the data is sufficient to determine compliance and the need for engineering controls. If in doubt, a competent, trained individual can be consulted to assist in the evaluation of worker exposure.

Engineering surveys can be accomplished using an SLM or an octave band analyzer. The latter identifies specific frequencies that contribute to the total noise arising from individual noise sources. The engineering control solution often depends on the frequencies measured. Materials specified for controlling noise at the source must be matched to the frequency of the noise generated by the source.

It is important to remember that the decibel scale is logarithmic, so one cannot simply add decibel values when combining values from multiple noise sources. Two noises of exactly the same level, measured independently of one another, have a combined noise level that is 3 dBA higher than the individual values. The greater the difference between two individual noise sources, the lower the combined noise level.

Table 15.7-4 shows a simple way to determine combined noise levels. If more than two sources contribute to the overall noise level, then this process should be repeated iteratively to calculate the combined noise level. The *OSHA Technical Manual* (1999) contains a good description of this technique. When dealing with multiple noise sources, the best approach is to determine which source generates the most noise, and this source should be the first priority in deciding where to apply engineering controls.

Table 15.7-4 Adding decibels for two noise sources (in decibels)

Difference in dB Values	Add to Higher Level
0–1	3
2–3	2
4–9	1
≥10	0

Source: OSHA 1999.

Whether a survey is for health or engineering purposes, it is essential to document survey results so that one can understand the sources of noise and ensure that data conforms to regulatory requirements. Documentation must be kept in a form suitable for comparison with future survey results. Such comparisons are useful in evaluating engineering controls and equipment performance over time. In survey documentation, the following information should be recorded:

- Date and time
- Instrumentation details (model, serial number, settings, calibration)
- Workplace descriptions
- Task descriptions
- Unusual conditions

BASIC NOISE-CONTROL CONCEPTS

Noise control is complicated at best. The preferred method for reducing noise is to control it at the source so that it does not create an exposure hazard for workers. Mining operations present difficult challenges for the application of engineering controls, including underground equipment operation, special considerations for environment/climate, and type and age of equipment.

When obtaining new or replacement equipment, appropriate language must be included in the contract or purchase order ensuring that the purchased equipment will generate the lowest noise level that is technically feasible. Obtaining new equipment with noise-control features included reduces the need to retrofit controls in the work environment. The manufacturer is in the best position to reduce noise at the source by means of improved design and integration of noise controls.

As most mining equipment has a relatively long service life, noise controls may need to be added to current equipment to reduce exposure risk. Three basic types of control can be considered:

1. Substitution
2. Modification of the noise source
3. Modification of the sound wave

Substitution is a long-term effort to replace existing equipment or processes with alternatives that generate less noise. Although substitution may ultimately reduce noise levels, replacement schedules for most equipment in mining operations are too long to accommodate near-term improvements. Substitution should always be considered, but the need to protect workers and meet regulations means that alternative controls must be implemented sooner.

Modification of the noise source usually involves changing the way equipment works so as to reduce noise levels. Although some changes can be made to mobile equipment, changes may be more effective when made to stationary equipment such as screening and crushing equipment, motors, pumps, dryers, and air compressors. Typical control measures for stationary equipment include installation of isolation dampers and vibration-damping materials for vibrating equipment or surfaces, and use of exhaust mufflers and rerouting of intake/exhaust points on pneumatic equipment. Additional measures include improved bracing, increased stiffness, or application of damping materials to vibrating surfaces.

Modification of the sound wave is frequently the easiest method to implement. This can be accomplished by confining sound waves in acoustical enclosures, absorbing sound waves within a room or space, and absorbing sound along its transmission path. Effective enclosure design involves many factors, including the following:

- Selection of construction materials for the enclosure wall
- Selection of acoustical linings based on the frequency distribution of the noise
- Sealing of enclosure joints
- Mounting of the enclosure so as to minimize vibration
- Addition of access panels (doors) for maintenance and other operational needs
- Consideration of the effect that the enclosure design will have on the operating equipment
- Ventilation by means of inlets/outlets designed to prevent the escape of noise

It is important to focus on controlling the frequencies that most adversely affect human hearing in the speech range. Those frequencies are 500, 1,000, 2,000, 3,000, 4,000, and 6,000 Hz. The higher frequencies are usually easier to control.

If full enclosure is not possible, partial enclosures, noise barriers, or shields that separate the noise from the worker should be considered to reduce noise levels. Barriers are most effective for high- or medium-frequency noise. Barriers should be as high as practical and as close to the noise source as possible.

Ventilation ducts and conveyor systems can be important transmitters of noise. Ducts can be lined to attenuate sound traveling through them. Similarly, noise in conveyor systems can be attenuated by selection of appropriate lining materials or application of vibration-damping materials.

In addition to enclosing the noise source, it is also possible to reduce worker exposure to noise by enclosing the worker. This option is particularly effective for mobile-equipment operators. Many manufacturers offer original-equipment enclosures that reduce noise as well as minimize exposure to airborne contaminants and temperature extremes. Enclosed sound-reducing stationary-equipment control rooms can minimize worker exposure to noise at fixed locations. These enclosures can also serve as quiet areas where other workers can gain relief from noise during their work shifts.

It is important to consider the flammability of materials selected for engineering controls in mining operations. In the United States, MSHA references ASTM International's ASTM E162-87 in its regulations for product flammability in underground mines. Understanding the flammability of acoustical materials selected is particularly important where burning, welding, or other hot work is done in the vicinity of the selected materials.

More specific information regarding the design and construction of noise controls in mining is available from a variety of sources. In the United States, MSHA has a number of publications providing detailed guidance on applying engineering controls in mining operations (MSHA 1999a, 1999b, 1999c), many of which are available on the MSHA Web site at www.msha.gov. The U.S. National Institute for Occupational Safety and Health (NIOSH) has specific programs for mining that are available on their Web site at www.cdc.gov/niosh/mining, including an area devoted to noise. The U.K. HSE lists case studies and sound solutions on their Web site at www.hse.gov.uk. Other countries may have similar information available. These resources can provide up-to-date information concerning control of noise in the mining industry.

Administrative controls are changes in work schedule or operations that reduce worker exposure to noise. Administrative controls should be used as adjuncts to—and not in lieu of—engineering controls. They may also have a place in minimizing worker exposure to noise while engineering controls are being designed and installed. As noise dose is a function of both noise level and duration of exposure, work can be organized into tasks that involve worker exposure to noise for only limited amounts of time during the work shift such that the noise dose remains below the regulatory or recommended limit. The balance of the work shift would include tasks where the noise is below the level that would contribute to the overall noise dose. Such rotational work schedules must be carefully designed so that worker exposures remain below acceptable levels. Other administrative controls could include remote operation of equipment or provision of quiet areas where employees can gain relief from noise. Improved maintenance of equipment and engineering controls can help to keep noise levels at the lowest level. In practice, however, administrative controls are of limited use because workers may not be allowed or able to switch among jobs, or work schedules can become too complicated to manage.

VIBRATION

Minimizing vibration, especially induced vibration, can contribute significantly to noise reduction. Many sources of noise that affect worker hearing are related to the vibration of equipment or surfaces. Vibration can be caused by a number of types of action, including rotation and reciprocation of devices. Examples of devices that tend to vibrate include compressors, pumps, fans, electric motors, pulleys, transfer belts, bin surfaces, gasoline and diesel engines, screening devices, and many more too numerous to mention. Vibration can result directly when rotating-equipment parts are unbalanced, misaligned, loose, or eccentric, or indirectly when induced in another piece of equipment.

The measurement and control of vibration serves two purposes: (1) prevention of premature wear and failure due to structural damage; and (2) reduction of noise levels. Measurement of vibration requires specialized equipment and experience in data interpretation. Many preventive-maintenance programs make use of vibration analysis to predict maintenance scheduling and equipment replacement.

Health effects beyond those related to hearing are also related to vibration. These additional effects are broadly considered to be ergonomics or human-factors issues related to whole-body vibration from activities such as operating mobile or stationary equipment. Both acute and chronic effects are of concern. Possible acute effects include loss of control of vibrating equipment due to mechanically induced decoupling from the hand. Possible chronic effects relate to the link between whole-body vibration and musculoskeletal problems including back pain and degenerative spinal disk disease.

The ACGIH, in its TLV recommendations and documentation, includes a TLV for whole-body vibration (ACGIH 2001). Whole-body vibration hazards and controls are not covered in this chapter but should be considered when vibration related to noise control is evaluated.

Vibration is controlled principally by reducing the mechanical disturbance that causes the vibration, isolating the disturbance from a radiating surface, and reducing the response of the radiating surface by applying damping or stiffening materials. Techniques such as balancing rotating equipment and installing isolation dampers may be obvious from initial evaluation. Increased maintenance to ensure that all equipment operates as designed and to observe early changes that may indicate increased vibration can help minimize vibration. Vibration control can be an iterative process involving incremental improvements as vibrating components are identified and managed.

PERSONAL PROTECTION

Even with the design and installation of engineering controls in mining operations, workers are still likely to be exposed to levels of noise that exceed regulatory or recommended limits. Even at levels below regulatory limits, noise-induced hearing loss can still occur. Most noise regulations established in recent years incorporate the concept of an action level. An action level is lower than the regulatory limit allowed for work-shift exposure, but still at a level where some hearing loss can occur. At this level, most regulations require that a number of measures be taken to monitor worker hearing and ensure that additional protective measures are implemented.

Most government regulations regarding noise in the workplace now require implementation of hearing-conservation programs for workers exposed above the action level. A hearing-conservation program should contain the following elements:

- Periodic assessment of worker exposure to noise (risk assessment)
- Health surveillance including audiometric testing
- Training
- Hearing protection

Periodic assessment of worker exposure to noise provides information about the continued effectiveness of installed engineering controls. Engineering controls over time may become damaged or require other maintenance to retain maximum effectiveness. Exposure data taken over time can also reveal changes in equipment noise levels caused by wear and tear. Continued compliance with regulatory or recommended exposure limits can be confirmed through periodic monitoring, and corrective actions can be taken as necessary. Worker use of engineering controls and hearing protection can be reinforced during periodic exposure assessments.

Health surveillance of workers exposed to high noise levels can identify workers with early signs of hearing loss so that further preventive measures can be taken. Usually health surveillance is conducted under the guidance of a physician who ensures that worker hearing is tested using approved audiometric test equipment and procedures and that test results are compared to normal hearing levels. If hearing loss is detected, additional testing can be undertaken and a specialist can be consulted to confirm the test results. If hearing loss is confirmed, the risk assessment should be reviewed and appropriate intervention taken for workers with confirmed hearing loss. The assessment should also explore whether workers have significant exposure to noise from activities outside of their employment.

Training, instruction, and information are key components to a successful hearing-conservation program. Most workers who understand the risks will take precautionary measures designed to prevent hearing loss. Information on the following subjects should be covered at a level at which the workers can understand:

- Nature of the hazards from exposure to noise
- Exposure limits compared to the exposure levels measured
- Risk assessment results
- Engineering controls in place
- Health surveillance program, including audiometric testing
- Purpose and value of wearing hearing protectors
- Types of hearing protection provided
- Government regulations as required

Computer software simulators have recently demonstrated the effects of noise-induced hearing loss. Use of these tools in training efforts can clearly demonstrate the effects of overexposure to noise. NIOSH has developed a hearing-loss simulator called HLSim, available from their Web site at www.cdc.gov/niosh/mining/products. The simulator demonstrates normal hearing and presents examples showing the effect of hearing loss on the ability to understand speech.

Hearing protection is critical for the prevention of worker hearing loss and is the last defense against the risks of noise. However, when properly selected and worn, hearing protectors are very effective in preventing hearing loss. Various types of protector are widely available, including ear plugs, ear caps, and ear muffs. The appropriate type for a particular situation depends on the level and frequency of noise measured during the risk assessment. In rare cases, double protection may be required. In the United States, MSHA has established a dual hearing-protection level of 105 dBA or greater for an 8-hour work shift. If necessary, a combination of ear plugs and ear muffs can provide added protection for very high noise exposure. Training programs should cover the various types of protector selected, the effectiveness of each type, and how to wear the protector properly.

Most hearing protectors have a noise-reduction rating (NRR) that indicates relative effectiveness. The NRR must be used with adjustment to compensate for variations in actual use. Several methods exist for calculating the adequacy of a protector. The easiest is to start with the NRR for a specific protector and calculate the attenuation value according to an approved method. NIOSH has published three methods for determining protector adequacy that vary by type of noise measurement data used (NIOSH 1975). OSHA has published a modified method for calculating adequacy based on the NRR value (OSHA 2004a, 2004b). Any of these methods can help ensure that the noise reaching the worker's ear has been reduced to protective levels.

SUMMARY

Noise exposure in mining operations remains a hazard, and continued diligence is required to protect worker hearing. Regulations to reduce the acceptable levels of exposure permitted for workers require the use of the following hierarchy of controls to reduce risk:

- Elimination
- Substitution
- Engineering controls
- Administrative controls
- Personal protective equipment

The controls at the top of this list are potentially more effective and protective than those at the bottom of the list, and result in the implementation of inherently safer systems that depend less on individual action. Thus, in the design of new processes and equipment, elimination and substitution, although the most difficult to implement, have the best chance for success. Administrative controls and personal protective equipment should be reserved for existing situations while engineering controls designed to remove the hazard are investigated and installed. If engineering controls do not reduce risk to acceptable levels, administrative controls and personal protective equipment may, by default, become long-term solutions.

The engineer plays an important role in the design and installation of engineering controls to reduce noise levels as much as reasonably achievable with the ultimate goal of eliminating noise as a risk in the work environment. There is broad recognition that this goal can be difficult to achieve. Understanding the risks of noise exposure and the importance of designing equipment and processes to minimize noise can help protect workers and minimize risk.

REFERENCES

ACGIH (American Conference of Governmental Industrial Hygienists). 2001. *Documentation of TLV—Whole Body Vibration: TLV Physical Agents*, 7th ed. Cincinnati: ACGIH.

ACGIH (American Conference of Governmental Industrial Hygienists). 2006. *Documentation of TLV—Noise*. Cincinnati: ACGIH.

ACGIH (American Conference of Governmental Industrial Hygienists). 2009. *Threshold Limit Values for Chemical Substances and Physical Agents*. Cincinnati: ACGIH.

ANSI S1.4-1983 (R2001). *Specification for Sound Level Meters*. New York: ANSI. Available from www.ansi.org.

ANSI S1.11-2004. *Specification of Octave Band Filters*. New York: ANSI. Available from www.ansi.org.

ANSI S1.25-1991 (R1997). *Specification for Personal Noise Dosimeters*. New York. ANSI. Available from www.ansi.org.

ANSI S1.43-1997 (R2002). *Specifications for Integrating Averaging Sound Level Meters*. New York: ANSI. Available from www.ansi.org.

ASTM E162-87. *Standard Test Method for Surface Flammability of Materials Using a Radiant Heat Energy Source*. West Conshohocken, PA: ASTM International. Available from www.astm.org.

EU Directive 2003/10/EC. 2003. *Directive of the European Parliament and of the Council of 6 February 2003 on the Minimum Health and Safety Requirements Regarding the Exposure of Workers to the Risks Arising from Physical Agents (Noise) (Seventeenth Individual Directive Within the Meaning of Article 16(1) of Directive 89/391/EEC)*. In the Official Journal of the European Union, L 042 of 15/02/2003. pp. 0038–0044.

HSE (Health and Safety Executive). 2005. *Control of Noise at Work Regulations 2005*. United Kingdom: HSE. Available from www.tsoshop.co.uk/bookstore.asp.

ISO 1999:1990. *Acoustics—Determination of Occupational Noise Exposure and Estimation of Noise-Induced Hearing Impairment*. Geneva: ISO. Available from www.iso.org.

MSHA (Mine Safety and Health Administration). 1999a. *Noise Control Resource Guide—Surface Mining*. Washington, DC: MSHA. www.msha.gov/1999noise/Guides/SurftotalFinal.pdf. Accessed November 2009.

MSHA (Mine Safety and Health Administration). 1999b. *Noise Control Resource Guide—Underground Mining*. Washington, DC: MSHA. www.msha.gov/1999noise/Guides/ugtotalFinal.pdf. Accessed November 2009.

MSHA (Mine Safety and Health Administration). 1999c. *Noise Control Resource Guide—Mills and Preparation Plants*. Washington, DC: MSHA. www.msha.gov/1999noise/Guides/MillPreptotalFinal.pdf. Accessed November 2009.

MSHA (Mine Safety and Health Administration). 2006. 30 CFR Part 62. *Occupational Noise Exposure*. Washington, DC: MSHA. Available from www.msha.gov.

NIOSH (National Institute for Occupational Safety and Health). 1975. *List of Personal Hearing Protectors and Attenuation Data*. Technical Report Publication No. 76-120. Cincinnati: NIOSH. pp 21–37. Available from www.cdc.gov/niosh.

OSHA (Occupational Health and Safety Administration). 1999. Physics of sound. In *OSHA Technical Manual*. Washington, DC: OSHA. www.osha.gov/dts/osta/otm/noise/health_effects/physics.html. Accessed November 2009.

OSHA (Occupational Health and Safety Administration). 2004a. 29 CFR 1910.95. *Occupational Noise Exposure*. Washington, DC: OSHA. www.osha.gov/pls/oshaweb/owadisp.show_document?p_table=STANDARDS&p_id=9735. Accessed November 2009.

OSHA (Occupational Health and Safety Administration). 2004b. 29 CFR 1910.95. *Methods for Estimating the Adequacy of Hearing Protector Attenuation*. Washington, DC: OSHA. www.osha.gov/pls/oshaweb/owadisp.show_document?p_table=STANDARDS&p_id=9737. Accessed November 2009.

Peterson, A.P.G. 1980. *Handbook of Noise Measurement*, 9th ed. Concord, MA: GenRad. www.ietlabs.com/pdf/Handbook%20of%20Noise%20Measurment.pdf. Accessed November 2009.

PART 16

Environmental Issues

CHAPTER 16.1

Site Environmental Considerations

Michael G. Nelson

INTRODUCTION

Activities associated with mining have direct and lasting effects on both the physical and the human environments. The social, economic, and political effects in the latter category are often considered sustainability issues and are discussed briefly in this chapter. Also in the latter category are issues of worker health and safety, which are discussed in Part 15 of this handbook.

Mining includes the following activities, all of which have distinct impacts on the environment:

- **Exploration.** Economic deposits are identified and their characteristics are determined to allow recovery.
- **Development.** Preparations are made for mining.
- **Extraction.** Valuable material is removed for sale or processing.
- **Reclamation.** Disturbances caused by any of the preceding activities are corrected or ameliorated.
- **Closure.** Activity ceases and the area is abandoned or returned to another use.

This chapter will address the environmental considerations incumbent in the development and extraction in some detail, including the permitting process required (in almost all jurisdictions) before mining activities begin.

Environmental regulations vary among the countries of the world. Most large mining companies endeavor to follow uniform practices at all their mines, regardless of location, and have publicly committed to follow well-articulated standards of sustainability, such as those described under "Sustainable Practices" in this chapter. However, because of the variations in regulations, it is difficult to give a detailed description of environmental requirements and practices that applies worldwide. Thus, the focus of this chapter is to give a general description of environmentally sound practices and procedures. Some of the most important are illustrated by specific examples. The first example shows the approach to environmental risk assessment taken by CODELCO (Corporación Nacional del Cobre de Chile), the national copper corporation of Chile, as part of its environmental management plan under the International Organization for Standardization (ISO) Standard 14001. The second example gives a description of surface coal mining regulations in the United States, probably some of the most detailed and specific regulations in existence. The third example describes the permitting and approval process for the Safford mine, located in a historic mining area in southern Arizona (United States), and includes descriptions of the public input to this process. The fourth example describes a successful approach to early involvement of local communities in mine planning and permitting at the Diavik diamond mine, which operates in a pristine northern environment where no mining had previously occurred.

HISTORIC MINING PRACTICES

In the recent past, mining was done with little knowledge of its effects on the environment and, from a modern perspective, with little concern for the effects that were known. The consequences of this approach often resulted in significant damage to the natural environment, including (but not limited to)

- Unreclaimed mine pits, shafts, tunnels, and waste piles that may result in landslides and large amounts of blowing dust;
- Surface and groundwater that may be contaminated by solid particulates and chemical contaminants released by active and abandoned workings, or by waste piles;
- Abandoned pits, shafts, and tunnels that may create potential falling hazards to humans, livestock, and wildlife;
- Similarly, release of particulates and chemical contaminants from mine workings or waste and spoil piles that may cause direct injury to or damage the health of humans or animals;
- Erosion, with its consequent loss of soil and vegetation in and around unreclaimed workings, that may be a significant problem; and
- Underground workings, current or abandoned, that may cause surface subsidence, which can result in damaged surface structures, and in fissures or escarpments that are hazardous to humans and animals (Craig et al. 2001).

Examples of these kinds of mining-induced damages are readily found in any historic mining district. In fact, water

---

Michael G. Nelson, Department Chair, Mining Engineering, College of Mines & Earth Sciences, University of Utah, Salt Lake City, Utah, USA contamination and abandoned waste piles left by ancient Roman mining activity in Spain resulted in rediscovery of the ore body on which Rio Tinto, one of the largest mining companies in the world, was founded (Raymond 1986).

In most cases, the environmental damage caused by mining was not well understood. There were some notable exceptions, such as the lawsuit in 1884 by downstream farmers against mining companies in California's "Mother Lode" district. By that time (well after the initial gold rush of 1849) those companies used a method called hydraulicking, in which entire hillsides were washed away with a powerful stream of water so the gold-bearing gravels could be processed. The resulting debris clogged streams and rivers and flooded meadows and fields, causing serious damage to agriculture (Hill et al. 2001). Other early environmental lawsuits related to the mining industry were directed at smelter operators by farmers who alleged that smelter gases were damaging their crops. Such suits were filed in England in 1865 (Brubaker 1995), and in the United States in Kansas in the 1880s (Junge and Bean 2006) and in Utah in 1903 (Lamborn and Peterson 1985). While these lawsuits may have been driven more by economic concerns than by a pure concern for the environment in the abstract, they certainly addressed what today would be considered environmental issues.

The public's expectations of the mining industry began to change in the 1950s, and by the end of the 1970s, governments in developed countries had enacted broad environmental laws that had direct bearing on all industrial activities, including mining (Kaas and Parr 1992). Although environmental standards still vary among countries, almost all major mining companies now state as policy that they will operate all their mines, regardless of location, to first-world standards of environmental protection and worker health and safety.

SUSTAINABLE PRACTICES

Leading mining companies have recently formulated, and pledged to follow, standards and principles for sustainable development of mineral resources worldwide. While not all of these principles relate directly to environmental practices, they represent a significant change in approach for the mining industry as a whole, a change that has already affected environmental practices in the industry. For that reason, the principles of sustainable development (as applied to mineral extraction) are discussed here.

In 1999, nine of the largest mining companies decided to embark on a new initiative intended to achieve a serious change in the way industry approached today's problems. They called this the Global Mining Initiative. It included a program of internal reform, a review of the various associations the companies belonged to, and a rigorous study of the societal issues they had to face. As a result, the International Institute for Environment and Development was commissioned to undertake the Mining, Minerals and Sustainable Development (MMSD) project.

Between 2000 and 2002, the MMSD project identified critical issues associated with development of mineral resources in four "spheres":

Economic sphere

- Maximize human well-being.
- Ensure efficient use of all resources, natural and otherwise, by maximizing rents.
- Seek to identify and internalize environmental and social costs.
- Maintain and enhance the conditions for viable enterprise.

Social sphere

- Ensure a fair distribution of the costs and benefits of development for all those alive today.
- Respect and reinforce the fundamental rights of human beings, including civil and political liberties, cultural autonomy, social and economic freedoms, and personal security.
- Seek to sustain improvements over time; ensure that depletion of natural resources will not deprive future generations through replacement with other forms of capital.

Environmental sphere

- Promote responsible stewardship of natural resources and the environment, including remediation of past damage.
- Minimize waste and environmental damage along the whole of the supply chain.
- Exercise prudence where impacts are unknown or uncertain.
- Operate within ecological limits and protect critical natural capital.

Governance sphere

- Support representative democracy, including participatory decision making.
- Encourage free enterprise within a system of clear and fair rules and incentives.
- Avoid excessive concentration of power through appropriate checks and balances.
- Ensure transparency through providing all stakeholders with access to relevant and accurate information.
- Ensure accountability for decisions and actions, which are based on comprehensive and reliable analysis.
- Encourage cooperation in order to build trust and shared goals and values.
- Ensure that decisions are made at the appropriate level, adhering to the principle of subsidiarity where possible.

In 2001, the board of the metals industry's representative organization, the International Council on Metals and the Environment agreed to broaden the group's mandate and transform itself into the International Council on Mining and Metals (ICMM).

In 2002, ICMM member companies signed the Toronto Declaration committing ICMM to continue the work started by the MMSD project and engage in constructive dialogue with key stakeholders, and in 2003, the International Council on Mining and Metals committed corporate members to implement and measure their performance against the ten principles shown in Table 16.1-1 (ICMM 2006).

Initially there was considerable debate in the mining community regarding the concepts of sustainability as applied to mineral extraction (NWMA 2002). Some argued that, because mineral resources are by nature finite, mineral extraction can never be truly sustainable. However, these concepts have in general been adopted by most of the mining industry, and they influence corporate practices in all areas of mining activity.

Table 16.1-1 ICMM's ten principles

1. **Implement and maintain ethical business practices and sound systems of corporate governance.**
   - Develop and implement company statements of ethical business principles, and practices that management is committed to enforcing.
   - Implement policies and practices that seek to prevent bribery and corruption.
   - Comply with or exceed the requirements of host-country laws and regulations.
   - Work with governments, industry and other stakeholders to achieve appropriate and effective public policy, laws, regulations and procedures that facilitate the mining, minerals and metals sector's contribution to sustainable development within national sustainable development strategies.

2. **Integrate sustainable development considerations within the corporate decision-making process.**
   - Integrate sustainable development principles into company policies and practices.
   - Plan, design, operate and close operations in a manner that enhances sustainable development.
   - Implement good practice and innovate to improve social, environmental and economic performance while enhancing shareholder value.
   - Encourage customers, business partners and suppliers of goods and services to adopt principles and practices that are comparable to our own.
   - Provide sustainable development training to ensure adequate competency at all levels among our own employees and those of contractors.
   - Support public policies and practices that foster open and competitive markets.

3. **Uphold fundamental human rights and respect cultures, customs and values in dealings with employees and others who are affected by our activities.**
   - Ensure fair remuneration and work conditions for all employees and do not use forced, compulsory or child labor.
   - Provide for the constructive engagement of employees on matters of mutual concern.
   - Implement policies and practices designed to eliminate harassment and unfair discrimination in all aspects of our activities.
   - Ensure that all relevant staff, including security personnel, are provided with appropriate cultural and human rights training and guidance.
   - Minimize involuntary resettlement, and compensate fairly for adverse effects on the community where they cannot be avoided.
   - Respect the culture and heritage of local communities, including indigenous peoples.

4. **Implement risk management strategies based on valid data and sound science.**
   - Consult with interested and affected parties in the identification, assessment and management of all significant social, health, safety, environmental and economic impacts associated with our activities.
   - Ensure regular review and updating of risk management systems.
   - Inform potentially affected parties of significant risks from mining, minerals and metals operations and of the measures that will be taken to manage the potential risks effectively.
   - Develop, maintain and test effective emergency response procedures in collaboration with potentially affected parties.

5. **Seek continual improvement of our health and safety performance.**
   - Implement a management system focused on continual improvement of all aspects of operations that could have a significant impact on the health and safety of our own employees, those of contractors and the communities where we operate.
   - Take all practical and reasonable measures to eliminate workplace fatalities, injuries and diseases among our own employees and those of contractors.
   - Provide all employees with health and safety training, and require employees of contractors to have undergone such training.
   - Implement regular health surveillance and risk-based monitoring of employees.
   - Rehabilitate and reintegrate employees into operations following illness or injury, where feasible.

6. **Seek continual improvement of our environmental performance.**
   - Assess the positive and negative, the direct and indirect, and the cumulative environmental impacts of new projects—from exploration through closure.
   - Implement an environmental management system focused on continual improvement to review, prevent, mitigate or ameliorate adverse environmental impacts.
   - Rehabilitate land disturbed or occupied by operations in accordance with appropriate post-mining land uses.
   - Provide for safe storage and disposal of residual wastes and process residues.
   - Design and plan all operations so that adequate resources are available to meet the closure requirements of all operations.

7. **Contribute to conservation of biodiversity and integrated approaches to land use planning.**
   - Respect legally designated protected areas.
   - Disseminate scientific data on and promote practices and experiences in biodiversity assessment and management.
   - Support the development and implementation of scientifically sound, inclusive and transparent procedures for integrated approaches to land use planning, biodiversity, conservation and mining.

8. **Facilitate and encourage responsible product design, use, reuse, recycling and disposal of our products.**
   - Advance understanding of the properties of metals and minerals and their life-cycle effects on human health and the environment.
   - Conduct or support research and innovation that promotes the use of products and technologies that are safe and efficient in their use of energy, natural resources and other materials.
   - Develop and promote the concept of integrated materials management throughout the metals and minerals value chain.
   - Provide regulators and other stakeholders with scientifically sound data and analysis regarding our products and operations as a basis for regulatory decisions.
   - Support the development of scientifically sound policies, regulations, product standards and material choice decisions that encourage the safe use of mineral and metal products.

9. **Contribute to the social, economic and institutional development of the communities in which we operate.**
   - Engage at the earliest practical stage with likely affected parties to discuss and respond to issues and conflicts concerning the management of social impacts.
   - Ensure that appropriate systems are in place for ongoing interaction with affected parties, making sure that minorities and other marginalized groups have equitable and culturally appropriate means of engagement.
   - Contribute to community development from project development through closure in collaboration with host communities and their representatives.
   - Encourage partnerships with governments and non-governmental organizations to ensure that programs (such as community health, education, local business development) are well designed and effectively delivered.
   - Enhance social and economic development by seeking opportunities to address poverty.

10. **Implement effective and transparent engagement, communication and independently verified reporting arrangements with our stakeholders.**
    - Report on our economic, social and environmental performance and contribution to sustainable development.
    - Provide information that is timely, accurate and relevant.
    - Engage with and respond to stakeholders through open consultation processes.

Source: ICMM 2006, © International Council on Mining and Metals.

Table 16.1-2 ISO 14001 standards

Systemic requirements
  a. Establish, document, implement, maintain, and continually improve an environmental management system in accordance with the standard.

Policy requirements
  a. Establish, define, document, implement, maintain, and communicate the organization's environmental policy.

Planning requirements
  a. Establish, implement, document, and maintain procedures to identify the environmental aspects of all activities, products, and services.
  b. Establish, implement, and maintain procedures to identify and clarify the legal and other requirements that apply to the organization's environmental aspects.
  c. Establish, implement, and maintain environmental objectives and targets.
  d. Establish, implement, and maintain programs to achieve environmental objectives and targets.

Operational requirements
  a. Personnel and resources
      i. Provide the resources needed to establish, implement, maintain, and improve the environmental management system.
      ii. Define, document, and communicate environmental management roles, responsibilities, and authorities.
      iii. Appoint someone to assume the role of management representative.
  b. Training
      i. Ensure competency of those who perform tasks that could have a significant environmental impact by identifying training needs, delivering training programs, and maintaining records of training activities.
      ii. Establish, implement, and maintain a procedure to make people aware of the organization's environmental management system.
  c. Communication
      i. Establish, implement, and maintain procedures to control the organization's internal and external environmental communications.
      ii. Document the organization's environmental policy, objectives, and targets, and the main parts of its environmental management system.
      iii. Describe how the parts of the organization's environmental management system interact.
  d. Documents
      i. Control all environmental management documents and records required by the ISO 14001:2004 standard.
      ii. Establish, document, implement, and maintain procedures to manage and control operational situations that could have significant environmental impacts.
      iii. Establish, document, implement, and maintain procedures to control significant environmental aspects of goods and services provided by suppliers and contractors.
  e. Emergency situations
      i. Establish, implement, and maintain procedures for potential emergency situations and accidents that could have an impact on the environment.
      ii. Establish, implement, and maintain procedures to respond to actual emergency situations and accidents that have an impact on the environment.
      iii. Test environmental emergency response procedures.
      iv. Respond to actual environmental emergencies and accidents.
      v. Prevent or mitigate the adverse environmental impacts that emergencies and accidents can and do cause.
      vi. Review and revise environmental emergency preparedness and response procedures.

Compliance requirements
  a. Measurement and monitoring
      i. Establish, implement, and maintain procedures to monitor and measure the operational characteristics that could have a significant impact on the environment.
      ii. Use and maintain calibrated or verified environmental monitoring and measuring equipment.
      iii. Keep a record of environmental monitoring and measuring activities.
  b. Documentation
      i. Establish, implement, and maintain a procedure to periodically evaluate how well the organization complies with all relevant legal and other environmental requirements.
      ii. Record the results of the organization's legal and other environmental compliance evaluations.
  c. Dealing with nonconformities
      i. Establish, implement, and maintain nonconformance management procedures.
      ii. Change documents when nonconformities make it necessary.
  d. Controlling records
      i. Establish environmental records for the organization.
      ii. Establish, implement, and maintain procedures to control the organization's environmental records.
  e. Internal audits
      i. Establish, implement, and maintain an environmental management audit program and procedures.
      ii. Conduct internal audits of environmental management system.
      iii. Report internal audit results to management.
  f. Environmental management reviews.
      i. Review the suitability, adequacy, and effectiveness of the environmental management system.
      ii. Assess whether or not the organization's environmental management system, policy, objectives, or targets should be changed.
      iii. Keep a record of all environmental reviews.

ENVIRONMENTAL MANAGEMENT SYSTEMS

ISO 14001, issued in 2004, provides standards by which a community or organization may put in place and implement a series of practices and procedures that, when taken together, result in an environmental management system (EMS).

An EMS is one part of an organization's larger management system. The EMS is used to establish an environmental policy and to manage the environmental aspects of the organization's activities, products, and services. A management system is a network of interrelated elements that include responsibilities, authorities, relationships, functions, processes, procedures, practices, and resources. A management system uses these elements to establish policies and objectives and to develop ways of applying these policies and achieving these objectives.

ISO 14001 is a detailed document comprising many pages. It describes systemic, policy, planning, operational, and compliance requirements. Table 16.1-2 outlines those requirements.

ISO 14001 is not a technical standard and thus does not in any way replace technical requirements embodied in statutes or regulations. It also does not set prescribed standards of performance for organizations. However, ISO certification or equivalent has become a critical component to acceptance by environmental agencies, nongovernmental organizations, and funding organizations.

The following example shows how a large mining company (CODELCO) complies with part of ISO 14001.

Example 1. Identification and Evaluation of Environmental Consequences

CODELCO adopted a sustainable development policy in June 2003 (CODELCO 2009). In December 2006, the company issued the Corporate Directive for Identification of Environmental Aspects and Evaluation of the Risk of Their Impacts (CODELCO 2006), providing a detailed and systematic method for assessing environmental risk, which is applied to all projects and activities in the corporation. The directive includes the following steps:

1. Identification of unit operations or activities, routine and nonroutine, past, present, and future.

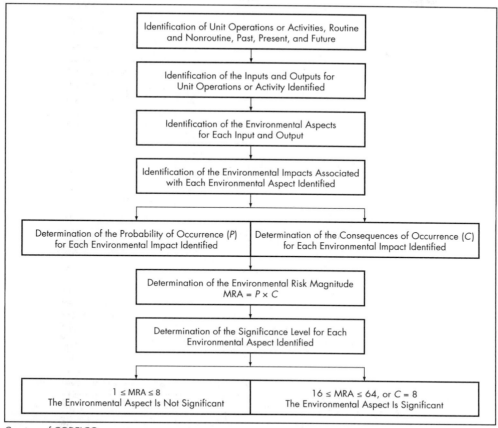

Courtesy of CODELCO.

Figure 16.1-1 Process for identifying significant environmental aspects

2. Identification of all the inputs and outputs for each unit operation or identified activity.
3. Identification of the environmental aspects associated with the inputs and outputs of each unit operation or identified activity.
4. Identification of the potential associated environmental impacts of each identified environmental aspect.
5. Evaluation of the environmental risk, as determined by the probability of occurrence of each potential identified impact, as a function of its frequency, and the consequences of that impact on the environment, as a function of its reversibility and extent.
6. Determination of the magnitude of environmental risk (MRA, for the Spanish *magnitud del riesgo ambiental*), according to the probability and consequences, for each potential impact identified.
7. Determination of the degree of significance of each environmental aspect, according to the value of the MRA for that aspect.

The directive follows ISO 14001:2004 and distinguishes between environmental aspects and environmental impacts. An environmental aspect is an element of the activities, products, or services of one organization that can interact with the environment; environmental impact is any change in the environment, whether adverse or beneficial, wholly or partially resulting from environmental aspects of an organization.

The process begins with generation of the process map for each level in a given work center, using the environmental impacts as the main elements. This allows the direction of the process and any associated equipment to achieve common objectives aligned with the corporate standards. Here, the term *process* includes any activity of production, service, or administration and is defined as a sequence of activities that effect the transformation of certain inputs, with the use of certain resources, labor, goods, and services, to satisfy those who receive or use the process output. A process map is defined as a diagrammatic representation of a process in the abstract, showing connections among constituent activities and flows of material, energy, or information, and designed to facilitate understanding and analysis of the processes. Figure 16.1-1 shows a process map for the identification of significant environmental aspects.

In the identification of unit operations and activities, when there are areas or activities that are not defined within a specific organizational unit or may be subject to overlapping jurisdictions, the environmental management representative in charge of environmental affairs, together with the leaders of those areas or activities, define the allocation of responsibility for assessment of environmental risk.

In the identification of inputs and outputs, the responsible personnel examine each unit operation or activity (routine and nonroutine, past, present, and future). They identify all the inputs, including raw materials, resources, consumables, intermediate or recycled products, energy, services, and in general everything that the unit operation or activity receives from the preceding unit operation. Similarly, they identify all

Table 16.1-3 Evaluation criteria for the probability of occurrence of an environmental impact

	Criterion: Occurrence of the Impact, as a Function of Its Frequency	Value
Probability (P)	The impact always occurs in association with a given activity, or the impact happens often or frequently.	High (8)
	The impact is expected to occur sometimes, or the impact is known to have occurred on some occasions.	Medium (4)
	The impact is expected to occur on one occasion, or the impact is known to have occurred on one occasion.	Low (2)
	Occurrence of the impact is highly unlikely, or the impact has never occurred.	Negligible (1)

Courtesy of CODELCO.

Table 16.1-4 Evaluation criteria for consequences of the occurrence of an environmental impact

	Criterion: Description of the Impact Based on Its Reversibility and Extent	Value
Consequence (C)	The alteration of the original environmental conditions is considered irrecoverable, or the extension of the impact exceeds the limits for the area of direct and indirect influence of the project activity, or it significantly alters environmental conditions, and it cannot be attenuated.	High (8)
	The alteration of the original environmental conditions is considered to be 50%–90% recoverable, or the extent of the impact does not exceed the limit of the area of direct and indirect influence of the project of activity, or the impact significantly alters or has the potential to alter environmental conditions, but the effects can be mitigated by 50%–90%.	Medium (4)
	The alteration of the original environmental conditions is expected to be minor or transitory, or the possible alteration can, with appropriate control measures, be transitory or reduced to minor levels, or the environmental impacts occur only within the company property.	Low (2)
	The alterations in the original condition will not result in any noticeable changes to the environment, or it will be possible to minimize any alterations so they will be imperceptible, or controlled completely.	Negligible (1)

Courtesy of CODELCO.

the outputs, including products sent to market, intermediate products, services to third parties, wastes and residues, and in general everything that the unit operation or activity delivers to subsequent unit operations or activities. The output of one unit operation or activity may often be the input to the next operation. In such cases, the line executive and the environmental management coordinators of both areas must define which operation is responsible for the identification of environmental aspects and the potential environmental impacts associated with this process flow, including the discharge, transport, required surge storage, and so on.

When identifying the environmental aspects, it is important to consider that different aspects can be associated with each operation or activity, not just for those performing the immediate task but for third parties. An example is the incorrect execution of a maintenance task, which may have effects on both the maintenance worker and on those who use the equipment after maintenance. The identification of these environmental aspects should consider (1) activities of all personnel who have access to the workplace, including contractors, suppliers, and visitors; (2) work in areas provided by the company and by third parties; and (3) activities that interact with other resources or protected areas, including tourist attractions, landscapes, or archaeological, anthropological, historical, or cultural heritage sites.

The probability of occurrence of an environmental impact, P, is evaluated according to Table 16.1-3, and the consequences of an environmental impact are evaluated according to Table 16.1-4. Next, for each identified environmental impact, the corresponding value in Table 16.1-5 is calculated as the product of the probability and the consequence for that impact.

Finally, the significance of the environmental aspects that give rise to each environmental impact is determined from Table 16.1-6. There are two conditions under which an environmental aspect is classified as significant: when the MRA is 16 or higher, or when the consequences of an associated environmental impact are high.

Figure 16.1-2 shows an example of a worksheet used to assess environmental risk by CODELCO's vice president of corporate projects, Office of Sustainable Development (E. Córdova, personal communication). The first column of the worksheet is prepared by listing every function or activity associated with a given project, in this case, the development of the Pilar Norte (the North Pillar) sector of the El Teniente deposit. The second column is completed with the environmental aspects and affected areas for each function or activity as identified by project personnel.

Table 16.1-5 Values for environmental risk magnitude

		Consequences (C)			
	MRA	Negligible (1)	Low (2)	Medium (4)	High (8)
Probability (P)	Negligible (1)	1	2	4	8
	Low (2)	2	4	8	16
	Medium (4)	4	8	16	32
	High (8)	8	16	32	64

Courtesy of CODELCO.

Table 16.1-6 Significance criteria for environmental aspects

MRA = P × C	Classification of Environmental Impact	Significance of Environmental Aspect Leading to Environmental Impact
1–8	Insignificant	Insignificant
16–64	Significant	Significant
Consequence 8	Significant	Significant

Courtesy of CODELCO.

The sheet, with its first two columns completed, is circulated to experts within the company, each of whom assigns a value of 1, 2, 4, or 8 to the probability and the consequence of each environmental aspect in the second column. This procedure is similar to that used for hazard identification and risk assessment in the workplace (Chapanis 1986; Colling 1990). The product of the probability and the consequence agreed on for each aspect determines whether the aspect is significant or insignificant, according to the procedures and criteria described previously. Aspects with significant environmental aspects are addressed as priorities in project planning and management.

Function or Activity	Environmental Aspect	Affected Area	Probability	Consequence	Magnitude	Classification
Operation of diesel equipment	Particulate emissions	Work area	8	1	8	Not significant
	Gas emissions	Work area	8	1	8	Not significant
	Fuel consumption	Work area	8	1	8	Not significant
Handling hazardous materials	Potential for spills	Ground & air/Work area	4	4	16	Significant
	Generation of hazardous solid residues	Work area	8	2	16	Significant
Equipment maintenance and repair	Welding gas discharge	Work area	8	1	8	Not significant
	Generation of any solid residues	Work area	8	2	16	Significant
	Potential for spills of hazardous materials	Ground & air/Work area	4	4	16	Significant
Installation of workplace utilities and structures	Electrical energy consumption	Natural resources	4	1	4	No Significant
	Generation of non-process wastewater	Agua o Suelo	1	4	4	Not significant
	Generation of trash and other wastes	Work area	4	1	4	Not significant
	Generation of any solid residues	Work area	4	2	8	Not significant
	Potable water consumption	Natural resources	4	1	4	Not significant
Civil, mechanical, and electrical area control rooms; emergency drifts; main electrical room; transformer room; air shaft connections at level; extraction drift connections at level; hydraulic breaker foundations and control rooms; fan rooms	Welding gas discharge	Work area	4	1	4	Not significant
	Generation of any solid residues	Work area	8	2	16	Significant
	Electrical energy consumption	Natural resources	4	1	4	Not significant
	Water consumption	Natural resources	8	1	8	Not significant
	Potential for spills of hazardous materials	Ground & air/Work area	4	4	16	Significant
Electical cables, cableways, and lights	Generation of any solid residues	Work area	8	2	16	Significant
	Electrical energy consumption	Natural resources	4	1	4	Not significant
Installation and assembly of ventilation doors	Generation of any solid residues	Work area	8	2	16	Significant
	Electrical energy consumption	Natural resources	4	1	4	Not significant
	Water consumption	Natural resources	8	1	8	Not significant
Installation of compressed air system	Potential for spills of non-hazardous materials	Ground & air/Work area	4	4	16	Significant
	Generation of hazardous solid residues	Work area	8	2	16	Significant
	Electrical energy consumption	Natural resources	4	1	4	Not significant
Foundations and other concrete structures for hydraulic breakers	Generation of any solid residues	Work area	8	2	16	Significant
	Electrical energy consumption	Natural resources	4	1	4	**Not significant**
	Water consumption	Natural resources	8	1	8	**Not significant**
	Potential for spills of non-hazardous materials	Ground & air/Work area	4	4	16	Significant
Installation of hydraulic breakers with hydraulic power systems and control panels	Generation of any solid residues	Work area	8	2	16	Significant
	Electrical energy consumption	Natural resources	4	1	4	**Not significant**
	Water consumption	Natural resources	8	1	8	**Not significant**
	Potential for spills of hazardous materials	Ground & air/Work area	4	4	16	Significant
Installation of fire prevention system	Potential for spills of non-hazardous materials	Work area	4	2	8	Not significant
	Electrical energy consumption	Natural resources	4	1	4	Not significant
	Water consumption	Natural resources	8	1	8	Not significant
Installation of industrial water system	Emission of welding gases	Work area	4	1	4	Not significant
	Potential for spills of non-hazardous materials	Work area	8	2	16	Significant
	Electrical energy consumption	Natural resources	4	1	4	Not significant
Installation of water system for fire control	Welding gas discharge	Work area	4	1	4	Not significant
	Potential for spills of non-hazardous materials	Work area	4	2	8	Not significant
	Electrical energy consumption	Natural resources	4	1	4	Not significant
Installation of hydraulic power system	Generation of any solid residues	Work area	8	2	16	Significant
	Potential for spills of hazardous materials	Ground & air/Work area	4	4	16	Significant
Installation of ventilation fans	Potential for spills of non-hazardous materials	Work area	4	2	8	Not significant
Installation of steel arch-type roof supports	Potential for spills of non-hazardous materials	Work area	8	2	16	Significant
	Electrical energy consumption	Natural resources	4	1	4	Not significant
Modification and reconditioning of equipment washdown areas	Generation of any solid residues	Work area	8	2	16	Significant
	Electrical energy consumption	Natural resources	4	1	4	Not significant
	Water consumption	Natural resources	8	1	8	Not significant
	Potential for spills of hazardous materials	Ground & air/Work area	4	4	16	Significant
Construction of civil and electrical maintenance shops	Generation of any solid residues	Work area	8	2	16	Significant
	Electrical energy consumption	Natural resources	4	1	4	Not significant
	Potential for spills of hazardous materials	Ground & air/Work area	4	4	16	Significant
Installation of bridge crane	Generation of any solid residues	Work area	8	2	16	Significant
	Electrical energy consumption	Natural resources	4	1	4	Not significant
	Potential for spills of hazardous materials	Ground & air/Work area	4	4	16	Significant
Installation of water treatment (de-acidification) plant	Emission of welding gases	Work area	4	1	4	Not significant
	Generation of any solid residues	Work area	8	2	16	Significant
	Electrical energy consumption	Ground & air/Work area	4	1	4	Not significant
	Water consumption	Work area	8	1	8	Not significant
	Potential for spills of hazardous materials	Work area	4	4	16	Significant
Installation of air compressors and lubrication systems	Generation of any solid residues	Work area	8	2	16	Significant
	Electrical energy consumption	Natural resources	4	1	4	Not significant
	Potential for spills of hazardous materials	Work area	4	4	16	Significant

Courtesy of CODELCO.

Figure 16.1-2 Identifying and evaluating environmental effects of projects

ENVIRONMENTAL REPORTING

The conveyance of information is critical at all phases of a mining project. All of the people and institutions with an interest in how the project is conducted should be well informed of the ongoing plans for and progress with the project. MMSD addressed the issue of access to information in a report titled *Breaking New Ground* (MMSD 2002):

Information flow is essential in a sustainable stakeholder society. Information comes in different forms and is of variable quality. Information about a company and its operations is used by a range of actors, such as communities, investors, employees, lenders, suppliers, and customers, often through appropriate accounting and reporting procedures based on defined indicators and measurement techniques. Within the industry, information is used by management to monitor performance efficiency and the impacts of operations. At the exploration phase, accurate geoscientific data and maps are crucial.

The working group that prepared *Breaking New Ground* held a workshop in 2001 in Vancouver. The workshop provided the opportunity to discuss the information used in preparing the report and examine how that information was viewed and handled by the mining industry. It was concluded that information often fails to flow to communities in a timely and transparent fashion, that disclosure practices often fall short of current best practice, and that one-size-fits-all systems of public reporting or a global reporting standard would be an extremely difficult initiative to develop. The distinct nature of specific mines, projects, companies, locations, and communities means that a different mix of indicators, metrics, and evaluations is needed. One effective way of scoping the need for information around any project is to ask the community what they need to know in considering project proposals.

The ISO 14001 standard includes disclosure as an outcome of the environmental auditing and management systems. The standard mainly addresses internal environmental management systems. Certain items within the standard refer in general terms to external communications as part of an organization's EMSs.

There are also voluntary initiatives to standardize the way corporations convey information. Most closely examined by the MMSD was the Global Reporting Initiative (GRI). It was convened in 1997 by the Coalition for Environmentally Responsible Economies "to make sustainability reporting as routine and credible as financial reporting in terms of comparability, rigour and verifiability" through "designing, disseminating and promoting standardized reporting practices, core measurements and customized sector specific measurements." The GRI Sustainability Reporting Guidelines suggest that reports include a CEO statement; key indicators; a profile of the reporting entity; policies, organization, and management systems; management performance; operational and product performance; and a sustainability overview (GRI 2009).

SOUND ENVIRONMENTAL PRACTICES

Past publications have provided detailed descriptions of environmental regulations and permitting procedures for a given location—often the United States. In this chapter, a more general approach is taken by describing the principles of sound environmental practice, which can and should be applied in any political jurisdiction. However, some examples of regulations and accepted practice are given for illustrative purposes.

Before discussing the individual activities involved in mining, it is important to point out that, in general, mining activities should be conducted with the future in mind. This will not only minimize the environmental effects of each activity, but will also result in significant cost savings. This approach will lead to some considerations that are very broad and apply to almost all projects. For example,

- Government officials and all residents in the area of the proposed mine should be well informed of and directly involved in all project plans from the beginning. Such involvement will help address the concerns of local people as they occur and may well obviate formal objections when permit applications are filed; and
- The overall mine plan should include consideration of the requirements of reclamation and restoration. Placement of waste piles, tailings ponds, and similar materials should be carefully planned to minimize rehandling of material.

At the same time, other considerations will be specific to a given project. For example,

- Design and location of roads developed for exploration should consider future needs for mining and processing. This will minimize unnecessary road building and decrease effects on the local ecosystem; and
- When containment ponds are required for drilling fluids or cuttings, consideration should be given to the future use of those ponds for tailings impoundment, storm water catchment, or settling ponds.

The mining operation should be thoroughly planned, in consultation with government agencies and people living in the region. Historic, cultural, and biological resources should be identified, and plans made for their protection. This is especially important in areas where indigenous people have little exposure to the technologies used in mining. Mining companies should take the appropriate steps to ensure that the indigenous people understand the mining plans. The indigenous peoples' values and land-use practices must be understood, and steps must be taken to protect them. In some locations, this will necessitate the involvement of anthropologists, sociologists, and other experts. It may also require a planning and approval process that differs from those to which the company is accustomed. For example, some indigenous peoples make decisions by group consensus, necessitating large community meetings that may last for several days.

Mine Development

Development is the preparation of the facilities, equipment, and infrastructure required for extraction of the valuable mineral material. It includes land acquisition, equipment selection and specification, infrastructure and surface facilities design and construction, environmental planning and permitting, and initial mine planning.

Hunt (1992) provided a useful checklist for mine development projects in the form of a questionnaire (see Table 16.1-7). While some aspects of this list are specific to operations in the United States, its breadth and detail make it a worthwhile resource for almost any mining project. It includes considerations for projects in areas where no mining has occurred, for projects in previously mined areas, and for expansion projects at operating mines.

Infrastructure and Surface Facilities Design and Construction

Infrastructure and surface facilities may include all of the following:

- Roads and railroads
- Power plants, electric power lines, and substations
- Fuel supply lines and tank farms
- Dams, diversions, reservoirs, groundwater well fields, water supply lines, water tanks, and water treatment plants
- Sewage lines and sewage treatment plants
- Maintenance shops
- Storage sheds and warehouses
- Office buildings, parking lots, shower facilities, and changehouses

- Worker accommodations (housing, cafeteria, infirmary, and recreation facilities)
- Houses for pumps, fans, and hoists
- Waste piles and impoundments
- Ponds for catchment of surface and groundwater, and mine drainage
- Industrial and household waste landfills

Six general environmental considerations apply to the design and construction of surface facilities and infrastructure:

1. Select locations of surface structures and infrastructure to minimize the effects of their construction and use on surface water, groundwater, plant and animal ecosystems, and nearby human habitation. Surface structures and infrastructure include roads, railroads, and power lines, for example.
2. Remove and store topsoil for future use from areas that will be later reclaimed.
3. Control runoff so that water exiting the boundaries of the permitted mine site is captured and treated as required to meet applicable discharge standards. Many jurisdictions permit mines as zero-discharge facilities, meaning that all water or solid waste is contained within the permitted mine area.
4. Protect surface water and wetlands. Because these features may exist directly over mineral reserves, their relocation may be necessary.
5. Revegetate disturbed areas. Recontour the land to a defined standard and revegetate with an approved mix of seeds and plantings.
6. Control emissions of dust and noise to meet the requirements of local regulations and the reasonable expectations of persons living nearby. In particular, if explosives are used, control air blast and ground vibration as required.

Specific considerations for some surface facilities or infrastructure include

- Maintenance shops should be designed to avoid contamination of soil and water by spilled fuel and lubricants;
- Surface facilities should be designed to minimize energy consumption, using solar water heating and alternative electric power generation wherever possible; and
- Surface facilities should also be designed architecturally to harmonize with the natural surroundings in the locale.

Environmental Planning and Permitting

The forward-thinking approach mentioned earlier is especially important in environmental planning and permitting. In the permitting process, careful planning is imperative, so that information is gathered efficiently, the applications submitted to regulators meet all requirements, including timely submission, and mine planning costs are minimized by avoiding redesign required after agency review. As pointed out by Hunt (1992), "One of the most serious deficiencies encountered by many operators in the permitting process is to fail to adequately plan the time required to collect baseline data and to provide for public and agency review of applications." Hunt goes on to list the major tasks and activities required during early phases of environmental planning:

- Identification of the regulatory agencies that will be responsible for permitting the proposed mining operation
- Acquisition of all necessary permitting forms and copies of applicable regulations
- Identification of possible legal or technical restrictions that may make approval difficult or require special attention in the permitting process
- Identification of environmental resources information that will be required and the development of a plan and schedule for collection of the required data

Development of a comprehensive permit plan during the initial stages of mine development is critical. The plan should identify predecessor information that must be generated at each stage in the permitting, thereby identifying critical path issues and potential bottlenecks. A Gantt chart is often used for this purpose. For example, if the groundwater discharge permit for the tailing impoundment takes 2 years, and the design must be submitted with the application, the mine may need to move the "design tailing impoundment" task up on the schedule. It is recommended that an environmental permitting specialist be tasked with preparing the comprehensive permit plan.

After preliminary planning, permit applications must be completed. Many jurisdictions require a detailed environmental audit, which is then described in a report, often called an environmental impact statement, or EIS. Depending on local regulations, the EIS may be issued for public review and comment. Topics covered in an EIS usually include the following 19 areas:

1. Permits required and permitting status
2. Ownership of surface and mineral rights
3. Ownership of the company or companies that will be operating the mine, including contractors
4. Previous history of the entities described in Item 3
5. Information on archaeological, historical, and cultural resources within the planned mine site and adjacent areas
6. Hydrology of the mine site and adjacent areas, including the presence of potential pollutants and plans for dealing with them
7. Potential for dust and air pollution, and plans for mitigation
8. Plans for use and control of explosives
9. Plans for disposal of mine waste, including methods for controlling dust generation, slope stability, and seepage of contaminated water
10. Plans for impoundments, and methods for preventing failure or overtopping
11. Plans for construction of backfilled areas during mining, and for backfilling of mine excavations when mining is finished, including methods for preventing slides or other instability
12. Potential for surface subsidence, both immediate and long term, and proposed methods of preventing subsidence or compensating for its negative effects
13. Potential for fires in mine workings, at outcrops, and in spoil piles, and methods for prevention and control of such fires
14. Planned use of injection wells or other underground pumping of fluids
15. Information on previous mine workings, plans for reclaiming previous workings, and description of how the planned workings will interact with previous workings
16. Socioeconomic impact analysis
17. Visual impact analysis
18. Noise impact analysis
19. Biological assessment and impact analysis

Table 16.1-7 Environmental audit checklist

Category	Checklist Items
Permits	☐ Provide a list of all permits existing, applied for, and in preparation for the mine. Indicate the current status of each permit. ☐ If the mine has National Pollutant Discharge Elimination System permits, air permits, waste permits, water withdrawal permits, or other permits, include a listing of each applicable permit. ☐ Are any additional permits required for the facility but not already obtained? ☐ Were any citizen protests or comments filed on any of the permits? If so, describe the issues raised. ☐ What is the status of the bond on each permit? Indicate the amount of bonding liability under each permit.
Surface and mineral rights information	☐ Is the surface property owned or controlled by the mining company? ☐ Does the mine have a specific written agreement with each surface owner pertaining to and providing applicable rights to the surface? ☐ If the mine is an underground mine, do the surface rights specifically include language that says the company has the right to subside? ☐ Has the mine ever received any complaints from a surface owner or any of the surrounding landowners? Indicate the nature of any complaint received. ☐ With respect to mineral rights, has there ever been a claim made suggesting that the mining company does not have complete and full mineral rights? ☐ Have any other entities been granted access rights to the property that may conflict with the rights to extract the mineral (e.g., oil wells, pipelines, utility easements, etc.)?
Ownership and violation information	☐ Is the mine permit held by any company not owned and controlled 100% by the operator? ☐ Is the mine permit being held by an independent contractor? ☐ Is the mine operation being conducted by any company not owned and controlled 100% by the permittee? ☐ Is the mine being operated by an independent contractor? Give a a brief summary of applicable relation. ☐ If the mine is permitted by or being operated by an independent contractor, has a review been completed of the contractor's violation history with regard to any other mines potentially affiliated with the contractor? ☐ If the mine is permitted by or being operated by an entity only partially owned, has a review been completed of the contractor's violation history with regard to any other mines potentially affiliated with the other parties? ☐ How is the relationship between mine personnel and the mine inspectors (excellent, good, poor)? If poor, please describe the problems encountered. ☐ Have there been any citizen complaints filed on the mine? If so, please describe the complaints filed.
Information on archaeological sites, cemeteries, etc.	☐ Within the permit area and adjacent areas, including the area over existing, past, and proposed underground workings, does the mine contain any of the following features? • Are there any federal lands? Indicate the type of federal land involved. • Is there any archaeological or historical site that has been previously identified? Indicate the type of site involved. • Are there public roads where there is not a specific written approval for mining through or under the public road by the agency that has jurisdiction over the road? • Are there occupied dwellings where there is not a specific written waiver from the owner of the occupied dwelling? • Are there any public buildings, schools, churches, community, or institutional buildings? • Are there any public parks? • Are there any public or private cemeteries?
Hydrology	☐ Has the mine obtained all necessary water discharge permits? ☐ Is the mineral being mined associated with potential pollutants, such as high-sulfur-bearing strata? ☐ Has the mine ever experienced any difficulties with poor-quality mine drainage? Briefly describe the situation encountered. ☐ Does any of the mine water need to be treated to meet effluent limitations or other applicable standards? ☐ Has any poor-quality mine drainage been observed on any part of the permit area, in any of the surrounding areas, whether associated with the mine or not? ☐ Have any of the surrounding landowners or occupants ever complained about loss of water or the mine contaminating their water supplies? ☐ Is the water used by surrounding landowners of poor quality? For example, is the water discolored, does it have a bad odor or bad taste, or are there other problems with water quality in the area? ☐ Has the mine ever had any complaints about the water level in streams increasing or decreasing, or the fish population in streams increasing or decreasing? ☐ Are there any in-stream treatment facilities or in-stream sediment ponds associated with the mine? ☐ Are there any wetland areas, wet areas with cattails, or similar aquatic vegetation that could be considered wetlands, within the vicinity of the mine?
Air pollution	☐ Has the mine obtained all the necessary quality permits? ☐ Have there been complaints about dust or air quality discharges from the mine? ☐ Are mine haul trucks covered with tarps to minimize dust and to prevent material from falling off? ☐ Is there a program for controlling dust and air pollution from the mine?
Explosives	☐ Has the mine had any complaints from landowners pertaining to blasting or the use of explosives? ☐ Is the mine access limited because of topographic conditions, or is it possible for local people to obtain unauthorized access to the mine site? ☐ If it is possible for unauthorized persons to obtain access to the site, what measures are currently taken to control this access? ☐ Are any personnel other than trained mine personnel allowed within the mine pit or the mine property in an area that may ever be subject to blasting?

(continues)

Table 16.1-7 Environmental audit checklist (continued)

Category	Checklist Items
Waste disposal	☐ Does the operation include a facility for the disposal of mine waste? ☐ Have there ever been any discolored or bad-quality water seeps from the waste pile? ☐ Does the groundwater indicate any levels of pollution from the waste pile? ☐ What is the worst-case chemical analysis for the waste material being disposed of at the mine? ☐ Have there ever been any signs of slumping or instability of the waste pile? ☐ Does the waste pile have any public roads, homes, occupied dwellings, or other structures downstream that could be affected by failure? ☐ Is the waste pile routinely inspected and certified by a professional engineer to verify that it is being built in accordance with the approved plan? ☐ Are there any abandoned mine waste piles within the vicinity of the mine, including the area overlying underground workings? ☐ Do any of the abandoned mine waste piles show any signs of instability or poor mine drainage? ☐ Does the disposal of any domestic or other nonmine waste occur within the property boundaries, including the disposal of garbage by mine or nonmine personnel? ☐ Has a survey been done to determine whether there is any nonmine waste on the property? Are there any abandoned nonmine waste piles within the mine property? ☐ What does the mine do with trash from the mine?
Impoundments	☐ Are there any large impoundments within the vicinity of the mine or included in the mine permit? ☐ Are there any public roads, facilities, dwellings, or other such facilities downstream of an impoundment that could be affected by the impoundment's failure? If so, briefly describe. ☐ Are all impoundments routinely inspected and certified by a professional engineer to verify that they are being built in accordance with the approved plans? ☐ Have any of the impoundments ever experienced overtopping?
Backfilling and slides	☐ Has the mine ever experienced a slide or instability of backfilled material or of outslope material? ☐ Does the mine have an excess spoil fill? Does the mine have a durable rock fill? Were any of the fills constructed in a location where there was a stream or running water prior to mining? ☐ If the mine is a surface mine, what is the maximum size pit and the size of the total pit currently open for the mine? How many cubic meters (cubic yards) of material would be required if the mine was closed today and the pit had to be backfilled? ☐ What is the maximum amount of highwall currently open on the operation, including total length and height? How many cubic yards (cubic meters) of material would be required if the mine was closed today and the highwall had to be completely eliminated?
Subsidence	☐ Does the mine include longwall or pillar removal mining? ☐ Has the mine ever identified any substance subsidence? ☐ Have there ever been any complaints by surface owners about surface subsidence? ☐ Does the mine have any significant problems with roof or floor control?
Mine fires	☐ Have there ever been any mine fires, including outcrop fires, identified on the mine property?
Underground injections	☐ Does the mine have any injection wells or pump any fluids or other material of any type underground? ☐ If the mine is an underground mine, has an evaluation ever been completed to determine what will occur after the mine is closed and the mine workings become flooded? If yes, briefly describe the resulting conclusion.
Previously mined areas	☐ Does the area affected by the mine include a previously mined area? ☐ Are there any of the following problems on the previously mined area? • Bad water drainage? • Unstable spoil? • Spoil on the outslope? • Refuse material on the outslope? • Unreclaimed areas where revegetation will be difficult? • Unreclaimed highwall?
Miscellaneous	☐ How close is the nearest house? ☐ Have there been complaints about the use of roads in the area? ☐ Has the mine ever received a closure or cessation order? ☐ Are there cleaning or other chemical solvents located at the mine? Have all necessary forms been filled out and provided in regard to "right to know" requirements for chemical exposure? ☐ Have all transformers or other electrical equipment been checked for polychlorinated biphenyl contamination? ☐ Is oil or fuel stored on the property? What types of facilities are provided for oil or fuel storage? Are there any underground storage tanks? Have all such tanks been tested? ☐ Are there any oil or gas wells, or similar facilities on the property that could conflict with mining?

Source: Hunt 1992.

Some countries require one comprehensive permit that addresses all environmental aspects, but some countries, such as the United States, require many individual permits covering myriad aspects. The following example illustrates the planning and permitting process for a surface coal mine in the United States.

Example 2. Premine Planning and Permitting for Surface Coal Mines in the United States

Table 16.1-8 describes reclamation requirements for surface coal mining based on requirements given in the Surface Mining Control and Reclamation Act (SMCRA) of 1977. Under SMCRA, specific requirements are given for obtaining a mining permit, and formal documents must be filed (CFR 2009).

A large amount of information is required for preparation of the reclamation and operations plan. Table 16.1-9 summarizes those requirements.

Government Agency Review of Permit Applications

The processing of a permit application begins with the submission of a completed application for a new permit, permit revision, or permit renewal to the regulatory authority and ends with the final decision to approve or deny the application. Agency review begins when the applicant officially files the permit application. For most permits, after the agency determines the application is complete, they begin the formal review process, notify affected parties, and begin a public comment period. In most cases, members of the public will be able to both submit written comments and request a hearing on the merits of the particular application. If the permit is granted, the operator may be required to post a bond. Upon approval of the bond, mining operations may commence.

Most agencies will conduct a detailed technical review of each application received to ensure that all applicable regulatory requirements are met. As part of this technical review, the applicant is notified of any deficiencies identified in the application as submitted and provided an opportunity to respond to the agency comments and to correct deficiencies. Failure to provide an adequate response or to correct any permit deficiency could result in denial of the application.

The time required for agency reviews will range from a few months to 2–3 years. This time can be even longer in areas where opposition to a project is well coordinated and well funded. The permitting for the Safford mine in Arizona, described in detail in this section, took 13 years. Thus it is extremely important that permit applications be prepared with great care and reviewed meticulously to minimize delays by the agencies.

In some countries, efforts have been made to speed up the permit review process. For example, in Chile, when environmental regulations were promulgated in 1994, specific approval times were specified. The maximum time for approval of an Environmental Impact Declaration is 90 days, not including the days required for response to questions from regulatory agencies. For the more complex environmental impact study, which requires establishment of baseline standards and community participation, the maximum is 180 days (República de Chile 2002).

Example 3. Freeport-McMoRan's Safford Mine

Exploration of the Dos Pobres/San Juan area of the Safford Mining District began in about 1940. The district is in Graham County, Arizona, about 12.9 km (8 miles) north of Safford. It comprises four porphyry copper deposits—Dos Pobres, San Juan, Lone Star, and Sanchez—located along the southwestern slope of the Gila Mountains north of the Gila River.

The Phelps Dodge Corporation (PD; now a part of Freeport-McMoRan) acquired an interest in the property in 1957 and began underground development in 1968. Underground activity stopped in 1982, and by 1992 interest had shifted to the oxide mineralization in the area. By 1994, PD owned most of these mineral deposits, but some extended onto public lands administered by the U.S. Bureau of Land Management (BLM). The portions of the deposits on BLM land were controlled by PD through mining claims filed under the U.S. General Mining Law of 1872 (Cooper 2008).

In an effort to consolidate its holdings in the Safford Mining District, PD proposed a land exchange to the BLM, in which PD would acquire public lands (referred to as the selected lands) within and adjacent to its existing private property in the mining district in trade for other lands (the offered lands) in Arizona currently owned by PD. Both parties signed an Agreement to Initiate (ATI), which began formal consideration of the land exchange.

The administration of public lands in the United States is often controversial, especially in the western states. In most areas, groups of citizens have organized to oppose almost any development on those lands. These groups are often supported by large, well-funded environmental organizations, such as the Sierra Club. Before making a decision about the land exchange, the BLM decided to prepare an EIS to comply with provisions of the U.S. National Environmental Policy Act (NEPA). After the parties agreed to formally consider a land exchange, public scoping took place in late fall of 1994 and baseline studies commenced.

Under the U.S. Clean Water Act, the U.S. Army Corps of Engineers (ACE) issues permits for the discharge of dredged or fill material into the navigable waters at specified disposal sites. Review of these permits usually requires several months after notice and the opportunity for public hearings. In late 1995, ACE stated that it would likely require an EIS as part of its environmental review for a Section 404 permit to implement the foreseeable mining uses if the land exchange was authorized. At about the same time, PD accelerated the planning and development schedules for the Dos Pobres/San Juan project.

PD submitted a mining plan of operations (MPO) in May 1996, which allowed the BLM, ACE, and EPA (U.S. Environmental Protection Agency), as cooperating agencies, to consolidate their respective environmental reviews for the proposed mining activities. The MPO alternative provided agency employees and the general public with more information regarding the potential effects of the foreseeable mining uses of the selected lands and also conformed to regulations of the Council on Environmental Quality and the guidelines published in *NEPA's Forty Most Asked Questions* (NEPA 2009).

In response to PD's submittal of an MPO, the BLM determined that the EIS should reflect the fact that an MPO was now the proposed action and made the land exchange proposal one alternative to the MPO. The BLM also involved ACE and EPA as cooperating agencies in the EIS process. The BLM then reinitiated the scoping process begun earlier because submittal of the MPO represented a significant change in scope from the original land exchange proposal. In December 1997, The U.S. Bureau of Indian Affairs (BIA) became a cooperating

Table 16.1-8 SMCRA reclamation requirements

Requirement	Description
Operations plan	• The type and method of coal mining, procedures, and proposed engineering techniques; anticipated annual and total production of coal, by tonnage; and the major equipment to be used for all aspects of those operations • The construction, modification, use, maintenance, and removal of the following facilities: retention dams, embankments, and other impoundments; overburden and topsoil handling and storage areas and structures; coal removal, handling, storage, cleaning, and transportation areas and structures; spoil, coal processing waste, and noncoal waste removal, handling, storage, transportation, and disposal areas and structures; mine facilities; and water and air • Existing structures proposed to be used in connection with or to facilitate the surface coal mining and reclamation operation • Blasting practices and blast monitoring system
Maps	Locations of • Lands proposed to be affected throughout the operation and any change in a facility or feature to be caused by the proposed operations • Buildings, utility corridors, and facilities to be used • The area of land to be affected within the proposed permit area, according to the sequence of mining and reclamation • Each area of land for which a performance bond or other equivalent guarantee will be posted • Each coal storage, cleaning, and loading area • Each topsoil, spoil, coal waste, and noncoal waste storage area • Each water diversion, collection, conveyance, treatment, storage, and discharge facility to be used • Each air pollution collection and control facility • Each source of waste and each waste disposal facility relating to coal processing or pollution control • Each facility to be used to protect and enhance fish and wildlife and related environmental values • Each explosives storage and handling facility • Each sedimentation pond, permanent water impoundment, coal processing waste bank, and coal processing waste dam and embankment, and fill area for the disposal of excess spoil
Air pollution control plan	Description of fugitive dust control practices and a monitoring program to provide sufficient data to evaluate the effectiveness of the fugitive dust control practices proposed to comply with federal and state air quality standards
Fish and wildlife information	All listed or endangered species, and all unusual habitats; protective measures to be undertaken during mining and reclamation; and plans for habitat enhancement
Reclamation plan	• A detailed timetable for the completion of each major step in the reclamation plan • A detailed estimate of the cost of reclamation of the proposed operations required to be covered by a performance bond, with supporting calculations for the estimates • A plan for backfilling, soil stabilization, compacting, and grading, with contour maps or cross sections that show the anticipated final surface configuration of the proposed permit area • A plan for removal, storage, and redistribution of topsoil, subsoil, and other material • A demonstration of the suitability of topsoil substitutes or supplements, based on analysis of the thickness of soil horizons, total depth, texture, percentage of coarse fragments, pH, and areal extent of the different kinds of soils • A plan for revegetation including, but not limited to, descriptions of the revegetation schedule; the species and amounts per acre of seeds and seedlings to be used; methods to be used in planting and seeding, mulching techniques, and irrigation (if appropriate); pest and disease control measures, if any; measures proposed to be used to determine the success of revegetation; a soil testing plan for evaluation of the topsoil results; and the handling and reclamation procedures related to revegetation • A description of the measures to be used to maximize the use and conservation of the coal resource • A description of measures to be employed to ensure that all debris, acid-forming and toxic-forming materials, and materials constituting a fire hazard are disposed of properly, and a description of the contingency plans that have been developed to preclude sustained combustion of such materials • A description, including appropriate cross sections and maps, of the measures to be used to seal or manage mine openings, and to plug, case, or manage exploration holes, other boreholes, wells, and other openings within the proposed permit area • A description of steps to be taken to comply with the requirements of the Clean Air Act, the Clean Water Act, and other applicable air and water quality laws and regulations, and health and safety standards • Description of the land use within the proposed mining area before mining, and of the proposed land use after mining and reclamation are complete • A general plan and a detailed design plan for each proposed siltation structure, water impoundment, and coal processing waste bank, dam, or embankment within the proposed permit area, and of the plans for removing each such structure
Hydrology	• Sampling and analysis methodology • Baseline information • Groundwater information, including the location and ownership for the permit and adjacent areas of existing wells, springs, and other groundwater resources; seasonal quality and quantity of groundwater and usage. Water quality descriptions shall include, at a minimum, total dissolved solids or specific conductance, pH, total iron, and total manganese. • Surface water information, including the name, location, ownership, and description of all surface water bodies such as streams, lakes, and impoundments; the location of any discharge into any surface water body in the proposed permit and adjacent areas; and information on surface water quality and quantity sufficient to demonstrate seasonal variation and water usage • Baseline cumulative impact area information • Alternative water source information • Probable hydrologic consequences determination • Cumulative hydrologic impact assessment • Hydrologic reclamation plan • Groundwater monitoring plan • Surface water monitoring plan

(continues)

Table 16.1-8 SMCRA reclamation requirements (continued)

Requirement	Description
Geology	• Sufficient detail to assist in determining the probable hydrologic consequences of the operation on the quality and quantity of surface and groundwater in the permit and adjacent areas, including the extent to which surface water and groundwater monitoring is necessary; all potentially acid- or toxic-forming strata down to and including the stratum immediately below the lowest coal seam to be mined; and whether reclamation can be accomplished and whether the proposed operation has been designed to prevent material damage to the hydrologic balance outside the permit area, including, at a minimum, the following: A description of the geology of the proposed permit and adjacent areas down to and including the deeper of either the stratum immediately below the lowest coal seam to be mined or any aquifer below the lowest coal seam to be mined that may be adversely impacted by mining. The description shall include the areal and structural geology of the permit and adjacent areas, and other parameters that influence the required reclamation and the occurrence, availability, movement, quantity, and quality of potentially impacted surface and groundwaters. • Analyses of samples collected from test borings, drill cores, or fresh, unweathered, uncontaminated samples from rock outcrops from the permit area, down to and including the deeper of either the stratum immediately below the lowest coal seam to be mined or any aquifer below the lowest seam to be mined that may be adversely impacted by mining. The analyses shall result in the following: (a) logs showing the lithologic characteristics including physical properties and thickness of each stratum and location of groundwater where occurring; (b) chemical analyses identifying those strata that may contain acid- or toxic-forming or alkalinity-producing materials and to determine their content, except that the regulatory authority may find that the analysis for alkalinity-producing materials is unnecessary; and (c) chemical analyses of the coal seam for acid- or toxic-forming materials, including the total sulfur and pyritic sulfur, except that the regulatory authority may find that the analysis of pyritic sulfur content is unnecessary.
Diversions	Stream channel and other diversions to be constructed within the proposed permit area, including maps and cross sections
Protection of area	Measures that will be taken to protect any publicly owned parks and historic places within the proposed permit area
Public and private interests	Description, including appropriate maps and cross sections, of measures that will be taken to ensure that the interests of the public and landowners affected are protected if during mining the operator undertakes the relocation or use of public roads
Disposal	Description, including appropriate maps and cross-sectional drawings, of methods for disposal of excess spoil, including the proposed disposal site, the design of the spoil disposal structures, the removal, if appropriate, of the site and structures, and any requisite geotechnical investigation
Roads	Description, including plans and drawings, for each road to be constructed, used, or maintained within the proposed permit area

Source: CFR 2009.

agency in the preparation of this EIS, specifically to contribute its expertise in the areas of Indian trust resources and tribal consultation. Protection of groundwater resources was of particular interest to the Indian (indigenous) people in the project area.

In September 1998, the draft EIS was published. It was more than 400 pages long (not including seven appendices), and included 108 tables and 73 figures. It addressed the scoping issues shown in Table 16.1-10. The EIS also identified the laws, regulations, and other government actions applicable to the project, including nine federal acts, five federal executive orders, six federal land-use or management plans, and twelve state and local laws.

A groundwater model and the Model, Monitor, and Mitigate program for groundwater protection were prepared. Both documents were reviewed at length by the BLM's hydrologist, the BIA, and a consultant retained by the BIA. However, the agencies disagreed on these and other issues, and the BIA withdrew as a cooperator in June 2000, without facilitating consultations with Indian tribes regarding potential impacts to trust resources. In April 2001, BLM reinitiated direct consultations with the Gila River Indian Community and the San Carlos Apache Tribe, which continued throughout the preparation and evaluation of the EIS.

Publication of the draft EIS was followed by a 60-day public comment period that was extended twice, and during which public and tribal open house meetings were held in four locations. Notices were published in the *Federal Register* and in the *Eastern Arizona Courier*.

The draft EIS informed readers that the 60-day comment period would end on November 25, 1998. During this period, BLM received three requests by electronic or regular mail for an extension of the comment period. BLM notified these three requesters via electronic mail and the public at large by a notice published in the *Federal Register* on November 30, 1998, that the comment period was being extended until December 18, 1998. The public was notified of a second extension of the comment period until January 29, 1999, through a notice printed in the *Federal Register* on December 18, 1998. Thus, the public comment period lasted a total of 127 days.

Three public open house meetings lasting 4 hours each were held at BLM field offices in Safford, Tucson, and Phoenix on, respectively, October 27, 28, and 29, 1998. Representatives from the BLM, ACE, and EPA were present in each meeting. A total of 212 members of the public attended the three meetings. A tribal open house meeting was held on December 4, 1998, on the San Carlos Apache reservation and was attended by 23 persons representing the tribe. Graphic depictions of the mine layout and selected and offered lands, a summary of the impacts of the project, and self-addressed comment sheets were available to the public at these meetings.

Within the comment period, BLM received 269 letters on the draft EIS including two letters originally sent to ACE. Of the total, 127 letters were from private individuals and 142 were from persons representing organizations, groups, businesses, or agencies. The BLM identified 650 comments in these letters.

All comments made in the letters were addressed by BLM. Those comments, and the responses to them, were

Table 16.1-9 Information needed to prepare reclamation and operations plans for surface coal mining in the United States

Natural Factors		Cultural Factors	

Topography
  Relief
  Slope
Climate
  Precipitation
  Wind patterns and intensity
  Humidity
  Temperature
  Climate type
  Growing season
  Microclimatic characteristics
Altitude
Exposure (aspect)
Hydrology
  Surface hydrology
    • Watershed considerations
    • Flood plain delineations
    • Surface drainage patterns
    • Runoff amounts and qualities
  Groundwater hydrology
    • Groundwater table
    • Aquifers
    • Groundwater flow amounts and qualities
    • Recharge potential
	Geology
	  Stratigraphy
	  Structure
	  Geomorphology
	  Chemical nature of overburden
	  Coal characterization
	Soils
	  Agricultural characteristics
	    • Texture
	    • Structure
	    • Organic matter content
	    • Moisture content
	    • Permeability
	    • pH
	    • Depth to bedrock
	    • Color
	  Engineering characteristics
	    • Shrink–swell potential
	    • Wetness
	    • Depth to bedrock
	    • Erodibility
	    • Slope
	    • Bearing capacity
	    • Organic layers
	Terrestrial ecology
	  Natural vegetation
	    • Characterization
	    • Uses and survival needs
	  Crops
	  Game animals
	  Resident and migratory birds
	  Rare and endangered species
	Aquatic ecology
	  Aquatic animals
	  Aquatic plants
	  Aquatic life systems
	    • Characterization
	    • Uses and survival needs
		Location
		Accessibility
		  Travel distances
		  Travel times
		  Transportation networks
		Site size and shape
		Surrounding land use
		  Current
		  Historical
		  Land-use plans
		  Zoning ordinances
		Land ownership
		  Public
		  Commercial or industrial
		  Private or residential
		Type, intensity, and value of use
		  Agricultural
		  Forestry
		  Recreational
		  Residential
		  Commercial
		  Industrial
		  Institutional
			Type, intensity, and value of use
			  Agricultural
			  Forestry
			  Recreational
			  Residential
			  Commercial
			  Industrial
			  Institutional
			  Transportation/utilities
			  Water
			Population characteristics
			  Population
			  Population shifts
			  Population density
			  Age distribution
			  Number of households
			  Household size
			  Average income
			  Employment
			  Educational levels

Source: Adapted from Ramani et al. 1990.

organized into 14 categories. Discussion of the categories and the responses to the letters required 95 pages in the EIS. A review of that material showed that those letters came from the following sources:

- Fifteen from other government agencies, requesting clarification or further response to specific issues
- Seven from individuals, objecting to specific aspects of the project or to the project in general
- Five from nongovernmental organizations, objecting to specific aspects of the project or to the project in general
- Five from Indian tribes or related organizations, objecting to specific aspects of the project or to the project in general
- One from an Indian tribe stating no objection to the project
- One from PD, requesting clarification or further response to specific issues

The remaining 235 letters expressed support for the project or for the proposed land exchange.

In addition to the 269 letters received during the comment period, 24 letters were received after the close of the extended comment period on January 29, 1999. These letters are not included in the official administrative record for this project; however, all were read by BLM and analyzed for comments. The BLM determined that all the issues and comments raised in the late letters had already been raised by other commenters and were being considered in the preparation of the final EIS.

In August 2001, PD submitted an updated MPO that addressed several concerns raised during the draft EIS public comment period. Specifically, PD modified the crushing, pretreatment, and material handling elements of the MPO to reduce impacts and increase the efficiencies of the ore production processes. As a result of PD's continuing optimization efforts, the projected water usage, truck haulage and associated air emissions, tank storage, and sulfuric acid truck deliveries were meaningfully reduced.

In July 2004, the BLM issued a Record of Decision (ROD) in favor of the land exchange, allowing the project to move forward (BLM 2004). In August the San Carlos Apache Tribe and Western Mining Action Project (WMAP) claimed that the land exchange between BLM and PD did not benefit the public interest and should not take place, and both parties filed protests to the ROD. WMAP claimed that BLM failed to meet its responsibility to ensure that springs and water holes will not be dewatered, instead saying, "Groundwater pumping and aquifer drawdown from the proposed mine will significantly reduce and/or eliminate a number of springs and water holes in the area" (Jones 2004a). The same groups also questioned the validity of PD's mining claims on selected lands

Table 16.1-10 Scoping issues addressed in the Dos Pobres/San Juan Environmental Impact Statement

Land Use	Physical Resources	Biological Resources	Cultural Resources	Socioeconomic Resources	Indian Trust Resources
Public lands management	Climate	Vegetation	Archaeological resources	Population and demographics	Indian trust assets
Access and recreation	Air quality	Wildlife resources	Traditional cultural properties	Local and regional economy	
Encumbrances	Geology	Special interest species/critical habitat		Infrastructure	
Agriculture and grazing	Soil	Biodiversity		Transportation	
Mineral rights	Groundwater quality/quantity				
Surface water rights	Surface water quality/quantity (including legally constituted surface waters of the United States)				
Noise and vibrations					
Visual resources					
Hazardous materials					

Source: Data from BLM 2004.

in the Gila Mountains, saying the validity of the claims was assumed by BLM when it prepared the EIS and the ROD. The WMAP protest accused BLM of basing the EIS and ROD on the assumption that the mine was going to be built "without determining if the projects would comply with federal law" (Jones 2004b).

By September, BLM had responded to the protests and allowed the ROD to stand. PD proceeded to file for air quality and aquifer protection permits and began the final feasibility study for the project (Stockton 2005). The feasibility study was completed in 2006. Construction at the site began in June of 2007, and the first cathode copper was produced on December 31, 2007 (Cooper 2008).

Public Participation in the Review of Permits

The environmental regulations in most jurisdictions provide an opportunity for public participation in the permitting process. Notice of the submittal of the permit application is usually advertised locally, with information provided on the schedule for public comment, the availability of the application for public review, and the location where comments may be submitted and hearings requested.

Interested persons or organizations, particularly those who may be adversely affected by a proposed mine, have the right to submit comments and objections to the issuance of the permit during the public comment period. Requests for hearings or conferences on the application must also be submitted within the time provided for public comment. In most jurisdictions, affected persons who file comments on the application, and the applicant if the permit is denied, may appeal the agency's decision through an administrative or judicial hearing. In an administrative hearing, evidence will be accepted, and the administrative law judge or review panel will render a decision to uphold or overrule the agency determination. Additional appeals from a final administrative decision can be brought to appropriate state or federal courts if desired by an adversely affected party.

Recently, there has been an increase in opposition to mining projects in developing countries, from individuals and environmental organizations in developed countries. These organizations are well funded and expert in using legal maneuvering to delay or prevent projects they oppose. Legal delays are but one tactic used to prevent development projects (Olpin 1991).

One example of these organizations is Earthworks, which is "a non-profit organization dedicated to protecting communities and the environment from the destructive impacts of mineral development, in the U.S. and worldwide" (Earthworks 2009). Earthworks is the sponsor of the No Dirty Gold campaign, which has organized opposition to mines proposed in Canada, Costa Rica, Ghana, Papua New Guinea, Peru, Romania, and the United States (No Dirty Gold 2009). The No Dirty Gold campaign has been particularly effective in persuading prominent retail jewelers to pledge not to use gold produced from mines that the campaign judges to be noncompliant with its "Golden Rules," which are as follows:

1. Respect basic human rights outlined in international conventions and law.
2. Obtain the free, prior, and informed consent of affected communities.
3. Respect workers' rights and labor standards, including safe working conditions.
4. Ensure that operations are not located in areas of armed or militarized conflict.
5. Ensure that projects do not force communities off their lands.
6. Ensure that projects are not located in protected areas, fragile ecosystems, or other areas of high conservation or ecological value.
7. Refrain from dumping mine wastes into the ocean, rivers, lakes, or streams.
8. Ensure that projects do not contaminate water, soil, or air with sulfuric acid drainage or other toxic chemicals.
9. Cover all costs of closing down and cleaning up mine sites.
10. Fully disclose information about social and environmental effects of projects.
11. Allow independent verification of the above (No Dirty Gold 2009).

Clearly, these rules are quite similar to the principles of sustainability adopted by ICMM. However, some of the rules are particularly difficult for mine operators. For example, Rules 4, 5, and 6 offer no consideration of the fact that ore deposits are located without respect to political conditions, protected areas, or human communities. Similarly, the independent verification in Rule 11 may be difficult to attain. Career environmentalists are often mistrustful of government agencies and mining company personnel, so "independent verification" implies inspection and approval by someone selected by the environmental organizations. Unfortunately, these organizations do not get increased financial support by giving approval to mining and other resource development projects; quite the contrary, a mining company may be reluctant to allow "independent verification" of this type.

The direct involvement of organizations like Earthworks, Greenpeace, Oxfam, and others in the permitting process makes it even more important for mining companies to form good relationships with local people and governments well before permit applications are filed. An excellent example of a project where this was done successfully is the Diavik diamond mine, described in the following example.

Example 4. Diavik Diamond Mines Inc.

The Diavik diamond mine operates in the Northwest Territories of Canada, one of the most untouched and ecologically sensitive environments in the world. The mine is surrounded by tundra and is home to bears, wolverines, and migrating caribou. The water of nearby Lac de Gras is exceptionally pure and hosts many species of fish and birds.

Development of the Diavik mine required about 10 years. The first claims were staked in 1991 and 1992. Sampling and exploration continued, and Diavik Diamond Mines Inc. was formed in 1996. In 1997, baseline environmental studies were completed. As required by Canadian law, a project description was submitted in March 1998, and the environmental assessment report followed in September of the same year. Regulatory and financial approvals were given in 1998 and production began in 2003.

In planning the mine, Diavik consulted extensively with local communities about its operation and effects. Diavik is committed to providing training, employment, and business opportunities to residents in local communities of the Northwest Territories and the West Kitikmeot region of Nunavut, ensuring that it leaves a legacy of economically and socially stable local communities in the region. The process followed in developing agreements with indigenous peoples is described by Prest (2002):

> At the start, we developed a shared vision with our communities with regards to project development. As part of that vision, right from the beginning we integrated social and environmental dimensions into our project plans.
>
> We had regular meetings with our all neighboring aboriginal and other communities. We incorporated traditional knowledge learned from community elders and members. We are proud to say that information learned from our aboriginal communities has definitely led to a better designed project.
>
> As a third step in the engagement process, we entered into a number of formal agreements for the project. There are three legs to this framework, the Environmental Agreement, the Socio-Economic Monitoring Agreement, and, separate Participation Agreements with each of our five aboriginal groups.
>
> [Diavik's] Participation Agreements are legally binding, private contracts which contain mutually beneficial provisions for both the company and individual aboriginal communities. Parts of these agreements address shared responsibility for training, employment and business participation.

The negotiations with indigenous communities were conducted according to the traditions of those communities. Diavik's representatives met with each community and discussed all the issues until a consensus was reached. Reportedly, this sometimes took several days (R. Davey, personal communication).

Diavik's commitment to community development and investment is evident in the Diavik Socio-Economic Monitoring Agreement (SEMA), which was formalized in 1999 between Diavik Diamond Mines Inc. and the Government of Northwest Territories. The agreement was ratified by local indigenous communities and outlines Diavik's commitment to provide training, employment, and business opportunities to northerners and, more specifically, indigenous northerners. Diavik individualized these commitments through participation agreements negotiated with each of the same five aboriginal groups (Diavik 2009b):

1. Tlicho Government
2. Yellowknives Dene First Nation
3. North Slave Metis Alliance
4. Kitikmeot Inuit Association
5. Lutsel K'e Dene First Nation

Commitments under SEMA are reported publicly twice a year. Diavik has consistently exceeded expectations outlined in SEMA:

- Diavik committed to supporting a 40% northern work force during construction. At its conclusion, Diavik had reached 44% northern employment.
- Diavik committed to reaching levels of northern purchasing of 38% during construction. At its conclusion, Diavik had actually reached 74%, representing about Can$900 million (US$853 million) in contracts with northern companies, of which approximately $600 million was with indigenous companies.
- During operations, Diavik had committed to purchasing 70% of its annual requirements for goods and services from northern companies.
- For operations, Diavik committed to 66% northern employment and 40% indigenous employment for its operations. In 2007, Diavik's work force averaged 785 people, 67% of whom were northern, with one-half being indigenous.

The mine meets its environmental protection commitments through a comprehensive management system. This system is ISO 14001:2004 certified and is responsible for managing environmental protection activities, ensuring that all employees are properly trained, anticipating and avoiding environmental problems, ensuring regulatory compliance and due diligence, and ensuring consistency with the corporate environmental policy of the company (Diavik 2009a).

Diavik has several active programs that contribute to environmental protection and sustainable development values. In some cases, the measures taken in these programs go beyond the strict legal requirements that apply to the project area. The programs include the following:

- **Caribou management.** A small portion of the Bathurst caribou herd passes through the Lac de Gras region during

spring and fall migrations. To protect caribou passing near the mine, haul roads display advisory signs to ensure that caribou and other wildlife have the right of way. Diavik annually monitors the caribou within the region, with the assistance of elders from local indigenous communities.
- **Water quality management.** During early meetings with local communities, the indigenous people emphasized the importance of preserving water quality in the lakes and drainage. Diavik constructed an extensive water collection system to help protect the surrounding lake waters. Sumps, pipe networks, storage ponds, and reservoirs collect runoff water, which is used in processing and can then be treated before release to the environment. In addition, the Aquatic Effects Monitoring Program ensures that lake water is regularly sampled and analyzed at set locations over the complete range of depth, in all seasons of the year.
- **Fish habitat.** The mine constructed rock-fill dikes to access the three ore bodies under the lake, so that mining operations occupy less than 0.5% of the area of Lac de Gras. However, even this small area removed some fish habitat, so Diavik constructed a coarse rock-fill area on the outside of one dike to provide new fish habitat. During mining, rock shoals are constructed inside the dike area; when the mine closes, these will become fish habitat. The area behind the dikes will be flooded, the dikes will be breached to locate islands, and the area returned as part of Lac de Gras. Thus there will be no net loss of fish habitat, as required by Canadian fisheries regulations.
- **Progressive reclamation.** The progressive reclamation practiced at Diavik uses knowledge gained through community consultation to prepare the mine site for eventual closure. Some examples include the contouring of country rock piles to create smooth hills that allow caribou safe access, and the creation of new fish habitat. Revegetation studies and waste rock management projects are also being conducted.
- **Closure security.** Diavik is committed to complete closure of the mine site. It has committed a substantial amount of financial security to guarantee full closure when mining ceases, as established in the environmental agreement. The value of this security is revised regularly to adjust for the projected environmental liability and its associated cost of closure versus efforts the company has made to progressively reclaim the site.
- **Sustainable development research.** Diavik is cooperating with Canadian universities and researchers in numerous scientific studies of the environment and geology at the mine site. Recent studies have included research into effects of mine blasts on fish, evaluations of potential plant species for reclamation, and monitoring of dust distribution using lichen as a bioindicator. The results of these research projects are published regularly and are available on the Internet (Diavik 2009b). Research results are routinely incorporated into operating and reclamation practices at the site. Diavik is also examining the effect of the arctic climate on diamond waste rock piles, an area of which little is known. This research will facilitate long-term protection of the environment and has long-term applications throughout the world.

Mine Planning

All of the development activities described to this point constitute *mine planning*. However, the term as used here refers specifically to the planning of the mine workings, including gaining access to the mineral deposit, removing the valuable mineral, and handling any waste material produced. The activity as comprehended here also includes tasks described as *mine design*. Although the details of these processes are described in the next section, planning for them is begun in the development stage, and is thus discussed at this point.

In mine planning, the specific day-to-day operations of the mine are set out, analyzed, and documented. Mine planning continues throughout the life of every mine to account for changing geologic and economic conditions. The objective of mine planning at each point in time is to determine the manner in which the mine should be operated to optimize the goal(s) of the operating entity, usually the return on investment. Although mine planning varies with site-specific conditions, the following points must be determined in the initial plan, and continually reevaluated throughout the life of the mine:

- The extent of the mineral deposit
- The market for anticipated products of the mine
- The manner in which the deposit will be accessed for extraction
- The equipment and personnel that will be used for extraction
- The operating sequence for extraction
- The disposition of extracted material, both waste and valuable mineral
- The interactions of mine workings with adjacent or overlying areas and properties

Initial mine planning will obviously take place concurrently with the environmental planning and permitting, and the consideration of environmental consequences of the initial mine plan will be included in this process.

Extraction

Mining methods may be classified according to a variety of schemes. Here it is convenient to distinguish surface mining, underground mining, aquatic or marine mining, and solution mining. Solution mining is the removal of valuable minerals from in-place deposits by dissolving the mineral in a suitable liquid that is then removed for recovery of the desired constituent. It may also be called in-situ mining or in-situ leaching. Solution mining uses techniques similar to those used in the extraction of petroleum and natural gas, and is thus not discussed further in this chapter. The operational details and environmental considerations of surface mining, underground mining, and aquatic or marine mining are discussed separately.

Surface Mining

Surface mining is the removal of material from the earth in excavations that are open to the surface. In some cases, material is removed directly from the earth's surface; for example, sand and gravel may be removed directly from deposits of those materials put in place by ancient lakes or rivers. In other cases, material that has no economic value (overburden) covers the valuable material and must be removed.

A typical operational cycle for surface mining, described in detail by Saperstein (1992), is as follows:

1. Install erosion and sedimentation controls.
2. Remove topsoil from areas to be mined.

3. Prepare the first drill bench by leveling the bench with a bulldozer, inspecting and scaling the highwall as required, and laying out the blastholes.
4. Drill the blastholes.
5. Blast the rock.
6. Load the fragmented material.
7. Haul the fragmented material—waste to the waste dumps and product to the load-out for sale or to subsequent processing.
8. Manage the waste dumps as required by contouring waste piles to a stable configuration or returning waste to mine workings for use in reclamation.
9. Prepare mine workings for reclamation by
    a. Recontouring to the original contour, or to another approved and stable configuration.
    b. Returning the stored topsoil to the mine site and spreading it uniformly on the recontoured surfaces.
10. Reclaim the prepared surfaces by
    a. Revegetating with an approved mixture of seeds and plantings.
    b. Irrigate and maintain revegetated areas as required until a stable condition is reached.
11. Remove temporary drainage controls and stream diversions.

The environmental considerations that must be addressed during surface mining include the following:

- Control drainage from mine workings to avoid contamination of surface water, groundwater, and the existing ecosystem. Treatment may be required to remove particulates or chemical contamination.
- Control inflow or surface water and storm runoff so these waters are not contaminated by passing through the mine workings.
- Carefully analyze the groundwater regime in and around the mine workings, and provide monitoring wells as required. In some cases, it may be advisable to minimize the interaction of the mine workings with groundwater by pumping out the aquifers in and around the mine. There are many potential uses for this water. For example, it might be used in the processing of ore, discharged directly to surface drainages or discharged to constructed wetlands on the surface.
- Consider the formation of "pit lakes" when mined-out surface workings fill with water. This water, which may originate from surface runoff or flow from underground aquifers, may become acidified or otherwise contaminated, and the effects of such contamination on the environment (including wildlife, plant life, and humans) must be managed.
- Design and construct waste dumps and tailing impoundments to minimize erosion and protect surface and groundwater.
- Control emissions of dust and noise to meet the requirements of local regulations and the reasonable expectations of persons living nearby. In particular, if explosives are used, control air blast and ground vibration as required.
- When mining is done at night, consider the effects of artificial light on wildlife and humans living near the mine.
- Control slope stability and landslides. Of course, this is also necessary for successful mining operations, but inadequate control of slope stability may also lead to environmental damage. Slope failure may block or contaminate streams, damage wildlife habitat, and cause flooding that endangers wildlife and humans living near the mine.
- Consider the effects of increased traffic to and from the mine site, especially when product haulage will result in a marked increase in heavy truck traffic.

Underground Mining

Underground mining is the removal of material from the earth in excavations below the earth's surface. Access to such underground workings may be gained through a drift or adit, a shaft, or a slope. Drifts and adits are horizontal tunnels, usually in a hillside, that connect to the mineral deposit. In the case of minerals like coal that occur in seams (near-horizontal deposits of fairly uniform thickness and considerable areal extent), the drift may be developed in the mineral itself. Travel in drifts is by rail or rubber-tired vehicles. Shafts are vertical tunnels developed from the surface to access mineral bodies below. Travel in shafts is by cages or cars, which are lowered and raised by a mechanism on the surface, similar to elevators in tall buildings. Slopes are tunnels neither vertical nor horizontal and may be lineal or spiral. Travel in lineal slopes may be by rubber-tired vehicles or hoists but rarely by rail. Travel in spiral slopes (also called ramps) is by rubber-tired vehicles.

Saperstein (1992) has described in detail a typical operational cycle for underground mining in which material is fragmented by drilling and blasting, as summarized:

1. Enter the workplace after the previous blasting round is detonated.
2. Ensure that the workplace is in good condition and safe for continued work by checking that ventilation is adequate and that blasting fumes have been removed; providing for dust suppression; checking for the presence of hazardous gases; and inspecting for and removing loose material.
3. Load the fragmented material.
4. Haul the fragmented material to the appropriate location.
5. Install ground support as required.
6. Extend utilities as required: ventilation, power (electricity or compressed air), and transportation.
7. Survey and drill blastholes for the next round.
8. Load the explosives and connect the detonation system.
9. Leave the workplace and detonate the round.

This cycle is typical for the mining of narrow, steeply dipping veins (typical of metal ores), and for massive deposits of limestone, salt, and dimension stone, where the vertical extent of the valuable mineral is 6 m (6.6 yd) or higher.

The operational cycle for methods in which material is mechanically fragmented and removed in a continuous process may be similarly summarized:

1. Enter the workplace after the required ground support has been installed.
2. Ensure that the workplace is in good condition and safe for continued work by checking that ventilation is adequate and that blasting fumes have been removed; providing for dust suppression; checking for the presence of hazardous gases; and inspecting for and removing loose material.
3. Cut and load the fragmented material until the limit for advance is reached.
4. Concurrently, haul the fragmented material to the appropriate location.

5. Remove fragmentation, loading, and hauling equipment from the workplace.
6. Install ground support as required.
7. Extend utilities as required: ventilation, power (electricity or compressed air), and transportation.
8. Survey as required to ensure that the mined opening is maintaining the required dimensions and directional orientation.

This cycle is typical for mining in near-level seams of relatively soft material that have considerable areal extent and are thinner than 6 m (6.6 yd), such as coal, trona, phosphate, and potash.

Seven environmental considerations must be addressed during underground mining:

1. Locate ventilation fans and hoist houses to minimize the effects of noise on wildlife and humans living nearby.
2. Whenever possible, dispose of mine waste in underground mine workings.
3. When waste dumps are constructed on the surface, design them to minimize erosion and protect surface and groundwater.
4. Control drainage from mine workings to avoid contamination of surface water, groundwater, and the existing ecosystem. Treatment may be required to remove particulates or chemical contamination.
5. Control inflow or surface water and storm runoff so these waters are not contaminated by passing through the mine workings.
6. Carefully analyze the groundwater regime in and around the mine workings, and provide monitoring wells as required. In some cases, it may be advisable to minimize the interaction of the mine workings with groundwater by pumping out the aquifers in and around the mine, and discharging the pumped water to constructed wetlands on the surface.
7. Analyze and predict the subsidence likely to result from mine workings. Design workings to minimize the effects of subsidence on surface structure, utilities, and important natural features such as lakes, streams and rivers, and wildlife habitat.

Aquatic or Marine Mining

Aquatic or marine mining is the removal of unconsolidated minerals that are near or under water, with processes in which the extracted mineral is moved by or processed in the associated water. This type of mining may also be referred to as alluvial mining or placer mining. The two major types of aquatic or marine mining are dredging and hydraulicking.

Materials typically recovered by aquatic or marine mining have usually been deposited in fluvial, aeolian, or glacial environments, and are thus unconsolidated. They include aggregate (sand and gravel) and materials deposited because of their relatively high specific gravities. The latter deposits are called placers and include native (naturally occurring) precious metals, tin (as the oxide cassiterite), heavy mineral sands (oxides of zirconium, hafnium, titanium, and others), and precious stones. In some cases, placer deposits are mined by methods described in the section on surface mining; in such cases, the environmental precautions given for those methods apply.

Dredging is the use of a powered mechanism to remove unconsolidated material from a body of water. The mechanism is almost always a type of bucket or shovel. In the simplest case, it may be a metal bucket moved by chains or steel cables that are attached to a pole. The bucket is dropped through the water and into the solid material on the bottom. As the bucket is retracted, its weight and trajectory force its leading edge into the solid material, and the bucket fills. When the bucket comes to the surface, it is emptied. This type of dredge is usually installed on the shore near a body of water that covers a valuable mineral deposit.

More complex dredge mechanisms attach several buckets to a wheel or a ladder. A ladder is a structure designed to support a series of buckets attached to a chain, which moves continuously in a loop. Both mechanisms are usually installed in a floating vessel, the dredge. While moving, the bucket wheel or the dredge ladder is lowered into the unconsolidated material below the surface, picking up that material and returning it to the dredge where it is either further processed or transferred to the shore for further usage. The dredge vessel may operate in a natural body of water or in an artificial body called a dredge pond. A typical operational cycle for dredging is comprised of the following steps:

1. If dredging in a dredge pond, complete subitems a–g; if not, proceed to Step 3.
    a. Locate a source of water for filling the pond.
    b. Install erosion and sedimentation controls, and divert surface water as required.
    c. Prepare dikes or dams required for the dredge pond. The perimeter of the dredge pond must extend beyond the extent of the mineral to be recovered to allow for deposition of overburden or tailings removed in the first pass of the dredge.
    d. Fill the dredge pond.
    e. Remove vegetation.
    f. Remove and stockpile topsoil.
    g. If necessary, remove overburden with excavating equipment, and place overburden in stable piles.
    h. Proceed to Step 3.
2. If dredging from the shore, prepare the site for installation of the dredge mechanism.
3. Install the dredge.
4. Remove by dredging any overburden that was not removed in Step 1g.
    a. When dredging in a natural body of water, the overburden will be deposited under the water in an area from which the valuable mineral has already been removed or in an area under which there is no valuable mineral.
    b. When dredging in a dredge pond or from the shore, the overburden will be deposited beside or behind the dredge in an area from which the valuable mineral has already been removed or in an area under which there is no valuable mineral. When mining in a dredge pond, the overburden will for a time be at least partially under water; when mining from the shore, it will not. However, in both cases, the overburden will over time drain and be left exposed. It will thus require reclamation as described in Steps 12 and 13.
5. Remove the valuable mineral by dredging.
6. Process the valuable mineral as required to concentrate the valuable constituent(s). This processing is almost always done on the dredge and is integrated with the dredging of the material.

a. When dredging in a natural body of water, the tailings will be deposited under the water in an area from which the valuable mineral has already been removed or in an area under which there is no valuable mineral.
b. When dredging in a dredge pond or from the shore, the tailings will be deposited beside or behind the dredge in an area from which the valuable mineral has already been removed, or in an area under which there is no valuable mineral. When mining in a pond, the tailings will for a time be at least partially under water; when mining from the shore, it will not. However, in both cases, the tailings will over time drain and be left exposed. They will thus require reclamation, as described in Steps 12 and 13.
7. Remove the valuable constituent(s) for sale or further processing off-site.
8. Deposit the tailings under the water in an area from which the valuable mineral has already been removed or in an area under which there is no valuable mineral.
9. Continue dredging until reaching the limits of the dredge's cables and other connections, the limits of the pond, or the boundary of the deposit.
10. If dredging in an artificial pond, prepare and fill the next pond as in Step 1, and transfer the dredge. If dredging in a natural body of water, move the dredge to the next location.
11. Begin dredge operation in the new location by returning to Step 3. Simultaneously, and in accordance with the approved reclamation plan, reclaim overburden and tailings piles from previous work, according to Steps 12 and 13.
12. Prepare mine workings for reclamation:
    a. Recontour to the original, stable contour, or to another, approved and stable configuration.
    b. Return the stored topsoil to the mine site, and spread it uniformly on the recontoured surfaces.
13. Reclaim the prepared surfaces:
    a. Revegetate with an approved mixture of seeds and plantings.
    b. Irrigate and maintain revegetated areas as required until a stable condition is reached.
14. Remove temporary drainage controls and stream diversions.

There may be considerable variation in the cycle, depending on how the dredge is transported between sites.

Hydraulicking is a method of mining placer deposits that was used extensively in the past but has fallen into disfavor because of the potential for serious effects on the stability of the remaining surface and on nearby surface waters.

Hydraulicking required a large deposit of auriferous alluvium and a source of water that would provide sufficient volume and pressure head. Overburden and pay gravel were both removed by high-pressure water that flowed through a monitor—a large nozzle that could be rotated horizontally and vertically. Material removed by the flow from the monitor moved to the bottom of the valley, where the flow was controlled so that overburden passed directly into the stream channel and pay gravel flowed through a sluice or similar recovery device for recovery of the ore. Tailings from the sluice also flowed into the stream channel.

As mentioned earlier, hydraulicking of the placer gold deposits in the famous California Gold Rush of 1849 led to severe contamination of the rivers that drained the area of the gold deposit. The solids from the hydraulicking operations filled rivers with solids, which were eventually deposited downstream with serious consequences for agriculture in the downstream areas.

Hydraulicking is still a very inexpensive method for moving large volumes of unconsolidated material and is still used on a limited basis with five precautions:

1. Material discharge from a hydraulicking operation must be captured and treated to remove solids, including, in particular, any fine, suspended solids. This will require a settling pond, and chemical flocculants may also be necessary. Even water that appears to be clear must be tested by appropriate methods and treated on the basis of those tests.
2. Tailings, overburden, and deactivated settling ponds must be reclaimed appropriately, usually by recontouring and revegetation. It may be necessary to remove and stockpile topsoil for use in reclamation. Coarse gravel, which may accumulate in separate piles, should receive special attention to ensure it is reclaimed properly.
3. The area of the mining operation should be isolated from the flow of surface streams and runoff to prevent contamination of those waters.
4. Discharge of water from the hydraulicking operation should be managed to prevent interference with the existing flow regimes in the drainage. The quantities and velocities of discharge should not modify the existing stream flow in a manner that will cause erosion, undercutting of banks, or flooding.
5. Highwalls and embankments produced by hydraulicking should be reclaimed, again by recontouring and revegetation. Even when topsoil is removed and stockpiled, there may not be enough topsoil for the reclamation required. In such cases hydroseeding will likely be required.

CONCLUSION

Activities associated with mining have direct and lasting effects on both the physical and the human environments. In the recent past, mining was done with little concern for its effects on the environment. The consequences of this approach often resulted in significant damage to the natural environment. However, as political and cultural norms changed, and new legal requirements were enacted, almost all major mining companies adopted rigorous policies and procedures for sustainability, community engagement, and environmental risk assessment and mitigation. These companies apply such policies throughout their operations, many of which are worldwide. In addition, many mining companies work actively to remediate environmental damage caused by historic mining operations in areas where they have past or current operations.

Thus, environmental considerations are an important part of the modern mining industry. These considerations include preproject planning, permitting, compliant operation, reclamation, and closure. These considerations must be included in all project planning, and feasibility studies must account for the influence of environmental considerations on project schedules and costs.

REFERENCES

BLM (U.S. Bureau of Land Management). 2004. Dos Pobres/San Juan Project Record of Decision. www.blm.gov/az/st/en/info/nepa/environmental_library/arizona_resource_management/dos_pobres.print.html. Accessed October 2009.

Brubaker, E. 1995. *Property Rights in the Defence of Nature*. Environment Probe. www.environment.probeinternational.org/books/property-rights-defence-nature. Accessed January 2010.

CFR (Code of Federal Regulations). 2009. Surface Mining Permit Applications—Minimum Requirements for Reclamation and Operation Plan. Title 30—Mineral Resources, Part 780. Surface Mining Permit Applications—Minimum Requirements for Reclamation and Operation Plan. Available from http://ecfr.gpoaccess.gov/cgi/t/text/text-idx?c=ecfr&tpl=/ecfrbrowse/Title30/30cfr780_main_02.tpl. Accessed January 2010.

Chapanis, A. 1986. To err is human, to forgive design. In *Proceedings of the Annual Professional Development Conference*. New Orleans, LA: American Society of Safety Engineers.

CODELCO (Corporación Nacional del Cobre de Chile). 2006. Directriz corporativa para identificar aspectos ambientales y evaluar el riesgo de sus impactos. [Corporate directive for identification of environmental aspects and evaluation of the risk of their impacts]. Código SGASS-90, Versión 1, December.

CODELCO (Corporación Nacional del Cobre de Chile). 2009. Sustainable development. www.codelco.com/english/desarrollo/fr_desarrollo.html. Accessed September 2009.

Colling, D.A. 1990. *Industrial Safety: Management and Technology*. Englewood Cliffs, NJ: Prentice Hall.

Cooper, G. 2008. *Heap Leach Start-Up and Operation at Freeport-McMoRan Safford, Inc.* www.minexpo.com/Presentations/Cooper.pdf. Accessed August 2009.

Craig, J.R., Vaughan, D.J., and Skinner, B.J. 2001. *Resources of the Earth: Origin, Use, and Environmental Impact*. Upper Saddle River, NJ: Prentice Hall.

Diavik. 2009a. Our approach: Environment. www.diavik.ca/ENG/ourapproach/environment.asp. Accessed August 2009.

Diavik. 2009b. Our approach: Community investment. www.diavik.ca/ENG/ourapproach/community_investment.asp. Accessed August 2009.

Earthworks. 2009. About us. www.earthworksaction.org/about_us.cfm. Accessed August 2009.

GRI (Global Reporting Initiative). 2009. What is GRI? www.globalreporting.org/AboutGRI/WhatIsGRI/. Accessed January 2010.

Hill, R.L., Kohler, S.L., Higgins, C.T., and Youngs, L.G. 2001. The influence of gold-mining on the development of California. http://gsa.confex.com/gsa/2001CD/finalprogram/abstract_3915.htm. Accessed August 2009.

Hunt, D.K. 1992. Environmental protection and permitting. In *SME Mining Engineering Handbook*, 2nd ed. Edited by H.L. Hartman. Littleton, CO: SME.

ICMM (International Council on Mining and Metals). 2006. 10 Principles. www.icmm.com/our-work/sustainable-development-framework/10-principles#01. Accessed August 2008.

ISO 14001:2004. *Environmental Management Systems—Requirements with Guidance for Use*. Geneva: International Organization for Standardization.

Jones, G. 2004a. Land exchange protests similar. *Eastern Arizona Courier*. September 8.

Jones, G. 2004b. Groups question validity of PD's mining claim. *Eastern Arizona Courier*. September 13.

Junge, A., and Bean, R. 2006. *A Short History of the Zinc Smelting Industry in Kansas*. Kansas Department of Health and Environment. www.kdheks.gov/remedial/articles/smelterhistory.pdf. Accessed August 2009.

Kaas, L.M., and Parr, C.J. 1992. Environmental consequences. In *SME Mining Engineering Handbook*, 2nd ed. Edited by H.L. Hartman. Littleton, CO: SME.

Lamborn, J.E., and Peterson, C.S. 1985. The substance of the land: Agriculture vs. industry in the smelter cases of 1904 and 1906. *Utah Hist. Q.* 53(Fall): 308–325.

MMSD (Mining, Minerals and Sustainable Development Project). 2002. *Breaking New Ground: Mining, Minerals, and Sustainable Development*. London: Earthscan Publications.

NEPA (National Environmental Policy Act). 2009. *NEPA's Forty Most Asked Questions*. www.nepa.gov/nepa/regs/40/40p3.htm. Accessed August 2009.

No Dirty Gold. 2009. www.nodirtygold.org/. Accessed August 2009.

NWMA (Northwest Mining Association). 2002. Various articles. *NW Min. Assoc. Bull.* 108(1):4–5.

Olpin, O. 1991. The ethics of representing clients on environmental matters. *Resource Law Notes*, newsletter of the Natural Resource Law Center, University of Colorado, Boulder, 24 (January): 5.

Prest, S.F., and Kurtz, B. 2002. Building a Sustainable Development Framework: Community Relations. Presented at The Global Mining Initiative Conference, Resourcing the Future, Toronto, May 11–14. www.diavik.ca/documents/may_2002_global_mining_initiative_conf_may_6.pdf. Accessed August 2009.

Ramani, R.V., Sweigard, R.J., and Clar, M.L. 1990. *Reclamation planning. In Surface Mining*, 2nd ed. Littleton, CO: SME.

Raymond, R. 1986. *Out of the Fiery Furnace*. University Park, PA: Pennsylvania State University Press.

República de Chile. 2002. Reglamento del sistema de evaluación de impacto ambiental. *Diario Oficial*. December 7.

Saperstein, L.W. 1992. Basic tasks in the production cycle. In *SME Mining Engineering Handbook*, 2nd ed. Edited by H.L. Hartman. Littleton, CO: SME.

SMCRA (Surface Mining Control and Reclamation Act). 1977. Surface mining law. www.osmre.gov/topic/SMCRA/SMCRA.shtm. Accessed January 2010.

Stockton, L. 2005. Officials glad land exchange is complete. *Eastern Arizona Courier*. September 28.

CHAPTER 16.2

Mining and Sustainability

R. Anthony Hodge

INTRODUCTION

The 1987 publication of the report, *Our Common Future* by the United Nations (UN) World Commission on Environment and Development (WCED), brought the concept of sustainable development into the limelight. Chaired by the former Prime Minister of Norway, Gro Harlem Brundtland, and following hearings held across the world, the commission proposed an agenda for world development that would enhance security and reduce North–South disparities. It would be development "which meets the needs of the present without compromising the ability of future generations to meet their own needs" (WCED 1987).

Since then, a rich debate has ensued about what this means in practical terms. Though many other sets of words have been suggested for defining the phrase *sustainable development*, the Brundtland Commission definition has stood the test of time and remains the anchor. For a rich discussion, see the "definitions" portal of the International Institute for Sustainable Development, or IISD (SD Gateway n.d.).

In recent years, the word *sustainability* has also found its way into common use. The idea is simple. Sustainability is the persistence over a long time—indefinitely—of certain necessary and/or desired characteristics of both human society and the enveloping ecosystem (Robinson et al. 1990). These characteristics range from primary needs such as air, water, food, clothing, shelter, and basic human rights to a host of conditions that would collectively be called *quality of life*, not only for people but for other life forms as well.

It is here that the definitional issue becomes difficult for some, because the choice of which characteristics are to be sustained and the degree to which they will be sustained depends on the particular values that are applied. In turn, these depend on who is doing the applying. In other words, it is not a closed definition. What a company CEO chooses as important may be different than a politician, doctor, or librarian; what a Mexican chooses may be different from, for example, a Tanzanian or Australian. Because of these potential differences, a fair and effective process of interaction and seeking consensus is critical to the practical application of these ideas. Herein lies the rationale for why it is essential to always address not only the *what*, or the *substantive* part of human action, but also the *how*, or the *process* part. In other words, in the practical application of sustainable development concepts, not only what we do is important but how we do it.

Thus, the ideas of sustainable development and sustainability are different but synchronous. *Sustainability* is a more general term that captures the idea that we need to maintain certain important aspects of the world over the long term. *Sustainable development* is the human or action part of this set of ideas: As a society, we want to make choices about our actions that allow us to provide for the present without undermining the possibility for future generations to provide for themselves.

Together, these ideas are very appealing. However, their translation to practical action remains much debated. This is not surprising. Human society is complex. There are about 10,000 cells in the standard industrial classification—our way of classifying human activities within the market economy. This does not account for many more activities outside the market economy. There are about 200 countries across the world, and the global ecosystem is complex and not fully understood.

For its part, the mining, minerals, and metals industry has been a particularly active locus of sustainability-related policy and practice innovations because

- The potential implications—both positive and negative—of mining activities and the minerals and metals that result are significant;
- Many interests are touched by mining;
- The role of many of these interests in decision making is growing (e.g., communities and indigenous people);
- The nature of contemporary communications systems has brought the often dramatic nature of mining operations into the public eye; and
- Industry, governments, civil society organizations, and the public, in general, are all anxious to ensure mining makes a positive contribution that is fairly shared.

Importantly, the concept of sustainable development has not disappeared like so many "flavor of the month" ideas. Rather,

---

R. Anthony Hodge, President, International Council on Mining & Metals; Professor, Mining & Sustainability, Queen's University, Kingston, Ontario, Canada

it has grown in prominence and is now deeply entrenched in legislation, government, and corporate policy and practice. Sustainability is the subject of university curricula, given as a label to vice presidents and departments of mining companies, incorporated into the names of service providers, and included as an element of key performance indicators.

This chapter provides an overview that links the ideas of sustainable development and sustainability to the mining, minerals, and metals industry. To do so, a particular template is used for organizing the many interlinked bodies of knowledge that must be brought together: the Seven Questions to Sustainability (7QS) (MMSD North America 2002).

This template is pragmatic though much informed by the theoretical foundation of systems theory. It recognizes that to bring sustainability ideas into practice for the mining, minerals, and metals industry, the hard (well-defined) and the soft (ill-defined) systems of the real world must both be addressed, as well as the objective (independent of judgment) and the subjective (dependent on judgment). In doing so, all must be treated, if not exactly scientifically (which is not always possible), at least in a way that is characterized by intellectual rigor (see discussion of systems theory and sustainability in Hodge 1995 and 1996). The 7QS template weaves together ideas from many disciplines but recognizes that deeper exploration is often warranted depending on site-specific conditions. In this context, the objective of this chapter is to open a door as a first step to practical application.

MINING AND METALS INDUSTRY RESPONSE TO SUSTAINABLE DEVELOPMENT

The 1970s and 1980s were a time of reaction to dramatic change for mining. Echoing increasing concern for the environment across society, the late 1980s saw a number of leading mining companies publish "state-of-environment" reports related to their operations. Taking another important step, 30 leading mining and metals companies from across the world came together in 1991 to create the International Council on Metals and the Environment (ICME). ICME would give the industry an international voice on environmental matters.

Also in the early 1990s, many of these same companies joined with senior governments, labor unions, aboriginal peoples, and environmental nongovernmental organizations (NGOs) in a broad review of mining practices in Canada. The Whitehorse Mining Initiative turned out to be a precursor of a number of initiatives convened to bring sustainability ideas to practical application around the world. The resulting Leadership Accord (Whitehorse Mining Initiative Leadership Council Accord 1993) is a summons to change, framed within the context of a commitment to social and environmental goals. It seeks a sustainable mining industry within the framework of an evolving and sustainable society. The ideas it champions and the multi-interest process it uses elegantly capture sustainable development in practice.

For the mining industry, the decade of the 1990s was a bleak period. Commodity prices dropped while public criticism skyrocketed, much driven by a civil society that was quick to take advantage of newly available and quickly evolving computer-based communications. As a whole, the industry found itself under attack and in a defensive posture. Its *social license* to operate was threatened (though that particular label was to come later). In the late 1990s and faced with growing concern about access to capital, land, and human resources, the chief executive officers of nine of the world's largest mining companies took an unprecedented step.

Working through the World Business Council for Sustainable Development, they initiated the Global Mining Initiative (GMI). As part of GMI, they commissioned the International Institute for Environment and Development (London) to undertake a global review that would lead to the identification of how mining and minerals can best contribute to the global transition to sustainable development. The resulting project, Mining, Minerals, and Sustainable Development (MMSD), sparked a large and rich literature, including the project's final report, *Breaking New Ground: Mining, Minerals and Sustainable Development* (MMSD 2002).

Before the GMI was completed, participants moved to create an organization that would carry the resulting recommendations forward to implementation. Thus, in 2001 and building on the foundation established by ICME, the International Council on Mining and Metals (ICMM) was created. Many of the ideas summarized in this chapter have emerged from or been refined through subsequent work of ICMM.

Almost simultaneously with these events, NGO pressure on the World Bank Group led to the initiation in 2001 of a multi-interest review of the group's involvement in extractive industries. The Extractive Industries Review sought to test whether or not industry projects could be compatible with the World Bank Group's goal of sustainable development and poverty reduction. In its final report, *Striking a Better Balance* (Salim 2003), the review concluded in the affirmative but only if three enabling conditions were in place:

1. Public and corporate governance advocacy for the poor, including proactive planning and management to maximize poverty alleviation through sustainable development
2. Much more effective social and environmental policies
3. Respect for human rights

The resulting refocused World Bank policy emphasizes strengthened governance and transparency, ensuring that benefits reach the poor, mitigating environmental and social risks, protecting human rights, and promoting renewable energy and efficiency to combat climate change. This refocusing has in turn influenced mining approaches to implementing sustainable development on the ground. In sum, this process has served to elucidate and reinforce the concepts of sustainable development addressed in this discussion.

MINING'S CONTRIBUTION TO SUSTAINABLE DEVELOPMENT

At the base of the interlinked ideas of sustainability and sustainable development lies the simple idea that any human activity—including mining—should be undertaken in such a way that the activity itself and the products produced together provide a net contribution to human and ecosystem well-being over the long term.

From an engineering design perspective in general and a mine design perspective in particular, this simple idea gives rise to an overarching two-dimensional design criterion. That is, mining activity (or any human endeavour for that matter) should be designed to achieve (through the activity itself and the products that result) a net contribution to both human and ecosystem well-being over the long term. The achievement of design success should, in turn, be tested against this design criterion.

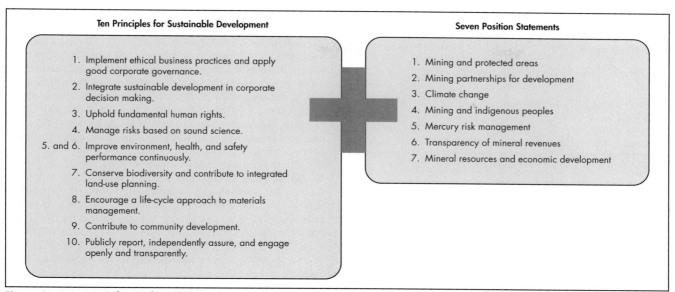

Figure 16.2-1 Principles and position statements that comprise the sustainable development framework of ICMM

Thus, the focus here is not on how mining can be sustainable—any mining project as a discrete activity cannot continue indefinitely—but on how mining, minerals, and metals can contribute to sustainable development. This is a conceptual shift away from a singular analysis and mitigation of *impacts* to a more comprehensive analysis and encouragement of *contribution*.

The focus on contribution is a tougher but fairer approach. It demands consideration of both the good and the bad. (The idea that the mining and metals industry should be designed for and tested against its contribution to the well-being of people and the environment—to sustainable development—was first proposed by Professor Robert Gibson in 2000 and later taken up in development of the 7QS template [MMSD North America 2002; Hodge 2004].) That a mining activity might be challenged to make a positive contribution to the ecosystem over the long term may seem a tough, even impossible challenge to some. However, articulating explicit engineering design criteria in this way sets a design objective that is essential if mining and metals-related activity and the resulting products are to achieve the sustainable development contribution that is being demanded by society.

Application of these ideas is not simply a *greening* phenomenon; it is related as much to well-being and security of people as the environment. And, interestingly, the mining industry's capacity to deal with the environmental aspects is currently stronger than its ability to address the full range of social aspects. This is likely because many environmental issues can be addressed through *hard* scientific and technical solutions, whereas social issues often require *soft* behavioral-type solutions, which can be much harder to design and implement—and which often fall outside the engineer's training.

These ideas also veer sharply away from thinking in terms of a trade-off between human and ecosystem well-being—it is not a balancing act that pits people against the environment. There are obviously many small trade-offs in any practical application: between interests, between components of the ecosystem, across time, and across space. This is particularly the case for mining and the related process of concentrating and refining. However, in an overarching sense, the ideas of sustainability and sustainable development call for both human and ecosystem well-being to be maintained or improved over the long term. Doing one at the expense of the other is not acceptable because, either way, the foundation of life is undermined.

PRINCIPLES AND FRAMEWORKS

The nature of applied sustainability is evident from the many attempts to articulate sustainable development principles. For a listing of more than 100 such principles, see the the "principles" portal of the IISD (n.d.). In short, there is no one-size-fits-all approach to defining, framing, and characterizing the ideas of sustainable development and sustainability. One set of principles of particular relevance to the mining industry is the 10-part set developed by the ICMM, along with its complementary set of position statements. ICMM brings together 19 of the largest mining companies of the world and, on an annual basis, member companies assess performance against these commitments using a procedure that includes third-party independent assurance (ICMM 2010a).

The ICMM principles and position statements as of 2010 are summarized in Figure 16.2-1. Just as many definitions and principles have been proposed, so too have many organizing frameworks been designed to bring theory to action. A comparative analysis of about 30 such characterizations or frameworks is found in Hodge 1997.

Almost three decades before the popularization of sustainable development ideas, geographer Walter Firey pointed out that three broad groupings of knowledge were pertinent to natural resource use (Firey 1960; and see discussion in Hodge 1997):

1. Ecological (environmental)
2. Ethnological (social/culture)
3. Economic

In the late 1980s, Firey's three-part model of natural resource use was adopted by a number of those attempting to operationalize the concept of sustainable development (Mitchell 1991). Since then, the three-part environmental–social–

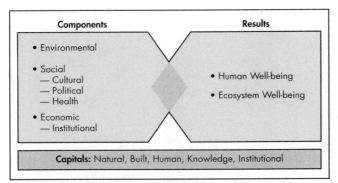

Figure 16.2-2 Different characterizations of sustainable development

economic characterization of sustainable development has gained broad usage. A significant limitation of this approach, however, arises because of the lack of common treatment for the social element.

Many other frameworks have also emerged that make sense for the particular application: population health, healthy communities, sustainable communities, foreign aid, urban design, agriculture, forestry, fisheries, indigenous peoples' needs, and so forth. From one particularly important perspective, the World Bank Group has used the lens and insights of economics to offer a "capitals" model to apply sustainable development ideas: natural, built, human, knowledge, and institutional. Figure 16.2-2 is a synthesis that brings a number of the alternatives together.

Another perspective that offers practical insight comes from an analysis of mine closure options that was undertaken for the Faro mine project in the Yukon Territory, Canada (Hodge and Merkhofer 2008). In this case, a sophisticated multi-attribute utility analysis—driven by a multistakeholder process—was used to assess the alternatives. Multi-attribute utility analysis is a form of decision analysis in which a set of objectives is articulated and each alternative's performance is assessed against those objectives. The underlying principle is that the alternative that best performs against the objectives is the best one. The foundation of this kind of analysis is the articulation of objectives that can be translated to a scale, which can be assessed in terms of a direction (e.g., more is better, less is worse) and magnitude (e.g., how much better or worse).

The rigorous scaling and multi-interest process of assessing and judging are complex. In the Faro analysis, two time horizons were used in the assessment: (1) short, 15–40 years; and (2) long, 500–1,000 years. Of relevance to this discussion is the framework of closure objectives, because it was set to reflect an overarching government policy of sustainable development. Eight objectives were used for assessing closure alternative at the Faro mine (Yukon Territory, Canada):

1. Maximize public health and safety
2. Maximize worker health and safety
3. Maximize restoration, protection and enhancement of the environment
4. Maximize local socioeconomic benefits
5. Maximize Yukon socioeconomic benefits
6. Minimize cost
7. Minimize restrictions on traditional land use
8. Minimize restrictions on local land use

As with all sustainable development frameworks, a broad range of topics is captured. However, this example is particularly useful for another reason. It is clear that the simultaneous treatment of all objectives sets up tensions that must ultimately be resolved: In a comparative analysis of alternatives, different value sets might well judge performance differently and, at the same time, place greater weight on certain objectives. This is a poignant example of applied sustainability in practice. Here again the process of finding common ground can be seen as critical to practical implementation of sustainable development ideas.

From a different but important perspective, as part of the World Bank Group response to the Extractive Industries Review, the International Finance Corporation (IFC) prepared and adopted a sustainability framework that consists of (1) the Policy on Social and Environmental Sustainability, (2) the Policy on Disclosure of Information; and (3) a set of performance standards on social and environmental sustainability.

In its performance standards, the IFC addresses eight topics:

1. Social and environmental assessment and management systems
2. Labor and working conditions
3. Pollution prevention and abatement
4. Community health, safety, and security
5. Land acquisition and involuntary resettlement
6. Biodiversity conservation and sustainable natural resource management
7. Indigenous peoples
8. Cultural heritage

Here again is another definition of what is appropriate to include in addressing application of sustainability concepts to mining. The IFC sustainability framework is currently under review with the expectation that a revised package is targeted for release in 2011. The IFC has also published a number of Good Practice Notes/Handbooks (IFC 2006) relevant to the practical issues of sustainability. Most importantly, the performance standards along with the World Bank Group's Environmental, Health, and Safety Guidelines form the basis of the Equator Principles, which have been adopted by 67 lending institutions worldwide (the Equator Principles lending institutions), many of whom provide financing for mining activities. The Equator Principles are a voluntary set of standards for determining, assessing, and managing social and environmental risk in project financing. They are considered the financial industry gold standard for sustainable project finance.

Another important framework is the Global Compact, a principle-based framework for businesses, which was established by the United Nations in 2000. The Global Compact is the world's largest voluntary corporate citizenship initiative and states 10 principles in the areas of human rights, labor, the environment, and anticorruption. Members are committed to aligning their operations and strategies with the 10 principles (United Nations Global Compact n.d.).

For their part, the Millennium Development Goals are eight international development goals that all 192 UN member states and at least 23 international organizations have agreed to achieve by the year 2015:

1. Eradicating extreme poverty and hunger
2. Achieving universal primary education

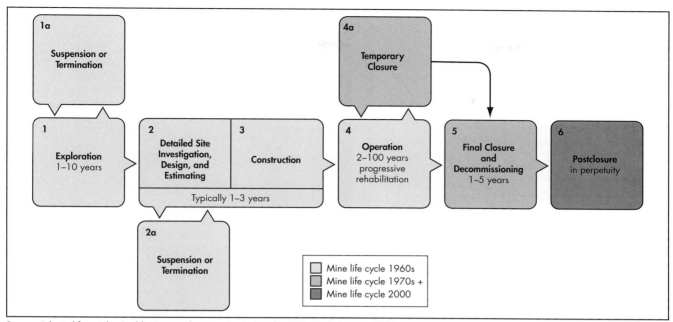

Figure 16.2-3 Mine project life cycle

Source: Adapted from John Gadsby (personal communication) and MMSD North America 2002.

3. Promoting gender equity and empowering women
4. Reducing child mortality
5. Improving maternal health
6. Combating HIV/AIDS, malaria, and other diseases
7. Ensuring environmental sustainability
8. Developing a global partnership for development

Each of the approaches to capturing what sustainability means in terms of principles and frameworks is useful for the relevant driving application. There is no single panacea that applies in all cases. However, taken together they capture the breadth of issues and topics that must be considered.

NONRENEWABLE NATURE OF MINING, MINERALS, AND METALS

Much of the literature through the 1980s and 1990s focused on renewable resource management and the idea of living off the interest of a continuing core stock. For some, nonrenewable resource-related activities such as mining simply did not fit into the sustainability concept, although the hat might be tipped toward recycling and reuse of nonrenewable materials as helpful strategies.

One result of this early emphasis was the marginalization, to a great extent, of concerns and perspectives about nonrenewable resources. Interestingly, the fact that minerals are nonrenewable (or stock) resources and, in some sense, fixed in absolute quantity turns out to be relatively unimportant from a sustainability perspective—at least at the macro scale (MMSD North America 2002). The nonrenewable character of minerals received a great deal of attention in the literature from 1950 into the 1970s. However, the long statistical record of continued output at relatively constant prices, together with growing understanding first of environmental issues and then of sustainability, has served to deemphasize this concern. At the same time, recognition is growing that the products of mining are needed both to provide for the world's population and, even more so, to support approaches that will allow society to walk more lightly on the earth. For example, all the strategies needed for development, transmission, and use of renewable energy sources depend on mined metals and minerals. Similarly, strategies to move to a carbon-reduced economy are only possible through creative uses of mined minerals and metals.

Thus, the sustainability-related focus is now appropriately on mining as an activity and its implications for the communities and ecosystem within which minerals are embedded. At any given site, whether a mining, smelting, refining, primary metals manufacturing, or recycling operation, there is a beginning and end: No mining/mineral activity can be expected to have an indefinite life span. However, the implications of that activity (not only as a direct result of the activity but also through the product that is produced) go on indefinitely.

In that sense, mining/mineral activities serve as a bridge to the future. The sustainability challenge is to ensure that the implications of mining activities and the products that result are net positive for people and ecosystems over the long term: It is the well-being of human society and the enveloping ecosystems that need sustaining. Limited-term mining projects can serve sustainability objectives if they are designed and implemented in ways that ensure they meet that challenge.

BOUNDARY CONDITIONS FOR APPLYING SUSTAINABILITY TO MINING

Bringing sustainability ideas into mine design has a significant impact on mine design boundary conditions. Four aspects of boundary conditions apply in this case, each of which has implications for setting the time and space dimensions and helps to identify the system components that must be considered:

1. Mine project life cycle (Figure 16.2-3)
2. Mineral life cycle (Figure 16.2-4)
3. Time horizon (Table 16.2-1)
4. Communities of interest

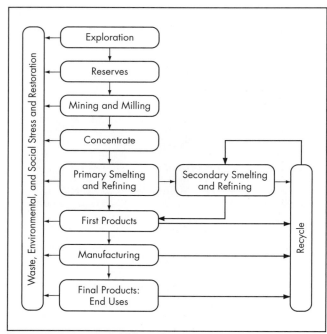

Figure 16.2-4 Mineral life cycle

Table 16.2-1 Perspectives on the time horizons for applied sustainability

Perspective	Typical Time Horizon
Financial/tax cycle	Annual or quarterly
Election cycle	2, 4, or 5 years
Standard engineering design life	Up to 50 years
Social time horizon (seven generations)	Up to 200 years
Environmental time horizon	200 to 10,000+ years

Source: Adapted from Freeze 1987.

and civil society organizations—and is not yet fully understood or appreciated.

The fourth and final aspect of defining the boundary conditions for applying sustainability relates to identifying the communities of interest that must be considered in the mine design, operation, and closure process. In times past, a company and government would simply come to a bilateral agreement on conditions that would govern mine activities. Today that is no longer the case. Many interests play active roles, and the definition of those roles itself requires great care and attention. Interests important to mining are

- Industry (investors, employees, industry associations, other companies);
- Support services (financial, consultants, contractors, suppliers);
- Government (federal, state/provincial, county/regional district, local);
- Indigenous people and their organizations;
- Organized labor;
- Mining-affected communities (by economic, social, and/or environmental [e.g., watershed] dependency);
- NGOs or civil society organizations; and
- Academic, learning, and research and development support (universities, technical schools, private and public research centers).

SEVEN QUESTIONS TO SUSTAINABILITY

One product of the mining industry's MMSD project was a template aimed at assessing the compatibility between mining operations and sustainability criteria (MMSD North America 2002; Hodge 2004, 2006). The 7QS offers seven queries for consideration in the mine design and assessment process. Each question is the interrogative form of a goal statement. Seen in another light, these seven goals define the application of sustainability concepts. The technique of using the interrogative form in this way is drawn from the accounting profession's approach to auditing and assuring the validity and accuracy of financial statements.

The focus on applying the 7QS approach is not on how a given mine can be sustainable—mining as a discrete activity cannot continue indefinitely—but on how the process of mining and the products it produces can best contribute to sustainable development. Thus it enshrines the conceptual shift away from analysis and mitigation of *impacts* to analysis and encouragement of *contribution*.

The seven questions are summarized in Figure 16.2-6. In this section, the 7QS template is used to systematically organize the parts of applied sustainability, rendering their application more practical. (The material presented in this section is modified from MMSD North America 2002.) In applying this

First, the mine design process must take into consideration the full life cycle of a mine project. Many of the greatest problems facing the mining industry today stem from the fact that this has not been done in the past. In the 1960s, the mine design process was limited to the end of operations. In the 1970s and driven by rising environmental consciousness, the need for mine reclamation gave rise to concerns about closure, which at the time focused on cleaning and grooming a site, with revegetation being a key activity. It is only in the last decade that the realization has set in that a postclosure phase can in some cases extend indefinitely because of the geochemical processes at work in waste rock piles, tailings, and exposed workings. These give rise to potential liabilities that must be factored into annual financial statements and the calculation of share value.

Second, the mine design process must consider the mineral life cycle. The mining industry has come to learn that it must not only consider the production of minerals and metals but also their use. Too often in the past, the mine design process has limited its perspective in time and space to the immediate mine operation. In fact, many significant implications extend across space and time in a kind of ripple effect (Figure 16.2-5). Some of these—both positive and negative—are significant. And all must be taken into consideration to ensure that the full contribution is accounted for.

Third, the time horizon to be used needs particular mention. Table 16.2-1 lists different perspectives on the time horizons relevant to applied sustainability. This element of the boundary conditions for the mine design, operation, and closure processes has changed dramatically in the last few decades. The mine project life cycle shown in Figure 16.2-3 illustrates how the time horizon evolved after the 1970s to eventually take into considerations the postclosure phase, which can sometimes stretch environmental and social obligations of mining operations into perpetuity. This is new terrain for all interests—industry, government, host communities,

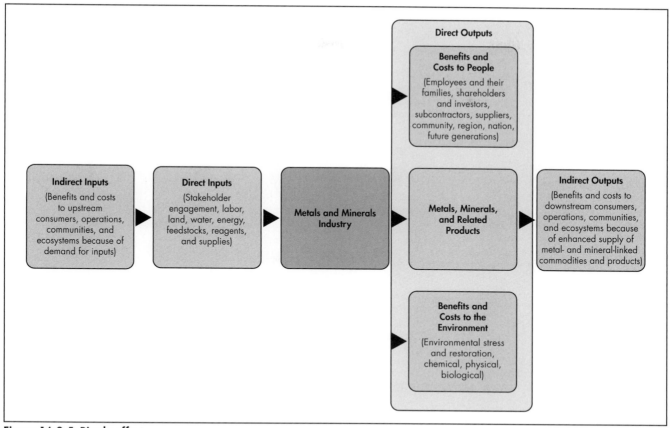

Figure 16.2-5 Ripple effect

template, attention must be paid to the compilation and analysis of both quantitative and qualitative insights. Ultimately in assessing effectiveness, efficiency, and progress, four distinct steps are essential (Hodge 2007):

1. Drawing the qualitative insight that comes through understanding the **story** that is relevant to the object(s) of the assessment
2. Undertaking and compiling the relevant quantitative **measurement**
3. Applying a systemic approach to synthesizing, setting criteria, and **judging** significance
4. Effectively **communicating** the results to different key interests

The 7QS approach encompasses four categories of insight:

1. **Relationships.** Question 1 in Figure 16.2-6 (engagement) deals with the state of relationships that are important to any given project or within any region that is being assessed (see Thomson and Joyce 2000 for a succinct discussion of this topic). A key issue facing many projects is the sense that the distribution of costs, benefits, and risk is unfair. This sense can cause stakeholder reactions that range from feelings of discontent to outright civil disobedience and damage to persons and property. Although detailed procedures will vary from site to site, the 7QS approach calls for addressing this issue early and in a way that facilitates constructive relationships achieved over the full project life cycle.
2. **Ends.** Questions 2 (people) and 3 (environment) focus on the end results that must be achieved and against which the success of any project must be tested—human and ecosystem well-being over the long term.
3. **Means to achieving ends.** Questions 4 (economy), 5 (traditional and nonmarket activities), and 6 (governance and institutions) cover the various means of achieving human and ecosystem well-being.
4. **Feedback.** Lastly, Question 7 (synthesis and continuous learning) provides the feedback mechanism that allows managers and others to ensure accountability and to learn from the inevitable mistakes, adapt and improve designs as necessary, and celebrate the successes, giving credit where due.

Effective Engagement

If relationships with those important to a mining/minerals/metals project are unhealthy, the chance of achieving a successful project—one that contributes to sustainability—is greatly reduced. Although this is a simple idea and one that is key to many successful nonmining businesses around the world, application of this idea is only now gaining momentum across the industry.

At any point in time, mining activities must align with the norms and values of society as a whole. When it does, a social license to operate is the result—an unwritten approval. If that alignment is not apparent, the social license will be challenged. Within the mining industry, elucidation of this concept has been led by Ian Thomson and Susan Joyce (2008).

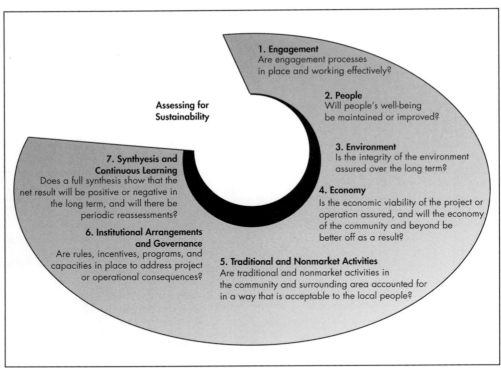

Courtesy of the International Institute for Sustainable Development.
Figure 16.2-6 Seven Questions to Sustainability

Thomson and Joyce (2006) have also played a lead role in recognizing the central role of the explorationist and junior companies (that are the first to enter an area) in either creating lasting positive relationships or a climate of long-term tension. Working to entrench such acceptance is an insurance policy that includes both the ability to recognize meaningful change and the ability to consider accommodating these changes.

Building a constructive engagement with the local community involves a series of challenging steps to (1) identify key interests, (2) learn how to listen to each interest's concerns amid all the noise of existing pressures, and (3) develop a way forward for mining and metals operations based on mutual respect, trust, and integrity. It requires effort and resources, sometimes just as much as many technical aspects of a project. And today, with the changing and growing role of many interests in society, if care is not taken to build the needed relationships, implications for proceeding effectively and efficiently on a project can be seriously undermined. At worst, physical conflict can occur.

A leading example of addressing community relationship issues has been completed by Newmont Gold Corporation (2009). In this case, concerns over conflict led an ethical investor to ask for a complete review of the relationships between Newmont operations and its host communities. The initial work, completed over 2 years, included an independent review panel. Building on existing and past attempts to build effective relationships, it led to a series of company–host community interactions that, in turn, changed Newmont's own internal management system. One result was that, in 2010, Newmont was ranked 16th in *Corporate Responsibility* magazine's 11th Annual List of 100 Best Corporate Citizens in the United States (*Corporate Responsibility* 2010), the only mining company in the top 20 and joining such businesses as Hewlett-Packard, Intel, Gap, IBM, and Microsoft.

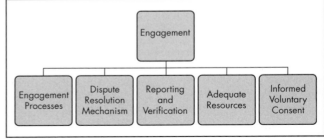

Courtesy of the International Institute for Sustainable Development.
Figure 16.2-7 Assessing the effectiveness of relationships between operations and other interests

A particularly important issue for the mining industry is building effective relationships with indigenous people. A useful overview of issues relevant to this topic can be found in Render 2006. A current perspective on how to best achieve effective working relationships with indigenous people is provided in ICMM 2010b. This is especially relevant for countries where the legal system does not provide strong protections for indigenous peoples. In the last decade, indigenous peoples' concerns have led to the emergence of special impact benefit agreements that formally entrench special arrangements for participation. A current tool kit addressing such agreements is now available (Gibson and O'Faircheallaigh 2010).

Considerations important to building effective relationships are shown in Figure 16.2-7. Table 16.2-2 is modified from the 7QS work and lists the ideal answer to the question, "Are engagement processes in place and working effectively?," and offers example indicators that can be considered for compilation in order to assess how close a given mining or metals

Table 16.2-2 Engagement: Are engagement processes in place and working effectively?

Question (Goal)	Ideal Answer (Objectives)	Example Indicators
Are processes of engagement committed to, designed, and implemented that • Ensure all affected communities of interest (including vulnerable or disadvantaged subpopulations due to, for example, minority status, gender, ethnicity, or poverty) are well informed and have the opportunity to participate in decisions that influence their own future; • Are understood and agreed upon by implicated communities of interest; and • Are consistent with the legal, institutional, and cultural characteristics of the community and country where the project is located?	Satisfactory processes of engagement have been designed and implemented that • Ensure all affected communities of interest (including vulnerable or disadvantaged subpopulations due to, for example, gender, ethnicity, or poverty) are well informed and have the opportunity to participate in the decisions that influence their own future; and • Are understood, agreed upon, and consistent with the legal, institutional, and cultural characteristics of the community and country where the project is located. As indicated by:	Input ⇓ Output ⇓ Result
	Engagement processes. Engagement processes are in place for all phases of the project/operation life cycle to serve as a mechanism for • Collaboratively identifying desired objectives, best approaches for gathering evidence in support of achieving objectives (quantitative and qualitative), assessment criteria, trade-offs and the bases for judging trade-offs; and • Overseeing the application of the approach to assessing the contribution to sustainability articulated here.	• Comprehensive mapping of interests completed. • Design of engagement strategy completed, including guidelines that are agreed upon by all interests • Full and satisfactory disclosure of project-related information • Effective implementation as signaled by participant satisfaction
	Dispute resolution mechanism. An agreed upon, affordable dispute resolution mechanism (or set) exists and is understood by and accessible to all communities of interest.	• Dispute resolution mechanism(s) • Effective mechanisms as signaled by participant satisfaction
	Reporting and verification. Appropriate systems of reporting and verification are in place.	• Systems in place • Systems working effectively from perspective of various interests
	Adequate resources. Adequate resources have been made available to ensure that all communities of interest can effectively participate as needed. Note: Responsibility for ensuring that this capacity is in place rests with a mix of government, company, and the local community itself. The exact distribution of this responsibility should be worked out collaboratively.	• Adequate resources available • Satisfaction with level of support • Effective participation achieved as assessed by company, community, indigenous peoples, and government
	Informed and voluntary consent. The informed and voluntary consent of those affected by the project or operation has been given. Note: Inclusion of this factor does not imply that consent be given as a requirement for a project to proceed. The responsibility for approval lies with the relevant regulatory agency that is mandated by the laws of the country. Rather, this factor is included as a means to assess the extent of concurrence of those affected by a project. If that concurrence is high, the potential for achieving a net positive contribution to sustainability is greatly enhanced. In contrast, if negative feeling toward a project or operation is extensive, that potential is greatly reduced.	• Broad community support

Source: Adapted from MMSD North America 2002.

operation is to achieving the ideal answer. In turn, the breadth of knowledge that Table 16.2-2 spans provides an indication of what any mining operation might be expected to take into consideration when addressing this aspect of sustainable development.

Contributing to Human Well-Being

Most inside the mining industry take it for granted that mining activities contribute to society in general, to investors that risk their capital, to management and workers who are gainfully employed as a result, and to the many host communities who experience the secondary and tertiary benefits that ripple out from any mining activity. However, following hard on the heels of the rise of the contemporary environmental movement came close examination of the social and economic implications of mining activities, particularly in emerging nations. One observation that emerged was that examples of developing economies suggested a link between mining activity and ongoing poverty. In these instances, was the presence of natural resources in fact a "resource curse" resulting from a combination of poor governance, corruption, and civil war? There is a large amount of literature on this subject (e.g., see Davis 2009, Crowson 2009, Auty 1993, and Sachs and Warner 1995).

More recently, work completed by the ICMM (McPhail 2008) has demonstrated that the resource curse needn't occur if appropriate collaborative action is taken. The ICMM work has identified six key areas of focus: (1) poverty reduction,

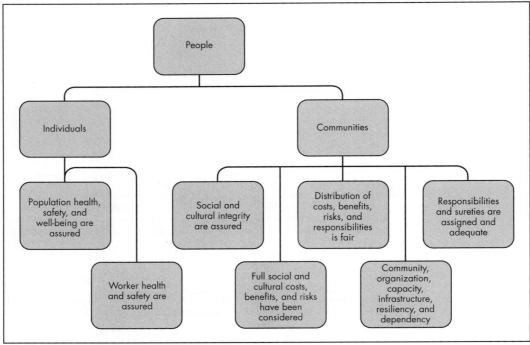

Courtesy of the International Institute for Sustainable Development.
Figure 16.2-8 Example factors to consider in assessing the contributions of mining and/or metals operations to human well-being

(2) revenue management, (3) regional development, (4) local content (the use of local labor and locally derived services and supplies), (5) social investment, and (6) dispute resolution. The key to success is the use of collaborative approaches involving host governments, companies, civil society, and donor agencies.

One spin-off of the emergence of sustainability ideas starting in the mid-1980s, the resource curse debate of the 1990s and 2000s, and the ongoing challenging of the mining industry by NGOs and in some cases host governments is a realization that much greater effort is required to understand, capture, and share the nature and extent of mining and metal's contribution to human well-being. No longer can it simply be taken for granted.

A key issue requiring attention in emerging economies is a documented gender bias in mining whereby benefits (such as employment, income, training, education, and health care) accrue largely to men and the local elite while most risks (such as family and social disruption, domestic violence, alcoholism, HIV/AIDS, increased prostitution, loss of gardens for subsistence agriculture, pollution, and water losses) fall on the poorer women, the less advantaged, and the families they care for (see Eftimie et al. 2009a and 2009b). A company purchasing the bulk or all of its produce needs locally can be beneficial for local farmers and landowners who see prices increase for their produce, but this can also have the unintended effect of worsening the situation for the poor community women who may be unable to afford the higher food prices. These are complex issues that require careful management approaches for any mining company.

Figure 16.2-8 offers example factors to be considered in answering the question—"Will people's well-being be maintained or improved" by a given mining, minerals or metals operation? These factors draw on the foundation provided by socioeconomic impact analysis but use the lens of contribution. Table 16.2-3 then provides both the ideal answer as well as examples of the kind of indicators that might be considered in assessing how close a given operation is to meeting the ideal answer. As in the previous section, the range of topics provides a sense of what sustainable development means to mining from this particular perspective.

Contributing to Ecosystem Well-Being

The relationship between a mining operation and its host environment is the focus of the environmental impact assessment and the resulting environmental management plan and related environmental management systems. The approach used here builds on this foundation.

Figure 16.2-9 shows one way of conceptualizing the environmental implications of a mining operation. It looks at chemical, physical, and biological implications from both the generation of environmental stress and the linked potential for ecosystem restoration. The concept of restoration that is used here is *not* the idea that everything needs to be returned to a chemical, physical, and biological premining state. Rather, a robust ecosystem is sought that is naturally reproducing and sustainable. This is the domain of the emerging science and art of *restoration ecology*.

A chemical, physical, and biological characterization of environmental implications is traditional and helpful from several perspectives, including (1) often regulatory requirements are organized in this manner; (2) the professional expertise that must be drawn on is often organized in this way; and (3) traditional education is offered in this way and therefore management's understanding is often aided by this approach.

Table 16.2-3 People: Will people's well-being be maintained or improved?

Question (Goal)	Ideal Answer (Objectives)	Example Indicators
Will the project/operation lead directly or indirectly to maintenance of people's well-being, preferably an improvement • During the life of the project or operation, or • In postclosure?	The project or operation will lead directly or indirectly to the maintenance or improvement of people's well-being • During the life of the project or operation or • In postclosure As indicated by:	Input ⇓ Output ⇓ Result
	Community organization and capacity. Effective and representative community organization and capacity (knowledge, skills, and resources) are in place in the local community, including representation of women and the disadvantaged in community leadership and decision making. Note: Responsibility for ensuring this capacity is in place rests with a mix of government, company, and the local community itself. The exact distribution of this responsibility should be worked out collaboratively.	• Presence of an organizational structure that links and represents the community in project-related decision-making processes • Training facilities in place • Local education/skills level to serve project needs and provide basis for postclosure activities • Community access to the information and expertise needed to ensure that properly informed decisions are made
	Social/cultural integrity. All communities of interest have a reasonable degree of confidence that social and cultural integrity will be maintained or preferably improved in a way that is consistent with the vision and aspirations of the community. Involuntary resettlement and other interventions will be undertaken in such as way as to maintain or preferably improve social and cultural integrity. Note: This category is particularly dynamic and will change as a project proceeds.	• Existence of community and regional visions expressed explicitly in development and land-use plans • Presence of key indicative social structures and their states • Sense of satisfaction signaled by all interests that social and cultural integrity will be maintained or improved (including separate consultations with community women and representatives of disadvantaged groups in the community) • Social and cultural indicators identified as significant by the community • Preparation and implementation of community-supported and project-induced migration management plans to help manage the impacts of the inflow of outsiders that occurs when a mining development is expected or announced
	Worker and population health, safety, and well-being. Improvement of indicators of worker and population health, safety, and well-being are maintained or improved. Note: Responsibility for gathering this data and information lies with a mix of company (in terms of workers), community, and government. However, statistics on population health, training and education, jobs, income, poverty, debt, community resiliency, and community dependency typically fall to government.	• Baseline studies completed that include basic demographics to track population change (birth rate, infant mortality, morbidity rates, in/out-migration), household incomes, and so forth, including gender disaggregated data • Worker health and safety • Population health • Training and education • Jobs, income, poverty, debt • Crime and security • Community resiliency • Community dependency
	Availability of basic infrastructure. The infrastructure to meet basic needs is available to workers and residents.	• Water supply, sewage and wastewater treatment, power, communications, transportation, education, health services
	Consideration of all direct, indirect, and induced or diffuse effects. All communities of interest have a reasonable degree of confidence that all direct, indirect, and induced or diffuse effects have been considered and addressed. Note: Requirements will change through the project.	• Direct, indirect, and induced or diffuse economic, social, and cultural effects of project • Changes in social behavior as a result of the project
	Full social/cultural costs, benefits, and risks. All communities of interest have a reasonable degree of confidence that the full costs, benefits, and risks to people have been identified and factored into project or operation-related decision making (as it applies throughout the full project or operation life cycle).	• Satisfaction that all social/cultural costs, benefits, and risks found across the full life cycle from exploration through postclosure have been identified and addressed

Source: Adapted from MMSD North America 2002.

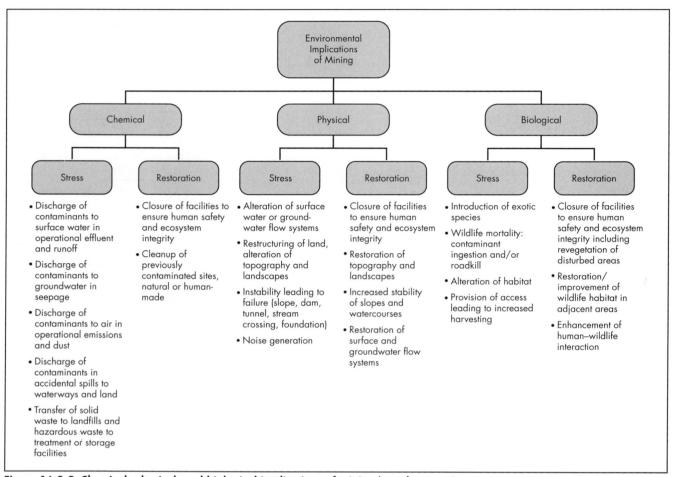

Figure 16.2-9 Chemical, physical, and biological implications of mining/metals operations

However, the ecological system does not function in such a compartmentalized way, and those expert in ecology in general and ecological restoration in particular have offered alternative ways of addressing this challenge. Figure 16.2-10 offers such an approach, and Table 16.2-4 addresses the question—"Is the integrity of the environment assured over the long term?"—and provides both an ideal answer to this question as well as an example set of indicators that, if compiled, would serve to facilitate an assessment of how compatible a mining/metals operation would be with sustainable development. An increasingly important emerging issue is understanding the full life-cycle environmental contribution of the metals. Much remains to be learned about the technique of life-cycle analysis.

Economic Viability: Project, Community, Nation
Understanding the economics of a project lies at the heart of successful mining from a management and investment perspective. Two key factors govern project economic stability: (1) the licensing and fiscal regimes under which mining takes place—that they are efficient, noncorrupt, and result in security of tenure; and (2) an equitable sharing of costs, benefits, risks, and responsibilities between the host country and the investor.

For any project to be successful, the deposit, the company, and industry must all be economically viable. The typical economics of a mine is such that the first few years (generally, years 1 to 5) are usually very profitable, which is necessary to serve investors who require a return on invested capital. Middle years are moderately profitable (generally, years 6 to 15), and older mines are usually marginally viable. Company managers strategize from this perspective because it is their legal responsibility to protect the investors. If a large number of operating mines within their system are in the older phase, a company can be vulnerable. However, from the perspective of an employee or service provider, the mine is meaningful as long as it operates and provides employment. And the concern from the local, regional, or national government is that the mine leads to a lasting contribution—however that is achieved. These different perspectives all factor into the sustainable development equation.

However, they do not imply that a project/company should be assuming responsibility for the local, regional, or national economy. Rather, it is often possible that by working in collaboration with in-country partners, benefits can be achieved for the host community, region, or country that involve little cost to the company but have significant benefits for the host. Further, if unintended economic consequences related to this broader perspective arise and are not recognized, they may result in a liability to the company over the long term that can have grave consequences, including the possibility of the national/local business environment deteriorating to the point

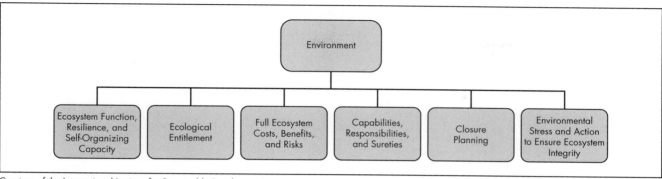

Courtesy of the International Institute for Sustainable Development.

Figure 16.2-10 Elements for assessing the contribution of mining/metals operations to the environment over the long term

Table 16.2-4 Environment: Is the integrity of the environment assured over the long term?

Question (Goal)	Ideal Answer (Objectives)	Example Indicators
Will the project or operation lead directly or indirectly to the maintenance or strengthening of the integrity of biophysical systems so that they can continue in postclosure to provide the needed support for the well-being of people and other life forms?	The project or operation will lead directly or indirectly to the maintenance or strengthening of the integrity of biophysical systems, as indicated by:	Input ⇓ Output ⇓ Result
	Ecosystem function, resilience, and self-organizing capacity. All communities of interest hold a reasonable degree of confidence that ecosystem function, resilience, and self-organizing capacity will be maintained or improved over the long term.	• Baseline studies completed • Monitoring systems in place • Projected effects of project on indicator species of aquatic and terrestrial flora and fauna (identified through both scientific assessment and traditional ecological knowledge studies) • Projected long-term well-being of water systems and renewable resources in the area of the mine/mineral project • Tracking rapid geological change
	Ecological entitlement. All communities of interest have a reasonable degree of confidence that the capacity of project-affected renewable and nonrenewable resources will be maintained or improved such that the needs of both present and future generations will be met.	• Degree of confidence
	Full ecosystem costs, benefits, and risks. All communities of interest have a reasonable degree of confidence that the full costs, benefits, and risks to the ecosystem have been identified and factored into project/operation-related decision making (as it applies throughout the full project/operation life cycle).	• Full or total cost-accounting tools to assess the implications of the project • Satisfaction that all social/cultural costs, benefits, and risks related to the full life cycle from exploration to postclosure have been identified and addressed
	Capabilities, responsibilities, and sureties. All communities of interest have a reasonable degree of confidence that the responsibilities and sureties for ensuring both short- and long-term ecosystem well-being have been fully and fairly assigned and accepted and that those responsible have the capacity and willingness to fulfill their roles (including responsibilities attached to company, community, government, or NGO).	• Financial sureties and mechanisms to address potential present and future environmental liabilities and effort required to ensure the continuing integrity of biophysical systems • Satisfaction that sureties and mechanisms will provide adequate bridging to postclosure state.
	Closure planning. All communities of interest are satisfied that closure plans meeting best industry practice have been completed; mine design and implementation are driven by closure requirements to ensure that long-term financial, environmental, and social costs are minimized.	• Closure plans to best industry practice are in place and fully integrated into mine design and implementation

(continues)

Table 16.2-4 Environment: Is the integrity of the environment assured over the long term? (continued)

Question (Goal)	Ideal Answer (Objectives)	Example Indicators

Will the project or operation lead directly or indirectly to the maintenance or strengthening of the integrity of biophysical systems so that they can continue in postclosure to provide the needed support for the well-being of people and other life forms?	**Environmental stress and action to ensure ecosystem integrity.** Physical, chemical, and biological stress imposed on the ecosystem by the project or operation does not threaten the function, resilience, and self-organizing capacity of the biophysical system; and appropriate actions are taken to ensure the continuing integrity of biophysical systems.	• Material inputs/flows—water (surface, ground), energy (by form and source), reagents (e.g., cyanide), fuels, solvents, lubricants, other supplies
		• Recycling
		  – Material recovery, oil, solvents, lubricants, batteries, tires, etc.
		  – Waste
		  – Hazardous and solid waste generated and discharged
		• Water quality
		  – Surface water: effluent chemistry, ambient water quality downstream, stream sediment storage and load
		  – Groundwater: contaminant plume, plume chemistry, ambient quality downstream
		• Water quantity—maintenance of surface and groundwater supplies
		• Acid rock drainage and metal leaching—tailings, waste rock, workings; short- and long-term treatment required
		• Soils
		  – Chemical, biological, and physical change
		  – Erosion
		• Rapid and long-term landscape change
		  – Deep open pits, mountaintop removal
		  – Landslides and avalanches
		• Air quality
		  – Emissions chemistry
		  – Indoor air quality
		  – Greenhouse gas emissions
		• Noise, on- and off-site
		• Environmental incidents, on- and off-site
		• Reclamation and restoration, on- and off-site
		• Environmental effects monitoring system in place

Source: Adapted from MMSD North America 2002.

where the company can no longer operate efficiently or, in the worst case, not at all. These are significant risks.

Figure 16.2-11 is a simple characterization of the economic aspects that require consideration. Table 16.2-5 responds to the question—"Is the economic viability of the project assured; will the community and broader economy be better off as a result?" It then offers an ideal answer to that question and a set of example indicators that can be compiled to assess how close a given operation is to the ideal condition.

Traditional and Nonmarket Activities

Since the 1980s, recognition has been growing of the significance of traditional and nonmarket activities—human activities that are not part of the market economy and therefore not captured in the accounting systems that underlie our systems of taxation or the national accounts.

National accounts are used to track the flow of goods and services in the economy. Contributing data are organized in an indtustry classification system, which provides a useful indication of human activity. The compiled data are used to calculate such indicators as national income and gross domestic product or gross national product. However, many activities important from a sustainability perspective lie outside the formal market economy. One way of getting a sense of these is to consider the vast array of organizations of civil society, those engaged in not-for-profit work. Table 16.2-6 is a listing of the general categories of a classification of non-profit institutions, which was developed at Johns Hopkins University and subsequently adopted by the United Nations (2003).

Often, reacting to public criticism, mining companies focus on environmental or social advocacy groups. A more strategic approach would be to consider all aspects of civil society that are important in a given community, ensuring that the contribution of the mining/metals activity to nonmarket and traditional activities is as efficiently and effectively pursued as possible. Traditional culture-based activities of indigenous people lie within this aspect of the sustainability equation and are of particular importance to mining/metals operations in many parts of the world.

Figure 16.2-12 is a simple characterization of the traditional and nonmarket aspects that require consideration. Table 16.2-7 responds to the question—"Are traditional and nonmarket activities in the community and surrounding area accounted for in a way that is acceptable to the local people?" It then offers an ideal answer to that question and a set of example indicators that can be compiled to assess how close a given operation is to the ideal condition.

Institutional Arrangements and Governance

This category addresses the effectiveness of the formal and informal rules that society creates to govern activities such

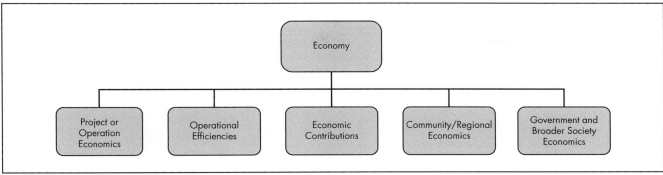

Courtesy of the International Institute for Sustainable Development.

Figure 16.2-11 Elements for assessing the economic viability of mining/metals operations over the long term

Table 16.2-5 Economy: Is the economic viability of the project assured?

Question (Goal)	Ideal Answer (Objectives)	Example Indicators
Is the financial health of the project/company assured and will the project or operation contribute (through planning, evaluations, decision making, and action) to the long-term viability of the local and regional economy in ways that will help ensure sufficiency for all and provide specific opportunities for the less advantaged?	The financial health of the project/company is assured and the project or operation will contribute (through planning, evaluations, decision making, and action) to the long-term viability of the local, regional, and global economy, as indicated by:	Input ⇓ Output ⇓ Result
	Project or operation economics. Project or operation economic success targets are achieved.	• Demonstrated positive project economics as calculated in the feasibility study; economic targets met as the project proceeds
	Operational efficiencies	• Mining efficiency • Processing efficiency • Refining efficiency
	Economic contributions: annual/total	Contributions • To the local economy • To the regional economy • To the national economy • To the global economy • Total
	Community/regional economics. Community and adjacent regional economic success targets are met, including direct employment, indirect employment, and small and medium business development.	• Demonstrated positive project economics as calculated in the feasibility study; economic targets met as project proceeds • Distribution of economic benefits resulting from the project between the company, government, and local community, and within the local community between men and women and between the wealthy and the disadvantaged
	Government/broader society economics. A net economic contribution to governments and broader society is achieved.	• Net financial gain for government as the project proceeds and meets economic targets, which contributes to economic growth and increased standards of living

Source: Adapted from MMSD North America 2002.

as mining/mineral projects and operations. This is a complex area that includes internal corporate management systems as well those found in broader society. Although some rules are written, others are unwritten and simply accepted as part of culture—internal corporate culture and culture of the host community and country.

Internal Management Systems
Adaptation of internal management systems to sustainability is well under way. A useful overview of many related issues is provided by Botvin 2009. An early foundation was provided by the work of the International Organization for Standardization (ISO), which has developed widely accepted quality management systems (ISO 9001), environmental management systems (ISO 14001), and risk management principles and guidelines (ISO 31001).

Reporting and Assurance
Since the mid-1990s, there have been a number of significant developments regarding how a mining/metals company publicly reports its sustainability-related performance and how that reported progress is assured using independent third-party assurance. Three initiatives are particularly relevant to the mining industry:

Table 16.2-6 United Nations classification of nonprofit institutions

Group	Subgroups
1. Culture and Recreation	Culture and Arts, Sports, Other
2. Education and Research	Primary and Secondary, Higher, Other, Research
3. Health	Hospitals and Rehabilitation Nursing Homes, Mental Health and Crisis Intervention, Other
4. Social Services	Social Services, Emergency Relief, Income Support and Maintenance
5. Environment	Environment, Animal Protection
6. Development and Housing	Economic, Social and Community Development, Housing, Employment and Training
7. Law, Advocacy, and Politics	Civic and Advocacy Organizations, Law and Legal Services, Political Organizations
8. Philanthropic Intermediaries and Volunteerism Promotion	Grant-Making Foundations, Other
9. International	International Activities
10. Religion	Religious Congregations and Associations
11. Business and Professional Associations, Unions	Business Associations, Professional Associations, Unions
12. Not elsewhere classified	Not elsewhere classified

Source: United Nations 2003.

1. Mining and Metals Sector Supplement (MMSS) of the Global Reporting Initiative (GRI)
2. ICMM Reporting and Assurance System
3. Towards Sustainable Mining (TSM) system of the Mining Association of Canada (MAC)

All of these initiatives recognize that for the industry to improve its reputation and credibility, improved performance linked to an alignment with public values is essential. From a company perspective, there is a close link to risk management. All three of these initiatives are cognizant of the need to synchronize.

GRI MMSS. The GRI is a network-based organization that over the past decade has pioneered the development of the world's most widely used sustainability reporting framework. The GRI framework sets out principles and indicators that organizations can use to measure and report their economic, environmental, and social performances. At its heart sits the generic 2006 G3 Guidelines that refer to all applications. In addition, sector supplements and national annexes are being developed over time. As of 2010, 15 sector supplements have been developed, one of which is focused on mining.

Following 6 years of intense multistakeholder development, the GRI MMSS was released in March 2010 (ICMM 2010c). MMSS addresses the following nine main sector topics:

1. Biodiversity
2. Emissions, effluents, and waste
3. Labor
4. Indigenous rights
5. Community
6. Artisanal and small-scale mining
7. Resettlement
8. Closure planning
9. Materials stewardship

ICMM's reporting and assurance system. ICMM's system serves as a means to demonstrate that members have delivered on the commitments contained in the 10 ICMM principles and the related position statements (Figure 16.2-1). Members must meet GRI reporting requirements, and independent third-party assurance is required on an overall company system basis.

MAC's TSM initiative. MAC describes TSM as "a strategy for improving the mining industry's performance by aligning its actions with the priorities and values of Canadians; and

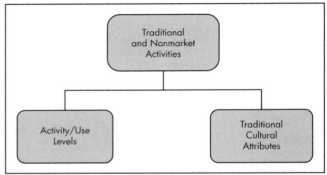

Courtesy of the International Institute for Sustainable Development.

Figure 16.2-12 Elements for considering implications of mining/metals operation on traditional and nonmarket activity

a process for finding common ground with our communities of interest, to build a better mining industry today and in the future" (MAC 2010). TSM focuses on company management systems and provides a means of evaluating their quality, comprehensiveness, and robustness. As of 2010, four performance elements were finalized: (1) tailings management, (2) energy use and greenhouse gas emission management, (3) external outreach, and (4) crisis management planning. In addition, work on aboriginal relations, biodiversity conservation, safety and health, and mine closure is in varying states of implementation.

Nation-Based Mining Law

In all cases, host countries own the natural resources. Thus complex systems of laws and regulations are put in place to guide exploration, production, closure, use, reuse and recycling. Every company is responsible for developing an understanding of the particular regulatory regime that is in place where it is active.

Important International Agreements

A large number of international agreements have significant implications for the mining, minerals, and metals industry. These range from the transport of hazardous materials to human rights and chemicals management. Here, too, it is the responsibility of companies to understand this complex part of the operating environment.

Table 16.2-7 Traditional and nonmarket activities: Are traditional and nonmarket activities accounted for in a way that is acceptable to the local people?

Question (Goal)	Ideal Answer (Objectives)	Example Indicators
Will the project/operation contribute to the long-term viability of traditional and nonmarket activities in the implicated community and region?	The project/operation will contribute to the long-term viability of traditional and nonmarket activities in the implicated community and region, as indicated by:	Input ⇓ Output ⇓ Result
	Activity/use levels. Maintenance of activity/use levels is designated by the community in question.	• Baseline study of traditional and nonmarket activities completed • Dependency levels on traditional and nonmarket activities
	Traditional cultural attributes. Maintenance of traditional cultural attributes is designated by the community in question.	• Use of indigenous language • See also Table 16.2-3

Source: Adapted from MMSD North America 2002.

Assessing the Effectiveness of Institutional Arrangements and Governance

Elements that could contribute to an assessment of the effectiveness of institutional arrangements and governance are shown in Figure 16.2-13 and are described in greater detail in Table 16.2-8.

Synthesis and Continuous Learning

Many companies employ an internal system of planning, decision making, implementation, performance assessment, and adjustment as a matter of course. This element of the 7QS framework calls for the same kind of approach to be taken but using the lens of sustainability as the driver. Example elements that could contribute to an assessment of the effectiveness of an overall integrated evaluation and capacity for continuous learning and improvement are summarized in Figure 16.2-14 and described in greater detail in Table 16.2-9.

PERSPECTIVES BY PROJECT LIFE-CYCLE PHASES

Application of sustainability concepts to the mining, minerals, and metals industry requires attention paid to the full project and mineral life cycles (see earlier discussion in the "Boundary Conditions for Applying Sustainability Ideas to Mining" section). A cursory summary of phase-by-phase observations follows. (This discussion is modified from MMSD 2002.)

Exploration

Exploration is the starting point of the entire mineral project life cycle. However, only 1 out of 1,000 exploration projects ever evolves to the next level of activity (often more exploration). Explorationists generally have the objective of getting into and out of areas quickly and moving on to the next evaluation. Exploration is highly competitive and therefore secretive by design. Without positive relationships with the local community, serious tensions can occur. For their part, local residents want to know what is happening and how their lives might be affected. Left to ferment, anger and/or unintended expectations can grow—either of which can subsequently cause difficulty. Although the extent of the human and environmental implications is less than in later phases of activity, they can be profound. Oftentimes concerns center on access routes, which can open up large areas of untouched territory.

Exploration is the first time that the desirability or appropriateness of mining in a particular area is considered. Because exploration sets the stage for all that follows, the nature of the relationship established between the exploration team and affected communities of interest sets the tone going forward. Only in the last few years have models for appropriate engagement during exploration begun to emerge (e.g., see Thomson and Joyce 2006 and 2008).

The leading node of guidance on exploration and sustainability comes from the Prospectors and Developers Association of Canada (PDAC) and its e3 Plus Framework for Responsible Exploration (PDAC 2010). The program is aimed at helping exploration companies continuously improve their social, environmental, and health and safety performances and to comprehensively integrate these three aspects into all their exploration programs. It includes a set of overarching principles and explanatory guidance notes; and three comprehensive tool kits addressing social responsibility, environmental stewardship, and health and safety. For their part, the principles deal with

- Responsible governance and management,
- Ethical business practices,
- Human rights,
- Project due diligence and risk assessment,
- Engagement with host communities and other affected and interested parties,
- Contributing to community development and social well-being
- Protecting the environment, and
- Safeguarding the health and safety of workers and the local population.

The e3 Plus program is currently examining options for reporting and assurance that will echo those described. A significant driver is the demand by senior firms that the conditions they might inherit through acquisition of advanced exploration projects take into consideration the full range of sustainability components, in particular the need for effective engagement with host communities.

Design and Construction

Relative to other phases of activity, the design and construction phase is short. However, this intense pulse of activities and related social and environmental implications can be destructive if not carefully managed. This phase of activity

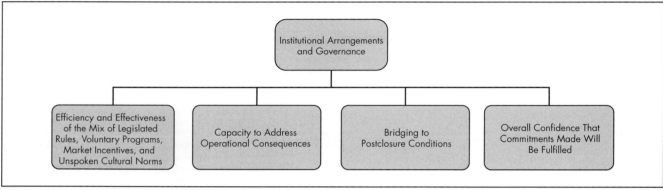

Courtesy of the International Institute for Sustainable Development.

Figure 16.2-13 Example elements for assessing the effectiveness of institutional arrangements and governance

Table 16.2-8 Institutional arrangements and governance: Are capacities in place to address project or operational consequences?

Question (Goal)	Ideal Answer (Objectives)	Example Indicators
Are the institutional arrangements and systems of governance in place to provide a reasonable degree of confidence that the capacity to address project or operation consequences will continue to exist through the full life cycle, including postclosure?	Satisfactory institutional arrangements and governance mechanisms are in place, as indicated by:	Input ⇓ Output ⇓ Result
	Mix of rules, market incentives, voluntary programs, and cultural norms. An effective mix of legislated rules, market incentives, voluntary programs, and cultural norms is in place for governing project activities.	• Satisfaction with mix from the perspective of the various communities of interest, including company, community, indigenous peoples, NGOs, government
	Capacity. A reasonable degree of confidence is held by all communities of interest that the capacity to address project or operation consequences is in place now and will continue to exist throughout the full project/operation life cycle, including postclosure.	• Capacity of community support infrastructure to meet the needs of residents and workers of the region • Monitoring and enforcement programs in place with adequate resources committed for the full project life cycle
	Bridging. A reasonable degree of confidence is held by all communities of interest that sufficient local capacity and commitment will be developed and an adequate level of resources will be accumulated and set aside throughout the life of the project/operation to ensure a smooth transition to an acceptable postclosure condition (ecological, social, cultural, economic) for the community that remains.	• Existence of community-based *nonmining* economic and social development initiatives
	Confidence that commitments made will be fulfilled. A reasonable degree of confidence is held by all communities of interest that commitments that have been made will be fulfilled.	• System of publicly reported reporting and assurance in place • Level of funding to cover rehabilitation/reclamation costs during operations and at closure

Source: Adapted from MMSD North America 2002.

also marks the time that formal approvals are sought and received. During approvals processes, key opportunities exist for engagement with communities of interest (see previous discussion of effective engagement in the "Effective Engagement" section).

The influx of outsiders seeking employment and new opportunities can create significant changes within the community—often for the worse—and preparation and implementation of a project-induced migration management plan that is acceptable to the established community population can help manage the impacts. Many of the activities are undertaken by subcontractors who may or may not follow the same practices as the project owner or manager. Effective systems (policy, management, oversight, incentives) for ensuring good practices are essential.

Operation

The operation phase typically receives the greatest amount of attention. When the general public conjures an image of a mining/mineral project or operation, it is the operation phase that is imagined, whether the activity be mining, smelting, refining, metals manufacturing, or recycling. Economic, social, cultural, and environmental implications are relatively well understood, although their treatment is uneven in practice.

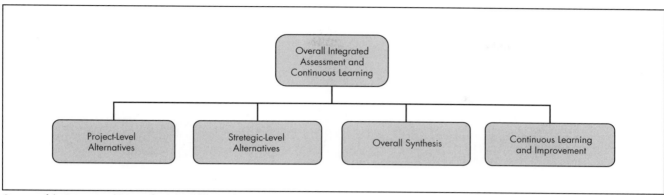

Courtesy of the International Institute for Sustainable Development.

Figure 16.2-14 Synthesis and continuous learning—elements to consider

Table 16.2-9 Synthesis and continuous learning: Does synthesis show that the net result will be positive or negative?

Question (Goal)	Ideal Answer (Objectives)	Example Indicators
Has an overall evaluation been made and is a system in place for periodic revaluation based on • At the project level, consideration of all reasonable alternative configurations and designs (including the no-go option in the initial evaluation); • At the overarching strategic level, consideration of all reasonable alternatives for supplying the commodity and the services it provides for meeting society's needs; and • A synthesis of all the factors raised in this list of questions, leading to an overall judgment that the contribution to people and ecosystems will be net positive in the long term?	An overall evaluation been made and a system is in place for periodic revaluation based on consideration of • At the project level, consideration of all reasonable alternative configurations including the no-go option; • At the overarching strategic level, consideration of all reasonable alternatives for supplying the commodity and the services it provides for meeting society's needs; and • Synthesis of all the factors raised in this list of questions, leading to an overall judgment that the contribution to people and ecosystems will be net positive in the long term; As indicated by:	Input ⇓ Output ⇓ Result
	Project-level alternatives. All reasonable project alternatives have been considered.	• All key alternatives considered, for example: – Access – Transportation – Energy supply – Water supply – Local infrastructure – Tailings and effluent management – Mineral processing options
	Strategic-level alternatives	• Strategic level review that confirms project need
	Overall synthesis. A synthesis has been completed, and the system is in place to for periodic reassessment.	• Synthesis undertaken
	Continuous learning and improvement. A commitment to continuous learning and improvement is held by all interests, including company, community, government, and others.	• Mechanism(s) and resources in place to periodically repeat the overall sustainability assessment and report the results publicly

Source: Adapted from MMSD North America 2002.

Closure: Temporary, Final, Post

Temporary closure (due to changing metal prices, accident, disaster, or labor strife) rarely receives the pre-event planning and consideration that it should. Though often a temporary closure can run a few short days or weeks, sometimes so-called temporary closures can stretch into many years' duration.

While concepts of *design-for-closure* date from the 1970s (J. Gadsby, personal communication), concepts of sustainability now demand *design-for-postclosure*. Huge benefits can be realized by ensuring that closure and postclosure activities are conceived of and implementation is begun during the design and construction and operational phases of activity. Today, design-for-closure includes (1) physical environmental aspects, (2) social and community sustainability aspects, and, importantly, (3) financial assurance that resources will be available for needed closure and postclosure activities.

Design-for-postclosure involves a significant increase in the time horizon governing project design criteria, whether the focus be social or environmental in nature. Furthermore, successful design-for-postclosure identifies a need for involvement of those affected by postclosure conditions from the earliest phases of any project. Fortunately, in the case of closure and postclosure, research and experience have produced a number of successful models. One thing emerges from all

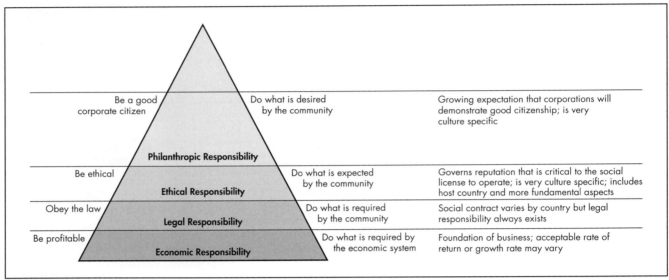

Source: Carroll 1991.

Figure 16.2-15 Carroll's pyramid of corporate responsibility and performance

of them: *succeeding custodians*, the people who will remain nearby long after mining operations cease, need to be at the table. It is only with their presence that their values can be factored into project implementation and the bridging role of a mine/mineral project or operation can be realized.

One delicate issue that the industry is now examining, which emerges because of the longer time horizons now being considered combined with the societal demand for greater transparency, is the whole question of estimating long-term closure liability and related financial assurance.

CORPORATE SOCIAL RESPONSIBILITY

Application of sustainability concepts to the mining, minerals and metals industry evokes yet another perspective that is growing in importance—the issue of corporate social responsibility (CSR). Establishing the nature and extent of a company's social responsibility is of deep importance in today's world. CSR provides a critical and natural overarching supplement to sustainable development and sustainability ideas and provides another direct connection to values through the addition of an ethical lens. At the same time, contemporary ideas of applied sustainability provide a foundation for CSR; they are closely linked but not the same.

Values are "something which is prized as of great worth and desirability: that which is respected and which motivates action...concepts which motivate actors in a general way to more specific goals" (Chadwick 1978). The described treatment of applied sustainability starts with a values-based proposition, one that says that we care for both people and the environment. In contrast, *ethics* is the science of right or wrong. It provides rules of behavior that signal what is considered decent and honest as opposed to misleading, dishonest, or unscrupulous.

Values and ethics combine to establish the boundaries of appropriate CSR. Over the last decade, there has been growing emphasis on CSR. The events in the financial services industry that sparked the 2008 recession have added fuel to the debate. Carroll (1991) offers an interesting and helpful pyramid of corporate responsibility and performance (Figure 16.2-15) and serves to illustrate a kind of hierarchy that applies to corporate responsibility.

Revenue Transparency and the Extractive Industries Transparency Initiative

One dimension of CSR that has been in the limelight has been the question of the revenues that flow from an active operation to the government. Concerns that these revenues were being intercepted for personal gain led to the creation of the Extractive Industries Transparency Initiative (EITI). Launched at the World Summit on Sustainable Development in Johannesburg, South Africa, in September 2002, EITI has grown to become a broad coalition of governments, major organizations of civil society, institutional investors, international organizations, international financial institutions, and companies (oil and gas, mining). EITI is active in 47 countries in which mining plays a significant role in the economy.

EITI represents a global standard for (1) extractive companies to publish the revenues they pay to governments in implementing countries (taxes, royalties, licenses, fees); (2) national governments to disclose what they receive; and (3) the payments and receipts to be reconciled in a third-party assured process. Each national application is overseen by an in-country tripartite multistakeholder committee consisting of representatives of industry, government, and civil society.

Bridging Issue of Human Rights

Respecting human rights is clearly part of corporate responsibility. For the 19 corporate members of ICMM and the 30 affiliated mining associations that, in turn, link to another 1,500 companies, respect for human rights is a key aspect of sustainable development.

In 2005, the UN secretary-general appointed John Ruggie, Berthold Beitz professor in human rights and international affairs at the Kennedy School of Government, Harvard University, as his special representative on business and human rights. Five thousand companies are now engaged worldwide in this initiative (Business and Human Rights Resource Centre n.d.). Ruggie played a central role

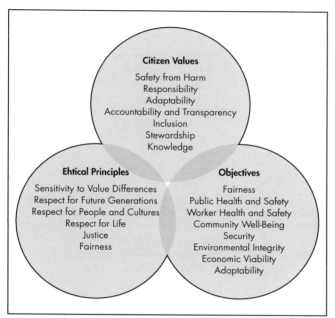

Source: Data from Joanna Facella, Nuclear Waste Management Organization, Toronto, Ontario, Canada.

Figure 16.2-16 Toward an ethical lens for mining

in preparing the UN Millennium Summit in 2000, including drafting the summit's final "Declaration," which adopted the Millennium Development Goals that have brought renewed energy and focus to the fight against global poverty. In 2008, Ruggie produced the now well-accepted "Protect, Respect, and Remedy" framework for corporate responsibility, which outlined in clear terms the distinctive but complementary roles of government and corporations regarding human rights: While governments have the primary responsibility for the protection of human rights, companies have a responsibility for respecting them. The reporting and assurance systems described previously include treatment of company respect for human rights.

Toward an Ethical Lens for Mining

It is inevitable that the material described in this chapter, taken together, signals that the world is moving toward developing a kind of *ethical lens* for assessing the mining, minerals, and metals industry. During the period 2002–2005, a Canadian initiative undertook a comprehensive review of alternatives for managing used nuclear fuel over the long term. Figure 16.2-16 shows the results of an extensive discussion with Canadians about the values (Watling et al. 2004), ethical principles (NWMO 2005), and related objectives that were used to assess options for long-term management of used nuclear fuel (NWMO 2004). Together they serve as an innovative ethical lens that is as applicable to mining in general as it is to the management of nuclear waste.

LOOKING FORWARD

Ongoing Issues

A number of issues remain to be fully addressed (modified from MMSD North America 2002 and Hodge 2004):

- **Fair distribution:** Detailed site-specific procedures for describing and addressing the distribution of costs, benefits, risks, and responsibilities among governments, men and women, rich and poor, and present and future generations.
- **Decision making, trade-offs:** Designing and implementing optimum decision making systems and approaches that effectively and fairly address trade-offs in any given site application.
- **Need and alternatives:** Effectively and fairly assessing the need for a given project and/or commodity in light of considerations and alternatives that span local to global implications.
- **Achieving a whole system perspective:** Seeing, understanding, and factoring in a sense of the whole system, not just the small component parts.
- **Uncertainty, precaution, and adaptive management:** Dealing with uncertainty using an appropriate level of precaution and an adaptive management approach.
- **Attribution problem and dealing with cumulative impacts:** Effectively addressing the common situation where a project is one of a number of contributors to social, cultural, economic, and environmental change or stress—establishing the cumulative implications, apportioning responsibility, and deciding who should take responsibility for the analysis.

These issues are not new. Together, they are a testament to some of the complexities that must be faced in bringing the ideas of sustainability from theory to practice. In addition to the previous list, the following practical issues remain to be resolved to improve the contribution of the mining industry to sustainability:

- **Capacity building:** Creating the necessary national and local governance systems that can facilitate effective and principled engagement among industry, government, and civil society.
- **Voluntary codes of conduct:** Increasing participation in and enhancement of values-based, principled codes of conduct that lift broad industry performance above legislated minimums.
- **Climate change:** Establishing how the mining industry can most effectively contribute to addressing the issue of global warming.
- **Artisanal and small-scale mining:** Establishing how such mining can be best managed in various countries and within the context of sustainable development, taking into account environmental and social implications as well as links to large-scale mining (for a comprehensive treatment of these issues, see CASM n.d.).
- **Consistent application of performance standards:** Achieving consistency in the application of sustainable development ideas within countries, between countries, and across the industry from juniors to majors.
- **Effective human resource management:** Effectively creating the foundation of education, training, and support that is required to ensure the needed flow of skilled personnel (see Freemen and Miller 2009).

All the issues listed are critical to effective application of sustainable development concepts to the mining and metals industry.

New Perspectives

In current public debate, many challenge the consumptive nature of modern society. Professor Saleem Ali has offered an alternative perspective on this divisive topic that has significant implications for the mining, minerals, and metals industry (Ali 2009). He argues that simply disavowing consumption of materials is not likely to help in planning for a resource-scarce future, given the extent of inequity in today's world, development imperatives, and our goals for a democratic global society. Rather than suppress the creativity and desire to discover—which he calls "the treasure impulse"—Ali proposes a new environmental paradigm, one that accepts our need to consume "treasure" for cultural and developmental reasons but warns of our concomitant need to conserve.

In evaluating the impact of treasure consumption on resource-rich countries, he argues that a way exists to consume responsibly and alleviate global poverty. Ali's thesis brings a much needed additional perspective relevant to applying sustainability ideas to the mining industry.

Change and Change Management
The role of metals and minerals in providing for both human and ecosystem well-being is critical, and the mining and metals industry enjoys a deep and rich heritage. However, this same heritage brings with it an innate resistance to change that can impede needed change. There are many contributing factors to this resistance, including

- Tradition: "We've always done it this way and it has worked, why change now?" or "If it's not broken, why fix it?";
- A number of economic factors that tend to favor risk-averse behaviors, including the volatility and uncertainty inherent in commodities markets, the scale and complexity of operations, the highly competitive nature of the business, and the resulting thin margins (Peterson et al. 2001); and
- A range of factors that can come into play at a personal level and apply not only in mining but in any industry. Examples include uncertainty, lack of training, lack of understanding about how change might affect a given job or activity, loss of job status, loss of security, and timing (summarized from Stanislao and Stanislao 1983).

To overcome this resistance requires well-thought-out and appropriate change strategies. Sometimes this calls for a catalytic role, sometimes a role in bringing people together, sometimes the provision of resources, and sometimes the design of a new and innovative solution. The key is to be strategic.

Since the 1980s, a number of factors have emerged to provide a vital and exciting climate of change in the mining industry. These factors include

- Increased public scrutiny of mining activities and the rise in the activity level, sophistication, and significance of civil society organizations;
- Public recognition of the significance of environmental and social risks and liabilities associated with mining practices and, in both regulatory and financial services industry practices, a shift that clearly assigns those liabilities to project owners and proponents;
- Changing practices in the financial services industry that are increasingly linking access to capital to a demonstration of corporate responsibility using sustainability criteria in the assessment process—a factor much reinforced by the 2008–2009 financial crisis;
- Consolidation among both operating firms and technology suppliers in response to commodity-price pressures and the desire to achieve economies of scale;
- Globalization of mining activities and the increased role of developing regions in mineral production and of mining in the developing countries' economies;
- Conflict in weak governance zones;
- Increasing worldwide consumption of key mined commodities with increasing demand in emerging economies;
- Increased dependence on metals to support a shift to a green economy; and
- Growing formal recognition that no one party can do it alone.

These drivers of change are not going to disappear; if anything, their significance will increase. The result is that the mining industry is embracing change in an unprecedented manner, and the concepts that have emerged to guide that change in a strategic and systematic way are those of sustainability and sustainable development.

ACKNOWLEDGMENTS
This chapter greatly benefited from reviews by Laeeque Daneshmend, head of the Department of Mining Engineering, Queen's University, Kingston, Ontario, Canada; Dirk van Zyl, professor of mining, University of British Columbia, Vancouver, Canada; Leigh Freeman, principal, Downing Teal, Denver, Colorado, United States; and John Strongman and colleagues in the Mining Group at the World Bank, Washington, D.C., United States. Peter Darling provided ongoing useful guidance throughout. Gemma Lee proofread and caught the many inevitable small mistakes. All of these inputs are greatly appreciated. However, while the comments received added much richness, responsibility for the substance as it stands and, in particular, any remaining errors rest with the author.

REFERENCES
Ali, S. 2009. *Treasures of the Earth Need, Greed, and a Sustainable Future*. New Haven, CT: Yale University Press.

Auty, R.M. 1993. *Sustaining Development in Mineral Economies: The Resource Curse Thesis*. London: Routledge.

Botvin, J.A., ed. 2009. *Sustainable Management of Mining Operations*. Littleton, CO: SME.

Business and Human Rights Resource Centre. n.d. www.business-humanrights.org. Accessed July 2010.

Carroll, A.B. 1991. The pyramid of corporate social responsibility: Toward the moral management of organizational stakeholders. *Bus. Horiz.* 34(4):39–48.

CASM (Communities and Small Scale Mining). n.d. Web site hosted by the World Bank at www.artisanalmining.org/index.cfm. Accessed July 2010.

Chadwick, G. 1978. *A Systems View of Planning—Towards a Theory of the Urban and Regional Planning Process*, 2nd ed. Vol. 1, Urban and Regional Planning Series. Oxford, England: Pergamon Press. p. 125.

Corporate Responsibility. 2010. Corporate Responsibility magazine releases 11th annual 100 best corporate citizens list. www.docstoc.com/docs/27223712/Corporate-Responsibility-Magazine-Releases-11th-Annual-100-Best-Corporate-Citizens-List. Accessed July 2010.

Crowson, P. 2009. The resource curse: A modern myth? In *Mining, Society and a Sustainable World*. Edited by J.P. Richards. Berlin: Springer. pp. 3–36.

Davis, G.A. 2009. Extractive economies, growth and the poor. In *Mining, Society and a Sustainable World*. Edited by J.P. Richards. Berlin: Springer. pp. 37–60.

Eftimie, A., Heller, C., and Strongman, J. 2009a. *Gender Dimensions of the Extractive Industries*. Washington, DC: World Bank.

Eftimie, A., Heller, C., and Strongman, J. 2009b. *Mainstreaming Gender into Extractive Industries Projects*. Washington, DC: World Bank.

Firey, W. 1960. *Man, Mind and Land—A Theory of Resource Use*. Glencoe, IL: Free Press. p. 20.

Freeman, L.W., and Miller, H.B. 2009. Human resource management. In *Sustainable Management of Mining Operations*. Littleton, CO: SME. pp. 133–176.

Freeze, R.A. 1987. *Technical Analysis and Social Decision-Making: The 1987 Hagey Lecture*. Waterloo, ON: University of Waterloo Press. p. 34.

Gibson, G., and O'Faircheallaigh, C. 2010. *IBA Community Toolkit—Negotiation and Implementation of Impact and Benefit Agreements*. www.ibacommunitytoolkit.ca/pdf/IBA_toolkit_March_2010_high_resolution.pdf. Accessed July 2010.

Gibson, R.B. 2000. Favouring the higher test: Contribution to sustainability as the central criterion for reviews and decisions under the Canadian Environmental Assessment Act. *J. Environ. Law Pract*. 10(1):39–54.

Hodge, R.A. 1995. Assessing progress toward sustainability: Development of a systemic framework and reporting structure. Ph.D. dissertation, School of Urban Planning, Faculty of Engineering, McGill University, Montreal, QC.

Hodge, R.A. 1996. A systemic approach to assessing progress toward sustainability. In *Achieving Sustainable Development*. Edited by A. Dale and J.B. Robinson. Vancouver, BC: University of British Columbia Press.

Hodge, R.A. 1997. Toward a conceptual framework for assessing progress toward sustainability. In *Social Indicators Research*, Vol. 40. Dordrecht: Kluwer Academic Press. pp. 5–98.

Hodge, R.A. 2004. Mining's seven questions to sustainability: From mitigating impacts to encouraging contribution. *Episodes* 27(3):177–185.

Hodge, R.A. 2006. Mining, minerals and sustainability. In *Linking Industry and Ecology—A Question of Design*. Edited by R. Cote, J. Tansey, and A. Dale. Vancouver, BC: University of British Columbia Press. pp. 151–175.

Hodge, R.A. 2007. Tracking progress toward sustainability: Linking the power of measurement and story. In *Mining Engineering*. Littleton, CO: SME.

Hodge, R.A., and Merkhofer, M.W. 2008. *Faro Mine Closure, Assessing the Alternatives: An Application of Multi-Attribute Utility Analysis*. Final Report of the Faro Closure Assessment Team. Whitehorse, YT: Assessment and Abandoned Mines Branch, Department of Energy Mines and Resources, Government of Yukon.

ICMM (International Council on Mining and Metals). 2010a. *Making a Difference: Annual Review of the International Council on Mining and Metals*. London: ICMM.

ICMM (International Council on Mining and Metals). 2010b. *Good Practice Guide: Indigenous People and Mining*. London: ICMM.

ICMM (International Council on Mining and Metals). 2010c. *Sustainability Reporting Guidelines and Mining and Metals Sector Supplement*. London: ICMM.

IFC (International Finance Corporation). 2006. *IFC Sustainability Framework: Policy on Social and Environmental Sustainability, Policy on Disclosure, and Performance Standards on Social and Environmental Sustainability*. www.ifc.org/ifcext/sustainability.nsf/AttachmentsByTitle/pol_SocEnvSustainability2006/$FILE/SustainabilityPolicy.pdf. Accessed July 2010.

IISD (International Institute for Sustainable Development). n.d. Sustainable development principles. www.iisd.org/sd/principle.aspx. Accessed October 2010.

ISO 9001. 2008. Quality Management Systems. Available from www.iso.org/iso/iso_catalogue/catalogue_tc/catalogue_detail.htm?csnumber=46486.

ISO 14001. 2004/2009. Environmental Management Systems. Available from www.iso.org/iso/iso_catalogue/catalogue_ics/catalogue_detail_ics.htm?ics1=13&ics2=20&ics3=10&csnumber=54536.

ISO 31001. 2009. Risk Management. Available from www.iso.org/iso/iso_catalogue/catalogue_tc/catalogue_detail.htm?csnumber=43170.

MAC (Mining Association of Canada). 2010. *Towards Sustainable Mining 101: A Primer*. www.mining.ca/www/media_lib/TSM_101/MAC_TSM_101_Primer_February_2010_FINAL.pdf. Accessed July 2010.

McPhail, K. 2008. Sustainable development in the mining and metals sector: The case for partnership at local, national and global levels. Essay prepared for the ICMM, London.

Mitchell, B. 1991. "Beating" conflict and uncertainty in resource management and development. In *Resource Management and Development*. Edited by B. Mitchell. Toronto, ON: Oxford University Press. pp. 268–285.

MMSD (Mining Minerals and Sustainable Development). 2002. *Breaking New Ground: Mining, Minerals and Sustainable Development*. London: International Institute for Environment and Development.

MMSD North America. 2002. *Seven Questions to Sustainability—Assessing How a Mine/Mineral Project or Operation Contributes to Sustainability*. Winnipeg, MB: International Institute for Sustainable Development.

Newmont Gold Corporation. 2009. Community Relationships Review. www.beyondthemine.com/2008/?pid=470. Accessed July 2010.

NWMO (Nuclear Waste Management Organization). 2004. *Assessing the Options—Future Management of Used Nuclear Fuel in Canada*. NWMO Assessment Team Report. www.nwmo.ca/uploads_managed/MediaFiles/1092_9-1assessingtheoptions_nwmoass.pdf. Accessed July 2010.

NWMO (Nuclear Waste Management Organization). 2005. *Ethical and Social Framework*. Roundtable on Ethics, Nuclear Waste Management Organization. www.nwmo.ca/ethicalandsocialframework. Accessed July 2010.

PDAC (Prospectors and Developers Association of Canada). 2010. e3 Plus, a framework for responsible exploration. www.pdac.ca/e3plus/index.aspx. Accessed July 2010.

Peterson, D.J., LaTourrettte, T., and Bartis, J.T. 2001. *New Forces at Work in Mining—Industry Views of Critical Technology*. Santa Monica, CA; Arlington, VA: RAND Science and Technology Policy Institute. p. 11.

Render, J.M. 2006. *Mining and Indigenous Peoples' Issue Review*. London: International Council on Mining and Metals.

Robinson, J.G., Francis, G., Legge, R., and Lerner, S. 1990. Defining a sustainable society: values, principles, and definitions. *Alternatives* 17(2):36–46.

Sachs, J.D., and Warner, A.M. 1995. Natural resource abundance and economic growth. Working paper, Cambridge, MA: Center for International Development and Harvard Institute for International Development Harvard University.

Salim, E. 2003. *Striking a Better Balance—The Final Report of the Extractive Industries Review*. Washington, DC: Extractive Industries Review. www.eireview.org. Accessed July 2010.

SD Gateway. n.d. Introduction to sustainable development: Definitions. http://sdgateway.net/introsd/definitions.htm. Accessed July 2010.

Stanislao, J., and Stanislao, B.C. 1983. Dealing with resistance to change. *Bus. Horiz.* July-August.

Thomson, I., and Joyce, S.A. 2000. Changing expectations: Future social and economic realities for mineral exploration. In *Proceedings: Mining Millennium 2000*. Toronto, ON: Canadian Institute of Mining, Metallurgy and Petroleum; Montreal, QC: Prospectors and Developers Association of Canada.

Thomson, I., and Joyce, S.A. 2006. Changing mineral exploration industry approaches to sustainability. In *Wealth Creation in the Minerals Industry: Integrating Science, Business and Education*. Edited by M.E. Doggett and J.R. Parry. Special Publication No. 12. Littleton, CO: Society of Economic Geologists.

Thomson, I., and Joyce, S.A. 2008. The social licence to operate: What it is and why it seems so hard to obtain. In *Proceedings, PDAC Convention*. Toronto, ON: Prospectors and Developers Association of Canada.

United Nations. 2003. *Handbook on Non-profit Institutions in the System of National Accounts*. www.jhu.edu/~ccss/publications/pdf/icnpo.pdf. Accessed July 2010.

United Nations Global Compact. n.d. Corporate Citizenship in the World Economy. www.unglobalcompact.org/docs/news_events/8.1/GC_brochure_FINAL.pdf. Accessed July 2010.

Watling, J., Maxwell, J., Saxena, N., and Taschereau, S. 2004. *Responsible Action—Citizens' Dialogue on the Long-Term Management of Used Nuclear Fuel*. Toronto, ON: Nuclear Waste Management Organization. www.nwmo.ca/uploads_managed/MediaFiles/877_13-1ResponsibleAction-CitizensDialogueontheLong-termManagementofUsedNuclearFuel.pdf. Accessed July 2010.

WCED (UN World Commission on Environment and Development). 1987. *Our Common Future—Report of the World Commission on Environment and Development*. London: Oxford University Press. p. 8.

Whitehorse Mining Initiative Leadership Council Accord. 1993. www.nrcan-rncan.gc.ca/mms-smm/poli-poli/gov-gov/wmi-imw-eng.htm. Accessed July 2010.

CHAPTER 16.3

Impacts and Control of Blasting

Charles Dowding

Control of blast effects near critical rock masses or constructed facilities depends on two main considerations:

1. The amount of explosives detonated per volume of rock and the shot pattern must be adjusted to ensure adequate fragmentation and relief.
2. Shot designs must reduce the amount of explosives detonated at any instant and adjust the initiation sequence to reduce resulting ground and airborne disturbances.

Therefore, the initiation sequence must be simultaneously concentrated appropriately in space but spread out in time. There is an optimum design that achieves both objectives: adequate fragmentation and control of disturbances. This optimum can be reached only through an understanding of the physics, rock mechanics, and geology of rock masses, structural and rock mass response to blast disturbance, and the interaction between rock fragmentation and shot design.

This chapter summarizes the state of the art in vibration measurement and structural response to facilitate optimum blast design. It also summarizes shot design for fragmentation. In addition, it applies the principles of earthquake engineering and nuclear blast protective design to blast-vibration monitoring and summarizes scientifically acceptable experimental observations of mining-induced ground motions and structural response. These technologies are integrated with the summary of the state of the art to mitigate several trends described in the following paragraphs.

There has been a general downward trend in regulatory limits on allowable blast-induced vibrations. This drift can be attributed, in part, to the tendency to take the limit of the last study and divide it in half to be safe. Unfortunately, too many studies whose limits were divided were themselves only summaries of past work that had also been divided past limits. This chapter presents the background of and references to original experiments conducted to determine safe blasting controls and therefore enables the reader to set appropriate limits based on original past work within the framework of existing regulations.

Another trend is the misapplication of peak particle-velocity limits that were determined for cosmetic cracking of residential structures. These limits have been applied to concrete tunnel liners, welded steel pipelines, radio towers, slabs on grade, and the curing of concrete, to name just a few instances. This chapter draws attention to studies that determined limits specifically for these and other cases. Where no studies exist, it presents methods based on response spectra or ground strains that allow setting of appropriate criteria or limits.

Frequency of vibration and ground strain form the foundation for the presentation. The importance of frequency cannot be overemphasized, as it is as critical as peak particle velocity in determining the response of aboveground structures. For belowground structures, frequency, in combination with propagation velocity, controls the response of the rock mass and embedded structures. In both cases, cracking results from induced strains, where particle velocity is used as an index of strain.

In addition, this chapter describes computerized monitoring instrumentation. Computerization simultaneously increases monitoring efficiency and decreases costs, both original capital costs and costs associated with record keeping. The labor-saving efficiencies associated with automated record keeping continue to be undervalued by many mining operations.

RANGE OF BLAST EFFECTS

Blast effects on surrounding earth materials and structures can be divided into two categories: permanent and transient displacements. Although the focus of this chapter is transient displacements, effects of permanent displacements are presented because they are associated with significant transient effects at relatively small distances.

Permanent Degradation and Displacement of Adjacent Rock

Permanent effects, with the exception of fly rock, are described within only a few hundred feet. They can be divided into two categories: degradation and displacement.

Degradation

Degradation is normally described by cracking intensity. Such blast-induced cracking has been observed experimentally to

---

Charles Dowding, Professor of Civil and Environmental Engineering, Northwestern University, Evanston, Illinois, USA

vary with hole diameter and rock type (Siskind and Fumanti 1974; Holmberg and Persson 1978). Small-hole-diameter construction blasting has induced cracking at distances of 3.5 to 7 ft (1 to 2 m). Likewise, large-hole-diameter mining blasting has induced cracking at distances of 35 to 70 ft (10 to 20 m). Careful blast design can dramatically reduce these maximum distances.

Displacement

Displacement can be produced by either delayed gas pressures (those that accumulate during detonation) or vibration-induced shaking. Delayed gas pressure has dislocated blocks as large as 1,300 yd^3 (1,000 m^3) during construction blasting (Dowding 1985). Such movement is unusual but is associated with isolated blocks, gas leaks along open joints, and poor shot design with large burdens. Vibratory or shaking-induced displacement is normally associated with unstable blocks in rock slopes and can occur wherever static factors of safety are low and ground motions produce permanent displacements that are greater than the first-order asperity wavelength (Dowding and Gilbert 1988). Gas-pressure-related displacement can occur out to several hundred feet.

Two special cases of permanent displacement exist:

1. **Fly rock.** Fly rock is a special case of permanent displacement of rock by explosive expulsion from the top of the blasthole. Rock has been observed to have been propelled as far as 330 to 3,300 ft (100 to 1,000 m) (Roth 1979). Statistical studies have shown that the probability of these extreme events is quite low under normal circumstances, 1 in 10,000,000 at 2,000 ft (600 m) (Lundborg 1981). Since the probability increases with decreasing distance, blasting mats are required in any construction blasting in an urban environment to prevent all fly rock.
2. **Soil densification or compaction.** Another special case of permanent displacement is the vibratory densification of a nearby mass of loose, clean sand. The propensity for such densification is a function of the soil's density, mineralogy, grain-size distribution, and saturation. Soils that are densifiable are loose sands with <5% of silt-size particles found below the water table. For example, such sands were densified out to distances of 70 ft (20 m) after detonation of single 11-lb (5-kg) charges within the loose sand mass itself (Ivanov 1967). Soils that are slightly cemented, contain >5% fines, or are above the water table are less subject to vibratory densification from typical ground motions.

Transient Structural Response

Transient effects result from the vibratory nature of ground and airborne disturbances that propagate outward from a blast. In this discussion, it is assumed that no permanent displacements are produced. Thus the only effects are those associated with the vibratory response of facilities in or on the rock or soil mass surrounding the blast. Transient means that the peak displacement is only temporary and lasts <1/100th of a second, and the structure returns to its original position afterward.

Transient structural effects can be arranged to reflect the expected distance from a blast. Beginning with the closest, transient effects are in this order: structural distortion, faulted or displaced cracks, falling objects, cosmetic cracking of wall coverings, excessive instrument and machinery response, human response, and microdisturbance. The first four effects, those that relate to structural response, are normally grouped together for experimental observation as structural response and do not normally occur when vibration levels are regulated to prevent cosmetic cracking.

Excessive structural response has been separated into three categories (Northwood et al. 1963; Siskind et al. 1980b). In order of declining severity and increasing distance of occurrence, beginning with effects that occur closest to the blast, the categories are as follows:

1. Major (permanent distortion): Seriously weaken the structure (e.g., large cracks or shifting of foundations or bearing walls, major settlement resulting in distortion or weakening of the superstructure, and walls out of plumb)
2. Minor (displaced cracks): Are surficial, and do not affect the strength of the structures (e.g., broken windows, loosened or fallen plaster, hairline cracks in masonry)
3. Threshold (cosmetic cracking): Open old cracks and formation of new plaster cracks, dislodging of loose objects (e.g., loose bricks in chimneys)

These specific definitions of response should not be described collectively as "damage." To do so blurs the distinction between threshold or cosmetic cracking and major response or structural distress.

Regulatory controls in North America are based on the occurrence of threshold cracking of plaster and gypsum wallboard in residential structures (Siskind et al. 1980a; Dowding 1985, 1996). Observed cracking is cosmetic in nature and does not affect structural stability. These cosmetic cracks are hair-sized and are similar to cracks that occur during the natural aging of structures. In fact, they are indistinguishable from cracks that result from natural aging. Control limits are based on direct observations of test homes immediately before and immediately after blast events, to avoid confusion with similar cracks that might occur from natural processes. These controls do not apply to engineered structures that are constructed of steel and concrete, buried structures, or adjacent rock.

Distinction Between Blast-Induced Cracking and Natural Cracking

Control of blast-induced transient effects to prevent threshold or cosmetic cracking reduces blast-induced displacement or strains in structures to or below that caused by everyday human activities and weather changes (Stagg et al. 1984; Dowding 1988). These cosmetic cracks in many cases are smaller than cracks caused by other natural or occupant-initiated processes that are active in all constructed facilities. Thus blast-induced threshold cracks can be scientifically observed only with visual inspection immediately before and after each blast. Observations made under less stringently controlled conditions have little scientific merit because of the high probability of environmentally produced cracks occurring between or before visual inspections.

Several institutional references present excellent summaries of the multiple origins of cracks (BRE 1977; Association of Casualty and Surety Companies 1956; Thoenen and Windes 1942). Basically, cracks are caused by the following:

- Differential thermal expansion
- Structural overloading
- Chemical changes in mortar, bricks, plaster, and stucco
- Shrinkage and swelling of wood
- Fatigue and aging of wall coverings
- Differential foundation settlement

Table 16.3-1 Strain levels induced by daily environmental changes and household activities with their corresponding blast levels

Loading Phenomena	Site*	Microstrain Induced by Phenomena, μin./in.	Corresponding Blast Level†	
			in./s	mm/s
Daily environmental changes	K_1	149	1.2	30.0
	K_2	385	3.0	76.0
Household activities				
Walking	S_2	9.1	0.03	0.8
Heel drops	S_2	16.0	0.03	0.8
Jumping	S_2	37.3	0.28	7.1
Door slams	S_1	48.8	0.50	12.7
Nail pounding	$S_{1,2}$	88.7	0.88	22.4

Source: Stagg et al. 1984.
*K_1 and K_2 were placed across a tape joint between two sheets of gypsum wallboard.
†Blast equivalent based on envelope line of strain vs. ground vibration.

Over time, all of the causes listed are likely to crack walls, whether or not blasting occurs. This list has three important implications: (1) Structures expand and contract preferentially along existing weaknesses (cracks). Seasonal expansion and contraction along these cracks return patching and repainting to the original cracked state within several years. This persistent cracking is annoying to owners who are unaware of the difficulty of patching existing cracks of any kind. (2) The distortion that caused the cracking also creates stress concentrations that may lower the resistance of a wall covering to vibration cracking; however, current regulatory limits already implicitly include these distortion effects as explained in the paragraphs that follow. (3) Natural cracks continue to occur over time. Therefore, any postblast inspection at low vibration levels is likely to find new cracks from natural aging unless preblast inspection is conducted immediately before the blast.

Response of Structures to Everyday Activities

A comparison of relative levels of strain gives perspective to the observation of cracking at low particle velocities. Table 16.3-1 compares strain levels from daily environmental changes (temperature and humidity) and from household activities measured in the U.S. Bureau of Mines (USBM) test house, and lists their corresponding blast levels; a door was slammed adjacent to the wall on which the strains were measured.

Apparently, in the course of daily life, an active family produces strains in walls similar to the strains produced by blasting vibrations of 0.1 to 0.5 in./s (2.5 to 12 mm/s). Most astonishing are the measurements, in a wood-framed house, of relatively enormous strains from daily changes in temperature and humidity. These strains alone are great enough to crack plaster.

Blast-Induced Air Overpressures

Blast-induced air overpressures are the air pressure waves generated by explosions. The higher-frequency portion of the pressure wave is audible and is the sound that accompanies a blast; the lower-frequency portion is not audible, but it excites structures and in turn causes a secondary and audible rattle within a structure.

Overpressure waves are of interest for three reasons: (1) the audible portion produces direct noise; (2) the inaudible portion by itself or in combination with ground motion can produce structural motions that in turn produce noise; and (3) they may crack windows, although air-blast pressure alone would have to be unusually high for such cracking. Response noise within a structure is the source of many complaints. Blast-induced air overpressures often generate sounds by means of the response of components such as loose windows and doors or highly flexible structural panels. Delicately balanced loose objects may rattle or fall at low levels of motion, generating even further concern.

Human Response

Humans are quite sensitive to motion and noise that accompany blast-induced ground and airborne disturbances. Therefore, human response is significant in the reporting of blast-induced cracking. Motion and noise from blasting can be startling and lead to a search for some physical manifestation of the phenomena. Frequently, a previously unnoticed crack provides such confirmation of the event. Furthermore, if a person is worried and observes a crack that was not noticed before, the crack's perceived significance increases over that of one noticed in the absence of any startling activity. These concerns are real and, in the mind of the observer, are sincere.

In typical mining situations, significant blast-induced inaudible air overpressure and audible noise immediately follow the ground motion and intensify human response. Both ground and airborne disturbances excite walls, rattle dishes, and together tend to produce more noise inside a structure than outside. Thus both the audible noise and the wall rattle produced by inaudible pressures contribute to human response. To complicate matters even more, inaudible air overpressures can vibrate walls to produce audible noise at large distances, and are inaccurately reported by occupants as ground motions because the audible effects are so similar.

CHARACTER OF BLAST EXCITATION AND STRUCTURAL RESPONSE

Figure 16.3-1 shows blast excitations by ground and airborne disturbances and the response of the walls and superstructure of a residential structure. Measurements were made about 2,000 ft (600 m) from a typical surface coal-mining blast. Both ground and airborne disturbances (the upper four time histories) produce structure responses (the lower four time histories). Because of the importance of frequency, the full waveform or time history should be recorded. When a critical location is known, blast response is best described by the strain at that location. Alternatively, particle velocity (shown in the figure) can be measured outside the structure of concern, as many recent cracking studies have correlated cracking with excitation particle velocity measured in the ground.

Ground Motion

Ground motion can be described by three mutually perpendicular components: longitudinal (L), transverse (T), and vertical (V) (Figure 16.3-1). The L and T directions are oriented in the horizontal plane with L directed along the line between the blast and recording transducer. When a study focuses on structural response, axes can be labeled H1, H2, and V, with H1 and H2 oriented parallel to the structure's principal axis.

Variation of peak motions in each component (L, V, and T in Figure 16.3-1) has led to difficulty in determining which is most important. Horizontal motions seem to control the horizontal response of walls and superstructures; vertical motions

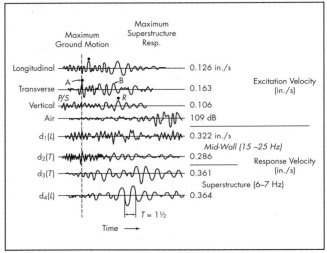

Source: Adapted from Dowding 1996.

Figure 16.3-1 Blast excitations by ground and airborne disturbances and the response of the walls and superstructure of a residential structure

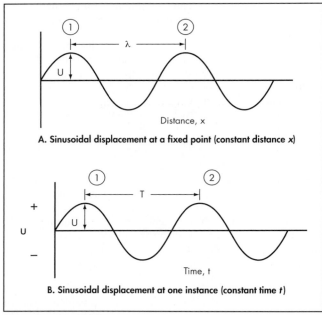

Source: Dowding 1985.

Figure 16.3-2 Sinusoidal approximations

seem to control the vertical response of floors. In an absolute sense, the peak ground motion is actually the maximum vector sum of the three components, which usually occurs at the largest peak of the three components, the dashed line in the figure. This *true* maximum vector sum is not the *false* maximum vector sum calculated with the maxima for each component (dots in the figure), no matter their time of occurrence. The false maximum vector sum may be as much as 40% greater than the true maximum vector sum, which is normally 5% to 10% greater than the maximum single-component peak.

In general, experimental observations of threshold or cosmetic cracking, which form the basis of blasting controls in North America, have been correlated with the maximum single component regardless of direction. Therefore, use of the false maximum vector sum, for control, provides a large unaccounted-for factor of safety.

Two wave types are produced by blasting, *body* and *surface*, and are illustrated by the ground motion in Figure 16.3-1. Body waves travel through earth materials; surface waves travel along surfaces and interfaces of earth materials. The most important surface wave is the Rayleigh surface wave, denoted R on the vertical trace in the figure. Body waves can be further subdivided into *compressive* (compression/tension) or soundlike waves and *distortional* or *shear* waves, denoted P/S on the vertical trace in the figure. Explosions produce predominantly body waves at small distances, which propagate outward in a spherical manner until they intersect a boundary such as another rock layer, soil, or the ground surface. At this intersection, shear and surface waves are produced. Rayleigh surface waves become important at large transmission distances, as shown in the vertical trace by the relatively large amplitude of R compared to that of P/S.

Sinusoidal Approximations

Typical blast vibrations, no matter the wave type, can be approximated as sinusoidally varying in either time or distance along the radial or longitudinal line, as shown by the time variations in Figure 16.3-2. This approximation is useful because it makes calculations for strain and for acceleration from particle velocity much simpler than they are for irregular pulses. Ground motion from a blast is similar to the motion of a floating cork caused by a passing water wave produced by dropping a stone in a pond. Displacement of the cork from its at-rest position is similar to the displacement u of a particle in the ground from its at-rest position. Similarly, the cork's velocity u as it bobs up and down is analogous to that of a particle in the ground; hence the term *particle velocity*.

The water wave that excites the cork can be described by its wavelength γ, which is the distance between wave crests; the wave speed or propagation velocity c at which it travels outward from the stone's impact; and the frequency f, which is how many times the cork bobs up and down in 1 second. Frequency f is equal to $1/T$, or the reciprocal of the period (time that it takes the cork to complete one cycle of motion). Frequency is measured in cycles per second or Hertz (Hz). Propagation velocity c should not be confused with particle velocity $\mathring{u}$, as c is the speed at which the water wave passes by the cork and $\mathring{u}$ is the speed at which the cork moves up and down while the wave passes. In the same fashion as for the water wave, blast-vibration waves can be described by their wavelength, propagation velocity, and frequency.

Kinemetric Relationships of Ground Motion

The general form for the sinusoidal approximation is best understood by beginning with the equation for sinusoidal displacement u:

$$u = U \sin(2\pi f t) \qquad (16.3\text{-}1)$$

where
  U = maximum displacement
  f = frequency
  t = time

The relationship between maximum particle displacement u, maximum particle velocity $\mathring{u}_{\max}$, and maximum acceleration $\ddot{u}_{max}$ is also greatly simplified by the sinusoidal approximation and is found by differentiation with respect to time whenever the sin/cos function maximizes at l:

$$u_{max} = U$$
$$\dot{u}_{max} = U(2\pi f) = 2\pi f u_{max}$$
$$\ddot{u}_{max} = U(4\pi^2 f^2) = 2\pi f \dot{u}_{max}$$
(16.3-2)

Usually, acceleration is normalized (i.e., divided by gravitational acceleration, 386.4 in./s² [9,814 mm/s²]). Therefore, an acceleration of 79 in./s² (2,000 mm/s²) is 2,000/9,814 = 79/386 = 0.2 g, or two-tenths that of gravity.

Kinemetric relations among particle displacement, velocity, and acceleration for complex waveforms are exactly related by integration or differentiation of any of the waveforms. For instance, integration of an acceleration time history gives a velocity time history; integration of the latter in turn gives a displacement time history. Even though a particle-velocity record can be differentiated to find acceleration, it is not recommended, as the procedure is sensitive to small changes in the slope of the velocity time history. Further discussion of the inaccuracies of differentiation and integration can be found in Dowding (1985) and in texts devoted to interpretation of time histories (e.g., Hudson 1979).

Transient Nature of Blast Motions

Great care should be taken not to confuse the effects of steady-state, single-frequency motions with those of transient blast motions. Most vibration studies conducted by personnel trained in mechanical and electrical engineering and in geophysics implicitly assume that the motions are continuous (last many cycles) and steady-state (have constant frequency and amplitude). However, as is evident in Figure 16.3-1, blast-induced motions last only one or two cycles at a relatively constant amplitude and frequency. Such conditions are not similar enough to steady-state motion to allow specific application of steady-state approximations such as resonance.

Estimation of Dominant Frequency

The use of frequency-based vibration criteria has made the estimation and calculation of dominant frequency an important concern. Dominant frequency can be estimated by visual inspection of the time history or calculated from Fourier frequency spectra or, alternatively, response spectra.

The accuracy or difficulty of visually estimating the dominant frequency depends on the complexity of the time history. The easiest type of time-history record for estimating frequency is one with a single dominant pulse like that shown in the inset of Figure 16.3-3. This dominant frequency can be determined by hand measurement of the time of the two zero crossings on either side of the peak. The difference between these times is one-half of the period, which is the inverse of twice the frequency of the dominant peak, as show in the inset.

Figure 16.3-3 shows that the relatively large blasts produced by surface coal mining, when measured at typically distant structures, tend to produce vibrations with lower principal frequencies than do construction blasts. Construction blasts involve smaller explosions, but the typically small distances between a structure and a blast, as well as rock-to-rock transmission paths, tend to produce high frequencies. The high-frequency motions associated with construction blasts are less apt to crack adjacent structures (Dowding 1985).

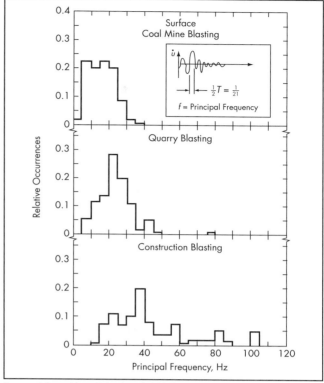

Source: Adapted from Siskind et al. 1980b.

Figure 16.3-3 Dominant-frequency histograms of blasts for three types of industry, measured at structures adjacent to the blast (the inset defines dominant frequency)

The most difficult type of record to interpret is one that contains nearly equal peaks at two dominant frequencies, such as that shown Figure 16.3-1. The two dominant frequencies are the initial 15- to 20-Hz portion (peak A) and the later 5- to 10-Hz portion (peak B). The initial portion produces the highest wall response; the second portion produces the greatest superstructure response. For the best frequency correlation of both types of response, both frequencies should be calculated.

The best computational approach to determining the dominant frequency involves the response spectrum. The response spectrum is preferred over the Fourier frequency spectrum because it can be related to structural strains (Dowding 1985). A compromise approach is to calculate the dominant frequency associated with each peak by the zero-crossing method described previously.

Since histories do not often contain as broad a range of dominant frequencies as that in Figure 16.3-1, most approaches require only the calculation of the frequency associated with the maximum particle velocity for blasts that produce small particle velocities. More complex frequency analyses are needed only when peak particle velocities approach control limits.

Propagation Effects

Ground motion always decreases with increasing distance. The effects of constructive and destructive interference and geology are included in the scatter of data for the mean trend of the decay of amplitude with distance. Although this scatter is

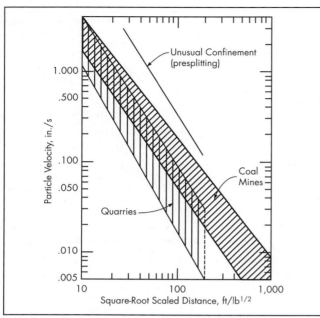

Source: Adapted from Siskind et al. 1980b.

Figure 16.3-4 Attenuation relationships showing scatter from geological and blast design effects as well as high expected velocities from confined shots such as presplitting

large, the associated decay with distance is observed in all blast-vibration studies. Typical examples of this decay are shown in Figure 16.3-4 where maximum particle velocity is plotted as a function of square-root-scaled distance from the blast.

Square-root scaling, or plotting peak particle velocity as a function of the distance R divided by the square root of the charge weight, $R/W^{1/2}$, is more traditional than cube-root scaling, which incorporates energy considerations (Hendron 1977). Both square- and cube-root scaling can be used to compare field data and predict the attenuation or decay of peak particle velocity; however, square-root scaling is more popular.

Several square-root attenuation relationships used in the United States are shown in Figure 16.3-4. They are banded to reflect scatter, which is typical of blasting operation. The upper (unusual confinement) line should be used for presplitting, cratering, and beginning new bench levels. It is also the basis for U.S. Office of Surface Mining Reclamation and Enforcement (OSM) regulations for conservative shot design when monitoring instruments are not used.

Dominant frequencies tend to decline with increasing distance and increasing importance of surface waves. At larger distances typical for mining, higher-frequency body waves begin to have relatively lower peak amplitudes than do lower-frequency surface waves, as shown in Figure 16.3-1. Since lower frequencies can induce greater structural response (Medearis 1976), OSM scaled-distance limits decline with increasing absolute distance.

Blast-Induced Air Overpressures

As for ground motions, the waves from blast-induced air over-pressures can be described with time histories as shown in Figure 16.3-1. The higher-frequency portion of the pressure wave is audible sound. The lower-frequency portion, although not audible, excites structures, causing a secondary and audible rattle within the structure. Air-blast excitation of the walls is evident from a comparison of the last one-quarter of the time histories of air blast and wall response in Figure 16.3-1. Unlike ground motions, air overpressures can be described completely with only one transducer, since at any one point air pressure is equal in all three orthogonal directions.

Propagation of blast-induced air overpressures has been studied by numerous investigators and is generally reported with cube-root-scaled rather than square-root-scaled distances. Peak pressures are reported in terms of decibels (dB), defined as

$$dB = 20 \log_{10}(P/P_0) \qquad (16.3\text{-}3)$$

where
P = measured peak sound pressure
P_0 = reference pressure of 2.9×10^{-9} psi (20×10^{-6} (P_a)

Figure 16.3-5 summarizes the effect of two important instrumentation and shot variables:

1. The effect of the weighting scales in the sound-pressure-level meter is dramatically evident. C weighting greatly reduces the recorded peak pressure at any scaled distance. This does not mean that the peak is reduced by changing instruments, but rather that the C weighting system does not respond to low-frequency pressure pulses. However, low-frequency pressure peaks excite structures and occupants whether or not they are sensed by measurement instruments. The other (5- and 0.1-Hz) labels denote the lower-frequency bounds of the recording capabilities of "linear" systems.
2. The effect of venting caused by inadequate stemming is also evident from the higher average pressures produced by the parting shots at any scaled distance. Parting shots are detonated in thin rock layers between coal strata in surface mines. Consequently, there is less hole height available for stemming, and these shots frequently eject the stemming and thereby produce abnormally high air overpressures. The unconfined relationship should be used for the demolition of structures after modification for the effects of weather and ground reflection.

Various effects of wind have been reported and should be added to the average relations presented in Figure 16.3-5. Wiss and Linehan's (1978) study of air overpressures produced by surface coal mining show that, in moderate winds, the typical 7.7-dB reduction for each doubling of distance is reduced by

$$7.7 - 1.6\, V_{mph} \cos\theta \text{ dB} \qquad (16.3\text{-}4)$$

where
V_{mph} = wind velocity, in miles per hour
θ = the angle between the line connecting the blast and transducer and the wind direction

An air-temperature inversion causes the sound-pressure wave to be refracted back to the ground and at times to be amplified at small 16-acre- (0.65-km^2-) sized locations. Such inversions occur when the normal decrease in temperature with altitude is reversed because of the presence of a warmer upper layer. Schomer et al. (1976) showed that, for propagation distances of 2 to 40 mi (3 to 60 km), inversions produce zones of intensification of up to 3 times the average attenuated or low air overpressures at those (but not smaller) distances;

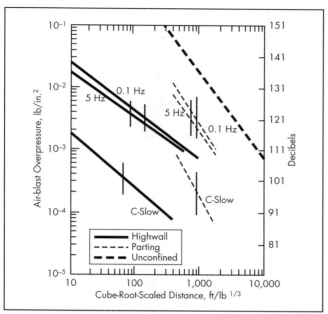

Source: Adapted from Siskind et al. 1980a.

Figure 16.3-5 Attenuation relationships for air overpressures produced by confined (highwall) and partially confined (parting) surface coal-mining blasts and unconfined blasts

the average increase is 1.8 times (5.1 dB). At distances of <2 mi (3 km), where high air overpressures are likely to occur, his measurements show no inversion effects.

Structural Strains vs. Particle Velocity

Although particle velocity is the traditional measurement of choice, structural strains control cracking. They should be measured directly from relative displacements on structures or within rock masses when critical locations are known, and can be obtained with a variety of strain and relative displacement gauges (Stagg et al. 1984). Unfortunately, critical locations may be either unknown or too many in number to measure economically. Therefore, some means of estimation is necessary.

Ground-motion and air-overpressure time histories can be used to calculate the relative displacement of structural components with a knowledge of the responding structure's dynamic-response characteristics (Dowding 1985, 1996). These relative displacements can in turn be used to calculate strains. The accuracy of these estimates is limited by the degree to which the structure behaves as a single-degree-of-freedom system and the accuracy of the estimate of dynamic response characteristics.

Measurement of Particle Velocity

Although any of the three kinematric descriptors (displacement, velocity, or acceleration) can be used to describe ground motion, particle velocity is the most preferable. It has the best correlation with scientific observation of blast-induced cracking, which forms the basis of vibration control. Furthermore, it can be integrated to calculate displacement. If acceleration is desired, it should be measured directly to avoid differentiation of the particle-velocity time history.

The location for measurement varies throughout the world. In North America, excitation or ground motion is measured on the ground adjacent to the structure of interest.

In Europe, it is measured on the structure's foundation. The difference stems from historical precedent and location of transducers during scientific observation of cracking rather than from difference in philosophy. In North America, it is frequently impossible to place transducers on adjacent property owned by a party not involved in the project. Excitation motions should be measured outside of and not on the structure. Structural-response motions should be measured on the most responsive structural members, which are not the basement or foundation walls because of the restraint provided by the ground.

Time histories of the three components of motion should be measured because of the importance of excitation frequency. Recording of peak motions alone does not yield information about the dominant frequency and time history details that control structural response. Peak motions and dominant frequency can be used to describe low-level, noncritical motions. Therefore, machines used to monitor critical motions should be capable of recording time histories of selected critical motions. Machines that record only peak motions can be used with those that record time histories to provide redundant measurement where frequency content does not vary widely.

USE OF MEASUREMENT INSTRUMENTS

This section describes the characteristics of instruments that measure ground motions (acceleration, velocity, displacement) and air blasts (air overpressure). Since there are many excellent sources for information on instruments, the principal characteristics of available systems are summarized rather than exhaustively reviewed. The most complete single reference for detailed instrumentation information is the *Shock and Vibration Handbook* (Harris and Crede 1976).

Almost all field-portable vibration-monitoring systems are now built around a digital central processing unit (sometimes called a data logger) and an analog-to-digital (A/D) converter. Geophone or air-pressure transducers convert time-varying velocity ground motion or air pressure into an analog (or continuous) voltage signal. The A/D converter translates this analog signal into digital (or discrete) form at a constant rate of sampling (typically 1,000–2,000 samples per second) and saves the data to a file in system memory. The onboard computer or central processing system analyzes the data and generates reports for download to and display on a portable computer. Typical A/D converters subdivide the transducer voltage range into 2^{12} (4,096) or more divisions; some new systems are capable of 2^{24} subdivisions. Every year, sampling speed increases and A/D converters become more precise. Some of these systems are self-contained and are the size of a brick. Others connect to exterior transducers and are the size of a thick notebook computer. In development are systems the size of a deck of playing cards that may eventually replace current instruments, provided that smaller batteries become available.

Transducers

Transducers are one of the weaker links in the measurement system because they must translate kinematric motions or pressures into electrical signals. The remaining components transform electrical signals or light beams and are not restricted by mechanical displacement. The main characteristics of transducers that affect their performance are sensitivity and frequency response.

For energy-converting transducers (i.e., those that do not require an energy source), instrument sensitivity is the ratio of the electrical output to the kinemetric displacement, velocity, and acceleration or overpressure. Since allowable limits are specified in terms of ground-particle velocity, all blast monitors come equipped with velocity gauges.

Frequency response is the frequency range over which electrical output is constant with constant mechanical motion. This constancy is normally expressed in terms of decibels. For instance, "linear within 3 dB between 5 and 200 Hz" means that the transducer produces a voltage output that is constant within 30% between 5 and 200 Hz, and up to 450 Hz on some European machines. Generally, it is better to look at the transducer's response spectra (such as those shown in Figure 16.3-6) to determine the frequencies where this difference occurs. For example, the difference occurs at low frequencies for the velocity transducers in the figure. At 70% of critical damping, the velocity transducer in the figure is ±3 dB (±30%) down at 1 Hz.

One of the most critical aspects of vibration monitoring is how the transducer is mounted in the field, which depends on the particle acceleration of the wave train being monitored. When the measurement surface is horizontal and the vertical maximum particle acceleration is <0.3 g, the type of mounting is not particularly critical, as the possibility of rocking is small and the transducer needs no device to hold it. When the measurement surface consists of soil and vertical maximum particle acceleration is in the range 0.3–1.0 g, the transducer should be buried completely (Johnson 1962). When the measurement surface consists of rock, asphalt, or concrete, the transducer should be fastened to the measurement surface with double-sided tape, epoxy, or quick-setting cement (Hydrocal or other gypsum-based cements set in 15–30 minutes). If these methods are unsatisfactory or accelerations exceed 1.0 g, only cement or bolts are sufficient to hold the transducer to a hard surface.

All transducers mounted on vertical surfaces should be bolted in place. Air-overpressure transducers should be placed at least 3 ft (1 m) aboveground, pointed downward to prevent rain damage and fitted with a windscreen to reduce wind-excitation-induced false events.

Instrument Calibration

It is obvious that the entire vibration-measurement system should be calibrated, as it is futile to record data if they cannot be exploited because of a lack of reference. Manufacturers supply calibration curves with their instruments that are similar to the response spectra for transducers shown in Figure 16.3-6. Recalibration or checking requires special vibrating platforms where frequency and displacement are controlled, and, in the field, a calibrating circuit to pulse the magnetic core of the geophone (Stagg and Engler 1980).

Number of Instruments

Although the obvious irreducible number of instruments for each blast is one, two instruments provide more thorough documentation of the spatial distribution of effects. If only one instrument is used, it should be located at the nearest or most critical receiver. This single instrument should be what we term a Type I instrument and should record time histories of the three axes of particle velocity as well as air overpressure. Since it must monitor continuously, it must trigger (begin recording) automatically and be capable of monitoring even while printing or communicating results.

When blasting will occur at more than one general location (i.e., will involve nearest structures separated by

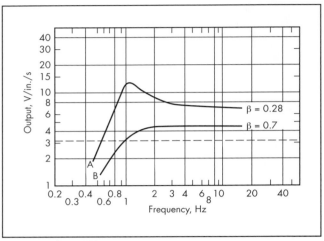

Source: Dowding 1985.

Figure 16.3-6 Sample response spectra for velocity transducers with differing percentages of damping

hundreds of feet or meters), then minimally two and optimally four instruments should be used, including a spare to ensure continuous coverage in case of failure.

Additional instruments other than the spare can be Type II, which provide a lesser level of service. They must at least continuously record the peak particle velocity in one axis and may or may not measure air overpressure. The best axis is the vertical, since no horizontal-direction decision is required and surface waves usually involve a significant vertical component regardless of the direction of the maximum horizontal component. These instruments should be located at distances greater than the nearest structure to monitor a large area.

The spare instrument can be either Type I or II. Where air overpressures are problematic or frequencies are critical, the spare should be Type I. The spare instrument can also be used to monitor sites where complaints develop. Such public-relations work is essential in North America where lawsuits arise even when all blast effects comply with regulatory guidelines.

This approach describes the least number of instruments. Applicable regulations and mining schedules may require a larger number.

Instrument Deployment During Test Blasts

When blasting projects begin, when geological conditions change radically, or when new initiation systems are introduced, test blasts should be conducted to minimize the number of instruments necessary to monitor production blasts. The test blasts should produce project-specific attenuation relationships for both air overpressures and ground motion. Such relationships vary from project to project because of changes in geology and blasting practices. Additionally, the test blasts should enable determination of the frequency of motions at different scaled and absolute distances. Frequency is important in estimating structural response by response-spectrum analyses.

The attenuation relationship is not solely a site property. Although it depends on geology, it also depends heavily on blast geometry and timing. For instance, with the same charge per delay, a blast with a larger burden produces an attenuation relationship with a similar slope or decay with distance but a larger intercept. Furthermore, differing initiation timing changes the time history in terms of both length and frequency.

During test blasts, at least four instruments should be used to measure peak particle velocities at widely differing scaled distances for the same blast. Therefore, for any one blast design, parameters and initiation sequences are constant, and the resulting attenuation relationship shows the effect of only distance, direction, and/or geology. Seismographs and/or transducers should be placed along a single line with constant geology to best determine the attenuation relationship, or at all critical structures to determine the effects of direction and variable geology. Ideally, linear orientation should be along a path with constant soil thickness and not cross any large geologic discontinuities such as faults. If geology changes radically, then two such attenuation lines are necessary, but not necessarily with each blast.

EVALUATION OF MEASUREMENTS

Documentation of blast effects involves two radically different endeavors: measurement of ground and air disturbances as well as observation of cosmetic cracking. Measurement can now be accomplished remotely with computers to eliminate completely human interaction, whereas scientific observation must involve meticulous human inspection immediately before and after a blast. While the focus of this section is instrumental monitoring, the alleged appearance of cracking by neighboring property owners is nonetheless a very serious consideration.

Principal problems in the evaluation of measured effects involve (1) accounting for geologic and weather effects on the overall attenuation with a small number of instruments and (2) incorporating structural response and frequency effects. Principal problems with the observation of blast-induced cracks involve (1) discriminating between blast-induced cracks and environmentally and human-induced cracks and (2) reducing the enormous amount of time necessary for direct observation. Observational problems are normally overcome by using instrumentally measurable blasting controls at levels that are low enough to avoid the threshold of cosmetic cracking even for old, degraded structures or to eliminate cracking altogether. Otherwise blast-induced cracks can be observed only by immediate before-and-after blast inspection.

Structural Response and Frequency Effects

Structures respond to both ground and airborne disturbances, as shown by the bottom four time histories in Figure 16.3-1. Walls respond more to the higher-frequency (15–20 Hz) waves that occur early during ground motion; the superstructure or overall skeleton of the structure responds more to the lower-frequency (5-Hz) waves that occur late during ground motion. Walls are again excited by the arrival of the air-pressure wave. Structural response can be calculated from ground motion if the natural frequency and damping of structural components are known or estimated.

Langan (1980) showed that measured structural response has a higher correlation with calculated single-degree-of-freedom (SDF) response than with peak ground motion. Therefore, structural motions can be estimated more accurately by assuming that they are proportional to response-spectrum values at a particular structure's natural frequency than by assuming that they are proportional to the peak ground motion. This improved correlation is largely a result of the consideration of frequency in the response spectrum, which is calculated from the SDF response.

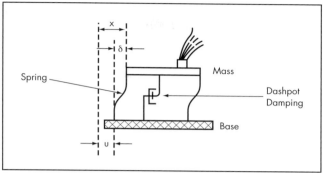

Source: Dowding 1985.

Figure 16.3-7 Single-degree-of-freedom model of a house, showing the relative displacement of the walls δ, and the analogy between model mass and roof mass, model stiffness, and wall stiffness

Origin of the SDF Model

One of the critical structural response factors is the amount of differential displacement (δ in Figure 16.3-7) that occurs between or along structural members, because it is proportional to strain, which, in turn, causes cracking. Such displacements can be computed by mathematical idealization of the SDF model shown in the figure.

To make calculations practical, it is necessary to simplify most structures. The fundamental characteristics of a structure that govern its behavior under vibratory or dynamic loading are (1) the masses of the main components (analogous to floor and roof masses), (2) the spring stiffness of the main components (analogous to wall stiffness), and (3) the amount of damping or energy dissipation (analogous to differential movement in cracks, joints, and connections). The behavior of one- and two-story buildings is directly analogous to the behavior of an SDF system when movement in only one direction is considered. For multiple-story structures, it is necessary to model the structure as multiple-degrees-of-freedom systems. However, even such systems can be idealized as SDF systems to enable calculation of the fundamental mode of response.

If a structure's damped natural frequency f_d and its fraction of critical damping β are known, the values of dynamic properties f_d and β can be accurately measured from a free-vibration time history of the building response. These measured parameters automatically account for factors that are difficult to quantify, such as the degree of fixity of the columns and the damping coefficient. As shown in Figure 16.3-1, these parameters can be measured from the structure's free response. The time between peaks is the period $T = 1/f_d$ and the decay of free oscillation is proportional to the damping β.

Estimation of Dynamic Response Properties

The fundamental natural frequency f_d of the superstructure of any tall building can be estimated from compilations of work in earthquake engineering (Newmark and Hall 1982):

$$f_d = 1/0.1N \qquad (16.3\text{-}5)$$

where N is the number of stories. Substitution for N of 1 and 2 for residential structures yields f values of 10 and 5 Hz for one- and two-story structures, respectively, which compares favorably with the results of actual measurements.

Table 16.3-2 Natural frequencies for unusual structures

Type of Structure	Height, m	f, Hz
Radio tower	30	3.8
Petroleum distillation tower	21	1.2
Coal silo	60	0.6
Bryce Canyon rock pinnacle	27	3

Source: Medearis 1975; Dowding and Kendorski 1983.

Damping β is a function of building construction and, to some extent, the intensity of vibration. Thus it cannot be simplified as easily as can the natural frequency. Measurements reveal a wide range of damping values for residential structures, with an average of 5% (Dowding et al. 1981). This value is appropriate for initial estimates for taller engineered structures. Further details for engineered structures are available in Newmark and Hall (1982).

Walls and floors vibrate independently of the superstructure and have their own but similar fundamental frequencies of vibration in the range 12–20 Hz and averaging 15 Hz (Dowding et al. 1981). Floors in office buildings with large floor spans tend to have natural frequencies that are lower than those of floors in residential structures, but similar to those of walls in residential structures.

Dynamic-response properties of some unusual tall structures cannot be estimated with the $1/0.1N$ equation (Equation 16.3-5). Field-measured natural frequencies for these types of structures are given in Table 16.3-2.

Response Spectra

The pseudo-velocity response spectrum of a single ground motion, such as that of the seven-delay quarry blast in Figure 16.3-8, is generated from the relative displacement δ_{max} values of a number of different SDF systems when excited by that motion. In the figure, the time history is operated on by the SDF equation to produce a computed relative displacement δ, which is then multiplied by the circular natural frequency ($2\pi f$ or $2\pi 10$ for point 1) to produce the pseudo-velocity response PV.

Consider two different components of the same structure: a 10-Hz superstructure and a 20-Hz wall. If the ground motions $\ddot{u}(t)$ of the quarry blast are processed twice by the SDF equation with $f = 10$ and 20 Hz and β is held constant at 3%, two δ_{max} values result. For the 10-Hz system, the associated values are

$$p = 2\pi(f) = 2\pi(10)$$

$$\delta_{max} = 0.25 \text{ mm } (0.01 \text{ in.}) \quad (16.3\text{-}6)$$

$$PV_{10} = \delta_{max} = 2\pi(10)(0.25) = 15.7 \text{ mm/s } (0.62 \text{ in./s})$$

For the 20-Hz system, the associated values are

$$p = 2\pi(f) = 2\pi(20)$$

$$\delta_{max} = 0.5 \text{ mm } (0.02 \text{ in.}) \quad (16.3\text{-}7)$$

$$PV_{20} = 2\pi(20)(0.5) = 63.5 \text{ mm/s } (2.5 \text{ in./s})$$

The PV values for the two systems are shown in Figure 16.3-8A as points 1 (PV_{10}) and 2 (PV_{20}). If the same ground-motion time history in Figure 16.3-8B is processed a number of times for a variety of f values and β is held constant, the resulting pseudo-velocities form the solid line in Figure 16.3-8A.

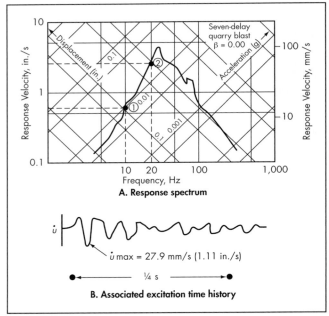

Source: Dowding 1985, 1986.

Figure 16.3-8 Construction of a pseudo-velocity response spectrum

Computers enable easy calculation of spectra from time histories. The two most commonly used spectra are the Fourier frequency and the pseudo-velocity response spectra. Although they are essentially different in meaning and typical use, they are similar for undamped response where the maximum motion occurs near the end of the time history (Dowding 1985; Hudson 1979). Since response spectra are calculated for damped response and peaks normally occur in the middle as well as the beginning of the time history, the two spectra are not usually the same.

Only the pseudo-velocity response spectrum can be used to directly calculate structural response. Because of the similarity of Fourier frequency and response spectra, either can be used to determine the dominant frequency in the ground motion.

Case Histories

Figure 16.3-9 compares time histories and response spectra from the longitudinal components for an urban construction blast and a surface coal mine blast. Although the peak particle velocities are similar—0.15 in./s (3.8 mm/s) for the construction blast (Figure 16.3-9A), and 0.13 in./s (3.3 mm/s) for the coal-mine blast (Figure 16.3-9B)—the response spectra in the frequency range 5–20 Hz differ greatly. This difference is greatest in the range of natural frequencies for residential structures and their components, 5–20 Hz. In this range, surface-mining motions produce response velocities that are 10 times greater than those for construction motions.

Urban Construction Blast

Ground motion induced by the construction blast was produced by a much smaller shot than that induced by the surface-mining blast. Five holes (12 ft [3.6 m] deep, 1.5 in. [38 mm] in diameter) were arranged in a single row. Each hole was charged with a stick of gelatin dynamite and initiated

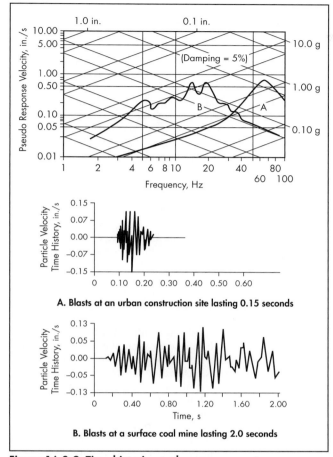

Figure 16.3-9 Time histories and response spectra

separately after a delay held constant at 25 ms. The burdens and hole spacings were small, approximately 2–3 ft (0.6–0.9 m). The total charge was 20 lb (9 kg), and the maximum charge per delay was 5 lb (2.3 kg).

The structure of concern, a historic theater, and the recording transducers were located 50 ft (15 m) away. The rock being fragmented consisted of granitized biotite schist. The shot produced peak particle velocities along the L, T, and V axes of 0.15, 0.16, and 0.28 in./s (3.8, 4.1, and 7.1 mm/s), respectively, with dominant frequencies between 75 and 125 Hz.

Surface Mine Blast

Ground motion induced by the surface-mining blast was produced by a multiple-row blast. Some 60 holes (83 ft [25 m] deep, 15 in. [380 mm] in diameter) were arranged in a four-row pattern. The burden between rows was 20 ft (6.1 m) and the hole spacing was 25 ft (7.6 m). Each hole contained four decks (charges that are detonated after an interval held constant at ≥17 ms). The ammonium nitrate and fuel oil (ANFO) charge weight per deck ranged from 100 to 130 lb (45 to 60 kg). Therefore, the largest charge per delay was 130 lb (60 kg), and the total charge was 27,700 lb (12,600 kg).

The geology between the blast and the transducer, which was located at the nearest residence, consisted of sedimentary rock with 10–30 ft (3–10 m) of overlying silty glacial till. A small gully (33 ft [10 m] deep) was located 1,300 ft (400 m) north of and between the blast and the transducer. The soil at the transducer was 10–15 ft (3–4 m) deep. Some 2,750 ft (825 m) separated the shot from the transducer, where the longitudinal velocity time history in Figure 16.3-9 with a peak of 0.13 in./s (3.3 mm/s) was recorded. The accompanying transverse and vertical peak particle velocities were 0.18 and 0.09 in./s (4.6 and 2.3 mm/s) with dominant frequencies between 11 and 13 Hz. Other similarly designed shots with distances between 1,900 and 2,700 ft (580 and 825 m) produced peak particle velocities between 0.17 and 0.23 in./s (4.3 and 5.8 mm/s) with dominant frequencies between 13 and 17 Hz.

These particle velocities are high at 2,700 ft (825 m), according to scaled-distance relationships. The high values may result from unusual confinement (too large a burden) of delay overlap. For instance, if two delays overlap so as to add, then the maximum charge per delay doubles and the square-root-scaled distance declines by 30%. Particle velocities measured at smaller-scaled distances are closer to expected levels.

Restrained Structures and Rock Masses

Aboveground structures such as homes and rock pinnacles, due to their capacity for free response, can selectively amplify incoming ground motions. In contrast, buried or restrained structures such as pipelines and rock masses cannot respond freely. Regardless of whether the response to strain is restrained or free, however, cracks are still initiated by strains. Strains in freely responding structures are proportional to the relative displacement between the ground and the superstructure, as shown in Figure 16.3-7; strains in restrained structures such as pipelines are usually those of the surrounding ground and can be approximated as the strain produced by plane-wave propagation (Dowding 1985, 1996):

$$\varepsilon = \mathring{u}_c/c_c \text{ and } \gamma = \mathring{u}_s/c_s \quad (16.3\text{-}8)$$

where

ε = axial strain
$\mathring{u}_c$ = maximum compressive-wave particle velocity
c_c = compressive-wave propagation velocity
γ = shear strain
$\mathring{u}_s$ = maximum shear-wave particle velocity
c_s = shear-wave propagation velocity

For cases involving one critical location along a pipeline, pipe strains should be measured directly on the metal. For cases involving tunnels and/or cavern liners, critical strains can be estimated by calculation of the relative flexibility of the rock and liner (Hendron and Fernandez 1983).

CONTROL OF BLAST EFFECTS

When discussing blast design, it is important to mention the significance of burden. The two most familiar blast design parameters, charge per delay and powder factor, do not optimize the absolute burden between holes. Charge per delay (or instant of time) distributes the detonation process in time; powder factor distributes the detonation process in space. Unfortunately, powder factor alone cannot be used to calculate the maximum or optimum burden, which often controls external blast effects such as ground motion. Burden is often "designed" by general rules such as "k times borehole diameter." Proportioning maximum burden by borehole diameter tends to promote larger-diameter holes for a given powder factor, since the larger the hole diameter, the larger the burden and thus the fewer the number of holes per unit volume and the lower the drilling cost. Unfortunately, large burdens for a given geology and rock type can lead to too large a separation of blastholes, which can lead to excessive back break and vibration.

One simple factor that can be used to check shot designs for excessively large burdens is the so-called drill factor, equal to (holes per unit surface area times hole depth)/unit volume. Experience with pipeline trench blasting reveals that shots designed with powder factors of 5.4 and 3.4 lb/yd^3 (3.2 and 2.0 kg/m^3), which does not seem like much of a difference, can produce vastly different results. These two shots involved 25 and 5 lb (11.4 and 2.3 kg) of explosive per running foot (0.3 m) of 16- and 10-ft- (5- and 3-m-) deep trenches and 76 and 10 lb (34.5 and 4.5 kg) of explosive per hole, respectively. The larger shot cratered the rock and damaged pipe 21 ft (6.4 m) away, whereas the more optimally designed shot did not crater or damage pipe 9 ft (2.7 m) away. The drill factors for these two shots are 1.1 and 3.4 ft/yd^3 (1.3 m/m^3), respectively.

Direct regulation or specification of effects, rather than shot design, is the most effective control from a regulatory viewpoint because effects are so dependent on the details of shot geometry and initiation sequence. Such dependency renders control impossible by simple regulatory specification of two- or three-shot design parameters. For instance, consider control by specification of the maximum charge weight detonated per instant at given distances from the nearest structure. Even with such detailed specification, intended vibration limits at the structure may be exceeded because of poor choice in the location of holes and/or the relative times of initiation.

Present regulatory control limits in many countries are below the levels at which cosmetic cracking may appear. There are two principal reasons for such tight restrictions. First, regulatory limits are influenced heavily by human response to blast-induced vibration and noise. Since humans are approximately 10 times more sensitive than structures to vibration, low regulatory limits are understandable. Second, many regulations appear to have been adopted without the documented scientific experimentation necessary to determine the vibration levels that cause cracking.

Statistical Analysis of Data by Pre- and Postblast Inspection

Unmeasurables in observation can be taken into account indirectly by considering the appearance of cosmetic cracks as a probabilistic event. As an aid in determining the effects of certain data sets on overall conclusions, the probabilities of cracking at various different particle-velocity levels have been calculated several times (Siskind et al. 1980b; Siskind 1981). These studies involve both immediate pre- and postblast inspection of walls in residential structures, many of which were old and distorted, and whose walls were covered with plaster. The definitions of observed cracking in each study are described previously in the "Range of Blast Effects" section.

Data from various sets of observations were analyzed with cracking points and the assumption that every cracking point excludes the possibility of noncracking at a higher particle velocity (Siskind et al. 1980b). If the probability of cracking is calculated as the percentage of points at lower levels of velocity, the result is the log-normal scaled plot of the probability of cracking vs. particle velocity in Figure 16.3-10, where threshold damage is the occurrence of hair-sized cosmetic cracks similar to those caused by natural expansions and contraction. This approach seems conservative as low particle-velocity observations do not count noncracking at higher levels.

According to Figure 16.3-10, there appears to be a lower limit of particle velocity of 0.5 in./s (12 mm/s) below which no cosmetic or threshold cracking (extension of hairline cracks) has been observed from blasting anywhere in the world. This observation includes the data with unusually low frequencies that were collected by Dvorak (1962). His data are those that tend to populate the lower region of the figure. High-frequency data (>40 Hz) show that a 5% probability of minor cracking does not occur until particle velocities reach 3 in./s (75 mm/s) (Siskind et al. 1980b).

The admissibility of Dvorak's data has been questioned because of the absence of time histories; some other studies, such as that by Langefors et al. (1958), are plagued by the same problem. To resolve this difficulty, only new USBM observations have been included in a recomputation of probabilities in Figure 16.3-11. The observations include low-frequency motions associated with surface mining. Again, there is a particle velocity, 0.79 in./s (20 mm/s), below which no blast-induced cracking was observed.

Comparison of Blast and Environmental Effects

Changes in crack width caused by ground motions of <1 in./s (25 mm/s) are smaller than those caused by the passage of weekly weather fronts (Dowding 1988). This conclusion was reached after measuring, for 8 months, the displacement response on a poorly built house located close to surface-coal-mining vibrations. Displacements were measured at 10 different wall positions that included cracked and uncracked wall coverings. Weather- and blast-induced crack displacements across the most dynamically responsive wall-covering cracks are compared in Figure 16.3-12. The continuous and highly cyclical curve is that of displacements produced by environmental changes. The small circles are the maximum zero-to-peak dynamic displacements recorded by the same gauge. Even though the maximum recorded particle velocity is as high as 0.95 in./s (24 mm/s), the maximum weather-induced displacements are three times higher than those produced by blasting. On other gauges, weather changes produced displacements that were 10 times higher than those produced by blasting. This crack response is typical of those for the more than 30 cracks in over 20 studies reviewed by Dowding (2008).

Special Considerations

Statistically determined control limits are too low for basement walls and engineered structures. They are based on the response of residential structures and on lower-limit cases involving cracking of aboveground plaster or gypsum wallboard coverings in older distorted structures.

Engineered Structures

Concrete is much stronger than plaster. Therefore, engineered structures constructed of concrete can withstand maximum particle velocities of at least 4 in./s (100 mm/s) without cracking (Crawford and Ward 1965). Furthermore, buried structures such as pipelines and tunnel linings are not free to respond as are aboveground residential structures whose response provides the data from which most limits are chosen. Therefore, underground structures are able to withstand even greater excitation motions (Dowding 1985).

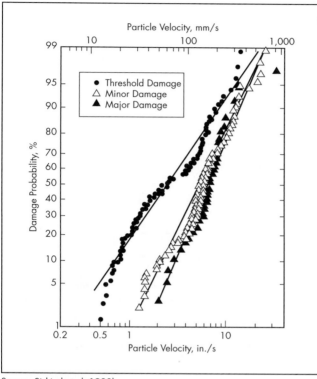

Source: Siskind et al. 1980b.
Figure 16.3-10 Probability analysis of worldwide blast cracking data

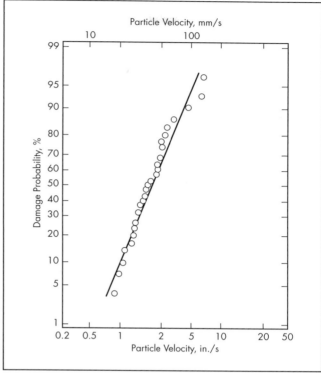

Source: Siskind 1981.
Figure 16.3-11 Probability analysis of blast-induced threshold cracks observed by the USBM

Specific engineered structures should be analyzed in terms of the strain that can be withstood by critical elements and the strain should be measured. This approach is particularly appropriate for singular structures with isolated portions near the blast source, such as buried pipelines. Often the ground and pipe strains can be confirmed by directly measuring the strains on the buried pipes, as was done by the USBM (Siskind et al. 1994). Their measurements confirmed that welded steel pipelines can withstand ground motions considerably above the typical control limit of 50 mm/s. Oriard (2002) believes that even the high allowable peak particle velocities reported by Siskind et al. are conservative for welded steel pipelines. Recently, use of measured strains at critical locations within engineered structures has been adopted for blasting immediately beneath steel-frame structures. Strain gauges attached to the steel H-beam cores of the building columns confirmed that high peak particle velocities were not causing high strains.

Fatigue or Repeated Events

Since current regulatory limits are so low as to restrain blast-induced displacements below those caused by the passage of weekly weather fronts, the question of repeated events becomes moot. Weather by itself over the years produces greater repeated-event effects than does blasting.

A repeated-event experiment conducted at the USBM test house (Stagg et al. 1984) confirms that current regulatory controls for fatigue cracking are low enough. The test house was framed in wood and had paper-backed gypsum-board interior walls. When continuously vibrated at an equivalent ground particle velocity of 0.5 in./s (12 mm/s), no response

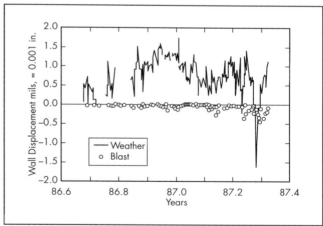

Figure 16.3-12 Displacements produced by weather-induced changes in humidity and temperature (continuous line) and by surface-coal-mine-induced ground motions (circles)

was observed until 52,000 cycles, when a taped joint between sheets of gypsum board cracked. These taped joints are the weakest and most compliant zones in a house with paper-backed gypsum-board walls.

Rock Mass Displacement and Cracking

Cracking rock immediately adjacent to a blast can be controlled by limiting particle velocities to 27 in./s (700 mm/s) in the volume of rock to be protected (Holmberg and Persson 1978).

Rock displacement by forces produced from delayed gas pressure cannot be controlled by specifying an allowable particle velocity. Fortunately, these displacements occur only very close to a blast, within 100 to 165 ft (30 to 50 m) and are associated with blocks that are unconstrained by other surrounding rock.

The sliding instability of individual rock blocks must be evaluated on a case-by-case basis. Each block must have an adequate factor of safety to prevent static failure (Dowding and Gilbert 1988). Since wavelengths of blast-induced shearing motions in rock might be 100–200 ft (roughly 30–60 m) for a 50-Hz wave, only blocks 23–46 ft (7–14 m) in size (~one-fourth the wavelength) can accelerate uniformly. Thus large slopes are unlikely to be destabilized by blasting where the peak particle velocity is high.

Frequency-Based Control with Dominant Frequency

Figure 16.3-13 shows the limit adopted by the OSM, based on a suggested but not rigorously validated proposal by the USBM (Siskind et al. 1980b). Corner 1 of the figure is the lowest particle velocity at which USBM personnel observed cosmetic cracking; corners 2 and 3 are unconfirmed. The dotted line shows the limit for safe construction blasting near engineered structures; the dashed line shows the limit for safe construction blasting in urban areas near older homes and historic buildings.

The dominant frequency that is consistent with Figure 16.3-13 is the frequency that is associated with peaks in the time history whose amplitudes are >50% of the peak or maximum particle velocity. The frequency of these peaks was calculated by the zero-crossing method, as shown in the inset for Figure 16.3-3. The frequency associated with the peak-particle-velocity pulse is a good first approximation and eliminates the need for sophisticated Fourier frequency analysis or response-spectrum analysis. Response-spectrum analyses are the most precise approach to account for the frequency effects of structural response and should be used in singular cases where exacting analysis is required.

Regulatory Compliance for Air Overpressure

Although broken glass is normally associated with excessive air-blast overpressures, limits in the United States are based on the wall response necessary to produce wall strains equivalent to those produced by surface-coal-mining-induced ground motions with a peak particle velocity of 0.75 in./s (19 mm/s). These limits are listed in Table 16.3-3. If a wall-strain level equivalent to that produced by a particle velocity of 1.0 in./s (25 mm/s) measured in the ground were chosen, the allowable overpressure would increase by 3 dB. Most cases of broken glass are reported to have been observed at air overpressures of 136–140 dB (as measured with a linear transducer). Wiggins (1969) reports that sonic-boom air overpressures of >140 dB were necessary to produce cosmetic cracking in military barracks and windows. This observation underscores the conservatism of the 133-dB regulatory limit for cosmetic cracking.

Because of the different sound-weighting scales used by monitoring instruments, the recommended levels in the table differ by instrument system. Since structures are most sensitive to low-frequency motions and the greatest air pressures occur at these inaudible frequencies, A-weighted scales cannot be used at all. Since C-weighted scales are least sensitive at low frequencies, their use requires the most restrictive limits.

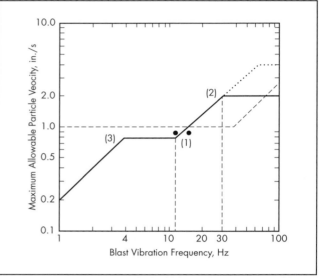

Figure 16.3-13 Frequency-based blast-vibration control limit to protect residential structures, modified from suggestion by USBM

Table 16.3-3 Air-overpressure control limits for various instrument lower-frequency limits

Lower-frequency limit, Hz (–3 dB)	Maximum level, dB
≤0.1-Hz flat response*	134 peak
≤2-Hz flat response	133 peak
≤6-Hz flat response	129 peak
C-weighted slow response*	105 peak dBC

Source: Stachura et al. 1981.
*Only when approved by the regulatory authority.

DIGITAL-AGE IMPROVEMENTS IN BLAST CONTROL

Significant digital-age improvements in blast control have accrued from the use of electronic detonators and signature hole analysis.

Electronic Detonators

Electronic detonators contain a microprocessor that improves the precision of firing times by almost two orders of magnitude (Bartley et al. 2000). This increase now makes it possible to precisely control the firing time of each blasthole, even for shots with several hundred holes. Although electronic detonators are more expensive than pyrotechnic detonators, they offer a number of advantages because of their increased control of detonation times as well as increased safety and flexibility of timing.

The increase in firing-time precision is dramatic. Several decades ago, single batches of pyrotechnic detonators had variations (coefficient of variation or 1 standard deviation divided by the average) in delay times as high as 5% to 10% (Winzer et al. 1979); today these values are on the order of 2%–5% (Cunningham 2000). Although 5% may not be important for shots with a small number of holes (<30), it is important for a four-row, 100-hole shot, designed, for example, with surface delays of 17–25 ms between holes and 42 ms between rows. If a nonelectric system is used for this shot, then all of the downhole delays must be at least 500 ms to allow the surface signal to traverse all of the holes and reach the last hole before the first hole detonates. Thus the last hole will fire between 475 and 525 ms 64% of the time, between 450 and 550 ms 95% of the time, and between 425 and 575 ms 99% of the time. Thus 64% of the delay times fall within ±1 coefficient of variation (5%) of the expected or average delay time of 500 ms. For every shot, there is a 5% possibility of a delay overlap with three (i.e., 50/17) of the previous holes. These increases in unintended external effects occur if overlap results in simultaneous detonations (less than about 8 ms apart), detonation of a rear-row hole before the front-row holes complete, or detonation of holes out of sequence along a row. Electronic detonators now eliminate these potential overlaps and associated increases in back break and peak particle velocity.

The downhole microprocessor in the electronic detonator allows complete freedom in programming firing times. It is now no longer necessary to carry an inventory of preprogrammed delays for all anticipated blast designs. This freedom allows designs to accommodate changes in inter-row and inter-hole delay times as a function of the changing free face at corners and other challenging shot geometries.

Electronic detonators not only reduce peak particle velocities, they also improve safety and in-hole diagnostics (Bartley and McClure 2003). They offer increased immunity to radio-frequency transmissions, electromagnetic interference, and stray current. They are completely testable and can be automated to test and disarm. Most important, in these days of threats of terrorism, they require a specific blasting machine as well as specific identification and password.

Signature-Hole Analysis
Production of synthetic seismograms (ground-motion time histories) by superposition of time-lagged waveforms is another product of the digital age. This superposition is popularly called *signature-hole analysis*, as it requires a waveform measured at the position of interest generated from a single (i.e., a signature) hole at the blast location. The time history recorded for the signature hole (which automatically accounts for the geological irregularities in the travel path) is superposed and time-lagged for each hole's designed detonation time (and travel path in more sophisticated versions) (Anderson 2008). Like electronic detonators, this analytical technique has benefited from the digital age, as it would not be possible without digital waveforms and computerized manipulation.

Signature-hole analysis provides a means to estimate the peak-particle-velocity waveform that results from a change in blast timing. Thus it enables estimates of both the resulting peak particle velocity and the frequency content to match frequency-dependent control limits. Scaled-distance relationships enable only the peak particle velocity, but not the frequency content, to be estimated. Regardless of the genesis of the changes in design, be it the need to improve fragmentation or ground motions or both, signature-hole analysis can be helpful in predicting the changes in amplitude and frequency of the ground-motion time history produced by blasting.

Signature-hole analysis has been the subject of both support and skepticism over the years. It was hampered in its early days by wide differences between design and actual delay firing times for pyrotechnic delays. Yet even its skeptics believe that, when applied correctly, it is a useful design tool. Increased precision of electronic detonators as described previously now eliminates mistiming as a cause of differences between values that are predicted by signature-hole analysis and values that are measured.

REFERENCES

Note: Blasting-related publications from the U.S. Bureau of Mines (closed in 1995) are still available from the International Society of Explosives Engineers (ISEE) at www.isee.org. In addition, the U.S. Office of Surface Mining Reclamation and Enforcement, which regulates surface coal mining in the United States, sponsors a modest blasting-related research program, the results of which are published at www.arblast.osmre.gov.

Anderson, D.A. 2008. Signature hole blast vibration control, twenty years hence and beyond. In *Proceedings of the 34th Annual Conference on Explosives and Blasting Technique*, New Orleans, LA, USA, January 27–30. Cleveland: International Society of Explosives Engineers.

Association of Casualty and Surety Companies. 1956. *Blasting Claims: A Guide for Adjusters*. New York: National Board of Fire Underwriters and the Association of Casualty and Surety Companies.

Bartley, D.A., and McClure, R. 2003. Further field applications of electronic detonator technology. *Fragblast* 7(1):13–22.

Bartley, D.A., McClure, R., Wingfield, R., and Trouselle, R. 2000. Electronic detonator technology: Field application and safety approach. In *Explosives and Blasting Technique: Proceedings of the 1st World Conference on Explosives and Blasting Technique*, Munich, Germany, September 6–8. Edited by R. Holmberg. Rotterdam: A.A. Balkema.

BRE (Building Research Establishment). 1977. Cracking in buildings. In *Building Research Establishment Digest*, Vol. 75. Garston, UK: Building Research Station.

Crawford, R., and Ward, H.S. 1965. *Dynamic Strains in Concrete and Masonry Walls*. Building Research Note 54. Ottawa: National Research Council, Division of Building Research.

Cunningham, C. 2000. The effect of timing precision on control of blasting effects. In *Explosives and Blasting Technique: Proceedings of the 1st World Conference on Explosives and Blasting Technique*, Munich, Germany, September 6–8. Edited by R. Holmberg. Rotterdam: A.A. Balkema.

Dowding, C.H. 1985. *Blast Vibration Monitoring and Control*. Englewood Cliffs, NJ: Prentice Hall.

Dowding, C.H. 1988. Comparison of environmental and blast induced effects through computerized surveillance. In *The Art and Science of Geotechnical Engineering at the Dawn of the 21st Century: A Volume Honoring Ralph B. Peck*. Edited by W.J. Hall. Englewood Cliffs, NJ: Prentice Hall. pp. 143–160.

Dowding, C.H. 1996. *Construction Vibrations*. Upper Saddle River, NJ: Prentice Hall.

Dowding, C.H. 2008. *Micrometer Response of Cracks to Weather and Vibrations*. Cleveland: International Society of Explosives Engineers.

Dowding, C.H., and Gilbert, C. 1988. Dynamic stability of rock slopes and high frequency traveling waves. *J. Geotech. Eng. ASCE* 114 GTIO:1069–1088.

Dowding, C.H., and Kendorski, E.S. 1983. Response of rock pinnacles to blasting vibrations and airblasts. *Bull. Assoc. Eng. Geol.* 20(J):271–281.

Dowding, C.H., Murray, P.D., and Atmatzidis, D.K. 1981. Dynamic response properties of residential structures subjected to blasting vibrations. *J. Struct. Eng.* 107(STI):1233–1249.

Dvorak, A. 1962. Seismic effects of blasting on brick houses. *Prace Geofyrikenina, Ustance,* Ceskoslovenski Akademie, Ved., No. 159, Geogysikalni, Sbornik.

Harris, C.M., and Crede, C.E., eds. 1976. *Shock and Vibration Handbook.* New York: McGraw-Hill.

Hendron, A.J. 1977. Engineering of rock blasting on civil projects. In *Structural and Geotechnical Mechanics: A Volume Honoring Professor Nathan M. Newmark.* Edited by W.J. Hall. Englewood Cliffs, NJ: Prentice Hall.

Hendron, A.J., and Fernandez, G. 1983. Dynamic and static design considerations for underground chambers. In *Seismic Design of Embankments and Caverns: Proceedings of a Symposium.* Edited by T.R. Howard. New York: American Society of Civil Engineers.

Holmberg; R., and Persson, P.A. 1978. The Swedish approach to contour blasting. In *Proceedings of the 4th Conference on Explosives and Blasting Techniques,* Solon, OH: Society of Explosives Engineers. pp. 113–127.

Hudson, D.E. 1979. *Reading and Interpreting Strong Motion Accelerograms.* Berkeley, CA: Earthquake Engineering Research Institute.

Ivanov, P.L. 1967. *Compaction of Noncohesive Soils by Explosions.* TA 710193. Translated from Russian. Denver, CO: National Science Foundation, U.S. Water and Power Resources Services.

Johnson, C.F. 1962. Coupling small vibration gauges to soil. *Earthquake Notes* 33(3):40–47.

Langan, R.T. 1980. Adequacy of single-degree-of-freedom system modeling of structural response to blasting vibrations. M.S. thesis, Northwestern University, Evanston, IL.

Langefors, U., Kihlstrom, B., and Westerberg, H. 1958. Ground vibrations in blasting. *Water Power* (February): 335–338, 390–395, 421–424.

Lundborg, N. 1981. *The Probability of Flyrock.* Report DS 1981:5. Stockholm, Sweden: Swedish Detonic Research Foundation.

Medearis, K. 1975. *Structural Response to Explosion Induced Ground Motions.* New York: American Society of Civil Engineers.

Medearis, K. 1976. *The Development of a Rational Damage Criteria for Low Rise Structures Subjected to Blasting Vibrations.* Report to the National Crushed Stone Association. Washington, DC.

Newmark, N.M., and Hall, W.J. 1982. *Earthquake Spectra and Design.* Berkeley, CA: Earthquake Engineering Research Institute.

Northwood, T.D., Crawford, R., and Edwards, A.T. 1963. Blasting vibrations and building damage. *The Engineer* 215(5601):973–978.

Oriard, L.L. 2002. *Explosives Engineering, Construction Vibrations and Geotechnology.* Cleveland: International Society of Explosives Engineers.

Roth, J. 1979. *A Model for the Determination of Flyrock Range as a Function of Shot Conditions.* Report NTIS, PB81 16.358. Los Altos, CA: Management Services Association, U.S. Bureau of Mines.

Schomer, P.D., Goff, R.J., and Little, M. 1976. *The Statistics of Amplitude and Spectrum of Blasts Propagated in the Atmosphere.* Technical Report N-13. Champaign, IL: U.S. Army Construction Engineering Research Laboratory.

Siskind, D.E. 1981. *Open-File Report of Responses to Questions Raised by RI 8507.* Minneapolis, MN: U.S. Bureau of Mines.

Siskind, D.E., and Fumanti, R. 1974. *Blast-Produced Fractures in Lithonia Granite.* Report of Investigations 7901. Washington, DC: U.S. Bureau of Mines (now available from ISEE).

Siskind, D.E., Stachura, V.J., Stagg, M.S., and Kopp, J.W. 1980a. *Structures Response and Damage Produced by Airblast from Surface Mining.* Report of Investigations 8485. Washington, DC: U.S. Bureau of Mines (now available from ISEE).

Siskind, D.E., Stagg, M.S., Kopp, J.W., and Dowding, C.H. 1980b. *Structure Response and Damage Produced by Ground Vibrations from Surface Blasting.* Report of Investigations 8507. Washington, DC: U.S. Bureau of Mines (now available from ISEE).

Siskind, D.E., Stagg, M.S., Wiegand, J.E., and Schultz, D.J. 1994. *Surface Mine Blasting Near Pressurized Transmission Pipelines.* Report of Investigations 9523. Washington, DC: U.S. Bureau of Mines (now available from ISEE).

Stachura, V.J., Siskind, D.E., and Engler, A. 1981. *Airblast Instrumentation and Measurement Techniques for Surface Mining.* Report of Investigations 8508. Washington, DC: U.S. Bureau of Mines (now available from ISEE).

Stagg, M.S., and Engler, A.J. 1980. *Measurement of Blast-Induced Ground Vibrations and Seismograph Calibrations.* Report of Investigations 8506. Washington, DC: U.S. Bureau of Mines (now available from ISEE).

Stagg, M.S., Siskind, D.E., Stevens, M.G., and Dowding, C.H. 1984. *Effects of Repeated Blasting on a Wood-Frame House.* Report of Investigations 8896. Washington, DC: U.S. Bureau of Mines (now available from ISEE).

Thoenen, J.R., and Windes, S.L. 1942. *Seismic Effects of Quarry Blasting.* Bulletin 441. Washington, DC: U.S. Bureau of Mines (now available from ISEE).

Wiggins, J.H. 1969. Effect of sonic boom on structural behavior. *Mater. Res. Stand.* 7(6).

Winzer, S.R., Furth, W., and Ritter, A. 1979. Initiator firing times and their relationship to blasting performance. In *Proceedings of the 20th Symposium on Rock Mechanics.* Austin, TX: University of Texas at Austin. pp. 461–470.

Wiss, J.F., and Linehan, P.W. 1978. *Control of Vibration and Blast Noise from Surface Coal Mining.* Research Report Contract 10255022. Washington, DC: U.S. Bureau of Mines (now available from ISEE).

CHAPTER 16.4

Water and Sediment Control Systems

Christopher D. Lidstone and Abby Korte

INTEGRATING ENVIRONMENTAL COSTS

Watershed plans have been integrated into public policy in response to the complex requirements of water resources and land-use planning. The permitting of a new mine and the expansion of an existing mine both face public scrutiny, especially in relation to the effects of mining on the local watershed or groundwater basin. Geographic information system (GIS) databases, which are available on numerous public Web sites, typically identify the presence of a mine as a potential point source for pollution.

The mining industry is responsible for much of this perception because of its historic failure to address drainage pollution of all types and because of its historic failure to include environmental costs as part of its public stewardship of the land. In the United States, new sand and gravel operations are often first reviewed by local planning and zoning boards. These planning and zoning boards consist of volunteers or individuals serving by local appointment rather than technical experts or regulators experienced with mining and the impacts related to mining. It is this local board of citizens that first addresses a mine's potential to have an adverse impact on area water wells or nearby flood levels.

The quality and quantity of the surface water runoff that leaves a mine site is often the public's first indication of the presence of a mining operation and the environmental behavior of the corporate citizen. Unlike industrial pollution, which is often generated from an undesirable or unforeseen by-product of the manufacturing process and which is short lived, water pollution from mining can continue long after the mine is shut down and the mineral resource has been fully exhausted. Underground mining can impact groundwater by changing its flow direction, by mixing multiple groundwater sources, and by changing water quality through the physiochemical processes of oxidation and reduction. Runoff from inadequately controlled or poorly reclaimed surface mine disturbances can result in large volumes of sediment-laden waters exiting the mine site and delivering toxic and acid-forming substances to off-site drainages. The act of surface mining will disrupt vegetation, parent material (bedrock), and soil beds. As a surface drainage system trends towards geomorphic instability, excessive erosion will occur and deliver sediment and turbid water off-site.

Cost Benefit and Probability of Failure

The design and implementation of an adequate drainage system before, during, and after surface and underground mining operations is essential to minimize adverse and costly environmental impacts. In addition, the mining industry is faced with the challenge of designing hydrologic and hydraulic control under a myriad of climatic conditions worldwide. For example, surface mines in Sumatra, Indonesia, need to control 3 m (10 ft) of rainfall over a 6-month period during the wet season, while surface mines in western New South Wales, Australia, need to control 0.3 m (12 in.) of rainfall in 60 minutes. Both situations create hydrologic challenges, and one universal truth can be identified—the larger the rainfall volume or the greater the rainfall intensity, the larger and more costly the hydrologic control structure.

Life of the Structure

It can be said that all engineered systems are designed to fail; it is the probability of failure that governs the degree of environmental protection. Each mining company must make this design decision either in response to regulatory criteria or as a cost–benefit analysis made during the planning stages. The probability of failure relationship can be simply stated as

$$P = 1 - q^n$$

where
P = probability of failure
$q = 1 - (1/T_R)$, where T_R is the return period of the design storm
n = anticipated life of the structure

For the purposes of this handbook, the four classes of structures (Class A through Class D) shown in Table 16.4-1 should be considered.

Christopher D. Lidstone, President, Lidstone and Associates, Inc., Fort Collins, Colorado, USA
Abby Korte, Project Hydrologist, Lidstone and Associates, Inc., Fort Collins, Colorado, USA

Table 16.4-1 Design life of structures

Type and Life of Structure	Return Period of Design Storm, T_R
Class A: 10 years	25 years or less
Class B: 11 to 20 years	50 years or less
Class C: Greater than 20 years	100 years or less
Class D: Enhanced protection need	Greater than 100 years

Many temporary haul roads, with a design life of less than 10 years, would dictate the use of a Class A structure (e.g., culverts or diversions). Railroad access structures that will be present for the life of the mine may dictate the use of Class C or Class D structures. A tailings dam or a diversion around a tailings dam will certainly dictate the use of a Class D structure design, including a structure designed for a probable maximum flood (PMF).

DESIGN CONSIDERATIONS RELATED TO MINING

Geomorphology

Surface drainage systems represent a delicate balance between supply and demand. There exists a delicate balance between the amount of material to be carried by a particular stream system and the amount of power available to move that material downstream. The relation can be presented as follows (Lane 1955):

$$QS \propto Q_s D_{50}$$

where
- Q = discharge
- S = channel slope
- Q_s = sediment discharge
- D_{50} = median grain size of the bed material

This equation shows that the discharge, Q, and channel slope, S, are proportional to the sediment discharge, Q_s. This relationship is presented graphically in Figure 16.4-1 and can be used to predict how a channel system might respond to changes in various parameters (sediment load, channel geometry, channel slope, discharge, etc.).

Whether constructing a diversion around a surface mineral mine or determining the appropriate setback distances for a floodplain sand and gravel mine, knowing how that system behaves geomorphically is critical. Historic aerial photo analysis can be a useful tool in determining a river system's channel migration zone and determining the risk of avulsive (versus accretionary) changes in the channel bank location. In addition, it is important to understand how a natural system behaves when designing a surface water diversion so as to avoid additional degradation or aggradation of the channel.

Groundwater Hydrology

It is critical for the engineer or mining company to put wells into the ground to evaluate both groundwater hydrology and chemistry before opening up the ground to extraction activities. Putting wells in the ground allows for the collection of data relating to groundwater levels, groundwater chemistry (oxidizing versus reducing environment), and the direction and rate of groundwater flow. If a proper evaluation of site conditions is not completed prior to the commencement of mining, the site can be inundated with more inflow than the pumping system can handle, resulting in flooding and possibly

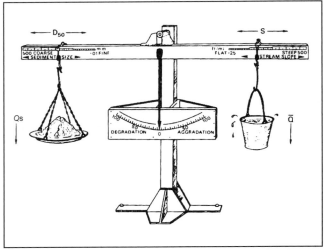

Source: Lane 1955, with permission from the American Society of Civil Engineers.

Figure 16.4-1 Lane's balance diagram

environmental excursions as mine personnel must handle higher than anticipated levels of "dirty" water.

Water and Sediment Management

Diversion and Interception of Undisturbed Runoff

It has been said that an ounce of prevention is worth a pound of cure. Treating disturbed sediment or nutrient-laden runoff from a mine site can be a complicated and costly endeavor. Diverting undisturbed runoff or upgradient groundwater around the mining operation via a constructed channel or drain will eliminate the need for postprocess water treatment. Diversions take on many design forms, depending on the quantity of runoff and the slope of the channel. Classically, as the quantity of runoff increases, there is a progression from bare earthen to grassed waterways to rock riprap channels. In the design of diversions, the development of a stable channel for the design storm appears to be the primary consideration. Often diversions can accommodate the large design storm event, but failure to consider potential sediment deposition from intermediate events and the subsequent loss of conveyance capacity sometimes leads to failure of the control structure.

If undisturbed runoff cannot be diverted around a mine site, it can sometimes be detained in one or a series of interceptor ponds. Interceptor ponds are designed to store and infiltrate all runoff associated with a given design storm, so that undisturbed runoff does not interact with land disturbances.

Treatment of Disturbed Land Runoff

In the event that surface runoff can be neither diverted around the mine site nor stored and infiltrated back into the groundwater via the use of interceptor ponds, then all runoff, which mixes with the mining process, must be treated to prevent the contamination of downstream water resources. One of the primary pollutants associated with surface mine operations is sediment. Because there are other chapters in this handbook dedicated to the treatment of acid mine drainage and various other chemical constituents, this chapter will deal exclusively with the removal of sediment from disturbed runoff.

Increased sediment inputs to stream systems can have a series of adverse ecological and economic impacts, ranging from increased sediment loading in reservoirs, degradation of fish-spawning habitat, and increased nutrient loading that can lead to eutrophication in stagnant water bodies.

The basis for sediment removal is settlement under quiescent flow conditions. Stokes' law describes the settling velocity of a spherical object in a viscous liquid as a result of the pull of gravity. The settling velocity of a small spherical particle (i.e., sediment) can be described by the following equation:

$$V_s = \frac{2(\rho_p - \rho_f)}{9\mu} gR^2$$

where
- V_s = downward settling velocity, m/s
- ρ_p = mass density of the particles, kg/m^3
- ρ_f = mass density of the fluid, kg/m^3
- μ = fluid's dynamic viscosity, Pa s
- g = acceleration due to gravity, m/s^2
- R = radius of the object, m

When the required sediment settling velocity is determined, a sediment detention pond can be designed such that turbid water is retained long enough for the design particle size to settle out of solution.

In some cases, the detention of turbid waters is not enough to remove clay-size particles, which are too small to be removed by settling. In these cases, the treatment of disturbed runoff by the use of flocculants may be necessary. Most flocculants are composed of heavy organic polymers with a strong positive charge that attracts the negatively charged clay particles in a "bundle," allowing them (collectively) to settle out faster than single particles might settle.

Flocculants tend to be cost prohibitive and are generally used only when at least one of the following conditions are present: there is a limitation on the amount of space and, therefore, time available to allow particulates to settle out; there is a particularly large proportion of the sediment load that is silt or clay in size; or the water quality of the receiving water body is of a pristine nature. If a flocculant is used, the most important component of the distribution system is to ensure that the flocculant is well mixed with the discharge.

SURFACE WATER HYDROLOGY

An accurate analysis of surface water hydrology and runoff processes is a critical first step in the design and management of water and sediment control systems. It is important not only for the water resources professional to determine whether the project cost, timing, and scope warrant the collection of surface water data, but also to recognize the limitations and errors associated with the collection and analysis of a time-limited data set. Often on-site data must be augmented with data from published sources. For any given project location and scope, there is no absolute best method. However, the only universal truism in surface water hydrology is that, if available, there is no substitution for locally collected data.

Selection of Analysis Method

Many methods are available for estimating design rainfall depths and for constructing synthetic rainfall distributions or a rainfall hyetograph. The construction of the rainfall hyetograph leads to the calculation and simulation of a runoff hydrograph. For the design of some surface water controls like culverts and some diversions, a peak runoff rate is sufficient. The design of storage components such as detention ponds and certain conveyance channels require the determination of the entire storm hyetograph and runoff hydrograph, as the design must take into account conveyance and residency time calculations.

Design Rainfall Event

The design rainfall event is typically mathematically generated from volume-duration-frequency data or intensity-duration-frequency data for a specific region or project location (Seybert 2006). The design rainfall event is often established by local permitting regulations and is the foundation of the runoff modeling and design calculations for hydraulic structures associated with surface mining operations. The most comprehensive source of rainfall data for the United States and its territories is provided by the Hydrometeorological Design Studies Center and the Office of Hydrological Development of the National Oceanic and Atmospheric Administration's National Weather Service through its Precipitation Frequency Data Server, which can be accessed through the Internet at www.weather.gov/ohd/hdsc/ (NOAA 2009). Through the listed Web site, one can access state-specific rainfall distribution data in the form of rainfall atlases (NWS 1961) and distribution tables (Arkell and Richards 1986). International organizations maintain similar rainfall information databases, and these can be researched in advance of mine planning. If it is important to know the temporal distribution of rainfall for a particular event, and real-time local data are unavailable. The Soil Conservation Service (SCS) and local agricultural services provide synthetic rainfall distribution curves (SCS 1973).

Peak Flow Determination

Rational Method

The rational method is the most commonly used technique for determining peak flow rates. The rational method formula is

$$Q_{PK} = CiAk$$

where
- Q_{PK} = peak flow rate, m^3/s (ft^3/s)
- C = dimensionless runoff coefficient
- i = rainfall intensity, mm/h (in./h)
- A = drainage area, ha (acres)
- k = conversion factor equal to 0.00278 in metric units (1.008 in English units)

The dimensionless runoff coefficient, C, is designed to incorporate differences of rainfall interception/infiltration, surface detention, and/or antecedent moisture conditions. It is nearly impossible for a single factor to incorporate all of these. Therefore, more commonly, the C factor is based on land-use type and imperviousness. A simplified list of C factor values is shown in Table 16.4-2. The C factor can also be area weight averaged throughout the basin of study (Haan and Barfield 1978). The rational method is a simple peak flow calculation based on several assumptions. Included in the assumptions is that the rainfall occurs uniformly throughout the basin and that the rainfall frequency equals the runoff frequency. Typically, this can be successfully used for smaller basins. The method is well described in numerous publications, including Haan and Barfield (1978) and Simons, Li and Associates (1982b).

Table 16.4-2 Simplified table of rational Method C factors

Groundcover	Runoff Coefficient, C
Lawns	0.05–0.35
Forest	0.05–0.25
Cultivated land	0.08–0.41
Meadow	0.1–0.5
Parks, cemeteries	0.1–0.25
Unimproved areas	0.1–0.3
Pasture	0.12–0.62
Residential areas	0.3–0.75
Business areas	0.5–0.95
Industrial areas	0.5–0.9
Asphalt streets	0.7–0.95
Concrete streets	0.7–0.95
Roofs	0.75–0.95

SCS Curve Number Method

An alternative is the SCS curve number (SCS-CN) method, which was first published in the *National Engineering Handbook* (SCS 1982) in 1956 by the Soil Conservation Service, now known as the Natural Resources Conservation Service. The SCS-CN method was developed from extensive field investigations and has been revised several times. The SCS-CN formulas are as follows:

$$Q = \frac{(P - I_a)^2}{(P - I + S)}$$

$$I_a = 0.2S$$

$$S = \frac{1,000}{CN} - 10$$

where
Q = runoff, in.
P = precipitation, in.
I_a = initial abstraction, in.
S = potential maximum retention, in.
CN = curve number

Initial abstraction generally consists of interception losses, surface storage, and soil infiltration. Because of the dimensional incongruity of these formulae, there is no direct coefficient conversion between English and metric units, and each variable must be converted separately. This method requires the water resources professional to collect information pertaining to land use, hydrologic condition, and hydrologic soil type. A more detailed explanation of the development and use of the SCS-CN method is beyond the scope of this chapter, but many excellent texts and papers outline the methodology and provide the necessary tabular data (Seybert 2006).

Time of Concentration

The time of concentration, t_c, is the length of time it takes for precipitation to travel from the most remote part of a watershed to the point of interest, often the basin outlet. Several methods exist for estimating t_c, including, but not limited to, the SCS upland method and the kinematic wave method (Kirpich 1940; Ragan and Duru 1972). The Ragan and Duru equation is

$$t_c = \frac{C_1 (nL)^{0.6}}{(i_e^{0.4} S^{0.3})}$$

where
t_c = time of concentration, min
C_1 = a constant equal to 1.977 in metric units (0.928 in English units)
n = Manning's roughness coefficient
L = length, m (ft)
i_e = excess rainfall intensity, mm/h (in./h)
S = slope (vertical/horizontal), m/m (ft/ft)

This equation is only valid when the product of i_e and L is greater than 152 m (500 ft).

The Manning's roughness coefficient, n, can be assigned as a single unit for the reach of interest or can be composited to better represent a channel with varying degrees of resistance to flow. Several sources of information are available for Manning's n values, including books with photographs of channel sections with measured n values (Chow 1959; Acrement and Schneider 1984).

The Kirpich equation for estimating t_c is

$$t_c = C_2 L_1^{0.77} \left(\frac{L_1}{H}\right)^{0.385}$$

where
C_2 = a constant equal to 0.0195 in metric units (0.0078 in English units)
L_1 = maximum flow length, m (ft)
H = elevation difference between the hydraulically most remote point in the watershed and the point of interest, m (ft)

SURFACE WATER HYDRAULICS

Designing a diversion ditch or reclamation channel requires that the mining professional must have a good understanding of the hydraulic and sediment transport characteristics, as well as channel geometry and morphometric characteristics of the native system. Flow conditions in open channels are complex. The position of the free water surface and channel geometry can change with respect to time and space. Depth of flow, velocity, discharge, and slope of the channel bottom and water surface are interdependent.

The basic hydraulic theories that relate flow velocity, depth, discharge, and slope are as follows:

- Conservation of mass (continuity)
- Conservation of linear momentum
- Conservation of energy

Conservation of mass states that matter can neither be created nor destroyed. Based on this it can be concluded that flow discharge in each section of a river reach is identical. Conservation of momentum concludes that the summation of all external forces (pressure, body, and shear forces) acting on the fluid is equal to the rate of a flux of momentum through the boundary of the fluid. The law of conservation of energy (the first law of thermodynamics) states that energy can neither be created nor destroyed. For steady flow, the total energy (kinematic and potential energy) at any point in a stream is equal to the sum of the total energy at any downstream point and the loss of energy between the two points.

The continuity, momentum, and energy equations can be used to determine the elevation of the water surface in an open channel for a given discharge. For steady flow, the energy equation is written in the following form to compute the water surface profile:

$$\delta L = \frac{H_2 - H_1}{S_o - S_f}$$

where

δL = distance between sections 1 and 2, m (ft)
H = specific head, m (ft)
S_o = bed slope (vertical/horizontal), m/m (ft/ft)
S_f = friction or energy slope (vertical/horizontal), m/m (ft/ft)

The specific head, H, is defined as

$$H = \frac{V^2}{2g} + y$$

where

V = average velocity, m/s (ft/s)
g = acceleration due to gravity, m^3/s (ft^3/s)
y = flow depth, m (ft)

The energy slope, S_f, is the slope of the energy line representing the elevation of the total energy of flow. It is related to the velocity and depth by the following empirical relationship, which is known as Manning's equation:

$$Q = VA = \left(\frac{k}{n}\right) AR^{2/3} S^{1/2}$$

where

Q = flow rate, m^3/s (ft^3/s)
V = velocity, m/s (ft/s)
A = flow area, m^2 (ft^2)
k = conversion factor equal to 1.00 in metric units (1.486 in English units)
n = Manning's roughness coefficient
R = hydraulic radius, m (ft)
S = channel slope (vertical/horizontal), m/m (ft/ft)

R is calculated by the cross-sectional area of the channel divided by the wetted perimeter. Basic channel hydraulics is well summarized in Simons, Li and Associates (1982b).

SEDIMENT PRODUCTION

Soil Loss Equations

The Universal Soil Loss Equation (USLE) calculates sediment production via a combination of sheet and rill erosion and the associated runoff for the landscape profile. The soil loss is an average erosion rate for the landscape profile. Erosion can vary widely, even on a uniform slope, depending on slope position and configuration of the slope profile. USLE does not estimate the amount of sediment leaving a field or watershed but estimates soil movement at a particular site. The quantity of sediment eroded from a slope depends on the erosive power of rainfall, soil characteristics, slope length, and gradient, as well as type and amount of soil cover and conservation practices. These parameters have been combined into the USLE (Wischmeier and Smith 1978).

$$A = R \times K \times LS \times C \times P$$

where

A = soil loss, kg/ha (st/acre)
R = rainfall factor
K = soil erodibility, t/ha (st/acre) per unit R
LS = dimensionless length–slope factor that accounts for the actual length and slope compared to a standard slope of 9% and 22.1 m (72.6 ft) in length
C = a factor that accounts for the effectiveness of vegetative cover, mulches, etc.
P = a factor that accounts for the effectiveness of conservation practices such as terraces or fallow cover

Values for these parameters (i.e., R, LS, C, and P) may be found in many sources and have been modified for the mining industry (Chow 1959; Wischmeier and Smith 1978).

Contemporary to the USLE, the Revised Universal Soil Loss Equation (RUSLE) computes sheet and rill erosion and the associated runoff from untilled lands, as the original USLE was derived and intended for use in agricultural settings. RUSLE contains several updates and changes from USLE, including new and improved soil erodibility indexes for some areas, time-variant soil erodibility to reflect freeze–thaw cycles, new equations to account for slope length and steepness, and additional C and P factors to account for cover and conservation practices on untilled lands.

Regression Studies

In some cases, particularly when there are anthropogenic or human-related sources of sediment present, it is possible to pull real data on sediment contributions from previously conducted field research. In many cases, real data have been collected on sediment production rates from given land-use types, and regression (i.e., retrospective) analyses have been performed to model the rate of sediment production as a function of several landscape variables. For example, several studies have been completed quantifying sediment production from various unsealed road surfaces. One such study has quantified and modeled sediment production from all sources, focusing on forest roads in the southern Sierra Nevada, California, United States (Korte and MacDonald 2005).

Other Methods

In the international setting, predictive models developed in the United States may not be applicable. International mining efforts typically require the site-specific application of empirical methods and observations. Denudation rates can be calculated from observation of past mining practices. One technique that is uniformly applicable includes the volumetric measurement of accumulated sediment within a pond over a known period of time. In an effort to develop an appropriate empirical model, it is important for the mining professional to evaluate observed conditions within the watershed that delivered water and sediment to that pond.

HYDROLOGIC STRUCTURES

Hydrologic control structures are an integral part of any mining operation. Such structures are generally referred to in the context of surface mines and their associated disturbances, but are becoming increasingly more important for both underground and in-situ recovery (ISR) mining operations. All mining operations include such surface disturbances associated with roads, buildings, power lines and substations, stockpile areas, loadout and water treatment facilities. Such facilities require water and sediment control during mining and reclamation at the end of mining. Surface mining operations are further complicated by the associated disruption of the existing topography to extract the ore or mineral commodity from the ground. Such operations will require a more sophisticated drainage control and a far more complicated final reclamation plan than underground or ISR operations.

To adequately control pollution, it is necessary to establish a complete drainage design plan for the entire permit area using sound engineering practices. It is often important to separate the hydrologic control plan on a temporal basis, that is, short term (1 to 5 years) and long term or "life of mine," depending on the size and dynamic nature of the operation. The short-term plan will include design details, engineering calculations, and likely all required construction permits from the various regulatory agencies. The long-term or life-of-mine plan is simply a concept plan, which will dictate the progression and type of hydrologic control features but will allow flexibility as mine production and mining methods inevitably change. This long-term plan is completed without actual engineering design or regulatory permitting.

In addressing the overall hydrologic control plan, it is essential that the design engineer recognize that the drainage basin or basins within the mining area are only part of a larger, more complex system. Mining within a portion of the drainage basin will elicit a complex response both upstream and downstream from the area of actual disturbance. The design engineer must consider the active mine controls and, even more importantly, the postmining restoration of this drainage system, such that the reclaimed land surface will properly function as part of a larger system. Hydrologic control structures are summarized in Table 16.4-3.

Diversion Channels and Ditches

Diversions are used throughout mining to intercept and convey runoff. Typically they are used to convey runoff from undisturbed areas around active mining sites and preclude mixing of runoff waters and reduce the quantity of runoff requiring treatment. In some cases, diversions are necessary to maintain water flow for downstream water users, including farmers, recreation, wildlife, and fisheries. Because the construction of a diversion of an intermittent or perennial stream is generally a high capital-cost item, mine management must closely address the question of the proposed design life and determine value engineering. As the size of the drainage basin increases, it can be assumed that the means to stabilize each diversion will transition from a simple earthen cut/fill at a graded slope to a well-established vegetated channel to a channel stabilized with geotextile, concrete, or riprap. Drop structures are typically constructed at locations where the design or as-constructed channel gradient is too steep, resulting in excessive flow velocities and the need for channel protection (Jones and Lidstone 1996). A diversion structure that will remain in place after mining is completed must also incorporate geomorphic stability and may include aquatic habitat features (e.g., pools, riffles, large woody debris, and riparian vegetation).

The diversion hydrology is determined after the design life is established. Permanent diversions or life-of-mine features are typically designed for the 100-year discharge event, with additional capacity to accommodate freeboard.

Table 16.4-3 Hydrologic control structures

Type of Structure	Design Life	Purpose of Structure
Diversion channels and ditches	Temporary or permanent	Divert undisturbed runoff around the mining operation in a stable fashion.
Culverts	Temporary	Divert runoff below mine access or haul road.
Sediment ponds	Temporary	Treat disturbed land runoff.
Interceptor ponds	Temporary or permanent	Prevent undisturbed runoff from entering a mine pit or mine area. Can serve as a stock pond or detention pond.
Small area sediment controls (best management practices)	Temporary	Treat disturbed land runoff from small acreage disturbance.
Dry wells	Temporary or permanent	Capture and treat disturbed land runoff.

Regulatory programs may dictate the use of the PMF or some proportion of the PMF, where mine tailings or hazardous wastes are involved. The diversion hydraulics can be simplified using normal depth procedures (Manning's equation) or using more complicated software programs such as SEDCAD, from Civil Software Design LLC, and HEC-RAS and HEC-HMS, which were developed by the Hydrologic Engineering Center of the U.S. Army Corps of Engineers. For the purposes of this section, conveyance capacity of a diversion can be summarized by

$$Q = VA$$

where

Q = design discharge, m³/s (ft³/s)
V = velocity, m/s (ft/s)
A = cross-sectional area, m² (ft²)

Design hydraulics can be simply calculated using Manning's equation as follows:

$$V = (1.49/n)R^{2/3}S^{1/2}$$

where

V = average velocity, m/s (ft/s)
n = roughness coefficient of the channel, which is empirically determined
$R = A/P$, or the cross-sectional area of the channel divided by the wetted perimeter, m (ft)
S = slope (vertical/horizontal) of the channel or hydraulic gradient, m/m (ft/ft)

The channel must be proportioned to carry the design runoff at average velocities less than or equal to the maximum permissible velocity. This is accomplished by applications of the Manning's formula with iterative adjustments in channel parameters until the maximum permissible velocity is achieved.

Numerous studies address the maximum permissible velocity of a channel, and the most commonly accepted is in Chow (1959). Channel type or channel lining must be established in the determination of maximum permissible velocity. Noncohesive (sand bed) channels are governed by initiation of motion and upper limits of 0.6 to 0.9 m/s (2 to 3 ft/s). Well-vegetated channels can withstand velocities of 1.5 to 2.1 m/s (5 to 7 ft/s) before erosion will take place. Cohesive bed (clay) channels are typically not governed by the principles of initiation of motion but rather by tractive force as follows:

$$\tau_c = \gamma DS$$

where

τ_c = critical shear
γ = unit weight of water, which is 1,000 kg/m³ (62.4 lb/ft³)
D = flow depth, m (ft)
S = channel slope (vertical/horizontal) or hydraulic gradient, m/m (ft/ft)

Allowable shear is dependent on the strength characteristics of the clay, plasticity, and electrical conductivity or salinity. Typical critical shear, τ_c, limits range from 0.73 to 3.12 kg/m² (0.15 to 0.64 lb/ft²).

Where erosion problems are anticipated in design, the engineer will need to employ hydrologic control structures such as riprap or gabion drop structures to reduce overall channel slope; riprap or geotextiles to protect the underlying soils from erosion; or interlocking concrete blocks to stabilize the problematic reach of the channel. Numerous publications assist in the design of such structures, including Haan and Barfield (1978); Simons, Li and Associates (1982a); Ferris (1996); and Jones and Lidstone (1996). The design and construction of a riprap drop structure or grade control is an important means to reduce the overall channel slope and thus the erosive energy within a stream channel or diversion.

Permanent diversions require forward planning and a dedication of design and construction resources. Such diversions, which are used for the reestablishment of a drainage system affected by surface mining, must perform adequately without the assured benefit of periodic maintenance. The designer must be familiar with the inherent stability of the upstream and downstream undisturbed watersheds, including channel geometry, channel slope, channel bed and bank materials, and geomorphic characteristics. Although the designer may not necessarily want to duplicate these natural conditions, in recognition that the act of mining has permanently changed the landform, such preexisting conditions will lend themselves to design concepts. Channel design should incorporate a pilot channel to convey the annual channel forming event, typically the 1.5- to 2-year flood event, and a floodplain that at a minimum can convey the 100-year/24-hour flood event. Channel meanders, pools, and riffles may be incorporated into the ultimate channel design. Implementation of final design must incorporate biological and riparian considerations. Typically, this includes the presence of randomly placed boulders, large woody debris, and bank stabilization with woody vegetation. Numerous publications can assist in the design of such permanent diversions (Simons, Li and Associates 1982b).

Diversion ditches are a smaller-scale diversion structure, but they have become increasingly important in areas disturbed by underground mines and ISR-type mining operations. Diversion ditches can protect a well field from overland flow, as well as protecting the face-up or main portal area from off-site erosion. The design event for such ditches are governed by the life of the structure to be protected or, in some cases, the length of time for the site to be successfully revegetated and become stable in its own right. The ditches control a small basin area and often deliver water to a culvert and ultimately to native, undisturbed receiving water. Similar to the drainages, adequate freeboard must be allowed, and the water should be passed via the diversion ditch in a nonerosive fashion.

Culverts

Culverts are used throughout mining to intercept and convey runoff under access roads, stockpile areas, and structures. The selection of a properly sized culvert is dependent on the life of the structure (design discharge), type of culvert, entrance configuration, pipe length and slope, headwater constraints (the available elevation above the top of pipe to pond water), and tail-water condition (the degree to which the downstream portion of the culvert is submerged). Design procedures have been somewhat simplified by the development of culvert-capacity charts or nomographs. Culvert capacity, Q, is governed by the following general equation:

$$Q = \frac{a\sqrt{2gH}}{\sqrt{1 + K_e + K_b + K_c L}}$$

where

a = cross-sectional area of the pipe, m² (ft²)
g = gravitational acceleration, which is 10 m/s² (32 ft/s²)
H = head loss through the pipe, m (ft)
K_b, K_c, and K_e = coefficients that reflect various head loss coefficients that are pipe dependent
L = length of the pipe, m (ft)

Most manufacturers provide design nomographs for their type of pipe, and the Federal Highway Administration (FHWA) presents an extensive culvert-design publication called HEC-5. Publicly available software, HY-8, can be downloaded from the FHWA Web site (FHWA 2009).

Several types and shapes of culverts are commonly used in modern mines. These include concrete pipe, reinforced concrete pipe, corrugated metal pipe, corrugated steel pipe, corrugated plastic pipe, and smooth-walled plastic pipe. Each pipe type has certain advantages in cost, transportability, loading characteristics, and durability. The shape of the culvert will often dictate its capacity and, given modern environmental concerns, its ability to allow sediment and fish to pass. Barrel culverts, which are often the most common, will typically maintain the highest velocities, but they can be a barrier to fish. Box culverts provide greater flow capacity, but can require more frequent maintenance and may not provide the depth of flow to allow fish passage. Arch culverts and, more preferably, bottomless arch culverts (i.e., those where there is neither a steel underbody nor an engineered lower section) can be used. Bottomless arch culverts can be founded on bedrock or on the natural alluvial material, which comprises the actual streambed. The use of such culverts can balance flow-capacity demands while allowing fish passage with a natural substrate. The development of a natural substrate within the floor of the culvert will allow the development of a low-flow channel for the movement of fish. However, such culverts can be sediment sinks, given that the floor of the culvert will often behave as a natural channel. The geomorphic position of bottomless arch culverts is a critical design consideration.

A fundamental mistake at many mine sites is the use of an undersized culvert or the placement of a culvert at too flat a grade. Under no circumstances should a culvert less than 0.45 m (18 in.) in diameter be used. The ability of the culvert to pass sediment and debris must be considered in all designs. Where regular maintenance is available, the use of a trash rack can prevent culvert clogging. In an area of high sediment yield, too flat a culvert grade will result in plugging. In this case, culverts should be set at a minimum grade of 2%. Where erosion may occur at the inlet or outlet or if a road overtopping design must be considered, the placement of erosion protection is a critical design concern. A fundamental error in culvert design and placement is the purchase of too short a pipe length. Where erosion control is required at the inlet or outlet portion, the pipe must project a sufficient length to allow the adequate construction of erosion control. Usually this requires an additional 3 to 6 m (10 to 20 ft) of culvert length, depending on the burial depth and diameter of the culvert. This erosion control could be riprap, gabion structures, or a concrete headwall. Riprap is the most commonly used. Riprap should be placed on a slope no steeper than a 1.5:1 (horizontal/vertical) slope. So, if a 1.2-m- (4-ft-) diameter culvert is placed under a haul road with 0.6-m (2-ft) burial depth, then the culvert will need to project 2.7 m (9 ft) on the upstream and downstream sides. For a typical 7.4-m- (24-ft-) wide haul road, this would require the purchase of a 13-m- (42-ft-) long culvert. Tail-water protection is equally important to ensure that the culvert is not undermined by erosion or, in the case of overtopping flows, that adequate protection is available. The tail-water protection is generally loose riprap and should extend a minimum of 3 m (10 ft) beyond the end of the culvert.

Water and Sediment Control Ponds

Water and sediment control ponds have been used extensively by the mining industry for stormwater management and sediment control. For simplicity purposes, three basic types of ponds are used in mining: sediment ponds, interceptor ponds, and infiltration basins. Sediment ponds are used to reduce peak flow, trap sediment, and meet effluent limits (Ferris 1996). In many countries, mine discharge must meet certain governmental standards. In the United States, these standards are dictated by both state and federal programs under the guise of the National Pollutant Discharge Elimination System, Water Pollution Control Facility, and/or a general stormwater permit. The key distinction in treatment requirements is to determine whether or not the source water mixes with process water or has the potential for being contaminated by some other pollutant besides soil. In general practice, it is important to keep stormwater from commingling with the following: truck traffic (oil and grease), process or wash water, and chemical leachate sources.

For this reason, interceptor ponds may be constructed. Interceptor ponds are used to prevent runoff from entering the mining pit and impacting mining operations. Many mine operators will construct a pond on the upstream side of a mine pit to ensure that at least a portion of the native stormwater is captured prior to entering the mining operations. This water can then be diverted or pumped around the mining operation or can be left to evaporate or infiltrate away. The design capacities of such ponds are not dictated by regulatory or treatment requirements but are more likely based on mining economics. A larger pond (100-year/24-hour capacity) can provide the pit with a high degree (0.01 annual probability) of protection. A smaller pond will provide incrementally less protection.

Interceptor ponds can also include small, elongated ponds, which are constructed on mine benches to capture groundwater and/or surface water. Such ponds will prevent stormwater from mixing with the mining operation, thereby eliminating the need for water treatment.

Finally, an infiltration basin is a depression in the earth that is designed to percolate the water to the subsurface. Such basins are designed on either highly pervious soils/rock, or alternatively, they may be overexcavated and filled with pervious rock. Infiltration basins are most common on aggregate mines, rock quarries, and placer mines where such pervious soils or fractured rock are most likely to occur. The advantage of such basins is that there will be a minimal water-handling requirement, and water treatment regulations typically do not apply.

Design Considerations

Sediment pond design must address placement or location standards, hydrology, sediment production from the upstream watershed, and the upstream watershed's efficiency to trap sediment. The location of this sediment pond within the watershed must consider the size and location of the mine-related disturbance, sizing criteria to ensure that the dam height and volumetric capacity will meet regulatory criteria, pond shape to ensure sediment trapping efficiency, and overall shape to allow sediment cleanout as required.

Sediment ponds are generally located downstream of the mining disturbance, so as to capture and treat disturbed land runoff. The mining engineer must compute design hydrology to permit the sizing of the structure and, as applicable, its embankment and spillway. Design hydrology is directly related to the discussion of the design life of the structure (see the "Life of the Structure" section earlier in this chapter). In the United States, design hydrology may be dictated by federal or state regulation. For example, a coal mining company in the United States must establish sediment control "so as to prevent to the extent possible additional contributions of suspended solids to stream flow, or runoff outside the permit area, but in no event shall contributions be in excess of requirements set by applicable State or Federal law" (30USC1265). Various state guidelines interpret sediment pond containment for the water and sediment yield for the 10-year/24-hour storm event. The volume associated with the design event is directly related to the position of the structure within the watershed and the runoff of the soils, as discussed earlier in the "Surface Water Hydrology" section. A larger water or sediment containment volume could translate into a larger embankment structure and may dictate a more complex permitting/design or construction process. Therefore, the selection of the embankment's location within the watershed (i.e., upstream contributing watershed) may be critical in an effort to avoid such complexities.

In the case of process water ponds, where the generation of sediment and dirty water is the result of material processing operations, basin or pond shape is critical. These ponds are often required to discharge, and as such they must meet discharge standards. The length-to-width ratio is important in that the longer the flow length, the greater the opportunity to deposit or settle suspended solids. Such ponds may incorporate baffles to increase flow length (settling time) and to allow periodic cleaning of the facility. Where sedimentation rates need to be accelerated and additional flow length cannot be achieved, flocculants can be added to water to enhance the sedimentation process.

Sediment pond design is well documented (Haan and Barfield 1978; Simons, Li and Associates 1982a; Ferris 1996).

Dams and Incised Structures

A variety of dam/impoundment types have been utilized for these structures, including embankment dams and excavated basins. The most common pond consists of a dam embankment that may range from 3.6 to 6 m (12 to 20 ft) in height and a collection basin. The dam height requirement of 6 m (20 ft) may be dictated by certain regulatory requirements. For example, the Mine Safety and Health Administration's Safety of Dam requirements will influence the determination of a sediment-pond dam height in the United States (30CFR 77.216; MSHA 1975). On an international basis, dam height consideration may be governed by local rules, but more importantly, dam height is governed by the talents of the designer and by quality control during construction. Large dams that exceed 6 m (20 ft) should be designed by an experienced engineer and will require foundation preparation, material selection and compaction control, among other requirements. The safest ponds are those with no embankment. Such pond basins are constructed as incised or below-grade basins. These basins can be of an unlimited depth and may include the mine pit itself. Such basins are not subject to rules and regulations dictating dam stability, but they can be subject to rules that require thorough drainage or passage of water for downstream users.

Inlet Structures

Inlet structures are an important design consideration in all sediment ponds. Such structures should be located to provide the hydraulically longest flow path between the inlet and outlet. This will help to reduce short-circuiting and provide a higher potential for mixing the inflow with the permanent pool, thereby reducing sediment concentration through dilution and providing a longer residence time needed for settling sediment.

Stabilizing the inlet structure is particularly important in an excavated basin where the sediment pond is incised below the elevation of the natural ground surface. In this case, the inlet structure conveys water from the natural stream's bed elevation to the bottom of the pond or to the elevation of the permanent pool. Without an effort to stabilize the inlet structure, erosion and excess generation of sediment will occur. Inlet structures can be stabilized by riprap, concrete, vegetation, or numerous erosion-control products.

Sediment Storage Capacity

Sediment storage can be addressed by numerous methods, including computer programs or a guideline such as x-acre-feet of storage per disturbed contributing watershed acreage (Ferris 1996). The most common method is an adaptation of the USLE or one of the numerous variations available in the literature, discussed previously in the "Sediment Production" section and later in the "Small-Area Sediment Controls" section. Depending on cleanout opportunities, each sediment pond should be designed to hold a minimum of 3 years of sediment accumulation.

Permanent-Pool Storage Capacity

The size of the permanent pool affects peak flow reduction, sediment trap efficiency, and effluent concentration. A permanent pool can be generated from groundwater inflow or surface water inflow where seasonal precipitation and runoff exceeds evaporation and infiltration. A permanent pool volume is calculated through the development of a water balance within the watershed and at the sediment pond site. A permanent pool can provide significant dilution opportunities; it will also result in less available storage and detention time for the influent waters.

Spillway Design

Spillway location, type, and size directly influence the performance of a sediment pond. Spillways consist of primary spillways, which are designed to allow smaller flows out of the impoundment, and emergency spillways, which are designed to pass a peak flow and to ensure the stability of the embankment. Most treatment-type reservoirs are designed with both a primary and an emergency spillway, so that treated water can be released on a regular basis while protecting the embankment. Detention-type reservoirs may only have an emergency spillway and are either dewatered by a pump or through evaporation and seepage.

Primary spillways are often gated culverts or pipes, which can be controlled to allow periodic discharge and draining of the reservoir. Where treatment is required, a drop inlet or riser may be attached to the culvert to remove water from the upper portion of the pond. All valve assemblies should be equipped so that they can be operated from above the surface of the water. All inlets should be protected by a screening device that prevents floating objects, brush, turtles, fish, and so forth from entering and clogging the pipe (Schwab et al. 1966). The design size and capacity of the primary spillway is dictated by dewatering needs and treatment time requirements. In some instances it may be dictated by downstream water users via regulated water rights or irrigation needs.

All sediment ponds and most interceptor ponds should have an emergency spillway. In the case of the incised pond or excavated basin, the emergency spillway is generally the downstream drainage channel. Typically, an incised pond has no embankment to protect. Good design practice, as well as environmental rules and regulations, dictate the design and construction of an emergency spillway to protect an embankment from overtopping and thus leading to ultimate failure. The emergency spillway will typically be sized for a peak runoff event greater than the actual design event for the pond. For example, if the pond is designed for storage of the 10-year/24-hour storm event, the emergency spillway should be designed to pass the 25-year/24-hour storm event at a minimum. If the design life of the structure is greater than 10 years, good engineering practices may dictate a more conservative design (e.g., 50-year/24-hour). The emergency spillway should be designed to be nonerosive at the design event. This may be accomplished with a vegetated spillway, or in some cases it can be stabilized by riprap, concrete, or numerous erosion-control products. This will be discussed in the following section. The emergency spillway should be constructed such that no flow will be conveyed against the embankment, so that the affected parts of the embankment are protected.

SMALL-AREA SEDIMENT CONTROLS

Small-area sediment controls or best management practices (BMPs) allow an area of land to be subdivided such that the volume of runoff coming from a given parcel is bifurcated and reduced. If the area of land draining to a single given point can be reduced, then the volume of surface runoff traveling to that given point will be reduced and the erosion potential will thereby be mitigated.

Best Management Practices

As is the case when designing hydrologic controls or control structures, the design life of the erosion-control structure is often the driving force when determining the type of material used for BMPs. In addition, the severity of the slope, runoff hydraulics, and potential for sediment production may also influence the determination of the type of BMP. Proper installation of erosion-control structures is critical because when overland flow has either bypassed or undermined the structure, failure is imminent. The mining professional must match the method to the site condition (i.e., straw bales may not be effective in a steep draw). In general, many of the discussed erosion-control structures (especially those designed to stabilize slopes and promote the development of vegetation) require very regular maintenance and have a short life span.

Dry Wells or French Drains

A dry well or French drain can be used to treat contaminated runoff and promote infiltration of surface water back into an aquifer. French drains are only effective at removing sediment from contaminated waters and are not designed to treat hazardous wastes. Both structures are passive, relying on gravity to pass runoff through a filter comprised of rock or gravel before it mixes with groundwater (Figure 16.4-2). These systems are often lined with filter fabric, and in general, the deeper the well or drain, the more effectively it can remove particulate pollution.

Revegetation

Steep, smooth, and nonvegetated slopes and spoil mounds are subject to high rates of water and wind erosion. High erosion rates decrease slope stability, result in the loss of valuable topsoil, and can be very costly to repair. Adding vegetation cover to an exposed soil surface can intercept and reduce raindrop impact and increase infiltration, which slows and redirects overland flow (Kleinman 1996). Several immediate revegetation establishment options exist, including drill seeding, hydroseeding, broadcast seeding, and transplanting entire live plants or plant cuttings. In addition, the placement of mulch can increase soil moisture, provide a temporary cover to reduce erosion risk, moderate soil temperature, and increase the likelihood of seed establishment (Kleinman 1996).

Natural and Biodegradable

Straw or hay bale check dams can be installed for temporary sediment or erosion control for one season or less. Straw or hay bale check dams are placed end to end on contour to reduce the run length of any exposed slope (Table 16.4-4). The most critical factor in proper installation of a straw bale check dam is the surface slope, which should be less than 20%. In addition to straw bale check dams, there are a variety of biodegradable matrix fabric structures that can be used both to protect steep slopes from erosion and to aid in the establishment of living vegetation (Barrett et al. 1998; North American Green 2009; RoLanka International 2009).

Artificial and Synthetic Materials

The most commonly used synthetic erosion-control product is sediment fencing. Sediment fences act as a dam, trapping sediment on the upslope side and allowing water to leave the

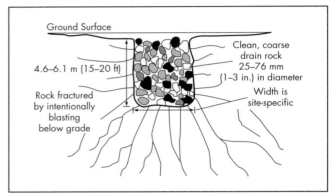

Source: Norman et al. 1998.

Figure 16.4-2 Dry well or French drain construction details

Table 16.4-4 Straw or hay bale check dam spacing

Slope, %	Maximum Slope Length Above a Single-Bale Check Dam or Between Successive Bale Check Dams, ft
≤2	250
5	100
10	50
15	35
20	25

site (Figure 16.4-3). They are effective in retaining suspended solids coarser than 0.02 mm (0.00079 in.). They are simple to construct, relatively inexpensive, and easily moved as development proceeds.

MECHANICAL TREATMENT (SEDIMENT SEPARATORS)

The mining industry has utilized mechanical treatment for sediment removal or sediment separation for decades. Mechanical separators can be incorporated into the mineral-processing stream (placer or aggregate mines, process mills) or can be used solely for environmental purposes. For the purposes of this chapter, only environmental use of this equipment will be discussed. The principal advantage of mechanical separators is their ability to effectively remove sediment from water where space or time is limited. With the addition of flocculants, these separators can generally meet discharge effluent standards relatively quickly without requiring several acres of treatment ponds. Mechanical separators will also reduce cleanout costs because sediment traps and particulate removal is integrated within the equipment process stream.

Mechanical separators can include a swirl concentrator or inertial separator and will intercept sediment-laden flow. The swirl concentrator separates sediment from the incoming flow through the centrifugal force generated by the inherent inertia of the flow (Warner and Dysart 1983). In many ways they operate in the same way as centrifugal cyclones used in mineral processing. Effluent with a high sediment load is transmitted to a small sediment trap, while the clearer flow is discharged directly to a stream or receiving water. The inertial separator is designed to retain incoming sediment in a tank between a series of V-cut bottom-slotted troughs (Sterling and Warner 1984). The sediment-laden influent is transferred from influent troughs into the tank, and clearer flow is withdrawn by deep troughs (Warner 1992). The concept behind these mechanical separators has been transferred to stormwater

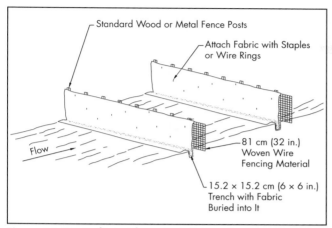

Figure 16.4-3 Sediment fence installation details

control agencies and has manifested as design guidelines for BMPs in the control of sediment-laden runoff. An example of this is presented in the Washington County (Oregon, United States) Clean Water Services design and construction standards, one of which is presented as Figure 16.4-4.

The large-scale use of mechanical separation in the mining industry may include hydrocyclone separators, which are prevalent in the gold mining industry, and water clarifiers, which are utilized extensively in the aggregate industry. Both systems allow operators to take a dirty stream of water and produce clean water immediately, concentrating the fines or solids to a thick state. Hydrocyclones are based on the centrifuge principal, where the particles are spun against the outside wall of the separator and gravitate toward the bottom and are collected in the sedimentation tank. A water clarifier is a similarly self-contained unit but uses a flocculant-driven process with a combination of settlement time and available space. It includes a dry polymer feed system, a hydraulic package, control panels, and a mud-discharge pump system. As the sediment is removed from the water, it is pumped as slurry out of the clarifier to its interim or final resting space. The clean water is pumped back into the process stream or to the receiving water. Hydrocyclones and clarifiers can typically handle greater flow volumes of up to 252 L/s (4,000 gpm) and greater sediment concentrations (up to 20% solids).

GROUNDWATER HYDROLOGY

Both surface and underground mines must develop a strategy to handle groundwater both for water quantity and quality. If the volume of intercepted groundwater exceeds an operation's ability to pump, flooding of the mine operations will result. If groundwater commingles with the mining operation, contamination of this water by the operation may ensue. Such contamination may result from interaction with equipment and haul traffic (oil, grease, sediment) or from an interaction with a by-product of mining, which could include the leaching of metals as the groundwater moves from a reduced to an oxidized state. Excess groundwater must be handled in a similar fashion as excess surface water and must be treated in advance of discharge.

The purpose of this section is to outline a strategy for designing a groundwater monitoring program that provides useful information throughout the life of the mining operation. Like surface water, if groundwater is successfully intercepted and diverted around the mining operation, the need for treatment is eliminated. Unlike surface water, groundwater interception requires extensive subsurface capital construction, which may include slurry walls, grout curtains, or a dewatering well field.

Predictions of Groundwater Inflow

Predictions of groundwater inflow require a detailed knowledge of the geologic framework, including structure, stratigraphy, and a definition of the aquifer properties. Such properties typically include an understanding of hydrologic boundaries (recharge and discharge), an aquifer's potentiometric surface, the presence of aquitards (low-permeability strata) and aquicludes (very low permeability strata), and areas of aquifer communication. Supporting data can be derived from the geologic exploration program, which may include drill-hole logs, geotechnical studies, geophysical logs, geological cross sections, monitor wells, and structure contour maps. Aquifer hydraulic characteristics must be derived from actual pump tests, which have been conducted on or near the mine property.

Aquifer Testing

Aquifer tests should be used to determine transmissivity, storage coefficient, hydrologic boundaries, leakage, aquifer homogeneity, and isotropy. For example, an aquifer pump test can be used to determine these groundwater properties in accordance with methods described by Theis (1935), Cooper and Jacob (1946), Boulton (1954), and Lohman (1979). The location and number of aquifer tests should be sufficient to characterize the different aquifers or the variability within the aquifer in the vicinity of the mine. The purpose of the data collection effort will dictate the degree of investment. For example, an in-situ mine requires an extensive network of wells and aquifer tests to not only define the hydrologic unit from which resource extraction is taking place but also to define the overlying and underlying aquifer and confining units. A coal or metal mine may require an aquifer testing program to establish baseline data and to ensure that the mine workings remain dry during and in some cases following the extraction process. Geochemical change, where groundwater passes through spent mine workings, is a long-term surface and underground mining concern.

Aquifer tests include single-well and multiple-well tests. A single-well test can allow the specialist to establish permeability or transmissivity of an aquifer on a local scale. A multiple-well test can allow for the calculation of storativity, as well as provide data for the degree of communication between a production zone and its overlying and underlying counterparts. A typical aquifer test includes the placement and testing of a pump in the pumping well and monitoring water level changes in that well and any other well in the immediate area. Well completion and aquifer testing are well presented in the literature (Sterrett 2007). The length of an aquifer test may change depending on its relative importance in the design process. A short-term test (1 to 4 hours) typically provides information in the immediate area of the well bore. An intermediate-length test (4 to 24 hours) provides good information relative to the aquifer in general. A long-term test (24 hours to 30 days) provides excellent information relative to the aquifer, aquifer sustainability, and area boundary conditions. A good guideline is that the longer the test, the greater the aquifer area tested. When designing an extensive underground mine operation, dependent on dewatering, a long-term multiple-well test is

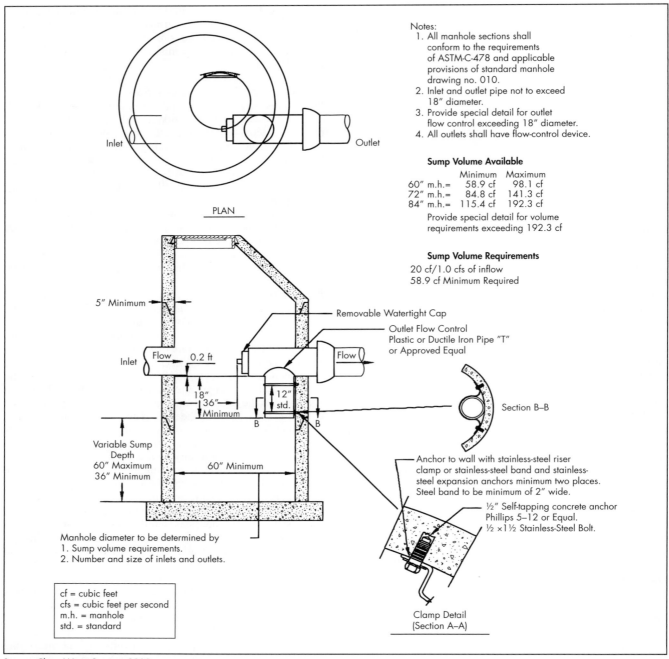

Source: Clean Water Services 2009.

Figure 16.4-4 Stormwater manhole and sediment separator

recommended. It is important to collect and maintain meteorological data throughout the course of the test.

Methods to Handle Groundwater Inflow

Groundwater can be handled before mine interaction, on interception by the mine, or after it passes through the mine. On a water quality and treatment basis, it is far better to handle groundwater before it interacts with the mine or mine workings. This can be done simply using a dewatering well field, or it can be done using slightly more complex and far more permanent slurry walls or a grout curtain. A dewatering well field requires closely spaced wells that are pumped continuously, and the respective cone of depression of each well will overlap and ensure that no groundwater passes through the well field. This is an effective means of controlling groundwater flow, but it requires nearly constant operation and maintenance. The advantages are its relatively low capital cost, its system design flexibility, and its short-term effectiveness. The disadvantages are the operational costs of keeping pumps and well screens operational, the electrical costs, and the need to move the system as the mine operation advances.

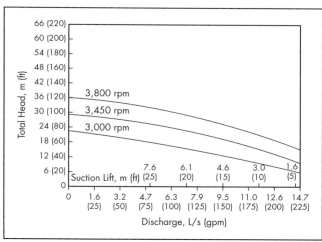

Source: Warner 1992.

Figure 16.4-5 Pump characteristic curves for engine-driven pump

A far more effective and long-term method is the construction of a grout curtain, which includes the drilling of exploration boreholes into a lower confining layer and emplacement (injection) of a concrete grout or, in limited cases, a freezing agent such as liquid nitrogen. Drill-hole spacing must be adequate to ensure the overlapping of the grout to guarantee an impermeable seal. In the case of the concrete grout curtain, its advantage is its relative permanent characteristic such that after the mine closes, groundwater will continue to be diverted around the mining operation, thereby avoiding the long-term geochemical impacts. The disadvantages are the initial capital cost and its overall ineffectiveness in some fractured aquifer systems.

A recent advancement in groundwater interception for mining is the use of slurry walls. This technique is far more effective than a grout wall in that it is constructed as a continuous trench that is backfilled with a bentonite/sand mix. Slurry walls are keyed into a lower confining layer, but may be limited by equipment reach and depth. Recent advances in slurry-wall construction allow depths up to 24 m (70 ft). The advantages of slurry walls are their relatively lower capital cost compared to grout walls and their relatively permanent behavior, even after the mine closes, so that groundwater will continue to be diverted around the mining operation. The disadvantages of such slurry walls are that they are limited to relatively shallow applications and they require a nearby source of clean sand. They are particularly effective in sand and gravel mining near or adjacent to the floodplain of a river.

When groundwater enters the mine operation, it must be collected in a sump (at a low spot within the mine workings) and removed from the mine operation by mechanical means. This is generally accomplished with a pump or a series of pumps. Pump and pipeline redundancy is absolutely critical if there is a pump or pipeline failure. Another critical concern is electrical failure and protection of the dewatering controls. Backup generators should be available at all mine sites, and the system of controls should be protected from a rising water level and/or placed at a location that is above the water table if all pumps were shut off (i.e., the recovered potentiometric surface).

Ideally, groundwater should be intercepted before it has an opportunity to be contaminated by mine equipment or oxidized waste rock. For this reason, sumps can be constructed on highwall benches or at the leading edge of an underground mine. If the groundwater is unaffected, it can be pumped directly to the surface streams. If it is affected, it must be handled in the same fashion as surface water and treated through sediment or treatment ponds. In most cases, additional treatment for mobilized metals will be required beyond pure sediment control, and this may include, but not be limited to, the addition of lime (pH control), organic matter (chelation), "bugs" (sulfate-reducing agents), or barium chloride (radium control).

Pumps

Pumps are widely used throughout the mining industry for dewatering underground mine operations and active pits, and for predraining areas adjacent to the pit prior to excavation (Warner 1992). Having the ability to control the presence of groundwater or surface water can alleviate numerous production problems and provide safer working conditions.

Centrifugal Pump

The basic components of a centrifugal pump are the inlet, the eye of the rotating impeller, the curved impeller vanes, the volute (i.e., casing), and the discharge connection. A centrifugal pump is a simple device with the only moving part being the impeller, which is attached to the shaft of a motor or engine. Assuming that the suction pipe and pump housing (volute) is filled with water to the level of the impeller eye (i.e., it is primed), then as the impellers rotate, a partial vacuum is created that allows atmospheric pressure to lift more water into the pump housing. The water, which enters the inlet opening at the eye of the impeller, is set in rotation by the impeller and creates centrifugal force, resulting in pressure at the outer perimeter of the impeller. Water moves outward from the impeller at a high velocity and pressure into an expanding volute and is then discharged.

Calculating Dynamic Head

The essential information needed to properly select a pump is the required flow rate in liters per second (gallons per minute) and the total dynamic head in meters (feet). Flow rate is dependent primarily on operational factors such as the anticipated rate of flow into a pit or underground mine. Determination of the total dynamic head requires consideration of the suction lift; the elevation difference between the pump and the discharge point; the pressure requirements, if any, at the discharge outlet; the friction along the pipes and fittings; and the velocity head.

Representative pump characteristic curves are illustrated in Figure 16.4-5. The curves show the relationship between the discharge and the total dynamic head. Also shown along the x-axis is the suction lift or net positive suction head. The discharge points along the x-axis indicate the maximum flows that the pump is capable of delivering at designated suction lifts.

Pump Troubleshooting

A properly designed and installed pump will reduce potential operational problems. A vacuum gauge, installed a minimum of one pipe diameter from the pump inlet, can solve the majority of problems. A high gauge reading indicates a partially blocked suction line, whereas a low reading often indicates air leakage through fittings on the suction side. A vacuum gauge can easily detect common problems, such as a nonprimed pump; too high a dynamic lift; excessive air in the water due

Table 16.4-5 Annotated list of computer modeling software

Use	Software	Source*
Surface water hydrology	WinTR55	USDA NRCS
	HEC-1	USACE
Surface water hydraulics	HEC-RAS	USACE
	HSPF	USGS, USEPA
	Flowmaster	Bentley
Culvert hydraulics	HY-8	USDOT, FHWA
Site grading, geomorphology	Carlson Natural Regrade with Geofluv	Carlson Software
Groundwater, aquifer characteristics	AQTESOLV	Hydrosolve
Groundwater modeling	Visual MODFLOW	Waterloo Hydrogeologic
Hydrology, sediment and structures	SEDCAD	Civil Software Design LLC
Sediment yield and transport	SIMSED99	USACE
Sediment transport	HEC-HMS	USACE
	GSTARS	USBR
	SMS	Scientific Software Group
Erosion	WEPP	USDA-ARS
Water quality	WATEQ4F	USGS
	AquaChem	Schlumberger Water Services
	QUAL2E	USEPA

*FHWA = Federal Highway Administration
NRCS = Natural Resources Conservation Service
USACE = U.S. Army Corps of Engineers
USBR = U.S. Bureau of Reclamation
USDA = U.S. Department of Agriculture
USDA-ARS = U.S. Department of Agriculture–Agricultural Research Service
USDOT = U.S. Department of Transportation
USEPA = U.S. Environmental Protection Agency
USGS = U.S. Geological Survey

to the intake being near the water surface; air leakage in the inlet pipe, fittings, or stuffing box; or the strainer, foot valve, or suction pipe that is too small or restricted by debris. If a discharge pressure gauge reads too low, the pump may not be primed, the revolutions per minute may be too low, the total dynamic head may be too high, the impeller may be rotating in the wrong direction, there may be excessive air in the water, there may be mechanical defects, and so forth. Excess pump vibration may indicate a misalignment, worn bearings, a bent shaft, or that an obstruction has lodged in one side of the impeller.

COMPUTER MODELS AND REFERENCES

Utilization of computer applications and software can offer readily available professional assistance to the mining engineer. Simulation models and software are very important tools that are used to evaluate both surface and groundwater hydrology as well as sediment and erosion processes. Much of this software is publicly available and can be downloaded for free or at a nominal charge. Table 16.4-5 identifies some of the currently available software that has been successfully used by the environmental professional, with a certain bias toward publicly available and free software. An excellent reference (including downloadable technology transfer) for design manuals and software used by the mining industry is the U.S. Department of the Interior, Office of Surface Mining Web site (DOI 2009).

A variety of software programs exist that allow the mining professional to evaluate both existing and proposed conditions. Not only are the modeling programs valuable on their own, but they can also be interfaced or joined with AutoCAD design software by Autodesk or with GIS software by ESRI.

REFERENCES

Acrement, G.J., and Schneider, V.R. 1984. *Guide for Selecting Manning's Roughness Coefficients for Natural Channels and Flood Plains.* Report No. FHWA-TS-84-204. Springfield, VA: Federal Highway Administration, U.S. Department of Transportation, National Technical Information Service.

Arkell, R.E., and Richards, F. 1986. Short duration rainfall relations for the Western United States. Reprinted from the preprint volume of the Conference on Climate and Water Management—A Critical Era and Conference on the Human Consequences of 1985s Climate, Asheville, NC, Aug. 4–7. Boston: American Meteorological Society.

Barrett, M.E., Malina, J.F., and Charbeneau, R.J. 1998. An evaluation of geotextiles for temporary sediment control. *Water Environ. Res.* 70(3):283–290.

Boulton, N.S. 1954. Drawdown of the water table under nonsteady conditions near a pumped well in an unconfined formation. In *Proceedings of the Institution of Civil Engineers*, Vol. 3. London: Institution of Civil Engineers. pp. 564–579.

Chow, V.T. 1959. *Open Channel Hydraulics.* Tokyo: McGraw-Hill.

Clean Water Services. 2009. Drawing 240. Washington County, Oregon. www.cleanwaterservices.org. Accessed 2009.

Cooper, H.H., Jr., and Jacob, C.E. 1946. A generalized graphical method for evaluating formation constants and summarizing well-field history. In *Transactions American Geophysical Union*, Vol. 27. Washington DC: American Geophysical Union. pp. 526–534.

DOI (U.S. Department of the Interior). 2009. Office of Surface Mining Reclamation and Enforcement (OSM), National Technology Development and Transfer. www.techtransfer.osmre.gov/. Accessed July 2009.

Ferris, F.K. 1996. *Handbook of Western Reclamation Techniques.* Washington, DC: U.S. Department of Interior, Office of Surface Mining.

FHWA (Federal Highway Administration). 2009. HY-8 software. Available at www.fhwa.dot.gov.

Haan, C.T., and Barfield, B.J. 1978. *Hydrology and Sedimentology of Surface Mined Lands.* Lexington, KY: University of Kentucky Office of Continuing Education and Extension.

Jones, C.M., and Lidstone, C.D. 1996. Drop structures. In *Handbook of Western Reclamation Techniques.* Edited by F.K. Ferris. Washington, DC: U.S. Department of the Interior, Office of Surface Mining. pp. II 21–32.

Kirpich, P.Z. 1940. Time of concentration of small agricultural watersheds. *Civil Eng.* 10(6):362.

Kleinman, L.H. 1996. Vegetative surface stabilization. In *Handbook of Western Reclamation Techniques.* Edited by F.K. Ferris. Washington, DC: U.S. Department of the Interior, Office of Surface Mining. pp. V 83–89.

Korte, A., and MacDonald, L.H. 2005. Road sediment production and delivery in the southern Sierra Nevada, California. Abstract H51E-0416. *Trans. Am. Geophys. Union* 86(52).

Lane, E.W. 1955. The importance of fluvial morphology in hydraulic engineering. In *Proceedings of the American Society of Civil Engineering*. Paper 745. pp. 1–17.

Lohman, S.W. 1979. *Ground-Water Hydraulics*. U.S. Geological Survey Professional Paper 708. Washington, DC: Government Printing Office.

MSHA (Mine Safety and Health Administration). 1975. 30CFR77.216. *Water, Sediment, or Slurry Impoundments and Impounding Structures; General*. Arlington, VA: MSHA. Available from www.msha.gov.

NOAA (National Oceanic and Atmospheric Administration). 2009. National Hydrometeorological Design Studies Center. www.weather.gov/ohd/hdsc/. Accessed July 2009.

Norman, D., Wampler, P., Throop, A., Schnitzer, F., and Roloff, J. 1998. *Best Management Practices for Reclaiming Surface Mines in Washington and Oregon*. Open-File Report 0-96-02. Portland, OR: Oregon Department of Geology and Mineral Industries.

North American Green. 2009. Comprehensive Erosion/Sediment Control and Turf Reinforcement Solutions. www.nagreen.com. Accessed July 2009.

NWS (National Weather Service). 1961. Rainfall Frequency Atlas of the United States. Technical Paper 40. Washington, DC: U.S. Department of Commerce.

Ragan, R.M., and Duru, J.O. 1972. Kinematic wave nomograph for times of concentration. In *Proceedings of the American Society of Civil Engineers*. 98(HY10):1765–1771.

RoLanka International. 2009. Resources for Effective Erosion and Sediment Control. www.geo-naturals.com/GN/index.html. Accessed July 2009.

Schwab, G.O., Frevert, R.K., Edminster, T.W., and Barnes, K.K. 1966. *Soil and Water Conservation Engineering*. New York: John Wiley and Sons.

SCS (Soil Conservation Service). 1973. *A Method for Estimating Volume and Rate of Runoff in Small Watersheds*. Publication SC-TP-149. Washington DC: Department of Agriculture.

SCS (Soil Conservation Service). 1982. Ponds—Planning, design, construction. In *U.S. Soil Conservation Service Agricultural Handbook 590*. Washington DC: Department of Agriculture.

Seybert, T.A. 2006. *Stormwater Management for Land Development: Methods and Calculations for Quality Control*. Hoboken, NJ: John Wiley and Sons.

Simons, Li and Associates. 1982a. *Design Manual for Sedimentation Control Through Sedimentation Ponds and Other Physical and Chemical Treatment*. Washington, DC: U.S. Department of Interior, Office of Surface Mining.

Simons, Li and Associates. 1982b. *Design Manual for Water Diversions on Surface Mine Operations*. Washington, DC: U.S. Department of the Interior, Office of Surface Mining.

Sterling, H.J., and Warner, R.C. 1984. The inertial separator as a sediment control device. In *Proceedings, 1984 Symposium on Surface Mining, Hydrology, Sedimentology, and Reclamation*. Lexington, KY: University of Kentucky.

Sterrett, R.J. 2007. *Groundwater and Wells*, 3rd ed. New Brighton, MN: Johnson Screens.

Theis, C.V. 1935. The relation between the lowering of the piezometric surface and the rate and duration of discharge of a well using groundwater storage. In *Trans. Am. Geophys. Union* 2:519–524.

30USC1265. Section 515(b)(10)B). Title 30: Mineral Lands and Mining, Environmental Protection Standards.

Warner, R.C. 1992. Design and management of water and sediment control systems. In *SME Mining Engineering Handbook*, 2nd ed. Littleton, CO: SME.

Warner, R.C., and Dysart, B.C., III. 1983. Potential use of the swirl concentrator for sediment control on surface mined lands. In *Proceedings, 1984 Symposium on Surface Mining, Hydrology, Sedimentology, and Reclamation*. Lexington, KY: University of Kentucky.

Wischmeier, W.H., and Smith, D.D. 1978. *Predicting rainfall erosion losses—A guide to conservation planning*. Agricultural Handbook No. 537. Washington, DC: U.S. Department of Agriculture.

CHAPTER 16.5

Mitigating Acid Rock Drainage

Rens Verburg

INTRODUCTION

Sustainable mining requires the prevention, mitigation, management, and control of mining impacts on the environment. Acid rock drainage (ARD) continues to be one of the most serious and visible environmental issues facing the mining industry because it is often the transport medium for a range of pollutants, which may affect on-site and off-site water resources, and associated human and ecological receptors. The impacts of ARD on near and distant water resources and receptors can also be long term and persist after mine closure. Therefore, ARD prevention, mitigation, and treatment are important components of overall mine water management over the entire life of a mining operation.

This chapter addresses the prediction, prevention, and treatment of ARD. A comprehensive approach to ARD management reduces the environmental risks and subsequent costs for the mining industry and governments, reduces adverse environmental impacts, and promotes public support for mining. The extent and particular elements of the ARD management approach that should be implemented at a certain operation will vary based on many site-specific factors, not limited to the project's potential to generate ARD.

Acid rock drainage is formed by the natural oxidation of sulfide minerals when exposed to air and water. Activities that involve the excavation of rock with sulfide minerals, such as metal and coal mining, accelerate the process. ARD results from a series of reactions and stages that typically proceed from near-neutral to more acidic pH conditions. When sufficient base minerals are present to neutralize the ARD, neutral mine drainage or saline drainage may result from the oxidation process. Neutral mine drainage (NMD) is characterized by elevated metals in solution at approximately neutral pH, whereas saline drainage (SD) contains high levels of sulfate at neutral pH without significant dissolved metal concentrations. Figure 16.5-1 illustrates the various types of drainage schematically.

A risk-based planning and design approach forms the basis for prevention and mitigation. This approach is applied throughout the mine life cycle but primarily in the assessment and design phases. The risk-based process aims to quantify the long-term impacts of alternatives and to use this knowledge to select the option that has the most desirable combination of attributes (e.g., protectiveness, regulatory acceptance, community approval, cost). Mitigation measures implemented as part of an effective control strategy should require minimal active intervention and management.

Proper mine characterization, drainage-quality prediction, and mine waste management can prevent ARD formation in most cases and minimize ARD formation in all cases. Prevention of ARD must commence at exploration and continue throughout the mine life cycle. Ongoing ARD planning and management is critical to the successful prevention of ARD.

Stopping ARD formation, once initiated, may be challenging because it is a process that, left unimpeded, will continue (and may accelerate) until one or more of the reactants (sulfide minerals, oxygen, water) are exhausted or excluded from reaction. The ARD formation process can continue to produce impacted drainage for decades or centuries after mining has ceased, as is illustrated by the portal in Spain shown in Figure 16.5-2, which dates from the Roman era.

The cost of ARD remediation at orphaned mines in North America alone has been estimated in the tens of billions of U.S. dollars. Individual mines can face postclosure liabilities of tens to hundreds of million dollars for ARD remediation and treatment if the sulfide oxidation process is not properly managed during the mine's life.

This chapter draws heavily from and follows the general structure of the *Global Acid Rock Drainage Guide* (*GARD Guide*), a state-of-practice summary of the best practices and technologies. It was developed under the auspices of the International Network for Acid Prevention (INAP) to assist ARD stakeholders, such as mine operators, regulators, communities, and consultants, with addressing issues related to sulfide mineral oxidation (INAP 2009). Readers are encouraged to make use of the *GARD Guide* and its references for further detail on the subjects covered in this chapter.

FORMATION OF ACID ROCK DRAINAGE

The process of sulfide oxidation and formation of ARD is very complex and involves a multitude of chemical and biological processes that can vary significantly depending on

---

Rens Verburg, Principal Geochemist, Golder Associates, Inc., Redmond, Washington, USA

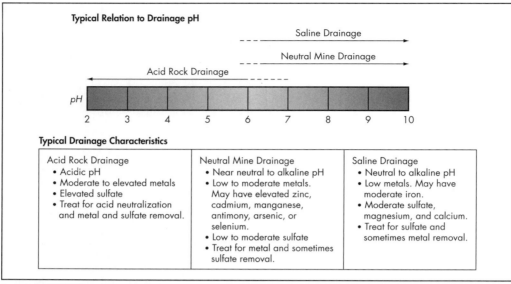

Source: INAP 2009.
Figure 16.5-1 Types of drainage produced by sulfide oxidation

Figure 16.5-2 Roman portal with acid rock drainage—Spain

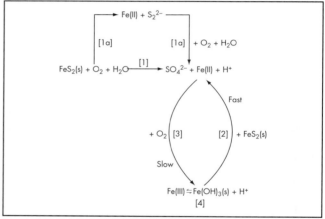

Source: Stumm and Morgan 1981.
Figure 16.5-3 Model for the oxidation of pyrite

environmental, geological, and climate conditions (Nordstrom and Alpers 1999). Sulfide minerals in ore deposits are formed under reducing conditions in the absence of oxygen. When exposed to atmospheric oxygen or oxygenated waters due to mining, mineral processing, excavation, or other earth-moving processes, sulfide minerals can become unstable and oxidize.

Figure 16.5-3 presents a simplified model that describes the oxidation of pyrite, the sulfide mineral responsible for the majority of ARD (Stumm and Morgan 1981). The reactions shown are schematic and may not represent the exact mechanisms or reaction stoichiometry, but the illustration is a useful visual aid for understanding sulfide oxidation.

The chemical reaction representing pyrite oxidation (Reaction 1 in Figure 16.5-3) requires three basic ingredients: pyrite, oxygen, and water. This reaction can occur either abiotically or biotically (i.e., mediated through microorganisms). In the latter case, bacteria such as *Acidithiobacillus ferrooxidans*, which derive their metabolic energy from oxidizing ferrous to ferric iron, can accelerate the oxidation reaction rate by many orders of magnitude relative to abiotic rates (Nordstrom 2003). In addition to direct oxidation, pyrite can also be dissolved and then oxidized (Reaction 1a).

Under the majority of circumstances, atmospheric oxygen acts as the oxidant. However, aqueous ferric iron can oxidize pyrite as well according to Reaction 2. This reaction is

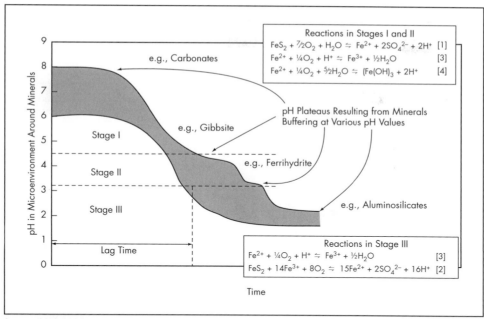

Source: INAP 2009.

Figure 16.5-4 Stages in the formation of ARD

considerably faster (two to three orders of magnitude) than the reaction with oxygen, and generates substantially more acidity per mole of pyrite oxidized. However, this reaction is limited to conditions in which significant amounts of dissolved ferric iron occur (i.e., acidic conditions: pH 4.5 and lower). Oxidation of ferrous iron by oxygen (Reaction 3) is required to generate and replenish ferric iron, and acidic conditions are required for the latter to remain in solution and participate in the ARD production process. As indicated by this reaction, oxygen is needed to generate ferric iron from ferrous iron. Also, the bacteria that may catalyze this reaction (primarily members of the *Acidithiobacillus* genus) demand oxygen for aerobic cellular respiration. Therefore, some nominal amount of oxygen is needed for this process to be effective, even when catalyzed by bacteria, although the oxygen requirement is considerably less than for abiotic oxidation.

A process of environmental importance related to ARD generation pertains to the fate of ferrous iron resulting from Reaction 1. Ferrous iron can be removed from solution under slightly acidic to alkaline conditions through oxidation and subsequent hydrolysis and the formation of a relatively insoluble iron (hydr)oxide (Reaction 4). When Reactions 1 and 4 are combined, as is generally the case when conditions are not acidic (i.e., pH > 4.5), oxidation of pyrite produces twice the amount of acidity relative to Reaction 1 as follows:

$$FeS_2 + 15/4 O_2 + 7/2 H_2O = Fe(OH)_3 + 2SO_4^{2-} + 4H^+$$

which is the overall reaction most commonly used to describe pyrite oxidation.

Although pyrite is by far the dominant sulfide responsible for the generation of acidity, different ore deposits contain different types of sulfide minerals. Not all of these sulfide minerals generate acidity when being oxidized. As a general rule, iron sulfides (pyrite, marcasite, pyrrhotite), sulfides with molar metal/sulfur ratios less than 1, and sulfosalts (e.g., enargite) generate acid when they react with oxygen and water. Sulfides with metal/sulfur ratios equal to 1 (e.g., sphalerite, galena, chalcopyrite) tend not to produce acidity when oxygen is the oxidant. Therefore, the acid generation potential of an ore deposit or mine waste generally depends on the amount of iron sulfide present. However, when aqueous ferric iron is the oxidant, all sulfides are capable of generating acidity.

Neutralization reactions also play a key role in determining the compositional characteristics of drainage originating from sulfide oxidation. As for sulfide minerals, the reactivity, and accordingly the effectiveness with which neutralizing minerals are able to buffer any acid being generated, can vary widely. Most carbonate minerals are capable of dissolving rapidly, making them effective acid consumers. However, hydrolysis of dissolved Fe or Mn following dissolution of their respective carbonates and subsequent precipitation of a secondary mineral may generate acidity. Although generally more common than carbonate phases, aluminosilicate minerals tend to be less reactive, and their buffering may only succeed in stabilizing the pH when rather acidic conditions have been achieved. Calcium–magnesium silicates have been known to buffer mine effluents at neutral pH when sulfide oxidation rates were very low (Jambor 2003).

The combination of acid generation and acid neutralization reactions typically leads to a stepwise development of ARD (Figure 16.5-4). Over time, pH decreases along a series of pH plateaus governed by the buffering of a range of mineral assemblages. The lag time to acid generation is a very important consideration in ARD prevention. It is far more effective (and generally far less costly in the long term) to control ARD generation during its early stages. The lag time also has significant ramifications for interpretation of test results. Because the first stage of ARD generation may last for a very long time, even for materials that will eventually be highly acid generating, it is critical to recognize the stage of oxidation when predicting ARD potential. The early results

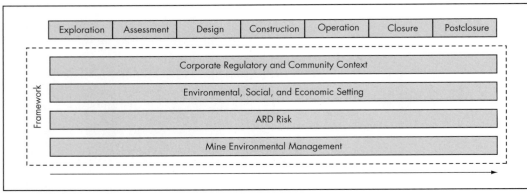

Source: INAP 2009.
Figure 16.5-5 Conceptual ARD management framework

of geochemical testing, therefore, may not be representative of long-term environmental stability and associated discharge quality. However, early test results provide valuable data to assess future conditions such as consumption rates of available neutralizing minerals.

A common corollary of sulfide oxidation is metal leaching (ML), leading to the frequent use of the acronyms *ARD/ML* or *ML/ARD* to describe the nature of acidic mine discharges more accurately. Major and trace metals in ARD, NMD, and SD originate from the oxidizing sulfides and dissolving acid-consuming minerals. In the case of ARD, Fe and Al are usually the principal major dissolved metals, while trace metals such as Cu, Pb, Zn, Cd, Mn, Co, and Ni can also achieve elevated concentrations. In mine discharges with a more circumneutral character, trace metal concentrations tend to be lower as the result of formation of secondary mineral phases and increased sorption. However, certain parameters remain in solution as the pH increases, in particular the metalloids As, Se, and Sb, as well as other trace metals (e.g., Cd, Cr, Mn, Mo, and Zn).

FRAMEWORK FOR ACID ROCK DRAINAGE MANAGEMENT

The issues and approaches to ARD prevention and management are the same around the world. However, the specific techniques used for ARD prediction, interpretation of ARD test results, and ARD management may differ depending on the local, regional, or country context and are adapted to climate, topography, and other site conditions.

Therefore, despite the global similarities of ARD issues, there is no "one size fits all" approach to address ARD management. The setting of each mine is unique and requires a carefully considered assessment to find a management strategy within the broader corporate, regulatory, and community framework that applies to the project in question. The site-specific setting comprises the social, economic, and environmental situation within which the mine is located; the framework comprises the applicable corporate, regulatory norms and standards and community-specific requirements and expectations. This framework applies over the complete life cycle of the mine and is illustrated conceptually in Figure 16.5-5.

All mining companies, regardless of size, need to comply with the national legislation and regulations pertaining to ARD for the countries within which they operate. It is considered good corporate practice to adhere to global ARD guidance as well; in many cases, such adherence is a condition of funding.

Many mining companies have established clear corporate guidelines that represent the company's view of the priorities to be addressed and their interpretation of generally accepted best practice related to ARD. Caution is needed to ensure that all specifics of the country's regulations are met, as corporate ARD guidelines cannot be a substitute for country regulations.

Mining companies operate within the constraints of a "social license" that, ideally, is based on a broad consensus from all stakeholders. This consensus tends to cover a broad range of social, economic, environmental, and governance elements (sustainable development). ARD plays an important part in the mine's social license because it tends to be one of the more visible environmental consequences of mining.

The costs of closure and postclosure management of ARD are increasingly recognized as a fundamental component of all proposed and operating mining operations. Some form of financial assurance is now required in many jurisdictions.

Characterization

The generation, release, transport, and attenuation of ARD are intricate processes governed by a combination of physical, chemical, and biological factors. Whether ARD becomes an environmental concern depends largely on the characteristics of the sources, pathways, and receptors involved. Characterization of these aspects is therefore crucial to the prediction, prevention, and management of ARD.

Environmental characterization programs are designed to collect sufficient data to answer the following questions:

1. Is ARD likely to occur?
2. What are the sources of ARD?
3. How much ARD will be generated and when?
4. What are the significant pathways that transport contaminants to the receiving environment?
5. What human and ecological receptors have contact with the receiving environment?
6. What level of risk would the anticipated concentrations of contaminants in the receiving environment pose to the receptors?
7. What can be done to prevent or mitigate/manage ARD?

The geologic and mineralogical characteristics of the ore body and host rock are the principal controls on the type of drainage that will be generated as a result of mining. Subsequently, the site climatic and hydrologic/hydrogeologic characteristics define how mine drainage and its constituents

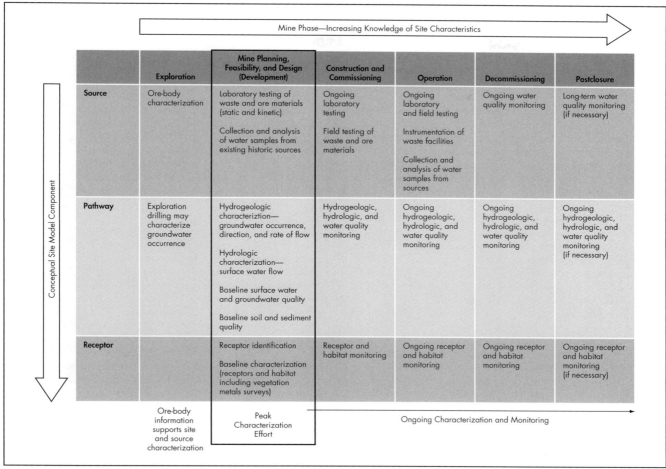

Figure 16.5-6 Overview of ARD characterization program by mine phase

are transported through the receiving environment to receptors. To evaluate these issues, expertise from multiple disciplines is required, including geology, mineralogy, hydrology, hydrogeology, geochemistry, (micro)biology, meteorology, and engineering.

The geologic characteristics of mineral deposits exert important and predictable controls on the environmental signature of mineralized areas (Plumlee 1999). Therefore, a preliminary assessment of the ARD potential should be made based on a review of geologic data collected during exploration. Baseline characterization of metal concentrations in various environmental media (i.e., water, soils, vegetation, animal tissue) may also provide an indication of ARD potential and serves to document the potential for naturally elevated metal concentrations.

During mine development and operation, the initial assessment of ARD potential is refined through detailed characterization data on the environmental stability of the waste and ore materials. The magnitude and location of mine discharges to the environment also are identified during mine development. Meteorological, hydrological, and hydrogeological investigations are conducted to characterize the amount and direction of water movement within the mine watershed(s) to evaluate transport pathways for constituents of interest. Identification of potential human and ecological receptors within the watershed boundary requires expertise in field biology and public health.

Over the mine life, the focus of the ARD characterization program evolves from establishing baseline conditions to predicting drainage release and transport and to monitoring of the environmental conditions, receptors, and impacts. Despite inherent differences at mine sites (e.g., based on commodity type, climate, mine phase, regulatory framework), the general approach to site characterization is similar.

Figures 16.5-6 and 16.5-7 present the chronology of an ARD characterization program and identify the data collection activities typically executed during each mine phase. The bulk of the characterization effort occurs prior to mining during the mine planning, assessment, and design (sometimes collectively referred to as the development phase). In addition, potential environmental impacts are identified and appropriate prevention and mitigation measures, intended to minimize environmental impacts and human and ecological risk, are incorporated. During the commissioning/construction and operation phases, a transition from site characterization to monitoring occurs, which is continued throughout the decommissioning/closure and postclosure phases. Ongoing monitoring helps refine the understanding of the site, which allows for adjustment of remedial measures, in turn resulting in reduced closure costs and improved risk management.

Waste or Facility Type—Potential ARD/NMD/SD Sources		Mine Phase—Increasing Knowledge of Source Material Characterization					
		Exploration	Mine Planning, (Pre-) Feasibility and Design	Construction and Commissioning	Operation	Decommissioning	Postclosure (Care and Maintenance)
	Waste Rock	Drill core descriptions and assay data (petrology and mineralogy) Block model (quantity of ore and waste) Review of any historical data	Laboratory testing of drill core samples—sample selection targets waste*	Ongoing laboratory testing of drill core or development rock samples* Field leach testing (barrels, test pads)	Ongoing laboratory testing* Ongoing field leach testing Collection and analysis of runoff and seepage samples from waste rock facility	Collection and analysis of runoff and seepage samples from waste rock facility	Collection and analysis of runoff and seepage samples from waste rock facility (if necessary)
	Tailings		Laboratory testing of pilot plant tailings* Analysis of pilot testing supernatant	Ongoing laboratory testing of pilot-plant tailings*	Ongoing laboratory testing of tailings discharge* Collection and analysis of supernatant and seepage samples from tailings storage facility	Collection and analysis of supernatant and seepage samples from tailings storage facility	Collection and analysis of supernatant and seepage samples from tailings storage facility (if necessary)
	Ore		Laboratory testing of drill core samples*		Ongoing laboratory testing*	If ore stockpiles exist, collection and analysis of runoff and seepage samples	
	Pit		Laboratory testing of drill core samples—sample selection targets pit walls*		Field scale leach testing (e.g., wall washing) Collection and analysis of water samples (i.e., runoff, sumps)	Collection and analysis of pit water and pit inflow(s) water samples	Collection and analysis of pit water samples (if necessary)
	Underground Workings		Laboratory testing of drill core samples—sample selection targets mine walls*		Collection and analysis of water samples (i.e., sumps, dewatering wells)	Collection and analysis of mine pool water samples	Collection and analysis of mine pool water samples (if necessary)

*Typical laboratory testing components: particle size, whole rock analysis, mineralogy, acid–base accounting, static and kinetic leach testing.

Source: INAP 2009.

Figure 16.5-7 ARD characterization program for individual source materials by mine phase

Prediction

One of the main objectives of site characterization is prediction of ARD potential and drainage chemistry. Because prediction is directly linked to mine planning, in particular with regard to water and mine waste management, the characterization effort needs to be phased in step with overall project planning. Early characterization tends to be generic and generally avoids presumptions about the future engineering/mine design, whereas later characterization and modeling must consider and be integrated with the specifics of engineering/mine design. Iteration may be required as evaluation of the ARD potential may result in the realization that a reassessment of the overall mine plan is needed. Integration of the characterization and prediction effort into the mine operation is a key element for successful ARD management.

Accurate prediction of future mine discharges requires an understanding of the sampling, testing, and analytical procedures used, consideration of the future physical and geochemical conditions, and the identity, location, and reactivity of the contributing minerals. All mine sites are unique for reasons related to geology, geochemistry, climate, commodity, processing method, regulations, and stakeholders. Prediction programs therefore need to be tailored to the mine in question. Also, the objectives of a prediction program can be variable. For instance, they can include definition of water treatment requirements, selection of mitigation methods, assessment of water quality impact, or determination of reclamation bond amounts.

Predictions of drainage quality are made in a qualitative and quantitative sense. Qualitative predictions are focused on assessing whether acidic conditions might develop in mine wastes, with the corresponding release of metals and acidity to mine drainage. Where qualitative predictions indicate a high probability of ARD generation, attention turns to review of alternatives to prevent ARD, and the prediction program is refocused to assist in the design and evaluation of these alternatives.

Significant advances in the understanding of ARD have been made over the last several decades, with parallel advances in mine water quality prediction and use of prevention techniques. However, quantitative mine water quality prediction can be challenging because of the wide array of the reactions involved and potentially very long time periods over which these reactions take place. Despite these uncertainties, quantitative predictions that have been developed using realistic assumptions (while recognizing associated limitations) have proven to be of significant value for identification of ARD management options and assessment of potential environmental impacts.

Prediction of mine water quality generally is based on one or more of the following:

- Test leachability of waste materials in the laboratory
- Test leachability of waste materials under field conditions
- Geological, hydrological, chemical, and mineralogical characterization of waste materials
- Geochemical and other modeling

Analog operating or historic sites are also valuable in ARD prediction, especially those that have been thoroughly characterized and monitored. The development of geoenvironmental models is one of the more prominent examples of the "analog" methodology. Geo-environmental models, which are constructs that interpret the environmental characteristics of an ore deposit in a geologic context, provide a very useful way to interpret and summarize the environmental signatures of mining and mineral deposits in a systematic geologic framework; they can be applied to anticipate potential environmental problems at future mines, operating mines, and orphan sites (Plumlee et al. 1999). A generic overall approach for ARD prediction is illustrated in Figure 16.5-8.

Prevention and Mitigation

The fundamental principle of ARD prevention is to apply a planning and design process to prevent, inhibit, retard, or stop the hydrological, chemical, physical, or microbiological processes that result in the impacts to water resources. Prevention should occur at, or as close to, the point where the deterioration in water quality originates (i.e., source reduction), or through implementation of measures to prevent or retard the transport of the ARD to the water resource (i.e., recycling, treatment, and/or secure disposal). This principle is universally applicable but methods of implementation are site specific.

Prevention is a proactive strategy that obviates the need for the reactive approach to mitigation. For an existing case of ARD that is adversely impacting the environment, mitigation will usually be the initial course of action. Despite this initial action, subsequent preventive measures are often considered with the objective of reducing future contaminant loadings, thus reducing the ongoing need for mitigation controls. Integration of the prevention and mitigation effort into the mine operation is a key element for successful ARD management.

Prior to identification or evaluation of prevention and mitigation measures, the strategic objectives must be identified. That process should consider assessment of the following:

- Quantifiable risks to ecological systems, human health, and other receptors
- Site-specific discharge water quality criteria
- Capital, operating, and maintenance costs of mitigation or preventive measures
- Logistics of long-term operations and maintenance
- Required longevity and anticipated failure modes

Prevention is the key to avoiding costly mitigation. The primary objective is to apply methods that minimize sulfide reaction rates, metal leaching, and the subsequent migration of weathering products that result from sulfide oxidation. Such methods involve the following:

- Minimizing oxygen supply
- Minimizing water infiltration and leaching
- Reducing, removing, or isolating sulfide minerals
- Controlling pore water solution pH
- Controlling bacteria and biogeochemical processes

Factors influencing selection of the above methods include the following:

- Geochemistry of source materials and the potential of source materials to produce ARD
- Type and physical characteristics of the source, including water flow and oxygen transport
- Mine development stage (more options are available at early stages)
- Phase of oxidation (more options are available at early stages when pH is still near neutral and oxidation products have not significantly accumulated)
- Time period for which the control measure is required to be effective
- Site conditions (i.e., location, topography and available mining voids, climate, geology, hydrology and hydrogeology, availability of materials and vegetation)
- Water quality criteria for discharge
- Risk acceptance by company and other stakeholders

More than one measure, or a combination of measures, may be required to achieve the desired objective.

Typical objectives for ARD control are to satisfy environmental criteria using the most cost-effective technique. Technology selection should consider predictions for discharge water chemistry, advantages and disadvantages of treatment options, risk to receptors, and the regulatory context related to mine discharges. Figure 16.5-9 provides a generic overview of the most common ARD prevention and mitigation measures available during the various stages of the mine life cycle.

Treatment

The objectives of mine drainage treatment are varied. Recovery and reuse of mine water within the mining operations may be desirable or required for processing of ores and minerals, conveyance of materials, operational use (dust suppression, mine cooling, irrigation of rehabilitated land), and so forth. Mine drainage treatment, in this case, is aimed at modifying the water quality, so that it is fit for the intended use on or off the mine site.

A more public objective of mine water treatment is the protection of human and ecological health in cases where people or ecological receptors may come in contact with the impacted mine water through indirect or direct use of on-site and off-site water resources. Water treatment then aims to remove the pollutants contained in mine drainage to prevent or mitigate environmental impacts.

In the majority of jurisdictions, any discharge of mine drainage into a public stream or aquifer must be approved by the relevant regulatory authorities; regulatory requirements stipulate a certain mine water discharge quality or associated discharge pollutant loads. Although discharge quality standards may not be available for many developing countries in which mining occurs, internationally acceptable environmental quality standards generally still apply as stipulated by project financiers and mining company policies.

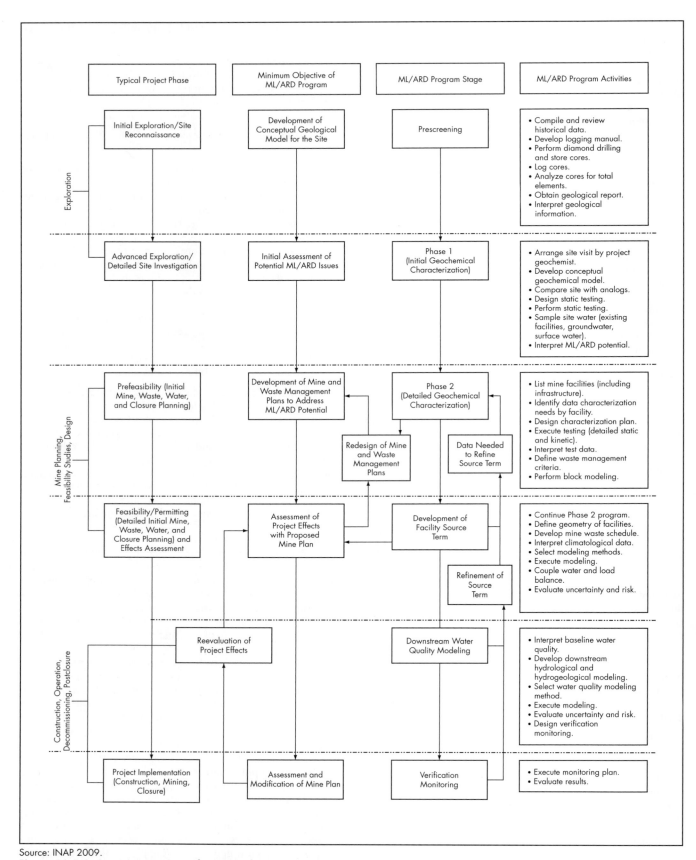

Figure 16.5-8 Generic overview of ARD prediction approach

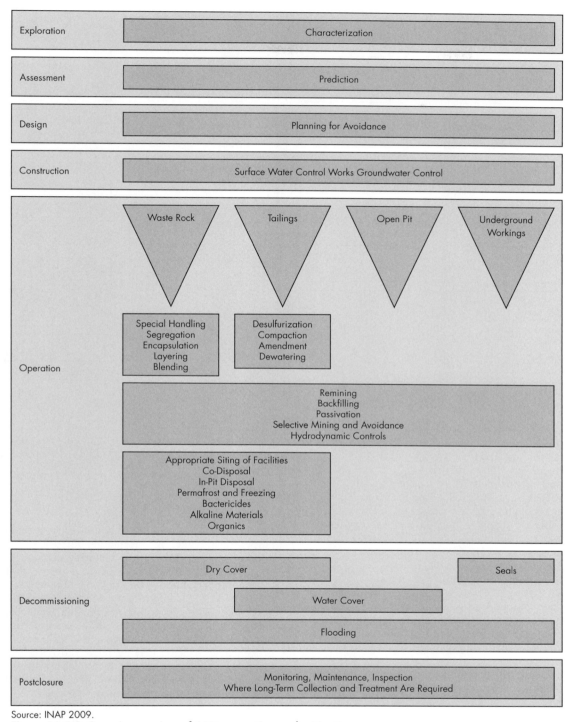

Source: INAP 2009.
Figure 16.5-9 Generic overview of ARD prevention and mitigation measures

The approach to selection of a mine drainage treatment method is premised on a thorough understanding of the integrated mine water system and circuits and the specific objective(s) to be achieved. The approach adopted for mine drainage treatment will be influenced by a number of considerations.

Prior to selecting the treatment process, a clear statement and understanding of the objectives of treatment should be prepared. Mine drainage treatment must always be evaluated and implemented within the context of the integrated mine water system. Treatment will have an impact on the flow and quality profile in the water system; hence, a treatment system is selected based on mine water flow, water quality, cost, and water use(s) and receptors.

Characterization of the mine drainage in terms of flow and chemical characteristics should include consideration

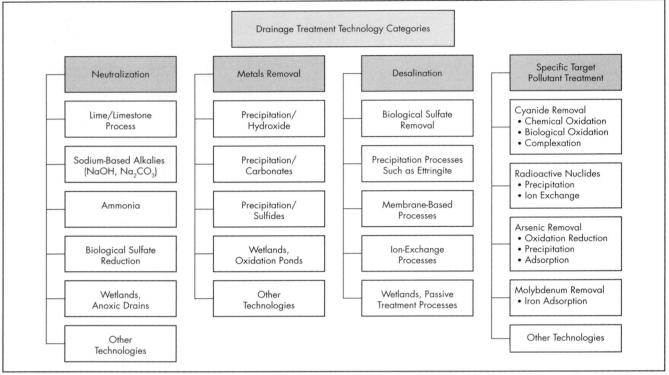

Source: INAP 2009.
Figure 16.5-10 Generic overview of ARD treatment alternatives

of temporal and seasonal changes. Flow data are especially important, as this information is required to size any treatment system properly. Of particular importance are extreme precipitation and snowmelt events that require adequate sizing of collection ponds and related piping and ditches. The key chemical properties of mine drainage relate to acidity/alkalinity, sulfate content, salinity, metal content, and the presence of specific compounds associated with specific mining operations, such as cyanide, ammonia, nitrate, arsenic, selenium, molybdenum, and radionuclides. There are also a number of mine drainage constituents (e.g., hardness, sulfate, silica), which may not be of regulatory or environmental concern in all jurisdictions but that could affect the selection of the preferred water treatment technology. Handling and disposal of treatment plant waste and residues such as sludges and brines and their chemical characteristics must also be factored into any treatment decisions.

A mine drainage treatment facility must have the flexibility to deal with increasing/decreasing water flows, changing water qualities, and regulatory requirements over the life of mine. This may dictate phased implementation and modular design and construction. Additionally, the postclosure phase may place specific constraints on the continued operation and maintenance of a treatment facility.

A variety of practical considerations related to mine site features will influence the construction, operation, and maintenance of a mine drainage treatment facility:

- Mine layout and topography
- Space
- Climate
- Sources of mine drainage feeding the treatment facility
- Nature and location of treated water receptors

Various ARD treatment alternatives are presented in Figure 16.5-10.

Monitoring
Monitoring is the process of routinely, systematically, and purposefully gathering information for use in management decision making. Mine site monitoring aims to identify and characterize any environmental changes from mining activities to assess conditions on the site and potential impacts to receptors both on- and off-site. Monitoring consists of both observation (e.g., recording information about the environment) and investigation (e.g., studies such as toxicity tests where environmental conditions are controlled). Monitoring is critical in decision making related to ARD management, for instance through assessing the effectiveness of mitigation measures and the subsequent implementation of adjustments to mitigation measures as required.

Development of an ARD monitoring program starts with a review of the mine plan, the geographical location, and the geological setting. The mine plan provides information on the location and magnitude of surface and subsurface disturbances, ore processing and milling procedures, waste disposal areas, effluent discharge locations, groundwater withdrawals, and surface water diversions. This information is used to identify potential sources of ARD, potential pathways for release of ARD to the receiving environment, receptors that may be impacted by these releases, and potential mitigation that may be required. Because the spatial extent of a monitoring

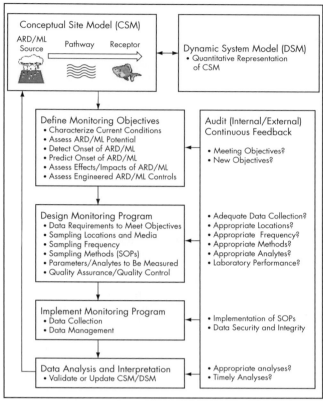

Source: INAP 2009.
Figure 16.5-11 Development of an ARD monitoring program

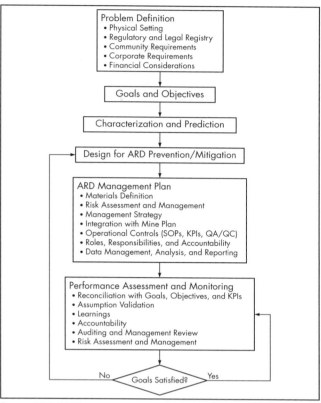

Source: INAP 2009.
Figure 16.5-12 Flow chart for ARD performance assessment and management review

program must include all these components, a watershed approach to ARD monitoring (including groundwater) is often required. Monitoring occurs at all stages of project development, from preoperational through postclosure. However, over the life of a mine, the objectives, components, and intensity of the monitoring activities will change. The development and components of a generic ARD monitoring program are presented in Figure 16.5-11.

Management and Performance Assessment

The management of ARD and the assessment of its performance are usually described within the site environmental management plan or in a site-specific ARD management plan. The ARD management plan represents the integration of the concepts and technologies described earlier in this chapter. It also references the engineering design processes and operational management systems employed by mining companies.

The need for a formal ARD management plan is usually triggered by the results of an ARD characterization and prediction program or the results of site monitoring. The development, implementation, assessment, and continuous improvement of an ARD management plan are ongoing processes throughout the life of a mine, which will typically follow the sequence of steps illustrated in Figure 16.5-12.

As shown in Figure 16.5-12, the development of an ARD management plan starts with establishment of clear goals and objectives, such as preventing ARD or achieving compliance with specific water quality criteria. This includes consideration of the physical setting, regulatory and legal registry, community and corporate requirements, and financial considerations. Characterization and prediction programs identify the potential magnitude of the ARD issue and provide the basis for the selection and design of appropriate ARD prevention and mitigation technologies. The design process includes an iterative series of steps in which ARD control technologies are assessed and then combined into a robust system of management and controls (i.e., the ARD management plan) for the specific site. The initial mine design may be used to develop the ARD management plan needed for an environmental assessment. The final design is usually developed in parallel with project permitting.

The ARD management plan identifies the materials and mine wastes that require special management. Risk assessment and management are included in the plan to refine strategies and implementation steps. To be effective, the ARD management plan must be fully integrated with the mine plan. Operational controls such as standard operating procedures (SOPs), key performance indicators (KPIs), and quality assurance/quality control (QA/QC) programs are established to guide its implementation. The ARD management plan identifies roles, responsibilities, and accountabilities for mine operating staff. Data management, analysis, and reporting schemes are included to track progress of the plan.

In the next step, monitoring is conducted to compare field performance against the design goals and objectives of the management plan. Assumptions made in the characterization and prediction programs and design of the prevention/

mitigation measures are tested and revised or validated. *Learnings,* or lessons learned during monitoring and assessment, are evaluated and incorporated into the plan as part of continuous improvement. Accountability for implementing the management plan is checked to ensure that those responsible are meeting the requirements stipulated in the plan. Internal and external reviews or audits should be conducted to gauge performance of personnel, management systems, and technical components to provide additional perspectives on the implementation of the ARD management plan. Review by site and corporate management of the entire plan is necessary to ensure that the plan continues to adhere to site and corporate policies. Finally, results are assessed against the goals. If the objectives are met, performance assessment and monitoring continues throughout the mine life with periodic rechecks against the goals. If the objectives are not met, then redesign and reevaluation of the management plan and performance assessment and monitoring systems for ARD prevention/ mitigation are required. This additional effort might also require further characterization and ARD prediction.

The process described in Figure 16.5-12 results in continuous improvement of the ARD management plan and its implementation, and accommodates possible modifications in the mine plan. If the initial ARD management plan is robust, it can be more readily adapted to mine plan changes.

Implementing the ARD management plan relies on a hierarchy of management tools. Corporate policies help define corporate or site standards, which lead to SOPs and KPIs that are specific to the site and guide operators in implementing the ARD management plan. Where corporate policies or standards do not exist, projects and operations should rely on industry best practice.

Communication and Consultation

The level of knowledge of ARD generation and mitigation has increased dramatically over the last few decades within the mining industry, academia, and regulatory agencies. However, for this knowledge to be meaningful to the wide range of stakeholders generally involved with a mining project, it needs to be translated into a format that can be readily understood. This communication should convey the predictions of future drainage quality and the effectiveness of mitigation plans, their degree of uncertainty, and contingency measures to address that uncertainty. An open dialogue on what is known, and what can be predicted with varying levels of confidence, helps build understanding and trust, and ultimately results in a better ARD management plan.

Communicating and consulting with stakeholders about ARD issues is essential to the company's social license to operate. Because of the generally highly visible nature of ARD, skilled people are needed to communicate effectively, and the involvement of representatives from all relevant technical disciplines may be required.

SUMMARY

A thorough evaluation of ARD potential should be conducted prior to mining and continued throughout the life of the mine. Consistent with sustainability principles, strategies for dealing with ARD should focus on prevention or minimization, rather than control or treatment. These strategies are formulated within an ARD management plan, to be developed in the early phases of the project together with monitoring requirements to assess its performance. Integration of the ARD management plan with the mine operation plan is critical to the success of ARD prevention or mitigation. ARD management practices continue to evolve but tend to be site specific and require special expertise.

REFERENCES

INAP (International Network for Acid Prevention). 2009. Main page. *GARD Guide: The Global Acid Rock Drainage Guide.* www.gardguide.com/index.php/Main_Page. Accessed January 2009.

Jambor, J.L. 2003. Mine-waste mineralogy and mineralogical perspectives of acid–base accounting. In *Environmental Aspects of Mine Wastes.* Edited by J.L. Jambor, D.W. Blowes, and A.I.M. Ritchie. Short Course Series Volume 31. Quebec, Canada: Mineralogical Association of Canada.

Nordstrom, D.K. 2003. Effects of microbiological and geochemical interactions in mine drainage. In *Environmental Aspects of Mine Wastes.* Edited by J.L. Jambor, D.W. Blowes, and A.I.M. Ritchie. Short Course Series, Vol. 31. Quebec, Canada: Mineralogical Association of Canada.

Nordstrom, D.K., and Alpers, C.N. 1999. Geochemistry of acid mine waters. In *The Environmental Geochemistry of Mineral Deposits, Part A: Processes, Techniques, and Health Issues.* Edited by G.S. Plumlee and M.J. Logsdon. Reviews in Economic Geology, Vol. 6A. Littleton, CO: Society of Economic Geologists.

Plumlee, G.S. 1999. The environmental geology of mineral deposits. In *The Environmental Geochemistry of Mineral Deposits, Part A: Processes, Techniques, and Health Issues.* Edited by G.S. Plumlee and M.J. Logsdon. Reviews in Economic Geology, Vol. 6A. Littleton, CO: Society of Economic Geologists.

Plumlee, G.S., Smith, K.S., Montour, M.R., Ficklin, W.H., and Mosier, E.L. 1999. Geologic controls on the composition of natural waters and mine waters draining diverse mineral-deposit types. In *The Environmental Geochemistry of Mineral Deposits, Part B: Case Studies and Research Topics.* Edited by L.H. Filipek and G.S. Plumlee. Reviews in Economic Geology, Vol. 6B. Littleton, CO: Society of Economic Geologists.

Stumm, W., and Morgan, J.J. 1981. *Aquatic Chemistry,* 2nd ed. New York: John Wiley and Sons.

CHAPTER 16.6

Waste Disposal and Contamination Management

Richard K. Borden

INTRODUCTION

Wastes generated by mining and mineral processing operations can have profound environmental impacts on land, water, and air resources if improperly managed. Mining and mineral processing generate large volumes of mineral waste, such as tailings and waste rock. At large mines the mass of mineral waste generated can commonly be measured in the tens of millions to billions of tons. Similarly, the surface area that must be disturbed for mineral waste disposal is often measured in the tens to thousands of hectares and may account for the majority of disturbance. If the mineral waste is also chemically reactive, the potential environmental impacts and the complexity of waste management increase dramatically.

Public concerns during project permitting are commonly centered on potential exposure risks and water quality impacts from chemically reactive mineral wastes and can result in project delays and costly permitting requirements. If mishandled, chemically reactive mineral waste can degrade water quality for centuries, creating large environmental and financial liabilities. Waste-related perpetual water management and treatment can account for more than half of the total closure cost at some mines. Fortunately, significant advances in mineral waste characterization and management have been made over the past several decades. Proactive mineral waste management can significantly reduce the intensity, footprint, and duration of environmental impacts. Companies that practice proactive management can reduce their financial liabilities, improve their reputations, and become miners of choice, helping ensure access to new mining opportunities.

Mining and processing operations also can generate significant volumes of nonmineral waste. Common hazardous and industrial wastes include solvents, used oil, oily debris, spent reagents, coolants, greases, batteries, and used paint. These wastes typically are sent to off-site recycling, treatment, or disposal facilities. Other nonhazardous wastes such as construction debris, packing materials, general office waste, and used tires may be recycled, sent to an off-site municipal landfill, or placed in permitted on-site landfills. Both hazardous waste and materials such as gasoline, strong acids, and some reagents must be carefully managed on-site to comply with legal requirements, protect worker health and safety, and avoid spills. Areas that have been impacted by historic hazardous material spills and releases may have contaminated soils and groundwater that can threaten human health or the environment. Active management or remediation of contaminated sites is commonly required at older mining operations.

MINERAL WASTE

The most common mineral wastes include overburden and waste rock, tailings, spent heap leach ore, block cave rubble, and slag. Waste rock, overburden, and interburden are unmineralized or weakly mineralized rock and unconsolidated sediments that must be removed to access the underlying ore. Waste rock is typically a poorly sorted mixture of clay, silt, sand, gravel, and boulder-sized material. The most common disposal method for waste rock is placement within dumps and stockpiles, although in-pit disposal is common in strip mines. Tailings are the fine-grained waste that remains after the minerals or elements of economic interest have been removed from the ore. Rejects from coal-washing facilities, fines generated by bauxite and iron ore washing, and red mud generated by aluminum refining also have many of the same physical characteristics and management requirements as tailings. Tailings are composed of the gangue minerals in the ore and residual minerals of economic interest that were not recovered, along with process water and any reagents that were added during the milling and beneficiation processes. Tailings are usually disposed of in specially engineered repositories that are capable of containing the fine-grained and often saturated tailings mass without risk of geotechnical failure.

Spent heap leach ore is the residual rock that is left after the elements of economic interest have been removed by leaching. It may have the same physical characteristics as waste rock or it may have undergone additional crushing, grinding, and/or agglomeration steps. Chemically, the spent heap leach ore has usually been altered by the addition of strongly acidic, saline, and/or basic leach solutions and the addition of reagents such as cyanide. Spent ore will also contain residual leach solution held within its pores, even after the initial drain down has been completed. In most circumstances

Richard K. Borden, Principal Advisor Environment, Rio Tinto, South Jordan, Utah, USA

spent ore is left in place on the heap leach pad after closure, but sometimes it may be actively cycled on and off the pads.

The block cave underground mining method can create a large mass of highly fractured and disturbed rubble overlying the underground workings. This rubble can have many of the same physical and chemical characteristics as waste rock, even though it has been generated in situ. Slag is the glassy waste residue generated by the smelting of metallic ores, typically at much lower volumes than other categories of mineral wastes.

Mineral waste disposal can be responsible for much of the environmental impact caused by mining. Potential impacts that must be assessed, minimized, and mitigated during mine design, operation, and closure include the following:

- **Direct land disturbance.** Construction of out-of-pit mineral waste storage facilities will typically require burial of the premining surface, its soils, and ecosystems beneath tens to hundreds of meters of waste. The surface area impacted by waste disposal is commonly larger than the disturbance required to access the ore body via open pits and underground workings.
- **Geotechnical instability.** Unless properly designed, thick waste piles may be prone to geotechnical instability and failure. Instability may range from excessive surface erosion to large deep-seated slope failures. Geotechnical risks are generally highest for tailings and other fine-grained waste materials that are saturated when deposited. Geotechnical issues and requirements related to waste management are detailed in Chapters 8.10 and 8.11.
- **Erosion and sediment release.** Erosive wastes are prone to the formation of gullies and other erosion features. The release of sediment at much higher rates than surrounding natural landforms can have negative impacts on down-gradient surface water bodies and aquatic ecosystems.
- **Visual impacts.** Large-scale mineral waste transport and placement can significantly modify the landscape, creating landforms that are taller than the surrounding topography, truncating valleys and drainage lines, and creating unnatural uniform planar landforms that do not blend in with the surrounding natural topography. Mineral wastes may inhibit vegetation establishment, creating surfaces that look different from the surrounding undisturbed lands. Visual impacts are likely to be a particular concern near population centers and recreational or protected areas.
- **Direct exposure risks and phytotoxicity.** As detailed in the next section, chemically reactive mineral waste may pose direct chemical exposure risks to people, plants, and animals that live on or near the waste. The pH, salinity, or metals content of the waste can also inhibit vegetation establishment and prevent successful rehabilitation of waste surfaces.
- **Water quality degradation.** Chemically reactive mineral wastes can degrade the quality of water that runs off of or seeps through the waste material. Unless properly managed, this can cause degradation of surface and groundwater quality, impacts to aquatic ecosystems, and loss of the beneficial use of water resources far from the point of initial waste placement.
- **Dust release.** Wind erosion and dust release can degrade air quality because of increases in suspended particulate matter. If the dust is derived from chemically reactive waste, wind transport may disperse potential contaminants over a broad area.

Chemically Reactive Mineral Waste

Chemically reactive mineral waste has chemical or mineral constituents that could now or in the future create an environmental hazard above and beyond the risks posed by inert dust in air or inert total suspended solids in water. Several types of chemical hazards are associated with mineral waste and are discussed in the following paragraphs.

Acid Rock Drainage

The oxidation of naturally occurring sulfide minerals such as pyrite and chalcopyrite releases acidity, metals, and sulfate. Sulfide-bearing mineral waste can generate drainage with low pH, high salinity, and high metals concentrations, and can prevent the establishment of vegetation. Acid rock drainage (ARD) is most commonly associated with ore deposits that formed as a result of sulfide mineralization such as precious (gold, silver) and base metal (copper, nickel, lead, zinc) ore bodies. However, ARD is possible at any ore body or associated country rock that formed or exists under reducing conditions such as those typically found below the premining water table. For example, ARD has also been observed at coal, diamond, iron ore, and uranium mines. Operations that use acid for processing (such as uranium or copper heap leach operations) may also create acidic solutions and wastes with properties similar to ARD. ARD represents the most significant and most common chemical hazard posed by mineral waste and is discussed in more detail in Chapter 16.5 Good overviews of ARD chemistry, prediction, and management are provided in the *Global Acid Rock Drainage Guide* (INAP 2009) and the Mine Environment Neutral Drainage Program manuals (MEND 2001).

Metals Soluble or Bioavailable at Neutral pH

Many elements are soluble at neutral pH and can leach from mineral waste at concentrations that may be of environmental concern. Once in solution, these elements may be toxic to aquatic life or may make the water unfit for use as drinking, irrigation, or stock water. Many of these elements are also bioavailable at neutral pH and can inhibit vegetation establishment or pose direct exposure risks to animals and people via uptake by vegetation growing on the waste or incidental ingestion. These elements may include boron, cadmium, cobalt, chromium, fluoride, mercury, manganese, molybdenum, nickel, selenium, thallium, and zinc (Smith and Huyck 1999). Neutral drainage problems may be associated with many types of ore bodies and processes but are particularly common at sulfide-bearing ore bodies, which also contain significant neutralization potential. In this case the acidity released by sulfide oxidation is neutralized in situ, but the metals released from the sulfide minerals are soluble at neutral pH and can be mobilized out of the waste material.

Salinity and Sodicity

Some mineral wastes contain readily soluble salts, which can impact water quality and inhibit vegetation establishment. Major ions typically associated with salinity impacts include sodium, magnesium, potassium, calcium, chloride, sulfate, carbonate, and bicarbonate (Richards 1954). Elevated salinity can cause environmental impacts through general osmotic effects or specific ion effects. Saline soils can inhibit vegetation establishment, and solute loading to surface water can degrade water quality, harming aquatic life and ecosystems. Saline soils with a high proportion of exchangeable sodium relative to calcium, potassium, and magnesium are defined as sodic soils. Sodic soils and wastes can have poor soil structure, are prone to erosion, and have poor drainage. Salt loading can make water unfit for use as drinking, stock, or irrigation water. Salinity guidelines for soils and water are usually expressed in terms of total dissolved solids concentration or electrical conductivity. However, water quality guidelines for specific major ions such as sulfate have also been promulgated in many jurisdictions. Many sectors of the mining industry may have salinity issues, including coal, borax, salt, potash, and any sulfide-bearing ore bodies. Operations that use seawater in the process can create tailings salinity issues. The evapoconcentration of zero-discharge or low-discharge water bodies and process circuits such as pit lakes and tailings impoundments may also result in high salinity.

Radionuclides
Radionuclides such as uranium, thorium, radium, and radon can pose exposure risks because of toxicity and/or radiological hazards. The release of uranium and its daughter products are an issue at uranium mines. However, radionuclides may also pose hazards at heavy mineral, coal, and rare earth ore bodies and at ore bodies associated with granitic rocks. Radon exposure may be a particular concern at some underground mining operations in granitic terrains.

Cyanide and Other Reagents
Cyanide is most commonly used in the mining industry to extract gold from ore and can also be used at low concentrations as a flotation reagent. As such it may be present in spent heap leach ore and tailings. Because of the toxicity of cyanide, airborne and aqueous releases to the environment are highly regulated in most jurisdictions. A good overview of cyanide geochemistry and environmental behavior is provided by Smith and Mudder (1999). Best practice for management of cyanide and cyanide-bearing wastes is provided by the International Cyanide Management Institute (ICMI 2009a, 2009b). Many other organic and inorganic reagents may be used in the flotation process or other ore-processing techniques. These reagents are commonly highly diluted within the resultant processing mineral waste but under some circumstances could pose risks to the environment and human health.

Nitrogen Compounds
Nitrogen compounds (ammonium, nitrate, and nitrite) are most commonly associated with incomplete combustion of blasting agents at hard-rock mines and can be present in both waste rock and tailings. Nitrate is also a common cyanide degradation product. Ammonium is also commonly used in uranium processing and is likely to be present in some uranium tailings. Nitrogen compounds have the potential to be directly toxic to aquatic life and can degrade water quality so that it cannot be used as drinking or stock water. The release of nitrogen and/or other nutrients such as phosphorus can cause algal blooms and eutrophication of surface water bodies.

Naturally Occurring Asbestos and Other Mineral Dust Health Issues
Although little commercial asbestos mining is still performed, asbestiform minerals may be present at low concentrations (as naturally occurring accessory or gangue minerals) in other types of ore bodies. These may include vermiculite, talc, and enriched banded iron formation ore bodies, and ore bodies associated with skarns, carbonatite intrusions, ultramafic igneous rocks, mafic alkaline intrusions, serpentinite, and metamorphosed carbonates (Virta 2002). Gangue asbestos in waste rock and tailings can pose airborne exposure risks to workers and nearby communities. Worker exposure to silica minerals, amorphous silica, and other fine mineral dusts can pose health risks. A good summary of the potential health effects of mineral dusts is provided by Ross (1999).

High pH
Many jurisdictions specify a maximum discharge limit of 8.5 to 9.0 for pH to protect aquatic ecosystems and drinking water resources. Aluminum refining generates bauxite residues (red mud), which typically have pH values of greater than 12. Alkaline leaching operations for the recovery of uranium, gold, and other commodities may also generate mineral waste with a high pH.

Spontaneous Combustion
Spontaneous combustion may pose a risk at operations that disturb or expose carbonaceous and/or highly reactive sulfide-bearing waste. Oxidation of the material may release heat faster than can be dissipated, until temperatures become high enough to trigger combustion of the waste. This is most likely to occur at coal mines but may also be a risk at operations that disturb carbonaceous and/or reactive pyritic shale.

Compared to inert waste, the presence of chemically reactive mineral waste can significantly increase the complexity and cost of waste management. Successful management of chemically reactive mineral waste requires a thorough understanding of pertinent regulations, well-designed characterization programs, careful site selection, good facility design, and rigorous ongoing management and monitoring (Figure 16.6-1).

Waste Characterization and Impact Prediction
When developing a waste characterization program, operations must identify and understand the physical and chemical characteristics and hazards of all mineral wastes that will be disturbed, exposed, produced, or imported over the life of the operation. The characterization program must be rigorous enough to provide reliable predictions of the long-term physical and chemical behavior of the waste. Ultimately, the program will be used to select appropriate management strategies that comply with pertinent regulations for each waste type; ensure that all repositories are physically and chemically safe and stable; and allow for successful rehabilitation and closure.

Physical hazards include materials that are highly erodable and may be prone to shallow or deep-seated geotechnical failure under local conditions. Very blocky or poorly graded coarse-grained materials may also inhibit vegetation establishment. The potential existence of each of the chemical hazards described in the previous section must be assessed. The characterization work should be implemented in a phased approach, beginning with a broad-based sampling program combined with simple laboratory tests, followed by more complex testing if required. Initial testing should be designed to determine if a potential problem exists, while later testing may be focused on defining geologic distribution, release rates, and ecological or health effects.

For most geochemical hazards, laboratory analysis typically progresses from

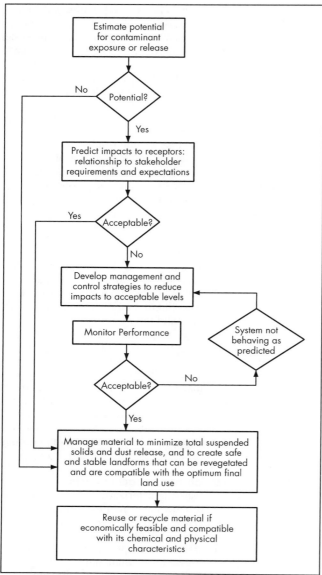

Courtesy of Rio Tinto.

Figure 16.6-1 Overarching management strategy for chemically reactive mineral waste

management (e.g., if the waste contains negligible neutralization potential and is located in a sensitive environment with little buffering capacity).

If materials that pose an ARD risk are identified by the preliminary testing, increased sampling density coupled with more detailed acid–base accounting, net acid generation testing, and mineralogical analysis should be implemented to further define the geochemical behavior of the material under surface weathering conditions. Acid–base accounting compares the acid-generating potential of the rock sample (as measured by sulfide sulfur) against its acid-neutralization potential (most commonly related to the calcium and magnesium carbonate content of the sample). Rock that contains excess neutralization potential will theoretically maintain a neutral pH as it weathers, even if it contains abundant sulfide minerals. Because of the uncertainties created by differential reaction kinetics, leaching rates, and mineral distribution in the rock, a commonly used screening criterion assumes that rocks with excess neutralization potential may still be acid generating unless the neutralization potential is at least twice the acid potential (Price et al. 1997).

Long-term kinetic leach tests should be carried out on selected samples if contact water quality and contaminant release rates need to be predicted. Kinetic leach testing should be performed in a manner consistent with the conditions to which the material will be exposed (e.g., will it be deposited above the water table or under water?). The assessment must also consider the effects of processing and deposition on the materials (e.g., will sulfides become concentrated during processing?). More detail on ARD characterization and prediction is presented in Chapter 16.5.

Preliminary physical analysis commonly consists of standard laboratory classification tests such as

- Grain size distribution and unified soils classification,
- Plasticity (Atterberg limits),
- Estimated placement density, and
- Moisture content.

More sophisticated tests may be required to assess weathering, strength, permeability, and erosion potential. Similarly, revegetation tests for mineral wastes may progress from

- Nutrient analyses and grain size distribution, to
- Greenhouse trials, to
- Field revegetation test plots and plant tissue sampling (to determine metals uptake).

Materials characterization data should be integrated with mine plans, waste inventories, and environmental baseline data such as climatic setting, hydrology, hydrogeology, and receiving environment sensitivity to develop an integrated conceptual understanding of how contaminants may be released, migrate, and impact receiving environments, biota, and people. Transport and migration pathways to be considered include direct runoff, wind erosion and dispersion, infiltration and water discharge from the facility's toe, and infiltration and migration to groundwater. Receiving environments and resources that may be impacted include surrounding terrestrial ecosystems; nearby communities; downstream aquatic ecosystems; groundwater-dependent ecosystems; and surface and groundwater resources used for drinking, stock, and/or irrigation water.

It is usually useful and often important to develop a conceptual model that provides a unified overview of how the

- Total element analyses, paste pH and conductivity, and short-term static leach tests; through
- Longer-term kinetic or sequential leach tests, mineralogical analysis, and ecotoxicity tests; to
- Field scale tests and trials.

For example, a typical phased testing program for ARD may begin with simple total sulfur analyses performed on a suite of well-distributed samples that represent all major lithologies, alteration types, ore grades, and waste types. If all samples contain very low sulfur values, then the risks posed by ARD are likely negligible, and more complex testing is unlikely to be required. At most sites, ARD risks are likely to be negligible if all total sulfur values are less than 0.1%, but under some geologic and environmental conditions, material with total sulfur values as low as 0.05% may require special

waste behaves in the environment and how it may impact key receptors. Although a model may be as simple as a schematic diagram, flow chart, or summary text, in some cases, it will need to be supplemented with numerical computer models to predict geochemical, hydrogeologic, and geotechnical behavior. Conceptual and numeric models should be validated and periodically updated in response to changes in operations and new monitoring and characterization data.

Site Selection and Disposal Area Footprint

Waste disposal facilities should be located in areas that minimize environmental impacts and long-term environmental liabilities. The selection process should include a review of site regulatory requirements, baseline conditions, and environmental considerations (see Chapter 16.1), environmental consequences, and the direct surface impacts caused by disposal. In general, the following factors should be considered when selecting locations for waste disposal facilities (modified from Rio Tinto 2008a):

- Only place waste within legally permitted areas.
- When practicable, preferentially place waste within inactive open pits, underground workings, or existing disturbed areas. This not only minimizes direct land disturbance but may allow rehabilitation of open pits that otherwise could not be returned to productive postmining use and can provide significant benefits for the management of some chemically reactive mineral waste.
- Avoid permanent disruption of drainage systems.
- Tie waste repositories into the surrounding topography to maintain natural, free-draining landforms and to reduce visual impacts.
- Avoid placement on land with high biodiversity or ecosystem services values.
- Avoid placement in areas with significant archaeological or social value.
- Avoid placement in close proximity to local communities.
- Preferentially place chemically reactive wastes in drainage basins that already contain reactive waste (thereby avoiding placement in undisturbed drainages).
- Limit the footprint of chemically reactive mineral waste to the maximum extent practicable. This will limit the volume of precipitation that may ultimately contact the waste, mobilizing contaminants and degrading water quality. It will also limit the surface area that may ultimately need to be covered or capped, often at significant cost.
- Avoid placement in areas with poor foundation conditions due to topography, underlying geology, or hydrology.
- Avoid placement of chemically reactive mineral waste over significant aquifers or groundwater recharge zones. Preferentially place waste over aquitards and aquicludes and in areas where hydrogeologic containment can be demonstrated.
- When the choice is available, such as in some mountainous terrains, preferentially place chemically reactive waste in areas with significantly dryer climates.
- Balance economic considerations such as haul profiles, potential resource sterilization, and pumping costs with environmental, social, and closure considerations.
- Avoid placement in or near perennial surface water bodies or in large ephemeral drainage lines when practicable, unless this represents the preferred environmental alternative.

Uncontrolled river or shallow marine disposal, which could allow rapid transport of the mineral waste away from the point of initial placement, should almost always be avoided. This does not necessarily apply to the return of mining-related inert dredge spoils into the river or shallow marine environments from which they were derived. River flow, wave action, and long-shore currents can transport wastes far from the point of initial discharge, making it difficult to control where the wastes are ultimately deposited. Aquatic environments can also be severely damaged by the smothering of benthic organisms, degradation of water quality, metals uptake by organisms contacting the waste, and increases in turbidity, which inhibits photosynthesis. However, controlled deposition into lakes and the deep marine environment may represent a preferred environmental alternative under some circumstances. For example, lake disposal may be preferred if

- ARD or geotechnical stability concerns need to be managed;
- It is restricted to well-defined, permanent deposition areas;
- Disturbance of sensitive terrestrial habitats needs to be avoided; and
- Placement can be controlled so that significant fisheries, high-biodiversity freshwater habitats, and down-gradient water resources are protected.

Deep marine mineral waste disposal should generally be considered only if the waste can be discharged and deposited below the maximum euphotic, upwelling, and mixed-layer depths in the surrounding ocean. The actual maximum depths of these three zones must be determined though detailed field studies. Coral reefs, kelp forests, and almost all other high-biodiversity marine environments are restricted to the euphotic zone (where light penetration allows photosynthesis), which typically extends to maximum depths of anywhere from 30 to 200 m in the ocean. Within this marine zone, productivity, fishery resources, ecosystem services, and biodiversity are highest, and waste can be routinely transported by wave action and currents. In most cases, deep marine disposal involves pretreatment, discharge from a pipe more than 100 m deep, and final tailings deposition in water greater than 1,000 m deep.

Waste Management Strategy and Facility Environmental Design

The design and engineering of waste disposal facilities should take into account local legal requirements; the physical, chemical, and mineralogical properties of the waste; and the environmental setting of the disposal area, including considerations of climate, water balance, underlying geology, hydrogeology and foundation conditions, potential seismic events, and nearby receiving environments and communities. Priority should be given to the proactive management of chemical hazards so that exposure risks are controlled and long-term liabilities such as perpetual water treatment are minimized. The selection of appropriate management strategies typically begins by comparing the impacts as predicted by conceptual or numeric models to environmental compliance and performance objectives. If needed, a strategy is selected to reduce the potential impacts and ensure that all compliance and performance objectives will be met during start-up, operation, and closure.

The selected strategy does not have to completely prevent any solute or contaminant release but must ensure that release rates meet regulatory requirements and are low enough to be assimilated by the receiving environment without causing harm to people, ecosystems, organisms, or resources. Numerous geochemical prevention and control strategies are available for consideration (Table 16.6-1). Based on the chemical hazard, waste characteristics, and environmental setting, many of the control strategies listed in Table 16.6-1 will be impractical or impossible to implement at a specific site. Final selection among the remaining viable control strategies should be based on a rigorous comparison of the potential options (or combination of options) and giving preference to strategies that

- Will consistently meet or exceed local regulatory or permit requirements;
- Pose the lowest risk of short- or long-term failure;
- Can be realistically implemented under actual field conditions (the simpler the better);
- Are compatible with life-of-mine and short-term mine plans including short-, intermediate, and long-term material balances;
- Are proactive preventive measures rather than reactive remedial measures;
- Minimize postclosure liabilities and most closely approach a "walk away" solution at closure; and
- Meet the surface discharge limits, groundwater quality requirements, air quality standards, and/or receiving environment objectives in the most cost-effective manner.

Not all strategies listed in Table 16.6-1 are applicable to each chemical hazard. Strategies that depend on subaqueous disposal to limit oxygen flux into the waste are extremely effective when attempting to control ARD, sulfate-dominated salinity, and metals release related to sulfide oxidation. The maximum concentration of dissolved oxygen in surface water is approximately 30 times less than in the atmosphere. More importantly, the diffusive transfer of oxygen in water is on the order of 10,000 times slower than in air, whether the water is fully oxygenated or not (MEND 2001). Subaqueous disposal in a nonerosive landform, such as a flooded inactive open pit or underground workings, usually represents the most effective and secure disposal option for sulfide-bearing mineral waste. However, subaqueous disposal is not appropriate for waste-containing metals and salts that are already in a soluble form, because the constituents will go into solution where they can be easily transported out of the waste. Subaqueous disposal may also not be appropriate for oxidized wastes, which may contain metals that are stable under oxidizing conditions but which would be mobilized under reducing conditions. Almost any kind of nonerosive inert cover will control dust release for hazards associated with naturally occurring asbestos and other contaminants that may be entrained in the dust. However, to minimize contaminant seepage from a waste repository, a cover must be designed to limit infiltration of precipitation or surface water into the waste.

Figure 16.6-2 provides an example of an integrated mineral waste management strategy at the Rio Tinto Barneys Canyon gold mine in the western United States. About 5% of the 150 Mt of waste rock produced at the mine was unoxidized, enriched in some metals, contained abundant pyrite, and was prone to ARD generation. Proactive control of this potential ARD source required implementation of several of the strategies from Table 16.6-1.

- **Selective handling:** Based on field inspection of color and periodic confirmatory sampling, the unoxidized sulfide-bearing rock was segregated from the benign oxide rock so that it could be specially handled.
- **In-pit disposal and selective waste placement:** The sulfide-bearing rock was placed in two small repositories: one an inactive open pit and another an area already used for oxide waste rock disposal. The repositories both had a small footprint (limiting incident precipitation), were located on previously disturbed land, were not in contact with significant drainage lines or springs, and were located away from property boundaries and groundwater resources.
- **Waste removal and consolidation:** Some sulfide-bearing, low-grade ore stockpiles were also moved into the repositories after it was decided that the material would not be processed.
- **Recontouring to promote runoff:** The repositories were constructed to tie into surrounding native hillsides and to have a final slope of at least 10°. This configuration minimizes run-on and promotes runoff during heavy precipitation events.
- **Infiltration limiting covers:** The mine is located in a semiarid climate, so a store-and-release infiltration limiting cover was constructed on the repositories. Evapotranspiration was maximized by construction of a 4-m-thick cover with good moisture retention properties and planting a mixture of deep-rooting native perennial grasses, forbs, and shrubs to maximize plant water usage. Both repository surfaces were also sloped to the south to maximize sun exposure, growing season, and evapotranspiration.

Store-and-release cover systems, which seek to limit water contact with the underlying mineral waste, are seldom viable in regions with high rainfall, particularly with distinct wet and dry seasons. When managing sulfide-bearing wastes in humid climates, it is commonly better to try and minimize oxidation of the waste by ensuring it remains saturated in perpetuity or is capped with a fine-grained, compacted cover that remains near saturation. Figure 16.6-3, the Rio Tinto Ridgeway gold mine in the eastern United States, is an example of a typical flooded, chemically reactive mineral waste repository designed to limit oxygen ingress into the waste.

Waste storage facilities should be designed and constructed so that postclosure environmental impacts and financial liabilities are minimized. For example, landforms can be constructed to minimize costly recontouring and earthmoving requirements at closure. In many jurisdictions, maximum allowable final slopes for landforms constructed of mineral waste can be no steeper than 2:1 or 3:1 (horizontal/vertical). Rehabilitation costs can be significantly decreased by benching final outer waste rock dump faces in short stair-stepped lifts to match the required final slope angle. Constructing final lifts out of benign waste material—whose physical and chemical properties can support vegetation, will not pose direct exposure risks, will not degrade water quality, and is not prone to erosion—can prevent the need to import costly cover material to meet these objectives at closure. Constructing waste disposal facilities in a manner that allows progressive rehabilitation can also allow operations to minimize the footprint of unreclaimed land at closure. Progressive rehabilitation and

Table 16.6-1 Common management strategies for chemically reactive mineral waste

Strategy	Features
	Proactive Preventive Measures
Selective mining	Early identification of rock masses that pose the greatest geochemical risk allows the mine plan to be designed to avoid exposing or disturbing these materials. The additional handling, closure, and water treatment costs associated with acid-generating or other chemically reactive materials should be included in any economic analysis when determining ore reserves and cutoff grades, open-pit geometry, and the layout of underground workings. Early characterization of relatively benign rock masses is also needed so that materials that could be used for clean fill, construction activities, encapsulation, mixing, or covering are identified and can be incorporated into the mine plan as appropriate.
Improved recoveries	Improving recovery rates during the beneficiation process means that less sulfide ore minerals (such as chalcopyrite and bornite) remain in the resultant tailings or spent ore, where they produce no economic benefit, can contribute to acid potential, and may increase metals concentrations in the waste. Similarly, improving leaching and recovery rates of elements with economic value can lower their concentration in leached tailings and spent heap leach ore where they may be potential contaminants.
Lowering cutoff grade	Anytime cutoff grade can be lowered, the volume of waste rock produced will be reduced. Because the highest-grade waste material is converted to ore when cutoff grades are lowered, this process commonly removes rock with the highest potential contaminant concentrations from waste rock production.
Reagent selection and management	Potential environmental impacts caused by reagents added during ore processing can be minimized by selecting less toxic but equally effective reagents or ensuring reagents such as cyanide are added at, but not above, rates that are optimum for processing. Ensuring that blasting techniques are optimized so that blasting agents are efficiently used can lessen the risks posed by residual nitrogen compounds in waste rock and tailings.
Chemical treatment of tailings	Tailings can be treated before they are discharged to waste repositories. For example, acid-leached tailings can be neutralized to raise the pH and lower metals solubility; rinsing and cyanide destruction can be used to reduce the hazard posed by cyanide leached materials; and red mud generated from aluminum refining can be neutralized with seawater addition.
Selective waste placement	Waste repositories should be designed so that transport pathways that may allow contaminants to migrate out of the waste are minimized. Issues that should be considered include avoiding placement of reactive wastes in areas where they may contact surface water flows or in areas where groundwater discharges to the surface, over highly permeable near-surface aquifers, over or immediately up-gradient of sensitive environments (such as lands with high biodiversity value) or facility boundaries, in unstable or highly erosive landforms, and in watersheds or in areas that have not already been impacted by previous chemically reactive mine waste disposal; and minimizing the ultimate footprint of the chemically reactive waste disposal areas.
Up-gradient water diversion	In locations where surface or groundwater comes into contact with chemically reactive waste materials or wall rock, it is beneficial to capture the clean up-gradient water before it contacts the reactive material. This will reduce the likelihood of contaminants being mobilized and will protect water quality from being degraded. A surface water control plan should be developed as part of the mine plan for all disturbed areas.
Selective handling/encapsulation	Chemically reactive mineral waste may be selectively handled and surrounded with benign waste to limit the contact of runoff water, infiltration, and groundwater; reduce direct plant uptake and animal exposure; reduce the risk posed by wind erosion; provide a growth media over phytotoxic waste; provide alkalinity to underlying acidic waste; or limit the access of oxygen to sulfidic waste. Benign materials can also be used for foundations in areas where water may perch at the base of the waste or in areas where groundwater may discharge into the base of the waste pile via seeps and springs.

(continues)

Table 16.6-1 Common management strategies for chemically reactive mineral waste (continued)

Strategy	Features
In-pit or underground disposal	This method is viable where mined-out pits or underground workings of sufficient size are available. With effective mine planning, the early closure of one of a series of pits or workings may allow for the effective disposal of chemically reactive wastes. Depending on the wall-rock permeability and the location of the water table, waste placed in open pits may need to be encapsulated. For pits and underground workings located below the postmining water table, this disposal method may be used in conjunction with subaqueous disposal to prevent sulfide oxidation but may not be appropriate for wastes containing soluble contaminants. Backfilling open pits above the postmining water table can also be used to prevent evapo-concentration and the creation of hypersaline conditions in arid or semiarid climates and the formation of pit lakes that may provide a direct exposure pathway to terrestrial animals.
Subaqueous disposal	The placement of acid-generating wastes below the water table or in a permanent surface water body and the permanent flooding of underground workings or pits will greatly limit the flux of oxygen into sulfide-bearing materials. This can lower the rate of acid generation and metals release by several orders of magnitude. This technique is most effective when flooding or covering is accomplished rapidly—before the material has time to oxidize. Subaqueous disposal is generally not a viable control option for wastes containing contaminants that are already in a soluble form and whose rate of release is not dependent on the supply of oxygen.
Blending/mixing	The blending and mixing of acid-generating wastes with net neutralizing materials may be used to ensure that bulk neutral pH conditions are maintained within the waste mass as a whole, either in perpetuity or for a sufficiently long period to allow another long-term management technique to be implemented. This control technique may be difficult to implement because of preferential flow paths established within the waste, armoring of neutralizing rock surfaces, and acidified pockets that are likely to form within the waste column. Proper implementation of the technique requires good ore and waste characterization and control, coupled with sequential mining and good dispatch control to ensure that acid-generating and acid-neutralizing wastes are intimately mixed at the appropriate ratios.
Enhanced tailings consolidation	Techniques that enhance tailings consolidation such as the construction of underdrains, ensuring subaerial placement, controlling the rate of rise, active removal of pore water with extraction bores, or initial placement as thickened, dry stacked, or paste tailings will create a more stable material with lower permeability. If solutes within the pore water pose an environmental risk, enhanced consolidation will also reduce the total contaminant mass and lower the potential flux of contaminants that could discharge from the tailings.
Codisposal	Materials with different physical characteristics may be mixed for codisposal to enhance stability and resistance to erosion or to reduce permeability. This technique may be applied to the codisposal of tailings with waste rock. The mixing of tailings and waste rock may create a well-graded material that typically has lower hydraulic conductivity and higher water storage capacity, and is relatively resistant to erosion. The geochemical compatibility and the physical behavior of the mixtures require careful evaluation. Techniques to deposit whole tailings (such as dry stacking, paste, or thickened tailings) that are not strongly segregated into coarse and fine fractions, or segregated into sulfide-depleted and sulfide-enriched fractions, may also provide lower hydraulic conductivity, maintain saturation and, if the parent material is net neutralizing, may prevent the creation of net acid-generating zones within the tailings.
Sulfide flotation	Gangue sulfides such as pyrite may be separated from the bulk of the tailings in order to reduce the tailings' acid potential and ensure they are net neutralizing. The relatively small volume of concentrated waste sulfide can then be specially handled to minimize its environmental risks. Metals that are associated with the sulfides may also be preferentially removed from the tailings by this technique.

(continues)

Table 16.6-1 Common management strategies for chemically reactive mineral waste (continued)

Strategy	Features
Compaction during placement	Chemically reactive waste rock or overlying benign material may be compacted during placement to lower the hydraulic conductivity of the material and limit the potential flux of water through the compacted layer. Compaction may also aid in the establishment of capillary breaks and/or perched water tables associated with the compacted layers, which will inhibit the flux of oxygen. Although engineered compacted layers should usually be specially constructed, some compaction of the waste can also be accomplished by dumping in shorter lifts to enhance the compaction that occurs naturally as mine vehicles travel over the surface.
Liners	Liners can be constructed under reactive mineral wastes to allow capture of contact water that would normally discharge to the surface or to groundwater. Liners can be simple and relatively inexpensive or complex and expensive, depending on the required performance criteria (generally the allowable design leakage rate). Complex liners are generally required for heap leach pads containing very hazardous leach solutions. For mineral wastes, complex liners are usually only practicable for use with small volumes of very reactive mineral waste. If hydrogeologic containment can be demonstrated because of underlying aquicludes, aquitards, and/or strong upward vertical gradients, then liner systems may not be needed, or less complex systems may be adequate.
Temporary synthetic covers	Synthetic covers such as high-density polyethylene (HDPE) sheeting may be used to temporarily cover reactive mineral wastes or ore stockpiles. This isolates the material from precipitation and greatly limits the volume of contact water that may discharge from the material. This is generally not a viable long-term strategy to limit water contact because the synthetic cover will degrade with time, but it can be used on a short-term basis until the ore is processed or the final closure strategy for the mineral waste is implemented. However, when temporary covers are removed, sulfide oxidation products and other soluble constituents may dissolve and produce a large spike in contaminant release.
Micro-encapsulation (passivation)	This process involves the coating of mine wastes or pit walls with products that encapsulate sulfide surfaces and inhibit oxidation reactions. The mechanism involves mixing the reagents with waste or spraying pit walls with treatment solutions. Treatments that are being investigated include phosphate, amorphous silica, polymer, and potassium permanganate application. Encouraging bioshrouding of sulfide minerals by promoting the growth of microbe biofilms with nutrient addition is also under investigation. However, these techniques are likely to be costly on a per-ton basis; the potential exists for environmental problems associated with the use of some reagents; retroactive treatment of large waste storage facilities is unlikely to be successful; and the long-term viability of the techniques has not been proven at an operational scale.
Bactericide application	Abiotic sulfide oxidation rates are typically several times slower than biologically controlled oxidation rates. Bactericides such as thiocyanate that inhibit the growth of sulfide-oxidizing microorganisms can substantially slow ARD generation rates. However, these techniques are likely to be costly on a per-ton basis; the potential exists for environmental problems associated with the use of some reagents; retroactive treatment of large waste storage facilities is unlikely to be successful; and the long-term viability of the techniques has not been proven at an operational scale.
Incorporation of organic matter and/or nutrients	Incorporation of organic matter or nutrients into wastes can promote the establishment of reducing conditions within the waste material (inhibiting sulfide oxidation and promoting sulfide formation) and may reduce metals solubility by providing sorption sites. Organic matter and nutrients can also promote the establishment of sulfate-reducing bacteria, potentially removing dissolved sulfate and metals from pore waters. Such systems may not be appropriate for use with some oxidized wastes because reductive dissolution of iron or manganese oxides may release adsorbed trace metals such as arsenic.
Accelerated sulfide oxidation	In specially designed heap leach pads, it may be possible to accelerate sulfide oxidation processes, depleting all the reactive gangue sulfide minerals before closure. This would help reduce the risk of long-term contaminant release from the mineral waste. This technique would likely only be viable for copper heap leach pads and other operations that are already designed to maximize the oxidation of sulfide ore minerals. However, costs are likely to be high and the viability of this option has not been proven at an operational scale.

(continues)

Table 16.6-1 Common management strategies for chemically reactive mineral waste (continued)

Strategy	Features
Reactive Remedial Measures	
Recontouring to promote runoff and inhibit infiltration	Waste surfaces can be recontoured to prevent precipitation from pooling and infiltrating at low points on the surface and/or to allow precipitation to be more efficiently shed from the surface. The surface can be contoured to naturally shed precipitation via runoff, or low points can be established from which runoff is collected and piped off the surface.
Direct revegetation	Some waste surfaces may be directly revegetated after minor physical or chemical modification, such as ripping to reduce compaction, addition of alkaline materials to increase the pH to near neutral, or the addition of organic matter. The establishment of vegetation can reduce erosion, significantly increase evapotranspiration, and reduce the amount of water that infiltrates the underlying waste material. Direct revegetation may allow many of the benefits of a store-and-release cover to be realized without the need to import large volumes of cover material.
General purpose covers	Covers can be constructed over reactive wastes using inert mineral waste, imported fill, or soil to reduce infiltration, preserve runoff water quality, allow vegetation to become established on phytotoxic wastes, limit plant uptake of bioavailable metals in the underlying waste, limit wind transport of reactive wastes, limit direct exposure pathways for animals and humans to metals and radionuclides in the underlying wastes, limit oxygen ingress into the underlying mineral waste, and/or provide alkalinity to the underlying waste.
Alkaline covers	Under certain climatic and geochemical conditions it may be beneficial to construct covers with relatively reactive carbonate material. The carbonate can provide alkalinity to the underlying waste via infiltrating precipitation to control ARD from weakly acid-generating wastes or to increase the thickness of the underlying neutralized waste layer, allowing plants to root more deeply.
Infiltration or oxygen-limiting covers	Two general types of covers to limit oxygen and/or water ingress may be constructed on top of existing mineral waste: (1) low-permeability covers to limit the flux of oxygen and water into the underlying waste or (2) store-and-release covers that limit the amount of water that may infiltrate the underlying waste. In general, low-permeability covers are best suited to wet climates, and store-and-release covers are best suited to dry climates (where evaporation is at least twice precipitation). Low-permeability covers that inhibit the entry of oxygen, using natural materials such as compacted clay, generally require that nearly saturated conditions be maintained within the cover at all times. Low-permeability covers are often subjected to natural processes that may cause degradation in performance over time such as biointrusion by plants and animals and increasing permeability because of natural freeze/thaw and wetting/drying cycles. It is important for compacted layers to be protected from these processes by a thick overlying cover layer. The performance of most store-and-release cover systems is strongly dependent on the thickness and texture of the cover material (i.e., high-moisture storage capacity and low susceptibility to cracking), as well as the density, rooting characteristics, and species composition of vegetation that is established on the cover to maximize evapotranspiration rates.
Infiltration or oxygen-limiting covers (continues)	Covers can be simple, thin, and relatively cheap or complex, thick, and relatively expensive depending on the required performance criteria (generally the design allowable flux of water or oxygen through the cover and into the underlying reactive mineral waste). Covers designed to limit infiltration may also include capillary breaks to inhibit vertical water movement.
Rinsing	Rinsing can be used to flush soluble constituents from mineral wastes such as spent heap leach ore. Generally, several pore volumes must be exchanged to reduce concentrations to acceptable levels. If all the potential contaminants are not present in a readily soluble form, then rinsing is unlikely to be viable. For example, rinsing would not be a viable alternative for sulfide-bearing wastes until almost all of the sulfides have been oxidized.

(continues)

Table 16.6-1 Common management strategies for chemically reactive mineral waste (continued)

Strategy	Features
Flooding	The permanent flooding of existing wastes, open pits, and workings can limit the access of oxygen to the waste material and greatly restrict the rate of sulfide oxidation. Ideally, flooding should occur as soon as possible after the waste material is exposed. Waste materials that have undergone long periods of subaerial storage prior to flooding may contain abundant soluble sulfide oxidation products that may be released when the material is initially contacted by water. It may be necessary to add neutralizing agents to weathered materials that have acidified before they are flooded or to add alkalinity to the flooding waters. Flooding is generally not a viable control strategy unless contamination release rates are dependent on the flux of oxygen into the waste. Large fluctuations in water level so that wastes are periodically exposed and resubmerged must be avoided.
Down-gradient collection and treatment of contact water	This is required for sites where contaminated water is already being released or where it cannot be avoided and involves the capture of impacted surface and groundwater for treatment and disposal. Potential collection and recovery systems may include gravity flow into catchment ponds, drains, and trenches, as well as pumping from pits, underground workings, and extraction wells. Management options include active water treatment such as the addition of alkaline materials or the use of reverse osmosis, internal mixing, and dilution with controlled release to receiving water bodies, evaporative disposal, reuse in the process, reactive barriers, settling ponds, and wetlands treatment for polishing.
In-pit water treatment	The water chemistry of pit lakes may be manipulated in situ to promote physical or chemical conditions and biological processes that improve or maintain water quality. Some examples include controlling the depth of flooding to completely cover acid-generating wall-rock zones, to prevent water from contacting chemically reactive zones, or to create lakes that are unlikely to turn over; the direct addition of alkaline materials to maintain a neutral pH and inhibit sulfide oxidation below the water surface; the addition of nutrients to promote biological activity that removes metals from the water column and helps establish anaerobic conditions at depth; and the addition of sorbing agents such as iron to remove dissolved metals from solution.
Demonstrating hydrogeologic containment	For open pits that intersect the water table in arid and semiarid climates it may be possible to demonstrate hydrogeologic containment of solutes and contaminated water within the pit. Evaporative losses from the pit lake may be sufficient to maintain radial groundwater flow toward the pit in perpetuity. Containment can be enhanced by reducing the volume of runoff water that reaches the pit floor or by pumping and treating enough pit water to keep the lake level below water levels in the surrounding bedrock. Contaminated groundwater underlying waste rock dumps and tailings impoundments located within the zone of groundwater capture of the open pit may also be permanently contained. In some arid climates it may be possible to preferentially direct contaminated surface water into the pit for disposal by evaporation.
Waste removal and consolidation	In some cases it may be beneficial to move tailings and waste rock from one location to another if they are broadly dispersed or located in a particularly sensitive location. The selective removal and consolidation of highly reactive wastes into a smaller footprint may allow for better mitigation, containment, and control of the waste.

Source: Adapted from Rio Tinto 2008b.

revegetation of tailings impoundments can be particularly important to prevent dust generation from inactive tailings surfaces as they dry out. Other closure requirements and considerations that are pertinent to mineral waste disposal facilities design and construction are detailed in Chapter 16.7.

Ongoing Management and Monitoring

After a mineral waste management strategy is selected and waste storage facilities have been designed, they must be constructed and successfully managed on an ongoing, long-term basis. If benign and chemically reactive wastes are to be segregated and specially handled, the complexity of mine planning and management is increased. Waste management strategies need to be fully integrated into mine plans and truck control and dispatch systems. Strategies and costs should also be fully integrated into any economic models used by mine planners. Standard operating procedures need to be developed to ensure that different types of waste are identified, hauled to the correct location, and then placed and managed appropriately. Successful field implementation of strategies and designs will typically require cooperation between environmental personnel, senior management, geologists, short- and long-term mine planners, field supervisors, equipment operators, and surveyors. At many operations, development of a formal mineral waste management plan may be justified. These plans typically include a brief description of management objectives and strategies, waste storage facility design criteria, detailed field procedures, clear assignment of responsibility for different tasks and actions, and ongoing sampling and monitoring requirements.

Monitoring is required to ensure successful implementation of the mineral waste management plan and to ensure that the strategy is leading to the intended results. Physical monitoring programs for waste disposal facilities typically will include at a minimum

- Regular visual inspections of surface structures and facilities such as spillways, piping, dykes, ditches, and other water management systems;
- Regular visual inspections for signs of excessive surface erosion and shallow or deep-seated failure on the outer slopes of waste repositories; and
- Monitoring of water levels and pore pressure within embankments and the waste.

Programs to monitor the geochemical behavior of waste disposal facilities will typically include

- Periodic sampling of runoff water and water discharging from the facility's toe in order to monitor flow volumes, solute concentrations, and the solute mass that is being released from the waste;
- Periodic sampling of down-gradient monitoring wells and surface water bodies to ensure that seepage from the waste is not adversely impacting receiving environment water quality; and
- Periodic assessment of revegetation success such as total cover, species composition, and plant health.

If needed, more specialized monitoring data may be collected, including

- Paste pH, conductivity, and leach testing of soils forming on the surface of the waste or on covers constructed over the waste;

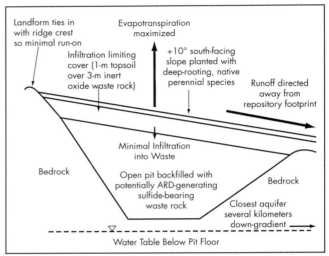

Figure 16.6-2 Example of a chemically reactive mineral waste repository in a semiarid climate

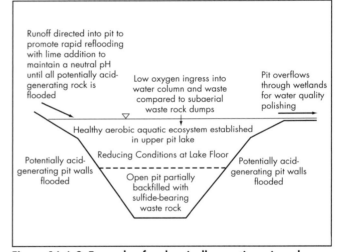

Figure 16.6-3 Example of a chemically reactive mineral waste repository in a humid climate

- Tissue sampling and metals analysis for plants growing on the waste;
- Monitoring of infiltration rates into the waste;
- Internal monitoring of the waste disposal facility for pertinent parameters such as temperature, oxygen content, moisture content, and pore water chemistry;
- Radiological measurements at the surface of the facility; and
- Monitoring of ecosystem health.

Monitoring data should be reviewed regularly, and historical trends should be examined so the longer-term chemical behavior of the mineral waste can be assessed. Time series of monitoring data should be maintained so that long-term changes in water quality, flow rate, or other key parameters can be tracked and significant changes can be identified. Significant unexpected changes should trigger a review of predictions, performance objectives, management strategy, and/or field implementation procedures. Appropriate feedback mechanisms should be in place to ensure timely management response to out-of-compliance, unexpected, or upset conditions. It is usually beneficial to maintain internal monitoring points in addition to the monitoring required at mandated external compliance points. Internal monitoring can allow early warning of potential problems before out-of-compliance conditions develop.

Closure

Waste disposal facility closure should create safe and stable landforms and allow land and water resources to be returned to productive postmining uses. Geochemical exposure hazards should be controlled, and long-term water management liabilities should be minimized. General reclamation and rehabilitation requirements are detailed in Chapter 16.7. Typical closure activities that are specific to mineral waste disposal facilities include the following:

- Removal of any surface debris or infrastructure.
- Large-scale recontouring to create natural-looking, safe, and stable landforms; to reestablish drainage lines; to reduce angle of repose slopes; and to comply with any other mandated closure criteria.
- Construction of long-term erosion and runoff control structures (Chapter 16.4).
- Placement of any cover material needed to prevent wind erosion, allow revegetation, preserve runoff water quality, limit infiltration, or control oxygen ingress. Cover material may be composed of benign mineral waste or imported borrow material with the appropriate physical and chemical characteristics. If erosion is a concern, a well-graded material should generally be used to support vegetation while also being resistant to erosion. If mineral waste is used as a growth medium, it should generally have the following characteristics:
  – Circumneutral pH (ideally between about 5.5 and 8.0) (Buckman and Brady 1969; Tucker et al. 1987),
  – Low salinity (typically an electrical conductivity of less than 2 dS/m [deciSiemens/meter] for moderately salt-sensitive species) (Maas 1990; ANZECC/ARMCANZ 2000),
  – Excess neutralization potential and low sulfide sulfur content (ideally a neutralization potential ratio of greater than 2:1 [Price et al. 1997] and total sulfur less than about 0.5%);
  – Metals concentrations that will not pose an exposure risk to plants or animals, and
  – Good moisture retention characteristics (adequate sand, silt, and clay in coarse wastes).
- Ripping or scarifying compacted surfaces to remove physical barriers to plant growth.
- Placement of topsoil and/or any soil amendments needed to promote plant growth. Topsoil that is free of weed seed will almost always enhance desired plant growth, and good topsoil management is an important part of most reclamation programs. Soil amendments may be needed when direct planting into mineral waste in order to modify the pH, increase water-holding capacity, improve soil structure, increase organic matter content, and add nutrients. Care should be taken not to overapply biosolids or other nutrients which may promote long-term weed dominance.
- Seed application and/or the planting of seedlings. The species selected for planting are likely to be based on the planned final land use, regulatory requirements, and stakeholder expectations.

Postclosure water management is commonly required for disposal facilities that contain chemically reactive mineral waste. Contaminated water that discharges from the waste material will typically need to be collected and treated before it can be discharged to the environment. Depending on the rates at which sulfide minerals oxidize and at which soluble constituents migrate out of the waste, water management may be required for decades or even centuries after closure. Spent heap leach pads containing hazardous process fluids may need to be flushed with many pore volumes of clean water to meet water quality discharge requirements. Long-term water management costs can be high, in many cases exceeding the cost of the initial rehabilitation. Postclosure monitoring programs are also usually required to ensure that surfaces are not eroding, the mandated vegetation is permanently established, and chemical exposure risks and impacts remain within acceptable ranges. If only inert waste is present, postclosure monitoring may be completed within a few years of closure, but if chemically reactive waste is present, monitoring may be required for many decades.

NONMINERAL WASTE

Common hazardous and industrial wastes generated at mining and mineral processing operations include used oil, oily debris, antifreeze, greases, batteries, solvents, coolants, spent reagents, used paint, and contaminated soils. These wastes are typically sent to off-site recycling, treatment, or disposal facilities. Other nonhazardous wastes, such as construction debris, packing materials, general office waste, and used tires, may be recycled, sent to an off-site municipal landfill, or placed in permitted on-site landfills. Remote operations that are not connected to municipal sewage systems must also construct and manage on-site sewage treatment systems.

Each waste stream needs to be characterized so that its hazards are understood, legal requirements are defined, and appropriate recycling, treatment, or disposal options are identified. Wastes need to be safely accumulated on-site until they are shipped off-site or placed into permanent on-site storage facilities. The volume of waste generated may be reduced by improving process efficiency, substituting hazardous materials with less hazardous but equally effective products, and by identifying on-site recycling and reuse opportunities.

The degree of effort required to characterize each waste stream will generally depend on its potential risks and management requirements. Some waste streams may only require detailed assessment during initial characterization and in response to significant changes in the generation process. Other wastes may require much more frequent analytical testing if the characteristics of the waste are highly variable or the final treatment or disposal method has strict physical or chemical requirements. For example, used oil burned for energy recovery typically has restrictions on metals and total halogens content and flash point. Municipal landfills and many industrial waste landfills are only permitted to receive nonhazardous wastes, so waste characteristics must be defined in accordance with local regulatory requirements to identify appropriate disposal facilities. Depending on the waste stream, key sources of information are likely to be

- Material safety data sheets;
- Other manufacturer and/or supplier information;

- Process and engineering knowledge and materials balances;
- Local regulations, which may list wastes generated by certain processes and/or industries and mandate how they must be handled;
- Analytical testing such as
  - Primary chemical constituents;
  - Trace chemical constituents and impurities such as metals, organic compounds, asbestos, and carcinogens;
  - Physiochemical characteristics such as flash point (typically considered hazardous if <60°C), corrosiveness (typically considered hazardous if pH <2 or >12.5), and reactivity (unstable when heated, compressed, or in contact with water or other wastes);
  - Toxicity leach tests such as the toxicity characteristic leaching procedure (Method 1311, EPA 2008) or the synthetic precipitation leaching procedure (Method 1312, EPA 2008); and
  - Radiological analysis.

Temporary on-site waste accumulation and storage areas should be clearly defined, utilizing fencing and appropriate signage as required by local regulations. Dispersed short-term accumulation areas typically will be located at or near the point of waste generation. Wherever possible, wastes should be segregated at source, which allows for improved recycling or recovery of materials, may increase the number of recycling or disposal options available for the waste, and may reduce the cost of waste disposal. In particular, disposal costs for mixed wastes may also be significantly higher if hazardous wastes are mixed with nonhazardous wastes.

Waste storage and containment practices must meet local, national, or international standards as applicable and must be commensurate with the risk posed by the waste. In many jurisdictions mines are required to have a central waste accumulation site where hazardous and industrial wastes are consolidated and stored before being shipped off-site. Central accumulation areas should typically be located in a secure location, on an impermeable surface to contain spills, and with a roof to minimize rainwater contact. New or used storage containers can be used, but they should be in good condition and must be compatible with the contained waste. Chemically incompatible wastes, which could react with each other, should not be stored together. Waste containers should be appropriately labeled so the contents and their primary hazards can be identified. In some jurisdictions, hazardous wastes have strict limits on the time they can be stored on-site before being shipped to a permitted disposal facility.

All waste must be permanently disposed in accordance with applicable laws, regulations, and permit conditions. Where local or national laws are weak or do not exist, internationally recognized good practices should generally be followed. The design and ongoing management of off-site waste disposal facilities must also be consistent with the regulatory classification and the hazards posed by the waste. Waste shipped off-site should be tracked and/or manifested in order to document its ultimate disposal location. It is often prudent to periodically audit off-site treatment, storage, and disposal facilities that are handling large volumes of hazardous waste to ensure that it is being appropriately managed. There are many examples of off-site contractors who have not correctly handled wastes, creating significant financial liabilities for the original waste generators.

Permanent on-site waste disposal should only occur within designed facilities with engineering and management controls commensurate with the health, safety, and environmental risks posed by the waste. In most cases the on-site disposal facilities will need to be permitted with the appropriate regulatory agencies. To ensure adequate levels of engineering design and management, active on-site landfills or waste storage facilities will generally require

- Clear restrictions on what types of waste materials can and can't be placed in the facility;
- Clear boundaries and limited access to the facility to ensure waste delivery is controlled and to minimize exposure risks to animals and people; and
- Operating procedures describing waste delivery and placement practices, inspection frequency, and covering requirements.

For active storage facilities and landfills that accept chemically reactive wastes capable of posing direct exposure risks or degrading surface and groundwater quality, additional controls are likely to be required, including

- Leachate and stormwater management systems consistent with the chemical hazards posed by the waste, for example:
  - Composite liners with leachate collection systems for chemically reactive wastes that may degrade groundwater quality,
  - Covered and enclosed structures for long-term storage of wastes that are unstable in contact with water, and
  - Stormwater collection systems for wastes that may degrade runoff water quality;
- Closure plans detailing how the facility will be closed—in particular, final cover designs needed to limit direct exposure risks to animals and people, to preserve runoff water quality, to limit infiltration, and to control wind dispersion of the waste; and
- Monitoring programs for groundwater, surface water, air quality, landfill gas, and windblown debris to ensure that potential impacts associated with the waste are being adequately controlled.

Remote sites may find it economic to operate dedicated recycling and/or incineration facilities for waste materials such as antifreeze, used oil, solvents, and medical waste. Again, these facilities must be designed and operated in accordance with all application waste management regulations and air quality standards.

HAZARDOUS MATERIALS AND CONTAMINATION CONTROL

Hazardous material is broadly defined as any substance (solid, liquid, or gas) that could pose a significant threat to human health or the environment if improperly handled or accidentally released. Typically, hazardous materials are toxic, corrosive, flammable, explosive, strongly oxidizing, or radioactive. Hazardous materials that are commonly managed at mining operations include cyanide, strong acids, caustic soda, lime, chlorine, pesticides, herbicides, explosives, solvents, gasoline, and hazardous waste. Other substances that can be considered hazardous under some circumstances include paints, diesel fuel, coolants, and lubricants. The actual risks posed by the handling of these materials depend on their innate hazards, the volumes that are present, potential receiving environments, and transport pathways that could connect the point of release with potential receptors.

The mishandling or unintentional release of hazardous materials can create acute or chronic health problems for workers and nearby communities, damage ecosystems, degrade land productivity and water quality, and create costly cleanup liabilities. Hazardous materials should be managed to minimize the frequency and severity of any releases. Pipes and tanks should be designed to safely transport and store hazardous materials with minimal risk of failure. Pipelines and tanks should be designed for the pressures used and the material being transported, be well maintained, and be regularly inspected. In general, the use of underground bulk storage tanks should be avoided. If the risks posed by contaminant release are high, additional safeguards such as secondary containments, leak detection systems, and double-walled pipelines may need to be considered.

Secondary containment systems such as bunds, berms, and liners for bulk storage tanks need to be built and maintained in compliance with all local and national requirements and permit conditions, or, in their absence, internationally recognized standards. Secondary containment systems may be designed for individual tanks or for entire tank farms and drum storage areas. Containment and capture systems (such as concrete pads sloped to collection sumps) may also be constructed in areas prone to recurring spills such as haul truck refueling areas.

Secondary containment facilities should be sized to hold a volume greater than that of the largest tank in the containment, along with any piping that drains back into the tank or containment and additional capacity for the design storm event (typically 110% of the largest foreseeable leak into the containment is sufficient). Stand alone double-walled tanks can provide valid secondary containment if designed to contain all leakage from the internal tank. The impermeable material used to construct the secondary containment must have a design leakage rate that is consistent with the environmental hazards posed by the material. In most cases, the secondary containment should be constructed with highly impermeable materials such as concrete, asphalt, or synthetic liners like HDPE. However, if risks are relatively low, more permeable materials such as earth berms and compacted fine-grained soils may be acceptable.

Liner design for the storage of hazardous materials should also always meet or exceed local legal and regulatory requirements. Leach pads and process water ponds that contain hazardous fluids, such as cyanide-bearing solutions or extremely acidic, basic, and/or saline solutions, will generally need to be underlain by a composite liner. Tailings impoundments that contain hazardous pore fluids may also need to be underlain by a liner system unless sufficient hydrogeologic containment can be demonstrated along with adequate seepage collection and monitoring systems. Liner systems may also sometimes be required for chemically reactive waste rock dumps in wetter climates unless sufficient hydrogeologic containment can be demonstrated along with adequate seepage collection and monitoring systems. Many types of nonmineral waste landfills will also require liners to allow capture and collection of leachate and landfill gases.

Composite liners are typically constructed with an underlying low-permeability, compacted clay layer and an overlying synthetic liner. Again, the actual design leakage rate of the liner system must be based on a risk assessment that takes into account the hazards posed by the liquid and its potential environmental impacts. If a higher leakage or failure rate is determined to be acceptable, then a composite liner may not be needed. Conversely, if the hazards posed by the fluid and the environmental setting are high, then a double or triple liner system with intervening leak detection and pressure relief layers (to minimize head on the lower liners) is likely to be needed.

Leach pads and other process circuits that contain large volumes of hazardous fluids should be designed for gravity flow into engineered process ponds and containment structures during power outages and other upset conditions. The capacity of these structures should be large enough to hold the maximum credible volume of the fluid that would be released by drain down before the upset condition could be corrected. In many cases, backup emergency power systems for pumps may be required.

If there is an uncontained spill, rapid spill response can minimize the environmental damage by limiting the volume of material released, the surface area impacted, and the exposure duration. Emergency response procedures for spill containment and cleanup should be developed and appropriate spill response equipment stored on-site. Typical spill response kits may include sorbent materials, neutralizing agents, drums, shovels, and appropriate personal protective equipment. At port facilities where spills onto open water are possible, spill response boats and booms may be maintained. Most operations should also maintain a trained spill-response team.

CONTAMINATED WATER MANAGEMENT

Mining operations can impact surrounding water resources and water-dependent ecosystems by (1) water withdrawal and dewatering impacts, and (2) the discharge of contaminated water. Surface and groundwater withdrawals to dewater the ore body or to supply operations can lower surrounding groundwater water tables, causing seeps, springs, and wells to dry up; harming groundwater-dependent vegetation and ecosystems; and reducing in-stream flow. If necessary, dewatering impacts must be predicted, monitored, and mitigated. Mitigation strategies may include

- Improving water efficiency through process and management improvements so that less water needs to be withdrawn,
- Intentional surface water discharge at key locations to maintain in-stream flow,
- Providing alternative water resources for impacted communities,
- Intentional recharge of groundwater to minimize drawdown impacts, and
- Construction of slurry walls and other subsurface flow barriers to minimize hydrogeologic connections.

Water may be discharged to the environment through centralized, engineered, and permitted discharge points; more dispersed, uncontrolled stormwater discharge locations; and/or uncontrolled subsurface flow. If the released water is contaminated, it may degrade down-gradient water quality, harming ecosystems, human health, and water resources. Water that contacts chemically reactive mineral waste is commonly contaminated, but poor-quality water can also be generated by ore-body dewatering, mineral processing, sewage discharge, recurring hazardous material releases, and outflows from pit lakes and underground workings.

As discussed in the earlier section on chemically reactive mineral waste, mine-impacted water typically poses risks because of ARD, salinity, metals soluble at neutral pH, radionuclides, nitrogen compounds, and added reagents such as cyanide. In many jurisdictions, discharge limits are established by law, regulation, or permit for potential contaminants of concern. In such cases, water may usually only be released from designated outfall points and must consistently meet the mandated discharge limits. These limits are typically based on maximum concentration, average concentration weighted by flow and/or time, and/or mass loading. In other cases, discharge limits must be selected and justified based on site-specific data and conditions. In these circumstances it is important to understand not only the potential discharge water quality but also migration pathways and the assimilative capacity and sensitivity of the receiving environment. It is important to differentiate between impacts to receiving environment water quality and environmental harm. Discharged mine water may result in measurable changes in solute concentrations in receiving water bodies without causing environmental harm. For example, metals concentration may measurably increase, while the ambient concentration remains well below threshold values for protection of aquatic ecosystems and quality of drinking, stock, and irrigation water. It is also important to consider cumulative impacts to the receiving environment. If multiple mines exist in a single drainage basin, the cumulative impact of all mine discharges must sometimes be considered.

Discharge water quality and volumes partly depend on the innate geology and climatic setting of the ore body being mined. However, mine and process design and management can also have a profound influence on the quality and volume of water that must be managed. The water and solute balance should be predicted and monitored during mine design, operation, and closure. These data can be used to select appropriate water management strategies to meet discharge limits and protect down-gradient water quality.

Water treatment before discharge can be costly, sometimes exceeding several dollars per 1,000 L. At large mines with significant ARD flows, cumulative treatment costs can be measured in the tens to hundreds of millions of dollars. Implementation of internal proactive management strategies that reduce the volume of water that must be treated and/or reduce the solute load in the water can be cost-effective, as well as ultimately more protective of the environment. The strategies discussed previously to minimize water contact with, and solute loading from, mineral wastes are a key component of overall site water quality management. Other broad internal water management strategies include the following.

- **Water diversion:** Capture and diversion of clean surface and groundwater flows up-gradient of the operation can limit the volume of water that may be contaminated by contact with the operational footprint.
- **Water segregation:** Within the footprint of the operation, careful segregation and separate handling of waters with different qualities can limit the volume of water that ultimately requires treatment. Clean stormwater can often be discharged without treatment if its quality is preserved. Unless needed as makeup water, precipitation and stormwater should generally be isolated from process water circuits, reactive mineral waste, hazardous materials, and contaminated areas.
- **Recycling and reuse:** Internal recycling and reuse of poor-quality water can reduce the volume of makeup water that must be withdrawn from the environment and allow beneficial use of poor-quality water, thereby reducing the volume of water that may need to be treated before discharge.
- **Improved water use efficiency:** Improvements in water use efficiency can also reduce the volume of water that must be imported into the operation and that ultimately must be discharged.
- **Reagent management:** Process water quality can be improved by the efficient use of reagents and/or replacement of hazardous reagents with less hazardous but equally effective substitutes.
- **Safe on-site storage and controlled release:** If sufficient safe water storage capacity exists on-site, mine water can sometimes be discharged without treatment at rates and times that can be assimilated by the receiving environment. For example, saline waters can be stored on-site during the dry season and discharged during peak storm-related runoff events, when natural salinity is low and in-stream flow is highest.
- **On-site evaporation:** Evaporative losses within the footprint of the operation will reduce the volume of water that must be discharged. Natural evaporation from storage ponds and tailings dams can be a significant source of water loss in some climates. Evaporation can be enhanced with simple sprinkler systems, and in some extreme situations energy intensive crystallizers may be used. Although evaporation can be effective for reducing the volume of water that must be managed, it does not remove soluble salts from the system and will ultimately produce saline solutions or soluble solids that must be managed.

In humid environments it is likely these proactive water management strategies will reduce but not eliminate the need to discharge water. However, mines in arid and semiarid climates may maintain a negative water balance because of evaporative losses, entrainment losses to shipped product, and long-term storage within tailings. The water balance implications of any significant operational changes such as new ore-body development, new processes, or footprint expansion must be carefully assessed. It is also important to distinguish between operational water balances when mining and processing activities may consume water, and postclosure conditions when these water sinks are no longer available.

In many cases, water treatment before discharge will be required. At a minimum, most mining operations will need to control sediment discharge via on-site erosion control and the use of settling ponds. Sewage treatment facilities will usually be required unless operations are connected to a municipal treatment system. Oil water separators are also likely to be needed to capture fuel, oil, and grease from refueling stations and maintenance shops. Mine waters that have been impacted by ARD, salinity, metals soluble at neutral pH, reagent addition, or other contaminants may require additional treatment before they can be discharged. Numerous mine-impacted water treatment technologies exist for different contaminants of concern. These technologies are too numerous to discuss in detail here, but many good summaries are available in the literature (Brown et al. 2002; Younger et al. 2002; INAP 2009). The various treatment technologies are most commonly designed to neutralize acidity and raise the pH, remove metals from solution via precipitation or adsorption, or remove salts from solution by membrane-based or ion exchange technologies.

Neutralization and metals precipitation is most commonly accomplished by the addition of alkaline materials such as lime, limestone, and soda ash, but many other viable technologies are available. In some situations, multiple technologies must be used in sequence such as neutralization to raise the pH and remove metals from solution, followed by flocculation and filtering to remove fine total suspended solids and then reverse osmosis to remove dissolved salts. Treatment systems may be either active or passive. Active systems usually require extensive infrastructure, electrical power, and continuous staff involvement for operation and maintenance. They are usually required for large flows of poor-quality water that must be discharged from active mines. Passive systems generally do not require continuous reagent and energy inputs or operator involvement and so are appropriate at some closed operations and remote sites. However, passive systems are commonly only viable for smaller flows (typically less than 10 L/s), for water that is only weakly contaminated, or as a polishing step for water pretreated with an active system. Common passive treatment systems include wetlands, aeration and settling ponds, and anoxic limestone drains.

The water treatment systems that are selected and used should reliably meet the discharge water quality criteria in the most cost-effective manner. Keys variables that should be considered during the selection process include

- Initial water quality and its variability;
- Required discharge water quality;
- Volume and variability of flow that must be treated;
- Operational versus closure requirements;
- Capital, operations, and maintenance costs;
- Regulatory requirements;
- Reliability of the technology;
- Power, labor, and footprint constraints; and
- Solid waste (generally sludge) management requirements.

Contaminated water with high solute concentrations can generate large volumes of sludge. Although commonly overlooked during initial analysis, the costs and technical challenges to manage water treatment sludges can be significant.

CONTAMINATED SITES MANAGEMENT

Unplanned releases of hazardous materials will typically lead to the contamination of soils, sediment, groundwater, and/or surface water. Historic waste and hazardous materials management was generally less rigorous than current practices, so contaminated sites are most likely to exist at older mines and at new mines in historic mining districts. Common historic contaminated sites at mining operations include leaking underground storage tanks; locations of recurring spills such as refueling areas; historic landfills and waste storage areas; old mineral processing, smelting, and refining sites; and chemically reactive mineral waste repositories. Common contaminants of concern include hydrocarbons, chlorinated solvents, PCBs (polychlorinated biphenyls), asbestos, acids, and metals such as arsenic, cadmium, chromium, copper, lead, mercury, nickel, selenium, and zinc.

Many jurisdictions have clear guidance on acceptable levels for these contaminants in soils, sediment, and water. When available, the appropriate local screening or investigation levels should be used. These criteria are generally derived from simple exposure models with conservative assumptions to protect human health, plants, and animals. Concentrations above these screening levels will generally trigger further remedial, management, or administrative actions to control the risks to human health and the environment. The appropriate criteria must be selected based on the media, the environmental setting, and the potential receptors requiring protection. Examples include standards for drinking water and the protection of aquatic life; and guidelines for stock and irrigation water, animal feed, and soil and sediment cleanup. Because the screening levels tend to be conservative, in some cases the cost and time required to derive site-specific criteria through a formal ecological or human health risk assessment process may be justified.

If current or potential future exposure risks are determined to be unacceptable, then remediation or management strategies (corrective actions) will need to be developed and implemented. Consideration should be given to stakeholder expectations, legal requirements, and the land's planned final use when defining remedial objectives. Corrective actions should then be selected to meet the environmental objectives in the most reliable and cost-effective manner. Only strategies that will consistently meet or exceed local regulatory or permit requirements should be considered. In some cases, more aggressive but more costly remediation strategies may be justified if they will raise the final value of the land or provide significant additional environmental or reputational benefit. Corrective action selection and refinement should be an iterative process with a range of potentially viable strategies assessed before the optimum strategy or combination of strategies is selected. Common general remediation and management strategies include those discussed in the following paragraphs (modified from Rio Tinto 2008c).

Removal Actions

Contaminated soils and sediments can be removed and transported to an engineered repository or treatment facility where the contaminants are neutralized or destroyed. Removal actions typically have a cleanup target in which all contaminated soils or sediments over a certain concentration are removed. The cleanup goals are typically based on the final assumed land use, such as industrial/commercial uses, which typically allow the highest residual concentrations to be left in place, and residential land uses, which usually require the lowest. Open space, wildlife habitat, or grazing and agricultural lands will typically (but not always) have intermediate values.

In-Situ Treatment

Some exposed or near-surface contaminated soils may be remediated in place without excavation and transport. Examples of such in-situ treatment techniques include

- Lime application to neutralize acidity and limit metals bioavailability and mobility,
- Application of nutrients or microbiota and mixing to accelerate the biological degradation of organic contaminants such as hydrocarbons,
- Air sparging and soil vapor extraction for removal of volatile compounds,
- Application of oxidizing agents to treat cyanide spills, and
- Bioremediation with plants that extract metals from the soil and can then be harvested.

Capping in Place

Contaminated soils can be capped with clean fill, soils, and impermeable layers (such as compacted soils, synthetic liners, asphalt, and concrete). Capping can serve several purposes, including

- Preserving runoff water quality;
- Controlling the dispersion of contaminated soils by wind, runoff, or tracking;

- Eliminating direct exposure risks to animals and people from contaminants exposed at the surface;
- Limiting plant contaminant uptake and subsequent exposure risks posed to browsing and grazing animals; and
- Limiting infiltration and subsequent contaminant migration to groundwater.

The cap must usually be designed to meet the environmental objectives for the long term because the contamination is left in place.

Groundwater Remediation
Contaminated groundwater may be remediated through a variety of pump and treatment systems, by in-situ treatment technologies, or by other subsurface control systems such as reactive barriers to control contaminant migration. Pump and treat systems typically involve either removal of the most contaminated portion of the plume (to remove the contaminant source) or capture of the leading edge of a migrating plume (to control its movement). Pump and treat systems for hydrocarbon contamination are usually focused on the removal of free phase product that is floating on the surface of the water table. Examples of in-situ treatment technologies for groundwater include injection of alkalinity to neutralize acidity; air sparging and vapor extraction to remove volatile organic compounds; and the addition of oxygen, microorganisms, and/or nutrients to promote biological degradation. Groundwater remediation will usually continue until water quality meets applicable criteria for drinking water, aquatic life, irrigation water, or stock water.

Groundwater Flow Control
The flux of contaminated groundwater may be reduced so that contaminant loading into and concentrations within down-gradient aquifers or surface water bodies remain within acceptable levels. Controls may include

- Impermeable covers that limit infiltration to groundwater through contamination in the vadose zone,
- Impermeable subsurface barriers that divert up-gradient flow around the contamination, or
- Impermeable subsurface barriers that divert down-gradient flows away from sensitive receptors.

Land-Use Restrictions
Exposure risks to people and animals may be controlled by establishing long-term land-use restrictions such as prohibitions against residential development and farming or fencing to keep even incidental visitors out. Areas where contaminated soils are capped in place will typically require restrictions against development or uses that could breach the cap and re-expose the underlying contamination. In some countries, long-term land-use restrictions may be difficult to enforce.

Natural Attenuation
Many organic contaminants such as solvents, hydrocarbons, and cyanide will degrade with time after they are released to soils, sediment, or water due to oxidation and biological activity. Acidic waters may be neutralized through mixing with more-alkaline waters or reaction with aquifer materials, and dissolved metals may be removed from solution by sorption and precipitation. Down-gradient mixing and dilution will also limit contaminant concentrations below the point of release. If these natural attenuation processes are identified and quantified, it may be possible to justify no further action other than long-term monitoring, if allowed in the local jurisdiction. However, generally, a rigorous evaluation will be needed to demonstrate that impacts to important receiving environments and receptors will remain within acceptable limits throughout the natural attenuation process.

In all cases, some form of post-remediation sampling or monitoring is generally required to ensure that environmental objectives are being met. For removal actions, this typically involves short-term sampling immediately after the excavation has been completed to confirm residual contaminant concentrations are below the target values. If natural attenuation is selected, then long-term monitoring will generally be required to ensure that contaminant concentrations in the affected media are declining at the rates predicted and that receiving environment criteria are being met.

REFERENCES
ANZECC/ARMCANZ (Australian and New Zealand Environment and Conservation Council and Agriculture and Resource Management Council of Australia and New Zealand). 2000. Australian and New Zealand Guidelines for Fresh and Marine Water Quality. October 2000.

Brown, M., Barley, B., and Wood, H. 2002. *Minewater Treatment: Technology, Application and Policy*. London: IWA Publishing.

Buckman, H.O., and Brady, N.C. 1969. *The Nature and Properties of Soils*. New York: MacMillan.

EPA (U.S. Environmental Protection Agency). 2008. SW846 Test Methods for Evaluating Solid Waste, Physical/Chemical Methods. www.epa.gov/epawaste/hazard/testmethods/SW846. Accessed February 2010.

ICMI (International Cyanide Management Institute). 2009a. International Cyanide Management Code. www.cyanidecode.org. Accessed January 2010.

ICMI (International Cyanide Management Institute). 2009b. Implementation Guidance for the International Cyanide Management Code. www.cyanidecode.org. Accessed January 2010.

INAP (International Network for Acid Prevention) 2009. *The Global Acid Rock Drainage Guide*. www.gardguide.com. Accessed January 2010.

Maas, E.V. 1990. Crop salt tolerance. In *Agricultural Salinity Assessment and Management*. Edited by K.K. Tanji. New York: American Society of Civil Engineers.

MEND (Mine Environment Neutral Drainage). 2001. MEND Manuals 5.4.2 a–f, Volumes 1–6. Ottawa, ON: Natural Resources Canada.

Price, W.A., Morin, K., and Hutt, N. 1997. Guidelines for the prediction of acid rock drainage and metal leaching for mines in British Columbia: Part II, Recommended procedures for static and kinetic testing. In *Proceedings of the Fourth International Conference on Acid Rock Drainage*, Vancouver, BC, May 31–June 6. Ottawa, ON: Mine Environment Neutral Drainage Program.

Richards, L.A. 1954. *Diagnosis and Improvement of Saline and Alkaline Soils*. Agricultural Handbook 60. Washington, DC: U.S. Department of Agriculture.

Rio Tinto. 2008a. Mineral Waste Management Standard (Version 2) and Guidance Note (Version 2). www.riotinto.com. Accessed April 2010.

Rio Tinto. 2008b. Acid Rock Drainage Prediction and Control Standard (Version 2) and Guidance Note (Version 3). www.riotinto.com. Accessed April 2010.

Rio Tinto. 2008c. Hazardous Materials and Contamination Control Standard (Version 2) and Guidance Note (Version 2). www.riotinto.com. Accessed April 2010.

Ross, M. 1999. The health effects of mineral dusts. In *The Environmental Geochemistry of Mineral Deposits, Part A: Processes, Techniques and Health Issues. Reviews in Economic Geology* 6A. Edited by G.S. Plumlee and M.J. Logsdon. Littleton, CO: Society of Economic Geologists.

Smith, A.C.S., and Mudder, T.I. 1999. The environmental geochemistry of cyanide. In *The Environmental Geochemistry of Mineral Deposits, Part A: Processes, Techniques and Health Issues. Reviews in Economic Geology* 6A. Edited by G.S. Plumlee and M.J. Logsdon. Littleton, CO: Society of Economic Geologists.

Smith, K.S., and Huyck, H.L.O. 1999. An overview of the abundance, relative mobility, bioavailability, and human toxicity of metals. In *The Environmental Geochemistry of Mineral Deposits, Part A: Processes, Techniques and Health Issues. Reviews in Economic Geology* 6A. Edited by G.S. Plumlee and M.J. Logsdon. Littleton, CO: Society of Economic Geologists.

Tucker, G.B., Berg, W.A., and Gentz, D.H. 1987. pH. In *Reclaiming Mine Soils and Overburden in the Western United States*. Edited by R.D. Williams and G.E. Schuman. Ankeny, Iowa: Soil Conservation Society of America.

Virta, R.L. 2002. *Asbestos: Geology, Mineralogy, Mining and Uses*. USGS Open-File Report 02-149. http://pubs.usgs.gov/of/2002. Accessed April 2010.

Younger, P.L., Banwart, S.A., and Hedin, R.S. 2002. *Mine Water Hydrology, Pollution, Remediation*. Dordrecht, Netherlands: Kluwer Academic Publishers.

CHAPTER 16.7

Closure Planning

Evelyn L. Jessup Bingham

INTRODUCTION
In recent years the minerals industry has recognized the importance of sustainable, integrated closure planning. Because of the finite nature of ore bodies, all mine and related facilities will eventually close, and the reputation of the industry is very much dependent on the legacy it leaves. Whether a facility has 10, 20, or 50 years of operational life remaining, integrating closure planning into the mining and metal processing business can result in value for both the company and the wider community.

Activities associated with mineral resource development have a significant impact on the social and environmental aspects of a site. Any site disturbance, even in the early stages of a project, creates a legal and financial closure obligation for the mining company. Communities, governments, and lending institutions are more and more aware of the benefit of good closure planning and demand it as part of the permitting process for new or expanded operations.

Closure of an asset, be it a mine or mineral processing facility, is generally not straightforward and requires dealing with the uncertainty of future conditions and events concerning commodity prices, stakeholder demands, regulatory requirements, environmental conditions, and ore grade. Ideally, closure is carefully planned and occurs with the exhaustion of the mineral resource, but it is not uncommon for closure announcements to come suddenly with the fall of commodity prices or some other unplanned event. Risk-based closure planning is inclusive of such uncertainty.

Closure of an asset is a process that works toward an end point that can be referred to as completion. Planning for a successful closure is a complex, multi-disciplinary task that is essential to minimizing long-term risk for the mining company, the environment, and affected stakeholders.

CHAPTER ORGANIZATION AND APPLICABILITY
This chapter is intended to provide basic practical guidance on closure planning. The information presented should be considered just a first glimpse of the wide-ranging discipline of closure planning.

In recent years a number of comprehensive and specialty documents have been created by governments, industry associations, and nongovernmental organizations (NGOs) on the subject of asset closure planning and financial assurance, such as those available through the International Council on Mining and Metals at www.icmm.com.

The core information provided in this chapter includes a number of practical tables and lists of information that have been found to be useful for closure planning practitioners who are creating or reviewing best-practice integrated closure plans.

After a brief overview of closure planning issues and a description of what integrated closure planning means, a set of principles for creating a framework for closure planning is presented. Following this, the steps involved in closure planning are outlined, and detailed components of an effective closure plan are described according to the closure planning steps. Also presented are a number of common risks to be considered in closure planning. Lists of this type are often useful in preparing for closure risk assessment workshops with a multi-disciplinary team to assist in the completeness of the review. Finally, a checklist is offered, showing how the detail and contents of a closure plan can change through the life cycle of a mineral resource development project.

The chapter content is applicable to a wide range of facilities and activities including mines, processing facilities, and greenfields and brownfields projects. The concepts and tools are appropriate for creating a closure plan for the first time or reviewing an existing closure plan.

COMPLETING THE ASSET LIFE CYCLE—CLOSURE PLAN OBJECTIVES
A successfully completed closure process is one in which the land affected by the asset has been rehabilitated to a sustainable condition. Ideally, in this completed state, the mining lease ownership has been relinquished by the mining company to another land user with minimized continued liability. This sustainable completed state can only be achieved through careful and informed planning over the long term. Closure planning is the responsibility of the mining company, but it cannot be done in a vacuum. Mining companies must involve governments, the communities of which they are part, and

---

Evelyn L. Jessup Bingham, Group Manager, Closure & Waste, BHP Billiton, Melbourne, Victoria, Australia

other stakeholders in closure planning to achieve a successful closure outcome.

Increasingly, financial institutions lending capital to mineral resource development projects set a high standard for asset closure planning as they seek to maximize value from the project and minimize risk. These institutions have learned hard lessons from the past where poor operational closure practices have resulted in ongoing legacies of water pollution, safety and health hazards, and the resulting public outrage and continuing liabilities. Many countries have specific regulations concerning closure planning and completion. Currently, however, there is no overarching world governance body covering mine and mineral processing facility closure, and there are still areas in the world where closure regulation is minimal or nonexistent.

External mining stakeholders such as local communities, conservation groups, and biodiversity advocates are becoming more and more sophisticated about the outcomes of good and bad closure planning practices. Given the opportunity, external stakeholders can become partners in closure planning, ultimately to the benefit of all.

Good closure planning toward the outcome of sustainable mine completion requires communication and engagement with stakeholders to develop agreed-on closure plan objectives, which should include the following:

- Comply with regulatory requirements or industry-leading practice where local regulatory requirements are insufficient to manage risk.
- Protect public health, safety, and welfare over the long term.
- Maximize environmental sustainability.
- Mitigate socioeconomic impacts during and after closure.
- Provide a reasonable basis for estimating and managing closure costs.
- Minimize costs for the mining company, the government, and the public.
- Achieve sustainable land-use conditions as agreed with stakeholders.
- Enhance the reputation of the mining company as a responsible corporate citizen, and ensure that shareholder value is preserved.

Timing and how these objectives are achieved through the life of the asset life cycle is very site specific. Certain types of mines (such as porphyry copper deposits and other hard-rock mines) and mineral processing centers may not allow for any concurrent or progressive rehabilitation or reclamation (for the purposes of this chapter, *reclamation* and *rehabilitation* are used interchangeably) during the operating phase because the disturbed areas (such as the open pit, tailings impoundment, waste dumps, and concentrator buildings) are in constant use during the operations. Assets of this type may spend many decades planning for closure activities that will occur at the end of the asset life.

Coal mines, on the other hand, generally have the opportunity to execute and obtain sign-off on rehabilitation through the life of the asset. However, even projects conducive to progressive closure will face challenges associated with life-of-asset (LOA) plans changing as commodity prices go up and down. It is not unheard of for coal mines to have to dig up or cover over well-established rehabilitation because coal seams once thought uneconomic are now scheduled to be mined.

Economic conditions going in the other direction more commonly present a serious obstacle to sustainable closure planning and ultimate completion of the closure process. A downturn in commodity prices usually results in a lack of resources for progressive rehabilitation (if the mine continues to operate) and for closure execution if sudden closure of the asset is announced.

To minimize the ups and downs of economic cycles and resource availability, effort on the part of the mining company must be made to understand and forecast changing conditions that might affect closure planning and execution as well as possible changes in stakeholder needs. Closure plans must be adjusted through time and adequate resources allocated for difficult economic conditions. Postmining land use and other objectives may need to change according to changes in the LOA plan and external factors.

INTEGRATING CLOSURE PLANNING FOR EFFECTIVENESS

A mining company's commitment to good closure planning practices will pay off in reduction of liabilities for the company and improved reputation with stakeholders. In the past, closure planning has often been a stand-alone activity in the business, usually done as a one-time effort to meet a regulatory obligation.

It is not uncommon for the asset closure plan to be the responsibility of the environmental function at a site while the LOA plan is the responsibility of the mine planners and the updates to the closure financial provision are handled by the accounting department, whose focus is on meeting accounting guidelines. A lack of communication between these various departments can cause the closure plan scope and details to fall well behind the inevitable updates to the LOA plan and the closure provision carried on the books to be out of date with both.

To address closure planning issues and meet business objectives to manage risk, it is necessary to ensure that closure planning is fully integrated in the core business of the asset. As business conditions change, LOA plans change, and so must the closure plan change. Additionally, the detail and accuracy of closure plans must change through the life cycle of the asset—starting out as conceptual and progressively becoming more detailed over time. Finally, as regulations and stakeholder expectations change, the closure plan needs to change. Effective closure plans must be integrated in a number of levels including the following:

- Integration with the LOA and long-term business plan. The closure plan necessarily links past disturbance, current operations, progressive rehabilitation, and future disturbance planned until the end of asset life.
- Integration with external stakeholder affairs. A focus of closure planning should be engagement with the external stakeholders affected by closure. This will include working with the government, local community, and NGOs to understand their needs and concerns. Interaction with external stakeholders during closure planning not only results in compliance with regulations but also affords the development of robust closure criteria to support an acceptable end land-use and communities that are better prepared to support the company when it comes time for completion and lease relinquishment.

- Integration with all the relevant functional disciplines on-site. Closure planning and ultimately the execution of the closure project toward completion will involve senior management, human resources, project engineering, business improvement specialists, environmental managers, operations, planning engineers, community and external affairs, legal advisors, risk management, and finance; all are necessary for effectiveness. Commonly, the following key roles and responsibilities will be essential to the multidisciplinary team:
  - Technical services provide the engineering design, LOA plan, and rehabilitation plan.
  - Project engineers and cost specialists are responsible for cost estimates, design of necessary pilot tests, and execution by project management.
  - Environmental specialists are responsible for developing closure criteria and performance requirements working with government representatives to meet regulations.
  - Community specialists are responsible for stakeholder engagement and closure scenario building.
  - Finance and accounting personnel are responsible for provisioning and maintaining closure risk registers.
  - Management is responsible for setting key performance indicators and milestones for closure.

Integrated closure plans are developed, reviewed, and updated starting in the LOA concept phase and moving through prefeasibility, feasibility, execution, operation, and closure phases of a mineral development project. Throughout this lengthy period, integration in the key areas will increase the probability of a closure plan meeting the business objectives upon execution.

CREATING A MANAGEMENT FRAMEWORK FOR INTEGRATED CLOSURE PLANNING

In order to fully integrate closure planning in the business, a long-term commitment to a set of basic principles is necessary from senior management. Closure planning is not a one-time endeavor but one that takes place over many years or decades. The people involved in closure planning will come and go; however, if the company has adopted a framework of principles, closure planning can proceed and improve through the life cycle of the mineral development project.

The following basic list of principles should be considered and adjusted as required for adoption as a company "closure policy" to form a solid framework from which the closure planning team can work over time.

- **Framework Principle 1: Regular Review and Updates**
  - Closure plans shall be maintained throughout the life cycle of all mineral development assets.
  - Closure plans shall be reviewed annually in conjunction with the LOA planning process. Closure plans and associated cost estimates shall be revised when there is a significant change to the LOA plan, and when there are regulations or factors significantly affecting the closure plan and associated cost estimates.
  - Closure planning processes and closure plans and cost estimates shall be subject to reporting, audit, and governance procedures, particularly where there is a risk of material misstatement on financial reports or a significant effect on business decisions.

Table 16.7-1 Closure plan upgrades across life-cycle phases

Life-Cycle Phase	Closure Plan Class
Projects in study phase	Conceptual-level closure plan and cost estimate ±50% to 30%
Assets with >10 years life remaining	Conceptual-level closure plan and cost estimates, e.g., ±30%. Expectation of progressive improvement in uncertainty and cost estimates annual review.
Assets 10–3 years life remaining	Prefeasibility-level closure plan and cost estimates, e.g., ±20%. Expectation of progressive improvement in uncertainty and cost estimates annual review.
Assets with 3–1 years life remaining	Feasibility-level closure plan and cost estimates, e.g., ±15%.
Assets in closure execution or postclosure	Execution-level closure plan and cost estimate, e.g., ±10%

- **Framework Principle 2: Management of Business Risk**
  - Closure planning shall include an assessment of all risks associated with closure including, but not limited to, risks associated with
    - Postclosure liabilities;
    - External stakeholder needs, aspirations, concerns;
    - Deferral or delay in progressive rehabilitation during operations;
    - Uncertainty in environmental, regulatory, or community aspects;
    - Uncertainty in cost estimations;
    - Uncertainty in technical aspects of closure; and
    - Delay in closure execution.
  - The management of closure risk shall be developed in accordance with formal risk management standards and associated guidance approved by company management.
- **Framework Principle 3: Accurate Estimates of Cost**
  - Closure plans shall be prepared in accordance with project management principles and contain sufficient detail to determine the cost of closure to the extent necessary to manage closure risk.
  - Closure plans will be progressively upgraded according to the life-cycle stage of the mineral development project as shown in Table 16.7-1.
  - Environmental and engineering studies will be performed to increase the certainty of closure cost estimates as an investment comes closer to the need to execute the required closure activity.
  - All closure plans shall include an estimate of the provision for closure, in accordance with international accounting guidelines.

THE CLOSURE PLANNING PROCESS

Whether creating a closure plan from scratch for a greenfield project, exploration project, or simply reviewing an existing closure plan for a large, mature mining property, the steps in the closure planning process are essentially the same. The level of detail and effort required to develop the closure plan through this process will vary as a function of site-specific issues and the life-cycle phase of the project.

Sudden or unplanned closure due to a downturn in commodity price or other adverse business decision may call for an acceleration of this process and require such things as interim care and maintenance plans to be implemented while the closure execution plan is being finalized.

Workshop Approach

The process of developing and updating an integrated closure plan should take as much thought and effort as that dedicated to design and development of any significant project. The result of the closure planning, even in the early phases, has the potential to significantly affect the business and stakeholders. Closure planning sets the course for years of activities, and it involves the expenditure of material funds for studies and, ultimately, closure execution.

Because of the multi-disciplinary nature of closure planning, facilitated workshop sessions are useful at intervals in the closure planning process. An internal closure risk assessment workshop is useful in the beginning to set the context, agree on strategy, and explore risks.

Workshop sessions with external stakeholders can be effective in building relationships and obtaining input about expectations. If available, an expert in workshop facilitation can add significant value, keeping the participants on point, while enabling constructive brainstorming, debate, discussion, and challenge.

A single point of accountability, a team organizational chart, and a work breakdown structure and schedule should be developed for the closure planning team to refer to. This should be updated annually during the regular closure plan review.

Basic Steps in Closure Planning

Recommended steps for closure planning are as follows:

1. Set the context by reviewing background data regulations, LOA plans, environmental baselines, and previous closure plans. Review stakeholders and engage as needed to understand their issues. Review the current status of the progressive rehabilitation plan and implementation schedule to determine whether the LOA plan, progressive rehabilitation plan, and closure plan align.
2. Establish closure plan strategy and criteria including scope, objectives, assumptions, design criteria, and completion criteria. This sets the basis for all subsequent steps. It is essential that the closure team reaches agreement on the standards and strategy to efficiently complete the planning.
3. Assess risks and evaluate closure options. The closure risk assessment process forms the backbone of integrated closure planning, and therefore a standardized risk assessment process, such as that developed by Australia and New Zealand (AS.NZS 4360:2004), should be used. Involving the proper set of people with the detailed knowledge of the site is essential to identifying all the risk issues. Standard options analysis techniques that would be used on any large project involving significant costs should be applied to ensure sufficient rigor in providing the basis of the preferred closure plan.
4. Select a preferred closure plan from the option analyses to control closure risks at an acceptable level. This step is usually an iterative process with Steps 3 and 5, as cost estimate and the risk profile necessarily enter into the decision-making process. The preferred closure plan will be adjusted with changes to the LOA plan and other changes to the context and risk profile of the site as the annual reviews proceed over time.
5. Generate robust cost estimates for the preferred closure plan in accordance with sound engineering and project management principles. The engineering cost estimate serves as the basis for a number of subset estimates, including internal business investment valuation, accounting provision for financial statements and government financial assurance or bonding. Both deterministic and probabilistic estimates may be generated depending on the level of uncertainty. Probabilistic techniques such as Monte Carlo simulations may require specialist effort but can very effectively incorporate the uncertainty of such items as undefined contamination and cost fluctuation. As the site nears closure, specialist demolition, salvage value, and remedial experts will be required to establish the cost of an executable closure plan.
6. Establish an action plan for improvement. Effective integrated closure planning requires improvement in the knowledge base in preparation for the next review cycle and beyond. This will include further stakeholder consultation, engineering and environmental studies, site characterization studies, definition of existing contamination, and rehabilitation pilot testing. These actions may require allocation of significant resources, and therefore a formal action plan for improvement establishing schedule and budget is necessary.
7. Update the closure document and record management system. The closure plan process represents a significant investment in human and financial resources over many years. A readily retrievable archive of the process, the information used, and the decisions made is essential for annual reviews and audits.

COMPONENTS OF THE INTEGRATED CLOSURE PLAN

Table 16.7-2 provides a guide to the components of a typical closure plan organized by the high-level steps taken by the closure planning team. Because of the site-specific nature of closure planning, components will vary according to many factors. The list in Table 16.7-2 can be used in a number of ways: as a closure planning checklist, a table of contents for a closure plan, an update list during the annual closure plan review, or a closure plan audit checklist.

CLOSURE RISK

Assessing closure risk requires a systematic, structured evaluation. As mentioned, the risk assessment forms the basis of the closure plan and cost estimates. The focus of the closure risk assessment will be to establish an acceptable risk profile for the company and all other stakeholders upon completion of the closure project.

Materiality of risks will be an important consideration for the closure team to prevent the closure planning process from becoming bogged down in unnecessary detail. It is useful for the team to establish a materiality threshold at the beginning of the closure planning process as an assumption. This will depend a great deal on where the asset is in the life cycle, the class of closure plan (concept, prefeasibility, feasibility, etc.), and the overall magnitude of the closure plan.

For example, a closure team may decide that risk issues identified which have a worst-case cost of addressing less than $200,000 are not material to the process of updating a concept-level closure plan that will total more than $100,000,000. The closure planning team may record these lesser risks or review them in subsequent updates but not spend valuable time putting them through the full risk assessment and option evaluation process.

Table 16.7-2 Key components of an integrated closure plan

	Plan Component	Description and Comment
▶	**Closure Planning Step 1**	Set the context.
1a.	Life-of-asset (LOA) plan summary and map	The LOA plan is the basis of the closure plan. Additionally, the closure plan can affect the LOA plan when the cost and risk profile of the closure plan is considered. The LOA plan establishes the impacted area and scope of works to be decommissioned, rehabilitated, closed, and released. Summaries of the final configuration of pits, dumps, and plant areas are included in the closure plan. LOA plans necessarily change with changing economic conditions and changing technology. Closure plans should be fully integrated and updated with the LOA plan.
1b.	Progressive rehabilitation plan summary and map	Plans to rehabilitate areas of disturbance back to final configuration during the operating phase (prior to closure execution) are integrated in the closure plan. Progressive rehabilitation plans are integrated with the LOA plan and closure plan, and are reviewed annually. Expenditures for progressive rehabilitation are part of the operations budget and commonly are excluded from the closure plan and cost estimate. A significant closure planning risk is that progressive rehabilitation during operations falls behind schedule and unexpectedly needs to be completed as part of closure execution. Similarly, sudden closure (e.g., due to commodity price downturn) can result in a breakdown in the progressive rehabilitation schedule and an underfunded closure plan.
1c.	Studies register—environmental, cultural, heritage, and biodiversity conditions	A reference list of studies documenting conditions for the LOA plan footprint, pollutant management area, and regional area impacted must be established. Baseline studies and recent reviews of conditions are important in establishing closure criteria. In particular, incorporating the ongoing results of groundwater monitoring, surface water monitoring, and vegetation studies into closure planning is critical. Basis for remedial projects (current and forecasted) and their expected outcome should be reviewed regularly and factored into closure planning.
1d.	Legal and corporate requirement register	A register of regulatory/legislative mandatory government requirements that cover the environmental, social, financial, and legal aspects of the closure plan must be kept up-to-date and reviewed annually. These requirements vary significantly from country to country and site to site, and must be well understood as they become the basis for the absolute minimum criteria for the closure plan to meet. Site-specific permit requirements are specified. Legal commitments to joint-venture partners, financial institutions, or other entities affecting closure planning must be met. A register of company requirements, such as internal, environmental, or community standards and industry-leading closure criteria and practices, must be kept. Distinctions must be made as to which are considered mandatory (as in the case where a constructive obligation has been created by publicly indicating a standard higher than legislative requirements) and which are discretionary.
1e.	Land and property ownership register	A register summarizing the site land information impacted by the site disturbance should be created. Property boundaries, geographic information system data, lease arrangements, deeds, and related obligations must be clearly recorded and reviewed for their impact on the closure plan.
1f.	Stakeholder analysis register	External and internal closure stakeholders must be identified; known stakeholders' needs, concerns, and expectations must be assessed and considered. Requirements and community expectations in developing overall closure objectives and criteria for postclosure land use and cultural and heritage values must be evaluated. This register is updated as closure planning proceeds through the years to reflect current demographics and changing views as the stakeholder consultation and external affairs management plan (discussed further in step 6b) is implemented.
▶	**Closure Planning Step 2**	Establish closure plan strategy and criteria.
2a.	Closure boundary and scope summary	This may be a simple list and diagram of the key elements of the site, established by the multifunctional closure team after review of existing information. Establishing the boundaries of the closure plan may not be completely straightforward. The closure plan may or may not include such elements as a joint venture or contractor-run coal processing plant. A rail spur on-site may be included in the closure plan at one site, while another site's third-party-owned railway passing through the site is not included. Intangibles such as the human resources severance plans or a company-sponsored community support trust are to be key elements recorded and clearly included, or not included, in the closure plan to avoid confusion and errors in closure provisions as time goes on.
2b.	Closure strategy, objectives, and assumption documentation	Site-specific high-level vision, strategy, and objectives, followed by a more detailed listing of supporting assumptions (e.g., target closure date, costing basis in-house or contractor), should be established. These are updated as needed during the annual review to reflect any changes in stakeholder and company requirements. The materiality of risks should be established for consideration in the risk assessment for the rest of the closure process.
2c.	Closure criteria register—design criteria and performance indicators	Determining closure design criteria and performance indicators is an interactive process that takes form as the closure team proceeds through the risk assessment and closure options analysis. Examples of specific criteria include depth of topsoil, minimum requirement for slopes to control erosion, and storm criteria for sizing water management structures. The specifics of these items will vary depending on where in the life cycle the asset is, starting out as conceptual when closure is many years away and becoming very specific during the closure execution and postclosure monitoring phase. The closure criteria register should provide a high-level summary of rationale and references for how the criteria were determined (e.g., reference to international engineering standards, site-specific risk control basis, local regulatory mandate).

(continues)

Table 16.7-2 Key components of an integrated closure plan (continued)

	Plan Component	Description and Comment
▶	**Closure Planning Step 3**	Assess risks and evaluate closure options.
3a.	**Closure risk register**	The closure plan can effectively be thought of as a set of risk control measures. Documenting the output of the formal detailed assessment of all business, environmental, and social sustainability aspects and risks of closure is critical to the closure planning process. It is essential that risk assessments are reviewed and updated annually or whenever there are major changes in the LOA plan or regulatory requirements. The risk register should document the full range of closure risk, risk rating, and enough background information to enable efficient review and communication for audit. The risk register also serves as the basis of the annual closure plan review. Table 16.7-3 provides a list of examples of common closure risks and the associated closure plan controls.
3b.	**Closure opportunities register**	While risks associated with closure are usually the primary focus of the closure planning team, it is important to identify and develop any closure opportunities. Postclosure land uses for the large expanses of land typically involved in a mining property should be explored. Industrial parks, emerging carbon sequestration techniques, alternative energy (such as solar and wind power generation stations) may present economic development opportunities. Timber, agriculture, and grazing may be appropriate. Biodiversity preserves and wildlife conservation areas are postclosure land uses that may present long-term opportunities. The community or regional government may have an interest in using infrastructure such as roads, bridges, power plants, and warehouses after operations have ceased. The capacity of a third party to manage these facilities needs to be carefully gauged prior to the release of the infrastructure. It is not uncommon for the mining company to find the liability coming back unexpectedly if the recipient of the infrastructure finds itself unable to bear the maintenance or operating costs. Other common opportunities to evaluate include third-party reprocessing of low-grade ore stockpiles prior to rehabilitation, recovery of metals from postclosure water treatment, and revenue created by salvage value during demolition. Closure opportunities assessment should be evaluated and documented in the same manner as downside risk.
3c.	**Closure option analysis summary**	As the closure team works through the evaluation of risks and opportunities, various closure plan options will be generated. Formal and informal options analysis will be required to choose the preferred set of activities that become the updated closure plan. As options are considered, objectives and criteria may need to be adjusted, data gaps may be identified, and the interactive nature of closure planning becomes apparent. Concisely documenting options considered will make subsequent reviews and updates more efficient.
▶	**Closure Planning Step 4**	Select preferred closure plan.
4a.	**Postclosure land use and final landform plan**	Postclosure land use will be developed and agreed upon with relevant stakeholders (both internal and external) and as the result of risk assessment and options analysis. The postclosure land use will have a significant effect on the final landform plan, and both are tied to developing an acceptable postclosure risk profile. Examples of this interaction include reducing erosion risk, where final landform slope angles are minimized; establishing self-sustaining vegetation, where landforms are designed to promote or discourage access for timber or grazing activities; or achieving flora and fauna biodiversity goals. More than one postclosure use may be compatible at the same site, such as a solar panel installation on a rehabilitated waste dump, which may work well with managed grazing. Conversely, postclosure land uses may be at odds with each other or pose unacceptable risk, such as managed grazing vs. biodiversity uses or access to water for grazing animals vs. public safety issues. Landform plans should reflect the postclosure public access situation. Re-contouring of land to deter public access where necessary may be possible in lieu of fencing that requires maintenance. Public safety and long-term environmental sustainability should always be an overriding consideration and must be worked through with applicable stakeholders. The final landform plan should include maps and other documentation that clearly show the final landforms and how they support the postclosure land use.
4b.	**Closure execution plan**	An adequate description and schedule of closure actions and activities are required. The level of detail included in the closure execution plan will depend on the class of closure plan (conceptual, feasibility, etc.), regulatory requirements, and the company's internal project management requirements. This plan is sometimes referred to as the annual closure plan update or preferred closure plan and commonly addresses all activities required for closure including • Infrastructure decommissioning; • Remediation of contaminated soils; • Water remediation; • Earthmoving, topsoiling, revegetation; • Social management—community and work force (training, economic development); and • Project management and administration. Attention to completeness is required, as this will provide the basis for the cost estimates. Supporting documentation may include organizational charts, work breakdown structures, and contingency plans. It is not uncommon for an asset to develop and mange multiple versions of a closure execution plan that contains varying levels of detail depending on the whether the audience is community stakeholders who want a high-level summary or the engineering functions needing the detailed plan.

(continues)

Table 16.7-2 Key components of an integrated closure plan (continued)

	Plan Component	Description and Comment
4c.	Postclosure plan	It is critical that a schedule of postclosure actions and activities be developed and updated. Following the initial phases of decommissioning and rehabilitation, the postclosure period continues until the mining lease is relinquished or the company is otherwise released of liabilities. This postclosure phase needs to be considered in the risk assessment process. Indeed, it can be a surprisingly lengthy and expensive phase, perhaps continuing into perpetuity at some older more problematic operation sites. Postclosure activities include long-term water monitoring, data analysis, and reporting to company and government. They also include maintenance of remaining facilities, such as fences, water treatment plants, and water conveyance structures; actions required to repair erosion; and other rehabilitation repairs. The project management and administrative activities required for the postclosure period must be included in the plan.
4d.	Completion plan/mining lease relinquishment plan	In most cases, one of the prime objectives of a closure plan is to meet the performance indicators set by government and other stakeholders and to obtain formal relinquishment of the mining lease and sign off that all company obligations are complete. Unfortunately, there are very few examples where this has been successfully accomplished. Developing a mining lease relinquishment plan in the early phases of mine life is intended to give clarity to the process and mechanisms required for completion of closure. Ideally, the plan is developed in conjunction with the formal approval body such as the state or federal government. Attention should be given to the fact that most governments around the world (in both developed and undeveloped countries) do not have adequate procedures to sign off on the relinquishment plan or officially release the company, even when it is clearly demonstrated that the closure criteria have been met.
▶	Closure Planning Step 5	**Generate cost estimates with sound principles.**
5a.	Deterministic cost estimate and range analysis	Cost estimates are made for the activities that constitute the selected closure execution plan. Typically, a deterministic range analysis estimating the range of costs of each activity is captured as line items in a spreadsheet. Cost estimates should follow the standard project management and engineering estimate procedures that would be undertaken for any large-scale, multimillion-dollar projects. It is useful to forecast expenditures for a 30-year period following the targeted closure date to see the larger, long-term picture and to ensure that all costs are captured to avoid underestimating the cost of closure. The closure team will need to review applicable government, accounting, and corporate requirements to decide if a deterministic cost estimate is a sufficient end point or if probabilistic modeling is required.
5b.	Probabilistic modeling for expected cost of closure	Probabilistic cost modeling, using various techniques including the Monte Carlo simulation method, can be very effective in managing the inherent uncertainty of closure cost estimates. Probabilities can be assigned to the cost range analysis, event risk occurrence, and timing of key aspects affecting present value calculations and the like. The results of a probabilistic cost model for closure may satisfy a number of government, accounting, and corporate requirements to report the "expected value" of closure that demonstrates an incorporation of risk management into to the cost estimate. The closure team will want to include the expert advice of probabilistic financial modelers as would be done for any large-project costing effort.
5c.	Closure cost summary report	The results of the base cost estimate (whether deterministic or probabilistic) will be manipulated for various purposes according to specific guidelines. The three most common types of closure cost estimates required are the following: • Business valuation. This all-inclusive cost estimate is used primarily for internal business decisions and follows company-specific investment standards. The closure valuation looks at the LOA disturbance (rather than existing disturbance), integrates progressive rehabilitation, and is usually adjusted to account for salvage value and tax benefits. • Accounting provision. International accounting standards are very specific for determining an acceptable provision to be declared in the mining company's financial statements. Review of current accounting standards by an accounting professional is necessary to ensure compliance with corporate government requirements such as the Sarbanes–Oxley Act of 2002 in the United States. Generally, provision calculations do not include salvage value, do not allow tax benefits to be recognized, and only address current reporting year current disturbance (rather than LOA disturbance). • Government financial assurance. Guidelines vary greatly country to country and state to state. It is critical that the closure team adequately reviews the site-specific regulatory requirements for calculating financial assurance. Progressive rehabilitation is most often treated as an operating cost and is not included in any of these closure cost estimates. Progressive rehabilitation is budgeted and tracked as any other operational expenditure according to local accounting guidelines. Discount rates are generally applied to all of the above cost-of-closure calculations. Assumptions made to determine present value may vary and should be carefully documented to avoid errors and confusion. Documenting assumptions made and guidelines followed to determine the reported costs of closure is important to avoid confusion in subsequent reviews and to provide a basis for internal and external auditing.
5d.	Closure plan review and audit plan	Regular, formal review of the closure plan and associated cost estimates are critical to keeping the closure plan accurate and effective for managing risk and liabilities and for making sound business decisions. An annual review by the closure planning team, integrated with the LOA plan update, should be scheduled and results documented. Review by an entity outside the closure team can be very helpful due to the complex nature of closure planning and the potential for errors in the risk-assessment process. This review can take many forms, including expert peer review, internal audit, audit by joint-venture partners, or consultant review. The scope and details of the internal and external peer reviews will vary with the level of site-specific business risk and where the mining venture is in the life cycle. As the site nears closure execution, detail reviews and audit of the closure plan and cost estimates become critical to managing business risk.

(continues)

Table 16.7-2 Key components of an integrated closure plan (continued)

	Plan Component	Description and Comment
▶	Closure Planning Step 6	Establish an action plan for improving the closure plan.
6a.	Closure study plan	A schedule of environmental and engineering studies and projects is required to systematically reduce the uncertainty in the closure plan. Examples include piloting wast dump covers, vegetation trials, groundwater studies, waste characterization, and community surveys. The closure study plan should contain sufficient detail on goals and objectives to be effective. The study plan should include estimates of human and cost resources required and should identify where the resources are budgeted.
6b.	Stakeholder consultation and external affairs management plan	A plan for engaging and informing stakeholders around closure issues should be prepared. Economic development for the community postclosure must take place over many years to be effective and should be explicitly addressed in the external affairs management plan. Additionally, planning is necessary to receive sufficient input on postclosure land use and sustainable development. Frequency and range of activity will change over time depending on the life cycle of the asset. Human resources and budgets required should be identified and scheduled.
▶	Closure Planning Step 7	Update closure document and record management system.
7a.	Closure document and record retention management system	Closure planning updates will take place over many years or decades and will be subject to audit and regular review. A formal system with accountability will need to be established for the short and long term. Information gathered during the closure planning process should be documented in a manner that allows easy access and change management. Particular attention to effective document retention is required in the postclosure period and following completion and lease relinquishment.

Table 16.7-3 Common closure planning risks

Type	Closure Risk Issue	Possible Causes	Potential Impacts	Example Controls
Safety and health	Worker injury during closure execution	Asbestos, exposure to hazardous materials, inexperienced demolition crew	Injury, loss of life, company reputation, financial loss, regulatory penalties	Effective health and safety, safety and change management program, use of qualified people
	Public injury postclosure	Subsidence of underground workings, open pits, high-wall failure, unsecured open voids, standing water	Injury, loss of life, company reputation, financial loss, regulatory penalties	Engineering design eliminating hazard, controlled access, public education, monitoring program, maintenance plan
Environment	Water contamination	Acid rock drainage, metal and salt seepage from waste rock/tailings/pit/underground workings, hydrocarbon soil/water contamination from poor operations practices, water catchment overflow, erosion, pit lake water quality deterioration	Third-party legal action, extensive remediation postclosure financial loss, relinquishment delay, loss of company reputation	Long-term water monitoring programs, baseline data collection, preventive actions during operations, adequate closure design criteria
	Water balance, water management	Inadequate quantitative water balance, poor storm design criteria, climate change	Uncontrolled discharge, legal action, penalties, contamination flow off site, overflow or failure of dams, catchments, injury or loss of life, financial loss, loss of reputation	Baseline climate and water data studies, quantitative water balance inclusive of water quality issues, understanding of surface/groundwater interaction, proper storm design criteria, climate change risk analysis
	Air	Improper air pollution control during operation (dust/stack emissions), decommissioning activities, inadequate cover/vegetation, improper void management	Soil contamination off property, financial loss, third-party lawsuits, loss of company reputation, community complaints	Effective air control programs during operations and closure execution, monitoring cover design criteria

(continues)

Table 16.7-3 Common closure planning risks (continued)

Type	Closure Risk Issue	Possible Causes	Potential Impacts	Example Controls
Environment (continued)	Tailings and landform failure	Slope failure, visual impairment, soil loss, storm erosion, vegetation failure, fauna habitat impaired, unsustainable ecosystem, seismic stability	Water contamination, financial loss, nongovernmental organization and community complaints, loss of company reputation, delay in relinquishment	Adequate baseline studies, rigorous engineering criteria development inclusive of ecosystem considerations and stakeholder desires
	Tailings and waste disposal	Geochemistry of rock dumps, tailings. On-site disposal of hazardous waste, long-term stability compromised by seismic or erosion events, radiation isolation, unknown hazards in landfill	Long-term water treatment, financial loss	Adequate environmental impact studies, qualified expert consultation, engineering criteria development
	Failure of remedial plan or closure activity	Poor execution by third-party contractor, soil or water contamination more serious than anticipated, vegetation failure	Financial loss, regulatory penalty, delay in relinquishment	Due diligence of third-party contractors, appropriate indemnities and insurance, independent review of high-risk plans, adequate environmental studies, and pilot testing
Community	Employee	Union strife, management retention during decommissioning, rehabilitation, and postclosure	Company reputation, access to qualified employees for closure project and future projects	Redundancy/severance planning, training programs, incentives for retention/hiring qualified personnel
	Local community outrage	Loss of income, real estate devaluation, population decline, loss of tax base, unrealistic expectations of mining company viewed as "deep pockets," failure of third party to maintain infrastructure handed over postclosure, safety hazard postclosure	Socioeconomic disruption, poverty; relocation of population; loss of emergency, medical, and educational services; company reputation; access to new projects; financial loss; long-term liability returns to company	Proactive community planning over entire life cycle, economic development partnerships, education and regular information updates to external stakeholders on LOA plan
Financial and legal	Government requirements change	Current regulations not clear, changes over time, sudden changes with change in government party. Common considerations: pit backfilling, sulfate limits for water discharge, return to original contour, unacceptable postclosure land use	Closure plan inadequate, financial loss, inability to achieve completion	Engagement of government and nongovernmental organizations, monitoring of regulations, obtaining government sign-off incrementally, participating in industry forum, adjusting LOA plan as required (including consideration of early closure)
	Loss of reputation/third-party actions	Contamination of surface water, groundwater, air; relinquishment of closed property to third party unable to provide proper care and maintenance; exposure to workers and public during operations and closure	Financial loss through payment of compensation, return of property and liabilities after relinquishment, loss of reputation, environmental penalties, and other statutory penalties	Thorough risk-based remediation, external review of third parties and appropriate sign-off, monitoring programs during operations in closure phase
	Inadequate provisioning	Failure to consider all possible risks/contingencies including tax consequences. Common shortfalls: long-term water treatment costs, soil remediation costs, lack of topsoil and cover availability, inadequate estimate of postclosure monitoring time period, inadequate project management, overestimation of salvage value	Financial loss, misstatement of financial report, poor business decisions	Use of standardized risk assessment processes throughout the asset life cycle, audit and other assurance activities, review by legal and tax experts, closure project planned and executed by project management professionals
	Sterilization of resource	Untimely execution of closure	Resource not available when commodity price rises	Use of standardized risk assessment to evaluate closure decision

Table 16.7-4 Life-cycle closure plan comparison checklist

Life-Cycle Phase	Concept Phase and Prefeasibility Study	Feasibility Planning and Execution/Project Construction	Operating Phase	Closure Execution Phase—Decommissioning and Postclosure
General considerations	Closure options are considered in these early study phases of an asset. If a study phase involves pilot testing, drilling, or other actual land disturbance, a closure plan is prepared to address any liabilities created.	A preferred conceptual closure plan for the asset is generally selected during the feasibility phase. As soon as construction of the mining asset begins, land disturbance is created and a closure plan and closure cost estimate should be in place and regularly updated to accurately reflect the closure liability as it is created.	Operation is usually the longest phase in the asset life cycle. A life-of-asset (LOA) plan is updated annually and addresses all closure liabilities by integrating the asset business plan, progressive rehabilitation plans, and closure activities to be executed at cessation of operations.	Following the announcement by management that operations have ceased, the closure plan becomes a project to be managed according to standard project management principles. The closure project does not end until relinquishment of the property has occurred.
Closure plan class	Preparation of a conceptual class of closure plan and cost estimate ±50% to 30% to support project evaluation. Expectation of progressive improvement in uncertainty and annual review of cost estimates. When asset has less than 10 years of life remaining, upgrade closure plan and cost estimate class to prefeasibility/feasibility level.		Transition to prefeasibility/feasibility-level closure plan and cost estimates, e.g., ±20% to 15%	Execution level closure plan and cost estimate, e.g., ±10%
1. Basic closure planning actions	The following applies to closure planning in all phases of the life cycle: ☐ Identify fatal flaws in operation plan that would prevent compliance as a result of closure. ☐ Identify information gaps, update plan, and schedule to rectify. ☐ Establish current government and other stakeholder requirements.			
2. Completion criteria and exit strategy	☐ Identify likely completion criteria. ☐ Outline completion strategy.		☐ Establish completion criteria and preliminary relinquishment sign-off process through studies and stakeholder engagement.	☐ Refine and implement relinquishment sign-off process.
3. Stakeholder engagement	☐ Establish stakeholder identification register. ☐ Identify stakeholder issues. ☐ Outline concept of how stakeholders are to be engaged.		☐ Prepare, maintain, and implement community relations plan that includes community consultation with respect to closure issues.	☐ Implement community relations plan focused on decommissioning, postclosure, and completion issues. ☐ Confirm community acceptance of closure actions for completion.
4. Closure risk management	☐ List closure risks identified in order of priority. Formally assess top closure risks to avoid fatal flaw issues.		☐ Conduct formal qualitative closure risk assessments. ☐ Establish and update a documented closure risk register.	
5. Closure plan cost estimate	☐ Document basis of closure cost estimates. ☐ Document use of deterministic and probabilistic methods used. ☐ Confirm that closure cost estimates are risk based and clearly tied to closure plan, i.e., expected value at the appropriate accuracy level for risk management.			
6. Government	☐ Prepare an overview of requirements.	☐ Generate a register listing requirements and high-level plan for compliance.	☐ Generate a register listing completion criteria requirements with plan, schedule, and cost for compliance.	☐ Obtain formal acceptance for and implement the relinquishment sign-off process.
7. Environmental baseline for issues affecting closure	☐ Identify baseline issues including preexisting contamination, sensitive ecosystems, and biodiversity.	☐ Implement baseline study plans fit for purpose. Compile information and use to plan for additional work as needed.		
8. Postclosure land use	☐ Describe current land use. ☐ Conceptualize postclosure land use for project and alternatives. ☐ Identify land tenure risks and issues for closure. ☐ Plan for additional information needed. ☐ Plan for stakeholder engagement on postclosure land use.	☐ Determine detailed land capability/postmining/postclosure land-use alternatives. ☐ Discuss postclosure land-use alternatives with stakeholder engagement. ☐ Prepare a preliminary LOA plan that includes postclosure land-use issues.	☐ Include postclosure land-use goals when LOA and closure plan are integrated into business. ☐ Progressively evaluate rehabilitation against final land-use status (species regrowth, fauna assemblages, etc.). ☐ Validate land-use status against closure completion criteria.	☐ Document how postclosure land-use goals are met.

(continues)

Table 16.7-4. Life-cycle closure plan comparison checklist (continued)

Life-Cycle Phase	Concept Phase and Prefeasibility Study	Feasibility Planning and Execution/Project Construction	Operating Phase	Closure Execution Phase—Decommissioning and Postclosure
9. Water management	☐ Describe current surface and groundwater status. ☐ List broad closure water management goals. ☐ Identify on-site and off-site water-related risks. ☐ Establish storm design criteria.	☐ Prepare a conceptual water management plan to address on- and off-site risks including • Surface and groundwater, • Impact on quantity, and • Impact on quality. ☐ Finalize landform drainage plan. ☐ Generate a preliminary monitoring plan. ☐ Assess on- and off-site risks. ☐ Determine postclosure goals including final use and completion criteria for dams and voids.	☐ Clearly define and document water completion criteria. ☐ Progressively update current water status report including • Surface runoff patterns, • Storm sizing criteria, • Aquifer delineation, • Surface and groundwater quality and quantity. ☐ Evaluate and validate status against closure completion criteria. ☐ Schedule and cost pollution prevention and remediation plans. ☐ Schedule and cost monitoring plans. ☐ Quantify final landform and drainage patterns (including creek diversions, dams, future flooding impacts).	☐ Implement closure water management, remediation plans. ☐ Decommission and/or rehabilitate any water facilities (wells, boreholes, dams, canals, diversions, etc.) that are not part of postclosure land use. ☐ Monitor and evaluate progress toward final hydrology and water quality completion criteria.
10. Plant/infrastructure	☐ Identify major infrastructure and define footprint. ☐ Prepare a conceptual decommissioning plan including what infrastructure will remain in place postclosure.	☐ Summarize demolition and decommissioning quantities. ☐ Identify off- and on-site disposal needs. ☐ Establish rehabilitation plan for footprint and adjacent areas.	☐ Progressively update decommissioning plans and cost estimates. ☐ Monitor salvage value and adjust closure cost estimate as required.	☐ Perform decommissioning. ☐ Perform remediation if necessary. ☐ Monitor and evaluate progress toward completion.
11. Overburden/rock piles, tailings dams, open-pit features	☐ Describe quantity and quality of material. ☐ Conceptualize alternative locations and forms identified. ☐ Conceptualize disposal and closure methods described. ☐ Quantify volumes. ☐ Quantify types including acid generation potential. ☐ Prepare topsoil stripping and management plan. ☐ Prepare preliminary LOA plan for dumps, tailings, pit features.	☐ Generate LOA plan to include • Final slope angles; • Structural safety defined for tailings and rock/overburden piles; • Water and sediment control features; • Structural stability including settlement, subsidence program; • Facility closure and rehabilitation methodology (impoundment capping, revegetation, final void); • Topsoil and other cover material plan; • Rehabilitation schedule; and • Configuration.	☐ Revise and update LOA plan in conjunction with development of closure plan annually. ☐ Progressively monitor final voids (including borrow pits and quarries), and report against closure goals. ☐ Regularly address gaps found in integrated LOA plan, progressive rehabilitation plan, and closure plan.	☐ Perform rehabilitation. ☐ Perform remediation if necessary. ☐ Monitor and evaluate progress toward completion.
12. Waste and chemical management	☐ Identify types of chemicals used. ☐ Identify types of waste to be generated. ☐ Describe general management and disposal plan. ☐ Quantify chemicals. ☐ Quantify waste generated.	☐ Generate chemical spill prevention and management plan. ☐ Develop waste management plan. ☐ Develop spill prevention and response plan. ☐ Develop rehabilitation plan for on-site waste disposal facilities.	☐ Track waste generation and annual disposal capacity against plan. ☐ Quantify accumulation of waste on-site. ☐ Develop soil decontamination plans. ☐ Document disposal plans for cumulative waste over operating life and at closure.	☐ Implement soil decontamination plans. ☐ Implement closure waste disposal plan for project. ☐ Monitor and evaluate progress toward completion.

(continues)

Table 16.7-4 Life-cycle closure plan comparison checklist (continued)

Life-Cycle Phase	Concept Phase and Prefeasibility Study	Feasibility Planning and Execution/Project Construction	Operating Phase	Closure Execution Phase—Decommissioning and Postclosure
12. Waste and chemical management (continued)	☐ Identify gaps between generation and disposal capacity. ☐ Identify off-site disposal facilities.		☐ Within 3 and 5 years of projected management decision to close, finalize detailed closure waste disposal plan.	
13. Haul roads and access roads	☐ Identify road system. ☐ Identify roads to remain at closure and likely completion criteria. ☐ Conceptualize rehabilitation plan for roads.		☐ Formalize completion criteria for roads to remain at completion of closure. ☐ Develop progressively more detailed rehabilitation plan for roads. ☐ Perform progressive decommissioning of roads where feasible.	☐ Perform decommissioning and rehabilitation of roads that will not be retained. ☐ Configure roads to be left upon completion of closure for long-term use. ☐ Perform remediation if necessary. ☐ Monitor and evaluate progress toward completion.
14. Air quality management	☐ Identify baseline air quality status. ☐ Describe project risks including air contaminant impact off- and on-site at closure.	☐ Conduct baseline air monitoring to establish air quality key performance indicators. ☐ Prepare air-quality management plan for minimizing on- and off-site impact.	☐ Develop a mitigation plan for surface treatments and revegetation to reduce dust and odor in accordance with key performance indicators at completion.	☐ Complete surface treatments and revegetation to meet air quality key performance indicators at completion.
15. Underground workings	☐ Describe subsidence-control issues. ☐ Describe water management issues.	☐ Identify underground workings areas. ☐ Prepare LOA subsidence management and mitigation plans. ☐ Prepare LOA underground water management plans.	☐ Develop underground decommissioning plan. ☐ Develop long-term subsidence management plan.	☐ Implement project plan for decommissioning underground workings.

Closure risks are very site specific and must be assessed by a cross-functional team familiar with the site and issues. However, there are a number of risks issues that are commonly encountered. Table 16.7-3 may serve as a helpful guide to brainstorming site-specific closure risks.

LIFE-CYCLE CLOSURE PLAN COMPARISON

Table 16.7-4 provides a comparison of how the closure plan should change as the mineral development project moves through the life cycle from a study project and feasibility plan into the lengthy operating phase and ultimately to the closure phase. The table is not meant to be exhaustive but can be used to prompt discussion and develop a high-level checklist to ensure that closure plans contain the correct level of detail as the mineral development project progresses.

ACKNOWLEDGMENT

This chapter was drafted with the assistance of Gary Bentel, geotechnical manager, BHP Billiton.

REFERENCE

AS/NZS 4360:2004. *Risk Management*. Sydney, NSW: Standards Australia International; Wellington: Standards New Zealand.

PART 17

Community and Social Issues

CHAPTER 17.1

Community Issues

Robin Evans and Deanna Kemp

INTRODUCTION

The excavation, refining and shipping of this ore to the smelters of Japan could bring great profit over the next 20 years to the shareholders of Rio Tinto–Zinc—at the cost of damage to the physical, social and spiritual well-being of Bougainville, which, until the mine came, was a peaceful and prosperous island. Moreover there is a danger that arguments over the ownership of the mine could cause political strife, even civil war, in this part of the South Pacific (West 1972).

In 1989, following a series of increasingly violent protests that included sabotage of power supplies and attacks on mine workers, Bougainville Copper Ltd. evacuated its work force from the Panguna mine located on Bougainville Island in Papua New Guinea. The operation was shut down at relatively short notice, with most equipment left in place, and has not operated since. In the year prior to the shutdown, Bougainville Copper was capitalized at US$1.5 billion (Humphreys 2000), and the operation represented one of the world's largest open-pit mines. In the ensuing years, the civil unrest developed to a full-scale conflict between the Bougainville Revolutionary Army and the Papua New Guinea Defence Force that devastated the island, with several thousand deaths and approximately 50,000 people (a third of the island's population) displaced from their homes (Regan 2001). A peace process that commenced in 1998 between the government and local communities has returned some calm to the island, but the events that surround the abandoned Panguna mine remain the most vivid and tragic example of community conflict surrounding a major mining operation.

The underlying causes of the conflict that erupted on the island were many and complex: they included ethnic differences and the emergence of a secessionist movement prior to the transition of Papua New Guinea from Australian administration to full independence in 1975. The mine became a catalyst, with community concerns about the distribution of economic benefits and the environmental impacts of mine waste on the local river system featuring prominently. According to Denoon (2000), for many of the landowners "Panguna was a social, economic and environmental disaster, and a spur to militant protest."

Mines and the communities they are associated with have always been inextricably linked via a complex network of relationships and issues such as these. Local community members usually form part of the work force at a mining operation, while others in the area supply goods and services. At the same time, individuals, families, and sometimes whole communities can be displaced by the development of a mining lease, while some may be affected by environmental impacts associated with an operation. Community livelihoods can be impacted both directly by land-use changes, and also by changes within local social structures. While safety standards in the industry have improved significantly in most areas, there is a history of workplace accidents and health issues that remains an important factor in the relationship between mining companies, work force, and communities. In some locations, operations can be in competition with parts of the community for scarce resources such as productive land and/or water. In others, mining companies are welcomed as an agent of development that can bring infrastructure, essential services, and economic opportunities to a region. In most cases there is a continuum of views held about proposed or existing mining operations.

The balancing of the benefits and costs of mining operations for local and regional communities has attracted debate for centuries: "Thus it is said, it is clear to all that there is greater detriment from mining than the value of the metals which the mining produces" (Agricola 1556). While Agricola himself went on to staunchly argue the case for the 16th-century mining industry, others have taken more critical positions when considering modern operations. Notwithstanding the resilience and intrinsic strengths of some mine-affected communities for dealing with changes and transformations brought about by mining, there are often power disparities between mines and many remote, rural, and/or indigenous communities (e.g., Banerjee 2001). Compared to companies,

---

Robin Evans, Senior Research Fellow, University of Queensland, Sustainable Minerals Institute, Centre for Social Responsibility in Mining, Brisbane, Australia
Deanna Kemp, Senior Research Fellow, University of Queensland, Sustainable Minerals Institute, Centre for Social Responsibility in Mining, Brisbane, Australia

communities usually have more limited access to information, knowledge, technology, and capital that can be leveraged to shape the nature and pace of mining development. Such disparities have been the primary driver for the increasing involvement of rights-based nongovernmental organizations (NGOs) that have seen international campaigns launched against companies or particular projects when grievances escalate. The boom/bust nature of the industry adds another dimension to the debate about whether mining brings positive or negative change, with mining's contribution to development challenged by "resource curse" theories.

In short, the issues associated with interactions between mining companies and communities have become more prominent, and have increasingly required more attention from those associated with managing resource companies and mineral operations. Those working in the industry are increasingly required to respond to "community issues" in ways that their predecessors were never expected to. It is also likely that many technical staff who get involved in social aspects of mining have little training or prior experience in these issues. The purpose of this chapter is to provide an introduction to the types of social and community issues involved with mining, and the contexts in which they could emerge. It explores what is meant by the term *community*, and also reviews how community issues have increasingly been investigated in the last 20 years, particularly through projects such as the Global Mining Initiative (GMI). Subsequent sections explore more systematically the contextual factors that influence community relationships, and also the types of impacts that attract most attention. More detailed chapters follow this overview and address specific themes in more detail, including indigenous peoples and mining projects, and specific processes to assess and manage social impacts.

WHAT IS "COMMUNITY"?

The concept of "community" is usually used in the minerals industry to describe those who live in the geographic region of an operation, either in defined settlements or dispersed settings. However, there are other equally valid ways to consider the term, especially as modern industry practice has moved to include a greater incidence of fly-in/fly-out arrangements whereby workers and their families live in a distant location. Another relevant example is where traditional owners of the land associated with the mining development have maintained their links to the land but reside elsewhere. In such cases, community impacts can occur many hundreds of miles from an operation. Mining projects often include transport infrastructure and supply chains that span large distances, connecting networks of mines, processing centers, and ports, significantly increasing the range and types of other potential community impacts. There are also many different definitions of community that are not geographically based, such as communities of practice and spiritual communities. However, the mining industry tends to emphasize physicality and proximity to an operation, either spatially or by issue. Leading industry practice acknowledges that communities are complex, evolving, political, and heterogeneous entities (DITR 2006). In 2005, the Australian Ministerial Council on Mineral and Petroleum Resources (MCMPR) defined community as

> a group of people living in a particular area or region. In mining industry terms, community is generally applied to the inhabitants of immediate and surrounding areas who are affected by a company's activities.

The term *local* or *host* community is usually applied to those living in the immediate vicinity of an operation, being indigenous or nonindigenous people, who may have cultural affinity, claim, or direct ownership of an area in which a company has an interest.

Affected community refers to the members of the community affected by a company's activities. The effects are most commonly social (resettlement, changed services such as education and health), economic (compensation, job prospects, creation of local wealth), environmental, and political.

A community is usually a diverse group of people with some common bonds. Diversity can come in the form of gender, ethnicity, religion, race, age, economic or social status, wealth, education, language, class, or caste. As a result, members of any community are likely to hold diverse opinions about a mining operation and its activities, as well as most other subjects. As mentioned earlier, individuals within a community will have different and sometimes overlapping associations with the mine as neighbors, employees, suppliers, and so on. It is not uncommon for disagreement and sometimes conflict to develop between different sections of a community in relation to mining operations. To different degrees, conflict will also exist within a community prior to the start of mining, from low-level tension to violent conflict. Some companies choose to operate in conflict or postconflict zones, which will involve yet another layer of complexity to the process of understanding the local community.

More recently, the term *stakeholder* has become a common term that is related to but distinct from *community*. The idea that business has responsibilities broader than its traditional role of generating a return for shareholders is reflected in stakeholder theory, popularized by Freeman (1984) in his seminal work, *Strategic Management: A Stakeholder Approach*. This theory holds that successful companies recognize that they have responsibilities to *stakeholders*, a term referring to any individual or groups who can affect or are affected by a corporation's activities (and where corporate responsibility extends beyond maximizing a financial return to shareholders). Stakeholder theory has not only been a powerful force in academic circles in terms of developing its own research tradition, but also in encouraging the corporate sector to see community concerns and aspirations as key considerations.

A common definition of stakeholders is "those who have an interest in a particular decision, either as individuals or representatives of a group. This includes people who influence a decision, or can influence it, as well as those affected by it" (MCMPR 2005).

Stakeholders might therefore include local community members, NGOs, governments, shareholders, and employees. The use of the term *stakeholder* has been contested on the basis that some communities are in fact *rights holders*—a stronger term than stakeholder—due to rights defined by relevant national or state law. For example, indigenous peoples in Australia with a recognized claim to an area have the right to negotiate under the Native Title Act (Commonwealth of Australia 1993). While local communities are usually viewed as key stakeholders, the potential range of all stakeholders is considerably broader. Both terms are relevant to the discussion that will occur in the next few chapters, although the main focus will be on the narrower group of "community" as outlined in its definition.

MINING AND SUSTAINABLE DEVELOPMENT

The increased focus on the mineral industry's impact on the environment and society, including the local community, can be attributed to both global trends associated with attitudes to private enterprise and large multinational corporations, and also to the mineral industry's poor track record with high-profile cases such as the Bougainville crisis and various other environmental incidents (e.g., Ok Tedi in Papua New Guinea, Marcopper in the Philippines, and Baia Mare in Romania). During the 1990s the minerals industry came under increasing challenge from various quarters, with the result that it found itself losing its "social license to operate." This term has now become a popular way of describing the influence that society in general has over the ability of an organization to carry out its activities, above and beyond the legal license issued by governments, which govern the extraction of mineral resources. Social license is variously described but commonly considered an ongoing process of approval from the community that is given at a point in time, and not necessarily for the future (AccountAbility 2004).

In response to this increased societal pressure, there were several early initiatives in different countries as local industry bodies sought to engage with both governments and their critics. One example is the Whitehorse Mining Initiative in Canada (Cooney 2008), a multistakeholder initiative developed in 1994 involving industry, government, and NGOs, which aimed to develop general principles for responsible mining. In Australia the Minerals Council of Australia developed the Australian Minerals Industry Code for Environmental Management, modified in the late 1990s to include additional requirements focusing on social and community issues. These types of initiatives were often linked to the concept of sustainable development, a term that has emerged over the last 20 years as a key organizing framework for the global community to consider the links between the development and environmental protection agendas. Dresner (2008) outlines both the history and the politics of this process, including key milestones such as the Brundtland report (WCED 1987) and the United Nations Rio Earth Summit in 1992. Sustainable development remains a contested concept, with many competing definitions and sets of principles, but its popularity has meant that many organizations have chosen to use it to frame their own activities in social and environmental areas.

In 1998, a dialogue between a small group of senior mining industry CEOs including Hugh Morgan of Western Mining Corporation and Sir Robert Wilson from Rio Tinto led to the formation of the GMI, an industry-led process that expanded to involve many of the world's largest mining companies. The initiative had three elements: a 3-year research project to investigate the activities of the industry through the lens of sustainable development; a major conference held in Toronto in 2002 to review the outcomes of the project; and the creation of a new global industry body, the International Council on Mining and Metals (ICMM), charged with implementation of the industry response to the outcomes of the project. The Mining, Minerals and Sustainable Development (MMSD) research project was managed by an independent research group, the International Institute for Environment and Development, and resulted in the *Breaking New Ground* report produced in 2002 in time for the Toronto conference (IIED 2002). It is beyond the scope of this chapter to explore the full range of issues covered by the MMSD report (many are touched on in other chapters in this handbook), but it is noteworthy that one of the nine key challenges identified for the industry to address was the area of mines and communities. A number of supporting research reports addressed specific issues under this theme such as indigenous peoples' rights, social impact analysis, and socioeconomic development. Danielson (2006) provides a comprehensive account of the origins and progress of the MMSD project, including a range of different stakeholder perspectives on various elements of the process. There is little doubt that the GMI and its associated research has been the most influential process to date of all those designed to examine the social and environmental aspects of mining industry activities.

In reviewing the progress and outcomes of both the Whitehorse Mining Initiative and the GMI, Cooney (2008) suggested that the companies who launched the two initiatives were primarily concerned about the public image of mining. Public criticism was being driven by misinformation about the actual impacts of mining, and improved communication would help address this. However, he suggested that through the course of both initiatives and subsequent processes such as the World Bank's *Extractive Industries Review*,

> the mining industry learned matters both of process and of substance: from the engagement process, mining companies have learned different models of comprehensive dialogue and consensus building with critics; by listening to their critics, the companies have learned different approaches to analyzing and managing critical issues. Self-education was not the mining industry's initial purpose in either the Whitehorse Mining Initiative or the Global Mining Initiative, but it was to be the outcome.

In addition to these industry-driven initiatives, many other stakeholder groups have initiated reviews and developed frameworks for reviewing environmental and social aspects of projects, some specific to the resources sector but others more generally focused. Such nonregulatory drivers are pushing minerals companies to focus on local-level social and community issues. Particularly for publicly listed companies, the screening process for socially responsible investment or ethical investment funds, other indexes such as the Dow Jones Sustainability Index, and public ratings are also influencing behavior. Most of these indexes and rating agencies require that organizations establish a systematic approach to managing the social dimensions of their projects, as they would environmental and economic aspects. Many funding agencies, such as the International Finance Corporation (IFC) also require this as a condition of finance. In the absence of mandatory legislation, pressure from third parties has helped to sustain attention on community relations in the mining industry. This growth of "soft" regulation has been significant in the past 10 years—examples that are relevant and referenced by the minerals industry include the following:

- **ICMM Sustainable Development Framework**—Ten principles and various elements that provide guidance on applying sustainable development principles to mineral

operations. Reporting against this framework is obligatory for members of the ICMM.
- **Kimberley Process Certification Scheme**—A multi-stakeholder initiative that aims to provide product certification for diamonds and reduce the trade in "blood diamonds."
- **Extractive Industries Transparency Initiative**—A voluntary initiative focused on ensuring transparency of payments associated with resource projects.
- **Equator Principles**—A set of environmental and social benchmarks developed by a group of major international banks for addressing environmental and social issues in development project finance.
- **IFC Environmental and Social Standards**—A standards framework developed by the IFC to apply to projects in which they invest World Bank funds.

As well as pressure from this type of "soft" regulation to improve social and environmental performance, many governments are becoming increasingly involved in regulating the community aspects of the development, operation, and closure of mines (Brereton 2002). The focus of governments was initially on environmental issues, but social dimensions are increasingly being regulated in developed and developing countries alike. In Australia, for example, most states have made basic community consultation mandatory for major new development projects, including mining, often as part of environmental and social impact assessments. The Australian Native Title Act (AustLII 1993) has also become a central part of the Australian regulatory regime, providing indigenous groups with the right to negotiate and a potential vehicle to deliver both social and economic benefits for indigenous communities. South Africa has a regulatory framework in place to progress black economic empowerment in the mining and petroleum industries. The Philippine Mining Act of 1995 requires that proposed projects undergo a comprehensive environmental assessment, which must consider socioeconomic as well as environmental impacts and provide evidence of broad social acceptability. In Canada, the requirement to incorporate sustainability considerations in environmental assessment processes has seen community issues come to the fore in several high profile and controversial mining cases such as the Voisey's Bay project, as described by Gibson (2006), and the more recent Kemess North review.

CONTEXT IS CRUCIAL

Before considering some examples of the types of community issues that emerge, it is important to emphasize the influence and importance of context. Many factors can have a significant impact on the interactions and relationships between mining operations and communities, including various social and political aspects, as well as the stage of the mining life cycle involved. Mining is a truly global activity, involving many different types of organizations and communities in settings that range from arid mountains in parts of the Andes, to remote areas within the Arctic Circle, to established agricultural regions in developed countries, and to tropical rainforest settings in developing economies in Asia. In addition to the obvious geographical differences, other contextual factors can be very important.

The history of mining in a country and region can influence community attitudes to mining projects. In some cases mining may be a relatively well-accepted activity, as, for example, in certain well-developed coal-mining areas of the United States and Australia. A number of well-known mining towns around the world have been created close to major mining deposits whose economies continue to be based around the exploitation of these resources. Examples include Kalgoorlie in Australia, Sudbury in Canada, and Cerro de Pasco in Peru. However, in some of these areas, the cumulative and historical social and environmental impacts associated with the mining operations can still be the subject of community debate. As the industry increases its activities in less-developed countries with little or no experience of large-scale mining, community attitudes can vary widely. A good example is Mongolia, which has only recently opened its industry to direct foreign investment. Large-scale mining by global companies is a new phenomenon, and the management of mineral revenues was the principal issue in the 2009 election held in that country.

Existing community land uses can include broad-scale farming, intensive agriculture, open grazing of livestock, and hunting and fishing (among many others). Although the physical footprint of mining operations is usually relatively low, its interaction with other land uses in terms of impacts on other resources such as water, labor, and infrastructure can be significant. In some locations, small-scale or artisanal mining (particularly of gemstones and precious metals) may already be a significant community activity—some estimates put the number of people involved in this type of mining worldwide as high as 20 million, including large numbers of women and children. In this sense, the displacement of artisanal activities by large-scale mining projects can have significant negative ramifications for certain groups of people. By the same token, mining companies have worked with artisanal mining communities to address health and environmental issues and to find opportunities for artisanal miners to find alternative sources of income.

In many parts of the world, elements of the physical landscape play an important role in the culture of local communities. A desert clay pan may represent a physical feature to be managed by a mining company, but it is a significant site to a local indigenous community. In some parts of Australia, red ochre is used by male members of indigenous groups in cultural ceremonies, and therefore its presence in mine overburden becomes a major logistical issue that requires an appropriate response. There are cases where the development of mining projects has resulted in the destruction of sites of major cultural significance, such as at the Argyle mine in Western Australia, for example. There are also cases where the presence of such sites have been the main reason for mining projects not going ahead, and some sites where mining projects have been designed to accommodate cultural heritage considerations.

Political and legal frameworks within a country will have a significant impact on the scale and nature of the mining industry and can also often be the subject of intense community focus. Government capacity to regulate the minerals industry and manage the benefits of mining for the local communities has been identified as a crucial aspect by recent studies (e.g., ICMM 2006) and has been the subject of recent World Bank projects in several developing countries, such as through the provision of technical assistance to the government in Laos and other countries in Asia to strengthen their ability to manage the burgeoning mining sector. Also of interest are legal and customary rights concerning land management, especially in cases where mining occurs on lands claimed by indigenous peoples. In Papua New Guinea, there is a well-developed system of customary land ownership that is linked to family descent, whereas in some parts of Africa communal tribal ownership and shared land-use arrangements make negotiations for land access for mining purposes considerably more complex. All around the world, many rural or traditional communities have operated on a system of land use and ownership that has not required formal land ownership. If a company compensates only those people with formal land ownership, there is a high potential that it will negatively affect livelihoods, culture, and social traditions that have been in place for generations. Where land titles are in place, it is also men who often hold title and receive compensation for land, highlighting the potential for women to be negatively impacted unless strategies are in place to ensure equitable distribution of benefits.

Other important contextual factors include the nature and scale of the mining operation itself. Large-scale open-pit and strip mines can result in more visible manifestations of mining activity in the form of spoil piles and waste dumps and can be more disruptive to other land uses such as agriculture. Underground mines generally employ more selective mining methods and produce less waste, but subsidence effects in longwall mines can result in impacts on surface environments and water resources. In some countries, the safety record of underground mining is significantly worse than surface mines, as, for example, in the underground coal industry in China (which includes many smaller, informal operations), where the reported fatalities in 2007 from a series of methane and dust explosions and cave-ins numbered nearly 3,800. It is believed that many fatalities and injuries go unreported in China and other countries.

The nature, size, and reputation of the company involved can also be influential. The larger global multinational companies tend to be engaged in the sustainable development debate and signatories to many of the frameworks and conventions that deal with community issues, whereas many of the smaller mining and exploration companies are less active in these areas. Size does not, however, always correspond to enhanced social performance. Not all operations within a large company perform to the same standard. Small or junior companies with a single asset may find creative ways to work with the local community to address social concerns, although they usually have significantly less capacity to adequately resource a community relations function and therefore do not always give social aspects the attention that is warranted. Smaller companies can also have a short-term outlook, as their focus is on discovering and developing the asset to the extent they can sell it, rather than operating the mine themselves. This tends to lead to an avoidance of social investment. Whether a company is from the country of origin can also determine the extent of community opposition. In several parts of the world, there has been community opposition to the involvement of foreign companies in mining and exporting valuable minerals when much of the profit is perceived to go offshore.

Finally, the stage of the mining life cycle can also have a significant bearing on the development of community issues. Some issues are specific to certain phases. For example, the impacts associated with accessing hitherto virgin jungle for the purposes of exploration or with a community hosting a large, temporary construction work force are very different from those that occur during actual mining operations. Closure is often associated with community concerns over the withdrawal of the mining work force and related economic activity. Some closures are planned, where the mine comes to a natural end because the resource has been exhausted, but other mines close suddenly because of changes in commodity prices, which means the mine becomes unprofitable. Mines can also be disrupted or closed because of community protest or conflict, as in the case of Bougainville discussed earlier. Community issues are not fixed; they evolve throughout the life of the operation.

Consider the example of Newmont Waihi Gold's mine located in Waihi, New Zealand, shown in Figure 17.1-1. At this historic mining town, the conversion of old underground workings into an open-pit development during the 1980s has brought the upper benches of the mine to within feet of residences and the town's main street. Although mining has been an activity in the area for more than 100 years, the impacts associated with the mine have affected many residents living close to the edge of the pit, and also the community more broadly. The community relations landscape has been dominated by these issues of amenity, including the impacts of noise, dust, and blast vibrations. However, there are many other contextual issues that also influence how the community perceives and interacts with the operation. The mine and its work force make up about 25% of the town's economy, with many local people employed at the operation. The development of the open pit removed a small hill that was of cultural significance to some Maori tribes of the area, who continue to oppose the presence of the mine on the basis of their traditional beliefs and values but remain in discussion with the company on the management of cultural issues. The mine operates in a developed country with strong environmental protection legislation, and is located in a farming area close to a major tourist region. Recently, the company worked with the community to develop a vision for the town after mine closure, which has been imminent for the past few years, and it is the potential closure of the operation and its impacts that are the current focus for many in the community. A good understanding of the overall context—economic, cultural, political, social, and environmental—as well as an understanding of the issues particular to each community and each group within the community, is therefore crucial in identifying and addressing the many issues that emerge from the closure process.

What Are Community Issues?

The introduction of the term *triple bottom line* (Elkington 1997) and its application in the context of sustainability reporting has seen social or community issues identified as a separate category to environmental or economic issues within the sustainable development framework. In reality, however, most communities are extremely focused on all aspects of mining development and do not necessarily separate out these issues into neat categories. Consider a remote traditional community that relies on the local waterway for catching fish as well as for spiritual worship and ceremony. They may explain the water as important to their survival, their traditions, their family, and their future, seeing these aspects as interconnected rather than separating them. Although there are different ways of understanding these dynamics, it is often environmental or economic aspects that are the main focus of attention. Figure 17.1-2, while not intended to be comprehensive, illustrates both the breadth of issues under consideration and their interrelated nature.

Courtesy of Newmont Waihi Gold.
Figure 17.1-1 Newmont Waihi Gold's operation in Waihi, New Zealand

Some themes have become particularly prominent in the last decade, partly as a result of such initiatives as the MMSD project but also due to a range of other drivers. Some of these have been mentioned previously, such as nascent "social regulation" in the form of legislation, voluntary initiatives, and pressure from NGOs and civil society. In recent years some investment funds have been deliberately disinvested from major companies in the minerals industry because of concerns about social risk. Several of these issues, which have become the focus of various groups and organizations, are explored briefly in the following sections.

Economic Development

The positive influence of mining projects on local, regional, and national economies has always been an argument used by proponents to support new developments. In contrast, the resource curse hypothesis suggests that in fact countries with high levels of natural resources suffer lower rates of economic growth than those with more diversified economies, in effect suffering a paradox of plenty. Much research supports arguments both for and against this proposition, one example being the Resource Endowment Initiative involving the ICMM, the United Nations Conference on Trade and Development, and the World Bank (ICMM 2006). This study concluded that mining investment does provide opportunities for economic growth, poverty reduction, and engagement in the global economy, pointing to specific examples such as Chile and Botswana where increased mining investment has coincided with an upturn in national economic growth. However, it also emphasized the need for effective and transparent governance regimes for the management of mineral wealth. Transfer of some of the benefits from taxes and royalty streams back to the regions where mines are located has been an issue of some contention in several countries such as Peru, for example, where several problems have developed as a result of changes to legislation (Arellano-Yanguas 2008).

In recent years more research has been focused on economic impacts at the local and regional levels. Companies are now starting to report on how much of the added value associated with wages and purchase of goods and services for mining operations stays at the local and regional levels. For example, see the economic indicators in the *Sustainability Reporting Guidelines and Mining and Metals Sector Supplement* (Global Reporting Initiative 2009). In addition, a company's contribution to community development activities are directed at growing local economic activity, often with a focus on non-mining-related businesses, in order to provide for a postclosure future. However, the additional cash flows injected into local economies can have negative impacts as well, as in the form of disproportionate inflation, for example. This can apply equally in less-developed contexts where market economies may be significantly changed by the introduction of mining industry wages, as well as developed economies where industry expansions can result in distortions in real estate and labor markets, with consequential impacts for other sections of the community.

Water and Mining

Access to fresh water represents an essential human need. Water is also fundamental to other ecosystem services required to sustain human life and is high on the political agendas of all levels of government, including the United Nations. It was recognized as a key theme in the MMSD project, with mines operating in the driest and the wettest regions on the planet. While not extracting as large a quantity of water as the agricultural sector in most countries, individual mines are often large consumers in their local context, and their impacts can be significant. Mining companies compete for water use within a range of market and nonmarket jurisdictions, and can often afford to pay considerably more than others, thereby running the risk of reducing the viability of other industries. Companies can also affect access to water if mining development does not consider the usage patterns of the local community. For example, by building roads or operations, some mining developments can inadvertently make access to water sources in developing countries more difficult. At the same time, mining companies are often responsible for developing water infrastructure used by other industries and communities (e.g., Brereton and Parmenter 2006).

Water use in many processing activities results in bodies of contaminated water in tailings storages and flooded pits that, if incorrectly managed, pose risks for downstream users. Many environmental legacies of mining have involved pollution of water systems, including the frequent incidence of acid mine drainage at closed or abandoned mines on many continents. Potential impacts of mining operations on both the quantity and quality of surface and groundwater resources are increasingly being raised as concerns by local communities, and in several cases have been the principal reason why some projects have not gone ahead.

Community Health

There are many ways in which mining operations can impact the area of community health. Population changes including in-migration in developing countries such as Papua New Guinea can be responsible for the spread of diseases such as human immunodeficiency virus (HIV) and tuberculosis. The ICMM recently released a report titled *Good Practice Guidance on HIV/AIDS, TB and Malaria* (ICMM 2008) for its members, reflecting the incidence of HIV in mining work forces and communities in different parts of the world. Other direct health impacts can come from emissions from processing operations, such as high blood lead levels found in communities located near older lead smelters on several continents. Local controversies have developed over the potential health impacts of riverine and marine tailings disposal processes

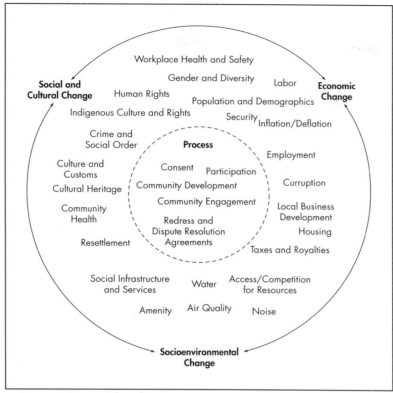

Figure 17.1-2 Examples of community issues in mining

such as those at Marinduque in the Philippines and Minahasa in Indonesia, for example. Both of these cases attracted significant international attention.

On the other hand, mining developments are often responsible for the establishment of health infrastructure and services in remote areas. Despite the environmental impacts associated with riverine waste and tailings disposal at the Ok Tedi mine in Papua New Guinea, the antimalarial and health clinic campaign introduced by the company to isolated communities near the mine resulted in extremely significant improvements in infant mortality and life expectancy statistics in those areas. A mining company–initiated campaign to eliminate filariasis from the island of Misima in the same country realized similarly impressive results, and there are many examples in Africa of similar involvement in regional antimalarial campaigns. While such initiatives at one level represent a risk reduction measure for the company's own work force, they often extend well beyond the level required for pure mitigation.

Resettlement and In-Migration

Mining often requires or results in movements of people, either out of or into a mining area. When minerals are located where people are living, resettlement is often required in order for the resource to be exploited. Resettlement may involve one family or a whole community. The IFC's resettlement standard has, by default, set the industry standard for resettlement. Aside from physical resettlement, mining can result in social, economic, and cultural dislocation, often stemming from physical resettlement, but not necessarily. For example, a haul road may inadvertently divide an otherwise connected group of people, or may impact access to resources such as productive land or water. But while it may serve to dislocate in some instances, a new road can also increase people's mobility and access to new markets for expanded economic opportunity, although these opportunities may not be experienced across the board.

There is a vigorous debate around whether mining companies should ever undertake resettlement involuntarily. There are some sites where communities have been involuntarily settled in the past, and as a result harbor resentment, particularly when relocation was facilitated by force or violence. In some cases, new owners have been faced with legacy issues from prior resettlement that cannot be ignored. While some communities are moved to enable resource extraction, there are other circumstances that result in migration into an area as a result of people seeking economic benefit through direct or indirect employment or other business opportunities, sometimes known as the honey pot effect. In-migration can be just as challenging for a company as resettlement, particularly if it is uncontrolled. Often companies expect governments to take responsibility for community planning around mine settlements or squatter camps, which, if not done adequately, can often cause great resentment.

Security and Human Rights

Human rights discourse has recently grown in strength, as have calls for business, and in particular multinational companies, to ensure that their activities do not harm the rights of others. The key international reference for human rights is the Bill of Rights, which includes the United Nations' Universal Declaration of Human Rights, proclaimed in 1948 by the UN General Assembly, and the two covenants—the International

Covenant on Civil and Political Rights, and the International Covenant on Economic, Social and Cultural Rights. As elements of the bill are ratified by governments, not companies, it has not been clear how human rights responsibilities confer to companies, particularly those operating on a transnational basis, including whether and how companies should be held accountable for their actions should they abuse human rights. The nature of global businesses has meant that it is often difficult to apply home country laws to companies operating abroad, although there have been attempts in both Canada and Australia to introduce legislation that specifically focuses on mining companies in this area. Further, the laws of some host countries may not have enshrined human rights into law, or may have weak or ineffective legislative frameworks that fail to hold companies to account for human rights violations. While the debate about the human rights responsibilities of companies continues, a recent report from the UN special representative of the secretary-general clarified the different yet complementary roles of governments and companies with respect to human rights (Ruggie 2008). The three-part framework outlines three key principles: the state duty to protect, the corporate responsibility to respect, and access to remedies. Under the recently extended mandate, "operationalizing" the framework will gain particular attention.

In the mining industry, human rights are most often raised in the context of the use of security forces to protect mining operations, which has been particularly controversial in militarized regions. Safety and working conditions have also been the subject of intense scrutiny. Attention is also being called to the human rights issues associated with environmental impacts, particularly if mines impact on the ability of local communities to establish sustainable livelihoods, as well as a number of other rights-related impacts, such as on culture (e.g., if sacred sites are impacted by mining activities), discrimination in the workplace, and the right to an adequate standard of living. While attention is often called to the negative impacts that mines have on local communities, it also needs to be recognized that companies help uphold a variety of human rights in their day-to-day business. Their contribution to economic growth, for example, can provide the support necessary for fulfillments of various economic, social, and cultural rights. Responsible mitigation of environmental impacts, close consultation with local communities, and employment procedures also help uphold different sorts of rights (ICMM 2009).

Mining, like many other economic activities, has often been conducted with little regard to the rights and interests of indigenous peoples on whose lands resources were located. Over the last few decades, indigenous peoples have been recognized as a distinct category of human societies under international law and, to varying degrees, in national law as well. The adoption in 2007 of the United Nations Declaration on the Rights of Indigenous Peoples has brought a sharper focus to this area. Harms that indigenous peoples suffer as a consequence of poor practices have included dispossession and forced relocation, destruction of culturally significant sites, loss of livelihoods, exposure to disease and "social vices" such as alcohol and prostitution, and in extreme cases, total cultural and social breakdown. It is vitally important that the rights of indigenous peoples are taken into account in any mining project. However, as human rights assessments are not yet common in the industry, particularly those that are publicly disclosed, the degree to which companies respect or otherwise impact human rights—indigenous or otherwise—is not always clear.

COMMUNITY ENGAGEMENT AND DEVELOPMENT

The common terminology for the processes undertaken by mining companies to understand and address community issues include both *community engagement* and *community development*. Community engagement is usually undertaken so that the company can better understand community perspectives. Community engagement is increasingly required by legislation, most commonly as part of project approval processes. It can also be undertaken voluntarily, as part of developing good relationships with local or host communities. Community engagement is not new in the sense that mining companies have always interacted with a range of groups around mining operations. However, in recent times, the focal point and rationale for community engagement has increasingly fallen under the banner of sustainable development and been linked to a broader range of issues than previously considered.

Community development is focused on the needs and aspirations of the community. Community development is concerned with issues of social justice, human rights, and empowerment of all groups in a community, including the most vulnerable. Community development can contribute to the management of the social impacts of a mining project, but this is not the core focus. Effective engagement is essential for community development, but community development does not automatically flow from engagement. The mining industry tends to present community development as a "mature" form of stakeholder engagement practice. A common reference in this area is the spectrum of public participation developed by the International Association for Public Participation (IAP2 2009), which shows a progression from informing and consulting to collaboration and empowerment of local communities. Under this model, community development is "high end" participation compared with public relations, which focuses more on corporate reputation through formal communication and information dissemination.

There are several stages of mine life, from exploration and project development to construction, extraction operations, and closure. In different ways, depending on the context, community engagement and development is important and relevant. Exploration is important in terms of establishing early relationships. This stage sets the scene for the future. By the time project development commences, there should have been sustained dialogue between the company and the community, various studies and assessments undertaken to determine potential impacts and benefits that might occur, and strategies for either impact mitigation or maximization discussed and agreed. The construction phase can result in significant changes and impacts to local communities. Regular dialogue remains important in these phases. It is often only after operations commence that profits are reaped and significant money starts to flow. Community development often becomes a central focus at this point, but the foundations for good development are ideally laid from the very beginning of the project life cycle. Closure and postclosure considerations should be discussed from the very outset. If community development has succeeded, a community will be well placed to deal with the changes that closure may bring.

The mining industry tends to use community development to describe activities undertaken directly or indirectly with communities in the geographic proximity of operations that aim to achieve positive economic, environmental, and/or social outcomes for those communities. Some companies also use terms such as *community programs, community support,* and *social investment*. The sphere of activity varies from site to site, depending on the context, size of operation, commodity, impact of the operation, and various social and political expectations. Activities typically included in sustainability reports under community development tend to include local employment (direct or indirect through the supply of goods and services), training and skills development, provision of infrastructure (such as roads, water, and sanitation facilities), service delivery (such as health and education), employee volunteerism, donations, and nonmining-related opportunities (such as capacity building and empowerment programs). However, it is often not the activities themselves that denote community development, but the processes used. An employment program can serve company goals, but if community-directed and empowering, it can also meet the needs and aspirations of the local community. Process is key to successful community development.

Of particular interest to this chapter is the increasing alignment of various mining companies with the international development and poverty agenda as expressed, for example, by the CEO of Anglo American Corporation at a mining industry conference in South Africa: "Mining companies are not development agencies, but we are important development actors.... I believe that the mining sector will—and must—play an increasingly important role in development and poverty alleviation in the continent" (Carroll 2008). Several multinational mining companies have publicly aligned themselves to the Millenium Development Goals, a set of ambitious targets released by the United Nations to address pressing global development needs. This move toward community development is unsurprising, given that mining companies increasingly operate in some of the poorest and most marginalized communities of the world. Many of the world's leading companies, particularly those operating within a sustainable development framework, are becoming involved in the provision of infrastructure and services, such as for health and education, and economic opportunities through compensation, royalties, direct or indirect employment, and small business enterprise, as well as initiating capacity building and other community development programs. However, while the core business of mining can have adverse impacts, so too can development projects initiated to secure social license to operate, and companies should always analyze the potential for perverse consequences of development projects, including the issue of community dependency.

ADDRESSING COMMUNITY ISSUES

As a testament to the increasing voice that society has over company affairs, and the continued emphasis on the social aspects of mining, in 2009 Newmont Mining Corporation released a major community relationships review. This was undertaken in response to a shareholder resolution at its 2007 annual general meeting put forward by Christian Brothers Investment Services Inc., endorsed by the board, and supported by almost 92% of shareholders. The aims of the review were defined as follows:

- To better understand Newmont's current community relationships and their contexts;
- To assess future risks and opportunities to Newmont with regard to these relationships;
- To analyze the relevance and effectiveness of Newmont's policies, systems, and controls as they relate to community relationships; and
- To identify the impact of resources, capacity and governance on the implementation of these policies and controls (Newmont Mining Corporation 2009).

The project was undertaken by an independent working group and reviewed by an external advisory panel whose members included some who had been publicly critical of the industry's approach to dealing with communities. It was designed around case studies of five Newmont operating sites located in different continents and with vastly different contexts. In reviewing the final report, the Environment and Social Responsibility Board Committee commented on the project outcomes:

> The Company has learned much about the need to foster and maintain good relationships with governments, communities and other stakeholders, not just the ones who support the Company in its mining ventures, but also those who object to mining in general or the Company in particular. We firmly believe that the future viability and sustainability of the Company's business requires that the Company manage our community relationships more effectively and with consistency. The Company must ensure that community engagement, community relations and conflict management become a more integral component of the Company's business (Newmont Mining Corporation 2009).

In responding to the emerging agenda around mines and communities, the industry has made some significant changes at a number of levels, including both policy and planning activities to address community issues. Typically, corporate policies of leading companies in the minerals industry now explicitly address a range of broader social justice objectives, which include such aspects as local and indigenous employment, security and human rights, sustainable livelihoods, culture and heritage, ethical procurement, and stakeholder and/or community consultation. Increasingly, these policies are focused not only on mitigating the negative impacts of mining on the environment and people, but also on delivering sustainable benefits for local and regional communities. However, "good deeds" undertaken in one area in order to maximize benefits does not compensate for social harms. At a minimum, companies must focus on avoiding or mitigating social harm that occurs as a result of their activities.

At the level of process, there is a growing emphasis on the need to adopt a more participatory and inclusive approach to interacting with stakeholders, including local communities. In order to achieve this, there are a series of essential processes that should be undertaken from the outset and repeated throughout the mine life, including inclusive engagement and relationship building, social mapping and baseline studies, social impact assessment, and social risk assessment and community planning. These processes are explored in more detail in a subsequent chapter. In an effort to improve their performance in community relations, some minerals companies are now investing considerable resources in developing and implementing management systems for this area. This entails adaptation and extension of the approach being taken in other dimensions of sustainable development, such as occupational health and safety and environmental management. The elements of these systems typically include an annual planning process, detailed documentation of procedures, regular reviews and audits against defined corporate standards, and a strong focus on information management. Examples that illustrate the "systematization" of community issues include cultural awareness protocols and training, and procedures for managing community complaints or grievances.

Structural arrangements for community relations vary according to contextual and organizational factors. Some operations include community relations departments within the communications, public relations, or external affairs functions, whereas others position community relations as part of the environmental or sustainable development departments. Many large mine sites, particularly those in developing countries or where there is a large indigenous population, have dedicated community development units or departments. Some companies detach or semidetach community development through a dedicated foundation or trust. At the corporate level, there is often a senior executive or separate department that has responsibility for social policy, as well as a board subcommittee in the area of sustainable development or social and environmental policy.

There are no known studies, either regionally or globally, that provide a comprehensive profile of organizational arrangements for community work within the global minerals industry. A notable contribution, however, was a study undertaken by Reichardt and Moshoeshoe (2003) that considered the structural arrangements and capacity for sustainable and community development undertaken in the context of the African mining industry. In the Australian context, a practitioner survey undertaken by Kemp (2004) found that people involved in community relations work were generally well educated and had considerable industry experience, but the majority did not hold qualifications in the social sciences. In addition, the group had a low level of prior experience in community relations–type work, either within or outside the industry. Subsequent anecdotal evidence suggests that the industry has recently targeted people with social science backgrounds from sectors such as international development, aid, social services, and community development while also building the social science and relational capacities of those from technical or operational backgrounds (Kemp 2009). The policy, structural, recruitment, and professional development arrangements outlined above speak to an ever-increasing focus on the institutionalization and professionalization of community relations in mining.

SUMMARY

Although many of the issues explored in this chapter are not new to the industry, and indeed have been the subject of public debate for decades, there is no doubt that the focus on community issues in mining has increased substantially in recent years. There are many reasons for this, some of which have been identified earlier, but in general the change does reflect an underlying shift in societal attitudes toward business and society, and the extractives industry in particular. For those working in the industry, it means that significant attention needs to be paid to this area, and not just by those working in dedicated community roles. In particular, engineers should recognize the diversity and complexity of communities, rather than viewing them as a collective entity; understand the direct and indirect ways in which mining operations impact communities; and appreciate the significance of contextual and cultural factors that might affect community responses to mining activities. They should also understand that the nature of community issues will change over the life cycle of a project, and they should appreciate the need for early and continuous engagement with the communities associated with their operations.

The mining industry has, to a great extent, accepted aspects of the moral arguments being made in this area. The industry also appears to be persuaded that investing in community relations activities to address these types of issues makes good business sense. A former chief economist for Rio Tinto identified two main sets of business case arguments, which he termed "show-stopping pressures" and "competitive pressures" (Humphreys 2000). The former includes disruptions and delays to projects that threaten their existence, while the latter focuses on the desire to be the "developer of choice" and the ability to move faster in the project establishment process. However, in the same paper the author adopted a less-instrumental argument: "How companies behave reflects underlying currents in the value systems of the world in which they operate. They and their employees are not something apart from civil society. They are an integral part of it and, as such, need to be sensitive to changes in its priorities and perceptions." The maintenance of the industry's "social license to operate" has become an imperative in an increasingly connected and challenging global environment.

The Panguna mine on Bougainville referred to at the beginning of this chapter has not operated since 1989. There have been suggestions in recent years that some in the local communities would like to investigate the possibility of reopening the mine. Other landowners remain bitterly opposed because of the environmental and social legacies left by the operation and subsequent conflict. As in many other locations, the possibility of using the economic benefits of resource development to provide opportunities for local communities remains attractive. Amidst the complexity, one thing is clear—in negotiating the future of any renewed mining activity on the island, the approach taken by mining companies, governments, and local stakeholders in addressing "community issues" will need to be very different from the past. To be successful, mining companies must reflect a better understanding of the needs, aspirations, strengths, and perspectives of local communities and focus on these social aspects as much as on the engineering design of any new mining proposal.

REFERENCES

AccountAbility. 2004. *Business and Economic Development: Mining Sector Report*. London: AccountAbility.

Agricola, G. 1556. *De Re Metallica*. Translated by Herbert C. Hoover and Lou H. Hoover. New York: Dover Publications (1950 edition).

Arellano-Yanguas, J. 2008. A thoroughly modern resource curse? The new natural resource policy agenda and the mining revival in Peru. IDS Working Paper 300. Brighton, UK: Institute of Development Studies.

AustLII (Australasian Legal Information Institute). 1993. Commonwealth Consolidated Acts—Native Title Act 1993. www.austlii.edu.au/au/legis/cth/consol_act/nta1993147/. Accessed February 2010.

Banerjee, S.B. 2001. Corporate citizenship and indigenous stakeholders: Exploring a new dynamic of organizational stakeholder relationships. *J. Corp. Citizen.* no. 1:39–55.

Brereton, D. 2002. The role of self-regulation in improving corporate social performance: The case of the mining industry. Presented at the Australian Institute of Criminology Conference on Current Issues in Regulation: Enforcement and Compliance, Melbourne, September.

Brereton, D., and Parmenter, J. 2006. Water, communities and mineral resource development: Understanding the risks and opportunities. In *Proceedings of the Water in Mining Conference*, Brisbane, November. Melbourne: Australasian Institute of Mining and Metallurgy.

Carroll, C. 2008. Keynote address presented at the African Mining Indaba, Capetown, February 5.

Commonwealth of Australia. 1993. *Native Title Act.* www.comlaw.gov.au/ComLaw/Legislation/ActCompilation1.nsf/0/906779FFBAA54B97CA257348001877EB/$file/NativeTitle1993_WD02.pdf. Accessed December 2009.

Cooney, J. 2008. Sustainable mining and the oil sands. Keynote address presented to the Alberta Environment Conference, Edmonton, April 23.

Danielson, L. 2006. *Architecture for change: An account of the mining, minerals and sustainable development project.* Berlin: Global Public Policy Institute.

Denoon, D. 2000. *Getting Under the Skin: The Bougainville Copper Agreement and the Creation of the Panguna Mine.* Melbourne: Melbourne University Press.

DITR (Department of Industry, Tourism and Resources). 2006. *Community Engagement and Development.* Canberra: DITR.

Dresner, S. 2008. *The Principles of Sustainability.* London: Earthscan.

Elkington, J. 1997. *Cannibals with Forks: The Triple Bottom Line of 21st Century Business.* Oxford: Capstone.

Freeman, R.E. 1984. *Strategic Management: A Stakeholder Approach.* Boston: Pitman.

Gibson, R.B. 2006. Sustainability assessment and conflict resolution: Reaching agreement to proceed with the Voisey's Bay nickel mine. *J. Cleaner Prod.* no. 14:334–348.

Global Reporting Initiative. 2009. *Sustainability Reporting Guidelines and Mining and Metals Sector Supplement* (draft for public comment). Amsterdam: Global Reporting Initiative.

Humphreys, D. 2000. A business perspective on community relations in mining. *Resour. Policy* no. 26:127–131.

IAP2 (International Association for Public Participation). 2009. Spectrum of Public Participation. www.iap2.org. Accessed June 2009.

ICMM (International Council on Mining and Metals). 2006. *The Challenge of Mineral Wealth: Using Resource Endowments to Foster Sustainable Development.* London: ICMM.

ICMM (International Council on Mining and Metals). 2008. *Good Practice Guidance on HIV/AIDS, TB and Malaria.* London: ICMM.

ICMM (International Council on Mining and Metals). 2009. *Human Rights in the Mining and Metals Industry: Overview, Management, Approach and Issues.* London: ICMM.

IIED (International Institute for Environment and Development). 2002. *Breaking New Ground: Mining, Minerals, and Sustainable Development: The Report of the MMSD Project.* London: Earthscan.

Kemp, D. 2004. The emerging field of community relations: Profiling the practitioner perspective. Presented at the Inaugural Minerals Council of Australia Global Sustainable Development Conference, Melbourne, November.

Kemp, D. 2009. Mining and community development: Problems and possibilities of local-level practice. *Community Dev. J.* http://cdj.oxfordjournals.org/cgi/content/abstract/bsp006. Accessed December 2009.

MCMPR (Ministerial Council on Mineral and Petroleum Resources). 2005. Principles for engagement with communities and stakeholders. www.ret.gov.au/resources/mcmpr/Pages/PrinciplesforEngagementwithCommunitiesandStakeholders.aspx. Accessed June 2009.

Newmont Mining Corporation. 2009. *Report of the Environmental and Social Responsibility Committee of the Board of Directors.* www.beyondthemine.com/2008/pdf/CRR_Board_Response_March_2009.pdf. Accessed June 2009.

Philippine Mining Act of 1995. Republic Act No. 7942. Enacted by the Senate and House of Representatives of the Philippines in Congress, March 3, 1995.

Regan, A. 2001. Why a neutral peace monitoring force? The Bougainville conflict and the peace process. In *Without a Gun: Australians' Experiences Monitoring Peace in Bougainville, 1997–2001.* Edited by M. Wehner and D. Denoon. Canberra: Pandanus Books.

Reichardt, M., and Moshoeshoe, M. 2003. *Pioneering New Approaches in Support of Sustainable Development in the Extractive Sector.* Report of the joint project of the International Council on Mining and Metals and the World Bank. Johannesburg: African Institute of Corporate Citizenship.

Ruggie, J. 2008. *Protect, Respect and Remedy: A Framework for Business and Human Rights.* Report of the Special Representative of the Secretary-General on the issue of human rights and transnational corporations and other business enterprises to the Human Rights Council. UN Doc. A/HRC/8/5, 7 April. New York: United Nations.

WCED (World Commission on Environment and Development). 1987. *Our Common Future.* Oxford: Oxford University Press.

West, R. 1972. *River of Tears: The Rise of the Rio Tinto–Zinc Mining Corporation.* London: Earth Island.

CHAPTER 17.2

Social License to Operate

Ian Thomson and Robert G. Boutilier

ORIGIN OF THE TERM

Mining is accepted by the public at large because of the role it plays in society as a provider of minerals and metals for the public's needs and general well-being. There can be no doubt as to the historic role that mineral exploitation has played in the advance of societies, and in more recent times in the economic growth and industrialization of specific countries such as Australia, Canada, Chile, South Africa, and the United States. At the level of individual mining projects, however, this acceptance is neither automatic nor unconditional, and since 1990 has become increasingly tenuous.

During the 1990s, the mining industry found itself under close public scrutiny following a series of well-publicized chemical spills, tailings dam failures, and increasing conflict with local communities around exploration and development projects (Thomson and Joyce 2006 provide a review of this time period). In 1996, a Roper opinion poll showed mining to rate last among 24 U.S. industries in terms of public popularity, behind the tobacco industry (Prager 1997). Internationally, mining became a pejorative term in many circles and widely regarded as a problem industry that was the cause of unwanted pollution and undesirable social impacts. This pervasively negative reputation constituted a liability to the industry.

At a meeting with World Bank personnel in Washington in early 1997, Jim Cooney, then director of international and public affairs with Placer Dome, proposed that the industry had to act positively to recover its reputation and gain a "social license to operate" in a process that, beginning at the level of individual mines and projects, would, over time, create a new culture and public profile for the mining industry. The concept and terminology surfaced in May 1997 in discussions within a conference on mining and the community in Quito, Ecuador, sponsored by the World Bank, and soon entered the vocabulary of the industry, civil society, and the communities that host mines and mining projects.

CHARACTER AND DEFINITION

At the level of an individual mine or mining project (including exploration projects), the social license is rooted in the beliefs, perceptions, and opinions held by the local population and other stakeholders about the mine or project. It is therefore "granted" by the community. It is also intangible, unless effort is made to measure these beliefs, opinions, and perceptions. Finally, it is dynamic and nonpermanent because beliefs, opinions, and perceptions are subject to change as new information is acquired. Hence the social license has to be earned and then maintained.

The social license has been defined as existing when a mine or project has the ongoing approval within the local community and other stakeholders (Business for Social Responsibility 2003b; AccountAbility and Business for Social Responsibility 2004), ongoing approval or broad social acceptance (Joyce and Thomson 2000), and most frequently as ongoing acceptance.

The differentiation of approval (having favorable regard, agreeing to, or being pleased with) and acceptance (disposition to tolerate, agree, or consent to) can be shown to be real and indicative of two levels of the social license—a lower level of acceptance and a higher level of approval. Although the lower level is sufficient to allow a project to proceed and a mine to enjoy a quiet relationship with its neighbors, the higher level is more beneficial for all concerned, including the industry as a whole.

On occasion, the social license can transcend approval when a substantial portion of the community and other stakeholders incorporate the mine or project into their collective identity. At this level of relationship, it is not uncommon for the community to become advocates or defenders of the mine or project since they consider themselves to be co-owners and emotionally vested in the future of the mine or project.

EXAMPLES FROM REAL COMMUNITIES

In the spring of 2008, a tourist traveling across southern British Columbia, Canada, checked into a motel in the community of Trail and chatted with the receptionist. As part of the exchange, the receptionist mentioned how excited she was about "our" new project. Thinking this to be some improvement to the motel, the visitor asked for more information and was interested to learn that the new project had captured the imagination of the whole community. It was, in fact, a process

---

Ian Thomson, Principal, On Common Ground Consultants, Inc., Vancouver, British Columbia, Canada
Robert G. Boutilier, President, Boutilier & Associates, Vancouver, British Columbia, Canada, and Cuernavaca, Mexico

to recover recyclable scrap metals at the nearby Teck Cominco smelter. Clearly, the receptionist and her neighbors had very positive feelings toward the project, the company, and the smelter; they not only approved of it, they identified with it, and in their collective minds the project was fully "licensed" to go ahead.

The same type of community identification with a company's operations can be found in small cities and towns throughout the world. In the information economy, San José, California, United States, and the surrounding vicinity identifies itself with the high-tech industry, dubbing itself "Silicon Valley." With this level of support, community and private sector organizations see their interests as highly aligned, both politically and economically.

PROBLEMS OF INTERPRETATION

The concept of an informal social license is comfortably compatible with legal norms in countries that operate under the principles of common law. However, the concept runs into difficulties in countries such as those in Latin America that operate under the principles of civil law, whereby only an official authority can grant a license. As a consequence, while communities and civil society are eager to see the social license in terms of a dynamic, ongoing relationship between the company and its stakeholders, regulators (and in turn many companies) see the license in terms of a formal permission linked to specific tasks and events in which the regulator plays the central role in granting the license.

In parallel with the above, there have been attempts in Canada and the United States to link the social license to specific tasks, and also to reformulate the concept as something that is established by the company and, once gained, becomes permanent. For example, Shepard (2008), suggests that the social license be defined as "a comprehensive and thoroughly documented process to have local stakeholders and other vested interests identify their values and beliefs as they participate in scoping the environmental impact assessment and in identifying alternative plans of operation for the project." (Notably this does not stipulate that the community, stakeholders, and other groups accept, approve, or support the project).

Neither of these capture the vision of the social license as being dynamic, granted by the community (or at the higher level, society in general), descriptive of the quality of the relationship between company and stakeholders, or involving the reputation of the company and hence industry. Moreover, the words *to operate* are sometimes confused with the strictly operational phase of a mine's life cycle when ore is extracted for processing. A better sense of the term *to operate* is to continue the project, no matter where in the mine life cycle, from the initiation of exploration through to closure.

COMPARISON WITH OTHER FORMS OF CONSULTATION AND CONSENT

FPIC is the acronym used to describe free, prior, and informed consultation and free, prior, and informed consent. Both involve close interactions with communities around resource development projects and have become critical issues for the mining industry. Superficially, there appears to be much in common between a social license and FPIC. From an operational perspective, however, while there is convergence, there are also significant differences between them, with FPIC limited in scope and time to gaining permission to enter land and/or initiate a mining project. More specific definitions can clarify the distinctions.

The free, prior, and informed consultation version of FPIC is perhaps best expressed in the Performance Standards on Social and Environmental Sustainability of the International Finance Corporation (IFC), although it is widely incorporated into national regulations for environmental impact assessment and the permitting of mining projects. In the IFC performance standards, free, prior, and informed consultation is defined as "an obligation of private sector project proponents to engage with project affected populations in a process of consultation that is 'Free' (free of intimidation or coercion), 'Prior' (timely disclosure of information—in effect before any decision is made), and 'Informed' (relevant, understandable, and accessible information)" (IFC 2006). Guidelines accompanying the IFC standards indicate that the process of FPIC should lead to "broad community support," an aspect that is substantially similar to gaining a social license. However, the FPIC process is a specific action or series of actions limited to the period before a one-time decision is taken about a project and focuses on application to the provision of information about a project and its potential consequences.

The free, prior, and informed consent version of FPIC is rooted in two international instruments: the International Labour Organization (ILO) Convention 169, also known as ILO 169, and the United Nations Declaration on the Rights of Indigenous Peoples, or UNDRIP (ILO 1989; UNGA 2007). It is important to note that these instruments are directed to governments, not the private sector, and are limited to interactions with indigenous and tribal peoples. Nevertheless, they are of concern to the private sector because companies are typically the advocates for development projects and frequently introduce the idea of a mining project before any such discussion has taken place between government and the indigenous people. Therefore, as the initiators, companies should take care that discussions with indigenous communities do not proceed ahead of consultation between government and the indigenous people, neither in the timing nor direction of the content.

ILO 169, which has been adopted by 20 nations (the majority in Latin America), requires governments to "consult the peoples concerned…whenever consideration is being given to legislative or administrative measures which may affect them directly;…establish means by which these peoples can freely participate…at all levels of decision-making"; and stipulates that "the consultations…shall be undertaken, in good faith…with the objective of achieving agreement or consent to the proposed measures" (Article 6). With respect to mineral resources and mining projects, ILO 169 is quite specific in requiring governments to consult indigenous and tribal people "with a view to ascertaining whether and to what degree their interests would be prejudiced, before undertaking or permitting any programs for the exploration or exploitation of such resources pertaining to their lands" (Article 15).

UNDRIP includes a much stronger statement on free, prior, and informed consent in requiring governments to "consult and cooperate in good faith with the indigenous peoples concerned through their own representative institutions in order to obtain their free and informed consent prior to the approval of any project affecting their lands or territories and other resources, particularly in connection with the development, utilization or exploitation of mineral, water or other resources" (Article 32.2).

In summary, FPIC is limited in scope and timing to a period before exploration or development can take place, is linked to a one-time decision-making process, and, in the case of ILO 169 and UNDRIP, is an obligation of governments. In contrast, the social license to operate is an expression of the quality of the relationship between a private sector project/ company and its neighbors. In terms of mining, this begins with first contact at the initiation of exploration and continues through the entire life of a project, which, if successful, includes mine construction, mine operation, closure, and increasingly into postclosure.

WHO GETS SOCIAL LICENSE AND WHO GRANTS IT

A social license is usually granted on a site-specific basis. Hence a company may have a social license for one mine or operation, but not for another. Usually the company has to have an operation that affects other groups, organizations, or aggregates of individuals. The bigger the effects, the more difficult it becomes to get the social license. For example, an independent Web-site designer working from a computer in his bedroom will normally get an automatic social license in most communities. A mining company wanting to relocate an entire village faces a much bigger challenge.

The license is granted by the *community*, a term used generically throughout this chapter to describe the network of stakeholders that share a common interest in a mining or exploration project and make up the granting entity. Identifying that the entity is a *network* makes salient the participation of individuals, groups, or organizations that might not necessarily be part of a geographic community, and recognizes the reality that communities are inherently heterogeneous. Calling them *stakeholders* means the network includes individuals, groups, and organizations that are either affected by the operation or that can affect the operation. For example, ranchers who would have to accept a land swap involving part of their pastureland would be affected by a proposed mining operation, without having much effect on it, provided they accepted the deal. By contrast, a paramilitary group of insurgents or an international environmental group that might attack the mine site, each in their own way, would *affect* the operation, without being affected much by it. They would be stakeholders too.

Use of the terms *community* and *stakeholder network* implies that the license is not granted by a single group or organization. It is a collective approval granted by a network of groups and individuals. Therefore, the existence of a handful of supporters amid a larger network of opponents would mean that the license has not been granted.

The requirement that the license be a sentiment shared across a network of groups and individuals introduces considerable complexity into the process. It invites the question about whether a coherent community or stakeholder network even exists. If one exists, how capable is it of reaching a consensus? What are the prerequisites a community must have before it becomes politically capable of granting a social license? These complexities make it more difficult to know when a social license has truly been earned. The complexities of the granting process are discussed later in this chapter.

COMPONENTS OF THE SOCIAL LICENSE

Accumulated experience supports the proposition made by Thomson and Joyce (2008) that the normative components of the social license comprise the community/stakeholder perceptions of the legitimacy and credibility of the mine or project and the presence or absence of true trust. These elements are acquired sequentially and are cumulative in building toward the social license. The mine or project must be seen as legitimate before credibility is of value in the relationship, and both must be in place before meaningful trust can develop.

THE BUSINESS CASE FOR INVESTING IN IT

Earning a social license to operate plays havoc with financial and engineering schedules. It involves a fuzzy relationship-building process that does not fit comfortably into technical planning frameworks that are built around concepts like material objectives, deadlines, and deliverables. So, why is this necessary?

The same question could be asked about safety. Doing safety training does not seem to directly contribute to the bottom line, so why bother? Several decades ago, health and safety was viewed in the same way that many people view social responsibility today. It was the responsibility of a specialist in the company. Accidents were accepted as the aspect of the job that justified higher pay. Today, safety is everyone's responsibility. The leading companies have systems in place to identify dangerous situations so that accidents can be prevented before they happen. The same will eventually be true for the social license. The leading companies will establish positive stakeholder relations before complaints and controversies erupt. Seeing the earning of a social license as an essential part of the project requires a higher-level perspective that is usually only the concern of executives and members of the board. However, project planners need to understand this bigger picture too, in order to allocate sufficient time and resources to earning a social license. The next two sections deal with this bigger picture.

Dependence on Resources Controlled by Stakeholders

It is an axiom of the mining industry that mines are made where mineral deposits are found. This truism reflects the fact that economically viable mineral deposits occur in relatively rare situations. As a result, the mining company has to adapt to the location of the deposit and engage with the people who live there and who consider the location part of their "backyard." By way of contrast, manufacturing industries can select the location for a new plant or facility based on a variety of factors, including the willingness of the local community to accept them. Indeed, during the first part of the decade, there was open competition between communities in North America to host new manufacturing and assembly plants and wholesale distribution depots.

Mineral deposits always have neighbors. Sometimes they have a village right on top of them. Roads always go through someone's land or affect someone's traffic or hunting ground or watershed. Even the most remote mines have to rely on the closest towns as supply centers and transportation links. The many ways that mines affect their neighbors make those neighbors stakeholders. However, the effects go in the opposite direction too. The mine can be affected by road closures, road use regulations, water rights negotiations, protests, blockades, and many other tactics that restrict access to essential resources. Even with the legal right to mine from a government, a company can still be stopped by a community determined to withhold the social license to operate. For example, in Peru alone, at the time of this writing, there were more than 80 sites where communities had prevented mining companies from exploring or exploiting mineral deposits.

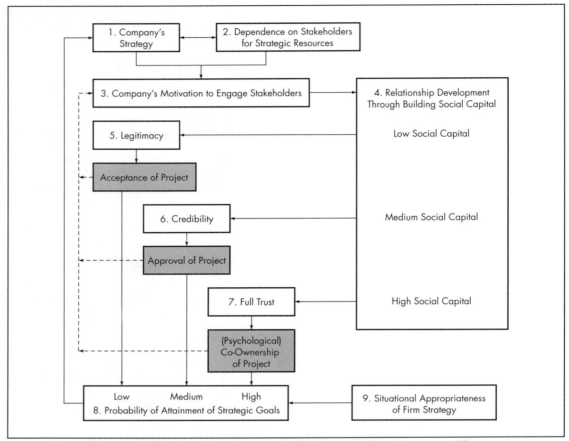

Figure 17.2-1 Adaptation of resource-based view to include process of gaining a social license to operate

The management theory that covers this type of challenge is called the *resource dependence* view of competitive advantage and the firm (Barney et al. 2001 provide a review). It explains why some companies thrived while others struggled. The theory emphasizes that companies need good relationships with those who control essential resources that are scarce, non-substitutable, or imperfectly imitable. That can include things like labor skills, some materials, risk financing, access to some markets, special legal permits, and, for mining, access to land and mineral deposits.

The Role of Strategy

Figure 17.2-1, shows the resource-dependence view augmented with acknowledgment of the role of strategy in setting priorities among resources. The chosen strategy (box 1) determines which resource-controlling stakeholder (box 2) must be dealt with. At the same time, because strategy is often based on existing strengths, the stakeholders the company can deal with best or worst influence what strategy will be chosen. Together, these factors identify the stakeholders with whom the company must engage (box 3).

Different Strategies at Different Stages of the Mine Life Cycle

Companies have different strategies for financial success at different stages in the life cycle of a mine. In the earlier stages of exploration and feasibility, the strategy is to buy the rights inexpensively, prove the richness of the deposit, and sell the property to a mine operating company. The financial gain depends on demonstrating the quality of the ore deposit (grade, tons, mineralogy, feasibility, etc.) and finding a motivated buyer who has the capital and knowledge needed to develop the property into an operating mine.

In the later stages, operating companies pursue a strategy of getting the property into production as quickly as possible in order to reap the financial rewards of selling the minerals from the site. Financial gain depends much more directly on the prices that mineral processors and industrial manufacturers are willing to pay for metals.

Mine builders want to buy properties that are fully permitted. They avoid properties at which stakeholders threaten to block construction or access to the deposits or other vital resources (e.g., water, power). Unfortunately, too many exploration companies think they can ignore or hide these kinds of stakeholder problems.

For example, from late 2003 to late 2004, the stock price for Monterrico Metals, a mineral exploration company, rose from around 150 to 600 pence per share on the London Stock Exchange. Drilling results had shown that the company had a world-class copper deposit at its Rio Blanco property in northern Peru, and the company was actively seeking a buyer to develop a mine. During 2005 and 2006, however, local peasants demonstrated against the development. A group of academics and British parliamentarians known as the Peru Support Group went to Rio Blanco to investigate. Their findings noted a permitting dispute between two Peruvian government bodies and, more particularly, widespread community opposition (Bebbington et al. 2007). The news of the apparent absence of a social license was followed by fall in share prices with their value ultimately dropping almost 70%. In early 2007, a Chinese mining company bought control of the company amid complaints from some Monterrico investors about the low price being paid. At the time of this writing, the project continues to be mired in conflict and there is no indication that a mine can be developed in the near future. In retrospect, the Monterrico investors should be satisfied with the price paid by the Chinese, but they have also learned the hard way that a social license to operate has a financial value when a buyer is sought to develop the property.

Identifying Stakeholders Based on Strategy

Boxes 1 and 2 in Figure 17.2-1 cover the criteria for identifying stakeholders. Those who will be *affected by the company* are implicated by the company's strategy (box 1). Those who have the potential to *affect the company* are covered by the company's dependence on others for resources (box 2). The combined set represents the stakeholders with whom the company should establish relationships.

When an exploration team determines that there is an interesting deposit, the activities in the first three boxes transpire. The team must identify by name local stakeholders who make up the community and then determine who speaks for each stakeholder group as an opinion leader in relation to the project. When the exploration team first introduces itself to those stakeholders, the social capital in the firm-stakeholder relationship is very low (box 4). The social capital dimension of stakeholder engagement and relationship is explored later in this chapter. For now it is sufficient to think of social capital as the collaborative capacity in the relationship.

The pivotal role of relationship building (box 4) implies that a policy aimed at keeping a low profile raises the sociopolitical risks associated with the project. This is especially true during the earlier stages of the project. If the company does not engage with all elements of the community, there is no way the community can collectively grant a social license. If relationship development is delayed, the social license will also be delayed, usually at a high financial cost to the company. Thus, anyone who argues that there is no time for a lengthy relationship-building process with the community is unwittingly advocating for lengthy delays to be imposed by the community.

Building Social Capital and Earning Higher Levels of Social License

Engagement entails the building of a relationship with the stakeholders, who are often already organized into a geographically based network (the community). If the relationship proceeds well, the community may view the company as legitimate (box 5 in Figure 17.2-1) and may grant a provisional/conditional social license (gray box, "acceptance of project").

Acceptance of the project can heighten the company's motivation to engage with the stakeholders, provided the company realizes that this is an investment in raising the probability that corporate strategic goals will be attained. Because not all companies realize this, the line feeding back to a company's motivation (box 3) may be dashed. Assuming the company does choose to build more social capital in the relationship, the community will come to see the company as credible (box 6). Consequently, it will grant full approval of the project (gray box, "approval of project"). Ideally, the relationships should evolve to this point before the company embarks on the feasibility and permitting stages of a project.

Probabilities of Success

If the company continues to put effort into the relationship (dashed line to the left in Figure 17.2-1), it can raise the probability of realizing its strategy and lower the inverse probability of sociopolitical risk. Taking the relationship to the level of full trust (box 7) yields exponentially more benefits for the project and the parent company. At that point, the social license reaches the level of "psychological co-ownership." This will result in the success of the project, provided that the company chose a winning strategy in the first place. Outside factors, however, like the price of metals, can always stop a project (box 9).

Figure 17.2-2 shows how the process of earning a social license to operate impacts a company's bottom line. *No access* means "red ink." Good access makes "black ink" possible, provided all other aspects of the project are positive. Although the relationship-building process might sound at first like a nonessential nicety, it is in fact a make-or-break factor for the successful development of a project. Every continent has its "off limits" mineral deposits that represent projects which failed at the relationship-building stage.

Dynamics of Stakeholder Influence in Networks

Some stakeholders have legal powers to affect the company. Governments, for example, can often act unilaterally to impose conditions or restrictions on mining activities. Most stakeholders, however, have their effects through political maneuvering. Some have direct control over the resources the company needs. They can exert influence by restricting access to those resources through tactics like prohibitive pricing, blockades, and boycotts.

Still other stakeholders have no legal power and no direct control over essential resources. They can, nonetheless, form alliances that pressure those with such direct power to act in solidarity with them. Environmental and human rights groups, for example, usually fall into this category. They often use tactics such as putting pressure on senior governments to change mining regulations or on municipal governments to withhold resource access (e.g., water) or on consumers to boycott the product (e.g., the blood diamonds campaign).

Because political alliances among stakeholders often determine their level of influence, it is important to know about those networks in order to identify the stakeholders who deserve more attention. Methods for acquiring this kind of information are discussed later in this chapter. For the moment, it is simply important to recognize that the sociopolitical dynamics in the stakeholder network do affect the project's support among stakeholders. Thus, acquiring a social license to operate is partially a sociopolitical task.

Resource Access Restriction at the Corporate Level

Stakeholders can also have influence over a project in less direct ways. For example, they can exert pressure through control of access to local resources; they can also affect access to the resources that the parent company needs. The local project may or may not be constituted as a subsidiary company of a larger company. In any case, the "parent(s)" sponsoring the project also needs resources. These include equity or debt financing, new highly trained employees, access to minerals at other sites in the future, and willing customers.

Stakeholders sometimes restrict the parent company's access to those resources by launching national or international campaigns that damage the company's reputation. The effect on share value, for instance, can restrict access to equity financing, which is particularly damaging to junior exploration companies, but can also affect operating companies. A company with a negative reputation will also have more trouble attracting talented employees. If the international campaign includes a consumer boycott (e.g., the dirty gold campaign), the pool of willing customers will shrink and income could be depressed. A bad reputation in community relations can also limit the number of sites a company can have access to in the future, no matter whether they are greenfield sites or acquisitions.

At the international level, the absence of a world government to regulate mining has led to the creation of a world governance regime instead. This network of international agreements, standards, and principles is supported by organizations that frequently finance mining projects, including the IFC and the banks that have adopted the Equator Principles (EPFIs 2006) as criteria for project financing. They stand ready to legitimize the complaints of local stakeholders and impose a variety of sanctions against the offending company. Again, these affect the parent company's reputation and access to money, talent, willing customers, and even further national legal permits.

In summary, earning a social license to operate is necessary for protecting the financial viability of the project and the company. It is as deserving of time, money, and career prestige as drilling to estimate ore reserves or preparing quarterly financial statements.

PHASES OF EARNING A SOCIAL LICENSE

A social license has distinguishable levels. At the same time, the process of moving from one level to another can be thought of as a smooth gradient of continuous relationship improvement through increasing social capital.

Figure 17.2-2 shows the four levels of social license and the three boundary criteria that separate them. The levels represent how the community treats the company. The boundary criteria represent how the community views the company, mostly based on the company's behavior. These transition criteria were derived from language repeatedly heard from communities themselves. Although the terms *legitimacy*, *credibility*, and *trust* emerged out of years of conversations while consulting with mining communities, social science and management literature has been shown to support these common-sense views.

The levels and boundary criteria are arranged in a hierarchy. It is possible to go both up and down the hierarchy. For example, if a company loses credibility, approval will be withdrawn and the project will hobble along on acceptance only. If a company loses legitimacy, the project will be shut down. If full trust is gained, the community will support and protect the project as its own.

Withholding/Withdrawal Level

Starting from the base, the rejection level of a social license is the worst-case scenario. This is when the community stops progress on the project. Many mineral deposits cannot be exploited because the community does not grant any level of social license to proceed.

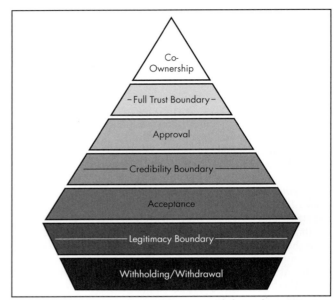

Figure 17.2-2 Levels of social license with boundary criteria between them

The withholding/withdrawal level is shown as narrower than the acceptance level above it in order to symbolize the possibility that, globally, more projects are accepted than rejected. However, at present, this is a supposition. The empirical studies to determine the correct width of each level in Figure 17.2-2 have yet to be done.

Legitimacy Boundary Criterion

The academic meaning of the term *legitimacy* is evolving as interest grows in how organizations gain or lose it, and what they can do with it once they have it (Deephouse and Suchman 2008 provide a review). Knoke (1985) defines legitimacy in the context of stakeholders and politics as "the acceptance by the general public and by relevant elite organizations of an association's right to exist and to pursue its affairs in its chosen manner." This adequately summarizes the bare minimum of legitimacy even when the company has no social license. Suchman (1995) offered a typology in his definition, which has become a touchstone for scholarship on legitimacy. He proposed a three-category typology of "legitimacies." They were (1) pragmatic, based on audience self-interest; (2) moral, based on normative approval; and (3) cognitive, based on comprehensibility and taken-for-grantedness. Each of these has aspects that continue to be important through higher levels of social license. For each type, the community asks itself questions such as the following:

- What do they want, and what is in it for us? How will the consequences of their actions affect us? How will the project affect the environmental resources we absolutely depend on for survival? If we cannot know that for sure, then can we at least discern whether they will be responsive to our concerns, or even share decision making with us? (Pragmatic legitimacy)
- Does anyone in authority recognize/respect us? Are they conforming to our social, cultural, or political norms? Have they followed the specific norms for approaching us with their proposal? Will the consequences of their

activity promote the general welfare of the community, according to our (i.e., the community's) own values? (Moral legitimacy)
- Does what they say make sense, or is it confusing or strange? Has this been done anywhere else? Are their proposals routine practice, or is this uncharted territory? Does that company have the capacity to do what they say they could do? In the case of an expansion of an existing operation, it is precisely the absence of questioning that indicates cognitive legitimacy. This occurs when the presence of the company and its activities are already taken for granted. They are seen as an inevitable part of the community economy. (Cognitive legitimacy)

In addition, particularly for mining exploration projects, the community may also ask questions about legal legitimacy. Do they have the legal permits and permission to do what they want to do?

The behaviors that lead to a company's gaining legitimacy are associated with spreading awareness about the company and what it does, listening to community concerns, and following the official and unofficial local norms, customs, and practices. The company should also have legal status nationally, inform the general public about how the proposed approach has worked for the benefit of other communities elsewhere, and solicit participation in planning and decision making in order to allay fears about the company implementing an arbitrary, uninformed, or high-handed development process.

Failure to engage all segments of the community (e.g., young, old, men, women), to inform them and to solicit their opinions, is often seen as evidence of illegitimacy by those excluded. It is normally important to communicate directly with the bases and not rely solely on leaders. At the same time, following local norms and protocols requires an understanding of, respect for, and use of local social structures and decision-making processes.

Acceptance Level

When legitimacy is established, the community response is that they will listen to the company and consider its proposals. If, by their own standards, they have no reason to doubt the company's credibility, they may allow the project to tentatively proceed. This constitutes the acceptance level of social license. It is a minimal objective for any company.

As can be seen in Figure 17.2-2, the acceptance level is bounded by the legitimacy criterion and the credibility criterion. This represents how acceptance requires that the company's legitimacy must be firmly established and its credibility should at least not be damaged.

Credibility Boundary Criterion

Credibility is the foundation of trust. When a company is regarded as credible, it is seen as following through on promises and dealing honestly with everyone. There is little danger of the company saying one thing one year and a different thing the next. Moreover, the company's policies are the same for everyone that it deals with.

The promise-keeping aspect of credibility is related to the company's responsiveness to community concerns and requests. Legitimacy can be earned by just listening; credibility requires doing something about what has been heard. Community members need to see action following their discussions.

An essential component of credibility comes from openness and transparency in the provision of information and decision making that demonstrates the company to be consistent in the way it treats different groups. Alternatively, transparency reveals the principles that determine why one group might be treated differently than another, reducing the risk of feelings of discrimination or marginalization.

The community asks itself questions such as these:

- Will they deliver on their promises? Are they living up to their responsibilities? Are they acting yet on what we have already said concerns us? Do their promises sound unrealistic?
- Do we understand why they treat some groups differently and how that is consistent behavior?
- Are they keeping any secrets? Do they avoid contact or avoid answering certain questions? Do they acknowledge difficulties or do they often sound glib?

Establishing credibility involves the cycle of listening to the community, responding with information and proposals, and implementing approved proposals. Demonstrations of responsiveness and principled action on a number of quick, short-cycle initiatives help to build a reputation for credibility in the early stages. There is pressure to show immediate benefits, as opposed to long-term promises. The planning horizons in rural communities seldom reach beyond one agricultural cycle. By contrast, exploration teams may make promises to be fulfilled by the future buyers of their properties.

In communities, credibility rests in a person's willingness to keep one's word. This is a significant challenge for the credibility of an exploration company that does not expect to be around when its promises are to be fulfilled. Many communities are suspicious of anyone who intends to transfer responsibility for stated promises to some unspecified, hypothetical person who may or may not even appear in the intangible marketplace for mineral deposits.

To earn credibility, the company should make and keep short-term promises. This is best done by using participatory processes to identify community priorities that the company can help make real. It helps to bring in third parties to verify the truth of company statements and to empower the community to be the watchdog on the company's activities (e.g., community-based environmental committees). Formal agreements and contracts give structure; manage expectations; avoid misunderstandings; and define rules, roles, and responsibilities. Providing false or incomplete information or failing to deliver on promises will quickly destroy credibility and lead to questions as to whether the company is even legitimate. Acceptance can be rapidly withdrawn.

Because companies usually carry so much economic clout, they are often expected to provide the kinds of services that are provided by the government in the developed world (e.g., education and health care). In contexts where governments have not provided these, a company's reluctance to talk about them may be taken as a sign that the company cannot be trusted to make an effort around basic community needs. The best way to address these expectations is with programs to help governments develop the capacity to shoulder such responsibilities. This shows a concern for the problems and a respect for local institutions. It avoids the trap that occurs when governments decide they do not need to do anything because the company is taking care of the community.

Even when relations with the community are good, there may still be attacks on the credibility of exploration companies from civil society sources that warn about the broken promises or social or environmental problems experienced by other communities that accepted mines. Even if some of these stories are false or exaggerated, the sources are credible because they do not appear on the surface to have anything to gain from misrepresenting the facts. When it is their word against a company's word, the company appears to have the greater motive to distort the truth.

Approval Level

When a company has established both legitimacy and credibility, a community is likely to grant approval of the project. This means the company has secure access to the resources it needs. The community regards the project favorably and is pleased with it. The community is now resistant to the "us-versus-them" rhetoric of antimining groups.

This level of social license represents the absence of sociopolitical risk. However, it is really only at the threshold of opportunity. Moving toward the top of this level, as represented in Figure 17.2-2, a company can begin to reap some positive benefit from their relationship building, such as strong community support for expansion of operations.

Full-Trust Boundary Criterion

In management research, trust has been shown to be important in relationships between and within organizations. It is just as important in relationships between organizations and stakeholders. Trust is especially important when bridging the boundary between businesses and civic sector organizations, which include many community groups.

Many typologies of trust have been proposed, some of which range from weak or superficial trust to deeper, more comprehensive trust. As used here, the term *full trust* means a broader and deeper trust.

Credibility is a basic level of trust related to honesty and reliability. A full trust relationship is one where there is a willingness to be vulnerable to the actions of others. Communities that have a full level of trust in a company believe that the company will always act in the community's best interests. Before moving to that level of trust, a community asks itself questions such as these:

- Have they fulfilled their promises repeatedly and consistently?
- Did they handle the unexpected problems in a way that showed they had our best interests at heart?
- Did they share power in a partnership approach?

In order to gain full trust, a company has to go beyond fulfilling its promises to jointly envisioning new development goals with the community. The company will have initiated activities to strengthen the community's ability to plan and achieve its goals for the future. For example, this can involve training in project management for local nongovernmental organizations (NGOs) or technical training for government employees. The company has managed to explain its decision-making principles to the degree that people accept rejection of proposals calmly because they trust the decision makers.

When fulfillment of a promise will be delayed, the company explains why and asks the community for ideas about how to overcome the obstacles to fulfillment. Whenever possible, the company involves community groups in decision making to ensure that no opportunity for capacity building and economic development is missed. The company hands over responsibility for specific aspects of the project to the community so that it takes ownership of both the opportunities and the risks.

Trust is the much desired quality in a relationship that is only earned over time and typically emerges as the product of shared experiences that have positive outcomes for all parties. Trust is hard to earn, easy to lose, and very difficult to recover once lost.

Co-Ownership Level

When the community sees the company as having full trust in it, the community takes responsibility for the project's success. Psychologically, both parties come to view it as a co-ownership arrangement. The limits of the responsibilities of each party are clear, as are ultimate decision criteria.

At this co-ownership level of social license, the company becomes an insider in the community social network. In social sciences, social identity theory describes this level of relationship as the dissolution of the us–them boundary (Williams 2001). Working closely together, the company and community often develop creative solutions to all types of challenges. If outside stakeholders, like the national government or an international NGO, move against the interests of the company, the community will mount a campaign in defense of the company. There have been cases where community members have traveled to foreign countries to challenge false information being promoted by NGO critics. In another case, community members marched on the national capital to make a point to politicians who were proposing a new tax that would affect the mine.

Few companies have taken their community relations to the co-ownership level. Many have difficulty seeing beyond the immediate transactions to the much greater benefits of establishing strong collaborative relationships. Nonetheless, as awareness of the potential benefits grows, more companies are attempting to win a higher level of social license.

MINE LIFE CYCLE AND FIRST IMPRESSIONS

The level of social license that a company enjoys can fluctuate throughout the life cycle of a project. Figure 17.2-3 shows the six stages of the mine life cycle: (1) exploration, (2) feasibility, (3) construction, (4) operation, (5) closure, and (6) postclosure. Obviously, the earning of an initial social license to operate has to be complete before construction begins (stage 3). However, in the past many companies have treated it as an afterthought sometime during the feasibility stage (2). Experience shows that ignoring the social license matter during exploration only creates problems in later stages.

In the first two stages, the key issue affecting the social license is access to land for exploration and feasibility studies. The exploration stage is especially important because that is when first impressions are made. It is a challenging period that can affect community relations during the whole mine life cycle. A positive relationship can lead to the early acquisition of a social license. If that is maintained, it can create the tolerance and mutual understanding needed to deal with conflicts and different interests during the whole life of the mine. Conversely, bad relationships during exploration can lead to social tensions, conflict, and premature shutdown of the project.

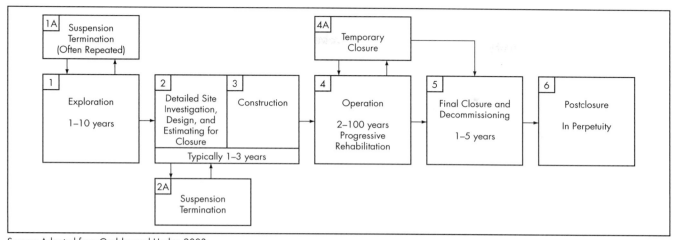

Source: Adapted from Gadsby and Hodge 2003.
Figure 17.2-3 Stages of the mine life cycle

A DIFFERENT STARTING POINT IN EVERY COMMUNITY

First impressions are important, but the preexisting disposition to trust also matters. Some societies are less trusting than others. Figure 17.2-4 shows percentages of generalized trust by population in selected countries from a survey by Delhey and Newton (2005). Obviously, earning a social license in Brazil would be much more difficult than in Norway. The reasons for these differences are not entirely clear, but there are suggestions that lower trust of people in general might be associated with fractionalization in terms of income inequality and political diversity and with higher levels of commitment to family, as opposed to the collective society.

In addition to these cultural factors, each community has its own history. Some have had negative experiences with previous operators of a mine or exploration project. Some have had bad experiences of a more general nature, such as exploitation by foreign companies or governments during colonial times. In many developing countries, there is still suspicion of corporations from developed countries as a result of the Third World debt crisis that followed World Bank policies aimed at encouraging developing countries to imitate rich countries.

All of these factors affect how long it will take to earn a social license in any particular community. Just because it took 18 months to earn enough of a social license to buy land in one community, it cannot be assumed that the next community will require the same amount of time.

DIFFERENT CULTURES, DIFFERENT EXPECTATIONS

There are profound cultural differences between corporations and communities, especially communities that are rural and operate at subsistence level. Companies tend to expect communities to be coherent, cohesive, rational economic actors with a keen sense of judicial impartiality regarding the issuing or revoking of a social license. However, cultural differences can radically redefine the meaning of these qualities. The community's perceptions are filtered through their world view. For example, the Dongria Kondh tribe, in the state of Orissa, India, consider their mountain sacred and view the bauxite mining proposal by Vedanta Resources as an assault on all that is good and holy (Survival International 2008). In these cases, the company must decide if it should withdraw, and avoid the risk of conflict, or invest in working with the community and other stakeholders over a longer period with the hope that a creative compromise can be found.

Many rural communities work on the tribal principles of personal authority and obligations to relatives. Such communities may see profit making as proper for the benefit of a family or community but improper for the benefit of any other body, from individual to company to state. In these societies, all companies, except family and community-owned businesses, start out as morally illegitimate because of the way they treat money and resources.

In many rural communities, nepotism is seen as commendable, not condemnable. Being seen as legitimate entails a willingness to treat community members like family. The company is expected to hire on the basis of local residence, rather than qualifications. Becoming qualified is seen as something that comes after being hired, rather than before.

Time is a common point of difference. Companies want things to happen according to a firm schedule, usually related to a budget. Communities, on the other hand, react organically, take time to reach decisions, and typically have no concern for the schedule set by the company.

THE LOW-PROFILE FALLACY

Earning a social license is complicated work. Occasionally companies are tempted to opt for a strategy that seems easier. They decide to keep a low profile or "stick to their knitting," especially in the exploration stage. This, of course, makes establishing legitimacy, credibility, and full trust quite impossible. The following anecdote describes the consequences.

> In July 2002, Meridian Gold, Inc., completed its acquisition of Brancote Holdings and took possession of the Esquel gold deposit, an exploration project located in the Cordón de Esquel in northwestern Chubut, Argentina. At the time, the concept of developing a mine at the Esquel gold deposit enjoyed general acceptance in the nearby town of Esquel as a result of successful efforts by Brancote to develop a positive relationship with the community. However,

popular sentiment changed very quickly as Meridian failed to respond to requests for information and assistance, and as local permitting authorities expressed a lack of confidence in the company and its actions. Over the following months, the situation became more polarized with the formation of community-based groups organized to seek information and clarification from the company on various issues. In the absence of replies that satisfied these community groups, an organized opposition to the idea of a mine formed in the town. Various protests, marches, and meetings took place with a common theme of "No to the Mine." Attempts to reach a solution through dialogue led by the Church failed to gain any momentum and, under increasing pressure from the local population and growing national and international opposition to the project, the mayor of Esquel authorized a plebiscite on the future of the proposed mine development. On March 23, 2003, the population of Esquel voted 81% (of a 75% turnout) against the mine. In a subsequent independent investigation by Business for Social Responsibility (2003a), it was noted that the Esquel gold mine project was characterized by a lack of engagement by Meridian with the community. This lack of engagement and the inability to have a meaningful dialogue about potential risks (and benefits) was the dominant factor influencing the community not to support the mine. In practice, the company did not provide timely and useful information, and on occasion made it difficult to obtain information, for example when the company refused to provide copies of the environmental impact study, a public document, available in CD format. What community members wanted was to feel informed and listened to, to participate in a real dialogue, and for the company to be responsive to their concerns. In short, the community wanted a partnership, something that the company never attempted to deliver (Business for Social Responsibility 2003a). The collapse of the Esquel project was financially disastrous for Meridian, with the company obliged to take a write-down of US$542.8 million. As of June 2009, the project remained frozen with a majority in the community firmly opposed to any form of mineral exploration or mine development.

Such scenarios are becoming more frequent. They are not always caused by a failure to talk to the community stakeholders. In most of these types of cases, the exploration teams have held community meetings. They have explained their plans. The community representatives showed approval for the projects. So what went wrong? Why do things so often fall apart?

Practical experience points to a number of problems created by the exploration team themselves. The most common problems are as follows:

- A failure to invest fully in relationship building, an approach generally justified by the knowledge that most exploration projects fail. Explorers often choose to delay full engagement with stakeholders until the "right moment" or when there is certainty that the project will go ahead.
- Selective engagement. Talking only to those who are already friendly toward the project creates a group

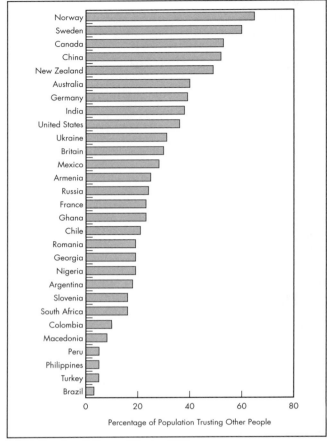

Source: Data from Delhey and Newton 2005.

Figure 17.2-4 Percentages of generalized trust for selected countries

marginalized from information and participation that may well respond by rejecting the project.
- Undermining the company's credibility by failing to deliver on promises already made or by providing false or incomplete information.
- Failing to understand the internal structure of the community and existing relationships between stakeholders and thereby unwittingly creating or amplifying rivalries or divisions that lead to conflict.
- Failing to respect and listen to the community and relying on legal positions and permits to operate.
- Failing to maintain close contact with the community when there is a change in project management or ownership, thereby weakening the continuity needed to sustain the social license.
- Misunderstanding, usually overestimating, the quality of the relationship with the community by confusing acceptance with approval, and cooperation with trust.

The last pitfall is particularly pernicious. Individuals and groups will cooperate with the company for many reasons, including courtesy, a desire for gain, a perception of having no alternative, or, as is often the case with authorities, a sense of obligation. Cooperation for these motives does not necessarily require a trusting relationship.

MAINTAINING A SOCIAL LICENSE

The four-level-phase view of earning a social license is simplistic. It helps mark progress and breaks the process into a practical, phase-like approach. It also highlights important pitfalls. However, every community and stakeholder network is unique in some ways. There is no one-size-fits-all process for earning a social license. There is always a need to adapt and adjust to the specifics of each case. For that reason, it is also important to understand the theoretical process underlying the phases. Any necessary adjustments and adaptations can then be derived from first principles.

Continuous Building of Social Capital

The movement from legitimacy through credibility to full trust is a process of building and balancing the social capital in the relationships between the company and the members of the network.

Definition of Social Capital

Adler and Kwon's (2002) definition of *social capital* says, "Social capital is the goodwill available to individuals or groups. Its source lies in the structure and content of the actor's social relations. Its effects flow from the information, influence, and solidarity it makes available to the actor." In other words, the effects of social capital flow from what it is, namely, information, influence, and solidarity.

Examined more closely, social capital can be seen as a set of benefits that network members can enjoy because of their unique patterns of connections in a network. (In social network analysis, the technical terms for connections in a network is *ties*.) The benefits fall into three general categories. First, social capital is access to the *information* that flows through the network. People who act as bridges between otherwise unconnected clusters of network members are especially likely to enjoy this type of benefit. They have "bridging" social capital. Second, social capital is *solidarity* in support of group norms and shared identity. For example, members of immigrant communities often stick together and police one another's behavior for mutual protection and benefit. This is an aspect of "bonding" social capital. Third, social capital is *influence* in the network. Those who have a good balance of bridging and bonding often enjoy a greater capacity to solicit assistance and resources from others, whether it is for a sociopolitical purpose or a more personal goal.

Sources of Social Capital

Adler and Kwon (2002) say the source of social capital is in the structure and content of the social relations of network members. The structure refers to the network pattern—a bridging or bonding pattern, for example. The content refers to qualities of the relationships between pairs of network actors. Adler and Kwon drew upon a set of distinctions proposed earlier by Nahapiet and Ghoshal (1998), who also identified a *structural* source of social capital, but divided the content into two distinct sources, the *relational* and the *cognitive*. The relational source encompasses qualities like reciprocity, mutual trust, and shared identity. The cognitive source covers shared language, shared problem-solving paradigms, common strategies, and even shared goals and values.

By cultivating these three sources of social capital simultaneously, a company can raise social capital in the community. Then it can "cash in" the social capital for a social license to proceed.

Tapping Sources of Social Capital Across the Phases

Developing the Structural Source of Social Capital

The structural source of social capital deals mostly with the patterns of connections in a network. For example, many networks have a core-periphery structure. The members at the core are well connected and influential. The peripheral members are more isolated with less access to resources. The structural dimension also deals with the strength of each tie. In a community network, for example, the groups or individuals with many weak ties will have access to a greater variety of information than others who have only a few strong ties. However, those with strong ties will be more capable of collaborating with one another to get things done.

When company representatives first arrive in a remote community, they begin forming ties with the network of community stakeholders. In the process of everyday transactions, they form very weak network ties with the hotel desk clerk, the restaurant owner, and the taxi driver. Since these people are all well embedded in the community, these ties represent bridges between the community network and the company's internal network. The leftmost panel of Figure 17.2-5 shows one thin line going from the company (i.e., the solid circle) to one rather central member of the community network. This is meant to represent a bridging relationship between the company and community. With nothing more than a weak bridging relationship in the community, the company is a peripheral member of the community network.

As the project proceeds, the company's presence grows in terms of numbers of representatives and in numbers of contacts with community members. Some of these relationships deepen. The company goes from having bridge relationships, to actually being a semiperipheral member of the community network. At this point, the company may also have created enough trust (relational source of social capital) and understanding (cognitive source of social capital) to be able to use its social capital to get an acceptance level of social license. The network structure would look something like the middle panel of Figure 17.2-5, where the original line representing the relationship formed in the leftmost panel becomes thicker in order to represent a stronger relationship.

The rightmost panel of Figure 17.2-5 shows the company as a member of the core of the community network. In this position it has considerable influence in the community and much more ready access to community-controlled resources. This would be a typical pattern for a company with an approval level of social license.

Developing the Relational Source of Social Capital

The relational source of social capital includes related qualities of social interactions like reciprocity, shared identity, and trust.

Reciprocity. Reciprocity begins with the simplest of courtesies in polite interaction. In many cultures this quickly involves mutual gift giving. As relationships develop, it is most prominent as various kinds of mutuality, including mutual respect and mutual self-disclosure. Reciprocity shades into generalized trust when the parties forgo immediate advantages to themselves in favor of actions that support the long-term interests of the group or of the bilateral relationship. This shows reciprocity because it requires a faith that such sacrifices will be reciprocated by others. In this way, reciprocity contributes to the solidarity benefit of social capital.

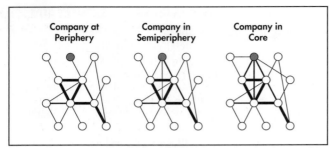

Figure 17.2-5 Structural dimension of social capital at each stage of the social license

Shared identity. Shared identity is important in preventing us–them dynamics from taking root. Members of stakeholder groups all share an identity by virtue of their membership in that group. Likewise, company representatives share an identity, especially in their roles as representatives. Two strategies can be used for avoiding intergroup conflict in such situations.

First, ties that cut across groups insulate members of each group from viewing members of the other group as "them." Crosscutting links either bilateral relationships or group memberships (that include people from both the stakeholder group and the company). For example, some company employees and stakeholder group members might both attend the same church and participate cooperatively in events there.

The second strategy is to heighten the salience of group memberships that completely encompass both groups. For example, both stakeholder group members and company employees might be citizens of the same country. Appeals to nationalistic sentiments can dissolve intergroup rancor and focus people on their shared interests.

At the highest level of social license, psychological co-ownership, the community, and the company share an identity centered around the industry itself. This happens, for example, when a community starts describing itself as a "mining town."

Trust. Trust is a very important aspect of the relational source of social capital. As Figure 17.2-5 shows, a company will start at different levels of trust even before any relationship begins. Early in relationships, trust is based more on calculations of risks and benefits and is limited to specific short-term transactions. Later on, trust comes to be based more on knowledge of the other party gained from experience over a period of time. Trust grows as the other party consistently keeps promises and demonstrably acts in the interests of the perceiver. This knowledge-based trust is important for gaining the acceptance level of social license. Companies can foster this kind of trust early by taking every opportunity to listen to the community and respond quickly (Svendsen et al. 2003).

Developing the Cognitive Source of Social Capital
The cognitive source of social capital comes from mutual understanding and agreement between the parties in a relationship. At the most basic level, speaking the same language creates a modicum of social capital. The level of language, however, is just as important. Companies must always use nontechnical language in their communications with stakeholders, even when it means using more words or losing some nonessential precision in meaning.

At a more abstract level, parties might use the same cognitive tools for approaching problems. For example, if both community stakeholders and the company habitually use strategic planning frameworks like cost–benefit analysis, strength-weakness-opportunity-threat analysis, critical-path analysis, and decentralized decision making, then they will enjoy more social capital together. More often, the stakeholders do not share these cognitive frameworks with companies. In those cases, both sides need to learn each other's ways of approaching collective action tasks. As the community learns the company's analytic techniques, it gains capabilities that it can use to achieve whatever goals it sets for itself, whether those goals are poverty reduction, ecological footprint reduction, or general community development.

Quite often the issues on which firms and stakeholders must develop a mutual understanding concern the general direction of future development in the community. Community stakeholders want to know what life will be like after the mine opens or after the mine closes. They need to know about downsides such as population influx and the possibility of more crime. For its part, the company wants to know what the community's priorities are for future development. Balancing short- and long-term interests here means not promising too much just to get an acceptance level of social license. In these cases, a future visioning process is in order. The process helps set planning priorities and taps the cognitive source of social capital.

How the Sources Work Together
Although the sources of social capital can be described as distinct, in practice they are intertwined in every action and activity. Likewise, the stakeholders' perceptions of legitimacy, credibility, and full trustworthiness develop organically along with the structural, relational, and cognitive aspects of the relationships.

Structural with Cognitive
Speaking the same language allows more communication to occur in the same amount of time. Inversely, communicating more frequently tends to create a shared set of understandings and frameworks. Shared understanding contributes to credibility because the stakeholders understand more thoroughly how mining works. Understanding, of course, also contributes to the cognitive aspect of legitimacy.

Structural with Relational
The gradual integration of the company into the stakeholder network (e.g., Figure 17.2-5) begins to create a shared identity. The community gradually sees the company as a fixture in the network. As crosscutting ties increase in number, a shared identity can emerge, thus tapping into the relational source of social capital. Likewise, any preexisting shared identity can facilitate the company's integration into the local network. For example, a local or domestic company might be able to create and strengthen ties more easily within the network.

Integration into the network builds legitimacy. Early in the process it creates the awareness and helps spread understanding needed for cognitive legitimacy. Later it contributes to the taken-for-grantedness of the company's presence, a more advanced form of cognitive legitimacy. Similarly, building trust, reciprocity, and shared identity contributes to moral legitimacy. Cognitive understanding contributes to pragmatic legitimacy, and vice versa.

Relational with Cognitive
A future visioning process affords an opportunity for the company to demonstrate honesty, transparency, concern for the community's priorities, respect for the community's leaders, responsiveness, and a willingness to share decision making on matters of mutual concern. All of these, in various proportions, contribute to stakeholder perceptions of legitimacy, credibility, and trustworthiness.

Often the process of coming to hold shared goals and visions for the future entails putting oneself in the other party's shoes. This can be difficult, but social capital from the relational source can make it much easier and often overlaps with the mutual perspective-taking feat (e.g., shared identity).

Dynamism in the Process

The levels of social capital in the firm–stakeholder relationships can fluctuate through the life cycle of the mine. Some of the determining factors may be beyond the control of either or both parties. More often, the actions of one or both parties raise or lower the level of social capital, with predictable effects on the level of social license. An example from an actual mine illustrates the point.

Figure 17.2-6 provides a graphical representation of the condition of the social license for the San Cristobal project over a 14-year time period from 1994 to 2008. Today, San Cristobal is a large zinc, lead, and silver mine in the southern Altiplano of Bolivia, operated by Minera San Cristobal (MSC), which is 65% owned by Apex Silver Mines Limited and 35% by Sumitomo Corporation. The quality of the social license was estimated initially from historical documents and the experience of persons who had been present throughout the life of the project using a basket of indicators and then verified through interviews with community members.

As can be seen in Figure 17.2-6 the quality of the relationship changed over time in response to various factors, which may be summarized as follows with reference to the numbers beside the graphical trace in the figure.

1. Rights to the mining concession are acquired by Mintec, a Bolivian mining consulting firm, which, together with permission from the community to access the surface, establishes formal legal status and permission to start work on the ground. The company rapidly builds social legitimacy with the local community by providing information and employment.
2. Geological indicators of widespread mineralization are identified. Mintec, and later ASC Bolivia LDC (ASC), an Apex Silver Mines Limited subsidiary that had acquired Mintec, steps up a dialogue with the community and provides additional benefits. Social and environmental baseline studies for an environmental impact assessment are initiated.
3. Clear indications of a very large mineral system are demonstrated through drilling. ASC believes that a mine can be developed in the near future. Accordingly, the company begins a process of consultation with the community of San Cristobal to relocate the population away from the mineral deposit. ASC empowers the community to manage essential aspects of the relocation such as selection of the new town site, design of houses and infrastructure, eligibility, and a benefit package. The community comes to feel that they are partners (co-owners) in the project—ASC has the mineral deposit, but the community has given its land to make the mine possible. Negotiations begin with the sister community of Culpina K for land for the tailings facility.
4. A comprehensive resettlement agreement is signed with the San Cristobal community. A land purchase agreement is reached with Culpina K. Trust reaches an all-time high.
5. The community of San Cristobal is relocated to the new town site together with the colonial church and cemetery.
6. The women, who were largely excluded from the negotiations and planning for the relocation, voice complaints about the houses: "They are not what we wanted." Culpina K residents start to think that they have made a bad deal.
7. The project is then transferred by ASC to MSC. Trust is eroded because the company has failed to deliver on commitments made in the resettlement agreement. Further, although the company has obtained necessary permits to construct and operate a mine, there are now doubts as to the feasibility of the project, and the company has drastically reduced the number of employees.
8. The project fails feasibility because of low metal prices. MSC closes all field operations. The communities are frustrated and trust is lost. Contacts between the company and the communities become infrequent. The company remains noncompliant with the terms of the resettlement agreement. Credibility is lost. The project remains legitimate in the minds of community members because they want the employment and the better future they hope it will bring.
9. Realizing the need to stabilize and strengthen the relationship with the community, the company initiates a program of assistance centered on employment for local people, agriculture, and tourism. Credibility is restored with delivery of the program.
10. A highly innovative program to encourage tourism is launched. Culpina K becomes heavily involved; San Cristobal, less so because it would rather have the mine.
11. An upturn in market conditions renders the project positive, and MSC announces the start of construction. Credibility peaks as the communities welcome the construction of "their" mine and the potential for employment during construction.
12. New management is installed to supervise construction that has no knowledge of the social history of the commitments to the communities. Contact between the company and the communities breaks down as the company drops meetings with the communities that involve top management. The communities feel deserted and disenfranchised because of a perception that they are not being respected and commitments for employment and training are not being honored. Credibility is lost almost immediately. Despite employment for almost all local people, social legitimacy drains away as the communities feel they have been overlooked in the construction phase and commitments dating back to 1999 remain unfulfilled. They still believe in "their" mine and mourn the loss of the partnership that existed at the time of the relocation of San Cristobal.
13. A contractor opens a road though community gardens outside of the agreed area of construction operations. For the community of Culpina K, this is an illegal act. Social legitimacy is compromised, and community relations collapse for a period with demonstrations and confrontation

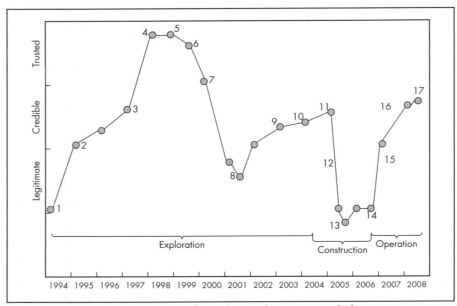

Figure 17.2-6 The ups and downs of social capital at San Cristobal mine

with the company. MSC is able to negotiate a letter of agreement that allows work to continue and begins a protracted process of negotiation for the acquisition of additional land for mine infrastructure facilities. Culpina K remains at a distance because of rising concerns that they will be adversely affected by the tailings facility.

14. Management comes to recognize the problems and risks created by the damaged relationship and takes action to improve the situation. The company proposes the formation of a community-based process for the design and management of community-development programs. The communities see this as an opportunity to take control of their own future and feel empowered. At the same time, MSC begins an accelerated program to comply with all prior commitments. The communities see progress and feel reassured. Construction ends; training and full-time employment is available to all. Dialogue is established with Culpina K around the issue of management of the tailings. Legitimacy is regained.

15. A management team is appointed to run the mine, replacing the construction team, which brings stability and quickly establishes a positive dialogue with the communities. All prior commitments have been met or are in a visible process of being met. An accelerated local employment program is initiated. Community-based planning for social and economic development is under way. Credibility is restored.

16. MSC responds to community concerns regarding the management of tailings by forming a joint monitoring committee with the community. Essential infrastructure improvements are made in San Cristobal and Culpina K. Credibility is strengthened. As "co-owners" of the mine, the communities make representations to the national government in support of the company in response to statements of increased taxes and threats of nationalization. Trust, however, remains elusive.

17. With the mine in operation and the communities fully involved in the management of their own development, in collaboration with MSC and local and regional government, there are early indications that trust could return to the relationship.

RELATIONSHIP BUILDING CHANGES BOTH SIDES

The Fiction of "Community" as an Entity

Up to this point, the term *community* has been used in a reified way, that is, in a way that grants it an identity and singleness of purpose that does not always exist. Many so-called communities are really aggregations of communities and/or kinship and interest groups. However, the concept of the social license to operate presupposes that all the families, clans, interest groups, and institutions in a geographic area have arrived at a shared vision for their future, or at least an agreed-upon set of priorities for today. This kind of cohesion is frequently absent. In such cases, it has to be built. That is why earning a social license to operate often involves the company in building social capital, also known as community building, capacity building, and institution strengthening.

Industries, and the companies in them, have to learn how to participate in the kind of community partnerships that earn a social license. Reciprocally, communities that want the benefits of participating in the mainstream economy have to develop the social structure that makes them capable of issuing a legitimate, credible, and trustworthy social license.

Communities Not Qualified to Issue a Social License to Operate

The key to a community's capacity to issue a meaningful social license is the pattern of social capital it has in its network structure. Without the right patterns of social capital, a community cannot stick to its commitment through the inevitable changes in local politics. Companies want a license that reduces the risk of investing in mining infrastructure. Figure 17.2-7A, B, and C illustrate three examples of how communities can have a social capital network structure that makes them unlikely to be capable of issuing the social license to operate that companies want.

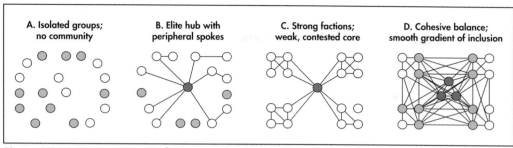

Figure 17.2-7 Four prototypical patterns of stakeholder network structure

Figure 17.2-7A shows a community with a severe lack of social capital. It has no core. It is an aggregate of groups that do not talk to each other. This is the pattern one would expect in a region after a natural disaster or war. There may be pockets of groups that deal with each other, but over the whole geographic region, or across all the stakeholders in the issue, there is little communication. No one can speak for the whole community, and therefore the community can neither issue nor revoke a social license to proceed. In this circumstance, companies usually fall back on the legal licensing process at the national level. Unfortunately, if the community subsequently comes together and finds a voice, that legal license can become worthless.

Another common hindrance to being capable of issuing a social license is the existence of an elite that purports to speak for everyone (see Figure 17.2-7B). The core has numerous relations of bridging social capital to peripheral groups. The peripheral groups have little social capital among themselves. Although dealing with this kind of structure is administratively convenient, there are dangers related to the elite's own legitimacy. Such elites are often tempted to put their own interests ahead of the interests of the whole community. Sometimes they stay in power by keeping everyone else divided, ignorant, or intimidated. Multinationals often fall prey to such groups when only the elites know the native language of the company representatives. Often they have been educated in the developed world and know the norms for doing business in the company's home country. The elites will say they can issue a social license, but once development gets underway, the objections of the excluded groups come to the surface. The company can then be lumped together with the elite as part of an illegitimate and corrupt establishment. This, of course, calls the social license into question.

Figure 17.2-7C shows another dysfunctional community structure. There is plenty of bonding social capital, but it is confined to tight-knit factions. To the extent that a core exists to speak for everyone, it is beholden to the leaders of the factions, be they mayors, war lords, clerics, or members of parliament. Because the factions are continually creating and destroying alliances in attempts to dominate the core, there is never a community with enough cohesion to ensure that any social license will be durable. The company can easily be drawn into the political machinations, either as a benevolent ally of one faction or an evil villain against which another faction rallies support.

Diagnosing Community Readiness to Issue

Companies that want a social license clearly need to know the patterns of the social capital in the network from which they expect to get the social license. Figure 17.2-7D is the ideal social capital structure for a stakeholder network. This type of community can make decisions that include everyone and that will work for everyone's benefit. It can stand behind the decisions for the long term but also has the flexibility to adapt its goals to changing circumstances.

Fortunately, social capital has been studied quite thoroughly in the social sciences for more than 15 years. Ways have been developed to quickly assess its patterns in a network, whether that is a geographic community or a wider set of stakeholders that includes national government departments and international NGOs. The basic approach is to interview the stakeholders about four areas of concern: (1) what their concerns are, (2) which other groups they consider to be stakeholders, (3) the quality of their relationship with the company, and (4) the quality of their relationships with other stakeholders. Data from interviews of this type provide a well-rounded basis for a stakeholder engagement strategy leading to a social license.

Ways to Approach Identified Problems

After a map of the social capital structure in the community is at hand, it is possible to determine where the effort should be focused. If the community looks like Figure 17.2-7A, it is necessary to start with basic communications among groups. If the community looks like Figure 17.2-7B, the first step is to make connections with the periphery groups and connect them with each other. If it looks like Figure 17.2-7C, efforts to strengthen the core are needed. A group of leaders from each faction must work together on a project long enough to develop a sense of shared purpose that is community-wide.

Each community has its own specific issues and problems that provide plenty of opportunities for bringing people together. Often the initial focus has to be setting community priorities and then identifying those that can be achieved in collaboration with the company. Wilson and Wilson (2006) offer numerous examples of initiatives that build social capital. These span diverse areas of operations from security to catering. Activities that build social capital rarely cost very much money. Usually they simply involve redirecting funds from an approach that ignores or damages the company's social license to an approach that improves and secures the social license. Sometimes they actually reduce costs, as occurs when the work to improve the capacities of local suppliers creates a stable of lower-cost local sources for goods and services previously imported from elsewhere.

Social License, Poverty Reduction, and Sustainable Development

The process of helping a community acquire the balance of social capital needed to issue a social license amounts to strengthening the community to the point where it can be a strong, reliable partner in achieving whatever goals it sets for itself. This happens to be the same process that community development specialists follow in initiatives to reduce poverty. Social capital building and balancing initiatives have been shown to reduce poverty, but as Figure 17.2-7 indicates, the pattern of social capital has to be balanced (e.g., Figure 17.2-7D). In addition to a strong core, the overall pattern should integrate all subgroups to avoid isolation of the periphery. When these conditions are met, communities can benefit from the presence of a national or international corporation. The company can create or broker links to the larger economy outside the community. Such links are called *linking* social capital (Woolcock 1998). Through linking social capital, a community can gain access to more distant markets, for both its goods and its services (i.e., remittance income). Therefore, gaining and maintaining a social license employs the same set of activities that promote sustainable community development. Both goals depend on strengthening the community to the point where it can be a stable, committed partner in development initiatives.

TOOLS AND GUIDELINES

Companies use numerous tools and guidelines to help them do the things that earn a social license. These tools often bear labels related to "corporate social responsibility," "sustainability," or various more specific issues like corruption, transparency, or the environment. Leipziger (2003) reviewed hundreds of such codes, lists of principles, and guidelines. She reviewed more than two dozen of them in detail. Some of them reiterate many of the principles discussed above. Those that contain recommendations for specific steps and activities will not always work. Communities and circumstances are too varied to be amenable to any one-size-fits-all program. That is why the more general principles must take precedence over concrete recommendations. Although some guidelines sound too vague to be practical, it is actually the more specific elements that are less practical because they help only in some circumstances but not in others. Ultimately the social, cultural, and political dynamics of the individual community or communities that are neighbors to mining projects have to be understood, and tools and guidelines must be adapted to meet the characteristics encountered on the ground.

Exploration Practices

For explorers, *The Principles and Guidance for Responsible Exploration*, produced by the Prospectors and Developers Association of Canada (PDAC 2009a) together with the *e3Plus* on-line reference manual for good practice (PDAC 2009b) are invaluable aides to gaining legitimacy and credibility with local communities and hence acceptance and a conditional social license.

For projects that advance to feasibility and construction, the key sources of information for the management of social, socio-environmental, socioeconomic, and certain cultural matters are the Performance Standards on Social and Environmental Sustainability of the International Finance Corporation (IFC) of the World Bank (IFC 2006) and the Equator Principles (EPFIs 2006), a parallel set of guidelines created by the private-sector banks. While not created specifically for the mining sector, the IFC standards and a series of accompanying manuals are based on extensive practical experience that is of immediate relevance to mining projects. The IFC standards emphasize disclosure of information, consultation, participatory processes of investigation, and decision making with the community and other stakeholders; inclusion and protection of vulnerable and marginalized groups; respect for rights; participation in the benefits of mining; and so forth, all of which are actions that can create social legitimacy and, properly implemented, credibility for the company and hence a base for building trust.

Mine Operation from Construction Through Postclosure

Business for Social Responsibility (2003b) and AccountAbility and Business for Social Responsibility (2004) provide case histories and discussion of the social license together with practical advice on how a social license may be gained.

The principles and guidelines of the International Council on Mining and Metals (ICMM) provide a point of reference at the global level for good practice that aids acquisition of the social license (ICMM 2003). In addition, the ICMM has prepared a series of good-practice handbooks that offer practical advice on the management of various aspects of mining, including social and community development. At the same time, mining associations in a number of countries have developed guidelines optimized to national legal, cultural, and social norms; both Canada and Australia, for example, have developed such instruments.

The Mining, Minerals and Sustainable Development (MMSD) project identified a portfolio of initiatives that support gaining and maintaining a social license and are critical if mining projects are to make a contribution to sustainable development (MMSD 2002a). They include

- Supporting people (social capital) and investing in people (human capital),
- Supporting and strengthening existing governance structures (social capital),
- Facilitating continuous learning and improvement (human capital), and
- Helping local populations anticipate and take advantage of mine development to improve their quality of life (as defined by them) beyond the life of the project or mine (balancing human and social capital).

The MMSD project also created a framework for assessing a mining project's contribution to sustainable development, which includes material that can be used to design projects in a way that helps in gaining the social license to operate (MMSD 2002b).

As with the examples above, full implementation of the recommendations contained in these documents in ways that are sensitive to local social and cultural norms can significantly assist in gaining and maintaining the social license.

General Guidelines for All Types of Businesses

In the developed world, various existing political forces came together to create an ad hoc governance regime to regulate global corporations. The result of the ad hoc, piecemeal collaboration has been the emergence of some mechanisms for influencing corporate behavior. The influence strategies are each centered on a different set of stakeholders.

One of the most important stakeholder groups has been investors. The socially responsible investment movement applies investment screening criteria that have been promoted by the aforementioned loose coalition. In 2007, ethically screened investments accounted for US$4.93 trillion in assets in United States and Europe alone (Social Investment Forum 2009). Prominent rating systems in the ethical investment field include the Calvert Social Index, the Dow Jones Sustainability Index, the FTSE4Good Index, and KLD's Domini 400 Social Index.

The same general influence mechanism has been called into play with lenders. The IFC has developed Performance Standards on Social and Environmental Sustainability, which are incorporated into the Equator Principles that have been adopted by most of the major commercial banks that finance mines. Companies seeking financing must pass independent assessments of their compliance with these standards for social and environmental performance.

In terms of employees and suppliers, there have been campaigns against child labor and sweatshops. The ILO has been very active in the area of mine safety. Its standards apply to governments, which in turn are supposed to enforce them on companies to whom it grants licenses.

The ILO 169 gives explicit instructions to governments to ensure that indigenous people are consulted before exploration and development of natural resources and that they share in the benefit that comes from resource development. This convention has a strong influence on corporate behavior even where governments are weak or absent, as is the case in many rural areas. The UNDRIP gives an even stronger instruction to governments to obtain the consent of indigenous people before any exploration or development of mineral resources.

Another approach is centered on consumers as stakeholders. Green and socially responsible consumer movements usually focus on specific product classes. The "dirty gold" and "blood diamonds" campaigns are aimed directly at the minerals industry.

The development of these influence channels has included the emergence of general codes and standards that spell out what is expected of companies. These expectations provide general guidance regarding how to earn a social license. Popular codes and standards that apply to all industries include the AA1000, the Global Reporting Initiative (GRI), and the venerable United Nations Universal Declaration of Human Rights (UNGA 1948).

The AA1000 was developed by the Institute for Social and Ethical AccountAbility (AccountAbility 2009). It consists of checklist-style framework for implementing changes in a company and a separate assurance process that verifies the company's adherence.

The GRI deals with standards for transparent reporting on nonfinancial matters. The most recent version was developed with input from thousands of companies from all parts of the world. It is a comprehensive checklist that was intended to apply to all industries. (GRI 2006). To help mining companies reduce the burdensome paperwork, the ICMM has published a guide for focusing on the parts of the GRI that are relevant to its sustainable development principles (ICMM 2005).

SUMMARY

The fundamental tools for building relationships between mining projects and communities are the values that permeate the actions of the company. These include

- Respect and inclusiveness;
- Transparency and honesty;
- Listening and empathy;
- Responsiveness and promise keeping;
- Goodwill, care, and protecting the interests of the other; and
- Clear rules and principled actions.

These are "soft" attributes that many mining company personnel find difficult to work with since they are not amenable to quantification in the same way as engineering and accounting. Nevertheless, they are keys to the creation of good, stable relationships and in many ways parallel the attributes of lasting personal relationships, friendships, and, indeed, a marriage.

ACKNOWLEDGMENTS

The authors thank Apex Silver for permission to present the history of the San Cristobal project; Juan Mamani, Javier Diez de Medina, and Margarita de Castro for providing the information on which the graphic history was first constructed; and the unknown community members who verified the data. The authors also acknowledge the support and intellectual contributions of Susan Joyce and Myriam Cabrera that led to the recognition of social legitimacy, credibility, and trust as the normative components of the social license to operate.

REFERENCES

AccountAbility. 2009. AA1000 Series. www.accountability.org.uk. Accessed September 2009.

AccountAbility and Business for Social Responsibility. 2004. *Business and Economic Development: Mining Sector Report*. London and San Francisco: AccountAbility and Business for Social Responsibility.

Adler, P.S., and Kwon, S.W. 2002. Social capital: Prospects for a new concept. *Acad. Manage. Rev.* 27(1):17–40.

Barney, J.B., Wright, M., and Ketchen, D.J., Jr. 2001. The resource-based view of the firm: Ten years after 1991. *J. Manage.* 27(6):625–641.

Bebbington, A., Connarty, M., Coxshall, W., O'Shaughessy, H., and Williams, M. 2007. *Mining Development in Peru*. London: Peru Support Group. www.perusupportgroup.org.uk/archive/news_120.html. Accessed September 2009.

Business for Social Responsibility. 2003a. *Minera El Disquite Report, Esquel, Argentina*. San Francisco: Business for Social Responsibility.

Business for Social Responsibility. 2003b. *The Social License to Operate*. San Francisco: Business for Social Responsibility.

Deephouse, D.L., and Suchman, M.C. 2008. Legitimacy in organizational institutionalism. In *The Sage Handbook of Organizational Institutionalism*. Edited by R. Greenwood, C. Oliver, K. Sahlin, and R. Suddaby. Oxford, UK: Sage.

Delhey, J., and Newton, K. 2005. Predicting cross-national levels of social trust: Global pattern or Nordic exceptionalism? *Eur. Sociol. Rev.* 21(4):311–327.

EPFIs (Equator Principles Financial Institutions). 2006. *The Equator Principles: A Financial Industry Benchmark for Determining, Assessing and Managing Social and Environmental Risk in Project Finance*. www.equator-principles.com. Accessed September 2009.

Gadsby, J., and Hodge, A. 2003. *Managing Environmental and Social Concerns for Mining*. Notes to accompany short course. Royal Roads University, Victoria, British Columbia, June, 2003.

GRI (Global Reporting Initiative). 2006. *G3 Guidelines*. Amsterdam: GRI. www.globalreporting.org/Reporting Framework/G3Guidelines/. Accessed October 2009.

ICMM (International Council on Mining and Metals). 2003. *Principles for Performance in Sustainable Development*. London: ICMM. www.icmm.com/our-work/sustainable-development-framework. Accessed October 2009.

ICMM (International Council on Mining and Metals). 2005. *Reporting Against the ICMM Sustainable Development Principles: Resource Guide to Assist ICMM Members Meet Their Reporting Commitments*. London: ICMM.

IFC (International Finance Corporation). 2006. Performance Standards on Social and Environmental Sustainability. www.ifc.org/ifcext/enviro.nsf/AttachmentsByTitle/pol_PerformanceStandards2006_PSIntro_HTML/$FILE/PS_Intro.pdf. Accessed September 2009.

ILO (International Labour Organization). 1989. Convention Concerning Indigenous and Tribal Peoples in Independent Countries. www.ilo.org/ilolex/cgi-lex/convde.pl?C169. Accessed September 2009.

Joyce, S., and Thomson, I. 2000. Earning a social license to operate: Social acceptability and resource development in Latin America. *Can. Inst. Min. Metall. Bull.* 93:49–53.

Knoke, D. 1985. The political economies of associations. In *Research in Political Sociology*. Edited by R.G. Braungart and M.M. Braungart. Greenwich, CT: JAI Press. p. 222.

Leipziger, D. 2003. *The Corporate Responsibility Code Book*. Sheffield, UK: Greenleaf.

MMSD (Mining, Minerals and Sustainable Development). 2002a. *Breaking New Ground: Mining, Minerals, and Sustainable Development—The Report of the MMSD Project*. London: Earthscan.

MMSD (Mining, Minerals and Sustainable Development). 2002b. *Seven Questions for Sustainability: How to Assess the Contribution of Mining and Minerals*. Winnipeg, MB: International Institute for Sustainable Development.

Nahapiet, J., and Ghoshal, S. 1998. Social capital, intellectual capital, and the organizational advantage. *Acad. Manage. Rev.* 23(2):242–266.

PDAC (Prospectors and Developers Association of Canada). 2009a. *Principles and Guidance for Responsible Exploration*. Toronto, ON: PDAC.

PDAC (Prospectors and Developers Association of Canada). 2009b. e3Plus: A Framework for Responsible Exploration. www.pdac.ca/e3plus/index.html. Accessed September 2009.

Prager, S. 1997. Changing North American mindset about mining. *Eng. Min. J.* 88(2):36–42.

Shepard, R.B. 2008. Gaining a social license to mine. *Mining.com* (April-June): 20–23.

Social Investment Forum. 2009. Socially responsible investing facts. www.socialinvest.org/resources/sriguide/srifacts.cfm. Accessed September 2009.

Suchman, M.C. 1995. Managing legitimacy: Strategic and institutional approaches. *Acad. Manage. Rev.* 20(3):571–610.

Survival International. 2008. India deals "devastating blow" to Orissa tribe. OneWorld.net. http://us.oneworld.net/article/supreme-court-deals-devastating-blow-indias-tribes. Accessed August 2008.

Svendsen, A.C., Boutilier, R.G., and Wheeler, D. 2003. *Stakeholder Relationships, Social Capital and Business Value Creation*. Toronto, ON: Canadian Institute of Chartered Accountants.

Thomson, I., and Joyce, S. 2006. Changing mineral industry approaches to sustainability. In *Wealth Creation in the Minerals Industry: Integrating Science, Business and Education*. Special Publication No. 12. Edited by M.D. Doggett and J.R. Parry. Littleton, CO: Society of Economic Geologists.

Thomson, I., and Joyce, S. 2008. The social license to operate: What it is and why it seems so hard to obtain. In *Proceedings of the 2008 Prospectors and Developers Association of Canada Convention*. Toronto, ON: Prospectors and Developers Association of Canada.

UNGA (United Nations General Assembly). 1948. United Nations Universal Declaration of Human Rights. General Assembly resolution 217 A (III). www.ohchr.org/EN/UDHR/pages/Introduction.aspx. Accessed September 2009.

UNGA (United Nations General Assembly). 2007. United Nations Declaration on the Rights of Indigenous Peoples. A/Res/61.295. www.un.org/esa/socdev/unpfii/en/declaration.html. Accessed September 2009.

Williams, M. 2001. In whom we trust: Group membership as an affective context for trust development. *Acad. Manage. Rev.* 26(3):377–396.

Wilson, C., and Wilson, P. 2006. *Make Poverty Business: Increase Profits and Reduce Risk by Engaging with the Poor*. Sheffield, UK: Greenleaf.

Woolcock, M. 1998. Social capital and economic development: Toward a theoretical synthesis and policy framework. *Theory Soc.* 27(2):151–208.

Cultural Considerations for Mining and Indigenous Communities

Ginger Gibson, Alistair MacDonald, and Ciaran O'Faircheallaigh

THE GROWING IMPORTANCE OF INDIGENOUS CULTURAL CONSIDERATIONS

Article 1 of the International Labour Organization's (ILO's) Convention 169 defines indigenous peoples as

(a) Tribal peoples in independent countries whose social, cultural and economic conditions distinguish them from other sections of the national community, and whose status is regulated wholly or partially by their own customs or traditions or by special laws or regulations; (b) peoples in independent countries who are regarded as indigenous on account of their descent from the populations which inhabited the country, or a geographical region to which the country belongs, at the time of conquest or colonisation or the establishment of present state boundaries and who, irrespective of their legal status, retain some or all of their own social, economic, cultural and political institutions (ILO 1989).

A cornerstone definition for the United Nations has been the "Cobo definition" of indigeneity, which highlights sacred attachment to and traditional ownership of land that has been appropriated by outsiders. This stands in contrast to people who self-identify as indigenous, which is the basis of the ILO definition.

The mining industry has not historically been required to closely analyze its potential impacts on indigenous cultures. This is no longer the case. The industry cannot afford the risks associated with the damage its activities can cause to indigenous culture. Nor can mining companies, in a highly competitive environment, ignore the potential of indigenous culture to add value to the enterprise.

More and more exploration and mining projects are located near indigenous communities whose culture is already under threat from dominant settler societies. The culture of indigenous communities, traditionally reliant on the land and its renewable resources, can be further threatened as they become exposed to and engaged with the mining industry (IFC 2007). Mining often becomes the new field of contestation between settler and indigenous cultures, as well as creating new relationships between the indigenous group and the distinct corporate culture of the modern mining firm.

Indigenous groups are not without levers to both shape these relationships and to impose sanctions on mining companies that refuse to acknowledge and respect their cultures. Many indigenous communities have gained enough power in recent decades to vigorously assert their rights to land, and have had these rights recognized through national law and international declarations and conventions such as the ILO's *Convention 169 Concerning Indigenous and Tribal Peoples* (1989) and the *United Nations Declaration on the Rights of Indigenous Peoples* (2007). This recognition has partly been driven by successful litigation and direct political action by indigenous peoples, often with the support of civil society groups. As part of this recognition of indigenous rights, cultural impact assessment of individual projects is now required in many regions. The Mackenzie Valley Resource Management Act in Canada's Northwest Territories, for example, requires environmental impact assessments (EIAs) to have regard to "the protection of the social, cultural and economic well-being of residents and communities of the Mackenzie Valley" (Government of Canada 1998).

Many of these indigenous cultures are resilient, vibrant, and resurgent. For example, in recent years, thousands of Aboriginal people from throughout Australia's Pilbara iron ore region have participated in annual ceremonial activities. Innu and Dene people of Canada are recapturing their cultural heritage through bushwalks and mass canoe trips (Zoe 2007). Indigenous communities have united to protect special cultural places and spaces. Another indicator of cultural resilience is that indigenous groups have increasingly challenged companies to make space for traditional cultural activities and values to be expressed at mine sites, to assist in protecting sensitive valued cultural components, and to adapt the corporate work culture of mining to the needs of indigenous people

---

Ginger Gibson, Adjunct Professor, University of British Columbia, Vancouver, British Columbia, Canada
Alistair MacDonald, Environmental Assessment Specialist, SENES Consultants Limited, Edmonton, Alberta, Canada
Ciaran O'Faircheallaigh, Professor, Politics and Public Policy, Griffith University, Brisbane, Queensland, Australia

(e.g., through alternative work rotations or respect for communal grieving) (Argyle Diamonds Ltd., Traditional Owners, and Kimberley Land Council Aboriginal Corporation 2004).

When cultural issues are not given appropriate consideration, the sanctions that indigenous groups impose on companies have become increasingly onerous. Companies who cannot show that they will not cause irrevocable damage may be denied the permits and licenses required to establish new projects. They may also be unable to secure mine financing if potential lenders perceive they are unable to manage cultural and social risks. Here are a few examples of downside risks resulting from inadequate community engagement or the presence of significant cultural impacts:

- Innu and Inuit landowners in Labrador, Canada, delayed the Voisey's Bay nickel project through court cases and activism because of the company's failure to prepare an environmental and cultural protection plan before starting exploration on indigenous lands (R. Gibson 2006).
- At Esquel in Argentina and Tambogrande in Peru (MacDonald and Gibson 2006), there was a fundamental lack of early and effective community engagement, and, as a result, insufficient effort to deal with indigenous concerns about adverse environmental, social, and cultural impacts. This led to a refusal by indigenous groups to grant companies a "social license to operate" (Joyce and Thompson 2000).
- Canada's federal government ultimately rejected a proposal for a modest drilling program in the remote Upper Thelon River basin when it became clear to the co-management board in charge of environmental assessment that the area held great cultural significance to the indigenous residents (MVEIRB 2007a).

(Note: The bulk of examples in this text are taken from the authors' experiences in Australasia and North America, but the tone is one focusing on expressing general concepts that emerge between indigenous communities and mining projects around the globe, and many of the general concepts, tips, and tools that can be applied in a variety of cultural contexts. However, there is no substitute for learning the site-specific cultural context and engaging culture holders in the cultural impact assessment process.)

Examples like these have done little to alleviate industry fears that dealing with indigenous cultures will consume valuable time, create new project costs, and open a Pandora's box of new risks for mining projects. Geologists, mining engineers, and other project managers have historically lacked training in cross-cultural engagement. This lack of experience can heighten a perception that cultural engagement is the first step in either a total rejection of their proposed project, a drastic reduction in its operating scope, or the imposition of restrictions that cost a great deal of money and reduce the life of any potential project.

Such a perception is both inaccurate and damaging to mining companies and their shareholders. In reality, the majority of potential cultural impacts are manageable at modest cost if they are clearly identified and appropriately addressed, and particularly if this occurs early in project life. Further, cross-cultural engagement can generate substantial value for the enterprise. For example, Canadian diamond mines have worked with indigenous groups to devise a 2-week-in/2-week-out work rotation schedule that maximizes the amount of time hunters can be out on the land. Any rotation greater than 2 weeks was considered too long, given that a herd of caribou can move through an area in weeks and be too far from communities to harvest by the time workers returned. In this case, the cultural practices and communal desires of the potentially affected culture groups were integrated into project planning. The result has been a win–win situation for the companies and the communities, with a more-committed work force for the companies, and an indigenous work force able to balance traditional and wage economies (G. Gibson 2008).

In areas largely populated by indigenous people, cross-culturally sensitive workplaces can add value by increasing indigenous employee retention rates. Unlike many nonindigenous mining employees who migrate or "commute" to remote mine sites, local indigenous people have a long-term commitment to live in the regions where mines are located. If they can be recruited and retained by mining companies, this can significantly reduce work-force turnover, which in turn can reduce company expenditure on staff recruitment, training, relocation, and severance, and enhance work-force cohesion and morale. In addition, resources committed to upgrading the skills of indigenous workers can generate higher returns for the corporation, as their ties to their "home place" make them less likely to move to other operations elsewhere, taking their new skills with them. Use of local labor can also reduce the costs involved in transporting workers into remote sites and providing them with camp accommodations. Companies may incur additional up-front costs in recruiting and training indigenous workers but, especially if a longer time frame is adopted in assessing returns, this investment can generate significant net gains (Barker and Brereton 2005; Barker 2008; Harvey and Gawler 2003; NWT and Nunavut Chamber of Mines 2005; SIWGMI 1996; G. Gibson 2008). Cross-culturally sensitive work environments can also minimize downtime and costs associated with turnover among indigenous and nonindigenous employees, reducing management–employee conflict and minimizing costly disruptions to the work environment.

Respectful cross-cultural engagement is essential if companies are to gain access to traditional indigenous knowledge, an outcome that can generate considerable value for the firm. Traditional knowledge is sometimes envisioned by developers as an input useful only during initial project planning, and as a legal requirement of the formal EIA process. Experience shows that these resilient cultures and their knowledge constitute a valuable resource that companies can draw from throughout the life of the mine. Cultural knowledge can be a valuable environmental planning and early warning tool for mining companies. Indigenous peoples are often in a unique position to assist mining companies in minimizing negative environmental impacts and in returning mine sites to acceptable environmental standards (O'Faircheallaigh 2010). One critical resource possessed by indigenous people is time depth in relation to information on the existing environment ("baseline data"). Typically, mining companies only start collecting baseline data some 2 to 3 years before project construction is planned to commence. Because environmental conditions can be highly variable over time, a few years' data may not provide a sufficient basis on which to develop an understanding of environmental dynamics. Indigenous peoples draw not only on decades of experience in observing environmental conditions on their traditional lands, an understanding of which is vital to their subsistence activities, but also on generations of knowledge handed down to them. Thus, their participation is essential in achieving a full and accurate understanding of existing environmental dynamics, which in turn is the foundation for effective environmental protection (Nadasdy 2003; Usher 2000).

The intimate and ongoing contact that many indigenous people have with their ancestral lands and waters also allows them to quickly detect ecological changes that may signal problems with a mine's environmental management system, helping to avoid potential damage and quickly take remedial actions. A range of more specific indigenous knowledge can also greatly assist effective environmental management. This can, for example, include understanding of animal behavior as essential in designing effective wildlife monitoring and management regimes.

The knowledge and skill of indigenous people can also be mobilized to make major technological shifts that increase productivity. As an example, using elders' knowledge of sea ice re-freeze conditions and time requirements to build a floating bridge, Inuit workers decreased wait times for shipping of ore from the Voisey's Bay mine in Labrador (I. Paine, personal communication).

In all of these cases, the requisite traditional knowledge may only be readily accessible if a respectful and trusting relationship, including strong ground rules for the protection of cultural knowledge, is established between the developer and knowledge holders.

Positive cross-cultural engagement can also allow companies to

- Demonstrate their corporate social responsibility, a major benefit in an industry with an often problematic public relations image;
- Develop the effective communication and understanding of the indigenous party (and vice versa) essential for negotiating, interpreting, and implementing agreements between indigenous communities and mining companies; and
- Obtain certainty of land tenure and permits in a timely manner.

Against this background, mining companies, small and large, have to learn how to deal with indigenous expectations that impacts on culture will be managed, and with the increased leverage exercised by indigenous people. Avoidance of cultural impact identification will only delay risks until a time in the future when greater financial losses will accrue because increased project investment has occurred. Industry associations are certainly taking heed that cultural issues merit closer consideration to both take advantage of upside benefits and avoid downside risks. As an example, the International Council on Mining and Metals has published ten principles of sustainable development, including the requirement that operations should "uphold fundamental human rights and respect cultures, customs and values in dealings with employees and others who are affected by our activities" (ICMM 2003). Also, the Prospectors and Developers Association of Canada has developed e3 Plus: A Framework for Responsible Exploration that includes expectations of "respecting and protecting local culture and traditions" (PDAC 2009).

Cultural impact assessment and the expectation that companies will achieve effective cross-cultural engagement with indigenous people are thus both "on the table." Despite the importance of cultural factors—their growing legal recognition, the demands of indigenous peoples around the world that they be considered before developments proceed, and the business case for positive engagement—culture quite often remains the "elephant in the living room" of mining project development. Everyone knows it is there and it is awfully hard to ignore, but dealing with it effectively is another matter. While some companies are making serious efforts to respectfully engage with indigenous cultures, common barriers to effective cross-cultural engagement and doing good cultural impact assessment still include

- Narrow interpretations of "culture," usually to the detriment of "intangible" dimensions of culture;
- Weak initial cross-cultural engagement often associated with lack of preparation and lack of in-house expertise;
- Power differentials and incompatible cultural values between settler society, indigenous groups, and mining companies;
- Emphasis on the technical/scientific/quantitative over the oral/traditional/qualitative in formal EIAs;
- The complex variety of contributing factors to cultural change;
- Lack of agreement on what constitute appropriate indicators of cultural impact;
- Lack of preexisting thresholds of acceptable change for cultural indicators;
- Companies not wanting to take on pseudo-governmental sociocultural protection roles; and
- Lack of formal legal requirements for implementation of cultural protections.

These are not insurmountable challenges; they can be overcome by changing attitudes with additional guidance and more effective approaches to cultural engagement and impact assessment, both at the corporate head-office level and by company managers "on the ground."

The purpose of this chapter is to identify some of the key issues that must be addressed, and principles and methods that can be used, during engagement between indigenous communities and mining companies. It seeks to help ensure that the current gaps in cultural impact assessment and community engagement practice can be overcome, relationships are respectful and appropriate, the benefits of cross-cultural reciprocity can be fully realized, and adverse impacts on indigenous culture can be avoided throughout the life cycle of an operation and at the various sites of cross-cultural interaction.

The sections that follow define culture, discussing intangible and tangible cultural assets; scoping, which requires engaging culture holders and understanding the potential impacts a project may have on culture; the collection and analysis of appropriate data and prediction of likely impacts and their significance; and the identification and implementation of appropriate mitigation, the means by which potential adverse impacts can be avoided or minimized and, equally importantly, benefits to (and of) culture can be maximized.

CONCEPTS OF CULTURE

Before exploring how mining development can impact indigenous culture and vice versa, it is necessary to know what culture is, who holds it and where, and how the values and priorities of indigenous cultures and mining's corporate culture tend to differ.

Tangible and Intangible Cultural Assets

The study of indigenous cultures as they relate to mining has often focused on the protection of historic artifacts that may be

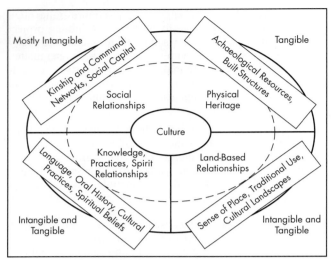

Source: Adapted from Candler 2008.

Figure 17.3-1 Tangible and intangible cultural assets

physically damaged by development. This "stones and bones" approach to cultural impact assessment ("heritage resources impact assessment") is practiced largely by archaeologists and other physical anthropologists, and has helped identify and protect the material manifestations of past occupation—features such as historic sites, gravesites, middens (refuse caches), rock and cave paintings, and scatterings of stone tools as well as dwellings or ceremonial sites and trails (King 2002; O'Faircheallaigh 2008). Each of these features leaves a physical clue as to how the earlier generations lived and traveled. They focus attention on the *tangible* signs of a culture's past.

Culture is, of course, much more than stones and bones; it is a living, continually adaptive system, not a remnant of the past. It is also highly complex, which makes precise or exhaustive definition of the concept impossible. A simple, general definition of culture is "a way of life; a system of knowledge, values, belief, and behavior, passed down between generations." As early as 1871, Sir Edward Tylor defined culture as "that complex whole which includes knowledge, belief, art, morals, law, custom, and any other capabilities and habits acquired by man as a member of society" (Tylor 1920).

Although it may be difficult to define it precisely, it is possible to identify some key elements of culture. Aside from the physical manifestations most often analyzed and protected during a heritage resources impact assessment, culture also includes values and worldviews, and the tools through which they are communicated—symbols, language, and behavior. It includes communally held knowledge, spiritual practices and beliefs, traditions and traditional pursuits, social relationships, and a common understanding of humankind's connection to the natural environment. These are among the *intangible* assets of a culture group. Manifestations of intangible cultural heritage include oral traditions, language, performing arts, ritual and festive events, knowledge concerning nature and the universe, traditional craftsmanship skills, traditional social practices, spirituality, and sense of place (UNESCO 2003; Andrews and Buggey 2008).

Common elements of culture, and particularly of indigenous cultures, are identified in Figure 17.3-1.

Common languages, stories, myths, songs, conversation, traditional activities, and art are used to communicate and transmit culture from generation to generation. Often in indigenous communities, the stories, knowledge, and practices that are critical to living well are transmitted as people are out on the land, engaged in the traditional economy as they pick berries and plants, hunt, trap, and fish. In these primarily oral societies, historical and mythological events are often recorded in landscape features. Literally for them, "the land is like a book" (Dene Elder Harry Simpson, quoted in G. Gibson 2008). As young people travel on the land to significant places, they engage with their elders, leaders, and families, strengthening their bonds and knowledge, while they come to know their history through storytelling at the significant sites of their people. The shared stories ensure that the younger generation acquires appropriate cultural information about traditions, way of life, and values.

The key thing to remember for those involved in cross-cultural engagement is that impacts on culture can come not only from changes to "things you can touch" or from physical effects on tangible and visible heritage resources, but they can also come from changes to "things that touch you"—intangible cultural assets that while perhaps not visible from without, are felt from within and are vital to cultural retention and maintenance.

Domains of Culture: Who Holds It and Where?

It is essential to understand the "domains" of culture—the places, spaces, and other planes of interaction where culture is transmitted, regenerated, and altered. Culture is acted out, transmitted, and received between the self, the family, and social networks; within and between communities; between different cultural groups; and within the context of an increasingly global popular culture. In many indigenous cultures there is an ongoing struggle between the norms and values realized through the practice of traditional ways such as subsistence harvesting and the values of a dominant settler society.

Culture has a spatial component as well. All cultures have a "home place" and members have a shared sense of place. For the mining company's corporate culture, the home place may be the corporate head office or the mine site. Indigenous cultures have ties to the physical habitations they reside in, just like settler communities do, but indigenous cultures may hold a much stronger, even symbiotic, relationship with their wider traditional lands.

For indigenous people the land itself is often a key repository of knowledge, an interpretive tool, and a key locator of meaning and history. There is growing international recognition (MacKay 2004; UNDESA 2004) that the ability to live on, care for, and use resources from their ancestral lands is central not only to the economic and social well-being of indigenous people but also to their very survival. Land is critical

- To physical sustenance;
- For social relationships that are bound up with relations to land;
- For law and culture, which are interwoven with use of the land and its resources; and
- For spirituality and religion, which have as their basis beliefs about the creation of the land, the way in which creation spirits continue to occupy the land and influence contemporary life, and the way in which ancestors and future generations are tied to the current generation through the land.

Indigenous knowledge of the land is tied to knowledge of past events related to people first by their ancestors and then passed by them to their own descendants. Quite literally for indigenous people, "wisdom sits in *places*" (Basso 1996, emphasis added).

Wisdom resides in spaces too. Places are sites. Spaces are routes, relational lines between points on the map. They are larger than a single place, and they can also represent a larger meaning as *cultural landscapes*. Cultural landscapes have been defined as "landscapes that are lived in" and that bring attention "to the way people within the landscape live, their traditions and everyday life" (NWT Cultural Places Program 2007). Even where there are no visible signs of human usage, there may be historic events, culture hero stories, or significant usages of an area.

> These kinds of activities can give cultural meaning to natural features in the landscape, such as hills, rocks, caves, water, trees, animals and plants. Natural features remind people of stories about events in their history and of how important animals, plants and spirits have always been for the people of their community (NWT Cultural Places Program 2007).

In a cultural values assessment for the proposed Sahyoue-Edacho National Historic Site in Canada's Northwest Territories, the importance of this cultural landscape was described as follows:

> These areas represent the relationship between the Sahtugotine and their land; the landscapes of Sahyoue and Edacho blend the natural and spiritual worlds of the Sahtugotine, creating a living web of myth and memory…[they] remain important places not only for subsistence, but also for spiritual renewal and a sense of cultural identity (Janes and Hanks 2003).

Cultural landscapes are the embodiment of cultural heritage at a grand scale, incorporating individual sites of heritage resources into a broader network of stories and collective memory tied to the land, essential to defining the people similarly tied to that land (Rose 1996). Buggey (2004) argues that a cultural landscape expresses an indigenous group's "unity with the natural and spiritual environment. It embodies their traditional knowledge of spirits, places, land uses, and ecology. Material remains of the association may be prominent, but will often be minimal or absent."

It is this lack of tangible presence that can be a problem for the outsider, both in coming to understand the values and priorities of an indigenous group and in determining the significance of impacts on cultural assets during impact assessment. The value of different cultural assets cannot be discerned from without; it must be determined from within, at least with the assistance of the indigenous peoples themselves.

In general, three types of "values" are associated with cultural assets:

1. *Use value* is associated with continuing utilization of an area. For example, an area may be important because it is used currently by culture holders on a day-to-day, seasonal, or annual basis for practice of the traditional economy.
2. *Bequest value* is associated with any cultural assets the holder desires to pass on unchanged to future generations and the satisfaction and peace of mind associated with this knowledge.
3. *Existence value* (sometimes known as wilderness value) is the benefit attributable to knowledge that an environmental resource exists, even in areas that are not used and likely will never be used by the individual. These values can be important in situations where indigenous groups are unable to access parts of their traditional lands for extended periods of time because, for instance, they are distant from settlements.

Often places carry multiple layers of meaning, deepening their value to culture holders. Cultural landscapes such as the Upper Thelon River basin in the Northwest Territories may be all the more important and sensitive because they hold use, bequest, and existence/wilderness values. For the Denesoline people, the area is "the place where God began." It is critical habitat for wildlife including the vitally important barren ground caribou herds. It is viewed as a place with which the Denesoline desire continued connection in its current, relatively pristine status to sustain their cultural history, traditional activities, and spiritual base into future generations (Lutsel K'e Dene First Nation 2008). The Upper Thelon River basin is also a highly valued conservation and ecotourism area (one of the largest forested areas north of the tree line), valued for its existence/wilderness value both among Canadians and internationally. This serves as a reminder that while impacts from mining may be primarily felt by those closest to the development (indigenous peoples), there may be impacts on other cultures associated with special places.

The existence and bequest values associated with special places and spaces belie using quantitative indicators like the existence of known physical heritage resources or intensity of current use for traditional activities as the only measuring sticks. Some places take much of their value from the peace of mind people enjoy from knowing they are not being disturbed. Other places are important not (or not only) because they house valuable renewable resources, but because they hold power and resources that should be avoided except in times of great need or in very specific circumstances, as, for instance, in a particular ceremonial context (NWT Cultural Places Program 2007). Some places may require avoidance by all people at all times, or should be visited only by particular people who hold specific knowledge, because they can be dangerous to the ignorant or the uninitiated. In many parts of Australia, specific places are variously considered dangerous to men or to women, or to any individual who has not completed specific initiation ceremonies, or to anyone who is not accompanied by custodians who can perform rituals to render the visitors safe. Aboriginal attempts to restrict access to such places are designed not only to protect them from damage, but also to protect non-Aboriginal people who would otherwise suffer serious injury by visiting them inappropriately (O'Faircheallaigh 2008).

Indigenous culture holders take their guardianship of spiritual and cultural places and spaces very seriously, and there may be severe repercussions for their failure to protect them. For instance, the destruction of an important women's cultural site by the Argyle diamond mine in Western Australia in the mid-1990s had serious social consequences for its female custodians, although they fought strenuously to stop the development (Dixon and Dillon 1990).

Culture is not only practiced on the land, it is practiced in the home and in the community. Cultural change is not limited to alterations on the land, but also occurs when social relations change, or when practice of language and other tools of cultural maintenance or one's sense of self, place, and community are altered (Basso 1996; G. Gibson 2008).

Understanding Cultural Differences
The work sites of operating mines, although they are often envisioned as "culture free" areas (Trigger 2005), are themselves one of many cross-cultural "mixing zones" associated with mining in indigenous communities (see Figure 17.3-2). Cultural differences are also felt during initial community engagement, around the negotiating table when agreements are being developed, and during the mine planning and formal cultural impact assessment processes.

Indigenous culture interacts with the institutional culture that prevails in mining camps and companies (Heyman 2004). The mining company can be distinguished by its different sites of operation: the operating mine site is quite distinct from the parent company in managerial style and strategic priorities. A parent company may watch over many operating mines and shepherd old mines out of existence. The management and corporate culture of the individual mine is concerned with the ore bodies on site, the understanding of which changes frequently and often expands as the ore body is newly delineated and made real through development and extraction.

The operating mine is characterized by key values, hierarchical relationships, strict lines of authority and communication, and prescribed policies for managerial and human resource systems. Key values of an operating mine are maximizing productivity and site safety (G. Gibson 2008). A head office's primary and overall responsibility is to maximize economic returns to company shareholders. The means to these culturally prioritized ends may conflict with the primary driver of indigenous culture: to protect the values imbued in the land under which the mineral resource lies.

"People" priorities may differ as well. Cross-cultural leadership qualities, social science skills, and human relations acumen are not necessarily valued prerequisites in mine management, as long as the individual has a track record of productivity. This can and has led to conflicts between indigenous workers and their settler culture managers (G. Gibson 2007, 2008). Indigenous workers are expected to assimilate into the occupational site, discarding, at least temporarily, their own practices and values in favor of those of, primarily, the institutional culture, and secondly, the settler society. As one senior manager put it, workers have to "take off their kimonos" and participate in the "industrial bubble" (Trigger 2005). Participation in the mine site can powerfully affect how individuals are able to reproduce and express their own values and practices (G. Gibson 2008). Exposure to a new workplace in an isolated fly-in/fly-out work camp where the dominant culture is defined by the employer not only exposes the indigenous person to new values and world views of people from settler cultures, but also to the institutional culture of the modern mining workplace. Common differences between the corporate culture of the mining work site and indigenous groups are outlined in Figure 17.3-3.

The difficulties for indigenous workers in adapting back and forth between the industrial production culture and the home community can create major challenges

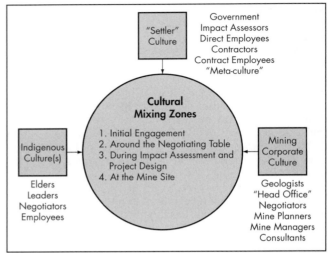

Figure 17.3-2 Cultural "mixing zones" associated with mining activities

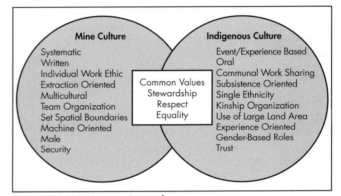

Figure 17.3-3 Comparing values

(O'Faircheallaigh 1995; Barker and Brereton 2005). Attempts to divorce the "workplace" culture from the "home place" culture lead to problems such as cross-cultural disputes and high turnover rates, often associated with increased familial tensions as well as workplace conflict. G. Gibson (2007, 2008) identified the top 10 cross-cultural conflicts encountered between indigenous people and mine managers at the diamond mines in northern Canada, a few of which are discussed here:

- In an area such as the open pit, it takes constant shifting, training, and machine jumping to gain the skills necessary to understand the full range of machines in the area. People who move up in the mines have been afforded these opportunities by their pit managers, given the time to train, and often have a few characteristics the manager sees as necessary. They may be aggressive and self-promoting. These characteristics and values may run contrary to the nature of an indigenous person.
- When a non-kin member of a family dies, corporate-leave policy may constrain managers from allowing a worker to attend the funerals of other than immediate relatives. In a society where kinship is defined by different rules, an indigenous person may need to attend funerals as part of a communal grieving custom.

Table 17.3-1 Elements of cultural difference that may be exposed during a cultural impact assessment

Factor	Mining Culture	Indigenous Culture
"On-the-ground" evidence of culture in area proposed for mining	Found archaeological sites	Stories, trails, use patterns (current and historic), density of place names, archaeological sites and potential for unfound sites, cultural landscapes
Prioritized cultural impact consideration	"Bones and stones"	Tools/images/objects/stories in combined place memory; innate values of the land
Value focus in cultural impact assessment	Current use value and bequest value associated with physical heritage resources	Historic and current use; bequest value and sometimes intrinsic existence values
Geographic scope of cultural impact assessment	Places and sites that will be physically altered or otherwise directly affected by the proposed development	Places, sites, spaces, cultural landscapes and relationships between them and people, home community, lived experience
What constitutes "heritage resources"?	Heritage resources as physical artifacts only	Heritage resources also include culture holders' shared history and experiences, knowledge of the land and resources, and ways of knowing and transmitting knowledge
Role of intangible cultural resources in cultural impact assessment	Minimal; view that vulnerability of things like language and time spent on the land is linked to cultural change beyond scope of project-specific cultural assessment	Project-specific cultural impact assessment should consider contribution of proposed development to potential erosion of cultural maintenance tools in context of cumulative change

Cultural differences must first be identified before there can be any successful coexistence and cross-cultural engagement of the many cultures on site. Only then can human resources policy be adapted to fit these unique cultural needs (work-site cross-cultural conflict mitigation strategies are discussed later in this chapter).

Obviously, the work site is not the only place where mining companies and indigenous communities interact. Common sources of tension between corporate/settler culture and indigenous cultures in other cultural mixing zones include

- Prioritization by companies of the written word and technical expertise over oral discussions in plain language, often characterized by company use of overly wordy and complex documents, presentations, and other materials about the development;
- Different canons of proof, with Western, "scientific" knowledge being privileged, and characterization of traditional knowledge and oral history, often provided by elders, as "hearsay" or "anecdotal" information without scientific support;
- Access to and control over information, including sensitive traditional knowledge;
- Misinterpretation of cultural signals (e.g., silence in settler society may indicate agreement, while for an indigenous society it may mean nonsupport);
- Conflict between settler laws and traditional indigenous laws, with corporations seeking to rely exclusively on the former;
- Lack of incorporation of ceremonies required prior to, during, and after meetings, breaking the land and agreements; and
- Different understandings of formal arrangements like impact benefit agreements as either final "set in stone" legal settlements (from the mining companies' perspective) or as the starting point for an ongoing, open, and adaptable relationship (from the viewpoint of the indigenous people) (G. Gibson 2008).

Many of these cultural differences are exposed during early community engagement and subsequent cultural impact assessment. Cultural differences not only govern how people interact but also how they perceive the environment and potential for impacts from mining projects. Table 17.3-1 provides a comparison of values in relation to key factors involved in a cultural impact assessment.

Some understanding of the cultural attributes of each other is essential prior to entering a formal relationship. Gaining a contextual knowledge of the community is essential so that relationships are based on mutual understandings of priorities and reasons behind actions. Without this in-depth knowledge, misinterpretation and a lack of trust can emerge. For example, concerns are sometimes raised about whether indigenous culture holders invent, fictionalize, or otherwise augment their histories to make potential cultural impacts appear more severe. This concern surfaces when an impact emerges in the last moment in an environmental assessment. This misperception is linked to the fact that the nature of information sought in order to understand culture is often highly sensitive. Depending on the cultural values of the potentially affected group, it may be hurtful or improper to custodians and to the wider indigenous community for cultural information in the form of traditional or local knowledge to be divulged, and so information may only be released at the very last moment when it becomes clear that to do so is the only way to protect a place. A conflict between the desire of formal impact assessment to quantify cultural assets and their value, and the desire for indigenous groups to retain control over the sensitive cultural knowledge and expression of value of their cultural resources often emerges. Sometimes cultural resource issues are simply too sensitive for public disclosure (King 2004; O'Faircheallaigh 2008). This is often the case with indigenous people who are concerned with the impact of mining activities on spiritual or religious sites. To divulge information about a site may actually put knowledge holders or the place itself, as well as recipients of the knowledge, at risk. This explains the reluctance to divulge it unless knowledge holders become convinced that protecting sites requires releasing information about them, even though revealing their location or significance may reduce their future safety.

EFFECTIVE EARLY COMMUNITY ENGAGEMENT

Early and appropriate community engagement can create a respectful relationship between parties and assist in proper identification of cultural issues that may arise from a project in a specific location, a process referred to as *scoping*. To scope properly, the developer needs to understand the unique cultural context in which the project is proposed and how mining activities have tended to impact indigenous cultures in this or other similar environments.

There is no broad-brush approach that can be used to understand indigenous peoples' lives. Each indigenous community will be unique, with issues, concerns, and an underlying context that will emerge only through engagement. At the outset of a project, developers need to develop a basic understanding of the cultural context in which they are operating and the unique vulnerabilities and resiliencies of the people involved. Developers need to be aware of

- The historic background of the community;
- The relationship between community members and the environment;
- Demographic characteristics of the community;
- Demographic subgroups within the community who are particularly vulnerable to adverse cultural impacts and are underrepresented;
- Internal (traditional laws and governance) and external (state) political and legal structures and customs;
- The community's relationship with regional, territorial, and federal governments;
- Existing community goals and aspirations for social and cultural well-being, as expressed, for instance, in development plans or proposals for protected areas;
- The cultural values that shape the perspectives of community members; and
- Vulnerabilities and strengths of the community, (e.g., strong language and other cultural programs) and trends in levels of subsistence activity.

Developing this contextual knowledge requires a substantial and early commitment of resources but has important benefits. It assists in accurate issue identification, which allows identification of the proper scope and scale for the required impact assessment. It also helps to identify locally and regionally relevant mechanisms and processes that are already in play to support and protect valued cultural components, such as the regulatory and legislative context for identifying and managing cultural impacts. Regulations on burial sites, cultural items, sacred sites, religious freedoms, and archaeological sites will also need to be understood (King 2004; Parker and King 1990).

Sources of contextual cultural information include

- Land-use plans and supporting documentation;
- Literature on protected areas in the vicinity, which often include some description of cultural values and specific spaces identified as requiring special protection;
- Documentation in relation to prior EIAs in the region in question;
- Published anthropological or sociological literature about the people and places involved;
- State and federal government departments responsible for indigenous populations' health and well-being, and cultural protection;
- Community-specific engagement protocols;
- Museums or cultural institutes dedicated to the study of the indigenous culture, which among other resources often house archaeological databases identifying physical heritage resource locations and areas of high heritage resource potential;

Table 17.3-2 Checklist of recommended activities prior to and during early community engagement

- ☐ Identify what land access agreements will be required in the area of the proposed development, and identify any overlapping land claims by different culture groups.
- ☐ Identify relevant community plans, regional land-use plans, and other planning documents. Identify whether the proposed development conforms to these plans.
- ☐ Use community engagement handbooks and talk to community liaison specialists before developing a consultation strategy. Consider whether a consultant or local expert from the cultural region is required.
- ☐ Identify which potentially affected communities and culture groups should be consulted initially.
- ☐ Identify whether each identified community, or the region as a whole, has specific policies or protocols that dictate how developers should conduct early community engagement, or for the collection and protection of traditional knowledge.
- ☐ Conduct desktop research on the social, economic, and cultural environment of communities that may be affected by the proposed development.
- ☐ Identify appropriate community contacts.
- ☐ Produce a preliminary in-house list of potential impacts and public concerns that may be raised (see the "Impacts of Mining on Indigenous Cultures" section later in this chapter). Use information about the proposed development, research on past developments in the potentially affected region, case studies of similar developments, and existing developments in the region as source material for this list.
- ☐ Distribute a plain-language description of the proposed development to potentially affected communities.
- ☐ Be familiar with cultural impact issues commonly brought up by indigenous community members, be ready to listen to whether and how these concerns relate to the proposed development, and be prepared to respectfully consider issues that may not seem to you to be directly relevant to the proposed development.
- ☐ Take responsibility for keeping detailed meeting minutes, be ready to provide them to the community consulted for verification, and establish a flexible and appropriate mechanism for follow-up communications with communities.

Source: Adapted from MVEIRB 2007b.

- Ethical guidelines for traditional knowledge collection and dissemination (e.g., MVEIRB 2005a); and
- Existing traditional knowledge reports, which may have certain confidentiality restrictions.

It is prudent to identify and use all available secondary (desktop) sources of information before committing to any dedicated primary (fieldwork) cultural impact assessment. Besides creating a useful contextual understanding to help avoid misunderstandings during initial engagement, this can also help avoid the "consultation fatigue" often cited by indigenous communities. Despite the obvious differences among indigenous communities, there are also remarkable consistencies in the issues they raise in relation to proposed developments. Therefore, it may be useful when determining the "cultural risk profile" of a development to consider case studies of similar developments in other indigenous environments. Table 17.3-2 identifies some of the activities a developer may want to undertake prior to formal engagement with a community.

Early community engagement can be essential in minimizing the type of misunderstandings that can damage, even irrevocably, a project in traditional indigenous territory. Important contextual information about the cultural group in question may in some cases only be unearthed during discussions with key contacts in the communities. The key here is to make sure that contact is "appropriate," meaning it is made through the proper channels according to relevant community engagement protocols. To bypass these protocols, which are set up by the culture holders to protect their interests, can get the engagement process off to a bad start.

Rio Tinto's (Rio's) Marandoo iron ore mine in the Pilbara region of Western Australia offers an example of the costs associated with failure to address cultural heritage protection early and in cooperation with indigenous custodians. The developer complied with the provisions of Western Australia's Aboriginal Heritage Act, but traditional owners believed this legislation offered little by way of effective protection and wished to negotiate stronger protective provisions. After Rio failed to establish a productive engagement with them during exploration for the project, the traditional owners undertook a number of legal actions designed to frustrate Rio's plans and slow development. While Rio eventually overcame these hurdles and secured project approvals, the action of traditional owners resulted in a delay of more than 2 years. This not only led to higher capital costs but meant that Marandoo came onstream too late to take advantage of an expected spike in iron ore prices, and Rio incurred significant losses as a result. Another result was, according to a Rio executive, "a low point in Aboriginal relationships for the company…[and] a legacy of deep distrust and bitterness in the [Aboriginal] communities" (Eggleston 2002). The experience contributed to a fundamental policy shift at Rio Tinto, with the company undertaking a major initiative to establish productive relationships with all indigenous communities located adjacent to its mine sites (Eggleston 2002; O'Faircheallaigh 2006).

IMPACTS OF MINING ON INDIGENOUS CULTURES

A second key task for managers is to understand how a project might impact culture. Many developers enter an area without fully understanding the likely impact of their activities on indigenous culture. This requires both a basic understanding of how mining projects have been known to impact indigenous culture, and specific knowledge of both the proposed development and the cultural context it is proposed to be situated in.

Common Impacts of Mining on Indigenous Cultures

At a workshop in Perth, Australia, in May 2008, EIA practitioners from around the world identified the cultural impacts that most concerned them in relation to mining. The results are indicated, along with the authors' own experiences, in Table 17.3-3 (updated from Gibson et al. 2008). The list is by no means is definitive.

Cultural impacts from mining activities occur to places and people. For example, the first layer of impact may be direct and tangible damage to heritage resource sites through construction, blasting, and earthmoving. As a result, the physical record of a culture group's past occupation may no longer be marked into the landscape (Angelbeck 2008). This destruction can in turn indirectly reduce the ability of people to practice rituals, ceremonies, and traditions, as well as add to disconnection from nature (Pilgrim et al. 2009). The space might also be a fundamental source for food, shelter, clothing,

Table 17.3-3 Common cultural impacts related to mining projects

Valued Cultural Component(s)	Impact Concern
Cultural heritage resources	• Physical loss of heritage resources and associated spiritual loss
Cultural landscapes and other special spiritual spaces/places	• Visual impacts redefining the way a place/landscape is "seen" in culture (inappropriate use or alteration of landscape can create loss of value and meaning regarding what places on the land can teach us about life)
Culture and land tenure	• Social organization may be tied closely to the system of land tenure, which may also reflect transmission of rights, property, and gender relationships. Changes to land tenure through mining or changes in property ownership may interfere with these relationships.
Overall relationship to land and traditional activities on the land (including practice of traditional economy)	• Sense of disconnection from traditional lands and wellspring of cultural well-being • Less time on the land conducting traditional practices can cause cascading impacts on well-being across a variety of categories (everything from poor diet to erosion of language and ways of learning) • Often linked to perceived and real physical degradation of habitat and lower harvesting success for key wildlife species • Less time in kinship relationships that are critical to transmission of knowledge
Values and belief systems	• "Culture free" working conditions can erode cultural values or lead to alienation • Large-scale in-migration can alter cultural norms and values • Spiritual/religious change, loss of faith in traditions
Methods of cultural transmission (including language, oral history, intergeneration relations, social networks)	• Contributions to changes in social structures leading to cultural loss (e.g., decline of intergenerational culture transmission) • Contribution to cumulative loss of indigenous language, even to extinction • Linked to changing political structures and power relations (e.g., elders vs. politicians or businesspeople as leaders)
Sense of self; sense of place; overall well-being	• Loss of sense of control over one's own fate • Health impact outcomes caused by cultural impacts (e.g., increased suicide, alcohol and drug addiction, poor diet)

medicine power, or a wellspring for much less tangible but equally important cultural resources (Angelbeck 2008). These changes can have cascading, potentially devastating impacts on a group's cultural history, traditional practices, and sense of place. These cascading impacts are more likely to occur if there are cumulative effects from other human activities contributing to pressures on indigenous cultural resources. Thus, destruction of the land can damage and even destroy cultures that remain closely tied to their environments, which in turn can lead to mental, psychological, and social pathologies (Pilgrim et al. 2009).

Impacts on culture, like all impacts, occur through these direct, indirect, and cumulative processes. Direct impacts are ones specifically caused by the development in question, such as the aforementioned mine construction requiring destruction or removal of physical heritage resources. Indirect impacts are changes caused or contributed to by a development through a chain of events. For example, increased employment of indigenous people at a large mine may reduce their reliance on, and interest in, spending time on the land conducting traditional pursuits. This can in some cases contribute to loss of connection to the land, reduction in intergenerational knowledge transfer, and other damage to maintenance of traditional culture. Alternatively, increased employment at a fly-in/fly-out mine may provide the added income and free time required to spend time on the land, creating an increase in practice of the traditional economy and beneficial cultural outcomes (O'Faircheallaigh 1995). The lesson here is twofold:

1. The term *impact* simply means "change." Impact outcomes are not inherently unidirectional. They can be beneficial or adverse, or either for different people, depending on the cultural context and the way people react to change.
2. Pathways by which impacts occur may be simple and involve single-chain cause–effect relationships (e.g., a bulldozer physically destroying an artifact), or they may involve complex causal chains with a variety of contributing factors along multiple pathways (e.g., less practice of traditional economic activity caused by a combination of increasing involvement in wage economy, fewer harvestable animals, changing demand for country food, and decreasing time on the land). The incorporation of culture holders in predicting both initial changes and their eventual outcomes may be required to create a realistic picture of the most likely impact outcomes in a specific cultural context.

Mine developments do not occur in cultural, economic, or social vacuums where they are the only cause of change. Cultural change is always occurring, and for multiple reasons. A common argument is that the most dominant forces of change in modern industrial society are beyond the capacity of individual developers to manage. This is true. Television, for example, brings into the homes of indigenous people the values and world views of dominant societies. Developers are not being asked to take full responsibility for the cumulative effects of these multiple contributors to cultural change. They are expected instead to recognize the pressures these cumulative effects may be placing on valued cultural components, as part of understanding the local operating environment to which the developer needs to adapt. The "weight of recent history" (Gibson et al. 2008) may need to be considered, as many indigenous communities have been bombarded with the negative effects associated with intrusions by settler cultures. Cumulative effects are often measured through trends in language retention rates, time spent on the land, and traditional harvesting levels. Equally important may be the more-qualitative measures of change that can only be captured by talking to the culture holders about their sense of place, their attachment to special sites and cultural landscapes, and their hopes and fears for the future.

It can be argued that impacts must be interpreted through the filter of the culture holders themselves. Impact significance is socially defined, and these social perceptions are linked to the values of the culture holder. Consider a situation where a defined cultural landscape of importance to an indigenous group is to be used for an open-pit mine. In absolute terms, the physical footprint of the mine and associated infrastructure may account for 1% of the cultural landscape. But that 1% may represent some of the most important hunting grounds and encompass an important spiritual site, and its importance is therefore much greater than its area would suggest. In addition, impacts on that 1% may affect the perceived value of a much wider area. Indeed, in some cases the destruction of one site can impact the values associated with the entire cultural landscape and have even further ramifications—the inability of traditional owners to protect a site may change relationships within the community. In addition, if certain sacred sites are destroyed, rituals and ceremonies may no longer be practiced, which will have far-reaching ramifications for cultural transmission.

The nature of cultural impacts may change over the mine life cycle. During initial exploration, concerns may focus on the cultural appropriateness of using a certain area for any sort of industrial development (MVEIRB 2007b). During mine development, increased activity on the ground may cause increased concern about immediate destruction or damage to specific physical heritage resources, spiritually important landscape features, and key wildlife species. There may also be increased concern about the effects on culture of increased in-migration, of increased indigenous involvement in the wage economy, and of associated changes to the traditional economy. During mining operations, concerns may focus on changing social network structures and familial tensions associated with long-distance commuting (Barker and Brereton 2005) or on workplace cross-cultural conflict (G. Gibson 2008).

Determining a Project's Potential for Cultural Impacts

Determining the potential cultural impacts of a particular mining development requires project- and culture group-specific analysis. Only by examining both the "site" (the project) and the "situation" (the cultural context within which it is embedded) can the developer answer these important questions:

- How much further work is required to fully understand potential cultural impacts?
- What should that work focus on?

The proper identification of impact issues allows a refining of the level of detail required and of the focus of cultural impact assessment work. For example, it makes no sense to conduct additional heritage resource assessments in areas where they have already been conducted or where all parties agree they are unlikely to be found. The exercise becomes especially fruitless if the real issue for the community is the potential for high levels of in-migration to exacerbate pressures on traditional culture. Therefore, although scoping for cultural impact assessment can be done in-house by the mining company, it is preferable to involve indigenous culture holders (e.g., through key contact interviews, focus groups, scoping sessions, or meeting with leaders or other groups) in the initial identification of potential impacts.

A range of warning signs is related both to the nature and scale of the proposed development and the cultural context. The following lists identify some attributes to consider when determining the likelihood that a project may cause or contribute to adverse cultural impacts. The first list identifies characteristics of a proposed project that are indicative of a higher impact potential on culture. The second focuses on aspects of the physical and human environmental context surrounding the proposed development. Again, these lists are far from exhaustive, and context-specific variables will emerge during dialogue between parties.

Site, or project, attributes associated with higher cultural impact potential may include situations such as the following:

- The physical and impact footprint of the project is large (e.g., in terms of the extent to which land, air, water, and wildlife may be adversely impacted).
- New or improved transportation access routes are involved.
- The land management status of an area would need to be changed.
- New, complex, or controversial technology is necessary.
- A high level of in-migration to a region is required.
- Long-distance commuting for all workers is required.
- The land use will lead to irrevocable visible changes (e.g., large-scale open-pit mining).
- It is a brand new development (known as a greenfield).

Situational, or cultural, contexts include the following:

- Indigenous communities/traditional lands/trails are nearby.
- Proposed or existing protected areas are in the vicinity.
- There is known or high potential for physical heritage resources or spiritual sites.
- More than one indigenous group is using the area.
- Wildlife intensity is high in the area, traditional pursuits are widely undertaken, and traditional harvesting is an important part of the economy.
- The area contains unique/rare landscape forms.
- The location is important to indigenous oral histories; that is, there is a high density of indigenous place names in the area.
- People practice agriculture in the region and have local forms of land tenure that is not recognized by the government.
- Other past, present, or reasonably foreseeable future developments contribute to cumulative effects in the area.
- Potentially affected communities are already dealing with the intensity of preexisting social, economic, or cultural problems.

The likelihood of major impacts and the level of effort required to manage them will grow as the number of attributes increases.

DIGGING DEEPER: DATA AND METHODS FOR DETERMINING CULTURAL IMPACTS

This section deals with data collection, impact identification, and predicting the significance of the determined impact. While each cultural impact assessment needs to be developed around the specific cultural context and expectations of the case at hand, there may be guidelines for social and cultural impact assessment from the jurisdiction where the work is being done or from prospective lending agencies (e.g., MVEIRB 2007b; MVEIRB forthcoming; IFC 2007).

In considering levels of effort for cultural impact assessment, a distinction is made between basic, moderate, and comprehensive. Consider, for example, a situation where the only potential cultural impacts identified during scoping are the physical disruption of heritage resources.

- Where there is a low risk of adverse impacts, a *basic* cultural impact assessment may require a summary of existing data on heritage resources and heritage resources potential, and inclusion of respected local knowledge holders to understand the impacts.
- Where there are no known heritage resources but medium-to-high potential for unfound heritage resources, a *moderate* impact assessment may require archaeological studies on the project site and development of a site-specific heritage resource protection plan(s).
- Where there are known heritage resources and high potential for unfound resources, a *comprehensive* impact assessment may require incorporation of traditional knowledge studies alongside local archaeological studies, a regional heritage resources cumulative effects assessment, and identification of a regional heritage resource protection plan, in concert with the responsible government authorities and indigenous groups.

For projects with potential impacts on less-tangible elements of culture, for example, whether they be effects on broader cultural landscapes, key wildlife species and traditional harvesting, or contributions to effects on language retention, a similar distinction from basic to comprehensive can be determined based on the results of scoping. In essence, a basic assessment looks to confirm that significant adverse impacts are unlikely, through secondary data collation and dialogue with potentially affected parties. A moderate assessment is used where there might be significant adverse impacts that require, at a minimum, adaptive management plans to be in place in case they do occur. A comprehensive assessment is based on the premise that adverse cultural impacts are likely, that there may be cumulative effects beyond the scope of the project that need to be factored in, and that aggressive mitigation and a dedicated monitoring system for project effects must be set up by the parties involved.

The discussion in this section assumes that the project in question has at least a moderate additional cultural impact assessment requirement.

Involving Cultural Resource Experts and the Culture Holders

Many approaches can be used to help understand impacts on indigenous cultures, including

- Traditional land-use studies (e.g., Tobias 2000);
- Traditional ecological knowledge studies;
- Physical anthropology or archaeology studies;
- Collection of oral histories from elders and other traditional knowledge holders;
- Linguistic and kinship studies;
- Place names research and other ethnogeographic studies;
- Cultural landscape studies;
- Land-use planning, including proposals for protected areas;
- Focus groups, interviews with key contacts, and town hall sessions geared toward identifying valued cultural components and concerns for them;
- Analysis of statistical trends in appropriate cultural indicators, usually collected by a bureau of statistics or similar body (e.g., land usage, language proficiency); and
- Community wellness surveys including indicators of cultural resilience and vulnerability.

Effective cross-cultural engagement and cultural impact assessments require input from trained practitioners, just as mining requires application of engineering and geological skills. The concept of a cultural impact assessment itself is often perceived as being "soft" or "fuzzy," a mistake compounded by the fact that this emboldens nonexperts to try to become practitioners without any training. As a result, the right professionals are often not brought on board to assist with community engagement and cultural impact assessment. Often, the engineer or geologist who seems best suited to consulting with communities—in other words, those with some empathy and communication skills—are given the added task of understanding culture, writing about it in an impact assessment, and finding mitigation strategies to avoid damaging it. The lack of engagement of anthropologists, linguists, or traditional knowledge specialists can lead to real problems down the road through misunderstandings and lack of rigor.

Technical experts alone are not enough. When it comes to documenting archaeological and other tangible impacts, there is no replacement for the expertise of archaeologists and cultural anthropologists, but their work should incorporate (or itself be used to augment) the local expertise of elders and traditional knowledge holders. Although archaeologists can determine the presence of ancestors via marks of their physical passing, they may not be aware of mythological events that have been worked into indigenous imaginations and cultural landscapes. When defining the cultural importance of a place or space, there is no substitute for the words of the culture holders themselves.

Despite this, companies often commission consultants to do desktop studies that include little or no engagement of culture holders themselves and rely entirely on archaeological evidence. This focus on desktop studies has led to an emphasis on summarizing available statistical and physical data to the neglect of the rich oral and indigenous narratives that can emerge locally. These narratives are laden with context and predictive power. Elders, leaders, spiritual healers, and cultural resource experts within a region can help to identify sacred areas, or values that are at stake in a region, and only local people can verify the meaning and extent of impacts on the cultural environment.

The developer should keep in mind that most indigenous communities want to maximize the control they have over cultural impact studies, especially ones that deal with sensitive traditional knowledge about land use, spiritual places and spaces, and other aspects of their culture. Research shows that control over cultural maintenance is linked to population health among indigenous groups. Chandler and Lalonde (2007) show how Aboriginal communities in British Columbia that have strong showings in eight indicators of control over "cultural continuity" (e.g., control over education, cultural facilities) have lower suicide rates than communities that lack control over cultural continuity. Lack of control over aspects of culture can be an impact in itself, one the developer should avoid exacerbating. However, there are often few financial or technical resources available for potentially affected communities to properly engage in studies of cultural impacts during EIA processes. The developer's role may be to assist in funding such studies rather than imposing them on the group from outside.

The parties should, early in their relationship, establish mechanisms for ensuring substantial indigenous control over research processes and cultural information. A basic start might be to give communities the authority to select archaeologists and other nonindigenous technical staff. Further, communities should be in charge of the information flow surrounding cultural heritage identification and protection. In Australia, for example, it is standard practice for indigenous organizations to enter into "cultural studies agreements" or "cultural heritage protocols" that

- Establish agreed processes for cultural heritage work;
- Allow for recognition of indigenous intellectual property in cultural heritage and indigenous control of sensitive information; and
- Provide for provision of reports that meet the developer's need for project permitting, yet avoid inappropriate disclosure or use of information.

Public consultation that occurs during a formal EIA requires specific engagement with expert knowledge holders, but these people may not be able to share knowledge with all indigenous community members, let alone the nonindigenous public. For example, women may only be able to pass knowledge on to young women during their puberty rites, and to pass knowledge to any men would be to violate norms and rules (O'Faircheallaigh 2008). Given the potential for information to be sacred or secret, early consultation should be undertaken with communities before public consultation, so that protocols can be arranged for control, content, and flow of sensitive information (King 2004).

Counting What Counts: Collecting and Analyzing the Right Data

The developer should have identified valued cultural components, the cultural assets people most want to protect and enhance. Appropriate ways to study changes to these cultural assets now need to be determined. Table 17.3-4 identifies some potential indicators for consideration when conducting a cultural impact assessment on mining projects.

Quantitative tools should be used where appropriate to augment qualitative approaches and vice versa. While quantitative indicators (e.g., percentage of language speakers) can surface some trends, the meaning and culturally relevant issues will only emerge from locally situated narratives. Community experts and outside experts can tend to weigh significance differently, prioritize certain issues, and account for change in culturally specific ways. As a result, it is important to include local experts in the definition of the appropriate issues, indicators, and narratives of impact and change. In some cases, quantitative indicators will be developed with and vetted by an appropriate cross section of the residents of potentially affected communities prior to the developer adopting them during any formal impact assessment process. Thresholds of acceptable change for all valued cultural components must also be established, and they cannot be effectively developed without community input. The same applies to qualitative narratives of the impacts and significance of these impacts.

As mentioned earlier, impacts on tangible culture, especially physical heritage resources, are the ones most often studied during a formal environmental impact assessment, and impact assessments for mining projects have largely tackled the concept of culture through a technical, "scientific" lens. As a result, only a limited range of indicators (almost entirely quantitative) have emerged to "capture" the concept of what constitutes culture and to assess potential cultural impacts (e.g., number of historic sites, numbers of traplines, language retention rates). Qualitative means of measuring cultural change have largely been neglected, often because they are difficult and time-consuming to measure, in part because relevant data are not available from published sources. Another reason for the limited use of qualitative data is a bias in the EIA business (of which cultural impact assessment is but a part) against what some professionals and EIA professionals

Table 17.3-4 Example valued cultural components, criteria, and indicators

Valued Component(s)	Indigenous Group Goal/Priority	Criteria or Indicator
Heritage resources	Protecting heritage and cultural resources	• Historic properties, including buildings, sites, structures, and signs • Cultural items, such as sacred objects and funerary items • Archaeological resources • Sites, features, or artifacts that are old (the specific required age is often determined by archaeological resource protection acts the developer needs to be familiar with) • Archaeological, historic, and scientific data that is usually found in association with archaeological resources and historic properties but may exist independently • Sacred sites
Cultural landscapes and other special spiritual spaces/places	The aesthetic, cultural, archaeological, and/or spiritual value attributed to places	• Reported time spent on the land for transmission of practices, language, oral histories • Types of cultural uses of and associations with the potentially affected area • Place names • Continuity of traditions • Relationship of site to other cultural sites in the landscape
Land tenure	Maintenance of traditional economic structures and forms of land tenure	• Landownership or traditional patterns of agriculture
Overall relationship to the land and traditional activities on the land (including practice of traditional economy)	Continuity of knowledge and practice	• Recreational and traditional economy that involves access to the land • Valuation of alternative land uses (i.e., tourism vs. hunting vs. industry) • Percentage of country food in diet • Number of traplines, or traditional use registered or known in area
Values and beliefs	Protection of indigenous culture in occupational sites and at home site	• Cultural norms, values, and beliefs that might be impacted at work sites
Methods of cultural transmission (including language, oral history, inter-generation relations, social networks)	Maintenance of traditional language, education, laws, and traditions; maintenance of cultural ties	• Practices: rituals, religious rites, and ceremonial activities • Percentage speaking indigenous language at home • Percentage retention of education, laws, and traditions • Density of relationships and networks in the culture
Sense of self, sense of place, overall well-being	Ability to influence decisions on critical cultural issues	• Percentage of gambling, reported crimes, substance abuse • Reports of sense of control over decision making

regard as "anecdotal" evidence. The stories, oral histories, and words of the culture holders are often discounted in favor of evidence that is easier to quantify.

When cultural practices are quantified (e.g., language, hunting, trapping, and food consumption), the indicators may be used as inappropriate proxies for what culture *is*, so that when one is not practicing this activity, one is not a "cultural" being. Local accounts of culture are usually about much more than participation in "traditional" practices. Elders will often tell complex stories that relate to values, roles, and relationships that are reinforced in meetings, at church, and in daily activities. These things cannot be reduced to numbers. They are complex concepts that are only reflected and reinforced through talk and practice. What holds the group together is more than participating in activities and speaking the language; it is also engaging and reengaging relationships to other people and animals, both past and present. Forces that interfere with unity, relationships, and the continuity of roles are what elders often define as threats or impacts (G. Gibson 2008). This reveals the importance of surfacing both qualitative and quantitative measures and narratives of impact.

A downside of focusing exclusively on quantitative data is that it can tell you *what* is happening, but rarely if ever can it tell you *why* change is occurring, especially on intangible assets of culture. Researchers tend to use available quantitative data on valued cultural components to get a sense of status and trends, looking for cultural "hot spots"—those valued cultural components that are in decline. Researchers then dig deeper using qualitative assessment tools to get at the reasons behind the phenomena.

A community's issues and concerns are often best surfaced through in-depth interviews, focus groups, or community meetings. Some training and knowledge of local context and politics is required to be effective. Talking to the wrong person about sensitive issues and places (e.g., a member of a family who does not use a certain area) can result in inaccurate information and inadvertently even raise tensions within the community.

In identifying appropriate criteria and indicators of cultural impacts and cultural resilience, it is important to consider not only readily available statistical data on culture collected by settler government agencies, but also community-based data and analysis. Figure 17.3-4 reveals the kinds of criteria for cultural transmission that an elder (in this case, a group of Tlicho elders from Canada's Northwest Territories) might raise with respect to the possible impacts from a proposed development. These are the criteria of cultural preservation that are often raised in a collective setting—the next step is connecting these criteria to the key issues and concerns that emerge from the proposed project. For example, if these are the areas the community most values and wants to enhance, the question of how mining can embrace these aspects might be addressed. Thus, Canadian mines operating in indigenous territory can bring in elders to tell stories, encouraging spiritual and cultural celebrations on site.

While these criteria do not lend themselves to quantification (e.g., it makes little sense to count the number of stories told over a year), the collective reflection in community settings on the issues and practices helps to imagine mitigation measures for a proposed development. Such community-based initiatives are more likely to be focused on criteria, stories, and indicators deemed most relevant to cultural health and maintenance by the culture holders themselves (G. Gibson 2008). Community-based initiatives may also have the added

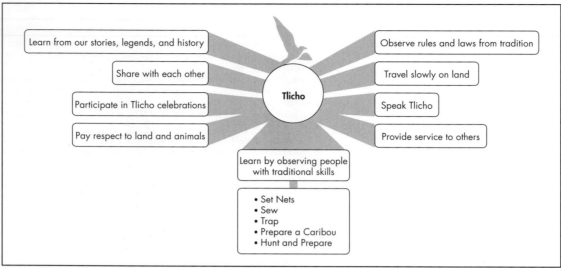

Source: G. Gibson 2008.
Figure 17.3-4 Community-defined criteria of cultural preservation

benefit of linking data collected to goal statements or thresholds of acceptable change that can assist in the determination of whether additional impacts will be significant. Where thresholds are discussed, there is the possibility of then defining when and how action will be taken, if significant changes come to pass.

Impact Prediction and Significance Considerations

A robust and defensible picture of potential impacts must be developed before effective mitigation strategies can be identified. Consider a situation where the question at hand is, What is the potential for impacts on heritage resources from Development X?, yet no on-the-ground archaeological studies have yet been conducted. How can a credible answer be derived?

The absence of physical evidence cannot be assumed in this case to be "evidence of absence" in terms of a location's cultural importance. In this case, different sources on the area's cultural resource potential can be consulted. Government officials responsible for assessing heritage resource potential across the jurisdiction, traditional land users, elders, and other traditional knowledge holders can be consulted to establish warning signs of the potential presence of unfound heritage resources or other cultural assets, which may include identification of

- The prevalence of trails, important hunting locations, meeting places, or historic habitations in the vicinity of the proposed development;
- Historic and current use of the area;
- Identification of heritage resources found around, rather than within, the project area;
- The role of the place and paths to it in the oral history of the culture, including density of place names;
- The ranges of important wildlife;
- Types of landforms, water bodies, and other physical characteristics that make it more likely to be used for cultural purposes;
- Whether the location fits into a larger cultural landscape, and how it relates to the whole; and

- The expressed desire by the culture holders to protect the location as it is, as seen, for instance, in public statements of concern, or attempts to withdraw the area from development or find other legal protections.

Using multiple indicators of cultural impact potential triangulated by several sources allows for an estimation of the importance of the area to a culture including, but not limited to, the presence of physical heritage resources. And this initial estimation can be accomplished in some cases without requiring a costly heritage resource impact assessment.

After cultural impact potential is identified, determination of cultural significance must be embraced as a social rather than a technical process. Because culture is a complex and always evolving human phenomenon, it is essential to focus not only on what impacts culture can absorb (a technically estimated threshold of *manageable* change), but also on what the culture holders are willing to take (a socially derived threshold of *acceptable* change). Therefore, when the time comes to assign significance to an impact on culture, a central voice should be that of the culture holders (Winds and Voices Environmental Services Inc. 2000). They are the primary experts on what will change or potentially be affected, because the significance of a site or space may not be apparent to the outsider, to a developer, assessment body, or regulatory agency. As the MVEIRB put it when deciding that a de facto burial ground under Great Slave Lake was not an appropriate location for exploration drilling:

> The Review Board relied on the most reliable and accurate experts it could when making its determination about cultural impacts—the Aboriginal cultural groups themselves. Cultural impacts are best identified and addressed when relayed by the holders of the cultural knowledge, the community members themselves (MVEIRB 2005b).

In indigenous cultures where oral history is the canon of proof (Andrews and Buggey 2008), the evidence of traditional knowledge holders such as elders and other land users,

Table 17.3-5 Factors to consider when determining cultural impact significance

Significance Factor	Questions the Assessors May Want to Consider
Nature of impact	• Which valued cultural components does the impact threaten? How does it threaten them? What pathway would it occur by? Knowledge of pathways of impact that can be blocked or managed by mitigation is essential to minimizing or eliminating impacts.
Situational context	• What is the evidence for use, bequest, and existence values associated with the potentially affected area? • How sensitive are the valued cultural components to change—will a little bit of change fundamentally alter the values associated with the place? • Is the place or landscape one of many similar important places or is it the "one and only"? • Is there the potential for cascading impacts from this cultural change? • Are there cumulative effects that need to be considered from other developments? • Does the area that will be impacted have multiple uses by many different groups? • How vulnerable or resilient are the valued cultural components that may be affected? What is the current status and trends for indicators related to them? Less-resilient elements of culture, if they are deemed important by culture holders, should be prioritized for mitigation.
Trade-offs	• Will there be beneficial effects to offset adverse effects? Will the beneficial effects be on the same valued components or different? For example, economic growth at the expense of cultural decline may require mitigation that takes financial capital and converts it to cultural assets. • Are there any activities or values associated with the location to be impacted that will be inevitably constrained, altered, or eliminated if the development proceeds?
Capacity to manage and social acceptability	• Will any identified thresholds of acceptable change be approached or breached if the development proceeds? Are there thresholds already breached that increase the significance of any additional changes? Does the predicted change exceed the existing capacity of the community and government services to absorb the change? • Are there management plans in place and can they handle the likely increased change?
Geographic area and distribution	• How many communities, culture groups, areas of cultural importance, and valued cultural components will be impacted by this development?
Likelihood	• Is there a greater than 50% chance the impact will occur? Does the estimated magnitude of the change make a 50% likelihood too high to consider? What is the estimate based on and what assumptions were used? Who made the estimation? Are there people who disagree with it?
Impact equity	• Are certain cultural groups or subgroups within a culture more likely to be impacted? For example, elders, people who rely on the traditional economy and live on the land, women, and youth, in that order, are often more likely to be impacted by adverse cultural change.
Public concern	• Is there a high level and wide cross section of the culture group (and others) expressing concern about the proposed activity?

affirmed by other culture holders, is not anecdotal evidence but rather the lived reality.

In assessing the authenticity of a claim about cultural impact, it is telling to look at the range of culture holders who will attest to the significance of the affected cultural site, space, or activity. Thus, significance is clearly established if an impact surfaces again and again in interviews, community meetings, focus groups, or public hearings.

A range of factors may be considered in weighting cultural impact significance, as identified in Table 17.3-5.

The situational context is often more important than the project context for determining the potential for significant adverse cultural impacts. The scale and meaning of cultural impacts cannot be foretold solely on the basis of project size. Just as severe water quality issues can emerge from the smallest of mines if acid mine drainage occurs, so too can a cultural impact from a small drilling program be significant if it is in a location of high sensitivity. Further, impacts on one part of a cultural landscape may have wider and more profound impacts on the whole, regardless of the scope or scale of the activity that causes the impact. Both of these issues were addressed in an MVEIRB decision to recommend rejection of an exploration drilling program in the Upper Thelon River basin of Canada's Northwest Territories:

> The Review Board is of the view that the degree of biophysical impact on the area is not always commensurate with the magnitude of the cultural impact experienced by the people who value it. The scale of the project is not the main consideration in this case. The project is small, but the issues are much bigger because the proposed development is located in a landscape of such vital cultural importance. In this case, the Review Board heard that this development, regardless of its size, cannot be reconciled with the values placed on the area where it is proposed (MVEIRB 2007a).

Because of rare instances like those noted here where developments are not allowed to proceed, the term *significance* can create fear in the minds of developers because there is an impression that significance is all about *whether* a project should proceed ("go" versus "no-go" zones). It is not always possible to mitigate, and as a result, indigenous groups will in some cases accept nothing but complete rejection of a project. Nonetheless, such outright rejection is the extremely rare exception rather than the rule. Significance determination is almost always more about helping to determine *how* a project should proceed, under what conditions, and with what level of cultural protection. Far from being "project breakers," most adverse impacts on culture initially deemed significant can be dealt with through proactive, cost-effective mitigation. The next section examines how appropriate mitigation measures can both minimize adverse impacts and maximize benefits to all parties.

CULTURAL IMPACT MITIGATION AND CULTURAL ENCHANCEMENT STRATEGIES

General Considerations

Mitigation measures to avoid, reduce, or compensate for adverse cultural impacts and to enhance cultural resilience run the gamut from complex re-routing of the proposed activity to small, day-to-day accommodations at the work site. For example, the type of food, the norms for recreation, the support for workers unfamiliar with the wage economy, the involvement of community-based elders and spiritual leaders and healers, and the naming of features within the site can all become a focus of cultural enhancement measures (Barker and Brereton 2005; G. Gibson 2008).

In many jurisdictions legislation nominally designed to protect indigenous heritage affords few if any enforceable tools to protect the complex and varied constituents of culture. This is particularly true for impacts on less-tangible elements of culture, such as cultural landscapes, which are often only given commemorative designations without attendant substantive land-use restrictions (NWT Cultural Places Program 2007), and impacts on cultural maintenance factors like language preservation. In the absence of formal statutory legal requirements, developers face two choices: They can negotiate effective mitigation measures with indigenous groups or they can face the risks of project delay and ultimately abandonment created by these groups as they use direct action and alternative legal or regulatory mechanisms, such as EIAs, to help make up for the absence of legislative protection.

It is in the interest of developers to prioritize cultural impact mitigation, not only because it helps companies gain the required social license to operate from the cultural group, but because successful mitigation, and cross-cultural engagement more generally, offer other substantial benefits. For instance, it can help make the work environment a welcoming one for indigenous people alongside the settler culture work force, which minimizes costly turnover and time spent on workplace conflict resolution. It can also help to demonstrate the corporate social responsibility expected by shareholders, funders, and the general public alike.

Finding appropriate means to minimize impacts on culture from a mining development requires asking

- What is the likely impact outcome?
- What are the pathways/triggers by which this impact might occur?
- How can this outcome be avoided, minimized, or compensated for (in that order of prioritization)?
- Does the proposed mitigation deal with the root of the problem (pathway/trigger)?
- What alternative mitigations are available and what are the advantages and disadvantages of each? What is the best mitigation option?
- Is the proposed mitigation acceptable to the impacted group?
- What have the cultural holders themselves proposed for avoidance, minimization, or compensation?

Mitigation should focus on altering the triggers of adverse change so that valued cultural resources are, at a minimum, not eroded and, at best, enhanced. Effective mitigation prioritizes elimination of the impact, but if this cannot be done, then impacts are controlled or minimized. Compensation should always be the last resort. One of the main reasons why impact avoidance or minimization is much preferred to compensation is that it is extremely hard and usually inappropriate to place a monetary value on cultural assets. Another is that monetary compensation can in some cases actually exacerbate other social, economic, or cultural impacts on indigenous communities rather than minimize them (Gibson and Klinck 2005). If the pathways by which culture is eroding are sought out, however, targeted investments of compensatory funds can go directly into avoiding root causes or bolstering cultural resilience strategies, such as bush camps, funding for traditional harvesters to get out on the land, documentation and dissemination of elders' knowledge and stories, and grade school cultural curriculum development.

Looking beyond adverse impact mitigation, developers can obtain substantial benefits by seeking to incorporate indigenous culture into the everyday life of mining projects. In particular, this can help create a workplace conducive to indigenous recruitment and retention, with major economic benefits for the mining company (O'Faircheallaigh 2010).

Specific Examples
Different types of mitigation measures should be established and managed at different stages of project development. For example, early consultation may serve to identify areas that are sacred, and the developer may remove these areas from its development plans. Cultural protection provisions may also be recommended in EIA reports issued by regulators and attached to licenses and permits. Mitigation commitments should be sought early and redefined as necessary as the development and relations between parties mature, and when potential impacts emerge via the monitoring systems established.

Table 17.3-6 comprises a range of mitigations that have been used in projects. However, mitigation will always have a site-specific element, and so new measures will emerge tailored to the local context.

In evaluating the strength of cultural heritage protections of a large number of agreements, O'Faircheallaigh (2008) identified two key criteria: the level of protection sought; and the activities, processes, and resources applied to ensure that this level of protection is actually realized on the ground. In relation to the first, he identified five distinct and alternative levels of protection. At one extreme, sites or areas of significance may be damaged or destroyed by project development without any reference to indigenous people. At the other, there is an unqualified requirement to avoid damage to cultural heritage, even if this involves major changes to project design and/or a decision not to develop specific ore bodies. In relation to activities, processes, and resources, he identified six key areas. In this case the criteria do not represent alternatives but are cumulative; that is, the more of them that are present, the more likely it is that the level of protection sought will actually be realized. The six key areas are

1. Provisions that maximize indigenous control of site clearance and heritage management processes;
2. Provision of financial and other resources to support cultural heritage identification and management;
3. Measures to enhance an indigenous community's general capacity for cultural heritage protection;
4. Explicit protection of any cultural knowledge provided by indigenous people as part of the cultural heritage protection regime;
5. Provisions that allow traditional owners to temporarily stop project activities to protect previously unknown sites; and
6. More general measures designed to support a *system* of cultural heritage protection.

O'Faircheallaigh (2008) found that the highest levels of cultural heritage protection were required in only one-fifth of

Table 17.3-6 Cultural and social measures from environmental assessments and contractual agreements

Area	Principles and Options
Cultural heritage protection	• Principle of avoiding damage, then minimize and compensate as last resort • Measures to avoid damage to cultural sites • Description of meaning of damage to, or interference with, site/object • Site clearance guidelines including time frame for site clearance reports • Prohibitions on interference with cultural spaces and archaeological sites • Indigenous access to areas that contain sacred sites (e.g., traditional owners at the Rio Tinto Argyle diamond mine in Western Australia have access to areas of the mining lease that contain such sites) • Employment of local cultural heritage protection monitors • Monitoring guidelines • Protocols for site or object management • Confidentiality of culturally sensitive information • Consequences if mining party defaults • Process or compensation for site/object damage
Cultural landscapes and other special spiritual spaces/places	• Protection of sacred sites • Continued practice of rituals to maintain site or protect visitors from harm
Culture and land tenure	• Protection or creation of alternative sites or livelihoods for agriculturalists
Overall relationship to the land and traditional activities on the land (including practice of traditional economy)	• Compensation for lost revenues from trapping and fishing caused by damage to equipment, loss of animals, including direct loss if animals are not in area or if there is increased cost associated with additional travel • Assistance in location of new areas • Rotation change such as moving from a 3-day rotation schedule to a 2-week rotation schedule to allow for time out harvesting on the land
Values	• Measures to control interactions of "outside" workers from large accommodation camps with small, primarily indigenous communities • Family and community support (e.g., money management, training) • Counseling in locally appropriate way and in own language for worker and for entire family • Flexible human resource policies that acknowledge different models of kinship • "Country" or "bush" food kitchens, and the ability to prepare food in the culturally appropriate style • Provisions for country food on the main dining menu • Possibilities or spaces for the practices, rituals, and spirituality of the indigenous peoples
Methods of cultural transmission (including language, oral history, intergeneration relations)	• Site visits by elders and conduct of ceremonial activities on site (e.g., the cultural practice of mantha, a ritual done to protect workers at the Argyle diamond mine or naming of landscape features using Aboriginal place names) • Special leave for culture and for family or community crisis • Creation of community–corporation relationship mechanisms such as family visits and out-on-the-land trips • Language-promotion activities • Community involvement in defining, monitoring, and analyzing cultural impact
Social networks	• Family accommodations and spousal visits • Develop family separation, life skills, drug and alcohol or gambling addictions programs; prepare people for lifestyle associated with development or rotation schedule management • Policies on consumption and use of alcohol and drugs • Training for and access to computers and availability of phone and computer access to family
Sense of self, sense of place, overall well-being	• Cultural exchanges of senior managers and senior indigenous leaders • Cultural sensitivity training for all incoming senior managers to ensure awareness and sensitivity to culture • Funding traditional knowledge studies—information collection and dissemination in an appropriate manner and led by the cultural group • Broader support for cultural activities (e.g., support for cultural activities) • Antiracism training and gender sensitivity training • Flexible work schedules and equal time on and off of rotations • Cross-cultural training workshops

the agreements he evaluated. This suggests that negotiated agreements are only able to achieve protection in some cases, given that the bargaining power of some indigenous communities is limited. Where the community is well-resourced and politically powerful, there is often a strong ability to use agreements to achieve cultural heritage protection. In other cases, indigenous peoples will need to use all available avenues for protection.

O'Faircheallaigh's research also highlighted the fact that even where a system is created for cultural heritage protection, there is often a systematic failure to implement it effectively. There are many reasons for this failure to implement agreements, such as changing circumstances, lack of resources, failure to review and renegotiate mechanisms over time, failure to anticipate all the issues that emerge, and inappropriate allocation of responsibility or authority for cultural heritage protection (O'Faircheallaigh 2003). This highlights the need to focus carefully on the mechanisms and resources needed to ensure that cultural heritage protection provisions continue to operate and work as previously envisioned.

Mining companies must develop, in collaboration with affected indigenous groups, specific measures to mitigate cultural impacts and support cross-cultural engagement, and ensure that these are not put in place solely to secure project approval but are maintained over the longer term. This requires substantial and ongoing commitment by senior management of the time and financial resources required to support indigenous participation and implementation of protective measures.

Workplace-specific mitigation is an essential element of any cultural impact management plan. Some of the measures that increase workplace satisfaction include antiracism training, gender sensitivity training, and policies on harassment, racism, and cultural leave for bereavement (G. Gibson 2008). In addition, the nature and length of a rotation will be of critical importance (Shrimpton and Storey 2001, 1989); an equal rotation (e.g., 2 weeks on and 2 weeks off) is often the most sustainable for families and workers. Exit interviews with Aboriginal miners at the Century mine in north Queensland found that appropriate support and mentoring for workers was essential (Barker and Brereton 2005). The commitment involved in providing this support is well justified in terms of the role it plays in securing an ongoing license to operate, in avoiding the serious investment risks that can occur if indigenous heritage is damaged, and in terms of the value that can accrue to the enterprise from the positive effects of cross-cultural engagement.

CONCLUSION

Using the tools outlined in this chapter will help developers get past the fear of the unknown and embrace the opportunities presented by cultural impact assessment and cross-cultural engagement. A mindset change is at least as important as a new toolkit. It is critical to remember that finding an adverse cultural impact does not in most cases mean a project will not proceed, and that avoidance of thorough and early cultural impact assessment is much more likely to create a high-risk situation than avoid one.

A good cultural impact assessment can do more than merely provide evidence for the EIA processes. Corporations can gain public prestige both by effectively engaging with indigenous communities and by supporting strong community-based identification, management, and protection of cultural assets. Cross-cultural engagement can also generate substantial value for enterprises by, for instance, allowing them to gain access to stable, skilled work forces that are committed to the regions in which minerals are developed.

Cooperative indigenous–company action to enhance culture can also generate benefits for indigenous culture holders. It can provide opportunities for indigenous people to document and appreciate their own history and culture. Where potential impacts on sacred land are being studied, the stories, songs, and practices that are central to this place may be archived and rejuvenated for younger generations of indigenous people.

The following general rules can make cross-cultural engagement and cultural impact assessments effective, risk-minimizing, and benefit-maximizing processes for all parties.

Key considerations for effective cross-cultural engagement:

- Proceed with caution and prudence. Don't rush in.
- Follow community engagement protocols.
- Be ready to put on the shoes of the "other"; identify and respect differences between your culture and that of the group you are interacting with. Suspend your disbelief early—just because a place has no cultural value or meaning to you, it does not follow that it has no meaning to others.
- Talk to indigenous people and their organizations well in advance of actions that may have impacts; provide information early, fully, and in a form that makes sense to people. Recognize that withholding information is itself a major cause of negative sociocultural impacts.
- Don't assume your own interpretation of what will or will not generate impacts is soundly based in the indigenous context.
- Share control and work cooperatively. For most indigenous people, a process someone else controls is a process they don't trust.
- Provide resources for communities to engage; communities are not in a position to do so, and it is an investment that will yield dividends.
- Use all available avenues, not just the formal EIA process, to engage with communities on cultural issues.
- Do not view cross-cultural engagement as a discrete stage in project life, or simply as one of a series of regulatory hurdles. See it as an integral part of all activities, like health and safety.

Key considerations for effective cultural impact assessment:

- Do early work—acclimatize to the indigenous culture, and understand what mining can do to culture.
- Include traditional knowledge as much as possible, but respect culture-group-defined protocols around gathering, control, and distribution of culturally sensitive information.
- Avoid the "fear factor"; realize that project-threatening impacts are relatively rare. Treat cultural impact assessment instead as an opportunity for proactive relationship building, not as a regulatory hurdle.
- Get beyond "bones and stones"; identify both tangible and intangible valued cultural components.
- See beyond the site to the situation; learn about cultural context.
- Recognize that wisdom sits in places, spaces, and people—all merit attention.
- Adopt the proper scope and scale for your project's cultural impact assessment based on robust scoping.
- Use experts where necessary and appropriate degrees of rigor.
- Don't study for study's sake; exhaust your analysis of existing data sources before starting additional fieldwork.
- Gradually ramp up effort, investment, and specialist advice as the project develops and potential impacts expand.
- Project size is only one factor in the equation—do not assume that the biophysical impact footprint is the only driver of cultural impact potential.
- Support and augment qualitative data with quantitative data, and vice versa.
- The culture holders themselves must play the key role in defining cultural impact significance.
- Consider whether community-defined thresholds of acceptable change are likely to be exceeded for valued cultural components, and work with other responsible parties to minimize additional impact loads.
- Recognize that the work site itself is not culture-free, and develop mitigation to avoid cross-cultural conflict.

ACKNOWLEDGMENTS

Alistair MacDonald acknowledges the Mackenzie Valley Environmental Impact Review Board for supporting his research for this chapter during its development of *Cultural Impact Assessment Guidelines* (MVEIRB).

REFERENCES

Andrews, T., and Buggey, S. 2008. Authenticity in Aboriginal cultural landscapes. *APT Bull.* 39(2-3):65–71.

Angelbeck, B. 2008. Archaeological heritage and traditional forests within the logging economy of British Columbia: An opportunity for corporate social responsibility. In *Earth Matters: Indigenous Peoples, the Extractive Industries and Corporate Social Responsibility*, Edited by C. O'Faircheallaigh and S. Ali. Sheffield, U.K.: Greenleaf Publishing.

Argyle Diamonds Ltd., Traditional Owners, and Kimberley Land Council Aboriginal Corporation. 2004. Argyle Diamond Mine Participation Agreement—Indigenous Land Use Agreement. www.atns.net.au/agreement.asp?EntityID=2591. Accessed January 2009.

Barker, T. 2008. Indigenous employment outcomes in the Australian mining industry. In *Earth Matters: Indigenous Peoples, the Extractive Industries and Corporate Social Responsibility*. Edited by C. O'Faircheallaigh and S. Ali. Sheffield, U.K.: Greenleaf Publishing.

Barker, T., and Brereton, D. 2005. Survey of local Aboriginal people formerly employed at Century mine: Identifying factors that contribute to voluntary turnover. Centre for Social Responsibility in Mining Research Paper No. 4. Brisbane: University of Queensland.

Basso, K. 1996. *Wisdom Sits in Places*. Santa Fe, NM: University of New Mexico Press.

Buggey, S. 2004. An approach to Aboriginal cultural landscapes in Canada. In *Northern Ethnographic Landscapes: Perspectives from Circumpolar Nations*. Edited by I. Krupnik, R. Mason, and T. Horton. Washington, D.C.: Smithsonian Institution.

Candler, C. 2008. Making a difference: Quality in a new era of cultural impact assessment and traditional use studies. Presented at the Western and Northern Canadian Chapter, International Association for Impact Assessment Conference, Cultural Impact Assessment: Beyond the Biophysical, February 28–29.

Chandler, M.J., and Lalonde, C.E. 2007. Cultural continuity as a moderator of suicide risk among Canada's First Nations. In *The Mental Health of Canadian Aboriginal Peoples: Transformations of Identity and Community*. Edited by L. Kirmayer and G. Valaskakis. Vancouver, BC: University of British Columbia Press.

Dixon, R., and Dillon, M. 1990. *Aborigines and Diamond Mining: The Politics of Resource Development in the East Kimberley, Western Australia*. Crawley: University of Western Australia Press.

Eggleston, P. 2002. Gaining Aboriginal community support for a new mine development and making a contribution to sustainable development. Presented at the Resources Law Conference, Edinburgh, Scotland, April 14–19.

Gibson, G. 2007. Give and take: Adaptation of worksite culture in Canada's diamond mines. Presented at the Energy and Mines Conference, Montreal, April 29–May 2.

Gibson, G. 2008. Negotiated spaces: Work, home and community in the Dene diamond economy. Ph.D. dissertation, University of British Columbia.

Gibson, G., and Klinck, J. 2005. Canada's resilient north: The impact of mining on Aboriginal communities. *Pimatziwin: Int. J. Aborig. Indig. Commut. Health* 3(1):116–139.

Gibson, G., O'Faircheallaigh, C., and MacDonald, A. 2008. Integrating cultural impact assessment into development planning. Workshop materials presented at the International Association for Impact Assessment's Annual Meeting, Perth, Australia, May 4.

Gibson, R. 2006. Sustainability assessment and conflict resolution: Reaching agreement to proceed with the Voisey's Bay nickel mine. *J. Cleaner Prod.* 14:334–348.

Government of Canada. 1998. Mackenzie Valley Resource Management Act. Ottawa, ON: Government of Canada.

Harvey, B., and Gawler, J. 2003. Aboriginal employment diversity in Rio Tinto. *Int. J. Divers. Organis. Commun. Nations* 3:197–209.

Heyman, J. 2004. The anthropology of power-wielding bureaucracies. *Hum. Organiz.* 63(4):487–500.

ICMM (International Council on Mining and Metals). 2003. 10 Principles to sustainable development. www.icmm.com/our-work/sustainable-development-framework/10-principles. Accessed May 2009.

IFC (International Finance Corporation). 2007. Guidance Note 7: Indigenous peoples. www.ifc.org/ifcext/sustainability.nsf/AttachmentsByTitle/pol_GuidanceNote2007_7/$FILE/2007+Updated+Guidance+Note_7.pdf. Accessed November 2008.

ILO (International Labour Organization). 1989. Convention 169 concerning indigenous and tribal peoples in independent countries, Geneva, June 27.

Janes, E., and Hanks, C. 2003. *Report on Cultural Values for Sahyoue-Edacho National Historic Site*. Unpublished report. Yellowknife, NT: GeoNorth Limited and Hanks Heritage Consulting.

Joyce, S., and Thomson, I. 2000. Earning a social license to operate: Social acceptability and resource development in Latin America. *Can. Min. Metall. Bull.* 93(1037):February.

King, T. 2002. *Thinking About Cultural Resource Management: Essays from the Edge*. Walnut Creek, CA: AltaMira Press.

King, T. 2004. *Cultural Resource Laws and Practice: An Introductory Guide*, 2nd ed. Walnut Creek, CA: AltaMira Press.

Lutsel K'e Dene First Nation. 2008. Correspondence to Minister of Indian and Northern Affairs Canada re: Reports of Environmental Assessment, Uravan Mineral Inc. Boomerang Claims. November 17, 2008. www.reviewboard.ca/registry/. Accessed April 2009.

MacDonald, A., and Gibson, G. 2006. The rise of sustainability: Changing public concerns and governance approaches toward exploration. *Soc. Econ. Geol. Spec. Pub.* 12:1–22.

MacKay, F. 2004. Indigenous Peoples' Right to Free, Prior and Informed Consent and the World Bank's Extractive Industries Review. Forest Peoples Programme. Unpublished report.

MVEIRB (Mackenzie Valley Environmental Impact Review Board). 2005a. *Guidelines for Incorporating Traditional Knowledge in Environmental Impact Assessment*. Yellowknife, NT: MVEIRB.

MVEIRB (Mackenzie Valley Environmental Impact Review Board). 2005b. *Letter from MVEIRB to the Minister of Indian and Northern Affairs re: New Shoshoni Ventures Preliminary Diamond Exploration*. Yellowknife, NT: MVEIRB.

MVEIRB (Mackenzie Valley Environmental Impact Review Board). 2007a. *Report of Environmental Assessment and Reasons for Decision on Ur Energy Inc.'s Screech Lake Uranium Exploration Project*. EA 0607-003. Yellowknife, NT: MVEIRB.

MVEIRB (Mackenzie Valley Environmental Impact Review Board). 2007b. *Socio-Economic Impact Assessment Guidelines*. Yellowknife, NT: MVEIRB.

MVEIRB (Mackenzie Valley Environmental Impact Review Board). Forthcoming. *Cultural Impact Assessment Guidelines*. Yellowknife, NT: MVEIRB.

Nadasdy, P. 2003. *Hunters and Bureaucrats: Power, Knowledge, and Aboriginal–State Relations in the Southwest Yukon*. Vancouver: University of British Columbia Press.

NWT Cultural Places Program. 2007. *Living with the Land: A Manual for Documenting Cultural Landscapes in the Northwest Territories*. Yellowknife, NT: Government of the Northwest Territories.

NWT and Nunavut Chamber of Mines. 2005. Sustainable Economies: Aboriginal participation in the Northwest Territories Mining Industry, 1990–2004. www.miningnorth.com/docs/Aboriginal Participation 2005 (2).pdf. Accessed November 2009.

O'Faircheallaigh, C. 1995. Long distance commuting in resource industries: Implications for native peoples in Australia and Canada. *Hum. Organiz.* 54(2):205–213.

O'Faircheallaigh, C. 2003. Implementing agreements between Indigenous peoples and resource developers in Australia and Canada. In *Aboriginal Politics and Public Sector Management*. Research Paper No. 13. January.

O'Faircheallaigh, C. 2006. Aborigines, mining companies and the state in contemporary Australia: A new political economy or "business as usual?" *Aust. J. Polit. Sci.* 41(1):1–22.

O'Faircheallaigh, C. 2008. Negotiating cultural heritage? Aboriginal-mining company agreements in Australia. *Dev. Change* 39(1):25–51.

O'Faircheallaigh, C. 2010. Corporate social responsibility, the mining industry and indigenous peoples: From cost and risk inimization to value creation and sustainable development. In *Innovative Corporate Social Responsibility: From Risk Management to Value Creation*. Edited by C. Louche, S.O. Idowu, and W.L. Filho. Sheffield, U.K.: Greenleaf Publishing.

Parker, P., and King, T. 1990. *Guidelines for Evaluating and Documenting Traditional Cultural Properties*. National Register Bulletin. Washington, DC: U.S. Department of the Interior, National Park Service.

PDAC (Prospectors and Developers Association of Canada). 2009. *e3 Plus: A Framework for Responsible Exploration Principles and Guidance*. Toronto: PDAC. www.pdac.ca/e3plus/. Accessed March 2009.

Pilgrim, S., Samson, C., and Pretty, J. 2009. *Rebuilding Lost Connections: How Revitalisation Projects Contribute to Cultural Continuity and Improve the Environment*. iCES Occasional Paper 2009-01. Colchester, U.K.: Interdisciplinary Centre for Environment and Society, University of Essex.

Rose, D.B. 1996. *Nourishing Terrains: Australian Aboriginal Views of Landscape and Wilderness*. Canberra: Australian Heritage Commission.

Shrimpton, M., and Storey, K. 2001. The Effects of Offshore Employment in the Petroleum Industry: A Cross-National Perspective. OCS Study MMS 2001-041. Herndon, VA: U.S. Department of the Interior Minerals Management Service Environmental Studies Program.

Shrimpton, M., and Storey, K. 1989. Fly-in mining and the future of the Canadian north. In *At the End of the Shift: Mines and Single-Industry Towns in Northern Ontario*. Edited by M. Bray and A. Thomson. Toronto: Dundurn Press.

SIWGMI (Sub-Committee of the Intergovernmental Working Group on the Mineral Industry). 1996. *Aboriginal Participation in the Mining Industry in Canada 1996: Seventh Annual Report*. www.publications.gov.sk.ca/details.cfm?p=7169. Accessed January 2009.

Tobias, T. 2000. *Chief Kerry's Moose: A Guide to Land Use and Occupancy Mapping, Research Design and Data Collection*. Vancouver: Ecotrust Canada.

Trigger, D. 2005. Mining projects in remote Aboriginal Australia: Sites for the articulation and contesting of economic and cultural futures. In *Culture, Economy and Governance in Aboriginal Australia*. Edited by D. Austin-Broos and G. MacDonald. Sydney: Sydney University Press.

Tylor, E. 1920 (1871). *Primitive Culture*. New York: J.P. Putnam's Sons.

UNDESA (United Nations Department of Economic and Social Affairs). 2004. An Overview of the Principle of Free, Prior and Informed Consent and Indigenous Peoples in International and Domestic Law and Practice. PFII/2004/WS.2/8. New York: UNDESA.

UNESCO (United Nations Educational, Scientific and Cultural Organization). 2003. Convention for the Safeguarding of the Intangible Cultural Heritage. www.unesco.org/culture/ich/index.php?pg=00006. Accessed December 2008.

United Nations. 2007. *United Nations Declaration on the Rights of Indigenous Peoples*. http://www.un.org/esa/socdev/unpfii/documents/DRIPS_en.pdf. Accessed February 2010.

Usher, P. 2000. Traditional ecological knowledge in environmental assessment and management. *Arctic* 53(2):183–193.

Winds and Voices Environmental Services, Inc. 2000. *Determining Significance of Environmental Effects: An Aboriginal Perspective*. Research and Development Monograph Series. Hull, QC: Canadian Environmental Assessment Agency.

Zoe, J.B., ed. 2007. *Trails of our Ancestors: Building a Nation*. Behchoko, NT: Tlicho Community Services Agency.

CHAPTER 17.4

Management of the Social Impacts of Mining

Daniel M. Franks

INTRODUCTION

The value of a "social license to operate" is increasingly recognized within the mining industry. Unmitigated negative social impacts have the potential to result in negative publicity, increased litigation, and reputational damage, or to delay, prevent, or close down mining in existing and prospective areas as a result of community concerns. On the other hand, the positive impacts associated with mining projects can be welcomed by communities and governments. Assessment and ongoing management of environmental impacts are relatively common fixtures in the mining industry, as evidenced through formal environmental impact assessment processes and environmental management plans. Only more recently have approaches emerged to link social impact assessment with ongoing management and to proactively respond to social and community issues.

This chapter outlines techniques and processes that assist first in identifying and responding to social issues during planning and then in guiding and monitoring projects during operation through to postclosure. The term *management* is used in this chapter to refer to the coordination of activities in responding to social impacts and social risks. Effective management requires an understanding of social issues, which can be gained through ongoing assessment. Through both assessment and management, the design and implementation of mining activities can be shaped to enhance environmental and community outcomes.

Social impact assessment and management are the responsibility of dedicated community relations practitioners at most mining operations. In addition, however, there is a need for mining engineering professionals to be familiar with such approaches, since effective management requires integration across all aspects of the operation. In recent years, mining companies have increasingly come to recognize the value of using management systems to manage all aspects of their operations. Commencing with the introduction of quality systems and environmental management systems based around key international standards (ISO 9000 and ISO 14000, respectively), the trend now is to bring these specialized areas together into integrated management systems. It is essential that the management of social issues be part of this trend and not considered in isolation from the other parts of the business.

This chapter begins with a discussion of the social impacts of mining and then looks at each of the phases of social impact assessment and management that make up current practice in the field (Figure 17.4-1). These phases are consistent with an adaptive management framework or the plan-do-check-act cycle. Adaptive management emphasizes continuous improvement through an iterative process of planning, implementation, monitoring, and adjustment. The phases of social impact assessment and management include (1) the scoping and formulation of alternatives, (2) profiling and baseline studies, (3) predictive assessment and revision of alternatives, (4) management strategies to avoid and mitigate negative social impacts and enhance positive impacts, (5) monitoring and reporting, and (6) evaluation and review. The chapter concludes with lessons that can inform technical and management roles. Through greater awareness and consideration of social impacts and risks, mining engineering professionals will find their projects are more acceptable to communities and have a greater prospect of success.

SOCIAL IMPACTS AND RISKS

A social impact is something that is experienced or felt (real or perceived) by an individual, social group, or economic unit. Social impacts are the effect of an action (or lack of action) and can be both positive and negative. Social impacts can vary in type and intensity, and over space and time. Examples of social impacts associated with mining operations include employment effects; changes to social services, such as health and childcare or the availability and cost of housing; and cultural change, such as changes in traditional family roles as a result of the demands of mining employment, or even the breakdown of traditional economies due to the introduction of a cash economy. Environmental impacts also have social implications. Mining activities can result in changes to community amenities, health, or the availability and quality of water and land (Table 17.4-1).

Impacts can be direct, such as the impact of noise and dust, or result from indirect pathways, such as road fatalities

---

Daniel M. Franks, Research Fellow, University of Queensland, Sustainable Minerals Institute, Centre for Social Responsibility in Mining, Brisbane, Australia

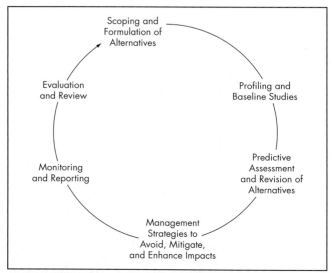

Figure 17.4-1 Phases of social impact assessment and management. The phases should be applied to each stage of the mining life cycle as part of an adaptive management cycle.

resulting from increased traffic in a nearby town servicing a mine. Impacts often accumulate and interact such that they trigger or become associated with other impacts. Cumulative impacts arise from the compounding activities of a single operation or multiple mining and processing operations, as well as from the interaction of mining impacts with other past, current, and future activities that may not be related to mining (Franks et al. 2009a).

A social risk is the potential for an existing or planned project to have an impact on individuals or groups or, conversely, to be impacted by them. Like impacts, social risks are both positive and negative because of the potential for mining to generate social and economic opportunities, such as economic and community development and employment.

The fear of an action can often be as important a generator of impact as the action itself. Perceptions of impact, opportunity, or risk are subjective, and public understandings do not necessarily correspond with scientific perspectives. There are a number of reasons for this. Nontechnical factors, such as personal value systems, previous experience, levels of trust in information sources and methods, and openness to change, all influence the way individuals and communities perceive and respond to change.

PHASES OF SOCIAL IMPACT ASSESSMENT AND MANAGEMENT

Social impact assessment (SIA) is a process for understanding the social issues associated with development. SIA is focused on how to identify, avoid, mitigate, and enhance outcomes for communities and is most effective as an iterative process, rather than a one-off activity at the outset of mining (Vanclay 2003; Becker and Vanclay 2006). Social impact management refers to systems and strategies undertaken during the implementation phases of a development (including exploration) to monitor, report, evaluate, review, and proactively respond to change. Together, SIA and impact management provide a very effective tool to address impacts, if the team conducting the work are well integrated with the overall management of the entire operation.

Not all social impacts are predictable. Because communities and the external environment are dynamic, an element of uncertainty will always be present. Similarly, not all impacts can be avoided or mitigated. However, SIA can provide insights, focus attention, and identify key issues as perceived by stakeholders to predict and anticipate change, and social impact management can assist in proactively responding to the intended and unintended consequences of mining developments.

SIA and impact management are most effective when undertaken across the life cycle of mining and resource processing, encompassing all of the activities from exploration, construction, extraction, and processing, through to postclosure, as well as including recycling and waste management. The varied social impacts across the mine life-cycle phases and the extraction and resource processing stages demand a varied range of approaches to assessment and management. For example, the exploration phase often entails a need to carefully manage community expectations, a challenge in maintaining continuity, and a risk that the outcomes at this early stage can greatly influence the future of relations. At the construction phase, there is invariably a large influx of temporary contract workers into a region that may pose challenges for communities; at closure, impacts are generated by the withdrawal of economic activity and employment, and management requires long-term planning to support alternate futures. Similarly the nature and scale of impacts, and thus the scale and scope of assessment and management, will vary with the type of extraction. Opencut pits involve the removal, processing, storage, and rehabilitation of large amounts of material that can pose risks to waterways or affect the visual amenity of a region; underground mines can create subsidence at the surface, which can affect other land users; heap leach and in-situ leach operations can pose risks to groundwater that may be used by other people and industries; and the various stages of processing will demand varied inputs and create by-products and hazardous materials that can pose risks to communities. The form and level of assessment and management is determined by the significance and scale of the action, as well as by the sensitivity of the community, the location, and its environment.

SIA is a common requirement of regulatory approval processes at the project approvals phase for mining and processing stages in many jurisdictions. These SIAs may be independent or integrated within environmental impact assessments. Beyond this regulatory requirement, SIAs are undertaken by mining operations in response to an actual (or potential) major change in the project or impact on a community—for example, in preparation for closure or for a major expansion. Social impact assessment can also address any particular issues faced during exploration, resettlement, temporary scale-down of operations, or a community development program. Assessments are most useful when maintained and periodically updated as "living" documents to inform decision making. Social impact management approaches, such as community reference panels, social management plans, grievance mechanisms, and community development initiatives, are increasingly required by government legislation, as well as by industry and corporate policies/standards.

The following sections introduce and describe each of the phases of SIA and impact management. Consistent with an adaptive management approach, many of the techniques mentioned may be useful in several of the phases. For example,

Table 17.4-1 Common changes induced by mining activities that can lead to social impacts and risks

Social and Cultural Change

Population and demographics	In-migration, out-migration, workers' camps, social inclusion, growth or decline of towns, conflict and tensions between social groups
Social infrastructure and services	Demands on and investment in housing, skills (shortages and staff retention), childcare, health, education, and training
Crime and social order	Corruption, domestic violence, sexual violence, substance abuse and trafficking, prostitution, change in social norms, pace of change for vulnerable communities
Culture and customs	Change in traditional family roles, changing production and employment base, effect of cash economy, reduced participation in civil society, community cohesion, sense of place, community leadership, cultural heritage
Community health and safety	Disease, vehicle accidents, spills, alcohol and substance abuse, pollution, interruption to traditional food supply, awareness and treatment programs
Labor	Health and safety, working conditions, remuneration, right to assemble, representation in unions, labor force participation for women
Gender and vulnerable groups	Disproportionate experience of impact and marginalization of vulnerable groups (e.g., women, disabled, aged, ethnic minorities, indigenous, and young), equity in participation and employment
Human rights and security	Abuses by security personnel (government, contractor, company), social disorder in camps, suppression of demonstrations, targeting of activists, rights awareness programs

Economic Change

Distribution of benefits	Employment, flow of profits, royalties and taxes, training, local business spending, community development and social programs, compensation, managing expectations, equitable distribution across state/regional/local/ethnic/family groups, cash economy
Inflation/deflation	Housing (ownership and rents), food, access to social services
Infrastructure	Demands on, and investment in, roads, rail, ports, sewerage, telecommunications, power and water supplies

Socio-Environmental Change

Pollution and amenity	Air (e.g., dust), water (e.g., acid and metalliferous drainage, cyanide, riverine and submarine waste disposal), noise, scenic amenity, vibration, radiation, traffic, government capacity to monitor and regulate
Resources (access/competition)	Land, mobility, water (groundwater, river, ocean), mineral resources (artisanal and small-scale mining), cultural heritage, forest resources, human, postmining land use
Resettlement	Consent and consultation for resettlement, compensation, ties to land, adequacy of resettlement housing and facilities, equity, postsettlement conditions, livelihoods
Disturbance	Disruption to economic and social activities (including by exploration), consultation for land access, frequency and timing, compensation

The Process of Change

Community engagement	Consultation, communication, participation, empowerment, access to decision makers, transparency, timing, inclusiveness—particularly for vulnerable and marginalized groups—respect of customs and authority structures, reporting
Consent	Indigenous sovereignty/title (free, prior, and informed consent), community consent
Participation	Planning, development of programs, monitoring, selection of alternatives and technologies, operational aspects
Remedy	Grievance and dispute resolution, acknowledgment of issues, compensation, mitigation
Agreements	Equity, timely honoring of commitments, issues with delivery, duress, clarity of obligations, capacity and governance (including government capacity to respond to and manage change)
Community development	Participation, adequacy, appropriateness, capacity to facilitate, consistency, prioritization

management approaches such as community reference groups can be effectively used to identify issues during scoping. To avoid duplication, the techniques will be discussed in the phase in which they are most commonly used. Table 17.4-2 presents a variety of social research methods and impact assessment techniques that are often, but not exclusively, applied at the scoping and predictive assessment phases.

Scoping and Formulation of Alternatives

When developing a new project, planners are usually faced with a number of alternative options. Impact assessments provide an opportunity to investigate these alternatives in greater detail to assist decision makers in choosing the most appropriate option.

The scoping phase sets the parameters for the later phases of assessment and management by determining the scale, timing, and focus of the assessment, ascertaining who is likely to be impacted and identifying the actions that are likely to result in impacts.

Scoping will begin by defining the purpose of the assessment and identifying background material that may influence the assessment. This includes regulations and legislation, government and corporate policy and programs, standards, and operating procedures. After these have been taken into account, the next step is to identify project stakeholders.

Early public involvement is crucial to identifying attitudes and perspectives, harnessing local knowledge, and defining processes for further public involvement (more information on the identification of stakeholders appears in "Profiling and Baseline Studies"). With the assistance of stakeholders, the actions that are likely to result in impacts and the extent to which these impacts may be felt can be initially identified. Case studies of related actions may assist in this regard, and any gaps in knowledge should be recorded.

Alternative options should be formulated for later analysis and an initial appraisal of the impacts of these alternatives undertaken. Impacts can be prioritized, in consultation with stakeholders, to narrow down the analysis

Table 17.4-2 Common social and economic research methods and assessment techniques

Method	Description
Literature review	A critical summary of current knowledge on a topic drawn from an existing body of literature.
Interview	A method of primary data collection that consists of in-depth questioning. Interviews may vary according to the type of informant, the type of medium (telephone or face-to-face, individual or group), the setting and recording, and the type of questioning (structured, semistructured, and unstructured).
Focus group	A group interview method where a facilitator poses questions to generate discussion among participants.
Survey/questionnaire	A form with questions used to solicit information from a statistically significant group of respondents. Surveys may be used in SIA to provide data on the characteristics and opinions of a population and may vary according to the choice and wording of questions, the type of instrument (e.g., mail-out, telephone, or face-to-face), the sample size, and sample frame.
Stakeholder analysis	A stakeholder is anyone who can affect, or is affected by, an action. Stakeholder analysis consists of the identification of stakeholders, analysis of their underlying attitudes and motivations, a determination of which stakeholders are most significant, an understanding of their networks and relationships, and the development and implementation of an engagement plan.
Social and economic profile	A process to collect relevant primary and secondary data about a community. The profile is a detailed description of the community, environment, and economy of a region and provides insight into values, priorities, and trends.
Social baseline	An appraisal of the current state of a community or social group including a consideration of trends.
Case study/comparative analysis	Detailed analysis of an example, often used to identify patterns and causal relationships. Case studies can be developed from multiple sources of information and can be supplemented by data gathering.
Cultural heritage mapping	A process for identifying and recording the meaning ascribed to landscapes by cultural groups.
Ethnography	The description of human societies, usually informed by field methods that include interviews, participant observation, and surveys.
Impact pathway analysis	A process-mapping exercise used to predict the pathway of impacts resulting from an action. The method prompts insights into the direct and indirect impacts of actions and their interaction. Also known as change mapping.
Social risk assessment	A participatory technique to identify, prioritize, and respond to the social risks and opportunities faced by an organization or communities. Through a facilitated workshop, key stakeholders determine the consequence and likelihood of each identified risk and develop controls to avoid, mitigate, or enhance priority risks. Also known as social risk and opportunity analysis.
Scenario analysis	A tool used to anticipate change under different plausible future situations. Scenario analysis assists the development of a proactive policy response through the testing of assumptions.
Trend analysis	The collection and analysis of historical and contemporary data to inform the prediction of the future.
Cost–benefit analysis	An economic technique that compares the costs and benefits, usually quantified in monetary terms, for scenarios with and without an action.
Cost-effectiveness analysis	An economic technique that compares the cost-effectiveness of alternative options for achieving an outcome. It is used to identify the alternative with the lowest direct financial cost.
Input–output analysis	An investigation of the relationships and interdependences of an economy through an analysis of the flow of resources. It considers the inputs to industry, transfers between sectors, household consumption, and the outputs of goods produced.
Choice modeling	An experimental technique used in economics to frame trade-offs between different options. The technique estimates the value of options by revealing how respondents are willing to trade them. Information is usually gathered through surveys.

and investigation of the later phases to address key issues of relevance. Cumulative impacts should be considered during this prioritization stage. Although an issue may appear acceptable when considered in isolation, the impact may have greater significance when combined with the impacts of other overlapping activities.

Future and simultaneous proposals should be considered during the scoping phase. Methods range from collation of known projects (announced, approved, or under construction) and government forecasting to industry surveys of potential future activities. Efforts to forecast potential future developments can demonstrate that the proposal is serious about (and investing in) regional futures. Issues of commercial confidentiality have been overcome in other resource provinces through the use of anonymous industry forecasting surveys, usually undertaken by a peak industry body. Such information can also help the industry negotiate with government on future infrastructure needs and priorities.

The output of the scoping phase may be the definition of the objective, scope, scale, priority issues, and terms of reference for the phases of assessment and management to follow.

Profiling and Baseline Studies

Social profiling consists of understanding the communities and stakeholders potentially impacted by the activity through social and economic research. Profiling involves analysis of the social and economic characteristics of a region at a given point of time. Baselines are an appraisal of the state of a community or social group before an activity takes place. Baseline studies provide a benchmark against which potential impacts can be anticipated and change measured. They are also valuable for building mutual understanding. The foundational information and understandings provided by profiling are useful across all of the phases of extractive projects. Leading companies now routinely require their operations to undertake such studies and update them at regular intervals, particularly when there is any significant change to the scale or shape of a project or community. For example, Anglo American PLC has developed its own Socio-Economic Assessment Toolbox process to assist and encourage their operations to regularly review social impacts at different stages through the life of the project. Through regular review of profiles and baseline studies, or by approaching profiling as a "living" output, longitudinal trends may be observed over time and a more accurate picture of the change processes will be developed.

Knowing the community assists in anticipating how people might respond to change. Actions, even well-meaning ones, that are not socially, culturally, or environmentally grounded in the people and places (potentially) affected by a development, invariably result in poor outcomes. Understanding communities involves an analysis of their relationships and networks and the values that may shape attitudes and behaviors.

Stakeholder analysis is a common profiling technique that may be useful across a number of phases. A stakeholder is someone that has, or potentially has, an interest in an issue—any entity that is affected by, or can affect, a project. Stakeholders can be individuals, groups, neighborhoods, or organizations. Stakeholder analysis consists of the identification of stakeholders, analysis of their underlying attitudes and motivations, a determination of which stakeholders are most significant, an understanding of their networks and relationships, and the development and implementation of an engagement plan. Targeted consultations (or focused plans) may need to be developed for each stakeholder, especially vulnerable groups. Stakeholders have varying degrees of power, legitimacy, and interest in an issue or a project. They may include communities located in the vicinity of mining operations, employees, shareholders, financial institutions, indigenous peoples, nongovernmental organizations, trade unions, and governments and their departments. Stakeholders also include people within resource companies that may be important to the planning, development, and implementation of the action.

Profiling can be assisted by establishing meaningful avenues of two-way dialogue with stakeholders. Dialogue can reveal stakeholder histories and decode cultural meanings and symbols. Knowing the operations or proposed activity is a crucial, but often missing, element of profiling. This step requires an understanding of how the various elements of the proposal fit together and how the work groups involved in planning and implementation make decisions.

Profiling includes analysis of demographic patterns and trends; population characteristics; ethnicity and culture; the local economy; the labor market; land-use and ownership patterns; social and political organization; family and community organization; matters involving health, nutrition, and disease; community infrastructure and services (housing, health, childcare, etc.); expectations and concerns community members have about the project; community needs and desired futures and the capacity to meet these needs; and the vulnerability of social groups.

After a review of secondary information, and the identification of knowledge gaps, a program for the collection of primary data is developed. It is important to know the purpose for the collection of each piece of data. The accumulation of vast amounts of information of marginal use can be overwhelming and leave little time to do analytical work. Methods of primary data collection include quantitative, qualitative, participatory, and technical methods, such as interviews, focus groups, literature reviews, and surveys and case studies, among others.

Predictive Assessment and Revision of Alternatives

In this phase, likely impacts are identified and predicted, and their scale and significance evaluated, using technical and participatory methods. Impact prediction is an opportunity to analyze issues in more detail and to undertake broader consultation and engagement. A number of techniques can be used in social impact assessment to inform this phase (Table 17.4-2). The choice of methods will depend on the nature of the activity and the phase within the mining life cycle. The focus of the analysis will be informed by the prior scoping and profiling phases, and the baseline data will form a foundation for anticipating and predicting risks and impacts. Predictive assessment usually requires the collection of additional data.

Where possible, analysis should consider associated facilities, policies, or programs, such as roads, power transmission lines, and community development initiatives. Predictions should both describe how the proposed activity will contribute to the existing situation and assess the capacity of environmental and social systems to absorb impacts. The methodology and level of confidence should also be clearly explained, as well as any gaps in knowledge.

The outcomes of predictive assessment and analysis are usually prioritized by their scale and level of significance (sometimes as a matrix). They are used to provide feedback to engineers and project developers in order to modify and revise the project, as well as to enable them to decide which proposed project alternative best achieves the objectives of the project while still enhancing social outcomes and avoiding negative impacts.

As part of the trend toward integrated management systems, most mining companies have now adopted extensive risk-management processes that encompass many different elements. It is common to find consequence tables in risk-assessment procedures that include both environmental and community dimensions (Barclay et al. 2009). Social risk assessment, also known as social risk and opportunity analysis, is a technique that can be undertaken during the predictive assessment phase, but it may also be periodically employed to identify and update key social risks and opportunities, and to respond with appropriate strategies, during the impact management phases. Social risk assessments usually involve a workshop of key stakeholders (internal and/or external to the organization) and may be focused on the social risks and opportunities faced by the company or communities. The key tasks are to

1. List the potential social risks and opportunities that may affect (a) the company, and (b) the community;
2. Think about the causal factors for each of the risks/opportunities identified;
3. Make an assessment of the likelihood of the risk/opportunity occurring;
4. Make an assessment of the consequence if the risk/opportunity were to materialize;
5. Prioritize the risks and opportunities based on the assessment; and

6. Develop controls (ways to mitigate or enhance) for the most significant risks and opportunities (Evans et al. 2007).

The outcomes of such a social risk-assessment process should be captured as part of the overall "risk register" for the project. Social risks can be among the highest priority category for mining projects.

Management Strategies to Avoid, Mitigate, and Enhance Impacts

In many circumstances, there is a lack of integration between SIA and the ongoing management of social and economic issues after a project commences or after an operation closes. This can happen for many reasons. SIAs are often conducted by external consultants to serve the regulatory need for independent analysis. In such cases, there can be a lack of continuity between impact assessment and relationship development with community and stakeholders. Another possible reason for a lack of integration is that impact assessment might be the responsibility of a small team, most likely in the community relations section, whereas management requires coordination across all aspects of the operation.

It is important that the outcomes of SIA be embedded across all aspects of the business, similar to the way that health and safety have been embraced at a corporate level in recent years. Social impact management can be formalized into management systems, typified by various series of the International Organization for Standardization, site plans, agreements, development of standard operating procedures for high-risk issues, and systems to handle complaints and grievances. Examples of management procedures to address social issues include formal complaint handling systems, cultural heritage management plans, human rights and cultural awareness training (linked to human resources systems), and local sourcing and purchasing policies.

For some impacts, particularly cumulative impacts, the most effective approach may not be to target a particular impact generated from mining but to invest "off-site" to ameliorate or enhance impacts generated by other activities. At the broader level, there are four approaches to the management of impacts. These are to (1) mitigate or enhance the impacts of past and existing activities; (2) mitigate or enhance the impacts of the project or activity under consideration; (3) mitigate or enhance the impacts of potential future projects; or (4) consider whether and how these projects or activities should proceed (Duinker and Greig 2006).

This section has introduced a number of strategies for ongoing management. There is a great deal of overlap between the strategies outlined here and the practice of community relations, community development, and community engagement. The following discussion details a sample of approaches including social management plans, scenario planning, community reference groups, community development and social investments, networking and working groups, and complaints and grievance handling mechanisms.

Social Management Plans
Social management plans (also known as "environmental and social management plans," "social and labor plans" or "environmental and social action plans") summarize the findings of the impact assessment; they outline the measures adopted to enhance positive impacts and to avoid, mitigate, and (as a last resort) offset and compensate negative impacts (Franks et al. 2009b). In addition, the plans provide estimates of the timing, frequency, duration, and cost of management measures. They also establish monitoring and reporting procedures (discussed later in greater depth).

Social management plans are usually developed in partnership with regulatory agencies, investors, and the community; they identify the responsibilities of each party in the management of impacts, opportunities, and risks. Management plans also provide an opportunity to link activities with local and regional planning processes and, if developed with reference to the management plans of other operations, can assist in addressing cumulative impacts. They also provide the facility to coordinate project activities with service and infrastructure planning by government.

The plans may explicitly refer to capacity-building activities, where the institutional or community capacity to undertake such activities is lacking, and may include details of community development and social investments. Finally, social management plans outline the procedures for how social issues will be addressed in site management systems and plans, the processes for ongoing public participation and information disclosure, and the mechanisms for handling community grievances and feedback. Measures that are the responsibility of other parties are recorded and form a basis for ongoing partnerships (Franks et al. 2009b).

Scenario Planning
Scenario planning can assist organizations in preparing for unplanned activities. Scenario analysis is a tool to anticipate change under different plausible future situations. It assists the development of a proactive policy response through the testing of assumptions. If conducted with communities, scenario planning can help to inform the public of risks and manage expectations. For example, the boom-and-bust nature of the industry can increase the risk of premature or temporary closure and downsizing of operations. The effects of mine closures (where impacts are generated by the absence of activities) can be a significant challenge for regional communities and economies. Planning for closure should proceed well before an operation starts, and measures should be put in place to prepare communities and companies for such an eventuality, which will most definitely occur (however large the deposit or valuable its product).

Community Reference Groups
Community engagement is an important component of social impact management. It involves activities such as the communication of the project proposal to stakeholders; the incorporation of stakeholder views in order to modify project details; the ongoing involvement of stakeholders in community boards and reference groups; and stakeholder participation in the submission of ideas for, and implementation of, community projects.

Regardless of the form of engagement, it is important to be upfront and straightforward with stakeholders when communicating the potential impacts of a project. False impressions can distort expectations and become an ongoing point of contention and a breach of trust when the true nature of impacts becomes clear (Franks 2009).

Community reference groups provide a forum for ongoing consultation and engagement. Representation can include groups such as youth and aged organizations, local business, tourism, health, welfare, policing, and education, in addition to environment, government, and community groups. Broad representation helps to ensure that a range of issues are covered, (although reference groups should be kept to a manageable size). For community reference groups to be at their most effective, there needs to be strong mechanisms for feedback to the broader community to provide an opportunity for input and to report on outcomes.

Community Development and Social Investments

A key strategy adopted by mining companies to manage social impacts involves programs to support community development. These may include health and education programs or support of organizations such as schools, clubs, and societies. It is important to focus and coordinate investments to target community priorities and identified needs. Community development may be prioritized by the outcomes of the scoping, profiling, and predictive assessment phases and, more importantly, through community participation. Partnerships are often the best way to facilitate local capacity-building and development programs, social services, and infrastructure (Kemp 2003, 2010). Partnerships with organizations, service providers, governments, other mining companies, and peak industry bodies can be effective in mobilizing greater resources, leveraging investments, and coordinating activities. Many mining organizations prefer an approach where community-led initiatives are supported by industry. This approach seeks to build the capacity of communities and their organizations to undertake activities and avoid overdependence on mining companies. Examples of community trusts include BHP Billiton's Minera Escondida Foundation in Chile and Rio Tinto's Coal and Allied Community Trust in Australia.

Networking and Multistakeholder Working Groups

Informal and formal networks can provide important opportunities to exchange experiences at the operational and strategic level to better manage the impacts of activities. Professional networking is an opportunity to exchange ideas and advice, as well as to communicate approaches (both successes and failures). Formal networking arrangements, such as forums of mine managers and professional staff, provide an ongoing opportunity to discuss common issues and coordinate activities. Examples include the Central Queensland Mining Rehabilitation Group, which is a forum for sharing experiences about environmental management on mine sites in central Queensland, Australia, and the Muswellbrook Mine Managers forum, which is a regular meeting between the general managers of multiple coal mines and local government in the Hunter Valley, Australia.

Multistakeholder working groups are an opportunity to facilitate partnerships around a particular goal. Working groups can share strategic information, develop and coordinate solutions, undertake research into best practice, and facilitate cross-sector communication. Multistakeholder working groups are well placed to focus on the management of social issues at a regional scale.

Complaints and Grievance Handling Mechanisms

Complaints and grievance handling mechanisms are becoming more common at the operational level within the mining industry as a means to actively respond to community concerns. A grievance is a concern, issue, or problem that is usually expressed through a complaint or protest by individuals or groups. According to Kemp and Gotzman (2009) grievance mechanisms include

- A dedicated pathway (or pathways) and processes of engagement for handling grievances;
- Procedural elements, such as a documented procedure outlining steps to be taken to handle community grievances;
- Records, such as complaints/grievance logs and data, evidence of communication about the process, and documentation of outcomes;
- Dedicated resources, such as human and financial resources, as well as formally defined responsibilities for grievance handling;
- Evidence of dialogue with aggrieved parties and/or use of alternative dispute-resolution techniques (negotiation, mediation, arbitration, etc.) where direct dialogue is not possible or has little potential (through normal channels) of leading to a resolution of issues; and
- Substantive outcomes, such as improved organizational practice and relationships, and conflict resolution (validated by aggrieved parties).

Monitoring and Reporting

The monitoring and reporting phase involves collection, analysis, and dissemination of information on social impacts, opportunities, and risks over time. This phase can (1) assist in refining assessments, (2) track the progress of social impact management approaches and identify changes needed (adaptive management), (3) assess the return that a company is getting on its community investments, (4) report to communities on how they are being impacted, and (5) facilitate an informed dialogue around these issues.

Monitoring

Social impact monitoring can be challenging for many reasons: time lags between actions and outcomes, difficulty in isolating impacts, difficulties in obtaining and comparing data, differences in the way people experience impacts, the risk of information overload for managers, and the complexity of the way impacts interact. Not all impacts can be separated and analyzed independently because they do not exist in isolation of the social and ecological context. It is important, therefore, to have ongoing points of intervention and monitoring.

Monitoring should be undertaken over meaningful time scales and spatial extent. Regional monitoring can help to address the cumulative impacts of multiple actions (Franks et al. 2009a). Monitoring should also be designed to facilitate community participation. The participation of community members can assist many aspects of monitoring, including the collection of data, participation in the development of indicator frameworks, or multistakeholder monitoring organizations. Meaningful participation can assist in building public confidence and trust in the monitoring and resolution process.

Monitoring consists of a number of discrete activities. These include the following:

1. Decide what to monitor. Prioritization can be based on community engagement and the profiling, assessment, and impact management phases. It is important to monitor outcomes, not just effort and activity.
2. Define targets, limits, and thresholds. What are the desired and undesired outcomes? Thresholds refer to

scientifically defined points where undesirable changes result if exceeded. Limits consider what may be acceptable to the community as determined through consultation and participation, with targets being designed as the desired future outcomes.
3. Select indicators. What would indicate that change is occurring or has occurred?
4. Establish measures. What data are preferred, what data are available, how will data be collected, and how often will it be collected?
5. Report and interpret data. Identify and verify trends (or nontrends), interpret trends (attribute change), and communicate results.

The collection of irrelevant information can be avoided through planning and prioritization, and the measurement load can be minimized through the appropriate design of administrative systems and the use of existing data whenever possible. Qualitative data can provide exceptionally useful information if the methodology employed is consistent and robust.

Reporting
Sustainability reporting has become a significant activity for the mining industry to communicate performance at the operational and corporate levels. Reporting consists of the documentation and communication of information on numerous activities and outcomes. In recent years there has been a trend toward standardized reporting requirements, such as the Global Reporting Initiative (GRI). The GRI is a sustainability reporting framework that requires reporting on the economic impacts on project stakeholders and systems; environmental inputs, outputs, and expenditure; labor practices; human rights; and social risks to communities. The GRI has developed, in collaboration with the International Council on Mining and Metals, a mining and metals sector supplement that details specific disclosures and indicators for the industry (GRI 2010). The Extractives Industry Transparency Initiative (EITI) is another such approach. The EITI is a global standard for transparency in oil, gas, and mining that is implemented by both businesses and governments. EITI requires industry to publish all of the payments made to governments, such as through royalties and taxes, and for governments to disclose the revenues they receive from resource developments (EITI 2009).

The documentation of social decisions and agreements (internal and external to the operations) is a key aspect of reporting that is often overlooked. The nondelivery of agreed outcomes is a salient issue at many operations. Nondelivery can create unaccounted-for future liabilities, breach trust with communities, and damage the reputation of the operation.

For some operations and issues, reporting is best addressed at a regional scale. In circumstances where multiple mining operations are located in close proximity to a single town or community, there is often an absence of information that provides a comprehensive overview of industry investments, activities, aggregate impacts, and the state of the environment. Collective reporting to the community on the economic, social, and environmental performance of the industry may be more effective at communicating the overall contribution of the industry and the totality of activities and impacts (Brereton et al. 2008; Franks et al. 2009a). Regional organizations and industry bodies are best placed to coordinate such efforts; however, the absence of a representative organization is not necessarily prohibitive.

Evaluation and Review
The final phase of social impact assessment and management is to evaluate and review the assessment and management processes. An active process of evaluation and review—and importantly, the adjustment of actions—are fundamental features of adaptive management and should be integrated into each of the previous phases. The reconciliation of impacts predicted during the assessment phase with the actual impacts experienced during implementation will assist in refining and improving future approaches. Structured review should be built into ongoing programs, policies, projects, and agreements in order to reflect contemporary conditions.

LESSONS FOR THE EFFECTIVE MANAGEMENT OF SOCIAL IMPACTS
This section summarizes key lessons that can inform mine managers, engineers, and technical professionals to effectively manage the social impacts of mining. It is important to note that communities are complex and, at times, difficult to predict, that local context is paramount, and that the transferability of approaches is often not straightforward. At the same time, consideration of these issues will assist mining professionals to design and implement projects that are more acceptable to communities and therefore have a greater prospect of success.

1. Engage with community stakeholders and build relationships through understanding and goodwill. Be upfront and straightforward about potential risks. Establish meaningful and timely avenues for two-way dialogue, and use this to understand stakeholder histories, relationships, and networks, as well as the values that shape attitudes and behaviors. Assist stakeholders to articulate community concerns and visions. Listen to and design the project or activity within the parameters of such perspectives. Be responsive and adaptive; respect customs and political and authority structures; and, where appropriate, gain informed consent.
2. Align and integrate the outcomes of assessment and monitoring into site management systems and plans. Ensure company-wide understanding and respect of community expectations, concerns, and future visions. SIA and impact management are only effective to the extent that the responsible staff and the outcomes are well integrated into the overall management of the entire operation.
3. Draw on specialist skills and social and economic research to build knowledge and understanding. The use of specialists, however, should not come at the expense of integrating understandings into the organization, nor should it confuse accountability or responsibilities. Relationships with community stakeholders are most effective when stakeholders feel the people they are dealing with have the authority to make decisions and changes.
4. Build the capacity of stakeholders. Tailor community development activities to enhance independence and postmining legacies.
5. Monitor what matters. Design monitoring in such a way that the community's concerns are identified and acted on. Acknowledge and remedy (past) issues and grievances,

and fulfill outstanding commitments. Use appropriate systems to record issues and to coordinate and deliver outcomes.
6. Continuously review activities and programs, and periodically update assessments as "living" documents.
7. Assess and manage social impacts and risks across the life cycle of mining and resource processing activities, including all of the activities from exploration through to postclosure, as well as recycling and waste management. The scale and focus of activities will vary depending on the stage, and local and operational context.

The processes of social impact assessment and impact management detailed in this chapter will assist in identifying key stakeholder issues, predicting and anticipating change, and embedding these understandings into ongoing systems and strategies to proactively respond to the consequences of mineral exploitation. Trust is an important feature of managing social impacts. Mining companies may not originate from the region, or be familiar with the local culture, customs, and lifestyles, but nevertheless they have the power to transform the environment and society. As outsiders, mining companies may be viewed with suspicion by communities and must earn community trust. Even well-meaning actions, when not socially, culturally, or environmentally relevant, may result in poor outcomes. However, by proactively responding to community issues, facilitating meaningful participation, and shaping mutually beneficial futures, mining companies can avoid conflict with communities and the associated costs.

ACKNOWLEDGMENTS

The author acknowledges the contributions of David Brereton, Catherine Pattenden, Deanna Kemp, Robin Evans, Warwick Browne, and Julia Keenan, who developed aspects of the research presented in this chapter and/or provided comments on earlier drafts.

REFERENCES

Barclay, M., Franks, D., and Pattenden, C. 2009. Risk communication: A framework for technology development and implementation in the mining and minerals processing industries. Parker Cooperative Research Centre for Integrated Hydrometallurgy Solutions final report. Brisbane, Australia: Centre for Social Responsibility in Mining, Sustainable Minerals Institute, University of Queensland.

Becker, H., and Vanclay, F., eds. 2006. *The International Handbook of Social Impact Assessment: Conceptual and Methodological Advances*. Cheltenham, UK: Edward Elgar. pp. 74–91.

Brereton, D., Moran, C.J., McIlwain, G., McIntosh, J., and Parkinson, K. 2008. Assessing the cumulative impacts of mining on regional communities: An exploratory study of coal mining in the Muswellbrook area of New South Wales. ACARP Project C14047. Brisbane, Australia: Centre for Social Responsibility in Mining, Centre for Water in the Minerals Industry, and Australian Coal Association Research Program.

Duinker, P., and Greig, L. 2006. The impotence of cumulative effects assessment in Canada: Ailments and ideas for redeployment. *Environ. Manage.* 37(2):153–161.

EITI (Extractive Industries Transparency Initiative). 2009. Principles and criteria. www.eiti.org/eiti/principles. Accessed September 2009.

Evans, R., Brereton, D., and Joy, J. 2007. Risk assessment as a tool to explore sustainable development issues: Lessons from the Australian coal industry. *Int. J. Risk Assess. Manage.* 7(5):607–619.

Franks, D. 2009. Avoiding mine-community conflict: From dialogue to shared futures. In *Proceedings of the First International Seminar on Environmental Issues in the Mining Industry (Enviromine 2009)*, Santiago, Chile, September 30–October 2. Edited by J. Wiertz and C.J. Moran. Santiago, Chile: Gecamin.

Franks, D., Brereton, D., and C.J. Moran. 2009a. Surrounded by Change—Collective strategies for managing the cumulative impacts of multiple mines. In *Proceedings of the International Conference on Sustainable Development Indicators in the Minerals Industry*, Gold Coast, Queensland, Australia, July 6–8. Victoria, Australia: Australasian Institute of Mining and Metallurgy.

Franks, D., Fidler, C., Brereton, D., Vanclay, F., and Clark, P. 2009b. Leading practice strategies for addressing the social impacts of resource developments. Briefing paper for the Department of Employment, Economic Development and Innovation, Queensland Government. Brisbane, Australia: Centre for Social Responsibility in Mining, Sustainable Minerals Institute, University of Queensland.

GRI (Global Reporting Initiative). 2009. Sustainability reporting guidelines and mining and metals sector supplement. Draft sector supplement for public comment. January 28–April 29, 2009. Version 6. www.globalreporting.org/NR/rdonlyres/E75BAED5-F176-477E-A78E-DC2E434E1FB2/2454/DraftFinalMiningandMetalsSectorSupplment.pdf. Accessed December 2009.

Kemp, D. 2003. Discovering participatory development through corporate-NGO collaboration: A mining industry case study. Research Paper No. 2. Brisbane, Australia: Centre for Social Responsibility in Mining.

Kemp, D. 2010. Community relations in the global mining industry: Exploring the internal dimensions of externally orientated work. *Corp. Soc. Resp. Environ. Manage.* 17(1):1–14.

Kemp, D., and Gotzmann, N. 2009. Community grievance mechanisms and the Australian minerals industry: An industry discussion paper. Brisbane, Australia: Centre for Social Responsibility in Mining, Sustainable Minerals Institute, University of Queensland.

Vanclay, F. 2003. International Principles for Social Impact Assessment. *Impact Assess. Proj. Apprais.* 21(1):5–11.

APPENDIX A

Web Sites Related to Mining

UNITED STATES PROFESSIONAL ORGANIZATIONS

American Coal Ash Association (ACAA): www.acaa-usa.org
American Coal Foundation: www.teachcoal.org
American Geological Institute: www.agiweb.org
American Geophysical Union: www.agu.org
American Institute of Mining, Metallurgical, and Petroleum Engineers (AIME): www.aimehq.org
American Institute of Professional Geologists (AIPG): www.aipg.org
American Rock Mechanics Association (ARMA): www.armarocks.org
American Society for Testing and Materials (ASTM): www.astm.org
Association of American State Geologists (AASG): www.stategeologists.org
Association of Environmental and Engineering Geologists (AEG): www.aegweb.org
China Clay Producers Association (CCPA): www.kaolin.com
Coal Operators and Associates, Inc.: www.miningusa.com/coa
Electric Power Research Institute (EPRI): www.epri.com
Energy and Mineral Law Foundation (EMLF): www.emlf.org
(The) Environmental and Engineering Geophysical Society (EEGS): www.eegs.org
Gems, Rocks and Minerals: www.gemsrocks.com
GEOindex: www.geoindex.com/geoindex
Geo-Institute: www.geoinstitute.org
(The) Geological Society of America (GSA): www.geosociety.org
Illinois Clean Coal Institute (ICCI): www.icci.org
International Center for Aggregates Research (ICAR): www.icar.utexas.edu
International Society for Soil Mechanics and Geotechnical Engineering (ISSMGE): www.issmge.org
International Society of Explosives Engineers (ISEE): www.isee.org
Kentucky Coal Education: www.coaleducation.org
Kentucky Mining Institute (KMI): www.miningusa.com/kmi
Lignite Energy Council: www.lignite.com
Mineral Economics and Management Society: www.minecon.com
Minerals and Geotechnical Logging Society (MGLS): http://mgls.spwla.org
(The) Minerals, Metals and Materials Society (TMS): www.tms.org
Mining Industry Council of Missouri: www.momic.com
Mississippi Valley Trade and Transport Council (MVTTC): www.mvttc.com
National Association of Geoscience Teachers (NAGT): www.nagt.org
National Council of Examiners for Engineering and Surveying (NCEES): www.ncees.org
National Mine Land Reclamation Center: http://wvwri.nrcce.wvu.edu/programs/nmlrc/index.cfm
(The) National Mining Association (NMA): www.nma.org

National Research Center for Coal and Energy (NRCCE): www.nrcce.wvu.edu
National Stone, Sand, and Gravel Association (NSSGA): www.nssga.org
North Carolina Coal Institute (NCCI): www.nccoal.org
Nuclear Energy Institute (NEI): www.nei.org
Rocky Mountain Association of Geologists (RMAG): www.rmag.org
Rocky Mountain Mineral Law Foundation (RMMLF): www.rmmlf.org
Seismological Society of America (SSA): www.seismosoc.org
Society for Mining, Metallurgy, and Exploration, Inc. (SME): www.smenet.org
Society for Sedimentary Geology (SEPM): www.sepm.org
Society of Economic Geologists (SEG): www.segweb.org
Society of Exploration Geophysicists (SEG): www.seg.org
Society of Independent Professional Earth Scientists (SIPES): www.sipes.org
Society of Petroleum Engineers (SPE): www.spe.org
Soil Science Society of America (SSSA): www.soils.org
Solution Mining Research Institute (SMRI): www.solutionmining.org
United States Universities Council on Geotechnical Education and Research (USUCGER): www.usucger.org

UNITED STATES REFERENCE, INFORMATION, AND PUBLICATIONS
Aggregates and Roadbuilding: www.rocktoroad.com
Argus: www.energyargus.com
Chemical Engineering: www.che.com
Coal Age: http://www.coalage.com
Coal Information Network: www.coalinfo.org
Dredging News Online: www.sandandgravel.com
Earth: www.earthmagazine.org
Electronic Journal of Geotechnical Engineering: www.ejge.com/Index_ejge.htm
Energy Market Report: www.econ.com/emrindex.html
Engineering and Mining Journal (E&MJ): www.e-mj.com
Engineering News-Record (ENR): www.enr.com
GoldAndSilverMines.com: www.goldandsilvermines.com
Goldsheet Mining Directory: www.goldsheetlinks.com
ICMJ's Prospecting and Mining Journal: www.icmj.com
InfoMine: www.infomine.com
Journal of Metals (JOM): www.tms.org/jom.html
Metal World: www.metalworld.com
Mine-On-Line: www.mine-on-line.com
Mineral Information Institute: www.mii.org
Miners News: www.minersnews.com
Minesite.*com:* www.minesite.com
Mining Engineering: http://me.smenet.org
Mining.com: www.mining.com
(The) Mining Record: www.miningrecord.com
(The) Northern Miner: www.northernminer.com
Pit and Quarry: www.pitandquarry.com
Platts: www.platts.com
Rock and Dirt: www.rockanddirt.com
Skillings Mining Review: www.skillings.net
World-Wide Web Virtual Library of Geotechnical Engineering: www.ejge.com/GVL

INTERNATIONAL MINING ECONOMICS
AME Mineral Economics (Australia): www.ame.com.au
American Metal Market: www.amm.com
Bloomberg: www.bloomberg.com
Bloomsbury Minerals Economics (United Kingdom): www.bloomsburyminerals.com
Chicago Stock Exchange: www.chx.com

CME Group: www.cmegroup.com
Coal Trading Association: www.coaltrade.org
Commodity Research Bureau: www.crbtrader.com
Denver Gold Group: www.denvergold.org
Dow Jones: www.dowjones.com
Financial Times Energy (United Kingdom): www.ftenergy.com.companies/energy
London Metal Exchange (United Kingdom): www.lme.co.uk
Metals Economics Group (United Kingdom): www.metalseconomics.com
MineMarket.com: www.minemarket.com
NASDAQ: www.nasdaq.com
New York Mercantile Exchange (NYMEX/COMEX): www.nymex.com/index.aspx
New York Stock Exchange (NYSE Euronext): www.nyse.com
OTC Bulletin Board: www.otcbb.com
Reuters: www.reuters.com
Standard and Poor's (S&P): www.standardandpoors.com/home/en/us
Wall Street Journal: www.wsj.com
World Mine Cost Data Exchange, Inc.: www.minecost.com

MINING COMPUTING ORGANIZATIONS
Engineering Software Center: www.engsoftwarecenter.com
Geotechnical and Geoenvironmental Software Directory: www.ggsd.com
Macintosh Geological Software: http://geowww.geo.tcu.edu/faculty/geosoftware.html
Mining Internet Services, Inc.: www.miningusa.com

UNITED STATES GOVERNMENT

Federal Agencies
Army Corps of Engineers (ACE): www.usace.army.mil/pages/default.aspx
Bureau of Land Management (BLM): www.blm.gov/wo/st/en.html
Bureau of Reclamation: www.usbr.gov
Commodity Futures Trading Commission: www.cftc.gov
Council on Environmental Quality (CEQ): www.whitehouse.gov/administration/eop/ceq
Department of Commerce: www.commerce.gov
Department of Energy (DOE): www.energy.gov
Department of the Interior (DOI): www.doi.gov
Energy Information Administration (EIA): www.eia.doe.gov
Environmental Protection Agency (EPA): www.epa.gov
Fossil Energy: www.fe.doe.gov
Geological Survey (USGS): www.usgs.gov
Mine Safety and Health Administration (MSHA): www.msha.gov
Minerals Information: http://minerals.usgs.gov/minerals
Minerals Management Service (MMS): www.mms.gov
National Geophysical Data Center: www.ngdc.noaa.gov
National Institute for Occupational Safety and Health (NIOSH): www.cdc.gov/niosh/mining
National Science Foundation (NSF): www.nsf.gov
National Technical Information Service (NTIS): www.ntis.gov
Nuclear Regulatory Commission (NRC): www.nrc.gov
Office of Surface Mining (OSM): www.osmre.gov
State Minerals Statistics and Information: http://minerals.usgs.gov/minerals/pubs/state

Mining Agencies and Associations (by State)
Alabama Surface Mining Commission: www.surface-mining.state.al.us
Alaska Department of Natural Resources, Division of Mining, Land and Water: http://dnr.alaska.gov/mlw
Alaska Miners Association: www.alaskaminers.org
Arizona Department of Mines and Mineral Resources: www.admmr.state.az.us

Arizona State Mine Inspector: www.asmi.state.az.us
Arkansas Department of Environmental Quality: www.adeq.state.ar.us/mining
California Mining Association: www.calmining.org
California Office of Mine Reclamation: www.conservation.ca.gov/omr
California State Mining and Geology Board: www.conservation.ca.gov/smgb
Colorado Division of Reclamation Mining and Safety: http://mining.state.co.us
Colorado Mining Association: www.coloradomining.org
Connecticut Department of Environmental Protection: www.ct.gov/dep/sit/default.asp
Delaware Geologic Survey: www.dgs.udel.edu
Florida Mining and Minerals Regulation: www.dep.state.fl.us/water/mines/index.htm
Georgia Environmental Protection Division: www.gaepd.org
Hawaii Department of Lands and Natural Resources: http://hawaii.gov/dlnr/land
Idaho Department of Lands: www.idl.idaho.gov
Idaho Mining Association: www.idahomining.org/ima
Illinois Office of Mines and Minerals: http://dnr.state.il.us/mines
Indiana Department of Natural Resources, Division of Reclamation: www.in.gov/dnr/reclamation
Iowa Soil Conservation Division: www.agriculture.state.ia.us/soilConservation.asp
Kansas Department of Health and Environment: www.kdheks.gov/mining
Kentucky Coal Association: www.kentuckycoal.org
Kentucky Division of Mine Reclamation and Enforcement: www.dmre.ky.gov
Kentucky Office of Mine Safety and Licensing: www.omsl.ky.gov
Louisiana Office of Conservation, Surface Mining Section: http://dnr.louisiana.gov/CONS/CONSERIN/Surfmine.ssi
Maine Bureau of Land and Water Quality: www.maine.gov/dep/blwq
Maryland Bureau of Mines: www.mde.state.md.us/Programs/WaterPrograms/MiningInMaryland/BOM
Massachusetts Department of Conservation and Recreation: www.mass.gov/dcr
Michigan Department of Natural Resources: www.michigan.gov/dnr
Minnesota Division of Lands and Minerals: www.dnr.state.mn.us/lands-minerals/index.html
(Minnesota) Iron Mining Association of Minnesota: www.taconite.org
Mississippi Mining and Reclamation Division: www.deq.state.ms.us/mdeq.nsf/page/geology-mining-and-reclamation
Missouri Department of Natural Resources: www.dnr.mo.gov
Montana Department of Environmental Quality: www.deq.state.mt.us
Nebraska School of Natural Resources, Conservation and Survey Division: http://snr.unl.edu/csd/
Nevada Bureau of Mines and Geology: www.nbmg.unr.edu
Nevada Division of Minerals: http://minerals.state.nv.us
Nevada Mining Association: www.nevadamining.org
New Hampshire Department of Environmental Services: http://des.nh.gv
New Jersey Geological Survey: www.state.nj.us/dep/njgs
New Mexico Bureau of Mine Safety: www.bmi.state.nm.us
New Mexico Mining and Minerals Division: www.emnrd.state.nm.us/mmd
New York Division of Mineral Resources: www.dec.ny.gov/about/636.html
North Carolina Division of Land Resources: www.dlr.enr.state.nc.us
North Dakota Geological Survey: www.dmr.nd.gov/ndgs
North Dakota Public Service Commission, Coal Mine Reclamation: www.psc.state.nd.us/jurisdiction/reclamation.html
North Dakota State Land Department: www.land.nd.gov
Northwest Mining Association: www.nwma.org
Ohio Aggregates and Industrial Minerals Association: www.oaima.org
Ohio Coal Association: www.ohiocoal.com
Ohio Division of Mineral Resources Management: www.dnr.state.oh.us/mineral/default/tabid/10352/Default.aspx
Oklahoma Department of Mines: www.ok.gov/mines
Oregon Department of Geology and Mineral Industries: www.oregongeology.org/sub/default.htm

Pennsylvania Office of Mineral Resources Management: www.dep.state.pa.us/dep/deputate/minres/minres.htm
Rhode Island Bureau of Natural Resources: www.dem.ri.gov/programs/bnatres
South Carolina Geological Survey: www.dnr.sc.gov/geology
South Dakota Minerals and Mining Program: http://denr.sd.gov/des/mm/mmprogram.aspx
Tennessee Department of Environment and Conservation: www.state.tn.us/environment
Texas Bureau of Economic Geology: www.beg.utexas.edu
(Texas) Railroad Commission of Texas: www.rrc.state.tx.us
Utah Department of Natural Resources: http://naturalresources.utah.gov
Utah Division of Oil, Gas, and Mining: http://ogm.utah.gov
Utah Mining Association: www.utahmining.org
Vermont Agency of Natural Resources: www.anr.state.vt.us
Virginia Department of Mines, Minerals and Energy: www.mme.state.va.us
Washington Geology and Earth Resources Division: www.dnr.wa.gov/ResearchScience/GeologyEarthSciences
West Virginia Coal Association: www.wvcoal.com
West Virginia Department of Environmental Protection: www.wvdep.org
West Virginia Office of Miners' Health Safety and Training: www.state.wv.us/mhst
Wisconsin Waste and Materials Management: www.dnr.state.wi.us.org/aw/wm
Wyoming Department of Environmental Quality: http://deq.state.wy.us
Wyoming Mining Association: www.wma-minelife.com

INTERNATIONAL PUBLICATIONS AND ASSOCIATIONS

Africa
Chamber of Mines of South Africa: www.bullion.org.za
MBendi (South Africa): www.mbendi.com

Asia
Asian Journal of Mining: www.asianmining.com
Chamber of Mining Engineers of Turkey: www.maden.org.tr/index_en.php
China Coal Information Network: www.coalinfo.net.cnChina Coal Research Institute, Xi'an Branch: http://xianccri.en.alibaba.com/
Japan Coal Energy Center: www.jcoal.or.jp/index-en.html
Mineral Resources Authority of Mongolia: www.mram.gov.mn/en/index.php
Mining India: www.miningindia.com

Australia and New Zealand
Association of Mining and Exploration Companies: www.amec.asn.au
Australasian Institute of Mining and Metallurgy (AusIMM): www.ausimm.com.au
Australian Centre for Mining Environmental Research (ACMER): www.acmer.uq.edu.au
Australian Coal Association Research Program: www.acarp.com.au
Australian Institute of Geoscientists: http://aig.org.au
Australian Journal of Mining: www.theajmonline.com.au
Australian Mines and Metals Association.: www.amma.org.au
Chamber of Mines and Energy of Western Australia, Inc.: www.cmewa.com
New South Wales Minerals Council: www.nswmin.com.au
New Zealand Minerals Industry Association: www.minerals.co.nz/html/index.html
OzGold Database Technology: www.comcen.com.au/~ozgold/main.html
Queensland Resources Council: www.qrc.org.au/default.asp

Canada
Association for Mineral Exploration British Columbia: www.amebc.ca/Home.aspx
CAMESE (Canadian Association of Mining Equipment and Services for Export): www.camese.org
Canadian Diamond Drilling Association: www.canadiandrilling.com
Canadian Institute of Mining, Metallurgy and Petroleum (CIM): www.cim.org/splash

Canadian Mining Industry Research Organization (CAMIRO): www.camiro.org
Canadian Mining Journal: www.canadianminingjournal.com
Coal Association of Canada: www.coal.ca/content
International Council on Mining and Metals: www.icmm.com
Mining Association of Canada: www.mining.ca
Natural Resources Canada: www.nrcan-rncan.gc.ca.com
NWT and Nunavut Chamber of Mines: www.miningnorth.com
Ontario Mining Association: www.oma.on.ca/en
Ontario Ministry of Northern Development, Mines and Forestry: www.mndm.gov.on.ca
Prospectors and Developers Association of Canada: www.pdac.ca
Yukon Chamber of Mines: www.ycmines.ca

Europe
Coal Services International (Ireland): www.coalservices.com
EuSalt (European Salt Producers' Association) (France): www.eusalt.com
German Brown Coal Association: www.braunkohle.de
German Coal Importers Association: www.verein-kohlenimporteure.de
GInfoServer: www.geo.uni-bonn.de/members/haack/gisinfo.html
IEA Coal Industry Advisory Board (France): www.iea.org/ciab
International Energy Agency (IEA): www.iea.org
International Organization for Standardization (ISO)(Switzerland): www.iso.org
International Society for Rock Mechanics (Portugal): www.isrm.net

Latin America
Editec (Chile): www.editec.cl
La Camara de Mineria del Ecuador: www.cme.org.ec/portal
National Society of Mining Petroleum and Energy (Peru): www.snmpe.org.pe
Panorama Minero (Argentina): www.panoramaminero.com.ar

United Kingdom
Association of Geotechnical and Geoenvironmental Specialists: www.ags.org.uk/site/home/index.cfm
British Geological Survey: www.bgs.ac.uk
(The) Coal Authority: www.coal.gov.uk
Coal International: www.quarryworld.co.uk/coalinternational.php
(The) Geological Society: www.geolsoc.org.uk/index.html
Geopages: www.geopages.co.uk/about/aboutus.html
globalCOAL: www.globalcoal.com
IEA Clean Coal Centre: www.iea-coal.org.uk
Industrial Minerals: www.mineralnet.co.uk
Institute of Materials, Minerals and Mining: www.iom3.org
Institute of Quarrying: www.inst-of-quarrying.org/iq
International Association of Hydrogeologists: www.iah.org
International Mining: www.im-mining.com
(The) McCloskey Group: www.mccloskeycoal.com
Minerals and Energy—Raw Materials Report (Sweden): www.tandf.co.uk/journals/SMIN
Mining and Quarry World: www.quarryworld.co.uk/miningquarryworld.php
Mining Journal Online: www.mining-journal.com
Mining Magazine: www.miningmagazine.com
World Coal: www.energyglobal.com/magazine/latestissue/world-coal.aspx
World Coal Institute: www.wci-coal.com
World Mining Equipment: www.wme.com
World Nuclear Association: www.world-nuclear.org

APPENDIX B

Coal Mine Gas Chart

Coal Mine Gas Chart

Name of Gas	Symbol	Specific Gravity	Effect on Humans	Explosive Limits	Characteristics
Hydrogen	H_2	0.07	Nonpoisonous, but will not support life	LEL, 4%; UEL, 74% (NP: 5.1% O_2, 4.3% H_2)	Colorless, odorless, tasteless
Methane	CH_4	0.55	Nonpoisonous (slight narcotic effect at high concentrations)	LEL, 5%; UEL, 14% (NP: 12.2% O_2, 5.9% CH_4)	Colorless, odorless, tasteless
Carbon monoxide	CO	0.97	Poisonous chemical; asphyxiant (2,000 ppm, possibly fatal exposure)	LEL, 12.5%; UEL, 74% (NP: 6.1% O_2, 13.8% CO)	Colorless, odorless, tasteless
Nitrogen	N_2	0.97	Nonpoisonous (can cause nitrogen narcosis under pressure)	Inert (extinguishing agent)	Colorless, odorless, tasteless
Air	Air	1	20.93% O_2; 78.11% N_2; 0.03% CO_2; 0.93% others	NA	Colorless, odorless, tasteless
Oxygen	O_2	1.11	Essential to life (becomes toxic at >1 atmosphere pressure)	Nonflammable of itself (but supports and greatly increases combustion of all flammable substances)	Colorless, odorless, tasteless
Hydrogen sulfide	H_2S	1.19	Poisonous (possibly fatal at 500 ppm; instant death at 1,000 ppm)	LEL, 4.5%; UEL, 45%	Colorless, sweet taste, rotten egg odor
Carbon dioxide	CO_2	1.53	Increases respiration 50% at 2%, 100% at 3%, toxic at >5% (possibly fatal at >15%)	Inert (extinguishing agent)	Colorless, pungent smell, soda water taste
Nitrogen dioxide	NO_2	1.6	Poisonous (200 ppm possibly fatal)	Nonflammable of itself (but supports combustion)	Reddish brown, acrid smell, acidic taste
Sulfur dioxide	SO_2	2.26	Poisonous (500 ppm fatal)	Inert	Colorless, acid taste, burning sulfur smell

(continues)

Coal Mine Gas Chart (continued)

Name of Gas	Where Found	Detection
Hydrogen	Spontaneous combustion, behind seals, battery charging	Gas chromatograph
Methane	Seam gas, cut coal, roof cavities, face areas, sealed areas, heatings	Catalytic combustion, thermal conductivity, infrared analyzer, gas chromatograph
Carbon monoxide	Heatings, explosion, fires, product of incomplete combustion, water gas, producer gas, diesels, shot-firing, seals	Multitube detectors, infrared analyzer, electrochemical sensor, gas chromatograph
Nitrogen	Atmosphere, 78.11% of air	Gas chromatograph
Air	All airways	NA
Oxygen	Air, battery charging, oxidizing chemical storage, cylinder storage	Gas chromatograph, electrochemical sensor (partial pressure and volumetric), paramagnetic cells
Hydrogen sulfide	Stagnant water, heating in coal-containing pyrites, seam gas	Multitube detectors, electrochemical sensor, odor
Carbon dioxide	Seam gas, breathing, oxidation, after explosions, combustion	Multitube detectors, infrared analyzer, electrochemical sensor
Nitrogen dioxide	Shot-firing, diesels, afterdamp	Tube detectors, electrochemical, color, odor, taste
Sulfur dioxide	Heating, combustion of high-sulfur coals	Tube detector, electrochemical taste, odor

Courtesy of Queensland Mines Rescue Service.
Notes: LEL = lower explosive limit; MAC = maximum allowable concentration; NA = not applicable; NP = nose point; STEL = short-term exposure limit; UEL = upper explosive limit.

APPENDIX C

Conversion Chart

Conversion of English (Imperial) to SI (metric) units

To convert from	to	Multiply by
abampere	ampere (A)	1.000 000 10^1
abcoulomb	coulomb (C)	1.000 000* 10^1
abfarad	farad (F)	1.000 000* 10^9
abhenry	henry (H)	1.000 000* 10^{-9}
abmho	siemens (S)	1.000 000* 10^9
abohm	ohm (Ω)	1.000 000* 10^{-9}
abvolt	volt (V)	1.000 000* 10^{-8}
acre foot (U.S. survey)†	cubic meter (m^3)	1.233 489 10^3
acre (U.S. survey)†	square meter (m^2)	4.046 873 10^3
acre	hectare (ha)	4.046 873 10^{-1}
ampere hour	coulomb (C)	3.600 000* 10^3
are	square meter (m^2)	1.000 000 10^2
angstrom	meter (m)	1.000 000* 10^{-10}
astronomical unit	meter (m)	1.495 979 10^{11}
atmosphere (standard)	pascal (Pa)	1.013 250* 10^5
atmosphere (technical = 1 kgf/cm^2)	pascal (Pa)	9.806 650* 10^4
bar	pascal (Pa)	1.000 000* 10^5
barn	square meter (m^2)	1.000 000* 10^{-28}
barrel (for petroleum, 42 gal)	cubic meter (m^3)	1.589 873 10^{-1}
board foot	cubic meter (m^3)	2.359 737 10^{-3}
British thermal unit (International Table)	joule (J)	1.055 056 10^3
Btu (International Table)·ft/h·ft^2·°F (k, thermal conductivity)	watt per meter kelvin (W/m·K)	1.730 735 10^0
Btu (International Table) in./h·ft^2·°F (k, thermal conductivity)	watt per meter kelvin (W/m·K)	1.442 279 10^{-1}
Btu (International Table) in./s·ft^2·°F (k, thermal conductivity)	watt per meter kelvin (W/m·K)	5.192 204 10^2
Btu (International Table)/h	watt (W)	2.930 711 10^{-1}
Btu (International Table)/ft^2	joule per square meter (J/m^2)	1.135 653 10^4
Btu (International Table)/h·ft^2·°F (C, thermal conductance)	watt per square meter kelvin (W/m^2·K)	5.678 263 10^0
Btu (International Table)/s·ft^2·°F	watt per square meter kelvin (W/m^2·K)	2.044 175 10^4
Btu (International Table)/lb	joule per kilogram (J/kg)	2.326 000* 10^3
Btu (International Table)/lb·°F (c, heat capacity)	joule per kilogram kelvin (J/kg·K)	4.186 800* 10^3
bushel (U.S.)	cubic meter (m^3)	3.523 907 10^{-2}
caliber (inch)	meter (m)	2.540 000* 10^{-2}
calorie (International Table)	joule (J)	4.186 800* 10^0
calorie (kilogram, International Table)	joule (J)	4.186 800* 10^3
cal (International Table)/g	joule per kilogram (J/kg)	4.186 800* 10^3
cal (International Table)/g·°C	joule per kilogram kelvin (J/kg·K)	4.186 800* 10^3
cal (thermochemical)/min	watt (W)	6.973 333 10^{-2}
cal (thermochemical)/cm^2·min	watt per square meter (W/m^2)	6.973 333 10^2
carat (metric)	kilogram (kg)	2.000 000* 10^{-4}

(continues)

Conversion of English (Imperial) to SI (metric) units (continued)

To convert from	to	Multiply by	
centimeter of mercury (0°C)	pascal (Pa)	1.333 22	10^3
centimeter of water (4°C)	pascal (Pa)	9.806 38	10^1
centipoise	pascal second (Pa·s)	1.000 000*	10^{-3}
centistokes	square meter per second (m²/s)	1.000 000*	10^{-6}
circular mil	square meter (m²)	5.067 075	10^{-10}
clo	kelvin square meter per watt (K·m²/W)	2.003 712	10^{-1}
cup	cubic meter (m³)	2.365 882	10^{-4}
curie	becquerel (Bq)	3.700 000*	10^{10}
day (mean solar)	second (s)	8.640 000	10^4
day (sidereal)	second (s)	8.616 409	10^4
degree (angle)	radian (rad)	1.745 329	10^{-2}
degree Celsius	kelvin (K)	$t_K = t°_C + 273.15$	
degree Fahrenheit	degree Celsius	$t°_C = (t°_F - 32)/1.8$	
degree Fahrenheit	kelvin (K)	$t_K = (t°_F + 459.67)/1.8$	
degree Rankine	kelvin (K)	$t_K = t°_R/1.8$	
°F·h·ft²/Btu (International Table) (R, thermal resistance)	kelvin square meter per watt (K·m²/W)	1.761 102	10^{-1}
denier	kilogram per meter (kg/m)	1.111 111	10^{-7}
dyne	newton (N)	1.000 000*	10^{-5}
dyne-cm	newton meter (N·m)	1.000 000*	10^{-7}
dyne·cm²	pascal (Pa)	1.000 000*	10^{-1}
electronvolt	joule (J)	1.602 19	10^{-19}
EMU of capacitance	farad (F)	1.000 000*	10^9
EMU of current	ampere (A)	1.000 000*	10^1
EMU of electric potential	volt (V)	1.000 000*	10^{-8}
EMU of inductance	henry (H)	1.000 000*	10^{-9}
EMU of resistance	ohm (Ω)	1.000 000*	10^{-9}
ESU of capacitance	farad (F)	1.112 650	10^{-12}
ESU of current	ampere (A)	3.335 6	10^{-10}
ESU of electric potential	volt (V)	2.997 9	10^2
ESU of inductance	henry (H)	8.987 554	10^{11}
ESU of resistance	ohm (Ω)	8.987 554	10^{11}
erg	joule (J)	1.000 000*	10^{-7}
erg/cm²·s	watt per square meter (W/m²)	1.000 000*	10^{-3}
erg/s	watt (W)	1.000 000*	10^{-7}
faraday (chemical)	coulomb (C)	9.649 57	10^4
faraday (physical)	coulomb (C)	9.652 19	10^4
fathom	meter (m)	1.828 8	10^0
fermi (femtometer)	meter (m)	1.000 000*	10^{-15}
fluid ounce (U.S.)	cubic meter (m³)	2.957 353	10^{-5}
foot	meter (m)	3.048 000*	10^{-1}
foot (U.S. survey)†	meter (m)	3.048 006	10^{-1}
foot of water (39.2°F)	pascal (Pa)	2.988 98	10^3
ft²	square meter (m²)	9.290 304*	10^{-2}
ft²/h (thermal diffusivity)	square meter per second (m²/s)	2.580 640*	10^{-5}
ft²/s	square meter per second (m²/s)	9.290 304*	10^{-2}
ft³ (volume; section modulus)	cubic meter (m³)	2.831 685	10^{-2}
ft³/min	cubic meter per second (m³/s)	4.719 474	10^{-4}
ft³/s	cubic meter per second (m³/s)	2.831 685	10^{-2}
ft³/ton (tonnage factor)	cubic meter per metric ton (m³/t)	3.121 39	10^{-2}
ft⁴ (moment of section)‡	meter⁴ (m⁴)	8.630 975	10^{-3}
ft/h	meter per second (m/s)	8.466 667	10^{-5}
ft/min	meter per second (m/s)	5.080 000*	10^{-3}
ft/s	meter per second (m/s)	3.048 000*	10^{-1}
ft/s²	meter per square second (m/s²)	3.048 000*	10^{-1}
ft/ton (drilling factor)	meter per metric ton (m/t)	3.359 8	10^{-1}

(continues)

Conversion of English (Imperial) to SI (metric) units (continued)

To convert from	to	Multiply by	
footcandle	lux (lx)	1.076 391	10^1
footlambert	candela per square meter (cd/m²)	3.426 259	10^0
ft·lbf	joule (J)	1.355 818	10^0
ft·lbf/h	watt (W)	3.766 161	10^{-4}
ft·lbf/min	watt (W)	2.259 697	10^{-2}
ft·lbf/s	watt (W)	1.355 818	10^0
ft·poundal	joule (J)	4.214 011	10^{-2}
ft·ton (moment)	meter·metric ton (m·t)	2.765 1	10^{-1}
free fall, standard (g)	meter per square second (m/s²)	9.806 650*	10^0
gallon	meter per square second (m/s²)	1.000 000*	10^{-2}
gallon (Canadian liquid)	cubic meter (m³)	4.546 090	10^{-3}
gallon (U.K. liquid)	cubic meter (m³)	4.546 092	10^{-3}
gallon (U.S. dry)	cubic meter (m³)	4.404 884	10^{-3}
gallon (U.S. liquid)	cubic meter (m³)	3.785 412	10^{-3}
gallon (U.S. liquid)/day	cubic meter per second (m³/s)	4.381 264	10^{-8}
gallon (U.S. liquid)/min	cubic meter per second (m³/s)	6.309 020	10^{-5}
gallon (U.S. liquid)/hp·h (SFC, specific fuel consumption)	cubic meter per joule (m³/J)	1.410 089	10^{-9}
gamma	tesla (T)	1.000 000*	10^{-9}
gauss	tesla (T)	1.000 000*	10^{-4}
gilbert	ampere (A)	7.957 747	10^{-1}
gill (U.K.)	cubic meter (m³)	1.420 654	10^{-4}
gill (U.S.)	cubic meter (m³)	1.182 941	10^{-4}
grad	degree (angular)	9.000 000*	10^{-1}
grad	radian (rad)	1.570 796	10^{-2}
grain (1/7,000 lb avoirdupois)	kilogram (kg)	6.479 891*	10^{-5}
grain (lb avoirdupois/7,000)/gal (U.S. liquid)	kilogram per cubic meter (kg/m³)	1.711 806	10^{-2}
gram	kilogram (kg)	1.000 000*	10^{-3}
g/cm³	kilogram per cubic meter (kg/m³)	1.000 000*	10^3
gram-force/cm²	pascal (Pa)	9.806 650*	10^1
hectare	square meter (m²)	1.000 000*	10^4
horsepower (550 ft·lbf/s)	watt (W)	7.456 999	10^2
horsepower (boiler)	watt (W)	9.809 50	10^3
horsepower (electric)	watt (W)	7.460 000*	10^2
horsepower (metric)	watt (W)	7.354 99	10^2
horsepower (water)	watt (W)	7.460 43	10^2
horsepower (U.K.)	watt (W)	7.457 0	10^2
hour (mean solar)	second (s)	3.600 000	10^3
hour (sidereal)	second (s)	3.590 170	10^3
hundredweight (long)	kilogram (kg)	5.080 235	10^1
hundredweight (short)	kilogram (kg)	4.535 924	10^1
inch (in.)	meter (m)	2.540 000*	10^{-2}
inch of mercury (32°F)	pascal (Pa)	3.386 38	10^3
inch of mercury (60°F)	pascal (Pa)	3.376 85	10^3
inch of water (39.2°F)	pascal (Pa)	2.490 82	10^2
inch of water (60°F)	pascal (Pa)	2.488 4	10^2
in.²	square meter (m²)	6.451 600*	10^{-4}
in.³ (volume; section modulus)	cubic meter (m³)	1.638 706	10^{-5}
in.³/min	cubic meter per second (m³/s)	2.731 177	10^{-7}
in.⁴ (moment of section)‡	meter⁴ (m⁴)	4.162 314	10^{-7}
in./s	meter per second (m/s)	2.540 000*	10^{-2}
in./s²	meter per square second (m/s²)	2.540 000*	10^{-2}
kayser	1 per meter (1/m)	1.000 000*	10^2
kelvin	degree Celsius	$t°_C = t_K - 273.15$	
kilocalorie (International Table)	joule (J)	4.186 800*	10^3
kilocalorie (thermochemical)/min	watt (W)	6.973 333	10^1
kilogram-force (kgf)	newton (N)	9.806 650*	10^0

(continues)

Conversion of English (Imperial) to SI (metric) units (continued)

To convert from	to	Multiply by
kgf·m	newton meter (N·m)	$9.806\ 650^* \times 10^0$
kgf·s²/m (mass)	kilogram (kg)	$9.806\ 650^* \times 10^0$
kgf/cm²	pascal (Pa)	$9.806\ 650^* \times 10^4$
kgf/m²	pascal (Pa)	$9.806\ 650^* \times 10^0$
kgf/mm²	pascal (Pa)	$9.806\ 650^* \times 10^6$
kilometer per hour (km/h)	meter per second (m/s)	$2.777\ 778 \times 10^1$
kilopond	newton (N)	$9.806\ 650^* \times 10^0$
kWh	joule (J)	$3.600\ 000^* \times 10^6$
kip (1,000 lbf)	newton (N)	$4.448\ 222 \times 10^3$
kip/in.² (ksi)	pascal (Pa)	$6.894\ 757 \times 10^6$
knot (international)	meter per second (m/s)	$5.144\ 444 \times 10^{-1}$
lambert	candela per square meter (cd/m²)	$1/\pi^* \times 10^4$
lambert	candela per square meter (cd/m²)	$3.183\ 099 \times 10^3$
langley	joule per square meter (J/m²)	$4.184\ 000^* \times 10^4$
league	meter (m)	†
light year	meter (m)	$9.460\ 55 \times 10^{15}$
liter	cubic meter (m³)	$1.000\ 000^* \times 10^{-3}$
maxwell	weber (Wb)	$1.000\ 000^* \times 10^{-8}$
mho	siemens (S)	$1.000\ 000^* \times 10^0$
microinch	meter (m)	$2.540\ 000^* \times 10^{-8}$
micrometer (μm)	meter (m)	$1.000\ 000^* \times 10^{-6}$
mil	meter (m)	$2.540\ 000^* \times 10^{-5}$
mile (international)	meter (m)	$1.609\ 344^* \times 10^3$
mile (statute)	meter (m)	$1.609\ 3 \times 10^3$
mile (U.S. survey)†	meter (m)	$1.609\ 347 \times 10^3$
mile (international nautical)	meter (m)	$1.852\ 000^* \times 10^3$
mile (U.S. nautical)	meter (m)	$1.852\ 000^* \times 10^3$
mile² (international)	square meter (m²)	$2.589\ 988 \times 10^6$
mile² (U.S. survey)†	square meter (m²)	$2.589\ 998 \times 10^6$
mile/h (international)	meter per second (m/s)	$4.470\ 400^* \times 10^{-1}$
mile/h (international)	kilometer per hour (km/h)	$1.609\ 344^* \times 10^0$
mile/min (international)	meter per second (m/s)	$2.682\ 240^* \times 10^1$
mile/s (international)	meter per second (m/s)	$1.609\ 344^* \times 10^3$
millibar	pascal (Pa)	$1.000\ 000^* \times 10^2$
millimeter of mercury (0°C)	pascal (Pa)	$1.333\ 22 \times 10^2$
minute (angle)	radian (rad)	$2.908\ 882 \times 10^{-4}$
minute (mean solar)	second (s)	$6.000\ 000 \times 10^1$
minute (sidereal)	second (s)	$5.983\ 617 \times 10^1$
month (mean calendar)	second (s)	$2.628\ 000 \times 10^6$
oersted	ampere per meter (A/m)	$7.957\ 747 \times 10^1$
ohm centimeter	ohm meter (Ω·m)	$1.000\ 000^* \times 10^{-2}$
ohm circular-mil per foot	ohm square millimeter per meter (Ω·mm²/m)	$1.662\ 426 \times 10^{-3}$
ounce (avoirdupois)	kilogram (kg)	$2.834\ 952 \times 10^{-2}$
ounce (troy or apothecary)	kilogram (kg)	$3.110\ 348 \times 10^{-2}$
ounce (U.K. fluid)	cubic meter (m³)	$2.841\ 307 \times 10^{-5}$
ounce (U.S. fluid)	cubic meter (m³)	$2.957\ 353 \times 10^{-5}$
ounce-force	newton (N)	$2.780\ 139 \times 10^{-1}$
oz-in.	newton meter (N·m)	$7.061\ 552 \times 10^{-3}$
oz (avoirdupois)/gal (U.K. liquid)	kilogram per cubic meter (kg/m³)	$6.236\ 021 \times 10^0$
oz (avoirdupois)/gal (U.S. liquid)	kilogram per cubic meter (kg/m³)	$7.489\ 152 \times 10^0$
oz (avoirdupois)/in.³	kilogram per cubic meter (kg/m³)	$1.729\ 994 \times 10^3$
oz (avoirdupois)/ft²	kilogram per square meter (kg/m²)	$3.051\ 517 \times 10^{-1}$
oz (avoirdupois)/yd²	kilogram per square meter (kg/m²)	$3.390\ 575 \times 10^{-2}$
oz/ton (ore grade)	grams per metric ton (g/t)	3.125×10^1
parsec	meter (m)	$3.085\ 678 \times 10^{16}$

(continues)

Conversion of English (Imperial) to SI (metric) units (continued)

To convert from	to	Multiply by	
peck (U.S.)	cubic meter (m³)	8.809 768	10⁻³
pennyweight	kilogram (kg)	1.555 174	10⁻³
perm (0°C)	kilogram per pascal second square meter (kg/Pa·s·m²)	5.721 35	10⁻¹¹
perm (23°C)	kilogram per pascal second square meter (kg/Pa·s·m²)	5.745 25	10⁻¹¹
perm in. (0°C)	kilogram per pascal second meter (kg/Pa·s·m)	1.453 22	10⁻¹²
perm in. (23°C)	kilogram per pascal second meter (kg/Pa·s·m)	1.459 29	10⁻¹²
phot	lumen per square meter (lm/m²)	1.000 000*	10⁴
pica (printer's)	meter (m)	4.217 518	10⁻³
pint (U.S. dry)	cubic meter (m³)	5.506 105	10⁻¹
pint (U.S. liquid)	cubic meter (m³)	4.731 765	10⁻⁴
point (printer's)	meter (m)	3.514 598*	10⁻⁴
poise (absolute viscosity)	pascal second (Pa·s)	1.000 000*	10⁻¹
pound (lb avoirdupois)	kilogram (kg)	4.535 924	10⁻¹
pound (troy or apothecary)	kilogram (kg)	3.732 417	10⁻¹
lb·ft² (moment of inertia)	kilogram square meter (kg·m²)	4.214 011	10⁻²
lb·in.² (moment of inertia)	kilogram square meter (kg·m²)	2.926 397	10⁻⁴
lb·min²/ft⁴ (air friction factor, K)	kilogram per cubic meter (kg/m³)	1.855	10⁶
lb/ft·h	pascal second (Pa·s)	4.133 789	10⁻⁴
lb/ft·s	pascal second (Pa·s)	1.488 164	10⁰
lb/ft (loading factor)	kilogram per meter (kg/m)	1.488 156	10⁰
lb/ft²	kilogram per square meter (kg/m²)	0.882 428	10⁰
lb/ft³ (specific weights)	kilogram per cubic meter (kg/m³)	1.601 846	10¹
lb/gal (U.K. liquid)	kilogram per cubic meter (kg/m³)	9.977 633	10¹
lb/gal (U.S. liquid)	kilogram per cubic meter (kg/m³)	1.198 264	10²
lb/h	kilogram per second (kg/s)	1.259 979	10⁻⁴
lb/hp·h (SFC, specific fuel consumption)	kilogram per joule (kg/J)	1.689 659	10⁻⁷
lb/in.³	kilogram per cubic meter (kg/m³)	2.767 990	10⁴
lb/min	kilogram per second (kg/s)	7.559 873	10⁻³
lb/s	kilogram per second (kg/s)	4.535 924	10⁻¹
lb/ton (powder factor)	kilogram per metric ton (kg/t)	5.000	10⁻¹
lb/yd³ (powder factor)	kilogram per cubic meter (kg/m³)	5.932 764	10⁻¹
poundal	newton (N)	1.382 550	10⁻¹
poundal/ft²	pascal (Pa)	1 488 164	10⁰
poundal·s/ft²	pascal second (Pa·s)	1.488 164	10⁰
pound-force (lbf)	newton (N)	4.448 222	10⁰
lbf·ft	newton meter (N·m)	1.355 818	10⁰
lbf·ft/in.	newton meter per meter (N·m/m)	5.337 866	10¹
lbf·in.	newton meter (N·m)	1.129 848	10⁻¹
lbf·in./in.	newton meter per meter (N·m/m)	4.448 222	10⁰
lbf·s/ft²	pascal second (Pa·s)	4.788 026	10¹
lbf/ft	newton per meter (N/m)	1.459 390	10¹
lbf/ft²	pascal (Pa)	1.788 026	10¹
lbf/in.	newton per meter (N/m)	1.751 268	10²
lbf/in.² (psi)	pascal (Pa)	6.894 757	10³
lbf/lb (thrust/weight [mass] ratio)	newton per kilogram (N/kg)	9.806 650	10⁰
quart (U.S. dry)	cubic meter (m³)	1.101 221	10⁻³
quart (U.S. liquid)	cubic meter (m³)	9.463 529	10⁻⁴
rad (radiation dose absorbed)	gray (Gy)	1.000 000*	10⁻²
rhe	1 per pascal second (1/Pa·s)	1.000 000*	10¹
rod	meter (m)	†	
roentgen	coulomb per kilogram (C/kg)	2.58	10⁻⁴
second (angle)	radian (rad)	4.818 137	10⁻⁶

(continues)

Conversion of English (Imperial) to SI (metric) units (continued)

To convert from	to	Multiply by
second (sidereal)	second (s)	$9.972\ 696 \times 10^{-1}$
section	square meter (m^2)	†
shake	second (s)	$1.000\ 000^* \times 10^{-8}$
slug	kilogram (kg)	$1.459\ 390 \times 10^1$
slug/ft·s	pascal second (Pa·s)	$4.788\ 026 \times 10^1$
slug/ft^3	kilogram per cubic meter (kg/m^3)	$5.153\ 788 \times 10^2$
statampere	ampere (A)	$3.335\ 640 \times 10^{-10}$
statcoulomb	coulomb (C)	$3.335\ 640 \times 10^{-10}$
statfarad	farad (F)	$1.112\ 650 \times 10^{-12}$
stathenry	henry (H)	$8.987\ 554 \times 10^{11}$
statmho	siemens (S)	$1.112\ 650 \times 10^{-12}$
statohm	ohm (Ω)	$8.987\ 554 \times 10^{11}$
statvolt	volt (V)	$2.997\ 925 \times 10^2$
stere	cubic meter (m^3)	$1.000\ 000^* \times 10^0$
stilb	candela per square meter (cd/m^2)	$1.000\ 000^* \times 10^4$
stokes (kinematic viscosity)	square meter per second (m^2/s)	$1.000\ 000^* \times 10^{-4}$
tablespoon	cubic meter (m^3)	$1.178\ 676 \times 10^{-5}$
teaspoon	cubic meter (m^3)	$4.928\ 922 \times 10^{-6}$
tex	kilogram per meter (kg/m)	$1.000\ 000^* \times 10^{-6}$
therm	joule (J)	$1.055\ 056 \times 10^8$
ton (assay)	kilogram (kg)	$2.916\ 667 \times 10^{-2}$
ton (long, 2,240 lb)	kilogram (kg)	$1.016\ 017 \times 10^3$
ton (metric)	kilogram (kg)	$1.000\ 000^* \times 10^3$
ton (nuclear equivalent of TNT)	joule (J)	4.184×10^9
ton (refrigeration)	watt (W)	$3.516\ 800 \times 10^3$
ton (register)	cubic meter (m^3)	$2.831\ 685 \times 10^0$
ton (short, 2,000 lb)	kilogram (kg)	$9.071\ 817 \times 10^2$
ton/yd^3 (specific weight)	metric ton per cubic meter (t/m^3)	$1.186\ 66 \times 10^0$
ton (long)/yd^3 (specific weight)	kilogram per cubic meter (kg/m^3)	$1.328\ 939 \times 10^3$
ton (short)/h	kilogram per second (kg/s)	$2.519\ 958 \times 10^{-1}$
ton-force (2,000 lbf)	newton (N)	$8.896\ 111 \times 10^3$
ton/yd^3 (specific weight)	metric ton per cubic meter (t/m^3)	$1.186\ 55 \times 10^0$
torr (mm Hg, 0°C)	pascal (Pa)	$1.333\ 22 \times 10^2$
township	square meter (m^2)	†
unit pole	weber (Wb)	$1.256\ 637 \times 10^{-7}$
W·h	joule (J)	$3.600\ 000^* \times 10^3$
W·s	joule (J)	$1.000\ 000^* \times 10^0$
W/cm^2	watt per square meter (W/m^2)	$1.000\ 000^* \times 10^4$
W/in.^{2}	watt per square meter (W/m^2)	$1.550\ 003 \times 10^3$
yard	meter (m)	$9.144\ 000^* \times 10^{-1}$
yd^2	square meter (m^2)	$8.361\ 274 \times 10^{-1}$
yd^3	cubic meter (m^3)	$7.645\ 549 \times 10^{-1}$
yd^3/min	cubic meter per second (m^3/s)	$1.274\ 258 \times 10^{-2}$
yd^3/ton (stripping ratio)	cubic meter per metric ton (m^3/t)	$8.427\ 8 \times 10^{-1}$
year (calendar)	second (s)	$3.153\ 600 \times 10^7$
year (sidereal)	second (s)	$3.155\ 815 \times 10^7$

*Exact value.
†Conversion factors for land measure may be determined from the following relationships:
  1 league = 3 miles (exactly)
  1 rod = 16½ feet (exactly)
  1 section = 1 square mile (exactly)
  1 township = 36 square miles (exactly)
‡This is sometimes called the moment of inertia of a plane section about a specified axis.

Index

Note: *f.* indicates figure; *t.* indicates table.
Pages 1–1046 are found in Volume 1; pages 1047–1840 are found in Volume 2.

AA1000, 1795
AAPG Computer Applications in Geology, No. 4, 158
ABAQUS (finite element analysis) software, 609
Abrasion test, 1474
Acid rock drainage (ARD), 1721, 1722*f.*, 1734
    acid generation and neutralization, 1723–1724, 1723*f.*
    chemical characterization, 676
    and ferrous iron, 1723
    formation of, 1721–1724
    long-term, 1721, 1722*f.*
    and metal leaching (ARD/ML), 1724
    pyrite oxidation, 1722, 1722*f.*
    and stormwater management, 676–677
    and sustainability, 33
    and waste piles and dumps, 676–677
Acid rock drainage management, 1721, 1732
    characterization, 1724–1725, 1725*f.*, 1726*f.*
    communication and consultation, 1732
    framework, 1724–1732, 1724*f.*
    Global Acid Rock Drainage Guide (GARD Guide), 1721, 1734
    management and performance assessment, 1731–1732, 1731*f.*
    monitoring, 1730–1731, 1731*f.*
    prediction, 1726–1727, 1728*f.*
    prevention and mitigation, 1727, 1729*f.*
    treatment, 1727–1730, 1730*f.*
Acid sulfate, 1088
Acid–base accounting (ABA) of tailings, 655–656
Acidic chloride, 1088
ADDCAR system, 1030
Advance rate prediction (rock breakage)
    based on cutterhead design, 432
    based on specific energy, 430–432, 430*f.*, 431*t.*
AFCs. *See* Armored face conveyors
Africa
    mining history, 4
    publications and associations (Web sites), 1831
AG/SAG. *See* Autogenous grinding/semiautogenous grinding
Aggregates
    in shotcrete, 580, 580*t.*
    See also Quarrying

Agricola, Georgius (Georg Bauer), 3, 127, 227, 627
    on detriments and advantages of mining, 1767
    on heap leaching, 1073
Air conditioning, 1611
Aker Wirth continuous mining machine, 428, 429*f.*
Alaska, University of, frozen placer research, 1062
Alcoa bauxite mines (Australia), and IPCC, 941
Algoma BIF deposits, 97
Alimak raise mining, 1193, 1283, 1284*f.*, 1362, 1362*f.*
Alkaline ammonia-based lixiviant, 1088
Alternative Investment Market (AIM), 65
Amax Gold, 299
AME Mineral Economics, 62
American Bureau of Metal Statistics, 62
American Concrete Institute, 589
American Conference of Governmental Industrial Hygienists (ACGIH)
    and noise regulations, 1633, 1634
    on whole-body vibration, 1638
American Iron and Steel Institute, 744
American Society of Heating, Refrigerating, and Air-Conditioning Engineers (ASHRAE), 1613
    psychrometric chart, 1612*f.*
Ammonium nitrate-fuel oil (ANFO), 866, 455–456, 458–459
    in quarrying, 1035
Ampere's law, 1534
Anaconda mine (Montana)
    and air conditioning, 1611
    geologic mapping system, 157
Analysis of Retreat Mining Pillar Stability (ARMPS) method, 389
Anglo American, 78
Annual Conference on Ground Control in Mining, 597
Annual value, 302
ARD. *See* Acid rock drainage
Argentina
    Meridian Gold, Inc., and failed social license approach in Esquel gold deposit, 1787–1788
    mining history, 4

Argyle diamond mine (Australia)
    drill-and-blast tunneling, 1198
    and interaction with indigenous culture, 1801
Armored face conveyors (AFCs)
    in longwall mining, 1399
    in soft-rock underground mining, 1159–1160, 1161*f.*, 1162*f.*
Artificial intelligence, 842
Assaying, 187
    digestion techniques, 190
    fire, 108, 190
    geochemical, 189
    instrumental measurement, 190
    methods, 135, 189–191
    neutron activation analysis (NAA), 190–191
    nondestructive analysis, 190–191
    quantitative, 189
    wet chemistry, 190
    X-ray, 190
Asset management, 781, 803–804
    ARP base case and uncertainties, 796–797, 797*f.*
    ARP contents, 794–795, 795*f.*
    ARP development process, 796, 796*f.*
    asset, defined, 783
    asset hierarchy, 782, 782*f.*
    asset information management system, defined, 784
    asset life cycle, defined, 783
    asset management, defined, 783
    asset management information system, defined, 783–784
    asset management objectives, defined, 784
    asset management plan, defined, 784
    asset management plan implementation, 791
    asset management policy, 788
    asset management policy, defined, 784
    asset management strategy, defined, 784
    asset management strategy and plans, 788–789
    asset management system, defined, 783
    asset management system structure, 784–786, 785*f.*
    asset manager, 793
    asset reference plan (ARP), 793–794, 794*f.*
    asset system, defined, 784
    auditing, 792
    benefits of whole life-cycle approach, 782
    business model, 792–797

concept and terminology, 781–782, 783–784
connections between organization goals and plan, 787, 788f.
cost/risk/performance decision making and optimization, 801–802, 802f., 803f.
defining mining assets, 792–793
Deming cycle (plan-do-check-act), 787, 787f.
emerging standards of, 782–783
enablers, 785, 789–791
failure modes and effects analysis (FMEA), 798, 799
human factors, 791
identification of appropriate solutions or improvement actions, 792
identification of failures and improvement opportunities, 792
implementation of corrective or preventive action, 792
information management, 789–790
key attributes, 785–786, 786f.
life-cycle costing, 797–798, 799f.
life-cycle planning, 797, 798f.
management review, 792
managing aging assets, 802–803, 803f., 804f.
methodologies and techniques, 797–803
monitoring and improvement, 791–792
optimization, defined, 784, 785
overall equipment effectiveness (OEE), 800
PAS 55, 781–782, 783–784, 787, 802f.
performance and condition monitoring, 792
processes, 785, 786f.
questions for ARP to answer, 794
record-keeping and documentation updates, 792
reliability centered maintenance (RCM), 798, 799, 800f., 825, 826, 826f., 827, 828f., 835
requirements, 787–792, 787f.
risk and reliability engineering, 798–799, 799f.
risk management, 789, 790f., 791f.
risk-based inspection (RBI), 800
root cause analysis, 801
shamrock diagram, 789, 791f.
Six Sigma, 800–801
strengths, weaknesses, opportunities, threats (SWOT), 793
team, 793
terminology, 783–784
total productive maintenance (TPM), 800
total quality management (TQM), 800–801
Association of Exploration Geochemists, 128
Associations, U.S. (Web sites), 1827–1828
ASTM Standard Methods for coal and coke sample preparation, 191
Atlas Copco, 707, 709, 711
Atomic absorption, 108
Atomic absorption spectrophotometry (AAS), 133, 135, 190
Atterberg limits, 654
Auction pricing, 13
The Australasian Code for Reporting of Exploration Results, Mineral Resources and Ore Reserves, 464
Australasian Institute of Mining and Metallurgy Joint Ore Reserves Committee (JORC) and Code, 14, 193, 203, 231, 298–299, 882
VALMIN code, 219–220
Australia
aboriginal people, 1797
Central Queensland Mining Rehabilitation Group, 1823
and community involvement, 1768, 1769, 1770
Cooperative Research Centre (CRC), 433
Cooperative Research Centre for Mining (CRCMining), 813
Department of Industry, Tourism and Resources, 646
Geochemistry Research Centre of Australia, 108
leadership in safety risk management, 303
lost-time injury frequency rate (2000–2007), 1558, 1558f.
mining history, 4
mining industry fatalities (2000–2007), 1558, 1558f.
Ministerial Council on Mineral and Petroleum Resources, 1768
Muswellbrook Mine Managers forum, 1823
Native Title Act (1993), 1768, 1770
New South Wales Coal Mine Health and Safety Regulation 2006, 466, 468–469
Occupational Health and Safety Act (New South Wales, 1983, 2000), 1560
publications and associations (Web sites), 1831
relative performance of Australian dollar vs. CRB Index, 77, 79f.
underground mine fatality rates 2004–2006, 303, 304t.
use of R&P in coal mining, 1339
Workplace Health and Safety Act (Tasmania, 1995), 1560
See also Alcoa bauxite mines; Boundary Hill pit; Goonyella Riverside mine; Groote Eylandt mine; Moura coal mine; Mt. Arthur North mine; Northparkes mine; Norwich Park mine; Rio Tinto; Tarong coal mine; Wagerup alumina refinery; West Angelas iron ore mine
Australia/China Coal Mine Safety Demonstration Project, 465
Australian Bureau of Agricultural and Resource Economics (ABARE), 63
Australian Coal Association Research Program (ACARP), 813
Australian Minerals Industry Code for Environmental Management, 1769
Autogenous grinding (AG), ball charges, 1466–1467
Autogenous grinding/semiautogenous grinding (AG/SAG), 1461, 1466
ball charges, 1466–1467
feed hardness, 1468–1469
feed sizes, 1469
gearless drives, 1466
high- and low-aspect-ratio mills, 1466
mill liner designs, 1469
pebble crushing, 1467–1468
single-stage, 1466
unit power, 1476–1477
See also Semiautogenous grinding
Automatic control systems, 728–729
and relevancy of process inputs, 729
and sensor accuracy, 729
and validity of assumptions, 729
Automation, 5–6, 26, 805
and accurate mine and machine model, 819
advanced sensing technologies, 817
automated blasthole drilling (case study), 820–822, 820t., 821t., 822f.
automated blasthole drills (ABDs), 809–810
automated digging, 811–813, 812f.
automated dozers, 809
automated drill sensors, 820, 820t.
automated drilling compared with manual, 821t., 822f.
automated dump trucks (ADTs), 808
automated horizon control, 814–815, 185f.
automated underground loading and haulage, 808–809
automatic, defined, 805
automatic face alignment and creep control, 813–814
autonomous, defined, 806
autonomous haulage systems, 806–808
autonomous mining, 806
and avoidance of mixed fleets, 820
benefits of, 26, 805
challenges, 806
and computing power, 27
data fusion, 27
defined, 806
dragline automation, 816–817
driverless trains, 27
duty meters, 817
and efficiency, 818–819
efficiency gains through process control, 819
face alignment control system (example), 814
and fatigue elimination, 818
future challenges, 817
future possibilities, 26
and improved information system maintenance, 820
and information system improvement, 819
integrated site awareness, 818
intelligent, defined, 806
"islands of" (mine sites), 26–27
longwall, 813–816
and machine health and maintenance, 820
machine-level haulage automation, 807–808
mechanized, defined, 805
mine-wide information systems, 817
and minimization of operator error, 818
onboard sensors, 817
of operations centers, 27
and planning and scheduling, 25–26
and predictable working environment, 820
and remote locations, 818
remote operations centers (ROCs), 810–811
required operational changes, 820
requirements for full implementation, 819–820
safety advantages, 817–818
and scarcity of skilled operators, 818
semiautomatic, defined, 805
site-level haulage automation, 806–807
"state-based" shearer automation, 815
terminology and definitions, 805–806
turnkey full-mine systems, 27
vision of future, 817–818
and workers, 27–28
See also Robotics
Autonomous operation, 15
and planning and scheduling, 25–26
Average grade, 196
evaluating differences in, 207, 207t.
Avoca method, 366, 367t., 395, 395f.

Backfill mining, 1375
automated systems, 1383
backfill types, 1376–1383
cemented hydraulic backfill and backfill plant requirements, 1380

cemented hydraulic backfill and cement
 content vs. strength and deformation
 properties, 1379, 1379f.
cemented hydraulic backfill and cement
 content vs. use, 1379, 1380t.
cemented hydraulic backfill and
 freestanding height, 1379–1380
cemented hydraulic backfill and pump
 requirements, 1380
cemented hydraulic backfill system design,
 1379–1380
cemented rock fill and binder placement,
 1380
cemented rock fill and rock fill size
 considerations, 1380
cemented rock fill system design, 1380
and dry sand and rock fill, 1376–1377
effect of cemented fill in mill feed on
 processing plant recovery, 1383
feeder and blower considerations,
 1381–1382
flowable fill equipment requirements, 1383
flowable fill system design, 1382–1383,
 1383t.
heat generation of sulfide backfills, 1383
history, 1375–1376
hydraulic backfill, 1375
information sources, 1383
particle size distribution, 1381–1382
paste fill and backfill plant requirements,
 1381
paste fill and paste particle size
 requirements, 1380–1381
paste fill and pipe wear, 1381
paste fill and pump requirements, 1381
paste fill system design, 1380–1381
pipe wear, 1382
pneumatic stowing, 1376, 1381
pneumatic stowing transport considerations,
 1381
power requirements, 1382
remote placement of backfills, 1383
solids loading and system pressure drop,
 1382
thickened tailings, 1376
transport of slurry solids content,
 1377–1378
uncemented hydraulic backfill and backfill
 plant requirements, 1378–1379
uncemented hydraulic backfill and pipe
 material selection, 1378
uncemented hydraulic backfill and wear in
 pipes and line fittings, 1378, 1378t.
uncemented hydraulic backfill drainage
 requirements, 1377
uncemented hydraulic backfill pipe friction
 losses and pump requirements,
 1377–1378
uncemented hydraulic backfill system
 design, 1377–1379
uncemented hydraulic backfill transport
 pipe size requirements, 1377, 1378t.
Backwardation, 58
Bad Grund mine (Germany), and paste fill,
 1376
Balance sheets, 80
Ball mills, 1470–1471
Banded iron formation (BIF) deposits, 97
Bandura, Albert, 328
Bauxite deposits, 100
Baxter, George, 297
Beam stage loader, crusher, and boot end (in
 longwall mining), 1160–1161

Behavior modeling, 328, 329t.
Benefit–cost ratio, 302
Bernoulli equation, 753, 1578
BHP Billiton
 dragline automation trials, 816
 Fatal Risk Control Protocols, 465
Bickford, William, and safety fuse, 3
Bingham Canyon copper mine (U.S.), 15, 721
 planning process, 29–30
Bishop's method, 499
Black lung disease. *See* Coal workers'
 pneumoconiosis
Blasthole drilling, 435
 automated blasthole drills (ABDs),
  809–810, 820–822, 820t., 821t., 822f.
 cluster drills, 436, 436f., 437f.
 coil tubing, 440
 computer-controlled systems, 440
 contractors, 459
 directional, 440
 down-the-hole (DTH) hammer, 435, 436f.
 drag-type bits, 439
 drill jumbos, 436, 437–438, 437f.
 face drilling rigs, 436
 future trends, 24
 and Global Positioning Systems, 459
 in-the-hole (ITH) hammer, 435, 437, 438
 jackleg drills, 436
 new developments, 440
 penetration rate (percussive drilling), 438,
  439t.
 percussion bits, 435–436, 436f.
 percussive, 435–438
 percussive drilling equipment, 436–438
 percussive drilling equipment for surface
  use, 438
 polycrystalline diamond compact (PDC)
  bits, 439
 production (long-hole) drilling rigs, 436
 rate of advance (rotary drilling), 439–440,
  440t.
 ring-drilling drills, 436–437, 437f.
 rotary, 435, 438–440, 439f.
 stoper drills, 436
 tricone bits, 438–439
 in wet conditions, 748
 See also Drilling and blasting
Blasthole stoping (end slicing), 391, 391f.
Blasting, 443
 and air blast, 458
 alternatives to, 24
 with ANFO, 455–456, 458–459
 bench blast design nomenclature, 447, 448f.
 bench blasting, 447
 blast design, 447–454
 burden and spacing, 448, 449f.
 burn cut, 451–452, 452f.
 controlled, and slope design, 515–516, 516f.
 costs, 455–456, 456t., 457f.
 decking, 450, 458
 delays, 450–451, 451f., 452f.
 design for underground mining, 451–454
 detonators in holes, 454
 and dust control, 1604
 with dynamite, 458–459
 effects on environment, 874
 face-height-to-burden ratio, 454–455, 455f.
 fan cut (slashing), 451, 452, 452f., 453f.
 and flyrock, 457–458
 and free faces, 447, 447f.
 future trends, 24
 gas fumes from, 458
 and ground vibration, 458

 headings, 451–452, 452f.
 high-gas (higher-than-average-gas), 454
 hole diameter and burden vs. face height,
  447–448, 448f., 449f.
 holes and double-priming, 454, 455f.
 improper stemming, 458
 improper timing, 458
 long-hole stope blasts, 452–453
 mass blasts, 453, 453f., 454f.
 number of rows, 450
 and overbreak, 458
 patterns, 450, 450f.
 pillar recovery, 453–454
 powder column length, 449
 powder factors, 446, 447, 447f., 450
 repump emulsion systems, 454, 454f.
 restrictions in surface mining, 406
 shot service contractors, 458, 459
 specialized techniques in strip mining,
  1022–1024, 1023f.
 stemming, 449
 subdrilling, 448–449
 surface blast design, 447–451
 underground blast design, 451–454
 V-cut, 451, 452, 452f.
 See also Drilling and blasting; Explosives
Blasting impacts and control, 1689
 blast-induced air overpressures, 1691,
  1694–1695, 1695f.
 character of blast excitation and structural
  response, 1691–1695
 comparison of blast and environmental
  effects, 1700, 1701f.
 control of blast effects, 1699–1702
 digital-age improvements in blast control,
  1702–1703
 distinction between blast-induced and
  natural cracking, 1690–1691
 and electronic detonators, 1702–1703
 and engineered structures, 1700–1701
 estimation of dominant frequency, 1693,
  1693f.
 estimation of dynamic response properties,
  1697–1698, 1698t.
 evaluation of measurements, 1697–1699
 fatigue or repeated events, 1701
 frequency-based control with dominant
  frequency, 1702, 1702f.
 ground motion, 1691–1692, 1692f.
 human response, 1691
 instrument calibration, 1696, 1696f.
 instrument deployment during test blasts,
  1696–1697
 kinemetric relationships of ground motion,
  1692–1693
 measurement of particle velocity, 1695
 number of measurement instruments, 1696
 permanent degradation of adjacent rock,
  1689–1690
 permanent displacement of adjacent rock,
  1690
 propagation effects, 1693–1694, 1694f.
 range of blast effects, 1689–1691
 regulatory compliance for air overpressure,
  1702, 1702t.
 response of structures to everyday activities,
  1691, 1691t.
 response spectra, 1698–1699, 1698f.
 response spectra in surface mine blast,
  1699, 1699f.
 response spectra in urban construction blast,
  1698–1699, 1699f.
 restrained structures and rock masses, 1699

rock mass displacement and cracking, 1701–1702
Shock and Vibration Handbook, 1695
and signature-hole analysis, 1703
single-degree-of-freedom (SDF) model, 1697, 1697f.
sinusoidal approximations, 1692, 1692f.
statistical analysis of data by pre- and postblast inspection, 1700, 1701f.
structural response and frequency effects, 1697–1698
structural strains vs. particle velocity, 1695
and transducers, 1695–1696
transient nature of blast motions, 1693
transient structural response, 1690
use of measurement instruments, 1695–1696
Block caving, 28, 1437
   advance undercutting, 1442–1443, 1442f.
   air blasts, 1449
   automation, 1446
   capital cost considerations, 1448, 1448t.
   capital development, 1448
   cavability, 1438, 1450
   cave management, 1447–1448
   cave progress tracking, 1447
   cave shape and risk, 1450
   caving rate and risk, 1450
   conceptual development, 28
   concrete/shotcrete delivery systems, 1447
   current and planned mines, 1437, 1438f.
   design, 28
   draw control, 1447–1448
   drawpoint hang-ups, 1445, 1445f.
   drawpoint opening rates, 1445
   drawpoint opening sequences, 1445
   El Teniente (straight-through) extraction level layout, 1439–1440, 1440f.
   environmental impacts and permitting, 1450
   excavation stability and risk, 1450
   extraction level, 1439–1440
   fixed facilities design and risk, 1450
   forecasts, 1447
   fragmentation characteristics, 1438–1439
   full-gravity (grizzly) system, 1441, 1443f.
   general mining systems, 1441
   geotechnical characteristics, 1438–1439
   in hard-rock mining, 364–365, 365f.
   haulage level, 1440–1441
   herringbone extraction level layout, 1439, 1440f.
   high hang-ups, 1446
   hydraulic radius, 1438
   inrushes, 1449–1450
   in-situ fragmentation, 1439
   in-situ stress and risk, 1450
   labor, 1448–1449
   labor and risk, 1450
   labor distribution by category, 1449, 449f.
   as least expensive underground mining method, 1437
   LHD (rubber-tired) system, 1441
   and LHDs, 1439–1440, 1446
   locomotives, 1447
   long-hole drills, 1446–1447
   low hang-ups, 1445
   major operational hazards, 1449–1450
   major risk categories, 1450
   medium-height hang-ups, 1445–1446
   mine development, 1213
   mine layout, 1439–1441
   mining methods, 1441–1443
   and modeling, 30
   monitoring, 1439
   movement of workers, materials, and utilities, 1447
   mud rushes, 1449–1450
   operating cost considerations, 1448, 1449t.
   and operations, 30
   ore-body characteristics, 1437–1438
   orepasses, 1443–1444
   output rates, 29
   planning, 29–30
   post- or conventional undercutting, 1442, 1442f.
   preproduction development and construction, 1443–1444, 1443f., 144f.
   pre-undercutting, 1442, 1442f.
   primary fragmentation, 1439
   procurement and risk, 1450
   production haulage (trucks and trains), 1446
   production issues, 1445–1446
   qualitative and quantitative risk assessments, 1449
   reliability in constructability and construction, 28–29
   reporting, 1447
   reserves, 1447
   risk considerations, 1449–1450
   rock bursts, 1450
   rock mass characterization, 1438
   safety and risk, 1450
   secondary drills and rock breakers, 1447
   secondary fragmentation, 1439
   and sensing technologies, 30
   service/ventilation level, 1440
   stope development for gravity systems, 1213, 1214f.
   stope development for LHD systems, 1213, 1213f.
   stope development for slusher systems, 1213, 1214f.
   stress-induced collapse, 1450
   super caves, 1437
   sustaining capital, 1448
   typical equipment, 1446–1447
   undercut level, 1439
   undercut sequence design and risk, 1450
   undercutting methods and issues, 1441–1443
   and underground infrastructure, 29
   underground stability issues, 1450
   water inrushes, 1450
   See also Cave mining
Block Caving Geomechanics, 1449
The Blue Book, 468
Boards of directors, 18
Bobcats, 1173
Boliden raise cage, 1209
Bond abrasion test, 1474
Bond equation, 1470, 1476
Borates, 1110
   in-situ solution mining of, 1110
   principal minerals, 1110t.
   processing of, 1110–1111
Bord-and-pillar mining. *See* Room-and-pillar mining
Borer miners, 381–382, 425, 425f.
Boshkov and Wright classification system, 369, 370t.
Bougainville Copper Ltd., and community conflicts, 1767
Boundary Hill pit (Australia), automated dragline trial, 816
Box–Cox transformation, 843–844, 844t.
Boyle, R.W., 127
Brazil, mining history, 4
Breaking New Ground, 1649–1650, 1666, 1769
Breathe and Live, 1578
Brewery Creek (Alaska) heap leach installation, 1079, 1079f.
Bre-X project (Indonesia), 299
BRIC countries (Brazil, Russia, India, China), 11
Bridge conveyors and carriers, 381
British Coal Corporation, 606
British Columbia Securities Commission, 225
British Medical Research Council (BMRC)/Johannesburg, respirable size-selective curves, 1601
Bronze Age, 3
Brundtland, Gro Harlem, and Brundtland Commission, 32, 1665, 1769
Bucket fill factor, 866–867
Bucket-wheel excavator systems, 911–912, 911f., 912f.
Bucyrus, 937
Bullion markets, 59
The Burrowers, 1199

Cable electric shuttle cars (in R&P mining), 1167, 1168f.
Canada
   and community issues, 1770
   Innu, Inuit, and Dene people, 1797, 1798
   internal responsibility system (IRS) for health and safety, 1560
   National Instruments 43-101, 203, 231
   publications and associations (Web sites), 1831–1832
   and tar sands, 6
   Whitehorse Mining Initiative, 1666, 1769
Canada Cement Lafarge Ltd., 1376
Canada Centre for Mineral and Energy Technology (CANMET), 433, 720–721
Canadian Institute of Mining, Metallurgy and Petroleum (CIM), 146
CIMVAL code, 219–220
Canadian Rockburst Support Handbook, 620
Carbonatite Cu, rare earth element, Nd, Fe, P deposits, 91–92
Carnegie Mellon University (CMU), 806
Casapalca mine (Peru), and shrinkage stoping, 1351
Cash flow
   defined, 80
   discount rates applied to projections, 70
   external factors affecting, 68
   internal factors affecting, 69–70
   See also Discounted cash-flow; *and under* Market capitalization
Cassiterite, and placer mining, 1057
Caterpillar, 806, 917, 920, 925, 937
Caterpillar Performance Handbook, 285, 938, 1019
Cave mining, 1385, 1395–1397
   anomalies in quoted performance of caving operations, 1385, 1385t.
   and cavability, 1386–1388
   cost models for block-cave mining, 275, 275f., 275t.–277t.
   daily production from (2008), 1385, 1385t.
   dilution, 1392, 1393f., 1394f., 1395f.
   draw control, 1391, 1392f., 1392t., 1393f., 1394f.
   drawpoint productivity, 1389–1390, 1390f.
   drawzone spacing, 1390–1391, 1390f.
   flow lines and stresses around drawzones, 1390, 1391f.

geomechanics classification, 1388, 1388*f.*
in-situ rock mass ratings (IRMRs), 1386
lateral extension, 1386–1388
layouts, 1392–1393, 1395*f.*
maximum/minimum spacing of drawzones, 1390, 1391*f.*
modified rock mass ratings (MRMRs), 1386, 1386*f.*
parameters to be considered before implementation, 1386, 1387*t.*
and particle size distribution, 1388–1390
primary fragmentation, 1388, 1389*f.*
rate of caving, 1388
rate of damage, 1388
rate of undercutting, 1388
role of, 1385
sand model tests, 1390, 1390*f.*
secondary fragmentation, 1388–1389, 1389*f.*
size distribution of cave fragmentation, 1389, 1389*f.*
stability diagram, 1388, 1388*t.*
support requirements, 1394–1395, 1396*f.*, 1397*t.*
undercutting, 1393–1394, 1395*f.*, 1396*f.*
vertical extension, 1386
See also Block caving; Longwall top coal caving; Panel caving; Sublevel caving
Caving, 364–365, 387, 387*f. See also* Stoping
Cayuga mine (New York), 721
Centre for Mining Technology and Equipment (CMTE), 433
CES. *See* U.S. Bureau of Mines Cost Estimating System
Chemical sediment deposits, 100
Chezy–Darcy law for frictional pressure drop in a pipe, 1578
Chicago Board of Trade, 58
China
    as leading coal producer, 1339
    as number one consumer of mined commodities, 77, 78*f.*
    projected increase in ore consumption, 21, 22*f.*, 22*t.*
Christian Brothers Investment Services Inc., 1775
Christmas-tree method (coal), 389, 389*f.*
Chuquicamata copper mine (Chile), 15
Chuquicamata IPCC plant (Chile), 954
CIF. *See* Cost, insurance, and freight
CIMVAL code, 219–220
Civil engineers, and dewatering, 744
Classification, 1481. *See also* Hydrocyclones; Screens and screening
Clastic sediment-hosted copper deposits, 96–97
Climatic change, 6, 17–18
    and energy usage, 33–34
CLIMSIM software, 1611, 1621–1623, 1623*f.*
Closure
    analysis of options (Faro mine), 1668
    design-for-postclosure, 1683–1684
    environmental issues, 346
    of heap leaching facilities, 1083–1084
    and mining method selection, 346
    open-pit mining, 875, 893, 893*t.*
    security, 1660
    site environmental considerations, 1643
    strip mining, 997–998
    and supply, 44
    and sustainability, 1683–1684
    and waste disposal and contamination management, 1745
    waste piles and dumps, 677
    See also Environmental issues

Closure planning, 1753
    accurate estimates of cost, 1755
    action plan for improvement, 1756
    and asset life cycle, 1754
    changes as project moves through life cycle, 1762*t.*–1764*t.*, 1764
    community and stakeholder involvement, 1753–1754
    components of integrated plan, 1756, 1757*t.*–1760*t.*
    cost estimates, 1756
    documentation and record management, 1756
    establishing strategy, 1756
    evaluating options, 1756
    information sources, 1753
    integrating into life-of-asset planning, 1754–1755
    management framework, 1755, 1755*t.*
    management of business risk, 1755
    objectives, 1753–1754
    process, 1755–1756
    regular review and updates, 1755
    rehabilitation, 1754
    risk assessment, 1756–1764, 1760*t.*–1761*t.*
    selecting a plan from option analysis, 1756
    setting the context, 1756
    steps in, 1756
    workshop approach, 1756
Cluster drills, 436, 436*f.*, 437*f.*
Coal Information, 62–63
Coal, 377
    alternative uses of, 6
    bottom-dump coal haulers, 925, 925*f.*
    Christmas-tree method (room-and-pillar mining), 389, 389*f.*
    clean, 35
    clearance equipment, 1170–1171, 1175*t.*
    combination methods (room-and-pillar mining), 389, 389*f.*
    continuous room-and-pillar mining operation, 388
    conventional room-and-pillar mining operation, 388
    data collection on geology of, 158
    dust control, 1603–1605
    gasification, 35, 366
    geological data collection, 148, 158
    mine gas chart, 1833–1834
    outside-lift method (room-and-pillar mining), 389–390, 389*f.*, 390*f.*
    percentage mined using R&P among leading coal producers, 1339
    pillar-extraction methods, 389, 389*f.*
    pricing of steam coal, 60–61
    productivity of surface vs. underground mines (U.S.), 343, 345*t.*
    reclamation requirements for surface mining of (U.S.), 1654, 1655*t.*–1656*t.*, 1657*t.*
    room-and-pillar mining, 387–390
    room-and-pillar mining methods, 388
    room-and-pillar retreat methods, 388–390
    sample preparation, 191
    split-and-fender method (room-and-pillar mining), 389, 389*f.*, 390, 390*f.*
    strip mining, 996
    subsidence contours above longwall panels, 629, 630*f.*
    top coal caving, 1161–1162, 1162*f.*, 1163*f.*
    underground coal gasification, 1130*f.*, 1131
    U.S. DOE underground coal gasification experiments, 1131
    U.S. EPA and underground coal gasification, 1131

    See also Longwall mining; Methane (coal-bed production); Room-and-pillar mining in coal; *and under* Gas control; Room-and-pillar mining; Soft-rock equipment selection (underground mining)
Coal Mine Roof Rating (CMRR) system, 157–158
Coal seam gasification, 366
Coal seam methane drainage, 366, 1345
Coal workers' pneumoconiosis (CWP; aka black lung disease), 1414, 1600, 1601, 1601*f.*, 1602*f.*
Coalition for Environmentally Responsible Economies, 1650
Cobalt, Ontario, 705
CODELCO (Corporación Nacional del Cobre de Chile), 1643
    environmental risk assessment, 1646–1648
    environmental risk assessment worksheet, 1648, 1649*f.*
    evaluation criteria for consequences of environmental impact occurrence, 1648, 1648*t.*
    evaluation criteria for probability of environmental impact occurrence, 1648, 1648*t.*
    significance criteria for environmental aspects, 1648, 1648*t.*
    values for environmental risk magnitude, 1648, 1648*t.*
Coefficient of variation, analyzing, 207, 207*t.*
Coil sensors, 109*t.*, 110
Coke, sample preparation of, 191
Collahuasi IPCC plant (Chile), 954
Colorimetric method, 108–109
Comminution, 1461
    application of power, 1475–1477
    beneficiation during, 1473
    and Bond equation, 1470, 1476
    and choice of mining method, 1461
    circuit design, 1474–1477
    circuit selection, 1477
    effects of mining methods on, 1461–1464
    and energy usage, 30, 31*f.*, 34–35
    future, 1479
    grindability, 1474
    and knowledge of deposit and mining method, 1473
    liners, 1478
    media sizes for various mill types, 1477–1478, 1478*t.*
    ore characterization tests, 1474, 1475*t.*
    Pennsylvania (Bond) abrasion test, 1474
    pre-comminution blending of samples, 1473–1474
    selected circuit configurations, 1461, 1463*t.*
    simulation-based design, 1476
    single-particle breakage, 1474
    size range of common comminution machines, 1461, 1462*f.*
    standard materials, 1474
    unit operations for stages of, 1461, 1462*t.*
    wear rates, 1474
    See also Crushing; Grinding; Milling
Commodity stewardship, 8
Commonwealth Scientific and Industrial Research Organisation (CSIRO), 433
    digital terrain map, 816, 817*f.*
    dragline automation research, 816–817
    hollow inclusion cell, 511
Communication systems. *See* Mine communication systems

Community issues, 1767–1768, 1776
　addressing, 1775–1776
　affected community, 1768
　community, defined, 1768–1769
　community development, 1774–1775
　community engagement, 1774
　community involvement in closure planning, 1753–1754
　context, 1770–1774
　economic development, 1772
　health, 1772–1773
　local or host community, 1768
　and nature and scale of mining operation, 1771
　and nature, size, and reputation of company, 1771
　and physical landscape, 1770
　and political and legal frameworks, 1770–1771
　resettlement and in-migration, 1773
　rights holders, defined, 1768–1769
　scope of, and examples, 1771–1772, 1773f.
　security and human rights, 1773–1774
　and stage of mining life cycle, 1771
　stakeholder, defined, 1768–1769
　and sustainable development, 1769–1770
　water and mining, 1772
　See also Indigenous cultural considerations; Social impact assessment and management; Social license to operate; Sustainability
Compensation pressure, 566–567
Compressed air
　aftercoolers, 714
　in air flushing, 707–708
　applications, 706
　blockage prevention, 713
　boosters, 709–710
　compressor rental, 711, 712f.
　compressor types, 708–710
　and contaminant concentration, 713, 713f.
　contaminants and treatment, 712
　in COPROD system, 707
　corrosion from moisture, 713–714, 714f.
　design criteria for air purity, 712
　determining air pressure and flows, 708
　dewpoint as function of ambient temperature, relative humidity, and working pressure, 714, 715f.
　in down-the-hole (DTH) drilling, 706–707
　in drilling, 706–707
　dynamic (centrifugal) type, 708
　economies of drilling, 710–711
　emission regulations, 716
　and energy savings, 706, 710
　filters, 712–713
　flushing velocity, 707, 707f.
　free air, defined, 706
　free air delivery (FAD), 706, 710
　and hammer drills, 706
　heavy-duty systems, 709
　history of, 705–706
　humidity reduction, 714
　and increased drilling efficiency, 707
　leakage, 713–714, 714t.
　life-cycle costs, 710, 711f.
　medium pressure, 706
　moisture reduction, 713–714
　network distribution, 714–716
　noise directives, 716
　oil-lubricated vs. oil-free compressors, 709
　portable systems, 708, 709
　positive displacement type, 708
　prefilters, 713
　preventive maintenance of equipment, 711
　and processing, 706
　refrigerant drying, 714
　relative costs over time of different production activities, 708, 708f.
　remote monitoring and control, 716
　rotary screw type, 708
　safety of, 706
　stationary systems, 708–709
　and Swedish method, 706
　theory of, 705
　in top hammer drilling, 706
　variable-speed drive (VSD) systems, 709, 710
　in ventilation, 709
　VSDs in energy efficiency and cost savings, 710–711
Computational fluid dynamics (CFD), 743
　and hydrocyclones, 1505
Computing organizations (Web sites), 1829
Contango, 58
Continuous haulage systems, 381
Continuous miners, 381, 381f., 417, 426–427
　Aker Wirth continuous mining machine, 428, 429f.
　full-face, 425, 425f.
　in in-pit crushing and conveying (IPCC), 942–943
　place-change (cut-and-flit), 1166, 1167f.
　in R&P coal mining, 1341–1342, 1341f., 1345
　in R&P mining, 1166
　sequential cut-and-bolt, 1166
　simultaneous cut-and-bolt, 1166, 1167f.
　in soft-rock underground mining, 1158f.
Continuous room-and-pillar mining (coal), 388
Control systems. See Automatic control systems
Conventional room-and-pillar mining (coal), 388
Conversion chart (English to SI units), 1835–1840
Conveyor Equipment Manufacturers Association (CEMA), 982
Conveyors, 1288, 1288f.
　armored face conveyors (AFCs), 1159–1160, 1161f., 1162f.
　bridge type, 381
　conveyor stacking in heap leaching, 1082, 1083f., 1084f.
　costs, 1289
　and electric power distribution and utilization, 692–693, 693f.
　flexible conveyor trains (FCTs) in R&P coal mining, 1345
　future designs, 1289
　integration aspects, 1289
　limits of application, 1289
　mobile conveyors and loaders in ship loading, 980
　noise attenuation, 1637
　in ore emplacement (solution mining), 1095
　safety considerations, 1288
　selection criteria, 1288–1289
　technology, 1289
　underground, 1178
　See also Hauling and conveying; In-pit crushing and conveying
Cooling, 1618, 1623
　airflow increase, 1618
　airflow requirements, 1622–1623, 1623f.
　airway, ventilation, and heat flow parameters, 1622, 1622t.
　bulk cool air underground, 1619
　bulk cool intake air on surface, 1619
　cabs, 1620
　chillers (aka district cooling), 1619
　chilling service water, 1618–1619
　and CLIMSIM software, 1621–1623, 1623f.
　controlled recirculation, 1621
　and costs, 1621
　energy recovery, 1620–1621
　and governmental regulation, 1621
　and heat sources, 1621
　ice cooling, 1620
　integrated refrigeration systems, 1620
　large spray coolers, 1620
　maintenance, 1621
　and mining methods, 1621
　and mining rate, 1621
　and ore-body, 1621
　planning, 1621–1623
　and seasonal surface ambient conditions, 1621
　and size and condition of major airways, 1621
　small spray chambers, 1620
　spot coolers, 1619
　system selection, 1621
　thermal storage systems, 1621
　towers, 1619–1620
　and ventilation schematic, 1622, 1622f.
　vests, 1620
　See also Heat; Mine ventilation
Cooney, Jim, 1779
Cooperative Research Centre (CRC; Australia), 433
Cooperative Research Centre for Mining (CRCMining; Australia), 813
Coos Bay mine (Oregon), placer mining, 1060, 1061f.
Copper
　cutoff grade (solution mining), 1092
　heap leaching and chemistry of deposits, 1076–1077
　heap leaching and geology of deposits, 1075–1076
　lixiviants, 1088–1089
　oxide ores and heap leaching, 1076
　price trend, 11, 12f.
　recovery (solution mining), 1099
　sulfide ores and heap leaching, 1076
Coriolis force, 1316
Cornish pump, 3
Corporate social responsibility, 1684–1685, 1684f.
Cost estimating, 263, 281–282
　capital costs in feasibility studies, 301t.
　operating costs in feasibility studies, 301t.
Cost estimating (surface mines)
　ancillary systems, 287
　broad-brush methods, 281
　capital costs, 287
　comparative approach, 281
　cost models, 281, 288–289, 290t.–293t.
　dangers of averaging parameters or combining cost components, 282
　drilling and blasting, 283–284
　engineering-based and itemized, 281–282
　equipment component, 287, 288
　excavating and hauling, 284–286
　factored approach, 281
　labor component, 287–288, 289, 289t.
　mine producing 1,000 metric tons per day, 290t.–291t.
　mine producing 10,000 metric tons per day, 291t.–292t.

mine producing 80,000 metric tons per day, 292t.–293t.
operating costs, 287
ore and waste haul profiles, 282
parametric method, 281
powder factor, 282
professional salaries, 289, 289t.
scaling factors, 281
stripping ratio, 282
supply component, 287, 288, 288t.
target production rate, 282
Cost estimating (underground mines), 263–264
amount of rock to be removed for openings, 266
capital costs, 270
contingency fund, 270
continuous-flow calculatons, 267–268
cost models for block-cave mining, 275, 275f., 275t.–277t.
cost models for mechanized cut-and-fill mining, 277, 277f., 277t.–279t.
cost models for room-and-pillar mining, 273, 273f., 273t.–274t.
cost parameters, 265–269
critical access distances, 264
cycle-time calculations, 266–267
drilling requirements, 269
economic evaluation, 270–271
engineering, management, and administrative costs, 270, 271t.
equipment, 265
equipment operating costs, 269
estimation process, 269–270
labor, 265
labor costs (including benefits), 269, 270t., 271t., 272
model construction, 271–272
models, 271
operating costs, 270–271
ore haul distances, 266
powder factor, 266, 269
power and pressure equations (pumping), 268
and preliminary mine design, 264–265
production rate, 265–266
project life, 265
resource size, 265
rock characteristics, 272, 272t.
size of hauling equipment, 266
size of openings, 266
stope design parameters, 264–265
supply, 265
supply consumption, 268–269
supply costs, 269–270, 271t.
underground development openings, 264
unit costs, 272
ventilation calculations, 268
Cost, insurance, and freight (CIF), 50, 60
Cost Reference Guide, 268–269
Costs
average and marginal cost curves, 311, 312f.
and change in, as reserves are depleted, 310
externalities, 311
fixed, 310, 311
marginal, 311–313, 312f.
operating, 311
opportunity, 311
real, 310
recoverable, 310
sunk, 310
variable, 311
Council for Scientific and Industrial Research (CSIR; South Africa), 433

Cove prospect (Nevada), 299
Crandall Canyon mine (Utah) disaster, 717
Creighton mine (Canada), and air conditioning, 1611
Cross, Hardy, 1587–1588
Crosscuts, 1209–1210
CRU, 62
Crushers, 1285, 1286f., 1287f.
Allis design, 1465, 1466f.
alternatives to, 1463–1464
cone type, 188, 1039–1040
design and construction issues, 1285–1286
and electric power distribution and utilization, 695
equipment overview, 1286–1287
and grizzlies, 1286
gyratory, 1039, 1039f., 1463f.
impact type, 1040–1041, 1040f.
integration aspects, 1287
jaw type, 188, 1039
and life of mine/expansion plans, 1286
location, 1286
optional equipment for, 1287
and rock breakers, 1286–1287
roll type, 188, 1041
rooms, 1226–1227, 1226f., 1240
safety considerations, 1285
in sample preparation, 188
selection, 1286
and soft-rock underground mining, 1160–1161
and special particle shapes, 1473
Symons design, 1465, 1466f.
technology, 1287–1288
See also In-pit crushing and conveying
Crushing, 1461
apron feeders, 1463
bins, 1463
feeding considerations, 1463, 1464f.
location of primary crushers, 1462
staged crushing plants for low-tonnage operations, 1477
See also Comminution; In-pit crushing and conveying
Cummins, 710
Current trends, 19
alternative uses of coal, 6
automation, 5–6
climate change, 6, 17–18
economic uncertainty, 5
exploration technology, 14
fly-in/fly-out (FIFO) operations, 18
increased governance of mining, 18, 19
larger companies, 19
low-grade ore bodies, 4
marketing, 13
in mine operations, 15–16
in mineral exploration, 13–14
mining engineering knowledge, 5
ore reserves and valuation, 14–15
safety, 18
skilled employee shortage, 18
social and environmental issues, 4–5, 8, 17, 19
supply and demand, 11–13, 12f.
sustainability, 17–18
tar sands, 6
vertical integration, 16–17
See also Future trends
Customer/supplier engagement, 33
Cut-and-fill mining, 394, 1365
and access, infrastructure, and development cost, 1366

and amenability to mechanization, 1366
and availability of fill material, 1367
Avoca method, 395, 395f.
back stoping, 1366
bench-and-fill, 1366, 1371–1372, 1372f.
blasthole stoping, 1366
blasthole stoping with delayed backfill, 1367
cemented fill, mines using, 1372–1373
cemented fill stopes, 1367–1369
cemented rock fill, 1366, 1373
concurrent, 1366
current trends, 1373
delayed, 1366, 1367
and dilution, 1366–1367
drift-and-fill, 396, 396f., 1366, 1367–1368
extraction principles, 394–395
favorable conditions for, 1365
fill types, 1366
and geotechnical constraints, 1367
limitations, 1365–1366
longitudinal blasthole stoping with delayed backfill, 1367, 1368f.
mechanized, cost models for, 277, 277f., 277t.–279t.
method selection criteria, 1366–1367
and ore-body morphology, 1366
overhand, 365, 367t., 368f., 394, 394f., 395, 396, 1366
overhand drift-and-fill, 1368, 1369f.
overhand stoping, 365, 367t., 368f.
paste fill, 1366, 1373
postpillar, 365–366, 396, 397f., 1366, 1370, 1370f.
resuing, 1370–1371, 1372f.
sand fill, 1366
stope mechanization, 1373
stoping in flat, thick, wide ore bodies, 1212
stoping in steeply dipping, narrow ore bodies, 1212, 1212f.
transverse blasthole stoping with delayed backfill, 1367, 1368f.
two-pass, 394–395, 394f.
types, 1366
uncemented fill, mines using, 1373
uncemented fill stopes, 1369–1372
uncemented rock fill, 1366
undercut, 396, 397f.
underhand, 367t., 395, 396, 396f., 397f., 1366
underhand drift-and-fill, 1358–1359, 1359f.
uphole, flat-back slicing, 1366
vertical or near-vertical stoping, 1370, 1371f.
Cutoff grade optimization, 846
algorithm for single constraint problem, 850
heuristic algorithms, 846
heuristic techniques, 847–849, 847t., 848f., 848t., 849t.
Lane's approach, 849–850, 849t.
and mixed-integer linear programming (MILP), 850–852, 851f.
traditional cutoff grades in open-pit mining, 846–847, 846t.
Cyanide leaching, 1088
Cyclones, 1496
air separators, 1497, 1498f.
heavy medium cyclones, 1513
in placer ore sizing, 1068
See also Hydrocyclones

Darcy's equation, 654
Darcy–Weisbach equation, 1378

Data fusion, 27
DCF. *See* Discounted cash-flow
De Re Metallica, 3, 627, 1073
Debt/EBITDA ratio, 78
Declines, 1195, 1197t., 1198t.
Defense Advanced Research Projects Agency (DARPA) Grand Challenge, 806
Demand, 39
   current trends, 11–13, 12f.
   effect of supply on prices, 40
   future trends, 21
   and intensity of use, 40
   and markets for mineral products, 40
   and per capita incomes, 40
   and recycling, 40–41
   and substitutions and prices, 40
Dene, 1797
Denesoline, 1801
Depletion, 45
   and rents, 45
   and supply, 41, 42
   and sustainability, 47
   and technological change, 42–43
Deposits. *See* Mineral deposits
Design basis report, 234
   need for, 234
   Vol. 1: Management summary, 234–235, 256–257
   Vol. 2: Project economics, 235, 257
   Vol. 3: Technical narrative, 235, 257–259
   Vol. 4: Project execution plan, 235, 259–260
   Vol. 5: Operating plan, 235, 260–261
Deutsche Institut für Normung (DIN) conveyor standards, 982
Dewatering
   as future water source, 8
   in open-pit mining, 871, 873–874
   in quarrying, 1035
   wells, 518
Dewatering (liquid–solid separation), 1457, 1547, 1553
   centrifuges, 1552
   clarifiers, 1550
   and closed water circuits, 1547
   conventional thickeners, 1550
   costs, 1547–1548
   countercurrent decantation (CCD) thickeners, 1550–1551, 1551f.
   cyclones, 1549
   disk filters, 1547
   drag coefficient and related functions for spherical particles, 1548, 1549t.
   equipment, 1549–1553
   filters, 1551–1552
   high-density thickeners, 1550, 1550f.
   high-rate thickeners, 1550
   high-tonnage (continuous) pressure filters, 1552, 1553t.
   low-tonnage pressure filters, 1551–1552, 1552f.
   major influences on, 1548–1549
   and particle size and shape, 1548–1549, 1549t.
   and Reynolds number, 1548
   screens, 1549
   steps in, 1548
   and Stokes' law, 1548
   tailings ponds, 1552–1553
   thermal drying, 1547
   thickeners, 1550–1551
   vacuum filters, 1551, 1551f.
   and water discharge, 1547, 1548
   and water reclamation, 1547

Dewatering surface operations, 743, 873–874
   access safety and maintenance planning, 745–746
   air locking pipelines (troubleshooting), 759–760
   anchors, 755
   arctic conditions and permafrost, pit pumping in, 749–750
   and barges, 746
   basin runoff calculation, 744
   booster stations, 758–759
   cavitation (troubleshooting), 760
   centrifugal pumps, 751
   command and control system, 757
   costs, 762
   deserts, pit pumping in, 750
   diesel-driven pumps, 758
   electrical motor-driven pumps, 758
   emergency planning, 745
   field flow measurements, 755–756
   floating barges, 759
   flow measurement, 755–756
   and groundwater, 747–749
   haulage road runoff control, 746–747
   hauling wet ore (direct cost example), 762
   head loss in pipes, 835
   high-density polyethylene (HDPE) pipe, 754–755
   impoundment safety, 745
   indirect costs of water in pit, 746
   in-pit water control, 746–747
   insufficient suction head (troubleshooting), 760–761, 760f.
   locating booster pumps, 753–754, 754f.
   maintenance, 835–837
   maintenance strategy, 836–837, 836t.
   mine development, and groundwater, 747
   mine operation, and groundwater, 748
   minimum flow (troubleshooting), 761
   modeling, 756–757, 757t.
   open channel design, 750–751, 751f.
   pipe discharge drop method, 756, 756f.
   pipe materials and characteristics, 754–755
   pipeline material selection, 754
   pipelines, 753–755
   pipelines and confined stress due to temperature changes, 755
   pit pumping (direct cost example), 762
   pit pumping problems (special conditions), 749–750
   planning and managing, 835–836
   point loading, 755
   point precipitation frequency (PPF), 744
   precipitation and storm protection calculations, 744
   premining feasibility studies (re groundwater), 747
   pressure bleed through check valves (troubleshooting), 761
   pressure measurement, 755–756
   priming (troubleshooting), 760
   pump affinity laws, 753, 753f.
   pump curves, 751–752, 752f.
   pump selection, 751–753, 835
   pump station design, 757–759
   pump system operating point, 752, 752f.
   pump throttling, 752–753
   safety aspects, 744–746
   sediment, debris, and screens (troubleshooting), 761
   site access to lift equipment, 746
   slope failures into impoundments, 745
   sumps, 747

   surface hydraulic calculations, 744
   team, 743–744
   thrust blocks, 755
   trapezoidal channels, 751, 751f.
   troubleshooting pumping systems, 759–762
   Valdez Creek groundwater control (case study), 748–749
   water hammer (troubleshooting), 761–762
   wet drilling and blasting, 748
Dewatering underground operations, 765
   air lifts, 777
   backfilling, 770
   borehole flowmeter surveys, 768
   bulkheads and plugs, 778–779
   cavitation, 774
   clarification, 772
   clarification and settling systems, 772–774
   contingency planning, 772
   control and collection of water, 769–772
   control of surface water, 769
   dewatering pump stations, 1178
   drainage system design, 772
   draining underground water, 770–771
   and drilling management, 769–770, 770t.
   effect of barrier surrounding a shaft being deepened, 767, 768f.
   effective porosity (coefficient of storage), 767
   emergency storage required for 50-year storm, 767t.
   flow rate, 774–775
   flow to wells in dewatered and saturated aquifers, 767, 767f.
   ground freezing, 772
   grouting, 769, 771–772
   head or pressure relationship, 775
   high-speed centrifugal pumps, 774
   horizontal multistage centrifugal pumps, 777, 777f.
   horizontal sumps, 772, 773f.
   hydrological studies and modeling, 766–769
   impact of water on operations, 765
   inrushes, causes of, 779
   inrushes, defined, 779
   inrushes, frequency and location of, 779–780
   inrushes, prevention of, 770
   lamella plate settlers, 773
   mud-handling systems, 773–774, 774f.
   net positive suction head (NPSH), 774
   panel dewatering (R&P mining), 1169
   performance testing of pumps, 777–778
   pipework, 778
   positive displacement pumps, 774
   predicting precipitation and runoff, 766
   predicting water inflows, 766–769
   pressure tests, 768, 768f.
   progressive cavity pumps, 777, 777f.
   pump costs, 778
   pump selection, 776–777
   pump station design, 776
   pumping system design, 776
   pumping systems, 774–778
   pumping tests, 768
   reciprocating positive displacement pumps, 776, 776f.
   single-stage horizontally split centrifugal pumps, 776–777
   small portable or semiportable pumps, 776
   suction and losses for single-stage centrifugal pumps, 774, 775f.
   sumps, 772, 773f.
   surface water disposal, 780

and tracers, 769
vee settlers, 773, 773f.
vertical centrifugal pumps, 777
vertical sumps, 772, 773f.
vertical turbine (deep well) pumps, 777
water sources, 765
and water table (piezometric surface), 767
Diagenetic hydrothermal formation processes, 83, 86
Diamonds
   core sampling, 106
   drill-and-blast tunneling, 1198
   and interaction with indigenous culture (Australia), 1801
   magmatic deposits, 85t., 89
   pricing of gem diamonds, 49
Diamond-wire saws, 429–430
Diavik diamond mine (Canada), 1643
   community involvement in mine planning and permitting, 1659–1660
Diesel engines, heat from, 1618
Diesel exhaust control, 1605
   combusting and collecting generated DPM, 1606–1607
   and diesel equipment maintenance, 1607
   diesel particulate matter (DPM), 1605
   DPM control, 1606–1607
   DPM dilution, 1607
   DPM standards, 1605, 1606t.
   monitoring DPM emissions, 1607
   reducing DPM generation, 1606, 1606t.
Diesel-driven pumps, 758
Diesel–electric trucks, 869
Disc cutters, 419, 420f.
Discount rate, 302
Discounted cash-flow (DCF), 302
   analysis, 314–315
   in hard-rock mining method selection, 374
Discrete event simulation (DES), 841–842
Dispatch software, 1279
Dividend yield, defined, 80
Doe Run lead operations (Missouri), 8
DOZ mine (Indonesia), 1443f., 1445f.
Dragline systems, 903–905, 904f.
   advance benching, 905, 906f.
   allowable load, 910
   bank volumes, 910
   bucket capacity, 908
   bucket rigging, 908, 908f.
   buckets, 903
   cycle time, 908–909
   design, 904f.
   digging positions, 905, 905f.
   dragline selection, 909–910
   dump radius, 909–910, 910f.
   effective management of rehandle, 909
   extended benching, 906–907, 906f., 907f.
   fill factor, and buckets, 908
   hoist dependent cycles, 908
   lateral positions, 905
   as loaders, 910–911
   loading draglines, 939
   maximum suspended load, 910
   operating efficiency factor, 909, 909t.
   operating hours and calendar hours, 909, 909f.
   operating methods, 905–907
   pads, 904
   pits, 904, 904f.
   prime bank cubic meters or yards (PBCM/PBCY), 910
   prime volumes, 910
   production, 907–909
   production factor, 910
   productive digging, 909
   quality operation planning and coordination, 909
   rated suspended load, 910
   selection, 931–932
   sets (blocks), 904
   simple side casting, 905
   spoil-side benching, 904f., 907, 907f.
   stand-off distance, 909, 910f.
   swell factor, and buckets, 908
   swing dependent cycles, 908
   total bank cubic meters or yards (TBCM/TBCY), 910
   walking draglines, 931–932
   See also Excavation; Hauling and conveying
Drag-type bits, 439
Dredgers of the World, 1066
Dredging
   in hydraulic mining, 1049
   in surface mining, 405t., 407
   See also Placer mining and dredging
Drift-and-fill cut-and-fill mining, 396, 396f., 1366
Drifts, 1209–1210
Drill jumbos, 436, 437–438, 437f.
Drilling and blasting
   compared with TBMs, 1260–1263
   cost estimates, 1220, 1220t.
   cost estimating for surface mines, 283–284
   and leaching, 1092
   in open-pit mining, 863–866, 864f., 865f.
   in quarrying, 1035
   in R&P coal mining, 1342, 1342f.
   in shaft sinking, 1183–1184, 1183f.
   in strip mining, 990–991
   in sublevel caving, 1428–1429, 1428f.
   in subsurface mine development, 1209, 1220, 1220t.
   in tunneling, 1198–1199, 1199f.
   wet, dewatering surface operations for, 748
   See also Blasthole drilling; Blasting; Blasting impacts and control
Drilling for solution mining, 440
Drills, 3
   cluster type, 436, 436f., 437f.
   and compressed air, 706–707
   down-the-hole (DTH), 706–707, 865, 866
   down-the-hole (DTH) hammer, 435, 436f.
   drag-type bits, 439
   hammer type, 706
   handheld drills in hard-rock underground mining, 1155
   in-the-hole (ITH) hammer, 435, 437, 438
   jackleg, 436
   jumbo drills in hard-rock underground mining, 1152–1153, 1152f.
   jumbos, 436, 437–438, 437f.
   laser, 417
   percussion bits, 435–436, 436f.
   percussive equipment, 436–438
   polycrystalline diamond compact (PDC) bits, 439
   production drills in hard-rock underground mining, 1154–1155, 1155f.
   raise drilling machines, 423–424, 424f.
   ring-drilling, 436–437, 437f.
   rotary, 865
   shaft drilling machines, 422, 423f.
   stoper, 436
   top hammer drills in hard-rock underground mining, 1154
   top hammer type, 706, 865–866
   tricone bits, 438–439
Drum shearers. *See* Longwall drum shearers
Ducktown, Tennessee, 705
Dumps. *See* Waste piles and dumps
Dust control, 1595
   air helmets, 1605
   alveolar retention of particles of unit density, 1600, 1600f.
   and blasting, 1604
   in coal mining (continuous mining sections), 1603
   in coal mining (longwall sections), 1603, 1603t.
   in coal mining (surface), 1603–1604
   and coal processing, 1605
   and coal workers' pneumoconiosis (CWP), 1414, 1600, 1601, 1601f., 1602f.
   contain and capture, 1605
   and conveyor belts, 1604
   on drill rigs, 1603, 1604
   dust classification, 1599, 1599t.
   dust measuring instruments, 1601–1602
   dust sampling criteria, 1601
   dust suppressants, 1604–1605
   engineering control, 1602–1605
   GCA Ram, 1601
   and haul roads, 859, 969–971, 970t., 971f., 1604
   instantaneous dust monitors, 1602
   in metal mining, 1604
   and mine ventilation, 1578
   and mineral processing, 1605
   and minerals in transport, 1604–1605
   MINIRAM (miniature real-time aerosol monitor), 1602
   mobile equipment, 1603–1604
   MRE gravimetric dust sampler, 1601
   open-pit mining dust effects, 874–875
   and ore transport, 980, 987
   in orepasses, 1604
   permissible exposure limits for mine air contaminants, 1595, 1596t.
   personal exposure measurement, 1600–1602
   personal protection equipment, 1605
   prevention of dust generation, 1604
   R&P mining in coal and dust displacement, 1345
   and radon, 1628
   replaceable filter respirators, 1605
   respirable dust permissible exposure limits, 1600, 1600t.
   and roadheaders, 1604
   Simslin, 1601
   surfactants, 1604
   and tailings impoundments and dams, 663–664
   TEOMB (tapered-element oscillating microbalance), 1602, 1602f.
   and transfer points, 1604
   Tyndallometer, 1601
   tyndalloscope, 1601
   U.S. personal gravimetric sampler, 1601–1602
   ventilation air in continuous coal mining, 1603
   in waste passes, 1604
   water infusion, 1603
   water scrubbers, 1603
   water spray systems, 1603, 1604
Dust Control Handbook for Mineral Processing, 1605

Duval Corporation copper-molybdenum mine (Arizona), and IPCC, 942
Dynamite, invention of, 3

Eagle River gold mine (Ontario), 1352
    shrinkage stoping case study, 1352–1353
Earthworks, 1658
EBITDA, defined, 80
Eco-efficiency, 33
Economic decision making, 309
    average and marginal cost curves, 311, 312f.
    and change in costs as reserves are depleted, 310
    and competitors' knowledge, 310
    and conventional economic models, 309
    and cost externalities, 311
    cost perspective, 309
    and customers' knowledge, 310
    and fixed costs, 310, 311
    and high sunk costs, 310
    and interconnected global marketplace, 309
    and marginal costs, 311–313, 312f.
    and marginal revenue calculation, 312–313, 313t.
    and operating costs, 311
    and opportunity costs, 311
    and real costs, 310
    and recoverable costs, 310
    and risk/return trade-off, 309–310
    strategic perspective, 309
    and sunk costs, 310
    and time value of money, 309, 313–315
    types of costs, 310–313
    and variable costs, 311
Economic uncertainty, 5
Economics, 39
    boundaries of mining industry, 39
    cost estimation, 300–301, 301t.
    criteria for investment analysis, 302
    demand, 39–41
    Dutch disease, 46
    investment analysis, 301–302
    investors and speculators, 52–53
    mine management, 302–304
    mineral property evaluation, 298–299
    mineral property feasibility studies, 299–302
    mining law, 304–306
    overestimated properties, 299
    political factors, 44–47
    rents, 45–46
    resource and reserve reporting guides, 298–299
    supply, 41–44
    value added, 39, 45
    See also Costs; Investment analysis; Market capitalization; Pricing
Efficiency, 22
EGCH. See Exploration geochemistry
EITI. See Extractive Industries Transparency Initiative
Ejector-tray trucks, 1150
El Teniente (Chile)
    block caving production, 1437
    extraction level layout, 1439–1440
Electric power distribution and utilization, 683
    active filters, 690–691
    air-insulated switchgear (AIS) systems, 687, 688f.
    alternating current, 683–684
    ampacity, 699
    and ball mill drives, 692, 693f.
    basic distribution arrangements, 700
    and bucket-wheel and bucket-chain excavators, 695–696, 697f., 698f.
    cable couplers, 700
    cable insulation, 699
    cable temperature class, 699
    cable voltage ratings, 699, 699t.
    cables, 699–700, 699t.
    converter transformers, 691–692, 692f., 693f.
    and conveyor belts, 692–693, 693f.
    costs, 685
    and cutters, 683
    and cyclone feed pumps, 693–694
    demand, 684
    demand factor, 684
    design criteria, 684–685
    design for earthquakes, 696–698
    design for high altitude, 696, 699t.
    design for power self-generation, 698–699
    dewatering pumps, maintenance of, 834
    direct current, 683
    direct current usage, 685–686, 686f.
    distribution arrangements, 700–703
    diversity factor, 684
    dragline excavators, 694–695, 696f.
    drives and motors, 691–696
    earth continuity protection, 834–835
    earth leakage protection, 833, 834–835
    effect of current exposure to human body, 833–834, 834f.
    equipment solutions, 685–690
    filter and compensation (F&C) units, 690, 690f.
    gas-insulated switchgear (GIS) systems, 687, 689f.
    and high-pressure grinding rolls (HPGRs), 692, 693f.
    history in mining, 683–684
    HV and MV technologies, 687–688
    HV substations, 686–687, 687t.
    increased voltage in lines, 685, 686f.
    load factor, 684
    and loaders, 683
    and location of equipment, 685
    LV technologies, 688–690, 690t.
    maintenance of infrastructure, 832–835
    maximum demand, 684
    mean time between failures (MTBF), 685
    mean time to repair, 685
    and mine hoists, 694, 694f., 695f.
    motor starters, maintenance of, 834–835
    and motors, 684
    multiple-drive configuration, 684
    overhead power transmission, 685–686
    parallel lines, 685, 686f.
    peak load, 684
    point of common coupling (PCC), 687
    portable cables, 699
    power factor, 833
    power production directly at plant, 686
    power quality support equipment, 690–691
    power terminology, 684
    and primary crushers, 695
    protective equipment maintenance, 834
    rock drills, maintenance of, 834
    and safety, 685
    and SAG mill drives, 691–692, 692f.
    and shovels, 695
    and shuttle cars, 683
    stages of power distribution, 833
    static VAR compensators (SVCs), 690, 691f.
    surface mine distribution arrangements, 700–702, 701f., 702f.
    surface overhead-line distribution arrangements, 700
    surge arrestors, 687
    switchgear maintenance, 834
    switchgear systems, 687, 688f., 689f., 689t.
    synchronous condensers, 691, 691f.
    and system efficiency, 685
    tap changers, 687
    total connected load, 684
    trailing cables, 699
    trailing cables and maintenance, 834
    transformer maintenance, 834
    typical power distribution configuration for mine and plant, 687, 688f.
    underground cables, 687
    underground mine distribution arrangements, 703
    ventilation fans, maintenance of, 834
    voltage, 684
    voltage level, 686, 686t.
    and voltage regulation, 685
Electrical engineers, and dewatering, 744
Electrical resistivity, 109t., 110
Electrical self-potential, 109t., 110
Electricity generation, 6–7
Electronic cone penetrometer testing (CPT), 652
Electrostatic separation, 1533, 1543–1544
    conductive induction, 1544, 1544f.
    industrial roll-type device, 1544–1545, 1545f.
    ion bombardment, 1544–1545, 1545f.
    minerals and their magnetic and electrostatic responses, 1534, 1535t.
    operating principle of roll-type device, 1544, 1545f.
    of placer ores, 1070
    triboelectrification, 1544
    work function, 1544, 1544f.
Emma mine (Utah), 297–298
Energy
    alternative, for trucks, 25
    clean coal, 35
    and climate change, 33–34
    coal gasification, 35
    conspicuous consumption, 7
    emerging innovations, 34–35
    future usage trends, 32
    greenhouse gas emissions for selected metals mining, 34, 34f.
    and HIsmelt process, 35
    hydrogen, 35
    minerals industry consumption, 34, 35f.
    more efficient comminution, 34–35
    and sustainability, 33–35
    usage in comminution, 30, 31f.
    See also Compressed air; Electric power distribution and utilization
English units, conversion to SI units (chart), 1835–1840
Enterprise value (EV), defined, 80
Entries, 1204, 1205f.
    decline (slope), 1205
    drift (adit), 1204–1205
    vertical shaft, 1205–1207, 1206f.
Environmental engineers, and dewatering, 744
Environmental issues, 4–5, 8, 17, 19, 22
    blasting effects on environment, 874
    and block caving, 1450
    environmental and closure requirements, 346
    environmental monitoring, 724
    and mine subsidence, 636–641

and mineral property valuation, 222
mining law and environmental laws, 336
mining law and environmental protection, 304–305
and mining method selection, 346
and open-pit mining, 872–875
and ore transport and recovery, 986–987
and soft-rock mining method selection, 380
and solution mining, 1098–1099
and surface mining, 406
zero harm, 380
  See also Acid rock drainage; Acid rock drainage management; Community issues; Indigenous cultural considerations; International Council on Metals and the Environment; ISO 14000; ISO 14001; Site environmental considerations; Social impact assessment and management; Social license to operate; Sustainability; Waste disposal and contamination management
Epigenetic mineral deposits, 83–84
Epithermal gold and silver deposits, 93–94
Equator Principles, 1668, 1770
Escondida copper mine (Chile)
  as example of large-scale operations, 15, 15t.
  and SX-EW, 16
Estimation of Mineral Resources and Mineral Reserves Best Practice Guidelines, 146
European publications and associations (Web sites), 1832
European Union
  mining equipment safety directive, 607
  noise regulations, 1633, 1634
EV. *See* Enterprise value
EV/EBITDA (enterprise value/earnings before interest, taxes, depreciation, and amortization), 65, 75
  defined, 80
Evaporite deposits, 100
Excavation. *See* Open-pit mining; Subsurface mine development; Tunnel-boring machines; Underground development openings and infrastructure
Excavation equipment selection, 931
  bucket fill factor, 932, 933f.
  bucket-chain excavators, 935
  bucket-wheel excavators, 935
  bulldozers, 939
  continuous excavators, 935
  cost estimation, 933, 934t.
  final selection, 932–933
  job operating efficiency, 932
  load-and-carry wheel loaders, 938
  loading draglines, 939
  machine geometry, 932
  mechanical availability, 932
  size selection, 932, 933t.
  stripping shovels, 932
  surface miners, 939
  theoretical cycle time, 932, 933t.
  tractor scrapers, 938–939
  walking draglines, 931–932
Exploration
  advances in technology for, 14
  and aerial geophysics, 106
  and aerial photography, 106
  analysis of survey data, 106
  competitors in, 14
  and computer databases, 106
  core sampling, 106

discovery risk vs. political risk, 13
electrical methods, 109t., 110
electromagnetic methods, 109t., 110
fluid inclusion studies, 111–112
future trends, 23
geochemical zoning, 111
and geographic information systems (GISs), 106
and geological surveys, 106
geometallurgy, 112
and geophysical exploration, 106
geophysical methods, 109–110, 109t.
by global mining companies, 14
gravimetric methods, 109, 109t.
isotopic studies, 112
by junior exploration companies, 14
local geological studies, 106
by local or parastatal companies, 14
magnetic surveys, 109, 109t.
mineralogical zoning, 111
and multispectral data, 106
objectives, 105
ore deposit models and classification, 111
ore deposit zoning, 111, 112f.
and ore deposits, 106
and post-mineralization cover, 106–107
and preexisting data, 106
radioactive methods, 109t., 110
remote sensing, 109t., 110
reverse circulation drilling, 106
seismic surveys, 109t., 110
shift to less-developed countries, 13–14
sound methods, 105
by state-owned enterprises (SOEs), 14
and statistical analytical methods, 106
and wall-rock alteration, 111
Exploration and Mining Geology, 174
Exploration geochemistry (EGCH), 106, 107, 127
  analytical methods, 108–109, 133–135, 134t.
  and animal actions, 108
  and anomalous areas, 106, 107
  anomalous lows, 136
  aqueous fluid dispersions, 107
  assay methods, 135
  baseline studies, 140
  chemical analysis, 139
  communication with other professionals, 140
  data interpretation, 107, 109
  deposit zoning and mine sequencing, 139
  digestion methods, 134–135, 134f.
  dispersion models, 107–108
  dispersion patterns, 136
  duplicate sampling, 135–136, 136f.
  and element mobility, 108
  elements as pathfinders of ore deposits, 107, 107t.
  gaseous dispersions, 107
  geochemical methods, 135
  geochemical provinces, 107
  glacial deposit collection, 131
  glacier sediments, 108
  gold analyses, 135
  heavy mineral separations, 133
  history of, 127
  humus sampling, 132
  infrared analyzers, 139
  interpretation principles, 136–138
  lake water and sediment collection, 131
  methods, 127–128
  mine site sampling, 139

  model of formation of geochemical anomalies, 136, 137f.
  and multivariate statistical analysis, 137, 138
  orientation surveys, 128–130, 129t.
  overburden drilling, 131
  panning, 138
  primary dispersion halos, 108
  program planning, 128
  quality control and assurance, 135–136, 135f., 139
  regular fire assay, 108
  rock sample preparation, 133
  rock sampling, 131–132
  sample collection and handling, 130–132
  sample drying, 132–133
  sample preparation, 132–133
  secondary dispersion halos, 107–108
  sequential extractions, 135
  sieving and crushing of samples, 132–133
  and soil analysis, 108
  soil collection, 130
  soil gas sampling, 132
  soil grid showing anomalous gold population, 138, 139f.
  soil profile illustrating anomalies related to geological and geomorphological complexities, 136–137, 137f.
  spectrometric methods, 108
  statistical methods, 137–138, 138f.
  stream sediment collection, 130–131
  and stream sediments, 108
  surface water collection, 131
  threshold of background population, 136
  and univariate statistics, 137–138, 138f.
  UV light method, 138
  and vegetation, 108
  vegetation sample preparation, 133
  vegetation sampling, 132
  and weathering, 107–108
Exploration geology, 145
Explosives, 443
  ammonium nitrate, 1035
  bulk emulsion, 1035
  charge-up machines, 1155
  classifications, 443–444, 444t.
  detonators and initiation systems, 444–446
  electric detonators, 444–446, 445f.
  electronic detonators, 445, 446, 459
  fracturing, 446
  fragmentation, 446–447
  nitroglycerine explosives, 1035
  nonelectric detonators, 445–446, 445f.
  nonelectric shock tube (nonel), 1035
  pentaerythritol tetranitrate (PETN), 443
  in quarrying, 1035
  shock-tube initiators, 444, 445
  storage areas in subsurface mining, 1218
  transportation classifications, 444, 444t.
  See also Blasthole drilling; Blasting; Drilling and blasting
Extractive Industries Review, 1666, 1769
Extractive Industries Transparency Initiative (EITI), 1684, 1770, 1824

Face drilling rigs, 436
Fair market value (FMV), in mineral property valuation, 219
Falconbridge Nickel Mines Ltd., 1375–1376
Falun copper mine (Sweden), 3
Faraday's law, 119
Faro mine project (Yukon Territory), analysis of closure options, 1668

Federal Highway Administration (U.S.)
  culvert-design publication, 1711
  visual impact studies, 670
Feeder breakers (in R&P mining), 1168
Felsic magmas, mineral deposits associated with, 85t., 89–91
Finance
  institutions, 18
  modeling, 74–75
  See also Cost estimating (surface mines); Cost estimating (underground mines); Economic decision making; Economics; Fair market value; International Finance Corporation; Market capitalization; Mineral property valuation; Net present value; World Bank; World Bank Group
Firey, Walter, 1667–1668
FLAC (Fast Lagrangian Analysis of Continua) software, 609, 609f.
Flame jets, 417
Flat-blade dilatometer test, 475
Flood basalt association deposits, 88
Flotation
  development in Australia, 4
  and larger particles, 30–31
Flushing velocity, 707, 707f.
Fly-in/fly-out (FIFO) operations, 18
FMV. *See* Fair market value
FOB. *See* Free on board
Four "C's," 49
Four-wheel-drive vehicles, in hard-rock underground mining, 1155
Frasch, Herman, 1115–1116
Frasch sulfur mining, 1115–1116
  basic requirements for, 1116
  facilities needed, 1118
  field development, 1116–1118, 1117f.
  ore deposit, 1116
  well and development sequence, 1117, 1117f., 1118, 1118f.
Fraser Institute indexes of policy and mineral potential, 305–306
Free on board (FOB), 50, 60
Freeport-McMoRan
  IPCC plant (Chile), 954
  Safford mine (Arizona) permitting and approval process, 1654–1658, 1658t.
Froth flotation, 1517
  acids in pH control, 1524
  activators, 1524–1525
  adsorption of anionic collectors, 1521, 1521f.
  alkalis in pH control, 1524
  alkyl sulfates, 1522
  anionic collectors, 1521–1522, 1521f.
  anionic collectors for oxide minerals, 1522
  anionic collectors for sulfide minerals, 1521–1522
  axial mixing, 1527–1528, 1528f.
  bubble generation, 1527
  cationic collectors, 1522–1523, 1523f.
  cell selection, 1529–1530
  cell selection calculations, 1529–1530
  chemisorption in collection, 1520
  circuit design, 1529
  collection in froth layer, 1520
  collectors, 1520–1523, 1521f.
  conventional cells, 1526, 1526f.
  corrosive and wear properties of ore, 1529
  cyanide as depressant, 1525, 1525f.
  depressants, 1525
  design software, 1530
  dithiophosphates, 1521
  dithiophosphinates, 1521
  effect of wash water on entrainment of feed into froth, 1527, 1527f.
  effects of relative bubble and particle sizes, 1527, 1527f.
  enrichment ratio, 1518
  equipment, 1525–1528
  flotation columns, 1526–1528, 1526f.
  flotation time needed, 1529, 1530t.
  frother function, 1523
  frothers, 1523–1524
  grade–recovery curves, 1518–1519, 1519f.
  hydrophobicity/hydrophilicity, 1519, 1519f., 1520f.
  hydroxamates, 1522, 1522f.
  interrelated components of systems, 1517, 1518f.
  laboratory and plant flotation kinetics in screening, 1528–1529
  laboratory reagent screening, 1528–1529
  lime as depressant, 1525
  mercaptans, 1522
  mercaptobenzothiazole, 1521
  mineral applications, 1517
  modifiers, 1524–1525
  monothiophosphates, 1522
  nonionic collectors, 1520–1521, 1521f.
  nonmineral applications, 1517
  oleic acid, 1522, 1522f.
  optimum grind size of ore, 1529
  optimum pulp density, 1529
  organic depressants, 1525
  particle–bubble contact, 1519–1520, 1520f.
  percent metal loss, 1518
  percent metal recovery, 1518
  percent weight recovery, 1518
  performance calculations, 1517–1519, 1519f.
  performance parameters, 1517–1518
  pH control, 1524, 1524f.
  phosphonic acid, 1522, 1523f.
  phosphoric acid esters, 1522, 1523f.
  physisorption in collection, 1520
  ratio of concentration, 1517–1518
  reagent addition, 1529
  reagents, 1520–1525
  recovery of gangue particles by entrainment and hydrophobicity, 1525
  release analysis screening method, 1528, 1528f.
  sulfonates, 1522
  sulfosuccinamates, 1522, 1522f.
  synthetic and natural frothers, 1524
  thionocarbamates, 1521
  trithiocarbonate, 1521
  variability of ore, 1529
  xanthates, 1521, 1521f.
  xanthogen formates, 1522
FTSE 350 Mining Index, 65
FTSE Med 100, 65
Full-face continuous miners, 425, 425f.
Full-face machine cutters, 424–425
Future trends, 21
  advanced processing, 30–32
  automation, 26–28
  climate change, 6, 17–18
  commodity stewardship, 8
  and conspicuous power consumption, 7
  demand, 21
  electricity generation, 6–7
  environmental and social issues, 8, 22
  exploration, 23
  geological characterization, 23–24
  hauling and conveying, 24–25
  marine mining, 8
  mining in space, 8
  planning and scheduling, 25–26
  supply, 21–23
  surface mining, 24–26
  underground mining, 28–30
  value of water, 7–8
  See also Current trends
Future value, 302

Gas control, 1595
  coal seam degasification, 1597–1599
  combusting and collecting generated DPM, 1606–1607
  cross-measure boreholes, 1599, 1599f.
  and diesel equipment maintenance, 1607
  diesel exhaust, 1605–1607
  diesel particulate matter (DPM), 1605
  DPM control, 1606–1607
  DPM dilution, 1607
  DPM standards, 1605, 1606t.
  gas emission reservoir, 1596–1597, 1597f.
  gas fumes from blasting, 458
  gassiness of coal seams, 1596, 1596t.
  greenhouse gas emissions for selected metals mining, 34, 34f.
  horizontal boreholes drilled from surface, 1598, 1598f.
  in-mine horizontal drilling, 1597, 1598f.
  methane control in coal mines, 1595–1599
  methane explosions in U.S. coal mines (1970–2006), 1596, 1596t.
  monitoring DPM emissions, 1607
  permissible exposure limits for mine air contaminants, 1595, 1596t.
  postmining coal seam degasification, 1598–1599
  premining coal seam degasification, 1597–1598, 1598f.
  radon gas, 1607–1608, 1608t.
  reducing DPM generation, 1606, 1606t.
  strategy, 1595–1599
  vertical gob wells, 1599, 1599f.
  vertical hydrofraced well, 1597–1598
Geochemical prospecting. *See* Exploration geochemistry (EGCH)
Geochemistry Research Centre of Australia, 108
Geographic information systems (GISs)
  and exploration, 106
  Web-based services, 567
Geologic interpretation, 173
Geologic modeling, 173–174
  and cores, 176
  data analysis, review and verification, 175–176
  empirical, 174
  empirical geologic VMS model, 177, 178–179, 182f.
  empirical porphyry model showing distribution of alteration, mineralization zones, and sulfide occurrences in copper system, 177–179, 183f.
  genetic (conceptual), 174
  gold deposits, 174
  information to include, 174
  and maps and cross sections, 176–177, 178f., 179f.
  methodology, 175–179
  and ongoing data acquisition, 176–177

ore-body representation, 174–175, 175f.
plan (level, slice) maps, 177, 177f., 180f., 181f.
porphyry-related deposits, 174
purpose, 179–181
and sound geologic interpretation, 180
volcanogenic massive sulfide (VMS) deposits, 174
Geologic representation, 181–184
Geological characterization
future trends, 23–24
improved measurement techniques, 23
improved predictive modeling, 23
Geological data collection, 146–148
abbreviations, 153, 154t.–156t.
acetate overlays on base map or air photo, 157, 159f.
alteration data, 147
Anaconda mapping system, 157
bench mapping, 160, 161f.
bulk density (gamma-gamma density) logging, 165
caliper log, 166
coal data, 148
coal geology, 158
compass and tape in mapping, 156–157
computer-based systems, 146, 157
conventional method, 146, 157
core/cutting logging process, 160–163, 163f., 164f., 165f.
culture symbols (surface or underground), 152t.–153t.
data on miscellaneous features, 148
database access and restriction, 147
digitization of data, 147
evaluation support, 166–169
exploration, project, and mining geology, distinguished, 145
field note sheets, 157, 158f., 159f., 161f.
geological symbols, 150t.–151t.
geophysical logging, 163–166
laterolog (electrical conductivity, induction log), 166
lithologic data, 147
lithological symbols, 149t.
location data, 147
log types, 165–166, 167f., 168f.
mapping, 153–160
mapping and resource evaluation, 153
mapping process (steps in), 153–156
mineralization data, 147–148
natural gamma ray logging, 165
neutron density logging, 165
pit mapping, 160
quality assurance and control, 146
quantification, 146
required data, 147
reserves and resources, 145–146
resistivity (electrical) logging, 165–166
sedimentological symbols, 151t.–152t.
software, 158–159
sonic (acoustic) velocity logging, 165
spontaneous potential logging, 165
standardization, 146
structural data, 147
surface mapping, 159–160
symbols (mapping), 148, 149t.–153t.
underground mapping, 160
X-ray analyzers, 157
Geological Survey of Canada, 127
Geological surveys, 106
Geology of Ore Deposits, 174

Geomechanics, 463, 469
and availability of adequate information, 464
and "competent persons," 464–465
and concept studies, 464
demonstrating geological similarity, 464
feasibility assessment, 464–465
and mining feasibility, 463–465
and mining operations, 465
prefeasibility assessment, 464
and risk, 465
and safety, 465, 466
and strata management, 465, 468–469
and use of appropriate assessment methods, 464
See also Rock mechanics; Soil mechanics
Geophysical exploration, 106
Geophysics prospecting, 113
absolute gravity, 118
active (human-made) transmitting sources, 113, 114t.
active sources in electromagnetic methods (principle of induction), 119–120
active-source methods, 120
airborne electromagnetic methods, 120, 121f.
and airborne gravity systems, 118
best at prefeasibility stage, 113
brownfields exploration, 119
chemical remanent magnetization (CRM), 116
data compilation and interpretation, 117–118
defined, 113
depositional or detritus remanent magnetization (DRM), 116
drill-hole resistivity arrays, 123
electromagnetic fields from natural/passive sources, 119–123
electromagnetic methods, 123, 124t.
EM field functions and units, 120
factors affecting electrolytic conductivity of rocks, 121
and Faraday's law, 119
ferrimagnetic materials, 116
ferromagnetic materials, 114–115
galvanic methods, 120–121
and galvanic resistivity arrays, 122f., 121–123
and Global Positioning Systems (GPSs), 117
gravity gradiometer response of ore bodies, 118, 119f.
and gravity instruments, 118–119
and gravity meters, 118, 118f.
and induced polarization, 123, 123t.
isothermal remanent magnetization (IRM), 116
and Lenz's law, 119–120
magnetic field strength (H), 114
magnetic moment per unit volume (M), 114
magnetic survey guidelines for scope of work, 117
magnetic surveys and survey design, 117–118, 117t.
and magnetite, 116
and magnetometers, 116–117
magnetotellurics, 119
magnetotellurics and skin depth, 119, 119f.
methods, 113, 114t.
natural source audio-frequency magnetotellurics (NSAMT), 119

natural transmitting sources, 113, 114t.
normal remanent magnetization (NRM), 116
potential fields (gravity), 118–119
potential fields (magnetics), 113–118
potential fields by objective, 113, 115t.–116t.
pressure or piezo-remanent magnetization (PRM), 116
regional mapping arrays, 122, 122f.
relative gravity, 118
remanent magnetization, 116
right-hand rule, 119–120, 120f.
seismic methods, 123–125, 125t.
seismic reflection survey checklist, 125, 125t.
and Snell's law, 123, 125f.
and superconducting quantum interference devices (SQUIDs), 117
survey costs, 118
survey equipment specifications, 117
survey flying specifications, 117
susceptibility (k), 114
targeting for drill-hole positioning arrays, 122–123, 122f.
thermo-remanent magnetization (TRM), 116
and time domain electromagnetic (TDEM) systems, 120, 121f.
time domain wave shape, 123f.
types of magnetism, 114–116
units (Tesla, centimeter-gram-second), 113
wire-line and downhole geophysical logging, 125–126
Geostatistics, 194
and drilling and sampling, 195
and grade estimation, 195
kriging, 194, 195
model validation, 195–196
and ore-body modeling, 195–196
and resource evaluation, 196
summary statistics, 194
variance, 194
variograms, 194, 194f.
and volumetric modeling (block model), 195, 197–197
Geotechnical design, 465–466
analytical, 467
based on local experience, 467
based on rules of thumb, 467
empirical, 467–468
FoS concept and formula, 467
implementation of, 468–469
input data, 468
and legislation, 466–467, 469
and non-geotechnical matters, 468
numerical, 467
objectives, 466
requirements of, 466
and risk, 467
and safety, 466
Geotechnical instrumentation, 551
accuracy, 551
borehole displacements, 561–564
borehole extensometers, 562, 563f.
borehole geophysics, 553–554, 554f.
borehole pressure cells, 566–567
borehole televiewers, 553, 553f.
conceptual information sources, 551–552
conformance, 551
convergence monitoring systems, 562–563
core orientation tools, 552–553, 552t., 553f.
crosshole tomography, 554
data acquisition and presentation, 567–568

data immersion and visualization, 568, 568f.
data integration and management, 567
data reliability, 567
digital photogrammetry in investigation, 555, 556f., 557f.
extensometers (in monitoring), 560–561, 561f.
fiber optical technology, 560t., 563–564
geodetic monitoring, 559–560
geophysical logs (wireline tools), 553–554, 554t., 555f.
Global Positioning Systems (GPSs), 560
ground characterization (borehole techniques), 552–554
ground characterization (remote sensing), 554–556
ground-penetrating radar (GPR), 556
groundwater characterization, 556–558
hydraulic fracturing (HF), 558–559, 558t., 559f.
hydraulic testing of preexisting fractures (HTPFs), 558, 558t., 559
in-place inclinometers, 561–562
in-situ stress measurement, 558–559, 558t.
integrated pit slope monitoring systems, 561
for investigation, 551, 552–559
LiDAR in investigation, 555, 555f.
LiDAR in monitoring, 566
microseismicity, 567
for monitoring, 551, 559–567, 559t., 560t.
overcoring, 558, 558t.
photogrammetry in monitoring, 566
piezometers, 556–557
pore pressure monitoring, 566–567
precision, 551
probe inclinometers, 561, 562f.
range, 551
reliability, 551
remote sensing of ground deformation, 564–566
resolution, 551
robustness, 551
satellite InSAR, 564, 565f.
selection, 551
shaped accelerometers, 563, 564f.
stress change monitoring, 566–567
surface displacement monitoring, 559–561
surface geophysics, 556
surface radar, 564–566, 566f.
3-D surveys, 556
tiltmeters, 561
time domain reflectometry (TDR), 563
type of data required, 552
vertical profiling, 554, 554f.
Web GIS services, 567
wireless data transmission, 567–568
GFMS, 62
Gibson, Robert, 1667
GISs. See Geographic information systems
Global Acid Rock Drainage Guide (GARD Guide), 1721, 1734
Global Compact, 1668
Global Mining Initiative (GMI), 4, 1644, 1666, 1769
Global Navigation Satellite System (GNSS), 732
Global Positioning Systems (GPSs), 727
bandwidth, 728
in blasthole drilling, 459
challenges with, 727–728
and equipment dispatching and monitoring, 871–872
in geologic data collection, 146
in geophysics prospecting, 117
in geotechnical monitoring, 560
improving accuracy, 728
latency, 728
and mine surveying, 732
in proximity warning systems, 728
satellite availability, 728
vendor- or equipment-specific systems, 728
Global Reporting Initiative (GRI), 1650, 1795, 1824
Global Seismic Hazard Assessment Program, 652
globalCOAL, 60–61
Web site, 62
Globalization, 11, 19
Goaf, 387
Gob, 387
Gold
analyses, 135
clay-rich deposits and heap leaching, 1075
geologic modeling of deposits, 174
gravity-recoverable test, 201
heap leaching and chemistry of deposits, 1076
hydrothermal gold and silver deposits, 85t., 93–94
iron oxide copper-gold (with or without uranium) deposits, 85t., 92–93
Kolar Gold Fields (India), 1375
lixiviants, 1088
No Dirty Gold campaign, 1658
and placer mining, 1057
recovery systems (solution mining), 1099–1100
rushes (western U.S.), 4
shrinkage stoping, 1351, 1352–1353
silver-rich deposits and heap leaching, 1075
soil grid showing anomalous gold population, 138, 139f.
survey of gold-silver mineralizations, 128, 130f.
and uranium in conglomerates, 99
See also under Mineral deposits
Golder Paste Technology Ltd., and paste fill, 1376
Good Practice Guidance on HIV/AIDS, TB and Malaria, 1772
Goonyella Riverside mine (Australia), multiple-seam dragline operation with prestrip (case study), 1005–1006, 1006f.
Government agencies, and dewatering, 744
GPS. See Global Positioning Systems
Grade estimation, 195
Granite-hosted Sn-W deposits, 90–91
Grant, Baron, 297
Grasberg copper mine (Indonesia), 15
Gravity concentration and concentrators, 1507
absolute separation, 1507
air jigs, 1509
Bartles–Mozley tables, 1511
Batac jig, 1508
Baum jig, 1508, 1509f.
centrifugal (enhanced gravity) concentrators, 1511–1512
classification of devices, 1508, 1508f.
Cleveland placer jig, 1508
coal jigs, 1508, 1509f.
concentration criterion (C), 1507
Denver jig, 1508
equipment selection, 1515, 1515t.
Falcon concentrators, 1512, 1512f.
feed size ranges for devices, 1508, 1508t.
flowing film separators, 1507, 1509–1511, 1510f., 1511f.
and fluid media, 1507
Gemini tables, 1511
Harz jig, 1508
in-line pressure jigs, 1509, 1509f.
jigging devices, 1507, 1508–1509, 1509f., 1510f.
Kelsey centrifugal jig, 1509, 1510f.
Knelson concentrators, 1511–1512, 1512f.
mechanisms of movement, 1507–1508
mineral jigs, 1508, 1509f.
multi-G devices, 1507
Pan-American placer jig, 1508, 1509f.
performance curves, 1514–1515, 1514f., 1515t.
pinched sluices, 1510, 1510f.
Reichert cones, 1510, 1510f.
relative separation, 1507
riffled sluices, 1510
shaking devices, 1507, 1511, 1511f.
slurry feed, 1507
spiral concentrators, 1510–1511, 1511f.
values of concentration criterion required for effective separation, 1507, 1507t.
Gravity separation. See Gravity concentration and concentrators; Heavy medium separation
GRI. See Global Reporting Initiative
Grinding
dry, 1473
energies required, 1472
fine, 1472
See also Autogenous grinding; Autogenous grinding/semiautogenous grinding; Comminution; Semiautogenous grinding
Groote Eylandt mine (Australia), cast, doze, excavate strip mining operation (case study), 1000
Ground control, 611
defined, 595
See also Ground control (cable and rock bolting); Ground support; Hard-rock ground control; Shotcrete; Soft-rock ground control
Ground control (cable and rock bolting), 611
active support or reinforcement, defined, 611
auxiliary fittings, 613
cable bolt support, 618–619, 621f.
connectible Super Swellex, 619
continuous frictionally coupled (CFC) elements, 612, 612f.
continuous mechanically coupled (CMC) elements, 612, 612f.
deep-seated ground anchors (greater than 15 m), 620–621
discrete load transfer, 612, 612f.
discrete mechanically and frictionally coupled (DMFC) elements, 612–613, 612f.
dynamic support, 613–614
encapsulation techniques, 613
and fault slip bursts, 614
frictional support, 616–618
grouted mechanically anchored rock bolts, 615–616, 615f.
load transfer concepts for reinforcement, 612–613
medium-depth dynamic reinforcement elements, 619
medium-depth dynamic support design, 620, 621f., 622t.

medium-depth reinforcement design (3–15 m), 618–620
near-surface dynamic reinforcement elements, 618, 620f.
near-surface static design guidelines, 618, 619t., 620f.
near-surface static reinforcement (3 m or less), 614–618
passive support or reinforcement, defined, 611
and pillar bursts, 614
postexcavation reinforcement, 613
preexcavation reinforcement, 613
resin-anchored dowels, 616
resin-anchored pre-tensioned bolts, 616, 616f.
and rock bursts, 613–614, 614f., 615t.
rock reinforcement, 611–614
rock reinforcement, defined, 611
rock reinforcement support elements, 614–621
rock support, defined, 611
and seismic events, 613
spiling, 618
Split Set bolts, 616–617, 616f., 617t., 618
Stability Graph Method, 619
static support, 613
static support design, 619
and strain bursts, 614
strata control, defined, 611
support posttensioning, 613
support pre-tensioning, 613
Swellex bolts, 617–618, 617t.
ungrouted mechanically anchored rock bolts, 614–615
Ground mechanics. *See* Geomechanics
Ground support, 611
concrete liners, 622
and duty of care, 624
hydraulic props, 624
and mining legislation, 624
rigid steel sets, 623, 623f.
safety, 624
screen or mesh, 621, 622f.
shotcrete, 621–622
shotcrete pillars, 623
steel support, 623
straps, 621, 622f.
timber cribs, 623–624, 624f.
timber support, 623–624
work-force training, 624
yielding steel sets, 623, 623f.
See also Ground control
Groundwater
characterization, 556–558
contaminated flow control, 1750
control (Valdez Creek mine case study), 748–749
depression in shaft sinking, 1186–1188
and dewatering surface operations, 747–749
hydrology, 1706, 1715–1718
remediation, 1750
restoration and uranium, 1115, 1116t.
and surface mining, 406
and tailings impoundments and dams, 648
and waste piles and dumps, 673
Grouse Creek mine (Idaho), 299
A Guide to Health and Safety Regulation in Great Britain, 304
Guide to Shotcrete, 589
Gunpowder, introduction of, 3
Gy's sampling formula, 197–198

Handbook of Exploration and Environmental Geochemistry, 128
Handbook of Mining and Tunneling Machinery, 433
Handbook of Noise Measurement, 1635
Handbook on Mine Fill, 1383
Handy & Harman, 59
Hard rock, 595
mines, 595–596
Hard-rock equipment selection (underground mining), 1143
appropriate technology, 1144
articulated rear-dump trucks, 1147, 1147f.
average annual cost estimation, 1146–1147
bucket load (LHDs), 1151
choice of manufacturer or supplier, 1144
ejector-tray trucks, 1150
equipment matching, 1143–1144
equipment replacement, 1145
explosive charge-up machines, 1155
financial comparisons, 1145–1146
fleet description, 1143
four-wheel-drive vehicles, 1155
handheld drills, 1155
information required for financial analysis, 1146
and innovative models, 1144–1145
integrated tool carriers, 1155
jumbo drills, 1152–1153, 1152f.
leasing and contracting, 1146
LHD cycle, 1151
load-haul-dump machines (LHDs), 1150–1152, 1151f.
production drills, 1154–1155, 1155f.
rail equipment, 1155
remote control and automation of LHDs and trucks, 1152
rigid-body rear-dump trucks, 1147
road maintenance equipment, 1155
robust scalers, 1155
rock-bolting machines, 1153–1154
rope scrapers (slushers), 1155
shotcrete machines, 1155
sizing to stope development, 1143
top hammer drills, 1154
tractor–trailer truck units, 1147, 1148f.
trolley-wire electric trucks, 1150
truck capacity and operating cost, 1148
and truck clearance to sidewalls, 1148
and truck clearance to ventilation ducts, 1147–1148
truck cycle times, 1149
truck efficiency factors, 1149
truck load, 1149
truck productivity as function of haul distance, 1150, 1150f.
truck productivity as function of truck capacity, 1150, 1150f.
and truck turning radius, 1148
trucks, 1147–1150
trucks and development openings, 1147, 1148f.
trucks, key consideration for, 1148–1149
two-boom jumbo drill cycle time, 1153, 1154t.
two-boom jumbo drill development performance, 1153, 1154t.
two-boom jumbo drill physical development parameters, 1153, 1153t.
underground road trains, 1147, 1148, 1148f.
Hard-rock ground control, 573, 593
bolt plates, 576, 577f., 577t.
bolting machines, 574–575, 574f.
cable lacing, 578, 579f.
chain-link mesh, 575, 576f., 576t.
load transfer between mesh and reinforcement, 575–576
load vs. displacement for chain-link mesh, 575, 576f.
load vs. displacement for welded-wire mesh, 575, 575f.
mechanized installation of chain-link mesh, 575, 576f.
mesh aperture, 574, 574f.
mesh straps, 577, 578f.
Osro (Oslo) straps, 577–578, 579f.
pressure plates, 574f., 575
reinforcement, 573
and rockfall injuries, 573
steel mesh, 573–578
steel straps, 577, 578f.
straps, 576–578
surface support, 573
thin spray-on liners (TSLs), 591
typical welded-wire mesh dimensions, 574, 574f.
welded-wire mesh, 574–575, 574f.
wire thickness, 574
See also Ground control (cable and rock bolting); Shotcrete; Soft-rock ground control
Hard-rock mining method selection, 357
and advanced exploration concept studies, 374–375
Avoca method, 366, 367t.
block caving, 364–365, 365f.
Boshkov and Wright classification system, 369, 370t.
caving, 364–365
coal seam gasification, 366
coal seam methane drainage, 366
and deposit discovery, 374
economic analysis, 374
and final feasibility studies, 375
Hartman selection process, 369, 371f.
Horodiam method, 366
key influences on, 357–362
Laubscher selection process, 369, 372f.
longwall mining, 365
modified rock mass rating (MRMR), 360–361, 361f.
modified rock quality index (Q), 362
modified stability number (N'), 362
Morrison classification system, 369, 372f.
Nicholas classification system, 371, 372t., 373t., 374t.
numerical evaluation methodologies, 371–374
open-cut mining, 362–363, 363f.
overhand cut-and-fill stoping, 365, 367t., 368f.
postpillar cut-and-fill, 365–366
and preliminary feasibility studies, 375
qualitative and quantitative ranking systems, 369–371
risk assessment, 374
rock mass classification of underground excavations, 358–362
rock mass rating (RMR), 360
rock quality designation (RQD), 359–360, 360f.
rock structure rating (RSR), 360
rock tunneling quality index (Q), 361–362
room-and-pillar mining, 365, 367t.
selection and evaluation methodologies, 367–375

shrink stopes, 365, 367t., 368f.
square-set stoping, 367t.
stability in open-pit slopes, 358, 359f.
stages in application of process, 374–375
stoping, 365, 367t.
and strength and character of rock mass, 358–362
strip mining, 363–364, 363f.
and style of mineralization, 357–358
sublevel caving, 365, 366f.
sublevel open stopes, 367t.
sublevel retreat open stopes, 367t.
surface mining methods, 362–364
surface to underground transition, 366–367
Terzaghi's rock mass classification, 358–359
underground mining methods, 364–366, 364f.
underground retorting, 366
underhand cut-and-fill, 367t.
vertical crater retreat (VCR), 365, 369f.
See also Soft-rock mining method selection
Hartman selection process, 369, 371f.
Haul roads, 858–860, 858f., 869, 957, 974
categories, 958, 959f.
combined alignment, 962
cross slope or camber, 961, 963f.
curvature and switchbacks, 960
curve superelevation (banking), 961, 963t.
ditches and drainage, 963–964
dust palliation, 969–971, 970t., 971f.
functional design component, 957, 958, 958f., 966–971
functional design method specification notes, 968
geometric design component, 957, 958, 958–964, 958f., 960f., 961f.
horizontal (longitudinal) alignment, 960–962
integrated design components, 957–958, 958f.
intersection layout, 962, 963f.
and larger trucks, 957
maintenance cost model, 973, 974f.
maintenance management design component, 957, 958, 958f., 971–974
maintenance management system flow chart, 972, 974f.
maintenance management systems, 971–973, 973f., 973t.
mechanistic structural design procedure, 965–966, 966f., 967f.
optimal and maximum sustained grades, 960
in quarrying, 1036
and rolling resistance, 957
rolling resistance, estimating, 968, 969t.
rolling resistance, estimating from road visual assessments, 973–974, 975f., 975t.
run-out, 961–962
safety berms, 963, 964f.
sight distances, 959–960, 962f.
stopping distance limits, 959
structural design component, 957, 958, 958f., 964–966, 965f.
structural design method specification notes, 966
vehicle operating cost model, 973
vertical alignment, 959–960
wearing course material selection, 968, 968f., 969t.
width of road, 960, 962t.

Hauling and conveying, 912, 919
alternative energy for trucks, 25
articulated dump trucks (ADTs), 869, 1036
automated underground loading and haulage, 808–809
autonomous haulage systems, 806–808
belt conveyors, 982–984
belt conveyors in quarrying, 1038, 1038f., 1038t.
bottom-dump coal haulers, 925, 925f.
bottom-dump trucks, 978
bridge conveyors and carriers, 381
bulk flow simulation techniques, 982
cable belt conveyors, 982–983
continuous haulage systems, 381
conveyor belts, and electric power distribution and utilization, 692–693, 693f.
conveyor design guides, 982
conveyor sides and covers, 983–984
conveyor systems, future trends in, 25
cost estimating (surface mines), 282, 284–286
cost estimating (underground mines), 266
diesel–electric trucks, 869
downhill conveyors, 982
effect of mechanical cutters and new sorting technologies, 31
electric-drive trucks, 938
end-dump trucks, 978
equipment as cost drivers, 916
frameless trailers, 978
future trends, 24–25
haul distances for various haulers, 919, 919t.
haul roads, 858–860, 858f., 869
haulage road runoff control, 746–747
hauling wet ore, 762
in-pit crushing and conveying (IPCC), 869, 870–871, 928–929, 928f.
load-out stations, 978, 978f.
long-range, 25
machine-level haulage automation, 807–808
mechanical-drive trucks, 869, 938
mobile conveyors, 983, 984f.
off-highway articulated dump trucks, 920, 920f., 920t., 937, 937t.
off-highway rigid-frame trucks, 920–923, 921f., 922f., 923f.
on-highway trucks, 919–920, 937
in open-pit mining, 869–871
open-pit mining loading and hauling needs, 410–411, 410t.
overland conveyors, 982, 983f.
pipe conveyors, 983, 983f.
pneumatic trailers, 978
productivity estimation, 938
in quarrying, 1036
rigid-frame haul trucks, 869, 937–938, 937t., 1036
road trains (tractors with multiple trailers), 978
sandwich belt conveyors, 983
semitrailer belly-dump trucks, 919
semitrailer rear-dump trucks, 919
shiftable conveyors, 983
side-dump trucks, 978, 978f.
site-level haulage automation, 806–807
standard dump trucks, 919
strip mining loading and hauling needs, 410–411, 410t.
trolley-assisted haulage, 870
trolley-assisted mining trucks, 925–926, 926f., 977, 978f.

truck drive-train type, 923, 923f., 924t.
truck haulage, 977–978
truck selection, 937–938
truck size selection, 937, 937t.
truck tire management, 870
trucks, 869–870
wheel-tractor scraper, 926–927, 927f.
See also Conveyors; Dragline systems; Haul roads; In-pit crushing and conveying; Loaders; Load-haul-dump vehicles; Rail haulage systems
Hawkes, H.E., 127
Hayden Hill project (California), 299
Health, 1567, 1575
black lung disease (pneumoconiosis), 1414
community, 1571–1575
control of exposure to hazards, 1568, 1569t.
fatigue, 1571
as first priority of management, 317–318
fitness for work, 1569
heat stress, 1570–1571
HIV/AIDS, 1572, 1572t., 1573–1574, 1573f., 1574f.
infectious diseases, 1568
insurance, 1568
malaria, 1572–1573, 1572t., 1573f.
medical surveillance, 1569–1570
and mining law, 335–336
noise-induced hearing loss, 1570
occupational, 1568
occupational, medical side of, 1569–1571
occupational diseases, 1570–1571
personal, 1567–1568
and personal hygiene, 1568
and remote locations of mining, 1567
risk assessment, 1567, 1568–1569
silicosis, 1570
and soft-rock mining method selection, 380
and travel, 1567, 1568
tuberculosis, 1572, 1572t., 1574–1575, 1574f.
and underground ore handling, 1290
wellness, 1571
See also Cooling; Dust control; Gas control; Heat; Mine Safety and Health Administration; Mine ventilation; National Institute for Occupational Safety and Health; Occupational Safety and Health Administration; Noise hazards and controls; Radiation; Safety
Health and Safety at Work Act (U.K., 1974), 1559–1560
Heap leaching, 667, 668, 1085
acid production problems, 1085
agglomeration, 1081, 1081f.
capital and operating costs, 1085, 1086t.
and Carlin-type sedimentary gold ores, 1075
and chemistry of copper deposits, 1076–1077
and chemistry of gold deposits, 1076
and chemistry of nickel deposits, 1077
and chemistry of uranium deposits, 1077
and clay-rich gold deposits, 1075
configurations, 1074–1075
conveyor stacking, 1082, 1083f., 1084f.
drainage base, 1078
drip irrigation, 1079, 1079f.
dynamic heaps, 1074
efficiency of solution displacement, 1077–1078
and erosion, 676
flow sheet, 1458, 1458f.

future trends, 31
and geology of copper deposits, 1075–1076
and geology of gold deposits, 1075
and geology of nickel deposits, 1076
and geology of uranium deposits, 1076
geosynthetic raincoats, 675
and granitic uranium deposits, 1076
heap height, 1080
heap permeability and flow efficiency, 1077–1078
high-rate evaporative sprinklers, 1079
history of, 1073
intermediate liners, 1078
laboratory work, 1077
and laterites, 1075
leach pads, 1080
liner leak problems, 1085
liner systems, 673, 673f., 675, 676t.
list of operations, 1073
and low-sulfide acid volcanics or intrusives, 1075
metallurgical balance, 1083
mining for, 1080–1081
operations, 1073, 1074f.
and oxide copper ores, 1076
oxidized massive sulfides, 1075
permanent heaps, 1074
permeability problems, 1085
pH problems, 1085
poor recovery troubleshooting, 1085
principle of, 1073
reasons for, 1073–1074
reciprocating sprinklers, 1079
reclamation and closure, 1083–1084
recovery delay in multilift heaps, 1078
and saprolites, 1075
and sedimentary uranium deposits, 1076
and silver-rich gold deposits, 1075
solution application, 1079
solution application rate and leach time, 1078–1079
solution collection, 1083, 1084f.
stability problems, 1085
and sulfide copper ores, 1076
troubleshooting, 1084–1085
truck stacking, 1081–1082, 1082f.
valley-fill heaps, 1074–1075, 1083f.
water balance, 1080
wobbler sprinklers, 1079
See also Solution mining (in-situ techniques); Solution mining (surface techniques); Waste piles and dumps
Heat, 1611, 1623
and air conditioning, 1611
autocompression, 1617
autocompression, defined, 1611
and barometers, 1615, 1615f.
from broken rock, 1618
and CLIMSIM software, 1611
critical depth, defined, 1612
defined, 1611
from diesel engines, 1618
from electrical equipment, 1617–1618
enthalpy, defined, 1612
heat strain, defined, 1611
heat stress, defined, 1611
heat stress indices, 1614–1615
HotWork software, 1613–1614
instrumentation, 1615–1616, 1615f.
latent (hidden) heat, defined, 1612
miscellaneous sources, 1618
and psychrometers, 1615, 1615f., 1616f.
psychrometry, 1612f., 1613–1614, 1614f.
from rock, 1616–1617, 1617f., 1617t.
rock temperature measurement (resistance thermometers, thermocouples), 1616, 1616f.
sensible heat, defined, 1612
sources, 1616–1618
temperature, defined, 1612
thermal conductivity, 1616, 1617t.
thermal properties of common substances, 1616–1617, 1617t.
total heat load, 1618
See also Cooling; Mine ventilation
Heavy medium separation, 1512–1513
barite medium, 1513
Drewboy device, 1513
Dutch State Mines device, 1513
Dyna Whirlpool separators, 1513, 1514f.
equipment selection, 1515, 1515t.
feed preparation, 1513
ferrosilicon medium, 1513
galena medium, 1513
heavy medium cyclones, 1513
heavy medium solids, 1513
magnetite medium, 1513
medium recovery, 1513
Norwalt device, 1513
performance curves, 1514–1515, 1514f., 1515t.
separators, 1513, 1513f.
Wemco device, 1513
HEC-HMS software, 1710
HEC-RAS software, 1710
Heinrich Robert mine (Germany), roadheader usage, 1219
Henderson mine (Colorado)
block caving, 1439, 1440f., 1443f.
panel caving, 341–342, 342f.
High-density polyethylene (HDPE) pipe, 754–755
High-pressure grinding rolls (HPGRs), 1461, 1469–1470, 1470f., 1471f., 1471t.
conditions favored for, 1477
unit power, 1476–1477
Highwall mining, 1020, 1027
advantages and limitations, 1028
auger mining, 1027, 1028
case study (Jim Bridger mine, Wyoming), 1029–1030
costs per metric ton, 1028, 1029t.
dedicated systems, 1028, 1029, 1029f.
equipment and operation, 1028–1029
future developments, 1029
and highwall stability, 1027
modeling, 1028
pillar design formula, 1027–1028
pillar loading formula, 1028
pillar width calculation, 1028, 1028t.
recent developments, 1029
site requirements and conditions, 1027–1028
Hinsley, F.B., 1587
HIsmelt process, 35
Hitachi, 925, 937
HIV/AIDS, 1572, 1572t., 1573–1574, 1573f.
Good Practice Guidance on HIV/AIDS, TB and Malaria, 1772
stages of infection, 1574, 1574f.
workplace programs, 1574
Hoek–Brown
curvilinear failure envelope, 505, 505f., 506
failure criterion, 533–534, 534f.
m values, 531t.
Hoisting systems, 1295
aerodynamic and buffeting forces, 1316
annual shaft maintenance summary (shaft with two hoists), 1303, 1303t.
arrestor gears, 1306–1307
automatic loading, 1307–1308
bearings, 1297, 1298f.
bending ratio (friction hoists), 1311
bending-of-sheaves fatigue (hoist ropes), 831
bicylindrical conical drum hoists, 1296, 1296f.
brakes, 1298, 1298f., 1299f.
cage compartment size, 1314
cages, 1303, 1306
cappels, 1313
catch-gear rack and catch plate, 1307, 1307f.
clutches, 1299
clutching accuracy, 1322
combined modes fatigue (hoist ropes), 831–832
comparison of friction and drum hoists, 1297
compartment arrangement, 1314–1315
compartment size, 1314
concrete shaft lining, 1317
condition monitoring of guides, 830–831
conveyance terminology, 1303
Coriolis force, 1316
counterweight compartment size, 1315
counterweights, 1306
daily hoist examination and testing requirement (N. America), 1303, 1303t.
deflection sheaves, 1301
deterioration from movement, 830
deterioration from shaft operation, 829–830
differential rope loads, 1301
direct loading from orepass, 1307, 1308f.
disc brakes, 1298, 1299f.
divided single-drum hoists, 1296, 1296f.
double-drum hoist duty cycle, 1302f., 1303
double-drum hoists, both drums clutched, 1296, 1296f.
double-drum hoists, one drum clutched, 1296, 1296f.
drive motors and power conversion, 1299
drum brakes, 1298
drum hoist rope changing procedures, 1312
drum hoist ropes, 1308
drum hoist safety devices, 1306
drum hoist selection for production, 1299–1300
drum hoist selection for service, 1300
drum hoists, 1295–1296, 1296f.
drum/sheave diameter (drum hoist ropes), 1310
drums, 1297
dumps with mechanical scrolls, 1304, 1305f.
elastic rope stretch for skip, 1322
electromagnetic analysis of hoist ropes, 832
equipment monitoring facilities, 1320
fatigue damage, 830
fixed-body skips, 1304–1305, 1305f.
flattened strand, Lang lay ropes, 1309, 1310f., 1310t.
fleet angles (drum hoist ropes), 1310–1311
free bending fatigue (hoist ropes), 831
friction (Koepe) hoists, 1295, 1296–1297, 1296f.

friction hoist rope changing procedures, 1312
friction hoist ropes, 1308–1309
friction hoist safety devices, 1306–1307
friction hoist selection, 1300–1302
full locked coil ropes, 1309, 1310f., 1310t.
gearing, 1297–1298
ground-mounted friction hoists, 1296f., 1297
guide alignment, 1316
guide and bunton structural design, 1315–1316
guide rope design, 1316, 1317f.
guide rope location, 1316–1317, 1318f.
guide rope tensioning, 1317
guide ropes, 1311
headframe construction materials, 1318–1319
headframe design considerations, 1319–1322
headframe foundation, 1319
headframe heating and ventilation, 1321
headframe height, 1321
headframe loads, 1319
headframes, 1318–1323
headframes and conveyance and rope handling, 1320–1321, 1321f.
headframes and mine services, 1319–1320
headframes (temporary or permanent) and sinking provisions, 1320
hoist components, 1297–1299
hoist control systems, 1299
hoist duty cycles, 1302–1303
hoist rope inspection procedures, 832
hoist ropes, 831–832, 1309, 1310t.
hoisting distances for different ropes, 1311
hoisting plant, defined, 1295
horizontal ride, 830
hours of use per year, 1303
hydraulic dumps, 1304, 1305f.
inclined (slope) haulage, 1311–1312
inertial platform, 831
interior members, 1315–1317
jaw braking system, 1298, 1298f.
Koepe hoist duty cycle, 1302f., 1303
Lang lay ropes and hoisting distance, 1311
laser beams, 830
loading pockets, 1307–1308
loading via conveyor, 1307, 1308f.
loading via measuring flask, 1307, 1308f.
maintenance, 829–832
major components, 829, 829f.
manway compartment size, 1314
mechanical damage, 830
minimum clearance calculations, 1322–1323
and mining regulations, 1321
moving-beam systems, 831
multiple strand, nonspin ropes, 1309, 1310t.
multiple-rope drum hoists, blair type, 1296, 1296f.
multiple-stage, 1209
number of hoists, 1314
operating allowance, 1321
overturning (Kimberley) skips, 1303, 1304f.
parallel motion braking system, 1298, 1298f.
plumb lines, 830
rail guides, 1315
rope attachments, 1312–1313
rope construction, 1309, 1309f.
rope cuts, 1312
rope guides, 1315

rope harmonics and vibrations, 1321–1322, 1322f.
rope inspection procedures, 1312
rope installation and handling, 1312
rope lubrication, 1312
rope operating practices, 1312
rope sag, 1311
rope selection for drum hoists, 1309–1311
rope selection for friction hoists, 1311
rope selection requirements, 1309
rope storage, 1312
rope stretch allowance, 1321
rope tension, calculation of, 1312
rope tension equalization, 1321
rope torque, 1316
rope tread, 1301
ropes, 1308–1313
ropes and abrasive wear, 1312
round strand, regular lay ropes, 1309, 1310f., 1310t.
rubbing ropes, 1311
safety devices, 1306–1307
safety dogs, 1306
safety factor (drum hoist ropes), 1310
safety factor (friction hoists), 1311
safety factor for rope attachments, 1313
service compartment size, 1314
shaft and guide systems, 1313–1318
shaft buntons, 1315
shaft design procedure, 1313
shaft exterior shape, 1315
shaft guides, 1315
shaft inclination, 1314
shaft lining design considerations, 1318
shaft linings, 1317–1318
shaft location, 1314
shaft structure deterioration mechanisms, 829
shaft structure maintenance, 830, 831t.
shaft structures, 829
shaft systems, 1295
shafts, purpose of, 1313
sheave sizing, 1312, 1312t.
shotcrete shaft lining, 1317
single-drum hoists, 1295–1296, 1296f.
skip bail design, 1306
skip body design, 1305
skip compartment size, 1314
skip crosshead design, 1306
skip design methods, 1305
skip door design, 1305–1306
skip feed chute, 1307, 1308f.
skip guide roller design, 1306
skip guide shoe design, 1306
skip loading, 1307
skips, 1303–1305
South African Bureau of Standards on hoist ropes, 1311
spacing of Koepe wheel and deflection sheave, 1321
spill pockets, 1308
split differential diameter drum hoists, 1296, 1296f.
static drives, 1299
steel corrosion, 829–830
steel or cast-iron tubbing (shaft lining), 1317–1318
steel shaft guides, 1315
stopping allowance, 1321
stopping distance, 1323
swing-out-body skips, 1304, 1304f., 1305f.
swivels, 1313
T_1/T_2 ratio, 1300–1301

tail rope loop diameter, 1311
tail rope loop dividers, 1301–1302
tail rope orientation, 1311
tensile fatigue (hoist ropes), 831
thimble cappels, 1313
thimbles, 1313
torsion fatigue (hoist ropes), 831
tower-mounted friction hoists, 1296–1297, 1296f.
unequal drum and rope diameters, 1322
unequal rope lengths, 1323
ventilation characteristics, 1318
vertical closure and squeezing, 830
vertical strain, 830
Wabi single-tooth dog, 1306
wear, 830
wheels, 1297
wooden shaft guides, 1315
Holland formula, 1340, 1341f.
Homestake gold mine (South Dakota)
 and air conditioning, 1611
 and shrinkage stoping, 1351
Horodiam method, 366
HotWork software, 1613–1614
HSE. *See* Health and Safety Executive *under* United Kingdom
Hughes Tool Company, 438
Hunt, Brook, 62
Hydraulic breakers, 428–429
Hydraulic haulage systems (in underground ore handling), 1293
Hydraulic mining, 1049
 control hut safety, 1054
 design parameters, 1050
 dredging, 1049
 energy costs, 1049
 history of, 1049
 mobile plant operation, 1055
 monitor actuation (steering), 1052–1053, 1053f.
 monitor and control hut placement, 1054
 monitor control hut, 1053–1054, 1053f.
 monitor design, 1051–1054
 monitor operation, 1054–1055, 1055f.
 monitors, 1049, 1050f., 1052f., 1053f.
 nozzles, 1052, 1053f.
 operation, 1054–1055
 ore stream pumping vs. trucking, 1055
 pipelines, 1050–1051, 1051f.
 pumps, and arrangements of, 1050
 pumps, and net positive suction head, 1050
 and ring mains, 1051
 streamformers, 1052, 1053f.
 system design, 1050–1051
 water-jet cutting, 1049, 1050, 1051
Hydrocyclones, 1496, 1496f.
 apex size correlation, 1502, 1502f.
 capacity correlation, 1502, 1502f.
 common application classes, 1496, 1496f.
 and computational fluid dynamics (CFD), 1505
 conceptual model of operation, 1503, 1503f.
 correction factors for hydrocyclone feed solids volumetric concentration, 1501, 1501f.
 design enhancement, 1504, 1504f.
 effect of cone angle on performance, 1504, 1504f.
 experimental data for inefficient hydrocyclone, 1503, 1503t.
 instrumentation, 1505
 media choices, 1505
 mounting, 1499

new systems, 1504–1505
overflow, 1498
in parallel, 1496–1497, 1497f.
particle density, 1499, 1500f.
partition curve for inefficient hydrocyclone, 1503, 1503f.
performance assessment, 1503–1504
pressure drop correction factor, 1501, 1501f.
principal design elements, 1497, 1498f.
principles of operation, 1497–1499
in quarrying sand processing, 1043
recent developments, 1504–1506
ReCyclone, 1504–1505, 1505f.
regions in, 1498–1499, 1499f.
roping, 1499, 1499f.
selection, 1500–1503, 1500f., 1501f., 1501t., 1502f., 1503f.
specific gravity correction factor, 1501, 1502f.
spray flow, 1499, 1499f.
standard sizing and relative pricing, 1506, 1506f.
and Stokes' law, 1498
three-product, 1504
underflow, 1497–1498
vortex finder correction factor, 1502, 1502f.
water-only (automedium), 1497, 1497f.
Hydrofluoric acid, 134
Hydrogen Energy (company), 35
Hydrologists, 744
Hydrology, 743
Hydrometallurgy, 16
Hydrostatic lift, 743
Hydrothermal formation processes, 83, 84
Hydrothermal gold and silver deposits, 85t., 93–94
Hypogene (primary) processes, 84

ICME. See International Council on Metals and the Environment
ICMM. See International Council on Mining and Metals
Idarado mine (Colorado), and shrinkage stoping, 1351
IISD. See International Institute for Sustainable Development
Ilmenite, and placer mining, 1057
Impact hammers, 420, 428–429
  production rates, 432, 433f.
Imperial College of Science and Technology, Applied Geochemical Research Group, 127
Inclinometers
  borehole type in monitoring of tailings impoundments and dams, 660
  borehole type in slope management, 521
  in-place type, 561–562
  probe type, 561, 562f.
INCO Metals, and thickened tailings, 1376
India, and use of R&P in coal mining, 1339
Indicator methods of grade estimation, 195
Indigenous cultural considerations, 1797, 1814
  barriers to respectful engagement, 1799
  bequest value, 1801
  and Cobo definition, 1797
  collecting and analyzing right data, 1808–1810, 1809t., 1810f.
  comparing values, 1802, 1802f.
  concepts of culture, 1799–1803
  cultural assets, 1801
  cultural enhancement strategies, 1812–1813
  cultural impact assessments, 1803, 1803t.
  cultural landscapes, 1801
  cultural mixing zones, 1802–1803, 1802f.
  culture, defined, 1800
  data and methods for determining cultural impacts, 1807–1811
  determining a project's potential for cultural impacts, 1806–1807
  domains of culture, 1800–1802
  effective early community engagement, 1803–1805, 1804t.
  existence (wilderness) value, 1801
  growing importance of, 1797–1799
  and ILO Convention 169, 1797
  impact mitigation, 1811–1814, 1813t.
  impact prediction, 1810
  impact significance, 1810–1811, 1811t.
  impacts of mining on indigenous cultures, 1805–1807, 1805t.
  indigenous employee retention rates, 1798
  indigenous environmental knowledge, 1799
  involving cultural resource experts, 1807–1808
  involving culture holders, 1808
  land, importance of, 1800–1801
  scoping, 1803–1804
  tangible and intangible cultural assets, 1799–1800, 1800f.
  and technological shifts, 1799
  traditional indigenous knowledge, 1798–1799
  understanding cultural differences, 1802–1803, 1802f., 1803t.
  use value, 1801
  See also Community issues; Social license to operate
Induced polarization, 109t., 110
Inductively coupled plasma (ICP) spectrometry, 133
Inductively coupled plasma–emission spectrometry (ICP-ES), 190
Inductively coupled plasma–mass spectrometry (ICP-MS), 108
Inductively coupled plasma–optical emission spectrometry (ICP-OES), 108, 135
Industrial Minerals, 61, 63
Industrial Minerals and Rocks, 187
InfoMine USA, 269
Infrastructure maintenance. See Maintenance (of infrastructure)
Innu, 1797
  and Voisey's Bay nickel project, 1798
In-pit crushing and conveying (IPCC), 869, 870–871, 928–929, 928f., 941
  advantages of, 942–943
  capital costs, 944
  and continuous mining, 942–943
  current trends, 954–955
  design parameters, 943–945
  development of, 941–942
  disadvantages of, 943
  double-roll crushers, 947, 947f.
  fixed crushers, 948, 948f.
  fixed rim-mounted crusher operation, 950–951
  fixed rim-mounted crushing plants, 948, 948f.
  fully mobile continuous crushing system operation, 953–954
  fully mobile continuous crushing systems, 950, 951f.
  fully mobile crusher operation, 953, 953f.
  fully mobile crushing plants, 950, 951f.
  future trends, 955–956
  gyratory crusher and low-speed sizer compared, 956, 956f.
  gyratory crushers, 946–947, 946f.
  hybrid roll sizers, 947–948, 947f.
  in-pit crushing system types, 948–950
  and life-of-mine/expansion plans, 944
  low-speed sizers, 947, 947f.
  maintenance requirements, 945
  in open-pit mining, 869, 870–871
  operational considerations, 944–945
  operations, 950–954
  and ore characteristics, 944
  plant layout and design, 945, 945f.
  primary crusher selection, 945–946, 946t.
  primary crusher types (large capacity), 946–948
  process control, 954
  process design requirements and criteria, 944, 944t., 945
  production requirements, 944
  and project location, 944
  semifixed crusher operation, 952
  semifixed crushing plants, 948, 949f.
  semimobile direct-dump crushing plant operation, 952–953
  semimobile direct-dump crushing plants, 949–950, 950f.
  semimobile indirect feed crushing plants, 948–949, 949f.
  semimobile indirect feed crushing station operation, 952
  in South Africa, 941
  stationary in-ground crushing plants, 948, 948f.
  See also Conveyors; Hauling and conveying
In-place mining, 381
In-situ leaching, 31
Institute for Social and Ethical AccountAbility, 1795
Institution of Mining and Metallurgy (United Kingdom), 298–299
Instrumental neutron activation analysis (INAA), 133
Instrumentation engineers, and dewatering, 744
Interaction, 596–597
Internal rate of return (IRR), 302
International Autogenous and Semiautogenous Grinding Technology Conferences, 1461
International Building Code, 685, 698
International Commission on Large Dams Bulletin 121, 661
International Council on Metals and the Environment (ICME), 1644
  formation of, 1666
  statement on tailings management, 646
  tailings management guidelines, 662
International Council on Mining and Metals (ICMM), 17
  closure planning information, 1753
  formation of, 1666, 1769
  on site environmental considerations, 1644
  on sustainability, 1644, 1645t., 1667, 1667f., 1667t., 1680, 1769–1770, 1799
  tailings management guidelines, 662
  Toronto Declaration, 1644
International Covenant on Civil and Political Rights, 1773–1774
International Covenant on Economic, Social and Cultural Rights, 1773–1774
International Cyanide Management Institute, 1735
International Electrotechnical Commission (IEC), 825
International Federation of Chemical, Energy, Mine and General Workers' Unions, 318

International Finance Corporation (IFC)
  environmental and social standards, 1770
  Good Practice Notes/Handbooks, 1668
  Performance Standards on Social and Environmental Sustainability, 1780, 1795
  and social issues, 1769
  sustainability framework, 1668
  tailings management standards, 662
International Institute for Environment and Development (IIED), 1644
  tailings management standards, 662
International Institute for Sustainable Development (IISD), 1665
International Journal of Rock Mechanics, Mining Sciences, 433
International Labour Organization (ILO)
  Convention 169, 1780–1781, 1795, 1797
  on mine surveying, 732
  on occupational health, 1568
International mining economics (Web sites), 1828–1829
International Nickel Company (Canada), and backfilling, 1375–1376
International Organization for Standardization (ISO), 303
  mining truck design standards, 922
International publications and associations (Web sites), 1831–1832
International Rock Excavation Data Exchange Standard, 721
International Seabed Authority, 1068
International Society for Mine Surveying (ISM), 731
International Society for Rock Mechanics (ISRM), 468, 553
International System of Units (SI), 49
  units, conversion from English units (chart), 1835–1840
International Valuation Standards Council, 220
Intrusion-related gold deposits, 91
Inuit, and Voisey's Bay nickel project, 1798
Inverse-distance weighting, 195
Investment analysis, 73, 79–80
  analyst's skills, 76
  and asset's position on cash cost curve, 76, 77f.
  buy-side, 73
  "castles in the sky" theory, 73
  client contact, 75
  and company's management skills, 78–79
  earnings momentum and the cycle, 77, 77f.
  evaluation metrics, 75, 75t.
  financial modeling, 74–75
  and global commodities trade, 77, 78f.
  importance of, 73–74
  and independent research, 75
  intrinsic-value (firm-foundation) theory, 73
  investment skills, 75–76
  and leverage and liquidity, 78
  and operational leverage, 76
  and optionality, 76
  and political/legal environment, 79
  and quality of asset, 76
  relative market cap performance between Fortescue Metals and Mount Gibson (2002–2008), 73, 74f.
  relative performance of Australian dollar vs. CRB Index, 77, 79f.
  relative share performance between Xstrata and Anglo American (2002–2008), 73, 74f.
  research notes, 75, 76f.
  sell-side, 73
  sell-side analyst's skills, 76
  tasks, 74–75
  tools, 76–79
IPCC. *See* In-pit crushing and conveying
Irish-type Pb-Zn (Cu) deposits, 96
Iron
  automated blasthole drilling, 820–822, 820t., 821t., 822f.
  banded iron formation (BIF) deposits, 97
  ferrous, and acid rock drainage, 1723
  sediment-hosted deposits, 97
Iron oxide copper-gold (with or without uranium) deposits, 85t., 92–93
IRR. *See* Internal rate of return
ISO. *See* International Organization for Standardization
ISO 9000 series, 193, 1817
ISO 14000, 986–987, 1817
ISO 14001, 1643, 1646, 1646t.
ISO 17025, 193

Jackleg drills, 436
Janbu's method, 499
Jim Bridger mine (Wyoming), highwall mining case study, 1029–1030
John Deere, 710, 920
Joint roughness coefficients (JRCs), 530, 530t.
JORC code. *See under* Australasian Institute of Mining and Metallurgy

Kador Engineering, 925
Kidd Creek mine (Ontario)
  and air conditioning, 1611
  and thickened tailings, 1376
Kilburn, Lionel, 225
Kimberley Process Certification Scheme, 1770
Kirchhoff, Gustav R., 1585
Kirchhoff's laws, 756, 1585–1586, 1587
  direct application of, 1587, 1588–1589, 1588f.
Kirpich equation, 1708
Koepe, Frederick, 1296
Kolar Gold Fields (India), 1375
Komatsu, 917, 920, 925, 937
Kress, 925

La Quinta mine (Peru), tailings management scheme, 660
LAMODEL (highwall modeling), 1028, 1030
Land professionals, and dewatering, 744
Landriault, Dave, 1376
Laronde mine (Canada), and air conditioning, 1611
Laser drills, 417
Laterite deposits, 100
Laterites, 1075
Lateritic gold deposits, 101
Latin America
  mining history, 4
  publications and associations (Web sites), 1832
Laubscher rock mass classification system, 1386
Laubscher selection process, 369, 372f.
Law. *See* Mining law
Law of Conservation, 405
Layered chromite deposits, 87
LBMA. *See* London Bullion Market Association
Leaching, 4. *See also* Heap leaching; In-situ leaching
Legal professionals, and dewatering, 744
Legge, Thomas, 1568
Leica Automatic Deformation Monitoring System (GeoMoS), 736
Lenz's law, 119–120
Lerchs–Grossman pit optimization method, 889, 890
Liebherr, 937
Life-cycle assessment, 33
Liquid–solid separation. *See* Dewatering (liquid–solid separation)
Lithium, 1111–1112
  chemical composition of lithium brine lakes, 1112t.
  in-situ solution mining of, 1112
  processing of, 1112, 1113t.
Lithium metaborate, 134
LME. *See* London Metal Exchange
Load-haul-dump vehicles (LHDs)
  in hard-rock underground mining, 1150–1152, 1151f.
  in soft-rock underground mining, 1173, 1176f.
  in underground horizontal and inclined development, 1198
  See also Hauling and conveying
Loaders, 912, 914t., 1276, 1277f.
  automation, 1279
  bucket fill factor, 935, 936t.
  capacity, 913
  comparison, 936–937, 936t.
  costs, 1279
  digging profiles, 914
  dragline systems as, 910–911
  drawpoints for, 1277, 1278f.
  drive systems, 914
  drive-by loading, 915–916, 916f.
  electric cable shovels, 936t.
  and electric power distribution and utilization, 683
  fixed single-point ship loaders, 980
  flask train-loading system, 979
  flexibility, 913–914
  flood train-loading system, 979
  front-end, 936t.
  gradeability/speed/rimpull charts, 1277, 1278f.
  hydraulic excavators, 936t.
  hydraulic shovels, 912–913, 912f., 914t.
  job operating efficiency, 935
  life, 914
  linear ship loaders, 981
  load-and-carry wheel loaders, 938
  loading silos, 979
  luffing, slewing, and telescopic ship loaders, 980–981, 981f.
  maintenance, 915, 1280
  manufacturers and model specs, 917, 918t.
  matching to haulers, 916, 916f.
  material conditions, 915
  mechanical availability, 935–936
  mine design and construction for, 1277–1279
  mining shovels, 912–913, 913f., 914t.
  mobile conveyors and loaders in ship loading, 980
  mobility, 913
  one-sided loading, 915, 915f.
  operating methods, 915–916
  operational aspects, 1279
  in placer mining and dredging, 1060–1061, 1061f.
  as production drivers, 916
  and production management systems, 1279

quadrant ship loaders, 981
rubber-tired, 408
safety considerations, 1277
selection, 913–915, 935–937
ship loading, 980–981
size selection, 935
stockpile bays for loaders, 1278
stope mucking, 1279
support requirements, 914–915
swing angle, 913, 914t.
theoretical cycle time, 935
tipple accesses for, 1278
track dozer productivity, 918–919, 918t.
track dozers, 916–919, 917f., 918t.
tramming, 1279
tramming drifts (drives) for, 1277
traveling ship loaders, 980, 981f.
truck load-out stations, 978, 978f.
truck-loading bays, 1279
two-sided loading, 915, 915f.
wheel loader spotting, 915, 916f.
wheel loaders, 912–913, 912f., 914t., 1035–1036, 1036t.
See also Dragline systems; Hauling and conveying; Load-haul-dump vehicles; Unloaders
Lode (or orogenic) gold deposits, 94
London Bullion Market Association (LBMA), 59
  Web site, 62
London Metal Exchange (LME), 50, 57–59
  copper prices (1950–2008), 50, 51f.
  Web site, 62
London Platinum and Palladium Market (LPPM), 59
  Web site, 62
Longwall drum shearers, 417, 427, 427f.
Longwall mining, 1399, 1414–1415
  abrasivity, 1406–1407
  advantages and disadvantages, 384, 384t.
  AFC power demand calculations, 1166
  and armored face conveyors, 1399
  automation, 1162, 1163f.
  balance chambers, 1412, 1413f.
  beam stage loader, crusher, and boot end, 1160–1161
  and black lung disease (pneumoconiosis), 1414
  bleeder ventilation systems, 1412, 1412f.
  bleederless ventilation systems, 1412, 1413f.
  and carbon dioxide, 1402–1403
  and carbon monoxide, 1402
  and coal, 1399
  conceptual loading model for longwall shield capacity evaluation, 1404, 1405f.
  contemporary issues, 1414
  conventional layout, 1407, 1408f.
  and Coward's triangle, 1411–1412, 1411f.
  cross-measure boreholes, 1414
  cumulative probability distribution of loads required to stabilize longwall roof, 1404, 1406f.
  and deposit depth, 1399–1400
  and deposit dip, 1400
  and deposit thickness, 1400
  and deposit uniformity, 1400–1402
  districts, 1407
  dog bones (dumb bells), 1401
  electrical equipment, 1178
  electrical systems, 1161
  face length, 1409
  favorable characteristics, 384t.
  float time, 1407
  and fracturing or faulting, 1401
  gate roads, 1407–1409, 1409f.
  and geomechanics, 464
  gob ventilation boreholes (GVBs), 1413–1414
  in hard-rock mining, 365
  horizontal boreholes, 1414
  and hydrogen sulfide, 1403
  major components, 1399, 1400f.
  and methane, 1402
  mine layouts, 1407–1409, 1408f.
  monorails, 1161
  multiseam longwall, 1410–1411
  and nameplate capacity (NPC), 1162–1163, 1165
  panel length, 1409–1410
  periodic weighting behavior of longwall roof, 1404, 1405f.
  place-changing operations, 1408
  plow systems, 1410, 1411f.
  pump stations, 1161
  punch (highwall) layout, 1407
  roof and floor strata, 1403–1405
  roof bolting, 1408
  roof supports, 1158–1159, 1160f.
  as safe, efficient method, 1414
  and sand channels, 1401–1402, 1042f.
  seam topography, 1401, 1401f.
  as serially dependent processes, 1399
  set load, 1403
  shearers, 1159, 1161f., 1399
  and shields, 1403–1405
  soft-rock equipment selection, 1157, 1158–1166, 1158f., 1160f.
  in soft-rock mining, 377, 383–384, 383f., 597–599, 599f.
  specific energy of cutting, 1405–1406
  and spontaneous combustion, 1403
  stageloaders, 1399
  and strata gases, 1402–1403
  and support density, 1163
  and surface subsidence, 1414
  and thick seams, 1410
  and thin seams, 1410
  tube-bundle systems, 1412–1413
  ventilation, 1407, 1411–1414, 1412f., 1413f., 1430, 1430f., 1446
  and water, 1403
  yield load, 1404
Longwall top coal caving (LTCC), 1166, 1410, 1411f.
  equipment selection, 1161–1162, 1162f., 1163f., 1166
  See also Cave mining
LPPM. See London Platinum and Palladium Market
Lucky Friday mine (Idaho), undercut-and-fill stoping, 1212, 1212f.
Lundberg, H., 127

Mackenzie Valley Resource Management Act (Canada), 1797
Mafic and ultramafic intrusive association deposits, 89
Mafic magmas, mineral deposits in, 85t., 87–89
Magma Copper Company San Manuel mine (Arizona), 1267
  and air conditioning, 1611
  equipment selection, 1267–1268, 1268f., 1268t.
  mine geology, 1267
  mining setbacks and TBM modification, 1268–1269, 1269f., 1269t.
  TBM case study, 1267–1269
  TBM use, 1219
  tunneling scheme, 1267, 1267f.
Magmatic diamond deposits, 85t., 89
Magmatic hydrothermal formation processes, 83, 84–86
Magmatic platinum group metal (PGM) deposits, 87–88
Magnesium, 1108
  geologic conditions and in-situ solution mining of, 1108–1109
  principal magnesium minerals, 1109t.
  processing of magnesium chloride, 1109
  processing of magnesium hydroxide, 1109
Magnetic separation, 1533
  closed-cycle liquefier superconducting system, 1542
  conventional magnets, 1538–1540
  diamagnetic minerals, 1533, 1534–1538
  dry permanent magnetic separators, 1540
  eddy current separators, 1542
  ferromagnetic minerals, 1533
  flux density, 1534
  Frantz Isodynamic separators, 1540
  high-intensity magnetic separators, 1538–1540
  indirect-cooling superconduction system, 1543
  induced-roll magnetic separators, 1538–1539, 1539f., 1541–1542, 1542t.
  Jones separators, 1539–1540, 1540f.
  lift-type magnetic separators, 1539, 1540f.
  low-intensity magnetic separators, 1538
  low-loss superconducting system, 1542–1543
  magnetic drum separators, 1538, 1538f.
  magnetic field, 1534
  magnetic force, 1534
  magnetic pulleys, 1538, 1538f.
  magnetic theory, 1533–1538
  magnetization, 1534
  minerals and their magnetic and electrostatic responses, 1534, 1535t.–1537t.
  paramagnetic minerals, 1533, 1534
  permanent magnets, 1540–1542
  permeability, 1534
  protective magnets, 1538, 1538f.
  rare-earth drum (RED) separators, 1540–1542, 1541f., 1542f., 1542t.
  rare-earth roll (RER) separators, 1541–1542, 1541f., 1542f.
  superconducting dry open-gradient magnetic separators, 1543
  superconducting magnets, 1542–1543
  superconducting wet high-gradient magnetic separators, 1543, 1543f.
  susceptibility, 1534
  wet magnetic separators, 1538, 1539f.
  wet permanent magnetic separators, 1542, 1542f.
Maintenance (of infrastructure), 825, 837
  availability of infrastructure, 825–826, 826f.
  capacity of infrastructure, 826
  defined, 825
  dependability of infrastructure, 825, 826f.
  design-out maintenance, 826–827
  dewatering systems, 835–837
  electrical power distribution systems, 832–835
  hoisting systems, 829–832

infrastructure performance measures, 825–826
life-cycle reliability approach, 827, 828f., 835
maintenance performance indicator (MPI), 829
optimal frequency, 827, 828f.
primary functions of assets, 827
Quality Standard QS-9000 guidelines, 829
reliability of infrastructure, 825, 826, 826f., 827, 828f.
risk priority number (RPN), 829
roadway compaction, 832
roadway degradation mechanisms, 832
roadway drainage, 832
roadway grading profiles, 832
roadway resurfacing, 832
root-cause analysis, 827
secondary functions of assets, 827
synchronizing maintenance and production plans, 827–829
tactics, 826–827, 827f.
underground roadway maintenance, 832
workshop facilities, 837

Malaria, 1572, 1573f.
control programs, 1572–1573
Good Practice Guidance on HIV/AIDS, TB and Malaria, 1772

Management, 302
actions during contract negotiations, 320
avoidance of ethnocentrism, 322
basic principles of, 318
and behavior modeling, 328, 329t.
characteristics of effective manager, 319
competencies, 320
contemporary values, 302–303
control and cultural differences, 322
and cultural differences, 320–321, 321t.
decision making and cultural differences, 322
employee development, 324–325
employee recruitment and retention, 323–324
employee relations, 317
establishing and maintaining constructive relationships, 318, 319
focusing on the situation, not the person, 318
and global transfer of knowledge, 325–326
good-faith bargaining, 319–320
health and safety as first priority, 317–318
historic approaches, 302
and individualism vs. collectivism, 321
labor relations, 303
leadership and cultural differences, 321–322
leading by example, 318
learning from the outside, 326
and learning organizations, 324–325
and long- vs. short-term orientation, 321
maintaining self-esteem of employees, 318
mandatory bargaining issues, 320
and masculinity vs. femininity, 321
and mental models, 324–325
motivation and cultural differences, 322
in multinational environment, 317, 320–323
nonmandatory bargaining issues, 320
and personal mastery, 324
prohibition against bargaining individually with union employees, 320
prohibition against threats, interrogations, promises, and spying (TIPS), 320
safety risk, 303–304

and sequential vs. synchronous time concepts, 321
and shared vision, 325
and simulation training, 328–329
and skilled work force, 317
and small vs. large power distance, 320–321
and systems thinking, 325
taking initiative to make things better, 318
and teams, 325
training evaluation and record keeping, 328
training for work with other cultures, 322–323
and uncertainty avoidance, 321
and union contracts, 319
and unions, 318–320
See also Asset management; Closure; Finance; Health; Safety; Social impact assessment and management; Social license to operate; Training

Manganese
lixiviants, 1090
ore deposits, 98
sediment-hosted, 97
Manning's equation, 750, 1709, 1710
Manning's factor, 1261
Manning's roughness coefficient, 1708
Marchetti dilatometer test, 475
Marine mining, 8
Mark–Bieniawski pillar design formula, 1027–1028
Market capitalization, 65
and commodity prices, 68, 69f., 77
and cost escalation over life of mine, 70
and cost of capital, 70
and cost of extracting and treating ore, 69
and currency trends, 68
defined, 65, 80
differences with juniors, 71
discount rates applied to projected cash flows, 70
and diversity of company's asset portfolio, 70–71
and exploration and development costs, 69
and financing costs, 70
and growth potential, 71
and merger and acquisition activity, 71
multiple on theoretical NPV, 70–71
and net asset value,
and net present value (NPV), 65
and ore in the ground, 66–67
performance of mining indices vs. FTSE Med 100, 65, 66f.
and point in economic-and-commodity cycle, 71
and political risk, 70
and probability-weighted NPV, 67–68
and production capacity, 68
production cost standard, 69, 69t.
and project development risk, 70
and royalties and taxes, 70
and size and quality of mineral deposit, 69
and smelting and refining costs, 70
and spot vs. long-term prices, 68
valuation drivers through economic cycle, 65, 66f.
valuation of junior miners, 65–68
valuation of producing companies (external factors affecting cash flows), 68
valuation of producing companies (internal factors affecting cash flows), 69–70
value of mining project vs. investment risk, 65, 67f.

Market protection, 33
Material transporters and utility vehicles, 1173
Maximum credible earthquake (MCE), 652
Maxter-Atlas, 925
McCoy property (Nevada), 299
Mechanical engineers, and dewatering, 744
Mechanical rock breakage, 417
advance rate prediction based on cutterhead design, 432
advance rate prediction based on specific energy, 430–432, 430f., 431t.
advantages, 417
Aker Wirth continuous mining machine, 428, 429f.
bit materials, 419
bit selection, 419–420, 421f.
bit types, 419, 420f.
borer miners, 425, 425f.
continuous miners, 417, 426–427
diamond-wire saws, 429–430
disc cutters, 419, 420f.
drag-type tools, 418, 419
estimation of cutter life, 432–433
flame jets, 417
full-face continuous miners, 425, 425f.
full-face machine cutters, 424–425
full-face machines, 420, 421–425
hydraulic breakers, 428–429
impact hammer production rates, 432, 433f.
impact hammers, 420, 428–429
information sources, 433
laser drills, 417
longwall drum shearers, 417, 427, 427f.
machine categories, 420
by microwave, 417
partial-face machines, 420, 425–428
partial-face machines for hard-rock applications, 428, 429f.
performance estimates, 430–433
plow systems, 427
by projectile impact, 417
raise boring machines, 417, 423, 1283 1284f.
raise drilling machines, 423–424, 424f.
road milling machines, 427–428, 428f.
roadheader production estimation, 432
roadheaders, 417, 425–426, 426f.
Robbins Mobile Miner, 428, 429f.
rock breakers, 420
rock indentation process, 418, 418f., 419f.
rock mechanics tests, 418
rock saws, 429
rock/bit interaction, 418–419
roller-type tools, 418, 419
shaft boring machines, 421–422
shaft drilling machines, 422, 423f.
and specific energy (SE), 419
and specific penetration (SP), 419
stope boring machines, 424
strawberry cutters, 419, 420f.
TBM advance prediction, 432
trenchers, 429
tunnel boring machines, 417, 421, 422f.
Voest-Alpine surface miner, 429
water jets, 417, 430
water monitors, 430
Mega, 925
Meikle mine (Nevada), and air conditioning, 1611
Mergers and conglomerates, 5
Meridian Gold, Inc., and failed social license approach in Esquel gold deposit (Argentina), 1787–1788

Merriespruit (South Africa) tailings failure, 662
Metal Bulletin, 62
Metal exchanges, 57–59
Metallurgical efficiency, 1455
Metallurgical testing, 193
   accuracy, 194
   acid–base accounting analysis (ABA), 202
   advanced media competency test, 200
   agitated leach test, 201–202
   and assay pulps, 197
   autogenous and semiautogenous grinding (AG & SAG) milling tests, 200
   autogenous media competency test, 200
   average grade, 196
   beneficiation tests, 199
   bias, 194
   Bond abrasion index (Ai) test, 200
   Bond ball mill work index test, 200
   Bond low-energy crushing work index (Wi_c) test, 200
   Bond rod mill work index test, 200
   bottle rolls test, 201
   bulk density, 196
   carbon adsorption test, 202
   and channel samples, 197
   chemical analysis, 199
   column leach test, 202
   contained metal, 196
   density, 196
   detailed test program, 198–199
   diagnostic leach test, 201
   environmental testing, 202
   evaluation methods, 196–197
   flotation testing, 201
   and geostatistics, 194–196
   gold ore testing, 201
   grade–tonnage curves, 196, 197*f*.
   gravity testing, 200–201
   gravity-recoverable gold test, 201
   Gy's sampling formula, 197–198
   heavy media separation test, 200–201
   humidity cell test, 202
   JK Tech drop-weight test, 200
   laboratory accreditation and certification, 193
   laboratory test programs, 198–199
   leachate analysis, 202
   MacPherson autogenous work index test, 200
   mineralogical analysis, 199
   objectives, 197
   physical testing, 200
   pilot-plant tests, 199–200
   precision, 194
   preliminary test program, 198
   quality control, 193–194
   resource classification, 196
   resource evaluation, 196
   sample preparation, 199
   sample selection, 198
   sample types, 197
   scoping test program, 198
   specific gravity, 196–197
   spiral test, 201
   and split cores, 197
   stripping ratio, 196
   table test, 201
   tailings characterization test, 202
   test-program stages, 199
   and trenching and pitting samples, 197
Metamorphic hydrothermal formation processes, 83, 86
Meteorologists, 744
Methane (coal-bed production), 6, 1121
   carbon dioxide sequestration, 1130
   cavitation technique, 1127*f*.
   coal-bed methane reserves, 1121*t*.
   coal-bed reserves, 1121–1123, 1122*f*.
   coal-bed reservoir properties, 1123–1124, 1123*f*.
   in deep reservoirs, 1127, 1127*f*., 1128*f*.
   deep-well injection of produced water, 1129, 1129*f*.
   diffusivity of coal seams, 1124
   drainage from coal seams, 366
   gas composition, 1129
   gas gathering and measurement, 1129
   gas processing, 1129, 1129*f*.
   ground stresses around coal seam, 1124
   horizontal laterals drilled from the surface, 1128
   measurement of coal-bed gas contents, 1123, 1123*f*.
   in medium-depth reservoirs, 1125–1127, 1126*f*., 1127*f*.
   molecular gate separation system, 1130, 1130*f*.
   permeability of coal seam, 1123
   produced water disposal, 1128–1129, 1129*f*.
   production technology, 1124–1127
   reservoir characteristics, 1126*t*.
   reservoir pressure of coal seam, 1123–1124
   rock mechanics properties around coal seam, 1124
   in shallow reservoirs, 1124–1125, 1126*f*.
   surface disposal of produced water, 1128–1129, 1129*f*.
   underground coal gasification, 1130*f*., 1131
   vertical hydrofraced well, 1128
   water content and quality in coal seam, 1124
   well servicing and maintenance, 1127–1128, 1128*f*.
   wire-line logging, 1124, 1125*f*.
Metso, 1037
Michigan, and compressed air, 705
Micromine PITRAM software, 1279
Microsoft Excel, 302
Microwave mechanical rock breakage, 417
Middle Ages, 3
Millennium Development Goals, 1668–1669
Milling. *See* Primary milling; Secondary milling; Tertiary milling
Mine and Mill Equipment Operating Costs Estimator's Guide, 268–269
Mine Backfill, 1383
Mine communication systems, 717, 729
   accommodation of information source mobility, 721–722, 722*f*.
   accommodation of information sources and infrastructure, 721
   basics, 717, 718*f*.
   best-of-breed approach, 725
   client devices, 720
   client mesh networks, 720
   communication pathway or network, 717
   computer networks, 720
   data warehouses, 724, 725*f*.
   desired attributes, 721
   ease of maintenance, 723
   environmental monitoring, 724
   equipment status monitoring, 724
   Ethernet networks, 720
   expansion capability, 723
   fiber-optic systems, 720
   high frequency (HF) technologies, 718–719, 719*t*.
   implementation, 724–725
   information receiver, 717
   information source or transmitter, 717
   interfaces, 724
   low frequency (LF) technologies, 717–718, 719*t*.
   medium frequency (MF) technologies, 718, 719*t*.
   monitoring systems, 724–725
   pagers, 720
   pre-installation checklist, 723
   production monitoring, 724
   proximity warning systems (GPS- or sensor-based), 728
   redundant operation, 722–723, 723*f*., 726
   reporting systems, 724
   safety monitoring, 724
   self-healing, redundant-ring architecture, 722–723, 723*f*.
   sensors, 717
   strategic vendor approach, 725
   TCP (Transmission Control Protocol), 721
   TCP/IP (Transmission Control Protocol/Internet Protocol), 720
   telephones, 720
   transmission media, 719–721
   transmission technologies, 717–719, 719*t*.
   ultra high frequency (UHF) band, 718–719, 719*f*., 720
   ultra low frequency (ULF) band, 717, 719*t*.
   ultra wideband (UW) technologies, 719, 719*t*.
   very high frequency (VHF) band, 718–719, 719*f*.
   very low frequency (VLF) band, 717, 719*t*.
   wired networks, 719–720
   wireless mesh networks, 720, 721*f*., 726
   wireless networks, 719, 720–721
   See also Automatic control systems; Global Positioning Systems; Miner tracking systems
Mine Improvement and New Emergency Response (MINER) Act of 2006 (U.S.), 303, 717
   and miner tracking systems, 725, 726
Mine infrastructure, defined, 825
Mine infrastructure maintenance. *See* Maintenance (of infrastructure)
Mine rehabilitation. *See under* Closure planning; Open-pit mining; Open-pit planning and design; Quarrying; Strip mine planning and design
Mine Safety and Health Administration (MSHA)
   fatality information, 817, 1560
   mine ventilation video, 1578
   and miner tracking systems, 725
   noise regulations and engineering guidance, 1633, 1634, 1637
Mine subsidence, 627
   active, 635
   angle of break (angle of fracture), 629
   angle of draw, 628, 629*f*., 629*t*.
   architectural and structural considerations, 642
   area of influence, 628–629, 629*f*.
   backfilling in control of, 641
   and caving, 627–628
   chimney caving, 627, 628*f*.
   and competence of surrounding materials, 631
   and comprehensive planning, 643
   compressive, 629

continuous, 627, 632–635
contractile, 629
critical, 628, 629f.
crown holes, 627, 628f.
damage control and prevention, 641–643
damage to agricultural resources, 639
damage to airport runways, 637
damage to buildings and structures, 636–637, 638f., 638t., 639f.
damage to hydrological resources, 639, 640f.
damage to pipelines, 637–638
damage to railways, 637
damage to roads, 637
damage to subsurface structures, 638–639, 640f.
defined, 627
and degree of extraction, 631
development of, 627–639
discontinuous, 627–628, 632, 632f.
and elapsed time, 632
environmental effects, 636–641
extensile, 629
and extraction rate, 631
extraction rate in control of, 641
and extraction thickness, 630
factors affecting, 629–632
and geological discontinuities, 631, 631f.
graphical methods in prediction of, 635, 635f.
harmonic mining in control of, 641, 642f.
horizontal displacement, 629, 630f., 636
and hydrogeology, 632
and inclination of extraction horizon, 630
influence functions in prediction of, 633–634, 634f., 634t.
and in-situ stress state, 631
limit angle, 628, 629f.
limit of, 628
long-term effects, 635–636, 636t.
and longwall face, 627, 628f.
and longwall mining, 1414
measurement and monitoring, 632–633
and method of working, 631
mine layout changes in control of, 641
and mined area, 631
and mining depth, 630
and near-surface geology and surface topography, 631
noneffective width, 628
partial mining in control of, 641
phenomenological prediction, 633
pillar collapse, 635–636
plug, 627, 628f.
and postmining stabilization, 642
prediction of, 633–635
profile functions in prediction of, 634–635
pseudo-mining damage, 639–641
residual, 635–636
roof collapse, 635–636
scope of term, 627
structural deformation definitions, 636, 637f.
subcritical, 628, 629f.
and sublevel caving, 1433–1435, 1435f.
subsidence contours above longwall coal panels, 629, 630f.
supercritical, 628, 629f.
surface, 627
tensile, 629
and tolerable damage, 636
vertical displacement, 629, 636
vertical displacement measurement, 632–633

Mine surveying, 731–732
aerial and terrestrial photogrammetry surveys, 734
control benchmarks, 733, 733f.
coplaning or aligning, 739
data processing, 734
defined, 731
detail surveys, 740–741
detail surveys of massive deposits, 741
detail surveys of tabular deposits, 741
direction and grade lasers, 742
equipment and techniques, 732
and Global Navigation Satellite Systems (GNSS), 732, 734
GNSS surveys, 734
grade chains, 742
grade sticks, 741
Hause quadrilateral, 740
horizontal and vertical control, 736–737, 738f.
laser-scan surveys, 734, 735f.
laws and regulations, 732
mine-survey control network, 733, 733f.
open-pit mines, 732–736
position and direction transfer (surface to underground), 738–740, 740f., 740t., 741f.
and safety, 731–732
setting-out surveys, 741–742
slope-stability surveys, 734–736
software, 734
strings and grade pegs, 742
support surveys for equipment and machine control, 736, 737f.
surveyor's responsibilities, 732
3-D digital modeling, 734, 735f.
topographic surveys, 733–734
total station surveys, 733, 734f.
traversing with total stations in the backs (classical traversing), 737, 739f.
traversing with total stations in the walls (wall-station traversing), 737–738, 739f., 740f.
underground mines, 736–742
Weisbach triangle, 739–740
Mine ventilation, 1577
active devices, 1580–1582
air crossings, 1582
air doors, 1582
air movers, 1584
air quantity surveys, 1584–1585, 1584f.
airflow, 1578
airlocks, 1582
airway resistance (Atkinson's law), 1578–1579, 1580f.
analytical methods for solving ventilation networks, 1586–1587
auxiliary systems, 1583–1584
and Bernoulli equation, 1578
in block caving, 1440
booster fans, 1581–1582, 1582f.
and Chezy–Darcy law, 1578
in coal and metalliferous mines, 1591–1593
and compressed air, 709
cost estimating calculations, 268
deviations from square law, 1586
diesel dilution requirements, 1590, 1590t.
dilution of pollutants, 1577, 1578
and dust suppression systems, 1578
equivalent resistances, 1586–1587, 1587f.
face ventilation (R&P mining), 1169
factors in hazards and methods of control in, 1577, 1578f.
fan and duct systems, 1583–1584, 1583f.
fan maintenance, 834
fan pressure–quantity characteristics curves and operating points, 1581, 1581f.
friction factors, 1579, 1579t.
and frictional pressure drop, 1578–1579, 1587
and heating and cooling systems, 1578
and hoisting systems, 1318, 1321
jet fans, 1584
and Kirchhoff's laws, 1585–1586, 1587
Kirchhoff's laws, direct application of, 1587, 1588–1589, 1588f.
line brattices, 1583, 1583f.
in longwall mines, 1407, 1411–1414, 1412f., 1413f., 1430, 1430f., 1446, 1591–1592, 1592f.
main fans, 1580–1581, 1581f.
maximum air velocities (recommended), 1590, 1590t.
and methane drainage, 1578
minimum oxygen concentration, 1577
network analysis, 1585–1589
network simulation programs, 1589
numerical methods for solving ventilation networks, 1587–1589, 1588f.
in ore-body deposits, 1593, 1593f.
orepasses, 1283
parallel circuits, 1586
passive devices, 1582–1583
planning, 1589–1591, 1589f., 1590t.
pressure surveys, 1585
purpose of, 1577–1578
and radon, 1627, 1628f.
regulators, 1582
in room-and-pillar mines, 1592, 1592f., 1593f.
in room-and-pillar mining in hard rock, 1336, 1340f., 1342f., 1344–1345
seals, 1582–1583, 1582f.
series circuits, 1586
shock losses, 1579–1580
square law, 1578–1579, 1586
and steady-flow energy equation, 1578
stoppings, 1582
in stratified deposits, 1591–1592
surveys, 1584–1585
system schematic, 1580, 1580f.
systems, 1580–1584
TBMs and ventilation openings, 1259, 1260f., 1261
theory of, 1578–1580
through-flow systems, 1591, 1591f.
truck clearance to ducts, 1147–1148
U-tube systems, 1591, 1591f.
ventilation drifts, 1232, 1233f.
See also Cooling; Heat
Miner tracking systems, 725, 727
components, 725
fiber infrastructure systems, 726–727
integration options, 727
redundancy (survivability) requirements, 726, 727
RFID leaky feeder infrastructure systems, 726
and seam height, 726
serial infrastructure systems, 727
system layout, 727
system selection, 725–726
system types, 726–727
wired mesh systems, 726
wireless mesh systems, 726
Minera Escondida IPCC plant (Chile), 954

Mineral deposits, 83
　Algoma BIF, 97
　associated with felsic magmas, 85*t.*, 89–91
　associated with peralkaline/carbonatite magmas, 85*t.*, 91–92
　banded iron formation, 97
　bauxite, 100
　carbonatite Cu, rare earth element, Nd, Fe, P, 91–92
　chemical sediments, 100
　classification scheme, 85*t.*, 86
　clastic sediment-hosted copper, 96–97
　concealed (subsurface), 105
　defined, 83
　diagenetic hydrothermal formation processes, 83, 86
　epigenetic, 83–84
　epithermal gold and silver, 93–94
　evaporites, 100
　flood basalt, 88
　geological processes forming, 83
　gold and uranium in conglomerates, 99
　granite-hosted Sn-W, 90–91
　hydrothermal formation processes, 83, 84
　hydrothermal gold and silver, 85*t.*, 93–94
　hypogene (primary), 84
　intrusion-related gold, 91
　Irish-type Pb-Zn (Cu), 96
　iron oxide copper-gold (with or without uranium), 85*t.*, 92–93
　laterites, 100
　lateritic gold, 101
　layered chromite, 87
　lode (or orogenic) gold, 94
　in mafic magmas, 85*t.*, 87–89
　magmatic diamond, 85*t.*, 89
　magmatic hydrothermal formation processes, 83, 84–86
　magmatic platinum group metal (PGM), 87–88
　manganese ore, 98
　metamorphic hydrothermal formation processes, 83, 86
　Mississippi Valley type Pb-Zn, 95–96
　nickel (cobalt) laterite, 100–101
　nickel sulfide, 88
　ore genesis, 83–84
　ores related to weathering, 85*t.*, 100–102
　orthomagmatic formation processes, 83, 84
　other mafic and ultramafic intrusive associations, 89
　outcropping (surface), 105
　peralkaline Ta-Nb rare earth element, 91
　placer deposits, 85*t.*, 102
　plate tectonics, 83, 84*f.*
　podiform chromite, 87
　porphyry Cu-Mo-Au, 89–90
　porphyry Mo (W), 90
　Rapitan BIF, 98
　sandstone-hosted uranium, 99
　secondary copper, 101
　secondary zinc, 101–102
　sediment-hosted, 85*t.*, 95–100
　sediment-hosted gold, 93
　sediment-hosted iron and manganese, 97
　sediment-hosted sulfide, 95
　sedimentary exhalative Pb-Zn (Cu) in clastic sediments, 95
　sedimentary uranium, 98
　size and quality of, and market capitalization, 69
　skarn and carbonate replacement, 85*t.*, 92
　sudbury, 88
　supergene (secondary), 84
　supergene weathering, 101
　Superior BIF, 97–98
　surface hydrothermal or seafloor formation processes, 83, 86
　surface upgrade, 86
　surficial formation processes, 83
　syngenetic, 83
　titanomagnetite, 87
　ultramafic volcanic association, 88–89
　unconformity vein-type uranium, 98–99
　volcanic-hosted or volcanogenic massive sulfide, 85*t.*, 94–95
Mineral potential index, 305–306
Mineral PriceWatch, 61, 63
Mineral processing, 1455
　abundance of various metals in earth's crust, 1456, 1456*t.*
　aqueous dissolution, 1457
　comminution and energy usage, 30, 31*f.*
　concentration, 1457
　crushing and sizing flow sheet, 1457, 1457*f.*
　dewatering, 1457
　economic concerns, 1455–1456
　flotation and larger particles, 30–31
　flotation flow sheet, 1457–1458, 1458*f.*
　flow chart, extraction through processing, 1455, 1456*f.*
　froth flotation, 1457
　future trends, 30–32
　goals of, 1455
　grade, 1455
　gravity concentration, 1457
　gravity concentration flow sheet, 1457, 1458*f.*
　grinding and agitated leaching flow sheet, 1458–1459, 1458*f.*
　heap leaching flow sheet, 1458*f.*, 1459
　leaching, 1457
　magnetic and electrostatic concentration, 1457
　material balance for unit operation, 1455, 1456*f.*
　metallurgical efficiency, 1455
　percent recovery, 1455
　processing sequences for selected metals, 1455, 1456*t.*
　recovery, 1455
　size reduction, 1457
　size separation, 1457
　technology, 15–16
　total and recycled supply of selected metals (U.S., 1996), 1455–1456, 1456*t.*
　unit operations, 1456–1457
　See also Comminution; Dewatering (liquid–solid separation); Electrostatic separation; Froth flotation; Gravity concentration and concentrators; Heavy medium separation; Hydrocyclones; Magnetic separation; Screens
Mineral property feasibility studies, 227, 299–302
　additional testing activities, 249–250
　Classic Stage 1: Conceptual study, 228
　Classic Stage 2: Preliminary or prefeasibility study, 228
　Classic Stage 3: Feasibility study, 228
　classic three-phased approach, 228–229
　combining classic and recommended approaches, 234
　complete activity definitions, 237–256
　considerations, 228
　criteria for, 299–300, 300*t.*
　errors in, 300
　industry approach, 228–229
　minimizing risk in, 300
　need for test mine, 233
　objectives, 228
　problems with conceptual study, 228–229
　problems with final feasibility study, 229
　problems with prefeasibility study, 229
　project design basis report, 234–235, 256–261
　project team, 236–237
　recommended (engineered, systematic) approach, 229–231
　Recommended Phase 1: Preliminary (or conceptual) feasibility stage, 230, 231–232, 237–241
　Recommended Phase 2: Intermediate (prefeasibility) feasibility stage, 230, 232–233, 241–249
　Recommended Phase 3: Final feasibility stage, 230, 233–234, 250–256
　and stage of the project, 227–228
　timing and schedule, 235–236, 236*t.*
　who should perform them, 227
　work breakdown structure and numbering system, 230–231, 230*t.*, 231*t.*
　work breakdown structure of engineered, systematic approach, 231–234
Mineral property valuation, 219
　assumptions, 219
　capital asset pricing model for discount rate derivation, 221–222
　codes, 219–220
　commodity price selection, 221
　for developed or operating properties, 221–224
　discount rate determination, 221–222
　distinguished from evaluation, 219
　and environmental considerations, 222
　for exploration properties, 225
　and fair market value (FMV), 219
　geoscience matrix approach, 225, 225*t.*
　and highest and best use, 219
　income (cash-flow) approach, 220, 220*t.*, 221–223
　and infrastructure-related risk, 222
　and labor/management issues, 222
　market multiples approach, 220, 223–224
　market-related approach, 220, 220*t.*, 223
　market-related transactions and development or operating properties, 223
　market-related transactions and exploration properties, 223
　methods, 220–225, 220*t.*
　Monte Carlo simulations, 220–221, 220*t.*, 224
　and net present value (NPV), 220
　and operating and capital costs, 222
　option/real option pricing valuations, 220, 220*t.*, 224
　and political and social issues, 222
　and prices and markets, 222
　and processing-related risk, 222
　and project status, 222
　property types, 219
　qualifications for valuator, 225–226
　and quality of analytical data, 222
　replacement-cost approach, 220, 220*t.*, 224
　risk buildup discount rate derivation, 222, 223*t.*
　risk factors, 222

rule-of-thumb approach, 225, 226t.
for undeveloped properties, 224–225
weighted average cost of capital discount rate derivation, 221
Mineral resource estimation, 203
   anisotropies, 212
   area-averaging method, 211
   average grade, evaluating differences in, 207, 207t.
   basic statistics, 207
   block size, 212
   coefficient of variation, analyzing, 207, 207t.
   compositing, 206–207
   and cost, 214
   data collection and verification, 204
   and deposit geometry, 215–216
   dilution and ore losses, 214
   exponential variogram model, 210, 210f.
   and folding, 205
   and geologic features modeled, 205
   geologic interpretation, 204–206
   geometric methods, 211–212
   grade distribution, 207
   grade distribution and geologic parameters, 207–208, 208f.
   and grade variability, 216
   and grade zoning, 205–206
   inverse-distance estimation, 213
   kriging, 213–214
   linear variogram model, 210, 210f.
   methodology, 203–204
   modeling, 211–216
   and moderately complex geometry, 215, 216t.
   moving-average methods, 212–214
   and ore boundaries, 214–215
   ore reserve estimate, 203–204
   polygonal methods, 211–212
   reporting, 203
   resource model, 203
   sample data, 203
   sample selection criteria, 212–213
   selection of method, 214–216, 216t.
   and simple geometry, 215, 216t.
   spherical variogram model, 210, 210f.
   variogram modeling, 208–210, 209f.
   variograms, computing, 209–210, 209f.
   and very complex geometry, 216, 216t.
   wireframe models, 204–205, 205f.
Mining, defined, 146
Mining and Metallurgical Society of America, 219–220
Mining Association of Canada, on tailings management, 646, 659, 663
Mining computing organizations (Web sites), 1829
Mining Cost Service, 267, 269, 271, 272, 282
Mining economics (Web sites), 1828–1829
Mining Economics and Strategy, 314
Mining engineering
   and dewatering, 744
   knowledge, 5
Mining geology, 145
Mining in space, 8
Mining industry
   boundaries of, 39
   economics of, 39–47
   history, 3–4
   increased governance of, 18, 19
   management sophistication of, 18–19
   as mature and evolving, 8–9
   trend toward larger companies, 19
   vertical integration of, 16–17
   See also Current trends; Future trends
Mining law, 304, 331, 337
   abandonment procedures, 336
   accelerated depreciation, 336
   access to the land, 304, 332
   agreements, 335
   allowable expenses, 337
   amortization, 336–337
   auditing capacity, 337
   cancellation of title, 334
   carried interest of state, 335
   and common vs. civil law, 331
   cultural and social issues, 305
   definition of mineral, 333
   described, 331–332
   dividend withholding tax, 337
   and employment, 304
   and entry issues, 332
   and environmental laws, 336
   environmental protection, 304–305
   and global business, 305
   and health and safety, 335–336
   import/export duties, 337
   and land issues, 332
   land unsuitability for mining (legal sterilization of lands), 305
   leasing fees, 304
   licenses and leasing, 332–333
   loss carryforward, 337
   mineral potential index, 305–306
   and mining code, 331
   ownership of minerals (state or individual), 332
   participating interest of state, 335
   policy potential index, 305–306
   pollution penalties, 336
   powers of officials, 334
   reclamation, 337
   and reclamation plans, 336
   registration of titles, 333–334
   rentals and fees, 334
   right to apply for mineral rights, 333
   right to mine, 304
   royalties, 304, 337
   safety legislation, 1559–1560
   security of tenure, 333
   and small-scale mining, 334–335
   surrender of title, 334
   suspension of title, 334
   taxation, 304, 336–337
   termination of title, 334
   and work conditions, 304
   and worker compensation for injury or illness, 304
Mining Magazine, 158
Mining method classification system, 349
   and company capabilities, 349, 349t.
   deposit geometry (underground mining), 352
   deposits classified by depth, 350, 350t.
   deposits classified by geometry and type, 350, 350t.
   deposits related to geometry, genesis, and strength, 350, 351t.
   depth related to excavating technique and stripping ratio, 351
   depth related to inclination (surface mining), 351
   fill (underground mining), 353, 353t.
   ground control (underground mining), 352, 352t.
   input statement categories, 349, 349t.
   main roof (underground mining), 352
   and natural conditions, 349, 349t.
   pillars (underground mining), 353
   and public policy, 349, 349t.
   rocks classified by strength, 350, 351t.
   secondary factors, 353–355, 354t.
   spatial description, 349–350
   and state of the art, 349t.
   surface mining, 351–352, 351t.
   surface pit slopes related to rock strength and time, 350, 350t.
   tabular deposits classified by attitude, bulk handling, and rock strength, 350t.
   underground deposits classified by thickness, 350, 350t.
   underground mining, 352–353, 354t.
Mining method selection, 341, 348
   capital and operating cost comparisons, 345–346, 346f.
   decrease in underground mining production (U.S.), 342, 344t.
   development tasks and costs, 344–345
   environmental and closure requirements, 346
   and geology of deposit, 347
   and location of deposit, 346–347
   material produced (surface vs. underground mining, U.S.), 342, 344t.
   and political and social conditions, 347–348
   and processing requirements, 347
   productivity (tons per hour), 343, 345t.
   safety (surface vs. underground mines, U.S.), 343–344, 345t.
   surface mine size and tonnage, 343, 344t.
   surface mining, 341
   surface mining followed by underground mining, 341, 342
   underground mine size and tonnage, 343, 344t.
   underground mining, 341, 341–342
   and work force, 348
   See also Hard-rock mining method selection; Soft-rock mining method selection
Mining, Minerals and Sustainable Development (MMSD) project, 1644, 1772
   Breaking New Ground, 1649–1650, 1666, 1769
   economic, social, environmental, and governance spheres, 1644
   formation of, 1666
   and Global Reporting Initiative (GRI), 1650, 1795
   initiatives that support gaining and maintaining social license, 1794
   tailings management standards, 662
Mining Research Establishment, 606
Mining Source Book, 267, 269, 283, 284
Mississippi Valley type Pb-Zn deposits, 95–96
Mobile bolters (in R&P mining), 1168
Mobile Metal Ion, 108
Mobile miners, 1200
Modeling
   artificial intelligence, 842
   dewatering networks, 756–757, 757t.
   discrete element modeling in bulk flow simulation techniques, 982
   discrete event simulation (DES), 841–842
   financial, 74–75
   for highwall mining, 1028
   improved predictive capabilities, 23
   neural networks, 842–843
   resources, 1016

shotcrete characteristics and properties, 588
soft-rock ground control, 609–610, 609f.
statistical process control (SPC), 841, 841f.
in systems engineering, 841–843
and underground mining, 30
See also Geologic modeling; and under Cost estimating (surface mines); Cost estimating (underground mines); Geostatistics; Mineral resource estimation
Modified rock mass rating (MRMR), 360–361, 361f.
Modified rock quality index (Q), 362
Modified stability number (N'), 362, 363f.
Modular Mining Dispatch software, 1279
Mohr–Coulomb
 failure criteria, 506, 506f., 533
 strength envelope, 655
Monorails (in longwall mining), 1161
Montana, and backfilling, 1375
Monte Carlo simulations
 in mineral property valuation, 220–221, 220t., 224
 in slope stability assessment, 503
Morgan, Hugh, 1769
Morococha mine (Peru), and shrinkage stoping, 1351
Morrison classification system, 369, 372f.
Mount Cenis tunnel (Switzerland), 705
Mount Nansen (Yukon Territory) survey of gold-silver mineralizations, 128, 130f.
Moura coal mine (Australia), 1994 explosion, 303
Mt. Arthur North mine (Australia), truck and shovel operation (case study), 1006–1012, 1007f.
Multipurpose vehicles, 1173

NAA. See Neutron activation analysis
National Coal Board (United Kingdom), 632
National Engineering Handbook, 1708
National Institute for Occupational Safety and Health (NIOSH; U.S.), 157
 MF radio research funding, 723
 noise control guidance, 1637–1638
 Reports of Investigations of the Mining Research Group, 433
National Oceanic and Atmospheric Administration (NOAA; U.S.)
 and dewatering research, 744
 National Weather Service rainfall data sources, 1707
Natural Resources Canada, 63
Naturalis Historiae, 3
Nautilus Minerals, Inc., 1068
Nearest-neighbor (polygonal) weighting, 195
Neptune Minerals, 1068
Net asset value, 66
Net positive suction head (NPSH)
 and dewatering underground operations, 774
 and hydraulic mining, 1050
Net present value (NPV), 65, 302
 deriving, 73
 in hard-rock mining method selection, 374
 in mineral property valuation, 220
 multiple on theoretical, 70–71
 in open-pit planning, 882, 883, 884, 888–889, 899
 probability-weighted, 67–68
Neural networks, 842
 advanced concepts, 843
 data subdivision, 842

design of, 842–843
 number of inputs, 842
 number of neurons, 842
 size of hidden layer, 842
 training algorithm, 842–843
Neutral mine drainage (NMD), 1721, 1722f.
Neutron activation analysis (NAA), 190–191
Nevada, and heap leaching, 1073
New York Mercantile Exchange (NYMEX), 57–58
 Web site, 62
New Zealand
 placer mining and dredging, 1068
 publications and associations (Web sites), 1831
Newcomen, Thomas, and pumping engine, 3
Newmont Mining Corporation
 on community issues, 1775
 Waihi Gold mine (New Zealand), and community issues, 1771, 1772f.
Newton–Rittinger equation, 707, 1378
NGOs. See Nongovernmental organizations
Nicholas classification system, 371, 372t., 373t., 374t.
Nickel
 and backfilling, 1375–1376
 heap leaching and chemistry of deposits, 1077
 heap leaching and geology of deposits, 1076
 lixiviants, 1089–1090
 recovery systems (solution mining), 1100
 Voisey's Bay nickel project, 1798
Nickel (cobalt) laterite deposits, 100–101
Nickel sulfide deposits, 88
Nitroglycerine explosives, in quarrying, 1035
No Dirty Gold campaign, 1658
Nobel, Alfred, and dynamite, 3
Noise hazards and controls, 1633, 1639
 acoustical enclosures, 1637
 action level, 1633
 action level (MSHA), 1633
 administrative controls, 1638
 attenuation in conveyor systems, 1637
 basic noise-control concepts, 1637–1638
 combined noise levels, 1636, 1637t.
 compressed air noise directives, 716
 criterion level,1633
 dB, 1633
 dBA, 1634
 dosimeters, 1634, 1635–1636, 1636f.
 exchange rate, 1634
 exposure limits, 1634
 and flammability of acoustic materials, 1637
 Handbook of Noise Measurement, 1635
 instrumentation and survey techniques, 1634–1637
 lining ducts, 1637
 lower exposure action value (HSE), 1633
 modification of noise source, 1637
 modification of sound wave, 1637
 noise barriers, 1637
 noise dose, 1634
 noise shields, 1637
 noise-induced hearing loss, 1570
 noise-level equivalencies at various sound levels (MSHA and EU reference standards), 1634, 1635t.
 octave band analyzers, 1636, 1636t.
 open-pit mining noise effects, 874
 partial enclosures, 1637
 permissible exposure limit (PEL), 1634
 personal noise dosimeters, 1635–1636, 1636f.

 personal protection, 1638–1639
 regulations, 1633, 1634
 sound level meter, defined, 1634
 sound level meters (SLMs), 1635–1636, 1635f.
 substitution as noise control, 1637
 threshold, 1634
 threshold-limit value (ceiling), 1634
 threshold-limit value (time-weighted average) for physical agents, 1634
 upper exposure action value (HSE), 1633
 vibration minimization,1638
 worker enclosures, 1637
Nongovernmental organizations (NGOs), 18
North Antelope Rochelle mine (Wyoming), 341, 341t.
Northparkes mine (Australia), 342, 343f.
Norwich Park mine (Australia), single-seam dragline operation (case study), 1003–1004, 1004f.
NPSH. See Net positive suction head
NPV. See Net present value
Nuclear power, 7
Nuclear Regulatory Commission, on post-uranium mining groundwater restoration, 1115
Numerical evaluation methodologies, 371–374
NYMEX. See New York Mercantile Exchange

Obert–Duvall formula, 1340
Ocampo (Mexico) heap leach, 1083f.
Occupational Health and Safety Act (New South Wales, 1983, 2000), 1560
Occupational Safety and Health Administration (OSHA; U.S.)
 noise regulations, 1633, 1634
 OSHA Technical Manual, 1636
Ok Tedi mine (Papua New Guinea), 1769, 1773
Open-cut mining, 362–363, 363f.
Open-ending room-and-pillar mining, 390, 390f.
Open-pit mining, 405, 405t., 407–408, 857
 access, 863
 advantages and disadvantages, 413t.
 ancillary equipment and mine services, 871
 articulated dump trucks (ADTs), 869
 benches, 857–858, 858f.
 blast design, 863–865, 864f., 865f.
 blasting, 866
 blasting effects on environment, 874
 bucket fill factor, 866–867
 and cable shovels, 408, 409
 chemical contaminants, 874
 communication infrastructure, 871
 defined, 857
 deposit conditions favorable to, 413t.
 diesel–electric trucks, 869
 drilling, 865–866
 drilling and blasting, 863–866
 drive-by operation, 867, 868f.
 drop cuts, 867
 dust effects, 874–875
 dust management (haul roads), 859
 end-dumping (top-down), 860, 861f.
 environmental considerations, 872–875
 equipment choice and capital-vs.-operating-cost-analysis, 408–409
 equipment dispatching and monitoring, 871–872
 excavation, 866–869
 excavation equipment, 866–867
 excavation practices, 867
 frontal cuts, 867, 868f.

geometry (layout), 857–863, 858f.
geotechnical design, 872
grade control, 867
haul road maintenance, 859–860
haul roads, 858–860, 858f., 869
haulage systems, 869–871
hydraulic excavators, 866
and hydraulic shovels, 408, 409
in-pit crushing and conveying (IPCC), 869, 870–871
in-pit water management (dewatering), 871, 873–874
loading and hauling needs, 410–411, 410t.
loading equipment, 408
mechanical-drive trucks, 869
mine rehabilitation and closure, 875
noise effects, 874
operating shifts, 411–412, 411t., 412t.
overburden disposal, 860–861
overburden removal, 863
paddock dumping (bottom-up), 860–861, 861f.
phases, 409
and physical change of landform, 874
pit expansion, 861–862
power systems, 871
production planning, 872
pushbacks (cutbacks), 861–862, 862f.
reconciliation, 868, 869
rehabilitation, 875
rigid-frame haul trucks, 869
road traffic effects, 874, 875
rolling resistance (haul roads), 859–860
rope shovel buckets, 866
rope shovels, 866
and rubber-tired loaders, 408
selective mining, 867
slope monitoring, 872–873
stop-and-reverse parallel operation, 867, 868f.
topsoil removal, 863
transition to underground mining, 366–367, 862–863
trolley-assisted haulage, 870
truck tire management, 870
trucks, 869–870
typical deposits, 857
unit operations, 863–872
used vs. new equipment, 409
wearing course materials (haul roads), 860
Open-pit planning and design, 877
allowing for blending, 889–891, 889f., 890f., 891f.
allowing for mine closure and rehabilitation, 893, 893t.
allowing for mining width, 889, 889f.
allowing for safety berms and ramps, 886–887, 888f.
analyzing case for flexibility, 891
assessment, 878t., 881
assumption sets, 880–881
best-case plan, 881
blending stockpiles, 896
block models, 894–896, 895f., 896f., 896t.
buffer stockpiles, 897
building a model of highly uncertain data, 900–901, 900f.
capital costs, defined, 897
cost factor, 883
cost information, 880
cost models for optimization, 897–898
current operations information, 878
cutoff and stockpile optimization, 894t.
determining open-pit/underground interface, 892–893, 892t.
determining optimal production rate, 884, 885f.
determining options for preservation, 891–892, 892t.
determining processing cutoffs, 885–886, 886f., 887f.
evidence-based decision making, 882
explicit treatment of uncertainty and risk, 882
final plan, 881
fixed costs, defined, 897
framing, 878t., 881, 881t., 899
general preparation, 877–878, 878t.
grade stockpiles, 896
influence diagrams, 900, 900f.
Lerchs–Grossman pit optimization method, 889, 890
life-of-mine plans, 882, 883t.
maximal value, defined, 893
mine planning optimization, 893, 894t.
minimal value, defined, 893
mining costs, described, 897t.
mining direction, 884
models and tools, 893–901
optimal value, defined, 893
optimization, defined, 893
optimization model, 893
optimized plan, 881
option preservation actions, 881
pit optimization, 894t.
pit parameterization methods, 882–884
plan for final pit, 887–888
planning process, 877–881, 878t.
planning team and committee's roles and responsibilities, 877–878, 878t.
planning techniques, 882–893
preparation of inputs, 878–881, 878t.
price forecasts, 879–880
primary objectives, 882
principles, 881–882
project alternatives, 878–879
purposes and scopes, 878, 879t.
pushbacks, 888–889, 888f., 889f.
regular block models, 894
rehabilitation costs, described, 897t.
resource information, 878
resource models, 894–896
revenue factor, 883
risk identification and management, 891
risk management actions, 881
schedule optimization, 894t.
secondary objectives, 882, 883t.
selection, 878t., 881
selling costs, described, 897t.
separation of public reporting from planning, 882
stakeholder analysis, 877
Steering Committee, 878, 878t., 879t.
stockpile models, 896–897
and taxes, 899
terminal costs, defined, 897
theoretical best-case plan, 884
and topographical features, 886, 887f.
variable cost categories, 897t.
variable costs, defined, 897
vertical bounding, 883
Ordinary Method of Slices, 499
Ore bodies
depth of, 15, 16f.
low-grade, 4
low-grade, and mineral processing, 15–16
Ore deposits, defined, 83
Ore in the ground, 66–67
Ore movement. See Ore transport; Underground ore handling
Ore recovery, 985
blending, 986
bucket-wheel/stacker reclaimer, 986f.
circular blending stockpile, 986f.
dead capacity, 986
environmental considerations, 986–987
live capacity, 986
reclaiming systems, 986
Ore reserves, 145–146
reporting, 464
structure and metallurgy, 14
valuation of, 14–15
Ore resources, 145–146
Ore storage, 985
configurations, 985
covered, 985
sizing, 985
Ore transport, 977
Big Bags (Bulk Bags, Super Sacks), 982
bottom-dump hopper cars, 979
clamshell-type barge unloaders, 980
containerized, 982
continuous barge unloading systems, 980
dust control, 980, 987
environmental considerations, 986–987
flask train-loading system, 979
flexible intermediate bulk containers (FIBCs), 982
flood train-loading system, 979
hydraulic conveying (pipe systems), 984–985
loading silos, 979
marine transportation, 980–982
multimodal transportation, 982
pipe transport, 984–985
pneumatic barge unloaders, 980
pneumatic conveying (pipe systems), 984
rail transportation, 978–980
river barges, 980
rotary car dumpers, 979
shifting cars, 979–980
ship loading, 980–981
ship unloading, 980, 981–982
unit trains, 979
See also Haul roads; Hauling and conveying; In-pit crushing and conveying (IPCC); Loaders
Orepasses, 1228–1230, 1229f., 1247, 1282
advantages and disadvantages, 285
clearing of blockages, 1282
construction, 1283, 1284f.
construction and installation, 1248
construction risks, 1282
cost considerations, 1285
design, 1247–1248
design and construction issues, 1282–1283
excavation and ground support, 1247, 1249f.
fall prevention, 1282
inclination angle, 1282
length, 1282
lining, 1247
maintenance, 1248
muck flow control, 1229f., 1247–1248, 1248f.
and mud rushes, 1282
opening size and shape, 1247
operational aspects, 1283–1285
orientation, 1282

safety considerations, 1282
shape, 1282–1283
support systems, 1283
ventilation, 1283
width, 1283, 1283t.
Ores related to weathering, deposits of, 85t., 100–102
Organisation for Economic Co-operation and Development, International Energy Agency, 62–63
Organizations (Web sites)
international, 1831–1832
mining computing, 1829
U.S., 1827–1828
Orthomagmatic formation processes, 83, 84
OSHA Technical Manual, 1636
Our Common Future, 1665
Outside-lift method (coal), 389–390, 389f., 390f.
Overhand cut-and-fill mining, 394, 395, 394f., 396
Overhand cut-and-fill stoping, 365, 367t., 368f.

Pachuca mine (Mexico), and rill shrinkage, 1351
Palabora copper mine (South Africa), 15
Panel caving, 398f., 399
block size, 402
cave management, 402
extraction principles, 399–400, 400f.
undercutting and formation of extraction trough, 400–402, 401f.
See also Cave mining
Panguna mine (Papua New Guinea), and community conflicts, 1767
Papua New Guinea
and community conflicts over Panguna mine, 1767
placer mining and dredging, 1068
Park, Trenor W., 29
Paste and Thickened Tailings, 1383
Pennsylvania abrasion test, 1474
Peralkaline Ta-Nb rare earth element deposits, 91
Peralkaline/carbonatite magmas, mineral deposits associated with, 85t., 91–92
Percussion bits, 435–436, 436f.
Peru, mining history, 4
Philippine Mining Act of 1995, 1770
Photogypsum tailings, 677
Physical asset management. *See* Asset management
Piezometers, 556–557
in monitoring of tailings impoundments and dams, 659
Pillaring, 387, 387f.
Piper Alpha disaster, 783
PITRAM software, 1279
Place-change continuous mining, 381
Placer mining and dredging, 1057
alluvial fan placers, 1058
backfill sand, 1060–1061
beach placers, 1058, 1060, 1061f.
bucket ladders, 1064–1065, 1065f.
bucket wheels, 1065, 1065f.
bucket-wheel cutterheads, 1066
bulk sampling advantages and disadvantages, 1059
California (spud) bucket ladders, 1064
conventional equipment for dry unconsolidated placers, 1060–1061
debris flow, 1058
deposit type formation, 1058–1059
desert placers, 1058
by dozer and front-end loaders, 1060–1061, 1061f.
draglines, 1064
dredge head variations, 1065–1066, 1067f.
dredge location, 1063
dredge manufacturers, 1066
dredge types, 1064–1066
dredging costs, 1066–1067
dredging operations, 1062–1068
drift mining, 1060
drill sampling advantages and disadvantages, 1059–1060
eluvial (colluvial, hillside) placers, 1058
eolian placers, 1058
estimated cost, 1062, 1062t.
exploration and sampling techniques, 1059–1060
fixed-arm buckets, 1064
frozen placers, 1061–1062
future of industry, 1068
glacial placers, 1058–1059
grab buckets, 1064, 1064f.
grade control, 1063
gulch placers, 1058
horizontal auger heads, 1066
hydraulic dredges, 1062, 1065–1066, 1066f.
hydraulic mining of placers, 1062
hydrothermal alteration and weathering, 1058
manufacturers' information sources, 1066
marine placers, 1059
mechanical dredges, 1062, 1064–1065, 1064f., 1065f.
and natural hazards, 1063
New Zealand bucket ladders, 1064
nondredging methods, 1060–1062
physical release, 1058
placer, defined, 1057
placer classification, 1057
preparations and planning, 1063
processing, 1068–1071
pulse placers, 1058
Ramsey scraper bucket, 1060, 1060f.
release mechanism, 1058
residual placers, 1058
rotary-cutterhead dredges, 1066, 1067f.
of seafloor massive suflides (SMS), 1068
and silting, 1064
source material, 1057
straight suction heads, 1065
stream placers, 1058
subgrade hydraulic mining, 1062
suction hopper dredges, 1065, 1066f.
tailings disposal, 1063–1064, 1070
trailing drag heads, 1065, 1067f.
Placer ores
attribution scrubbers in sizing of, 1068
beneficiation costs, 1071, 1071t.
centrifugal jigs in beneficiation of, 1069
cyclones in sizing of, 1068
dry magnet separators, 1069
electrostatic plates (ESPs) in separation of, 1070
electrostatic separation, 1070
flotation, 1070
flow sheets, 1070–1071, 1071f.
gravity separation, 1069
high-tension rolls (HTRs) in separation of, 1070
hot acid leaching, 1070
hydraulic classifiers in sizing of, 1069
low-intensity magnetic separators (LIMS), 1069
magnetic separation, 1069
sizing, 1068–1069
spirals in beneficiation of, 1069
triboelectric separators, 1069
trommel screens in sizing of, 1068
vibrating screens in sizing of, 1069
wet high-intensity magnetic separators (WHIMS), 1069
wet magnet separators, 1069
wet shaking tables in beneficiation of, 1069
wet vs. dry beneficiation, 1068
Placers
classification, 1057
defined, 1057
deposits, 85t., 102
Planning and scheduling, future trends in, 25–26
Platinum, and placer mining, 1057
Platts Metals Week, 62
Pliny the Elder, 3
Plow systems, 427
Pocket-and-wing room-and-pillar mining, 390, 390f.
Podiform chromite deposits, 87
Poisson distribution, 840
Poisson's ratio, 503, 511, 965
Policy potential index, 305–306
Political factors, 44–45
depletion, 45
depletion and sustainability, 47
economic management, 46
effect of price volatility on developing countries, 46
and exploration, 45
institutional capacity for fiscal policy, 46–47
mineral rents, 45
political risk, 13, 70
taxation, 46
user cost, 45
Polycrystalline diamond compact (PDC) bits, 439
Porphyry Cu-Mo-Au deposits, 89–90
Porphyry Mo (W) deposits, 90
Postpillar cut-and-fill mining, 365–366, 396, 397f., 1366
Potash, 377, 1106
geologic conditions, 1106
in-situ solution mining of abandoned or flooded conventional mines, 1107
in-situ solution mining of impermeable salts, 1107
in-situ solution mining of permeable salts, 1106–1107
processing of, 1107–1108
production of, 1108
Power
application in comminution, 1475–1477
conspicuous consumption of, 7
power and pressure equations (pumping), 268
requirements in backfill mining, 1382
See also Electric power distribution and utilization; Nuclear power
Present value, 302
Price/earnings (P/E) multiples, 75
Price/earnings (P/E) ratio, defined, 80
Pricing, 49, 68t.
and bullion markets, 59
closing prices, 59
commodity, 68, 69f., 77
cost, insurance, and freight (CIF), 50, 60
and cost of entry, 56, 56f.
determining factors, 51–56

and exchange rates, 50–51, 52f.
ex-works (EXW), 50
four "C's," 49
free on board (FOB), 50, 60
of gem diamonds, 49
hedging, 58
and homogeneity (degree of standardization), 56, 57, 57f.
hybrid (terminal market/negotiated), 61
of industrial metals and minerals, 50, 61
and inflation or deflation, 50, 51f.
information sources, 62–63
of intermediate metallurgical products, 49
and location, 49–50
Loco London, 59
Loco Zurich, 59
long-term, 53–54, 54f.
of metal concentrates, 49
and metal exchanges, 57–59
of metals, 49
near-term, 51–53, 53f.
negotiated, 60–61
nominal terms, 50
official prices, 58
by producer, 59–60
of raw materials, 49
reference prices, 61
risk analysis, 49
and seaborne commodities, 60
spot vs. long-term, 68
of steam coal, 60–61
strike, 58
TC/RCs (treatment and refining charges), 61
and traders (intermediaries), 57
and transportability, 56, 57f.
trends, 55–56
and units of measure, 49–50
unofficial prices, 59
variability, 62, 63f.
where and how prices are determined, 56–62
Primary development, 1179
Primary milling, 1464
AG/SAG circuits, 1466–1469, 1466f., 1467f., 1468f.
crusher–HPGR circuits, 1469–1470, 1470f., 1471f., 1471t.
crushing circuits, 1464–1465, 1465f.
See also Comminution; Secondary milling; Tertiary milling
Processing. See Mineral processing
Product disclosure, 33
Product stewardship, 33
Production (long-hole) drilling rigs, 436
Professional societies, U.S. (Web sites), 1827–1828
Profit-and-loss (P&L) statement, defined, 80
Programmable logic controller (PLC) systems, in dewatering, 757
Project geology, 145
Projectile impact in mechanical rock breakage, 417
Prospecting, 105
Prospectors and Developers Association of Canada, e3 Plus Framework for Responsible Exploration, 1681, 1799
Pumps and pumping, 1717
centrifugal pumps, 1717
characteristic curves, 1717, 1717f.
Cornish, 3
cyclone feed type, 693–694
dewatering pump stations in soft-rock equipment underground mining, 1178
dewatering stations in underground operations, 1178
dynamic head calculation, 1717
in hydraulic mining, 1050
ore stream pumping vs. trucking in hydraulic mining, 1055
power and pressure equations, 268
repump emulsion systems, 454, 454f.
in solution mining, 1097
stations in soft-rock underground longwall mining, 1161
stations in subsurface mine development, 1217
sumps and pump rooms in underground development openings and infrastructure, 1223–1226, 1224f., 1225f., 1226f.
troubleshooting, 1717–1718
See also under Dewatering surface operations; Dewatering underground operations; Underground development openings and infrastructure

Quality control and assurance
in exploration geochemistry, 135–136, 135f., 139
in geological data collection, 146
in metallurgical testing, 193–194
in sample preparation, 191
Quarrying, 405, 405t., 407–408, 1031
and archaeological remains, 1033, 1034
backhoe excavators, 1036, 1036f., 1036t.
and belt conveyors, 1038, 1038f., 1038t.
cone crushers, 1039–1040
crushing, 1038–1041
dewatering, 1035
dimension stone, 1045
and distance to market, 1032, 1045–1046
draglines, 1035
drilling and blasting, 1035
dry screening (sand processing), 1043–1044
elutriators (sand processing), 1043
face loading, 1035–1036
face shovels, 1036
factors affecting location, 1032, 1032t.
free settling classifiers (sand processing), 1042
geology and location, 1031–1032
geotechnical design, 1034–1035
gyratory crushers, 1039, 1039f., 1463f.
haul roads, 1036
haul trucks, 1036
hauling to plant, 1036–1037
hindered settling classifiers (sand processing), 1042
horizontal current classifiers (sand processing), 1042
hydraulic backhoe excavators, 1036
hydrocyclones or hydroclones (sand processing), 1043
igneous rock, 1031
impact crushers, 1040–1041, 1040f.
industry, 1046
in-pit processing, 1037–1038, 1037f.
jaw crushers, 1039
lakes, 1045
lignite removal, 1043
log washers, 1042
management and disposal of silt, 1043
manufactured sand, 1044
marine aggregates, 1044
marketing, 1045–1046
metamorphic rock, 1031–1032

multibench granite operation, 1032f.
overburden stripping, 1034
permit to operate, 1032–1033
post-production landfill sites, 1045
quarry design, 1034
quarry face design, 1034
quarry face inspection, 1034
quarry infrastructure, 1033–1035
rehabilitation, 1045
reversion to agriculture, 1045
right-sized trucks, 1036
risk assessment, 1034
rock operations, 1031–1032
roll crushers, 1041
safety, 1034
sand and gravel washing, 1041–1044, 1042f.
sand processing methods, 1042–1043
sand/gravel sites, 1031–1032
screen types and media, 1041
sedimentary rock, 1031
site security, 1033
size of haul truck fleet, 1037
soil stripping, 1034
specialist operations, 1044–1045
stockpiles, 1034–1035
super quarries, 1044
trommels, 1041–1042
typical conditions for permit, 1033
underground quarries, 1044
and vertical integration, 1046
in wet conditions, 1035
water-handling system, 1034
wheel loaders, 1035–1036, 1036t.

R&P. See Room-and-pillar mining
Radiation, 1625
monitoring, 1628–1631
occupational exposure limits, 1626
units of, 1625–1626
Radon, 132, 677–678, 1625
activated carbon canisters (in sampling), 1630
alpha track detector systems (in sampling), 1631
alpha track detectors, 1630
coefficients of diffusion (various materials), 1627, 1627t.
continuous monitors, 1631
continuous progeny monitors, 1631
controlling progenies at underground uranium mines, 1627–1628
Cote method (sampling), 1630
diffusion-based ionization chamber (monitoring), 1631
and dust control, 1628
effect of progeny on human health, 1626
electret detector systems (in sampling), 1631
emanation at underground mines, 1627
grab sampling, 1628–1630
integrated sampling, 1630–1631
Kusnetz method (sampling), 1629
modified Tsivoglou method (sampling), 1629
progeny sampling, 1629–1631
rock method (sampling), 1630, 1630t.
Rolle method (sampling), 1629–1630, 1630t.
sampling with scintillation cells, 1628–1629
scintillation cell monitor, 1631
thermoluminescent dosimeter (TLD) in sampling, 1630–1631, 1631f.
and ventilation systems, 1627, 1628f.
and water control, 1627–1628

RAECOM (Radiation Attenuation Effectiveness and Cover Optimization with Moisture Effects) code, 678
Ragan and Duru equation, 1708
Rail haulage systems (in underground ore handling), 1289, 1291f.
  capacity, 1291
  capital costs, 1290
  design issues, 1291
  equipment, 1292–1293
  health and safety considerations, 1290
  and material properties and handling, 1292
  and mine plan, 1291
  operating cost, 1290
  and ore-body geometry, 1291
  and production growth, 1291
  production rate and scale of operations, 1290–1291
  reliability and operability, 1290
  selection criteria, 1290–1293
  system layout and geometry, 1292, 1292f.
  system life, 1291
  See also Hauling and conveying
Railroads
  driverless trains, 27
  future trends, 25
Raise boring machines, 417, 423, 1283 1284f.
Raise drilling machines, 423–424, 424f.
Raises, 1209, 1209f., 1210f.
  blind-hole raise drilling with "box hole" drill, 1209, 1209f.
  Boliden raise cage, 1209
  boring by upreaming, 1209, 1209f.
  bottom-up drill-and-blast drop raise, 1209
  conventional drilling and blasting, 1209
  defined, 1210
  raise climbers, 1209, 1209f.
Raising, 1192
  Alimak method, 1193, 1283, 1284f.
  cage or gig, 1193, 1193f.
  conventional, 1192
  longhole, 1193, 1194f.
  raise boring, 1193–1195, 1195f., 1196t.
Ram cars (in R&P mining), 1167–1168, 1168f.
Ramps, 1210
Rapid Excavation and Tunneling Conferences, 433
Rapitan BIF deposits, 98
ReCyclone, 1504–1505, 1505f.
Regulations and regulators, 18
Rehabilitation. *See under* Closure planning; Open-pit mining; Open-pit planning and design; Quarrying; Strip mine planning and design
Reliability-Centered Maintenance, 827, 828f.
Remote operation, 15
Rents, 45
  and depletion, 45
  and interest groups, 46
  and taxation, 46
Reserves. *See* Ore reserves
Resource characterization, defined, 173
Resources. *See* Mineral resource estimation; Ore resources
Reynolds number, 1548
Rights holders, defined, 1768–1769
Rimpull, 925
Ring-drilling drills, 436–437, 437f.
Rio Tinto
  acquisitions, 78
  ancient Spanish mine, 3
  automated operations center (Australia), 27

Barneys Canyon gold mine integrated waste management strategy, 1738, 1744f.
  and driverless trains (Australia), 27
  and heap leaching, 1073
  and HIsmelt process, 35
  and labor relations, 303
  management approach (*The Way We Work*), 302
  Marandoo iron ore mine and failed cultural protection (Australia), 1805
  and product stewardship, 33
  Ridgeway gold mine flooded, chemically reactive mineral waste repository, 1738, 1744f.
  and underground mining, 28
Road milling machines, 427–428, 428f.
Roadheaders, 417, 425–426, 426f.
  production estimation, 432
  in R&P mining, 1166–1167
  in underground horizontal and inclined development, 1200–1201, 1201f.
Robbins Mobile Miner, 428, 429f.
Robens Report (U.K.), 1559–1560
Robinsky, Professor, 1376
Robotics, 806. *See also* Automation
Rock breakers, 420
Rock bursts, 613
  anticipated damage, 615t.
  damage mechanisms and levels, 614, 614f., 615t.
  fault slip bursts, 614
  pillar bursts, 614
  strain bursts, 614
Rock mass rating (RMR), 360. *See also* Modified rock mass rating
Rock mechanics, 527, 548–549
  (an)isotropic rock masses, 528
  basic friction angles, 531t.
  behavior model, 529
  body stresses, 544, 546f.
  brittle criterion, 533, 534, 534f.
  CHILE rock mass, 528
  compressive strength testing terminology, 532, 532f.
  compressive stresses, 542
  confined strength, 533–534
  defined, 527
  deformation parameters (uniaxial testing), 532–533, 533f.
  design pathways for concept formulation and analysis, 528, 528f.
  deviatoric stress, 542–543, 543f.
  DIANE rock mass, 528
  discontinuities, 530
  (dis)continuous rock masses, 528
  distribution of stresses with circular excavation, 543, 543f.
  distribution of stresses with rectangular roadways, 543, 543f.
  elastic redistributions, 542–544, 543f.
  elastic stresses around surface excavation, 543–544, 545f.
  excavation design, 547, 547f., 548f.
  failure and collapse modes around excavations, 547–548
  failure and collapse modes around floors, 548
  failure and collapse modes around roofs, 547–548
  failure and collapse modes around sides (pit walls and underground), 548
  geological strength index (GSI), 535–539, 540f.

Hoek–Brown criterion, 533–534, 534f.
Hoek–Brown m values, 531t.
horizontal stress, 542, 543f.
(in)homogeneous rock masses, 528
in-situ stresses, 542
intact rock strength, 532–534
intermediate principal stresses, 542
joint roughness coefficients (JRCs), 530, 530t.
large excavations resulting in large changes in stress regime at adjacent smaller excavations, 543, 545f.
major principal stresses, 542
mean stress, 542
and mine design, 528–529, 546
minor principal stresses, 542
modular ratios, 531t.
Mohr–Coulomb criterion, 533
multiple pathways in design, 529, 529f.
nature of rock, 527–528
nonelastic behavior around excavations, 544
(non)linear elastic rock masses, 528
numerical approaches, 529
orientation, 530
and overall mine design, 546
persistence, 530, 530t.
post-peak (yielding) behavior, 534–535, 534f., 535t.
practical implementation of, 544–548
Q rock mass classification, 535, 537t.–539t.
RMR rock mass classification, 535, 536t.
rock mass classification, 535–541
rock properties, 529–535
and scale, 527–528, 528f.
shear strength, 530–532, 531f., 531t.
spacing, 530, 530t.
stress terminology and equations, 541–542
stresses, 541–544
surface conditions, 530, 530t.
tests, 418
uniaxial compressive strength (UCS), 532, 532t.
variation in stress concentration factors with K ratio and aspect ratio of rectangular roadways, 543, 544f.
vertical stress, 542, 543f.
yielding, 544, 546f.
See also Geomechanics
Rock Mechanics and Rock Engineering, 433
Rock quality designation (RQD), 359–360, 360f.
Rock reinforcement, defined, 611
Rock saws, 429
Rock structure rating (RSR), 360
Rock support, defined, 611
Rock tunneling quality index (Q), 361–362
Rock-bolting machines, in hard-rock underground mining, 1153–1154
Rock–soil transitional materials, 471
Room-and-pillar mining, 377, 381–383, 381f., 385, 599–600, 601f.
  access to R&P mine, 385–386
  advantages and disadvantages, 382t.
  cost models, 273, 273f., 273t.–274t.
  favorable characteristics, 382t.
  layouts, 387f., 388f.
  open-ending sequence, 390, 390f.
  pocket-and-wing sequence, 390, 390f.
  retreat methods using breaker posts or mechanized chocks, 390–391, 390f.
  in soft rock, 387–390

Room-and-pillar mining in coal, 387–390, 1339, 1346
  air horsepower (AHP) formula, 1344
  air quantity formula, 1344
  cable bolts, 1343
  Christmas-tree method, 389, 389f.
  coal production operation, 1339
  combination methods, 389, 389f.
  continuous mining method, 388, 1341–1342, 1341f., 1345
  conventional (drill-and-blast) mining method, 388, 1342, 1342f.
  crib support systems, 1343, 1343f.
  dust displacement, 1345
  entry development, 1339–1340
  and flexible conveyor trains (FCTs), 1345
  functions of pillars, 1339
  geotechnical design, 1340–1341
  ground control, 1342–1344, 1343f.
  haulage, 1167–1168, 1168f., 1345
  main entries, 1340, 1340f.
  methane drainage, 1345
  mining methods, 388
  orthogonal layout, 1339, 1340f.
  outside-lift method, 389–390, 389f., 390f.
  panel entries, 1340, 1340f.
  pillar design, 1340–1341, 1341f.
  pillar extraction, 389, 389f., 1345, 1346f.
  pillar extraction ratio, 1340, 1341f.
  pillar strength, 1340, 1341f.
  pulverized limestone or rock dust to prevent explosions, 1345
  retreat methods, 388–390
  roof bolting, 1342–1343, 1343f.
  roof trusses, 1343
  scrubbers, 1344, 1344f.
  split-and-fender method, 389, 389f., 390, 390f.
  steel plate supports, 1343, 1343f.
  steel sets, 1343
  submain entries, 1340, 1340f.
  system rationalization and optimization, 1345
  as unsupported, 1339
  ventilation, 1340f., 1342f., 1344–1345
  yieldable arches, 1344
Room-and-pillar mining in hard rock, 365, 367t., 1327
  barrier pillar design, 1330
  capital costs, 1336–1337, 1337t.
  effect of changing market conditions on mine planning, 1336
  equipment and operations planning from materials-moving simulation program, 1332, 1332f.
  mine haulage development, 1331–1333
  mine plan view, 1327, 1327f.
  moving ore from stope face to hoisting conveyance, 1332, 1333f.
  multiple-pass stoping, 1333, 1334–1335, 1334f.
  operating costs, 1336, 1337, 1337t.
  ore-body characteristics, 1327
  pillar extraction methods, 1335–1336
  pillar orientation due to in-situ stress, 1331
  pillar strength equations, 1328–1330, 1329t.
  pillar strength rating system, 1330, 1331f.
  pillar stress equations, 1328
  pillar width, 386
  room-and-pillar orientation due to dip, 1331
  room width, 386
  secondary extraction methods, 386–387
  single-pass stoping, 1333–1334, 1334f.
  stoping, 386–387
  traditional strength-based pillar design, 1327–1331
  typical room-and-pillar sizes for metal mines, 1330–1331
  ventilation, 1336
Rope scrapers (slushers), in hard-rock underground mining, 1155
Ruggie, John, 1684–1685
Rutile, and placer mining, 1057

Safety, 18, 1557–1558, 1565
  accidents in surface vs. underground mines (U.S.), 343–344, 345t.
  and appropriate equipment, 1562
  audits for tailings impoundments and dams, 660
  and automation, 817–818
  berms, and haul roads, 963, 964f.
  berms and ramps, and open-pit planning and design, 886–887, 888f.
  BHP Billiton Fatal Risk Control Protocols, 465
  and block caving, 1450
  and community acceptance of risk, 1559, 1559t., 1563
  and competent people, 1562
  and compressed air, 706
  and control huts in hydraulic mining, 1054
  and controlled work environment, 1562–1563
  and culture of denial, 1563
  defenses, 1563
  defining, 1558
  and dewatering surface operations, 744–746
  disasters, 717, 783
  disasters, avoiding, 1563–1565
  and due diligence, 1562
  and effective systems and practices, 1562
  and electric power distribution and utilization, 685
  factor for waste piles and dumps, 674
  factor in slope stability, 495
  fatal injury frequency rate, 1558–1559, 1558f.
  as first priority of management, 317–318
  and geomechanics, 465, 466
  and geotechnical design, 466
  ground support, 624
  and hoisting systems, 1306–1307, 1310, 1311, 1313
  and human error, 1560–1562, 1564t.
  information sharing, 1560
  latent conditions, 1563
  legislation, 1559–1560
  longwall mining as safe, efficient method, 1414
  lost-time injury frequency rate, 1558–1559, 1558f.
  and management reluctance to withdraw workers from danger, 1563
  measuring, 1558–1559
  and mine surveying, 731–732
  miner survey on causes of accidents and incidents, 1561, 1561f.
  miner survey on human error and accidents, 1562, 1562f.
  miner survey on reasons for human error, 1562, 1563f.
  mining equipment safety directive (European Union), 607
  and mining law, 335–336
  monitoring, communication systems in, 724
  and normalizing of evidence, 1563
  as part of sustainable mining practices, 1557, 1557f.
  and quarrying, 1034
  Reason's Swiss cheese model, 1563, 1565f.
  risk management, 303–304, 1557–1558, 1559, 1559f.
  safe system of work, 1562–1563, 1565f.
  shotcrete and safe reentry times, 583–584
  and shrinkage stoping, 1352
  and soft-rock mining method selection, 380
  and strip mine planning, 1025
  TBM record, 1261
  underground mine fatality rates 2004–2006 (U.S. & Australia), 303, 304t.
  and underground ore handling, 1273, 1275, 1277, 1280, 1282, 1285, 1288, 1290
  zero harm, 380, 1557
  See also Cooling; Dust control; Gas control; Health; Heat; Mine Safety and Health Administration; Mine ventilation; Miner tracking systems; National Institute for Occupational Safety and Health; Noise hazards and controls; Occupational Safety and Health Administration; Radiation
Safety fuse, 3
Safford mine (Arizona) permitting and approval process, 1654–1658, 1658t
Sago mine (West Virginia) disaster, 717
Sahuarita IPCC (Arizona), 942
Saline drainage (SD), 1721, 1722f.
Salt, 377, 1105
  geologic conditions and in-situ solution mining of, 1106
  processing and production of, 1106
Sample preparation, 187
  automated systems, 189
  blenders, 189
  coal and coke, 191
  cone crushers, 188
  crushers, 188
  dryers, 188
  equipment, 188–189
  hammer mills, 188
  jaw crushers, 188
  plate pulverizers, 188–189
  procedure selection, 187–188
  pulp splitters, 189
  pulverizers, 188–189
  quality control, 191
  riffle splitters, 188
  roll crushers, 188
  rotating sectorial splitters, 188
  screens, 188
  splitters, 188
  vibratory ring or swing mills, 189
SAMVAL code, 219–220
San Cristobal mine (Bolivia), and social license to operate, 1791–1792, 1792f.
San Manuel mine (Arizona). See Magma Copper Company San Manuel mine (Arizona)
Sandstone-hosted uranium deposits, 99
Sandvik, 1037
Saprolites, and heap leaching, 1075
Saskatchewan uranium mines, and tailings management scheme, 660
Savage Zinc Company mine (Tennessee), development for information, 1216
Savery, Thomas, and fire engine, 3
Saxony, 3
Schenck, Robert C., 297–298

Scott, D.R., 1587
Screens and screening, 1481
- aperture shapes, 1485, 1486*f.*
- aperture size and throughfall size, 1485, 1486*f.*
- application classes, 1481, 1482*f.*
- banana screen, 1486, 1486*f.*
- blinding (bridging), 1485, 1485*f.*
- capacity factor (C) as function of aperture size, 1488, 1488*f.*
- correction factor (K) as function of percent half size, 1488, 1488*f.*
- correction factor (M) as function of percent oversize, 1488, 1488*f.*
- deck, 1485
- deck motion, 1486–1487, 1487*f.*
- defined, 1481
- design and operating variables, 1485, 1485*t.*
- discrete element simulation methods, 1492, 1493*f.*
- efficiency, 1490, 1490*f.*
- fine screening, 1493–1496
- fixed types, 1481
- G-forces, 1486, 1486*t.*
- grizzly screens, 1481, 1483*f.*
- inclining deck, 1485–1486
- instrumentation, 1491–1492, 1491*f.*
- moving (linear) types, 1481
- moving (rotating) types, 1481
- open area, 1485, 1485*f.*
- operational types, 1481, 1482*t.*
- orbit motion, 1487
- particle motion, 1486, 1487*f.*
- partition curves for 254-mm hydrocyclone and Derrick Stack Sizer, 1495–1496, 1495*f.*
- partition curves for Derrick Stack Sizer with different screen apertures, 1496, 1496*f.*
- partition factors as function of position on deck, 1490, 1491*f.*
- pegging (plugging), 1485
- percent of rated capacity vs. E_u, 1489–1490, 1490*f.*
- performance assessment, 1488–1491, 1489*f.*, 1490*t.*
- polyurethane screen media, 1493, 1494*f.*, 1494*t.*
- population balance simulation methods, 1492, 1492*f.*, 1492*t.*
- principles of operation (vibrating screens), 1481–1487
- process matrix (vibrating screens), 1487, 1487*t.*
- Q factors, 1488, 1489*t.*
- real-time condition monitoring, 1491, 1491*f.*
- real-time deck motion modulation, 1491
- recent developments, 1491–1493
- relative mass flows and mean particle sizes (horizontal vibrating screen), 1484, 1484*f.*
- rubber screen media, 1493, 1494*f.*, 1494*t.*
- screen area, 1485
- screen media, 1493, 1493*f.*, 1494*f.*, 1494*t.*
- screen selection, 1487–1488, 1488*f.*, 1489*t.*
- sieve bends, 1481, 1484*f.*
- speed ranges, 1486
- Stack Sizers, 1481, 1484*f.*, 1494–1495
- standard screen sizing and relative pricing, 1493, 1495*t.*
- stroke angle, 1487
- strokes, 1486
- trommel screens, 1481, 1483*f.*
- vibrating screens, 1481, 1483*f.*
- wet or dry operation, 1481–1485
- wire mesh screen media, 1493, 1493*f.*, 1494*t.*

Secondary copper deposits, 101
Secondary development, 1179
Secondary milling, 1470. *See also* Comminution; Primary milling; Tertiary milling
Secondary zinc deposits, 101–102
SEDCAD software, 1710
Sediment-hosted deposits, 85*t.*, 95–100
- gold, 93
- iron and manganese, 97
- sulfide, 95

Sedimentary exhalative Pb-Zn (Cu) in clastic sediment deposits, 95
Sedimentary minerals, 377
Sedimentary uranium deposits, 98
Semiautogenous grinding (SAG), 1466, 1467*f.*
- ball charges, 1466–1467
- conditions favored for, 1477
- flow sheet, 1466, 1468*f.*
- makeup balls, 1478
- *See also* Autogenous grinding; Autogenous grinding/semiautogenous grinding

Separation. *See* Dewatering (liquid–solid separation); Electrostatic separation; Gravity concentration and concentrators; Heavy medium separation; Magnetic separation
Service facilities, developing, 1216–1218
Settling velocity, 707
Seven Questions to Sustainability (7QS), 1666, 1670–1671, 1672*f.*
Shaft boring machines, 421–422
Shaft drilling machines, 422, 423*f.*
Shafts, 1180
- blasthole drilling, 1184–1185
- circular drilling patterns for different rock types, 1184, 1184*f.*
- collars, 1181, 1182*f.*, 1207
- defined, 1180
- depression of groundwater level in sinking, 1186–1188
- depth of, 1207
- determination of stress coefficient in linings, 1181, 1182*f.*
- drill-and-blast sinking, 1183–1184, 1183*f.*
- drilling, 1189–1190, 1193*f.*
- freezing method (sinking), 1189, 1190*f.*, 1191*t.*–1192*t.*
- ground support, 1207
- grouting, 1188–1189, 1188*f.*
- incline, 1209
- insets and sumps, 1181
- internal arrangement of services, 1207
- jumbos, 1185
- lining construction, 1186
- linings, 1181, 1207
- mucking, 1185–1186, 1185*t.*
- parallel system (sinking), 1183
- series system (sinking), 1183
- shaft-sinking equipment, 1207
- shape and diameter, 1180–1181
- simultaneous system (sinking), 1183
- sinking stages, 1186
- sinking technology, 1181–1183, 1183*f.*
- vs. slopes, 1207–1209, 1208*f.*
- specialized sinking methods, 1186–1192, 1187*t.*

Shanghai Metal Exchange, 58
Shareholders, 18
Shear strength. *See under* Soil; Slope stability; Rock mechanics
Shearers, in soft-rock underground mining (longwall mining), 1159, 1161*f.*
Sheorey's equation, 505
Sherpa Cost Estimating Software for Surface Mines, 289
Shock and Vibration Handbook, 1695
Shotcrete, 578–579
- admixtures, 580
- aggregates in, 580, 580*t.*
- basic components, 580–581
- basket mechanism, 585, 586*f.*
- cement in, 580
- cementitious additions, 580–581
- in compression after application that terminates above tunnel floor, 584, 585*f.*
- in compression after applicaton to tunnel floor or all around tunnel, 584, 585*f.*
- damage classification, 591, 591*t.*
- design of liners, 586–588
- deterministic design approaches, 588, 589*f.*, 590*f.*, 590*t.*
- distributed surface loading, 585, 587*f.*
- dry-mix, 579
- dynamic testing of surface support, 584
- EFNARC test, 582
- empirical design approaches, 586–588, 588*f.*, 589*f.*
- extended faceplate, 585, 586*f.*
- failure mechanisms, 585–586, 587*f.*
- failure modes, 588, 590*f.*
- fiber-reinforced, 581–582, 581*f.*, 582*f.*, 582*t.*
- in ground support, 621–622
- guide, 589
- in hard-rock ground control, 578–591
- machines, 1155
- mesh-reinforced, 581
- numerical modeling approaches, 588
- performance, 588–591
- pillars, 592, 592*f.*, 623
- promotion of block interlock, 584–585, 586*f.*
- quality control, 588–591, 591*f.*
- rebound factor, 583, 584*t.*
- reinforced, 581–582
- reinforced arches, 592–593, 592*f.*, 592*t.*
- roughness factor, 583, 583*t.*, 584*t.*
- round determinate panel (RDP) test, 582, 583*f.*
- and safe reentry times, 583–584
- slab enhancement, 585, 586*f.*
- stabilization mechanisms, 584–585
- steel fiber types, 581, 582*f.*
- stress-induced loading, 585–586, 587*f.*
- structural arch, 585, 586*f.*
- support for rock wedges, 588
- surface preparation, 582
- surface protection, 585
- synthetic fiber types, 581, 582*f.*
- thickness, 582–583
- typical mixes, 582, 582*t.*
- water component, 580
- wedge and block loading, 585, 587*f.*
- wet-mix, 579

Shrinkage stoping, 1347, 1353
- advantages and disadvantages, 1352
- with chutes, 1349*f.*
- costs, 1352
- and deposit shape, dip, and size, 1347
- development and preparation, 1347–1348

development for production, 1211, 1211f.
with drawpoint extraction, 1348f.
drilling, 1349–1350
Eagle River mine (Ontario) case study, 1352
efficiencies, 1352
and ground control, 1350
in hard-rock mining, 365, 367t., 368f.
host rock characteristics, 1347
inclined shrinkage, 1351
isometric view with LHD drawpoints, 1350f.
Jackpot headpieces, 1350–1351
long-hole, 1352, 1353f.
ore characteristics for, 1347
and ore grade, 1347
production rates, 1347
raising, 1348
rill shrinkage, 1349f., 1351
safety, 1352
and shafts, winzes, etc. 1352
stope drawdown, 1351
stoping operations, 1348–1351
stulling, 1350–1351
in sublevel open-stoping, 393–394, 394f.
variations and applications, 1351–1352
SI. *See* International System of Units
Sierrita mine (Arizona), tailings management scheme, 660
Silliman, Professor, 297
Silver
epithermal deposits, 93–94
heap leaching and silver-rich gold deposits, 1075
hydrothermal deposits, 85t., 93–94
lixiviants, 1088
Mount Nansen (Yukon Territory) survey of gold-silver mineralizations, 128, 130f.
recovery systems, 1099–1100
Simulation methods of grade estimation, 195
Site environmental considerations, 1643, 1663
and aquatic or marine mining, 1662–1663
in closure, 1643
closure security, 1660
CODELCO compliance with ISO 14001 (case study), 1646–1648
community involvement in mine planning and permitting (Diavik diamond mine, Canada), 1659–1660
in development, 1643
environmental audit checklist, 1650, 1652t.–1653t.
environmental disclosure and reporting, 1649–1650
environmental impact statement, 1651, 1654–1657
environmental planning and permitting, 1651–1654
environmental risk assessment worksheet (CODELCO), 1648, 1649f.
evaluation criteria for consequences of environmental impact occurrence (CODELCO), 1648, 1648t.
evaluation criteria for probability of environmental impact occurrence (CODELCO), 1648, 1648t.
in exploration, 1643
and extraction, 1643, 1660–1663
Freeport-McMoRan Safford mine permitting and approval process, 1654–1658, 1658t.
and Global Mining Initiative, 1644
and Global Reporting Initiative (GRI), 1650, 1795

government agency review of permit applications, 1654
and historic mining practices, 1643–1644
infrastructure and surface facilities design and construction, 1650–1651
and International Council on Mining and Metals (ICMM), 1644
ISO 14001, environmental management systems (EMS), 1646, 1646t.
mine development practices, 1650
and Mining, Minerals and Sustainable Development (MMSD) project, 1644
planning of mine workings, 1660
progressive reclamation, 1660
public participation in review of permits, 1658–1659
in reclamation, 1643
reclamation requirements for surface coal mining per Surface Mining Control and Reclamation Act of 1877 (U.S.), 1654, 1655t.–1656t., 1657t.
significance criteria for environmental aspects (CODELCO), 1648, 1648t.
sound environmental practices, 1650–1663
and surface mining, 1660–1661
sustainable development research, 1660
sustainable practices, 1644
and underground mining, 1661–1662
values for environmental risk magnitude (CODELCO), 1648, 1648t.
water quality management, 1660
wildlife management, 1659–1660
See also Environmental issues; Sustainability
Six Sigma, 800–801, 839, 845, 845f.
Skarn and carbonate replacement deposits, 85t., 92
Skillings Mining Review, 62
Skinner, B.F., 328
Slope design, 511–512
back break, 514, 514f.
catch bench (berm) design, 512–514, 513f.
catch bench width vs. design bench width, 512, 512f.
controlled blasting, 515–516
cost of slope failure, 515
design sectors, 512
dewatering wells, 518
drain holes, 518
drainage galleries or adits, 518–519
failure volume estimation, 515
geotextiles, 516–517
grading, 515
interramp design, 514–515
interramp slope angle, 514, 514f.
line drilling, 516
major structures, 515
mechanical stabilization, 516–517
modified production blasts, 516
operational considerations, 512
overall slope, 515
pit plan with slope design sectors, 511–512, 512f.
presplit blasting, 516, 516f.
protective blankets, 516
shotcrete, 517
slope support and stabilization, 515–519
steel reinforcement members, 517, 518f.
steps in, 511
structural domain, 512
structural stabilization, 517
trim (cushion) blasting, 516
vegetative stabilization, 517–518

wall orientation, 512
water control, 518–519
wire net or mesh, 517, 517f.
Slope management, 520
borehole extensometers, 521
borehole inclinometers, 521
contingency planning, 523–524
data reduction and reporting, 522
detection and monitoring of instability, 520
detection and monitoring tips, 521–522
initial response to slope creation, 522–523
instruments for measuring components of rock deformation, 519, 520t.
monitoring schedule, 522
progressive condition, 523, 523f.
regressive condition, 523, 523f.
regressive/progressive condition, 523, 523f.
slide management, 523
slope-stability radar, 521, 521f.
subsurface displacement detection, 521
surface displacement detection, 519–521
survey monitoring, 520–521
tension crack mapping, 519
time-dependent slope movement characteristics, 522–523
wireline extensometers, 519–520
Slope stability, 495
assessment techniques, 500–504
and bench angles, 496, 497f.
block flow failure, 499–500
cell mapping, 510
contoured pole density plots, 510, 510f.
Coulomb wedge analysis applied to problems of, 486, 486f.
and creep, 496
criteria, 495–496
data collection, 508–510
defined, 495
detail line method, 509f., 510
detailed study, 496–497
discontinuity survey data sheet, 509f., 510
and discontinuities, 497–498, 498f., 499f.
factors contributing to changes in stress and rock strength, 507–508
field work, 496–497
fracture set mapping, 510
general failure modes in rock mass, 497–504
general surface failure, 499
geologic data presentation, 510
Hoek–Brown curvilinear failure envelope, 505, 505f., 506
hydrology data, 511
in-situ stress, 504–505
instability without failure, 495
and intact rock, 497, 498f.
and interramp angle, 496, 497f.
limit equilibrium assessment technique, 500–502, 501f., 503f.
and major rock structures, 497
and minor rock structures, 497
Mohr–Coulomb failure criteria, 506, 506f.
monitoring, 496, 872–873
nonplanar failure surfaces, 499–500
numerical models (continuum and discontinuum), 503
office and laboratory work, 497
in open-pit mining, 358, 359f., 496
oriented core method, 510–511
and overall slope angle, 496, 497f.
planar failure geometry, 498, 500f.
plane shear failure geometries, 498–499, 500f.

and pole plots, 498
primary failure modes, 497
probabilistic assessment method, 502–504, 504f., 505f.
raveling, 499, 500
reliability method, 495
rock components of slope, 497
rock fabric data mapping, 510–511
and rock mass, 497, 498f.
and rock mechanics, 495
rock strength property data, 511
rockfalls, 499, 500
rotational shear, 499
safety factor, 495
shear strength of discontinuities, 505–506
shear strength of rough discontinuities, 506, 507f.
shear strength of smooth discontinuities, 505f., 506, 506f.
slab failure geometry, 499
slope failure, 495–496
slope geometry and terminology, 496, 497f.
step path failure geometry, 498–499
step wedge failure geometry, 499
stress measurements, 511
stresses and strength, 504–508
and strip mine planning, 1025
and surface mining, 406
surveys, 734–736
and tension cracks, 495, 496f.
toppling, 499, 500, 501f.
two-block failure geometry, 499
wedge failure geometry, 499, 500f.
Slopes vs. shafts, 1207–1209, 1208f.
The SME Guide for Reporting Exploration Results, Mineral Resources, and Mineral Reserves (2007), 145–146, 299, 464
Snell's law, 123, 125f.
Social impact assessment and management, 1817, 1824–1825
baseline studies, 1820–1821
community development and social investments, 1823
community engagement and reference groups, 1822–1823, 1824
and community relations practitioners, 1817
complaint and grievance handling mechanisms, 1823
evaluation and review, 1824, 1825
formulation of alternatives, 1819–1820
integration throughout operation, 1817, 1824
key lessons, 1824–1825
management, defined, 1817
management strategies, 1822–1823
monitoring, 1823–1825
multistakeholder working groups, 1823, 1824
networking, 1823
phases of, 1817, 1818–1824
predictive assessment, 1821–1822
reporting, 1823, 1824
and risk assessment and management, 1821–1822
scenario planning, 1822
scoping phase, 1819–1820
social impact assessment (SIA), described, 1818–1819, 1820t., 1824
social impacts, 1817–1818, 1819t.
social management plans, 1822
social profiling, 1820–1821
social risks, 1818
stakeholder analysis, 1821
throughout mine life cycle, 1825

See also Indigenous cultural considerations
Social issues, 4–5, 8, 17, 19, 22
International Finance Corporation on, 1769
and mineral property valuation, 222
and mining law, 305
See also Indigenous cultural considerations; ISO 9000
Social license to operate, 5, 304, 1666
acceptance level, 1784f., 1785
acceptance vs. approval, 1779
approval level, 1784f., 1786
building social capital, 1782f., 1783
business case for investing in, 1781–1784
character of, 1779
communities not qualified to issue, 1792–1793, 1793f.
and community, 1781
community readiness to issue, 1793, 1793f.
components of, 1781
continuous building of social capital, 1789
co-ownership level, 1784f., 1786
credibility boundary criterion, 1784f., 1785–1786
and cultural and social issues, 305
defined, 1779
dependence on resources controlled by stakeholders, 1781–1782, 1782f.
developing cognitive source of social capital, 1790
developing relational source of social capital, 1789–1790
developing structural source of social capital, 1789, 1790f.
different cultures, different expectations, 1787
different starting point in every community, 1787, 1788f.
different strategies at different stages of mine life cycle, 1782–1783
dynamics of stakeholder influence in networks, 1783
dynamism in process, 1791–1792, 1792f.
earning higher levels of, 1783
examples, 1779–1780
and exploration practices, 1794
fiction of community as entity, 1792
formal and informal understandings of, 1780
and free, prior, and informed consent/consultaton (FPIC), 1780, 1781
full-trust boundary criterion, 1784f., 1786
general guidelines for all businesses, 1794–1795
grantors and grantees, 1781
identifying stakeholders based on strategy, 1782f., 1783
IFC Performance Standards on Social and Environmental Sustainability, 1780
and legal sterilization of lands, 305
legitimacy, defined, 1784–1785
legitimacy boundary criterion, 1784–1785, 1784f.
linking social capital, 1794
low-profile fallacy, 1787–1788
maintaining, 1789–1792
mine life cycle and first impressions, 1786, 1787f.
and mine operation through postclosure, 1794
and mineral property valuation, 222
origin of term, 1779
and other forms of consultation and consent, 1780–1781
phases of earning, 1784–1786, 1784f.

and poverty reduction, 1793–1794
probabilities of success, 1782f., 1783, 1784f.
and reciprocity, 1789–1790
relational source with cognitive, 1791
relationship building, 1792–1794
resource access restriction at corporate level, 1783–1784
resource dependence, 1782, 1782f.
and shared identity, 1790
social capital, defined, 1789
social capital sources, 1789
and stakeholders, 1781
strategy, 1782–1783, 1782f.
structural source with cognitive, 1790
structural source with relational, 1790
and sustainability, 32–33, 1794
"to operate," defined, 1780
tools and guidelines, 1794–1795
and trust, 1790
ways to approach identified problems, 1793, 1793f.
withholding/withdrawal level, 1784, 1784f.
See also Indigenous cultural considerations
Societe Grenobloise d'Etudes et d'Application Hydrauliques (France), 1375
Society for Mining, Metallurgy, and Exploration (SME), 145–146
Working Party No. 79, Ore Reserve Definition, 298
Society of Automotive Engineers, (SAE), 922
Soda ash, 1109
geologic conditions, 1109
in-situ solution mining of, 1109–1110
processing of, 1110
Sodium sulfate, 1111
geologic deposits and in-situ solution mining of, 1111
principal minerals, 1111t.
processing of, 1111
Soft rock, 595
classifications and ores, 377
index properties, 472
longwall mining, 597–599, 599f., 600f.
mines, 595–596
rock–soil transitional materials, 471
room-and-pillar mining, 599–600, 601f.
Soft-rock equipment selection (underground mining), 1157
actual advance rate (AAR; R&P mining), 1170
actual productivity, 1165
AFC power demand calculations (longwall mining), 1166
ancillary equipment, 1171–1174
ancillary equipment (R&P mining), 1169
armored face conveyors (AFCs), 1159–1160, 1161f., 1162f.
beam stage loader, crusher, and boot end (longwall mining), 1160–1161
bobcats, 1173
cable electric shuttle cars (R&P mining), 1167, 1168f.
coal clearance equipment, 1170–1171
coal haulage (R&P mining), 1167–1168, 1168f.
commercial prequalification, 1163
continuous miners, 1158f.
continuous miners (R&P mining), 1166
development process options flow chart (R&P mining), 1166, 1167f.
development system arrangement (R&P mining), 1169, 1170t.

dewatering pump stations, 1178
electrical equipment and distribution, 1174–1178, 1176t.
electrical equipment for development panels, 1177–1178, 1178t.
electrical equipment for longwall panels, 1178
electrical load flow study (longwall mining), 1166
electrical power (R&P mining), 1168–1169
electrical systems (longwall mining), 1161
equipment sizing (R&P mining), 1169–1170, 1171f., 1172t.–1173t., 1174t.
expression of interest, 1163
and face length (longwall mining), 1162
face ventilation (R&P mining), 1169
feeder breakers (R&P mining), 1168
hydraulic flow simulations (longwall mining), 1166
load-haul-dump vehicles (LHDs), 1173, 1176f.
longwall automation, 1162, 1163f.
longwall equipment sizing, 1165–1166
longwall equipment specifications, 1163, 1164t.
longwall equipment variations, 1161–1162
and longwall mining, 1157, 1158, 1158f., 1160f.
longwall mining equipment, 1158–1166
longwall ploughs, 1162, 1163f.
longwall top coal caving (LTCC), 1166
material transporters and utility vehicles, 1173
mobile bolters (R&P mining), 1168
mobile equipment, 1173
mobile roof supports (R&P mining), 1169
monorails (longwall mining), 1161
multipurpose vehicles, 1173
nameplate advance rate (NAR; R&P mining), 1169
nameplate capacity (NPC; longwall mining), 1162–1163, 1165
nameplate capacity (NPC; R&P mining), 1169
operational availability, 1165
operational availability (R&P mining), 1170
panel dewatering (R&P mining), 1169
personnel carriers, 1173
place-change (cut-and-flit) continuous miners (R&P mining), 1166, 1167f.
process advance rate (PAR; R&P mining), 1170
process cycle capacity (PCC), 1165, 1166
process reduction factor (PRF; R&P mining), 1170
productivity reduction factor (PRF), 1165
pump stations (longwall mining), 1161
ram cars (R&P mining), 1167, 1168, 1168f.
roadheaders (R&P mining), 1166–1167
roof supports (longwall mining), 1158–1159, 1160f.
and room-and-pillar mining, 1157–1158, 1158f.
room-and-pillar mining equipment, 1166–1170
sequential cut-and-bolt continuous miners (R&P mining), 1166
shearers (longwall mining), 1159, 1161f.
short listing, 1163
simultaneous cut-and-bolt continuous miners (R&P mining), 1166, 1167f.
special-purpose vehicles, 1173–1174, 1176f., 1176t.
specification (coal clearance), 1171, 1175t.
specification aspects and longwall equipment selection, 1162–1163
specifications (R&P mining), 1169, 1170t.
standard longwall equipment, 1158–1161
standard room-and-pillar mining equipment, 1166–1169
and support density (longwall mining), 1163
technical prequalification, 1163
and top coal caving, 1161–1162, 1162f., 1163f.
underground conveyors, 1178
underground lighting, 1178
underground reticulation systems, 1176–1177, 1177t.
and web depth (longwall mining), 1162
and Wongawilli extraction, 1158, 1159f.
Soft-rock ground control, 595
arch joints, 603, 604f.
British colliery arch, 602–603
cambered-top steel arches, 620, 603f.
concrete or cast steel segments, 601
design methodologies, 607–610
experience, 608
face support, 604–607
four-leg chock shields, 605, 605f.
H-section steel arches, 602, 603f.
inclined-leg shields, 605
injection of resins and grouts into fracture planes, 600
interaction, 596–597
lemniscate linkage, 605
and mean load density (MLD), 606
measurement, 608–609
numerical modeling, 609–610, 609f.
packs, 601
physical modeling, 609
published work on, 597
roadway support system strength, 604
rock bolts, 600, 601–602
roof support, 605, 605f.
setting load, 606
shotcrete, 601
standardization, 607
steel standing supports, 600–601, 602–604
stilts for arch legs, 603–604
stress and displacement around an excavation, 596, 596f.
stress contours below remnant pillar, 597, 598f.
struts for arch supports, 603
support requirements, 606–607
support selection, 600–601, 608
T-H yielding arches, 602, 603f., 604
theoretical background, 596–597
two-leg shields, 605–606, 606f.
yield load, 606
See also Ground control; Ground control (cable and rock bolting); Ground support; Hard-rock ground control
Soft-rock mining method selection, 377, 384
and capital cost, 380
definition stage, 381
and deposit depth, 379
and deposit dip, 378, 378t.
and deposit grade, 379, 379t.
and deposit shape, 378, 378t.
and deposit size, 379
and deposit thickness, 379, 379t.
and development timing, 380
and environmental effects, 380
and flexibility, 380
and health and safety, 380
and host rock strength, 378
identification stage, 381
longwall mining, 377, 383–384, 383f., 384t.
and mechanization, 377, 380
mine planning, 381
and operating cost, 379–380
and ore deposit characteristics, 377–379
and ore strength, 378, 378t.
and ore uniformity, 379
and production rate, 380
room-and-pillar mining, 377, 381–383, 381f., 382t.
selection, 381
and selectivity, 380
small-scale methods, 384
stope-and-pillar mining, 377
See also Hard-rock mining method selection
SoftRock Solutions, 736
Soil
Atterberg limits, 472
characteristics, 471–472
coarse grained, 472
coefficient of permeability, 475–476, 475t.
compressibility, 471, 476–477
cone penetration test (CPT), 474–475
consolidation, 472
creep, 481–482, 496
defined, 471
deterioration, 482–483
dilatometer test, 475
drainage conditions and shear strength parameters in the field, 481
effective stress, 471–472
equilibrium relative humidity (ERH), 483
expansibility, 476–477
field exploration, 472–475
field vane test, 475
fine grained, 472
fully softened shear strength, 479, 479f., 480f.
grain size, 472
index properties, 472
initial deformation, 477
liners, 667
modulus of deformation, 477
peak shear strength, 479
permeability, 475–476, 475t., 476f.
pore water pressure, 471, 472
pressuremeter test (PMT), 475
primary consolidation or swelling, 477–478, 477f.
properties, 471–481
residual shear strength, 479f., 479–480, 480f.
rock–soil transitional materials, 471
secondary consolidation or swelling, 478, 478t.
shale slake durability, 482, 483t.
shear strength, 471, 478–481
shear strength of fine-grained soils, 478
shear strength of sand, 478
shrinkage limit, 472
soil plasticity, 472
squeezing ground, 481, 490–491
standard penetration test (SPT), 472–474, 474t.
strength, 471
swelling, 472
unconfined compressive strength, 480
undrained condition, 477
undrained shear strength, 480–481, 481f.
Unified Soil Classification System, 472, 473t.

Soil mechanics, 471
  active and passive earth pressures, 484–486, 484f.
  bearing capacity of mine pillars on underclays, 488, 488f.
  circular failure surface for stability of cohesive material, 486–487, 486f.
  controlling ground movements in tunneling, 491–492
  Coulomb wedge analysis applied to slope-stability problems, 486, 486f.
  Coulomb wedge analysis of limiting conditions, 485–486, 485f.
  curved failure surfaces for passive earth pressure, 487, 487f.
  defined, 471
  estimating ground movements around shallow tunnels, 491, 491f.
  general bearing-capacity relationship, 487–488, 487f.
  limit equilibrium, 484–488
  limiting stress conditions, 483
  Rankine analysis of active and passive earth pressures, 484–485, 484f., 485f.
  settlement of foundations in sands, 488, 489
  settlement of foundations in soft clay, 488–489
  settlement of foundations in weak rock and stiff soils, 488, 489
  stress-deformation conditions, 483–484
  and tunneling, 489–492
  tunneling and squeezing ground, 490–491, 491t.
  and tunneling conditions, 489–490
  Tunnelman's Ground Classification System, 490, 490t.
  See also Geomechanics
Solution mining (in-situ techniques), 1103
  in abandoned or flooded conventional mines, 1105
  and alkali evaporite deposits, 1103
  of borates, 1110–1111, 1110t.
  deposit characteristics, 1103–1104
  drilling for, 440
  Frasch sulfur mining, 1115–1118, 1117f., 1118f.
  in hydraulic fracturing, 1105
  in impermeable salt deposits, 1104–1105, 1104f.
  of lithium, 1111–1112, 1112t.
  of magnesium, 1108–1109, 1109t.
  and marine evaporite deposits, 1103
  and metal deposits, 1104
  metals exploited, 1112–1115
  methods, 1104–1105
  and natural brine deposits, 1103
  in permeable metal deposits, 1105
  in permeable salt deposits, 1105
  of potash, 1106–1108
  of salt, 1105–1106
  salts exploited, 1105–1112
  of soda ash, 1109–1110, 1109t.
  of sodium sulfate, 1111, 1111t.
  of uranium, 1112–1115, 1113t., 1114f., 1115f., 1116f., 1116t.
  See also Heap leaching; Solution mining (surface techniques)
Solution mining (surface techniques), 1087
  acid sulfate, 1088
  acidic chloride, 1088
  agglomeration, 1093–1094
  alkaline ammonia-based lixiviant, 1088
  alternate and variant leach systems, 1093
  ancillary and infrastructure requirements, 1100
  application rate, 1095–1096, 1096t., 1097
  bioleaching, 1090
  and blockage, 1098
  commodity recovery systems, 1099–1100
  conveyor systems in ore emplacement, 1095
  copper cutoff grade, 1092
  copper lixiviants, 1088–1089
  copper recovery, 1099
  and coverage, 1098
  cyanide leaching, 1088
  and drilling and blasting, 1092
  and droplet impact, 1098
  environmental aspects, 1098–1099
  and evaporation rate, 1098
  expanding pads, 1099
  fluid flow phenomena, 1095–1097
  geotechnical design, 1098–1099
  gold lixiviants, 1088
  gold recovery systems, 1099–1100
  irrigation rate, 1097–1098
  labor requirements, 1100
  leach systems, 1092–1093
  leach/rest cycles, 1096–1097
  manganese lixiviants, 1090
  materials handling, 1092–1095
  metallurgical testing, 1090–1092, 1092f.
  nickel lixiviants, 1089–1090
  nickel recovery systems, 1100
  on/off single-lift leach systems, 1093
  and open-pit mining, 1092
  optimum application rate, 1096
  ore emplacement, 1094–1095
  ore preparation, 1093–1095
  ore-lixiviant systems, 1087–1090
  permanent multilift leach systems, 1092–1093
  pipeline requirements, 1097
  plug dumping in ore emplacement, 1095
  polymetallic system leaching, 1090
  power requirements, 1100
  pressure emitters, 1098
  production aspects, 1097–1098
  pulsed leaching, 1097
  pumping systems, 1097
  reusable pads, 1099
  sand-slime separation, 1093, 1094
  silver lixiviants, 1088
  silver recovery systems, 1099–1100
  site requirements, 1100
  solution application systems, 1097–1098
  solution conditioning, 1100
  solution containment, 1099
  solution management, 1095–1099
  solution retention and capillary rise, 1095, 1096f.
  sphalerite lixiviant, 1090
  sprinklers, 1098
  stability, 1099
  sulfuric acid, 1088
  surface acceptance rate, 1097
  truck dumping and dozing in ore emplacement, 1094–1095
  typical leach curve for a heap leach, 1092f.
  uranium lixiviants, 1089
  uranium recovery systems, 1100
  valley dumping, 1099
  water requirements, 1100
  and wind, 1098
  and winter weather, 1098
  zinc heap leaching, 1090
  See also Heap leaching; Solution mining (in-situ techniques)
South Africa
  and black economic empowerment, 1770
  Bureau of Standards on hoist ropes, 1311
  flask train-loading system, 979
  in-pit crushing and conveying (IPCC), 941
  mining history, 4
  R&P use in coal mining, 1339
  SAMREC Code, 203
  SAMVAL code, 219–220
Soviet Union, and development of exploration geochemistry, 127
Special-purpose vehicles, 1173–1174, 1176f., 1176t.
Specific energy (SE), 419
Specific penetration (SP), 419
Sphalerite lixiviant, 1090
Split-and-fender method (coal), 389, 389f., 390, 390f.
Square-set stoping, 367t.
Stakeholders
  analysis of, in open-pit planning and design, 877
  defined, 1768–1769
  involvement in closure planning, 1753–1754
  Strategic Management: A Stakeholder Approach, 1768
  See also Community issues; ISO 9000; Social impact assessment and management; Social license to operate
Standards of Disclosure for Mineral Projects, 146
State-owned enterprises (SOEs), 14
Statistical process control (SPC), 841, 841f.
Stava dam (Italy), tailings dam failure, 661–662
Steady-flow energy equation, 1578
Steel, intensity per capita, 11, 12f.
Stewardship. *See* Commodity stewardship; Product stewardship
Stewart, William Morris, 297
Stillwater Mining Company
  and long-hole shrinkage, 1352
  TBM use at Montana platinum mine, 1267
  TBMs in driving of long lateral drifts, 1219
  3200 Haulage System Trade-Off Study, 1218–1219
Stirred mills, 1472
Stokes' law, 1378, 1498, 1548
Stoper drills, 436
Stopes and stoping, 365, 367t.
  blasthole stoping (end slicing), 391, 391f.
  block caving (panel caving), 1213
  boring machines, 424
  cut-and-fill stoping in flat, thick, wide ore bodies, 1212
  cut-and-fill stoping in steeply dipping, narrow ore bodies, 1212, 1212f.
  in hard-rock mining, 365, 367t., 386–387
  hard-rock room-and-pillar development, 1210–1211
  in hard rock room-and-pillar mining, 386–387
  long-hole stope blasts, 452–453
  overhand cut-and-fill stoping, 365, 367t., 368f.
  sizing equipment to match stope development, 1143
  square-set stoping, 367t.
  stope design parameters, 264–265

stope development for gravity block caving systems, 1213, 1214f.
stope development for LHD block caving systems, 1213, 1213f.
stope development for slusher block caving, 1213, 1214f.
stope-and-pillar mining in soft rock, 377
stoper drills, 436
sublevel caving, 1212–1213
sublevel open stopes, 367t.
sublevel open-stope development, 1211–1212
sublevel open-stoping method, 391–394, 391f., 392f., 393f., 394f.
sublevel retreat open stopes, 367t.
in subsurface mine development, 1210–1213
vertical crater retreat stoping, 392–393, 393f.
See also Shrinkage stoping; Sublevel stoping; Sublevel open stoping method
Strata control, defined, 611
Strategic Management: A Stakeholder Approach, 1768
Strawberry cutters, 419, 420f.
Strike price, 58
Striking a Better Balance, 1666
Strip mine planning and design, 1013
    ancillary resources (oil, gas, etc.), 1017
    area mining, 1019
    base-case plan, 1014
    blasting with parallel tie-up, 1023, 1023f.
    box cut, 1021, 1021f.
    brownfield expansion, 1013
    contour mining, 1019
    conventional blasting profile, 1023, 1023f.
    depth-width ratio, 1023
    dragline sequence, 1021–1022, 1021f.
    drill-hole coverage, 1015–1016
    environmental baseline, 1017
    equipment selection, 1025
    existing infrastructure, 1019
    feasibility studies, 1013, 1014t.
    geological model, 1016
    greenfield development, 1013
    ground control, 1025
    haul access, 1020
    highwall ramps, 1020
    impact analysis and monitoring, 1017–1018
    infrastructure mapping, 1019
    land and mineral ownership, 1017
    land rehabilitation, 1025–1026
    long-term (life-of-mine) planning, 1013–1014
    mine infrastructure, 1019
    mining methods, 1019–1020
    mining sequence, 1020–1022
    mitigation plans, 1017
    orientation of pit to strike and dip of seam, 1020
    outlining reserves, 1014–1017
    permit preparation time span, 1018–1019
    permitting, 1017–1019, 1018t.
    pit design, 1019–1025
    pit geometry, 1020
    pit slope stability, 1025
    pit-end ramps, 1020
    planning process, 1013, 1015t.
    presplit blasting profile, 1023, 1023f.
    project mapping, 1014–1015
    rehandle, 1020
    reserve evaluation, 1016
    resource control, 1017
    risk and reward, 1014
    safety, 1025
    short-term planning, 1013
    single-seam, cross-pit, chop-down operation, 1024, 1024f.
    specialized blasting techniques, 1022–1024, 1023f.
    strip ratio, 1016–1017, 1027
    stripping previously worked deposits, 1025
    topographical map, 1016
    two-seam method, 1024
Strip mining, 363–364, 363f., 405, 405t., 407–408, 989
    advantages and disadvantages, 413t.
    box cut, 1021, 1021f.
    bucket-wheel excavators, 1019
    and cable shovels, 408, 409
    cast blasting, 1019, 1022–1024, 1022f., 1023f.
    cast, doze, excavate operation, 1000, 1008t.–1010t.
    cast–doze excavating, 991
    chop cut with in-pit bench method, 1001–1002, 1002f.
    clearing and topsoil removal, 990, 990f.
    coal mining, 996
    continuous system stripping, 994
    contract stripping, 409
    deposit conditions favorable to, 413t.
    dilution, 996
    double in-pit bench method, 1002, 1002f.
    dozer ripping, 990
    dozer-assisted dragline stripping, 992
    dragline sequence, 1021–1022, 1021f.
    dragline stripping, 990f., 991–992, 991f., 992f., 1008t.–1010t.
    draglines, 1019
    drilling and blasting, 990–991
    electric shovels, 1019
    equipment choice and capital-vs.-operating-cost analysis, 408–409
    equipment, 408
    extended key with in-pit bench method, 1001, 1002f., 1003
    fragmentation, 990–991, 990f.
    future trends, 1011t., 1012
    and highwall mining, 1020
    and hydraulic shovels, 408, 409
    interburden operation, 993
    key bridge method, 1001, 1002f., 1103
    landform strategy, 996
    loading and hauling needs, 410–411, 410t.
    loading equipment, 408
    loss, 996
    mine closure, 997–998
    multiple-seam dragline operation with prestrip, 1004–1006, 1006f.
    offset configuration, 1004–1005
    operating shifts, 411–412, 411t., 412t.
    parting operation, 993
    phases, 409
    prestrip operation, 993
    processes, 989–999, 990f.
    pullback technique, 1002, 1002f.
    rejects and tailings storage or disposal, 999
    relative economics by process and depth changes, 998, 998f.
    river and creek diversions, 999
    and rubber-tired loaders, 408
    side cast method, 1001, 1002f.
    single-seam, cross-pit, chop-down operation, 1024, 1024f.
    single-seam dragline operation with prestrip, 1000–1004, 1001f., 1002f., 1004f.
    spreader dumps, 996, 996f.
    stacked configuration, 1004
    strategies, 999–1000
    strategies compared by selection driver, 1008t.–1010t., 1012
    truck and shovel operation, 990f., 993, 993f., 1003, 1006–1007, 1007f., 1008t.–1010t.
    two-seam method, 1024
    used vs. new equipment, 409
    waste management, 999
    waste placement, 994–996, 995f., 996f.
    waste removal, 991–994
    See also Highwall mining
Structural engineers, and dewatering, 744
Sublevel caving, 365, 366f., 396–397, 1417–1418
    and abutment stresses, 1430–1431, 1432f.
    air gap, 1433
    and aquifers, streams, and dams, 1431
    back break, 1422–1423, 1424f., 1425f.
    cavability, 1433
    cave monitoring, 1433
    cave-front layouts, 1431, 1433f.
    chaotic and variable flow, 1423–1425, 1425f., 1426f., 1427f.
    communications systems, 1430
    and compressed air, 1430
    dead zones, 1429
    depth of draw, 1422, 1423f.
    draw strategy, 1431–1432
    extraction, 1432
    as factory method, 1417
    and fines migration, 1432
    geomechanical factors, 1430–1431
    geometry at Kiruna mine, 1419, 1420f.
    Grängesberg marker trials, 1419, 1420f.
    and gravity flow, 1418–1425, 1419f.
    hang-ups, 1429
    independent draw, 1431
    interactive draw, 1431
    Kiruna marker trials, 1419–1420, 1420f.
    layout, 397–399, 398f.
    layout dimensions, 1426–1428
    layout orientation, 1425–1426, 1427f.
    lead–lag rules, 1431–1432
    and LHDs, 1427–1428, 1428f.
    method, 1417–1418
    mine design, 1425–1431
    mine development, 1212–1213
    and minor structures, 1431
    and natural caving, 1430
    ore recovery over multiple levels, 1422, 1424f.
    ore-body characteristics, 1418
    overdraw, 1432
    Perseverance mine marker trials, 1421, 1422f.
    precharging, 1429
    preloading, 1429
    primary recovery, 1421
    production draw control, 1431–1432
    production drift shape, 1427–1428, 1428f.
    production drift spacing, 1427, 1428f.
    production drill-and-blast, 1428–1429, 1428f.
    production management, 1432
    recovery, 1432

recovery and dilution, 399
Ridgeway marker trials, 1420–1421, 1421f.
rings, 1417–1418, 1418f.
schematic, 1417f.
secondary recovery, 1421
services, 1430
smart markers, 1421
stages of cave propagation, 1433, 1434f.
stoping, 1212–1213
and stress field orientation, 1430
sublevel height, 1426
subsidence, 1433–1435, 1435f.
tertiary recovery, 1421
void monitoring, 1433
and water, 1430
width of draw, 1421–1422, 1421f.
See also Cave mining
Sublevel open stopes, 367t.
Sublevel open-stoping method, 391
   blasthole stoping (end slicing), 391, 391f.
   extraction principles, 391
   shrinkage stoping, 393–394, 394f.
   sublevel stoping, 391–392, 392f.
   vein mining, 393, 393f.
   vertical crater retreat stoping, 392–393, 393f.
Sublevel retreat open stopes, 367t.
Sublevel stoping, 391–392, 392f., 1355, 1356f., 1363
   Alimak raise mining, 1362, 1362f.
   composite plan section for design blast layout, 1358, 1358f.
   definitions, 1355, 1355f.
   design considerations, 1358–1362, 1357f.
   development considerations, 1359
   dilution, 1355, 1356f.
   dip, 1356
   drilling, 1355
   external (unplanned) dilution, defined, 1355
   geometric constraints, 1357f., 1358
   geotechnical variables, 1356
   height, defined, 1355, 1355f.
   intermediate levels, 1355–1356, 1356f.
   internal dilution, defined, 1355
   layout, 1356, 1357f.
   long-hole stoping, 1360–1361, 1360f., 1361f.
   longitudinal mining, 1363, 1363f.
   longitudinal pillar, defined, 1355, 1355f.
   longitudinal section of blasthole layout, 1358, 1358f.
   ore handling considerations, 1359
   piggyback stope, 1361, 1361f.
   pillar size, 1358
   rib pillar, defined, 1355, 1355f.
   ring section, 1358f., 1359
   selectivity, 1358
   shape, 1356
   sill pillar, defined, 1355
   size, 1356, 1357f.
   slot raise, 1358f., 1359
   span, defined, 1355, 1355f.
   stope spans, 1356–1358
   sublevel stoping, 1359–1360, 1359f.
   transverse open stoping, 1357f., 1362–1363
   variables, 1356–1358
   variations, 1356, 1362–1363
   vein mining, 1362, 1362f.
   vertical crater retreat, 1356, 1357f., 1361–1362, 1361f.
   width, defined, 1355, 1355f.
Subsidence. See Mine subsidence
Subsidence Engineers' Handbook, 635

Subsurface mine development, 1203
   blind-hole raise drilling with "box hole" drill, 1209, 1209f.
   for block caving (panel caving), 1213
   Boliden raise cage, 1209
   bottom-up drill-and-blast drop raise, 1209
   conventional raise drilling and blasting, 1209
   crosscuts, 1209–1210
   cut-and-fill stoping in flat, thick, wide ore bodies, 1212
   cut-and-fill stoping in steeply dipping, narrow ore bodies, 1212, 1212f.
   decision making, 1218–1219
   decline (slope) entry, 1205
   depth of shafts, 1207
   developing service facilities, 1216–1218
   developments driven for obtaining information, 1215–1216
   drift entry (adit), 1204–1205
   drifts, 1209–1210
   drill-and-blast cost estimates, 1220, 1220t.
   entries, 1204–1209, 1205t.
   excavation planning, 1213–1218
   exploration developments, 1215
   explosive storage area, 1218
   flat and vertical development spacing, 1203–1204
   fuel storage area, 1218
   ground support (shafts), 1207
   hard-rock room-and-pillar development, 1210–1211
   incline shafts, 1209
   influence of geology, 1204
   influence of stress condition, 1204
   internal arrangement of services (shafts), 1207
   lunch area, 1218
   mechanical excavation methods, 1219–1220
   multiple-stage hoisting, 1209
   office space, 1218
   orientation of adjacent excavation developments, 1204
   production accessways, 1204–1213
   pump station, 1217
   raise boring by upreaming, 1209, 1209f.
   raise climber, 1209, 1209f.
   raise developments, 1209, 1209f., 1210f.
   raises, defined, 1210
   ramps, 1210
   rapid-development incentives, 1213–1215
   rapid-development necessities, 1214–1215
   rapid-excavation payoff, 1215
   shaft collars, 1207
   shaft linings, 1207
   shaft-sinking equipment, 1207
   shop and storehouse developments, 1216–1217
   shrinkage stoping development for production, 1211, 1211f.
   size and configuration of shafts, 1207
   slopes vs. shafts, 1207–1209, 1208f.
   spacing and alignment of excavations, 1203–1204
   stope development, 1210–1213
   stope development for gravity block caving systems, 1213, 1214f.
   stope development for LHD block caving systems, 1213, 1213f.
   stope development for slusher block caving, 1213, 1214f.
   storage pocket (skip pocket) size, 1217–1218

   sublevel caving, 1212–1213
   sublevel open-stope development, 1211–1212
   successive turnouts (breakaways), 1204
   sump area, 1217
   TBM design recommendations for, 1219
   TBMs in hard-rock applications, 1219
   vertical shaft entry, 1205–1207, 1206f.
   winzes, 1210
Sudbury deposits, 88
Sulfur. See Frasch sulfur mining
Sulfuric acid, 1088
Sullivan mine (British Columbia), and pneumatic stowing, 1376
Sunshine mine (Idaho), and air conditioning, 1611
Superconductive quantum interference device (SQUID) sensors, 109t., 110, 117
Supercycles, 11–13, 19, 65
Supergene (secondary) processes, 84
Supergene weathering deposits, 101
Superior BIF deposits, 97–98
Supply, 41
   and barriers to entry, 44
   and capital expenditure requirements, 43
   and closures and cutbacks, 44
   current trends, 11–13, 12f.
   and depletion, 41, 42
   and economies of scale, 43
   effect of ore depletion and technological change, 42–43
   effect on prices, 40
   and exploration, 42
   future trends, 21–23
   and geographical location, 41
   and influences on costs, 42
   and ore grade, 42
   and pricing, 44
   ratio of ore to waste, 41
   and transport, 41–42
Surface mining, 405, 412–413
   advantages and disadvantages of methods, 413t.
   aqueous, 405t., 407
   auger mining, 405t., 407
   and benches, 407
   and blasting restrictions, 406
   deposit conditions favorable to, 413t.
   depth related to inclination, 351
   dredging, 405t., 407
   environmental and social considerations, 406
   and equipment selection, 407
   evaporite processing, 405t.
   future trends, 24–26
   and groundwater conditions, 406
   in hard-rock mining, 362–364
   hydraulic mining, 405t., 407
   in-situ leaching, 405t., 407
   and labor costs, 406
   mechanical, 405t.
   method selection, 341, 351–352, 351t.
   methods, 405t.
   open-pit mining, 405, 405t., 407–408
   pit limits, 405–406
   placer mining, 405t.
   quarrying, 405, 405t., 407–408
   and remote operations, 406
   and risk of investment, 406
   seam deposit mining, 405
   and seasonality, 406
   and size of operation, 407
   and slope stability, 406

solution mining, 405t., 407
strip mining, 405, 405t., 407–408
subdivisions, 407
surface to underground transition, 366–367, 862–863
and topography, 406
and waste disposal, 406–407
See also Cost estimating (surface mines); Hard-rock mining method selection; Mining method classification system; Mining method selection
Surface Mining, 1013
Surface Mining Control and Reclamation Act of 1877 (U.S.), 1654, 1655t.–1656t.
Surface or seafloor hydrothermal formation processes, 83, 86
Surficial formation processes, 83
Surveying. *See* Mine surveying
Sustainability Reporting Guidelines and Mining and Metals Sector Supplement, 1772
Sustainability, 32, 1644, 1665–1666
  and acid rock drainage, 33
  and artisanal and small-scale mining, 1685
  assessing effectiveness of institutional arrangements and governance, 1681, 1682f., 1682t.
  and attribution problem, 1685
  boundary conditions for application to mining, 1669–1670
  and capacity building, 1685
  and change management, 1686
  and climate change, 1685
  and closure, 1683–1684
  and communities of interest, 1669, 1670
  and community issues, 1769–1770
  and contributing to ecosystem well-being, 1674–1676, 1676f., 1677f., 1677t.–1678t.
  and contributing to human well-being, 1673–1674, 1674f., 1675t.
  and corporate social responsibility, 1684–1685, 1684f.
  and decision making and trade-offs, 1685
  and depletion, 47
  and design and construction, 1681–1682
  design-for-postclosure, 1683–1684
  and economic viability of project, community, and nation, 1676–1678, 1679f., 1679t.
  and effective engagement, 1671–1673, 1672f., 1673t.
  and effective human resource management, 1685
  and energy, 33–35
  Equator Principles, 1668, 1770
  ethical lens for mining, 1685, 1685f.
  and ethics, 1684
  e3 Plus Framework for Responsible Exploration, 1681, 1799
  and exploration phase, 1681
  Extractive Industries Transparency Initiative (EITI), 1684, 1770, 1824
  and fair distribution, 1685
  Faro mine project analysis of closure options, 1668
  Firey's groupings of knowledge in natural resource use, 1667–1668
  future, 1685–1686
  Global Compact, 1668
  and GRI MMSS, 1680
  and human rights, 1684–1685
  ICMM principles, 1644, 1645t.
  and ICMM reporting and assurance system, 1667f., 1680
  ICMM 7 position statements, 1667, 1667t.
  ICMM 10 principles, 1667, 1667t.
  important international agreements, 1680
  and institutional arrangements and governance, 1678–1681
  and internal management systems, 1679
  International Finance Corporation framework, 1668
  MAC TSM initiative, 1680
  Millennium Development Goals, 1668–1669
  and mine project life cycle, 1669, 1670, 1669f.
  and mineral life cycle, 1669, 1670, 1670f., 1671f.
  mining industry contribution to sustainable development, 1666–1667
  mining industry response to sustainable development, 1666
  and nation-based mining law, 1680
  and need and alternatives, 1685
  new perspectives, 1685–1686
  and nonrenewable nature of mining, minerals, and metals, 1669
  and operations, 1682
  and *Our Common Future*, 1665
  and performance standards, 1685
  perspectives by project life-cycle phases, 1681–1684
  and product stewardship, 33
  and reporting and assurance, 1679–1680
  and resource curse, 1673–1674
  revenue transparency, 1684
  and safety, 1557, 1557f.
  Seven Questions to Sustainability (7QS), 1666, 1670–1671, 1672f.
  and social license to operate, 32–33
  Sustainability Reporting Guidelines (Global Reporting Initiative), 1650, 1795
  and sustainable development, 1665
  synthesis and continuous learning, 1681, 1683f., 1683t.
  and time horizon, 1669, 1670, 1670t.
  and traditional and nonmarket activities, 1678, 1680f., 1680t., 1681t.
  and uncertainty, precaution, and adaptive management, 1685
  and values, 1684
  and voluntary codes of conduct, 1685
  and waste disposal, 33
  and whole system perspective, 1685
  World Bank Group capitals model, 1668, 1668f.
  See also Community issues; Environmental issues; Indigenous cultural considerations; Site environmental considerations; Social impact assessment and management; Social license to operate
SveBeFo (Sweden), 433
Swedish method, 706
SX-EW (solvent extraction electrowinning), 16
Syngenetic mineral deposits, 83
Systems engineering, 839
  algorithm for determining optimum cutoff grades for single constraint problem, 850
  analysis techniques, 843–845
  analyzing non-normal data, 843–844, 844t.
  artificial intelligence, 842
  Box–Cox transformation, 843–844, 844t.
  cutoff grade and scheduling optimizaton, 846–852
  cutoff grade optimization by Lane's approach, 849–850, 849t.
  data collection, 839–841
  determining sample size, 839
  discrete event simulation (DES), 841–842
  heuristic cutoff grade optimization algorithms, 846
  heuristic cutoff grade optimization techniques, 847–849, 847t., 848f., 848t., 849t.
  minimum number of samples to estimate within certain error range, 839
  modeling techniques, 841–843
  neural networks, 842–843
  optimization, 845–852
  Poisson distribution, 840
  sampling proportions, 840
  scheduling and cutoff grade optimization using mixed-integer linear programming (MILP), 850–852, 851f.
  Six Sigma, 839, 845, 845f.
  smaller samples, 840
  standard deviation, 839
  standard optimization techniques, 845–846
  statistical process control (SPC), 841, 841f.
  suggested data descriptors for common distributions, 843, 843t.
  time studies, 840–841
  traditional cutoff grades in open-pit mining, 846–847, 846t.

Tailings
  acid–base accounting (ABA), 655–656
  and acid-generating potential, 655
  Atterberg limits, 654
  bulk and dry density relationship (formula), 654
  bulk density (formula), 653
  characteristics, 653–656
  and contaminant/metal release, 655
  continuum, 650–651
  degree of saturation (formula), 654
  delivery and deposition, 658
  dry density (formula), 653
  earthquakes and liquefaction, 655
  effective stress, 655
  and effluent guidelines, 656
  geochemistry, 655–656
  hydraulic gradient, 654
  kinetic testing, 656
  liquefaction, 655
  liquid limit (LL), 654
  loading cases, 654–655
  management principles, 646
  and metal-leaching characteristics, 655
  moisture content (formula), 653
  nonplastic, 654
  paste consistency (nonsegregating), 650
  permeability coefficient, 654
  phases, 653
  plastic, 654
  plastic limit (PL), 654
  plasticity index (PI), 654
  quasi-static pore pressure (aka dynamic pore pressure), 655
  saturated, 653
  semisolid, 654
  shear strength, 654–655
  slurry density, 650

solid, 654
specific gravity (formula), 654
static loading and liquefaction, 655
static testing, 656
terminology, 653, 653t.
thickening, 650–651
total stress, 655
total volume, 653
total weight or mass, 653
viscous fluid, 654
void ratio (formula), 654
yield stress, 650
See also Uranium tailings; Waste disposal and contamination management
Tailings impoundments and dams, 645–646
    algae introduction into tailings stream, 660
    alternatives analysis and selection, 653
    basin liner systems, 648–649
    borehole inclinometers in monitoring, 660
    centerline embankment, 649
    and climate, 651
    composite liners, 649
    construction considerations, 658–659
    construction materials, 647–648
    current and future trends, 662–664
    decant barges, 658
    decant tower systems, 657–658
    design components, 645–646
    design objective confirmation, 659
    disposal of tailings within other mining facilities, 660
    downstream embankment, 649
    drains (subaerial deposition), 657
    and dust control, 663–664
    embankment types, 649–650
    and geology, 651
    geomembrane liners, 649
    geotechnical site investigation, 652
    and groundwater, 648
    guidance documents, 662
    historical sketch of, 660–661
    and hydrology, 651
    improved consistency of management practices in industry, 663
    incidents and lessons, 660–662
    increased application of dewatered tailings, 663
    integrating facilities with other mining and mine waste facilities, 664
    international standards and guidelines, 662–663
    leakage collection and recovery systems, 649
    liquid–solid separation, 650
    low-permeability soil liners, 649
    management technologies, 662
    Merriespruit tailings failure, 662
    metallurgical improvements, 662
    modified centerline embankment, 650
    monitoring, 659–660
    mud farming, 660
    natural materials in construction of, 648
    operation, maintenance, and surveillance (OMS), 659
    and piezometers, 659
    pit disposal, 660
    recognition of facilities as permanent features, 663
    recognition of uniqueness of each facility, 663
    role of design engineer, 658–659
    safety audits, 660
    site characterization, 651–653
    site seismicity, 652–653
    slope movement measurement, 659–660
    staged development, 648
    Stava dam incident, 661–662
    survey methods in monitoring, 659–660
    tailings delivery and deposition, 658
    terrain analysis, 651–652
    types and selection of facilities, 646–651, 647f.
    unique management schemes, 660
    upstream embankment, 649
    waste materials in construction of, 647–648
    water balance (incl. equation), 656–657
    water management, 656–658
    water removal systems, 657–658
    water storage, 657
Tar sands, 6
Tarong coal mine (Australia), automated dragline demonstration, 816
Taylor's rule, 265, 282–283
Tayoltita mine (Mexico), and shrinkage stoping, 1351
TC/RCs (treatment and refining charges), 61
Teck Cominco recycling plant (British Columbia), and community acceptance, 1779–1780
Terex, 920
Tertiary milling, 1471–1473
    defined, 1471
    and particle size analysis, 1472–1473
    See also Comminution; Primary milling; Secondary milling
Terzaghi's rock mass classification, 358–359
The Tex Report, 62
Theis solution, 767
Thiobacillus ferrooxidans, 1090
Thiobacillus thiooxidans, 1090
Threats, interrogations, promises, and spying (TIPS), 320
Time value of money, 309, 313
    discounted cash-flow analysis, 314–315
    valuation at constant point in time, 313–314
Tin mining (Cornwall), 3
TIPS. See Threats, interrogations, promises, and spying
Titanomagnetite deposits, 87
Tlicho, 1809, 1810f.
Tokyo Commodity Exchange, 58
Toronto Stock Exchange, 65
Tractor–trailer truck units, in hard-rock underground mining, 1147, 1148f.
Training
    annual refresher, 327–328
    behavior modeling, 328, 329t.
    methods, 328
    needs assessment, 324
    for new miners, 326–327
    for newly hired experienced miners, 327
    scope, 326–328
    for trainers, 328
    for work with other cultures, 322–323
Transport of ore. See Ore transport
Trenchers, 429
Trends. See Current trends; Future trends
Tricone bits, 438–439
Trolley-wire electric trucks, in hard-rock underground mining, 1150
Trona, 377
Trucks
    alternative energy for, 25
    articulated dump trucks (ADTs), 869, 1036, 1280, 1280f.
    articulated rear-dump trucks, 1147, 1147f.
    automated dump trucks (ADTs), 808
    bottom-dump trucks, 978
    capacity and operating cost, 1148
    chute-loading systems for, 1281
    clearance to sidewalls, 1148
    clearance to ventilation ducts, 1147–1148
    comparison of shaft, conveyor, and truck options in underground ore handling, 1275–1276
    costs, 1281
    cycle times, 1149
    decline haulage access for, 1280f., 1281, 1281f.
    and development openings, 1147, 1148f.
    design and construction issues for, 1281
    diesel–electric, 869
    dumping and dozing in ore emplacement, 1094–1095
    efficiency factors, 1149
    ejector-tray trucks, 1150
    electric-drive, 938
    end-dump trucks, 978
    and haul roads, 957
    haul trucks, 1036
    haulage, 977–978
    key consideration for, 1148–1149
    load, 1149
    loading bays, 1279,
    load-out stations, 978, 978f.
    mechanical-drive, 869, 938
    mine design and construction for, in underground ore handling, 1277–1279
    mining truck design standards, 922
    off-highway articulated dump trucks, 920, 920f., 920t., 937, 937t.
    off-highway rigid-frame trucks, 920–923, 921f., 922f., 923f.
    on-highway, 919–920, 937
    in open-pit mining, 869–870
    vs. ore stream pumping, 1055
    orepass dump points (tipples) for, 1291
    productivity as function of capacity, 1150, 1150f.
    productivity as function of haul distance, 1150, 1150f.
    remote control and automation of, 1152
    right-sized trucks, 1036
    rigid dump trucks, 1280
    rigid-body rear-dump trucks, 1147
    rigid-frame haul trucks, 869, 937–938, 937t., 1036
    safety considerations, 1280
    selection, 937–938
    selection (hard-rock underground mining), 1147–1150
    selection criteria, 1280–1281
    semitrailer belly-dump trucks, 919
    semitrailer rear-dump trucks, 919
    side-dump trucks, 978, 978f.
    side-loading bays, 1281
    size of haul truck fleet, 1037
    size selection, 937, 937t.
    stacking, 1081–1082, 1082f.
    standard dump trucks, 919
    in strip mining (truck and shovel operation), 990f., 993, 993f., 1003, 1006–1012, 1007f., 1008t.–1010t.
    tire management, 870
    tractor–trailer units, 1147, 1148f.
    trolley-assisted, 925–926, 926f., 977, 978f.
    trolley-wire electric trucks, 1150
    truck drive-train type, 923, 923f., 924t.
    turning radius, 1148
    in underground ore handling, 1280–1281

Tuberculosis, 1572, 1572t., 1574–1575, 1574f.
Good Practice Guidance on HIV/AIDS, TB and Malaria, 1772
Tungsten, and placer mining, 1057
Tunnel-boring machines (TBMs), 417, 421, 422f., 1255
   advance prediction, 432
   advantages compared with drill-and-blast, 1260–1262
   circular cross section, 1261
   classifications, 1255
   consumable costs in hard, abrasive rock, 1262
   and declines, 1260
   design and operating problems and solutions, 1263–1264, 1264t.
   design recommendations for subsurface mine development, 1219
   and development access openings, 1259, 1260f.
   disadvantages compared with drill-and-blast, 1262–1263
   double-shield, 1258–1259, 1258f.
   and exploration openings, 1259, 1259f.
   good safety record, 1261
   in hard-rock applications, 1219
   and haulage openings, 1259, 1259f., 1260f.
   high capital cost, 1262
   high mechanical and electrical skill levels required, 1263
   high rate of advance, 1260
   improved ventilation characteristics, 1261
   and inclines, 1260
   Kelly-drive, 1257, 1257f.
   large components that are difficult to transport, 1262
   large power supply required, 1263
   large value of spares and cutter inventory, 1262
   less concrete when lining is required, 1262
   limited range of application in varying ground conditions, 1262
   limited to circular cross section, 1262–1263
   limited turning radius, 1263
   list of mining projects and machine manufacturers, 1265t.–1266t.
   long lead time for delivery, 1262
   lower Manning's factor, 1261
   main-beam, 1256–1257, 1256f., 1257f.
   mining applications, 1255–1256, 1256f., 1259–1260
   and muck removal, 1259–1260
   and multiple-use openings, 1259, 1261f.
   no blast fumes or vibration, 1260
   open-gripper, 1257, 1257f.
   open-type, 1255
   operational and management problems and solutions, 1264–1267
   reduced ground disturbance and support, 1260
   reduced labor costs, 1261
   reduced lining costs, 1260
   shielded-type, 1255
   single-shield, 1257–1258, 1258f.
   soft-ground type, 1255
   and Stillwater Mining Company, 1267
   substantial crew training required, 1263
   substantial preparation at work site, 1263, 1263f.
   suitability for automation, 1260–1261
   types and design features, 421, 423f., 1256–1259
   in underground horizontal and inclined development, 1199–1200, 1200f.
   uniform muck size, 1260
   use by Magma Copper Company San Manuel mine (case study), 1267–1269, 1267f., 1268f., 1269f., 1269t.
   and varying excavation diameter, 1263
   and ventilation openings, 1259, 1260f.
   and water drainage openings, 1259, 1261f.
Tunneling
   drill-and-blast, 1198–1199, 1199f.
   full-face tunnel boring systems (TBMs), 1199–1200, 1200f.
   and load-haul-dump units (LHDs), 1198
   mechanized, 1199–1201
   rock tunneling quality index (Q), 361–362
   and soil mechanics, 489–492, 491t.
   in underground horizontal and inclined development, 1195–1197
Turnouts (breakaways), 1204
Two-pass cut-and-fill mining, 394–395, 394f.
Tylor, Edward, 1800

UDEC (highwall modeling), 1028, 1030
Ultramafic volcanic association deposits, 88–89
Unconformity vein-type uranium deposits, 98–99
Undercut cut-and-fill mining, 396, 397f.
Underground development openings and infrastructure, 1223
   access drifts, 1231–1232, 1233f.
   closeout and commissioning, 1237
   construction, 1234–1237
   construction procurement, 1235
   construction schedule, 1237
   conveyor drifts, 1231, 1232f.
   crusher room closeout and commissioning, 1242
   crusher room construction and installation, 1240–1242, 1241f., 1242f., 1243f.
   crusher room design, 1240, 1240f., 1241t.
   crusher room maintenance, 1242–1243
   crusher room preplanning, 1240
   crusher rooms, 1226–1227, 1226f., 1240
   design, 1232–1234
   emergency access, 1230, 1230f., 1248
   emergency access construction and installation, 1230f., 1249
   emergency access design, 1248–1249
   emergency access evacuation method, 1249
   emergency access location and layout, 1248–1249
   emergency access maintenance, 1249–1250
   emergency access size and cross section, 1248
   excavation methods, 1235–1236, 1236t.
   excavation sequence, 1236
   fit-out and installation of permanent equipment, 1236t.
   loading pocket construction and installation, 1251
   loading pocket design, 1250–1251
   loading pocket loading and materials-handling systems, 1250
   loading pocket maintenance, 1251
   loading pocket O&M costs, 1250
   loading pocket operating criteria, 1250
   loading pocket spill control, 1250–1251
   loading pockets, 1230–1231, 1231f., 1250
   maintenance, 1237
   major construction equipment, 1236–1237
   ore bin commissioning and closeout, 1246
   ore bin construction and installation, 1245–1246
   ore bin design, 1244–1245
   ore bin excavation and ground support design 1244–1245
   ore bin infrastructure, 1228f., 1229f., 1231f., 1245,
   ore bin location and access, 1244
   ore bin maintenance, 1246–1247
   ore bin preplanning, 1245–1246
   ore bin procurement, 1246
   ore bin shape, size, and capacity, 1244
   ore bins, 1227–1228, 1228f., 1229f., 1244
   orepass construction and installation, 1248
   orepass design, 1247–1248
   orepass excavation and ground support, 1247, 1249f.
   orepass lining, 1247
   orepass maintenance, 1248
   orepass muck flow control, 1229f., 1247–1248, 1248f.
   orepass opening size and shape, 1247
   orepasses, 1228–1230, 1229f., 1247
   project preplanning, 1235
   pump room arrangement, 1238, 1239f.
   special purpose drift construction and installation, 1252–1253, 1252f., 1253f.
   special purpose drift design, 1251
   special purpose drift excavation and ground support, 1252
   special purpose drift layout, 1251, 1252f.
   special purpose drift location, 1251
   special purpose drift maintenance, 1253
   special purpose drift preplanning, 1252
   special purpose drift procurement, 1252
   special purpose drift purpose, 1251
   special purpose drifts, 1231–1232, 1232f., 1233f., 1251
   sump and pump room closeout and commissioning, 1239
   sump and pump room construction and installation, 1237–1239, 1239f.
   sump and pump room design, 1237, 1238t.
   sump and pump room maintenance, 1239–1240
   sump and pump room preplanning, 1237–1238
   sump and pump room procurement, 1238
   sumps and pump rooms, 1223–1226, 1224f., 1225f., 1226f.
   underground maintenance shop closeout and commissioning, 1243
   underground maintenance shop construction and installation, 1243, 1245f.
   underground maintenance shop design, 1243, 1244t.
   underground maintenance shop maintenance, 1243
   underground maintenance shops, 1227, 1227f., 1243
   ventilation drifts, 1232, 1233f.
   waste pass construction and installation, 1248
   waste pass excavation and ground support, 1247
   waste pass lining, 1247
   waste pass maintenance, 1248
   waste pass muck flow control, 1229f., 1247–1248, 1248f.
   waste pass opening size and shape, 1247
   waste passes, 1228, 1247

Underground horizontal and inclined development methods, 1179
  circular shaft drilling patterns for different rock types, 1184, 1184f.
  declines, 1195, 1197t., 1198t.
  depression of groundwater level in shaft sinking, 1186–1188
  determination of stress coefficient in shaft linings, 1181, 1182f.
  drill-and-blast shaft sinking, 1183–1184, 1183f.
  freezing method (shaft sinking), 1189, 1190f., 1191t.–1192t.
  mobile miners, 1200
  parallel system (shaft sinking), 1183
  raise boring, 1193–1195, 1195f., 1196t.
  raising, 1192
  raising, Alimak method of, 1193, 1283, 1284f.
  raising, cage or gig, 1193, 1193f.
  raising, conventional, 1192
  raising, longhole, 1193, 1194f.
  roadheaders, 1200–1201, 1201f.
  series system (shaft sinking), 1183
  shaft blasthole drilling, 1184–1185
  shaft collars, 1181, 1182f.
  shaft drilling, 1189–1190, 1193f.
  shaft grouting, 1188–1189, 1188f.
  shaft insets and sumps, 1181
  shaft jumbos, 1185
  shaft lining construction, 1186
  shaft linings, 1181
  shaft mucking, 1185–1186, 1185t.
  shaft shape and diameter, 1180–1181
  shaft sinking stages, 1186
  shaft sinking technology, 1181–1183, 1183f.
  shafts, 1180
  shafts, defined, 1180
  simultaneous system (shaft sinking), 1183
  specialized shaft sinking methods, 1186–1192, 1187t.
  tunnel boring machines, 1199–1200, 1200f.
  tunneling, 1195–1197
  tunneling, drill-and-blast, 1198–1199, 1199f.
  tunneling, mechanized, 1199–1201
  underground mine infrastructure, 1180f.
  winzing, 1192
Underground mine planning, 1135
  application flexibility of equipment, 1140
  characteristics of extracted material, 1139
  equipment acceptance, selection, and versatility, 1140
  field-tested equipment, 1140
  geologic and mineralogic information needed, 1135–1136
  government considerations affecting mine size, 1138–1139
  land and water considerations, 1137
  microlevel geologic data, 1139
  organizational planning, 1139–1140
  and rock properties, 1136
  sizing the mine production, 1137–1138
  structural information needed, 1136
  technical information needed, 1135
  test mine, 1136
  and time allotted for development, 1138
  work force and production design, 1139–1140
Underground mining, 352–353, 354t., 385
  automated underground loading and haulage, 808–809
  basic infrastructures, 1180f.
  blast design, 451–454
  block caving, 28–30
  cut-and-fill method, 394–396, 394f., 395f., 396f.
  deposit geometry, 352
  electric power distribution and utilization., 703
  failure and collapse modes around sides, 548
  fatality rates 2004–2006 (U.S. & Australia), 303, 304t.
  fill, 353, 353t.
  future trends, 28–30
  ground control, 352, 352t.
  in hard-rock mining, 364–366, 364f.
  hauling and conveying cost estimating, 266
  main roof, 352
  mapping, 160
  methods, 364–366, 364f.
  mine surveying, 736–742
  modeling, 30
  panel caving mining method, 398f., 399–402, 400f., 401f.
  pillars, 353
  room-and-pillar method, 385–391, 387f., 388f., 389f., 390f.
  sublevel caving mining method, 396–399, 398f.
  sublevel open-stoping method, 391–394, 391f., 392f., 393f., 394f.
  surface to underground transition, 366–367
  transition from open-pit mining, 366–367, 862–863
  underground coal gasification, 1130f., 1131
  underground roadway maintenance, 832
  See also Cost estimating (underground mines); Dewatering underground operations; Hard-rock equipment selection (underground mining); Hard-rock mining method selection; Mining method classification system; Mining method selection; Soft-rock equipment selection (underground mining); Subsurface mine development; Underground development openings and infrastructure; Underground horizontal and inclined development methods; Underground ore handling
Underground ore handling, 1271
  articulated dump trucks, 1280, 1280f.
  and availability and reliability, 1273
  bottlenecks, 1271–1272
  buffers, 1272–1273, 1272f.
  and change in mining method, 1274
  and change in ore-handling system, 1274
  chute-loading systems for trucks, 1281
  comparison of shaft, conveyor, and truck options, 1275–1276
  contingency plans, 1273, 1274t.
  and continuous improvement, 1274
  conveyor costs, 1289
  conveyor integration aspects, 1289
  conveyor safety considerations, 1288
  conveyor selection criteria, 1288–1289
  conveyor technology, 1289
  conveyors, 1288–1289, 1288f.
  conveyors, and limits of application, 1289
  cost comparisons, 1275, 1276f.
  crusher design and construction issues, 1285–1286
  crusher equipment overview, 1286–1287
  crusher integration aspects, 1287
  crusher location, 1286
  crusher safety considerations, 1285
  crusher selection, 1286
  crusher technology, 1287–1288
  crushers, 1285–1288, 1286f., 1287f.
  crushers, and life of mine/expansion plans, 1286
  decline haulage access for trucks, 1280f., 1281, 1281f.
  defined, 1271
  design and construction for loaders and trucks, 1277–1279
  design and construction issues for trucks, 1281
  drawpoints for loaders, 1277, 1278f.
  future designs of conveyors, 1289
  and geographic footprint, 1273
  gradeability/speed/rimpull charts, 1277, 1278f.
  grizzlies for crushers, 1286
  and ground conditions, 1273
  hydraulic haulage systems, 1293
  and license to operate, 1273
  loader automation, 1279
  loader costs, 1279
  loader maintenance, 1280
  loader safety considerations, 1277
  loaders, 1276–1279, 1277f.
  maintenance plans, 1273, 1274t.
  and maintenance support, 1273
  miscellaneous haulage sysems, 1293
  and new technologies, 1274
  operational aspects of loaders, 1279
  optional equipment for crushers, 1287
  optionality comparisons, 1275–1276
  optionality plans, 1273, 1274t.
  orepass advantages and disadvantages, 285
  orepass construction, 1283, 1284f.
  orepass construction risks, 1282
  orepass cost considerations, 1285
  orepass design and construction issues, 1282–1283
  orepass dump points (tipples) for trucks, 1291
  orepass inclination angle, 1282
  orepass length, 1282
  orepass operational aspects, 1283–1285
  orepass orientation, 1282
  orepass safety considerations, 1282
  orepass shape, 1282–1283
  orepass support systems, 1283
  orepass ventilation, 1283
  orepass width, 1283, 1283t.
  orepasses, 1282–1285
  orepasses and clearing of blockages, 1282
  orepasses and fall prevention, 1282
  orepasses and mud rushes, 1282
  production management systems, 1279
  rail haulage and material properties and handling, 1292
  rail haulage and mine plan, 1291
  rail haulage and ore-body geometry, 1291
  rail haulage and production growth, 1291
  rail haulage capacity, 1291
  rail haulage capital costs, 1290
  rail haulage design issues, 1291
  rail haulage equipment, 1292–1293
  rail haulage health and safety considerations, 1290
  rail haulage operating cost, 1290
  rail haulage production rate and scale of operations, 1290–1291
  rail haulage reliability and operability, 1290
  rail haulage selection criteria, 1290–1293

rail haulage system layout and geometry, 1292, 1292f.
rail haulage system life, 1291
rail haulage systems, 1289–1293, 1291f.
and replacement or major maintenance, 1273
and revenue, 1273
rigid dump trucks, 1280
road trains, 1280, 1280f.
rock breakers for crushers, 1286–1287
and safety, 1273
safety comparisons, 1275
selection process, 1273–1275, 1275t.
and short-term operational variability, 1275
shuttle cars, 1280, 1281, 1281f.
side-loading bays for trucks, 1281
stockpile bays for loaders, 1278
stope mucking (loaders), 1279
strategy, 1271–1273, 1272f.
and system robustness, 1274
and throughput options, 1273
timing comparisons, 1275
tipple accesses for loaders, 1278
tramming (loaders), 1279
tramming drifts (drives) for loaders, 1277
truck safety considerations, 1280
truck selection criteria, 1280–1281
trucking costs, 1281
truck-loading bays, 1279,
trucks, 1280–1281
use of contractors to manage buffers, 1272, 1273f.
Underground retorting, 366
Underground road trains, in hard-rock underground mining, 1147, 1148, 1148f.
Underhand cut-and-fill mining, 367t., 395, 396, 396f., 397f.
Uniaxial compressive strength (UCS), 532, 532t.
Unions, 18
  contracts, 319
  membership as UN universal right, 318
  and strikes, 319
  transnational strategies, 318
United Kingdom
  Health and Safety Executive (HSE) noise regulations and solutions, 1633, 1634, 1638
  mining safety legislation, 1559–1560
  publications and associations (Web sites), 1832
United Mine Workers of America, 303
United Nations
  Brundtland Commission, 32, 1665, 1769
  Declaration on the Rights of Indigenous Peoples (UNDRIP), 1780–1781, 1795, 1797
  Environment Programme, 662
  Global Compact, 1668
  and indigenous cultural considerations, 1797
  International Labour Organization (ILO) on mine surveying, 732
  Millennium Development Goals, 1668–1669
  Rio Earth Summit (1992), 1769
  Universal Declaration of Human Rights, 1773, 1795
United States
  federal government agencies (Web sites), 1829
  professional societies, associations, and organizations (Web sites), 1827–1828
  references, information, and publications (Web sites), 1828
  state mining agencies and associations (Web sites), 1829–1831
  use of R&P in coal mining, 1339
  See also entries beginning with U.S.
Unloaders
  bottom-dump ore cars, 979
  bucket-chain continuous unloaders for ships, 980, 981
  bucket-wheel unloaders for ships, 980
  clamshell-type barge unloaders, 980
  clamshell-type ship unloaders, 980, 981
  continuous unloading systems for ships, 980, 981
  and fine materials, 979
  and frozen materials, 979
  pneumatic barge unloaders, 980
  pneumatic unloaders for ships, 981
  rotary train car dumpers, 979
  self-unloading ships, 982
  shifting locomotives, 979
  for ships, 980, 981–982
  for trains, 979–980
  trestle system for trains, 979
  See also Loaders; Load-haul-dump vehicles
Unwedge software, 588
Uranium, 1112–1114
  decay series, 1625, 1626f.
  with gold in conglomerates, 99
  granitic deposits and heap leaching, 1076
  and groundwater restoration, 1115, 1116t.
  and heap leaching, 1076–1077
  heap leaching and chemistry of deposits, 1077
  heap leaching and geology of deposits, 1076
  in-situ solution mining of, 1114
  iron oxide copper-gold (with or without uranium) deposits, 85t., 92–93
  ISR facility, 1114, 1116f.
  lixiviants, 1089
  processing of, 1114–1115
  production of, 1115
  and radiation, 1625
  recovery systems, 1100
  sandstone-hosted, 99
  sedimentary, 98
  sedimentary deposits and heap leaching, 1076
  solution mining (in-situ techniques), 1112–1115, 1113t., 1114f., 1115f., 1116f., 1116t.
  species conversion factors, 1113, 1113t.
  unconformity vein-type, 98–99
  wells and well patterns, 1114, 1114f., 1115f.
  See also Radon; and under Heap leaching; Mineral deposits; Solution mining (in-situ techniques)
Uranium Mill Tailings Radiation Control Act of 1978, 677
Uranium tailings, 677
  cover calculator, 677–678
  cover for, 677, 677f.
  management scheme, 660
  See also Tailings; Tailings impoundments and dams
U.S. Army Corps of Engineers, 1654
  Hydraulic Engineering Center, 744
U.S. Atomic Energy Commission, dust size-selection device, 1601
U.S. Bureau of Indian Affairs, 1654–1656
U.S. Bureau of Land Management (BLM), 1654–1658
U.S. Bureau of Mines
  Cost Estimating System (CES), 281
  and heap leaching, 1073
  highwall mining research, 1029
  on in-pit crushing and conveying (IPCC), 942
  Minerals Health and Safety Technology Division on haul road maintenance, 971
  and pneumatic stowing, 1376
U.S. Clean Water Act, 1654
U.S. Department of Energy, underground coal gasification experiments, 1131
U.S. Department of the Interior, Office of Surface Mining Reclamation and Enforcement, and flowable fill, 1376
U.S. Environmental Protection Agency (EPA), 1654
  toxicity characteristic leaching procedure, 676
  and underground coal gasification, 1131
U.S. Geological Survey, 63, 127
  and dewatering research, 744
  seismic hazard maps, 652
U.S. Metal and Industrial Mineral Mine Salaries, Wages and Benefits, 2009 Survey Results, 272, 288, 289
U.S. National Geodetic Survey (NGS), 736
U.S. Office of Surface Mining, and coal mining surface restoration, 669
U.S. Securities and Exchange Commission, 146
  Industry Guide 7, 203, 231

Valdez Creek mine (Alaska) groundwater control (case study), 748–749
VALMIN code, 219–220
Vein mining, 393, 393f.
Ventilation. See Mine ventilation
Vertical crater retreat (VCR), 365, 369f.
  stoping, 392–393, 393f.
Voest-Alpine surface miner, 429
Voisey's Bay nickel project, 1798
Volcanic-hosted or volcanogenic massive sulfide deposits, 85t., 94–95
Volvo, 920

Wabi Iron and Steel Corp., 1306
Wagerup alumina refinery (Australia), tailings management scheme, 660
Warren, H., 127
Waste disposal and contamination management, 33, 1733
  acid rock drainage, 1734
  asbestiform minerals, 1735
  capping in place, 1749–1750
  chemically reactive mineral waste, 1734–1735, 1736f.
  chemically reactive mineral waste repository in humid climate, 1738, 1744f.
  chemically reactive mineral waste repository in semiarid climate, 1738, 1744f.
  closure, 1745
  contaminated site management, 1749–1750
  contaminated water management, 1747–1749
  cyanide and other reagents, 1735
  and direct exposure risks and phytotoxicity, 1734
  and direct land disturbance, 1734
  disposal area footprint, 1737
  and dust release, 1734

and erosion and sediment release, 1734
facility environmental design, 1738–1744
and geotechnical instability, 1734
groundwater flow control, 1750
groundwater remediation, 1750
hazardous materials, 1746–1747
high-pH waste, 1735
improved water use efficiency, 1748
infiltration limiting covers, 1738
in-pit disposal and selective waste placement, 1738
in-situ treatment, 1749
land-use restrictions, 1750
liners for harzardous materials, 1747
metals soluble or bioavailable at neutral pH, 1734
mineral waste, 1733–1745
natural attenuation, 1750
nitrogen compounds, 1735
nonmineral waste, 1745–1746
ongoing management and monitoring, 1744–1745
on-site evaporation, 1748
radionuclides, 1735
reagent management, 1748
recontouring to promote runoff, 1738
Rio Tinto examples, 1738, 1744f.
safe on-site water storage and controlled release, 1748
salinity, 1734–1735
secondary containment, 1747
selective handling, 1738
site selection, 1737
sodicity, 1735
soil and sediment removal actions, 1749
spontaneous combustion, 1735
and visual impacts, 1734
waste characterization and impact prediction, 1735–1737
waste management strategy, 1737–1738, 1739t.–1743t., 1744f.
waste removal and consolidation, 1738
water diversion, 1748
and water quality degradation, 1734
water recycling and reuse, 1748
water segregation, 1748
See also Acid rock drainage management; Tailings; Tailings impoundments and dams

Waste piles and dumps, 667
and acid rock drainage, 676–677
base failure (spreading), 670, 671f.
block translation, 670, 671f.
closure, 677
configuratons, 668, 668f.
cross-valley type, 668, 668f.
defined, 667
design of, 670–676
diked embankment type, 668, 668f.
drainage, 675
and dump geometry, 671–672
and erosion, 675–676
factor of safety, 674
factors affecting slope stability, 671–674
failure modes, 670–671, 671f.
and geotechnical properties of foundation, 672
and geotechnical properties of geosynthetics, 673
and geotechnical properties of mine waste, 672
and groundwater, 673
historical sketch, 667

hydraulic conductivity function (HCF), 675
impacts of, 669–670
and land disturbance, 669
liquefaction, 670–671, 671f.
and phreatic surface, 673
placer waste and tailings deposits, 669
and radioactive waste rock, 677–678, 677f.
reclamation, 677
reliability, 674
ridge embankment type, 668, 668f.
rock drains, 675
rock dumps and erosion, 676
rotational circular failures, 670, 671f.
seepage, 675
and seismic forces, 673–674
settlement, 674–675
shallow flow slides, 670, 671f.
sidehill structure, 668, 668f.
simplified deformation analyses, 674
and site topography, 671
slope stability, 670–674
soil–water characteristic curves (SWCCs), 675
and stacking method, 671–672
stockpiles, 669
surface or edge slumping, 670, 671f.
types of, 667–669
valley-fill type, 668, 668f.
and visual impacts, 669–670
and water quality, 669
See also Heap leaching

Water
acid rock drainage (ARD) and stormwater management, 676–677
balance in heap leaching, 1080
in dust control, 1603, 1604
handling system in quarrying, 1034
inrushes and block caving, 1450
land water considerations in underground mine planning, 1137
and longwall mining, 1403
National Weather Service rainfall data sources, 1707
pore water pressure in soil, 471, 472
quality and waste piles and dumps, 669
requirements in solution mining, 1100
soil–water characteristic curves (SWCCs), 675
and sublevel caving, 1430
surface water collection in exploration geochemistry, 131
surface water hydraulics, 1708–1709
surface water hydrology, 1707–1708
value of, 7–8
See also Dewatering (liquid–solid separation); Dewatering surface operations; Dewatering underground operations; Groundwater; Pumps and pumping; Tailings impoundments and dams

Water and sediment control systems, 1705
aquifer testing, 1715–1716
artificial and synthetic materials, 1714, 1715f.
best management practices, 1714
centrifugal pumps, 1717
computer modeling, 1718, 1718t.
conservation of energy, 1708–1709
conservation of linear momentum, 1708–1709
conservation of mass, 1708–1709
control ponds, 1712–1713
cost benefit, 1705

culverts, 1711–1712
dams and incised structures, 1713
design considerations, 1706–1707
design rainfall event, 1707
diversion and interception of undisturbed runoff, 1706
diversion channels and ditches, 1710–1711
dry wells, 1714, 1714f.
dynamic head calculation, 1717
French drains, 1714, 1714f.
and geomorphology, 1706, 1706f.
and GIS, 1705
groundwater hydrology, 1706, 1715–1718
groundwater inflow management, 1716–1717
groundwater inflow prediction, 1715
hydraulics software, 1710
hydrologic structures, 1709–1713, 1710t.
inlet structures, 1713
Kirpich equation, 1708
Lane's balance diagram, 1706, 1706f.
Manning's equation, 1709, 1710
Manning's roughness coefficient, 1708
mechanical treatment (sediment separators), 1714–1715, 1716f.
natural and biodegradable sediment controls, 1714, 1714t.
permanent-pool storage capacity, 1713
probability of failure (design life of structures), 1705–1706, 1706t.
pump characteristic curves, 1717, 1717f.
pump troubleshooting, 1717–1718
pumps, 1717–1718
Ragan and Duru equation, 1708
rational method of peak flow determination, 1707, 1708t.
regression studies, 1709
revegetation, 1714
Revised Universal Soil Loss Equation (RUSLE), 1709
and runoff, 1705
SCS curve number (SCS-CN) method of peak flow determination, 1708
sediment production, 1709
sediment storage capacity, 1713
selection of analysis method, 1707
and slope design, 518–519
small-area sediment controls, 1713–1714
spillway design, 1713
surface water hydraulics, 1708–1709
surface water hydrology, 1707–1708
time of concentration, 1708
treatment of disturbed land runoff, 1706–1707
Universal Soil Loss Equation (USLE), 1709
watershed planning, 1705
Water jets, 417, 430
in hydraulic mining, 1049, 1050, 1051
Water monitors, 430
Water rights professionals, and dewatering, 744
Watt, James, and Cornish pump, 3
Web sites, 1827–1832
Webb, J.S., 127
West Angelas iron ore mine (Australia), automated blasthole drilling (case study), 820–822, 820t., 821t., 822f.
Wet ball mills, 1470
Whitehorse Mining Initiative (Canada), 1666, 1769
Wieliczka salt mine (Poland), 3
Wilson, Robert, 1769
Winzes and winzing, 1192, 1210
Witwatersrand (South Africa), 1375

Wongawilli extraction, 1158, 1159f.
Workplace Health and Safety Act (Tasmania, 1995), 1560
World Bank
   effluent guidelines, 656
   Extractive Industries Review, 1666, 1769
   and social license to operate, 1779
   tailings management standards, 662
World Bank Group
   capitals model, 1668, 1668f.
   Environmental, Health, and Safety Guidelines, 1668
   Equator Principles, 1668, 1770
   and International Finance Corporation sustainability framework, 1668
World Bureau of Metal Statistics, 62
World Business Council for Sustainable Development, 1666
World Information Service on Energy: Uranium Project, 1073

X-ray analyzers, 157
X-ray diffractometer, 190
X-ray fluorescence (XRF), 108, 133, 138
   in assaying, 190
Xstrata acquisitions, 78

Young's modulus, 503, 511

Zero harm, 380
Zinc, heap leaching of, 1090
Zircon, and placer mining, 1057
Zoning patterns, 173